General Information 1

Data Acquisition Circuits 2

Display Drivers 3

Line Drivers and Receivers 4

Peripheral Drivers/Actuators 5

Memory Interface Circuits 6

Speech Synthesis Circuits 7

Appendix A Power Derating Curves A

Appendix B Ordering Instructions, Mechanical Data, IC Sockets B

Appendix C Explanation of Logic Symbols C

Interface Circuits Data Book

Texas Instruments

IMPORTANT NOTICE

Texas Instruments (TI) reserves the right to make changes in the devices or the device specifications identified in this publication without notice. TI advises its customers to obtain the latest version of device specifications to verify, before placing orders, that the information being relied upon by the customer is current.

TI warrants performance of its semiconductor products to current specifications in accordance with TI's standard warranty. Testing and other quality control techniques are utilized to the extent TI deems such testing necessary to support this warranty. Unless mandated by government requirements, specific testing of all parameters of each device is not necessarily performed.

In the absence of written agreement to the contrary, TI assumes no liability for TI applications assistance, customer's product design, or infringement of patents or copyrights of third parties by or arising from use of semiconductor devices described herein. Nor does TI warrant or represent that any license, either express or implied, is granted under any patent right, copyright, or other intellectual property right of TI covering or relating to any combination, machine, or process in which such semiconductor devices might be or are used.

Specifications contained in this data book supersede all data for these products published by TI in the United States before September 1986.

ISBN 0-89512-199-9

INTRODUCTION

Texas Instruments offers a broad line of Interface and Speech Products serving analog signal conditioning/processing and interface applications that may require higher currents and/or higher voltages than can be achieved with conventional digital devices.

TI's Interface circuits represent technologies from classic bipolar through BIDFET, Advanced Low-Power Schottky (ALS), IMPACT™, LinCMOS™, and ADVANCED LinCMOS™ processes. The ALS and IMPACT™ oxide-isolated technologies provide the Interface family with improved speed-power characteristics. LinCMOS™ and ADVANCED LinCMOS™ technologies feature a step-function improvement in impedance, speed, power dissipation, and threshold stability.

TI's Interface products include such devices as data transmission circuits that tie computers and their associated peripherals together according to a set of industry (EIA) standards that prescribe line length, data rates, and propagation delays, among other things. With the recent growth of the flat panel display market, TI's high-voltage display drivers are providing cost-effective and reliable solutions to the AC-plasma, vacuum-fluorescent, and electroluminescent display markets. Analog-to-digital and digital-to-analog converters are offered as peripheral support chips in microprocessor-based systems and DSP (digital signal processing) related analog interfaces. TI's line of high-current motor/printhead and MOSFET drivers combine logic control and high current-drive capability on one IC.

During the last decade, TI has produced a wide range of speech-generating devices based on the technique of pitch-excited linear predictive coding (LPC). This technique extracts data from original recorded speech to define the control parameters for a mathematical model of the vocal tract. The speech generated from this model retains all the inflection and voice characteristics of the original spoken phrase while minimizing digitized data storage requirements; and it does not exhibit the robotic quality often associated with synthesis-by-rule systems.

This data book provides information on the following types of products:

- Analog Switches
- High-Current Actuators and Peripheral Drivers
- Switched-Capacitor General Purpose Filters
- A/D and D/A Converters
- High-Voltage Display Drivers
- IBM 360/370 I/O Line Drivers
- IEEE-488 (GPIB) Octal Bus Transceivers
- RS-422-A Line Drivers
- RS-422-A, RS-423-A, and RS-485 Line Receivers
- LPC10 and LPC12 Voice Synthesis Functions on a Chip
- One-Chip Speech System
- Auxiliary Speech Memories

These products cover the dynamic development of linear circuits from the classical operational amplifier to the high-performance A/D and D/A converters and speech-generating devices. New surface-mount packages (8 to 84 leads) include both ceramic and plastic chip carriers, and the small-outline (D) plastic packages that optimize board density with minimum impact on power-dissipation capability.

The Selection Guide includes a functional description of each product, and to assist the design engineer, the Guide is organized into sections containing information on key parameters and packaging. Ordering information and mechanical data are in Appendix B.

During the last decade, TI has produced one of the largest IC socket families. TI's sockets include every type and size socket in common use today and are available in a wide choice of contact materials and designs. Details on TI's sockets are presented in Appendix B.

While this volume offers design and specification data only for Interface and Speech components, complete technical data for any TI semiconductor product is available from your nearest TI Field Sales Office, local authorized TI distributor, or by writing directly to:

Texas Instruments Incorporated
LITERATURE RESPONSE CENTER
P. O. Box 809066
DALLAS, TEXAS 75380-9066

We sincerely feel that you will discover the new 1987 *Interface Circuits Data Book* to be a significant addition to your collection of technical literature.

General Information 1
Alphanumeric Index
Selection Guide

Data Acquisition Circuits 2
Cross-Reference Guide
Data Sheets

Display Drivers 3
Data Sheets

Line Drivers and Receivers 4
Cross-Reference Guide
Data Sheets

Peripheral Drivers/Actuators 5
Cross-Reference Guide
Data Sheets

Memory Interface Circuits 6
Data Sheets

Speech Synthesis Circuits 7
Data Sheets

Appendix A Power Derating Curves A

Appendix B Ordering Instructions, Mechanical Data, IC Sockets B

Appendix C Explanation of Logic Symbols C

DEVICE	PAGE NO.
ADC0803C	2-9
ADC0803I	2-9
ADC0804C	2-15
ADC0804I	2-15
ADC0805C	2-9
ADC0805I	2-9
ADC0808	2-21
ADC0808M	2-29
ADC0809	2-21
ADC0831A	2-37
ADC0831B	2-37
ADC0832A	2-37
ADC0832B	2-37
ADC0834A	2-45
ADC0834B	2-45
ADC0838A	2-45
ADC0838B	2-45
AM26LS31C	4-5
AM26LS31M	4-5
AM26LS32AC	4-13
AM26LS32AM	4-13
AM26LS33AC	4-13
AM26LS33AM	4-13
AM26S10C	4-23
AM26S10M	4-23
AM26S11C	4-23
AM26S11M	4-23
DS3680	5-5
L293	5-9
L293D	5-13
L298	5-17
MC3446	4-31
MC3450	4-35
MC3452	4-35
MC3453	4-43
MC3486	4-47
MC3487	4-53
N8T26	4-57
PBL3717A	5-19
SN55107A	4-73
SN55107B	4-73
SN55108A	4-73
SN55108B	4-73
SN55109A	4-89
SN55110A	4-89
SN55113	4-101
SN55114	4-113
SN55115	4-121
SN55116	4-131
SN55117	4-131
SN55118	4-131
SN55119	4-131
SN55121	4-143
SN55122	4-147
SN55138	4-181
SN55150	4-205
SN55152	4-223
SN55154	4-237
SN55157	4-255
SN55158	4-261
SN55182	4-377
SN55183	4-385
SN55188	4-391
SN55189	4-397
SN55189A	4-397
SN5520	6-3
SN5522	6-15
SN55234	6-35
SN5524	6-25
SN55325	6-45
SN55326	6-55
SN55327	6-55
SN55426B	3-3
SN55427B	3-3
SN55450B	5-81
SN55451B	5-81
SN55452B	5-81
SN55453B	5-81
SN55454B	5-81
SN55461	5-93
SN55462	5-93
SN55463	5-93
SN55464	5-93
SN55471	5-109
SN55472	5-109
SN55473	5-109
SN55474	5-109
SN55500E	3-17
SN55501E	3-29
SN55551	3-79
SN55552	3-79
SN55553	3-95
SN55554	3-95
SN55ALS192	4-481
SN65176B	4-351
SN65500E	3-23
SN65501E	3-35
SN65508	3-41
SN65509	3-47
SN65512B	3-53
SN65513B	3-59
SN65518	3-71
SN65551	3-87
SN65552	3-87
SN65553	3-101
SN65554	3-101
SN65555	3-109
SN65556	3-109
SN65557	3-117
SN65558	3-117
SN65559	3-125
SN65560	3-125
SN65563	3-133
SN65564	3-133
SN65567	3-141
SN65568	3-141
SN75061	4-63
SN75064	5-23
SN75065	5-23
SN75066	5-23
SN75067	5-23
SN75068	5-29
SN75069	5-29
SN75107A	4-73

ALPHANUMERIC INDEX

DEVICE	PAGE NO.
SN75107B	4-73
SN75108A	4-73
SN75108B	4-73
SN75109A	4-89
SN75110A	4-89
SN75111	4-97
SN75112	4-89
SN75113	4-101
SN75114	4-113
SN75115	4-121
SN75116	4-131
SN75117	4-131
SN75118	4-131
SN75119	4-131
SN75121	4-143
SN75122	4-147
SN75123	4-153
SN75124	4-157
SN75125	4-163
SN75127	4-163
SN75128	4-169
SN75129	4-169
SN75136	4-175
SN75138	4-181
SN75140	4-191
SN75141	4-191
SN75146	4-199
SN75150	4-205
SN751506	3-153
SN751508	3-161
SN75151	4-211
SN751516	3-153
SN751518	3-161
SN75152	4-223
SN75153	4-211
SN75154	4-237
SN75155	4-245
SN75157	4-255
SN75158	4-261
SN75159	4-269
SN75160B	4-281
SN75161B	4-289
SN75162B	4-289
SN75163B	4-301
SN75164B	4-309
SN75172	4-319
SN75173	4-327
SN75174	4-335
SN75175	4-343
SN75176B	4-351
SN75177B	4-361
SN75178B	4-361
SN75179B	4-371
SN75182	4-377
SN75183	4-385
SN75188	4-391
SN75189	4-397
SN75189A	4-397
SN75207	4-405
SN75207B	4-405
SN75208	4-405
SN75208B	4-405

DEVICE	PAGE NO.
SN75372	5-33
SN75374	5-43
SN75407	5-53
SN75408	5-53
SN75435	5-57
SN75436	5-63
SN75437A	5-63
SN75438	5-63
SN75440	5-69
SN754410	5-153
SN754411	5-159
SN75446	5-75
SN75447	5-75
SN75448	5-75
SN75449	5-75
SN75451B	5-81
SN75452B	5-81
SN75453B	5-81
SN75454B	5-81
SN75461	5-93
SN75462	5-93
SN75463	5-93
SN75465	5-101
SN75466	5-101
SN75467	5-101
SN75468	5-101
SN75469	5-101
SN75471	5-109
SN75472	5-109
SN75473	5-109
SN75476	5-117
SN75477	5-117
SN75478	5-117
SN75479	5-117
SN75491	3-7
SN75491A	3-7
SN75492	3-7
SN75492A	3-7
SN75494	3-15
SN75500E	3-23
SN75501E	3-35
SN75508	3-41
SN75509	3-47
SN75512B	3-53
SN75513B	3-59
SN75514	3-65
SN75518	3-71
SN75551	3-87
SN75552	3-87
SN75553	3-101
SN75554	3-101
SN75555	3-109
SN75556	3-109
SN75557	3-117
SN75558	3-117
SN75559	3-125
SN75560	3-125
SN75563	3-133
SN75564	3-133
SN75567	3-141
SN75568	3-141
SN75581	3-149

TEXAS INSTRUMENTS
POST OFFICE BOX 655012 • DALLAS, TEXAS 75265

DEVICE	PAGE NO.
SN75603	5-123
SN75604	5-123
SN75605	5-123
SN75608	5-133
SN75609	5-143
SN75ALS126	4-413
SN75ALS130	4-419
SN75ALS160	4-425
SN75ALS161	4-435
SN75ALS162	4-443
SN75ALS163	4-453
SN75ALS164	4-461
SN75ALS165	4-471
SN75ALS192	4-481
SN75ALS193	4-489
SN75ALS194	4-501
SN75ALS195	4-511
TL0808	2-57
TL0809	2-57
TL182C	2-65
TL182I	2-65
TL182M	2-65
TL185C	2-65
TL185I	2-65
TL185M	2-65
TL188C	2-65
TL188I	2-65
TL188M	2-65
TL191C	2-65
TL191I	2-65
TL191M	2-65
TL376C	5-165
TL4810B	3-171
TL4810BI	3-171
TL500C	2-71
TL501C	2-71
TL502C	2-71
TL503C	2-71
TL505C	2-85
TL507C	2-91
TL507I	2-91
TL5812	3-177
TL5812I	3-177
TL601C	2-97
TL601I	2-97
TL601M	2-97
TL604C	2-97
TL604I	2-97
TL604M	2-97
TL607C	2-97
TL607I	2-97
TL607M	2-97
TL610C	2-97
TL610I	2-97
TL610M	2-97
TLC04	2-103
TLC0820AC	2-113
TLC0820AI	2-113
TLC0820AM	2-113
TLC0820BC	2-113
TLC0820BI	2-113
TLC0820BM	2-113
TLC10	2-123
TLC1205AI	2-181
TLC1205BI	2-181
TLC1225AI	2-181
TLC1225BI	2-181
TLC14	2-103
TLC1540I	2-197
TLC1540M	2-197
TLC1541I	2-197
TLC1541M	2-197
TLC20	2-123
TLC32040I	2-271
TLC32040M	2-271
TLC4016I	2-205
TLC4016M	2-205
TLC4066I	2-213
TLC4066M	2-213
TLC532AI	2-139
TLC532AM	2-139
TLC533AI	2-139
TLC533AM	2-139
TLC540I	2-149
TLC540M	2-149
TLC541I	2-149
TLC541M	2-149
TLC543I	2-157
TLC543M	2-157
TLC544I	2-157
TLC544M	2-157
TLC545I	2-165
TLC545M	2-165
TLC546I	2-165
TLC546M	2-165
TLC548C	2-173
TLC548I	2-173
TLC548M	2-173
TLC549C	2-173
TLC549I	2-173
TLC549M	2-173
TLC7135	2-221
TLC7136C	2-233
TLC7524C	2-243
TLC7524I	2-243
TLC7528C	2-251
TLC7528I	2-251
TLC7533	2-263
TSP50C40A	7-3
TSP50C50	7-7
TSP5110A	7-11
TSP5220C	7-15
TSP60C20	7-19
TSP6100	7-23
uA9636AC	4-523
uA9637AC	4-529
uA9637AM	4-529
uA9638C	4-535
uA9639C	4-539
UCN4810A	3-183
UDN2841	5-169
UDN2845	5-169
ULN2001A	5-173
ULN2002A	5-173

ALPHANUMERIC INDEX

DEVICE	PAGE NO.
ULN2003A	5-173
ULN2004A	5-173
ULN2005A	5-173
ULN2064	5-181
ULN2065	5-181
ULN2066	5-181
ULN2067	5-181
ULN2068	5-187
ULN2069	5-187
ULN2074	5-193
ULN2075	5-193

Single-Slope and Dual-Slope A/D Converters

RESOLUTION	CONVERSION SPEED (ms)	FUNCTION	TYPE	PACKAGE	PAGE
4 1/2 Bits	80	Dual-Slope Analog Processors	TL500	J	2-71
8-10 Bits			TL501		2-71
4 1/2 Digits		Digital Processors with Seven-Segment Outputs	TL502	N	2-71
4 1/2 Digits		Digital Processors with BCD Outputs	TL503		2-71
10-Bits	50	Dual-Slope Analog	TL505		2-85
7-Bits	1	Pulse-Width Modulator for Single-Slope Converter	TL507	P	2-91
4 1/2 Digits	34	Dual-Slope ADC with BCD Output	TLC7135	N	2-221
3 1/2 Digits	333	Dual-Slope ADC with LCD Drivers	TLC7136		2-233

D/A Converters (5 to 15 Volts)

RESOLUTION	FUNCTION	TYPE	SETTLING TIME	PACKAGE	PAGE
8 Bits	Single Multiplying DAC	TLC7524	100 ns	D,N	2-243
8 Bits	Dual Multiplying DAC	TLC7528		N	2-251
10 Bits	Single Multiplying DAC	TLC7533	150 ns		2-263

Successive-Approximation A/D Converters

ADDRESS AND DATA I/O FORMAT	CONVERSION SPEED† (μs)	SIGNAL INPUTS ANALOG DEDICATED	SIGNAL INPUTS ANALOG DIGITAL‡	UNADJUSTED ERROR (MAX) ±LSB	TYPE	RESOLUTION BITS	POWER DISSIPATION (TYP)	PACKAGE	PAGE
PARALLEL	100	1§	0	0.5	ADC0803	8	10 mW	N	2-9
				1.0	ADC0804				2-15
					ADC0805				2-9
		8		0.75	ADC0808			FN,N	2-21
				0.75	ADC0808M			FK,JD	2-29
				1.25	ADC0809			FN,N	2-21
				0.75	TL0808		0.5 mW		2-57
				1.25	TL0809				2-57
	1	1		1.0	TLC0820A		35 mW	FN,N FK,J	2-113
				0.5	TLC0820B				
	15	5	6		TLC532A		6 mW	FN,N	2-139
	30				TLC533A				2-139
	10	1§	0	1.0	TLC1205A	12 Plus Sign	25 mW	N,J	2-181
				0.5	TLC1205B				2-181
				1.0	TLC1225A				2-181
				0.5	TLC1225B				2-181
SERIAL	84	1§	0	1.0	ADC0831A	8	10 mW	P	2-37
				0.5	ADC0831B				2-37
		2§		1.0	ADC0832A				2-37
				0.5	ADC0832B				2-37
		4§		1.0	ADC0834A			N	2-45
				0.5	ADC0834B				2-45
		8		1.0	ADC0838A			FN,N	2-45
					ADC0838B				2-45
	13	11		0.5	TLC540		6 mW		2-149
	25				TLC541				2-149
	22	5			TLC543			D,N	2-157
	25				TLC544				2-157
	13	19			TLC545			FN,N	2-165
	25				TLC546				2-165
	22	1			TLC548			D,P	2-173
	25				TLC549				2-173
	31	11			TLC1540	10		FN,N	2-197
	31			1.0	TLC1541				2-197
	52	2¶		#	TLC32040	14	200 mW	N	2-271

† Includes access time.

‡ Analog/digital inputs can be used either as digital logic inputs or inputs for analog to digital conversion. For example: The TLC532/3A can have 11 analog inputs, 5 analog inputs and 6 digital inputs, or any combination in between.

§ Differential Input.

¶ The TLC32040 has two differential inputs for the 14 bit A/D and a serial port input for the 14 bit D/A.

The A/D conversion accuracy for this device is measured in terms of signal-to-quantization distortion and also in LSB over certain converter ranges. Please refer to the data sheet.

TEXAS INSTRUMENTS
POST OFFICE BOX 655012 • DALLAS, TEXAS 75265

Analog Switches and Multiplexers

FUNCTION	TYPICAL IMPEDANCE (OHM)	VOLTAGE RANGE (V)	POWER SUPPLIES (V)	TYPE	PACKAGE	PAGE
TWIN SPDT	100	±10	±15	TL182	N	2-65
	150			TL185		2-65
DUAL SPST	100			TL188		2-65
TWIN DUAL SPST	150			TL191		2-65
SPDT	100	−17 to +25	±25	TL601	P	2-97
DUAL SPDT				TL604		2-97
SPST WITH ENABLE	100			TL607		2-97
SPST WITH LOGIC INPUTS	80			TL610		2-97
QUAD BILATERAL ANALOG SWITCH	50	2 to 12	12	TLC4016	N,D,J	2-205
	30			TLC4066		2-213

Switched-Capacitor Filter ICs

FUNCTION	FILTER ORDER	POWER SUPPLIES (V)	TYPE	PACKAGE	PAGE
DUAL FILTER, GENERAL PURPOSE	2	±4 to ±5	TLC10	FN,N	2-123
			TLC20		2-123
LOW PASS, BUTTERWORTH	4	±2.5 to ±6	TLC04	D,P	2-103
			TLC14		2-103

SELECTION GUIDE
DISPLAY DRIVERS

Electroluminescent Display Drivers

DESCRIPT	DRVRS PER PKG	INPUT COMPATIBILITY	POWER SUPPLY	PRODUCT FEATURES	TYPE	PKG	PAGE
ROW DRIVER	32	CMOS	V_{CC1} (logic) = 10.8 V to 15 V	• 225-V open-drain DMOS outputs • Serial-in, parallel-out architecture • 50-mA current sink output capability • Extremely low steady state power consumption • Left side (SNXX551) and right side (SNXX552) drivers enhance circuit layout	SN55551	FD	3-79
					SN65551 SN75551	FN,N	3-87
					SN55552	FD	3-79
					SN65552 SN75552	FN,N	3-87
	32			• Monolithic BIDFET integrated circuits • Very low steady-state power consumption • 300-mA output capability • High-voltage open-collector N-P-N outputs	SN65557 SN75557	FN	3-117
					SN65558 SN75558	FN	3-117
	34			• 225-V totem-pole BIDFET output structures • 70-mA output capability • Very low steady-state power consumption • 3-State capabilities • Selectable Open-Source or Open-Drain output	SN65563 SN75563	FN	3-133
					SN65564 SN75564	FN	3-133
COLUMN DRIVERS	32	CMOS	V_{CC1} (logic) = 10.8 V to 15 V	• 60-V totem-pole BIDFET output structures • Serial-in, parallel-out architecture • 15-mA sink or source output capability • Top (SNXX553) and bottom (SNXX554) drivers enhance circuit layout	SN55553	FD	3-95
					SN65553 SN75553	FN,N	3-101
					SN55554	FD	3-95
					SN65554 SN75554	FN,N	3-101
				• 90-V output voltage swing capability • 15-mA output source and sink current capability • High-speed serially-shifted data input • Totem-pole outputs • Latches on all driver outputs	SN65555 SN75555	FN,N	3-109
					SN65556 SN75556	FN,N	3-109
				• Energy recovery system compatible • 80-V totem-pole BIDFET output structures • Serial-in, parallel-out architecture • 15-mA sink or source output capability • Top (SNXX559) and bottom (SNXX560) drivers enhance circuit layout	SN65559 SN75559	FN,N	3-125
					SN65560 SN75560	FN,N	3-125
			V_{CC1} (logic) = 4.5 V to 5.5 V	• Energy recovery system compatible • 4.5-V to 5.5-V V_{CC1} operation at 5 MHz • Two Parallel high-speed 16-bit shift registers • 60-V totem-pole BIDFET output structures • 15-mA sink or source output capability • Top (SNXX567) and bottom (SNXX568) drivers enhance circuit layout	SN65567 SN75567	FN	3-141
					SN65568 SN75568	FN	3-141

TEXAS INSTRUMENTS
POST OFFICE BOX 655012 • DALLAS, TEXAS 75265

LED Display Drivers

DESCRIPT	DRVRS PER PKG	INPUT COMPATIBILITY	POWER SUPPLY	PRODUCT FEATURES	TYPE	PKG	PAGE
SEGMENT DRIVERS	4	MOS	10 V	• 50-mA source/sink capability	SN75491	N	3-7
			20 V		SN75491A		
DIGIT DRIVERS	6		10 V	• 250-mA sink capability	SN75492		3-7
			20 V		SN75492A		
			Variable from 3.2 V to 8.8 V	• 250-mA sink capability • Display blanking provisions	SN75494		3-15

Vacuum Fluorescent Display Drivers

DESCRIPT	DRVRS PER PKG	INPUT COMPATIBILITY	POWER SUPPLY	PRODUCT FEATURES	TYPE	PKG	PAGE
ANODE, GRID DRIVERS SEGMENT OR DOT MATRIX FORMATS	12	TTL	V_{CC1} (logic) =5 V to 15 V V_{CC2} (display) = 0 V to 60 V	• Serial-in, parallel-out architecture • 60-V totem-pole outputs • 25-mA current source output capability • On-board latches	SN65512B SN75512B	DW,N	3-53
			V_{CC1} (logic) = 5 V to 15 V, V_{CC2}, (display) = 0 V to 60 V	All features same as SN65512B except • Shift register reset replaces latches	SN65513B SN75513B	DW,N	3-59
		CMOS	V_{CC1} (logic) = 5 V to 15 V, V_{CC2}, V_{CC3}, (display) = 0 V to 60 V	All features same as SN65512B except • 125-V totem-pole output	SN75514	DW,N	3-65
	32	CMOS, TTL	V_{CC1} (logic) = 5 V to 15 V, V_{CC2}, (display) = 0 V to 60 V	All features same as SN65512B except • 32 bits for large format displays	SN65518 SN75518	FN,N	3-71
	10	CMOS, TTL	V_{CC1} (logic) = 5 V to 15 V V_{CC2} (display) = 0 V to 60 V	• Serial-in, parallel-out architecture • 60-V totem-pole outputs • 40-mA current source output capability • Second source to Sprague UCN-4810A	UCN4810A	N	3-183
	10	CMOS	V_{CC1} (logic) = 5 V to 15 V V_{CC2} (display) = 0 V to 60 V	• Serial-in, parallel-out architecture • 60-V totem-pole outputs • 40-mA current source output capability • Improved direct replacement for UCN4810A and TL4810A	TL4810B TL4810BI	DW,N	3-171
	20			• 70-V output voltage swing capability • Drives up to 20 lines • Direct replacement for Sprague UCN5812A	TL5812 TL5812I	FN,N	3-177

SELECTION GUIDE
DISPLAY DRIVERS

DC Plasma Display Drivers

DESCRIPT	DRVRS PER PKG	INPUT COMPATIBILITY	POWER SUPPLY	PRODUCT FEATURES	TYPE	PKG	PAGE
ROW DRIVERS	32	CMOS	V_{CC} (logic) = 4 V to 6 V	• 180-V open drain parallel output • 220-mA parallel output sink current • Left side (SN751506) and right side (SN751516) drivers enhance circuit layout	SN751506 SN751516	FT	3-153
COLUMN DRIVERS			V_{CC} (logic) = 4.5 V to 5.5 V	• −120-V open collector P-N-P parallel outputs • Two parallel high-speed 16-bit shift registers • Latches on all driver outputs • Top (SN751508) and bottom (SN751518) drivers enhance circuit layout	SN751508 SN751518		3-161
ANODE DRIVERS	7	TTL	V_{CC+} = 4.5 V to 5.5 V V_{CC-} = −10.8 V to −13.2 V	• Serial-in, parallel-out architecture • 100-V output capability • Alternative driver for VF	SN75581	J,N	3-149

AC Plasma Display Drivers

DESCRIPT	DRVRS PER PKG	INPUT COMPATIBILITY	POWER SUPPLY	PRODUCT FEATURES	TYPE	PKG	PAGE
AXIS DRIVERS	4	CMOS	V_{CC1} (logic) = 10 V to 14 V V_{CC2} (display) = 40 V to 90 V	• Independent addressing of each gate for serial and parallel applications • High input impedance 1 MΩ typically • 30-mA integral clamp diodes on outputs • Switches 70 V in 1.2 μs • 3-input AND function (SN55426B) NAND function (SN55427B)	SN55426B SN55427B	J	3-3
	32 (8-bits with 1 of 4 selectors)		V_{CC1} (logic) = 10.8 V to 13.2 V V_{CC2} (display) = 0 V to 100 V	• High-speed serial-in, parallel-out architecture (MHz) • Fast output transitions (< 150 ns) • 25-mA output current capability • X-axis driver (SNXX500) • Y-axis driver (SNXX501) • Military temperature packages available (SN55500, SN55501)	SN55500E	FD,JD	3-17
					SN65500E SN75500E	FN,N	3-23
	32 32 × 1				SN55501E	FD JD	3-29
					SN65501E SN75501E	FN,N	3-35
			V_{CC1} (logic) = 7.65 V to 9.35 V V_{CC2} (display) = V_{CC1} to 90 V	• High-speed serial-in, parallel-out • X-axis driver (SNCC508) • Y-axis driver (SNXX508)	SN65508 SN75508	FN	3-41
	32 (8 bits plus 2 select bits)		V_{CC1} (logic) = 8 V to 11.4 V V_{CC2} (display) = V_{CC1} to 90 V		SN65509 SN75509		3-47

Line Drivers

APPLICATION	OUTPUT	DRIVERS PER PACKAGE	DEVICE TYPE	PKG	PAGE NUMBER
EIA STANDARD RS-422-A	DIFFERENTIAL	2	SN55158	JG	4-261
			SN75158	D,JG,P	
			SN75159	D,J,N	4-269
			uA9638	D,JG,P	4-535
		4	AM26LS31	D,FK,J,N	4-5
			MC3487	D,J,N	4-53
			SN75151	DW,J,N	4-211
			SN75153	J,N	
			SN75172	J,N	4-319
			SN75174	J,N	4-335
			SN55ALS192	J,FK	4-481
			SN75ALS192	D,J,N	
			SN75ALS194	D,J,N	4-501
EIA STANDARD RS-485	DIFFERENTIAL	4	SN75172	J,N	4-319
			SN75174	J,N	4-335
EIA STANDARD RS-423-A	SINGLE-ENDED	2	uA9636A	D,JG,P	4-523
EIA STANDARD RS-232-C	SINGLE-ENDED	2	SN55150	JG,FK	4-205
			SN75150	D,JG,P	
			uA9636A	D,JG,P	4-523
		4	SN55188	J,FK	4-391
			SN75188	D,J	
IBM 360/370	SINGLE-ENDED	2	SN75123	D,J,N	4-153
		4	SN75ALS126	D,J,N	4-413
			SN75ALS130	D,J,N	4-419
GENERAL PURPOSE	SINGLE-ENDED	2	SN55121	FK,J	4-143
			SN75121	D,J,N	
GENERAL PURPOSE	DIFFERENTIAL	2	MC3453	D,J,N	4-43
			SN55109A	FK,J	4-89
			SN75109A	D,J,N	
			SN55110A	FK,J	
			SN75110A	D,J,N	
			SN75111	D,J,N	4-97
			SN75112	D,J,N	4-89
			SN55113	FK,J	4-101
			SN75113	D,J,N	
			SN55114	FK,J	4-113
			SN75114	D,J,N	
			SN55183	FK,J	4-385
			SN75183	D,J,N	

Line Receivers

APPLICATION	OUTPUT	RECEIVERS PER PACKAGE	DEVICE TYPE	PKG	PAGE NUMBER
EIA STANDARD RS-422-A	DIFFERENTIAL	2	SN75146	D,JG,P	4-199
			SN55157	JG	4-255
			SN75157	D,JG,P	
			uA9637A	D,JG,P	4-529
		4	uA9639	D,JG,P	4-539
			AM26LS32A	D,FK,J,N	4-13
		4	MC3486	D,J,N	4-47
			SN75173	D,J,N	4-327
			SN75175	D,J,N	4-343
			SN75ALS193	J	4-489
			SN75ALS195	J	4-511
EIA STANDARD RS-485	DIFFERENTIAL	4	SN75173	D,J,N	4-327
			SN75175	D,J,N	4-343
EIA STANDARD RS-423-A	SINGLE-ENDED	2	SN75146	D,JG,P	4-199
			SN75157	D,JG,P	4-255
			uA9637A	D,JG,P	4-529
			uA9639	D,JG,P	4-539
		4	AM26LS32A	D,FK,J,N	4-13
			MC3486	D,J,N	4-47
			SN75173	D,J,N	4-327
			SN75175	D,J,N	4-343
			SN75ALS193	J	4-489
			SN75ALS195	J	4-511
EIA STANDARD RS-232-C	SINGLE-ENDED	2	SN55152	J,FK	4-223
			SN75152	D,J,N	
		4	SN55154	J,FK	4-237
			SN75154	D,J,N	
			SN55189	J,FK	4-397
			SN75189	D,J,N	
			SN55189A	J,FK	
			SN75189A	D,J,N	
IBM 360/370	SINGLE-ENDED	3	SN75124	D,J,N	4-157
		7	SN75125	D,J,N	4-163
			SN75127		
		8	SN75128	DW,J,N	4-169
			SN75129		
GENERAL PURPOSE	SINGLE-ENDED	2	SN55122	FK,J	4-147
			SN75122	D,J,N	
			SN75140	D,JG,P	4-191
			SN75141	D,JG,P	

Line Receivers (Continued)

APPLICATION	OUTPUT	RECEIVERS PER PACKAGE	DEVICE TYPE	PKG	PAGE NUMBER
GENERAL PURPOSE	DIFFERENTIAL	2	SN55107A	FK,J	4-73
			SN75107A	D,J,N	
			SN55107B	FK,J	
			SN75107B	D,J,N	
			SN55108A	FK,J	
			SN75108A	D,J,N	
			SN55108B	FK,J	
			SN75108B	D,J,N	
			SN55115	FK,J	4-121
			SN75115	D,J,N	
			SN55182	FK,J	4-377
			SN75182	D,J,N	
			SN75207	D,J,N	4-405
			SN75207B	D,J,N	
			SN75208	D,J,N	
			SN75208B	D,J,N	
		4	AM26LS33A	D,FK,J,N	4-13
			MC3450	D,J,N	4-35
			MC3452	D,J,N	4-35

TEXAS INSTRUMENTS
POST OFFICE BOX 655012 • DALLAS, TEXAS 75265

Line Transceivers

APPLICATION	BUS I/O	TRANSCEIVERS PER PACKAGE	DEVICE TYPE	PKG	PAGE NUMBER
EIA STANDARD RS-232-C	SINGLE-ENDED	1	SN75155	D,JG,P	4-245
EIA STANDARD RS-422-A AND EIA STANDARD RS-485	DIFFERENTIAL	1	SN65176B	D,JG,P	4-351
			SN75176B	D,JG,P	
			SN75177B	D,JG,P	4-361
			SN75178B	D,JG,P	
			SN75179B	D,JG,P	4-371
EIA STANDARD 488 GPIB	SINGLE-ENDED	4	MC3446	D,J,N	4-31
		8	SN75160B	DW,J,N	4-281
			SN75ALS160	DW,J,N	4-425
			SN75161B	DW,J,N	4-289
			SN75ALS161	DW,J,N	4-435
			SN75162B	DW,N	4-289
			SN75ALS162	DW,N	4-443
			SN75164B	DW,N	4-309
			SN75ALS164	DW,N	4-461
			SN75ALS165	DW,J,N	4-471
IEEE 802.3 1BASE5	DIFFERENTIAL	1	SN75061	DW,J,N	4-63
GENERAL PURPOSE	SINGLE-ENDED	4	AM26S10C	D,J,N	4-23
			AM26S11C	D,J,N	4-23
			N8T26	D,J,N	4-57
			SN75136	D,J,N	4-175
			SN55138	FK,J	4-181
			SN75138	D,J,N	
		8	SN75163B	DW,J,N	4-301
			SN75ALS163	DW,J,N	4-453
	DIFFERENTIAL	1	SN55116	FK,J	4-131
			SN75116	D,J,N	
			SN55117	FK,JG	
			SN75117	D,JG,P	
			SN55118	J,FK	
			SN75118	D,J,N	
			SN55119	FK,JG	
			SN75119	D,JG,P	

General Purpose Drivers and Actuators

SWITCHING VOLTAGE MAX (V)	OFF-STATE VOLTAGE MAX (V)	OUTPUT CURRENT (mA)	DRIVERS PER PACKAGE	OUTPUT CLAMP DIODES	INPUT CAPABILITY	FUNCTION	DELAY TIME TYP (ns)	TYPE	PKG	PAGE
20	30	300	2	NO	TTL	AND	20	SN55450B	FK,J	5-81
20	30	300	2	NO	TTL	AND	18	SN55451B	FK,JG	5-81
20	30	300	2	NO	TTL	NAND	25	SN55452B	FK,JG	5-81
20	30	300	2	NO	TTL	OR	18	SN55453B	FK,JG	5-81
20	30	300	2	NO	TTL	NOR	26	SN55454B	FK,JG	5-81
20	30	300	2	NO	TTL	AND	18	SN75451B	D,P	5-81
20	30	300	2	NO	TTL	NAND	25	SN75452B	D,P	5-81
20	30	300	2	NO	TTL	OR	18	SN75453B	D,P	5-81
20	30	300	2	NO	TTL	NOR	26	SN75454B	D,P	5-81
20	30	300	2	NO	TTL	MOS DRIVER	35	SN75372	D,P	5-33
24	24	500	2	YES	TTL	MOS DRIVER	35	SN75374	D,N	5-33
24	24	500	4	YES	TTL	AND	28	SN55461	FK,JG	5-93
30	35	300	2	NO	TTL	NAND	38	SN55462	FK,JG	5-93
30	35	300	2	NO	TTL	OR	28	SN55463	FK,JG	5-93
30	35	300	2	NO	TTL	NOR	35	SN55464	FK,JG	5-93
30	35	300	2	NO	TTL	AND	28	SN75461	D,P	5-93
30	35	300	2	NO	TTL	NAND	38	SN75462	D,P	5-93
30	35	300	2	NO	TTL	OR	28	SN75463	D,P	5-93
30	35	300	2	NO	TTL	INVERT W ENAB	1050	SN75437A	NE	5-63
30	35	300	2	NO	TTL	INVERT W ENAB	750	SN75435	NE	5-57
35	70	500	4	YES	TTL,CMOS	BUFFER W ENAB	1450	SN75440	NE	5-69
35	70	600	4	YES	TTL,CMOS	INVERT W ENAB	1050	SN75438	NE	5-63
35	70	600	4	YES	CMOS,MOS,TTL	INVERT	500	SN75064	NE	5-23
35	70	1000	4	YES	TTL,CMOS	INVERT	500	SN75066	NE	5-23
35	50	1250	4	YES	TTL	INVERT	500	SN75068	NE	5-29
35	50	1250	4	YES	MOS	INVERT	500	UDN2841	NE	5-169
35	50	1250	4	YES	TTL,5 V MOS	INVERT	500	UDN2845	NE	5-169
35	50	1250	4	NO	TTL,5 V MOS	INVERT	500	ULN2064	NE	5-181
35	50	1500	4	NO	TTL,5 V MOS	INVERT	500	ULN2066	NE	5-181
35	50	1500	4	YES	TTL	INVERT	500	ULN2068	NE	5-187
35	50	1250	4	YES	MOS	INVERT	500	ULN2074	NE	5-193
35	50	1250	4	YES	TTL,CMOS	AND	300	SN75446	D,P	5-75
35	50	1250	4	NO	TTL,CMOS	NAND	300	SN75447	D,P	5-75
35	50	1250	4	YES	TTL,CMOS	OR	300	SN75448	D,P	5-75
55	70	350	2	YES	TTL,CMOS	NOR	300	SN75449	D,P	5-75
55	70	350	2	YES	TTL,CMOS	NAND	500	SN75407	D,P	5-53
55	70	350	2	YES	TTL,CMOS	OR	500	SN75408	D,P	5-53
55	70	350	2	YES	TTL,CMOS	INVERT W ENAB	1050	SN75436	NE	5-63
50	70	500	2	YES	TTL,CMOS	INVERT	250	ULN2001A	D,N	5-173
50	70	500	2	YES	TTL,CMOS	INVERT	250	ULN2002A	D,N	5-173
50	70	500	4	YES	TTL,CMOS	INVERT	250	ULN2003A	D,N	5-173
50	50	350	7	YES	TTL,CMOS,PMOS	INVERT	250	ULN2004A	D,N	5-173
50	50	350	7	YES	25 V PMOS	INVERT	250	ULN2005A	D,N	5-173
50	50	350	7	YES	TTL,CMOS					
50	50	350	7	YES	15 V MOS					
50	50	350	7	YES	TTL					

TEXAS INSTRUMENTS
POST OFFICE BOX 655012 • DALLAS, TEXAS 75265

SELECTION GUIDE
PERIPHERAL DRIVERS/ACTUATORS

General Purpose Drivers and Actuators (Continued)

SWITCHING VOLTAGE MAX (V)	OFF-STATE VOLTAGE MAX (V)	OUTPUT CURRENT (mA)	DRIVERS PER PACKAGE	OUTPUT CLAMP DIODES	INPUT CAPABILITY	FUNCTION	DELAY TIME TYP (ns)	TYPE	PKG	PAGE
50	80	1500	4	YES	TTL	INVERT	500	SN75065	NE	5-23
50	80	1500	4	YES	MOS	INVERT	500	SN75067	NE	5-23
50	80	1500	4	YES	TTL,5 V MOS	INVERT	500	SN75069	NE	5-29
50	80	1500	4	YES	TTL	INVERT	500	ULN2065	NE	5-181
50	80	1500	4	YES	MOS	INVERT	500	ULN2067	NE	5-181
50	80	1500	4	YES	TTL,5 V MOS	INVERT	500	ULN2069	NE	5-187
50	80	1500	4	NO	TTL,5 V MOS	INVERT	500	ULN2075	NE	5-193
55	70	300	2	NO	TTL	AND	28	SN55471	FK,JG	5-109
55	70	300	2	NO	TTL	NAND	38	SN55472	FK,JG	5-109
55	70	300	2	NO	TTL	OR	28	SN55473	FK,JG	5-109
55	70	300	2	NO	TTL	NOR	35	SN55474	FK,JG	5-109
55	70	300	2	NO	TTL	AND	28	SN75471	D,P	5-109
55	70	300	2	NO	TTL	NAND	38	SN75472	D,P	5-109
55	70	300	2	NO	TTL	OR	28	SN75473	D,P	5-109
55	70	300	2	YES	TTL,CMOS	AND	200	SN75476	D,P	5-117
55	70	300	2	YES	TTL,CMOS	NAND	200	SN75477	D,P	5-117
55	70	300	2	YES	TTL,CMOS	OR	200	SN75478	D,P	5-117
55	70	300	2	YES	TTL,CMOS	NOR	200	SN75479	D,P	5-117
60	60	100	4	YES	TTL,CMOS,MOS	TELECOM RY DRV	1000	DS3680	D,J,N	5-5
60	100	350	7	YES	TTL	INVERT	250	SN75465	D,N	5-101
60	100	350	7	YES	TTL,CMOS,PMOS	INVERT	250	SN75466	D,N	5-101
60	100	350	7	YES	25 V PMOS	INVERT	250	SN75467	D,N	5-101
60	100	350	7	YES	TTL,CMOS	INVERT	250	SN75468	D,N	5-101
60	100	350	7	YES	15 V MOS	INVERT	250	SN75469	D,N	5-101

Motor Drivers and Power Actuators

SWITCHING VOLTAGE MAX (V)	OFF-STATE VOLTAGE MAX (V)	OUTPUT CURRENT (mA)	DRIVERS PER PACKAGE	OUTPUT CLAMP DIODES	INPUT CAPABILITY	FUNCTION	DELAY TIME TYP (ns)	TYPE	PKG	PAGE
18	18	500	3	NO	TTL,MOS,CMOS	HALF-H DRIVER		TL376C	NE	5-165
36	36	600	4	YES	TTL	HALF-H DRIVER	600	L293D	NE	5-13
36	36	1000	4	NO	TTL	HALF-H DRIVER	600	L293	NE	5-9
36	36	1000	4	YES	TTL,CMOS	HALF-H DRIVER	600	SN754410	NE	5-153
36	36	1000	4	NO	TTL,CMOS	HALF-H DRIVER	600	SN754411	NE	5-159
40	40	2000	1	YES	TTL,CMOS	HALF-H DRIVER		SN75603	KC,KH,KV	5-123
40	40	2000	1	YES	TTL,CMOS	HALF-H DRIVER		SN75604	KC,KH,KV	5-123
40	40	2000	1	YES	TTL,CMOS	HALF-H DRIVER		SN75605	KC,KH,KV	5-123
46	46	1000	1	YES	TTL	STEPPER DRIVER		PBL3717	NE	5-19
46	46	2000	2	NO	TTL	FULL-H DRIVER		L298	KV	5-17
60	60	2500	2	YES	TTL,CMOS	ACTUATOR	800	SN75608	KV	5-133
60	60	2500	2	YES	TTL,CMOS	ACTUATOR	800	SN75609	KV	5-143

TEXAS INSTRUMENTS
POST OFFICE BOX 655012 • DALLAS, TEXAS 75265

Core-Memory Drivers

MAX OUTPUT CURRENT	t_{PD} TYP	POWER SUPPLIES	OUTPUTS	DEVICE TYPE	PKG	PAGE NUMBER
600 mA	45 ns	V_{CC1} = 5 V V_{CC2} = 4.5 V to 24 V	DUAL SOURCE, DUAL SINK	SN55325	FK,J	6-45
	40 ns	V_{CC} = 5 V	QUADRUPLE SINK	SN55326	J	6-55
	35 ns	V_{CC1} = 5 V V_{CC2} = 4.5 V to 24 V	QUADRUPLE SOURCE	SN55327	J	6-55

t_{PD} – Propagation Delay Time

Core-Memory Sense Amplifiers

THRESHOLD SENSITIVITY	t_{PD} TYP	UNITS PER PACKAGE	TYPE OF OUTPUT	DEVICE TYPE	PKG	PAGE NUMBER
±15 mV	35 ns	1	RESISTOR	SN5520	J	6-3
	30 ns	1	OPEN COLL OR RESISTOR	SN5522	J	6-15
	25 ns	2	RESISTOR	SN5524	J	6-25
		2	RESISTOR	SN55234	J	6-35
±7 mV	28 ns	2	TOTEM POLE	SN55236	WC	See Note 1
				SN75236		

t_{PD} – Propagation Delay Time
NOTE 1: For additional information, contact your nearest TI field sales office.

MOS-Memory Sense Amplifiers

THRESHOLD SENSITIVITY	t_{PD} TYP	UNITS PER PACKAGE	TYPE OF OUTPUT	DEVICE TYPE	PKG	PAGE NUMBER
±25 mV	17 ns	2	TOTEM POLE	SN55107A	FK,J	4-73
				SN75107A	D,J,N	4-73
	19 ns	2	OPEN COLLECTOR	SN55108A	FK,J	4-73
				SN75108A	D,J,N	4-73
±10 mV	25 ns	2	TOTEM POLE	SN75207	D,J,N	4-405
	25 ns	2	OPEN COLLECTOR	SN75208	D,J,N	4-405

t_{PD} – Propagation Delay Time

SELECTION GUIDE
SPEECH SYNTHESIS CIRCUITS

PROCESS	DESCRIPTION	PACKAGE	DEVICE TYPE	PAGE
PMOS	LPC-10 VOICE SYNTHESIZER, 4-BIT CONTROL BUS	N	TSP5110A	7-11
	LPC-10 VOICE SYNTHESIZER, 8-BIT CONTROL BUS		TSP5220C	7-15
	128K-BIT ROM FOR TSP5110A AND TSP5220C		TSP6100	7-23
CMOS	MICROPROCESSOR, SYNTHESIZER, 64K-BIT ROM	N	TSP50C40A	7-3
	256K-BIT ROM FOR TSP50C4X, TSP50C50 FAMILIES		TSP60C20	7-19
	LPC-12 HIGH-QUALITY VOICE SYNTHESIZER WITH 6-POLE LOW-PASS FILTER	J,N	TSP50C50	7-7

General Information	Alphanumeric Index Selection Guide	1
Data Acquisition Circuits	Cross-Reference Guide Data Sheets	2
Display Drivers	Data Sheets	3
Line Drivers and Receivers	Cross-Reference Guide Data Sheets	4
Peripheral Drivers/Actuators	Cross-Reference Guide Data Sheets	5
Memory Interface Circuits	Data Sheets	6
Speech Synthesis Circuits	Data Sheets	7
Appendix A	Power Derating Curves	A
Appendix B	Ordering Instructions Mechanical Data IC Sockets	B
Appendix C	Explanation of Logic Symbols	C

CROSS-REFERENCE GUIDE
(manufacturers arranged alphabetically)

Replacements were based on similarity of electrical and mechanical characteristics as shown in currently published data. Interchangeability in particular applications is not guaranteed. Before using a device as a substitute, the user should compare the specifications of the substitute device with the specifications of the original.

Texas Instruments makes no warranty as to the information furnished and buyer assumes all risk in the use thereof. No liability is assumed for damages resulting from the use of the information contained in this list.

ANALOG DEVICES	TI DIRECT REPLACEMENT	TI FUNCTIONAL REPLACEMENT	PAGE NO.
AD570JN		ADC0803CN	2-9
AD7512DIJN		TL182CN	2-65
AD7512DIJQ		TL182IN	2-65
AD7512DIKN		TL182CN	2-65
AD7512DIKQ		TL182IN	2-65
AD7512DISD		TL182MJ	2-65
AD7512DITD		TL182MJ	2-65
AD7533	TLC7533		2-263
AD7524JN	TLC7524CN		2-243
AD7524AD	TLC7524IN		2-243
AD7528LN	TLC7528CN		2-251
AD7528CQ	TLC7528IN		2-251
AD7820	TLC0820		2-113

BURR-BROWN	TI DIRECT REPLACEMENT	TI FUNCTIONAL REPLACEMENT	PAGE NO.
AD7533	TLC7533		2-263
AD7820	TLC0820		2-113
ADC82AG		TLC0820BIN	2-113
ADC82AM		TLC0820AIN	2-113

DATEL	TI DIRECT REPLACEMENT	TI FUNCTIONAL REPLACEMENT	PAGE NO.
ADC-830C	ADC0803CN		2-9
ADC-EK12DC		TLC7135CN or TLC7136CN or TL500/1/3CN	2-221 2-233 2-71
ADC-EK12DR		TLC7135CN or TLC7136CN or TL500/1/3CN	2-221 2-233 2-71

FUJITSU	TI FUNCTIONAL REPLACEMENT	PAGE NO.
MB4053P	TL507IN	2-91

DATA ACQUISITION CIRCUITS
CROSS-REFERENCE GUIDE

INTERSIL	TI DIRECT REPLACEMENT	TI FUNCTIONAL REPLACEMENT	PAGE NO.
ADC0803LCD	ADC0803IN		2-9
ADC0803LCN	ADC0803CN		2-9
ADC0804LCD	ADC0804IN		2-15
ADC0804LCN	ADC0804CN		2-15
DGM182AK	TL182MN	TL604MP	2-97
DGM182BJ	TL182CN/IN	TL604CP/IP	2-97
DGM185AK	TL185MN	TL604MP	2-97
DGM185BJ	TL185CN/IN	TL604CP/IP	2-97
DGM188AK	TL188MN	TL610MP	2-97
DGM188BJ	TL188CN/IN	TL610CP/IP	2-97
DGM191AK	TL191MN	TL610MP	2-97
DGM191BJ	TL191CN/IN	TL610CP/IP	2-97
ICL7106CPL		TLC7136CN	2-233
ICL7126CPL	TLC7136CN		2-233
ICL7135CPI	TLC7135CN		2-221
ICL7136CPL	TLC7136CN		2-233

HARRIS	TI DIRECT REPLACEMENT	PAGE NO.
HF10	TLC10	2-123

LINEAR TECHNOLOGY	TI DIRECT REPLACEMENT	PAGE NO.
LTC1060ACN	TLC10N	2-123
LTC1060CN	TLC20N	2-123

MAXIM	TI DIRECT REPLACEMENT	PAGE NO.
MF10BN	TLC10N	2-123
MF10CN	TLC20N	2-123
ICL7135	TLC7135	2-221

MICRO NETWORKS	TI FUNCTIONAL REPLACEMENT	PAGE NO.
MN5100/5101	TLC0820ACN	2-113
	TLC0820BCN	2-113
MN5120/5130/5140	TLC0820ACN	2-113
	TLC0820BCN	2-113

MICRO POWER SYSTEMS	TI FUNCTIONAL REPLACEMENT	PAGE NO.
MP7138AN	TLC7135CN	2-221
	TL500/1/3CN	2-71
MP7574AD/BD	ADC0805IN series	2-9
MP7574JN/KN	ADC0804CN or	2-15
	ADC0805CN series	2-9
MP7581/JN/KN/	ADC0808N/	2-21
AD/BD	ADC809N	2-21

MOTOROLA	TI DIRECT REPLACEMENT	TI FUNCTIONAL REPLACEMENT	PAGE NO.
MC1405L		TL500CN TL501CN TL505CN	2-71 2-71 2-85
MC14433P		TLC7135CN or TL500/1/3CN	2-221 2-71
MC14442L	TLC533AMJ	TLC532AMJ	2-139
MC14442P	TLC533AIN	TLC532AIN	2-139
MC14443P		TL507IN	2-91
MC14444P		TLC546IN	2-165
MC14447P		TL507IP	2-91
MC145040FN	TLC541MFN	TLC540MFN	2-149
MC145040L	TLC541MJ	TLC540MJ	2-149
MC145040P	TLC541MN	TLC540MN	2-149
MC54HC4016J	TLC4016MJ		2-205
MC74HC4016J	TLC4016IN		2-205
MC74HC4016N	TLC4016IN		2-205
MC54HC4066J	TLC4066MJ		2-213
MC74HC4066J	TLC4066IN		2-213
MC74HC4066N	TLC4066IN		2-213

NATIONAL	TI DIRECT REPLACEMENT	TI FUNCTIONAL REPLACEMENT	PAGE NO.
ADC0803LCD	ADC0803IN		2-9
ADC0803LCN	ADC0803IN		2-9
ADC0804LCD	ADC0804IN		2-15
ADC0804LCN	ADC0804CN		2-15
ADC0805LCN	ADC0805IN		2-9
ADC0808CCJ	ADC0808N TL0808N		2-21 2-57
ADC0808CCN	ADC0808N TL0808N		2-21 2-57
ADC0809CCN	ADC0809N TL0809N		2-21 2-57
ADC0811BCJ	TLC541IN	TLC540IN	2-149
ADC0811BCN	TLC541IN	TLC540IN	2-149
ADC0811BCV	TLC541IFN	TLC540IFN	2-149
ADC0811BJ	TLC541MJ	TLC540MJ	2-149
ADC0811CCJ	TLC541IN	TLC540IN	2-149
ADC0811CCN	TLC541IN	TLC540IN	2-149
ADC0811CCV	TLC541IFN	TLC540IFN	2-149
ADC0811CJ	TLC541MJ	TLC540MJ	2-149
ADC0820BCD	TLC0820BIN		2-113
ADC0820BCN	TLC0820BCN		2-113
ADC0820BD	TLC0820BMJ		2-113
ADC0820CCD	TLC0820AIN		2-113
ADC0820CCN	TLC0820ACN		2-113
ADC0820CD	TLC0820AMJ		2-113
ADC0829BCN	TLC533AIN	TLC532AIN	2-139
ADC0829CCN	TLC533AIN	TLC532AIN	2-139
ADC0830BCN		TLC546IN	2-165
ADC0830CCN		TLC546IN	2-165
ADC0831BCJ	ADC0831BIP	TLC549IN	2-173
ADC0831BCN	ADC0831BCP	TLC549IN	2-173
ADC0831CCJ	ADC0831AIP	TLC549IN	2-173
ADC0831CCN	ADC0831ACP	TLC549IN	2-173
ADC0832BCJ	ADC0832BIP	TLC544IN	2-157
ADC0832BCN	ADC0832BCP	TLC544IN	2-157

(continued)

NATIONAL	TI DIRECT REPLACEMENT	TI FUNCTIONAL REPLACEMENT	PAGE NO.
ADC0832CCJ	ADC0832AIP	TLC544IN	2-157
ADC0832CCN	ADC0832ACP	TLC544IN	2-157
ADC0834BCJ	ADC0834BIN		2-45
ADC0834BCN	ADC0834BCN		2-45
ADC0834CCJ	ADC0834AIN		2-45
ADC0834CCN	ADC0834ACN		2-45
ADC0838BCJ	ADC0838BIN		2-45
ADC0838BCN	ADC0838BCN		2-45
ADC0838CCJ	ADC0838AIN		2-45
ADC0838CCN	ADC0838ACN		2-45
ADC1001CCJ		TLC1541IN	2-197
ADC1005BCJ		TLC1541IN	2-197
ADC1005CCJ		TLC1541IN	2-197
ADC1205	TLC1205		2-181
ADC1225	TLC1225		2-181
ADC3511CCN		TLC7135CN or TL500/1/3CN	2-221 2-71
ADC3711CCN		TLC7135CN or TL500/1/3CN	2-221 2-71
ADD3501CCN		TLC7136CN or TL500/1/2CN	2-233 2-71
ADD3701CCN		TLC7136CN or TL500/1/2CN	2-233 2-71
MF10BN	TLC10CN		2-123
MF10CN	TLC20CN		2-123
MM54HC4016J	TLC4016MJ		2-205
MM54HC4066J	TLC4066MJ		2-213
MM74HC4016N/J	TLC4016IN		2-205
MM74HC4066N/J	TLC4066IN		2-213
MF4-50	TLC04		2-103
MF4-100	TLC14		2-103

PRECISION MONOLITHICS	TI FUNCTIONAL REPLACEMENT	PAGE NO.
PM7524HP	TLC7524CN	2-243
PM7524FQ	TLC7524IN	2-243
PM7528	TLC7528	2-251
PM7533	TLC7533	2-263

RCA	TI DIRECT REPLACEMENT	TI FUNCTIONAL REPLACEMENT	PAGE NO.
CD4016AD	TLC4016MJ		2-205
CD4016AE	TLC4016IN		2-205
CD4066AD	TLC4066MJ		2-213
CD4066AE	TLC4066IN		2-213
CA3162E		TL501CN/TL503CN	2-71

SIGNETICS	TI DIRECT REPLACEMENT	TI FUNCTIONAL REPLACEMENT	PAGE NO.
ADC0803/4/5-1LCN	ADC0803/4/5IN		2-9
ADC0804-1CN	ADC0804CN		2-15
NE5034F		TLC532AIN	2-139
NE5036FE/N/D		TLC549CN/CD	2-173
NE5037F/N/D		TLC549CN/CD	2-173

SILICONIX	TI DIRECT REPLACEMENT	TI FUNCTIONAL REPLACEMENT	PAGE NO.
DG182AP	TL182MN	TL610MP	2-97
DG182BP	TL182CN/IN	TL610CP/IP	2-97
DG185AP	TL185MN	TL604MP	2-97
DG185BP	TL185CN/IN	TL604CP/IP	2-97
DG188AP	TL188MN	TL604MP	2-97
DG188BP	TL188CN/IN	TL604CP/IP	2-97
DG191AP	TL191MN	TL604MP	2-97
DG191BP	TL191CN/IN	TL604CP/IP	2-97
LD110CJ		TL503CN or TLC7135CN	2-71 2-221
LLD111ACJ		TL501CN or TLC7135CN	2-71 2-221
LD120CJ		TL500CN or TLC7135CN	2-71 2-221
LD121ACJ		TL503CN or TLC7135CN	2-71 2-221
Si520DJ		ADC0808N ADC0809N	2-21 2-21
Si7135CJ	TLC7135CN		2-221

TELEDYNE	TI DIRECT REPLACEMENT	TI FUNCTIONAL REPLACEMENT	PAGE NO.
TSC7106CPL		TLC7136CN	2-233
TSC7126	TLC7136CN		2-233
TSC7126ACPL	TLC7136CN		2-233
TSC7135CPI	TLC7135CN		2-221
TSC8700		ADC0808N	2-21
TSC8701		TLC1541IN	2-197
TSC8703		ADC0808N	2-21
TSC8704		TLC1541IN	2-197
TSC14433CN		TLC7135CN	2-221

D2754, NOVEMBER 1983—REVISED SEPTEMBER 1986

- 8-Bit Resolution
- Ratiometric Conversion
- 100 μs Conversion Time
- 135 ns Access Time
- Guaranteed Monotonicity
- High Reference Ladder Impedance 8 kΩ Typical
- No Zero Adjust Requirement
- On-Chip Clock Generator
- Single 5-Volt Power Supply
- Operates with Microprocessor or as Stand-Alone
- Designed to be Interchangeable with National Semiconductor and Signetics ADC0803 and ADC0805

N DUAL-IN-LINE PACKAGE
(TOP VIEW)

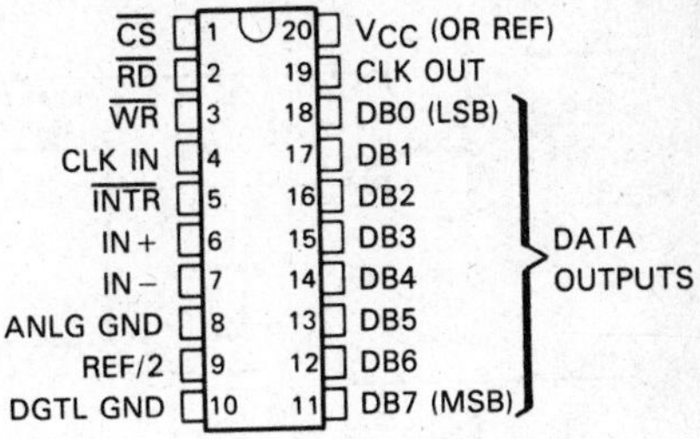
CS 1
RD 2
WR 3
CLK IN 4
INTR 5
IN+ 6
IN− 7
ANLG GND 8
REF/2 9
DGTL GND 10
20 VCC (OR REF)
19 CLK OUT
18 DB0 (LSB)
17 DB1
16 DB2
15 DB3
14 DB4
13 DB5
12 DB6
11 DB7 (MSB)
DATA OUTPUTS

description

The ADC0803 and ADC0805 are CMOS 8-bit successive-approximation analog-to-digital converters that use a modified potentiometric (256R) ladder. These devices are designed to operate from common microprocessor control buses, with the three-state output latches driving the data bus. The devices can be made to appear to the microprocessor as a memory location or an I/O port. Detailed information on interfacing to most popular microprocessors is readily available from the factory.

A differential analog voltage input allows increased common-mode rejection and offset of the zero-input analog voltage value. Although a reference input (REF/2) is available to allow 8-bit conversion over smaller analog voltage spans or to make use of an external reference, ratiometric conversion is possible with the REF/2 input open. Without an external reference, the conversion takes place over a span from V_{CC} to analog ground (ANLG GND). The devices can operate with an external clock signal or, with an additional resistor and capacitor, can operate using an on-chip clock generator.

The ADC0803I and ADC0805I are characterized for operation from −40°C to 85°C. The ADC0803C and ADC0805C are characterized from 0°C to 70°C.

functional block diagram (positive logic)

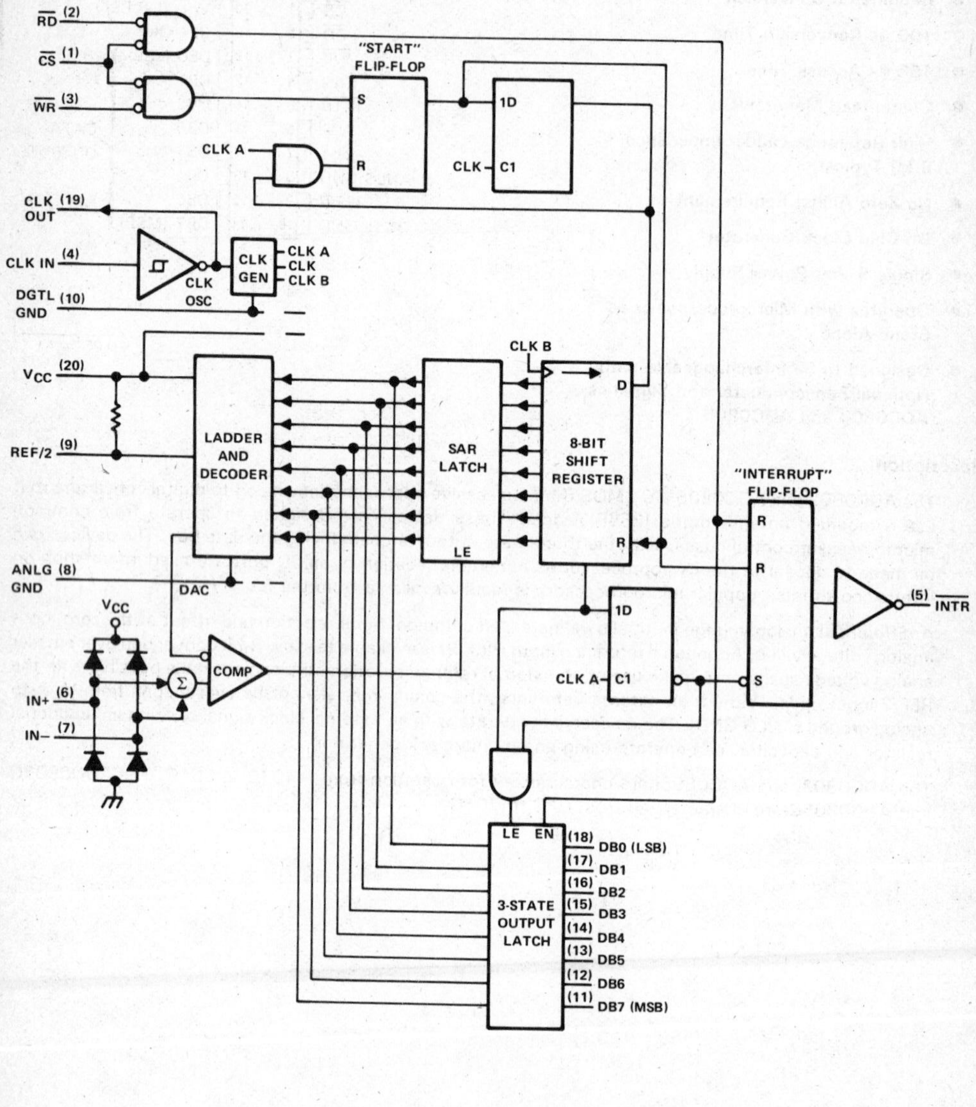
RD (2)
CS (1)
WR (3)
"START" FLIP-FLOP
S
R
1D
CLK C1
CLK A
CLK (19) OUT
CLK IN (4)
CLK OSC
CLK GEN
CLK A
CLK
CLK B
DGTL (10) GND
VCC (20)
REF/2 (9)
LADDER AND DECODER
SAR LATCH
LE
CLK B
8-BIT SHIFT REGISTER
D
R
"INTERRUPT" FLIP-FLOP
R
R
S
(5) INTR
ANLG (8) GND
DAC
VCC
COMP
IN+ (6)
IN− (7)
1D
CLK A C1
LE EN
3-STATE OUTPUT LATCH
(18) DB0 (LSB)
(17) DB1
(16) DB2
(15) DB3
(14) DB4
(13) DB5
(12) DB6
(11) DB7 (MSB)

TEXAS INSTRUMENTS
POST OFFICE BOX 655012 • DALLAS, TEXAS 75265

absolute maximum ratings over operating free-air temperature range (unless otherwise noted)

Supply voltage, V_{CC} (see Note 1) . . . . . 6.5 V
Input voltage range $\overline{CS}$, $\overline{RD}$, $\overline{WR}$ . . . . . −0.3 V to 18 V
Other inputs . . . . . −0.3 V to V_{CC} +0.3 V
Output voltage range . . . . . −0.3 V to V_{CC} +0.3 V
Operating free-air temperature range: ADC080_I . . . . . −40°C to 85°C
ADC080_C . . . . . 0°C to 70°C
Storage temperature range . . . . . −65°C to 150°C
Lead temperature 1,6 mm (1/16 inch) from case for 10 seconds . . . . . 260°C

NOTE 1: All voltage values are with respect to digital ground (DGTL GND) with DGTL GND and ANLG GND connected together unless otherwise noted.

recommended operating conditions

		MIN	NOM	MAX	UNIT
Supply voltage, V_{CC}		4.5	5	6.3	V
Analog input voltage (see Note 2)		−0.05		V_{CC}+0.05	V
Voltage at REF/2 (see Note 3), $V_{REF/2}$		0.25	2.5		V
High-level input voltage at $\overline{CS}$, $\overline{RD}$, or $\overline{WR}$, V_{IH}		2		15	V
Low-level input voltage at $\overline{CS}$, $\overline{RD}$, or $\overline{WR}$, V_{IL}				0.8	V
Analog ground voltage (see Note 4)		−0.05	0	1	V
Clock input frequency (see Note 5), f_{clock}		100	640	1460	kHz
Duty cycle for f_{clock} above 640 kHz (see Note 5)		40%		60%	
Pulse duration, clock input (high or low) for f_{clock} below 640 kHz, $t_{w(CLK)}$		275	781		ns
Pulse duration, $\overline{WR}$ input low, $t_{w(WR)}$		100			ns
Operating free-air temperature, T_A	ADC080_I	−40		85	°C
	ADC080_C	0		70	°C

NOTES: 2. When the differential input voltage ($V_{I+} - V_{I-}$) is less than or equal to 0 V, the output code is 0000 0000.
3. The internal reference voltage is equal to the voltage applied to REF/2 or approximately equal to one-half of the V_{CC} when REF/2 is left open. The voltage at REF/2 should be one-half the full-scale differential input voltage between the analog inputs. Thus, the differential input voltage range when REF/2 is open and V_{CC} = 5 V is 0 V to 5 V. $V_{REF/2}$ for an input voltage range from 0.5 V to 3.5 V (full-scale differential voltage of 3 V) is 1.5 V.
4. These values are with respect to DGTL GND.
5. Total unadjusted error is guaranteed only at an f_{clock} of 640 kHz with a duty cycle of 40% to 60% (pulse duration 625 ns to 937 ns). For frequencies above this limit or pulse duration below 625 ns, error may increase. The duty cycle limits should be observed for an f_{clock} greater than 640 kHz. Below 640 kHz, this duty cycle limit can be exceeded provided $t_{w(CLK)}$ remains within limits.

electrical characteristics over recommended operating free-air temperature range, V_{CC} = 5 V, f_{clock} = 640 kHz, $V_{REF/2}$ = 2.5 V (unless otherwise noted)

PARAMETER			TEST CONDITIONS		MIN	TYP†	MAX	UNIT
V_{OH}	High-level output voltage	All outputs	V_{CC} = 4.75 V,	I_{OH} = −360 μA	2.4			V
		DB and $\overline{INTR}$	V_{CC} = 4.75 V,	I_{OH} = −10 μA	4.5			
V_{OL}	Low-level output voltage	Data outputs	V_{CC} = 4.75 V,	I_{OL} = 1.6 mA			0.4	V
		$\overline{INTR}$ output	V_{CC} = 4.75 V,	I_{OL} = 1 mA			0.4	
		CLK OUT	V_{CC} = 4.75 V,	I_{OL} = 360 μA			0.4	
V_{T+}	Clock positive-going threshold voltage				2.7	3.1	3.5	V
V_{T-}	Clock negative-going threshold voltage				1.5	1.8	2.1	V
$V_{T+} - V_{T-}$	Clock input hysteresis				0.6	1.3	2	V
I_{IH}	High-level input current					0.005	1	μA
I_{IL}	Low-level input current					−0.005	−1	μA
I_{OZ}	Off-state output current		V_O = 0				−3	μA
			V_O = 5 V				3	
I_{OHS}	Short-current output current	Output high	V_O = 0,	T_A = 25°C	−4.5	−6		mA
I_{OLS}	Short-circuit output current	Output low	V_O = 5 V,	T_A = 25°C	9	16		mA
I_{CC}	Supply current plus reference current		$V_{REF/2}$ = open, $\overline{CS}$ = 5 V	T_A = 25°C,		1.1	1.8	mA
$R_{REF/2}$	Input resistance to reference ladder		See Note 6		2.5	8		kΩ
C_i	Input capacitance (control)					5	7.5	pF
C_o	Output capacitance (DB)					5	7.5	pF

NOTE 6: Resistance is calculated from the current drawn from a 5-volt supply applied to pins 8 and 9.

operating characteristics over recommended operating free-air temperature, V_{CC} = 5 V, $V_{REF/2}$ = 2.5 V, f_{clock} = 640 kHz (unless otherwise noted)

PARAMETER			TEST CONDITIONS		MIN	TYP†	MAX	UNIT
	Supply-voltage-variation error		V_{CC} = 4.5 V to 5.5 V,	See Note 7		±1/16	±1/8	LSB
	Total adjusted error	ADC0803	With full-scale adjust,	See Notes 7 and 8			±1/4	LSB
							±1/2	
	Total unadjusted error	ADC0805	$V_{REF/2}$ = 2.5 V,	See Notes 7 and 8			±1/2	LSB
			$V_{REF/2}$ open,	See Notes 7 and 8			±1	
	DC common-mode error		See Notes 7 and 8			±1/16	±1/8	LSB
t_{en}	Output enable time @ 25°C		T_A = 25°C,	C_L = 100 pF		135	200	ns
t_{dis}	Output disable time @ 25°C,		T_A = 25°C, C_L = 10 pF,	R_L = 10 kΩ		125	200	ns
$t_{d(INTR)}$	Delay time to reset $\overline{INTR}$ @ 25°C		T_A = 25°C			300	450	ns
t_{conv}	Conversion cycle time @ 25°C		f_{clock} = 100 kHz to 1.46 MHz, T_A = 25°C,	See Note 9	66		73	clock cycles
CR	Free-running conversion rate		$\overline{INTR}$ connected to $\overline{WR}$,	$\overline{CS}$ at 0 V			8770	conv/s

†All typical values are at T_A = 25°C.

NOTES: 7. These parameters are guaranteed over the recommended analog input voltage range.

8. All errors are measured with reference to an ideal straight line through the end-points of the analog-to-digital transfer characteristic.

9. Although internal conversion is completed in 64 clock periods, a $\overline{CS}$ or $\overline{WR}$ low-to-high transition is followed by 1 to 8 clock periods before conversion starts. After conversion is completed, part of another clock period is required before a high-to-low transition of $\overline{INTR}$ completes the cycle.

PARAMETER MEASUREMENT INFORMATION

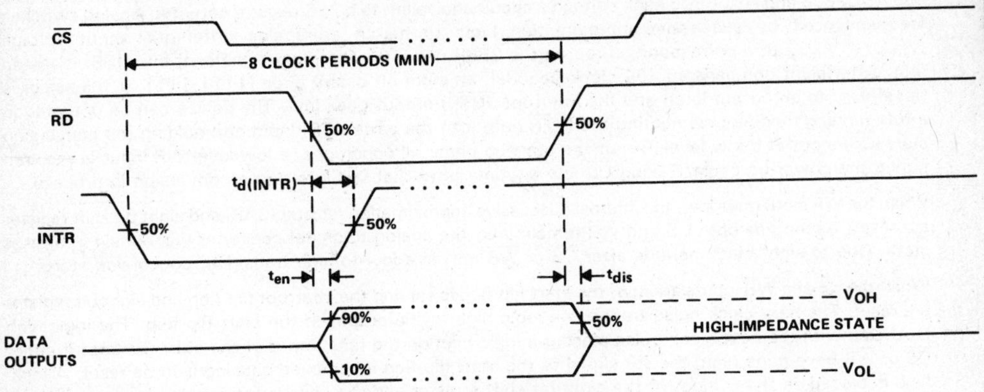
CS
8 CLOCK PERIODS (MIN)
RD
50%
50%
td(INTR)
INTR
50%
50%
ten
tdis
VOH
90%
50%
HIGH-IMPEDANCE STATE
DATA
OUTPUTS
10%
VOL

READ OPERATION TIMING DIAGRAM

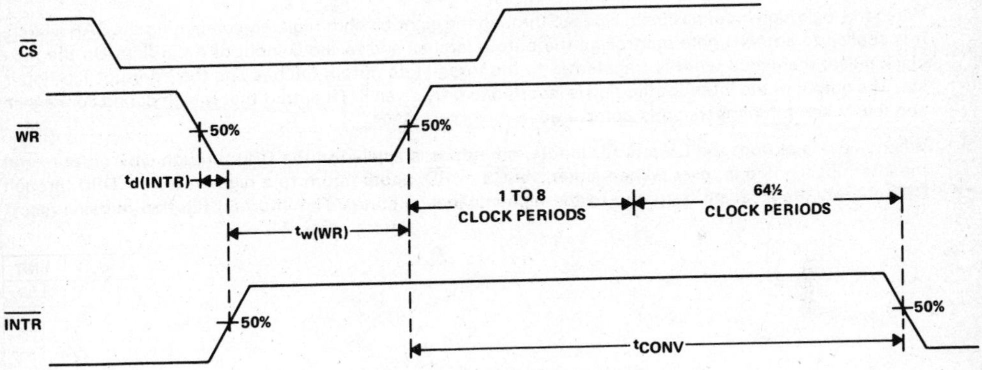
CS
WR
50%
50%
td(INTR)
1 TO 8
CLOCK PERIODS
64½
CLOCK PERIODS
tw(WR)
INTR
50%
50%
tCONV

WRITE OPERATION TIMING DIAGRAM

PRINCIPLES OF OPERATION

The ADC0803 and ADC0805 each contain a circuit equivalent to a 256-resistor network. Analog switches are sequenced by successive-approximation logic to match an analog differential input voltage ($V_{in+} - V_{in-}$) to a corresponding tap on the 256R network. The most significant bit (MSB) is tested first. After eight comparisons (64 clock periods), an eight-bit binary code (1111 1111 = full scale) is transferred to an output latch and the interrupt ($\overline{INTR}$) output goes low. The device can be operated in a free-running mode by connecting the $\overline{INTR}$ output to the write ($\overline{WR}$) input and holding the conversion start ($\overline{CS}$) input at a low level. To ensure start-up under all conditions, a low-level $\overline{WR}$ input is required during the power-up cycle. Taking $\overline{CS}$ low anytime after that will interrupt a conversion in process.

When the $\overline{WR}$ input goes low, the internal successive approximation register (SAR) and eight bit shift register are reset. As long as both $\overline{CS}$ and $\overline{WR}$ remain low, the analog-to-digital converter will remain in a reset state. One to eight clock periods after $\overline{CS}$ or $\overline{WR}$ makes a low-to-high transition, conversion starts.

When the $\overline{CS}$ and $\overline{WR}$ inputs are low, the start flip-flop is set and the interrupt flip-flop and eight bit register are reset. The next clock pulse transfers a logic high to the output of the start flip-flop. The logic high is ANDed with the next clock pulse placing a logic high on the reset input of the start flip-flop. If either $\overline{CS}$ or $\overline{WR}$ have gone high, the set signal to the start flip-flop is removed causing it to be reset. A logic high is placed on the D input of the eight-bit shift register and the conversion process is started. If the $\overline{CS}$ and $\overline{WR}$ inputs are still low, the start flip-flop, the eight-bit shift register, and the SAR remain reset. This action allows for wide $\overline{CS}$ and $\overline{WR}$ inputs with conversion starting from one to eight clock periods after one of the inputs goes high.

When the logic high input has been clocked through the eight-bit shift register, completing the SAR search, it is applied to an AND gate controlling the output latches and to the D input of a flip-flop. On the next clock pulse, the digital word is transferred to the three-state output latches and the interrupt flip-flop is set. The output of the interrupt flip-flop is inverted to provide an $\overline{INTR}$ output that is high during conversion and low when the conversion is completed.

When a low is at both the $\overline{CS}$ and $\overline{RD}$ inputs, an output is applied to the DB0 through DB7 outputs and the interrupt flip-flop is reset. When either the $\overline{CS}$ or $\overline{RD}$ inputs return to a high state, the DB0 through DB7 outputs are disabled (returned to the high-impedance state). The interrupt flip-flop remains reset.

TEXAS INSTRUMENTS
POST OFFICE BOX 655012 • DALLAS, TEXAS 75265

8-BIT ANALOG-TO-DIGITAL CONVERTER WITH DIFFERENTIAL INPUTS

D2755, OCTOBER 1983—REVISED SEPTEMBER 1986

- **8-Bit Resolution**
- **Ratiometric Conversion**
- **100 μs Conversion Time**
- **135 ns Access Time**
- **No Zero Adjust Requirement**
- **On-Chip Clock Generator**
- **Single 5-Volt Power Supply**
- **Operates with Microprocessor or as Stand-Alone**
- **Designed to be Interchangeable with National Semiconductor and Signetics ADC0804**

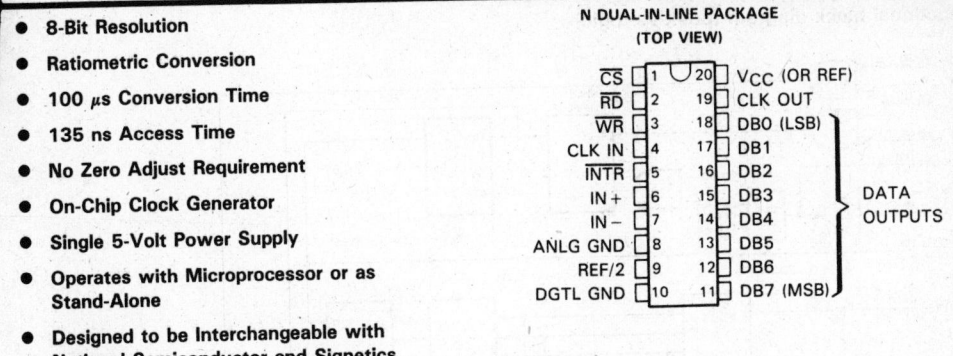

N DUAL-IN-LINE PACKAGE

(TOP VIEW)

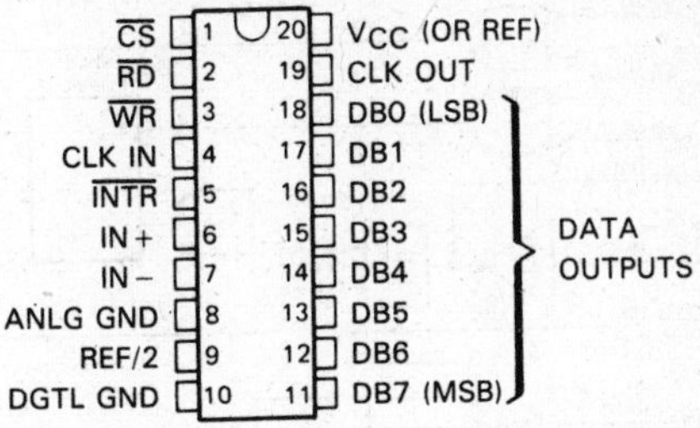
CS 1
RD 2
WR 3
CLK IN 4
INTR 5
IN+ 6
IN− 7
ANLG GND 8
REF/2 9
DGTL GND 10
20 VCC (OR REF)
19 CLK OUT
18 DB0 (LSB)
17 DB1
16 DB2
15 DB3
14 DB4
13 DB5
12 DB6
11 DB7 (MSB)
DATA OUTPUTS

description

The ADC0804 is a CMOS 8-bit successive-approximation analog-to-digital converter that uses a modified potentiometric (256R) ladder. The ADC0804 is designed to operate from common microprocessor control buses, with the three-state output latches driving the data bus. The ADC0804 can be made to appear to the microprocessor as a memory location or an I/O port. Detailed information on interfacing to most popular microprocessors is readily available from the factory.

A differential analog voltage input allows increased common-mode rejection and offset of the zero-input analog voltage value. Although a reference input (REF/2) is available to allow 8-bit conversion over smaller analog voltage spans or to make use of an external reference, ratiometric conversion is possible with the REF/2 input open. Without an external reference, the conversion takes place over a span from V_{CC} to analog ground (ANLG GND). The ADC0804 can operate with an external clock signal or, with an additional resistor and capacitor, can operate using an on-chip clock generator.

The ADC0804I is characterized for operation from −40 °C to 85 °C. The ADC0804C is characterized for operation from 0 °C to 70 °C.

functional block diagram (positive logic)

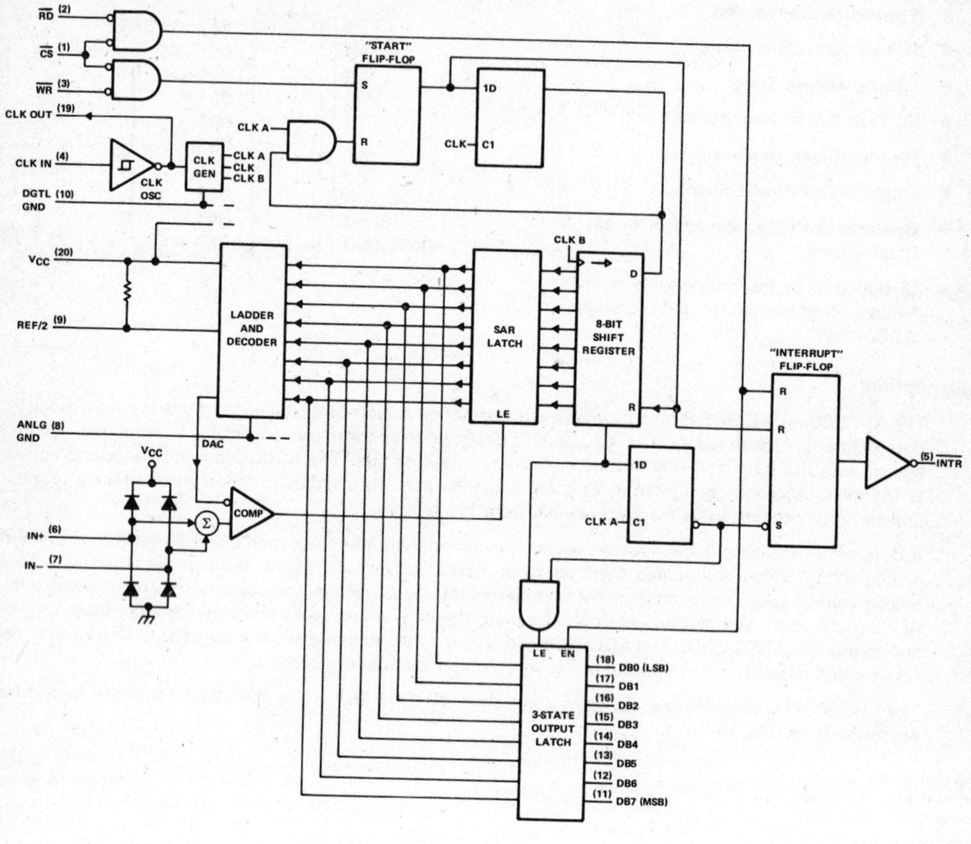
RD (2)
CS (1)
WR (3)
CLK OUT (19)
CLK IN (4)
DGTL (10) GND
VCC (20)
REF/2 (9)
ANLG (8) GND
IN+ (6)
IN− (7)
CLK OSC
CLK GEN
CLK A
CLK
CLK B
"START" FLIP-FLOP
S
R
1D
C1
LADDER AND DECODER
SAR LATCH
LE
8-BIT SHIFT REGISTER
D
DAC
COMP
Σ
"INTERRUPT" FLIP-FLOP
(5) INTR
LE EN
3-STATE OUTPUT LATCH
(18) DB0 (LSB)
(17) DB1
(16) DB2
(15) DB3
(14) DB4
(13) DB5
(12) DB6
(11) DB7 (MSB)

absolute maximum ratings over operating free-air temperature range (unless otherwise noted)

Supply voltage, V_{CC} (see Note 1)	6.5 V
Input voltage range $\overline{CS}$, $\overline{RD}$, $\overline{WR}$	−0.3 V to 18 V
other inputs	−0.3 V to V_{CC} + 0.3 V
Output voltage range	−0.3 V to V_{CC} + 0.3 V
Operating free-air temperature range: ADC0804I	−40°C to 85°C
ADC0804C	0°C to 70°C
Storage temperature range	−65°C to 150°C
Lead temperature 1,6 mm (1/16 inch) from case for 10 seconds	260°C

NOTE 1: All voltage values are with respect to digital ground (DGTL GND) with DGTL GND and ANLG GND connected together (unless otherwise noted).

recommended operating conditions

		MIN	NOM	MAX	UNIT
Supply voltage, V_{CC}		4.5	5	6.3	V
Voltage at REF/2, $V_{REF/2}$ (see Note 2)		0.25	2.5		V
High-level input voltage at $\overline{CS}$, $\overline{RD}$, or $\overline{WR}$, V_{IH}		2		15	V
Low-level input voltage at $\overline{CS}$, $\overline{RD}$, or $\overline{WR}$, V_{IL}				0.8	V
Analog ground voltage (see Note 3)		−0.05	0	1	V
Analog input voltage (see Note 4)		−0.05		V_{CC}+0.05	V
Clock input frequency, f_{clock} (see Note 5)		100	640	1460	kHz
Duty cycle for f_{clock} ≥ 640 kHz (see Note 5)		40		60	%
Pulse duration clock input (high or low) for f_{clock} < 640 kHz, $t_{w(CLK)}$ (see Note 5)		275	781		ns
Pulse duration, $\overline{WR}$ input low (start conversion), $t_{w(WR)}$		100			ns
Operating free-air temperature, T_A	ADC0804I	−40		85	°C
	ADC0804C	0		70	°C

NOTES: 2. The internal reference voltage is equal to the voltage applied to REF/2, or approximately equal to one-half of the V_{CC} when REF/2 is left open. The voltage at REF/2 should be one-half the full-scale differential input voltage between the analog inputs. Thus, the differential input voltage when REF/2 is open and V_{CC} = 5 V is 0 to 5 V. VREF/2 for an input voltage range from 0.5 V to 3.5 V (full-scale differential voltage of 3 V) is 1.5 V.

3. These values are with respect to DGTL GND.
4. When the differential input voltage ($V_{IN+} - V_{in-}$) is less than or equal to 0 V, the output code is 0000 0000.
5. Total unadjusted error is guaranteed only at an f_{clock} of 640 kHz with a duty cycle of 40% to 60% (pulse duration 625 ns to 937 ns). For frequencies above this limit or pulse duration below 625 ns, error may increase. The duty cycle limits should be observed for an f_{clock} greater than 640 kHz. Below 640 kHz, this duty cycle limit can be exceeded provided $t_{w(CLK)}$ remains within limits.

electrical characteristics over recommended operating free-air temperature range, V_{CC} = 5 V, f_{clock} = 640 kHz, REF/2 = 2.5 V (unless otherwise noted)

PARAMETER			TEST CONDITIONS	MIN	TYP†	MAX	UNIT
V_{OH}	High-level output voltage	All outputs	V_{CC} = 4.75 V, I_{OH} = −360 μA	2.4			V
		DB and $\overline{INTR}$	V_{CC} = 4.75 V, I_{OH} = −10 μA	4.5			V
V_{OL}	Low-level output voltage	Data outputs	V_{CC} = 4.75 V, I_{OL} = 1.6 mA			0.4	V
		$\overline{INTR}$ output	V_{CC} = 4.75 V, I_{OL} = 1 mA			0.4	V
		CLK OUT	V_{CC} = 4.75 V, I_{OL} = 360 μA			0.4	V
V_{T+}	Clock positive-going threshold voltage			2.7	3.1	3.5	V
V_{T-}	Clock negative-going threshold voltage			1.5	1.8	2.1	V
$V_{T+} - V_{T-}$	Clock input hysteresis			0.6	1.3	2	V
I_{IH}	High-level input current				0.005	1	μA
I_{IL}	Low-level input current				−0.005	−1	μA
I_{OZ}	Off-state output current		V_O = 0			−3	μA
			V_O = 5 V			3	μA
I_{OHS}	Short-circuit output current	Output high	V_O = 0, T_A = 25°C	−4.5	−6		mA
I_{OLS}	Short-circuit output current	Output low	V_O = 5 V, T_A = 25°C	9	16		mA
I_{CC}	Supply current plus reference current		REF/2 open, $\overline{CS}$ at 5 V, T_A = 25°C		1.9	2.5	mA
$R_{REF/2}$	Input resistance to reference ladder		See Note 6	1	1.3		kΩ
C_i	Input capacitance (control)				5	7.5	pF
C_o	Output capacitance (DB)				5	7.5	pF

NOTE 6: The resistance is calculated from the current drawn from a 5-V supply applied to pins 8 and 9.

operating characteristics over recommended operating free-air temperature, V_{CC} = 5 V, $V_{REF/2}$ = 2.5 V, f_{clock} = 640 kHz (unless otherwise noted)

PARAMETER		TEST CONDITIONS	MIN	TYP†	MAX	UNIT
	Supply-voltage-variation error (See Notes 2 and 7)	V_{CC} = 4.5 V to 5.5 V		±1/16	±1/8	LSB
	Total unadjusted error (See Notes 7 and 8)	$V_{REF/2}$ = 2.5 V			±1	LSB
	DC common-mode error (See Note 8)			±1/16	±1/8	LSB
t_{en}	Output enable time	C_L = 100 pF		135	200	ns
t_{dis}	Output disable time	C_L = 10 pF, R_L = 10 kΩ		125	200	ns
$t_{d(INTR)}$	Delay time to reset $\overline{INTR}$			300	450	ns
t_{conv}	Conversion cycle time (See Note 9)	f_{clock} = 100 kHz to 1.46 MHz	65½		72½	clock cycles
	Conversion time		103		114	μs
CR	Free-running conversion rate	$\overline{INTR}$ connected to $\overline{WR}$, $\overline{CS}$ at 0 V			8827	conv/s

† All typical values are at T_A = 25°C.

NOTES: 2. The internal reference voltage is equal to the voltage applied to REF/2, or approximately equal to one-half of the V_{CC} when REF/2 is left open. The voltage at REF/2 should be one-half the full-scale differential input voltage between the analog inputs. Thus, the differential input voltage when REF/2 is open and V_{CC} = 5 V is 0 to 5 V. $V_{REF/2}$ for an input voltage range from 0.5 V to 3.5 V (full-scale differential voltage of 3 V) is 1.5 V.

7. These parameters are guaranteed over the recommended analog input voltage range.
8. All errors are measured with reference to an ideal straight line through the end-points of the analog-to-digital transfer characteristic.
9. Although internal conversion is completed in 64 clock periods, a $\overline{CS}$ or $\overline{WR}$ low-to-high transition is followed by 1 to 8 clock periods before conversion starts. After conversion is completed, part of another clock period is required before a high-to-low transition of $\overline{INTR}$ completes the cycle.

timing diagrams

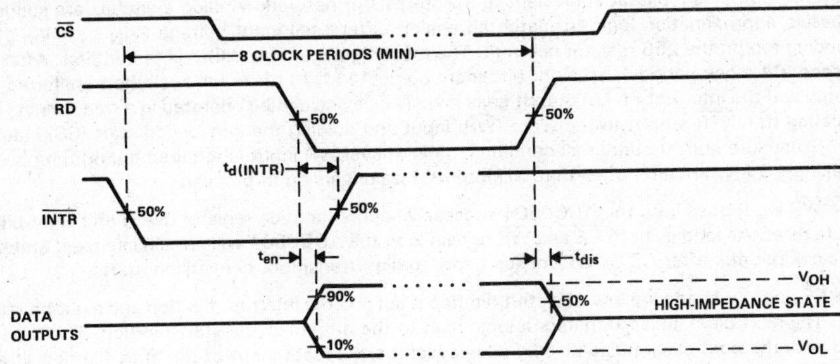
CS
8 CLOCK PERIODS (MIN)
RD
50%
50%
td(INTR)
INTR
50%
50%
ten
tdis
VOH
90%
50%
HIGH-IMPEDANCE STATE
DATA
OUTPUTS
10%
VOL

READ OPERATION TIMING DIAGRAM

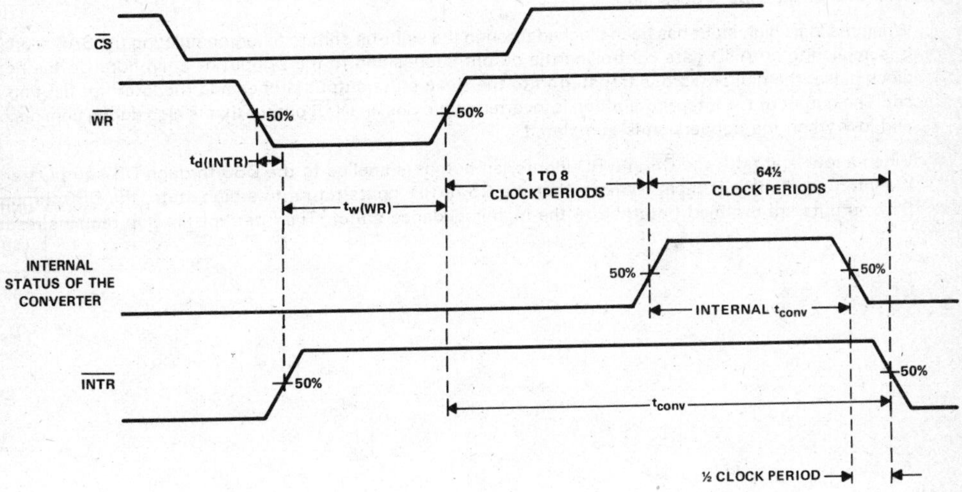
CS
WR
50%
50%
td(INTR)
1 TO 8
CLOCK PERIODS
64½
CLOCK PERIODS
tw(WR)
INTERNAL
STATUS OF THE
CONVERTER
50%
50%
INTERNAL tconv
INTR
50%
50%
tconv
½ CLOCK PERIOD

WRITE OPERATION TIMING DIAGRAM

PRINCIPLES OF OPERATION

The ADC0804 contains a circuit equivalent to a 256-resistor network. Analog switches are sequenced by successive approximation logic to match an analog differential input voltage ($V_{IN+} - V_{in-}$) to a corresponding tap on the 256-resistor network. The most-significant bit (MSB) is tested first. After eight comparisons (64 clock periods), an eight-bit binary code (1111 1111 = full scale) is transferred to an output latch and the interrupt ($\overline{\text{INTR}}$) output goes low. The device can be operated in a free-running mode by connecting the $\overline{\text{INTR}}$ output to the write ($\overline{\text{WR}}$) input and holding the conversion start ($\overline{\text{CS}}$) input at a low level. To ensure start-up under all conditions, a low-level $\overline{\text{WR}}$ input is required during the power-up cycle. Taking $\overline{\text{CS}}$ low anytime after that will interrupt a conversion in process.

When the $\overline{\text{WR}}$ input goes low, the ADC0804 successive approximation register (SAR) and eight-bit shift register are reset. As long as both $\overline{\text{CS}}$ and $\overline{\text{WR}}$ remain low, the ADC0804 will remain in a reset state. One to eight clock periods after $\overline{\text{CS}}$ or $\overline{\text{WR}}$ makes a low-to-high transition, conversion starts.

When the $\overline{\text{CS}}$ and $\overline{\text{WR}}$ inputs are low, the start flip-flop is set and the interrupt flip-flop and eight-bit register are reset. The next clock pulse transfers a logic high to the output of the start flip-flop. The logic high is ANDed with the next clock pulse placing a logic high on the reset input of the start flip-flop. If either $\overline{\text{CS}}$ or $\overline{\text{WR}}$ have gone high, the set signal to the start flip-flop is removed causing it to be reset. A logic high is placed on the D input of the eight-bit shift register and the conversion process is started. If the $\overline{\text{CS}}$ and $\overline{\text{WR}}$ inputs are still low, the start flip-flop, the eight-bit shift register, and the SAR remain reset. This action allows for wide $\overline{\text{CS}}$ and $\overline{\text{WR}}$ inputs with conversion starting from one to eight clock periods after one of the inputs goes high.

When the logic high input has been clocked through the eight-bit shift register, completing the SAR search, it is applied to an AND gate controlling the output latches and to the D input of a flip-flop. On the next clock pulse, the digital word is transferred to the three-state output latches and the interrupt flip-flop is set. The output of the interrupt flip-flop is inverted to provide an $\overline{\text{INTR}}$ output that is high during conversion and low when the conversion is completed.

When a low is at both the $\overline{\text{CS}}$ and $\overline{\text{RD}}$ inputs, an output is applied to the DB0 through DB7 outputs and the interrupt flip-flop is reset. When either the $\overline{\text{CS}}$ or $\overline{\text{RD}}$ inputs return to a high state, the DB0 through DB7 outputs are disabled (returned to the high-impedance state). The interrupt flip-flop remains reset.

- **Total Unadjusted Error . . . ±0.75 LSB Max for ADC0808 and ±1.25 LSB Max for ADC0809**
- **Resolution of 8 Bits**
- **100 μs Conversion Time**
- **Ratiometric Conversion**
- **Guaranteed Monotonicity**
- **No Missing Codes**
- **Easy Interface with Microprocessors**
- **Latched 3-State Outputs**
- **Latched Address Inputs**
- **Single 5-Volt Supply**
- **Low Power Consumption**
- **Designed to be Interchangeable with National Semiconductor ADC0808, ADC0809**

N
DUAL-IN-LINE PACKAGE
(TOP VIEW)

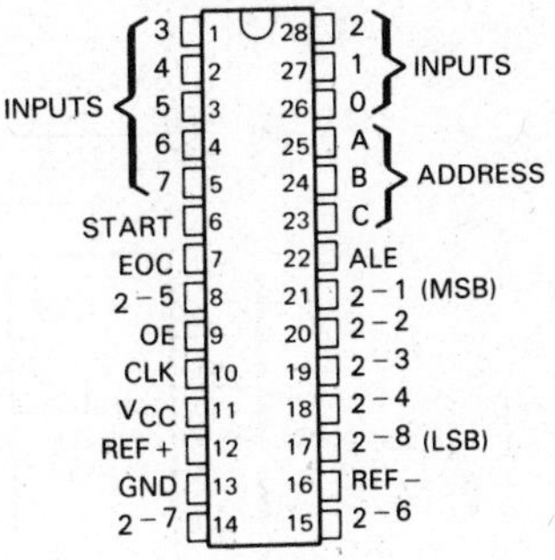
INPUTS
3
4
5
6
7
START
EOC
2 − 5
OE
CLK
V_{CC}
REF +
GND
2 − 7
1
2
3
4
5
6
7
8
9
10
11
12
13
14
28
27
26
25
24
23
22
21
20
19
18
17
16
15
2
1
0
INPUTS
A
B
C
ADDRESS
ALE
2 − 1 (MSB)
2 − 2
2 − 3
2 − 4
2 − 8 (LSB)
REF −
2 − 6

FN PACKAGE
(TOP VIEW)

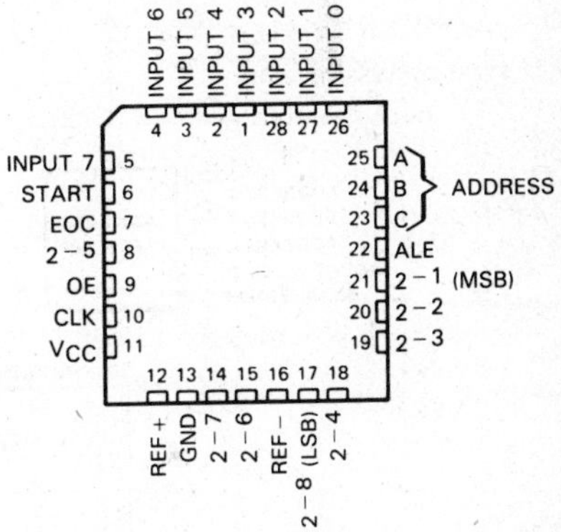
INPUT 6
INPUT 5
INPUT 4
INPUT 3
INPUT 2
INPUT 1
INPUT 0
4 3 2 1 28 27 26
INPUT 7 5
START 6
EOC 7
2 − 5 8
OE 9
CLK 10
V_{CC} 11
25 A
24 B
23 C
ADDRESS
22 ALE
21 2 − 1 (MSB)
20 2 − 2
19 2 − 3
12 13 14 15 16 17 18
REF +
GND
2 − 7
2 − 6
REF −
2 − 8 (LSB)
2 − 4

description

The ADC0808 and ADC0809 are monolithic CMOS devices with an 8-channel multiplexer, an 8-bit analog-to-digital (A/D) converter, and microprocessor-compatible control logic. The 8-channel multiplexer can be controlled by a microprocessor through a 3-bit address decoder with address load to select any one of eight single-ended analog switches connected directly to the comparator. The 8-bit A/D converter uses the successive-approximation conversion technique featuring a high-impedance threshold detector, a switched-capacitor array, a sample-and-hold, and a successive-approximation register (SAR). Detailed information on interfacing to most popular microprocessors is readily available from the factory.

The comparison and converting methods used eliminate the possibility of missing codes, nonmonotonicity, and the need for zero or full-scale adjustment. Also featured are latched 3-state outputs from the SAR and latched inputs to the multiplexer address decoder. The single 5-volt supply and low power requirements make the ADC0808 and ADC0809 especially useful for a wide variety of applications. Ratiometric conversion is made possible by access to the reference voltage input terminals.

The ADC0808 and ADC0809 are characterized for operation from −40 °C to 85 °C.

functional block diagram (positive logic)

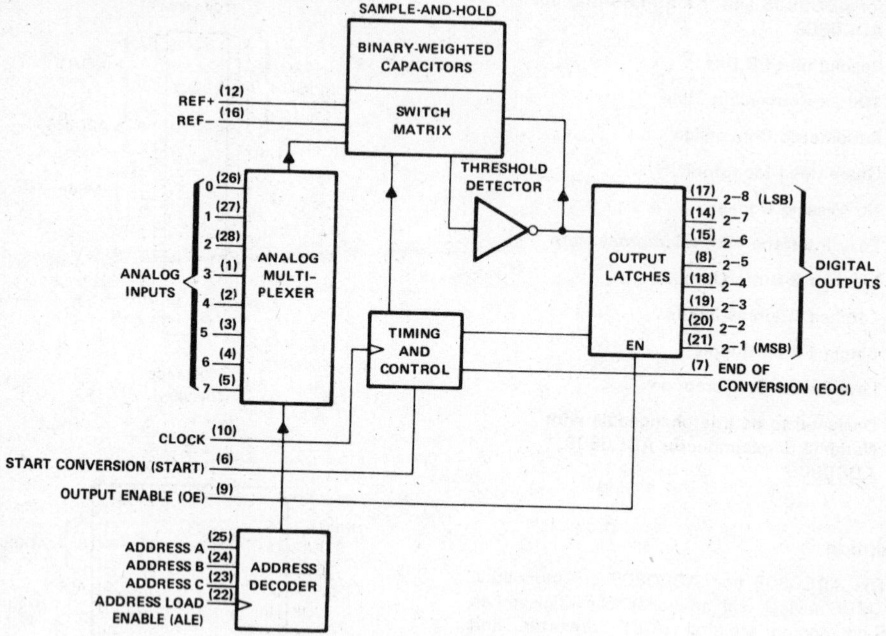
SAMPLE-AND-HOLD
BINARY-WEIGHTED CAPACITORS
SWITCH MATRIX
REF+ (12)
REF– (16)
THRESHOLD DETECTOR
ANALOG INPUTS
0 (26)
1 (27)
2 (28)
3 (1)
4 (2)
5 (3)
6 (4)
7 (5)
ANALOG MULTI-PLEXER
TIMING AND CONTROL
OUTPUT LATCHES
EN
(17) 2−8 (LSB)
(14) 2−7
(15) 2−6
(8) 2−5
(18) 2−4
(19) 2−3
(20) 2−2
(21) 2−1 (MSB)
DIGITAL OUTPUTS
(7) END OF CONVERSION (EOC)
CLOCK (10)
START CONVERSION (START) (6)
OUTPUT ENABLE (OE) (9)
ADDRESS A (25)
ADDRESS B (24)
ADDRESS C (23)
ADDRESS LOAD ENABLE (ALE) (22)
ADDRESS DECODER

MULTIPLEXER FUNCTION TABLE

INPUTS				SELECTED ANALOG CHANNEL
ADDRESS			ADDRESS STROBE	
C	B	A		
L	L	L	↑	0
L	L	H	↑	1
L	H	L	↑	2
L	H	H	↑	3
H	L	L	↑	4
H	L	H	↑	5
H	H	L	↑	6
H	H	H	↑	7

H = high level, L = low level
↑ = low-to-high transition

operating sequence

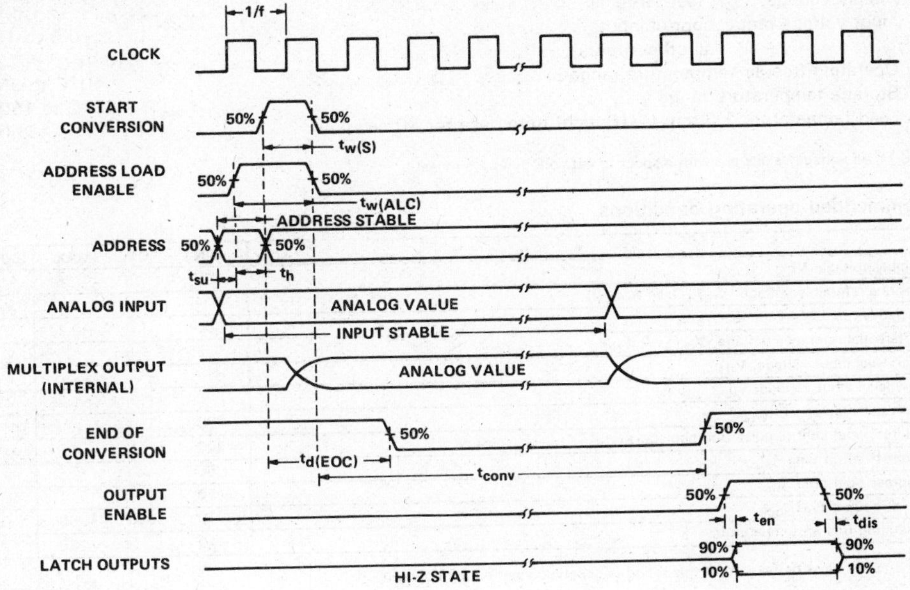
1/f
CLOCK
START CONVERSION
50%
50%
$t_{w(S)}$
ADDRESS LOAD ENABLE
50%
50%
$t_{w(ALC)}$
ADDRESS STABLE
ADDRESS
50%
50%
t_{su}
t_h
ANALOG INPUT
ANALOG VALUE
INPUT STABLE
MULTIPLEX OUTPUT (INTERNAL)
ANALOG VALUE
END OF CONVERSION
50%
50%
$t_{d(EOC)}$
t_{conv}
OUTPUT ENABLE
50%
50%
t_{en}
t_{dis}
LATCH OUTPUTS
90%
10%
90%
10%
HI-Z STATE

absolute maximum ratings over operating free-air temperature range (unless otherwise noted)

Supply voltage, V_{CC} (see Note 1) . . . . 6.5 V
Input voltage range: control inputs . . . . −0.3 to 15 V
all other inputs . . . . −0.3 V to V_{CC} + 0.3 V
Operating free-air temperature range . . . . −40 °C to 85 °C
Storage temperature range . . . . −65 °C to 150 °C
Lead temperature 1,6 mm (1/16 inch) from case for 10 seconds . . . . 260 °C

NOTE 1: All voltage values are with respect to network ground terminal.

recommended operating conditions

	MIN	NOM	MAX	UNIT
Supply voltage, V_{CC}	4.5	5	6	V
Positive reference voltage, V_{ref+} (see Note 2)		V_{CC}	$V_{CC}+0.1$	V
Negative reference voltage, V_{ref-}		0	−0.1	V
Differential reference voltage, $V_{ref+} - V_{ref-}$		5		V
High-level input voltage, V_{IH}	$V_{CC}-1.5$			V
Low-level input voltage, V_{IL}			1.5	V
Start pulse duration, $t_{w(S)}$	200			ns
Address load control pulse duration, $t_{w(ALC)}$	200			ns
Address setup time, t_{su}	50			ns
Address hold time, t_h	50			ns
Clock frequency, f_{clock}	10	640	1280	kHz
Operating free-air temperature, T_A	−40		85	°C

NOTE 2: Care must be taken that this rating is observed even during power-up.

electrical characteristics over recommended operating free-air temperature range, V_{CC} = 4.75 V to 5.25 V (unless otherwise noted)

total device

PARAMETER			TEST CONDITIONS	MIN	TYP†	MAX	UNIT
V_{OH}	High-level output voltage		I_O = −360 µA	V_{CC}−0.4			V
V_{OL}	Low-level output voltage	Data outputs	I_O = 1.6 mA			0.45	V
		End of conversion	I_O = 1.2 mA			0.45	
I_{OZ}	Off-state (high-impedance-state) output current		V_O = V_{CC}			3	µA
			V_O = 0			−3	
I_I	Control input current at maximum input voltage		V_I = 15 V			1	µA
I_{IL}	Low-level control input current		V_I = 0			−1	µA
I_{CC}	Supply current		f_{clock} = 640 kHz		0.3	3	mA
C_i	Input capacitance, control inputs		T_A = 25°C		10	15	pF
C_o	Output capacitance, data outputs		T_A = 25°C		10	15	pF
	Resistance from pin 12 to pin 16				1000		kΩ

analog multiplexer

PARAMETER		TEST CONDITIONS		MIN	TYP†	MAX	UNIT
I_{on}	Channel on-state current (see Note 3)	V_I = 5 V,	f_{clock} = 640 kHz			2	µA
		V_I = 0,	f_{clock} = 640 kHz			−2	
I_{off}	Channel off-state current	V_{CC} = 5 V, T_A = 25°C	V_I = 5 V		10	200	nA
			V_I = 0		−10	−200	
		V_{CC} = 5 V	V_I = 5 V			1	µA
			V_I = 0			−1	

†Typical values are at V_{CC} = 5 V and T_A = 25°C.

NOTE 3: Channel on-state current is primarily due to the bias current into or out of the threshold detector, and it varies directly with clock frequency.

operating characteristics, T_A = 25°C, V_{CC} = V_{REF+} = 5 V, V_{REF-} = 0 V, f_{clock} = 640 kHz (unless otherwise noted)

	PARAMETER	TEST CONDITIONS	ADC0808			ADC0809			UNIT
			MIN	TYP†	MAX	MIN	TYP†	MAX	
k_{SVS}	Supply voltage sensitivity	V_{CC} = V_{ref+} = 4.75 V to 5.25 V, T_A = −40°C to 85°C, See Note 4		±0.05			±0.05		%/V
	Linearity error (see Note 5)			±0.25			±0.5		LSB
	Zero error (see Note 6)			±0.25			±0.25		LSB
	Total unadjusted error (See Note 7)	T_A = 25°C		±0.25	±0.5		±0.5		LSB
		T_A = −40°C to 85°C			±0.75			±1.25	
		T_A = 0°C to 70°C						±1	
t_{en}	Output enable time	C_L = 50 pF, R_L = 10 kΩ		80	250		80	250	ns
t_{dis}	Output disable time	C_L = 10 pF, R_L = 10 kΩ		105	250		105	250	ns
t_{conv}	Conversion time	See Note 8	90	100	116	90	100	116	µs
$t_{d(EOC)}$	Delay time, end of conversion output	See Notes 8 and 9	0		14.5	0		14.5	µs

†Typical values for all except supply voltage sensitivity are at V_{CC} = 5 V, and all are at T_A = 25°C.

NOTES:
4. Supply voltage sensitivity relates to the ability of an analog-to-digital converter to maintain accuracy as the supply voltage varies. The supply and V_{ref+} are varied together and the change in accuracy is measured with respect to full-scale.
5. Linearity error is the maximum deviation from a straight line through the end points of the A/D transfer characteristic.
6. Zero error is the difference between 00000000 and the converted output for zero input voltage; full-scale error is the difference between 11111111 and the converted output for full-scale input voltage.
7. Total unadjusted error is the maximum sum of linearity error, zero error, and full-scale error.
8. Refer to the operating sequence diagram.
9. For clock frequencies other than 640 kHz, $t_{d(EOC)}$ maximum is 8 clock periods plus 2 µs.

PRINCIPLES OF OPERATION

The ADC0808 and ADC0809 each consists of an analog signal multiplexer, an 8-bit successive-approximation converter, and related control and output circuitry.

multiplexer

The analog multiplexer selects 1 of 8 single-ended input channels as determined by the address decoder. Address load control loads the address code into the decoder on a low-to-high transition. The output latch is reset by the positive-going edge of the start pulse. Sampling also starts with the positive-going edge of the start pulse and lasts for 32 clock periods. The conversion process may be interrupted by a new start pulse before the end of 64 clock periods. The previous data will be lost if a new start of conversion occurs before the 64th clock pulse. Continuous conversion may be accomplished by connecting the End-of-Conversion output to the start input. If used in this mode an external pulse should be applied after power up to assure start up.

converter

The CMOS threshold detector in the successive-approximation conversion system determines each bit by examining the charge on a series of binary-weighted capacitors (Figure 1). In the first phase of the conversion process, the analog input is sampled by closing switch S_C and all S_T switches, and by simultaneously charging all the capacitors to the input voltage.

In the next phase of the conversion process, all S_T and S_C switches are opened and the threshold detector begins identifying bits by identifying the charge (voltage) on each capacitor relative to the reference voltage. In the switching sequence, all eight capacitors are examined separately until all 8 bits are identified, and then the charge-convert sequence is repeated. In the first step of the conversion phase, the threshold detector looks at the first capacitor (weight = 128). Node 128 of this capacitor is switched to the reference voltage, and the equivalent nodes of all the other capacitors on the ladder are switched to REF−. If the voltage at the summing node is greater than the trip-point of the threshold detector (approximately one-half the V_{CC} voltage), a bit is placed in the output register, and the 128-weight capacitor is switched to REF−. If the voltage at the summing node is less than the trip point of the threshold detector, this 128-weight capacitor remains connected to REF+ through the remainder of the capacitor-sampling (bit-counting) process. The process is repeated for the 64-weight capacitor, the 32-weight capacitor, and so forth down the line, until all bits are counted.

With each step of the capacitor-sampling process, the initial charge is redistributed among the capacitors. The conversion process is successive approximation, but relies on charge redistribution rather than a successive-approximation register (and reference DAC) to count and weigh the bits from MSB to LSB.

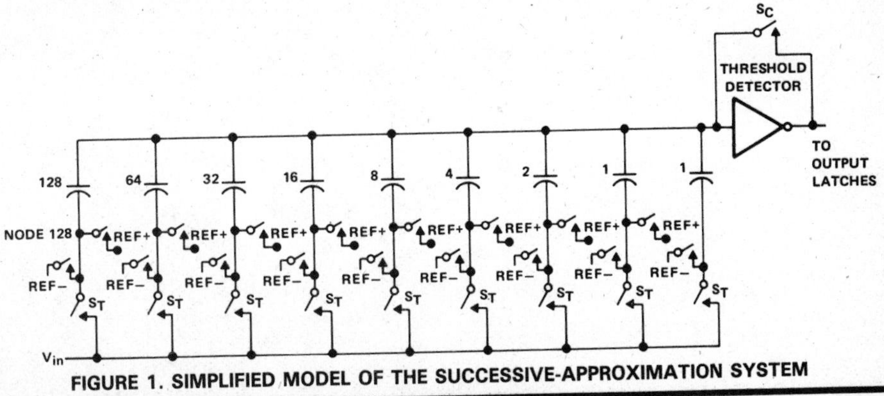
SC
THRESHOLD DETECTOR
TO OUTPUT LATCHES
128
64
32
16
8
4
2
1
1
NODE 128
REF+
REF−
ST
Vin

FIGURE 1. SIMPLIFIED MODEL OF THE SUCCESSIVE-APPROXIMATION SYSTEM

- Total Unadjusted Error . . . ±0.75 LSB Max
- Resolution of 8 Bits
- 100 μs Conversion Time
- Ratiometric Conversion
- Guaranteed Monotonicity
- No Missing Codes
- Easy Interface with Microprocessors
- Latched 3-State Outputs
- Latched Address Inputs
- Single 5-Volt Supply
- Low Power Consumption
- Designed to be Interchangeable with National Semiconductor ADC0808CJ

JD
DUAL-IN-LINE PACKAGE
(TOP VIEW)

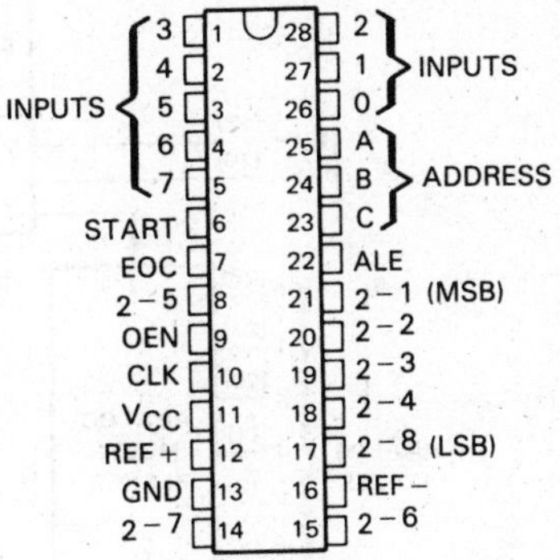
INPUTS
3 1
4 2
5 3
6 4
7 5
START 6
EOC 7
2−5 8
OEN 9
CLK 10
VCC 11
REF+ 12
GND 13
2−7 14
28 2
27 1
26 0
INPUTS
25 A
24 B
23 C
ADDRESS
22 ALE
21 2−1 (MSB)
20 2−2
19 2−3
18 2−4
17 2−8 (LSB)
16 REF−
15 2−6

FK PACKAGE
(TOP VIEW)

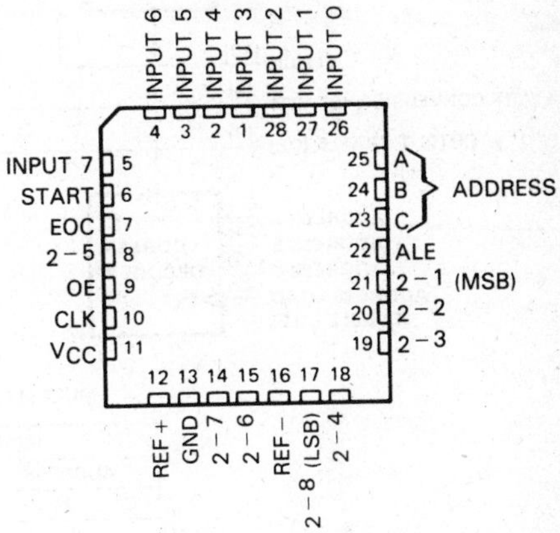
INPUT 6
INPUT 5
INPUT 4
INPUT 3
INPUT 2
INPUT 1
INPUT 0
4 3 2 1 28 27 26
INPUT 7 5
START 6
EOC 7
2−5 8
OE 9
CLK 10
VCC 11
25 A
24 B
23 C
ADDRESS
22 ALE
21 2−1 (MSB)
20 2−2
19 2−3
12 13 14 15 16 17 18
REF+
GND
2−7
2−6
REF−
2−8 (LSB)
2−4

description

The ADC0808M is a monolithic CMOS device with an 8-channel multiplexer, an 8-bit analog-to-digital (A/D) converter, and microprocessor-compatible control logic. The 8-channel multiplexer can be controlled by a microprocessor through a 3-bit address decoder with single-address load to select any one of eight single-ended analog switches connected directly to the comparator. The 8-bit A/D converter uses the successive-approximation conversion technique featuring a high-impedance threshold detector, a switched capacitor array, a sample-and-hold, and a successive-approximation register (SAR). Detailed information on interfacing to most popular microprocessors is readily available from the factory.

The comparison and converting methods used eliminate the possibility of missing codes, nonmonotonicity, and the need for zero or full-scale adjustment. Also featured are latched 3-state outputs from the SAR and latched inputs to the multiplexer address decoder. The single 5-volt supply and low power requirements make the ADC0808M especially useful for a wide variety of applications. Ratiometric conversion is made possible by access to the reference voltage input terminals.

The ADC0808M is characterized for operation over the full military temperature range of −55°C to 125°C.

functional block diagram (positive logic)

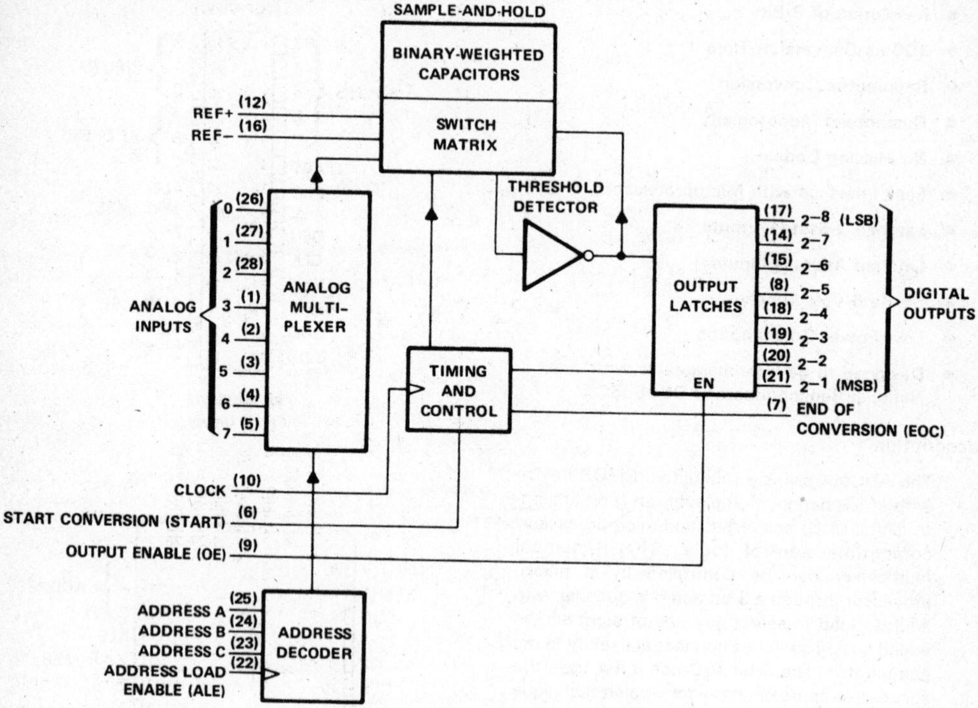
SAMPLE-AND-HOLD
BINARY-WEIGHTED CAPACITORS
SWITCH MATRIX
REF+ (12)
REF– (16)
THRESHOLD DETECTOR
ANALOG INPUTS
0 (26)
1 (27)
2 (28)
3 (1)
4 (2)
5 (3)
6 (4)
7 (5)
ANALOG MULTI-PLEXER
TIMING AND CONTROL
OUTPUT LATCHES
EN
(17) 2−8 (LSB)
(14) 2−7
(15) 2−6
(8) 2−5
(18) 2−4
(19) 2−3
(20) 2−2
(21) 2−1 (MSB)
DIGITAL OUTPUTS
(7) END OF CONVERSION (EOC)
CLOCK (10)
START CONVERSION (START) (6)
OUTPUT ENABLE (OE) (9)
ADDRESS A (25)
ADDRESS B (24)
ADDRESS C (23)
ADDRESS LOAD ENABLE (ALE) (22)
ADDRESS DECODER

MULTIPLEXER FUNCTION TABLE

INPUTS				SELECTED ANALOG CHANNEL
ADDRESS			ADDRESS STROBE	
C	B	A		
L	L	L	↑	0
L	L	H	↑	1
L	H	L	↑	2
L	H	H	↑	3
H	L	L	↑	4
H	L	H	↑	5
H	H	L	↑	6
H	H	H	↑	7

H = high level, L = low level
↑ = low-to-high transition

TEXAS INSTRUMENTS
POST OFFICE BOX 655012 • DALLAS, TEXAS 75265

operating sequence

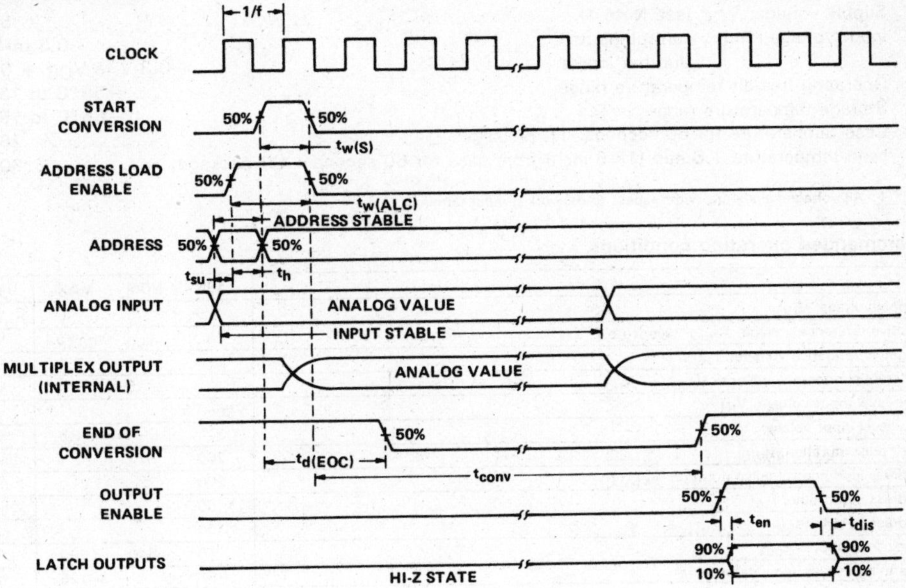
1/f
CLOCK
START CONVERSION
50%
50%
tw(S)
ADDRESS LOAD ENABLE
50%
50%
tw(ALC)
ADDRESS STABLE
ADDRESS
50%
50%
tsu
th
ANALOG INPUT
ANALOG VALUE
INPUT STABLE
MULTIPLEX OUTPUT (INTERNAL)
ANALOG VALUE
END OF CONVERSION
50%
50%
td(EOC)
tconv
OUTPUT ENABLE
50%
50%
ten
tdis
LATCH OUTPUTS
90%
10%
90%
10%
HI-Z STATE

absolute maximum ratings over operating free-air temperature range (unless otherwise noted)

Supply voltage, V_{CC} (see Note 1) .... 6.5 V
Input voltage range: control inputs .... −0.3 to 15 V
all other inputs .... −0.3 V to V_{CC} + 0.3 V
Operating free-air temperature range .... −55°C to 125°C
Storage temperature range .... −65°C to 150°C
Case temperature for 60 seconds: FK package .... 260°C
Lead temperature 1,6 mm (1/16 inch) from case for 60 seconds: JD package .... 300°C

NOTE 1: All voltage values are with respect to network ground terminal.

recommended operating conditions

	MIN	NOM	MAX	UNIT
Supply voltage, V_{CC}	4.5	5	6	V
Positive reference voltage, V_{ref+} (see Note 2)		V_{CC}	$V_{CC}+0.1$	V
Negative reference voltage, V_{ref-}		0	−0.1	V
Differential reference voltage, $V_{ref+} - V_{ref-}$		5		V
High-level input voltage, V_{IH}	$V_{CC}-1.5$			V
Low-level input voltage, V_{IL}			1.5	V
Start pulse duration, $t_{w(S)}$	200			ns
Address load control pulse duration, $t_{w(ALC)}$	200			ns
Address setup time, t_{su}	50			ns
Address hold time, t_h	50			ns
Clock frequency, f_{clock}	10	640	1280	kHz
Operating free-air temperature, T_A	−55		125	°C

NOTE 2: Care must be taken that this rating is observed even during power-up.

electrical characteristics over recommended operating free-air temperature range, V_{CC} = 4.5 V to 5.5 V (unless otherwise noted)

total device

PARAMETER			TEST CONDITIONS	MIN	TYP†	MAX	UNIT
V_{OH}	High-level output voltage		I_O = −360 μA	V_{CC}−0.4			V
V_{OL}	Low-level output voltage	Data outputs	I_O = 1.6 mA			0.45	V
		End of conversion	I_O = 1.2 mA			0.45	
I_{OZ}	Off-state (high-impedance-state) output current		V_O = V_{CC}			3	μA
			V_O = 0			−3	
I_I	Control input current at maximum input voltage		V_I = 15 V			1	μA
I_{IL}	Low-level control input current		V_I = 0			−1	μA
I_{CC}	Supply current		f_{clock} = 640 kHz		0.3	3	mA
C_i	Input capacitance, control inputs		T_A = 25°C		10		pF
C_o	Output capacitance, data outputs		T_A = 25°C		10		pF
	Resistance from pin 12 to pin 16				1000		kΩ

analog multiplexer

PARAMETER		TEST CONDITIONS		MIN	TYP†	MAX	UNIT
I_{on}	Channel on-state current (see Note 3)	V_I = V_{CC},	f_{clock} = 640 kHz			2	μA
		V_I = 0,	f_{clock} = 640 kHz			−2	
I_{off}	Channel off-state current	V_{CC} = 5 V, T_A = 25°C	V_I = 5 V		10	200	nA
			V_I = 0		−10	−200	
		V_{CC} = 5 V	V_I = 5 V			1	μA
			V_I = 0			−1	

† Typical values are at V_{CC} = 5 V and T_A = 25°C.

NOTE 3: Channel on-state current is primarily due to the bias current into or out of the threshold detector, and it varies directly with clock frequency.

operating characteristics, T_A = 25°C, V_{CC} = V_{REF+} = 5 V, V_{REF-} = 0 V, f_{clock} = 640 kHz (unless otherwise noted)

PARAMETER		TEST CONDITIONS	MIN	TYP†	MAX	UNIT
k_{SVS}	Supply voltage sensitivity	V_{CC} = V_{ref+} = 4.5 V to 5.5 V, T_A = −55°C to 125°C, See Note 4		±0.05		%/V
	Linearity error (see Note 5)			±0.25		LSB
	Zero error (see Note 6)			±0.25		LSB
	Total unadjusted error (see Note 7)	T_A = 25°C		±0.25	±0.5	LSB
		T_A = −55°C to 125°C			±0.75	
t_{PZL}	Output enable time to low level	See Figure 1		90	250	ns
t_{PZH}	Output enable time to high level	See Figure 1		150	360	ns
t_{dis}	Output disable time	See Figure 1		200	405	ns
t_{conv}	Conversion time	See Note 8 and 9 and Figure 1	90	100	116	μs
$t_{d(EOC)}$	Delay time, end of conversion output	See Notes 8 and 10 and Figure 1	0		14.5	μs

† Typical values for all except supply voltage sensitivity are at V_{CC} = 5 V, and all are at T_A = 25°C.

NOTES: 4. Supply voltage sensitivity relates to the ability of an analog-to-digital converter to maintain accuracy as the supply voltage varies. The supply and V_{ref+} are varied together and the change in accuracy is measured with respect to full-scale.
5. Linearity error is the maximum deviation from a straight line through the end points of the A/D transfer characteristic.
6. Zero error is the difference between 00000000 and the converted output for zero input voltage; full-scale error is the difference between 11111111 and the converted output for full-scale input voltage.
7. Total unadjusted error is the maximum sum of linearity error, zero error, and full-scale error.
8. Refer to the operating sequence diagram.
9. For clock frequencies other than 640 kHz, t_{conv} is 57 clock cycles minimum and 74 clock cycles maximum.
10. For clock frequencies other than 640 kHz, $t_{d(EOC)}$ maximum is 8 clock cycles plus 2 μs.

PARAMETER MEASUREMENT INFORMATION

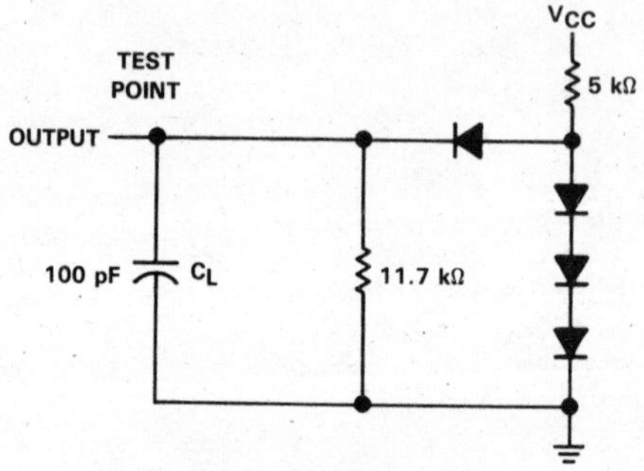
V_{CC}
TEST POINT
5 kΩ
OUTPUT
100 pF
C_L
11.7 kΩ

FIGURE 1. TEST CIRCUIT

PRINCIPLES OF OPERATION

The ADC0808M consists of an analog signal multiplexer, an 8-bit successive-approximation converter, and related control and output circuitry.

multiplexer

The analog multiplexer selects 1 of 8 single-ended input channels as determined by the address decoder. Address load control loads the address code into the decoder on a low-to-high transition. The output latch is reset by the positive-going edge of the start pulse. Sampling also starts with the positive-going edge of the start pulse and lasts for 32 clock periods. The conversion process may be interrupted by a new start pulse before the end of 64 clock periods. The previous data will be lost if a new start of conversion occurs before the 64th clock pulse. Continuous conversion may be accomplished by connecting the End-of-Conversion output to the start input. If used in this mode an external pulse should be applied after power up to assure start up.

converter

The CMOS threshold detector in the successive-approximation conversion system determines each bit by examining the charge on a series of binary-weighted capacitors (Figure 2). In the first phase of the conversion process, the analog input is sampled by closing switch S_C and all S_T switches, and by simultaneously charging all the capacitors to the input voltage.

In the next phase of the conversion process, all S_T and S_C switches are opened and the threshold detector begins identifying bits by identifying the charge (voltage) on each capacitor relative to the reference voltage. In the switching sequence, all eight capacitors are examined separately until all 8 bits are identified, and then the charge-convert sequence is repeated. In the first step of the conversion phase, the threshold detector looks at the first capacitor (weight = 128). Node 128 of this capacitor is switched to the reference voltage, and the equivalent nodes of all the other capacitors on the ladder are switched to REF−. If the voltage at the summing node is greater than the trip-point of the threshold detector (approximately one-half the V_{CC} voltage), a bit is placed in the output register, and the 128-weight capacitor is switched to REF−. If the voltage at the summing node is less than the trip point of the threshold detector, this 128-weight capacitor remains connected to REF+ through the remainder of the capacitor-sampling (bit-counting) process. The process is repeated for the 64-weight capacitor, the 32-weight capacitor, and so forth down the line, until all bits are counted.

With each step of the capacitor-sampling process, the initial charge is redistributed among the capacitors. The conversion process is successive approximation, but relies on charge redistribution rather than a successive-approximation register (and reference DAC) to count and weigh the bits from MSB to LSB.

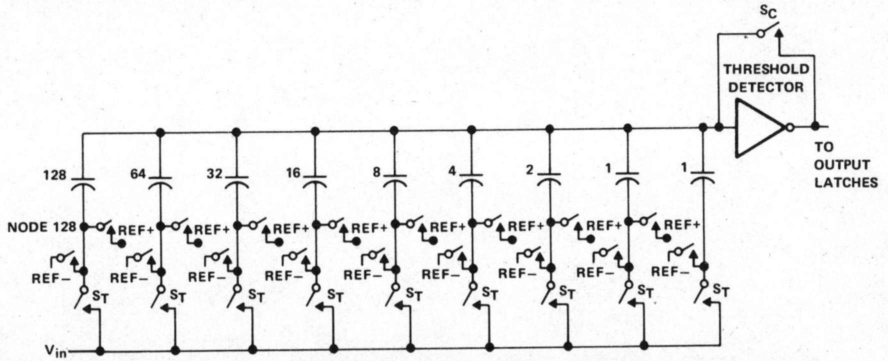
SC
THRESHOLD
DETECTOR
TO
OUTPUT
LATCHES
128
64
32
16
8
4
2
1
1
NODE 128
REF+
REF−
ST
Vin

FIGURE 2. SIMPLIFIED MODEL OF THE SUCCESSIVE-APPROXIMATION SYSTEM

ADC0831A, ADC0832A, ADC0831B, ADC0832B
A/D PERIPHERALS WITH SERIAL CONTROL

D2795, AUGUST 1985—REVISED JUNE 1986

- 8-Bit Resolution
- Easy Interface to Microprocessors or Stand-Alone Operation
- Operates Ratiometrically or with 5-V Reference
- Single Channel or Multiplexed Twin Channels with Single-Ended or Differential Input Options
- Input Range 0 to 5 V with Single 5-V Supply
- Inputs and Outputs are Compatible with TTL and MOS
- Conversion Time of 32 μs at CLK = 250 kHz
- Designed to be Interchangeable with National Semiconductor ADC0831 and ADC0832

DEVICE	TOTAL UNADJUSTED ERROR	
	A-SUFFIX	B-SUFFIX
ADC0831	± 1 LSB	± ½ LSB
ADC0832	± 1 LSB	± ½ LSB

ADC0831 . . . P DUAL-IN-LINE PACKAGE
(TOP VIEW)

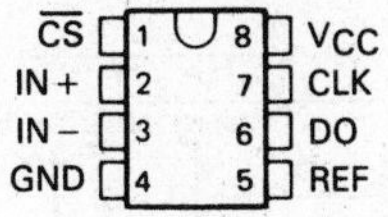
CS 1
IN+ 2
IN− 3
GND 4
8 VCC
7 CLK
6 DO
5 REF

ADC0832 . . . P DUAL-IN-LINE PACKAGE
(TOP VIEW)

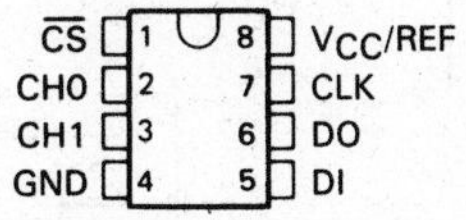
CS 1
CH0 2
CH1 3
GND 4
8 VCC/REF
7 CLK
6 DO
5 DI

description

These devices are 8-bit successive-approximation analog-to-digital converters. The ADC0831A and ADC0831B have single input channels; the ADC0832A and ADC0832B have multiplexed twin input channels. The serial output is configured to interface with standard shift registers or microprocessors. Detailed information on interfacing to most popular microprocessors is readily available from the factory.

The ADC0832 multiplexer is software configured for single-ended or differential inputs. The differential analog voltage input allows for common-mode rejection or offset of the analog zero input voltage value. In addition, the voltage reference input can be adjusted to allow encoding any smaller analog voltage span to the full 8 bits of resolution.

The operation of the ADC0831 and ADC0832 devices is very similar to the more complex ADC0834 and ADC0838 devices. Ratiometric conversion can be attained by setting the REF input equal to the maximum analog input signal value, which gives the highest possible conversion resolution. Typically, REF is set equal to V_{CC} (done internally on the ADC0832). For more detail on the operation of the ADC0831 and ADC0832 devices, refer to the ADC0834/ADC0838 data sheet.

The ADC0831AI, ADC0831BI, ADC0832AI, and ADC0832BI are characterized for operation from −40°C to 85°C. The ADC0831AC, ADC0831BC, ADC0832AC, and ADC0832BC are characterized for operation from 0°C to 70°C.

functional block diagram

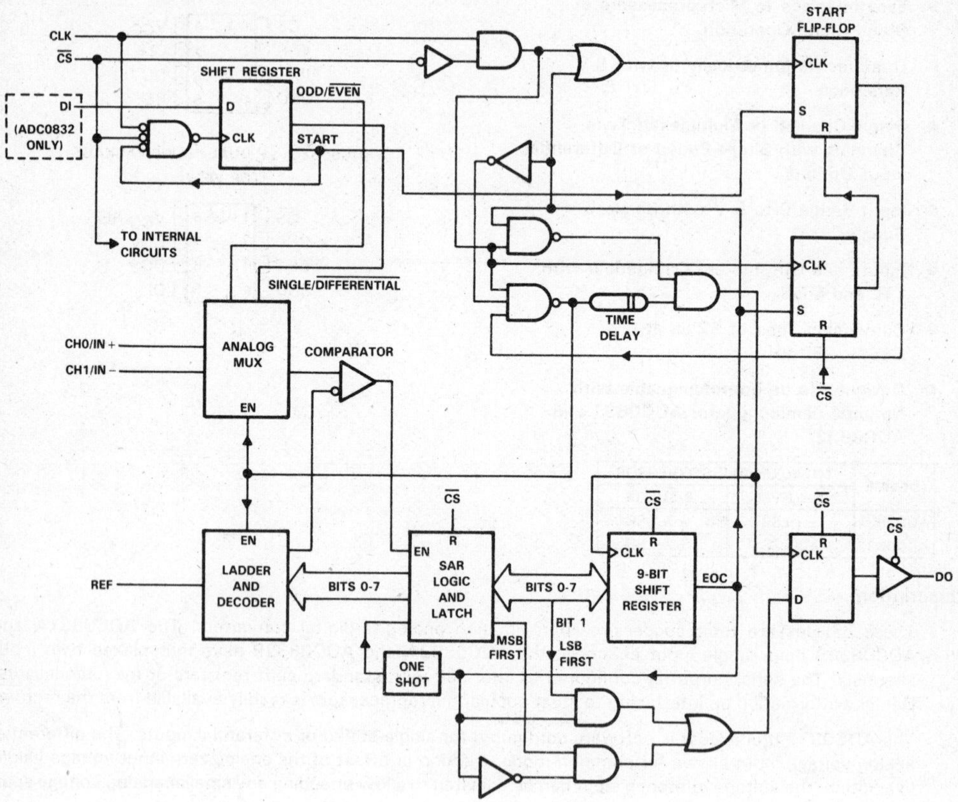
CLK
CS
DI
(ADC0832 ONLY)
SHIFT REGISTER
D
CLK
ODD/EVEN
START
START FLIP-FLOP
CLK
S
R
TO INTERNAL CIRCUITS
SINGLE/DIFFERENTIAL
CH0/IN +
CH1/IN −
ANALOG MUX
EN
COMPARATOR
TIME DELAY
CLK
S
R
CS
REF
EN
LADDER AND DECODER
BITS 0-7
CS
EN
R
SAR LOGIC AND LATCH
BITS 0-7
CS
CLK
R
9-BIT SHIFT REGISTER
EOC
CS
R
CLK
D
CS
DO
BIT 1
MSB FIRST
LSB FIRST
ONE SHOT

TEXAS INSTRUMENTS
POST OFFICE BOX 655012 • DALLAS, TEXAS 75265

sequence of operation

ADC0831

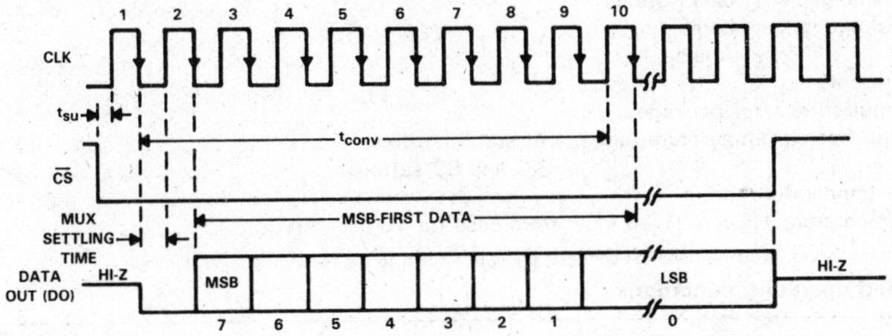
CLK
1
2
3
4
5
6
7
8
9
10
t_{su}
t_{conv}
$\overline{CS}$
MUX SETTLING TIME
MSB-FIRST DATA
DATA OUT (DO)
HI-Z
MSB
LSB
HI-Z
7
6
5
4
3
2
1
0

ADC0832

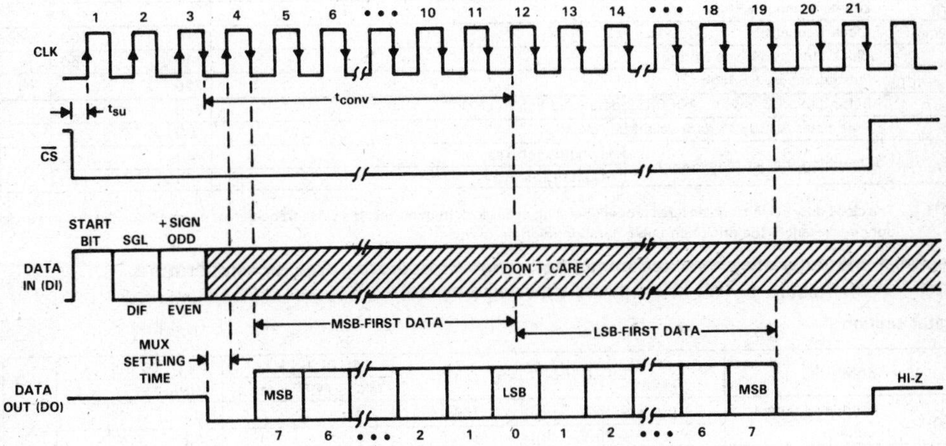
CLK
1
2
3
4
5
6
10
11
12
13
14
18
19
20
21
t_{su}
t_{conv}
$\overline{CS}$
START BIT
SGL
+SIGN ODD
DATA IN (DI)
DIF
EVEN
DON'T CARE
MSB-FIRST DATA
LSB-FIRST DATA
MUX SETTLING TIME
DATA OUT (DO)
MSB
LSB
MSB
HI-Z
7
6
2
1
0
1
2
6
7

ADC0832 MUX ADDRESS CONTROL LOGIC TABLE

MUX ADDRESS		CHANNEL NUMBER	
SGL/$\overline{DIF}$	ODD/$\overline{EVEN}$	0	1
L	L	+	−
L	H	−	+
H	L	+	
H	H		+

H = high level, L = low level, − or + = polarity of selected input pin

absolute maximum ratings over recommended operating free-air temperature range (unless otherwise noted)

Supply voltage, V_{CC} (see Note 1) .... 6.5 V
Input voltage range: Logic .... −0.3 V to 15 V
Analog .... −0.3 V to V_{CC}+0.3 V
Input current .... ±5 mA
Total input current for package .... ±20 mA
Operating free-air temperature range: AI and BI suffixes .... −40°C to 85°C
AC and BC suffixes .... 0°C to 70°C
Storage temperature range .... −65°C to 150°C
Lead temperature 1,6 mm (1/16 inch) from case for 10 seconds .... 260°C

NOTE 1: All voltage values, except differential voltages, are with respect to the network ground terminal.

recommended operating conditions

			MIN	NOM	MAX	UNIT
V_{CC}	Supply voltage		4.5	5	6.3	V
V_{IH}	High-level input voltage		2			V
V_{IL}	Low-level input voltage				0.8	V
f_{clock}	Clock frequency		10		400	kHz
	Clock duty cycle (see Note 2)		40		60	%
$t_{wH(CS)}$	Pulse duration, $\overline{CS}$ high		220			ns
t_{su}	Setup time, $\overline{CS}$ low or ADC0832 data valid before clock↑		350			ns
t_h	Hold time, ADC0832 data valid after clock↑		90			ns
T_A	Operating free-air temperature	AI and BI suffixes	−40		85	°C
		AC and BC suffixes	0		70	

NOTE 2: The clock duty cycle range ensures proper operation at all clock frequencies. If a clock frequency is used outside the recommended duty cycle range, the minimum pulse duration (high or low) is 1 μs.

electrical characteristics over recommended range of operating free-air temperature, V_{CC} = 5 V, f_{clock} = 250 kHz (unless otherwise noted)

digital section

PARAMETER		TEST CONDITIONS†		AI, BI SUFFIX			AC, BC SUFFIX			UNIT
				MIN	TYP‡	MAX	MIN	TYP‡	MAX	
V_{OH}	High-level output voltage	V_{CC} = 4.75 V,	I_{OH} = −360 μA	2.4			2.8			V
		V_{CC} = 4.75 V,	I_{OH} = −10 μA	4.5			4.6			
V_{OL}	Low-level output voltage	V_{CC} = 4.75 V,	I_{OL} = 1.6 mA	0.4			0.34			V
I_{IH}	High-level input current	V_{IH} = 5 V			0.005	1		0.005	1	μA
I_{IL}	Low-level input current	V_{IL} = 0			−0.005	−1		−0.005	−1	μA
I_{OH}	High-level output (source) current	V_{OH} = 0,	T_A = 25°C	−6.5	−14		−6.5	−14		mA
I_{OL}	Low-level output (sink) current	V_{OL} = V_{CC},	T_A = 25°C	8	16		8	16		mA
I_{OZ}	High-impedance-state output current (DO)	V_O = 5 V,	T_A = 25°C		0.01	3		0.01	3	μA
		V_O = 0,	T_A = 25°C		−0.01	−3		−0.01	−3	
C_i	Input capacitance				5			5		pF
C_o	Output capacitance				5			5		pF

† All parameters are measured under open-loop conditions with zero common-mode input voltage.
‡ All typical values are at V_{CC} = 5 V, T_A = 25°C.

electrical characteristics over recommended range of operating free-air temperature, V_{CC} = 5 V, f_{clock} = 250 kHz (unless otherwise noted)

analog and converter section

PARAMETER			TEST CONDITIONS†	MIN	TYP‡	MAX	UNIT
V_{ICR}	Common-mode input voltage range		See Note 3	−0.05 to V_{CC}+0.05			V
$I_{I(stdby)}$	Standby input current (see Note 4)	On-channel	V_I = 5 V at on-channel,			1	μA
		Off-channel	V_I = 0 at off-channel			−1	
		On-channel	V_I = 0 at on-channel,			−1	
		Off-channel	V_I = 5 V at off-channel			1	
$r_{i(REF)}$	Input resistance to reference ladder			1.3	2.4	5.9	kΩ

total device

PARAMETER			TEST CONDITIONS†	MIN	TYP‡	MAX	UNIT
I_{CC}	Supply current	ADC0831			1	2.5	mA
		ADC0832			3	5.2	

† All parameters are measured under open-loop conditions with zero common-mode input voltage.

‡ All typical values are at V_{CC} = 5 V, T_A = 25°C.

NOTES: 3. If channel IN− is more positive than channel IN+, the digital output code will be 0000 0000. Connected to each analog input are two on-chip diodes that will conduct forward current for analog input voltages one diode drop above V_{CC}. Care must be taken during testing at low V_{CC} levels (4.5 V) because high-level analog input voltage (5 V) can, especially at high temperatures, cause this input diode to conduct and cause errors for analog inputs that are near full-scale. As long as the analog voltage does not exceed the supply voltage by more than 50 mV, the output code will be correct. To achieve an absolute 0 V to 5 V input voltage range requires a minimum V_{CC} of 4.95 volts for all variations of temperature and load.

4. Standby input currents are currents going into or out of the on or off channels when the A/D converter is not performing conversion and the clock is in a high or low steady-state condition.

operating characteristics V_{CC} = REF = 5 V, f_{clock} = 250 kHz, t_r = t_f = 20 ns, T_A = 25°C (unless otherwise noted)

PARAMETER			TEST CONDITIONS†	BI, BC SUFFIX MIN	BI, BC SUFFIX TYP	BI, BC SUFFIX MAX	AI, AC SUFFIX MIN	AI, AC SUFFIX TYP	AI, AC SUFFIX MAX	UNIT
	Supply-voltage variation error		V_{CC} = 4.75 V to 5.25 V		±1/16	±1/4		±1/16	±1/4	LSB
	Total unadjusted error (see Note 5)		V_{ref} = 5 V, T_A = MIN to MAX			±1/2			±1	LSB
	Common-mode error		Differential mode		±1/16	±1/4		±1/16	±1/4	LSB
t_{pd}	Propagation delay time, output data after CLK↓ (see Note 6)	MSB-first data	C_L = 100 pF		650	1500		650	1500	ns
		LSB-first data			250	600		250	600	
t_{dis}	Output disable time, DO after $\overline{CS}$↑		C_L = 10 pF, R_L = 10 kΩ		125	250		125	250	ns
			C_L = 100 pF, R_L = 2 kΩ			500			500	
t_{conv}	Conversion time (multiplexer addressing time not included)					8			8	clock periods

† All parameters are measured under open-loop conditions with zero common-mode input voltage. For conditions shown as MIN or MAX, use the appropriate value specified under recommended operating conditions.

NOTES: 5. Total unadjusted error includes offset, full-scale, linearity, and multiplexer errors.

6. The most significant-bit-first data is output directly from the comparator and therefore requires additional delay to allow for comparator response time. Least-significant-bit-first data applies only to ADC0832.

PARAMETER MEASUREMENT INFORMATION

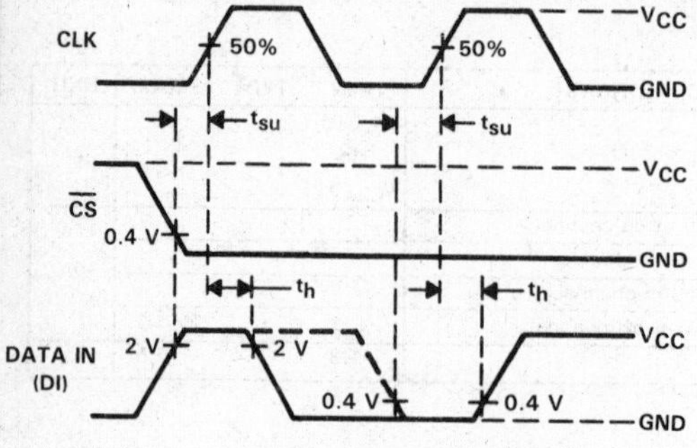
CLK
50%
50%
VCC
GND
tsu
tsu
CS
0.4 V
VCC
GND
th
th
DATA IN
(DI)
2 V
2 V
0.4 V
0.4 V
VCC
GND

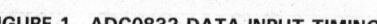

FIGURE 1. ADC0832 DATA INPUT TIMING

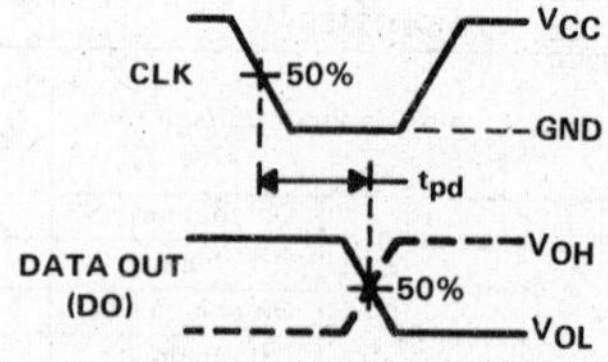
CLK
50%
VCC
GND
tpd
DATA OUT
(DO)
50%
VOH
VOL

FIGURE 2. DATA OUTPUT TIMING

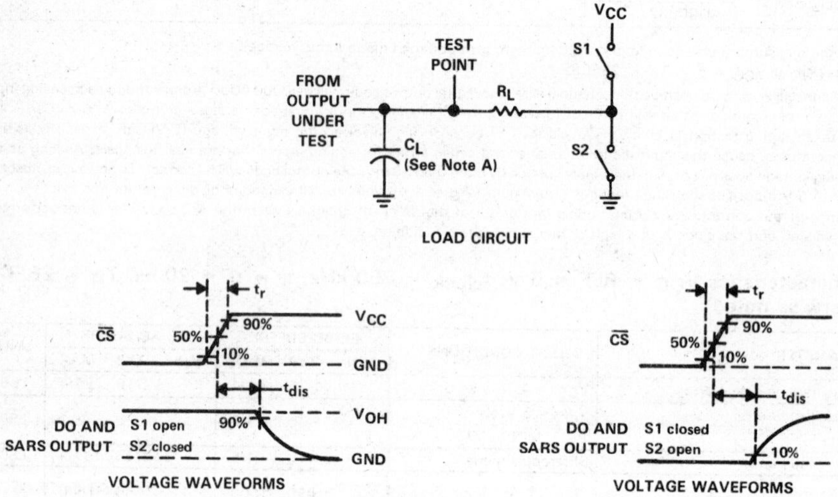
VCC
S1
TEST
POINT
FROM
OUTPUT
UNDER
TEST
RL
CL
(See Note A)
S2
LOAD CIRCUIT
tr
CS
50%
90%
10%
VCC
GND
tdis
DO AND
SARS OUTPUT
S1 open
S2 closed
90%
VOH
GND
VOLTAGE WAVEFORMS
tr
CS
50%
90%
10%
VCC
GND
tdis
DO AND
SARS OUTPUT
S1 closed
S2 open
10%
-VCC
GND
VOLTAGE WAVEFORMS

NOTE A: C_L includes probe and jig capacitance.

FIGURE 3. OUTPUT DISABLE TIME TEST CIRCUIT AND VOLTAGE WAVEFORMS

TEXAS INSTRUMENTS
POST OFFICE BOX 655012 • DALLAS, TEXAS 75265

TYPICAL CHARACTERISTICS

UNADJUSTED OFFSET ERROR
vs
REFERENCE VOLTAGE

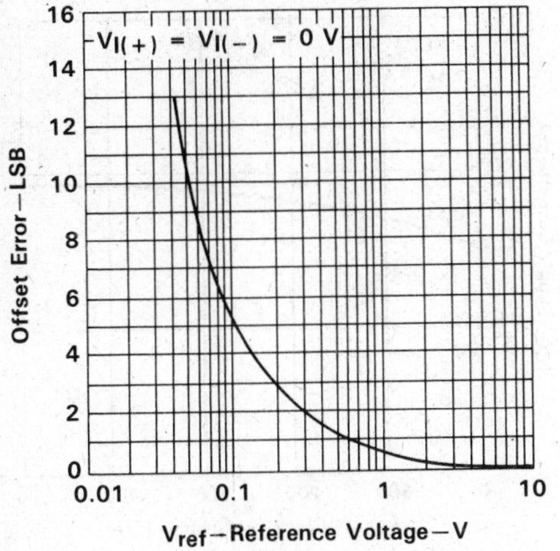
$V_{I(+)} = V_{I(-)} = 0$ V
Offset Error—LSB
16
14
12
10
8
6
4
2
0
0.01
0.1
1
10
V_{ref}—Reference Voltage—V

FIGURE 4

LINEARITY ERROR
vs
REFERENCE VOLTAGE

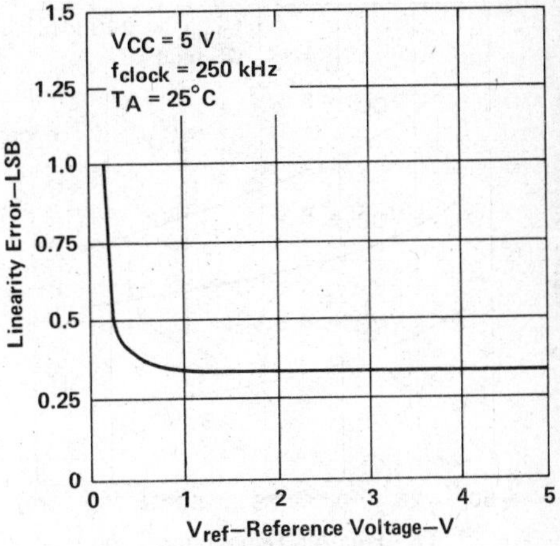
V_{CC} = 5 V
f_{clock} = 250 kHz
T_A = 25°C
Linearity Error—LSB
1.5
1.25
1.0
0.75
0.5
0.25
0
0
1
2
3
4
5
V_{ref}—Reference Voltage—V

FIGURE 5

LINEARITY ERROR
vs
FREE-AIR TEMPERATURE

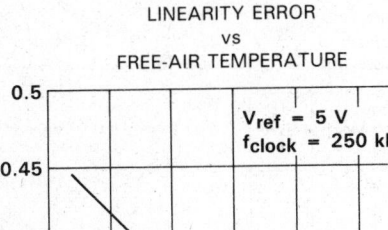

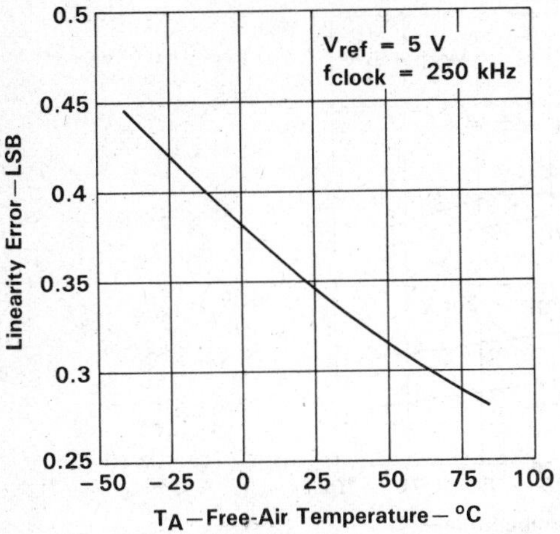
V_{ref} = 5 V
f_{clock} = 250 kHz
Linearity Error—LSB
0.5
0.45
0.4
0.35
0.3
0.25
−50
−25
0
25
50
75
100
T_A—Free-Air Temperature—°C

FIGURE 6

LINEARITY ERROR
vs
CLOCK FREQUENCY

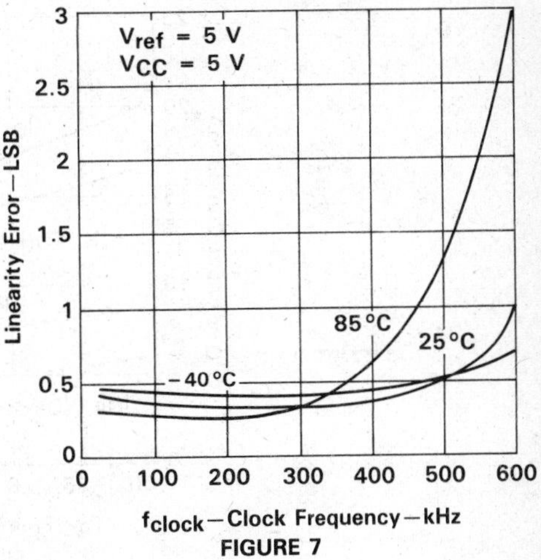
V_{ref} = 5 V
V_{CC} = 5 V
Linearity Error—LSB
3
2.5
2
1.5
1
0.5
0
85°C
25°C
−40°C
0
100
200
300
400
500
600
f_{clock}—Clock Frequency—kHz

FIGURE 7

TYPICAL CHARACTERISTICS

SUPPLY CURRENT
vs
FREE-AIR TEMPERATURE

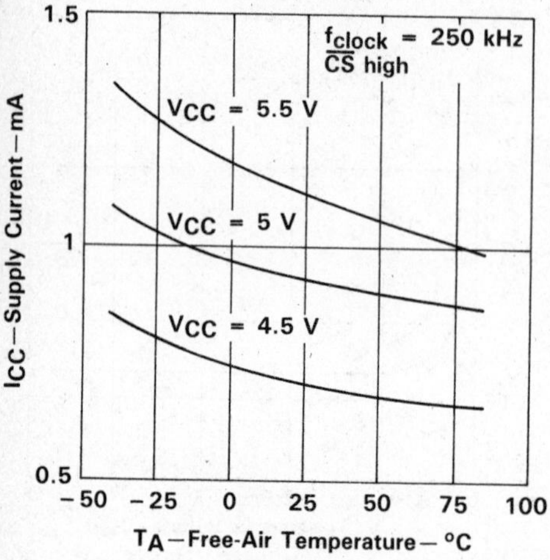
1.5
1
0.5
f_{clock} = 250 kHz
$\overline{CS}$ high
V_{CC} = 5.5 V
V_{CC} = 5 V
V_{CC} = 4.5 V
I_{CC}—Supply Current—mA
−50 −25 0 25 50 75 100
T_A—Free-Air Temperature—°C

FIGURE 8

SUPPLY CURRENT
vs
CLOCK FREQUENCY

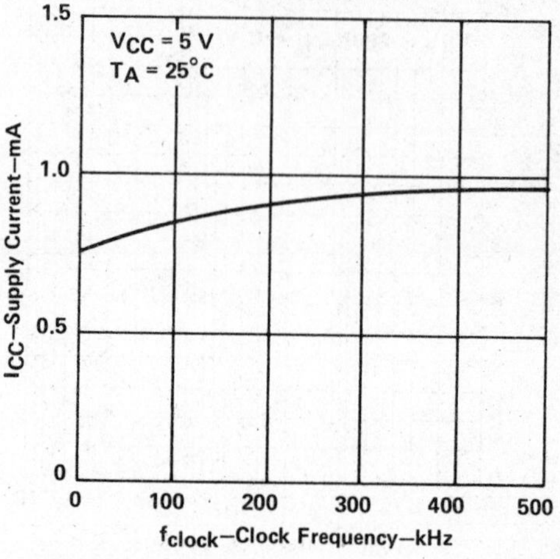
1.5
1.0
0.5
0
V_{CC} = 5 V
T_A = 25°C
I_{CC}—Supply Current—mA
0 100 200 300 400 500
f_{clock}—Clock Frequency—kHz

FIGURE 9

OUTPUT CURRENT
vs
FREE-AIR TEMPERATURE

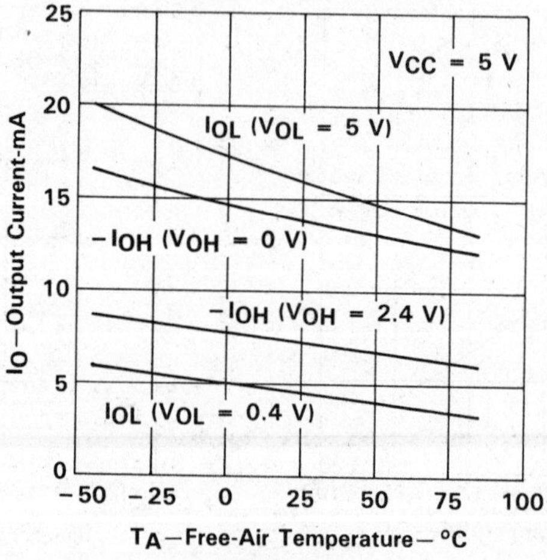
25
20
15
10
5
0
V_{CC} = 5 V
I_{OL} (V_{OL} = 5 V)
$-I_{OH}$ (V_{OH} = 0 V)
$-I_{OH}$ (V_{OH} = 2.4 V)
I_{OL} (V_{OL} = 0.4 V)
I_O—Output Current-mA
−50 −25 0 25 50 75 100
T_A—Free-Air Temperature—°C

FIGURE 10

TEXAS INSTRUMENTS
POST OFFICE BOX 655012 • DALLAS, TEXAS 75265

D2795, AUGUST 1985—REVISED OCTOBER 1986

- **8-Bit Resolution**
- **Easy Interface to Microprocessors or Stand-Alone Operation**
- **Operates Ratiometrically or with 5-V Reference**
- **4- or 8-Channel Multiplexer Options with Address Logic**
- **Shunt Regulator Allows Operation with High-Voltage Supplies**
- **Input Range 0 to 5 V with Single 5-V Supply**
- **Remote Operation with Serial Data Link**
- **Inputs and Outputs are Compatible with TTL and MOS**
- **Conversion Time of 32 μs at f_{clock} = 250 kHz**
- **Designed to be Interchangeable with National Semiconductor ADC0834 and ADC0838**

DEVICE	TOTAL UNADJUSTED ERROR	
	A SUFFIX	B SUFFIX
ADC0834	±1 LSB	±1/2 LSB
ADC0838	±1 LSB	±1/2 LSB

ADC0834 . . . N DUAL-IN-LINE PACKAGE
(TOP VIEW)

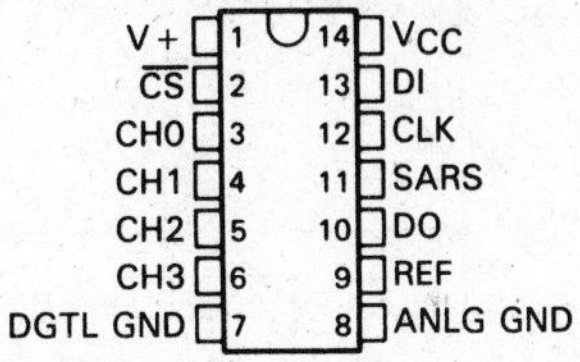
V+ 1
CS 2
CH0 3
CH1 4
CH2 5
CH3 6
DGTL GND 7
14 VCC
13 DI
12 CLK
11 SARS
10 DO
9 REF
8 ANLG GND

ADC0838 . . . N DUAL-IN-LINE PACKAGE
(TOP VIEW)

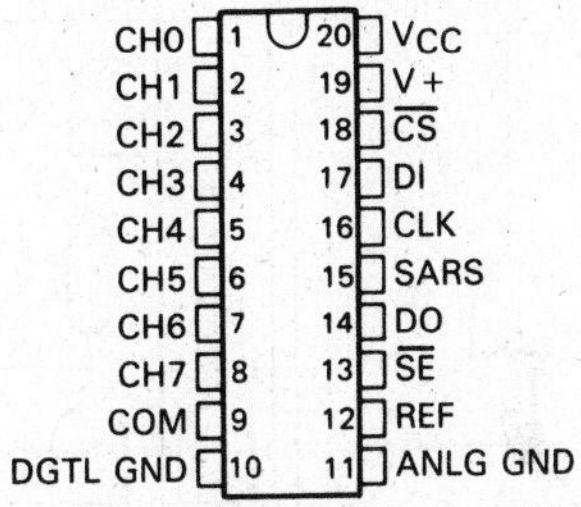
CH0 1
CH1 2
CH2 3
CH3 4
CH4 5
CH5 6
CH6 7
CH7 8
COM 9
DGTL GND 10
20 VCC
19 V+
18 CS
17 DI
16 CLK
15 SARS
14 DO
13 SE
12 REF
11 ANLG GND

ADC0838 . . . FN CHIP CARRIER PACKAGE
(TOP VIEW)

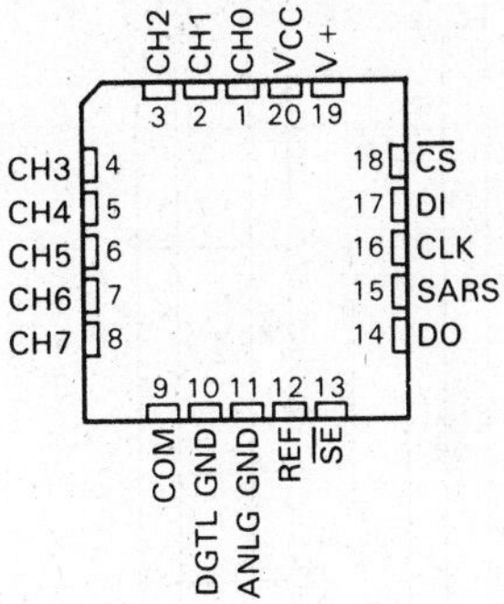
CH2 3
CH1 2
CH0 1
VCC 20
V+ 19
CH3 4
CH4 5
CH5 6
CH6 7
CH7 8
18 CS
17 DI
16 CLK
15 SARS
14 DO
9 COM
10 DGTL GND
11 ANLG GND
12 REF
13 SE

description

These devices are 8-bit successive-approximation analog-to-digital converters each with an input-configurable multichannel multiplexer and serial input/output. The serial input/output is configured to interface with standard shift registers or microprocessors. Detailed information on interfacing to most popular microprocessors is readily available from the factory.

The ADC0834 (4-channel) and ADC0838 (8-channel) multiplexer is software configured for single-ended or differential inputs as well as pseudo-differential input assignments. The differential analog voltage input allows for common-mode rejection or offset of the analog zero input voltage value. In addition, the voltage reference input can be adjusted to allow encoding any smaller analog voltage span to the full 8 bits of resolution.

The ADC0834AI, ADC0834BI, ADC0838AI, and ADC0838BI are characterized for operation from −40 °C to 85 °C. The ADC0834AC, ADC0834BC, ADC0838AC, and ADC0838BC are characterized for operation from 0 °C to 70 °C.

functional block diagram

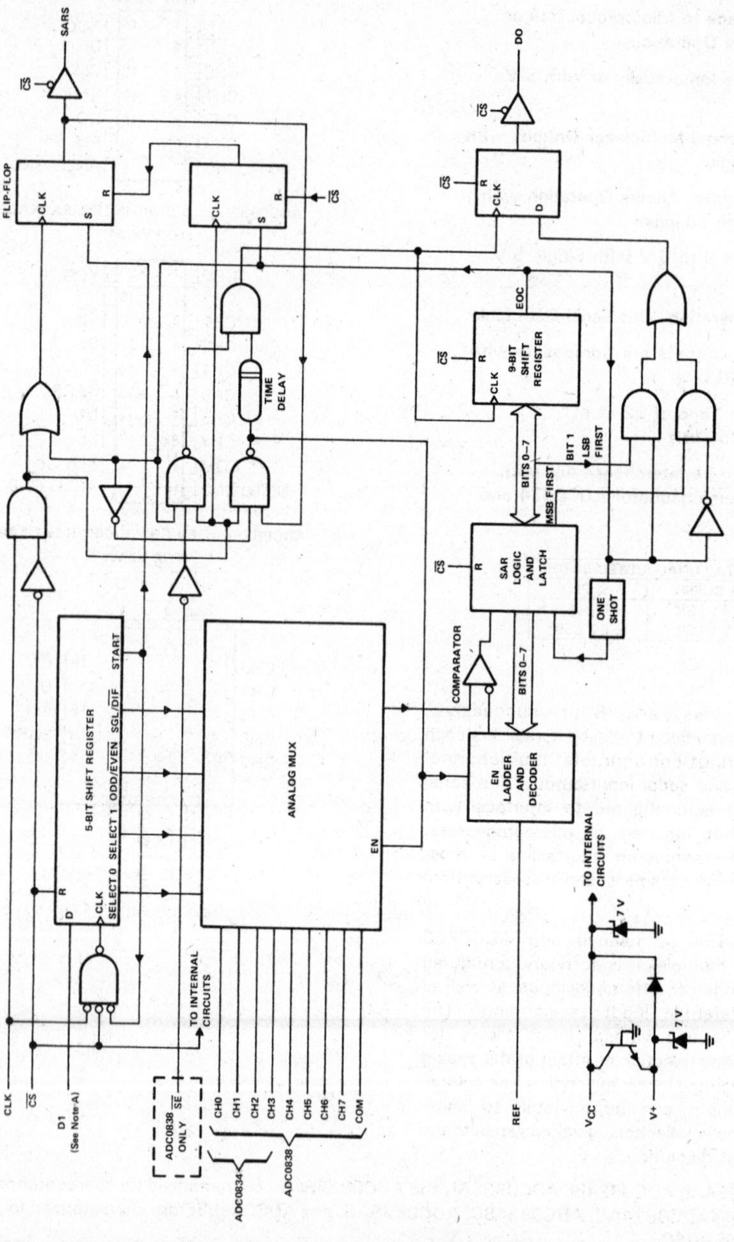
SARS
CS
START FLIP-FLOP
CLK
S
R
TIME DELAY
DO
EOC
9-BIT SHIFT REGISTER
BITS 0–7
MSB FIRST
BIT 1
LSB FIRST
SAR LOGIC AND LATCH
ONE SHOT
COMPARATOR
EN LADDER AND DECODER
TO INTERNAL CIRCUITS
7 V
5-BIT SHIFT REGISTER
START
SGL/DIF
ODD/EVEN
SELECT 1
SELECT 0
D
ANALOG MUX
EN
CLK
D1 (See Note A)
ADC0838 ONLY
SE
CH0
CH1
CH2
CH3
CH4
CH5
CH6
CH7
COM
ADC0834
ADC0838
REF
VCC
V+

NOTE A: For the ADC0834, DI is input directly to the D input of SELECT 1. SELECT 0 is forced to a high.

functional description

The ADC0834 and ADC0838 use a sample data comparator structure that converts differential analog inputs by a successive-approximation routine. Operation of both devices is similar with the exception of a select enable ($\overline{SE}$) input, analog common input, and multiplexer addressing. The input voltage to be converted is applied to a channel terminal and is compared to ground (single-ended), to an adjacent input (differential), or to a common terminal (pseudo-differential) that can be an arbitrary voltage. The input terminals are assigned a positive (+) or negative (−) polarity. If the signal inputs applied to the assigned positive terminal is less than the signal on the negative terminal, the converter output is all zeros.

Channel selection and input configuration are under software control using a serial data link from the controlling processor. A serial communication format allows more functions to be included in a converter package with no increase in size. In addition, it eliminates the transmission of low-level analog signals by locating the converter at the analog sensor and communicating serially with the controlling processor. This process returns noise-free digital data to the processor.

A particular input configuration is assigned during the multiplexer addressing sequence. The multiplexer address is shifted into the converter through the data input (DI) line. The multiplexer address selects the analog inputs to be enabled and determines whether the input is single-ended or differential. When the input is differential, the polarity of the channel input is assigned. Differential inputs are assigned to adjacent channel pairs. For example, channel 0 and channel 1 may be selected as a differential pair. These channels cannot act differentially with any other channel. In addition to selecting the differential mode, the polarity may also be selected. Either channel of the channel pair may be designated as the negative or positive input.

The common input on the ADC0838 can be used for a pseudo-differential input. In this mode, the voltage on the common input is considered to be the negative differential input for all channel inputs. This voltage can be any reference potential common to all channel inputs. Each channel input can then be selected as the positive differential input. This feature is useful when all analog circuits are biased to a potential other than ground.

A conversion is initiated by setting the chip select ($\overline{CS}$) input low. This enables all logic circuits. The $\overline{CS}$ input must be held low for the complete conversion process. A clock input is received from the processor. On each low-to-high transition of the clock input, the data on the DI input is clocked into the multiplexer address shift register. The first logic high on the input is the start bit. A 3- to 4-bit assignment word follows the start bit. On each successive low-to-high transition of the clock input, the start bit and assignment word are shifted through the shift register. When the start bit has been shifted into the start location of the multiplexer register, the input channel has been selected and conversion starts. The SAR Status output (SARS) goes high to indicate that a conversion is in progress and the DI input to the multiplexer shift register is disabled for the duration of the conversion.

An interval of one clock period is automatically inserted to allow for the selected multiplexed channel to settle. The data output DO comes out of the high-impedance state and provides a leading low for this one clock period of multiplexer settling time. The SAR comparator compares successive outputs from the resistive ladder with the incoming analog signal. The comparator output indicates whether the analog input is greater than or less than the resistive ladder output. As the conversion proceeds, conversion data is simultaneously output from the DO output pin with the most significant bit (MSB) first.

After eight clock periods the conversion is complete and the SAR Status (SARS) output goes low.

The ADC0834 outputs the least-significant-bit-first data after the MSB-first data stream. If the shift enable ($\overline{SE}$) line is held high on the ADC0838, the value of the least significant bit (LSB) will remain on the data line. When $\overline{SE}$ is forced low, the data is then clocked out as LSB-first data. (To output LSB first, the $\overline{SE}$ control input must first go low, then the data stored in the 9-bit shift register will output with LSB first.) When $\overline{CS}$ goes high, all internal registers are cleared. At this time the output circuits go to the high-impedance state. If another conversion is desired, the $\overline{CS}$ line must make a high-to-low transition followed by address information.

functional description (continued)

The DI and DO pins can be tied together and controlled by a bidirectional processor I/O bit received on a single wire. This is possible because the DI input is only examined during the multiplexer addressing interval and the DO output is still in a high-impedance state.

Detailed information on interfacing to most popular microprocessors is readily available from the factory.

sequence of operation

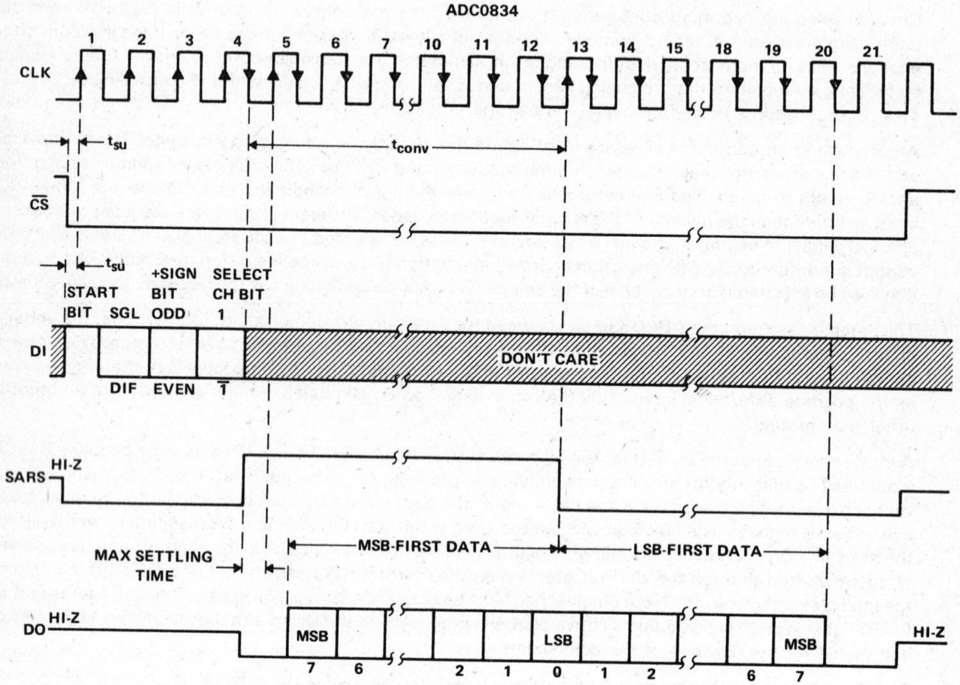
ADC0834
1 2 3 4 5 6 7 10 11 12 13 14 15 18 19 20 21
CLK
tsu
tconv
CS
tsu
START BIT
SGL
DIF
+SIGN BIT
ODD
EVEN
SELECT CH BIT 1
T
DI
DON'T CARE
HI-Z
SARS
MAX SETTLING TIME
MSB-FIRST DATA
LSB-FIRST DATA
HI-Z
DO
MSB
LSB
MSB
7 6 2 1 0 1 2 6 7
HI-Z

ADC0834 MUX ADDRESS CONTROL LOGIC TABLE

MUX ADDRESS			CHANNEL NUMBER			
SGL/$\overline{DIF}$	ODD/$\overline{EVEN}$	SELECT BIT 1	0	1	2	3
L	L	L	+	−		
L	L	H			+	−
L	H	L	−	+		
L	H	H			−	+
H	L	L	+			
H	L	H			+	
H	H	L		+		
H	H	H				+

H = high level, L = low level, − or + = polarity of selected input pin

sequence of operation

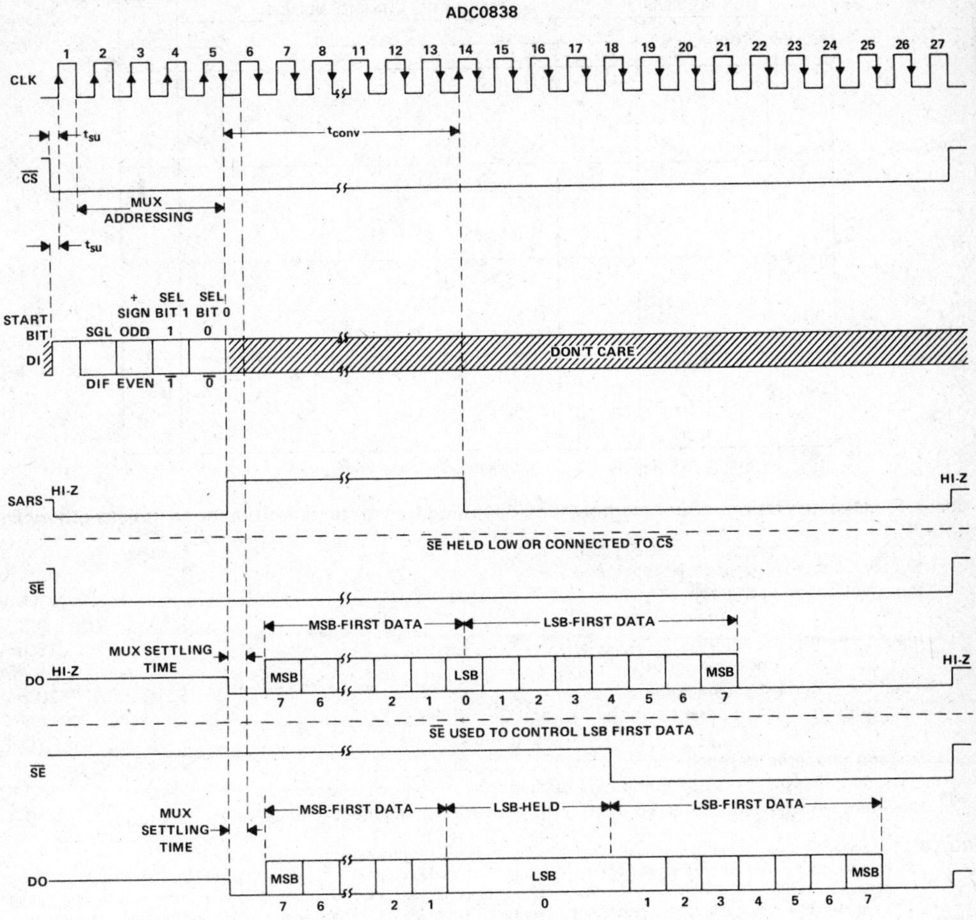
ADC0838
CLK
1 2 3 4 5 6 7 8 11 12 13 14 15 16 17 18 19 20 21 22 23 24 25 26 27
t_{su}
t_{conv}
CS
MUX ADDRESSING
t_{su}
START BIT
+ SIGN
SEL BIT 1
SEL BIT 0
SGL ODD 1 0
DI
DIF EVEN $\overline{1}$ $\overline{0}$
DON'T CARE
SARS
HI-Z
HI-Z
SE HELD LOW OR CONNECTED TO CS
SE
MSB-FIRST DATA
LSB-FIRST DATA
MUX SETTLING TIME
DO
HI-Z
MSB
LSB
MSB
7 6 2 1 0 1 2 3 4 5 6 7
HI-Z
SE USED TO CONTROL LSB FIRST DATA
SE
MSB-FIRST DATA
LSB-HELD
LSB-FIRST DATA
MUX SETTLING TIME
DO
MSB
LSB
MSB
7 6 2 1 0 1 2 3 4 5 6 7

ADC0838 MUX ADDRESS CONTROL LOGIC TABLE

MUX ADDRESS				SELECTED CHANNEL NUMBER								COM
SGL/$\overline{\text{DIF}}$	ODD/$\overline{\text{EVEN}}$	SELECT		0		1		2		3		
		1	0	0	1	2	3	4	5	6	7	
L	L	L	L	+	−							
L	L	L	H			+	−					
L	L	H	L					+	−			
L	L	H	H							+	−	
L	H	L	L	−	+							
L	H	L	H			−	+					
L	H	H	L					+	−			
L	H	H	H							−	+	
H	L	L	L	+								−
H	L	L	H			+						−
H	L	H	L					+				−
H	L	H	H							+		−
H	H	L	L		+							−
H	H	L	H				+					−
H	H	H	L						+			−
H	H	H	H								+	−

H = high level, L = low level, − or + = polarity of selected input

absolute maximum ratings over recommended operating free-air temperature range (unless otherwise noted)

Supply voltage, V_{CC} (see Notes 1 and 2)	6.5 V
Input voltage range: Logic	−0.3 V to 15 V
Analog	−0.3 V to V_{CC}+0.3 V
Input current: V+ input	15 mA
Any other input	±5 mA
Total input current for package	±20 mA
Operating free-air temperature range: AI and BI suffixes	−40°C to 85°C
AC and BC suffixes	0°C to 70°C
Storage temperature range	−65°C to 150°C
Case temperature for 10 seconds: FN package	260°C
Lead temperature 1,6 mm (1/16 inch) from case for 10 seconds: N package	260°C

NOTES: 1. All voltage values, except differential voltages, are with respect to the network ground terminal.
2. Internal zener diodes are connected from the V_{CC} input to ground and from the V+ input to ground. The breakdown voltage of each zener diode is approximately 7 volts. One zener diode can be used as a shunt regulator and connects to V_{CC} through a regular diode. When the voltage regulator powers the converter, this zener and regular diode combination ensures that the V_{CC} input (6.4 V) is less than the zener breakdown voltage. A series resistor is recommended to limit current into the V+ input.

recommended operating conditions

			MIN	NOM	MAX	UNIT
V_{CC}	Supply voltage		4.5	5	6.3	V
V_{IH}	High-level input voltage		2			V
V_{IL}	Low-level input voltage				0.8	V
f_{clock}	Clock frequency		10		400	kHz
	Clock duty cycle (see Note 3)		40		60	%
$t_{wH(CS)}$	Pulse duration, $\overline{CS}$ high		220			ns
t_{su}	Setup time, $\overline{CS}$ low, $\overline{SE}$ low, or data valid before clock↑		350			ns
t_h	Hold time, data valid after clock↑		90			ns
T_A	Operating free-air temperature	AI and BI suffixes	−40		85	°C
		AC and BC suffixes	0		70	

NOTE 3: The clock duty cycle range ensures proper operation at all clock frequencies. If a clock frequency is used outside the recommended duty cycle range, the minimum pulse duration (high or low) is 1 μs.

electrical characteristics over recommended range of operating free-air temperature, V_{CC} = V+ = 5 V, f_{clock} = 250 kHz (unless otherwise noted)

digital section

PARAMETER		TEST CONDITIONS†		AI, BI SUFFIX			AC, BC SUFFIX			UNIT
				MIN	TYP‡	MAX	MIN	TYP‡	MAX	
V_{OH}	High-level output voltage	V_{CC} = 4.75 V,	I_{OH} = −360 μA	2.4			2.8			V
		V_{CC} = 4.75 V,	I_{OH} = −10 μA	4.5			4.6			
V_{OL}	Low-level output voltage	V_{CC} = 5.25 V,	I_{OL} = 1.6 mA			0.4			0.34	V
I_{IH}	High-level input current	V_{IH} = 5 V			0.005	1		0.005	1	μA
I_{IL}	Low-level input current	V_{IL} = 0			−0.005	−1		−0.005	−1	μA
I_{OH}	High-level output (source) current	V_{OH} = 0,	T_A = 25°C	−6.5	−14		−6.5	−14		mA
I_{OL}	Low-level output (sink) current	V_{OL} = V_{CC},	T_A = 25°C	8	16		8	16		mA
I_{OZ}	High-impedance-state output current (DO or SARS)	V_O = 5 V,	T_A = 25°C		0.01	3		0.01	3	μA
		V_O = 0,	T_A = 25°C		−0.01	−3		−0.01	−3	
C_i	Input capacitance				5			5		pF
C_o	Output capacitance				5			5		pF

†All parameters are measured under open-loop conditions with zero common-mode input voltage (unless otherwise specified).
‡All typical values are at V_{CC} = V+ = 5 V, T_A = 25°C.

electrical characteristics over recommended range of operating free-air temperature, V_{CC} = V+ = 5 V, f_{clock} = 250 kHz (unless otherwise noted)

analog and converter section

PARAMETER			TEST CONDITIONS†	MIN	TYP‡	MAX	UNIT
V_{ICR}	Common-mode input voltage range		See Note 4	−0.05 to V_{CC}+0.05			V
$I_{I(stdby)}$	Standby input current (see Note 5)	On-channel	V_I = 5 V at on-channel,			1	μA
		Off-channel	V_I = 0 at off-channel			−1	μA
		On-channel	V_I = 0 at on-channel,			−1	μA
		Off-channel	V_I = 5 V at off-channel			1	μA
$r_{i(ref)}$	Input resistance to reference ladder			1.3	2.4	5.9	kΩ

total device

PARAMETER		TEST CONDITIONS†	MIN	TYP‡	MAX	UNIT
V_Z	Internal zener diode breakdown voltage	I_I = 15 mA at V+ pin, See Note 2	6.3	7	8.5	V
I_{CC}	Supply current			1	2.5	mA

†All parameters are measured under open-loop conditions with zero common-mode input voltage.
‡All typical values are at V_{CC} = 5 V, V+ = 5 V, T_A = 25°C.

NOTES: 2. Internal zener diodes are connected from the V_{CC} input to ground and from the V+ input to ground. The breakdown voltage of each zener diode is approximately 7 volts. One zener diode can be used as a shunt regulator and connects to V_{CC} through a regular diode. When the voltage regulator powers the converter, this zener and regular diode combination ensures that the V_{CC} input (6.4 V) is less than the zener breakdown voltage. A series resistor is recommended to limit current into the V+ input.

4. If channel IN− is more positive than channel IN+, the digital output code will be 0000 0000. Connected to each analog input are two on-chip diodes that will conduct forward current for analog input voltages one diode drop above V_{CC}. Care must be taken during testing at low V_{CC} levels (4.5 V) because high-level analog input voltage (5 V) can, especially at high temperatures, cause this input diode to conduct and cause errors for analog inputs that are near full-scale. As long as the analog voltage does not exceed the supply voltage by more than 50 mV, the output code will be correct. To achieve an absolute 0 V to 5 V input voltage range requires a minimum V_{CC} of 4.950 volts for all variations of temperature and load.

5. Standby input currents are currents going into or out of the on or off channels when the A/D converter is not performing conversion and the clock is in a high or low steady-state condition.

TEXAS INSTRUMENTS
POST OFFICE BOX 655012 • DALLAS, TEXAS 75265

operating characteristics V+ = V_{CC} = 5 V, f_{clock} = 250 kHz, t_r = t_f = 20 ns, T_A = 25°C (unless otherwise noted)

PARAMETER			TEST CONDITIONS†	BI, BC SUFFIX			AI, AC SUFFIX			UNIT
				MIN	TYP	MAX	MIN	TYP	MAX	
	Supply-voltage variation error		V_{CC} = 4.75 V to 5.25 V		±1/16	±1/4		±1/16	±1/4	LSB
	Total unadjusted error (see Note 6)		V_{ref} = 5 V, T_A = MIN to MAX			±1/2			±1	LSB
	Common-mode error		Differential mode		±1/16	±1/4		±1/16	±1/4	LSB
	Change in zero-error from V_{CC} = 5 V to internal zener diode operation (see Note 2)		I_I = 15 mA at V+ pin, V_{ref} = 5 V, V_{CC} open			1			1	LSB
t_{pd}	Propagation delay time, output data after CLK↓ (see Note 7)	MSB-first data	C_L = 100 pF		650	1500		650	1500	ns
		LSB-first data			250	600		250	600	
t_{dis}	Output disable time, DO or SARS after $\overline{CS}$↑		C_L = 10 pF, R_L = 10 kΩ		125	250		125	250	ns
			C_L = 100 pF, R_L = 2 kΩ			500			500	
t_{conv}	Conversion time (multiplexer addressing time not included)					8			8	clock periods

†All parameters are measured under open-loop conditions with zero common-mode input voltage. For conditions shown as MIN or MAX, use the appropriate value specified under recommended operating conditions.

NOTES: 2. Internal zener diodes are connected from the V_{CC} input to ground and from the V+ input to ground. The breakdown voltage of each zener diode is approximately 7 volts. One zener diode can be used as a shunt regulator and connects to V_{CC} through a regular diode. When the voltage regulator powers the converter, this zener and regular diode combination ensures that the V_{CC} input (6.4 V) is less than the zener breakdown voltage. A series resistor is recommended to limit current into the V+ input.
6. Total unadjusted error includes offset, full-scale, linearity, and multiplexer errors.
7. The most significant bit (MSB) data is output directly from the comparator and therefore requires additional delay to allow for comparator response time.

PARAMETER MEASUREMENT INFORMATION

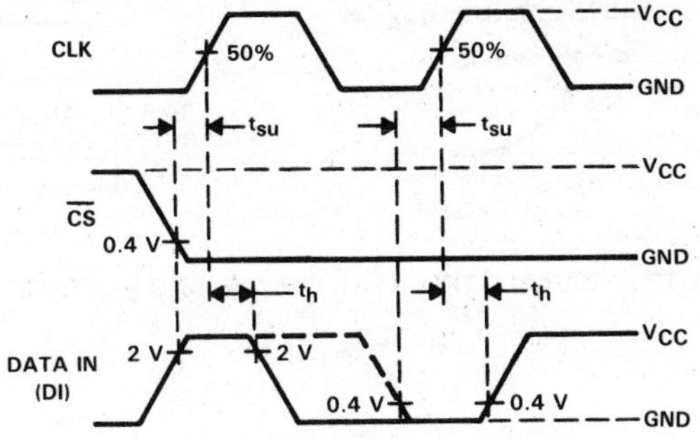
CLK
50%
50%
V_{CC}
GND
t_{su}
t_{su}
V_{CC}
$\overline{CS}$
0.4 V
GND
t_h
t_h
DATA IN
(DI)
2 V
2 V
V_{CC}
0.4 V
0.4 V
GND

FIGURE 1. DATA INPUT TIMING

PARAMETER MEASUREMENT INFORMATION

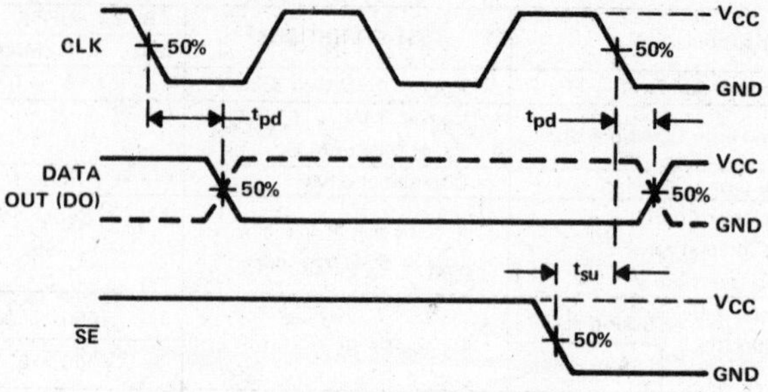
CLK
50%
50%
VCC
GND
tpd
tpd
DATA OUT (DO)
50%
50%
VCC
GND
tsu
SE
50%
VCC
GND

FIGURE 2. DATA OUTPUT TIMING

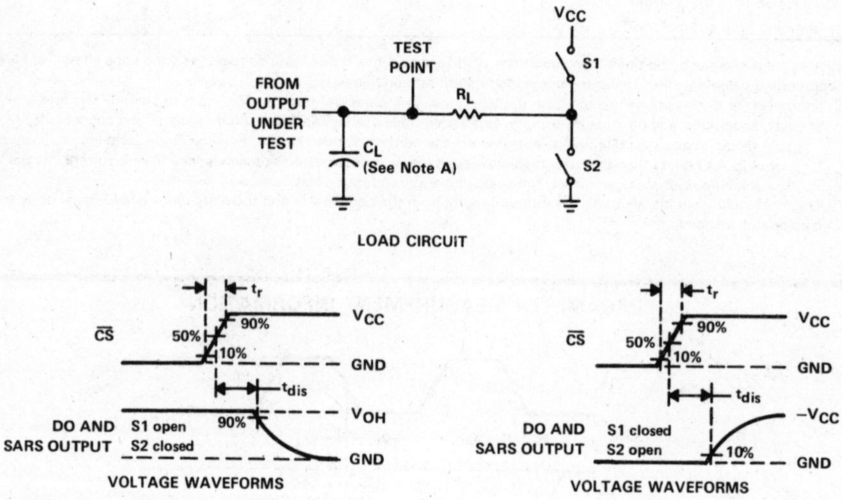
VCC
TEST POINT
S1
FROM OUTPUT UNDER TEST
RL
CL
(See Note A)
S2
LOAD CIRCUIT
tr
CS
50%
90%
10%
VCC
GND
tdis
DO AND SARS OUTPUT
S1 open
S2 closed
90%
VOH
GND
VOLTAGE WAVEFORMS
tr
CS
50%
90%
10%
VCC
GND
tdis
DO AND SARS OUTPUT
S1 closed
S2 open
10%
–VCC
GND
VOLTAGE WAVEFORMS

NOTE A: C_L includes probe and jig capacitance.

FIGURE 3. OUTPUT DISABLE TIME TEST CIRCUIT AND VOLTAGE WAVEFORMS

TYPICAL CHARACTERISTICS

UNADJUSTED OFFSET ERROR
vs
REFERENCE VOLTAGE

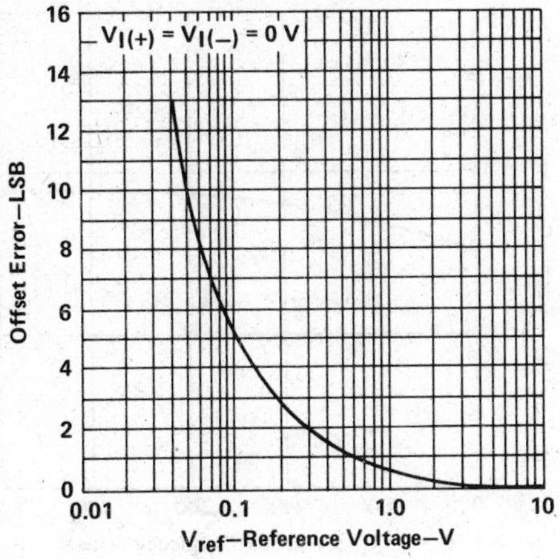
$V_{I(+)} = V_{I(-)} = 0$ V
Offset Error–LSB
V_{ref}–Reference Voltage–V
16
14
12
10
8
6
4
2
0
0.01
0.1
1.0
10

FIGURE 4

LINEARITY ERROR
vs
REFERENCE VOLTAGE

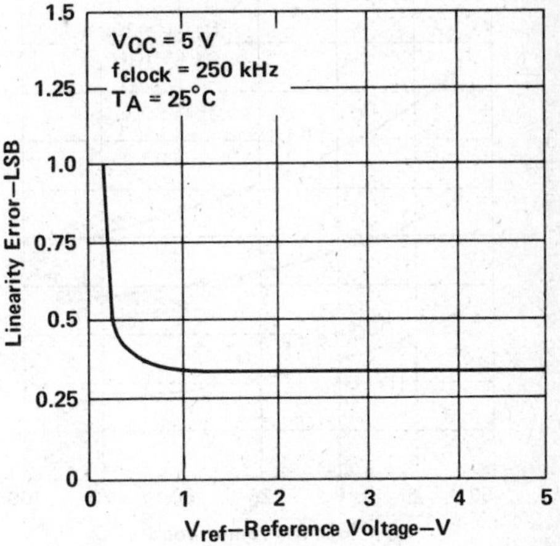
V_{CC} = 5 V
f_{clock} = 250 kHz
T_A = 25°C
Linearity Error–LSB
V_{ref}–Reference Voltage–V
1.5
1.25
1.0
0.75
0.5
0.25
0
0
1
2
3
4
5

FIGURE 5

LINEARITY ERROR
vs
FREE-AIR TEMPERATURE

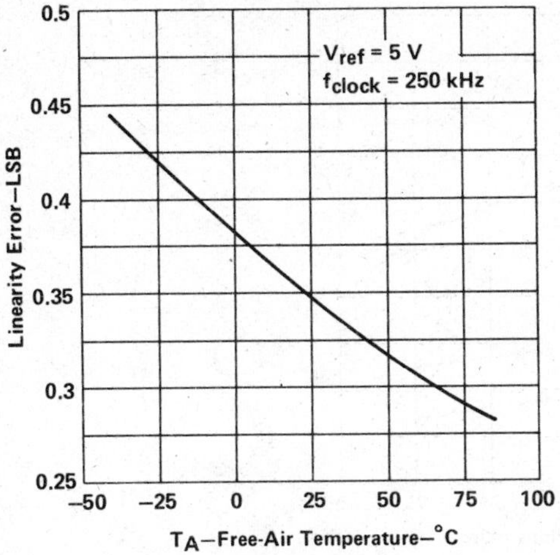
V_{ref} = 5 V
f_{clock} = 250 kHz
Linearity Error–LSB
T_A–Free-Air Temperature–°C
0.5
0.45
0.4
0.35
0.3
0.25
–50
–25
0
25
50
75
100

FIGURE 6

LINEARITY ERROR
vs
CLOCK FREQUENCY

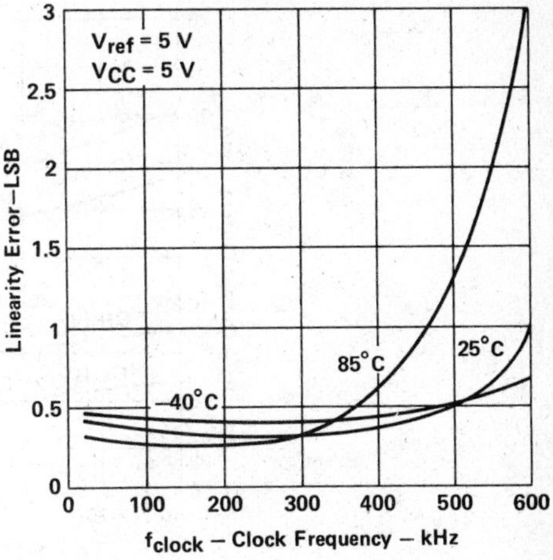
V_{ref} = 5 V
V_{CC} = 5 V
85°C
25°C
–40°C
Linearity Error–LSB
f_{clock} – Clock Frequency – kHz
3
2.5
2
1.5
1
0.5
0
0
100
200
300
400
500
600

FIGURE 7

TYPICAL CHARACTERISTICS

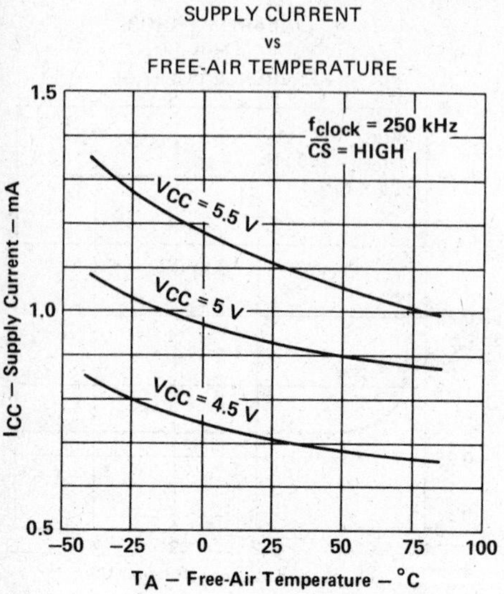
SUPPLY CURRENT
vs
FREE-AIR TEMPERATURE
fclock = 250 kHz
CS = HIGH
VCC = 5.5 V
VCC = 5 V
VCC = 4.5 V
ICC – Supply Current – mA
1.5
1.0
0.5
–50 –25 0 25 50 75 100
TA – Free-Air Temperature – °C

FIGURE 8

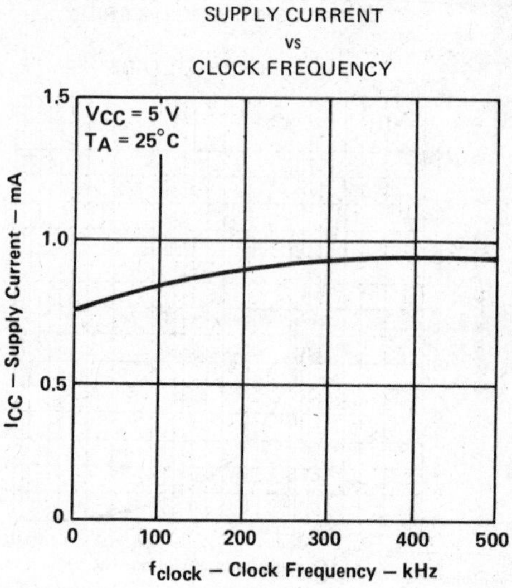
SUPPLY CURRENT
vs
CLOCK FREQUENCY
VCC = 5 V
TA = 25°C
ICC – Supply Current – mA
1.5
1.0
0.5
0
0 100 200 300 400 500
fclock – Clock Frequency – kHz

FIGURE 9

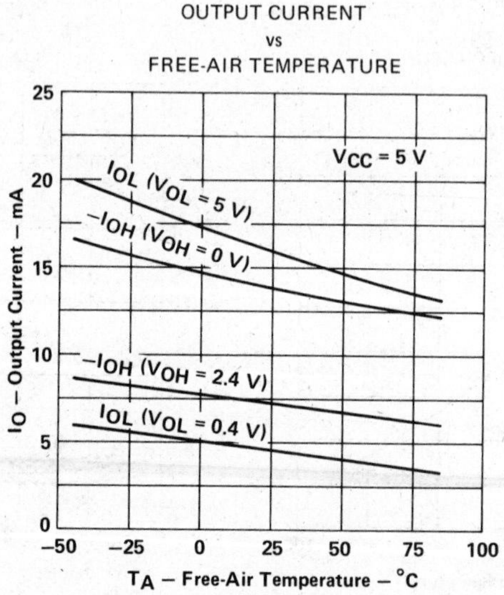
OUTPUT CURRENT
vs
FREE-AIR TEMPERATURE
VCC = 5 V
IOL (VOL = 5 V)
–IOH (VOH = 0 V)
–IOH (VOH = 2.4 V)
IOL (VOL = 0.4 V)
IO – Output Current – mA
25
20
15
10
5
0
–50 –25 0 25 50 75 100
TA – Free-Air Temperature – °C

FIGURE 10

TEXAS INSTRUMENTS
POST OFFICE BOX 655012 • DALLAS, TEXAS 75265

TL0808, TL0809
LOW-POWER CMOS ANALOG-TO-DIGITAL CONVERTERS WITH 8-CHANNEL MULTIPLEXERS

D2642, FEBRUARY 1986 – REVISED SEPTEMBER 1986

- **Total Unadjusted Error . . . ±0.75 LSB Max for TL0808 and ±1.25 LSB Max for TL0809 Over Temperature Range**
- **Ideal for Battery Operated, Portable Instrumentation Applications**
- **Resolution of 8 Bits**
- **100 μs Conversion Time**
- **Ratiometric Conversion**
- **Guaranteed Monotonicity**
- **No Missing Codes**
- **Easy Interface with Microprocessors**
- **Latched 3-State Outputs**
- **Latched Address Inputs**
- **Single 5-Volt Supply**
- **Extremely Low Power Consumption . . . 0.3 mW Typ**
- **Improved Direct Replacements for ADC0808, ADC0809**

N DUAL-IN-LINE PACKAGE
(TOP VIEW)

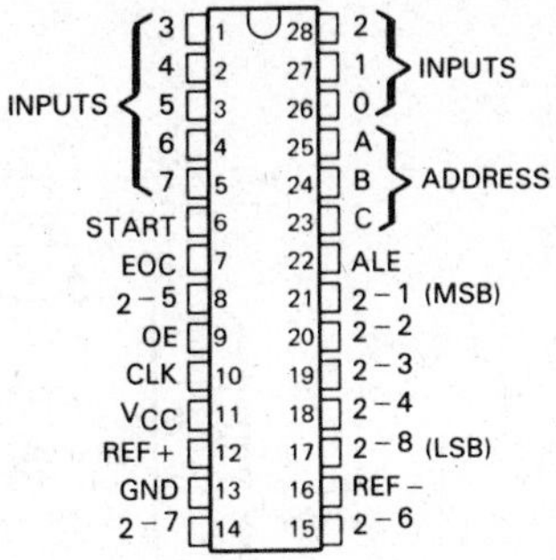
INPUTS
3 1
4 2
5 3
6 4
7 5
START 6
EOC 7
2 − 5 8
OE 9
CLK 10
VCC 11
REF + 12
GND 13
2 − 7 14
28 2
27 1
26 0
INPUTS
25 A
24 B
23 C
ADDRESS
22 ALE
21 2 − 1 (MSB)
20 2 − 2
19 2 − 3
18 2 − 4
17 2 − 8 (LSB)
16 REF −
15 2 − 6

FN PACKAGE
(TOP VIEW)

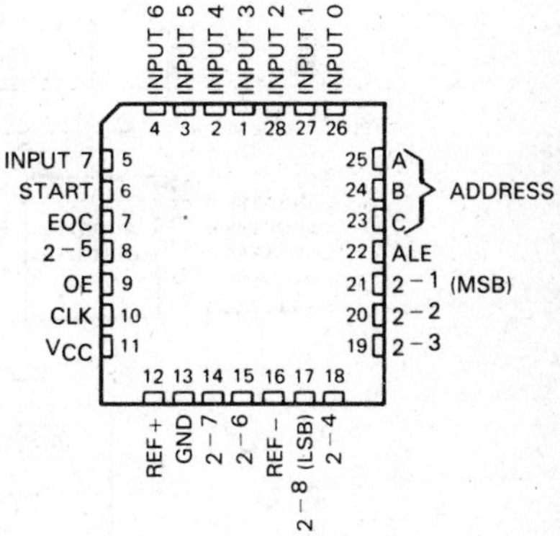
INPUT 6
INPUT 5
INPUT 4
INPUT 3
INPUT 2
INPUT 1
INPUT 0
4 3 2 1 28 27 26
INPUT 7 5
START 6
EOC 7
2 − 5 8
OE 9
CLK 10
VCC 11
25 A
24 B
23 C
ADDRESS
22 ALE
21 2 − 1 (MSB)
20 2 − 2
19 2 − 3
12 13 14 15 16 17 18
REF +
GND
2 − 7
2 − 6
REF −
2 − 8 (LSB)
2 − 4

description

The TL0808 and TL0809 are monolithic CMOS devices with an 8-channel multiplexer, an 8-bit analog-to-digital (A/D) converter, and microprocessor-compatible control logic. The 8-channel multiplexer can be controlled by a microprocessor through a 3-bit address decoder with address load to select any one of eight single-ended analog switches connected directly to the comparator. The 8-bit A/D converter uses the successive-approximation conversion technique featuring a high-impedance threshold detector, a switched-capacitor array, a sample-and-hold, and a successive-approximation register (SAR). Detailed information on interfacing to most popular microprocessors is readily available from the factory. These devices are designed to operate from common microprocessor control buses, with three-state output latches driving the data bus. The devices can be made to appear to the microprocessor as a memory location or an I/O port.

The comparison and converting methods used eliminate the possibility of missing codes, nonmonotonicity, and the need for zero or full-scale adjustment. Also featured are latched 3-state outputs from the SAR and latched inputs to the multiplexer address decoder. The single 3-volt supply and extremely low power requirements make the TL0808 and TL0809 especially useful for a wide variety of applications including portable battery and LCD applications. Ratiometric conversion is made possible by access to the reference voltage input terminals.

The TL0808 and TL0809 are characterized for operation from −40°C to 85°C.

functional block diagram (positive logic)

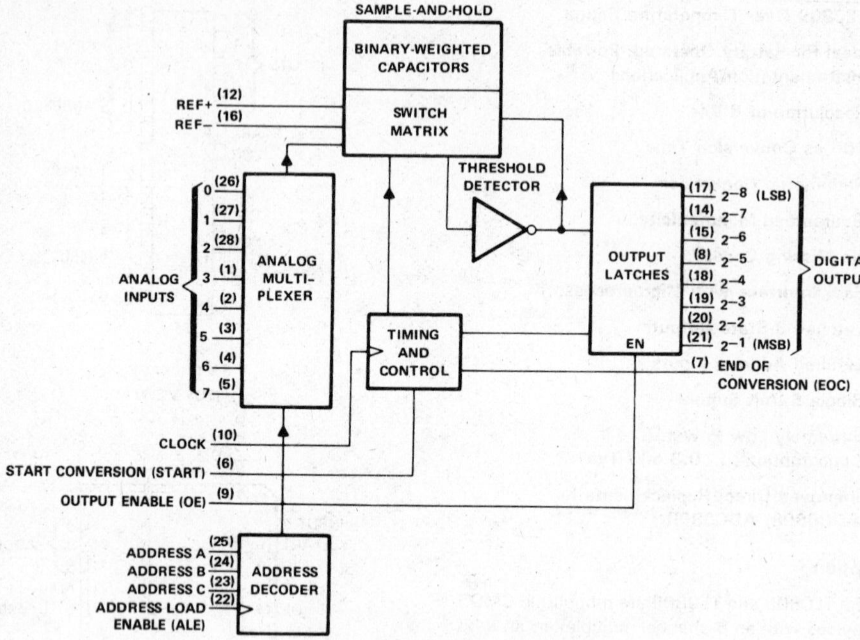
SAMPLE-AND-HOLD
BINARY-WEIGHTED CAPACITORS
SWITCH MATRIX
REF+ (12)
REF− (16)
THRESHOLD DETECTOR
ANALOG INPUTS
0 (26)
1 (27)
2 (28)
3 (1)
4 (2)
5 (3)
6 (4)
7 (5)
ANALOG MULTI-PLEXER
TIMING AND CONTROL
OUTPUT LATCHES
EN
(17) 2−8 (LSB)
(14) 2−7
(15) 2−6
(8) 2−5
(18) 2−4
(19) 2−3
(20) 2−2
(21) 2−1 (MSB)
DIGITAL OUTPUTS
(7) END OF CONVERSION (EOC)
CLOCK (10)
START CONVERSION (START) (6)
OUTPUT ENABLE (OE) (9)
ADDRESS A (25)
ADDRESS B (24)
ADDRESS C (23)
ADDRESS LOAD ENABLE (ALE) (22)
ADDRESS DECODER

MULTIPLEXER FUNCTION TABLE

INPUTS				SELECTED ANALOG CHANNEL
ADDRESS			ADDRESS STROBE	
C	B	A		
L	L	L	↑	0
L	L	H	↑	1
L	H	L	↑	2
L	H	H	↑	3
H	L	L	↑	4
H	L	H	↑	5
H	H	L	↑	6
H	H	H	↑	7

H = high level, L = low level
↑ = low-to-high transition

operating sequence

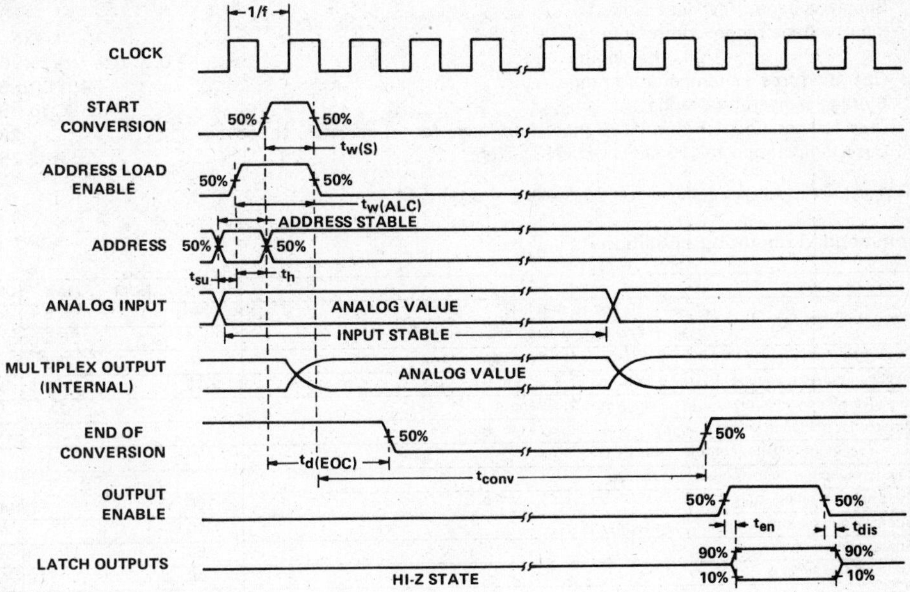
1/f
CLOCK
START CONVERSION
50%
50%
$t_{w(S)}$
ADDRESS LOAD ENABLE
50%
50%
$t_{w(ALC)}$
ADDRESS STABLE
ADDRESS
50%
50%
t_{su}
t_h
ANALOG INPUT
ANALOG VALUE
INPUT STABLE
MULTIPLEX OUTPUT (INTERNAL)
ANALOG VALUE
END OF CONVERSION
50%
50%
$t_{d(EOC)}$
t_{conv}
OUTPUT ENABLE
50%
50%
t_{en}
t_{dis}
LATCH OUTPUTS
HI-Z STATE
90%
10%
90%
10%

absolute maximum ratings over operating free-air temperature range (unless otherwise noted)

Supply voltage, V_{CC} (see Note 1) .... 6.5 V
Input voltage range: control inputs .... −0.3 to 15 V
all other inputs .... −0.3 V to V_{CC} + 0.3 V
Operating free-air temperature range .... −40°C to 85°C
Storage temperature range .... −65°C to 150°C
Lead temperature 1,6 mm (1/16 inch) from case for 10 seconds: N package .... 260°C
Case temperature for 10 seconds: FN package .... 260°C

NOTE 1: All voltage values are with respect to network ground terminal.

recommended operating conditions

		MIN	NOM	MAX	UNIT
Supply voltage, V_{CC}	f_{clock} = 10 kHz to 640 kHz	2.75		5.5	V
	f_{clock} = 640 kHz to 1280 kHz	4		5.5	
Clock frequency, f_{clock} (see supply voltage recommendation above)		10		1280	kHz
Positive reference voltage, V_{ref+} (see Notes 2, 3, and 4)		2.75	V_{CC}	V_{CC}+0.1	V
Negative reference voltage, V_{ref-} (see Notes 2, 3, and 4)			0	−0.1	V
Differential reference voltage, $V_{ref+} - V_{ref-}$ (see Note 4)			3		V
High-level input voltage, control inputs, V_{IH}		0.7 V_{CC}			V
Low-level input voltage, control inputs, V_{IL}				0.3 V_{CC}	V
Start pulse duration, $t_{w(S)}$		200			ns
Address load control pulse duration, $t_{w(ALC)}$		200			ns
Address setup time, t_{su}		50			ns
Address hold time, t_h		50			ns
Operating free-air temperature, T_A (see Note 4)		−40		85	°C

NOTES: 2. The accuracy of the conversion will depend on the stability of the reference voltages applied.
3. Analog voltages greater than or equal to V_{ref+} convert to all highs, and all voltages less than V_{ref-} convert to all lows.
4. For proper operation of the TL0808 and TL0809 at free-air temperatures below 0°C, V_{CC} and ($V_{ref+} - V_{ref-}$) should not be less than 3 volts.

electrical characteristics over recommended operating free-air temperature range, V_{CC} = 3 V to 5.25 V (unless otherwise noted)

total device

PARAMETER			TEST CONDITIONS	MIN	TYP†	MAX	UNIT
V_{OH}	High-level output voltage		$I_O = -360\ \mu A$	$V_{CC}-0.4$			V
V_{OL}	Low-level output voltage	Data outputs	I_O = 1.6 mA			0.4	V
		End of conversion	I_O = 1.2 mA			0.4	
I_{OZ}	Off-state (high-impedance-state) output current		$V_O = V_{CC}$			1	μA
			V_O = 0			−1	
I_I	Control input current at maximum input voltage		V_I = 15 V			1	μA
I_{IL}	Low-level control input current		V_I = 0			−1	μA
I_{CC}	Supply current		V_{CC} = 3 V, f_{clock} = 640 kHz		100	500	μA
			V_{CC} = 5 V, f_{clock} = 640 kHz		0.3	3	mA
C_i	Input capacitance, control inputs		T_A = 25°C		10	15	pF
C_o	Output capacitance, data outputs		T_A = 25°C		10	15	pF
	Resistance from pin 12 to pin 16			1	1000		kΩ

analog multiplexer

PARAMETER		TEST CONDITIONS		MIN	TYP†	MAX	UNIT
I_{on}	Channel on-state current (see Note 5)	V_I = 3 V,	f_{clock} = 640 kHz			2	μA
		V_I = 0,	f_{clock} = 640 kHz			−2	
I_{off}	Channel off-state current	V_{CC} = 3 V, T_A = 25°C	V_I = 3 V		10	200	nA
			V_I = 0		−10	−200	
		V_{CC} = 3 V	V_I = 3 V			1	μA
			V_I = 0			−1	

†Typical values are at V_{CC} = 3 V and T_A = 25°C.
NOTE 5: Channel on-state current is primarily due to the bias current into or out of the threshold detector, and it varies directly with clock frequency.

operating characteristics, T_A = 25 °C, V_{CC} = 3 V, V_{REF+} = 3 V, V_{REF-} = 0 V, f_{clock} = 640 kHz (unless otherwise noted)

PARAMETER		TEST CONDITIONS		TL0808 MIN	TL0808 TYP†	TL0808 MAX	TL0809 MIN	TL0809 TYP†	TL0809 MAX	UNIT
k_{SVS}	Supply voltage sensitivity	$V_{CC} = V_{ref+}$ = 3 V to 5.25 V, T_A = −40 °C to 85 °C, See Note 6			±0.05			±0.05		%/V
	Linearity error (see Note 7)				±0.5			±1		LSB
	Zero error (see Note 8)				±0.5			±0.5		LSB
	Total unadjusted error (See Note 9)	T_A = 25 °C			±0.25	±0.5		±0.5	±1	LSB
		T_A = −40 °C to 85 °C				±0.75			±1.25	LSB
t_{en}	Output enable time	C_L = 50 pF,	R_L = 10 kΩ		80	250		80	250	ns
t_{dis}	Output disable time	C_L = 10 pF,	R_L = 10 kΩ		105	250		105	250	ns
t_{conv}	Conversion time	See Note 10		90	100	116	90	110	116	μs
$t_{d(EOC)}$	Delay time, end of conversion output	See Notes 10 and 11		0		14.5	0		14.5	μs

†Typical values for all except supply voltage sensitivity are at V_{CC} = 3 V, and all are at T_A = 25 °C.

NOTES:
6. Supply voltage sensitivity relates to the ability of an analog-to-digital converter to maintain accuracy as the supply voltage varies. The supply and V_{ref+} are varied together and the change in accuracy is measured with respect to full-scale.
7. Linearity error is the maximum deviation from a straight line through the end points of the A/D transfer characteristic.
8. Zero error is the difference between 00000000 and the converted output for zero input voltage; full-scale error is the difference between 11111111 and the converted output for full-scale input voltage.
9. Total unadjusted error is the maximum sum of linearity error, zero error, and full-scale error.
10. Refer to the operating sequence diagram.
11. For clock frequencies other than 640 kHz, $t_{d(EOC)}$ maximum is 8 clock periods plus 2 μs.

PRINCIPLES OF OPERATION

The TL0808 and TL0809 each consists of an analog signal multiplexer, an 8-bit successive-approximation converter, and related control and output circuitry.

multiplexer

The analog multiplexer selects 1 of 8 single-ended input channels as determined by the address decoder. Address load control loads the address code into the decoder on a low-to-high transition. The output latch is reset by the positive-going edge of the start pulse. Sampling also starts with the positive-going edge of the start pulse and lasts for 32 clock periods. The conversion process may be interrupted by a new start pulse before the end of 64 clock periods. The previous data will be lost if a new start of conversion occurs before the 64th clock pulse. Continuous conversion may be accomplished by connecting the End-of-Conversion output to the start input. If used in this mode an external pulse should be applied after power up to assure start up.

converter

The CMOS threshold detector in the successive-approximation conversion system determines each bit by examining the charge on a series of binary-weighted capacitors (Figure 1). In the first phase of the conversion process, the analog input is sampled by closing switch S_C and all S_T switches, and by simultaneously charging all the capacitors to the input voltage.

In the next phase of the conversion process, all S_T and S_C switches are opened and the threshold detector begins identifying bits by identifying the charge (voltage) on each capacitor relative to the reference voltage. In the switching sequence, all eight capacitors are examined separately until all 8 bits are identified, and then the charge-convert sequence is repeated. In the first step of the conversion phase, the threshold detector looks at the first capacitor (weight = 128). Node 128 of this capacitor is switched to the reference voltage, and the equivalent nodes of all the other capacitors on the ladder are switched to REF−. If the voltage at the summing node is greater than the trip-point of the threshold detector (approximately one-half the V_{CC} voltage), a bit is placed in the output register, and the 128-weight capacitor is switched to REF−. If the voltage at the summing node is less than the trip point of the threshold detector, this 128-weight capacitor remains connected to REF+ through the remainder of the capacitor-sampling (bit-counting) process. The process is repeated for the 64-weight capacitor, the 32-weight capacitor, and so forth down the line, until all bits are counted.

With each step of the capacitor-sampling process, the initial charge is redistributed among the capacitors. The conversion process is successive approximation, but relies on charge redistribution rather than a successive-approximation register (and reference DAC) to count and weigh the bits from MSB to LSB.

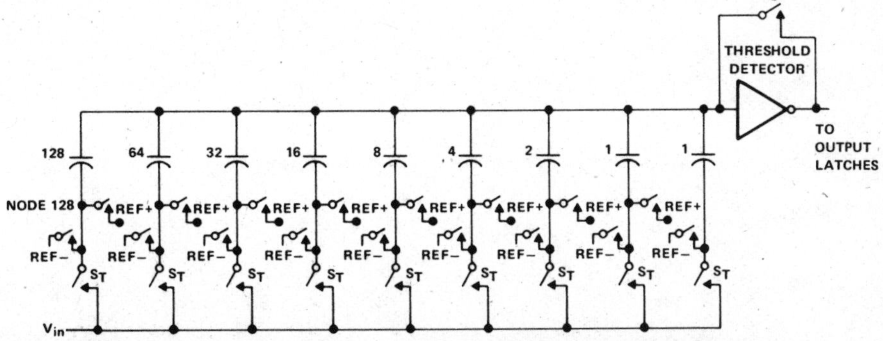
S_C
THRESHOLD DETECTOR
TO OUTPUT LATCHES
128
64
32
16
8
4
2
1
1
NODE 128
REF+
REF−
S_T
V_{in}

FIGURE 1. SIMPLIFIED MODEL OF THE SUCCESSIVE-APPROXIMATION SYSTEM

D2234, JUNE 1976—REVISED SEPTEMBER 1986

- Functionally Interchangeable with Siliconix DG182, DG185, DG188, DG191 with Same Terminal Assignments
- Monolithic Construction
- Adjustable Reference Voltage
- JFET Inputs
- Uniform On-State Resistance for Minimum Signal Distortion
- ±10-V Analog Voltage Range
- TTL, MOS, and CMOS Logic Control Compatibility

description

The TL182, TL185, TL188, and TL191 are monolithic high-speed analog switches using BI-MOS technology. They comprise JFET-input buffers, level translators, and output JFET switches. The TL182 switches are SPST; the TL185 switches are SPDT. The TL188 is a pair of complementary SPST switches as is each half of the TL191.

A high level at a control input of the TL182 turns the associated switch off. A high level at a control input of the TL185 turns the associated switch on. For the TL188, a high level at the control input turns the associated switches S1 on and S2 off.

The threshold of the input buffer is determined by the voltage applied to the reference input (V_{ref}). The input threshold is related to the reference input by the equation $V_{th} = V_{ref} + 1.4$ V. Thus, for TTL compatibility, the V_{ref} input is connected to ground. The JFET input makes the device compatible with bipolar, MOD, and CMOS logic families. Threshold compatibility may, again, be determined by $V_{th} = V_{ref} + 1.4$ V.

The output switches are junction field-effect transistors featuring low on-state resistance and high off-state resistance. The monolithic structure ensures uniform matching.

BI-MOS technology is a major breakthrough in linear integrated circuit processing. BI-MOS can have ion-implanted JFETs, p-channel MOS-FETs, plus the usual bipolar components all on the same chip. BI-MOS allows circuit designs that previously have been available only as expensive hybrids to be monolithic.

Devices with an "M" suffix are characterized for operation from −55°C to 125°C, those with an "I" suffix are characterized for operation from −25°C to 85°C, and those with a "C" suffix are characterized for operation from 0°C to 70°C.

TL182
N DUAL-IN-LINE PACKAGE
(TOP VIEW)

1S	1	14	2S
1D	2	13	2D
NC	3	12	NC
NC	4	11	NC
1A	5	10	2A
V_{CC}	6	9	V_{EE}
V_{LL}	7	8	V_{ref}

TL185
N DUAL-IN-LINE PACKAGE
(TOP VIEW)

1D1	1	16	1S1
NC	2	15	1A
1D2	3	14	V_{EE}
1S2	4	13	V_{ref}
2S1	5	12	V_{LL}
2D1	6	11	V_{CC}
NC	7	10	2A
2D2	8	9	2S2

TL188
N DUAL-IN-LINE PACKAGE
(TOP VIEW)

NC	1	14	NC
NC	2	13	NC
D1	3	12	D2
S1	4	11	S2
A	5	10	NC
V_{CC}	6	9	V_{EE}
V_{LL}	7	8	V_{ref}

TL191
N DUAL-IN-LINE PACKAGE
(TOP VIEW)

1D1	1	16	1S1
NC	2	15	1A
1D2	3	14	V_{EE}
1S2	4	13	V_{ref}
2S2	5	12	V_{LL}
2D2	6	11	V_{CC}
NC	7	10	2A
2D1	8	9	2S1

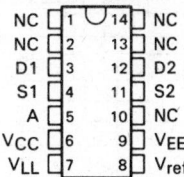

NC—No internal connection

TL182 TWIN SPST SWITCH

schematic (each channel)

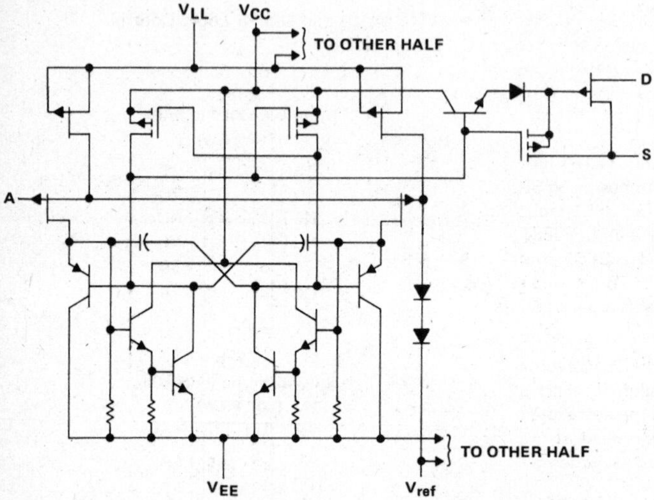
V_{LL}
V_{CC}
TO OTHER HALF
D
S
A
TO OTHER HALF
V_{EE}
V_{ref}

symbol

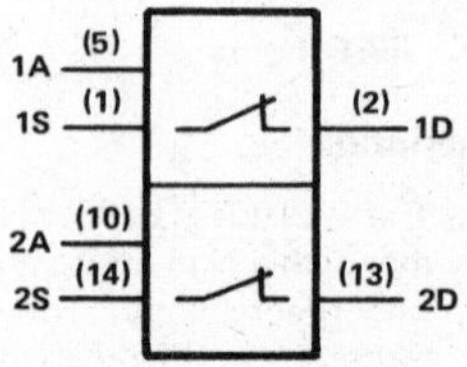
1A (5)
1S (1)
(2) 1D
2A (10)
2S (14)
(13) 2D

FUNCTION TABLE
(EACH HALF)

INPUT A	SWITCH S
L	ON (CLOSED)
H	OFF (OPEN)

TL185 TWIN DPST SWITCH

schematic (each channel)

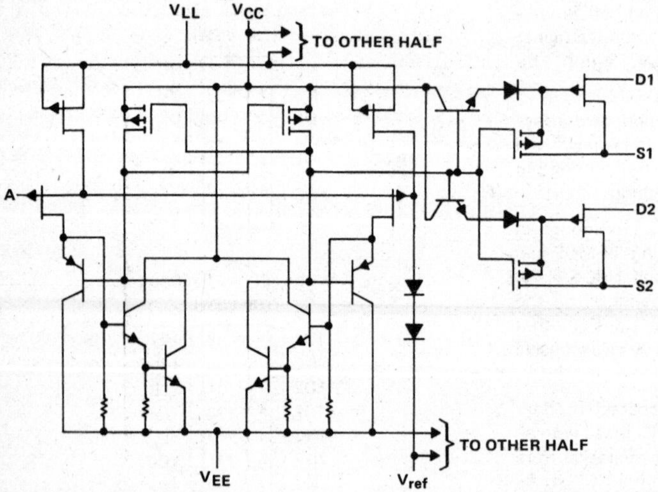
V_{LL}
V_{CC}
TO OTHER HALF
D1
S1
A
D2
S2
TO OTHER HALF
V_{EE}
V_{ref}

symbol

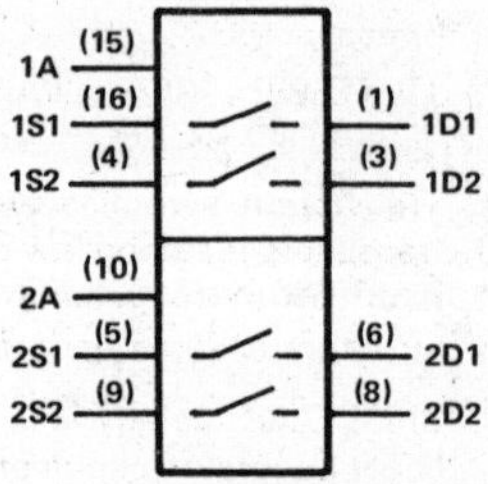
1A (15)
1S1 (16)
1S2 (4)
(1) 1D1
(3) 1D2
2A (10)
2S1 (5)
2S2 (9)
(6) 2D1
(8) 2D2

FUNCTION TABLE
(EACH HALF)

INPUT A	SWITCHES SW1 AND SW2
L	OFF (OPEN)
H	ON (CLOSED)

TEXAS INSTRUMENTS
POST OFFICE BOX 655012 • DALLAS, TEXAS 75265

TL188 DUAL COMPLEMENTARY SPST SWITCH

schematic

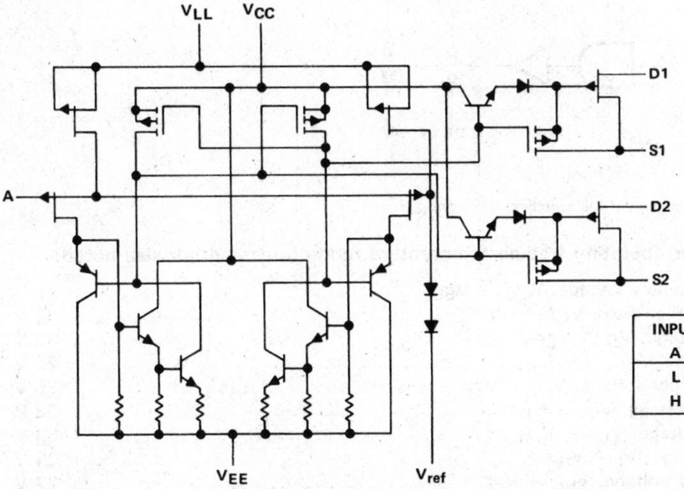
VLL
VCC
A
D1
S1
D2
S2
VEE
Vref

symbol

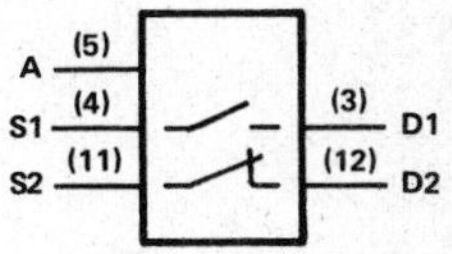
A (5)
S1 (4)
S2 (11)
(3) D1
(12) D2

FUNCTION TABLE

INPUT A	SWITCHES SW1	SW2
L	OFF (OPEN)	ON (CLOSED)
H	ON (CLOSED)	OFF (OPEN)

TL191 TWIN DUAL COMPLEMENTARY SPST SWITCH

schematic (each channel)

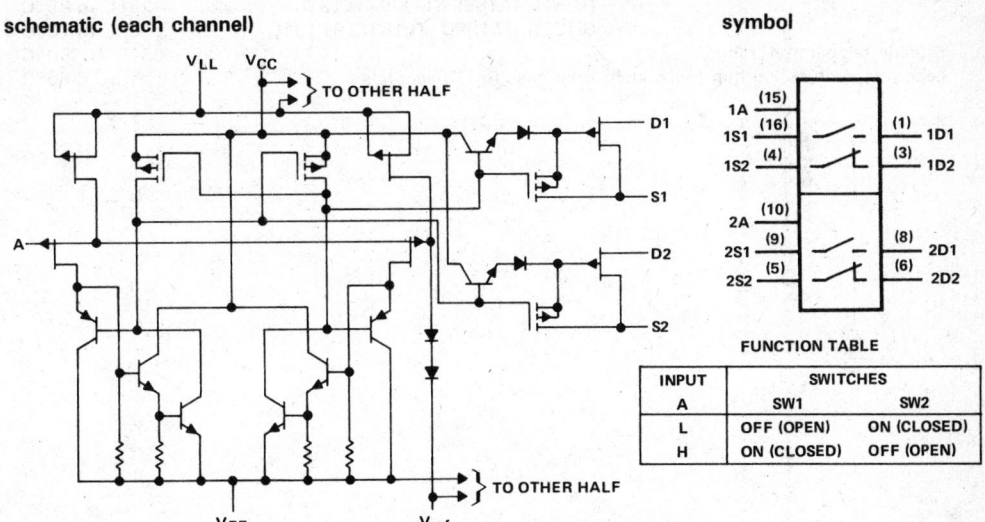

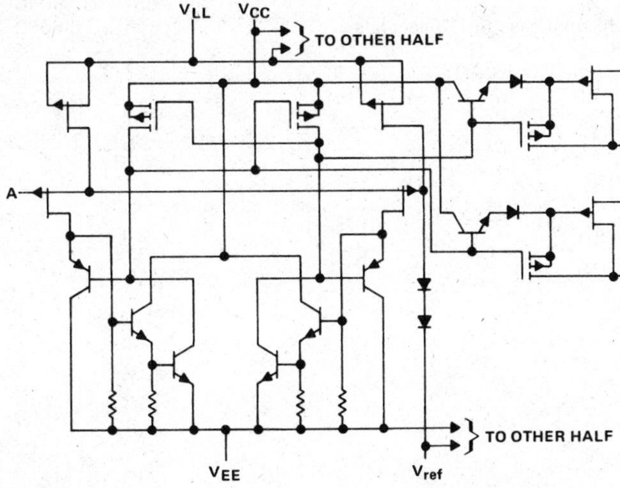
VLL
VCC
TO OTHER HALF
A
D1
S1
D2
S2
TO OTHER HALF
VEE
Vref

symbol

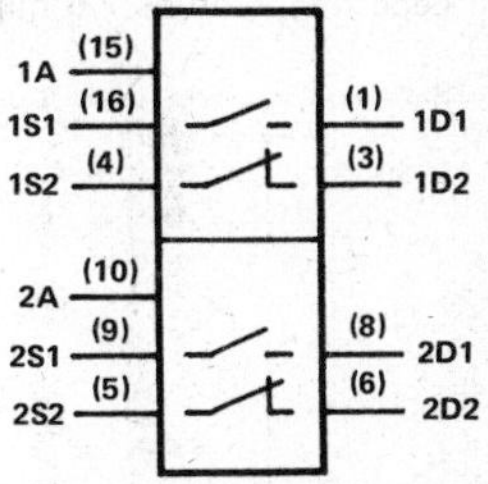
1A (15)
1S1 (16)
1S2 (4)
(1) 1D1
(3) 1D2
2A (10)
2S1 (9)
2S2 (5)
(8) 2D1
(6) 2D2

FUNCTION TABLE

INPUT A	SWITCHES SW1	SW2
L	OFF (OPEN)	ON (CLOSED)
H	ON (CLOSED)	OFF (OPEN)

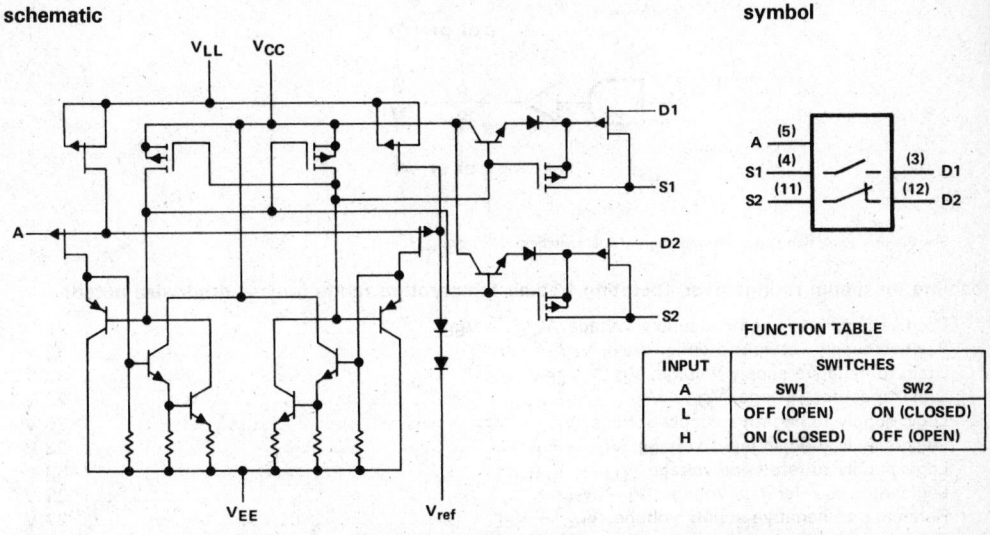

functional block diagram

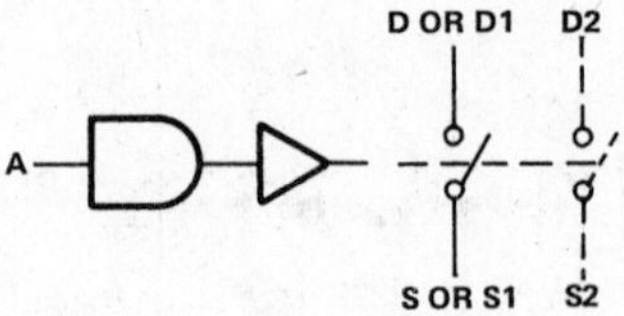
D OR D1
D2
A
S OR S1
S2

See the preceding two pages for operation of the switches.

absolute maximum ratings over operating free-air temperature range (unless otherwise noted)

Positive supply to negative supply voltage, $V_{CC} - V_{EE}$		36 V
Positive supply voltage to either drain, $V_{CC} - V_D$		33 V
Drain to negative supply voltage, $V_D - V_{EE}$		33 V
Drain to source voltage, $V_D - V_S$		±22 V
Logic supply to negative supply voltage, $V_{LL} - V_{EE}$		36 V
Logic supply to logic input voltage, $V_{LL} - V_I$		33 V
Logic supply to reference voltage, $V_{LL} - V_{ref}$		33 V
Logic input to reference voltage, $V_I - V_{ref}$		33 V
Reference to negative supply voltage, $V_{ref} - V_{EE}$		27 V
Reference to logic input voltage, $V_{ref} - V_I$		2 V
Current (any terminal)		30 mA
Operating free-air temperature range:	TL182M, TL185M, TL188M, TL191M	−55°C to 125°C
	TL182I, TL185I, TL188I, TL191I	−25°C to 85°C
	TL182C, TL185C, TL188C, TL191C	0°C to 70°C
Storage temperature range		−65°C to 150°C
Lead temperature 1,6 mm (1/16 inch) from case for 10 seconds		260°C

electrical characteristics, V_{CC} = 15 V, V_{EE} = −15 V, V_{LL} = 5 V, V_{ref} = 0 V

PARAMETER		TEST CONDITIONS		TL1_M MIN	TL1_M MAX	TL1_I MIN	TL1_I MAX	TL1_C MIN	TL1_C MAX	UNIT
V_{IH}	High-level control input voltage		T_A = MIN TO MAX	$V_{ref}+2$		$V_{ref}+2$		$V_{ref}+2$		V
V_{IL}	Low-level control input voltage		T_A = MIN to MAX		$V_{ref}+0.8$		$V_{ref}+0.8$		$V_{ref}+0.8$	V
I_{IH}	High-level control input current	V_I = 5 V	T_A = 25°C		10		10		20	μA
			T_A = MAX		20		20		20	
I_{IL}	Low-level control input current	V_I = 0	T_A = MIN to MAX		−250		−250		−250	μA
$I_{D(off)}$	Off-state drain current	V_D = 10 V, V_S = −10 V, V_{IH} = 2 V, V_{IL} = 0.8 V	T_A = 25°C				5		5	nA
			T_A = MAX		100		100		100	
$I_{S(off)}$	Off-state source current	V_D = −10 V, V_S = 10 V, V_{IH} = 2 V, V_{IL} = 0.8 V	T_A = 25°C				5		5	nA
			T_A = MAX		100		100		100	
$I_{D(on)}+I_{S(on)}$	On-state channel leakage current	V_D = −10 V, V_S = −10 V, V_{IH} = 2 V, V_{IL} = 0.8 V	T_A = 25°C				−10		−10	nA
			T_A = MAX		−200		−200		−200	
$r_{DS(on)}$	Drain-to-source on-state resistance	V_D = −10 V, I_S = 1 mA, V_{IH} = 2 V, V_{IL} = 0.8 V; TL182, TL188	T_A = MIN to 25°C		75		100		100	Ω
		TL182, TL188	T_A = MAX		100		150		150	
		TL185, TL191	T_A = MIN to 25°C		125		150		150	
		TL185, TL191	T_A = MAX		250		300		300	
I_{CC}	Supply current from V_{CC}	Both control inputs at 0 V	T_A = 25°C		1.5		1.5		1.5	mA
I_{EE}	Supply current from V_{EE}				−5		−5		−5	
I_{LL}	Supply current from V_{LL}				4.5		4.5		4.5	
I_{ref}	Reference current				−2		−2		−2	
I_{CC}	Supply current from V_{CC}	Both control inputs at 5 V	T_A = 25°C		1.5		1.5		1.5	mA
I_{EE}	Supply current from V_{EE}				−5		−5		−5	
I_{LL}	Supply current from V_{LL}				4.5		4.5		4.5	
I_{ref}	Reference current				−2		−2		−2	

switching characteristics, V_{CC} = 10 V, V_{EE} = −20 V, V_{LL} = 5 V, V_{ref} = 0 V, T_A = 25°C

PARAMETER		TEST CONDITIONS	TL1_M TYP	TL1_I TYP	TL1_C TYP	UNIT
t_{on}	Turn-on time	R_L = 300 Ω, C_L = 30 pF, Figure 1	175	175	175	ns
t_{off}	Turn-off time		350	350	350	

PARAMETER MEASUREMENT INFORMATION

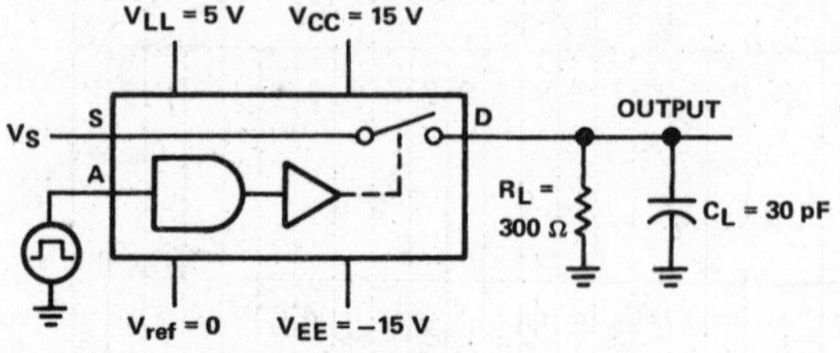
V_{LL} = 5 V
V_{CC} = 15 V
V_S
S
A
D
OUTPUT
R_L = 300 Ω
C_L = 30 pF
V_{ref} = 0
V_{EE} = −15 V

C_L includes probe and jig capacitance

V_S = 3 V for t_{on} and −3 V for t_{off}

$$V_O = V_S \frac{R_L}{R_L + r_{DS(on)}}$$

TEST CIRCUIT

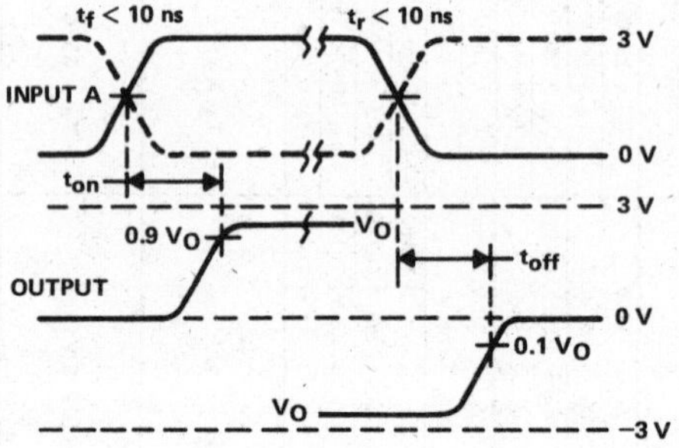
t_f < 10 ns
t_r < 10 ns
3 V
INPUT A
0 V
t_{on}
3 V
0.9 V_O
V_O
t_{off}
OUTPUT
0 V
0.1 V_O
V_O
−3 V

NOTE: A. The solid waveform applies for TL185 and SW1 of TL185 and TL191; the dashed waveform applies for TL182 and SW2 of TL185 and TL191.

B. V_O is the steady-state output with the switch on. Feed through via the gate capacitance may result in spikes (not shown) at the leading and trailing edges of the output waveform.

FIGURE 1. VOLTAGE WAVEFORMS

D2477, DECEMBER 1979—REVISED DECEMBER 1986

TL500C/TL501C ANALOG PROCESSORS

- True Differential Inputs
- Automatic Zero
- Automatic Polarity
- High Input Impedance . . . 10^9 Ohms Typically

TL500C CAPABILITIES

- Resolution . . . 14 Bits (with TL502C)
- Linearity Error . . . 0.001%
- 4 1/2-Digit Readout Accuracy with External Precision Reference

TL501C CAPABILITIES

- Resolution . . . 10-13 Bits (with TL502C)
- Linearity Error . . . 0.01%
- 3 1/2-Digit Readout Accuracy

TL502C/TL503C DIGITAL PROCESSORS

- Fast Display Scan Rates
- Internal Oscillator May Be Driven or Free-Running
- Interdigit Blanking
- Over-Range Blanking
- 4 1/2-Digit Display Circuitry
- High-Sink-Current Digit Driver for Large Displays

TL502C CAPABILITIES

- Compatible with Popular Seven-Segment Common-Anode Displays
- High-Sink-Current Segment Driver for Large Displays

TL503C CAPABILITIES

- Multiplexed BCD Outputs
- High-Sink-Current BCD Outputs

Caution. These devices have limited built-in gate protection. The leads should be shorted together or the device placed in conductive foam during storage or handling to prevent electrostatic damage to the MOS gates.

description

The TL500C and TL501C analog processors and TL502C and TL503C digital processors provide the basic functions for a dual-slope-integrating analog-to-digital converter.

The TL500C and TL501C contain the necessary analog switches and decoding circuits, reference voltage generator, buffer, integrator, and comparator. These devices may be controlled by the TL502C, TL503C, by discrete logic, or by a software routine in a microprocessor.

The TL502C and TL503C each includes oscillator, counter, control logic, and digit enable circuits. The TL502C provides multiplexed outputs for seven-segment displays, while the TL503C has multiplexed BCD outputs.

When used in complementary fashion, these devices form a system that features automatic zero-offset compensation, true differential inputs, high input impedance, and capability for 4 1/2-digit accuracy. Applications include the conversion of analog data from high-impedance sensors of pressure, temperature, light, moisture, and position. Analog-to-digital-logic conversion provides display and control signals for weight scales, industrial controllers, thermometers, light-level indicators, and many other applications.

principles of operation

The basic principle of dual-slope-integrating converters is relatively simple. A capacitor, C_X, is charged through the integrator from V_{CT} for a fixed period of time at a rate determined by the value of the unknown voltage input. Then the capacitor is discharged at a fixed rate (determined by the reference voltage) back to V_{CT} where the discharge time is measured precisely. The relationship of the charge and discharge values are shown below (see Figure 1).

$$V_{CX} = V_{CT} - \frac{V_I t_1}{R_X C_X} \quad \text{Charge} \qquad (1)$$

$$V_{CT} = V_{CX} - \frac{V_{ref} t_2}{R_X C_X} \quad \text{Discharge} \qquad (2)$$

Combining equations 1 and 2 results in:

$$\frac{V_I}{V_{ref}} = -\frac{t_2}{t_1} \qquad (3)$$

where:

V_{CT} = Comparator (offset) threshold voltage
V_{CX} = Voltage change across C_X during t_1 and during t_2 (equal in magnitude)
V_I = Average value of input voltage during t_1
t_1 = Time period over which unknown voltage is integrated
t_2 = Unknown time period over which a known reference voltage is integrated.

Equation (3) illustrates the major advantages of a dual-slop converter:

a. Accuracy is not dependent on absolute values of t_1 and t_2, but is dependent on their ratios. Long-term clock frequency variations will not affect the accuracy.
b. Offset values, V_{CT}, are not important.

The BCD counter in the digital processor (see Figure 2) and the control logic divide each measurement cycle into three phases. The BCD counter changes at a rate equal to one-half the oscillator frequency.

auto-zero phase

The cycle begins at the end of the integrate-reference phase when the digital processor applies low levels to inputs A and B of the analog processor. If the trigger input is at a high level, a free-running condition exists and continuous conversions are made. However, if the trigger input is low, the digital processor stops the counter at 20,000, entering a hold mode. In this mode, the processor samples the trigger input every 4000 oscillator pulses until a high level is detected. When this occurs, the counter is started again and is carried to completion at 30,000. The reference voltage is stored on reference capacitor C_{ref}, comparator offset voltage is stored on integration capacitor C_X, and the sum of the buffer and integrator offset voltages is stored on zero capacitor C_Z. During the auto-zero phase, the comparator output is characterized by an oscillation (limit cycle) of indeterminate waveform and frequency that is filtered and d-c shifted by the level shifter.

integrate-input phase

The auto-zero phase is completed at a BCD count of 30,000, and high levels are applied to both control inputs to initiate the integrate-input phase. The integrator charges C_X for a fixed time of 10,000 BCD counts at a rate determined by the input voltage. Note that during this phase, the analog inputs see only the high impedance of the noninverting operational amplifier input. Therefore, the integrator responds only to the difference between the analog input terminals, thus providing true differential inputs.

integrate-reference phase

At a BCD count of 39,999 + 1 = 40,000 or 0, the integrate-input phase is terminated and the integrate-reference phase is begun by sampling the comparator output. If the comparator output is low corresponding to a negative average analog input voltage, the digital processor applies a low and a high to inputs A and B, respectively, to apply the reference voltage stored on C_{ref} to the buffer. If the comparator output is high corresponding to a positive input, inputs A and B are made high and low, respectively, and the negative of the stored reference voltage is applied to the buffer. In either case, the processor automatically selects the proper logic state to cause the integrator to ramp back toward zero at a rate proportional to the reference voltage. The time required to return to zero is measured by the counter in the digital processor. The phase is terminated when the integrator output crosses zero and the counter contents are transferred to the register, or when the BCD counter reaches 20,000 and the over-range indication is activated. When activated, the over-range indication blanks all but the most significant digit and sign.

Seventeen parallel bits (4-1/2 digits) of information are strobed into the buffer register at the end of the integration phase. Information for each digit is multiplexed out to the BCD outputs (TL503C) or the seven-segment drivers (TL502C) at a rate equal to the oscillator frequency divided by 400.

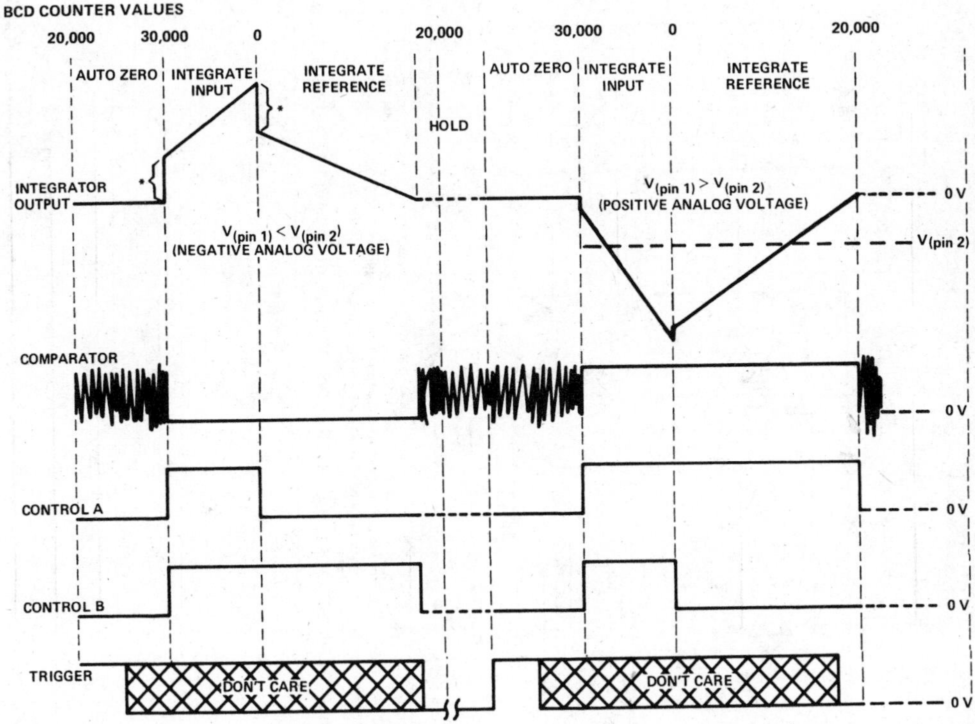
BCD COUNTER VALUES
20,000
30,000
0
20,000
30,000
0
20,000
AUTO ZERO
INTEGRATE INPUT
INTEGRATE REFERENCE
HOLD
AUTO ZERO
INTEGRATE INPUT
INTEGRATE REFERENCE
INTEGRATOR OUTPUT
V(pin 1) < V(pin 2)
(NEGATIVE ANALOG VOLTAGE)
V(pin 1) > V(pin 2)
(POSITIVE ANALOG VOLTAGE)
0 V
V(pin 2)
COMPARATOR
0 V
CONTROL A
0 V
CONTROL B
0 V
TRIGGER
DON'T CARE
DON'T CARE
0 V

*This step is the voltage at pin 2 with respect to analog ground.

FIGURE 1. VOLTAGE WAVEFORMS AND TIMING DIAGRAM

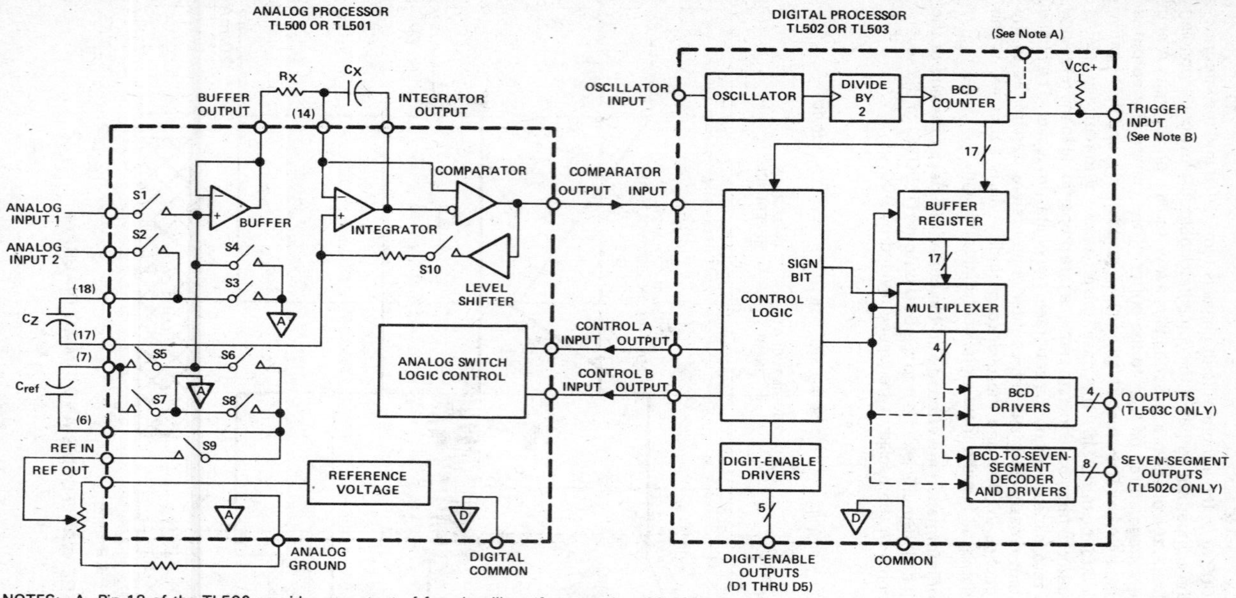
ANALOG PROCESSOR
TL500 OR TL501
DIGITAL PROCESSOR
TL502 OR TL503
(See Note A)
BUFFER OUTPUT
RX
CX
(14)
INTEGRATOR OUTPUT
OSCILLATOR INPUT
OSCILLATOR
DIVIDE BY 2
BCD COUNTER
VCC+
TRIGGER INPUT (See Note B)
17
COMPARATOR
COMPARATOR OUTPUT INPUT
ANALOG INPUT 1
S1
BUFFER
INTEGRATOR
BUFFER REGISTER
ANALOG INPUT 2
S2
S4
S10
SIGN BIT
S3
LEVEL SHIFTER
CONTROL LOGIC
MULTIPLEXER
(18)
CZ
(17)
(7)
Cref
(6)
S5
S6
S7
S8
S9
A
ANALOG SWITCH LOGIC CONTROL
CONTROL A INPUT OUTPUT
CONTROL B INPUT OUTPUT
4
BCD DRIVERS
Q OUTPUTS (TL503C ONLY)
REF IN
REF OUT
REFERENCE VOLTAGE
DIGIT-ENABLE DRIVERS
BCD-TO-SEVEN-SEGMENT DECODER AND DRIVERS
8
SEVEN-SEGMENT OUTPUTS (TL502C ONLY)
5
D
ANALOG GROUND
DIGITAL COMMON
DIGIT-ENABLE OUTPUTS (D1 THRU D5)
COMMON

NOTES: A. Pin 18 of the TL502 provides an output of f_{osc} (oscillator frequency) ÷ 20,000.
B. The trigger input assumes a high level if not externally connected.

FIGURE 2. BLOCK DIAGRAM OF BASIC ANALOG-TO-DIGITAL CONVERTER USING TL500C or TL501C and TL502C or TL503C

MODE	ANALOG INPUT	COMPARATOR	CONTROLS A AND B	ANALOG SWITCHES CLOSED
Auto Zero / Hold†	X	Oscillation	L L	S3, S4, S7, S9, S10
Integrate Input	Positive	H	H H	S1, S2
	Negative	L		
Integrate Reference	X	L‡	L H	S3, S6, S7
		H‡	H L	S3, S5, S8

H ≡ High, L ≡ low, X ≡ Irrelevant

† If the trigger input is low at the beginning of the auto-zero cycle, the system will enter the hold mode. A high level (or open circuit) will signal the digital processor to continue or resume normal operation.
‡ This is the state of the comparator output as determined by the polarity of the analog input during the integrate input phase.

TEXAS INSTRUMENTS
POST OFFICE BOX 655012 • DALLAS, TEXAS 75265

description of analog processors

J DUAL-IN-LINE PACKAGE
(TOP VIEW)

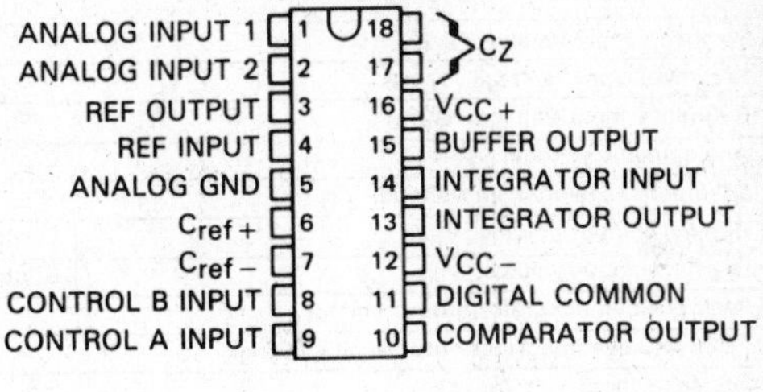
ANALOG INPUT 1
ANALOG INPUT 2
REF OUTPUT
REF INPUT
ANALOG GND
Cref +
Cref −
CONTROL B INPUT
CONTROL A INPUT
1
2
3
4
5
6
7
8
9
18
17
16
15
14
13
12
11
10
CZ
VCC+
BUFFER OUTPUT
INTEGRATOR INPUT
INTEGRATOR OUTPUT
VCC−
DIGITAL COMMON
COMPARATOR OUTPUT

The TL500C and TL501C analog processors are designed to automatically compensate for internal zero offsets, integrate a differential voltage at the analog inputs, integrate a voltage at the reference input in the opposite direction, and provide an indication of zero-voltage crossing. The external control mechanism may be a microcomputer and software routing, discrete logic, or a TL502C or TL503C controller. The TL500C and TL501C are designed primarily for simple, cost-effective, dual-slope analog-to-digital converters. Both devices feature true differential analog inputs, high input impedance, and an internal reference-voltage source. The TL500C provides 4-1/2-digit readout accuracy when used with a precision external reference voltage. The TL501C provides 100-ppm linearity error and 3-1/2-digit accuracy capability. These devices are manufactured using TI's advanced technology to produce JFET, MOSFET, and bipolar devices on the same chip. The TL500C and TL501C are intended for operation over the temperature range of 0°C to 70°C.

schematics of inputs and outputs

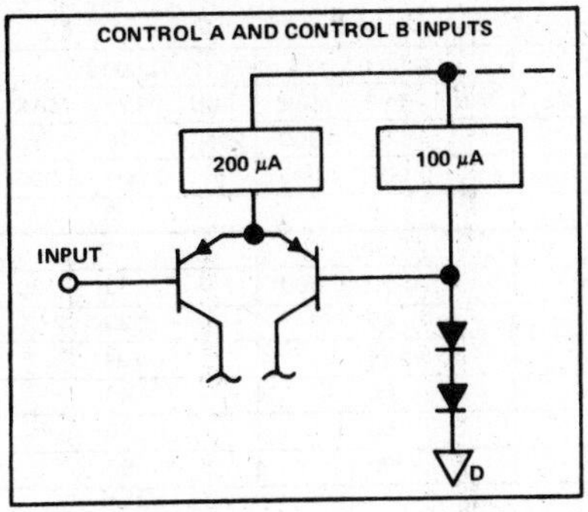
CONTROL A AND CONTROL B INPUTS
200 μA
100 μA
INPUT
D

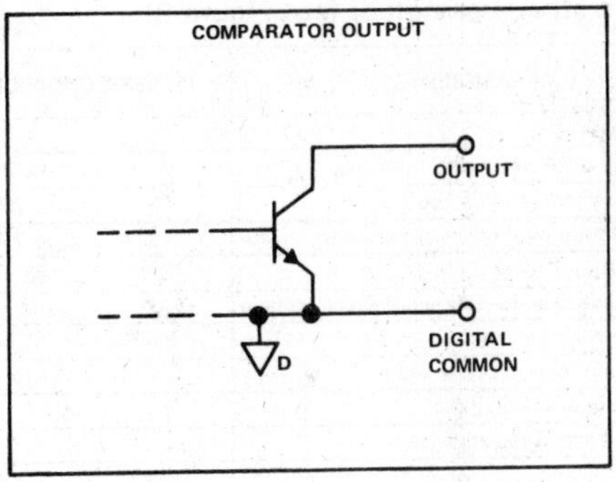
COMPARATOR OUTPUT
OUTPUT
DIGITAL
COMMON
D

absolute maximum ratings over operating free-air temperature range (unless otherwise noted)

Positive supply voltage, V_{CC+} (see Note 1)	+18 V
Negative supply voltage, V_{CC-}	−18 V
Input voltage, V_I	$\pm V_{CC}$
Comparator output voltage range (see Note 2)	0 V to V_{CC+}
Comparator output sink current (see Note 2)	20 mA
Buffer, reference, or integrator output source current (see Note 2)	10 mA
Total dissipation at (or below) 25°C free-air temperature (see Note 3)	1025 mW
Operating free-air temperature range	−0°C to 70°C
Storage temperature range	−65°C to 150°C
Lead temperature 1,6 mm (1/16 inch) from case for 60 seconds	300°C

NOTES: 1. Voltage values, except differential voltages, are with respect to the analog ground common pin tied together.
2. Buffer, integrator, and comparator outputs are not short-circuit protected.
3. For operation above 25°C free-air temperature, refer to Dissipation Derating Curves, Appendix A. TL500C and TL501C chips are glass mounted.

recommended operating conditions

		MIN	NOM	MAX	UNIT
Positive supply voltage, V_{CC+}		7	12	15	V
Negative supply voltage, V_{CC-}		−9	−12	−15	V
Reference input voltage, $V_{ref(I)}$		0.1		5	V
Analog input voltage, V_I				±5	V
Differential analog input voltage, V_{ID}				10	V
High-level input voltage, V_{IH}	Control inputs	2			V
Low-level input voltage, V_{IL}	Control inputs			0.8	V
Peak positive integrator output voltage, V_{OM+}		+9			V
Peak negative integrator output voltage, V_{OM-}		−5			V
Full scale input voltage				2 V_{ref}	
Autozero and reference capacitors, C_Z and C_{ref}		0.2			μF
Integrator capacitor, C_X		0.2			μF
Integrator resistor, R_X		15		100	kΩ
Integrator time constant, R_XC_X		See Note 4			
Free-air operating temperature, T_A		0		70	°C
Maximum conversion rate with TL502 or TL503			3	12.5	conv/sec

system electrical characteristics at $V_{CC\pm} = \pm 12$ V, $V_{ref} = 1{,}000 \pm 0.03$ mV, $T_A = 25°C$ (unless otherwise noted) (see Figure 3)

PARAMETER	TEST CONDITIONS	TL501C			TL500C			UNIT
		MIN	TYP	MAX	MIN	TYP	MAX	
Zero error			50	300		10	30	μV
Linearity error relative to full scale	$V_I = -2$ V to 2 V		0.005	0.05		0.001	0.005	%FS
Full scale temperature coefficient	$T_A = 0°C$ to $70°C$		6			6		ppm/°C
Temperature coefficient of zero error	$T_A = 0°C$ to $70°C$		4			1		μV/°C
Rollover error†			200	500		30	100	μV
Equivalent peak-to-peak input noise voltage			20			20		μV
Analog input resistance	Pin 1 or 2		10^9			10^9		Ω
Common-mode rejection ratio	$V_{IC} = -1$ V to +1 V		86			90		dB
Current into analog input	$V_I = \pm 5$ V		50			50		pA
Supply voltage rejection ratio			90			90		dB

†Rollover error is the voltage difference between the conversion results of the full-scale positive 2 volts and the full-scale negative 2 volts.

NOTE 4. The minimum integrator time constant may be found by use of the following formula:

$$\text{Minimum } R_XC_X = \frac{V_{ID}\text{ (full scale) } t_1}{|V_{OM-}| - V_{I(\text{pin }2)}}$$

where

V_{ID} = voltage at pin with respect to pin 2
$V_{I(\text{pin }2)}$ = voltage at pin 2 with respect to analog ground
t_1 = input integration time seconds

TEXAS INSTRUMENTS
POST OFFICE BOX 655012 • DALLAS, TEXAS 75265

electrical characteristics at $V_{CC\pm}$ = ±12 V, V_{ref} = 1 V, T_A = 25°C (see Figure 3)

integrator and buffer operational amplifiers

PARAMETER		TEST CONDITIONS	MIN	TYP	MAX	UNIT
V_{IO}	Input offset voltage			15		mV
I_{IB}	Input bias current			50		pA
V_{OM+}	Positive output voltage swing		9	11		V
V_{OM-}	Negative output voltage swing		−5	−7		V
A_{VD}	Voltage amplification			110		dB
B_1	Unity-gain bandwidth			3		MHz
CMRR	Common mode rejection	V_{IC} = −1 V to +1 V		100		dB
SR	Output slew rate			5		V/μs

comparator

PARAMETER		TEST CONDITIONS	MIN	TYP	MAX	UNIT
V_{IO}	Input offset voltage			15		mV
I_{IB}	Input bias current			50		pA
A_{VD}	Voltage amplification			100		dB
V_{OL}	Low-level output voltage	I_{OL} = 1.6 mA		200	400	mV
I_{OH}	High-level output current	V_{OH} = 3 V		5	20	nA

voltage reference output

PARAMETER		TEST CONDITIONS	MIN	TYP	MAX	UNIT
$V_{ref(0)}$	Reference voltage		1.12	1.22	1.32	V
α_{Vref}	Reference-voltage temperature coefficient	T_A = 0°C to 70°C		80		ppm/°C
r_o	Reference output resistance			3		Ω

logic control section

PARAMETER		TEST CONDITIONS	MIN	TYP	MAX	UNIT
I_{IH}	High-level input current	V_{IH} = 2 V		1	10	μA
I_{IL}	Low-level input current	V_{IL} = 0.8 V		−40	−300	μA

total device

PARAMETER		TEST CONDITIONS	MIN	TYP	MAX	UNIT
I_{CC+}	Positive supply current			15	20	mA
I_{CC-}	Negative supply current			12	18	mA

PARAMETER MEASUREMENT INFORMATION

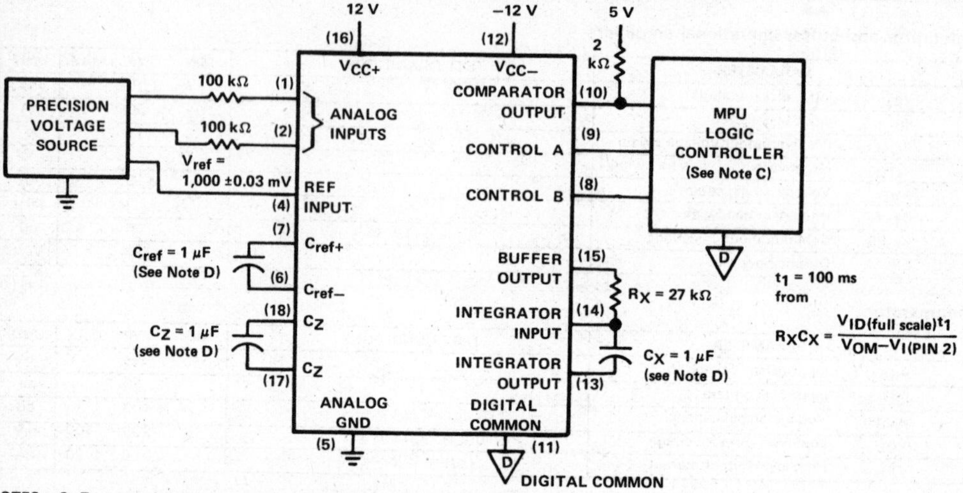
12 V
−12 V
5 V
(16)
(12)
2 kΩ
VCC+
VCC−
PRECISION VOLTAGE SOURCE
100 kΩ
(1)
100 kΩ
(2)
ANALOG INPUTS
COMPARATOR OUTPUT
(10)
(9)
CONTROL A
MPU LOGIC CONTROLLER (See Note C)
Vref = 1,000 ±0.03 mV
REF INPUT
(4)
(8)
CONTROL B
(7)
Cref+
Cref = 1 μF (See Note D)
(6)
Cref−
BUFFER OUTPUT
(15)
RX = 27 kΩ
t1 = 100 ms from
RXCX = VID(full scale)t1 / VOM−VI(PIN 2)
(18)
CZ
CZ = 1 μF (see Note D)
INTEGRATOR INPUT
(14)
CX = 1 μF (see Note D)
CZ
INTEGRATOR OUTPUT
(13)
(17)
ANALOG GND
DIGITAL COMMON
(5)
(11)
D
DIGITAL COMMON

NOTES: C. Tests are started approximately 5 seconds after power-on.

D. Capacitors used are TRW's X363UW polypropylene or equivalent for C_X, C_{ref}, and C_Z; however for C_{ref} and C_Z film-dielectric capacitors may be substituted.

FIGURE 3. TEST CIRCUIT CONFIGURATION

external-component selection guide

The autozero capacitor C_Z and reference capacitor C_{ref} should be within the recommended range of operating conditions and should have low-leakage characteristics. Most film-dielectric capacitors and some tantalum capacitors provide acceptable results. Ceramic and aluminum capacitors are not recommended because of their relatively high-leakage characteristics.

The integrator capacitor C_X should also be within the recommended range and must have good voltage linearity and low dielectric absorption. A polypropylene-dielectric capacitor similar to TRW's X363UW is recommended for 4-1/2-digit accuracy. For 3-1/2-digit applications, polyester, polycarbonate, and other film dielectrics are usually suitable. Ceramic and electrolytic capacitors are not recommended.

Stray coupling from the comparator output to any analog pin (in order of importance 17, 18, 14, 7, 6, 13, 1, 2, 15) must be minimized to avoid oscillations. In addition, all power supply pins should be bypassed at the package, for example, by a 0.01-μF ceramic capacitor.

Analog and digital common are internally isolated and may be at different potentials. Digital common can be within 4 volts of positive or negative supply with the logic decode still functioning properly.

The time constant R_XC_X should be kept as near the minimum value as possible and is given by the formula:

$$\text{Minimum } R_XC_X = \frac{V_{ID}\text{ (full scale) } t_1}{|V_{OM-}| - V_{I(pin2)}}$$

where:

$V_{ID(full\ scale)}$ = Voltage on pin 1 with respect to pin 2
t_1 = Input integration time in seconds
$V_{I(pin\ 2)}$ = Voltage on pin 2 with resepct to analog ground

description of digital processors

The TL502C and TL503C are control logic devices designed to complement the TL500C and TL501C analog processors. They feature interdigit blanking, over-range blanking, an internal oscillator, and a fast display scan rate. The internal-oscillator input is a Schmitt trigger circuit that can be driven by an external clock pulse or provide its own time base with the addition of a capacitor. The typical oscillator frequency is 120 kHz with a 470-picofarad capacitor connected between the oscillator input and ground.

The TL502C provides seven-segment-display output drivers capable of sinking 100 milliamperes and compatible with popular common-anode displays. The TL503C has four BCD output drivers capable of 100-milliampere sink currents. The code (see next page and Figure 4) for each digit is multiplexed to the output drivers in phase with a pulse on the appropriate digit-enable line at a digit rate equal to $f_{osc.}$ divided by 200. Each digit-enable output is capable of sinking 20-milliamperes.

The comparator input of each device, in addition to monitoring the output of the zero-crossing detector in the analog processor, may be used in the display test mode to check for wiring and display faults. A high logic level (2 to 6.5 volts) at the trigger input with the comparator input at or below 6.5 volts starts the integrate-input phase. Voltage levels equal to or greater than 7.9 volts on both the trigger and comparator inputs clear the system and set the BCD counter to 20,000. When normal operation resumes, the conversion cycle is restarted at the auto zero phase.

These devices are manufactured using I^2L and bipolar techniques. The TL502C and TL503C are intended for operation from 0°C to 70°C.

TL502 . . . N DUAL-IN-LINE PACKAGE
(TOP VIEW)

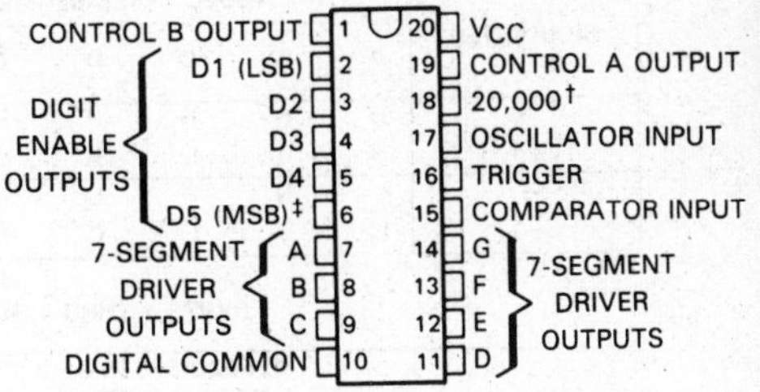
CONTROL B OUTPUT 1
D1 (LSB) 2
D2 3
D3 4
D4 5
D5 (MSB)‡ 6
DIGIT ENABLE OUTPUTS
7-SEGMENT DRIVER OUTPUTS
A 7
B 8
C 9
DIGITAL COMMON 10
20 VCC
19 CONTROL A OUTPUT
18 20,000†
17 OSCILLATOR INPUT
16 TRIGGER
15 COMPARATOR INPUT
14 G
13 F
12 E
11 D
7-SEGMENT DRIVER OUTPUTS

TL503 . . . N DUAL-IN-LINE PACKAGE
(TOP VIEW)

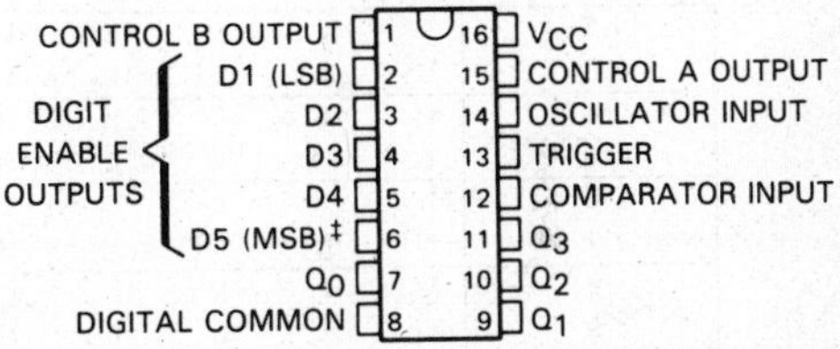
CONTROL B OUTPUT 1
D1 (LSB) 2
D2 3
D3 4
D4 5
D5 (MSB)‡ 6
DIGIT ENABLE OUTPUTS
Q0 7
DIGITAL COMMON 8
16 VCC
15 CONTROL A OUTPUT
14 OSCILLATOR INPUT
13 TRIGGER
12 COMPARATOR INPUT
11 Q3
10 Q2
9 Q1

† Pin 18 of TL502 provides an output of f_{osc} (oscillator frequencies) ÷20,000.
‡D5, the most significant bit, is also the sign bit.

TABLE OF SPECIAL FUNCTIONS
V_{CC} = 5 V ±10%

TRIGGER INPUT	COMPARATOR INPUT	FUNCTION
$V_I \leq 0.8$ V	$V_I \leq 6.5$ V	Hold at auto-zero cycle after completion of conversion
2 V $\leq V_I \leq$ 6.5 V	$V_I \leq 6.5$ V	Normal operation (continuous conversion)
$V_I \leq 6.5$ V	$V_I \geq 7.9$ V	Display Test: All BCD outputs high
$V_I \geq 7.9$ V	$V_I \leq 6.5$ V	Internal Test
Both inputs to go $V_I \geq 7.9$ V simultaneously		System clear: Sets BCD counter to 20,000. When normal operation is resumed, cycle begins with Auto Zero.

DIGIT 5 (MOST SIGNIFICANT DIGIT) CHARACTER CODES

CHARACTER	TL502C SEVEN-SEGMENT LINES							TL503C BCD OUTPUT LINES			
	A	B	C	D	E	F	G	Q3 8	Q2 4	Q1 2	Q0 1
+	H	H	H	H	L	L	L	H	L	H	L
+1	H	L	L	H	L	L	L	H	H	H	L
−	L	H	H	L	H	H	L	H	L	H	H
−1	L	L	L	L	H	H	L	H	H	H	H

DIGITS 1 THRU 4 NUMERIC CODE (See Figure 4)

NUMBER	TL502C SEVEN-SEGMENT LINES							TL503C BCD OUTPUT LINES			
	A	B	C	D	E	F	G	Q3 8	Q2 4	Q1 2	Q0 1
0	L	L	L	L	L	L	H	L	L	L	L
1	H	L	L	H	H	H	H	L	L	L	H
2	L	L	H	L	L	H	L	L	L	H	L
3	L	L	L	L	H	H	L	L	L	H	H
4	H	L	L	H	H	L	L	L	H	L	L
5	L	H	L	L	H	L	L	L	H	L	H
6	L	H	L	L	L	L	L	L	H	H	L
7	L	L	L	H	H	H	H	L	H	H	H
8	L	L	L	L	L	L	L	H	L	L	L
9	L	L	L	L	H	L	L	H	L	L	H

H = high level, L = low level

schematics of inputs and outputs

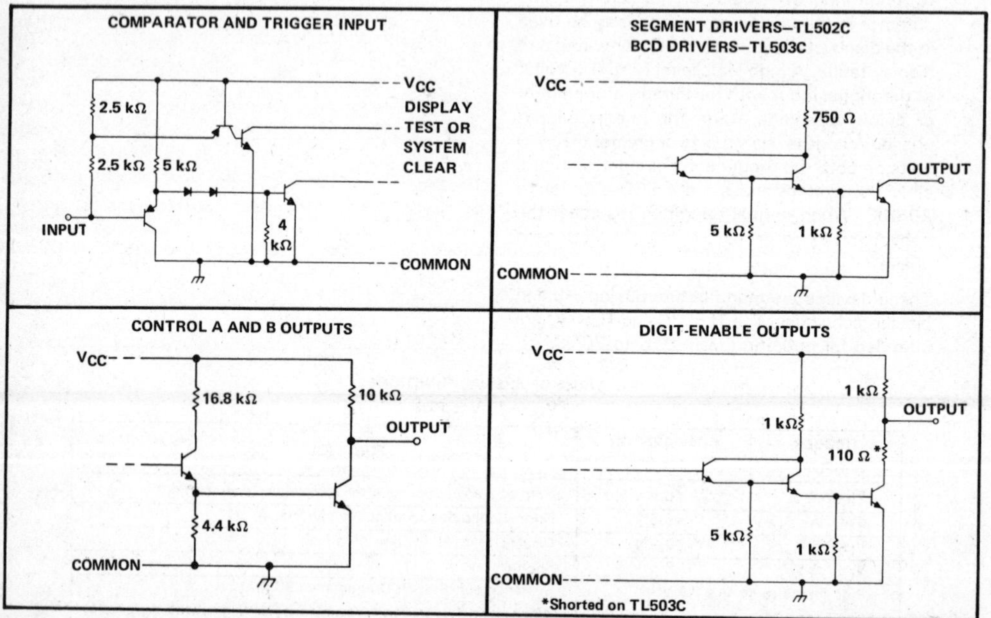
COMPARATOR AND TRIGGER INPUT
VCC
DISPLAY TEST OR SYSTEM CLEAR
2.5 kΩ
2.5 kΩ
5 kΩ
4 kΩ
INPUT
COMMON
SEGMENT DRIVERS–TL502C
BCD DRIVERS–TL503C
VCC
750 Ω
OUTPUT
5 kΩ
1 kΩ
COMMON
CONTROL A AND B OUTPUTS
VCC
16.8 kΩ
10 kΩ
OUTPUT
4.4 kΩ
COMMON
DIGIT-ENABLE OUTPUTS
VCC
1 kΩ
OUTPUT
1 kΩ
110 Ω *
5 kΩ
1 kΩ
COMMON
*Shorted on TL503C

absolute maximum ratings

Supply voltage, V_{CC} (see Note 5)		7	V
Input voltage, V_I	Oscillator	5.5	V
	Comparator or Trigger	9	
Output current	BCD or Segment drivers	120	mA
	Digit-enable outputs	40	
	Pin 18 (TLC502 only)	20	
Total power dissipation at (or below) 30 °C free-air temperature (see Note 6)		1100	mW
Operating free-air temperature range		0 to 70	°C
Storage temperature range		−65 to 150	°C
Lead temperaturee 1,6 mm (1/16 inch) from case for 10 seconds		260	°C

NOTES: 5. Voltage values are with respect to the network ground terminal.
6. For operation above 30 °C free-air temperature, derate linearly at the rate of 9.2 mW/°C.

recommended operating conditions

		MIN	NOM	MAX	UNIT
Supply voltage, V_{CC}		4.5	5	5.5	V
High-level input voltage, V_{IH}	Comparator and trigger inputs	2			V
Low-level input voltage, V_{IL}	Comparator and trigger inputs			0.8	V
Operating free-air temperature		0		70	°C

electrical characteristics at 25°C free-air temperature

	PARAMETER	TERMINAL	TEST CONDITIONS	TL502C MIN	TL502C TYP	TL502C MAX	TL503C MIN	TL503C TYP	TL503C MAX	UNIT
V_{IK}	Input clamp voltage	All inputs	V_{CC} = 4.5 V, I_I = −12 mA		−0.8	−1.5		−0.8	−1.5	V
V_{T+}	Positive-going input threshold voltage	Oscillator	V_{CC} = 5 V		1.5			1.5		V
V_{T-}	Negative-going input threshold voltage	Oscillator	V_{CC} = 5 V		0.9			0.9		V
$V_{T+} - V_{T-}$	Hysteresis	Oscillator	V_{CC} = 5 V	0.4	0.6	0.8	0.4	0.6	0.8	
I_{T+}	Input current at positive-going input threshold voltage	Oscillator	V_{CC} = 5 V	−40	−94	−170	−40	−94	−170	μA
I_{T-}	Input current at negative-going input threshold voltage	Oscillator	V_{CC} = 5 V	40	117	170	40	117	170	μA
V_{OH}	High-level output voltage	Digit enable	V_{CC} = 4.5 V, I_{OH} = 0	4.15	4.4		4.15	4.4		V
		Pin 18 (TL502C only)		4.25	4.4					
		Control A and B		4.25	4.4		4.25	4.4		
V_{OL}	Low-level output voltage	Digit enable	V_{CC} = 4.5 V, I_{OL} = 20 mA					0.2	0.5	V
		Pin 18 (TL502C only)	V_{CC} = 4.5 V, I_{OL} = 10 mA		0.15	0.4				
		Control A and B	V_{CC} = 4.5 V, I_{OL} = 2 mA		0.088	0.4		0.088	0.4	
		Segment drivers	V_{CC} = 4.5 V, I_{OL} = 100 mA		0.17	0.3				
		BCD drivers	V_{CC} = 4.5 V, I_{OL} = 100 mA					0.17	0.3	
I_I	Input current	Comparator, Trigger	V_{CC} = 5.5 V, V_I = 5.5 V		65	100		65	100	μA
		Oscillator				1			1	mA
I_{IH}	High-level input current	Comparator, Trigger	V_{CC} = 5.5 V, V_I = 2.4 V		−0.6	−1		−0.6	−1	mA
		Oscillator				0.5			0.5	
I_{IL}	Low-level input voltage	Oscillator	V_{CC} = 5.5 V, V_I = 0.4 V		−0.1	−0.17		−0.1	−0.17	mA
		Comparator, Trigger			−1	−1.6		−1	−1.6	
I_{OH}	High-level output current (Output transistor off)	Digit enable	V_{CC} = 4.5 V, V_O = 0.5 V,	−2.5	−4		−2.5	−4		mA
		Pin 18 (TL502C only)	V_{CC} = 4.5 V, V_O = 0.5 V	−0.5	−0.9					
		Control A and B	V_{CC} = 4.5 V, V_O = 0.5 V	−0.25	−0.4		−0.25	−0.4		
		Segment drivers	V_{CC} = 4.5 V, V_O = 5.5 V			0.25				
		BCD drivers	V_{CC} = 4.5 V, V_O = 5.5 V						0.25	
I_{OL}	Low-level output current (Output transistor on)	Digit enable	V_{CC} = 4.5 V, V_O = 3.55 V	18	23					mA
I_{CC}	Supply current	V_{CC}	V_{CC} = 5.5 V		73	110		73	110	mA

special functions† operating characteristics at 25 °C free-air temperature

PARAMETER		TEST CONDITIONS	MIN	TYP	MAX	UNIT
I_I	Input current into comparator or trigger inputs	V_{CC} = 5.5 V, V_I = 8.55 V		1.2	1.8	mA
		V_{CC} = 5.5 V, V_I = 6.25 V			0.5	mA

†The comparator and trigger inputs may be used in the normal mode or to perform special functions. See the Table of Special Functions.

TYPICAL APPLICATION DATA

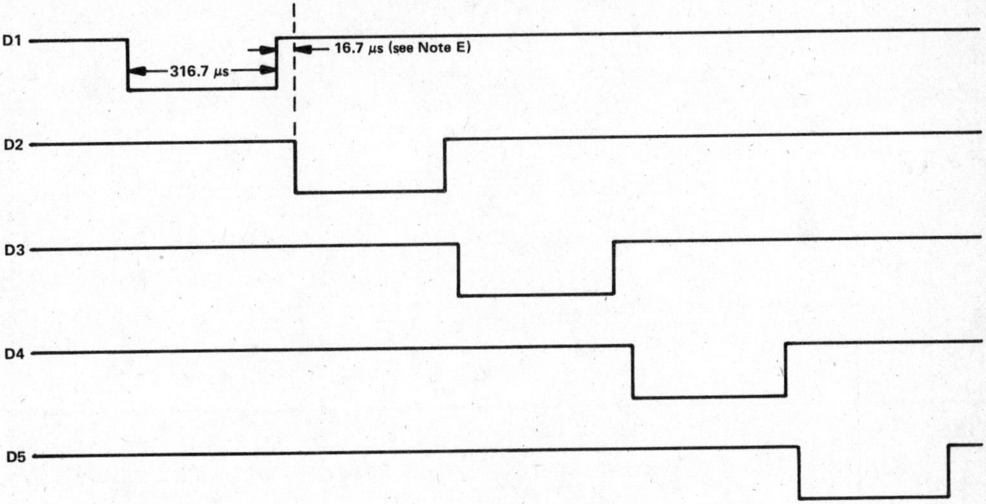
D1
316.7 μs
16.7 μs (see Note E)
D2
D3
D4
D5

NOTE E: The BCD or seven-segment driver outputs are present for a particular digit slightly before the falling edge of that digit enable.

FIGURE 4. TL502C, TL503C DIGIT TIMING WITH 120-kHz CLOCK SIGNAL AT OSCILLATOR INPUT

D2366, OCTOBER 1977—REVISED SEPTEMBER 1986

- 3-Digit Accuracy (0.1%)
- 10-Bit Resolution
- Automatic Zero
- Internal Reference Voltage
- Single-Supply Operation
- High-Impedance MOS Input
- Designed for Use with TMS1000 Type Microprocessors for Cost-Effective High-Volume Applications
- BI-MOS Technology
- Only 40 mW Typical Power Consumption

N DUAL-IN-LINE PACKAGE
(TOP VIEW)

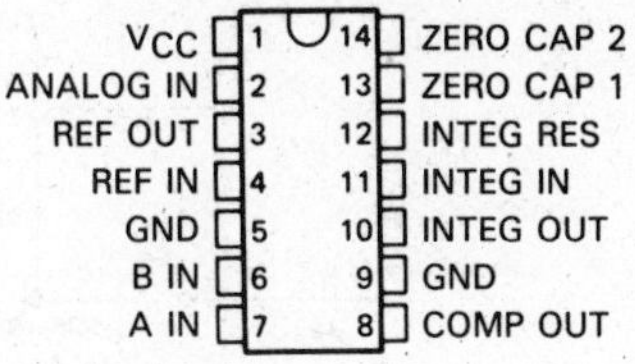
VCC 1
ANALOG IN 2
REF OUT 3
REF IN 4
GND 5
B IN 6
A IN 7
14 ZERO CAP 2
13 ZERO CAP 1
12 INTEG RES
11 INTEG IN
10 INTEG OUT
9 GND
8 COMP OUT

Caution. This device has limited built-in gate protection. The leads should be shorted together or the device placed in conductive foam during storage or handling to prevent electrostatic damage to the MOS gates.

description

The TL505C is an analog-to-digital converter building block designed for use with TMS1000 type microprocessors. It contains the analog elements (operational amplifier, comparator, voltage reference, analog switches, and switch drivers) necessary for a unipolar automatic-zeroing dual-slope converter. The logic for the dual-slope conversion can be performed by the associated MPU as a software routine or it can be implemented with other components such as the TL502 logic-control device.

The high-impedance MOS inputs permit the use of less expensive, lower value capacitors for the integration and offset capacitors and permit conversion speeds from 20 per second to 0.05 per second.

The TL505C is a product of TI's BI-MOS process, which incorporates bipolar and MOSFET transistors on the same monolithic circuit. The TL505C is characterized for operation from 0°C to 70°C.

functional block diagram

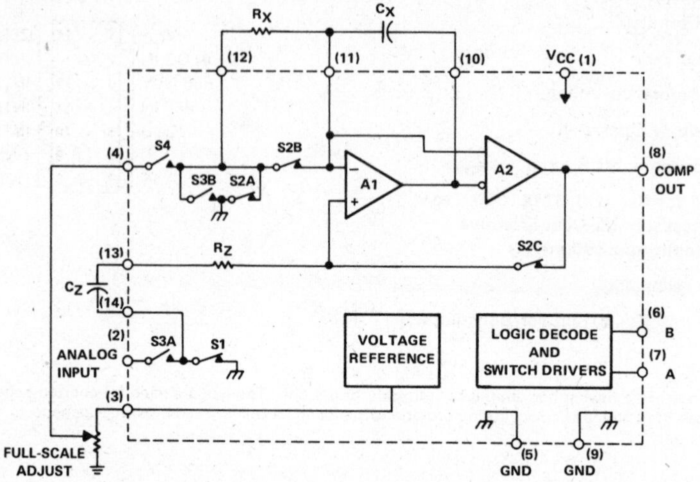
RX
CX
(12)
(11)
(10)
VCC (1)
(4)
S4
S2B
S3B
S2A
A1
A2
(8)
COMP
OUT
(13)
RZ
S2C
CZ
(14)
(2)
ANALOG
INPUT
S3A
S1
VOLTAGE
REFERENCE
LOGIC DECODE
AND
SWITCH DRIVERS
(6)
B
(7)
A
(3)
FULL-SCALE
ADJUST
(5)
GND
(9)
GND

NOTE: Analog and digital GND are internally connected together.

absolute maximum ratings over operating free-air temperature range (unless otherwise noted)

Supply voltage, V_{CC} (see Note 1) .... 18 V
Input voltage, pins 2, 4, 6, and 7 .... V_{CC}
Continuous total dissipation at (or below) 25°C free-air temperature (see Note 2) .... 875 mW
Operating free-air temperature range .... 0°C to 70°C
Storage temperature range .... −65°C to 150°C
Lead temperature 1,6 mm (1/16 inch) from case for 10 seconds .... 260°C

NOTES: 1. Voltage values are with respect to the two ground terminals connected together.
2. For operation above 25°C free-air temperature, derate linearly to 560 mW at 70°C at the rate of 7.0 mW/°C.

recommended operating conditions

	MIN	NOM	MAX	UNIT
Supply voltage, V_{CC}	7	9	15	V
Analog input voltage, V_I	0		4	V
Reference input voltage, $V_{ref(I)}$	0.5		3	V
High-level input voltage at A or B, V_{IH}	3.6		$V_{CC}+1$	V
Low-level input voltage at A or B, V_{IL}	0.2		1.8	V
Integrator capacitor, C_X	See "component selection"			
Integrator resistor, R_X	0.5		2	MΩ
Integration time, t_1	16.6		500	ms
Operating free-air temperature, T_A	0		70	°C

electrical characteristics, V_{CC} = 9 V, $V_{ref(I)}$ = 1 V, T_A = 25°C, connected as shown in Figure 1 (unless otherwise noted)

PARAMETER		TEST CONDITIONS	MIN	TYP	MAX	UNIT
V_{OH}	High-level output voltage at pin 8	I_{OH} = 0	7.5	8.5		V
I_{OH}	High-level output current at pin 8	V_{OH} = 7.5 V		−100		μA
V_{OL}	Low-level output voltage at pin 8	I_{OL} = 1.6 mA		200	400	mV
V_{OM}	Maximum peak output voltage swing at integrator output	$R_X \geq$ 500 kΩ	$V_{CC}-2$	$V_{CC}-1$		V
$V_{ref(0)}$	Reference output voltage	I_{ref} = −100 μA	1.15	1.22	1.35	V
α_{Vref}	Temperature coefficient of reference output voltage	T_A = 0°C to 70°C		±100		ppm/°C
I_{IH}	High-level input current into A or B	V_I = 9 V		1	10	μA
I_{IL}	Low-level input current into A or B	V_I = 1 V		10	200	μA
I_I	Current into analog input	V_I = 0 to 4 V, A input at 0 V		±10	±200	pA
I_{IB}	Total integrator input bias current			±10		pA
I_{CC}	Supply current	No load		4.5	8	mA

system electrical characteristics, V_{CC} = 9 V, $V_{ref(I)}$ = 1 V, T_A = 25°C, connected as shown in Figure 1 (unless otherwise noted)

PARAMETER	TEST CONDITIONS	MIN	TYP	MAX	UNIT
Zero error	V_I = 0		0.1	0.4	mV
Linearity error	V_I = 0 to 4 V		0.02	0.1	%FS
Ratiometric reading	V_I = $V_{ref(I)} \approx$ 1 V	0.998	1.000	1.002	
Temperature coefficient of ratiometric reading	$V_{ref(I)}$ constant and ≈ 1 V, T_A = 0°C to 70°C		±10		ppm/°C

DEFINITION OF TERMS

Zero Error

The intercept (b) of the anolog-to-digital converter system transfer function y = mx + b, where y is the digital output, x is the analog input, and m is the slope of the transfer function, which is approximated by the ratiometric reading.

Linearity Error

The maximum magnitude of the deviation from a straight line between the end points of the transfer function.

Ratiometric Reading

The ratio of negative integration time (t_2) to positive time (t_1).

PRINCIPLES OF OPERATION

A block diagram of an MPU system utilizing the TL505C is shown in Figure 1. The TL505C operates in a modified positive-integration three-step dual-slope conversion mode. The A/D converter waveforms during the conversion process are illustrated in Figure 2.

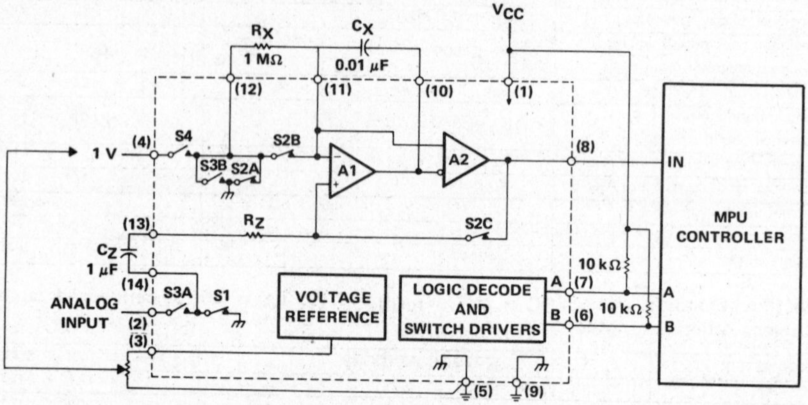
VCC
RX
1 MΩ
CX
0.01 μF
(12)
(11)
(10)
(1)
(4)
1 V
S4
S2B
S3B
S2A
A1
A2
(8)
IN
(13)
RZ
S2C
MPU
CONTROLLER
CZ
1 μF
10 kΩ
(14)
A (7)
S3A
S1
ANALOG
INPUT
(2)
(3)
VOLTAGE
REFERENCE
LOGIC DECODE
AND
SWITCH DRIVERS
B (6)
10 kΩ
A
B
(5)
(9)

FIGURE 1. FUNCTIONAL BLOCK DIAGRAM OF TL505C INTERFACE WITH A MICROPROCESSOR SYSTEM

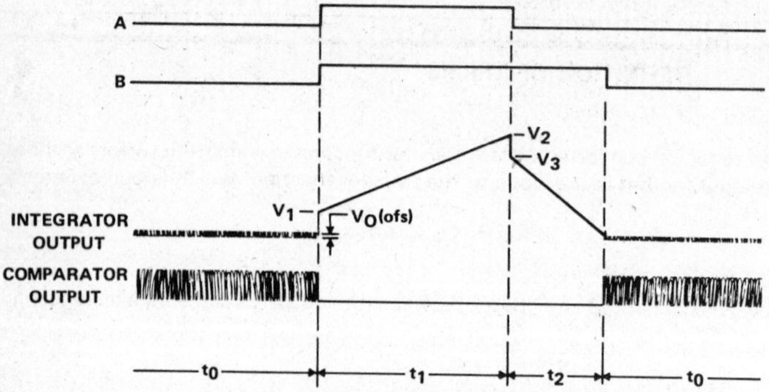
A
B
V2
V3
V1
VO(ofs)
INTEGRATOR
OUTPUT
COMPARATOR
OUTPUT
t0
t1
t2
t0

$V_1 = V_2 - V_3 = V_I + V_{O(ofs)}$

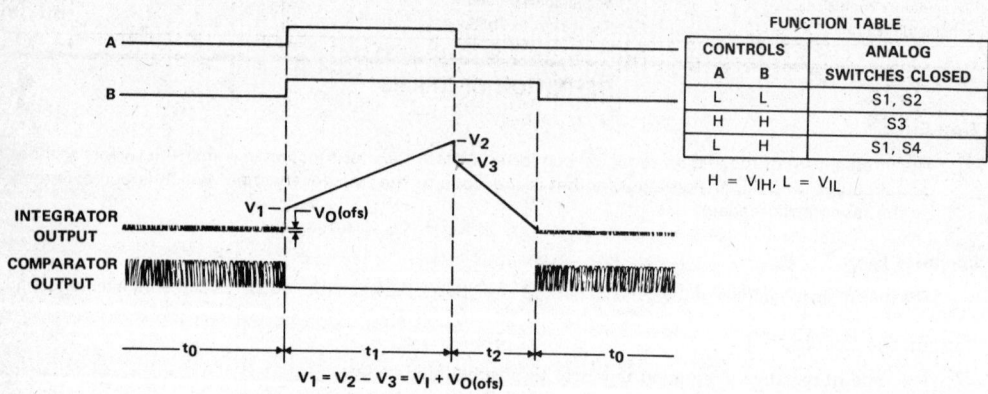

FUNCTION TABLE

CONTROLS		ANALOG
A	B	SWITCHES CLOSED
L	L	S1, S2
H	H	S3
L	H	S1, S4

$H = V_{IH}$, $L = V_{IL}$

FIGURE 2. CONVERSION PROCESS TIMING DIAGRAMS

TEXAS INSTRUMENTS
POST OFFICE BOX 655012 • DALLAS, TEXAS 75265

PRINCIPLES OF OPERATION

The first step of the conversion cycle is the auto-zero period t_0 during which the integrator offset is stored in the auto-zero capacitor and the offset of the comparator is stored in the integrator capacitor. To accomplish this, the MPU takes the A and B inputs both low. This is decoded by the switch drivers, which close S1 and S2. The output of the comparator is connected to the input of the integrator through the low-pass filter consisting of R_Z and C_Z. The closed loop of A1 and A2 will seek a null condition where the offsets of the integrator and comparator are stored in C_Z and C_X, respectively. This null condition is characterized by a high-frequency oscillation at the output of the comparator. The purpose of S2B is to shorten the amount of time required to reach the null condition.

At the conclusion of t_0, the MPU takes the A and B inputs both high closing S3 and opening all other switches. The input signal V_I is applied to the noninverting input of A1 through C_Z. V_I is then positively integrated by A1. Since the offset of A1 is stored in C_Z, the change in voltage across C_X will be due to only the input voltage. It should be noted that since the input is integrated in a positive integration during t_1, the output of A1 will be the sum of the input voltage, the integral of the input voltage, and the comparator offset, as shown in Figure 2. The change in voltage across capacitor C_X (V_{CX}) during t_1 is given by

$$\Delta V_{CX(1)} = \frac{V_I t_1}{R_1 C_X} \tag{1}$$

where $R_1 = R_X + R_{S3B}$ and

R_{S3B} is the resistance of switch S3B.

At the end of t_1, the MPU takes the A input low and the B input high closing S1 and S4 and opening all other switches. In this state, the reference is integrated by A1 in a negative sense until the integrator output reaches the comparator threshold. At this point, the comparator output goes high. This change in state is sensed by the MPU, which terminates t_2 by again taking the A and B inputs both low. During t_2, the change in voltage across C_X is given by

$$\Delta V_{CX(2)} = \frac{V_{ref} t_2}{R_2 C_X} \tag{2}$$

where $R_2 = R_X + R_{S4} + R_{ref}$ and

R_{ref} is the equivalent resistance of the reference divider.

Since $\Delta V_{CX1} = -\Delta V_{CX2}$, equations (1) and (2) can be combined to give

$$V_I = V_{ref} \frac{R_1 \cdot t_2}{R_2 \cdot t_1} \tag{3}$$

This equation is a variation on the ideal dual-slope equation, which is

$$V_I = V_{ref} \frac{t_2}{t_1} \tag{4}$$

Ideally then, the ratio of R_1/R_2 would be exactly equal to one. In a typical TL505C system where $R_X = 1\ \text{M}\Omega$, the scaling error introduced by the difference in R_1 and R_2 is so small that it can be neglected and equation (3) reduces to (4).

TYPICAL APPLICATION DATA

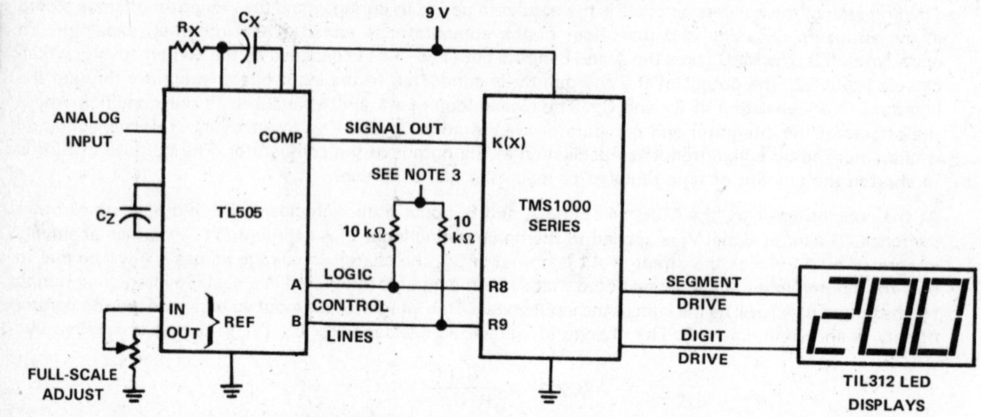
9 V
RX
CX
ANALOG
INPUT
COMP
SIGNAL OUT
K(X)
SEE NOTE 3
CZ
TL505
TMS1000
SERIES
10 kΩ
10
kΩ
LOGIC
A
R8
CONTROL
IN
OUT
REF
B
R9
LINES
SEGMENT
DRIVE
DIGIT
DRIVE
FULL-SCALE
ADJUST
TIL312 LED
DISPLAYS

NOTE 3: Connect to either 9 V or 0 V depending on which device in the TMS1000 series is used and how it is programmed.

FIGURE 3. TL505C IN CONJUNCTION WITH A TMS1000 SERIES MICROPROCESSOR FOR A 3-DIGITAL PANEL METER APPLICATION

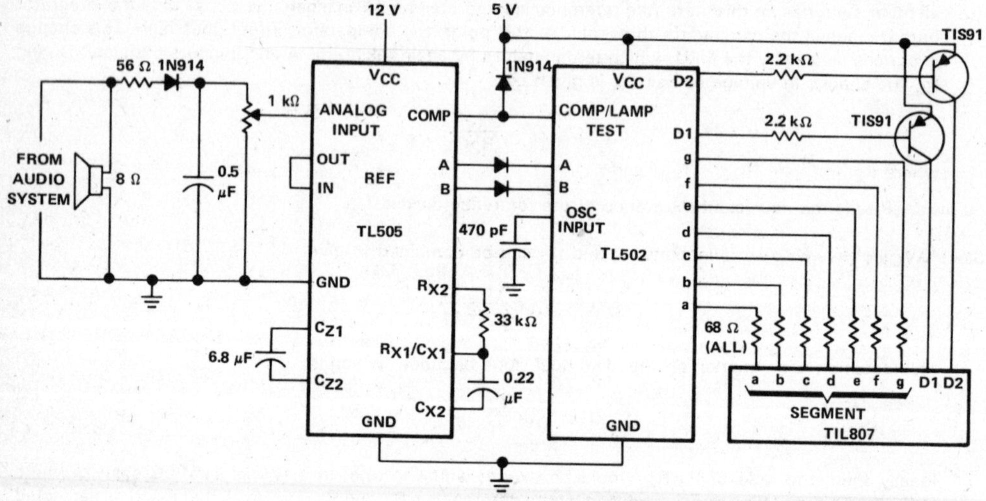
12 V
5 V
TIS91
56 Ω 1N914
VCC
1N914
VCC
D2
2.2 kΩ
1 kΩ
ANALOG
INPUT
COMP
COMP/LAMP
TEST
D1
2.2 kΩ
TIS91
FROM
AUDIO
SYSTEM
8 Ω
0.5
µF
OUT
IN
REF
A
B
g
f
e
d
c
b
a
OSC
INPUT
TL505
470 pF
TL502
GND
RX2
33 kΩ
68 Ω
(ALL)
CZ1
6.8 µF
RX1/CX1
CZ2
0.22
µF
a b c d e f g D1 D2
CX2
GND
GND
SEGMENT
TIL807

FIGURE 4. AUDIO PEAK POWER METER

TEXAS INSTRUMENTS
POST OFFICE BOX 655012 • DALLAS, TEXAS 75265

D2503, OCTOBER 1979—REVISED SEPTEMBER 1986

- Low Cost
- 7-Bit Resolution
- Guaranteed Monotonicity
- Ratiometric Conversion
- Conversion Speed . . . Approximately 1 ms
- Single-Supply Operation . . . Either Unregulated 8-V to 18-V (V_{CC2} Input), or Regulated 3.5-V to 6-V (V_{CC1} Input)
- I^2L Technology
- Power Consumption at 5 V . . . 25 mW Typ
- Regulated 5.5-V Output (≤1 mA)

P DUAL-IN-LINE PACKAGE
(TOP VIEW)

ENABLE	1	8	RESET
CLK	2	7	V_{CC2}
GND	3	6	V_{CC1}
OUTPUT	4	5	ANALOG INPUT

FUNCTION TABLE

ANALOG INPUT CONDITION	ENABLE	OUTPUT
X	L†	H
V_I < 200 mV	H	L
$V_{ramp} > V_I >$ 200 mV	H	H
$V_I > V_{ramp}$	H	L

†Low level on enable also inhibits the reset function.
H = high level, L = low level, X = irrelevant

A high level on the reset pin clears the counter to zero, which sets the internal ramp to 0.75 V_{CC}. Internal pull down resistors keep the reset and enable pins low when not connected.

description

The TL507 is a low-cost single-slope analog-to-digital converter designed to convert analog input voltages between 0.25 V_{CC1} and 0.75 V_{CC1} into a pulse-width-modulated output code. It contains a 7-bit synchronous counter, a binary weighted resistor ladder network, an operational amplifier, two comparators, a buffer amplifier, an internal regulator, and necessary logic circuitry. Integrated-injection logic (I^2L) technology makes it possible to offer this complex circuit at low cost in a small dual-in-line 8-pin package.

In continuous operation, it is possible to obtain conversion speeds up to 1000 per second. The TL507 requires external signals for clock, reset, and enable. Versatility and simplicity of operation, coupled with low cost, make this converter especially useful for a wide variety of applications.

The TL507C is characterized for operation from 0°C to 70°C, and the TL507I is characterized for operation from −40°C to 85°C.

functional block diagram (positive logic)

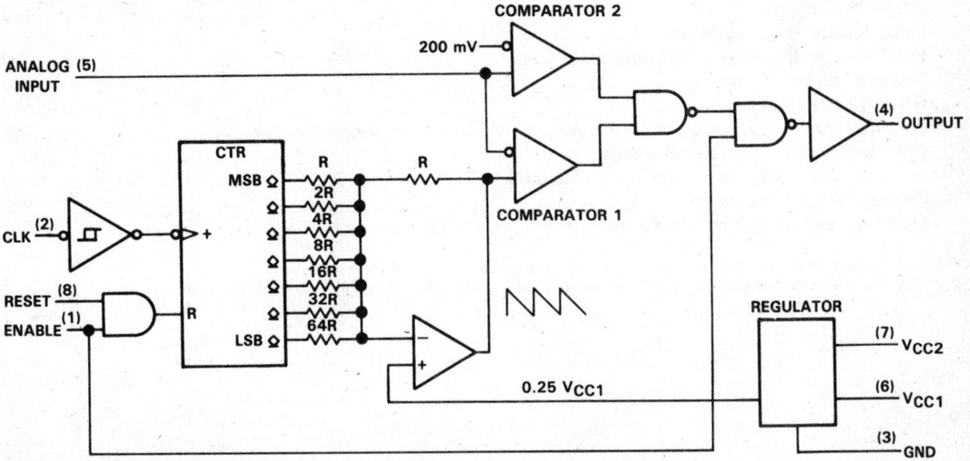
COMPARATOR 2
200 mV
ANALOG (5)
INPUT
(4)
OUTPUT
CTR
MSB
LSB
R
2R
4R
8R
16R
32R
64R
COMPARATOR 1
CLK (2)
RESET (8)
ENABLE (1)
REGULATOR
(7) VCC2
(6) VCC1
(3) GND
0.25 VCC1

◇ indicates an n-p-n open-collector output.

TEXAS INSTRUMENTS
POST OFFICE BOX 655012 • DALLAS, TEXAS 75265

schematics of inputs and outputs

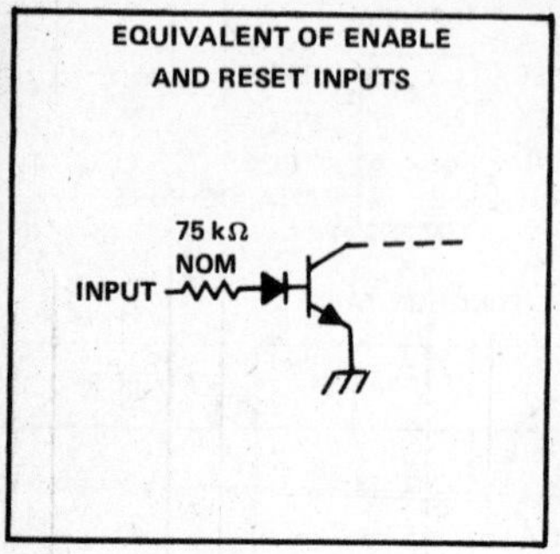
EQUIVALENT OF ENABLE
AND RESET INPUTS
75 kΩ
NOM
INPUT

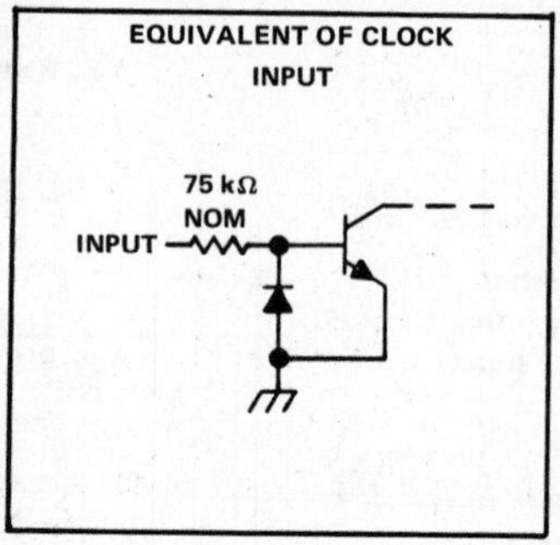
EQUIVALENT OF CLOCK
INPUT
75 kΩ
NOM
INPUT

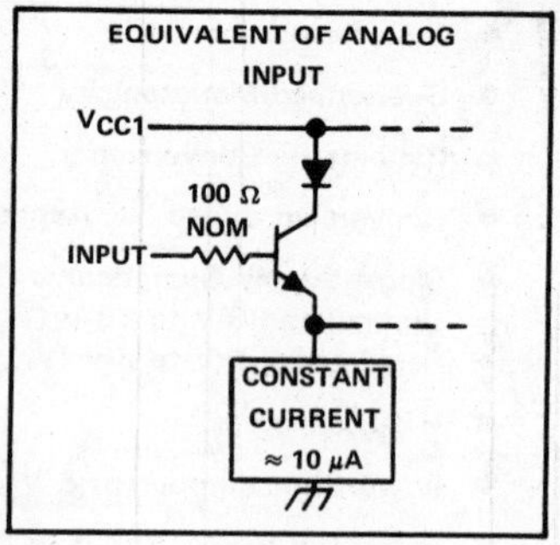
EQUIVALENT OF ANALOG
INPUT
VCC1
100 Ω
NOM
INPUT
CONSTANT
CURRENT
≈ 10 μA

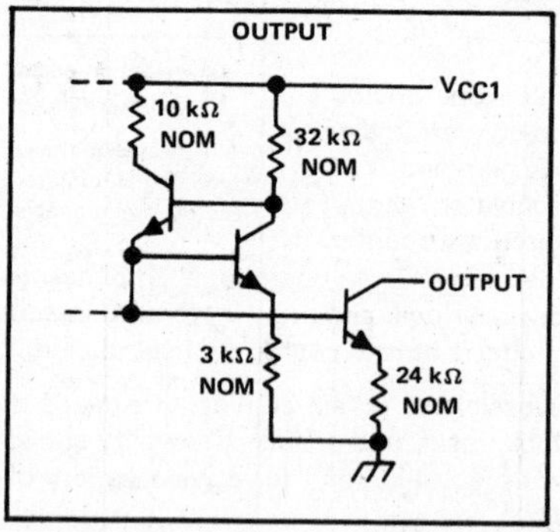
OUTPUT
VCC1
10 kΩ
NOM
32 kΩ
NOM
OUTPUT
3 kΩ
NOM
24 kΩ
NOM

absolute maximum ratings over operating free-air temperature range (unless otherwise noted)

Supply voltage, V_{CC1} (see Note 1)	6.5 V
Supply voltage, V_{CC2}	20 V
Input voltage at analog input	6.5 V
Input voltage at enable, clock, and reset inputs	±20 V
On-state output voltage	6 V
Off-state output voltage	20 V
Continuous total dissipation at (or below) 25°C free-air temperature (see Note 2)	725 mW
Operating free-air temperature range: TL507I	−40°C to 85°C
TL507C	−0°C to 70°C
Storage temperature range	−65°C to 150°C
Lead temperature 1,6 mm (1/16 inch) from case for 10 seconds	260°C

NOTES: 1. Voltage values are with respect to network ground terminal unless otherwise noted.
2. For operation above 25°C free-air temperature, refer to Dissipation Derating Curves, Appendix A.

recommended operating conditions

	MIN	NOM	MAX	UNIT
Supply voltage, V_{CC1}	3.5	5	6	V
Supply voltage, V_{CC2}	8	15	18	V
Input voltage at analog input	0		5.5	V
Input voltage at chip enable, clock, and reset inputs			±18	V
High-level input voltage, V_{IH}, reset and enable	2			V
Low-level input voltage, V_{IL}, reset and enable			0.8	V
On-state output voltage			5.5	V
Off-state output voltage			18	V
Clock frequency, f_{clock}	0	125	150	kHz

electrical characteristics over recommended operating free-air temperature range, $V_{CC1} = V_{CC2} = 5$ V (unless otherwise noted)

regulator section

PARAMETER		TEST CONDITIONS		MIN	TYP‡	MAX	UNIT
V_{CC1}	Supply voltage (output)	V_{CC2} = 10 to 18 V,	I_{CC1} = 0 to −1 mA	5	5.5	6	V
I_{CC1}	Supply current	V_{CC1} = 5 V,	V_{CC2} open		5	8	mA
I_{CC2}	Supply current	V_{CC2} = 15 V,	V_{CC1} open		7	10	mA

inputs

PARAMETER			TEST CONDITIONS	MIN	TYP‡	MAX	UNIT
V_{T+}	Positive-going threshold voltage§	Clock Input				4.5	V
V_{T-}	Negative-going threshold voltage§	Clock Input		0.4			V
V_{hys}	Hysteresis ($V_{T+} - V_{T-}$)	Clock Input		2	2.6	4	V
I_{IH}	High-level input current	Reset, Enable, and Clock	V_I = 2.4 V		17	35	μA
			V_I = 18 V	130	220	320	μA
I_{IL}	Low-level input current	Reset, Enable, and Clock	V_I = 0			±10	μA
I_I	Analog input current		V_I = 4 V		10	300	nA

output section

PARAMETER		TEST CONDITIONS	MIN	TYP‡	MAX	UNIT
I_{OH}	High-level output current	V_{OH} = 18 V		0.1	100	μA
I_{OL}	Low-level output current	V_{OL} = 5.5 V	5	10	15	mA
V_{OL}	Low-level output voltage	I_{OL} = 1.6 mA		80	400	mV

operating characteristics over recommended operating free-air temperature range, $V_{CC1} = V_{CC2} = 5.12$ V

PARAMETER	TEST CONDITIONS	MIN	TYP‡	MAX	UNIT
Overall error	See Figure 1			±80	mV
Differential nonlinearity	See Figure 1			±20	mV
Zero error§	Binary count = 0			±80	mV
Scale error	Binary count = 127			±80	mV
Full scale input voltage §	Binary count = 127	3.74	3.82	3.9	V
Propagation delay time from reset or enable			2		μs

‡ All typical values are at $T_A = 25°C$.
§ These parameters are linear functions of V_{CC1}.

definitions

zero error

The absolute value of the difference between the actual analog voltage at the 01H-to-00H transition and the ideal analog voltage at that transition.

overall error

The magnitude of the deviation from a straight line between the endpoints of the transfer function.

differential nonlinearity

The maximum deviation of an analog-value change associated with a 1-bit code change (1 clock pulse) from its theoretical value of 1 LSB.

PARAMETER MEASUREMENT INFORMATION

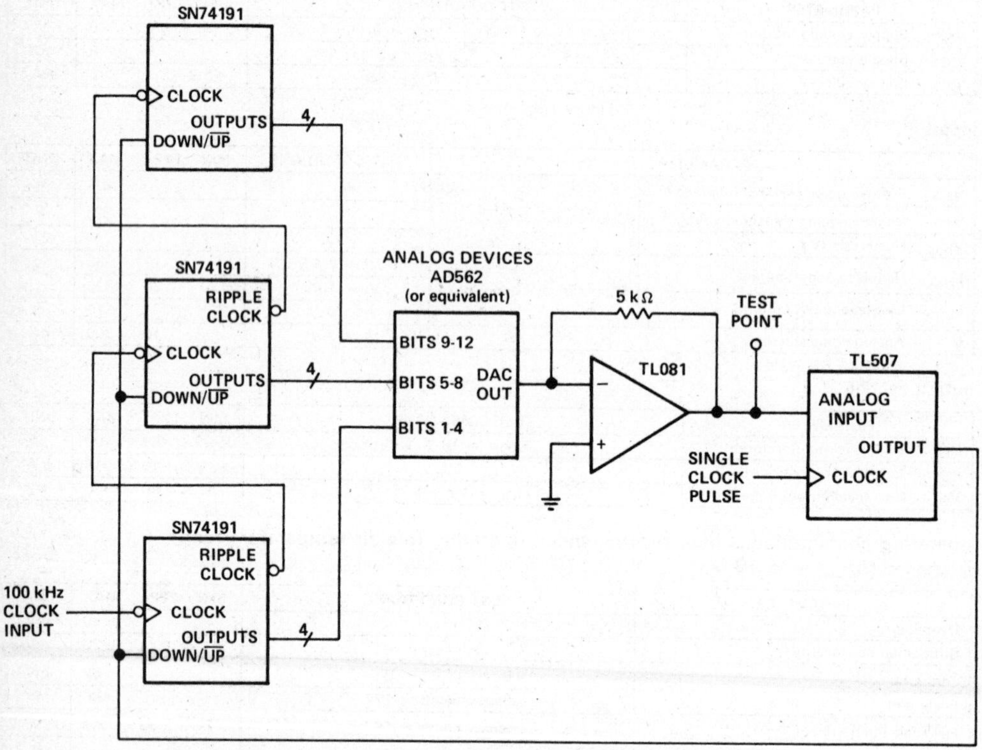
SN74191
CLOCK
OUTPUTS
DOWN/UP
4
SN74191
RIPPLE CLOCK
CLOCK
OUTPUTS
DOWN/UP
4
SN74191
RIPPLE CLOCK
100 kHz CLOCK INPUT
CLOCK
OUTPUTS
DOWN/UP
4
ANALOG DEVICES AD562 (or equivalent)
BITS 9-12
BITS 5-8
BITS 1-4
DAC OUT
5 kΩ
TL081
TEST POINT
TL507
ANALOG INPUT
OUTPUT
SINGLE CLOCK PULSE
CLOCK

FIGURE 1. MONOTONICITY AND NONLINEARITY TEST CIRCUIT

TEXAS INSTRUMENTS
POST OFFICE BOX 655012 • DALLAS, TEXAS 75265

PRINCIPLES OF OPERATION

The TL507 is a single-slope analog-to-digital converter. All single-slope converters are basically voltage-to-time or current-to-time converters. A study of the functional block diagram shows the versatility of the TL507.

An external clock signal is applied through a buffer to a negative-edge-triggered synchronous counter. Binary-weighted resistors from the counter are connected to an operational amplifier used as an adder. The operational amplifier generates a signal that ramps from 0.75 • V_{CC1} down to 0.25 • V_{CC1}. Comparator 1 compares the ramp signal to the analog input signal. Comparator 2 functions as a fault defector. With the analog input voltage in the range 0.25 • V_{CC1} to 0.75 • V_{CC1}, the duty cycle of the output signal is determined by the unknown analog input as shown in Figure 2 and the Function Table.

For illustration assume V_{CC1} = 5.12 V,

$$0.25 \cdot V_{CC1} = 1.28\ \text{V}$$

$$1\ \text{binary count} = \frac{(0.75 - 0.25)\ V_{CC1}}{128} = 20\ \text{mV}$$

$$0.75 \cdot V_{CC1} - 1\ \text{count} = 3.82\ \text{V}$$

The output is an open-collector n-p-n transistor capable of withstanding up to 18 volts in the off state. The output is current limited to the 8- to 12-milliampere range; however, care must be taken to ensure that the output does not exceed 5.5 volts in the on state.

The voltage regulator section allows operation from either an unregulated 8- to 18-volt V_{CC2} source or a regulated 3.5- to 6-volt V_{CC1} source. Regardless of which external power source is used, the internal circuitry operates at V_{CC1}. When operating from a V_{CC1} source, V_{CC2} may be connected to V_{CC1} or left open. When operating from a V_{CC2} source, V_{CC1} can be used as a reference voltage output.

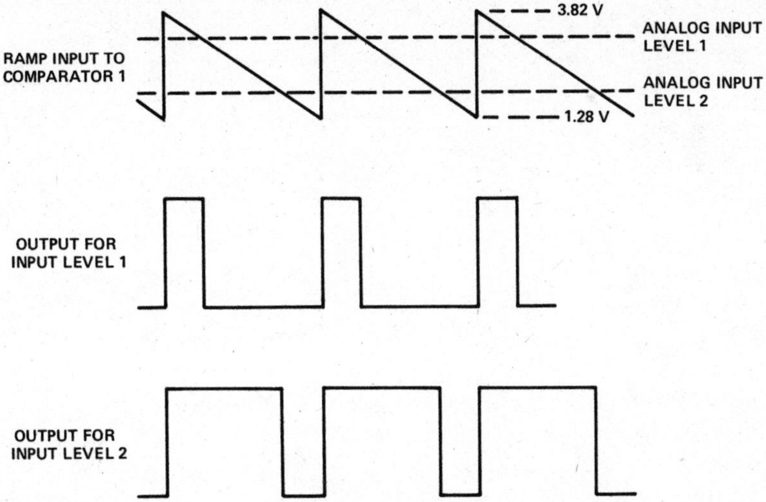
3.82 V
ANALOG INPUT LEVEL 1
RAMP INPUT TO COMPARATOR 1
ANALOG INPUT LEVEL 2
1.28 V
OUTPUT FOR INPUT LEVEL 1
OUTPUT FOR INPUT LEVEL 2

FIGURE 2

- Switches ±10-V Analog Signals
- TTL Logic Capability
- 5- to 30-V Supply Ranges
- Low (100 Ω) On-State Resistance
- High (10^{11} Ω) Off-State Resistance
- 8-Pin Functions

P PACKAGE
(TOP VIEW)

TL601

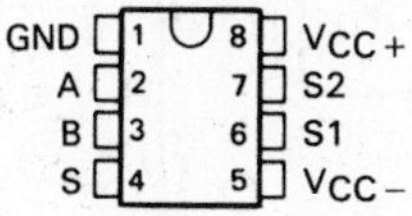
GND
A
B
S
1
2
3
4
8
7
6
5
VCC+
S2
S1
VCC−

TL604

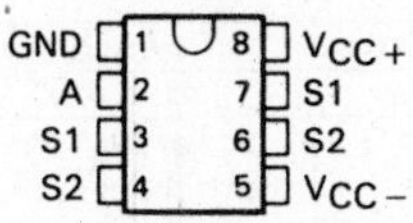
GND
A
S1
S2
1
2
3
4
8
7
6
5
VCC+
S1
S2
VCC−

TL607

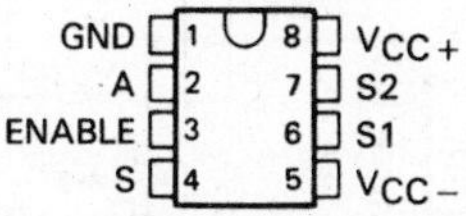
GND
A
ENABLE
S
1
2
3
4
8
7
6
5
VCC+
S2
S1
VCC−

TL610

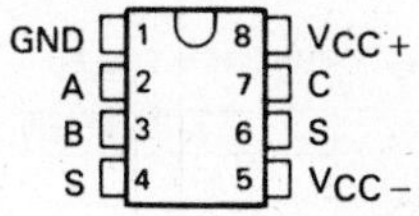
GND
A
B
S
1
2
3
4
8
7
6
5
VCC+
C
S
VCC−

description

The TL601, TL604, TL607, and TL610 are a family of monolithic P-MOS analog switches that provide fast switching speeds with high r_{off}/r_{on} ratio and no offset voltage. The p-channel enhancement-type MOS switches will accept analog signals up to ±10 volts and are controlled by TTL-compatible logic inputs. The monolithic structure is made possible by BI-MOS technology, which combines p-channel MOS with standard bipolar transistors.

These switches are particularly suited for use in military, industrial, and commercial applications such as data acquisition, multiplexers, A/D and D/A converters, MODEMS, sample-and-hold systems, signal multiplexing, integrators, programmable operational amplifiers, programmable voltage regulators, crosspoint switching networks, logic interface, and many other analog systems.

The TL601 is an SPDT switch with two logic control inputs. The TL604 is a dual complementary SPST switch with a single control input. The TL607 is an SPDT switch with one logic control input and one enable input. The TL610 is an SPST switch with three logic control inputs. The TL610 features a higher r_{off}/r_{on} ratio than the other members of the family.

The TL601M, TL604M, TL607M, and TL610M are characterized for operation over the full military temperature range of −55°C to 125°C, the TL601I, TL604I, TL607I, and TL610I are characterized for operation from −25°C to 85°C, and the TL601C, TL604C, TL607C, and TL610C are characterized for operation from 0°C to 70°C.

TYPICAL OF
ALL INPUTS

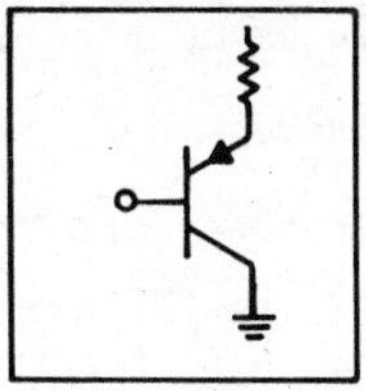

TYPICAL OF
ALL SWITCHES

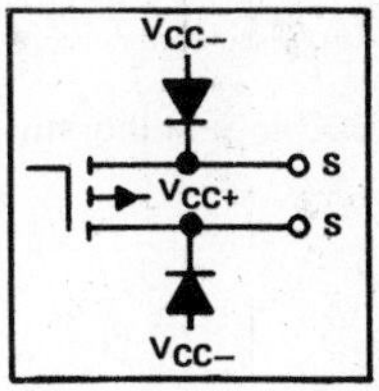
VCC−
S
VCC+
S
VCC−

TEXAS INSTRUMENTS
POST OFFICE BOX 655012 • DALLAS, TEXAS 75265

logic symbols† and switch diagrams

TL601

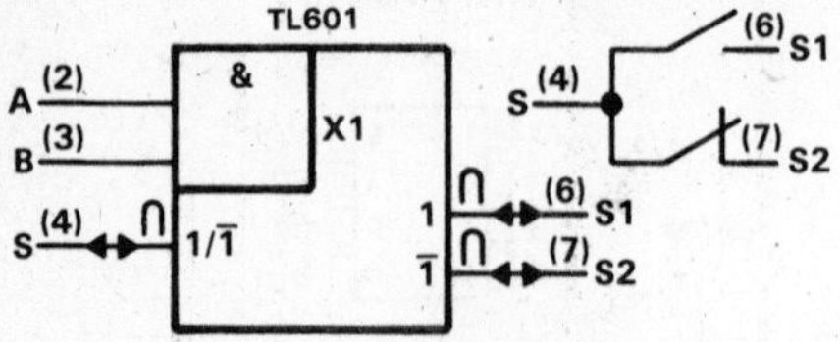
&
X1
1/1̄
A (2)
B (3)
S (4)
1
1̄
(6) S1
(7) S2
S (4)

TL602

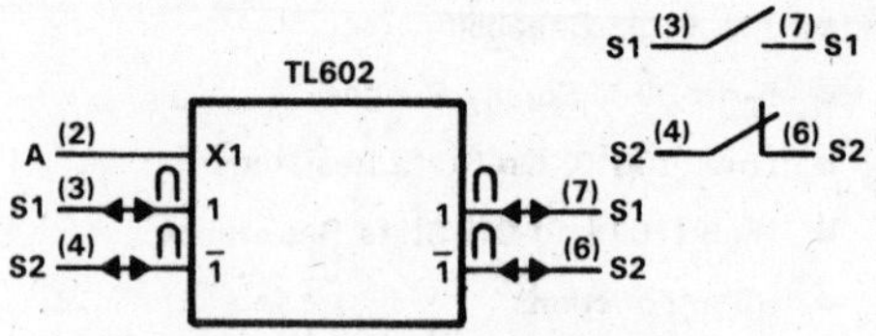
S1 (3) (7) S1
S2 (4) (6) S2
A (2)
S1 (3)
S2 (4)
X1
1
1̄
(7) S1
(6) S2

FUNCTION TABLE

LOGIC INPUTS		ANALOG SWITCH	
A	B	S1	S2
L	X	OFF (OPEN)	ON (CLOSED)
X	L	OFF (OPEN)	ON (CLOSED)
H	H	ON (CLOSED)	OFF (OPEN)

FUNCTION TABLE

LOGIC INPUT	ANALOG SWITCH	
A	S1	S2
H	ON (CLOSED)	OFF (OPEN)
L	OFF (OPEN)	ON (CLOSED)

TL607

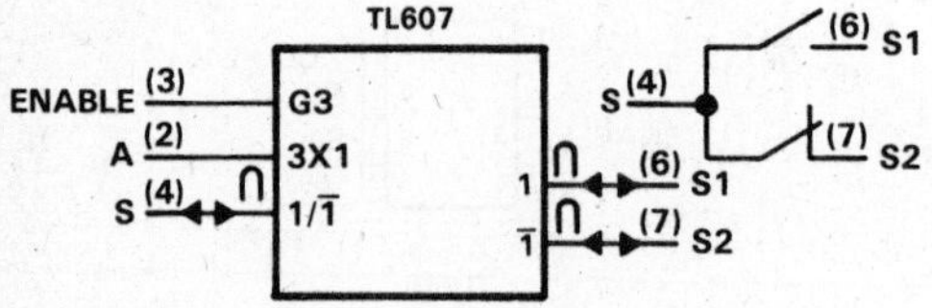
ENABLE (3)
A (2)
S (4)
G3
3X1
1/1̄
1
1̄
(6) S1
(7) S2
S (4)

TL610

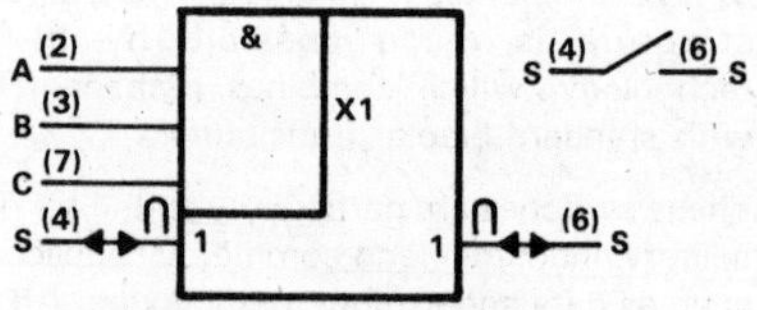
&
X1
A (2)
B (3)
C (7)
S (4)
1
(6) S
S (4) (6) S

FUNCTION TABLE

INPUTS		ANALOG SWITCH	
A	ENABLE	S1	S2
X	L	OFF (OPEN)	OFF (OPEN)
L	H	OFF (OPEN)	ON (CLOSED)
H	H	ON (CLOSED)	OFF (OPEN)

FUNCTION TABLE

INPUTS			ANALOG SWITCH
A	B	C	S
L	X	X	OFF (OPEN)
X	L	X	OFF (OPEN)
X	X	L	OFF (OPEN)
H	H	H	ON (CLOSED)

†These symbols are in accordance with ANSI/IEEE Std 91-1984.

TL607 logic diagram (positive logic)

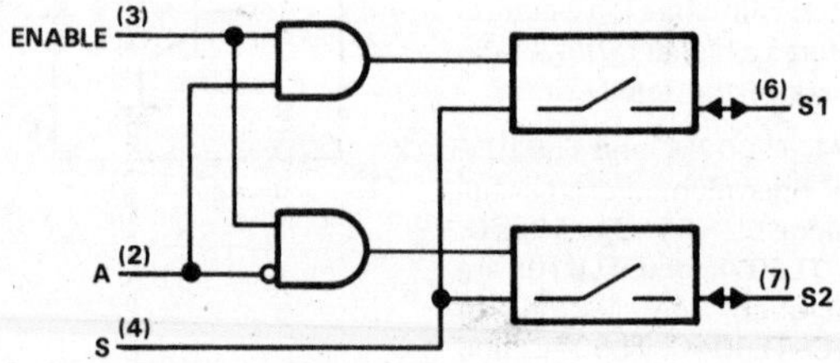
ENABLE (3)
A (2)
S (4)
(6) S1
(7) S2

absolute maximum ratings over operating free-air temperature range (unless otherwise noted)

Supply voltage, V_{CC+} (see Note 1)	30 V
Supply voltage, V_{CC-}	−30 V
V_{CC+} to V_{CC-} supply voltage differential	35 V
Control input voltage	V_{CC+}
Switch off-state voltage	30 V
Switch on-state current	10 mA
Operating free-air temperature range: TL601M, TL604M, TL607M, TL610M	−55 °C to 125 °C
TL601I, TL604I, TL607I, TL610I	−25 °C to 85 °C
TL601C, TL604C, TL607C, TL610C	0 °C to 70 °C
Storage temperature range	−65 °C to 150 °C
Lead temperature (1,6 mm) 1/16 inch from case for 60 seconds: JG package	300 °C
Lead temperature (1,6 mm) 1/16 inch from case for 10 seconds: P package	260 °C

NOTE 1: All voltage values are with respect to network ground terminal.

recommended operating conditions

		TL601M, TL604M TL607M, TL610M			TL601I, TL604I TL607I, TL610I			TL601C, TL604C TL607C, TL610C		UNIT
	MIN	NOM	MAX	MIN	NOM	MAX	MIN	NOM	MAX	
Supply voltage, V_{CC+} (see Figure 1)	5	10	25	5	10	25	5	10	25	V
Supply voltage, V_{CC-} (see Figure 1)	−5	−20	−25	−5	−20	−25	−5	−20	−25	V
V_{CC+} to V_{CC-} supply voltage differential (see Figure 1)	15		30	15		30	15		30	V
High-level control input voltage, V_{IH}	2		5.5	2		5.5	2		5.5	V
Low-level control input voltage, V_{IL} All inputs			0.8			0.8			0.8	
Voltage at any analog switch (S) terminal	$V_{CC-}+8$		V_{CC+}	$V_{CC-}+8$		V_{CC+}	$V_{CC-}+8$		V_{CC+}	V
Switch on-state current			10			10			10	mA
Operating free-air temperature, T_A	−55		125	−25		85	0		70	°C

PARAMETER MEASUREMENT INFORMATION

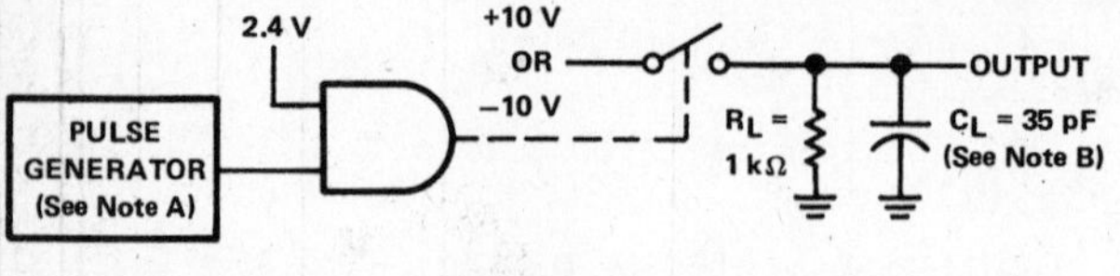
2.4 V
+10 V
OR
−10 V
PULSE
GENERATOR
(See Note A)
OUTPUT
R_L =
1 kΩ
C_L = 35 pF
(See Note B)

TEST CIRCUIT

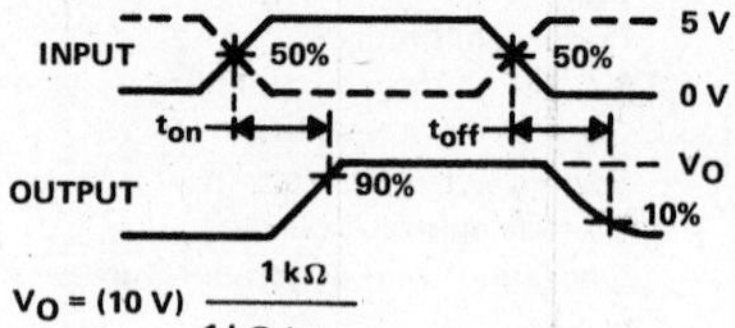
INPUT
50%
50%
5 V
0 V
t_{on}
t_{off}
OUTPUT
90%
10%
V_O

$V_O = (10\ V)\ \frac{1\ k\Omega}{1\ k\Omega + r_{on}}$

VOLTAGE WAVEFORMS

NOTES: A. The pulse generator has the following characteristics: $Z_{out} = 50\ \Omega$, $t_r \leq 15$ ns, $t_f \leq 15$ ns, $t_w = 500$ ns.
B. C_L includes probe and jig capacitance.

FIGURE 2

TYPICAL CHARACTERISTICS

SWITCH ON-STATE RESISTANCE
vs
FREE-AIR TEMPERATURE

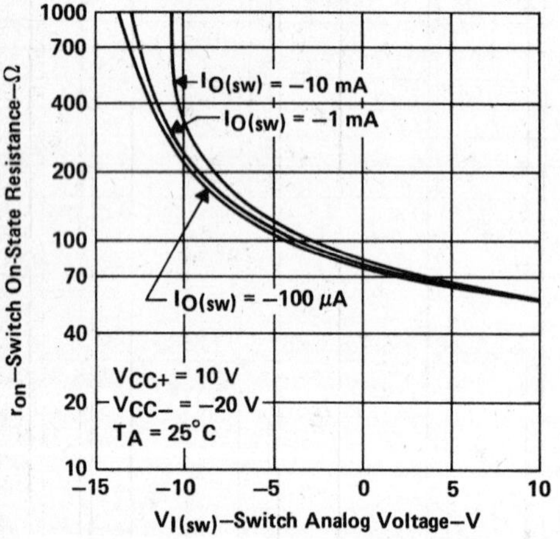
1000
700
400
200
100
70
40
20
10
−15
−10
−5
0
5
10
r_{on}—Switch On-State Resistance—Ω
$V_{I(sw)}$—Switch Analog Voltage—V
$I_{O(sw)}$ = −10 mA
$I_{O(sw)}$ = −1 mA
$I_{O(sw)}$ = −100 µA
V_{CC+} = 10 V
V_{CC-} = −20 V
T_A = 25°C

FIGURE 3

SWITCH ON-STATE RESISTANCE
vs
SWITCH ANALOG VOLTAGE

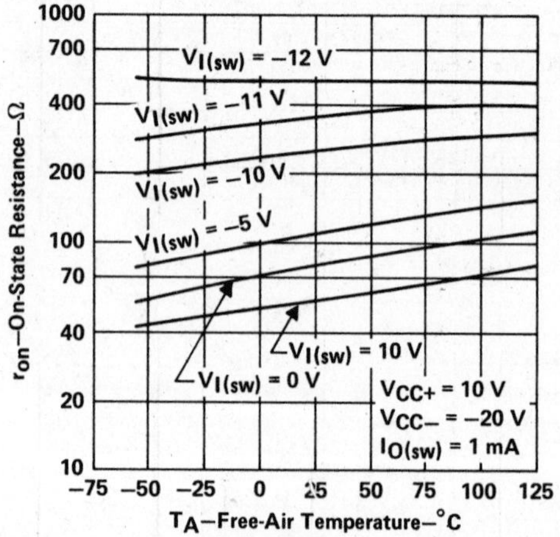
1000
700
400
200
100
70
40
20
10
−75
−50
−25
0
25
50
75
100
125
r_{on}—On-State Resistance—Ω
T_A—Free-Air Temperature—°C
$V_{I(sw)}$ = −12 V
$V_{I(sw)}$ = −11 V
$V_{I(sw)}$ = −10 V
$V_{I(sw)}$ = −5 V
$V_{I(sw)}$ = 10 V
$V_{I(sw)}$ = 0 V
V_{CC+} = 10 V
V_{CC-} = −20 V
$I_{O(sw)}$ = 1 mA

FIGURE 4

electrical characteristics over recommended operating free-air temperature range, V_{CC+} = 10 V, V_{CC-} = −20 V, analog switch test current = 1 mA (unless otherwise noted)

PARAMETER		TEST CONDITIONS†			TL6__M TL6__I MIN	TYP‡	MAX	TL6__C MIN	TYP‡	MAX	UNIT
I_{IH}	High-level input current	V_I = 5.5 V				0.5	10		0.5	10	μA
I_{IL}	Low-level input current	V_I = 0.4 V				−50	−250		−50	−250	μA
I_{off}	Switch off-state current	$V_{I(sw)}$ = −10 V, See Note 2		T_A = 25°C		−400			−500		pA
				T_A = MAX†		−50	−100		−10	−20	nA
r_{on}	Switch on-state resistance	$V_{I(sw)}$ = 10 V, $I_{O(sw)}$ = −1 mA		TL601, TL604, TL607		55	100		75	200	Ω
				TL610		40	80		40	100	
		$V_{I(sw)}$ = −10 V, $I_{O(sw)}$ = −1 mA		TL601, TL604, TL607		220	400		220	600	
				TL610		120	300		120	400	
r_{off}	Switch off-state resistance					25			20		GΩ
C_{on}	Switch on-state input capacitance	$V_{I(sw)}$ = 0 V, f = 1 MHz				16			16		pF
C_{off}	Switch off-state input capacitance	$V_{I(sw)}$ = 0 V, f = 1 MHz				8			8		pF
I_{CC+}	Supply current from V_{CC+}	Logic input(s) at 5.5 V, All switch terminals open		TL601, TL604		5	10		5	10	mA
			Enable input high	TL607		5	10		5	10	
			Enable input low	TL607		3	5		3	5	
				TL610		5	10		5	10	
I_{CC-}	Supply current from V_{CC-}	Logic input(s) at 5.5 V, All switch terminals open		TL601, TL604		−1.2	−2.5		−1.2	−2.5	mA
			Enable input high	TL607		−2.5	−5		−2.5	−5	
			Enable input low	TL607		−0.05	−0.5		−0.05	−0.5	
				TL610		−1.2	−2.5		−1.2	−2.5	

†MAX is 125°C for M-suffix types, 85°C for I-suffix types, and 70°C for C-suffix types.
‡All typical values are at T_A = 25°C except for I_{off} at T_A = MAX.
NOTE 2: The other terminal of the switch under test is at V_{CC+} = 10 V.

switching characteristics, V_{CC} = 10 V, V_{CC-} = −20 V, T_A = 25°C

PARAMETER		TEST CONDITIONS	MIN	TYP	MAX	UNIT
t_{off}	Switch turn-off time	R_L = 1 kΩ, C_L = 35 pF, See Figure 2		400	500	ns
t_{on}	Switch turn-on time			100	150	

Figure 1 shows power supply boundary conditions for proper operation of the TL601 Series. The range of operation for supply V_{CC+} from +5 V to +25 V is shown on the vertical axis. The range of V_{CC-} from −5 volts to −25 volts is shown on the horizontal axis. A recommended 30-volt maximum voltage differential from V_{CC+} to V_{CC-} governs the maximum V_{CC+} for a chosen V_{CC-} (or vice versa). A minimum recommended difference of 15 volts from V_{CC+} to V_{CC-} and the boundaries shown in Figure 1 allow the designer to select the proper combinations of the two supplies.

The designer-selected V_{CC+} for a chosen V_{CC-} supply values limit the maximum input voltage that can be applied to either switch terminal; that is, the input voltage should be between V_{CC-} +8 V and V_{CC+} to keep the on-state resistance within specified limits.

RECOMMENDED COMBINATIONS
OF SUPPLY VOLTAGES

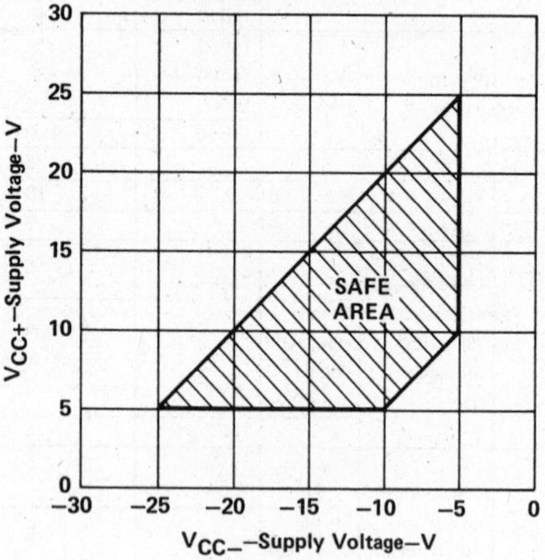
30
25
20
15
10
5
0
V_{CC+}—Supply Voltage—V
SAFE
AREA
−30
−25
−20
−15
−10
−5
0
V_{CC-}—Supply Voltage—V

FIGURE 1

TLC04, TLC14
BUTTERWORTH FOURTH-ORDER LOW-PASS SWITCHED-CAPACITOR FILTERS

D2970, NOVEMBER, 1986

- **Low Clock-to-Cutoff-Frequency Ratio Error**
  TLC04 . . . ±0.8%
  TLC14 . . . ±1%
- **Filter Cutoff Frequency Dependent Only on External-Clock Frequency Stability**
- **Minimum Filter Response Deviation Due to External Component Variations Over Time and Temperature**
- **Cutoff Frequency Range from 0.1 Hz to 20 kHz**
- **5-V to 12-V Operation**
- **Self Clocking or TTL-Compatible and CMOS-Compatible Clock Inputs**
- **Designed to be Interchangeable with National MF4-50 and MF4-100**

D OR N PACKAGE
(TOP VIEW)

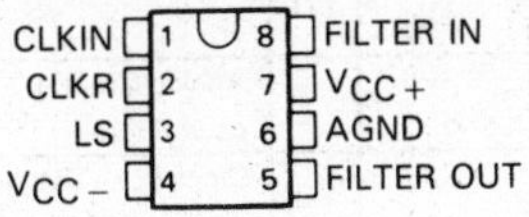
CLKIN 1
CLKR 2
LS 3
V_{CC-} 4
8 FILTER IN
7 V_{CC+}
6 AGND
5 FILTER OUT

description

The TLC04 and TLC14 are monolithic Butterworth low-pass switched-capacitor filters. Each is designed as a low-cost, easy-to-use device and to provide accurate fourth-order low-pass filter functions in circuit design configurations.

Each filter features cutoff frequency stability that is dependent only on the external-clock frequency stability. The cutoff frequency is clock tunable and has a clock-to-cutoff frequency ratio of 50:1 with less than ±0.8% error for the TLC04 and a clock-to-cutoff frequency ratio of 100:1 with less than ±1% error for the TLC14. The input clock features self-clocking or TTL- or CMOS-compatible options in conjunction with the level shift (LS) pin.

The TLC04 and TLC14 are characterized for operation from 0°C to 70°C.

functional block diagram

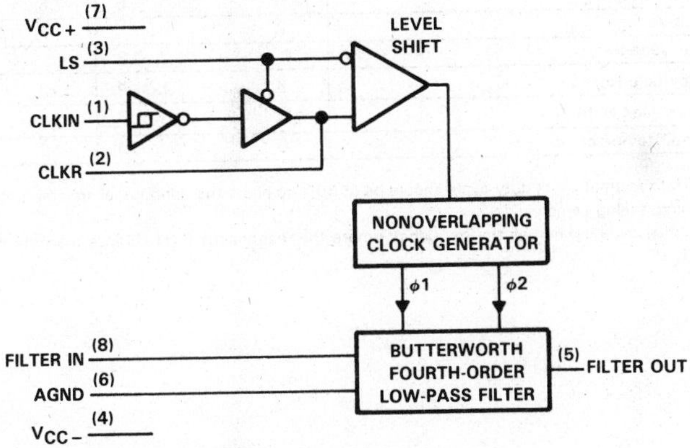
V_{CC+} (7)
LS (3)
CLKIN (1)
CLKR (2)
LEVEL SHIFT
NONOVERLAPPING CLOCK GENERATOR
φ1
φ2
FILTER IN (8)
AGND (6)
V_{CC-} (4)
BUTTERWORTH FOURTH-ORDER LOW-PASS FILTER
(5) FILTER OUT

TEXAS INSTRUMENTS
POST OFFICE BOX 655012 • DALLAS, TEXAS 75265

pin description

PIN NAME	PIN NO.	I/O	DESCRIPTION
AGND	6	I	Analog Ground — The noninverting input to the operational amplifiers of the Butterworth fourth-order low-pass filter.
CLKIN	1	I	Clock In — The clock input terminal for CMOS-compatible clock or self-clocking options. For either option, the Level Shift (LS) terminal is at V_{CC-}. For self-clocking, a resistor is connected between the CLKIN and CLKR terminal pins and a capacitor is connected from the CLKIN terminal pin to ground.
CLKR	2	I	Clock R — The clock input for a TTL-compatible clock. For a TTL clock, the level shift pin is connected to mid-supply and the CLKIN pin may be left open, but it is recommended that it be connected to either V_{CC+} or V_{CC-}.
FILTER IN	8	I	Filter Input
FILTER OUT	5	O	Butterworth fourth-order low-pass Filter Output
LS	3	I	Level Shift — This terminal accommodates the various input clocking options. For CMOS-compatible clocks or self-clocking, the level-shift terminal is at V_{CC-} and for TTL-compatible clocks, the level-shift terminal is at mid-supply.
V_{CC+}	7	I	Positive supply voltage terminal
V_{CC-}	4	I	Negative supply voltage terminal

absolute maximum ratings over operating free-air temperature range (unless otherwise noted)

Supply voltage, $V_{CC\pm}$ (see Note 1) .... ±7 V
Operating free-air temperature range .... 0°C to 70°C
Storage temperature range .... −65°C to 150°C
Lead temperature 1,6 mm (1/16 inch) from case for 10 seconds .... 260°C

NOTE 1: All voltage values are with respect to the AGND terminal.

recommended operating conditions

		TLC04		TLC14		UNIT
		MIN	MAX	MIN	MAX	
V_{CC+}	Positive supply voltage	2.5	6	2.5	6	V
V_{CC-}	Negative supply voltage	−2.5	−6	−2.5	−6	V
V_{IH}	High-level input voltage	2		2		V
V_{IL}	Low-level input voltage		0.8		0.8	V
f_{clock}	Clock frequency (see Note 2)	5	1×10^6	10	1×10^6	Hz
f_{co}	Cutoff frequency (see Note 3)	0.1	20×10^3	0.1	10×10^3	Hz
T_A	Operating free-air temperature	0	70	0	70	°C

NOTES: 2. Above 250 kHz, the input clock duty cycle should be at 50% to allow the operational amplifiers the maximum time to settle while processing analog samples.
3. The cutoff frequency is defined as the frequency where the response is 3.01 dB less than the dc gain of the filter.

electrical characteristics over recommended operating free-air temperature range, V_{CC+} = 2.5 V, V_{CC-} = −2.5 V, $f_{clock} \leq$ 250 kHz (unless otherwise noted)

filter section

PARAMETER			TEST CONDITIONS	TLC04 MIN	TLC04 TYP†	TLC04 MAX	TLC14 MIN	TLC14 TYP†	TLC14 MAX	UNIT
V_{OO}	Output voltage offset				−150			−300		mV
V_{OM}	Peak output voltages	V_{OM+}	R_L = 5 kΩ	2	2.3		2	2.3		V
		V_{OM-}		−1	−1.5		−1	−1.5		
I_{OS}	Short-circuit output current	Source	T_A = 25°C, See Note 4		−0.5			−0.5		mA
		Sink			28			28		
I_{CC}	Supply current		f_{clock} = 250 kHz		1.5	2.25		1.5	2.25	mA

NOTE 4: I_{OS} (source current) is measured by forcing the output to its maximum positive voltage and then shorting the output to the negative supply (V_{CC-}) terminal. I_{OS} (sink current) is measured by forcing the output to its maximum negative voltage and then shorting the output to the positive supply (V_{CC+}) terminal.

operating characteristics over recommended operating free-air temperature range, V_{CC+} = 2.5 V, V_{CC-} = −2.5 V (unless otherwise noted)

PARAMETER	TEST CONDITIONS		TLC04 MIN	TLC04 TYP†	TLC04 MAX	TLC14 MIN	TLC14 TYP†	TLC14 MAX	UNIT
Clock-to-cutoff-frequency ratio (f_{clock}/f_{co})	$f_{clock} \leq$ 250 kHz, T_A = 25°C		49.27	50.07	50.87	99	100	101	
Temperature coefficient of clock-to-cutoff frequency ratio	$f_{clock} \leq$ 250 kHz		−25	0	25	−25	0	25	ppm/°C
Frequency response above and below cutoff frequency (see Note 5)	f_{co} = 5 kHz, f_{clk} = 250 kHz, T_A = 25°C	f = 6 kHz	−8.11	−7.57	−7.03				dB
		f = 4.5 kHz	−1.7	−1.46	−1.22				
	f_{co} = 2.5 kHz, f_{clk} = 250 kHz, T_A = 25°C	f = 3 kHz				−7.92	−7.42	−6.92	dB
		f = 2.25 kHz				−1.77	−1.51	−1.25	
Dynamic range (see Note 6)	T_A = 25°C			80			78		dB
Stop-band frequency attentuation at 2 f_{co}	$f_{clock} \leq$ 250 kHz		24	25		24	25		dB
DC voltage amplification	$f_{clock} \leq$ 250 kHz, RS ≤ 2 kΩ		−0.15	0	0.15	−0.15	0	0.15	dB
Peak-to-peak clock feedthrough voltage	T_A = 25°C			15			15		mV

† All typical values are at T_A = 25°C.

NOTES: 5. The frequency responses at f are referenced to a dc gain of 0 dB.

6. The dynamic range is referenced to 2.82 V rms (4 V peak) where the wideband noise over a 20-kHz bandwidth is typically 282 μV rms for the TLC04 and 355 μV rms for the TLC14.

electrical characteristics over recommended operating free-air temperature range, V_{CC+} = 5 V, V_{CC-} = −5 V, $f_{clock} \leq$ 250 kHz, (unless otherwise noted)

filter section

PARAMETER			TEST CONDITIONS	TLC04 MIN	TLC04 TYP†	TLC04 MAX	TLC14 MIN	TLC14 TYP†	TLC14 MAX	UNIT
V_{OO}	Output voltage offset				−200			−400		mV
V_{OM}	Peak output voltages	V_{OM+}	R_L = 5 kΩ	4	4.5		4	4.5		V
		V_{OM-}		−4	−4.1		−4	−4.1		
I_{OS}	Short-circuit output current	Source	T_A = 25°C, See Note 4		−1.5			−1.5		mA
		Sink			50			50		
I_{CC}	Supply current		f_{clock} = 250 kHz		2.5	3.5		2.5	3.5	mA

NOTE 4: I_{OS} (source current) is measured by forcing the output to its maximum positive voltage and then shorting the output to the negative supply (V_{CC-}) terminal. I_{OS} (sink current) is measured by forcing the output to its maximum negative voltage and then shorting the output to the positive supply (V_{CC+}) terminal.

clocking section

PARAMETER			TEST CONDITIONS‡		MIN	TYP†	MAX	UNIT
V_{T+}	Positive-going input threshold voltage	CLKIN	V_{CC} = 10 V		6.1	7	8.9	V
			V_{CC} = 5 V		3.1	3.5	4.4	
V_{T-}	Negative-going input threshold voltage		V_{CC} = 10 V		1.3	3	3.8	V
			V_{CC} = 5 V		0.6	1.5	1.9	
V_{hys}	Hysteresis ($V_{T+} - V_{T-}$)		V_{CC} = 10 V		2.3	4	7.6	V
			V_{CC} = 5 V		1.2	2	3.8	
V_{OH}	High-level output voltage	CLKR	V_{CC} = 10 V	I_O = −10 μA	9			V
			V_{CC} = 5 V		4.5			
V_{OL}	Low-level output voltage		V_{CC} = 10 V	I_O = 10 μA			1	V
			V_{CC} = 5 V				0.5	
	Input leakage current		V_{CC} = 10 V	Level Shift pin at mid-supply, T_A = 25°C			2	μA
			V_{CC} = 5 V				2	
	Output current		V_{CC} = 10 V	CLKR shorted to V_{CC-}	−3	−6		mA
			V_{CC} = 5 V		−0.75	−1.5		
	Output current		V_{CC} = 10 V	CLKR shorted to V_{CC+}	2.5	5		mA
			V_{CC} = 5 V		0.65	1.3		

† All typical values are at T_A = 25°C.
‡ $V_{CC} = V_{CC+} - V_{CC-}$.

operating characteristics over recommended operating free-air temperature range, V_{CC+} = 5 V, V_{CC-} = −5 V (unless otherwise noted)

PARAMETER	TEST CONDITIONS		TLC04 MIN	TLC04 TYP†	TLC04 MAX	TLC14 MIN	TLC14 TYP†	TLC14 MAX	UNIT
Clock-to-cutoff-frequency ratio (f_{clock}/f_{co})	f_{clock} ≤ 250 kHz, T_A = 25°C		49.58	49.98	50.38	99	100	101	
Temperature coefficient of clock-to-cutoff frequency ratio	f_{clock} ≤ 250 kHz		−15	0	15	−15	0	15	ppm/°C
Frequency response above and below cutoff frequency (see Note 5)	f_{co} = 5 kHz, f_{clk} = 250 kHz, T_A = 25°C	f = 6 kHz	−7.84	−7.57	−7.3				dB
		f = 4.5 kHz	−1.56	−1.44	−1.32				
	f_{co} = 2.5 kHz, f_{clk} = 250 kHz, T_A = 25°C	f = 3 kHz				−7.67	−7.42	−7.17	dB
		f = 2.25 kHz				−1.64	−1.51	−1.38	
Dynamic range (see Note 7)	T_A = 25°C			80			78		dB
Stop-band frequency attentuation at 2 f_{co}	f_{clock} ≤ 250 kHz		24	25		24	25		dB
DC voltage amplification	f_{clock} ≤ 250 kHz, RS ≤ 2 kΩ		−0.15	0	0.15	−0.15	0	0.15	dB
Peak-to-peak clock feedthrough voltage	T_A = 25°C			25			25		mV

† All typical values are at T_A = 25°C.

NOTES: 5. The frequency responses at f are referenced to a dc gain of 0 dB.

7. The dynamic range is referenced to 2.82 V rms (4 V peak) where the wideband noise over a 20-kHz bandwidth is typically 282 μV rms for the TLC04 and 355 μV rms for the TLC14.

TYPICAL APPLICATION DATA

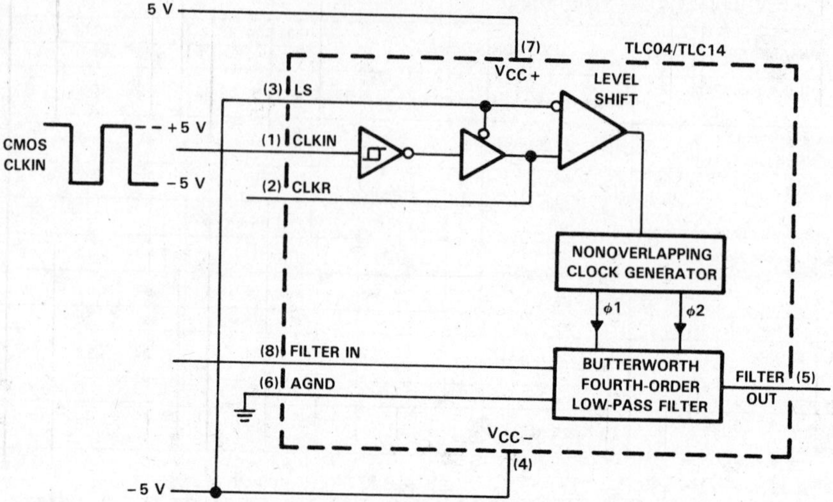
5 V
(7)
TLC04/TLC14
VCC+
LEVEL
SHIFT
(3) LS
CMOS
CLKIN
+5 V
−5 V
(1) CLKIN
(2) CLKR
NONOVERLAPPING
CLOCK GENERATOR
ϕ1
ϕ2
(8) FILTER IN
(6) AGND
BUTTERWORTH
FOURTH-ORDER
LOW-PASS FILTER
FILTER
OUT
(5)
VCC−
(4)
−5 V

FIGURE 1. CMOS-CLOCK-DRIVEN, DUAL-SUPPLY OPERATION

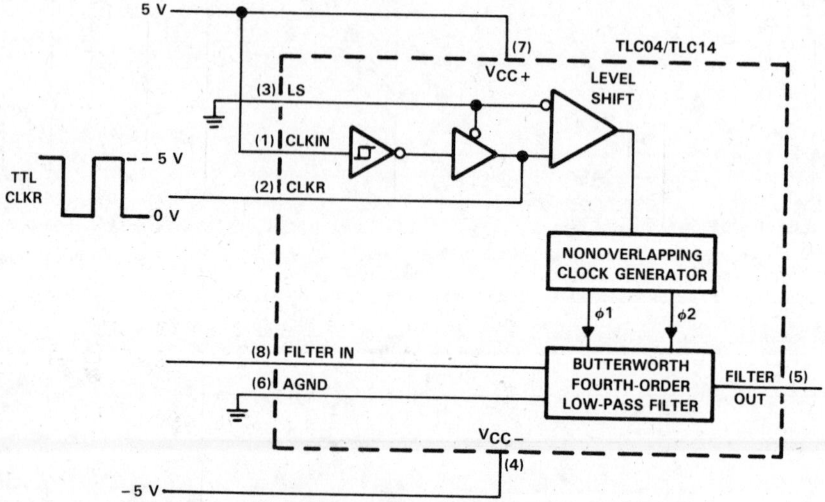
5 V
(7)
TLC04/TLC14
VCC+
LEVEL
SHIFT
(3) LS
(1) CLKIN
TTL
CLKR
5 V
0 V
(2) CLKR
NONOVERLAPPING
CLOCK GENERATOR
ϕ1
ϕ2
(8) FILTER IN
(6) AGND
BUTTERWORTH
FOURTH-ORDER
LOW-PASS FILTER
FILTER
OUT
(5)
VCC−
(4)
−5 V

FIGURE 2. TTL-CLOCK-DRIVEN, DUAL-SUPPLY OPERATION

TYPICAL APPLICATION DATA

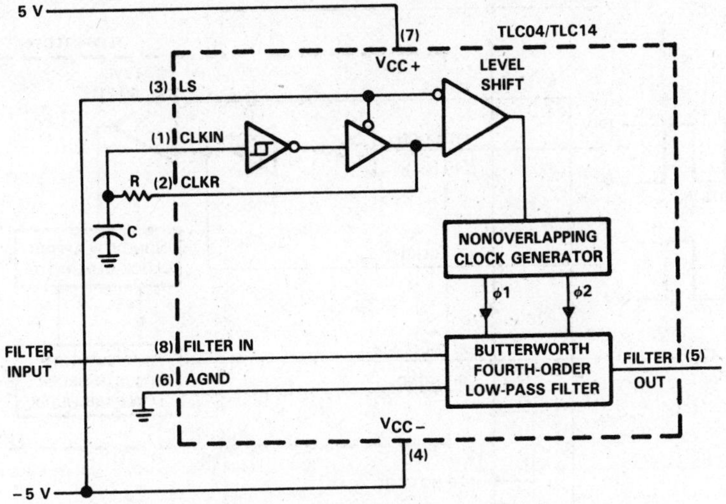
5 V
(7)
TLC04/TLC14
VCC+
LEVEL
SHIFT
(3) LS
(1) CLKIN
R
(2) CLKR
C
NONOVERLAPPING
CLOCK GENERATOR
ϕ1
ϕ2
FILTER
INPUT
(8) FILTER IN
(6) AGND
BUTTERWORTH
FOURTH-ORDER
LOW-PASS FILTER
FILTER
OUT
(5)
VCC−
(4)
−5 V

$$f_{clock} = \frac{1}{RC \times \ln\left[\left(\frac{V_{CC} - V_{T-}}{V_{CC} - V_{T+}}\right)\left(\frac{V_{T+}}{V_{T-}}\right)\right]}$$

For $V_{CC} = 10$ V,

$$f_{clock} = \frac{1}{1.69\ RC}$$

FIGURE 3. SELF-CLOCKING THROUGH SCHMITT TRIGGER OSCILLATOR, DUAL-SUPPLY OPERATION

TYPICAL APPLICATION DATA

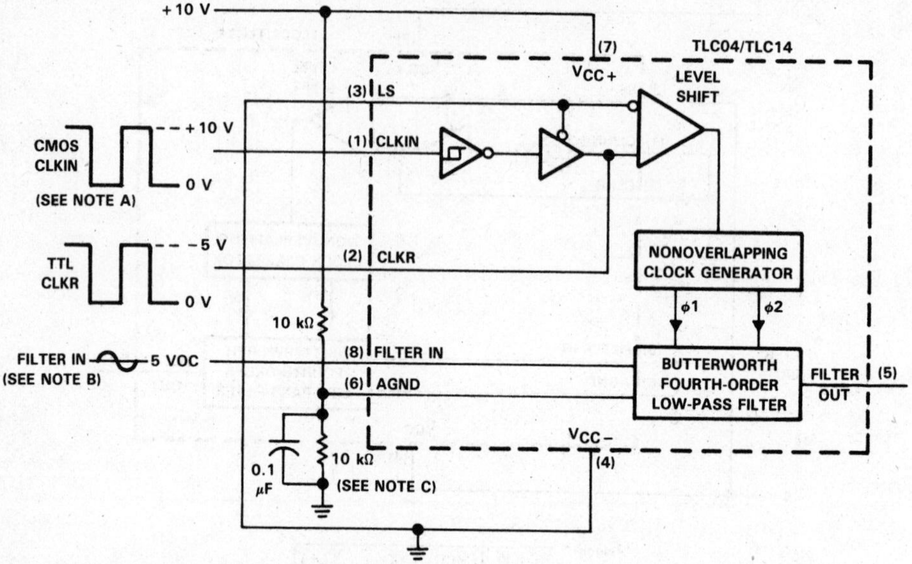
+10 V
(7)
TLC04/TLC14
VCC+
LEVEL SHIFT
(3) LS
CMOS CLKIN
+10 V
0 V
(SEE NOTE A)
(1) CLKIN
TTL CLKR
-5 V
0 V
(2) CLKR
NONOVERLAPPING CLOCK GENERATOR
φ1
φ2
10 kΩ
FILTER IN
5 VOC
(SEE NOTE B)
(8) FILTER IN
(6) AGND
BUTTERWORTH FOURTH-ORDER LOW-PASS FILTER
FILTER OUT
(5)
VCC−
(4)
0.1 μF
10 kΩ
(SEE NOTE C)

NOTES: A. The external clock used must be of CMOS level because the clock is input to a CMOS Schmitt trigger.
B. The Filter input signal should be dc-biased to mid-supply or ac-coupled to the terminal.
C. The AGND terminal must be biased to mid-supply.

FIGURE 4. EXTERNAL-CLOCK-DRIVEN SINGLE-SUPPLY OPERATION

TYPICAL APPLICATION DATA

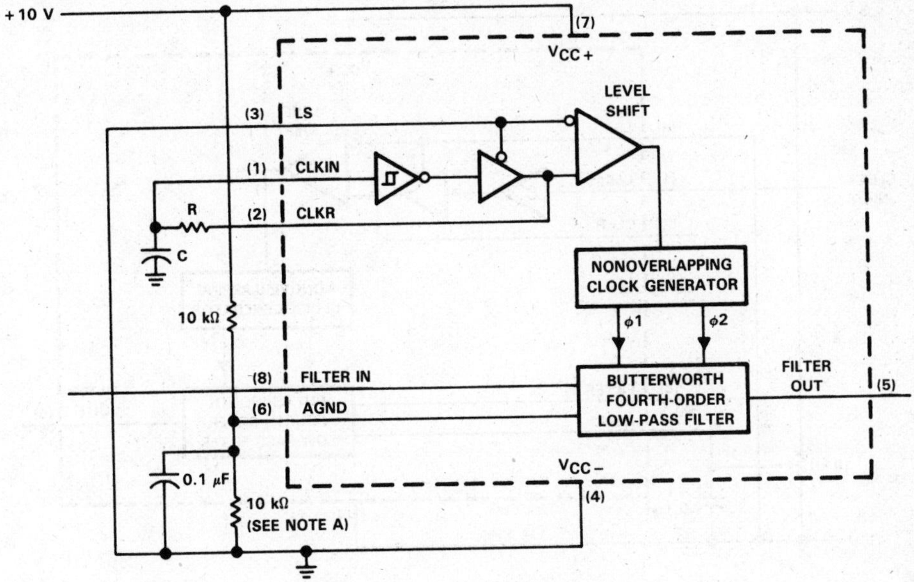
+10 V
(7)
VCC+
LEVEL
SHIFT
(3) LS
(1) CLKIN
R
(2) CLKR
C
NONOVERLAPPING
CLOCK GENERATOR
10 kΩ
φ1
φ2
FILTER
OUT
(8) FILTER IN
BUTTERWORTH
FOURTH-ORDER
LOW-PASS FILTER
(5)
(6) AGND
VCC−
0.1 μF
(4)
10 kΩ
(SEE NOTE A)

$$f_{clock} = \frac{1}{RC \times \ln\left[\left(\frac{V_{CC} - V_{T-}}{V_{CC} - V_{T+}}\right)\left(\frac{V_{T+}}{V_{T-}}\right)\right]}$$

For $V_{CC} = 10$ V,

$$f_{clock} = \frac{1}{1.69\ RC}$$

NOTE A: The AGND terminal must be biased to mid-supply.

FIGURE 5. SELF-CLOCKING THROUGH SCHMITT TRIGGER OSCILLATOR, SINGLE-SUPPLY OPERATION

TYPICAL APPLICATION DATA

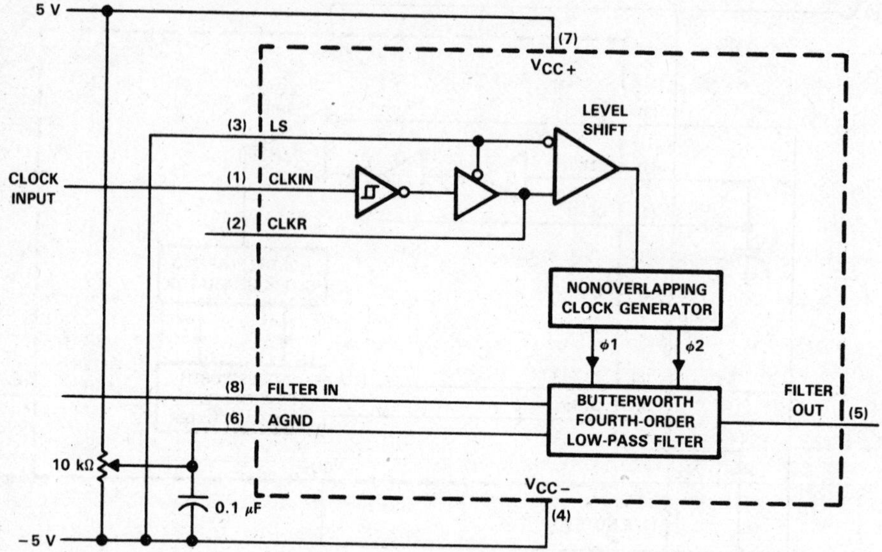
5 V
(7)
VCC+
LEVEL
SHIFT
(3) LS
CLOCK
INPUT
(1) CLKIN
(2) CLKR
NONOVERLAPPING
CLOCK GENERATOR
φ1
φ2
(8) FILTER IN
(6) AGND
BUTTERWORTH
FOURTH-ORDER
LOW-PASS FILTER
FILTER
OUT
(5)
10 kΩ
0.1 μF
VCC−
(4)
−5 V

FIGURE 6. DC OFFSET ADJUSTMENT

TLC0820A, TLC0820B
ADVANCED LinCMOS™ HIGH-SPEED 8-BIT ANALOG-TO-DIGITAL CONVERTERS USING MODIFIED "FLASH" TECHNIQUES

D2873, SEPTEMBER 1986—REVISED OCTOBER 1986

- Advanced LinCMOS™ Silicon-Gate Technology
- 8-Bit Resolution
- Differential Reference Inputs
- Parallel Microprocessor Interface
- Conversion Time
  Write-Read Mode . . . 0.9 μs and 1.1 μs
  Read Mode . . . 2.5 μs Max
- No External Clock or Oscillator Components Required
- On-Chip Track-and-Hold
- Low Power Consumption . . . 50 mW Typ
- Single 5-V Supply
- TLC0820B is Direct Replacement for National Semiconductor ADC0820B/BC and Analog Devices AD7820L/C/U; TLC0820A is Direct Replacement for National Semiconductor ADC0820C/CC and Analog Devices AD7820K/B/T

TLC0820AM, TLC0820BM . . . J OR N PACKAGE
TLC0820AI, TLC0820BI . . . N PACKAGE
TLC0820AC, TLC0820BC . . . N PACKAGE
(TOP VIEW)

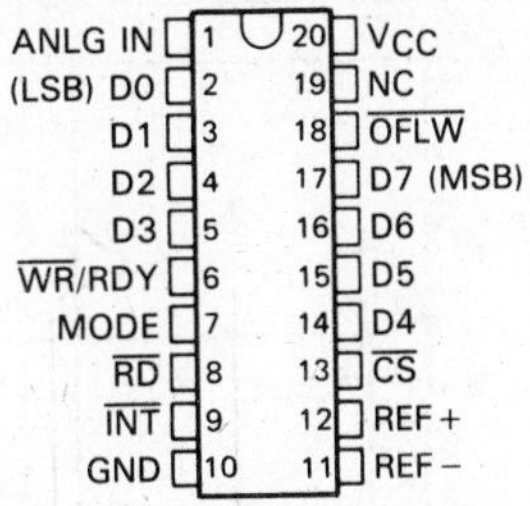
ANLG IN 1
(LSB) D0 2
D1 3
D2 4
D3 5
WR/RDY 6
MODE 7
RD 8
INT 9
GND 10
20 VCC
19 NC
18 OFLW
17 D7 (MSB)
16 D6
15 D5
14 D4
13 CS
12 REF+
11 REF−

TLC0820AM, TLC0820BM . . . FK PACKAGE
TLC0820AI, TLC0820BI . . . FN PACKAGE
TLC0820AC, TLC0820BC . . . FN PACKAGE
(TOP VIEW)

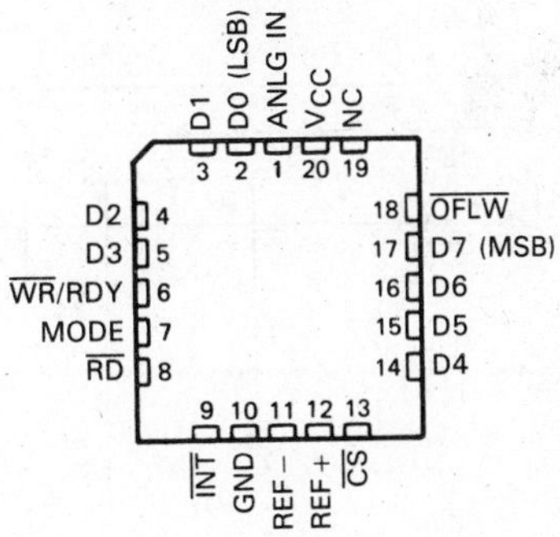
D1 3
D0 (LSB) 2
ANLG IN 1
VCC 20
NC 19
D2 4
D3 5
WR/RDY 6
MODE 7
RD 8
18 OFLW
17 D7 (MSB)
16 D6
15 D5
14 D4
INT 9
GND 10
REF− 11
REF+ 12
CS 13

NC—No internal connection

description

The TLC0820A and TLC0820B are Advanced LinCMOS™ 8-bit analog-to-digital converters each consisting of two 4-bit "flash" converters, a 4-bit digital-to-analog converter, a summing (error) amplifier, control logic, and a result latch circuit. The modified "flash" technique allows low-power integrated circuitry to complete an 8-bit conversion in 1.4 microseconds. The on-chip track-and-hold circuit has a 100-nanosecond sample window and allows the TLC0820A and TLC0820B to convert continuous analog signals having slew rates of up to 100 millivolts per microsecond without external sampling components. TTL-compatible three-state output drivers and two modes of operation allow interfacing to a variety of microprocessors. Detailed information on interfacing to most popular microprocessors is readily available from the factory.

The TLC0820AM and TLC0820BM are available in both the N plastic and the J ceramic packages and are characterized for operation over the full military temperature range of −55°C to 125°C. The TLC0820AI and TLC0820BI are characterized for operation from −40°C to 85°C. The TLC0820AC and TLC0820BC are characterized for operation from 0°C to 70°C.

Advanced LinCMOS is a trademark of Texas Instruments.

functional block diagram

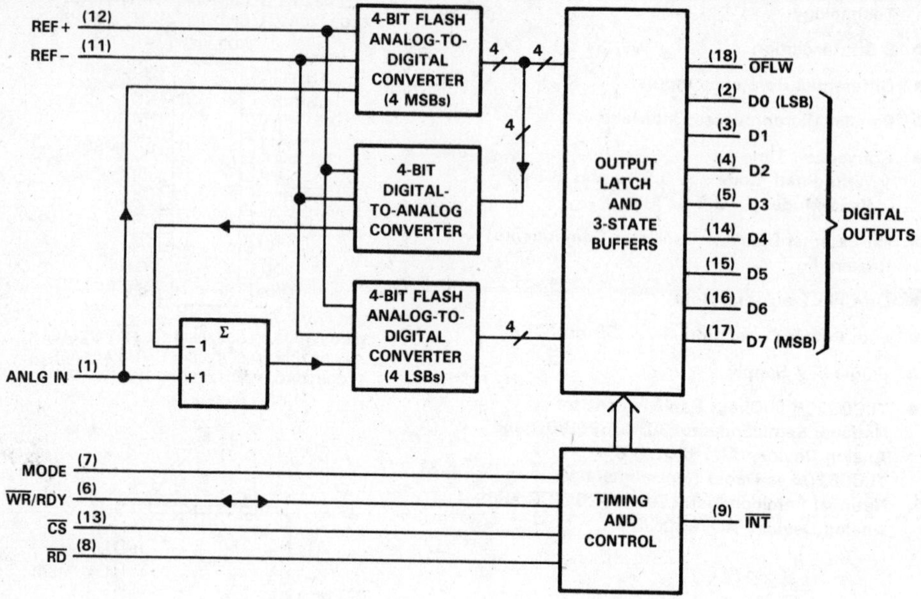
REF+ (12)
REF− (11)
ANLG IN (1)
4-BIT FLASH ANALOG-TO-DIGITAL CONVERTER (4 MSBs)
4-BIT DIGITAL-TO-ANALOG CONVERTER
4-BIT FLASH ANALOG-TO-DIGITAL CONVERTER (4 LSBs)
Σ
−1
+1
4
OUTPUT LATCH AND 3-STATE BUFFERS
(18) OFLW
(2) D0 (LSB)
(3) D1
(4) D2
(5) D3
(14) D4
(15) D5
(16) D6
(17) D7 (MSB)
DIGITAL OUTPUTS
MODE (7)
WR/RDY (6)
CS (13)
RD (8)
TIMING AND CONTROL
(9) INT

TEXAS INSTRUMENTS
POST OFFICE BOX 655012 • DALLAS, TEXAS 75265

PIN		DESCRIPTION
NAME	NUMBER	
ANLG IN	1	Analog input
$\overline{CS}$	13	This input must be low in order for $\overline{RD}$ or $\overline{WR}$ to be recognized by the ADC.
D0	2	Three-state data output, bit 1 (LSB)
D1	3	Three-state data output, bit 2
D2	4	Three-state data output, bit 3
D3	5	Three-state data output, bit 4
D4	14	Three-state data output, bit 5
D5	15	Three-state data output, bit 6
D6	16	Three-state data output, bit 7
D7	17	Three-state data output, bit 8 (MSB)
GND	10	Ground
$\overline{INT}$	9	In the WRITE-READ mode, the interrupt output, $\overline{INT}$, going low indicates that the internal count-down delay time, $t_{d(int)}$, is complete and the data result is in the output latch. $t_{d(int)}$ is typically 800 ns starting after the rising edge of the $\overline{WR}$ input (see operating characteristics and Figure 3). If $\overline{RD}$ goes low prior to the end of $t_{d(int)}$, $\overline{INT}$ goes low at the end of t_{dRIL} and the conversion results are available sooner (see Figure 2). $\overline{INT}$ is reset by the rising edge of either $\overline{RD}$ or $\overline{CS}$.
MODE	7	Mode-selection input. It is internally tied to GND through a 50-uA current source, which acts like a pull-down resistor. READ mode: Occurs when this input is low. WRITE-READ mode: Occurs when this input is high.
NC	19	No internal connection
$\overline{OFLW}$	18	Normally the $\overline{OFLW}$ output is a logical high. However, if the analog input is higher than the V_{REF+}, $\overline{OFLW}$ will be low at the end of conversion. It can be used to cascade 2 or more devices to improve resolution (9 or 10-bits).
$\overline{RD}$	8	In the WRITE-READ mode with $\overline{CS}$ low, the 3-state data outputs D0 through D7 are activated when $\overline{RD}$ goes low. $\overline{RD}$ can also be used to increase the conversion speed by reading data prior to the end of the internal count-down delay time. As a result, the data transferred to the output latch is latched after the falling edge of $\overline{RD}$. In the READ mode with $\overline{CS}$ low, the conversion starts with $\overline{RD}$ going low. $\overline{RD}$ also enables the three-state data outputs upon completion of the conversion. The RDY output going into the high-impedance state and $\overline{INT}$ going low indicates completion of the conversion.
REF −	11	This input voltage is placed on the bottom of the resistor ladder.
REF +	12	This input voltage is placed on the top of the resistor ladder.
V_{CC}	20	Power supply voltage
$\overline{WR}$/RDY	6	In the WRITE-READ mode with $\overline{CS}$ low, the conversion is started on the falling edge of the $\overline{WR}$ input signal. The result of the conversion is strobed into the output latch after the internal count-down delay time, $t_{d(int)}$, provided that the $\overline{RD}$ input does not go low prior to this time. $t_{d(int)}$ is approximately 800 ns. In the READ mode, RDY (an open-drain output) will go low after the falling edge of $\overline{CS}$, and will go into the high-impedance state when the conversion is strobed into the output latch. It is used to simplify the interface to a microprocessor system.

absolute maximum ratings over operating free-air temperature range (unless otherwise noted)

	TLC0820AM TLC0820BM	TLC0820AI TLC0820BI	TLC0820AC TLC0820BC	UNIT
Supply voltage, V_{CC} (see Note 1)	10	10	10	V
Input voltage range, all inputs (see Note 1)	−0.2 to $V_{CC}+0.2$	−0.2 to $V_{CC}+0.2$	−0.2 to $V_{CC}+0.2$	V
Output voltage range, all outputs (see Note 1)	−0.2 to $V_{CC}+0.2$	−0.2 to $V_{CC}+0.2$	−0.2 to $V_{CC}+0.2$	V
Operating free-air temperature range	−55 to 125	−40 to 85	0 to 70	°C
Storage temperature range	−65 to 150	−65 to 150	−65 to 150	°C
Case temperature for 60 seconds: FK package	260			°C
Case temperature for 10 seconds: FN package		260	260	°C
Lead temperature 1,6 mm (1/16 inch) from case for 60 seconds: J package	300			°C
Lead temperature 1,6 mm (1/16 inch) from case for 10 seconds: N package	260	260	260	°C

NOTE 1: All voltages are with respect to network ground terminal, pin 10.

recommended operating conditions

			TLC0820AM TLC0820BM			TLC0820AI TLC0820BI			TLC0820AC TLC0820BC			UNIT
			MIN	NOM	MAX	MIN	NOM	MAX	MIN	NOM	MAX	
Supply voltage, V_{CC}			4.5	5	8	4.5	5	8	4.5	5	8	V
Analog input voltage			−0.1		$V_{CC}+0.1$	−0.1		$V_{CC}+0.1$	−0.1		$V_{CC}+0.1$	V
Positive reference voltage, V_{REF+}			V_{REF-}		V_{CC}	V_{REF-}		V_{CC}	V_{REF-}		V_{CC}	V
Negative reference voltage, V_{REF-}			GND		V_{REF+}	GND		V_{REF+}	GND		V_{REF+}	V
High-level input voltage, V_{IH}	V_{CC} = 4.75 V to 5.25 V	$\overline{CS}$, $\overline{WR}$/RDY, $\overline{RD}$	2			2			2			V
		MODE	3.5			3.5			3.5			
Low-level input voltage, V_{IL}	V_{CC} = 4.75 V to 5.25 V	$\overline{CS}$, $\overline{WR}$/RDY, $\overline{RD}$			0.8			0.8			0.8	V
		MODE			1.5			1.5			1.5	
Delay time from $\overline{WR}$ to $\overline{RD}$ in write-read mode, t_{dWR} (see Figures 2 and 3)			0.6			0.6			0.6			μs
Write-pulse duration in write-read mode, t_{wW} (see Figures 2, 3, and 4)			0.6		50	0.6		50	0.6		50	μs
Operating free-air temperature, T_A			−55		125	−40		85	0		70	°C

electrical characteristics over recommended operating free-air temperature range, V_{CC} = 5 V (unless otherwise noted)

PARAMETER			TEST CONDITIONS	MIN	TYP†	MAX	UNIT
V_{OH}	High-level output voltage	Any D, $\overline{INT}$, or $\overline{OFLW}$	V_{CC} = 4.75 V, I_{OH} = −360 μA	2.4			V
			V_{CC} = 4.75 V, I_{OH} = −10 μA	4.5			
V_{OL}	Low-level output voltage	Any D, $\overline{OFLW}$, $\overline{INT}$, or $\overline{WR}$/RDY	V_{CC} = 5.25 V, I_{OL} = 1.6 mA			0.4	V
I_{IH}	High-level input current	$\overline{CS}$ or $\overline{RD}$	V_{IH} = 5 V		0.005	1	μA
		$\overline{WR}$/RDY			0.1	3	
		MODE			50	200	
I_{IL}	Low-level input current	$\overline{CS}$, $\overline{WR}$/RDY, $\overline{RD}$, or MODE	V_{IL} = 0		−0.005	−1	μA
I_{OZ}	Off-state (high-impedance state) output current	Any D or $\overline{WR}$/RDY	V_O = 5 V		0.1	3	μA
			V_O = 0		−0.1	−3	
I_I	Analog input current		$\overline{CS}$ at 5 V, V_I = 5 V			3	μA
			$\overline{CS}$ at 5 V, V_I = 0			−3	
I_{OS}	Short-circuit output current	Any D, $\overline{OFLW}$, $\overline{INT}$, or $\overline{WR}$/RDY	V_O = 5 V, T_A = 25°C	7	14		mA
		Any D or $\overline{OFLW}$	V_O = 0, T_A = 25°C	−6	−12		
		$\overline{INT}$		−4.5	−9		
R_{ref}	Reference resistance			1.25	2.3	6	kΩ
I_{CC}	Supply current		$\overline{CS}$, $\overline{WR}$/RDY, and $\overline{RD}$ at 0 V		7.5	15	mA
C_i	Input capacitance	Any digital			5		pF
		Analog (pin 1)			45		
C_o	Output capacitance	Any digital				5	pF

† All typical values are at T_A = 25°C.

operating characteristics, V_{CC} = 5 V, V_{REF+} = 5 V, V_{REF-} = 0, t_r = t_f = 20 ns, T_A = 25°C (unless otherwise noted)

PARAMETER		TEST CONDITIONS		TLC0820A MIN	TLC0820A TYP	TLC0820A MAX	TLC0820B MIN	TLC0820B TYP	TLC0820B MAX	UNIT
k_{SVS}	Supply voltage sensitivity	V_{CC} = 5 V ± 5%			±1/16	±1/4		±1/16	±1/4	LSB
	Total unadjusted error†	MODE pin at 0 V				1			1/2	LSB
t_{convR}	Read mode conversion time	MODE pin at 0 V, See Figure 1			1.6	2.5		1.6	2.5	μs
$t_{d(int)}$	Internal count-down delay time	MODE pin at 5 V, C_L = 50 pF, See Figures 3 and 4			800	1300		800	1300	ns
t_{aR}	Access time from $\overline{RD}$↓	MODE pin at 0 V, See Figure 1			t_{convR} +20	t_{convR} +50		t_{convR} +20	t_{convR} +50	ns
t_{aR1}	Access time from $\overline{RD}$↓	MODE pin at 5 V, $t_{dWR} < t_{d(int)}$, See Figure 2	C_L = 15 pF		190	280		190	280	ns
			C_L = 100 pF		210	320		210	320	
t_{aR2}	Access time from $\overline{RD}$↓	MODE pin at 5 V, $t_{dWR} > t_{d(int)}$ See Figure 3	C_L = 15 pF		70	120		70	120	ns
			C_L = 100 pF		90	150		90	150	
t_{aINT}	Access time from $\overline{INT}$↓	MODE pin at 5 V, See Figure 4			20	50		20	50	ns
t_{dis}	Disable time from $\overline{RD}$↑	R_L = 1 kΩ, C_L = 10 pF, See Figures 1, 2, 3, and 5			70	95		70	95	ns
t_{dRDY}	Delay time from $\overline{CS}$↓ to RDY↓	MODE pin at 0 V, C_L = 50 pF, See Figure 1			50	100		50	100	ns
t_{dRIH}	Delay time from $\overline{RD}$↑ to $\overline{INT}$↑	C_L = 50 pF, See Figures 1, 2, and 3			125	225		125	225	ns
t_{dRIL}	Delay time from $\overline{RD}$↓ to $\overline{INT}$↓	MODE pin at 5 V, $t_{dWR} < t_{d(int)}$, See Figure 2			200	290		200	290	ns
t_{dWIH}	Delay time from $\overline{WR}$↑ to $\overline{INT}$↑	MODE pin at 5 V, C_L = 50 pF, See Figure 4			175	270		175	270	ns
$t_{d(NC)}$	Delay to next conversion	See Figures 1, 2, 3, and 4				500			500	ns
	Slew rate tracking				0.1			0.1		V/μs

†Total unadjusted error includes offset, full-scale, and linearity errors.

TEXAS INSTRUMENTS
POST OFFICE BOX 655012 • DALLAS, TEXAS 75265

PARAMETER MEASUREMENT INFORMATION

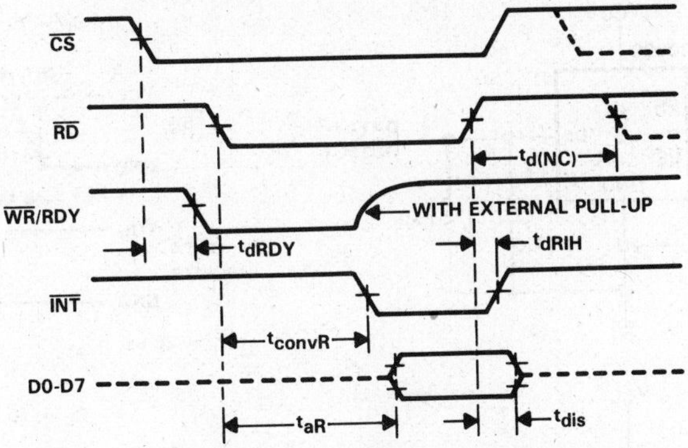
CS
RD
WR/RDY
INT
D0-D7
td(NC)
WITH EXTERNAL PULL-UP
tdRDY
tdRIH
tconvR
taR
tdis

FIGURE 1. READ MODE WAVEFORMS (MODE PIN LOW)

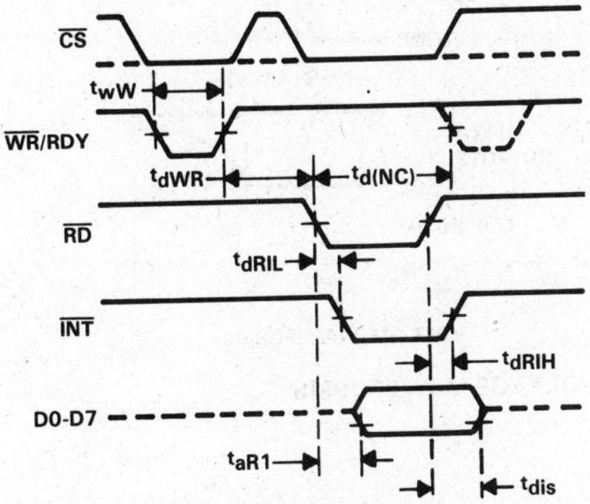
CS
twW
WR/RDY
tdWR
td(NC)
RD
tdRIL
INT
tdRIH
D0-D7
taR1
tdis

FIGURE 2. WRITE-READ MODE WAVEFORMS [MODE PIN HIGH AND $t_{dWR} < t_{d(int)}$]

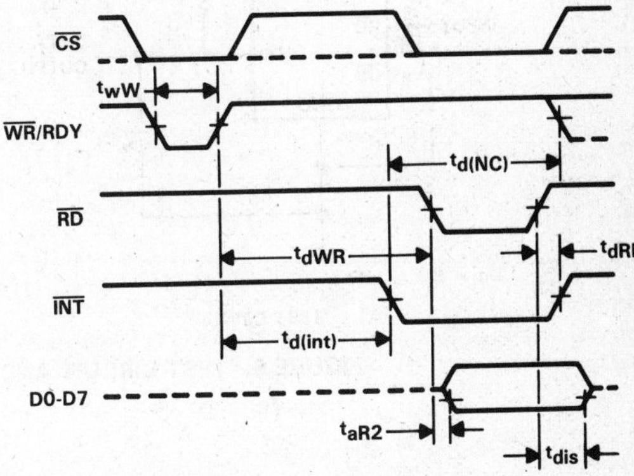
CS
twW
WR/RDY
td(NC)
RD
tdWR
tdRIH
INT
td(int)
D0-D7
taR2
tdis

FIGURE 3. WRITE-READ WAVEFORMS [MODE PIN HIGH AND $t_{dWR} > t_{d(int)}$]

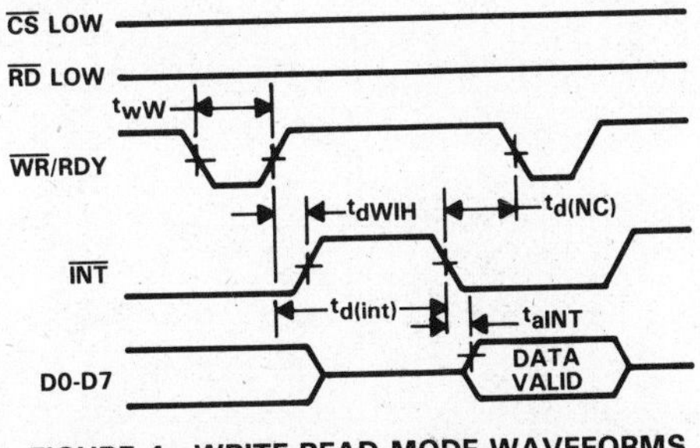
CS LOW
RD LOW
twW
WR/RDY
tdWIH
td(NC)
INT
td(int)
taINT
D0-D7
DATA VALID

FIGURE 4. WRITE-READ MODE WAVEFORMS (STAND-ALONE OPERATION, MODE PIN HIGH, AND RD LOW)

PARAMETER MEASUREMENT INFORMATION

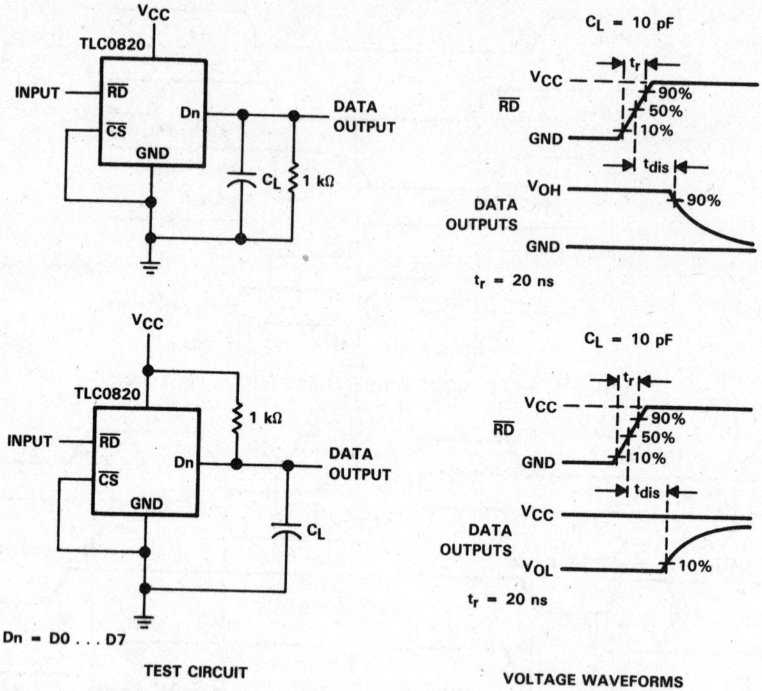
V_{CC}
TLC0820
INPUT
$\overline{RD}$
$\overline{CS}$
Dn
GND
DATA
OUTPUT
C_L
1 kΩ
C_L = 10 pF
t_r
t_{dis}
V_{CC}
$\overline{RD}$
GND
90%
50%
10%
V_{OH}
DATA
OUTPUTS
V_{OL}
t_r = 20 ns
Dn = D0 . . . D7
TEST CIRCUIT
VOLTAGE WAVEFORMS

FIGURE 5. TEST CIRCUIT AND VOLTAGE WAVEFORMS

PRINCIPLES OF OPERATION

The TLC0820A and TLC0820B each employ a combination of ''sampled-data'' comparator techniques and ''flash'' techniques common to many high-speed converters. Two 4-bit ''flash'' analog-to-digital conversions are used to give a full 8-bit output.

The recommended analog input voltage range for conversion is −0.1 V to V_{CC}+0.1 V. Analog input signals that are less than V_{REF-} + ½ LSB or greater than V_{REF+} − ½ LSB convert to 00000000 or 11111111 respectively. The reference inputs are fully differential with common-mode limits defined by the supply rails. The reference input values define the full-scale range of the analog input. This allows the gain of the ADC to be varied for ratiometric conversion by changing the V_{REF+} and V_{REF-} voltages.

The device operates in two modes, read (only) and write-read, which are selected by the MODE pin (pin 7). The converter is set to the read (only) mode when pin 7 is low. In the read mode, the $\overline{WR}$/RDY pin is used as an output and is referred to as the ''ready'' pin. In this mode, a low on the ''ready'' pin while $\overline{CS}$ is low indicates that the device is busy. Conversion starts on the falling edge of $\overline{RD}$ and is completed no more than 2.5 microseconds later when $\overline{INT}$ falls and the ''ready'' pin returns to a high-impedance state. Data outputs also change from high-impedance to active states at this time. After the data is read, $\overline{RD}$ is taken high, $\overline{INT}$ returns high, and the data outputs return to their high-impedance states.

The converter is set to the write-read mode when pin 7 is high and $\overline{WR}$/RDY is referred to as the ''write'' pin. Taking $\overline{CS}$ and the ''write'' pin low selects the converter and initiates measurement of the input signal. Approximately 600 nanoseconds after the ''write'' pin returns high, the conversion is completed. Conversion starts on the rising edge of $\overline{WR}$/RDY in the write-read mode.

The high-order 4-bit ''flash'' ADC measures the input by means of 16 comparators operating simultaneously. A high precision 4-bit DAC then generates a discrete analog voltage from the result of that conversion. After a time delay, a second bank of comparators does a low-order conversion on the analog difference between the input level and the high-order DAC output. The results from each of these conversions enter an 8-bit latch and are output to the three-state buffers on the falling edge of $\overline{RD}$.

TYPICAL APPLICATION DATA

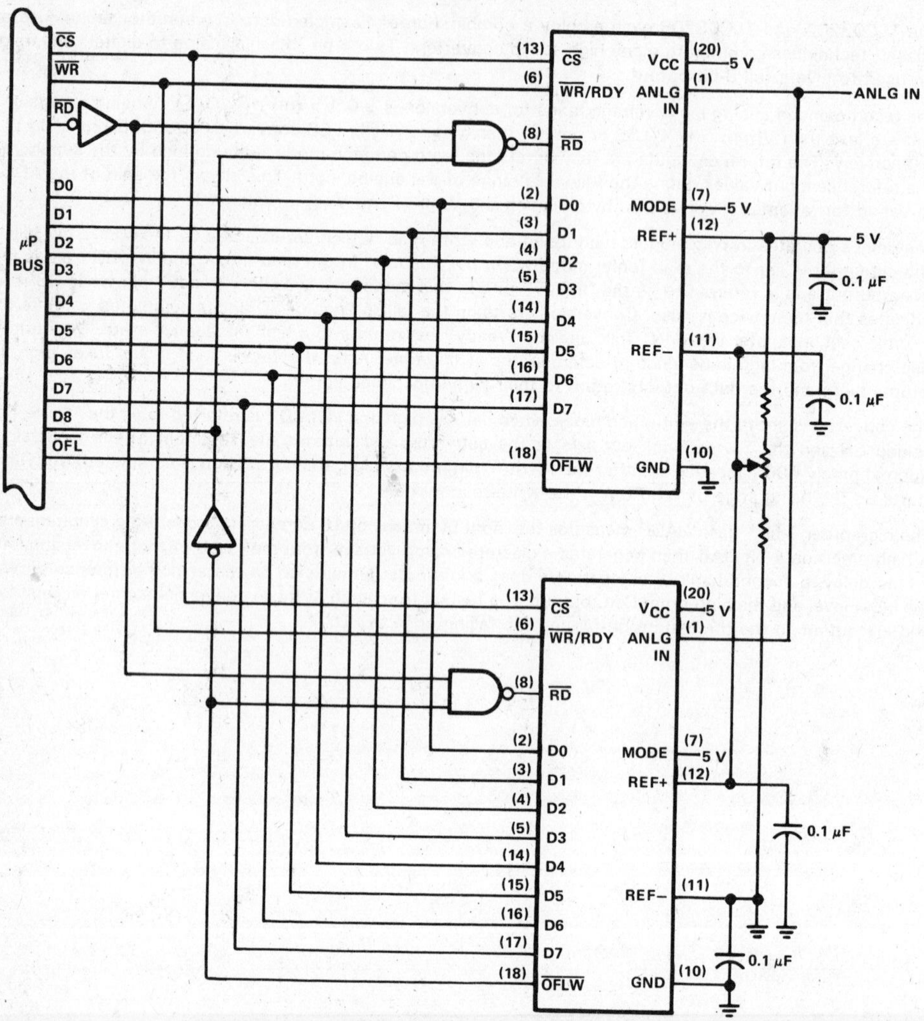
CS
WR
RD
D0
D1
μP
BUS
D2
D3
D4
D5
D6
D7
D8
OFL
(13)
(6)
(8)
(2)
(3)
(4)
(5)
(14)
(15)
(16)
(17)
(18)
CS
WR/RDY
RD
D0
D1
D2
D3
D4
D5
D6
D7
OFLW
VCC
ANLG IN
MODE
REF+
REF−
GND
(20)
(1)
(7)
(12)
(11)
(10)
5 V
ANLG IN
0.1 μF

FIGURE 6. CONFIGURATION FOR 9-BIT RESOLUTION

ADVANCE INFORMATION

UNIVERSAL DUAL SWITCHED-CAPACITOR FILTER

D2952, AUGUST 1986—REVISED SEPTEMBER 1986

- **Maximum Clock to Center-Frequency Ratio Error**
  TLC10 . . . ±0.6%
  TLC20 . . . ±1.5%
- **Filter Cutoff Frequency Stability Dependent Only on External-Clock Frequency Stability**
- **Minimum Filter Response Deviation Due to External Component Variations over Time and Temperature**
- **Critical-Frequency Times Q Factor Range Up to 200 kHz**
- **Critical-Frequency Operation Up to 30 kHz**
- **Designed to be Interchangeable with:**
  National MF10
  Maxim MF10
  Linear Technology LTC1060

N DUAL-IN-LINE PACKAGE
(TOP VIEW)

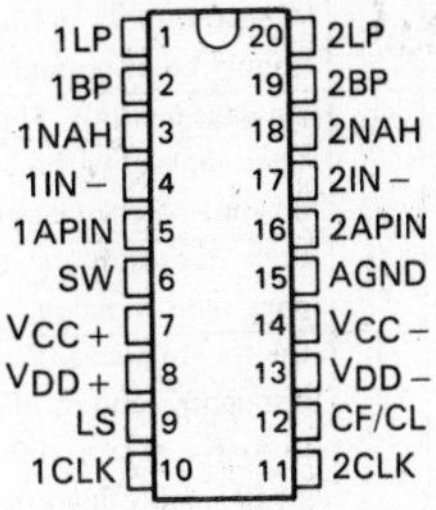
1LP 1 20 2LP
1BP 2 19 2BP
1NAH 3 18 2NAH
1IN− 4 17 2IN−
1APIN 5 16 2APIN
SW 6 15 AGND
V_{CC+} 7 14 V_{CC-}
V_{DD+} 8 13 V_{DD-}
LS 9 12 CF/CL
1CLK 10 11 2CLK

FN CHIP CARRIER PACKAGE
(TOP VIEW)

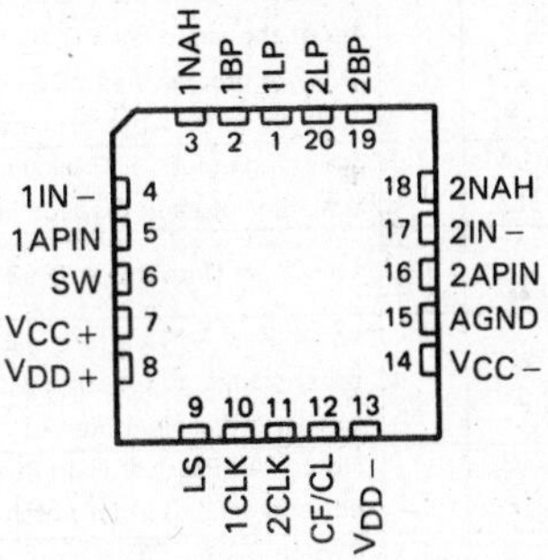
1NAH 1BP 1LP 2LP 2BP
3 2 1 20 19
1IN− 4
1APIN 5
SW 6
V_{CC+} 7
V_{DD+} 8
18 2NAH
17 2IN−
16 2APIN
15 AGND
14 V_{CC-}
9 10 11 12 13
LS 1CLK 2CLK CF/CL V_{DD-}

description

The TLC10 and TLC20 are monolithic general-purpose switched-capacitor CMOS filters each containing two independent active-filter sections. Each device facilitates configuration of Butterworth, Bessel, Cauer, or Chebyshev filter design.

Filter features include cutoff frequency stability that is dependent only on the external clock frequency stability and minimal response deviation over time and temperature. Features also include a critical-frequency times filter quality (Q) factor range of up to 200 kiloHertz.

With external clock and resistors, each filter section can be used independently to produce various second-order functions or both sections can be cascaded to produce fourth-order functions. For functions greater than fourth-order, ICs can be cascaded.

The TLC10 and TLC20 are characterized for operation from 0°C to 70°C.

TEXAS INSTRUMENTS
POST OFFICE BOX 655012 • DALLAS, TEXAS 75265

PIN NAME	PIN NO.	I/O	DESCRIPTION
AGND	15	I	Analog Ground — The noninverting inputs to the input operational amplifiers of both filter sections. This terminal should be at ground for dual supplies or at mid-supply level for single-supply operation.
1APIN 2APIN	5 16	I	All-Pass Inputs — The all-pass input to the summing amplifier of each respective filter section used for all-pass filter applications in configuration modes 1a, 4, 5, and 6. This terminal should be driven from a source having an impedance of less than 1 kilohm. In all other modes, this terminal is grounded. See Typical Application Data.
1BP 2BP	2 19	O	Band-Pass Outputs — The band-pass output of each respective filter section provides the second-order band-pass filter functions.
CF/CL	12	I	Center Frequency/Current Limit — This input terminal provides the option to select the input-clock-to-center-frequency ratio of 50:1 or 100:1 or to limit the current of the IC. For a 50:1 ratio, the CF/CL terminal is set to V_{DD+}. For a 100:1 ratio, the CF/CL terminal is set to ground for dual supplies or to mid-supply level for single-supply operation. For current limiting, the CF/CL terminal is set to V_{DD-}. This aborts filtering and limits the IC current to 0.5 milliamperes.
1CLK 2CLK	10 11	I	Clock Inputs — The clock input to the two-phase nonoverlapping generator of each respective filter section is used to generate the center frequency of the complex pole pair second-order function. Both clocks should be of the same level (TTL or CMOS) and have duty cycles close to 50%, especially when clock frequencies (f_{clock}) greater than 200 kiloHertz are used. At this duty cycle, the operational amplifiers have the maximum time to settle while processing analog samples.
1IN− 2IN−	4 17	I	Inverting Inputs — The inverting input side of the input operational amplifier whose output drives the summing amplifier of each respective filter section.
1LP 2LP	1 20	O	Low-Pass Outputs — The low-pass outputs of the second-order filters.
LS	9	I	Level Shift — This terminal accommodates various input clock levels of bipolar (CMOS) or unipolar (TTL or other clocks) to function with single or dual supplies. For CMOS (±5-volt) clocks, V_{DD-} or ground is applied to the LS terminal. For TTL and other clocks, ground is applied to the LS terminal.
1NAH 2NAH	3 18	O	Notch, All-Pass, or High-Pass Outputs — The output of each respective filter section can be used to provide either a second-order notch, all-pass, or high-pass output filter function, depending on circuit configuration.
SW	6	I	Switch Input — This input terminal is used to control internal switches to connect either the AGND input or the LP output to one of the inputs of the summing amplifier. The terminal controls both independent filter sections and places them in the same configuration simultaneously. If V_{CC-} is applied to the SW terminal, the AGND input terminal will be connected to one of the inputs of each summing amplifier. If V_{CC+} is applied to the SW terminal, the LP output will be connected to one of the inputs of the summing amplifier.
V_{CC+}	7		Analog positive supply voltage terminal
V_{CC-}	14		Analog negative supply voltage terminal
V_{DD+}	8		Digital positive supply voltage terminal
V_{DD-}	13		Digital negative supply voltage terminal

functional block diagram

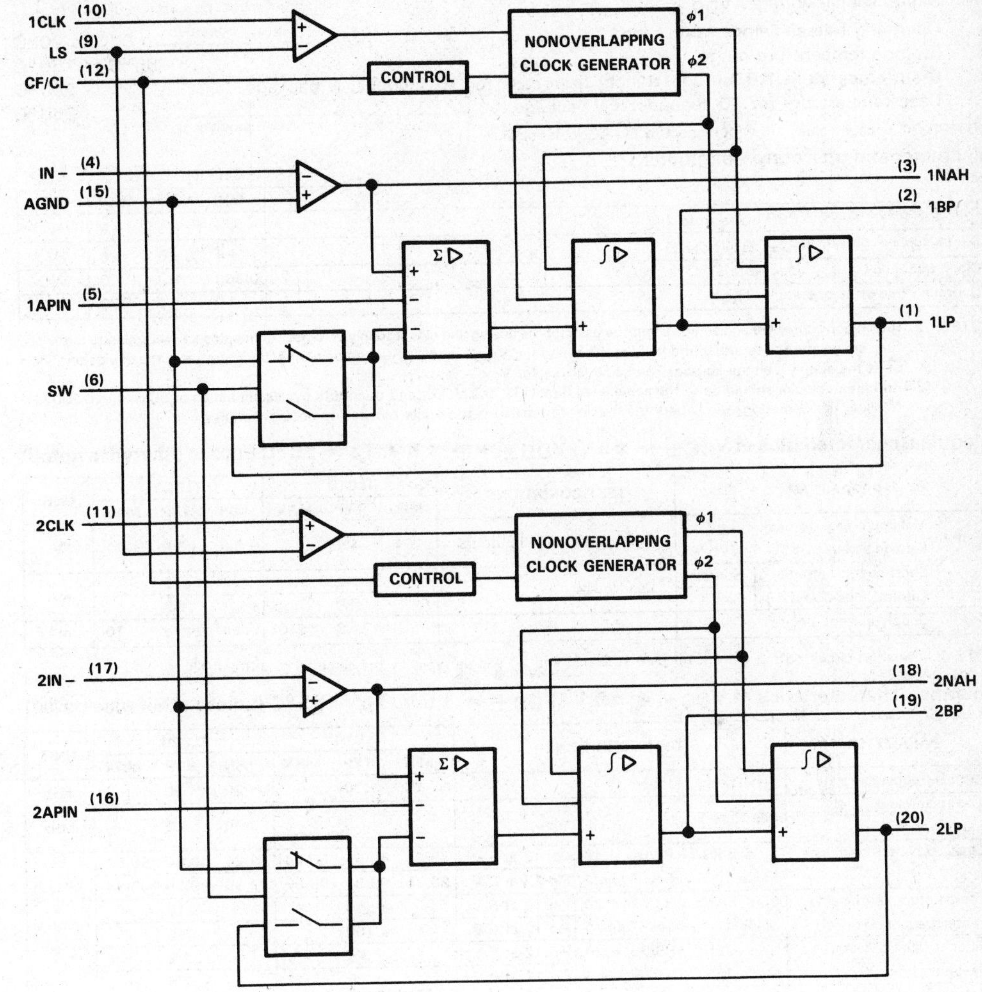
1CLK
(10)
LS
(9)
CF/CL
(12)
CONTROL
NONOVERLAPPING
CLOCK GENERATOR
ϕ1
ϕ2
IN –
(4)
AGND
(15)
(3)
1NAH
(2)
1BP
1APIN
(5)
SW
(6)
(1)
1LP
2CLK
(11)
2IN –
(17)
(18)
2NAH
(19)
2BP
2APIN
(16)
(20)
2LP

absolute maximum ratings over operating free-air temperature range (unless otherwise noted)

Analog supply voltage, $V_{CC\pm}$ (see Note 1)	±7 V
Digital supply voltage, $V_{DD\pm}$	±7 V
Operating free-air temperature range	0°C to 70°C
Storage temperature range	−65°C to 150°C
Lead temperature 1,6 mm (1/16 inch) from case for 10 seconds: N package	260°C
Case temperature for 10 seconds: FN package	260°C

NOTE 1: All voltage values are with respect to the AGND terminal.

recommended operating conditions

	MIN	NOM	MAX	UNIT
Analog supply voltage, $V_{CC\pm}$, (see Note 2)	±4	±5	±6	V
Digital supply voltage, $V_{DD\pm}$, (see Note 2)	±4	±5	±6	V
Clock frequency, f_{clock}, (see Note 3)	0.008		1.0	MHz
Operating free-air temperature, T_A	0		70	°C

NOTES: 2. A common supply voltage source should be used for the analog and digital supply voltages. Although each has separate terminals, they are connected together internally at the substrate. V_{CC+} and V_{DD+} can be connected together at the device terminals or at the supply voltage source. The same is true for V_{CC-} and V_{DD-}.

3. Both input clocks should be of the same level type (TTL or CMOS), and their duty cycles should be at 50% above 200 kHz to allow the operational amplifiers the maximum time to settle while processing analog samples.

electrical characteristics at $V_{CC\pm}$ = ±5 V, $V_{DD\pm}$ + = ±5 V, T_A = 25°C (unless otherwise noted)

PARAMETER			TEST CONDITIONS	TLC10			TLC20			UNIT
				MIN	TYP	MAX	MIN	TYP	MAX	
V_{OPP}	Maximum peak-to-peak output voltage swing		R_L = 3.5 kΩ at all outputs	±4	±4.1		±3.8	±3.9		V
I_{OS}	Short-circuit output current, Pins 3 and 18	Source	See Note 4		2			2		mA
		Sink			50			50		
I_{CC}	Supply current				8	10		8	10	mA

NOTE 4: The short-circuit output current for pins 1, 2, 19, and 20 will be typically the same as pins 3 and 18.

operating characteristics at $V_{CC\pm}$ = ±5 V, $V_{DD\pm}$ = ±5 V, T_A = 25°C (unless otherwise noted)

PARAMETER	TEST CONDITIONS		TLC10			TLC20			UNIT
			MIN	TYP	MAX	MIN	TYP	MAX	
Critical-frequency range	$f_o \times Q \leq 200$ kHz		20	30		20	30		kHz
Maximum clock frequency, f_{clock}	See Note 3		1	1.5		1	1.5		MHz
Clock to center-frequency ratio	$f_o \leq 5$ kHz, R3/R2 = 10, Mode 1, See Figure 1	Pin 12 at 5 V	49.64	49.94	50.24	49.24	49.94	50.64	
		Pin 12 at 0 V	98.75	99.35	99.95	97.86	99.35	100.84	
Temperature coefficient of center frequency	$f_o \leq 5$ kHz, R3/R2 = 20, Mode 1, See Figure 1	Pin 12 at 5 V		±10			±10		ppm/°C
		Pin 12 at 0 V		±100			±100		
Filter Q (quality factor) deviation from 20	$f_o \leq 5$ kHz, R3/R2 = 20, Mode 1, See Figure 1	Pin 12 at 5 V		±2%	±4%		±2%	±6%	
		Pin 12 at 0 V		±2%	±3%		±2%	±6%	
Temperature coefficient of measured filter Q	$f_o \leq 5$ kHz, R3/R2 = 20, Mode 1			±500			±500		ppm/°C
Low-pass output deviation from unity gain	R1 = R2 = 10 kΩ Mode 1, See Figure 1				±2%			±2%	
Crosstalk attenuation				60			60		dB
Clock feedthrough voltage				10			10		mV
Operational amplifier gain-bandwidth product				2.5			2.5		MHz
Operational amplifier slew rate				7			7		V/μs

TYPICAL APPLICATION DATA

modes of operation

The TLC10 and TLC20 are switched-capacitor (sampled-data) filters that closely approximate continuous filters. Each filter section is designed to approximate the response of a second-order variable filter. When the sampling frequency is much larger than the frequency band of interest, the sampled-data filter is a good approximation to its continuous time equivalent. In the case of the TLC10 and TLC20, the ratio is about 50:1 or 100:1. To fully describe their transfer function, a time domain approach would be appropriate. Since this may appear cumbersome, the following application examples are based on the well known frequency domain. It should be noted that in order to obtain the actual filter response, the filter's response must be examined in the z-domain.

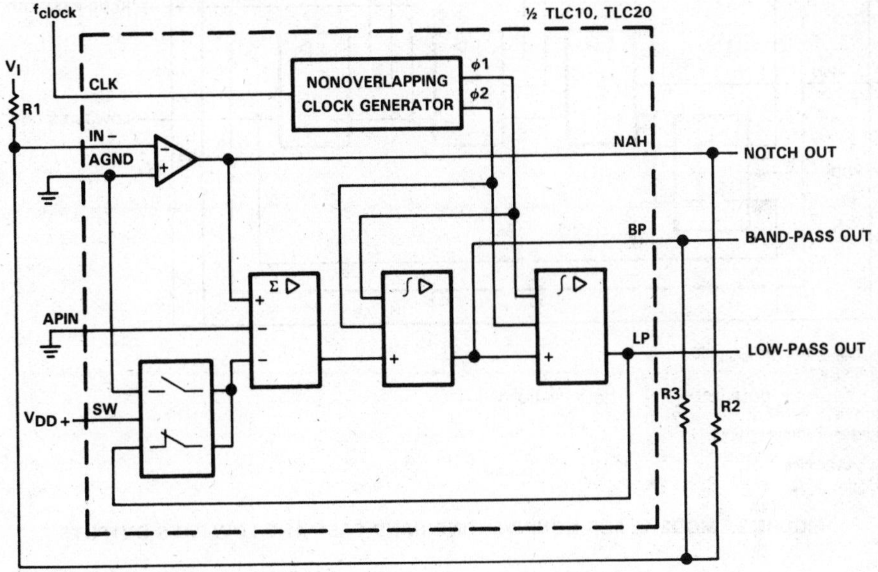
fclock
½ TLC10, TLC20
VI
R1
CLK
NONOVERLAPPING
CLOCK GENERATOR
ϕ1
ϕ2
IN−
AGND
NAH
NOTCH OUT
BP
BAND-PASS OUT
Σ
APIN
LP
LOW-PASS OUT
VDD+
SW
R3
R2

$f_o = f_{clock}/100$ or $f_{clock}/50$
$f_{notch} = f_o$
$H_{OLP} = -R2/R1$ (as $f \rightarrow 0$)
$H_{OBP} = -R3/R1$ (at $f = f_o$)
H_{ON} = notch gain $\begin{cases} \text{as f approaches 0} & -R2/R1 \\ \text{as f approaches 0.5 } f_{clock} & \end{cases}$
$Q = f_o/BW = R3/R2$

Circuit dynamics:
The following expressions determine the swing at each output as a function of the desired Q of the second-order function.
$H_{OLP} = H_{OBP}/Q$ or $H_{OLP} \times Q = H_{ON} \times Q$
H_{OLP} (peak) $= Q \times H_{OLP}$ (for high Qs)

FIGURE 1. MODE 1 FOR NOTCH, BAND-PASS, AND LOW-PASS OUTPUTS: $f_{notch} = f_o$

TYPICAL APPLICATION DATA

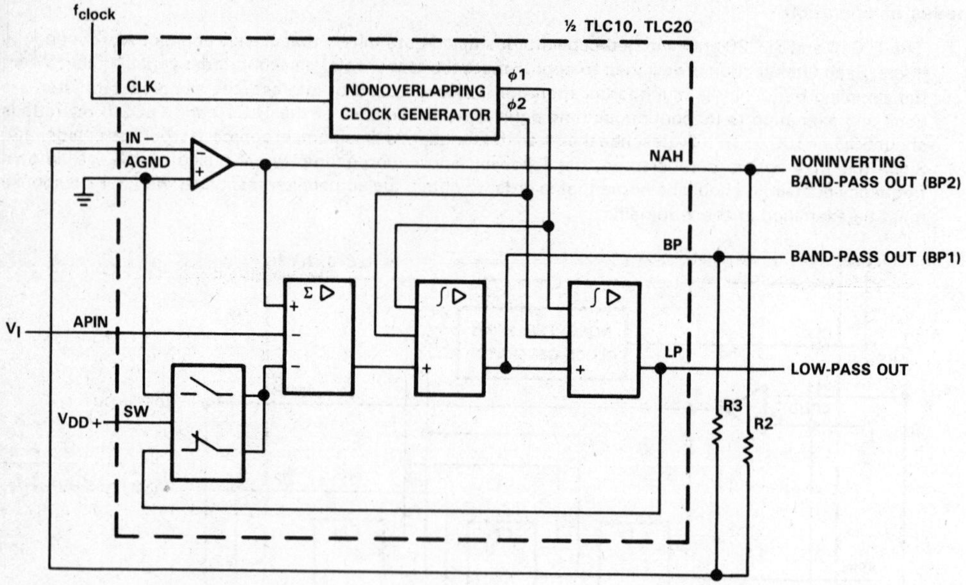
fclock
½ TLC10, TLC20
CLK
NONOVERLAPPING
CLOCK GENERATOR
ϕ1
ϕ2
IN –
AGND
NAH
NONINVERTING
BAND-PASS OUT (BP2)
BP
BAND-PASS OUT (BP1)
VI
APIN
LP
LOW-PASS OUT
VDD +
SW
R3
R2

$f_o = f_{clock}/100$ or $f_{clock}/50$
$Q = R3/R2$
$H_{OLP} = -1$ $\quad H_{OLP}$ (peak) $= Q \times H_{OLP}$ (for high Qs)
$H_{OBP1} = -R3/R2$
$H_{OBP2} = 1$ (noninverting)

Circuit dynamics:
$H_{OBP1} = Q$

FIGURE 2. MODE 1a FOR NONINVERTING BAND-PASS AND LOW-PASS OUTPUTS

TYPICAL APPLICATION DATA

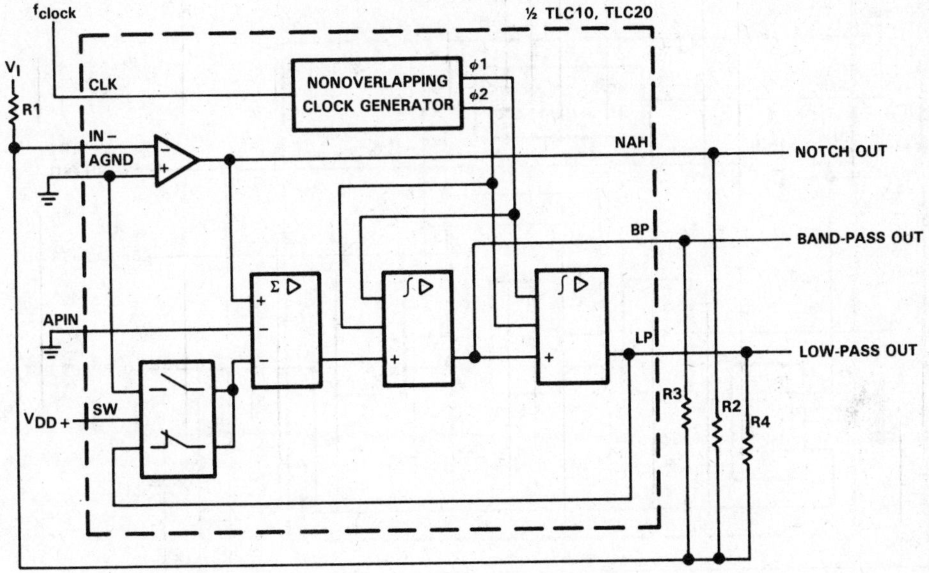
fclock
½ TLC10, TLC20
VI
R1
CLK
NONOVERLAPPING
CLOCK GENERATOR
φ1
φ2
IN−
AGND
NAH
NOTCH OUT
BP
BAND-PASS OUT
APIN
LP
LOW-PASS OUT
VDD +
SW
R3
R2
R4

$f_o = f_{notch} \times \sqrt{R2/R4 + 1}$

$f_{notch} = f_{clock}/100$ or $f_{clock}/50$

$$Q = \frac{\sqrt{R2/R4 + 1}}{R2/R3}$$

$$H_{OLP} \text{ (as f approaches 0)} = \frac{-R2/R1}{R2/R4 + 1}$$

$$H_{OBP} \text{ (at } f = f_o) = -R3/R1$$

$$H_{ON1} \text{ (as f approaches 0)} = \frac{-R2/R1}{R2/R4 + 1}$$

$$H_{ON2} \text{ (as f approaches 0.5 } f_{clock}) = -R2/R1$$

Circuit dynamics:

$$H_{OBP} = Q\sqrt{H_{OLP} \times H_{ON2}} = Q\sqrt{H_{ON1} \times H_{ON2}}$$

FIGURE 3. MODE 2 FOR NOTCH 2, BAND-PASS, AND LOW-PASS OUTPUTS: $f_{notch} < f_o$

TYPICAL APPLICATION DATA

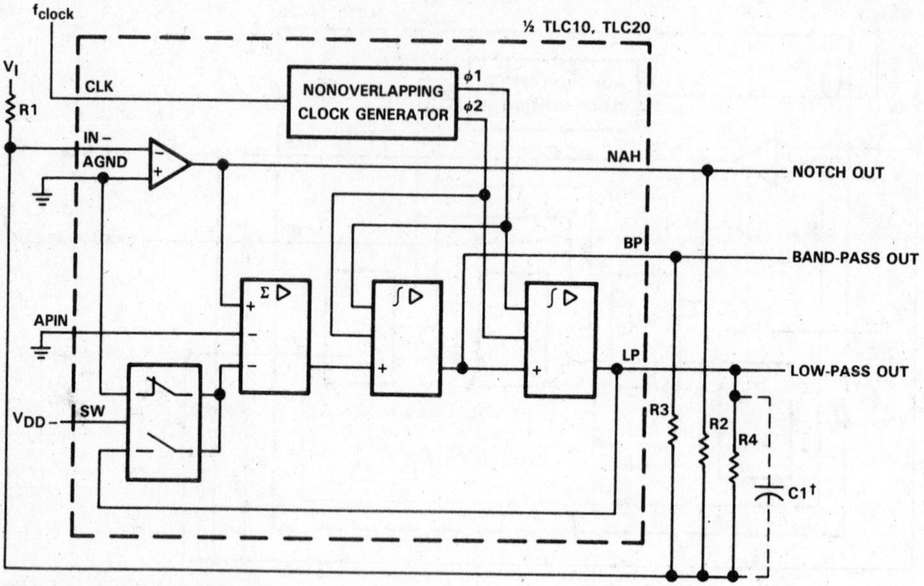
f clock
½ TLC10, TLC20
V I
R1
CLK
NONOVERLAPPING
CLOCK GENERATOR
φ1
φ2
IN –
AGND
NAH
NOTCH OUT
BP
BAND-PASS OUT
Σ
APIN
LP
LOW-PASS OUT
V DD
SW
R3
R2
R4
C1†

$f_o = (f_{clock}/100 \text{ or } f_{clock}/50)\ \sqrt{R2/R4}$

$Q = \sqrt{R2/R4} \times R3/R2$

H_{OHP} (as f approaches 0.5 f_{clock}) = $-R2/R1$

H_{OLP} (as f approaches 0) = $-R4/R1$

H_{OBP} (at $f = f_o$) = $-R3/R1$

Circuit dynamics:

$R2/R4 = H_{OHP}/H_{OLP}$: $H_{OBP} = \sqrt{H_{OHP} \times H_{OLP}} \times Q$

H_{OLP} (peak) = $Q \times H_{OLP}$ (for high Qs)

H_{OHP} (peak) = $Q \times H_{OHP}$ (for high Qs)

†In this mode, the feedback loop is closed around the input summing amplifier; the finite GBW product of this operational amplifier will cause a slight Q enhancement. If this is a problem, connect a low-value capacitor (10 pF to 100 pF) across R4 to provide some phase lead.

FIGURE 4. MODE 3 FOR HIGH-PASS, BAND-PASS, AND LOW-PASS OUTPUTS

TYPICAL APPLICATION DATA

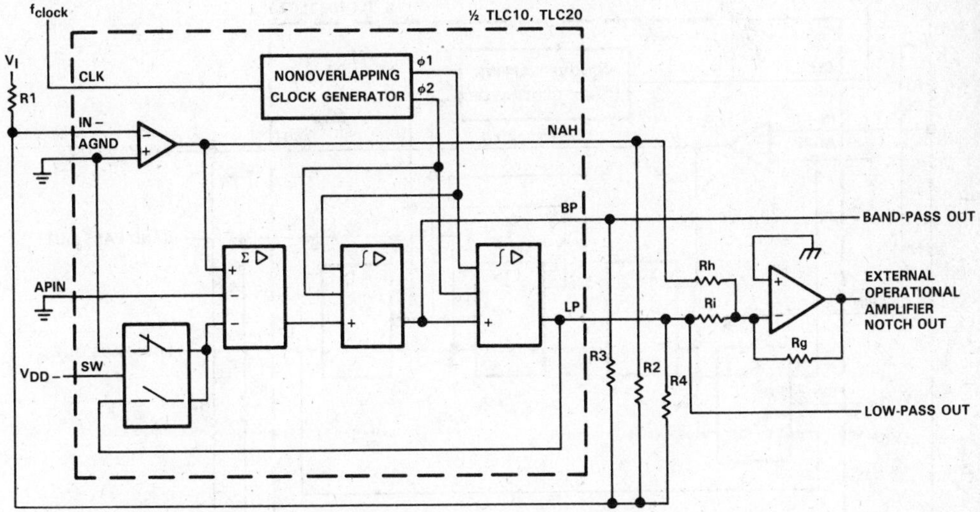

fclock
½ TLC10, TLC20
VI
R1
CLK
NONOVERLAPPING
CLOCK GENERATOR
φ1
φ2
IN −
AGND
NAH
BP
BAND-PASS OUT
APIN
Rh
Ri
EXTERNAL
OPERATIONAL
AMPLIFIER
NOTCH OUT
LP
Rg
R3
R2
R4
VDD
SW
LOW-PASS OUT

$f_o = (f_{clock}/100 \text{ or } f_{clock}/50)\ \sqrt{R2/R4}$

$Q = \sqrt{R2/R4} \times R3/R2$

$H_{OHP} = -R2/R1$

$H_{OBP} = -R3/R1$

$H_{OLP} = -R4/R1$

$f_{notch} = (f_{clock}/100 \text{ or } f_{clock}/50)\ \sqrt{Rh/Ri}$

H_{ON} (at $f = f_o$) = $|\ Q\ (Rg/Ri \times H_{OLP} - Rg/Rh \times HOHP)\ |$

H_{ON1} (as f approaches 0) = $Rg/Ri \times H_{OLP}$

H_{ON2} (as f approaches 0.5 f_{clock}) = $-Rg/Rh \times H_{OHP}$

FIGURE 5. MODE 3a FOR HIGH-PASS, BAND-PASS, LOW-PASS, AND NOTCH OUTPUTS WITH EXTERNAL OPERATIONAL AMPLIFIER

TYPICAL APPLICATION DATA

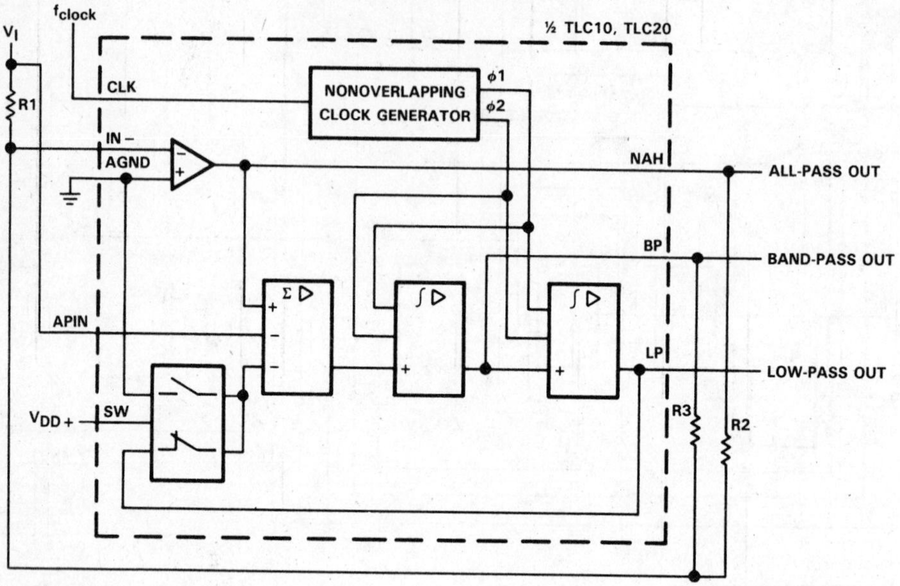
fclock
½ TLC10, TLC20
VI
R1
CLK
NONOVERLAPPING
CLOCK GENERATOR
φ1
φ2
IN −
AGND
NAH
ALL-PASS OUT
BP
BAND-PASS OUT
APIN
LP
LOW-PASS OUT
VDD +
SW
R3
R2

$f_o = f_{clock}/100$ or $f_{clock}/50$

$f_z = f_o$†

$Q = f_o/BW = R3/R2$

$Q_z = R3/R1$

H_{OAP} (at $0 \leq f \leq 0.5\ f_{clock}$) $= -R2/R1 = -1$
(for AP output R1 = R2)

H_{OLP} (as f approaches 0) $= -(R2/R1 + 1) = -2$

H_{OBP} (at $f = f_o$) $= -R3/R2\ (R2/R1 + 1) = -2\ (R3/R2)$

Circuit dynamics:

$H_{OBP} = H_{OLP} \times Q = (H_{OAP} + 1)\ Q$

†Due to the sampled-data nature of the filter, a slight mismatch of f_z and f_o occurs causing a 0.4-dB peaking around f_o of the all-pass filter amplitude response (which theoretically should be a straight line). If this is unacceptable, Mode 5 is recommended.

FIGURE 6. MODE 4 FOR ALL-PASS, BAND-PASS, AND LOW-PASS OUTPUTS

TYPICAL APPLICATION DATA

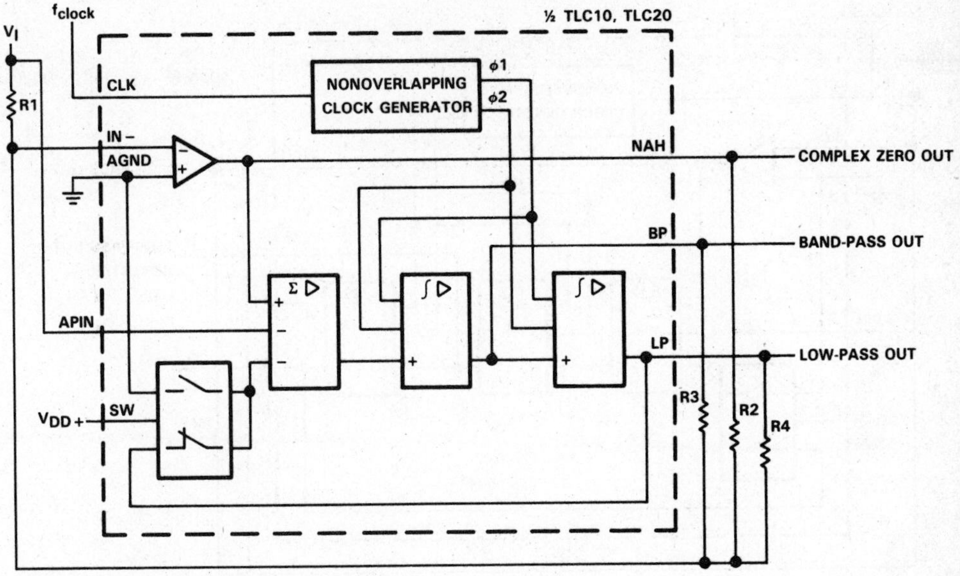
f_{clock}
½ TLC10, TLC20
V_I
R1
CLK
NONOVERLAPPING
CLOCK GENERATOR
φ1
φ2
IN −
AGND
NAH
COMPLEX ZERO OUT
BP
BAND-PASS OUT
Σ
APIN
LP
LOW-PASS OUT
V_{DD} +
SW
R3
R2
R4

$f_o = \sqrt{R2/R4 + 1} \times (f_{clock}/100 \text{ or } f_{clock}/50)$

$f_z = \sqrt{1 - R1/R4} \times (f_{clock}/100 \text{ or } f_{clock}/50)$

$Q = \sqrt{R2/R4 + 1} \times R3/R2$

$Q_z = \sqrt{1 - R1/R4} \times R3/R1$

H_{OZ1} (as f approaches 0) $= R2\,(R4 - R1)/R1\,(R2 + R4)$

H_{OZ2} (as f approaches 0.5 f_{clock}) $= R2/R1$

$H_{OBP} = (R2/R1 + 1) \times R3/R2$

$H_{OLP} = (R2 + R1)/(R2 + R4) \times R4/R1$

FIGURE 7. MODE 5 FOR NUMERATOR COMPLEX ZEROS, BAND-PASS, AND LOW-PASS OUTPUTS

TYPICAL APPLICATION DATA

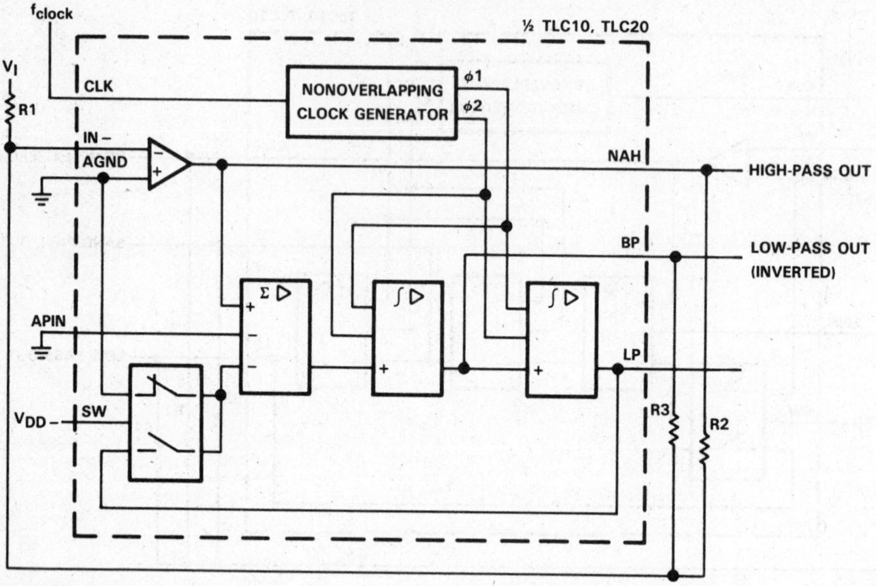
fclock
½ TLC10, TLC20
VI
R1
CLK
IN –
AGND
NONOVERLAPPING CLOCK GENERATOR
φ1
φ2
NAH
HIGH-PASS OUT
BP
LOW-PASS OUT (INVERTED)
APIN
LP
VDD
SW
R3
R2

f_c = R2/R3 (f_{clock}/100 or f_{clock}/50)
H_{OLP} = −R3/R1
H_{OHP} = −R2/R1

FIGURE 8. MODE 6 FOR SINGLE-POLE HIGH-PASS AND LOW-PASS OUTPUT

TYPICAL APPLICATION DATA

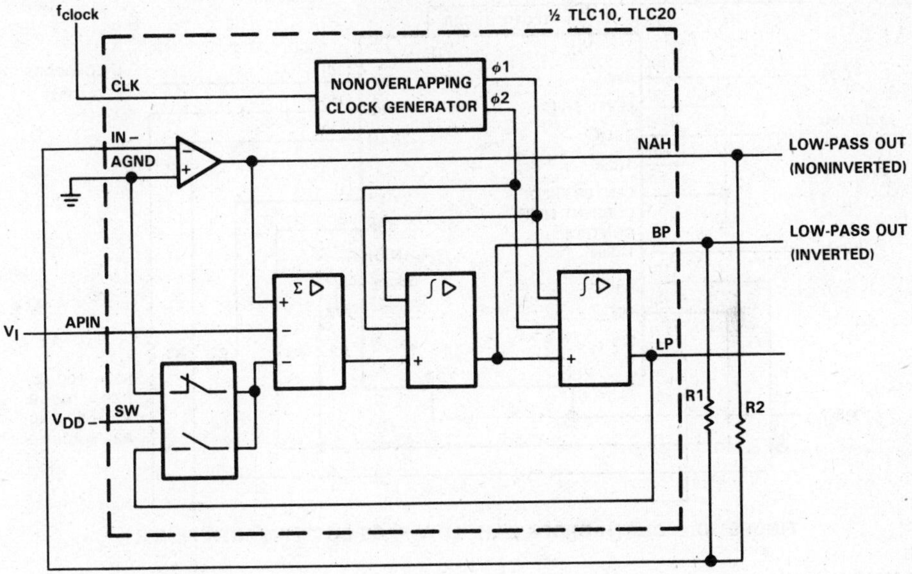
fclock
½ TLC10, TLC20
CLK
NONOVERLAPPING
CLOCK GENERATOR
φ1
φ2
IN−
AGND
NAH
LOW-PASS OUT
(NONINVERTED)
BP
LOW-PASS OUT
(INVERTED)
APIN
VI
LP
SW
VDD
R1
R2

$f_c = R2/R3 \times (f_{clock}/100 \text{ or } f_{clock}/50)$
$H_{OLP1} = 1$ (noninverting)
$H_{OLP2} = -R3/R2$

FIGURE 9. MODE 6a FOR SINGLE-POLE LOW-PASS OUTPUT (INVERTED AND NONINVERTED)

TYPICAL APPLICATION DATA

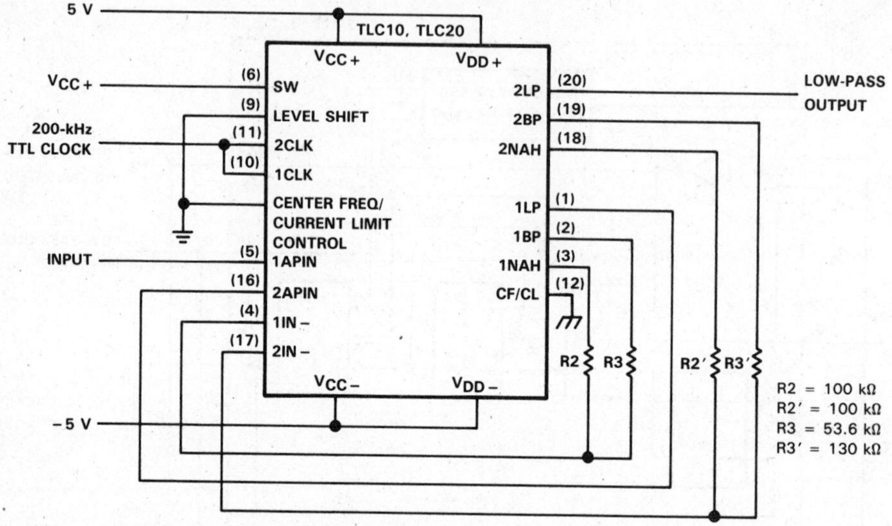
5 V
TLC10, TLC20
VCC+
VDD+
VCC+
(6) SW
(9) LEVEL SHIFT
200-kHz TTL CLOCK
(11) 2CLK
(10) 1CLK
CENTER FREQ/ CURRENT LIMIT CONTROL
INPUT
(5) 1APIN
(16) 2APIN
(4) 1IN−
(17) 2IN−
VCC−
VDD−
−5 V
2LP (20)
2BP (19)
2NAH (18)
1LP (1)
1BP (2)
1NAH (3)
CF/CL (12)
LOW-PASS OUTPUT
R2
R3
R2′
R3′
R2 = 100 kΩ
R2′ = 100 kΩ
R3 = 53.6 kΩ
R3′ = 130 kΩ

FIGURE 10. FOURTH-ORDER 2-kHz LOW-PASS BUTTERWORTH FILTER

filter terminology

f_c	The cutoff frequency of the low-pass or high-pass filter output
f_{clock}	The input clock frequency to the device
f_{notch}	The notch frequency of the notch output
f_o	The center frequency of the complex pole pair second-order function
f_z	The center frequency of the complex zero pair
H_{OBP}	The band-pass output voltage gain (V/V) at the band-pass center frequency
H_{OHP}	The high-pass output voltage gain (V/V) as the frequency approaches 0.5 f_{clock}
H_{OLP}	The low-pass output voltage gain (V/V) as the frequency approaches 0
HON	The notch output voltage gain (V/V) at the notch frequency
H_{ON1}	The low-side notch output voltage gain as the frequency approaches 0
H_{ON2}	The high-side notch output voltage gain as the frequency approaches 0.5 f_{clock}
H_{OZ1}	Gain at complex zero output (as $f \rightarrow 0$ Hz)
H_{OZ2}	Gain at complex zero output (as f approaches 0.5 f_{clock})
Q	The quality factor of the complex pole pair second-order function. Q is the ratio of f_o to the 3-dB bandwidth of the band-pass output. The value of Q also affects the possible peaking of the low-pass and high-pass outputs.
Q_z	The quality factor of the complex zero pair, if such a complex pair exists. This parameter is used when an all-pass filter output is desired.

TEXAS INSTRUMENTS
POST OFFICE BOX 655012 • DALLAS, TEXAS 75265

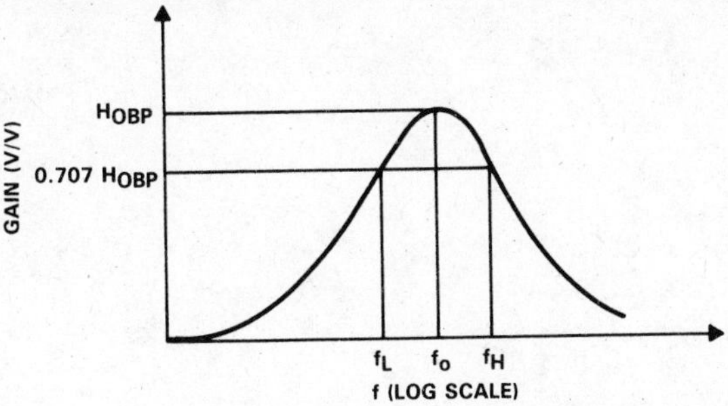
GAIN (V/V)
HOBP
0.707 HOBP
fL
fo
fH
f (LOG SCALE)

$$Q = \frac{f_o}{f_H - f_L} \ ; \ f_o = \sqrt{f_L f_H}$$

$$f_L = f_o \left(\frac{-1}{2Q} + \sqrt{\left(\frac{1}{2Q}\right)^2 + 1} \right)$$

$$f_H = f_o \left(\frac{1}{2Q} + \sqrt{\left(\frac{1}{2Q}\right)^2 + 1} \right)$$

FIGURE 11. BAND-PASS OUTPUT

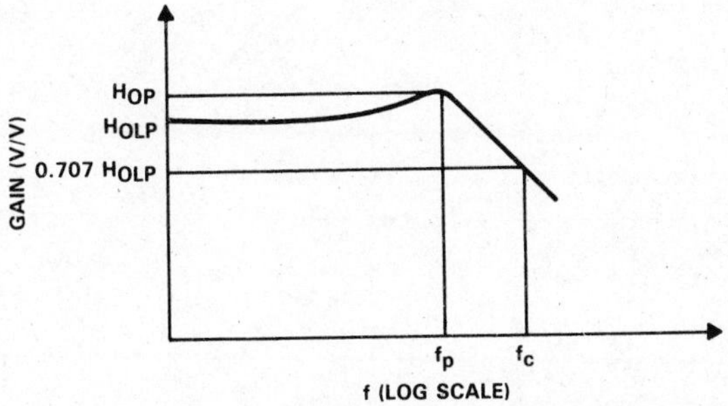
GAIN (V/V)
HOP
HOLP
0.707 HOLP
fp
fc
f (LOG SCALE)

$$f_c = f_o \times \sqrt{\left(1 - \frac{1}{2Q^2}\right) + \sqrt{\left(1 - \frac{1}{2Q^2}\right)^2 + 1}}$$

$$f_p = f_o \sqrt{1 - \frac{1}{2Q^2}}$$

$$H_{OP} = H_{OLP} \times \frac{1}{\frac{1}{Q}\sqrt{1 - \frac{1}{4Q^2}}}$$

FIGURE 12. LOW-PASS OUTPUT

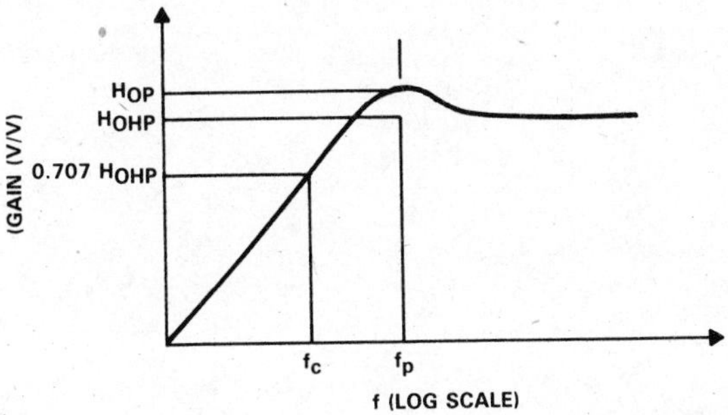
(GAIN (V/V)
HOP
HOHP
0.707 HOHP
fc
fp
f (LOG SCALE)

$$f_c = f_o \times \left[\sqrt{\left(1 - \frac{1}{2Q^2}\right) + \sqrt{\left(1 - \frac{1}{2Q^2}\right)^2 + 1}} \right]^{-1}$$

$$f_p = f_o \times \left[\sqrt{1 - \frac{1}{2Q^2}} \right]^{-1}$$

$$H_{OP} = H_{OHP} \times \frac{1}{\frac{1}{Q}\sqrt{1 - \frac{1}{4Q^2}}}$$

FIGURE 13. HIGH-PASS OUTPUT

TLC532AM, TLC532AI, TLC533AM, TLC533AI
LinCMOS™ 8-BIT ANALOG-TO-DIGITAL PERIPHERALS WITH 5 ANALOG AND 6 DUAL-PURPOSE INPUTS

D2819, NOVEMBER 1983—REVISED SEPTEMBER 1986

- LinCMOS™ Technology
- 8-Bit Resolution
- Total Unadjusted Error . . . ±0.5 LSB Max
- Ratiometric Conversion
- Access Plus Conversion Time:
  TLC532A . . . 15 μs Max
  TLC533A . . . 30 μs Max
- 3-State, Bidirectional I/O Data Bus
- 5 Analog and 6 Dual-Purpose Inputs
- On-Chip 12-Channel Analog Multiplexer
- Three On-Chip 16-Bit Data Registers
- Software Compatible with Larger TL530 and TL531 (21-Input Versions)
- On-Chip Sample-and-Hold Circuit
- Single 5-V Supply Operation
- Low Power Consumption . . . 6.5 mW Typ
- Improved Direct Replacements for Texas Instruments TL532 and TL533, National Semiconductor ADC0829, and Motorola MC14442

N DUAL-IN-LINE PACKAGE
(TOP VIEW)

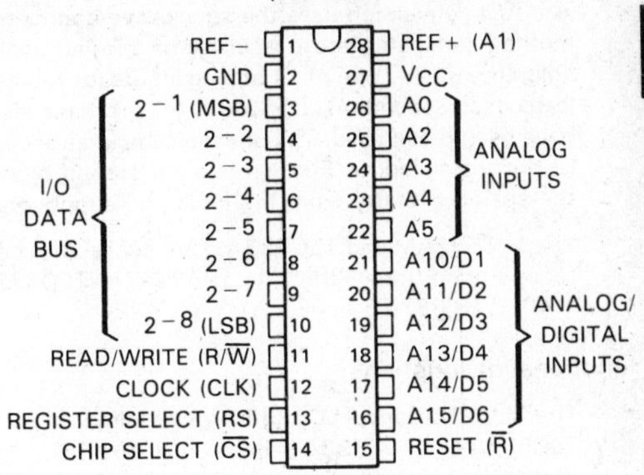
REF− 1
GND 2
2^{-1} (MSB) 3
2^{-2} 4
2^{-3} 5
2^{-4} 6
2^{-5} 7
2^{-6} 8
2^{-7} 9
2^{-8} (LSB) 10
I/O DATA BUS
READ/WRITE (R/$\overline{W}$) 11
CLOCK (CLK) 12
REGISTER SELECT (RS) 13
CHIP SELECT ($\overline{CS}$) 14
28 REF+ (A1)
27 V_{CC}
26 A0
25 A2
24 A3
23 A4
22 A5
ANALOG INPUTS
21 A10/D1
20 A11/D2
19 A12/D3
18 A13/D4
17 A14/D5
16 A15/D6
ANALOG/ DIGITAL INPUTS
15 RESET ($\overline{R}$)

FN CHIP CARRIER PACKAGE
(TOP VIEW)

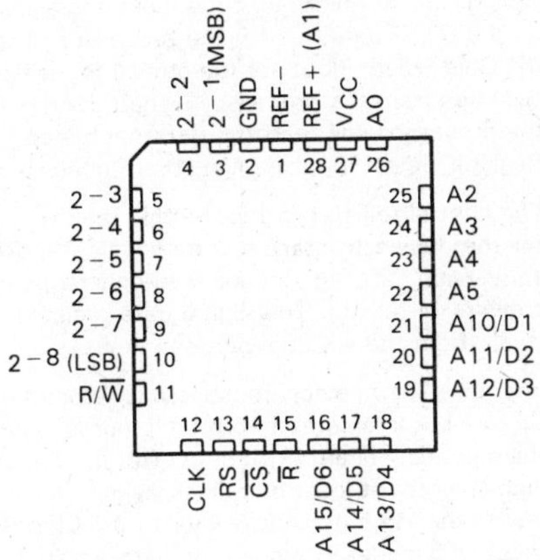
2^{-2}
2^{-1}(MSB)
GND
REF−
REF+ (A1)
VCC
A0
4 3 2 1 28 27 26
2^{-3} 5
2^{-4} 6
2^{-5} 7
2^{-6} 8
2^{-7} 9
2^{-8} (LSB) 10
R/$\overline{W}$ 11
25 A2
24 A3
23 A4
22 A5
21 A10/D1
20 A11/D2
19 A12/D3
12 13 14 15 16 17 18
CLK
RS
$\overline{CS}$
$\overline{R}$
A15/D6
A14/D5
A13/D4

description

The TLC532A and TLC533A are monolithic LinCMOS™ peripheral integrated circuits each designed to interface a microprocessor for analog data acquisition. These devices are complete peripheral data acquisition systems on a single chip and can convert analog signals to digital data from up to 11 external analog terminals. Each device features operation from a single 5-volt supply. Each contains a 12-channel analog multiplexer, an 8-bit ratiometric analog-to-digital (A/D) converter, a sample-and-hold, three 16-bit registers, and microprocessor-compatible control circuitry. Additional features include a built-in self-test, six multipurpose (analog or digital) inputs, five external analog inputs, and an 8-pin input/output (I/O) data port. The three on-chip data registers store the control data, the conversion results, and the input digital data that can be accesssed via the microprocessor data bus in two 8-bit bytes (most-significant byte first). In this manner, a microprocessor can access up to 11 external analog inputs or 6 digital signals and the positive reference voltage that may be used for self-test.

FUNCTION TABLE

ADDRESS/CONTROL					DESCRIPTION
R/$\overline{W}$	RS	$\overline{CS}$	$\overline{R}$	CLK	
X	X	X	L†		Reset
L	H	L	H	↓	Write bus data to control register
H	L	L	H	↑	Read data from analog conversion register
H	H	L	H	↑	Read data from ditigal data register
X	X	H	H	X	No response

H = High-level, L = Low-level, X = Irrelevant
↓ = High-to-low transition, ↑ = Low-to-high transition
†For proper operation, Reset must be low for at least three clock cycles.

TEXAS INSTRUMENTS
POST OFFICE BOX 655012 • DALLAS, TEXAS 75265

description (continued)

The A/D conversion uses the successive-approximation technique and switched-capacitor circuitry. This method eliminates the possibility of missing codes, nonmonotonicity, and a need for zero or full-scale adjustment. Any one of 11 analog inputs (or self-test) can be converted to an 8-bit digital word and stored in 10 microseconds (TLC532A) or 20 microseconds (TLC533A) after instructions from the microprocessor have been recognized. The on-chip sample-and-hold functions automatically to minimize errors due to noise on the analog inputs. Furthermore, differential high-impedance reference inputs are available to help isolate the analog circuitry from the logic and supply noises while easing ratiometric conversion and scaling.

The TLC532AM and TLC533AM are available in both the N and FN plastic packages and are characterized for operation from −55 °C to 125 °C. The TLC532AI and TLC533AI are characterized for operation from −40 °C to 85 °C.

functional description

The TLC532A and TLC533A provide direct interface to a microprocessor-based system. Control of the TLC532A and TLC533A is handled via the 8-line TTL-compatible 3-state data bus, the three control inputs (Read/Write, Register Select, and Chip Select), and the Clock input. Each device contains three 16-bit internal registers. These registers are the control register, the analog conversion data register, and the digital data register.

A high level at the Read/Write input and a low level at the Chip Select input set the device to output data on the 8-line data bus for the processor to read. A low level at the Read/Write input and a low level at the Chip Select input set the device to receive instructions into the internal control register on the 8-line data bus from the processor. When the device is in the read mode and the Register Select input is low, the processor will read the data contained in the analog conversion data register. However, when the Register Select input is high, the processor reads the data contained in the digital data register.

The control register is a write-only register into which the microprocessor writes command instructions for the device to start A/D conversion and to select the analog channel to be converted. The analog conversion data register is a read-only register that contains the current converter status and most recent conversion results. The digital data register is also a read-only register that holds the digital input logic levels from the six dual-purpose inputs.

Internally each device contains a byte pointer that selects the appropriate byte during two cycles of the Clock input in a normal 16-bit microprocessor instruction. The internal pointer will automatically point to the most-significant (MS) byte after the first complete clock cycle any time that the Chip Select is at the high level for at least one clock cycle. This causes the device to treat the next signal on the 8-line data bus as the MS byte. A low level at the Chip Select input activates the inputs and outputs and an internal function decoder. However, no data is transferred until the Clock goes high. The internal byte pointer first points to the MS byte of the selected register during the first clock cycle. After the first clock cycle in which the MS byte is accessed, the internal pointer switches to the LS byte and remains there for as long as Chip Select is low. The MS byte of any register may be accessed by either an 8-bit or a 16-bit microprocessor instruction; however, the LS byte may only be accessed by a 16-bit microprocessor instruction.

Normally, a two-byte word is written into or read from the controlling processor, but a single byte can be read by the processor by proper manipulation of the Chip Select input. This can be used to read conversion status from the analog conversion data register or the digital multipurpose input levels from the digital data register. The format and content of each two-byte word is shown in Figures 1 through 3.

functional description (continued)

A conversion cycle is started after a two-byte instruction is written into the control register and the start conversion (SC) bit is a logic high. This two-byte instruction also selects the input analog channel to be converted. The status (EOC) bit in the analog conversion data register is reset and it remains reset until the conversion is completed, at that time the status bit is then set again. After conversion, the results are loaded into the analog conversion data register. These results remain in the analog conversion data register until the next conversion cylce is completed. If a new conversion command is entered into the control register while the conversion cycle is in progress, the on-going conversion will be aborted and a new channel acquisition cycle will immediately begin.

The Reset input allows the device to be externally forced to a known state. When a low level is applied to the Reset input for a minimum of three clock periods, the start conversion bit is cleared. The A/D converter is then idled and all the outputs are placed in the high-impedance off-state. However, the content of the analog conversion data register is not affected by the Reset input going to a low level.

Detailed information on interfacing to most popular microprocessors is readily available from the factory.

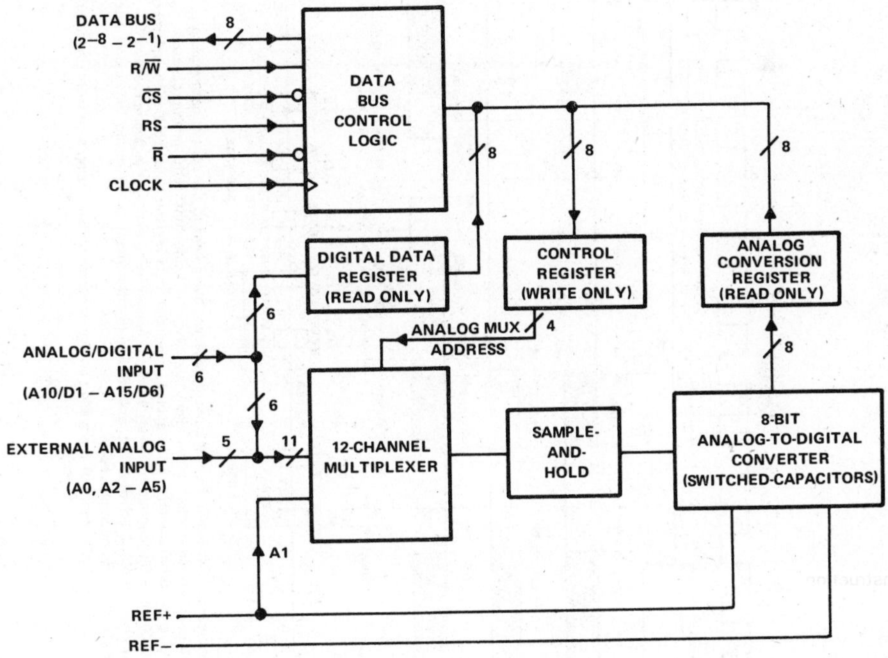
DATA BUS
(2−8 – 2−1)
8
R/W̄
CS̄
RS
R̄
CLOCK
DATA
BUS
CONTROL
LOGIC
8
8
8
DIGITAL DATA
REGISTER
(READ ONLY)
CONTROL
REGISTER
(WRITE ONLY)
ANALOG
CONVERSION
REGISTER
(READ ONLY)
6
ANALOG MUX
ADDRESS
4
8
ANALOG/DIGITAL
INPUT
(A10/D1 – A15/D6)
6
6
EXTERNAL ANALOG
INPUT
(A0, A2 – A5)
5
11
12-CHANNEL
MULTIPLEXER
SAMPLE-
AND-
HOLD
8-BIT
ANALOG-TO-DIGITAL
CONVERTER
(SWITCHED-CAPACITORS)
A1
REF+
REF−

typical operating sequence

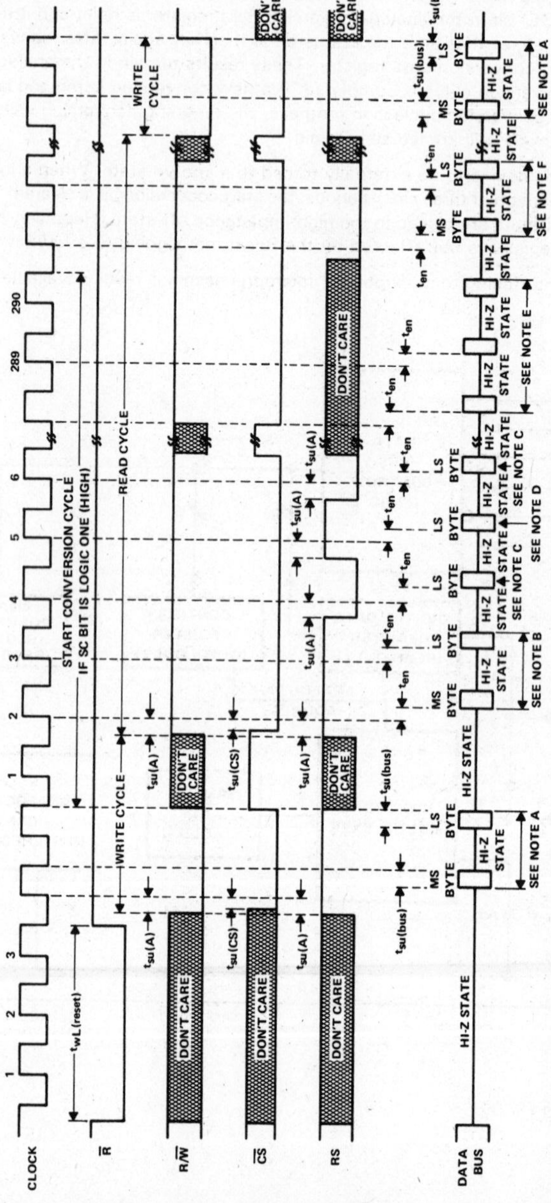
CLOCK
R̄
R/W̄
C̄S̄
RS
DATA BUS
DON'T CARE
HI-Z STATE
$t_{wL(reset)}$
WRITE CYCLE
START CONVERSION CYCLE IF SC BIT IS LOGIC ONE (HIGH)
READ CYCLE
$t_{su(A)}$
$t_{su(CS)}$
$t_{su(bus)}$
t_{en}
MS BYTE
LS BYTE
SEE NOTE A
SEE NOTE B
SEE NOTE C
SEE NOTE D
SEE NOTE E
SEE NOTE F

NOTES:
A. This is a 16-bit input instruction from the microprocessor being sent to the control data register.
B. This is the 2-byte (16-bit) content of the digital data register being sent to the microprocessor.
C. This is the LS byte (8-bit) content of the analog conversion data register being sent to the microprocessor.
D. This is the LS byte (8-bit) content of the digital data register being sent to the microprocessor.
E. These are MS byte (8-bit), LS byte (8-bit), and LS byte (8-bit) content of the analog conversion data register or digital data register being sent to the microprocessor.
F. This is the 2-byte (16-bit) content of the analog conversion data register being sent to the microprocessor.

read or write cycle time sequence

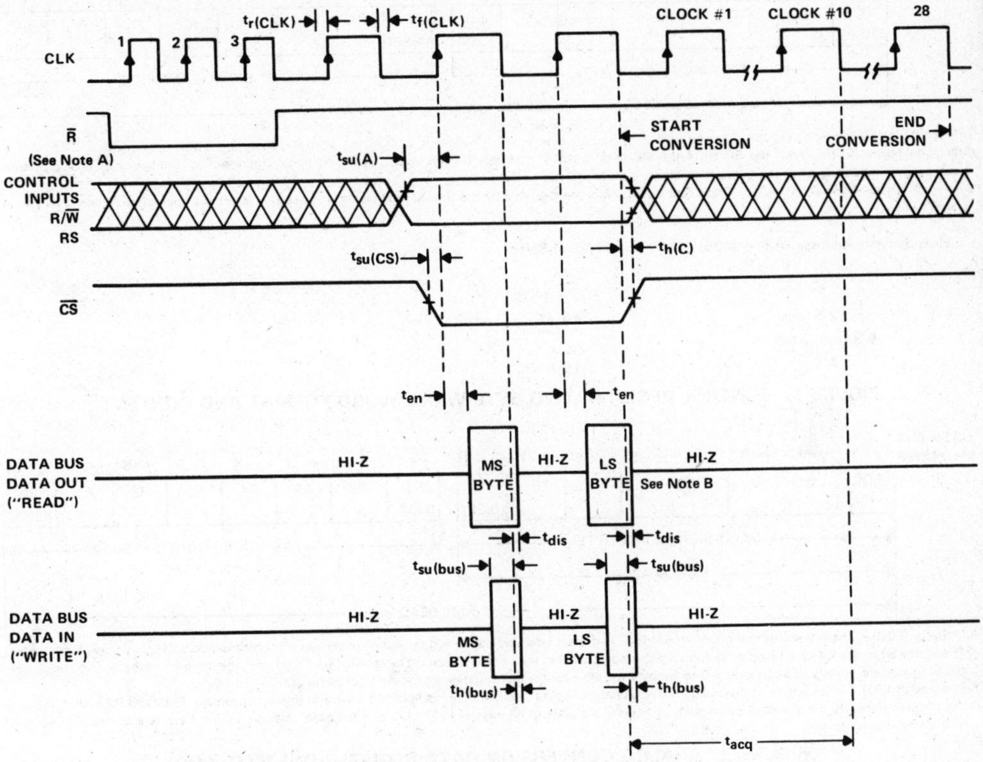
CLK
1
2
3
tr(CLK)
tf(CLK)
CLOCK #1
CLOCK #10
28
R̄
(See Note A)
START CONVERSION
END CONVERSION
tsu(A)
CONTROL INPUTS R/W̄ RS
th(C)
tsu(CS)
CS̄
ten
ten
DATA BUS DATA OUT ("READ")
HI-Z
MS BYTE
HI-Z
LS BYTE
See Note B
HI-Z
tdis
tdis
tsu(bus)
tsu(bus)
DATA BUS DATA IN ("WRITE")
HI-Z
MS BYTE
HI-Z
LS BYTE
HI-Z
th(bus)
th(bus)
tacq

NOTES: A. The reset pulse (R̄ low) is required only during power-up.

B. The most-significant byte output of Data Out occurs when CLK is high. When CLK is low, Data Out is in the high-impedance (off) state. When CLK goes high again, the least-significant byte is placed on the data bus. At this point, the least-significant byte will remain on the bus for as long as CLK is kept high.

DATA BUS LINES	2^{-1}	2^{-2}	2^{-3}	2^{-4}	2^{-5}	2^{-6}	2^{-7}	2^{-8}	2^{-1}	2^{-2}	2^{-3}	2^{-4}	2^{-5}	2^{-6}	2^{-7}	2^{-8}
	X (MSB)	X	X	X	X	X	X	SC (LSB)	X (MSB)	X	X	X	A3	A2	A1	A0 (LSB)
	MOST-SIGNIFICANT BYTE								LEAST-SIGNIFICANT BYTE							
	16-BIT WRITE															

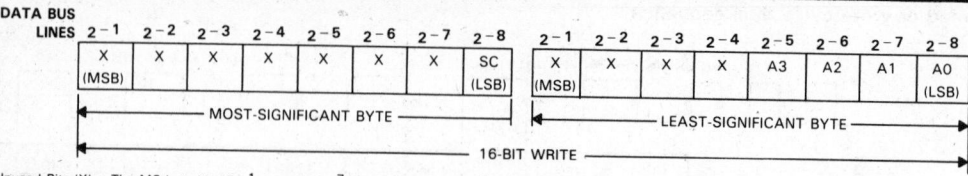

Unused Bits (X)— The MS byte bits 2^{-1} through 2^{-7} and LS byte bits 2^{-1} through 2^{-4} of the control register are not used internally.
Start Conversion (SC)— When the SC bit in the MS byte is set to a logical 1 (high level), analog-to-digital conversion of the specified analog channel will begin immediately after the completion of the control register write.
Analog Multiplex Address (A0-A3)— These four address bits are decoded by the analog multiplexer and used to select the appropriate analog channel as shown below:

Hexadecimal Address (A3 = MSB)	Channel Select
0	A0
1	REF+ (A1)
2-5	A2-A5
6-9 (not used)	
A-F	A10-A15

FIGURE 1. CONTROL REGISTER TWO-BYTE WRITE WORD FORMAT AND CONTENT

DATA BUS LINES	2^{-1}	2^{-2}	2^{-3}	2^{-4}	2^{-5}	2^{-6}	2^{-7}	2^{-8}	2^{-1}	2^{-2}	2^{-3}	2^{-4}	2^{-5}	2^{-6}	2^{-7}	2^{-8}
	EOC (MSB)	0	0	0	0	0	0	0 (LSB)	R7 (MSB)	R6	R5	R4	R3	R2	R1	R0 (LSB)
	MOST-SIGNIFICANT BYTE								LEAST-SIGNIFICANT BYTE							
	8-BIT READ															
	16-BIT READ															

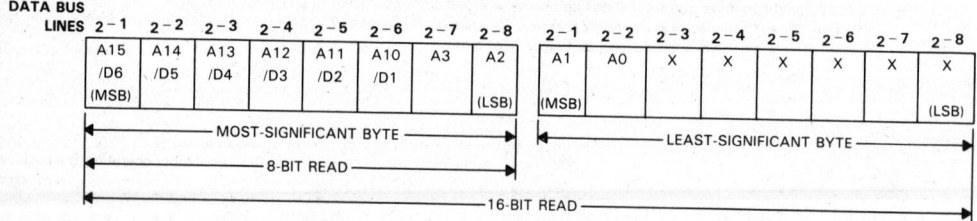

A/D Status (EOC)— The A/D status end-of-conversion (EOC) bit is set whenever an analog-to-digital conversion is successfully completed by the A/D converter. The status bit is cleared by a 16-bit write from the microprocessor to the control register. The remainder of the bits in the MS byte of the analog conversion data register are always reset to logical 0 to simplify microprocessor interrogation of the A/D converter status.
A/D Result (R0-R7)— The LS byte of the analog conversion data register contains the result of the analog-to-digital conversion. Result bit R7 is the MSB and the converter follows the standard convention of assigning a code of all ones (11111111) to a full-scale analog voltage. There are no special overflow or underflow indications.

FIGURE 2. ANALOG CONVERSION DATA REGISTER ONE-BYTE AND TWO-BYTE READ WORD FORMAT AND CONTENT

DATA BUS LINES	2^{-1}	2^{-2}	2^{-3}	2^{-4}	2^{-5}	2^{-6}	2^{-7}	2^{-8}	2^{-1}	2^{-2}	2^{-3}	2^{-4}	2^{-5}	2^{-6}	2^{-7}	2^{-8}
	A15 /D6 (MSB)	A14 /D5	A13 /D4	A12 /D3	A11 /D2	A10 /D1	A3	A2 (LSB)	A1 (MSB)	A0	X	X	X	X	X	X (LSB)
	MOST-SIGNIFICANT BYTE								LEAST-SIGNIFICANT BYTE							
	8-BIT READ															
	16-BIT READ															

Shared Digital Port (A10/D1-A15/D6)— The voltage present on these pins is interpreted as a digital signal and the corresponding states are read from these bits. A digital value will be given for each pin even if some or all of these pins are being used as analog inputs.
Analog Multiplexer Address (A0-A3)— The address of the selected analog channel presently addressed is given by these bits.
Unused Bits (X)— LS byte bits 2^{-3} through 2^{-8} of the digital data register are not used.

FIGURE 3. DIGITAL DATA REGISTER ONE-BYTE AND TWO-BYTE READ WORD FORMAT AND CONTENT

absolute maximum ratings over operating free-air temperature range (unless otherwise noted)

Supply voltage, V_{CC} (see Note 1) . . . . -0.3 V to 6.5 V
Input voltage range: Positive reference voltage . . . . V_{REF-} to V_{CC} + 0.3 V
Negative reference voltage . . . . -0.3 V to V_{REF+}
All other inputs . . . . -0.3 V to V_{CC} + 0.3 V
Input current, I_I (any input) . . . . ±10 mA
Total input current, (all inputs) . . . . ±20 mA
Operating free-air temperature range: TLC532AM, TLC533AM . . . . -55°C to 125°C
TLC532AI, TLC533AI . . . . -40°C to 85°C
Storage temperature range . . . . -65°C to 150°C
Lead temperature 1,6 mm (1/16 inch) from case for 10 seconds: N package . . . . 260°C
Case temperature for 10 seconds: FN package . . . . 260°C

NOTE 1: All voltage values are with respect to network ground terminal.

recommended operating conditions

		TLC532A			TLC533A			UNIT
		MIN	NOM	MAX	MIN	NOM	MAX	
Supply voltage, V_{CC}		4.75	5	5.5	4.75	5	5.5	V
Positive reference voltage, V_{REF+} (see Note 2)		2.5	V_{CC}	$V_{CC}+0.1$	2.5	V_{CC}	$V_{CC}+0.1$	V
Negative reference voltage, V_{REF-} (see Note 2)		-0.1	0	2.5	-0.1	0	2.5	V
Differential reference voltage, $V_{REF+} - V_{REF-}$		1	V_{CC}	$V_{CC}+0.2$	1	V_{CC}	$V_{CC}+0.2$	V
High-level input voltage, V_{IH}	Clock input	$V_{CC}-0.8$			$V_{CC}-0.8$			V
	All other digital inputs	2			2			
Low-level input voltage, V_{IL}	Any digital input			0.8			0.8	V
Clock frequency, f_{CLK}		0.1	2	2.048	0.1	1.048	1.06	MHz
CS setup time, $t_{su(CS)}$		75			100			ns
Address (R/W and RS) setup time, $t_{su(A)}$		100			145			ns
Data bus input setup time, $t_{su(bus)}$		140			185			ns
Control (R/W, RS, and CS) hold time, $t_{h(C)}$		10			20			ns
Data bus input hold time, $t_{h(bus)}$		15			20			ns
Pulse duration of control during read, $t_{w(C)}$		305			575			ns
Pulse duration, reset low, $t_{wL(reset)}$		3			3			Clock Cycles
Pulse duration of clock high, $t_{wH(CLK)}$		230			440			ns
Pulse duration of clock low, $t_{wL(CLK)}$		200			410			ns
Clock rise time, $t_{r(CLK)}$				15			25	ns
Clock fall time, $t_{f(CLK)}$				16			30	ns
Operating free-air temperature, T_A	TLC_ _AM	-55		125	-55		125	°C
	TLC_ _AI	-40		85	-40		85	

NOTE 2: Analog input voltages greater than or equal to that applied to the REF+ terminal convert to all ones (11111111), while input voltages equal to or less than that applied to the REF− terminal convert to all zeros (00000000). For proper operation, the positive reference voltage, V_{REF+}, must be at least 1-volt greater than the negative reference voltage, V_{REF-}. In addition, unadjusted errors may increase as the differential reference voltage, $V_{REF+} - V_{REF-}$, falls below 4.75 volts.

electrical characteristics over recommended operating free-air temperature range, $V_{REF+} = V_{CC}$, V_{REF-} at ground, f_{CLK} = 2 MHz (unless otherwise noted)

PARAMETER			TEST CONDITIONS	MIN	TYP†	MAX	UNIT
V_{OH}	High-level output voltage		$I_{OH} = -1.6$ mA	2.4			V
V_{OL}	low-level output voltage		$I_{OL} = 1.6$ mA			0.4	V
I_{IH}	High-level input current	Any digital or Clock input	$V_{IH} = 5.5$ V			10	μA
		Any control input				1	
I_{IL}	Low-level input current	Any digital or Clock input	$V_{IL} = 0$			−10	μA
		Any control input				−1	
I_{OZ}	Off-state (high impedance-state) output current		$V_O = V_{CC}$			10	μA
			$V_O = 0$			−10	
I_I	Analog input current (see Note 3)		$V_I = 0$ to V_{CC}			±500	nA
	Leakage current between selected channel and all other analog channels		$V_I = 0$ to V_{CC}, Clock input at 0 V			±400	nA
C_i	Input capacitance	Digital pins 3 thru 10			4	30	pF
		Any other input pin			2	15	
$I_{CC} + I_{REF+}$	Supply current plus reference current		$V_{CC} = V_{REF+} = 5.5$ V, Outputs open		1.5	3	mA
I_{CC}	Supply current		$V_{CC} = 5.5$ V		1.4	2	mA

NOTE 3: Analog input current is an average of the current flowing into a selected analog channel input during one full conversion cycle.

operating characteristics over recommended operating free-air temperature range, $V_{REF+} = V_{CC}$, V_{REF-} at ground, f_{CLK} = 2 MHz (unless otherwise noted)

PARAMETER			TEST CONDITIONS	MIN	TYP†	MAX	UNIT
	Linearity error		See Note 4			±0.5	LSB
	Zero error		See Note 5			±0.5	LSB
	Full-scale error		See Note 5			±0.5	LSB
	Total unadjusted error		See Note 6			±0.5	LSB
	Absolute accuracy error		See Note 7			±1	LSB
t_{conv}	Conversion time (including channel acquisition time)				30		Clock Cycles
t_{acq}	Channel acquisition time prior to starting conversion				10		Clock Cycles
t_{en}	Data output enable time (see Note 8)		$C_L = 50$ pF, $R_L = 3$ kΩ,			250	ns
t_{dis}	Data output disable time		$C_L = 50$ pF, $R_L = 3$ kΩ	10			ns
$t_{r(bus)}$	Data bus output rise time	High-impedance to high-level	$C_L = 50$ pF, $R_L = 3$ kΩ			150	ns
		Low to high-level				300	
$t_{f(bus)}$	Data bus output fall time	High-impedance to low-level	$C_L = 50$ pF, $R_L = 3$ kΩ			150	ns
		High to low-level				300	

†Typical values are at $V_{CC} = 5$ V, $T_A = 25$°C.

NOTES: 4. Linearity error is the deviation from the best straight line through the A/D transfer characteristics.

5. Zero error is the difference between 00000000 and the converted output for zero input voltage; full-scale error is the difference between 11111111 and the converted output for full-scale input voltage.

6. Total unadjusted error is the sum of linearity, zero, and full-scale errors.

7. Absolute accuracy error is the maximum difference between an analog value and the nominal midstep value within any step. This includes all errors including inherent quantization error, which is the ±0.5 LSB uncertainty caused by the A/D converters finite resolution.

8. If chip-select setup time, $t_{su(CS)}$, is less than 0.14 microseconds, the effective data output enable time, t_{en}, may extend such that $t_{su(CS)} + t_{en}$ is equal to a maximum of 0.475 microseconds.

electrical characteristics over recommended ranges V_{CC}, V_{REF+}, and operating free-air temperature, V_{REF-} at ground, f_{CLK} = 1.048 MHz (unless otherwise noted)

PARAMETER			TEST CONDITIONS	MIN	TYP†	MAX	UNIT
V_{OH}	High-level output voltage		I_{OH} = −1.6 mA	2.4			V
V_{OL}	Low-level output voltage		I_{OL} = 1.6 mA			0.4	V
I_{IH}	High-level input current	Any digital or Clock input	V_{IH} = 5.5 V			10	μA
		Any control input				1	
I_{IL}	Low-level input current	Any digital or Clock input	V_{IL} = 0			−10	μA
		Any control input				−1	
I_{OZ}	Off-state (high impedance-state) output current		$V_O = V_{CC}$			10	μA
			V_O = 0			−10	
I_I	Analog input current (see Note 3)		V_I = 0 to V_{CC}			±500	nA
	Leakage current between selected channel and all other analog channels		V_I = 0 to V_{CC}, Clock input at 0 V			±400	nA
C_i	Input capacitance	Digital pins 3 thru 10			4	30	pF
		Any other input pin			2	15	
$I_{CC} + I_{REF+}$	Supply current plus reference current		$V_{CC} = V_{REF+}$ = 5.5 V, Outputs open		1.3	3	mA
I_{CC}	Supply current		V_{CC} = 5.5 V		1.2	2	mA

NOTE 3: Analog input current is an average of the current flowing into a selected analog channel input during one full conversion cycle.

operating characteristics over recommended ranges V_{CC}, V_{REF+}, and operating free-air temperature, V_{REF-} at ground, f_{clock} = 1.048 MHz (unless otherwise noted)

PARAMETER			TEST CONDITIONS	MIN	TYP†	MAX	UNIT
	Linearity error		See Note 4			±0.5	LSB
	Zero error		See Note 5			±0.5	LSB
	Full-scale error		See Note 5			±0.5	LSB
	Total unadjusted error		See Note 6			±0.5	LSB
	Absolute accuracy error		See Note 7			±1	LSB
t_{conv}	Conversion time (including channel acquisition time)				30		Clock Cycles
t_{acq}	Channel acquisition time prior to starting conversion				10		Clock Cycles
t_{en}	Data output enable time (see Note 8)		C_L = 50 pF, R_L = 3 kΩ,			335	ns
t_{dis}	Data output disable time		C_L = 50 pF, R_L = 3 kΩ	10			ns
$t_{r(bus)}$	Data bus output rise time	High-impedance to high-level	C_L = 50 pF, R_L = 3 kΩ			150	ns
		Low to high-level				300	
$t_{f(bus)}$	Data bus output fall time	High-impedance to low-level	C_L = 50 pF, R_L = 3 kΩ			150	ns
		High to low-level				300	

†Typical values are at V_{CC} = 5 V, T_A = 25°C.

NOTES: 4. Linearity error is the deviation from the best straight line through the A/D transfer characteristics.
5. Zero error is the difference between 00000000 and the converted output for zero input voltage; full-scale error is the difference between 11111111 and the converted output for full-scale input voltage.
6. Total unadjusted error is the sum of linearity, zero, and full-scale errors.
7. Absolute accuracy error is the maximum difference between an analog value and the nominal midstep value within any step. This includes all errors including inherent quantization error, which is the ±0.5 LSB uncertainty caused by the A/D converters finite resolution.
8. If chip-select setup time, $t_{su(CS)}$, is less than 0.14 microseconds, the effective data output enable time, t_{en}, may extend such that $t_{su(CS)} + t_{en}$ is equal to a maximum of 0.475 microseconds.

LinCMOS™ 8-BIT ANALOG-TO-DIGITAL PERIPHERALS WITH SERIAL CONTROL AND 11 INPUTS

D2799, OCTOBER 1983—REVISED DECEMBER 1985

- LinCMOS™ Technology
- 8-Bit Resolution A/D Converter
- Microprocessor Peripheral or Stand-Alone Operation
- On-Chip 12-Channel Analog Multiplexer
- Built-In Self-Test Mode
- Software-Controllable Sample and Hold
- Total Unadjusted Error . . . ±0.5 LSB Max
- TLC541 is Direct Replacement for Motorola MC145040 and National Semiconductor ADC0811. TLC540 is Capable of Higher Speed
- Pinout and Control Signals Compatible with TLC1540 Family of 10-Bit A/D Converters

N DUAL-IN-LINE PACKAGE
(TOP VIEW)

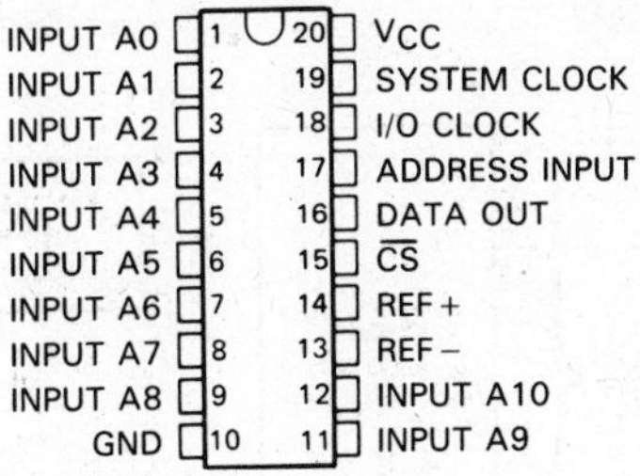
INPUT A0
INPUT A1
INPUT A2
INPUT A3
INPUT A4
INPUT A5
INPUT A6
INPUT A7
INPUT A8
GND
1
2
3
4
5
6
7
8
9
10
20
19
18
17
16
15
14
13
12
11
VCC
SYSTEM CLOCK
I/O CLOCK
ADDRESS INPUT
DATA OUT
CS
REF+
REF−
INPUT A10
INPUT A9

FN CHIP CARRIER PACKAGE
(TOP VIEW)

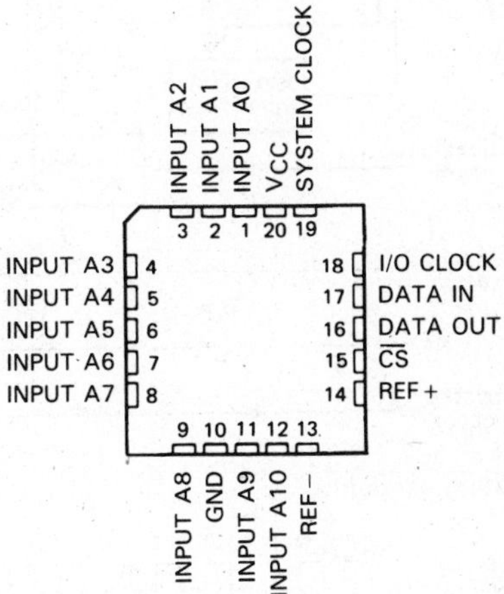
INPUT A2
INPUT A1
INPUT A0
VCC
SYSTEM CLOCK
3 2 1 20 19
INPUT A3
INPUT A4
INPUT A5
INPUT A6
INPUT A7
4
5
6
7
8
18
17
16
15
14
I/O CLOCK
DATA IN
DATA OUT
CS
REF+
9 10 11 12 13
INPUT A8
GND
INPUT A9
INPUT A10
REF−

TYPICAL PERFORMANCE	TLC540	TLC541
Channel Acquisition Sample Time	2 μs	3.6 μs
Conversion Time	9 μs	17 μs
Samples per Second	75×10^3	40×10^3
Power Dissipation	6 mW	6 mW

description

The TLC540 and TLC541 are LinCMOS™ A/D peripherals built around an 8-bit switched-capacitor successive-approximation A/D converter. They are designed for serial interface to a microprocessor or peripheral via a three-state output with up to four control inputs [including independent System Clock, I/O Clock, Chip Select (CS), and Address Input]. A 4-megahertz system clock for the TLC540 and a 2.1-megahertz system clock for the TLC541 with a design that includes simultaneous read/write operation allow high-speed data transfers and sample rates of up to 75,180 samples per second for the TLC540 and 40,000 samples per second for the TLC541. In addition to the high-speed converter and versatile control logic, there is an on-chip 12-channel analog multiplexer that can be used to sample any one of 11 inputs or an internal "self-test" voltage, and a sample-and-hold that can operate automatically or under microprocessor control. Detailed information on interfacing to most popular microprocessors is readily available from the factory.

The converters incorporated in the TLC540 and TLC541 feature dproofifferential high-impedance reference inputs that facilitate ratiometric conversion, scaling, and analog circuitry isolation from logic and supply noises. A switched-capacitor design allows guaranteed low-error (±0.5 LSB) conversion in 9 microseconds for the TLC540 and 17 microseconds for the TLC541 over the full operating temperature range.

The TLC540 and the TLC541 are available in both the N and FN plastic packages. The M-suffix versions are characterized for operation from −55°C to 125°C. The I-suffix versions are characterized for operation from −40°C to 85°C.

functional block diagram

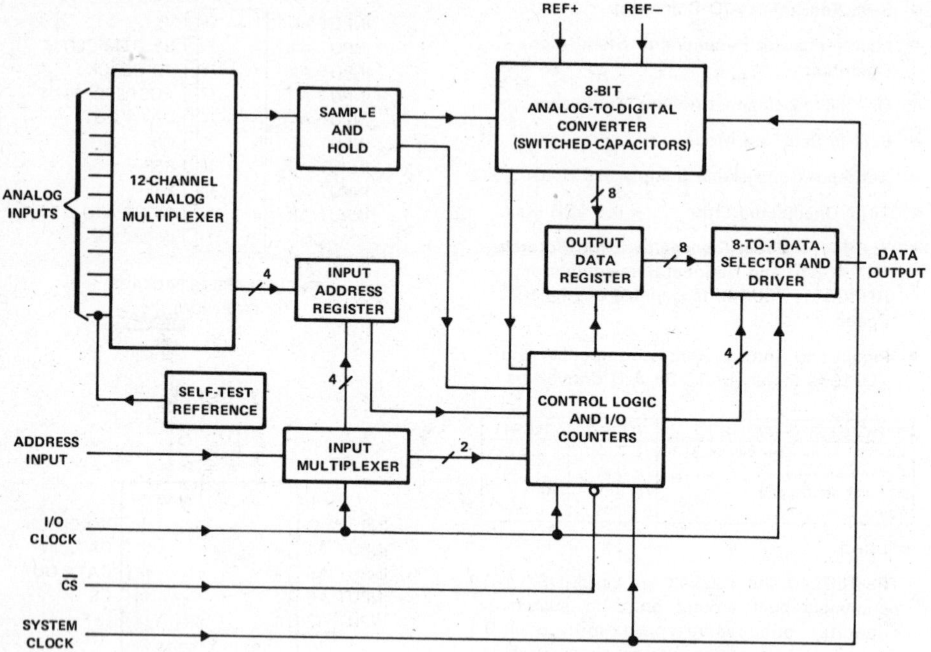
REF+
REF−
8-BIT ANALOG-TO-DIGITAL CONVERTER (SWITCHED-CAPACITORS)
SAMPLE AND HOLD
12-CHANNEL ANALOG MULTIPLEXER
ANALOG INPUTS
INPUT ADDRESS REGISTER
OUTPUT DATA REGISTER
8-TO-1 DATA SELECTOR AND DRIVER
DATA OUTPUT
SELF-TEST REFERENCE
CONTROL LOGIC AND I/O COUNTERS
INPUT MULTIPLEXER
ADDRESS INPUT
I/O CLOCK
CS
SYSTEM CLOCK
8
4
2

operating sequence

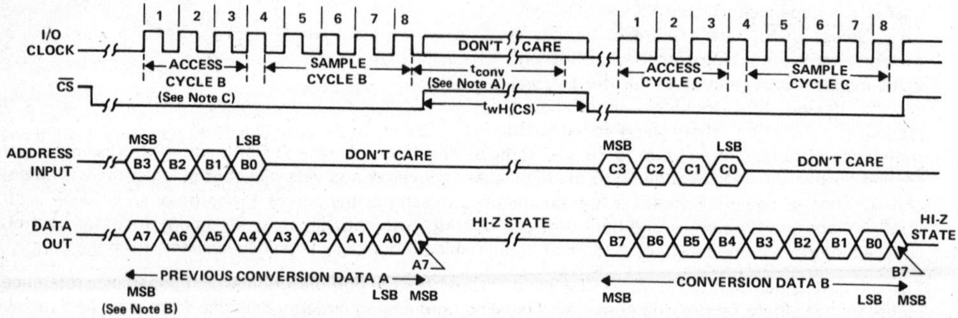
I/O CLOCK
CS
ADDRESS INPUT
DATA OUT
DON'T CARE
ACCESS CYCLE B
(See Note C)
SAMPLE CYCLE B
tconv
(See Note A)
twH(CS)
ACCESS CYCLE C
SAMPLE CYCLE C
MSB
LSB
B3 B2 B1 B0
C3 C2 C1 C0
A7 A6 A5 A4 A3 A2 A1 A0
HI-Z STATE
B7 B6 B5 B4 B3 B2 B1 B0
PREVIOUS CONVERSION DATA A
CONVERSION DATA B
(See Note B)

NOTES: A. The conversion cycle, which requires 36 System Clock periods, is initiated on the 8th falling edge of the I/O Clock after $\overline{CS}$ goes low for the channel whose address exists in memory at that time. If $\overline{CS}$ is kept low during conversion, the I/O Clock must remain low for at least 36 System Clock cycles to allow conversion to be completed.

B. The most significant bit (MSB) will automatically be placed on the DATA OUT bus after $\overline{CS}$ is brought low. The remaining seven bits (A6-A0) will be clocked out on the first seven I/O Clock falling edges.

C. To minimize errors caused by noise at the $\overline{CS}$ input, the internal circuitry waits for three System Clock cycles (or less) after a chip select falling edge is detected before responding to control input signals. Therefore, no attempt should be made to clock-in address data until the minimum chip-select setup time has elapsed.

TEXAS INSTRUMENTS
POST OFFICE BOX 655012 • DALLAS, TEXAS 75265

absolute maximum ratings over operating free-air temperature range (unless otherwise noted)

Supply voltage, V_{CC} (see Note 1) . . . . . 6.5 V
Input voltage range (any input) . . . . . −0.3 V to V_{CC} + 0.3 V
Output voltage range . . . . . −0.3 V to V_{CC} + 0.3 V
Peak input current range (any input) . . . . . ±10 mA
Peak total input current (all inputs) . . . . . ±30 mA
Operating free-air temperature range: TLC540I, TLC541I . . . . . −40°C to 85°C
TLC540M, TLC541M . . . . . −55°C to 125°C
Storage temperature range . . . . . −65°C to 150°C
Lead temperature 1,6 mm (1/16 inch) from case for 10 seconds: N package . . . . . 260°C
Case temperature for 10 seconds: FN package . . . . . 260°C

NOTE 1: All voltage values are with respect to digital ground with REF− and GND wired together (unless otherwise noted).

recommended operating conditions

			TLC540			TLC541			UNIT
			MIN	NOM	MAX	MIN	NOM	MAX	
Supply voltage, V_{CC}			4.75	5	5.5	4.75	5	5.5	V
Positive reference voltage, V_{REF+} (see Note 2)			2.5	V_{CC}	V_{CC}+0.1	2.5	V_{CC}	V_{CC}+0.1	V
Negative reference voltage, V_{REF-} (see Note 2)			−0.1	0	2.5	0.1	0	2.5	V
Differential reference voltage, $V_{REF+} - V_{REF-}$ (see Note 2)			1	V_{CC}	V_{CC}+0.2	1	V_{CC}	V_{CC}+0.2	V
Analog input voltage (see Note 2)			0		V_{CC}	0		V_{CC}	V
High-level control input voltage, V_{IH}			2			2			V
Low-level control input voltage, V_{IL}					0.8			0.8	V
Setup time, address bits at data input before I/O CLK↑, $t_{su(A)}$			200			400			ns
Hold time, address bits after I/O CLK↑, $t_{h(A)}$			0			0			ns
Setup time, $\overline{CS}$ low before clocking in first address bit, $t_{su(CS)}$ (see Note 3)			3			3			System clock cycles
$\overline{CS}$ high during conversion, $t_{wH(CS)}$			36			36			System clock cycles
Input/Output clock frequency, $f_{CLK(I/O)}$			0		2.048	0		1.1	MHz
System clock frequency, $f_{CLK(SYS)}$			$f_{CLK(I/O)}$		4	$f_{CLK(I/O)}$		2.1	MHz
System clock high, $t_{wH(SYS)}$			110			210			ns
System clock low, $t_{wL(SYS)}$			100			190			ns
Input/Output clock high, $t_{wH(I/O)}$			200			404			ns
Input/Output clock low, $t_{wL(I/O)}$			200			404			ns
Clock transition time (see Note 4)	System	$f_{CLK(SYS)} \le 1048$ kHz			30			30	ns
		$f_{CLK(SYS)} > 1048$ kHz			20			20	
	I/O	$f_{CLK(I/O)} \le 525$ kHz			100			100	ns
		$f_{CLK(I/O)} > 525$ kHz			40			40	
Operating free-air temperature, T_A		TLC540M, TLC541M	−55		125	−55		125	°C
		TLC540I, TLC541I	−40		85	−40		85	

NOTES: 2. Analog input voltages greater than that applied to REF+ convert as all "1"s (11111111), while input voltages less than that applied to REF− convert as all "0"s (00000000). For proper operation, REF+ voltage must be at least 1 volt higher than REF− voltage. Also, the total unadjusted error may increase as this differential reference voltage falls below 4.75 volts.

3. To minimize errors caused by noise at the chip select input, the internal circuitry waits for three System Clock cycles (or less) after a chip select falling edge is detected before responding to control input signals. Therefore, no attempt should be made to clock-in an address until the minimum chip select setup time has elapsed.

4. This is the time required for the clock input signal to fall from V_{IH} min to V_{IL} max or to rise from V_{IL} max to V_{IH} min. In the vicinity of normal room temperature, the devices function with input clock transition time as slow as 2 microseconds for remote data acquisition applications where the sensor and the A/D converter are placed several feet away from the controlling microprocessor.

electrical characteristics over recommended operating temperature range, $V_{CC} = V_{REF+} = 4.75$ V to 5.5 V (unless otherwise noted), $f_{CLK(I/O)} = 2.048$ MHz for TLC540 or $f_{CLK(I/O)} = 1.1$ MHz for TLC541

PARAMETER			TEST CONDITIONS		MIN	TYP†	MAX	UNIT
V_{OH}	High-level output voltage (pin 16)		$V_{CC} = 4.75$ V,	$I_{OH} = 360\ \mu A$	2.4			V
V_{OL}	Low-level output voltage		$V_{CC} = 4.75$ V,	$I_{OL} = 1.6$ mA			0.4	V
I_{OZ}	Off-state (high-impedance state) output current		$V_O = V_{CC}$,	$\overline{CS}$ at V_{CC}			10	μA
			$V_O = 0$,	$\overline{CS}$ at V_{CC}			−10	μA
I_{IH}	High-level input current		$V_I = V_{CC}$			0.005	2.5	μA
I_{IL}	Low-level input current		$V_I = 0$			−0.005	−2.5	μA
I_{CC}	Operating supply current		$\overline{CS}$ at 0 V			1.2	2.5	mA
	Selected channel leakage current		Selected channel at V_{CC}, Unselected channel at 0 V			0.4	1	μA
			Selected channel at 0 V, Unselected channel at V_{CC}			−0.4	−1	μA
$I_{CC} + I_{REF}$	Supply and reference current		$V_{REF+} = V_{CC}$,	$\overline{CS}$ at 0 V		1.3	3	mA
C_i	Input capacitance	Analog inputs				7	55	pF
		Control inputs				5	15	pF

†All typical values are at $T_A = 25°C$.

operating characteristics over recommended operating free-air temperature range, $V_{CC} = V_{REF+} = 4.75$ V to 5.5 V, $f_{CLK(I/O)} = 2.048$ MHz for TLC540 or 1.1 MHz for TLC541, $f_{CLK(SYS)} = 4$ MHz for TLC540 or 2.1 MHz for TLC541.

	PARAMETER	TEST CONDITIONS	TLC540			TLC541			UNIT
			MIN	TYP	MAX	MIN	TYP	MAX	
	Linearity error	See Note 5			±0.5			±0.5	LSB
	Zero error	See Notes 2 and 6			±0.5			±0.5	LSB
	Full-scale error	See Notes 2 and 6			±0.5			±0.5	LSB
	Total unadjusted error	See Note 7			±0.5			±0.5	LSB
	Self-test output code	Input A11 address = 1011 (See Note 8)	01111101 (125)		10000011 (131)	01111101 (125)		10000011 (131)	
t_{conv}	Conversion time	See Operating Sequence			9			17	µs
	Total access and conversion time	See Operating Sequence			13.3			25	µs
t_{acq}	Channel acquisition time (sample cycle)	See Operating Sequence			4			4	I/O clock cycles
t_v	Time output data remains valid after I/O clock↓		10			10			ns
t_d	Delay time, I/O clock↓ to data output valid	See Parameter Measurement Information			300			400	ns
t_{en}	Output enable time				150			150	ns
t_{dis}	Output disable time				150			150	ns
$t_{r(bus)}$	Data bus rise time				300			300	ns
$t_{f(bus)}$	Data bus fall time				300			300	ns

NOTES: 2. Analog input voltages greater than that applied to REF+ convert to all "1"s (11111111), while input voltages less than that applied to REF− convert to all "0"s (00000000). For proper operation, REF+ voltage must be at least 1 volt higher than REF− voltage. Also, the total unadjusted error may increase as this differential reference voltage falls below 4.75 volts.

5. Linearity error is the maximum deviation from the best straight line through the A/D transfer characteristics.
6. Zero error is the difference between 00000000 and the converted output for zero input voltage; full-scale error is the difference between 11111111 and the converted for full-scale input voltage.
7. Total unadjusted error is the sum of linearity, zero, and full-scale errors.
8. Both the input address and the output codes are expressed in positive logic. The A11 analog input signal is internally generated and is used for test purposes.

PARAMETER MEASUREMENT INFORMATION

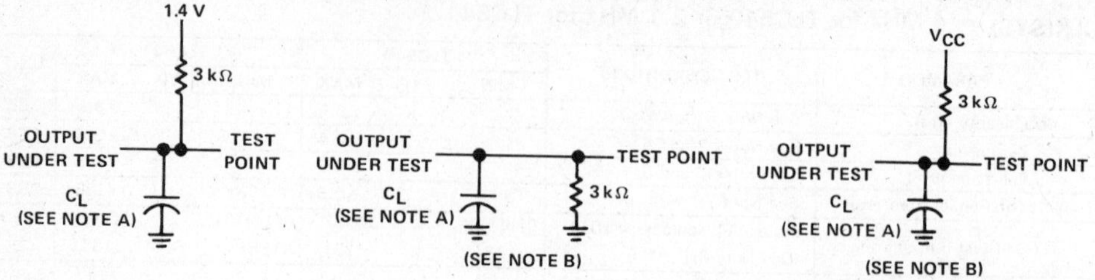
1.4 V
3 kΩ
OUTPUT UNDER TEST
TEST POINT
CL (SEE NOTE A)
OUTPUT UNDER TEST
TEST POINT
3 kΩ
CL (SEE NOTE A)
(SEE NOTE B)
VCC
3 kΩ
OUTPUT UNDER TEST
TEST POINT
CL (SEE NOTE A)
(SEE NOTE B)

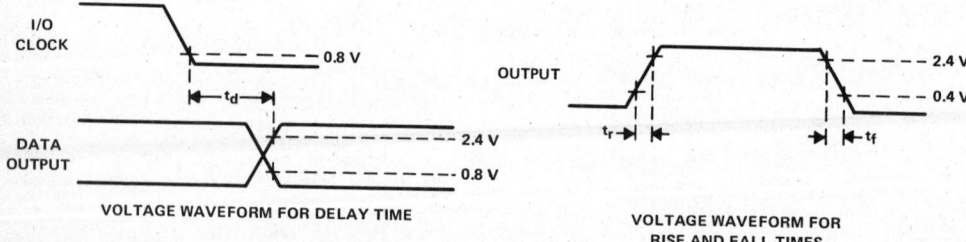

LOAD CIRCUIT FOR t_d, t_r, AND t_f

LOAD CIRCUIT FOR t_{PZH} AND t_{PHZ}

LOAD CIRCUIT FOR t_{PZL} AND t_{PLZ}

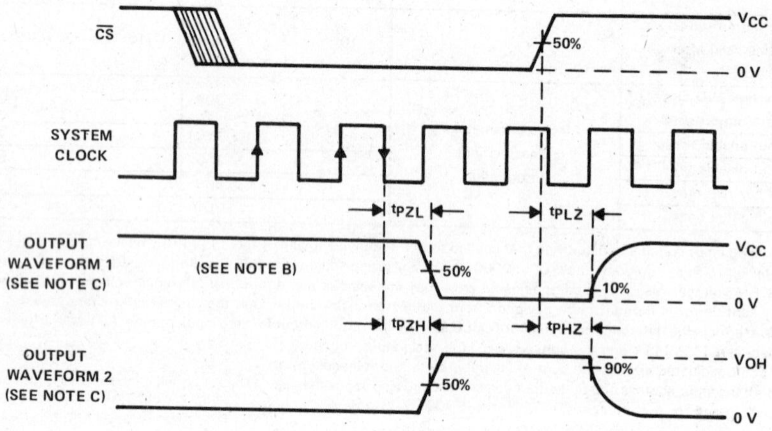
CS
50%
VCC
0 V
SYSTEM CLOCK
tPZL
tPLZ
OUTPUT WAVEFORM 1 (SEE NOTE C)
(SEE NOTE B)
50%
10%
VCC
0 V
tPZH
tPHZ
OUTPUT WAVEFORM 2 (SEE NOTE C)
50%
90%
VOH
0 V

VOLTAGE WAVEFORMS FOR ENABLE AND DISABLE TIMES

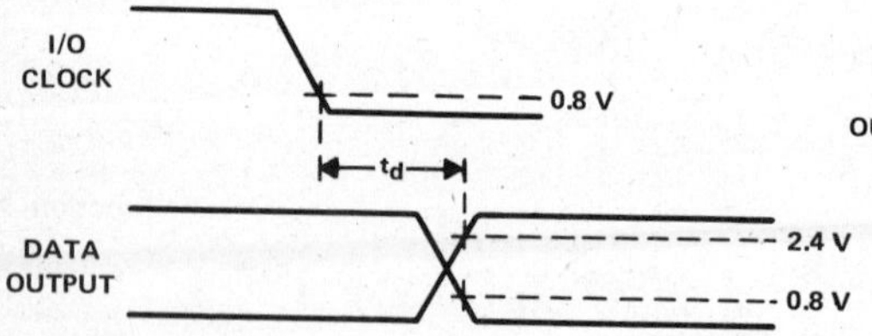
I/O CLOCK
0.8 V
td
DATA OUTPUT
2.4 V
0.8 V

VOLTAGE WAVEFORM FOR DELAY TIME

OUTPUT
2.4 V
0.4 V
tr
tf

VOLTAGE WAVEFORM FOR RISE AND FALL TIMES

NOTES: A. C_L = 50 pF for TLC540 and 100 pF for TLC541.
B. $t_{en} = t_{PZH}$ or t_{PZL}, $t_{dis} = t_{PHZ}$ or t_{PLZ}.
C. Waveform 1 is for an output with internal conditions such that the output is low except when disabled by the output control. Waveform 2 is for an output with internal conditions such that the output is high except when disabled by the output control.

principles of operation

The TLC540 and TLC541 are each complete data acquisition systems on a single chip. They include such functions as analog multiplexer, sample-and-hold, 8-bit A/D converter, data and control registers, and control logic. For flexibility and access speed, there are four control inputs [two clocks, chip select ($\overline{CS}$), and address]. These control inputs and a TTL-compatible 3-state output are intended for serial communications with a microprocessor or microcomputer. With judicious interface timing, with TLC540 a conversion can be completed in 9 microseconds, while complete input-conversion-output cycles can be repeated every 13 microseconds. With TLC541 a conversion can be completed in 17 microseconds, while complete input-conversion-output cycles are repeated every 25 microseconds. Furthermore, this fast conversion can be executed on any of 11 inputs or its built-in "self-test," and in any order desired by the controlling processor.

The System and I/O Clocks are normally used independently and do not require any special speed or phase relationships between them. This independence simplifies the hardware and software control tasks for the device. Once a clock signal within the specification range is applied to the System Clock input, the control hardware and software need only be concerned with addressing the desired analog channel, reading the previous conversion result, and starting the conversion by using the I/O Clock. The System Clock will drive the "conversion crunching" circuitry so that the control hardware and software need not be concerned with this task.

When $\overline{CS}$ is high, the Data Output pin is in a three-state condition and the Address Input and I/O Clock pins are disabled. This feature allows each of these pins, with the exception of the $\overline{CS}$ pin, to share a control logic point with their counterpart pins on additional A/D devices when additional TLC540/541 devices are used. In this way, the above feature serves to minimize the required control logic pins when using multiple A/D devices.

The control sequence has been designed to minimize the time and effort required to initiate conversion and obtain the conversion result. A normal control sequence is:

1. $\overline{CS}$ is brought low. To minimize errors caused by noise at the $\overline{CS}$ input, the internal circuitry waits for two rising edges and then a falling edge of the System Clock after a low $\overline{CS}$ transition, before the low transition is recognized. This technique is used to protect the device against noise when the device is used in a noisy environment. The MSB of the previous conversion result will automatically appear on the Data Out pin.
2. A new positive-logic multiplexer address is shifted in on the first four rising edges of the I/O Clock. The MSB of the address is shifted in first. The negative edges of these four I/O clock pulses shift out the second, third, fourth, and fifth most significant bits of the previous conversion result. The on-chip sample-and-hold begins sampling the newly addressed analog input after the fourth falling edge. The sampling operation basically involves the charging of internal capacitors to the level of the analog input voltage.
3. Three clock cycles are then applied to the I/O pin and the sixth, seventh, and eighth conversion bits are shifted out on the negative edges of these clock cycles.
4. The final eighth clock cycle is applied to the I/O Clock pin. The falling edge of this clock cycle completes the analog sampling process and initiates the hold function. Conversion is then performed during the next 36 System Clock cycles. After this final I/O Clock cycle, $\overline{CS}$ must go high or the I/O Clock must remain low for at least 36 System Clock cycles to allow for the conversion function.

$\overline{CS}$ can be kept low during periods of multiple conversion. When keeping $\overline{CS}$ low during periods of multiple conversion, special care must be exercised to prevent noise glitches on the I/O Clock line. If glitches occur on the I/O Clock line, the I/O sequence between the microprocessor/controller and the device will lose synchronization. Also, if $\overline{CS}$ is taken high, it must remain high until the end of the conversion. Otherwise, a valid falling edge of $\overline{CS}$ will cause a reset condition, which will abort the conversion in progress.

A new conversion may be started and the ongoing conversion simultaneously aborted by performing steps 1 through 4 before the 36 System Clock cycles occur. Such action will yield the conversion result of the previous conversion and not the ongoing conversion.

principles of operation (continued)

It is possible to connect the System and I/O Clock pins together in special situations in which controlling circuitry points must be minimized. In this case, the following special points must be considered in addition to the requirements of the normal control sequence previously described.

1. When $\overline{CS}$ is recognized by the device to be at a low level, the common clock signal is used as an I/O Clock. When $\overline{CS}$ is recognized by the device to be at a high level, the common clock signal is used to drive the ''conversion crunching'' circuitry.
2. The device will recognize a $\overline{CS}$ low transition only when the $\overline{CS}$ input changes and subsequently the System Clock pin receives two positive edges and then a negative edge. For this reason, after a $\overline{CS}$ negative edge, the first two clock cycles will not shift in the address because a low $\overline{CS}$ must be recognized before the I/O Clock can shift in an analog channel address. Also, upon shifting in the address, $\overline{CS}$ must be raised after the sixth I/O Clock pulse that has been recognized by the device, so that a $\overline{CS}$ low level will be recognized upon the lowering of the eighth I/O Clock signal that is recognized by the device. Otherwise, additional common clock cycles will be recognized as I/O Clock pulses and will shift in an erroneous address.

For certain applications, such as strobing applications, it is necessary to start conversion at a specific point in time. This device will accommodate these applications. Although the on-chip sample-and-hold begins sampling upon the negative edge of the fourth I/O Clock cycle, the hold function is not initiated until the negative edge of the eighth I/O Clock cycle. Thus, the control circuitry can leave the I/O Clock signal in its high state during the eighth I/O Clock cycle until the moment at which the analog signal must be converted. The TLC540/TLC541 will continue sampling the analog input until the eighth falling edge of the I/O Clock. The control circuitry or software will then immediately lower the I/O Clock signal and hold the analog signal at the desired point in time and start conversion.

Detailed information on interfacing to most popular microprocessors is readily available from the factory.

TLC543M, TLC543I, TLC544M, TLC544I 8-BIT ANALOG-TO-DIGITAL PERIPHERALS WITH SERIAL CONTROL AND 5 INPUTS

D2799, SEPTEMBER 1986

- LinCMOS™ Technology
- 8-Bit Resolution A/D Converter
- On-Chip 6-Channel Analog Multiplexer
- Built-In Self-Test Mode
- Software-Controllable Sample and Hold
- Total Unadjusted Error . . . ±0.5 LSB Max
- End-of-Conversion Output
- Conversion Time . . . 17 μs Max
- Internal System Clock . . . 4 MHz Typ
- Low Power Consumption . . . 6 mW Typ
- Total Access and Conversion Cycles:
  TLC543 . . . 45,500 c/s Min
  TLC544 . . . 40,000 c/s Min

D, J, OR N PACKAGE
(TOP VIEW)

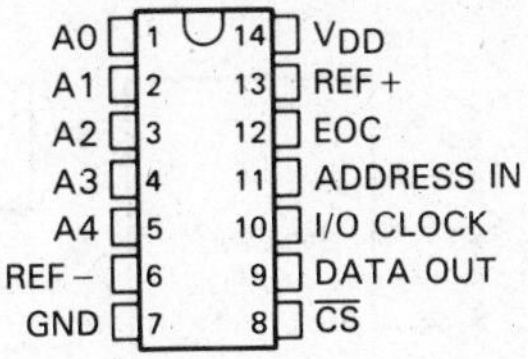
A0 1
A1 2
A2 3
A3 4
A4 5
REF− 6
GND 7
14 VDD
13 REF+
12 EOC
11 ADDRESS IN
10 I/O CLOCK
9 DATA OUT
8 CS

description

The TLC543 and TLC544 are LinCMOS™ A/D peripherals built around an 8-bit switched-capacitor successive-approximation A/D converter. They are designed for serial interface to a microprocessor or peripheral via a three-state output with up to four control lines that include I/O Clock, Chip Select ($\overline{CS}$), Address Input, and End-of-Conversion Output (EOC). A 4-megahertz on-chip system clock and simultaneous read/write operations permit high-speed data transfer and minimum sample rates of 45,500 cycles per second for TLC543 and 40,000 cycles for the TLC544. In addition to the high-speed converter and versatile control logic, there is an on-chip 6-channel analog multiplexer that can be used to sample any one of five inputs or an internal "self-test" voltage and a sample-and-hold that can operate automatically or under processor control.

The converters incorporated in the TLC543 and TLC544 feature differential high-impedance reference inputs that permit ratiometric conversion, scaling, and analog circuitry isolation from logic and supply noise.

A totally switched-capacitor design allows guaranteed low-error (±0.5 LSB) conversion in 17 microseconds maximum for the TLC543 and the TLC544 over the full operating temperature range. The TLC543M and TLC544M are characterized for operation over the full military temperature range of −55°C to 125°C. The TLC543I and TLC544I are characterized for operation from −40°C to 85°C.

LinCMOS is a trademark of Texas Instruments Incorporated

TEXAS INSTRUMENTS
POST OFFICE BOX 655012 • DALLAS, TEXAS 75265

functional block diagram

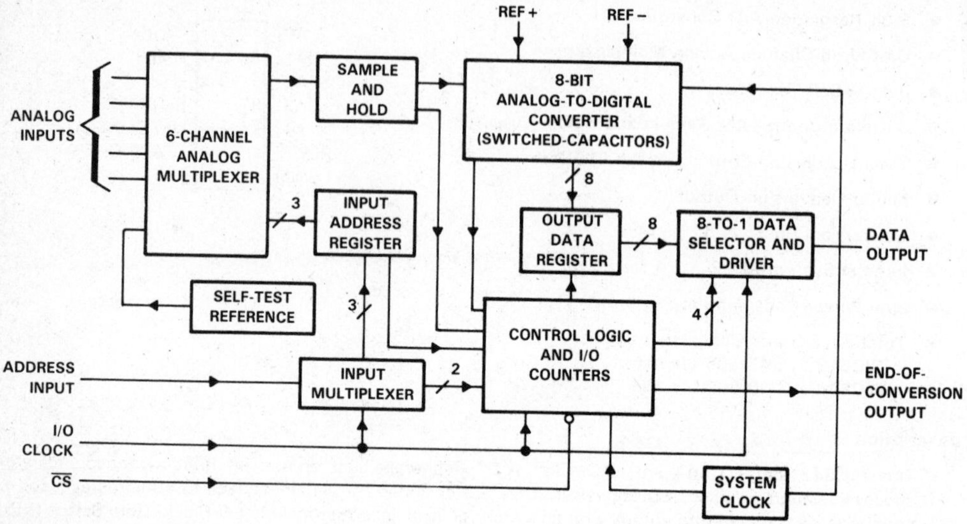
REF +
REF −
ANALOG INPUTS
6-CHANNEL ANALOG MULTIPLEXER
SAMPLE AND HOLD
8-BIT ANALOG-TO-DIGITAL CONVERTER (SWITCHED-CAPACITORS)
INPUT ADDRESS REGISTER
OUTPUT DATA REGISTER
8-TO-1 DATA SELECTOR AND DRIVER
DATA OUTPUT
SELF-TEST REFERENCE
CONTROL LOGIC AND I/O COUNTERS
ADDRESS INPUT
INPUT MULTIPLEXER
END-OF-CONVERSION OUTPUT
I/O CLOCK
CS
SYSTEM CLOCK

operating sequence

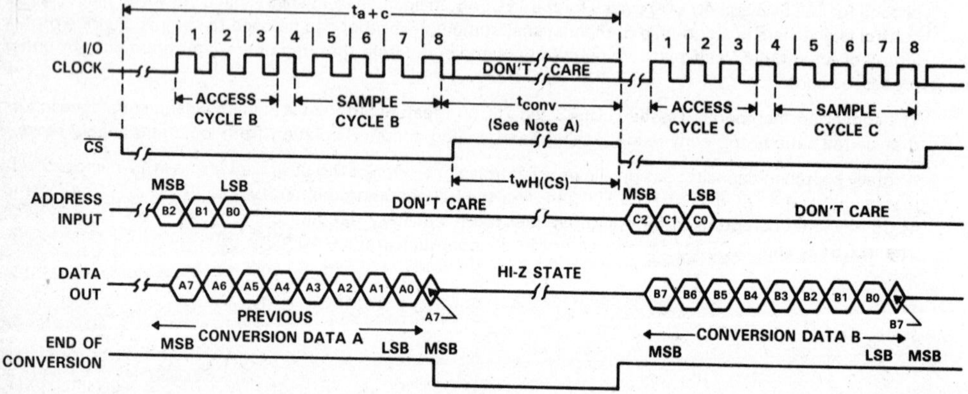
ta+c
I/O CLOCK
1 2 3 4 5 6 7 8
DON'T CARE
ACCESS CYCLE B
SAMPLE CYCLE B
tconv (See Note A)
ACCESS CYCLE C
SAMPLE CYCLE C
CS
twH(CS)
ADDRESS INPUT
MSB LSB
B2 B1 B0
DON'T CARE
C2 C1 C0
DATA OUT
A7 A6 A5 A4 A3 A2 A1 A0
A7
HI-Z STATE
B7 B6 B5 B4 B3 B2 B1 B0
B7
PREVIOUS CONVERSION DATA A
CONVERSION DATA B
END OF CONVERSION

NOTES: A. The conversion cycle, which requires 36 internal system clock periods, is initiated on the 8th falling edge of the I/O Clock after $\overline{CS}$ goes low for the channel whose address exists in memory at that time. If $\overline{CS}$ is kept low during conversion, the I/O clock must remain low for at least 36 system clock cycles to allow conversion to be completed.

B. The most significant bit (MSB) will automatically be placed on the DATA OUT bus after $\overline{CS}$ is brought low. The remaining seven bits (A6-A0) will be clocked out on the first seven I/O Clock falling edges.

C. To minimize errors caused by noise at the $\overline{CS}$ input, the internal circuitry waits for three internal system clock cycles (1.4 μs at 2 MHz) after a chip select transition before responding to control input signals. Therefore, no attempt should be made to clock-in address data until the minimum chip-select setup time has elapsed.

absolute maximum ratings over operating free-air temperature range (unless otherwise noted)

Supply voltage, V_{CC} (see Note 1) .... 6.5 V
Input voltage range (any input) .... −0.3 V to V_{CC} + 0.3 V
Output voltage range .... −0.3 V to V_{CC} + 0.3 V
Peak input current range (any input) .... ±10 mA
Peak total input current (all inputs) .... ±30 mA
Operating free-air temperature range: TLC543I, TLC544I .... −40 °C to 85 °C
TLC543M, TLC544M .... −55 °C to 125 °C
Storage temperature range .... −65 °C to 150 °C
Lead temperature 1,6 mm (1/16 inch) from case for 10 seconds: D or N package .... 260 °C
Lead temperature 1,6 mm (1/16 inch) from case for 60 seconds: J package .... 300 °C

NOTE 1: All voltages are with respect to ground (GND pin) with REF− and GND wired together (unless otherwise noted).

recommended operating conditions

		TLC543			TLC544			UNIT
		MIN	NOM	MAX	MIN	NOM	MAX	
Supply voltage, V_{CC}		3	5	6	3	5	6	V
Positive reference voltage, V_{REF+} (see Note 2)		2.5	V_{CC}	$V_{CC}+0.1$	2.5	V_{CC}	$V_{CC}+0.1$	V
Negative reference voltage, V_{REF-} (see Note 2)		−0.1	0	2.5	0.1	0	2.5	V
Differential reference voltage, $V_{REF+} - V_{REF-}$ (see Note 2)		1	V_{CC}	$V_{CC}+0.2$	1	V_{CC}	$V_{CC}+0.2$	V
Analog input voltage (see Note 2)		0		V_{CC}	0		V_{CC}	V
High-level control input voltage, V_{IH} (for V_{CC} = 4.75 to 5.5 V)		2			2			V
Low-level control input voltage, V_{IL} (for V_{CC} = 4.75 to 5.5 V)				0.8			0.8	V
Input/Output clock frequency, $f_{CLK(I/O)}$ (for V_{CC} = 4.75 to 5.5 V)		0		2.048	0		1.1	MHz
System clock frequency, $f_{CLK(I/O)}$ (for V_{CC} = 4.75 to 5.5 V)				4			2.1	MHz
Input/Output clock high, $t_{wH(I/O)}$		200			404			ns
Input/Output clock low, $t_{wL(I/O)}$		200			404			ns
I/O clock transition time (see Note 3)	$f_{CLK(I/O)} < 1.1$ MHz			100			100	ns
	$f_{CLK(I/O)} > 1.1$ MHz			40				
Duration of $\overline{CS}$ input high state during conversion, $t_{wH(CS)}$		17			17			μs
Setup time, address bits at data input before I/O CLOCK↑, $t_{su(A)}$		200			400			ns
Hold time, address bits after I/O CLOCK↑, $t_{h(A)}$		0			0			ns
Setup time, $\overline{CS}$ low before clocking in first address bits, $t_{su(CS)}$ (see Note 4)		1.4			1.4			μs
Operating free-air temperature, T_A	TLC543M, TLC544M	−55		125	−55		125	°C
	TLC543I, TLC544I	−40		85	−40		85	

NOTES: 2. Analog input voltages greater than that applied to REF+ convert as all "1"s (11111111) and input voltages less than that applied to REF− convert as all "0"s (00000000). For proper operation, REF+ voltage must be at least 1 volt higher than REF− voltage. Also, adjusted errors may increase as this differential reference voltage falls below 4.75 volts.

3. This is the time required for the clock input signal to fall from V_{IH} min to V_{IL} max or to rise from V_{IL} max to V_{IH} min. In the vicinity of normal room temperature, the devices function with input clock transitions as slow as 2 microseconds for remote data acquisition applications where the sensor and the A/D converter are placed several feet away from the controlling microprocessor.

4. To minimize errors caused by noise at the Chip Select input, the internal circuitry waits for three system clock cycles (1.4 μs at 2 MHz) after a chip select falling edge is detected before responding to control input signals. Therefore, no attempt should be made to clock-in address data until the minimum chip select setup time has elapsed.

electrical characteristics over recommended operating temperature range, $V_{CC} = V_{REF+} = 4.75$ V to 5.5 V (unless otherwise noted), $f_{CLK(I/O)} = 2.048$ MHz for TLC543 or $f_{CLK(I/O)} = 1.1$ MHz for TLC544

PARAMETER			TEST CONDITIONS		MIN	TYP†	MAX	UNIT
V_{OH}	High-level output voltage, Data out, EOC		$V_{CC} = 4.75$ V,	$I_{OH} = -360$ μA	2.4			V
V_{OL}	Low-level output voltage	Data out	$V_{CC} = 4.75$ V,	$I_{OL} = 3.2$ mA			0.4	V
		EOC	$V_{CC} = 4.75$ V,	$I_{OL} = 1.6$ mA			0.4	
I_{OZ}	Off-state (high-impedance state) output current		$V_O = V_{CC}$,	$\overline{CS}$ at V_{CC}			10	μA
			$V_O = 0$,	$\overline{CS}$ at V_{CC}			−10	
I_{IH}	High-level input current		$V_I = V_{CC} + 0.3$ V			0.005	2.5	μA
I_{IL}	Low-level input current		$V_I = 0$			−0.005	−2.5	μA
I_{CC}	Operating supply current		$\overline{CS}$ at 0 V			1.2	2	mA
I_{lkg}	Selected channel leakage current		Selected channel at V_{CC}, Unselected channel at 0 V	See Figure 1		0.4	1	μA
			Selected channel at 0 V, Unselected channel at V_{CC}			−0.4	−1	
I_{REF}	Reference current		$V_{REF+} = V_{CC}$,	$\overline{CS}$ at 0 V		0.1	1	mA
C_i	Input capacitance	Analog inputs				7	55	pF
		Control inputs				5	15	

† All typical values are at $V_{CC} = 5$ V, $T_A = 25$°C.

PARAMETER MEASUREMENT INFORMATION

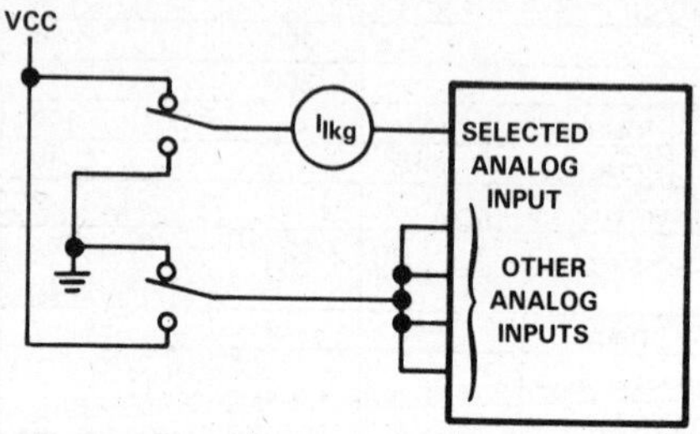
VCC
I_{lkg}
SELECTED ANALOG INPUT
OTHER ANALOG INPUTS

FIGURE 1. SELECTED CHANNEL LEAKAGE CURRENT

operating characteristics over recommended operating free-air temperature range, $V_{CC} = V_{REF+} = 4.75$ to 5.5 V, $f_{CLK(I/O)} = 2.048$ MHz for TLC543 or 1.1 MHz for TLC544

PARAMETER		TEST CONDITIONS	TLC543			TLC544			UNIT
			MIN	TYP	MAX	MIN	TYP	MAX	
	Linearity error	See Note 5			±0.5			±0.5	LSB
	Zero error	See Note 6			±0.5			±0.5	LSB
	Full-scale error	See Note 6			±0.5			±0.5	LSB
	Total unadjusted error	See Note 7			±0.5			±0.5	LSB
	Self-test output code	Input A5 address = 101_{10}, See Note 8	01111101 (125)		10000011 (131)	01111101 (125)		10000011 (131)	
t_{conv}	Conversion time	See Operating Sequence		8	17		12	17	µs
t_{a+c}	Total access and conversion time	See Operating Sequence		12	22		19	25	µs
t_{acq}	Channel acquisition time (sample cycle)	See Operating Sequence			4			4	I/O clock cycles
t_v	Time output data remains valid after I/O clock↓		10			10			ns
t_d	Delay time, I/O clock↓ to data output valid				300			400	ns
t_{en}	Output enable time	See Figure 2			1.4			1.4	ns
t_{dis}	Output disable time				150			150	ns
$t_{r(bus)}$	Data bus and EOC rise time				300			300	ns
$t_{f(bus)}$	Data bus and EOC fall time				300			300	ns
$t_{PHL(EOC)}$	Propagation delay, 8th I/O clock↓ to EOC				400			400	ns
$t_{d(EOC)}$	Delay time, EOC to DATA OUT (MSB) (see Note 9)			−1			−1		µs

NOTES: 5. Linearity error is the maximum deviation from the best straight line through the A/D transfer characteristics.
6. Zero error is the difference between the output of an ideal and an actual A/D converter for zero input voltage; full-scale error is that same difference for full-scale input voltage.
7. Total unadjusted error comprises linearity, zero, and full-scale errors.
8. Both the input address and the output codes are expressed in positive logic. The A5 analog input signal is internally generated and is used for test purposes.
9. The EOC signal is output after 40 internal clock cycles, while the data is available after 36 internal clock cycles. Thus, the delay time, EOC to DATA OUT, is a negative value equal to four internal system clock cycles less internal propagation delays.

PARAMETER MEASUREMENT INFORMATION

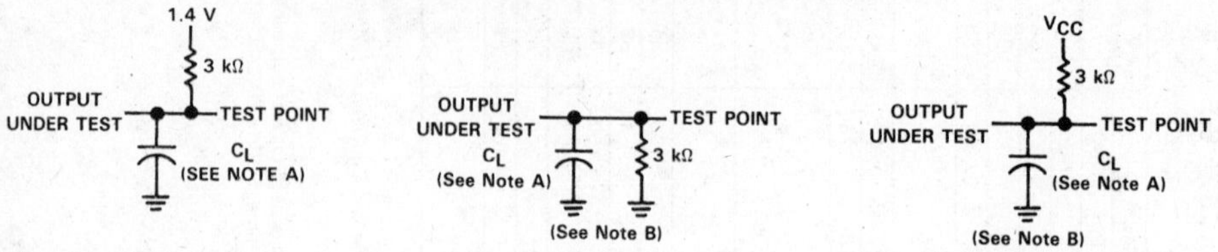
1.4 V
3 kΩ
OUTPUT UNDER TEST
TEST POINT
C_L (SEE NOTE A)
OUTPUT UNDER TEST
TEST POINT
C_L (See Note A)
3 kΩ
(See Note B)
V_{CC}
3 kΩ
OUTPUT UNDER TEST
TEST POINT
C_L (See Note A)
(See Note B)

LOAD CIRCUIT FOR t_d, t_r, and t_f

LOAD CIRCUIT FOR t_{PZH} AND t_{PHZ}

LOAD CIRCUIT FOR t_{PZL} AND t_{PLZ}

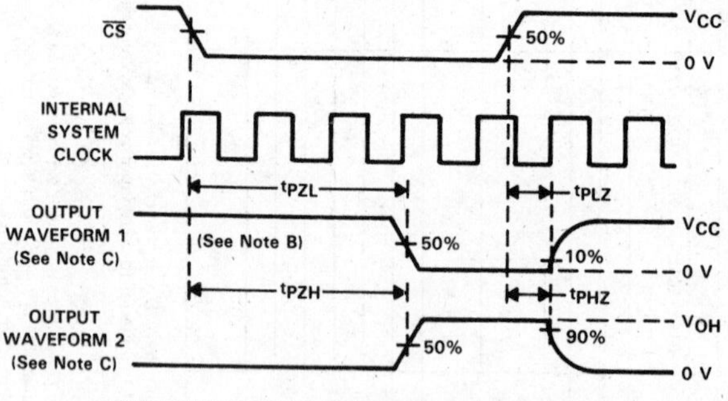
CS
V_{CC}
50%
0 V
INTERNAL SYSTEM CLOCK
t_{PZL}
t_{PLZ}
OUTPUT WAVEFORM 1 (See Note C)
(See Note B)
50%
10%
V_{CC}
0 V
t_{PZH}
t_{PHZ}
OUTPUT WAVEFORM 2 (See Note C)
50%
90%
V_{OH}
0 V

VOLTAGE WAVEFORMS FOR ENABLE AND DISABLE TIMES

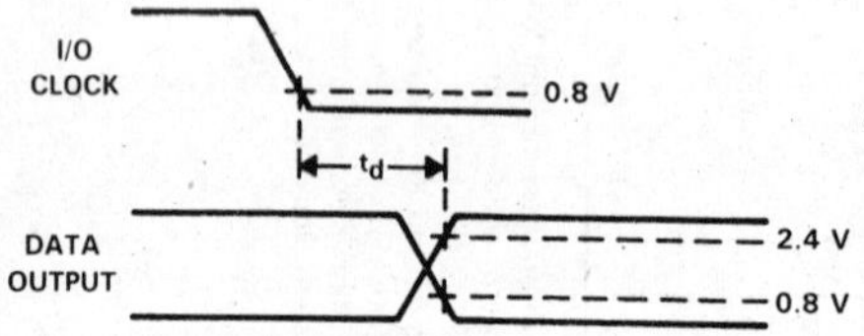
I/O CLOCK
0.8 V
t_d
DATA OUTPUT
2.4 V
0.8 V

VOLTAGE WAVEFORM FOR DELAY TIME

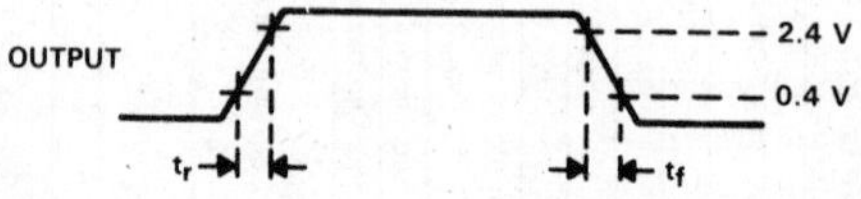
OUTPUT
2.4 V
0.4 V
t_r
t_f

VOLTAGE WAVEFORM FOR RISE AND FALL TIMES

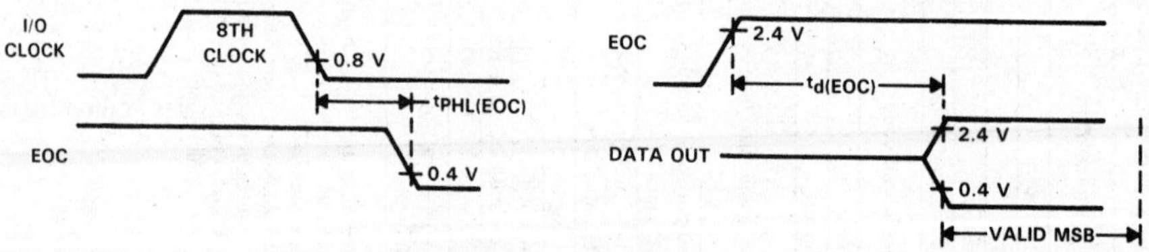
I/O CLOCK
8TH CLOCK
0.8 V
$t_{PHL(EOC)}$
EOC
0.4 V
EOC
2.4 V
$t_{d(EOC)}$
DATA OUT
2.4 V
0.4 V
VALID MSB

VOLTAGE WAVEFORMS FOR EOC TIMING

NOTES: A. C_L = 50 pF for TLC543 and 100 pF for TLC544.
B. $t_{en} = t_{PZH}$ or t_{PZL}, $t_{dis} = t_{PHZ}$ or t_{PLZ}.
C. Waveform 1 is for an output with internal conditions such that the output is low except when disabled by the output control. Waveform 2 is for an output with internal conditions such that the output is high except when disabled by the output control.

FIGURE 2. OPERATING CHARACTERISTICS

TEXAS INSTRUMENTS
POST OFFICE BOX 655012 • DALLAS, TEXAS 75265

PRINCIPLES OF OPERATION

introduction

TLC543 and TLC544 are each complete data acquisition systems on a single chip. They include the functions of analog multiplexer, sample and hold, 8-bit A/D converter, data and control registers, and control logic. Flexible serial communication is achieved with a microprocessor or microcomputer using a TTL-compatible three-state Data Out and four control lines: Chip Select ($\overline{CS}$), I/O Clock, Address Input, and End of Conversion (EOC) output.

To maximize access speed, the device simultaneously writes the previous conversion result, reads a new multiplexer address, and acquires the analog signal. This is followed by the A/D conversion, whose end is signalled by EOC output going high. These Total Access and Conversion Cycles are completed in a minimum of 22 μs for the TLC543 and 25 μs for the TLC544. Conversion can take place, in any order, on the five analog inputs or the built-in self-test system.

The system clock, which drives the control logic and the switched-capacitor successive approximation A/D converter, is internal to the device and typically runs at a frequency of 4 MHz. This internal system clock runs independently and there are no required phase or frequency relationships with other signals.

digital interface

The I/O clock controls the acquisition of the analog signal as well as all serial data communications between the TLC543 or TLC544 and the host processor. This I/O clock from the host consists of a burst of eight pulses separated by the conversion time. Timing may be achieved by chip select ($\overline{CS}$) synchronously gating a continuous I/O clock or directly from the host with $\overline{CS}$ held low continuously.

With $\overline{CS}$ high, Data Out is in a high-impedance condition with the Address Input and I/O clock input disabled. This feature allows the interface pins, with the exception of $\overline{CS}$ and EOC, to share a common bus with additional TLC543 or TLC544 devices or other members of the TLC543/544 family of devices.

typical operating sequence

Consider an access and conversion sequence where $\overline{CS}$ is being used: $\overline{CS}$ is brought low and recognized after the time out of the noise-rejection circuitry. The MSB of the previous conversion result appears at Data Out, whose three-state output is enabled. The MSB of the new multiplexer address should be present at the Address Input to conform with the setup time, $t_{su(A)}$, requirements before the first rising edge of the I/O clock. The multiplexer address is shifted in on the first three rising edges of the I/O clock.

The first seven falling edges of I/O CLOCK shift out the remaining seven bits of the previous conversion on DATA OUT. The eighth I/O clock falling edge returns the MSB to the Data Out. Optimum serial transfer takes place with the bit streams being read on the rising edges of the I/O clock for the respective devices and Data Out and Address In lines.

At the fourth falling edge of the I/O clock, the on-chip sample and hold begins to acquire the newly addressed analog input and continues until the eighth (and final) falling edge. A hold function is initiated by the eighth I/O clock pulse falling edge. If it is desired to start the conversion at a specific point in time (or lengthen the acquisition time), the host processor may leave the eighth I/O clock pulse in the high state until the moment at which the analog signal must be sampled. After bringing the eighth I/O pulse low, the A/D function is performed in the next 36 internal system clock cycles.

In applications where $\overline{CS}$ is held low continuously, the bursts of eight I/O clock pulses should be timed to be at least t_{conv} apart.

$\overline{CS}$ input

To minimize bus contention caused by noise enabling the three-state Data Out, when the $\overline{CS}$ input is brought low the device waits for two rising edges and a falling edge of the internal system clock before recognizing the $\overline{CS}$ transition. Hence, the setup time $t_{su(CS)}$ should be observed when using the $\overline{CS}$ input. This applies also to a $\overline{CS}$ high-to-low transition, except for disabling of DATA OUT, which goes into a high-impedance state immediately within the t_{dis} specification (see Figure 3). If this interruption of $\overline{CS}$ in the low state is less than 1.5 internal system clock cycles, and hence not recognized, DATA OUT will be immediately enabled with the return of $\overline{CS}$ to the low state. DATA OUT becomes enabled after a $\overline{CS}$ high-to-low transition in time t_{en} (equivalent to $t_{su(CS)}$ for this device, see Figure 3).

$\overline{CS}$ can be brought high during a conversion without affecting the ongoing conversion but must remain high until the end of conversion. Otherwise, a $\overline{CS}$ falling edge will cause a reset condition that will abort the conversion in progress. When a new access cycle is started, the previous conversion result will be output.

A new conversion may be restarted by toggling $\overline{CS}$ high-to-low at least $t_{su(CS)}$ before the eighth falling edge of the I/O clock. The ongoing access cycle will be aborted. Again, when a new access cycle is started, the previous conversion result will be output.

end of conversion output (EOC)

EOC goes low a propagation delay time $t_{PHL(EOC)}$ after the 8th falling edge of the I/O clock, and goes high when conversion is complete. At this time the MSB is available at Data Out; however, if $\overline{CS}$ is high it will be necessary to bring $\overline{CS}$ low and wait for the $\overline{CS}$ recognition time before Data Out is available, since Data Out is in a high-impedance state when $\overline{CS}$ is high. Delay time $t_{d(EOC)}$ of EOC to Data Out is a negative value of 4 internal system clock cycles less internal propagation delay, because the EOC signal is output after 40 internal system clock cycles whereas conversion is complete with data available after 36 cycles.

TLC545M, TLC545I, TLC546M, TLC546I LinCMOS™ 8-BIT ANALOG-TO-DIGITAL PERIPHERALS WITH SERIAL CONTROL AND 19 INPUTS

D2850, DECEMBER 1985

- LinCMOS™ Technology
- 8-Bit Resolution A/D Converter
- Microprocessor Peripheral or Stand-Alone Operation
- On-Chip 20-Channel Analog Multiplexer
- Built-In Self-Test Mode
- Software-Controllable Sample and Hold
- Total Unadjusted Error . . . ±0.5 LSB Max
- Timing and Control Signals Compatible with 8-Bit TLC540 and 10-Bit TLC1540 A/D Converter Families

TYPICAL PERFORMANCE	TL545	TL546
Channel Acquisition Time	1.5 μs	2.7 μs
Conversion Time	9 μs	17 μs
Sampling Rate	76 × 10^3	40 × 10^3
Power Dissipation	6 mW	6 mW

N DUAL-IN-LINE PACKAGE
(TOP VIEW)

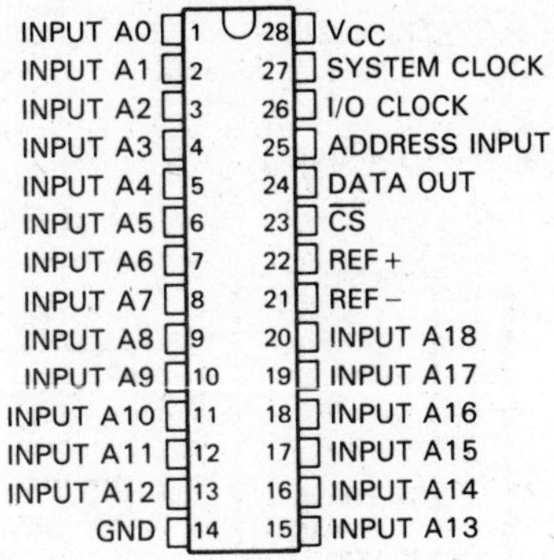
INPUT A0 1
INPUT A1 2
INPUT A2 3
INPUT A3 4
INPUT A4 5
INPUT A5 6
INPUT A6 7
INPUT A7 8
INPUT A8 9
INPUT A9 10
INPUT A10 11
INPUT A11 12
INPUT A12 13
GND 14
28 VCC
27 SYSTEM CLOCK
26 I/O CLOCK
25 ADDRESS INPUT
24 DATA OUT
23 CS
22 REF+
21 REF−
20 INPUT A18
19 INPUT A17
18 INPUT A16
17 INPUT A15
16 INPUT A14
15 INPUT A13

FN CHIP CARRIER PACKAGE
(TOP VIEW)

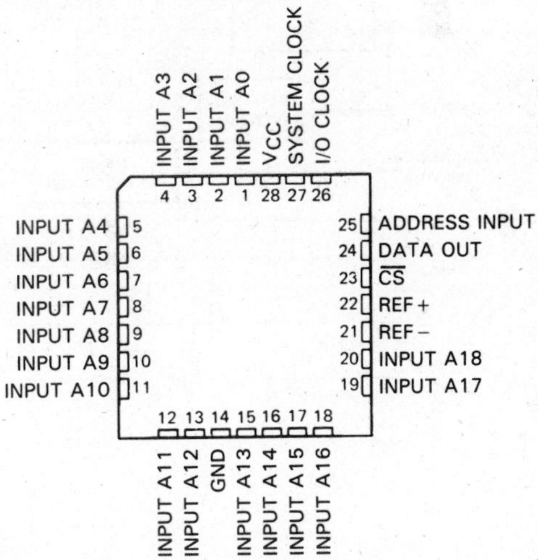
INPUT A3
INPUT A2
INPUT A1
INPUT A0
VCC
SYSTEM CLOCK
I/O CLOCK
4 3 2 1 28 27 26
INPUT A4 5
INPUT A5 6
INPUT A6 7
INPUT A7 8
INPUT A8 9
INPUT A9 10
INPUT A10 11
25 ADDRESS INPUT
24 DATA OUT
23 CS
22 REF+
21 REF−
20 INPUT A18
19 INPUT A17
12 13 14 15 16 17 18
INPUT A11
INPUT A12
GND
INPUT A13
INPUT A14
INPUT A15
INPUT A16

description

The TLC545 and TLC546 are LinCMOS™ A/D peripherals built around an 8-bit switched-capacitor successive-approximation A/D converter. They are designed for serial interface to a microprocesor or peripheral via a three-state output with up to four control inputs [including independent System Clock, I/O Clock, Chip Select ($\overline{CS}$), and Address Input]. A 4-megahertz system clock for the TLC545 and a 2.1-megahertz system clock for the TLC546 with a design that includes simultaneous read/write operation allow high-speed data transfers and sample rates of up to 76,923 samples per second for the TLC545, and 40,000 samples per second for the TLC546. In addition to the high-speed converter and versatile control logic, there is an on-chip 20-channel analog multiplexer that can be used to sample any one of 19 inputs or an internal "self-test" voltage, and a sample-and-hold that can operate automatically or under microprocessor control.

The converters incorporated in the TLC545 and TLC546 feature differential high-impedance reference inputs that facilitate ratiometric conversion, scaling, and analog circuitry isolation from logic and supply noises. A totally switched-capacitor design allows guaranteed low-error (±0.5 LSB) conversion in 9 microseconds for

LinCMOS is a trademark of Texas Instruments Incorporated

the TLC545, and 17 microseconds for the TLC546 over the full operating temperature range. Detailed information on interfacing to most popular microprocessors is readily available from the factory.

The TLC545M and the TLC546M are characterized for operation from −55 °C to 125 °C. The TLC545I and the TLC546I are characterized for operation from −40 °C to 85 °C.

functional block diagram

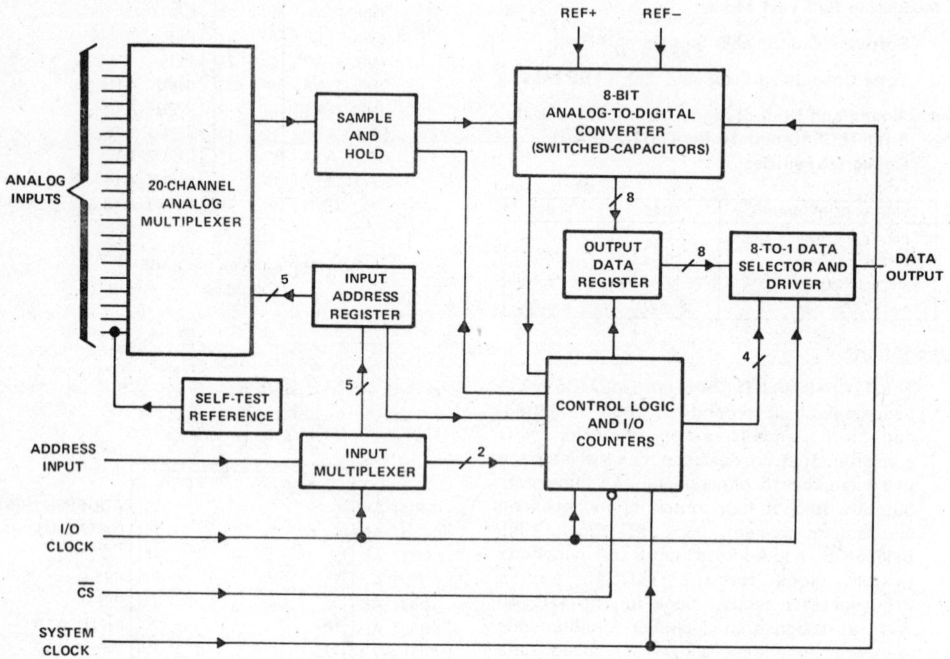
REF+
REF−
8-BIT ANALOG-TO-DIGITAL CONVERTER (SWITCHED-CAPACITORS)
SAMPLE AND HOLD
ANALOG INPUTS
20-CHANNEL ANALOG MULTIPLEXER
8
OUTPUT DATA REGISTER
8
8-TO-1 DATA SELECTOR AND DRIVER
DATA OUTPUT
5
INPUT ADDRESS REGISTER
4
5
SELF-TEST REFERENCE
CONTROL LOGIC AND I/O COUNTERS
ADDRESS INPUT
INPUT MULTIPLEXER
2
I/O CLOCK
CS
SYSTEM CLOCK

operating sequence

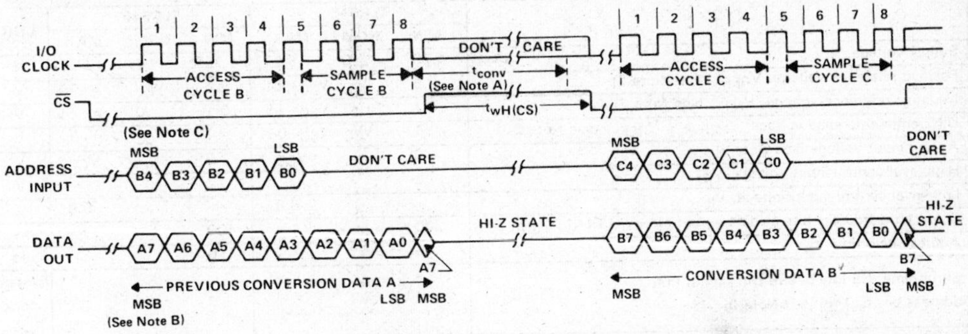
I/O CLOCK
1 2 3 4 5 6 7 8
DON'T CARE
ACCESS CYCLE B
SAMPLE CYCLE B
tconv (See Note A)
ACCESS CYCLE C
SAMPLE CYCLE C
CS
twH(CS)
(See Note C)
ADDRESS INPUT
MSB
LSB
B4 B3 B2 B1 B0
DON'T CARE
C4 C3 C2 C1 C0
DON'T CARE
DATA OUT
A7 A6 A5 A4 A3 A2 A1 A0
A7
HI-Z STATE
B7 B6 B5 B4 B3 B2 B1 B0
B7
PREVIOUS CONVERSION DATA A
CONVERSION DATA B
(See Note B)

NOTES: A. The conversion cycle, which requires 36 system clock periods, is initiated with the 8th I/O clock↓ after CS↓ for the channel whose address exists in memory at that time.

B. The most significant bit (MSB) will automatically be placed on the DATA OUT bus after $\overline{CS}$ is brought low. The remaining seven bits (A6-A0) will be clocked out on the first seven I/O clock falling edges.

C. To minimize errors caused by noise at the $\overline{CS}$ input, the internal circuitry waits for three system clock cycles (or less) after a chip select transition before responding to control input signals. Therefore, no attempt should be made to clock-in address data until the minimum chip-select setup time has elapsed.

absolute maximum ratings over operating free-air temperature range (unless otherwise noted)

Supply voltage, V_{CC} (see Note 1)	6.5 V
Input voltage range (any input)	−0.3 V to V_{CC} +0.3 V
Output voltage range	−0.3 V to V_{CC} +0.3 V
Peak input current range (any input)	±10 mA
Peak total input current (all inputs)	±30 mA
Operating free-air temperature range: TLC545I, TLC546I	−40°C to 85°C
TLC545M, TLC546M	−55°C to 125°C
Storage temperature range	−65°C to 150°C
Lead temperature 1,6 mm (1/16 inch) from case for 10 seconds: N package	260°C
Case temperature for 10 seconds: FN package	260°C

NOTE 1: All voltage values are with respect to network ground terminal.

recommended operating conditions

			TLC545 MIN	TLC545 NOM	TLC545 MAX	TLC546 MIN	TLC546 NOM	TLC546 MAX	UNIT
Supply voltage, V_{CC}			4.75	5	5.5	4.75	5	5.5	V
Positive reference voltage, V_{REF+} (see Note 2)			2.5	V_{CC}	$V_{CC}+0.1$	2.5	V_{CC}	$V_{CC}+0.1$	V
Negative reference voltage, V_{ref-} (see Note 2)			−0.1	0	$V_{CC}-2.5$	−0.1	0	$V_{CC}-2.5$	V
Differential reference voltage, $V_{REF+} - V_{REF-}$ (see Note 2)			1	V_{CC}	$V_{CC}+0.2$	1	V_{CC}	$V_{CC}+0.2$	V
Analog input voltage (see Note 2)			0		V_{CC}	0		V_{CC}	V
High-level control input voltage, V_{IH}			2			2			V
Low-level control input voltage, V_{IL}					0.8			0.8	V
Setup time, address bits at data input before I/O CLK↑, $t_{su(A)}$			200			400			ns
Address hold time, t_h			0			0			ns
Setup time, $\overline{CS}$ low before clocking in first address bit, $t_{su(CS)}$ (see Note 3)			3			3			System clock cycles
Chip select high during conversion, $t_{wH(CS)}$			36			36			System clock cycles
Input/Output clock frequency, $f_{CLK(I/O)}$			0		2.048	0		1.1	MHz
System clock frequency, $f_{CLK(SYS)}$			$f_{CLK(I/O)}$		4	$f_{CLK(I/O)}$		2.1	MHz
System clock high, $t_{wH(SYS)}$			110			210			ns
System clock low, $t_{wL(SYS)}$			100			190			ns
Input/Output clock high, $t_{wH(I/O)}$			200			404			ns
Input/Output clock low, $t_{wL(I/O)}$			200			404			ns
Clock transition time (see Note 4)	System	$f_{CLK(SYS)} \leq 1048$ kHz			30			30	ns
		$f_{CLK(SYS)} > 1048$ kHz			20			20	
	I/O	$f_{CLK(I/O)} \leq 525$ kHz			100			100	ns
		$f_{CLK(I/O)} > 525$ kHz			40			40	
Operating free-air temperature, T_A		TLC545M, TLC546M	−55		125	−55		125	°C
		TLC545I, TLC546I	−40		85	−40		85	

NOTES: 2. Analog input voltages greater than that applied to REF+ convert as all ''1''s (11111111), while input voltages less than that applied to REF− convert as all ''0''s (00000000). For proper operation, REF+ voltage must be at least 1 volt higher than REF− voltage. Also, total unadjusted errors may increase as this differential reference voltage falls below 4.75 volts.
3. To minimize errors caused by noise at the Chip Select input, the internal circuitry waits for three system clock cycles (or less) after a chip select falling edge or rising edge is detected before responding to control input signals. Therefore, no attempt should be made to clock-in address data until the minimum chip select setup time has elapsed.
4. This is the time required for the clock input signal to fall from V_{IH} min to V_{IL} max or to rise from V_{IL} max to V_{IH} min. In the vicinity of normal room temperature, the devices function with input clock transition time as slow as 2 microseconds for remote data acquisition applications where the sensor and the A/D converter are placed several feet away from the controlling microprocessor.

electrical characteristics over recommended operating temperature range, $V_{CC} = V_{REF+} = 4.75$ V to 5.5 V (unless otherwise noted), $f_{CLK(I/O)} = 2.048$ MHz for TLC545 or $f_{CLK(I/O)} = 1.1$ MHz for TLC546

PARAMETER			TEST CONDITIONS		MIN	TYP†	MAX	UNIT
V_{OH}	High-level output voltage (pin 24)		$V_{CC} = 4.75$ V,	$I_{OH} = -360$ μA	2.4			V
V_{OL}	Low-level output voltage		$V_{CC} = 4.75$ V,	$I_{OL} = 3.2$ mA			0.4	V
I_{OZ}	Off-state (high-impedance state) output current		$V_O = V_{CC}$,	$\overline{CS}$ at V_{CC}			10	μA
			$V_O = 0$,	$\overline{CS}$ at V_{CC}			−10	μA
I_{IH}	High-level input current		$V_I = V_{CC}$			0.005	2.5	μA
I_{IL}	Low-level input current		$V_I = 0$			−0.005	−2.5	μA
I_{CC}	Operating supply current		$\overline{CS}$ at 0 V			1.2	2.5	mA
	Selected channel leakage current		Selected channel at V_{CC}, Unselected channel at 0 V			0.4	1	μA
			Selected channel at 0 V, Unselected channel at V_{CC}			−0.4	−1	μA
$I_{CC} + I_{REF}$	Supply and reference current		$V_{REF+} = V_{CC}$,	$\overline{CS}$ at 0 V		1.3	3	mA
C_i	Input capacitance	Analog inputs				7	55	pF
		Control inputs				5	15	pF

†All typical values are at $T_A = 25°C$.

operating characteristics over recommended operating free-air temperature range, $V_{CC} = V_{REF+} = 4.75$ V to 5.5 V, $f_{CLK(I/O)} = 2.048$ MHz for TLC545 or 1.1 MHz for TLC546, $f_{CLK(SYS)} = 4$ MHz for TLC545 or 2.1 MHz for TLC546

PARAMETER		TEST CONDITIONS	TLC545			TLC546			UNIT
			MIN	TYP	MAX	MIN	TYP	MAX	
	Linearity error	See Note 5			±0.5			±0.5	LSB
	Zero error	See Note 6			±0.5			±0.5	LSB
	Full-scale error	See Note 6			±0.5			±0.5	LSB
	Total unadjusted error	See Note 7			±0.5			±0.5	LSB
	Self-test output code	Input A19 address = 10011 (See Note 8)	01111101 (125)		10000011 (131)	01111101 (125)		10000011 (131)	
t_{conv}	Conversion time	See Operating Sequence			9			17	μs
	Total access and conversion time	See Operating Sequence			13			25	μs
t_{acq}	Channel acquisition time (sample cycle)	See Operating Sequence			3			3	I/O clock cycles
t_v	Time output data remains valid after I/O clock↓		10			10			ns
t_d	Delay time, I/O clock↓ to data output valid	See Parameter Measurement Information			300			400	ns
t_{en}	Output enable time				150			150	ns
t_{dis}	Output disable time				150			150	ns
$t_{r(bus)}$	Data bus rise time				300			300	ns
$t_{f(bus)}$	Data bus fall time				300			300	ns

NOTES: 5. Linearity error is the maximum deviation from the best straight line through the A/D transfer characteristics.
6. Zero error is the difference between 00000000 and the converted output for zero input voltage; full-scale error is the difference between 11111111 and the converted for full-scale input voltage.
7. Total unadjusted error is the sum of linearity, zero, and full-scale errors.
8. Both the input address and the output codes are expressed in positive logic. The A19 analog input signal is internally generated and is used for test purposes.

PARAMETER MEASUREMENT INFORMATION

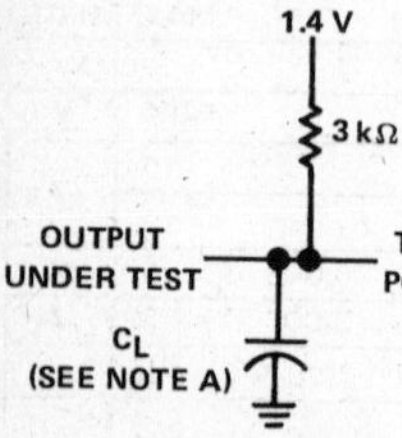
1.4 V
3 kΩ
OUTPUT
UNDER TEST
TEST
POINT
C_L
(SEE NOTE A)

LOAD CIRCUIT FOR
t_d, t_r, AND t_f

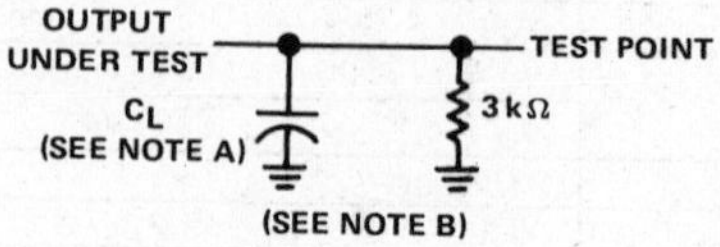
OUTPUT
UNDER TEST
TEST POINT
C_L
(SEE NOTE A)
3 kΩ
(SEE NOTE B)

LOAD CIRCUIT FOR
t_{PZH} AND t_{PHZ}

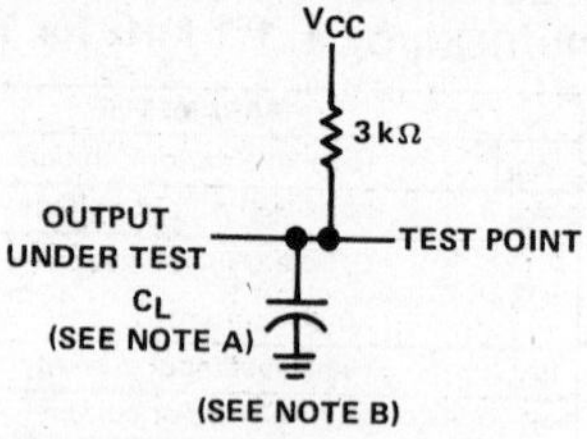
V_{CC}
3 kΩ
OUTPUT
UNDER TEST
TEST POINT
C_L
(SEE NOTE A)
(SEE NOTE B)

LOAD CIRCUIT FOR
t_{PZL} AND t_{PLZ}

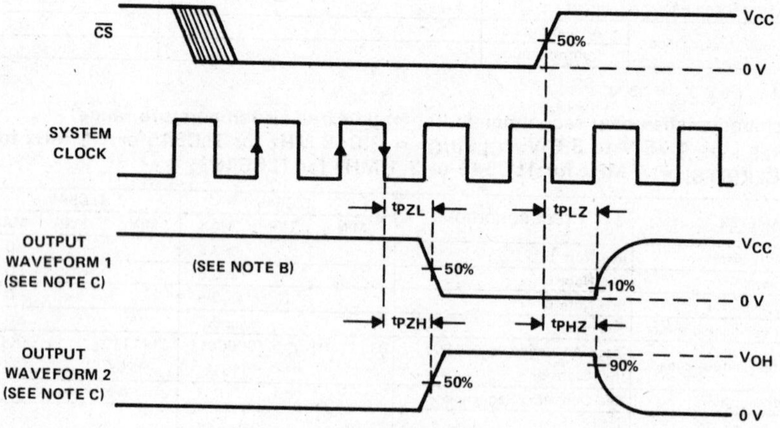
CS
SYSTEM
CLOCK
OUTPUT
WAVEFORM 1
(SEE NOTE C)
(SEE NOTE B)
OUTPUT
WAVEFORM 2
(SEE NOTE C)
V_{CC}
50%
0 V
t_{PZL}
t_{PLZ}
V_{CC}
50%
10%
0 V
t_{PZH}
t_{PHZ}
V_{OH}
90%
50%
0 V

VOLTAGE WAVEFORMS FOR ENABLE AND DISABLE TIMES

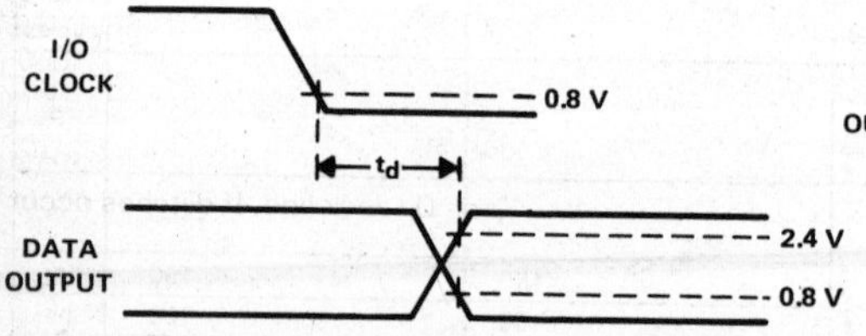
I/O
CLOCK
0.8 V
t_d
DATA
OUTPUT
2.4 V
0.8 V

VOLTAGE WAVEFORM FOR DELAY TIME

OUTPUT
2.4 V
0.4 V
t_r
t_f

VOLTAGE WAVEFORM FOR
RISE AND FALL TIMES

NOTES: A. C_L = 50 pF for TLC545 and 100 pF for TLC546
B. t_{en} = t_{PZH} or t_{PZL}, t_{dis} = t_{PHZ} or t_{PLZ}
C. Waveform 1 is for an output with internal conditions such that the output is low except when disabled by the output control.
Waveform 2 is for an output with internal conditions such that the output is high except when disabled by the output control.

principles of operation

The TLC545 and TLC546 are both complete data acquisition systems on single chips. Each includes such functions as system clock, sample-and-hold, 8-bit A/D converter, data and control registers, and control logic. For flexibility and access speed, there are four control inputs; Chip Select ($\overline{CS}$), Address Input, I/O clock, and System clock. These control inputs and a TTL-compatible 3-state output facilitate serial communications with a microprocessor or microcomputer. The TLC545 and TLC546 can complete conversions in a maximum of 9 and 17 microseconds respectively, while complete input-conversion-output cycles can be repeated at a miximum of 13 and 25 microseconds, respectively.

The System and I/O clocks are normally used independently and do not require any special speed or phase relationships between them. This independence simplifies the hardware and software control tasks for the device. Once a clock signal within the specification range is applied to the System clock input, the control hardware and software need only be concerned with addressing the desired analog channel, reading the previous conversion result, and starting the conversion by using the I/O clock. The System clock will drive the ''conversion crunching'' circuitry so that the control hardware and software need not be concerned with this task.

When $\overline{CS}$ is high, the Data Output pin is in a high-impedance condition, and the Address Input and I/O Clock pins are disabled. This feature allows each of these pins, with the exception of the $\overline{CS}$, to share a control logic point with their counterpart pins on additional A/D devices when additional TLC545/TLC546 devices are used. Thus, the above feature serves to minimize the required control logic pins when using multiple A/D devices.

The control sequence has been designed to minimize the time and effort required to initiate conversion and obtain the conversion result. A normal control sequence is:

1. $\overline{CS}$ is brought low. To minimize errors caused by noise at the $\overline{CS}$ input, the internal circuitry waits for two rising edges and then a falling edge of the System clock after a $\overline{CS}$ transition before the transition is recognized. The MSB of the previous conversion result will automatically appear on the Data Out pin.
2. A new positive-logic multiplexer address is shifted in on the first five rising edges of the I/O clock. The MSB of the address is shifted in first. The negative edges of these five I/O clocks shift out the 2nd, 3rd, 4th, 5th, and 6th most significant bits of the previous conversion result. The on-chip sample-and hold begins sampling the newly addressed analog input after the 5th falling edge. The sampling operation basically involves the charging of internal capacitors to the level of the analog input voltage.
3. Two clock cycles are then applied to the I/O pin and the 7th and 8th conversion bits are shifted out on the negative edges of these clock cycles.
4. The final 8th clock cycle is applied to the I/O clock pin. The falling edge of this clock cycle completes the analog sampling process and initiates the hold function. Conversion is then performed during the next 36 system clock cycles. After this final I/O clock cycle, $\overline{CS}$ must go high or the I/O clock must remain low for at least 36 system clock cycles to allow for the conversion function.

$\overline{CS}$ can be kept low during periods of multiple conversion. When keeping $\overline{CS}$ low during periods of multiple conversion, special care must be exercised to prevent noise glitches on the I/O Clock line. If glitches occur on the I/O Clock line, the I/O sequence between the microprocessor/controller and the device will lose synchronization. Also, if $\overline{CS}$ is taken high, it must remain high until the end of conversion. Otherwise, a valid falling edge of $\overline{CS}$ will cause a reset condition, which will abort the conversion in progress.

A new conversion may be started and the ongoing conversion simultaneously aborted by performing steps 1 through 4 before the 36 system clock cycles occur. Such action will yield the conversion result of the previous conversion and not the ongoing conversion.

It is possible to connect the system and I/O clocks together in special situations in which controlling circuitry points must be minimized. In this case, the following special points must be considered in addition to the requirements of the normal control sequence previously described.

1. When $\overline{CS}$ is recognized by the device to be at a low level, the common clock signal is used as an I/O clock. When the $\overline{CS}$ is recognized by the device to be at a high level, the common clock signal is used to drive the ''conversion crunching'' circuitry.
2. The device will recognize a $\overline{CS}$ transition only when the $\overline{CS}$ input changes and subsequently the system clock pin receives two positive edges and then a negative edge. For this reason, after a $\overline{CS}$ negative edge, the first two clock cycles will not shift in the address because a low $\overline{CS}$ must be recognized before the I/O clock can shift in an analog channel address. Also, upon shifting in the address, $\overline{CS}$ must be raised after the 6th I/O clock, which has been recognized by the device, so that a $\overline{CS}$ low level will be recognized upon the lowering of the 8th I/O clock signal recognized by the device. Otherwise, additional common clock cycles will be recognized as I/O clocks and will shift in an erroneous address.

For certain applications, such as strobing applications, it is necessary to start conversion at a specific point in time. This device will accommodate these applications. Although the on-chip sample-and-hold begins sampling upon the negative edge of the 5th I/O clock cycle, the hold function is not initiated until the negative edge of the 8th I/O clock cycle. Thus, the control circuitry can leave the I/O clock signal in its high state during the 8th I/O clock cycle, until the moment at which the analog signal must be converted. The TLC545/546 will continue sampling the analog input until the 8th falling edge of the I/O clock. The control circuitry or software must then immediately lower the I/O clock signal to initiate the hold function at the desired point in time and to start conversion.

Detailed information on interfacing to most popular microprocesors is readily available from the factory.

TLC548, TLC549
LinCMOS™ 8-BIT ANALOG-TO-DIGITAL PERIPHERAL WITH SERIAL CONTROL

D2816, NOVEMBER 1983—REVISED JULY 1986

TLC548M, TLC549M . . . D OR P PACKAGE
TLC548I, TLC549I . . . D OR P PACKAGE
TLC548C, TLC549C . . . D PACKAGE
(TOP VIEW)

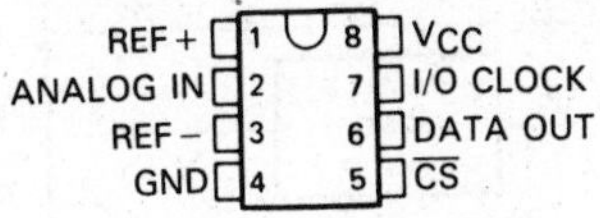
REF+ 1
ANALOG IN 2
REF− 3
GND 4
8 VCC
7 I/O CLOCK
6 DATA OUT
5 CS

- LinCMOS™ Technology
- Microprocessor Peripheral or Stand-Alone Operation
- 8-Bit Resolution A/D Converter
- Differential Reference Input Voltages
- Conversion Time . . . 17 μs Max
- Total Access and Conversion Cycles Per Second
  TLC548 . . . up to 45,500
  TLC549 . . . up to 40,000
- On-Chip Software-Controllable Sample-and-Hold
- Total Unadjusted Error . . . ±0.5 LSB Max
- 4-MHz Typical Internal System Clock
- Wide Supply Range . . . 3 V to 6 V
- Low Power Consumption . . . 6 mW Typ
- Ideal for Cost-Effective, High-Performance Applications Including Battery-Operated Portable Instrumentation
- Pinout and Control Signals Compatible with the TLC540 and TLC545 8-Bit A/D Converters and with the TLC1540 10-Bit A/D Converter

description

The TLC548 and TLC549 are LinCMOS™ A/D peripheral integrated circuits built around an 8-bit switched-capacitor successive-approximation ADC. They are designed for serial interface with a microprocessor or peripheral through a 3-state data output and an analog input. The TLC548 and TLC549 use only the Input/Output Clock (I/O Clock) input along with the Chip Select ($\overline{CS}$) input for data control. The maximum I/O clock input frequency of the TLC548 is guaranteed up to 2.048 megahertz, and the I/O clock input frequency of the TLC549 is guaranteed to 1.1 megahertz. Detailed information on interfacing to most popular microprocessors is readily available from the factory.

Operation of the TLC548 and the TLC549 is very similar to that of the more complex TLC540 and TLC541 devices; however, the TLC548 and TLC549 provide an on-chip system clock that operates typically at 4 megahertz and requires no external components. The on-chip system clock allows internal device operation to proceed independently of serial input/output data timing and permits manipulation of the TLC548 and TLC549 as desired for a wide range of software and hardware requirements. The I/O Clock together with the internal system clock allow high-speed data transfer and conversion rates of 45,500 conversions per second for the TLC548, and 40,000 conversions per second for the TLC549.

Additional TLC548 and TLC549 features include versatile control logic, an on-chip sample-and-hold circuit that can operate automatically or under microprocessor control, and a high-speed converter with differential high-impedance reference voltage inputs that ease ratiometric conversion, scaling, and circuit isolation from logic and supply noises. Design of the totally switched-capacitor successive-approximation converter circuit allows conversion with a maximum total error of ±0.5 least significant bit (LSB) in less than 17 microseconds.

The TLC548M and TLC549M are available in the D or P plastic package and are characterized for operation over the temperature range of −55°C to 125°C. The TLC548I and TLC549I are characterized for operation from −40°C to 85°C. The TLC548C and TLC549C are characterized for operation from 0°C to 70°C.

LinCMOS is a trademark of Texas Instruments.

functional block diagram

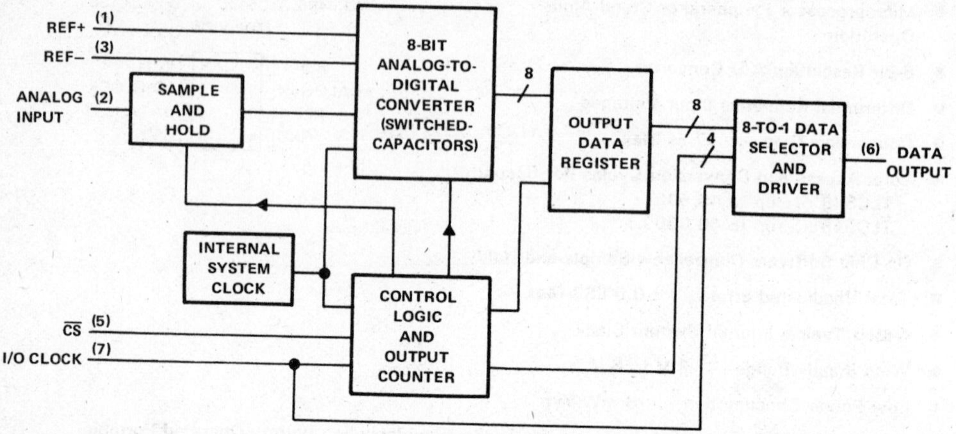
REF+ (1)
REF− (3)
ANALOG INPUT (2)
SAMPLE AND HOLD
8-BIT ANALOG-TO-DIGITAL CONVERTER (SWITCHED-CAPACITORS)
8
OUTPUT DATA REGISTER
8
4
8-TO-1 DATA SELECTOR AND DRIVER
(6) DATA OUTPUT
INTERNAL SYSTEM CLOCK
CONTROL LOGIC AND OUTPUT COUNTER
CS (5)
I/O CLOCK (7)

operating sequence

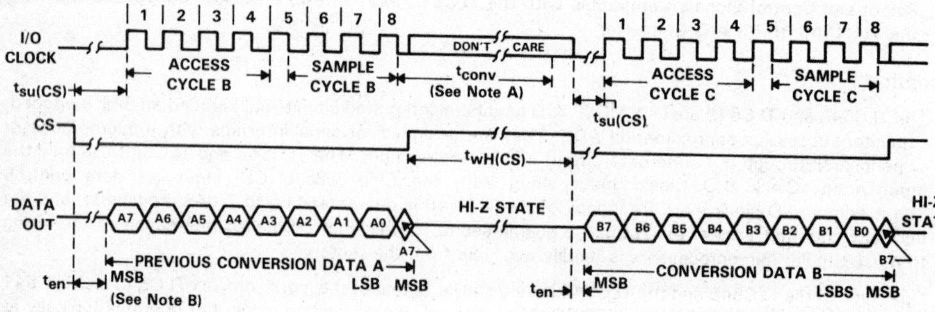
I/O CLOCK
1 2 3 4 5 6 7 8
DON'T CARE
ACCESS CYCLE B
SAMPLE CYCLE B
tconv (See Note A)
ACCESS CYCLE C
SAMPLE CYCLE C
tsu(CS)
CS
twH(CS)
DATA OUT
A7 A6 A5 A4 A3 A2 A1 A0
HI-Z STATE
B7 B6 B5 B4 B3 B2 B1 B0
PREVIOUS CONVERSION DATA A
A7
LSB MSB
ten
MSB (See Note B)
CONVERSION DATA B
B7
LSBS MSB

NOTES: A. The conversion cycle, which requires 36 internal system clock periods (17 μs maximum), is initiated with the 8th I/O clock pulse trailing edge after $\overline{CS}$ goes low for the channel whose address exists in memory at the time.

B. The most significant bit (A7) will automatically be placed on the DATA OUT bus after $\overline{CS}$ is brought low. The remaining seven bits (A6-A0) will be clocked out on the first seven I/O clock falling edges. B7-B0 will follow in the same manner.

TEXAS INSTRUMENTS
POST OFFICE BOX 655012 • DALLAS, TEXAS 75265

absolute maximum ratings over operating free-air temperature range (unless otherwise noted)

Supply voltage, V_{CC} (see Note 1) .... 6.5 V
Input voltage range at any input .... −0.3 V to V_{CC} + 0.3 V
Output voltage range .... −0.3 V to V_{CC} + 0.3 V
Peak input current range (any input) .... ±10 mA
Peak total input current range (all inputs) .... ±30 mA
Operating free-air temperature range (see Note 2): TLC548M, TLC549M .... −55°C to 125°C
TLC548I, TLC549I .... −40°C to 85°C
TLC548C, TLC549C .... 0°C to 70°C
Storage temperature range .... −65°C to 150°C
Lead temperature 1,6 mm (1/16 inch) from case for 10 seconds .... 260°C

NOTES: 1. All voltage values are with respect to the network ground terminal with the REF− and GND terminal pins connected together, unless otherwise noted.
2. The D package is not guaranteed below −40°C.

recommended operating conditions

		TLC548			TLC549			UNIT
		MIN	NOM	MAX	MIN	NOM	MAX	
Supply voltage, V_{CC}		3	5	6	3	5	6	V
Positive reference voltage, V_{REF+} (see Note 3)		2.5	V_{CC}	V_{CC}+0.1	2.5	V_{CC}	V_{CC}+0.1	V
Negative reference voltage, V_{REF-} (see Note 3)		−0.1	0	2.5	−0.1	0	2.5	V
Differential reference voltage, V_{REF+}, V_{REF-} (see Note 3)		1	V_{CC}	V_{CC}+0.2	1	V_{CC}	V_{CC}+0.2	V
Analog input voltage (see Note 3)		0		V_{CC}	0		V_{CC}	V
High-level control input voltage, V_{IH} (for V_{CC} = 4.75 V to 5.5 V)		2			2			V
Low-level control input voltage, V_{IL} (for V_{CC} = 4.75 V to 5.5 V)				0.8			0.8	V
Input/output clock frequency, $f_{CLK(I/O)}$ (for V_{CC} = 4.75 V to 5.5 V)		0		2.048	0		1.1	MHz
Input/output clock high, $t_{wH(I/O)}$ (for V_{CC} = 4.75 V to 5.5 V)		200			404			ns
Input/output clock low, $t_{wL(I/O)}$ (for V_{CC} = 4.75 V to 5.5 V)		200			404			ns
Input/output clock transition time, $t_{t(I/O)}$ (see Note 4) (for V_{CC} = 4.75 V to 5.5 V)				100			100	ns
Duration of $\overline{CS}$ input high state during conversion, $t_{wH(CS)}$ (for V_{CC} = 4.75 V to 5.5 V)		17			17			μs
Setup time, $\overline{CS}$ low before first I/O clock, $t_{su(CS)}$ (for V_{CC} = 4.75 V to 5.5 V) (see Note 5)		1.4			1.4			μs
Operating free-air temperature, T_A	TLC548M, TLC549M	−55		125	−55		125	°C
	TLC548I, TLC549I	−40		85	−40		85	°C
	TLC548C, TLC549C	0		70	0		70	°C

NOTES: 3. Analog input voltages greater than that applied to REF+ convert to all ones (11111111), while input voltages less than that applied to REF− convert to all zeros (00000000). For proper operation, the positive reference voltage V_{REF+}, must be at least 1 volt greater than the negative reference voltage V_{REF-}. In additon, unadjusted errors may increase as the differential reference voltage $V_{REF+} - V_{REF-}$ falls below 4.75 V.
4. This is the time required for the input/output clock input signal to fall from V_{IH} min to V_{IL} max or to rise from V_{IL} max to V_{IH} min. In the vicinity of normal room temperature, the devices function with input clock transition time as slow as 2 μs for remote data acquisition applications in which the sensor and the ADC are placed several feet away from the controlling microprocessor.
5. To minimize errors caused by noise at the $\overline{CS}$ input, the internal circuitry waits for two rising edges and one falling edge of internal system clock after $\overline{CS}$↓ before responding to control input signals. This $\overline{CS}$ set-up time is given by the t_{en} and $t_{su(CS)}$ specifications.

electrical characteristics over recommended operating free-air temperature range, $V_{CC} = V_{REF+} = 4.75$ V to 5.5 V (unless otherwise noted), $f_{CLK(I/O)} = 2.048$ MHz for TLC548 or 1.1 MHz for TLC549

PARAMETER			TEST CONDITIONS		MIN	TYP†	MAX	UNIT
V_{OH}	High-level output voltage		$V_{CC} = 4.75$ V,	$I_{OH} = -360\ \mu A$	2.4			V
V_{OL}	Low-level output voltage		$V_{CC} = 4.75$ V,	$I_{OL} = 3.2$ mA			0.4	V
I_{OZ}	Off-state (high-impedance state) output current		$V_O = V_{CC}$,	$\overline{CS}$ at V_{CC}			10	V
			$V_O = 0$,	$\overline{CS}$ at V_{CC}			−10	
I_{IH}	High-level input current, control inputs		$V_I = V_{CC}$			0.005	2.5	μA
I_{IL}	Low-level input current, control inputs		$V_I = 0$			−0.005	−2.5	μA
$I_{I(on)}$	Analog channel on-state input current, during sample cycle		Analog input at V_{CC}			0.4	1	μA
			Analog input at 0 V			−0.4	−1	
I_{CC}	Operating supply current		$\overline{CS}$ at 0 V			1.8	2.5	mA
$I_{CC}+I_{REF}$	Supply and reference current		$V_{REF+} = V_{CC}$			1.9	3	mA
C_i	Input capacitance	Analog inputs				7	55	pF
		Control inputs				5	15	

operating characteritics over recommended operating free-air temperature range, $V_{CC} = V_{REF+} = 4.75$ V to 5.5 V (unless otherwise noted), $f_{CLK(I/O)} = 2.048$ MHz for TLC548 or 1.1 MHz for TLC549

PARAMETER		TEST CONDITIONS	TLC548			TLC549			UNIT
			MIN	TYP†	MAX	MIN	TYP†	MAX	
	Linearity error	See Note 6			±0.5			±0.5	LSB
	Zero error	See Note 7			±0.5			±0.5	LSB
	Full-scale error	See Note 7			±0.5			±0.5	LSB
	Total unadjusted error	See Note 8			±0.5			±0.5	LSB
t_{conv}	Conversion time	See Operating Sequence		8	17		12	17	μs
	Total access and conversion time	See Operating Sequence		12	22		19	25	μs
t_{acq}	Channel acquisition time (sample cycle)	See Operating Sequence			4			4	I/O clock cycles
t_v	Time output data remains valid after I/O clock↓		10			10			ns
t_d	Delay time to data output valid	I/O clock↓			300			400	ns
t_{en}	Output enable time				1.4			1.4	μs
t_{dis}	Output disable time	See Parameter Measurement Information			150			150	ns
$t_{r(bus)}$	Data bus rise time				300			300	ns
$t_{f(bus)}$	Data bus fall time				300			300	ns

†All typicals are at $V_{CC} = 5$ V, $T_A = 25$ °C.

NOTES: 6. Linearity error is the deviation from the best straight line through the A/D transfer characteristics.
7. Zero error is the difference between 00000000 and the converted output for zero input voltage; full-scale error is the difference between 11111111 and the converted output for full-scale input voltage.
8. Total unadjusted error is the sum of linearity, zero, and full-scale errors.

PARAMETER MEASUREMENT INFORMATION

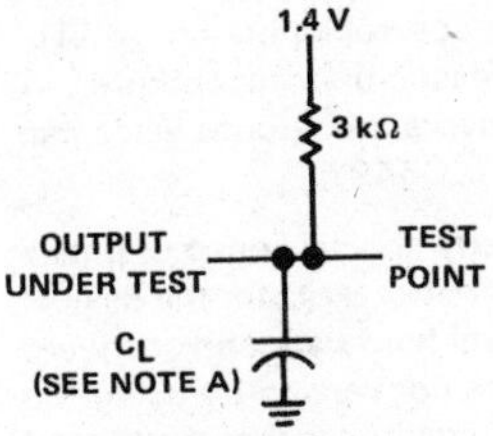
1.4 V
3 kΩ
OUTPUT
UNDER TEST
TEST
POINT
C_L
(SEE NOTE A)

LOAD CIRCUIT FOR t_d, t_r, AND t_f

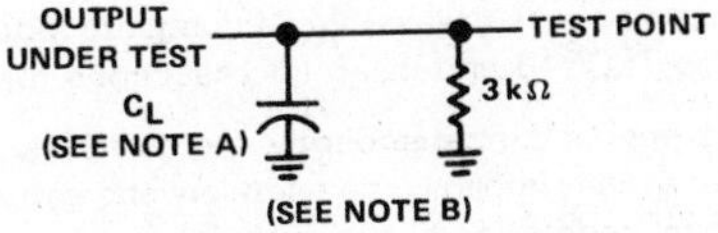
OUTPUT
UNDER TEST
TEST POINT
C_L
(SEE NOTE A)
3 kΩ
(SEE NOTE B)

LOAD CIRCUIT FOR t_{PZH} AND t_{PHZ}

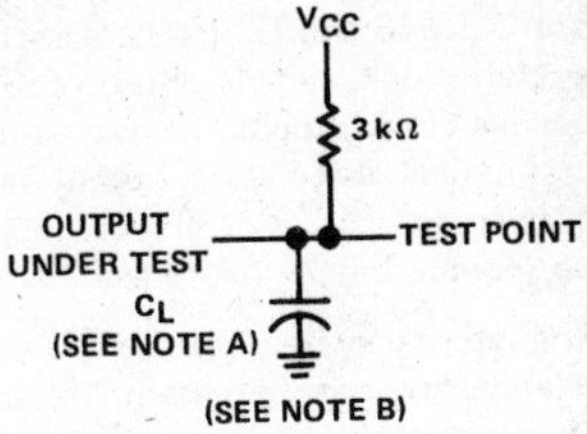
V_{CC}
3 kΩ
OUTPUT
UNDER TEST
TEST POINT
C_L
(SEE NOTE A)
(SEE NOTE B)

LOAD CIRCUIT FOR t_{PZL} AND t_{PLZ}

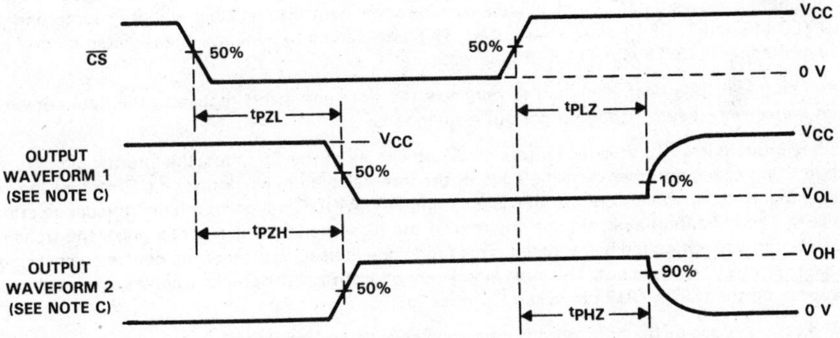
CS
50%
50%
V_{CC}
0 V
t_{PZL}
t_{PLZ}
OUTPUT
WAVEFORM 1
(SEE NOTE C)
V_{CC}
50%
10%
V_{OL}
t_{PZH}
V_{OH}
OUTPUT
WAVEFORM 2
(SEE NOTE C)
90%
50%
t_{PHZ}
0 V
(SEE NOTE B)

VOLTAGE WAVEFORMS FOR ENABLE AND DISABLE TIMES

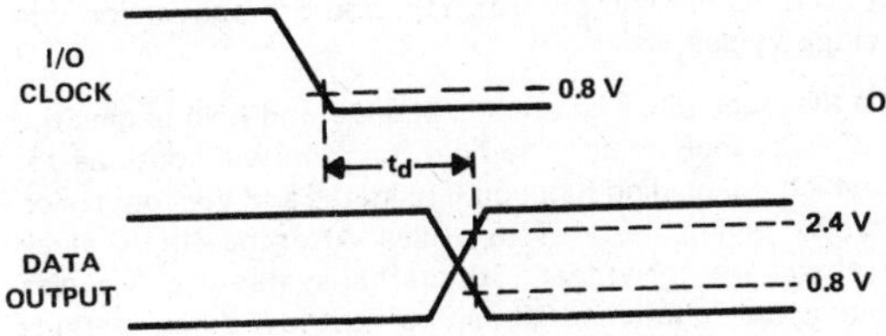
I/O
CLOCK
0.8 V
t_d
DATA
OUTPUT
2.4 V
0.8 V

VOLTAGE WAVEFORM FOR DELAY TIME

OUTPUT
2.4 V
0.4 V
t_r
t_f

VOLTAGE WAVEFORM FOR RISE AND FALL TIMES

NOTES: A. C_L = 50 pF for TLC548 and 100 pF for TLC549; C_L includes jig capacitance.
B. $t_{en} = t_{PZH}$ or t_{PZL}, $t_{dis} = t_{PHZ}$ or t_{PLZ}.
C. Waveform 1 is for an output with internal conditions such that the output is low except when disabled by the output control.
Waveform 2 is for an output with internal conditions such that the output is high except when disabled by the output control.

PRINCIPLES OF OPERATION

The TLC548 and TLC549 are each complete data acquisition systems on a single chip. Each contains an internal system clock, sample-and-hold, 8-bit A/D converter, data register, and control logic circuitry. For flexibility and access speed, there are two control inputs: I/O Clock and Chip Select ($\overline{CS}$). These control inputs and a TTL-compatible three-state output facilitate serial communications with a microprocessor or minicomputer. A conversion can be completed in 17 microseconds or less, while complete input-conversion-output cycles can be repeated in 22 microseconds for the TLC548 and in 25 microseconds for the TLC549.

The internal system clock and I/O clock are used independently and do not require any special speed or phase relationships between them. This independence simplifies the hardware and software control tasks for the device. Due to this independence and the internal generation of the system clock, the control hardware and software need only be concerned with reading the previous conversion result and starting the conversion by using the I/O clock. In this manner, the internal system clock drives the "conversion crunching" circuitry so that the control hardware and software need not be concerned with this task.

When $\overline{CS}$ is high, the data output pin is in a high-impedance condition and the I/O clock pin is disabled. This $\overline{CS}$ control function allows the I/O Clock pin to share the same control logic point with its counterpart pin when additional TLC548 and TLC549 devices are used. This also serves to minimize the required control logic pins when using multiple TLC548 and TLC549 devices.

The control sequence has been designed to minimize the time and effort required to initiate conversion and obtain the conversion result. A normal control sequence is:

1. $\overline{CS}$ is brought low. To minimize errors caused by noise at the $\overline{CS}$ input, the internal circuitry waits for two rising edges and then a falling edge of the internal system clock after a $\overline{CS}\downarrow$ before the transition is recognized. However, upon a $\overline{CS}$ rising edge, DATA OUT will go to a high-impedance state within the t_{dis} specification even though the rest of the IC's circuitry will not recognize the transition until the $t_{su(CS)}$ specification has elapsed. This technique is used to protect the device against noise when used in a noisy environment. The most significant bit (MSB) of the previous conversion result will initially appear on the DATA OUT pin when $\overline{CS}$ goes low.
2. The falling edges of the first four I/O clock cycles shift out the 2nd, 3rd, 4th, and 5th most significant bits of the previous conversion result. The on-chip sample-and-hold begins sampling the analog input after the 4th high-to-low transition of the I/O Clock. The sampling operation basically involves the charging of internal capacitors to the level of the analog input voltage.
3. Three more I/O clock cycles are then applied to the I/O pin and the 6th, 7th, and 8th conversion bits are shifted out on the falling edges of these clock cycles.
4. The final, (the 8th), clock cycle is applied to the I/O clock pin. The on-chip sample-and-hold begins the hold function upon the high-to-low transition of this clock cycle. The hold function will continue for the next four internal system clock cycles, after which the holding function terminates and the conversion is performed during the next 32 system clock cycles, giving a total of 36 cycles. After the 8th I/O clock cycle, $\overline{CS}$ must go high or the I/O clock must remain low for at least 36 internal system clock cycles to allow for the completion of the hold and conversion functions. $\overline{CS}$ can be kept low during periods of multiple conversion. When keeping $\overline{CS}$ low during periods of multiple conversion, special care must be exercised to prevent noise glitches on the I/O Clock line. If glitches occur on the I/O Clock line, the I/O sequence between the microprocessor/controller and the device will lose synchronization. If $\overline{CS}$ is taken high, it must remain high until the end of conversion. Otherwise, a valid high-to-low transition of $\overline{CS}$ will cause a reset condition, which will abort the conversion in progress.

A new conversion may be started and the ongoing conversion simultaneously aborted by performing steps 1 through 4 before the 36 internal system clock cycles occur. Such action will yield the conversion result of the previous conversion and not the ongoing conversion.

TEXAS INSTRUMENTS
POST OFFICE BOX 655012 • DALLAS, TEXAS 75265

PRINCIPLES OF OPERATION

For certain applications, such as strobing applications, it is necessary to start conversion at a specific point in time. This device will accommodate these applications. Although the on-chip sample-and-hold begins sampling upon the high-to-low transition of the 4th I/O clock cycle, the hold function does not begin until the high-to-low transition of the 8th I/O clock cycle, which should occur at the moment when the analog signal must be converted. The TLC548 and TLC549 will continue sampling the analog input until the high-to-low transition of the 8th I/O clock pulse. The control circuitry or software will then immediately lower the I/O clock signal and start the holding function to hold the analog signal at the desired point in time and start conversion.

Detailed information on interfacing to the most popular microprocessor is readily available from Texas Instruments.

TLC1205A, TLC1205B, TLC1225A, TLC1225B
SELF-CALIBRATING 12-BIT-PLUS-SIGN UNIPOLAR OR BIPOLAR ANALOG-TO-DIGITAL CONVERTERS

D2982, FEBRUARY 1987

- ADVANCED LinCMOS™ Technology
- Self-Calibration Eliminates Expensive Trimming at Factory and Offset Adjustment in the Field
- 12-Bit Plus Sign Unipolar or Bit Bipolar
- ± ½ and ± 1 LSB Linearity Error in Unipolar Configuration
- 10 μs Conversion Time (Mode 2) (clock = 2.6 MHz)
  20 μs Conversion Time (Mode 1) (clock = 2.6 MHz)
- Compatible with All Microprocessors
- True Differential Analog Voltage Inputs
- 0 to 5 V Analog Voltage Range with Single 5-V Supply (Unipolar Configuration)
- −5 V to 5 V Analog Voltage Range with ± 5-V Supplies (Bipolar Configuration)
- Low Power . . . 25 mW Maximum
- Replaces National Semiconductor ADC1205 and ADC1225 in Mode 1 Operation

TLC1205
J OR N DUAL-IN-LINE PACKAGE
(TOP VIEW)

Signal	Pin	Pin	Signal
ANLG V_{CC-}	1	24	DGTL V_{CC}
IN −	2	23	D12/D7/0 (status)
IN +	3	22	D12/D6/SARS
ANLG GND	4	21	D12/D5/0/DI5
REF	5	20	D12/D4/0/DI4
ANLG V_{CC+}	6	19	D11/D3/0/DI3
VOS	7	18	D10/D2/BYST/DI2
CLK IN	8	17	D9/D1/EOC/DI1
$\overline{WR}$	9	16	D8/D0/INT/DI0
$\overline{CS}$	10	15	$\overline{INT}$
$\overline{RD}$	11	14	READY OUT
DGTL GND	12	13	$\overline{STATUS}$

(Pins 23 through 16: I/O BUS)

TLC1225
J OR N DUAL-IN-LINE PACKAGE
(TOP VIEW)

Signal	Pin	Pin	Signal
ANLG V_{CC-}	1	28	DGTL V_{CC}
IN −	2	27	D12
IN +	3	26	D11
ANLG GND	4	25	D10
REF	5	24	D9
ANLG V_{CC+}	6	23	D8
VOD	7	22	D7
CLK IN	8	21	D6
$\overline{WR}$	9	20	D5/DI5
$\overline{CS}$	10	19	D4/DI4
$\overline{RD}$	11	18	D3/DI3
DGTL GND	12	17	D2/DI2
READY OUT	13	16	D1/DI1
$\overline{INT}$	14	15	D0/DI0

(Pins 27 through 15: I/O BUS)

description

The TLC1205 and TLC1225 converters are manufactured with Texas Instruments highly efficient ADVANCED LinCMOS™ technology. Either of the TLC1205 or TLC1225 CMOS analog-to digital converters can be operated as a unipolar or bipolar converter. A unipolar input (0 to 5 V) can be accommodated with a single 5-volt supply, while a bipolar input (−5 V to 5 V) requires the addition of a 5-volt negative supply. Conversion is performed via the successive-approximation method. The 24-pin TLC1205 outputs the converted data in two 8-bit bytes, while the TLC1225 outputs the converted data in a parallel word and interfaces directly to a 16-bit data bus. Negative numbers are given in the 2's complement data format. All digital signals are fully TTL and CMOS compatible.

These converters utilize a self-calibration technique by which seven of the internal capacitors in the capacitive ladder of the A/D conversion circuitry can be automatically or manually calibrated. If the converters are operated in Mode 1, one of the seven internal capacitors is calibrated during the first part of the conversion sequence. For example, one capacitor is calibrated during the first conversion. The next capacitor is calibrated during the second conversion. If the converters are operated in Mode 2, the internal capacitors are calibrated during a nonconversion, capacitor-calibrate cycle in which all seven of the internal capacitors are calibrated at the same time. A Mode 2 conversion requires only 10 μs (2.6 MHz clock) after the nonconversion, capacitor-calibrating cycle has been completed. The calibration or conversion cycle may be initiated at any time by issuing the proper address to the data bus. The self-calibrating techniques eliminate the need for expensive trimming of thin-film resistors at the factory and provide excellent performance at low cost.

ADVANCED LinCMOS™ is a trademark of Texas Instruments Incorporated

functional block diagram

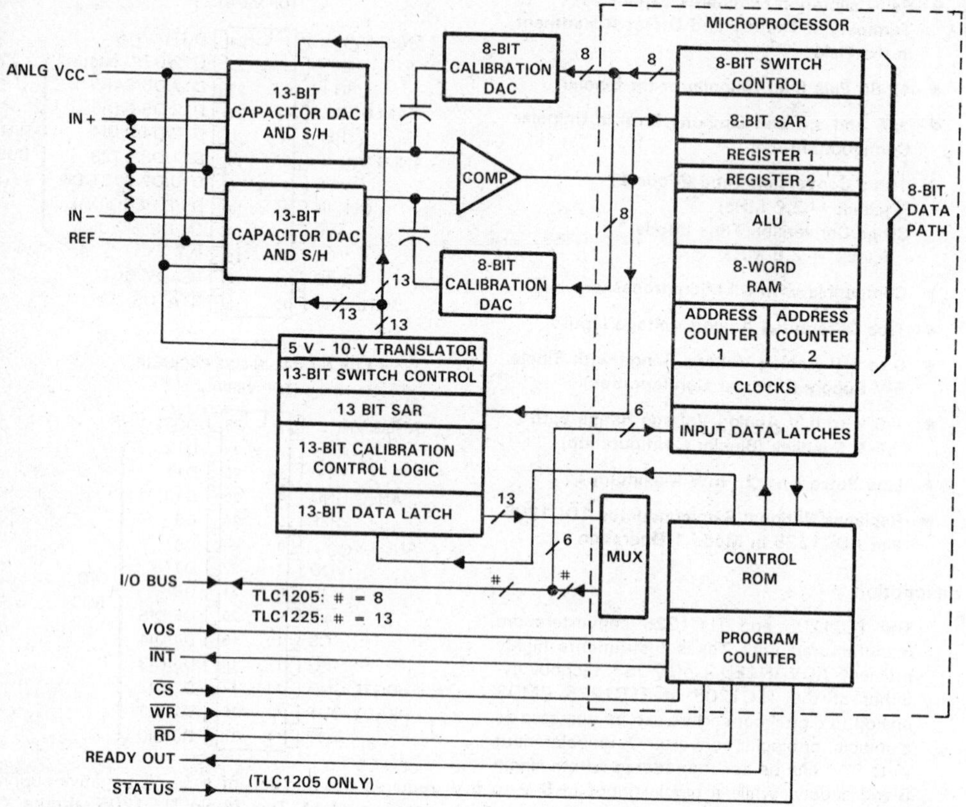
ANLG VCC
IN +
IN –
REF
13-BIT CAPACITOR DAC AND S/H
13-BIT CAPACITOR DAC AND S/H
8-BIT CALIBRATION DAC
8-BIT CALIBRATION DAC
COMP
5 V - 10 V TRANSLATOR
13-BIT SWITCH CONTROL
13 BIT SAR
13-BIT CALIBRATION CONTROL LOGIC
13-BIT DATA LATCH
MICROPROCESSOR
8-BIT SWITCH CONTROL
8-BIT SAR
REGISTER 1
REGISTER 2
ALU
8-WORD RAM
ADDRESS COUNTER 1
ADDRESS COUNTER 2
CLOCKS
INPUT DATA LATCHES
8-BIT DATA PATH
MUX
CONTROL ROM
PROGRAM COUNTER
I/O BUS
TLC1205: # = 8
TLC1225: # = 13
VOS
INT
CS
WR
RD
READY OUT
STATUS
(TLC1205 ONLY)

In Mode 1, these converters are replacements for National Semiconductor ADC1205 and ADC1225 integrated circuits. The Mode 1 conversion time for guaranteed accuracy is 51 clock cycles. In the Mode 2 operation, these devices are no longer true replacements. However, the Mode 2 conversion time for guaranteed accuracy is only 26 clock cycles.

The TLC1205AM, TLC1205BM, TLC1225AM, and TLC1225BM are characterized for operation over the full military temperature range of −55°C to 125°C. The TLC1205AI, TLC1205BI, TLC1225AI, and TLC1225BI are characterized for operation from −40°C to 85°C.

TEXAS INSTRUMENTS
POST OFFICE BOX 655012 • DALLAS, TEXAS 75265

operation description

calibration of comparator offset

The following actions are performed to calibrate the comparator offset:

1. The IN+ and IN− inputs are internally shorted together in order that the comparator input is zero. A course comparator offset calibration is performed by storing the offset voltages of the interconnecting comparator stages on the coupling capacitors, which connect the interconnecting stages. Refer to Figure 1. The storage of offset voltages is accomplished by closing all switches and then opening switches A and A′, then switches B and B′, and then C and C′. This process continues until all interconnecting stages of the comparator are calibrated. After this action, some of the comparator offset still remains uncalibrated.

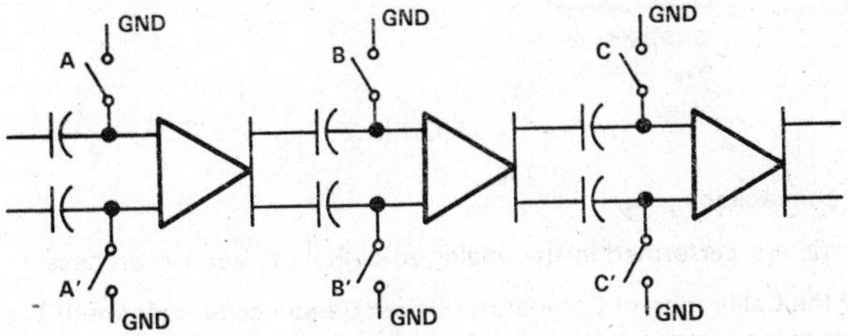
GND
A
GND
B
GND
C
A′
GND
B′
GND
C′
GND

FIGURE 1

2. An A/D conversion is done on the remaining offset with the 8-bit calibration DACs and 8-bit SAR and the result is stored in the RAM.

capacitor calibration of the ADC's Capacitive Ladder

The following actions are performed to calibrate capacitors in the 13-bit DAC's, which comprise the ADC's capacitive ladder:

1. The IN+ and IN− inputs are internally disconnected from the 13-bit capacitive DACs.
2. The most-significant-bit (MSB) capacitor is tied to REF, while the rest of the ladder capacitors are tied to GND. The A/D conversion result for the remaining comparator offset, obtained in step 2 above, is retrieved from the RAM and is input to the 8-bit DACs.
3. Step 1 of the Calibration of Comparator Offset sequence is performed. The 8-bit DAC input is returned to zero and the remaining comparator offset is then subtracted. Thus, the comparator offset is completely corrected.
4. Now the MSB capacitor is tied to GND, while the rest of the ladder capacitors, C_X, are tied to REF. An MSB capacitor voltage error (see Figure 2) on the comparator output will occur if the MSB capacitor does not equal the sum of the other capacitors in the capacitive ladder. This error voltage is converted to an 8-bit word from which a capacitor error is computed and stored in the RAM.
5. The capacitor voltage error for the next most significant capacitor is calibrated by keeping the MSB capacitor grounded and then performing the above Steps 1 - 4 while using the next most significant capacitor in lieu of the MSB capacitor. The seven most significant capacitors can be calibrated in this manner.

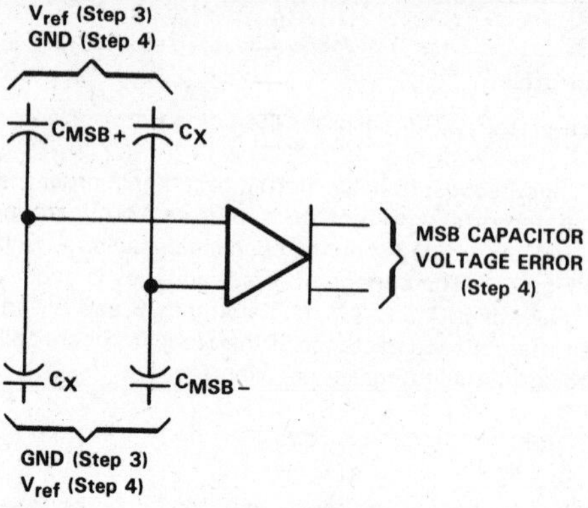
V_{ref} (Step 3)
GND (Step 4)
C_{MSB+}
C_X
MSB CAPACITOR VOLTAGE ERROR (Step 4)
C_X
C_{MSB-}
GND (Step 3)
V_{ref} (Step 4)

FIGURE 2

analog-to-digital conversion

The following steps are performed in the analog-to-digital conversion process:

1. Step 1 of the Calibration of Comparator Offset Sequence is performed. The A/D conversion result for the remaining comparator offset, which was obtained in Step 2 of the Calibration of Comparator Offset, is retrieved from the RAM and is input to the 8-bit DACs. Thus the comparator offset is completely corrected.
2. IN+ and IN− are sampled onto the 13-bit capacitive ladders.
3. The 13-bit analog-to-digital conversion is performed. As the successive-approximation conversion proceeds successively through the seven most significant capacitors, the error for each of these capacitors is recovered from the RAM and accumulated in a register. This register controls the 8-bit DACs so the total accumulated error for these capacitors is subtracted out during the conversion process.

absolute maximum ratings over operating free-air temperature range (unless otherwise noted)

Supply voltage (ANLG V_{CC+} and DGTL V_{CC}) (see Note 1)	15 V
Supply voltage, ANLG V_{CC-}	−15 V
Control and Clock input voltage range	−0.3 V to +15 V
Analog input (IN+, IN−) voltage range, V_{I+} and V_{I-}	ANLG V_{CC-} −0.3 V to ANLG V_{CC+} + 0.3 V
Reference voltage range, V_{ref}	−0.3 V to ANLG V_{CC+} + 0.3 V
Mode select voltage range, VOS	−0.3 V to ANLG V_{CC+} + 0.3 V
Output voltage range	−0.3 V to DGTL V_{CC} + 0.3 V
Input current (per pin)	±5 mA
Input current (per package)	±20 mA
Operating free-air temperature range:	
TLC1205AM, TLC1205BM, TLC1225AM, TL1225BM	−55 °C to 125 °C
TLC1205AI, TLC1205BI, TLC1225AI, TLC1225BI	−40 °C to 85 °C
Storage temperature range	−65 °C to 150 °C
Lead temperature 1,6 mm (1/16 inch) from the case for 60 seconds: J package	300 °C
Lead temperature 1,6 mm (1/16 inch) from the case for 10 seconds: N package	260 °C

Note 1: All analog voltages are referred to ANLG GND and all digital voltages are referred to DGTL GND.

TEXAS INSTRUMENTS
POST OFFICE BOX 655012 • DALLAS, TEXAS 75265

recommended operating conditions

		MIN	MAX	UNIT
Supply voltage	ANLG V_{CC+}	4.5	6	V
	ANLG V_{CC-}	−5.5	ANLG GND	
	DGTL V_{CC}	4.5	6	
High-level input voltage, V_{IH}, all digital inputs except CLK IN (V_{CC} = 4.75 V to 5.25 V)		2		V
Low level input voltage, V_{IL}, all digital inputs except CLK IN (V_{CC} = 4.75 V to 5.25 V)			0.8	V
Analog input voltage, V_{I+}, V_{I-}	Bipolar range	ANLG V_{CC-} − 0.05	ANLG V_{CC+} + 0.05	V
	Unipolar range	ANLG GND − 0.05	ANLG V_{CC+} + 0.05	
Clock input frequency, f_{clock}		0.3	2.6	MHz
Clock duty cycle		40%	60%	
Pulse duration, $\overline{CS}$ and $\overline{WR}$ both low, t_w ($\overline{CS}\cdot\overline{WR}$)		350		ns
Setup time before $\overline{WR}$↑ or $\overline{CS}$↑, t_{su}			100	ns
Hold time after $\overline{WR}$↑ or $\overline{CS}$↑, t_h			20	ns
Operating free-air temperature, T_A	TLC1205AM, TLC1225AM TLC1205BM, TLC1225BM	−55	125	°C
	TLC1205AI, TLC1225AI TLC1205BI, TLC1225BI	−40	85	

electrical characteristics over recommended operating free-air temperature range, ANLG V_{CC+} = DGTL V_{CC} = V_{ref} = 5 V, ANLG V_{CC-} = −5 V (for bipolar input range), ANLG V_{CC-} = ANLG GND (for unipolar input range) (unless otherwise noted) (see Note 1)

PARAMETER		TEST CONDITIONS		MIN	MAX	UNIT
V_{OH}	High-level output voltage	DGTL V_{CC} = 4.75 V	I_O = −1.8 mA	2.4		V
			I_O = −50 μA	4.5		
V_{OL}	Low-level output voltage	DGTL V_{CC} = 4.75 V,	I_O = 8 mA		0.4	V
V_{T+}	Clock positive-going threshold voltage			2.7	3.5	V
V_{T-}	Clock negative-going threshold voltage			1.4	2.1	V
V_{hys}	Clock input hysteresis	V_{T+}min − V_{T-}max		0.6		V
		V_{T+}max − V_{T-}min			2.1	
R_{ref}	Input resistance, REF terminal			1	10	MΩ
I_{IH}	High-level input current	V_I = 5 V			1	μA
I_{IL}	Low-level input current	V_I = 0			−1	μA
I_{OZ}	High-impedance-state output leakage current	V_O = 0			−3	μA
		V_O = 5 V			3	
I_O	Output current	V_O = 0			−6	mA
		V_O = 5 V			8	
DGTL I_{CC}	Supply current from DGTL V_{CC}	f_{clk} = 2.6 MHz,	$\overline{CS}$ high		3	mA
ANLG I_{CC+}	Supply current from ANLG V_{CC+}	f_{clk} = 2.6 MHz,	$\overline{CS}$ high		3	mA
ANLG I_{CC-}	Supply current from ANLG V_{CC-}	f_{clk} = 2.6 MHz,	$\overline{CS}$ high		−3	mA

NOTE 1: Bipolar input range is defined as: V_{I+} = −5.05 V to +5.05 V, V_{I-} = −5.05 V to +5.05 V, and $|V_{I+} - V_{I-}| \le 5.05$ V. The unipolar input voltage range is defined as: V_{I+} = −0.05 V to 5.05 V, V_{I-} = −0.05 V to 5.05 V, and $|V_{I+} - V_{I-}| \le 5.05$ V.

operating characteristics over recommended operating free-air temperature range, ANLG V_{CC+} = DGTL V_{CC} = V_{ref} = 5 V, ANLG V_{CC-} = −5 V (for bipolar input range), ANLG V_{CC-} = ANLG GND (for unipolar input range), f_{clock} = 2.6 MHz (unless otherwise noted)(see Note 1)

PARAMETER			TEST CONDITIONS		MIN	MAX	UNIT
	Linearity error		Unipolar input range	TLC1205A, TLC1225A		±1	LSB
				TLC1205B, TLC1225B		±0.5	
			Bipolar input range	TLC1205A, TLC1225A		±2	
				TLC1205B, TLC1225B		±1.5	
	Zero error					±0.5	LSB
	Adjusted positive and negative full-scale error (see Note 2)		Unipolar input range			±1	LSB
	Adjusted positive and negative full-scale error (see Note 3)		Bipolar input range			±1	LSB
	Temperature coefficient of gain					15	ppm/°C
	Temperature coefficient of offset point					1.5	ppm/°C
k_{SVS}	Supply voltage sensitivity	Zero error	ANLG V_{CC+} = 5 V ± 5%, ANLG V_{CC-} = −5 V ± 5%, DGTL V_{CC} = 5 V ± 5%			±0.75	LSB
		Positive and negative full-scale error				±0.75	
		Linearity error				±0.25	
t_c	Conversion time	Mode 1				51	$\frac{1}{f_{clk}}$
		Mode 2				26	
t_a	Access time (delay from falling edge of $\overline{CS}$•$\overline{RD}$ to data output)		C_L = 100 pF			210	ns
t_{dis}	Disable time, output (delay from rising edge of $\overline{RD}$ to high-impedance state		R_L = 10 kΩ,	C_L = 10 pF		260	ns
			R_L = 2 kΩ,	C_L = 100 pF		290	
$t_{d(READY)}$	$\overline{RD}$ or $\overline{WR}$ to READY OUT delay					400	ns
$t_{d(INT)}$	$\overline{RD}$ or $\overline{WR}$ to reset of $\overline{INT}$ delay					400	ns

NOTES: 1. Bipolar input range is defined as: V_{I+} = −5.05 V to +5.05 V, V_{I-} = −5.05 V to +5.05 V, and $|V_{I+} - V_{I-}| \le 5.05$ V. The unipolar input voltage range is defined as: V_{I+} = −0.05 V to 5.05 V, V_{I-} = −0.05 V to 5.05 V, and $|V_{I+} - V_{I-}| \le 5.05$ V.
2. See section — Positive and Negative Full-Scale Adjustment, Unipolar Inputs.
3. See section — Positive and Negative Full-Scale Adjustment, Bipolar Inputs.

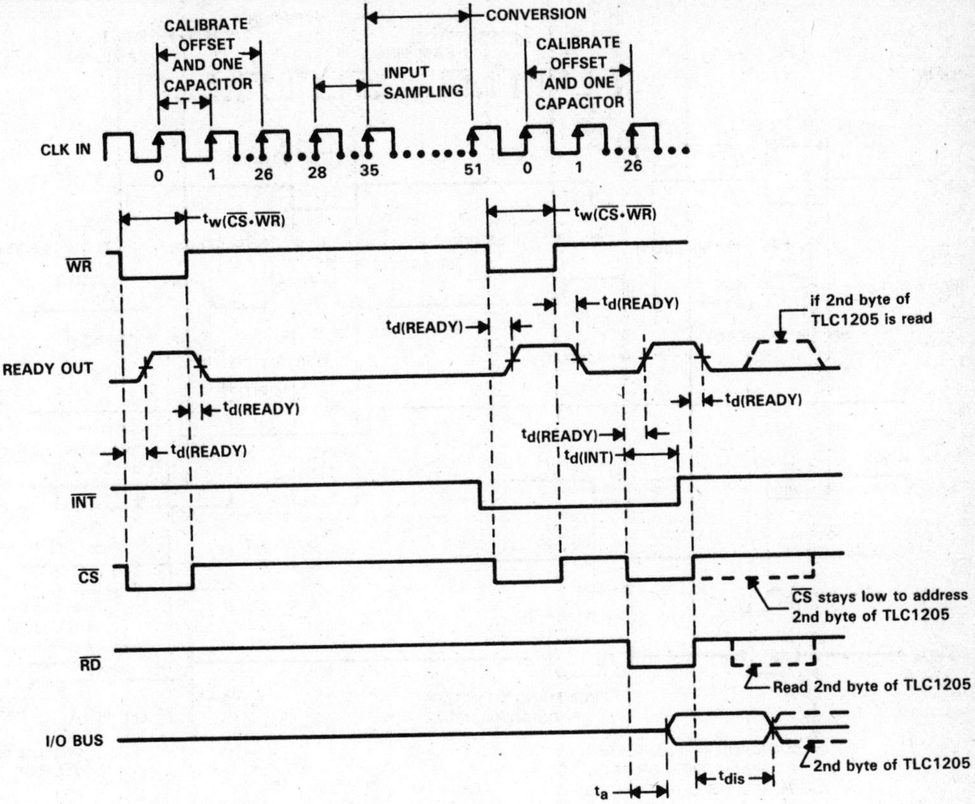
CONVERSION
CALIBRATE OFFSET AND ONE CAPACITOR
T
INPUT SAMPLING
CALIBRATE OFFSET AND ONE CAPACITOR
CLK IN
0
1
26
28
35
51
0
1
26
tw(CS·WR)
tw(CS·WR)
WR
td(READY)
td(READY)
if 2nd byte of TLC1205 is read
READY OUT
td(READY)
td(READY)
td(READY)
td(READY)
td(INT)
INT
CS
CS stays low to address 2nd byte of TLC1205
RD
Read 2nd byte of TLC1205
I/O BUS
2nd byte of TLC1205
ta
tdis

FIGURE 3. MODE 1 TIMING DIAGRAM

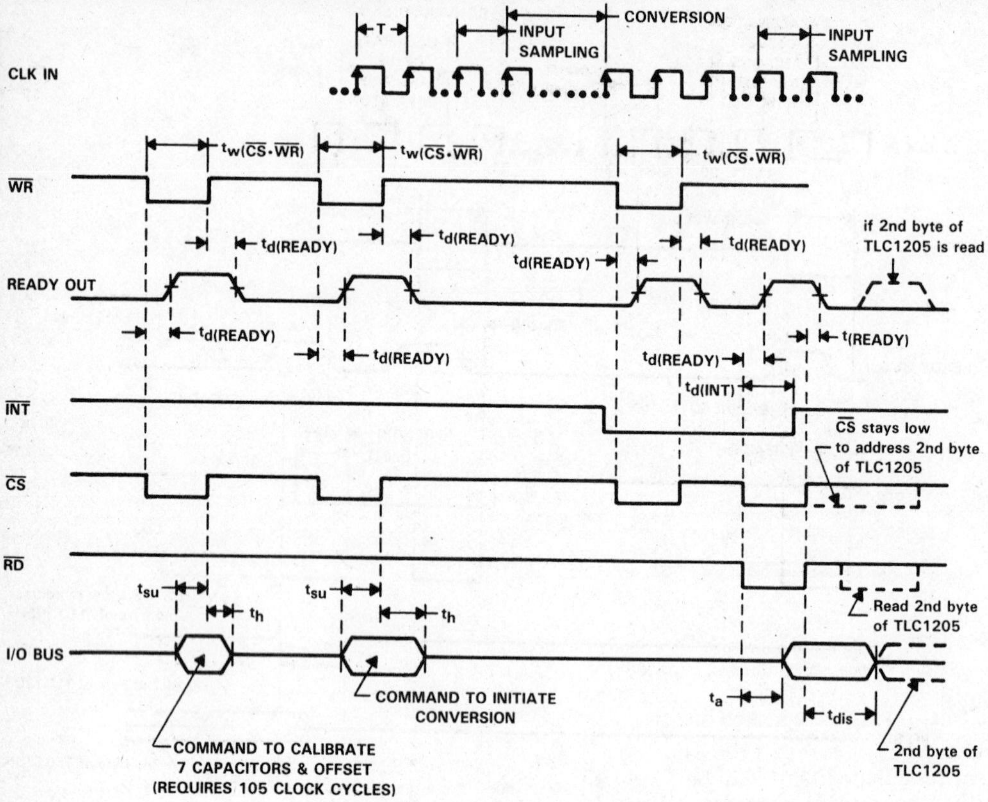
T
INPUT
SAMPLING
CONVERSION
INPUT
SAMPLING
CLK IN
$t_{w(\overline{CS}\cdot\overline{WR})}$
$\overline{WR}$
$t_{d(READY)}$
if 2nd byte of
TLC1205 is read
READY OUT
$t_{(READY)}$
$t_{d(INT)}$
$\overline{INT}$
$\overline{CS}$ stays low
to address 2nd byte
of TLC1205
$\overline{CS}$
$\overline{RD}$
t_{su}
t_h
Read 2nd byte
of TLC1205
I/O BUS
COMMAND TO INITIATE
CONVERSION
t_a
t_{dis}
COMMAND TO CALIBRATE
7 CAPACITORS & OFFSET
(REQUIRES 105 CLOCK CYCLES)
2nd byte of
TLC1205

FIGURE 4. MODE 2 TIMING DIAGRAM

PARAMETER MEASUREMENT INFORMATION

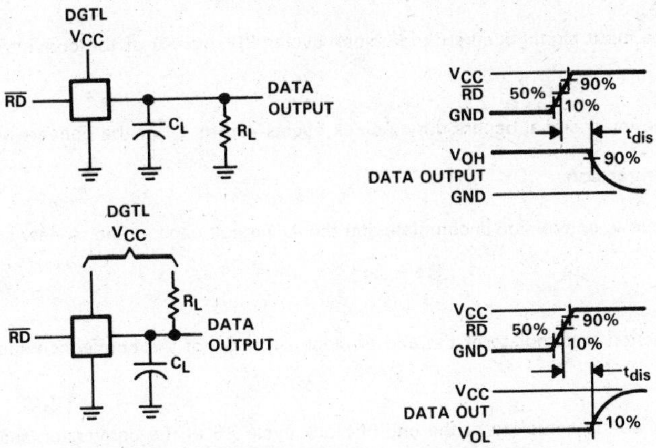
DGTL
VCC
RD
DATA
OUTPUT
CL
RL
VCC
RD
GND
50%
90%
10%
tdis
VOH
DATA OUTPUT
GND
90%
DGTL
VCC
RL
RD
DATA
OUTPUT
CL
VCC
RD
GND
50%
90%
10%
tdis
VCC
DATA OUT
VOL
10%

FIGURE 5. LOAD CIRCUITS AND WAVEFORMS

PRINCIPLES OF OPERATION

The following information is categorized into Mode 1 and Mode 2 groupings to allow the designer to concentrate on a particular mode of interest

power-up calibration sequence

Mode 1

When the chip is powered-up, the internal capacitors are automatically calibrated as part of the power-up sequence. This initial calibration sequence requires 105 clock cycles. The chip will not perform an A/D conversion during this calibration sequence.

Mode 2

Power-Up calibration is not automatic and calibration is initiated by writing control words to the six least significant bits of the data bus. If addressed or initiated, conversion can begin after the first clock cycle. However, full A/D conversion accuracy is not guaranteed until after internal capacitor calibration.

conversion start sequence

Mode 1

The conversion sequence is initiated when $\overline{CS}$ and $\overline{WR}$ are both low.

Mode 2

The writing of the conversion command word to the six least significant bits of the data bus, when either $\overline{CS}$ or $\overline{WR}$ goes high, initiates the conversion sequence.

analog sampling sequence

Mode 1

Sampling of the input signal occurs during clock cycles 29 thru 35 of the conversion sequence.

Mode 2

Sampling of the input signal occurs during clock cycles 4 thru 10 of the conversion sequence.

completed A/D conversion

When $\overline{INT}$ goes low, conversion is complete and the A/D result can be read. A new conversion can begin immediately.

Mode 1

The A/D conversion is complete at the end of clock cycle 51 of the conversion sequence.

Mode 2

The A/D conversion is complete at the end of clock cycle 26 of the conversion sequence.

aborting a conversion in process and beginning a new conversion

Mode 1 and Mode 2

If a conversion is initiated while a conversion sequence is in process, the ongoing conversion will be aborted and a new conversion sequence will begin.

Mode 1

If the new conversion is started before the Analog Sampling begins (see Analog Sampling Sequence section and the Mode 1 Timing Diagram), the particular internal capacitor that was being calibrated during the aborted conversion sequence will be calibrated during the new conversion sequence. Otherwise, the next internal capacitor will be calibrated during the new conversion sequence.

reading the conversion result

TLC1205

Upon activating the required control signals to read the conversion result or status information, the appropriate pins are brought out of a high-impedance state and drive the data bus with the proper information. These pins are D12/D7/0 through D8/DO/INT/DIO.

If $\overline{STATUS}$, $\overline{CS}$, and $\overline{RD}$ are all low, status information can be read. The format of the conversion result and status information and the respective pins for output are presented in Table 1.

TABLE 1

BYTES	$\overline{STATUS}$	$\overline{CS}$	$\overline{RD}$	I/O BUS D12/ D7/ 0	D12/ D6/ SARS	D12/ D5/ 0/ DI5	D12/ D4/ 0/ DI4	D11/ D3/ 0/ DI3	D10/ D2/ BYST/ DI2	D9/ D1/ EOC/ DI1	D8/ D0/ INT/ DI0
MSB	H	L	L	D12	D12	D12	D12	D11	D10	D9	D8
LSB	H	L	↑↓	D7	D6	D5	D4	D3	D2	D1	D0
STATUS	L	L	L	L	SARS	L	L	L	BYST	EOC	INT

The status information is described in Table 2.

TABLE 2

STATUS BIT	BIT DESCRIPTION	TO CLEAR BIT
L	The output has no meaning and is low.	
SARS	A high indicates that conversion is in progress.	
BYST	A low indicates that the next conversion result read will be the most significant conversion byte. A high indicates that the next conversion result read will be the least significant conversion byte. The BYST bit is toggled by reading the conversion result bytes. This bit can be cleared with a "status write" instruction.	By a "status write" or toggled by reading a conversion data byte
EOC	A high indicates that conversion is complete and the conversion data has been transferred to the output latch.	
INT	A high indicates that conversion is complete and the conversion data has been transferred to the output latch and is ready to read.	By reading a conversion data byte, reading the status byte, or a "status write"

With $\overline{STATUS}$ high, when $\overline{CS}$ and $\overline{RD}$ both go low, the most significant byte (MSB) of the conversion result can be read. Then by taking $\overline{RD}$ high and back low, the least significant byte (LSB) of the conversion result can be read. Subsequently taking $\overline{RD}$ high and low causes the alternate reading of the MSB and LSB of the conversion result.

The format of the output is extended sign with 2's complement, right justified data. For both unipolar and bipolar cases, the sign bit D12 is low if $V_{I+} - V_{I-}$ is positive and high if $V_{I+} - V_{I-}$ is negative. The format of the conversion result and the respective output pins are presented in Table 2. The format of the conversion result and the respective pins for output are presented in Table 1.

TLC1225

When both $\overline{CS}$ and $\overline{RD}$ go low, all 13 bits of conversion data are output to the I/O bus. The format of the output is extended sign with 2's complement, right justified data. Unlike the TLC1205, the TLC1225 does not have internal status information or a $\overline{STATUS}$ pin. For both unipolar and bipolar cases, the sign bit D12 is low if $V_{I+} - V_{I-}$ is positive and high if $V_{I+} - V_{I-}$ is negative.

general

reset INT

When reading the conversion data, the falling edge of the first low-going combination of $\overline{CS}$ and $\overline{RD}$ will reset $\overline{INT}$. The falling edge of the low-going combination of $\overline{CS}$ and $\overline{WR}$ will also reset $\overline{INT}$.

ready out

For high-speed microprocessors, READY OUT allows the TLC1205 and the TLC1225 to insert a wait state in the microprocessor's read cycle.

status write (TLC1205)

A status write resets the internal logic and status bits and aborts any conversion in process. A status write occurs when $\overline{CS}$, $\overline{WR}$, and $\overline{STATUS}$ are taken low.

reference voltage (V_{ref})

This voltage defines the range for $|V_{I+} - V_{I-}|$. When $|V_{I+} - V_{I-}|$ equals V_{ref}, the highest conversion data value results. When $|V_{I+} - V_{I-}|$ equals 0, the conversion data value is zero. Thus, for a given input, the conversion data changes ratiometrically with changes in V_{ref}.

VOS

This pin is a digital input and is used to select Mode 1 or Mode 2 operation. A logic low selects Mode 1; a logic high selects Mode 2.

In Mode 1, the ICs are true replacements for National Semiconductor's ADC1205 and ADC1225. The ADC1205 and ADC1225 use the VOS pin to adjust zero error. Since the zero error adjustment voltage is below the TLC1205's and TLC1225's maximum acceptable level for a logic low signal, the TLC1205 and TLC1225 ICs are true replacements. Even in Mode 1, the TLC1205's and TLC1225's converted data can be read earlier than the ADC1205's and ADC1225's.

calibration and conversion considerations

Mode 1

Calibration of the seven internal capacitors is an integral part of the A/D conversion. One of the seven internal capacitors is calibrated during the first part of the conversion sequence. For example, one of the capacitors is calibrated during the first conversion. The next capacitor is calibrated during the second conversion. After seven conversions, the pattern for calibrating the internal capacitors repeats. A conversion sequence requires 51 clock cycles.

A conversion is initiated by the low-going combination of $\overline{CS}$ and $\overline{WR}$. The conversion sequence is illustrated in the Mode 1 timing diagram.

Mode 2

Calibration of the internal capacitor and A/D conversion are two separate actions. Each action is independently initiated. Mode 2 conversion is much faster than Mode 1, since Mode 2 conversion is not accompanied by the calibration of internal capacitors. In Mode 2, a calibration command that calibrates all seven internal capacitors is normally issued first. A conversion command then initiates the A/D conversion without calibrating the internal capacitors. Subsequent conversions can be performed by issuing additional conversion commands. The calibration and conversion commands are totally independent from one another and can be initiated in any order. Calibration and conversion commands require 105 and 26 clock cycles, respectively.

The calibrate and conversion commands are initiated by writing control words on the six least significant bits of the data bus. These control words are written into the IC when either $\overline{CS}$ or $\overline{WR}$ goes high. The initiation of these commands is illustrated in the Mode 2 Timing Diagram. The bit patterns for the commands are shown in Table 3.

TABLE 3. MODE 2 CONVERSION COMMANDS

COMMAND	$\overline{CS}$ + $\overline{WR}$	I/O BUS DI5	DI4	DI3	DI2	DI1	DI0	Required number of clock cycles
Conversion	↑	H	L	X	X	X	L	26
Calibrate†	↑	L	X	L	L	L	L	105

†Calibration is lost when clock is stopped.

analog inputs

differential inputs provide common mode rejection

The differential inputs reduce common-mode noise. Common-mode noise is noise common to both IN+ and IN− inputs, such as 60-Hz noise. There is no time interval between the sampling of the IN+ and IN− so these inputs are truly differential. Thus, no conversion errors result from a time interval between the sampling of the IN+ and IN− inputs.

input bypass capacitors

Input bypass capacitors may be used for noise filtering. However, the charge on these bypass capacitors will be depleted during the input sampling sequence when the internal sampling capacitors are charged. Note that the charging of the bypass capacitors through the differential source resistances must keep pace with the charge depletion of the bypass capacitors during the input sampling sequence. Note that higher source resistances reduce the amount of charging current for the bypass capacitors. Also, note that fast, successive conversion will have the greatest charge depletion effect on the bypass capacitors. Therefore, the above phenomenon becomes more significant as source resistances and the conversion rate (i.e., higher clock frequency and conversion initiation rate) increase.

In addition, if the above phenomenon prevents the bypass capacitors from fully charging between conversions, voltage drops across the source resistances will result due to the ongoing bypass capacitor charging currents. The voltage drops will cause a conversion error. Also, the voltage drops increase with higher $|V_{I+} - V_{I-}|$ values, higher source resistances, and lower charge on the bypass capacitors (i.e., faster conversion rate).

For low-source-resistance applications ($R_{source} < 100\ \Omega$), a 0.001-μF bypass capacitor at the inputs will prevent pickup due to the series lead inductance of a long wire. A 100-ohm resistor can be placed between the capacitor and the output of an operational amplifier to isolate the capacitor from the operational amplifier.

input leads

The input leads should be kept as short as possible, since the coupling of noise and digital clock signals to the inputs can cause errors.

power supply considerations

Noise spikes on the V_{CC} lines can cause conversion error. Low-inductance tantalum capacitors (> 1 μF) with short leads should be used to bypass ANLG V_{CC} and DGTL V_{CC}. A separate regulator for the TLC1205 or TLC1225 and other analog circuitry will greatly reduce digital noise on the supply line.

positive and negative full-scale adjustment

unipolar inputs

Apply a differential input voltage that is 0.5 LSB below the desired analog full-scale voltage (V_{FS}) and adjust the magnitude of the REF input so that the output code is just changing from 0 1111 1111 1110 to 0 1111 1111 1111. If this transition is desired for a different input voltage, the reference voltage can be adjusted accordingly.

bipolar inputs

First, follow the procedure for the Unipolar case.

Second, apply a differential input voltage so that the digital output code is just changing from 1 0000 0000 0001 to 1 0000 0000 0000. Call this actual differential voltage V_X. The ideal differential voltage for this transition is:

$$-V_{FS} + \frac{V_{FS}}{8192} \qquad (1)$$

The difference between the actual and ideal differential voltages is:

$$\text{Delta} = V_X - \left(-V_{FS} + \frac{V_{FS}}{8192}\right) \qquad (2)$$

Then apply a differential input voltage of:

$$V_X - \frac{\text{Delta}}{2} \qquad (3)$$

and adjust V_{ref} so the digital output code is just changing from 1 0000 0000 0001 to 1 0000 0000 0000. This procedure produces positive and negative full-scale transitions with symmetrical minimum error.

TYPICAL APPLICATIONS

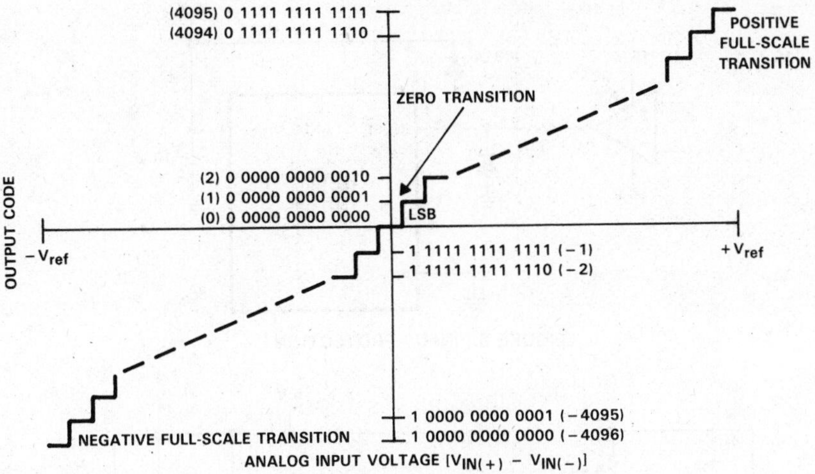
(4095) 0 1111 1111 1111
(4094) 0 1111 1111 1110
POSITIVE FULL-SCALE TRANSITION
ZERO TRANSITION
(2) 0 0000 0000 0010
(1) 0 0000 0000 0001
(0) 0 0000 0000 0000
LSB
OUTPUT CODE
$-V_{ref}$
$+V_{ref}$
1 1111 1111 1111 (−1)
1 1111 1111 1110 (−2)
1 0000 0000 0001 (−4095)
1 0000 0000 0000 (−4096)
NEGATIVE FULL-SCALE TRANSITION
ANALOG INPUT VOLTAGE $[V_{IN(+)} - V_{IN(-)}]$

FIGURE 6. TRANSFER CHARACTERISTIC

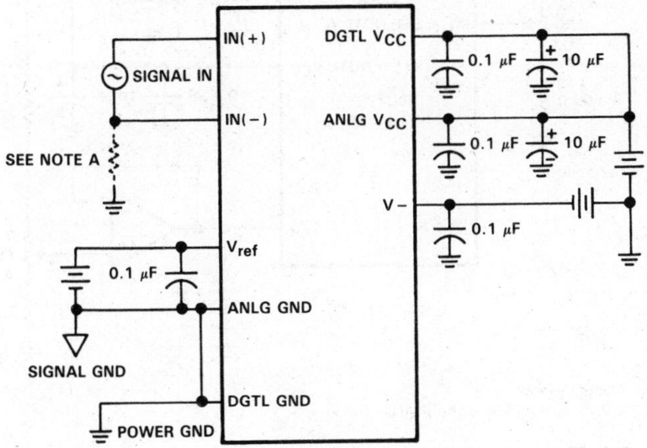
IN(+)
SIGNAL IN
IN(−)
SEE NOTE A
DGTL V_{CC}
0.1 μF
10 μF
ANLG V_{CC}
0.1 μF
10 μF
V−
0.1 μF
V_{ref}
0.1 μF
ANLG GND
SIGNAL GND
DGTL GND
POWER GND

NOTE: A. The analog input must have some current return path to ANALOG GND.
B. Bypass capacitor leads must be as short as possible.

FIGURE 7. ANALOG CONSIDERATIONS

TYPICAL APPLICATIONS (Continued)

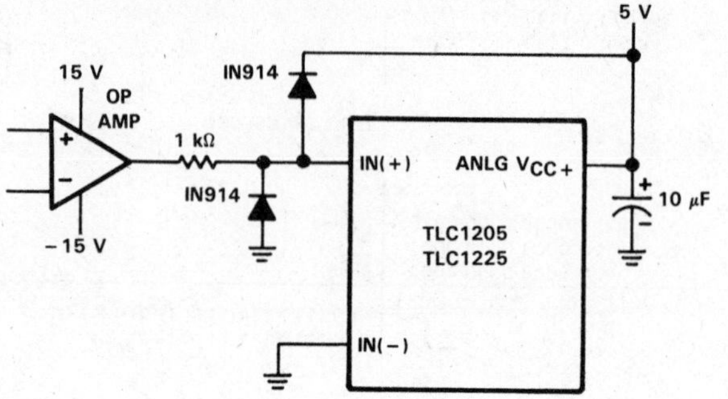
5 V
15 V
OP AMP
IN914
1 kΩ
IN914
−15 V
IN(+)
ANLG VCC+
10 µF
TLC1205
TLC1225
IN(−)

FIGURE 8. INPUT PROTECTION

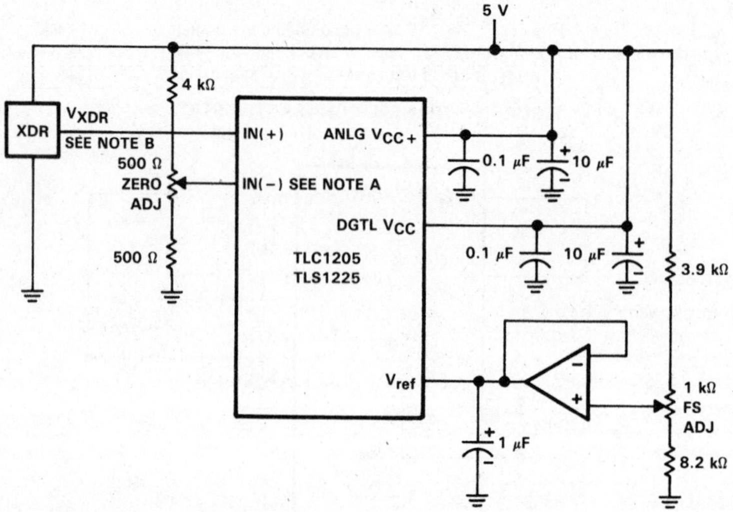
5 V
4 kΩ
XDR
VXDR
SEE NOTE B
500 Ω
ZERO
ADJ
500 Ω
IN(+)
ANLG VCC+
IN(−) SEE NOTE A
0.1 µF
10 µF
DGTL VCC
0.1 µF
10 µF
TLC1205
TLS1225
3.9 kΩ
Vref
1 kΩ
FS
ADJ
1 µF
8.2 kΩ

NOTE: A. $V_{I-} = 0.15 \times$ ANLG V_{CC+}.
B. 15% of ANALOG $V_{CC} \le V_{XDR} \le$ 85% of ANALOG V_{CC}.

FIGURE 9. OPERATING WITH RATIOMETRIC TRANSDUCERS

- LinCMOS™ Technology
- 10-Bit Resolution A/D Converter
- Microprocessor Peripheral or Stand-Alone Operation
- On-Chip 12-Channel Analog Multiplexer
- Built-In Self-Test Mode
- Software-Controllable Sample and Hold
- Total Unadjusted Error . . .
  TLC1540: ±0.5 LSB Max
  TLC1541: ±1.0 LSB Max
- Pinout and Control Signals Compatible with TLC540 and TLC549 Families of 8-Bit A/D Converters

TYPICAL PERFORMANCE	
Channel Acquisition Sample Time	5.5 μs
Conversion Time	21 μs
Samples per Second	32×10^3
Power Dissipation	6 mW

N DUAL-IN-LINE PACKAGE
(TOP VIEW)

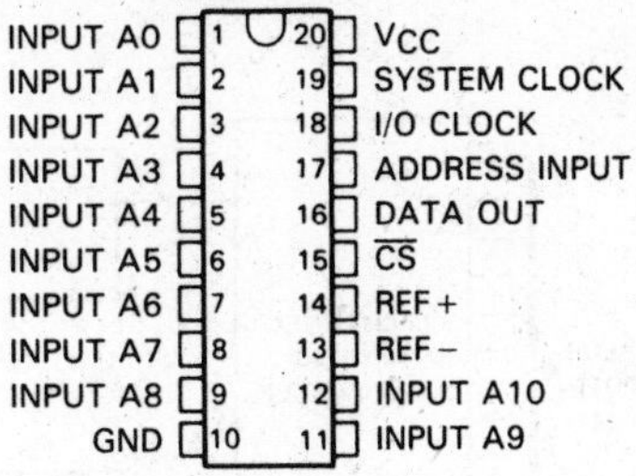
INPUT A0 1 20 VCC
INPUT A1 2 19 SYSTEM CLOCK
INPUT A2 3 18 I/O CLOCK
INPUT A3 4 17 ADDRESS INPUT
INPUT A4 5 16 DATA OUT
INPUT A5 6 15 CS
INPUT A6 7 14 REF+
INPUT A7 8 13 REF−
INPUT A8 9 12 INPUT A10
GND 10 11 INPUT A9

FN CHIP CARRIER PACKAGE
(TOP VIEW)

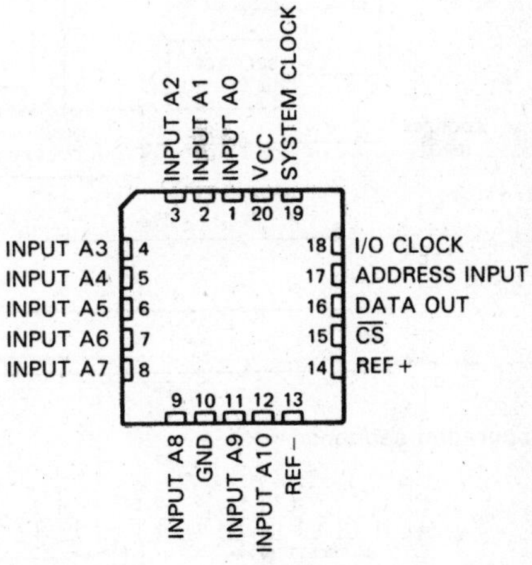
INPUT A2
INPUT A1
INPUT A0
VCC
SYSTEM CLOCK
3 2 1 20 19
INPUT A3 4
INPUT A4 5
INPUT A5 6
INPUT A6 7
INPUT A7 8
18 I/O CLOCK
17 ADDRESS INPUT
16 DATA OUT
15 CS
14 REF+
9 10 11 12 13
INPUT A8
GND
INPUT A9
INPUT A10
REF−

description

The TLC1540 and TLC1541 are LinCMOS™ A/D peripherals built around an 10-bit switched-capacitor successive-approximation A/D converter. They are designed for serial interface to a microprocessor or peripheral via a three-state output with up to four control inputs [including independent System Clock, I/O Clock, Chip Select ($\overline{CS}$), and Address Input]. A 2.1-megahertz system clock for the TLC1540 and TLC1541, with a design that includes simultaneous read/write operation, allows high-speed data transfers and sample rates of up to 32,258 samples per second. In addition to the high-speed converter and versatile control logic, there is an on-chip 12-channel analog multiplexer that can be used to sample any one of 11 inputs or an internal "self-test" voltage, and a sample-and-hold that can operate automatically or under microprocessor control. Detailed information on interfacing to most popular microprocessors is readily available from the factory.

The converters incorporated in the TLC1540 and TLC1541 feature differential high-impedance reference inputs that facilitate ratiometric conversion, scaling, and analog circuitry isolation from logic and supply noises. A totally switched-capacitor design allows guaranteed low-error conversion (±0.5 LSB for the TLC1540, ±1 LSB for the TLC1541) in 21 microseconds over the full operating temperature range.

The TLC1540 and the TLC1541 are available in both the N and FN plastic packages. The M-suffix versions are characterized for operation from −55°C to 125°C. The I-suffix versions are characterized for operation from −40°C to 85°C.

LinCMOS is a trademark of Texas Instruments Incorporated

functional block diagram

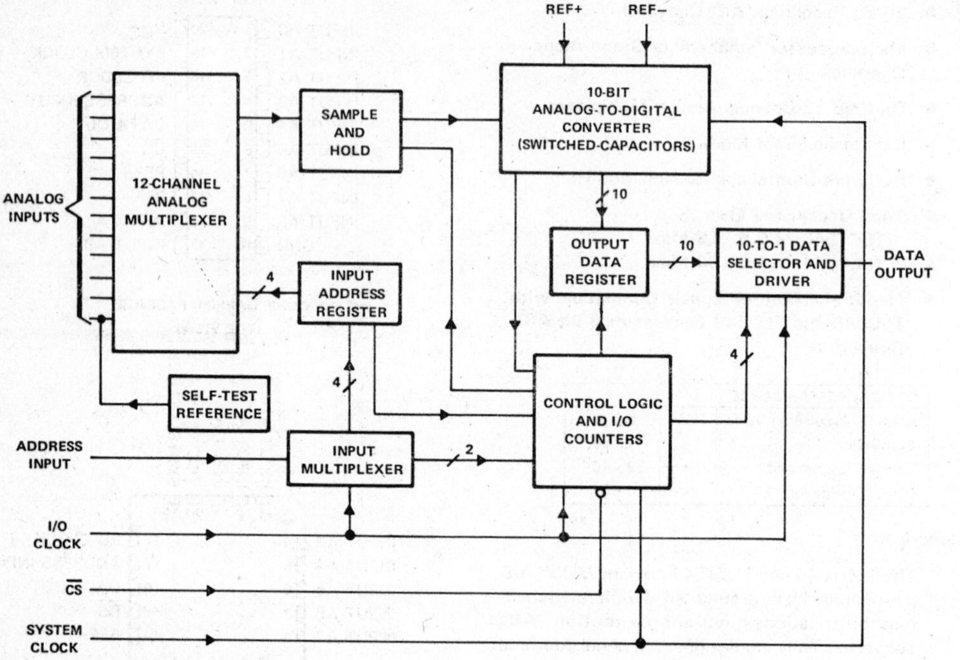
REF+
REF–
10-BIT ANALOG-TO-DIGITAL CONVERTER (SWITCHED-CAPACITORS)
SAMPLE AND HOLD
12-CHANNEL ANALOG MULTIPLEXER
ANALOG INPUTS
INPUT ADDRESS REGISTER
4
10
OUTPUT DATA REGISTER
10-TO-1 DATA SELECTOR AND DRIVER
DATA OUTPUT
SELF-TEST REFERENCE
CONTROL LOGIC AND I/O COUNTERS
ADDRESS INPUT
INPUT MULTIPLEXER
2
I/O CLOCK
CS
SYSTEM CLOCK

operating sequence

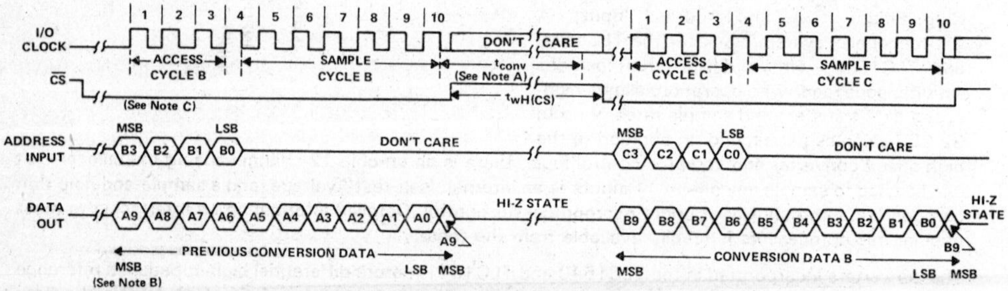
I/O CLOCK
1 2 3 4 5 6 7 8 9 10
DON'T CARE
ACCESS CYCLE B
SAMPLE CYCLE B
t_{conv} (See Note A)
ACCESS CYCLE C
SAMPLE CYCLE C
CS
(See Note C)
$t_{wH(CS)}$
ADDRESS INPUT
MSB
LSB
B3 B2 B1 B0
DON'T CARE
C3 C2 C1 C0
DATA OUT
A9 A8 A7 A6 A5 A4 A3 A2 A1 A0
A9
HI-Z STATE
B9 B8 B7 B6 B5 B4 B3 B2 B1 B0
B9
PREVIOUS CONVERSION DATA
CONVERSION DATA B
MSB (See Note B)
LSB MSB

NOTES: A. The conversion cycle, which requires 44 System Clock periods, is initiated on the 10th falling edge of the I/O Clock↓ after CS↓ goes low for the channel whose address exists in memory at that time. If $\overline{CS}$ is kept low during conversion, the I/O Clock must remain low for at least 44 System Clock cycles to allow conversion to be completed.

B. The most significant bit (MSB) will automatically be placed on the DATA OUT bus after $\overline{CS}$ is brought low. The remaining nine bits (A8-A0) will be clocked out on the first nine I/O Clock falling edges.

C. To minimize errors caused by noise at the CS input, the internal circuitry waits for three System Clock cycles (or less) after a chip-select falling edge is detected before responding to control input signals. Therefore, no attempt should be made to clock-in address data until the minimum chip-select setup time has elapsed.

absolute maximum ratings over operating free-air temperature range (unless otherwise noted)

Supply voltage, V_{CC} (see Note 1) . . . . 6.5 V
Input voltage range (any input) . . . . −0.3 V to V_{CC} + 0.3 V
Output voltage range . . . . −0.3 V to V_{CC} + 0.3 V
Peak input current range (any input) . . . . ±10 mA
Peak total input current (all inputs) . . . . ±30 mA
Operating free-air temperature range: TLC1540I, TLC1541I . . . . −40°C to 85°C
TLC1540M, TLC1541M . . . . −55°C to 125°C
Storage temperature range . . . . −65°C to 150°C
Case temperature for 10 seconds: FN package . . . . 260°C
Lead temperature 1,6 mm (1/16 inch) from the case for 10 seconds: N package . . . . 260°C

NOTE 1: All voltage values are with respect to digital ground with REF− and GND wired together (unless otherwise noted).

recommended operating conditions

			TLC1540, TLC1541			UNIT
			MIN	NOM	MAX	
Supply voltage, V_{CC}			4.75	5	5.5	V
Positive reference voltage, V_{REF+} (see Note 2)			2.5	V_{CC}	V_{CC}+0.1	V
Negative reference voltage, V_{REF-} (see Note 2)			−0.1	0	2.5	V
Differential reference voltage, $V_{REF+} - V_{REF-}$ (see Note 2)			1	V_{CC}	V_{CC}+0.2	V
Analog input voltage (see Note 2)			0		V_{CC}	V
High-level control input voltage, V_{IH}			2			V
Low-level control input voltage, V_{IL}					0.8	V
Setup time, address bits before I/O CLK↑, $t_{su(A)}$			400			ns
Hold time, address bits after I/O CLK↑, $t_{h(A)}$			0			ns
Setup time, $\overline{CS}$ low before clocking in first address bit, $t_{su(CS)}$ (see Note 3)			3			System clock cycles
$\overline{CS}$ high during conversion, $t_{wH(CS)}$			44			System clock cycles
Input/Output clock frequency, $f_{CLK(I/O)}$			0		1.1	MHz
System clock frequency, $f_{CLK(SYS)}$			$f_{CLK(I/O)}$		2.1	MHz
System clock high, $t_{wH(SYS)}$			210			ns
System clock low, $t_{wL(SYS)}$			190			ns
Input/Output clock high, $t_{wH(I/O)}$			404			ns
Input/Output clock low, $t_{wL(I/O)}$			404			ns
Clock transition time (see Note 4)	System	$f_{CLK(SYS)} \le$ 1048 kHz			30	ns
		$f_{CLK(SYS)} >$ 1048 kHz			20	
	I/O	$f_{CLK(I/O)} \le$ 525 kHz			100	ns
		$f_{CLK(I/O)} >$ 525 kHz			40	
Operating free-air temperature, T_A		TLC1540M, TLC1541M	−55		125	°C
		TLC1540I, TLC1541I	−40		85	

NOTES: 2. Analog input voltages greater than that applied to REF+ convert as all "1"s (11111111), while input voltages less than that applied to REF− convert as all "0"s (00000000). For proper operation, REF+ voltage must be at least 1 volt higher than REF− voltage. Also, the total unadjusted error may increase as this differential reference voltage falls below 4.75 volts.
3. To minimize errors caused by noise at the chip select input, the internal circuitry waits for three System Clock cycles (or less) after a chip select falling edge is detected before responding to control input signals. Therefore, no attempt should be made to clock-in an address until the minimum chip select setup time has elapsed.
4. This is the time required for the clock input signal to fall from V_{IH} min to V_{IL} max or to rise from V_{IL} max to V_{IH} min. In the vicinity of normal room temperature, the devices function with input clock transition time as slow as 2 microseconds for remote data acquisition applications where the sensor and the A/D converter are placed several feet away from the controlling microprocessor.

electrical characteristics over recommended operating temperature range, $V_{CC} = V_{REF+} = 4.75$ V to 5.5 V (unless otherwise noted), $f_{CLK(I/O)} = 1.1$ MHz, $f_{CLK(SYS)} = 2.1$ MHz

PARAMETER			TEST CONDITIONS		MIN	TYP†	MAX	UNIT
V_{OH}	High-level output voltage (pin 16)		$V_{CC} = 4.75$ V,	$I_{OH} = 360$ µA	2.4			V
V_{OL}	Low-level output voltage		$V_{CC} = 4.75$ V,	$I_{OL} = 3.2$ mA			0.4	V
I_{OZ}	Off-state (high-impedance state) output current		$V_O = V_{CC}$,	$\overline{CS}$ at V_{CC}			10	µA
			$V_O = 0$,	$\overline{CS}$ at V_{CC}			−10	µA
I_{IH}	High-level input current		$V_I = V_{CC}$			0.005	2.5	µA
I_{IL}	Low-level input current		$V_I = 0$			−0.005	−2.5	µA
I_{CC}	Operating supply current		$\overline{CS}$ at 0 V			1.2	2.5	mA
	Selected channel leakage current		Selected channel at V_{CC}, Unselected channel at 0 V			0.4	1	µA
			Selected channel at 0 V, Unselected channel at V_{CC}			−0.4	−1	µA
$I_{CC} + I_{REF}$	Supply and reference current		$V_{REF+} = V_{CC}$,	$\overline{CS}$ at 0 V		1.3	3	mA
C_i	Input capacitance	Analog inputs				7	55	pF
		Control inputs				5	15	pF

operating characteristics over recommended operating free-air temperature range, $V_{CC} = V_{REF+} = 4.75$ V to 5.5 V, $f_{CLK(I/O)} = 1.1$ MHz, $f_{CLK(SYS)} = 2.1$ MHz

PARAMETER		TEST CONDITIONS	TLC1540 MIN	TLC1540 MAX	UNIT
	Linearity error	See Note 5		±0.5	LSB
	Zero error	See Notes 2 and 6		±0.5	LSB
	Full-scale error	See Notes 2 and 6		±0.5	LSB
	Total unadjusted error	See Note 7		±0.5	LSB
	Self-test output code	Input A11 address = 1011 (See Note 8)	0111110100 (500)	1000001100 (524)	
t_{conv}	Conversion time	See Operating Sequence		21	µs
	Total access and conversion time	See Operating Sequence		31	µs
t_{acq}	Channel acquisition time (sample cycle)	See Operating Sequence		6	I/O clock cycles
t_v	Time output data remains valid after I/O clock↓		10		ns
t_d	Delay time, I/O clock↓ to data output valid	See Parameter Measurement Information		400	ns
t_{en}	Output enable time	See Parameter Measurement Information		150	ns
t_{dis}	Output disable time	See Parameter Measurement Information		150	ns
$t_{r(bus)}$	Data bus rise time	See Parameter Measurement Information		300	ns
$t_{f(bus)}$	Data bus fall time	See Parameter Measurement Information		300	ns

† All typical values are at $V_{CC} = 5$ V, $T_A = 25°C$.

NOTES: 2. Analog input voltages greater than that applied to REF+ convert to all ''1''s (11111111), while input voltages less than that applied to REF− convert to all''0''s (00000000). For proper operation, REF+ voltage must be at least 1 volt higher than REF− voltage. Also, the total unadjusted error may increase as this differential reference voltage falls below 4.75 volts.

5. Linearity error is the maximum deviation from the best straight line through the A/D transfer characteristics.
6. Zero error is the difference between 00000000 and the converted output for zero input voltage; full-scale error is the difference between 11111111 and the converted output for full-scale input voltage.
7. Total unadjusted error comprises linearity, zero, and full-scale errors.
8. Both the input address and the output codes are expressed in positive logic. The A11 analog input signal is internally generated and is used for test purposes.

TEXAS INSTRUMENTS
POST OFFICE BOX 655012 • DALLAS, TEXAS 75265

electrical characteristics over recommended operating temperature range, $V_{CC} = V_{REF+} = 4.75$ V to 5.5 V (unless otherwise noted), $f_{CLK(I/O)} = 1.1$ MHz, $f_{CLK(SYS)} = 2.1$ MHz

PARAMETER			TEST CONDITIONS		MIN	TYP†	MAX	UNIT
V_{OH}	High-level output voltage (pin 16)		$V_{CC} = 4.75$ V,	$I_{OH} = 360$ μA	2.4			V
V_{OL}	Low-level output voltage		$V_{CC} = 4.75$ V,	$I_{OL} = 3.2$ mA			0.4	V
I_{OZ}	Off-state (high-impedance state) output current		$V_O = V_{CC}$,	$\overline{CS}$ at V_{CC}			10	μA
			$V_O = 0$,	$\overline{CS}$ at V_{CC}			−10	
I_{IH}	High-level input current		$V_I = V_{CC}$			0.005	2.5	μA
I_{IL}	Low-level input current		$V_I = 0$			−0.005	−2.5	μA
I_{CC}	Operating supply current		$\overline{CS}$ at 0 V			1.2	2.5	mA
	Selected channel leakage current		Selected channel at V_{CC}, Unselected channel at 0 V			0.4	1	μA
			Selected channel at 0 V, Unselected channel at V_{CC}			−0.4	−1	
$I_{CC} + I_{REF}$	Supply and reference current		$V_{REF+} = V_{CC}$,	$\overline{CS}$ at 0 V		1.3	3	mA
C_i	Input capacitance	Analog inputs				7	55	pF
		Control inputs				5	15	

operating characteristics over recommended operating free-air temperature range, $V_{CC} = V_{REF+} = 4.75$ V to 5.5 V, $f_{CLK(I/O)} = 1.1$ MHz, $f_{CLK(SYS)} = 2.1$ MHz

PARAMETER		TEST CONDITIONS	TLC1541 MIN	TLC1541 MAX	UNIT
	Linearity error	See Note 5		±1	LSB
	Zero error	See Notes 2 and 6		±1	LSB
	Full-scale error	See Notes 2 and 6		±1	LSB
	Total unadjusted error	See Note 7		±1	LSB
	Self-test output code	Input A11 address = 1011 (See Note 8)	0111110100 (500)	1000001100 (524)	
t_{conv}	Conversion time	See Operating Sequence		21	μs
	Total access and conversion time	See Operating Sequence		31	μs
t_{acq}	Channel acquisition time (sample cycle)	See Operating Sequence		6	I/O clock cycles
t_v	Time output data remains valid after I/O clock↓		10		ns
t_d	Delay time, I/O clock↓ to data output valid	See Parameter Measurement Information		400	ns
t_{en}	Output enable time			150	ns
t_{dis}	Output disable time			150	ns
$t_{r(bus)}$	Data bus rise time			300	ns
$t_{f(bus)}$	Data bus fall time			300	ns

† All typical values are at $V_{CC} = 5$ V, $T_A = 25$°C.

NOTES: 2. Analog input voltages greater than that applied to REF+ convert to all "1"s (11111111), while input voltages less than that applied to REF− convert to all "0"s (00000000). For proper operation, REF+ voltage must be at least 1 volt higher than REF− voltage. Also, the total unadjusted error may increase as this differential reference voltage falls below 4.75 volts.

5. Linearity error is the maximum deviation from the best straight line through the A/D transfer characteristics.
6. Zero error is the difference between 00000000 and the converted output for zero input voltage; full-scale error is the difference between 11111111 and the converted output for full-scale input voltage.
7. Total unadjusted error comprises linearity, zero, and full-scale errors.
8. Both the input address and the output codes are expressed in positive logic. The A11 analog input signal is internally generated and is used for test purposes.

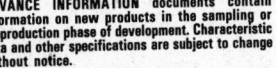

PARAMETER MEASUREMENT INFORMATION

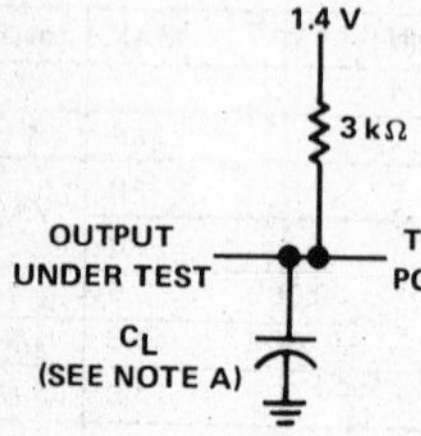
1.4 V
3 kΩ
OUTPUT
UNDER TEST
TEST
POINT
C_L
(SEE NOTE A)

LOAD CIRCUIT FOR
t_d, t_r, AND t_f

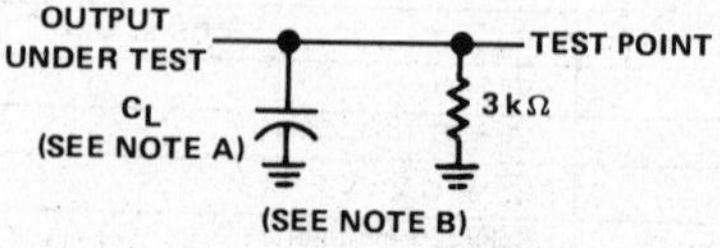
OUTPUT
UNDER TEST
TEST POINT
C_L
(SEE NOTE A)
3 kΩ
(SEE NOTE B)

LOAD CIRCUIT FOR
t_{PZH} AND t_{PHZ}

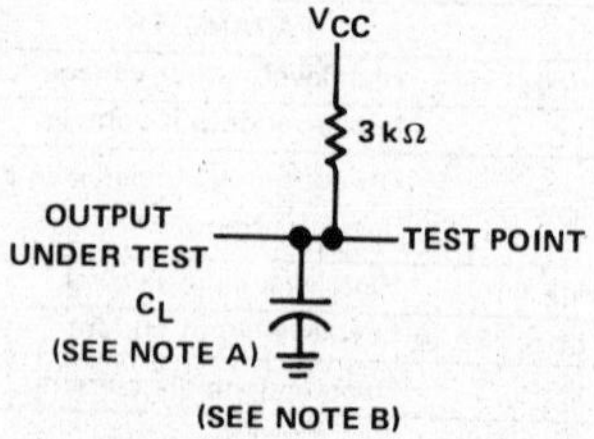
V_{CC}
3 kΩ
OUTPUT
UNDER TEST
TEST POINT
C_L
(SEE NOTE A)
(SEE NOTE B)

LOAD CIRCUIT FOR
t_{PZL} AND t_{PLZ}

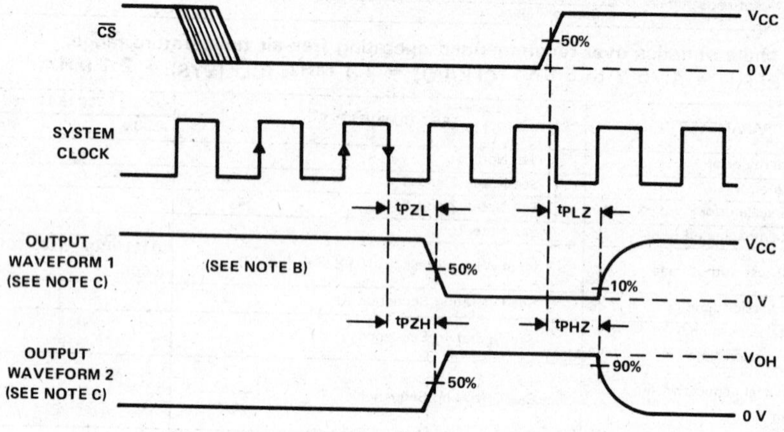
$\overline{CS}$
V_{CC}
50%
0 V
SYSTEM
CLOCK
t_{PZL}
t_{PLZ}
OUTPUT
WAVEFORM 1
(SEE NOTE C)
(SEE NOTE B)
V_{CC}
50%
10%
0 V
t_{PZH}
t_{PHZ}
OUTPUT
WAVEFORM 2
(SEE NOTE C)
V_{OH}
90%
50%
0 V

VOLTAGE WAVEFORMS FOR ENABLE AND DISABLE TIMES

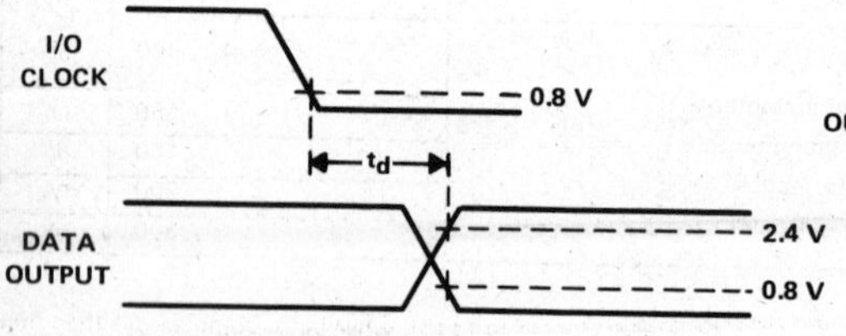
I/O
CLOCK
0.8 V
t_d
DATA
OUTPUT
2.4 V
0.8 V

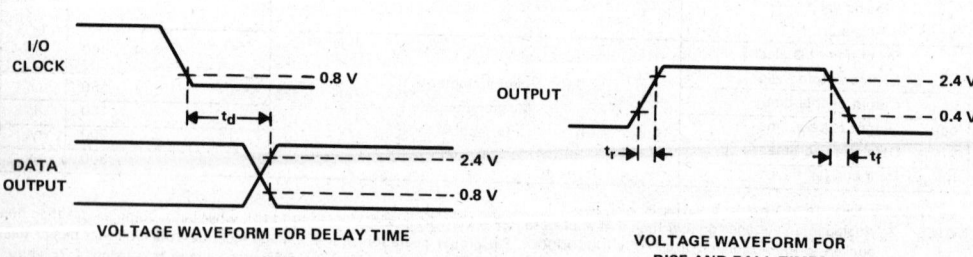

VOLTAGE WAVEFORM FOR DELAY TIME

OUTPUT
2.4 V
0.4 V
t_r
t_f

VOLTAGE WAVEFORM FOR
RISE AND FALL TIMES

NOTES: A. C_L = 50 pF
B. $t_{en} = t_{PZH}$ or t_{PZL}, $t_{dis} = t_{PHZ}$ or t_{PLZ}.
C. Waveform 1 is for an output with internal conditions such that the output is low except when disabled by the output control.
Waveform 2 is for an output with internal conditions such that the output is high except when disabled by the output control.

principles of operation

The TLC1540 and TLC1541 are both complete data acquisition systems on single chips. Each includes such functions as sample-and-hold, 10-bit A/D converter, data and control registers, and control logic. For flexibility and access speed, there are four control inputs; Chip Select ($\overline{CS}$), Address Input, I/O Clock, and System Clock. These control inputs and a TTL-compatible three-state output are intended for serial communications with a microprocessor or microcomputer. The TLC1540 and TLC1541 can complete conversions in a maximum of 21 microseconds, while complete input-conversion-output cycles can be repeated at a maximum of 31 microseconds.

The System and I/O Clocks are normally used independently and do not require any special speed or phase relationships between them. This independence simplifies the hardware and software control tasks for the device. Once a clock signal within the specification range is applied to the System Clock input, the control hardware and software need only be concerned with addressing the desired analog channel, reading the previous conversion result, and starting the conversion by using the I/O Clock. The System Clock will drive the ''conversion crunching'' circuitry so that the control hardware and software need not be concerned with this task.

When $\overline{CS}$ is high, the Data Output pin is in a three-state condition and the Address Input and I/O Clock pins are disabled. This feature allows each of these pins, with the exception of the $\overline{CS}$ pin, to share a control logic point with their counterpart pins on additional A/D devices when additional TLC1540/1541 devices are used. In this way, the above feature serves to minimize the required control logic pins when using multiple A/D devices.

The control sequence has been designed to minimize the time and effort required to initiate conversion and obtain the conversion result. A normal control sequence is:

1. $\overline{CS}$ is brought low. To minimize errors caused by noise at the $\overline{CS}$ input, the internal circuitry waits for two rising edges and then a falling edge of the System Clock after a low $\overline{CS}$ transition, before the low transition is recognized. This technique is used to protect the device against noise when the device is used in a noisy environment. The MSB of the previous conversion result will automatically appear on the Data Out pin.
2. A new positive-logic multiplexer address is shifted in on the first four rising edges of the I/O Clock. The MSB of the address is shifted in first. The negative edges of these four I/O Clock pulses shift out the second, third, fourth, and fifth most significant bits of the previous conversion result. The on-chip sample-and-hold begins sampling the newly addressed analog input after the fourth falling edge. The sampling operation basically involves the charging of internal capacitors to the level of the analog input voltage.
3. Five clock cycles are then applied to the I/O pin and the sixth, seventh, eighth, ninth, and tenth conversion bits are shifted out on the negative edges of these clock cycles.
4. The final tenth clock cycle is applied to the I/O Clock pin. The falling edge of this clock cycle completes the analog sampling process and initiates the hold function. Conversion is then performed during the next 44 System Clock cycles. After this final I/O Clock cycle, $\overline{CS}$ must go high or the I/O Clock must remain low for at least 44 System Clock cycles to allow for the conversion function.

$\overline{CS}$ can be kept low during periods of multiple conversion. When keeping $\overline{CS}$ low during periods of multiple conversion, special care must be exercised to prevent noise glitches on the I/O Clock line. If glitches occur on the I/O Clock line, the I/O sequence between the microprocessor/controller and the device will lose synchronization. Also, if $\overline{CS}$ is taken high, it must remain high until the end of the conversion. Otherwise, a valid falling edge of $\overline{CS}$ will cause a reset condition, which will abort the conversion in progress.

A new conversion may be started and the ongoing conversion simultaneously aborted by performing steps 1 through 4 before the 44 System Clock cycles occur. Such action will yield the conversion result of the previous conversion and not the ongoing conversion.

principles of operation (continued)

It is possible to connect the System and I/O Clock pins together in special situations in which controlling circuitry points must be minimized. In this case, the following special points must be considered in addition to the requirements of the normal control sequence previously described.

1. When $\overline{CS}$ is recognized by the device to be at a low level, the common clock signal is used as an I/O Clock. When $\overline{CS}$ is recognized by the device to be at a high level, the common clock signal is used to drive the "conversion crunching" circuitry.
2. The device will recognize a $\overline{CS}$ low transition only when the $\overline{CS}$ input changes and subsequently the System Clock pin receives two positive edges and then a negative edge. For this reason, after a $\overline{CS}$ negative edge, the first two clock cycles will not shift in the address because a low $\overline{CS}$ must be recognized before the I/O Clock can shift in an analog channel address. Also, upon shifting in the address, $\overline{CS}$ must be raised after the eighth I/O Clock that has been recognized by the device, so that a $\overline{CS}$ low level will be recognized upon the lowering of the tenth I/O Clock signal that is recognized by the device. Otherwise, additional common clock cycles will be recognized as I/O Clock pulses and will shift in an erroneous address.

For certain applications, such as strobing applications, it is necessary to start conversion at a specific point in time. This device will accommodate these applications. Although the on-chip sample-and-hold begins sampling upon the negative edge of the fourth I/O Clock cycle, the hold function is not initiated until the negative edge of the tenth I/O Clock cycle. Thus, the control circuitry can leave the I/O Clock signal in its high state during the tenth I/O Clock cycle until the moment at which the analog signal must be converted. The TLC1540/TLC1541 will continue sampling the analog input until the tenth falling edge of the I/O Clock. The control circuitry or software will then immediately lower the I/O Clock signal and hold the analog signal at the desired point in time and start conversion.

Detailed information on interfacing to most popular microprocessors is readily available from the factory.

D2922, JANUARY 1986

- High Degree of Linearity
- High On-Off Output Voltage Ratio
- Low Crosstalk Between Switches
- Low On-State Impedance of 50 Ohms Typ at V_{CC} = 9 V
- Individual Switch Controls
- Extremely Low Input Current

TLC4016M . . . J OR N PACKAGE
TLC4016I . . . D OR N PACKAGE
(TOP VIEW)

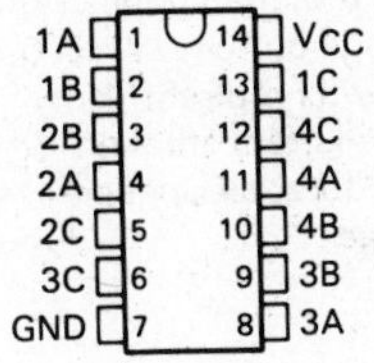
1A 1
1B 2
2B 3
2A 4
2C 5
3C 6
GND 7
14 VCC
13 1C
12 4C
11 4A
10 4B
9 3B
8 3A

description

The TLC4016 is a silicon-gate CMOS quadruple analog switch integrated circuit designed to handle both analog and digital signals. Each switch permits signals with amplitudes up to 12 volts peak to be transmitted in either direction.

Each switch section has its own enable input control. A high-level voltage applied to this control terminal turns on the associated switch section.

Applications include signal gating, chopping, modulation or demodulation (modem), and signal multiplexing for analog-to-digital and digital-to-analog conversion systems.

The TLC4016M is characterized for operation from −55°C to 125°C, and the TLC4016I is characterized from −40°C to 85°C.

logic symbol†

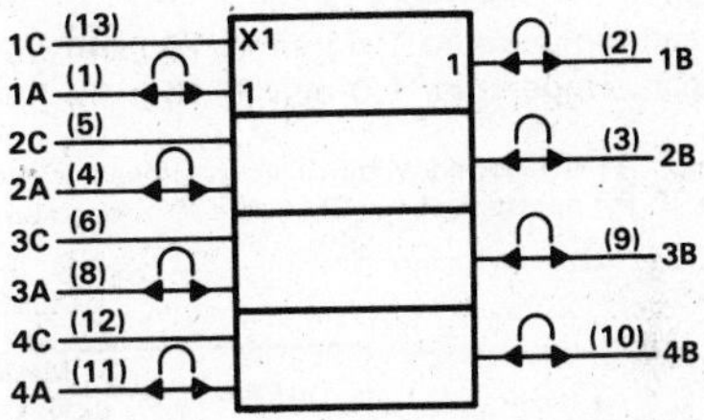
1C (13)
1A (1)
2C (5)
2A (4)
3C (6)
3A (8)
4C (12)
4A (11)
X1
1
1
(2) 1B
(3) 2B
(9) 3B
(10) 4B

† This symbol is in accordance with ANSI/IEEE Std 91-1984 and IEC Publication 617-12.

logic diagram (positive logic)

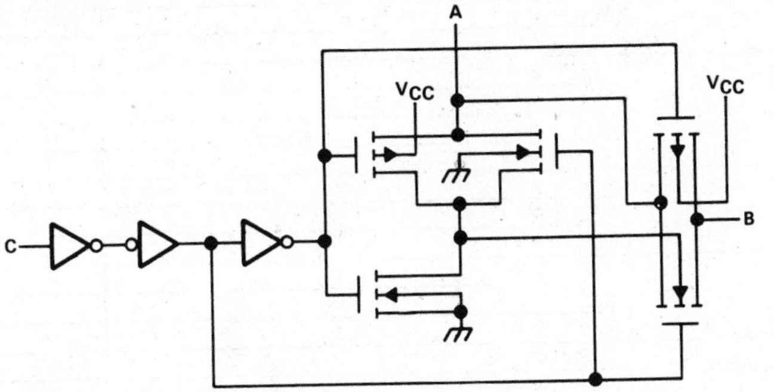
A
VCC
VCC
C
B

absolute maximum ratings over operating free-air temperature range (unless otherwise noted)

Supply voltage range (see Note 1) . . . . −0.5 V to 15 V
Control-input diode current ($V_I < 0$ or $V_I > V_{CC}$) . . . . ±20 mA
I/O port diode current ($V_I < 0$ or $V_{I/O} > V_{CC}$) . . . . ±20 mA
On-state switch current ($V_{I/O}$ = 0 to V_{CC}) . . . . ±25 mA
Continuous current through V_{CC} or GND pins . . . . ±50 mA
Continuous total dissipation at (or below) 25°C free-air temperature (see Note 2):
D package . . . . 950 mW
J package . . . . 1025 mW
N package . . . . 875 mW
Operating free-air temperature, T_A: TLC4016M . . . . −55°C to 125°C
TLC4016I . . . . −40°C to 85°C
Storage temperature range . . . . −65°C to 150°C
Lead temperature 1,6 mm (1/16 inch) from case for 10 seconds: D and N packages . . . . 260°C
Lead temperature 1,6 mm (1/16 inch) from case for 60 seconds: J package . . . . 300°C

NOTES: 1. All voltages are with respect to ground unless otherwise specified.
2. For operation above 25°C free-air temperature, see Dissipation Derating Table.

DISSIPATION DERATING TABLE

Package	Maximum Power Dissipation			Derating Factor
	25°C	85°C	125°C	
D	950 mW	494 mW		7.6 mW/°C
J	1025 mW	533 mW	205 mW	8.2 mW/°C
N	875 mW	455 mW	175 mW	7.0 mW/°C

recommended operating conditions

		MIN	NOM	MAX	UNIT
Supply voltage, V_{CC}		2†	5	12	V
I/O port voltage, $V_{I/O}$		0		V_{CC}	V
High-level input voltage, V_{IH}	V_{CC} = 2 V	1.5		V_{CC}	V
	V_{CC} = 4.5 V	3.15		V_{CC}	
	V_{CC} = 9 V	6.3		V_{CC}	
	V_{CC} = 12 V	8.4		V_{CC}	
Low-level input voltage, V_{IL}	V_{CC} = 2 V	0		0.3	V
	V_{CC} = 4.5 V	0		0.9	
	V_{CC} = 9 V	0		1.8	
	V_{CC} = 12 V	0		2.4	
Input rise time, t_r	V_{CC} = 2 V			1000	ns
	V_{CC} = 4.5 V			500	
	V_{CC} = 9 V			400	
Input fall time, t_f	V_{CC} = 2 V			1000	ns
	V_{CC} = 4.5 V			500	
	V_{CC} = 9 V			400	
Operating free-air temperature, T_A	TLC4016M	−55		125	°C
	TLC4016I	−40		85	

†With supply voltages at or near 2 volts, the analog switch on-state resistance becomes very nonlinear. It is recommended that only digital signals be transmitted at these low supply voltages.

electrical characteristics over recommended operating free-air temperature range (unless otherwise noted).

PARAMETER		TEST CONDITIONS	V_{CC}	TLC4016M			TLC4016I			UNIT
				MIN	TYP†	MAX	MIN	TYP†	MAX	
r_{Son}	On-state switch resistance	I_S = 1 mA, V_A = 0 to V_{CC}, See Figure 1	4.5 V		100	220		100	200	Ω
			9 V		50	120		50	105	
			12 V		30	100		30	85	
		I_S = 1 mA, V_A = 0 or V_{CC}, See Figure 1	2 V		120	240		120	215	
			4.5 V		50	120		50	100	
			9 V		35	80		35	75	
			12 V		20	70		20	60	
	On-state switch resistance matching	V_A = 0 to V_{CC}, See Figure 1	4.5 V		10	20		10	20	Ω
			9 V		5	15		5	15	
			12 V		5	15		5	15	
I_I	Control input current	V_I = 0 or V_{CC}	2 V to 6 V			±1			±1	μA
		V_I = 0 or V_{CC}, T_A = 25°C				±0.1			±0.1	
I_{Soff}	Off-state switch leakage current	V_S = ±V_{CC}, See Figure 2	5.5 V		±10	±600		±10	±600	nA
			9 V		±15	±800		±15	±800	
			12 V		±20	±1000		±20	±1000	
I_{Son}	On-state switch leakage current	V_A = 0 or V_{CC}, See Figure 3	5.5 V		±10	±150		±10	±150	nA
			9 V		±15	±200		±15	±200	
			12 V		±20	±300		±20	±300	
I_{CC}	Supply current	V_I = 0 or V_{CC}, I_O = 0	5.5 V		2	40		2	20	μA
			9 V		8	160		8	80	
			12 V		16	320		16	160	
C_i	Input capacitance	A or B	2 V to 12 V		15			15		pF
		C			5	10		5	10	
C_f	Feedthrough capacitance	A to B, V_I = 0	2 V to 12 V		5			5		pF

†All typical values are at T_A = 25°C.

switching characteristics over recommended operating free-air temperature range, C_L = 50 pF (unless otherwise noted)

PARAMETER		TEST CONDITIONS	V_{CC}	TLC4016M			TLC4016I			UNIT
				MIN	TYP†	MAX	MIN	TYP†	MAX	
t_{pd}	Propagation delay time, A to B or B to A	See Figure 4	2 V		25	75		25	62	ns
			4.5 V		5	15		5	13	
			9 V		4	14		4	12	
			12 V		3	13		3	11	
t_{on}	Switch turn-on time	R_L = 1 kΩ, See Figures 5 and 6	2 V		32	150		32	125	ns
			4.5 V		8	30		8	25	
			9 V		6	18		6	15	
			12 V		5	15		5	13	
t_{off}	Switch turn-off time	R_L = 1 kΩ, See Figures 5 and 6	2 V		45	252		45	210	ns
			4.5 V		15	54		15	45	
			9 V		10	48		10	40	
			12 V		8	45		8	38	
f_{co}	Switch cutoff frequency (channel loss = 3 dB)		4.5 V		100			100		MHz
			9 V		120			120		
$V_{OCF(PP)}$	Control feedthrough voltage to any switch, peak to peak	See Figure 7	4.5 V			180			180	mV
	Frequency at which crosstalk attenuation between any two switches equals 50 dB	See Figure 8	4.5 V		1			1		MHz

†All typical values are at T_A = 25°C.

PARAMETER MEASUREMENT INFORMATION

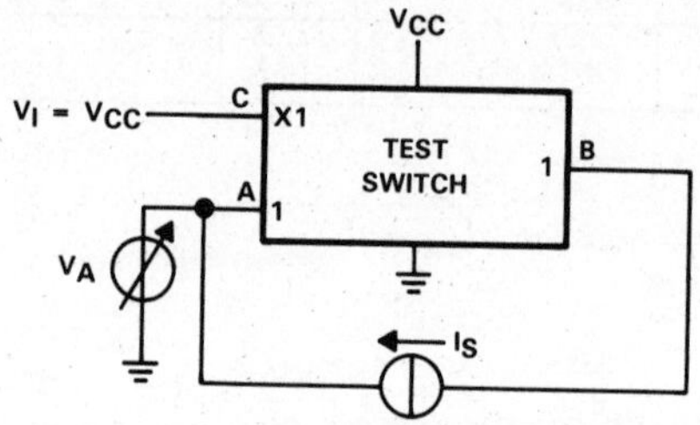
V_{CC}
$V_I = V_{CC}$
C
X1
TEST
SWITCH
1
B
A
1
V_A
I_S

FIGURE 1. ON-STATE RESISTANCE TEST CIRCUIT

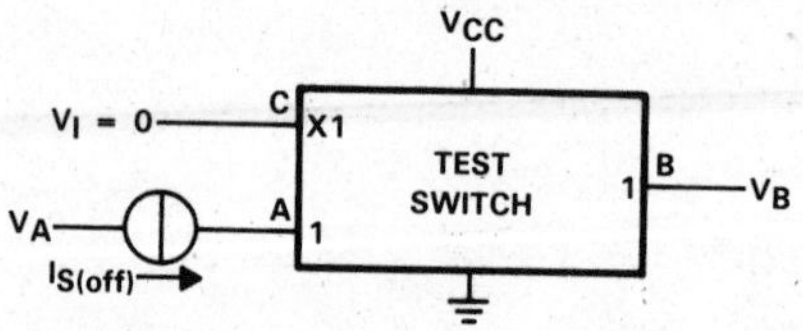
V_{CC}
$V_I = 0$
C
X1
TEST
SWITCH
1
B
V_B
V_A
A
1
$I_{S(off)}$

$V_S = V_A - V_B$

CONDITION 1: $V_A = 0$, $V_B = V_{CC}$

CONDITION 2: $V_A = V_{CC}$, $V_B = 0$

FIGURE 2. OFF-STATE SWITCH LEAKAGE CURRENT TEST CIRCUIT

PARAMETER MEASUREMENT INFORMATION

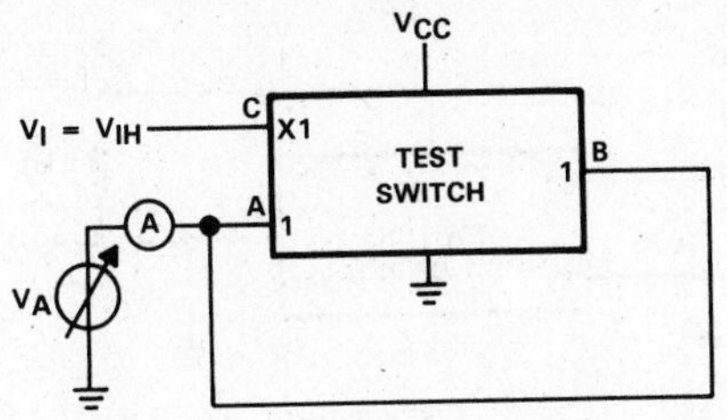
VCC
VI = VIH
C
X1
TEST
SWITCH
1
B
A
1
A
VA

FIGURE 3. ON-STATE SWITCH LEAKAGE CURRENT TEST CIRCUIT

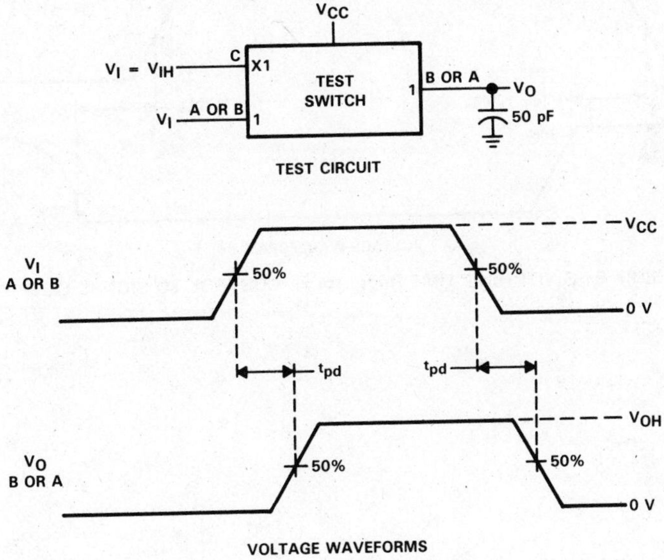
VCC
VI = VIH
C
X1
TEST
SWITCH
1
B OR A
VO
50 pF
VI
A OR B
1
TEST CIRCUIT
VCC
VI
A OR B
50%
50%
0 V
tpd
tpd
VOH
VO
B OR A
50%
50%
0 V
VOLTAGE WAVEFORMS

FIGURE 4. PROPAGATION DELAY TIME, SIGNAL INPUT TO SIGNAL OUTPUT

PARAMETER MEASUREMENT INFORMATION

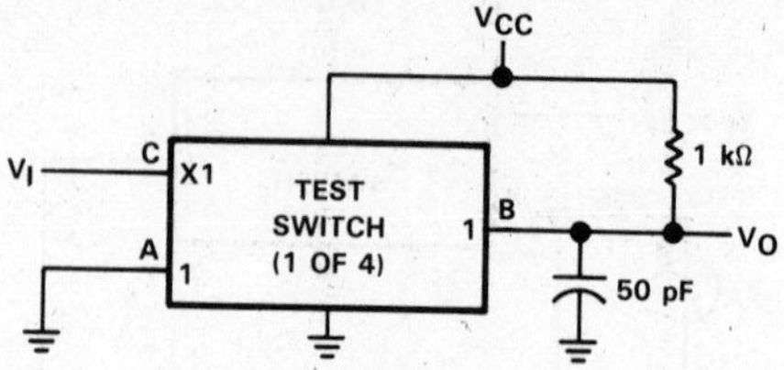
VCC
VI
C
X1
TEST
SWITCH
(1 OF 4)
A
1
1
B
1 kΩ
VO
50 pF

TEST CIRCUIT

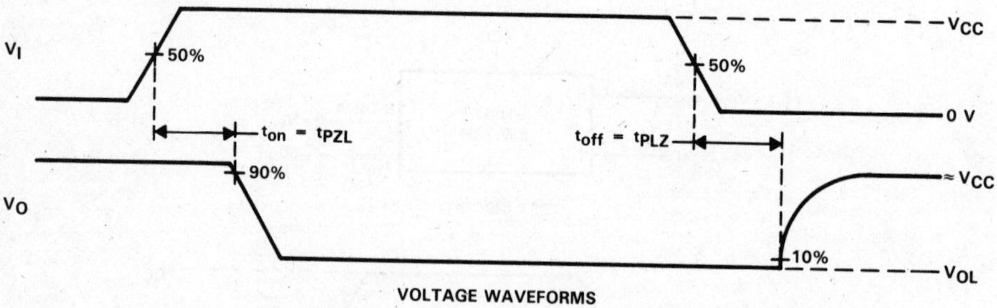
VI
50%
50%
VCC
0 V
ton = tPZL
toff = tPLZ
VO
90%
≈ VCC
10%
VOL

VOLTAGE WAVEFORMS

FIGURE 5. SWITCHING TIME (t_{PZL}, t_{PLZ}), CONTROL TO SIGNAL OUTPUT

PARAMETER MEASUREMENT INFORMATION

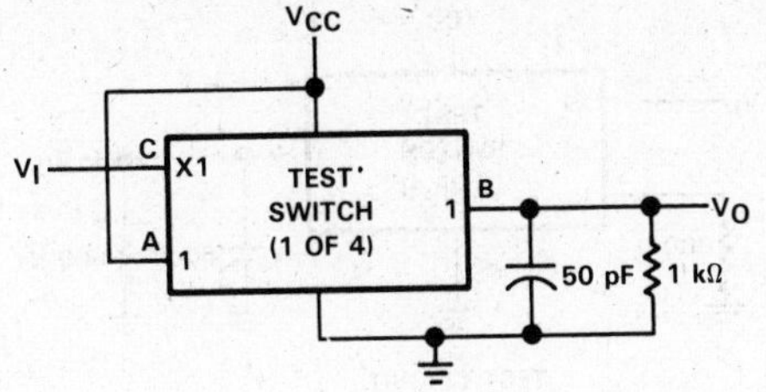
VCC
VI
C
X1
A
1
TEST
SWITCH
(1 OF 4)
1
B
VO
50 pF
1 kΩ

TEST CIRCUIT

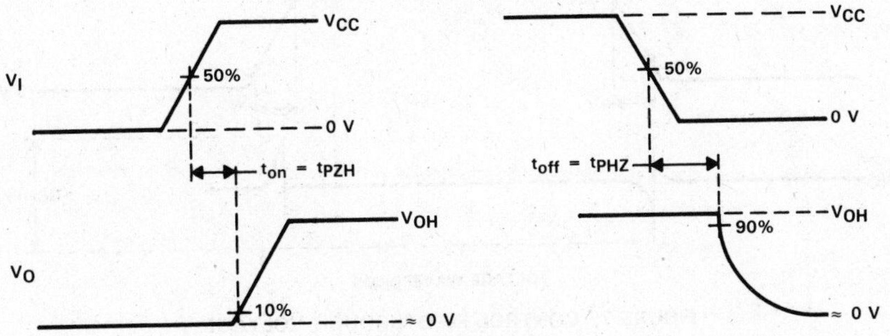
VI
VCC
50%
0 V
ton = tPZH
VOH
VO
10%
≈ 0 V
VCC
50%
0 V
toff = tPHZ
VOH
90%
≈ 0 V

VOLTAGE WAVEFORMS

FIGURE 6. SWITCHING TIME (t_{PZH}, t_{PHZ}), CONTROL TO SIGNAL OUTPUT

PARAMETER MEASUREMENT INFORMATION

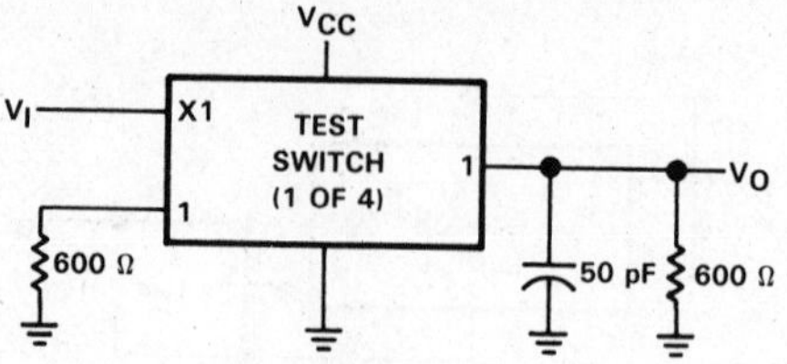
VCC
VI
X1
TEST
SWITCH
(1 OF 4)
1
1
600 Ω
50 pF
600 Ω
VO

TEST CIRCUIT

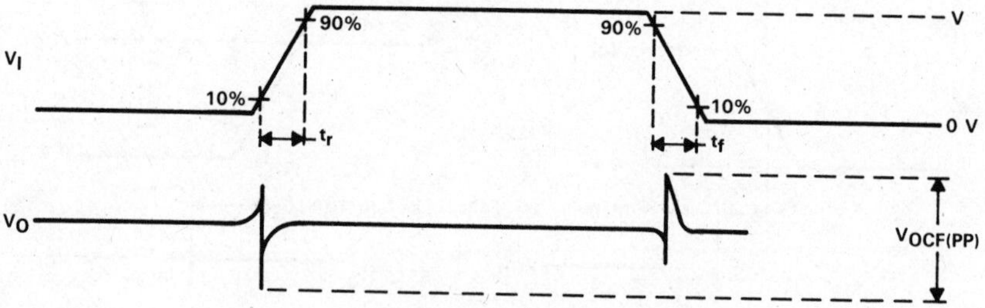
VI
90%
10%
90%
10%
V
0 V
tr
tf
VO
VOCF(PP)

VOLTAGE WAVEFORMS

FIGURE 7. CONTROL FEEDTHROUGH VOLTAGE

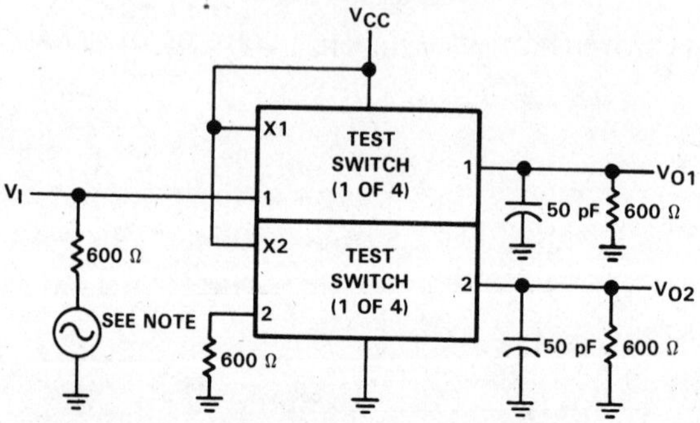
VCC
X1
TEST
SWITCH
(1 OF 4)
1
VO1
VI
1
50 pF
600 Ω
600 Ω
X2
TEST
SWITCH
(1 OF 4)
2
VO2
SEE NOTE
2
600 Ω
50 pF
600 Ω

NOTE: ADJUST f for $a_X = \frac{V_{O2}}{V_{O1}} = 50$ dB.

FIGURE 8. CROSSTALK BETWEEN ANY TWO SWITCHES, TEST CIRCUIT

SILICON-GATE CMOS QUADRUPLE BILATERAL ANALOG SWITCH

D2922, JANUARY 1986

- High Degree of Linearity
- High On-Off Output Voltage Ratio
- Low Crosstalk Between Switches
- Low On-State Impedance . . . Typically 30 Ohms at V_{CC} = 12 V
- Individual Switch Controls
- Extremely Low Input Current
- Functionally Interchangeable with National Semiconductor MM54/74HC4066, Motorola MC54/74HC4066, and RCA CD4066A

TLC4066M . . . J OR N PACKAGE
TLC4066I . . . D OR N PACKAGE
(TOP VIEW)

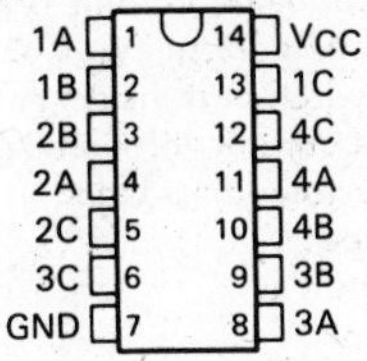
1A 1
1B 2
2B 3
2A 4
2C 5
3C 6
GND 7
14 VCC
13 1C
12 4C
11 4A
10 4B
9 3B
8 3A

logic symbol†

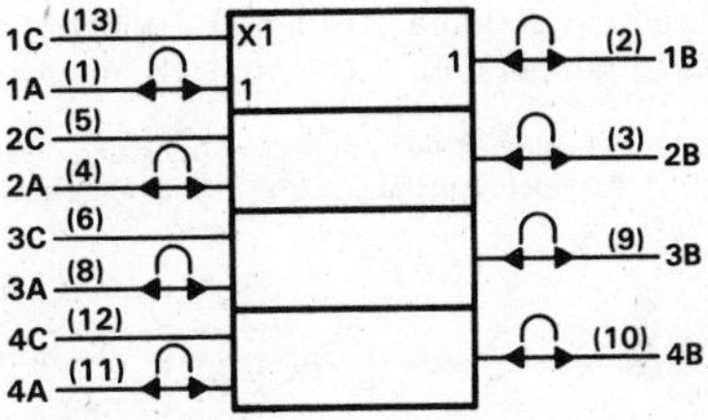
1C (13)
1A (1)
2C (5)
2A (4)
3C (6)
3A (8)
4C (12)
4A (11)
X1
1
1
(2) 1B
(3) 2B
(9) 3B
(10) 4B

† This symbol is in accordance with ANSI/IEEE Std 91-1984 and IEC Publication 617-12.

description

The TLC4066 is a silicon-gate CMOS quadruple analog switch integrated circuit designed to handle both analog and digital signals. Each switch permits signals with amplitudes up to 12 volts peak to be transmitted in either direction.

Each switch section has its own enable input control. A high-level voltage applied to this control terminal turns on the associated switch section.

Applications include signal gating, chopping, modulation or demodulation (modem), and signal multiplexing for analog-to-digital and digital-to-analog conversion systems.

The TLC4066M is characterized for operation from −55 °C to 125 °C. The TLC4066I is characterized from −40 °C to 85 °C.

logic diagram (positive logic)

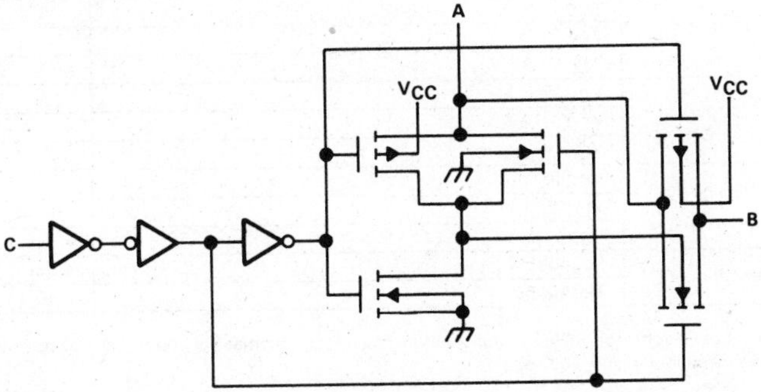
A
VCC
VCC
B
C

absolute maximum ratings over operating free-air temperature range (unless otherwise noted)

Supply voltage range (see Note 1) . . . . −0.5 V to 15 V
Control-input diode current ($V_I < 0$ or $V_I > V_{CC}$) . . . . ±20 mA
I/O port diode current ($V_I < 0$ or $V_{I/O} > V_{CC}$) . . . . ±20 mA
On-state switch current ($V_{I/O}$ = 0 to V_{CC}) . . . . ±25 mA
Continuous current through V_{CC} or GND pins . . . . ±50 mA
Continuous total dissipation at (or below) 25 °C free-air temperature (see Note 2):
  D package . . . . 950 mW
  J package . . . . 1025 mW
  N package . . . . 875 mW
Operating free-air temperature, T_A: TLC4066M . . . . −55 °C to 125 °C
  TLC4066I . . . . −40 °C to 85 °C
Storage temperature range . . . . −65 °C to 150 °C
Lead temperature 1,6 mm (1/16 inch) from case for 10 seconds: D and N packages . . . . 260 °C
Lead temperature 1,6 mm (1/16 inch) from case for 60 seconds: J package . . . . 300 °C

NOTES: 1. All voltages are with respect to ground unless otherwise specified.
2. For operation above 25 °C free-air temperature, see Dissipation Derating Table.

DISSIPATION DERATING TABLE

Package	Maximum Power Dissipation 25 °C	85 °C	125 °C	Derating Factor
D	950 mW	494 mW		7.6 mW/°C
J	1025 mW	533 mW	205 mW	8.2 mW/°C
N	875 mW	455 mW	175 mW	7.0 mW/°C

recommended operating conditions

		MIN	NOM	MAX	UNIT
Supply voltage, V_{CC}		2†	5	12	V
I/O port voltage, $V_{I/O}$		0		V_{CC}	V
High-level input voltage, V_{IH}	V_{CC} = 2 V	1.5		V_{CC}	V
	V_{CC} = 4.5 V	3.15		V_{CC}	
	V_{CC} = 9 V	6.3		V_{CC}	
	V_{CC} = 12 V	8.4		V_{CC}	
Low-level input voltage, V_{IL}	V_{CC} = 2 V	0		0.3	V
	V_{CC} = 4.5 V	0		0.9	
	V_{CC} = 9 V	0		1.8	
	V_{CC} = 12 V	0		2.4	
Input rise time, t_r	V_{CC} = 2 V			1000	ns
	V_{CC} = 4.5 V			500	
	V_{CC} = 9 V			400	
Input fall time, t_f	V_{CC} = 2 V			1000	ns
	V_{CC} = 4.5 V			500	
	V_{CC} = 9 V			400	
Operating free-air temperature, T_A	TLC4066M	−55		125	°C
	TLC4066I	−40		85	

†With supply voltages at or near 2 volts, the analog switch on-state resistance becomes very nonlinear. It is recommended that only digital signals be transmitted at these low supply voltages.

electrical characteristics over recommended operating free-air temperature range (unless otherwise noted)

PARAMETER		TEST CONDITIONS	V_{CC}	TLC4066M MIN	TLC4066M TYP†	TLC4066M MAX	TLC4066I MIN	TLC4066I TYP†	TLC4066I MAX	UNIT
r_{Son}	On-state switch resistance	I_S = 1 mA, V_A = 0 to V_{CC}, See Figure 1	4.5 V		100	220		100	200	Ω
			9 V		50	110		50	105	
			12 V		30	90		30	85	
		I_S = 1 mA, V_A = 0 or V_{CC}, See Figure 1	2 V		120	240		120	215	
			4.5 V		50	120		50	100	
			9 V		35	80		35	75	
			12 V		20	70		20	60	
	On-state switch resistance matching	V_A = 0 to V_{CC}, See Figure 1	4.5 V		10	20		10	20	Ω
			9 V		5	15		5	15	
			12 V		5	15		5	15	
I_I	Control input current	V_I = 0 or V_{CC}	2 V or 6 V			±1			±1	μA
I_{Soff}	Off-state switch leakage current	V_S = ±V_{CC}, See Figure 2	5.5 V		±10	±600		±10	±600	nA
			9 V		±15	±800		±15	±800	
			12 V		±20	±1000		±20	±1000	
I_{Son}	On-state switch leakage current	V_A = 0 or V_{CC}, See Figure 3	5.5 V		±10	±150		±10	±150	nA
			9 V		±15	±200		±15	±200	
			12 V		±20	±300		±20	±300	
I_{CC}	Supply current	V_I = 0 or V_{CC}, I_O = 0	5.5 V		2	40		2	20	μA
			9 V		8	160		8	80	
			12 V		16	320		16	160	
C_i	Input capacitance, A or B		2 V to 12 V		15			15		pF
	Input capacitance, C				5	10		5	10	
C_f	Feedthrough capacitance, A to B	V_I = 0	2 V to 12 V		5			5		pF

†All typical values are at T_A = 25°C.

switching characteristics over recommended operating free-air temperature range, C_L = 50 pF (unless otherwise noted)

PARAMETER		TEST CONDITIONS	V_{CC}	TLC4066M			TLC4066I			UNIT
				MIN	TYP†	MAX	MIN	TYP†	MAX	
t_{pd}	Propagation delay time, A to B or B to A	See Figure 4	2 V		25	75		15	30	ns
			4.5 V		5	15		5	13	
			9 V		4	12		4	10	
			12 V		3	13		3	11	
t_{on}	Switch turn-on time	R_L = 1 kΩ, See Figures 5 and 6	2 V		32	150		32	125	ns
			4.5 V		8	30		8	25	
			9 V		6	18		6	15	
			12 V		5	15		5	13	
t_{off}	Switch turn-off time	R_L = 1 kΩ, See Figures 5 and 6	2 V		45	252		45	210	ns
			4.5 V		15	54		15	45	
			9 V		10	48		10	40	
			12 V		8	45		8	38	
f_{co}	Switch cutoff frequency (channel loss = 3 dB)		4.5 V		100			100		MHz
			9 V		120			120		
$V_{OCF(PP)}$	Control feedthrough voltage to any switch, peak to peak	See Figure 7	4.5 V			180			180	mV
	Frequency at which crosstalk attenuation between any two switches equals 50 dB	See Figure 8	4.5 V		1			1		MHz

†All typical values are at T_A = 25°C.

PARAMETER MEASUREMENT INFORMATION

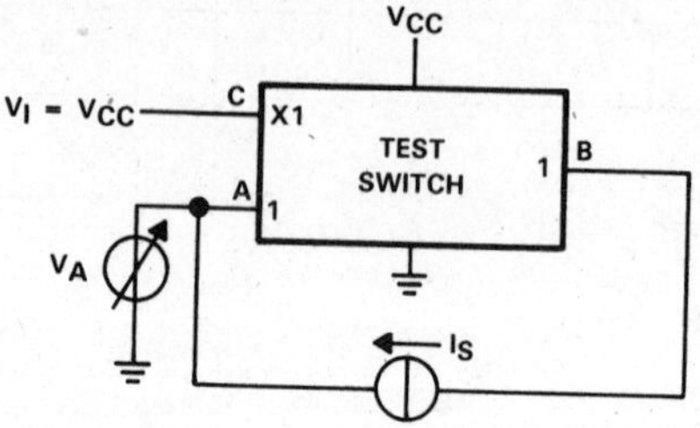
VCC
VI = VCC
C
X1
TEST
SWITCH
1
B
A
1
VA
IS

FIGURE 1. ON-STATE RESISTANCE TEST CIRCUIT

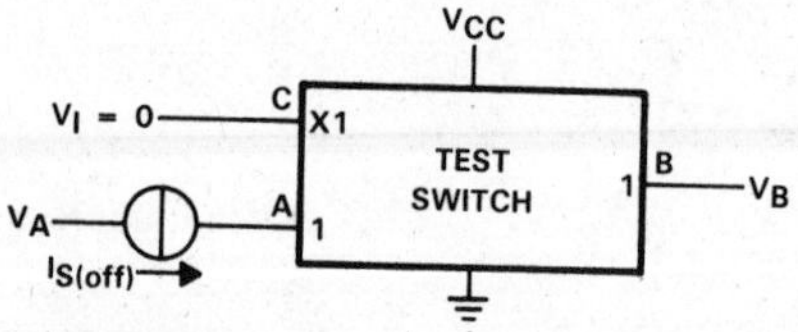
VCC
VI = 0
C
X1
TEST
SWITCH
1
B
VB
VA
A
1
IS(off)

$V_S = V_A - V_B$

CONDITION 1: $V_A = 0$, $V_B = V_{CC}$

CONDITION 2: $V_A = V_{CC}$, $V_B = 0$

FIGURE 2. OFF-STATE SWITCH LEAKAGE CURRENT TEST CIRCUIT

PARAMETER MEASUREMENT INFORMATION

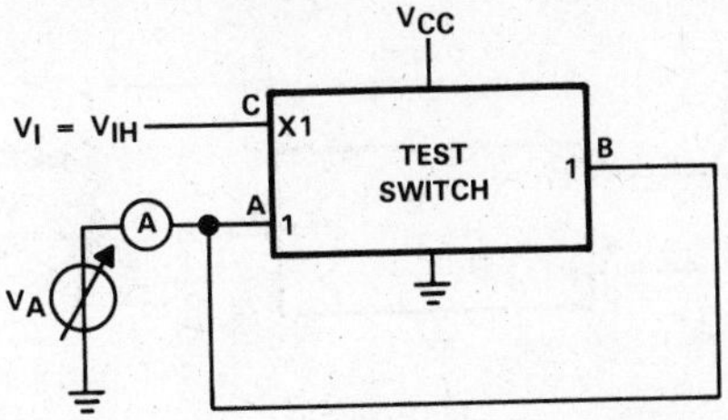
VCC
VI = VIH
C
X1
TEST
SWITCH
B
1
A
1
A
VA

FIGURE 3. ON-STATE SWITCH LEAKAGE CURRENT TEST CIRCUIT

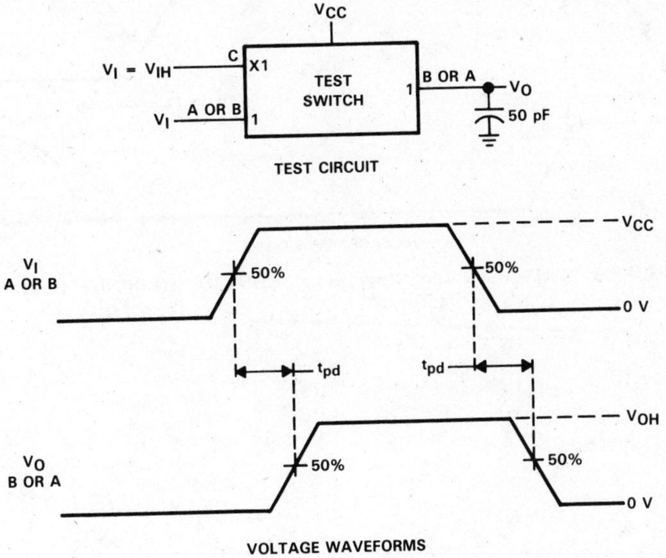
VCC
VI = VIH
C
X1
TEST
SWITCH
1
B OR A
VO
50 pF
VI
A OR B
1
TEST CIRCUIT
VCC
VI
A OR B
50%
50%
0 V
tpd
tpd
VOH
VO
B OR A
50%
50%
0 V
VOLTAGE WAVEFORMS

FIGURE 4. PROPAGATION DELAY TIME, SIGNAL INPUT TO SIGNAL OUTPUT

PARAMETER MEASUREMENT INFORMATION

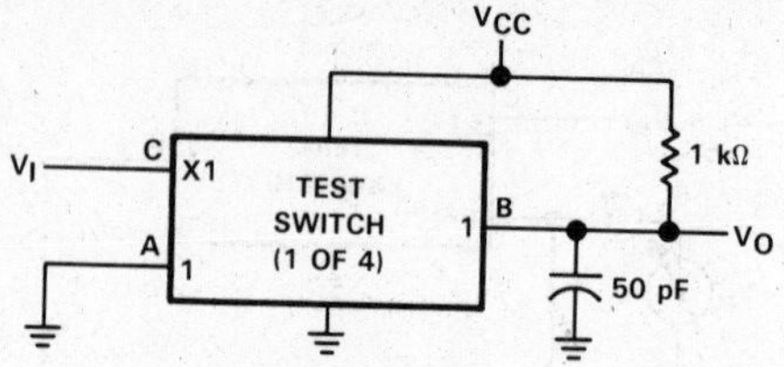
V_{CC}
V_I
C
X1
TEST
SWITCH
(1 OF 4)
A
1
1
B
1 kΩ
V_O
50 pF

TEST CIRCUIT

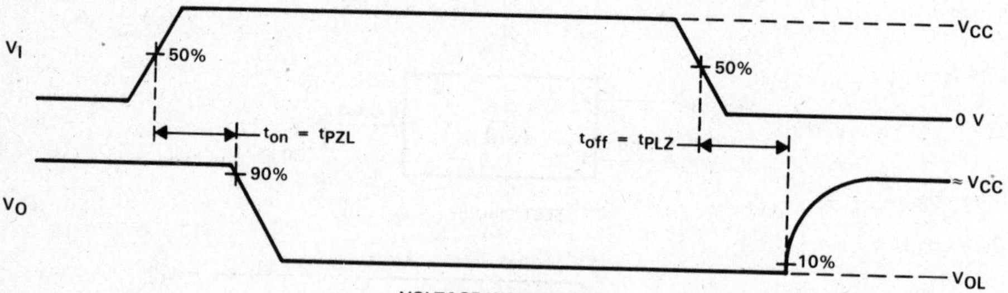
V_I
50%
50%
V_{CC}
0 V
t_{on} = t_{PZL}
t_{off} = t_{PLZ}
90%
V_O
≈ V_{CC}
10%
V_{OL}

VOLTAGE WAVEFORMS

FIGURE 5. SWITCHING TIME (t_{PZL}, t_{PLZ}), CONTROL TO SIGNAL OUTPUT

PARAMETER MEASUREMENT INFORMATION

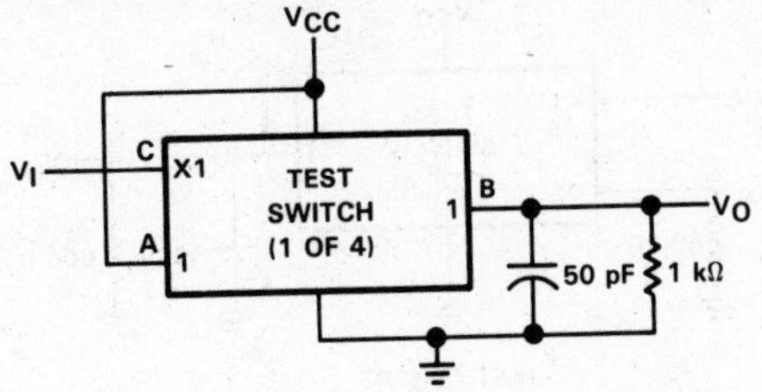
VCC
VI
C
X1
A
1
TEST
SWITCH
(1 OF 4)
1
B
VO
50 pF
1 kΩ

TEST CIRCUIT

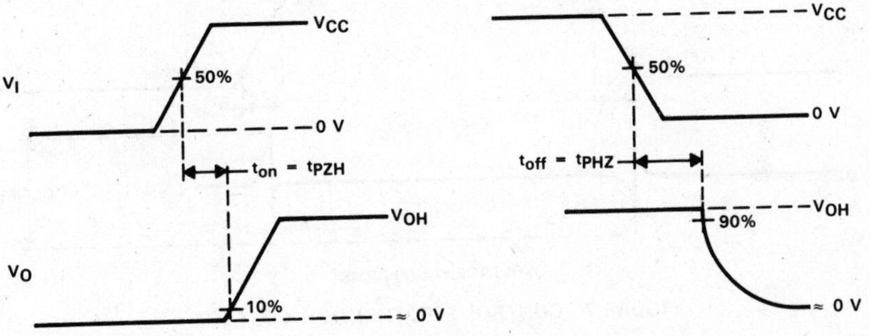
VCC
VI
50%
0 V
ton = tPZH
VOH
VO
10%
≈ 0 V
VCC
50%
0 V
toff = tPHZ
VOH
90%
≈ 0 V

VOLTAGE WAVEFORMS

FIGURE 6. SWITCHING TIME (t_{PZH}, t_{PHZ}), CONTROL TO SIGNAL OUTPUT

PARAMETER MEASUREMENT INFORMATION

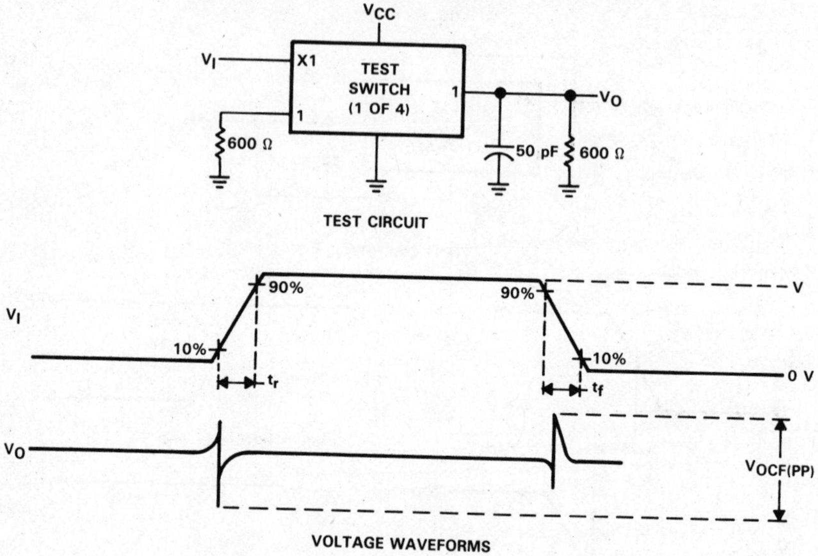
VCC
VI
X1
TEST
SWITCH
(1 OF 4)
1
1
600 Ω
50 pF
600 Ω
VO
TEST CIRCUIT
90%
90%
V
VI
10%
10%
0 V
tr
tf
VO
VOCF(PP)
VOLTAGE WAVEFORMS

FIGURE 7. CONTROL FEEDTHROUGH VOLTAGE

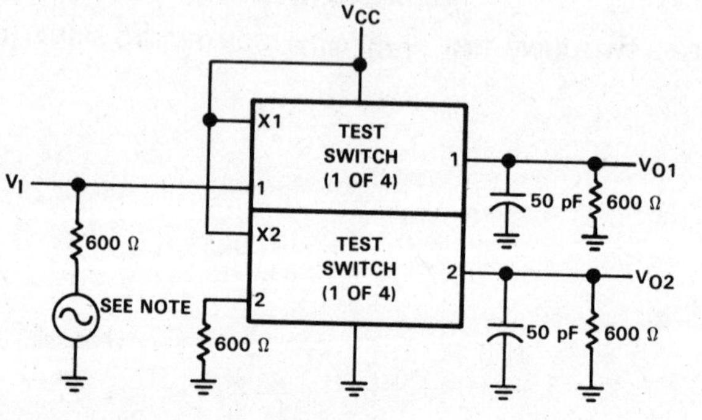
VCC
X1
TEST
SWITCH
(1 OF 4)
1
1
VO1
VI
50 pF
600 Ω
600 Ω
X2
TEST
SWITCH
(1 OF 4)
2
2
VO2
SEE NOTE
600 Ω
50 pF
600 Ω

NOTE: ADJUST f for $a_X = \frac{V_{O2}}{V_{O1}} = 50$ dB.

FIGURE 8. CROSSTALK BETWEEN ANY TWO SWITCHES, TEST CIRCUIT

TLC7135
Advanced LinCMOS™ 4 1/2-DIGIT PRECISION ANALOG-TO-DIGITAL CONVERTER

D2851, DECEMBER 1986

- ADVANCED LinCMOS™ Technology
- Zero Reading for 0-V Input
- Precision Null Detection with True Polarity at Zero
- 1-pA Typical Input Current
- True Differential Input
- Multiplexed Binary-Coded-Decimal Output
- Low Rollover Error: ±1 Count Maximum
- Control Signals Allow Interfacing with UARTs or Microprocessors
- Autoranging Capability with Over- and Under-Range Signals
- TTL-Compatible Outputs
- Direct Replacement for Teledyne TSC7135, Intersil ICL7135, Maxim ICL7135, and Siliconix Si7135

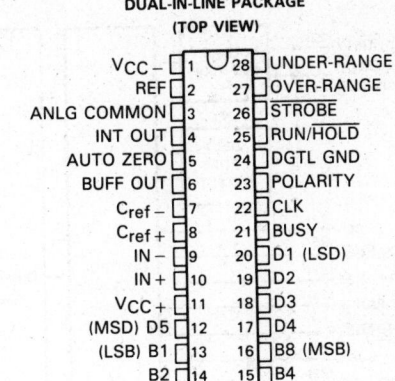

N
DUAL-IN-LINE PACKAGE
(TOP VIEW)

Signal	Pin	Pin	Signal
V_{CC-}	1	28	UNDER-RANGE
REF	2	27	OVER-RANGE
ANLG COMMON	3	26	$\overline{\text{STROBE}}$
INT OUT	4	25	RUN/$\overline{\text{HOLD}}$
AUTO ZERO	5	24	DGTL GND
BUFF OUT	6	23	POLARITY
C_{ref-}	7	22	CLK
C_{ref+}	8	21	BUSY
IN−	9	20	D1 (LSD)
IN+	10	19	D2
V_{CC+}	11	18	D3
(MSD) D5	12	17	D4
(LSB) B1	13	16	B8 (MSB)
B2	14	15	B4

Caution. This device has limited built-in gate protection. The leads should be shorted together or the device placed in conductive foam during storage or handling to prevent electrostatic damage.

description

The TLC7135 converter is manufactured with Texas Instruments highly efficient ADVANCED LinCMOS™ technology. This 4 1/2-digit dual-slope-integrating analog-to-digital converter is designed to provide interfaces to both a microprocessor and a visual display. The digit-drive outputs D1 through D4 and multiplexed binary-coded-decimal outputs, B1 through B4, provide an interface for LED or LCD decoder/drivers as well as microprocessors.

The TLC7135 offers 50-ppm (one part in 20,000) resolution with a maximum linearity error of one count. The zero error is less than 10 μV and zero drift is less than 0.5 μV/°C. Source-impedance errors are minimized by low input current (less than 10 pA). Rollover error is limited to ±1 count.

The TLC7135 BUSY, $\overline{\text{STROBE}}$, RUN/$\overline{\text{HOLD}}$, OVER-RANGE, and UNDER-RANGE control signals support microprocessor-based measurement systems. The control signals also can support remote data acquisition systems with data transfer via universal asynchronous receiver transmitters (UARTs).

The TLC7135 is characterized for operation from 0°C to 70°C.

ADVANCED LinCMOS™ is a trademark of Texas Instruments Incorporated.

functional block diagram

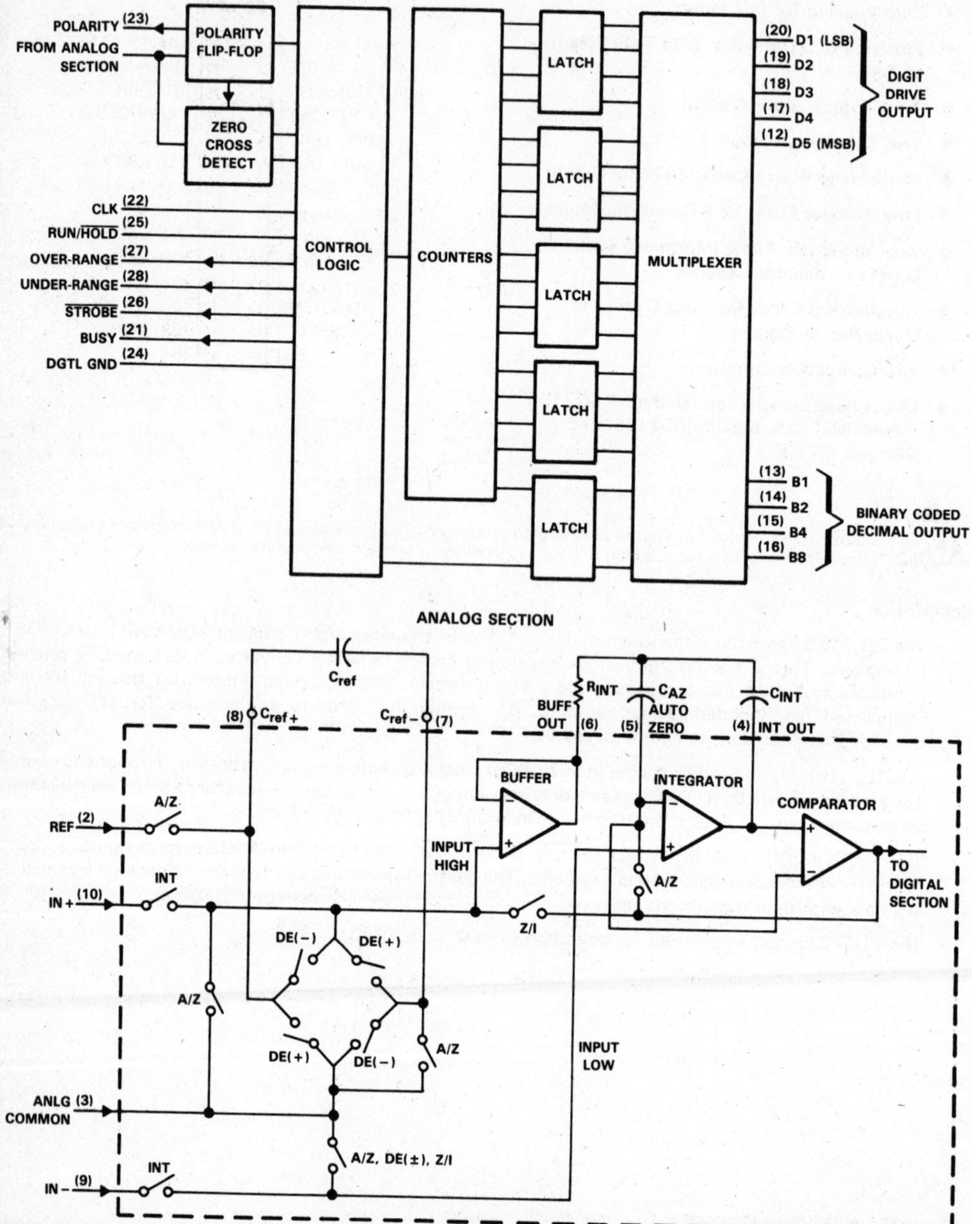
POLARITY (23)
FROM ANALOG SECTION
POLARITY FLIP-FLOP
ZERO CROSS DETECT
CLK (22)
RUN/HOLD (25)
OVER-RANGE (27)
UNDER-RANGE (28)
STROBE (26)
BUSY (21)
DGTL GND (24)
CONTROL LOGIC
COUNTERS
LATCH
MULTIPLEXER
(20) D1 (LSB)
(19) D2
(18) D3
(17) D4
(12) D5 (MSB)
DIGIT DRIVE OUTPUT
(13) B1
(14) B2
(15) B4
(16) B8
BINARY CODED DECIMAL OUTPUT
ANALOG SECTION
Cref
(8) Cref +
Cref − (7)
RINT
BUFF OUT (6)
CAZ
(5) AUTO ZERO
CINT
(4) INT OUT
BUFFER
INTEGRATOR
COMPARATOR
REF (2)
A/Z
IN + (10)
INT
INPUT HIGH
Z/I
TO DIGITAL SECTION
DE(−)
DE(+)
INPUT LOW
ANLG COMMON (3)
A/Z, DE(±), Z/I
IN − (9)

TEXAS INSTRUMENTS
POST OFFICE BOX 655012 • DALLAS, TEXAS 75265

absolute maximum ratings over operating free-air temperature range (unless otherwise noted)

Supply voltage (V_{CC+} with respect to V_{CC-}) .... 15 V
Analog input voltage (pin 9 or pin 10) .... V_{CC-} to V_{CC+}
Reference voltage range .... V_{CC-} to V_{CC+}
Clock input voltage range .... 0 V to V_{CC+}
Operating free-air temperature range .... 0°C to 70°C
Storage temperature range .... −65°C to 150°C
Lead temperature 1,6 mm (1/16 inch) from case for 10 seconds .... 260°C

recommended operating conditions

	MIN	NOM	MAX	UNIT
Supply voltage, V_{CC+}	4	5	6	V
Supply voltage, V_{CC-}	−3	−5	−8	V
Reference voltage, V_{ref}		1		V
High-level input voltage, CLK, RUN/$\overline{HOLD}$, V_{IH}	2.8			V
Low-level input voltage, CLK, RUN/$\overline{HOLD}$, V_{IL}			0.8	V
Differential input voltage, V_{ID}	V_{CC-} +1		V_{CC+} −0.5	V
Maximum operating frequency, f_{clock} (see Note 1)	1.2	2		MHz
Operating free-air temperature range, T_A	0		70	°C

NOTE 1: Clock frequency range extends down to 0 Hz.

electrical characteristics, V_{CC+} = 5 V, V_{CC-} = −5 V, V_{ref} = 1 V, f_{clock} = 120 kHz, T_A = 25°C (unless otherwise noted)

PARAMETER			TEST CONDITIONS		MIN	TYP	MAX	UNIT
V_{OH}	High-level output voltage	D1-D5,B1,B2,B4,B8	I_O = −1 mA		2.4		5	V
		Other outputs	I_O = −10 μA		4.9		5	
V_{OL}	Low-level output voltage		I_O = 1.6 mA				0.4	V
	Peak-to-peak output noise voltage (see Note 2)		V_{ID} = 0,	Full Scale = 2 V		15		μV
α_{VO}	Zero-reading temperature coefficient of output voltage		V_{ID} = 0,	0°C ≤ T_A ≤ 70°C		0.5	2	μV/°C
I_{IH}	High-level input current		V_I = 5 V,	0°C ≤ T_A ≤ 70°C		0.1	10	μA
I_{IL}	Low-level input current		V_I = 0 V,	0°C ≤ T_A ≤ 70°C		−0.02	−0.1	mA
I_I	Input leakage current, pins 9 and 10		V_{ID} = 0	T_A = 25°C		1	10	pA
				0°C ≤ T_A ≤ 70°C			250	
I_{CC+}	Positive supply current		f_{clock} = 0	T_A = 25°C		1	2	mA
				0°C ≤ T_A ≤ 70°C			3	
I_{CC-}	Negative supply current		f_{clock} = 0	T_A = 25°C		−0.8	−2	mA
				0°C ≤ T_A ≤ 70°C			−3	
C_{pd}	Power dissipation capacitance		See Note 3			40		pF

NOTES: 2. This is the peak-to-peak value that is not exceeded 95% of the time.
3. Factor relating clock-frequency to increase in supply current. At V_{CC+} = 5 V

$$I_{CC+} = I_{CC+}(f_{clock} = 0) + C_{pd} \times 5\ V \times f_{clock}$$

operating characteristics, V_{CC+} = 5 V, V_{CC-} = −5 V, V_{ref} = 1 V, f_{clock} = 120 kHz, T_A = 25°C (unless otherwise noted)

	PARAMETER	TEST CONDITIONS	MIN	TYP	MAX	UNIT
α_{FS}	Full-scale temperature coefficient (see Note 4)	V_{ID} = 2 V, 0°C ≤ T_A ≤ 70°C			5	ppm/°C
	Linearity error	−2 V ≤ V_{ID} ≤ 2 V		0.5	1	count
	Differential linearity error (see Note 5)	−2 V ≤ V_{ID} ≤ 2 V		0.01		LSB
	± Full-scale symmetry error (see Note 6) (rollover error)	V_{ID} = ±2 V		0.5	1	count
	Display reading with 0-V input	V_{ID} = 0, 0°C ≤ T_A ≤ 70°C	−.0000	±.0000	+.0000	Digital Reading
	Display reading in ratiometric operation	V_{ID} = V_{ref}, T_A = 25°C	+.9998	+.9999	+1.0000	Digital Reading
		0°C ≤ T_A ≤ 70°C	+.9995	+.9999	+1.0005	

NOTES: 4. This parameter is measured with an external reference having a temperature coefficient of less than 0.01 ppm/°C.
5. The magnitude of the difference between the worst case step of adjacent counts and the ideal step.
6. Rollover error is the difference between the absolute values of the conversion for 2 V and −2 V.

timing diagrams

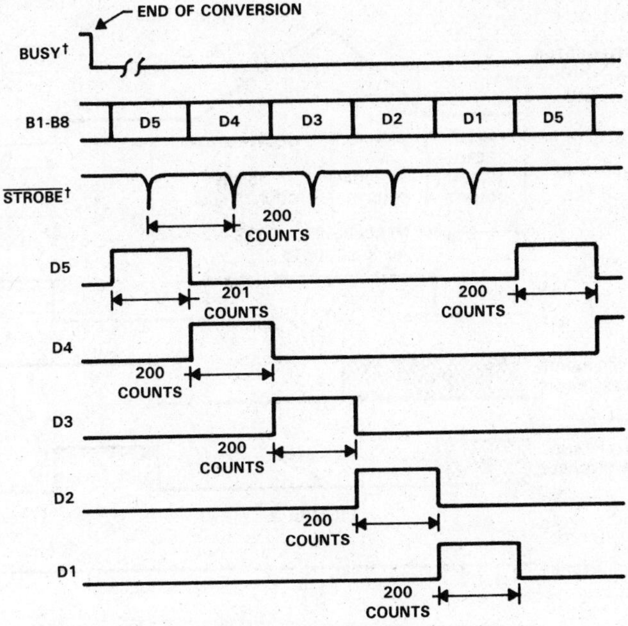
END OF CONVERSION
BUSY†
B1-B8
D5
D4
D3
D2
D1
D5
STROBE†
200
COUNTS
D5
201
COUNTS
200
COUNTS
D4
200
COUNTS
D3
200
COUNTS
D2
200
COUNTS
D1
200
COUNTS

† Delay between BUSY going low and the first $\overline{\text{STROBE}}$ pulse is dependent upon the analog input.

FIGURE 1

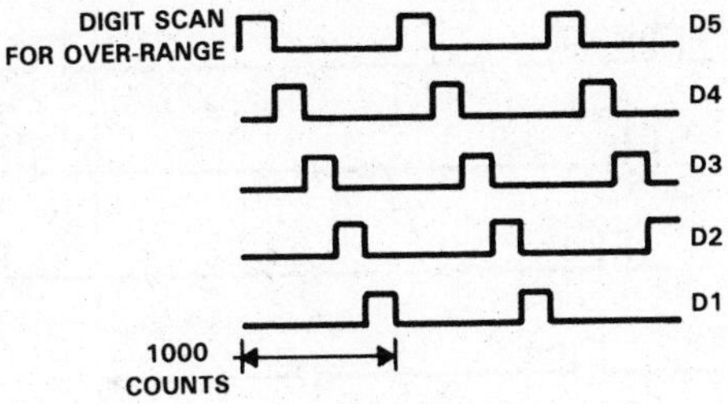
DIGIT SCAN
FOR OVER-RANGE
D5
D4
D3
D2
D1
1000
COUNTS

FIGURE 2

timing diagrams (continued)

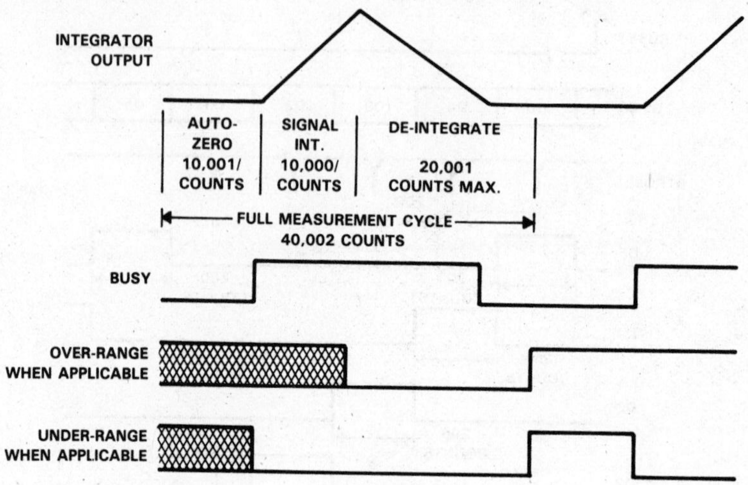
INTEGRATOR
OUTPUT
AUTO-
ZERO
10,001/
COUNTS
SIGNAL
INT.
10,000/
COUNTS
DE-INTEGRATE
20,001
COUNTS MAX.
FULL MEASUREMENT CYCLE
40,002 COUNTS
BUSY
OVER-RANGE
WHEN APPLICABLE
UNDER-RANGE
WHEN APPLICABLE

FIGURE 3

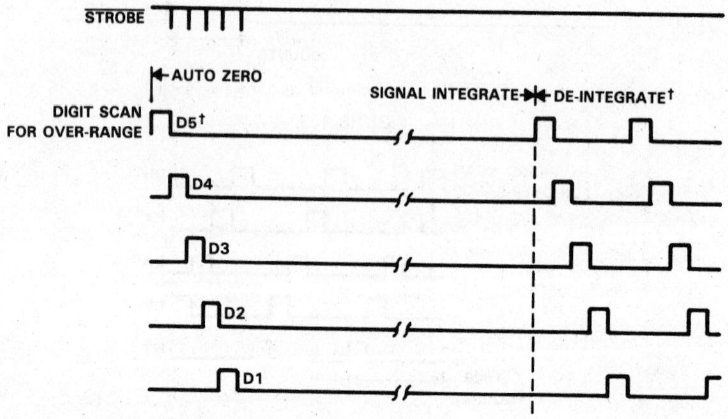
STROBE
AUTO ZERO
SIGNAL INTEGRATE
DE-INTEGRATE†
DIGIT SCAN
FOR OVER-RANGE
D5†
D4
D3
D2
D1

†First D5 of AUTO ZERO and DE-INTEGRATE is one count longer.

FIGURE 4

PRINCIPLES OF OPERATION

A measurement cycle for the TLC7135 consists of the following four phases.

1. Auto-Zero Phase. The internal IN+ and IN− inputs are disconnected from the pins and internally connected to ANLG COMMON. The reference capacitor is charged to the reference voltage. The system is configured in a closed loop and the auto-zero capacitor is charged to compensate for offset voltages in the buffer amplifier, integrator, and comparator. The auto-zero accuracy is limited only by the system noise, and the overall offset, as referred to the input, is less than 10 μV.
2. Signal Integrate Phase. The auto-zero loop is opened and the internal IN+ and IN− inputs are connected to the external pins. The differential voltage between these inputs is integrated for a fixed period of time. If the input signal has no return with respect to the converter power supply, IN− can be tied to ANLG COMMON to establish the correct common-mode voltage. Upon completion of this phase, the polarity of the input signal is recorded.
3. De-integrate Phase. The reference is used to perform the de-integrate task. The internal IN− is internally connected to ANLG COMMON and IN+ is connected across the previously charged reference capacitor. The recorded polarity of the input signal is used to ensure that the capacitor will be connected with the correct polarity so that the integrator output polarity will return to zero. The time, which is required for the output to return to zero, is proportional to the amplitude of the input signal. The return time is displayed as a digital reading and is determined by the equation 10,000 x (V_{ID}/V_{ref}). The maximum or full-scale conversion occurs when V_{ID} is two times V_{ref}.
4. Zero Integrator Phase. The internal IN− is connected to ANLG COMMON. The system is configured in a closed loop to cause the integrator output to return to zero. Typically this phase requires 100 to 200 clock pulses. However, after an over-range conversion, 6200 pulses are required.

description of analog circuits

input signal range

The common mode range of the input amplifier extends from 1 V above the negative supply to 1 V below the positive supply. Within this range, the common mode rejection ratio (CMRR) is typically 86 dB. Both differential and common mode voltages cause the integrator output to swing. Therefore, care must be exercised to assure the integrator output does not saturate.

analog common

Analog common (ANLG COMMON) is connected to the internal IN− during the auto-zero, de-integrate, and zero integrator phases. If IN− is connected to a voltage which is different than analog common during the signal integrate phase, the resulting common mode voltage will be rejected by the amplifier. However, in most applications, IN LO will be set at a known fixed voltage (power supply common for instance). In this application, analog common should be tied to the same point, thus removing the common mode voltage from the converter. Removing the common mode voltage in this manner will slightly increase conversion accuracy.

reference

The reference voltage is positive with respect to analog common. The accuracy of the conversion result is dependent upon the quality of the reference. Therefore, to obtain a high accuracy conversion, a high quality reference should be used.

description of digital circuits

RUN/$\overline{\text{HOLD}}$ input

When the RUN/$\overline{\text{HOLD}}$ input is high or open, the device will continuously perform measurement cycles every 40,002 clock pulses. If this input is taken low, the IC will continue to perform the ongoing measurement cycle and then hold the conversion reading for as long as the pin is held low. If the pin is held low after completion of a measurement cycle, a short positive pulse (greater than 300 ns) will initiate a new measurement cycle. If this positive pulse occurs before the completion of a measurement cycle, it will not be recognized. The first STROBE pulse, which occurs 101 counts after the end of a measurement cycle, is an indication of the completion of a measurement cycle. Thus, the positive pulse could be used to trigger the start of a new measurment after the first STROBE pulse.

$\overline{\text{STROBE}}$ input

Negative going pulses from this input are used to transfer the BCD conversion data to external latches, UARTS, or microprocesors. At the end of the measurement cycle, the digit-drive (D5) input goes high and remains high for 201 counts. The most significant digit (MSD) BCD bits are placed on the BCD pins. After the first 101 counts, halfway through the duration of output D1-D5 going high, the $\overline{\text{STROBE}}$ pin goes low for 1/2 clock pulse width. The placement of the $\overline{\text{STROBE}}$ pulse at the midpoint of the D5 high pulse allows the information to be latched into an external device on either a low-level or an edge. Such placement of the $\overline{\text{STROBE}}$ pulse also ensures that the BCD bits for the second MSD will not yet be competing for the BCD lines and latching of the correct bits is assured. The above process is repeated for the second MSD and the D4 output. Similarly, the process is repeated through the least significant digit (LSD). Subsequently, inputs D5 through D1 and the BCD lines will continue scanning without the inclusion of $\overline{\text{STROBE}}$ pulses. This subsequent continuous scanning causes the conversion results to be continuously displayed. Such subsequent scanning does not occur when an over-range condition occurs.

BUSY output

The BUSY output goes high at the beginning of the signal integrate phase and remains high until the first clock pulse after zero-crossing or at the end of the measurement cycle if an over-range condition occurs. It is possible to use the BUSY pin to serially transmit the conversion result. Serial transmission can be accomplished by ANDing the BUSY and CLOCK signals and transmitting the ANDed output. The transmitted output consists of 10,001 clock pulses, which occur during the signal integrate phase, and the number of clock pulses, which occur during the de-integrate phase. The conversion result can be obtained by subtracting 10,001 from the total number of clock pulses.

OVER-RANGE output

When an over-range condition occurs, this pin goes high after the BUSY signal goes low at the end of the measurement cycle. As previously noted, the BUSY signal remains high until the end of the measurement cycle when an over-range condition occurs. The OVER-RANGE output goes high at end of BUSY and goes low at the beginning of the de-integrate phase in the next measurement cycle.

UNDER-RANGE output

At the end of the BUSY signal, this pin goes high if the conversion result is less than or equal to 9% (count of 1800) of the full-scale range. The UNDER-RANGE output is brought low at the beginning of the signal integrate phase of the next measurement cycle.

PRINCIPLES OF OPERATION

POLARITY output

The POLARITY output is high for a positive input signal and is updated at the beginning of each de-integrate phase. The polarity output is valid for all inputs including ±0 and over-range signals.

digit-drive (D5, D4, D2 and D1) outputs

Each digit-drive output (D1 through D5) sequentially goes high for 200 clock pulses. This sequential process is continuous unless an over-range occurs. When an over-range occurs, all of the digit drive outputs are blanked from the end of the strobe sequence until the beginning of the de-integrate phase (when the sequential digit drive activation begins again). The blanking activity, during an over-range condition, may be used to cause the display to flash and indicate the over-range condition.

BCD outputs

The BCD bits (B8, B4, B2 and B1) for a given digit are sequentially activated on these outputs. Simultaneously, the appropriate Digit-drive line for the given digit is activated.

system aspects

integrating resistor

The value of the integrating resistor (R_{INT}) is determined by the full scale input voltage and the output current of the integrating amplifier. The integrating amplifier can supply 20 μA of current with negligible non-linearity. The equation for determining the value of this resistor is as follows:

$$R_{INT} = \frac{\text{FULL-SCALE VOLTAGE}}{I_{INT}}$$

Integrating amplifier current, I_{INT}, from 5 to 40 μA will yield good results. However, the nominal and recommended current is 20 μA.

integrating capacitor

The product of the integrating resistor and capacitor should be selected to give the maximum voltage swing without causing the integrating amplifier output to saturate and get too close to the power supply voltages. If the amplifier output is within 0.3 V of either supply, saturation will occur. With ±5-V supplies and ANLG COMMON connected to ground, the designer should design for a ±3.5-V to ±4-V integrating amplifier swing. A nominal capacitor value is 0.47 μF. The equation for determining the value of the integrating capacitor (C_{INT}) is as follows:

$$C_{INT} = \frac{10{,}000 \times \text{CLOCK PERIOD} \times I_{INT}}{\text{INTEGRATOR OUTPUT VOLTAGE SWING}}$$

where: I_{INT} is nominally 20 μA.

Capacitors with large tolerances and high dielectric absorption can induce conversion inaccuracies. A capacitor, which is too small could cause the integrating amplifier to saturate. High dielectric absorption causes the effective capacitor value to be different during the signal integrate and de-integrate phases. Polypropylene capacitors have very low dielectric absorption. Polystyrene and Polycarbonate capacitors have higher dielectric absorption, but also work well.

PRINCIPLES OF OPERATION

auto-zero and reference capacitor

Large capacitors will tend to reduce noise in the system. Dielectric absorption is unimportant except during power-up or overload recovery. Typical values are 1 μF.

reference voltage

For high-accuracy absolute measurements, a high quality reference should be used.

rollover resistor and diode

The TLC7135 has a small rollover error, however it can be corrected. The correction is to connect the cathode of any silicon diode to the INT OUT pin and the anode to a resistor. The other end of the resistor is connected to ANLG COMMON or ground. For the recommended operating conditions the resistor value is 100 kΩ. This value may be changed to correct any rollover error which has not been corrected. In many non-critical applications, the resistor and diode are not needed.

maximum clock frequency

For most dual-slope A/D converters, the maximum conversion rate is limited by the frequency response of the comparator. In this circuit, the comparator follows the integrator ramp with a 3 μs delay. Therefore, with a 160 kHz clock frequency (6 μs period), half of the first reference integrate clock period is lost in delay. Hence, the meter reading will change from 0 to 1 with a 50 μV input, 1 to 2 with a 150 μV input, 2 to 3 with a 250 μV input, etc. This transition at midpoint is desirable; however, if the clock frequency is increased appreciably above 160 kHz, the instrument will flash "1" on noise peaks even when the input is shorted. The above transition points assume a 2-V input range is equivalent to 20,000 clock cycles.

If the input signal is always of one polarity, comparator delay need not be a limitation. Clock rates of 1 MHz are possible since non-linearity and noise do not increase substantially with frequency. For a fixed clock frequency, the extra count or counts caused by comparator delay will be a constant and can be subtracted out digitally.

For signals with both polarities, the clock frequency can be extended above 160 kHz without error by using a low value resistor in series with the integrating capacitor. This resistor causes the integrator to jump slightly towards the zero-crossing level at the beginning of the de-integrate phase and thus, compensates for the comparator delay. This series resistor should be 10 to 50 ohms. This approach allows clock frequencies up to 480 kHz.

minimum clock frequency

The minimum clock frequency limitations result from capacitor leakage from the auto-zero and reference capacitors. Measurement cycles as high as 10 seconds are not influenced by leakage error.

rejection of 50 Hz or 60 Hz pickup

To maximize the rejection of 50 Hz or 60 Hz pickup, the clock frequency should be chosen so that an integral multiple of 50 Hz or 60 Hz periods occur during the signal integrate phase. To achieve rejection of these signals, some clock frequencies which could be used are as follows:

50 Hz: 250, 166.66, 125, 100 kHz, etc.
60 Hz: 300, 200, 150, 120, 100, 40, 33.33 kHz, etc.

TEXAS INSTRUMENTS
POST OFFICE BOX 655012 • DALLAS, TEXAS 75265

PRINCIPLES OF OPERATION

zero-crossing flip-flop

This flip-flop interrogates the comparator's zero-crossing status. The interrogation is performed after the previous clock cycle and the positive half of the ongoing clock cycle have occurred so that any comparator transients which result from the clock pulses do not affect the detection of a zero-crossing. This procedure delays the zero-crossing detection by one clock cycle. To eliminate the inaccuracy, which is caused by this delay, the counter is disabled for one clock cycle at the beginning of the de-integrate phase. Therefore, when the zero-crossing is detected one clock cycle later than the zero-crossing actually occurs, the correct number of counts is displayed.

noise

The peak-to-peak noise around zero is approximately 15 μV (peak-to-peak value not exceeded 95% of the time). Near full scale, this value increases to approximately 30 μV. Much of the noise originates in the auto-zero loop, and is proportional to the ratio of the input signal to the reference.

analog and digital grounds

For high-accuracy applications, ground loops must be avoided. Return currents from digital circuits must not be sent to the analog ground line.

power supplies

The TLC7135 is designed to work with ±5-V power supplies. However, 5-V operation is possible if the input signal does not vary more than ±1.5 V from mid-supply.

Advanced LinCMOS™ 3 1/2-DIGIT PRECISION ANALOG-TO-DIGITAL CONVERTER AND LCD DRIVER

D2849, OCTOBER 1986

- ADVANCED LinCMOS™ Technology
- Zero Reading for 0-V Input on All Scales
- Precision Null Detection with True Polarity at Zero
- 1-pA Typical Input Current
- True Differential Input and Reference
- Direct LCD Display Drive with No External Components
- Low Noise — 15 μVp-p Without Hysteresis or Overrange Hangover
- On-Chip Clock Oscillator and Reference
- Convenient 9-V Battery Operation with Low Power Dissipation, Less than 1 mW
- Direct Replacement for Intersil and Maxim ICL7136
- Pin Compatible with Intersil ICL7106; ICL7126 and Teledyne TSC7106, TSC7136

N DUAL-IN-LINE PACKAGE

(TOP VIEW)

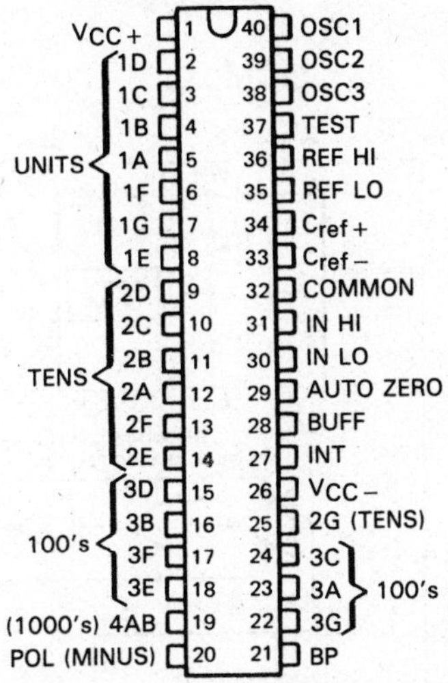
VCC+ 1
1D 2
1C 3
1B 4
UNITS 1A 5
1F 6
1G 7
1E 8
2D 9
2C 10
2B 11
TENS 2A 12
2F 13
2E 14
3D 15
3B 16
100's 3F 17
3E 18
(1000's) 4AB 19
POL (MINUS) 20
40 OSC1
39 OSC2
38 OSC3
37 TEST
36 REF HI
35 REF LO
34 Cref+
33 Cref−
32 COMMON
31 IN HI
30 IN LO
29 AUTO ZERO
28 BUFF
27 INT
26 VCC−
25 2G (TENS)
24 3C
23 3A 100's
22 3G
21 BP

Caution. These devices have limited built-in gate protection. The leads should be shorted together or the device placed in conductive foam during storage or handling to prevent electrostatic damage to the MOS gates.

description

The TLC7136C is a high-performance, very low-power 3 1/2-digit analog-to-digital converter (ADC). The TLC7136C contains seven-segment decoders, display drivers, a clock, and a reference. This device is designed to interface with liquid crystal displays and incorporates a backplane drive. The device can easily be powered with a 9-volt battery because the supply current is less than 100 microamperes.

The TLC7136C provides high accuracy and versatility and such features as auto-zeroing to less than 10 microvolts, zero drift of less than 1 μV/°C, maximum input bias current of 10 picoamperes, and rollover error of less than 1 count.

The differential input and on-chip reference are particularly useful when measuring load cells, strain gauges, and other bridge-type transducers. Single-supply operation provides economy in that a high-performance panel meter can be built with only seven passive components and a display. The TLC7136C is an improved version of the Intersil ICL7126 in that overrange hangover and hysteresis effects are eliminated.

The TLC7136C is characterized for operation from 0°C to 70°C.

ADVANCED LinCMOS™ is a trademark of Texas Instruments Incorporated.

functional block diagram (with external components)

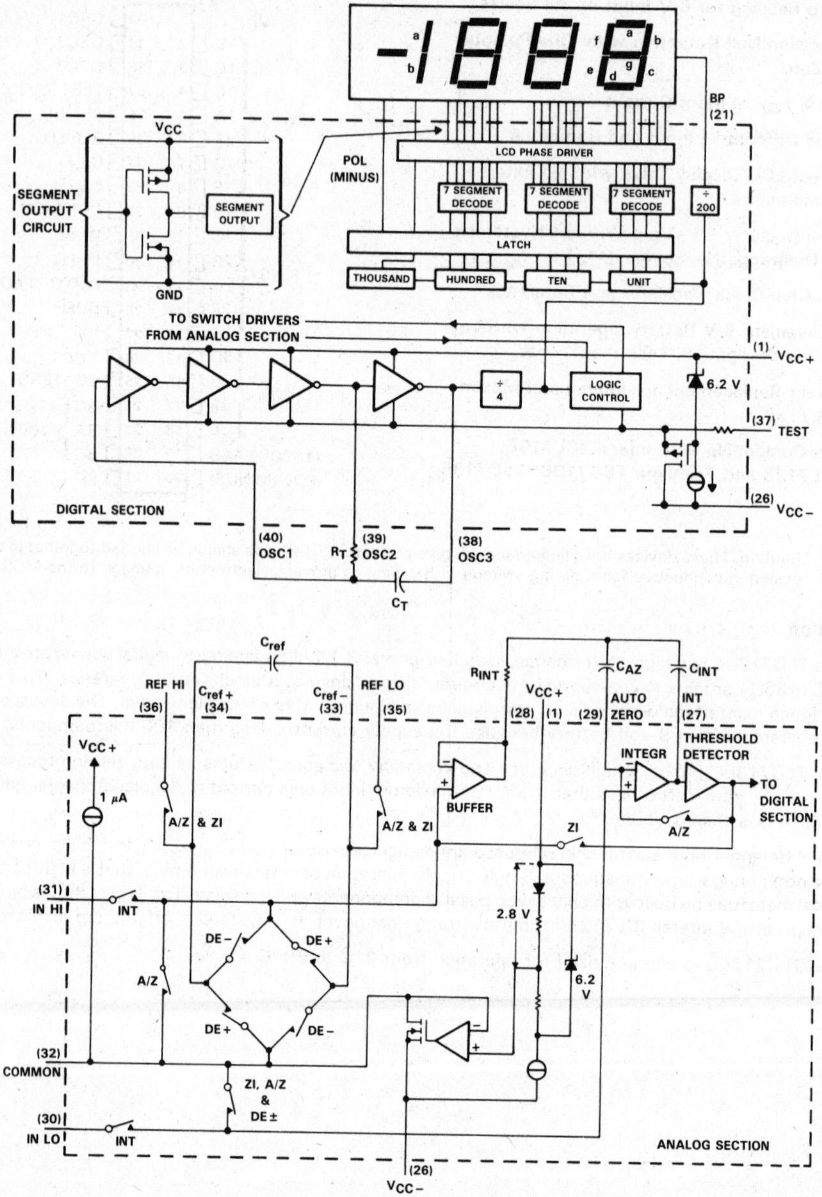
BP
(21)
VCC
GND
SEGMENT
OUTPUT
CIRCUIT
SEGMENT
OUTPUT
POL
(MINUS)
LCD PHASE DRIVER
7 SEGMENT
DECODE
7 SEGMENT
DECODE
7 SEGMENT
DECODE
÷
200
LATCH
THOUSAND
HUNDRED
TEN
UNIT
TO SWITCH DRIVERS
FROM ANALOG SECTION
(1)
VCC+
÷
4
LOGIC
CONTROL
6.2 V
(37)
TEST
(26)
VCC−
DIGITAL SECTION
(40)
OSC1
(39)
RT
OSC2
(38)
OSC3
CT
Cref
REF HI
(36)
Cref+
(34)
Cref−
(33)
REF LO
(35)
RINT
VCC+
(28)
(1)
CAZ
AUTO
(29)
ZERO
CINT
INT
(27)
INTEGR
THRESHOLD
DETECTOR
TO
DIGITAL
SECTION
VCC+
1 μA
A/Z & ZI
A/Z & ZI
BUFFER
ZI
A/Z
(31)
IN HI
INT
DE−
DE+
2.8 V
A/Z
6.2
V
DE+
DE−
(32)
COMMON
ZI, A/Z
&
DE±
(30)
IN LO
INT
ANALOG SECTION
(26)
VCC−

NOTE: Letters beside switches indicate state of conversion during switch closure.

TEXAS INSTRUMENTS
POST OFFICE BOX 655012 • DALLAS, TEXAS 75265

absolute maximum ratings over operating free-air temperature range (unless otherwise noted)

Supply voltage (V_{CC+} with respect to V_{CC-}), V_{CC} .... 15 V
Voltage range for any input except clock (see Note 1) .... V_{CC-} to V_{CC+}
Clock input voltage range .... V_{test} to V_{CC+}
Operating free-air temperature range .... 0°C to 70°C
Storage temperature range .... −65°C to 150°C
Lead temperature 1,6 mm (1/16 inch) from case for 60 seconds .... 260°C

NOTE 1: Input voltages may exceed the supply voltages provided the input current is limited to ±100 μA.

recommended operating conditions

			MIN	NOM	MAX	UNIT
V_{CC}	Supply voltage			9		V
V_{ref}	Reference input voltage	FS (full scale) V_{ID} = 200 mV, See Note 2		100		mV
		FS V_{ID} = 2 V		1		V
	Full-scale input voltage			2 V_{ref}		V
V_I	Input voltage at IN HI or IN LO		$V_{CC-}+1$		$V_{CC+}-0.5$	V
C_{ref}	Reference capacitor		0.1		1	μF
C_z	Auto-zero capacitor		0.033		0.47	μF
C_x	Integrator capacitor		0.047		0.15	μF
R_s	Integrator resistor	FS = 200 mV		180		kΩ
		FS = 2 V		1.8		MΩ
T_A	Operating free-air temperature		0		70	°C

electrical characteristics, V_{CC} = 9 V, f_{clock} = 16 kHz, T_A = 25°C (unless otherwise noted)

PARAMETER	TEST CONDITIONS	MIN	TYP	MAX	UNIT
Common-mode rejection ratio	V_{IC} = ±1 V, V_{ID} = 0, FS = 200 mV		50		μV/V
Noise voltage (peak-to-peak value not exceeded 95% of time)	V_{ID} = 0, FS = 200 mV		15		μV
Input leakage current	V_{ID} = 0		1	10	pA
Scale factor temperature coefficient	V_{ID} = 199 mV, T_A = 0 to 70°C, See Note 3		1	5	ppm/°C
Analog common voltage (with respect to V_{CC+})	250 kΩ between COMMON and V_{CC+}	−2.6	−3	−3.2	V
Temperature coefficient of analog common voltage (with respect to V_{CC+})	250 kΩ between COMMON and V_{CC+}		150		ppm/°C
Peak-to-peak segment drive voltage (see Note 4)		4	5	6	V
Peak-to-peak backplane drive voltage (see Note 4)		4	5	6	V
Supply current (see Note 5)	V_{ID} = 0		50	100	μA
Power dissipation capacitance	See Note 6		40		pF

NOTES: 2. V_{ID} is the voltage at IN HI with respect to IN LO.
3. This is measured using a fixed external reference voltage with 0-ppm/°C temperature coefficient.
4. Backplane drive is in phase with segment drive for a turned-off segment, 180° out of phase for a turned-on segment. Backplane frequency is 20 times the conversion rate. The average dc component is less than 50 mV.
5. This does not include current through the common terminal. During the auto-zero phase, current is 10 to 20 μA higher. Use of a 48-kHz oscillator increases current by typically 8 μA.
6. This can be used to determine the no-load dynamic power dissipation. $P_D = C_{pd} \cdot V_{CC}^2 \cdot f + I_{CC} \cdot V_{CC}$.

operating characteristics over recommended operating free-air temperature range, V_{CC} = 9 V

PARAMETER	TEST CONDITIONS	MIN	TYP	MAX	UNIT
Zero-input digital reading	V_{ID} = 0, FS = 200 mV	−0.000	±0.000	+0.000	
Ratiometric digital reading	V_{ID} = V_{ref} = 100 mV	999	999/1000	1000	
Rollover error (see Note 7)	V_{ID-} = V_{ID+} ≈ 200 mV or 2 V		±0.2	±1	Count
Linearity error	FS = 200 mV or 2 V		±0.2	±1	Count
Zero-reading temperature coefficient	V_{ID} = 0, T_A = 0°C to 70°C		0.2	1	μV/°C

NOTE 7: Rollover error is the difference between the magnitudes of the conversion results for equal positive and negative inputs near full scale.

PARAMETER MEASUREMENT INFORMATION

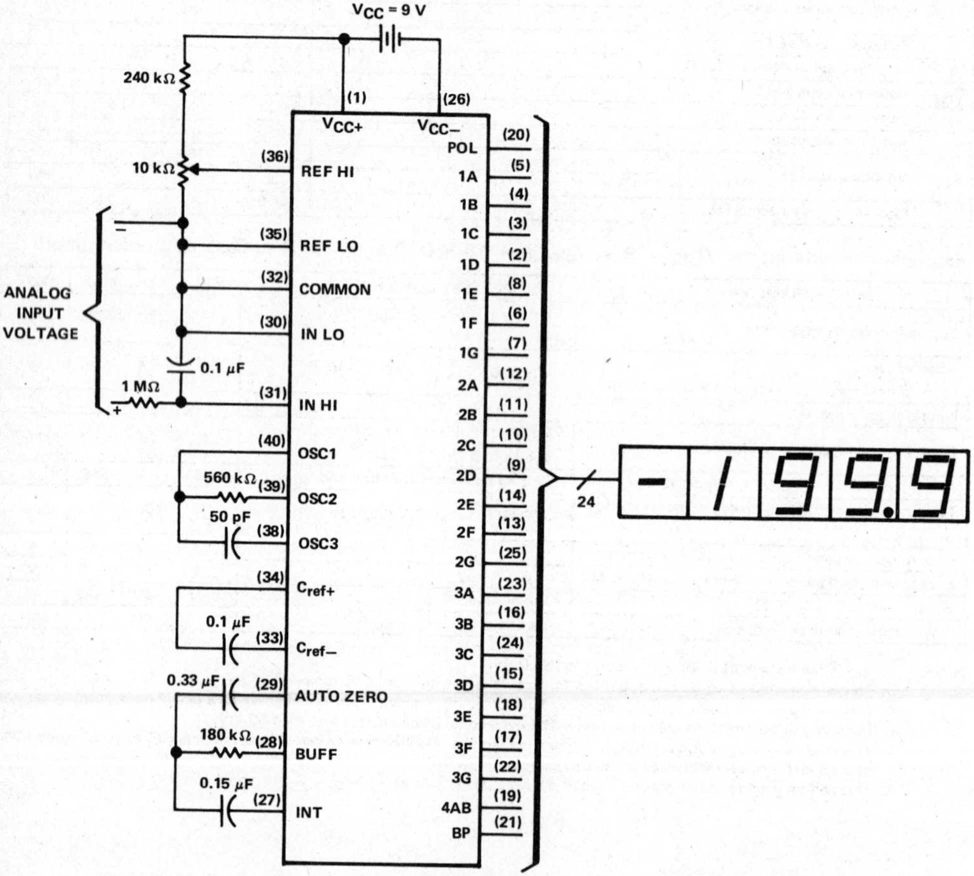
VCC = 9 V
240 kΩ
10 kΩ
ANALOG INPUT VOLTAGE
0.1 μF
1 MΩ
560 kΩ
50 pF
0.1 μF
0.33 μF
180 kΩ
0.15 μF
(1)
(26)
VCC+
VCC−
(36) REF HI
(35) REF LO
(32) COMMON
(30) IN LO
(31) IN HI
(40) OSC1
(39) OSC2
(38) OSC3
(34) Cref+
(33) Cref−
(29) AUTO ZERO
(28) BUFF
(27) INT
POL (20)
1A (5)
1B (4)
1C (3)
1D (2)
1E (8)
1F (6)
1G (7)
2A (12)
2B (11)
2C (10)
2D (9)
2E (14)
2F (13)
2G (25)
3A (23)
3B (16)
3C (24)
3D (15)
3E (18)
3F (17)
3G (22)
4AB (19)
BP (21)
24
-1999

FIGURE 1. TEST CIRCUIT (CLOCK FREQUENCY = 16 kHz, 1 READING PER SECOND)

PARAMETER MEASUREMENT INFORMATION

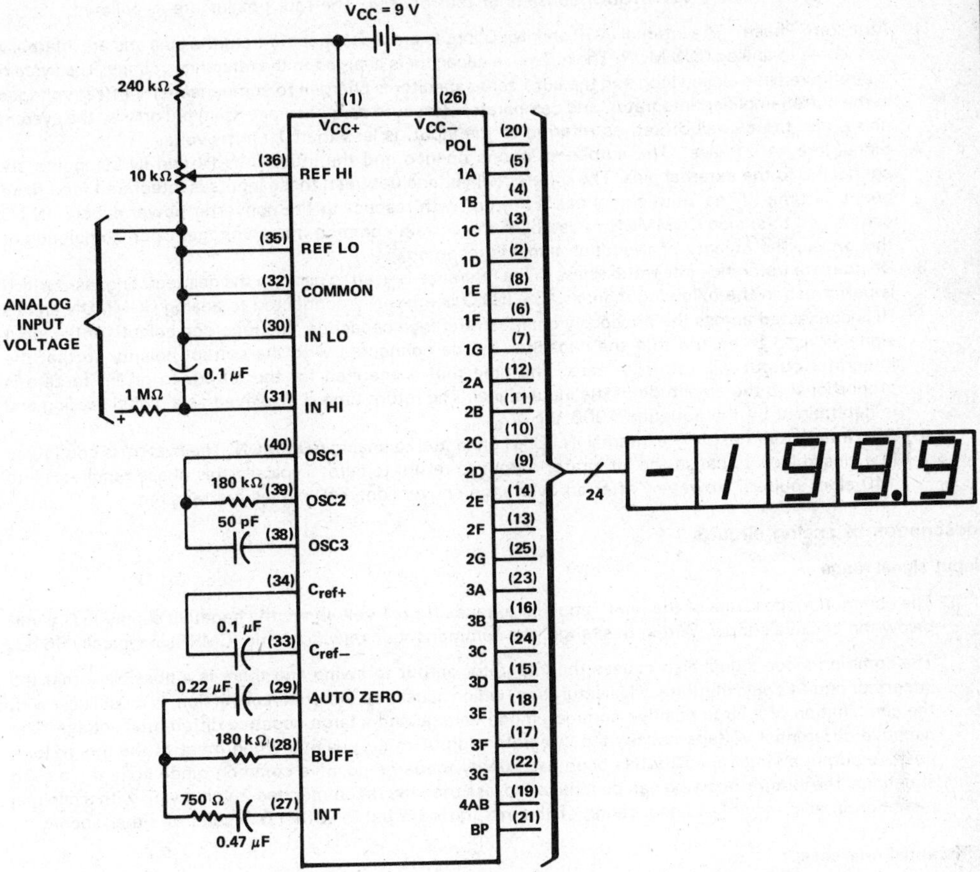
VCC = 9 V
240 kΩ
10 kΩ
ANALOG INPUT VOLTAGE
0.1 μF
1 MΩ
180 kΩ
50 pF
0.1 μF
0.22 μF
180 kΩ
750 Ω
0.47 μF
(1)
(26)
VCC+
VCC−
(36) REF HI
(35) REF LO
(32) COMMON
(30) IN LO
(31) IN HI
(40) OSC1
(39) OSC2
(38) OSC3
(34) Cref+
(33) Cref−
(29) AUTO ZERO
(28) BUFF
(27) INT
POL (20)
1A (5)
1B (4)
1C (3)
1D (2)
1E (8)
1F (6)
1G (7)
2A (12)
2B (11)
2C (10)
2D (9)
2E (14)
2F (13)
2G (25)
3A (23)
3B (16)
3C (24)
3D (15)
3E (18)
3F (17)
3G (22)
4AB (19)
BP (21)
24
-1 99.9

FIGURE 2. TEST CIRCUIT (CLOCK FREQUENCY = 48 kHz, 3 READINGS PER SECOND)

PRINCIPLES OF OPERATION

A measurement cycle, for the TLC7136C, consists of four phases. The four phases are as follows:

1. Auto-Zero Phase. The internal IN HI and IN LO inputs are disconnected from the pins and are internally connected to analog COMMON. The reference capacitor is charged to the reference voltage. The system is configured in a closed loop and the auto-zero capacitor is charged to compensate for offset voltages in the buffer amplifier, integrator, and comparator. The auto-zero accuracy is limited only by the system noise, and the overall offset, as referred to the input, is less than 10 microvolts.
2. Signal Integrate Phase. The auto-zero loop is opened and the internal IN HI and IN LO inputs are connected to the external pins. The differential voltage between these inputs is integrated for a fixed period of time. If the input signal has no return with respect to the converter power supply, IN LO can be tied to analog COMMON to establish the correct common-mode voltage. Upon completion of this phase, the polarity of the input signal is recorded.
3. Deintegrate (reference-integrate) Phase. The reference is used to perform the deintegrating task, which is performed in the following manner. The IN LO is internally connected to analog COMMON and IN HI is connected across the previously charged reference capacitor. The recorded polarity of the input signal is used to ensure that the capacitor will be connected with the correct polarity so that the integrator output will return to zero. The time that is required for the output to return to zero is proportional to the amplitude of the input signal. The return time is displayed as a digital reading and is determined by the equation 1000 V_{ID}/V_{ref}.
4. Zero Integrator Phase. The internal IN LO is connected to analog COMMON. The system is configured in a closed loop to cause the integrator output to return to zero. Typically this phase requires 11 to 140 clock pulses. However, after an overrange conversion, 740 pulses are required.

description of analog circuits

input signal range

The common-mode range of the input amplifier extends from 1 volt above the negative supply to 0.5 volt below the positive supply. Within this range, the common mode rejection ratio (CMRR) is typically 86 dB.

The common-mode signal also causes the integrator output to swing and there is a possibility that the integrator output could saturate. This saturation, which causes an incorrect conversion, is most likely with the combination of a large positive common mode voltage and a large negative differential voltage. The negative differential voltage causes the integrator output to go positive when most of the integrator's positive output swing capability has been used up by the large positive common mode voltage. In such situations, the integrator swing can be reduced to less than the recommended 2-volt swing with a minimal reduction in accuracy. The linear range of the integrator output is within 0.3 volts of either supply.

differential reference

The reference voltage must lie within the device power supply range. The major source of common-mode error is caused by the loss or gain of charge from the reference capacitor due to stray capacitances. With large common-mode voltages, the reference capacitor will gain charge or voltage while deintegrating a positive signal and, conversely, lose charge or voltage while deintegrating a negative signal. This gain or loss of reference capacitor voltage will cause a rollover error. The selection of a reference capacitor that is large in comparison to the stray capacitance will reduce the rollover error to less than 0.5 counts (see Component Value Selection).

analog common

For battery operation or when the inputs are floating with respect to the TLC7136C power supply, the analog COMMON pin is used to set the common mode voltage. The COMMON pin is preset by internal circuits to a voltage that is approximately 3 volts less than the TLC7136C positive supply. This preset voltage will give a 6-volt end-of-battery life.

PRINCIPLES OF OPERATION

When the power supply voltage is greater than 7 volts, the TLC7136C zener will be in a regulating mode and the preset voltage at the COMMON pin will have reference-like qualities. The preset voltage will then have a low 0.001%-per-volt voltage coefficient, a low output impedance of approximately 35 ohms, and a temperature coefficient of less than 80 ppm/°C. Therefore, the preset voltage could be used for an on-chip reference, however, there are some limitations. For 2°C to 8°C temperature changes, a scale factor of one count or more can result. Also, if the power supply voltage drops below 7 volts, the voltage coefficient will be poor since the zener will no longer be in a regulating mode.

Analog COMMON is connected to the internal IN LO during the auto-zero, deintegrate, and zero integrator phases. If IN LO is connected to a voltage that is different from analog common during the signal-integrate phase, the resulting common-mode voltage will be rejected by the amplifier. However, in certain applications, IN LO is set at a fixed known voltage, for example the power supply common voltage. For these applications, the COMMON pin should be tied to IN LO to eliminate the common-mode rejection error. The same consideration applies to the reference voltage. Referring the reference voltage to analog COMMON eliminates another common-mode error source. Referring the reference voltage to an analog common is accomplished by connecting COMMON to either REF LO or REF HI.

test

The TEST pin performs two functions. First, it is connected to the internally generated digital supply (negative side) through a 500-ohm resistor. This connection allows the TEST pin to be used as the negative supply for external segment drivers, such as decimal points or any LCD segment that requires up to 1-milliampere load current. Second, the pin performs a test function. When the TEST pin is pulled up to V_{CC+}, all segments will turn on and the display will read −1888. In this test mode, a constant DC voltage is applied to the segments, rather than a square wave, and the segments may be damaged if the test is prolonged.

description of digital circuits

An internal digital ground is generated with a 6-volt zener diode and a large P-channel source follower. This generated supply can handle the large capacitive currents that result when the backplane (BP) voltage is switched. Dividing the clock frequency by 800 gives the BP frequency. For 3 readings per second, the BP signal is a 5-volt, 60-Hz squarewave. The segments that are driven at the same frequency and amplitude are in phase with BP when off, and out of phase with BP when on. Except in the test mode, a negligible amount of DC voltage is placed across the segments. For negative-polarity inputs, the polarity indication will become active. Also, if the placement of IN LO and IN HI is switched, the polarity indication can be switched accordingly.

system timing

The TLC7136C clock circuit is shown in Figure 3. The three possible clock setups are pin 40 connected to an external oscillator, a crystal between pins 39 and 40, or an RC oscillator with connections to pins 38, 39, and 40.

The frequency of the clock oscillator is first divided by four and then the resulting clock signal is used to clock the decade counters. The divide-by-four clock signal is then further divided to form the four convert-cycle phases, which are as follows:

1. 1,000 counts for signal integration.
2. 0 to 2,000 counts for reference deintegration.

PRINCIPLES OF OPERATION

3. 11 to 140 counts for zero integration (with an overranged conversion of greater than 2,060 counts, the zero integrator phase will require 740 counts, and auto-zero will require 260 counts).
4. 910 to 2900 counts for auto-zero (for signals less than full-scale, auto-zero gets the unused portion of reference deintegration and zero integration).

The total measurement cycle requires 4,000 counts or 16,000 clock pulses. A 48-kilohertz oscillator would be required for three readings per second.

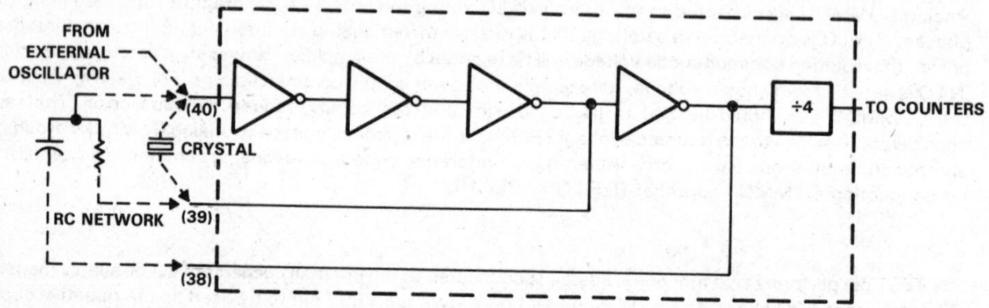
FROM
EXTERNAL
OSCILLATOR
(40)
CRYSTAL
RC NETWORK
(39)
(38)
÷4
TO COUNTERS

NOTE: This figure shows all three external control circuits connected; however, only one external circuit (crystal, RC network or external oscillator) is connected for proper operation.

FIGURE 3. CLOCK CIRCUITS

component value selection

integrating resistor

The buffer amplifier and integrator class A output stages require approximately 6 microamperes of quiescent current and can source −1 microampere of current without inducing any significant nonlinearity. The integrating resistor should be sufficiently large that the buffer amplifier and integrator will remain in this linear region. However, the resistor must also be small enough that PC board leakage remains insignificant. Values of 180 kilohms and 1.8 megohms are recommended for the respective 200-millivolt and 2-volt full-scale voltages.

integrating capacitor

The integrating capacitor should be chosen to give the maximum voltage swing, yet not allow the combined tolerances of the integrating resistor and capacitor to cause the integrator to saturate. The linear range of the integrator extends to within 0.3 volt of V_{CC-} or V_{CC+}. A +2-volt full-scale integrator swing works fine when analog common is used as the reference. Capacitor values of 0.047 microfarad and 0.15 microfarad are recommended for 3 (48-kilohertz oscillator) and 1 (16-kilohertz oscillator) readings per second respectively. As the oscillator frequency is increased, the capacitor value must be decreased to maintain the same output swing. Polypropylene capacitors are recommended because of their reasonable cost and low dielectric absorption, which produces low roll-over errors.

PRINCIPLES OF OPERATION

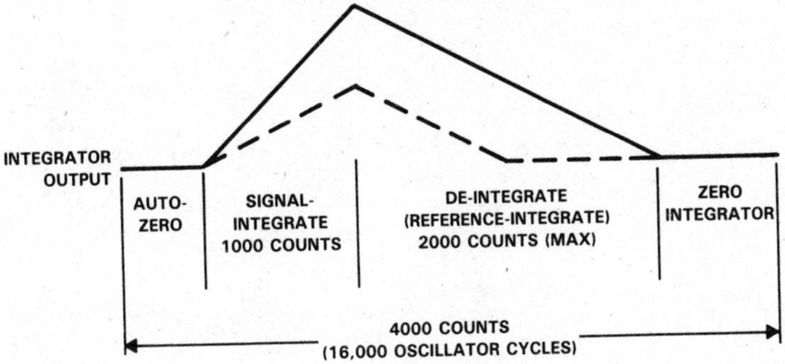
INTEGRATOR OUTPUT
AUTO-ZERO
SIGNAL-INTEGRATE 1000 COUNTS
DE-INTEGRATE (REFERENCE-INTEGRATE) 2000 COUNTS (MAX)
ZERO INTEGRATOR
4000 COUNTS (16,000 OSCILLATOR CYCLES)

FIGURE 4. TIMING DIAGRAM

auto-zero capacitor

The size of this capacitor has an effect upon the noise of the system. For 200-millivolt full-scale applications, in which noise must be kept to a minimum, a 0.47-microfarad capacitor is recommended. The zero-integrator phase allows the use of a large auto-zero capacitor without the accompanying hysteresis or overrange hangover problems that can occur with the ICL7126 or ICL7106.

reference capacitor

A 0.1-microfarad capacitor is fine for most applications. However, with large common-mode signals, if the REF LO pin is not connected to analog COMMON and the full-scale voltage is 200 millivolts, a larger capacitor is required to prevent rollover error. A 1-microfarad capacitor will hold this rollover error to 0.5 counts.

oscillator components

A 50-pF capacitor is recommended for all frequency ranges. The resistor can be selected from the equation:

$$f = 0.45/RC$$

where: R = 180 kilohms for 48-kilohertz oscillator (3 readings per second) and 560 kilohms for 16-kilohertz oscillator (1 reading per second).

reference voltage

An input voltage of 2 V_{ref} is required to obtain a full-scale reading of 2,000 counts. Therefore, for a full-scale of 200 millivolts and 2 volts, V_{ref} should be 100 millivolts and 1 volt respectively. In many situations, the designer might like to have a full-scale voltage other than 200 millivolts or 2 volts. In situations where the designer desires a full-scale voltage of X volts, the designer can select a V_{ref} of X/2 volts. The value of the integrating resistor can be determined by the following equation:

$$\frac{\text{X volts (desired full-scale)}}{\text{200 mV}} = \frac{\text{Value of Integrating Resistor for X volts full-scale}}{\text{180 k}\Omega\text{ (integrating resistor for 200 mV full-scale)}}$$

If X volts is greater than 200 millivolts, it is better to work with an X/2 volts reference, since dividing the X volts down to 200 millivolts will cause the input signal to be more susceptible to noise.

Sometimes a designer will want a digital reading of zero when V_I does not equal zero. This desire can be met by connecting V_I between IN HI and COMMON and the zero-reading V_I between COMMON and IN LO.

TLC7524
Advanced LinCMOS™ 8-BIT MULTIPLYING DIGITAL-TO-ANALOG CONVERTER

D3008, SEPTEMBER 1986

- Advanced LinCMOS™ Silicon-Gate Technology
- Easily Interfaced to Microprocessors
- On-Chip Data Latches
- Guaranteed Monotonicity
- Segmented High-Order Bits Ensure Low-Glitch Output
- Designed to be Interchangeable with Analog Devices AD7524, PMI PM-7524, and Micro Power Systems MP7524
- Fast Control Signaling for Digital Signal Processor Applications Including Interface with TMS320

D OR N PACKAGE
(TOP VIEW)

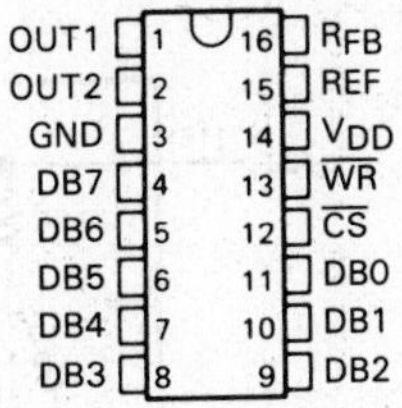
OUT1 1
OUT2 2
GND 3
DB7 4
DB6 5
DB5 6
DB4 7
DB3 8
16 RFB
15 REF
14 VDD
13 WR
12 CS
11 DB0
10 DB1
9 DB2

KEY PERFORMANCE SPECIFICATIONS	
Resolution	8 Bits
Linearity error	½ LSB Max
Power dissipation at V_{DD} = 5 V	5 mW Max
Settling time	100 ns Max
Propagation delay	80 ns Max

description

The TLC7524 is an Advanced LinCMOS™ 8-bit digital-to-analog converter (DAC) designed for easy interface to most popular microprocessors.

The TLC7524 is an 8-bit multiplying DAC with input latches and with a load cycle similar to the "write" cycle of a random access memory. Segmenting the high-order bits minimizes glitches during changes in the most-significant bits, which produce the highest glitch impulse. The TLC7524 provides accuracy to ½ LSB without the need for thin-film resistors or laser trimming, while dissipating less than 5 milliwatts typically.

Featuring operation from a 5-V to 15-V single supply, the TLC7524 interfaces easily to most microprocessor buses or output ports. Excellent multiplying (2 or 4 quadrant) makes the TLC7524 an ideal choice for many microprocessor-controlled gain-setting and signal-control applications.

The TLC7524I is characterized for operation from −25°C to 85°C, and the TLC7524C is characterized for operation from 0°C to 70°C.

TEXAS INSTRUMENTS
POST OFFICE BOX 655012 • DALLAS, TEXAS 75265

functional block diagram

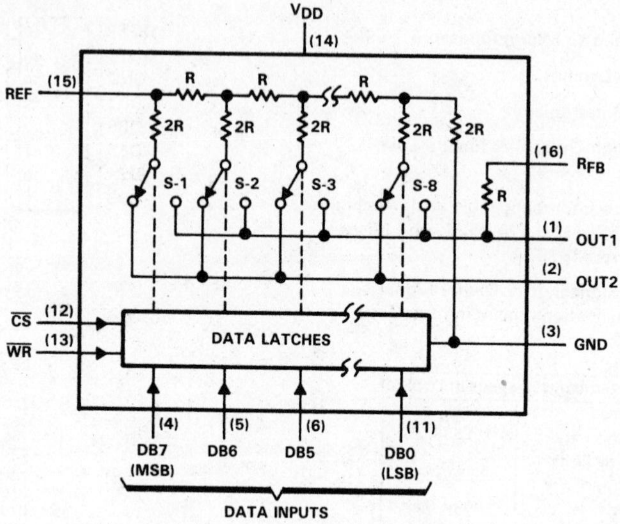
VDD
(14)
REF
(15)
R
R
R
2R
2R
2R
2R
2R
S-1
S-2
S-3
S-8
(16)
RFB
R
(1)
OUT1
(2)
OUT2
CS
(12)
WR
(13)
DATA LATCHES
(3)
GND
(4)
(5)
(6)
(11)
DB7
(MSB)
DB6
DB5
DB0
(LSB)
DATA INPUTS

operating sequence

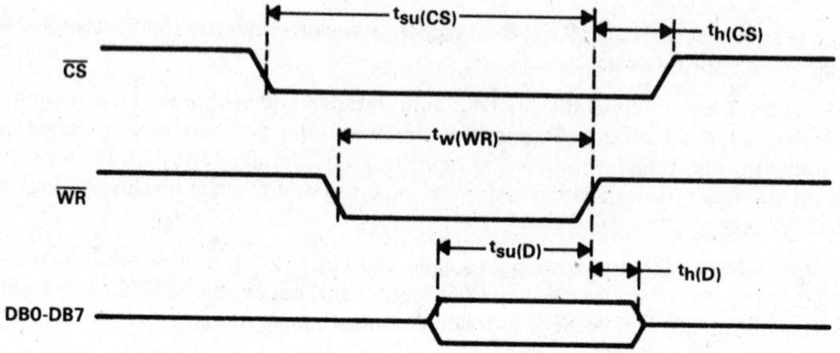
tsu(CS)
th(CS)
CS
tw(WR)
WR
tsu(D)
th(D)
DB0-DB7

absolute maximum ratings over operating free-air temperature range (unless otherwise noted)

Supply voltage, V_{DD} . . . . . −0.3 V to 16.5 V
Digital input voltage, V_I . . . . . −0.3 V to V_{DD} + 0.3 V
Reference voltage, V_{ref} . . . . . ±25 V
Peak digital input current, I_I . . . . . 10 μA
Operating free-air temperature range: TLC7524I . . . . . −25°C to 85°C
TLC7524C . . . . . 0°C to 70°C
Storage temperature range . . . . . −65°C to 150°C
Lead temperature 1,6 mm (1/16 inch) from case for 10 seconds . . . . . 260°C

recommended operating conditions

		V_{DD} = 5 V			V_{DD} = 15 V			UNIT
		MIN	NOM	MAX	MIN	NOM	MAX	
Supply voltage, V_{DD}		4.75	5	5.25	14.5	15	15.5	V
Reference voltage, V_{ref}			±10			±10		V
High-level input voltage, V_{IH}		2.4			13.5			V
Low-level input voltage, V_{IL}				0.8			1.5	V
CS setup time, $t_{su(CS)}$		40			40			ns
CS hold time, $t_{h(CS)}$		0			0			ns
Data bus input setup time, $t_{su(D)}$		25			25			ns
Data bus input hold time, $t_{h(D)}$		10			10			ns
Pulse duration, $\overline{WR}$ low, $t_{w(WR)}$		40			40			ns
Operating free-air temperature, T_A	TLC7524I	−25		85	−25		85	°C
	TLC7524C	0		70	0		70	

electrical characteristics over recommended operating free-air temperature range, V_{ref} = ±10 V, OUT1 and OUT2 at GND (unless otherwise noted)

PARAMETER			TEST CONDITIONS	V_{DD} = 5 V			V_{DD} = 15 V			UNIT
				MIN	TYP	MAX	MIN	TYP	MAX	
I_{IH}	High-level input current		$V_I = V_{DD}$			10			10	μA
I_{IL}	Low-level input current		$V_I = 0$			−10			−10	μA
	Output leakage current, OUT1 (Pin 1)		DB0-DB7 at 0 V, $\overline{WR}$, $\overline{CS}$ at 0 V, V_{ref} = ±10 V			±400			±200	nA
	Output leakage current, OUT2 (Pin 2)		DB0-DB7 at V_{DD}, $\overline{WR}$, $\overline{CS}$ at 0 V, V_{ref} = ±10 V			±400			±200	nA
I_{DD}	Supply current (quiescent)		DB0-DB7 at $V_{IH(min)}$ or $V_{IL(max)}$			1			2	mA
I_{DD}	Supply current (standby)		DB0-DB7 at 0 V or V_{DD}			500			500	μA
k_{SVS}	Supply voltage sensitivity, $\Delta gain/\Delta V_{DD}$		ΔV_{DD} = ±10%		0.01	0.16		0.005	0.04	%FSR
C_i	Input capacitance, DB0-DB7, $\overline{WR}$, $\overline{CS}$		$V_I = 0$			5			5	pF
C_o	Output capacitance,	OUT1	DB0-DB7 at 0 V, $\overline{WR}$ and $\overline{CS}$ at 0 V			30			30	pF
		OUT2				120			120	
C_o	Output capacitance	OUT1	DB0-DB7 at V_{DD}, $\overline{WR}$ and $\overline{CS}$ at 0 V			120			120	pF
		OUT2				30			30	
	Reference input impedance (Pin 15 to GND)			5		20	5		20	kΩ

operating characteristics over recommended operating free-air temperature range, V_{ref} = ±10 V, OUT1 and OUT2 at GND (unless otherwise noted)

PARAMETER	TEST CONDITIONS	V_{DD} = 5 V MIN	V_{DD} = 5 V TYP†	V_{DD} = 5 V MAX	V_{DD} = 15 V MIN	V_{DD} = 15 V TYP†	V_{DD} = 15 V MAX	UNIT
Linearity error				±0.5			±0.5	LSB
Gain error	See Note 1			±2.5			±2.5	LSB
Settling time (to ½ LSB)	See Note 2			100			100	ns
Propagation delay (from digital input to 90% of final analog output current)	See Note 2			80			80	ns
Feedthrough at OUT1 or OUT2	V_{ref} = ±10 V (100-kHz sinewave) $\overline{WR}$ and $\overline{CS}$ at 0 V, DB0-DB7 at 0 V			0.5			0.5	%FSR
Temperature coefficient of gain	T_A = 25°C to MAX		±0.004			±0.001		%FSR/°C

†Typical values at T_A = 25°C.

NOTES: 1. Gain error is measured using the internal feedback resistor. Nominal Full Scale Range (FSR) = V_{ref} − 1 LSB.
2. OUT1 load = 100 Ω, C_{ext} = 13 pF, $\overline{WR}$ at 0 V, $\overline{CS}$ at 0 V, DB0-DB7 at 0 V to V_{DD} or V_{DD} to 0 V.

principles of operation

The TLC7524 is an 8-bit multiplying D/A converter consisting of an inverted R-2R ladder, analog switches, and data input latches. Binary weighted currents are switched between the OUT1 and OUT2 bus lines, thus maintaining a constant current in each ladder leg independent of the switch state. The high-order bits are decoded and these decoded bits, through a modification in the R-2R ladder, control three equally weighted current sources. Most applications only require the addition of an external operational amplifier and a voltage reference.

The equivalent circuit for all digital inputs low is seen in Figure 1. With all digital inputs low, the entire reference current, I_{ref}, is switched to OUT2. The current source I/256 represents the constant current flowing through the termination resistor of the R-2R ladder, while the current source I_{lkg} represents leakage currents to the substrate. The capacitances appearing at OUT1 and OUT2 are dependent upon the digital input code. With all digital inputs high, the off-state switch capacitance (30 pF maximum) appears at OUT2 and the on-state switch capacitance (120 pF maximum) appears at OUT1. With all digital inputs low, the situation is reversed as shown in Figure 1. Analysis of the circuit for all digital inputs high is similar to Figure 1; however, in this case, I_{ref} would be switched to OUT1.

Interfacing the TLC7524 D/A converter to a microprocessor is accomplished via the data bus and the $\overline{CS}$ and $\overline{WR}$ control signals. When $\overline{CS}$ and $\overline{WR}$ are both low, the TLC7524 analog output responds to the data activity on the DB0-DB7 data bus inputs. In this mode, the input latches are transparent and input data directly affects the analog output. When either the $\overline{CS}$ signal or $\overline{WR}$ signal goes high, the data on the DB0-DB7 inputs are latched until the $\overline{CS}$ and $\overline{WR}$ signals go low again. When $\overline{CS}$ is high, the data inputs are disabled regardless of the state of the $\overline{WR}$ signal.

The TLC7524 is capable of performing 2-quadrant or full 4-quadrant multiplication. Circuit configurations for 2-quadrant or 4-quadrant multiplication are shown in Figures 2 and 3. Input coding for unipolar and bipolar operation are summarized in Tables 1 and 2, respectively.

principles of operation (continued)

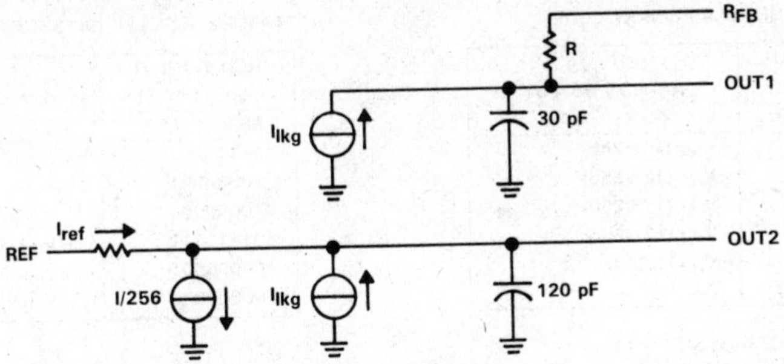
RFB
R
OUT1
Ilkg
30 pF
Iref
REF
OUT2
I/256
Ilkg
120 pF

FIGURE 1. TLC7524 EQUIVALENT CIRCUIT WITH ALL DIGITAL INPUTS LOW

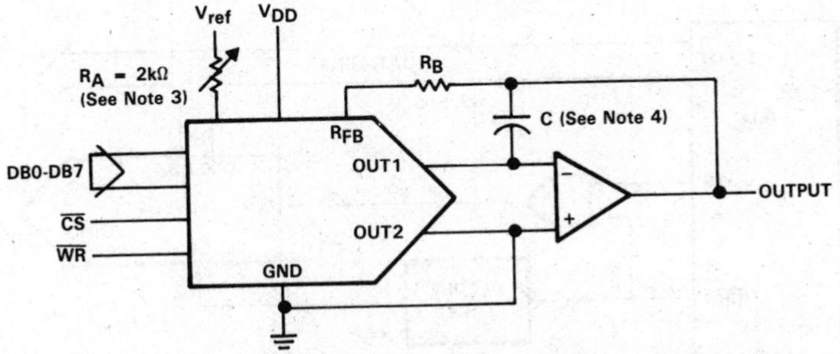
Vref
VDD
RA = 2kΩ
(See Note 3)
RB
C (See Note 4)
RFB
OUT1
DB0-DB7
OUTPUT
CS
OUT2
WR
GND

FIGURE 2. UNIPOLAR OPERATION (2-QUADRANT MULTIPLICATION)

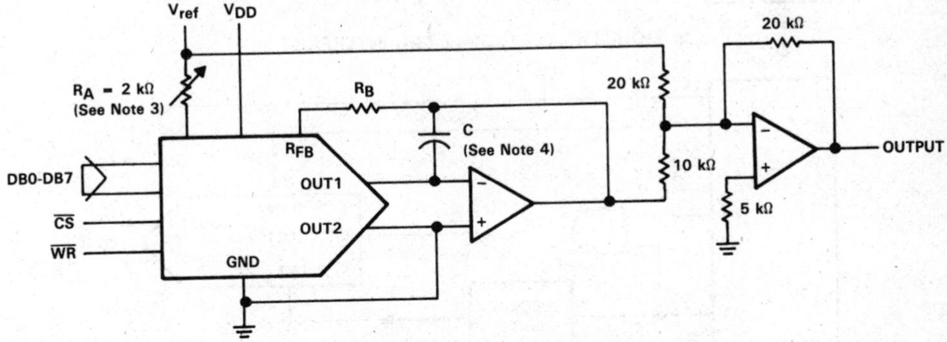
Vref
VDD
20 kΩ
RA = 2 kΩ
(See Note 3)
RB
20 kΩ
C
(See Note 4)
RFB
DB0-DB7
OUT1
10 kΩ
OUTPUT
CS
OUT2
5 kΩ
WR
GND

FIGURE 3. BIPOLAR OPERATION (4-QUADRANT OPERATION)

NOTES: 3. R_A and R_B used only if gain adjustment is required.
4. C phase compensation (10-15 pF) is required when using high-speed amplifiers to prevent ringing or oscillation.

principles of operation (continued)

TABLE 1. UNIPOLAR BINARY CODE

DIGITAL INPUT (SEE NOTE 5) MSB LSB	ANALOG OUTPUT
11111111	$-V_{ref}$ (255/256)
10000001	$-V_{ref}$ (129/256)
10000000	$-V_{ref}$ (128/256) = $-V_{ref}/2$
01111111	$-V_{ref}$ (127/256)
00000001	$-V_{ref}$ (1/256)
00000000	0

TABLE 2. BIPOLAR (OFFSET BINARY) CODE

DIGITAL INPUT (SEE NOTE 6) MSB LSB	ANALOG OUTPUT
11111111	V_{ref} (127/128)
10000001	V_{ref} (1/128)
10000000	0
01111111	$-V_{ref}$ (1/128)
00000001	$-V_{ref}$ (127/128)
00000000	$-V_{ref}$

NOTES: 5. LSB = 1/256 (V_{ref}).
6. LSB = 1/128 (V_{ref}).

microprocessor interfaces

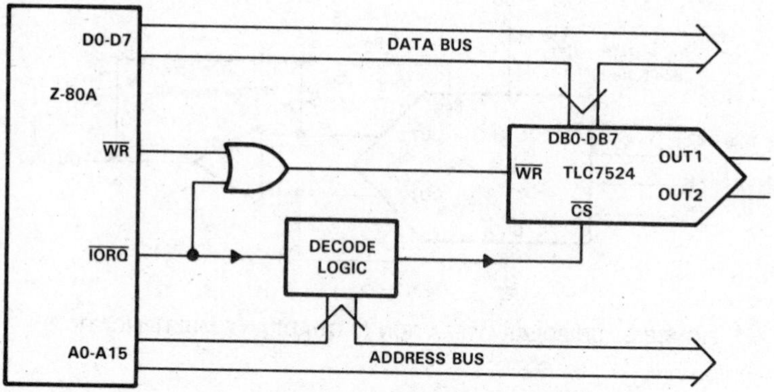
D0-D7
DATA BUS
Z-80A
WR
DB0-DB7
WR
TLC7524
OUT1
OUT2
CS
IORQ
DECODE
LOGIC
A0-A15
ADDRESS BUS

FIGURE 4. TLC7524—Z-80A INTERFACE

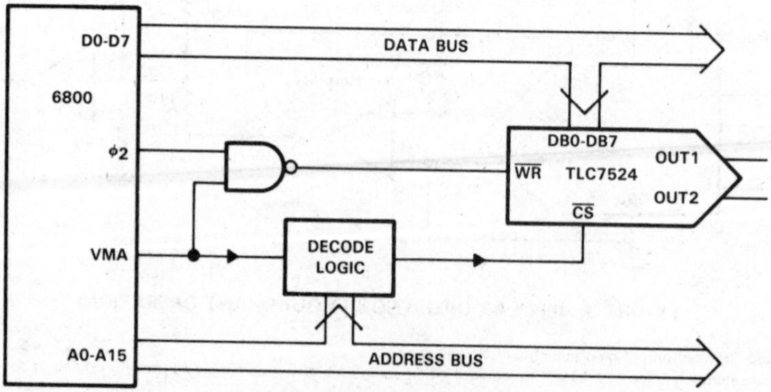
D0-D7
DATA BUS
6800
φ2
DB0-DB7
WR
TLC7524
OUT1
OUT2
CS
VMA
DECODE
LOGIC
A0-A15
ADDRESS BUS

FIGURE 5. TLC7524—6800 INTERFACE

TEXAS INSTRUMENTS
POST OFFICE BOX 655012 • DALLAS, TEXAS 75265

microprocessor interfaces (continued)

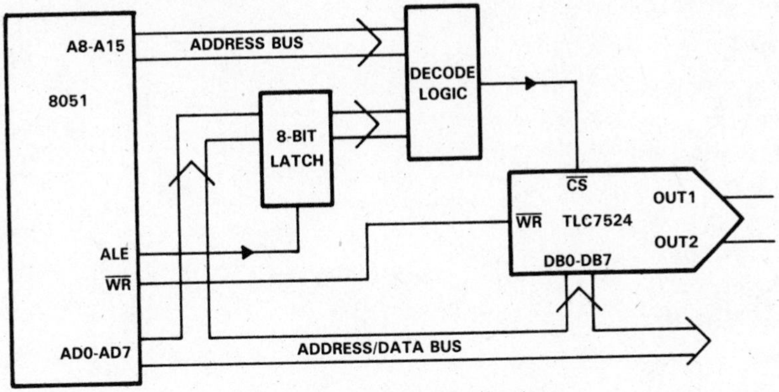
A8-A15
ADDRESS BUS
8051
DECODE LOGIC
8-BIT LATCH
CS
OUT1
WR
TLC7524
OUT2
DB0-DB7
ALE
WR
AD0-AD7
ADDRESS/DATA BUS

FIGURE 6. TLC7524—8051 INTERFACE

TLC7528
Advanced LinCMOS™ DUAL 8-BIT MULTIPLYING DIGITAL-TO-ANALOG CONVERTER

D2979, JANUARY 1987

- ADVANCED LinCMOS™ Silicon-Gate Technology
- Easily Interfaced to Microprocessors
- On-Chip Data Latches
- Guaranteed Monotonicity
- Designed to be Interchangeable with Analog Devices ADC7528 and PMI PM-7528
- Fast Control Signaling for Digital Signal Processor Applications Including Interface with TMS320

N DUAL-IN-LINE PACKAGE

(TOP VIEW)

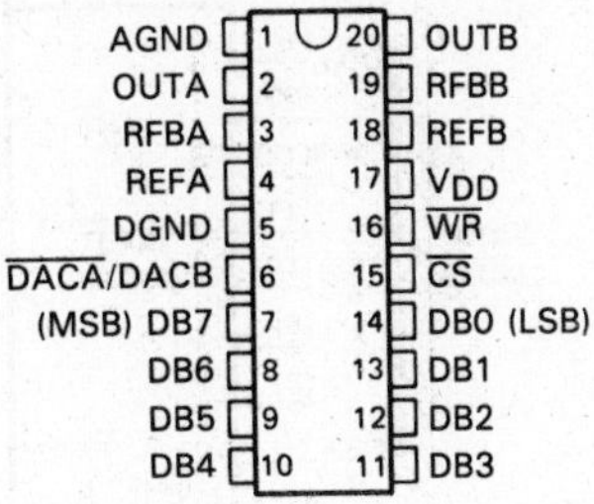
AGND 1
OUTA 2
RFBA 3
REFA 4
DGND 5
DACA/DACB 6
(MSB) DB7 7
DB6 8
DB5 9
DB4 10
20 OUTB
19 RFBB
18 REFB
17 VDD
16 WR
15 CS
14 DB0 (LSB)
13 DB1
12 DB2
11 DB3

KEY PERFORMANCE SPECIFICATIONS	
Resolution	8 bits
Linearity Error	1/2 LSB
Power Dissipation at V_{DD} = 5 V	5 mW
Settling Time at V_{DD} = 5 V	100 ns
Propagation Delay at V_{DD} = 5 V	80 ns

description

The TLC7528 is a dual 8-bit digital-to-analog converter designed with separate on-chip data latches and featuring excellent DAC-to-DAC matching. Data is transferred to either of the two DAC data latches via a common 8-bit input port. Control input $\overline{\text{DACA}}$/DACB determines which DAC is to be loaded. The ''load'' cycle of the TLC7528 is similar to the ''write'' cycle of a random-access memory, allowing easy interface to most popular microprocessor busses and output ports. Segmenting the high-order bits minimizes glitches during changes in the most significant bits, where glitch impulse is typically the strongest.

The TLC7528 operates from a 5-volt to 15-volt power supply and dissipates less than 15 mW (typical). Excellent 2- or 4-quadrant multiplying makes the TLC7528 a sound choice for many microprocessor-controlled gain-setting and signal-control applications.

The TLC7528I is characterized for operation from −25 to 85 °C. The TLC7528C is characterized for operation from 0 °C to 70 °C.

ADVANCED LinCMOS is a trademark of Texas Instruments Incorporated

TEXAS INSTRUMENTS
POST OFFICE BOX 655012 • DALLAS, TEXAS 75265

functional block diagram

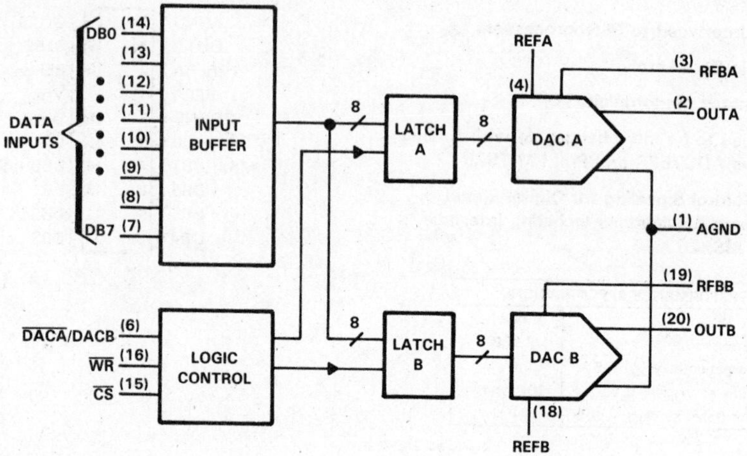
DATA
INPUTS
DB0
DB7
(14)
(13)
(12)
(11)
(10)
(9)
(8)
(7)
INPUT
BUFFER
DACA/DACB
WR
CS
(6)
(16)
(15)
LOGIC
CONTROL
8
LATCH
A
LATCH
B
DAC A
DAC B
REFA
(4)
(3)
RFBA
(2)
OUTA
(1)
AGND
(19)
RFBB
(20)
OUTB
(18)
REFB

operating sequence

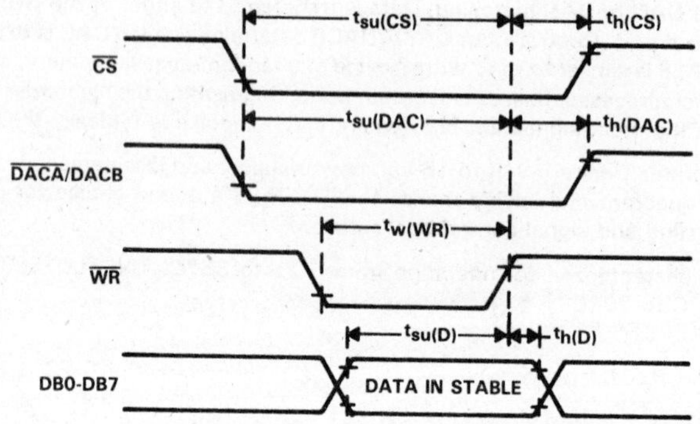
CS
DACA/DACB
WR
DB0-DB7
tsu(CS)
th(CS)
tsu(DAC)
th(DAC)
tw(WR)
tsu(D)
th(D)
DATA IN STABLE

absolute maximum ratings over operating free-air temperature range (unless otherwise noted)

Supply voltage, V_{DD} (to AGND or DGND) . . . . −0.3 V to 16.5 V
Voltage between AGND and DGND . . . . $\pm V_{DD}$
Input voltage, V_I (to DGND) . . . . -0.3 V to $V_{DD} + 0.3$
Reference voltage, V_{refA} or V_{refB} (to AGND) . . . . ±25 V
Output voltage, V_{OA} or V_{OB} (to AGND) . . . . ±25 V
Peak input current . . . . 10 μA
Operating free-air temperature range: TLC7528I . . . . −25°C to 85°C
TLC7528C . . . . 0°C to 70°C
Storage temperature range . . . . −65°C to 150°C
Lead temperature 1,6 mm (1/16 inch) from case for 10 seconds . . . . 260°C

recommended operating conditions

		V_{DD} = 4.75 V to 5.25 V			V_{DD} = 14.5 V to 15.5 V			UNIT
		MIN	NOM	MAX	MIN	NOM	MAX	
Reference voltage, V_{refA} or V_{refB}			±10			±10		V
High-level input voltage, V_{IH}		2.4			13.5			V
Low-level input voltage, V_{IL}				0.8			1.5	V
$\overline{CS}$ setup time, $t_{su(CS)}$		50			50			ns
$\overline{CS}$ hold time, $t_{h(CS)}$		0			0			ns
DAC select setup time, $t_{su(DAC)}$		50			50			ns
DAC select hold time, $t_{h(DAC)}$		10			10			ns
Data bus input setup time $t_{su(D)}$		25			25			ns
Data bus input hold time $t_{h(D)}$		0			0			ns
Pulse duration, $\overline{WR}$ low, $t_{w(WR)}$		50			50			ns
Operating free-air temperature, T_A	TLC7528I	−25		85	−25		85	°C
	TLC7528C	0		70	0		70	

electrical characteristics over recommended operating free-air temperature range, V_{refA} = V_{refB} = 10 V, V_{OA} and V_{OB} at 0 V (unless otherwise noted)

PARAMETER			TEST CONDITIONS	V_{DD} = 5 V MIN	V_{DD} = 5 V TYP†	V_{DD} = 5 V MAX	V_{DD} = 15 V MIN	V_{DD} = 15 V TYP†	V_{DD} = 15 V MAX	UNIT
I_{IH}	High-level input current		V_I = V_{DD}			10			10	μA
I_{IL}	Low-level input current		V_I = 0 V			−10			−10	μA
	Reference input impedance (Pin 15 to GND)			8	11	15	8	11	15	kΩ
I_{lkg}	Output leakage current	OUTA	DACA data latch loaded with 00000000, V_{refA} = ±10 V			±400			±200	nA
		OUTB	DACB data latch loaded with 00000000, V_{refB} = ±10 V			±400			±200	
	Input resistance match (REFA to REFB)					±1%			±1%	
	DC supply sensitivity, Δ gain/Δ V_{DD}		ΔV_{DD} = ±10%			0.04			0.02	%/%
I_{DD}	Supply current (quiescent)		DB0-DB7 at V_{IH}min or V_{IL}max			1			1	mA
I_{DD}	Supply current (standby)		DB0-DB7 at 0 V or V_{DD}			0.5			0.5	mA
C_i	Input capacitance	DB0-DB7				10			10	pF
		$\overline{WR}$, $\overline{CS}$ $\overline{DACA}$/DACB				15			15	
C_o	Output capacitance, (OUTA, OUTB)		DAC data latches loaded with 00000000			50			50	pF
			DAC data latches loaded with 11111111			120			120	

†All typical values are at T_A = 25°C.

TEXAS INSTRUMENTS
POST OFFICE BOX 655012 • DALLAS, TEXAS 75265

operating characteristics over recommended operating free-air temperature range, V_{refA} = V_{refB} = 10 V, V_{OA} and V_{OB} at 0 V (unless otherwise noted)

PARAMETER		TEST CONDITIONS	V_{DD} = 5 V MIN	V_{DD} = 5 V TYP	V_{DD} = 5 V MAX	V_{DD} = 15 V MIN	V_{DD} = 15 V TYP	V_{DD} = 15 V MAX	UNIT
Linearity error					±1/2			±1/2	LSB
Settling time (to 1/2 LSB)		See Note 1			100			100	ns
Gain error		See Note 2			2.5			2.5	LSB
AC feedthrough	REFA to OUTA	See Note 3			−65			−65	dB
	REFB to OUTB				−65			−65	
Temperature coefficient of gain		See Note 4			0.007			0.0035	%FSR/°C
Propagation delay (from digital input to 90% of final analog output current)		See Note 5			80			80	ns
Channel-to-channel isolation	REFA to OUTB	See Note 6		77			77		dB
	REFB to OUTA	See Note 7		77			77		
Digital-to-analog glitch impulse area		Measured for code transition from 00000000 to 11111111, T_A = 25°C		160			440		nVs
Digital crosstalk glitch impulse area		Measured for code transition from 00000000 to 11111111, T_A = 25°C		30			60		nVs
Harmonic distortion		V_i = 6 V rms, f = 1 kHz, T_A = 25°C		−85			−85		dB

NOTES: 1. OUTA, OUTB load = 100 Ω, C_{ext} = 13 pF; $\overline{WR}$ and $\overline{CS}$ at 0 V; DB0-DB7 at 0 V to V_{DD} or V_{DD} to 0 V.
2. Gain error is measured using an internal feedback resistor. Nominal Full Scale Range (FSR) = V_{ref} − 1 LSB.
3. V_{ref} = 20 V peak-to-peak, 100-kHz sine wave; DAC data latches loaded with 0000000.
4. Temperature coefficient of gain measured from 0°C to 25°C or from 25°C to 70°C.
5. V_{refA} = V_{refB} = 10 V; OUTA/OUTB load = 100 Ω, C_{ext} = 13 pF; $\overline{WR}$ and $\overline{CS}$ at 0 V; DB0-DB7 at 0 V to V_{DD} or V_{DD} to 0 V.
6. Both DAC latches loaded with 11111111; V_{refA} = 20 V peak-to-peak, 100-kHz sine wave; V_{refB} = 0; T_A = 25°C.
7. Both DAC latches loaded with 11111111; V_{refB} = 20 V peak-to-peak, 100-kHz sine wave; V_{refA} = 0; T_A = 25°C.

principles of operation

The TLC7528 contains two identical 8-bit multiplying D/A converters, DACA and DACB. Each DAC consists of an inverted R-2R ladder, analog switches, and input data latches. Binary-weighted currents are switched between DAC output and AGND, thus maintaining a constant current in each ladder leg independent of the switch state. Most applications require only the addition of an external operational amplifier and voltage reference. A simplified D/A circuit for DACA with all digital inputs low is shown in Figure 1.

Figure 2 shows the DACA equivalent circuit. A similar equivalent circuit can be drawn for DACB. Both DACs share the analog ground pin 1 (AGND). With all digital inputs high, the entire reference current flows to OUTA. A small leakage current (I_{lkg}) flows across internal junctions, and as with most semiconductor devices, doubles every 10°C. C_o is due to the parallel combination of the NMOS switches and has a value that depends on the number of switches connected to the output. The range of C_o is 50 pF to 120 pF maximum. The equivalent output resistance r_o varies with the input code from 0.8R to 3R where R is the nominal value of the ladder resistor in the R-2R network.

Interfacing the TLC7528 to a microprocessor is accomplished via the data bus, $\overline{CS}$, $\overline{WR}$, and $\overline{DACA}$/DACB control signals. When $\overline{CS}$ and $\overline{WR}$ are both low, the TLC7528 analog output, specified by the $\overline{DACA}$/DACB control line, responds to the activity on the DB0-DB7 data bus inputs. In this mode, the input latches are transparent and input data directly affects the analog output. When either the $\overline{CS}$ signal or $\overline{WR}$ signal goes high, the data on the DB0-DB7 inputs is latched until the $\overline{CS}$ and $\overline{WR}$ signals go low again. When $\overline{CS}$ is high, the data inputs are disabled regardless of the state of the $\overline{WR}$ signal.

The digital inputs of the TLC7528 provide TTL compatibility when operated from a supply voltage of 5 V. The TLC7528 may be operated with any supply voltage in the range from 5 V to 15 V, however, input logic levels are not TTL compatible above 5 V.

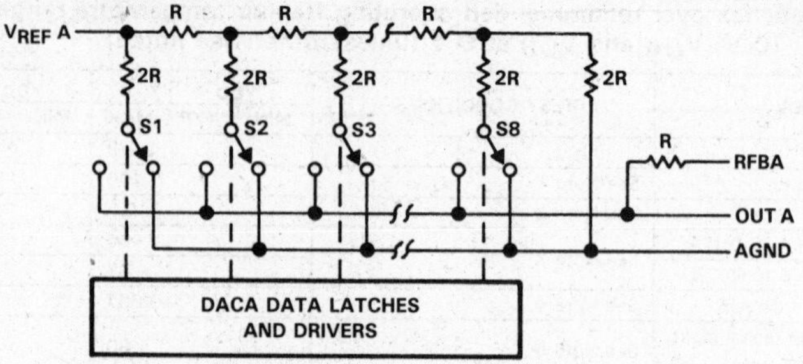
V_{REF} A
R
R
R
2R
2R
2R
2R
2R
S1
S2
S3
S8
R
RFBA
OUT A
AGND
DACA DATA LATCHES
AND DRIVERS

FIGURE 1. SIMPLIFIED FUNCTIONAL CIRCUIT FOR DACA

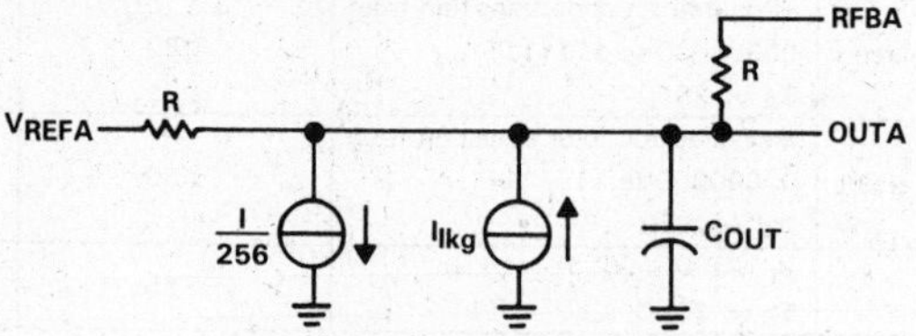
RFBA
R
V_{REFA}
R
OUTA
$\frac{I}{256}$
I_{lkg}
C_{OUT}

FIGURE 2. TLC7528 EQUIVALENT CIRCUIT, DACA LATCH LOADED WITH 11111111.

MODE SELECTION TABLE

$\overline{DACA}$/DACB	$\overline{CS}$	$\overline{WR}$	DACA	DACB
L	L	L	WRITE	HOLD
H	L	L	HOLD	WRITE
X	H	X	HOLD	HOLD
X	X	H	HOLD	HOLD

L = low level, H = high level, X = don't care

TYPICAL APPLICATION DATA

The TLC7528 is capable of performing 2-quadrant or full 4-quadrant multiplication. Circuit configurations for 2-quadrant and 4-quadrant multiplication are shown in Figures 3 and 4. Input coding for unipolar and bipolar operation are summarized in Tables 1 and 2, respectively.

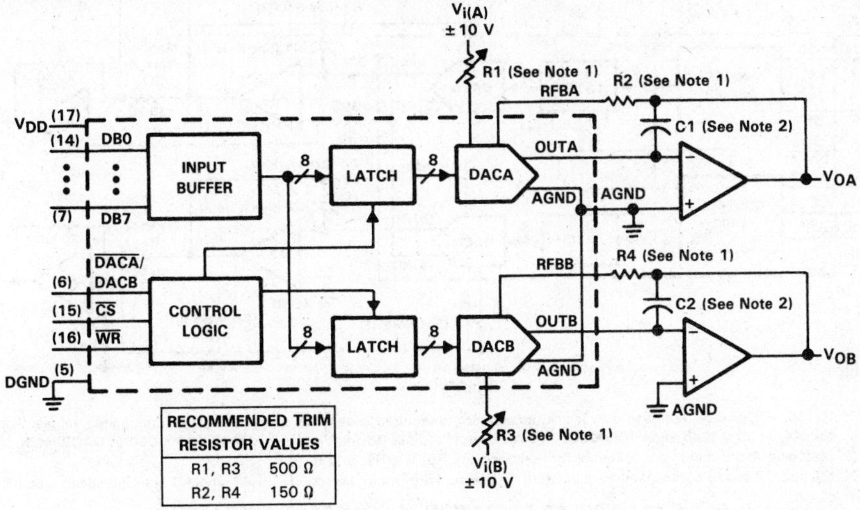
Vi(A)
±10 V
R1 (See Note 1)
RFBA
R2 (See Note 1)
VDD
(17)
(14)
DB0
(7)
DB7
INPUT BUFFER
8
LATCH
8
DACA
OUTA
AGND
AGND
C1 (See Note 2)
VOA
DACA/DACB
(6)
(15) CS
(16) WR
CONTROL LOGIC
(5)
DGND
RFBB
R4 (See Note 1)
C2 (See Note 2)
8
LATCH
8
DACB
OUTB
AGND
VOB
AGND
R3 (See Note 1)
Vi(B)
±10 V

RECOMMENDED TRIM RESISTOR VALUES	
R1, R3	500 Ω
R2, R4	150 Ω

NOTES: 1. R1, R2, R3, and R4 are used only if gain adjustment is required. See table for recommended values. Make gain adjustment with digital input of 255.
2. C1 and C2 phase compensation capacitors (10 pF to 15 pF) are required when using high-speed amplifiers to prevent ringing or oscillation.

FIGURE 3. UNIPOLAR OPERATION (2-QUADRANT MULTIPLICATION)

TYPICAL APPLICATION DATA

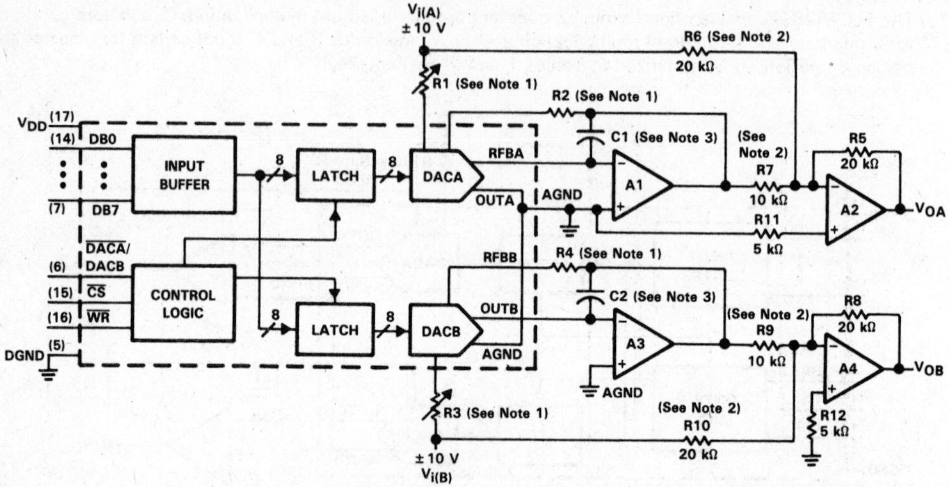
V_I(A)
±10 V
R6 (See Note 2)
20 kΩ
R1 (See Note 1)
R2 (See Note 1)
VDD (17)
(14) DB0
INPUT BUFFER
LATCH
DACA
RFBA
C1 (See Note 3)
(See Note 2)
R7
10 kΩ
R5
20 kΩ
A1
A2
VOA
(7) DB7
OUTA
AGND
R11
5 kΩ
DACA/DACB
(6)
(15) CS
(16) WR
CONTROL LOGIC
RFBB
R4 (See Note 1)
C2 (See Note 3)
LATCH
DACB
OUTB
(See Note 2)
R9
10 kΩ
R8
20 kΩ
A3
A4
VOB
DGND (5)
AGND
AGND
R3 (See Note 1)
(See Note 2)
R10
20 kΩ
R12
5 kΩ
±10 V
V_I(B)

NOTES: 1. R1, R2, R3, and R4 are used only if gain adjustment is required. See table in Figure 5 for recommended values. Adjust R1 for V_{OA} = 0 V with code 10000000 in DACA latch. Adjust R3 for V_{OB} = 0 V with 10000000 in DACB latch.
2. Matching and tracking are essential for resistor pairs R6, R7, R9, and R10.
3. C1 and C2 phase compensation capacitors (10 pF to 15 pF) may be required if A1 and A3 are high-speed amplifiers.

FIGURE 4. BIPOLAR OPERATION (4-QUADRANT OPERATION)

TABLE 1. UNIPOLAR BINARY CODE

DAC LATCH CONTENTS MSB LSB†	ANALOG OUTPUT
11111111	$-V_i$ (255/256)
10000001	$-V_i$ (129/256)
10000000	$-V_i$ (128/256) = $-V_i/2$
01111111	$-V_i$ (127/256)
00000001	$-V_i$ (1/256)
00000000	$-V_i$ (0/256) = 0

† 1 LSB = $(2^{-8})V_i$

TABLE 2. BIPOLAR (OFFSET BINARY) CODE

DAC LATCH CONTENTS MSB LSB‡	ANALOG OUTPUT
11111111	V_i (127/128)
10000001	V_i (1/128)
10000000	0 V
01111111	$-V_i$ (1/128)
00000001	$-V_i$ (127/128)
00000000	$-V_i$ (128/128)

‡ 1 LSB = $(2^{-7})V_i$

TEXAS INSTRUMENTS
POST OFFICE BOX 655012 • DALLAS, TEXAS 75265

TYPICAL APPLICATION DATA

microprocessor interface information

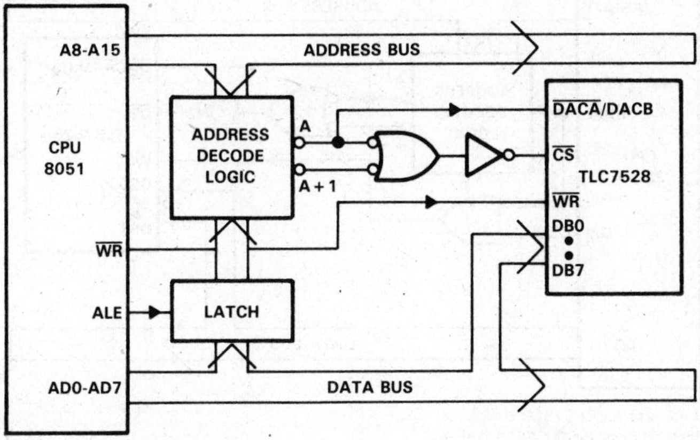
A8-A15
ADDRESS BUS
CPU
8051
ADDRESS
DECODE
LOGIC
A
A+1
DACA/DACB
CS
TLC7528
WR
DB0
DB7
WR
ALE
LATCH
AD0-AD7
DATA BUS

NOTE: A = decoded address for TLC7528 DACA.
A+1 = decoded address for TLC7528 DACB.

FIGURE 5. TLC7528 — INTEL 8051 INTERFACE

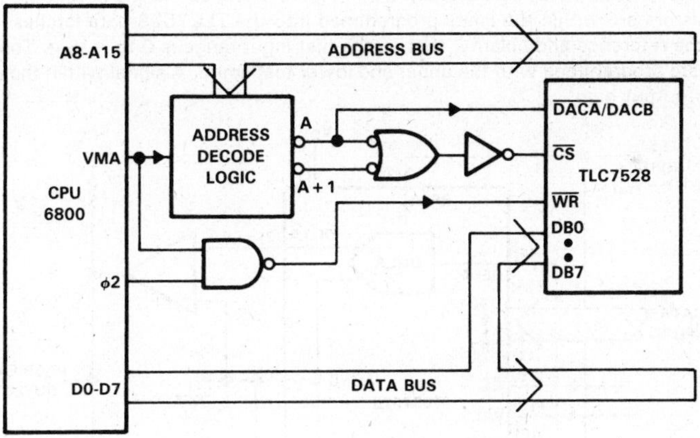
A8-A15
ADDRESS BUS
ADDRESS
DECODE
LOGIC
VMA
CPU
6800
A
A+1
DACA/DACB
CS
TLC7528
WR
DB0
DB7
φ2
D0-D7
DATA BUS

NOTE: A = decoded address for TLC7528 DACA.
A+1 = decoded address for TLC7528 DACB.

FIGURE 6. TLC7528 — 6800 INTERFACE

TYPICAL APPLICATION DATA

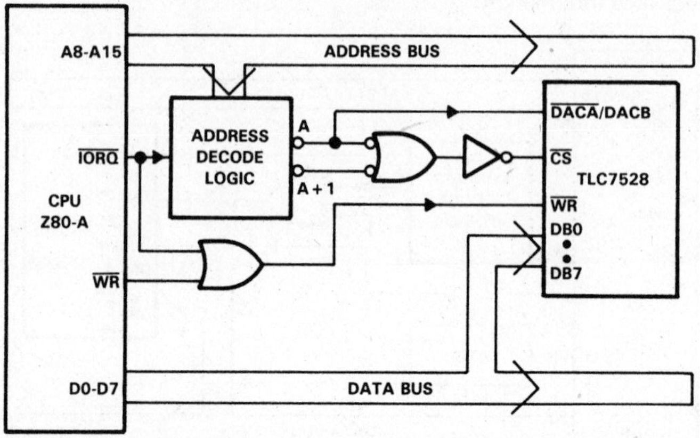
A8-A15
ADDRESS BUS
ADDRESS
DECODE
LOGIC
A
A + 1
IORQ
CPU
Z80-A
WR
D0-D7
DATA BUS
DACA/DACB
CS
TLC7528
WR
DB0
DB7

NOTE: A = decoded address for TLC7528 DACA.
A+1 = decoded address for TLC7528 DACB.

FIGURE 7. TLC7528 TO Z80-A INTERFACE

programmable window detector

The programmable window comparator shown in Figure 8 will determine if voltage applied to the DAC feedback resistors are within the limits programmed into the TLC7528 data latches. Input signal range depends on the reference and polarity, that is, the test input range is 0 to $-V_{ref}$. The DACA and DACB data latches are programmed with the upper and lower test limits. A signal within the programmed limits will drive the output high.

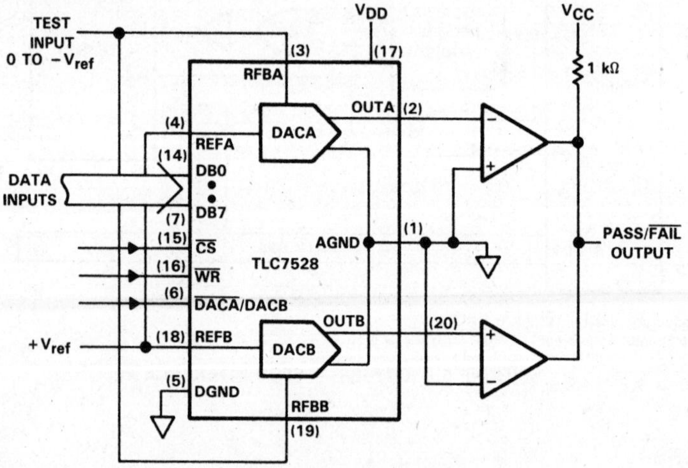
TEST
INPUT
0 TO −Vref
VDD
VCC
(3)
(17)
RFBA
1 kΩ
OUTA
(2)
(4)
REFA
DACA
(14)
DB0
DATA
INPUTS
DB7
(7)
(15)
CS
AGND
(1)
PASS/FAIL
OUTPUT
(16)
WR
TLC7528
(6)
DACA/DACB
OUTB
(20)
+Vref
(18)
REFB
DACB
(5)
DGND
RFBB
(19)

FIGURE 8. DIGITALLY PROGRAMMABLE WINDOW COMPARATOR (UPPER- AND LOWER-LIMIT TESTER)

TEXAS INSTRUMENTS
POST OFFICE BOX 655012 • DALLAS, TEXAS 75265

TYPICAL APPLICATION DATA

digitally controlled signal attenuator

Figure 9 shows the TLC7528 configured as a two-channel programmable attenuator. Applications include stereo audio and telephone signal level control. Table 3 shows input codes vs attenuation for a 0 to 15.5 dB range.

Attentuation db = $-20 \log_{10} D/256$, D = digital input code

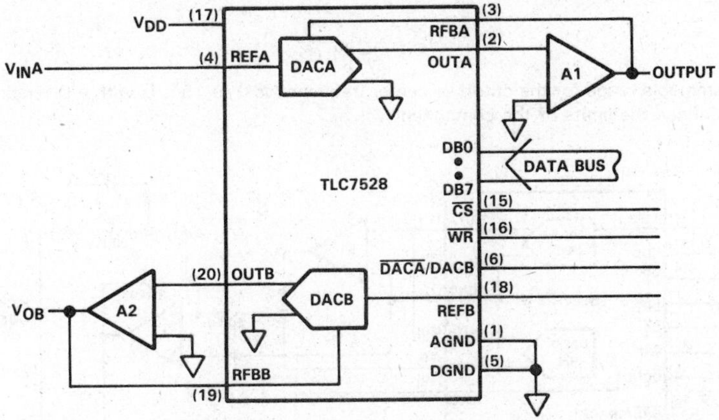
VDD
(17)
(3)
RFBA
(2)
VINA
(4)
REFA
DACA
OUTA
A1
OUTPUT
DB0
DATA BUS
TLC7528
DB7
(15)
CS
(16)
WR
(6)
DACA/DACB
(20)
OUTB
(18)
DACB
VOB
A2
REFB
(1)
AGND
(5)
RFBB
DGND
(19)

FIGURE 9. DIGITALLY CONTROLLED DUAL TELEPHONE ATTENUATOR

TABLE 3. ATTENUATION vs DACA, DACB CODE

ATTN(dB)	DAC INPUT CODE	CODE IN DECIMAL	ATTN(dB)	DAC INPUT CODE	CODE IN DECIMAL
0	11111111	255	8.0	01100110	102
0.5	11110010	242	8.5	01100000	96
1.0	11100100	228	9.0	01011011	91
1.5	11010111	215	9.5	01010110	86
2.0	11001011	203	10.0	01010001	81
2.5	11000000	192	10.5	01001100	76
3.0	10110101	181	11.0	01001000	72
3.5	10101011	171	11.5	01000100	68
4.0	10100010	162	12.0	01000000	64
4.5	10011000	152	12.5	00111101	61
5.0	10010000	144	13.0	00111001	57
5.5	10001000	136	13.5	00110110	54
6.0	10000000	128	14.0	00110011	51
6.5	01111001	121	14.5	00110000	48
7.0	01110010	114	15.0	00101110	46
7.5	01101100	108	15.5	00101011	43

TYPICAL APPLICATION DATA

programmable state-variable filter

This programmable state-variable or universal filter configuration provides low-pass, high-pass, and band-pass outputs, and is suitable for applications in which microprocessor control of filter parameters is required.

As shown in Figure 10, DACA1 and DACB1 control the gain and Q of the filter while DACA2 and DACB2 control the cutoff frequency. Both halves of the DACA2 and DACB2 must track accurately in order for the cutoff-frequency equation to be true. With the TLC7528, this is easily achieved.

$$f_c = \frac{1}{2\pi\ R1\ C1}$$

The programmable range for the cutoff or center frequency is 0 to 15 kHz with a Q ranging from 0.3 to 4.5. This defines the limits of the component values.

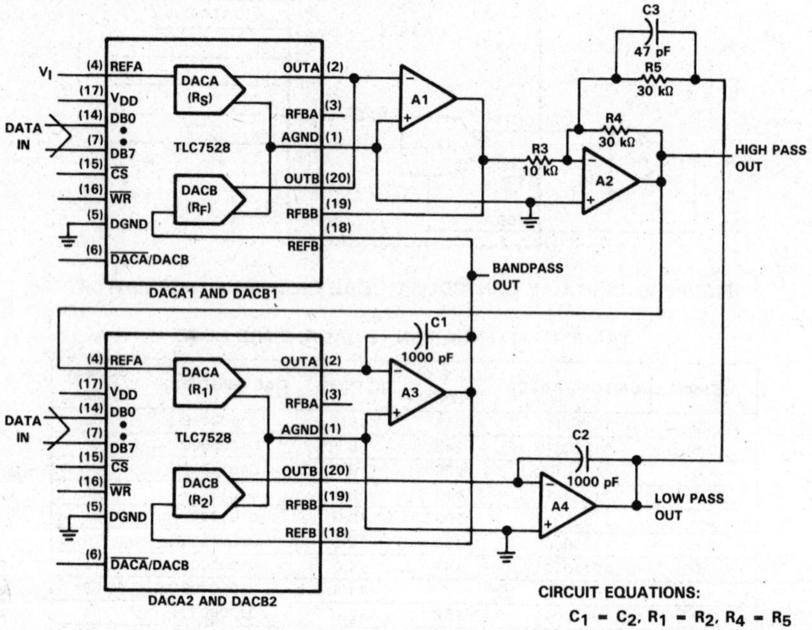
C3
47 pF
R5
30 kΩ
R4
30 kΩ
R3
10 kΩ
V_I
DATA IN
REFA
VDD
DB0
DB7
CS
WR
DGND
DACA/DACB
DACA (RS)
DACB (RF)
TLC7528
OUTA
RFBA
AGND
OUTB
RFBB
REFB
A1
A2
HIGH PASS OUT
BANDPASS OUT
DACA1 AND DACB1
C1
1000 pF
A3
DACA (R1)
DACB (R2)
C2
1000 pF
A4
LOW PASS OUT
DACA2 AND DACB2

CIRCUIT EQUATIONS:

$C_1 = C_2,\ R_1 = R_2,\ R_4 = R_5$

$$Q = \frac{R_3}{R_4} \cdot \frac{R_F}{R_{fb(DACB1)}}$$

$$A_o = -\frac{R_F}{R_S}$$

NOTES: A. Op-amps A1, A2, A3, and A4 are TL287.
B. C3 compensates for the op-amp gain-bandwidth limitations.
C. DAC equivalent resistance equals $\frac{256 \times \text{(DAC ladder resistance)}}{\text{DAC digital code}}$

FIGURE 10. DIGITALLY CONTROLLED STATE-VARIABLE FILTER

TLC7533I, TLC7533C
Advanced LinCMOS™ 10-BIT MULTIPLYING DIGITAL-TO-ANALOG CONVERTERS

D2166, OCTOBER 1986

- ADVANCED LinCMOS™ Silicon-Gate Technology
- Guaranteed Monotonicity
- Fast Settling Time
- CMOS/TTL Compatible
- Four-Quadrant Multiplication
- Designed to be Interchangeable with Analog Devices AD7533, AD7520, and PMI PM-7533

N DUAL-IN-LINE PACKAGE
(TOP VIEW)

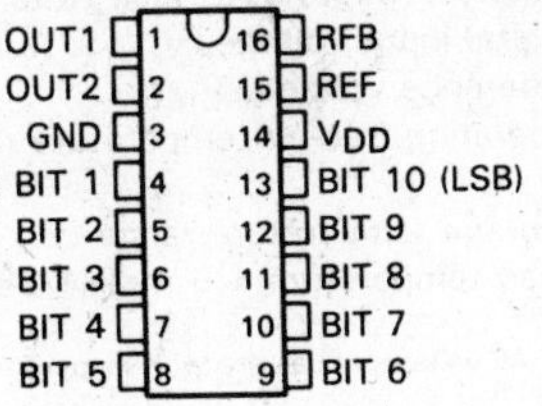
OUT1 1
16 RFB
OUT2 2
15 REF
GND 3
14 V_{DD}
(MSB) BIT 1 4
13 BIT 10 (LSB)
BIT 2 5
12 BIT 9
BIT 3 6
11 BIT 8
BIT 4 7
10 BIT 7
BIT 5 8
9 BIT 6

KEY PERFORMANCE SPECIFICATIONS	
Resolution	10 Bits
Linearity Error	1/2 LSB
Power Dissipation	30 mW
Settling Time	150 ns

description

The TLC7533 is an ADVANCED LinCMOS™ 10-bit digital-to-analog converter featuring two- and four-quadrant multiplication.

The TLC7533 is pin and functionally equivalent to the AD7520 and AD7533. Texas Instruments advanced thin-film-on-monolithic-CMOS fabrication process provides 10-bit linearity without laser trimming.

The TLC7533 features TTL or CMOS compatibility with low input leakage currents from 5-V to 15-V power supplies. Output scaling is provided by an internal feedback resistor and an external operational amplifier. Both positive and negative reference voltages can be utilized.

The TLC7533I is characterized for operation from −25°C to 85°C. The TLC7533C is characterized for operation from 0°C to 70°C.

ADVANCED LinCMOS is a trademark of Texas Instruments Incorporated

absolute maximum ratings over operating free-air temperature range (unless otherwise noted)

- Supply voltage, V_{DD} (see Note 1) .... −0.3 V to 16.5 V
- Digital input voltage, V_I .... −0.3 to V_{DD}+0.3 V
- Reference voltage, V_{ref} .... ±25 V
- Operating free-air temperature range: TLC7533I .... −25 °C to 85 °C
  - TLC7533C .... 0 °C to 70 °C
- Storage temperature range .... −65 °C to 150 °C
- Lead temperature 1,6 mm (1/16 inch) from case for 10 seconds .... 260 °C

NOTE 1: All voltage values are with respect to the network ground terminal.

recommended operating conditions

		MIN	NOM	MAX	UNIT
Supply voltage, V_{DD}		5		16.5	V
Reference voltage, V_{ref}			±10		V
High-level input voltage, V_{IH}		2.4			V
Low-level input voltage, V_{IL}				0.8	V
Operating free-air temperature, T_A	TLC7533I	−25		85	°C
	TLC7533C	0		70	°C

electrical characteristics over recommended operating temperature range, V_{DD} = 15 V, V_{ref} = ±10 V, OUT1 and OUT2 at 0 V (unless otherwise noted)

PARAMETER			TEST CONDITIONS	MIN	MAX	UNIT
I_I	Input leakage current, digital input		V_I = 0 or V_{DD}		±1	μA
r_I	Input resistance (pin 15) (see Note 2)			5	20	kΩ
I_{lkg}	Output leakage current	OUT1	Digital inputs at V_{IL}		±200	nA
		OUT2	Digital inputs at V_{IH}		±200	nA
k_{SVS}	Supply voltage sensitivity ($\Delta A_V/\Delta V_{DD}$) (see Note 3)		V_{DD} = 14 V to 16.5 V, Digital inputs at V_{IH}		0.008	%/%
I_{DD}	Supply current				2	mA
C_i	Input capacitance, digital input		V_I = V_{IL}		10	pF
C_o	Output capacitance	OUT1	Digital inputs at V_{IH}		100	pF
		OUT2	Digital inputs at V_{IH}		35	pF
		OUT1	Digital inputs at V_{IL}		35	pF
		OUT2	Digital inputs at V_{IL}		100	pF

NOTES: 2. Temperature coefficient is approximately −300 ppm/°C.
3. A_V is the ratio of the DAC's external operational amplifier output voltage to the REF input voltage when using the internal feedback resistor.

operating characteristics over recommended operating free-air temperature range, V_{DD} = 15 V, V_{ref} = 10 V, OUT1 and OUT2 at 0 V (unless otherwise noted)

PARAMETER	TEST CONDITIONS	MIN	MAX	UNIT
Relative accuracy	See Note 4		±0.05	%FSR
Gain error	Digital inputs at V_{IH}, See Notes 4 and 5		±1.5	%FS
Output current settling time	To ±0.05% FSR, R_L = 100 Ω, Digital inputs changing from V_{IH} to V_{IL}, or V_{IL} to V_{IH}		150	ns
Feedthrough error	Digital inputs at V_{IL}, V_{ref} = ±10 V sine wave at 100 kHz		±0.1	%FSR

NOTES: 4. Practical Full Scale Range (FSR) = V_{ref} − 1 LSB.
5. Gain error is measured using the internal feedback resistor. Full-Scale (FS) = $-V_{ref}$ (1023/1024). Maximum gain change from T_A = 25 °C to minimum or maximum temperature is 0.1% FSR.

PRINCIPLES OF OPERATION

The TLC7533 is a 10-bit multiplying D/A converter consisting of an inverted R-2R ladder and analog switches. Binary-weighted currents are switched between the OUT1 and OUT2 bus lines by NMOS current switches. The on-state resistances of these switches are binarily scaled so that the voltage drop across every switch is the same. The OUT1 and OUT2 bus lines should be maintained at the same potential so that the current in each ladder leg remains constant and is independent of the switch state. Most applications require only the addition of an external operational amplifier and a voltage reference.

The equivalent circuit for all digital inputs low is shown in Figure 1. With all of the digital inputs low, the entire reference current, I_{ref}, is switched to OUT2 as shown in Figure 2. The current source $I_{ref}/1024$ represents the constant current flowing through the termination resistor of the R-2R ladder; while the current source I_{lkg} represents leakage currents to the substrate. The output capacitances, $C_{o(1)}$ and $C_{o(2)}$, are due to the capacitance of the NMOS current switches and vary with the switch state. With all digital inputs low, all of the current switches and the entire resistor ladder are switched to the OUT2 bus line. The capacitance appearing at OUT2 is a maximum of 100 pF; at OUT1 there is a maximum of 35 pF. With all digital inputs high, all of the current switches are switched to OUT1, and 100 pF maximum appears at OUT1. A maximum of 35 pF appears at OUT2 as shown in Figure 3.

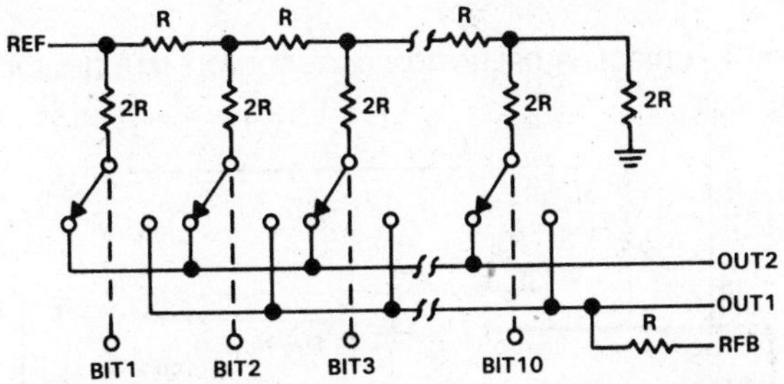
REF
R
R
R
2R
2R
2R
2R
2R
OUT2
OUT1
R
RFB
BIT1
BIT2
BIT3
BIT10

FIGURE 1. SIMPLIFIED DAC CIRCUIT — ALL DIGITAL INPUTS LOW

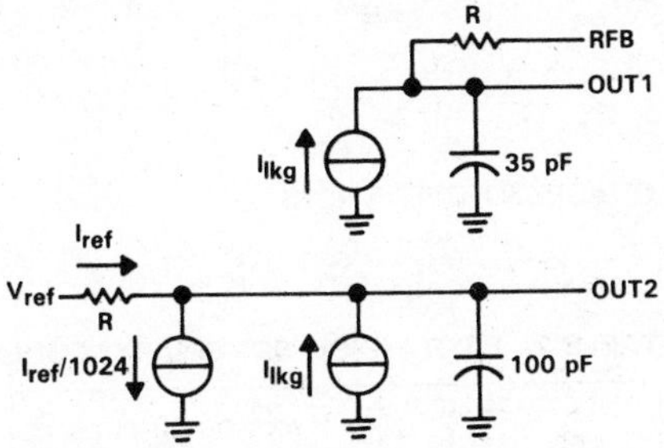
R
RFB
OUT1
Ilkg
35 pF
Iref
Vref
R
OUT2
Iref/1024
Ilkg
100 pF

FIGURE 2. DAC EQUIVALENT CIRCUIT — ALL DIGITAL INPUTS LOW

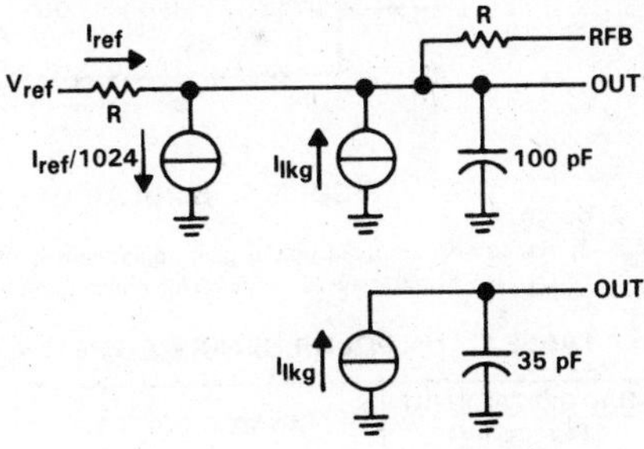
Iref
R
RFB
Vref
OUT1
R
Iref/1024
Ilkg
100 pF
OUT2
Ilkg
35 pF

FIGURE 3. DAC EQUIVALENT CIRCUIT — ALL DIGITAL INPUTS HIGH

TYPICAL APPLICATION DATA

The TLC7533 is capable of performing 2-quadrant or full 4-quadrant multiplication. Circuit configurations for 2-quadrant or 4-quadrant multiplication are shown in Figures 4 and 5. Input coding for unipolar and bipolar operation are summarized in Tables 1 and 2, respectively.

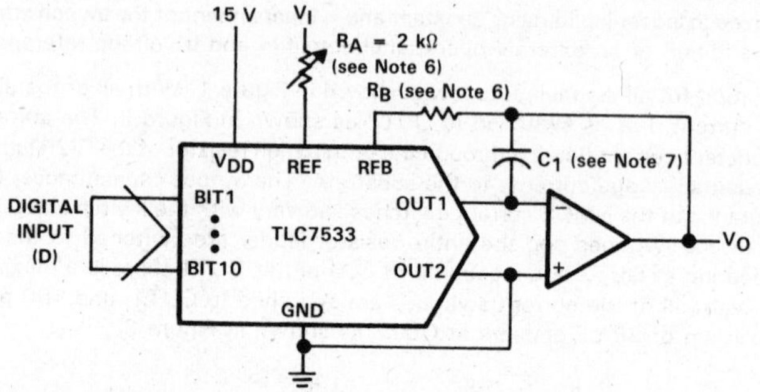
15 V
V_I
R_A = 2 kΩ (see Note 6)
R_B (see Note 6)
C_1 (see Note 7)
V_{DD}
REF
RFB
DIGITAL INPUT (D)
BIT1
BIT10
TLC7533
OUT1
OUT2
GND
V_O

FIGURE 4. UNIPOLAR OPERATION (2-QUADRANT MULTIPLICATION)

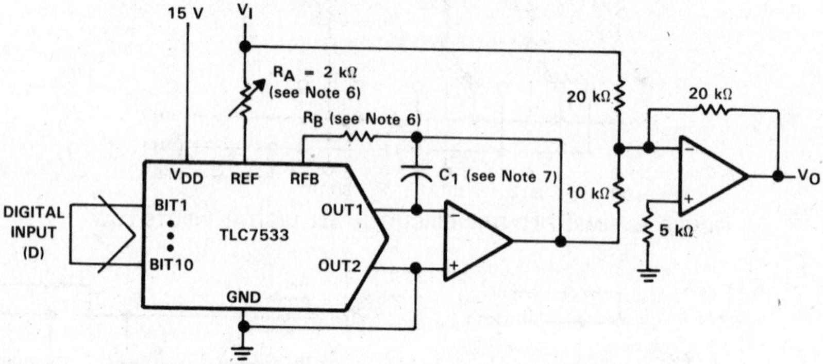
15 V
V_I
R_A = 2 kΩ (see Note 6)
R_B (see Note 6)
20 kΩ
20 kΩ
C_1 (see Note 7)
10 kΩ
5 kΩ
V_{DD}
REF
RFB
DIGITAL INPUT (D)
BIT1
BIT10
TLC7533
OUT1
OUT2
GND
V_O

FIGURE 5. BIPOLAR OPERATION (4-QUADRANT OPERATION)

NOTES: 6. R_A and R_B are used only if gain adjustment is required.
7. C_1 (10-33 pF) may be required for phase compensation when using high-speed op-amps.

TABLE 1. UNIPOLAR BINARY CODE

DAC DIGITAL INPUT MSB LSB†	ANALOG OUTPUT
1111111111	$-V_I$ (1023/1024)
1000000001	$-V_I$ (513/1024)
1000000000	$-V_I$ (512/1024) = $-V_{ref}/2$
0111111111	$-V_I$ (511/1024)
0000000001	$-V_I$ (1/1024)
0000000000	$-V_I$ (0/1024) = 0

†1 LSB = $(2^{-10})\ V_I$

TABLE 2. BIPOLAR (OFFSET BINARY) CODE

DAC DIGITAL INPUT MSB LSB‡	ANALOG OUTPUT
1111111111	$+V_I$ (511/512)
1000000001	$+V_I$ (1/512)
1000000000	0
0111111111	$-V_I$ (1/512)
0000000001	$-V_I$ (511/512)
0000000000	$-V_I$ (512/512) = $-V_I$

‡1 LSB = $(2^{-9})\ V_I$

TYPICAL APPLICATION DATA

The TLC7533 may be used in voltage output operation as shown in Figure 6. In this configuration, the input voltage is applied to the OUT1 terminal and the output voltage is taken from the REF terminal. The output voltage varies with the digital input code according to the equation shown. The output should be buffered to prevent loading errors due to the high output resistance of this circuit (typically 10 kilohms). The input voltage should not exceed 1.5 volts to ensure nonlinearity errors less than 1 LSB.

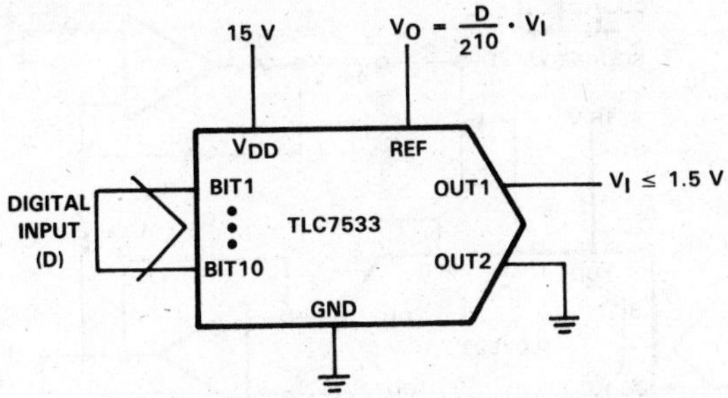
15 V
$V_O = \frac{D}{2^{10}} \cdot V_I$
VDD
REF
BIT1
OUT1
$V_I \leq 1.5$ V
DIGITAL
INPUT
(D)
TLC7533
BIT10
OUT2
GND

FIGURE 6. VOLTAGE OUTPUT OPERATION

By connecting the DAC in the feedback of an op-amp as shown in Figure 7, the circuit behaves as a programmable gain amplifier with the transfer function:

$$V_O = -V_I \left(\frac{1024}{D}\right)$$

where D = Digital Input Code (expressed as a decimal number)

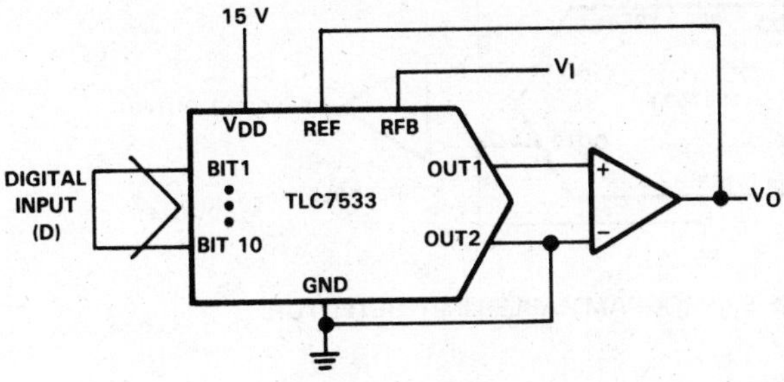
15 V
V_I
VDD
REF
RFB
BIT1
OUT1
DIGITAL
INPUT
(D)
TLC7533
BIT 10
OUT2
GND
V_O

GAIN TABLE	
D	V_O/V_I
1023	−1.00097
512	−2
256	−4
128	−8
2	−512
1	−1024
0	open loop

FIGURE 7. PROGRAMMABLE GAIN AMPLIFIER

TYPICAL APPLICATION DATA

The programmable function generator shown in Figure 8 produces both square and triangular wave output at a frequency determined by the digital input code. The digital input of the digitally programmable limit detector shown in Figure 9 determines the trip point of the PASS/FAIL output. For a digital input of 00000 00000, the threshold is 0 V, for 11111 11111, the threshold is $-V_{ref}$.

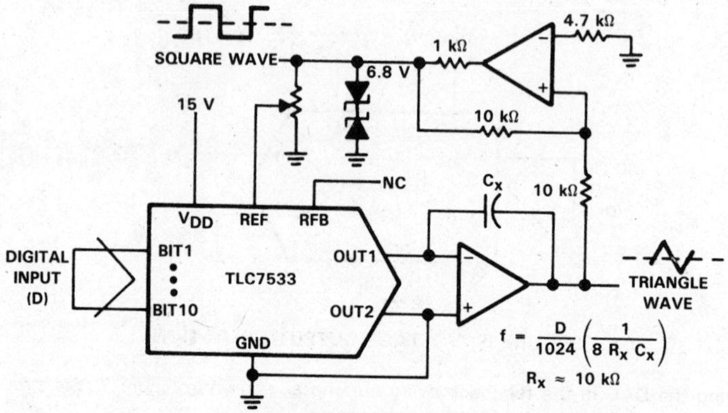
SQUARE WAVE
15 V
6.8 V
1 kΩ
4.7 kΩ
10 kΩ
NC
C_x
10 kΩ
V_{DD}
REF
RFB
DIGITAL
INPUT
(D)
BIT1
BIT10
TLC7533
OUT1
OUT2
GND
TRIANGLE
WAVE
$f = \frac{D}{1024}\left(\frac{1}{8 R_x C_x}\right)$
$R_x \approx 10\ k\Omega$

FIGURE 8. PROGRAMMABLE FUNCTION GENERATOR

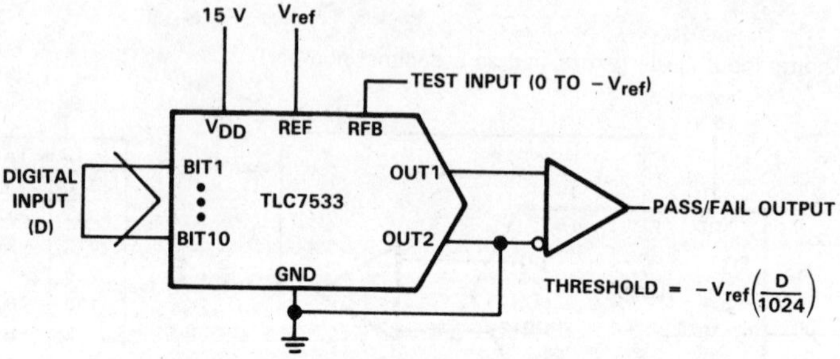
15 V
V_{ref}
TEST INPUT (0 TO $-V_{ref}$)
V_{DD}
REF
RFB
DIGITAL
INPUT
(D)
BIT1
BIT10
TLC7533
OUT1
OUT2
GND
PASS/FAIL OUTPUT
THRESHOLD = $-V_{ref}\left(\frac{D}{1024}\right)$

FIGURE 9. PROGRAMMABLE LIMIT DETECTOR

TYPICAL APPLICATION DATA

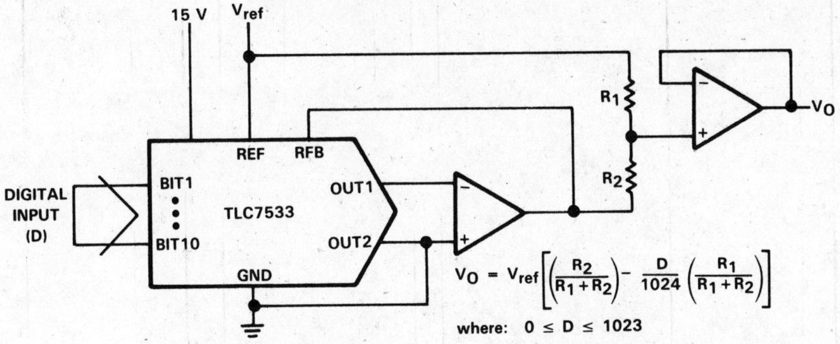
15 V
Vref
REF
RFB
BIT1
BIT10
DIGITAL
INPUT
(D)
TLC7533
OUT1
OUT2
GND
R1
R2
VO

$$V_O = V_{ref}\left[\left(\frac{R_2}{R_1 + R_2}\right) - \frac{D}{1024}\left(\frac{R_1}{R_1 + R_2}\right)\right]$$

where: $0 \leq D \leq 1023$

FIGURE 10. MODIFIED SCALE-FACTOR AND OFFSET

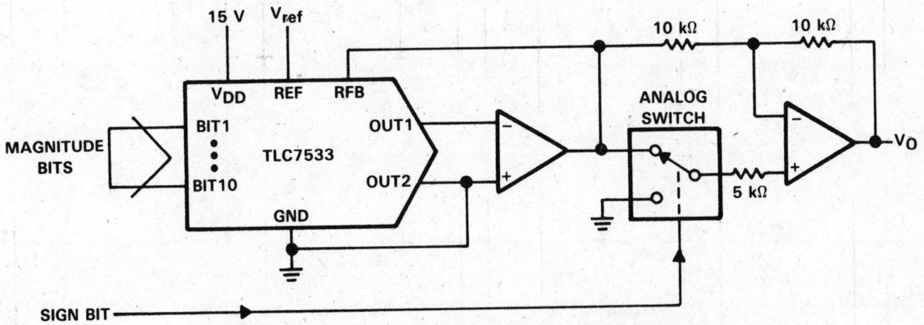
15 V
Vref
VDD
REF
RFB
MAGNITUDE
BITS
BIT1
BIT10
TLC7533
OUT1
OUT2
GND
10 kΩ
10 kΩ
ANALOG
SWITCH
5 kΩ
VO
SIGN BIT

FIGURE 11. 10-BIT AND SIGN MULTIPLYING DAC

- ADVANCED LinCMOS™ Silicon Gate Process Technology
- 14-Bit Dynamic Range ADC and DAC
- 10-Bit ADC and DAC Linearity Over Any 10-Bit Range
- Variable ADC and DAC Sampling Rate Up to 19,200 Samples per Second
- Switched-Capacitor Antialiasing Input Filter and Output-Reconstruction Filter
- Serial Port for Direct Interface to TMS32011, TMS32020, and TMS32025 Digital Processors
- Synchronous or Asynchronous ADC and DAC Conversion Rates with Programmable Incremental ADC and DAC Conversion Timing Adjustments
- Serial Port Interface to SN54299 or SN74299 Serial-to-Parallel Shift Registers for Parallel Interface to TMS32010 or Other Digital Processors

J OR N PACKAGE
(TOP VIEW)

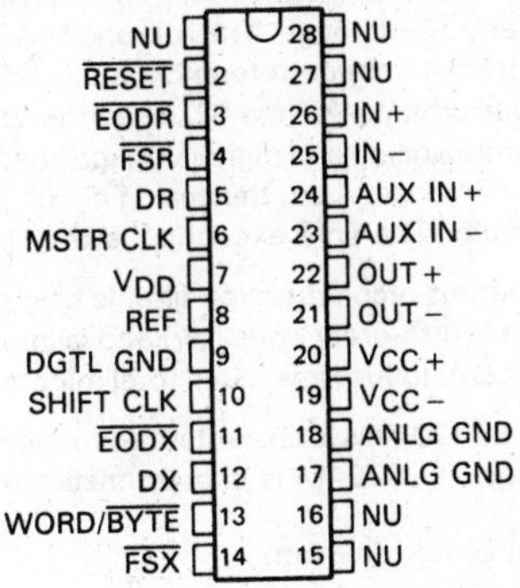
NU 1
RESET 2
EODR 3
FSR 4
DR 5
MSTR CLK 6
VDD 7
REF 8
DGTL GND 9
SHIFT CLK 10
EODX 11
DX 12
WORD/BYTE 13
FSX 14
28 NU
27 NU
26 IN+
25 IN−
24 AUX IN+
23 AUX IN−
22 OUT+
21 OUT−
20 VCC+
19 VCC−
18 ANLG GND
17 ANLG GND
16 NU
15 NU

NU—Nonusable; no external connection should be made to these pins

description

The TLC32040 is a complete analog-to-digital and digital-to-analog input/output system on a single monolithic CMOS chip. This device integrates a bandpass switched-capacitor antialiasing input filter, a 14-bit resolution A/D converter, four microprocessor-compatible serial port modes, a 14-bit resolution D/A converter, and a low-pass switched-capacitor output-reconstruction filter. The device offers numerous combinations of Master Clock input frequencies and conversion/sampling rates, which can be changed via digital processor control.

Typical applications for this IC include modems (7.2-, 8-, 9.6-, 14.4-, and 19.2-kHz sampling rate), analog interface for digital signal processors, speech recognition/storage systems, industrial process control, biomedical instrumentation, acoustical signal processing, spectral analysis, data acquisition, and instrumentation recorders. Four serial modes, which allow direct interface to the TMS32011, TMS32020, and TMS32025 digital signal processors, are provided. Also, when the transmit and receive sections of the Analog Interface Circuit (AIC) are operating synchronously, it will interface to two SN54299 or SN74299 serial-to-parallel shift registers. These serial-to-parallel shift registers can then interface in parallel to the TMS32010, other digital signal processors, or external FIFO circuitry. Output data pulses are emitted to inform the processor that data transmission is complete, or to allow the DSP to differentiate between two transmitted bytes. A flexible control scheme is provided so that the functions of the IC can be selected and adjusted coincidentally with signal processing via software control.

The antialiasing input filter comprises seventh-order and fourth-order CC-type (Chebyshev/elliptic transitional) low-pass and high-pass filters, respectively, and a fourth-order equalizer. The input filter is implemented in switched-capacitor technology and is preceded by a continuous time filter to eliminate any possibility of aliasing caused by sampled data filtering. When no filtering is desired, the entire composite filter can be switched out of the signal path. A selectable, auxiliary, differential analog input is provided for applications where more than one analog input is required.

ADVANCED LinCMOS™ is a trademark of Texas Instruments Incorporated

TEXAS INSTRUMENTS
POST OFFICE BOX 655012 • DALLAS, TEXAS 75265

description (continued)

The A/D and D/A converters each have 14 bits of resolution with 10 bits of integral linearity guaranteed over any 10-bit range. The A/D and D/A architectures guarantee no missing codes and monotonic operation. An internal voltage reference is provided to ease the design task and to provide complete control over the performance of the IC. The internal voltage is brought out to a pin and is available to the designer. Separate analog and digital voltage supplies and grounds are provided to minimize noise and ensure a wide dynamic range. Also, the analog circuit path contains only differential circuitry to keep noise to an absolute minimum. The only exception is the DAC sample-and-hold, which utilizes pseudo-differential circuitry.

The output-reconstruction filter is a seventh-order CC-type (Chebyshev/elliptic transitional low-pass filter with a fourth-order equalizer) and is implemented in switched-capacitor technology. This filter is followed by a continuous-time filter to eliminate images of the digitally encoded signal.

The TLC32040M is characterized for operation over the full military temperature range of −55°C to 125°C, and the TLC32040I is characterized for operation from −40°C to 85°C.

functional block diagram

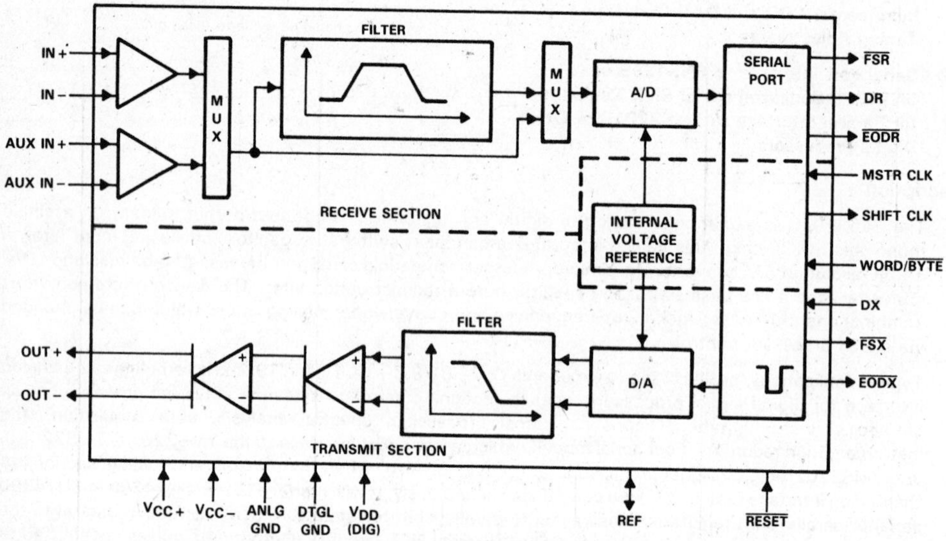
IN +
IN −
AUX IN +
AUX IN −
MUX
FILTER
MUX
A/D
SERIAL PORT
FSR
DR
EODR
MSTR CLK
SHIFT CLK
WORD/BYTE
DX
FSX
EODX
RECEIVE SECTION
INTERNAL VOLTAGE REFERENCE
FILTER
D/A
OUT +
OUT −
TRANSMIT SECTION
VCC+
VCC−
ANLG GND
DTGL GND
VDD (DIG)
REF
RESET

PRINCIPLES OF OPERATION

analog input

Two sets of analog inputs, IN+, IN−, and AUX IN+, AUX IN−, are provided. Each input set can be operated in either differential or single-ended modes, since sufficient common-mode range and rejection are provided. Normally, the IN+ and IN− inputs are used; however, the auxiliary inputs, AUX IN+ and AUX IN−, can be used if a second input is required. The gain for the IN+, IN−, and auxiliary AUX IN+ and AUX IN− inputs can be programmed to either 1, 2, or 4 (see the Gain Control Table). Either input circuit can be selected via software control. It is important to note that a wide dynamic range is assured by the differential internal analog architecture and by the separate analog and digital voltage supplies and grounds.

A/D bandpass filter, A/D bandpass filter clocking, and A/D conversion rate timing

The A/D bandpass filter can be selected or bypassed via software control. The frequency response of this filter is presented in the following pages. This response results when the switched-capacitor filter clock frequency is 288 kHz. Several possible options can be used to attain a 288-kHz switched-capacitor filter clock. When the filter clock frequency is not 288 kHz, the filter transfer function is frequency-scaled by the ratio of the actual clock frequency to 288 kHz. The low-frequency roll-off of the high-pass section is 300 kHz. However, the high-pass section low-frequency roll-off can be changed to 200 kHz with a metal mask option.

The Internal Timing Configuration and AIC DX Data Word Format sections of this data sheet indicate the many options for attaining a 288-kHz bandpass switched-capacitor filter clock. These sections indicate that the RX Counter A can be programmed to give a 288-kHz bandpass-switched capacitor filter clock for several Master Clock input frequencies.

The A/D conversion rate is then attained by frequency-dividing the 288-kHz bandpass switched-capacitor filter clock with the RX Counter B. Thus, unwanted aliasing is prevented because the A/D conversion rate is an integral submultiple of the bandpass switched-capacitor filter sampling rate, and the two rates are synchronously locked.

A/D converter performance specifications

Fundamental performance specifications for the A/D converter circuitry are presented in the A/D converter operating characteristics section of this data sheet. The realization of the A/D converter circuitry with switched-capacitor techniques provides an inherent sample-and-hold.

analog output

The analog output circuitry is an analog output power amplifier. Both noninverting and inverting amplifier outputs are brought out of the IC. This amplifier can drive transformer hybrids or low-impedance loads directly in either a differential or single-ended configuration.

D/A low-pass filter, D/A low-pass filter clocking, and D/A conversion rate timing

The frequency response of this filter is presented in the following pages. This response results when the low-pass switched-capacitor filter clock frequency is 288 kHz. Like the A/D filter, the transfer function of this filter is frequency-scaled when the clock frequency is not 288 kHz. A continuous-time filter is provided on the output of the D/A low-pass filter to greatly attenuate any switched-capacitor clock feedthrough.

The D/A conversion rate is then attained by frequency-dividing the 288-kHz switched-capacitor filter clock with TX Counter B. Thus, unwanted aliasing is prevented because the D/A conversion rate is an integral submultiple of the switched-capacitor low-pass filter sampling rate, and the two rates are synchronously locked.

PRINCIPLES OF OPERATION (continued)

asynchronous versus synchronous operation

If the transmit section of the AIC (low-pass filter and DAC) and receive section (bandpass filter and ADC) are operated asynchronously, the low-pass and band-pass filter clocks are independently generated from the Master Clock signal. Also, the D/A and A/D conversion rates are independently determined. If the transmit and receive sections are operated synchronously, the low-pass filter clock drives both low-pass and band-pass filters. In synchronous operation, the A/D conversion timing is derived from, and is equal to, the D/A conversion rate timing. (See description of the WORD/$\overline{\text{BYTE}}$ pin in the Pin Functional Description Section.)

D/A converter performance specifications

Fundamental performance specifications for the D/A converter circuitry are presented in the D/A converter operating characteristics section of the data sheet. The D/A converter has a sample-and-hold that is realized with a switched-capacitor ladder.

system frequency response correction

Sin x/x correction circuitry is performed in digital signal processor software. The system frequency response can be corrected via DSP software to ±0.1 dB accuracy to a band-edge of 3000 Hz for all sampling rates. This correction is accomplished with a first-order digital correction filter, which requires only seven TMS320 instruction cycles. With a 200-ns instruction cycle, seven instructions represent an overhead factor of only 1.1% and 1.3% for sampling rates of 8 and 9.6 kHz, respectively (see the sin x/x Correction Section for more details).

serial port

The serial port has four possible modes that are described in detail in the pin description section. These modes are briefly described below.

1. The transmit and receive sections of the AIC are operated asynchronously, and the AIC serial port interfaces directly with the TMS32011.
2. The transmit and receive sections of the AIC are operated asynchronously, and the AIC serial port interfaces directly with the TMS32020 and the TMS32025.
3. The transmit and receive sections of the AIC are operated synchronously, and the AIC serial port interfaces directly with the TMS32011.
4. The transmit and receive sections of the AIC are operated synchronously, and the AIC serial port interfaces directly with the TMS32020, TMS32025, or two SN54299 or SN74299 serial-to-parallel shift registers, which can then interface in parallel to the TMS32010, to any other digital signal processor, or to external FIFO circuitry.

testing

An addendum accompanying this data sheet fully describes the test capabilities of the IC, provided by the design.

internal voltage reference

The internal reference eliminates the need for an external voltage reference, and thus provides overall circuit cost reduction. Additionally, the internal reference makes the performance of the IC less susceptible to noise. Thus, the internal reference eases the design task and provides complete control over the performance of the IC. The internal reference is brought out to a pin and is available to the designer. To keep the amount of noise on the reference signal to a minimum, an external capacitor may be connected between REF and ANLG GND.

PRINCIPLES OF OPERATION (continued)

reset

A reset function is provided to initiate serial communications between the AIC and DSP and to allow fast, cost-effective testing during manufacturing. The reset function will initialize all AIC registers, including the control register. The reset pin has an internal pull-up resistor. After a negative-going pulse on the $\overline{\text{RESET}}$ pin, the AIC will be initialized. This initialization allows normal serial port communications activity to occur between AIC and DSP (see AIC DX Data Word Format section).

loopback

This feature allows the user to test the circuit remotely. In loopback, the OUT+ and OUT− pins are internally connected to the IN+ and IN− pins. Thus, the DAC bits (d15 to d2), which are transmitted to the DX pin, can be compared with the ADC bits (d15 to d2), which are received from the DR pin. An ideal comparison would be that the bits on the DR pin equal the bits on the DX pin. However, in practice there will be some difference in these bits due to the ADC and DAC output offsets.

The loopback feature is implemented with digital signal processor control by transmitting the appropriate serial port bit to the control register (see AIC Data Word Format section).

PIN NAME	PIN NO.	I/O	DESCRIPTION
ANLG GND	17,18		Analog ground return for all internal analog circuits. Not internally connected to DGTL GND.
AUX IN+	24	I	Noninverting auxiliary analog input stage. This input can be switched into the bandpass filter and A/D converter path via software control. If the appropriate bit in the Control register is a 1, the auxiliary inputs will replace the IN+ and IN− inputs. If the bit is a 0, the IN+ and IN− inputs will be used (see the AIC DX Data Word Format section).
AUX IN−	23	I	Inverting auxiliary analog input (see the above AUX IN+ pin description).
DGTL GND	9		Digital ground for all internal logic circuits. Not internally connected to ANLG GND.
DR	5	O	This pin is used to transmit the ADC output bits from the AIC to the TMS320 serial port. This transmission of bits from the AIC to the TMS320 serial port is synchronized with the SHIFT CLK signal.
DX	12	I	This pin is used to receive the DAC input bits and timing and control information from the TMS320. This serial transmission from the TMS320 serial port to the AIC is synchronized with the SHIFT CLK signal.
$\overline{\text{EODR}}$	2	O	(See the WORD/$\overline{\text{BYTE}}$ pin description and the Serial Port Timing Diagram.) During the word-mode timing, this signal is a low-going pulse that occurs immediately after the 16 bits of A/D information have been transmitted from the AIC to the TMS320 serial port. This signal can be used to interrupt a microprocessor upon completion of serial communications. Also, this signal can be used to strobe and enable external serial-to-parallel shift registers, latches, or external FIFO RAM, and to facilitate parallel data bus communications between the AIC and the serial-to-parallel shift registers. During the byte-mode timing, this signal goes low after the first byte has been transmitted from the AIC to the TMS320 serial port and is kept low until the second byte has been transmitted. The TMS32011 can use this low-going signal to differentiate between the two bytes as to which is first and which is second.

PIN NAME	PIN NO.	I/O	DESCRIPTION
$\overline{EODX}$	11	O	(See the WORD/$\overline{BYTE}$ pin description and the Serial Port Timing Diagram.) During the word-mode timing, this signal is a low-going pulse that occurs immediately after the 16 bits of D/A converter and control or register information have been transmitted from the TMS320 serial port to the AIC. This signal can be used to interrupt a microprocessor upon the completion of serial communications. Also, this signal can be used to strobe and enable external serial-to-parallel shift registers, latches, or an external FIFO RAM, and to facilitate parallel, data-bus communications between the AIC and the serial-to-parallel shift registers. During the byte-mode timing, this signal goes low after the first byte has been transmitted from the TMS320 serial port to the AIC and is kept low until the second byte has been transmitted. The TMS32011 can use this low-going signal to differentiate between the two bytes as to which is first and which is second.
$\overline{FSR}$	4	O	In the serial transmission modes, which are described in the WORD/$\overline{BYTE}$ pin description, the $\overline{FSR}$ pin is held low during bit transmission. When the $\overline{FSR}$ pin goes low, the TMS320 serial port will begin receiving bits from the AIC via the DR pin of the AIC. The most significant DR bit will be present on the DR pin before $\overline{FSR}$ goes low. (See Serial Port Timing and Internal Timing Configuration Diagrams.)
$\overline{FSX}$	14	O	When this pin goes low, the TMS320 serial port will begin transmitting bits to the AIC via the DX pin AIC. In all serial transmission modes, which are described in the WORD/$\overline{BYTE}$ pin description, the $\overline{FSX}$ pin is held low during bit transmission (see Serial Port Timing and Internal Timing Configuration Diagrams).
IN+	26	I	Noninverting input to analog input amplifier stage
IN−	25	I	Inverting input to analog input amplifier stage
MSTR CLK	6	I	The Master Clock signal is used to derive all the key logic signals of the AIC, such as the Shift Clock, the switched-capacitor filter clocks, and the A/D and D/A timing signals. The Internal Timing Configuration diagram shows how these key signals are derived. The frequencies of these key signals are synchronous submultiples of the Master Clock frequency to eliminate unwanted aliasing when the sampled analog signals are transferred between the switched-capacitor filters and the A/D and D/A converters (see the Internal Timing Configuration).
OUT+	22	O	Noninverting output of analog output power amplifier. Can drive transformer hybrids or high-impedance loads directly in either a differential or a single-ended configuration.
OUT−	21	O	Inverting output of analog output power amplifier; functionally identical with and complementary to OUT+.
REF	8		The internal voltage reference is brought out to this pin.
$\overline{RESET}$	2	I	A reset function is provided to initialize the TA, TA′, TB, RA, RA′, RB, and control registers. This reset function initiates serial communications between the AIC and DSP. The reset function will initialize all AIC registers including the control register. After a negative-going pulse on the $\overline{RESET}$ pin, the AIC registers will be initialized to provide an 8-kHz data conversion rate for a 5.184-MHz master clock input signal. The conversion rate adjust registers, TA′ and RA′, will be reset to 1. The CONTROL register bits will be reset as follows (see AIC DX Data Word Format section). d7 = 1, d6 = 1, d5 = 1, d4 = 0, d3 = 0, d2 = 1 This initialization allows normal serial-port communication to occur between AIC and DSP. This pin has an internal pull-up resistor and is set to a high logic level unless it is pulled to ground.
SHIFT CLK	10	O	The Shift Clock signal is obtained by dividing the Master Clock signal frequency by four. This signal is used to clock the serial data transfers of the AIC, described in the WORD/$\overline{BYTE}$ pin description below (see the Serial Port Timing and Internal Timing Configuration diagram).
V_{DD}	7		Digital supply voltage, 5 V ±5%
V_{CC+}	20		Positive analog supply voltage, 5 V ±5%
V_{CC-}	19		Negative analog supply voltage −5 V ±5%

PIN NAME	PIN NO.	I/O	DESCRIPTION
WORD/$\overline{\text{BYTE}}$	13	I	This pin, in conjunction with a bit in the CONTROL register, is used to establish one of four serial modes. These four serial modes are described below. This pin has an internal pull-up resistor and is set to a logic high unless it is pulled to ground., *AIC transmit and receive sections are operated asynchronously.* The following description applies when the AIC is configured to have asynchronous transmit and receive sections. If the appropriate data bit in the Control register is a 0 (see the AIC DX Data Word Format), the transmit and receive sections will be asynchronous. L Serial port will directly interface with the serial port of the TMS32011 and communicates in two 8-bit bytes. The operation sequence is as follows (see Serial Port Timing diagrams). 1. The $\overline{\text{FSX}}$ or $\overline{\text{FSR}}$ pin is brought low. 2. One 8-bit byte is transmitted or one 8-bit byte is received. 3. The $\overline{\text{EODX}}$ or $\overline{\text{EODR}}$ pin is brought low. 4. The $\overline{\text{FSX}}$ or $\overline{\text{FSR}}$ pin emits a positive frame-sync pulse that is four Shift Clock cycles wide. 5. One 8-bit byte is transmitted or one 8-bit byte is received. 6. The $\overline{\text{EODX}}$ or $\overline{\text{EODR}}$ pin is brought high. 7. The $\overline{\text{FSX}}$ or $\overline{\text{FSR}}$ pin is brought high. H Serial port will directly interface with the serial port of the TMS32020 and communicates in one 16-bit word. The operation sequence is as follows (see Serial Port Timing diagrams): 1. The $\overline{\text{FSX}}$ or $\overline{\text{FSR}}$ pin is brought low. 2. One 16-bit word is transmitted or one 16-bit word is received. 3. The $\overline{\text{FSX}}$ or $\overline{\text{FSR}}$ pin is brought high. 4. The $\overline{\text{EODX}}$ or $\overline{\text{EODR}}$ pin emits a low-going pulse. *AIC transmit and receive sections are operated synchronously.* If the appropriate data bit in the Control register is a 1, the transmit and receive sections will be configured to be synchronous. In this case, the bandpass switched-capacitor filter and the A/D conversion timing will be derived from the TX Counter A, TX Counter B, and TA, TA′, and TB registers, rather than the RX Counter A, RX Counter B, and RA, RA′, and RB registers. In this case, the AIC $\overline{\text{FSX}}$ and $\overline{\text{FSR}}$ timing will be identical, as will the $\overline{\text{EODX}}$ and $\overline{\text{EODR}}$ timing. The synchronous operation sequences are as follows (see Serial Port Timing diagrams). L Serial port will directly interface with the serial port of the TMS32011 and communicates in two 8-bit bytes. The operation sequence is as follows (see Serial Port Timing diagrams): 1. The $\overline{\text{FSX}}$ and $\overline{\text{FSR}}$ pins are brought low. 2. One 8-bit byte is transmitted and one 8-bit byte is received. 3. The $\overline{\text{EODX}}$ and $\overline{\text{EODR}}$ pins are brought low. 4. The $\overline{\text{FSX}}$ and $\overline{\text{FSR}}$ pins emit positive frame-sync pulses that are four Shift Clock cycles wide. 5. One 8-bit byte is transmitted and one 8-bit byte is received. 6. The $\overline{\text{EODX}}$ and $\overline{\text{EODR}}$ pins are brought high. 7. The $\overline{\text{FSX}}$ and $\overline{\text{FSR}}$ pins are brought high. H Serial port will directly interface with the serial port of the TMS32020 and communicates in one 16-bit word. The operation sequence is as follows (see Serial Port Timing diagrams): 1. The $\overline{\text{FSX}}$ and $\overline{\text{FSR}}$ pins are brought low. 2. One 16-bit word is transmitted and one 16-bit word is received. 3. The $\overline{\text{FSX}}$ and $\overline{\text{FSR}}$ pins are brought high. 4. The $\overline{\text{EODX}}$ or $\overline{\text{EODR}}$ pins emit low-going pulses. Since the transmit and receive sections of the AIC are now synchronous, the AIC serial port, with additional NOR and AND gates, will interface to two SN54299 or SN74299 serial-to-parallel shift registers. Interfacing the AIC to the SN54299 or SN74299 shift register allows the AIC to interface to an external FIFO RAM and facilitates parallel, data bus communications between the AIC and the digital signal processor. The operation sequence is the same as the above sequence (see Serial Port Timing diagrams).

INTERNAL TIMING CONFIGURATION

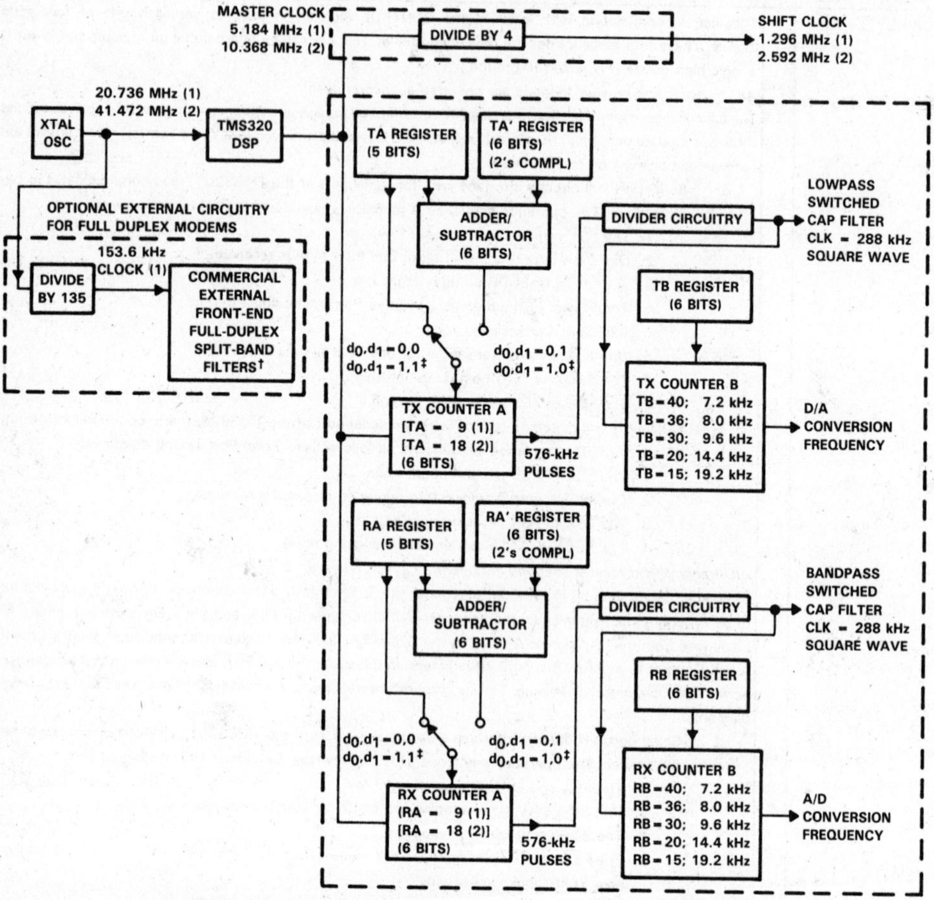
MASTER CLOCK
5.184 MHz (1)
10.368 MHz (2)
DIVIDE BY 4
SHIFT CLOCK
1.296 MHz (1)
2.592 MHz (2)
20.736 MHz (1)
41.472 MHz (2)
XTAL OSC
TMS320 DSP
OPTIONAL EXTERNAL CIRCUITRY FOR FULL DUPLEX MODEMS
153.6 kHz CLOCK (1)
DIVIDE BY 135
COMMERCIAL EXTERNAL FRONT-END FULL-DUPLEX SPLIT-BAND FILTERS†
TA REGISTER (5 BITS)
TA' REGISTER (6 BITS) (2's COMPL)
ADDER/ SUBTRACTOR (6 BITS)
DIVIDER CIRCUITRY
LOWPASS SWITCHED CAP FILTER CLK = 288 kHz SQUARE WAVE
TB REGISTER (6 BITS)
d0,d1 = 0,0
d0,d1 = 1,1‡
d0,d1 = 0,1
d0,d1 = 1,0‡
TX COUNTER A
[TA = 9 (1)]
[TA = 18 (2)]
(6 BITS)
576-kHz PULSES
TX COUNTER B
TB = 40; 7.2 kHz
TB = 36; 8.0 kHz
TB = 30; 9.6 kHz
TB = 20; 14.4 kHz
TB = 15; 19.2 kHz
D/A CONVERSION FREQUENCY
RA REGISTER (5 BITS)
RA' REGISTER (6 BITS) (2's COMPL)
ADDER/ SUBTRACTOR (6 BITS)
DIVIDER CIRCUITRY
BANDPASS SWITCHED CAP FILTER CLK = 288 kHz SQUARE WAVE
RB REGISTER (6 BITS)
d0,d1 = 0,0
d0,d1 = 1,1‡
d0,d1 = 0,1
d0,d1 = 1,0‡
RX COUNTER A
(RA = 9 (1))
[RA = 18 (2)]
(6 BITS)
576-kHz PULSES
RX COUNTER B
RB = 40; 7.2 kHz
RB = 36; 8.0 kHz
RB = 30; 9.6 kHz
RB = 20; 14.4 kHz
RB = 15; 19.2 kHz
A/D CONVERSION FREQUENCY

NOTE: Frequency 1, 20.736 MHz, is used to show how 153.6 kHz (for a commercially available modem split-band filter clock), popular speech and modem sampling signal frequencies, and an internal 288-kHz switched-capacitor filter clock can be derived synchronously and as submultiples of the crystal oscillator frequency. Since these derived frequencies are synchronous submultiples of the crystal frequency, aliasing does not occur as the sampled analog signal passes between the analog converter and switched-capacitor filter stages. Frequency 2, 41.472 MHz, is used to show that the AIC can work with high-frequency signals, which are used by high-speed digital signal processors.

†Split-band filtering can alternatively be performed after the analog input function via software in the TMS320.

‡These control bits are described in the AIC DX Data Word Format section.

explanation of internal timing configuration

All of the internal timing of the AIC is derived from the high-frequency clock signal that drives the Master Clock input pin. The Shift Clock signal, which strobes the serial port data between the AIC and DSP, is derived by dividing the Master Clock input signal frequency by four.

TX Counter A and TX Counter B, which are driven by the Master Clock signal, determine the D/A conversion period timing. Similarly, RX Counter A and RX Counter B determine the A/D conversion period timing. In order for the switched-capacitor low-pass and band-pass filters to meet their transfer function specifications, the frequency of the clock inputs of the switched-capacitor filter must be 288 kHz. If the frequencies of the clock inputs are not 288 kHz, the filter transfer function frequencies are scaled by the ratios of the clock frequencies to 288 kHz. Thus, to obtain the specified filter responses, the combination of Master Clock frequency and TX Counter A and RX Counter A values must yield 288-kHz switched-capacitor clock signals. These 288-kHz clock signals can then be divided by the TX Counter B and RX Counter B to establish the D/A and A/D conversion period timings.

TX Counter A and TX Counter B are reloaded every D/A conversion period, while RX Counter A and RX Counter B are reloaded every A/D conversion period. The TX Counter B and RX Counter B are loaded with the values in the TB and RB Registers respectively. Via software control, the TX Counter A can be loaded with either the TA Register, the TA Register less the TA′ Register, or the TA Register plus the TA′ Register. By selecting the TA Register less the TA′ Register option, the upcoming conversion period timing will occur earlier by an amount of time that equals TA′ times the signal period of the Master Clock. By selecting the TA Register plus the TA′ Register option, the upcoming conversion period timing will occur later by an amount of time that equals TA′ times the signal period of the Master Clock. Thus, the D/A conversion timing can be advanced or retarded. An identical ability to alter the A/D conversion timing is provided. In this case, however, the RX Counter A can be programmed via software control with the RA Register, the RA Register less the RA′ Register, or the RA Register plus the RA′ Register.

The above feature is particularly useful for modem applications. This feature allows controlled changes in the A/D and D/A conversion timing. This feature can be used to enhance signal-to-noise performance, to perform frequency-tracking functions, and to generate nonstandard modem frequencies.

If the transmit and receive sections are configured to be synchronous (see WORD/$\overline{\text{BYTE}}$ pin description), then both the low-pass and bandpass switched-capacitor filter clocks are derived from TX Counter A. Also, both the D/A and A/D conversion timing are derived from the TX Counter A and TX Counter B. When the transmit and receive sections are configured to be synchronous, the RX Counter A, RX Counter B, RA Register, RA′ Register, and RB Registers are not used.

AIC DR or DX word bit pattern

A/D or D/A MSB

1st bit sent ↓ (D15) — 1st bit sent of 2nd byte ↓ (D7) — A/D or D/A LSB ↓ (D2)

D15	D14	D13	D12	D11	D10	D9	D8	D7	D6	D5	D4	D3	D2	D1	D0

AIC DX data word format section

d15 d14 d13 d12 d11 d10 d9 d8 d7 d6 d5 d4 d2	d1	d0	COMMENTS
primary DX serial communication protocol			
← d15 (MSB) through d2 go to the D/A converter register →	0	0	The TX and RX Counter A's are loaded with the TA and RA register values. The TX and RX Counter B's are loaded with TB and RB register values.
← d15 (MSB) through d2 go to the D/A converter register →	0	1	The TX and RX Counter A's are loaded with the TA+TA′ and RA+RA′ register values. The TX and RX Counter B's are loaded with the TB and RB register values. NOTE: d1 = 0, d0 = 1 will cause the next D/A and A/D conversion periods to be changed by the addition of TA′ and RA′ Master Clock cycles, in which TA′ and RA′ can be positive or negative or zero. Please refer to the Conversion Period Adjustment Error Detection Table.
← d15 (MSB) through d2 go to the D/A converter register →	1	0	The TX and RX Counter A's are loaded with the TA−TA′ and RA−RA′ register values. The TX and RX Counter B's are loaded with the TB and RB register values. NOTE: d1 = 1, d0 = 0 will cause the next D/A and A/D conversion periods to be changed by the subtraction of TA′ and RA′ Master Clock cycles, in which TA′ and RA′ can be positive or negative or zero. Please refer to the Conversion Period Adjustment Error Detection Table.
← d15 (MSB) through d2 go to the D/A converter register →	1	1	The TX and RX Counter A's are loaded with the TA and RA register values. The TX and RX Counter B's are loaded with the TB and RB register values. After a delay of four Shift Clock cycles, a secondary transmission will immediately follow to program the AIC to operate in the desired configuration.

NOTE: Setting the two least significant bits to 1 in the normal transmission of DAC information (Primary Communications) to the AIC will initiate Secondary Communications upon completion of the Primary Communications.

Upon completion of the Primary Communication, $\overline{FSX}$ will remain high for four SHIFT CLOCK cycles and will then go low and initiate the Secondary Communication. The timing specifications for the Primary and Secondary Communications are identical. In this manner, the Secondary Communication, if initiated, is interleaved between successive Primary Communications. This interleaving prevents the Secondary Communication from interfering with the Primary Communications and DAC timing, thus preventing the AIC from skipping a DAC output.

secondary DX serial communication protocol

x x |← to TA register →| x x |← to RA register →|	0 0	d13 and d6 are MSBs
x |← to TA′ register →| x |← to RA′ register →|	0 1	d14 and d7 are 2′s complement sign bits
x |← to TB register →| x |← to RB register →|	1 0	d14 and d7 are MSBs
x x x x x x x x d7 d6 d5 d4 d3 d2 |← CONTROL REGISTER →|	1 1	d2 = 0/1 deletes/inserts the bandpass filter d3 = 0/1 disables/enables the loopback function d4 = 0/1 disables/enables the AUX IN+ and AUX IN− pins d5 = 0/1 asynchronous/synchronous transmit and receive sections d6 = 0/1 gain control bits (see Gain Control Section) d7 = 0/1 gain control bits (see Gain Control Section)

reset function

A reset function is provided to initiate serial communications between the AIC and DSP. The reset function will initialize all AIC registers, including the control register. After a negative-going pulse on the $\overline{\text{RESET}}$ pin, the AIC registers will be initialized to provide an 8-kHz A/D and D/A conversion rate for a 5.184 MHz master clock input signal. The AIC, excepting the CONTROL register, will be initialized as follows (see AIC DX Data Word Format section):

REGISTER	INITIALIZED REGISTER VALUE (HEX)
TA	9
TA′	1
TB	24
RA	9
RA′	1
RB	24

The CONTROL register bits will be reset as follows (see AIC DX Data Word Format section):

d7 = 1, d6 = 1, d5 = 1, d4 = 0, d3 = 0, d2 = 1

This initialization allows normal serial port communications to occur between AIC and DSP. If the transmit and receive sections are configured to operate synchronously and the user wishes to program different conversion rates, only the TA, TA′, and TB register need to be programmed, since both transmit and receive timing are synchronously derived from these registers (see the pin descriptions and AIC DX Word Format sections).

AIC responses to improper conditions

The AIC has provisions for responding to improper conditions. These improper conditions and the response of the AIC to these conditions are presented in Table 1 below:

AIC register constraints

The following constraints are placed on the contents of the AIC registers:

1. TA register must be > 1.
2. TA′ register can be either positive, negative, or zero.
3. RA register must be > 1.
4. RA′ register can be either positive, negative, or zero.
5. (TA register ± TA′ register) must be > 1.
6. (RA register ± RA′ register) must be > 1.
7. TB register must be > 1.

TABLE 1. AIC RESPONSES TO IMPROPER CONDITIONS

IMPROPER CONDITION	AIC RESPONSE
TA register + TA′ register = 0 or 1 TA register − TA′ register = 0 or 1	Reprogram TX Counter A with TA register value
TA register + TA′ register < 0	MOD 64 arithmetic is used to ensure that a positive value is loaded into the TX Counter A, i.e., TA register + TA′ register + 40 HEX is loaded into TX Counter A
RA register + RA′ register = 0 or 1 RA register − RA′ register = 0 or 1	Reprogram RX Counter A with RA register value
RA register + RA′ register = 0 or 1	MOD 64 arithmetic is used to ensure that a positive value is loaded into RX Counter A, i.e., RA register + RA′ register + 40 HEX is loaded into RX Counter A
TA register = 0 or 1 RA register = 0 or 1	AIC is shut down
TB register = 0 or 1	Reprogram TB register with 24 HEX
RB register = 0 or 1	Reprogram RB register with 24 HEX
AIC and DSP cannot communicate	Hold last DAC output

improper operation due to conversion times being too close together

If the difference between two successive D/A conversion frame syncs is less that 1/19.2 kHz, the AIC operates improperly. In this situation, the second D/A conversion frame sync occurs too quickly and there is not enough time for the ongoing conversion to be completed. This situation can occur if the A and B registers are improperly programmed or if the A + A′ register or A − A′ register result is too small. When incrementally adjusting the conversion period via the A + A′ register options, the designer should be very careful not to violate this requirement (see diagram below).

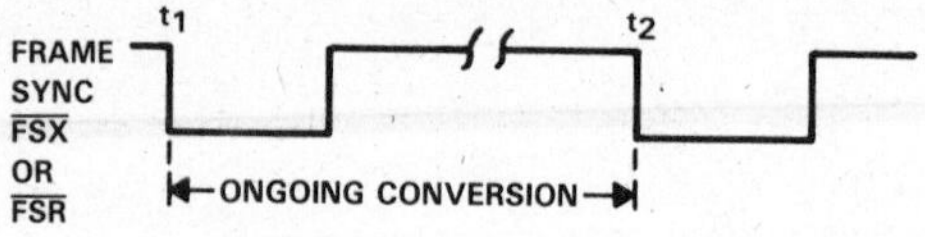
t1
t2
FRAME SYNC FSX OR FSR
ONGOING CONVERSION

$t_2 - t_1 \geq 1/19.2$ kHz

TEXAS INSTRUMENTS
POST OFFICE BOX 655012 • DALLAS, TEXAS 75265

asynchronous operation — more than one receive frame sync occurring between two transmit frame syncs

When incrementally adjusting the conversion period via the A + A′ or A − A′ register options, a specific protocol is followed. The command to use the incremental conversion period adjust option is sent to the AIC during a $\overline{FSX}$ frame sync. The ongoing conversion period is then adjusted. However, either Receive Conversion Period A or B may be adjusted. For both transmit and receive conversion periods, the incremental conversion period adjustment is performed near the end of the conversion period. Therefore, if there is sufficient time between t1 and t2, the receive conversion period adjustment will be performed during Receive Conversion Period A. Otherwise, the adjustment will be performed during Receive Conversion Period B. The adjustment command only adjusts one transmit conversion period and one receive conversion period. To adjust another pair of transmit and receive conversion periods, another command must be issued during a subsequent $\overline{FSX}$ frame (see figure below).

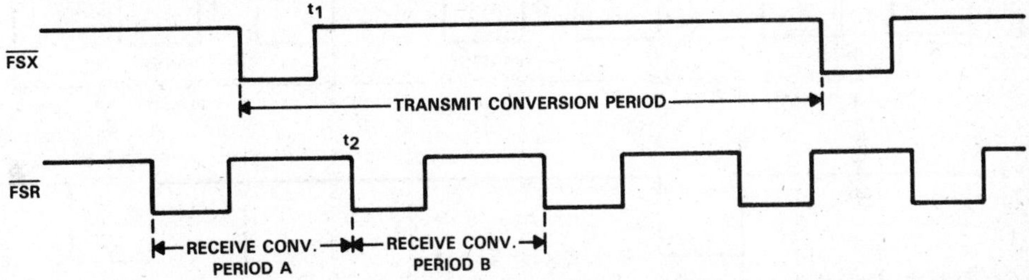
t1
FSX
TRANSMIT CONVERSION PERIOD
t2
FSR
RECEIVE CONV. PERIOD A
RECEIVE CONV. PERIOD B

asynchronous operation — more than one transmit frame sync occurring between two receive frame syncs

When incrementally adjusting the conversion period via the A + A′ or A − A′ register options, a specific protocol is followed. For both transmit and receive conversion periods, the incremental conversion period adjustment is performed near the end of the conversion period. The command to use the incremental conversion period adjust options is sent to the AIC during a $\overline{FSX}$ frame sync. The ongoing transmit conversion period is then adjusted. However, three possibilities exist for the receive conversion period adjustment in the diagram as shown in the figure below. If the adjustment command is issued during Transmit Conversion Period A, Receive Conversion Period A will be adjusted if there is sufficient time between t1 and t2. Or, if there is not sufficient time between t1 and t2, Receive Conversion Period B will be adjusted. Or, the receive portion of an adjustment command may be ignored if the adjustment command is sent during a receive conversion period, which is already being or will be adjusted due to a prior adjustment command. For example, if adjustment commands are issued during Transmit Conversion Periods A, B, and C, the first two commands may cause Receive Conversion Periods A and B to be adjusted, while the third receive adjustment command is ignored. The third adjustment command is ignored since it was issued during Receive Conversion Period B, which already will be adjusted via the Transmit Conversion Period B adjustment command.

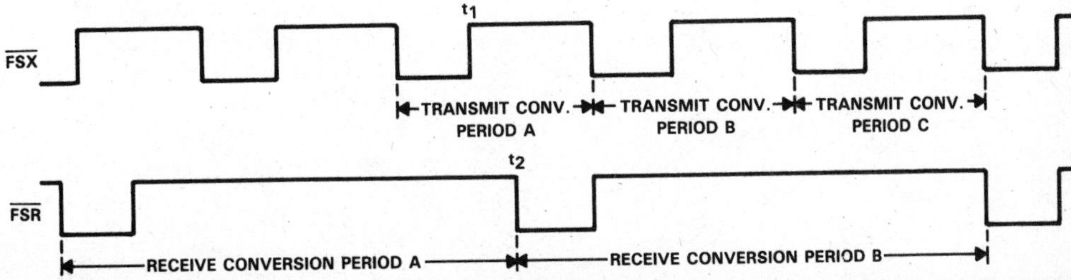
t1
FSX
TRANSMIT CONV. PERIOD A
TRANSMIT CONV. PERIOD B
TRANSMIT CONV. PERIOD C
t2
FSR
RECEIVE CONVERSION PERIOD A
RECEIVE CONVERSION PERIOD B

asynchronous operation — more than one set of primary and secondary DX serial communication occurring between two receive frame sync (see AIC DX Data Word Format section)

The TA, TA′, TB, and control register information that is transmitted in the secondary communications is always accepted and is applied during the ongoing transmit conversion period. If there is sufficient time between t1 and t2, the TA, RA′, and RB register information, which is sent during Transmit Conversion Period A, will be applied to Receive Conversion Period A. Otherwise, this information will be applied during Receive Conversion Period B. If RA, RA′, and RB register information has already been received and is being applied during an ongoing conversion period, any subsequent RA, RA′, or RB information that is received during this receive conversion period will be disregarded (see diagram below).

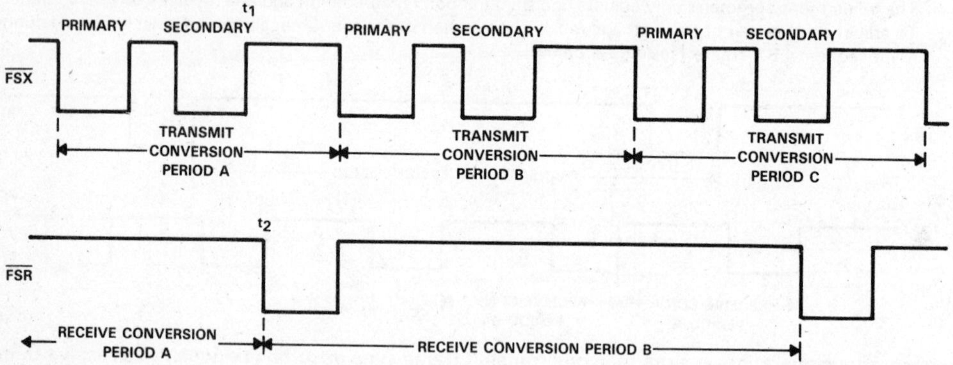
PRIMARY
SECONDARY
t_1
PRIMARY
SECONDARY
PRIMARY
SECONDARY
$\overline{FSX}$
TRANSMIT CONVERSION PERIOD A
TRANSMIT CONVERSION PERIOD B
TRANSMIT CONVERSION PERIOD C
t_2
$\overline{FSR}$
RECEIVE CONVERSION PERIOD A
RECEIVE CONVERSION PERIOD B

absolute maximum ratings over operating free-air temperature range (unless otherwise noted)

Supply voltage, V_{CC+} (see Note 1)	−0.3 V to 15 V
Supply voltage, V_{DD}	−0.3 V to 15 V
Output voltage, V_O	−0.3 V to 15 V
Input voltage, V_I	−0.3 V to 15 V
Digital ground voltage	−0.3 V to 15 V
Operating free-air temperature range: TLC32040M	−55°C to 125°C
TLC32040I	−40°C to 85°C
Storage temperature range	−65°C to 150°C
Lead temperature 1,6 mm (1/16 inch) from case for 60 seconds: J package	300°C
Lead temperature 1,6 mm (1/16 inch) from case for 10 seconds: N package	260°C

NOTE 1: Voltage values for maximum ratings are with respect to VCC−.

recommended operating conditions

PARAMETER		MIN	NOM	MAX	UNIT
Supply voltage, V_{CC+} (see Note 2)		4.75	5	5.25	V
Supply voltage, V_{CC-} (see Note 2)		−4.75	−5	−5.25	V
Digital supply voltage, V_{DD} (see Note 2)		4.75	5	5.25	V
Digital ground voltage with respect to ANLG GND, DGTL GND			0		V
High-level input voltage, V_{IH}		2		V_{DD}+0.3	V
Low-level input voltage, V_{IL} (see Note 3)		−0.3		0.8	V
Load resistance at OUT+ and/or OUT−, R_L		300			Ω
Load capacitance at OUT+ and/or OUT−, C_L				100	pF
MSTR CLK frequency (see Note 4)		0.075	5	10.368	MHz
Analog input amplifier common mode input voltage (see Note 5)				±1.5	V
A/D or D/A conversion rate				19.2	kHz
Operating free-air temperature, T_A	TLC32040M	−55		125	°C
	TLC32040I	−40		85	°C

NOTES: 2. Voltages at analog inputs and outputs, V_{CC+}, and V_{CC-}, are with respect to the ANLG GND terminal. Voltages at digital inputs and outputs and V_{CC+}(DIG) are with respect to the DGTL GND terminal.
3. The algebraic convention, in which the least positive (most negative) value is designated minimum, is used in this data sheet for logic voltage levels and temperature only.
4. The bandpass and low-pass switched-capacitor filter responses are only guaranteed when the switched-capacitor clock frequency is 288 kHz. For switched-capacitor filter clocks at frequencies other than 288 kHz, the filter response is shifted by the ratio of switched-capacitor filter clock frequency to 288 kHz.
5. This range applies when (IN+ − IN−) or (AUX+ − AUX−) equals ±6 V.

electrical characteristics over recommended operating free-air temperature range, V_{CC+} = 5 V, V_{CC-} = −5 V, V_{DD} = 5 V (unless otherwise noted)

total device, MSTR CLK frequency = 5.184 MHz, outputs not loaded

PARAMETER		TEST CONDITIONS	MIN	MAX	UNIT
V_{OH}	High-level output voltage	V_{DD} = 4.75 V, I_{OH} = −300 μA	2.4		V
V_{OL}	Low-level output voltage	V_{DD} = 4.75 V, I_{OL} = 2 mA		0.4	V
I_{CC+}	Supply current from V_{CC+}			25	mA
I_{CC-}	Supply current from V_{CC-}			−25	mA
$I_{CC(stdby)}$	Standby current (MSTR CLK, SHIFT CLK, or $\overline{FSR}$ SYNC in static state)			5	mA

receive amplifier input

PARAMETER		TEST CONDITIONS	MIN	TYP†	MAX	UNIT
	A/D converter offset error (filters bypassed)			5	20	mV
	A/D converter offset error (filters in)			50	140	mV
CMRR	Common-mode rejection ratio at IN+, IN−, or AUX+, AUX−	See Note 6		55		
r_I	Input resistance at IN+, IN− or AUX IN+, AUX IN−			100		kΩ

transmit filter output

PARAMETER		TEST CONDITIONS	MIN	TYP†	MAX	UNIT
V_{OO}	Output offset voltage at OUT+ or OUT− (single-ended relative to ANLG GND)			50		mV
V_{OM}	Maximum peak output voltage swing across R_L at OUT+ or OUT− (single-ended)	$R_L \geq 300\ \Omega$, Offset voltage = 0	±3			V
V_{OM}	Maximum peak output voltage swing between OUT+ and OUT− (differential output)	$R_L \geq 600\ \Omega$	±6			V

†All typical values are at T_A = 25°C.

NOTE 6: The test condition is a 0 dBm, 1-kHz input signal with an 8-kHz conversion rate.

electrical characteristics over recommended operating free-air temperature range, V_{CC+} = 5 V, V_{CC-} = −5 V, V_{DD} = 5 V (unless otherwise noted)

specific modem specifications, SCF clock frequency = 288 kHz

PARAMETER		TEST CONDITIONS	MIN	TYP†	MAX	UNIT
Attenuation of second harmonic of transmitted analog signal	single-ended	See Note 7	60	65		dB
	differential		60	65		
Attenuation of third and higher harmonics of transmitted analog signal	single-ended	See Note 7	60	65		dB
	differential		60	65		

gain and dynamic range

PARAMETER	TEST CONDITIONS	MIN	MAX	UNIT
Absolute transmit gain tracking error while transmitting into 600 Ω (see Note 8)	−50 to 0 dBm signal range		±1.0	dB
Absolute receive gain tracking error (see Note 8)	−50 to 0 dBm signal range		±1.0	dB

power supply rejection and crosstalk attenuation

PARAMETER		TEST CONDITIONS	MIN	TYP†	MAX	UNIT
V_{CC+} or V_{CC-} supply voltage rejection ratio, receive channel	f = 0 to 30 kHz	Idle channel, supply signal at 200 mV p-p measured at DR (ADC output)		30		dB
	f = 30 kHz to 50 kHz			45		
V_{CC+} or V_{CC-} supply voltage rejection ratio, transmit channel (single-ended)	f = 0 to 30 kHz	Idle channel, supply signal at 200 mV p-p measured at OUT+		30		dB
	f = 30 kHz to 50 kHz			45		
Crosstalk attenuation, transmit-to-receive (single-ended)				80		dB

†All typical values are at T_A = 25°C.

NOTES: 7. The test condition is a 0 dBm, 1-kHz input signal into 600 Ω with an 8-kHz conversion rate.
8. Gain tracking is relative to the absolute gain at 1-kHz.

delay distortion, SCF clock frequency = 288 kHz, input (IN+ − IN−) is ±3-V sinewave

Please refer to filter response graphs for delay distortion specifications.

bandpass filter transfer function with 300-Hz high-pass roll-off (see curves), SCF clock frequency = 288 kHz, input (IN+ − IN−) is a ±3-V sinewave (see Note 9)

PARAMETER	TEST CONDITIONS		MIN	MAX	UNIT
Gain relative to gain at 1 kHz	Input signal reference is 0 dB	f = 100 Hz		−45	dB
		f = 150 Hz		−33	
		300 Hz ≤ f ≤ 3.4 kHz	−0.5	0.5	
		f = 4 kHz		−16	
		f ≥ 4.6 kHz		−60	

bandpass filter transfer function with 200-Hz high-pass roll-off (see curves), SCF clock frequency = 288 kHz, input (IN+ − IN−) is a ±3-V sinewave (see Note 9)

PARAMETER	TEST CONDITIONS		MIN	MAX	UNIT
Gain relative to gain at 1 kHz	Input signal reference is 0 dB	f = 100 Hz		−37	dB
		f = 150 Hz		−12	
		300 Hz ≤ f ≤ 3.4 kHz	−0.5	0.5	
		f = 4 kHz		−16	
		f ≥ 4.6 kHz		−60	

low-pass filter transfer function, SCF clock frequency = 288 kHz (see Note 9)

PARAMETER	TEST CONDITIONS		MIN	MAX	UNIT
Gain relative to gain at 1 kHz	Output signal reference is 0 dB	f ≤ 3.4 kHz	−0.5	0.5	dB
		f = 3.6 kHz		−6	
		f = 4 kHz		−30	
		f ≥ 4.4 kHz		−60	

serial port

PARAMETER		TEST CONDITIONS	MIN	TYP†	MAX	UNIT
V_{OH}	High-level output voltage	$I_{OH} = -300\ \mu A$	2.4			V
V_{OL}	Low-level output voltage	$I_{OL} = 2$ mA			0.4	V
I_I	Input current				±10	μA
C_I	Input capacitance			15		pF
C_O	Output capacitance			15		pF

† All typical values are at $T_A = 25°C$.

NOTE 9: The above filter specifications are guaranteed for a switched-capacitor filter clock range of 288 kHz. For switched-capacitor filter clocks at frequencies other than 288 kHz, the filter response is shifted by the ratio of switched-capacitor filter clock frequency to 288 kHz.

operating characteristics over recommended operating free-air temperature range, $V_{CC+} = 5$ V, $V_{CC-} = -5$ V, $V_{DD} = 5$ V

A/D converter (2's complement output, 14-bit resolution)

PARAMETER		TEST CONDITIONS	MIN	TYP†	MAX	UNIT
Integral linearity, f = 4.5 kHz to 19.2 kHz (See Note 10)	bit 1 thru bit 10	Sixteenth full scale		± ½		bit 1
	bit 2 thru bit 11	Eighth full scale		± ½		bit 2
	bit 3 thru bit 12	Quarter full scale		± ½		bit 3
	bit 4 thru bit 13	Half full scale		± ½		bit 4
	bit 5 thru bit 14	Full scale		± ½		bit 5
Conversion rate			1		20	kHz
Signal-to-quantization distortion ratio (for input signals > −15 dBm in the 300 Hz to 3400 Hz band)			60			dB
Equivalent input noise (relative to 600 Ω) at the ADC input		Inputs grounded		75		μV rms

D/A converter (2's complement input, 14-bit resolution)

PARAMETER		TEST CONDITIONS	MIN	TYP†	MAX	UNIT
Integral linearity, f = 4.5 kHz to 19.2 kHz (See Note 10)	bit 1 thru bit 10	Sixteenth full scale		± ½		bit 1
	bit 2 thru bit 11	Eighth full scale		± ½		bit 2
	bit 3 thru bit 12	Quarter full scale		± ½		bit 3
	bit 4 thru bit 13	Half full scale		± ½		bit 4
	bit 5 thru bit 14	Full scale		± ½		bit 5
Settling time				10		μs
Conversion time			1		20	kHz

noise (measurement includes low-pass and bandpass switched-capacitor filters)

PARAMETER		TEST CONDITIONS	TYP	MAX	UNIT
Transmit noises	single-ended	DX input = 0000000000000, constant input code		125	μV rms
	differential			250	μV rms
Receive noise (see Note 11)		Inputs grounded, gain = 1		150	μV rms

timing requirements

serial port — AIC input signals

	PARAMETER	MIN	MAX	UNIT
$t_{c(MCLK)}$	Master clock cycle time	95		ns
$t_{r(MCLK)}$	Master clock rise time		10	ns
$t_{f(MCLK)}$	Master clock fall time		10	ns
	Master clock duty cycle	42%	58%	
$t_{su(DX)}$	DX setup time before SCLK↓	20		ns
$t_{h(DX)}$	DX hold time after SCLK↓	$t_{c(SCLK)}/2$		ns

†All typical values are at $T_A = 25$ °C.

NOTES: 10. Integral linearity for the A/D and D/A converters is guaranteed over the conversion frequency range of 4.5 kHz to 19.2 kHz. Over this range the slew rates of the A/D and D/A converters' sample-and-hold circuits are adequate to guarantee the above integral linearity specifications.

11. This noise is referred to the input with a buffer gain of one. If the buffer gain is two or four, the noise figure will be correspondingly reduced. The noise is computed by statistically evaluating the digital output of the A/D converter.

operating characteristics over recommended operating free-air temperature range, V_{CC+} = 5 V, V_{CC-} = −5 V, V_{DD} = 5 V (continued)

serial port — AIC output signals

PARAMETER		MIN	MAX	UNIT
$t_{c(SCLK)}$	Shift clock (SCLK) cycle time	38		ns
$t_{f(SCLK)}$	Shift clock (SCLK) fall time		50	ns
$t_{r(SCLK)}$	Shift clock (SCLK) rise time		50	ns
	Shift clock (SCLK) duty cycle	45	55	%
$t_{d(CH-FL)}$	Delay from SCLK↑ to FSR/FSX↓		90	ns
$t_{d(CH-FH)}$	Delay from SCLK↑ to FSR/FSK)↑		90	ns
$t_{d(CH-DR)}$	DR valid after SCLK↑		90	ns
$t_{dw(CH-EL)}$	Delay from SCLK↑ to EODX/EODR↓ in word mode		90	ns
$t_{dw(CH-EH)}$	Delay from SCLK↑ to EODX/EODR↑ in word mode		90	ns
$t_{f(EODX)}$	EODX fall time		15	ns
$t_{f(EODR)}$	EODR fall time		15	ns
$t_{db(CH-EL)}$	Delay from SCLK↑ to EODX/EODR↓ in byte mode		100	ns
$t_{db(CH-EH)}$	Delay from SCLK↑ to EODX/EODR↑ in byte mode		100	ns

analog input signal required for full-scale A/D conversion

INPUT CONFIGURATIONS	CONTROL REGISTER BITS		ANALOG INPUT	A/D CONVERSION RESULT
	d6	d7		
Differential configuration	1	1	±6 V	full-scale
Analog input = IN+ − IN−	0	0		
= AUX+ − AUX−	1	0	+3 V	full-scale
	0	1	±1.5 V	full-scale
Single-ended configuration	1	1	±3 V	half-scale
Analog input = IN+ − ANLG GND	0	0		
= AUX+ − ANLG GND	1	0	±3 V	full-scale
	0	1	±1.5 V	full-scale

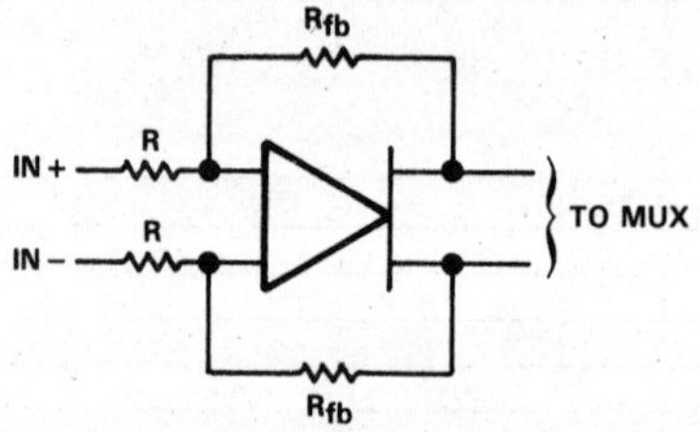
Rfb
R
IN+
R
IN−
TO MUX
Rfb

R_{fb} = R for d6 = 1, d7 = 1
d6 = 0, d7 = 0
R_{fb} = 2R for d6 = 1, d7 = 0
R_{fb} = 4R for d6 = 0, d7 = 1

FIGURE 1. IN+ AND IN− GAIN CONTROL CIRCUITRY

TEXAS INSTRUMENTS
POST OFFICE BOX 655012 • DALLAS, TEXAS 75265

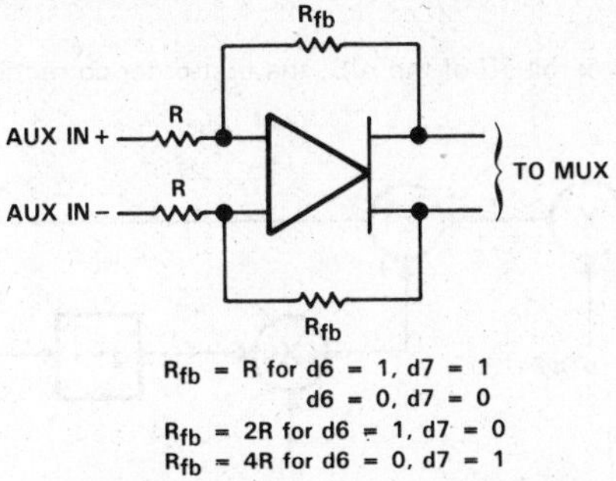
Rfb
R
AUX IN +
R
AUX IN −
TO MUX
Rfb
Rfb = R for d6 = 1, d7 = 1
d6 = 0, d7 = 0
Rfb = 2R for d6 = 1, d7 = 0
Rfb = 4R for d6 = 0, d7 = 1

FIGURE 2. AUXILIARY INPUT CIRCUITRY

sin x/x correction section

The AIC does not have sin x/x correction circuitry after the digital-to-analog converter. Sin x/x correction can be accomplished easily and efficiently in digital signal processor (DSP) software. Excellent correction accuracy can be achieved to a band edge of 3000 Hz by using a first-order digital correction filter. The results, which are shown below, are typical of the numerical correction accuracy that can be achieved for sample rates of interest. The filter requires only seven instruction cycles per sample on the TMS320 DSPs. With a 200-ns instruction cycle, nine instructions per sample represents an overhead factor of 1.4% and 1.7% for sampling rates of 8000 Hz and 9600 Hz, respectively. This correction will add a slight amount of group delay at the upper edge of the 300—3000-Hz band.

sin x/x roll-off for a zero-order hold function

The sin x/x roll-off for the AIC DAC zero-order hold function at a band-edge frequency of 3000 Hz for the various sampling rates is shown in the table below.

TABLE 2. sin x/x ROLL-OFF

f_s (Hz)	$20 \log \frac{\sin \pi f/f_s}{\pi f/f_s}$ (f = 3000 Hz) (dB)
7200	−2.64
8000	−2.11
9600	−1.44
14400	−0.63
19200	−0.35

Note that the actual AIC sin x/x roll-off will be slightly less than the above figures, because the AIC has less than a 100 percent duty cycle hold interval.

correction filter

To compensate for the sin x/x roll-off of the AIC, the first-order correction filter, which is shown below, is recommended.

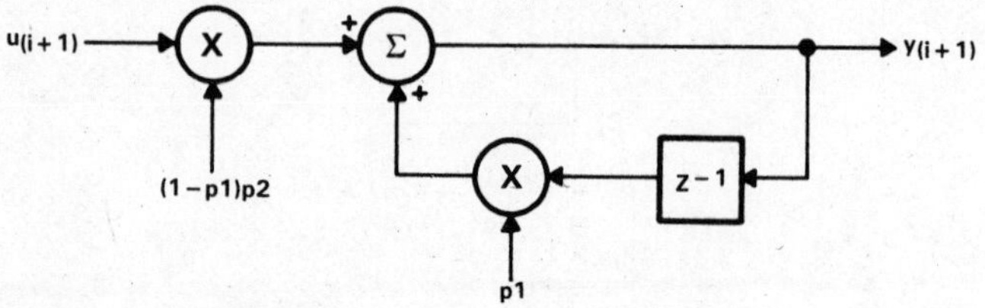
u(i+1)
X
Σ
+
+
y(i+1)
(1−p1)p2
X
z−1
p1

The difference equation for this correction filter is:

$$y_{i+1} = p2(1-p1)\ (u_{i+1}) + p1\ y1$$

where the constant p1 determines the pole locations.

The resulting squared magnitude transfer function is:

$$|H(f)|^2 = \frac{p2^2\ (1-p1)^2}{1\ -\ 2p1\ \cos(2\ \pi\ f/f_S)\ +\ p1^2}$$

correction results

Table 3 below shows the optimum p values and the corresponding correction results for 8000 Hz and 9600 Hz sampling rates.

TABLE 3

f (Hz)	ERROR (dB) f_s = 8000 Hz p1 = −0.14813 p2 = 0.9888	ERROR (dB) f_s = 9600 Hz p1 = −0.1307 p2 = 0.9951
300	−0.099	−0.043
600	−0.089	−0.043
900	−0.054	0
1200	−0.002	0
1500	0.041	0
1800	0.079	0.043
2100	0.100	0.043
2400	0.091	0.043
2700	−0.043	0
3000	−0.102	−0.043

TMS320 software requirements

The digital correction filter equation can be written in state variable form as follows:

$$Y = k1Y + k2U$$

where Y is the filter state and U is the next I/O sample. With the assumption that TMS processor page pointer and memory configuration are properly initialized, the equation can be executed in seven instructions or seven cycles with the following program:

```
ZAC
LT K2
MPY U
LTD K1
MPY Y
APAC
SACH (dma)
```

byte-mode timing

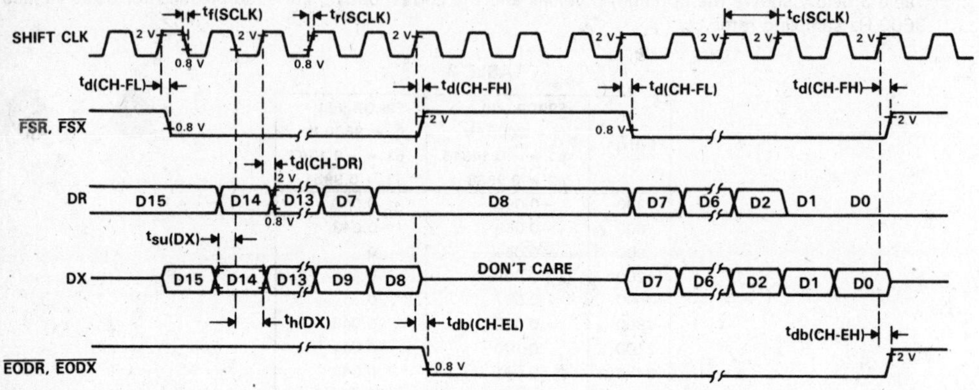
SHIFT CLK
$t_{f(SCLK)}$
$t_{r(SCLK)}$
$t_{c(SCLK)}$
2 V
0.8 V
$t_{d(CH-FL)}$
$t_{d(CH-FH)}$
$\overline{FSR}$, $\overline{FSX}$
$t_{d(CH-DR)}$
DR
D15
D14
D13
D7
D8
D6
D2
D1
D0
$t_{su(DX)}$
DX
D9
DON'T CARE
$t_{h(DX)}$
$t_{db(CH-EL)}$
$t_{db(CH-EH)}$
$\overline{EODR}$, $\overline{EODX}$

word-mode timing

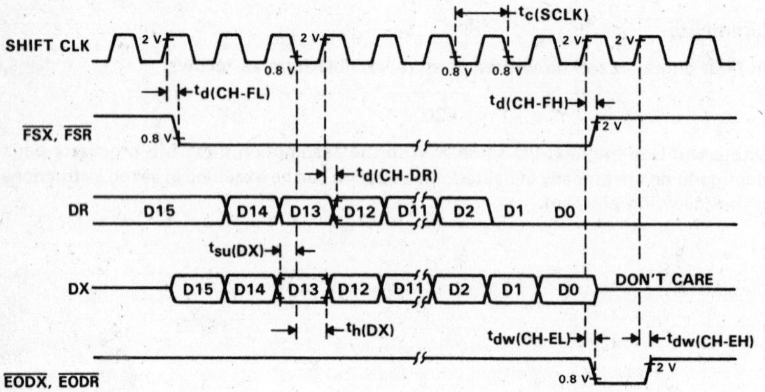
SHIFT CLK
$t_{c(SCLK)}$
2 V
0.8 V
$t_{d(CH-FL)}$
$t_{d(CH-FH)}$
$\overline{FSX}$, $\overline{FSR}$
$t_{d(CH-DR)}$
DR
D15
D14
D13
D12
D11
D2
D1
D0
$t_{su(DX)}$
DX
DON'T CARE
$t_{h(DX)}$
$t_{dw(CH-EL)}$
$t_{dw(CH-EH)}$
$\overline{EODX}$, $\overline{EODR}$

FIGURE 3. SERIAL PORT TIMING

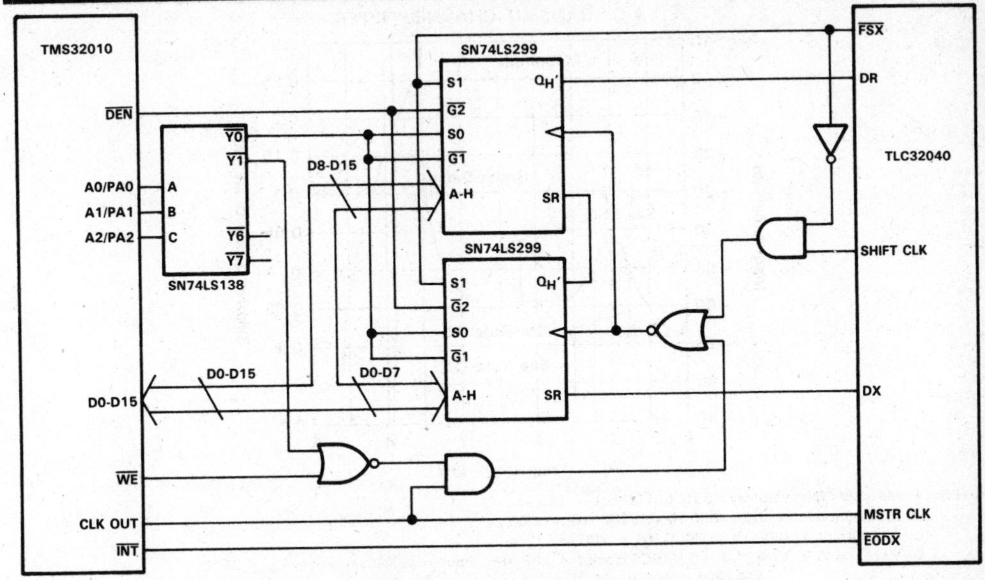
TMS32010
DEN
A0/PA0
A1/PA1
A2/PA2
D0-D15
WE
CLK OUT
INT
SN74LS138
A
B
C
Y0
Y1
Y6
Y7
SN74LS299
S1
G2
S0
G1
A-H
QH'
SR
D8-D15
D0-D15
D0-D7
TLC32040
FSX
DR
SHIFT CLK
DX
MSTR CLK
EODX

FIGURE 4. TLC32010—TLC32040 INTERFACE CIRCUIT

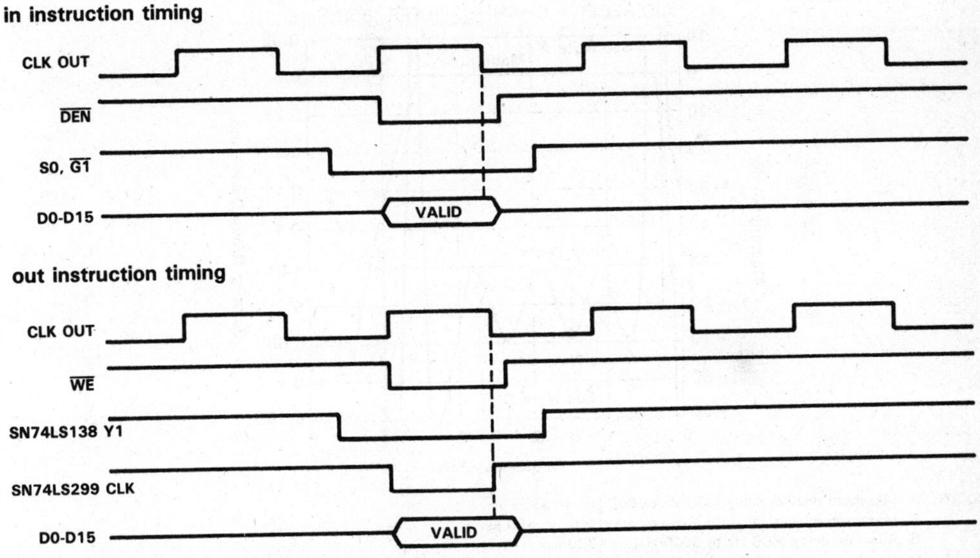
in instruction timing
CLK OUT
DEN
S0, G1
D0-D15
VALID
out instruction timing
CLK OUT
WE
SN74LS138 Y1
SN74LS299 CLK
D0-D15
VALID

FIGURE 5. TMS32010—TMS32040 INTERFACE TIMING

AIC TRANSMIT CHANNEL FILTER

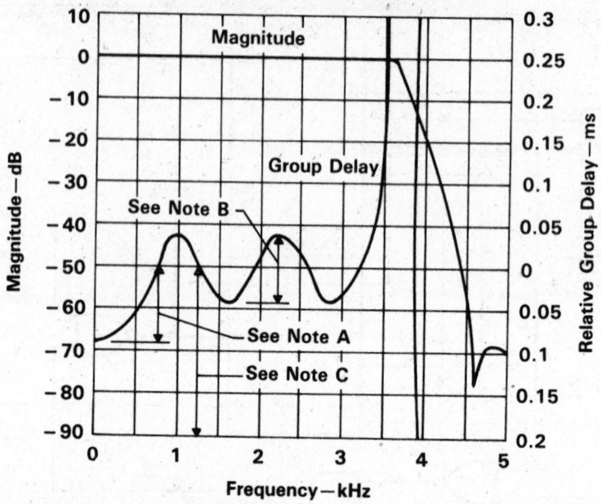
Magnitude
Group Delay
See Note B
See Note A
See Note C
Magnitude—dB
Relative Group Delay—ms
Frequency—kHz
10
0
−10
−20
−30
−40
−50
−60
−70
−80
−90
0.3
0.25
0.2
0.15
0.1
0.05
0
0.05
0.1
0.15
0.2
0
1
2
3
4
5

NOTES: A. Maximum relative delay (0 Hz to 600 Hz) = 125 μs.
B. Maximum relative delay (600 Hz to 3000 Hz) = ±50 μs.
C. Absolute delay (600 Hz to 3000 Hz) = 700 μs.
D. V_{CC+} = 5 V, V_{CC-} = −5 V, SCF clock f = 288 kHz, input = ±3-V sinewave, T_A = 25°C.

FIGURE 6

AIC RECEIVE CHANNEL FILTER (300 Hz)

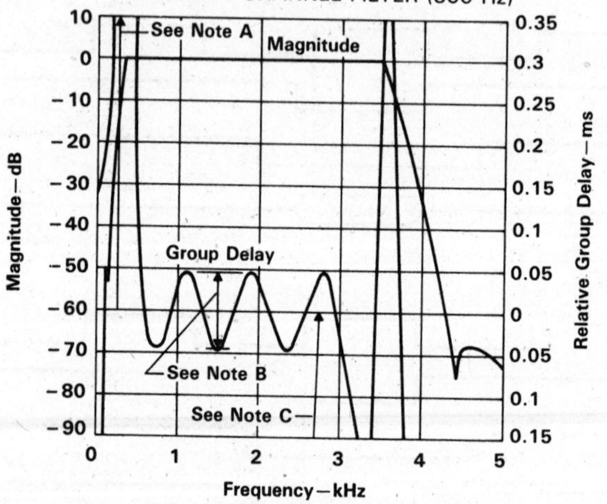
See Note A
Magnitude
Group Delay
See Note B
See Note C
Magnitude—dB
Relative Group Delay—ms
Frequency—kHz
10
0
−10
−20
−30
−40
−50
−60
−70
−80
−90
0.35
0.3
0.25
0.2
0.15
0.1
0.05
0
0.05
0.1
0.15
0
1
2
3
4
5

NOTES: A. Maximum relative delay (200 Hz to 600 Hz) = 3350 μs.
B. Maximum relative delay (600 Hz to 3000 Hz) = ±50 μs.
C. Absolute delay (600 Hz to 3000 Hz) = 1230 μs.
D. V_{CC+} = −5 V, V_{CC-} = −5 V, SCF clock f = 288 kHz, input = ±3-V sinewave, T_A = 25°C.

FIGURE 7

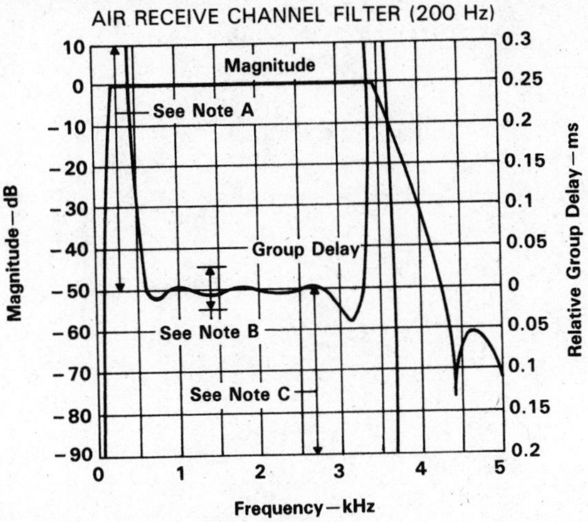
AIR RECEIVE CHANNEL FILTER (200 Hz)
Magnitude
See Note A
Group Delay
See Note B
See Note C
Magnitude—dB
Relative Group Delay—ms
Frequency—kHz
10
0
−10
−20
−30
−40
−50
−60
−70
−80
−90
0.3
0.25
0.2
0.15
0.1
0.05
0
0.05
0.1
0.15
0.2
0
1
2
3
4
5

NOTES: A. Maximum relative delay (200 Hz to 600 Hz) = 3350 μs.
B. Maximum relative delay (600 Hz to 3000 Hz) = ±50 μs.
C. Absolute delay (600 Hz to 3000 Hz) = 1080 μs.
D. V_{CC+} = −5 V, V_{CC-} = −5 V, SCF clock f = 288 kHz, input = ±3-V sinewave, T_A = 25°C.

FIGURE 8

General Information — 1
Alphanumeric Index
Selection Guide

Data Acquisition Circuits — 2
Cross-Reference Guide
Data Sheets

Display Drivers — 3
Data Sheets

Line Drivers and Receivers — 4
Cross-Reference Guide
Data Sheets

Peripheral Drivers/Actuators — 5
Cross-Reference Guide
Data Sheets

Memory Interface Circuits — 6
Data Sheets

Speech Synthesis Circuits — 7
Data Sheets

Appendix A Power Derating Curves — A

Appendix B Ordering Instructions, Mechanical Data, IC Sockets — B

Appendix C Explanation of Logic Symbols — C

D2520, MARCH 1979—REVISED OCTOBER 1986

- 90-V Output Swing
- CMOS-Compatible Inputs
- Quad Drivers with Independent Addressing of Each Gate for Serial or Parallel Applications
- High Data Input Impedance . . . 1 MΩ Typ
- 30-mA Clamp Diodes on Output

SN55426B . . . J
DUAL-IN-LINE PACKAGE
(TOP VIEW)

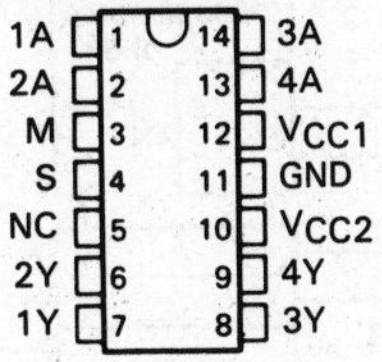
1A 1
2A 2
M 3
S 4
NC 5
2Y 6
1Y 7
14 3A
13 4A
12 V_{CC1}
11 GND
10 V_{CC2}
9 4Y
8 3Y

SN55427B . . . J
DUAL-IN-LINE PACKAGE
(TOP VIEW)

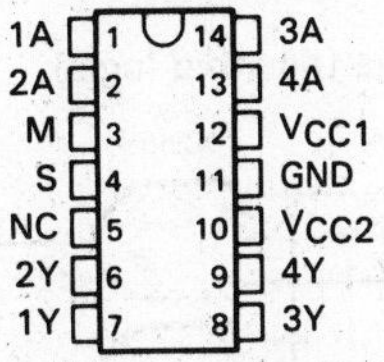
1A 1
2A 2
M 3
S 4
NC 5
2Y 6
1Y 7
14 3A
13 4A
12 V_{CC1}
11 GND
10 V_{CC2}
9 4Y
8 3Y

NC—No internal connection

description

The SN55426B and SN55427B are monolithic integrated-circuit plasma display drivers. The logic of the two drivers is complementary to permit controlled writing or erasing at a specified point on the display. The '426B noninverting pulser is normally near ground potential and is pulsed near V_{CC2}; the '427B inverting pulser is normally near V_{CC2} potential and is pulsed near ground potential. The devices are designed to accept CMOS logic input signals and drive one display line per output.

There are four gates per package with individual data inputs. Additionally, each device has a strobe and a multiplex input controlling all four gates. The devices require two power supplies: the logic section power supply V_{CC1}, and the high-voltage bias supply V_{CC2}. V_{CC2} controls the magnitude of the output swing.

Each output is designed to sustain 20-milliampere switching transients on the output. Each output is also protected by source and sink clamp diodes with 30-milliampere current capability. Each device is designed to be operated at 50 kilohertz but may be operated as high as 85 kilohertz.

The multiplex and strobe inputs (inputs M and S, respectively) act on all four gates simultaneously and aid in plasma panel design.

The SN55426B and SN55427B are characterized for operation over the full military temperature range of −55°C to 125°C.

FUNCTION TABLE

INPUTS			OUTPUTS	
A	M	S	'426B	'427B
L	X	X	L	H
X	L	X	L	H
X	X	L	L	H
H	H	H	H	L

H = high level, L = low level,
X = irrelevant

TEXAS INSTRUMENTS
POST OFFICE BOX 655012 • DALLAS, TEXAS 75265

logic symbols†

'426B

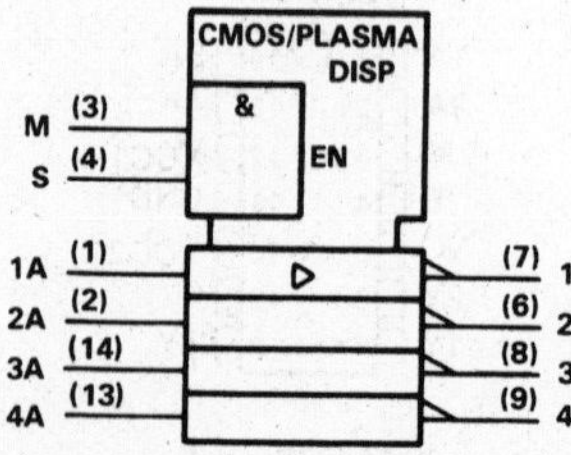
CMOS/PLASMA DISP
&
EN
M (3)
S (4)
1A (1)
2A (2)
3A (14)
4A (13)
(7) 1Y
(6) 2Y
(8) 3Y
(9) 4Y

'427B

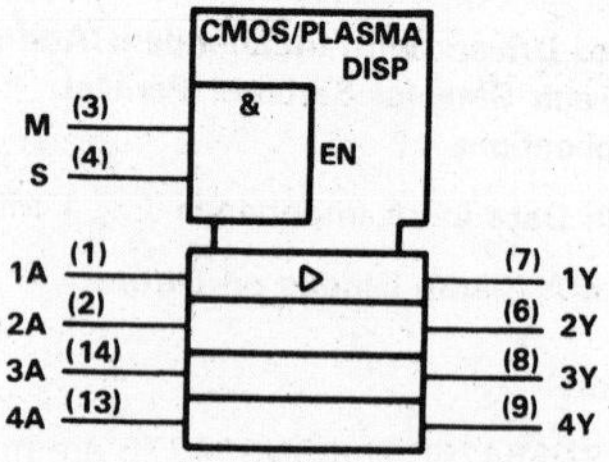
CMOS/PLASMA DISP
&
EN
M (3)
S (4)
1A (1)
2A (2)
3A (14)
4A (13)
(7) 1Y
(6) 2Y
(8) 3Y
(9) 4Y

† These symbols are in accordance with ANSI/IEEE Std 91-1984 and IEC Publication 617-12.

logic diagrams (positive logic)

'426B

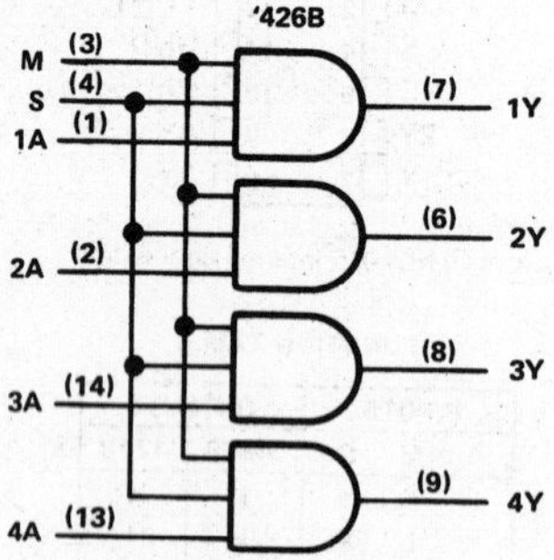
M (3)
S (4)
1A (1)
2A (2)
3A (14)
4A (13)
(7) 1Y
(6) 2Y
(8) 3Y
(9) 4Y

POSITIVE LOGIC: $Y = A \cdot M \cdot S$

'427B

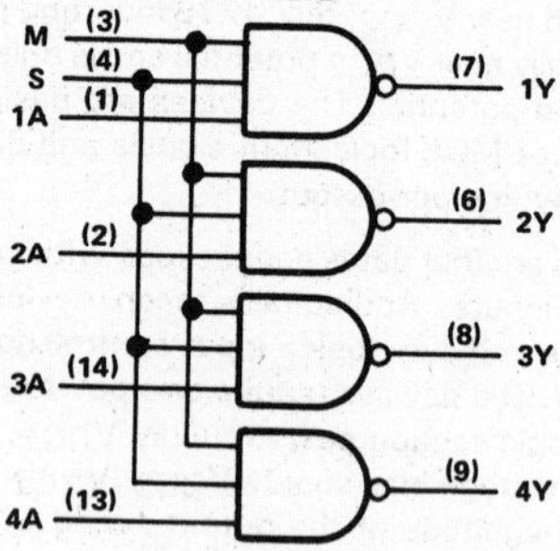
M (3)
S (4)
1A (1)
2A (2)
3A (14)
4A (13)
(7) 1Y
(6) 2Y
(8) 3Y
(9) 4Y

POSITIVE LOGIC: $Y = \overline{A \cdot M \cdot S}$

schematics of inputs and outputs

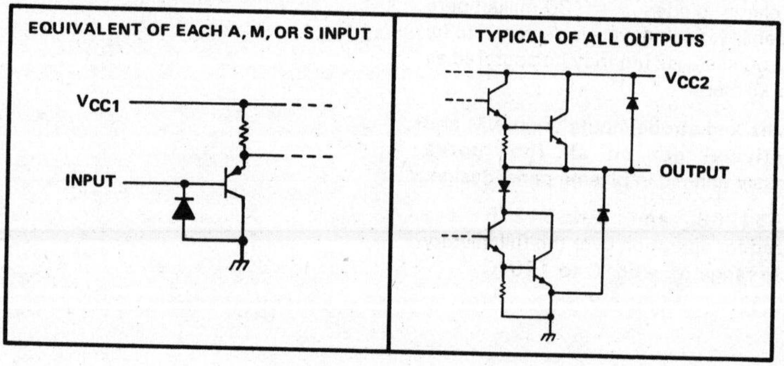
EQUIVALENT OF EACH A, M, OR S INPUT
VCC1
INPUT
TYPICAL OF ALL OUTPUTS
VCC2
OUTPUT

absolute maximum ratings over operating free-air temperature range (unless otherwise noted)

Supply voltage, V_{CC1} (see Note 1) . . . . . 15 V
Supply voltage, V_{CC2} . . . . . 95 V
Input voltage, V_I . . . . . 15 V
Continuous output current, I_O . . . . . 20 mA
Continuous total dissipation at (or below) 25°C free-air temperature (see Note 2) . . . . . 800 mW
Operating free-air temperature range . . . . . −55°C to 125°C
Storage temperature range . . . . . −65°C to 150°C
Lead temperature 1,6 mm (1/16 inch) from case for 60 seconds . . . . . 300°C

NOTES: 1. All voltage values are with respect to network ground terminal.
2. For operation above 25°C free-air temperature, refer to Dissipation Derating Curves in Appendix A. In the J package, SN55426B and SN55427B chips are alloy mounted.

recommended operating conditions

	MIN	NOM	MAX	UNIT
Supply voltage for logic section, V_{CC1}	10	12	14	V
Supply voltage for output section, V_{CC2}	40	70	90	V
High-level input voltage, V_{IH}	7			V
Low-level input voltage, V_{IL}			3	V
Strobe frequency	0		85	kHz
Data input frequency	0	50	85	kHz
Duration of strobe pulse	1.5	5		μs
Operating free-air temperature, T_A	−55		125	°C

electrical characteristics, V_{CC1} = 12 V, V_{CC2} = 70 V, T_A = −55°C to 125°C (unless otherwise noted)

PARAMETER			TEST CONDITIONS		MIN	TYP†	MAX	UNIT
V_{OH}	High-level output voltage		V_{IH} = 7 V,	I_O = −1 mA	$V_{CC2}-4$	$V_{CC2}-1$		V
			V_{IL} = 3 V	I_O = −15 mA	$V_{CC2}-8$	$V_{CC2}-1.8$		
V_{OL}	Low-level output voltage		V_{IH} = 7 V,	I_O = 1 mA		2	4	V
			V_{IL} = 3 V	I_O = 15 mA		3.5	8	
V_{OK}	Output clamp voltage		Output high,	I_O = 30 mA		$V_{CC2}+0.8$	$V_{CC2}+2$	V
			Output low,	I_O = −30 mA		−0.9	−2	
I_{IH}	High-level input current	A	V_{IH} = 12 V			12	60	μA
		M, S				50	200	
I_{CC1}	Supply current, logic section		V_{CC1} = 12 V,	All inputs at 12 V		10	15	mA
I_{CC2}	Supply current, output section		V_{CC2} = 90 V,	All outputs high		1.1	1.9	mA
			No load	All outputs low		0.1	0.6	
$I_{CC1(av)}$	Average supply current, logic section		t_w = 5 μs,	f = 50 kHz,		10		mA
$I_{CC2(av)}$	Average supply current, output section		No load			1.3		mA

† All typical values are at 25°C.

switching characteristics, V_{CC1} = 12 V, V_{CC2} = 70 V, T_A = 25°C

PARAMETER		TEST CONDITIONS	MIN	TYP	MAX	UNIT
t_{PLH}	Propagation delay time, low-to-high-level output	C_L = 20 pF, R_L = 100 kΩ,		0.7	1.2	μs
t_{PHL}	Propagation delay time, high-to-low-level output	See Figure 1		0.3	0.8	μs

PARAMETER MEASUREMENT INFORMATION

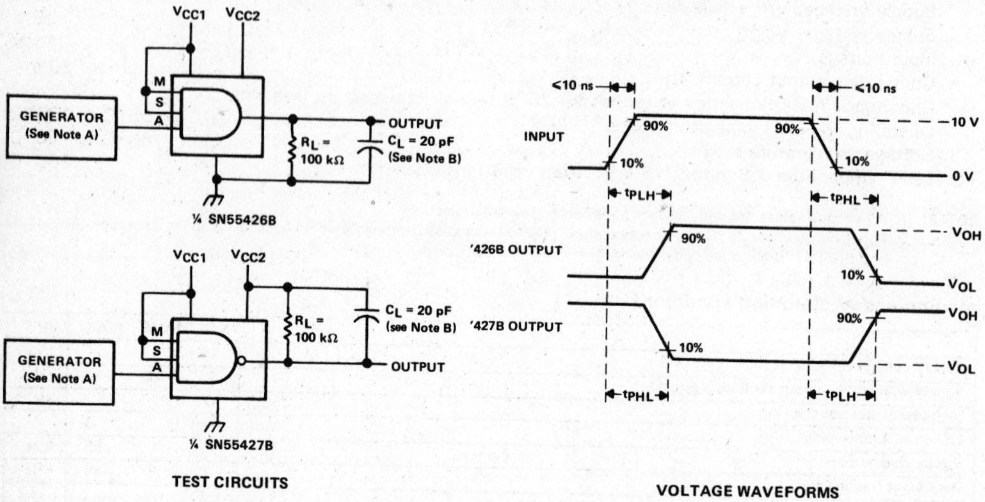
VCC1
VCC2
M
S
A
GENERATOR (See Note A)
OUTPUT
RL = 100 kΩ
CL = 20 pF (See Note B)
¼ SN55426B
CL = 20 pF (see Note B)
OUTPUT
¼ SN55427B
TEST CIRCUITS
≤10 ns
INPUT
90%
10%
10 V
0 V
tPLH
tPHL
'426B OUTPUT
VOH
VOL
'427B OUTPUT
VOLTAGE WAVEFORMS

NOTES: A. The pulse generator has the following characteristics: Z_O = 50 Ω, PRR ≤ 50 kHz, t_w = 5 μs.
B. C_L includes probe and jig capacitance.

FIGURE 1. SWITCHING TIMES

D2355, OCTOBER 1972—REVISED SEPTEMBER 1986

QUAD SEGMENT DRIVER AND HEX DIGIT DRIVER FOR INTERFACING BETWEEN MOS AND LIGHT-EMITTING-DIODE (LED) DISPLAYS

- 50-mA Source or Sink Capability ('491, '491A)
- 250-mA Sink Capability ('492, '492A)
- Rated for 10-V Operation ('491, '492)
- Rated for 20-V Operation ('491A, '492A)
- Low Input Current for MOS Compatability
- Low Standby Power
- High-Gain Darlington Circuits

SN75491, SN75491A
N DUAL-IN-LINE PACKAGE
(TOP VIEW)

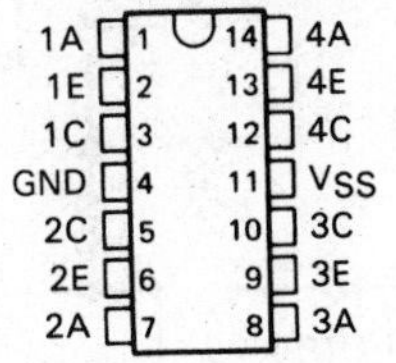
1A 1 14 4A
1E 2 13 4E
1C 3 12 4C
GND 4 11 VSS
2C 5 10 3C
2E 6 9 3E
2A 7 8 3A

SN75492, SN75492A
N DUAL-IN-LINE PACKAGE
(TOP VIEW)

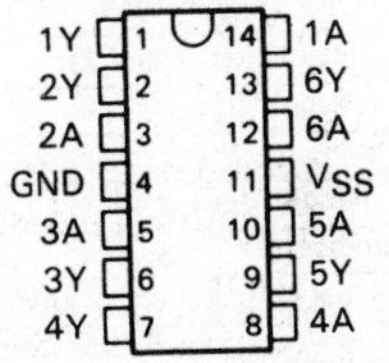
1Y 1 14 1A
2Y 2 13 6Y
2A 3 12 6A
GND 4 11 VSS
3A 5 10 5A
3Y 6 9 5Y
4Y 7 8 4A

description

The SN75491, SN75491A, SN75492, and SN75492A are monolithic integrated circuits designed to be used together with MOS integrated circuits and common-cathode LED's in serially addressed multi-digit displays. This time-multiplexed system, which uses a segment-address-and-digit-scan method of LED drive, minimizes the number of drivers required.

The SN75491 and SN75491A are quadruple segment drivers. The SN75492 and SN75492A are hex digit drivers. The SN75491 and SN75492 are characterized for operation to 10 volts. The SN75491A and SN75492A are characterized for operation to 20 volts.

The SN75491, SN75491A, SN75492, and SN75492A are characterized for operation from 0°C to 70°C.

logic diagram (each driver)

SN75491, SN75491A

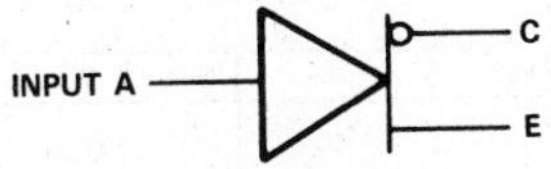
INPUT A
C
E

SN75492, SN75492A

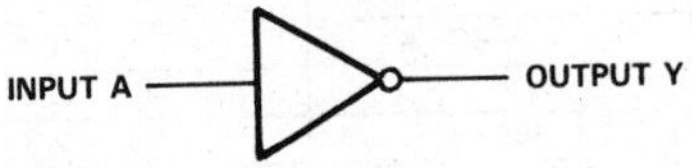
INPUT A
OUTPUT Y

logic symbols†

SN75491, SN75491A

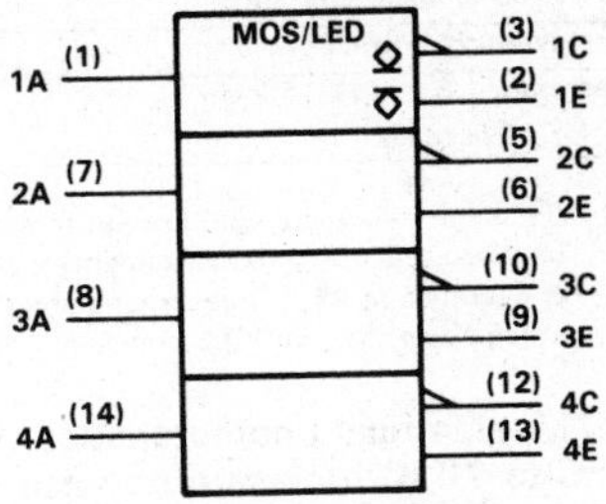
MOS/LED
1A (1)
(3) 1C
(2) 1E
2A (7)
(5) 2C
(6) 2E
3A (8)
(10) 3C
(9) 3E
4A (14)
(12) 4C
(13) 4E

SN75492, SN75492A

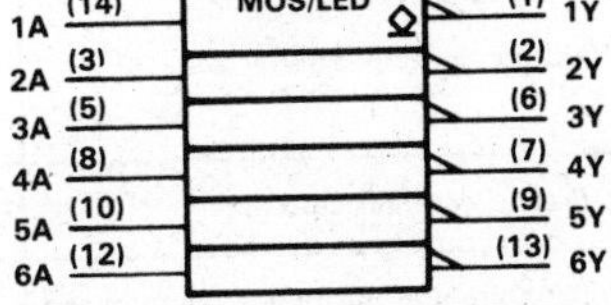
MOS/LED
1A (14)
2A (3)
3A (5)
4A (8)
5A (10)
6A (12)
(1) 1Y
(2) 2Y
(6) 3Y
(7) 4Y
(9) 5Y
(13) 6Y

† These symbols are in accordance with ANSI/IEEE Std 91-1984 and IEC Publication 617-12.

schematics

SN75491, SN75491A (each driver)

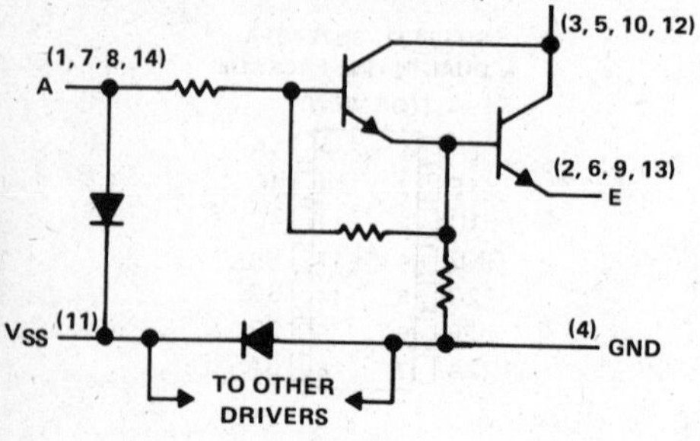
C
(3, 5, 10, 12)
(1, 7, 8, 14)
A
(2, 6, 9, 13)
E
V_{SS} (11)
(4)
GND
TO OTHER DRIVERS

SN75492, SN75492A (each driver)

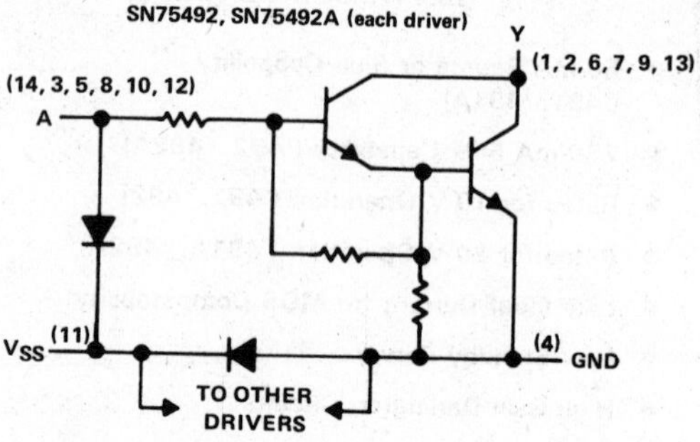
Y
(1, 2, 6, 7, 9, 13)
(14, 3, 5, 8, 10, 12)
A
V_{SS} (11)
(4)
GND
TO OTHER DRIVERS

absolute maximum ratings over operating free-air temperature range (unless otherwise noted)

		SN75491	SN75491A	SN75492	SN75492A	UNIT
Input voltage range (see Notes 1 and 2)		−5 V to V_{SS}	−5 V to V_{SS}	−5 V to V_{SS}	−5 V to V_{SS}	
Collector (output) voltage, V_C		10	20	10	20	V
Collector (output)-to-input voltage		10	20	10	20	V
Emitter-to-ground voltage ($V_I \geq 5$ V)		10	20			V
Emitter-to-input voltage		5	5			V
Voltage at V_{SS} terminal with respect to any other device terminal		10	20	10	20	V
Collector (output) current, I_C	Each collector (output)	50	50	250	250	mA
	All collectors (outputs)	200	200	600	600	
Continuous total dissipation at (or below) 25 °C free-air temperature (see Note 3)		875	875	875	875	mW
Operating free-air temperature range		0 to 70	0 to 70	0 to 70	0 to 70	°C
Storage temperature range		−65 to 150	−65 to 150	−65 to 150	−65 to 150	°C
Lead temperature 1,6 mm (1/16 inch) from case for 10 seconds		260	260	260	260	°C

NOTES: 1. All voltage values are with respect to network ground terminal.
2. The input is the only device terminal that may be negative with respect to ground.
3. For operation at 25 °C free-air temperature, refer to Dissipation Derating Curves in Appendix A. For these devices in the N package, use the 7-mW/°C curve.

'491, '491A electrical characteristics, V_{SS} = 10 V for SN75491, V_{SS} = 20 V for SN75491A, T_A = 0 °C to 70 °C (unless otherwise noted)

PARAMETER		TEST CONDITIONS		MIN	TYP†	MAX	UNIT
$V_{CE(on)}$	On-State collector-emitter voltage	Input = 8.5 V through 1 kΩ, V_E = 5 V, I_C = 50 mA, T_A = 25 °C			0.9	1.2	V
		Input = 8.5 V through 1 kΩ, V_E = 5 V, I_C = 50 mA				1.5	
$I_{C(off)}$	Off-state collector current	$V_C = V_{SS}$, V_E = 0, I_I = 40 μA				100	μA
		$V_C = V_{SS}$, V_E = 0, V_I = 0.7 V				100	
I_I	Input current at maximum input voltage	$V_I = V_{SS}$, V_E = 0, I_C = 20 mA	'491		2.2	3.3	mA
			'491A		4.7	6.5	
I_E	Emitter reverse current	V_I = 0, V_E = 5 V, I_C = 0				100	μA
I_{SS}	Current into V_{SS} terminal					1	mA

†All typical values are at T_A = 25 °C.

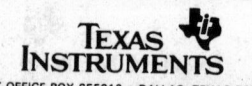

'492, '492A electrical characteristics, V_{SS} = 10 V for SN75492, V_{SS} = 20 V for SN75492A, T_A = 0°C to 70°C (unless otherwise noted)

PARAMETER		TEST CONDITIONS		MIN	TYP†	MAX	UNIT
V_{OL}	Low-level output voltage	Input = 6.5 V through 1 kΩ, I_{OL} = 250 mA, T_A = 25°C			0.9	1.2	V
		Input = 6.5 V through 1 kΩ, I_{OL} = 250 mA,				1.5	
I_{OH}	High-level output current	V_{OH} = V_{SS}, V_I = 40 μA				200	μA
		V_{OH} = V_{SS}, V_I = 0.5 V				200	
I_I	Input current at maximum input voltage	V_I = V_{SS}, I_{OL} = 20 mA	'492		2.2	3.3	mA
			'492A		4.7	6.5	
I_{SS}	Current into V_{SS} terminal					1	mA

†All typical values are at T_A = 25°C.

SN75491, SN75491A switching characteristics, V_{SS} = 7.5 V, T_A = 25°C

PARAMETER		TEST CONDITIONS	MIN	TYP	MAX	UNIT
t_{PLH}	Propagation delay time, low-to-high-level output (collector)	V_{IH} = 4.5 V, V_E = 0, R_L = 200 Ω, C_L = 15 pF		100		ns
t_{PHL}	Propagation delay time, high-to-low-level output (collector)			20		ns

SN75492, SN75492A switching characteristics, V_{SS} = 7.5 V, T_A = 25°C

PARAMETER		TEST CONDITIONS	MIN	TYP	MAX	UNIT
t_{PLH}	Propagation delay time, low-to-high-level output	V_{IH} = 7.5 V, R_L = 39 Ω, C_L = 15 pF		300		ns
t_{PHL}	Propagation delay time, high-to-low-level output			30		ns

PARAMETER MEASUREMENT INFORMATION

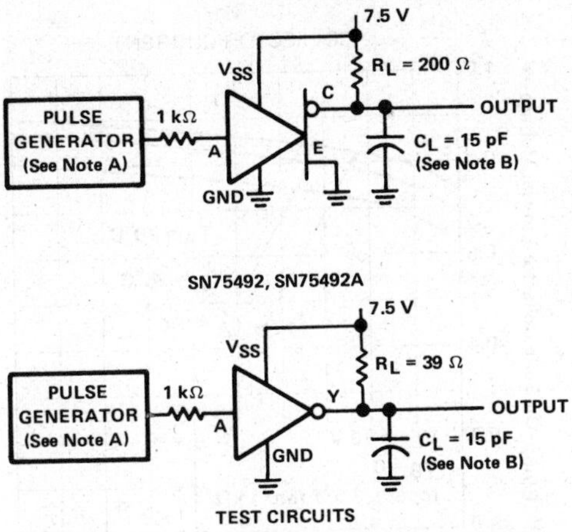
SN75491, SN75491A
7.5 V
VSS
RL = 200 Ω
C
PULSE GENERATOR (See Note A)
1 kΩ
A
E
OUTPUT
CL = 15 pF (See Note B)
GND
SN75492, SN75492A
7.5 V
VSS
RL = 39 Ω
1 kΩ
Y
OUTPUT
CL = 15 pF (See Note B)
GND
TEST CIRCUITS

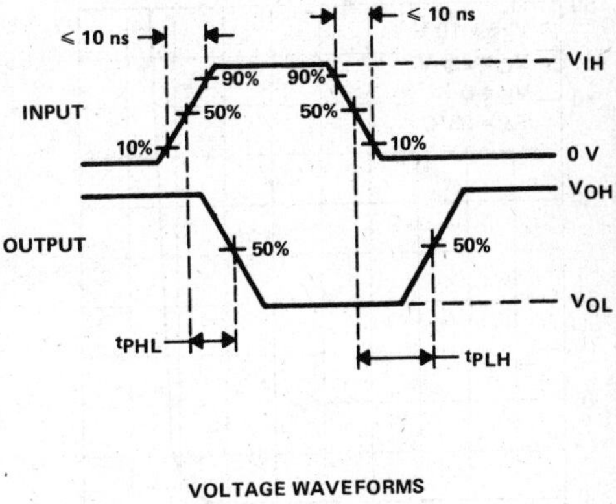
≤ 10 ns
≤ 10 ns
VIH
90%
90%
INPUT
50%
50%
10%
10%
0 V
VOH
OUTPUT
50%
50%
VOL
tPHL
tPLH
VOLTAGE WAVEFORMS

NOTES: A. The pulse generator has the following characteristics: Z_{out} = 50 Ω, PRR ≤ 100 kHz, t_w = 1 μs.
B. C_L includes probe and jig capacitance.

FIGURE 1. PROPAGATION DELAY TIMES

3 Display Drivers

TYPICAL CHARACTERISITCS

INPUT CURRENT
vs
INPUT VOLTAGE

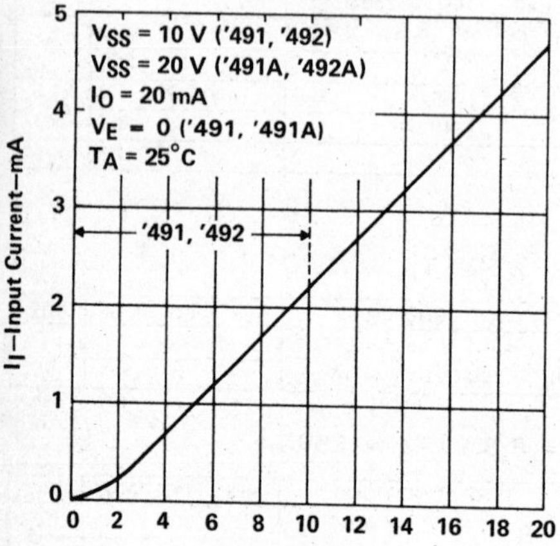
V_{SS} = 10 V ('491, '492)
V_{SS} = 20 V ('491A, '492A)
I_O = 20 mA
V_E = 0 ('491, '491A)
T_A = 25°C
'491, '492
I_I—Input Current—mA
V_I—Input Voltage—V

FIGURE 2

SN75491, SN75491A
COLLECTOR CURRENT
vs
INPUT CURRENT

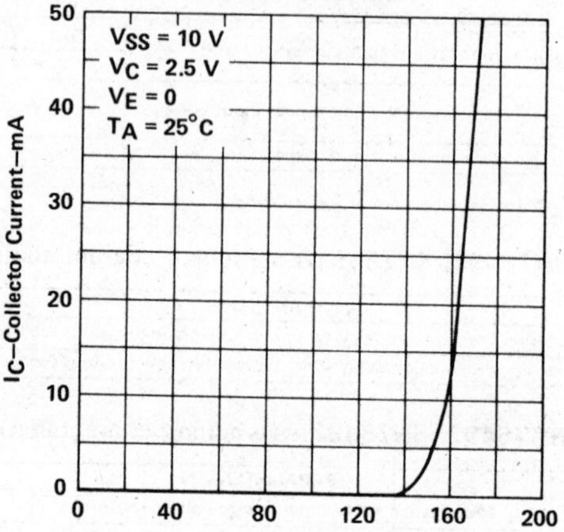
V_{SS} = 10 V
V_C = 2.5 V
V_E = 0
T_A = 25°C
I_C—Collector Current—mA
I_I—Input Current—μA

FIGURE 3

SN75491, SN75491A
COLLECTOR CURRENT
vs
INPUT VOLTAGE

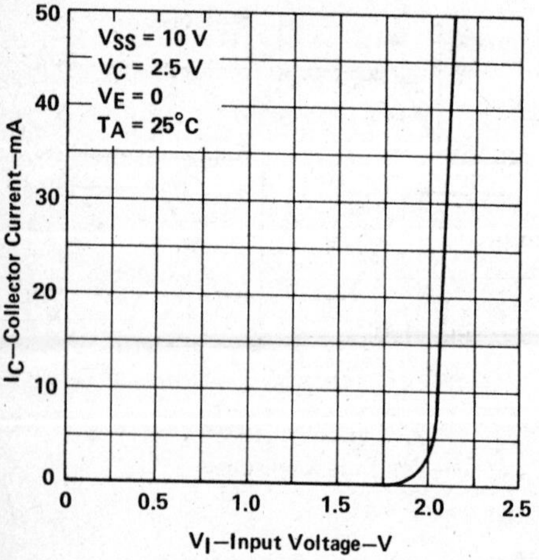
V_{SS} = 10 V
V_C = 2.5 V
V_E = 0
T_A = 25°C
I_C—Collector Current—mA
V_I—Input Voltage—V

FIGURE 4

SN75491, SN75491A
ON-STATE COLLECTOR-EMITTER VOLTAGE
vs
COLLECTOR CURRENT

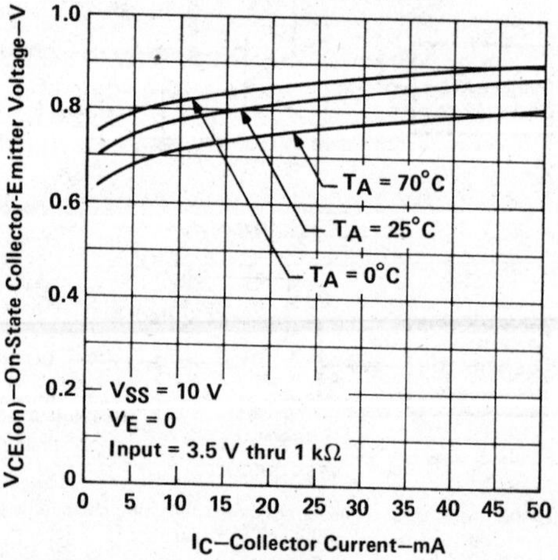
T_A = 70°C
T_A = 25°C
T_A = 0°C
V_{SS} = 10 V
V_E = 0
Input = 3.5 V thru 1 kΩ
$V_{CE(on)}$—On-State Collector-Emitter Voltage—V
I_C—Collector Current—mA

FIGURE 5

TYPICAL CHARACTERISTICS

SN75492, SN75492A
OUTPUT CURRENT
vs
INPUT CURRENT

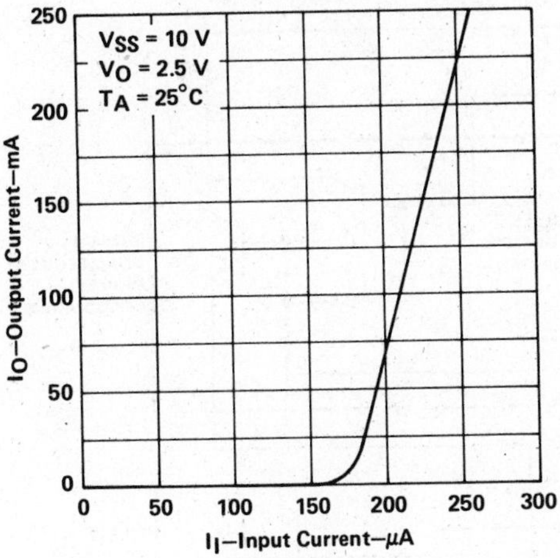
V_{SS} = 10 V
V_O = 2.5 V
T_A = 25°C
250
200
150
100
50
0
0
50
100
150
200
250
300
I_O—Output Current—mA
I_I—Input Current—μA

FIGURE 6

SN75492, SN75492A
OUTPUT CURRENT
vs
INPUT VOLTAGE

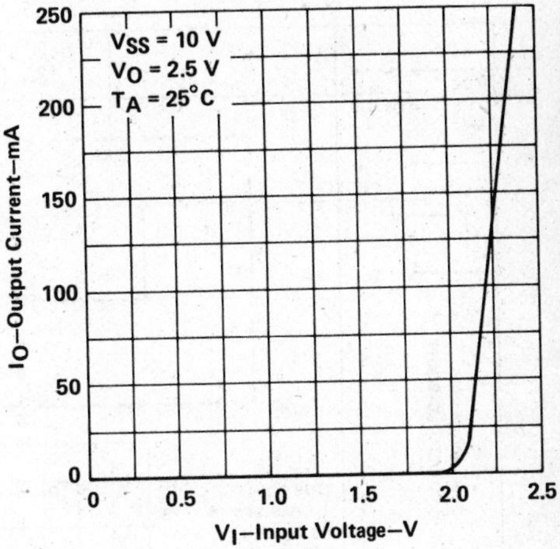
V_{SS} = 10 V
V_O = 2.5 V
T_A = 25°C
250
200
150
100
50
0
0
0.5
1.0
1.5
2.0
2.5
I_O—Output Current—mA
V_I—Input Voltage—V

FIGURE 7

SN75492, SN75492A
LOW-LEVEL OUTPUT VOLTAGE
vs
OUTPUT CURRENT

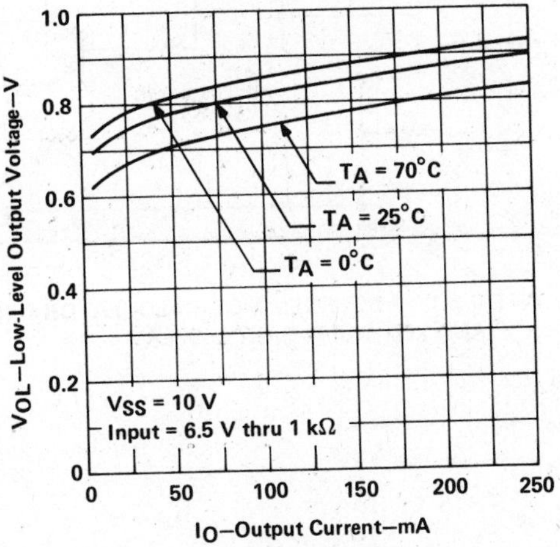
T_A = 70°C
T_A = 25°C
T_A = 0°C
V_{SS} = 10 V
Input = 6.5 V thru 1 kΩ
1.0
0.8
0.6
0.4
0.2
0
0
50
100
150
200
250
V_{OL}—Low-Level Output Voltage—V
I_O—Output Current—mA

FIGURE 8

3
Display Drivers

TYPICAL APPLICATION DATA

Figure 9 is an example of time multiplexing the individual digits in a display to minimize circuitry. Up to twelve digits, each of which use a seven-segment display with decimal point, may be displayed using only two SN75491 and two SN75492 drivers.

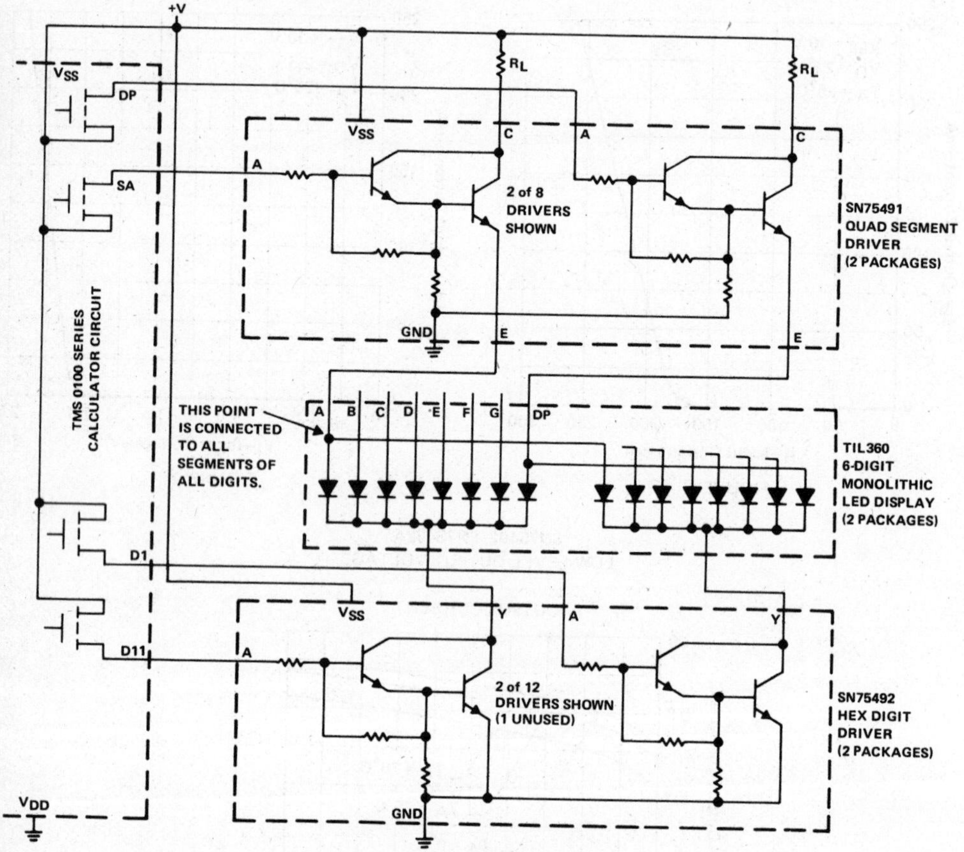
+V
V_{SS}
DP
SA
R_L
V_{SS}
C
A
C
A
2 of 8
DRIVERS
SHOWN
SN75491
QUAD SEGMENT
DRIVER
(2 PACKAGES)
GND
E
E
TMS 0100 SERIES
CALCULATOR CIRCUIT
THIS POINT
IS CONNECTED
TO ALL
SEGMENTS OF
ALL DIGITS.
A B C D E F G DP
TIL360
6-DIGIT
MONOLITHIC
LED DISPLAY
(2 PACKAGES)
D1
D11
V_{SS}
Y
A
Y
A
2 of 12
DRIVERS SHOWN
(1 UNUSED)
SN75492
HEX DIGIT
DRIVER
(2 PACKAGES)
V_{DD}
GND

FIGURE 9. INTERFACING BETWEEN MOS CALCULATOR CIRCUIT AND LED MULTI-DIGIT DISPLAY

TYPICAL APPLICATION DATA

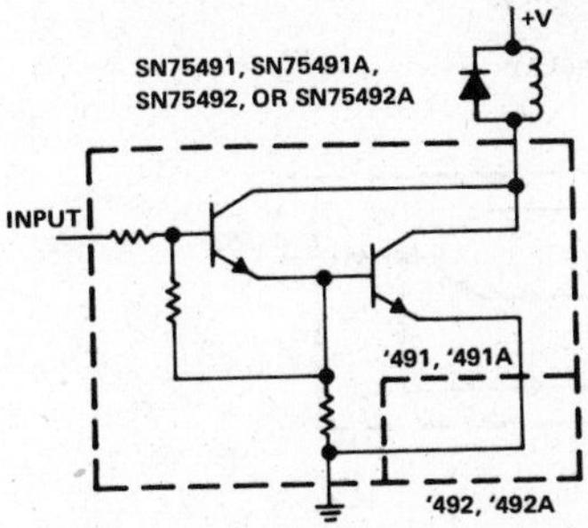
+V
SN75491, SN75491A,
SN75492, OR SN75492A
INPUT
'491, '491A
'492, '492A

FIGURE 10. QUAD OR HEX RELAY DRIVER

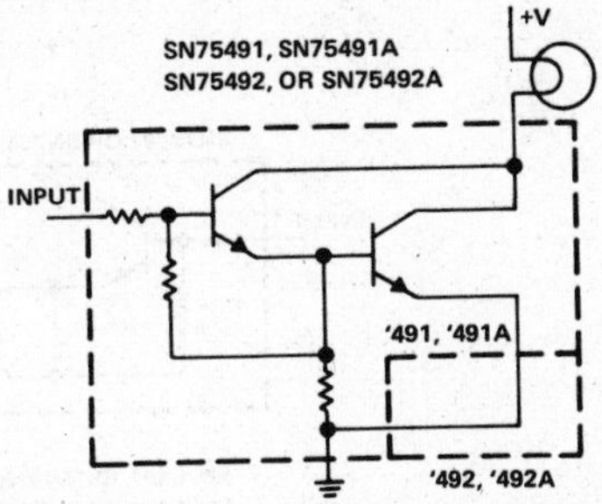
+V
SN75491, SN75491A
SN75492, OR SN75492A
INPUT
'491, '491A
'492, '492A

FIGURE 11. QUAD OR HEX LAMP DRIVER

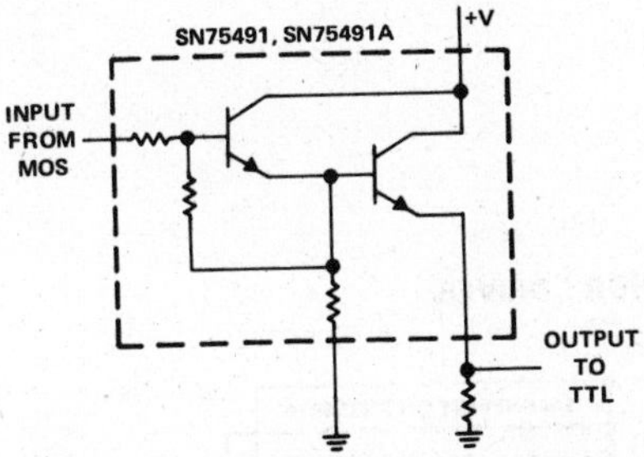
SN75491, SN75491A
+V
INPUT
FROM
MOS
OUTPUT
TO
TTL

FIGURE 12. MOS-TO-TTL LEVEL SHIFTER

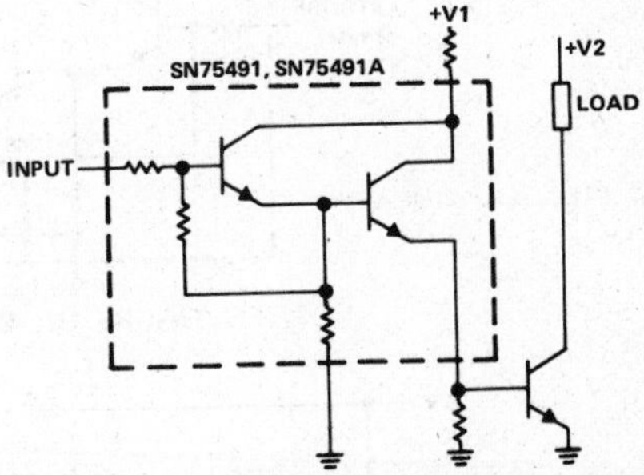
+V1
SN75491, SN75491A
+V2
LOAD
INPUT

FIGURE 13. QUAD HIGH-CURRENT N-P-N TRANSISTOR DRIVER

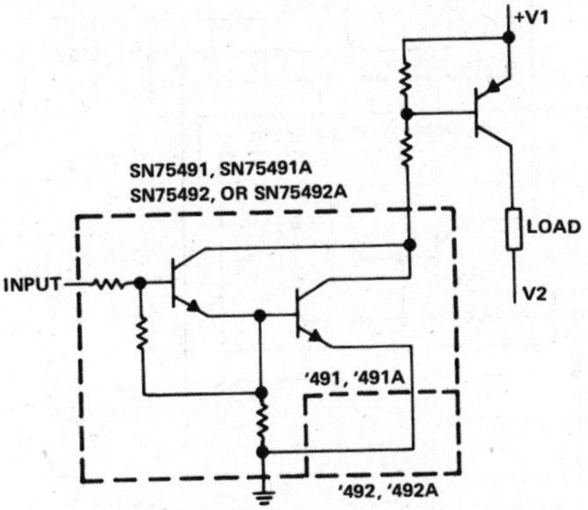
+V1
SN75491, SN75491A
SN75492, OR SN75492A
INPUT
LOAD
V2
'491, '491A
'492, '492A

NOTE A: This circuit may be used as a digit driver for common-mode LED displays.

FIGURE 14. QUAD OR HEX HIGH-CURRENT P-N-P TRANSISTOR DRIVER

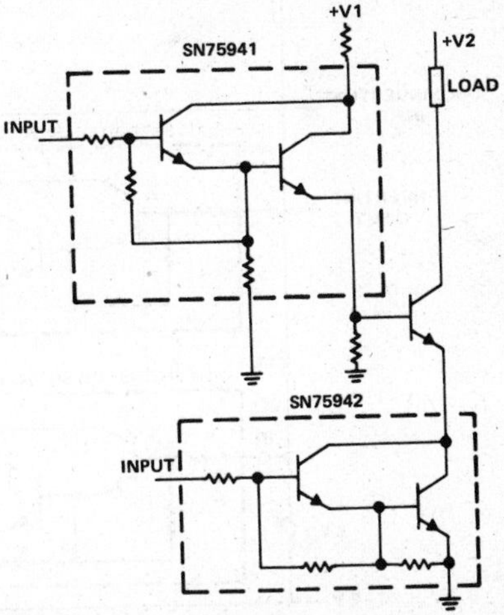
+V1
SN75941
+V2
LOAD
INPUT
SN75942
INPUT

FIGURE 15. BASE/EMITTER SELECT N-P-N TRANSISTOR DRIVER

TYPICAL APPLICATION DATA

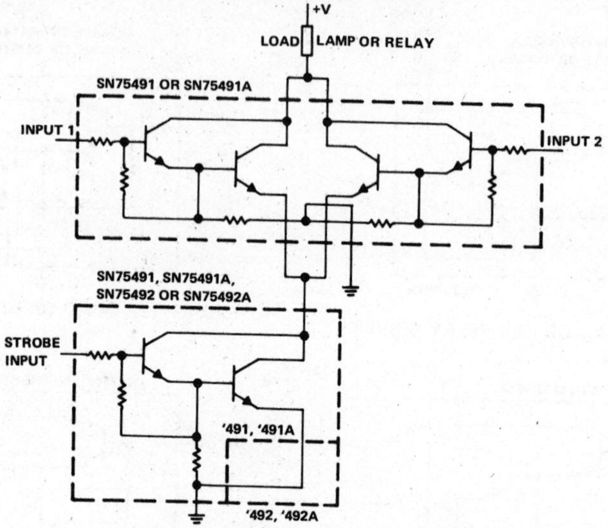
+V
LOAD
LAMP OR RELAY
SN75491 OR SN75491A
INPUT 1
INPUT 2
SN75491, SN75491A,
SN75492 OR SN75492A
STROBE
INPUT
'491, '491A
'492, '492A

FIGURE 16. STROBED "NOR" DRIVER

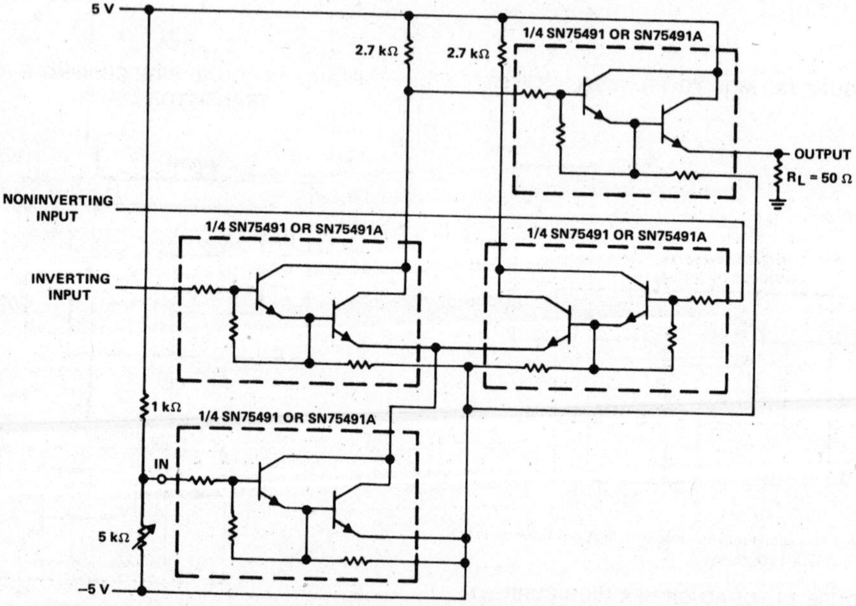
5 V
2.7 kΩ
2.7 kΩ
1/4 SN75491 OR SN75491A
OUTPUT
R_L = 50 Ω
NONINVERTING
INPUT
1/4 SN75491 OR SN75491A
1/4 SN75491 OR SN75491A
INVERTING
INPUT
1 kΩ
1/4 SN75491 OR SN75491A
IN
5 kΩ
−5 V

FIGURE 17. SN75491/SN75491A USED AS AN INTERFACE CIRCUIT BETWEEN THE BALANCED 30-MHz OUTPUT OF AN RF AMPLIFIER AND A COAXIAL CABLE

TEXAS INSTRUMENTS
POST OFFICE BOX 655012 • DALLAS, TEXAS 75265

D1932, MARCH 1983—REVISED JANUARY 1987

- Low Input Current for MOS Compatibility
- Low Voltage Operation
- Low Standby Power
- Display Blanking Capability
- 250-mA Sink Capability
- Low-Voltage Saturating Outputs
- High-Gain Circuits

N
DUAL-IN-LINE PACKAGE
(TOP VIEW)

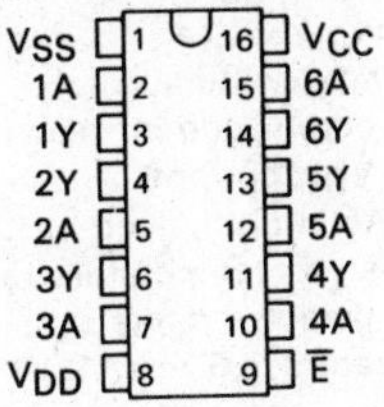
V_{SS} 1
1A 2
1Y 3
2Y 4
2A 5
3Y 6
3A 7
V_{DD} 8
16 V_{CC}
15 6A
14 6Y
13 5Y
12 5A
11 4Y
10 4A
9 $\overline{E}$

description

The SN75494 is designed to be used as an interface between MOS integrated circuits and LEDs in serially addressed multidigit displays. This device is similar in operation to the SN75492, but has several advantages over the earlier circuit. The SN75494 can be operated at lower supply voltages therefore, reducing power consumption. The enable ($\overline{E}$) input is used as a blanking input.

logic symbol†

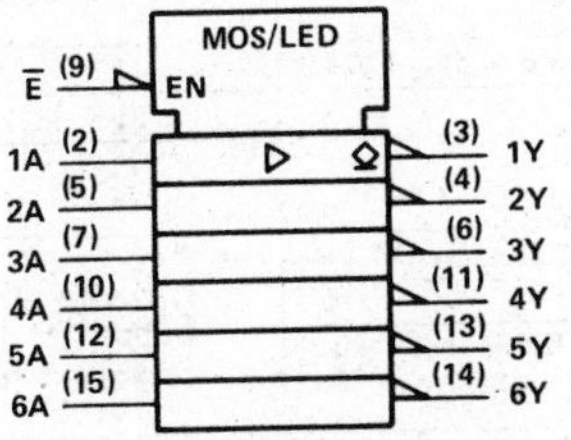
MOS/LED
$\overline{E}$ (9)
EN
1A (2)
2A (5)
3A (7)
4A (10)
5A (12)
6A (15)
(3) 1Y
(4) 2Y
(6) 3Y
(11) 4Y
(13) 5Y
(14) 6Y

†This symbol is in accordance with ANSI/IEEE Std 91-1984 and IEC Publication 617-12.

logic diagram (positive logic)

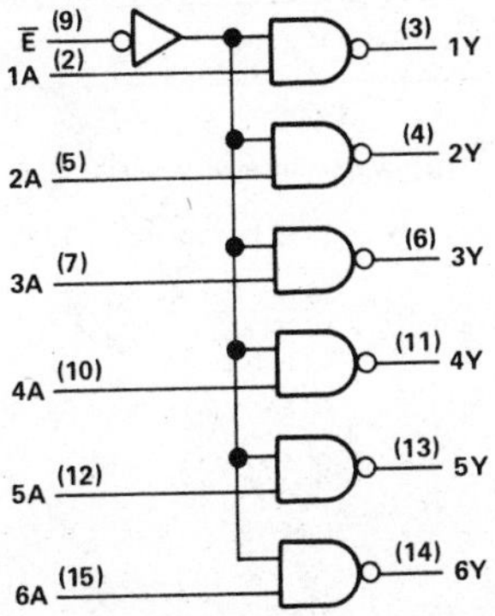
$\overline{E}$ (9)
1A (2)
2A (5)
3A (7)
4A (10)
5A (12)
6A (15)
(3) 1Y
(4) 2Y
(6) 3Y
(11) 4Y
(13) 5Y
(14) 6Y

schematic

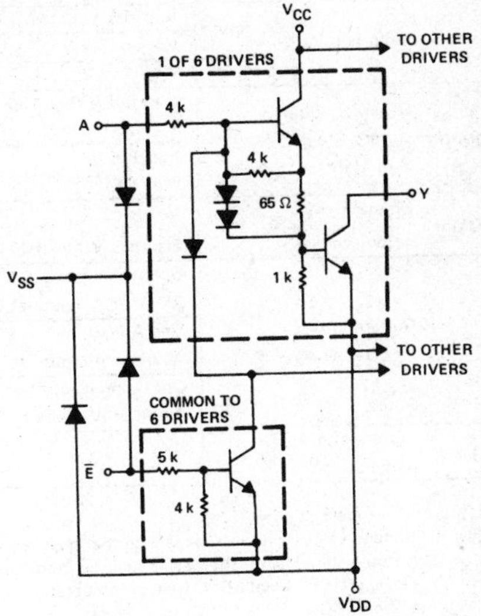
V_{CC}
TO OTHER DRIVERS
1 OF 6 DRIVERS
A
4 k
4 k
65 Ω
Y
1 k
V_{SS}
TO OTHER DRIVERS
COMMON TO 6 DRIVERS
$\overline{E}$
5 k
4 k
V_{DD}

NOTES: A. The V_{SS} terminal must be connected to the most positive voltage that is applied to the device.
B. Resistor values shown are nominal and in ohms.

TEXAS INSTRUMENTS
POST OFFICE BOX 655012 • DALLAS, TEXAS 75265

absolute maximum ratings over operating free-air temperature (unless otherwise noted)

Supply voltage, V_{CC} (see Note 1) .... 10 V
Supply voltage, V_{SS} (see Note 2) .... 10 V
Input voltage .... V_{SS}
Off-state output voltage .... 10 V
Continuous output current (each driver) .... 250 mA
Continuous V_{DD} current .... 600 mA
Continuous total dissipation at (or below) 25°C free-air temperature (see Note 3) .... 800 mW
Operating free-air temperature range .... 0°C to 70°C
Storage temperature range .... −65°C to 150°C
Lead temperature 1,6 mm (1/16 inch) from case for 10 seconds .... 260°C

recommended operating conditions

	MIN	MAX	UNIT
Supply voltage, V_{CC}	3.2	8.8	V
Supply voltage, V_{SS}	6.5	8.8	V
Operating free-air temperature, T_A	0	70	°C

electrical characteristics, V_{CC} = 8.8 V, V_{SS} = 8.8 V, T_A = 0°C to 70°C (unless otherwise noted)

PARAMETER			TEST CONDITIONS	MIN	TYP†	MAX	UNIT
I_I	Input current	A input	A at 8.8 V, $\bar{E}$ at 8.8 V		2	3	mA
			V_{CC} = 3.2 V, A at 8.8 V, $\bar{E}$ to 8.8 V thru 100 kΩ		1.8	2.5	
		$\bar{E}$ input	V_{CC} = 3.2 V, $\bar{E}$ at 8.8 V		1.6	2.5	
$I_{O(off)}$	Off-state output current (from Y to V_{DD})		A to 8.8 V thru 100 kΩ, $\bar{E}$ at 0 V, Y at 10 V		1	200	μA
			A at 8.8 V, $\bar{E}$ to 6.5 V thru 1 kΩ, Y at 10 V		1	100	
$V_{O(on)}$	On-state output voltage		V_{CC} = 3.2 V, V_{SS} = 6.5 V, A to 6.5 V thru 1 kΩ, $\bar{E}$ to 8.8 V thru 100 kΩ, I_{OL} = 250 mA		0.25	0.4	V
I_{CC}	Current into V_{CC} terminal		One A input to 8.8 V thru 100 kΩ, $\bar{E}$ at 0 V, All other A inputs at 0 V		10	500	μA
			One A input at 8.8 V, $\bar{E}$ to 6.5 V thru 1 kΩ, All other A inputs at 0 V		60	500	
			One A input at 8.8 V, $\bar{E}$ at 0 V, All other A inputs at 0 V		11	20	mA
I_{SS}	Current into V_{SS} terminal		V_{CC} = 3.2 V, $\bar{E}$ at 0 V, All A inputs at 0 V		10	500	μA

† All typical values are at T_A = 25°C.

NOTES: 1. Voltage values are with respect to the most negative device terminal, V_{DD}, unless otherwise noted.
2. No other terminal on the device may be more positive than V_{SS}.
3. For operation above 25°C free-air temperature, derate linearly from 800 mW at 63°C to 736 mW at 70°C at the rate of 9.2 mW/°C.

TEXAS INSTRUMENTS
POST OFFICE BOX 655012 • DALLAS, TEXAS 75265

D2471, DECEMBER 1984—REVISED NOVEMBER 1986

- Controls 32 Electrodes
- 100-V Totem-Pole Outputs
- Low Stand-by Power Consumption
- All Outputs Contain Sink and Source Clamp Diodes
- 15 mA Steady-State Output Current
- Rugged DMOS Outputs
- CMOS Inputs
- Dependable Texas Instruments Quality and Reliability
- Direct Replacement for SN55500D

JD PACKAGE
(TOP VIEW)

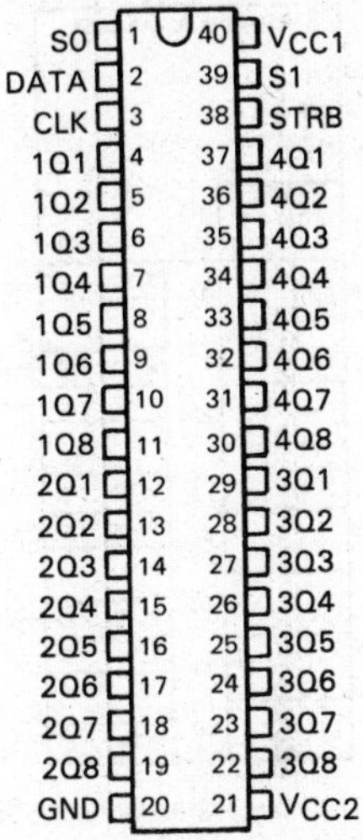
S0 1
DATA 2
CLK 3
1Q1 4
1Q2 5
1Q3 6
1Q4 7
1Q5 8
1Q6 9
1Q7 10
1Q8 11
2Q1 12
2Q2 13
2Q3 14
2Q4 15
2Q5 16
2Q6 17
2Q7 18
2Q8 19
GND 20
40 VCC1
39 S1
38 STRB
37 4Q1
36 4Q2
35 4Q3
34 4Q4
33 4Q5
32 4Q6
31 4Q7
30 4Q8
29 3Q1
28 3Q2
27 3Q3
26 3Q4
25 3Q5
24 3Q6
23 3Q7
22 3Q8
21 VCC2

FD PACKAGE
(TOP VIEW)

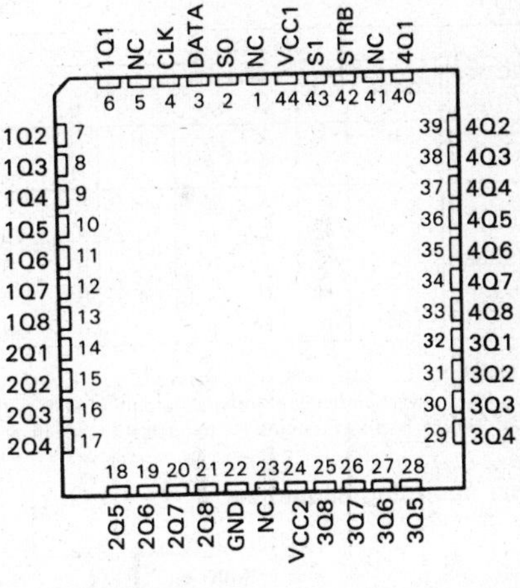
1Q1 NC CLK DATA S0 NC VCC1 S1 STRB NC 4Q1
6 5 4 3 2 1 44 43 42 41 40
1Q2 7
1Q3 8
1Q4 9
1Q5 10
1Q6 11
1Q7 12
1Q8 13
2Q1 14
2Q2 15
2Q3 16
2Q4 17
39 4Q2
38 4Q3
37 4Q4
36 4Q5
35 4Q6
34 4Q7
33 4Q8
32 3Q1
31 3Q2
30 3Q3
29 3Q4
18 19 20 21 22 23 24 25 26 27 28
2Q5 2Q6 2Q7 2Q8 GND NC VCC2 3Q8 3Q7 3Q6 3Q5

NC—No internal connection

description

The SN55500E is a monolithic BIDFET† integrated circuit designed to perform the line select operation of a matrix-addressable display. The device inputs are diode-clamped CMOS inputs.

The outputs of the driver are normally low and can be selectively switched high when the strobe input is low. Selection of the outputs is achieved through the data, S0, and S1 inputs. The 8-bit data stored internally in the serial register is inverted and sent to one of four output sections by the 2-line to 4-line decoder. All other outputs remain low. Internal circuits provide a high-current pulse to the level-shifting circuit during positive output transitions. When the output transition is complete, the low steady-state current reduces the circuits stand-by power consumption. All outputs contain clamp diodes to the V_{CC2} and GND supply inputs.

The SN55500E is characterized for operation over the full military temperature range of −55°C to 125°C.

† BIDFET—Bipolar, double-diffused, N-channel and P-channel MOS transistors on same chip — patented process.

TEXAS INSTRUMENTS
POST OFFICE BOX 655012 • DALLAS, TEXAS 75265

logic symbol†

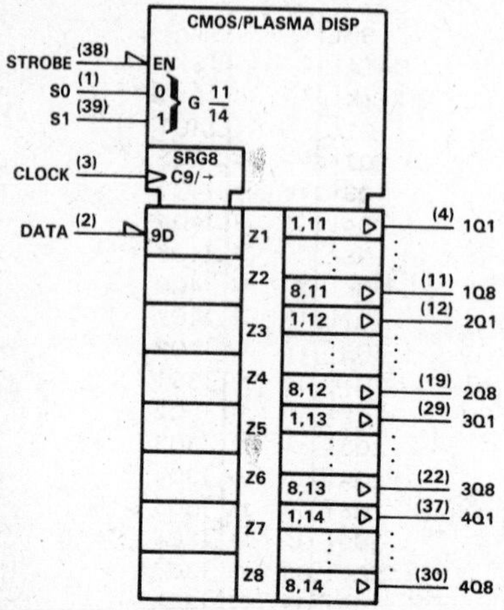
CMOS/PLASMA DISP
STROBE (38) EN
S0 (1) 0
S1 (39) 1
G 11/14
SRG8
CLOCK (3) C9/→
DATA (2) 9D
Z1
Z2
Z3
Z4
Z5
Z6
Z7
Z8
1,11
8,11
1,12
8,12
1,13
8,13
1,14
8,14
(4) 1Q1
(11) 1Q8
(12) 2Q1
(19) 2Q8
(29) 3Q1
(22) 3Q8
(37) 4Q1
(30) 4Q8

†This symbol is in accordance with ANSI/IEEE Std 91-1984 and IEC Publication 617-12.

Pin numbers shown are for the JD package.

functional block diagram (positive logic)

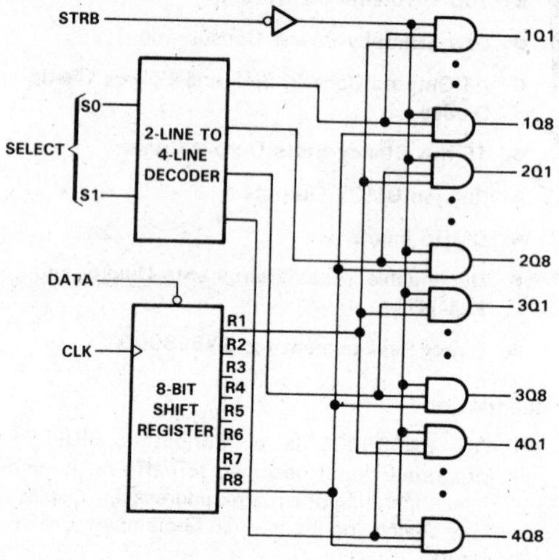
STRB
SELECT
S0
S1
2-LINE TO 4-LINE DECODER
DATA
CLK
8-BIT SHIFT REGISTER
R1
R2
R3
R4
R5
R6
R7
R8
1Q1
1Q8
2Q1
2Q8
3Q1
3Q8
4Q1
4Q8

FUNCTION TABLE

FUNCTION	INPUTS					OUTPUTS							
	DATA	CLK	SELECT		STRB	SHIFT REGISTER							
			S1	S0		R1	R2	R3 ... R8	1Q1 ... 1Q8	2Q1 ... 2Q8	3Q1 ... 3Q8	4Q1 ... 4Q8	
LOAD	H	↑	X	X	H	L	$R1_n$	$R2_n$ ... $R7_n$	L ... L	L ... L	L ... L	L ... L	
	L	↑	X	X	H	H	$R1_n$	$R2_n$ ... $R7_n$	L ... L	L ... L	L ... L	L ... L	
STROBE	X	X	X	X	H	$R1_n$	$R2_n$	$R3_n$ ... $R8_n$	L ... L	L ... L	L ... L	L ... L	
	X	H	L	L	L	$R1_n$	$R2_n$	$R3_n$ ... $R8_n$	R1 ... R8	L ... L	L ... L	L ... L	
	X	H	L	H	L	$R1_n$	$R2_n$	$R3_n$ ... $R8_n$	L ... L	R1 ... R8	L ... L	L ... L	
	X	H	H	L	L	$R1_n$	$R2_n$	$R3_n$ ... $R8_n$	L ... L	L ... L	R1 ... R8	L ... L	
	X	H	H	H	L	$R1_n$	$R2_n$	$R3_n$ ... $R8_n$	L ... L	L ... L	L ... L	R1 ... R8	

H = high level, L = low level, X = irrelevant, ↑ = low-to-high transition.
R1 . . . R8 = levels currently at internal outputs of shift registers one through eight, respectively.
$R1_n$ . . . $R8_n$ = levels at outputs R1 through R8 respectively, before the most recent ↑ transition of the clock.

typical operating sequence

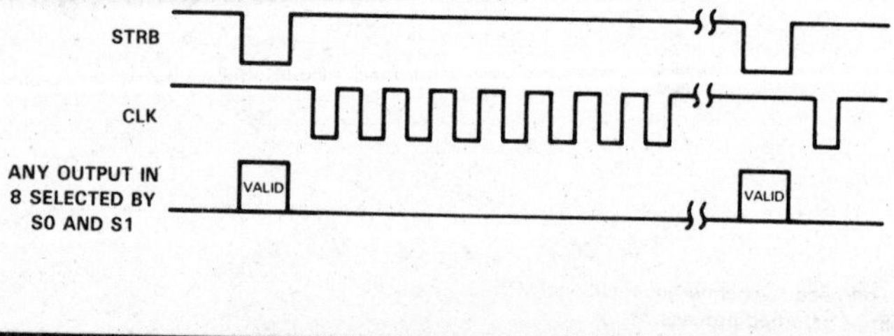
STRB
CLK
ANY OUTPUT IN 8 SELECTED BY S0 AND S1
VALID
VALID

schematics of inputs and outputs

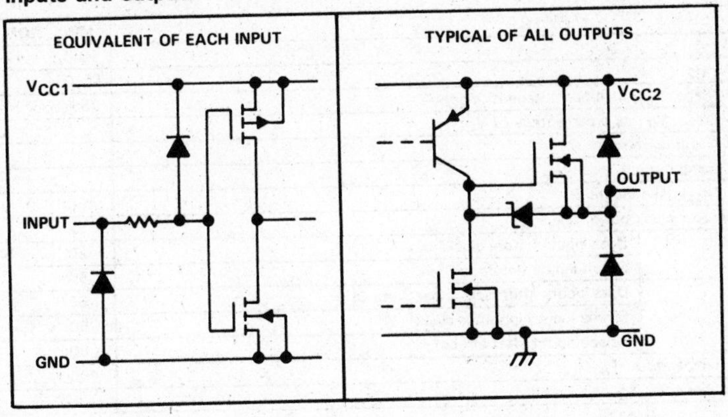
EQUIVALENT OF EACH INPUT
VCC1
INPUT
GND
TYPICAL OF ALL OUTPUTS
VCC2
OUTPUT
GND

absolute maximum ratings over operating free-air temperature range (unless otherwise noted)

Supply voltage, V_{CC1} (see Note 1)	13.8 V
Supply voltage, V_{CC2}	100 V
Input voltage	V_{CC1} + 0.3 V
Continuous total dissipation at (or below) 25 °C free-air temperature (see Note 2)	1825 mW
Operating free-air temperature range	−55 °C to 125 °C
Storage temperature range	−65 °C to 150 °C
Case temperature for 60 seconds: FD package	260 °C
Lead temperature 1,6 mm (1/16 inch) from case for 60 seconds: JD package	300 °C

NOTES: 1. Voltage values are with respect to network ground terminal.
2. For operation above 25 °C free-air temperature, see Dissipation Derating Table.

DISSIPATION DERATING TABLE

PACKAGE	POWER RATING	DERATING FACTOR	ABOVE T_A
FD	1825 mW	14.6 mW/°C	25 °C
JD	1825 mW	22 mW/°C	67 °C

recommended operating conditions

		MIN	NOM	MAX	UNIT
Supply voltage, V_{CC1}		10.8	12	13.2	V
Supply voltage, V_{CC2}		0		100	V
High-level input voltage, V_{IH}, as a percentage of V_{CC1}		75%			
Low-level input voltage, V_{IL}, as a percentage of V_{CC1}				25%	
High-level output clamp current				20	mA
Low-level output clamp current				−20	mA
Clock frequency, f_{clock} (see Figure 2)		0		8	MHz
Duration of high or low clock pulse, t_w		62			ns
Setup time, t_{su}	Data inputs before clock↑	20			ns
	Select inputs before strobe↓	50			
Hold time, t_h	Data inputs after clock↑ (see Note 3)	50			ns
	Strobe input high after clock↑	50			
	Select inputs after strobe↑	50			
Operating free-air temperature, T_A		−55			°C
Operating case temperature, T_C				125	°C

NOTE 3: For operation above 25°C junction temperature, refer to Figure 2.

electrical characteristics over recommended operating temperature range (unless otherwise noted)

PARAMETER		TEST CONDITIONS		MIN	TYP†	MAX	UNIT
V_{IK}	Input clamp voltage	V_{CC1} = 12 V,	I_I = −12 mA		−1	−1.5	V
V_{OH}	High-level output voltage	V_{CC1} = 13.2 V, V_{CC2} = 100 V	I_{OH} = −1 mA	94	97.5		V
			I_{OH} = −10 mA	92	94.5		
			I_{OH} = −15 mA	90	93.5		
V_{OL}	Low-level output voltage	V_{CC1} = 13.2 V, V_{CC2} = 100 V	I_{OL} = 1 mA		0.85	2	V
			I_{OL} = 10 mA		2	4	
			I_{OL} = 15 mA		2.75	5	
V_{OK}	Output clamp voltage	V_{CC2} = 0	I_O = 20 mA		1	2.5	V
			I_O = −20 mA		−1.2	−2.5	
I_{IH}	High-level input current	V_{CC1} = 13.2 V,	$V_I = V_{IH}$ min			1	μA
I_{IL}	Low-level input current	V_{CC1} = 13.2 V,	$V_I = V_{IL}$ max			−1	μA
I_{CC1}	Supply current	V_{CC1} = 13.2 V,	V_{CC2} = 100 V		0.05	1	mA
I_{CC2}	Supply current	V_{CC2} = 100 V			1	5	mA

†All typical values are at V_{CC} = 12 V, T_A = 25°C.

switching characteristics, V_{CC1} = 12 V, V_{CC2} = 100 V, T_A = 25°C

PARAMETER		TEST CONDITIONS	MIN	MAX	UNIT
t_{DHL}	Delay time, high-to-low-level output from strobe input	C_L = 30 pF, See Figure 1		250	ns
t_{DLH}	Delay time, low-to-high-level output from strobe input			450	ns
t_{THL}	Transition time, high-to-low-level output			200	ns
t_{TLH}	Transition time, low-to-high-level output			300	ns

TEXAS INSTRUMENTS
POST OFFICE BOX 655012 • DALLAS, TEXAS 75265

PARAMETER MEASUREMENT INFORMATION

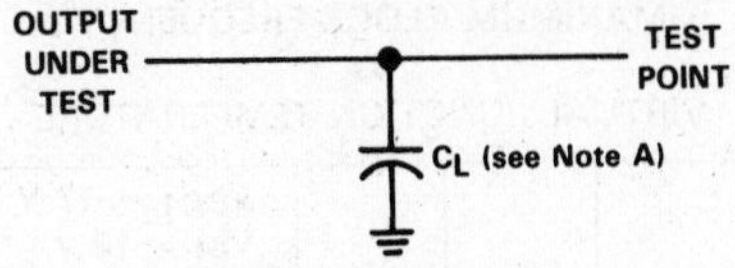
OUTPUT
UNDER
TEST
TEST
POINT
C_L (see Note A)

LOAD TEST CIRCUIT

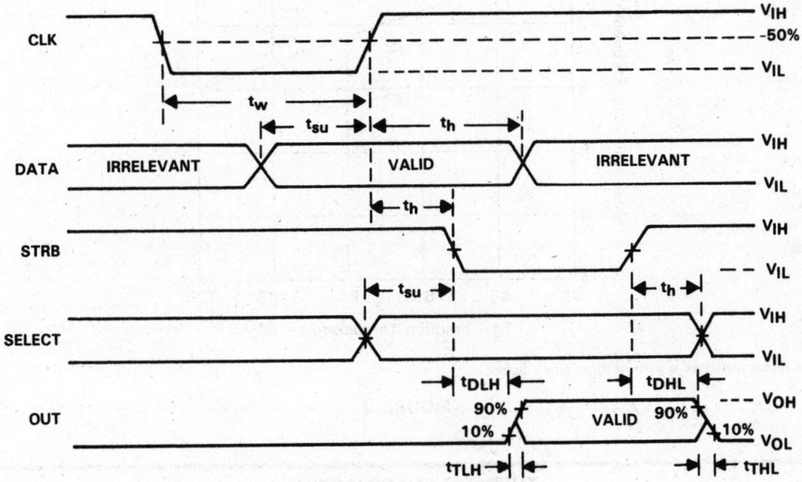
CLK
DATA
STRB
SELECT
OUT
V_{IH}
-50%
V_{IL}
t_w
t_{su}
t_h
IRRELEVANT
VALID
IRRELEVANT
V_{OH}
V_{OL}
t_{DLH}
t_{DHL}
90%
10%
VALID
90%
10%
t_{TLH}
t_{THL}

VOLTAGE WAVEFORMS

NOTE A. C_L includes probe and jig capacitance.

FIGURE 1. SWITCHING CHARACTERISTICS

TYPICAL CHARACTERISTICS

MAXIMUM CLOCK FREQUENCY
vs
VIRTUAL JUNCTION TEMPERATURE

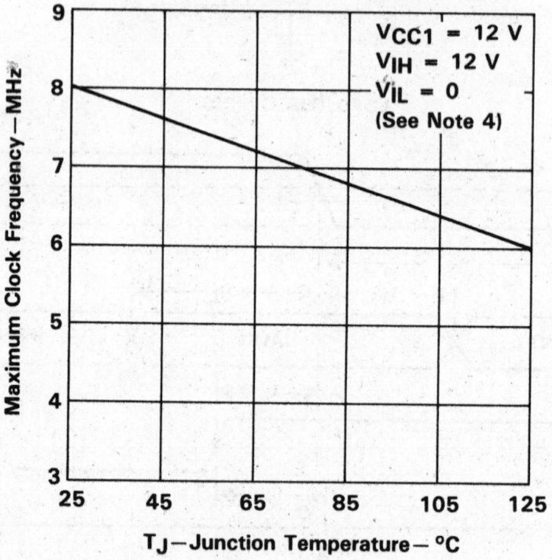
V_{CC1} = 12 V
V_{IH} = 12 V
V_{IL} = 0
(See Note 4)
Maximum Clock Frequency — MHz
9
8
7
6
5
4
3
25
45
65
85
105
125
T_J—Junction Temperature— °C

NOTE 4: This curve assumes a symmetrical clock pulse.

FIGURE 2

THERMAL INFORMATION

junction temperature formula

$$T_J = T_A + P_D R_{\theta JA}$$

$$T_J = T_C + P_D R_{\theta JC}$$

where

T_J = virtual junction temperature
T_A = free-air temperature
P_D = average device power dissipation
R_θ = thermal resistance (junction-to-air, $R_{\theta JA}$, or junction-to-case, $R_{\theta JC}$)

PACKAGE TYPE	$R_{\theta JA}$	$R_{\theta JC}$
FD 44-pin ceramic	68°C/W	20°C/W
JD 40-pin ceramic	45°C/W	12°C/W

SN65500E, SN75500E
AC PLASMA DISPLAY DRIVERS

D2471, DECEMBER 1985

- Controls 32 Electrodes
- 100-V Totem-Pole Outputs
- Low Stand-by Power Consumption
- All Outputs Contain Sink and Source Clamp Diodes
- 15 mA Steady-State Output Current
- Rugged DMOS Outputs
- CMOS Inputs
- Direct Replacement for SN75500A

N PACKAGE
(TOP VIEW)

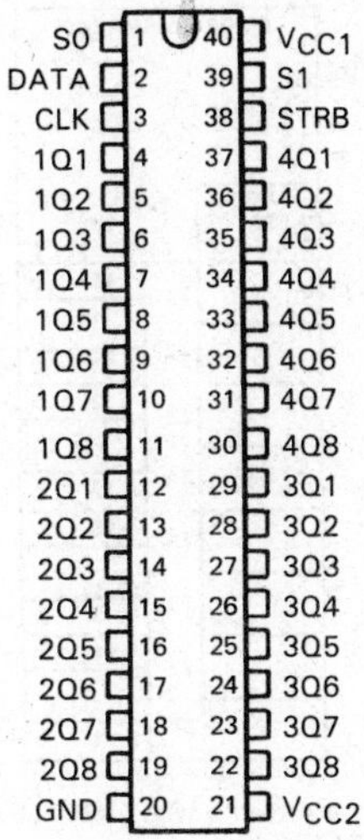
S0 1
DATA 2
CLK 3
1Q1 4
1Q2 5
1Q3 6
1Q4 7
1Q5 8
1Q6 9
1Q7 10
1Q8 11
2Q1 12
2Q2 13
2Q3 14
2Q4 15
2Q5 16
2Q6 17
2Q7 18
2Q8 19
GND 20
40 VCC1
39 S1
38 STRB
37 4Q1
36 4Q2
35 4Q3
34 4Q4
33 4Q5
32 4Q6
31 4Q7
30 4Q8
29 3Q1
28 3Q2
27 3Q3
26 3Q4
25 3Q5
24 3Q6
23 3Q7
22 3Q8
21 VCC2

description

The SN65500E and SN75500E are monolithic BIDFET† integrated circuits designed to perform the line select operation of a matrix-addressable display. The device inputs are diode-clamped CMOS inputs.

The outputs of these drivers are normally low and can be selectively switched high when the strobe input is low. Selection of the outputs is achieved through the data, S0, and S1 inputs. The 8-bit data stored internally in the serial register is inverted and sent to one of four output sections by the 2-line to 4-line decoder. All other outputs remain low. Internal circuits provide a high-current pulse to the level-shifting circuit during positive output transitions. When the output transition is complete, the low steady-state current reduces the circuit's standby power consumption. All outputs contain clamp diodes to the V_{CC2} and GND supply inputs.

The SN65500E is characterized for operation from −40°C to 85°C. The SN75500E is characterized for operation from 0°C to 70°C.

FN PACKAGE
(TOP VIEW)

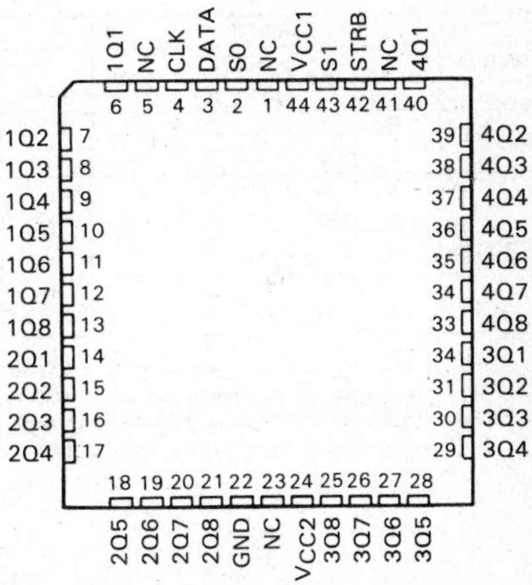
1Q1 NC CLK DATA S0 NC VCC1 S1 STRB NC 4Q1
6 5 4 3 2 1 44 43 42 41 40
1Q2 7
1Q3 8
1Q4 9
1Q5 10
1Q6 11
1Q7 12
1Q8 13
2Q1 14
2Q2 15
2Q3 16
2Q4 17
39 4Q2
38 4Q3
37 4Q4
36 4Q5
35 4Q6
34 4Q7
33 4Q8
34 3Q1
31 3Q2
30 3Q3
29 3Q4
18 19 20 21 22 23 24 25 26 27 28
2Q5 2Q6 2Q7 2Q8 GND NC VCC2 3Q8 3Q7 3Q6 3Q5

NC—No internal connection

†BIDFET—Bipolar, double-diffused, N-channel and P-channel MOS transistors on same chip — patented process.

logic symbol†

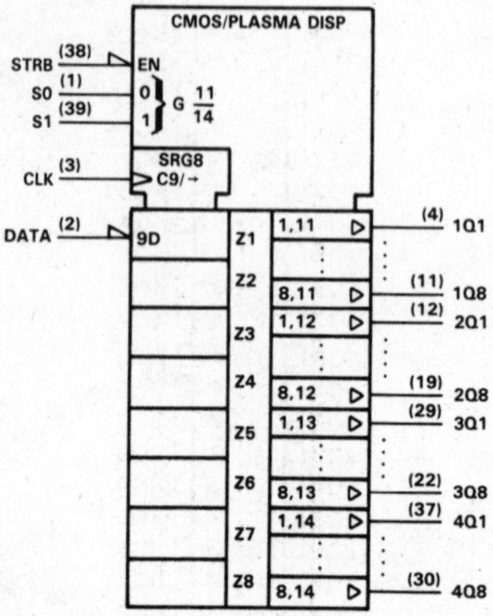
CMOS/PLASMA DISP
STRB
(38)
EN
S0
(1)
S1
(39)
G 11/14
CLK
(3)
SRG8
C9/→
DATA
(2)
9D
Z1
Z2
Z3
Z4
Z5
Z6
Z7
Z8
1,11
8,11
1,12
8,12
1,13
8,13
1,14
8,14
(4) 1Q1
(11) 1Q8
(12) 2Q1
(19) 2Q8
(29) 3Q1
(22) 3Q8
(37) 4Q1
(30) 4Q8

†This symbol is in accordance with ANSI/IEEE Std 91-1984 and IEC Publication 617-12.
Pin numbers shown are for the N package.

functional block diagram (positive logic)

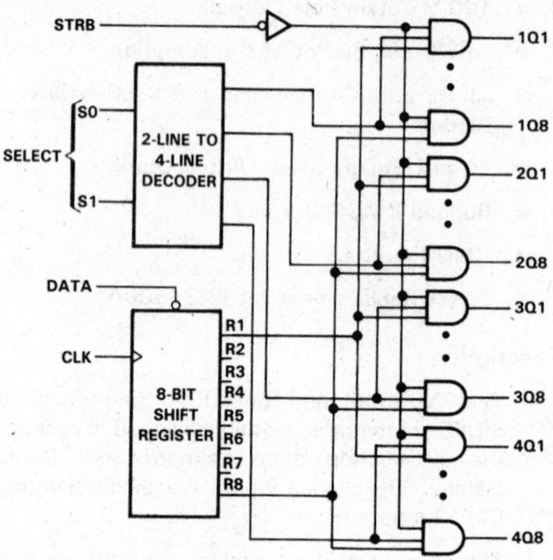
STRB
SELECT
S0
S1
2-LINE TO 4-LINE DECODER
DATA
CLK
8-BIT SHIFT REGISTER
R1
R2
R3
R4
R5
R6
R7
R8
1Q1
1Q8
2Q1
2Q8
3Q1
3Q8
4Q1
4Q8

FUNCTION TABLE

FUNCTION	INPUTS					OUTPUTS						
	DATA	CLK	SELECT S1	SELECT S0	STRB	SHIFT REGISTER R1	R2	R3 ... R8	1Q1 ... 1Q8	2Q1 ... 2Q8	3Q1 ... 3Q8	4Q1 ... 4Q8
LOAD	H	↑	X	X	H	L	$R1_n$	$R2_n$ ... $R7_n$	L ... L	L ... L	L ... L	L ... L
	L	↑	X	X	H	H	$R1_n$	$R2_n$ ... $R7_n$	L ... L	L ... L	L ... L	L ... L
STROBE	X	X	X	X	H	$R1_n$	$R2_n$	$R3_n$ ... $R8_n$	L ... L	L ... L	L ... L	L ... L
	X	H	L	L	L	$R1_n$	$R2_n$	$R3_n$ ... $R8_n$	R1 ... R8	L ... L	L ... L	L ... L
	X	H	L	H	L	$R1_n$	$R2_n$	$R3_n$ ... $R8_n$	L ... L	R1 ... R8	L ... L	L ... L
	X	H	H	L	L	$R1_n$	$R2_n$	$R3_n$ ... $R8_n$	L ... L	L ... L	R1 ... R8	L ... L
	X	H	H	H	L	$R1_n$	$R2_n$	$R3_n$ ... $R8_n$	L ... L	L ... L	L ... L	R1 ... R8

H = high level, L = low level, X = irrelevant, ↑ = low-to-high transition.
R1 . . . R8 = levels currently at internal outputs of shift registers one through eight, respectively.
$R1_n$ . . . $R8_n$ = levels at shift-register outputs R1 through R8, respectively, before the most recent ↑ transition of the clock.

TEXAS INSTRUMENTS
POST OFFICE BOX 655012 • DALLAS, TEXAS 75265

typical operating sequence

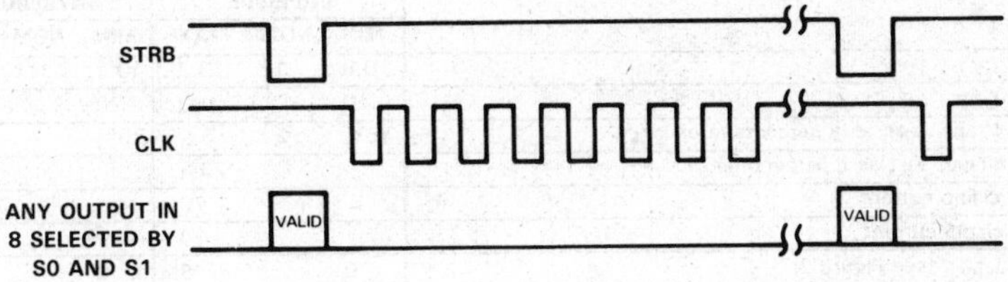
STRB
CLK
ANY OUTPUT IN 8 SELECTED BY S0 AND S1
VALID
VALID

schematics of inputs and outputs

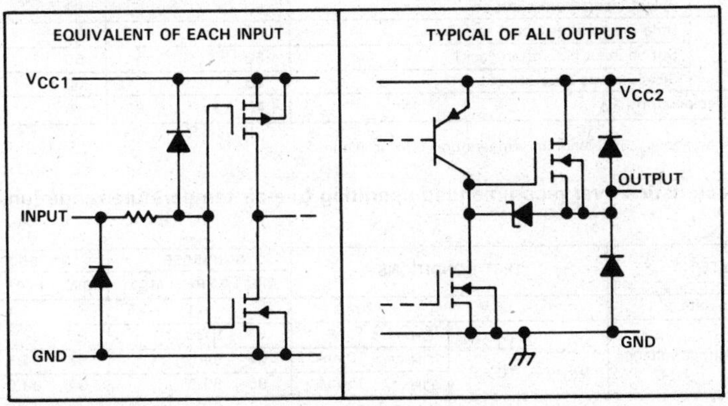
EQUIVALENT OF EACH INPUT
VCC1
INPUT
GND
TYPICAL OF ALL OUTPUTS
VCC2
OUTPUT
GND

absolute maximum ratings over operating free-air temperature range (unless otherwise noted)

Supply voltage, V_{CC1} (see Note 1) .... 15 V
Supply voltage, V_{CC2} .... 100 V
Input voltage .... V_{CC1} +0.3 V
Continuous total dissipation at (or below) 25 °C free-air temperature (see Note 2):
FN package .... 1775 mW
N package .... 1275 mW
Operating free-air temperature range: SN65500E .... −40 °C to 85 °C
SN75500E .... 0 °C to 70 °C
Storage temperature .... −65 °C to 150 °C
Lead temperature 1,6 mm (1/16 inch) from case for 10 seconds: N package .... 260 °C
Case temperature for 10 seconds: FN package .... 260 °C

NOTES: 1. Voltage values are with respect to network ground terminal.
2. For operation above 25 °C free-air temperature, see Dissipation Derating Table.

DISSIPATION DERATING TABLE

PACKAGE	POWER RATING	DERATING FACTOR	ABOVE T_A
FN	1775	14.2	25 °C
N	1275	10.2	25 °C

recommended operating conditions

		SN65500E			SN75500E			UNIT
		MIN	NOM	MAX	MIN	NOM	MAX	
Supply voltage, V_{CC1}		10.8	12	13.2	10.8	12	13.2	V
Supply voltage, V_{CC2}		0		100	0		100	V
High-level input voltage, V_{IH}, as a percentage of V_{CC1}		75%			75%			
Low-level input voltage, V_{IL}, as a percentage of V_{CC1}				25%			25%	
High-level output clamp current				20			20	mA
Low-level output clamp current				−20			−20	mA
Clock frequency, f_{clock} (see Figure 2)		0		8	0		8	MHz
Duration of high or low clock pulse, t_w		62			62			ns
Setup time, t_{su}	Data inputs before clock↑	20			20			ns
	Select inputs before strobe↓	50			50			
Hold time, t_h	Data inputs after clock↑ (see Note 3)	50			50			ns
	Strobe input high after clock↑	50			50			
	Select inputs after strobe↑	50			50			
Operating free-air temperature, T_A		−40		85	0		70	°C

NOTE 3: For operation above 25°C junction temperature, refer to Figure 2.

electrical characteristics over recommended operating free-air temperature range (unless otherwise noted)

PARAMETER		TEST CONDITIONS		SN65500E			SN75500E			UNIT
				MIN	TYP†	MAX	MIN	TYP†	MAX	
V_{IK}	Input clamp voltage	V_{CC1} = 12 V,	I_I = −12 mA		−1	−1.5		−1	−1.5	V
V_{OH}	High-level output voltage	V_{CC1} = 13.2 V, V_{CC2} = 100 V	I_{OH} = −1 mA	94	97.5		95	97.5		V
			I_{OH} = −10 mA	92	94.5		93	94.5		
			I_{OH} = −15 mA	90	93.5		91	93.5		
V_{OL}	Low-level output voltage	V_{CC1} = 13.2 V, V_{CC2} = 100 V	I_{OL} = 1 mA		0.85	2		0.85	2	V
			I_{OL} = 10 mA		2	4		2	4	
			I_{OL} = 15 mA		2.75	5		2.75	5	
V_{OK}	Output clamp voltage	V_{CC2} = 0	I_O = 20 mA		1	2.5		1	2.5	V
			I_O = −20 mA		−1.2	−2.5		−1.2	−2.5	
I_{IH}	High-level input current	V_{CC1} = 13.2 V,	V_I = V_{IH} min			1			1	μA
I_{IL}	Low-level input current	V_{CC1} = 13.2 V,	V_I = V_{IL} max			−1			−1	μA
I_{CC1}	Supply current	V_{CC1} = 13.2 V,	V_{CC2} = 100 V		0.05	1		0.05	1	mA
I_{CC2}	Supply current	V_{CC2} = 100 V			1	5		1	3	mA

†All typical values are at V_{CC1} = 12 V, T_A = 25°C.

switching characteristics, V_{CC1} = 12 V, V_{CC2} = 100 V, T_A = 25°C

PARAMETER		TEST CONDITIONS	MIN	MAX	UNIT
t_{DHL}	Delay time, high-to-low-level output from strobe input	C_L = 30 pF, See Figure 1		250	ns
t_{DLH}	Delay time, low-to-high-level output from strobe input			450	ns
t_{THL}	Transition time, high-to-low-level output			200	ns
t_{TLH}	Transition time, low-to-high-level output			300	ns

PARAMETER MEASUREMENT INFORMATION

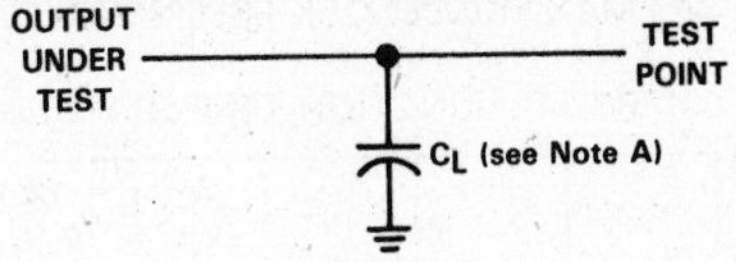
OUTPUT
UNDER
TEST
TEST
POINT
C_L (see Note A)

LOAD TEST CIRCUIT

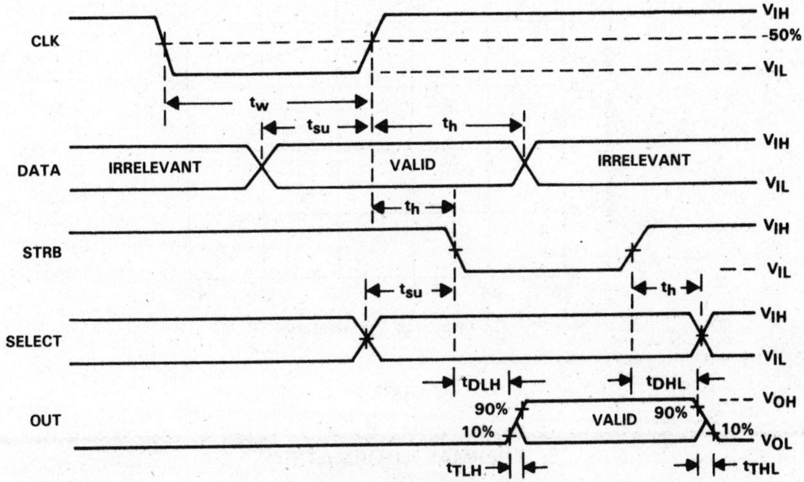
CLK
DATA
STRB
SELECT
OUT
V_{IH}
-50%
V_{IL}
t_w
t_{su}
t_h
IRRELEVANT
VALID
IRRELEVANT
t_{DLH}
t_{DHL}
90%
10%
VALID
V_{OH}
V_{OL}
t_{TLH}
t_{THL}

VOLTAGE WAVEFORMS

NOTE A. C_L includes probe and jig capacitance.

FIGURE 1. SWITCHING CHARACTERISTICS

TYPICAL CHARACTERISTICS

MAXIMUM CLOCK FREQUENCY
vs
VIRTUAL JUNCTION TEMPERATURE

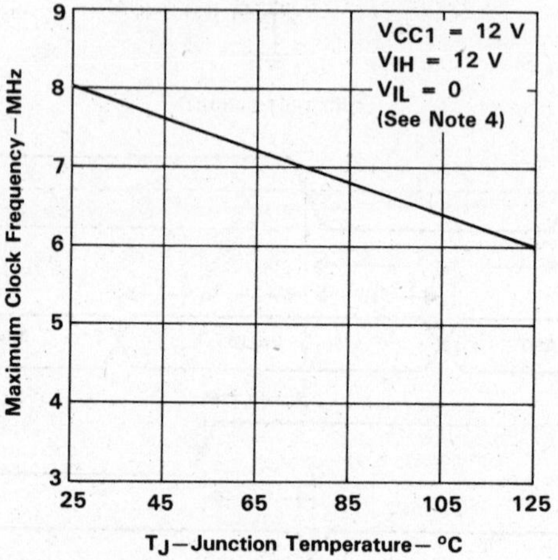

V_{CC1} = 12 V
V_{IH} = 12 V
V_{IL} = 0
(See Note 4)
Maximum Clock Frequency—MHz
9
8
7
6
5
4
3
25
45
65
85
105
125
T_J—Junction Temperature—°C

NOTE 4: This curve assumes a symmetrical clock pulse..

FIGURE 2

THERMAL INFORMATION

junction temperature formula

$$T_J = T_A + P_D R_{\theta JA}$$
$$T_J = T_C + P_D R_{\theta JC}$$

where

T_J = virtual junction temperature
T_A = free-air temperature
P_D = average device power dissipation
R_θ = thermal resistance (junction-to-air, $R_{\theta JA}$, or junction-to-case, $R_{\theta JC}$)

PACKAGE TYPE	$R_{\theta JA}$	$R_{\theta JC}$
FN 44-pin plastic	70°C/W	22°C/W
N 40-pin plastic	97°C/W	27°C/W

D2472, APRIL 1986

- Controls 32 Electrodes
- 100-V Totem-Pole Outputs
- Low Stand-by Power Consumption
- All Outputs Contain Sink and Source Clamp Diodes
- 15 mA Steady-State Output Current
- Rugged DMOS Outputs
- CMOS Inputs
- Direct Replacement for SN55501C, SN55501D

JD PACKAGE
(TOP VIEW)

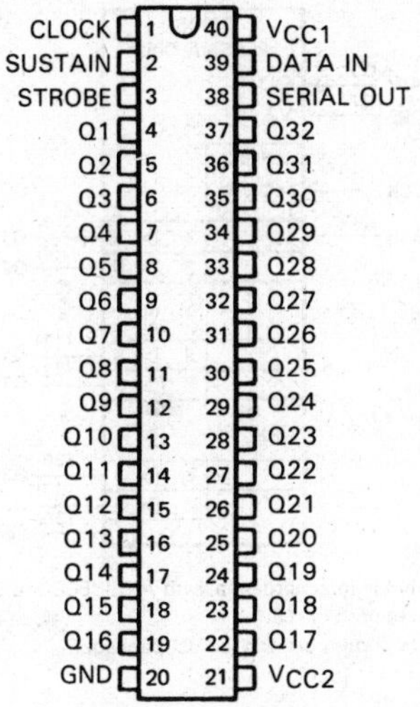
CLOCK 1
SUSTAIN 2
STROBE 3
Q1 4
Q2 5
Q3 6
Q4 7
Q5 8
Q6 9
Q7 10
Q8 11
Q9 12
Q10 13
Q11 14
Q12 15
Q13 16
Q14 17
Q15 18
Q16 19
GND 20
40 VCC1
39 DATA IN
38 SERIAL OUT
37 Q32
36 Q31
35 Q30
34 Q29
33 Q28
32 Q27
31 Q26
30 Q25
29 Q24
28 Q23
27 Q22
26 Q21
25 Q20
24 Q19
23 Q18
22 Q17
21 VCC2

FD PACKAGE
(TOP VIEW)

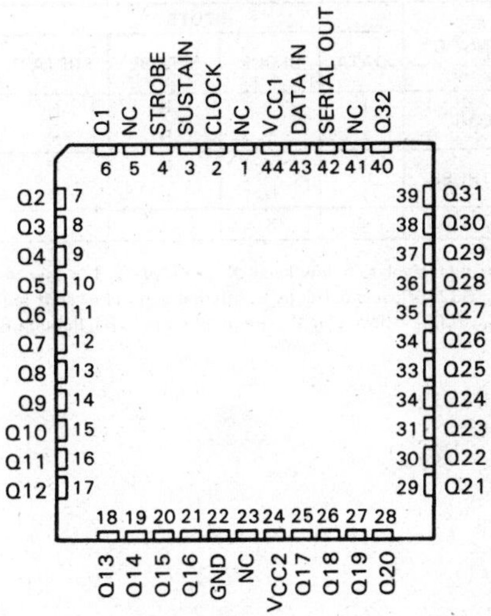
Q1 NC STROBE SUSTAIN CLOCK NC VCC1 DATA IN SERIAL OUT NC Q32
6 5 4 3 2 1 44 43 42 41 40
Q2 7
Q3 8
Q4 9
Q5 10
Q6 11
Q7 12
Q8 13
Q9 14
Q10 15
Q11 16
Q12 17
39 Q31
38 Q30
37 Q29
36 Q28
35 Q27
34 Q26
33 Q25
34 Q24
31 Q23
30 Q22
29 Q21
18 19 20 21 22 23 24 25 26 27 28
Q13 Q14 Q15 Q16 GND NC VCC2 Q17 Q18 Q19 Q20

NC—No internal connection

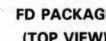

description

The SN55501E is a monolithic BIDFET† integrated circuit designed to provide the serial-to-parallel conversion and level translation of data in a matrix-addressable display. This device has diode-clamped CMOS inputs.

The Q outputs of these drivers are normally high and can be switched either selectively or together. Any output whose associated register bit (in the internal 32-bit serial register) contains a low will switch low when the strobe input is switched low if the sustain input is high. All other outputs remain high. When the sustain input is switched low, all outputs switch low independently of the data or strobe inputs. This feature can be used to generate a portion of the sustain pulse required in the operation of an AC plasma display. The internal level-shift circuits provide additional drive during the times that the outputs switch high to facilitate fast rise times while maintaining low standby power consumption. All outputs contain clamp diodes to the V_{CC2} and GND supply inputs.

The SN55501E is characterized for operation over the full military temperature range of −55°C to 125°C.

† BIDFET—Bipolar, double-diffused, N-channel and P-channel MOS transistors on same chip — patented process.

logic symbol†

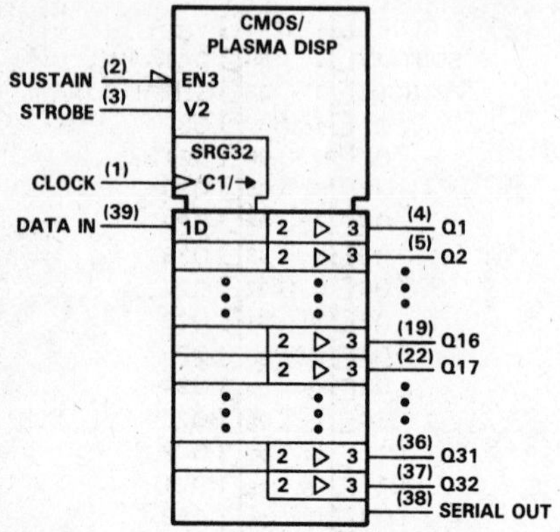
CMOS/
PLASMA DISP
SUSTAIN (2) EN3
STROBE (3) V2
SRG32
CLOCK (1) C1/→
DATA IN (39) 1D
2 ▷ 3 (4) Q1
2 ▷ 3 (5) Q2
2 ▷ 3 (19) Q16
2 ▷ 3 (22) Q17
2 ▷ 3 (36) Q31
2 ▷ 3 (37) Q32
(38) SERIAL OUT

† This symbol is in accordance with ANSI/IEEE Std 91-1984 and IEC Publication 617-12.

Pin numbers shown are for the JD package.

functional block diagram (positive logic)

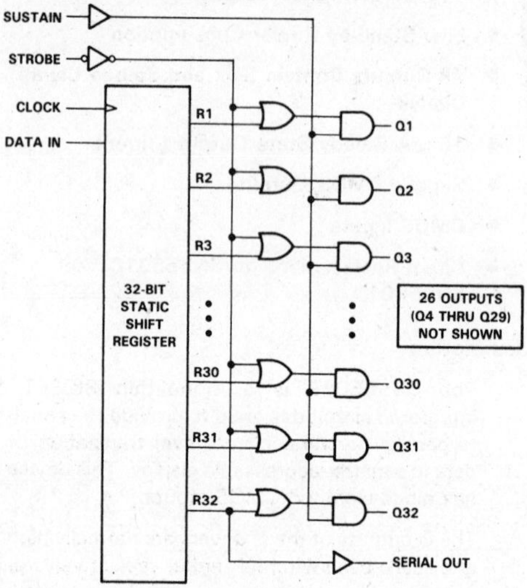
SUSTAIN
STROBE
CLOCK
DATA IN
32-BIT
STATIC
SHIFT
REGISTER
R1
R2
R3
R30
R31
R32
Q1
Q2
Q3
Q30
Q31
Q32
26 OUTPUTS
(Q4 THRU Q29)
NOT SHOWN
SERIAL OUT

FUNCTION TABLE

FUNCTION	INPUTS				OUTPUTS						
	DATA	CLOCK	STROBE	SUSTAIN	SHIFT REGISTER R1	R2	R3...R32	SERIAL DATA	Q1	Q2	Q3....Q32
LOAD	H	↑	H	H	H	$R1_n$	$R2_n...R31_n$	$R32_n$	H	H	H....H
	L	↑	H	H	L	$R1_n$	$R2_n...R31_n$	$R32_n$	H	H	H....H
STROBE	X	X	H	H	$R1_n$	$R2_n$	$R3_n...R32_n$	$R32_n$	H	H	H....H
	X	H	L	H	$R1_n$	$R2_n$	$R3_n...R32_n$	$R32_n$	R1	R2	R3....R32
SUSTAIN	X	X	X	L	$R1_n$	$R2_n$	$R3_n...R32_n$	$R32_n$	L	L	L....L

H = high level, L = low level, X = irrelevant, ↑ = low-to-high-level transition.
R1...R32 = levels currently at internal outputs of shift registers one through thirty-two, respectively.
$R1_n...R32_n$ = levels at shift-register outputs R1 through R32 respectively, before the most recent ↑ transition at the Clock input.

typical operating sequence

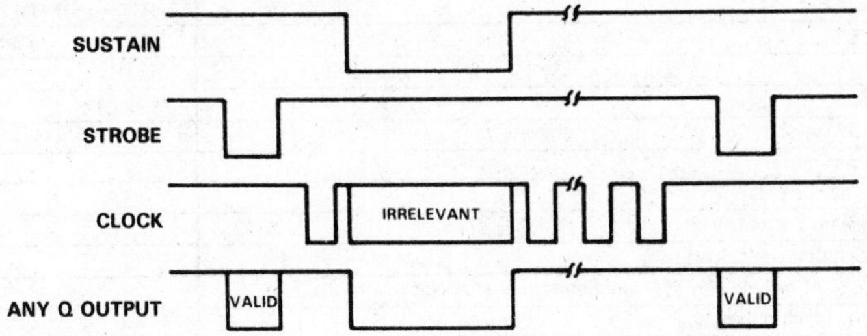
SUSTAIN
STROBE
CLOCK
IRRELEVANT
ANY Q OUTPUT
VALID
VALID

schematics of inputs and outputs

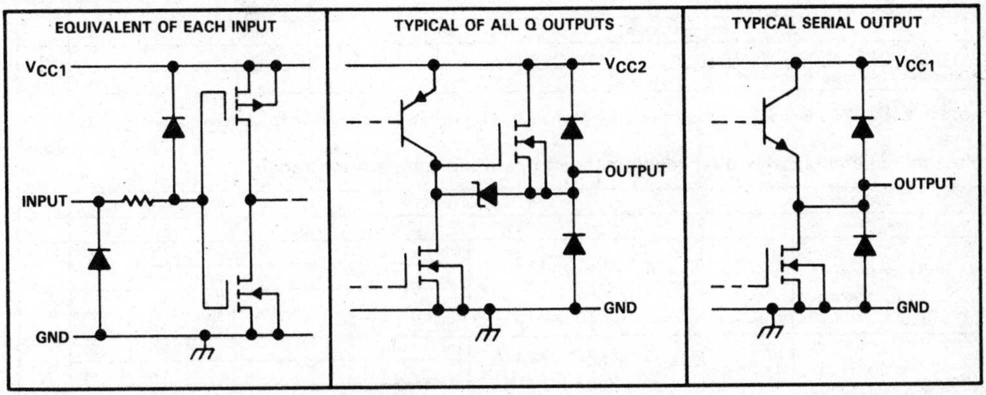
EQUIVALENT OF EACH INPUT
VCC1
INPUT
GND
TYPICAL OF ALL Q OUTPUTS
VCC2
OUTPUT
GND
TYPICAL SERIAL OUTPUT
VCC1
OUTPUT
GND

absolute maximum ratings over operating free-air temperature range (unless otherwise noted)

Supply voltage, V_{CC1} (see Note 1)	15 V
Supply voltage, V_{CC2}	100 V
Input voltage	V_{CC1} + 0.3 V
Continuous total dissipation at (or below) 25°C free-air (see Note 2)	1825 mW
Operating free-air temperature range	−55°C to 125°C
Storage temperature range	−65°C to 150°C
Case temperature for 60 seconds: FD package	260°C
Lead temperature 1,6 mm (1/16 inch) from case for 60 seconds: JD package	300°C

NOTES: 1. All voltage values are with respect to network ground terminal.
2. For operation above 25°C free-air temperature, see Dissipation Derating Table.

DISSIPATION DERATING TABLE

PACKAGE	POWER RATING	DERATING FACTOR	ABOVE T_A
FD	1825 mW	14.6 mW/°C	25°C
JD	1825 mW	22 mW/°C	67°C

recommended operating conditions

		MIN	NOM	MAX	UNIT
Supply voltage, V_{CC1}		10.8	12	13.2	V
Supply voltage, V_{CC2}		0		100	V
High-level input voltage, V_{IH}		0.75 V_{CC1}			
Low-level input voltage, V_{IL}				0.25 V_{CC1}	
Peak high-level Q output current, I_{OH}				−20	mA
Peak low-level Q output current, I_{OL}				20	mA
High-level Q output clamp current, I_{OKH}				20	mA
Low-level Q output clamp current, I_{OKL}				−20	mA
Clock frequency, f_{clock}, at or below, 25°C junction temperature (see Note 3)		0		8	MHz
Duration of high or low clock pulse, t_w		62			ns
Setup time, t_{su}	Data inputs before clock↑	20			ns
Hold time, t_h	Data hold time after clock↑	50			ns
	Strobe high after clock↑	150			
	Strobe high after sustain↑	250			
Operating free-air temperature, T_A		−55			°C
Operating case temperature, T_C				125	

NOTE 3: See Figure 3 for maximum clock frequency when devices are operated in cascade or for operation above T_J = 25°C.

electrical characteristics over recommended operating temperature range

PARAMETER			TEST CONDITIONS		MIN	TYP†	MAX	UNIT
V_{IK}	Input clamp voltage		V_{CC1} = 12 V,	I_I = 12 mA		−1	−1.5	V
V_{OH}	High-level output voltage	Q outputs	V_{CC1} = 13.2 V, V_{CC2} = 100 V	I_{OH} = −1 mA	94	97.5		V
				I_{OH} = −10 mA	92	94.5		
				I_{OH} = −15 mA	90	93.5		
		Serial out	V_{CC1} = 10.8 V,	I_{OH} = −100 μA	9	10		
V_{OL}	Low-level output voltage	Q outputs	V_{CC1} = 13.2 V, V_{CC2} = 100 V	I_{OL} = 1 mA		0.85	2	V
				I_{OL} = 10 mA		2	4	
				I_{OL} = 15 mA		2.75	5	
		Serial out	V_{CC1} = 10.8 V,	I_{OL} = 100 μA		0.1	1	
V_{OK}	Output clamp voltage	Q output	V_{CC2} = 0	I_{OK} = 20 mA		1	2.5	V
				I_{OK} = −20 mA		−1.2	−2.5	
I_{IH}	High-level input current		V_{CC1} = 13.2 V, V_{CC2} = 100 V	V_{IH} = V_{IH}min,			1	μA
I_{IL}	Low-level input current		V_{CC1} = 13.2 V, V_{CC2} = 100 V	V_{IL} = V_{IL}max,			−1	μA
I_{CC1}	Supply current from V_{CC1}		V_{CC1} = 13.2 V,	V_{CC2} = 100 V		0.05	1	mA
I_{CC2}	Supply current from V_{CC2}		V_{CC2} = 100 V	Outputs low		0.1	1	mA
				Outputs high		1	5	

†Typical values are at V_{CC1} = 12 V, T_A = 25°C.

switching characteristics, V_{CC1} = 12 V, V_{CC2} = 100 V, T_A = 25°C

PARAMETER			TEST CONDITIONS	MIN	TYP	MAX	UNIT
t_{DHL}	Delay time, high-to-low-level outputs	Strobe to Q outputs	C_L = 30 pF			250	ns
		Sustain to Q outputs	C_L = 30 pF			250	
		Clock to serial data output	C_L = 20 pF			147	
t_{DLH}	Delay time, low-to-high-level outputs	Strobe to Q outputs	C_L = 30 pF			450	ns
		Sustain to Q outputs	C_L = 30 pF			450	
		Clock to serial data output	C_L = 20 pF			147	
t_{THL}	Transition time, high-to-low-level Q output		C_L = 30 pF			200	ns
t_{TLH}	Transition time, low-to-high-level Q output		C_L = 30 pF			300	ns

PARAMETER MEASUREMENT INFORMATION

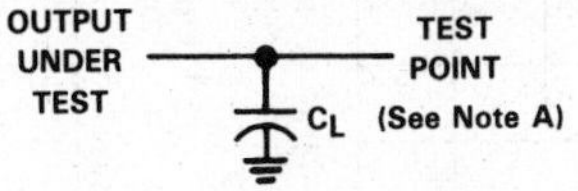
OUTPUT
UNDER
TEST
TEST
POINT
C_L (See Note A)

LOAD TEST CIRCUIT

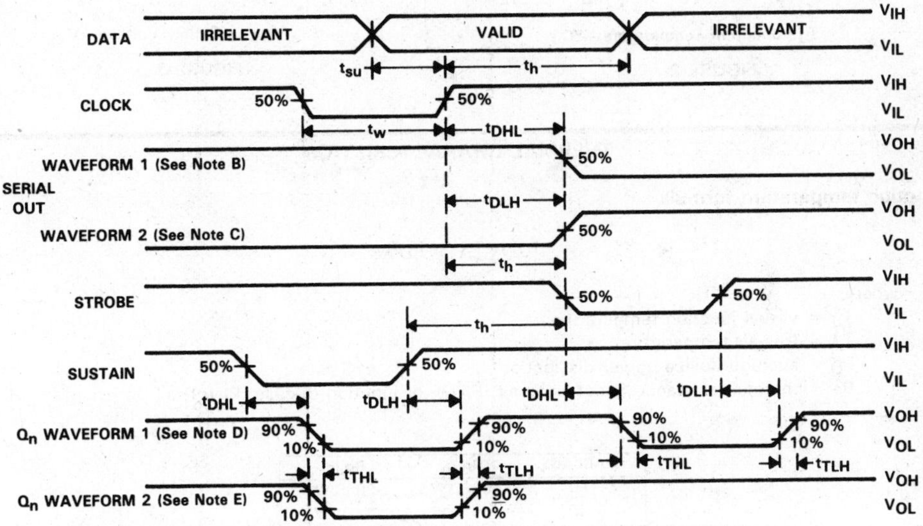
DATA
IRRELEVANT
VALID
IRRELEVANT
t_{su}
t_h
CLOCK
50%
50%
t_w
t_{DHL}
WAVEFORM 1 (See Note B)
SERIAL
OUT
50%
t_{DLH}
WAVEFORM 2 (See Note C)
50%
t_h
STROBE
50%
50%
t_h
SUSTAIN
50%
50%
t_{DHL}
t_{DLH}
t_{DHL}
t_{DLH}
Q_n WAVEFORM 1 (See Note D)
90%
10%
90%
10%
90%
10%
90%
10%
t_{THL}
t_{TLH}
t_{THL}
t_{TLH}
Q_n WAVEFORM 2 (See Note E)
90%
10%
90%
10%
V_{IH}
V_{IL}
V_{OH}
V_{OL}

NOTES: A. C_L includes probe and jig capacitance.
B. Serial out waveform for internal conditions such that a low is registered in R32.
C. Serial out waveform for internal conditions such that a high is registered in R32.
D. Q_n output with a low stored in associated register R_n.
E. Q_n output with a high stored in associated register R_n.

VOLTAGE WAVEFORMS

FIGURE 1. SWITCHING CHARACTERISTICS

RECOMMENDED OPERATING CONDITIONS

MAXIMUM CLOCK FREQUENCY
vs
JUNCTION TEMPERATURE

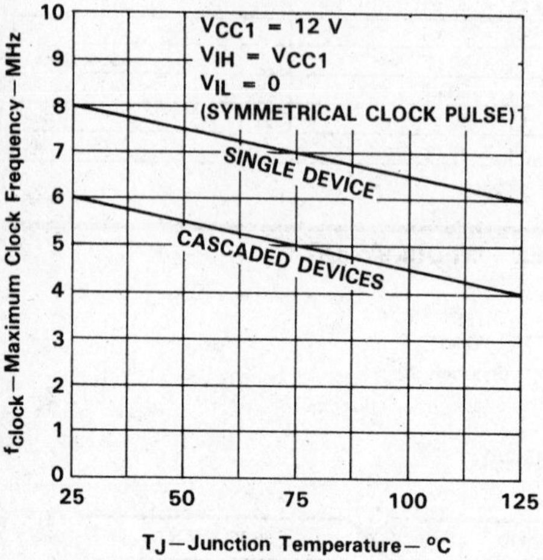
VCC1 = 12 V
VIH = VCC1
VIL = 0
(SYMMETRICAL CLOCK PULSE)
SINGLE DEVICE
CASCADED DEVICES
fclock – Maximum Clock Frequency – MHz
TJ – Junction Temperature – °C
0
1
2
3
4
5
6
7
8
9
10
25
50
75
100
125

FIGURE 2

INPUT VOLTAGE LOGIC LEVEL LIMITS
vs
V_{CC1} SUPPLY VOLTAGE

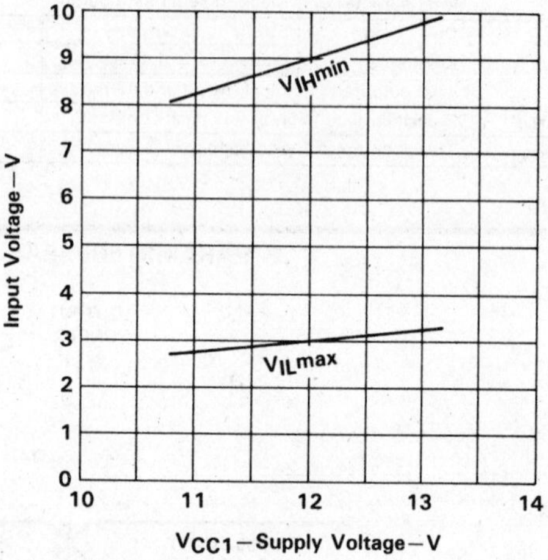
VIHmin
VILmax
Input Voltage – V
VCC1 – Supply Voltage – V
0
1
2
3
4
5
6
7
8
9
10
10
11
12
13
14

FIGURE 3

THERMAL CHARACTERISTICS

junction temperature formula

$$T_J = T_A + P_D R_\theta$$

where

T_J = virtual junction temperature
T_A = free-air temperature
P_D = average device power dissipation
R_θ = thermal resistance (junction-to-air, $R_{\theta JA}$, or junction-to-case, $R_{\theta JC}$)

PACKAGE	$R_{\theta JA}$	$R_{\theta JC}$
FD	68°C/W	20°C/W
JD	45°C/W	12°C/W

D2472, MARCH 1983—REVISED DECEMBER 1985

- **Controls 32 Electrodes**
- **100-V Totem-Pole Outputs**
- **Low Stand-by Power Consumption**
- **All Outputs Contain Sink and Source Clamp Diodes**
- **15 mA Steady-State Output Current**
- **Rugged DMOS Outputs**
- **CMOS Inputs**
- **Direct Replacement for SN75501C**

N PACKAGE
(TOP VIEW)

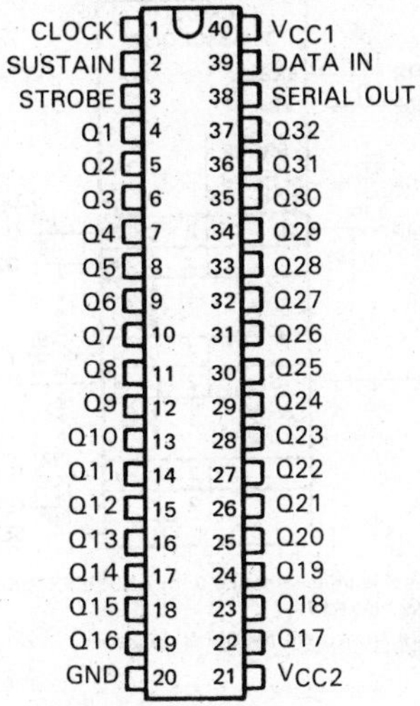
CLOCK 1
SUSTAIN 2
STROBE 3
Q1 4
Q2 5
Q3 6
Q4 7
Q5 8
Q6 9
Q7 10
Q8 11
Q9 12
Q10 13
Q11 14
Q12 15
Q13 16
Q14 17
Q15 18
Q16 19
GND 20
40 VCC1
39 DATA IN
38 SERIAL OUT
37 Q32
36 Q31
35 Q30
34 Q29
33 Q28
32 Q27
31 Q26
30 Q25
29 Q24
28 Q23
27 Q22
26 Q21
25 Q20
24 Q19
23 Q18
22 Q17
21 VCC2

FN PACKAGE
(TOP VIEW)

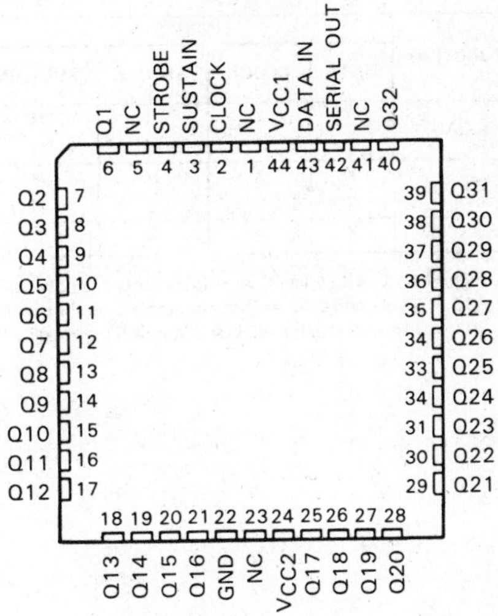
Q1 NC STROBE SUSTAIN CLOCK NC VCC1 DATA IN SERIAL OUT NC Q32
6 5 4 3 2 1 44 43 42 41 40
Q2 7
Q3 8
Q4 9
Q5 10
Q6 11
Q7 12
Q8 13
Q9 14
Q10 15
Q11 16
Q12 17
39 Q31
38 Q30
37 Q29
36 Q28
35 Q27
34 Q26
33 Q25
34 Q24
31 Q23
30 Q22
29 Q21
18 19 20 21 22 23 24 25 26 27 28
Q13 Q14 Q15 Q16 GND NC VCC2 Q17 Q18 Q19 Q20

NC—No internal connection

description

The SN65501E and SN75501E are monolithic BIDFET† integrated circuits designed to provide the serial-to-parallel conversion and level translation of data in a matrix-addressable display. The device inputs are diode-clamped CMOS inputs.

The Q outputs of these drivers are normally high and can be switched either selectively or together. Any output whose associated register bit (in the internal 32-bit serial register) contains a low will switch low when the strobe input is switched low if the sustain input is high. All other outputs remain high. When the sustain input is switched low, all outputs switch low independently of the data or strobe inputs. This feature can be used to generate a portion of the sustain pulse required in the operation of an AC plasma display. The internal level-shift circuits provide additional drive during the times that the outputs switch high to facilitate fast rise times while maintaining low standby power consumption. All outputs contain clamp diodes to the V_{CC2} and GND supply inputs.

The SN65501E is characterized for operation over the temperature range of −40 °C to 85 °C. The SN75501E is characterized for operation over the temperature range of 0 °C to 70 °C.

† BIDFET—Bipolar, double-diffused, N-channel and P-channel MOS transistors on same chip — patented process.

logic symbol†

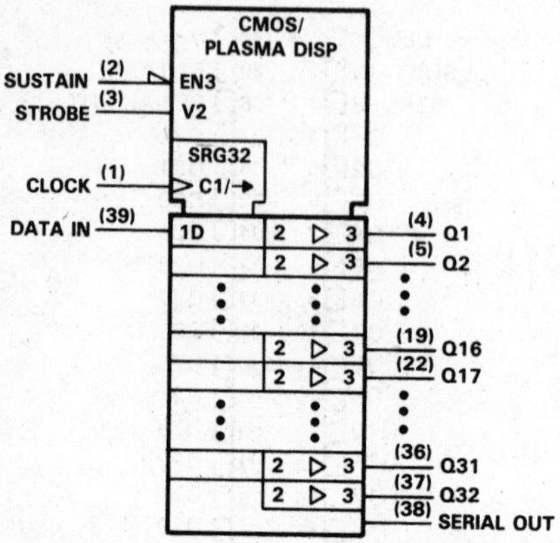
CMOS/
PLASMA DISP
SUSTAIN (2)
EN3
STROBE (3)
V2
SRG32
CLOCK (1)
C1/→
DATA IN (39)
1D
2 ▷ 3
(4) Q1
(5) Q2
(19) Q16
(22) Q17
(36) Q31
(37) Q32
(38) SERIAL OUT

† This symbol is in accordance with ANSI/IEEE Std 91-1984 and IEC Publication 617-12.
Pin numbers shown are for the N package.

functional block diagram (positive logic)

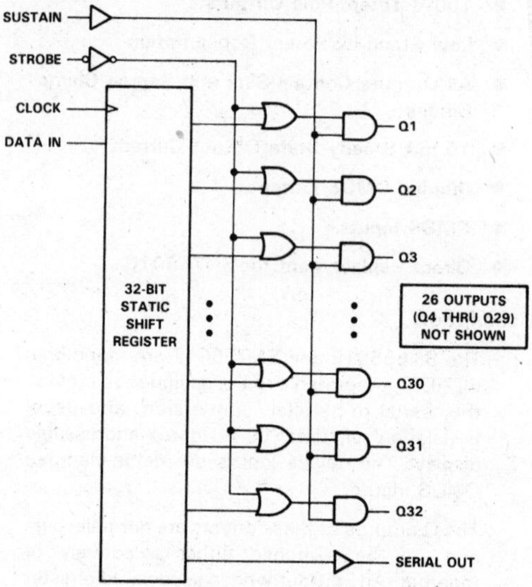
SUSTAIN
STROBE
CLOCK
DATA IN
32-BIT
STATIC
SHIFT
REGISTER
Q1
Q2
Q3
26 OUTPUTS
(Q4 THRU Q29)
NOT SHOWN
Q30
Q31
Q32
SERIAL OUT

FUNCTION TABLE

FUNCTION	INPUTS				OUTPUTS						
	DATA	CLOCK	STROBE	SUSTAIN	SHIFT REGISTER R1	SHIFT REGISTER R2	SHIFT REGISTER R3...R32	SERIAL DATA	Q1	Q2	Q3....Q32
LOAD	H	↑	H	H	H	$R1_n$	$R2_n...R31_n$	$R32_n$	H	H	H....H
	L	↑	H	H	L	$R1_n$	$R2_n...R31_n$	$R32_n$	H	H	H....H
STROBE	X	X	H	H	$R1_n$	$R2_n$	$R3_n...R32_n$	$R32_n$	H	H	H....H
	X	H	L	H	$R1_n$	$R2_n$	$R3_n...R32_n$	$R32_n$	R1	R2	R3....R32
SUSTAIN	X	X	X	L	$R1_n$	$R2_n$	$R3_n...R32_n$	$R32_n$	L	L	L....L

H = high level, L = low level, X = irrelevant, ↑ = low-to-high-level transition.
R1...R32 = levels currently at internal outputs of shift registers one through thirty-two, respectively.
$R1_n...R32_n$ = levels at shift-register outputs R1 through R32 respectively, before the most recent ↑ transition at the Clock input.

TEXAS INSTRUMENTS
POST OFFICE BOX 655012 • DALLAS, TEXAS 75265

typical operating sequence

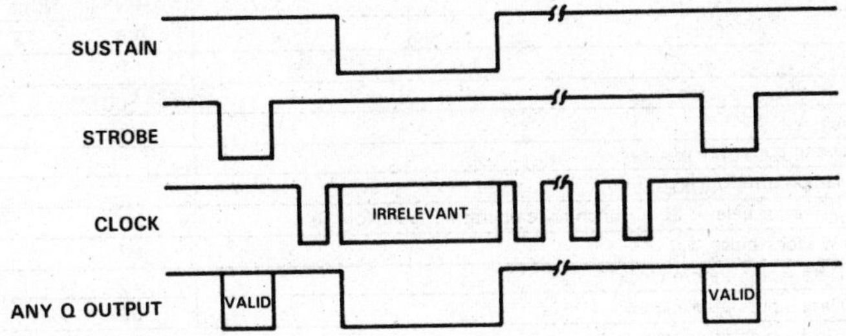
SUSTAIN
STROBE
CLOCK
IRRELEVANT
ANY Q OUTPUT
VALID
VALID

schematics of inputs and outputs

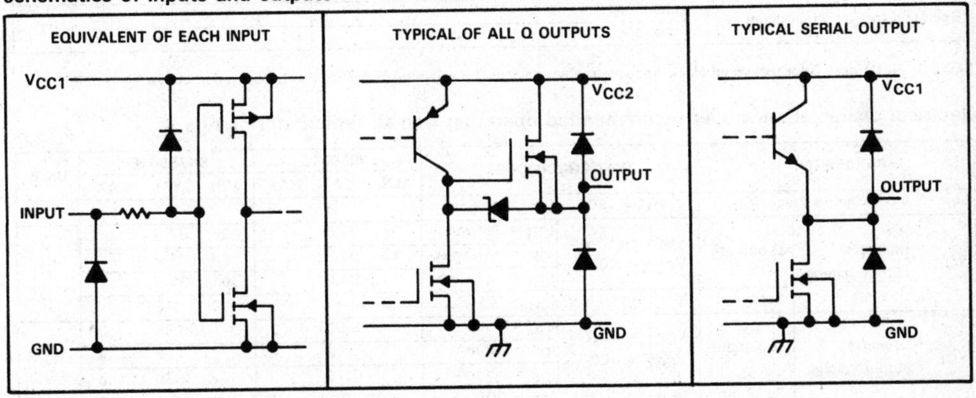
EQUIVALENT OF EACH INPUT
VCC1
INPUT
GND
TYPICAL OF ALL Q OUTPUTS
VCC2
OUTPUT
GND
TYPICAL SERIAL OUTPUT
VCC1
OUTPUT
GND

absolute maximum ratings over operating free-air temperature range (unless otherwise noted)

Supply voltage, V_{CC1} (see Note 1) . . . . . 15 V
Supply voltage, V_{CC2} . . . . . 100 V
Input voltage . . . . . V_{CC1} to 0.3 V
Continuous total dissipation at (or below) 25 °C free-air (see Note 2): FN package . . . . . 1775 mW
N package . . . . . 1250 mW
Operating free-air temperature range: SN65501E . . . . . −40 °C to 85 °C
SN75501E . . . . . 0 °C to 70 °C
Storage temperature range . . . . . −65 °C to 150 °C
Lead temperature 1,6 mm (1/16 inch) from case for 10 seconds: N package . . . . . 260 °C
Case temperature for 10 seconds: FN package . . . . . 260 °C

NOTES: 1. All voltage values are with respect to network ground terminal.
2. For operation above 25 °C free-air temperature, see Dissipation Derating Table.

DISSIPATION DERATING TABLE

PACKAGE	POWER RATING	DERATING FACTOR	ABOVE T_A
FN	1775 mW	14.2 mW/°C	25 °C
N	1250 mW	10.0 mW/°C	25 °C

recommended operating conditions

			MIN	NOM	MAX	UNIT
Supply voltage, V_{CC1}			10.8	12	13.2	V
Supply voltage, V_{CC2}			0		100	V
High-level input voltage, V_{IH}			0.75 V_{CC1}			
Low-level input voltage, V_{IL}					0.25 V_{CC1}	
High-level Q output clamp current, I_{OKH}					20	mA
Low-level Q output clamp current, I_{OKL}					−20	mA
Clock frequency, f_{clock}, at or below, 25°C junction temperature (see Note 3)			0		8	MHz
Duration of high or low clock pulse, t_w			62			ns
Setup time, t_{su}	Data inputs before clock↑		20			ns
	Data inputs after clock↑		50			
Hold time, t_h	Strobe high after clock↑		150			ns
	Strobe high after sustain↑		250			
Operating free-air temperature, T_A		SN65501E	−40		85	°C
		SN75501E	0		70	

NOTE 3: See Figure 3 for maximum clock frequency when devices are operated in cascade or for operation above T_J = 25°C.

electrical characteristics over recommended operating free-air temperature range

PARAMETER			TEST CONDITIONS		SN65501E MIN	SN65501E TYP†	SN65501E MAX	SN75501E MIN	SN75501E TYP†	SN75501E MAX	UNIT
V_{IK}	Input clamp voltage		V_{CC1} = 12 V,	I_I = 12 mA		−1	−1.5		−1	−1.5	V
V_{OH}	High-level output voltage	Q outputs	V_{CC1} = 13.2 V, V_{CC2} = 100 V	I_{OH} = −1 mA	94	97.5		95	97.5		V
				I_{OH} = −10 mA	92	94.5		93	94.5		
				I_{OH} = −15 mA	90	93.5		91	93.5		
		Serial out	V_{CC1} = 10.8 V,	I_{OH} = −100 μA	9	10		9	10		
V_{OL}	Low-level output voltage	Q outputs	V_{CC1} = 13.2 V, V_{CC2} = 100 V	I_{OL} = 1 mA		0.85	2		0.85	2	V
				I_{OL} = 10 mA		2	4		2	4	
				I_{OL} = 15 mA		2.75	5		2.75	5	
		Serial out	V_{CC1} = 10.8 V,	I_{OL} = 100 μA		0.1	1		0.1	1	
V_{OK}	Output clamp voltage	Q output	V_{CC2} = 0	I_{OK} = 20 mA		1	2.5		1	2.5	V
				I_{OK} = −20 mA		−1.2	−2.5		−1.2	−2.5	
I_{IH}	High-level input current		V_{CC1} = 13.2 V, V_{IH} = V_{IH}min, V_{CC2} = 100 V				1			1	μA
I_{IL}	Low-level input current		V_{CC1} = 13.2 V, V_{IL} = V_{IL}max, V_{CC2} = 100 V				−1			−1	μA
I_{CC1}	Supply current from V_{CC1}		V_{CC1} = 13.2 V, V_{CC2} = 100 V			0.05	1		0.05	1	mA
I_{CC2}	Supply current from V_{CC2}		V_{CC2} = 100 V			1	5		1	3	mA

† Typical values are at V_{CC1} = 12 V, T_A = 25°C.

switching characteristics, V_{CC1} = 12 V, V_{CC2} = 100 V, T_A = 25°C

PARAMETER			TEST CONDITIONS	MIN	TYP	MAX	UNIT
t_{DHL}	Delay time, high-to-low-level outputs	Strobe to Q outputs	C_L = 30 pF			250	ns
		Sustain to Q outputs	C_L = 30 pF			250	
		Clock to serial data output	C_L = 20 pF			147	
t_{DLH}	Delay time, low-to-high-level outputs	Strobe to Q outputs	C_L = 30 pF			450	ns
		Sustain to Q outputs	C_L = 30 pF			450	
		Clock to serial data output	C_L = 20 pF			147	
t_{THL}	Transition time, high-to-low-level Q output		C_L = 30 pF			200	ns
t_{TLH}	Transition time, low-to-high-level Q output		C_L = 30 pF			300	ns

PARAMETER MEASUREMENT INFORMATION

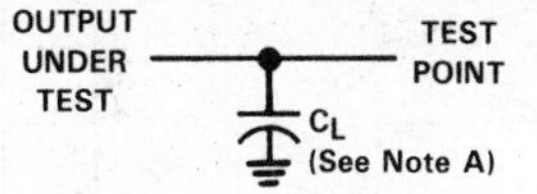
OUTPUT
UNDER
TEST
TEST
POINT
C_L
(See Note A)

LOAD TEST CIRCUIT

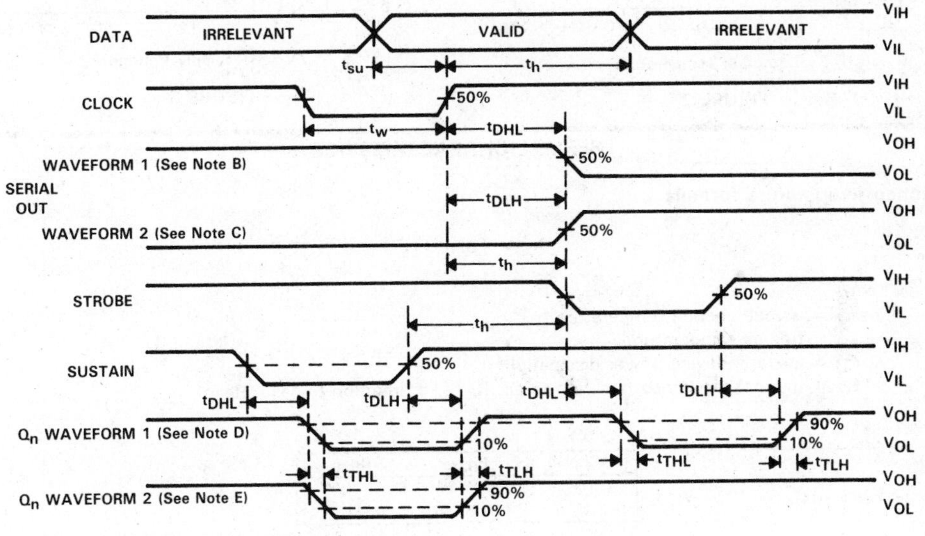
DATA
IRRELEVANT
VALID
IRRELEVANT
t_{su}
t_h
CLOCK
50%
t_w
t_{DHL}
WAVEFORM 1 (See Note B)
SERIAL
OUT
WAVEFORM 2 (See Note C)
t_{DLH}
t_h
STROBE
SUSTAIN
50%
t_{DHL}
t_{DLH}
Q_n WAVEFORM 1 (See Note D)
Q_n WAVEFORM 2 (See Note E)
90%
10%
t_{THL}
t_{TLH}
V_{IH}
V_{IL}
V_{OH}
V_{OL}

NOTES: A. C_L includes probe and jig capacitance.
B. Serial out waveform for internal conditions such that a low is registered in R32.
C. Serial out waveform for internal conditions such that a high is registered in R32.
D. Q_n output with a low stored in associated register R_n.
E. Q_n output with a high stored in associated register R_n.

VOLTAGE WAVEFORMS

FIGURE 1. SWITCHING CHARACTERISTICS

TYPICAL CHARACTERISTICS

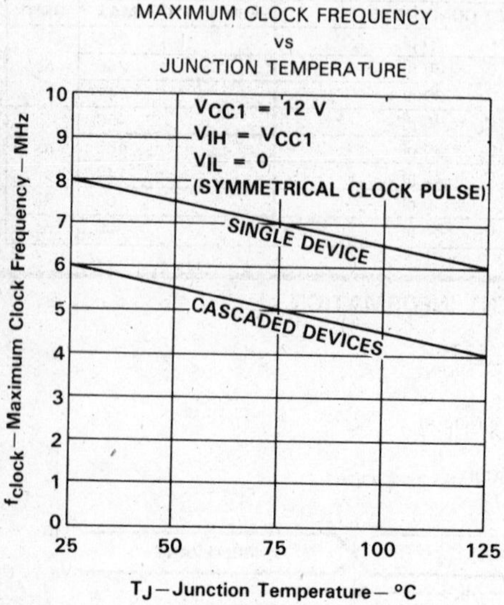
MAXIMUM CLOCK FREQUENCY
vs
JUNCTION TEMPERATURE
VCC1 = 12 V
VIH = VCC1
VIL = 0
(SYMMETRICAL CLOCK PULSE)
SINGLE DEVICE
CASCADED DEVICES
fclock—Maximum Clock Frequency—MHz
TJ—Junction Temperature—°C
0
1
2
3
4
5
6
7
8
9
10
25
50
75
100
125

FIGURE 2

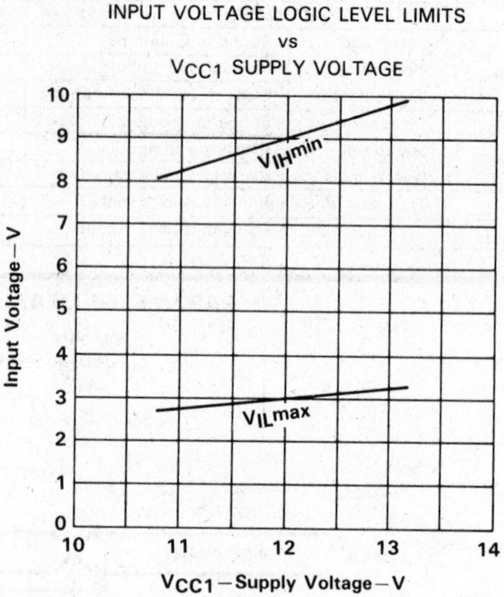
INPUT VOLTAGE LOGIC LEVEL LIMITS
vs
VCC1 SUPPLY VOLTAGE
VIHmin
VILmax
Input Voltage—V
VCC1—Supply Voltage—V
0
1
2
3
4
5
6
7
8
9
10
10
11
12
13
14

FIGURE 3

THERMAL CHARACTERISTICS

junction temperature formula

$$T_J = T_A + P_D R_\theta$$

where

T_J = virtual junction temperature
T_A = free-air temperature
P_D = average device power dissipation
R_θ = thermal resistance (junction-to-air, $R_{\theta JA}$, or junction-to-case, $R_{\theta JC}$)

PACKAGE	$R_{\theta JA}$	$R_{\theta JC}$
FN	70°C/W	22°C/W
N	100°C/W	27°C/W

D2924, DECEMBER 1985–REVISED OCTOBER 1986

- Controls 32 Electrodes
- Very Low Steady-State Power Consumption
- Rugged DMOS Outputs
- CMOS-Compatible Inputs
- Dependable Texas Instruments Quality and Reliability

FN PACKAGE
(TOP VIEW)

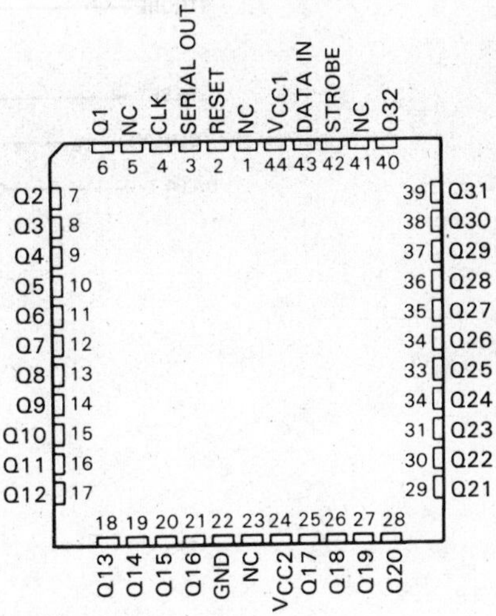
Q1 NC CLK SERIAL OUT RESET NC VCC1 DATA IN STROBE NC Q32
6 5 4 3 2 1 44 43 42 41 40
Q2 7
Q3 8
Q4 9
Q5 10
Q6 11
Q7 12
Q8 13
Q9 14
Q10 15
Q11 16
Q12 17
39 Q31
38 Q30
37 Q29
36 Q28
35 Q27
34 Q26
33 Q25
34 Q24
31 Q23
30 Q22
29 Q21
18 19 20 21 22 23 24 25 26 27 28
Q13 Q14 Q15 Q16 GND NC VCC2 Q17 Q18 Q19 Q20

NC–No internal connection

description

The SN65508 and SN75508 are monolithic BIDFET† integrated circuits designed to provide the serial-to-parallel conversion and level translation of data in a matrix-addressable display. All inputs are CMOS compatible and all outputs are totem-pole DMOS structures.

If the strobe input is at a high logic level, all outputs are high. When the strobe input goes low, any output whose associated register bit contains a low will go low. All outputs whose associated register bit contains a high will remain high. When the reset input is low, all register bits are low. In this condition, all outputs will go low when the strobe input goes low. The serial data output from the shift register may be used to cascade additional devices. This output is not affected by the Strobe input.

The SN65508 is characterized for operation from −40 °C to 85 °C. The SN75508 is characterized for operation from 0 °C to 70 °C.

logic symbol†

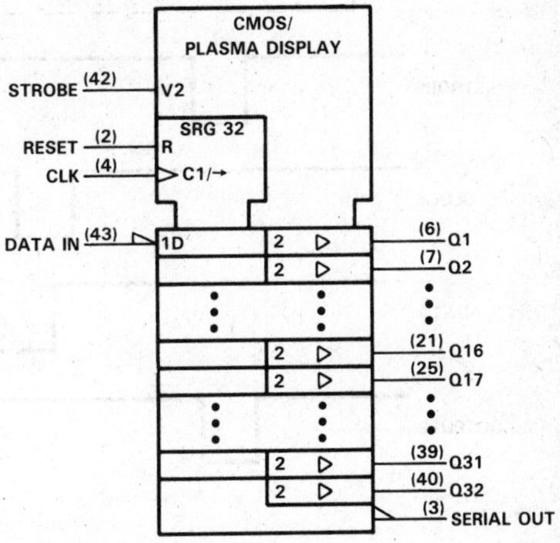
CMOS/
PLASMA DISPLAY
STROBE (42) V2
SRG 32
RESET (2) R
CLK (4) C1/→
DATA IN (43) 1D
2
(6) Q1
(7) Q2
(21) Q16
(25) Q17
(39) Q31
(40) Q32
(3) SERIAL OUT

†This symbol is in accordance with ANSI/IEEE Std 91-1094 and IEC Publication 617-12.

†BIDFET–Bipolar, double-diffused, N-channel and P-channel MOS transistors on same chip — patented process

functional block diagram

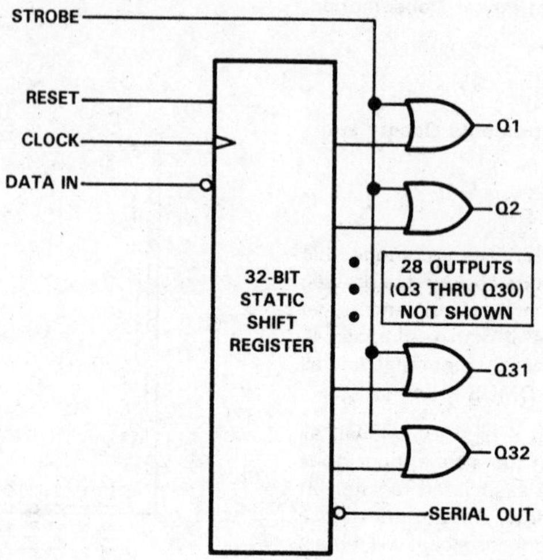
STROBE
RESET
CLOCK
DATA IN
32-BIT STATIC SHIFT REGISTER
Q1
Q2
28 OUTPUTS (Q3 THRU Q30) NOT SHOWN
Q31
Q32
SERIAL OUT

typical operating sequence

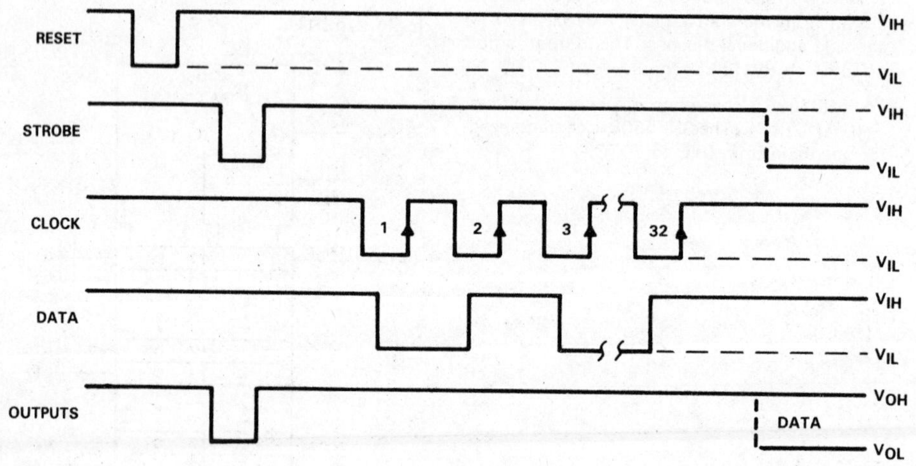
RESET
STROBE
CLOCK
DATA
OUTPUTS
1
2
3
32
DATA
V_{IH}
V_{IL}
V_{OH}
V_{OL}

TEXAS INSTRUMENTS
POST OFFICE BOX 655012 • DALLAS, TEXAS 75265

schematics of inputs and outputs

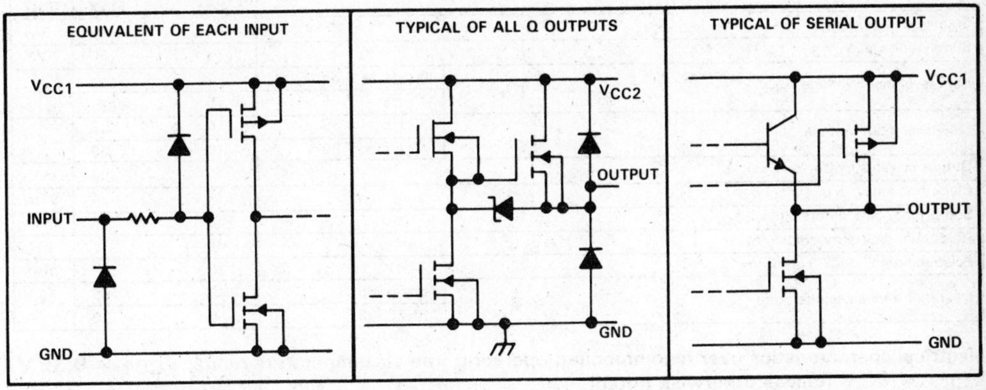
EQUIVALENT OF EACH INPUT
VCC1
INPUT
GND
TYPICAL OF ALL Q OUTPUTS
VCC2
OUTPUT
GND
TYPICAL OF SERIAL OUTPUT
VCC1
OUTPUT
GND

absolute maximum ratings over operating free-air temperature range (unless otherwise noted)

Supply voltage, V_{CC1} (see Note 1)		15 V
Supply voltage, V_{CC2}		95 V
Input voltage, V_I		−0.3 V to V_{CC} + 0.3 V
High-level output voltage, V_{OH}		95 V
High-level output current, I_{OH}		−3 mA
Continuous total power dissipation at (or below) 25°C free-air temperature (see Note 2)		1700 mW
Operating free-air temperature range:	SN65508	−40°C to 85°C
	SN75508	0°C to 70°C
Storage temperature range		−65°C to 150°C
Case temperature for 10 seconds		260°C

NOTES: 1. Voltage values are with respect to network ground terminal.
2. For operation above 25°C free-air temperature, derate linearly to 1088 mW at 70°C at the rate of 13.6 mW/°C.

recommended operating conditions

		MIN	MAX	UNIT
Supply voltage, V_{CC1}		7.65	9.35	V
Supply voltage, V_{CC2}		V_{CC1}	90	V
High-level input voltage, V_{IH}	V_{CC1} = 9.35 V	7	V_{CC1}	V
	V_{CC1} = 7.65 V	5.75	V_{CC1}	
Low-level input voltage, V_{IL}	V_{CC1} = 9.35 V	0	2.3	V
	V_{CC1} = 7.65 V	0	1.9	
Output current, I_O ($t_w \leq 1\ \mu s$)			80	mA
Clock frequency, f_{clock}			4	MHz
Setup time, data before clock↓, t_{su}		100		ns
Hold time, data after clock↓, t_h		62		ns
Pulse duration, clock high or low, t_{wCLK}		125		ns
Operating free-air temperature, T_A	SN65508	−40	85	°C
	SN75508	0	70	

electrical characteristics over recommended operating free-air temperature range, V_{CC1} = 9.35 V, V_{CC2} = 90 V (unless otherwise noted)

PARAMETER			TEST CONDITIONS		MIN	MAX	UNIT
V_{OH}	High-level output voltage	Q outputs	V_{CC1} = 7.65 V,	I_{OH} = −3 mA	83	87	V
		Serial output	V_{CC1} = 7.65 V,	I_{OH} = −50 μA	6.8	7.65	
V_{OL}	Low-level output voltage	Q outputs	V_{CC1} = 7.65 V,	I_{OL} = 10 mA	1.4	2.4	V
		Serial output	V_{CC1} = 7.65 V,	I_{OL} = 50 μA	0	0.8	
V_{OK}	Output clamp voltage		V_{CC2} = 0	I_O = 100 mA, $t_w \leq 1\ \mu s$		2.5	V
				I_O = −100 mA, $t_w \leq 1\ \mu s$		−2.7	
I_{IH}	High-level input current		V_I = 9.35 V			1	μA
I_{IL}	Low-level input current		V_I = 0.4 V			−1	μA
I_{OS}	Short-circuit output current		V_O = 0			−20	mA
I_{CC1}	Supply current from V_{CC1}					500	μA
I_{CC2}	Supply current from V_{CC2}		Output high			500	μA
			Output low			8.5	mA

switching characteristics, V_{CC1} = 7.65 V, T_A = 25°C

PARAMETER		TEST CONDITIONS	MIN	MAX	UNIT
t_{wRSTL}	Pulse duration, reset low		125		ns
t_{d1}	Delay time, V_{CC2} to Q outputs (10%−10%)	R_L = 100 kΩ, C_L = 100 pF		800	ns
t_{d2}	Delay time, V_{CC2} to Q outputs (90%−90%)	R_L = 100 kΩ, C_L = 100 pF		800	ns

PARAMETER MEASUREMENT INFORMATION

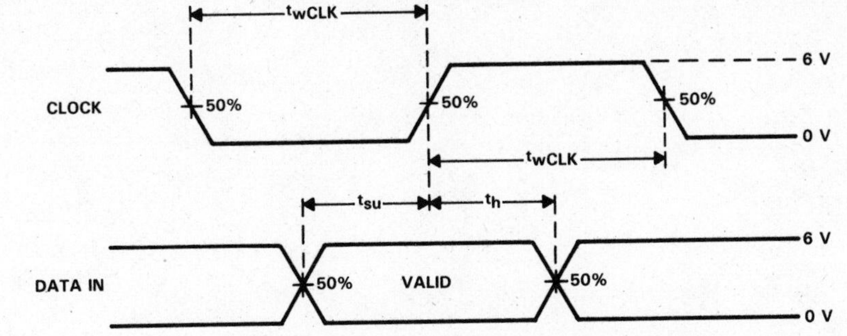
tWCLK
CLOCK
50%
50%
50%
6 V
0 V
tWCLK
tsu
th
DATA IN
50%
VALID
50%
6 V
0 V

FIGURE 1. INPUT TIMING VOLTAGE WAVEFORMS

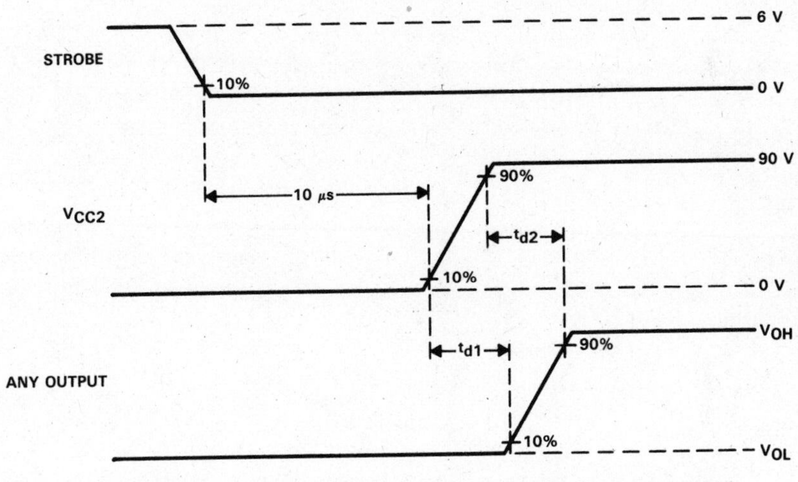
STROBE
6 V
10%
0 V
90 V
90%
10 μs
VCC2
td2
10%
0 V
VOH
td1
90%
ANY OUTPUT
10%
VOL

FIGURE 2. VOLTAGE WAVEFORMS FOR OUTPUT DELAY TIMES

D2923, DECEMBER 1985—REVISED OCTOBER 1986

- Controls 32 Electrodes
- Very Low Steady-State Power Consumption
- Rugged DMOS Outputs
- CMOS-Compatible Inputs
- Dependable Texas Instruments Quality and Reliability

FN PACKAGE
(TOP VIEW)

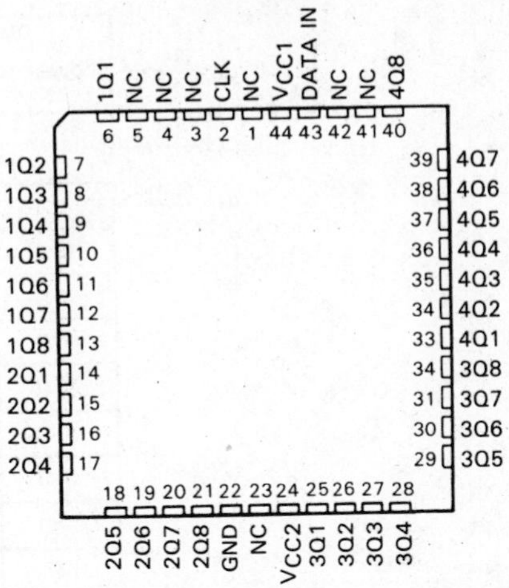
1Q1 NC NC NC CLK NC VCC1 DATA IN NC NC 4Q8
6 5 4 3 2 1 44 43 42 41 40
1Q2 7
1Q3 8
1Q4 9
1Q5 10
1Q6 11
1Q7 12
1Q8 13
2Q1 14
2Q2 15
2Q3 16
2Q4 17
39 4Q7
38 4Q6
37 4Q5
36 4Q4
35 4Q3
34 4Q2
33 4Q1
34 3Q8
31 3Q7
30 3Q6
29 3Q5
18 19 20 21 22 23 24 25 26 27 28
2Q5 2Q6 2Q7 2Q8 GND NC VCC2 3Q1 3Q2 3Q3 3Q4

NC—No internal connection

description

The SN65509 and SN75509 are monolithic BIDFET† integrated circuits designed to perform the line-select operation of a matrix-addressable display. All inputs are CMOS compatible and all outputs are totem-pole DMOS structures.

The 8-bit data stored internally in the serial register is transferred to one of four output sections selected by the last two bits entered into the 10-bit shift register. All 24 unselected outputs will remain at the high level while the state of the eight selected outputs will be set by the corresponding data in the shift register. V_{CC2} can be used as an output strobe as shown in typical operating sequence.

The SN65509 is characterized for operation from −40°C to 85°C. The SN75509 is characterized for operation from 0°C to 70°C.

FUNCTION TABLE

INPUT BITS			OUTPUTS			
FIRST ENTERED D8 D7 D6 D5 D4 D3 D2 D1	LAST ENTERED S1	S0	BYTE 4 4Q8−4Q1	BYTE 3 3Q8−3Q1	BYTE 2 2Q8−2Q1	BYTE 1 1Q8−1Q1
← D8−D1 →	L	L	All H	All H	All H	D8−D1
← D8−D1 →	L	H	All H	All H	D8−D1	All H
← D8−D1 →	H	L	All H	D8−D1	All H	All H
← D8−D1 →	H	H	D8−D1	All H	All H	All H

† BIDFET — Bipolar, double-diffused, N-channel and P-channel MOS transistors on same chip — patented process

logic symbol†

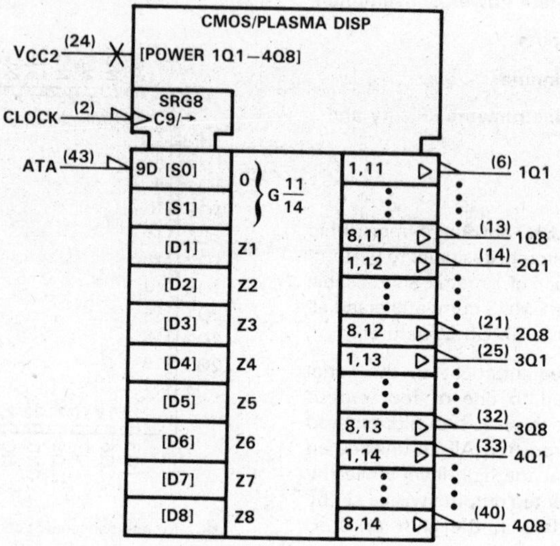
CMOS/PLASMA DISP
VCC2
(24)
[POWER 1Q1–4Q8]
SRG8
CLOCK
(2)
C9/→
ATA
(43)
9D [S0]
[S1]
[D1]
[D2]
[D3]
[D4]
[D5]
[D6]
[D7]
[D8]
0
G 11/14
Z1
Z2
Z3
Z4
Z5
Z6
Z7
Z8
1,11
8,11
1,12
8,12
1,13
8,13
1,14
8,14
(6) 1Q1
(13) 1Q8
(14) 2Q1
(21) 2Q8
(25) 3Q1
(32) 3Q8
(33) 4Q1
(40) 4Q8

† This symbol is in accordance with ANSI/IEEE Std 91-1984 and IEC Publication 617-12.

functional block diagram

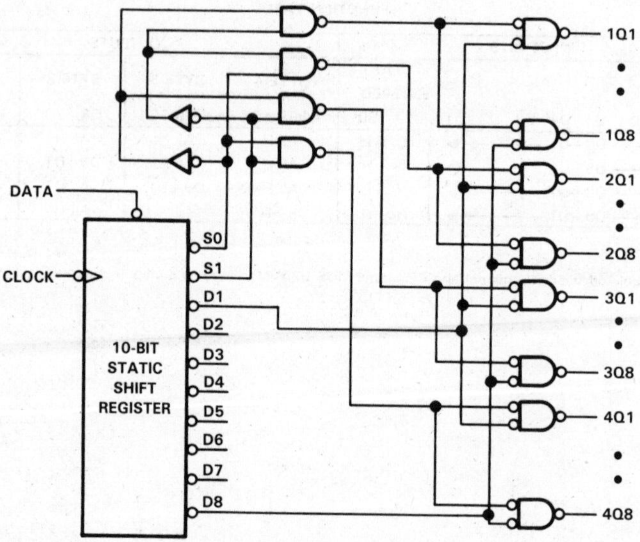
DATA
CLOCK
10-BIT
STATIC
SHIFT
REGISTER
S0
S1
D1
D2
D3
D4
D5
D6
D7
D8
1Q1
1Q8
2Q1
2Q8
3Q1
3Q8
4Q1
4Q8

typical operating sequence

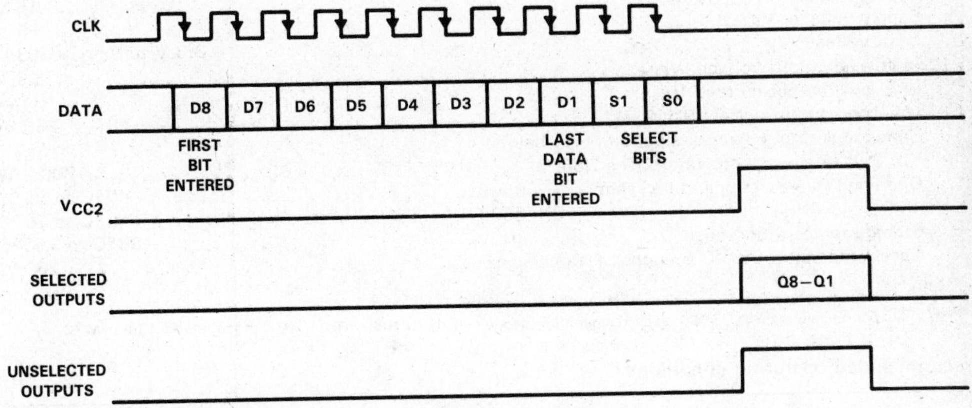
CLK
DATA
D8
D7
D6
D5
D4
D3
D2
D1
S1
S0
FIRST BIT ENTERED
LAST DATA BIT ENTERED
SELECT BITS
VCC2
SELECTED OUTPUTS
Q8–Q1
UNSELECTED OUTPUTS

schematic of inputs and outputs

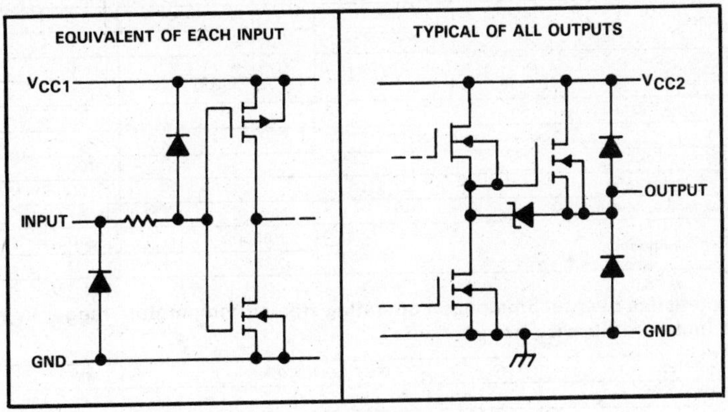
EQUIVALENT OF EACH INPUT
VCC1
INPUT
GND
TYPICAL OF ALL OUTPUTS
VCC2
OUTPUT
GND

absolute maximum ratings over operating free-air temperature range (unless otherwise noted)

Supply voltage, V_{CC1} (see Note 1)	15 V
Supply voltage, V_{CC2}	95 V
Input voltage, V_I	−0.3 V to V_{CC} + 0.3 V
High-level output voltage, V_{OH}	95 V
High-level output current, I_{OH}	−3 mA
Low-level output current, I_{OL}	25 mA
Continuous total power dissipation at (or below) 25°C free-air temperature (see Note 2)	1700 mW
Operating free-air temperature range: SN65509	−40°C to 85°C
SN75509	0°C to 70°C
Storage temperature range	−65°C to 150°C
Case temperature for 10 seconds: FN package	260°C

NOTES: 1. Voltage values are with respect to network ground terminal.
2. For operation above 25°C free-air temperature, derate linearly to 1088 mW at 70°C at the rate of 13.6 mW/°C.

recommended operating conditions

		MIN	MAX	UNIT
Supply voltage, V_{CC1}		8	11.4	V
Supply voltage, V_{CC2}		V_{CC1}	90	V
High-level input voltage, V_{IH}	V_{CC1} = 11.4 V	8.5	V_{CC1}	V
	V_{CC1} = 8 V	6	V_{CC1}	
Low-level input voltage, V_{IL}	V_{CC1} = 11.4 V	0	2.9	V
	V_{CC1} = 8 V	0	2	
Output current, I_O ($t_w \le 1\ \mu s$)			80	mA
Clock frequency, f_{clock}			3.1	MHz
Setup time, data before clock↓, t_{su}		100		ns
Hold time, data after clock↓, t_h		100		ns
Pulse duration, clock high or low, t_{wCLK}		161		ns
Operating free-air temperature, T_A	SN65509	−40	85	°C
	SN75509	0	70	

electrical characteristics over recommended operating free-air temperature range, V_{CC1} = 11.4 V, V_{CC2} = 90 V (unless otherwise noted)

PARAMETER		TEST CONDITIONS		MIN	MAX	UNIT
V_{OH}	High-level output voltage	V_{CC1} = 8 V,	I_{OH} = −3 mA	83	87	V
V_{OL}	Low-level output voltage	V_{CC1} = 8 V,	I_{OL} = 20 mA	1.4	2.4	V
V_{OK}	Output clamp voltage	V_{CC2} = 0	I_O = 100 mA, $t_w \le 1\ \mu s$		2.5	V
			I_O = −100 mA, $t_w \le 1\ \mu s$		−2.7	
I_{IH}	High-level input current	V_I = 11.4 V			1	μA
I_{IL}	Low-level input current	V_I = 0.4 V			−1	μA
I_{OS}	Short-circuit output current	V_O = 0			−20	mA
I_{CC1}	Supply current from V_{CC1}				500	μA
I_{CC2}	Supply current from V_{CC2}	All outputs high			500	μA
		Eight outputs low			5	mA

TEXAS INSTRUMENTS
POST OFFICE BOX 655012 • DALLAS, TEXAS 75265

switching characteristics, V_{CC1} = 8 V, T_A = 25°C

PARAMETER	TEST CONDITIONS	MIN	MAX	UNIT
t_{d1} Delay time, V_{CC2} to Q outputs (10% – 10%)	R_L = 100 kΩ, C_L = 100 pF		800	ns
t_{d2} Delay time, V_{CC2} to Q outputs (90% – 90%)	R_L = 100 kΩ, C_L = 100 pF		800	ns

PARAMETER MEASUREMENT INFORMATION

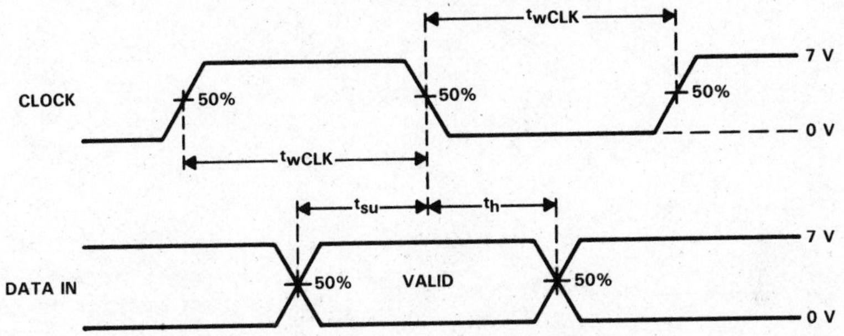

tWCLK
7 V
CLOCK
50%
50%
50%
0 V
tWCLK
tsu
th
7 V
DATA IN
50%
VALID
50%
0 V

FIGURE 1. INPUT TIMING VOLTAGE WAVEFORMS

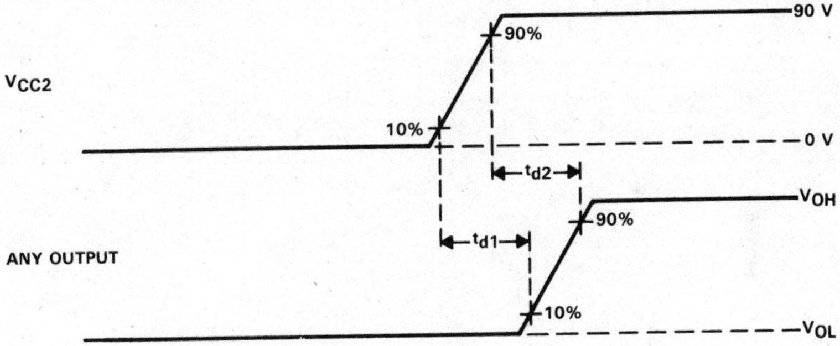

90 V
90%
VCC2
10%
0 V
td2
VOH
90%
td1
ANY OUTPUT
10%
VOL

FIGURE 2. VOLTAGE WAVEFORMS FOR OUTPUT DELAY

SN65512B, SN75512B
VACUUM FLUORESCENT DISPLAY DRIVERS

D2654, DECEMBER 1985

- Each Device Drives 12 Lines
- 60-V Output Voltage Swing Capability
- 25-mA Output Source Current Capability
- High-Speed Serially-Shifted Data Input
- TTL-Compatible Inputs
- Latches on All Driver Outputs

DW, N
DUAL-IN-LINE PACKAGE
(TOP VIEW)

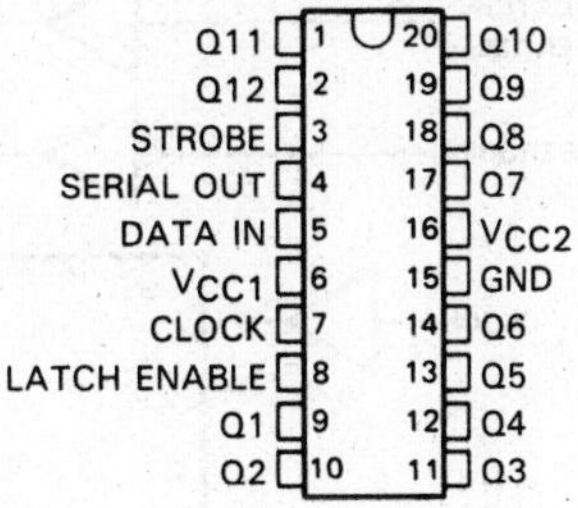
Q11 1 20 Q10
Q12 2 19 Q9
STROBE 3 18 Q8
SERIAL OUT 4 17 Q7
DATA IN 5 16 VCC2
VCC1 6 15 GND
CLOCK 7 14 Q6
LATCH ENABLE 8 13 Q5
Q1 9 12 Q4
Q2 10 11 Q3

description

The SN65512B and SN75512B are monolithic BIDFET† integrated circuits designed to drive a dot matrix or segmented vacuum fluorescent display.

All device inputs are diode-clamped p-n-p inputs and will assume a high logic level when open-circuited. The nominal input threshold is 1.5 volts. Outputs are totem-pole structures formed by an n-p-n emitter follower and double-diffused MOS (DMOS) transistors.

The device consists of a 12-bit shift register, 12 latches, and 12 output AND gates. Serial data is entered into the shift register on the low-to-high transition of the Clock. When high, the Latch Enable input transfers the shift register contents to the outputs of the 12 latches. The active-low strobe input enables all Q outputs. Serial data output from the shift register may be used to cascade shift registers. This output is not affected by the Latch Enable or Strobe inputs.

The SN65512B is characterized for operation from −40°C to 85°C. The SN75512B is characterized for operation from 0°C to 70°C.

logic symbol‡

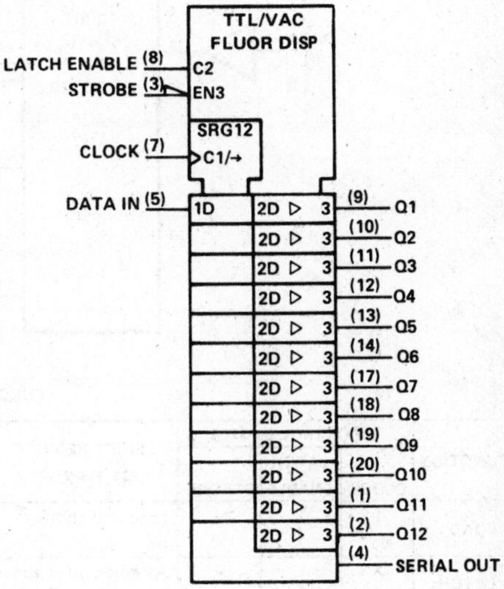
TTL/VAC
FLUOR DISP
LATCH ENABLE (8)
C2
STROBE (3)
EN3
SRG12
CLOCK (7)
C1/→
DATA IN (5)
1D
2D ▷ 3
(9) Q1
(10) Q2
(11) Q3
(12) Q4
(13) Q5
(14) Q6
(17) Q7
(18) Q8
(19) Q9
(20) Q10
(1) Q11
(2) Q12
(4) SERIAL OUT

‡ This symbol is in accordance with ANSI/IEEE Std 91-1984 and IEC Publication 617-12.

† BIDFET — Bipolar, double-diffused, N-channel and P-channel MOS transistors on same chip — patented process.

functional block diagram (positive logic)

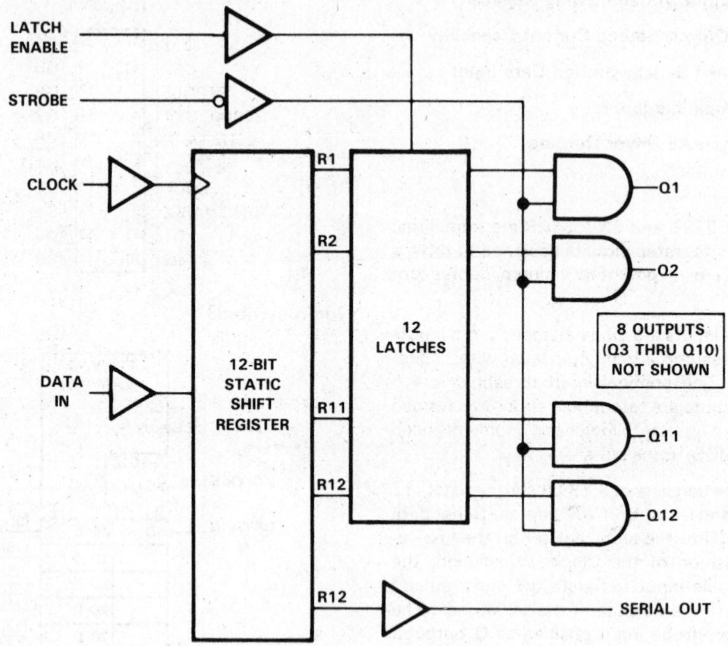
LATCH ENABLE
STROBE
CLOCK
DATA IN
R1
R2
R11
R12
R12
12-BIT STATIC SHIFT REGISTER
12 LATCHES
8 OUTPUTS (Q3 THRU Q10) NOT SHOWN
Q1
Q2
Q11
Q12
SERIAL OUT

FUNCTION TABLE

FUNCTION	CONTROL INPUTS			SHIFT REGISTER R1 THRU R12	LATCHES LC1 THRU LC12	OUTPUTS	
	CLOCK	LATCH ENABLE	STROBE			SERIAL	Q1 THRU Q12
LOAD	↑	X	X	Load and shift†	Determined by Latch Enable‡	R12	Determined by Strobe
	Not↑	X	X	No change	Determined by Latch Enable‡	R12	Determined by Strobe
LATCH	X	L	X	As determined above	Stored data	R12	Determined by Strobe
	X	H	X	As determined above	New data	R12	Determined by Strobe
STROBE	X	X	H	As determined above	Determined by Latch Enable‡	R12	All L
	X	X	L	As determined above	Determined by Latch Enable‡	R12	LC1 thru LC12, respectively

H = high level, L = low level, X = irrelevant, ↑ = low-to-high-level transition.

† R12 takes on the state of R11, R11 takes on the state of R10, . . . R2 takes on the state of R1, and R1 takes on the state of the data input.

‡ New data enter the latches while Latch Enable is high. These data are stored while Latch Enable is low.

typical operating sequence

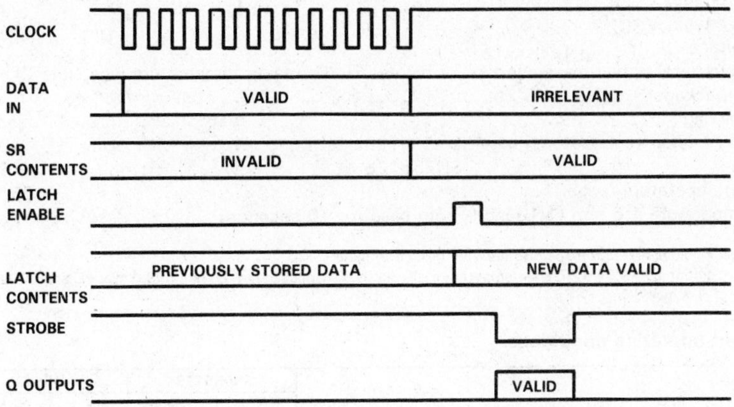
CLOCK
DATA IN
VALID
IRRELEVANT
SR CONTENTS
INVALID
VALID
LATCH ENABLE
LATCH CONTENTS
PREVIOUSLY STORED DATA
NEW DATA VALID
STROBE
Q OUTPUTS
VALID

schematics of inputs and outputs

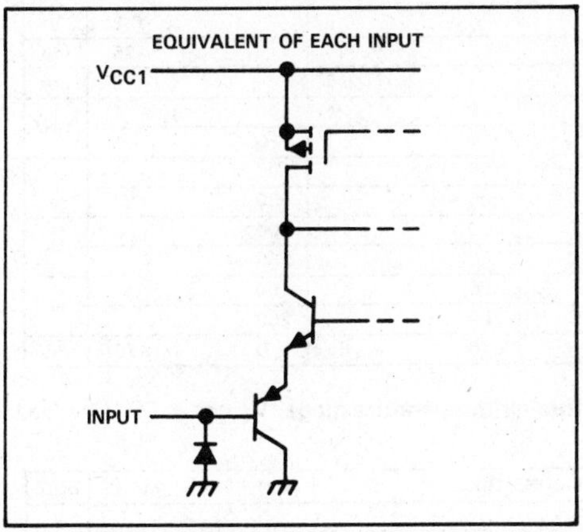
EQUIVALENT OF EACH INPUT
V_{CC1}
INPUT

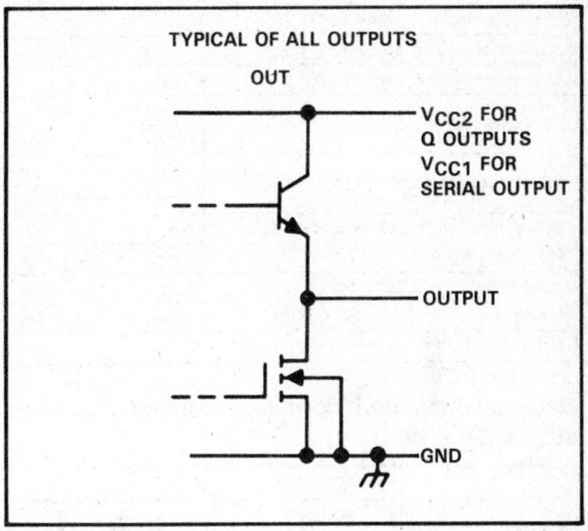
TYPICAL OF ALL OUTPUTS
OUT
V_{CC2} FOR Q OUTPUTS
V_{CC1} FOR SERIAL OUTPUT
OUTPUT
GND

3
Display Drivers

absolute maximum ratings over operating free-air temperature range (unless otherwise noted)

Supply voltage, V_{CC1} (see Note 1) . . . . 15 V
Supply voltage, V_{CC2} . . . . 70 V
Input voltage . . . . V_{CC1}
Continuous total power dissipation at (or below) 25°C free-air temperature (see Note 2)
  DW package . . . . 1125 mW
  N package . . . . 1150 mW
Operating free-air temperature range: SN65512B . . . . −40°C to 85°C
  SN75512B . . . . 0°C to 70°C
Storage temperature range . . . . −65°C to 150°C
Lead temperature 1,6 mm (1/16 inch) from case for 10 seconds . . . . 260°C

NOTES: 1. Voltage values are with respect to network ground terminal.
2. For operation above 25°C free-air temperature, derate the DW package at the rate of 9.0 mW/°C and the N package at the rate of 9.2 mW/°C.

recommended operating conditions

		SN65512B MIN	SN65512B MAX	SN75512B MIN	SN75512B MAX	UNIT
Supply voltage, V_{CC1}		5	15	5	15	V
Supply voltage, V_{CC2}		0	60	0	60	V
High-level input voltage, V_{IH}		2		2		V
Low-level input voltage, V_{IL}			0.8		0.8	V
High-level output current, I_{OH}			−25		−25	mA
Low-level output current, I_{OL}	V_{CC1} = 10 V		5		5	mA
Clock frequency, f_{clock}	V_{CC1} = 15 V, T_A = 25°C	0	4	0	4	MHz
	V_{CC1} = 5 V, T_A = 25°C	0	1	0	1	
Pulse duration, clock high or low, t_w	V_{CC1} = 15 V, T_A = 25°C	100		100		ns
	V_{CC1} = 5 V, T_A = 25°C	500		500		
Setup time, data before clock↑, t_{su} (see Figure 1)	V_{CC1} = 15 V, T_A = 25°C	100		100		ns
	V_{CC1} = 5 V, T_A = 25°C	250		250		
Hold time, data after clock↑, t_h (see Figure 1)	V_{CC1} = 15 V, T_A = 25°C	50		50		ns
	V_{CC1} = 5 V, T_A = 25°C	250		250		
Operating free-air temperature, T_A		−40	85	0	70	°C

electrical characteristics over recommended operating free-air temperature range, V_{CC2} = 60 V (unless otherwise noted)

PARAMETER			TEST CONDITIONS		MIN	TYP†	MAX	UNIT
V_{IK}	Input clamp voltage		I_I = −12 mA				−1.5	V
V_{OH}	High-level output voltage	Q outputs	I_{OH} = −25 mA		57.5	58		V
		Serial output	I_{OH} = −200 μA, V_{CC1} = 10 V		9	9.5		
V_{OL}	Low-level output voltage	Q outputs	I_{OL} = 5 mA, V_{CC1} = 10 V			2.6	5	V
		Serial output	I_{OL} = 200 μA, V_{CC1} = 10 V			0.05	0.2	
I_{IH}	High-level input current		V_{CC1} = 15 V, V_I = 5 V			0.01	1	μA
I_{IL}	Low-level input current		V_{CC1} = 15 V, V_I = 0.8 V			−25	−150	μA
I_{CC1}	Supply current from V_{CC1}		V_{CC1} = 15 V	V_I = 5 V		80	500	μA
				V_I = 0.8 V		2	6	mA
I_{CC2}	Supply current from V_{CC2}		V_{CC1} = 15 V	All outputs high		10	100	μA
				Strobe at 2 V		0.8	3	mA

† All typical values are at V_{CC1} = 10 V, T_A = 25°C.

switching characteristics, V_{CC1} = 10 V, V_{CC2} = 60 V, T_A = 25°C

PARAMETER		TEST CONDITIONS	MIN	MAX	UNIT
t_{DHL}	Delay time, high-to-low-level output	C_L = 30 pF, See Figure 2		300	ns
t_{DLH}	Delay time, low-to-high-level output			300	ns
t_{THL}	Transition time, high-to-low-level output			500	ns
t_{TLH}	Transition time, low-to-high-level output			500	ns

PARAMETER MEASUREMENT INFORMATION

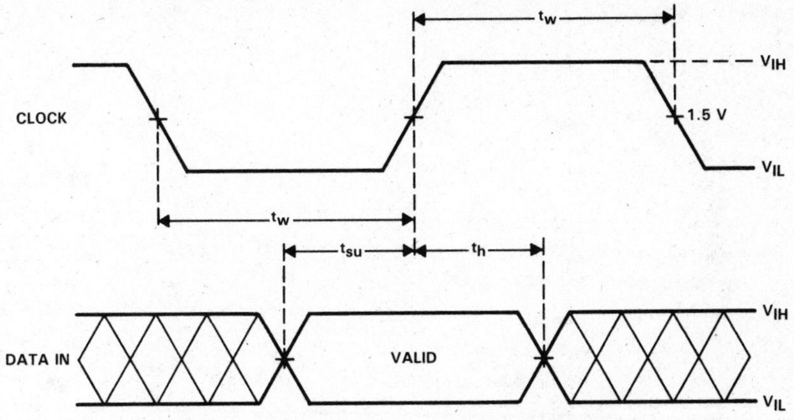
tw
VIH
CLOCK
1.5 V
VIL
tw
tsu
th
VIH
DATA IN
VALID
VIL

FIGURE 1. INPUT TIMING VOLTAGE WAVEFORMS

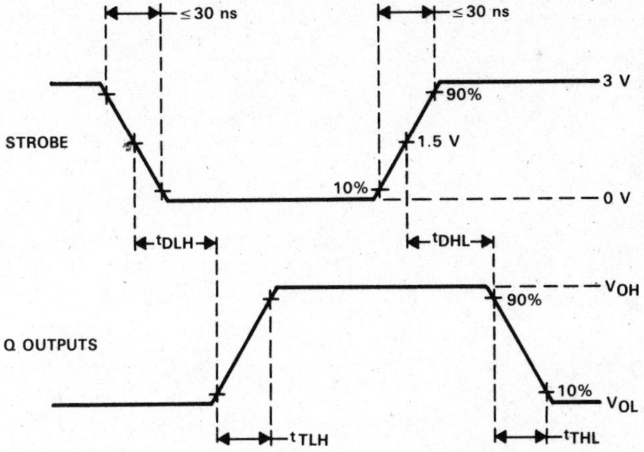
≤30 ns
≤30 ns
3 V
90%
STROBE
1.5 V
10%
0 V
tDLH
tDHL
VOH
90%
Q OUTPUTS
10%
VOL
tTLH
tTHL

FIGURE 2. SWITCHING-TIME VOLTAGE WAVEFORMS

D2721, MARCH 1983—REVISED SEPTEMBER 1986

- **Each Device Drives 12 Lines**
- **60-V Output Voltage Swing Capability**
- **25-mA Output Source Current Capability**
- **High-Speed Serially-Shifted Data Input**
- **TTL-Compatible Input**
- **Reset Input**

DW OR N PACKAGE
(TOP VIEW)

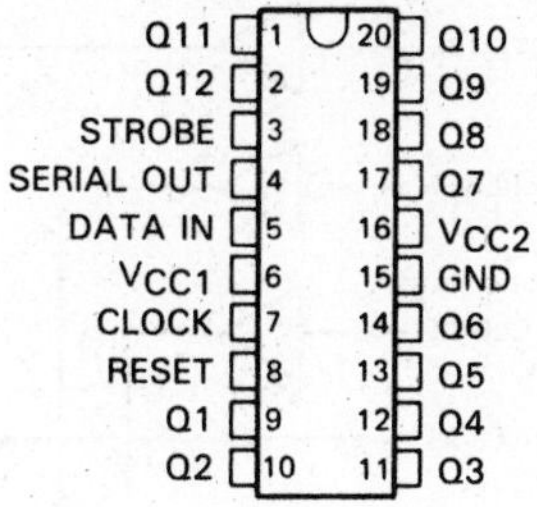
Q11 1
Q12 2
STROBE 3
SERIAL OUT 4
DATA IN 5
VCC1 6
CLOCK 7
RESET 8
Q1 9
Q2 10
20 Q10
19 Q9
18 Q8
17 Q7
16 VCC2
15 GND
14 Q6
13 Q5
12 Q4
11 Q3

description

The SN65513B and SN75513B are monolithic BIDFET† integrated circuits designed to drive a dot matrix or segmented vacuum fluorescent display.

All device inputs are diode-clamped p-n-p inputs and will assume a high logic level when left open. The nominal input threshold is 1.5 volts. Outputs are totem-pole structures formed by n-p-n emitter follower and double-diffused MOS (DMOS) transistors.

The device consists of a 12-bit shift register and 12 output AND gates. Data is entered into the shift register on the low-to-high transition of the Clock input. The active-low strobe input enables all Q outputs. The Reset input sets the shift register contents to all lows. The serial data output from the shift register may be used to cascade additional devices. This output is not affected by the strobe input.

The SN65513B is characterized for operation from −40°C to 85°C and the SN75513B is characterized for operation from 0°C to 70°C.

logic symbol‡

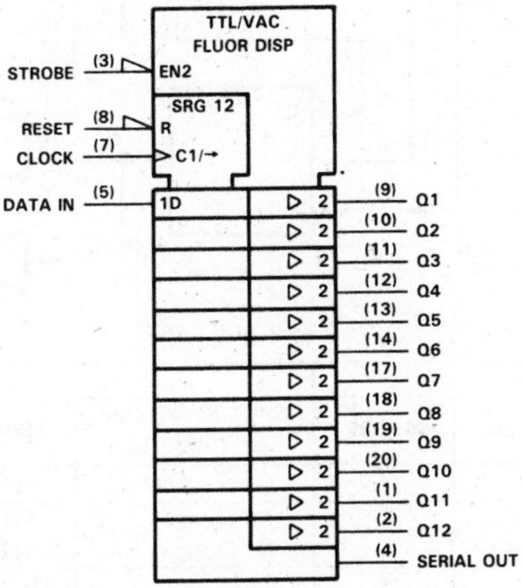
TTL/VAC
FLUOR DISP
STROBE (3) EN2
SRG 12
RESET (8) R
CLOCK (7) C1/→
DATA IN (5) 1D
(9) Q1
(10) Q2
(11) Q3
(12) Q4
(13) Q5
(14) Q6
(17) Q7
(18) Q8
(19) Q9
(20) Q10
(1) Q11
(2) Q12
(4) SERIAL OUT

† BIDFET—Bipolar, double-diffused, N-channel and P-channel MOS transistors on same chip—patented process.
‡ This symbol is in accordance with ANSI/IEEE Std 91-1984 and IEC Publication 617-12.

logic diagram (positive logic)

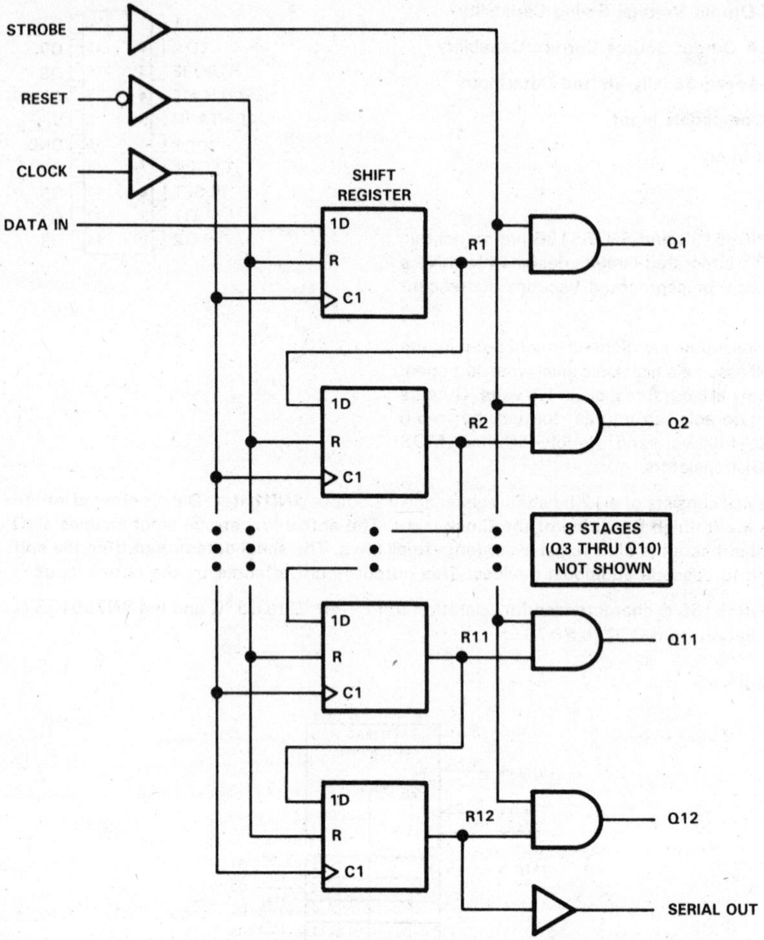
STROBE
RESET
CLOCK
DATA IN
SHIFT REGISTER
1D
R
C1
R1
R2
R11
R12
Q1
Q2
Q11
Q12
8 STAGES
(Q3 THRU Q10)
NOT SHOWN
SERIAL OUT

FUNCTION TABLE

FUNCTION	INPUTS				OUTPUTS	
	RESET	CLOCK	STROBE	SHIFT REGISTERS R1 THRU R12	SERIAL	Q1 THRU Q12
LOAD	H	↑	X	Load and Shift†	R12†	Determined by strobe
STROBE	H	Not↑	H	No Change	R12	All L
	H	Not↑	L	No Change	R12	R1 thru R12, respectively
RESET	L	H	X	All L	L	All L

H = high level, L = low level, X = irrelevant, ↑ = low-to-high transition.
† R12 and the serial output take on the state of R11, R11 takes on the state of R10 . . . R2 takes on the state of R1, and R1 takes on the state of the data input.

typical operating sequence

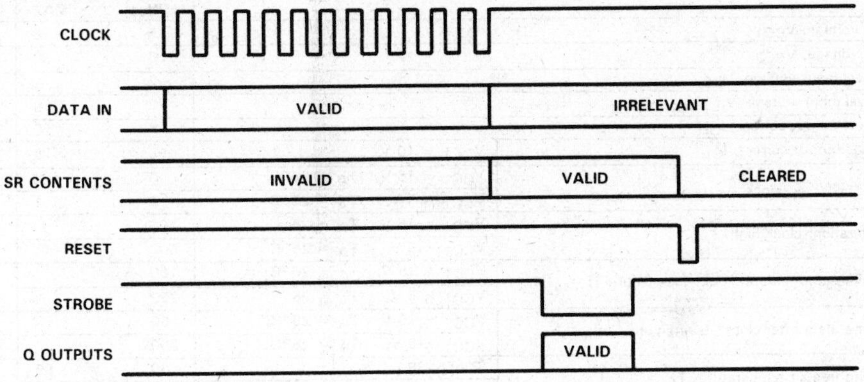
CLOCK
DATA IN
VALID
IRRELEVANT
SR CONTENTS
INVALID
VALID
CLEARED
RESET
STROBE
Q OUTPUTS
VALID

schematics of inputs and outputs

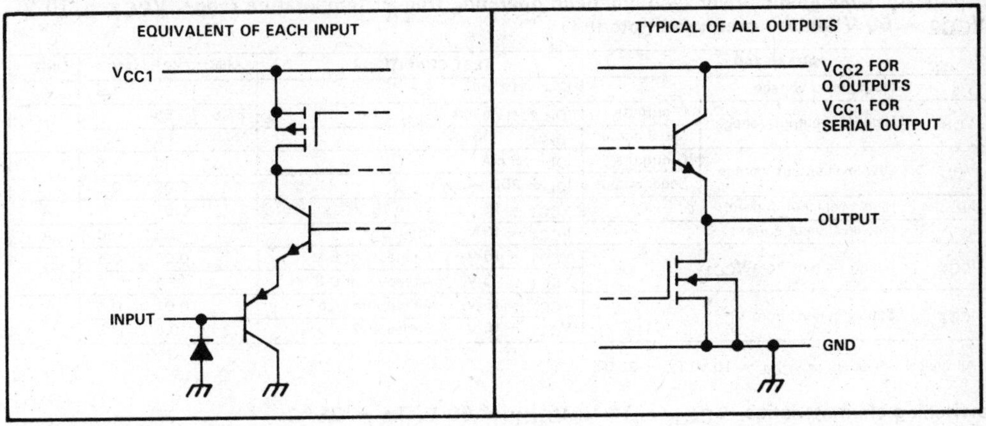
EQUIVALENT OF EACH INPUT
VCC1
INPUT
TYPICAL OF ALL OUTPUTS
VCC2 FOR
Q OUTPUTS
VCC1 FOR
SERIAL OUTPUT
OUTPUT
GND

absolute maximum ratings over operating free-air temperature range (unless otherwise noted)

Supply voltage, V_{CC1} (see Note 1)	15 V
Supply voltage, V_{CC2}	70 V
Input voltage	V_{CC1}
Continuous total dissipation at (or below) 25°C free-air temperature (see Note 2)	1150 mW
Operating free-air temperature range: SN65513B	−40°C to 85°C
SN75513B	0°C to 70°C
Storage temperature range	−65°C to 150°C
Lead temperature 1,6 mm (1/16 inch) from case for 10 seconds	260°C

NOTES: 1. Voltage values are with respect to network ground terminal.
2. For operation above 25°C free-air temperature, derate linearly to 598 mW at 85°C at the rate of 9.2 mW/°C.

recommended operating conditions, T_A = −40°C to 85°C (unless otherwise noted)

		MIN	MAX	UNIT
Supply voltage, V_{CC1}		5	15	V
Supply voltage, V_{CC2}		0	60	V
High-level input voltage, V_{IH}		2		V
Low-level input voltage, V_{IL}			0.8	V
High-level output current, I_{OH}			25	mA
Low-level output current, I_{OL}	V_{CC1} = 10 V		5	mA
Clock frequency, f_{clock}	V_{CC1} = 15 V, T_A = 25°C	0	4	MHz
	V_{CC1} = 5 V, T_A = 25°C	0	1	
Pulse duration, clock high, t_w	V_{CC1} = 15 V, T_A = 25°C	100		ns
	V_{CC1} = 5 V, T_A = 25°C	500		
Setup time, data before clock↑ (see Figure 1), t_{su}	V_{CC1} = 15 V, T_A = 25°C	100		ns
	V_{CC1} = 5 V, T_A = 25°C	250		
Hold time, data after clock↑ (see Figure 1), t_h	V_{CC1} = 15 V, T_A = 25°C	50		ns
	V_{CC1} = 5 V, T_A = 25°C	250		
Operating free-air temperature, T_A	SN65513B	−40	85	°C
	SN75513B	0	70	

electrical characteristics over recommended operating free-air temperature range, V_{CC1} = 10 V, V_{CC2} = 60 V (unless otherwise noted)

PARAMETER			TEST CONDITIONS		MIN	TYP†	MAX	UNIT
V_{IK}	Input clamp voltage		I_I = −12 mA				−1.5	V
V_{OH}	High-level output voltage	Q outputs	I_{OH} = −25 mA		57.5	58		V
		Serial output	I_{OH} = −200 μA		9	9.5		
V_{OL}	Low-level output voltage	Q outputs	I_{OL} = 5 mA			2.6	5	V
		Serial output	I_{OL} = 200 μA			0.05	0.2	
I_{IH}	High-level input current		V_{CC1} = 15 V,	V_I = 15 V		0.01	1	μA
I_{IL}	Low-level input current		V_{CC1} = 15 V,	V_I = 0 V		−25	−150	μA
I_{CC1}	Supply current from V_{CC1}		V_{CC1} = 15 V,	All inputs at 5 V		0.08	0.5	mA
			V_{CC1} = 15 V,	All inputs at 0.8 V		2	6	
I_{CC2}	Supply current from V_{CC2}		V_{CC1} = 15 V,	All outputs high		0.01	0.1	mA
			V_{CC1} = 15 V,	Strobe at 2 V		0.8	3	

† All typical values are at V_{CC1} = 10 V, T_A = 25°C.

switching characteristics, V_{CC1} = 10 V, V_{CC2} = 60 V, T_A = 25°C

PARAMETER		TEST CONDITIONS	MIN	MAX	UNIT
t_{DHL}	Delay time, high-to-low-level output	C_L = 30 pF, See Figure 2		300	ns
t_{DLH}	Delay time, low-to-high-level output			300	ns
t_{THL}	Transition time, high-to-low-level output			500	ns
t_{TLH}	Transition time, low-to-high-level output			500	ns

TEXAS INSTRUMENTS
POST OFFICE BOX 655012 • DALLAS, TEXAS 75265

PARAMETER MEASUREMENT INFORMATION

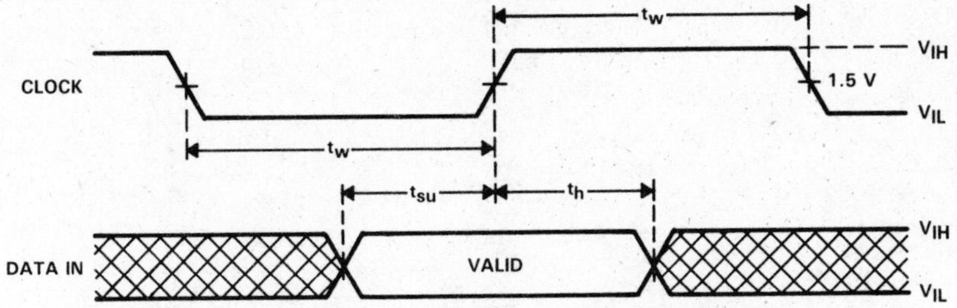
CLOCK
DATA IN
VALID
t_w
t_{su}
t_h
1.5 V
V_{IH}
V_{IL}

FIGURE 1. INPUT TIMING VOLTAGE WAVEFORMS

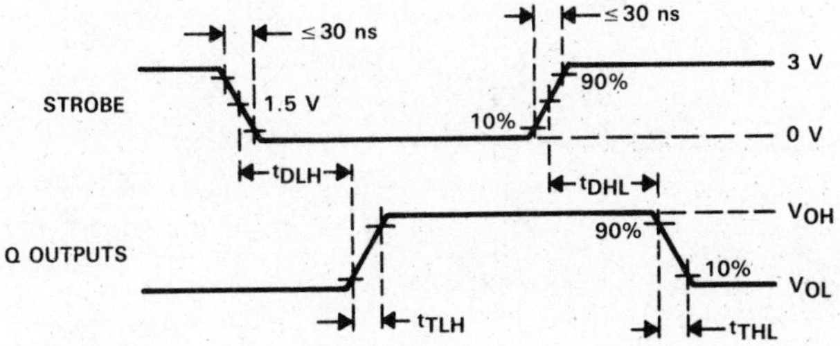
STROBE
Q OUTPUTS
≤ 30 ns
1.5 V
10%
90%
3 V
0 V
t_{DLH}
t_{DHL}
t_{TLH}
t_{THL}
V_{OH}
V_{OL}

FIGURE 2. SWITCHING-TIME VOLTAGE WAVEFORMS

ADVANCE INFORMATION

SN75514
VACUUM FLUORESCENT DISPLAY DRIVER

D2732, APRIL 1983—REVISED SEPTEMBER 1986

- Each Device Drives 12 Lines
- 125-V Output Voltage Swing Capability
- 25-mA Output Source Current Capability
- High-Speed Serially Shifted Data Input
- CMOS-Compatible Inputs
- Latches on All Driver Outputs

DW OR N PACKAGE
(TOP VIEW)

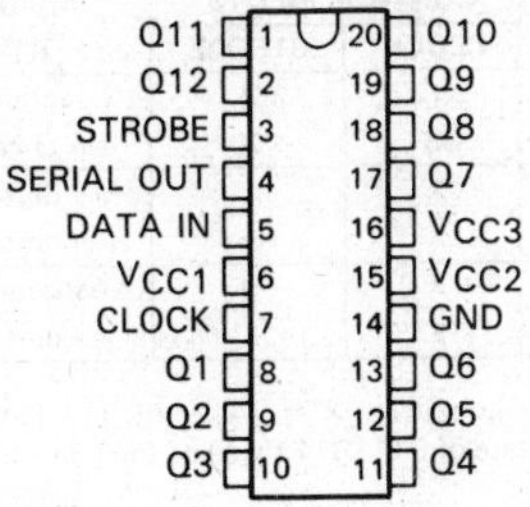
Q11 1 20 Q10
Q12 2 19 Q9
STROBE 3 18 Q8
SERIAL OUT 4 17 Q7
DATA IN 5 16 VCC3
VCC1 6 15 VCC2
CLOCK 7 14 GND
Q1 8 13 Q6
Q2 9 12 Q5
Q3 10 11 Q4

description

The SN75514 is a monolithic BIDFET† integrated circuit designed to drive a dot matrix or segmented vacuum fluorescent display. All device inputs are diode-clamped CMOS compatible inputs. The outputs are totem-pole structures formed with double-diffused MOS (DMOS) transistors.

The device consists of a 12-bit shift register, a 12-bit storage register, and 12 output AND gates. Serial data is entered into the shift register on the low-to-high transition of the clock input. On the high-to-low transition of the strobe input, data is transferred from the shift registers to the latches. When Strobe goes high, all Q outputs are enabled. Serial data output from the shift register may be used to cascade additional devices. Serial Out is not affected by the Strobe input.

Supply voltage V_{CC2} and V_{CC3} are used to provide 25-milliampere output source current capability at acceptable static device power dissipation. In this mode of operation V_{CC3} should be equal to V_{CC2} + 10 volts. It is possible to operate this device with $V_{CC3} = V_{CC2}$. However, the current capability will be reduced.

The SN75514 is characterized for operation from 0°C to 70°C.

logic symbol‡

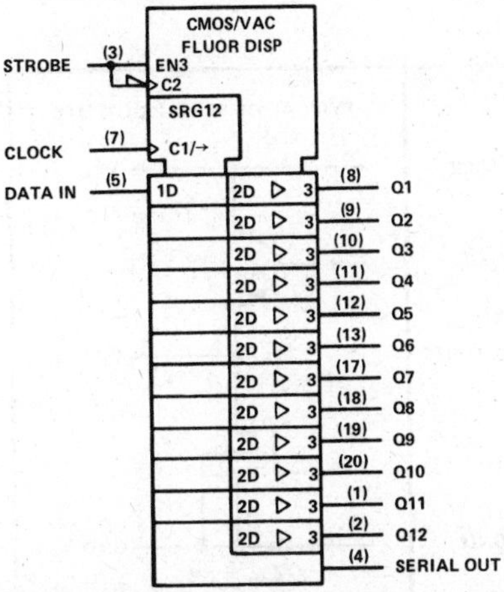
CMOS/VAC FLUOR DISP
STROBE (3) EN3 C2
SRG12
CLOCK (7) C1/→
DATA IN (5) 1D
2D ▷ 3
(8) Q1
(9) Q2
(10) Q3
(11) Q4
(12) Q5
(13) Q6
(17) Q7
(18) Q8
(19) Q9
(20) Q10
(1) Q11
(2) Q12
(4) SERIAL OUT

logic diagram (positive logic)

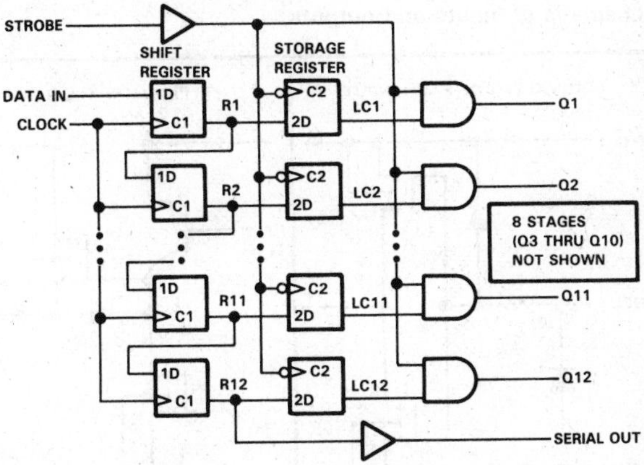
STROBE
SHIFT REGISTER
STORAGE REGISTER
DATA IN
CLOCK
1D
C1
C2
2D
R1
R2
R11
R12
LC1
LC2
LC11
LC12
Q1
Q2
Q11
Q12
8 STAGES (Q3 THRU Q10) NOT SHOWN
SERIAL OUT

†BIDFET—Bipolar, double-diffused, N-channel and P-channel MOS transistors on the same chip—patented process.
‡This symbol is in accordance with ANDI/IEEE Std 91-1984 IEC Publication 617-12.

FUNCTION TABLE

FUNCTION	CONTROL INPUTS		SHIFT REGISTERS	LATCHES	OUTPUTS	
	CLOCK	STROBE	R1 THRU R12	LC1 THRU LC12	SERIAL	Q1 THRU Q12
LOAD	↑	X	Load and shift*	Stored data	R12	Determined by Strobe
	No↑	X	No change	Stored data	R12	Determined by Strobe
LATCH	X	↓	As determined above	New data	R12	Determined by Strobe
	X	No↓	As determined above	Stored data	R12	Determined by Strobe
STROBE	X	H	As determined above	Stored data	R12	LC1 thru LC12, respectively
	X	L	As determined above	Stored data	R12	All L

H = high level, L = Low level, X = irrelevant, ↑ = low-to-high-level transition, ↓ = high-to-low-level transition.
*R12 takes on the state of R11, R11 takes on the state of R10 . . . R2 takes on the state of R1, and R1 takes on the state of the data input.

typical operating sequence

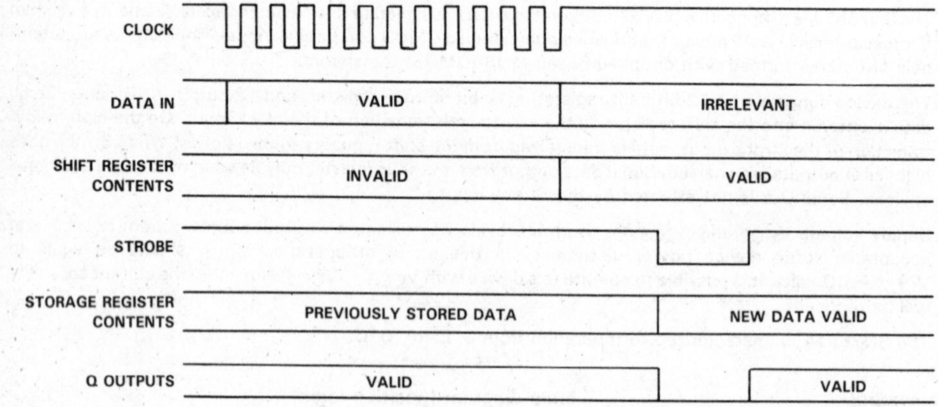
CLOCK
DATA IN
VALID
IRRELEVANT
SHIFT REGISTER CONTENTS
INVALID
VALID
STROBE
STORAGE REGISTER CONTENTS
PREVIOUSLY STORED DATA
NEW DATA VALID
Q OUTPUTS
VALID
VALID

schematic of inputs and outputs

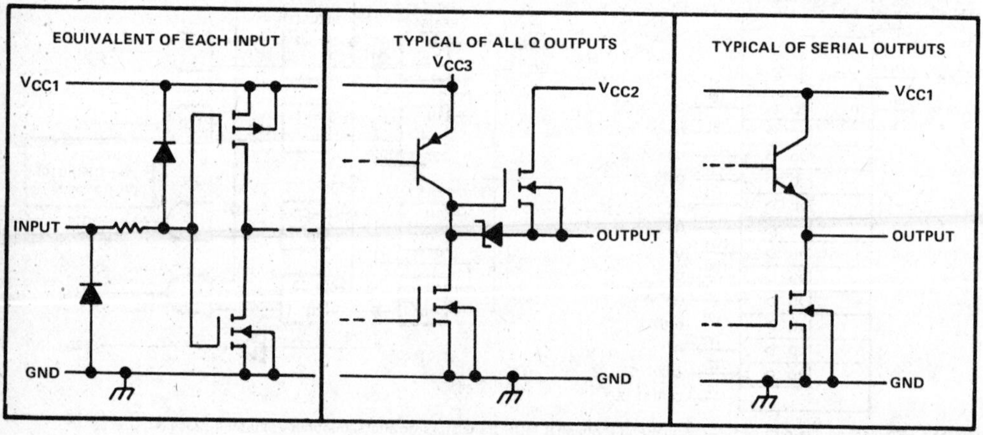
EQUIVALENT OF EACH INPUT
VCC1
INPUT
GND
TYPICAL OF ALL Q OUTPUTS
VCC3
VCC2
OUTPUT
GND
TYPICAL OF SERIAL OUTPUTS
VCC1
OUTPUT
GND

TEXAS INSTRUMENTS
POST OFFICE BOX 655012 • DALLAS, TEXAS 75265

absolute maximum ratings over operating free-air temperature range (unless otherwise noted)

Supply voltage, V_{CC1} (see Note 1) .... 15 V
Supply voltage, V_{CC2} .... 130 V
Supply voltage, V_{CC3} .... 140 V
Supply voltage difference, $V_{CC3} - V_{CC2}$ .... 75 V
Input voltage .... V_{CC1}
Continuous total dissipation at (or below) 25°C free-air temperature (see Note 2):
DW package .... 1125 mW
N package .... 1150 mW
Operating free-air temperature range .... 0°C to 70°C
Storage temperature range .... −65°C to 150°C
Lead temperature 1,6 mm (1/16 inch) from case for 10 seconds .... 260°C

NOTES: 1. Voltage values are with respect to network ground terminal.
2. For operation above 25°C free-air temperature, derate the DW package to 720 mW at 70°C at the rate of 9.0 mW/°C, and the N package to 736 mW at 70°C at the rate of 9.2 mW/°C.

recommended operating conditions

		MIN	NOM	MAX	UNIT
Supply voltage, V_{CC1}		5		15	V
Supply voltage, V_{CC2}		0		130	V
Supply voltage, V_{CC3}		V_{CC2}		$V_{CC2}+10$	V
High-level input voltage, V_{IH} (see Figure 1)	V_{CC1} = 5 V	4			V
	V_{CC1} = 15 V	11.25			
Low-level input voltage, V_{IL} (see Figure 1)	V_{CC1} = 5 V			1	V
	V_{CC1} = 15 V			3.75	
High-level output current, I_{OH} (T_A = 25°C)				−25	mA
Low-level output current, I_{OL}				2.5	mA
Clock frequency, f_{clock} (see Figure 2)		0		7.5	MHz
Data setup time before clock↑, t_{su} (see Figure 3)		150			ns
Data hold time after clock↑, t_h (see Figure 3)		150			ns
Delay time, strobe low to clock high, $t_{d(SL-CH)}$	V_{CC} = 5 V	1200			ns
	V_{CC} = 15 V	500			
Operating free-air temperature, T_A		0		70	°C

electrical characteristics over recommended operating free-air temperature range, V_{CC1} = 10 V (unless otherwise noted)

PARAMETER			TEST CONDITIONS		MIN	TYP†	MAX	UNIT
V_{IK}	Input clamp voltage		I_I = −1 mA				−1.5	V
V_{OH}	High-level output voltage	Q outputs	V_{CC2} = 130 V,	I_O = −25 mA	125	126		V
		Serial	I_{OH} = −200 μA		9	9.3		
V_{OL}	Low-level output voltage	Q outputs	I_{OL} = 2.5 mA			1.5	5	V
		Serial	I_{OL} = 200 μA				1	
I_{IH}	High-level input current		V_{CC1} = 10 V,	V_I = 10 V		0.01	1	μA
I_{IL}	Low-level input current		V_{CC1} = 10 V,	V_I = 0 V			−5	μA
I_{CC1}	Supply Current from V_{CC1}		V_{CC1} = 15 V				5	mA
			V_{CC1} = 5 V				5	
I_{CC2}	Supply Current from V_{CC2}		V_{CC1} = 15 V, V_{CC2} = 130 V, V_{CC3} = 140 V	All outputs high			−5	mA
				All outputs low			0.1	
I_{CC3}	Supply Current from I_{CC3}		V_{CC1} = 15 V, V_{CC2} = 130 V, V_{CC3} = 140 V	All outputs high			5	mA
				Strobe at 0 V			0.1	

†All typical values are at T_A = 25°C.

switching characteristics, V_{CC1} = 15 V, V_{CC2} = 130 V, T_A = 25°C

PARAMETER		TEST CONDITIONS	MIN	MAX	UNIT
t_{DHL}	Delay time, high-to-low-level output	C_L = 30 pF, See Figure 4		0.8	μs
t_{DLH}	Delay time, low-to-high-level output			0.8	μs
t_{THL}	Transition time, high-to-low-level output			1	μs
t_{TLH}	Transition time, low-to-high-level output			3	μs

TEXAS INSTRUMENTS
POST OFFICE BOX 655012 • DALLAS, TEXAS 75265

RECOMMENDED OPERATING CONDITIONS

INPUT THRESHOLD
vs
SUPPLY VOLTAGE V_{CC1}

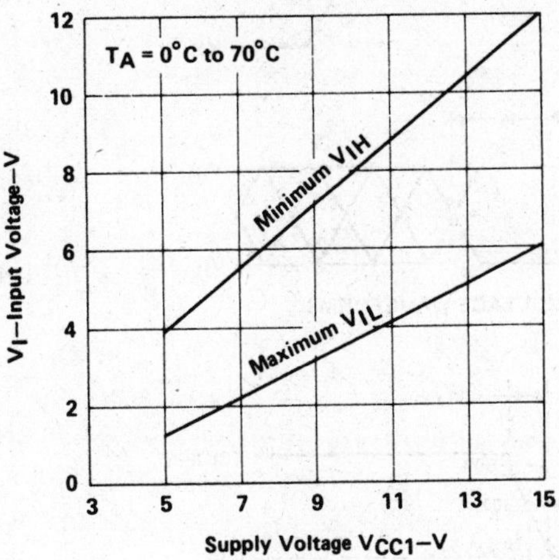
12
10
8
6
4
2
0
3
5
7
9
11
13
15
$T_A = 0°C$ to $70°C$
Minimum V_{IH}
Maximum V_{IL}
V_I—Input Voltage—V
Supply Voltage V_{CC1}—V

FIGURE 1

MAXIMUM INPUT DATA RATE
vs
SUPPLY VOLTAGE

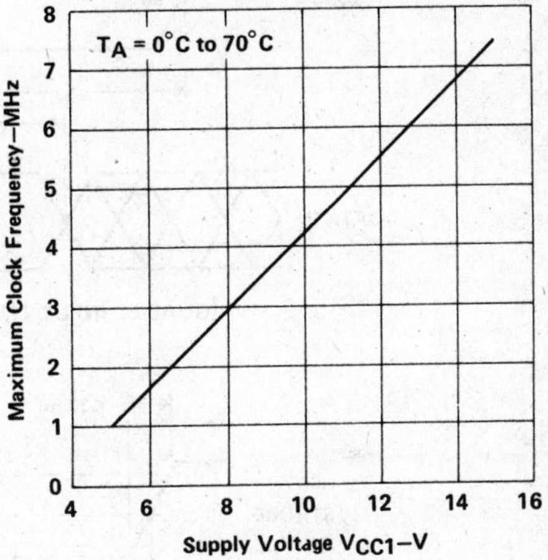
8
7
6
5
4
3
2
1
0
4
6
8
10
12
14
16
$T_A = 0°C$ to $70°C$
Maximum Clock Frequency—MHz
Supply Voltage V_{CC1}—V

FIGURE 2

PARAMETER MEASUREMENT INFORMATION

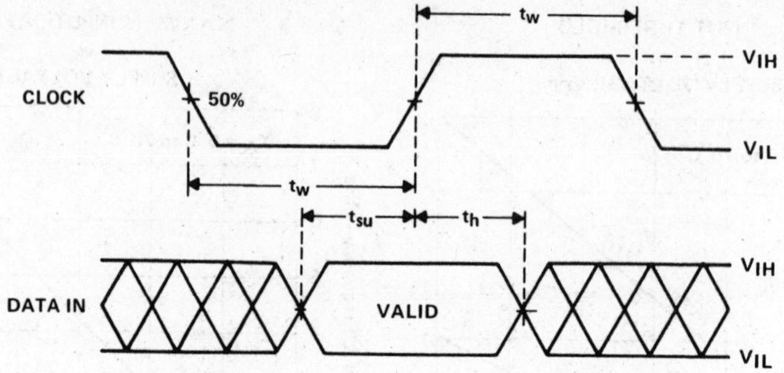
CLOCK
50%
t_w
t_w
t_{su}
t_h
V_{IH}
V_{IL}
DATA IN
VALID
V_{IH}
V_{IL}

FIGURE 3. INPUT TIMING VOLTAGE WAVEFORMS

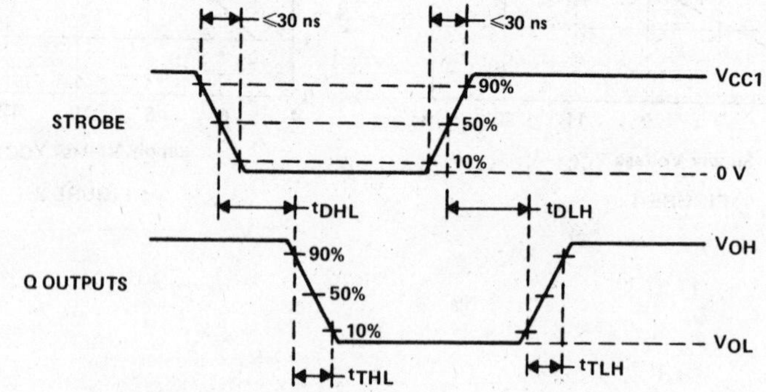
≤30 ns
≤30 ns
STROBE
90%
50%
10%
V_{CC1}
0 V
t_{DHL}
t_{DLH}
Q OUTPUTS
90%
50%
10%
V_{OH}
V_{OL}
t_{THL}
t_{TLH}

FIGURE 4. SWITCHING-TIME VOLTAGE WAVEFORMS

D2720, MARCH 1983—REVISED OCTOBER 1986

- **Each Device Drives 32 Lines**
- **60-V Output Voltage Swing Capability**
- **25-mA Output Source Current Capability**
- **High-Speed Serially Shifted Data Input**
- **Latches on All Driver Outputs**

N DUAL-IN-LINE PACKAGE
(TOP VIEW)

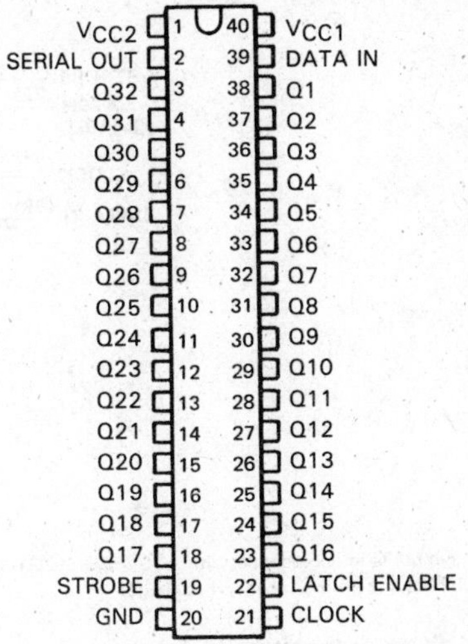
VCC2 1 40 VCC1
SERIAL OUT 2 39 DATA IN
Q32 3 38 Q1
Q31 4 37 Q2
Q30 5 36 Q3
Q29 6 35 Q4
Q28 7 34 Q5
Q27 8 33 Q6
Q26 9 32 Q7
Q25 10 31 Q8
Q24 11 30 Q9
Q23 12 29 Q10
Q22 13 28 Q11
Q21 14 27 Q12
Q20 15 26 Q13
Q19 16 25 Q14
Q18 17 24 Q15
Q17 18 23 Q16
STROBE 19 22 LATCH ENABLE
GND 20 21 CLOCK

description

The SN65518 and SN75518 are monolithic BIDFET† integrated circuits designed to drive a dot matrix or segmented vacuum fluorescent display.

The devices each consist of a 32-bit shift register, 32 latches, and 32 output AND gates. Serial data is entered into the shift register on the low-to-high transition of the clock input. While the Latch Enable input is high, parallel data is transferred to the output buffers through a 32-bit latch. Data present in the latch during the high-to-low transition of Latch Enable is latched. When the Strobe input is low, all Q outputs are enabled. When the Strobe input is high, all Q outputs are low.

Serial data output from the shift register may be used to cascade additional devices. This output is not affected by the Latch Enable or Strobe inputs.

The SN65518 is characterized for operation from −40°C to 85°C and the SN75518 is characterized for operation from 0°C to 70°C.

FN PACKAGE
(TOP VIEW)

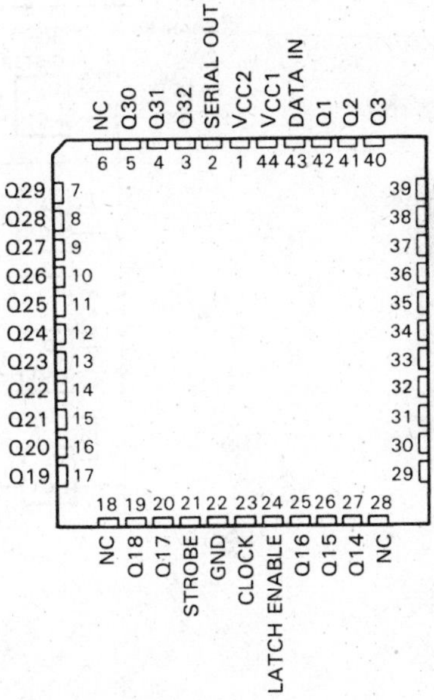
NC 6
Q30 5
Q31 4
Q32 3
SERIAL OUT 2
VCC2 1
VCC1 44
DATA IN 43
Q1 42
Q2 41
Q3 40
Q29 7
Q28 8
Q27 9
Q26 10
Q25 11
Q24 12
Q23 13
Q22 14
Q21 15
Q20 16
Q19 17
39 Q4
38 Q5
37 Q6
36 Q7
35 Q8
34 Q9
33 Q10
32 Q11
31 Q12
30 Q13
29 NC
18 NC
19 Q18
20 Q17
21 STROBE
22 GND
23 CLOCK
24 LATCH ENABLE
25 Q16
26 Q15
27 Q14
28 NC

†BIDFET—Bipolar, double-diffused, N-channel and P-channel MOS transistors on same chip—patented process.

NC—No internal connection

logic symbol†

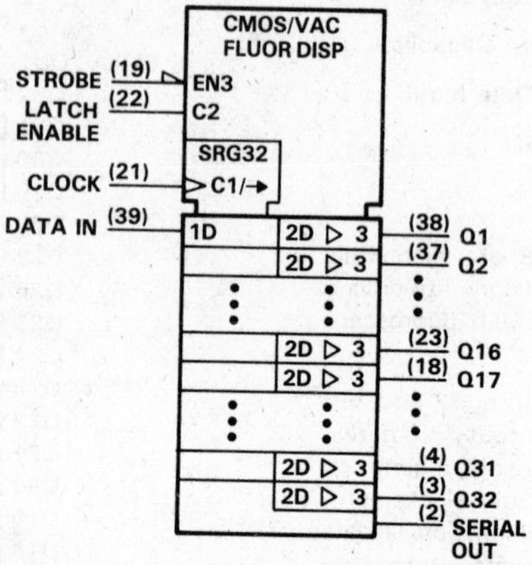
CMOS/VAC
FLUOR DISP
STROBE
(19)
EN3
LATCH
ENABLE
(22)
C2
SRG32
CLOCK
(21)
C1/→
DATA IN
(39)
1D
2D ▷ 3
(38)
Q1
(37)
Q2
(23)
Q16
(18)
Q17
(4)
Q31
(3)
Q32
(2)
SERIAL
OUT

† This symbol is in accordance with ANSI/IEEE Std 91-1984 and IEC Publication 617-12.
Pin numbers shown are for the N package.

logic diagram (positive logic)

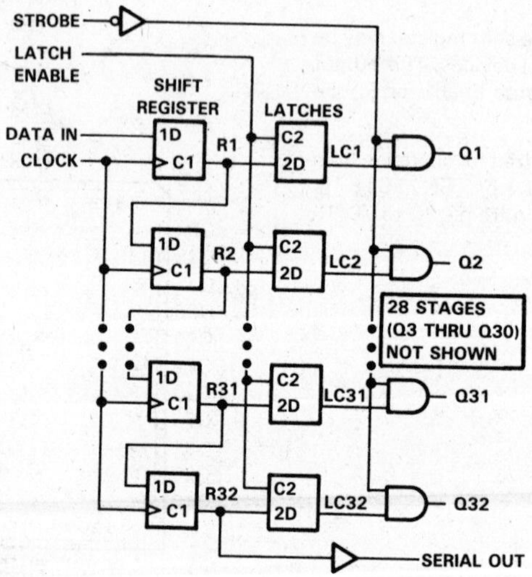
STROBE
LATCH
ENABLE
SHIFT
REGISTER
LATCHES
DATA IN
CLOCK
1D
C1
R1
C2
2D
LC1
Q1
R2
LC2
Q2
28 STAGES
(Q3 THRU Q30)
NOT SHOWN
R31
LC31
Q31
R32
LC32
Q32
SERIAL OUT

FUNCTION TABLE

FUNCTION	CONTROL INPUTS			SHIFT REGISTER R1 THRU R32	LATCHES LC1 THRU LC32	OUTPUTS	
	CLOCK	LATCH ENABLE	STROBE			SERIAL	Q1 THRU Q32
LOAD	↑	X	X	Load and shift*	Determined by Latch Enable§	R32	Determined by Strobe
	No↑	X	X	No change	Determined by Latch Enable§	R32	Determined by Strobe
LATCH	X	L	X	As determined above	Stored data	R32	Determined by Strobe
	X	H	X	As determined above	New data	R32	Determined by Strobe
STROBE	X	X	H	As determined above	Determined by Latch Enable§	R32	All L
	X	X	L	As determined above	Determined by Latch Enable§	R32	LC1 thru LC32, respectively

H = high level, L = low level, X = irrelevant, ↑ = low-to-high-level transition.

* R32 and the serial output take on the state of R31, R31 takes on the state of R30, . . . R2 takes on the state of R1, and R1 takes on the state of the data input.

§New data enter the latches while Latch Enable is high. These data are stored while Latch Enable is low.

typical operating sequence

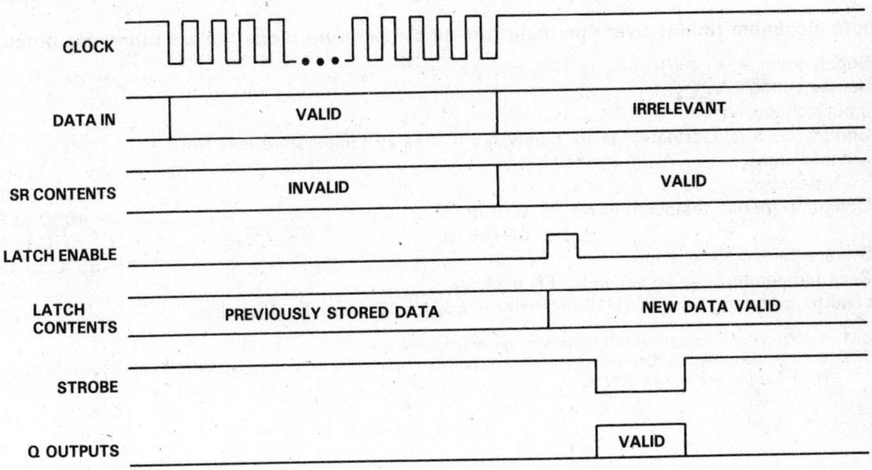
CLOCK
DATA IN
VALID
IRRELEVANT
SR CONTENTS
INVALID
VALID
LATCH ENABLE
LATCH CONTENTS
PREVIOUSLY STORED DATA
NEW DATA VALID
STROBE
Q OUTPUTS
VALID

schematic of inputs and outputs

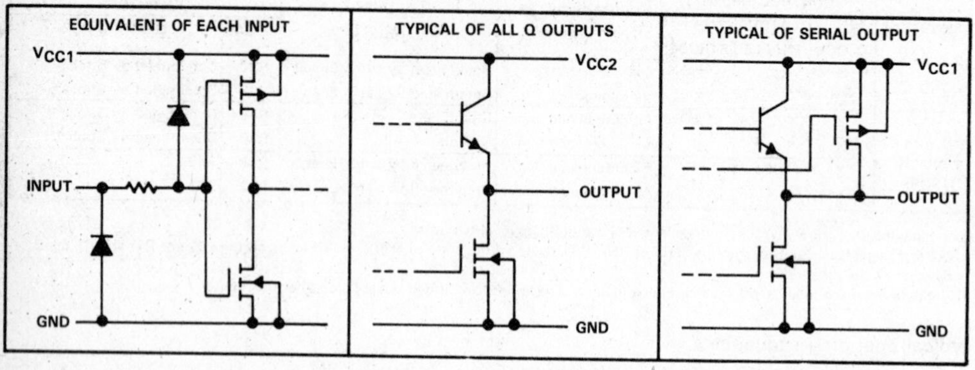
EQUIVALENT OF EACH INPUT
VCC1
INPUT
GND
TYPICAL OF ALL Q OUTPUTS
VCC2
OUTPUT
GND
TYPICAL OF SERIAL OUTPUT
VCC1
OUTPUT
GND

absolute maximum ratings over operating free-air temperature range (unless otherwise noted)

Supply voltage, V_{CC1} (see Note 1)	15 V
Supply voltage, V_{CC2}	70 V
Input voltage, V_I	V_{CC1}
Continuous total dissipation at (or below) 25°C free-air temperature (see Note 2):	
N package	1650 mW
FN package	1700 mW
Operating free-air temperature range: SN65518	−40°C to 85°C
SN75518	0°C to 70°C
Storage temperature range	−65°C to 150°C
Case temperature for 10 seconds: FN package	260°C
Lead temperature 1,6 mm (1/16 inch) from case for 10 seconds: N package	260°C

NOTES: 1. Voltage values are with respect to network ground terminal.
2. For operation above 25°C free-air temperature, derate the N package linearly at the rate of 13.2 mW/°C and the FN package linearly at the rate of 13.6 mW/°C.

recommended operating conditions, $T_A = 25°C$ (unless otherwise noted)

		MIN	MAX	UNIT
Supply voltage, V_{CC1}		4.5	15	V
Supply voltage, V_{CC2}		0	60	V
High-level input voltage, V_{IH} (see Figure 1)	$V_{CC1} = 4.5$ V	3.5		V
	$V_{CC1} = 15$ V	12		
Low-level input voltage, V_{IL} (see Figure 1)	$V_{CC1} = 4.5$ V		1	V
	$V_{CC1} = 15$ V		6	
High-level output current, I_{OH}			−25	mA
Low-level output current, I_{OL}			2	mA
Clock frequency, f_{clock} (see Figure 2)	$V_{CC1} = 10$ V to 15 V	0	5	MHz
	$V_{CC1} = 4.5$ V	0	1	
Pulse duration, clock high, $t_{w(CKH)}$	$V_{CC1} = 10$ V to 15 V	100		ns
	$V_{CC1} = 4.5$ V	500		
Pulse duration, clock low, $t_{w(CKL)}$	$V_{CC1} = 10$ V to 15 V	100		ns
	$V_{CC1} = 4.5$ V	500		
Setup time, data before clock↑, t_{su}	$V_{CC1} = 10$ V to 15 V	75		ns
	$V_{CC1} = 4.5$ V	150		
Hold time, data after clock↑, t_h	$V_{CC1} = 10$ V to 15 V	75		ns
	$V_{CC1} = 4.5$ V	150		
Operating free-air temperature, T_A	SN65518	−40	85	°C
	SN75518	0	70	

electrical characteristics over recommended ranges of operating free-air temperature and V_{CC1} (unless otherwise noted), $V_{CC2} = 60$ V

PARAMETER			TEST CONDITIONS	MIN	TYP†	MAX	UNIT
V_{IK}	Input clamp voltage		$I_I = -12$ mA			−1.5	V
V_{OH}	High-level output voltage	Q outputs	$I_{OH} = -25$ mA	57.5	58		V
		Serial output	$V_{CC1} = 5$ V, $I_{OH} = -20$ μA	4.5	4.9	5	
V_{OL}	Low-level output voltage	Q outputs	$I_{OL} = 1$ mA			5	V
		Serial output	$I_{OL} = 20$ μA		0.06	0.8	
I_{IH}	High-level input current		$V_{CC1} = 15$ V, $V_I = 15$ V		0.1	1	μA
I_{IL}	Low-level input current		$V_{CC1} = 15$ V, $V_I = 0$ V		−0.1	−1	μA
I_{CC1}	Supply current		$V_{CC1} = 4.5$ V		1.8	4	mA
			$V_{CC1} = 15$ V		2	5	
I_{CC2}	Supply current	SN65518	Outputs high, $T_A = -40°C$			12	mA
		SN65518,	Outputs high, $T_A = 0°C$ to MAX		7	10	
		SN75518	Outputs low		0.01	0.5	

† All typical values are at $T_A = 25°C$.

switching characteristics, V_{CC2} = 60 V, C_L = 50 pF, T_A = 25°C (unless otherwise noted)

PARAMETER				TEST CONDITIONS	MIN	MAX	UNIT
t_d	Delay time, Clock to data output		V_{CC1} = 4.5 V	C_L = 15 pF, See Figure 4		600	ns
			V_{CC1} = 15 V			150	
t_{DHL}	Delay time, high-to-low-level Q output	from latch enable	V_{CC1} = 4.5 V	See Figure 5		1.5	μs
		from strobe		See Figure 6		1	
		from latch enable	V_{CC1} = 15 V	See Figure 5		0.5	
		from strobe		See Figure 6		0.5	
t_{DLH}	Delay time, low-to-high-level Q output	from latch enable	V_{CC1} = 4.5 V	See Figure 5		1.5	μs
		from strobe		See Figure 6		1	
		from latch enable	V_{CC1} = 15 V	See Figure 5		0.25	
		from strobe		See Figure 6		0.25	
t_{THL}	Transition time, high-to-low-level Q output		V_{CC1} = 4.5 V	See Figure 6		3	μs
			V_{CC1} = 15 V			1.5	
t_{TLH}	Transition time, low-to-high-level Q output		V_{CC1} = 4.5 V	See Figure 6		2.5	μs
			V_{CC1} = 15 V			0.75	

RECOMMENDED OPERATING CONDITIONS

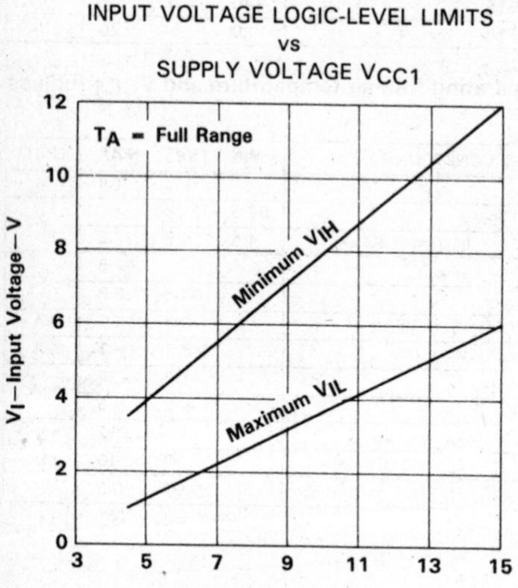
INPUT VOLTAGE LOGIC-LEVEL LIMITS
vs
SUPPLY VOLTAGE VCC1
TA = Full Range
Minimum VIH
Maximum VIL
VI—Input Voltage—V
Supply Voltage VCC1—V
0
2
4
6
8
10
12
3
5
7
9
11
13
15

FIGURE 1

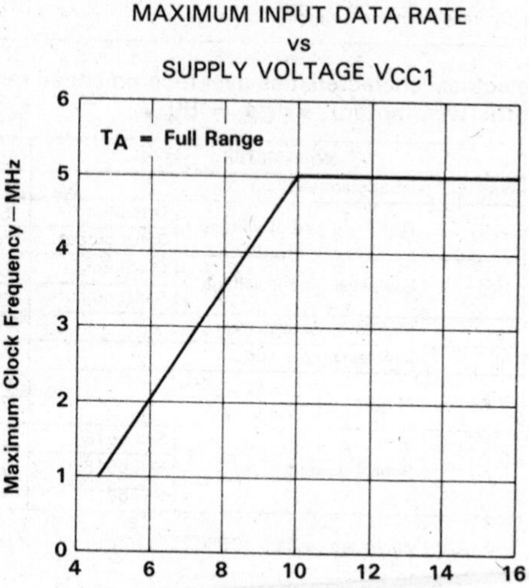
MAXIMUM INPUT DATA RATE
vs
SUPPLY VOLTAGE VCC1
TA = Full Range
Maximum Clock Frequency—MHz
Supply Voltage VCC1—V
0
1
2
3
4
5
6
4
6
8
10
12
14
16

FIGURE 2

PARAMETER MEASUREMENT INFORMATION

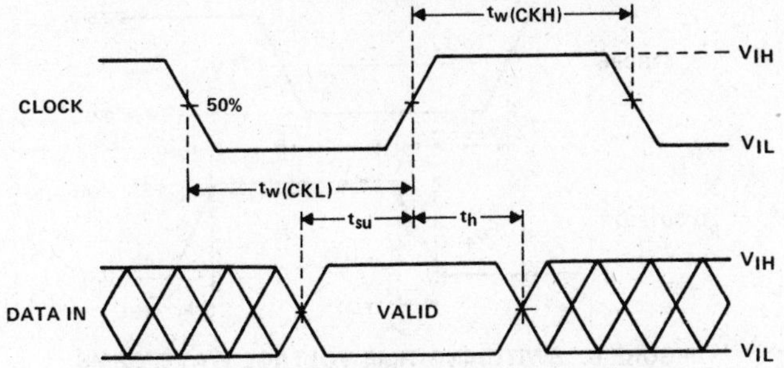
$t_{w(CKH)}$
V_{IH}
CLOCK
50%
V_{IL}
$t_{w(CKL)}$
t_{su}
t_h
V_{IH}
DATA IN
VALID
V_{IL}

FIGURE 3. INPUT TIMING VOLTAGE WAVEFORMS

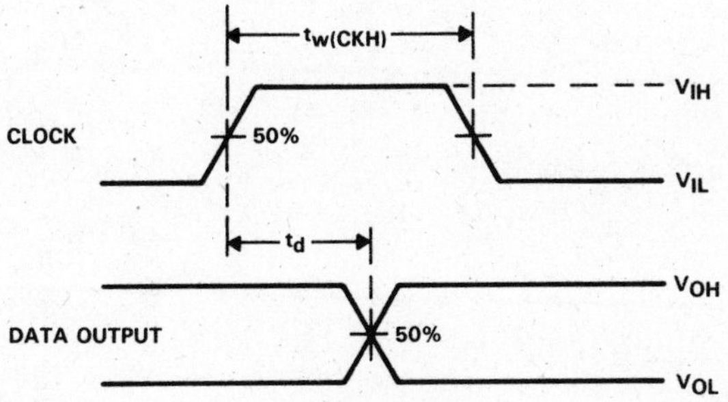
$t_{w(CKH)}$
V_{IH}
CLOCK
50%
V_{IL}
t_d
V_{OH}
DATA OUTPUT
50%
V_{OL}

FIGURE 4

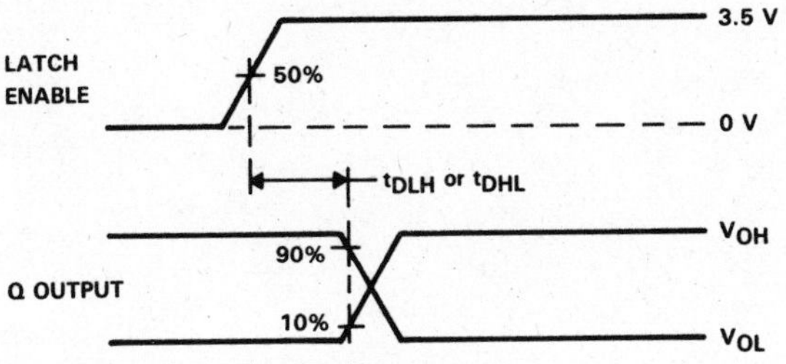
3.5 V
LATCH ENABLE
50%
0 V
t_{DLH} or t_{DHL}
V_{OH}
90%
Q OUTPUT
10%
V_{OL}

FIGURE 5

NOTE: For testing purposes, all input pulses have maximum rise and fall times of 30 ns.

PARAMETER MEASUREMENT INFORMATION

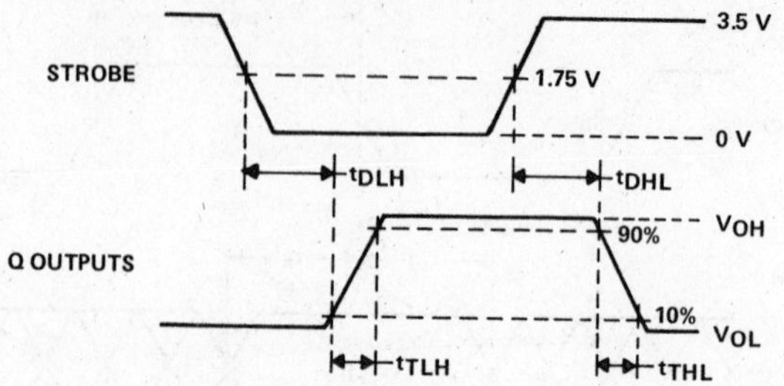
3.5 V
STROBE
1.75 V
0 V
tDLH
tDHL
VOH
90%
Q OUTPUTS
10%
VOL
tTLH
tTHL

FIGURE 6. SWITCHING-TIME VOLTAGE WAVEFORMS

NOTE: For testing purposes, all input pulses have maximum rise and fall times of 30 ns.

TEXAS INSTRUMENTS
POST OFFICE BOX 655012 • DALLAS, TEXAS 75265

- Each Device Drives 32 Electrodes
- High-Voltage Open-Drain DMOS Outputs
- 50-mA Output Current Capability
- CMOS-Compatible Inputs
- Very Low Steady-State Power Consumption

description

The SN55551 and SN55552 are monolithic BIDFET† integrated circuits designed to drive the row electrodes of an electroluminescent display. All inputs are CMOS-compatible and all outputs are high-voltage open-drain DMOS transistors. The SN55552 output sequence has been reversed from the SN55551 for ease in printed circuit board layout.

The devices consist of a 32-bit shift register, 32 AND gates, and 32 output OR gates. Typically, a composite row drive signal is externally generated by a high-voltage switching circuit and applied to the Substrate Common terminal. Serial data is entered into the shift register on the high-to-low transition of the clock input. A high Enable input allows those outputs with a high in their associated register to be turned on causing the corresponding row to be connected to the composite row drive signal. When the Strobe input is low, all output transistors are turned on. The Serial Data output from the shift register may be used to cascade additional devices. This output is not affected by the Enable or Strobe inputs.

The SN55551 and SN55552 are characterized for operation over the full military temperature range of −55°C to 125°C.

SN55551 . . . FD PACKAGE
(TOP VIEW)

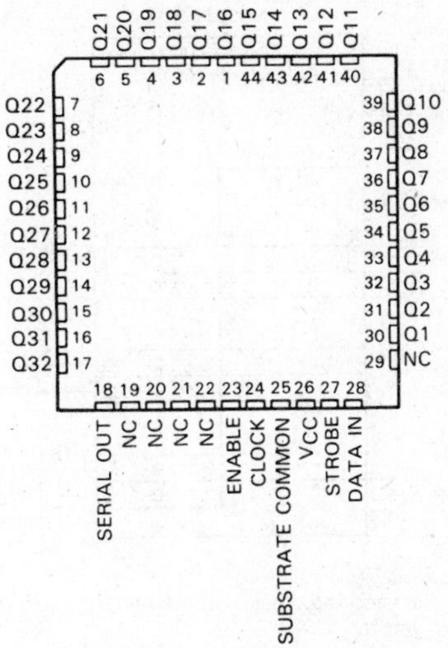
Q21 Q20 Q19 Q18 Q17 Q16 Q15 Q14 Q13 Q12 Q11
6 5 4 3 2 1 44 43 42 41 40
Q22 7
Q23 8
Q24 9
Q25 10
Q26 11
Q27 12
Q28 13
Q29 14
Q30 15
Q31 16
Q32 17
39 Q10
38 Q9
37 Q8
36 Q7
35 Q6
34 Q5
33 Q4
32 Q3
31 Q2
30 Q1
29 NC
18 19 20 21 22 23 24 25 26 27 28
SERIAL OUT NC NC NC NC ENABLE CLOCK SUBSTRATE COMMON VCC STROBE DATA IN

SN55552 . . . FD PACKAGE
(TOP VIEW)

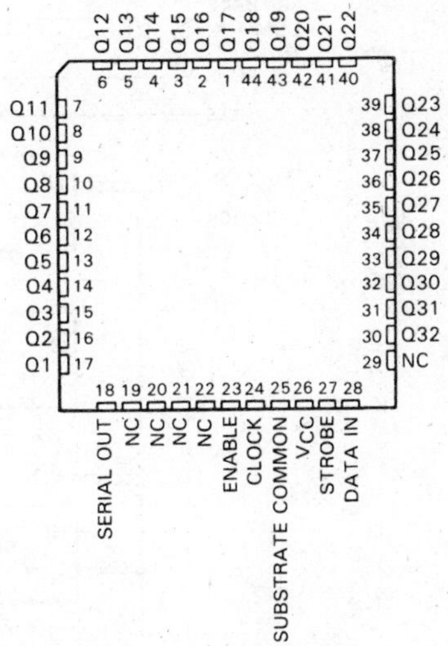
Q12 Q13 Q14 Q15 Q16 Q17 Q18 Q19 Q20 Q21 Q22
6 5 4 3 2 1 44 43 42 41 40
Q11 7
Q10 8
Q9 9
Q8 10
Q7 11
Q6 12
Q5 13
Q4 14
Q3 15
Q2 16
Q1 17
39 Q23
38 Q24
37 Q25
36 Q26
35 Q27
34 Q28
33 Q29
32 Q30
31 Q31
30 Q32
29 NC
18 19 20 21 22 23 24 25 26 27 28
SERIAL OUT NC NC NC NC ENABLE CLOCK SUBSTRATE COMMON VCC STROBE DATA IN

NC—No internal connection

† BIDFET — Bipolar, double-diffused, N-channel and P-channel MOS transistors on same chip — patented process.

logic symbols†

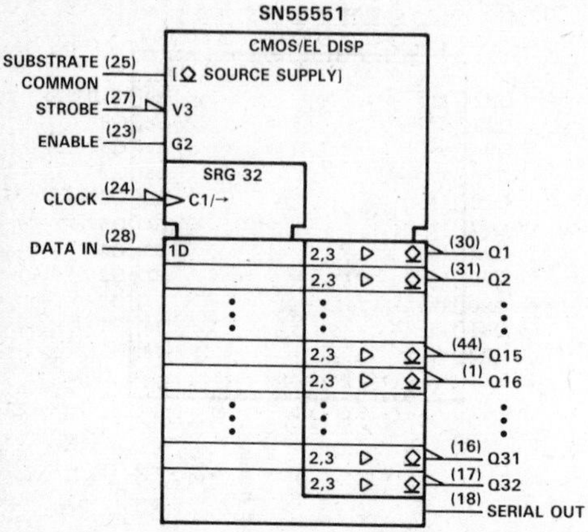
SN55551
CMOS/EL DISP
SUBSTRATE COMMON (25)
[SOURCE SUPPLY]
STROBE (27)
V3
ENABLE (23)
G2
SRG 32
CLOCK (24)
C1/→
DATA IN (28)
1D
2,3
(30) Q1
(31) Q2
(44) Q15
(1) Q16
(16) Q31
(17) Q32
(18) SERIAL OUT

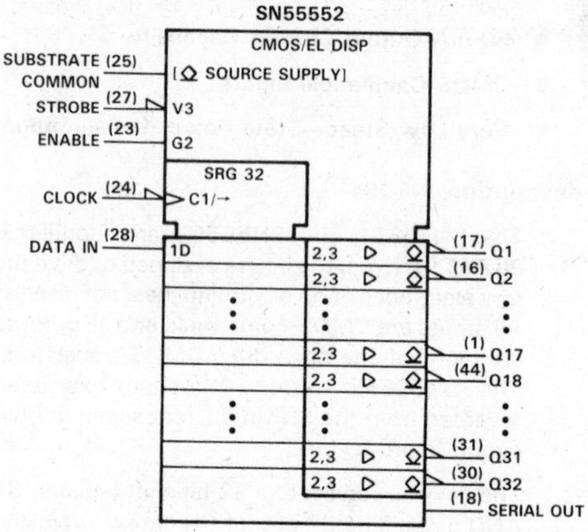
SN55552
CMOS/EL DISP
SUBSTRATE COMMON (25)
[SOURCE SUPPLY]
STROBE (27)
V3
ENABLE (23)
G2
SRG 32
CLOCK (24)
C1/→
DATA IN (28)
1D
2,3
(17) Q1
(16) Q2
(1) Q17
(44) Q18
(31) Q31
(30) Q32
(18) SERIAL OUT

† These symbols are in accordance with ANSI/IEEE Std 91-1984 and IEC Publication 617-12. The symbol ◇ here indicates an n-channel open-drain output.

logic diagram (positive logic)

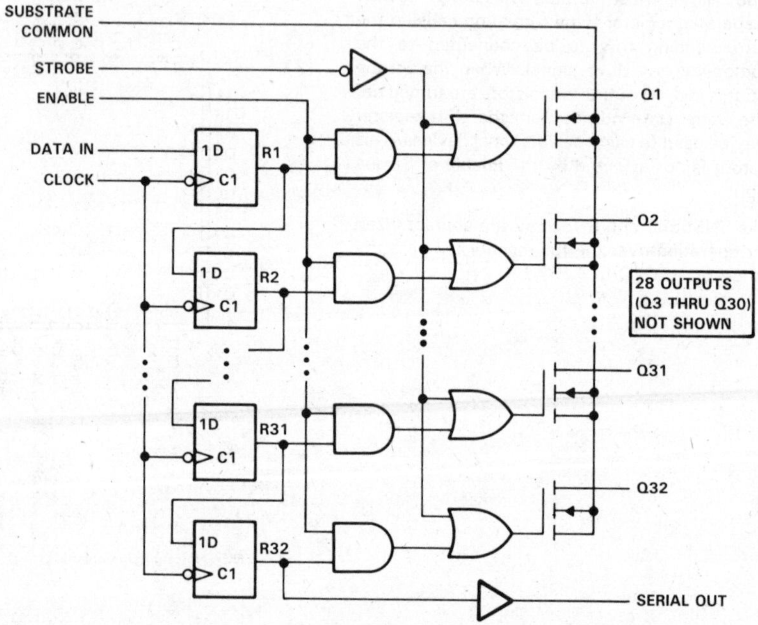
SUBSTRATE COMMON
STROBE
ENABLE
DATA IN
CLOCK
1D
C1
R1
R2
R31
R32
Q1
Q2
Q31
Q32
28 OUTPUTS (Q3 THRU Q30) NOT SHOWN
SERIAL OUT

FUNCTION TABLE

FUNCTION	CONTROL INPUTS			SHIFT REGISTERS R1 THRU R32	OUTPUTS	
	CLOCK	ENABLE	STROBE		SERIAL	Q1 THRU Q32
LOAD	↓	X	X	Load and Shift†	R32	Determined by Enable and Strobe
	No. ↓	X	X	No Change	R32	Determined by Enable and Strobe
ENABLE	X	L	H	As determined above	R32	All Q outputs off
	X	H	H	As determined above	R32	Determined by R1 through R32
STROBE	X	X	L	As determined above	R32	All Q outputs on

H = high level, L = low level, X = irrelevant, ↓ = high-to-low transition.
†Register R32 takes on the state of R31, R31 takes on the state of R30,. . .R2 takes on the state of R1, and R1 takes on the state of the data input.

typical operating sequence

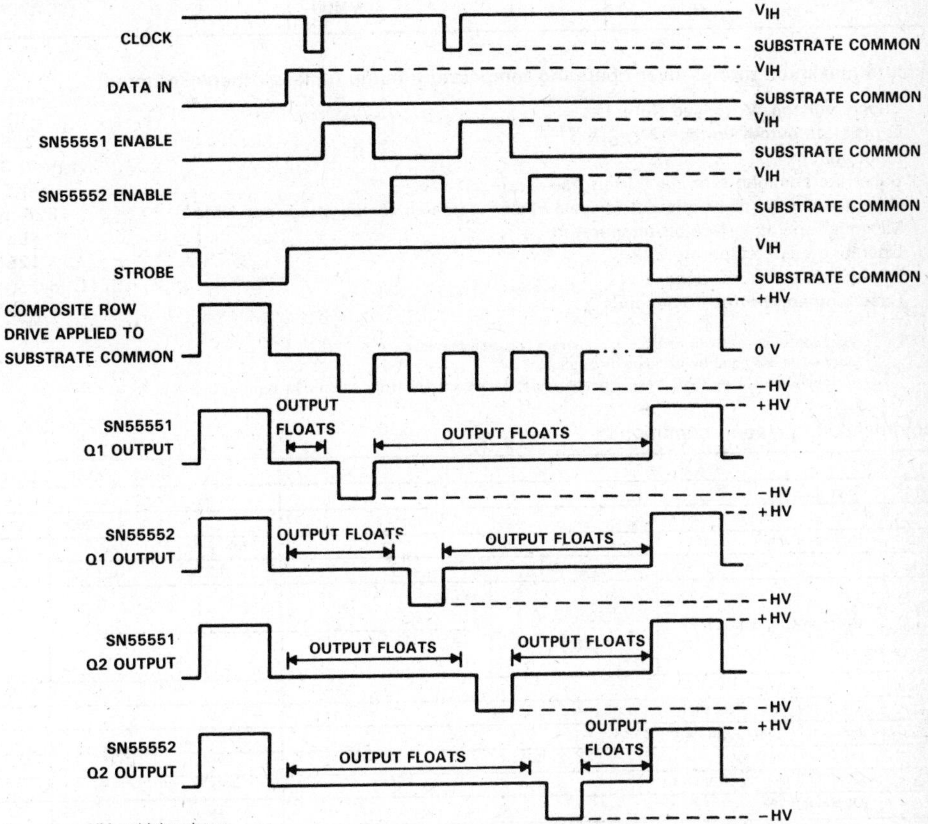
CLOCK
DATA IN
SN55551 ENABLE
SN55552 ENABLE
STROBE
COMPOSITE ROW DRIVE APPLIED TO SUBSTRATE COMMON
SN55551 Q1 OUTPUT
SN55552 Q1 OUTPUT
SN55551 Q2 OUTPUT
SN55552 Q2 OUTPUT
OUTPUT FLOATS
V_{IH}
SUBSTRATE COMMON
+HV
0 V
−HV

HV = high voltage

NOTE: During operation Clock, Data In, Enable, and Strobe are referenced to the Composite Row Drive signal received at the Substrate Common pin of the device.

schematic of inputs and outputs

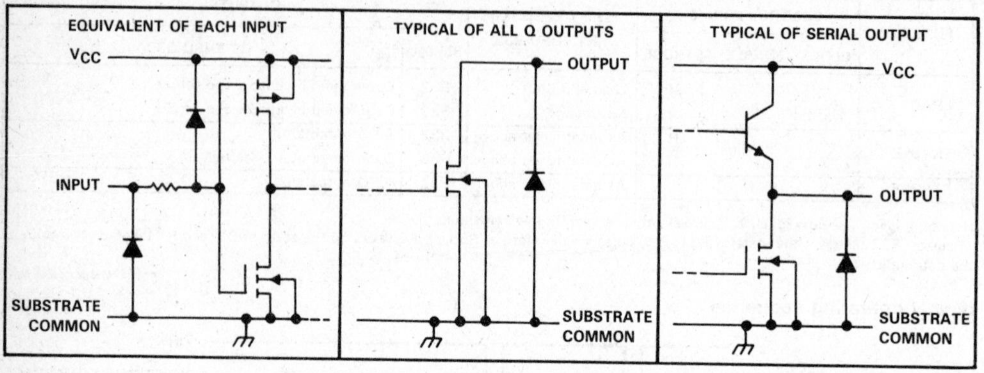
EQUIVALENT OF EACH INPUT
VCC
INPUT
SUBSTRATE COMMON
TYPICAL OF ALL Q OUTPUTS
OUTPUT
SUBSTRATE COMMON
TYPICAL OF SERIAL OUTPUT
VCC
OUTPUT
SUBSTRATE COMMON

absolute maximum ratings over operating temperature range (unless otherwise noted)

Supply voltage, V_{CC} (see Note 1) .... 18 V
Q off-state output voltage, $V_{O(off)}$ .... 225 V
Input voltage .... $V_{CC}+0.3$ V
Substrate common terminal current (see Note 2) .... 1.5 A
Continuous total dissipation at (or below) 25°C free-air temperature (see Note 3) .... 1825 mW
Minimum operating free-air temperature .... −55°C
Operating case temperature .... 125°C
Storage temperature range .... −65°C to 150°C
Case temperature for 60 seconds .... 260°C

NOTES: 1. Voltage values are with respect to substrate common terminal.
2. Duty cycle is limited by package dissipation.
3. For operation above 25°C free-air temperature, derate linearly at the rate of 14.6 mW/°C.

recommended operating conditions

				MIN	NOM	MAX	UNIT
V_{CC}	Supply voltage			10.8	12	15	V
$V_{O(off)}$	Off-state Q output voltage			0		200	V
V_{IH}	High-level input voltage			$0.75V_{CC}$		$V_{CC}+0.3$	V
V_{IL}	Low-level input voltage			−0.3		$0.25V_{CC}$	V
$I_{O(on)}$	On-state Q output current	V_{DD} = 80 V, Duty cycle ≤ 1%	V_{CC} = 10.8 V, T_C = 25°C			50	mA
			V_{CC} = 15 V, T_C = 25°C			80	
f_{clock}	Clock frequency, T_A = 25°C					6.25	MHz
t_w	Clock pulse duration, high or low, T_A = 25°C			80			ns
t_{su}	Setup time, data valid before clock↓, T_A = 25°C			20			ns
t_h	Hold time, data valid after clock↓, T_A = 25°C			110			ns
T_A	Operating free-air temperature			−55			°C
T_C	Operating case temperature					125	°C

electrical characteristics over recommended operating temperature range, V_{CC} = 12 V, substrate common at 0 V

PARAMETER			TEST CONDITIONS	MIN	MAX	UNIT
V_{OH}	High-level output voltage	Serial outputs	I_O = −100 μA	10		V
V_{OL}	Low-level output voltage	Q outputs	I_O = 50 mA		50	V
		Serial output	I_O = 100 μA		1.5	
I_{IH}	High-level input current		V_I = 12 V		5	μA
I_{IL}	Low-level input current		V_I = 0		−5	μA
$I_{O(off)}$	Off-state Q output current		V_O = 200 V		50	μA
I_{CC}	Supply current				500	μA

switching characteristics, V_{CC} = 12 V, T_C = 25°C

PARAMETER		TEST CONDITIONS	MIN	MAX	UNIT
t_{dLH}	Delay time, clock↓ to serial↓	C_L = 45 pF to common, See Figure 1		200	ns
t_{dHL}	Delay time, clock↓ to serial↑			200	ns
t_{dHL}	Delay time, enable to Q output↓	V_{DD} = 100 V, R_L = 2 kΩ, C_L = 45 pF to common, See Figure 1		500	ns

3 Display Drivers

PARAMETER MEASUREMENT INFORMATION

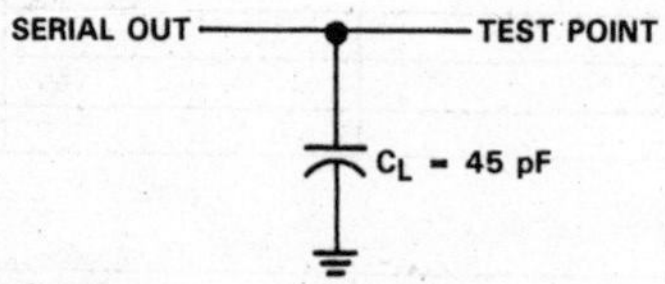
SERIAL OUT
TEST POINT
CL = 45 pF

SERIAL OUTPUT LOAD CIRCUIT

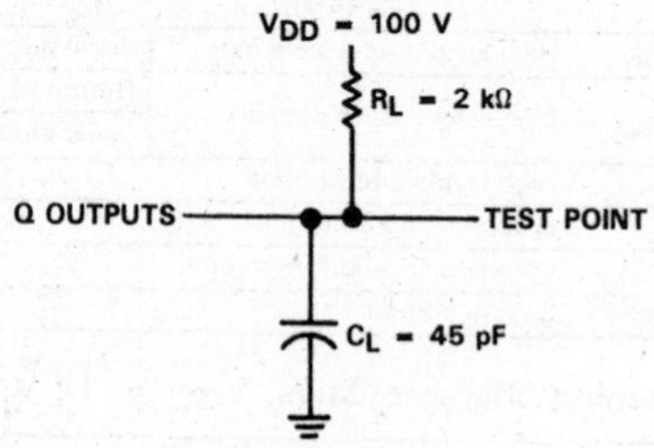
VDD = 100 V
RL = 2 kΩ
Q OUTPUTS
TEST POINT
CL = 45 pF

Q OUTPUT LOAD CIRCUIT

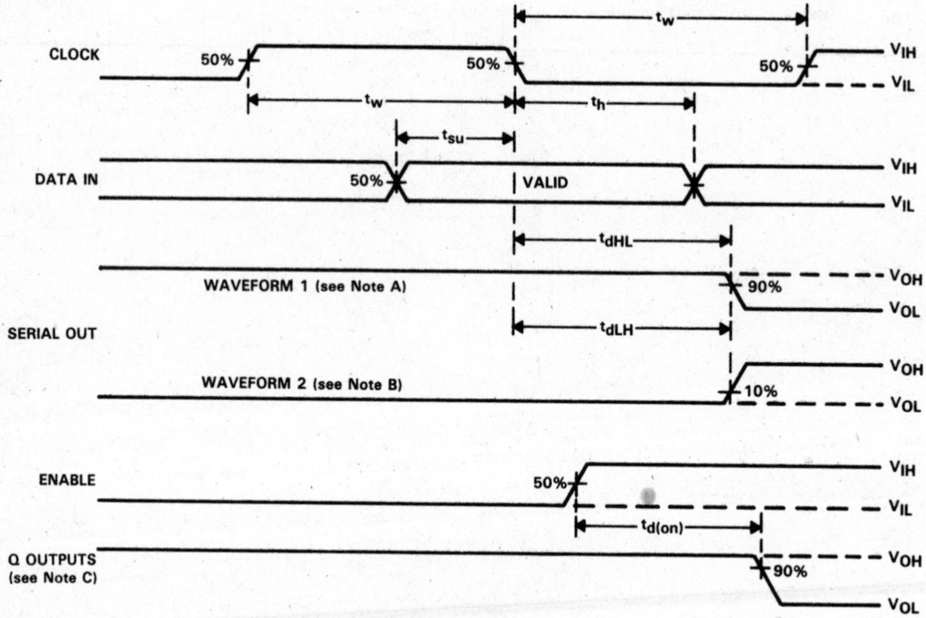
CLOCK
50%
50%
50%
tw
tw
th
tsu
VIH
VIL
DATA IN
50%
VALID
VIH
VIL
tdHL
WAVEFORM 1 (see Note A)
90%
VOH
VOL
SERIAL OUT
tdLH
WAVEFORM 2 (see Note B)
10%
VOH
VOL
ENABLE
50%
VIH
VIL
td(on)
Q OUTPUTS
(see Note C)
90%
VOH
VOL

VOLTAGE WAVEFORMS

NOTES: A. Waveform 1 is for internal conditions such that a low is clocked into R32.
B. Waveform 2 is for internal conditions such that a high is clocked into R32.
C. To measure $t_{d(on)}$, a high is stored in the associated register.

FIGURE 1. SWITCHING CHARACTERISTICS

RECOMMENDED OPERATING CONDITIONS

MAXIMUM ON-STATE Q OUTPUT CURRENT
vs
SUPPLY VOLTAGE

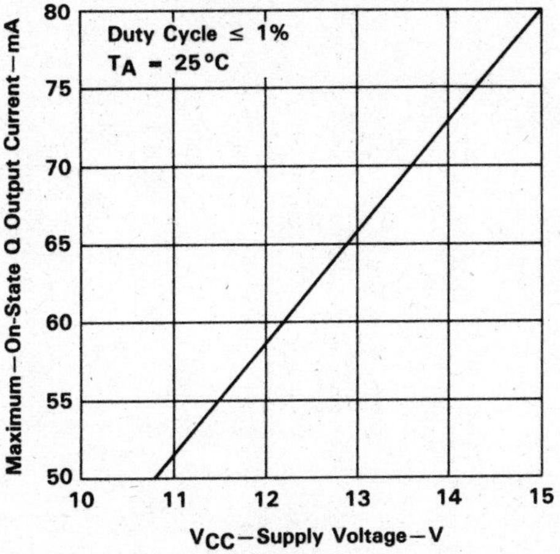
Duty Cycle ≤ 1%
T_A = 25°C
80
75
70
65
60
55
50
10
11
12
13
14
15
Maximum—On-State Q Output Current—mA
V_{CC}—Supply Voltage—V

FIGURE 2

TYPICAL CHARACTERISTICS

OUTPUT CHARACTERISTICS SHOWING
SAFE OPERATION AREA (SOA)

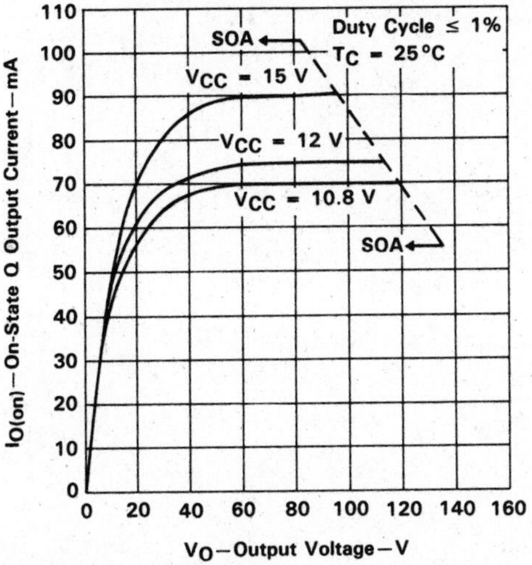
Duty Cycle ≤ 1%
T_C = 25°C
SOA
SOA
V_{CC} = 15 V
V_{CC} = 12 V
V_{CC} = 10.8 V
110
100
90
80
70
60
50
40
30
20
10
0
0
20
40
60
80
100
120
140
160
$I_{O(on)}$—On-State Q Output Current—mA
V_O—Output Voltage—V

FIGURE 3

OUTPUT SATURATION CURRENT
vs
JUNCTION TEMPERATURE

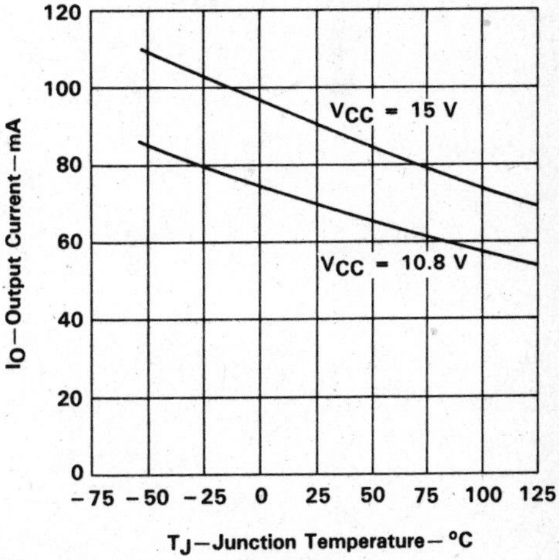
V_{CC} = 15 V
V_{CC} = 10.8 V
120
100
80
60
40
20
0
−75
−50
−25
0
25
50
75
100
125
I_O—Output Current—mA
T_J—Junction Temperature—°C

FIGURE 4

D2743, MARCH 1983—REVISED SEPTEMBER 1986

- Each Device Drives 32 Electrodes
- High-Voltage Open-Drain DMOS Outputs
- 50-mA Output Current Capability
- CMOS-Compatible Inputs
- Very Low Steady-State Power Consumption

N
DUAL-IN-LINE-PACKAGES
(TOP VIEW)

SN65551, SN75551

Signal	Pin	Pin	Signal
Q16	1	40	Q15
Q17	2	39	Q14
Q18	3	38	Q13
Q19	4	37	Q12
Q20	5	36	Q11
Q21	6	35	Q10
Q22	7	34	Q9
Q23	8	33	Q8
Q24	9	32	Q7
Q25	10	31	Q6
Q26	11	30	Q5
Q27	12	29	Q4
Q28	13	28	Q3
Q29	14	27	Q2
Q30	15	26	Q1
Q31	16	25	NC
Q32	17	24	DATA IN
SERIAL OUT	18	23	STROBE
ENABLE	19	22	V_{CC}
CLOCK	20	21	SUBSTRATE COMMON

SN65552, SN75552

Signal	Pin	Pin	Signal
Q17	1	40	Q18
Q16	2	39	Q19
Q15	3	38	Q20
Q14	4	37	Q21
Q13	5	36	Q22
Q12	6	35	Q23
Q11	7	34	Q24
Q10	8	33	Q25
Q9	9	32	Q26
Q8	10	31	Q27
Q7	11	30	Q28
Q6	12	29	Q29
Q5	13	28	Q30
Q4	14	27	Q31
Q3	15	26	Q32
Q2	16	25	NC
Q1	17	24	DATA IN
SERIAL OUT	18	23	STROBE
ENABLE	19	22	V_{CC}
CLOCK	20	21	SUBSTRATE COMMON

NC—No internal connection

description

The SN65551, SN65552, SN75551, and SN75552 are monolithic BIDFET† integrated circuits designed to drive the row electrodes of an electroluminescent display. All inputs are CMOS-compatible and all outputs are high-voltage open-drain DMOS transistors. The SN75552 output sequence has been reversed from the SN75551 for ease in printed circuit board layout.

The devices consist of a 32-bit shift register, 32 AND gates, and 32 output OR gates. Typically, a composite row drive signal is externally generated by a high-voltage switching circuit and applied to the Substrate Common terminal. Serial data is entered into the shift register on the high-to-low transition of the clock input. A high Enable input allows those outputs with a high in their associated register to be turned on causing the corresponding row to be connected to the composite row drive signal. When the Strobe input is low, all output transistors are turned on. The Serial Data output from the shift register may be used to cascade additional devices. This output is not affected by the Enable or Strobe inputs.

The SN65551 and SN65552 are characterized for operation from −40°C to 85°C. The SN75551 and SN75552 are characterized for operation from 0°C to 70°C.

† BIDFET — Bipolar, double-diffused, N-channel and P-channel MOS transistors on same chip — patented process.

SN65551, SN75551 . . . FN PACKAGE
(TOP VIEW)

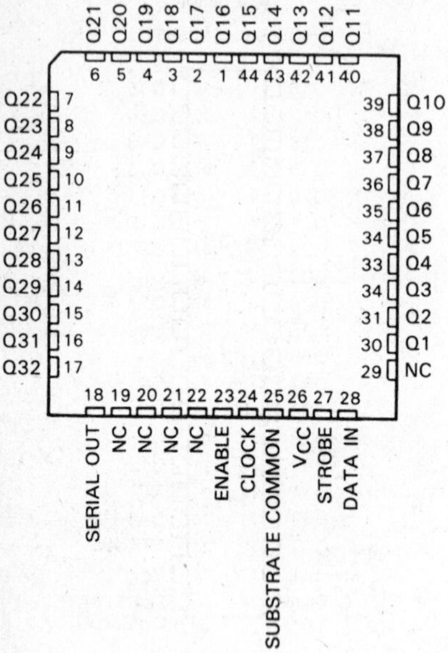
Q21 Q20 Q19 Q18 Q17 Q16 Q15 Q14 Q13 Q12 Q11
6 5 4 3 2 1 44 43 42 41 40
Q22 7
Q23 8
Q24 9
Q25 10
Q26 11
Q27 12
Q28 13
Q29 14
Q30 15
Q31 16
Q32 17
39 Q10
38 Q9
37 Q8
36 Q7
35 Q6
34 Q5
33 Q4
34 Q3
31 Q2
30 Q1
29 NC
18 19 20 21 22 23 24 25 26 27 28
SERIAL OUT NC NC NC NC ENABLE CLOCK SUBSTRATE COMMON VCC STROBE DATA IN

SN65552, SN75552 . . . FN PACKAGE
(TOP VIEW)

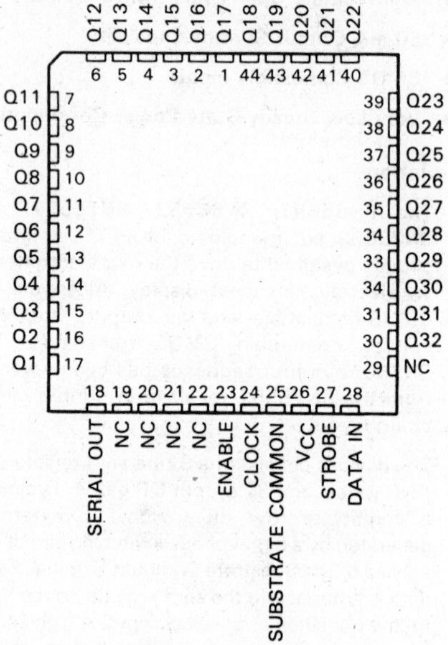
Q12 Q13 Q14 Q15 Q16 Q17 Q18 Q19 Q20 Q21 Q22
6 5 4 3 2 1 44 43 42 41 40
Q11 7
Q10 8
Q9 9
Q8 10
Q7 11
Q6 12
Q5 13
Q4 14
Q3 15
Q2 16
Q1 17
39 Q23
38 Q24
37 Q25
36 Q26
35 Q27
34 Q28
33 Q29
34 Q30
31 Q31
30 Q32
29 NC
18 19 20 21 22 23 24 25 26 27 28
SERIAL OUT NC NC NC NC ENABLE CLOCK SUBSTRATE COMMON VCC STROBE DATA IN

NC—No internal connection

logic symbols†

SN65551, SN75551

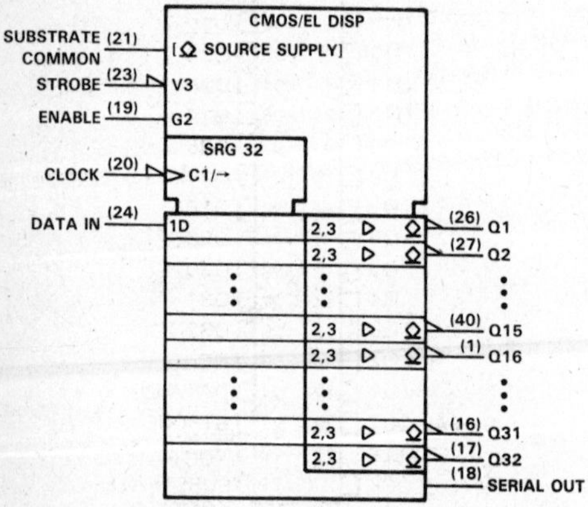
CMOS/EL DISP
SUBSTRATE COMMON (21)
[SOURCE SUPPLY]
STROBE (23)
V3
ENABLE (19)
G2
SRG 32
CLOCK (20)
C1/→
DATA IN (24)
1D
2,3
(26) Q1
(27) Q2
(40) Q15
(1) Q16
(16) Q31
(17) Q32
(18) SERIAL OUT

SN65552, SN75552

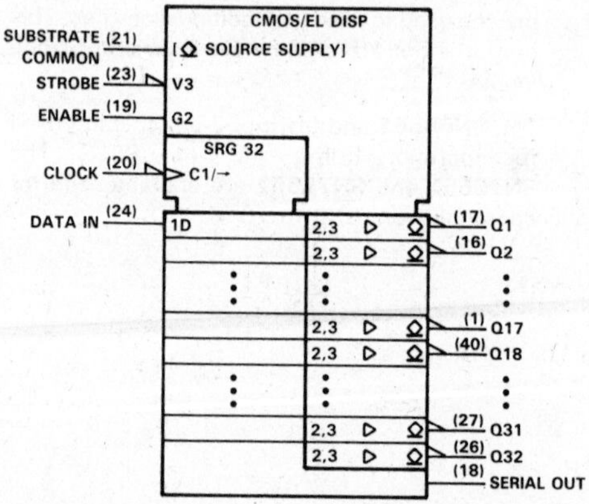
CMOS/EL DISP
SUBSTRATE COMMON (21)
[SOURCE SUPPLY]
STROBE (23)
V3
ENABLE (19)
G2
SRG 32
CLOCK (20)
C1/→
DATA IN (24)
1D
2,3
(17) Q1
(16) Q2
(1) Q17
(40) Q18
(27) Q31
(26) Q32
(18) SERIAL OUT

†These symbols are in accordance with ANSI/IEEE Std 91-1984 and IEC Publication 617-12. The symbol ◇ here indicates an n-channel open-drain output.
Pin numbers shown are for N package.

logic diagram (positive logic)

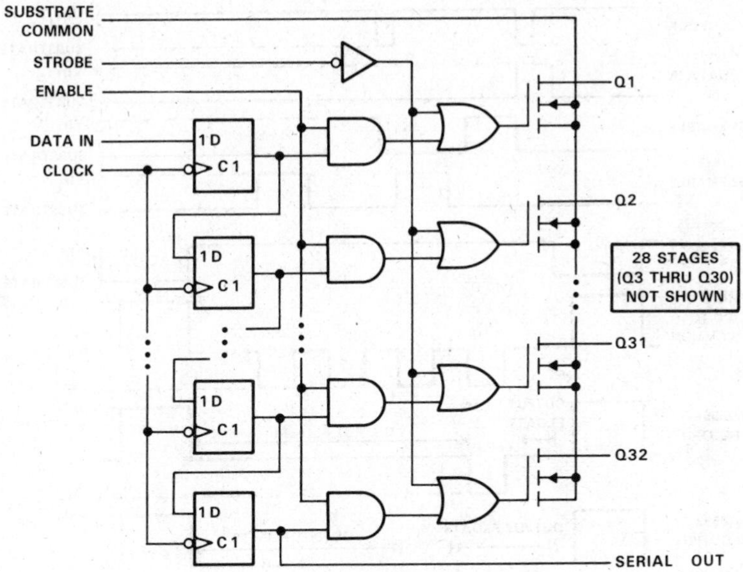
SUBSTRATE
COMMON
STROBE
ENABLE
DATA IN
CLOCK
1D
C1
Q1
Q2
28 STAGES
(Q3 THRU Q30)
NOT SHOWN
Q31
Q32
SERIAL OUT

FUNCTION TABLE

FUNCTION	CONTROL INPUTS			SHIFT REGISTERS R1 THRU R32	OUTPUTS	
	CLOCK	ENABLE	STROBE		SERIAL	Q1 THRU Q32
LOAD	↓	X	X	Load and Shift†	R32	Determined by Enable and Strobe
	No. ↓	X	X	No Change	R32	Determined by Enable and Strobe
ENABLE	X	L	H	As determined above	R32	All Q outputs off
	X	H	H	As determined above	R32	Determined by R1 through R32
STROBE	X	X	L	As determined above	R32	All Q outputs on

H = high level, L = Low level, X = irrelevant, ↓ = high-to-low transition.

†Register R32 takes on the state of R31, R31 takes on the state of R30,. . .R2 takes on the state of R1, and R1 takes on the state of the data input.

3

Display Drivers

typical operating sequence

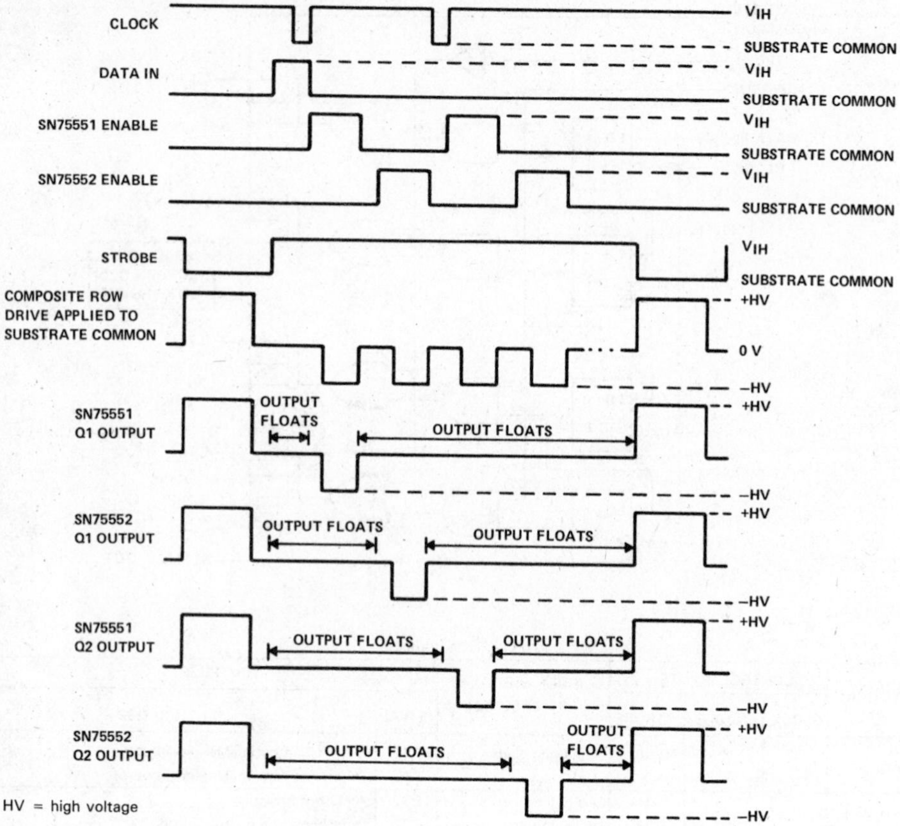
CLOCK
DATA IN
SN75551 ENABLE
SN75552 ENABLE
STROBE
COMPOSITE ROW DRIVE APPLIED TO SUBSTRATE COMMON
SN75551 Q1 OUTPUT
SN75552 Q1 OUTPUT
SN75551 Q2 OUTPUT
SN75552 Q2 OUTPUT
V_{IH}
SUBSTRATE COMMON
+HV
0 V
–HV
OUTPUT FLOATS

HV = high voltage

NOTE: During operation Clock, Data In, Enable, and Strobe are referenced to the Composite Row Drive signal received at the Substrate Common pin of the device.

TEXAS INSTRUMENTS
POST OFFICE BOX 655012 • DALLAS, TEXAS 75265

schematic of inputs and outputs

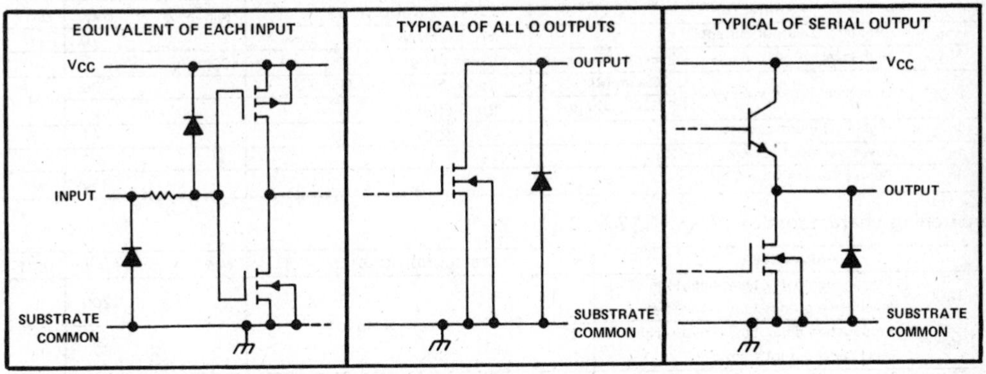
EQUIVALENT OF EACH INPUT
VCC
INPUT
SUBSTRATE COMMON
TYPICAL OF ALL Q OUTPUTS
OUTPUT
SUBSTRATE COMMON
TYPICAL OF SERIAL OUTPUT
VCC
OUTPUT
SUBSTRATE COMMON

absolute maximum ratings over operating free-air temperature range (unless otherwise noted)

Supply voltage, V_{CC} (see Note 1) . . . . 18 V
Q off-state output voltage, $V_{O(off)}$ . . . . 225 V
Input voltage . . . . $V_{CC}+0.3$ V
Substrate common terminal current (see Note 2) . . . . 1.5 A
Continuous total dissipation at (or below) 25°C free-air temperature
(see Note 3): FN package . . . . 1700 mW
N package . . . . 1250 mW
Operating free-air temperature range: SN65551, SN65552 . . . . −40°C to 85°C
SN75551, SN75552 . . . . 0°C to 70°C
Storage temperature range . . . . −65°C to 150°C
Lead temperature 1,6 mm (1/16 inch) from case for 10 seconds . . . . 260°C

NOTES: 1. Voltage values are with respect to substrate common terminal.
2. Duty cycle is limited by package dissipation.
3. For operation above 25°C free-air temperature, refer to Dissipation Derating Curves in Appendix A. In the N package, use the 10.0-mW/°C curve for these devices.

recommended operating conditions

			MIN	NOM	MAX	UNIT
V_{CC}	Supply voltage		10.8	12	15	V
V_{IH}	High-level input voltage (see Figure 1)	V_{CC} = 10.8 V	8.1		11.1	V
		V_{CC} = 15 V	11.25		15.3	
V_{IL}	Low-level input voltage (see Figure 1)	V_{CC} = 10.8 V	−0.3		2.7	V
		V_{CC} = 15 V	−0.3		3.75	
$V_{O(off)}$	Off-state Q output voltage		0		200	V
$I_{O(on)}$	On-state output current, duty cycle ≤ 1%, (see Figures 2, 3, and 4)	V_{CC} = 10.8 V, T_A = 25°C			50	mA
		V_{CC} = 15 V, T_A = 25°C			80	
I_{OK}	Output clamp current				−45	mA
f_{clock}	Clock frequency		0		4	MHz
t_w	Pulse duration, clock high or low		125			ns
t_{su}	Setup time, data before clock (see Figure 3)		50			ns
t_h	Hold time, data after clock (see Figure 3)		100			ns
T_A	Operating free-air temperature	SN65551, SN65552	−40		85	°C
		SN75551, SN75552	0		70	

electrical characteristics over recommended operating free-air temperature range

PARAMETER			TEST CONDITIONS	MIN	MAX	UNIT
$I_{O(off)}$	Off-state Q output current		$V_O = 200$ V		10	μA
V_{OH}	High-level output voltage	Serial outputs	$I_O = -100$ μA	$V_{CC}-1.5$		V
V_{OL}	Low-level output voltage	Q outputs	$I_{OL} = 50$ mA, See Figure 3		30	V
		Serial output	$I_{OL} = 100$ μA		1	
I_{IH}	High-level input current		V_I @ V_{CC}		1	μA
I_{IL}	Low-level input current		$V_I = 0$		−1	μA
I_{CC}	Supply current from V_{CC}				250	μA

switching characteristics, V_{CC} = 12 V, T_A = 25°C

PARAMETER		TEST CONDITIONS	MIN	MAX	UNIT
t_{PHL}	Propagation delay time, high-to-low level serial output from clock	C_L = 20 pF to ground, See Figure 7		200	ns
t_{PLH}	Propagation delay time, low-to-high level serial output from clock			200	ns
$t_{d(on)}$	Turn-on delay time, Q outputs from enable	I_{OL} = 50 mA, Strobe at V_{CC}, R_L = 1.4 kΩ to 100 V, See Figure 7		500	ns

RECOMMENDED OPERATING CONDITIONS

INPUT VOLTAGE LOGIC-LEVEL LIMITS
vs
SUPPLY VOLTAGE

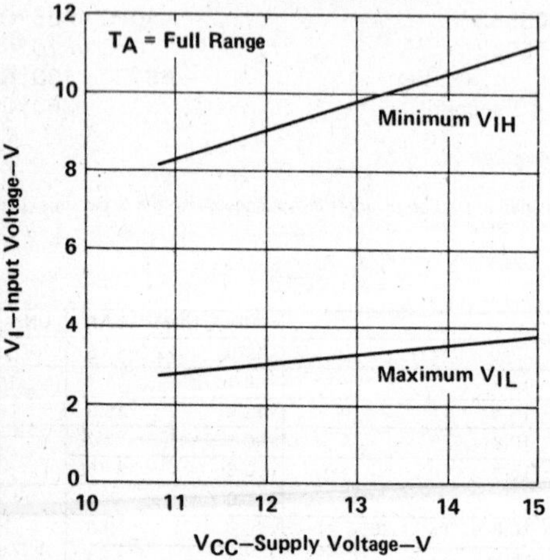

T_A = Full Range
Minimum V_{IH}
Maximum V_{IL}
V_I—Input Voltage—V
V_{CC}—Supply Voltage—V
0
2
4
6
8
10
12
10
11
12
13
14
15

FIGURE 1

MAXIMUM ON-STATE Q OUTPUT CURRENT
vs
SUPPLY VOLTAGE

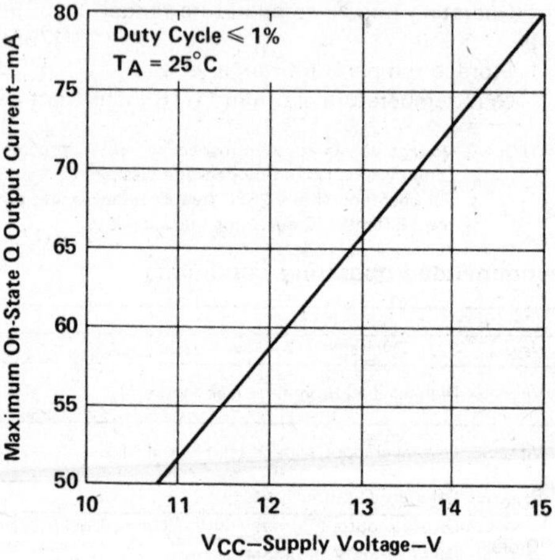

Duty Cycle ≤ 1%
T_A = 25°C
Maximum On-State Q Output Current—mA
V_{CC}—Supply Voltage—V
50
55
60
65
70
75
80
10
11
12
13
14
15

FIGURE 2

TYPICAL CHARACTERISTICS

ON-STATE Q OUTPUT CURRENT
vs
OUTPUT VOLTAGE

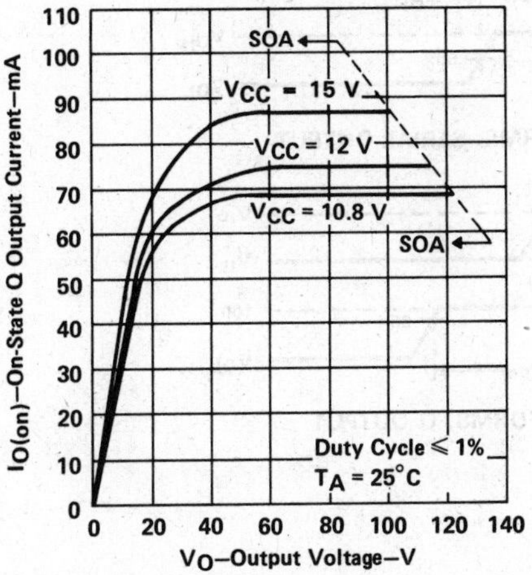
SOA
VCC = 15 V
VCC = 12 V
VCC = 10.8 V
SOA
Duty Cycle ≤ 1%
TA = 25°C
IO(on)—On-State Q Output Current—mA
VO—Output Voltage—V
0
10
20
30
40
50
60
70
80
90
100
110
0
20
40
60
80
100
120
140

SOA = Safe Operating Area

FIGURE 3

OUTPUT SATURATION CURRENT
vs
FREE-AIR TEMPERATURE†

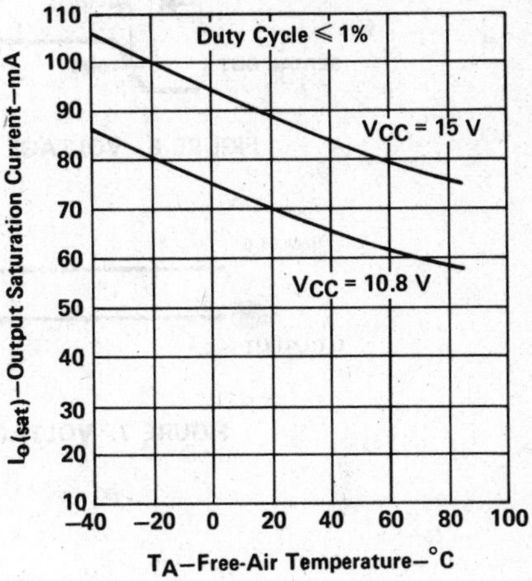
Duty Cycle ≤ 1%
VCC = 15 V
VCC = 10.8 V
Io(sat)—Output Saturation Current—mA
TA—Free-Air Temperature—°C
10
20
30
40
50
60
70
80
90
100
110
−40
−20
0
20
40
60
80
100

FIGURE 4

† Data for temperatures below 0°C and above 70°C apply only for SN65551 and SN65552.

PARAMETER MEASUREMENT INFORMATION

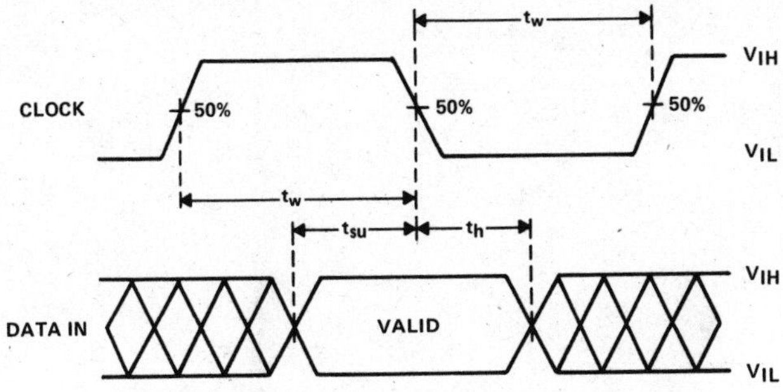
tw
VIH
CLOCK
50%
50%
50%
VIL
tw
tsu
th
VIH
DATA IN
VALID
VIL

FIGURE 5. INPUT TIMING VOLTAGE WAVEFORMS

3
Display Drivers

PARAMETER MEASUREMENT INFORMATION

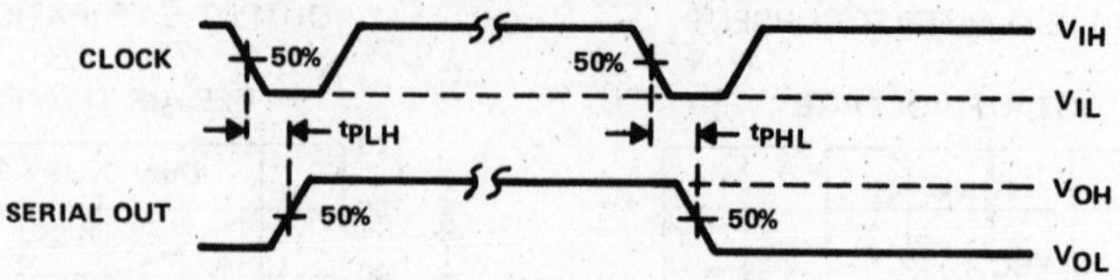
CLOCK
50%
50%
VIH
VIL
tPLH
tPHL
SERIAL OUT
50%
50%
VOH
VOL

FIGURE 6. VOLTAGE WAVEFORMS, SERIAL OUTPUT

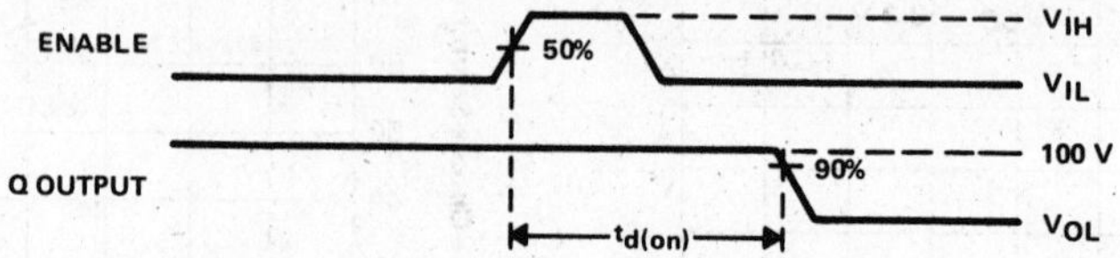
ENABLE
50%
VIH
VIL
Q OUTPUT
90%
100 V
VOL
td(on)

FIGURE 7. VOLTAGE WAVEFORMS, Q OUTPUT

TEXAS INSTRUMENTS
POST OFFICE BOX 655012 • DALLAS, TEXAS 75265

D2744, APRIL 1986

- Each Device Drives 32 Electrodes
- 60-V Output Voltage Swing Capability
- 15-mA Output Source and Sink Current Capability
- High-Speed Serially-Shifted Data Input
- Totem-Pole Outputs
- Latches on All Driver Outputs

description

The SN55553 and SN55554 are monolithic BIDFET† integrated circuits designed to drive the column electrodes of an electroluminescent display. The SN55554 output sequence has been reversed from the SN55553 for ease in printed circuit board layout.

The devices consist of a 32-bit shift register, 32 latches, and 32 output AND gates. Serial data is entered into the shift register on the low-to-high transition of the clock input. When high, the Latch Enable input transfers the shift register contents to the outputs of the 32 latches. When Output Enable is high, all Q outputs are enabled. Serial data output from the shift register may be used to cascade shift registers. This output is not affected by the Latch Enable or Output Enable inputs.

The SN55553 and SN55554 are characterized for operation over the full military temperature range of −55°C to 125°C.

SN55553 . . . FD PACKAGE
(TOP VIEW)

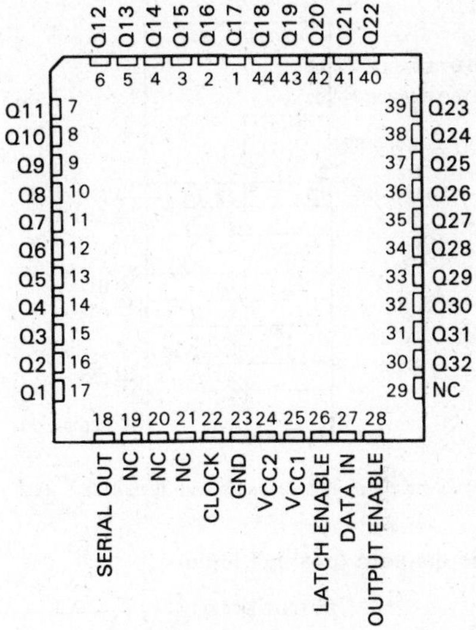
Q12 Q13 Q14 Q15 Q16 Q17 Q18 Q19 Q20 Q21 Q22
6 5 4 3 2 1 44 43 42 41 40
Q11 7
Q10 8
Q9 9
Q8 10
Q7 11
Q6 12
Q5 13
Q4 14
Q3 15
Q2 16
Q1 17
39 Q23
38 Q24
37 Q25
36 Q26
35 Q27
34 Q28
33 Q29
32 Q30
31 Q31
30 Q32
29 NC
18 19 20 21 22 23 24 25 26 27 28
SERIAL OUT NC NC NC CLOCK GND VCC2 VCC1 LATCH ENABLE DATA IN OUTPUT ENABLE

SN55554 . . . FD PACKAGE
(TOP VIEW)

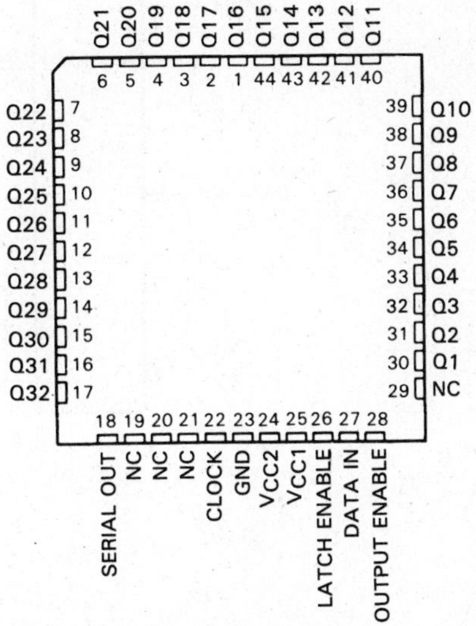
Q21 Q20 Q19 Q18 Q17 Q16 Q15 Q14 Q13 Q12 Q11
6 5 4 3 2 1 44 43 42 41 40
Q22 7
Q23 8
Q24 9
Q25 10
Q26 11
Q27 12
Q28 13
Q29 14
Q30 15
Q31 16
Q32 17
39 Q10
38 Q9
37 Q8
36 Q7
35 Q6
34 Q5
33 Q4
32 Q3
31 Q2
30 Q1
29 NC
18 19 20 21 22 23 24 25 26 27 28
SERIAL OUT NC NC NC CLOCK GND VCC2 VCC1 LATCH ENABLE DATA IN OUTPUT ENABLE

NC—No internal connection

†BIDFET — Bipolar, double-diffused, N-channel and P-channel MOS transistors on same chip — patented process.

logic symbols†

SN55553

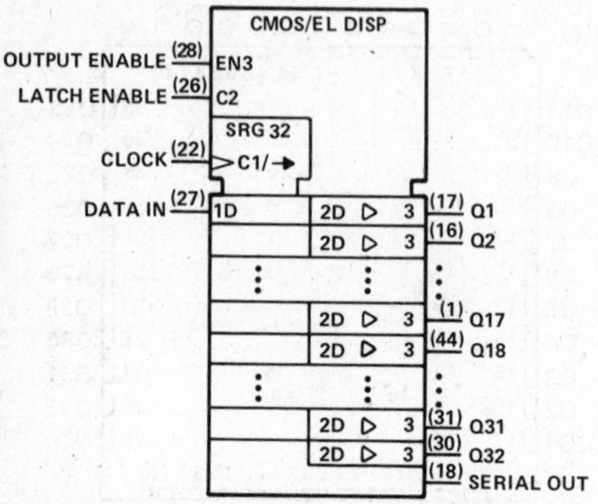
CMOS/EL DISP
OUTPUT ENABLE (28) EN3
LATCH ENABLE (26) C2
SRG 32
CLOCK (22) C1/→
DATA IN (27) 1D
2D 3 (17) Q1
2D 3 (16) Q2
2D 3 (1) Q17
2D 3 (44) Q18
2D 3 (31) Q31
2D 3 (30) Q32
(18) SERIAL OUT

SN55554

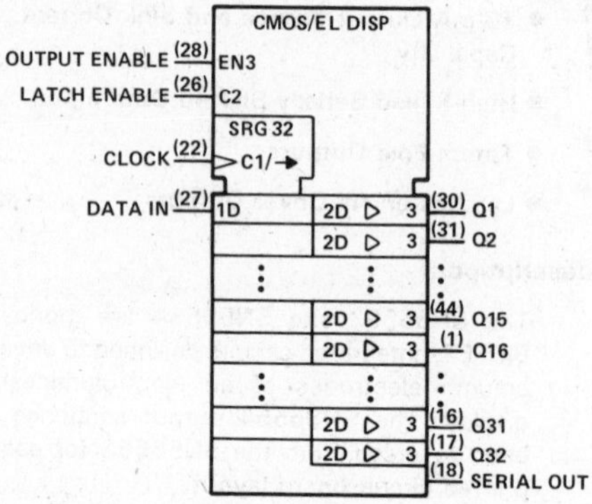
CMOS/EL DISP
OUTPUT ENABLE (28) EN3
LATCH ENABLE (26) C2
SRG 32
CLOCK (22) C1/→
DATA IN (27) 1D
2D 3 (30) Q1
2D 3 (31) Q2
2D 3 (44) Q15
2D 3 (1) Q16
2D 3 (16) Q31
2D 3 (17) Q32
(18) SERIAL OUT

†These symbols are in accordance with ANSI/IEEE Std 91-1984 and IEC Publication 617-12.

logic diagram (positive logic)

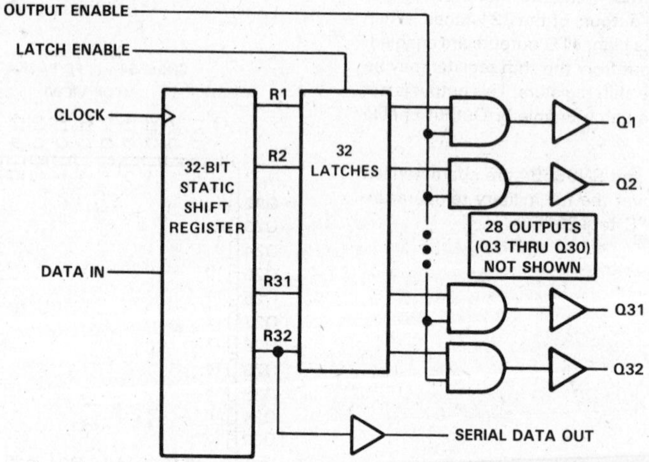
OUTPUT ENABLE
LATCH ENABLE
CLOCK
DATA IN
32-BIT STATIC SHIFT REGISTER
R1
R2
R31
R32
32 LATCHES
Q1
Q2
28 OUTPUTS (Q3 THRU Q30) NOT SHOWN
Q31
Q32
SERIAL DATA OUT

FUNCTION TABLE

FUNCTION	CONTROL INPUTS			SHIFT REGISTER R1 THRU R32	LATCHES LC1 THRU LC32	OUTPUTS	
	CLOCK	LATCH ENABLE	OUTPUT ENABLE			SERIAL	Q1 THRU Q32
LOAD	↑	X	X	Load and shift†	Determined by	R32	Determined by
	No↑	X	X	No change	Latch Enable‡	R32	Output Enable
LATCH	X	L	X	As determined above	Stored data	R32	Determined by
	X	H	X	As determined above	New data	R32	Output Enable
OUTPUT ENABLE	X	X	L	As determined above	Determined by	R32	All L
	X	X	H	As determined above	Latch Enable‡	R32	LC1 thru LC32, respectively

H = high level, L = low level, X = irrelevant, ↑ = low-to-high-level transition.
†R32 and the serial output take on the state of R31, R31 takes on the state of R30,. . .R2 takes on the state of R1, and R1 takes on the state of the data input.
‡New data enter the latches while Latch Enable is high. These data are stored while Latch Enable is low.

typical operating sequence

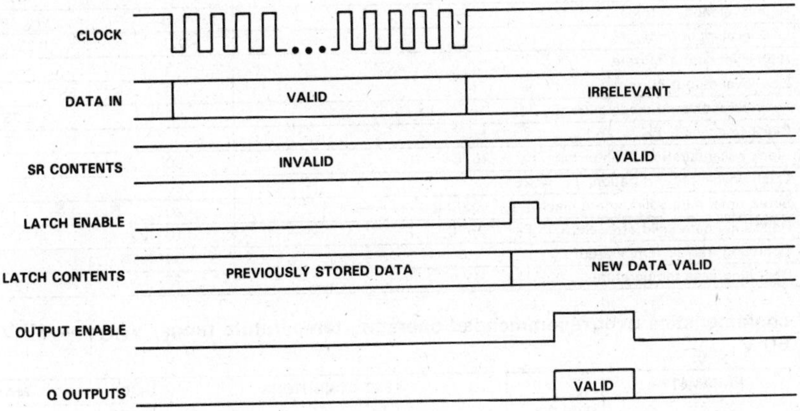
CLOCK
DATA IN
VALID
IRRELEVANT
SR CONTENTS
INVALID
VALID
LATCH ENABLE
LATCH CONTENTS
PREVIOUSLY STORED DATA
NEW DATA VALID
OUTPUT ENABLE
Q OUTPUTS
VALID

schematic of inputs and outputs

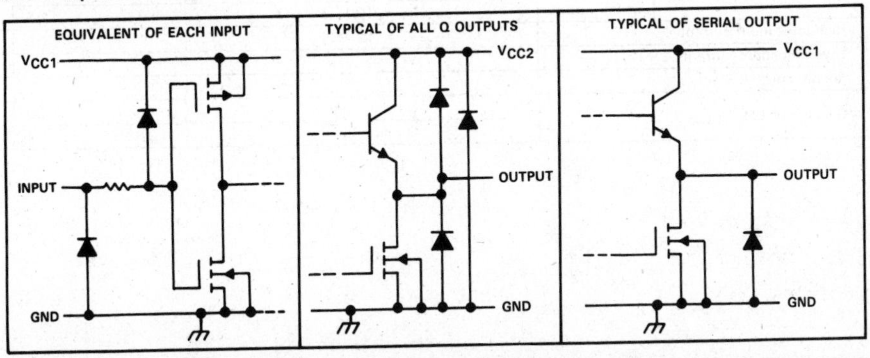
EQUIVALENT OF EACH INPUT
VCC1
INPUT
GND
TYPICAL OF ALL Q OUTPUTS
VCC2
OUTPUT
GND
TYPICAL OF SERIAL OUTPUT
VCC1
OUTPUT
GND

absolute maximum ratings over operating temperature range (unless otherwise noted)

Supply voltage, V_{CC1} (see Note 1)	18 V
Supply voltage, V_{CC2}	70 V
Input voltage	V_{CC1} + 0.3 V
Ground current	700 mA
Continuous total dissipation at (or below) 25°C free-air temperature (see Note 2)	1825 mW
Minimum operating free-air temperature	−55°C
Operating case temperature	125°C
Storage temperature range	−65°C to 150°C
Case temperature for 60 seconds	260°C

NOTES: 1. Voltage values are with respect to network ground terminal.
2. For operation above 25°C free-air temperature, derate linearly at the rate of 14.6 mW/°C.

recommended operating conditions

		MIN	NOM	MAX	UNIT
V_{CC1}	Supply voltage	10.8	12	13.2	V
V_{CC2}	Supply voltage	0		60	V
V_{IH}	High-level input voltage	$0.75V_{CC}$		$V_{CC}+0.3$	V
V_{IL}	Low-level input voltage	−0.3		$0.25V_{CC}$	
I_{OH}	High-level output current	−15			mA
I_{OL}	Low-level output current	15			mA
I_{OK}	Peak output clamp diode current			±20	mA
f_{clock}	Clock frequency, T_A = 25°C			6.25	MHz
$t_{w(CLK)}$	Clock pulse duration, high or low, T_A = 25°C	80			ns
$t_{w(LE)}$	Latch enable pulse duration, T_A = 25°C	80			
t_{su}	Setup time, data valid before clock↑, T_A = 25°C	20			ns
t_h	Hold time, data valid after clock ↑, T_A = 25°C	110			ns
T_A	Operating free-air temperature	−55			
T_C	Operating case temperature			125	

electrical characteristics over recommended operating temperature range, V_{CC1} = 12 V, V_{CC2} = 60 V

PARAMETER			TEST CONDITIONS	MIN	MAX	UNIT
V_{OH}	High-level output voltage	Q outputs	I_O = −15 mA	55		V
		Serial output	I_O = −100 μA	10		
V_{OL}	Low-level output voltage	Q outputs	I_O = 15 mA		10	V
		Serial output	I_O = 100 μA		1.5	
I_{IH}	High-level input current		V_I = 12 V		5	μA
I_{IL}	Low-level input current		V_I = 0		−5	μA
I_{CC1}	Supply current, V_{CC1}				7	mA
I_{CC2}	Supply current, V_{CC2}		Outputs high		20	mA
			Outputs low		2	

switching characteristics, V_{CC1} = 12 V, V_{CC2} = 60 V, T_C = 25°C

PARAMETER		TEST CONDITIONS	MIN	MAX	UNIT
t_{dLH}	Delay time, clock↑ to serial↑	C_L = 45 pF to ground, See Figures 1 and 2		200	ns
t_{dHL}	Delay time, clock↑ to serial↓			200	ns
t_{dLH}	Delay time, LE to Q output↑	C_L = 45 pF to ground, See Figures 1 and 3		1000	ns
t_{dHL}	Delay time, LE to Q output↓			500	ns

PARAMETER MEASUREMENT INFORMATION

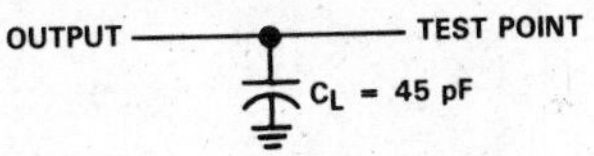
OUTPUT
TEST POINT
C_L = 45 pF

FIGURE 1. OUTPUT LOAD CIRCUIT

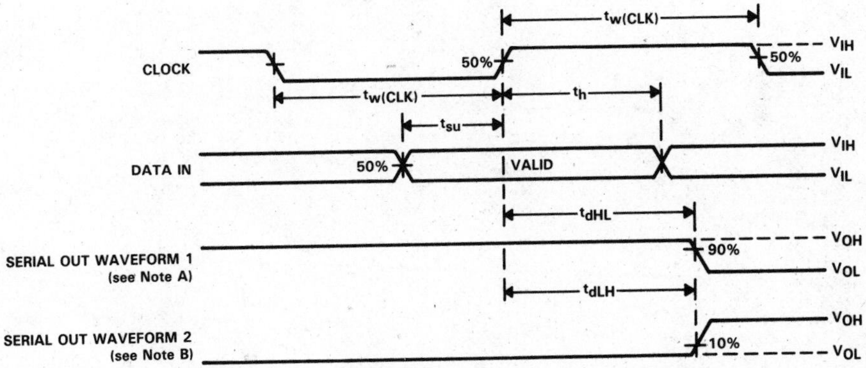
CLOCK
t_w(CLK)
50%
V_IH
V_IL
t_h
t_su
DATA IN
VALID
t_dHL
SERIAL OUT WAVEFORM 1
(see Note A)
90%
V_OH
V_OL
t_dLH
SERIAL OUT WAVEFORM 2
(see Note B)
10%

FIGURE 2. VOLTAGE WAVEFORMS FOR SERIAL OUTPUT

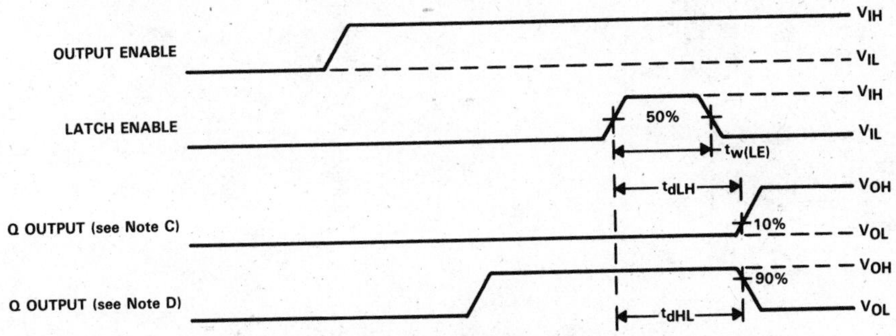
OUTPUT ENABLE
V_IH
V_IL
LATCH ENABLE
50%
t_w(LE)
t_dLH
Q OUTPUT (see Note C)
10%
V_OH
V_OL
Q OUTPUT (see Note D)
90%
t_dHL

FIGURE 3. VOLTAGE WAVEFORMS FOR Q OUTPUTS

NOTES: A. Waveform 1 is for internal conditions such that a low is clocked into R32.
B. Waveform 2 is for internal conditions such that a high is clocked into R32.
C. To measure t_{dLH}, initially a low is stored in the latch and a high is stored in the shift register.
D. To measure t_{dHL}, initially a high is stored in the latch and a low is stored in the shift register.

D2744, MARCH 1983—REVISED SEPTEMBER 1986

- Each Device Drives 32 Electrodes
- 60-V Output Voltage Swing Capability
- 15-mA Output Source and Sink Current Capability
- High-Speed Serially-Shifted Data Input
- Totem-Pole Outputs
- Latches on All Driver Outputs

description

The SN65553, SN65554, SN75553, and SN75554 are monolithic BIDFET† integrated circuits designed to drive the column electrodes of an electroluminescent display. The SN65554 and SN75554 output sequence has been reversed from the SN65553 and SN75553 for ease in printed circuit board layout.

The devices consist of a 32-bit shift register, 32 latches, and 32 output AND gates. Serial data is entered into the shift register on the low-to-high transition of the clock input. When high, the Latch Enable input transfers the shift register contents to the outputs of the 32 latches. When Output Enable is high, all Q outputs are enabled. Serial data output from the shift register may be used to cascade shift registers. This output is not affected by the Latch Enable or Output Enable inputs.

The SN65553 and SN65554 are characterized for operation from −40°C to 85°C. The SN75553 and SN75554 are characterized for operation from 0°C to 70°C.

N
DUAL-IN-LINE PLASTIC PACKAGE
(TOP VIEW)
SN65553, SN75553

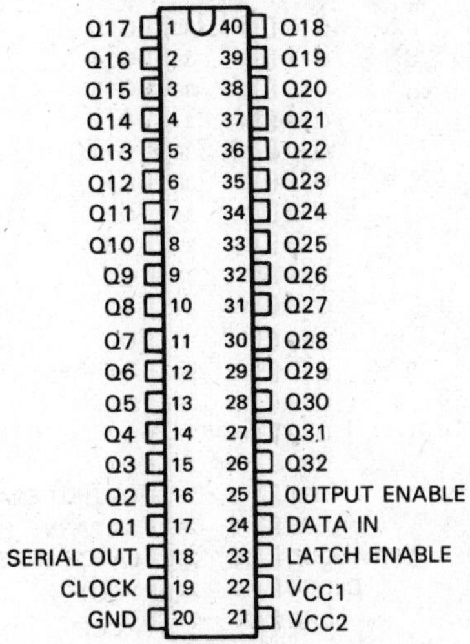
Q17 1 40 Q18
Q16 2 39 Q19
Q15 3 38 Q20
Q14 4 37 Q21
Q13 5 36 Q22
Q12 6 35 Q23
Q11 7 34 Q24
Q10 8 33 Q25
Q9 9 32 Q26
Q8 10 31 Q27
Q7 11 30 Q28
Q6 12 29 Q29
Q5 13 28 Q30
Q4 14 27 Q31
Q3 15 26 Q32
Q2 16 25 OUTPUT ENABLE
Q1 17 24 DATA IN
SERIAL OUT 18 23 LATCH ENABLE
CLOCK 19 22 VCC1
GND 20 21 VCC2

FN PLASTIC CHIP CARRIER PACKAGE
(TOP VIEW)
SN65553, SN75553

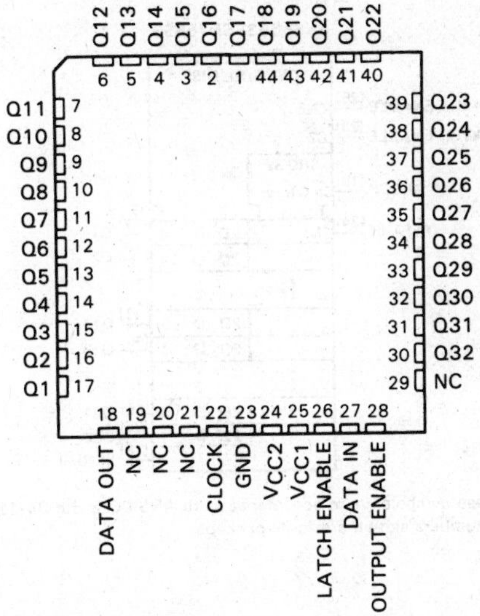
Q12 Q13 Q14 Q15 Q16 Q17 Q18 Q19 Q20 Q21 Q22
6 5 4 3 2 1 44 43 42 41 40
Q11 7
Q10 8
Q9 9
Q8 10
Q7 11
Q6 12
Q5 13
Q4 14
Q3 15
Q2 16
Q1 17
39 Q23
38 Q24
37 Q25
36 Q26
35 Q27
34 Q28
33 Q29
32 Q30
31 Q31
30 Q32
29 NC
18 19 20 21 22 23 24 25 26 27 28
DATA OUT NC NC NC CLOCK GND VCC2 VCC1 LATCH ENABLE DATA IN OUTPUT ENABLE

NC—No internal connection

†BIDFET — Bipolar, double-diffused, N-channel and P-channel MOS transistors on same chip — patented process.

N
DUAL-IN-LINE PLASTIC PACKAGE
(TOP VIEW)
SN65554, SN75554

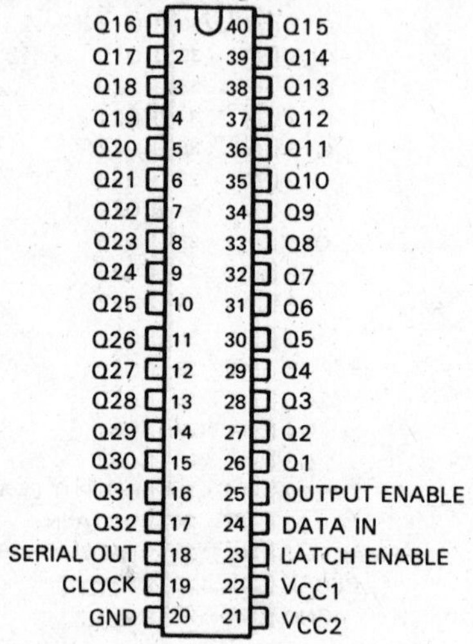
Q16 1 40 Q15
Q17 2 39 Q14
Q18 3 38 Q13
Q19 4 37 Q12
Q20 5 36 Q11
Q21 6 35 Q10
Q22 7 34 Q9
Q23 8 33 Q8
Q24 9 32 Q7
Q25 10 31 Q6
Q26 11 30 Q5
Q27 12 29 Q4
Q28 13 28 Q3
Q29 14 27 Q2
Q30 15 26 Q1
Q31 16 25 OUTPUT ENABLE
Q32 17 24 DATA IN
SERIAL OUT 18 23 LATCH ENABLE
CLOCK 19 22 VCC1
GND 20 21 VCC2

FN PLASTIC CHIP CARRIER PACKAGE
(TOP VIEW)
SN65554, SN75554

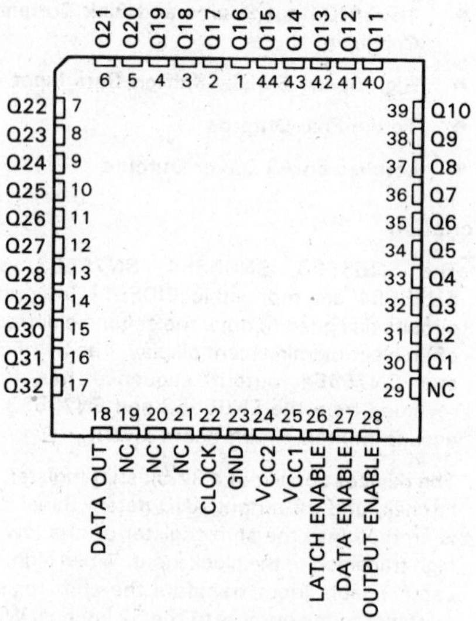
Q21 Q20 Q19 Q18 Q17 Q16 Q15 Q14 Q13 Q12 Q11
6 5 4 3 2 1 44 43 42 41 40
Q22 7
Q23 8
Q24 9
Q25 10
Q26 11
Q27 12
Q28 13
Q29 14
Q30 15
Q31 16
Q32 17
39 Q10
38 Q9
37 Q8
36 Q7
35 Q6
34 Q5
33 Q4
32 Q3
31 Q2
30 Q1
29 NC
18 19 20 21 22 23 24 25 26 27 28
DATA OUT NC NC NC CLOCK GND VCC2 VCC1 LATCH ENABLE DATA ENABLE OUTPUT ENABLE

NC—No internal connection.

logic symbols†

SN65553, SN75553

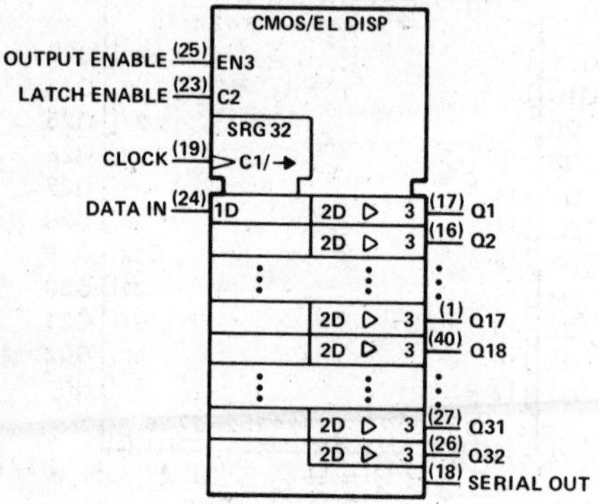
CMOS/EL DISP
OUTPUT ENABLE (25) EN3
LATCH ENABLE (23) C2
SRG 32
CLOCK (19) C1/→
DATA IN (24) 1D
2D ▷ 3 (17) Q1
2D ▷ 3 (16) Q2
2D ▷ 3 (1) Q17
2D ▷ 3 (40) Q18
2D ▷ 3 (27) Q31
2D ▷ 3 (26) Q32
(18) SERIAL OUT

SN65554, SN75554

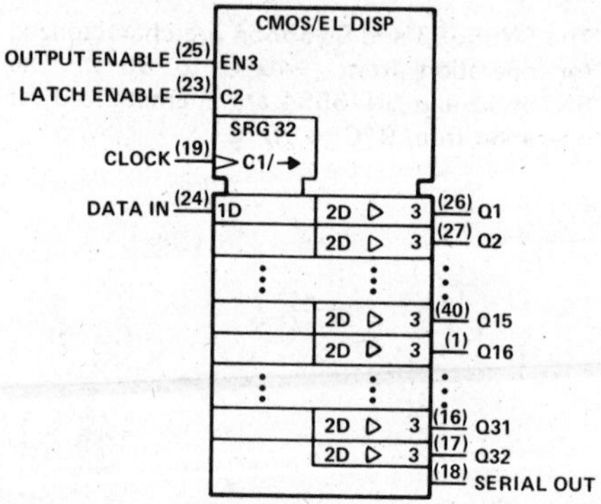
CMOS/EL DISP
OUTPUT ENABLE (25) EN3
LATCH ENABLE (23) C2
SRG 32
CLOCK (19) C1/→
DATA IN (24) 1D
2D ▷ 3 (26) Q1
2D ▷ 3 (27) Q2
2D ▷ 3 (40) Q15
2D ▷ 3 (1) Q16
2D ▷ 3 (16) Q31
2D ▷ 3 (17) Q32
(18) SERIAL OUT

†These symbols are in accordance with ANSI/IEEE Std 91-1984 and IEC Publication 617-12.
Pin numbers shown are for N packages.

logic diagram (positive logic)

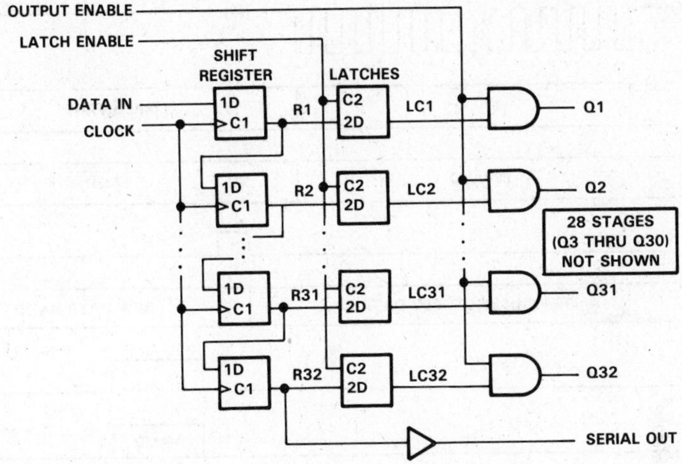
OUTPUT ENABLE
LATCH ENABLE
SHIFT REGISTER
LATCHES
DATA IN
CLOCK
1D
C1
R1
C2
2D
LC1
Q1
R2
LC2
Q2
28 STAGES (Q3 THRU Q30) NOT SHOWN
R31
LC31
Q31
R32
LC32
Q32
SERIAL OUT

FUNCTION TABLE

FUNCTION	CONTROL INPUTS			SHIFT REGISTER R1 THRU R32	LATCHES LC1 THRU LC32	OUTPUTS	
	CLOCK	LATCH ENABLE	OUTPUT ENABLE			SERIAL	Q1 THRU Q32
LOAD	↑	X	X	Load and shift†	Determined by	R32	Determined by
	No↑	X	X	No change	Latch Enable‡	R32	Output Enable
LATCH	X	L	X	As determined above	Stored data	R32	Determined by
	X	H	X	As determined above	New data	R32	Output Enable
OUTPUT ENABLE	X	X	L	As determined above	Determined by	R32	All L
	X	X	H	As determined above	Latch Enable‡	R32	LC1 thru LC32, respectively

H = high level, L = low level, X = irrelevant, ↑ = low-to-high-level transition.
†R32 and the serial output take on the state of R31, R31 takes on the state of R30,. . .R2 takes on the state of R1, and R1 takes on the state of the data input.
‡New data enter the latches while Latch Enable is high. These data are stored while Latch Enable is low.

typical operating sequence

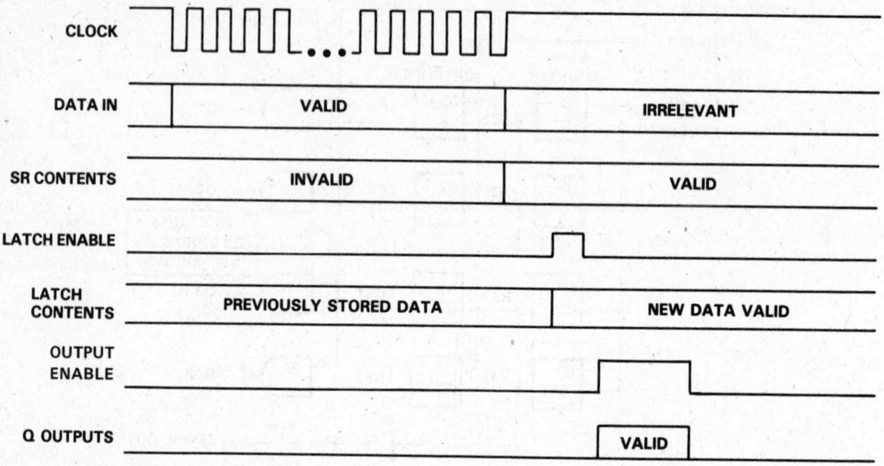
CLOCK
DATA IN
VALID
IRRELEVANT
SR CONTENTS
INVALID
VALID
LATCH ENABLE
LATCH CONTENTS
PREVIOUSLY STORED DATA
NEW DATA VALID
OUTPUT ENABLE
Q OUTPUTS
VALID

schematic of inputs and outputs

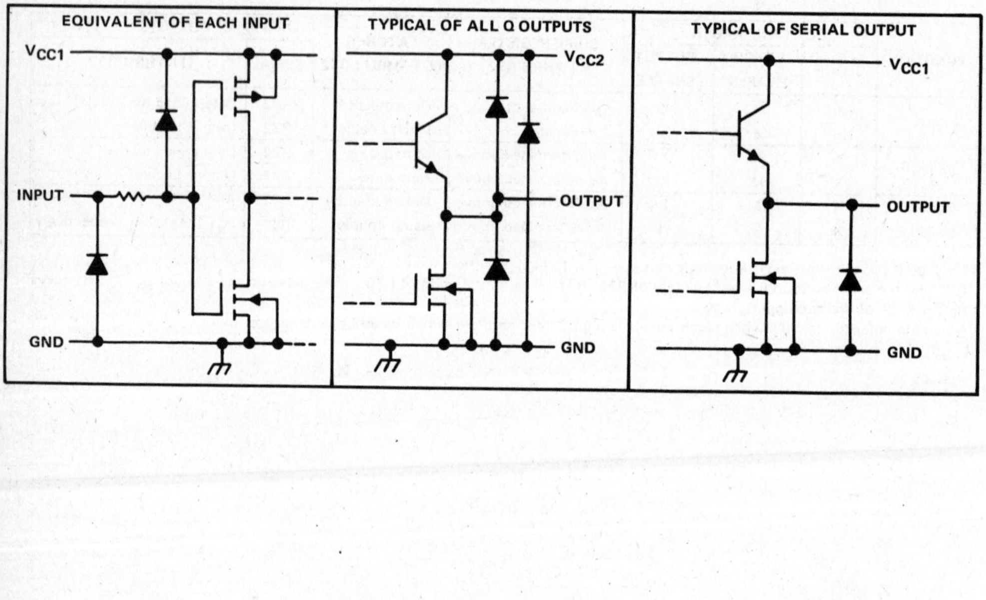
EQUIVALENT OF EACH INPUT
VCC1
INPUT
GND
TYPICAL OF ALL Q OUTPUTS
VCC2
OUTPUT
GND
TYPICAL OF SERIAL OUTPUT
VCC1
OUTPUT
GND

TEXAS INSTRUMENTS
POST OFFICE BOX 655012 • DALLAS, TEXAS 75265

absolute maximum ratings over operating free-air temperature range (unless otherwise noted)

Supply voltage, V_{CC1} (see Note 1) . . . . 18 V
Supply voltage, V_{CC2} . . . . 70 V
Input voltage . . . . V_{CC1} + 0.3 V
Ground current . . . . 700 mA
Continuous total dissipation at (or below) 25°C free-air temperature (see Note 2)
FN package . . . . 1700 mW
N package . . . . 1250 mW
Operating free-air temperature range: SN65553, SN65554 . . . . −40°C to 85°C
SN75553, SN75554 . . . . 0°C to 70°C
Storage temperature range . . . . −65°C to 150°C
Lead temperature 1,6 mm (1/16 inch) from case for 10 seconds: N package . . . . 260°C
Case temperature for 10 seconds: FN package . . . . 260°C

NOTES: 1. Voltage values are with respect to network ground terminal.
2. For operation above 25°C free-air temperature, refer to Dissipation Derating Curves in Appendix A. In the N package, use the 10.0-mW/°C curve for these devices.

3 Display Drivers

recommended operating conditions

		MIN	NOM	MAX	UNIT
Supply voltage, V_{CC1}		10.8	12	15	V
Supply voltage, V_{CC2}		0		60	V
High-level input voltage, V_{IH} (see Figure 1)	V_{CC1} = 10.8 V	8.1		11.1	V
	V_{CC1} = 15 V	11.25		15.3	
Low-level input voltage, V_{IL} (see Figure 1)	V_{CC1} = 10.8 V	−0.3		2.7	V
	V_{CC1} = 15 V	−0.3		3.75	
High-level output current, I_{OH}		−15			mA
Low-level output current, I_{OL}		15			mA
Output clamp current, I_{OK}				20	mA
Clock frequency, f_{clock}		0		6.25	MHz
Pulse duration, clock high or low, $t_{w(CLK)}$ (see Figure 2)		80			ns
Pulse duration, latch enable, $t_{w(LE)}$ (see Figure 4)		80			ns
Data setup time before clock ↑, t_{su} (see Figure 2)		20			ns
Data hold time after clock ↑, t_h (see Figure 2)		80			ns
Operating free-air temperature, T_A	SN65553, SN65554	−40		85	°C
	SN75553, SN75554	0		70	

electrical characteristics over recommended ranges of V_{CC1} and operating free-air temperature, V_{CC2} = 60 V (unless otherwise noted)

PARAMETER			TEST CONDITIONS	MIN	MAX	UNIT
V_{OH}	High-level output voltage	Q outputs	I_O = −15 mA	57		V
		Serial output	I_O = −100 μA	V_{CC1} − 1.5		
V_{OL}	Low-level output voltage	Q outputs	I_{OL} = 15 mA		8	V
		Serial output	I_{OL} = 100 μA		1	
I_{IH}	High-level input current		V_I = V_{CC1}		1	μA
I_{IL}	Low-level input current		V_I = 0		−1	μA
I_{CC1}	Supply current from V_{CC1}				5	mA
I_{CC2}	Supply current from V_{CC2}	SN65553, SN65554			12	mA
		SN75553, SN75554			10	

switching characteristics, V_{CC1} = 12 V, V_{CC2} = 60 V, T_A = 25°C

PARAMETER		TEST CONDITIONS	MIN	MAX	UNIT
t_{PHL}	Propagation delay time, high-to-low-level Serial output from Clock	C_L = 20 pF to ground, See Figure 3		140	ns
t_{PLH}	Propagation delay time, low-to-high-level Serial output from Clock			140	ns
t_{DHL}	Delay time, high-to-low-level Q output from Latch Enable	C_L = 20 pF to ground, See Figure 4		500	ns
t_{DLH}	Delay time, low-to-high-level Q output from Latch Enable	C_L = 20 pF to ground, See Figure 4		1	μs

RECOMMENDED OPERATION CONDITIONS

INPUT VOLTAGE LOGIC-LEVEL LIMITS
vs
SUPPLY VOLTAGE V_{CC1}

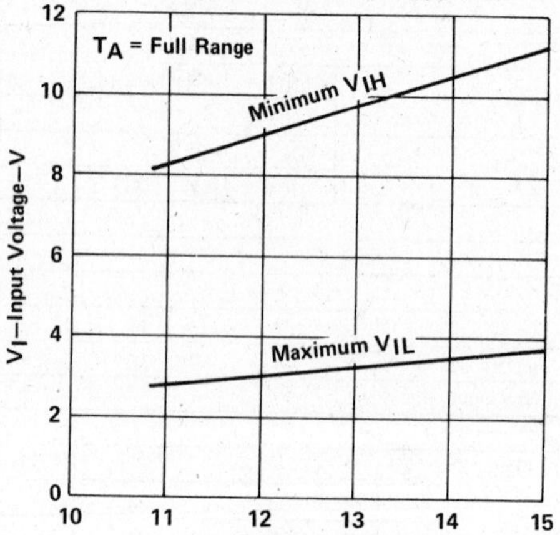
TA = Full Range
Minimum VIH
Maximum VIL
VI—Input Voltage—V
VCC—Supply Voltage—V
0
2
4
6
8
10
12
10
11
12
13
14
15

FIGURE 1

TEXAS INSTRUMENTS
POST OFFICE BOX 655012 • DALLAS, TEXAS 75265

PARAMETER MEASUREMENT INFORMATION

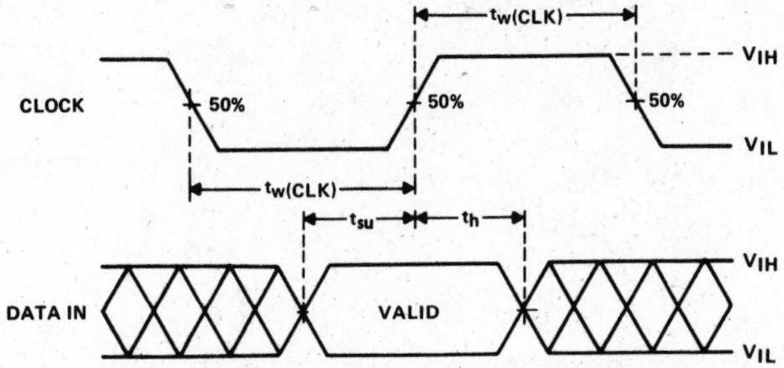
tw(CLK)
VIH
CLOCK
50%
50%
50%
VIL
tw(CLK)
tsu
th
VIH
DATA IN
VALID
VIL

FIGURE 2. INPUT TIMING VOLTAGE WAVEFORMS

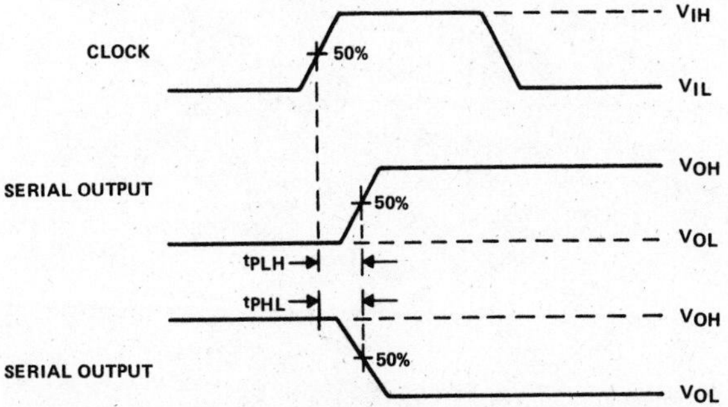
VIH
CLOCK
50%
VIL
VOH
SERIAL OUTPUT
50%
VOL
tPLH
tPHL
VOH
50%
SERIAL OUTPUT
VOL

FIGURE 3. VOLTAGE WAVEFORMS FOR PROPAGATION DELAY CLOCK TO SERIAL OUTPUT

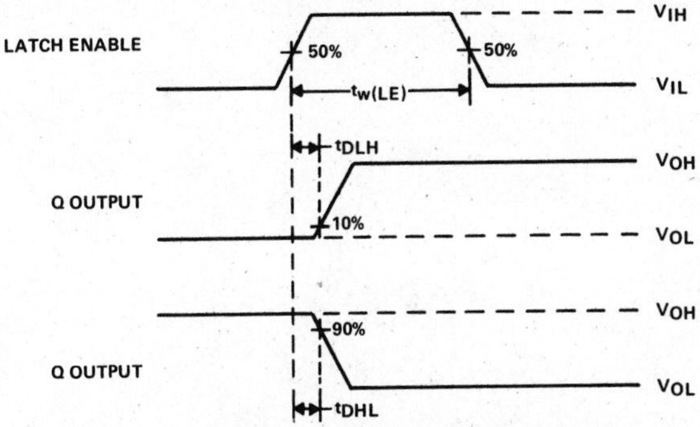
VIH
LATCH ENABLE
50%
50%
VIL
tw(LE)
tDLH
VOH
Q OUTPUT
10%
VOL
VOH
90%
Q OUTPUT
VOL
tDHL

FIGURE 4. VOLTAGE WAVEFORMS FOR DELAY TIMES, LATCH ENABLE TO Q OUTPUTS

D2744, APRIL 1985—REVISED SEPTEMBER 1986

- Each Device Drives 32 Electrodes
- 90-V Output Voltage Swing Capability Using Ramped Supply
- 15-mA Output Source and Sink Current Capability
- High-Speed Serially-Shifted Data Input
- Totem-Pole Outputs
- Latches on All Driver Outputs

SN65555, SN75555
N DUAL-IN-LINE PACKAGE
(TOP VIEW)

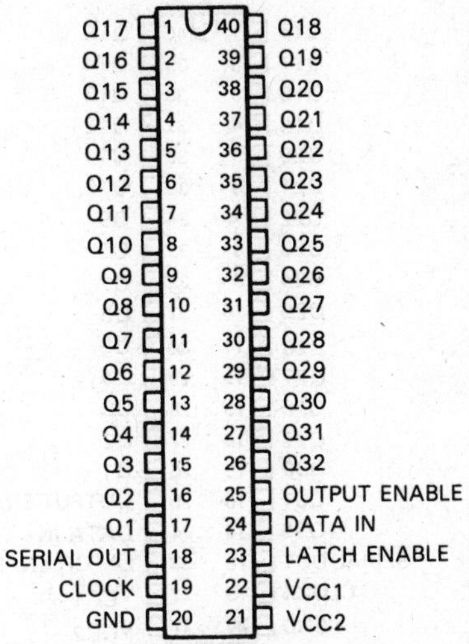
Q17 1 40 Q18
Q16 2 39 Q19
Q15 3 38 Q20
Q14 4 37 Q21
Q13 5 36 Q22
Q12 6 35 Q23
Q11 7 34 Q24
Q10 8 33 Q25
Q9 9 32 Q26
Q8 10 31 Q27
Q7 11 30 Q28
Q6 12 29 Q29
Q5 13 28 Q30
Q4 14 27 Q31
Q3 15 26 Q32
Q2 16 25 OUTPUT ENABLE
Q1 17 24 DATA IN
SERIAL OUT 18 23 LATCH ENABLE
CLOCK 19 22 VCC1
GND 20 21 VCC2

description

The SN65555, SN65556, and SN75556 are monolithic BIDFET† integrated circuits designed to drive the column electrodes of an electroluminescent display. The SN65556 and SN75556 output sequence has been reversed from the SN65555 and SN75555 for ease in printed circuit board layout.

The devices consist of a 32-bit shift register, 32 latches, and 32 output AND gates. Serial data is entered into the shift register on the low-to-high transition of the clock input. When high, the Latch Enable input transfers the shift register contents to the outputs of the 32 latches. When Output Enable is high, all Q outputs are enabled. Data must be loaded into the latches and Output Enable must be high before supply voltage V_{CC2} is ramped up.

Serial data output from the shift register may be used to cascade shift registers. This output is not affected by the Latch Enable or Output Enable inputs.

The SN65555 and SN65556 are characterized for operation from −40°C to 85°C. The SN75555 and SN75556 are characterized for operation from 0°C to 70°C.

SN65555, SN75555
FN PLASTIC CHIP CARRIER PACKAGE
(TOP VIEW)

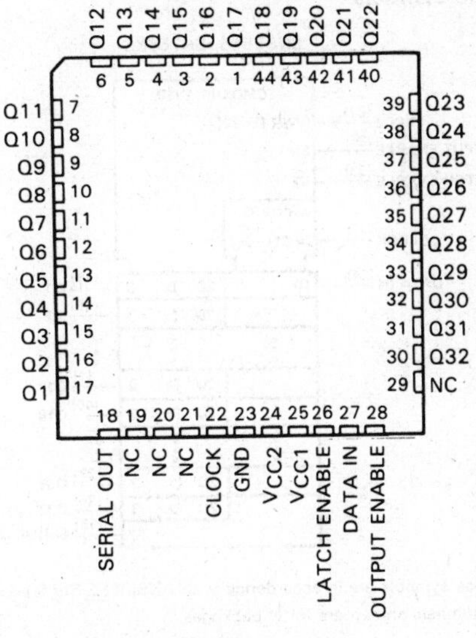
Q12 Q13 Q14 Q15 Q16 Q17 Q18 Q19 Q20 Q21 Q22
6 5 4 3 2 1 44 43 42 41 40
Q11 7
Q10 8
Q9 9
Q8 10
Q7 11
Q6 12
Q5 13
Q4 14
Q3 15
Q2 16
Q1 17
39 Q23
38 Q24
37 Q25
36 Q26
35 Q27
34 Q28
33 Q29
32 Q30
31 Q31
30 Q32
29 NC
18 19 20 21 22 23 24 25 26 27 28
SERIAL OUT NC NC NC CLOCK GND VCC2 VCC1 LATCH ENABLE DATA IN OUTPUT ENABLE

NC—No internal connection

†BIDFET — Bipolar, double-diffused, N-channel and P-channel MOS transistors on same chip — patented process.

TEXAS INSTRUMENTS
POST OFFICE BOX 655012 • DALLAS, TEXAS 75265

SN65556, SN75556
N DUAL-IN-LINE PACKAGE
(TOP VIEW)

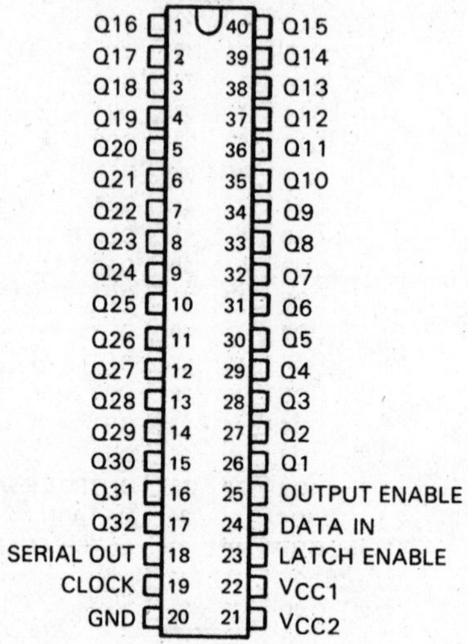
Q16 1 40 Q15
Q17 2 39 Q14
Q18 3 38 Q13
Q19 4 37 Q12
Q20 5 36 Q11
Q21 6 35 Q10
Q22 7 34 Q9
Q23 8 33 Q8
Q24 9 32 Q7
Q25 10 31 Q6
Q26 11 30 Q5
Q27 12 29 Q4
Q28 13 28 Q3
Q29 14 27 Q2
Q30 15 26 Q1
Q31 16 25 OUTPUT ENABLE
Q32 17 24 DATA IN
SERIAL OUT 18 23 LATCH ENABLE
CLOCK 19 22 VCC1
GND 20 21 VCC2

SN65556, SN75556
FN PLASTIC CHIP CARRIER PACKAGE
(TOP VIEW)

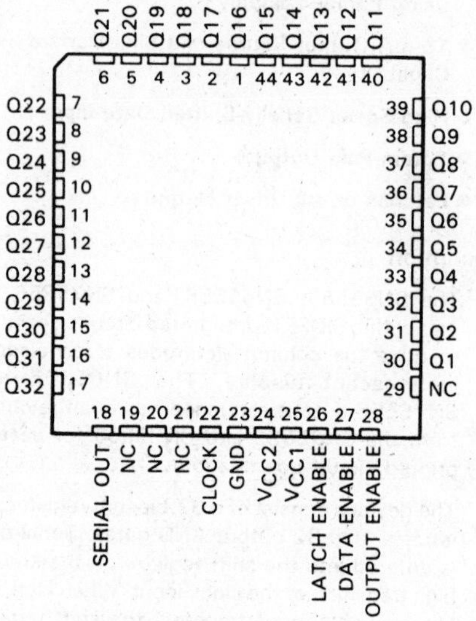
Q21 Q20 Q19 Q18 Q17 Q16 Q15 Q14 Q13 Q12 Q11
6 5 4 3 2 1 44 43 42 41 40
Q22 7
Q23 8
Q24 9
Q25 10
Q26 11
Q27 12
Q28 13
Q29 14
Q30 15
Q31 16
Q32 17
39 Q10
38 Q9
37 Q8
36 Q7
35 Q6
34 Q5
33 Q4
32 Q3
31 Q2
30 Q1
29 NC
18 19 20 21 22 23 24 25 26 27 28
SERIAL OUT NC NC NC CLOCK GND VCC2 VCC1 LATCH ENABLE DATA ENABLE OUTPUT ENABLE

NC—No internal connection

logic symbols†

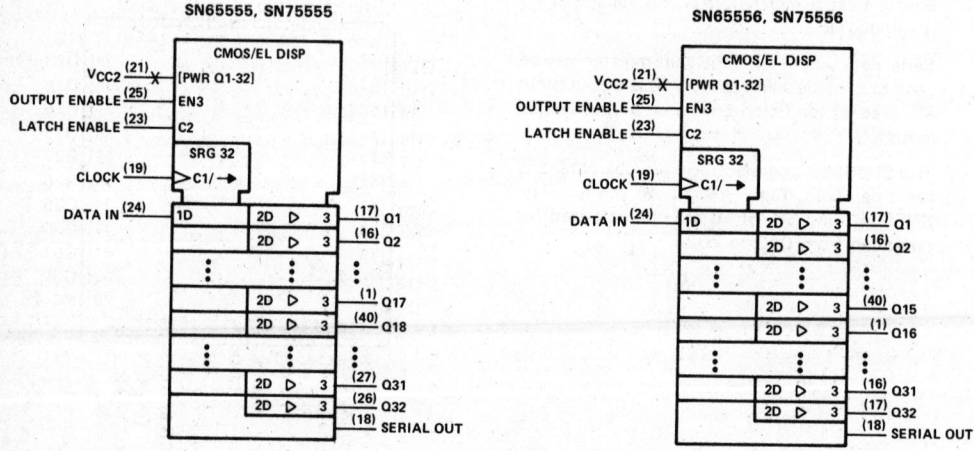

SN65555, SN75555

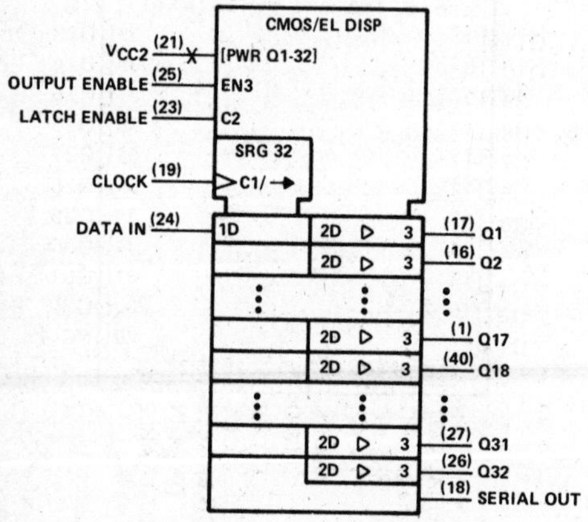
CMOS/EL DISP
VCC2 (21) [PWR Q1-32]
OUTPUT ENABLE (25) EN3
LATCH ENABLE (23) C2
SRG 32
CLOCK (19) C1/→
DATA IN (24) 1D
2D 3 (17) Q1
2D 3 (16) Q2
2D 3 (1) Q17
2D 3 (40) Q18
2D 3 (27) Q31
2D 3 (26) Q32
(18) SERIAL OUT

SN65556, SN75556

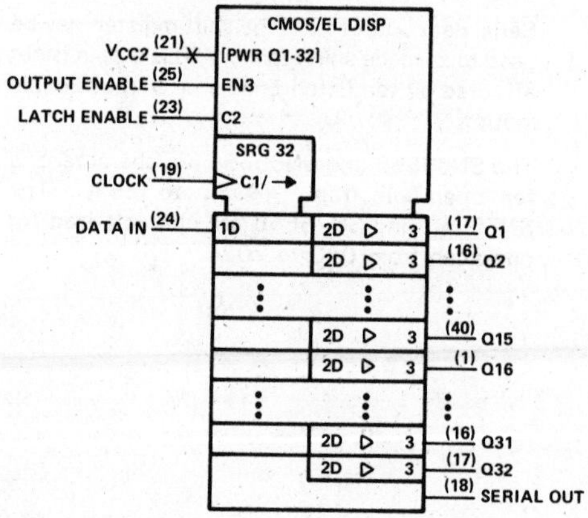
CMOS/EL DISP
VCC2 (21) [PWR Q1-32]
OUTPUT ENABLE (25) EN3
LATCH ENABLE (23) C2
SRG 32
CLOCK (19) C1/→
DATA IN (24) 1D
2D 3 (17) Q1
2D 3 (16) Q2
2D 3 (40) Q15
2D 3 (1) Q16
2D 3 (16) Q31
2D 3 (17) Q32
(18) SERIAL OUT

†These symbols are in accordance with ANSI/IEEE Std 91-1984 and IEC Publication 617-12.
Pin numbers shown are for N packages.

TEXAS INSTRUMENTS
POST OFFICE BOX 655012 • DALLAS, TEXAS 75265

logic diagram (positive logic)

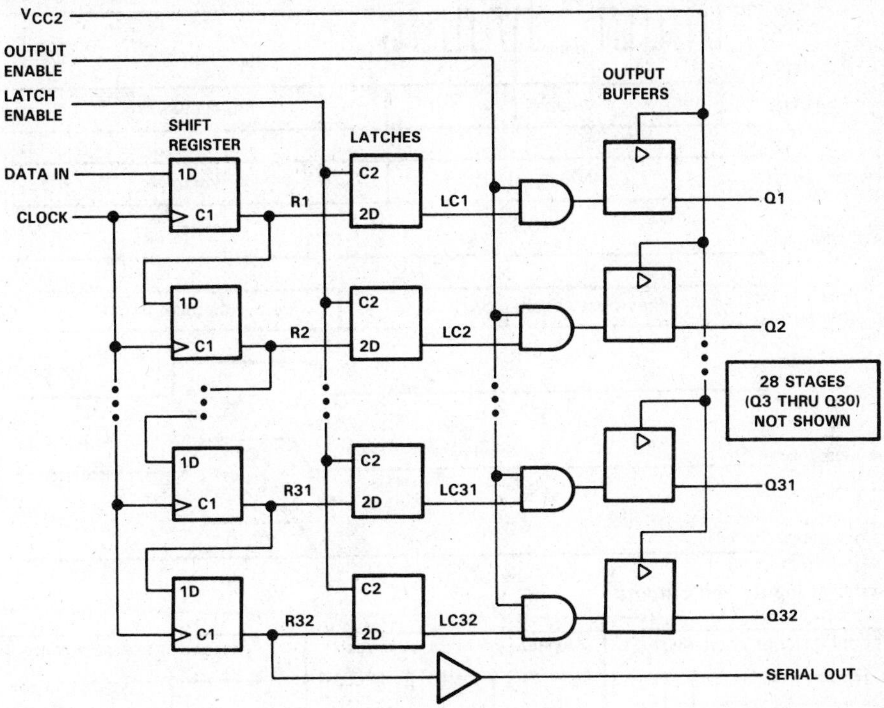
VCC2
OUTPUT ENABLE
LATCH ENABLE
SHIFT REGISTER
LATCHES
OUTPUT BUFFERS
DATA IN
CLOCK
1D
C1
C2
2D
R1
R2
R31
R32
LC1
LC2
LC31
LC32
Q1
Q2
Q31
Q32
28 STAGES (Q3 THRU Q30) NOT SHOWN
SERIAL OUT

FUNCTION TABLE

FUNCTION	CONTROL INPUTS			SHIFT REGISTER R1 THRU R32	LATCHES LC1 THRU LC32	OUTPUTS	
	CLOCK	LATCH ENABLE	OUTPUT ENABLE			SERIAL	Q1 THRU Q32
LOAD	↑	X	X	Load and shift†	Determined by Latch Enable‡	R32	Determined by Output Enable
	Not↑	X	X	No change	Determined by Latch Enable‡	R32	Determined by Output Enable
LATCH	X	L	X	As determined above	Stored data	R32	Determined by Output Enable
	X	H	X	As determined above	New data	R32	Determined by Output Enable
OUTPUT ENABLE	X	X	L	As determined above	Determined by Latch Enable‡	R32	All L
	X	X	H	As determined above	Determined by Latch Enable‡	R32	LC1 thru LC32, respectively

H = high level, L = low level, X = irrelevant, ↑ = low-to-high-level transition.

†R32 and the serial output take on the state of R31, R31 takes on the state of R30,. . .R2 takes on the state of R1, and R1 takes on the state of the data input.

‡New data enter the latches while Latch Enable is high. These data are stored while Latch Enable is low.

typical operating sequence

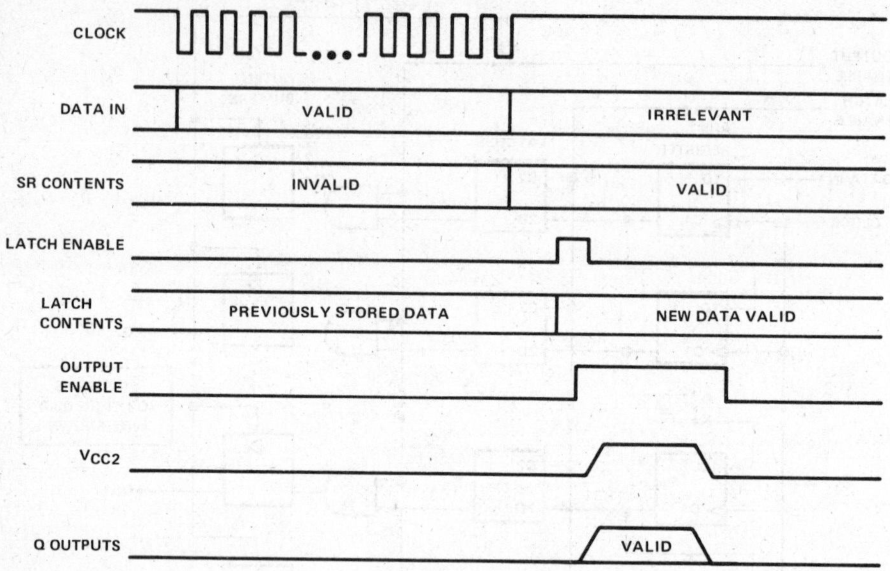
CLOCK
DATA IN
VALID
IRRELEVANT
SR CONTENTS
INVALID
VALID
LATCH ENABLE
LATCH CONTENTS
PREVIOUSLY STORED DATA
NEW DATA VALID
OUTPUT ENABLE
VCC2
Q OUTPUTS
VALID

schematic of inputs and outputs

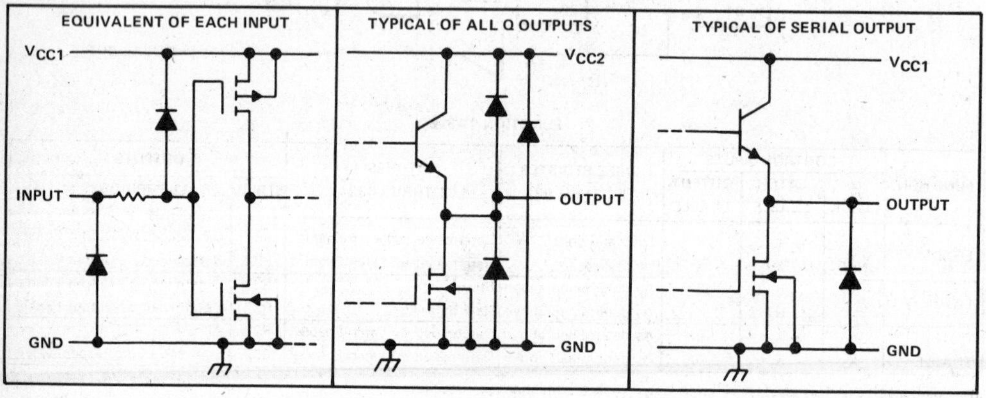
EQUIVALENT OF EACH INPUT
VCC1
INPUT
GND
TYPICAL OF ALL Q OUTPUTS
VCC2
OUTPUT
GND
TYPICAL OF SERIAL OUTPUT
VCC1
OUTPUT
GND

TEXAS INSTRUMENTS
POST OFFICE BOX 655012 • DALLAS, TEXAS 75265

absolute maximum ratings over operating free-air temperature range (unless otherwise noted)

Supply voltage, V_{CC1} (see Note 1)	18 V
Supply voltage, V_{CC2} (see Note 2)	90 V
Input voltage	V_{CC1} + 0.3 V
Ground current	700 mA
Continuous total dissipation at (or below) 25°C free-air temperature (see Note 3):	
N package	1250 mW
FN package	1700 mW
Operating free-air temperature range: SN65555, SN65556	−40°C to 85°C
SN75555, SN75556	0°C to 70°C
Storage temperature range	−65°C to 150°C
Case temperature for 10 seconds: FN package	260°C
Lead temperature 1,6 mm (1/16 inch) from case for 10 seconds: N package	260°C

NOTES: 1. Voltage values are with respect to network ground terminal.
2. These devices have been designed to be used in applications in which the high-voltage supply, V_{CC2}, is switched to ground before changing the state of the outputs.
3. For operation above 25°C free-air temperature, derate the N package at the rate of 10 mW/°C and the FN package at the rate of 13.6 mW/°C.

recommended operating conditions

			MIN	NOM	MAX	UNIT
V_{CC1}	Supply voltage		10.8	12	15	V
V_{CC2}	Supply voltage		0		80	V
V_{IH}	High-level input voltage (see Figure 1)	V_{CC1} = 10.8 V	8.1		11.1	V
		V_{CC1} = 15 V	11.25		15.3	
V_{IL}	Low-level input voltage (see Figure 1)	V_{CC1} = 10.8 V	−0.3†		2.7	V
		V_{CC1} = 15 V	−0.3†		3.75	
I_{OH}	High-level output current				−15	mA
I_{OL}	Low-level output current				15	mA
I_{OK}	Output clamp current				20	mA
f_{clock}	Clock frequency		0		6.25	MHz
$t_{w(CLK)}$	Pulse duration, clock high or low (see Figure 2)		80			ns
$t_{w(LE)}$	Pulse duration, latch enable (see Figure 4)		80			ns
t_{su}	Setup time	Data before clock ↑ (see Figure 2)	20			ns
		Output enable before V_{CC}↑ (see Figure 4)	500			
t_h	Hold time	Data after clock ↑ (see Figure 2)	80			ns
		Output enable after V_{CC}↑ (see Figure 4)	100			
dv/dt	Rate of rise for V_{CC2} (see Figure 4)				80	V/µs
T_A	Operating free-air temperature	SN65555, SN65556	−40		85	°C
		SN75555, SN75556	0		70	

†The algebraic convention, in which the least positive (most negative) value is designated minimum, is used in this data sheet for logic voltage levels.

electrical characteristics over recommended operating free-air temperature range, V_{CC1} = 12 V, V_{CC2} = 80 V

PARAMETER			TEST CONDITIONS	MIN	MAX	UNIT
V_{OH}	High-level output voltage	Q outputs	I_O = −15 mA	77		V
		Serial output	I_O = −100 μA	10.5		V
V_{OL}	Low-level output voltage	Q outputs	I_{OL} = 15 mA		8	V
		Serial output	I_{OL} = 100 μA		1	V
I_{IH}	High-level input current		V_I = 12 V		1	μA
I_{IL}	Low-level input current		V_I = 0		−1	μA
I_{CC1}	Supply current from V_{CC1}				2	mA
I_{CC2}	Supply current from V_{CC2}				5	mA

switching characteristics, V_{CC1} = 12 V, T_A = 25°C

PARAMETER		TEST CONDITIONS	MIN	MAX	UNIT
t_{PHL}	Propagation delay time, high-to-low-level Serial output from Clock	C_L = 20 pF to ground, V_{CC2} = 0, See Figure 3		140	ns
t_{PLH}	Propagation delay time, low-to-high-level Serial output from Clock	C_L = 20 pF to ground, V_{CC2} = 0, See Figure 3		140	ns
t_d	Delay time, V_{CC2} to Q outputs	dv/dt = 80 V/μs, See Figure 4		100	ns

RECOMMENDED OPERATION CONDITIONS

INPUT VOLTAGE LOGIC-LEVEL LIMITS
vs
SUPPLY VOLTAGE V_{CC1}

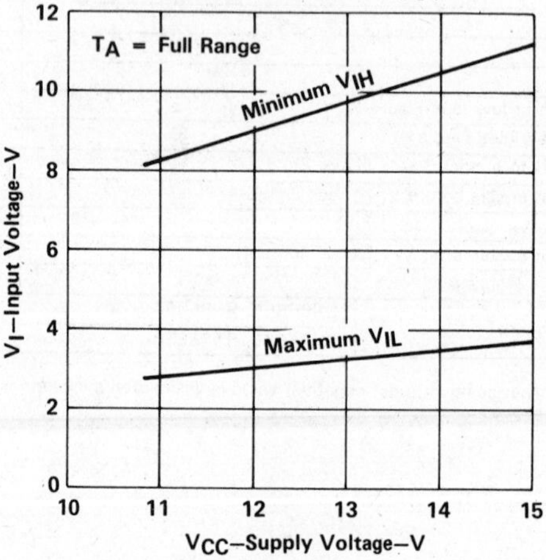
T_A = Full Range
Minimum V_IH
Maximum V_IL
V_I—Input Voltage—V
0
2
4
6
8
10
12
10
11
12
13
14
15
V_CC—Supply Voltage—V

FIGURE 1

TEXAS INSTRUMENTS
POST OFFICE BOX 655012 • DALLAS, TEXAS 75265

PARAMETER MEASUREMENT INFORMATION

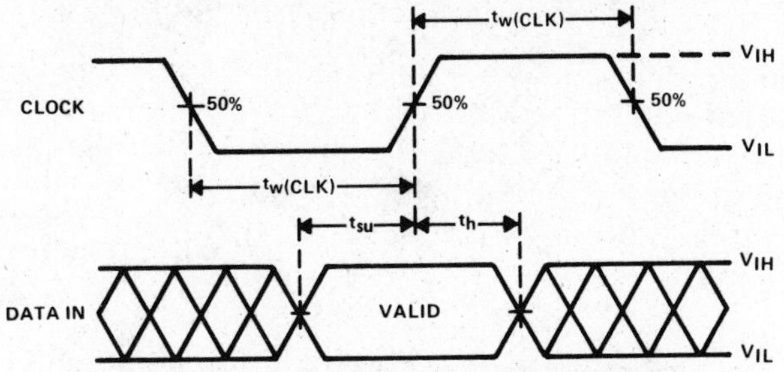
tw(CLK)
VIH
CLOCK
50%
50%
50%
VIL
tw(CLK)
tsu
th
VIH
DATA IN
VALID
VIL

FIGURE 2. INPUT TIMING VOLTAGE WAVEFORMS

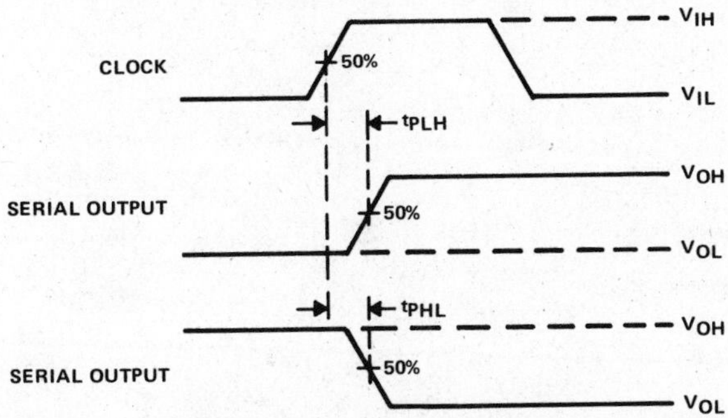
VIH
CLOCK
50%
VIL
tPLH
VOH
SERIAL OUTPUT
50%
VOL
tPHL
VOH
SERIAL OUTPUT
50%
VOL

FIGURE 3. VOLTAGE WAVEFORMS FOR PROPAGATION DELAY
CLOCK TO SERIAL OUTPUT

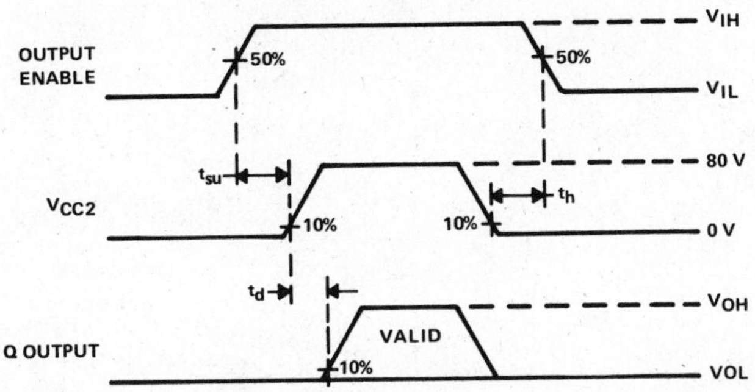
VIH
OUTPUT ENABLE
50%
50%
VIL
80 V
tsu
th
VCC2
10%
10%
0 V
td
VOH
Q OUTPUT
VALID
10%
VOL

FIGURE 4. VOLTAGE WAVEFORMS FOR DELAY TIMES, LATCH ENABLE TO Q OUTPUTS

- Each Device Drives 32 Electrodes
- High-Voltage Open-Collector N-P-N Outputs Using Ramped Supply
- 300-mA Output Current Capability
- CMOS-Compatible Inputs
- Very Low Steady-State Power Consumption

description

These devices are monolithic BIDFET† integrated circuits designed to drive the row electrodes of an electroluminescent display. All inputs are CMOS-compatible and all outputs are high-voltage open-collector n-p-n transistors. The SN65558 and SN75558 output sequences have been reversed from the SN65557 and SN75557 for ease in printed circuit board layout.

The devices consist of a 32-bit shift register, 32 AND gates, and 32 output OR gates. Typically, a composite row drive signal is externally generated by a high-voltage switching circuit and applied to the Substrate Common terminal. Serial data is entered into the shift register on the high-to-low transition of the clock input. A high Enable input allows those outputs with a high in their associated register to be turned on causing the corresponding row to be connected to the composite row drive signal. When the Strobe input is low, all output transistors are turned on. The Serial output from the shift register may be used to cascade additional devices. This output is not affected by the Enable or Strobe inputs.

The SN65557 and SN65558 are characterized for operation from −40°C to 85°C. The SN75557 and SN75558 are characterized for operation from 0°C to 70°C.

SN65557, SN75557 . . . FN PACKAGE
(TOP VIEW)

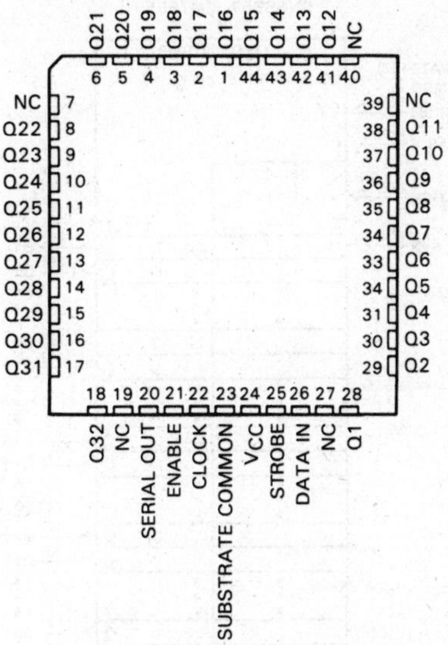
Q21 Q20 Q19 Q18 Q17 Q16 Q15 Q14 Q13 Q12 NC
6 5 4 3 2 1 44 43 42 41 40
NC 7
Q22 8
Q23 9
Q24 10
Q25 11
Q26 12
Q27 13
Q28 14
Q29 15
Q30 16
Q31 17
39 NC
38 Q11
37 Q10
36 Q9
35 Q8
34 Q7
33 Q6
34 Q5
31 Q4
30 Q3
29 Q2
18 19 20 21 22 23 24 25 26 27 28
Q32 NC SERIAL OUT ENABLE CLOCK SUBSTRATE COMMON VCC STROBE DATA IN NC Q1

SN65558, SN75558 . . . FN PACKAGE
(TOP VIEW)

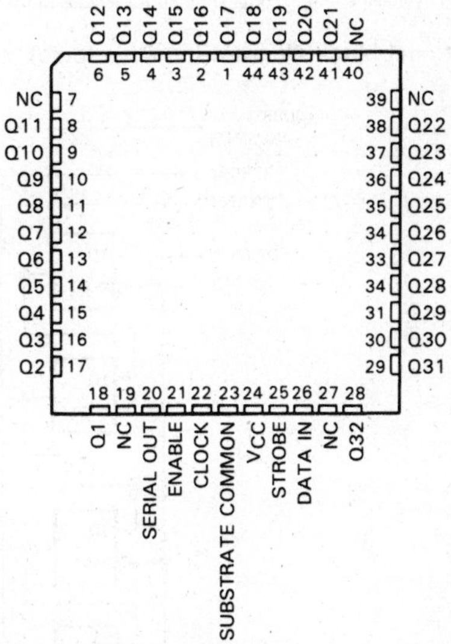
Q12 Q13 Q14 Q15 Q16 Q17 Q18 Q19 Q20 Q21 NC
6 5 4 3 2 1 44 43 42 41 40
NC 7
Q11 8
Q10 9
Q9 10
Q8 11
Q7 12
Q6 13
Q5 14
Q4 15
Q3 16
Q2 17
39 NC
38 Q22
37 Q23
36 Q24
35 Q25
34 Q26
33 Q27
34 Q28
31 Q29
30 Q30
29 Q31
18 19 20 21 22 23 24 25 26 27 28
Q1 NC SERIAL OUT ENABLE CLOCK SUBSTRATE COMMON VCC STROBE DATA IN NC Q32

NC—No internal connection

† BIDFET — Bipolar, double-diffused, N-channel and P-channel MOS transistors on same chip — patented process

logic symbols†

SN65557, SN75557

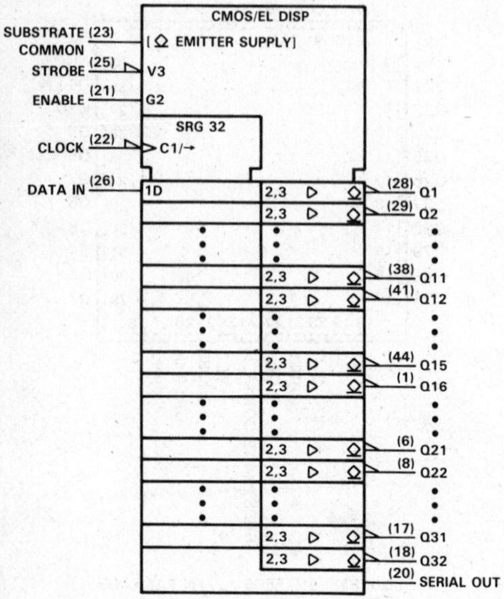
CMOS/EL DISP
SUBSTRATE COMMON (23)
[EMITTER SUPPLY]
STROBE (25)
V3
ENABLE (21)
G2
SRG 32
CLOCK (22)
C1/→
DATA IN (26)
1D
2,3
(28) Q1
(29) Q2
(38) Q11
(41) Q12
(44) Q15
(1) Q16
(6) Q21
(8) Q22
(17) Q31
(18) Q32
(20) SERIAL OUT

SN75558, SN75558

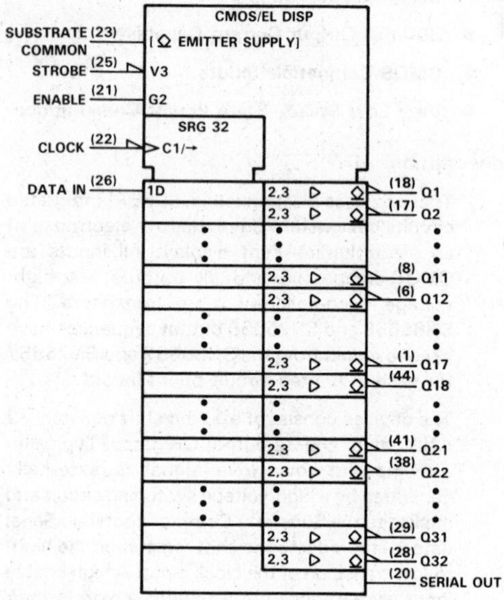
CMOS/EL DISP
SUBSTRATE COMMON (23)
[EMITTER SUPPLY]
STROBE (25)
V3
ENABLE (21)
G2
SRG 32
CLOCK (22)
C1/→
DATA IN (26)
1D
2,3
(18) Q1
(17) Q2
(8) Q11
(6) Q12
(1) Q17
(44) Q18
(41) Q21
(38) Q22
(29) Q31
(28) Q32
(20) SERIAL OUT

† These symbols are in accordance with ANSI/IEEE Std 91-1984 and IEC Publication 617-12.

functional block diagram (positive logic)

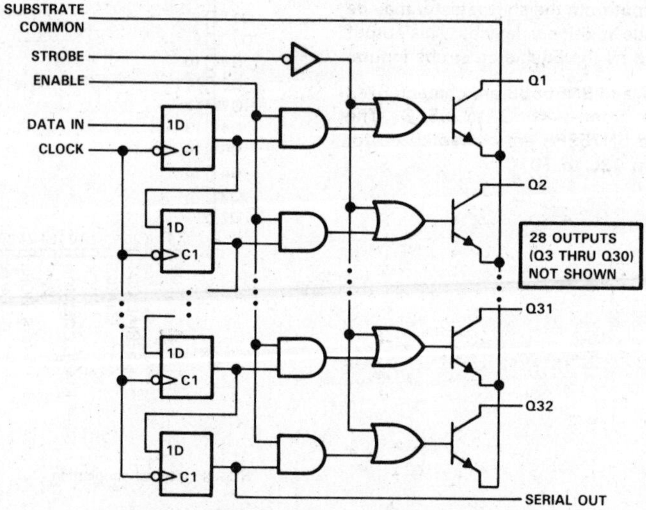
SUBSTRATE COMMON
STROBE
ENABLE
DATA IN
CLOCK
1D
C1
Q1
Q2
Q31
Q32
28 OUTPUTS (Q3 THRU Q30) NOT SHOWN
SERIAL OUT

FUNCTION TABLE

FUNCTION	CONTROL INPUTS			SHIFT REGISTERS R1 THRU R32	OUTPUTS	
	CLOCK	ENABLE	STROBE		SERIAL	Q1 THRU Q32
LOAD	↓	X	X	Load and Shift†	R32	Determined by Enable and Strobe
	No ↓	X	X	No Change	R32	Determined by Enable and Strobe
ENABLE	X	L	H	As determined above	R32	All Q outputs off
	X	H	H	As determined above	R32	Determined by R1 through R32
STROBE	X	X	L	As determined above	R32	All Q outputs on

H = high level, L = low level, X = irrelevant, ↓ = high-to-low transition.

†Register R32 takes on the state of R31, R31 takes on the state of R30,. . .R2 takes on the state of R1, and R1 takes on the state of the data input.

schematics of inputs and outputs

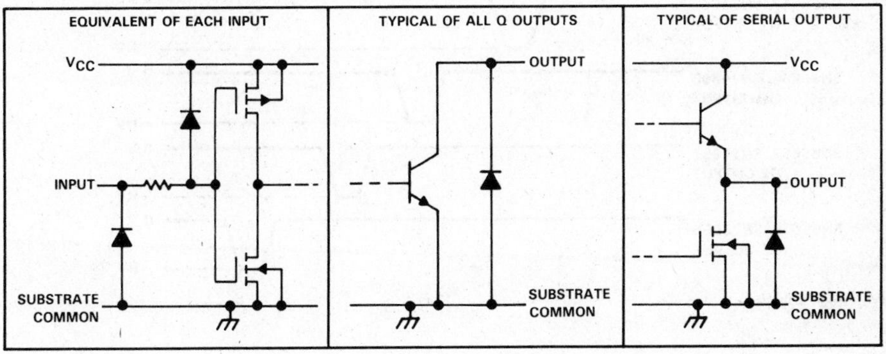
EQUIVALENT OF EACH INPUT
V_{CC}
INPUT
SUBSTRATE COMMON
TYPICAL OF ALL Q OUTPUTS
OUTPUT
SUBSTRATE COMMON
TYPICAL OF SERIAL OUTPUT
V_{CC}
OUTPUT
SUBSTRATE COMMON

typical operating sequence

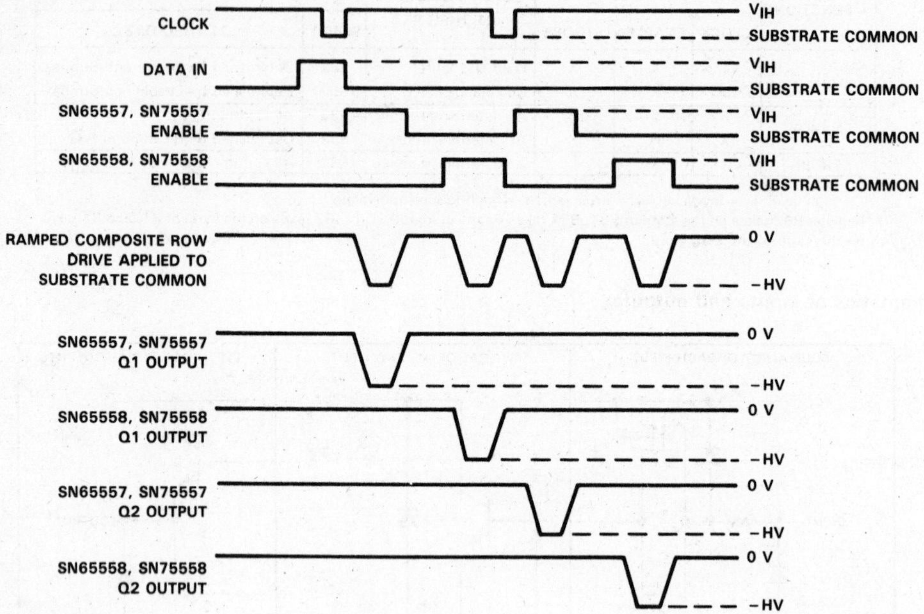
CLOCK
DATA IN
SN65557, SN75557 ENABLE
SN65558, SN75558 ENABLE
RAMPED COMPOSITE ROW DRIVE APPLIED TO SUBSTRATE COMMON
SN65557, SN75557 Q1 OUTPUT
SN65558, SN75558 Q1 OUTPUT
SN65557, SN75557 Q2 OUTPUT
SN65558, SN75558 Q2 OUTPUT
V_{IH}
SUBSTRATE COMMON
0 V
−HV

HV = High voltage

absolute maximum ratings over operating free-air temperature range (unless otherwise noted)

Supply voltage, V_{CC} (see Note 1)	18 V
Off-state output voltage, $V_{O(off)}$ (see Note 2)	110 V
Input voltage	V_{CC}+0.3 V
Substrate common terminal current (see Note 3)	750 mA
Continuous total dissipation at (or below) 25°C free-air temperature (see Note 4):	1700 mW
Operating free-air temperature range: SN65557, SN65558	−40°C to 85°C
SN75557, SN75558	0°C to 70°C
Storage temperature range	−65°C to 150°C
Case temperature for 10 seconds	260°C

NOTES: 1. Voltage values are with respect to substrate common terminal.
2. Data must be clocked into the shift register and Q outputs enabled prior to ramping substrate common to −HV (see typical operating sequence).
3. Duty cycle is limited by package dissipation.
4. For operation above 25°C free-air temperature, derate linearly to 1088 mW at the rate of 13.6 mW/°C.

recommended operating conditions

		MIN	NOM	MAX	UNIT
Supply voltage, V_{CC}		10.8	12	15	V
High-level input voltage, V_{IH} (see Figure 1)	V_{CC} = 10.8 V	8.1		11.1	V
	V_{CC} = 15 V	11.25		15.3	
Low-level input voltage, V_{IL} (see Figure 1)	V_{CC} = 10.8 V	−0.3		2.7	V
	V_{CC} = 15 V	−0.3		3.75	
Off-state Q output voltage, $V_{O(off)}$		−0.3		100	V
On-state Q output current, $I_{O(on)}$, duty cycle ≤ 1%, V_{CC} = 15 V				300	mA
Rate of rise for substrate common, dV/dt (see Figure 4)				100	V/μs
Clock frequency, f_{clock}		0		4	MHz
Pulse duration, clock high or low, t_w		125			ns
Setup time, t_{su}	Data before clock↓ (see Figure 3)	50			ns
	Enable before substrate common↓ (see Figure 4)	500			
Hold time, t_h, data after clock↓ (see Figure 3)		100			ns
Operating free-air temperature, T_A	SN65557, SN65558	−40		85	°C
	SN75557, SN75558	0		70	

electrical characteristics over recommended operating free-air temperature range, V_{CC} = 12 V (unless otherwise noted)

PARAMETER			TEST CONDITIONS	SN65557 SN65558		SN75557 SN75558		UNIT
				MIN	MAX	MIN	MAX	
$I_{O(off)}$	Off-state Q output current		V_O = 100 V		20		10	μA
V_{OH}	High-level output voltage	Serial outputs	I_O = −100 μA	10.5		10.5		V
V_{OL}	Low-level output voltage	Q outputs	I_{OL} = 300 mA		20		10	V
		Serial output	I_{OL} = 100 μA		1		1	
I_{IH}	High-level input current		V_I = 12 V		1		1	μA
I_{IL}	Low-level input current		V_I = 0		−1		−1	μA
I_{CC}	Supply current from V_{CC}				250		250	μA

switching characteristics, V_{CC} = 12 V, T_A = 25°C

PARAMETER		TEST CONDITIONS	MIN	MAX	UNIT
t_{PHL}	Propagation delay time, high-to-low-level serial output from clock	C_L = 20 pF to substrate common (see Figure 4)		200	ns
t_{PLH}	Propagation delay time, low-to-high-level serial output from clock			200	ns
$t_{d(on)}$	Turn-on delay time, Q outputs from enable	dV/dt = 100 V/μs, Strobe at V_{CC}, R_L = 2 kΩ to 60 V (see Figure 4)		500	ns

RECOMMENDED OPERATING CONDITIONS

INPUT VOLTAGE LOGIC-LEVEL LIMITS
vs
SUPPLY VOLTAGE

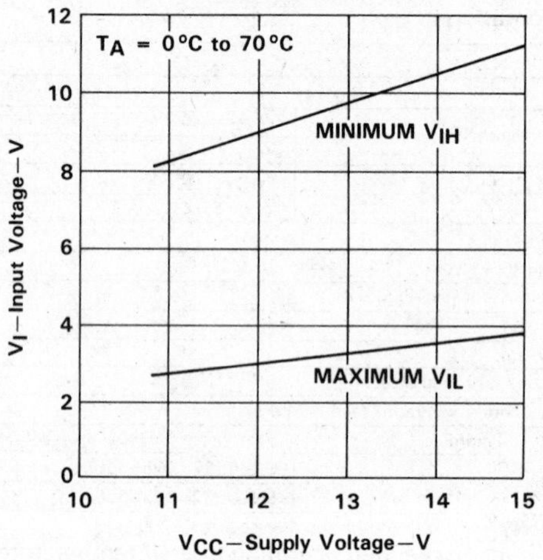

12
10
8
6
4
2
0
10
11
12
13
14
15
TA = 0°C to 70°C
MINIMUM VIH
MAXIMUM VIL
VI—Input Voltage—V
VCC—Supply Voltage—V

FIGURE 1

PARAMETER MEASUREMENT INFORMATION

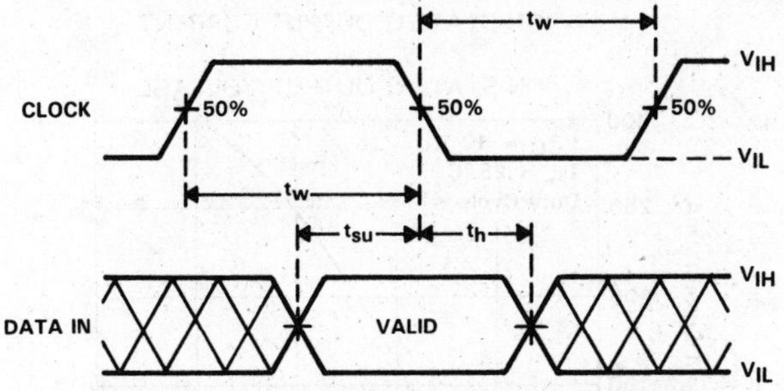
tw
VIH
CLOCK
50%
50%
50%
VIL
tw
tsu
th
VIH
DATA IN
VALID
VIL

FIGURE 2. INPUT TIMING VOLTAGE WAVEFORMS

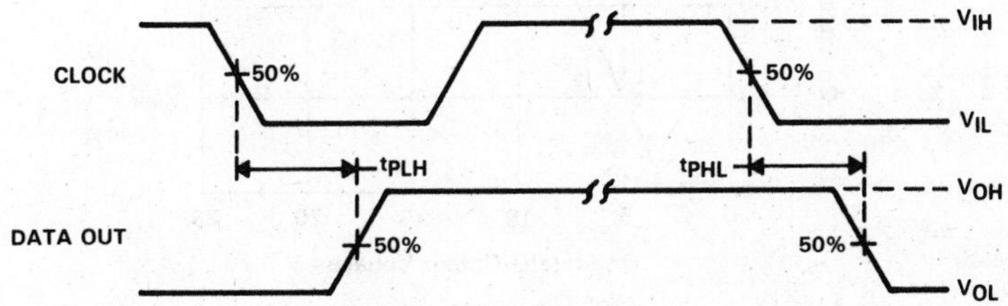
VIH
CLOCK
50%
50%
VIL
tPLH
tPHL
VOH
DATA OUT
50%
50%
VOL

FIGURE 3. VOLTAGE WAVEFORMS FOR PROPAGATION DELAY TIMES, CLOCK TO DATA OUT

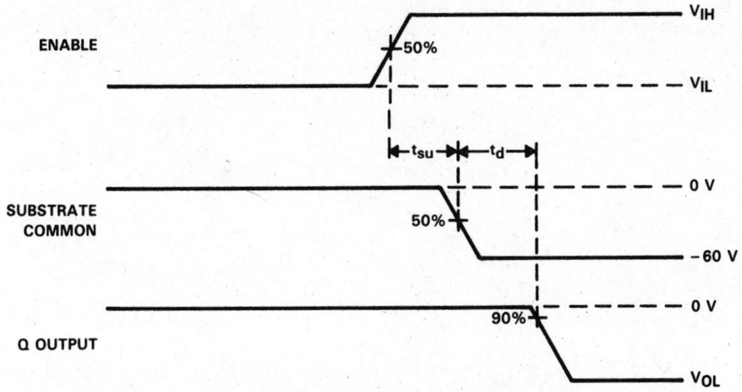
VIH
ENABLE
50%
VIL
tsu
td
0 V
SUBSTRATE
COMMON
50%
-60 V
0 V
90%
Q OUTPUT
VOL

FIGURE 4. VOLTAGE WAVEFORMS FOR TURN ON DELAY TIME,
SUBSTRATE COMMON TO Q OUTPUT

TYPICAL CHARACTERISTICS

ON-STATE Q OUTPUT CURRENT
vs
ON-STATE Q OUTPUT VOLTAGE

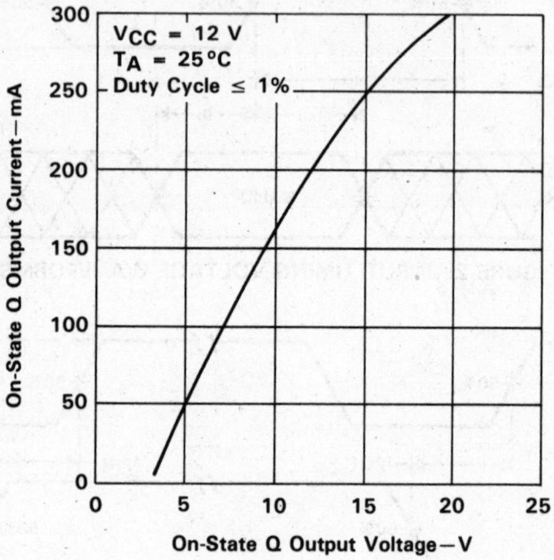
V_{CC} = 12 V
T_A = 25°C
Duty Cycle ≤ 1%
300
250
200
150
100
50
0
0
5
10
15
20
25
On-State Q Output Current—mA
On-State Q Output Voltage—V

FIGURE 5

D2947, APRIL 1986—REVISED OCTOBER 1986

- Controls 32 Electrodes
- 80-V (Ramped V_{CC2}) Totem-Pole Outputs
- Low CMOS Stand-By Power Consumption
- Energy Recovery System Compatible
- 15-mA Source and Sink Compatibility
- High-Speed Serially-Shifted Data Input

description

The SN65559, SN65560, SN75559, and SN75560 are monolithic BIDFET† integrated circuits designed to provide the serial-to-parallel conversion and level translation of data in a matrix-addressable electroluminescent display. The device inputs are diode-clamped CMOS inputs. The SN65560 and SN75560 output sequences are reversed from the SN65559 and SN75559 for ease in printed circuit board layout.

These column drivers consist of a 32-bit static shift register, 32 latches, and 32 high-voltage outputs. Serial data is entered into the shift register on the low-to-high transition of the clock signal. A logic high signal on the Latch Enable input transfers the data from the shift register to the latches while the V_{CC2} bus is low. Once stable in the latch circuits, the V_{CC2} rail is ramped up to allow the data to appear at the high-voltage outputs. By limiting V_{CC2} to a maximum of 60 volts, these devices may be safely operated in a non-ramped V_{CC2} mode. Drivers may be cascaded via the serial data output of the static shift register. This output is not affected by the Latch Enable input.

The SN65559 and SN65560 are characterized for operation from −40°C to 85°C. The SN75559 and SN75560 are characterized for operation from 0°C to 70°C.

SN65559, SN75559
N DUAL-IN-LINE PACKAGE
(TOP VIEW)

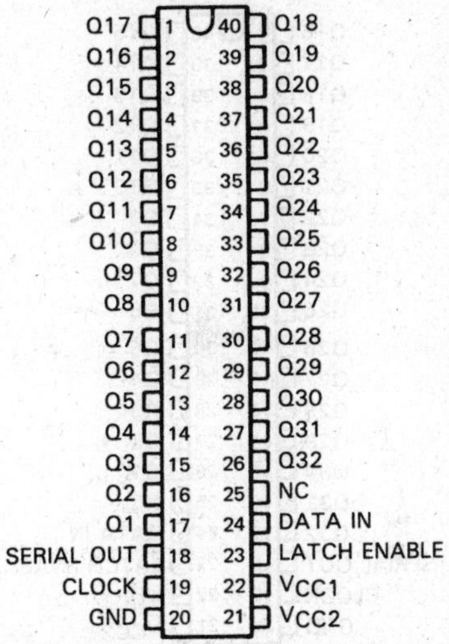
Q17 1
Q16 2
Q15 3
Q14 4
Q13 5
Q12 6
Q11 7
Q10 8
Q9 9
Q8 10
Q7 11
Q6 12
Q5 13
Q4 14
Q3 15
Q2 16
Q1 17
SERIAL OUT 18
CLOCK 19
GND 20
40 Q18
39 Q19
38 Q20
37 Q21
36 Q22
35 Q23
34 Q24
33 Q25
32 Q26
31 Q27
30 Q28
29 Q29
28 Q30
27 Q31
26 Q32
25 NC
24 DATA IN
23 LATCH ENABLE
22 VCC1
21 VCC2

SN65559, SN75559 . . . FN PACKAGE
(TOP VIEW)

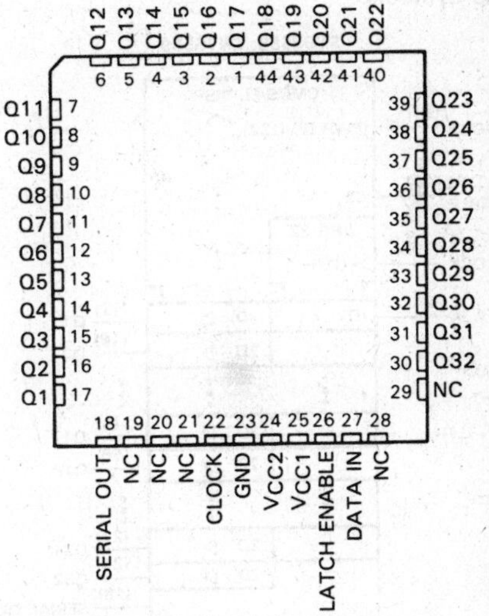
Q12 Q13 Q14 Q15 Q16 Q17 Q18 Q19 Q20 Q21 Q22
6 5 4 3 2 1 44 43 42 41 40
Q11 7
Q10 8
Q9 9
Q8 10
Q7 11
Q6 12
Q5 13
Q4 14
Q3 15
Q2 16
Q1 17
39 Q23
38 Q24
37 Q25
36 Q26
35 Q27
34 Q28
33 Q29
32 Q30
31 Q31
30 Q32
29 NC
18 19 20 21 22 23 24 25 26 27 28
SERIAL OUT NC NC NC CLOCK GND VCC2 VCC1 LATCH ENABLE DATA IN NC

NC—No internal connection

†BIDFET—Bipolar, double-diffused, N-channel and P-channel MOS transistors on the same chip—Patented Process

TEXAS INSTRUMENTS
POST OFFICE BOX 655012 • DALLAS, TEXAS 75265

ADVANCE INFORMATION

SN65560, SN75560
N DUAL-IN-LINE PACKAGE
(TOP VIEW)

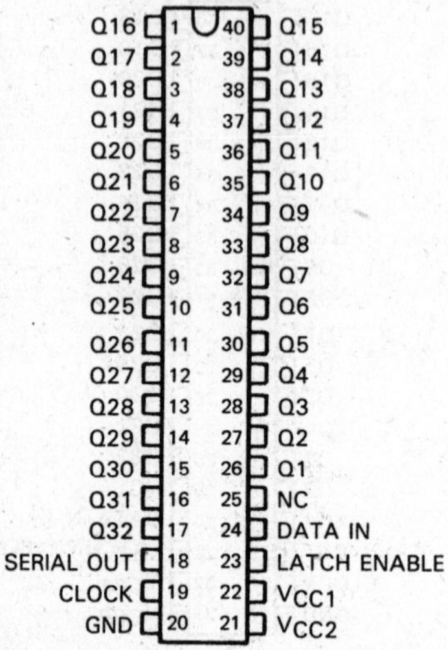
Q16 1 40 Q15
Q17 2 39 Q14
Q18 3 38 Q13
Q19 4 37 Q12
Q20 5 36 Q11
Q21 6 35 Q10
Q22 7 34 Q9
Q23 8 33 Q8
Q24 9 32 Q7
Q25 10 31 Q6
Q26 11 30 Q5
Q27 12 29 Q4
Q28 13 28 Q3
Q29 14 27 Q2
Q30 15 26 Q1
Q31 16 25 NC
Q32 17 24 DATA IN
SERIAL OUT 18 23 LATCH ENABLE
CLOCK 19 22 VCC1
GND 20 21 VCC2

SN65560, SN75560 . . . FN PACKAGE
(TOP VIEW)

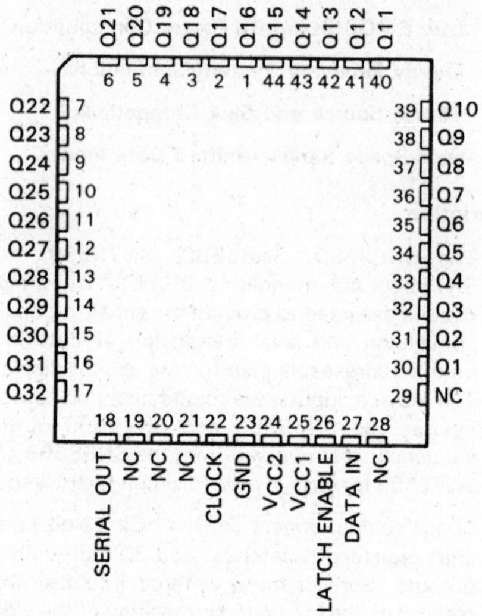
Q21 Q20 Q19 Q18 Q17 Q16 Q15 Q14 Q13 Q12 Q11
6 5 4 3 2 1 44 43 42 41 40
Q22 7
Q23 8
Q24 9
Q25 10
Q26 11
Q27 12
Q28 13
Q29 14
Q30 15
Q31 16
Q32 17
39 Q10
38 Q9
37 Q8
36 Q7
35 Q6
34 Q5
33 Q4
32 Q3
31 Q2
30 Q1
29 NC
18 19 20 21 22 23 24 25 26 27 28
SERIAL OUT NC NC NC CLOCK GND VCC2 VCC1 LATCH ENABLE DATA IN NC

NC—No internal connection

logic symbols†

SN65559, SN75559

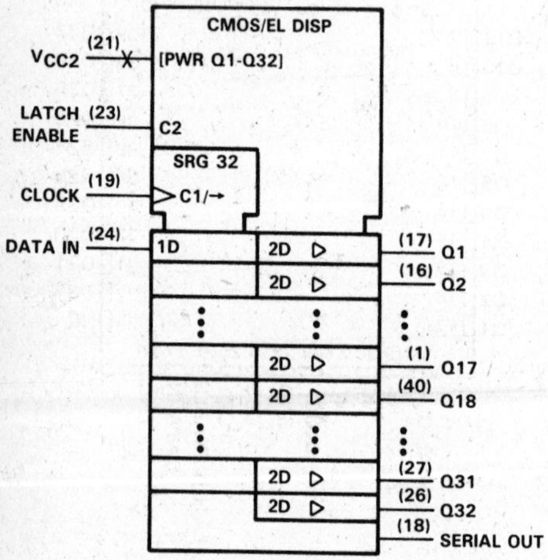
CMOS/EL DISP
VCC2 (21) [PWR Q1-Q32]
LATCH ENABLE (23) C2
SRG 32
CLOCK (19) C1/→
DATA IN (24) 1D
2D (17) Q1
2D (16) Q2
2D (1) Q17
2D (40) Q18
2D (27) Q31
2D (26) Q32
(18) SERIAL OUT

SN65560, SN75560

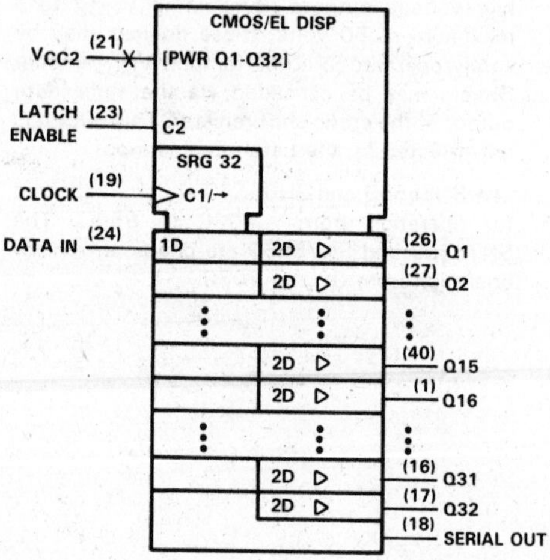
CMOS/EL DISP
VCC2 (21) [PWR Q1-Q32]
LATCH ENABLE (23) C2
SRG 32
CLOCK (19) C1/→
DATA IN (24) 1D
2D (26) Q1
2D (27) Q2
2D (40) Q15
2D (1) Q16
2D (16) Q31
2D (17) Q32
(18) SERIAL OUT

†These symbols are in accordance with ANSI/IEEE Std 91-1984 and IEC Publication 617-12.
Pin numbers shown are for N packages.

TEXAS INSTRUMENTS
POST OFFICE BOX 655012 • DALLAS, TEXAS 75265

FUNCTION TABLE

FUNCTION	CONTROL INPUTS		SHIFT REGISTERS R1 THRU R32	LATCHES LC1 THRU LC32	OUTPUTS	
	CLOCK	LATCH ENABLE†			SERIAL	Q1 THRU Q32
LOAD	↑	X	Load and Shift	Determined by Latch Enable	R32‡	LC1 thru LC32 respectively
	No↑	X	No change		R32	
LATCH	X	L	As determined above	Stored data	R32	LC1 thru LC32 respectively
	X	H		New data	R32	

H = high level; L = low level; X = irrelevant; ↑ = low-to-high-level transition

† New data enters the latches while Latch Enable is high. These data are stored while Latch Enable is low.

‡ R32 and the serial output take on the state of R31, R31 takes on the state of R30 . . . R2 takes on the state of R1 and R1 takes on the state of the data input.

logic diagram (positive logic)

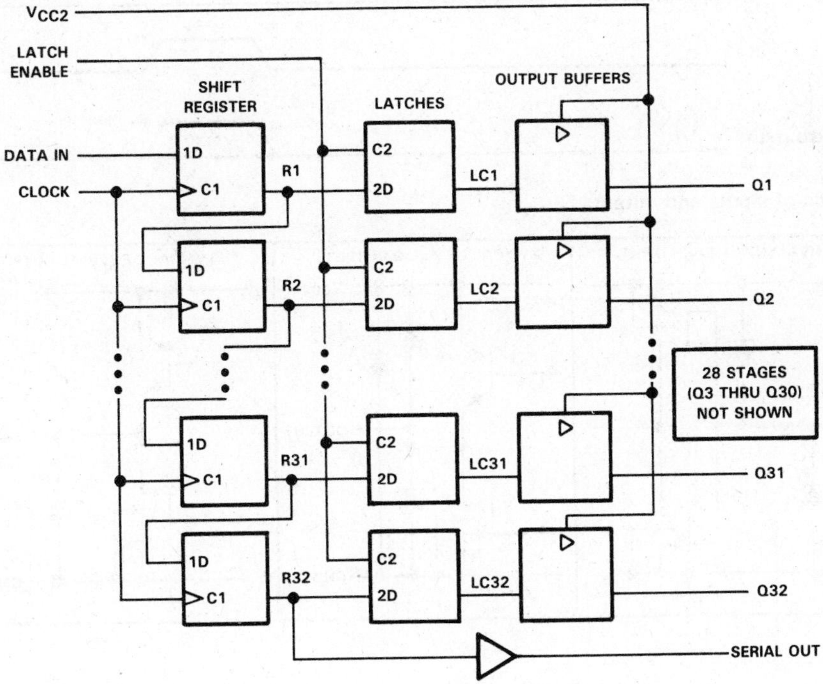
VCC2
LATCH ENABLE
SHIFT REGISTER
LATCHES
OUTPUT BUFFERS
DATA IN
CLOCK
1D
C1
R1
C2
2D
LC1
Q1
R2
LC2
Q2
28 STAGES (Q3 THRU Q30) NOT SHOWN
R31
LC31
Q31
R32
LC32
Q32
SERIAL OUT

typical operating sequence

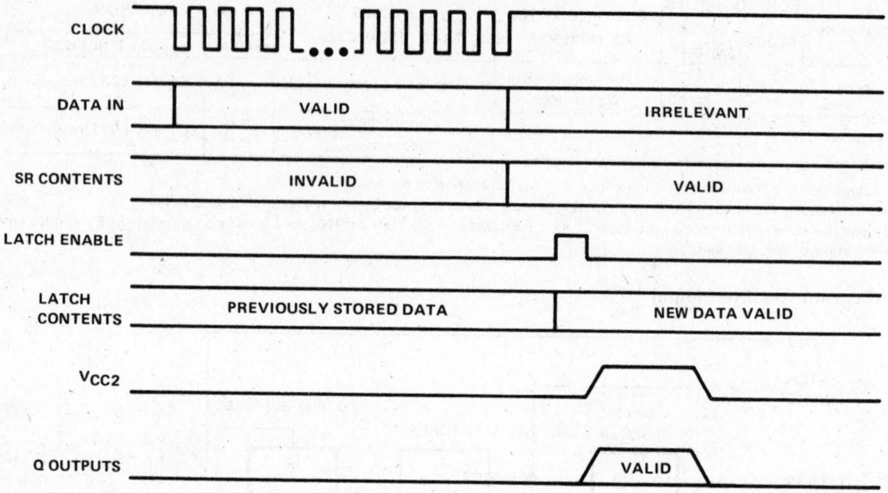
CLOCK
DATA IN
VALID
IRRELEVANT
SR CONTENTS
INVALID
VALID
LATCH ENABLE
LATCH
CONTENTS
PREVIOUSLY STORED DATA
NEW DATA VALID
VCC2
Q OUTPUTS
VALID

schematic of inputs and outputs

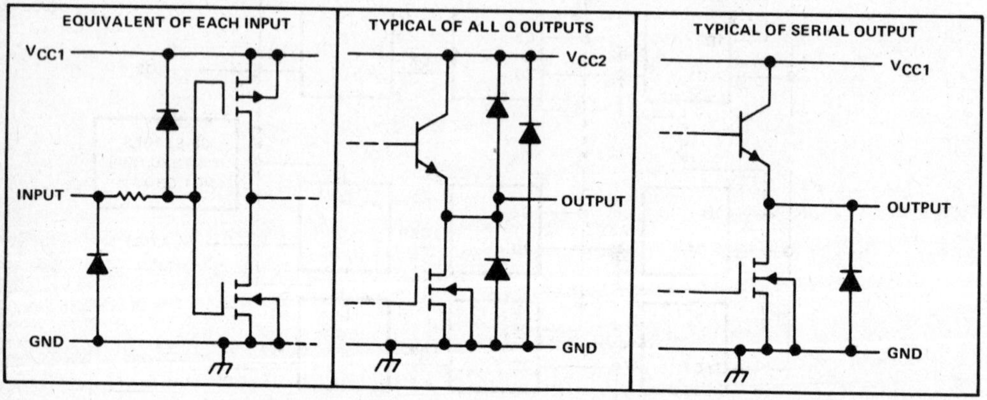
EQUIVALENT OF EACH INPUT
VCC1
INPUT
GND
TYPICAL OF ALL Q OUTPUTS
VCC2
OUTPUT
GND
TYPICAL OF SERIAL OUTPUT
VCC1
OUTPUT
GND

TEXAS INSTRUMENTS
POST OFFICE BOX 655012 • DALLAS, TEXAS 75265

absolute maximum ratings over operating free-air temperature range (unless otherwise noted)

Supply voltage, V_{CC1} (see Note 1)	18 V
Supply voltage, V_{CC2}	90 V
Input voltage, V_I	V_{CC1} + 0.3 V
Continuous total dissipation at (or below) 25 °C free-air temperature (see Note 2): N package	1250 mW
FN package	1700 mW
Operating free-air temperature range: SN65559, SN65560	−40 °C to 85 °C
SN75559, SN75560	0 °C to 70 °C
Storage temperature range	−65 °C to 150 °C
Case temperature for 10 seconds: FN package	260 °C
Lead temperature 1,6 mm (1/16 inch) from case for 10 seconds: N package	260 °C

NOTES: 1. Voltage values are with respect to network ground terminal.
2. For operation above 25 °C free-air temperature, derate the N package linearly at the rate of 10 mW/°C and the FN package at the rate of 13.6 mW/°C.

recommended operating conditions

		MIN	NOM	MAX	UNIT
Supply voltage, V_{CC1}		10.8	12	15	V
Supply voltage, V_{CC2}		0		80	V
High-level input voltage, V_{IH} (see Figure 1)	V_{CC1} = 10.8 V	8.1		11.1	V
	V_{CC1} = 15 V	11.25		15.3	
Low-level input voltage, V_{IL} (see Figure 1)	V_{CC1} = 10.8 V	−0.3		2.7	V
	V_{CC1} = 15 V	−0.3		3.75	
High-level Q output current, I_{OH}		−15			mA
Low-level Q output current, I_{OL}		15			mA
Q output clamp current, I_{OK}				20	mA
Clock frequency, f_{clock}		0		8	MHz
Pulse duration, clock high, $t_{w(CLK)}$		62			ns
Pulse duration, latch enable high, $t_{w(LE)}$		62			ns
Setup time, data before clock ↑, t_{su}		20			ns
Hold time, data after clock ↑, t_h		50			ns
Rate of rise of V_{CC2}, dv/dt (see Figure 4)				80	V/μs
Operating free-air temperature, T_A	SN65559, SN65560	−40		85	°C
	SN75559, SN75560	0		70	

NOTE 3: V_{CC2} must be ramped only when data within the latches is stable.

electrical characteristics over recommended operating free-air temperature range, V_{CC1} = 12 V

PARAMETER			TEST CONDITIONS		MIN	MAX	UNIT
V_{OH}	High-level output voltage	Q outputs	V_{CC2} = 80 V,	I_{OH} = −15 mA	77		V
		Serial output	I_{OH} = 100 μA		10.5		
V_{OL}	Low-level output voltage	Q outputs	I_{OL} = 15 mA			8	V
		Serial output	I_{OL} = 100 μA			1	
I_{IH}	High-level input current		V_{IH} = V_{CC1}			1	μA
I_{IL}	Low-level input current		V_{IL} = GND			−1	μA
I_{CC1}	Supply current from V_{CC1}		V_I = V_{CC1}			500	μA
I_{CC2}	Supply current from V_{CC2}		V_{CC2} = 80 V	Outputs low		3	mA
				Outputs high		0.5	

switching characteristics, V_{CC1} = 12 V, V_{CC2} = 0, T_A = 25°C

PARAMETER		FROM INPUT	TO OUTPUT	TEST CONDITIONS	MIN	MAX	UNIT
t_{PHL}	Propagation delay time, high-to-low level	Clock	Serial Out	C_L = 20 pF, See Figure 3		140	ns
t_{PLH}	Propagation delay time, low-to-high level	Clock	Serial Out	C_L = 20 pF, See Figure 3		140	ns
t_d	Delay time, V_{CC2} to Q output	V_{CC2}	Q	dv/dt = 80 V/μs, C_L = 100 pF, See Figure 4		100	ns

RECOMMENDED OPERATING CONDITIONS

INPUT VOLTAGE LOGIC-LEVEL LIMITS
vs
SUPPLY VOLTAGE V_{CC1}

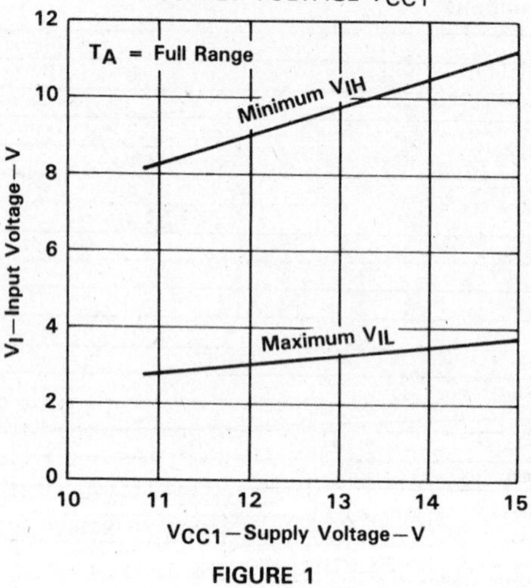
12
10
8
6
4
2
0
10
11
12
13
14
15
T_A = Full Range
Minimum V_IH
Maximum V_IL
V_I—Input Voltage—V
V_CC1—Supply Voltage—V

FIGURE 1

PARAMETER MEASUREMENT INFORMATION

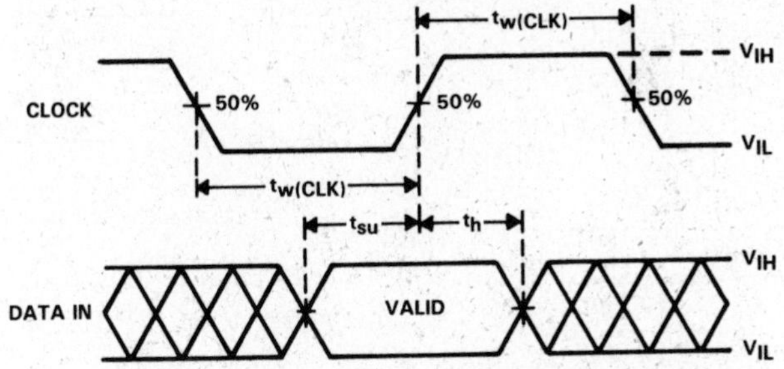
CLOCK
DATA IN
VALID
50%
50%
50%
$t_{w(CLK)}$
$t_{w(CLK)}$
t_{su}
t_h
V_{IH}
V_{IL}
V_{IH}
V_{IL}

FIGURE 2. INPUT TIMING VOLTAGE WAVEFORMS

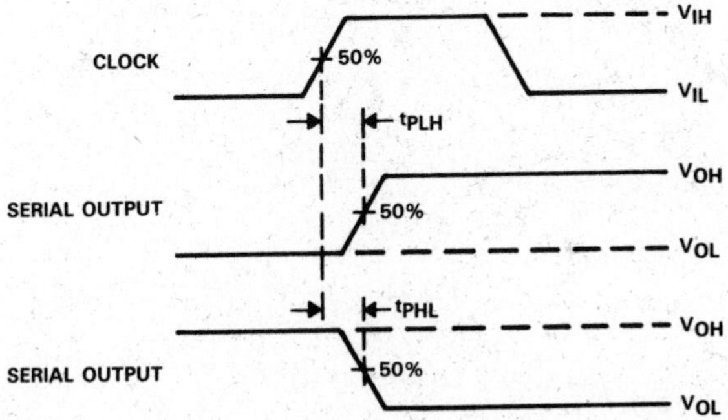
CLOCK
SERIAL OUTPUT
SERIAL OUTPUT
50%
50%
50%
t_{PLH}
t_{PHL}
V_{IH}
V_{IL}
V_{OH}
V_{OL}
V_{OH}
V_{OL}

FIGURE 3. VOLTAGE WAVEFORMS FOR PROPAGATION DELAY CLOCK TO SERIAL OUTPUT

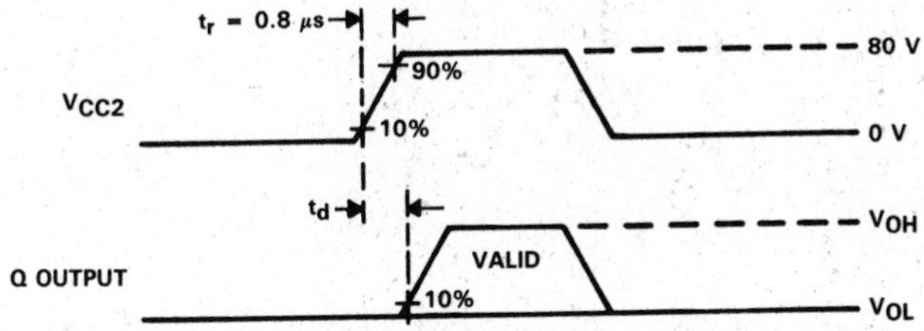
t_r = 0.8 μs
V_{CC2}
90%
10%
80 V
0 V
t_d
Q OUTPUT
VALID
10%
V_{OH}
V_{OL}

FIGURE 4. VOLTAGE WAVEFORMS FOR DELAY TIMES, LATCH ENABLE TO Q OUTPUTS

D2945, MAY 1986 REVISED NOVEMBER 1986

- **Each Device Drives 34 Electrodes**
- **Selectable Open-Source or Open-Drain Output**
- **Outputs Rated at 225 V**
- **±70-mA Output Current Capability**
- **CMOS-Compatible Inputs**
- **Very Low Steady-State Power Consumption**

description

The SN65563, SN65564, SN75563, and SN75564 are monolithic BIDFET† integrated circuits designed to drive the row electrodes of an electroluminescent display. All inputs are CMOS compatible. If the Positive Write input is high, the Q outputs act like open-source outputs and output data is not inverted with respect to input data. If the Positive Write input is low, the Q outputs act like open-drain outputs and output data is inverted with respect to input data. The SN65564 and SN75564 output sequences have been reversed from the SN65563 and SN75563 for ease in printed circuit board layout.

Typically, composite V_{CC2} and ground signals are externally generated by a high-voltage switching circuit. Serial data is entered into the shift register on the high-to-low transition of the clock input. A high Enable input allows those outputs with a high in their associated register to be turned on causing the corresponding row to be connected to V_{CC2} when Positive Write is high or to ground when Positive Write is low. The Serial Output from the shift register may be used to cascade additional devices. This output is not affected by the Enable or Positive Write inputs.

The SN65563 and SN65564 are characterized for operation over the full automotive operating temperature range of −40°C to 85°C. The SN75563 and SN75564 are characterized for operation from 0°C to 70°C.

SN65563, SN75563 . . . FN PACKAGE
(TOP VIEW)

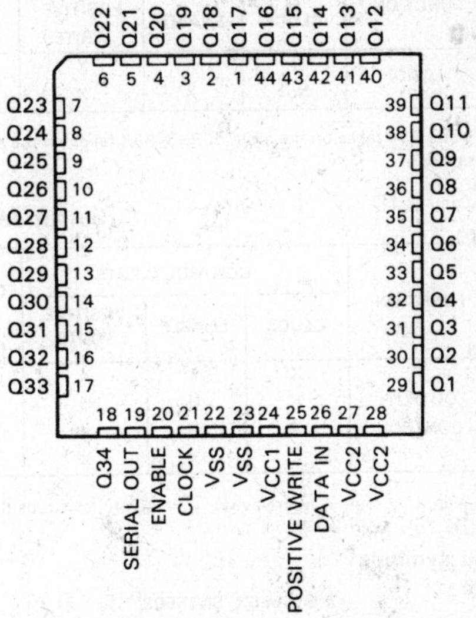
Q22 Q21 Q20 Q19 Q18 Q17 Q16 Q15 Q14 Q13 Q12
6 5 4 3 2 1 44 43 42 41 40
Q23 7
Q24 8
Q25 9
Q26 10
Q27 11
Q28 12
Q29 13
Q30 14
Q31 15
Q32 16
Q33 17
39 Q11
38 Q10
37 Q9
36 Q8
35 Q7
34 Q6
33 Q5
32 Q4
31 Q3
30 Q2
29 Q1
18 19 20 21 22 23 24 25 26 27 28
Q34 SERIAL OUT ENABLE CLOCK VSS VSS VCC1 POSITIVE WRITE DATA IN VCC2 VCC2

SN65564, SN75564 . . . FN PACKAGE
(TOP VIEW)

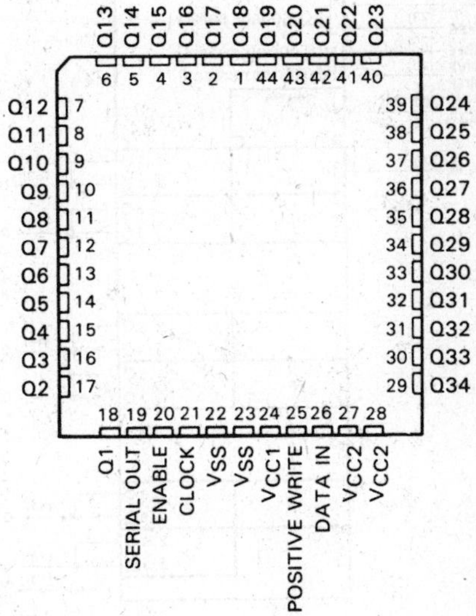
Q13 Q14 Q15 Q16 Q17 Q18 Q19 Q20 Q21 Q22 Q23
6 5 4 3 2 1 44 43 42 41 40
Q12 7
Q11 8
Q10 9
Q9 10
Q8 11
Q7 12
Q6 13
Q5 14
Q4 15
Q3 16
Q2 17
39 Q24
38 Q25
37 Q26
36 Q27
35 Q28
34 Q29
33 Q30
32 Q31
31 Q32
30 Q33
29 Q34
18 19 20 21 22 23 24 25 26 27 28
Q1 SERIAL OUT ENABLE CLOCK VSS VSS VCC1 POSITIVE WRITE DATA IN VCC2 VCC2

†BIDFET-Bipolar, double-diffused, N-channel and P-channel MOS transistors on the same chip — Patented Process

TEXAS INSTRUMENTS
POST OFFICE BOX 655012 • DALLAS, TEXAS 75265

LOAD FUNCTION TABLE

FUNCTION	CONTROL INPUTS			SHIFT REGISTER R1 THRU R34	OUTPUTS	
	CLOCK	ENABLE	POSITIVE WRITE		SERIAL	Q1 THRU Q34
LOAD	↓	X	X	Load and Shift†	R34	Determined by Enable and Positive Write
	No↓	X	X	No Change	R34	Determined by Enable and Positive Write

† Register R34 takes on the state of R33, R33 takes on the state of R32, . . . R2 takes on the state of R1, R1 takes on the state of the data input.

OUTPUT CONTROL FUNCTION TABLE

FUNCTION	CONTROL INPUTS			SHIFT REGISTER CONTENTS Rn FOR R1 THRU R34 (Determined Above)	OUTPUTS	
	CLOCK	ENABLE	POSITIVE WRITE		SERIAL	Q1 THRU Q34
OUTPUT CONTROL	X	L	X	X	R34	High-Impedance
	X	H	H	H	R34	H
	X	H	L	H	R34	L
	X	X	X	L	R34	High-Impedance

H = high, L = low, X = irrelevant, ↓ = high-to-low transition

logic symbols‡

SN65563, SN75563

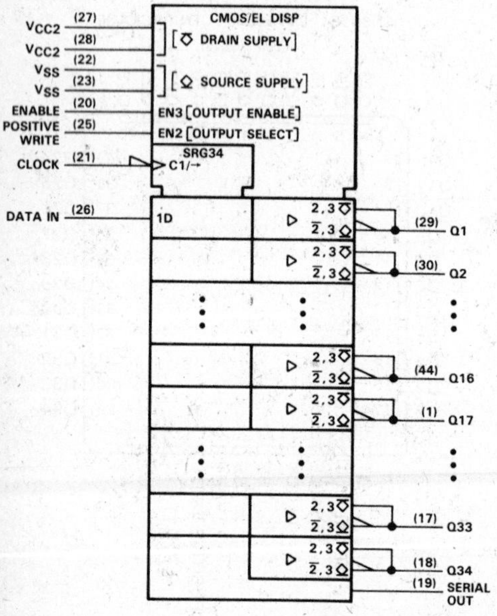
CMOS/EL DISP
VCC2 (27)
VCC2 (28)
[DRAIN SUPPLY]
VSS (22)
VSS (23)
[SOURCE SUPPLY]
ENABLE (20)
EN3 [OUTPUT ENABLE]
POSITIVE WRITE (25)
EN2 [OUTPUT SELECT]
SRG34
CLOCK (21)
C1/→
DATA IN (26)
1D
(29) Q1
(30) Q2
(44) Q16
(1) Q17
(17) Q33
(18) Q34
(19) SERIAL OUT

SN65564, SN75564

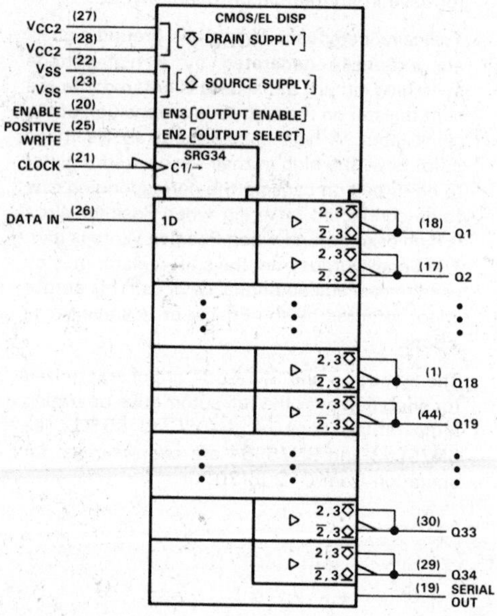
CMOS/EL DISP
VCC2 (27)
VCC2 (28)
[DRAIN SUPPLY]
VSS (22)
VSS (23)
[SOURCE SUPPLY]
ENABLE (20)
EN3 [OUTPUT ENABLE]
POSITIVE WRITE (25)
EN2 [OUTPUT SELECT]
SRG34
CLOCK (21)
C1/→
DATA IN (26)
1D
(18) Q1
(17) Q2
(1) Q18
(44) Q19
(30) Q33
(29) Q34
(19) SERIAL OUT

‡These symbols are in accordance with ANSI/IEEE Std 91-1984 and IEC Publication 617-12.

TEXAS INSTRUMENTS
POST OFFICE BOX 655012 • DALLAS, TEXAS 75265

logic diagram (positive logic)

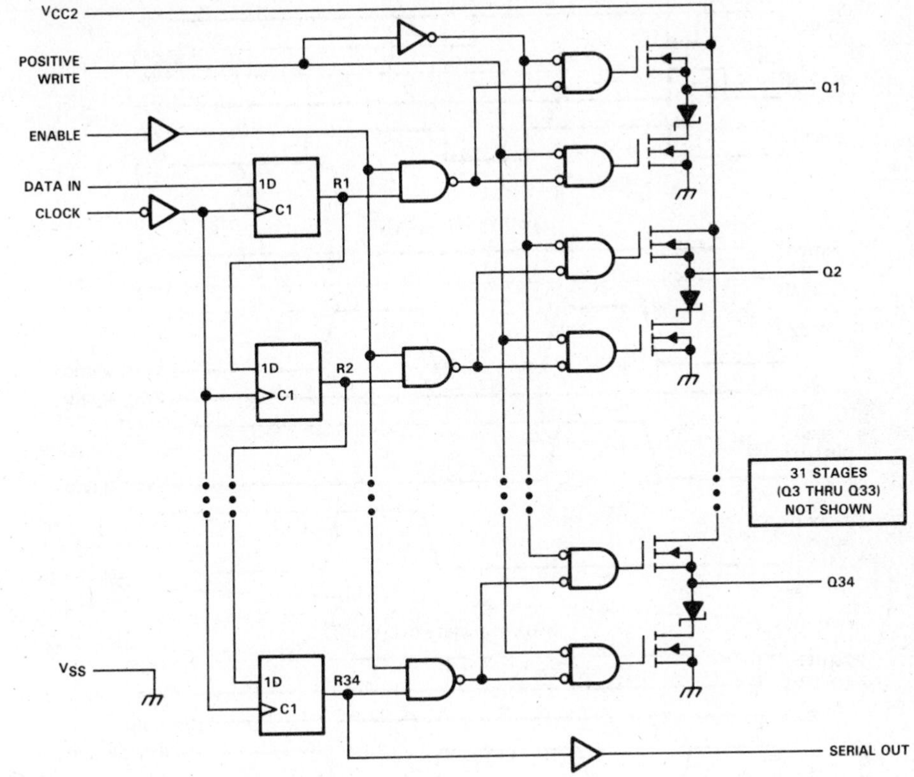
VCC2
POSITIVE
WRITE
ENABLE
DATA IN
CLOCK
1D
C1
R1
R2
R34
Q1
Q2
Q34
31 STAGES
(Q3 THRU Q33)
NOT SHOWN
VSS
SERIAL OUT

schematics of inputs and outputs

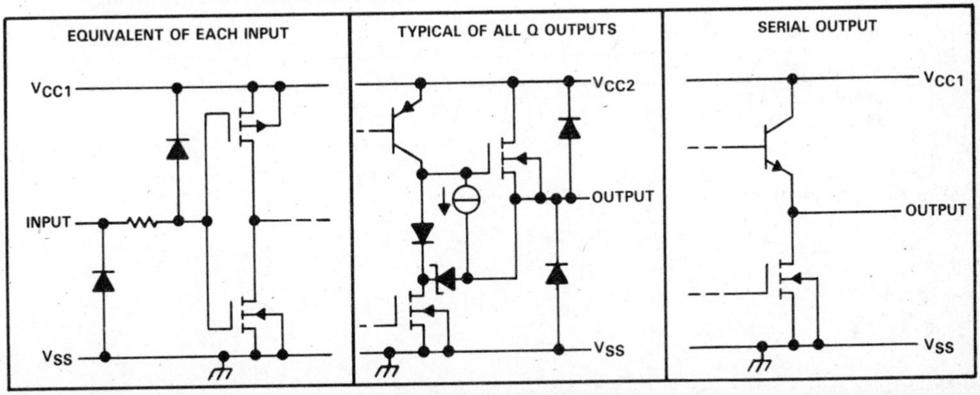
EQUIVALENT OF EACH INPUT
VCC1
INPUT
VSS
TYPICAL OF ALL Q OUTPUTS
VCC2
OUTPUT
VSS
SERIAL OUTPUT
VCC1
OUTPUT
VSS

3

Display Drivers

typical operating sequence

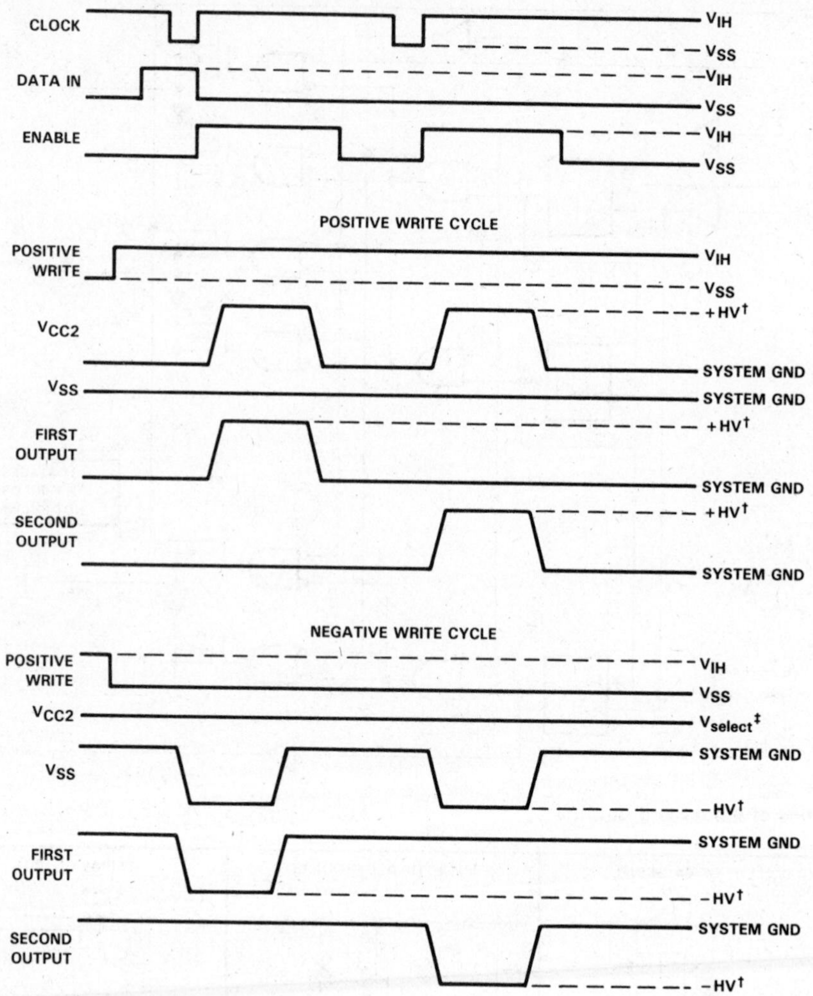
CLOCK
DATA IN
ENABLE
V_{IH}
V_{SS}
V_{IH}
V_{SS}
V_{IH}
V_{SS}
POSITIVE WRITE CYCLE
POSITIVE WRITE
V_{CC2}
V_{SS}
FIRST OUTPUT
SECOND OUTPUT
V_{IH}
V_{SS}
+HV†
SYSTEM GND
SYSTEM GND
+HV†
SYSTEM GND
+HV†
SYSTEM GND
NEGATIVE WRITE CYCLE
POSITIVE WRITE
V_{CC2}
V_{SS}
FIRST OUTPUT
SECOND OUTPUT
V_{IH}
V_{SS}
V_{select}‡
SYSTEM GND
−HV†
SYSTEM GND
−HV†
SYSTEM GND
−HV†

†HV = high voltage
‡V_{select} is a voltage level typically equal to V_{CC2} of the column driver.

ADVANCE INFORMATION

TEXAS INSTRUMENTS
POST OFFICE BOX 655012 • DALLAS, TEXAS 75265

absolute maximum ratings over operating free-air temperature range (unless otherwise noted)

Supply voltage, V_{CC1} (see Note 1) .... 15 V
Supply voltage, V_{CC2} .... 230 V
Supply voltage, V_{SS} .... −230 V
Input voltage .... −0.3 V to V_{CC1} + 0.3 V
Continuous total power dissipation at (or below) 25°C free-air temperature (see Note 2) .... 1700 mW
Operating free-air temperature range: SN65563, SN65564 .... −40°C to 85°C
SN75563, SN75564 .... 0°C to 70°C
Storage temperature range .... −65°C to 150°C
Lead temperature 1,6 mm (1/16 inch) from case for 10 seconds .... 260°C

NOTES: 1. Voltage values are with respect to V_{SS}.
2. For operation above 25°C free-air temperature, derate to 1088 mW at 70°C at the rate of 13.6 mW/°C.

recommended operating conditions (see Note 1, Figure 1, and Figure 2)

		MIN	NOM	MAX	UNIT
Supply voltage, V_{CC1}		10.8	12	13.2	V
Supply voltage, V_{CC2}		0		225	V
Supply voltage, V_{SS}		0		−225	V
High-level input voltage, V_{IH}		$0.75V_{CC1}$		$V_{CC1}+0.3$	V
Low-level input voltage, V_{IL}†		−0.3		$0.25V_{CC1}$	V
High-level output current, I_{OH}				−70	mA
Low-level output current, I_{OL}				70	mA
Output clamp current, I_{OK}				±70	mA
Clock frequency, f_{clock}				4	MHz
Pulse duration, Clock high or low, t_{wCLK}		125			ns
Setup time, data high or low before clock↓, t_{su1}		100			ns
Setup time, Clock low before V_{CC2}↑ or V_{SS}↓, t_{su2}		300			ns
Setup time, Enable high before V_{CC2}↑ or V_{SS}↓, t_{su3}		300			ns
Setup time, Positive Write high or low before V_{CC2}↑ or V_{SS}↓, t_{su4}		300			ns
Hold time, data high or low after clock↓, t_{h1}		100			ns
Hold time, Clock high after V_{CC2}↓ or V_{SS}↑, t_{h2}		500			ns
Hold time, Enable high after V_{CC2}↓ or V_{SS}↑, t_{h3}		300			ns
Hold time, Positive Write after V_{CC2}↓ or V_{SS}↑, t_{h4}		300			ns
Slew rate, V_{CC2} or V_{SS} with one active output driving a 4.7-nF load to V_{SS} or V_{CC2}, dv/dt				45	V/μs
Rest time, period between successive rampings of V_{CC2} or V_{SS}				5	μs
Operating free-air temperature, T_A	SN65563, SN65564	−40		85	°C
	SN75563, SN75564	0		70	°C

†The algebraic convention, in which the less positive (more negative) limit is designated as minimum, is used in this data sheet for logic voltage levels only.

electrical characteristics over recommended operating ranges of V_{CC1} and free-air temperature, V_{CC2} = 225 V, V_{SS} = 0 (unless otherwise noted)

PARAMETER			TEST CONDITIONS	MIN	MAX	UNIT
$I_{O(off)}$	Off-state Q output current		V_O = 225 V		20	μA
			V_O = 0		−20	
V_{OH}	High-level output voltage	Q outputs	I_O = −70 mA	V_{CC2}−30		V
		Serial Out	I_O = −100 μA, V_{CC1} = 12 V	10.5		
V_{OL}	Low-level output voltage	Q outputs	I_O = 70 mA		30	V
		Serial Out	I_O = 100 μA		1	
I_{IH}	High-level input current		V_{IH} = V_{CC1}		1	μA
I_{IL}	Low-level input current		V_{IL} = 0		−1	μA
I_{CC1}	Supply current from V_{CC1}				500	μA
I_{CC2}	Supply current from V_{CC2}		One Q output high		5	mA
			All Q outputs low		200	μA

switching characteristics operating range of V_{CC1}, T_A = 25°C

PARAMETER		TEST CONDITIONS	MIN	MAX	UNIT
t_{PLH}	Propagation delay time, low-to-high level serial output from clock	C_L = 50 pF to V_{SS}, See Figures 3 and 5		400	ns
t_{PHL}	Propagation delay time, high-to-low level serial output from clock			400	ns
t_{PLH}	Propagation delay time, low-to-high level Q output from V_{CC2}	dv/dt = 45 V/μs, One output on with C_L = 4.7 nF to V_{SS}, See Figures 4 and 5		6	μs
t_{PHL}	Propagation delay time, high-to-low level Q output from V_{CC2}			6	μs
t_{PLH}	Propagation delay time, low-to-high level Q output from V_{SS}	dv/dt = 45 V/μs, One output on with C_L = 4.7 nF to V_{CC2}, See Figures 4 and 6		6	μs
t_{PHL}	Propagation delay time, high-to-low level Q output from V_{SS}			6	μs

PARAMETER MEASUREMENT INFORMATION

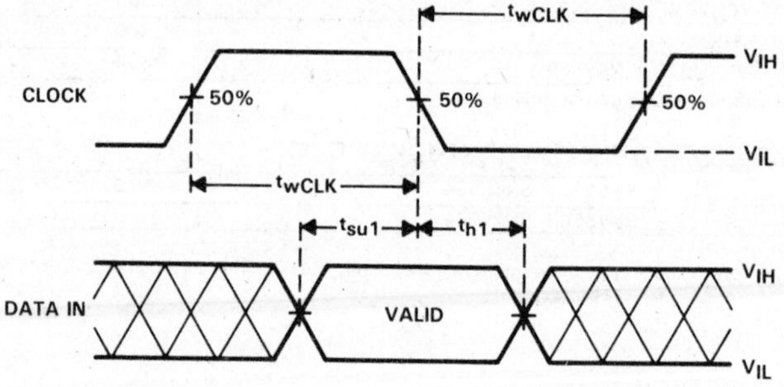
t_wCLK
CLOCK
50%
50%
50%
V_IH
V_IL
t_wCLK
t_su1
t_h1
DATA IN
VALID
V_IH
V_IL

FIGURE 1. INPUT TIMING VOLTAGE WAVEFORMS

PARAMETER MEASUREMENT INFORMATION

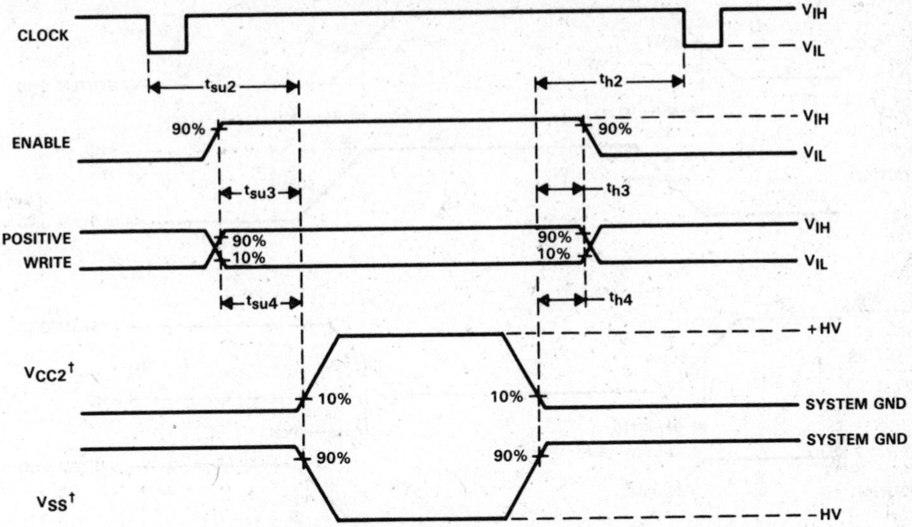
CLOCK
V_{IH}
V_{IL}
t_{su2}
t_{h2}
ENABLE
90%
90%
t_{su3}
t_{h3}
POSITIVE
WRITE
90%
10%
t_{su4}
t_{h4}
+HV
V_{CC2}†
10%
SYSTEM GND
SYSTEM GND
90%
V_{SS}†
-HV

†Timing waveforms are with respect to V_{CC2} or V_{SS}, as appropriate.

FIGURE 2. CONTROL INPUT TIMING VOLTAGE WAVEFORMS

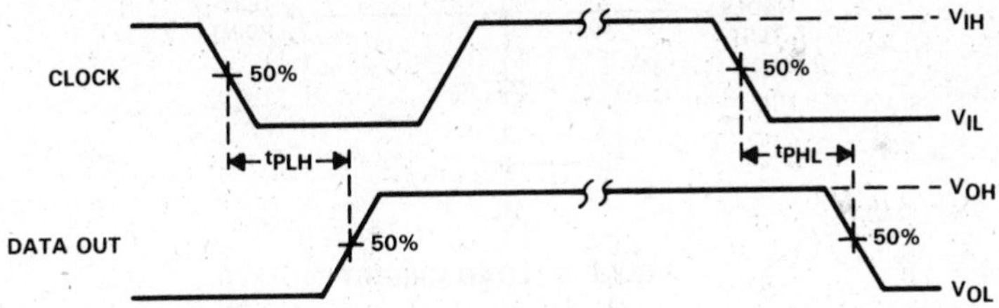
CLOCK
50%
V_{IH}
V_{IL}
t_{PLH}
t_{PHL}
DATA OUT
50%
V_{OH}
V_{OL}

FIGURE 3. VOLTAGE WAVEFORMS FOR PROPAGATION DELAY TIMES, CLOCK TO DATA OUT

PARAMETER MEASUREMENT INFORMATION

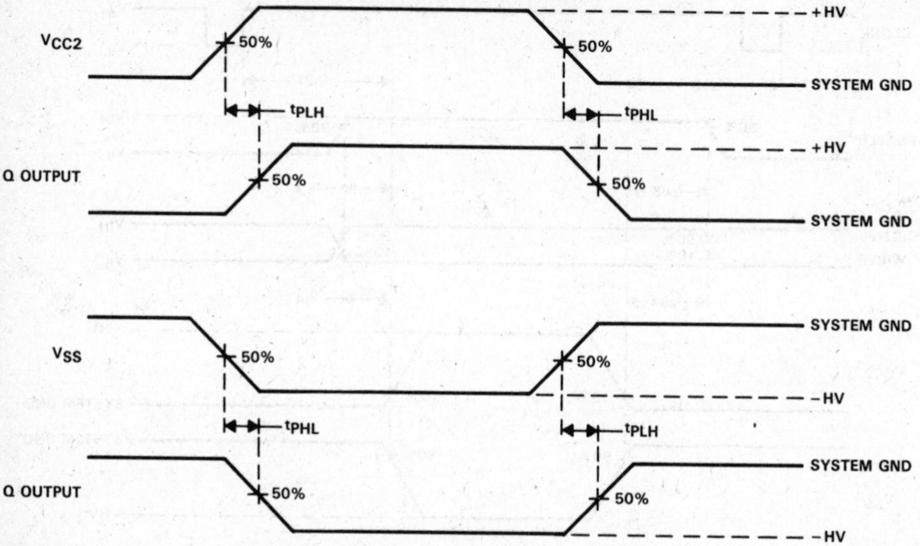
VCC2
50%
50%
+HV
SYSTEM GND
tPLH
tPHL
Q OUTPUT
50%
50%
+HV
SYSTEM GND
VSS
50%
50%
SYSTEM GND
−HV
tPHL
tPLH
Q OUTPUT
50%
50%
SYSTEM GND
−HV

FIGURE 4. VOLTAGE WAVEFORMS FOR PROPAGATION DELAY TIMES, V_{CC2} (V_{SS}) TO Q OUTPUT

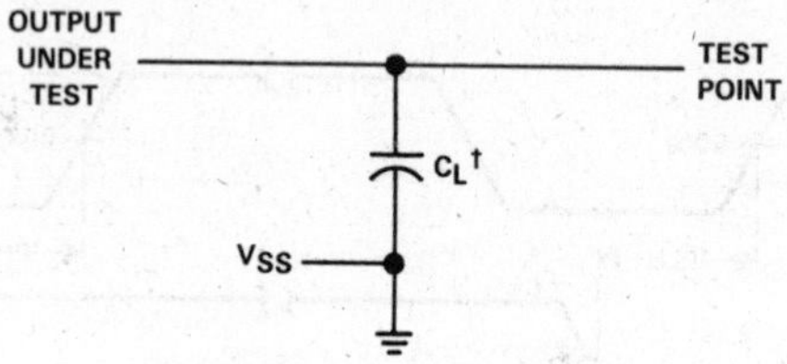
OUTPUT UNDER TEST
TEST POINT
CL†
VSS

FIGURE 5. LOAD CIRCUIT

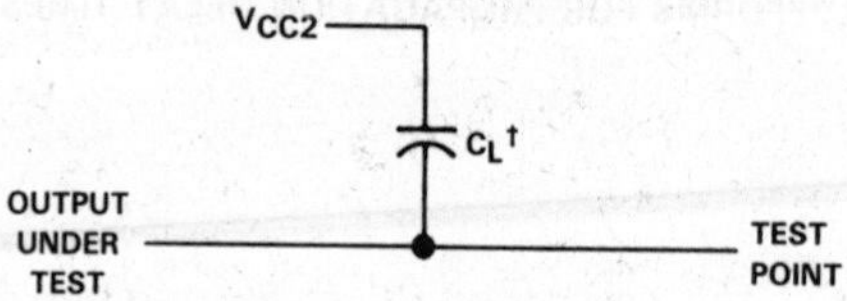
VCC2
CL†
OUTPUT UNDER TEST
TEST POINT

FIGURE 6. LOAD CIRCUIT

†C_L includes probe and jig capacitance.

TEXAS INSTRUMENTS
POST OFFICE BOX 655012 • DALLAS, TEXAS 75265

- **Controls 32 Electrodes**
- **60-V (Ramped V_{CC2}) Totem-Pole Outputs**
- **Low CMOS Stand-By Power Consumption**
- **Energy Recovery System Compatible**
- **15-mA Source and Sink Compatibility**
- **High-Speed Serially-Shifted Data Input**

SN65567, SN75567 . . . FN PACKAGE

(TOP VIEW)

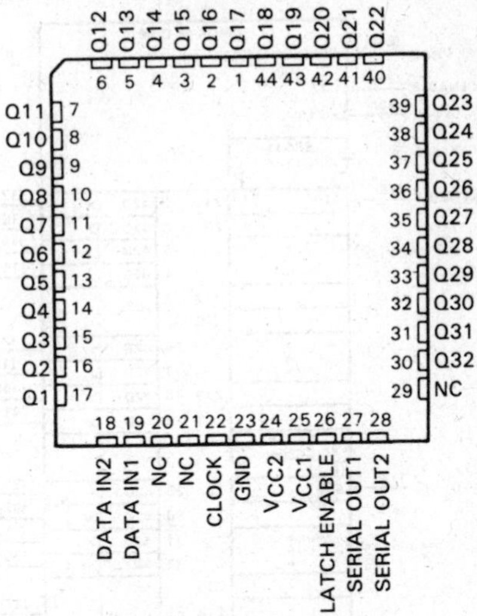
Q12 Q13 Q14 Q15 Q16 Q17 Q18 Q19 Q20 Q21 Q22
6 5 4 3 2 1 44 43 42 41 40
Q11 7
Q10 8
Q9 9
Q8 10
Q7 11
Q6 12
Q5 13
Q4 14
Q3 15
Q2 16
Q1 17
39 Q23
38 Q24
37 Q25
36 Q26
35 Q27
34 Q28
33 Q29
32 Q30
31 Q31
30 Q32
29 NC
18 19 20 21 22 23 24 25 26 27 28
DATA IN2 DATA IN1 NC NC CLOCK GND VCC2 VCC1 LATCH ENABLE SERIAL OUT1 SERIAL OUT2

description

The SN65567, SN65568, SN75567, and SN75568 are monolithic BIDFET† integrated circuits designed to provide the serial-to-parallel conversion and level translation of data in a matrix-addressable electroluminescent display. The device inputs are diode-clamped CMOS inputs. The SN65568 and SN75568 output sequences are reversed from the SN65567 and SN75567 for ease in printed circuit board layout.

These column drivers consist of two 16-bit static shift registers, 32 latches, and 32 high-voltage outputs. Typically, a 32-bit data string is split into two 16-bit data strings externally and then entered in parallel into the shift registers on the low-to-high transition of the clock signal. The register associated with Data Input 1 loads the odd bits while the shift register associated with Data Input 2 loads the even bits of the 32 latches. This method of entering data effectively doubles the clock frequency of a 32-bit shift register. A logic high signal on the latch enable input transfers the data from the shift register to the latches while the V_{CC2} bus is low. Once stable in the latch circuits, the V_{CC2} rail is ramped up to allow the data to appear at the high-voltage outputs. By limiting V_{CC2} to a maximum of 50 volts, these devices may be safely operated in a non-ramped V_{CC2} mode. Drivers may be cascaded via the serial data outputs of the static shift registers. These outputs are not affected by the latch enable input.

The SN65567 and SN65568 are characterized for operation from −40°C to 85°C. The SN75567 and SN75568 are characterized for operation from 0°C to 70°C.

SN65568, SN75568 . . . FN PACKAGE

(TOP VIEW)

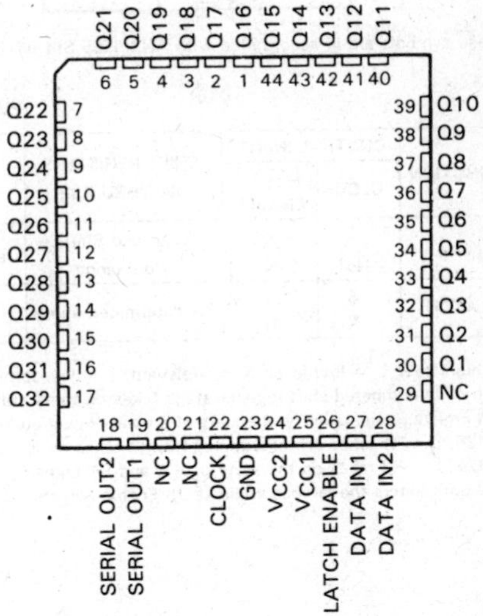
Q21 Q20 Q19 Q18 Q17 Q16 Q15 Q14 Q13 Q12 Q11
6 5 4 3 2 1 44 43 42 41 40
Q22 7
Q23 8
Q24 9
Q25 10
Q26 11
Q27 12
Q28 13
Q29 14
Q30 15
Q31 16
Q32 17
39 Q10
38 Q9
37 Q8
36 Q7
35 Q6
34 Q5
33 Q4
32 Q3
31 Q2
30 Q1
29 NC
18 19 20 21 22 23 24 25 26 27 28
SERIAL OUT2 SERIAL OUT1 NC NC CLOCK GND VCC2 VCC1 LATCH ENABLE DATA IN1 DATA IN2

NC—No internal connection

†BIDFET—Bipolar, double-diffused, N-channel and P-channel MOS transistors on the same chip—Patented Process

3

Display Drivers

ADVANCAE INFORMATION

logic symbols†

SN65567, SN75567

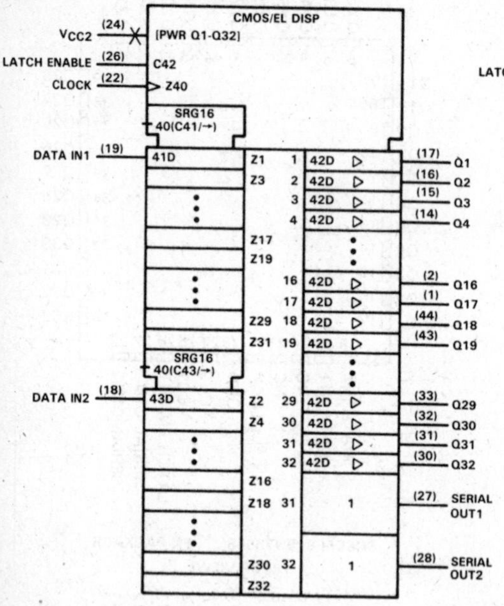
CMOS/EL DISP
VCC2 (24) [PWR Q1-Q32]
LATCH ENABLE (26) C42
CLOCK (22) Z40
SRG16
40(C41/→)
DATA IN1 (19) 41D
Z1 1 42D (17) Q1
Z3 2 42D (16) Q2
3 42D (15) Q3
4 42D (14) Q4
Z17
Z19
16 42D (2) Q16
17 42D (1) Q17
Z29 18 42D (44) Q18
Z31 19 42D (43) Q19
SRG16
40(C43/→)
DATA IN2 (18) 43D
Z2 29 42D (33) Q29
Z4 30 42D (32) Q30
31 42D (31) Q31
32 42D (30) Q32
Z16
Z18 31 1 (27) SERIAL OUT1
Z30 32 1 (28) SERIAL OUT2
Z32

SN65568, SN75568

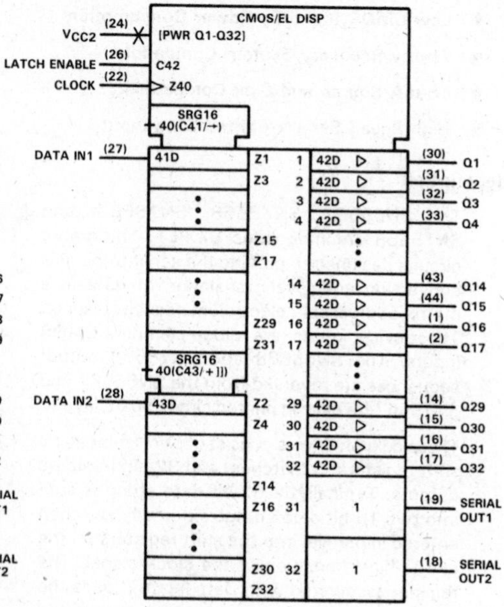
CMOS/EL DISP
VCC2 (24) [PWR Q1-Q32]
LATCH ENABLE (26) C42
CLOCK (22) Z40
SRG16
40(C41/→)
DATA IN1 (27) 41D
Z1 1 42D (30) Q1
Z3 2 42D (31) Q2
3 42D (32) Q3
4 42D (33) Q4
Z15
Z17
14 42D (43) Q14
15 42D (44) Q15
Z29 16 42D (1) Q16
Z31 17 42D (2) Q17
SRG16
40(C43/+]])
DATA IN2 (28) 43D
Z2 29 42D (14) Q29
Z4 30 42D (15) Q30
31 42D (16) Q31
32 42D (17) Q32
Z14
Z16 31 1 (19) SERIAL OUT1
Z30 32 1 (18) SERIAL OUT2
Z32

†These symbols are in accordance with ANSI/IEEE Std 91-1984 and IEC Publication 617-12.

FUNCTION TABLE

FUNCTION	CONTROL INPUTS: CLOCK	CONTROL INPUTS: LATCH ENABLE	SHIFT REGISTERS R1 THRU R32	LATCHES LC1 THRU LC32	OUTPUTS SERIAL: SO1	OUTPUTS SERIAL: SO2	OUTPUTS PARALLEL: Q1 THRU Q32
LOAD	↑	X	Load and Shift†	Determined by Latch Enable‡	R31	R32	LC1 thru LC32 respectively
	Not↑	X	No change		R31	R32	
LATCH	X	L	As determined above	Stored data	R31	R32	LC1 thru LC32 respectively
	X	H		New data	R31	R32	

H = high level; L = low level; X = irrelevant; ↑ = low-to-high-level transition

†Each even-numbered shift register stage takes on the state of the next-lower even-numbered stage and likewise each odd-numbered shift register stage takes on the state of the next-lower odd-numbered state; i.e., R32 takes on the state of R30, R30 takes on the state of R28, . . . R4 takes on the state of R2, R2 takes on the state of Data In 2, R31 takes on the state of R29, R29 takes on the state of R27, . . . R3 takes on the state of R1, and R1 takes on the state of Data In 1.

‡New data enters the latches while Latch Enable is high. These data are stored while Latch Enable is low.

logic diagram (positive logic)

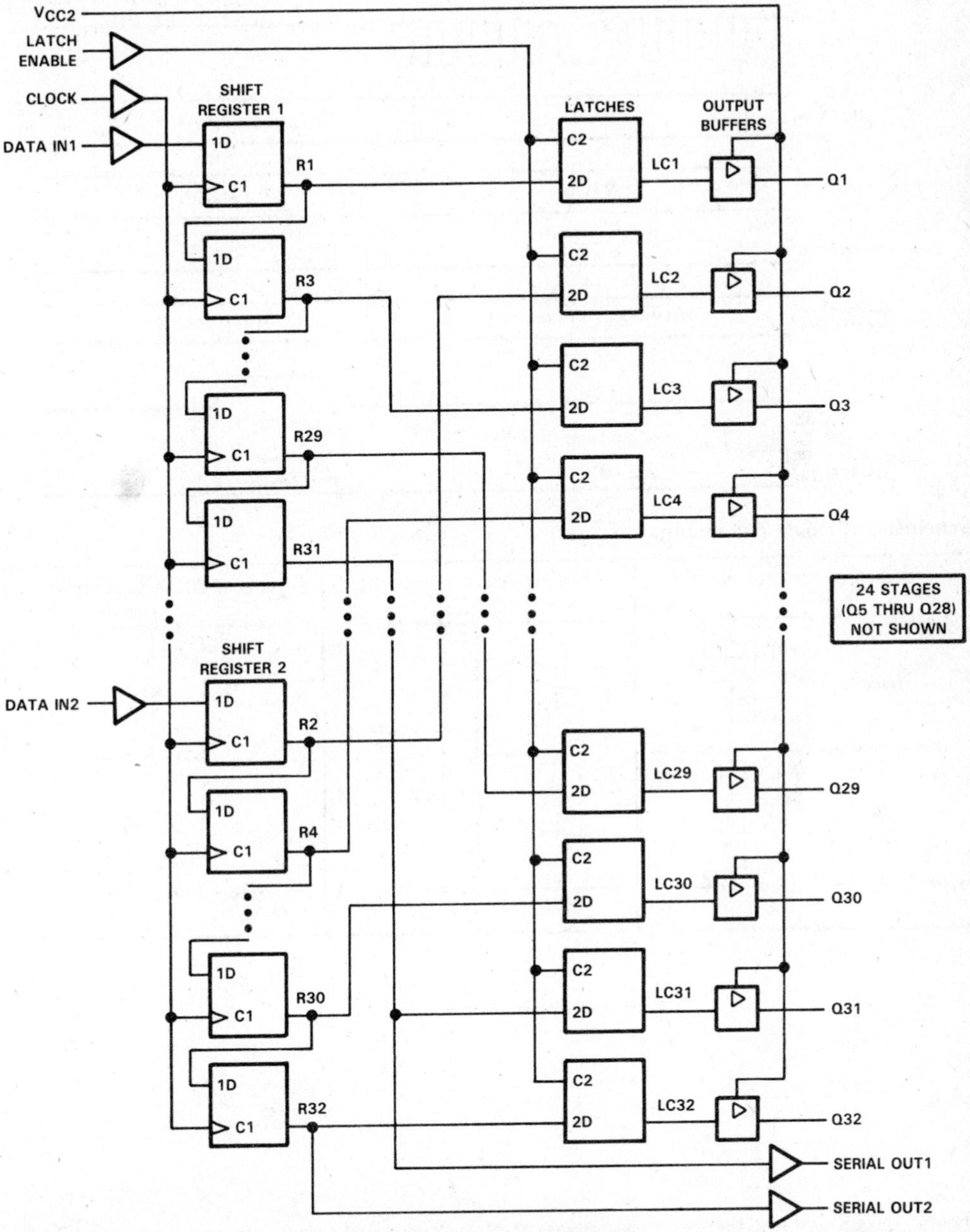

V_{CC2}
LATCH ENABLE
CLOCK
DATA IN1
SHIFT REGISTER 1
1D
C1
R1
R3
R29
R31
SHIFT REGISTER 2
DATA IN2
R2
R4
R30
R32
LATCHES
C2
2D
LC1
LC2
LC3
LC4
LC29
LC30
LC31
LC32
OUTPUT BUFFERS
Q1
Q2
Q3
Q4
Q29
Q30
Q31
Q32
24 STAGES (Q5 THRU Q28) NOT SHOWN
SERIAL OUT1
SERIAL OUT2

typical operating sequence

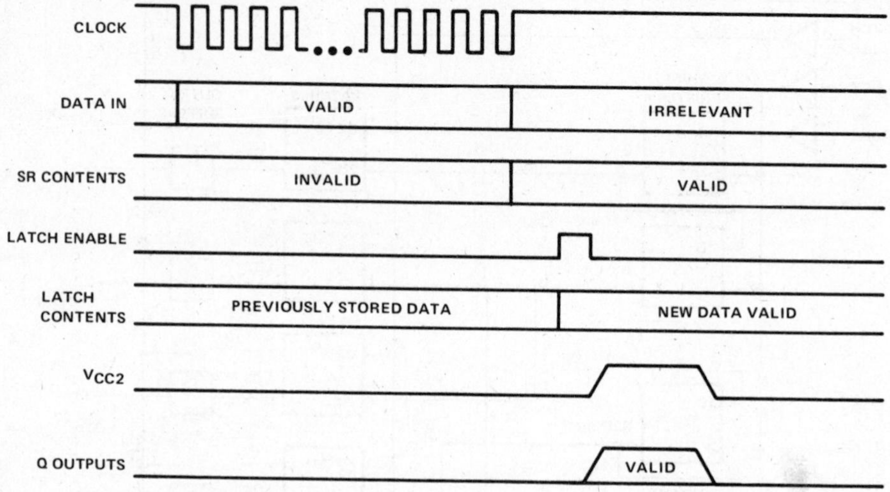
CLOCK
DATA IN
VALID
IRRELEVANT
SR CONTENTS
INVALID
VALID
LATCH ENABLE
LATCH
CONTENTS
PREVIOUSLY STORED DATA
NEW DATA VALID
VCC2
Q OUTPUTS
VALID

schematic of inputs and outputs

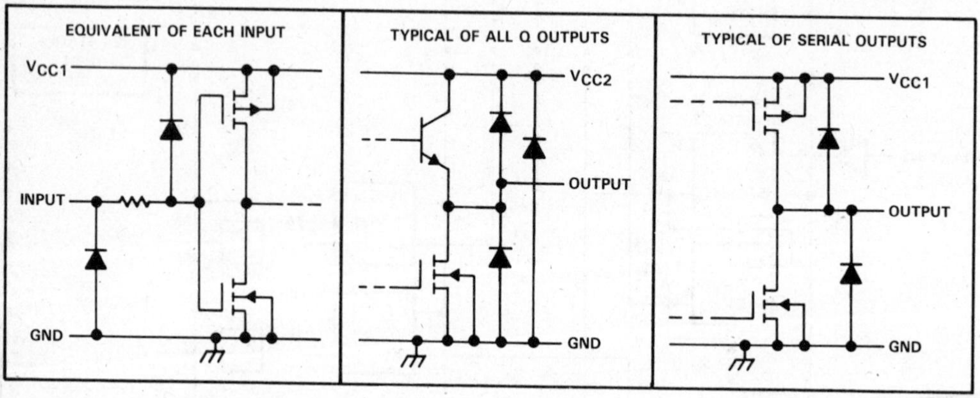
EQUIVALENT OF EACH INPUT
VCC1
INPUT
GND
TYPICAL OF ALL Q OUTPUTS
VCC2
OUTPUT
GND
TYPICAL OF SERIAL OUTPUTS
VCC1
OUTPUT
GND

absolute maximum ratings over operating free-air temperature range (unless otherwise noted)

Supply voltage, V_{CC1} (see Note 1) .... 8 V
Supply voltage, V_{CC2} .... 70 V
Input voltage, V_I .... V_{CC1} +0.3 V
Continuous total dissipation at (or below) 25°C free-air temperature (see Note 2) .... 1700 mW
Operating free-air temperature range: SN65567, SN65568 .... −40°C to 85°C
SN75567, SN75568 .... 0°C to 70°C
Storage temperature range .... −65°C to 150°C
Case temperature for 10 seconds .... 260°C

NOTES: 1. Voltage values are with respect to network ground terminal.
2. For operation above 25°C free-air temperature, derate linearly at the rate of 13.6 mW/°C.

recommended operating conditions

		MIN	NOM	MAX	UNIT
Supply voltage, V_{CC1}		4.5	5	5.5	V
Supply voltage, V_{CC2}		0		60	V
High-level input voltage, V_{IH}	V_{CC1} = 4.5 V	3.4		4.8	V
	V_{CC1} = 5.5 V	4.2		5.8	
Low-level input voltage, V_{IL}	V_{CC1} = 4.5 V	−0.3		1.1	V
	V_{CC1} = 5.5 V	−0.3		1.3	
High-level Q output current, I_{OH}		−15			mA
Low-level Q output current, I_{OL}		15			mA
Q output clamp current, I_{OK}				20	mA
Clock frequency, f_{clock}		0		5	MHz
Pulse duration, clock high, $t_{w(CLK)}$		100			ns
Pulse duration, latch enable high, $t_{w(LE)}$		100			ns
Setup time, data before clock ↑, t_{su}		50			ns
Hold time, data after clock ↑, t_h		50			ns
Rate of rise of V_{CC2}, dv/dt (see Figure 3)				60	V/μs
Operating free-air temperature, T_A	SN65567, SN65568	−40		85	°C
	SN75567, SN75568	0		70	

NOTE 3: V_{CC2} must be ramped only when data within the latches is stable.

electrical characteristics over recommended operating free-air temperature range, V_{CC1} = 5 V

PARAMETER			TEST CONDITIONS		MIN	MAX	UNIT
V_{OH}	High-level output voltage	Q outputs	V_{CC2} = 60 V, I_{OH} = −15 mA		57		V
		Serial output	I_{OH} = 100 μA		3.8		
V_{OL}	Low-level output voltage	Q outputs	I_{OL} = 15 mA			8	V
		Serial output	I_{OL} = 100 μA			1	
I_{IH}	High-level input current		V_{IH} = V_{CC1}			1	μA
I_{IL}	Low-level input current		V_{IL} = 0			−1	μA
I_{CC1}	Supply current from V_{CC1}		V_I = V_{CC1}			500	μA
I_{CC2}	Supply current from V_{CC2}		V_{CC2} = 60 V	Outputs low		0.5	mA
				Outputs high		0.5	

switching characteristics, V_{CC1} = 5 V, V_{CC2} = 0, T_A = 25°C

PARAMETER		FROM INPUT	TO OUTPUT	TEST CONDITIONS	MIN	MAX	UNIT
t_{PHL}	Propagation delay time, high-to-low level	Clock	Serial Out	C_L = 20 pF, See Figure 2		140	ns
t_{PLH}	Propagation delay time, low-to-high level	Clock	Serial Out	C_L = 20 pF, See Figure 2		140	ns
t_d	Delay time, V_{CC2} to Q output	V_{CC2}	Q	dv/dt = 60 V/µs, C_L = 100 pF, See Figure 3		100	ns

PARAMETER MEASUREMENT INFORMATION

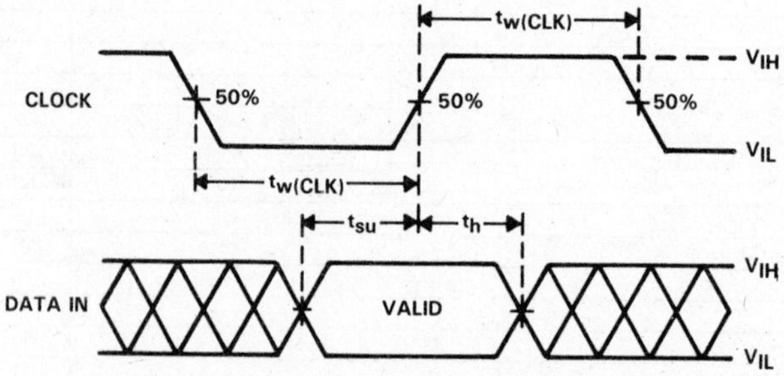
CLOCK
DATA IN
VALID
50%
50%
50%
$t_{w(CLK)}$
$t_{w(CLK)}$
t_{su}
t_h
V_{IH}
V_{IL}
V_{IH}
V_{IL}

FIGURE 1. INPUT TIMING VOLTAGE WAVEFORMS

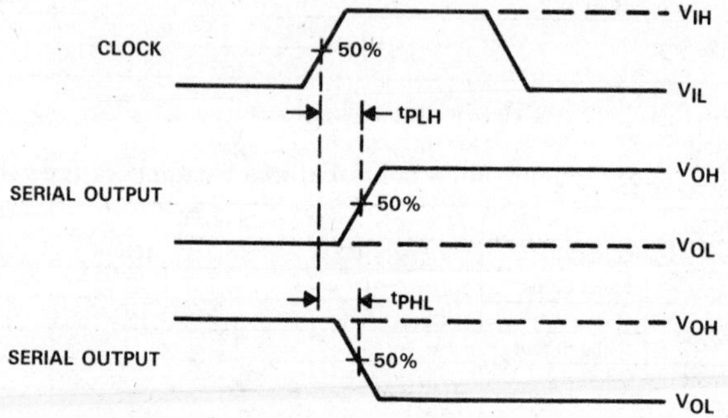
CLOCK
SERIAL OUTPUT
SERIAL OUTPUT
50%
50%
50%
t_{PLH}
t_{PHL}
V_{IH}
V_{IL}
V_{OH}
V_{OL}
V_{OH}
V_{OL}

FIGURE 2. VOLTAGE WAVEFORMS FOR PROPAGATION DELAY CLOCK TO SERIAL OUTPUT

TEXAS INSTRUMENTS
POST OFFICE BOX 655012 • DALLAS, TEXAS 75265

PARAMETER MEASUREMENT INFORMATION

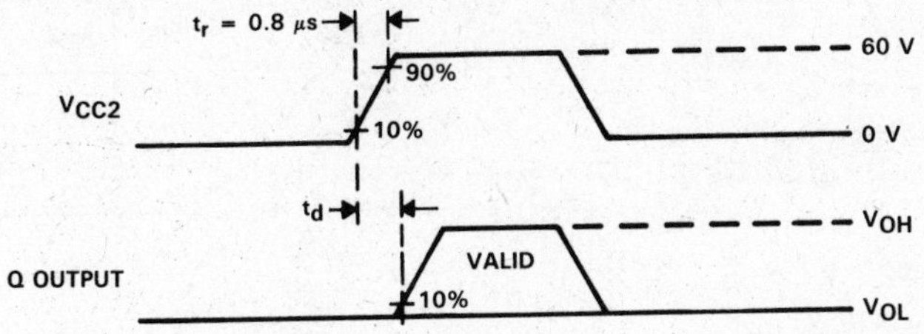
t_r = 0.8 μs
90%
10%
V_{CC2}
60 V
0 V
t_d
VALID
Q OUTPUT
10%
V_{OH}
V_{OL}

FIGURE 3. VOLTAGE WAVEFORMS FOR DELAY TIMES, LATCH ENABLE TO Q OUTPUTS

- Each Device Drives 7 Lines
- 150-V Output Voltage Swing Capability
- TTL Compatible Inputs
- Latches on All Driver Outputs
- High-Speed Serially Shifted Data Input
- Output Enable/Disable Function
- Serial Data Output for Cascade Operation
- Shift Register Has Synchronous Clear Function

J AND N
DUAL-IN-LINE PACKAGES
(TOP VIEW)

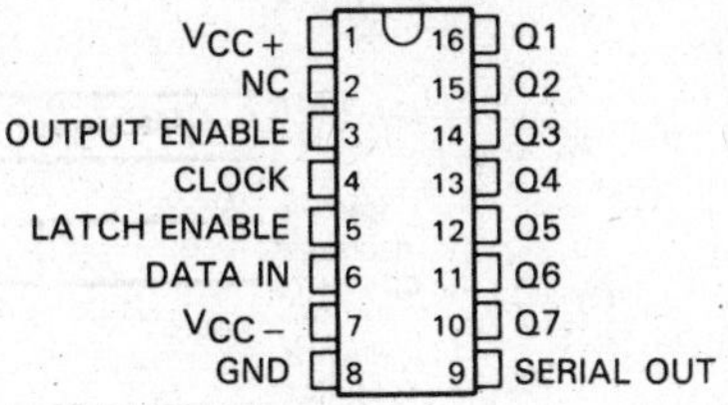
VCC+ 1
NC 2
OUTPUT ENABLE 3
CLOCK 4
LATCH ENABLE 5
DATA IN 6
VCC− 7
GND 8
16 Q1
15 Q2
14 Q3
13 Q4
12 Q5
11 Q6
10 Q7
9 SERIAL OUT

NC—No internal connection

description

The SN75581 is a monolithic BIDFET† integrated circuit designed to drive a dot matrix or segmented display. The output characteristics of this driver make it compatible to several display types including VF and DC plasma displays.

All device inputs are diode-clamped p-n-p inputs and, when left open, assume a high-logic level. The nominal input threshold is 1.5 volts. Outputs are open-source DMOS transistors for excellent high-voltage characteristics and reliability.

The device consists of a 7-bit shift register, seven latches, and seven output AND gates. Serial data is entered into the shift register on the low-to-high transition of the Clock input. When the Latch Enable input is high, data is transferred from the shift registers to the latch outputs. When Latch Enable makes a high-to-low transition with the Clock input high, the shift register is cleared. Taking the Output Enable input high enables all Q outputs simultaneously. The Serial Output is not affected by the Output Enable input.

The SN75581 is characterized for operation from 0°C to 70°C.

logic symbol‡

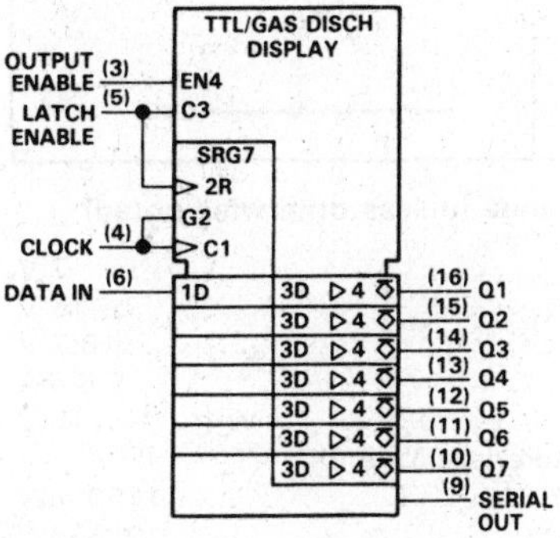
TTL/GAS DISCH DISPLAY
OUTPUT ENABLE (3) EN4
LATCH ENABLE (5) C3
SRG7
2R
G2
CLOCK (4) C1
DATA IN (6) 1D
3D ▷4 (16) Q1
3D ▷4 (15) Q2
3D ▷4 (14) Q3
3D ▷4 (13) Q4
3D ▷4 (12) Q5
3D ▷4 (11) Q6
3D ▷4 (10) Q7
(9) SERIAL OUT

logic diagram (positive logic)

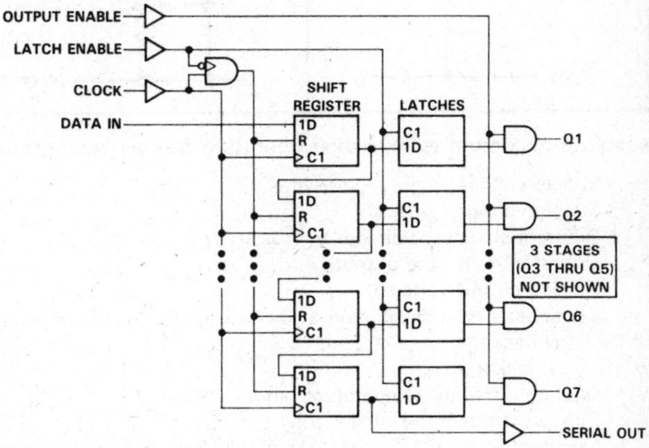
OUTPUT ENABLE
LATCH ENABLE
CLOCK
DATA IN
SHIFT REGISTER
LATCHES
1D
R
C1
C1
1D
Q1
Q2
3 STAGES (Q3 THRU Q5) NOT SHOWN
Q6
Q7
SERIAL OUT

† BIDFET—Bipolar, double-diffused, N-channel and P-channel MOS transistors on same chip—patented process.
‡ This symbol is in accordance with ANSI/IEEE Std 91-1984 and IEC Publication 617-12.

typical operating sequence

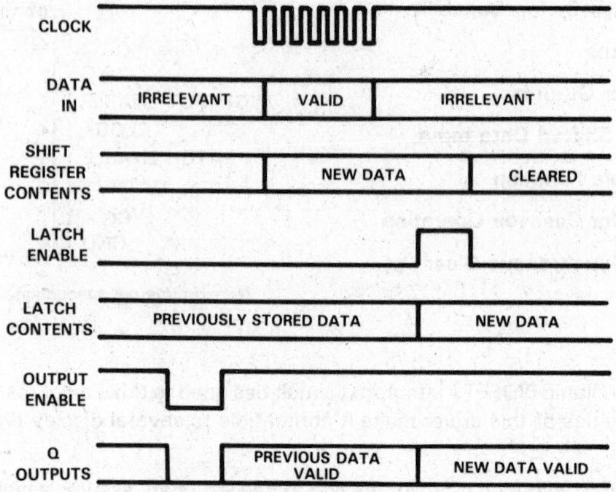
CLOCK
DATA IN
IRRELEVANT
VALID
IRRELEVANT
SHIFT REGISTER CONTENTS
NEW DATA
CLEARED
LATCH ENABLE
LATCH CONTENTS
PREVIOUSLY STORED DATA
NEW DATA
OUTPUT ENABLE
Q OUTPUTS
PREVIOUS DATA VALID
NEW DATA VALID

schematics of inputs and outputs

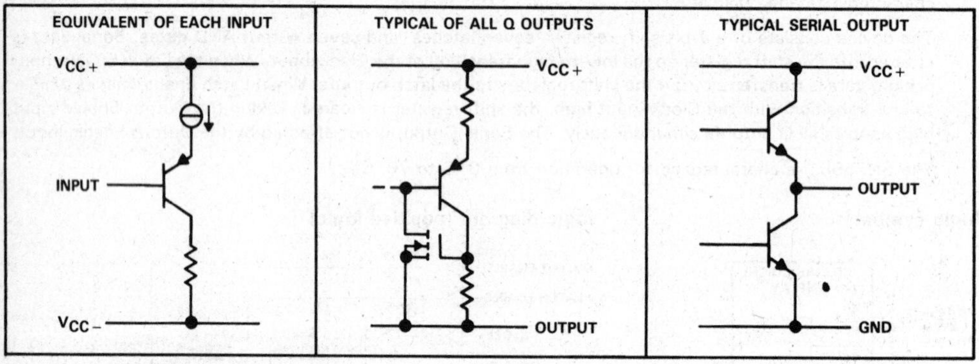
EQUIVALENT OF EACH INPUT
VCC+
INPUT
VCC−
TYPICAL OF ALL Q OUTPUTS
VCC+
OUTPUT
TYPICAL SERIAL OUTPUT
VCC+
OUTPUT
GND

absolute maximum ratings over operating free-air temperature range (unless otherwise noted)

Supply voltage, V_{CC+} (see Note 1)	7 V
Supply voltage, V_{CC-}	−15 V
Differential supply voltage, $V_{CC+} - V_{CC-}$	18.7 V
Output current (one output)	−5.5 mA
Applied output voltage	$V_{CC+} - 145$ V
Continuous total dissipation at (or below) 25°C free-air temperature (see Note 2):	
N package	1150 mW
J package	1025 mW
Operating free-air temperature range	0°C to 70°C
Storage temperature range	−65°C to 150°C
Lead temperature 1,6 mm (1/16 inch) from case for 60 seconds: J package	300°C
Lead temperature 1,6 mm (1/16 inch) from case for 10 seconds: N package	260°C

NOTES: 1. Voltage values are with respect to network ground terminal.
2. For operation above 25°C free-air temperature, derate the J package linearly to 656 mW at 70°C at the rate of 8.2 mW/°C and the N package to 736 mW at 70°C at the rate of 9.2 mW/°C.

recommended operating conditions

		MIN	NOM	MAX	UNIT
Supply voltage, V_{CC+}		4.5	5	5.5	V
Supply voltage, V_{CC-}		−10.8	−12	−13.2	V
High-level input voltage, V_{IH}		2			V
Low-level input voltage, V_{IL}				0.8	V
Clock frequency, f_{clock}				2	MHz
Pulse duration, clock high, $t_{w(CKH)}$		140			ns
Pulse duration, clock low, $t_{w(CKL)}$		320			ns
Pulse duration, latch enable high, $t_{w(LEH)}$		250			ns
Pulse duration, output enable low, $t_{w(OEL)}$		3			μs
Setup time, t_{su}	Data before clock↑	70			ns
	Clock high before latch enable↑	75			
Hold time, t_h	Data after clock↑	70			ns
	Clock high after latch enable↓	500			
Operating free-air temperature, T_A		0		70	°C

electrical characteristics over recommended operating free-air temperature range, V_{CC+} = 4.5 V to 5.5 V (unless otherwise noted)

PARAMETER		TEST CONDITIONS	MIN	TYP†	MAX	UNIT
V_{OH}	High-level output voltage, serial output	I_{OH} = −500 μA	2.4	4.7		V
V_{OL}	Low-level output voltage, serial output	V_{CC+} = 5.5 V, I_{OL} = 1.6 mA		0.15	0.4	V
$I_{O(on)}$	On-state output current, Q outputs	V_{OH} = V_{CC+} − 10 V	−2	−5.5		mA
$I_{O(off)}$	Off-state output current, Q outputs	V_{CC+} = 5.5 V, V_O = −140 V			5	μA
I_{IH}	High-level input current	V_I = 5.5 V			5	μA
I_{IL}	Low-level input current	V_I = 0.4 V			50	μA
I_{CC+}	Supply current from V_{CC+}	V_{CC+} = 5.5 V, V_{CC-} = −13.2 V		12	30	mA
I_{CC-}	Supply current from V_{CC-}	V_{CC+} = 5.5 V, V_{CC-} = −13.2 V		−11	−28	mA

switching characteristics, C_L = 20 pF, T_A = 25°C

PARAMETER		TEST CONDITIONS	MIN	TYP†	MAX	UNIT
t_{PHL}	Propagation delay time, high-to-low-level Q output from latch enable or output enable	R_L = 25 kΩ, See Figure 4		2.2	3	μs
t_{PLH}	Propagation delay time, low-to-high-level Q output from latch enable or output enable			0.75	2	
t_{PHL}	Propagation delay time, high-to-low-level serial data from clock	R_L = 3 kΩ, See Figure 5		200	350	ns
t_{PLH}	Propagation delay time, low-to-high-level serial data from clock			180	350	

† All typical values are at V_{CC+} = 5 V, V_{CC-} = −12 V, T_A = 25°C.

PARAMETER MEASUREMENT INFORMATION

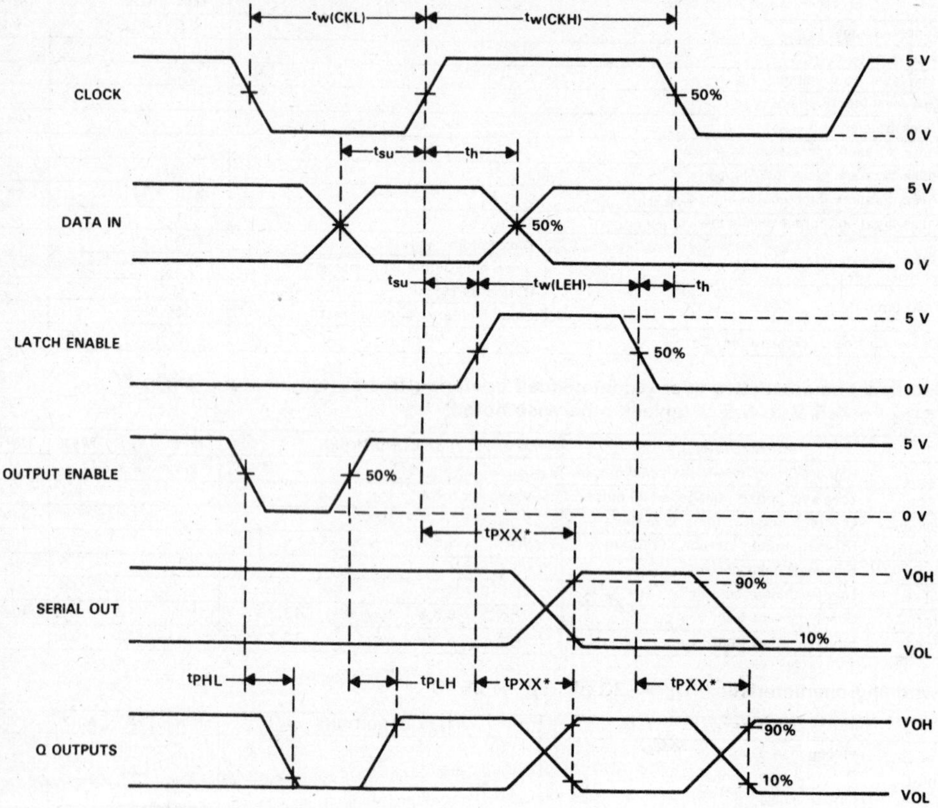
tw(CKL)
tw(CKH)
CLOCK
5 V
50%
0 V
tsu
th
DATA IN
50%
LATCH ENABLE
tw(LEH)
OUTPUT ENABLE
tPXX*
SERIAL OUT
VOH
90%
10%
VOL
tPHL
tPLH
Q OUTPUTS

*tPXX is tPHL or tPLH (whichever is appropriate)

FIGURE 3. VOLTAGE WAVEFORMS

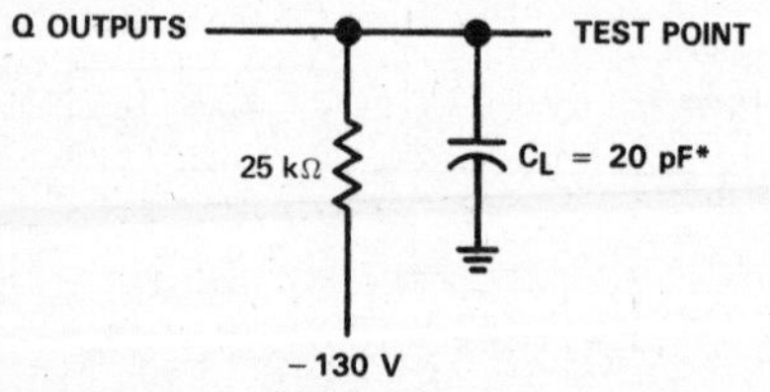
Q OUTPUTS
TEST POINT
25 kΩ
CL = 20 pF*
−130 V

*Includes probe and jig capacitance

FIGURE 4. Q OUTPUT LOAD CONDITIONS

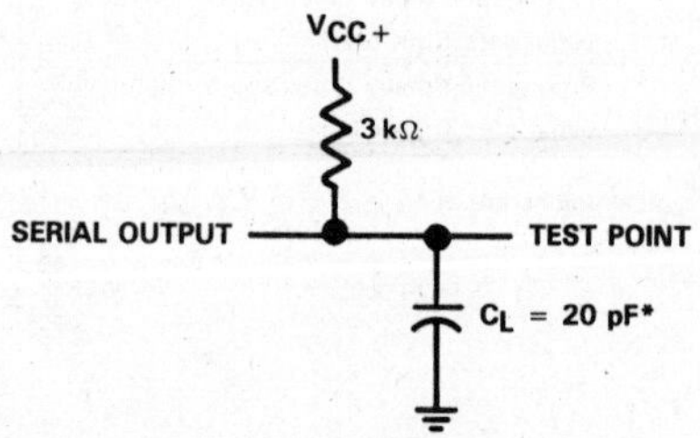
VCC+
3 kΩ
SERIAL OUTPUT
TEST POINT
CL = 20 pF*

*Includes probe and jig capacitance

FIGURE 5. SERIAL OUTPUT LOAD CONDITIONS

SN751506, SN751516
DC PLASMA DISPLAY DRIVERS

D3005, DECEMBER 1986

- Each Device Drives 32 Lines
- 180-V Open Drain Parallel Outputs
- 220-mA Parallel Output Sink Current Capability
- CMOS-Compatible Inputs
- Strobe Input Provided
- Serial Data Output for Cascade Operation
- Inputs Have Built-in Electrostatic Discharge Protection

SN751506 . . . FT PACKAGE
(TOP VIEW)

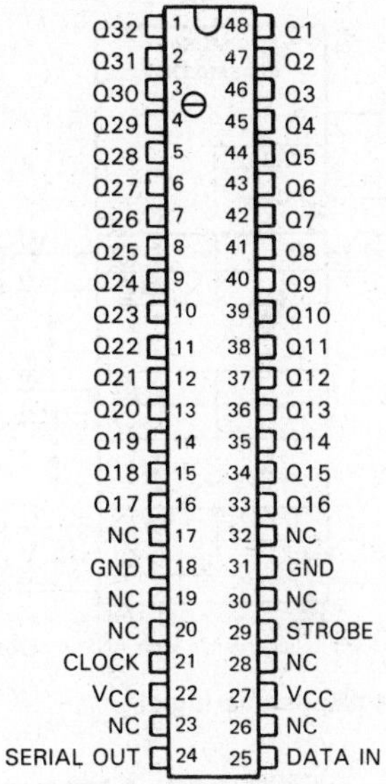
Q32 1 48 Q1
Q31 2 47 Q2
Q30 3 46 Q3
Q29 4 45 Q4
Q28 5 44 Q5
Q27 6 43 Q6
Q26 7 42 Q7
Q25 8 41 Q8
Q24 9 40 Q9
Q23 10 39 Q10
Q22 11 38 Q11
Q21 12 37 Q12
Q20 13 36 Q13
Q19 14 35 Q14
Q18 15 34 Q15
Q17 16 33 Q16
NC 17 32 NC
GND 18 31 GND
NC 19 30 NC
NC 20 29 STROBE
CLOCK 21 28 NC
VCC 22 27 VCC
NC 23 26 NC
SERIAL OUT 24 25 DATA IN

SN751516 . . . FT PACKAGE
(TOP VIEW)

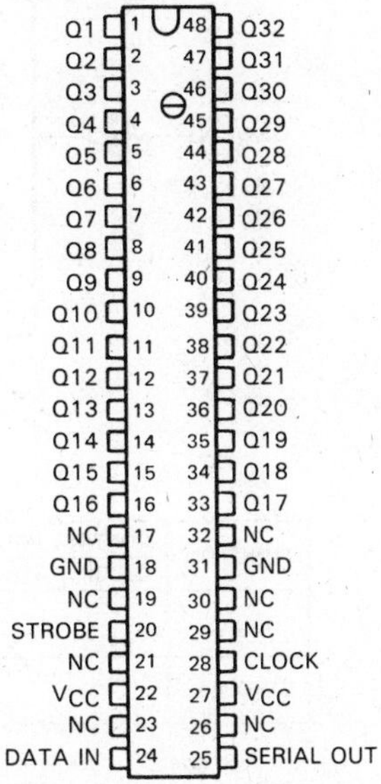
Q1 1 48 Q32
Q2 2 47 Q31
Q3 3 46 Q30
Q4 4 45 Q29
Q5 5 44 Q28
Q6 6 43 Q27
Q7 7 42 Q26
Q8 8 41 Q25
Q9 9 40 Q24
Q10 10 39 Q23
Q11 11 38 Q22
Q12 12 37 Q21
Q13 13 36 Q20
Q14 14 35 Q19
Q15 15 34 Q18
Q16 16 33 Q17
NC 17 32 NC
GND 18 31 GND
NC 19 30 NC
STROBE 20 29 NC
NC 21 28 CLOCK
VCC 22 27 VCC
NC 23 26 NC
DATA IN 24 25 SERIAL OUT

description

The SN751506 and the SN751516 are monolithic integrated circuits designed to drive the scan lines of a dc plasma panel display. The SN751516 pin sequence is reversed from the SN751506 for ease in printed circuit board layout.

Each device consists of a 32-bit shift register and 32 OR gates. Serial data is entered into the shift register on the high-to-low transition of the clock input. When the strobe input is low, all Q outputs are in the off-state. Outputs are open-drain JFET transistors with a breakdown voltage in excess of 180 volts. The outputs have a 220-milliampere sink current capability in the on state. Only one Q output should be allowed to be in the on state at a time.

Serial data output from the shift register may be used to cascade shift registers. This output is not affected by the strobe input. All inputs are CMOS compatible with ESD protection built in.

The SN751506 and SN751516 are characterized for operation from 0°C to 70°C.

TEXAS INSTRUMENTS
POST OFFICE BOX 655012 • DALLAS, TEXAS 75265

logic symbols†

SN751506

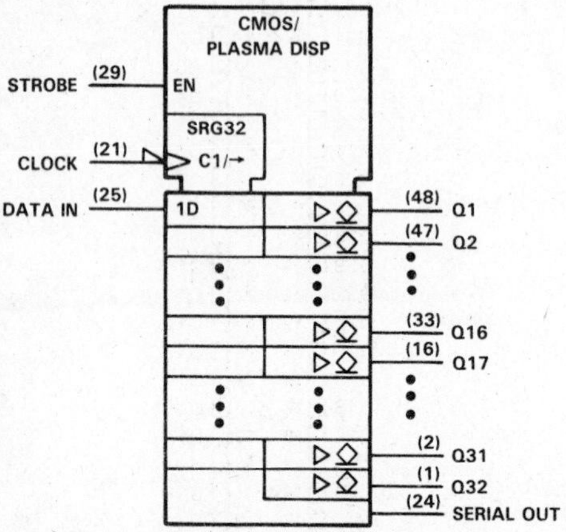
CMOS/
PLASMA DISP
STROBE (29) EN
SRG32
CLOCK (21) C1/→
DATA IN (25) 1D
(48) Q1
(47) Q2
(33) Q16
(16) Q17
(2) Q31
(1) Q32
(24) SERIAL OUT

SN751516

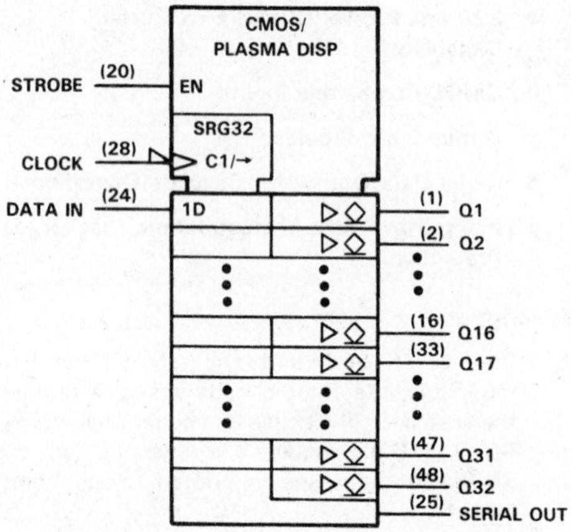
CMOS/
PLASMA DISP
STROBE (20) EN
SRG32
CLOCK (28) C1/→
DATA IN (24) 1D
(1) Q1
(2) Q2
(16) Q16
(33) Q17
(47) Q31
(48) Q32
(25) SERIAL OUT

† These symbols are in accordance with ANSI/IEEE Std 91-1984 and IEC Publication 617-12.

logic diagram (positive logic)

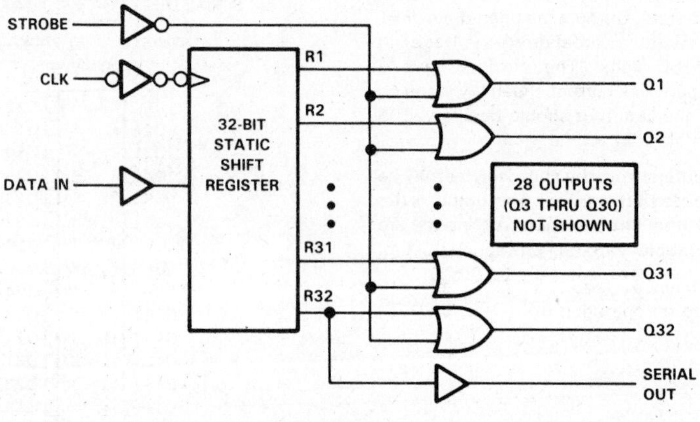
STROBE
CLK
DATA IN
32-BIT
STATIC
SHIFT
REGISTER
R1
R2
R31
R32
Q1
Q2
28 OUTPUTS
(Q3 THRU Q30)
NOT SHOWN
Q31
Q32
SERIAL
OUT

FUNCTION TABLE

FUNCTION	CONTROL INPUTS		SHIFT REGISTERS	OUTPUTS	
	CLOCK	STROBE	R1 THRU R32	SERIAL	Q1 THRU Q32
LOAD	↓	X	Load and shift‡	R32	Determined by STROBE
	No↓	X	No change	R32	
STROBE	X	L	As determined above	R32	All high impedance
	X	H		R32	R1 thru R32

H = high level, L = low level, X = irrelevant, ↓ = high to low transition.

‡ R32 takes on the state of R31, R31 takes on the state of R30, . . . R2 takes on the state of R1, and R1 takes on the state of the data input.

TEXAS INSTRUMENTS
POST OFFICE BOX 655012 • DALLAS, TEXAS 75265

typical operating sequence

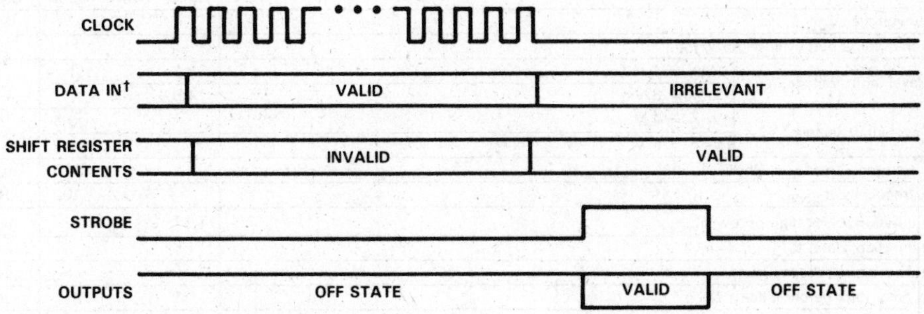
CLOCK
DATA IN†
VALID
IRRELEVANT
SHIFT REGISTER
CONTENTS
INVALID
VALID
STROBE
OUTPUTS
OFF STATE
VALID
OFF STATE

†Only 1 bit in 32 should be low in the input data.

schematics of inputs and outputs

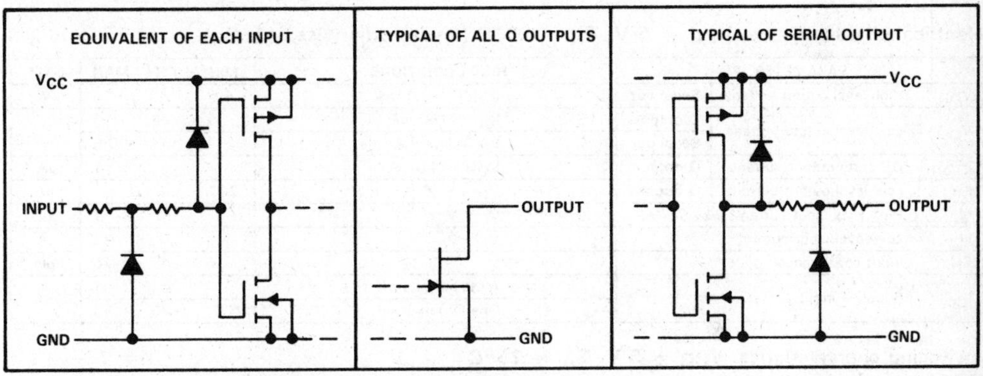
EQUIVALENT OF EACH INPUT
VCC
INPUT
GND
TYPICAL OF ALL Q OUTPUTS
OUTPUT
GND
TYPICAL OF SERIAL OUTPUT
VCC
OUTPUT
GND

absolute maximum ratings over operating free-air temperature range (unless otherwise noted)

Supply voltage, V_{CC} (see Note 1)	−0.4 V to 7 V
On-state Q output voltage, V_O	−0.4 V to 125 V
Off-state Q output voltage, V_O	−0.4 V to 180 V
Input voltage	−0.4 V to V_{CC}+0.4 V
Serial output voltage	−0.4 V to V_{CC}+0.4 V
Q output on-state time duration (see Note 2)	100 μs
Q output duty cycle (see Note 2)	1/200
Continuous total power dissipation at (or below) 25°C free-air temperature (see Note 3)	1025 mW
Operating free-air temperature range	0°C to 70°C
Storage temperature range	−55°C to 150°C
Lead temperature 1,6 mm (1/16 inch) from case for 10 seconds	260°C

NOTES: 1. Voltage values are with respect to GND.
2. Only one Q output should be on at a time.
3. For operation above 25°C free-air temperature, derate linearly to 656 mW at 70°C at the rate of 8.2 mW/°C.

3

Display Drivers

ADVANCE INFORMATION

recommended operating conditions

		MIN	NOM	MAX	UNIT
Supply voltage, V_{CC}		4	5	6	V
Peak on-state Q output voltage, $V_{O(on)}$				110	V
High-level input voltage, V_{IH}	V_{CC} = 4 V	3.2			V
	V_{CC} = 6 V	4.8			
Low-level input voltage, V_{IL}	V_{CC} = 4 V			0.8	V
	V_{CC} = 6 V			1.2	
Output current, I_O (T_A = 25°C)				220	mA
Clock frequency, f_{clock}				200	kHz
Pulse duration, clock high or low, t_{wCLK}		1.5†			μs
Pulse duration, data, t_{wD}		5			μs
Pulse duration, strobe, t_{wSTRB}		2			μs
Setup time, data before clock↓, t_{su}		1			μs
Hold time, data after clock↓, t_h		1.2			μs
Operating free-air temperature, T_A		0		70	°C

† The minimum clock period is 5 μs.

electrical characteristics, V_{CC} = 5 V, T_A = 25°C (unless otherwise noted)

PARAMETER			TEST CONDITIONS	MIN	TYP	MAX	UNIT
V_{OH}	High-level output voltage	Serial out	I_{OH} = −0.1 mA	4.5			V
V_{OL}	Low-level output voltage	Q outputs	I_{OL} = 180 mA		6	10	V
		Serial out	I_{OL} = 0.1 mA			0.5	
$I_{O(off)}$	Off-state output current	Q outputs	V_{OH} = 110 V			1	μA
I_{OL}	Low-level output current	Q outputs	V_{OL} = 16 V	220			mA
I_{IH}	High-level input current		$V_I = V_{CC}$			1	μA
I_{IL}	Low-level input current		V_I = 0			−1	μA
C_i	Input capacitance					15	pF
I_{CC}	Supply current		All Q outputs off			1	mA
			One Q output on		20	40	

switching characteristics, V_{CC} = 5 V, T_A = 25°C

PARAMETER		TEST CONDITIONS	MIN	TYP	MAX	UNIT
t_{pd}	Propagation delay time, clock to serial output	C_L = 15 pF		0.2	0.5	μs
t_{DHL}	Delay time, high-to-low-level Q output from strobe or clock inputs	C_L = 150 pF, R_L = 470 Ω, See Figures 2 and 3		0.2‡	0.6	μs
t_{DLH}	Delay time, low-to-high-level Q output from strobe or clock inputs			0.35‡	1	μs
t_{THL}	Transition time, high-to-low-level Q output			0.1	0.3	μs
t_{TLH}	Transition time, low-to-high-level Q output			0.35	1	μs

‡ Typical values are for clock inputs. Typical times from strobe inputs will be less.

TEXAS INSTRUMENTS
POST OFFICE BOX 655012 • DALLAS, TEXAS 75265

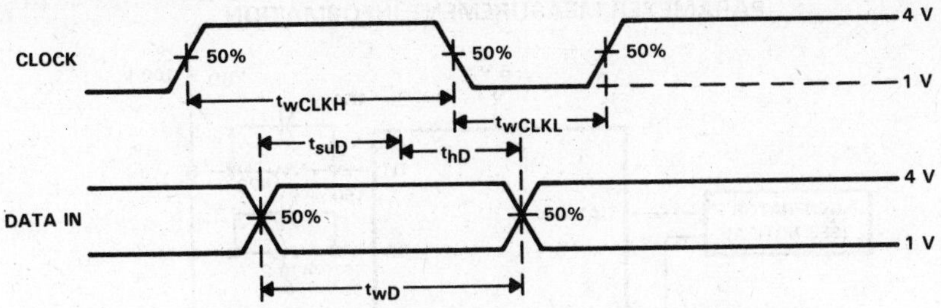
CLOCK
50%
50%
50%
4 V
1 V
t_{wCLKH}
t_{wCLKL}
t_{suD}
t_{hD}
DATA IN
50%
50%
4 V
1 V
t_{wD}

FIGURE 1. INPUT TIMING VOLTAGE WAVEFORMS

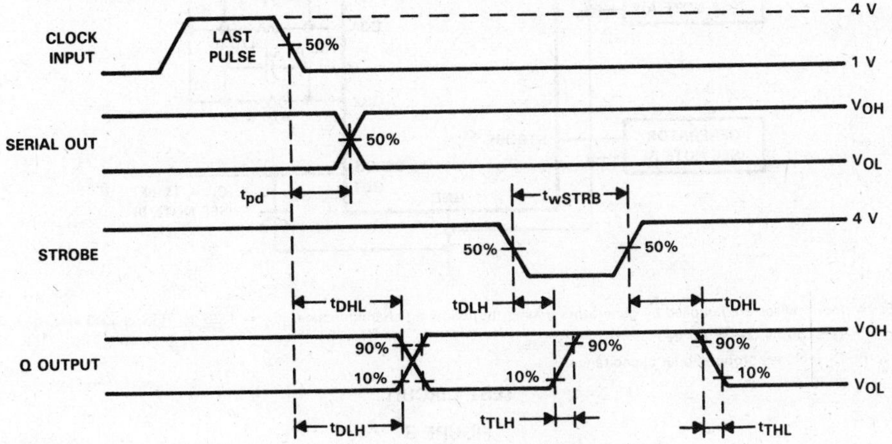
CLOCK INPUT
LAST PULSE
50%
4 V
1 V
SERIAL OUT
50%
V_{OH}
V_{OL}
t_{pd}
t_{wSTRB}
STROBE
50%
50%
4 V
t_{DHL}
t_{DLH}
t_{DHL}
Q OUTPUT
90%
10%
10%
90%
90%
10%
V_{OH}
V_{OL}
t_{DLH}
t_{TLH}
t_{THL}

FIGURE 2. SWITCHING CHARACTERISTICS

PARAMETER MEASUREMENT INFORMATION

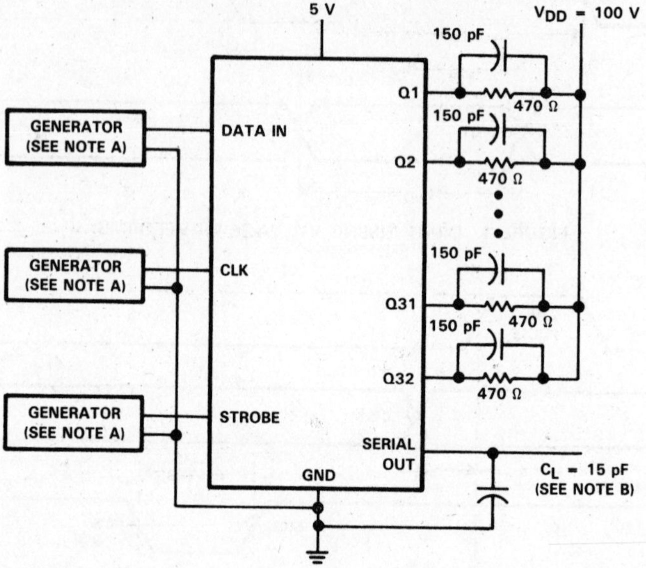
5 V
V_{DD} = 100 V
150 pF
Q1
470 Ω
150 pF
Q2
470 Ω
150 pF
Q31
470 Ω
150 pF
Q32
470 Ω
GENERATOR
(SEE NOTE A)
DATA IN
GENERATOR
(SEE NOTE A)
CLK
GENERATOR
(SEE NOTE A)
STROBE
SERIAL
OUT
GND
C_L = 15 pF
(SEE NOTE B)

NOTES: A. Input pulses are supplied by generators having the following characteristics: t_w = 1.25 μs, PRR ≤ 200 kHz, t_r ≤ 30 ns, t_f ≤ 30 ns, Z_o = 50 Ω.

B. C_L includes probe and jig capacitance.

TEST CIRCUIT

FIGURE 3

TYPICAL CHARACTERISTICS

LOW-LEVEL Q OUTPUT VOLTAGE
vs
FREE-AIR TEMPERATURE

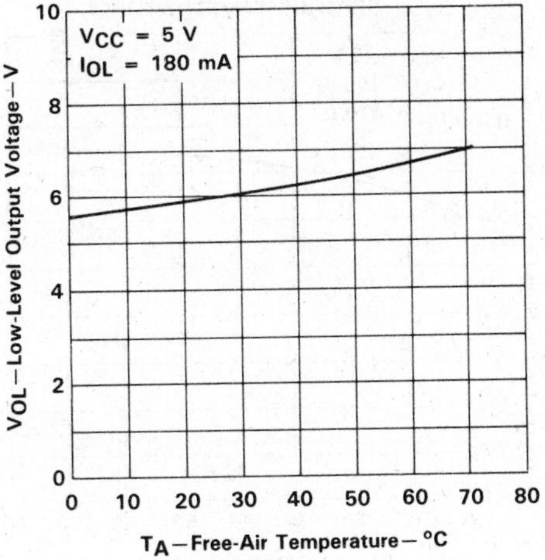
V_{CC} = 5 V
I_{OL} = 180 mA
V_{OL}—Low-Level Output Voltage—V
10
8
6
4
2
0
0 10 20 30 40 50 60 70 80
T_A—Free-Air Temperature—°C

FIGURE 4

LOW-LEVEL OUTPUT CURRENT Q OUTPUTS
vs
FREE-AIR TEMPERATURE

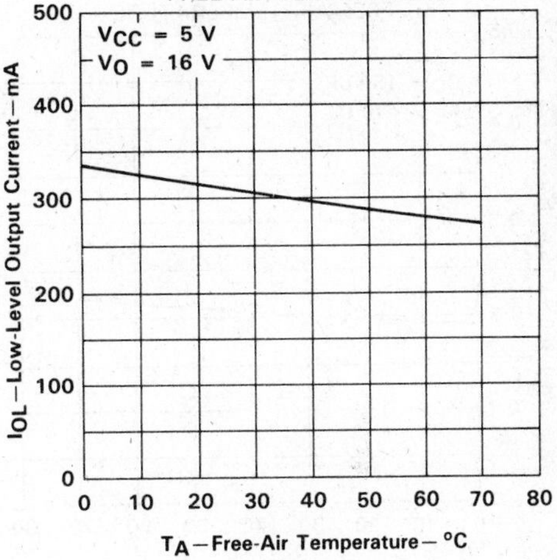
V_{CC} = 5 V
V_O = 16 V
I_{OL}—Low-Level Output Current—mA
500
400
300
200
100
0
0 10 20 30 40 50 60 70 80
T_A—Free-Air Temperature—°C

FIGURE 5

SUPPLY CURRENT
vs
FREE-AIR TEMPERATURE

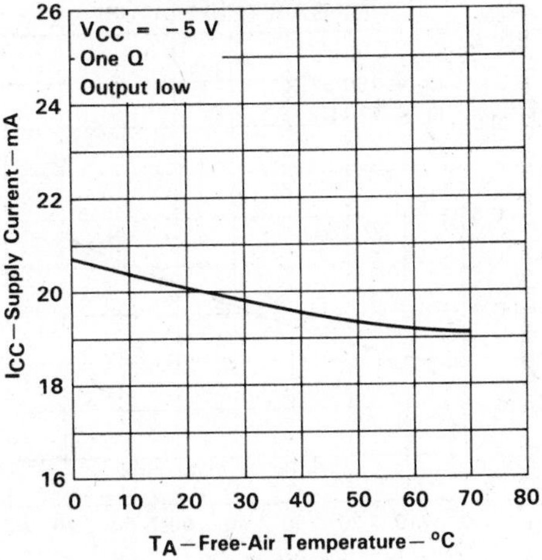
V_{CC} = −5 V
One Q
Output low
I_{CC}—Supply Current—mA
26
24
22
20
18
16
0 10 20 30 40 50 60 70 80
T_A—Free-Air Temperature—°C

FIGURE 6

PROPAGATION DELAY TIME,
CLOCK TO SERIAL OUTPUT
vs
FREE-AIR TEMPERATURE

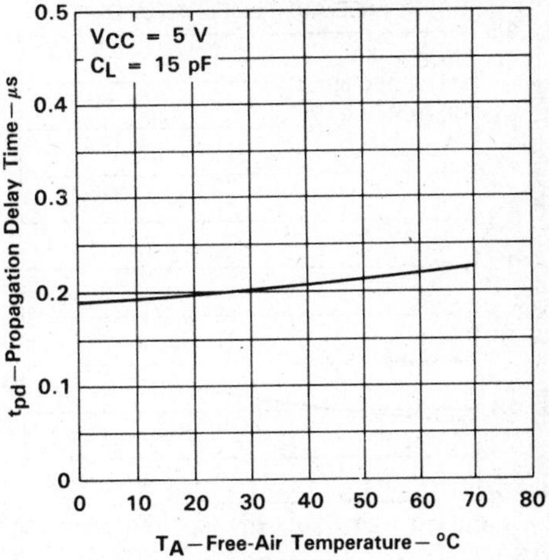
V_{CC} = 5 V
C_L = 15 pF
t_{pd}—Propagation Delay Time—μs
0.5
0.4
0.3
0.2
0.1
0
0 10 20 30 40 50 60 70 80
T_A—Free-Air Temperature—°C

FIGURE 7

3

Display Drivers

ADVANCE INFORMATION

TYPICAL CHARACTERISTICS

DELAY TIME,
HIGH-TO-LOW-LEVEL Q OUTPUT
vs
FREE-AIR TEMPERATURE

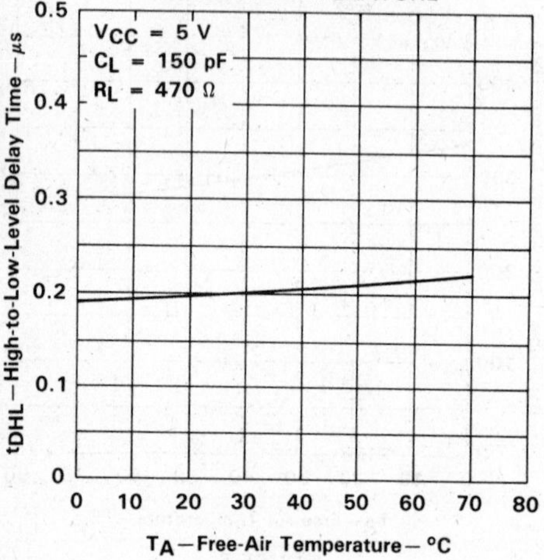
V_{CC} = 5 V
C_L = 150 pF
R_L = 470 Ω
t_{DHL}—High-to-Low-Level Delay Time—µs
0.5
0.4
0.3
0.2
0.1
0
0 10 20 30 40 50 60 70 80
T_A—Free-Air Temperature—°C

FIGURE 8

DELAY TIME,
LOW-TO-HIGH-LEVEL Q OUTPUT
vs
FREE-AIR TEMPERATURE

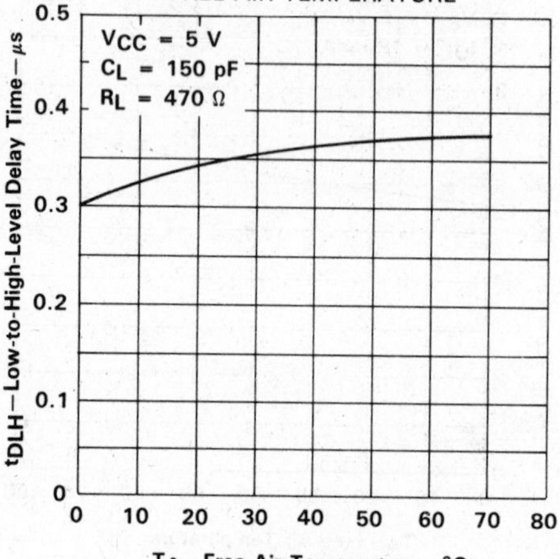
V_{CC} = 5 V
C_L = 150 pF
R_L = 470 Ω
t_{DLH}—Low-to-High-Level Delay Time—µs
0.5
0.4
0.3
0.2
0.1
0
0 10 20 30 40 50 60 70 80
T_A—Free-Air Temperature—°C

FIGURE 9

TRANSITION TIME,
HIGH-TO-LOW-LEVEL Q OUTPUT
vs
FREE-AIR TEMPERATURE

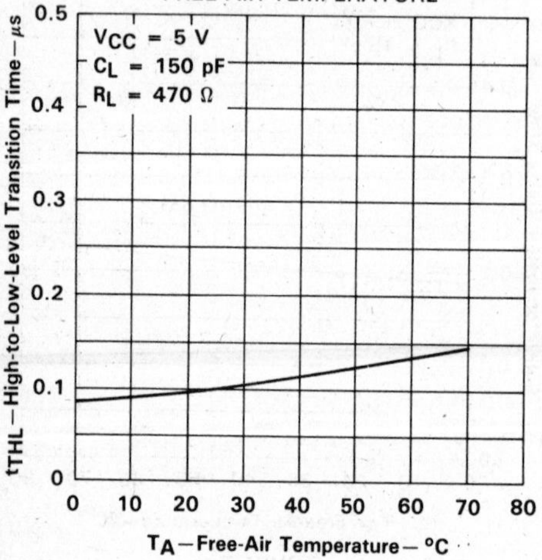
V_{CC} = 5 V
C_L = 150 pF
R_L = 470 Ω
t_{THL}—High-to-Low-Level Transition Time—µs
0.5
0.4
0.3
0.2
0.1
0
0 10 20 30 40 50 60 70 80
T_A—Free-Air Temperature—°C

FIGURE 10

TRANSITION TIME,
LOW-TO-HIGH-LEVEL Q OUTPUT
vs
FREE-AIR TEMPERATURE

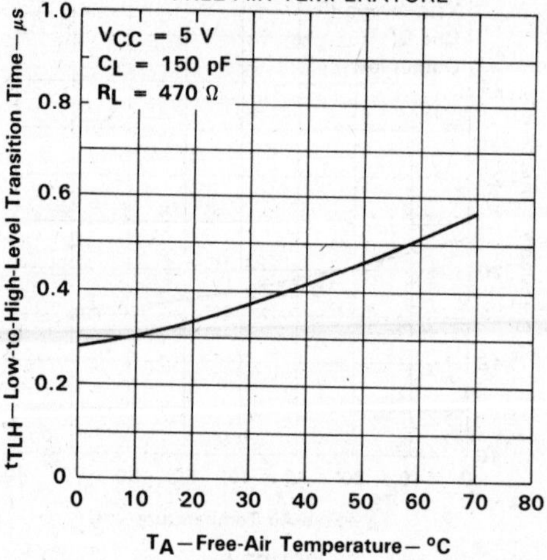
V_{CC} = 5 V
C_L = 150 pF
R_L = 470 Ω
t_{TLH}—Low-to-High-Level Transition Time—µs
1.0
0.8
0.6
0.4
0.2
0
0 10 20 30 40 50 60 70 80
T_A—Free-Air Temperature—°C

FIGURE 11

SN751508, SN751518 DC PLASMA DISPLAY DRIVERS

- Each Device Drives 32 Lines
- −120-V P-N-P Open-Collector Parallel Outputs
- High-Speed Serially Shifted Data Inputs
- CMOS-Compatible Inputs
- Strobe and Sustain Inputs Provided
- Serial Data Output for Cascade Operation

description

The SN751508 and SN751518 are monolithic integrated circuits designed to drive the data lines of a dc plasma panel display. The SN751518 pin sequence is reversed from the SN751508 for ease in printed circuit board layout.

Each device consists of two 16-bit shift registers, 32 latches, 32 OR gates, and 32 P-N-P open-collector output AND gates. Typically, a 32-bit data string is split into two 16-bit data strings externally and then entered in parallel into the shift registers on the high-to-low transition of the clock signal. A logic high signal on the Latch Enable input transfers the data from the shift registers to the inputs of 32 OR gates through the latches. Data present in the latch during the high-to-low transition of Latch Enable is stored. When the Strobe input is high, the latch is masked and a high will be placed on the data input of the output AND gates. When the Strobe input is low, and the Sustain input is high, data from the latches is reflected at the outputs. A logic low signal on the Sustain input will force all outputs to their off state. Drivers may be cascaded via the serial data outputs of the static shift registers. These outputs are not affected by the Latch Enable, Strobe, or Sustain inputs.

The SN751508 and the SN751518 are characterized from 0°C to 70°C.

SN751508 . . . FT PACKAGE
(TOP VIEW)

Q32	1	48	Q1
Q31	2	47	Q2
Q30	3	46	Q3
Q29	4	45	Q4
Q28	5	44	Q5
Q27	6	43	Q6
Q26	7	42	Q7
Q25	8	41	Q8
Q24	9	40	Q9
Q23	10	39	Q10
Q22	11	38	Q11
Q21	12	37	Q12
Q20	13	36	Q13
Q19	14	35	Q14
Q18	15	34	Q15
Q17	16	33	Q16
GND	17	32	GND
NC	18	31	SUSTAIN
STROBE	19	30	NC
NC	20	29	LATCH ENABLE
CLOCK	21	28	NC
V_{CC}	22	27	V_{CC}
SERIAL OUT2	23	26	DATA IN2
SERIAL OUT1	24	25	DATA IN1

SN751518 . . . FT PACKAGE
(TOP VIEW)

Q1	1	48	Q32
Q2	2	47	Q31
Q3	3	46	Q30
Q4	4	45	Q29
Q5	5	44	Q28
Q6	6	43	Q27
Q7	7	42	Q26
Q8	8	41	Q25
Q9	9	40	Q24
Q10	10	39	Q23
Q11	11	38	Q22
Q12	12	37	Q21
Q13	13	36	Q20
Q14	14	35	Q19
Q15	15	34	Q18
Q16	16	33	Q17
GND	17	32	GND
SUSTAIN	18	31	NC
NC	19	30	STROBE
LATCH ENABLE	20	29	NC
NC	21	28	CLOCK
V_{CC}	22	27	V_{CC}
DATA IN2	23	26	SERIAL OUT2
DATAIN1	24	25	SERIAL OUT1

NC—No internal connection.

logic symbols†

SN751508

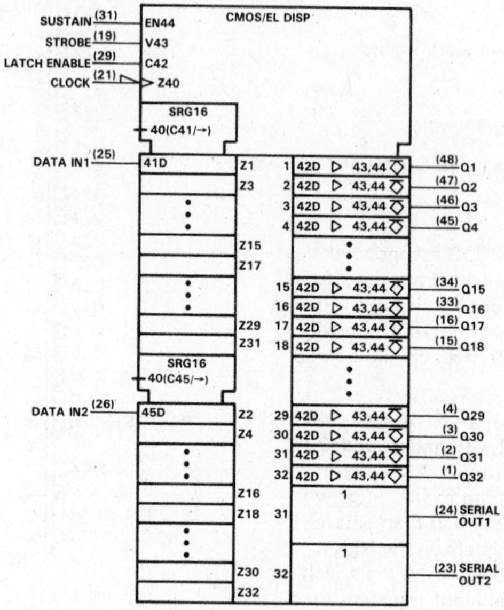
SUSTAIN (31) EN44
STROBE (19) V43
LATCH ENABLE (29) C42
CLOCK (21) Z40
CMOS/EL DISP
SRG16
40(C41/→)
DATA IN1 (25) 41D
SRG16
40(C45/→)
DATA IN2 (26) 45D
Q1 (48), Q2 (47), Q3 (46), Q4 (45), Q15 (34), Q16 (33), Q17 (16), Q18 (15), Q29 (4), Q30 (3), Q31 (2), Q32 (1)
42D ▷ 43,44
(24) SERIAL OUT1
(23) SERIAL OUT2

SN751518

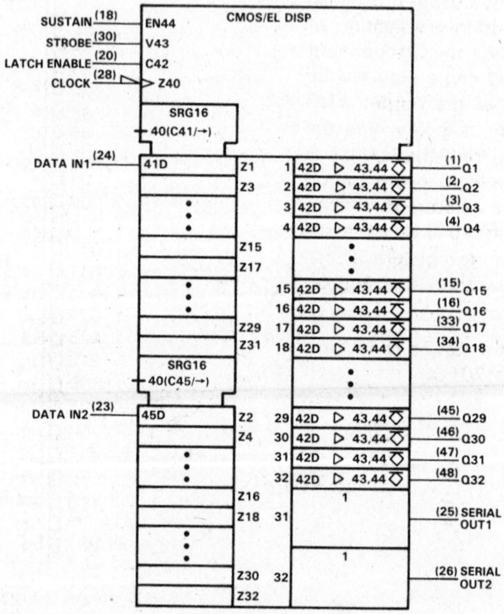
SUSTAIN (18) EN44
STROBE (30) V43
LATCH ENABLE (20) C42
CLOCK (28) Z40
CMOS/EL DISP
SRG16
40(C41/→)
DATA IN1 (24) 41D
SRG16
40(C45/→)
DATA IN2 (23) 45D
Q1 (1), Q2 (2), Q3 (3), Q4 (4), Q15 (15), Q16 (16), Q17 (33), Q18 (34), Q29 (45), Q30 (46), Q31 (47), Q32 (48)
42D ▷ 43,44
(25) SERIAL OUT1
(26) SERIAL OUT2

†These symbols are in accordance with ANSI/IEEE Std 91-1984 and IEC Publication 617-12.

logic diagram (positive logic)

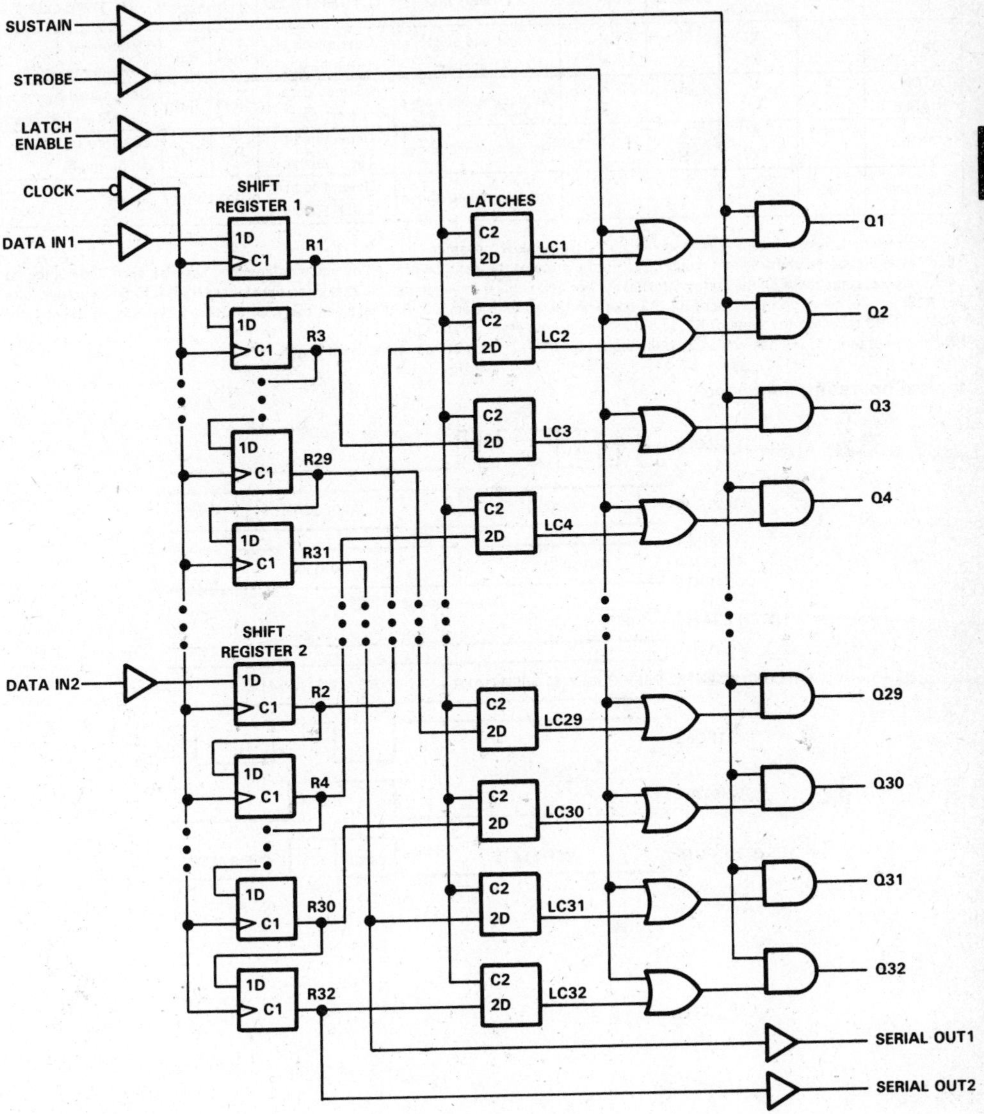
SUSTAIN
STROBE
LATCH
ENABLE
CLOCK
DATA IN1
SHIFT
REGISTER 1
LATCHES
1D
C1
R1
C2
2D
LC1
Q1
R3
LC2
Q2
LC3
Q3
R29
LC4
Q4
R31
SHIFT
REGISTER 2
DATA IN2
R2
LC29
Q29
R4
LC30
Q30
R30
LC31
Q31
R32
LC32
Q32
SERIAL OUT1
SERIAL OUT2

FUNCTION	CONTROL INPUTS				SHIFT REGISTERS R1 THRU R32	LATCHES LC1 THRU LC32	OUTPUTS		
	CLOCK	LATCH ENABLE	STROBE	SUSTAIN			SERIAL		Q1 THRU Q32
							SO1	SO2	
LOAD	↓ No ↓	X X	X X	X X	Load and shift† No change	Determined by Latch Enable‡	R31	R32	Determined by Sustain and Strobe
LATCH ENABLE	X X	L H	X X	X X	As determined above	Stored data New data	R31	R32	Determined by Sustain and Strobe
STROBE	X X	X X	L H	H H	As determined above	Determined by Latch Enable‡	R31	R32	LC1 thru LC32 All on (high)
SUSTAIN	X	X	X	L	As determined above	Determined by Latch Enable‡	R31	R32	All off

H = high level, L = low level, X = irrelevant, ↓ = high-to-low transition

† Each even-numbered shift register stage takes on the state of the next-lower even-numbered stage, and likewise each odd-numbered shift register stage takes on the state of the next-lower odd-numbered stage; i.e., R32 takes on the state of R30, R30 takes on the state of R28, . . . R4 takes on the state of R2, R2 takes on the state of Data In2, R31 takes on the state of R29, R29 takes on the state of R27, . . . R3 takes on the state of R1, and R1 takes on the state on Data In1.

‡ New data enters the latches while Latch Enable is high. This data is stored while Latch Enable is low.

typical operating sequence

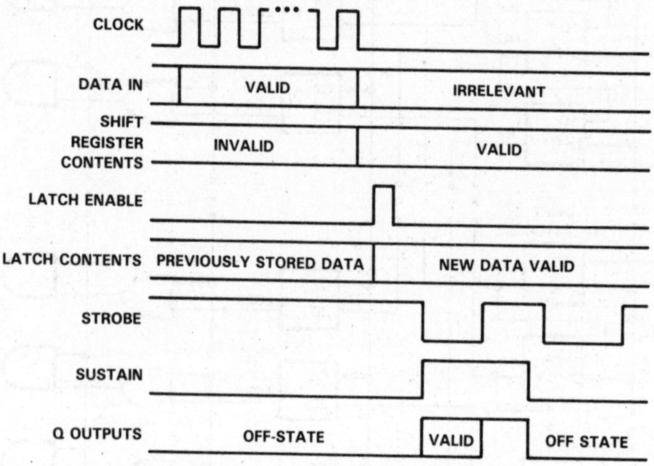
CLOCK
DATA IN
VALID
IRRELEVANT
SHIFT REGISTER CONTENTS
INVALID
VALID
LATCH ENABLE
LATCH CONTENTS
PREVIOUSLY STORED DATA
NEW DATA VALID
STROBE
SUSTAIN
Q OUTPUTS
OFF-STATE
VALID
OFF STATE

TEXAS INSTRUMENTS
POST OFFICE BOX 655012 • DALLAS, TEXAS 75265

schematics of inputs and outputs

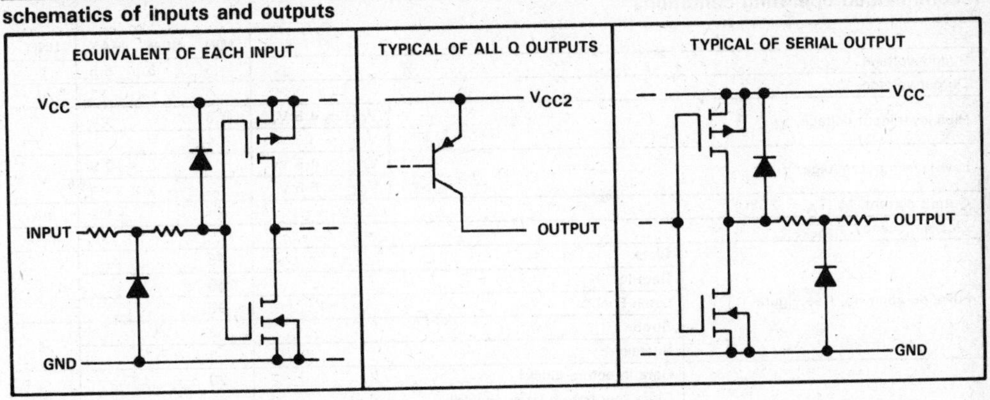
EQUIVALENT OF EACH INPUT
VCC
INPUT
GND
TYPICAL OF ALL Q OUTPUTS
VCC2
OUTPUT
TYPICAL OF SERIAL OUTPUT
VCC
OUTPUT
GND

absolute maximum ratings over operating free-air temperature (unless otherwise noted)

Supply voltage, V_{CC} (see Note 1)	−0.4 to 7 V
On-state Q output voltage, V_O	−120 V to V_{CC}+0.4 V
Input voltage	−0.4 V to V_{CC}+0.4 V
Serial output voltage	−0.4 V to V_{CC}+0.4 V
Continuous total power dissipation at (or below) 25°C free-air temperature (see Note 2)	1025 mW
Operating free-air temperature range	0°C to 70°C
Storage temperature range	−65°C to 150°C
Lead temperature 1,6 mm (1/16 inch) from case for 10 seconds	260°C

NOTES: 1. Voltages values are with respect to GND.
2. For operation above 25°C free-air temperature, derate linearly to 656 mW at 70°C at the rate of 8.2 mW/°C.

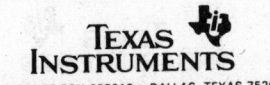

recommended operating conditions

		MIN	NOM	MAX	UNIT
Supply voltage, V_{CC}		4.5	5	5.5	V
Output voltage, V_O				−75	V
High-level input voltage, V_{IH}	V_{CC} = 4.5 V	3.6			V
	V_{CC} = 5.5 V	4.4			
Low-level input voltage, V_{IL}	V_{CC} = 4.5 V			0.9	V
	V_{CC} = 5.5 V			1	
Output current, I_O (T_A = 25°C)				−1.2	mA
Clock frequency, f_{clock}				5	MHz
Pulse duration, t_w (see Figure 1)	Clock	75			ns
	Data In	160			
	Latch Enable	90			
	Strobe	2			μs
	Sustain	2			
Setup time, t_{su} (see Figure 1)	Data In before clock↓	20			ns
	Clock low before latch enable↑	50			
	Latch-Enable low before clock↓	0			
	Latch-Enable high before strobe↓	0			
	Latch-Enable high before sustain↑	0			
Hold time, Data In after clock↓, t_h (see Figure 1)		50			μs
Operating free-air temperature, T_A		0		70	°C

electrical characteristics, V_{CC} = 5 V, T_A = 0°C to 70°C (unless otherwise noted)

PARAMETER		TEST CONDITIONS		MIN	TYP†	MAX	UNIT
V_{OH} High-level output voltage	Q out	I_{OH} = −0.5 mA		4	4.5		V
	Serial Out	V_{CC} = 5.5 V	I_{OH} = −100 μA	4.3	4.6		
			I_{OH} = −20 μA	4.4			
		V_{CC} = 4.5 V	I_{OH} = −100 μA	3.4	3.6		
			I_{OH} = −20 μA	3.6			
V_{OL} Low-level output voltage	Serial Out	V_{CC} = 5.5 V	I_{OL} = 100 μA		0.9	1.2	V
			I_{OL} = 20 μA			1.1	
		V_{CC} = 4.5 V	I_{OL} = 100 μA		0.9	1.1	
			I_{OL} = 20 μA			0.9	
I_{OH} High-level Q output current		T_A = 25°C,	V_O = 3 V	−1.2			mA
I_{OL} Low-level Q output current		T_A = 25°C,	V_O = −75 V			−500	μA
I_{IH} High-level input current		T_A = 25°C,	V_I = V_{CC}			1	μA
I_{IL} Low-level input current		T_A = 25°C,	V_I = 0			−1	μA
I_{CC} Supply current		All Q outputs high,	V_{CC} = 5.5 V		17	25	mA
		All Q outputs low				3	
C_i Input capacitance						15	pF

†All typical values are at T_A = 25°C.

switching characteristics V_{CC} = 5 V, T_A = 25°C (unless otherwise noted)

PARAMETER		TEST CONDITIONS	MIN	TYP	MAX	UNIT
t_{pd}	Propagation delay time, Clock to Serial Out	C_L = 15 pF		100	150	ns
t_{DLH}	Delay time, low-to-high-level Q output from Sustain or Strobe	C_L = 15 pF, R_L = 91 kΩ, See Figures 1 and 2		0.3‡	1	μs
t_{DHL}	Delay time, high-to-low-level Q output from Sustain or Strobe			1‡	2.5	μs
t_{TLH}	Transition time, low-to-high-level Q output			2	5	μs
t_{THL}	Transition time, high-to-low-level Q output			11	18	μs

‡Typical values for delay times are measured from the Sustain input.

PARAMETER MEASUREMENT INFORMATION

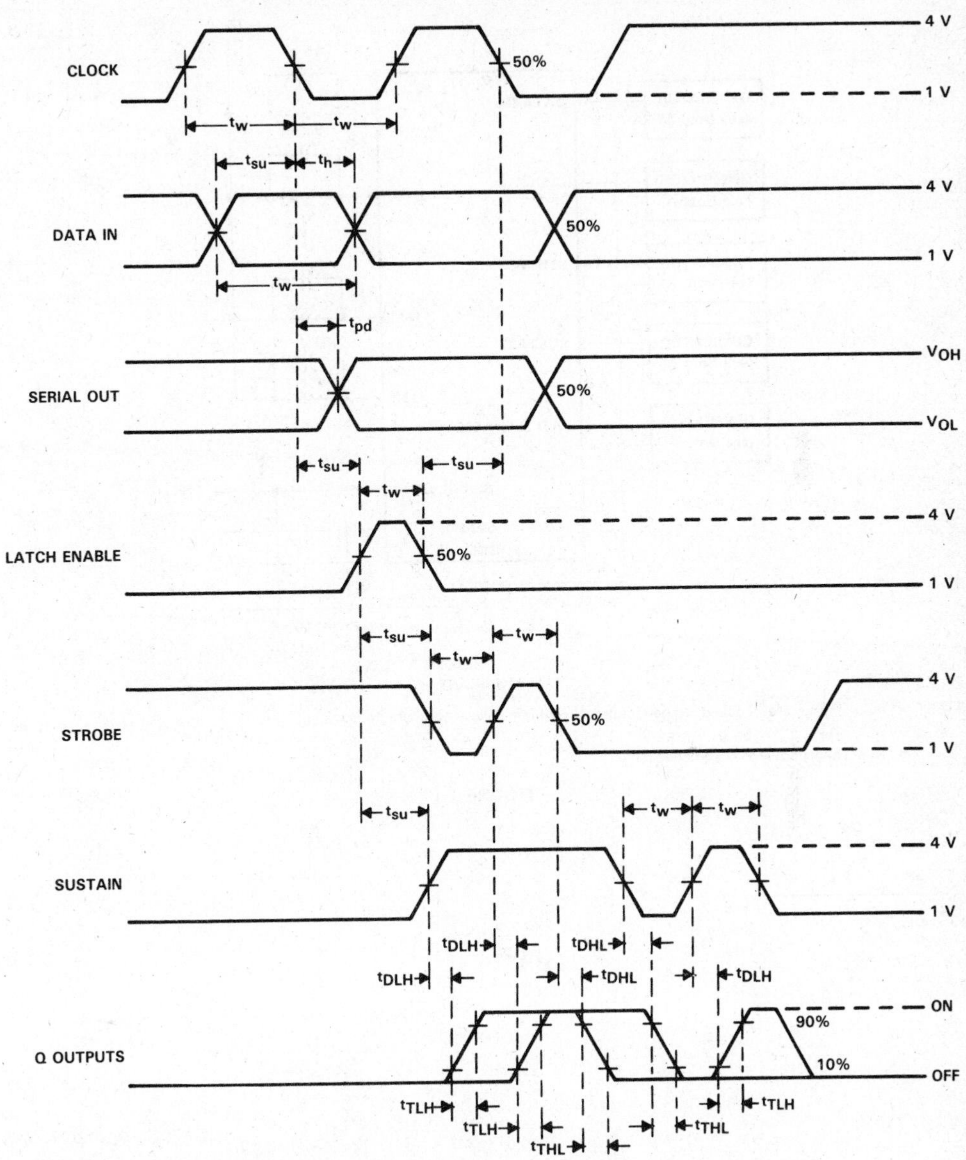
CLOCK
DATA IN
SERIAL OUT
LATCH ENABLE
STROBE
SUSTAIN
Q OUTPUTS
4 V
1 V
50%
VOH
VOL
ON
OFF
90%
10%
tw
tsu
th
tpd
tDLH
tDHL
tTLH
tTHL

NOTE: Input t_r and t_f are less than or equal to 10 ns.

FIGURE 1. INPUT TIMING AND SWITCHING TIME VOLTAGE WAVEFORMS

PARAMETER MEASUREMENT INFORMATION

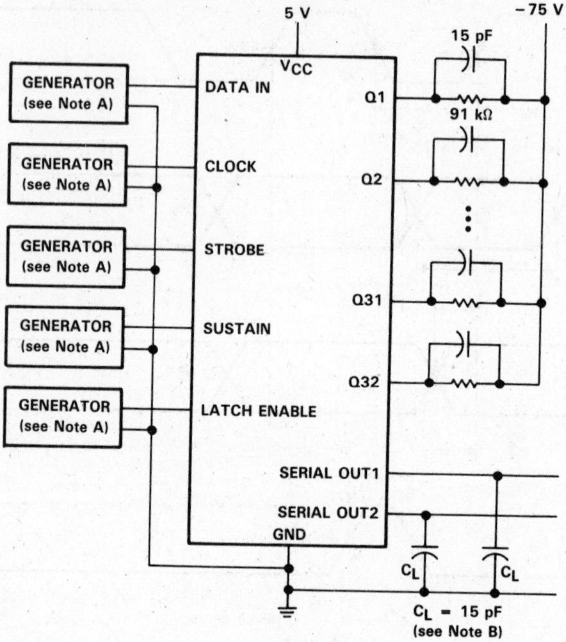
5 V
−75 V
15 pF
V_{CC}
GENERATOR (see Note A)
DATA IN
Q1
91 kΩ
GENERATOR (see Note A)
CLOCK
Q2
GENERATOR (see Note A)
STROBE
Q31
GENERATOR (see Note A)
SUSTAIN
Q32
GENERATOR (see Note A)
LATCH ENABLE
SERIAL OUT1
SERIAL OUT2
GND
C_L
C_L
C_L = 15 pF
(see Note B)

TEST CIRCUIT

NOTES: A. Input pulses are supplied by generators having the following characteristics: t_w = 100 ns, PRR ≤ 5 MHz, t_r ≤ 10 ns, t_f ≤ 10 ns, Z_o = 50 Ω.

B. C_L includes probe and jig capacitance.

FIGURE 2

TEXAS INSTRUMENTS
POST OFFICE BOX 655012 • DALLAS, TEXAS 75265

TYPICAL CHARACTERISTICS

SUPPLY CURRENT
vs
FREE-AIR TEMPERATURE

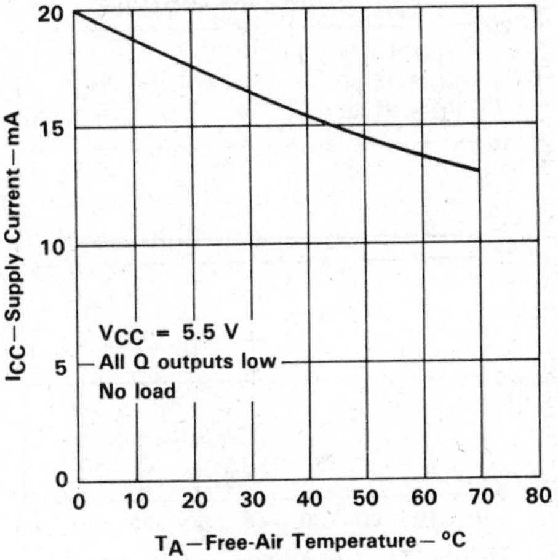
I_{CC}—Supply Current—mA
20
15
10
5
0
V_{CC} = 5.5 V
All Q outputs low
No load
0 10 20 30 40 50 60 70 80
T_A—Free-Air Temperature—°C

FIGURE 3

DELAY TIME, CLOCK TO SERIAL OUTPUT
vs
FREE-AIR TEMPERATURE

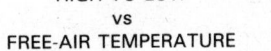

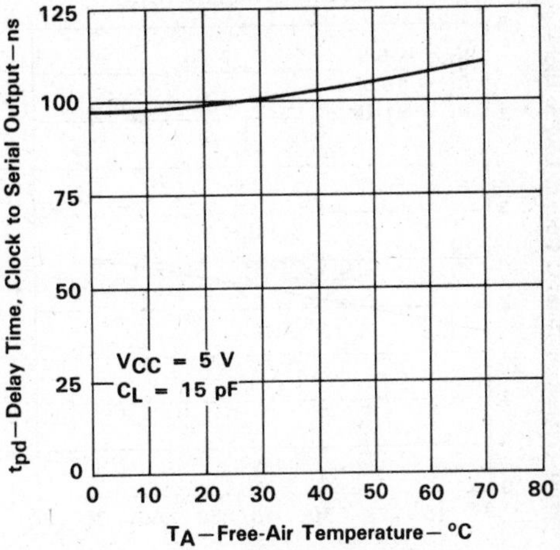
t_{pd}—Delay Time, Clock to Serial Output—ns
125
100
75
50
25
0
V_{CC} = 5 V
C_L = 15 pF
0 10 20 30 40 50 60 70 80
T_A—Free-Air Temperature—°C

FIGURE 4

DELAY TIME, SUSTAIN INPUT TO Q OUTPUT,
LOW TO HIGH
vs
FREE-AIR TEMPERATURE

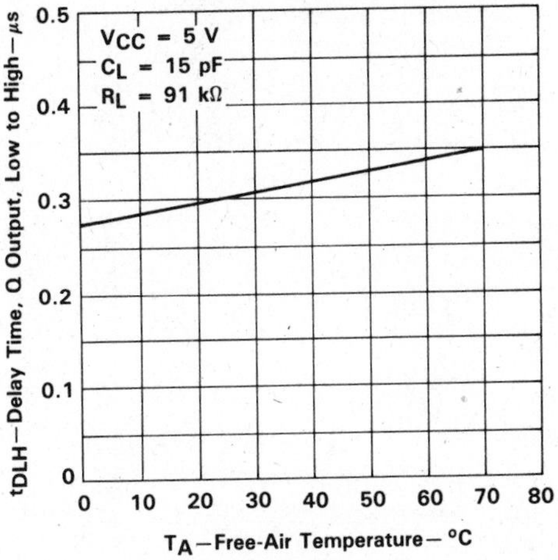
t_{DLH}—Delay Time, Q Output, Low to High—μs
0.5
0.4
0.3
0.2
0.1
0
V_{CC} = 5 V
C_L = 15 pF
R_L = 91 kΩ
0 10 20 30 40 50 60 70 80
T_A—Free-Air Temperature—°C

FIGURE 5

DELAY TIME, SUSTAIN INPUT TO Q OUTPUT,
HIGH TO LOW
vs
FREE-AIR TEMPERATURE

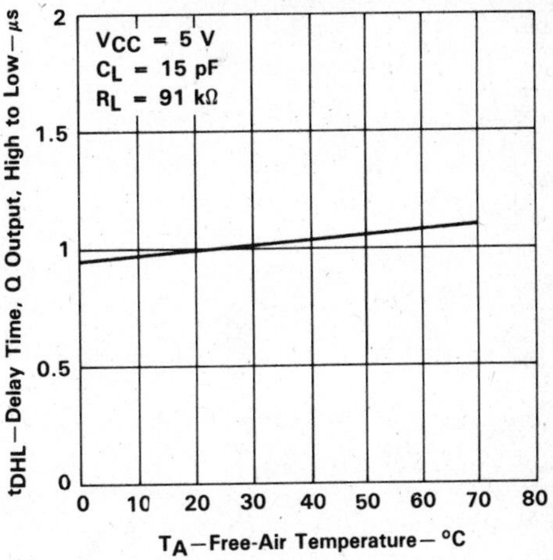
t_{DHL}—Delay Time, Q Output, High to Low—μs
2
1.5
1
0.5
0
V_{CC} = 5 V
C_L = 15 pF
R_L = 91 kΩ
0 10 20 30 40 50 60 70 80
T_A—Free-Air Temperature—°C

FIGURE 6

TYPICAL CHARACTERISTICS

TRANSITION TIME, Q OUTPUT,
LOW TO HIGH
vs
FREE-AIR TEMPERATURE

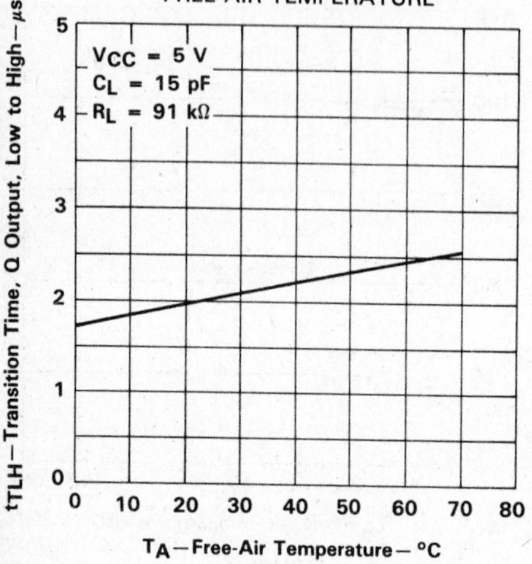
VCC = 5 V
CL = 15 pF
RL = 91 kΩ
tTLH—Transition Time, Q Output, Low to High—μs
TA—Free-Air Temperature—°C

FIGURE 7

TRANSITION TIME, Q OUTPUT,
HIGH TO LOW
vs
FREE-AIR TEMPERATURE

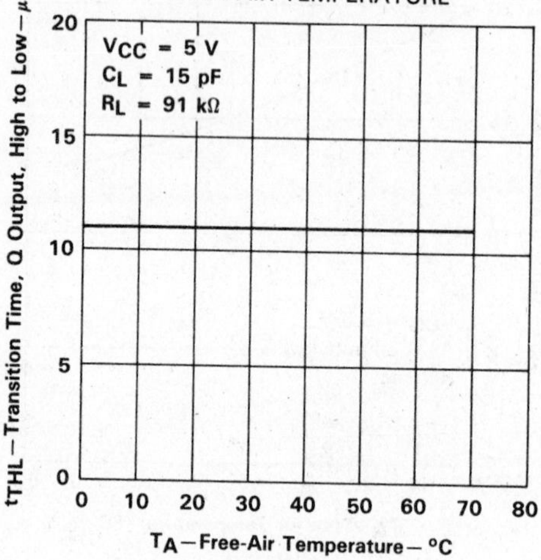
VCC = 5 V
CL = 15 pF
RL = 91 kΩ
tTHL—Transition Time, Q Output, High to Low—μs
TA—Free-Air Temperature—°C

FIGURE 8

D2715, DECEMBER 1984—REVISED FEBRUARY 1986

- **Each Device Drives 10 Lines**
- **60-V Output Voltage Rating**
- **40-mA Output Source Current**
- **High-Speed Serially-Shifted Data Input**
- **CMOS-Compatible Inputs**
- **Latches on All Driver Outputs**
- **Improved Direct Replacement for UCN4810A and TL4810A**

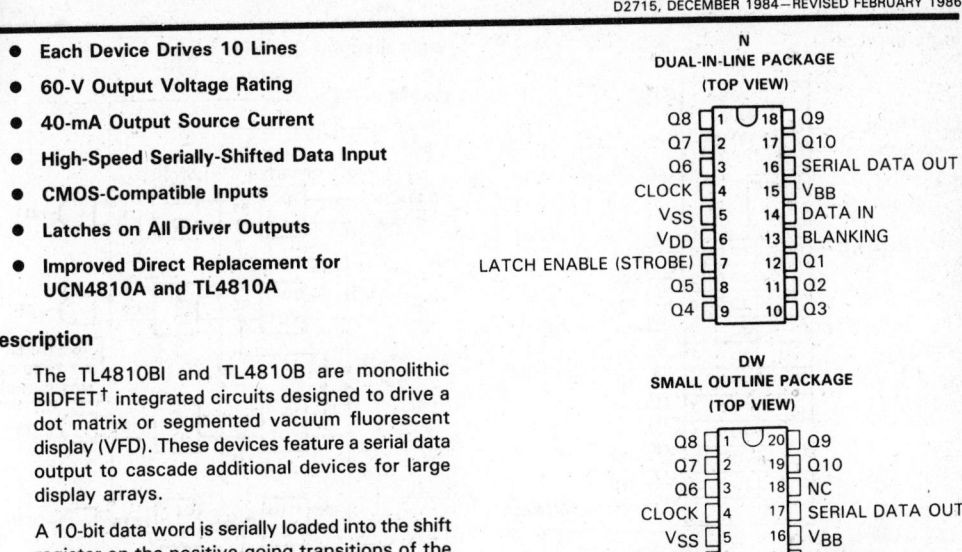

N
DUAL-IN-LINE PACKAGE
(TOP VIEW)

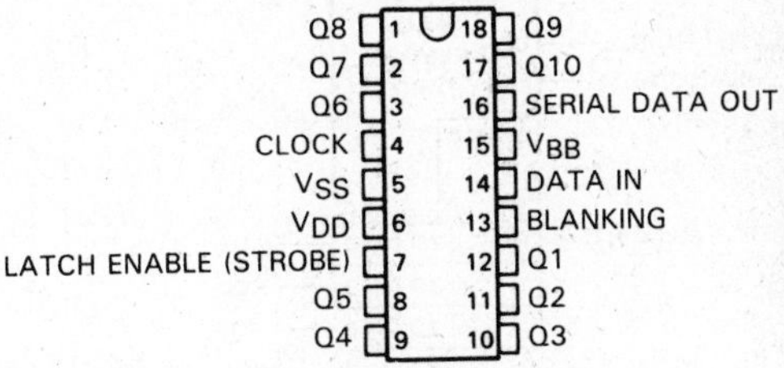
Q8 1
Q7 2
Q6 3
CLOCK 4
VSS 5
VDD 6
LATCH ENABLE (STROBE) 7
Q5 8
Q4 9
18 Q9
17 Q10
16 SERIAL DATA OUT
15 VBB
14 DATA IN
13 BLANKING
12 Q1
11 Q2
10 Q3

DW
SMALL OUTLINE PACKAGE
(TOP VIEW)

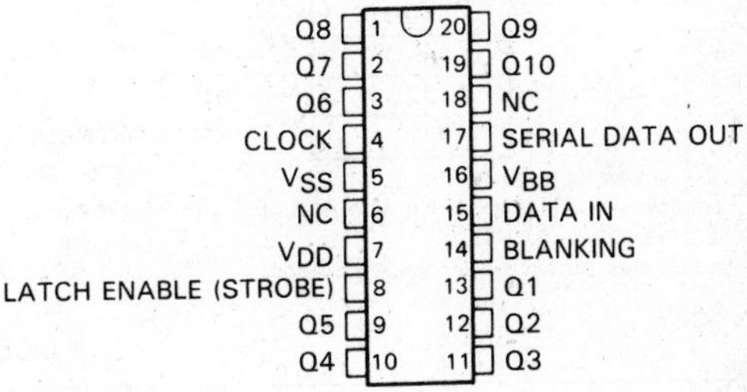
Q8 1
Q7 2
Q6 3
CLOCK 4
VSS 5
NC 6
VDD 7
LATCH ENABLE (STROBE) 8
Q5 9
Q4 10
20 Q9
19 Q10
18 NC
17 SERIAL DATA OUT
16 VBB
15 DATA IN
14 BLANKING
13 Q1
12 Q2
11 Q3

NC—No internal connection

description

The TL4810BI and TL4810B are monolithic BIDFET† integrated circuits designed to drive a dot matrix or segmented vacuum fluorescent display (VFD). These devices feature a serial data output to cascade additional devices for large display arrays.

A 10-bit data word is serially loaded into the shift register on the positive-going transitions of the clock. Parallel data is transferred to the output buffers through a 10-bit D-type latch while the latch enable input is high and is latched when the latch enable is low. When the blanking input is high, all outputs are low.

Outputs are totem-pole structures formed by n-p-n emitter-follower and double-diffused MOS (DMOS) transistors with output voltage ratings of 70 volts and 40 milliamperes source-current capability. All inputs are compatible with CMOS and TTL levels, but each requires the addition of a pull-up resistor to V_{DD} when driven by TTL logic.

The TL4810BI is characterized for operation from −40°C to 85°C. The TL4810B is characterized for operation from 0°C to 70°C.

3
Display Drivers

† BIDFET—Bipolar, Double-Diffused, N-Channel and P-Channel MOS transistors on same chip—patented process.

logic symbol†

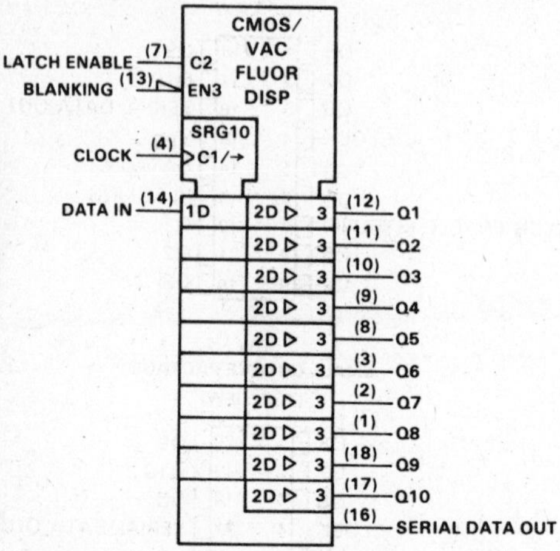
CMOS/
VAC
FLUOR
DISP
LATCH ENABLE (7) C2
BLANKING (13) EN3
CLOCK (4) C1/→
SRG10
DATA IN (14) 1D
2D ▷ 3 (12) Q1
2D ▷ 3 (11) Q2
2D ▷ 3 (10) Q3
2D ▷ 3 (9) Q4
2D ▷ 3 (8) Q5
2D ▷ 3 (3) Q6
2D ▷ 3 (2) Q7
2D ▷ 3 (1) Q8
2D ▷ 3 (18) Q9
2D ▷ 3 (17) Q10
(16) SERIAL DATA OUT

† This symbol is in accordance with ANSI/IEEE Std 91-1984 and IEC Publication 617-12.
Pin numbers shown are for the N package.

logic diagram (positive logic)

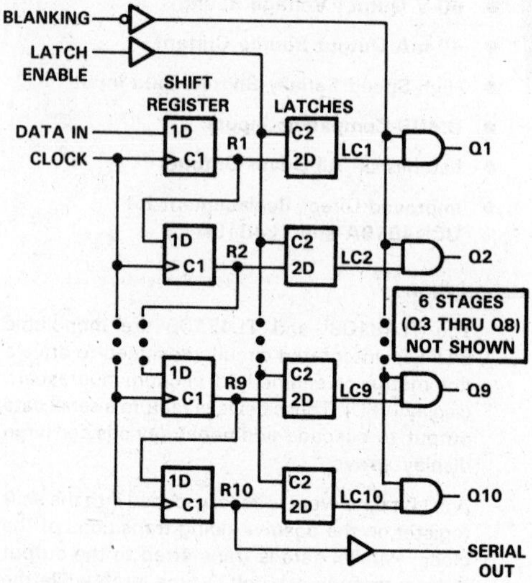
BLANKING
LATCH ENABLE
SHIFT REGISTER
LATCHES
DATA IN
CLOCK
1D
C1
R1
R2
R9
R10
C2
2D
LC1
LC2
LC9
LC10
Q1
Q2
Q9
Q10
6 STAGES (Q3 THRU Q8) NOT SHOWN
SERIAL OUT

FUNCTION TABLE

FUNCTION	CONTROL INPUTS			SHIFT REGISTERS R1 THRU R10‡	LATCHES LC1 THRU LC10	OUTPUTS	
	CLOCK	LATCH ENABLE	BLANKING			SERIAL	Q1 THRU Q10
LOAD	↑	X	X	Load and shift‡	Determined by Latch Enable§	R10	Determined by Blanking
	Not↑	X	X	No change	Determined by Latch Enable§	R10	Determined by Blanking
LATCH	X	L	X	As determined above	Stored data	R10	Determined by Blanking
	X	H	X	As determined above	New data	R10	Determined by Blanking
BLANK	X	X	H	As determined above	Determined by Latch Enable§	R10	All L
	X	X	L	As determined above	Determined by Latch Enable§	R10	LC1 thru LC10 respectively

H = high level, L = low level, X = irrelevant, ↑ = low-to-high-level transition.
‡ Register R10 takes on the state of R9, R9 takes on the state of R8 . . . R2 takes on the state of R1, and R1 takes on the state of the data input.
§ New data enter the latches while Latch Enable is high. These data are stored while Latch Enable is low.

TEXAS INSTRUMENTS
POST OFFICE BOX 655012 • DALLAS, TEXAS 75265

typical operating sequence

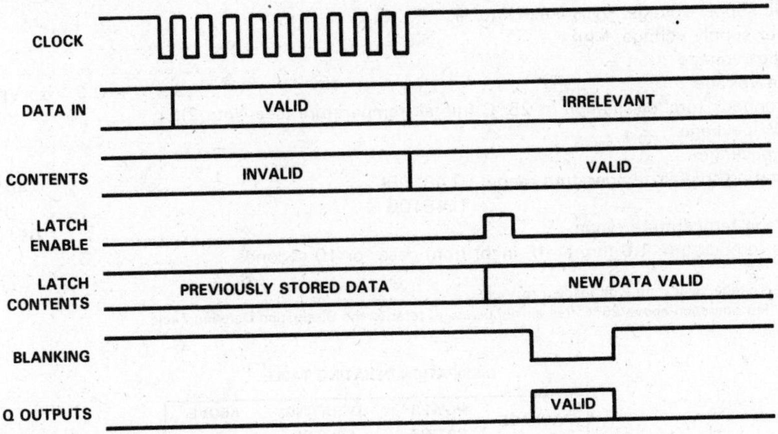
CLOCK
DATA IN
VALID
IRRELEVANT
SR CONTENTS
INVALID
VALID
LATCH ENABLE
LATCH CONTENTS
PREVIOUSLY STORED DATA
NEW DATA VALID
BLANKING
Q OUTPUTS
VALID

schematics of inputs and outputs

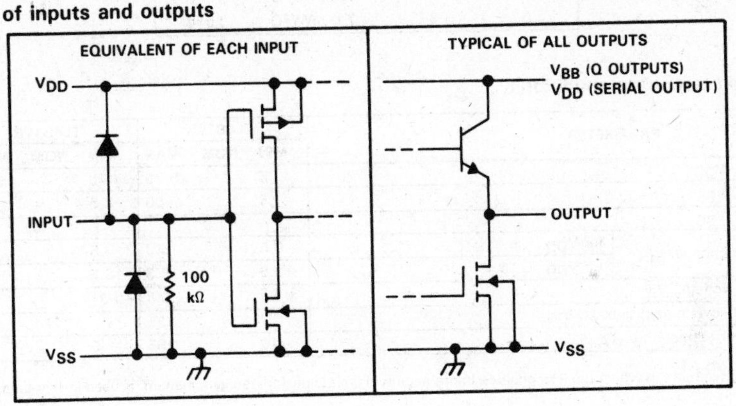
EQUIVALENT OF EACH INPUT
V_{DD}
INPUT
100 kΩ
V_{SS}
TYPICAL OF ALL OUTPUTS
V_{BB} (Q OUTPUTS)
V_{DD} (SERIAL OUTPUT)
OUTPUT
V_{SS}

absolute maximum ratings over operating free-air temperature range (unless otherwise noted)

Logic supply voltage, V_{DD} (see Note 1) .... 18 V
Driver supply voltage, V_{BB} .... 70 V
Output voltage .... 70 V
Input voltage .... −0.3 V to V_{DD} + 0.3 V
Continuous total dissipation at 25°C free air-temperature (see Note 2):
DW package .... 1150 mW
N package .... 875 mW
Operating free-air temperature range: TL4810BI .... −40°C to 85°C
TL4810B .... 0°C to 70°C
Storage temperature range .... −65°C to 150°C
Lead temperature 1,6 mm (1/16 inch) from case for 10 seconds .... 260°C

NOTES: 1. Voltage values are with respect to V_{SS}.
2. For operation above 25°C free-air temperature, refer to the Dissipation Derating Table.

DISSIPATION DERATING TABLE

PACKAGE	POWER RATING	DERATING FACTOR	ABOVE T_A
DW	1150	9.2 mW/°C	25°C
N	875	7.0 mW/°C	25°C

recommended operating conditions

PARAMETER		TL4810BI			TL4810B			UNIT
		MIN	NOM	MAX	MIN	NOM	MAX	
Supply voltage, V_{DD}		4.75		15.75	4.75		15.75	V
Supply voltage, V_{BB}		5		60	5		60	V
Supply voltage, V_{SS}			0			0		V
High-level input voltage, V_{IH}	for V_{DD} = 5 V	3.5		5.3	3.5		5.3	V
	for V_{DD} = 15 V	13.5		15.3	13.5		15.3	
Low-level input voltage, V_{IL}		−0.3†		0.8	−0.3†		0.8	V
Continuous high-level output current, I_{OH}				−25			−25	mA
Operating free-air temperature, T_A		−40		85	0		70	°C

† The algebraic convention, in which the less positive (more negative) limit is designated as minimum, is used in this data sheet for logic voltages only.

electrical characteristics over recommended operating free-air temperature range, V_{DD} = 5 V to 15 V, V_{BB} = 60 V, V_{SS} = 0 (unless otherwise noted)

PARAMETER			TEST CONDITIONS†		TL4810BI MIN	TL4810BI TYP‡	TL4810BI MAX	TL4810B MIN	TL4810B TYP‡	TL4810B MAX	UNIT
V_{OH}	High-level output voltage	Q outputs	I_{OH} = −25 mA		57.5	58		57.5	58		V
		Serial output	V_{DD} = 5 V,	I_{OH} = −100 μA	4	4.5		4	4.5		
			V_{DD} = 15 V,	I_{OH} = −100 μA	14	14.7		14	14.7		
V_{OL}	Low-level output voltage	Q outputs	I_{OH} = 1 μA,	Blanking input at V_{DD}		0.5	1		0.5	1	V
		Serial output	V_{DD} = 5 V,	I_{OL} = 100 μA		0.05	0.1		0.05	0.1	
			V_{DD} = 15 V,	I_{OL} = 100 μA		0.02	0.1		0.02	0.1	
I_{OL}	Low-level Q output current (pull-down current)		V_O = 60 V, T_A = MIN to 70°C	Blanking input at V_{DD},	2.5	3.7		2.5	3.7		mA
			V_O = 60 V, T_A = 85°C	Blanking input at V_{DD},	2						
$I_{O(off)}$	Off-state output current		V_O = 0, T_A = MAX	Blanking input at V_{DD},		−1	−15		−1	−15	μA
I_H	High-level input current		V_I = V_{DD}			30	50		30	50	μA
I_{BB}	Supply current from V_{BB}		All outputs low			0.5	1		0.5	1	mA
			All outputs high, T_A = 0°C to MAX			2.7	4		2.7	4	
			All outputs high, T_A = −40°C				5				
I_{DD}	Supply current from V_{DD}		All inputs at 0 V, One Q output high	V_{DD} = 5 V		10	50		10	50	μA
				V_{DD} = 15 V		10	100		10	100	
			All inputs at 0 V, All outputs low	V_{DD} = 5 V		10	50		10	50	
				V_{DD} = 15 V		10	100		10	100	

† For conditions shown as MIN or MAX, use the appropriate value specified under recommended operating conditions.
‡ All typical values are at T_A = 25°C, except for I_O.

timing requirements over recommended operating free-air temperature range

PARAMETER		V_{DD} = 5 V MIN	V_{DD} = 5 V MAX	V_{DD} = 15 V MIN	V_{DD} = 15 V MAX	UNIT
$t_{w(CKH)}$	Pulse duration, clock high	250		50		ns
$t_{w(LEH)}$	Pulse duration, latch enable high	250		50		ns
$t_{su(D)}$	Setup time, data before clock↑	125		25		ns
$t_{h(D)}$	Hold time, data after clock↑	125		25		ns
$t_{CKH-LEH}$	Delay time, clock ↑ to latch enable high	125		25		ns

switching characteristics, V_{BB} = 60 V, T_A = 25°C

PARAMETER			MIN	TYP	MAX	UNIT
t_{pd}	Propagation delay time, latch enable to output	V_{DD} = 5 V		1		μs
		V_{DD} = 15 V		0.5		

PARAMETER MEASUREMENT INFORMATION

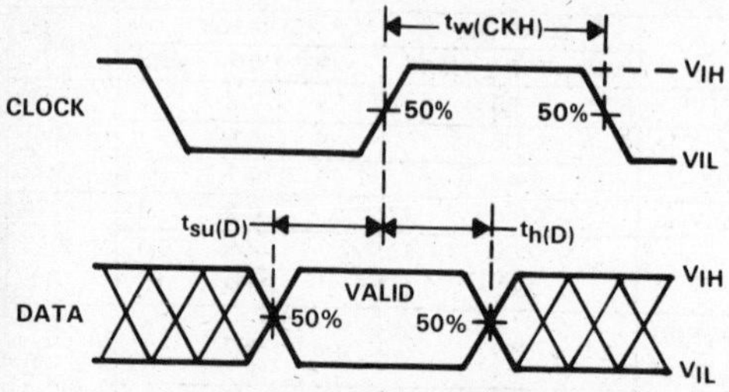
$t_{w(CKH)}$
CLOCK
50%
50%
V_{IH}
V_{IL}
$t_{su(D)}$
$t_{h(D)}$
DATA
VALID
50%
50%
V_{IH}
V_{IL}

FIGURE 1. INPUT TIMING

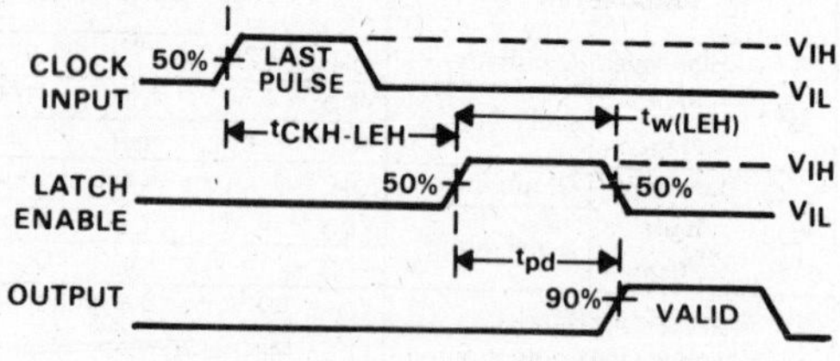
CLOCK INPUT
50%
LAST PULSE
V_{IH}
V_{IL}
$t_{CKH-LEH}$
$t_{w(LEH)}$
LATCH ENABLE
50%
50%
V_{IH}
V_{IL}
t_{pd}
OUTPUT
90%
VALID

FIGURE 2. OUTPUT SWITCHING TIMES

THERMAL INFORMATION

DW PACKAGE DUTY CYCLE
vs
FREE-AIR TEMPERATURE

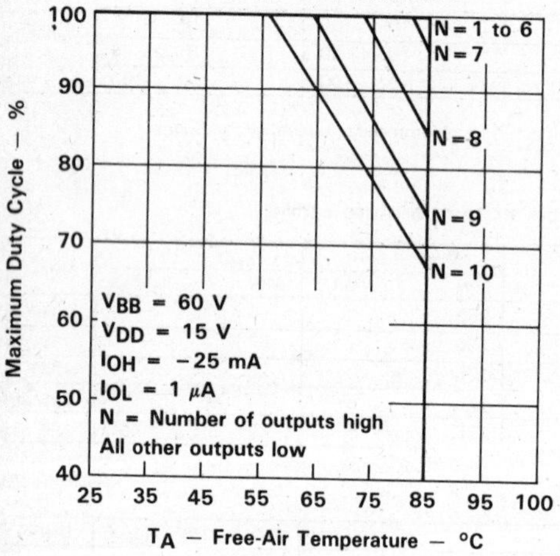
Maximum Duty Cycle — %
100
90
80
70
60
50
40
25 35 45 55 65 75 85 95 100
N = 1 to 6
N = 7
N = 8
N = 9
N = 10
V_{BB} = 60 V
V_{DD} = 15 V
I_{OH} = −25 mA
I_{OL} = 1 μA
N = Number of outputs high
All other outputs low
T_A — Free-Air Temperature — °C

FIGURE 3

N PACKAGE DUTY CYCLE
vs
FREE-AIR TEMPERATURE

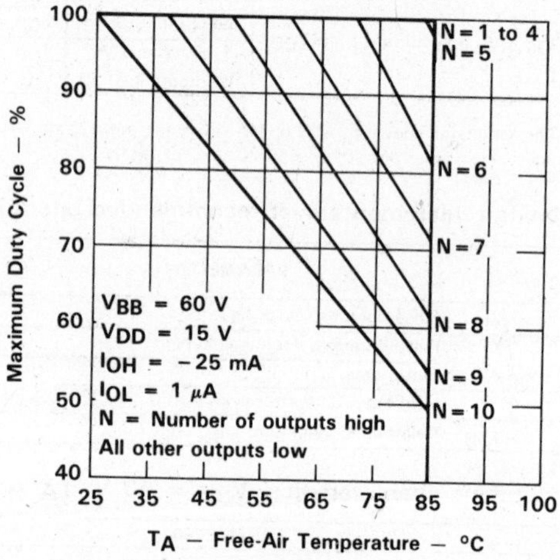
Maximum Duty Cycle — %
100
90
80
70
60
50
40
25 35 45 55 65 75 85 95 100
N = 1 to 4
N = 5
N = 6
N = 7
N = 8
N = 9
N = 10
V_{BB} = 60 V
V_{DD} = 15 V
I_{OH} = −25 mA
I_{OL} = 1 μA
N = Number of outputs high
All other outputs low
T_A — Free-Air Temperature — °C

FIGURE 4

D2914, OCTOBER 1985—REVISED AUGUST 1986

- Drives Up to 20 Lines
- 70-V Output Voltage Swing Capability
- 40-mA Output Source Current Capability
- High-Speed Serially-Shifted Data Input
- CMOS-Compatible Inputs
- Direct Replacement for Sprague UCN5812A

N
DUAL-IN-LINE PACKAGE
(TOP VIEW)

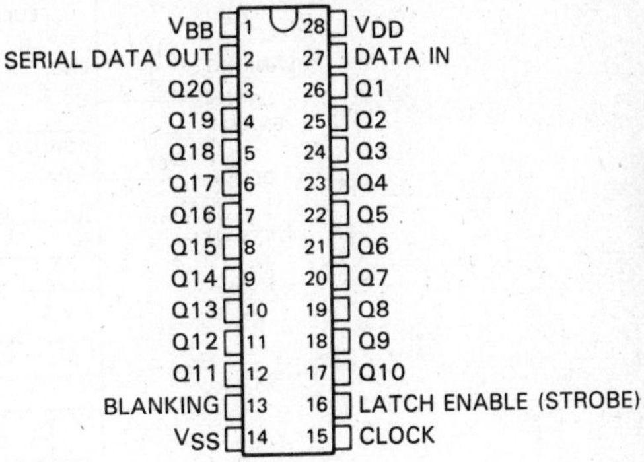
V_{BB} 1
SERIAL DATA OUT 2
Q20 3
Q19 4
Q18 5
Q17 6
Q16 7
Q15 8
Q14 9
Q13 10
Q12 11
Q11 12
BLANKING 13
V_{SS} 14
28 V_{DD}
27 DATA IN
26 Q1
25 Q2
24 Q3
23 Q4
22 Q5
21 Q6
20 Q7
19 Q8
18 Q9
17 Q10
16 LATCH ENABLE (STROBE)
15 CLOCK

FN PACKAGE
(TOP VIEW)

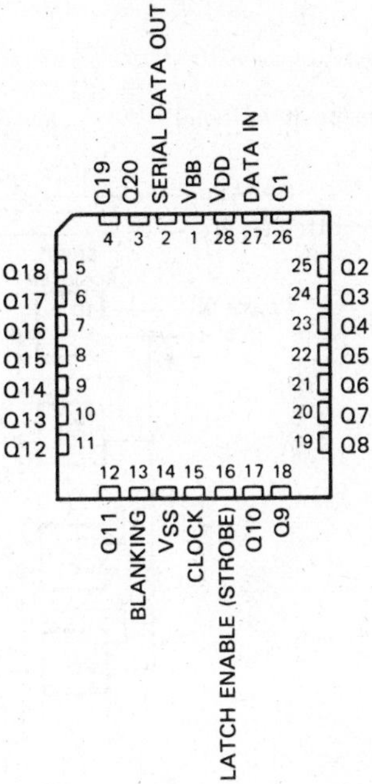
Q19
Q20
SERIAL DATA OUT
V_{BB}
V_{DD}
DATA IN
Q1
4 3 2 1 28 27 26
Q18 5
Q17 6
Q16 7
Q15 8
Q14 9
Q13 10
Q12 11
25 Q2
24 Q3
23 Q4
22 Q5
21 Q6
20 Q7
19 Q8
12 13 14 15 16 17 18
Q11
BLANKING
V_{SS}
CLOCK
LATCH ENABLE (STROBE)
Q10
Q9

description

The TL5812I and TL5812 are monolithic BIDFET† integrated circuits designed to drive a dot matrix or segmented vacuum fluorescent display (VFD). Each device features a serial data output to cascade additional devices for large display arrays.

A 20-bit data word is serially loaded into the shift register on the low-to-high transition of the clock input. Parallel data is transferred to the output buffers through a 20-bit D-type latch while the Latch Enable input is high and is latched when the Latch Enable input is low. When the blanking input is high, all outputs are low.

The outputs are totem-pole structures formed by n-p-n emitter-follower and double-diffused MOS (DMOS) transistors with output voltage ratings of 70 volts and a source-current capability of 40 milliamperes. All inputs are CMOS compatible.

The TL5812I is characterized for operation from −40 °C to 85 °C. The TL5812 is characterized for operation from 0 °C to 70 °C.

† BIDFET — Bipolar, double-diffused, N-channel and P-channel MOS transistors on the same chip — patented process.

TEXAS INSTRUMENTS
POST OFFICE BOX 655012 • DALLAS, TEXAS 75265

logic symbol†

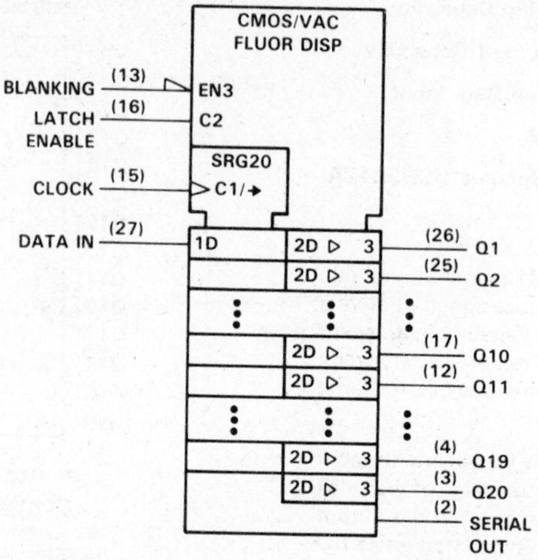
CMOS/VAC
FLUOR DISP
BLANKING
(13)
EN3
LATCH ENABLE
(16)
C2
SRG20
CLOCK
(15)
C1/→
DATA IN
(27)
1D
2D ▷ 3
(26) Q1
(25) Q2
(17) Q10
(12) Q11
(4) Q19
(3) Q20
(2) SERIAL OUT

† This symbol is in accordance with ANSI/IEEE Std 91-1984 and IEC Publication 617-12.

logic diagram (positive logic)

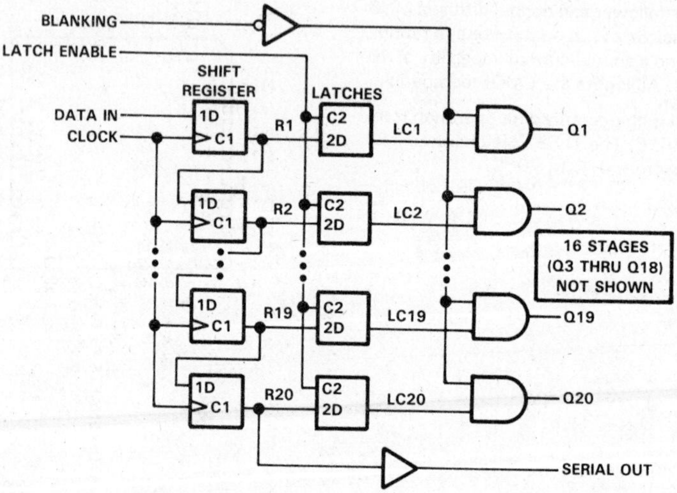
BLANKING
LATCH ENABLE
SHIFT
REGISTER
LATCHES
DATA IN
CLOCK
1D
C1
C2
2D
R1
R2
R19
R20
LC1
LC2
LC19
LC20
Q1
Q2
Q19
Q20
16 STAGES
(Q3 THRU Q18)
NOT SHOWN
SERIAL OUT

TEXAS INSTRUMENTS
POST OFFICE BOX 655012 • DALLAS, TEXAS 75265

FUNCTION TABLE

FUNCTION	CONTROL INPUTS			SHIFT REGISTER R1 THRU R20	LATCHES LC1 THRU LC20	OUTPUTS	
	CLOCK	LATCH ENABLE	BLANK-ING			SERIAL	Q1 THRU Q20
LOAD	↑	X	X	Load and shift‡	Determined by	R20	Determined by
	No↑	X	X	No change	Latch Enable§	R20	Blanking
LATCH	X	L	X	As determined above	Stored data	R20	Determined by
	X	H	X	As determined above	New data	R20	Blanking
BLANK	X	X	H	As determined above	Determined by	R20	All L
	X	X	L	As determined above	Latch Enable§	R20	LC1 thru LC20, respectively

H = high level, L = low level, X = irrelevant, ↑ = low-to-high-level transition.

‡R20 takes on the state of R19, R19 takes on the state of R18,. . .R2 takes on the state of R1, and R1 takes on the state of the data input.

§New data enter the latches while Latch Enable is high. These data are stored while Latch Enable is low.

typical operating sequence

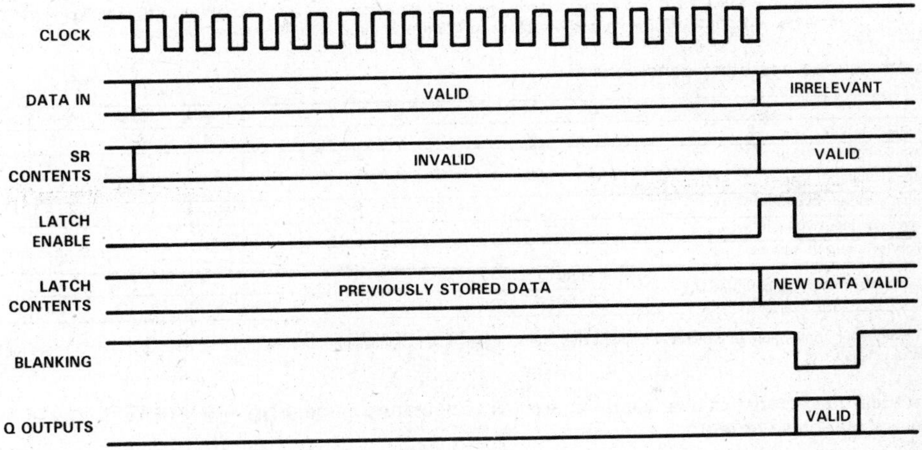
CLOCK
DATA IN
VALID
IRRELEVANT
SR CONTENTS
INVALID
VALID
LATCH ENABLE
LATCH CONTENTS
PREVIOUSLY STORED DATA
NEW DATA VALID
BLANKING
Q OUTPUTS
VALID

schematics of inputs and outputs

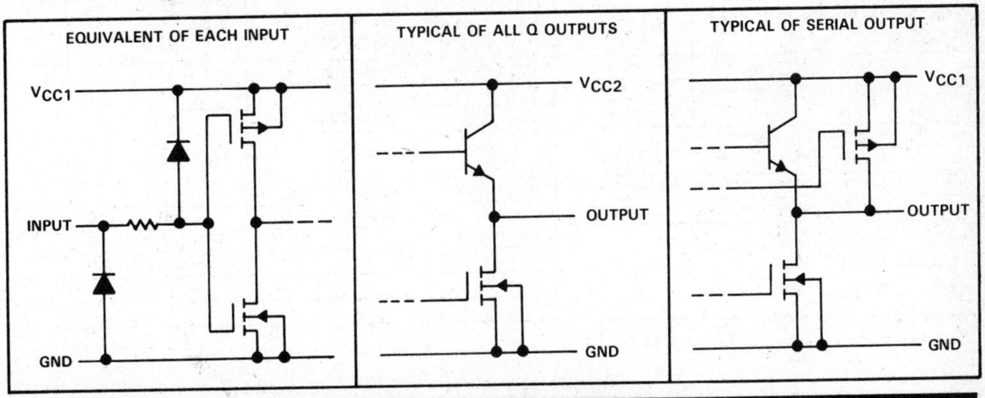
EQUIVALENT OF EACH INPUT
VCC1
INPUT
GND
TYPICAL OF ALL Q OUTPUTS
VCC2
OUTPUT
GND
TYPICAL OF SERIAL OUTPUT
VCC1
OUTPUT
GND

absolute maximum ratings over free-air operating temperature range (unless otherwise noted)

Supply voltage, V_{DD} (see Note 1) .... 15 V
Supply voltage, V_{BB} .... 70 V
Output voltage, V_O .... 70 V
Input voltage, V_I .... −0.3 V to V_{DD}+0.3 V
Output current, I_O .... −40 mA
Continuous total power dissipation at (or below) 25 °C free-air temperature (see Note 2):
  N package .... 1150 mW
  FN package .... 1400 mW
Operating free-air temperature range:
  TL5812I .... −40 °C to 85 °C
  TL5812 .... 0 °C to 70 °C
Storage temperature range .... −65 °C to 150 °C
Lead temperature 1,6 mm (1/16 inch) from case for 10 seconds .... 260 °C

NOTES 1. All voltage values are with respect to V_{SS}.
2. For operation above 25 °C free-air temperature, derate the N package linearly to 598 mW at 85 °C or to 736 mW at 70 °C at the rate of 9.2 mW/°C. Derate the FN package to 728 mW at 85 °C or to 896 mW at 70 °C at the rate of 11.2 mW/°C.

recommended operating conditions

		MIN	NOM	MAX	UNIT
Supply voltage, V_{DD}		4.5		15	V
Supply voltage, V_{BB}		0		60	V
Supply voltage, V_{SS}			0		V
High-level input voltage, V_{IH}		V_{DD} − 1.5		V_{DD} + 0.3	V
Low-level input voltage, V_{IL}		−0.3†		0.8	V
High-level output current, I_{OH}				−40	mA
Operating free-air temperature, T_A	TL5812I	−40		85	°C
	TL5812	0		70	°C

† The algebraic convention, in which the less positive (more negative) limit is designated as minimum, is used in this data sheet for logic voltage levels.

electrical characteristics over operating free-air temperature range, V_{DD} = 5 V to 15 V, V_{BB} = 60 V (unless otherwise noted)

PARAMETER			TEST CONDITIONS		MIN	TYP†	MAX	UNIT
V_{OH}	High-level output voltage	Q outputs	I_{OH} = −25 mA		57.5	58.2		V
		Serial outputs	V_{DD} = 5 V,	I_{OH} = −20 μA	4.5	4.9		
			V_{DD} = 15 V,	I_{OH} = −20 μA	14.5	14.9		
V_{OL}	Low-level output voltage	Q outputs	I_{OL} = 1 mA,	Blanking at V_{DD}		0.7	1.5	V
		Serial outputs	V_{DD} = 5 V,	I_{OL} = 20 μA		0.06	0.3	
			V_{DD} = 15 V,	I_{OL} = 20 μA		0.03	0.3	
I_{IH}	High-level input current		V_I = V_{DD}			0.3	1	μA
I_{IL}	Low-level input current		V_I = 0			−0.3	−1	μA
I_{OL}	Low-level output current (pull down current)		V_O = 60 V,	Blanking at V_{DD}	2.5	3.2		mA
$I_{O(off)}$	Off-state output current		V_O = 0,	Blanking at V_{DD}		< −1	−15	μA
I_{BB}	Supply current from V_{BB}		Outputs high			3.5	8	mA
			Outputs low			0.02	0.5	
I_{DD}	Supply current from V_{DD}		V_{DD} = 5 V			1.5	3	mA
			V_{DD} = 15 V			1.7	4	

† All typical characteristics are at T_A = 25 °C.

TEXAS INSTRUMENTS
POST OFFICE BOX 655012 • DALLAS, TEXAS 75265

timing requirements over operating free-air temperature range

PARAMETER			MIN	MAX	UNIT
t_{wCKH}	Pulse duration, clock high	V_{DD} = 5 V	500		ns
		V_{DD} = 15 V	100		
t_{wLEH}	Pulse duration, latch enable high	V_{DD} = 5 V	500		ns
		V_{DD} = 15 V	100		
t_{suD}	Setup time, data before clock↑	V_{DD} = 5 V	150		ns
		V_{DD} = 15 V	75		
t_{hD}	Hold time, data after clock↑	V_{DD} = 5 V	150		ns
		V_{DD} = 15 V	75		
$t_{CKH-LEH}$	Delay time, clock↑ to latch enable high	V_{DD} = 5 V	150		ns
		V_{DD} = 15 V	75		

switching characteristics, V_{BB} = 60 V, T_A = 25°C

PARAMETER			MIN	TYP	MAX	UNIT
t_{pd}	Propagation delay time, latch enable to output	V_{DD} = 5 V		2.2		μs
		V_{DD} = 15 V		0.8		

PARAMETER MEASUREMENT INFORMATION

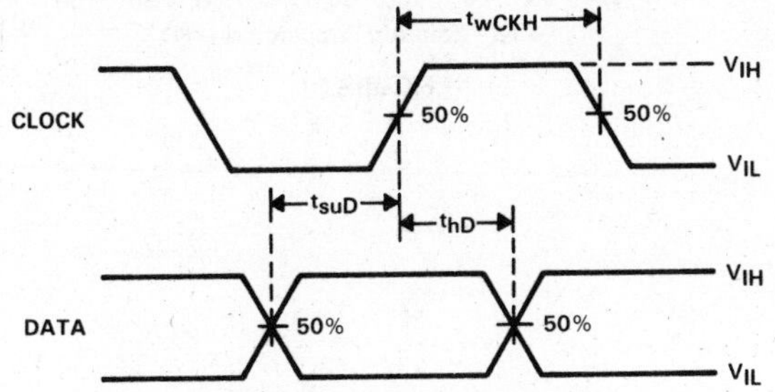

CLOCK
DATA
t_{wCKH}
t_{suD}
t_{hD}
50%
50%
50%
50%
V_{IH}
V_{IL}
V_{IH}
V_{IL}

FIGURE 1. INPUT TIMING

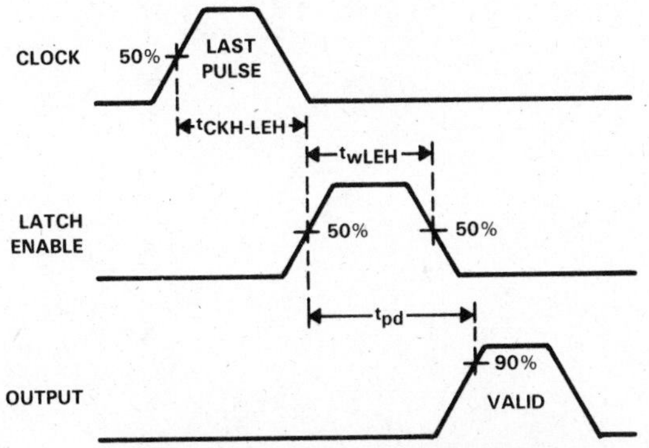

CLOCK
50%
LAST
PULSE
$t_{CKH-LEH}$
t_{wLEH}
LATCH
ENABLE
50%
50%
t_{pd}
OUTPUT
90%
VALID

FIGURE 2. OUTPUT SWITCHING TIMES

THERMAL INFORMATION

DUTY CYCLE
vs
FREE-AIR TEMPERATURE

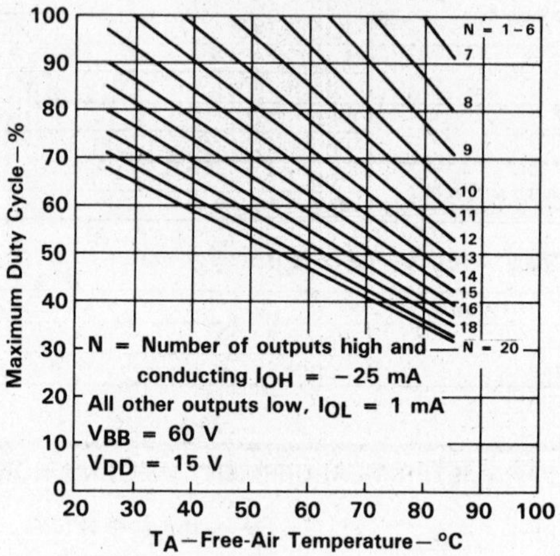
100
90
80
70
60
50
40
30
20
10
0
Maximum Duty Cycle — %
N = 1–6
7
8
9
10
11
12
13
14
15
16
18
N = 20
N = Number of outputs high and conducting IOH = −25 mA
All other outputs low, IOL = 1 mA
VBB = 60 V
VDD = 15 V
20 30 40 50 60 70 80 90 100
TA — Free-Air Temperature — °C

FIGURE 3

D2676, OCTOBER 1982–REVISED NOVEMBER 1986

- Each Device Drives 10 Lines
- 60-V Output Voltage Rating
- 40-mA Output Source Current
- High-Speed Serially-Shifted Data Input
- CMOS-Compatible Inputs
- Latches on All Driver Outputs
- Designed to be Interchangeable with Sprague UCN4810A

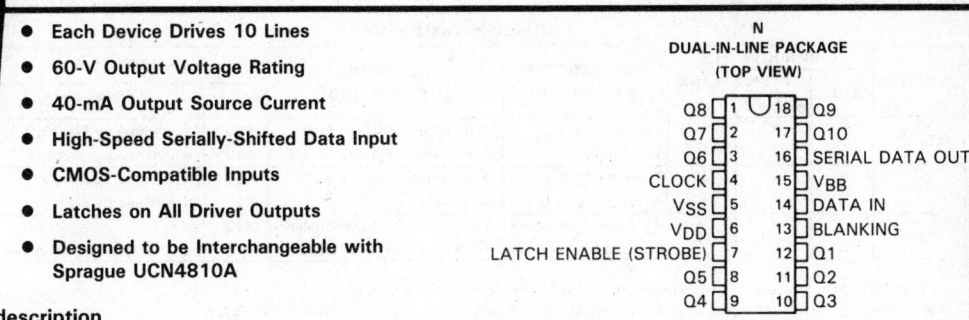

N
DUAL-IN-LINE PACKAGE
(TOP VIEW)

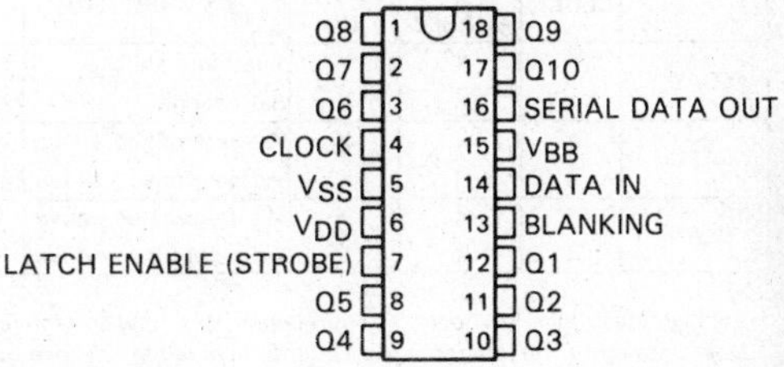
Q8 1
Q7 2
Q6 3
CLOCK 4
VSS 5
VDD 6
LATCH ENABLE (STROBE) 7
Q5 8
Q4 9
18 Q9
17 Q10
16 SERIAL DATA OUT
15 VBB
14 DATA IN
13 BLANKING
12 Q1
11 Q2
10 Q3

description

The UCN4810A is a monolithic BIDFET† integrated circuit designed to drive a dot matrix or segmented vacuum fluorescent display (VFD). This device features a serial data output to cascade additional devices for large display arrays.

A 10-bit data word is serially loaded into the shift register on the positive-going transitions of the clock. Parallel data is transferred to the output buffers through a 10-bit D-type latch while the latch enable input is high and will be latched when the latch enable is low. When the blanking input is high, all outputs are low.

Outputs are totem-pole structures formed by n-p-n emitter-follower and double-diffused MOS (DMOS) transistors with output voltage ratings of 60 volts, and 40 milliamperes source-current capability. All inputs are compatible with CMOS and TTL levels, but each requires the addition of a pull-up resistor to V_{DD} when driven by TTL logic.

The UCN4810A is characterized for operation from 0°C to 70°C.

logic symbol‡

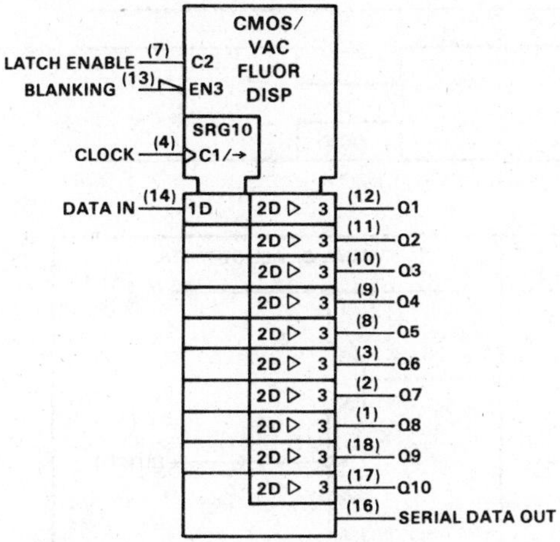
CMOS/
VAC
FLUOR
DISP
LATCH ENABLE (7) C2
BLANKING (13) EN3
SRG10
CLOCK (4) C1/→
DATA IN (14) 1D
2D ▷ 3 (12) Q1
2D ▷ 3 (11) Q2
2D ▷ 3 (10) Q3
2D ▷ 3 (9) Q4
2D ▷ 3 (8) Q5
2D ▷ 3 (3) Q6
2D ▷ 3 (2) Q7
2D ▷ 3 (1) Q8
2D ▷ 3 (18) Q9
2D ▷ 3 (17) Q10
(16) SERIAL DATA OUT

logic diagram (positive logic)

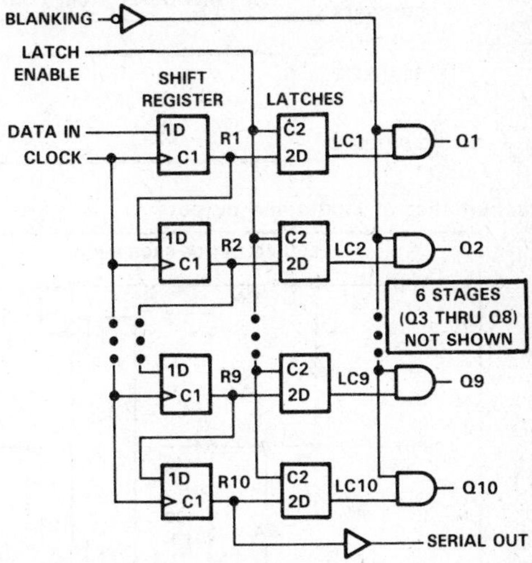
BLANKING
LATCH
ENABLE
SHIFT
REGISTER
LATCHES
DATA IN
CLOCK
1D
C1
R1
C2
2D
LC1
Q1
R2
LC2
Q2
6 STAGES
(Q3 THRU Q8)
NOT SHOWN
R9
LC9
Q9
R10
LC10
Q10
SERIAL OUT

† BIDFET – Bipolar Double-Diffused, N-Channel and P-Channel MOS transistors on same chip – patented process.
‡ This symbol is in accordance with ANSI/IEEE Std 91-1984 and IEC Publication 617-12.

TEXAS INSTRUMENTS
POST OFFICE BOX 655012 • DALLAS, TEXAS 75265

FUNCTION TABLE

FUNCTION	CONTROL INPUTS			SHIFT REGISTERS R1 THRU R10	LATCHES LC1 THRU LC10†	OUTPUTS	
	CLOCK	LATCH ENABLE	BLANKING			SERIAL	Q1 THRU Q10
LOAD	↑	X	X	Load and shift*	Determined by Latch Enable†	R10*	Determined by Blanking
	No ↑	X	X	No change	Determined by Latch Enable†	R10	Determined by Blanking
LATCH	X	L	X	As determined above	Stored data	R10	Determined by Blanking
	X	H	X	As determined above	New data	R10	Determined by Blanking
BLANK	X	X	H	As determined above	Determined by Latch Enable†	R10	All L
	X	X	L	As determined above	Determined by Latch Enable†	R10	LC1 thru LC12 respectively

H = high level, L = low level, X = irrelevant, ↑ = low-to-high-level transition.
† New data enter the latches while Latch Enable is high. These data are stored while Latch Enable is low.
* R10 takes on the state of R9, R9 takes on the state of R8 . . . R2 takes on the state of R1, and R1 takes on the state of the data input.

typical operating sequence

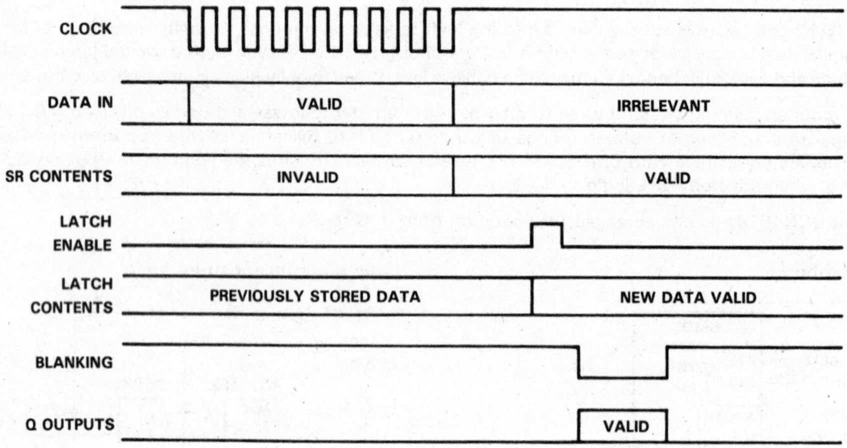
CLOCK
DATA IN
VALID
IRRELEVANT
SR CONTENTS
INVALID
VALID
LATCH ENABLE
LATCH CONTENTS
PREVIOUSLY STORED DATA
NEW DATA VALID
BLANKING
Q OUTPUTS
VALID

schematics of inputs and outputs

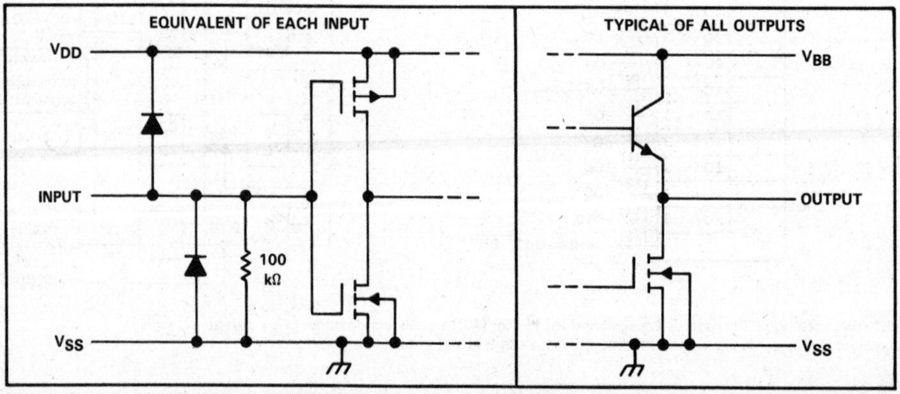
EQUIVALENT OF EACH INPUT
VDD
INPUT
100 kΩ
VSS
TYPICAL OF ALL OUTPUTS
VBB
OUTPUT
VSS

TEXAS INSTRUMENTS
POST OFFICE BOX 655012 • DALLAS, TEXAS 75265

absolute maximum ratings over operating free-air temperature range (unless otherwise noted)

Logic supply voltage, V_{DD} (see Note 1)	18 V
Driver supply voltage, V_{BB}	60 V
Output voltage	60 V
Input voltage	−0.3 V to V_{DD}+0.3 V
Continuous output current	−40 mA
Continuous total dissipation at 25°C free-air temperature (see Note 2)	1150 mW
Operating free-air temperature range	0°C to 70°C
Storage temperature range	−65°C to 150°C
Lead temperature 1,6 mm (1/16 inch) from case for 10 seconds	260°C

NOTES: 1. Voltage values are with respect to V_{SS}.
2. For operation above 25°C free-air temperature, derate linearly to 736 mW at 70°C at the rate of 9.2 mW/°C.

recommended operating conditions

		MIN	NOM	MAX	UNIT
Supply voltage, V_{DD}		4.75		15.75	V
Supply voltage, V_{BB}		5		60	V
Supply voltage, V_{SS}			0		V
High-level input voltage, V_{IH}	for V_{DD} = 5 V	3.5		5.3	V
	for V_{DD} = 15 V	13.5		15.3	
Low-level input voltage, V_{IL}		−0.3†		0.8	V
Continuous high-level output current, I_{OH}				−25	mA
Operating free-air temperature, T_A		0		70	°C

† The algebraic convention, in which the less positive (more negative) limit is designated as minimum, is used in this data sheet for logic voltage levels only.

electrical characteristics, V_{DD} = 4.75 V to 15.75 V, V_{BB} = 60 V, V_{SS} = 0, T_A = 25°C (unless otherwise noted)

PARAMETER		TEST CONDITIONS		MIN	MAX	UNIT
V_{OH}	High-level output voltage	I_{OH} = −25 mA		57.5		V
V_{OL}	Low-level output voltage	I_{OL} = 1 μA,	Blanking input at V_{DD}		1	V
I_{OL}	Low-level output current (pull-down current)	V_O = 60 V,	Blanking input at V_{DD}	0.4	0.85	mA
$I_{O(off)}$	Off-state output current	V_O = 60 V, All other terminals open,	V_{SS} = 0 V, T_A = 70°C		15	μA
I_{IH}	High-level input current	V_{DD} = 5 V,	V_I = 5 V		0.1	mA
		V_{DD} = 15 V,	V_I = 15 V		0.3	
r_i	Input resistance	V_{DD} = 5 V		50		kΩ
r_o	Output resistance	V_{DD} = 5 V			20	kΩ
		V_{DD} = 15 V			6	
I_{BB}	Supply current from V_{BB}	All outputs high			13	mA
		All outputs low			1.3	
I_{DD}	Supply current from V_{DD}	All inputs at 0 V, One output high	V_{DD} = 5 V		1	mA
			V_{DD} = 15 V		3	
		All inputs at 0 V, All outputs low	V_{DD} = 5 V		0.1	
			V_{DD} = 15 V		0.2	

timing requirements for V_{DD} = 5 V and V_{DD} = 15 V, T_A = 0°C to 70°C

PARAMETER	V_{DD} = 5 V MIN	V_{DD} = 5 V MAX	V_{DD} = 15 V MIN	V_{DD} = 15 V MAX	UNIT
Pulse duration, clock high, $t_{w(CKH)}$	1000		250		ns
Pulse duration, latch enable high, $t_{w(LEH)}$	500		300		ns
Setup time, data before clock ↑, $t_{su(D)}$	250		150		ns
Hold time, data after clock ↑, $t_{h(D)}$	250		150		ns
Delay time, clock ↑ to latch enable high, $t_{CKH\text{-}LEH}$	1000		400		ns

switching characteristics, V_{DD} = 5 V or 15 V, T_A = 25°C

	PARAMETER	MIN	TYP	MAX	UNIT
t_{pd}	Propagation delay time, latch enable to output		1		μs

PARAMETER MEASUREMENT INFORMATION

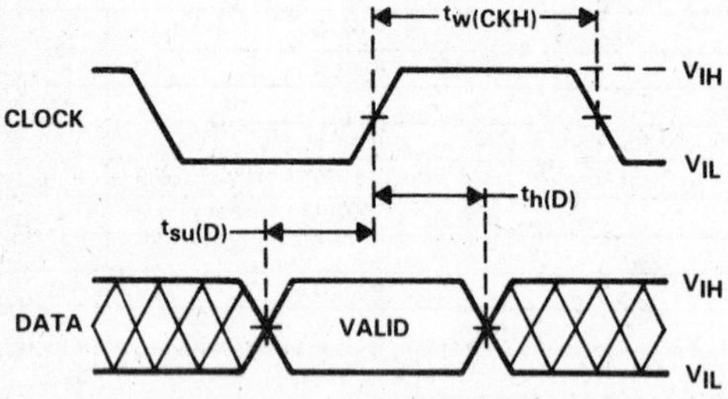
tw(CKH)
CLOCK
VIH
VIL
th(D)
tsu(D)
DATA
VALID
VIH
VIL

FIGURE 1. INPUT TIMING

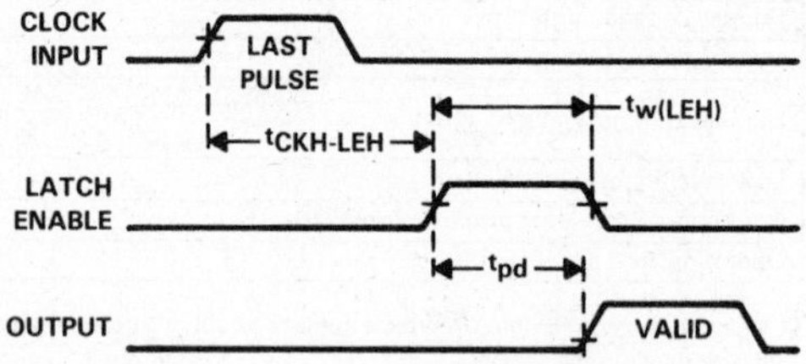
CLOCK INPUT
LAST PULSE
tw(LEH)
tCKH-LEH
LATCH ENABLE
tpd
OUTPUT
VALID

FIGURE 2. OUTPUT SWITCHING TIMES

THERMAL INFORMATION

DUTY CYCLE
vs
FREE-AIR TEMPERATURE

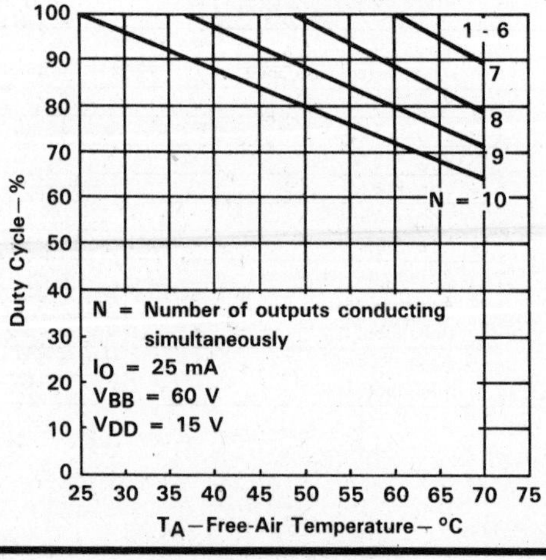
1 - 6
7
8
9
N = 10
N = Number of outputs conducting simultaneously
IO = 25 mA
VBB = 60 V
VDD = 15 V
Duty Cycle — %
100
90
80
70
60
50
40
30
20
10
0
25
30
35
40
45
50
55
60
65
70
75
TA — Free-Air Temperature — °C

TEXAS INSTRUMENTS
POST OFFICE BOX 655012 • DALLAS, TEXAS 75265

General Information 1

Alphanumeric Index
Selection Guide

Data Acquisition Circuits 2

Cross-Reference Guide
Data Sheets

Display Drivers 3

Data Sheets

Line Drivers and Receivers 4

Cross-Reference Guide
Data Sheets

Peripheral Drivers/Actuators 5

Cross-Reference Guide
Data Sheets

Memory Interface Circuits 6

Data Sheets

Speech Synthesis Circuits 7

Data Sheets

Appendix A Power Derating Curves A

Appendix B Ordering Instructions, Mechanical Data, IC Sockets B

Appendix C Explanation of Logic Symbols C

CROSS-REFERENCE GUIDE
(manufacturers arranged alphabetically)

Replacements were based on similarity of electrical and mechanical characteristics as shown in currently published data. Interchangeability in particular applications is not guaranteed. Before using a device as a substitute, the user should compare the specifications of the substitute device with the specifications of the original.

Texas Instruments makes no warranty as to the information furnished and buyer assumes all risk in the use thereof. No liability is assumed for damages resulting from the use of the information contained in this list.

AMD	SUGGESTED TI REPLACEMENT	PAGE NO.
AM26LS31C	AM26LS31C	4-5
AM26LS32C	AM26LS32AC	4-13
AM26LS33C	AM26LS33AC	4-13
AM26S10C	AM26S10C	4-23
AM26S10M	AM26S10M	4-23
AM26S11C	AM26S11C	4-23
AM26S11M	AM26S11M	4-23

FAIRCHILD	SUGGESTED TI REPLACEMENT	PAGE NO.
µA1488C	SN75188	4-391
µA1489AC	SN75189A	4-397
µA1489C	SN75189	4-397
µA26LS31C	AM26LS31C	4-5
µA26LS32C	AM26LS32AC	4-13
µA3486C	MC3486	4-47
µA3487C	MC3487	4-53
µA55107AM	SN55107A	4-73
µA55107BM	SN55107B	4-73
µA55108AM	SN55108A	4-73
µA55108M	SN55108B	4-73
µA55110M	SN55110A	4-89
µA55121M	SN55121	4-143
µA55122M	SN55122	4-147
µA75107AC	SN75107A	4-73
µA75108AC	SN75108A	4-73
µA75108BC	SN75108B	4-73
µA75108C	SN75107B	4-73
µA75110C	SN75110A	4-89
µA75150C	SN75150	4-205
µA75154C	SN75154	4-237
µA8T13C	SN75121	4-143
µA8T13M	SN55121	4-143
µA8T14C	SN75122	4-147
µA8T14M	SN55122	4-147
µA8T23C	SN75123	4-153
µA8T24C	SN75124	4-157
µA9614C	SN75114	4-113
µA9614M	SN55114	4-113
µA9615C	SN75115	4-121
µA9615M	SN55115	4-121
µA96172C	SN75172	4-319
µA96173C	SN75173	4-327
µA96174C	SN75174	4-335
µA96175C	SN75175	4-343
µA96176	SN75176B	4-351
µA96177	SN75177B	4-361
µA96178	SN75178B	4-361
µA9636AC	uA9636AC	4-523

FAIRCHILD	SUGGESTED TI REPLACEMENT	PAGE NO.
µA9637AC	uA9637AC	4-529
µA9637AM	uA9637AM	4-529
µA9638C	uA9638C	4-535
µA9639AC	uA9639C	4-539
µA9640C	AM26S10C	4-23
µA9640M	AM26S10M	4-23
µA9641C	AM26S11C	4-23
µA9641M	AM26S11M	4-23

MOTOROLA	SUGGESTED TI REPLACEMENT	PAGE NO.
AM26LS31	AM26LS31C	4-5
AM26LS32	AM26LS32AC	4-13
MC1488	SN75188	4-391
MC1489	SN75189	4-397
MC1489A	SN75189A	4-397
MC26S10	AM26S10C	4-23
MC26S11	AM26S11C	4-23
MC3446A	MC3446	4-31
MC3450	MC3450	4-35
MC3452	MC3452	4-35
MC3453	MC3453	4-43
MC3481	SN75ALS126	4-43
MC3485	SN75ALS130	4-43
MC3486	MC3486	4-47
MC3487	MC3487	4-53
MC55107	SN55107A	4-73
MC55108	SN55108A4-73	
MC75107	SN75107A	4-73
MC75108	SN75108A	4-73
MC75125	SN75125	4-163
MC75127	SN75127	4-163
MC75128	SN75128	4-169
MC75129	SN75129	4-169
MC75140	SN75140	4-191
MC75S110	SN75110A	4-89
SN75172	SN75172	4-319
SN75173	SN75173	4-327
SN75174	SN75174	4-335
SN75175	SN75175	4-343
SN75176	SN75176B	4-351
SN75177	SN75177B	4-361
SN75178	SN75178B	4-361

LINE DRIVERS AND RECEIVERS CROSS-REFERENCE GUIDE

NATIONAL	SUGGESTED TI REPLACEMENT	PAGE NO.
DS1488	SN75188	4-391
DS1489	SN75189	4-397
DS1489A	SN75189A	4-397
DS26LS31	AM26LS31C	4-5
DS26LS32	AM26LS32AC	4-13
DS26LS32M	AM26LS32AM	4-13
DS26LS33C	AM26LS33AC	4-13
DS26LS33M	AM26LS33AM	4-13
DS26S10C	AM26S10C	4-23
DS26S10M	AM26A10M	4-23
DS26S11C	AM26S11C	4-23
DS26S11M	AM26S11M	4-23
DS3486	MC3486	4-47
DS3487	MC3487	4-53
DS55107	SN55107B	4-73
DS55108	SN55108	4-73
DS55109	SN55109A	4-89
DS55110	SN55110A	4-89
DS55113	SN55113	4-101
DS55114	SN55114	4-113
DS55115	SN55115	4-121
DS55121	SN55121	4-143
DS55122	SN55122	4-147
DS75107	SN75107B	4-73
DS75108	SN75108B	4-73
DS75109	SN75109A	4-89
DS75110	SN75110A	4-89
DS75113	SN75113	4-101
DS75114	SN75114	4-113
DS75115	SN75115	4-121
DS75121	SN75121	4-143
DS75122	SN75122	4-147
DS75123	SN75123	4-153
DS75124	SN75124	4-157
DS75125	SN75125	4-163
DS75127	SN75127	4-163
DS75128	SN75128	4-169
DS75129	SN75129	4-169
DS75150	SN75150	4-205
DS75154	SN75154	4-237
DS75207	SN75207B	4-405
DS75207	SN75207	4-405
DS75208	SN75208	4-405
DS75208	SN75208B	4-405
DS75108	SN75108B	4-73
DS7820A	SN55182	4-377
DS78220	SN55182	4-377
DS7830	SN55183	4-385
DS8820	SN75182	4-377
DS8820A	SN75182	4-377
DS8830	SN75183	4-385

SIGNETICS	SUGGESTED TI REPLACEMENT	PAGE NO.
8T125	SN75125	4-163
8T126	SN75ALS126	4-163
8T127	SN75127	4-163
8T128	SN751284-169	
8T129	SN75129	4-169
8T13	SN75121	4-143
8T14	SN75122	4-147
8T23	SN75123	4-153
8T24	SN75124	4-157
8T26	N8T26	4-57
DM7820	SN55182	4-377
DM7830	SN55183	4-385
DM8820	SN75182	4-377
DM8830	SN75183	4-385
MC1488	SN75188	4-391
MC1489	SN75189	4-397
MC1489A	SN75189A	4-397

D2433, JANUARY 1979—REVISED OCTOBER 1986

- Meets EIA Standard RS-422-A
- Operates from a Single 5-V Supply
- TTL Compatible
- Complementary Outputs
- High Output Impedance in Power-Off Conditions
- Complementary Output Enable Inputs

AM26LS31M . . . J PACKAGE
AM26LS31C . . . D, J, OR N PACKAGE
(TOP VIEW)

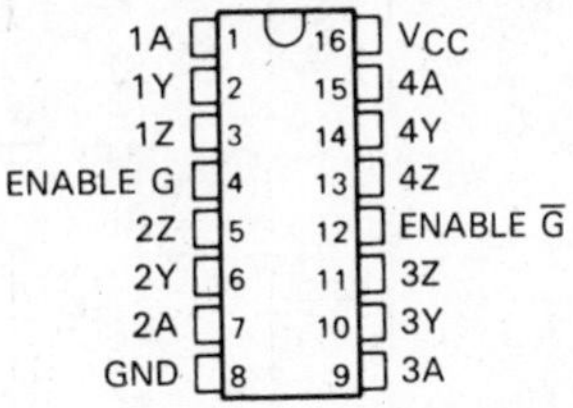
1A
1Y
1Z
ENABLE G
2Z
2Y
2A
GND
VCC
4A
4Y
4Z
ENABLE G̅
3Z
3Y
3A

AM26LS31M . . . FK PACKAGE
(TOP VIEW)

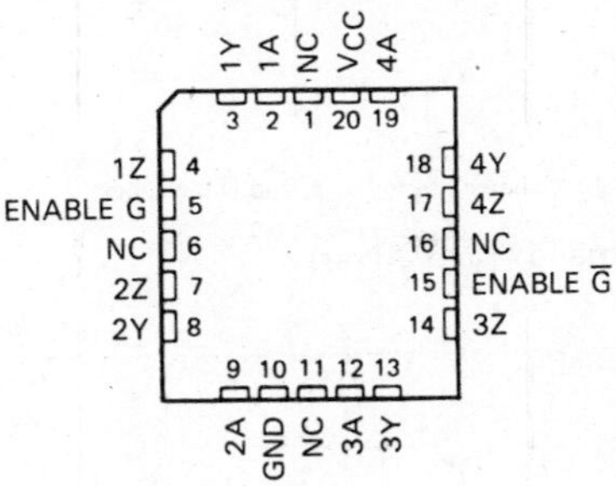
1Y 1A NC VCC 4A
1Z
ENABLE G
NC
2Z
2Y
4Y
4Z
NC
ENABLE G̅
3Z
2A GND NC 3A 3Y

NC—No internal connection

description

The AM26LS31M and AM26LS31C are quadruple complementary-output line drivers designed to meet the requirements of EIA Standard RS-422-A and Federal Standard 1020. The three-state outputs have high-current capability for driving balanced lines such as twisted-pair or parallel-wire transmission lines, and they provide a high-impedance state in the power-off condition. The enable function is common to all four drivers and offers the choice of an active-high or active-low enable input. Low-power Schottky circuitry reduces power consumption without sacrificing speed.

The AM26LS31M is characterized for operation over the full military temperature range of −55 °C to 125 °C. The AM26LS31C is characterized for operation from 0 °C to 70 °C.

FUNCTION TABLE (EACH DRIVER)

INPUT	ENABLES		OUTPUTS	
A	G	G̅	Y	Z
H	H	X	H	L
L	H	X	L	H
H	X	L	H	L
L	X	L	L	H
X	L	H	Z	Z

H = high level
L = low level
X = irrelevant
Z = high impedance (off)

logic symbol†

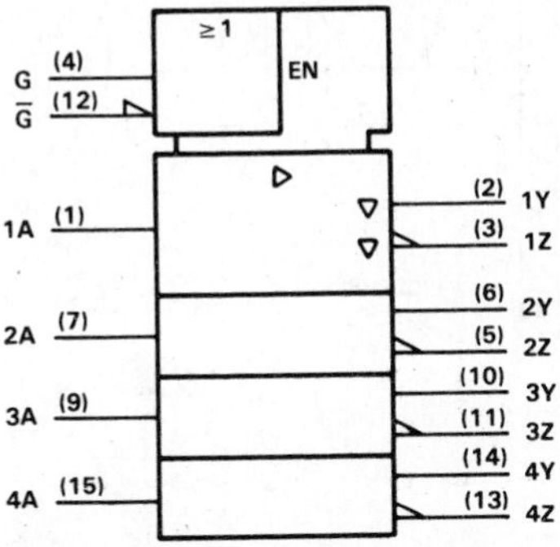
≥1
EN
G (4)
G̅ (12)
1A (1)
2A (7)
3A (9)
4A (15)
(2) 1Y
(3) 1Z
(6) 2Y
(5) 2Z
(10) 3Y
(11) 3Z
(14) 4Y
(13) 4Z

† This symbol is in accordance with ANSI/IEEE Std 91-1984 and IEC Publication 617-12.
Pin numbers shown are for D, J, and N packages.

logic diagram (positive logic)

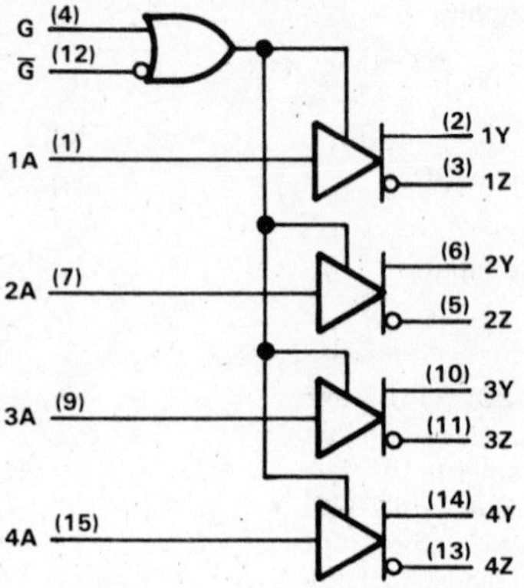
G (4)
$\overline{G}$ (12)
1A (1)
(2) 1Y
(3) 1Z
2A (7)
(6) 2Y
(5) 2Z
3A (9)
(10) 3Y
(11) 3Z
4A (15)
(14) 4Y
(13) 4Z

Pin numbers shown are for D, J, and N packages.

schematic (each driver)

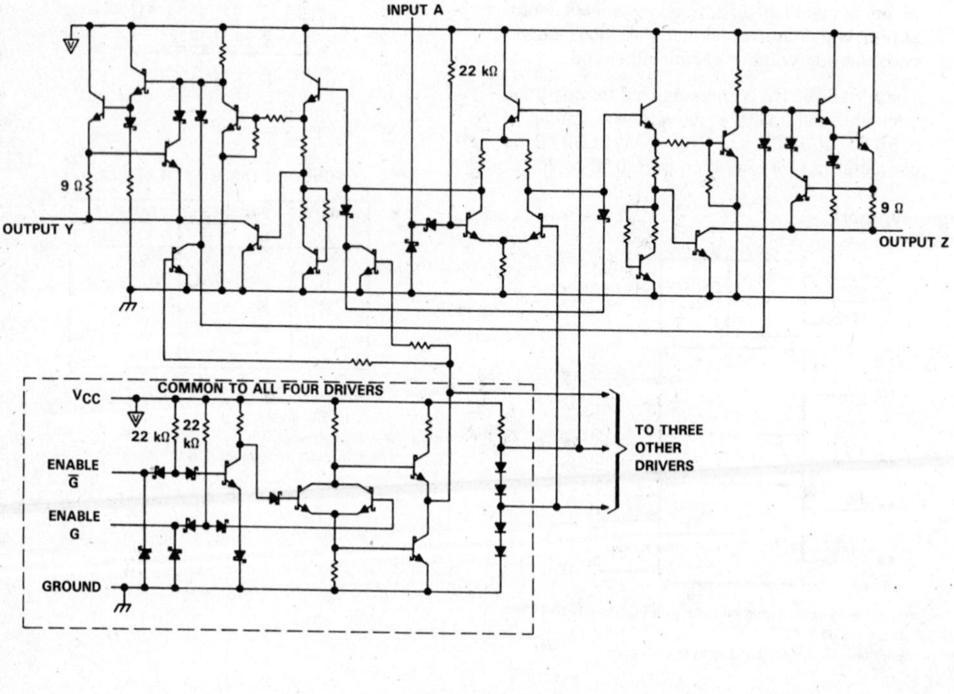
INPUT A
22 kΩ
9 Ω
OUTPUT Y
9 Ω
OUTPUT Z
COMMON TO ALL FOUR DRIVERS
V_{CC}
22 kΩ
22 kΩ
ENABLE $\overline{G}$
ENABLE G
GROUND
TO THREE OTHER DRIVERS

TEXAS INSTRUMENTS
POST OFFICE BOX 655012 • DALLAS, TEXAS 75265

absolute maximum ratings over operating free-air temperature range (unless otherwise noted)

		AM26LS31M	AM26LS31C	UNIT
Supply voltage, V_{CC} (see Note 1)		7	7	V
Input voltage		7	7	V
Output off-state voltage		5.5	5.5	V
Continuous total dissipation at (or below) 25°C free-air temperature (see Note 2)	D package		950	mW
	FK package	1375		
	J package	1375	1025	
	N package		1150	
Operating free-air temperature range		−55 to 125	0 to 70	°C
Storage temperature range		−65 to 150	−65 to 150	°C
Case temperature for 60 seconds: FK package		260		°C
Lead temperature 1,6 mm (1/16 inch) from case for 60 seconds: J package		300	300	°C
Lead temperature 1,6 mm (1/16 inch) from case for 10 seconds: D or N package			260	°C

NOTES: 1. All voltage values, except differential output voltage V_{OD}, are with respect to network ground terminal.
2. For operation above 25°C free-air temperature, refer to the Dissipation Derating Curves in Appendix A. In the J package, AM26LS31M chips are alloy mounted and AM26LS31C chips are glass mounted. In the N package, use the 9.2-mW/°C curve for these devices.

recommended operating conditions

	AM26LS31M			AM26LS31C			UNIT
	MIN	NOM	MAX	MIN	NOM	MAX	
Supply voltage, V_{CC}	4.5	5	5.5	4.75	5	5.5	V
High-level input voltage, V_{IH}	2			2			V
Low-level input voltage, V_{IL}			0.8			0.8	V
High-level output current, I_{OH}			−20			−20	mA
Low-level output current, I_{OL}			20			20	mA
Operating free-air temperature, T_A	−55		125	0		70	°C

electrical charcteristics over operating free-air temperature range (unless otherwise noted)

PARAMETER		TEST CONDITIONS†		MIN	TYP‡	MAX	UNIT
V_{IK}	Input clamp voltage	V_{CC} = MIN,	I_I = −18 mA			−1.5	V
V_{OH}	High-level output voltage	V_{CC} = MIN,	I_{OH} = −20 mA	2.5			V
V_{OL}	Low-level output voltage	V_{CC} = MIN,	I_{OL} = 20 mA			0.5	V
I_{OZ}	Off-state (high-impedance-state) output current	V_{CC} = MAX	V_O = 0.5 V			−20	μA
			V_O = 2.5 V			20	
I_I	Input current at maximum input voltage	V_{CC} = MAX,	V_I = 7 V			0.1	mA
I_{IH}	High-level input current	V_{CC} = MAX,	V_I = 2.7 V			20	μA
I_{IL}	Low-level input current	V_{CC} = MAX,	V_I = 0.4 V			−0.36	mA
I_{OS}	Short-circuit output current§	V_{CC} = MAX		−30		−150	mA
I_{CC}	Supply current (both drivers)	V_{CC} = MAX,	All outputs disabled		32	80	mA

† For conditions shown as MIN or MAX, use the appropriate value specified under recommended operating conditions.
‡ All typical values are at V_{CC} = 5 V and T_A = 25°C.
§ Not more than one output should be shorted at a time, and duration of the short-circuit should not exceed one second.

switching characteristics, V_{CC} = 5 V, T_A = 25°C

PARAMETER		TEST CONDITIONS	MIN	TYP	MAX	UNIT
t_{PLH}	Propagation delay time, low-to-high-level output	C_L = 30 pF, See Figure 1, S1 and S2 open		14	20	ns
t_{PHL}	Propagation delay time, high-to-low-level output			14	20	ns
	Output-to-output skew			1	6	ns
t_{PZH}	Output enable time to high level	C_L = 30 pF, R_L = 75 Ω, See Figure 1		25	40	ns
t_{PZL}	Output enable time to low level	C_L = 30 pF, R_L = 180 Ω, See Figure 1		37	45	ns
t_{PHZ}	Output disable time from high level	C_L = 10 pF, See Figure 1, S1 and S2 closed		21	30	ns
t_{PLZ}	Output disable time from low level			23	35	ns

TEXAS INSTRUMENTS
POST OFFICE BOX 655012 • DALLAS, TEXAS 75265

PARAMETER MEASUREMENT INFORMATION

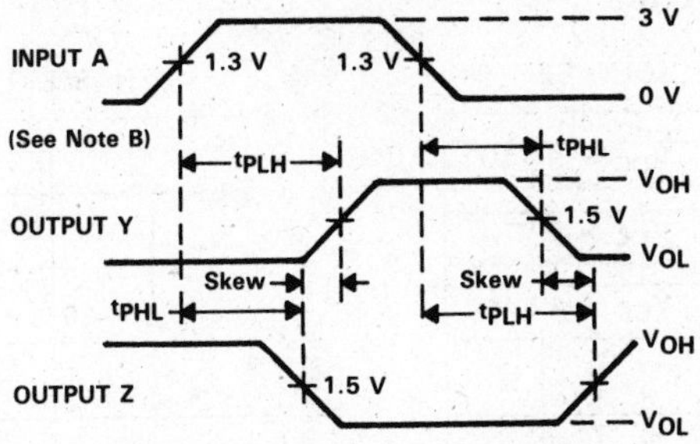
INPUT A
(See Note B)
1.3 V
1.3 V
3 V
0 V
tPLH
tPHL
VOH
OUTPUT Y
1.5 V
VOL
Skew
Skew
tPHL
tPLH
VOH
OUTPUT Z
1.5 V
VOL

PROPAGATION DELAY TIMES AND SKEW

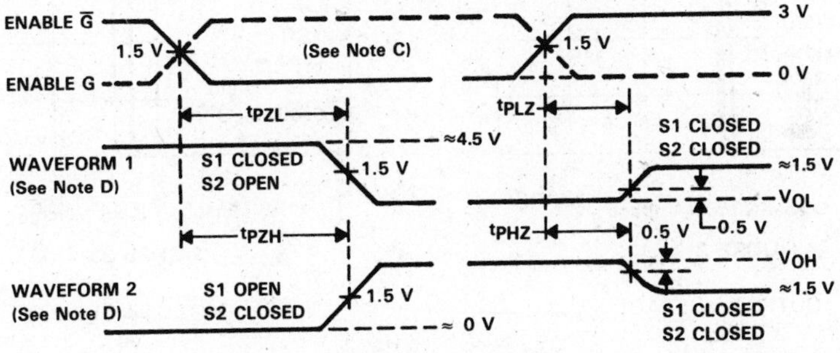
ENABLE G̅
1.5 V
(See Note C)
1.5 V
3 V
ENABLE G
0 V
tPZL
tPLZ
≈4.5 V
S1 CLOSED
S2 CLOSED
WAVEFORM 1
(See Note D)
S1 CLOSED
S2 OPEN
1.5 V
≈1.5 V
VOL
tPZH
tPHZ
0.5 V
0.5 V
VOH
WAVEFORM 2
(See Note D)
S1 OPEN
S2 CLOSED
1.5 V
≈1.5 V
≈ 0 V
S1 CLOSED
S2 CLOSED

ENABLE AND DISABLE TIMES

VOLTAGE WAVEFORMS

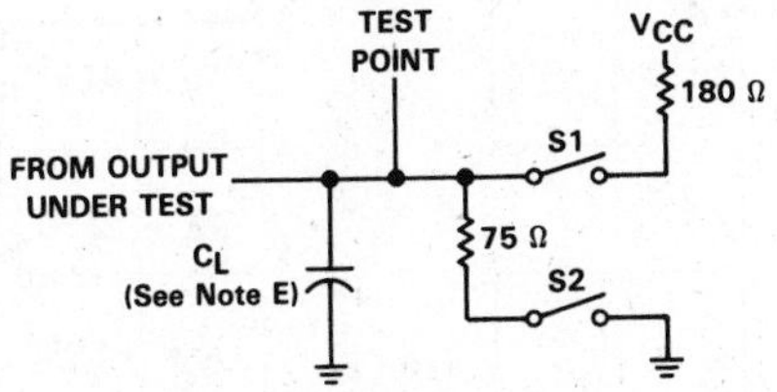
TEST
POINT
VCC
180 Ω
S1
FROM OUTPUT
UNDER TEST
75 Ω
CL
(See Note E)
S2

TEST CIRCUIT

NOTES: A. All input pulses are supplied by generators having the following characteristics: PRR ≤ 1 MHz, Z_{out} ≈ 50 Ω, t_r ≤ 15 ns, and t_f ≤ 6 ns.

B. When measuring propagation delay times and skew, switches S1 and S2 are open.

C. Each enable is tested separately.

D. Waveform 1 is for an output with internal conditions such that the output is low except when disabled by the output control. Waveform 2 is for an output with internal conditions such that the output is high except when disabled by the output control.

E. C_L includes probe and jig capacitance.

FIGURE 1. SWITCHING TIMES

TYPICAL CHARACTERISTICS†

OUTPUT VOLTAGE
vs
ENABLE G INPUT VOLTAGE

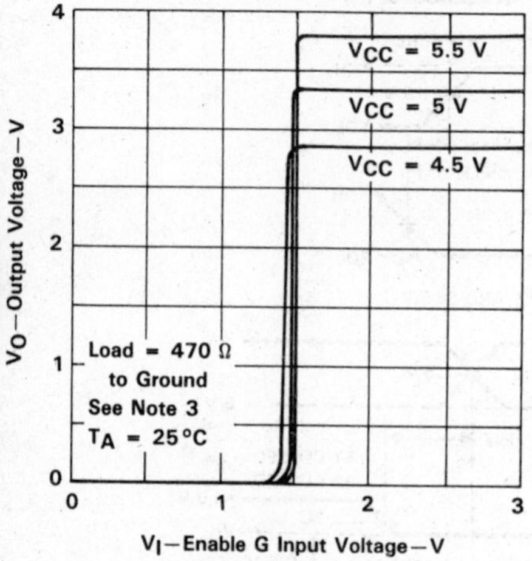
V_{CC} = 5.5 V
V_{CC} = 5 V
V_{CC} = 4.5 V
Load = 470 Ω
to Ground
See Note 3
T_A = 25°C
V_O—Output Voltage—V
V_I—Enable G Input Voltage—V

FIGURE 2

OUTPUT VOLTAGE
vs
ENABLE G INPUT VOLTAGE

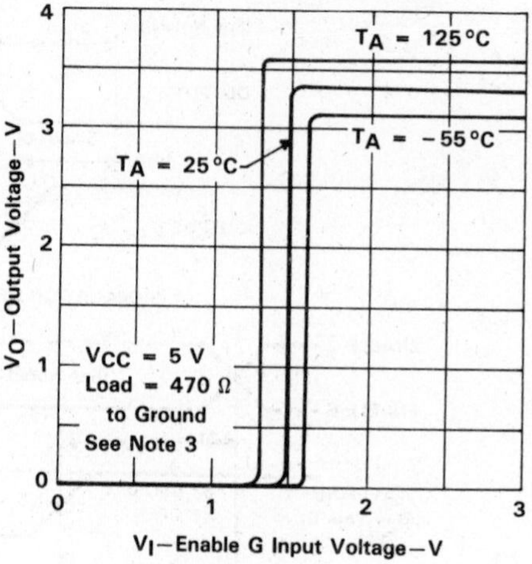
T_A = 125°C
T_A = −55°C
T_A = 25°C
V_{CC} = 5 V
Load = 470 Ω
to Ground
See Note 3
V_O—Output Voltage—V
V_I—Enable G Input Voltage—V

FIGURE 3

OUTPUT VOLTAGE
vs
ENABLE G INPUT VOLTAGE

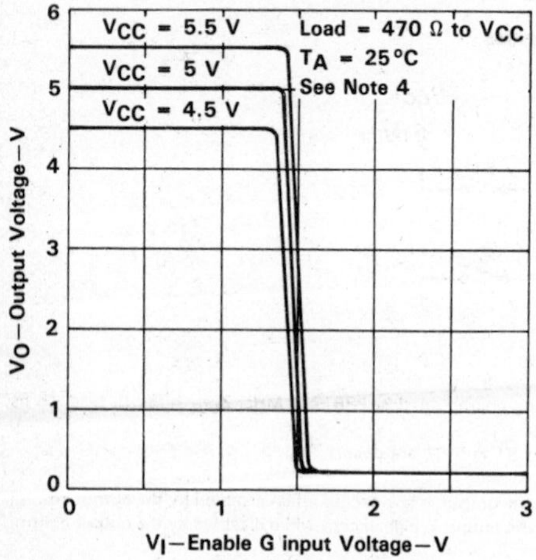
V_{CC} = 5.5 V
V_{CC} = 5 V
V_{CC} = 4.5 V
Load = 470 Ω to V_{CC}
T_A = 25°C
See Note 4
V_O—Output Voltage—V
V_I—Enable G input Voltage—V

FIGURE 4

OUTPUT VOLTAGE
vs
ENABLE G INPUT VOLTAGE

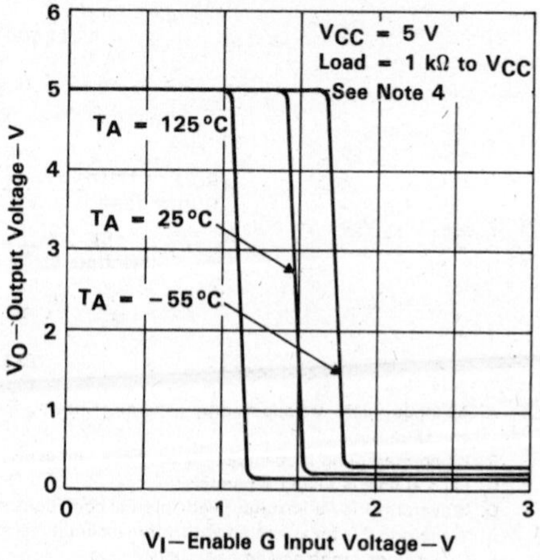
V_{CC} = 5 V
Load = 1 kΩ to V_{CC}
See Note 4
T_A = 125°C
T_A = 25°C
T_A = −55°C
V_O—Output Voltage—V
V_I—Enable G Input Voltage—V

FIGURE 5

† Data for temperature below 0°C and above 70°C are applicable to AM26LS31M circuits only.

NOTES: 3. The A input is connected to V_{CC} during the testing of the Y outputs and to ground during testing of the Z outputs.
4. The A input is connected to ground during the testing of the Y outputs and to V_{CC} during the testing of the Z outputs.

TYPICAL CHARACTERISTICS†

HIGH-LEVEL OUTPUT VOLTAGE
vs
FREE-AIR TEMPERATURE

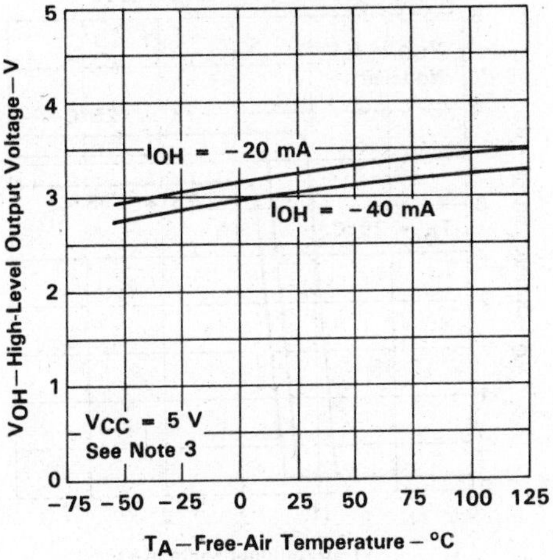
V_{OH}—High-Level Output Voltage—V
5
4
3
2
1
0
I_{OH} = −20 mA
I_{OH} = −40 mA
V_{CC} = 5 V
See Note 3
−75 −50 −25 0 25 50 75 100 125
T_A—Free-Air Temperature – °C

FIGURE 6

HIGH-LEVEL OUTPUT VOLTAGE
vs
OUTPUT CURRENT

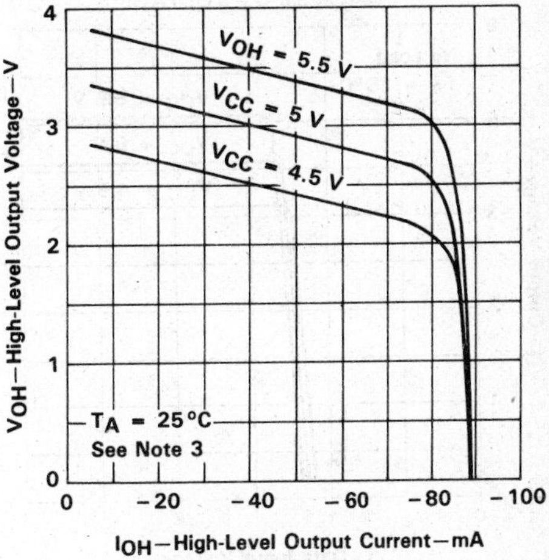
V_{OH}—High-Level Output Voltage—V
4
3
2
1
0
V_{OH} = 5.5 V
V_{CC} = 5 V
V_{CC} = 4.5 V
T_A = 25°C
See Note 3
0 −20 −40 −60 −80 −100
I_{OH}—High-Level Output Current—mA

FIGURE 7

LOW-LEVEL OUTPUT VOLTAGE
vs
FREE-AIR TEMPERATURE

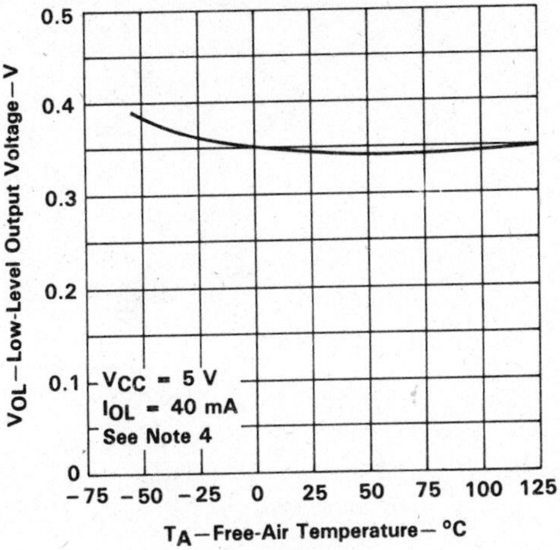
V_{OL}—Low-Level Output Voltage—V
0.5
0.4
0.3
0.2
0.1
0
V_{CC} = 5 V
I_{OL} = 40 mA
See Note 4
−75 −50 −25 0 25 50 75 100 125
T_A—Free-Air Temperature— °C

FIGURE 8

LOW-LEVEL OUTPUT VOLTAGE
vs
OUTPUT CURRENT

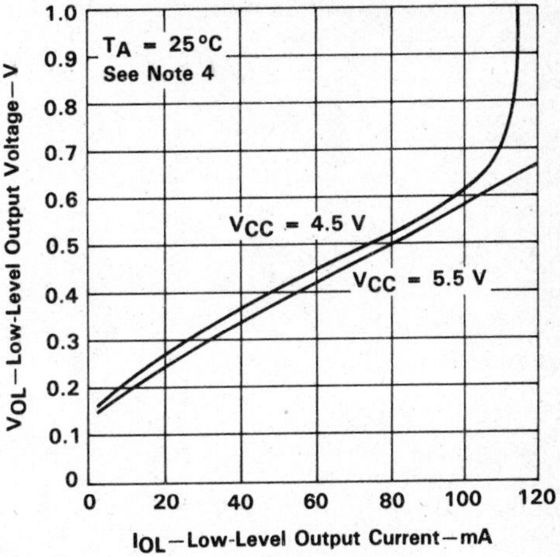
V_{OL}—Low-Level Output Voltage—V
1.0
0.9
0.8
0.7
0.6
0.5
0.4
0.3
0.2
0.1
0
T_A = 25°C
See Note 4
V_{CC} = 4.5 V
V_{CC} = 5.5 V
0 20 40 60 80 100 120
I_{OL}—Low-Level Output Current—mA

FIGURE 9

†Data for temperature below 0°C and above 70°C are applicable to AM26LS31M circuits only.

NOTES: 3. The A input is connected to V_{CC} during the testing of the Y outputs and to ground during testing of the Z outputs.
4. The A input is connected to ground during the testing of the Y outputs and to V_{CC} during the testing of the Z outputs.

TYPICAL CHARACTERISTICS†

Y OUTPUT VOLTAGE
vs
DATA INPUT VOLTAGE

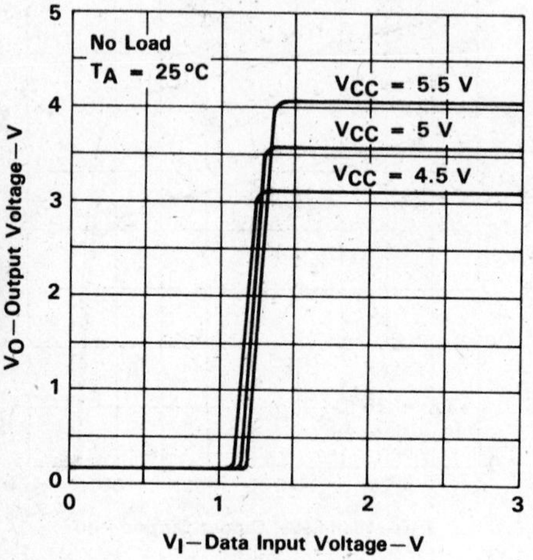
No Load
T_A = 25°C
V_{CC} = 5.5 V
V_{CC} = 5 V
V_{CC} = 4.5 V
V_O—Output Voltage—V
V_I—Data Input Voltage—V
0
1
2
3
4
5

FIGURE 10

Y OUTPUT VOLTAGE
vs
DATA INPUT VOLTAGE

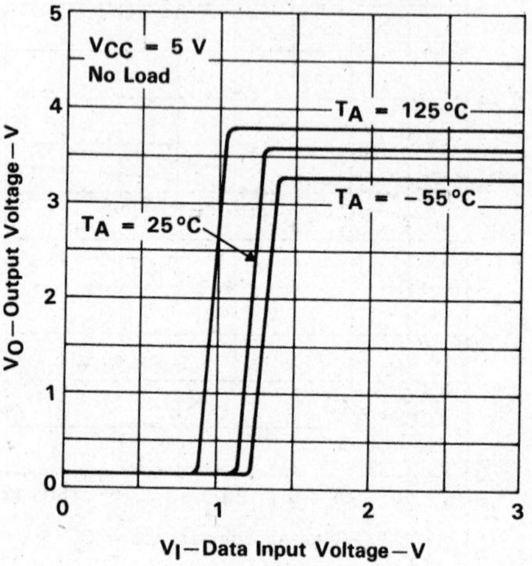
V_{CC} = 5 V
No Load
T_A = 125°C
T_A = −55°C
T_A = 25°C
V_O—Output Voltage—V
V_I—Data Input Voltage—V
0
1
2
3
4
5

FIGURE 11

†Data for temperature below 0°C and above 70°C are applicable to AM26LS31M circuits only.

TEXAS INSTRUMENTS
POST OFFICE BOX 655012 • DALLAS, TEXAS 75265

D2434, OCTOBER 1980—REVISED SEPTEMBER 1986

- AM26LS32A Meets EIA Standards RS-422-A and RS-423-A
- AM26LS32A has ±7-V Common-Mode Range with ±200-mV Sensitivity
- AM26LS33A has ±15-V Common-Mode Range with ±500 mV Sensitivity
- Input Hysteresis . . . 50 mV Typical
- Operates from a Single 5-V Supply
- Low-Power Schottky Circuitry
- 3-State Outputs
- Complementary Output Enable Inputs
- Input Impedance . . . 12 kΩ Min
- Designed to be Interchangeable with Advanced Micro Devices AM26LS32C and AM26LS33C

AM26LS32AM, AM26LS33AM . . . J PACKAGE
AM26LS32AC, AM26LS33AC . . . D, J, OR N PACKAGE
(TOP VIEW)

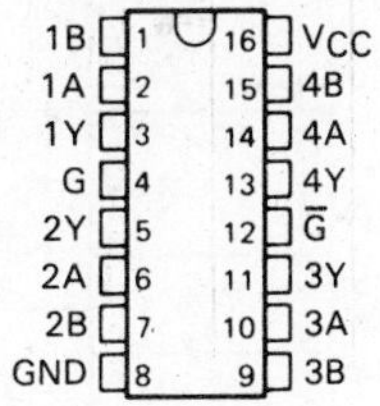
1B 1
1A 2
1Y 3
G 4
2Y 5
2A 6
2B 7
GND 8
16 VCC
15 4B
14 4A
13 4Y
12 G̅
11 3Y
10 3A
9 3B

AM26LS32AM, AM26LS33AM . . . FK PACKAGE
(TOP VIEW)

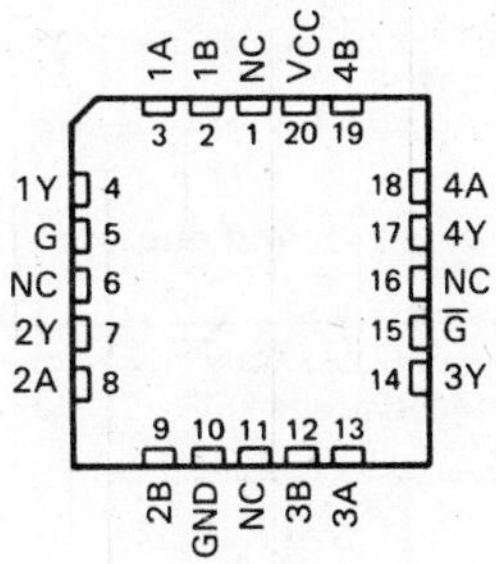
1A 1B NC VCC 4B
3 2 1 20 19
1Y 4
G 5
NC 6
2Y 7
2A 8
18 4A
17 4Y
16 NC
15 G̅
14 3Y
9 10 11 12 13
2B GND NC 3B 3A

NC→No internal connection

description

The AM26LS32A and AM26LS33A are quadruple line receivers for balanced and unbalanced digital data transmission. The enable function is common to all four receivers and offers a choice of active-high or active-low input. Three-state outputs permit connection directly to a bus-organized system. Fail-safe design ensures that if the inputs are open, the outputs will always be high.

Compared to the AM26LS32C and the AM26LS33C, the AM26LS32A and AM26LS33A incorporate an additional stage of amplification to improve sensitivity. The input impedance has been increased resulting in less loading of the bus line. The additional stage has increased propagation delay; however, this will not affect interchangeability in most applications.

The AM26LS32AM and the AM26LS33AM are characterized for operation over the full military temperature range of −55°C to 125°C. The AM26LS32AC and AM26LS33AC are characterized for operation from 0°C to 70°C.

FUNCTION TABLE (EACH RECEIVER)

DIFFERENTIAL INPUT	ENABLES G	ENABLES $\overline{G}$	OUTPUT
$V_{ID} \geq V_{TH}$	H	X	H
	X	L	H
$V_{TL} \leq V_{ID} \leq V_{TH}$	H	X	?
	X	L	?
$V_{ID} \leq V_{TL}$	H	X	L
	X	L	L
X	L	H	Z

H = high level, L = low level, X = irrelevant
Z = high impedance (off), ? = indeterminate

TEXAS INSTRUMENTS
POST OFFICE BOX 655012 • DALLAS, TEXAS 75265

logic symbol†

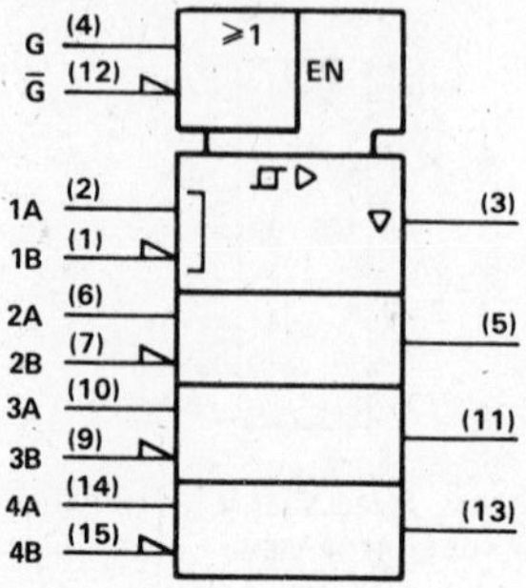
G (4)
G̅ (12)
≥1
EN
1A (2)
1B (1)
(3) 1Y
2A (6)
2B (7)
(5) 2Y
3A (10)
3B (9)
(11) 3Y
4A (14)
4B (15)
(13) 4Y

† This symbol is in accordance with ANSI/IEEE Std 91-1984 and IEC Publication 617-12.

Pin numbers shown are for D, J, and N packages.

logic diagram (positive logic)

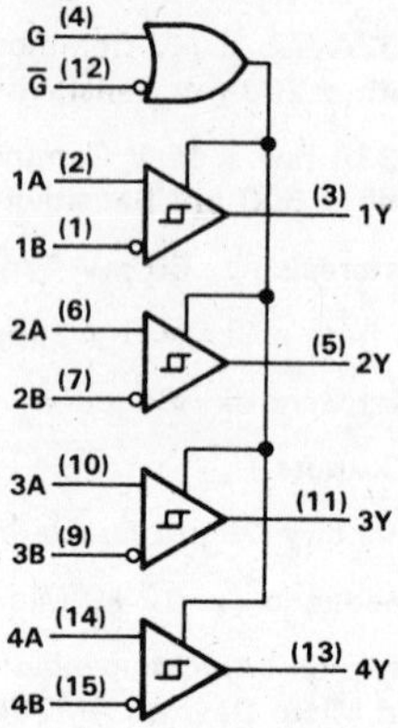
G (4)
G̅ (12)
1A (2)
1B (1)
(3) 1Y
2A (6)
2B (7)
(5) 2Y
3A (10)
3B (9)
(11) 3Y
4A (14)
4B (15)
(13) 4Y

schematics of inputs and outputs

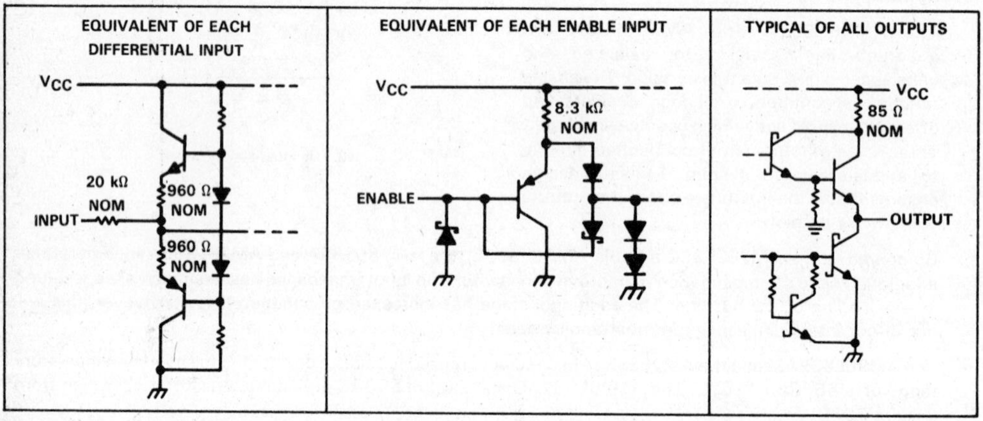
EQUIVALENT OF EACH DIFFERENTIAL INPUT
VCC
20 kΩ NOM
INPUT
960 Ω NOM
960 Ω NOM
EQUIVALENT OF EACH ENABLE INPUT
VCC
8.3 kΩ NOM
ENABLE
TYPICAL OF ALL OUTPUTS
VCC
85 Ω NOM
OUTPUT

absolute maximum ratings over operating free-air temperature range (unless otherwise noted)

		AM26LS32AM AM26LS33AM	AM26LS32AC AM26LS33AC	UNIT
Supply voltage, V_{CC} (see Note 1)		7	7	V
Input voltage, any differential input		±25	±25	V
Differential input voltage (see Note 2)		±25	±25	V
Continuous total dissipation at (or below) 25°C free-air temperaure (see Note 3)	D package		950	mW
	FK package	1375		mW
	J package	1375	1025	mW
	N package		1150	mW
Operating free-air temperature range		−55 to 125	0 to 70	°C
Storage temperature range		−65 to 150	−65 to 150	°C
Lead temperature 1,6 mm (1/16 inch) from case for 10 seconds	D or N package		260	°C
Case temperature for 60 seconds	FK package	260		°C
Lead temperature 1,6 mm (1/16 inch) from case for 60 seconds	J package	300	300	°C

NOTES: 1. All voltage values, except differential voltages, are with respect to the network ground terminal.
2. Differential voltage values are at the noninverting (A) input terminals with respect to the inverting (B) input terminals.
3. For operation above 25°C free-air temperature, refer to Dissipation Derating Curves in Appendix A. In the J package, AM26LS32AM and AM26LS33AM chips are alloy mounted and AM26LS32AC and AM26LS33AC chips are glass mounted. In the N package, use the 9.2 mW/°C curve.

recommended operating conditions

		AM26LS32AM AM26LS33AM			AM26LS32AC AM26LS33AC			UNIT
		MIN	NOM	MAX	MIN	NOM	MAX	
Supply voltage, V_{CC}		4.5	5	5.5	4.75	5	5.25	V
High-level input voltage, V_{IH}		2			2			V
Low-level input voltage, V_{IL}				0.8			0.8	V
Common-mode input voltage, V_{IC}	AM26LS32AM, AM26LS32AC			±7			±7	V
	AM26LS33AM, AM26LS33AC			±15			±15	V
High-level output current, I_{OH}				−440			−440	μA
Low-level output current, I_{OL}				8			8	mA
Operating free-air temperature, T_A		−55		125	0		70	°C

4

Line Drivers/Receivers

electrical characteristics over recommended ranges of V_{CC}, V_{IC}, and operating free-air temperature (unless otherwise noted)

PARAMETER		TEST CONDITIONS		MIN	TYP†	MAX	UNIT
V_{TH}	Differential input high-threshold voltage	$V_O = V_{OH}$min, $I_{OH} = -440\ \mu A$	AM26LS32A			0.2	V
			AM26LS33A			0.5	
V_{TL}	Differential input low-threshold voltage	$V_O = 0.45$ V, $I_{OL} = 8$ mA	AM26LS32A	−0.2‡			V
			AM26LS33A	−0.5‡			
V_{hys}	Hysteresis, $V_{T+} - V_{T-}$ §				50		mV
V_{IK}	Enable input clamp voltage	V_{CC} = MIN, $I_I = -18$ mA				−1.5	V
V_{OH}	High-level output voltage	V_{CC} = MIN, $V_{ID} = 1$ V, $V_{I(\overline{G})} = 0.8$ V, $I_{OH} = -440\ \mu A$	'32AM, '33AM	2.5			V
			'32AC, '33AC	2.7			
V_{OL}	Low-level output voltage	V_{CC} = MIN, $V_{ID} = -1$ V, $V_{I(\overline{G})} = 0.8$ V	$I_{OL} = 4$ mA			0.4	V
			$I_{OL} = 8$ mA			0.45	
I_{OZ}	Off-state (high-impedance-state) output current	V_{CC} = MAX	$V_O = 2.4$ V			20	μA
			$V_O = 0.4$ V			−20	
I_I	Line input current	$V_I = 15$ V, Other input at −10 V to 15 V				1.2	mA
		$V_I = -15$ V, Other input at −15 V to 10 V				−1.7	
$I_{I(EN)}$	Enable input current	$V_I = 5.5$ V				100	μA
I_{IH}	High-level enable current	$V_I = 2.7$ V				20	μA
I_{IL}	Low-level enable current	$V_I = 0.4$ V				−0.36	mA
r_i	Input resistance	$V_{IC} = -15$ V to 15 V, One input to AC ground		12	15		kΩ
I_{OS}	Short-circuit output current¶	V_{CC} = MAX		−15		−85	mA
I_{CC}	Supply current	V_{CC} = MAX, All outputs disabled			52	70	mA

† All typical values are at $V_{CC} = 5$ V, $T_A = 25°C$, and $V_{IC} = 0$.
‡ The algebraic convention, where the less positive (more negative) limit is designated as minimum, is used in this data sheet for threshold levels only.
§ Hysteresis is the difference between the positive-going input threshold voltage, V_{T+}, and the negative-going input threshold voltage, V_{T-}. See Figures 10 and 11.
¶ Not more than one output should be shorted at a time.

switching characteristics, $V_{CC} = 5$ V, $T_A = 25°C$

PARAMETER		TEST CONDITIONS	MIN	TYP	MAX	UNIT
t_{PLH}	Propagation delay time, low-to-high-level output	$C_L = 15$ pF, See Figure 1		20	35	ns
t_{PHL}	Propagation delay time, high-to-low-level output			22	35	ns
t_{PZH}	Output enable time to high level	$C_L = 15$ pF, See Figure 1		17	22	ns
t_{PZL}	Output enable time to low level			20	25	ns
t_{PHZ}	Output disable time from high level	$C_L = 5$ pF, See Figure 1		21	30	ns
t_{PLZ}	Output disable time from low level			30	40	ns

PARAMETER MEASUREMENT INFORMATION

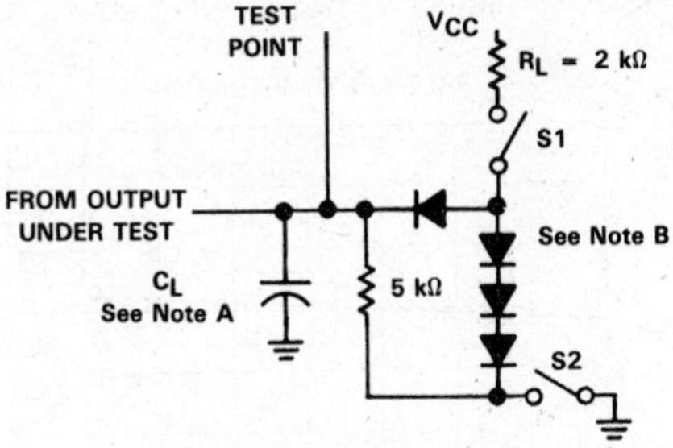
TEST POINT
VCC
RL = 2 kΩ
S1
FROM OUTPUT UNDER TEST
See Note B
CL
See Note A
5 kΩ
S2

TEST CIRCUIT

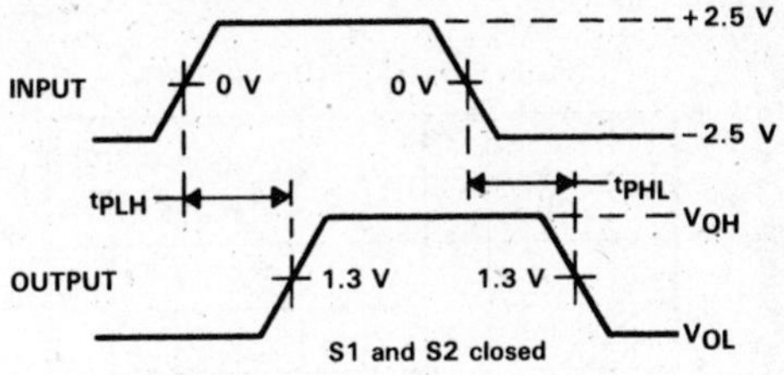
+2.5 V
INPUT
0 V
0 V
−2.5 V
tPLH
tPHL
VOH
OUTPUT
1.3 V
1.3 V
VOL
S1 and S2 closed

VOLTAGE WAVEFORMS FOR t_{PLH}, t_{PHL}

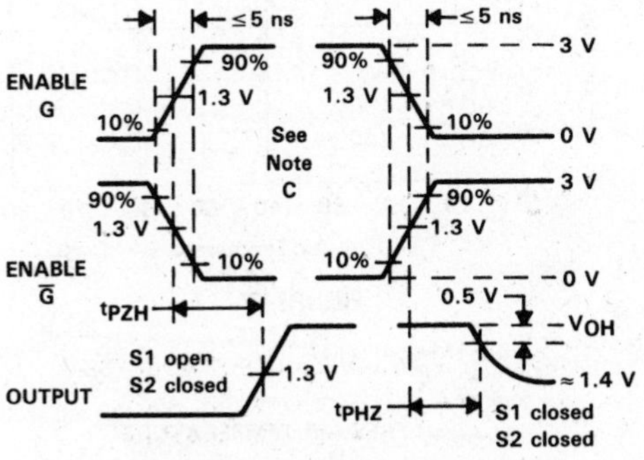
≤5 ns
≤5 ns
3 V
ENABLE G
90%
90%
1.3 V
1.3 V
10%
10%
0 V
See Note C
3 V
90%
90%
1.3 V
1.3 V
ENABLE G̅
10%
10%
0 V
0.5 V
VOH
tPZH
S1 open
S2 closed
1.3 V
≈ 1.4 V
OUTPUT
tPHZ
S1 closed
S2 closed

VOLTAGE WAVEFORMS FOR t_{PHZ}, t_{PZH}

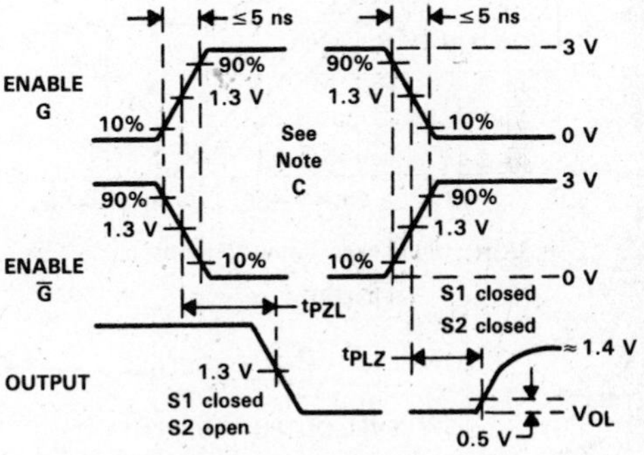
≤5 ns
≤5 ns
3 V
ENABLE G
90%
90%
1.3 V
1.3 V
10%
10%
0 V
See Note C
3 V
90%
90%
1.3 V
1.3 V
ENABLE G̅
10%
10%
0 V
S1 closed
S2 closed
tPZL
tPLZ
≈ 1.4 V
1.3 V
OUTPUT
S1 closed
S2 open
VOL
0.5 V

VOLTAGE WAVEFORMS FOR t_{PLZ}, t_{PZL}

NOTES: A. C_L includes probe and jig capacitance.
B. All diodes are 1N3064 or equivalent.
C. Enable G is tested with $\overline{G}$ high; $\overline{G}$ is tested with G low.

FIGURE 1

TYPICAL CHARACTERISTICS

HIGH-LEVEL OUTPUT VOLTAGE
vs
HIGH-LEVEL OUTPUT CURRENT

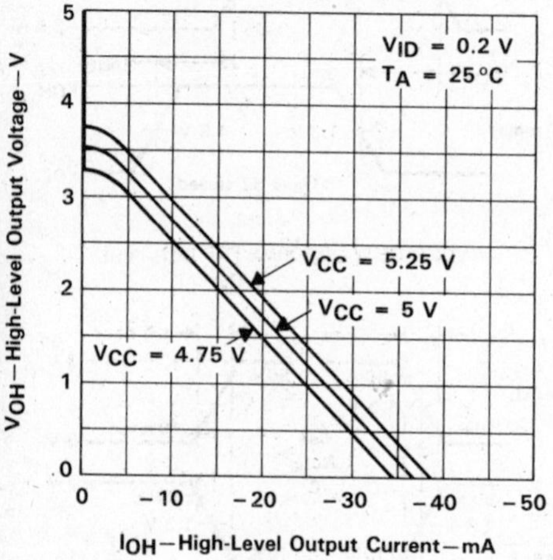
V_{ID} = 0.2 V
T_A = 25°C
V_{CC} = 5.25 V
V_{CC} = 5 V
V_{CC} = 4.75 V
V_{OH}—High-Level Output Voltage—V
I_{OH}—High-Level Output Current—mA
0 1 2 3 4 5
0 −10 −20 −30 −40 −50

FIGURE 2

HIGH-LEVEL OUTPUT VOLTAGE
vs
FREE-AIR TEMPERATURE

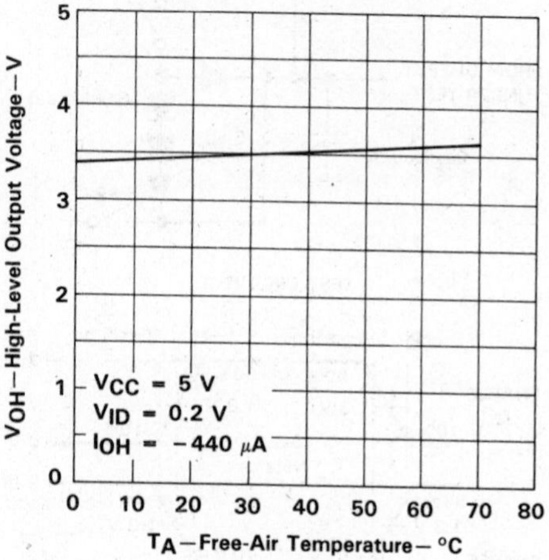
V_{CC} = 5 V
V_{ID} = 0.2 V
I_{OH} = −440 μA
V_{OH}—High-Level Output Voltage—V
T_A—Free-Air Temperature—°C
0 1 2 3 4 5
0 10 20 30 40 50 60 70 80

FIGURE 3

LOW-LEVEL OUTPUT VOLTAGE
vs
LOW-LEVEL OUTPUT CURRENT

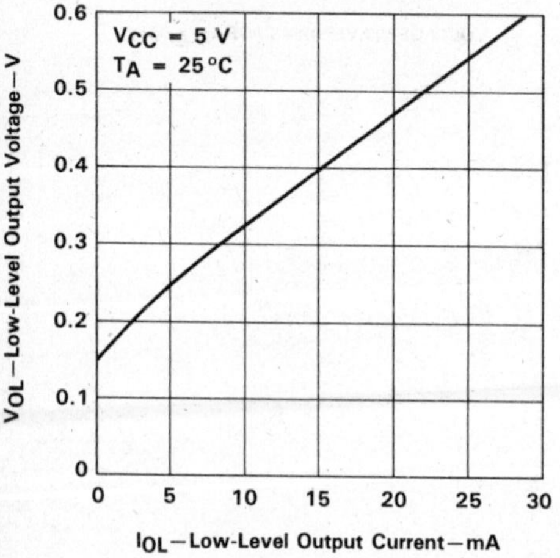
V_{CC} = 5 V
T_A = 25°C
V_{OL}—Low-Level Output Voltage—V
I_{OL}—Low-Level Output Current—mA
0 0.1 0.2 0.3 0.4 0.5 0.6
0 5 10 15 20 25 30

FIGURE 4

LOW-LEVEL OUTPUT VOLTAGE
vs
FREE-AIR TEMPERATURE

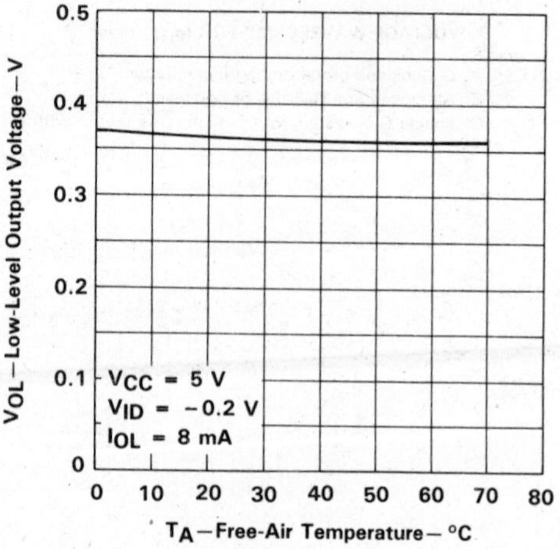
V_{CC} = 5 V
V_{ID} = −0.2 V
I_{OL} = 8 mA
V_{OL}—Low-Level Output Voltage—V
T_A—Free-Air Temperature—°C
0 0.1 0.2 0.3 0.4 0.5
0 10 20 30 40 50 60 70 80

FIGURE 5

TYPICAL CHARACTERISTICS

OUTPUT VOLTAGE
vs
ENABLE G VOLTAGE

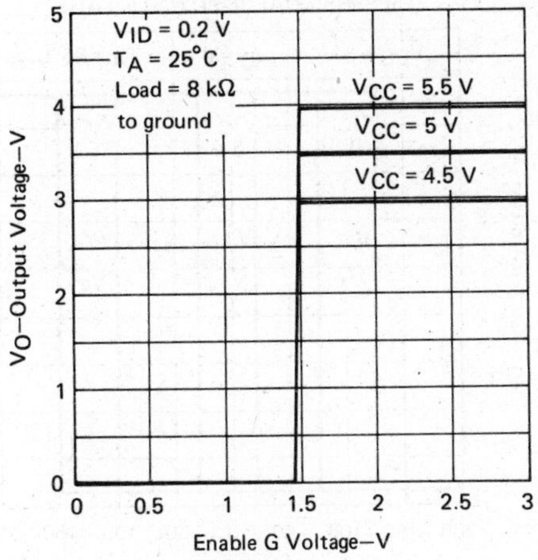
V_{ID} = 0.2 V
T_A = 25°C
Load = 8 kΩ to ground
V_{CC} = 5.5 V
V_{CC} = 5 V
V_{CC} = 4.5 V
V_O—Output Voltage—V
Enable G Voltage—V

FIGURE 6

OUTPUT VOLTAGE
vs
ENABLE G VOLTAGE

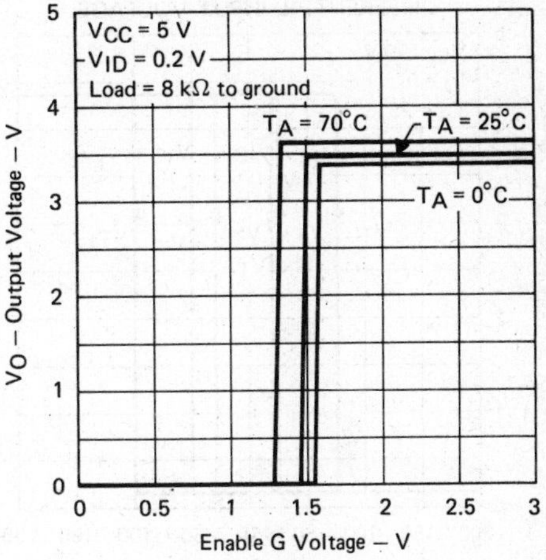
V_{CC} = 5 V
V_{ID} = 0.2 V
Load = 8 kΩ to ground
T_A = 70°C
T_A = 25°C
T_A = 0°C
V_O – Output Voltage – V
Enable G Voltage – V

FIGURE 7

OUTPUT VOLTAGE
vs
ENABLE G VOLTAGE

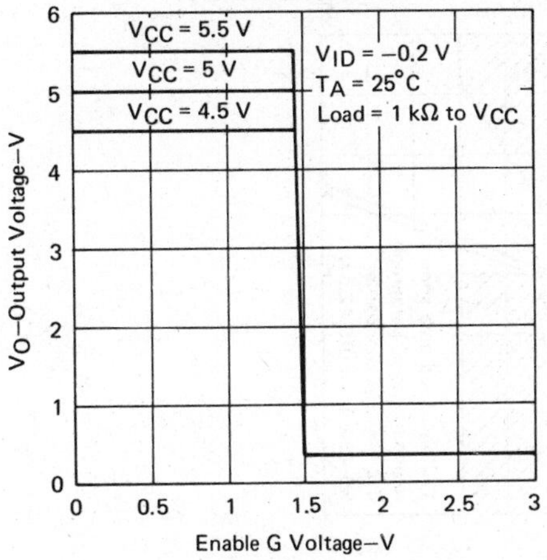
V_{CC} = 5.5 V
V_{CC} = 5 V
V_{CC} = 4.5 V
V_{ID} = –0.2 V
T_A = 25°C
Load = 1 kΩ to V_{CC}
V_O—Output Voltage—V
Enable G Voltage—V

FIGURE 8

OUTPUT VOLTAGE
vs
ENABLE G VOLTAGE

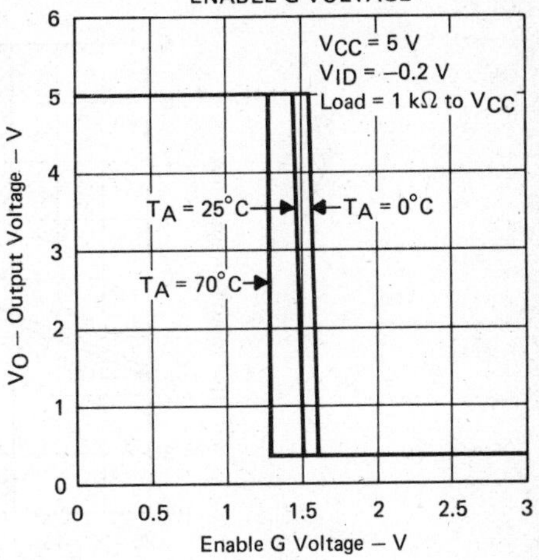
V_{CC} = 5 V
V_{ID} = –0.2 V
Load = 1 kΩ to V_{CC}
T_A = 25°C
T_A = 0°C
T_A = 70°C
V_O – Output Voltage – V
Enable G Voltage – V

FIGURE 9

4

Line Drivers/Receivers

TYPICAL CHARACTERISTICS

AM26LS32A
OUTPUT VOLTAGE
vs
DIFFERENTIAL INPUT VOLTAGE

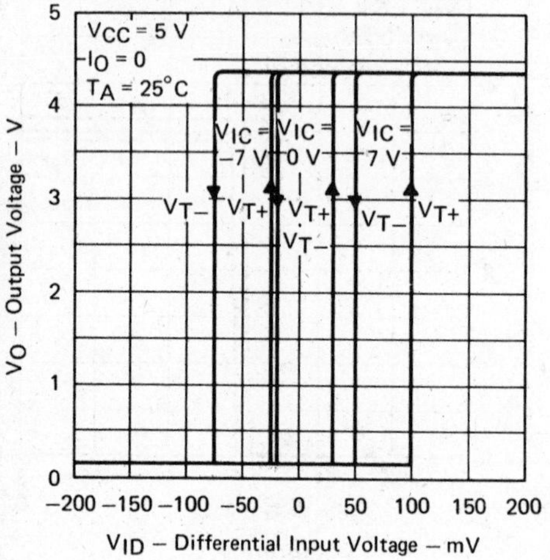
VCC = 5 V
IO = 0
TA = 25°C
VIC = −7 V
VIC = 0 V
VIC = 7 V
VT−
VT+
VT+
VT−
VT−
VT+
VO – Output Voltage – V
VID – Differential Input Voltage – mV
5
4
3
2
1
0
−200 −150 −100 −50 0 50 100 150 200

FIGURE 10

AM26LS33A
OUTPUT VOLTAGE
vs
DIFFERENTIAL INPUT VOLTAGE

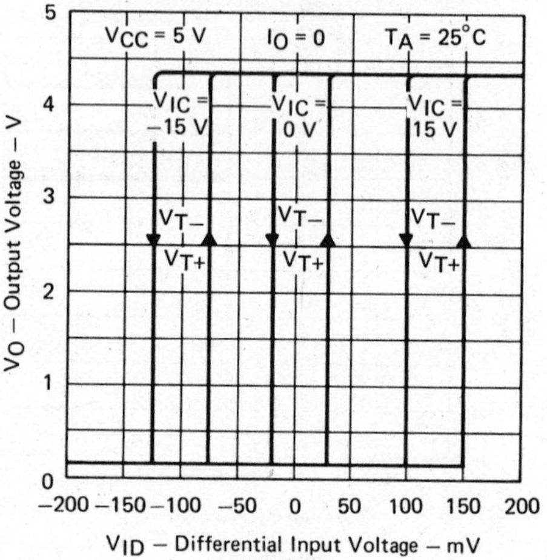
VCC = 5 V
IO = 0
TA = 25°C
VIC = −15 V
VIC = 0 V
VIC = 15 V
VT−
VT+
VO – Output Voltage – V
VID – Differential Input Voltage – mV
5
4
3
2
1
0
−200 −150 −100 −50 0 50 100 150 200

FIGURE 11

INPUT CURRENT
vs
INPUT VOLTAGE

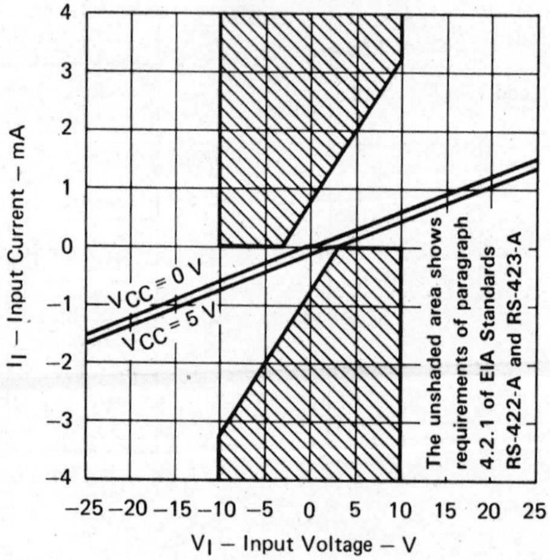
VCC = 0 V
VCC = 5 V
The unshaded area shows requirements of paragraph 4.2.1 of EIA Standards RS-422-A and RS-423-A
II – Input Current – mA
VI – Input Voltage – V
4
3
2
1
0
−1
−2
−3
−4
−25 −20 −15 −10 −5 0 5 10 15 20 25

FIGURE 12

TYPICAL APPLICATION

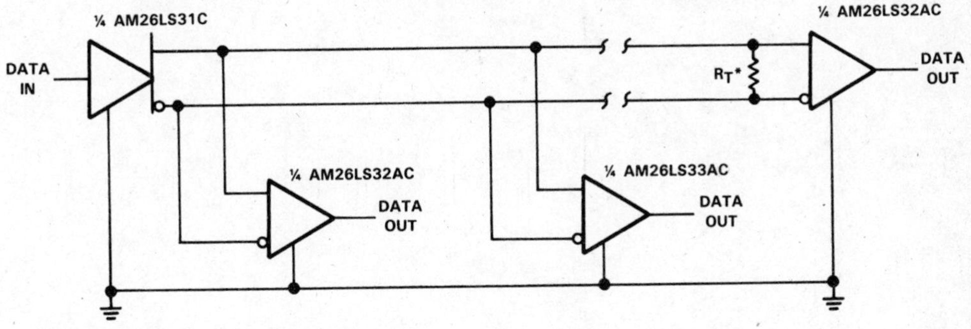
¼ AM26LS31C
DATA IN
¼ AM26LS32AC
DATA OUT
¼ AM26LS32AC
DATA OUT
¼ AM26LS33AC
DATA OUT
R_T*

*R_T equals the characteristic impedance of the line.

FIGURE 13. CIRCUIT WITH MULTIPLE RECEIVERS

D2298, JANUARY 1977—REVISED SEPTEMBER 1986

- **Schottky Circuitry for High Speed, Typical Propagation Delay Time . . . 12 ns**
- **Drivers Feature Open-Collector Outputs for Party-Line (Data Bus) Operation**
- **Driver Outputs Can Sink 100 mA at 0.8 V Maximum**
- **P-N-P Inputs for Minimal Input Loading**
- **Designed to be Interchangeable with Advanced Micro Devices AM26S10 and AM26S11**

AM26S10M, AM26S11M . . . J PACKAGE
AM26S10C, AM26S11C . . . D, J, OR N PACKAGE
(TOP VIEW)

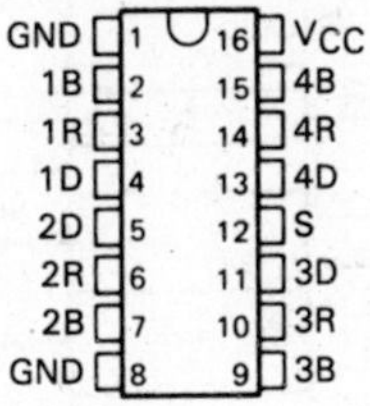
GND 1 16 VCC
1B 2 15 4B
1R 3 14 4R
1D 4 13 4D
2D 5 12 S
2R 6 11 3D
2B 7 10 3R
GND 8 9 3B

AM26S10M, AM26S11M . . . FK PACKAGE
(TOP VIEW)

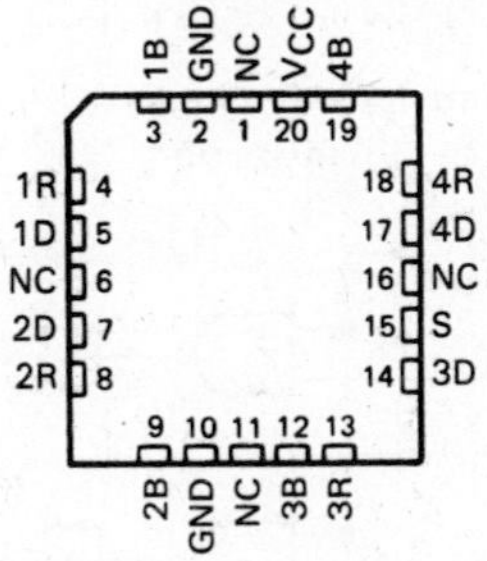
1B GND NC VCC 4B
3 2 1 20 19
1R 4
1D 5
NC 6
2D 7
2R 8
18 4R
17 4D
16 NC
15 S
14 3D
9 10 11 12 13
2B GND NC 3B 3R

NC—No internal connection

description

The AM26S10 and AM26S11 are quadruple bus transceivers utilizing Schottky-diode-clamped transistors for high speed. The drivers feature open-collector outputs capable of sinking 100 mA at 0.8 V maximum. The driver and strobe inputs use p-n-p transistors to reduce the input loading.

The driver of the AM26S10 is inverting; the driver of the AM26S11 is noninverting. Each device has two ground connections for improved ground current-handling capability. For proper operation, the ground pins should be tied together.

The AM26S10M and AM26S11M are characterized for operation over the full military temperature range of −55°C to 125°C. The AM26S10C and AM26S11C are characterized for operation over the temperature range of 0°C to 70°C.

AM26S10
FUNCTION TABLE
(TRANSMITTING)

INPUTS		OUTPUTS	
S	D	B	R
L	H	L	H
L	L	H	L

AM26S11
FUNCTION TABLE
(TRANSMITTING)

INPUTS		OUTPUTS	
S	D	B	R
L	H	H	L
L	L	L	H

AM26S10 AND AM26S11
FUNCTION TABLE
(RECEIVING)

INPUTS			OUTPUT
S	B	D	R
H	H	X	L
H	L	X	H

H = high level, L = low level, X = irrelevant

TEXAS INSTRUMENTS
POST OFFICE BOX 655012 • DALLAS, TEXAS 75265

logic symbols†

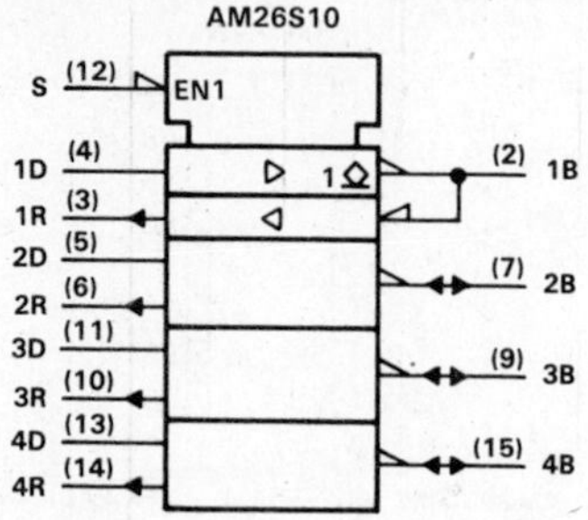
AM26S10
S
(12)
EN1
1D
(4)
1R
(3)
2D
(5)
2R
(6)
3D
(11)
3R
(10)
4D
(13)
4R
(14)
(2)
1B
(7)
2B
(9)
3B
(15)
4B

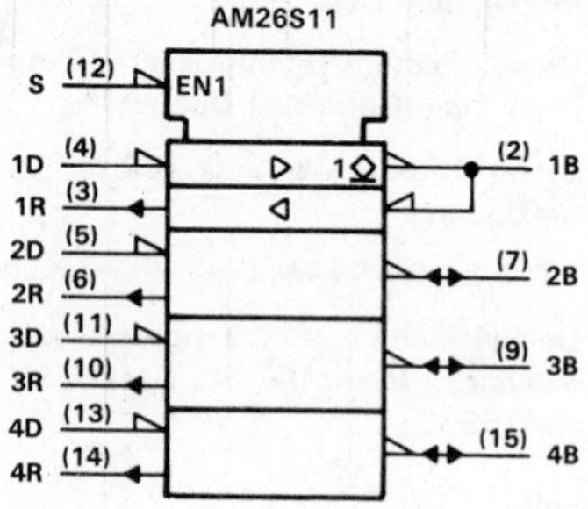
AM26S11
S
(12)
EN1
1D
(4)
1R
(3)
2D
(5)
2R
(6)
3D
(11)
3R
(10)
4D
(13)
4R
(14)
(2)
1B
(7)
2B
(9)
3B
(15)
4B

†These symbols are in accordance with ANSI/IEEE Std 91-1984 and IEC Publication 617-12.
Pin numbers shown are for D, J, and N packages.

logic diagrams (positive logic)

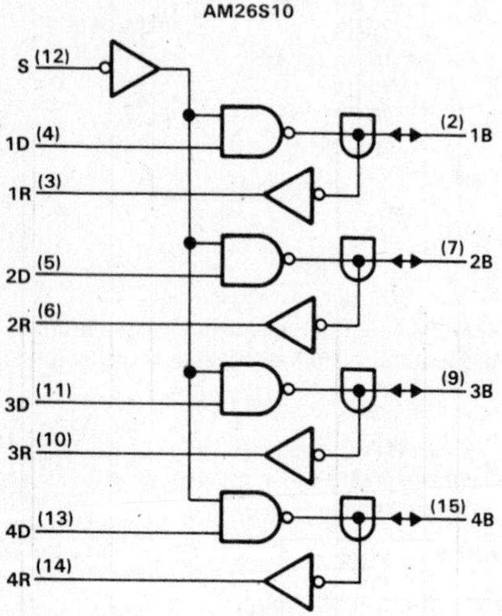
AM26S10
S
(12)
1D
(4)
1R
(3)
2D
(5)
2R
(6)
3D
(11)
3R
(10)
4D
(13)
4R
(14)
(2)
1B
(7)
2B
(9)
3B
(15)
4B

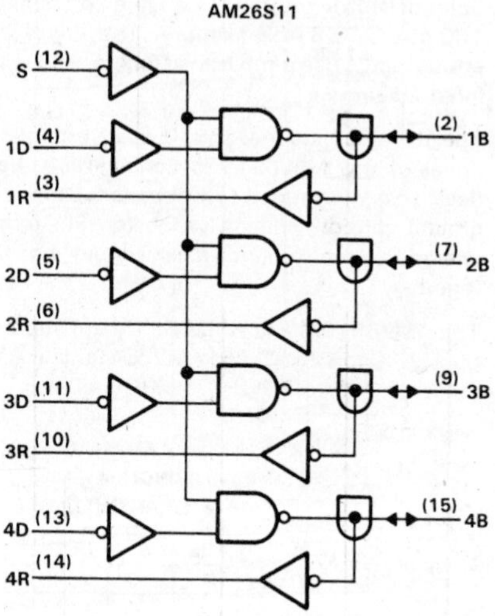
AM26S11
S
(12)
1D
(4)
1R
(3)
2D
(5)
2R
(6)
3D
(11)
3R
(10)
4D
(13)
4R
(14)
(2)
1B
(7)
2B
(9)
3B
(15)
4B

Pin numbers shown are for D, J, and N packages.

schematic (each transceiver)

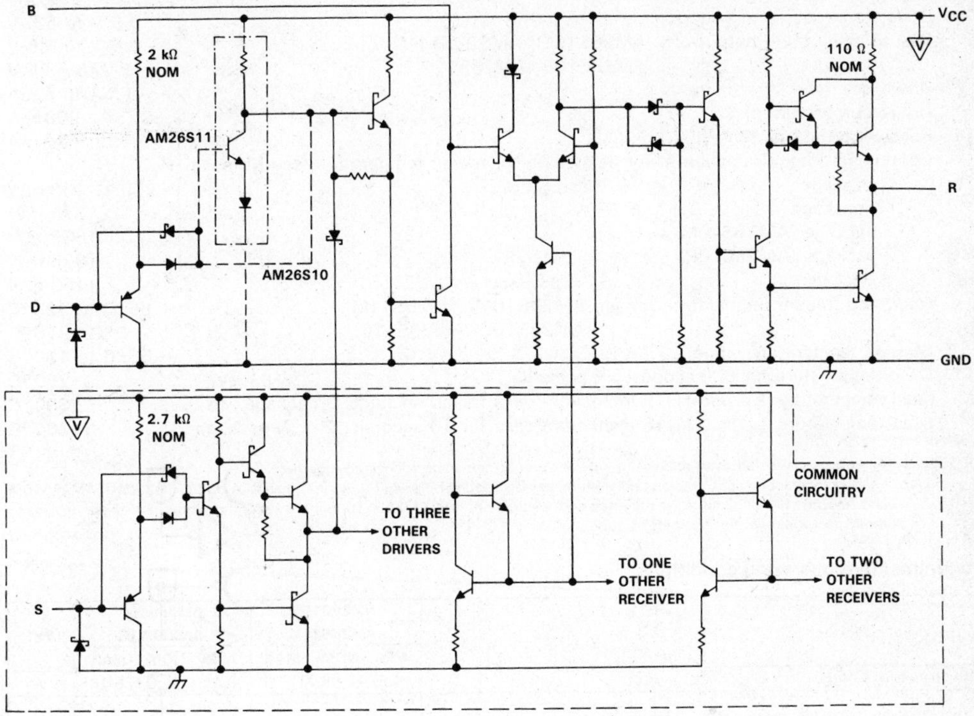
B
VCC
2 kΩ
NOM
110 Ω
NOM
AM26S11
R
AM26S10
D
GND
2.7 kΩ
NOM
COMMON
CIRCUITRY
TO THREE
OTHER
DRIVERS
TO ONE
OTHER
RECEIVER
TO TWO
OTHER
RECEIVERS
S

absolute maximum ratings over operating free-air temperature range (unless otherwise noted)

Supply voltage, V_{CC} (see Note 1) . . . . . . −0.5 V to 7 V
Driver or strobe input voltage . . . . . . −0.5 V to 5.5 V
Bus voltage, driver output off: AM26S10M, AMS26S11M . . . . . . −0.5 V to 5.5 V
AM26S10C, AM26S11C . . . . . . −0.5 V to 5.25 V
Driver or strobe input current . . . . . . −30 mA to 5 mA
Driver output current . . . . . . 200 mA
Receiver output current . . . . . . 30 mA
Continuous total dissipation at (or below) 25°C free-air temperature (see Note 2):
D package . . . . . . 950 mW
FK package . . . . . . 1375 mW
J package (AM26S10M) . . . . . . 1375 mW
J package (AM26S10C) . . . . . . 1025 mW
N package . . . . . . 1150 mW
Operating free-air temperature range: AM26S10M, AM26S11M . . . . . . −55°C to 125°C
AM26S10C, AM26S11C . . . . . . 0°C to 70°C
Storage temperature range . . . . . . −65°C to 150°C
Case temperature for 60 seconds: FK package . . . . . . 260°C
Lead temperature 1,6 mm (1/16 inch) from case for 60 seconds: J package . . . . . . 300°C
Lead temperature 1,6 mm (1/16 inch) from case for 10 seconds: D or N package . . . . . . 260°C

NOTES: 1. All voltage values are with respect to network ground terminals connected together.
2. For operation above 25°C free-air temperature, see Dissipation Derating Curves in Appendix A. In the J package, AM26S10M and AM26S11M chips are alloy mounted and AM26S10C and AM26S11C chips are glass mounted. For these devices in the N package, use the 9.2-mW/°C curve.

recommended operating conditions

		AM26S10M AM26S11M			AM26S10C AM26S11C			UNIT
		MIN	NOM	MAX	MIN	NOM	MAX	
Supply voltage, V_{CC}		4.5	5	5.5	4.75	5	5.25	V
High-level input voltage, V_{IH}	D or S	2			2			V
	B	2.4			2.25			
Low-level input voltage, V_{IL}	D or S			0.8			0.8	V
	B			1.6			1.75	
Receiver high-level output current, I_{OH}				−1			−1	mA
Low-level output current, I_{OL}	Driver			100			100	mA
	Receiver			20			20	
Operating free-air temperature, T_A		−55		125	0		70	°C

electrical characteristics over recommended operating free-air temperature range (unless otherwise noted)

PARAMETER			TEST CONDITIONS†	AM26S10M AM26S11M MIN	TYP‡	MAX	AM26S10C AM26S11C MIN	TYP‡	MAX	UNIT
V_{IK}	Input clamp voltage	D or S	V_{CC} = MIN, I_I = −18 mA			−1.2			−1.2	V
V_{OH}	High-level output voltage	R	V_{CC} = MIN, V_{IH} = 2 V, V_{IL} = V_{IL} max, I_{OH} = −1 mA	2.5	3.4		2.7	3.4		V
V_{OL}	Low-level output voltage	R	V_{CC} = MIN, V_{IH} = V_{IHMIN}, V_{IL} = 0.8 V; I_{OL} = 20 mA			0.5			0.5	V
		B	I_{OL} = 40 mA		0.33	0.5		0.33	0.5	
			I_{OL} = 70 mA		0.42	0.7		0.42	0.7	
			I_{OL} = 100 mA		0.51	0.8		0.51	0.8	
$I_{O(off)}$	Off-state output current	B	V_{IH} = 2 V, V_{IL} = 0.8 V; V_{CC} = MAX, V_O = 0.8 V			−50			−50	µA
			V_{CC} = MAX, V_O = 4.5 V			200			100	
			V_{CC} = 0, V_O = 4.5 V			100			100	
I_{IH}	High-level input current	D	V_{CC} = MAX, V_I = 2.7 V			30			30	µA
		S				20			20	
I_I	Input current at maximum input voltage	D or S	V_{CC} = MAX, V_I = 5.5 V			100			100	µA
I_{IL}	Low-level input current	D	V_{CC} = MAX, V_I = 0.4 V			−0.54			−0.54	mA
		S				−0.36			−0.36	
I_{OS}	Short-circuit output current§	R	V_{CC} = MAX	−20		−55	−18		−60	mA
I_{CC}	Supply current AM26S10		V_{CC} = MAX, Strobe at 0 V, No load, All driver outputs low		45	70		45	70	mA
	Supply current AM26S11					80			80	

† For conditions shown as MIN or MAX, use the appropriate value shown under recommended operating conditions.
‡ All typical values are at T_A = 25°C and V_{CC} = 5 V.
§ Not more than one output should be shorted to ground at a time, and duration of the short circuit should not exceed one second.

switching characteristics, V_{CC} = 5 V, T_A = 25°C

PARAMETER	FROM	TO	TEST CONDITIONS	AM26S10 MIN	TYP	MAX	AM26S11 MIN	TYP	MAX	UNIT
t_{PLH} Propagation delay time, low-to-high-level output	D	B	See Figure 1		10	15		12	19	ns
t_{PHL} Propagation delay time, high-to-low-level output					10	15		12	19	
t_{PLH} Propagation delay time, low-to-high-level output	S	B			14	18		15	20	ns
t_{PHL} Propagation delay time, high-to-low-level output					13	18		14	20	
t_{PLH} Propagation delay time, low-to-high-level output	B	R			10	15		10	15	ns
t_{PHL} Propagation delay time, high-to-low-level output					10	15		10	15	
t_{TLH} Transition time, low-to-high-level output		B		4	10		4	10		ns
t_{THL} Transition time, high-to-low-level output				2	4		2	4		

PARAMETER MEASUREMENT INFORMATION

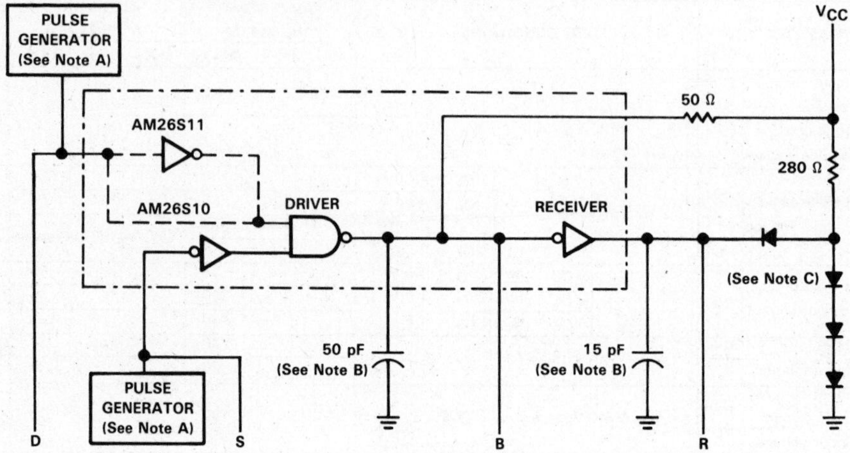
PULSE GENERATOR (See Note A)
VCC
50 Ω
AM26S11
280 Ω
AM26S10
DRIVER
RECEIVER
(See Note C)
50 pF (See Note B)
15 pF (See Note B)
PULSE GENERATOR (See Note A)
D
S
B
R

TEST CIRCUIT

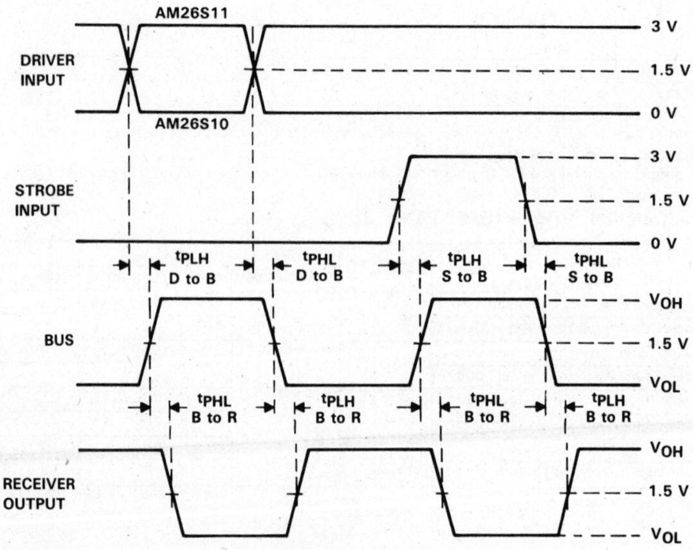
AM26S11
DRIVER INPUT
AM26S10
3 V
1.5 V
0 V
STROBE INPUT
3 V
1.5 V
0 V
tPLH D to B
tPHL D to B
tPLH S to B
tPHL S to B
BUS
VOH
1.5 V
VOL
tPHL B to R
tPLH B to R
tPHL B to R
tPLH B to R
RECEIVER OUTPUT
VOH
1.5 V
VOL

VOLTAGE WAVEFORMS

NOTES: A. The pulse generators have the following characteristics: Z_{out} = 50 Ω, t_r = 10 ± 5 ns.
B. Includes probe and jig capacitance.
C. All diodes are 1N916 or equivalent.

FIGURE 1

TEXAS INSTRUMENTS
POST OFFICE BOX 655012 • DALLAS, TEXAS 75265

TYPICAL APPLICATION

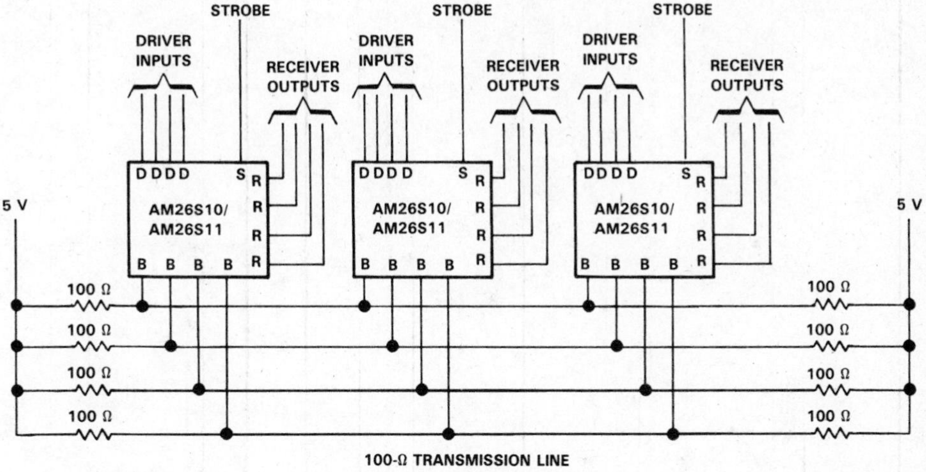
STROBE
DRIVER INPUTS
RECEIVER OUTPUTS
D D D D
S
R
R
R
R
AM26S10/
AM26S11
B B B B
5 V
100 Ω
100-Ω TRANSMISSION LINE

FIGURE 2. PARTY-LINE SYSTEM

D2290, JANUARY 1977—REVISED SEPTEMBER 1986

- **Driver Inputs Compatible with TTL and MOS Circuitry**
- **Driver Outputs Stay Off During Power Up and Power Down**
- **Drivers Feature Open-Collector Outputs for Party-Line Operation**
- **Designed for Interchangeability with Motorola MC3446**
- **Meet IEEE Standard 488-1975**

D, J, OR N DUAL-IN-LINE PACKAGE
(TOP VIEW)

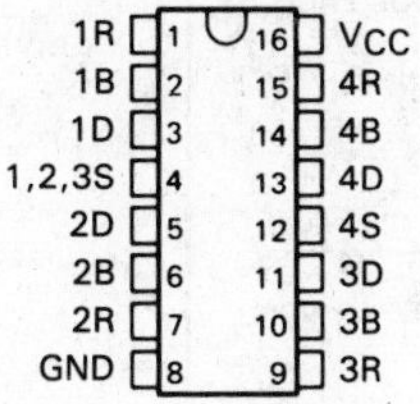
1R 1
1B 2
1D 3
1,2,3S 4
2D 5
2B 6
2R 7
GND 8
9 3R
10 3B
11 3D
12 4S
13 4D
14 4B
15 4R
16 VCC

description

These circuits are quadruple single-ended line transceivers designed for bidirectional flow of data and instructions. The bus terminal characteristic complies with paragraph 3.5.3 of IEEE Standard 488 (see Figure 3). Each driver output is tied to the junction of an internal voltage divider that sets the no-load output voltage and provides bus termination. The driver outputs are guaranteed to be "off" during power up and power down if either input is high. The receivers feature 950 millivolts typical hysteresis for noise immunity.

The MC3446 is characterized for operation from 0°C to 70°C.

FUNCTION TABLE
(TRANSMITTING)

INPUTS		OUTPUT	
S	D	B	R
L	H	H	H
L	L	L	L

FUNCTION TABLE
(RECEIVING)

INPUTS			OUTPUT
S	B	D	R
H	H	X	H
H	L	X	L

logic diagram (positive logic)

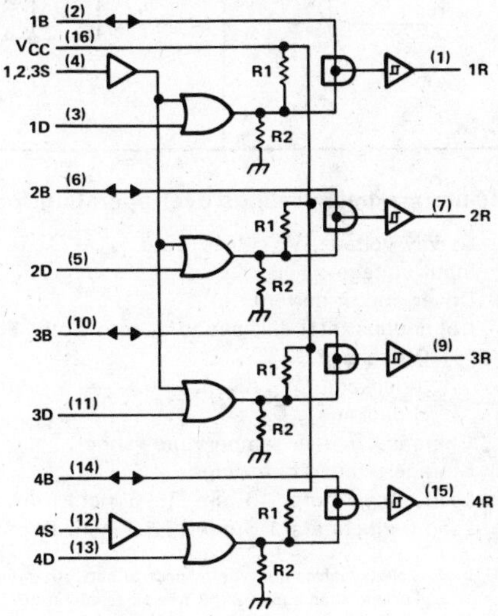
1B (2)
VCC (16)
1,2,3S (4)
1D (3)
R1
R2
(1) 1R
2B (6)
2D (5)
(7) 2R
3B (10)
3D (11)
(9) 3R
4B (14)
4S (12)
4D (13)
(15) 4R

R1 = 2.4 kΩ NOM, R2 = 5 kΩ NOM

logic symbol†

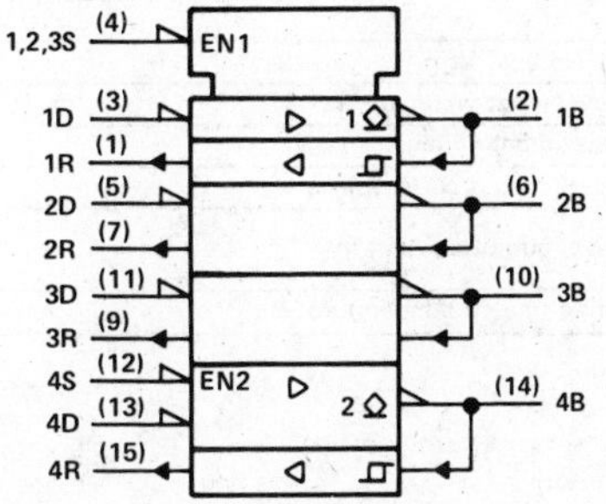
1,2,3S (4)
EN1
1D (3)
1R (1)
2D (5)
2R (7)
3D (11)
3R (9)
4S (12)
4D (13)
4R (15)
EN2
(2) 1B
(6) 2B
(10) 3B
(14) 4B

† This symbol is in accordance with ANSI/IEEE Std 91-1984 and IEC Publication 617-12.

schematics of inputs and outputs

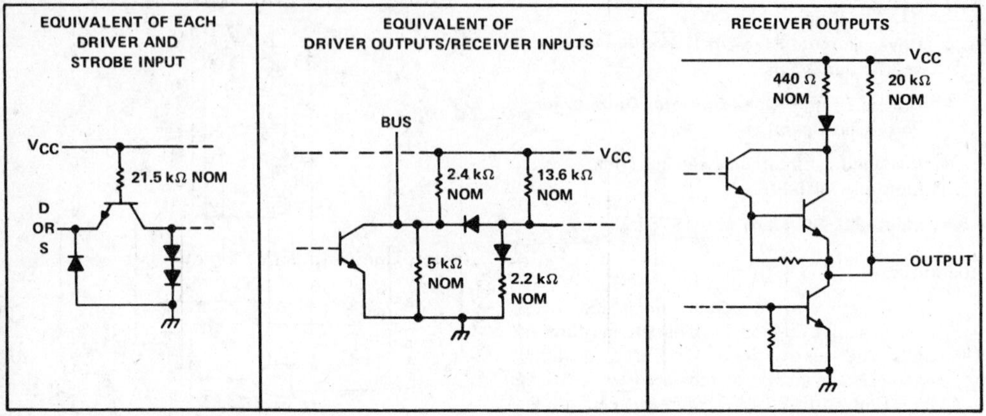
EQUIVALENT OF EACH DRIVER AND STROBE INPUT
VCC
21.5 kΩ NOM
D OR S
EQUIVALENT OF DRIVER OUTPUTS/RECEIVER INPUTS
BUS
VCC
2.4 kΩ NOM
13.6 kΩ NOM
5 kΩ NOM
2.2 kΩ NOM
RECEIVER OUTPUTS
VCC
440 Ω NOM
20 kΩ NOM
OUTPUT

absolute maximum ratings over operating free-air temperature range (unless otherwise noted)

Supply voltage, V_{CC} (see Note 1) . . . . 7 V
Input voltage . . . . 5.5 V
Driver output current . . . . 150 mA
Continuous total dissipation at (or below) 25 °C free-air temperature (see Note 2):
- D package . . . . 950 mW
- J package . . . . 1025 mW
- N package . . . . 1050 mW

Operating free-air temperature range . . . . 0 °C to 70 °C
Storage temperature range . . . . −65 °C to 150 °C
Lead temperature 1,6 mm (1/16 inch) from case for 60 seconds: J package . . . . . . 300 °C
Lead temperature 1,6 mm (1/16 inch) from case for 10 seconds: D and N package . . . . . . 260 °C

NOTES: 1. Voltage values are with respect to network ground terminal.
2. For operation above 25 °C free-air temperature, refer to Dissipation Derating Curves in Appendix A. In the J package, use the 8.2 mW/°C curve, in the D package, use the 7.6 mW/°C curve, and in the N package, use the 9.2-mW/°C curve.

recommended operating conditions

		MIN	NOM	MAX	UNIT
Supply voltage, V_{CC}		4.75	5	5.25	V
High-level input voltage, V_{IH}	D or S	2			V
Low-level input voltage, V_{IL}	D or S			0.8	V
High-level output current, I_{OH}	Receiver			−0.4	mA
Low-level output current, I_{OL}	Driver			48	mA
	Receiver			8	
Operating free-air temperature, T_A		0		70	°C

electrical characteristics over recommended ranges of V_{CC} and operating free-air temperature (unless otherwise noted)

PARAMETER			TEST CONDITIONS	MIN	TYP†	MAX	UNIT
V_{IK}	Input clamp voltage	D or S	$I_I = -12$ mA			−1.5	V
V_{T+}	Positive-going input threshold voltage	B		1.5	1.8	2	V
V_{T-}	Negative-going input threshold voltage	B		0.6	0.85	1.1	V
V_{hys}	Input hysteresis, $(V_{T+} - V_{T-})$	B		400	950		mV
V_{OH}	High-level output voltage	B	$V_{IH} = 2.4$ V, $I_{OH} = 0$	2.5	3.3	3.7	V
		R	$V_{IH} = 2$ V, $I_{OH} = -400$ μA	2.4			
V_{OL}	Low-level output voltage	B	$V_{IL} = 0.8$ V, $I_{OL} = 48$ mA			0.4	V
		R	$V_{IL} = 0.8$ V, $I_{OL} = 8$ mA			0.4	
$I_{O(bus)}$	Bus current	B	$V_{IH} = 2.4$ V, $V_O = 5.5$ V			2.5	mA
			$V_{IH} = 2.4$ V, $V_O = 5$ V	0.7			
			$V_{IH} = 2.4$ V, $V_O = 0.4$ V	−1.3		−3.2	
V_{OK}	Output clamp voltage	B	$I_O = -12$ mA			−1.5	V
I_I	Input current at maximum input voltage	D or S	$V_I = 5.5$ V			1	mA
I_{IH}	High-level input current	D or S	$V_{IH} = 2.4$ V		5	20	μA
I_{IL}	Low-level input current	D or S	$V_{CC} = 5$ V, $V_{IL} = 0.4$ V, $T_A = 25°C$		0.2	0.36	mA
I_{OS}	Short-circuit output current	R	$V_{IH} = 2$ V	4		14	mA
I_{CCH}	Supply current, all outputs high		No load		10	19	mA
I_{CCL}	Supply current, all outputs low		No load		32	39	mA

†All typical values are at $V_{CC} = 5$ V, $T_A = 25°C$.

switching characteristics, $V_{CC} = 5$ V, $T_A = 25°C$

PARAMETER		FROM	TO	TEST CONDITIONS	MIN	MAX	UNIT
t_{PLH}	Propagation delay time, low-to-high-level output	D	B	See Figure 1		40	ns
t_{PHL}	Propagation delay time, high-to-low-level output					50	
t_{PLH}	Propagation delay time, low-to-high-level output	S	B			50	ns
t_{PHL}	Propagation delay time, high-to-low-level output					50	
t_{PLH}	Propagation delay time, low-to-high-level output	B	R	See Figure 2		50	ns
t_{PHL}	Propagation delay time, high-to-low-level output					40	

PARAMETER MEASUREMENT INFORMATION

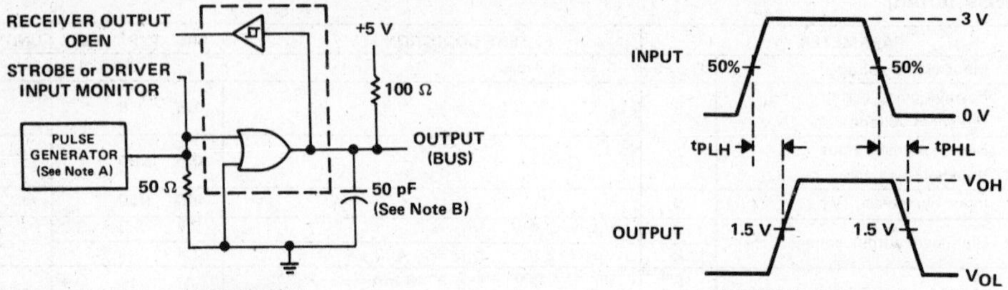
RECEIVER OUTPUT OPEN
STROBE or DRIVER INPUT MONITOR
PULSE GENERATOR (See Note A)
50 Ω
+5 V
100 Ω
OUTPUT (BUS)
50 pF (See Note B)
INPUT
50%
50%
3 V
0 V
tPLH
tPHL
VOH
OUTPUT
1.5 V
1.5 V
VOL

TEST CIRCUIT — VOLTAGE WAVEFORMS

FIGURE 1

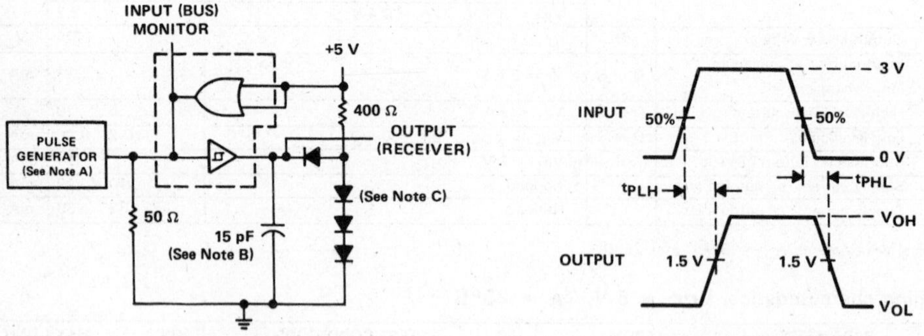
INPUT (BUS) MONITOR
+5 V
400 Ω
OUTPUT (RECEIVER)
PULSE GENERATOR (See Note A)
(See Note C)
50 Ω
15 pF (See Note B)
INPUT
50%
50%
3 V
0 V
tPLH
tPHL
VOH
OUTPUT
1.5 V
1.5 V
VOL

TEST CIRCUIT — VOLTAGE WAVEFORMS

FIGURE 2

NOTES: A. The input pulse is supplied by a generator having the following characteristics: t_w = 100 ns, PRR ≤ 1 MHz, t_r ≤ 10 ns, t_f ≤ 10 ns, Z_{out} ≈ 50 Ω.
B. This value includes probe and jig capacitance.
C. All diodes are 1N916 or 1N3064.

TYPICAL CHARACTERISTICS

DRIVER OUTPUT CHARACTERISTCS†

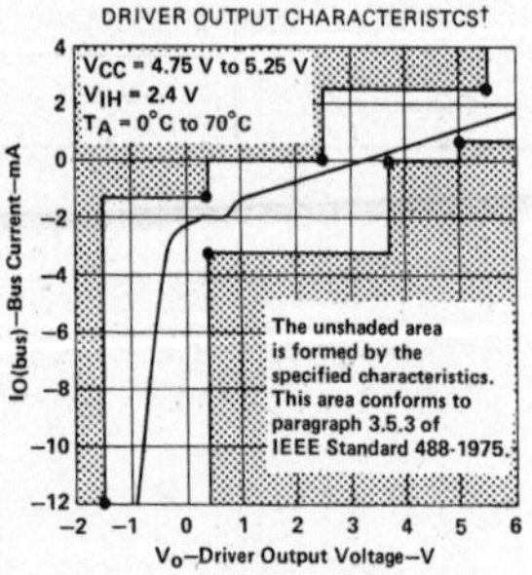
VCC = 4.75 V to 5.25 V
VIH = 2.4 V
TA = 0°C to 70°C
The unshaded area is formed by the specified characteristics. This area conforms to paragraph 3.5.3 of IEEE Standard 488-1975.
IO(bus)–Bus Current–mA
VO–Driver Output Voltage–V
4
2
0
−2
−4
−6
−8
−10
−12
−2 −1 0 1 2 3 4 5 6

FIGURE 3

RECEIVER TRANSFER CHARACTERISTICS

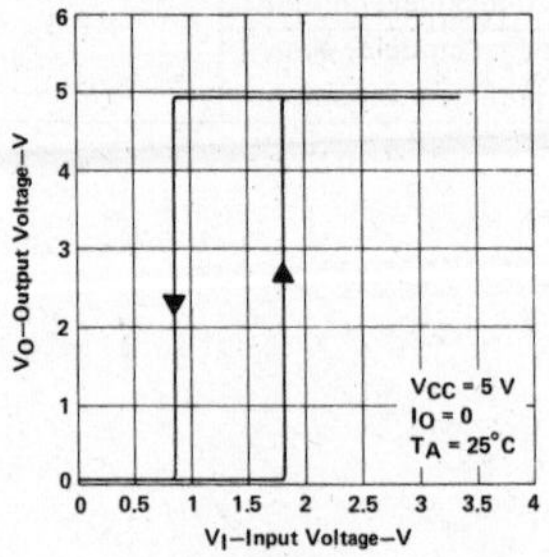
VCC = 5 V
IO = 0
TA = 25°C
VO–Output Voltage–V
VI–Input Voltage–V
6
5
4
3
2
1
0
0 0.5 1 1.5 2 2.5 3 3.5 4

FIGURE 4

†Conditions for typical curve are V_{CC} = 5 V, T_A = 25°C.

TEXAS INSTRUMENTS
POST OFFICE BOX 655012 • DALLAS, TEXAS 75265

D3006, FEBRUARY 1986—REVISED OCTOBER 1986

- Four Independent Receivers with Common Enable Input
- High Input Sensitivity . . . 25 mV Max
- High Input Impedance
- MC3450 has Three-State Outputs
- MC3452 has Open-Collector Outputs
- Glitch-Free Power-Up/Power-Down Operation

D, J, OR N PACKAGE
(TOP VIEW)

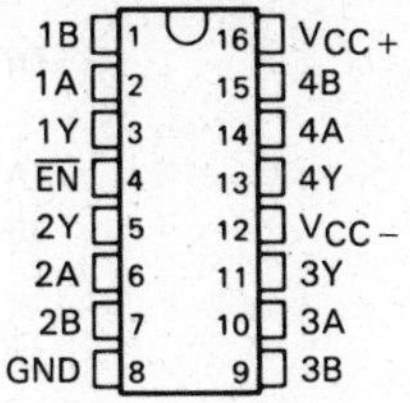
1B 1 16 VCC+
1A 2 15 4B
1Y 3 14 4A
EN 4 13 4Y
2Y 5 12 VCC−
2A 6 11 3Y
2B 7 10 3A
GND 8 9 3B

description

The MC3450 and MC3452 are quadruple differential line receivers designed for use in balanced and unbalanced digital data transmission. The MC3450 and MC3452 are the same except that the MC3450 has three-state ouputs whereas the MC3452 has open-collector outputs, which permit the wire-AND function with similar output devices. Three-state and open-collector outputs permit connection directly to a bus-organized system.

The MC3450 and MC3452 are designed for optimum performance when used with either the MC3453 quadruple differential line driver or SN75109A, SN75110A, and SN75112 dual differential drivers.

The MC3450 and MC3452 are characterized for operation from 0°C to 70°C.

FUNCTION TABLE

DIFFERENTIAL INPUTS A-B	ENABLE $\overline{EN}$	OUTPUT Y
$V_{ID} \geq 25$ mV	L	H
-25 mV $< V_{ID} < 25$ mV	L	?
$V_{ID} \leq 25$ mV	L	L
X	H	Z

H = high level, L = low level, ? = indeterminate,
Z = impedance (off)

logic symbols†

MC3450

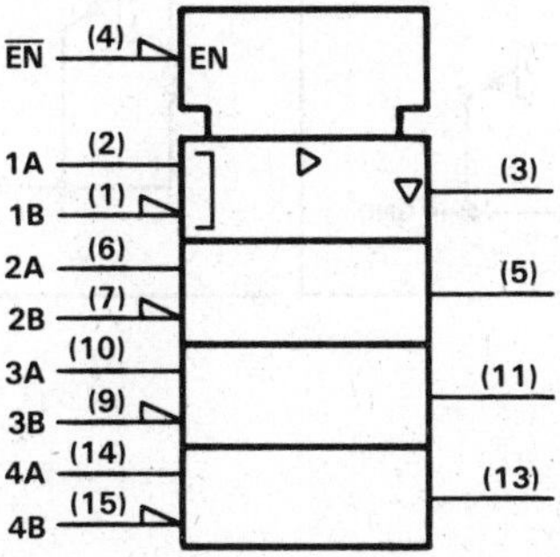
EN
(4)
EN
1A
(2)
1B
(1)
(3)
1Y
2A
(6)
2B
(7)
(5)
2Y
3A
(10)
3B
(9)
(11)
3Y
4A
(14)
4B
(15)
(13)
4Y

MC3452

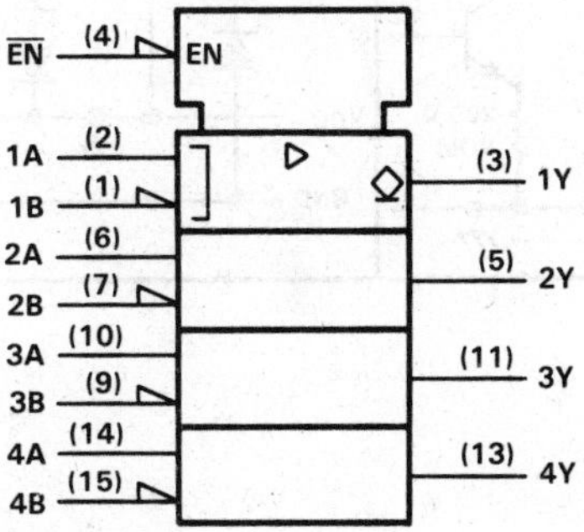
EN
(4)
EN
1A
(2)
1B
(1)
(3)
1Y
2A
(6)
2B
(7)
(5)
2Y
3A
(10)
3B
(9)
(11)
3Y
4A
(14)
4B
(15)
(13)
4Y

† These symbols are in accordance with ANSI/IEEE Std 91-1984 and IEC Publication 617-12.

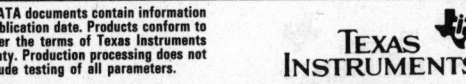

logic diagram (positive logic)

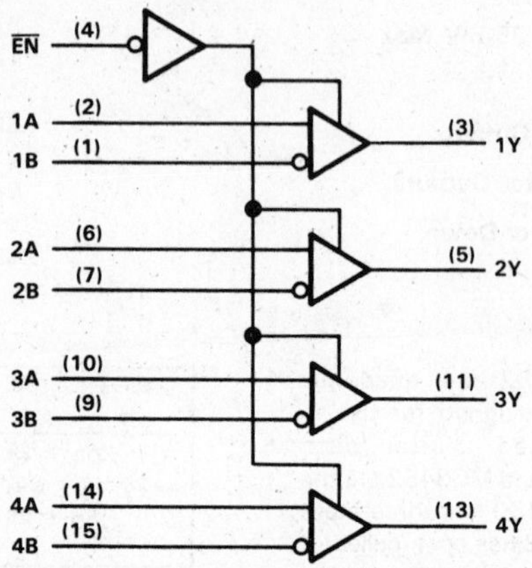
EN
(4)
1A
(2)
1B
(1)
(3)
1Y
2A
(6)
2B
(7)
(5)
2Y
3A
(10)
3B
(9)
(11)
3Y
4A
(14)
4B
(15)
(13)
4Y

schematics of inputs and outputs

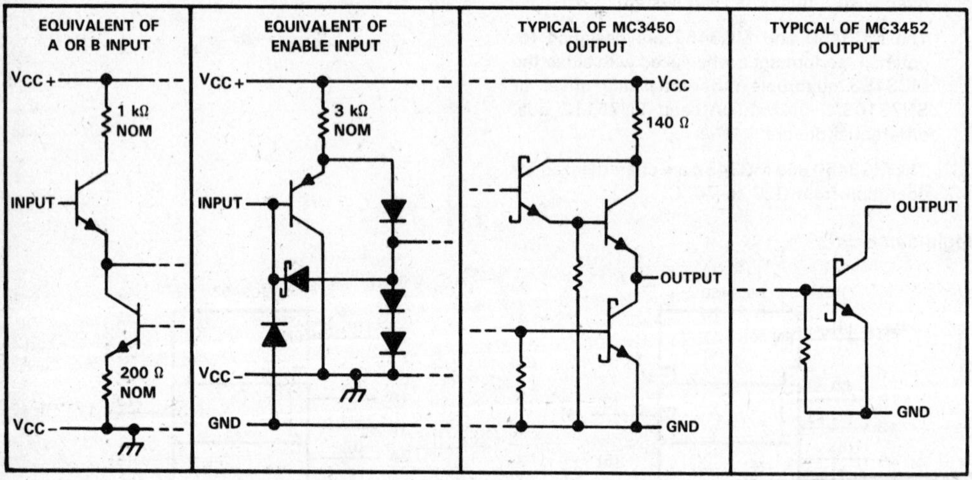
EQUIVALENT OF
A OR B INPUT
VCC+
1 kΩ
NOM
INPUT
200 Ω
NOM
VCC−
EQUIVALENT OF
ENABLE INPUT
VCC+
3 kΩ
NOM
INPUT
VCC−
GND
TYPICAL OF MC3450
OUTPUT
VCC
140 Ω
OUTPUT
GND
TYPICAL OF MC3452
OUTPUT
OUTPUT
GND

absolute maximum ratings over operating free-air temperature range (unless otherwise noted)

Supply voltage, V_{CC+} (see Note 1)	7 V
Supply voltage, V_{CC-}	−7 V
Differential input voltage (see Note 2)	±6 V
Common-mode input voltage (see Note 3)	±5 V
Enable input voltage	5.5 V
Continuous total dissipation at (or below) 25 °C free-air temperature (see Note 4):	
D package	950 mW
J package	1025 mW
N package	1150 mW
Operating free-air temperature range	0 °C to 70 °C
Storage temperature range	−65 °C to 150 °C
Lead temperature 1,6 mm (1/16 inch) from case for 10 seconds: D or N package	260 °C
Lead temperature 1,6 mm (1/16 inch) from case for 60 seconds: J package	300 °C

NOTES: 1. All voltage values, except differential input voltage, are with respect to network ground terminal.
2. Differential input voltage is measured at the noninverting input with respect to the corresponding inverting input.
3. Common-mode input voltage is the average of the voltages at the A and B inputs.
4. For operation above 25 °C free-air temperature, derate the D package to 608 mW at 70 °C at the rate of 7.6 mW/ °C, the J package to 656 mW at 70 °C at the rate of 8.2 mW/ °C, and the N package to 736 mW at 70 °C at the rate of 9.2 mW/ °C. In the J package, MC3450 and MC3452 chips are glass mounted.

recommended operating conditions

	MIN	NOM	MAX	UNIT
Supply voltage, V_{CC+}	4.75	5	5.25	V
Supply voltage, V_{CC-}	−4.75	−5	−5.25	V
High-level enable input voltage, V_{IH}	2			V
Low-level enable input voltage, V_{IL}			0.8	V
Low-level output current, I_{OL}			−16	mA
Differential input voltage, V_{ID} (see Note 5)	−5†		5	V
Common-mode input voltage, V_{IC} (see Note 5)	−3†		3	V
Input voltage range, any differential input to ground	−5†		3	V
Operating free-air temperature, T_A	0		70	°C

† The algebraic convention, in which the less positive (more negative) limit is designated minimum, is used in this data sheet for common-mode input voltage.

NOTE 5: The recommended combinations of input voltages fall within the shaded area of Figure 1.

RECOMMENDED COMBINATIONS OF INPUT VOLTAGES

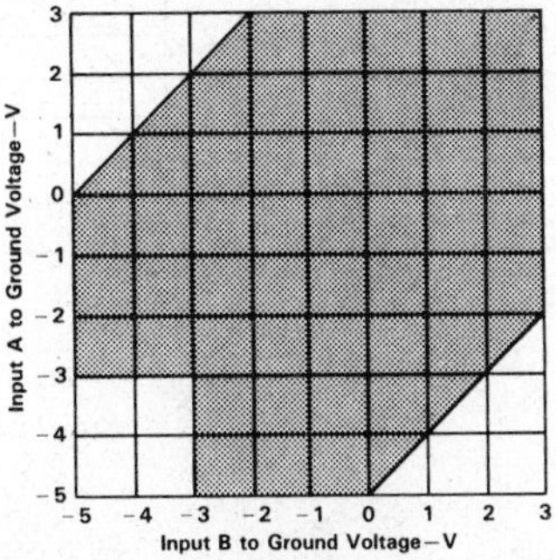
Input A to Ground Voltage—V
Input B to Ground Voltage—V
3
2
1
0
−1
−2
−3
−4
−5
−5 −4 −3 −2 −1 0 1 2 3

FIGURE 1

electrical characteristics over recommended operating free-air temperature range, $V_{CC\pm}$ = ±5.25 V (unless otherwise noted)

PARAMETER			TEST CONDITIONS	MC3450			MC3452			UNIT
				MIN	TYP†	MAX	MIN	TYP†	MAX	
I_{IH}	High-level input current	A inputs	V_{ID} = −2 V		30	75		30	75	μA
		B inputs	V_{ID} = −2 V		30	75		30	75	
		$\overline{EN}$	V_{IH} = 2.4 V			40			40	μA
			V_{IH} = 5.25 V			1			1	mA
I_{IL}	Low-level input current	A inputs	V_{ID} = 2 V			−10			−10	μA
		B inputs	V_{ID} = 2 V			−10			−10	
		$\overline{EN}$	V_{IL} = 0.4 V			−1.6			−1.6	mA
V_{OH}	High-level output voltage		$V_{CC\pm}$ = ±4.75 V, V_{ID} = 25 mV, $\overline{EN}$ at 0.8 V, I_{OH} = −400 μA, V_{IC} = −3 V to 3 V	2.4						V
I_{OH}	High-level output current		$V_{CC\pm}$ = ±4.75 V, V_{OH} = 5.25 V						250	μA
V_{OL}	Low-level output voltage		$V_{CC\pm}$ = ±4.75 V, V_{ID} = −25 mV, $\overline{EN}$ at 2 V, I_{OL} = 16 mA, V_{IC} = −3 V to 3 V			0.5			0.5	V
I_{OZ}	High-impedance-state output current		V_O = 2.4 V			40				μA
			V_O = 0.4 V			−40				
I_{OS}	Short-circuit output current‡		V_{ID} = 25 mV, V_O = 0, $\overline{EN}$ at 0.8 V	−18		−70				mA
I_{CCH+}	Supply current from V_{CC+}, outputs high					60			60	mA
I_{CCH-}	Supply current from V_{CC-}, outputs high					−30			−30	mA

† All typical values are at V_{CC+} = 5 V, V_{CC-} = −5 V, T_A = 25°C.
‡ Not more one output should be shorted at a time.

switching characteristics, $V_{CC\pm}$ = ±5 V, T_A = 25°C

PARAMETER	FROM (INPUT)	TO (OUTPUT)	TEST CONDITIONS	MC3450			MC3452			UNIT
				MIN	TYP†	MAX	MIN	TYP†	MAX	
t_{PLH}	A and B	Y	C_L = 50 pF, See Figure 2		17	25				ns
			C_L = 15 pF, See Figure 2					19	25	
t_{PHL}	A and B	Y	C_L = 50 pF, See Figure 2		17	25				ns
			C_L = 15 pF, See Figure 2					19	25	
t_{PZH}	$\overline{EN}$	Y	C_L = 50 pF, See Figure 2			21				ns
t_{PZL}	$\overline{EN}$	Y				27				
t_{PHZ}	$\overline{EN}$	Y	C_L = 15 pF, See Figure 3			18				ns
t_{PLZ}	$\overline{EN}$	Y				29				
t_{PLH}	$\overline{EN}$	Y	C_L = 15 pF, See Figure 4						25	ns
t_{PHL}	$\overline{EN}$	Y	C_L = 15 pF, See Figure 4						25	ns

† All typical values are at V_{CC+} = 5 V, V_{CC-} = −5 V, T_A = 25°C.

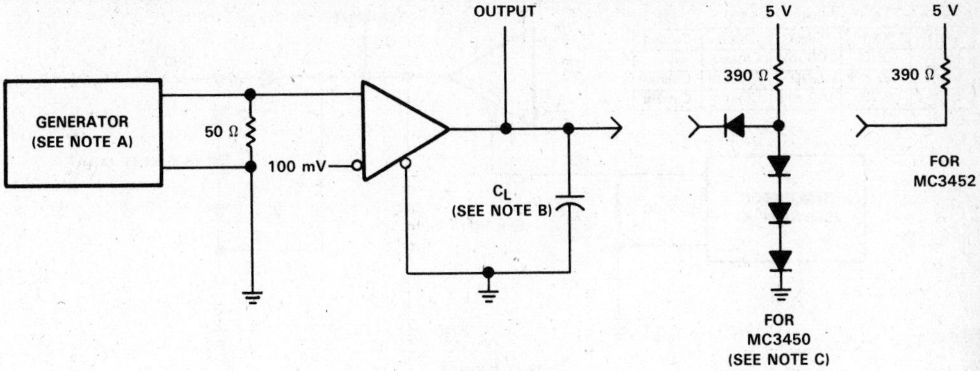
OUTPUT
5 V
5 V
390 Ω
390 Ω
GENERATOR
(SEE NOTE A)
50 Ω
100 mV
C_L
(SEE NOTE B)
FOR
MC3452
FOR
MC3450
(SEE NOTE C)

TEST CIRCUIT

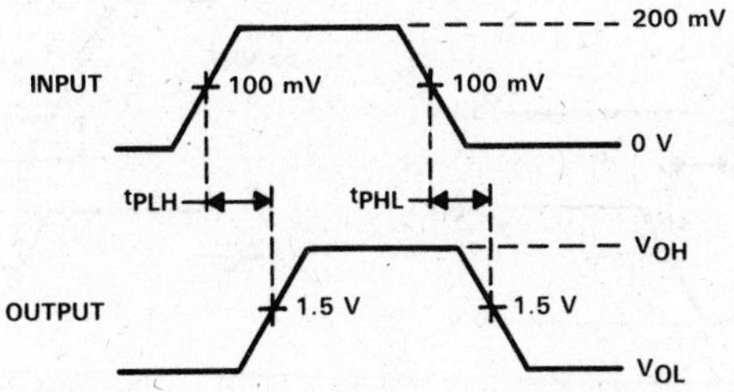
200 mV
INPUT
100 mV
100 mV
0 V
t_{PLH}
t_{PHL}
V_{OH}
OUTPUT
1.5 V
1.5 V
V_{OL}

NOTES: A. The input pulse is supplied by a generator having the following characteristics: PRR ≤ 1 MHz, duty cycle = 50%, $t_r \leq 6$ ns, $t_f \leq 6$ ns.
B. C_L includes probe and jig capacitance.
C. All diodes are 1N916 or equivalent.

VOLTAGE WAVEFORMS

FIGURE 2. PROPAGATION DELAY TIMES

	A	B	S1	S2
t_{PZH}	100 mV	GND	Open	Closed
t_{PZL}	GND	100 mV	Closed	Open
t_{PHZ}	100 mV	GND	Closed	Closed
t_{PLZ}	GND	100 mV	Closed	Closed

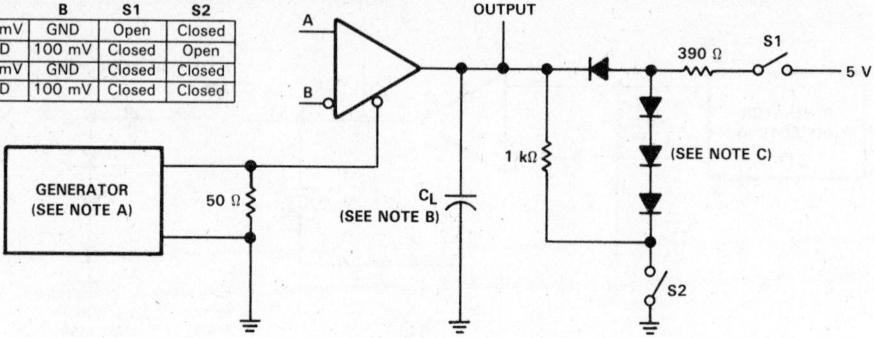
A
B
OUTPUT
390 Ω
S1
5 V
1 kΩ
(SEE NOTE C)
GENERATOR
(SEE NOTE A)
50 Ω
C_L
(SEE NOTE B)
S2

TEST CIRCUIT

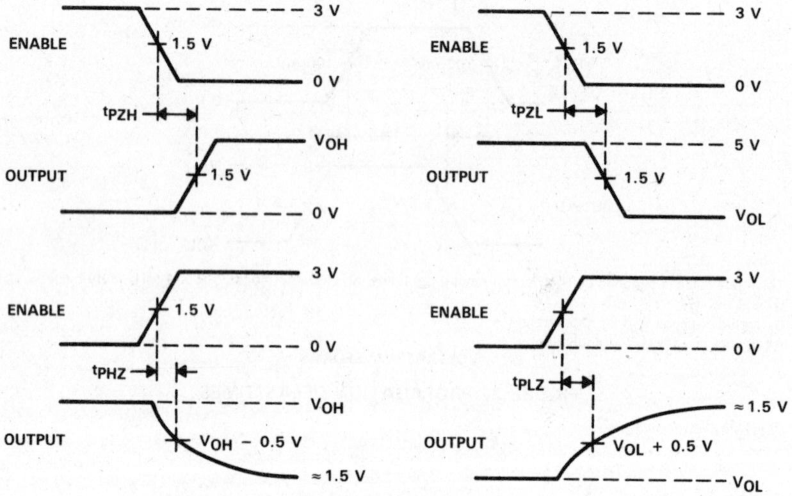
ENABLE
3 V
1.5 V
0 V
tPZH
OUTPUT
VOH
1.5 V
0 V
ENABLE
3 V
1.5 V
0 V
tPZL
OUTPUT
5 V
1.5 V
VOL
ENABLE
3 V
1.5 V
0 V
tPHZ
OUTPUT
VOH
VOH − 0.5 V
≈1.5 V
ENABLE
3 V
0 V
tPLZ
OUTPUT
≈1.5 V
VOL + 0.5 V
VOL

VOLTAGE WAVEFORMS

NOTES: A. The input pulse is supplied by a generator having the following characteristics: PRR ≤ 1 MHz, duty cycle = 50%, $t_r \leq 6$ ns, $t_f \leq 6$ ns.

B. C_L includes probe and jig capacitance.

C. All diodes are 1N916 or equivalent.

FIGURE 3. MC3450 ENABLE AND DISABLE TIMES

TEXAS INSTRUMENTS
POST OFFICE BOX 655012 • DALLAS, TEXAS 75265

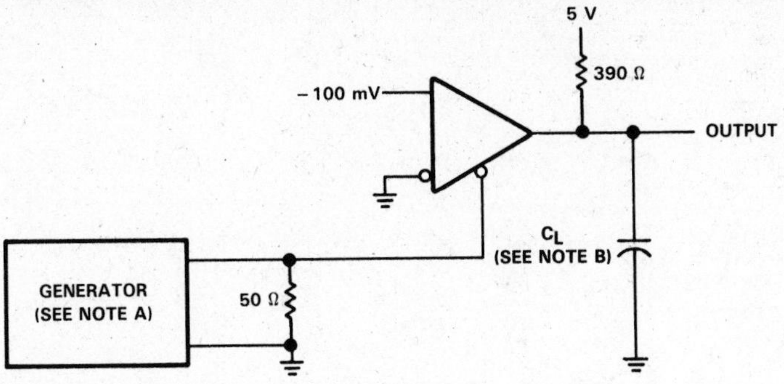
5 V
390 Ω
−100 mV
OUTPUT
C_L
(SEE NOTE B)
GENERATOR
(SEE NOTE A)
50 Ω

TEST CIRCUIT

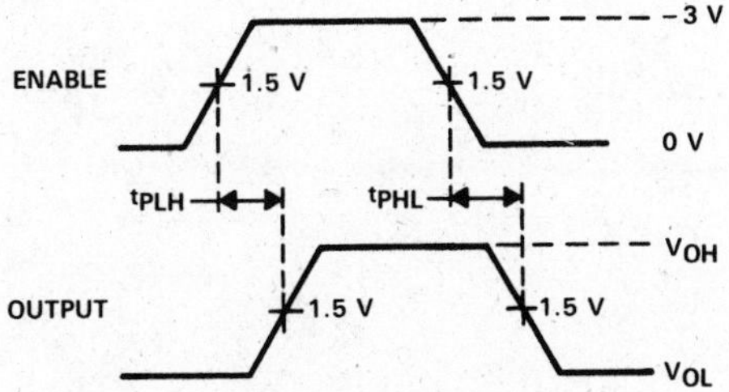
3 V
ENABLE
1.5 V
1.5 V
0 V
t_{PLH}
t_{PHL}
V_{OH}
OUTPUT
1.5 V
1.5 V
V_{OL}

VOLTAGE WAVEFORMS

NOTES: A. The input pulse is supplied by a generator having the following characteristics: PRR ≤ 1 MHz, duty cycle = 50%, $t_r \le 6$ ns, $t_f \le 6$ ns.
B. C_L includes probe and jig capacitance.

FIGURE 4. MC3452 PROPAGATION DELAY TIMES FROM ENABLE

D3000, FEBRUARY 1986

- **Similar to a Dual Version of SN75110A Line Driver**
- **Improved Stability Over Supply Voltage and Temperature Ranges**
- **Constant-Current Outputs**
- **High Output Impedance**
- **High Common-Mode Output Voltage Range (−3 V to 10 V)**
- **Glitch-Free Power-Up/Power-Down Operation**
- **TTL Input Compatibility**
- **Common Enable Circuit**
- **Designed to be Interchangeable with Motorola MC3453**

D, J, OR N
DUAL-IN-LINE PACKAGE
(TOP VIEW)

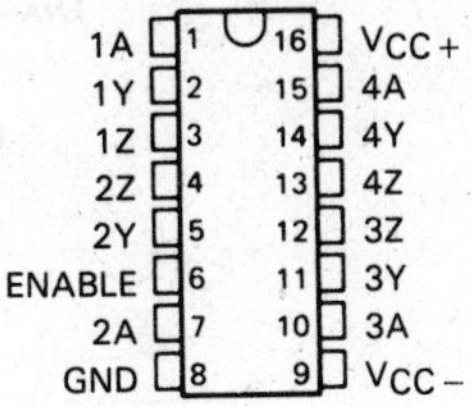
1A 1
1Y 2
1Z 3
2Z 4
2Y 5
ENABLE 6
2A 7
GND 8
16 VCC+
15 4A
14 4Y
13 4Z
12 3Z
11 3Y
10 3A
9 VCC−

FUNCTION TABLE

LOGIC INPUT	ENABLE INPUT	OUTPUT CURRENT Z	OUTPUT CURRENT Y
H	H	ON	OFF
L	H	OFF	ON
H	L	OFF	OFF
L	L	OFF	OFF

L = low logic level
H = high logic level

description

The MC3453 features four line drivers with a common enable input. When the enable input is high, a constant output current is switched between each pair of output terminals in response to the logic level at that channel's input. When the enable is low, all channel outputs are nonconductive (transistors biased to cutoff). This minimizes loading in party-line systems where a large number of drivers share the same line.

The driver outputs have a common-mode voltage range of −3 volts to 10 volts, allowing common-mode voltages on the line without affecting driver performance.

All inputs are diode clamped and are designed to satisfy TTL-system requirements. The inputs are tested at 2 volts for high-logic-level input conditions and 0.8 volt for low-logic-level input conditions. These tests guarantee 400 millivolts of noise margin when interfaced with Series 54/74 TTL.

The MC3453 is characterized for operation from 0°C to 70°C.

logic symbol†

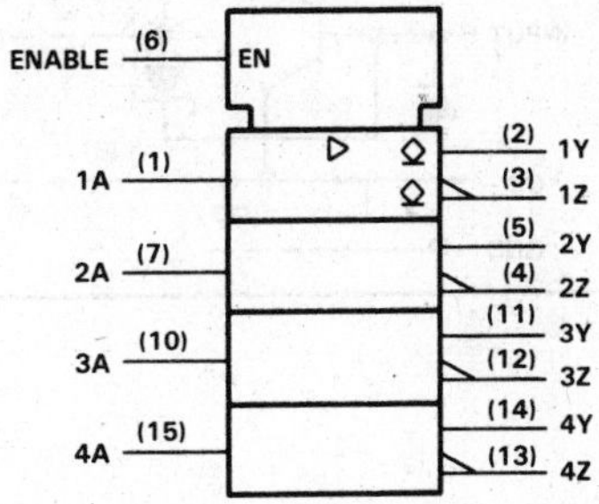
ENABLE
(6)
EN
1A
(1)
2A
(7)
3A
(10)
4A
(15)
(2) 1Y
(3) 1Z
(5) 2Y
(4) 2Z
(11) 3Y
(12) 3Z
(14) 4Y
(13) 4Z

† This symbol is in accordance with ANSI/IEEE Std 91-1984 and IEC Publication 617-12.

TEXAS INSTRUMENTS
POST OFFICE BOX 655012 • DALLAS, TEXAS 75265

logic diagram (positive logic)

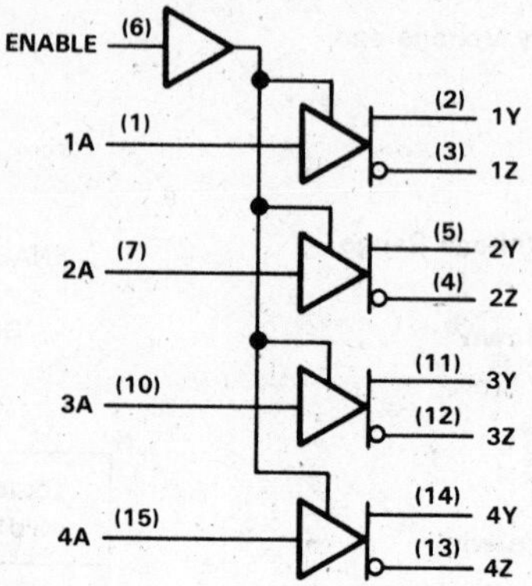
ENABLE
(6)
1A
(1)
(2)
1Y
(3)
1Z
2A
(7)
(5)
2Y
(4)
2Z
3A
(10)
(11)
3Y
(12)
3Z
4A
(15)
(14)
4Y
(13)
4Z

schematics of inputs and outputs

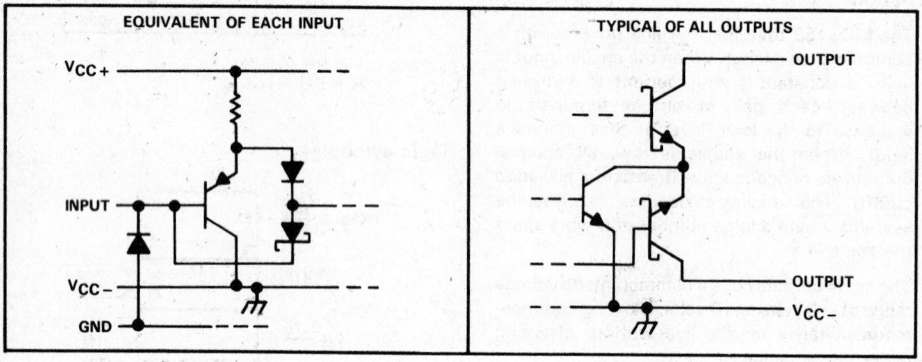
EQUIVALENT OF EACH INPUT
VCC+
INPUT
VCC−
GND
TYPICAL OF ALL OUTPUTS
OUTPUT
OUTPUT
VCC−

TEXAS INSTRUMENTS
POST OFFICE BOX 655012 • DALLAS, TEXAS 75265

absolute maximum ratings over operating free-air temperature range (unless otherwise noted)

Supply voltage, V_{CC+} (see Note 1) .... 7 V
Supply voltage, V_{CC-} .... −7 V
Input voltage (any input) .... 5.5 V
Output voltage range (any output) .... −5 V to 12 V
Continuous total dissipation at (or below) 25 °C free-air temperature (see Note 2):
- D package .... 950 mW
- J package .... 1025 mW
- N package .... 1150 mW

Operating free-air temperature range .... 0 °C to 70 °C
Storage temperature range .... −65 °C to 150 °C
Lead temperature 1,6 mm (1/16 inch) from case for 10 seconds: D, N package .... 260 °C
Lead temperature 1,6 mm (1/16 inch) from case for 60 seconds: J package .... 300 °C

NOTES: 1. All voltage values are with respect to network ground terminal.
2. For operation above 25 °C free-air temperature, derate the D package to 608 mW at 70 °C at the rate of 7.6 mW/°C, derate the J package to 656 mW at 70 °C at the rate of 8.2 mW/°C, and the N package to 736 mW at 70 °C at the rate of 9.2 mW/°C. In the J package the MC3453 is glass mounted.

recommended operating conditions

		MIN	NOM	MAX	UNIT
Supply voltage, V_{CC+}		4.75	5	5.25	V
Supply voltage, V_{CC-}		−4.75	−5	−5.25	V
High-level input voltage, V_{IH}		2		5.5	V
Low-level input voltage, V_{IL}		0		0.8	V
Common-mode output voltage range	V_{OCR+}	0		10	V
	V_{OCR-}	0		−3	V
Operating free-air temperature, T_A		0		70	°C

NOTE 3: All unused outputs must be grounded.

electrical characteristics over recommended operating free-air temperature range, V_{CC+} = 5.25 V, V_{CC-} = −5.25 V (unless otherwise noted)

PARAMETER		TEST CONDITIONS		MIN	TYP†	MAX	UNIT
V_{IK}	Input clamp voltage	I_I = −12 mA			−0.9	−1.5	V
$I_{O(on)}$	On-state output current	V_{CC+} = 5.25 V,	V_{CC-} = −5.25 V		11	15	mA
		V_{CC+} = 4.75 V,	V_{CC-} = − 4.75 V	6.5	11		
$I_{O(off)}$	Off-state output current	V_{CC+} = 4.75 V,	V_{CC-} = −4.75 V			100	μA
I_{IH}	High-level input current	V_I = 2.4 V				40	μA
		V_I = 5.25 V				1	mA
I_{IL}	Low-level input current	V_I = 0.4 V				−1.6	mA
I_{CC+}	Supply current from V_{CC+}	A inputs at 0.4 V	Enable at 2 V		33	50	mA
			Enable at 0.4 V		33	50	
I_{CC-}	Supply current from V_{CC-}	A inputs at 0.4 V	Enable at 2 V		−68	−90	mA
			Enable at 0.4 V		−31	−40	

† All typical values are at V_{CC+} = 5 V, V_{CC-} = −5 V, and T_A = 25 °C.

switching characteristics, V_{CC+} = 5 V, V_{CC-} = −5 V, R_L = 50 Ω, C_L = 40 pF, T_A = 25°C

PARAMETER		FROM (INPUT)	TO (OUTPUT)	TEST CONDITIONS	MIN	TYP	MAX	UNIT
t_{PLH}	Propagation delay time, low-to-high-level output	A	Y or Z	See Figure 1		9	15	ns
t_{PHL}	Propagation delay time, high-to-low-level output	A	Y or Z			7	15	ns
t_{PLH}	Propagation delay time, low-to-high-level output	Enable	Y or Z			14	25	ns
t_{PHL}	Propagation delay time, high-to-low-level output	Enable	Y or Z			15	25	ns

PARAMETER MEASUREMENT INFORMATION

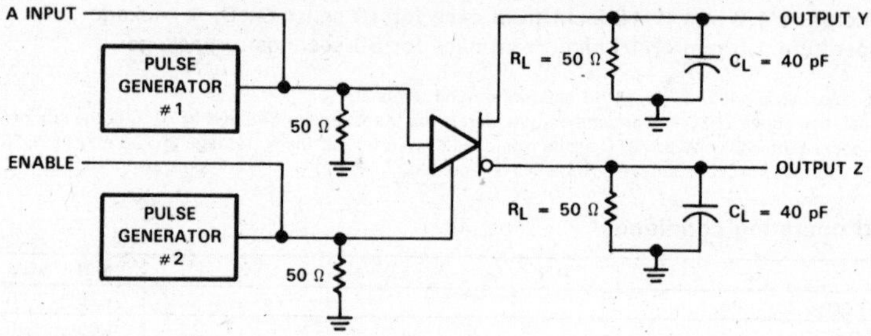
A INPUT
PULSE GENERATOR #1
50 Ω
ENABLE
PULSE GENERATOR #2
50 Ω
R_L = 50 Ω
C_L = 40 pF
OUTPUT Y
OUTPUT Z
R_L = 50 Ω
C_L = 40 pF

TEST CIRCUIT

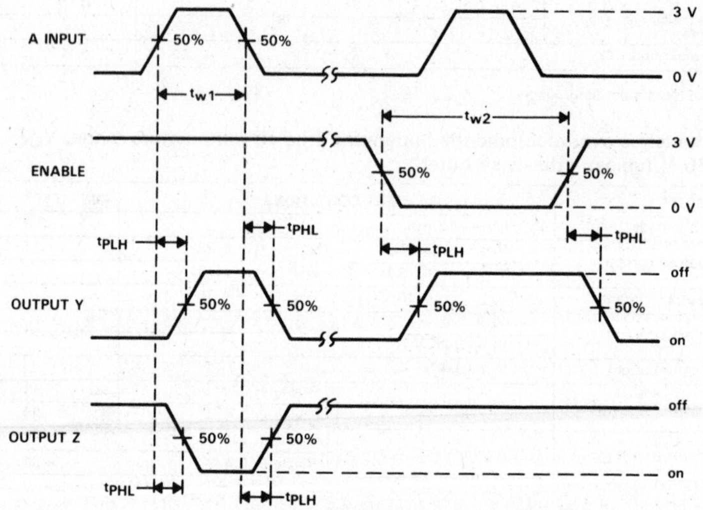
A INPUT
50%
50%
3 V
0 V
t_{w1}
t_{w2}
ENABLE
50%
50%
3 V
0 V
t_{PLH}
t_{PHL}
t_{PLH}
t_{PHL}
OUTPUT Y
50%
50%
50%
50%
off
on
OUTPUT Z
50%
50%
off
on
t_{PHL}
t_{PLH}

VOLTAGE WAVEFORMS

NOTES: A. The pulse generators have the following characteristics: Z_O = 50 Ω, t_r = t_f = 10 ±5 ns, t_{w1} = 200 ns, PRR ≤ 1 MHz, t_{w2} = 1 μs, PRR ≤ 500 kHz.
B. C_L includes probe and jig capacitance.

FIGURE 1. PROPAGATION DELAY TIMES

TEXAS INSTRUMENTS
POST OFFICE BOX 655012 • DALLAS, TEXAS 75265

D2434, JUNE 1980—REVISED SEPTEMBER 1986

- **Meets EIA Standards RS-422-A and RS423-A and Federal Standards 1020 and 1030**
- **Three-State, TTL-Compatible Outputs**
- **Fast Transition Times**
- **Operates from Single 5-Volt Supply**
- **Designed to be Interchangeable with Motorola MC3486**

D, J OR N PACKAGE
(TOP VIEW)

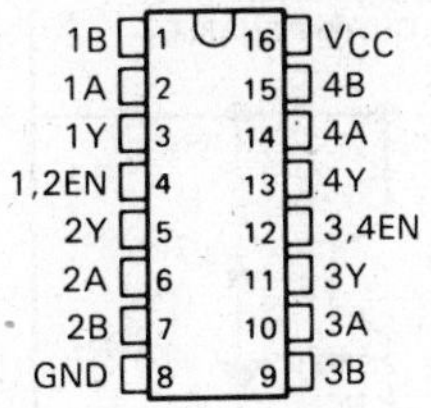
1B 1 16 VCC
1A 2 15 4B
1Y 3 14 4A
1,2EN 4 13 4Y
2Y 5 12 3,4EN
2A 6 11 3Y
2B 7 10 3A
GND 8 9 3B

description

The MC3486 is a monolithic quadruple differential line receiver designed to meet the specifications of EIA Standards RS-422-A and RS-423-A and Federal Standards 1020 and 1030. The MC3486 offers four independent differential-input line receivers that have TTL-compatible outputs. The outputs utilize three-state circuitry to provide a high-impedance state at any output when the appropriate output enable is at a low logic level.

The MC3486 is designed for optimum performance when used with the MC3487 quadruple differential line driver. It is supplied in a 16-pin package and operates from a single 5-volt supply.

The MC3486 is characterized for operation from 0°C to 70°C.

FUNCTION TABLE (EACH RECEIVER)

DIFFERENTIAL INPUTS A-B	ENABLE	OUTPUT Y
$V_{ID} \geq 0.2$ V	H	H
-0.2 V $< V_{ID} < 0.2$ V	H	?
$V_{ID} \leq -0.2$ V	H	L
Irrelevant	L	Z

H = high level, L = low level, Z = high-impedance (off), ? = indeterminate

logic symbol†

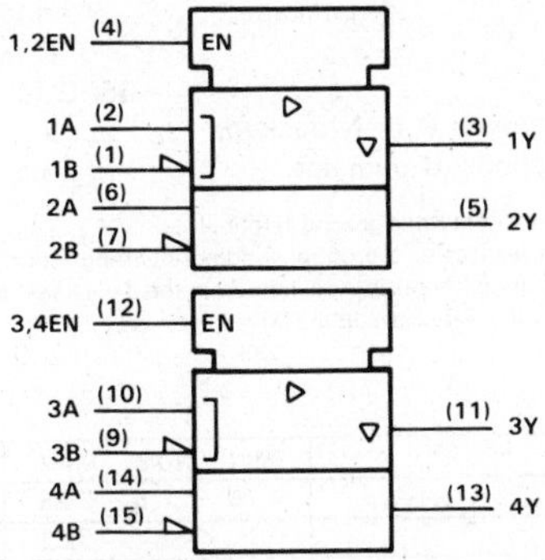
1,2EN (4)
EN
1A (2)
1B (1)
(3) 1Y
2A (6)
2B (7)
(5) 2Y
3,4EN (12)
EN
3A (10)
3B (9)
(11) 3Y
4A (14)
4B (15)
(13) 4Y

†This symbol is in accordance with ANSI/IEEE Std 91-1984 and IEC Publication 617-12.

logic diagram (positive logic)

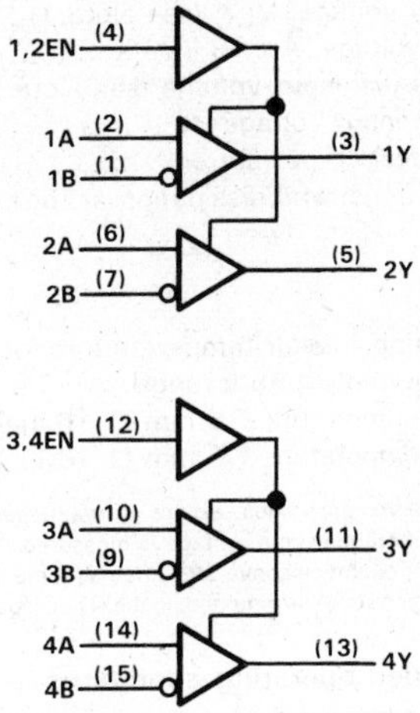
1,2EN (4)
1A (2)
1B (1)
(3) 1Y
2A (6)
2B (7)
(5) 2Y
3,4EN (12)
3A (10)
3B (9)
(11) 3Y
4A (14)
4B (15)
(13) 4Y

schematics of inputs and outputs

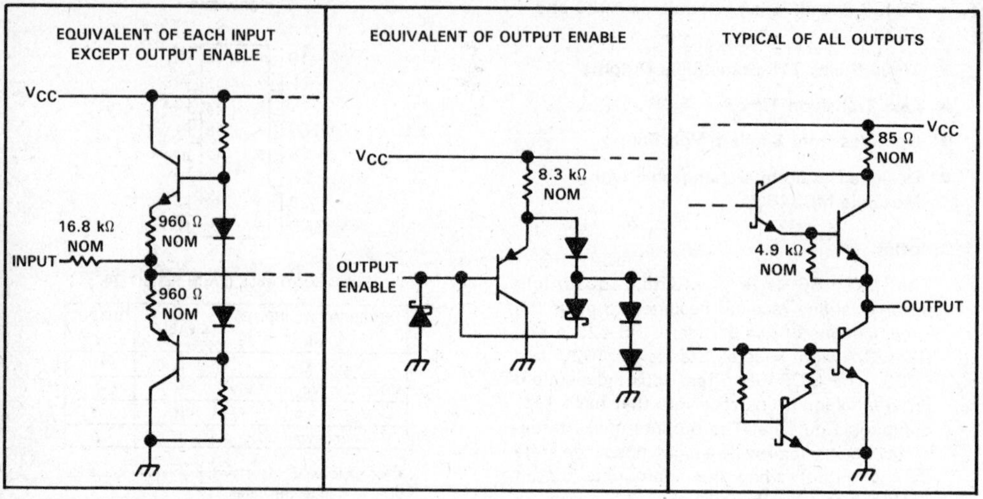
EQUIVALENT OF EACH INPUT EXCEPT OUTPUT ENABLE
VCC
16.8 kΩ NOM
INPUT
960 Ω NOM
960 Ω NOM
EQUIVALENT OF OUTPUT ENABLE
VCC
8.3 kΩ NOM
OUTPUT ENABLE
TYPICAL OF ALL OUTPUTS
VCC
85 Ω NOM
4.9 kΩ NOM
OUTPUT

absolute maximum ratings over operating free-air temperature range (unless otherwise noted)

Supply voltage, V_{CC} (see Note 1)	8 V
Input voltage, A or B inputs	±15 V
Differential input voltage (see Note 2)	±25 V
Enable input voltage	8 V
Low-level output current	50 mA
Continuous total dissipation at (or below) 25°C free-air temperature (see Note 3): D package	950 mW
J package	1025 mW
N package	1150 mW
Operating free-air temperature range	0°C to 70°C
Storage temperature range	−65°C to 150°C
Lead temperature 1,6 mm (1/16 inch) from case for 10 seconds: D or N package	260°C
Lead temperature 1,6 mm (1/16 inch) from case for 60 seconds: J package	300°C

NOTES: 1. All voltage values, except differential-input voltage, are with respect to network ground terminal.
2. Differential-input voltage is measured at the noninverting input with respect to the corresponding inverting input.
3. For operation above 25°C free-air temperature, refer to Dissipation Derating Curves in Appendix A. In the J package, MC3486 chips are glass mounted. In the N package, use the 9.2-mW/°C curve for these devices. In the D package, use 7.6 mW/°C curve.

recommended operating conditions

	MIN	NOM	MAX	UNIT
Supply voltage, V_{CC}	4.75	5	5.25	V
Common-mode input voltage, V_{IC}			±7	V
Differential input voltage, V_{ID}			±6	V
High-level enable input voltage, V_{IH}	2			V
Low-level enable input voltage, V_{IL}			0.8	V
Operating free-air temperature, T_A	0		70	°C

TEXAS INSTRUMENTS
POST OFFICE BOX 655012 • DALLAS, TEXAS 75265

electrical characteristics over recommended ranges of common-mode input voltage, supply voltage, and operating free-air temperature (unless otherwise noted)

PARAMETER		TEST CONDITIONS		MIN	MAX	UNIT
V_{TH}	Differential-input high-threshold voltage	$V_O = 2.7$ V, $I_O = -0.4$ mA			0.2	V
V_{TL}	Differential-input low-threshold voltage	$V_O = 0.5$ V, $I_O = 8$ mA		−0.2†		V
V_{IK}	Enable-input clamp voltage	$I_I = -10$ mA			−1.5	V
V_{OH}	High-level output voltage	V_{ID}★ = 0.4 V, $I_O = -0.4$ mA, See Note 4 and Figure 1		2.7		V
V_{OL}	Low-level output voltage	V_{ID}★ = −0.4 V, $I_O = 8$ mA, See Note 4 and Figure 1			0.5	V
I_{OZ}	High-impedance-state output current	$V_{IL} = 0.8$ V, $V_{ID} = -3$ V, $V_O = 2.7$ V			40	μA
		$V_{IL} = 0.8$ V, $V_{ID} = 3$ V, $V_O = 0.5$ V			−40	
I_{IB}	Differential-input bias current	$V_{CC} = 0$ V or 5.25 V, Other inputs at 0 V	$V_I = -10$ V		−3.25	mA
			$V_I = -3$ V		−1.5	
			$V_I = 3$ V		1.5	
			$V_I = 10$ V		3.25	
I_{IH}	High-level enable input current	$V_I = 5.25$ V			100	μA
		$V_I = 2.7$ V			20	
I_{IL}	Low-level enable input current	$V_I = 0.5$ V			−100	μA
I_{OS}	Short-circuit output current	$V_{ID} = 3$ V, $V_O = 0$, See Note 5		−15	−100	mA
I_{CC}	Supply current	$V_{IL} = 0$			85	mA

† The algebraic convention, in which the least positive (most negative) limit is designated as minimum, is used in this data sheet for threshold voltages only.

NOTES: 4. Refer to EIA Standards RS-422-A and RS-423-A for exact conditions.
5. Only one output at a time should be shorted.

switching characteristics, V_{CC} = 5 V, T_A = 25°C

PARAMETER		TEST CONDITIONS	MIN	TYP	MAX	UNIT
t_{PHL}	Propagation delay time, high-to-low-level output	$C_L = 15$ pF, See Figure 2		28	35	ns
t_{PLH}	Propagation delay time, low-to-high-level output			27	30	ns
t_{PZH}	Output enable time to high level	$C_L = 15$ pF, See Figure 3		13	30	ns
t_{PZL}	Output enable time to low level			20	30	ns
t_{PHZ}	Output disable time from high level			26	35	ns
t_{PLZ}	Output disable time from low level			27	35	ns

PARAMETER MEASUREMENT INFORMATION

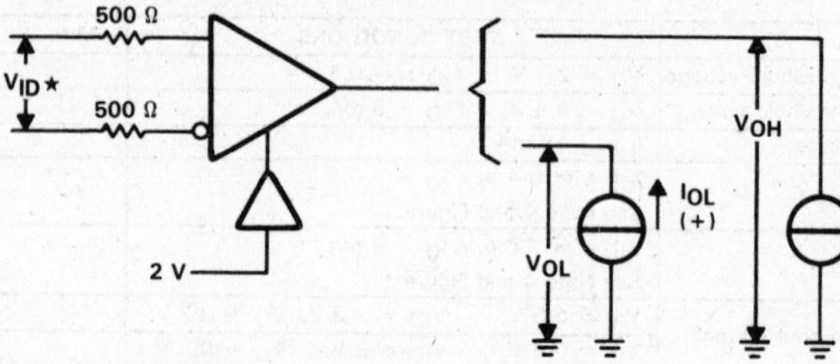
500 Ω
VID ★
500 Ω
2 V
VOH
IOL (+)
IOH (−)
VOL

FIGURE 1. V_{OH}, V_{OL}

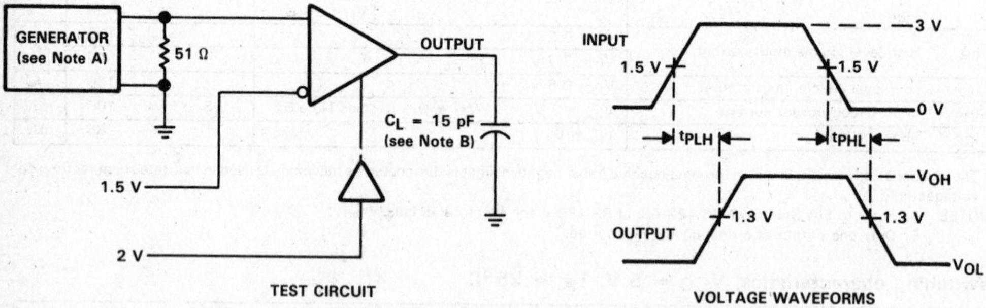
GENERATOR (see Note A)
51 Ω
OUTPUT
CL = 15 pF (see Note B)
1.5 V
2 V
TEST CIRCUIT
INPUT
3 V
1.5 V
1.5 V
0 V
tPLH
tPHL
VOH
1.3 V
1.3 V
OUTPUT
VOL
VOLTAGE WAVEFORMS

FIGURE 2. PROPAGATION DELAY TIMES

NOTES: A. The input pulse is supplied by a generator having the following characteristics: PRR ≤ 1 MHz, duty cycle ≈ 50%, $t_r \leq 6$ ns, $t_f \leq 6$ ns.

B. C_L includes probe and stray capacitance.

TEXAS INSTRUMENTS
POST OFFICE BOX 655012 • DALLAS, TEXAS 75265

PARAMETER MEASUREMENT INFORMATION

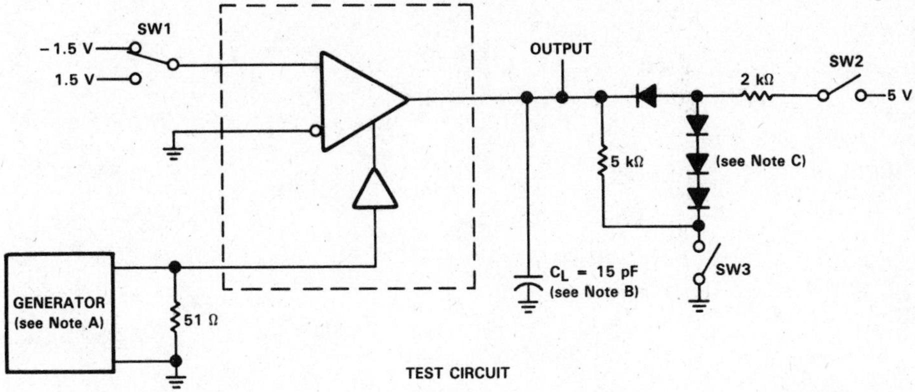
SW1
−1.5 V
1.5 V
OUTPUT
SW2
2 kΩ
5 V
5 kΩ
(see Note C)
C_L = 15 pF
(see Note B)
SW3
GENERATOR
(see Note A)
51 Ω

TEST CIRCUIT

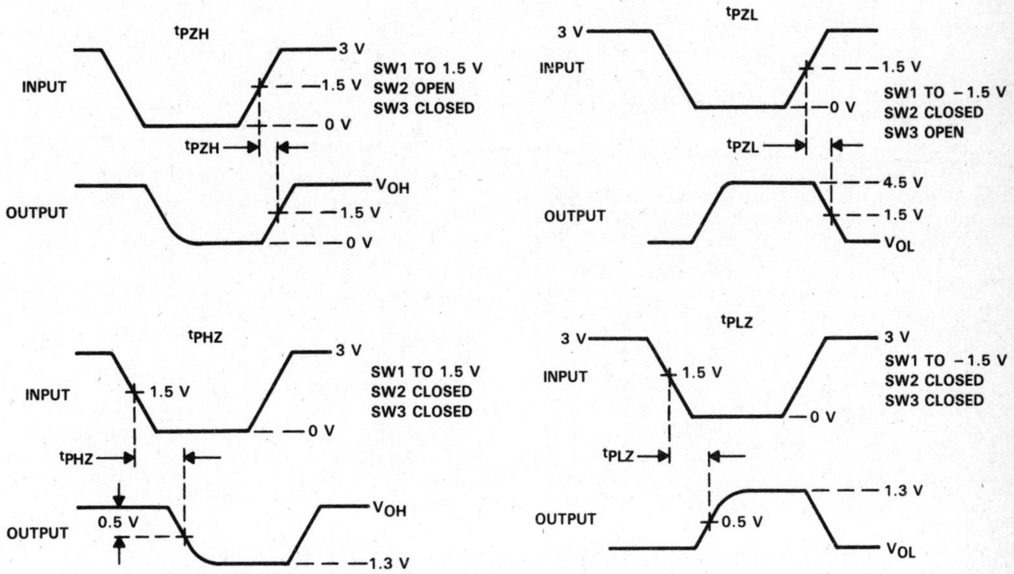
t_{PZH}
INPUT
3 V
1.5 V
0 V
SW1 TO 1.5 V
SW2 OPEN
SW3 CLOSED
OUTPUT
V_{OH}
1.5 V
0 V
t_{PZL}
3 V
INPUT
1.5 V
0 V
SW1 TO −1.5 V
SW2 CLOSED
SW3 OPEN
OUTPUT
4.5 V
1.5 V
V_{OL}
t_{PHZ}
INPUT
3 V
1.5 V
0 V
SW1 TO 1.5 V
SW2 CLOSED
SW3 CLOSED
OUTPUT
0.5 V
V_{OH}
1.3 V
t_{PLZ}
3 V
INPUT
1.5 V
0 V
SW1 TO −1.5 V
SW2 CLOSED
SW3 CLOSED
OUTPUT
1.3 V
0.5 V
V_{OL}

VOLTAGE WAVEFORMS

FIGURE 3. ENABLE AND DISABLE TIMES

NOTES: A. The input pulse is supplied by a generator having the following characteristics: PRR ≤ 1 MHz, duty cycle ≈ 50%, t_r ≤ 6 ns, t_f ≤ 6 ns.
B. C_L includes probe and stray capacitance.
C. All diodes are 1N916 or equivalent.

D2578, MAY 1980—REVISED SEPTEMBER 1986

- **Meets EIA Standard RS-422-A and Federal Standard 1020**
- **Three-State, TTL-Compatible Outputs**
- **Fast Transition Times**
- **High-Impedance Inputs**
- **Single 5-Volt Supply**
- **Power-Up and Power-Down Protection**
- **Designed to be Interchangeable with Motorola MC3487**

D, J, OR N DUAL-IN-LINE PACKAGE
(TOP VIEW)

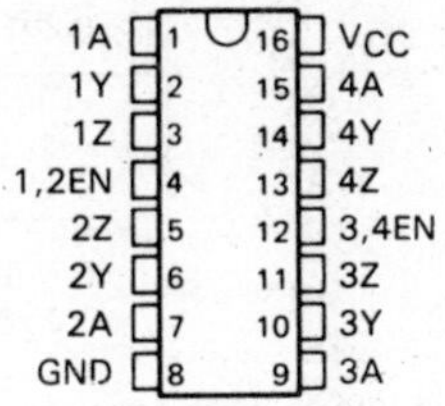
1A 1
1Y 2
1Z 3
1,2EN 4
2Z 5
2Y 6
2A 7
GND 8
16 VCC
15 4A
14 4Y
13 4Z
12 3,4EN
11 3Z
10 3Y
9 3A

description

The MC3487 offers four independent differential line drivers designed to meet the specifications of EIA Standard RS-422-A and Federal Standard 1020. Each driver has a TTL-compatible input buffered to reduce current and minimize loading.

The driver outputs utilize 3-state circuitry to provide high-impedance states at any pair of differential outputs when the appropriate output enable is at a low logic level. Internal circuitry is provided to ensure a high-impedance state at the differential outputs during power-up and power-down transition times, provided the output enable is low. The outputs are capable of source or sink currents of 48 milliamperes.

The MC3487 is designed for optimum performance when used with the MC3486 quadruple line receiver. It is supplied in a 16-pin dual-in-line package and operates from a single 5-volt supply.

The MC3487 is characterized for operation from 0°C to 70°C.

logic symbol†

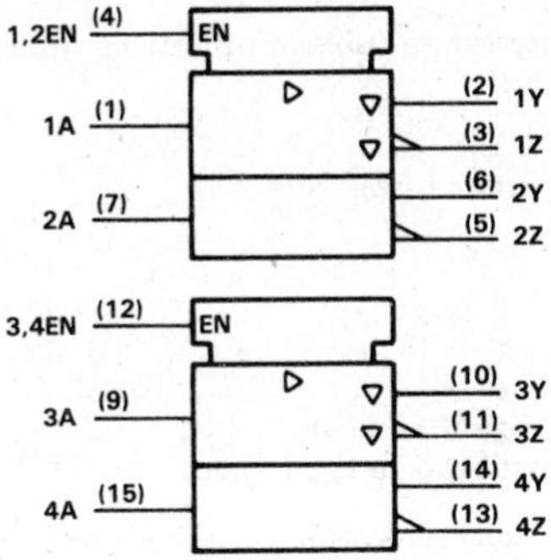
1,2EN (4)
EN
1A (1)
2A (7)
(2) 1Y
(3) 1Z
(6) 2Y
(5) 2Z
3,4EN (12)
EN
3A (9)
4A (15)
(10) 3Y
(11) 3Z
(14) 4Y
(13) 4Z

logic diagram (positive logic)

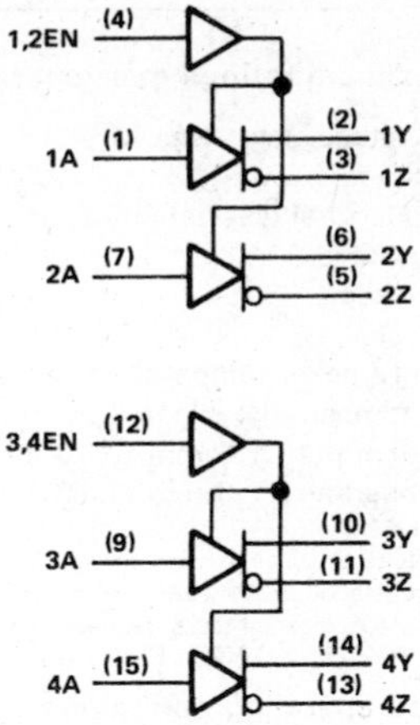
1,2EN (4)
1A (1)
(2) 1Y
(3) 1Z
2A (7)
(6) 2Y
(5) 2Z
3,4EN (12)
3A (9)
(10) 3Y
(11) 3Z
4A (15)
(14) 4Y
(13) 4Z

† This symbol is in accordance with ANSI/IEEE Std 91-1984 and IEC Publication 617-12.

TEXAS INSTRUMENTS
POST OFFICE BOX 655012 • DALLAS, TEXAS 75265

FUNCTION TABLE (EACH DRIVER)

INPUT	OUTPUT ENABLE	OUTPUTS Y	OUTPUTS Z
H	H	H	L
L	H	L	H
X	L	High-Impedance	High-Impedance

H = TTL high level X = irrelevant
L = TTL low level

schematics of inputs and outputs

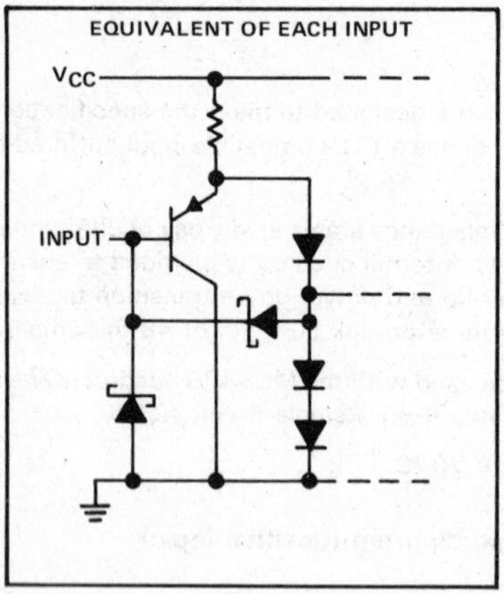
EQUIVALENT OF EACH INPUT
V_{CC}
INPUT

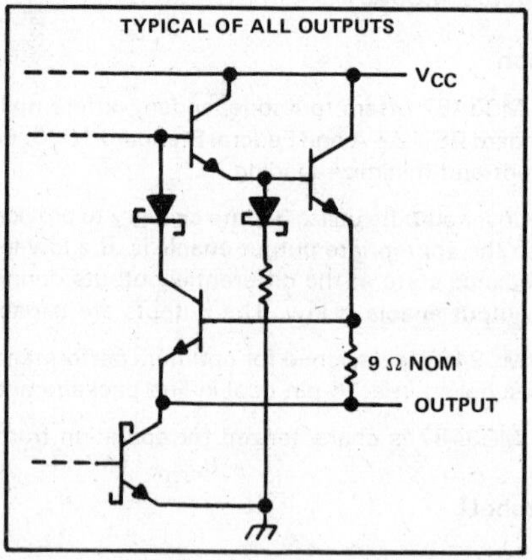
TYPICAL OF ALL OUTPUTS
V_{CC}
9 Ω NOM
OUTPUT

absolute maximum ratings over operating free-air temperature range (unless otherwise noted)

Supply voltage, V_{CC} (see Note 1)	8 V
Input voltage	5.5 V
Continuous total dissipation at (or below) 25 °C free-air temperature (see Note 2):	
D package	950 mW
J package	1025 mW
N package	1150 mW
Operating free-air temperature range	0 °C to 70 °C
Storage temperature range	−65 °C to 150 °C
Lead temperature 1,6 mm (1/16 inch) from case for 60 seconds: J package	300 °C
Lead temperature 1,6 mm (1/16 inch) from case for 10 seconds: D and N packages	260 °C

NOTES: 1. All voltage values, except differential output voltage, V_{OD}, are with respect to the network ground terminal.
2. For operation above 25 °C free-air temperature, refer to the Dissipation Derating Curves in Appendix A. In the J package, MC3487 chips are glass mounted. In the N package, use the 9.2-mW/°C curve for these devices.

recommended operating conditions

	MIN	NOM	MAX	UNIT
Supply voltage, V_{CC}	4.75	5	5.25	V
High-level input voltage, V_{IH}	2			V
Low-level input voltage, V_{IL}			0.8	V
Operating free-air temperature, T_A	0		70	°C

TEXAS INSTRUMENTS
POST OFFICE BOX 655012 • DALLAS, TEXAS 75265

electrical characteristics over recommended ranges of supply voltage and operating free-air temperature (unless otherwise noted)

PARAMETER		TEST CONDITIONS			MIN	MAX	UNIT
V_{IK}	Input clamp voltage	$I_I = -18$ mA				−1.5	V
V_{OH}	High-level output voltage	$V_{IL} = 0.8$ V,	$V_{IH} = 2$ V,	$I_{OH} = -20$ mA	2.5		V
V_{OL}	Low-level output voltage	$V_{IL} = 0.8$ V,	$V_{IH} = 2$ V,	$I_{OL} = 48$ mA		0.5	V
$\|V_{OD}\|$	Differential output voltage	$R_L = 100\ \Omega$,	See Figure 1		2		V
$\Delta\|V_{OD}\|$	Change in magnitude of differential output voltage†	$R_L = 100\ \Omega$,	See Figure 1			±0.4	V
V_{OC}	Common-mode output voltage‡	$R_L = 100\ \Omega$,	See Figure 1			3	V
$\Delta\|V_{OC}\|$	Change in magnitude of common-mode output voltage†	$R_L = 100\ \Omega$,	See Figure 1			±0.4	V
I_O	Output current with power off	$V_{CC} = 0$	$V_O = 6$ V			100	μA
			$V_O = -0.25$ V			−100	
I_{OZ}	High-impedance-state output current	Output enables at 0.8 V	$V_O = 2.7$ V			100	μA
			$V_O = 0.5$ V			−100	
I_I	Input current at maximum input voltage	$V_I = 5.5$ V				100	μA
I_{IH}	High-level input current	$V_I = 2.7$ V				50	μA
I_{IL}	Low-level input current	$V_I = 0.5$ V				−400	μA
I_{OS}	Short-circuit output current§	$V_I = 2$ V			−40	−140	mA
I_{CC}	Supply current (all drivers)	Outputs disabled				105	mA
		Outputs enabled,	No load			85	

†$\Delta|V_{OD}|$ and $\Delta|V_{OC}|$ are the changes in magnitude of V_{OD} and V_{OC}, respectively, that occur when the input is changed from a high level to a low level.
‡In EIA Standard RS-422-A, V_{OC}, which is the average of the two output voltages with respect to ground, is called output offset voltage, V_{OS}.
§Only one output at a time should be shorted and duration of the short-circuit should not exceed one second.

switching characteristics over recommended range of operating free-air temperature, $V_{CC} = 5$ V

PARAMETER		TEST CONDITION		MIN	MAX	UNIT
t_{PLH}	Propagation delay time, low-to-high-level output	$C_L = 15$ pF,	See Figure 2		20	ns
t_{PHL}	Propagation delay time, high-to-low-level output				20	ns
	Skew				6	ns
t_{TD}	Differential-output transition time	$C_L = 15$ pF,	See Figure 3		20	ns
t_{PZH}	Output enable time to high level	$C_L = 50$ pF,	See Figure 4		30	ns
t_{PZL}	Output enable time to low level				30	ns
t_{PHZ}	Output disable time from high level				25	ns
t_{PLZ}	Output disable time from low level				30	ns

PARAMETER MEASUREMENT INFORMATION

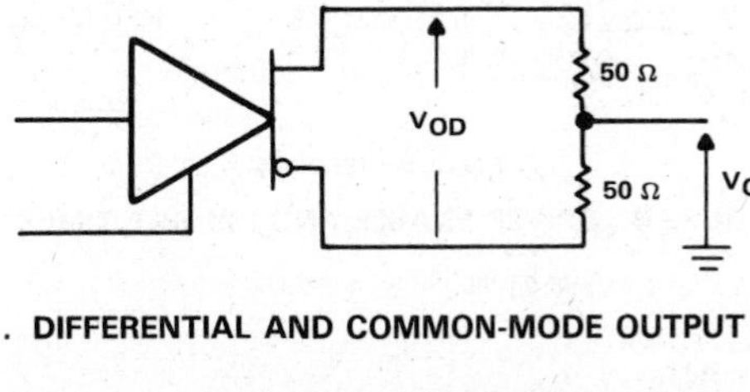
50 Ω
V_{OD}
50 Ω
V_{OC}

FIGURE 1. DIFFERENTIAL AND COMMON-MODE OUTPUT VOLTAGES

Line Drivers/Receivers 4

PARAMETER MEASUREMENT INFORMATION

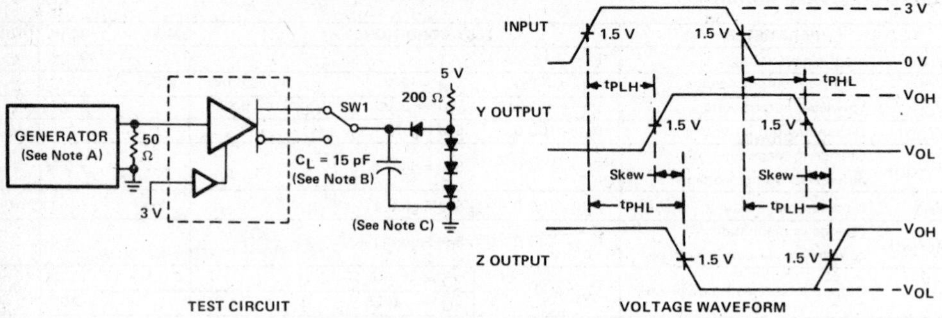
GENERATOR
(See Note A)
50 Ω
3 V
SW1
5 V
200 Ω
C_L = 15 pF
(See Note B)
(See Note C)
TEST CIRCUIT
INPUT
1.5 V
1.5 V
3 V
0 V
t_{PLH}
t_{PHL}
V_{OH}
Y OUTPUT
V_{OL}
Skew
Skew
Z OUTPUT
VOLTAGE WAVEFORM

FIGURE 2. PROPAGATION DELAY TIMES

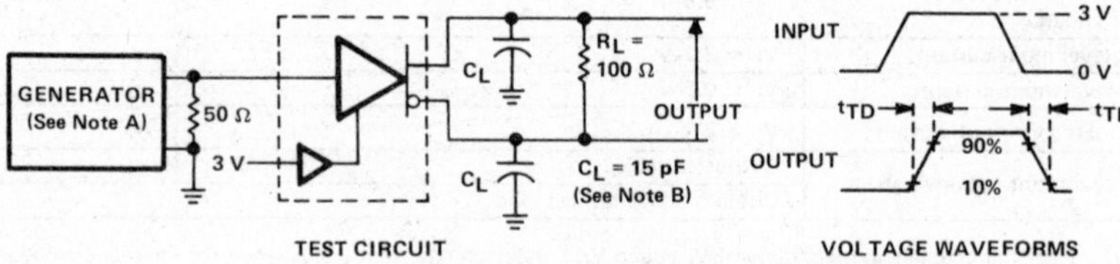
GENERATOR
(See Note A)
50 Ω
3 V
C_L
R_L = 100 Ω
OUTPUT
C_L = 15 pF
(See Note B)
TEST CIRCUIT
INPUT
3 V
0 V
t_{TD}
t_{TD}
OUTPUT
90%
10%
VOLTAGE WAVEFORMS

FIGURE 3. DIFFERENTIAL-OUTPUT TRANSITION TIMES

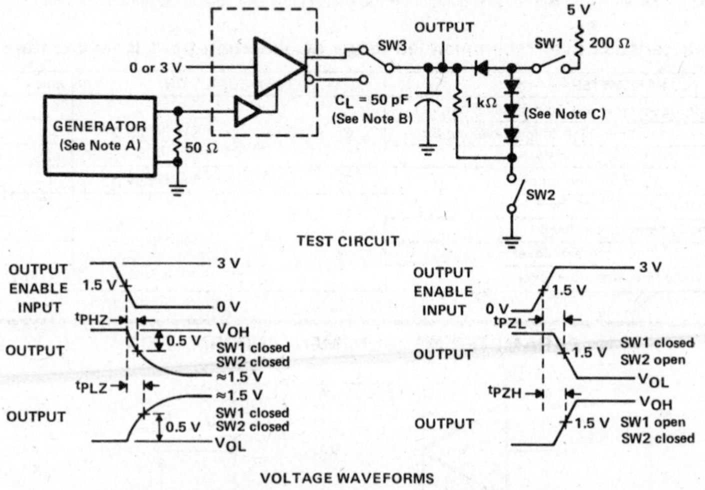
0 or 3 V
GENERATOR
(See Note A)
50 Ω
OUTPUT
SW3
5 V
SW1
200 Ω
C_L = 50 pF
(See Note B)
1 kΩ
(See Note C)
SW2
TEST CIRCUIT
OUTPUT ENABLE INPUT
1.5 V
3 V
0 V
t_{PHZ}
V_{OH}
0.5 V
OUTPUT
SW1 closed
SW2 closed
≈1.5 V
t_{PLZ}
≈1.5 V
OUTPUT
0.5 V
SW1 closed
SW2 closed
V_{OL}
OUTPUT ENABLE INPUT
0 V
1.5 V
3 V
t_{PZL}
OUTPUT
1.5 V
SW1 closed
SW2 open
V_{OL}
t_{PZH}
V_{OH}
OUTPUT
1.5 V
SW1 open
SW2 closed
VOLTAGE WAVEFORMS

FIGURE 4. DRIVER ENABLE AND DISABLE TIMES

NOTES: A. The input pulse is supplied by a generator having the following characteristics: $t_r \le 5$ ns, $t_f \le 5$ ns, PRR ≤ 1 MHz, duty cycle = 50%, Z_O = 50 Ω.
B. C_L includes probe and stray capacitance.
C. All diodes are 1N916 or 1N3064.

QUADRUPLE BUS TRANSCEIVER WITH 3-STATE OUTPUTS

D2462, MAY 1978—REVISED SEPTEMBER 1986

- **P-N-P Inputs for Minimal Input Loading (200 μA Maximum)**
- **High-Speed Schottky Circuitry**
- **3-State Outputs for Driver and Receiver**
- **Party-Line (Data-Bus) Operation**
- **Single 5-V Supply**
- **Designed to be Interchangeable with Signetics N8T26, also Called 8T26**

D, J, OR N PACKAGE
(TOP VIEW)

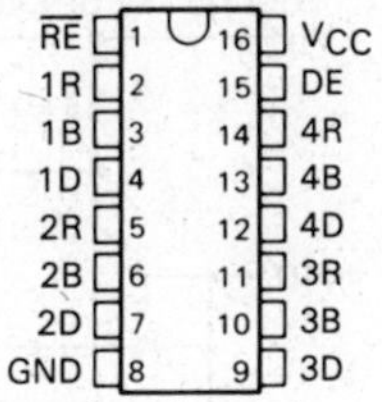
RE 1
1R 2
1B 3
1D 4
2R 5
2B 6
2D 7
GND 8
16 VCC
15 DE
14 4R
13 4B
12 4D
11 3R
10 3B
9 3D

description

The N8T26 is a quadruple transceiver utilizing Schottky-diode-clamped transistors. Both the driver and receiver have three-state outputs. With p-n-p inputs, the input loading is reduced to a maximum input current of 200 micro-amperes. This device is capable of high switching rates into high-capacitance loads and are suitable for driving long bus lines.

The N8T26 is characterized for operation from 0°C to 70°C.

FUNCTION TABLE (DRIVER)

INPUT		OUTPUT
DE	D	B
H	L	H
H	H	L
L	X	Z

FUNCTION TABLE (RECEIVER)

INPUT		OUTPUT
RE	B	R
L	L	H
L	H	L
H	X	Z

H = high level
L = low level
X = irrelevant
Z = high impedance

logic symbol†

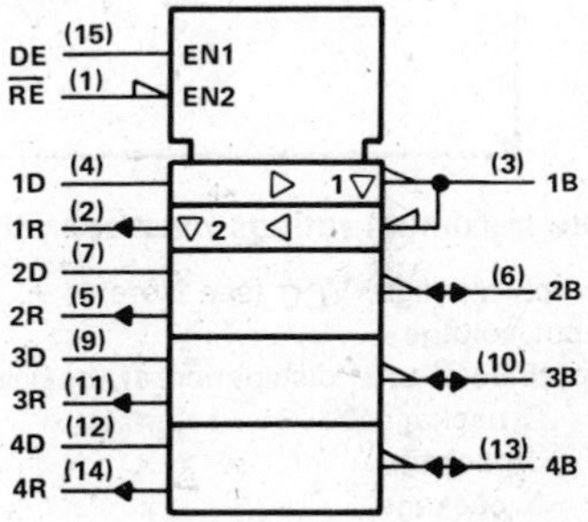
DE (15)
RE (1)
EN1
EN2
1D (4)
1R (2)
2D (7)
2R (5)
3D (9)
3R (11)
4D (12)
4R (14)
1∇
∇2
(3) 1B
(6) 2B
(10) 3B
(13) 4B

† This symbol is in accordance with ANSI/IEEE Std 91-1984 and IEC Publication 617-12.

logic diagram (positive logic)

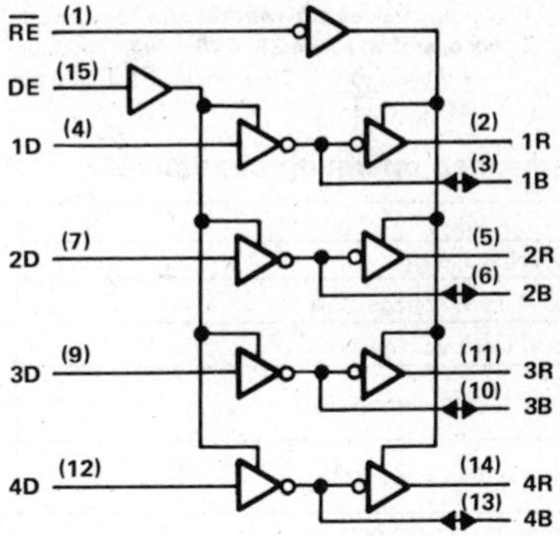
RE (1)
DE (15)
1D (4)
2D (7)
3D (9)
4D (12)
(2) 1R
(3) 1B
(5) 2R
(6) 2B
(11) 3R
(10) 3B
(14) 4R
(13) 4B

schematics of inputs and outputs

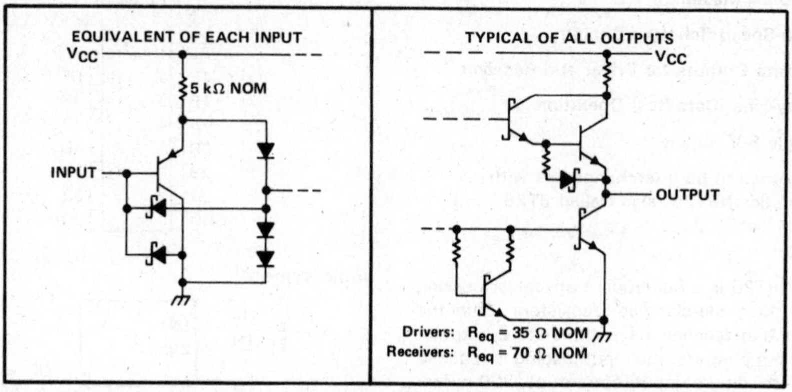
EQUIVALENT OF EACH INPUT
VCC
5 kΩ NOM
INPUT
TYPICAL OF ALL OUTPUTS
VCC
OUTPUT
Drivers: Req = 35 Ω NOM
Receivers: Req = 70 Ω NOM

absolute maximum ratings over operating free-air temperature range (unless otherwise noted)

Supply voltage, V_{CC} (see Note 1) .... 7 V
Input voltage .... 5.5 V
Continuous total dissipation at (or below) 25°C free-air temperature (see Note 2):
D package .... 950 mW
J package .... 1025 mW
N package .... 1150 mW
Operating free-air temperature range .... 0°C to 70°C
Storage temperature range .... −65°C to 150°C
Lead temperature 1,6 mm (1/16 inch) from case for 60 seconds: J package .... 300°C
Lead temperature 1,6 mm (1/16 inch) from case for 10 seconds: D or N package .... 260°C

NOTES: 1. Voltage values are with respect to network ground terminal.
2. For operation above 25°C free-air temperature, refer to Dissipation Derating Curves in Appendix A. For N8T26 in the N package, use the 9.2-mW/°C curve. In the J package, N8T26 chips are glass mounted.

recommended operating conditions

		MIN	NOM	MAX	UNIT
Supply voltage, V_{CC}		4.75	5	5.25	V
High-level input voltage, V_{IH}	B, D, DE, $\overline{RE}$	2			V
Low-level input voltage, V_{IL}	B, D, DE, $\overline{RE}$			0.85	V
High-level output current, I_{OH}	Driver, B			−10	mA
	Receiver, R			−2	
Low-level output current, I_{OL}	Driver, B			40	mA
	Receiver, R			16	
Operating free-air temperature, T_A		0		70	°C

electrical characteristics over recommended operating free-air temperature and supply voltage range (unless otherwise noted)

PARAMETER			TEST CONDITIONS	MIN	TYP†	MAX	UNIT
V_{IK}	Input clamp voltage	B,D,DE,$\overline{RE}$	I_I = −5 mA			−1	V
V_{OH}	High-level output voltage	B	V_{IH} = 2 V, V_{IL} = 0.85 V, I_{OH} = −10 mA	2.6	3.1		V
		R	V_{IL} = 0.85 V I_{OH} −2 mA	2.6	3.1		
V_{OL}	Low-level output voltage	B	V_{IH} = 2 V, I_{OL} = 40 mA			0.5	V
		R	V_{IH} = 2 V, V_{IL} = 0.85 V, I_{OL} = 16 mA			0.5	
I_{OZ}	Off-state (high-impedance state) output current	B,R	DE at 0.85 V RE at 2 V, V_O = 2.6 V			100	μA
		R	RE at 2 V, V_O = 0.5 V			−100	
I_{IH}	High-level input current	D,DE,$\overline{RE}$	V_I = 5.25 V			25	μA
I_{IL}	Low-level input current	B,D,DE,$\overline{RE}$	V_I = 0.4 V			−200	μA
I_{OS}	Short-circuit output current‡	B	V_{CC} = 5.25 V	−50		−150	mA
		R		−30		−75	
I_{CC}	Supply current		V_{CC} = 5.25 V, No load			87	mA

†All typical values are at T_A = 25°C and V_{CC} = 5 V.
‡Only one output should be shorted to ground at a time, and duration of the short circuit should not exceed one second.

switching characteristics, V_{CC} = 5 V, T_A = 25°C

PARAMETER		FROM	TO	TEST CONDITIONS	MIN	TYP	MAX	UNIT
t_{PLH}	Propagation delay time, low-to-high-level output	B	R	C_L = 30 pF, See Figure 1		8	18	ns
t_{PHL}	Propagation delay time, high-to-low-level output					7	10	
t_{PLH}	Propagation delay time, low-to-high-level output	D	B	C_L = 300 pF, See Figure 2		14	20	ns
t_{PHL}	Propagation delay time, high-to-low-level output					12	20	
t_{PLZ}	Output disable time from low level	RE	R	C_L = 30 pF, See Figure 3		9	17	ns
t_{PZL}	Output enable time to low level					15	30	
t_{PLZ}	Output disable time from low level	DE	B	C_L = 300 pF, See Figure 4		20	43	ns
t_{PZL}	Output enable time to low level					20	38	

PARAMETER MEASUREMENT INFORMATION

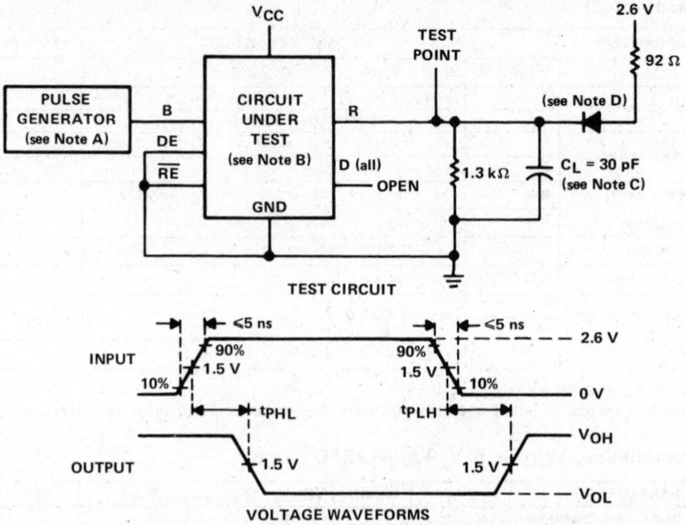
VCC
TEST POINT
2.6 V
92 Ω
PULSE GENERATOR (see Note A)
B
DE
RE
CIRCUIT UNDER TEST (see Note B)
R
D (all)
OPEN
GND
(see Note D)
1.3 kΩ
CL = 30 pF (see Note C)
TEST CIRCUIT
≤5 ns
≤5 ns
2.6 V
INPUT
90%
1.5 V
10%
0 V
tPHL
tPLH
VOH
OUTPUT
1.5 V
VOL
VOLTAGE WAVEFORMS

FIGURE 1. PROPAGATION DELAY TIMES FROM BUS TO RECEIVER OUTPUT

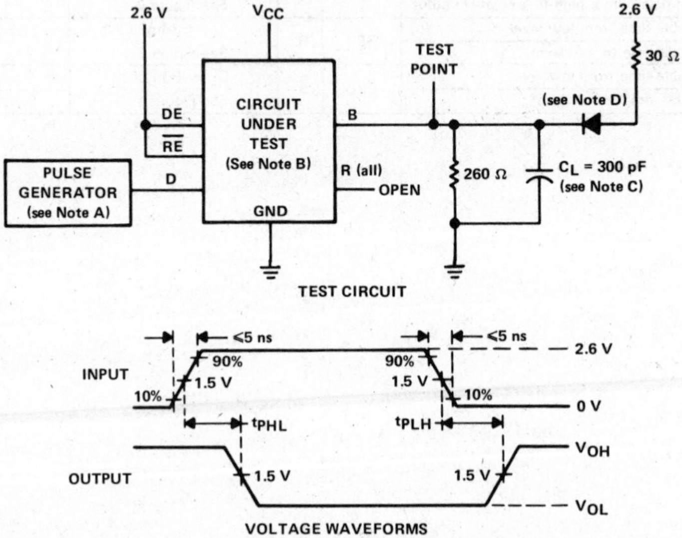
2.6 V
VCC
TEST POINT
2.6 V
30 Ω
DE
RE
CIRCUIT UNDER TEST (See Note B)
B
(see Note D)
PULSE GENERATOR (see Note A)
D
R (all)
OPEN
GND
260 Ω
CL = 300 pF (see Note C)
TEST CIRCUIT
≤5 ns
≤5 ns
2.6 V
INPUT
90%
1.5 V
10%
0 V
tPHL
tPLH
VOH
OUTPUT
1.5 V
VOL
VOLTAGE WAVEFORMS

FIGURE 2. PROPAGATION DELAY TIMES FROM DRIVER INPUT TO BUS

NOTES: A. The pulse generator in Figures 1 and 2 has the following characteristics: PRR ≤ 10 MHz, duty cycle = 50%, Z_{out} ≈ 50 Ω.
B. All inputs and outputs not shown are open.
C. C_L includes probe and jig capacitance.
D. All diodes are 1N916 or 1N3064.

TEXAS INSTRUMENTS
POST OFFICE BOX 655012 • DALLAS, TEXAS 75265

PARAMETER MEASUREMENT INFORMATION

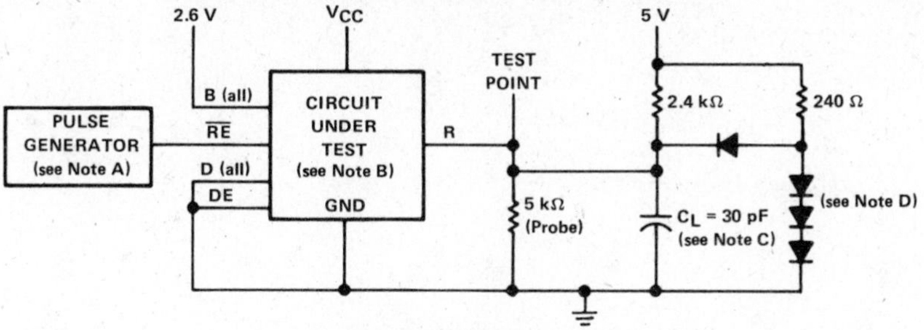
2.6 V
VCC
5 V
TEST POINT
B (all)
RE
D (all)
DE
CIRCUIT UNDER TEST (see Note B)
GND
R
PULSE GENERATOR (see Note A)
2.4 kΩ
240 Ω
5 kΩ (Probe)
CL = 30 pF (see Note C)
(see Note D)

TEST CIRCUIT

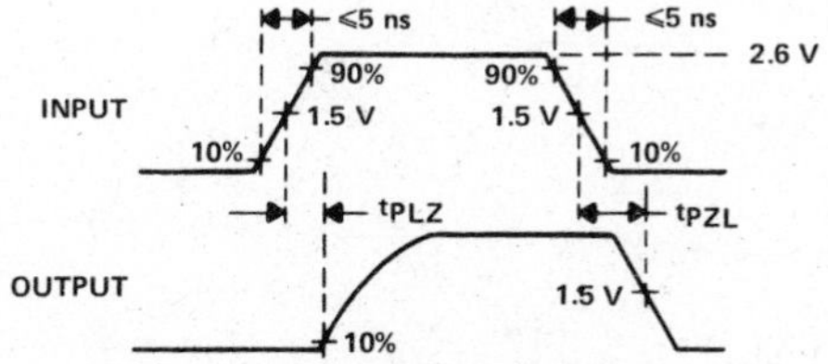
≤5 ns
≤5 ns
2.6 V
INPUT
90%
90%
1.5 V
1.5 V
10%
10%
tPLZ
tPZL
OUTPUT
1.5 V
10%

VOLTAGE WAVEFORMS

FIGURE 3. RECEIVER ENABLE AND DISABLE TIMES

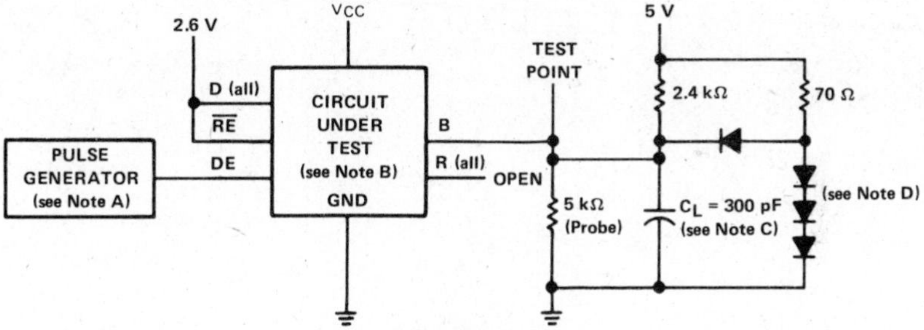
2.6 V
VCC
5 V
TEST POINT
D (all)
RE
DE
CIRCUIT UNDER TEST (see Note B)
GND
B
R (all)
OPEN
PULSE GENERATOR (see Note A)
2.4 kΩ
70 Ω
5 kΩ (Probe)
CL = 300 pF (see Note C)
(see Note D)

TEST CIRCUIT

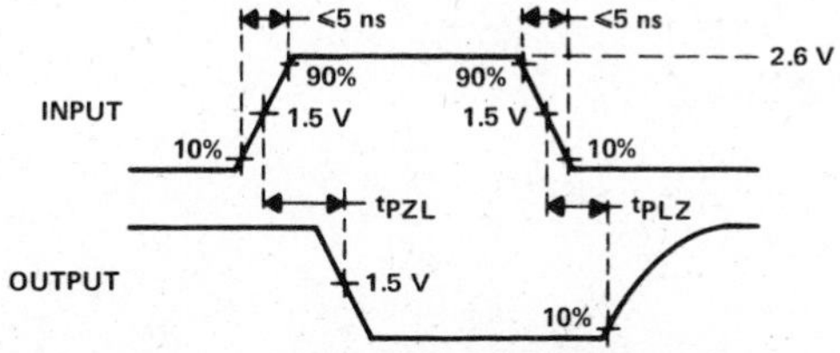
≤5 ns
≤5 ns
2.6 V
INPUT
90%
90%
1.5 V
1.5 V
10%
10%
tPZL
tPLZ
OUTPUT
1.5 V
10%

VOLTAGE WAVEFORMS

FIGURE 4. DRIVER ENABLE AND DISABLE TIMES

NOTES: A. The pulse generator in Figures 3 and 4 has the following characteristics: PRR ≤ 5 MHz, duty cycle = 50%, Z_{out} ≈ 50 Ω.
B. All inputs and outputs not shown are open.
C. C_L includes probe and jig capacitance.
D. All diodes are 1N916 or 1N3064.

D2959, JANUARY 1987

- **IEEE 802.3 1BASE5 Driver and Receiver**
- **On-Chip Receiver Squelch with Adjustable Threshold**
- **Adjustable Squelch Delay**
- **Direct TTL-Level Squelch Output**
- **Squelch Circuit Allows for External Noise Filtering**
- **Two Driver-Enable Options**
- **On-Chip Start-of-Idle Detection and Disable**
- **Driver Guarantees 2.0 Volts Minimum into a 50-Ohm Differential Load to Allow for Use with Doubly-Terminated Lines and Multipoint Architectures**
- **On-Chip Driver Slew-Rate Control for Very Closely Matched Output Rise and Fall Times**

DW, J, OR N PACKAGE
(TOP VIEW)

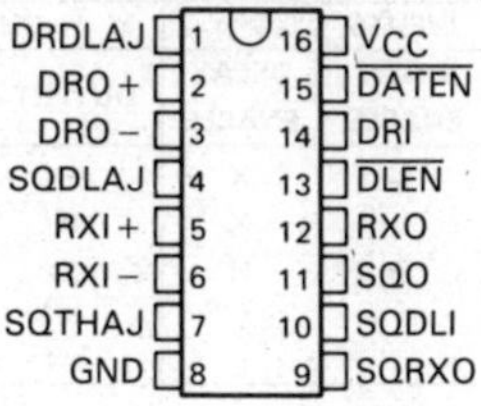
DRDLAJ 1
DRO+ 2
DRO− 3
SQDLAJ 4
RXI+ 5
RXI− 6
SQTHAJ 7
GND 8
16 VCC
15 DATEN
14 DRI
13 DLEN
12 RXO
11 SQO
10 SQDLI
9 SQRXO

PIN NAME	PIN NUMBER	DESCRIPTION
$\overline{\text{DATEN}}$	15	Driver Data Enable. When low, places driver outputs in an active state. When high, the driver outputs are in a high-impedance state if $\overline{\text{DLEN}}$ is also high.
$\overline{\text{DLEN}}$	13	Driver Delay Enable. When this signal is low and $\overline{\text{DATEN}}$ is high, the driver outputs are active for a period of time set by DRDLAJ after a positive-going transition on DRI. If there is no active data on DRI, the outputs are in a high-impedance state.
DRDLAJ	1	Driver Delay Adjust is a connection for the external R-C combination that determines the duration of the driver output active state after a positive transition on DRI when $\overline{\text{DLEN}}$ is low and $\overline{\text{DATEN}}$ is high.
DRI	14	Driver Data Input
DRO+	2	Noninverting Driver Output
DRO−	3	Inverting Driver Output
GND	8	Ground. Common for all voltages
RXI+	5	Noninverting Receiver Input
RXI−	6	Inverting Receiver Input
RXO	12	Main Receiver Output
SQDLAJ	4	Squelch Delay Adjust is a connection for an external R-C combination that determines the duration of the receiver unsquelch after a negative-going transition on SQDLI.
SQDLI	10	Squelch Delay Input is the input to the one-shot that controls the duration of the receiver unsquelch period. The main receiver output remains unsquelched as long as SQDLI is held high. Timing of the unsquelch period begins on the high-to-low transition of SQDLI.
SQO	11	Squelch Output is high while the receiver is squelched.
SQRXO	9	Squelch Receiver Output is high only when the differential receiver input exceeds the threshold set by SQTHAJ.
SQTHAJ	7	Squelch Receiver Threshold Adjust. The voltage at this input determines the threshold of the squelch receiver in a ratio of −2, SQTHAJ to threshold. If left open, the squelch receiver threshold defaults to −600 mV.
V_{CC}	16	Supply voltage input

FUNCTION TABLES

DRIVER				
INPUTS			OUTPUTS	
DRIVER IN	DATA ENABLE	DELAY ENABLE	OUTPUT +	OUTPUT −
L	L	X	L	H
H	L	X	H	L
X	H	H	Z	Z
H	H	L	H†	L†
L	H	L	L‡	H‡

RECEIVER§				
CONDITION	INPUTS		OUTPUTS	
	IN +	IN −	RECEIVER OUT	SQUELCH THRESHOLD
No active signal¶	X	X	H	H
Active signal¶	L	H	L	L
	H	L	H	L

† This condition is valid during the time period set by Driver Delay Adjust following a rising transition on Driver In. Following this, if no subsequent positive transition occurs on Driver In, the outputs will go to the high impedance state.
‡ This condition is valid if it occurs within the enable time set by Driver Delay Adjust after a rising transition on Driver In. Otherwise the outputs will be in the high-impedance state.
§ Pins 9 and 10 are tied together.
¶ An active signal is one that has an amplitude greater than the threshold level set by Squelch Threshold Adjust.

logic diagram (positive logic)

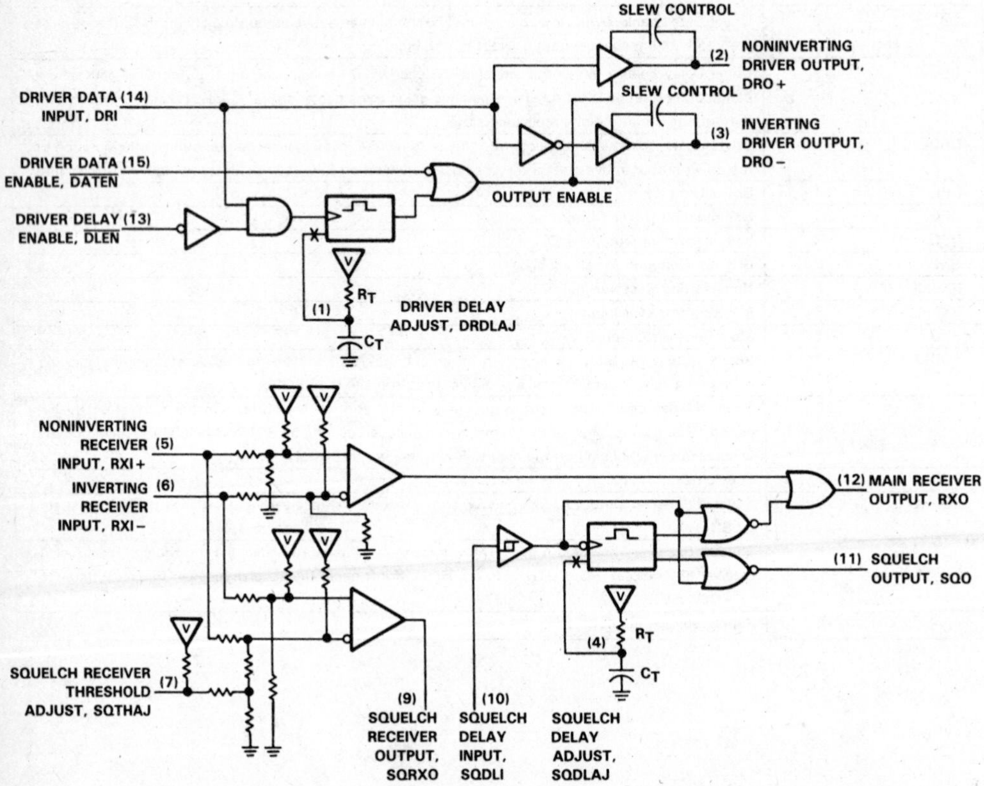
SLEW CONTROL
(2) NONINVERTING DRIVER OUTPUT, DRO+
SLEW CONTROL
(3) INVERTING DRIVER OUTPUT, DRO−
DRIVER DATA (14) INPUT, DRI
DRIVER DATA (15) ENABLE, DATEN
OUTPUT ENABLE
DRIVER DELAY (13) ENABLE, DLEN
V
RT
(1)
DRIVER DELAY ADJUST, DRDLAJ
CT
NONINVERTING RECEIVER (5) INPUT, RXI+
INVERTING (6) RECEIVER INPUT, RXI−
(12) MAIN RECEIVER OUTPUT, RXO
(11) SQUELCH OUTPUT, SQO
(4)
SQUELCH RECEIVER THRESHOLD (7) ADJUST, SQTHAJ
(9) SQUELCH RECEIVER OUTPUT, SQRXO
(10) SQUELCH DELAY INPUT, SQDLI
SQUELCH DELAY ADJUST, SQDLAJ

description

The SN75061 is a single-channel driver/receiver pair designed for use in IEEE 802.3, 1BASE5 applications as well as other general data communications circuits. The SN75061 offers the system designer both a driver and a receiver that are easily configured for use with a variety of controllers and data encoder/decoders.

The receiver features a full analog squelch circuit with an adjustable threshold and a programmable squelch delay. Internal nodes of the squelch circuitry are brought out to external connections to allow for the insertion of noise filtering circuitry of the designer's choice.

As with the receiver, the driver offers the user a variety of implementation options. Driver enabling may be controlled directly by an external logic input, or by use of an on-chip one-shot that is retriggered as long as data is being sent to the driver. The driver will then automatically go to the high-impedance state when end-of-packet occurs. The driver features internal slew-rate control for optimal matching of rise and fall times allowing for reduction of driver-induced jitter.

receiver

The SN75061 receiver implements full analog squelch functions by integrating both a separate, parallel squelch receiver with an externally programmable threshold, and a programmable one-shot. The output of the squelch receiver and the input to the high-level dc-triggered one-shot are brought out to external connections. These pins can be shorted for direct implementation, or used for the insertion of noise-filtering circuitry of the implementer's design. The receiver one-shot can be effectively bypassed by applying a high logic level to Squelch Delay In. The squelch threshold may be set externally by applying an external voltage set to a level that is −2 times the desired threshold voltage. If Squelch Threshold Adjust is left open, the squelch receiver will default to its internal preset value of −600 millivolts. The receiver also outputs a high logic ''squelch'' signal when there is no active data present at the receiver inputs. When no data is present on the transmission line, the receiver output assumes a high level. The ''unsquelch'' duration is set externally with an R-C combination at Squelch Delay Adjust.

driver

The driver offers the user a variety of implementation options. Driver enabling may be controlled directly by an active-low external logic input on Data Enable, or by use of another on-chip one-shot that retriggers with positive-going transitions on the driver input line. If no positive transition occurs within the pulse duration set by an external R-C combination, the one-shot times out and the driver is automatically put into a high-impedance state. When operating in the delay-enable mode, the 2-bit-time high-level start-of-idle pulse prescribed by IEEE 802.3 1BASE5 causes the one-shot to time out and automatically place the driver outputs in the high-impedance state. This delay time is also adjustable for use in other applications.

The driver implements an output slew-rate control that is internally set for nominally 40 mV/ns. (This is roughly a 100-ns peak-to-peak differential transition time.) The driver outputs are capable of driving a 50-ohm differential load with a guaranteed minimum output level of 2 volts. Short-circuit output current is guaranteed to be greater than 100 milliamperes.

absolute maximum ratings over operating free-air temperature range (unless otherwise noted)

Supply voltage, V_{CC} . . . . . 7 V
Input voltage (any logic input) . . . . . 7 V
Receiver differential input voltage . . . . . ±25 V
Receiver input voltage . . . . . ±15 V
Driver output voltage . . . . . −0.5 V to 15 V
Continuous total dissipation at (or below) 25°C free-air temperature (see Note 1):
  DW or J package . . . . . 1025 mW
  N package . . . . . 1150 mW
Operating free-air temperature range . . . . . 0°C to 70°C
Storage temperature range . . . . . −65°C to 150°C
Lead temperature 1,6 mm (1/16 inch) from case for 60 seconds: J package . . . . . 300°C
Lead temperature 1,6 mm (1/16 inch) from case for 10 seconds: DW or N package . . . . . 260°C

NOTE 1: For operation above 25°C free-air temperature, derate the DW and J packages to 656 mW at 70°C at the rate of 8.2 mW/°C, and the N package to 736 mW at 70°C at the rate of 9.2 mW/°C.

recommended operating conditions

	MIN	NOM	MAX	UNIT
Supply voltage, V_{CC}	4.75	5	5.25	V
Driver high-level input voltage, V_{IH}	2			V
Driver low-level input voltage, V_{IL}			0.8	V
Driver high-level output current, I_{OH}			−150	mA
Driver low-level output current, I_{OL}			150	mA
Receiver common-mode input voltage, V_{IC} (see Note 2)	−2.5		5	V
External timing resistance, R_{ext}	5		260	kΩ
External timing capacitance, C_{ext}	No restriction			
Operating free-air temperature, T_A	0		70	°C

NOTE 2: The algebraic convention, in which the less positive (more negative) limit is designated as minimum, is used in this data sheet for common-mode input voltage V_{IC} and threshold levels V_{TH} and V_{TL}.

electrical characteristics over recommended operating free-air and supply voltage range (unless otherwise noted)

driver

PARAMETER		TEST CONDITIONS	MIN	TYP†	MAX	UNIT
V_{IK}	Input clamp voltage	$I_I = -18$ mA			−1.5	V
V_{OD}	Differential-output voltage	$R_L = 50\ \Omega$	2	2.4	3.3	V
		$R_L = 115\ \Omega$			3.65	
ΔV_{OD}	Change in differential-output voltage for a change in logic input state				50	mV
I_{IH}	High-level input current	$V_I = 2.4$ V			20	μA
I_{IL}	Low-level input current	$V_I = 0.5$ V			−35	μA
I_{OS}	Short-circuit output current	$V_O = 0$ V or 6 V, $V_I = 0.8$ V or 2.5 V	±100		±300	mA
I_{OZ}	High-impedance output current	$V_{CC} = 5.25$ V, $V_{OC} = 10$ V			100	μA
		$V_{CC} = 5.25$ V, $V_{OC} = 0$			−100	

receiver

PARAMETER			TEST CONDITIONS	MIN	TYP†	MAX	UNIT
V_{IK}	Input clamp voltage, squelch delay		$I_I = -18$ mA			−1.5	V
V_{TH}	Differential-input high-threshold voltage		$V_O = 2.7$ V, $I_O = -0.4$ mA, $V_{IC} = 5$ V			50	mV
V_{TL}	Differential-input low-threshold voltage (see Note 2)		$V_O = 0.5$ V, $I_O = 16$ mA, $V_{IC} = 5$ V	−50			mV
V_{hys}	Hysteresis ($V_{TH} - V_{TL}$)				50		mV
V_{IC}	Common-mode input voltage					5	V
V_{OH}	High-level output voltage	RXO	$V_{CC} = 4.75$ V, $I_{OH} = -400$ μA, SQDLAJ at 0.8 V	2.7			V
		SQO	$V_{CC} = 4.75$ V, $I_{OH} = -400$ μA, SQDLAJ at 0.8 V	2.7	3.5		
		SQRXO	$V_{CC} = 4.75$ V, $I_{OH} = -20$ μA, $V_{ID(RXI)} = -0.7$ V, SQDLAJ open	2.7	4.65		
V_{OL}	Low-level output voltage	RXO	$V_{CC} = 4.75$ V, SQDLAJ at 2 V, $I_{OL} = 8$ mA			0.45	V
			$V_{CC} = 4.75$ V, SQDLAJ at 2 V, $I_{OL} = 16$ mA			0.5	
		SQO	$V_{CC} = 4.75$ V, SQDLAJ at 2 V, $I_{OL} = 8$ mA		0.35	0.5	
		SQRXO	$V_{CC} = 4.75$ V, $V_{ID(RXI)} = 50$ mV, $I_{OL} = 8$ mA			0.45	
			$V_{CC} = 4.75$ V, $V_{ID(RXI)} = 50$ mV, $I_{OL} = 16$ mA			0.5	
I_{IH}	High-level input current	SQDLI	$V_I = 2.4$ V			20	μA
I_{IL}	Low-level input current	SQDLI	$V_I = 0.5$ V			−35	μA
I_{OS}	Short-circuit output current	RXO	$V_{CC} = 5.25$ V, $V_O = 0$	−15		−85	mA
		SQO	$V_{CC} = 5.25$ V, $V_O = 0$	−15		−100	
		SQRXO	$V_{CC} = 5$ V, $V_O = 0$	−0.8	−1	−1.2	
r_I	Input resistance				10		kΩ
	Squelch preset threshold voltage			−570	−600	−630	mV
	Ratio of Squelch Threshold Adjust input voltage to actual squelch threshold voltage		SQTHAJ at 200 mV to 4 V	−1.9		−2.1	

driver and receiver

PARAMETER		TEST CONDITIONS	MIN	TYP†	MAX	UNIT
I_{CC}	Supply current	$V_{CC} = 5.25$ V, Driver outputs disabled, No loads			70	mA

† All typical values are at $V_{CC} = 5$ V, $T_A = 25°C$.

NOTE 2: The algebraic convention, in which the less-positive (more negative) limit is designated as minimum, is used in this data sheet for common-mode input voltage V_{IC} and threshold levels V_{TH} and V_{TL}.

switching characteristics, V_{CC} = 5 V, T_A = 25°C

driver

PARAMETER		TEST CONDITIONS	MIN	TYP	MAX	UNIT
SR	Differential-output slew rate	V_o = −2 V to 2 V, R_1 = 100 Ω (differential), See Figure 1	28	40	52	mV/ns
t_{DD}	Differential-output delay time (t_{DD+} and t_{DD-})	C_1 = 15 pF, R_1 = 100 Ω (differential), See Figure 2		128	140	ns
$t_{DD+} - t_{DD-}$	Differential-output delay time difference	R_1 = 100 Ω (differential), See Figure 2			5	ns
t_{PHZ}	Disable time from $\overline{DATEN}$	See Figures 3, 4, and 5			220	ns
t_{PLZ}					250	ns
t_{PZH}	Enable time from $\overline{DATEN}$				220	ns
t_{PZL}					290	ns
t_{PZH}	Enable time from $\overline{DLEN}$				250	ns
$t_{w(en)}$	Enable duration time (with $\overline{DLEN}$ low)	C_{ext} = 100 pF, R_{ext} = 62 kΩ, See Figure 6	2	2.5	3	μs

receiver

PARAMETER		TEST CONDITIONS	MIN	TYP	MAX	UNIT
$t_{en(RX)}$	Receiver enable time	Squelch off, See Figure 7		56		ns
t_{PLH}	Propagation delay time, low-to-high-level output	Squelch off, See Figure 8		20	35	ns
t_{PHL}	Propagation delay time, high-to-low-level output	Squelch off, See Figure 8		22	35	ns
t_{unsq}	Unsquelch duration time	C_{ext} = 50 pF, R_{ext} = 51 kΩ, See Figure 9	1	1.2	1.45	μs
		C_{ext} = 0, R_{ext} = 6.8 kΩ, See Figure 9			180	ns

PARAMETER MEASUREMENT INFORMATION

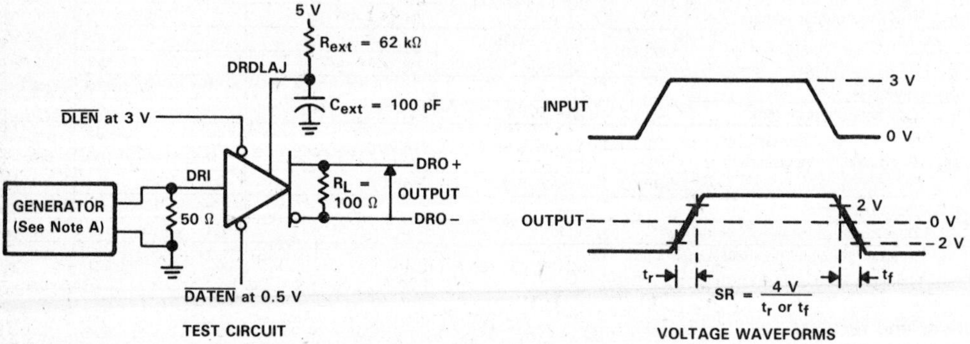
5 V
R_{ext} = 62 kΩ
DRDLAJ
C_{ext} = 100 pF
$\overline{DLEN}$ at 3 V
DRI
GENERATOR (See Note A)
50 Ω
R_L = 100 Ω
DRO +
OUTPUT
DRO −
$\overline{DATEN}$ at 0.5 V
TEST CIRCUIT
INPUT
3 V
0 V
OUTPUT
2 V
0 V
−2 V
t_r
t_f
SR = $\frac{4 V}{t_r \text{ or } t_f}$
VOLTAGE WAVEFORMS

FIGURE 1. DRIVER SLEW RATE MEASUREMENTS

NOTE A: The input pulse is supplied by a generator having the following characteristics: PRR ≤ 1 MHz, Duty Cycle ≤ 50%, t_r ≤6 ns, t_f≤ 6 ns, Z_{out} = 50 Ω.

PARAMETER MEASUREMENT INFORMATION

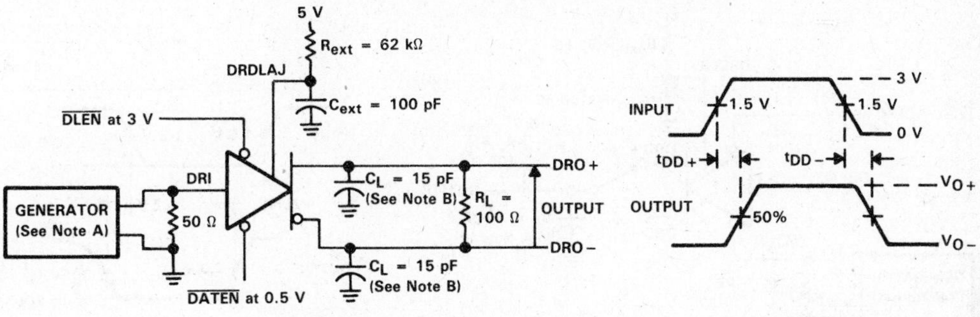
5 V
Rext = 62 kΩ
DRDLAJ
Cext = 100 pF
DLEN at 3 V
DRI
GENERATOR (See Note A)
50 Ω
CL = 15 pF (See Note B)
RL = 100 Ω
DRO+
OUTPUT
DRO−
DATEN at 0.5 V
INPUT
1.5 V
3 V
0 V
tDD+
tDD−
OUTPUT
50%
VO+
VO−

TEST CIRCUIT VOLTAGE WAVEFORMS

FIGURE 2. DRIVER DIFFERENTIAL DELAY TIMES

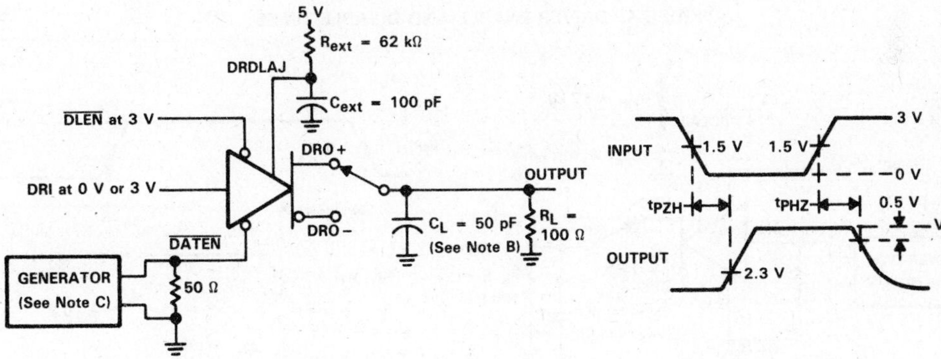
5 V
Rext = 62 kΩ
DRDLAJ
Cext = 100 pF
DLEN at 3 V
DRO+
DRI at 0 V or 3 V
OUTPUT
DRO−
CL = 50 pF (See Note B)
RL = 100 Ω
DATEN
GENERATOR (See Note C)
50 Ω
INPUT
1.5 V
3 V
0 V
tPZH
tPHZ
0.5 V
VOH
OUTPUT
2.3 V

TEST CIRCUIT VOLTAGE WAVEFORMS

FIGURE 3. DRIVER ENABLE AND DISABLE TIMES

NOTES: A. The input pulse is supplied by a generator having the following characteristics: PRR ≤ 1 MHz, Duty Cycle ≤ 50%, $t_r \leq 6$ ns, $t_f \leq 6$ ns, $Z_{out} = 50\ \Omega$.

B. C_L includes probe and jig capacitance.

C. The input pulse is supplied by a generator having the following characteristics: PRR ≤ 500 kHz, Duty Cycle ≤ 50%, $t_r \leq 6$ ns, $t_f \leq 6$ ns, $Z_{out} = 50\ \Omega$.

PARAMETER MEASUREMENT INFORMATION

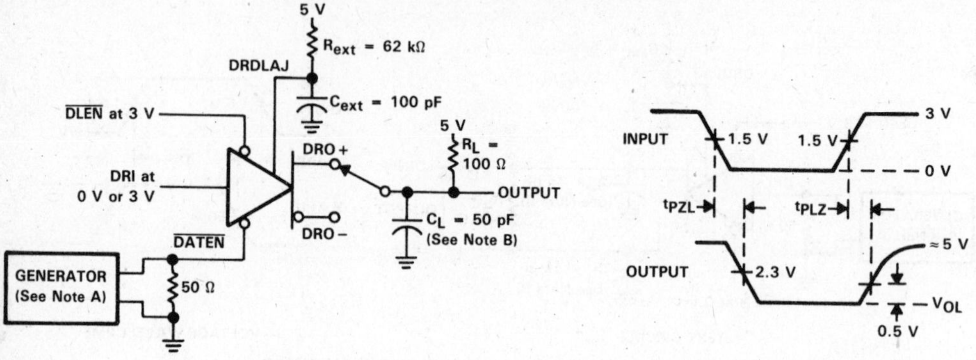
5 V
R_{ext} = 62 kΩ
DRDLAJ
C_{ext} = 100 pF
$\overline{DLEN}$ at 3 V
DRI at
0 V or 3 V
DRO +
DRO −
5 V
R_L =
100 Ω
OUTPUT
C_L = 50 pF
(See Note B)
$\overline{DATEN}$
GENERATOR
(See Note A)
50 Ω
INPUT
1.5 V
1.5 V
3 V
0 V
t_{PZL}
t_{PLZ}
OUTPUT
2.3 V
≈5 V
V_{OL}
0.5 V

TEST CIRCUIT

VOLTAGE WAVEFORMS

FIGURE 4. DRIVER ENABLE AND DISABLE TIMES

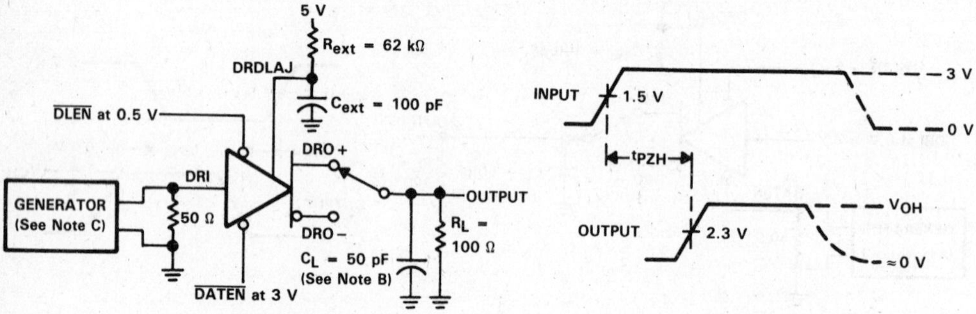
5 V
R_{ext} = 62 kΩ
DRDLAJ
C_{ext} = 100 pF
$\overline{DLEN}$ at 0.5 V
DRO +
DRI
GENERATOR
(See Note C)
50 Ω
OUTPUT
R_L =
100 Ω
DRO −
C_L = 50 pF
(See Note B)
$\overline{DATEN}$ at 3 V
INPUT
1.5 V
3 V
0 V
t_{PZH}
OUTPUT
2.3 V
V_{OH}
≈0 V

TEST CIRCUIT

VOLTAGE WAVEFORMS

FIGURE 5. ENABLE TIMES FROM DELAY ENABLE

NOTES: A. The input pulse is supplied by a generator having the following characteristics: PRR ≤ 500 kHz, Duty Cycle ≤ 50%, t_r ≤ 6 ns, t_f ≤ 6 ns, Z_{out} = 50 Ω.

B. C_L includes probe and jig capacitance.

C. The input pulse is supplied by a generator having the following characteristics: PRR ≤ 1 MHz, Duty Cycle ≤ 50%, t_r ≤6 ns, t_f≤ 6 ns, Z_{out} = 50 Ω.

PARAMETER MEASUREMENT INFORMATION

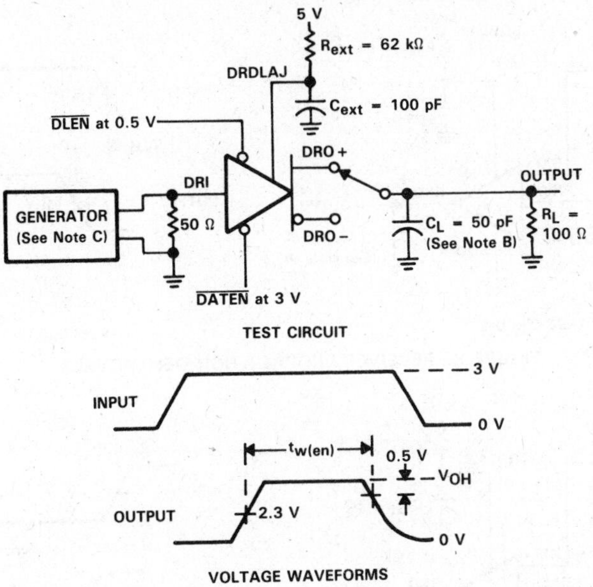
5 V
Rext = 62 kΩ
DRDLAJ
Cext = 100 pF
DLEN at 0.5 V
DRO +
DRI
OUTPUT
GENERATOR
(See Note C)
50 Ω
DRO −
CL = 50 pF
(See Note B)
RL =
100 Ω
DATEN at 3 V
TEST CIRCUIT
3 V
INPUT
0 V
tw(en)
0.5 V
VOH
OUTPUT
2.3 V
0 V
VOLTAGE WAVEFORMS

FIGURE 6. ENABLE DURATION TIME WITH DELAY ENABLE LOW

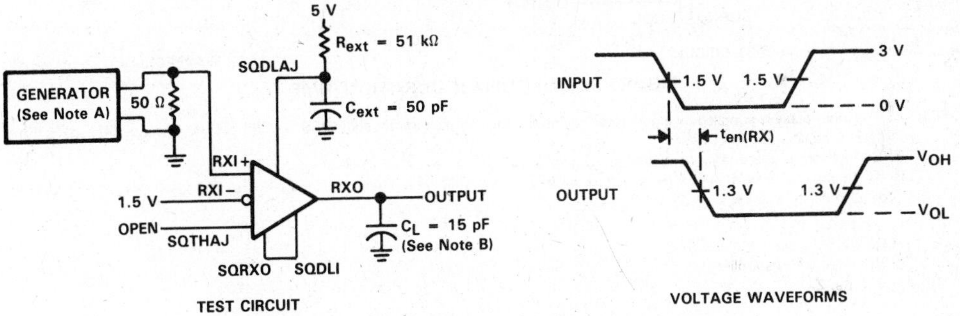
5 V
Rext = 51 kΩ
SQDLAJ
GENERATOR
(See Note A)
50 Ω
Cext = 50 pF
RXI +
RXI −
1.5 V
RXO
OUTPUT
OPEN
SQTHAJ
CL = 15 pF
(See Note B)
SQRXO
SQDLI
TEST CIRCUIT
INPUT
1.5 V
1.5 V
3 V
0 V
ten(RX)
OUTPUT
1.3 V
1.3 V
VOH
VOL
VOLTAGE WAVEFORMS

FIGURE 7. RECEIVER ENABLE (UNSQUELCH) TIMES

NOTES: A. The input pulse is supplied by a generator having the following characteristics: PRR ≤ 500 MHz, Duty Cycle ≤ 50%, $t_r \leq 6$ ns, $t_f \leq 6$ ns, Z_{out} = 50 Ω.

B. C_L includes probe and jig capacitance.

C. The input pulse is supplied by a generator having the following characteristics: PRR ≤ 200 kHz, Duty Cycle ≤ 50%, $t_r \leq 6$ ns, $t_f \leq 6$ ns, Z_{out} = 50 Ω.

PARAMETER MEASUREMENT INFORMATION

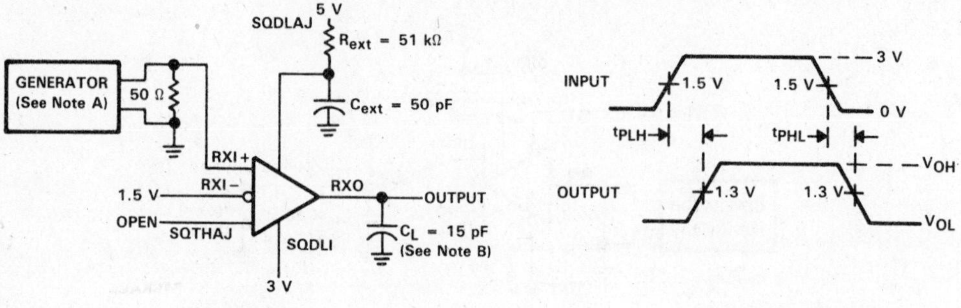
5 V
SQDLAJ
R_{ext} = 51 kΩ
GENERATOR (See Note A)
50 Ω
C_{ext} = 50 pF
RXI+
RXI−
1.5 V
OPEN
SQTHAJ
RXO
OUTPUT
SQDLI
C_L = 15 pF (See Note B)
3 V
INPUT
3 V
1.5 V
1.5 V
0 V
t_{PLH}
t_{PHL}
OUTPUT
1.3 V
1.3 V
V_{OH}
V_{OL}

TEST CIRCUIT

VOLTAGE WAVEFORMS

FIGURE 8. RECEIVER PROPAGATION DELAY TIMES

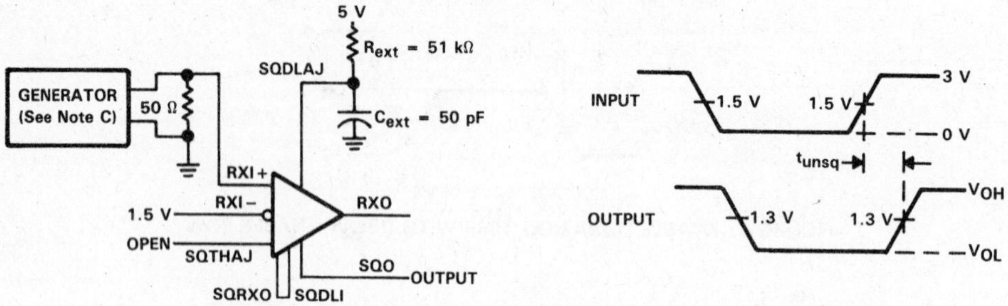
5 V
R_{ext} = 51 kΩ
SQDLAJ
GENERATOR (See Note C)
50 Ω
C_{ext} = 50 pF
RXI+
RXI−
1.5 V
OPEN
SQTHAJ
RXO
SQO
OUTPUT
SQRXO
SQDLI
INPUT
3 V
1.5 V
1.5 V
0 V
t_{unsq}
OUTPUT
1.3 V
1.3 V
V_{OH}
V_{OL}

TEST CIRCUIT

VOLTAGE WAVEFORMS

FIGURE 9. UNSQUELCH DURATION TIME

NOTES: A. The input pulse is supplied by a generator having the following characteristics: PRR ≤ 1 MHz, Duty Cycle ≤ 50%, t_r ≤ 6 ns, t_f ≤ 6 ns, Z_{out} = 50 Ω.

B. C_L includes probe and jig capacitance.

C. The input pulse is supplied by a generator having the following characteristics: PRR ≤ 100 kHz, Duty Cycle ≤ 50%, t_r ≤ 6 ns, t_f ≤ 6 ns, Z_{out} = 50 Ω.

TEXAS INSTRUMENTS
POST OFFICE BOX 655012 • DALLAS, TEXAS 75265

DUAL LINE RECEIVERS

D2304, JANUARY 1977—REVISED OCTOBER 1986

- High Speed
- Standard Supply Voltage
- Dual Channels
- High Common-Mode Rejection Ratio
- High Input Impedance
- High Input Sensitivity
- Differential Input Common-Mode Range of ±3 V
- Strobe Inputs for Receiver Selection
- Gate Inputs for Logic Versatility
- TTL Drive Capability
- High DC Noise Margin
- '107A and '107B Have Totem-Pole Outputs
- '108A and '108B Have Open-Collector Outputs
- "B" Versions Have Diode-Protected Input for Power-Off Condition

SN55107A, SN55107B, SN55108A
SN55108B . . . J PACKAGE
SN75107A, SN75107B, SN75108A
SN75108B . . . D, J, OR N PACKAGE
(TOP VIEW)

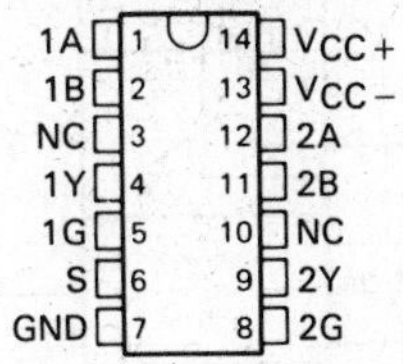
1A 1
1B 2
NC 3
1Y 4
1G 5
S 6
GND 7
14 VCC+
13 VCC−
12 2A
11 2B
10 NC
9 2Y
8 2G

SN55107A, SN55107B, SN55108A,
SN55108B . . . FK PACKAGE
(TOP VIEW)

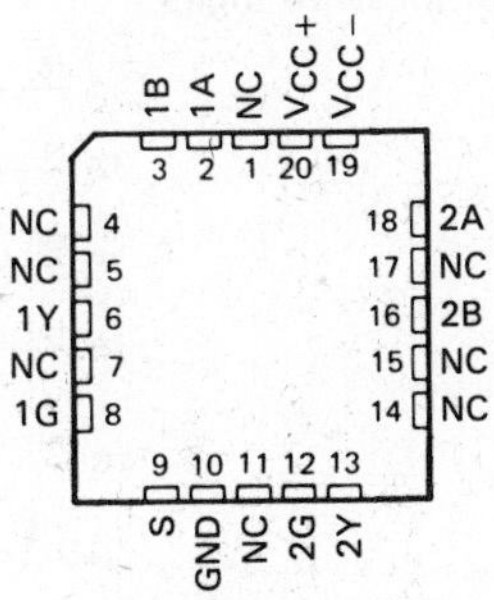
1B 1A NC VCC+ VCC−
3 2 1 20 19
NC 4
NC 5
1Y 6
NC 7
1G 8
18 2A
17 NC
16 2B
15 NC
14 NC
9 10 11 12 13
S GND NC 2G 2Y

NC—No internal connection

description

These circuits are TTL-compatible high-speed line receivers. Each is a monolithic dual circuit featuring two independent channels. They are designed for general use as well as such specific applications as data comparators and balanced, unbalanced, and party-line transmission systems. These devices are unilaterally interchangeable with and are replacements for the SN55107, SN55108, SN75107, and SN75108, but offer diode-clamped strobe inputs to simplify circuit design.

The essential difference between the "A" and "B" versions can be seen in the schematics. Input-protection diodes are in series with the collectors of the differential-input transistors of the "B" versions. These diodes are useful in certain "party-line" systems that may have multiple V_{CC+} power supplies and may be operated with some of the V_{CC+} supplies turned off. In such a system, if a supply is turned off and allowed to go to ground, the equivalent input circuit connected to that supply would be as follows:

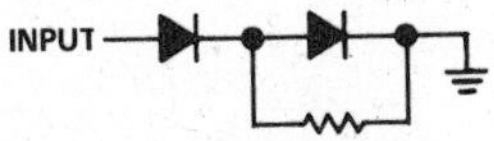
INPUT

"A" VERSION

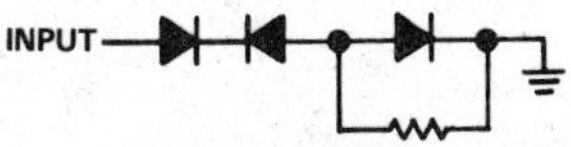
INPUT

"B" VERSION

This would be a problem in specific systems that might possibly have the transmission lines biased to some potential greater than 1.4 volts.

The SN55107A, SN55107B, SN55108A, and SN55108B are characterized for operation over the full military temperature range of −55 °C to 125 °C. The SN75107A, SN75107B, SN75108A, and SN75108B are characterized for operation from 0 °C to 70 °C.

TEXAS INSTRUMENTS
POST OFFICE BOX 655012 • DALLAS, TEXAS 75265

logic symbols†

SN55107
SN75107

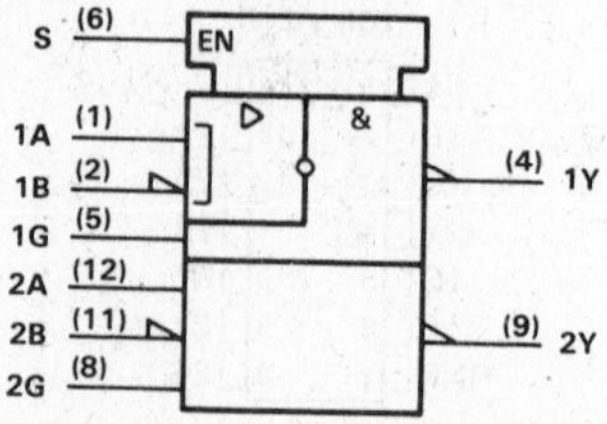
S (6)
EN
&
1A (1)
1B (2)
1G (5)
2A (12)
2B (11)
2G (8)
(4) 1Y
(9) 2Y

SN55108
SN75108

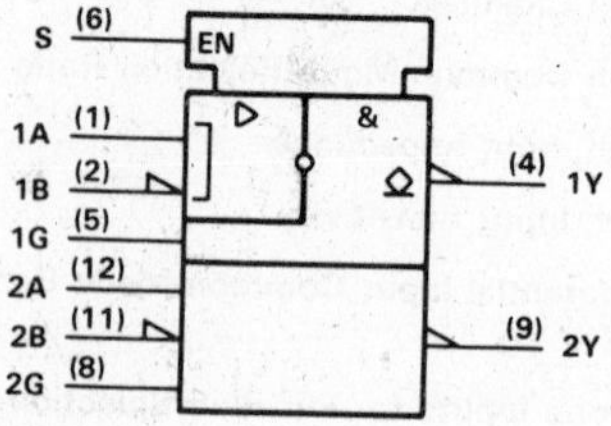
S (6)
EN
&
1A (1)
1B (2)
1G (5)
2A (12)
2B (11)
2G (8)
(4) 1Y
(9) 2Y

†These symbols are in accordance with ANSI/IEEE Std 91-1984 and IEC Publication 617-12.
Pin numbers shown are for D, J, and N packages.

logic diagram (positive logic)

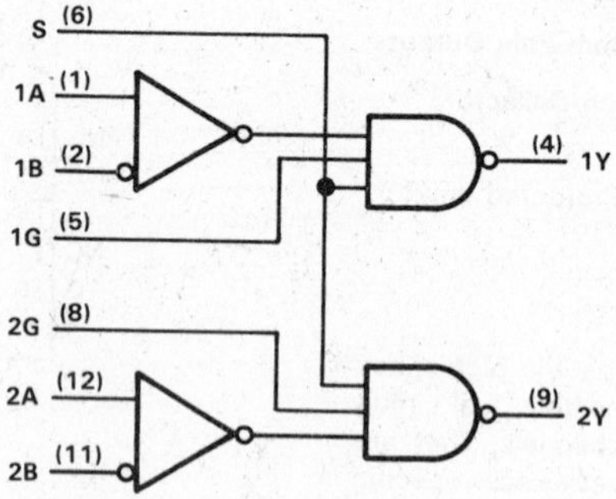
S (6)
1A (1)
1B (2)
1G (5)
2G (8)
2A (12)
2B (11)
(4) 1Y
(9) 2Y

FUNCTION TABLE

DIFFERENTIAL INPUTS A-B	STROBES G	STROBES S	OUTPUT Y
$V_{ID} \geq 25$ mV	X	X	H
-25 mV $< V_{ID} < 25$ mV	X	L	H
	L	X	H
	H	H	Indeterminate
$V_{ID} \leq -25$ mV	X	L	H
	L	X	H
	H	H	L

H = high level, L = low level, X = irrelevant

schematic (each receiver)

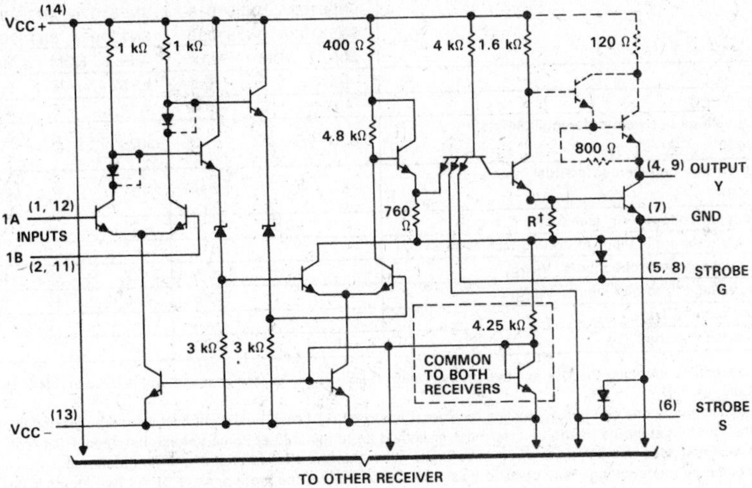
(14)
VCC+
1 kΩ
1 kΩ
400 Ω
4 kΩ
1.6 kΩ
120 Ω
4.8 kΩ
800 Ω
(4, 9) OUTPUT Y
(1, 12)
1A
INPUTS
1B
(2, 11)
760 Ω
R†
(7) GND
(5, 8) STROBE G
4.25 kΩ
3 kΩ
3 kΩ
COMMON TO BOTH RECEIVERS
(6) STROBE S
(13)
VCC−
TO OTHER RECEIVER

†R = 1 kΩ for '107A and '107B, 750 Ω for '108A and '108B.

NOTES: 1. Resistor values shown are nominal.
2. Components shown with dashed lines in the output circuitry are applicable to the '107A and '107B only. Diodes in series with the collectors of the differential input transistors are short-circuited on '107A and '108A.

absolute maximum ratings over operating free-air temperature range (unless otherwise noted)

Supply voltage, V_{CC+} (see Note 3)	7 V
Supply voltage, V_{CC-}	−7 V
Differential input voltage (see Note 4)	±6 V
Common-mode input voltage (see Note 5)	±5 V
Strobe input voltage	5.5 V
Continuous total dissipation at (or below) 25°C free-air temperature (see Note 6):	
D package	950 mW
FK package	1375 mW
J package	1025 mW
N package	1150 mW
Operating free-air temperature range: Series 55	−55°C to 125°C
Series 75	0°C to 70°C
Storage temperature range	−65°C to 150°C
Case temperature for 60 seconds: FK package	260°C
Lead temperature 1,6 mm (1/16 inch) from case for 60 seconds: J package	300°C
Lead temperature 1,6 mm (1/16 inch) from case for 10 seconds: D or N package	260°C

NOTES: 3. All voltage values, except differential voltages, are with respect to network ground terminal.
4. Differential voltage values are at the noninverting (A) terminal with respect to the inverting (B) terminal.
5. Common-mode input voltage is the average of the voltages at the A and B inputs.
6. For operation above 25°C free-air temperature, derate linearly at the following rates: 7.6 mW/°C for the D package, 11.0 mW/°C for the FK package, 8.2 mW/°C for the J package, and 9.2 mW/°C for the N package.

recommended operating conditions (see Note 7)

	SN55107A, SN55107B SN55108A, SN55108B			SN75107A, SN75107B SN75108A, SN75108B			UNIT
	MIN	NOM	MAX	MIN	NOM	MAX	
Supply voltage, V_{CC+}	4.5	5	5.5	4.75	5	5.25	V
Supply voltage, V_{CC-}	−4.5	−5	−5.5	−4.75	−5	−5.25	V
High-level input voltage between differential inputs, V_{IDH} (see Note 8)	0.025		5	0.025		5	V
Low-level input voltage between differential inputs, V_{IDL} (see Note 8)	−5†		−0.025	−5†		−0.025	V
Common-mode input voltage, V_{IC} (see Notes 8 and 9)	−3†		3	−3†		3	V
Input voltage, any differential input to ground (see Note 8)	−5†		3	−5†		3	V
High-level input voltage at strobe inputs, $V_{IH(S)}$	2		5.5	2		5.5	V
Low-level input voltage at strobe inputs, $V_{IL(S)}$	0		0.8	0		0.8	V
Low-level output current, I_{OL}			−16			−16	mA
Operating free-air temperature, T_A	−55		125	0		70	°C

† The algebraic convention, in which the less positive (more negative) limit is designated as minimum, is used in this data sheet for input voltage levels only.

NOTES: 7. When using only one channel of the line receiver, the strobe G of the unused channel should be grounded and at least one of the differential inputs of the unused receiver should be terminated at some voltage between −3 V and 3 V.

8. The recommended combinations of input voltages fall within the shaded area of the figure shown.

9. The common-mode voltage may be as low as −4 V provided that the more positive of the two inputs is not more negative than −3 V.

RECOMMENDED COMBINATIONS OF INPUT VOLTAGES

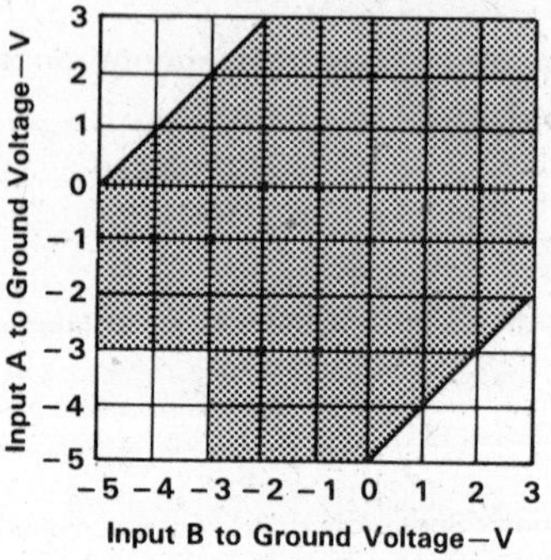
Input A to Ground Voltage—V
3
2
1
0
−1
−2
−3
−4
−5
−5 −4 −3 −2 −1 0 1 2 3
Input B to Ground Voltage—V

electrical characteristics over recommended free-air temperature range (unless otherwise noted)

PARAMETER			TEST CONDITIONS†		'107A, '107B MIN	TYP‡	MAX	'108A, '108B MIN	TYP‡	MAX	UNIT
I_{IH}	High-level input current	A	$V_{CC\pm}$ = MAX	V_{ID} = 5 V		30	75		30	75	μA
		B		V_{ID} = −5 V		30	75		30	75	
I_{IL}	Low-level input current	A	$V_{CC\pm}$ = MAX	V_{ID} = −5 V			−10			−10	μA
		B		V_{ID} = 5 V			−10			−10	
I_{IH}	High-level input current into 1G or 2G		$V_{CC\pm}$ = MAX, $V_{IH(S)}$ = 2.4 V				40			40	μA
			$V_{CC\pm}$ = MAX, $V_{IH(S)}$ = MAX V_{CC+}				1			1	mA
I_{IL}	Low-level input current into 1G or 2G		$V_{CC\pm}$ = MAX, $V_{IL(S)}$ = 0.4 V				−1.6			−1.6	mA
I_{IH}	High-level input current into S		$V_{CC\pm}$ = MAX, $V_{IH(S)}$ = 2.4 V				80			80	μA
			$V_{CC\pm}$ = MAX, $V_{IH(S)}$ = MAX $V_{CC\pm}$				2			2	mA
I_{IL}	Low-level input current into S		$V_{CC\pm}$ = MAX, $V_{IL(S)}$ = 0.4 V				−3.2			−3.2	mA
V_{OH}	High-level output voltage		$V_{CC\pm}$ = MIN, $V_{IL(S)}$ = 0.8 V, V_{IDH} = 25 mV, I_{OH} = −400 μA, V_{IC} = −3 V to 3 V		2.4						V
V_{OL}	Low-level output voltage		$V_{CC\pm}$ = MIN, $V_{IH(S)}$ = 2 V, V_{IDL} = −25 mV, I_{OL} = 16 mA, V_{IC} = −3 V to 3 V				0.4			0.4	V
I_{OH}	High-level output current		$V_{CC\pm}$ = MIN, V_{OH} = MAX V_{CC+}							250	μA
I_{OS}	Short-circuit output current§		$V_{CC\pm}$ = MAX		−18		−70				mA
I_{CCH+}	Supply current from V_{CC+}, outputs high		$V_{CC\pm}$ = MAX, T_A = 25°C			18	30		18	30	mA
I_{CCH-}	Supply current form V_{CC-}, outputs high		$V_{CC\pm}$ = MAX, T_A = 25°C			−8.4	−15		−8.4	15	mA

† For conditions shown as MIN or MAX, use the appropriate value specified under recommended operating conditions.
‡ All typical values are at V_{CC+} = 5 V, V_{CC-} = −5 V, T_A = 25°C.
§ Not more than one output should be shorted at a time.

switching characteristics, $V_{CC\pm}$ = ±5 V, T_A = 25°C, see Figure 1

PARAMETER		TEST CONDITIONS	'107A, '107B MIN	TYP	MAX	'108A, '108B MIN	TYP	MAX	UNIT
$t_{PLH(D)}$	Propagation delay time, low-to-high-level output, from differential inputs A and B	R_L = 390 Ω, C_L = 50 pF		17	25				ns
		R_L = 390 Ω, C_L = 15 pF					19	25	
$t_{PHL(D)}$	Propagation delay time, high-to-low-level output, from differential inputs A and B	R_L = 390 Ω, C_L = 50 pF		17	25				ns
		R_L = 390 Ω, C_L = 15 pF					19	25	
$t_{PLH(S)}$	Propagation delay time, low-to-high-level output, from strobe input G or S	R_L = 390 Ω, C_L = 50 pF		10	15				ns
		R_L = 390 Ω, C_L = 15 pF					13	20	
$t_{PHL(S)}$	Propagation delay time, high-to-low-level output, from strobe input G or S	R_L = 390 Ω, C_L = 50 pF		8	15				ns
		R_L = 390 Ω, C_L = 15 pF					13	20	

PARAMETER MEASUREMENT INFORMATION

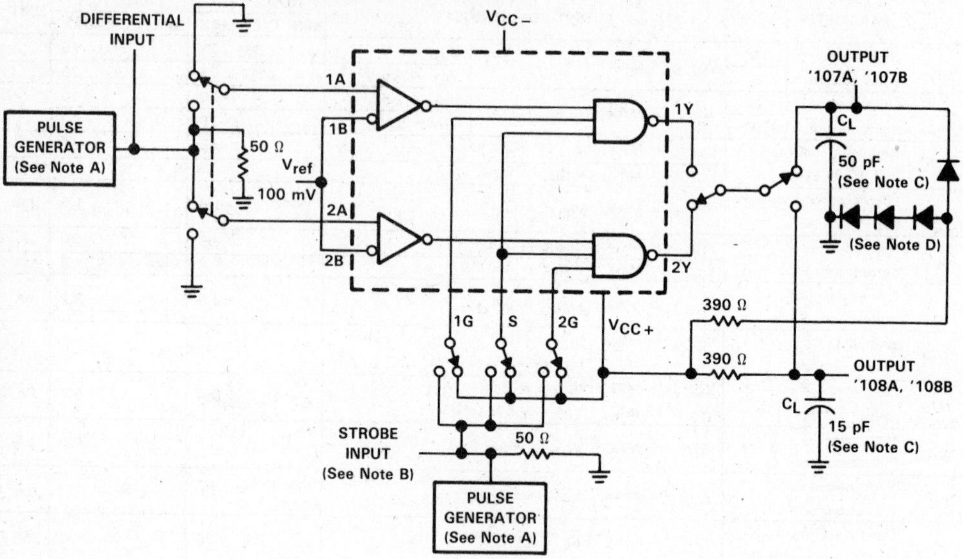
DIFFERENTIAL
INPUT
V_{CC-}
1A
1B
2A
2B
50 Ω
V_{ref}
100 mV
PULSE
GENERATOR
(See Note A)
1Y
2Y
OUTPUT
'107A, '107B
C_L
50 pF
(See Note C)
(See Note D)
390 Ω
390 Ω
1G
S
2G
V_{CC+}
OUTPUT
'108A, '108B
C_L
15 pF
(See Note C)
STROBE
INPUT
(See Note B)
50 Ω
PULSE
GENERATOR
(See Note A)

TEST CIRCUIT

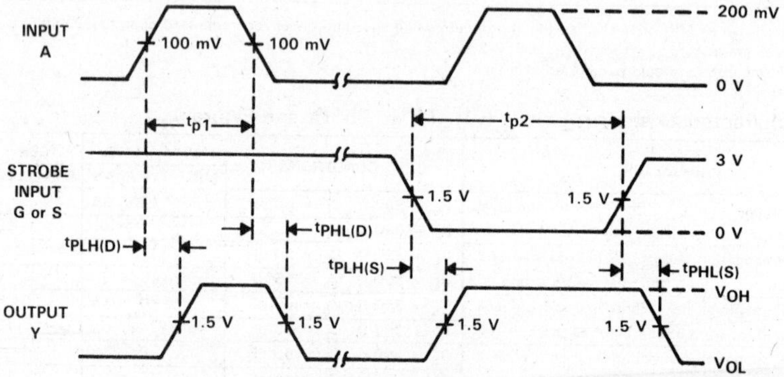
INPUT
A
100 mV
100 mV
200 mV
0 V
t_{p1}
t_{p2}
STROBE
INPUT
G or S
1.5 V
1.5 V
3 V
0 V
$t_{PLH(D)}$
$t_{PHL(D)}$
$t_{PLH(S)}$
$t_{PHL(S)}$
OUTPUT
Y
1.5 V
1.5 V
1.5 V
1.5 V
V_{OH}
V_{OL}

VOLTAGE WAVEFORMS

FIGURE 1. PROPAGATION DELAY TIMES

NOTES: A. The pulse generators have the following characteristics: Z_{out} = 50 Ω, t_r = 10 ±5 ns, t_f = 10 ±5 ns, t_{pd1} = 500 ns, PRR ≤ 1 MHz, t_{pd2} = 1 μs, PRR ≤ 500 kHz.

B. Strobe input pulse is applied to Strobe 1G when inputs 1A-1B are being tested, to Strobe S when inputs 1A-1B or 2A-2B are being tested, and to Strobe 2G when inputs 2A-2B are being tested.

C. C_L includes probe and jig capacitance.

D. All diodes are 1N916.

TEXAS INSTRUMENTS
POST OFFICE BOX 655012 • DALLAS, TEXAS 75265

TYPICAL CHARACTERISTICS†

OUTPUT VOLTAGE
vs
DIFFERENTIAL INPUT VOLTAGE

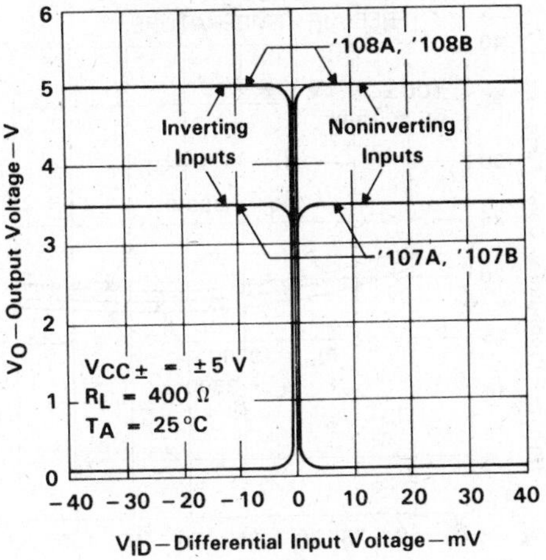
'108A, '108B
Inverting Inputs
Noninverting Inputs
'107A, '107B
VCC± = ±5 V
RL = 400 Ω
TA = 25°C
VO—Output Voltage—V
VID—Differential Input Voltage—mV

FIGURE 2

HIGH-LEVEL INPUT CURRENT INTO 1A or 2A
vs
FREE-AIR TEMPERATURE

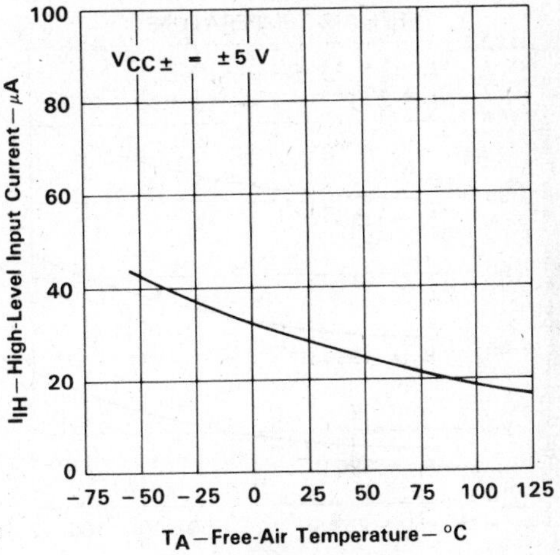
VCC± = ±5 V
IIH—High-Level Input Current—μA
TA—Free-Air Temperature—°C

FIGURE 3

SUPPLY CURRENT, OUTPUTS HIGH
vs
FREE-AIR TEMPERATURE

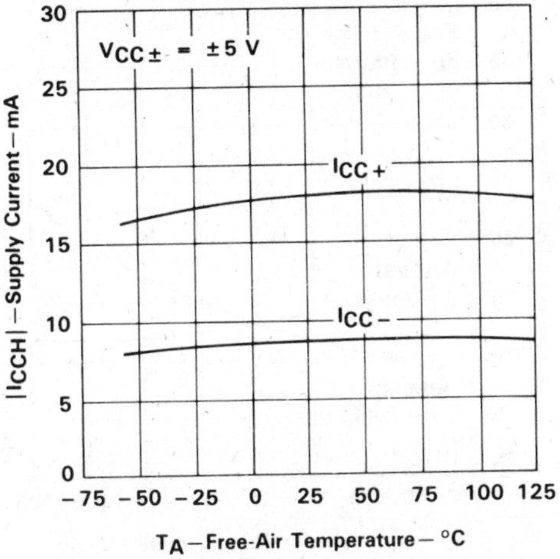
VCC± = ±5 V
ICC+
ICC−
|ICCH|—Supply Current—mA
TA—Free-Air Temperature—°C

FIGURE 4

'107A, '107B
PROPAGATION DELAY TIME
(DIFFERENTIAL INPUTS)
vs
FREE-AIR TEMPERATURE

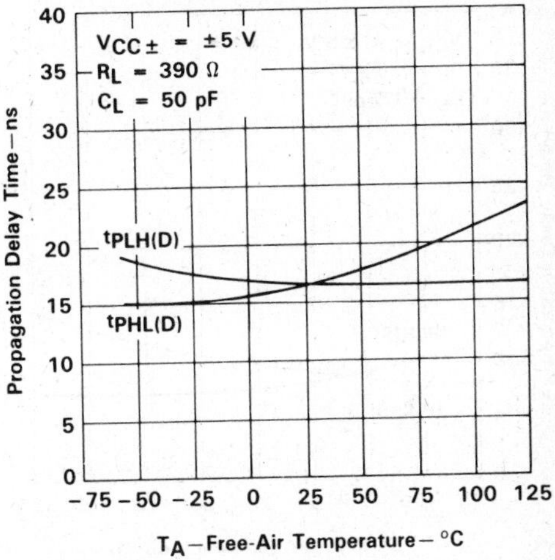
VCC± = ±5 V
RL = 390 Ω
CL = 50 pF
tPLH(D)
tPHL(D)
Propagation Delay Time—ns
TA—Free-Air Temperature—°C

FIGURE 5

† Values below 0°C and above 70°C apply to SN55 Series only.

TYPICAL CHARACTERISTICS†

'108A, '108B
PROPAGATION DELAY TIME, LOW-TO-HIGH LEVEL (DIFFERENTIAL INPUTS) vs FREE-AIR TEMPERATURE

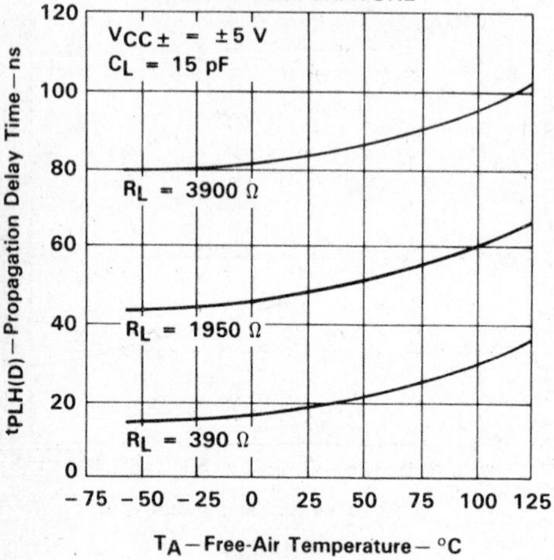

VCC± = ±5 V
CL = 15 pF
RL = 3900 Ω
RL = 1950 Ω
RL = 390 Ω
tPLH(D) – Propagation Delay Time – ns
TA – Free-Air Temperature – °C

FIGURE 6

'108A, '108B
PROPAGATION DELAY TIME, HIGH-TO-LOW LEVEL (DIFFERENTIAL INPUTS) vs FREE-AIR TEMPERATURE

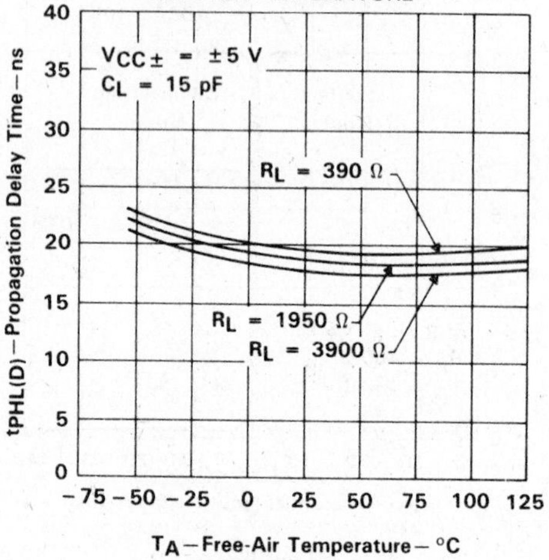

VCC± = ±5 V
CL = 15 pF
RL = 390 Ω
RL = 1950 Ω
RL = 3900 Ω
tPHL(D) – Propagation Delay Time – ns
TA – Free-Air Temperature – °C

FIGURE 7

'107A, '107B
PROPAGATION DELAY TIME (STROBE INPUTS) vs FREE-AIR TEMPERATURE

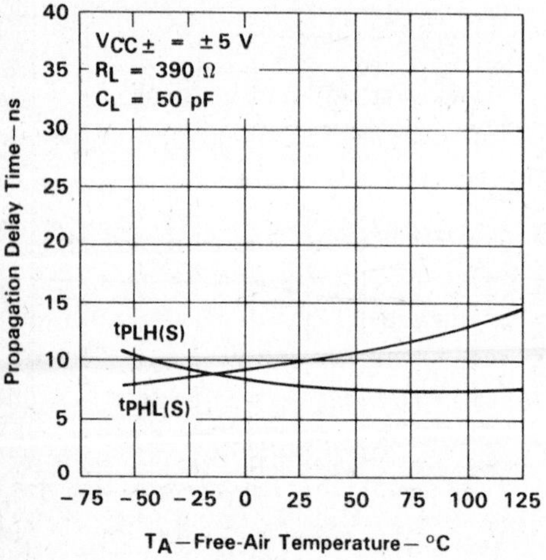

VCC± = ±5 V
RL = 390 Ω
CL = 50 pF
tPLH(S)
tPHL(S)
Propagation Delay Time – ns
TA – Free-Air Temperature – °C

FIGURE 8

'108A, '108B
PROPAGATION DELAY TIME (STROBE INPUTS) vs FREE-AIR TEMPERATURE

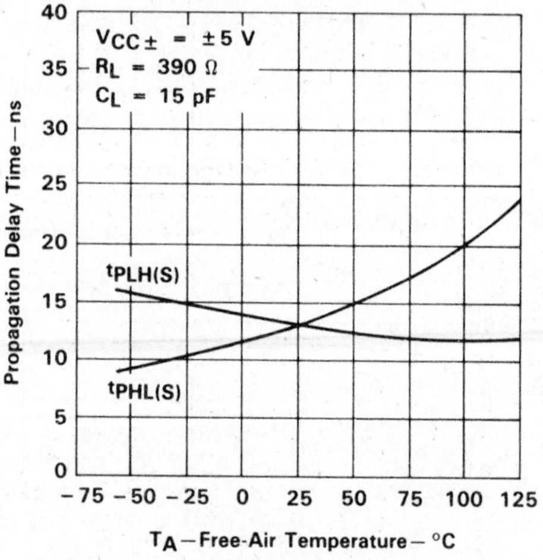

VCC± = ±5 V
RL = 390 Ω
CL = 15 pF
tPLH(S)
tPHL(S)
Propagation Delay Time – ns
TA – Free-Air Temperature – °C

FIGURE 9

† Values below 0°C and above 70°C apply to SN55 Series only.

4 Line Drivers/Receivers

TYPICAL APPLICATION DATA

basic balanced-line transmission system

The '107A, '107B, '108A, and '108B dual line circuits are designed specifically for use in high-speed data transmission systems that utilize balanced, terminated transmission lines such as twisted-pair lines. The system operates in the balanced mode, so noise induced on one line is also induced on the other. The noise appears common-mode at the receiver input terminals where it is rejected. The ground connection between the line driver and receiver is not part of the signal circuit so that system performance is not affected by circulating ground currents.

The unique driver-output circuit allows terminated transmission lines to be driven at normal line impedances. High-speed system operation is ensured since line reflections are virtually eliminated when terminated lines are used. Crosstalk is minimized by low signal amplitudes and low line impedances.

The typical data delay in a system is approximately (30 + 1.3 L) nanoseconds, where L is the distance in feet separating the driver and receiver. This delay includes one gate delay in both the driver and receiver.

Data is impressed on the balanced-line system by unbalancing the line voltages with the driver output current. The driven line is selected by appropriate driver-input logic levels. The voltage difference is approximately:

$$V_{DIFF} \approx 1/2 I_{O(on)} \cdot R_T.$$

High series line resistance will cause degradation of the signal. The receivers, however, will detect signals as low as 25 mV (or less). For normal line resistances, data may be recovered from lines of several thousand feet in length.

Line-termination resistors (R_T) are required only at the extreme ends of the line. For short lines, termination resistors at the receiver only may prove adequate. The signal amplitude will then be approximately:

$$V_{DIFF} \approx I_{O(on)} \cdot R_T.$$

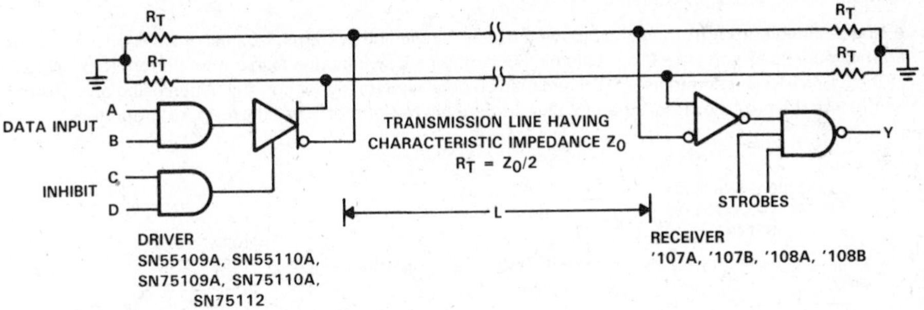
RT
RT
RT
RT
DATA INPUT
A
B
INHIBIT
C
D
TRANSMISSION LINE HAVING
CHARACTERISTIC IMPEDANCE Z0
RT = Z0/2
L
Y
STROBES
DRIVER
SN55109A, SN55110A,
SN75109A, SN75110A,
SN75112
RECEIVER
'107A, '107B, '108A, '108B

FIGURE 10

data-bus or party-line system

The strobe feature of the receivers and the inhibit feature of the drivers allow these dual line circuits to be used in data-bus or party-line systems. In these applications, several drivers and receivers may share a common transmission line. An enabled driver transmits data to all enabled receivers on the line while other drivers and receivers are disabled. Data is thus time-multiplexed on the transmission line. The device specifications allow widely varying thermal and electrical environments at the various driver and receiver locations. The data-bus system offers maximum performance at minimum cost.

TYPICAL APPLICATION DATA

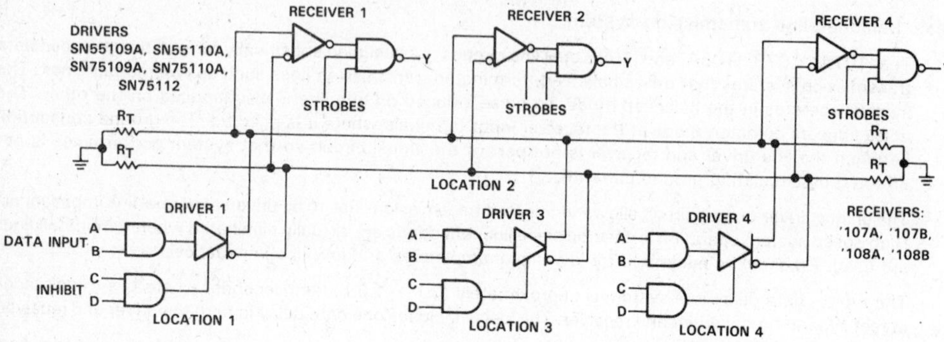
DRIVERS
SN55109A, SN55110A,
SN75109A, SN75110A,
SN75112
RECEIVER 1
RECEIVER 2
RECEIVER 4
STROBES
Y
RT
LOCATION 2
DRIVER 1
DRIVER 3
DRIVER 4
DATA INPUT
INHIBIT
A
B
C
D
LOCATION 1
LOCATION 3
LOCATION 4
RECEIVERS:
'107A, '107B,
'108A, '108B

FIGURE 11

unbalanced or single-line systems

These dual line circuits may also be used in unbalanced or single-line systems. Although these systems do not offer the same performance as balanced systems for long lines, they are adequate for very short lines where environmental noise is not severe.

The receiver threshold level is established by applying a dc reference voltage to one receiver input terminal. The signal from the transmission line is applied to the remaining input. The reference voltage should be optimized so that signal swing is symmetrical about it for maximum noise margin. The reference voltage should be in the range of −3 volts to 3 volts. It can be provided by a voltage supply or by a voltage divider from an available supply voltage.

A single-ended output from a driver may be used in single-line systems. Coaxial or shielded line is preferred for minimum noise and crosstalk problems. For large signal swings, the high output current (typically 27 mA) of the SN75112 is recommended. Drivers may be paralleled for higher current. When using only one channel of the line drivers, the other channel should be inhibited and/or have its outputs grounded.

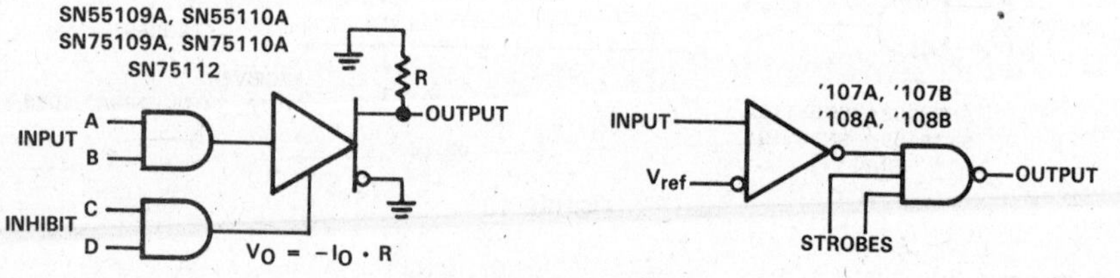
SN55109A, SN55110A
SN75109A, SN75110A
SN75112
R
OUTPUT
INPUT
A
B
INHIBIT
C
D
$V_O = -I_O \cdot R$
'107A, '107B
'108A, '108B
INPUT
V_{ref}
OUTPUT
STROBES

FIGURE 12

TEXAS INSTRUMENTS
POST OFFICE BOX 655012 • DALLAS, TEXAS 75265

TYPICAL APPLICATION DATA

'108A, '108B dot-AND output connections

The '108A, '108B line receivers feature an open-collector-output circuit that can be connected in the dot-AND logic configuration with other similar open-collector outputs. This allows a level of logic to be implemented without additional logic delay.

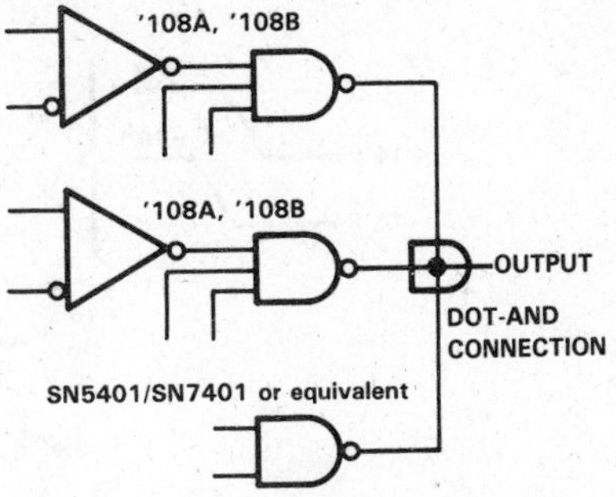
'108A, '108B
'108A, '108B
OUTPUT
DOT-AND
CONNECTION
SN5401/SN7401 or equivalent

FIGURE 13

increasing common-mode input voltage range of receiver

The common-mode voltage range or CMVR is defined as the range of voltage applied simultaneously to both input terminals that if exceeded does not allow normal operation of the receiver.

The recommended operating CMVR is ±3 volts, making it useful in all but the noisiest environments. In extremely noisy environments, common-mode voltage can easily reach ±10 V to ±15 V if some precautions are not taken to reduce ground and power supply noise, as well as crosstalk problems. When the receiver must operate in such conditions, input attenuators should be used to decrease the system common-mode noise to a tolerable level at the receiver inputs. Differential noise is also reduced by the same ratio.

These attenuators have been intentionally omitted from the receiver input terminals so the designer may select resistors that will be compatible with his particular application or environment. Furthermore, the use of attenuators adversely affects the input sensitivity, the propagation delay time, the power dissipation, and in some cases (depending on the selected resistor values) the input impedance, therefore reducing the versatility of the receiver.

The ability of the receiver to operate with approximately ±15 volts common-mode voltage at the inputs has been checked using the circuit shown in Figure 14. The resistors R1 and R2 provide a voltage divider network. Dividers with three different values presenting a 5-to-1 attenuation were used so as to operate the differential inputs at approximately ±3 volts common-mode voltage. Careful matching of the two attenuators is needed so as to balance the overdrive at the input stage. The resistors used are shown in Table A.

TABLE A

Attenuator 1:	R1 = 2 kΩ, R2 = 0.5 kΩ
Attenuator 2:	R1 = 6 kΩ, R2 = 1.5 kΩ
Attenuator 3:	R1 = 12 kΩ, R2 = 3 kΩ

Table B shows some of the typical switching results obtained under such conditions.

TABLE B. TYPICAL PROPAGATION DELAYS FOR RECEIVER WITH ATTENUATOR TEST CIRCUIT SHOWN IN FIGURE 14

DEVICE	PARAMETERS	INPUT ATTENUATOR	TYPICAL (ns)
'107A, '107B	t_{PLH}	1	20
		2	32
		3	42
	t_{PHL}	1	22
		2	31
		3	33
'108A, '108B	t_{PLH}	1	36
		2	47
		3	57
	t_{PHL}	1	29
		2	38
		3	41

TYPICAL APPLICATION DATA

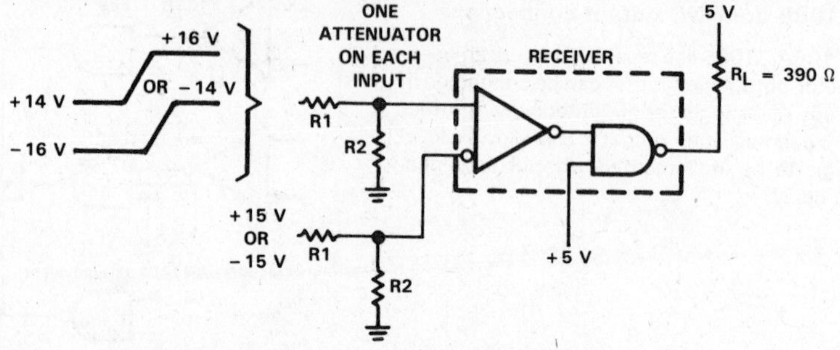
+16 V
OR −14 V
+14 V
−16 V
ONE ATTENUATOR ON EACH INPUT
R1
R2
+15 V
OR
−15 V
RECEIVER
5 V
R_L = 390 Ω
+5 V

FIGURE 14. COMMON-MODE CIRCUIT FOR TESTING INPUT ATTENUATORS, WITH RESULTS SHOWN IN TABLE B

Two methods of terminating a transmission line to reduce reflections are:

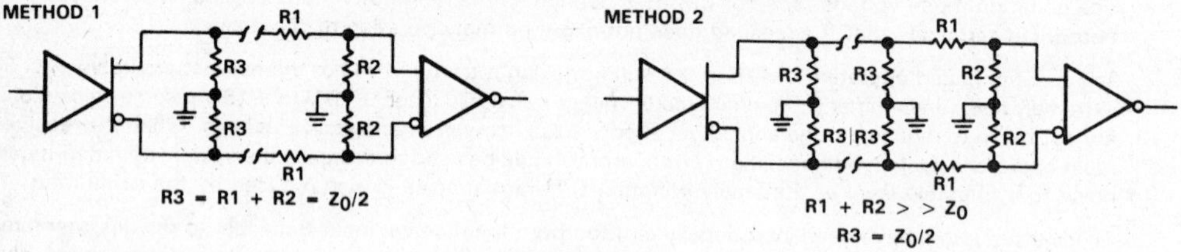
METHOD 1
R1
R3
R2
$R3 = R1 + R2 = Z_0/2$
METHOD 2
R1
R3
R2
$R1 + R2 >> Z_0$
$R3 = Z_0/2$

FIGURE 15

The first method uses the resistors as the attenuation network and line termination. The second method uses two additional resistors for the line terminations.

TEXAS INSTRUMENTS
POST OFFICE BOX 655012 • DALLAS, TEXAS 75265

TYPICAL APPLICATION DATA

For party-line operation, method 2 should be used as follows:

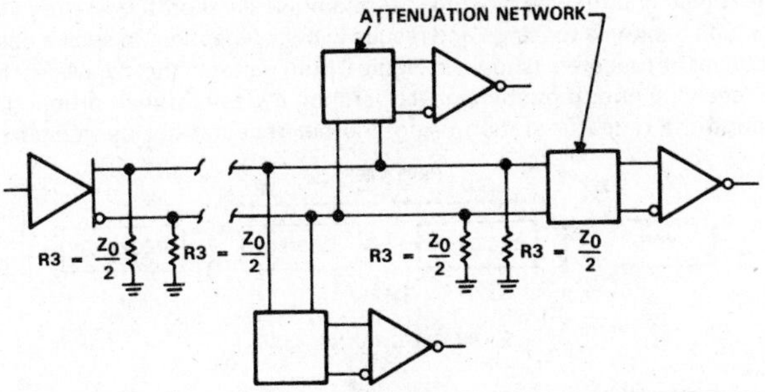
ATTENUATION NETWORK
R3 = Z0/2
R3 = Z0/2
R3 = Z0/2
R3 = Z0/2

FIGURE 16

To minimize the loading, the values of R1 and R2 should be fairly large. Examples of possible values are shown in Table A.

furnace control using the SN75108A

The furnace control circuit in Figure 17 is an example of the possible use of the SN55107A Series in areas other than what would normally be considered electronic systems. Basically the operation of this control is as follows. When the room temperature is below the desired level, the resistance of the room temperature sensor is high and channel 1 noninverting input is below (less positive than) the reference level set on the input differential amplifier. This situation causes a low output, operating the "heat on" relay and turning on the heat. The channel 2 noninverting input is below the reference level when the bonnet temperature of the furnace reaches the desired level. This causes a low output, thus operating the blower relay. Normally the furnace is shut down when the room temperature reaches the desired level and the channel 1 output goes high, turning the heat off. The blower remains on as long as the bonnet temperature is high, even after the "heat on" relay is off. There is also a safety switch in the bonnet that shuts the furnace down if the temperature there exceeds desired limitations. The types of temperature-sensing devices and bias-resistor values used are determined by the particular operating conditions encountered.

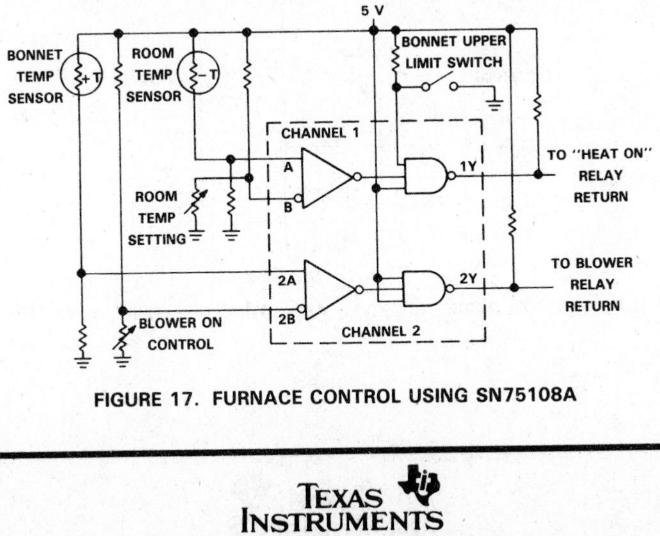
5 V
BONNET TEMP SENSOR
+T
ROOM TEMP SENSOR
−T
BONNET UPPER LIMIT SWITCH
CHANNEL 1
A
B
1Y
TO "HEAT ON" RELAY RETURN
ROOM TEMP SETTING
2A
2B
2Y
TO BLOWER RELAY RETURN
BLOWER ON CONTROL
CHANNEL 2

FIGURE 17. FURNACE CONTROL USING SN75108A

4

Line Drivers/Receivers

TYPICAL APPLICATION DATA

repeaters for long lines

In some cases, the driven line may be so long that the noise level on the line reaches the common-mode limits or the attenuation becomes too large and results in poor reception. In such a case, a simple application of a receiver and a driver as repeaters [shown in Figure 18(a)] restores the signal level and allows an adequate signal level at the receiving end. If multichannel operation is desired, then proper gating for each channel must be sent through the repeater station using another repeater set as in Figure 18(b).

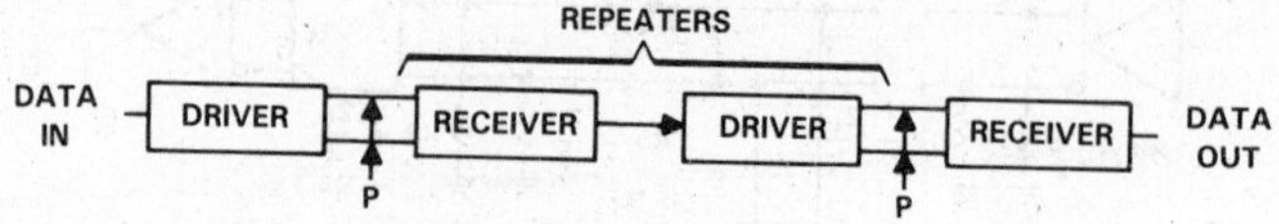
REPEATERS
DATA IN
DRIVER
RECEIVER
DRIVER
RECEIVER
DATA OUT
P
P

a. SINGLE-CHANNEL LINE

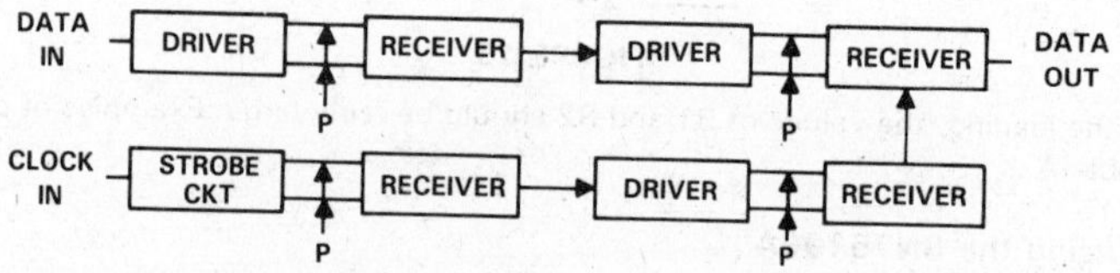
DATA IN
DRIVER
RECEIVER
DRIVER
RECEIVER
DATA OUT
CLOCK IN
STROBE CKT
RECEIVER
DRIVER
RECEIVER
P
P
P
P

b. MULTICHANNEL LINE WITH STROBE

FIGURE 18. RECEIVER-DRIVER REPEATERS

receiver as dual differential comparator

There are many applications for differential comparators, such as voltage comparison, threshold detection, controlled Schmitt triggering, and pulse width control.

As a differential comparator, a '107A or '108A may be connected so as to compare the noninverting input terminal with the inverting input as shown in Figure 19. Thus the output will be high or low resulting from the A input being greater or less than the reference. The strobe inputs allow additional control over the circuit so that either output or both may be inhibited.

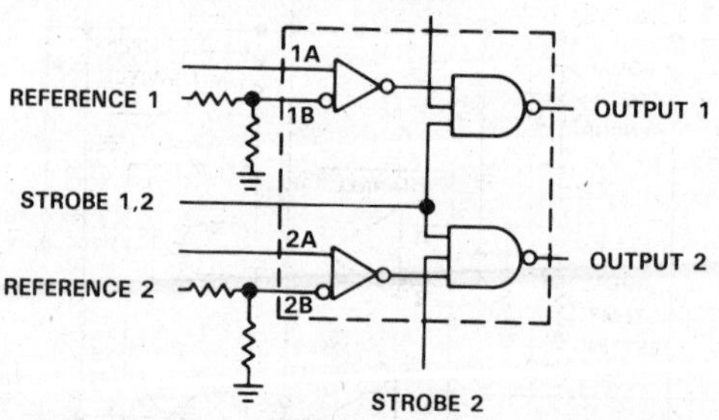
STROBE 1
1A
REFERENCE 1
1B
OUTPUT 1
STROBE 1,2
2A
REFERENCE 2
2B
OUTPUT 2
STROBE 2

FIGURE 19. SN55107A SERIES RECEIVER AS A DUAL DIFFERENTIAL COMPARATOR

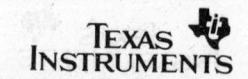

TEXAS INSTRUMENTS
POST OFFICE BOX 655012 • DALLAS, TEXAS 75265

TYPICAL APPLICATION DATA

window detector

The window detector circuit in Figure 20 has a large number of applications in test equipment and in determining upper limits, lower limits, or both at the same time — such as detecting whether a voltage or signal has exceeded its limits or ''window''. Illumination of the upper-limit (lower-limit) indicator shows that the input voltage is above (below) the selected upper (lower) limit. A mode selector is provided for selecting the desired test. For window detecting, the ''upper and lower limits'' test position is used.

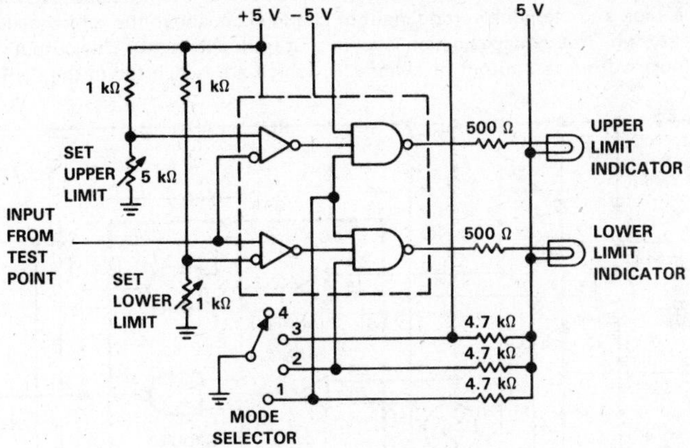
+5 V
−5 V
5 V
1 kΩ
1 kΩ
SET UPPER LIMIT
5 kΩ
INPUT FROM TEST POINT
SET LOWER LIMIT
1 kΩ
500 Ω
UPPER LIMIT INDICATOR
500 Ω
LOWER LIMIT INDICATOR
4.7 kΩ
4.7 kΩ
4.7 kΩ
4
3
2
1
MODE SELECTOR

MODE SELECTOR LEGEND

POSITION	CONDITION
1	OFF
2	TEST FOR UPPER LIMIT
3	TEST FOR LOWER LIMIT
4	TEST FOR UPPER AND LOWER LIMITS

FIGURE 20. WINDOW DETECTOR USING SN75108A

TYPICAL APPLICATION DATA

temperature controller with zero-voltage switching

The circuit in Figure 21 switches an electric resistive heater on or off by providing negative-going pulses to the gate of a triac during the time interval when the line voltage is passing through zero. The pulse generator is the 2N5447 and four diodes. This portion of the circuit provides negative-going pulses during the short time (approximately 100 μs) when the line voltage is near zero. These pulses are fed to the inverting input of one channel of the '108A. If the room temperature is below the desired level, the resistance of the thermistor is high and the noninverting input of channel 2 is above the reference level determined by the thermostat setting. This provides a high-level output from channel 2. This output is AND'ed with the positive-going pulses from the output of channel 1, which are reinverted in the 2N5449.

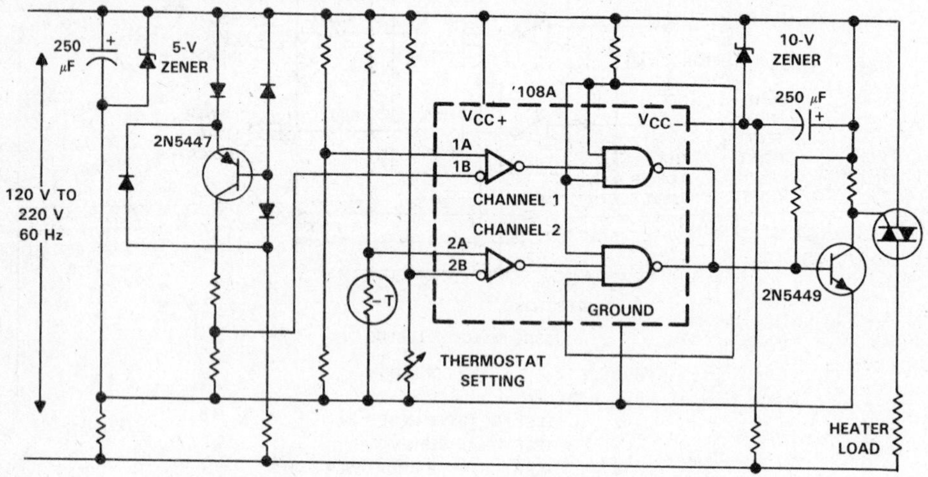
250 μF
5-V ZENER
2N5447
120 V TO 220 V 60 Hz
'108A
VCC+
VCC−
1A
1B
CHANNEL 1
CHANNEL 2
2A
2B
−T
GROUND
THERMOSTAT SETTING
10-V ZENER
250 μF
2N5449
HEATER LOAD

FIGURE 21. ZERO-VOLTAGE SWITCHING TEMPERATURE CONTROLLER

TEXAS INSTRUMENTS
POST OFFICE BOX 655012 • DALLAS, TEXAS 75265

D2106, DECEMBER 1975—REVISED SEPTEMBER 1986

- Improved Stability over Supply Voltage and Temperature Ranges
- Constant-Current Outputs
- High Speed
- Standard Supply Voltages
- High Output Impedance
- High Common-Mode Output Voltage Range (−3 V to 10 V)
- TTL Input Compatibility
- Inhibitor Available for Driver Selection

−55°C to 125°C J or FK PACKAGE	0°C to 70°C J or N PACKAGE	OUTPUT FUNCTION
SN55109A	SN75109A	6-mA Current Switch
SN55110A	SN75110A	12-mA Current Switch
	SN75112	27-mA Current Switch

SN55109A, SN55110A, . . . J PACKAGE
SN75109A, SN75110A, SN75112 . . .
D, J, OR N PACKAGE
(TOP VIEW)

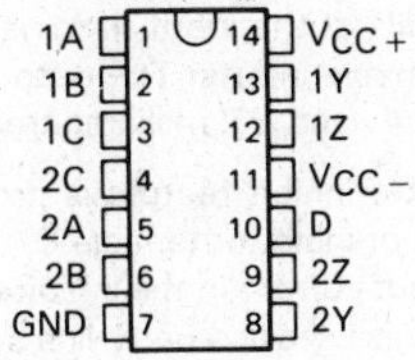
1A 1
1B 2
1C 3
2C 4
2A 5
2B 6
GND 7
14 VCC+
13 1Y
12 1Z
11 VCC−
10 D
9 2Z
8 2Y

SN55109A, SN55110A . . . FK PACKAGE
(TOP VIEW)

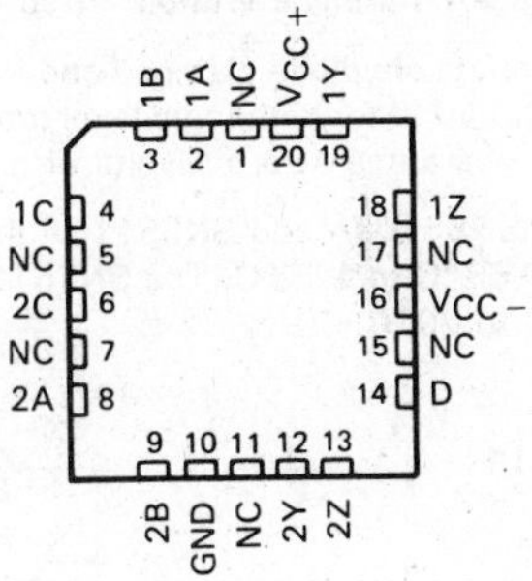
1B 1A NC VCC+ 1Y
3 2 1 20 19
1C 4
NC 5
2C 6
NC 7
2A 8
18 1Z
17 NC
16 VCC−
15 NC
14 D
9 10 11 12 13
2B GND NC 2Y 2Z

NC—No internal connection

description

The SN55109A, SN55110A, SN75109A, SN75110A, and SN75112 have improved output current regulation with supply voltage and temperature variations. In addition, the higher current of the SN75112 (27 mA) allows data to be transmitted over longer lines. These drivers offer optimum performance when used with the SN55107A, SN55108A, SN75107A, and SN75108A line receivers.

logic symbol†

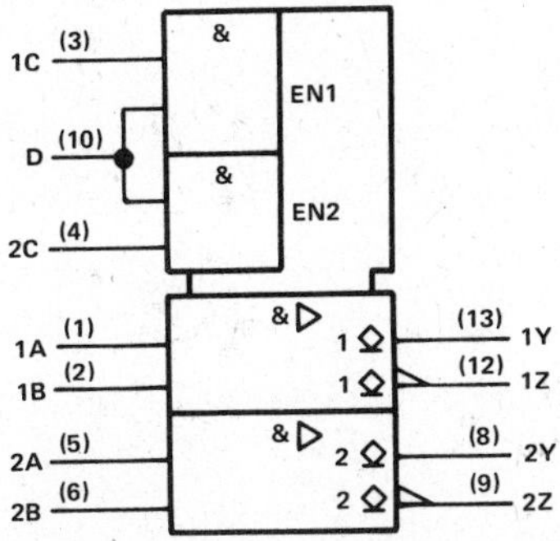
1C (3)
D (10)
2C (4)
&
EN1
&
EN2
1A (1)
1B (2)
2A (5)
2B (6)
(13) 1Y
(12) 1Z
(8) 2Y
(9) 2Z

† This symbol is in accordance with ANSI/IEEE Std 91-1984 and IED Publication 617-12.

logic diagram (positive logic)

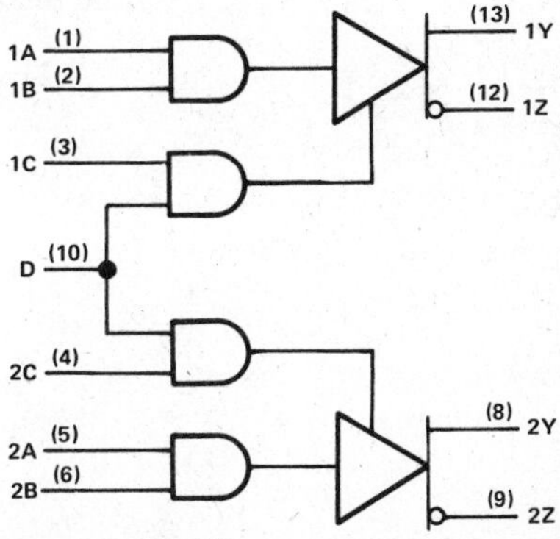
1A (1)
1B (2)
1C (3)
D (10)
2C (4)
2A (5)
2B (6)
(13) 1Y
(12) 1Z
(8) 2Y
(9) 2Z

Pin numbers shown are for D, J, and N packages.

description (continued)

These drivers feature independent channels with common voltage supply and ground terminals. The significant difference between the three drivers is in the output current specification. The driver circuits feature a constant output current that is switched to either of two output terminals by the appropriate logic levels at the input terminals. The output current can be switched off (inhibited) by low logic levels on the enable inputs. The output current is nominally 6 milliamperes for the '109A, 12 milliamperes for the '110A, and 27 milliamperes for the SN75112.

The enable/inhibit feature is provided so the circuits can be used in party-line or data-bus applications. A strobe or inhibitor (enable D), common to both drivers, is included for increased driver-logic versatility. The output current in the inhibited mode, $I_{O(off)}$, is specified so that minimum line loading is induced when the driver is used in a party-line system with other drivers. The output impedance of the driver in the inhibited mode is very high—the output impedance of a transistor biased to cutoff.

The driver outputs have a common-mode voltage range of −3 volts to 10 volts, allowing common-mode voltage on the line without affecting driver performance.

All inputs are diode clamped and are designed to satisfy TTL-system requirements. The inputs are tested at 2.0 volts for high-logic-level input conditions and 0.8 volt for low-logic-level input conditions. These test guarantee 400 millivolts of noise margin when interfaced with Series 54/74 TTL.

The SN55109A and SN55110A are characterized for operation over the full military temperature range of −55°C to 125°C. The SN75109A, SN75110A, and SN75112 are characterized for operation from 0°C to 70°C.

FUNCTION TABLE (EACH DRIVER)

LOGIC INPUTS		ENABLE INPUTS		OUTPUTS†	
A	B	C	D	Y	Z
X	X	L	X	OFF	OFF
X	X	X	L	OFF	OFF
L	X	H	H	ON	OFF
X	L	H	H	ON	OFF
H	H	H	H	OFF	ON

H = high level, L = low level, X = irrelevant

† When using only one channel of the line drivers, the other channel should be inhibited and/or have its outputs grounded.

schematic (each driver)

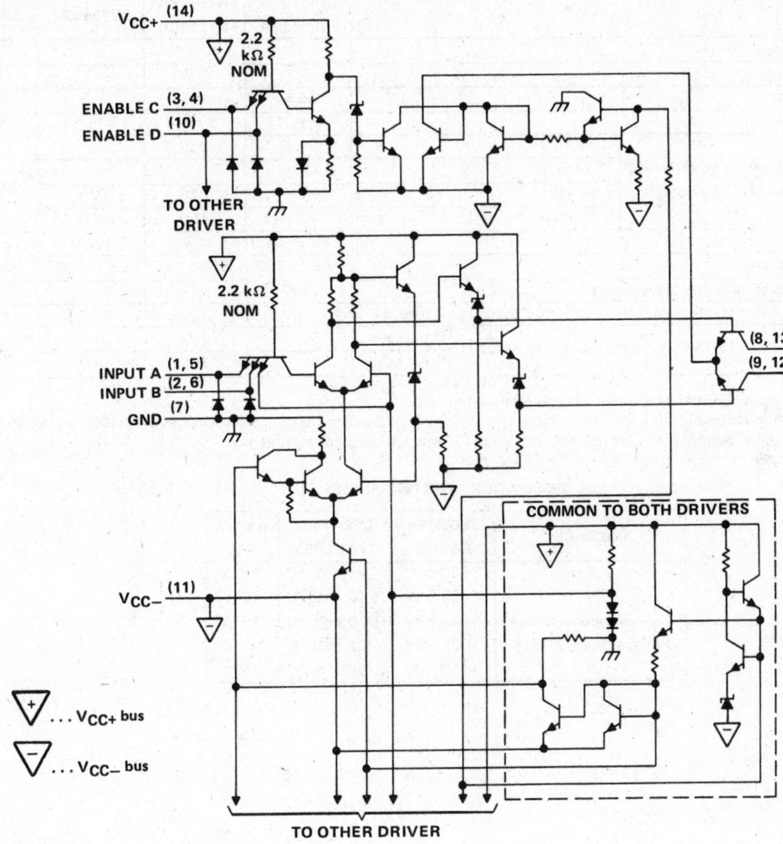
VCC+ (14)
2.2 kΩ NOM
ENABLE C (3, 4)
ENABLE D (10)
TO OTHER DRIVER
2.2 kΩ NOM
INPUT A (1, 5)
INPUT B (2, 6)
GND (7)
(8, 13) OUTPUT Y
(9, 12) OUTPUT Z
COMMON TO BOTH DRIVERS
VCC− (11)
. . . VCC+ bus
. . . VCC− bus
TO OTHER DRIVER

Pin numbers shown are for D, J, and N packages.

absolute maximum ratings over operating free-air temperature range (unless otherwise noted)

		SN55109A SN55110A	SN75109A SN75110A	SN75112	UNIT
V_{CC+} Supply voltage (see Note 1)		7	7	7	V
V_{CC-} Supply voltage		−7	−7	−7	V
V_I Input voltage		5.5	5.5	5.5	V
Output voltage range		−5 to 12	−5 to 12	−5 to 12	V
Continuous total dissipation at (or below) 25°C free-air temperature (see Note 2)	D package		950		mW
	FK package	1375			mW
	J package	1375	1025	1025	mW
	N package		1150	1150	mW
Operating free-air temperature range		−55 to 125	0 to 70	0 to 70	°C
Storage temperature range		−65 to 150	−65 to 150	−65 to 150	°C
Case temperature for 60 seconds: FK package		260			°C
Lead temperature 1,6 mm (1/16 inch) from case for 60 seconds: J package		300	300	300	°C
Lead temperature 1,6 mm (1/16 inch) from case for 10 seconds	D package		260	260	°C
	N package		260	260	°C

NOTES: 1. Voltage values are with respect to network ground terminal.
2. For operation above 25°C free-air temperature, refer to Dissipation Derating Table. In the J package, SN55109A and SN55110A chips are alloy mounted, and SN75109A and SN75110A chips are glass mounted.

DISSIPATION DERATING TABLE

PACKAGE	POWER RATING	DERATING FACTOR	ABOVE T_A
D	950 mW	7.6 mW/°C	25°C
FK	1375 mW	11.0 mW/°C	25°C
J (Alloy-mounted chip)	1375 mW	11.0 mW/°C	25°C
J (Glass-mounted chip)	1025 mW	8.2 mW/°C	25°C
N	1150 mW	9.2 mW/°C	25°C

recommended operating conditions (see Note 3)

		SN55109A, SN55110A MIN	NOM	MAX	SN75109A, SN75110A SN75112 MIN	NOM	MAX	UNIT
Supply Voltage V_{CC+}	$T_A \geq 0°C$	4.5	5	5.5	4.75	5	5.25	V
	$T_A < 0°C$	4.75	5	5.5				
Supply voltage V_{CC-}	$T_A \geq 0°C$	−4.5	−5	−5.5	−4.75	−5	−5.25	V
	$T_A < 0°C$	−4.75	−5	−5.5				
Positive common-mode output voltage		0		10	0		10	V
Negative common-mode output voltage		0		−3	0		−3	V
High-level input voltage, V_{IH}		2			2			V
Low-level input voltage, V_{IL}				0.8			0.8	V
Operating free-air temperature, T_A		−55		125	0		70	°C

NOTE 3: When using only one channel of the line drivers, the other channel should be inhibited and/or have its outputs grounded.

electrical characteristics over recommended operating free-air temperature range (unless otherwise noted)

PARAMETER			TEST CONDITIONS†	SN55109A, SN75109A MIN	TYP‡	MAX	SN55110A, SN75110A MIN	TYP‡	MAX	SN75112 MIN	TYP‡	MAX	UNIT
V_{IK}	Input clamp voltage		$V_{CC\pm}$ = MIN, I_I = −12 mA		−0.9	−1.5		−0.9	−1.5		−0.9	−1.5	V
$I_{O(on)}$	On-state output current		$V_{CC\pm}$ = MAX, V_O = 10 V		6	7		12	15		27	36	mA
			$V_{CC\pm}$ = MIN, V_O = −3 V	3.5	6		6.5	12		18	27		
$I_{O(off)}$	Off-state output current		$V_{CC\pm}$ = MIN, V_O = 10 V			100			100			100	μA
I_I	Input current at maximum input voltage	A, B, or C inputs	$V_{CC\pm}$ = MAX, V_I = 5.5 V			1			1			1	mA
		D input				2			2			2	
I_{IH}	High-level input current	A, B, or C inputs	$V_{CC\pm}$ = MAX, V_I = 2.4 V			40			40			40	μA
		D input				80			80			80	
I_{IL}	Low-level input current	A, B, or C inputs	$V_{CC\pm}$ = MAX, V_I = 0.4 V			−3			−3			−3	mA
		D input				−6			−6			−6	
$I_{CC+(on)}$	Supply current from V_{CC+} with driver enabled		$V_{CC\pm}$ = MAX, A and B inputs at 0.4 V, C and D inputs at 2 V		18	30		23	35		25	40	mA
$I_{CC-(on)}$	Supply current from V_{CC-} with driver enabled				−18	−30		−34	−50		−65	−100	
$I_{CC+(off)}$	Supply current from V_{CC+} with driver inhibited		$V_{CC\pm}$ = MAX, A, B, C, and D inputs at 0.4 V		18			21			30		mA
$I_{CC-(off)}$	Supply current from V_{CC-} with driver inhibited				−10			−17			−32		

†For conditions shown as MIN or MAX, use appropriate value specified under recommended operating conditions.
‡All typical values are at V_{CC+} = 5 V, V_{CC-} = −5 V, T_A = 25°C.

switching characteristics, V_{CC+} = 5 V, V_{CC-} = −5 V, T_A = 25°C

PARAMETER†	FROM (INPUT)	TO (OUTPUT)	TEST CONDITIONS	MIN	TYP	MAX	UNIT
t_{PLH}	A or B	Y or Z	C_L = 40 pF, R_L = 50 Ω, See Figure 1		9	15	ns
t_{PHL}					9	15	ns
t_{PLH}	C or D	Y or Z			16	25	ns
t_{PHL}					13	25	ns

† t_{PLH} = Propagation delay time, low-to-high-level output.
t_{PHL} = Propagation delay time, high-to-low-level output.

PARAMETER MEASUREMENT INFORMATION

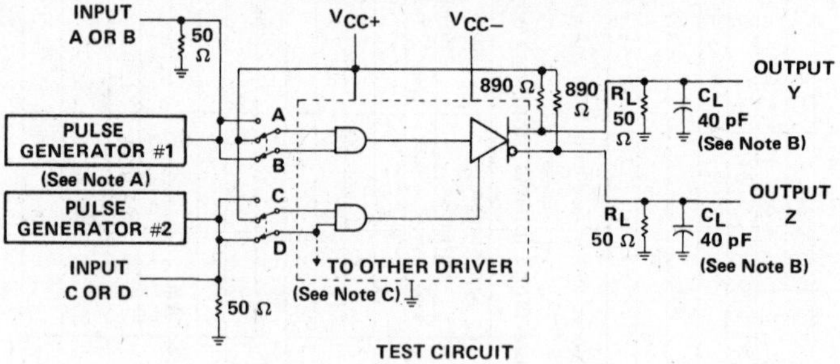
INPUT
A OR B
50 Ω
VCC+
VCC−
890 Ω
890 Ω
OUTPUT
Y
RL
50 Ω
CL
40 pF
(See Note B)
PULSE
GENERATOR #1
(See Note A)
A
B
C
D
PULSE
GENERATOR #2
OUTPUT
Z
RL
50 Ω
CL
40 pF
(See Note B)
INPUT
C OR D
TO OTHER DRIVER
(See Note C)
50 Ω

TEST CIRCUIT

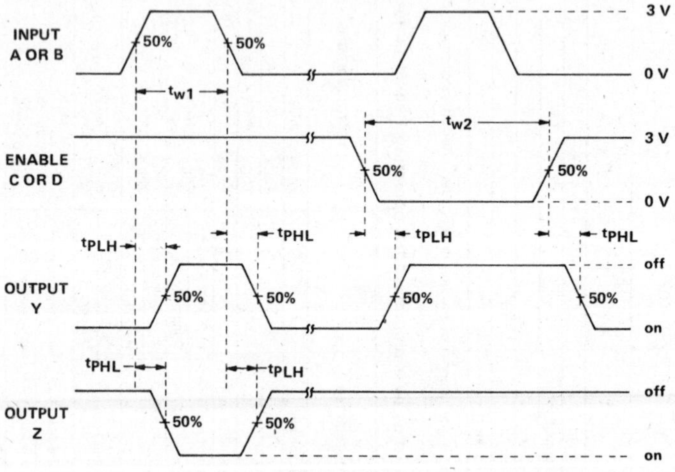
INPUT
A OR B
50%
50%
3 V
0 V
tw1
tw2
ENABLE
C OR D
50%
50%
3 V
0 V
tPLH
tPHL
tPLH
tPHL
OUTPUT
Y
50%
50%
50%
50%
off
on
tPHL
tPLH
OUTPUT
Z
50%
50%
off
on

VOLTAGE WAVEFORMS

NOTES: A. The pulse generators have the following characteristics: Z_{out} = 50 Ω, t_r = t_f = 10 ±5 ns, t_{w1} = 500 ns, PRR ≤ 1 MHz, t_{w2} = 1 μs, PRR ≤ 500 kHz.
B. C_L includes probe and jig capacitance.
C. For simplicity, only one channel and the enable connections are shown.

FIGURE 1. PROPAGATION DELAY TIMES

TEXAS INSTRUMENTS
POST OFFICE BOX 655012 • DALLAS, TEXAS 75265

TYPICAL CHARACTERISTICS

ON-STATE OUTPUT CURRENT
vs
NEGATIVE SUPPLY VOLTAGE

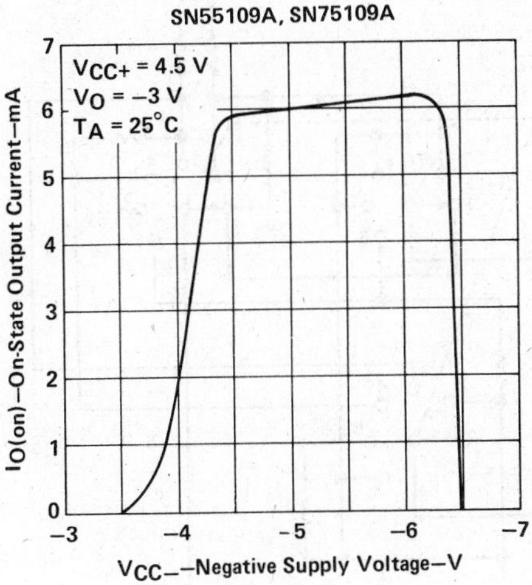
SN55109A, SN75109A
VCC+ = 4.5 V
VO = −3 V
TA = 25°C
IO(on)—On-State Output Current—mA
7
6
5
4
3
2
1
0
−3
−4
−5
−6
−7
VCC−—Negative Supply Voltage—V

FIGURE 2

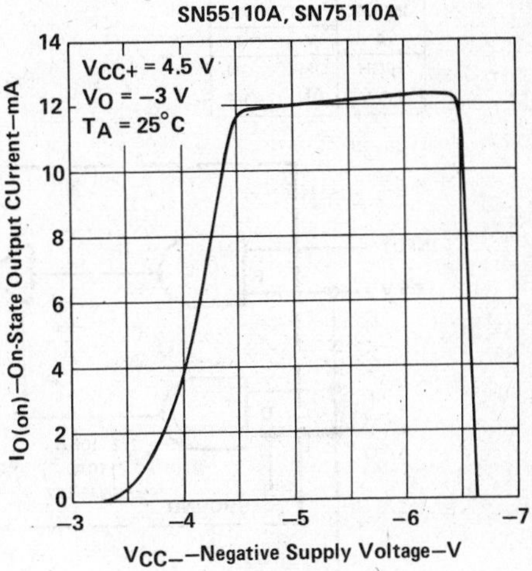
SN55110A, SN75110A
VCC+ = 4.5 V
VO = −3 V
TA = 25°C
IO(on)—On-State Output CUrrent—mA
14
12
10
8
6
4
2
0
−3
−4
−5
−6
−7
VCC−—Negative Supply Voltage—V

FIGURE 3

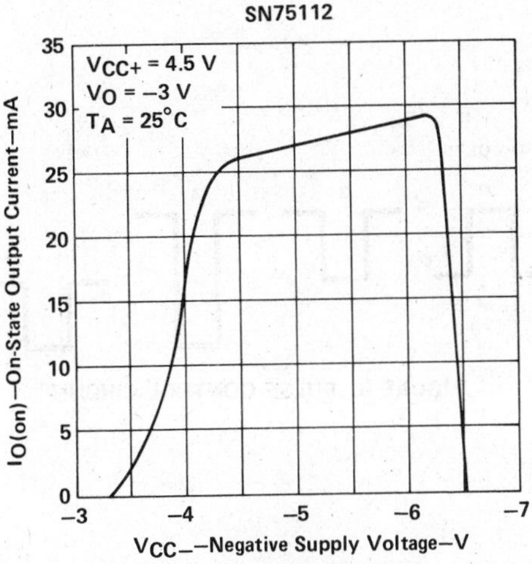
SN75112
VCC+ = 4.5 V
VO = −3 V
TA = 25°C
IO(on)—On-State Output Current—mA
35
30
25
20
15
10
5
0
−3
−4
−5
−6
−7
VCC−—Negative Supply Voltage—V

FIGURE 4

TYPICAL APPLICATION DATA

special pulse-control circuit

Figure 5 shows a circuit that may be used as a pulse generator output or in many other testing applications.

INPUT	OUTPUTS	
A	Y	Z
HIGH	OFF	ON
LOW	ON	OFF

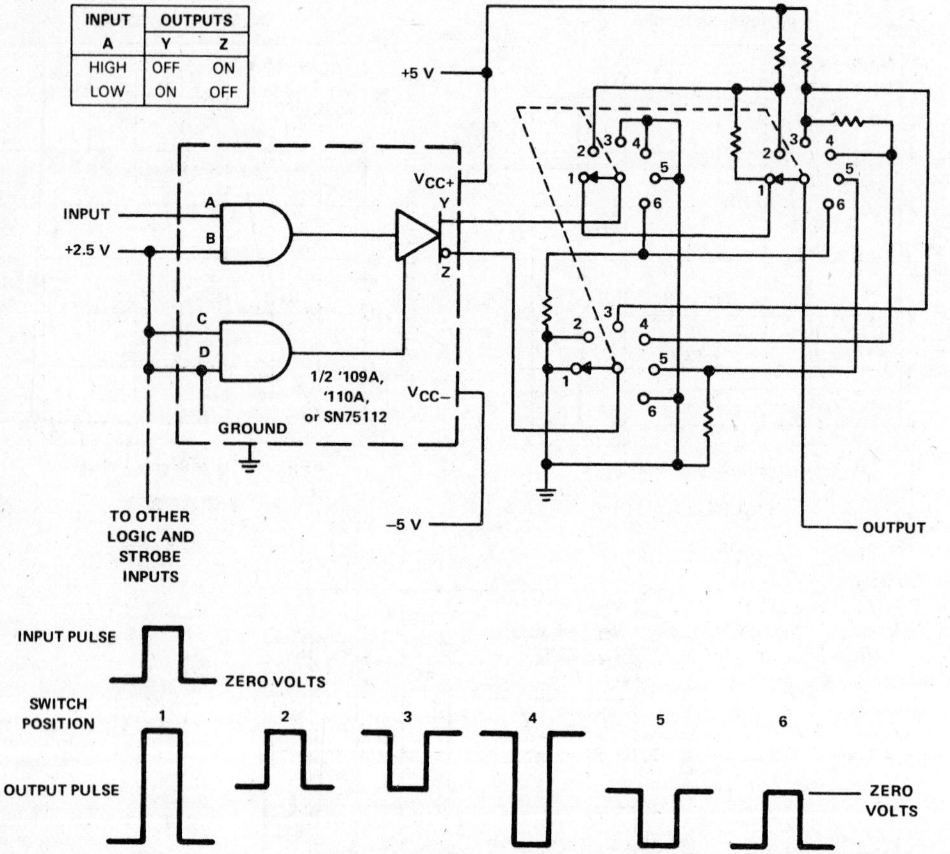
+5 V
INPUT
+2.5 V
A
B
C
D
V_{CC+}
Y
Z
1/2 '109A,
'110A,
or SN75112
V_{CC-}
GROUND
TO OTHER
LOGIC AND
STROBE
INPUTS
–5 V
OUTPUT
INPUT PULSE
ZERO VOLTS
SWITCH
POSITION
1
2
3
4
5
6
OUTPUT PULSE
ZERO
VOLTS

FIGURE 5. PULSE CONTROL CIRCUIT

- Similar to a Dual Version of SN75109A Line Driver
- Improved Stability Over Supply Voltage and Temperature Ranges
- Constant-Current Outputs
- High Output Impedance
- High Common-Mode Output Voltage Range (−3 V to 10 V)
- Glitch-Free Power-Up/Power-Down Operation
- TTL Input Compatibility
- Common Enable Circuit

D, J, OR N
DUAL-IN-LINE PACKAGE
(TOP VIEW)

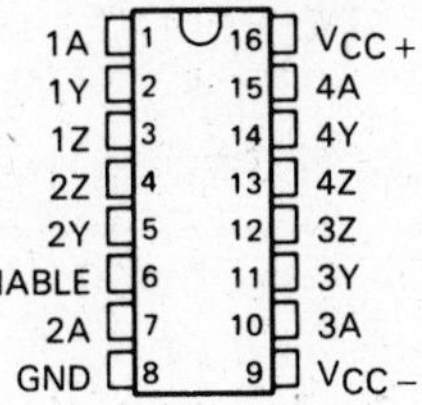
1A 1
1Y 2
1Z 3
2Z 4
2Y 5
ENABLE 6
2A 7
GND 8
16 VCC+
15 4A
14 4Y
13 4Z
12 3Z
11 3Y
10 3A
9 VCC−

FUNCTION TABLE

LOGIC INPUT	ENABLE INPUT	OUTPUT CURRENT Z	Y
H	H	ON	OFF
L	H	OFF	ON
H	L	OFF	OFF
L	L	OFF	OFF

L = low logic level
H = high logic level

description

The SN75111 features four line drivers with a common enable input. When the enable input is high, a constant output current is switched between each pair of output terminals in response to the logic level at that channel's input. When the enable is low, all channel outputs are nonconductive (transistors biased to cutoff). This minimizes loading in party-line systems where a large number of drivers share the same line.

The driver outputs have a common-mode voltage range of −3 volts to 10 volts, allowing common-mode voltages on the line without affecting driver performance.

All inputs are diode clamped and are designed to satisfy TTL-system requirements. The inputs are tested at 2 volts for high-logic-level input conditions and 0.8 volt for low-logic-level input conditions. These tests guarantee 400 millivolts of noise margin when interfaced with Series 54/74 TTL.

The SN75111 is characterized for operation from 0°C to 70°C.

logic symbol†

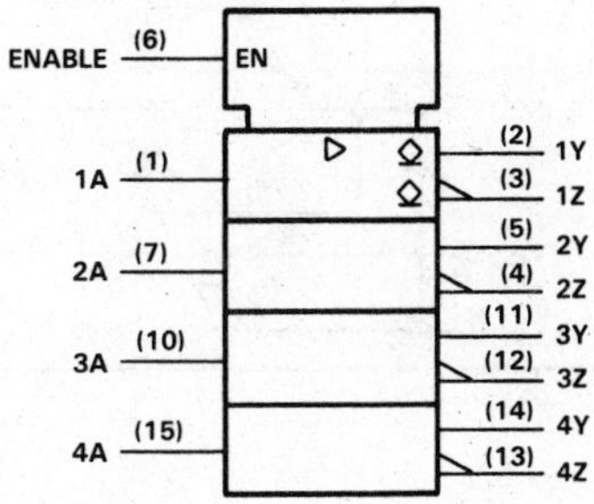
ENABLE (6)
EN
1A (1)
2A (7)
3A (10)
4A (15)
(2) 1Y
(3) 1Z
(5) 2Y
(4) 2Z
(11) 3Y
(12) 3Z
(14) 4Y
(13) 4Z

† This symbol is in accordance with ANSI/IEEE Std 91-1984 and IEC Publication 617-12.

logic diagram (positive logic)

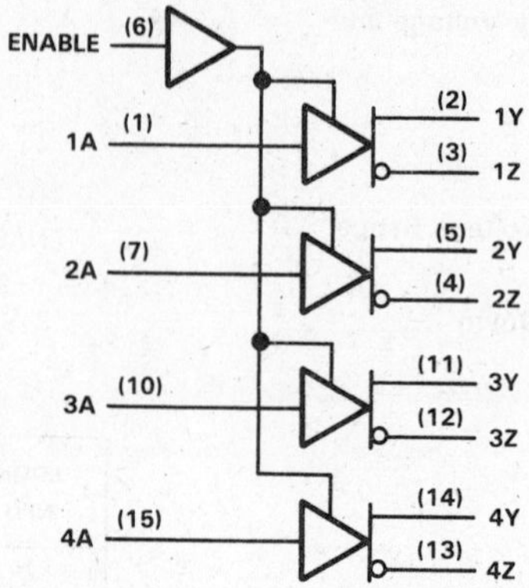
ENABLE
(6)
1A
(1)
(2)
1Y
(3)
1Z
2A
(7)
(5)
2Y
(4)
2Z
3A
(10)
(11)
3Y
(12)
3Z
4A
(15)
(14)
4Y
(13)
4Z

schematics of inputs and outputs

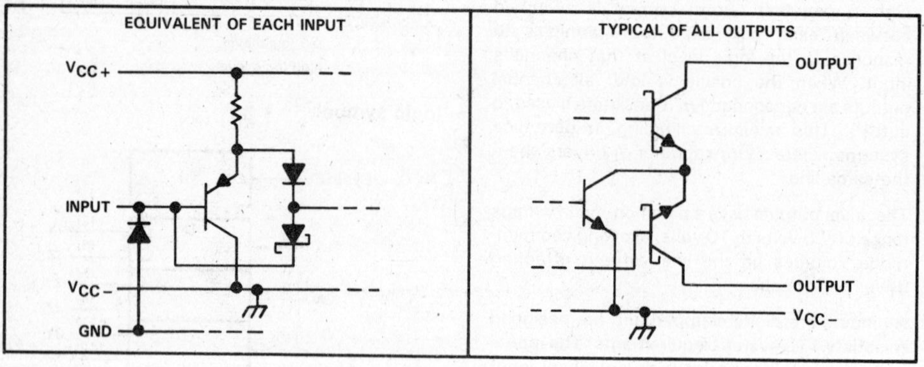
EQUIVALENT OF EACH INPUT
VCC+
INPUT
VCC−
GND
TYPICAL OF ALL OUTPUTS
OUTPUT
OUTPUT
VCC−

4 Line Drivers/Receivers

absolute maximum ratings over operating free-air temperature range (unless otherwise noted)

Supply voltage, V_{CC+} (see Note 1) .... 7 V
Supply voltage, V_{CC-} .... −7 V
Input voltage (any input) .... 5.5 V
Output voltage range (any output) .... −5 V to 12 V
Continuous total dissipation at (or below) 25°C free-air temperature (see Note 2):
 D package .... 950 mW
 J package .... 1025 mW
 N package .... 1150 mW
Operating free-air temperature range .... 0°C to 70°C
Storage temperature range .... −65°C to 150°C
Lead temperature 1,6 mm (1/16 inch) from case for 10 seconds: D, N package .... 260°C
Lead temperature 1,6 mm (1/16 inch) from case for 60 seconds: J package .... 300°C

NOTES: 1. All voltage values are with respect to network ground terminal.
2. For operation above 25°C free-air temperature, derate the D package to 608 mW at 70°C at the rate of 7.6 mW/°C, derate the J package to 656 mW at 70°C at the rate of 8.2 mW/°C, and the N package to 736 mW at 70°C at the rate of 9.2 mW/°C. In the J package the SN75111 is glass mounted.

recommended operating conditions

		MIN	NOM	MAX	UNIT
Supply voltage, V_{CC+}		4.75	5	5.25	V
Supply voltage, V_{CC-}		−4.75	−5	−5.25	V
High-level input voltage, V_{IH}		2		5.5	V
Low-level input voltage, V_{IL}		0		0.8	V
Common-mode output voltage range	V_{OCR+}	0		10	V
	V_{OCR-}	0		−3	V
Operating free-air temperature, T_A		0		70	°C

NOTE 3: All unused outputs must be grounded.

electrical characteristics over recommended operating free-air temperature range, V_{CC+} = 5.25 V, V_{CC-} = −5.25 V (unless otherwise noted)

PARAMETER		TEST CONDITIONS		MIN	TYP†	MAX	UNIT
V_{IK}	Input clamp voltage	I_I = −12 mA			−0.9	−1.5	V
$I_{O(on)}$	On-state output current	V_{CC+} = 5.25 V,	V_{CC-} = −5.25 V		5.5	7	mA
		V_{CC+} = 4.75 V,	V_{CC-} = −4.75 V	3.5	5.5		
$I_{O(off)}$	Off-state output current	V_{CC+} = 4.75 V,	V_{CC-} = −4.75 V			100	µA
I_{IH}	High-level input current	V_I = 2.4 V				40	µA
		V_I = 5.25 V				1	mA
I_{IL}	Low-level input current	V_I = 0.4 V				−1.6	mA
I_{CC+}	Supply current from V_{CC+}	A inputs at 0.4 V	Enable at 2 V		28	40	mA
			Enable at 0.4 V		27	40	
I_{CC-}	Supply current from V_{CC-}	A inputs at 0.4 V	Enable at 2 V		−43	−55	mA
			Enable at 0.4 V		−25	−35	

† All typical values are at V_{CC+} = 5 V, V_{CC-} = −5 V, and T_A = 25°C.

switching characteristics, V_{CC+} = 5 V, V_{CC-} = −5 V, R_L = 50 Ω, C_L = 40 pF, T_A = 25°C

PARAMETER		FROM (INPUT)	TO (OUTPUT)	TEST CONDITIONS	MIN	TYP	MAX	UNIT
t_{PLH}	Propagation delay time, low-to-high-level output	A	Y or Z	See Figure 1		9	15	ns
t_{PHL}	Propagation delay time, high-to-low-level output	A	Y or Z			7	15	ns
t_{PLH}	Propagation delay time, low-to-high-level output	Enable	Y or Z			14	25	ns
t_{PHL}	Propagation delay time, high-to-low-level output	Enable	Y or Z			15	25	ns

PARAMETER MEASUREMENT INFORMATION

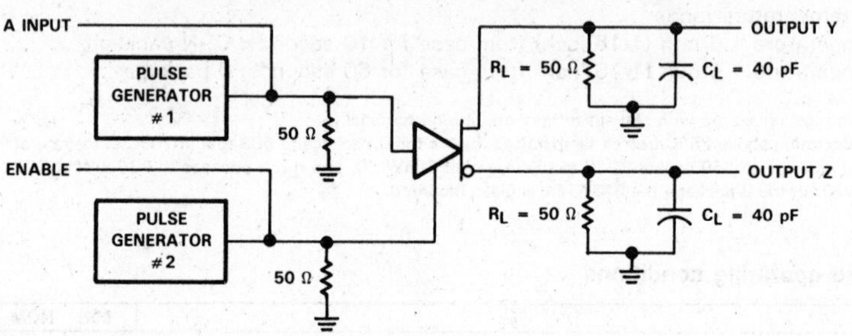
A INPUT
PULSE GENERATOR #1
50 Ω
ENABLE
PULSE GENERATOR #2
50 Ω
OUTPUT Y
RL = 50 Ω
CL = 40 pF
OUTPUT Z
RL = 50 Ω
CL = 40 pF

TEST CIRCUIT

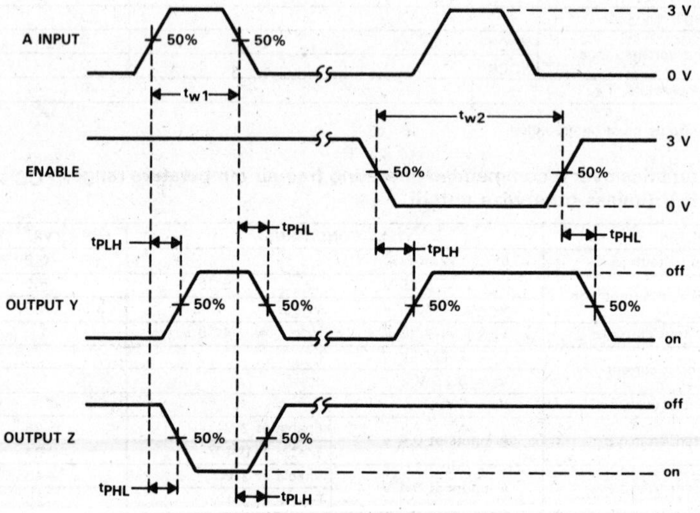
A INPUT
50%
50%
3 V
0 V
tw1
tw2
ENABLE
50%
50%
3 V
0 V
tPLH
tPHL
tPLH
tPHL
OUTPUT Y
50%
50%
50%
50%
off
on
OUTPUT Z
50%
50%
off
on
tPHL
tPLH

VOLTAGE WAVEFORMS

NOTES: A. The pulse generators have the following characteristics: Z_O = 50 Ω, t_r = t_f = 10 ±5 ns, t_{w1} = 200 ns, PRR ≤ 1 MHz, t_{w2} = 1 μs, PRR ≤ 500 kHz.
B. C_L includes probe and jig capacitance.

FIGURE 1. PROPAGATION DELAY TIMES

D1315, SEPTEMBER 1973—REVISED SEPTEMBER 1986

- Choice of Open-Collector, Open-Emitter, or 3-State Outputs
- High-Impedance Output State for Party-Line Applications
- Single-Ended or Differential AND/NAND Outputs
- Single 5-V Supply
- Dual Channel Operation
- Compatible with TTL
- Short-Circuit Protection
- High-Current Outputs
- Common and Individual Output Controls
- Clamp Diodes at Inputs and Outputs
- Easily Adaptable to SN55114 and SN75114 Applications
- Designed for Use with SN55115 and SN75115

SN55113 . . . J PACKAGE
SN75113 . . . D, J, OR N PACKAGE
(TOP VIEW)

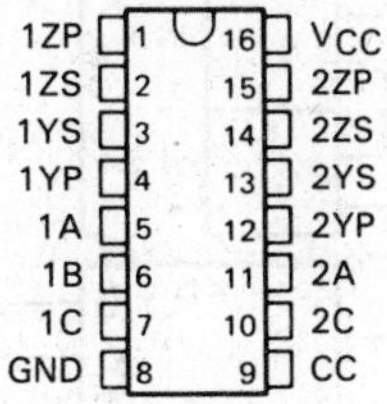
1ZP 1
1ZS 2
1YS 3
1YP 4
1A 5
1B 6
1C 7
GND 8
16 VCC
15 2ZP
14 2ZS
13 2YS
12 2YP
11 2A
10 2C
9 CC

SN55113 . . . FK PACKAGE
(TOP VIEW)

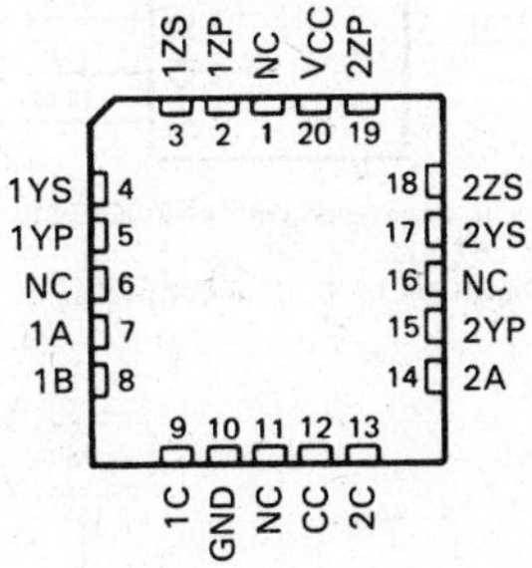
1ZS 3
1ZP 2
NC 1
VCC 20
2ZP 19
1YS 4
1YP 5
NC 6
1A 7
1B 8
18 2ZS
17 2YS
16 NC
15 2YP
14 2A
9 1C
10 GND
11 NC
12 CC
13 2C

NC—No internal connection

description

The SN55113 and SN75113 dual differential line drivers with three-state outputs are designed to provide all the features of the SN55114 and SN75114 line drivers with the added feature of driver output controls. Individual controls are provided for each output pair, as well as a common control for both output pairs. If any output is low, the associated output is in a high-impedance state and the output can neither drive nor load the bus. This permits many devices to be connected together on the same transmission line for party-line applications.

The output stages are similar to TTL totem-pole outputs, but with the sink outputs, YS and ZS, and the corresponding active pull-up terminals, YP and ZP, available on adjacent package pins.

The SN55113 is characterized for operation over the full military temperature range of −55°C to 125°C. The SN75113 is characterized for operation over the temperature range of 0°C to 70°C.

FUNCTION TABLE

INPUTS				OUTPUTS	
OUTPUT C	CONTROL CC	DATA A	DATA B†	AND Y	NAND Z
L	X	X	X	Z	Z
X	L	X	X	Z	Z
H	H	L	X	L	H
H	H	X	L	L	H
H	H	H	H	H	L

H = high level, L = low level, X = irrelevant,
Z = high impedance (off)
†B input and 4th line of function table are applicable only to driver number 1.

4

Line Drivers/Receivers

PRODUCTION DATA documents contain information current as of publication date. Products conform to specifications per the terms of Texas Instruments standard warranty. Production processing does not necessarily include testing of all parameters.

TEXAS INSTRUMENTS
POST OFFICE BOX 655012 • DALLAS, TEXAS 75265

logic symbol†

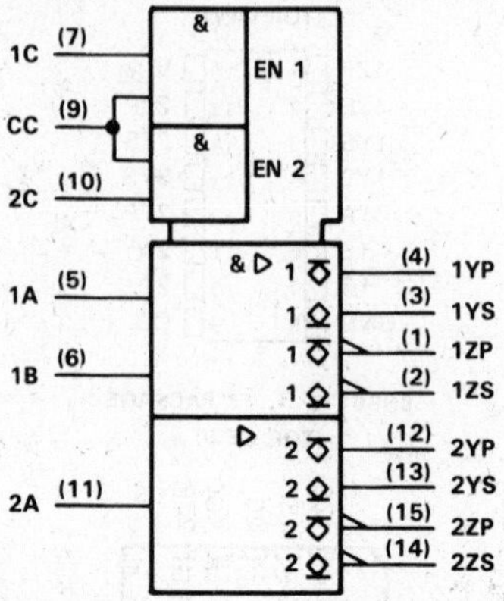
1C (7)
CC (9)
2C (10)
&
EN 1
EN 2
1A (5)
1B (6)
2A (11)
(4) 1YP
(3) 1YS
(1) 1ZP
(2) 1ZS
(12) 2YP
(13) 2YS
(15) 2ZP
(14) 2ZS

†This symbol is in accordance with ANSI/IEEE Std 91-1984 and IEC Publication 617-12.
Pin numbers shown are for D, J, and N packages.

logic diagram (positive logic)

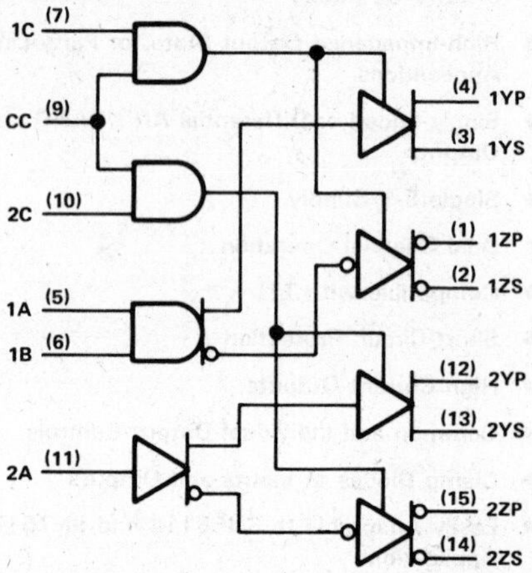
1C (7)
CC (9)
2C (10)
1A (5)
1B (6)
2A (11)
(4) 1YP
(3) 1YS
(1) 1ZP
(2) 1ZS
(12) 2YP
(13) 2YS
(15) 2ZP
(14) 2ZS

schematic

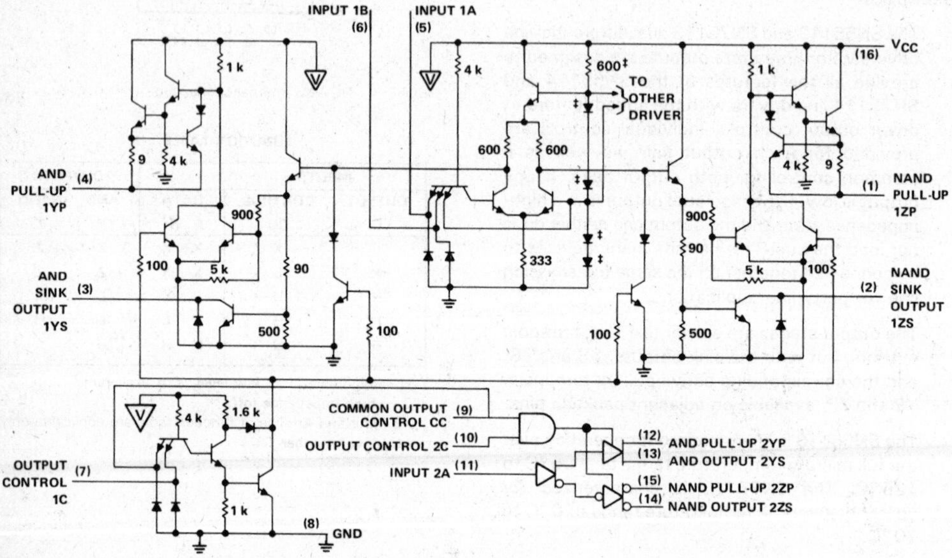
INPUT 1B (6)
INPUT 1A (5)
VCC (16)
TO OTHER DRIVER
AND PULL-UP (4) 1YP
AND SINK OUTPUT (3) 1YS
NAND PULL-UP (1) 1ZP
NAND SINK OUTPUT (2) 1ZS
COMMON OUTPUT CONTROL CC (9)
OUTPUT CONTROL 2C (10)
INPUT 2A (11)
OUTPUT CONTROL 1C (7)
(8) GND
AND PULL-UP 2YP (12)
AND OUTPUT 2YS (13)
NAND PULL-UP 2ZP (15)
NAND OUTPUT 2ZS (14)

▽ . . . V_{CC} bus

‡These components common to both drivers.
Resistor values shown are nominal and in ohms.

TEXAS INSTRUMENTS
POST OFFICE BOX 655012 • DALLAS, TEXAS 75265

absolute maximum ratings over operating free-air temperature range (unless otherwise noted)

Supply voltage, V_{CC} (see Note 1) .... 7 V
Input voltage .... 5.5 V
Off-state voltage applied to open-collector outputs .... 12 V
Continuous total dissipation at (or below) 25°C free-air temperature (see Note 2):
- D package .... 950 mW
- FK or J package .... 1000 mW
- N package .... 1150 mW

Operating free-air temperature range: SN55113 .... −55°C to 125°C
SN75113 .... 0°C to 70°C
Storage temperature range .... −65°C to 150°C
Lead temperature 1,6 mm (1/16 inch) from case for 10 seconds: D or N package .... 260°C
Case temperature for 60 seconds: FK package .... 260°C
Lead temperature 1,6 mm (1/16 inch) from case for 60 seconds: J package .... 300°C

NOTES: 1. All voltage values are with respect to network ground terminal.
2. For operation above 25°C free-air temperature, see Dissipation Derating Curves in Appendix A. In the J and FK packages, SN55113 chips are alloy mounted; SN75113 chips are glass mounted. In the N package, use the 9.2-mW/°C curve for these devices.

recommended operating conditions

	SN55113			SN75113			UNIT
	MIN	NOM	MAX	MIN	NOM	MAX	
Supply voltage, V_{CC}	4.5	5	5.5	4.75	5	5.25	V
High-level input voltage, V_{IH}	2			2			V
Low-level input voltage, V_{IL}			0.8			0.8	V
High-level output current, I_{OH}			−40			−40	mA
Low-level output current, I_{OL}			40			40	mA
Operating free-air temperature, T_A	−55		125	0		70	°C

electrical characteristics over recommended operating free-air temperature range (unless otherwise noted)

PARAMETER		TEST CONDITIONS†	SN55113 MIN	SN55113 TYP‡	SN55113 MAX	SN75113 MIN	SN75113 TYP‡	SN75113 MAX	UNIT
V_{IK}	Input clamp voltage	V_{CC} = MIN, I_I = −12 mA		−0.9	−1.5		−0.9	−1.5	V
V_{OH}	High-level output voltage	V_{CC} = MIN, V_{IH} = 2 V, V_{IL} = 0.8 V, I_{OH} = −10 mA	2.4	3.4		2.4	3.4		V
		V_{CC} = MIN, V_{IH} = 2 V, V_{IL} = 0.8 V, I_{OH} = −40 mA	2	3.0		2	3.0		
V_{OL}	Low-level output voltage	V_{CC} = MIN, V_{IH} = 2 V, V_{IL} = 0.8 V, I_{OL} = 40 mA		0.23	0.4		0.23	0.4	V
V_{OK}	Output clamp voltage	V_{CC} = MAX, I_O = −40 mA		−1.1	−1.5		−1.1	−1.5	V
$I_{O(off)}$	Off-state open-collector output current	V_{CC} = MAX, V_{OH} = 12 V, T_A = 25°C		1	10				μA
		V_{CC} = MAX, V_{OH} = 12 V, T_A = 125°C			200				
		V_{CC} = MAX, V_{OH} = 5.25 V, T_A = 25°C					1	10	
		V_{CC} = MAX, V_{OH} = 5.25 V, T_A = 70°C						20	
I_{OZ}	Off-state (high-impedance-state) output current	V_{CC} = MAX, Output controls at 0.8 V, T_A = 25°C, V_O = 0 to V_{CC}			±10			±10	μA
		V_{CC} = MAX, Output controls at 0.8 V, T_A = MAX, V_O = 0			−150			−20	
		V_{CC} = MAX, Output controls at 0.8 V, T_A = MAX, V_O = 0.4 V			±80			±20	
		V_{CC} = MAX, Output controls at 0.8 V, T_A = MAX, V_O = 2.4 V			±80			±20	
		V_{CC} = MAX, Output controls at 0.8 V, T_A = MAX, V_O = V_{CC}			80			20	
I_I	Input current at maximum input voltage — A, B, C	V_{CC} = MAX, V_I = 5.5 V			1			1	mA
	Input current at maximum input voltage — CC	V_{CC} = MAX, V_I = 5.5 V			2			2	
I_{IH}	High-level input current — A, B, C	V_{CC} = MAX, V_I = 2.4 V			40			40	μA
	High-level input current — CC	V_{CC} = MAX, V_I = 2.4 V			80			80	
I_{IL}	Low-level input current — A, B, C	V_{CC} = MAX, V_I = 0.4 V			−1.6			−1.6	mA
	Low-level input current — CC	V_{CC} = MAX, V_I = 0.4 V			−3.2			−3.2	
I_{OS}	Short-circuit output current§	V_{CC} = MAX, V_O = 0, T_A = 25°C	−40	−90	−120	−40	−90	−120	mA
I_{CC}	Supply current (both drivers)	All inputs at 0 V, No load, T_A = 25°C, V_{CC} = MAX		47	65		47	65	mA
		All inputs at 0 V, No load, T_A = 25°C, V_{CC} = 7 V		65	85		65	85	

†All parameters with the exception of off-state open-collector output current are measured with the active pull-up connected to the sink output.
‡All typical values are at T_A = 25°C and V_{CC} = 5 V, with the exception of I_{CC} at 7 V.
§Only one output should be shorted at a time, and duration of the short-circuit should not exceed one second.

switching characteristics, V_{CC} = 5 V, C_L = 30 pF, T_A = 25°C

PARAMETER		TEST CONDITIONS	SN55113 MIN	SN55113 TYP	SN55113 MAX	SN75113 MIN	SN75113 TYP	SN75113 MAX	UNIT
t_{PLH}	Propagation delay time, low-to-high-level output	See Figure 1		13	20		13	30	ns
t_{PHL}	Propagation delay time, high-to-low-level output	See Figure 1		12	20		12	30	ns
t_{PZH}	Output enable time to high level	R_L = 180 Ω, See Figure 2		7	15		7	20	ns
t_{PZL}	Output enable time to low level	R_L = 250 Ω, See Figure 3		14	30		14	40	ns
t_{PHZ}	Output disable time from high level	R_L = 180 Ω, See Figure 2		10	20		10	30	ns
t_{PLZ}	Output disable time from low level	R_L = 250 Ω, See Figure 3		17	35		17	35	ns

TEXAS INSTRUMENTS
POST OFFICE BOX 655012 • DALLAS, TEXAS 75265

PARAMETER MEASUREMENT INFORMATION

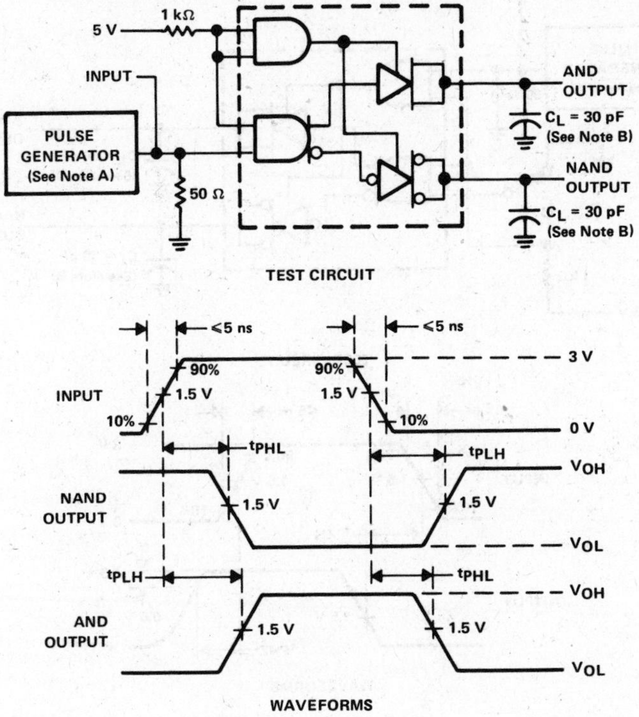
5 V
1 kΩ
INPUT
PULSE GENERATOR (See Note A)
50 Ω
AND OUTPUT
C_L = 30 pF (See Note B)
NAND OUTPUT
C_L = 30 pF (See Note B)
TEST CIRCUIT
≤5 ns
≤5 ns
3 V
90%
90%
1.5 V
1.5 V
INPUT
10%
10%
0 V
t_{PHL}
t_{PLH}
V_{OH}
NAND OUTPUT
1.5 V
1.5 V
V_{OL}
t_{PLH}
t_{PHL}
V_{OH}
AND OUTPUT
1.5 V
1.5 V
V_{OL}
WAVEFORMS

FIGURE 1. t_{PLH} and t_{PHL}

NOTES: A. The pulse generator has the following characteristics: Z_{out} = 50 Ω, PRR ≤ 500 kHz, t_w = 100 ns.
B. C_L includes probe and jig capacitance.

PARAMETER MEASUREMENT INFORMATION

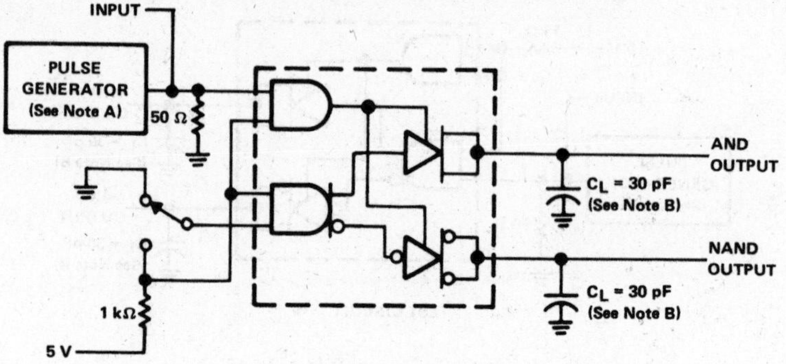
INPUT
PULSE
GENERATOR
(See Note A)
50 Ω
1 kΩ
5 V
AND
OUTPUT
C_L = 30 pF
(See Note B)
NAND
OUTPUT
C_L = 30 pF
(See Note B)

TEST CIRCUIT

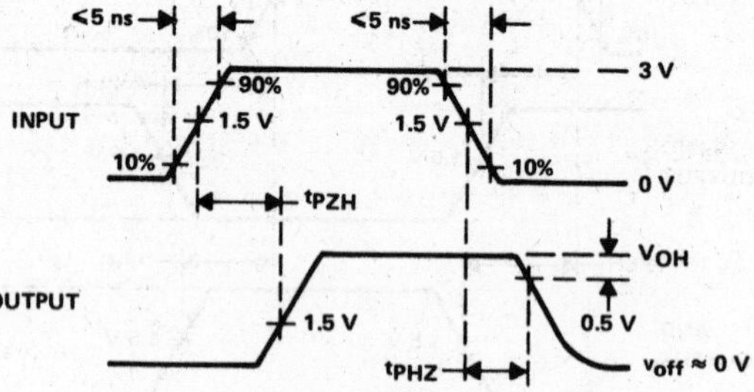
≤5 ns
≤5 ns
3 V
90%
90%
INPUT
1.5 V
1.5 V
10%
10%
0 V
t_{PZH}
V_{OH}
OUTPUT
1.5 V
0.5 V
t_{PHZ}
v_{off} ≈ 0 V

WAVEFORMS

FIGURE 2. t_{PZH} and t_{PHZ}

NOTES: A. The pulse generator has the following characteristics: Z_{out} = 50 Ω, PRR ≤ 500 kHz, t_w = 100 ns.
B. C_L includes probe and jig capacitance.

TEXAS INSTRUMENTS
POST OFFICE BOX 655012 • DALLAS, TEXAS 75265

PARAMETER MEASUREMENT INFORMATION

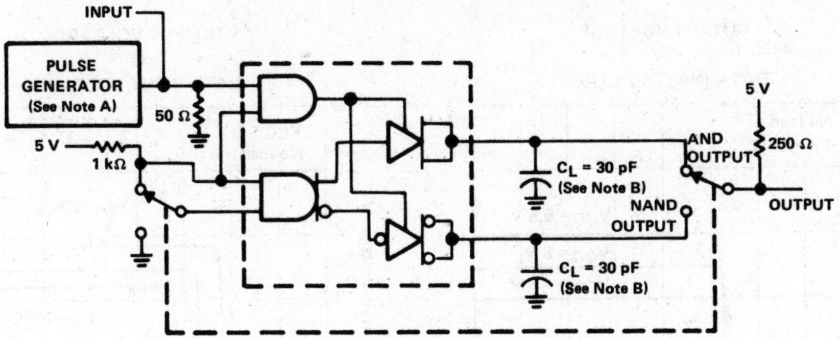
INPUT
PULSE GENERATOR (See Note A)
50 Ω
5 V
1 kΩ
5 V
250 Ω
AND OUTPUT
C_L = 30 pF (See Note B)
NAND OUTPUT
C_L = 30 pF (See Note B)
OUTPUT

TEST CIRCUIT

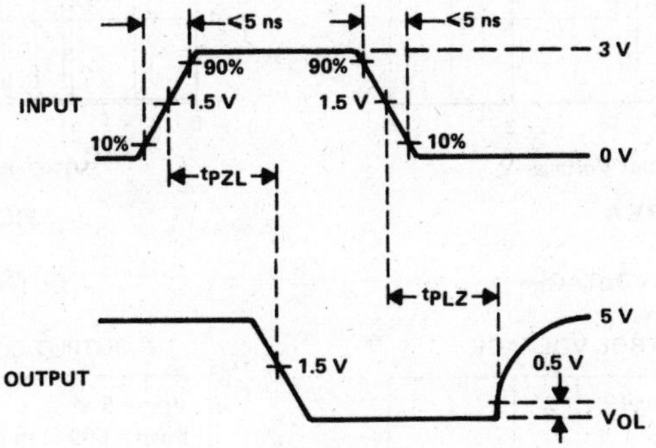
≤5 ns
≤5 ns
3 V
90%
90%
INPUT
1.5 V
1.5 V
10%
10%
0 V
tPZL
tPLZ
5 V
OUTPUT
1.5 V
0.5 V
VOL

WAVEFORMS

FIGURE 3. t_{PZL} and t_{PLZ}

NOTES: A. The pulse generator has the following characteristics: Z_{out} = 50 Ω, PRR ≤ 500 kHz, t_w = 100 ns.
B. C_L includes probe and jig capacitance.

TYPICAL CHARACTERISTICS†

OUTPUT VOLTAGE
vs
DATA INPUT VOLTAGE

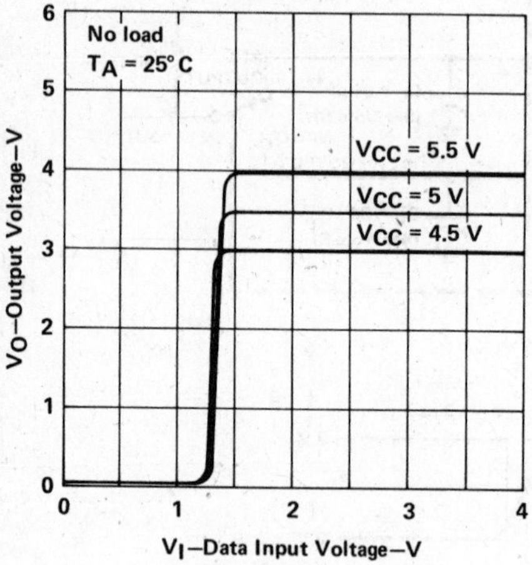
No load
TA = 25°C
VCC = 5.5 V
VCC = 5 V
VCC = 4.5 V
VO—Output Voltage—V
VI—Data Input Voltage—V

FIGURE 4

OUTPUT VOLTAGE
vs
DATA INPUT VOLTAGE

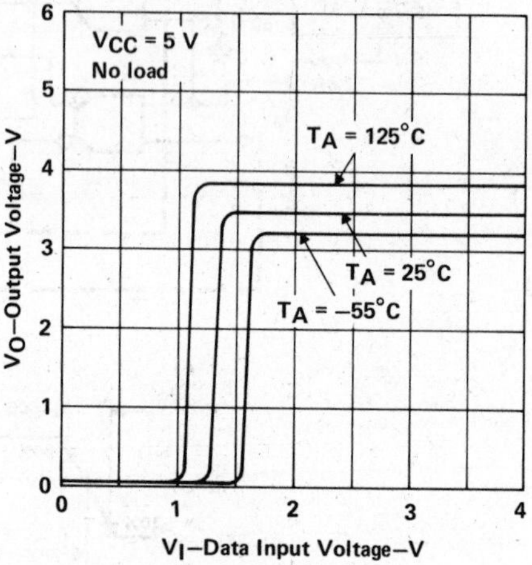
VCC = 5 V
No load
TA = 125°C
TA = 25°C
TA = −55°C
VO—Output Voltage—V
VI—Data Input Voltage—V

FIGURE 5

OUTPUT VOLTAGE
vs
OUTPUT CONTROL VOLTAGE

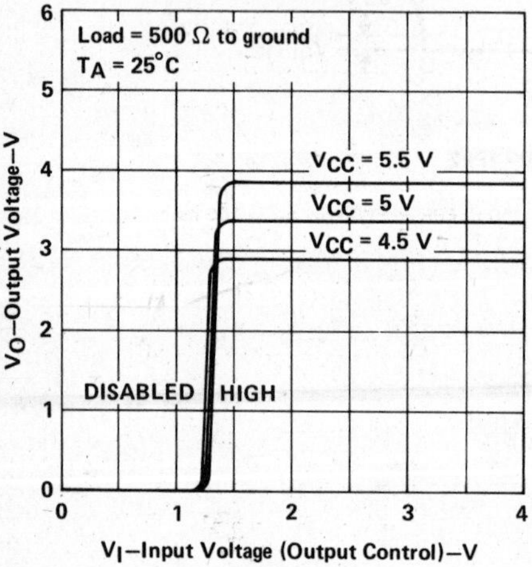
Load = 500 Ω to ground
TA = 25°C
VCC = 5.5 V
VCC = 5 V
VCC = 4.5 V
DISABLED
HIGH
VO—Output Voltage—V
VI—Input Voltage (Output Control)—V

FIGURE 6

OUTPUT VOLTAGE
vs
OUTPUT CONTROL VOLTAGE

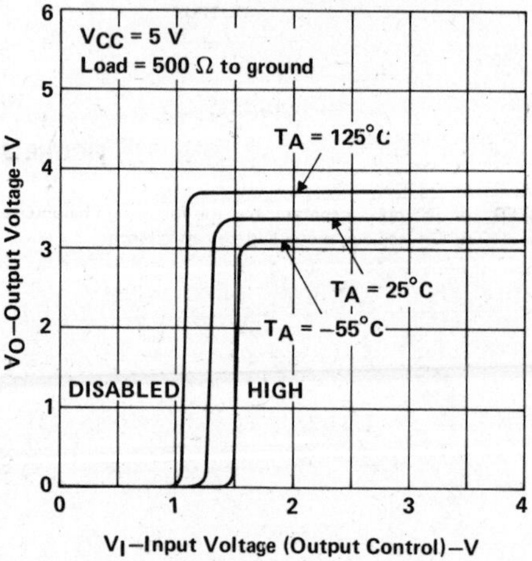
VCC = 5 V
Load = 500 Ω to ground
TA = 125°C
TA = 25°C
TA = −55°C
DISABLED
HIGH
VO—Output Voltage—V
VI—Input Voltage (Output Control)—V

FIGURE 7

†Data for temperatures below 0 °C and above 70 °C and for supply voltages below 4.75 V and above 5.25 V are applicable to SN55113 circuits only. These parameters were measured with the active pull-up connected to the sink output.

TEXAS INSTRUMENTS
POST OFFICE BOX 655012 • DALLAS, TEXAS 75265

TYPICAL CHARACTERISTICS†

OUTPUT VOLTAGE
vs
OUTPUT CONTROL VOLTAGE

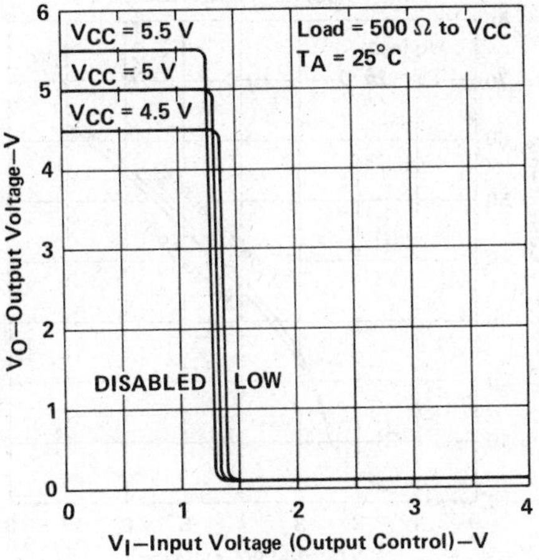
VCC = 5.5 V
VCC = 5 V
VCC = 4.5 V
Load = 500 Ω to VCC
TA = 25°C
DISABLED
LOW
VO–Output Voltage–V
VI–Input Voltage (Output Control)–V

FIGURE 8

OUTPUT VOLTAGE
vs
OUTPUT CONTROL VOLTAGE

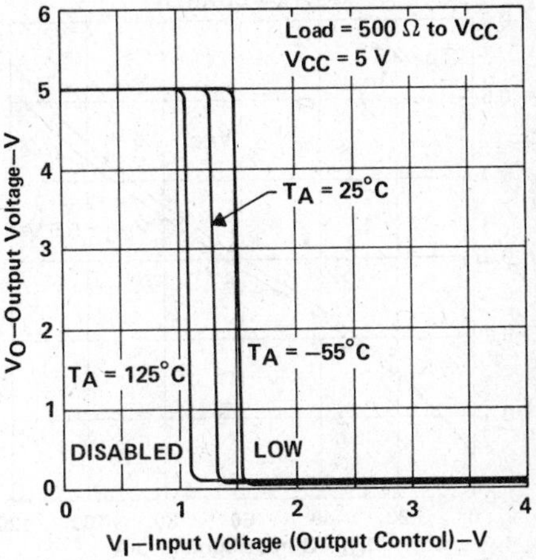
Load = 500 Ω to VCC
VCC = 5 V
TA = 25°C
TA = –55°C
TA = 125°C
DISABLED
LOW
VO–Output Voltage–V
VI–Input Voltage (Output Control)–V

FIGURE 9

OUTPUT VOLTAGE
vs
FREE-AIR TEMPERATURE

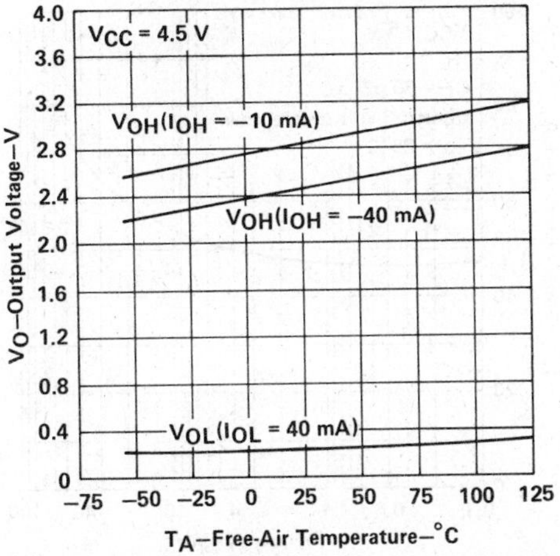
VCC = 4.5 V
VOH(IOH = –10 mA)
VOH(IOH = –40 mA)
VOL(IOL = 40 mA)
VO–Output Voltage–V
TA–Free-Air Temperature–°C

FIGURE 10

HIGH-LEVEL OUTPUT VOLTAGE
vs
OUTPUT CURRENT

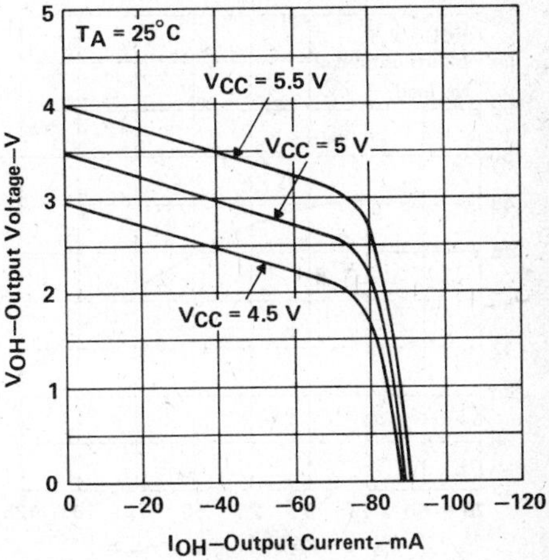
TA = 25°C
VCC = 5.5 V
VCC = 5 V
VCC = 4.5 V
VOH–Output Voltage–V
IOH–Output Current–mA

FIGURE 11

†Data for temperatures below 0°C and above 70°C and for supply voltages below 4.75 V and above 5.25 V are applicable to SN55113 circuits only. These parameters were measured with the active pull-up connected to the sink output.

TYPICAL CHARACTERISTICS†

LOW-LEVEL OUTPUT VOLTAGE
vs
OUTPUT CURRENT

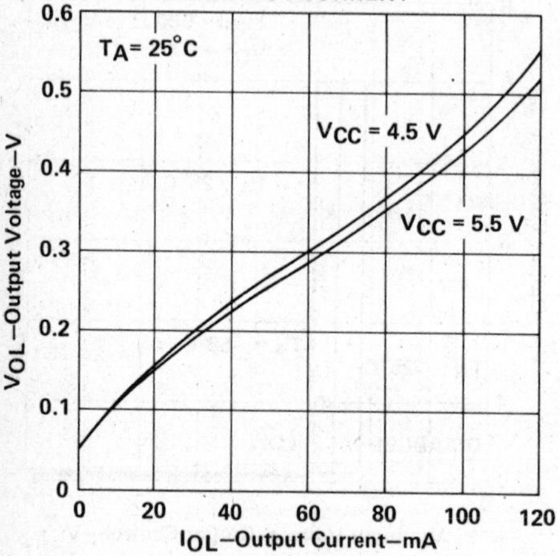
0.6
0.5
0.4
0.3
0.2
0.1
0
T_A = 25°C
V_{CC} = 4.5 V
V_{CC} = 5.5 V
V_{OL}—Output Voltage—V
0
20
40
60
80
100
120
I_{OL}—Output Current—mA

FIGURE 12

SUPPLY CURRENT
(BOTH DRIVERS)
vs
SUPPLY VOLTAGE

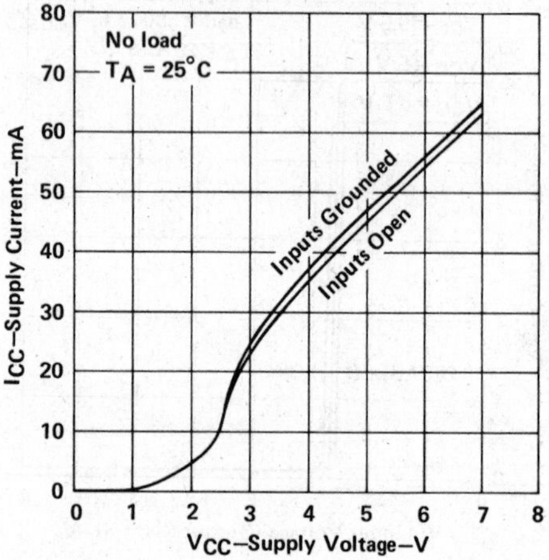
80
70
60
50
40
30
20
10
0
No load
T_A = 25°C
Inputs Grounded
Inputs Open
I_{CC}—Supply Current—mA
0
1
2
3
4
5
6
7
8
V_{CC}—Supply Voltage—V

FIGURE 13

SUPPLY CURRENT
(BOTH DRIVERS)
vs
FREE-AIR TEMPERATURE

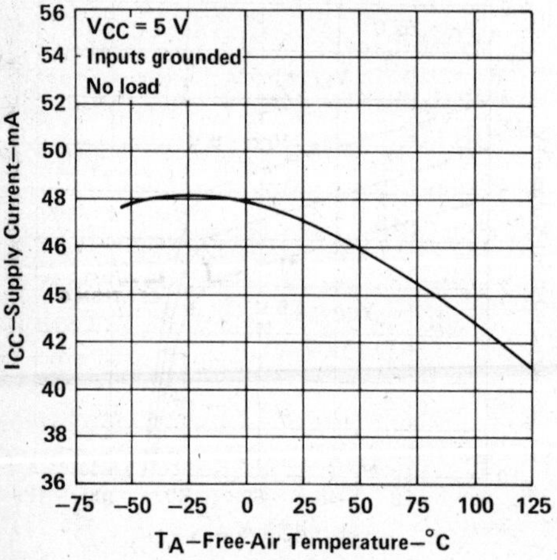
56
54
52
50
48
46
45
42
40
38
36
V_{CC} = 5 V
Inputs grounded
No load
I_{CC}—Supply Current—mA
−75
−50
−25
0
25
50
75
100
125
T_A—Free-Air Temperature—°C

FIGURE 14

SUPPLY CURRENT
(BOTH DRIVERS)
vs
FREQUENCY

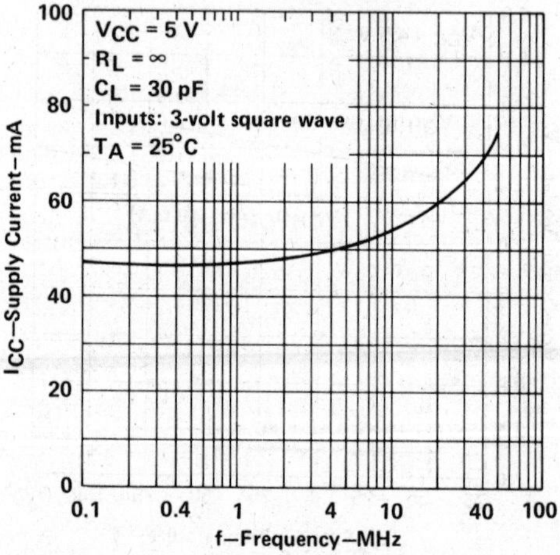
100
80
60
40
20
0
V_{CC} = 5 V
R_L = ∞
C_L = 30 pF
Inputs: 3-volt square wave
T_A = 25°C
I_{CC}—Supply Current—mA
0.1
0.4
1
4
10
40
100
f—Frequency—MHz

FIGURE 15

†Data for temperature below 0°C and above 70°CC and for supply voltages below 4.75 V and above 5.25 V are applicable to SN55113 circuits only. These parameters were measured with the active pull-up connected to the sink output.

TEXAS INSTRUMENTS
POST OFFICE BOX 655012 • DALLAS, TEXAS 75265

TYPICAL CHARACTERISTICS†

PROPAGATION DELAY TIMES
FROM DATA INPUTS
vs
FREE-AIR TEMPERATURE

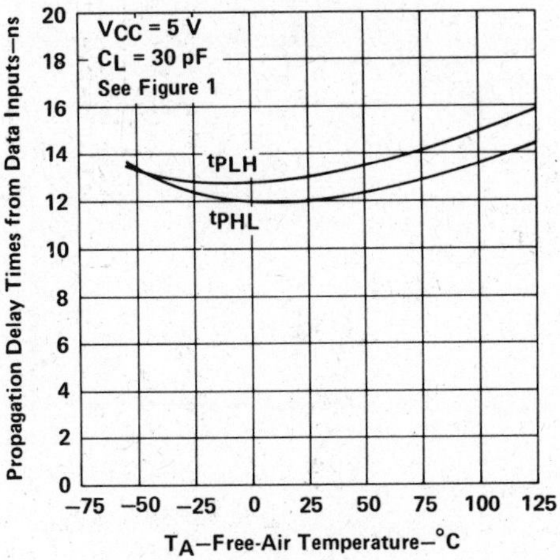
Propagation Delay Times from Data Inputs–ns
VCC = 5 V
CL = 30 pF
See Figure 1
tPLH
tPHL
20
18
16
14
12
10
8
6
4
2
0
–75 –50 –25 0 25 50 75 100 125
TA–Free-Air Temperature–°C

FIGURE 16

OUTPUT ENABLE AND DISABLE TIMES
vs
FREE-AIR TEMPERATURE

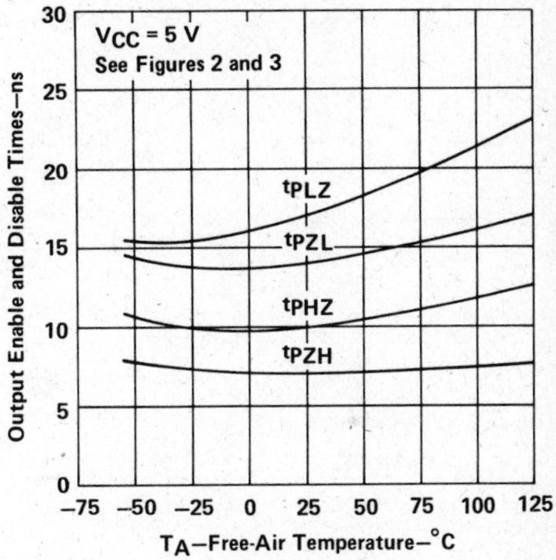
Output Enable and Disable Times–ns
VCC = 5 V
See Figures 2 and 3
tPLZ
tPZL
tPHZ
tPZH
30
25
20
15
10
5
0
–75 –50 –25 0 25 50 75 100 125
TA–Free-Air Temperature–°C

FIGURE 17

†Data for temperature below 0°C and above 70°CC and for supply voltages below 4.75 V and above 5.25 V are applicable to SN55113 circuits only. These parameters were measured with the active pull-up connected to the sink output.

TYPICAL APPLICATION DATA

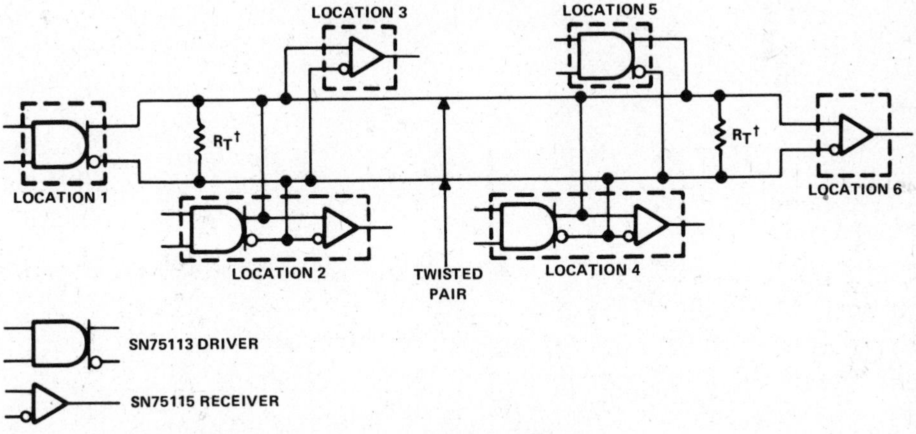
LOCATION 3
LOCATION 5
LOCATION 1
RT†
RT†
LOCATION 6
LOCATION 2
TWISTED PAIR
LOCATION 4
SN75113 DRIVER
SN75115 RECEIVER

†$R_T = Z_O$. A capacitor may be connected in series with R_T to reduce power dissipation.

FIGURE 18. BASIC PARTY-LINE OR DATA-BUS DIFFERENTIAL DATA TRANSMISSION

D1315, SEPTEMBER 1973—REVISED SEPTEMBER 1986

- Choice of Open-Collector, Open-Emitter, or Totem-Pole Outputs
- Single-Ended or Differential AND/NAND Outputs
- Single 5-V Supply
- Dual Channel Operation
- TTL-Compatible
- Short-Circuit Protection
- High-Current Outputs
- Triple Inputs
- Clamp Diodes at Inputs and Outputs
- Designed for Use with SN55115 and SN75115 Differential Line Receivers
- Designed to be Interchangeable with Fairchild 9614 Line Driver

SN55114 . . . J PACKAGE
SN75114 . . . D, J, OR N PACKAGE
(TOP VIEW)

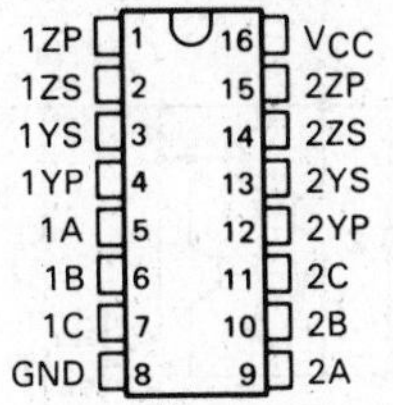
1ZP 1 16 VCC
1ZS 2 15 2ZP
1YS 3 14 2ZS
1YP 4 13 2YS
1A 5 12 2YP
1B 6 11 2C
1C 7 10 2B
GND 8 9 2A

SN55114 . . . FK PACKAGE
(TOP VIEW)

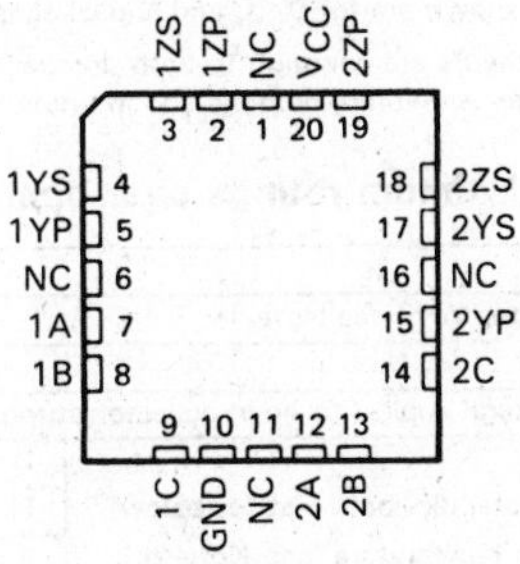
1ZS 1ZP NC VCC 2ZP
3 2 1 20 19
1YS 4
1YP 5
NC 6
1A 7
1B 8
18 2ZS
17 2YS
16 NC
15 2YP
14 2C
9 10 11 12 13
1C GND NC 2A 2B

NC—No internal connection

description

The SN55114 and SN75114 dual differential line drivers are designed to provide differential output signals with the high-current capability for driving balanced lines, such as twisted pair, at normal line impedances without high power dissipation. The output stages are similar to TTL totem-pole outputs, but with the sink outputs, YS and ZS, and the corresponding active pull-up terminals, YP and ZP, available on adjacent package pins. Since the output stages provide TTL-compatible output levels, these devices may also be used as TTL expanders or phase splitters.

The SN55114 is characterized for operation over the full military temperature range of −55°C to 125°C. The SN75114 is characterized for operation from 0°C to 70°C.

FUNCTION TABLE

INPUTS			OUTPUTS	
A	B	C	Y	Z
H	H	H	H	L
ALL OTHER INPUT COMBINATIONS			L	H

H = high level, L = low level

logic symbol†

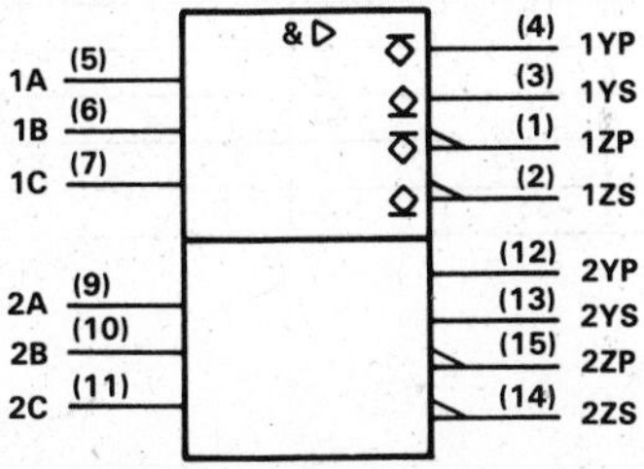
& ▷
1A (5)
1B (6)
1C (7)
(4) 1YP
(3) 1YS
(1) 1ZP
(2) 1ZS
2A (9)
2B (10)
2C (11)
(12) 2YP
(13) 2YS
(15) 2ZP
(14) 2ZS

†This symbol is in accordance with ANSI/IEEE Std 91-1984 and IEC Publication 617-12.

logic diagram (positive logic)

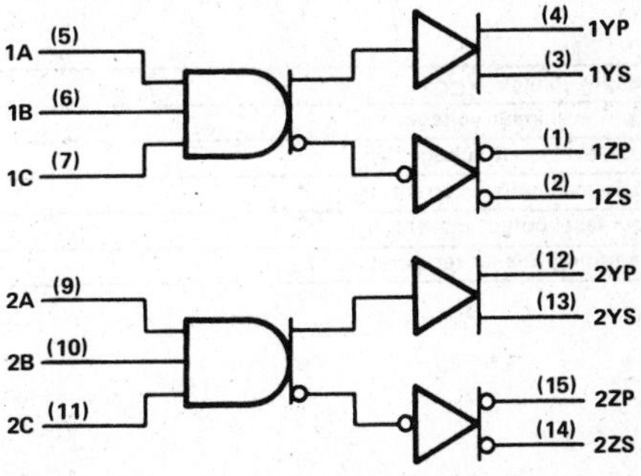
1A (5)
1B (6)
1C (7)
(4) 1YP
(3) 1YS
(1) 1ZP
(2) 1ZS
2A (9)
2B (10)
2C (11)
(12) 2YP
(13) 2YS
(15) 2ZP
(14) 2ZS

Pin numbers shown are for D, J, and N packages.

schematic (each driver)

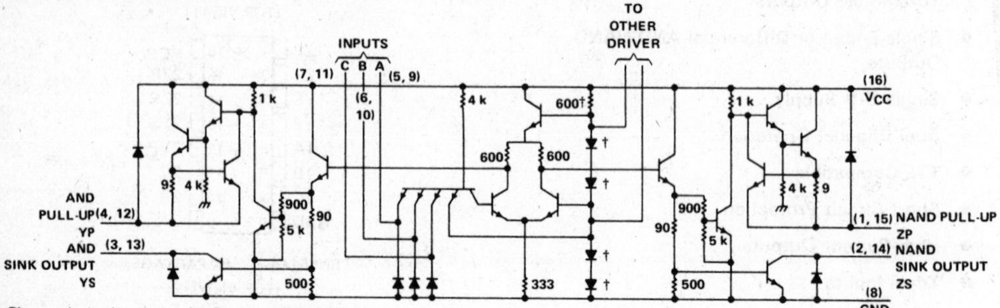
INPUTS
C B A
(7, 11)
(5, 9)
(6, 10)
TO OTHER DRIVER
(16) VCC
1 k
4 k
600†
600
600
9
4 k
900
90
5 k
AND PULL-UP (4, 12)
YP
AND (3, 13)
SINK OUTPUT
YS
500
333
†
1 k
4 k
9
900
90
5 k
(1, 15) NAND PULL-UP
ZP
(2, 14) NAND
SINK OUTPUT
ZS
500
(8)
GND

Pin numbers shown are for D, J, and N packages

†These components are common to both drivers.

Resistor values shown are nominal and in ohms.

absolute maximum ratings over operating free-air temperature range (unless otherwise noted)

		SN55114	SN75114	UNIT
Supply voltage, V_{CC} (see Note 1)		7	7	V
Input voltage		5.5	5.5	V
Off-state voltage applied to open-collector outputs		12	12	V
Continuous total dissipation at (or below) 25 °C free-air temperature (see Note 2)	D package		950	mW
	FK package	1375		mW
	J package	1375	1025	mW
	N package		1150	mW
Operating free-air temperature range		−55 to 125	0 to 70	°C
Storage temperature range		−65 to 150	−65 to 150	°C
Case temperature for 60 seconds: FK package		260		°C
Lead temperature 1,6 mm (1/16 inch) from case for 60 seconds: J package		300		°C
Lead temperature 1,6 mm (1/16 inch) from case for 10 seconds: D or N package			260	°C

NOTES: 1. Voltage values are with respect to network ground terminal.
2. For operation above 25 °C free-air temperature, refer to Dissipation Derating Curves in Appendix A. In the FK and J packages, SN55114 chips are alloy mounted. In the J package, SN75114 chips are glass mounted.

recommended operating conditions

	SN55114			SN75114			UNIT
	MIN	NOM	MAX	MIN	NOM	MAX	
Supply voltage, V_{CC1}	4.5	5	5.5	4.75	5	5.25	V
High-level input voltage, V_{IH}	2			2			V
Low-level input voltage, V_{IL}			0.8			0.8	V
High-level output current, I_{OH}			−40			−40	mA
Low-level output current, I_{OL}			40			40	mA
Operating free-air temperature, T_A	−55		125	0		70	°C

TEXAS INSTRUMENTS
POST OFFICE BOX 655012 • DALLAS, TEXAS 75265

electrical characteristics over recommended operating free-air temperature range (unless otherwise noted)

PARAMETER		TEST CONDITIONS†			SN55114			SN75114			UNIT
					MIN	TYP‡	MAX	MIN	TYP‡	MAX	
V_{IK}	Input clamp voltage	V_{CC} = MIN, I_I = −12 mA				−0.9	−1.5		−0.9	−1.5	V
V_{OH}	High-level output voltage	V_{CC} = MIN, V_{IH} = 2 V, V_{IL} = 0.8 V,		I_{OH} = −10 mA	2.4	3.4		2.4	3.4		V
				I_{OH} = −40 mA	2	3.0		2	3.0		
V_{OL}	Low-level output voltage	V_{CC} = MIN, V_{IH} = 2 V, V_{IL} = 0.8 V, I_{OL} = 40 mA				0.2	0.4		0.2	0.45	V
V_{OK}	Output clamp voltage	V_{CC} = 5 V, I_O = 40 mA,		T_A = 25°C		6.1	6.5		6.1	6.5	V
		V_{CC} = MAX, I_O = −40 mA,		T_A = 25°C		−1.1	−1.5		−1.1	−1.5	
$I_{O(off)}$	Off-state open-collector output current	V_{CC} = MAX	V_{OH} = 12 V	T_A = 25°C		1	100				µA
				T_A = 125°C			200				
			V_{OH} = 5.25 V	T_A = 25°C					1	100	
				T_A = 70°C						200	
I_I	Input current at maximum input voltage	V_{CC} = MAX, V_I = 5.5 V					1			1	mA
I_{IH}	High-level input current	V_{CC} = MAX, V_I = 2.4 V					40			40	µA
I_{IL}	Low-level input current	V_{CC} = MAX, V_I = 0.4 V				−1.1	−1.6		−1.1	−1.6	mA
I_{OS}	Short-circuit output current§	V_{CC} = MAX, V_O = 0,		T_A = 25°C	−40	−90	−120	−40	−90	−120	mA
I_{CC}	Supply current (both drivers)	All inputs at 0 V, No load, T_A = 25°C		V_{CC} = MAX		37	50		37	50	mA
				V_{CC} = 7 V		47	65		47	70	

† All parameters with the exception of off-state open-collector output current are measured with the active pullup connected to the sink output. For conditions shown as MIN or MAX, use the appropriate value specified under recommended operating conditions.
‡ All typical values are at T_A = 25°C and V_{CC} = 5 V, with the exception of I_{CC} at 7 V.
§ Only one output should be shorted at a time, and duration of the short-circuit should not exceed one second.

switching characteristics, V_{CC} = 5 V, T_A = 25°C

PARAMETER		TEST CONDITIONS	SN55114			SN75114			UNIT
			MIN	TYP	MAX	MIN	TYP	MAX	
t_{PLH}	Propagation delay time, low-to-high-level output	C_L = 30 pF, See Figure 1		15	20		15	30	ns
t_{PHL}	Propagation delay time, high-to-low-level output			11	20		11	30	ns

PARAMETER MEASUREMENT INFORMATION

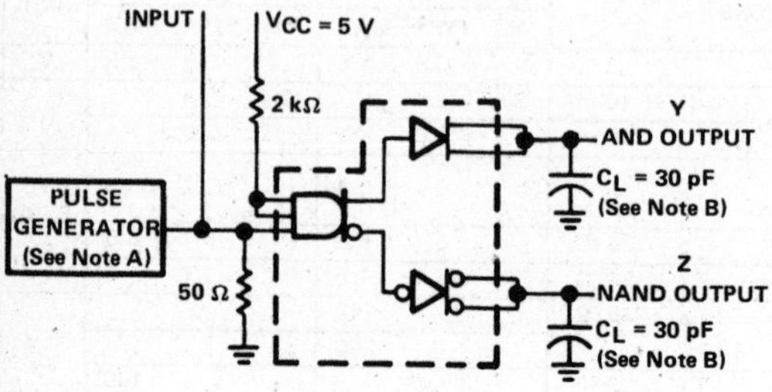
INPUT
V_{CC} = 5 V
2 kΩ
PULSE GENERATOR (See Note A)
50 Ω
Y
AND OUTPUT
C_L = 30 pF (See Note B)
Z
NAND OUTPUT
C_L = 30 pF (See Note B)

TEST CIRCUIT

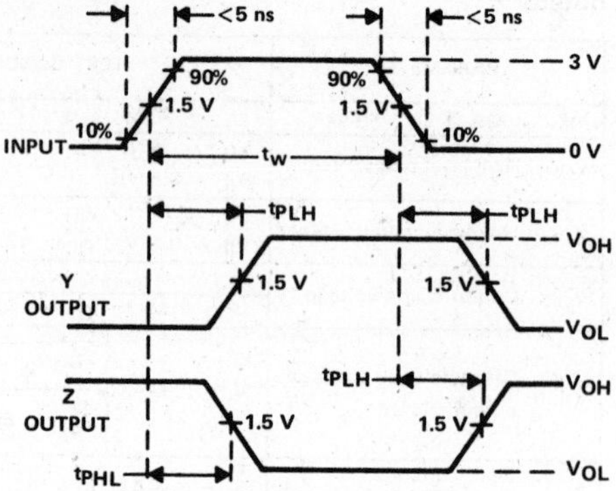
<5 ns
<5 ns
3 V
90%
90%
1.5 V
1.5 V
10%
10%
INPUT
0 V
t_w
t_{PLH}
t_{PLH}
V_{OH}
Y OUTPUT
1.5 V
1.5 V
V_{OL}
t_{PLH}
V_{OH}
Z OUTPUT
1.5 V
1.5 V
t_{PHL}
V_{OL}

VOLTAGE WAVEFORMS

NOTES: A. The pulse generator has the following characteristics: Z_{out} = 500 Ω, PRR ≤ 500 kHz, t_w ≥ 100 ns.
B. C_L includes probe and jig capacitance.

FIGURE 1. PROPAGATION DELAY TIMES

TYPICAL CHARACTERISTICS†

OUTPUT VOLTAGE
vs
DATA INPUT VOLTAGE

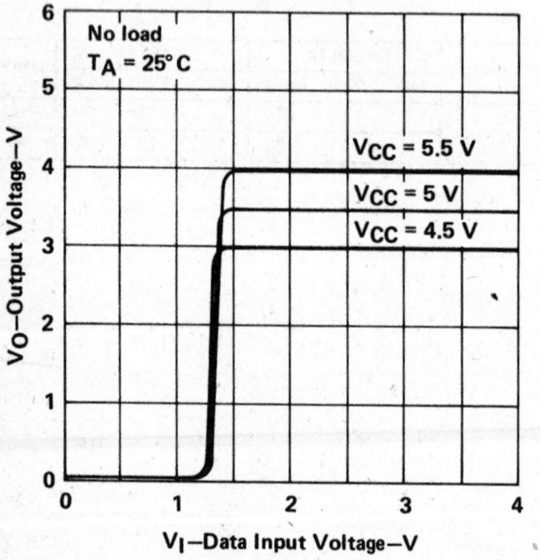
No load
T_A = 25°C
V_{CC} = 5.5 V
V_{CC} = 5 V
V_{CC} = 4.5 V
V_O—Output Voltage—V
V_I—Data Input Voltage—V
0
1
2
3
4
5
6

FIGURE 2

OUTPUT VOLTAGE
vs
DATA INPUT VOLTAGE

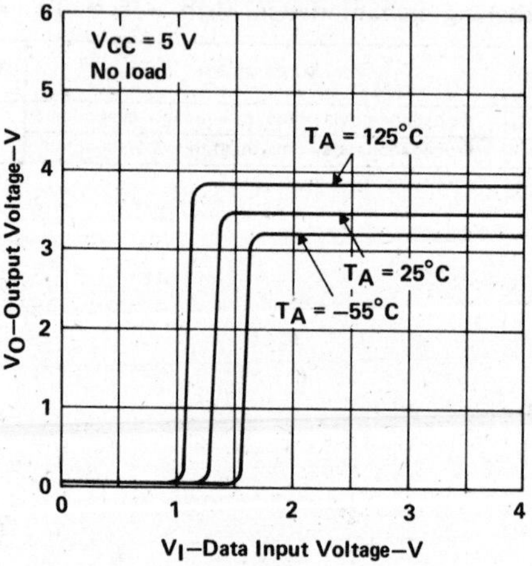
V_{CC} = 5 V
No load
T_A = 125°C
T_A = 25°C
T_A = −55°C
V_O—Output Voltage—V
V_I—Data Input Voltage—V
0
1
2
3
4
5
6

FIGURE 3

† Data for temperatures below 0°C and above 70°C and for supply voltages below 4.75 V and above 5.25 V are applicable to SN55114 circuits only. These parameters were measured with the active pullup connected to the sink output.

TYPICAL CHARACTERISTICS†

HIGH-LEVEL OUTPUT VOLTAGE
vs
OUTPUT CURRENT

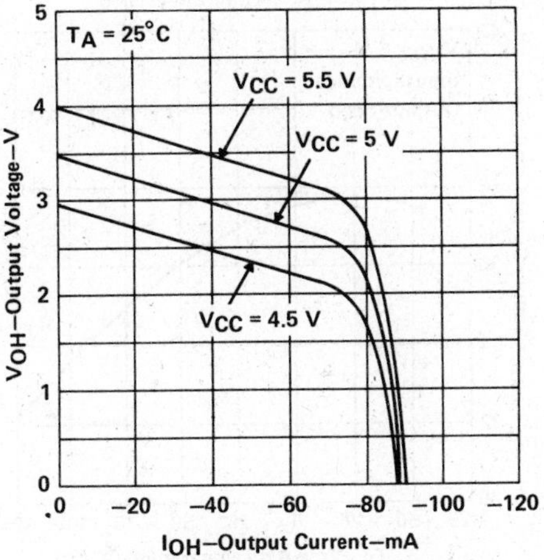
TA = 25°C
VCC = 5.5 V
VCC = 5 V
VCC = 4.5 V
VOH—Output Voltage—V
IOH—Output Current—mA
0
1
2
3
4
5
0
−20
−40
−60
−80
−100
−120

FIGURE 4

LOW-LEVEL OUTPUT VOLTAGE
vs
OUTPUT CURRENT

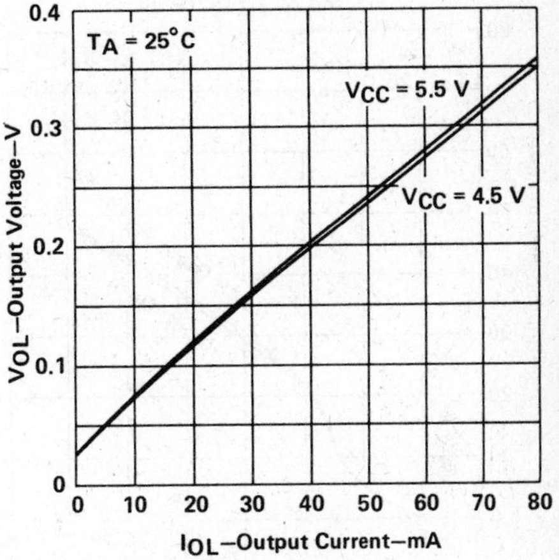
TA = 25°C
VCC = 5.5 V
VCC = 4.5 V
VOL—Output Voltage—V
IOL—Output Current—mA
0
0.1
0.2
0.3
0.4
0
10
20
30
40
50
60
70
80

FIGURE 5

OUTPUT VOLTAGE
vs
FREE-AIR TEMPERATURE

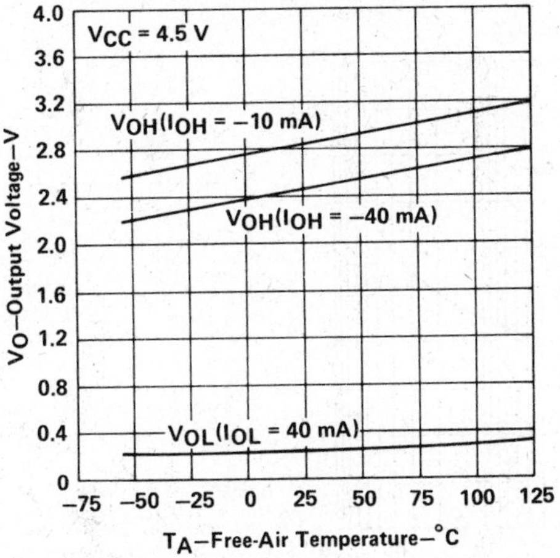
VCC = 4.5 V
VOH(IOH = −10 mA)
VOH(IOH = −40 mA)
VOL(IOL = 40 mA)
VO—Output Voltage—V
TA—Free-Air Temperature—°C
0
0.4
0.8
1.2
1.6
2.0
2.4
2.8
3.2
3.6
4.0
−75
−50
−25
0
25
50
75
100
125

FIGURE 6

PROPAGATION DELAY TIMES
vs
FREE-AIR TEMPERATURE

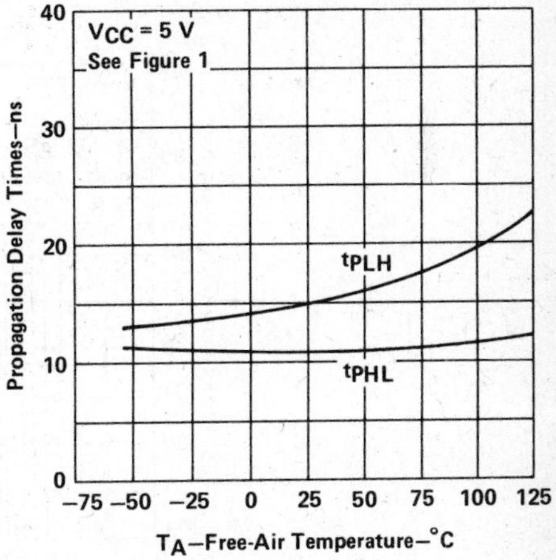
VCC = 5 V
See Figure 1
tPLH
tPHL
Propagation Delay Times—ns
TA—Free-Air Temperature—°C
0
10
20
30
40
−75
−50
−25
0
25
50
75
100
125

FIGURE 7

† Data for temperatures below 0 °C and above 70 °C are applicable to SN55114 circuits only. These parameters were measured with the active pullup connected to the sink output.

4

Line Drivers/Receivers

TYPICAL CHARACTERISTICS†

SUPPLY CURRENT
(BOTH DRIVERS)
vs
SUPPLY VOLTAGE

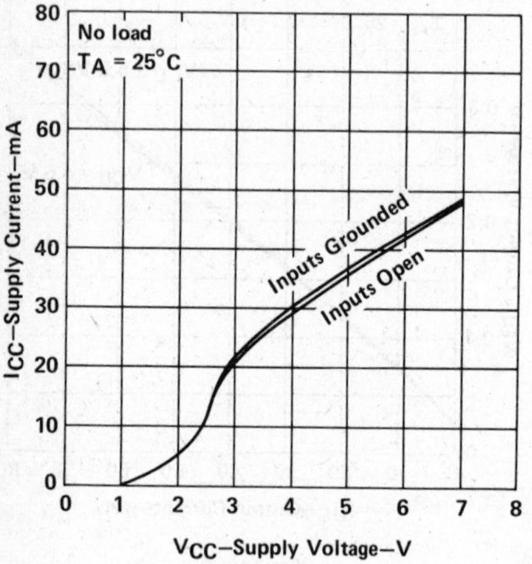
No load
T_A = 25°C
Inputs Grounded
Inputs Open
I_{CC}—Supply Current—mA
V_{CC}—Supply Voltage—V
0
10
20
30
40
50
60
70
80
1
2
3
4
5
6
7
8

FIGURE 8

SUPPLY CURRENT
(BOTH DRIVERS)
vs
FREE-AIR TEMPERATURE

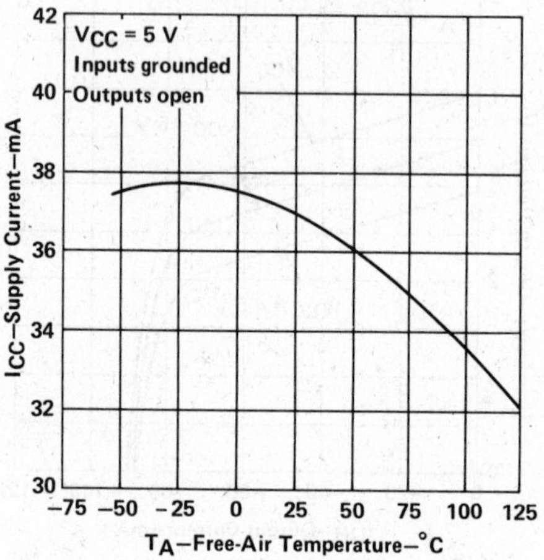
V_{CC} = 5 V
Inputs grounded
Outputs open
I_{CC}—Supply Current—mA
T_A—Free-Air Temperature—°C
30
32
34
36
38
40
42
−75 −50 −25 0 25 50 75 100 125

FIGURE 9

SUPPLY CURRENT
(BOTH DRIVERS)
vs
FREQUENCY

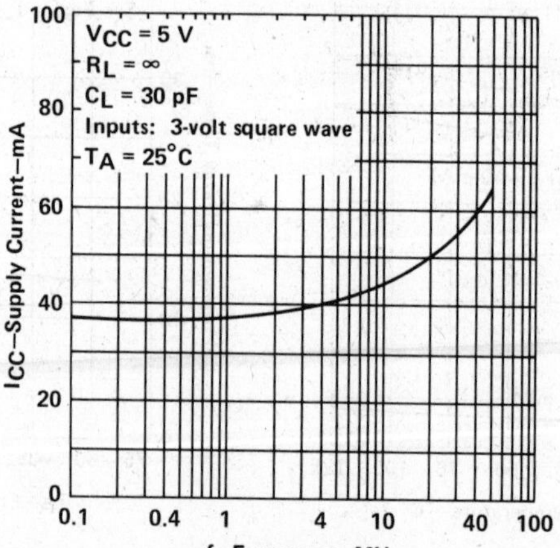
V_{CC} = 5 V
R_L = ∞
C_L = 30 pF
Inputs: 3-volt square wave
T_A = 25°C
I_{CC}—Supply Current—mA
f—Frequency—MHz
0
20
40
60
80
100
0.1
0.4
1
4
10
40
100

FIGURE 10

† Data for temperatures below 0°C and above 70°C are applicable to SN55114 circuits only. These parameters were measured with the active pullup connected to the sink output.

TYPICAL CHARACTERISTICS†

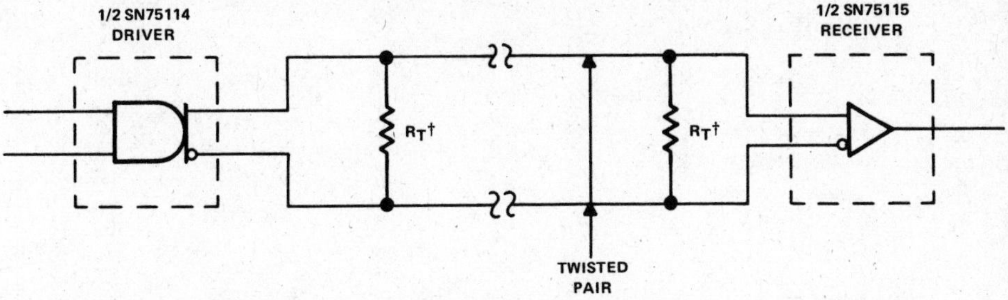
1/2 SN75114
DRIVER
R_T†
R_T†
1/2 SN75115
RECEIVER
TWISTED
PAIR

† $R_T = Z_O$. A capacitor may be connected in series with R_T to reduce power dissipation.

FIGURE 11. BASIC PARTY-LINE OR DATA-BUS DIFFERENTIAL DATA TRANSMISSION

D1315, SEPTEMBER 1973—REVISED OCTOBER 1986

- **Choice of Open-Collector or Active Pull-Up (Totem-Pole) Outputs**
- **Single 5-V Supply**
- **Differential Line Operation**
- **Dual-Channel Operation**
- **TTL Compatible**
- **± 15 V Common-Mode Input Voltage Range**
- **Optional-Use Built-In 130-Ω Line-Terminating Resistor**
- **Individual Frequency Response Controls**
- **Individual Channel Strobes**
- **Designed for Use with SN55113, SN75113, SN55114, and SN75114 Drivers**
- **Designed to be Interchangeable with Fairchild 9615 Line Receivers**

description

The SN55115 and SN75115 dual differential line receivers are designed to sense small differential signals in the presence of large common-mode noise. These devices give TTL-compatible output signals as a function of the differential input voltage. The open-collector output configuration permits the wire-ANDing of similar TTL outputs (such as SN5401/SN7401) or other SN55115/SN75115 line receivers. This permits a level of logic to be implemented without extra delay. The output stages are similar to TTL totem-pole outputs, but with sink outputs, 1YS and 2YS, and the corresponding active pull-up terminals, 1YP and 2YP, available on adjacent package pins. The frequency response and noise immunity may be provided by a single external capacitor. A strobe input is provided for each channel. With the strobe in the low level, the receiver is disabled and the outputs are forced to a high level.

The SN55115 is characterized for operation over the full military range of −55°C to 125°C. The SN75115 is characterized for operation from 0°C to 70°C.

SN55115 . . . J DUAL-IN-LINE PACKAGE
SN75115 . . . D, J, OR N PACKAGE
(TOP VIEW)

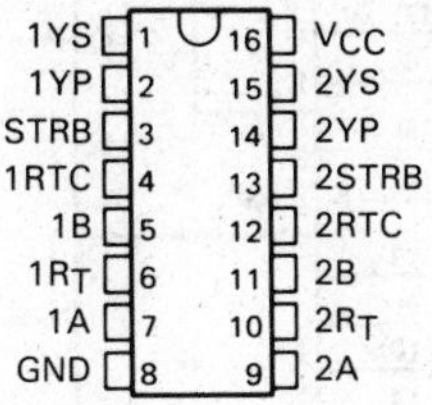
1YS 1
1YP 2
1STRB 3
1RTC 4
1B 5
1RT 6
1A 7
GND 8
16 VCC
15 2YS
14 2YP
13 2STRB
12 2RTC
11 2B
10 2RT
9 2A

SN55115 . . . FK PACKAGE
(TOP VIEW)

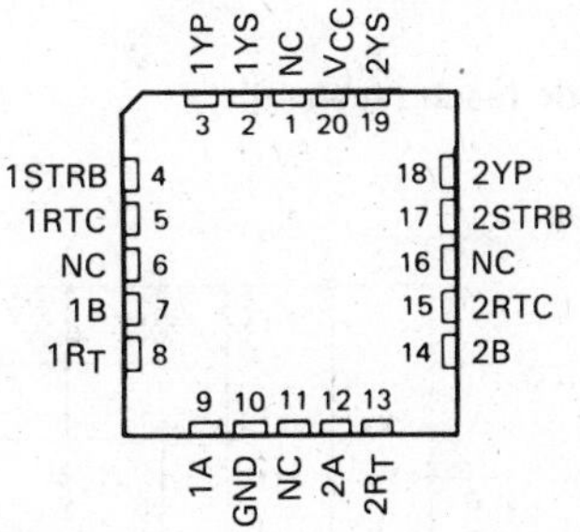
1YP 3
1YS 2
NC 1
VCC 20
2YS 19
1STRB 4
1RTC 5
NC 6
1B 7
1RT 8
18 2YP
17 2STRB
16 NC
15 2RTC
14 2B
1A 9
GND 10
NC 11
2A 12
2RT 13

NC—No internal connection

FUNCTION TABLE

STROBE	DIFF INPUT	OUTPUT (YP AND YS TIED TOGETHER)
L	X	H
H	L	H
H	H	L

H = $V_I \geq V_{IH}$ min or V_{ID} more positive than V_{TH} max
L = $V_I \leq V_{IL}$ max or V_{ID} more negative than V_{TL} max
X = irrelevant

4

Line Drivers/Receivers

logic symbol†

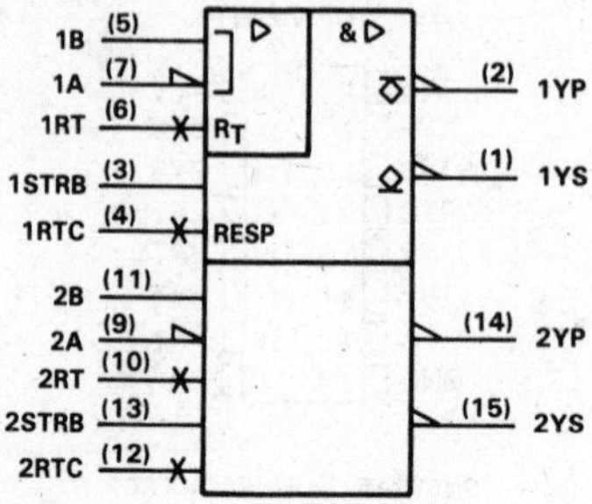
1B (5)
1A (7)
1RT (6)
1STRB (3)
1RTC (4)
2B (11)
2A (9)
2RT (10)
2STRB (13)
2RTC (12)
&
RT
RESP
(2) 1YP
(1) 1YS
(14) 2YP
(15) 2YS

† This symbol is in accordance with ANSI/IEEE Std 91-1984 and IEC Publication 617-12.

logic diagram (positive logic)

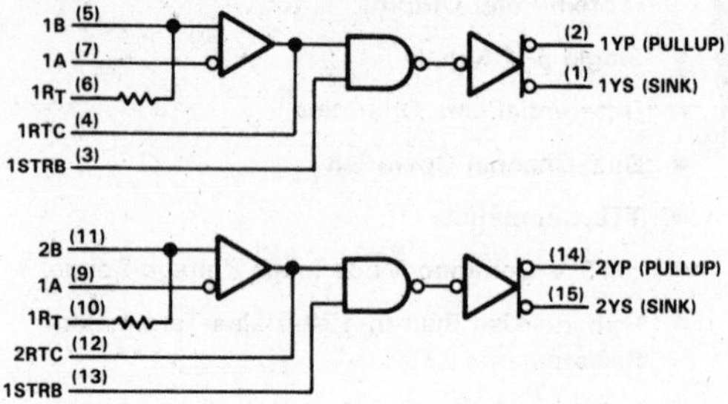
1B (5)
1A (7)
1RT (6)
1RTC (4)
1STRB (3)
(2) 1YP (PULLUP)
(1) 1YS (SINK)
2B (11)
1A (9)
1RT (10)
2RTC (12)
1STRB (13)
(14) 2YP (PULLUP)
(15) 2YS (SINK)

schematic (each receiver)

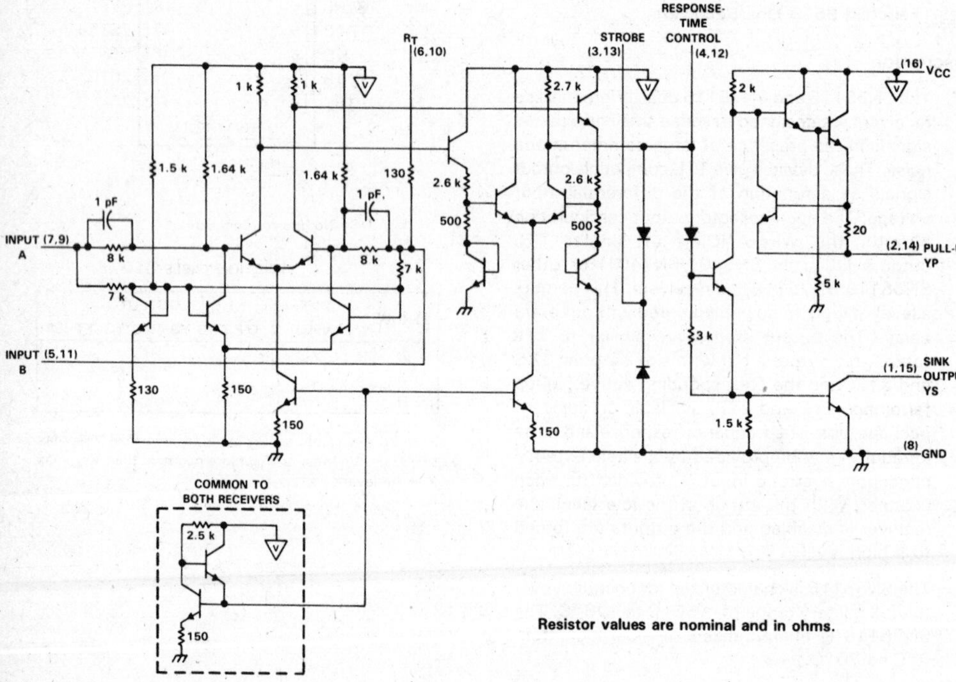
RT (6,10)
STROBE (3,13)
RESPONSE-TIME CONTROL (4,12)
(16) VCC
INPUT (7,9) A
INPUT (5,11) B
1 pF
8 k
7 k
1.5 k
1.64 k
1 k
1.64 k
130
1 pF,
8 k
7 k
130
150
150
2.6 k
500
2.7 k
2.6 k
500
2 k
20
(2,14) PULL-UP YP
5 k
3 k
(1,15) SINK OUTPUT YS
1.5 k
150
(8) GND
COMMON TO BOTH RECEIVERS
2.5 k
150

Resistor values are nominal and in ohms.

Pin numbers shown are for D, J, and N packages.

TEXAS INSTRUMENTS
POST OFFICE BOX 655012 • DALLAS, TEXAS 75265

absolute maximum ratings over operating free-air temperature range (unless otherwise noted)

		SN55115	SN75115	UNIT
Supply voltage, V_{CC} (see Note 1)		7	7	V
Input voltage at A, B, and R_T inputs		±25	±25	V
Input voltage at strobe input		5.5	5.5	V
Off-state voltage applied to open-collector outputs		14	14	V
Continuous total dissipation at (or below) 25°C free-air temperature (see Note 2)	D package		950	mW
	FK package	1375		
	J package	1375	1025	
	N package		1150	
Operating free-air temperature range		−55 to 125	0 to 70	°C
Storage temperature range		−65 to 150	−65 to 150	°C
Case temperature for 60 seconds: FK package		260		°C
Lead temperature 1,6 mm (1/16 inch) from case for 60 seconds: J package		300		°C
Lead temperature 1,6 mm (1/16 inch) from case for 10 seconds: D or N package			260	°C

NOTES: 1. All voltage values, except differential input voltage, are with respect to network ground terminal.
2. For operation above 25°C free-air temperature, refer to Dissipation Derating Curves in Appendix A. In the FK and J packages, SN55115 chips are alloy mounted and SN75115 chips are glass mounted. For these devices in the N package, use the 7.0-mW/°C curve. For the D package, use the 8.2 mW/°C curve.

recommended operating conditions

	SN55115			SN75115			UNIT
	MIN	NOM	MAX	MIN	NOM	MAX	
Supply voltage, V_{CC}	4.5	5	5.5	4.75	5	5.25	V
High-level (strobe) input voltage, V_{IH}	2.4			2.4			V
Low-level (strobe) input voltage, V_{IL}			0.4			0.4	V
High-level output current, I_{OH}			−5			−5	mA
Low-level output current, I_{OL}			15			15	mA
Operating free-air temperature, T_A	−55		125	0		70	°C

electrical characteristics over recommended operating free-air temperature range (unless otherwise noted)

PARAMETER		TEST CONDITIONS†		SN55115 MIN	SN55115 TYP‡	SN55115 MAX	SN75115 MIN	SN75115 TYP‡	SN75115 MAX	UNIT
V_{TH}§	Differential input high-threshold voltage	V_O = 0.4 V, I_{OL} = 15 mA, V_{IC} = 0				500			500	mV
V_{TL}§	Differential input low-threshold voltage	V_O = 2.4 V, I_{OH} = −5 mA, V_{IC} = 0		−500¶			−500¶			mV
V_{ICR}	Common-mode input voltage range	V_{ID} = ±1 V		+15 to −15	+24 to −19		+15 to −15	+24 to −19		V
V_{OH}	High-level output voltage	V_{CC} = MIN, V_{ID} = −0.5 V, I_{OH} = −5 mA	T_A = MIN	2.2			2.4			V
			T_A = 25°C	2.4	3.4		2.4	3.4		
			T_A = MAX	2.4			2.4			
V_{OL}	Low-level output voltage	V_{CC} = MIN, V_{ID} = 0.5 V, I_{OL} = 15 mA			0.22	0.4		0.22	0.45	V
I_{IL}	Low-level input current	V_{CC} = MAX, V_I = 0.4 V, Other input at 5.5 V	T_A = MIN			−0.9			−0.9	mA
			T_A = 25°C		−0.5	−0.7		−0.5	−0.7	
			T_A = MAX			−0.7			−0.7	
I_{SH}	High-level strobe current	V_{CC} = MIN, V_{ID} = −0.5 V, V_{strobe} = 4.5 V	T_A = 25°C			2			5	µA
			T_A = MAX			5			10	
I_{SL}	Low-level strobe current	V_{CC} = MAX, V_{ID} = 0.5 V, V_{strobe} = 0.4 V	T_A = 25°C		−1.15	−2.4		−1.15	−2.4	mA
$I_{(RTC)}$	Response-time-control current	V_{CC} = MAX, V_{ID} = 0.5 V, V_{RC} = 0	T_A = 25°C	−1.2	−3.4		−1.2	−3.4		mA
$I_{O(off)}$	Off-state open-collector output current	V_{CC} = MIN, V_{OH} = 12 V, V_{ID} = −4.5 V	T_A = 25°C			100				µA
			T_A = MAX			200				
		V_{CC} = MIN, V_{OH} = 5.25 V, V_{ID} = −4.75 V	T_A = 25°C						100	
			T_A = MAX						200	
R_T	Line-terminating resistance	V_{CC} = 5 V	T_A = 25°C	77	130	167	74	130	179	Ω
I_{OS}	Short-circuit output current‖	V_{CC} = MAX, V_O = 0, V_{ID} = −0.5 V	T_A = 25°C	−15	−40	−80	−14	−40	−100	mA
I_{CC}	Supply current (both receivers)	V_{CC} = MAX, V_{ID} = 0.5 V, V_{IC} = 0	T_A = 25°C		32	50		32	50	mA

† Unless otherwise noted V_{strobe} = 2.4 V. All parameters with the exception of off-state open-collector output current are measured with the active pull-up connected to the sink output.
‡ All typical values are at V_{CC} = 5 V, T_A = 25°C, and V_{IC} = 0.
§ Differential voltages are at the B input terminal with respect to the A input terminal.
¶ The algebraic convention, in which the less positive (more negative) limit is designated as minimum, is used in this data sheet for threshold voltages only.
‖ Only one output should be shorted to ground at a time, and duration of the short-circuit should not exceed one second.

switching characteristics, V_{CC} = 5 V, C_L = 30 pF, T_A = 25°C

PARAMETER		TEST CONDITIONS	SN55115 MIN	SN55115 TYP	SN55115 MAX	SN75115 MIN	SN75115 TYP	SN75115 MAX	UNIT
t_{PLH}	Propagation delay time, low-to-high-level output	R_L = 3.9 kΩ, See Figure 1		18	50		18	75	ns
t_{PHL}	Propagation delay time, high-to-low-level output	R_L = 390 Ω, See Figure 1		20	50		20	75	ns

PARAMETER MEASUREMENT INFORMATION

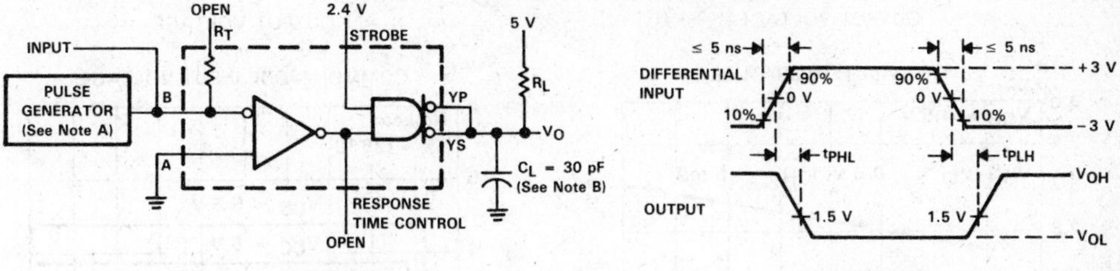
OPEN
R_T
2.4 V
STROBE
5 V
R_L
INPUT
PULSE GENERATOR (See Note A)
B
A
YP
YS
V_O
C_L = 30 pF (See Note B)
RESPONSE TIME CONTROL
OPEN
≤ 5 ns
≤ 5 ns
DIFFERENTIAL INPUT
90%
90%
0 V
0 V
10%
10%
+3 V
−3 V
t_{PHL}
t_{PLH}
V_{OH}
OUTPUT
1.5 V
1.5 V
V_{OL}

TEST CIRCUIT | WAVEFORMS

NOTES: A. The pulse generator has the following characteristics: Z_{out} = 50 Ω, PRR ≤ 500 kHz, t_w = 100 ns, duty cycle = 50%.
B. C_L includes probe and jig capacitance.

FIGURE 1. PROPAGATION DELAY TIMES

TYPICAL CHARACTERISTICS

INPUT CURRENT
vs
INPUT VOLTAGE

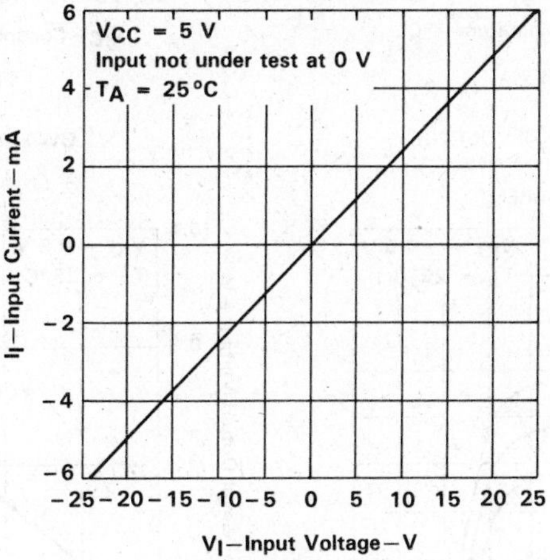
V_{CC} = 5 V
Input not under test at 0 V
T_A = 25°C
6
4
2
0
−2
−4
−6
I_I—Input Current—mA
−25 −20 −15 −10 −5 0 5 10 15 20 25
V_I—Input Voltage—V

FIGURE 2

TYPICAL CHARACTERISTICS†

OUTPUT VOLTAGE
vs
FREE-AIR TEMPERATURE

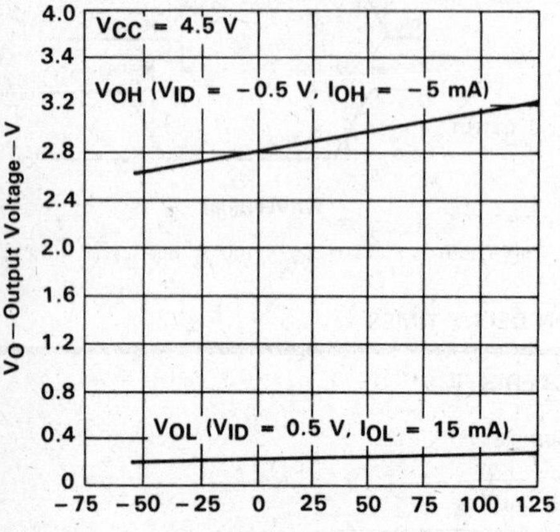
VCC = 4.5 V
VOH (VID = −0.5 V, IOH = −5 mA)
VOL (VID = 0.5 V, IOL = 15 mA)
VO—Output Voltage—V
TA—Free-Air Temperature—°C
4.0
3.4
3.2
2.8
2.4
2.0
1.6
1.2
0.8
0.4
0
−75 −50 −25 0 25 50 75 100 125

FIGURE 3

OUTPUT VOLTAGE
vs
COMMON-MODE INPUT VOLTAGE

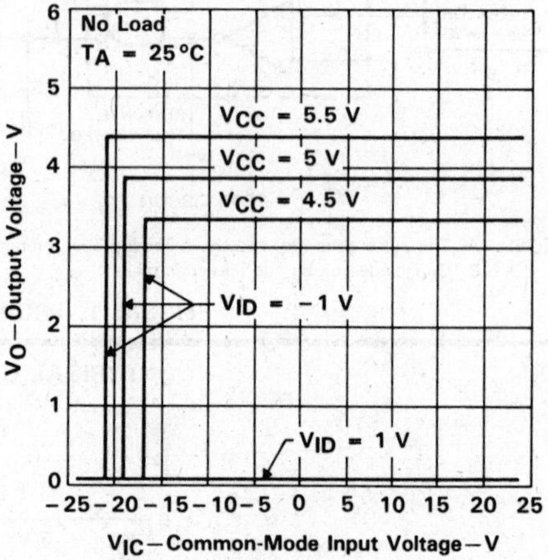
No Load
TA = 25°C
VCC = 5.5 V
VCC = 5 V
VCC = 4.5 V
VID = −1 V
VID = 1 V
VO—Output Voltage—V
VIC—Common-Mode Input Voltage—V
6
5
4
3
2
1
0
−25 −20 −15 −10 −5 0 5 10 15 20 25

FIGURE 4

HIGH-LEVEL OUTPUT VOLTAGE
vs
OUTPUT CURRENT

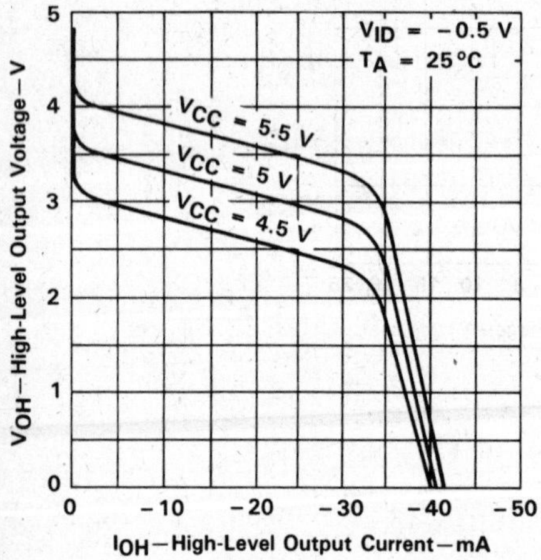
VID = −0.5 V
TA = 25°C
VCC = 5.5 V
VCC = 5 V
VCC = 4.5 V
VOH—High-Level Output Voltage—V
IOH—High-Level Output Current—mA
5
4
3
2
1
0
0 −10 −20 −30 −40 −50

FIGURE 5

LOW-LEVEL OUTPUT VOLTAGE
vs
OUTPUT CURRENT

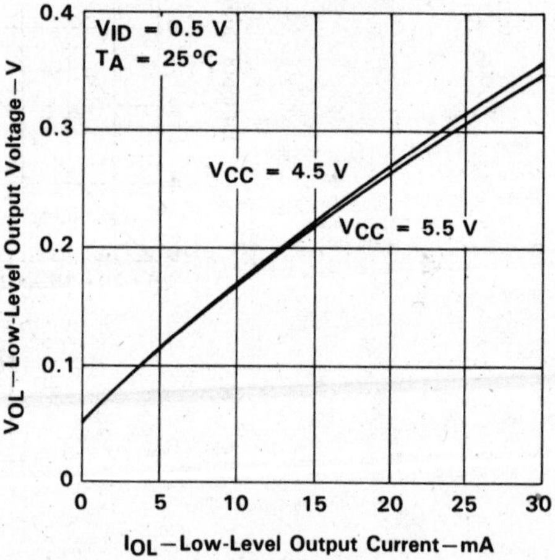
VID = 0.5 V
TA = 25°C
VCC = 4.5 V
VCC = 5.5 V
VOL—Low-Level Output Voltage—V
IOL—Low-Level Output Current—mA
0.4
0.3
0.2
0.1
0
0 5 10 15 20 25 30

FIGURE 6

† Data for temperatures below 0°C and above 70°C and for supply voltages below 4.75 V and above 5.25 V are applicable to SN55115 circuits only. These parameters were measured with the active pull-up connected to the sink output.

TEXAS INSTRUMENTS
POST OFFICE BOX 655012 • DALLAS, TEXAS 75265

TYPICAL CHARACTERISTICS†

OUTPUT VOLTAGE
vs
DIFFERENTIAL INPUT VOLTAGE

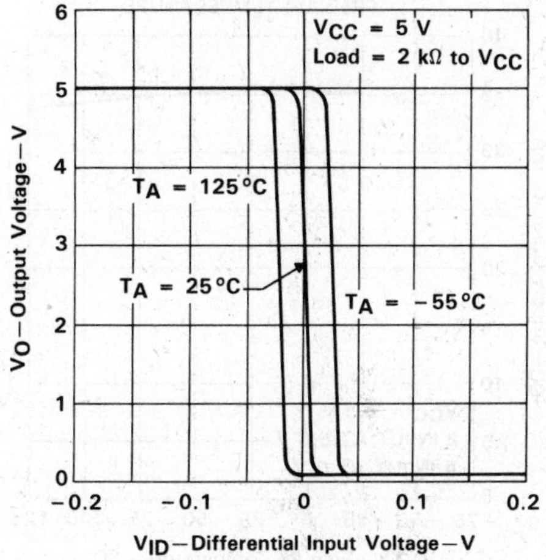
VCC = 5 V
Load = 2 kΩ to VCC
TA = 125°C
TA = 25°C
TA = −55°C
VO—Output Voltage—V
VID—Differential Input Voltage—V

FIGURE 7

OUTPUT VOLTAGE
vs
DIFFERENTIAL INPUT VOLTAGE

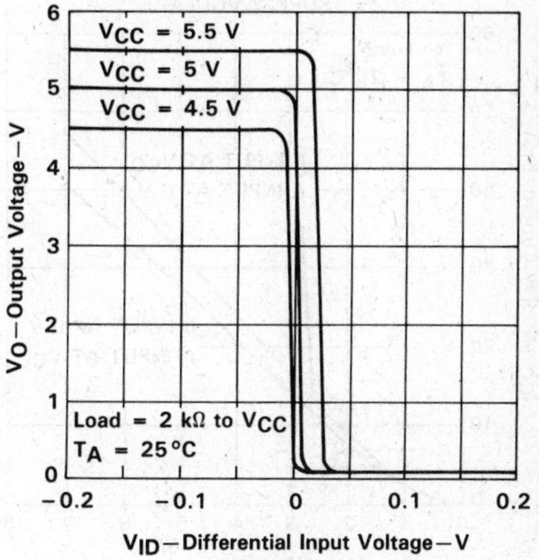
VCC = 5.5 V
VCC = 5 V
VCC = 4.5 V
Load = 2 kΩ to VCC
TA = 25°C
VO—Output Voltage—V
VID—Differential Input Voltage—V

FIGURE 8

OUTPUT VOLTAGE
vs
STROBE INPUT VOLTAGE

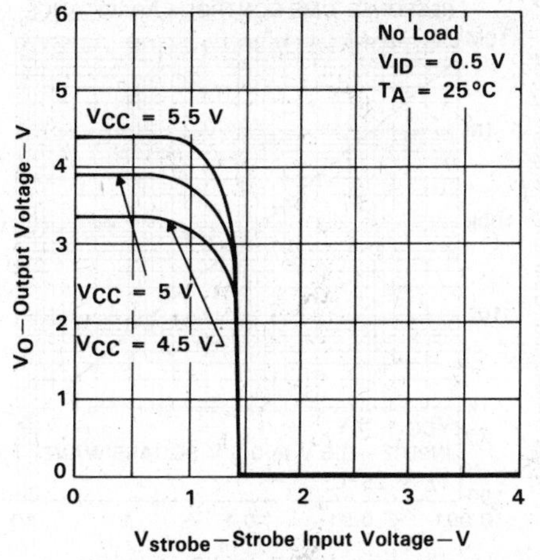
No Load
VID = 0.5 V
TA = 25°C
VCC = 5.5 V
VCC = 5 V
VCC = 4.5 V
VO—Output Voltage—V
Vstrobe—Strobe Input Voltage—V

FIGURE 9

OUTPUT VOLTAGE
vs
STROBE INPUT VOLTAGE

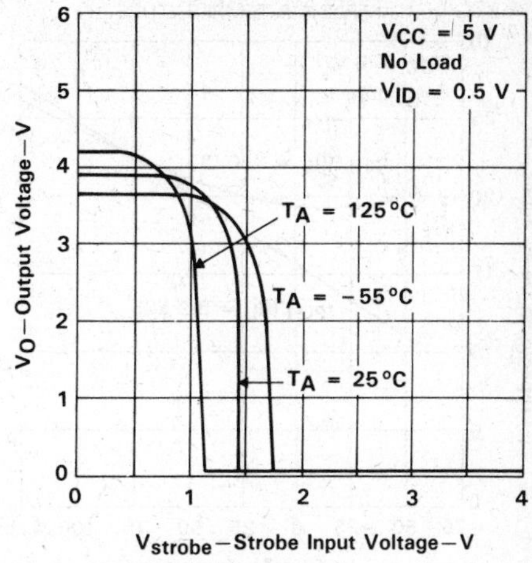
VCC = 5 V
No Load
VID = 0.5 V
TA = 125°C
TA = −55°C
TA = 25°C
VO—Output Voltage—V
Vstrobe—Strobe Input Voltage—V

FIGURE 10

† Data for temperatures below 0°C and above 70°C and for supply voltages below 4.75 V and above 5.25 V are applicable to SN55115 circuits only. These parameters were measured with the active pull up connected to the sink output.

TYPICAL CHARACTERISTICS†

SUPPLY CURRENT
(BOTH RECEIVERS)
vs
SUPPLY VOLTAGE

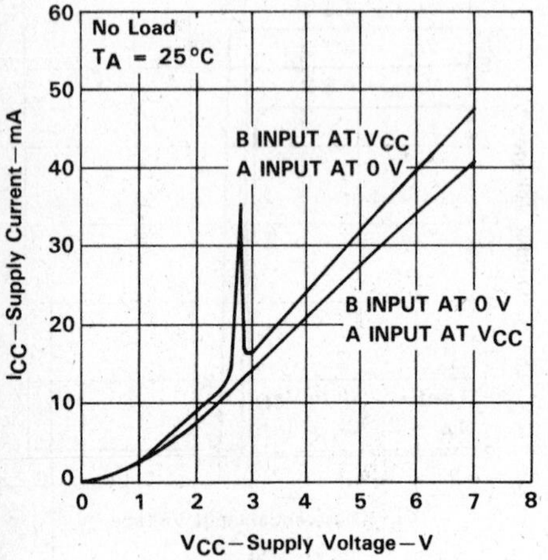
No Load
T_A = 25°C
B INPUT AT V_{CC}
A INPUT AT 0 V
B INPUT AT 0 V
A INPUT AT V_{CC}
I_{CC}—Supply Current—mA
V_{CC}—Supply Voltage—V
0
10
20
30
40
50
60
1
2
3
4
5
6
7
8

FIGURE 11

SUPPLY CURRENT
(BOTH RECEIVERS)
vs
FREE-AIR TEMPERATURE

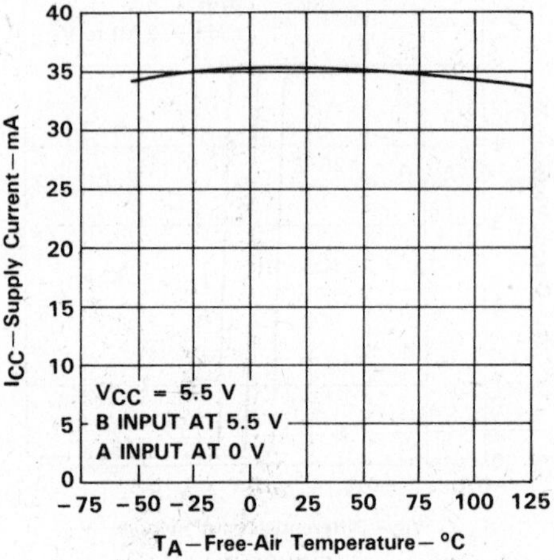
V_{CC} = 5.5 V
B INPUT AT 5.5 V
A INPUT AT 0 V
I_{CC}—Supply Current—mA
T_A—Free-Air Temperature—°C
0
5
10
15
20
25
30
35
40
−75
−50
−25
0
25
50
75
100
125

FIGURE 12

PROPAGATION DELAY TIMES
vs
FREE-AIR TEMPERATURE

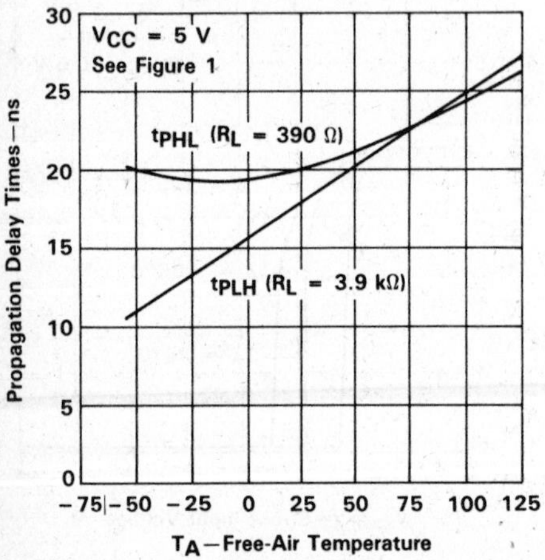
V_{CC} = 5 V
See Figure 1
t_{PHL} (R_L = 390 Ω)
t_{PLH} (R_L = 3.9 kΩ)
Propagation Delay Times—ns
T_A—Free-Air Temperature
0
5
10
15
20
25
30
−75
−50
−25
0
25
50
75
100
125

FIGURE 13

MAXIMUM OPERATING FREQUENCY
vs
RESPONSE-TIME-CONTROL CAPACITANCE

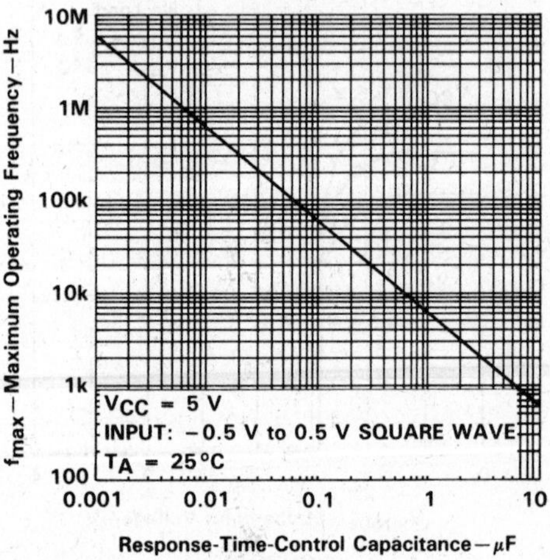
V_{CC} = 5 V
INPUT: −0.5 V to 0.5 V SQUARE WAVE
T_A = 25°C
f_{max}—Maximum Operating Frequency—Hz
Response-Time-Control Capacitance—μF
100
1k
10k
100k
1M
10M
0.001
0.01
0.1
1
10

FIGURE 14

† Data for temperatures below 0°C and above 70°C and for supply voltages below 4.75 V and above 5.25 V are applicable to SN55115 circuits only. These parameters were measured with the active pull-up connected to the sink output.

TEXAS INSTRUMENTS
POST OFFICE BOX 655012 • DALLAS, TEXAS 75265

TYPICAL APPLICATION DATA

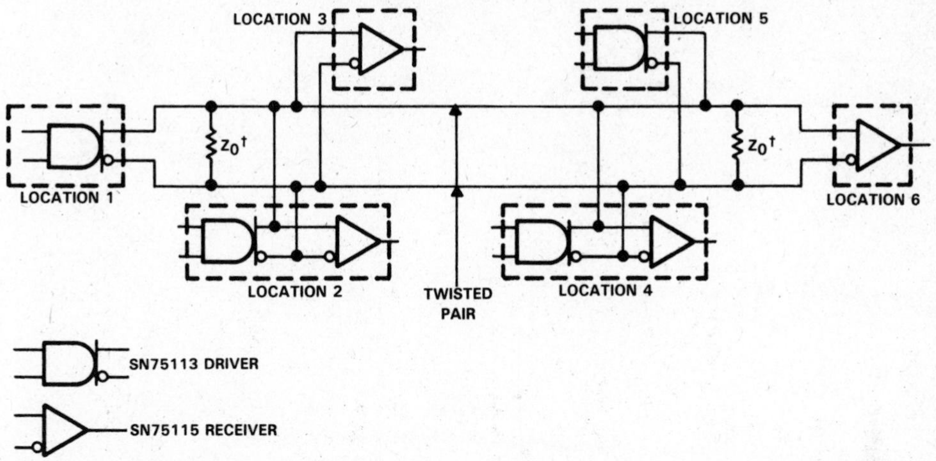
LOCATION 3
LOCATION 5
LOCATION 1
LOCATION 6
LOCATION 2
LOCATION 4
TWISTED PAIR
Z_O†
Z_O†
SN75113 DRIVER
SN75115 RECEIVER

† A capacitor may be connected in series with Z_O to reduce power dissipation.

FIGURE 15. BASIC PARTY-LINE OR DATA-BUS DIFFERENTIAL DATA TRANSMISSION

DIFFERENTIAL LINE TRANSCEIVERS

D2143, MAY 1976—REVISED SEPTEMBER 1986

features common to all types

- Single 5-V Supply
- 3-State Driver Output Circuitry
- TTL-Compatible Driver Inputs
- TTL-Compatible Receiver Output
- Differential Line Operation
- Receiver Output Strobe ('116, '117) or Enable ('118, '119)
- Designed for Party-Line (Data-Bus) Applications
- Choice of Ceramic or Plastic Packages

additional features of the SN55116/SN75116

- Independent Driver and Receiver
- Choice of Open-Collector or Totem-Pole Outputs on Both Driver and Receiver
- Dual Data Inputs on Driver
- Optional Line-Termination Resistor in Receiver
- ± 15-V Receiver Common-Mode Capability
- Receiver Frequency Response Control

additional features of the SN55117/SN75117

- Driver Output Internally Connected to Receiver Input

The SN55118/SN75118 is an SN55116/SN75116 with 3-State Receiver Output Circuitry
The SN55119/SN75119 is an SN55117/SN75117 with 3-State Receiver Output Circuitry

description

These integrated circuits are designed for use in interfacing between TTL-type digital systems and differential data transmission lines. They are especially useful for party-line (data-bus) applications. Each of these circuit types combine in one package a three-state differential line driver and a differential-input line receiver, both of which operate from a single 5-volt power supply. The driver inputs and receiver outputs are TTL compatible. The driver employed is similar to the SN55113/SN75113 three-state line driver, and the receiver is similar to the SN55115/SN75115 line receiver.

The '116 and '118 circuits offer all the features of the SN55113/SN75113 driver and the SN55115/SN75115 receiver combined. The driver performs the dual input AND and NAND functions when enabled, or presents a high impedance to the load when in the disabled state. The driver output stages are similar to TTL totem-pole outputs, but have the current-sink portion separated from the current-sourcing portion and both are brought out to adjacent package pins. This feature allows the user the option of using the driver in the open-collector output configuration, or, by connecting the adjacent source and sink pins together, of using the driver in the normal totem-pole output configuration.

The receiver portion of the '116 and '118 features a differential-input circuit having a common-mode voltage range of ±15 volts. An internal 130-ohm resistor is also provided, which may optionally be used for terminating the transmission line. A frequency response control pin allows the user to reduce the speed of the receiver or to improve differential noise immunity. The receiver of the '116 also has an output strobe and a split totem-pole output. The receiver of the '118 has an output-enable for the three-state split totem-pole output. The receiver section of either circuit is independent of the driver section except for the V_{CC} and ground pins.

The '117 and '119 circuits provide the basic driver and receiver functions of the '116 and '118, but use a package that is only half as large. The '117 and '119 are intended primarily for party-line or bus-organized systems as the driver outputs are internally connected to the receiver inputs. The driver has a single data input and a single enable input, and the '117 receiver has an output strobe while the '119 receiver has a three-state-output enable. These devices do not, however, provide output connection options, line termination resistors, or receiver frequency-response controls.

The SN55116, SN55117, SN55118, and SN55119 are characterized for operation over the full military temperature range of −55°C to 125°C; the SN75116, SN75117, SN75118, and SN75119 are characterized for operation from 0°C to 70°C.

TEXAS INSTRUMENTS
POST OFFICE BOX 655012 • DALLAS, TEXAS 75265

SN55116 . . . J PACKAGE
SN75116 . . . D, J, OR N PACKAGE
(TOP VIEW)

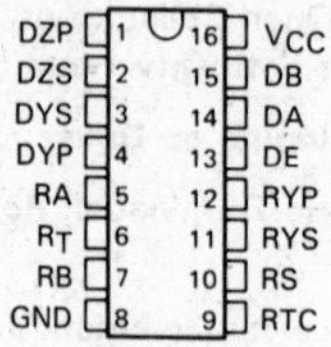
DZP 1 16 VCC
DZS 2 15 DB
DYS 3 14 DA
DYP 4 13 DE
RA 5 12 RYP
RT 6 11 RYS
RB 7 10 RS
GND 8 9 RTC

SN55118 . . . J PACKAGE
SN75118 . . . D, J, OR N PACKAGE
(TOP VIEW)

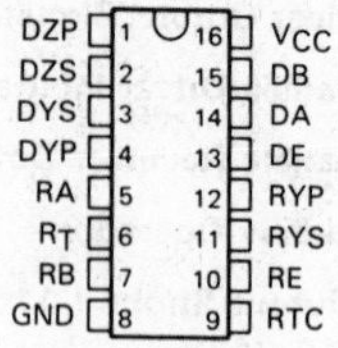
DZP 1 16 VCC
DZS 2 15 DB
DYS 3 14 DA
DYP 4 13 DE
RA 5 12 RYP
RT 6 11 RYS
RB 7 10 RE
GND 8 9 RTC

SN55116
FK PACKAGE
(TOP VIEW)

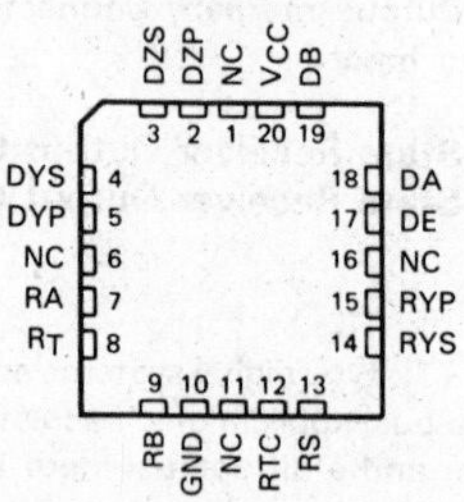
DZS DZP NC VCC DB
3 2 1 20 19
DYS 4 18 DA
DYP 5 17 DE
NC 6 16 NC
RA 7 15 RYP
RT 8 14 RYS
9 10 11 12 13
RB GND NC RTC RS

SN55118
FK PACKAGE
(TOP VIEW)

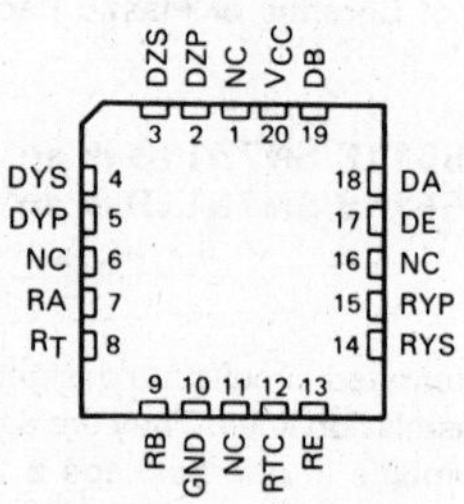
DZS DZP NC VCC DB
3 2 1 20 19
DYS 4 18 DA
DYP 5 17 DE
NC 6 16 NC
RA 7 15 RYP
RT 8 14 RYS
9 10 11 12 13
RB GND NC RTC RE

SN55117 . . . JG PACKAGE
SN75117 . . . D, JG, OR P PACKAGE
(TOP VIEW)

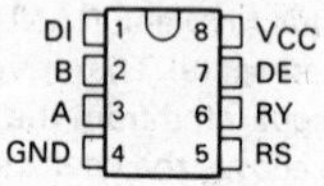
DI 1 8 VCC
B 2 7 DE
A 3 6 RY
GND 4 5 RS

SN55119 . . . JG PACKAGE
SN75119 . . . D, JG, OR P PACKAGE
(TOP VIEW)

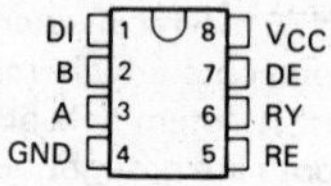
DI 1 8 VCC
B 2 7 DE
A 3 6 RY
GND 4 5 RE

SN55117
FK PACKAGE
(TOP VIEW)

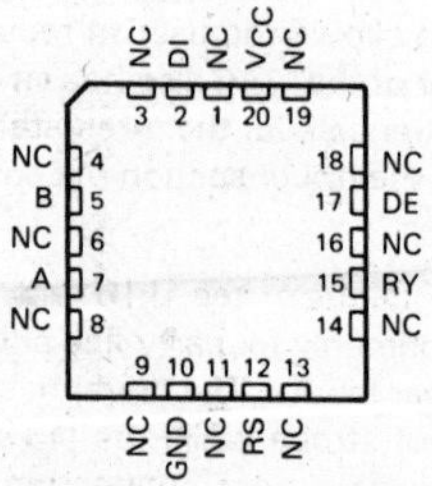
NC DI NC VCC NC
3 2 1 20 19
NC 4 18 NC
B 5 17 DE
NC 6 16 NC
A 7 15 RY
NC 8 14 NC
9 10 11 12 13
NC GND NC RS NC

SN55119
FK PACKAGE
(TOP VIEW)

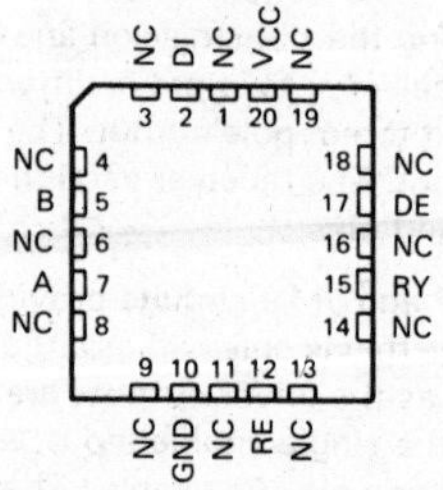
NC DI NC VCC NC
3 2 1 20 19
NC 4 18 NC
B 5 17 DE
NC 6 16 NC
A 7 15 RY
NC 8 14 NC
9 10 11 12 13
NC GND NC RE NC

NC—No internal connection.

4

Line Drivers/Receivers

'116, '118
FUNCTION TABLE
OF DRIVER

INPUTS			OUTPUTS	
DE	DA	DB	DY	DZ
L	X	X	Z	Z
H	L	X	L	H
H	X	L	L	H
H	H	H	H	L

'117, '119
FUNCTION TABLE
OF DRIVER

INPUTS		OUTPUTS	
DI	DE	A	B
H	H	H	L
L	H	L	H
X	L	Z	Z

'116, '118
FUNCTION TABLE OF RECEIVER

RS/RE	DIFF INPUT	OUTPUT RY	
		'116	'118
L	X	H	Z
H	L	H	H
H	H	L	L

'117, '119
FUNCTION TABLE OF RECEIVER

INPUTS			OUTPUT RY	
A	B	RS/RE	'117	'119
H	L	H	H	H
L	H	H	L	L
X	X	L	H	Z

H = high level ($V_I \geq V_{IH}$ min or V_{ID} more positive than V_{TH} max)
L = low level ($V_I \leq V_{IL}$ max or V_{ID} more negative than V_{TL} max)
X = irrelevant
Z = high impedance (off)

schematics of inputs and outputs

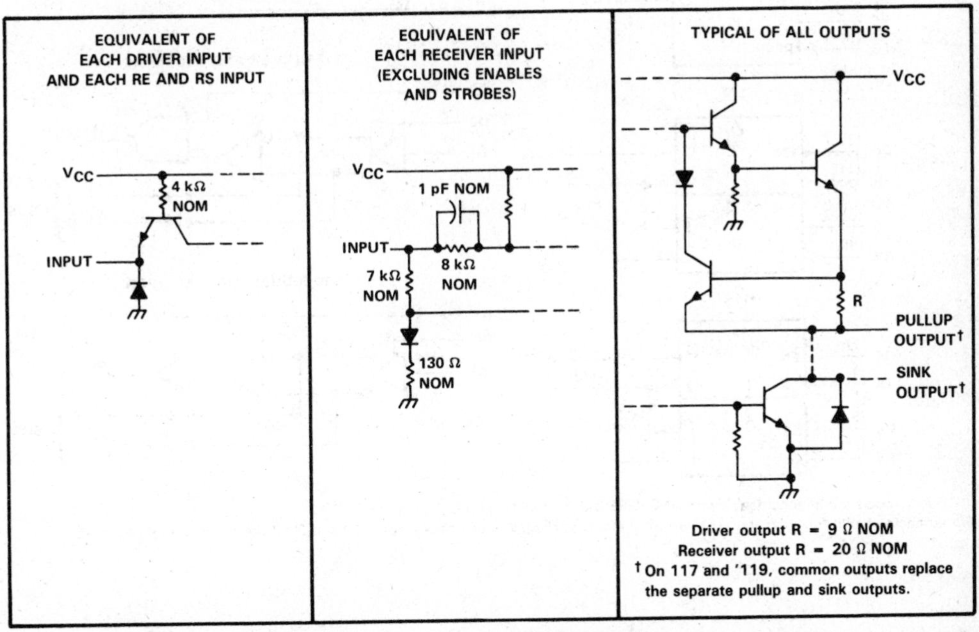
EQUIVALENT OF EACH DRIVER INPUT AND EACH RE AND RS INPUT
VCC
4 kΩ NOM
INPUT
EQUIVALENT OF EACH RECEIVER INPUT (EXCLUDING ENABLES AND STROBES)
VCC
1 pF NOM
INPUT
8 kΩ NOM
7 kΩ NOM
130 Ω NOM
TYPICAL OF ALL OUTPUTS
VCC
R
PULLUP OUTPUT†
SINK OUTPUT†
Driver output R = 9 Ω NOM
Receiver output R = 20 Ω NOM
† On 117 and '119, common outputs replace the separate pullup and sink outputs.

logic symbols†

'116

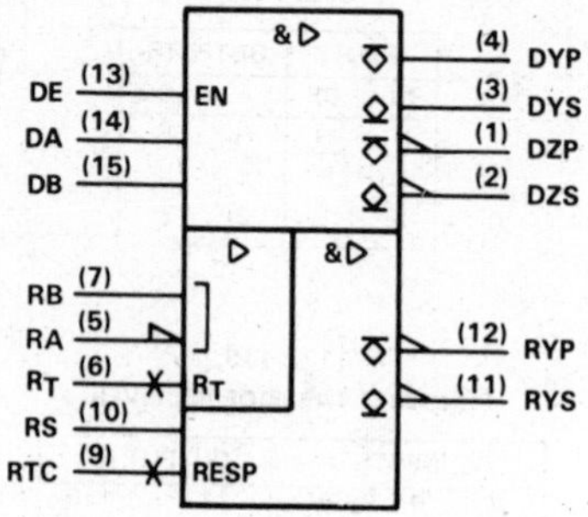
DE (13)
DA (14)
DB (15)
EN
&▷
(4) DYP
(3) DYS
(1) DZP
(2) DZS
RB (7)
RA (5)
RT (6)
RS (10)
RTC (9)
RT
RESP
(12) RYP
(11) RYS

'118

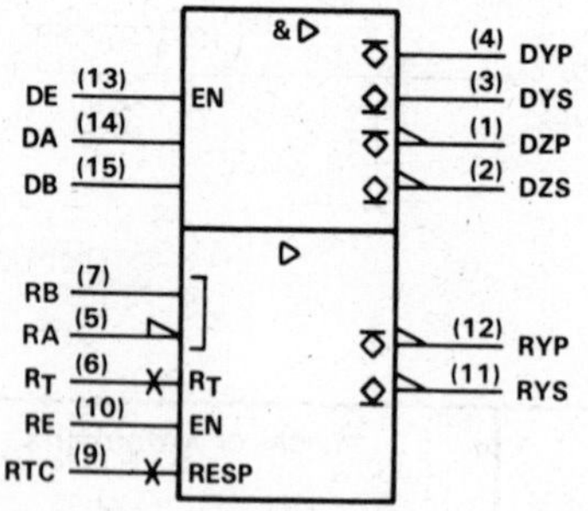
DE (13)
DA (14)
DB (15)
EN
(4) DYP
(3) DYS
(1) DZP
(2) DZS
RB (7)
RA (5)
RT (6)
RE (10)
RTC (9)
RT
EN
RESP
(12) RYP
(11) RYS

'117

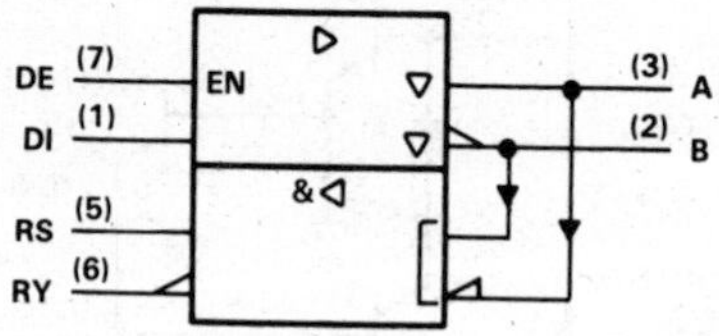
DE (7)
DI (1)
EN
RS (5)
RY (6)
(3) A
(2) B

'119

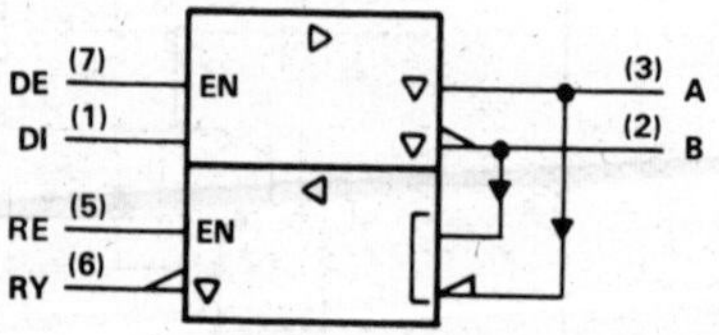
DE (7)
DI (1)
EN
RE (5)
RY (6)
EN
(3) A
(2) B

logic diagrams (positive logic)

'116 AND '118 DRIVER

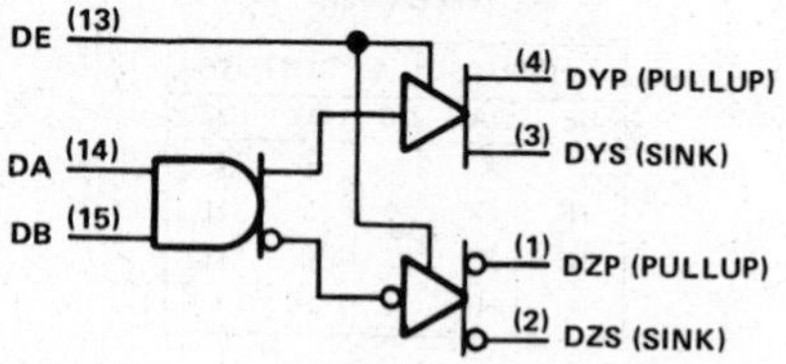
DE (13)
DA (14)
DB (15)
(4) DYP (PULLUP)
(3) DYS (SINK)
(1) DZP (PULLUP)
(2) DZS (SINK)

'116 RECEIVER

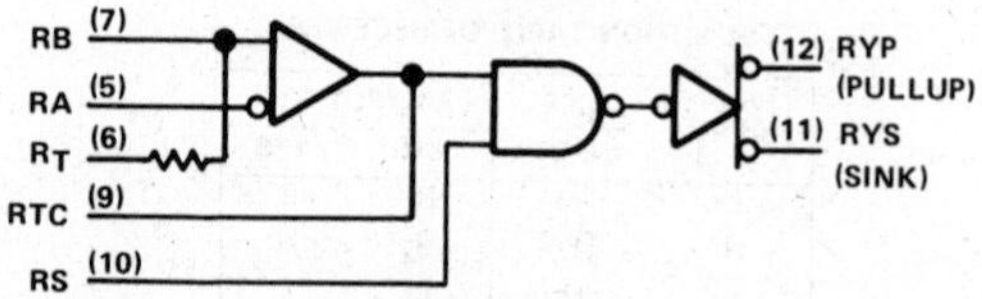
RB (7)
RA (5)
RT (6)
RTC (9)
RS (10)
(12) RYP (PULLUP)
(11) RYS (SINK)

'118 RECEIVER

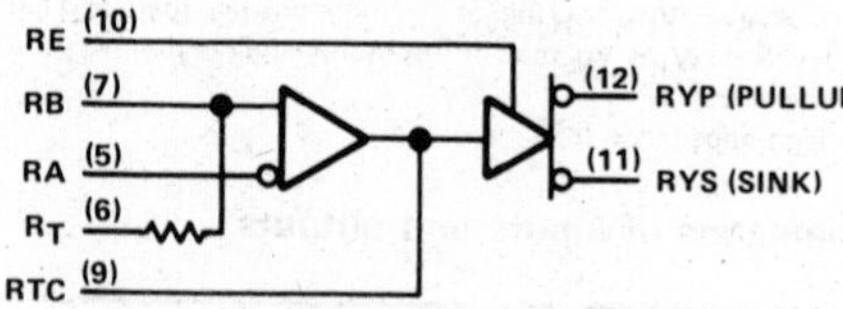
RE (10)
RB (7)
RA (5)
RT (6)
RTC (9)
(12) RYP (PULLUP)
(11) RYS (SINK)

'117 DRIVER AND RECEIVER

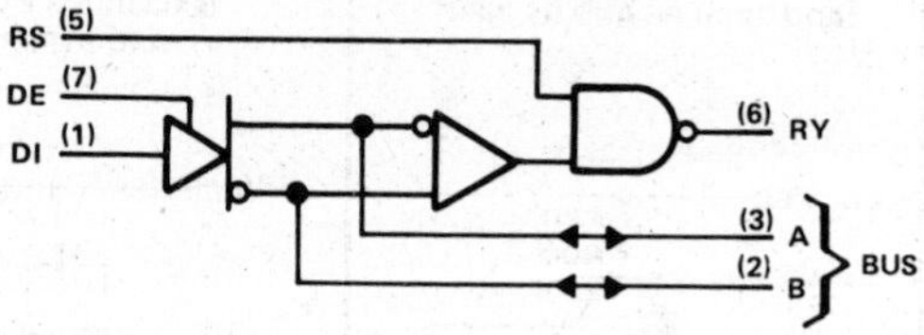
RS (5)
DE (7)
DI (1)
(6) RY
(3) A
(2) B
BUS

'119 DRIVER AND RECEIVER

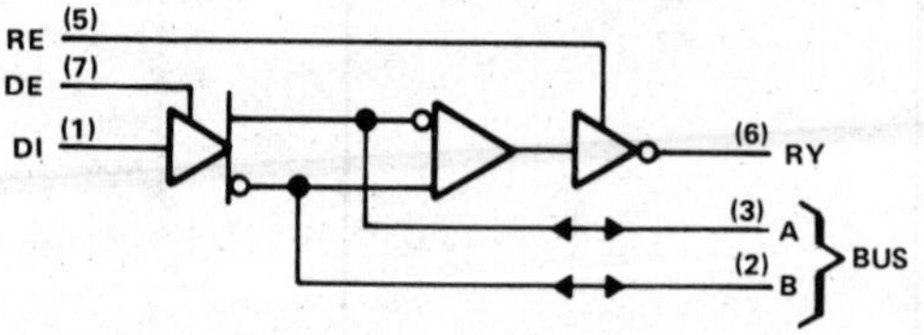
RE (5)
DE (7)
DI (1)
(6) RY
(3) A
(2) B
BUS

† These symbols are in accordance with ANSI/IEEE Std 91-1984 and IEC Publication 617-12.
Pin numbers shown for '116 and '118 are for J and N packages; those shown for '117 and '119 are for JG and P packages.

absolute maximum ratings over operating free-air temperature range (unless otherwise noted)

		'116, '118	'117, '119	UNIT
Supply voltage, V_{CC}		7	7	V
Input voltage, V_I	DA, DB, DE, DI, RE, RS	5.5	5.5	V
	RA, RB, RT	±25		
	A and B		0 to 6	
Off-state voltage applied to open-collector outputs		12		V

		SN55116 THRU SN55119	SN75116 THRU SN75119	UNIT
Continuous total dissipation at (or below) 25 °C free-air temperature (see Note 2)	D package		950	mW
	FK package	1375		
	J package	1375	1025	
	JG package	1050	825	
	N package		1150	
	P package		1000	
Operating free-air temperature range		−55 to 125	0 to 70	°C
Storage temperature range		−65 to 150	−65 to 150	°C
Case temperature for 60 seconds: FK package		260		°C
Lead temperature 1,6 mm (1/16 inch) from case for 60 seconds: J and JG packages		300	300	°C
Lead temperature 1,6 mm (1/16 inch) from case for 10 seconds: D, N, or P package			260	°C

NOTES: 1. All voltage values are with respect to network ground terminal.
2. For operation above 25 °C free-air temperature, refer to Dissipation Derating Curves in Appendix A. In the FK and J packages, SN55116 through SN55119 chips are alloy mounted and SN75116 through SN75119 chips are glass mounted. In the JG package, SN55117 and SN55119 are alloy mounted and SN75117 and SN75119 chips are glass mounted. In the N package, use the 9.2 mW/°C curve for these devices. In the P package, use the 8-mW/°C curve for these devices.

recommended operating conditions

PARAMETER		SN55' MIN	SN55' TYP	SN55' MAX	SN75' MIN	SN75' TYP	SN75' MAX	UNIT
Supply voltage, V_{CC}		4.5	5	5.5	4.75	5	5.25	V
High-level input voltage, V_{IH}	All inputs except differential inputs	2			2			V
Low-level input voltage, V_{IL}				0.8			0.8	V
High-level output current, I_{OH}	Drivers			−40			−40	mA
	Receivers			−5			−5	
Low-level output current, I_{OL}	Drivers			40			40	mA
	Receivers			15			15	
Receiver input voltage, V_I	'116, '118			±15			±15	V
	'117, '119	0		6	0		6	
Operating free-air temperature, T_A		−55		125	0		70	°C

electrical characteristics over recommended operating free-air temperature range (unless otherwise noted)

driver section

PARAMETER		TEST CONDITIONS†	'116, '118 MIN	'116, '118 TYP‡	'116, '118 MAX	'117, '119 MIN	'117, '119 TYP‡	'117, '119 MAX	UNIT
V_{IK}	Input clamp voltage	V_{CC} = MIN, I_I = −12 mA		−0.9	−1.5		−0.9	−1.5	V
V_{OH}	High-level output voltage	V_{CC} = MIN, V_{IL} = 0.8 V, V_{IH} = 2 V, T_A = 25°C (SN55′), I_{OH} = −10 mA	2.4	3.4		2.4	3.4		V
		V_{CC} = MIN, V_{IL} = 0.8 V, V_{IH} = 2 V, T_A = 0°C to 70°C (SN75′), I_{OH} = −40 mA	2	3		2	3		
		V_{CC} = MIN, V_{IL} = 0.8 V, V_{IH} = 2 V, T_A = −55°C to 125°C (SN55′), I_{OH} = −10 mA	2			2			
		V_{CC} = MIN, V_{IL} = 0.8 V, V_{IH} = 2 V, T_A = −55°C to 125°C (SN55′), I_{OH} = −40 mA	1.8			1.8			
V_{OL}	Low-level output voltage	V_{CC} = MIN, V_{IH} = 2 V, V_{IL} = 0.8 V, I_{OL} = 40 mA			0.4			0.4	V
V_{OK}	Output clamp voltage	V_{CC} = MAX, I_O = −40 mA, DE at 0.8 V			−1.5			−1.5	V
$I_{O(off)}$	Off-state open-collector output current	V_{CC} = MAX, V_O = 12 V, T_A = 25°C		1	10				μA
		V_{CC} = MAX, V_O = 12 V, T_A = MAX, SN55′			200				
		V_{CC} = MAX, V_O = 12 V, T_A = MAX, SN75′			20				
I_{OZ}	Off-state (high-impedance-state) output current	V_{CC} = MAX, V_O = 0 to V_{CC}, DE at 0.8 V, T_A = 25°C			±10				μA
		V_{CC} = MAX, DE at 0.8 V, T_A = MAX, V_O = 0, SN55′			−300				
		V_{CC} = MAX, DE at 0.8 V, T_A = MAX, V_O = 0.4 V to V_{CC}, SN55′			±150				
		V_{CC} = MAX, DE at 0.8 V, T_A = MAX, V_O = 0 to V_{CC}, SN75′			±20				
I_I	Input current at maximum input voltage (Driver or enable input)	V_{CC} = MAX, V_I = 5.5 V			1			1	mA
I_{IH}	High-level input current (Driver or enable input)	V_{CC} = MAX, V_I = 2.4 V			40			40	μA
I_{IL}	Low-level input current (Driver or enable input)	V_{CC} = MAX, V_I = 0.4 V			−1.6			−1.6	mA
I_{OS}	Short-circuit output current§	V_{CC} = MAX, V_O = 0, T_A = 25°C	−40		−120	−40		−120	mA
I_{CC}	Supply current (driver and receiver combined)	V_{CC} = MAX, T_A = 25°C		42	60		42	60	mA

† All parameters with the exception of off-state open-collector output current are measured with the active pull-up connected to the sink output. For conditions shown as MIN or MAX, use the appropriate value specified under recommended operating conditions.
‡ All typical values are at V_{CC} = 5 V and T_A = 25°C.
§ Not more than one output should be shorted at a time, and duration of the short circuit should not exceed one second.

switching characteristics, V_{CC} = 5 V, C_L = 30 pF, T_A = 25°C

driver section

PARAMETER		TEST CONDITIONS	MIN	TYP	MAX	UNIT
t_{PLH}	Propagation delay time, low-to-high-level output	See Figure 13		14	30	ns
t_{PHL}	Propagation delay time, high-to-low-level output	See Figure 13		12	30	ns
t_{PZH}	Output enable time to high level	R_L = 180 Ω, See Figure 14		8	20	ns
t_{PZL}	Output enable time to low level	R_L = 250 Ω, See Figure 15		17	40	ns
t_{PHZ}	Output disable time from high level	R_L = 180 Ω, See Figure 14		16	30	ns
t_{PLZ}	Output disable time from low level	R_L = 250 Ω, See Figure 15		20	35	ns

electrical characteristics over recommended operating free-air temperature range (unless otherwise noted)

receiver section

	PARAMETER		TEST CONDITIONS†			'116, '118 MIN	'116, '118 TYP‡	'116, '118 MAX	'117, '119 MIN	'117, '119 TYP‡	'117, '119 MAX	UNIT
V_{TH}	Differential input high-threshold voltage§		V_O = 0.4 V, I_{OL} = 15 mA, See Note 3	V_{CC} = MIN, See Note 4				0.5			0.5	V
				V_{CC} = 5 V, See Note 5				1			1	
V_{TL}	Differential input low-threshold voltage§		V_O = 2.4 V, I_{OH} = −5 mA, See Note 3	V_{CC} = MIN, See Note 4		−0.5¶			−0.5¶			V
				V_{CC} = 5 V, See Note 5		−1¶			−1¶			
V_I	Input voltage range#		V_{CC} = 5 V, V_{ID} = −1 V or 1 V, See Note 3			15 to −15			6 to 0			V
V_{OH}	High-level output voltage		I_{OH} = −5 mA, See Note 3	V_{CC} = MIN, V_{ID} = −0.5 V, See Note 4		2.4			2.4			V
				V_{CC} = 5 V, V_{ID} = −1 V, See Note 5		2.4			2.4			
V_{OL}	Low-level output voltage		I_{OL} = 15 mA, See Note 3	V_{CC} = MIN, V_{ID} = 0.5 V, See Note 4				0.4			0.4	V
				V_{CC} = 5 V, V_{ID} = 1 V, See Note 5				0.4			0.4	
$I_{I(rec)}$	Receiver input current		V_{CC} = MAX, See Note 3	V_I = 0 V,	Other input at 0 V		−0.5	−0.9		−0.5	−1	mA
				V_I = 0.4 V,	Other input at 2.4 V		−0.4	−0.7		−0.4	−0.8	
				V_I = 2.4 V,	Other input at 0.4 V		0.1	0.3		0.1	0.4	
I_I	Input current at maximum input voltage	Strobe	V_{CC} = MIN, V_{ID} = −0.5 V, V_{strobe} = 4.5 V		'116, '117			5			5	μA
		Enable	V_{CC} = MAX, V_I = 5.5 V		'118, '119			1			1	mA
I_{IH}	High-level input current	Enable	V_{CC} = MAX, V_I = 2.4 V		'118, '119			40			40	μA
I_{IL}	Low-level input current	Strobe	V_{CC} = MAX, V_{ID} = 0.5 V, V_{strobe} = 0.4 V, See Note 4		'116, '117			−2.4			−2.4	mA
		Enable	V_{CC} = MAX, V_I = 0.4 V		'118, '119			−1.6			−1.6	
$I_{(RC)}$	Response-time-control current (Pin 9)		V_{CC} = MAX, V_{ID} = 0.5 V, RC at 0 V, See Note 4		T_A = 25°C	−1.2						mA
$I_{O(off)}$	Off-state open-collector output current		V_{CC} = MAX, V_O = 12 V, V_{ID} = −1 V	T_A = 25°C			1	10				μA
				T_A = MAX	SN55'			200				
					SN75'			20				
I_{OZ}	Off-state (high-impedance state) output current		V_{CC} = MAX, V_O = 0 to V_{CC}, RE at 0.4 V	T_A = 25°C	'118, '119			±10			±10	μA
				T_A = MAX	SN55118			±150				
					SN55119						±150	
					SN75118			±20				
					SN75119						±20	
R_T	Line-terminating resistance		V_{CC} = 5 V		T_A = 25°C	77		167				Ω
I_{OS}	Short-circuit output current‖		V_{CC} = MAX, V_O = 0, V_{ID} = −0.5 V, See Note 4		T_A = 25°C	−15		−80	−15		−80	mA
I_{CC}	Supply current (driver and receiver combined)		V_{CC} = MAX, V_{ID} = 0.5 V, See Note 4		T_A = 25°C		42	60		42	60	mA

† Unless otherwise noted V_{strobe} = 2.4 V. All parameters with the exception of off-state open-collector output current are measured with the active pull-up connected to the sink output. For conditions shown as MIN or MAX, use the appropriate value specified under recommended operating conditions.

‡ All typical values are at V_{CC} = 5 V, T_A = 25°C, and V_{IC} = 0.

§ Differential voltages are at the B input terminal with respect to the A input terminal. Neither receiver input of the '117 or '119 should be taken negative with respect to GND.

¶ The algebraic convention, where the less positive (more negative) limit is designated as minimum, is used in this data sheet for threshold voltages only.

Input voltage range is the voltage range that, if exceeded at either input, will cause the receiver to cease functioning properly.

‖ Not more than one output should be shorted at a time.

NOTES: 3. Measurement of these characteristics on the '117 and '119 requires the driver to be disabled with the driver enable at 0.8 V.

4. This applies with the less positive receiver input grounded.

5. For '116 and '118, this applies with the more positive receiver input at 15 V or the more negative receiver input at −15 V. For '117 and '119, this applies with the more positive receiver input at 6 V.

switching characteristics, V_{CC} = 5 V, C_L = 30 pF, T_A = 25°C

receiver section

PARAMETER			TEST CONDITIONS	MIN	TYP	MAX	UNIT
t_{PLH}	Propagation delay time, low-to-high-level output		R_L = 400 Ω, See Figure 16		20	75	ns
t_{PHL}	Propagation delay time, high-to-low-level output				17	75	ns
t_{PZH}	Output enable time to high level	'118 and '119 only	R_L = 480 Ω, See Figure 14		9	20	ns
t_{PZL}	Output enable time to low level		R_L = 250 Ω, See Figure 15		16	35	ns
t_{PHZ}	Output disable time from high level		R_L = 480 Ω, See Figure 14		12	30	ns
t_{PLZ}	Output disable time from low level		R_L = 250 Ω, See Figure 15		17	35	ns

TYPICAL CHARACTERISTICS

DRIVER OUTPUT VOLTAGE
vs
DRIVER INPUT VOLTAGE

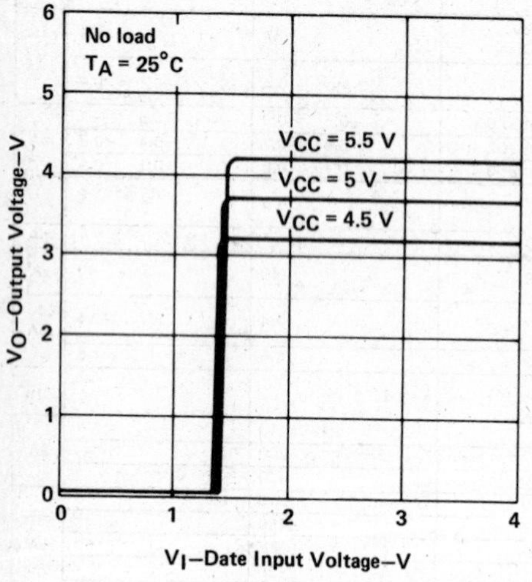
No load
T_A = 25°C
V_{CC} = 5.5 V
V_{CC} = 5 V
V_{CC} = 4.5 V
V_O–Output Voltage–V
V_I–Date Input Voltage–V
0
1
2
3
4
5
6

FIGURE 1

DRIVER OUTPUT VOLTAGE
vs
DRIVER INPUT VOLTAGE

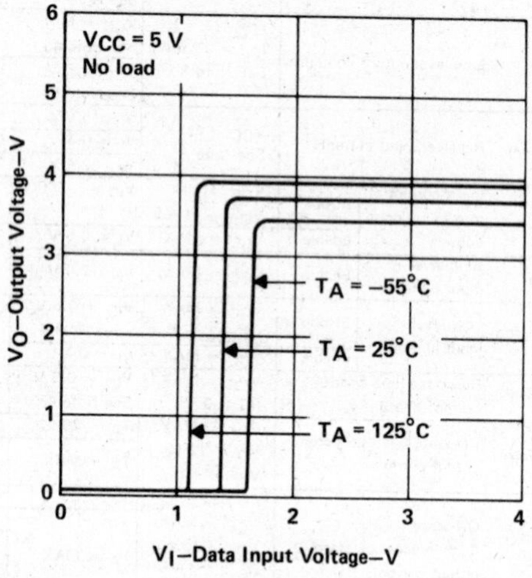
V_{CC} = 5 V
No load
T_A = –55°C
T_A = 25°C
T_A = 125°C
V_O–Output Voltage–V
V_I–Data Input Voltage–V
0
1
2
3
4
5
6

FIGURE 2

TYPICAL CHARACTERISTICS

DRIVER HIGH-LEVEL OUTPUT VOLTAGE
vs
OUTPUT CURRENT

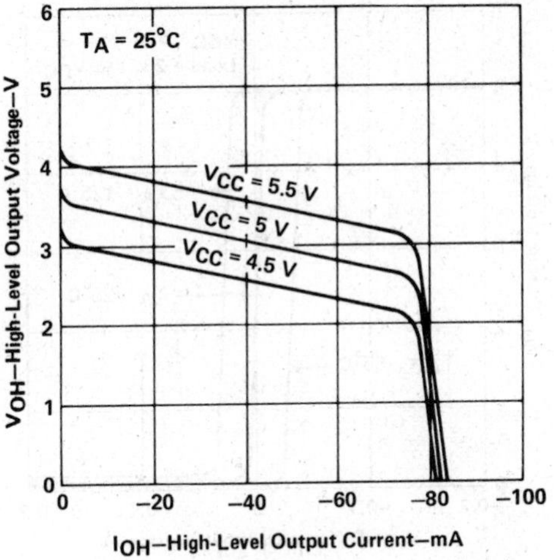
$T_A = 25°C$
$V_{CC} = 5.5$ V
$V_{CC} = 5$ V
$V_{CC} = 4.5$ V
V_{OH}—High-Level Output Voltage—V
I_{OH}—High-Level Output Current—mA

FIGURE 3

DRIVER LOW-LEVEL OUTPUT VOLTAGE
vs
OUTPUT CURRENT

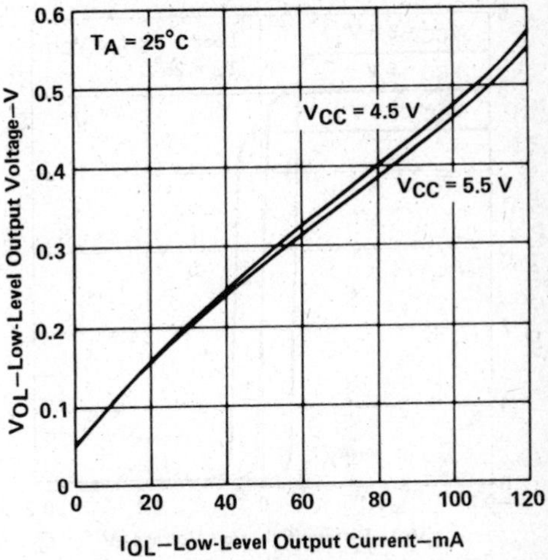
$T_A = 25°C$
$V_{CC} = 4.5$ V
$V_{CC} = 5.5$ V
V_{OL}—Low-Level Output Voltage—V
I_{OL}—Low-Level Output Current—mA

FIGURE 4

DRIVER PROPAGATION DELAY TIMES
vs
FREE-AIR TEMPERATURE†

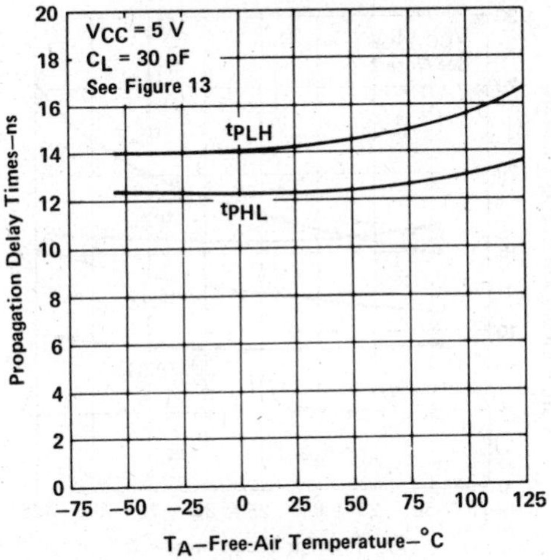
$V_{CC} = 5$ V
$C_L = 30$ pF
See Figure 13
t_{PLH}
t_{PHL}
Propagation Delay Times—ns
T_A—Free-Air Temperature—°C

FIGURE 5

DRIVER OUTPUT ENABLE AND DISABLE TIMES
vs
FREE-AIR TEMPERATURE†

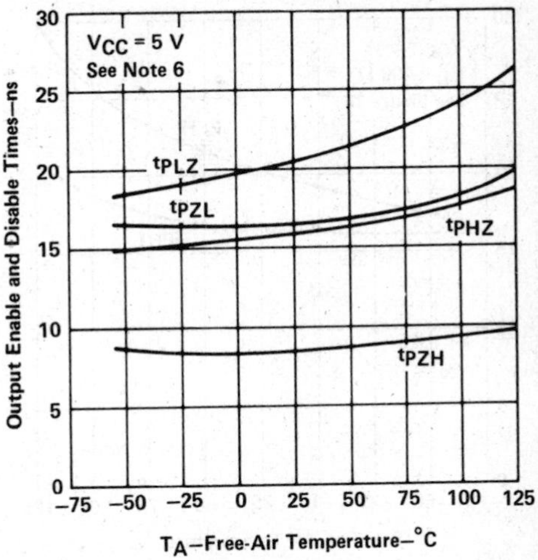
$V_{CC} = 5$ V
See Note 6
t_{PLZ}
t_{PZL}
t_{PHZ}
t_{PZH}
Output Enable and Disable Times—ns
T_A—Free-Air Temperature—°C

FIGURE 6

† Data for temperatures below 0°C and above 70°C are applicable to SN55116 through SN55119 devices only.

NOTE 6: For t_{PZH} and t_{PHZ}: $R_L = 180\ \Omega$, see Figure 14. For t_{PZL} and t_{PLZ}: $R_L = 250\ \Omega$, see Figure 15.

TYPICAL CHARACTERISTICS

RECEIVER OUTPUT VOLTAGE
vs
DIFFERENTIAL INPUT VOLTAGE

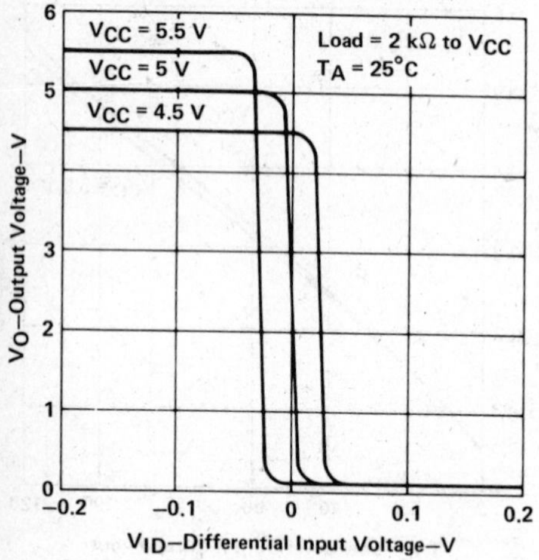
V_{CC} = 5.5 V
V_{CC} = 5 V
V_{CC} = 4.5 V
Load = 2 kΩ to V_{CC}
T_A = 25°C
V_O—Output Voltage—V
V_{ID}—Differential Input Voltage—V

FIGURE 7

RECEIVER OUTPUT VOLTAGE
vs
DIFFERENTIAL INPUT VOLTAGE†

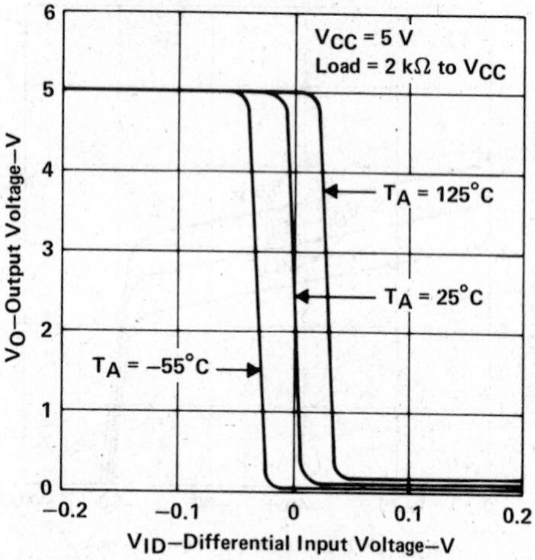
V_{CC} = 5 V
Load = 2 kΩ to V_{CC}
T_A = 125°C
T_A = 25°C
T_A = −55°C
V_O—Output Voltage—V
V_{ID}—Differential Input Voltage—V

FIGURE 8

RECEIVER PROPAGATION DELAY TIMES
vs
FREE-AIR TEMPERATURE†

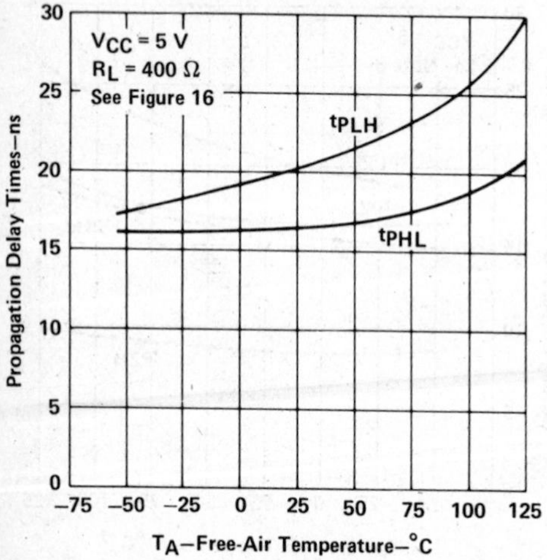
V_{CC} = 5 V
R_L = 400 Ω
See Figure 16
t_{PLH}
t_{PHL}
Propagation Delay Times—ns
T_A—Free-Air Temperature—°C

FIGURE 9

RECEIVER OUTPUT ENABLE AND DISABLE TIMES
vs
FREE-AIR TEMPERATURE†

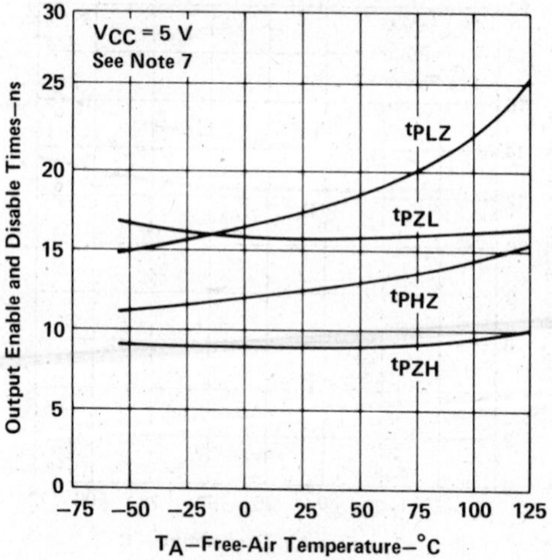
V_{CC} = 5 V
See Note 7
t_{PLZ}
t_{PZL}
t_{PHZ}
t_{PZH}
Output Enable and Disable Times—ns
T_A—Free-Air Temperature—°C

FIGURE 10

† Data for temperatures below 0°C and above 70°C are applicable to SN55116 through SN55119 devices only.

NOTE 7: For t_{PZH} and t_{PHZ}: R_L = 480 Ω, see Figure 14. For t_{PZL} and t_{PLZ}: R_L = 250 Ω, see Figure 15.

TYPICAL CHARACTERISTICS

SUPPLY CURRENT (DRIVER AND RECEIVER)
vs
SUPPLY VOLTAGE

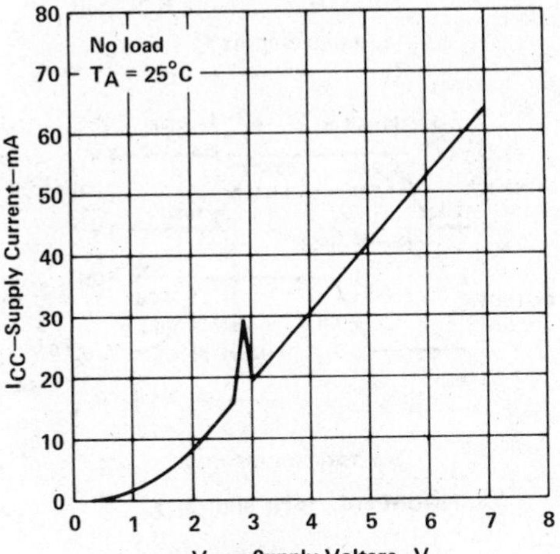
80
70
60
50
40
30
20
10
0
No load
T_A = 25°C
I_{CC}—Supply Current—mA
0
1
2
3
4
5
6
7
8
V_{CC}—Supply Voltage—V

FIGURE 11

SUPPLY CURRENT (DRIVER & RECEIVER)
vs
FREE-AIR TEMPERATURE†

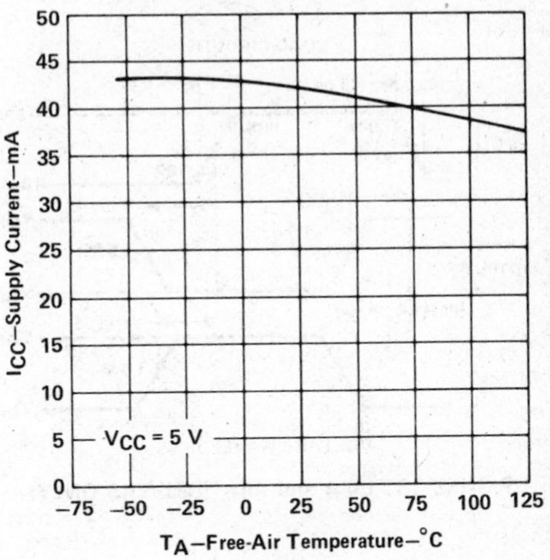
50
45
40
35
30
25
20
15
10
5
0
V_{CC} = 5 V
I_{CC}—Supply Current—mA
−75
−50
−25
0
25
50
75
100
125
T_A—Free-Air Temperature—°C

FIGURE 12

† Data for temperatures below 0°C and above 70°C are applicable to SN55116 through SN55119 devices only.

PARAMETER MEASUREMENT INFORMATION

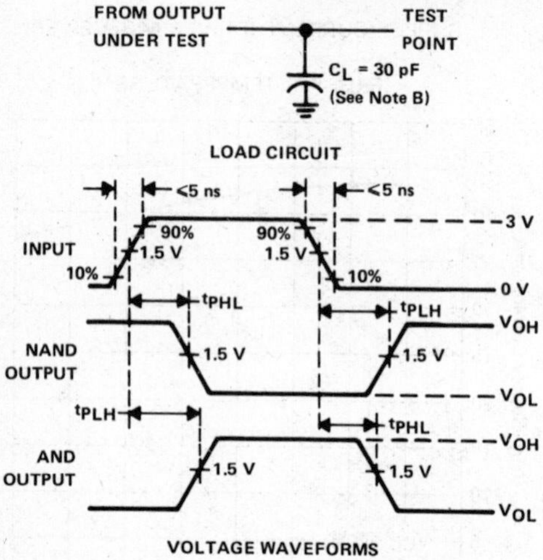
FROM OUTPUT UNDER TEST
TEST POINT
C_L = 30 pF (See Note B)
LOAD CIRCUIT
<5 ns
3 V
INPUT
90%
1.5 V
10%
0 V
t_{PHL}
t_{PLH}
V_{OH}
NAND OUTPUT
V_{OL}
AND OUTPUT
VOLTAGE WAVEFORMS

FIGURE 13. t_{PLH} and t_{PHL} (DRIVERS ONLY)

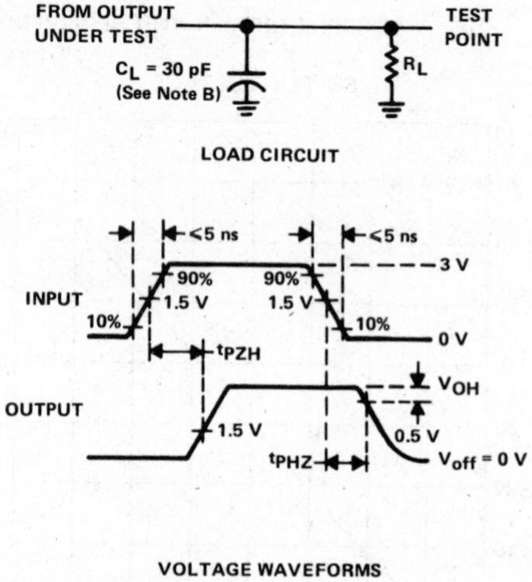
FROM OUTPUT UNDER TEST
TEST POINT
C_L = 30 pF (See Note B)
R_L
LOAD CIRCUIT
<5 ns
3 V
INPUT
90%
1.5 V
10%
0 V
t_{PZH}
V_{OH}
OUTPUT
0.5 V
t_{PHZ}
V_{off} = 0 V
VOLTAGE WAVEFORMS

FIGURE 14. t_{PZH} and t_{PHZ}

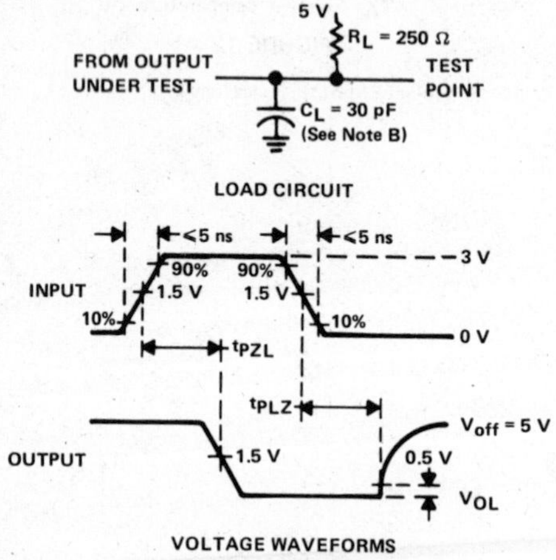
5 V
R_L = 250 Ω
FROM OUTPUT UNDER TEST
TEST POINT
C_L = 30 pF (See Note B)
LOAD CIRCUIT
<5 ns
3 V
INPUT
90%
1.5 V
10%
0 V
t_{PZL}
t_{PLZ}
V_{off} = 5 V
OUTPUT
0.5 V
V_{OL}
VOLTAGE WAVEFORMS

FIGURE 15. t_{PZL} and t_{PLZ}

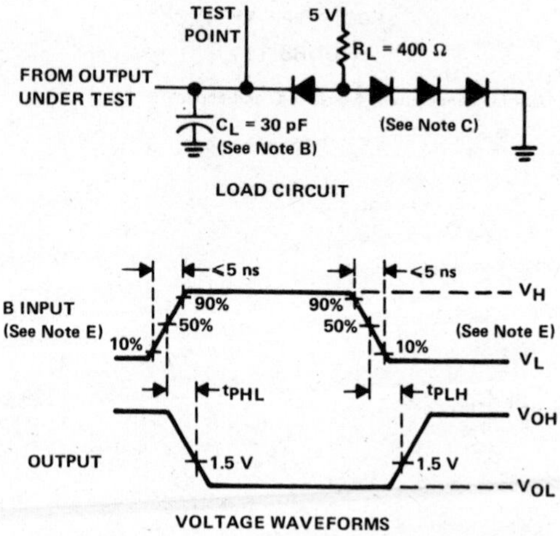
TEST POINT
5 V
R_L = 400 Ω
FROM OUTPUT UNDER TEST
C_L = 30 pF (See Note B)
(See Note C)
LOAD CIRCUIT
<5 ns
B INPUT (See Note E)
90%
50%
10%
V_H
(See Note E)
V_L
t_{PHL}
t_{PLH}
V_{OH}
OUTPUT
1.5 V
V_{OL}
VOLTAGE WAVEFORMS

FIGURE 16. t_{PLH} and t_{PHL} (RECEIVERS ONLY)

NOTES:
- A. Input pulses are supplied by generators having the following characteristics Z_{out} = 50 Ω, PRR ≤ 500 kHz, t_w = 100 ns.
- B. C_L includes probe and jig capacitance.
- C. All diodes are 1N3064 or equivalent.
- D. When testing the '116 and '118 receiver sections, the response-time control and the termination resistor pins are left open.
- E. For '116 and '118, V_H = 3 V, V_L = −3 V, the A input is at 0 V.
  For '117 and '119, V_H = 3 V, V_L = 0 V, the A input is at 1.5 V.

TEXAS INSTRUMENTS
POST OFFICE BOX 655012 • DALLAS, TEXAS 75265

D1334, SEPTEMBER 1973—REVISED SEPTEMBER 1986

- **Designed for Digital Data Transmission over 50-Ω to 500-Ω Coaxial Cable, Strip Line, or Twisted Pair**
- **High-Speed . . . t_{pd} = 20 ns Max at C_L = 15 pF**
- **TTL Compatible with Single 5-V Supply**
- **2.4-V Output at I_{OH} = −75 mA**
- **Uncommitted Emitter-Follower Output Structure for Party-Line Operation**
- **Short-Circuit Protection**
- **AND-OR Logic Configuration**
- **Designed for Use with Triple Line Receivers SN55122, SN75122**
- **Designed to be Interchangeable with Signetics N8T13**

SN55121 . . . J PACKAGE
SN75121 . . . D, J, OR N PACKAGE
(TOP VIEW)

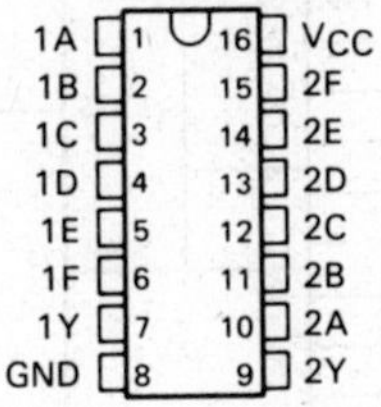
1A 1
1B 2
1C 3
1D 4
1E 5
1F 6
1Y 7
GND 8
16 VCC
15 2F
14 2E
13 2D
12 2C
11 2B
10 2A
9 2Y

SN55121 . . . FK PACKAGE
(TOP VIEW)

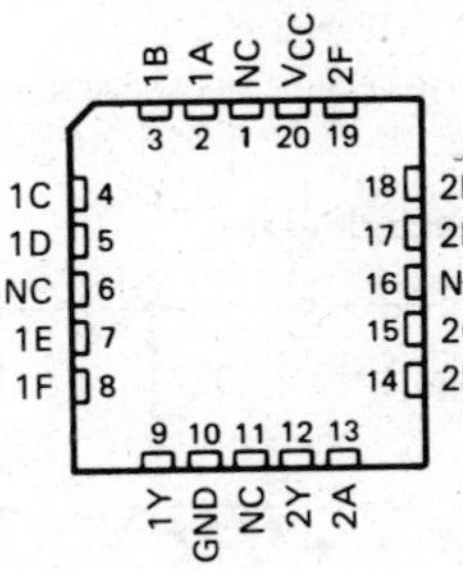
1B 3
1A 2
NC 1
VCC 20
2F 19
1C 4
1D 5
NC 6
1E 7
1F 8
18 2E
17 2D
16 NC
15 2C
14 2B
9 1Y
10 GND
11 NC
12 2Y
13 2A

NC—No internal connection

description

The SN55121 and SN75121 dual line drivers are designed for digital data transmission over lines having impedances from 50 to 500 Ω. They are also compatible with standard TTL logic and supply voltage levels.

The low-impedance emitter-follower outputs of the SN55121 and SN75121 will drive terminated lines such as coaxial cable or twisted pairs. Having the outputs uncommitted allows wired-OR logic to be performed in party-line applications. Output short-circuit protection is provided by an internal clamping network that turns on when the output voltage drops below approximately 1.5 volts. All of the inputs are in conventional TTL configuration and the gating can be used during power-up and power-down sequences to ensure that no noise is introduced to the line.

The SN55121 is characterized for operation over the full military temperature range of −55°C to 125°C. The SN75121 is characterized for operation from 0°C to 70°C.

FUNCTION TABLE

INPUTS						OUTPUT
A	B	C	D	E	F	Y
H	H	H	H	X	X	H
X	X	X	X	H	H	H
All other input combinations						L

H = high level
L = low level
X = irrelevant

logic symbol†

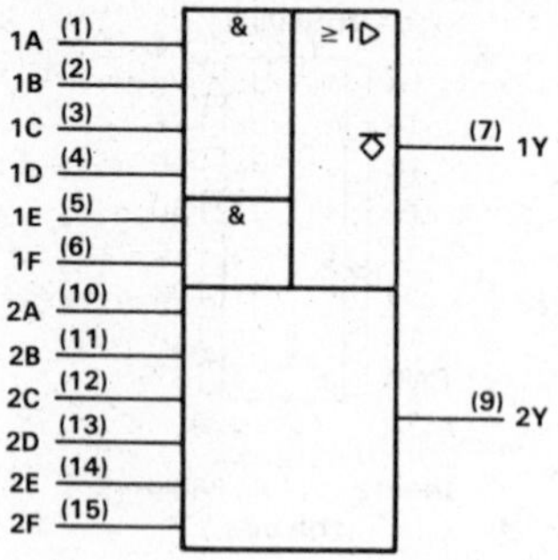
1A (1)
1B (2)
1C (3)
1D (4)
1E (5)
1F (6)
2A (10)
2B (11)
2C (12)
2D (13)
2E (14)
2F (15)
&
&
≥1
(7) 1Y
(9) 2Y

logic diagram (positive logic)

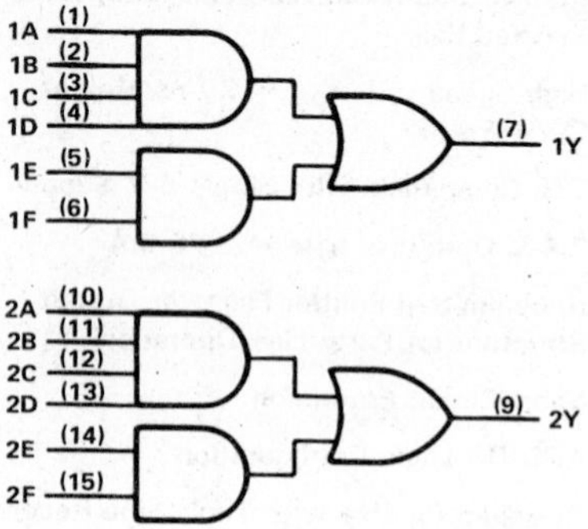
1A (1)
1B (2)
1C (3)
1D (4)
1E (5)
1F (6)
(7) 1Y
2A (10)
2B (11)
2C (12)
2D (13)
2E (14)
2F (15)
(9) 2Y

†This symbol is in accordance with ANSI/IEEE Std 91-1984 and IEC Publication 617-12.
Pin numbers shown are for J and N packages.

schematic (each driver)

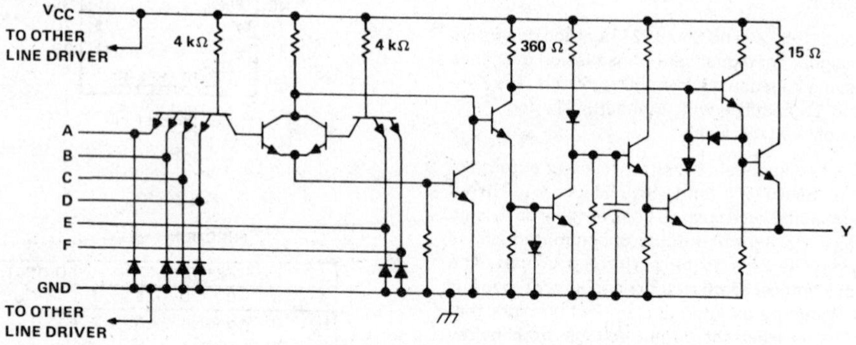
VCC
TO OTHER LINE DRIVER
4 kΩ
4 kΩ
360 Ω
15 Ω
A
B
C
D
E
F
Y
GND
TO OTHER LINE DRIVER

absolute maximum ratings over operating free-air temperature range (unless otherwise noted)

		SN55121	SN75121	UNIT
Supply voltage, V_{CC} (see Note 1)		6	6	V
Input voltage		6	6	V
Output voltage		6	6	V
Continuous total dissipation at (or below) 25 °C free air temperature (see Note 2)	D package		950	mW
	FK or J package	1375		mW
	J package		1025	mW
	N package		1150	mW
Operating free-air temperature range		−55 to 125	0 to 70	°C
Storage temperature range		−65 to 150	−65 to 150	°C
Case temperature for 60 seconds FK package		260		°C
Lead temperature 1,6 mm (1/16 inch) from case for 60 seconds: J package		300	300	°C
Lead temperature 1,6 mm (1/16 inch) from case for 10 seconds: D or N package			260	°C

NOTES: 1. All voltage values are with respect to both ground terminals connected together.
2. For operation above 25 °C free-air temperature, refer to Dissipation Derating Curves in Appendix A. In the J package, SN55121 chips are alloy mounted and SN75121 chips are glass mounted. In the N package, use the 9.2 mW/°C curve for these devices.

recommended operating conditions

	SN55121			SN75121			UNIT
	MIN	NOM	MAX	MIN	NOM	MAX	
Supply voltage, V_{CC}	4.75	5	5.25	4.75	5	5.25	V
High-level input voltage, V_{IH}	2			2			V
Low-level input voltage, V_{IL}			0.8			0.8	V
High-level output current, I_{OH}			−75			−75	mA
Operating free-air temperature, T_A	−55		125	0		70	°C

electrical characteristics over recommended ranges of supply voltage and operating free-air temperature (unless otherwise noted)

PARAMETER		TEST CONDITIONS	MIN	MAX	UNIT
V_{IK}	Input clamp voltage	V_{CC} = 5 V, I_I = −12 mA		−1.5	V
$V_{(BR)I}$	Input breakdown voltage	V_{CC} = 5 V, I_I = 10 mA	5.5		V
V_{OH}	High-level output voltage	V_{IH} = 2 V, I_{OH} = −75 mA, See Note 3	2.4		V
I_{OH}	High-level output current	V_{CC} = 5 V, V_{IH} = 4.5 V, V_{OH} = 2 V, T_A = 25°C, See Note 3	−100	−250	mA
I_{OL}	Low-level output current	V_{IL} = 0.8 V, V_{OL} = 0.4 V, See Note 3		−800	μA
$I_{O(off)}$	Off-state output current	V_{CC} = 3 V, V_O = 3 V		500	μA
I_{IH}	High-level input current	V_I = 4.5 V		40	μA
I_{IL}	Low-level input current	V_I = 0.4 V	−0.1	−1.6	mA
I_{OS}	Short-circuit output current†	V_{CC} = 5 V, T_A = 25°C		−30	mA
I_{CCH}	Supply current, outputs high	V_{CC} = 5.25 V, All inputs at 2 V, Outputs open		28	mA
I_{CCL}	Supply current, outputs low	V_{CC} = 5.25 V, All inputs at 0.8 V, Outputs open		60	mA

†Not more than one output should be shorted at a time.

NOTE 3. The output voltage and current limits are guaranteed for any appropriate combination of high and low inputs specified by the function table for the desired output.

switching characteristics, V_{CC} = 5 V, T_A = 25°C

PARAMETER		TEST CONDITIONS	MIN	TYP	MAX	UNIT
t_{PLH}	Propagation delay time, low-to-high-level output	R_L = 37 Ω, C_L = 15 pF, See Figure 1		11	20	ns
t_{PHL}	Propagation delay time, high-to-low-level output			8	20	
t_{PLH}	Propagation delay time, low-to-high-level output	R_L = 37 Ω, C_L = 1000 pF, See Figure 1		22	50	ns
t_{PHL}	Propagation delay time, high-to-low-level output			20	50	

PARAMETER MEASUREMENT INFORMATION

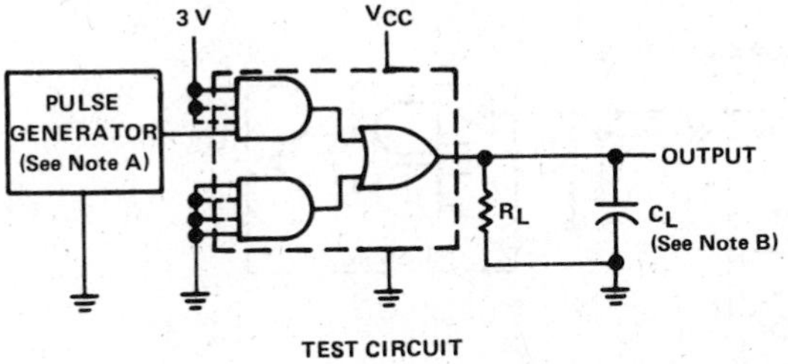
3 V
VCC
PULSE GENERATOR (See Note A)
OUTPUT
RL
CL (See Note B)
TEST CIRCUIT

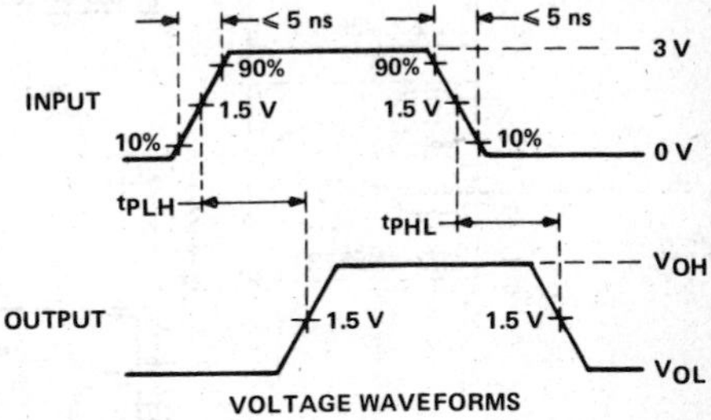
≤ 5 ns
≤ 5 ns
3 V
90%
90%
INPUT
1.5 V
1.5 V
10%
10%
0 V
tPLH
tPHL
VOH
OUTPUT
1.5 V
1.5 V
VOL
VOLTAGE WAVEFORMS

FIGURE 1. SWITCHING TIMES

NOTES: A. The pulse generators have the following characteristics: $Z_{out} \approx 50\ \Omega$, t_w = 200 ns, duty cycle ≤ 50%.
B. C_L includes probe and jig capacitance.

TYPICAL CHARACTERISTICS

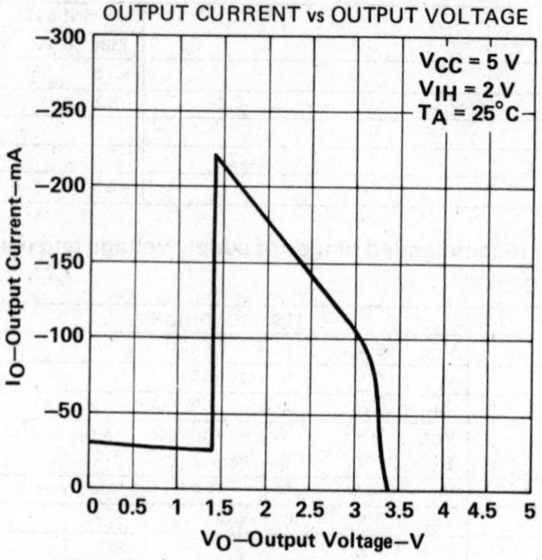
OUTPUT CURRENT vs OUTPUT VOLTAGE
V_{CC} = 5 V
V_{IH} = 2 V
T_A = 25°C
I_O—Output Current—mA
−300
−250
−200
−150
−100
−50
0
0 0.5 1 1.5 2 2.5 3 3.5 4 4.5 5
V_O—Output Voltage—V

FIGURE 2

TYPICAL APPLICATION DATA

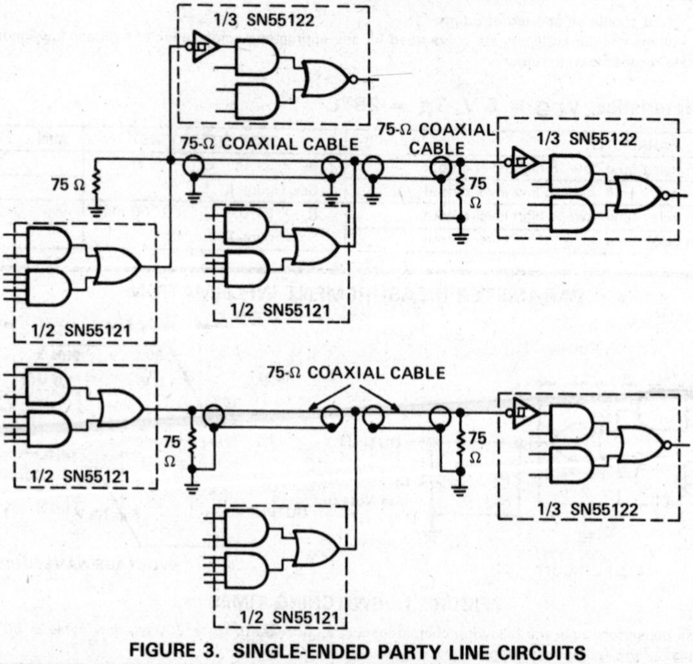
1/3 SN55122
75-Ω COAXIAL CABLE
75-Ω COAXIAL CABLE
75 Ω
75 Ω
1/3 SN55122
1/2 SN55121
1/2 SN55121
1/2 SN55121
75-Ω COAXIAL CABLE
75 Ω
75 Ω
1/3 SN55122
1/2 SN55121

FIGURE 3. SINGLE-ENDED PARTY LINE CIRCUITS

TEXAS INSTRUMENTS
POST OFFICE BOX 655012 • DALLAS, TEXAS 75265

D1334, SEPTEMBER 1973—REVISED SEPTEMBER 1986

- Designed for Digital Data Transmission Over Coaxial Cable, Strip Line, or Twisted Pair
- Designed for Operation with 50-Ω to 500-Ω Transmission Lines
- TTL Compatible
- Single 5-V Supply
- Built-In Input Threshold Hysteresis
- High Speed . . . Typical Propagation Delay Time = 20 ns
- Independent Channel Strobes
- Input Gating Increases Application Flexibility
- Fanout to 10 Series 54/74 Standard Loads
- Can be Used with Dual Line-Drivers SN55121 and SN75121
- Interchangeable with Signetics N8T14

SN55122 . . . J PACKAGE
SN75122 . . . D, J, OR N PACKAGE
(TOP VIEW)

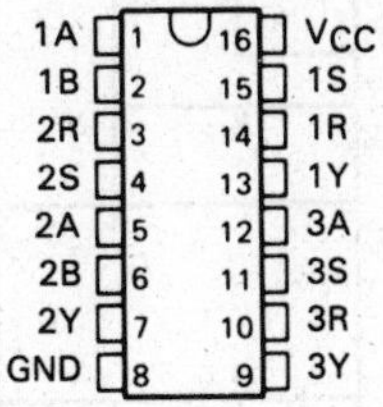
1A 1
1B 2
2R 3
2S 4
2A 5
2B 6
2Y 7
GND 8
16 VCC
15 1S
14 1R
13 1Y
12 3A
11 3S
10 3R
9 3Y

SN55122 . . . FK PACKAGE
(TOP VIEW)

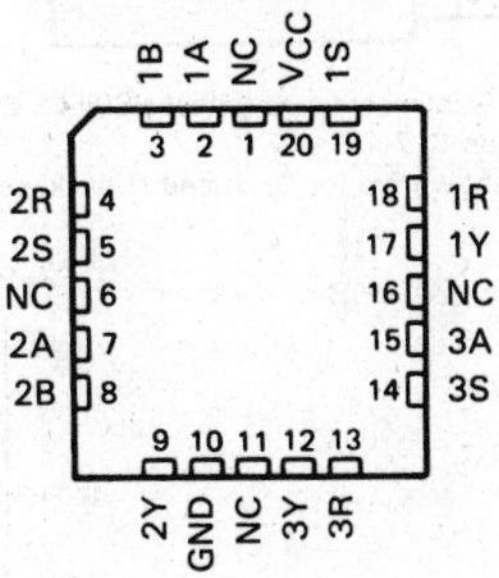
1B
1A
NC
VCC
1S
3 2 1 20 19
2R 4
2S 5
NC 6
2A 7
2B 8
18 1R
17 1Y
16 NC
15 3A
14 3S
9 10 11 12 13
2Y
GND
NC
3Y
3R

NC—No internal connection

description

The SN55122 and SN75122 are triple line-receivers that are designed for digital data transmission over lines having impedances from 50 to 500 ohms. They are also compatible with standard TTL logic and supply voltage levels.

The SN55122 and SN75122 have receiver inputs with built-in hysteresis to provide increased noise margin for single-ended systems. The high impedance of this input presents a minimum load to the driver and allows termination of the transmission line in its characteristic impedance to minimize line reflection. An open line will affect the receiver input as would a low-level voltage. The receiver can withstand a level of −0.15 volt with power on or off. The other inputs are in TTL configuration. The S input must be high to enable the receiver input. Two of the line receivers have A and B inputs that, if both are high, will hold the output low. The third receiver has only an A input that, if high, will hold the output low.

The SN55122 is characterized for operation over the full military temperature range of −55°C to 125°C. The SN75122 is characterized for operation from 0°C to 70°C.

logic symbol†

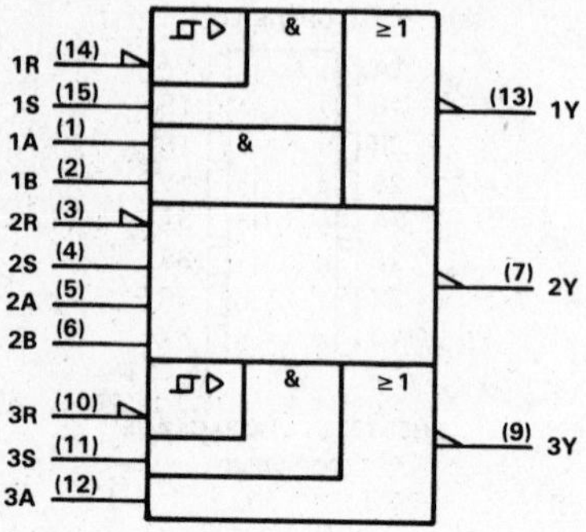
1R
1S
1A
1B
2R
2S
2A
2B
3R
3S
3A
(14)
(15)
(1)
(2)
(3)
(4)
(5)
(6)
(10)
(11)
(12)
&
&
≥1
&
≥1
(13)
1Y
(7)
2Y
(9)
3Y

†This symbol is in accordance with ANSI/IEEE Std 91-1984 and IEC Publication 617-12.

Pin numbers shown are for D, J, and N packages.

logic diagram

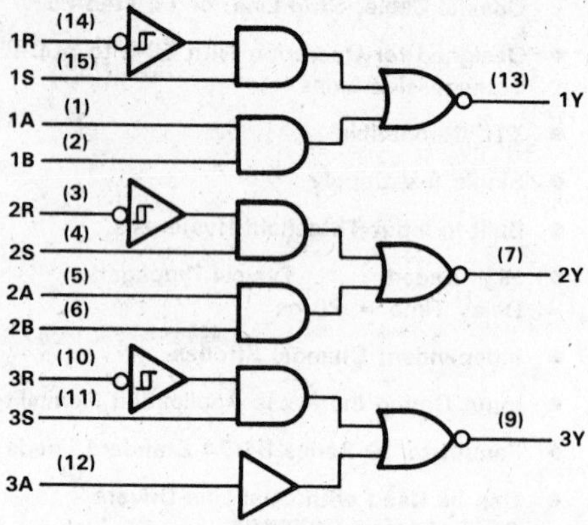
1R
1S
1A
1B
2R
2S
2A
2B
3R
3S
3A
(14)
(15)
(1)
(2)
(3)
(4)
(5)
(6)
(10)
(11)
(12)
(13)
1Y
(7)
2Y
(9)
3Y

FUNCTION TABLE

INPUTS				OUTPUT
A	B‡	R	S	Y
H	H	X	X	L
X	X	L	H	L
L	X	H	X	H
L	X	X	L	H
X	L	H	X	H
X	L	X	L	H

‡ B input and last two lines of the function table are applicable to receivers 1 and 2 only.

H = high level
L = low level
X = irrelevant

TEXAS INSTRUMENTS
POST OFFICE BOX 655012 • DALLAS, TEXAS 75265

schematic diagram (each receiver)

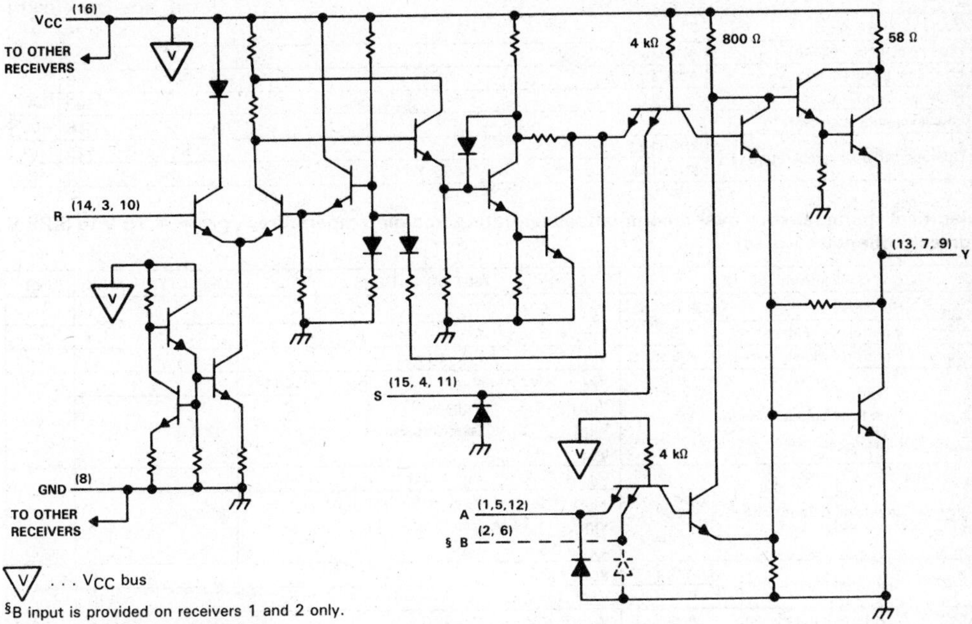
(16)
VCC
TO OTHER RECEIVERS
(14, 3, 10)
R
4 kΩ
800 Ω
58 Ω
(13, 7, 9)
Y
(15, 4, 11)
S
4 kΩ
(1,5,12)
A
(2, 6)
§ B
(8)
GND
TO OTHER RECEIVERS
V . . . VCC bus

§B input is provided on receivers 1 and 2 only.
Resistor values shown are nominal.

absolute maximum ratings over operating free-air temperature range (unless otherwise noted)

Supply voltage, V_{CC} (see Note 1)	6 V
Input voltage: R input	6 V
A, B, or S input	5.5 V
Output voltage	6 V
Output current	±100 mA
Continuous total power dissipation at (or below) 25 °C case temperature (see Note 2):	
D package	950 mW
J or FK package	1375 mW
N package	1150 mW
Operating free-air temperature range: SN55122	−55 °C to 125 °C
SN75122	0 °C to 70 °C
Storage temperature range	−65 °C to 150 °C
Case temperature for 60 seconds: FK package	260 °C
Lead temperature 1,6 mm (1/16 inch) from case for 60 seconds: J package	300 °C
Lead temperature 1,6 mm (1/16 inch) from case for 10 seconds: D or N package	260 °C

NOTES: 1. Voltage values are with respect to network ground terminal.
2. For operation above 25 °C free-air temperature, refer to the Dissipation Derating Curves in Appendix A. In the FK and J package, SN55122 chips are alloy mounted and in the J package, SN75122 chips are glass mounted. For derating the N package, use the 9.2-mW/°C curve and for the D package, use the 7.6-mW/°C curve.

recommended operating conditions

		MIN	NOM	MAX	UNIT
Supply voltage, V_{CC}		4.75	5	5.25	V
High-level input voltage, V_{IH}	A, B, R, or S	2			V
Low-level input voltage, V_{IL}	A, B, R, or S			0.8	V
High-level output current, I_{OH}				−500	μA
Low-level output current, I_{OL}				16	mA
Operating free-air temperature, T_A	SN55122	−55		125	°C
	SN75122	0		70	°C

electrical characteristics over recommended operating free-air temperature, V_{CC} = 4.75 V to 5.25 V (unless otherwise noted)

PARAMETER			TEST CONDITIONS	MIN	TYP†	MAX	UNIT
V_{hys}‡	Hysteresis	R	V_{CC} = 5 V, T_A = 25°C	0.3	0.6		V
V_{IK}	Input clamp voltage	A,B, or S	V_{CC} = 5 V, I_I = −12 mA			−1.5	V
$V_{I(BR)}$	Input breakdown voltage	A,B, or S	V_{CC} = 5 V, I_I = 10 mA	5.5			V
V_{OH}	High-level output voltage		V_{IH} = 2 V, V_{IL} = 0.8 V, I_{OH} = −500 μA	2.6			V
			$V_{I(A)}$ = 0, $V_{I(B)}$ = 0, $V_{I(S)}$ = 2 V, $V_{I(R)}$ = 1.45 V (see Note 3), I_{OH} = −500 μA	2.6			
V_{OL}	Low-level output voltage		V_{IH} = 2 V, V_{IL} = 0.8 V, I_{OL} = 16 mA			0.4	V
			$V_{I(A)}$ = 0, $V_{I(B)}$ = 0, $V_{I(S)}$ = 2 V, $V_{I(R)}$ = 1.45 V (see Note 4), I_{OL} = 16 mA			0.4	
I_{IH}	High-level input current	A,B, or S	V_I = 4.5 V			40	μA
		R	V_I = 3.8 V			170	
I_{IL}	Low-level input current	A,B, or S	V_I = 0.4 V, V_{IR} = 0.8 V	−0.1		−1.6	mA
I_{OS}§	Short-circuit output current		V_{CC} = 5 V, T_A = 25°C	−50		−100	mA
I_{CCH}	High-level supply current		V_{CC} = 5.25 V, All inputs at 0.8 V, Outputs open			72	mA
I_{CCL}	Low-level supply current		V_{CC} = 5.25 V, All inputs at 2 V, Outputs open			100	mA

† All typical values are at V_{CC} = 5 V and T_A = 25°C.
‡ Hysteresis is the difference between the positive-going input threshold voltage, V_{T+}, and the negative-going input threshold voltage, V_{T-}. See Figure 4.
§ Not more than one output should be shorted at a time and duration of the short circuit should not exceed one second.
NOTES: 3. The receiver input was high immediately before being reduced to 1.45 V.
4. The receiver input was low immediately before being increased to 1.45 V.

switching characteristics, V_{CC} = 5 V, T_A = 25°C

PARAMETER		TEST CONDITIONS	MIN	TYP	MAX	UNIT
t_{PLH}	Propagation delay time, low-to-high-level output from R input	See Figure 1		20	30	ns
t_{PHL}	Propagation delay time, high-to-low-level output from R input	See Figure 1		20	30	ns

TEXAS INSTRUMENTS
POST OFFICE BOX 655012 • DALLAS, TEXAS 75265

PARAMETER MEASUREMENT INFORMATION

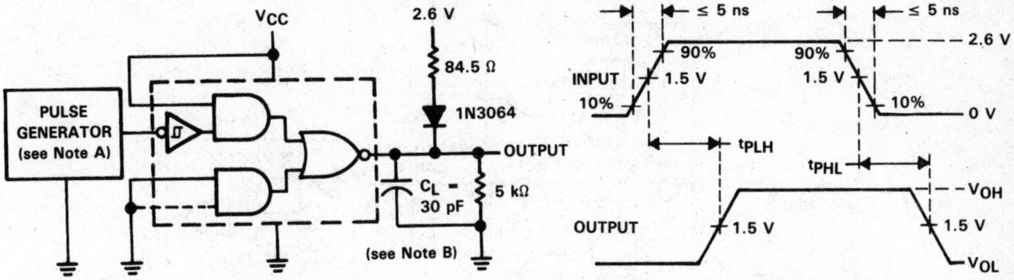
VCC
2.6 V
84.5 Ω
PULSE GENERATOR (see Note A)
1N3064
OUTPUT
CL = 30 pF
5 kΩ
(see Note B)
≤ 5 ns
≤ 5 ns
2.6 V
90%
90%
INPUT
1.5 V
1.5 V
10%
10%
0 V
tPLH
tPHL
VOH
OUTPUT
1.5 V
1.5 V
VOL

TEST CIRCUIT

VOLTAGE WAVEFORMS

NOTES: A. The pulse generator has the following characteristics: $Z_{out} \approx 50\ \Omega$, $t_w = 200$ ns, duty cycle = 50%.
B. C_L includes probe and jig capacitance.

FIGURE 1. SWITCHING TIMES

TYPICAL CHARACTERISTICS

OUTPUT VOLTAGE
vs
INPUT VOLTAGE

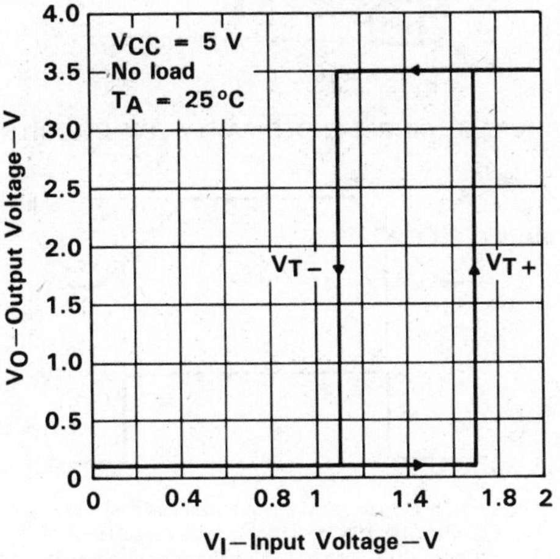
VCC = 5 V
No load
TA = 25°C
VO—Output Voltage—V
VT−
VT+
VI—Input Voltage—V
4.0
3.5
3.0
2.5
2.0
1.5
1.0
0.5
0
0
0.4
0.8
1
1.4
1.8
2

FIGURE 2

TYPICAL APPLICATION DATA

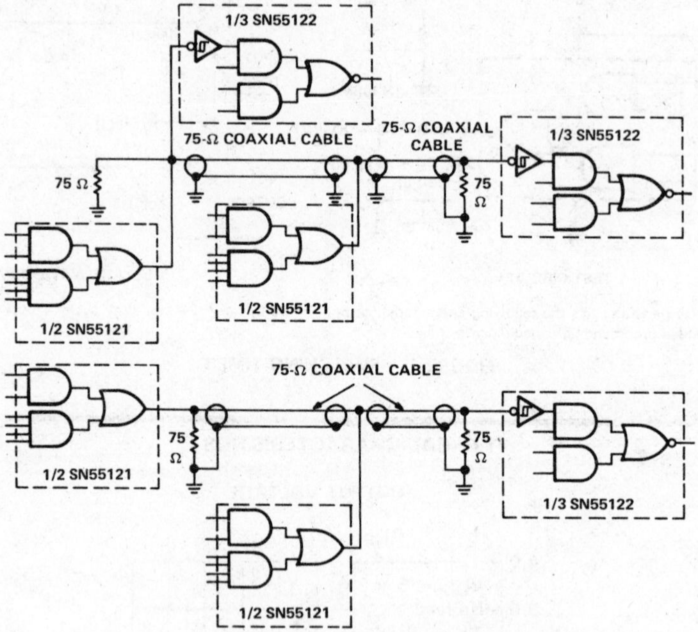
1/3 SN55122
75-Ω COAXIAL CABLE
75-Ω COAXIAL CABLE
1/3 SN55122
75 Ω
75 Ω
1/2 SN55121
1/2 SN55121
75-Ω COAXIAL CABLE
75 Ω
75 Ω
1/2 SN55121
1/3 SN55122
1/2 SN55121

FIGURE 3. SINGLE-ENDED PARTY LINE CIRCUITS

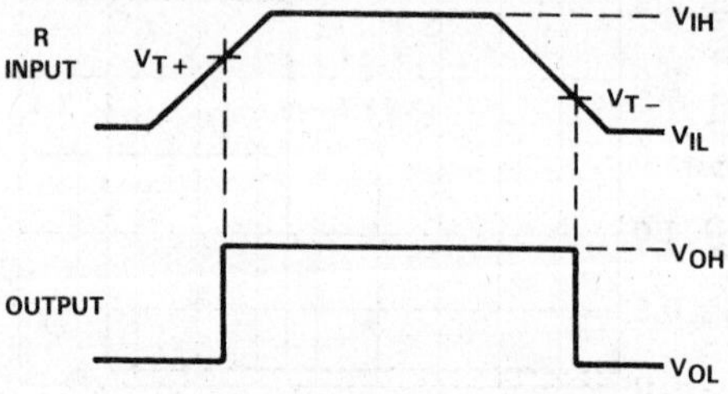
R INPUT
VT+
VIH
VT−
VIL
OUTPUT
VOH
VOL

The high gain and built-in hysteresis of the SN55122 and SN75122 line receivers enable them to be used as Schmitt triggers in squaring pulses.

FIGURE 4. PULSE SQUARING

TEXAS INSTRUMENTS
POST OFFICE BOX 655012 • DALLAS, TEXAS 75265

D1322, SEPTEMBER 1973—REVISED SEPTEMBER 1986

- Meets IBM System 360 Input/Output Interface Specifications
- Operates from Single 5-V Supply
- TTL Compatible
- 3.11 V Output at I_{OH} = −59.3 mA
- Uncommitted Emitter-Follower Output Structure for Party-Line Operation
- Short-Circuit Protection
- AND-OR Logic Configuration
- Designed for Use with Triple Line Receiver SN75124
- Designed to be Interchangeable with Signetics N8T23

D, J, OR N PACKAGE
(TOP VIEW)

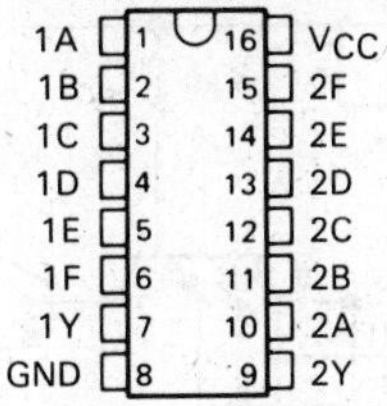
1A 1 16 VCC
1B 2 15 2F
1C 3 14 2E
1D 4 13 2D
1E 5 12 2C
1F 6 11 2B
1Y 7 10 2A
GND 8 9 2Y

FUNCTION TABLE

INPUTS						OUTPUT
A	B	C	D	E	F	Y
H	H	H	H	X	X	H
X	X	X	X	H	H	H
All other input combinations						L

H = high level
L = low level
X = irrelevant

description

The SN75123 dual line driver is specifically designed to meet the input/output interface specifications for IBM System 360. It is also compatible with standard TTL logic and supply voltage levels.

The low-impedance emitter-follower outputs of the SN75123 will drive terminated lines such as coaxial cable or twisted pair. Having the outputs uncommitted allows wired-OR logic to be performed in party-line applications. Output short-circuit protection is provided by an internal clamping network that turns on when the output voltage drops below approximately 1.5 volts. All the inputs are in conventional TTL configuration and the gating can be used during power-up and power-down sequences to ensure that no noise is introduced to the line.

The SN75123 is characterized for operation from 0°C to 70°C.

logic symbol†

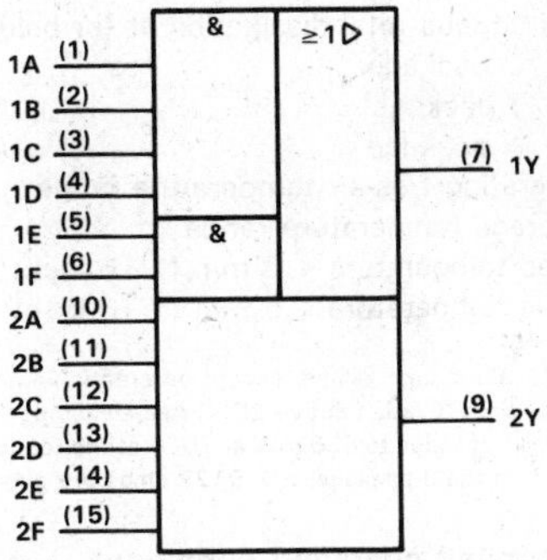
&
≥1
&
1A (1)
1B (2)
1C (3)
1D (4)
1E (5)
1F (6)
2A (10)
2B (11)
2C (12)
2D (13)
2E (14)
2F (15)
(7) 1Y
(9) 2Y

†This symbol is in accordance with ANSI/IEEE Std 91-1984 and IEC Publication 617-12.

logic diagram, each driver (positive logic)

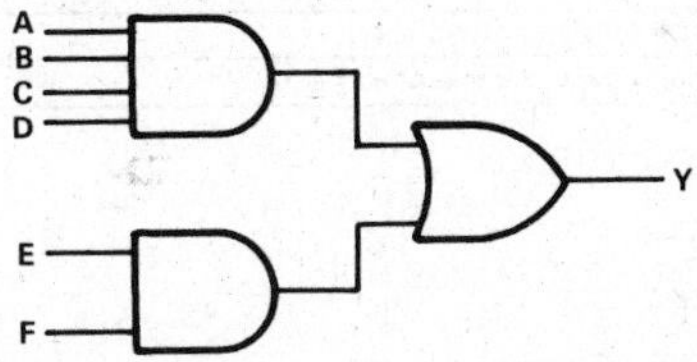
A
B
C
D
E
F
Y

PRODUCTION DATA documents contain information current as of publication date. Products conform to specifications per the terms of Texas Instruments standard warranty. Production processing does not necessarily include testing of all parameters.

TEXAS INSTRUMENTS
POST OFFICE BOX 655012 • DALLAS, TEXAS 75265

schematic (each driver)

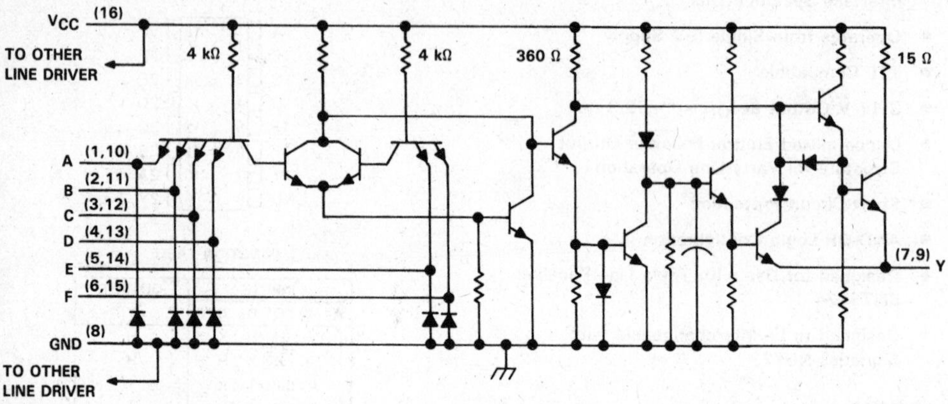
VCC
(16)
TO OTHER
LINE DRIVER
4 kΩ
4 kΩ
360 Ω
15 Ω
A (1,10)
B (2,11)
C (3,12)
D (4,13)
E (5,14)
F (6,15)
GND (8)
TO OTHER
LINE DRIVER
(7,9)
Y

Resistor values shown are nominal.

absolute maximum ratings over operating free-air temperature range (unless otherwise noted)

Supply voltage, V_{CC} (see Note 1) . . . . 7 V
Input voltage . . . . 5.5 V
Output voltage . . . . 7 V
Continuous total dissipation at (or below) 25°C free-air temperature (see Note 2):
  D package . . . . 950 mW
  J package . . . . 1025 mW
  N package . . . . 1150 mW
Operating free-air temperature range . . . . 0°C to 70°C
Storage temperature range . . . . −65°C to 150°C
Lead temperature 1,6 mm (1/16 inch) from case for 10 seconds: D or N package . . . . 260°C
Lead temperature 1,6 mm (1/16 inch) from case for 60 seconds: J package . . . . 300°C

NOTES: 1. All voltage values, except differential input voltage, are with respect to network ground terminal.
2. For operation above 25°C free-air temperature, derate the D package to 608 mW at 70°C at the rate of 7.6 mW/°C, the J package to 656 mW at 70°C at the rate of 8.2 mW/°C, and the N package to 736 mW at 70°C at the rate of 9.2 mW/°C. In the J package, SN75123 chips are glass mounted.

recommended operating conditions

	MIN	NOM	MAX	UNIT
Supply voltage, V_{CC}	4.75	5	5.25	V
High-level input voltage, V_{IH}	2			V
Low-level input voltage, V_{IL}			0.8	V
High-level output current, I_{OH}			−100	mA
Operating free-air tempeature, T_A	0		70	°C

electrical characteristics, V_{CC} = 4.75 V to 5.25 V, T_A = 0°C to 70°C (unless otherwise noted)

PARAMETER		TEST CONDITIONS	MIN	MAX	UNIT
V_{IK}	Input clamp voltage	V_{CC} = 5 V, I_I = −12 mA		−1.5	V
$V_{(BR)I}$	Input breakdown voltage	V_{CC} = 5 V, I_I = 10 mA	5.5		V
V_{OH}	High-level output voltage	V_{CC} = 5 V, V_{IH} = 2 V, I_{OH} = −59.3 mA, See Note 3, T_A = 25°C	3.11		V
		V_{CC} = 5 V, V_{IH} = 2 V, I_{OH} = −59.3 mA, See Note 3, T_A = 0°C to 70°C	2.9		
I_{OH}	High-level output current	V_{CC} = 5 V, V_{IH} = 4.5 V, V_{OH} = 2 V, T_A = 25°C, See Note 3	−100	−250	mA
V_{OL}	Low-level output voltage	V_{IL} = 0.8 V, I_{OL} = −240 μA, See Note 3		0.15	V
$I_{O(off)}$	Off-state output current	V_{CC} = 0, V_O = 3 V		40	μA
I_{IH}	High-level input current	V_I = 4.5 V		40	μA
I_{IL}	Low-level input current	V_I = 0.4 V	−0.1	−1.6	mA
I_{OS}	Short-circuit output current†	V_{CC} = 5 V, T_A = 25°C		−30	mA
I_{CCH}	Supply current, outputs high	V_{CC} = 5.25 V, All inputs at 2 V, Outputs open		28	mA
I_{CCL}	Supply current, outputs low	V_{CC} = 5.25 V, All inputs at 0.8 V, Outputs open		60	mA

†Not more than one output should be shorted at a time.

NOTE 3: The output voltage and current limits are guaranteed for any appropriate combination of high and low inputs specified by the function table for the desired output.

switching characteristics, V_{CC} = 5 V, T_A = 25°C

PARAMETER		TEST CONDITIONS	MIN	TYP	MAX	UNIT
t_{PLH}	Propagation delay time, low-to-high-level output	R_L = 50 Ω, C_L = 15 pF, See Figure 1		12	20	ns
t_{PHL}	Propagation delay time, high-to-low-level output			12	20	
t_{PLH}	Propagation delay time, low-to-high-level output	R_L = 50 Ω, C_L = 100 pF, See Figure 1		20	35	ns
t_{PHL}	Propagation delay time, high-to-low-level output			15	25	

PARAMETER MEASUREMENT INFORMATION

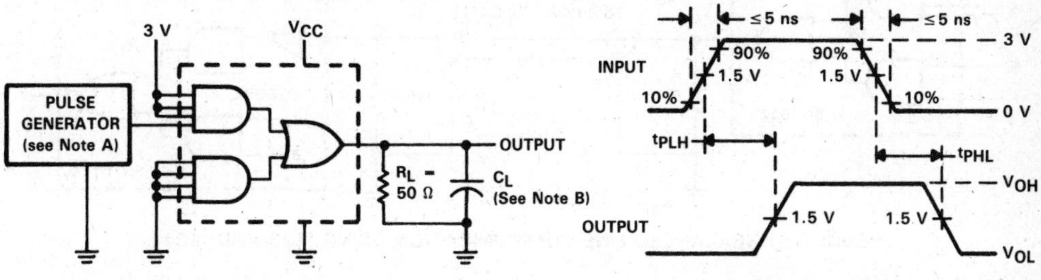
3 V
VCC
PULSE GENERATOR (see Note A)
OUTPUT
RL = 50 Ω
CL (See Note B)
≤5 ns
≤5 ns
3 V
INPUT
90%
90%
1.5 V
1.5 V
10%
10%
0 V
tPLH
tPHL
VOH
OUTPUT
1.5 V
1.5 V
VOL

TEST CIRCUIT　　　VOLTAGE WAVEFORMS

NOTES: A. The pulse generator has the following characteristics: Z_{out} ≈ 50 Ω; t_w = 200 ns, duty cycle = 50%.
B. C_L includes probe and jig capacitance.

FIGURE 1. SN75123 SWITCHING TIMES

TYPICAL CHARACTERISTICS

OUTPUT CURRENT
vs
OUTPUT VOLTAGE

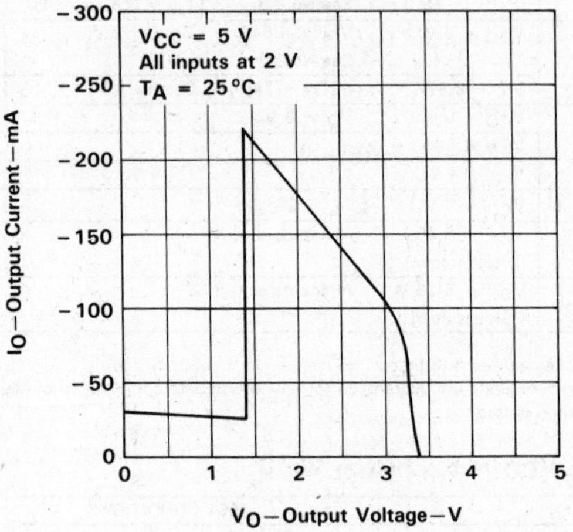
V_{CC} = 5 V
All inputs at 2 V
T_A = 25°C
I_O—Output Current—mA
−300
−250
−200
−150
−100
−50
0
0
1
2
3
4
5
V_O—Output Voltage—V

FIGURE 2

TYPICAL APPLICATION DATA

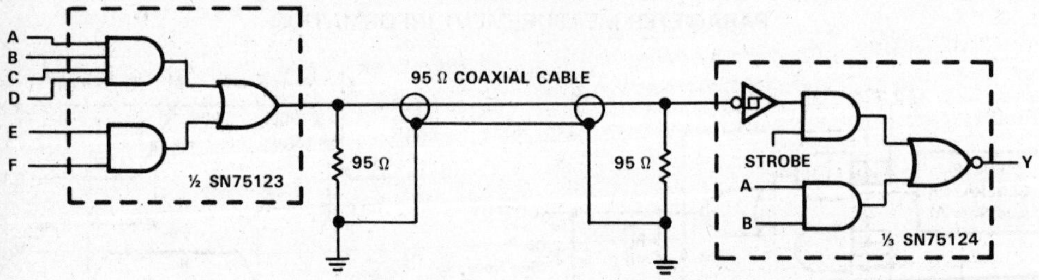
A
B
C
D
E
F
½ SN75123
95 Ω COAXIAL CABLE
95 Ω
95 Ω
STROBE
A
B
Y
⅓ SN75124

FIGURE 3. UNBALANCED LINE COMMUNICATION USING '123 AND '124

TEXAS INSTRUMENTS
POST OFFICE BOX 655012 • DALLAS, TEXAS 75265

D1322, SEPTEMBER 1973—REVISED SEPTEMBER 1986

- Meets IBM System 360 Input/Output Interface Specifications
- Operates from Single 5-V Supply
- TTL Compatible
- Built-In Input Threshold Hysteresis
- High Speed . . . Typical Propagation Delay Time = 20 ns
- Independent Channel Strobes
- Input Gating Increases Application Flexibility
- Designed for Use with Dual Line Driver SN75123
- Designed to be Interchangeable with Signetics N8T24

D, J, OR N PACKAGE
(TOP VIEW)

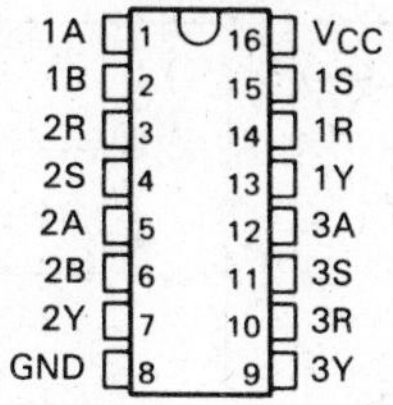
1A
1B
2R
2S
2A
2B
2Y
GND
VCC
1S
1R
1Y
3A
3S
3R
3Y
1
2
3
4
5
6
7
8
16
15
14
13
12
11
10
9

description

The SN75124 triple line receiver is specifically designed to meet the input/output interface specifications for IBM System 360. It is also compatible with standard TTL logic and supply voltage levels.

The SN75124 has receiver inputs with built-in hysteresis to provide increased noise margin for single-ended systems. An open line will affect the receiver input as would a low-level input voltage and the receiver input can withstand a level of −0.15 volt with power on or off. The other inputs are in TTL configuration. The S input must be high to enable the receiver input. Two of the line receivers have A and B inputs that, if both are high, will hold the output low. The third receiver has only an A input that, if high, will hold the output low.

The SN75124 is characterized for operation from 0°C to 70°C.

FUNCTION TABLE

INPUTS				OUTPUT
A	B‡	R	S	Y
H	H	X	X	L
X	X	L	H	L
L	X	H	X	H
L	X	X	L	H
X	L	H	X	H
X	L	X	L	H

‡B input and last two lines of the function table are applicable to receivers 1 and 2 only.

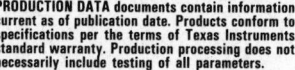

logic symbol†

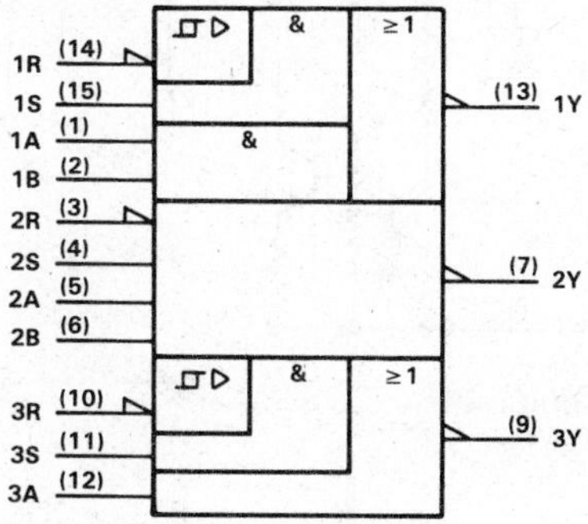
1R
1S
1A
1B
2R
2S
2A
2B
3R
3S
3A
(14)
(15)
(1)
(2)
(3)
(4)
(5)
(6)
(10)
(11)
(12)
&
≥1
(13)
1Y
(7)
2Y
(9)
3Y

†This symbol is in accordance with ANSI/IEEE Std 91-1984 and IEC Publication 617-12.

logic diagram (positive logic)

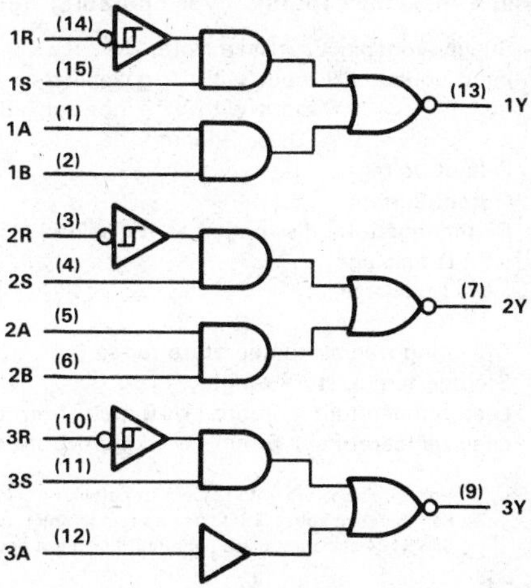
1R
1S
1A
1B
2R
2S
2A
2B
3R
3S
3A
(14)
(15)
(1)
(2)
(3)
(4)
(5)
(6)
(10)
(11)
(12)
(13)
1Y
(7)
2Y
(9)
3Y

schematic (each receiver)

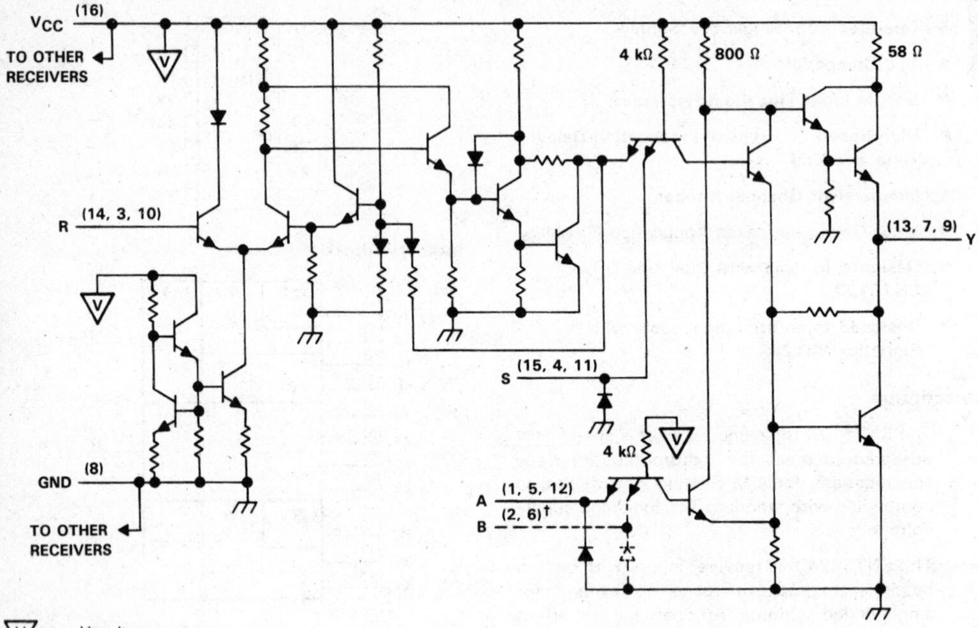
VCC
(16)
TO OTHER RECEIVERS
4 kΩ
800 Ω
58 Ω
(14, 3, 10)
R
(13, 7, 9)
Y
(15, 4, 11)
S
4 kΩ
(8)
GND
TO OTHER RECEIVERS
(1, 5, 12)
A
(2, 6)†
B

V . . . V_{CC} bus

†B input is provided on receivers 1 and 2 only.
Resistor values shown are nominal.

absolute maximum ratings over operating free-air temperature range (unless otherwise noted)

Supply voltage, V_{CC} (see Note 1)	7 V
Input voltage: R input with V_{CC} applied	7 V
R input with V_{CC} not applied	6 V
A, B, or S input	5.5 V
Output voltage	7 V
Output current	±100 mA
Continuous total dissipation at (or below) 25°C free-air temperature (see Note 2):	
D package	950 mW
J package	1025 mW
N package	1150 mW
Operating free-air temperature range	0°C to 70°C
Storage temperature range	−65°C to 150°C
Lead temperature 1,6 mm (1/16 inch) from case for 60 seconds: J package	300°C
Lead temperature 1,6 mm (1/16 inch) from case for 10 seconds: D or N package	260°C

NOTES: 1. Voltage values are with respect to network ground terminal.
2. For operation above 25°C free-air temperature, refer to the Dissipation Derating Curves in Appendix A. In the J package, SN75124 chips are glass mounted. For these devices in the N package, use the 9.2-mW/°C curve.

TEXAS INSTRUMENTS
POST OFFICE BOX 655012 • DALLAS, TEXAS 75265

recommended operating conditions

		MIN	NOM	MAX	UNIT
Supply voltage, V_{CC}		4.75	5	5.25	V
High-level input voltage, V_{IH}	A, B, or S	2			V
	R	1.7			
Low-level input voltage, V_{IL}	A, B, or S			0.8	V
	R			0.7	
High-level output current, I_{OH}				−800	μA
Low-level output current, I_{OL}				16	mA
Operating free-air temperature, T_A		0		70	°C

electrical characteristics, V_{CC} = 4.75 V to 5.25 V, T_A = 0°C to 70°C (unless otherwise noted)

PARAMETER			TEST CONDITIONS	MIN	TYP†	MAX	UNIT
V_{hys}	Hysteresis ($V_{T+} - V_{T-}$)	R	V_{CC} = 5 V, T_A = 25°C	0.2	0.4		V
V_{IK}	Input clamp voltage	A,B, or S	V_{CC} = 5 V, I_I = −12 mA			−1.5	V
$V_{(BR)I}$	Input breakdown voltage	A,B, or S	V_{CC} = 5 V, I_I = 10 mA	5.5			V
V_{OH}	High-level output voltage		V_{IH} = V_{IH} min, V_{IL} = V_{IL} max, I_{OH} = −800 μA, See Note 3	2.6			V
V_{OL}	Low-level output voltage		V_{IH} = V_{IH} min, V_{IL} = V_{IL} max, I_{OL} = 16 mA, See Note 3			0.4	V
I_I	Input current at maximum input voltage	R	V_I = 7 V			5	mA
			V_I = 6 V, V_{CC} = 0			5	
I_{IH}	High-level input current	A,B, or S	V_I = 4.5 V			40	μA
		R	V_I = 3.11 V			170	
I_{IL}	Low-level input current	A,B, or S	V_I = 0.4 V	−0.1		−1.6	mA
I_{OS}	Short-circuit output current‡		V_{CC} = 5 V, T_A = 25°C	−50		−100	mA
I_{CC}	Supply current		V_{CC} = 5.25 V			72	mA

†Typical value is at V_{CC} = 5 V, T_A = 25°C.
‡Not more than one output should be shorted at a time, and duration of the short-circuit should not exceed one second.
NOTE 3: The output voltage and current limits are guaranteed for any appropriate combination of high and low inputs specified by the function table for the desired output.

switching characteristics, V_{CC} = 5 V, T_A = 25°C

PARAMETER		TEST CONDITIONS	MIN	TYP	MAX	UNIT
t_{PLH}	Propagation delay time, low-to-high-level output from R input	See Figure 1		20	30	ns
t_{PHL}	Propagation delay time, high-to-low-level output from R input			20	30	

PARAMETER MEASUREMENT INFORMATION

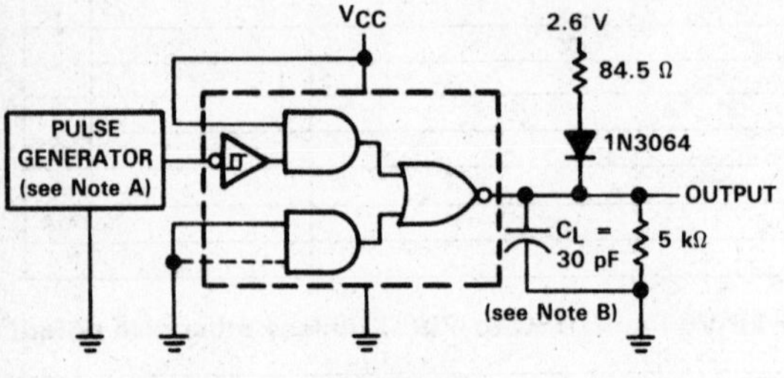
VCC
2.6 V
84.5 Ω
PULSE
GENERATOR
(see Note A)
1N3064
OUTPUT
CL =
30 pF
5 kΩ
(see Note B)
TEST CIRCUIT

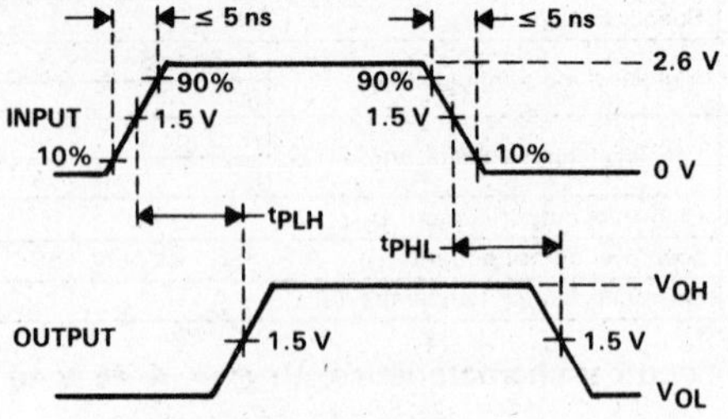
≤ 5 ns
≤ 5 ns
2.6 V
90%
90%
INPUT
1.5 V
1.5 V
10%
10%
0 V
tPLH
tPHL
VOH
OUTPUT
1.5 V
1.5 V
VOL
VOLTAGE WAVEFORMS

NOTES: A. The pulse generator has the following characteristics: Z_{out} ≈ 50 Ω, PRR ≤ 5 MHz, duty cycle = 50%.
B. C_L includes probe and jig capacitance.

FIGURE 1. SN75124 SWITCHING TIMES

TYPICAL CHARACTERISTICS

OUTPUT VOLTAGE
vs
RECEIVER INPUT VOLTAGE

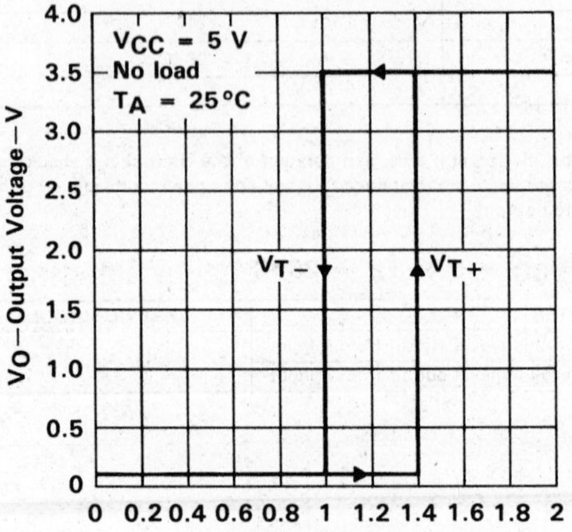
4.0
3.5
3.0
2.5
2.0
1.5
1.0
0.5
0
VCC = 5 V
No load
TA = 25°C
VT−
VT+
VO—Output Voltage—V
0 0.2 0.4 0.6 0.8 1 1.2 1.4 1.6 1.8 2
VI—Input Voltage—V

FIGURE 2

TEXAS INSTRUMENTS
POST OFFICE BOX 655012 • DALLAS, TEXAS 75265

TYPICAL APPLICATION DATA

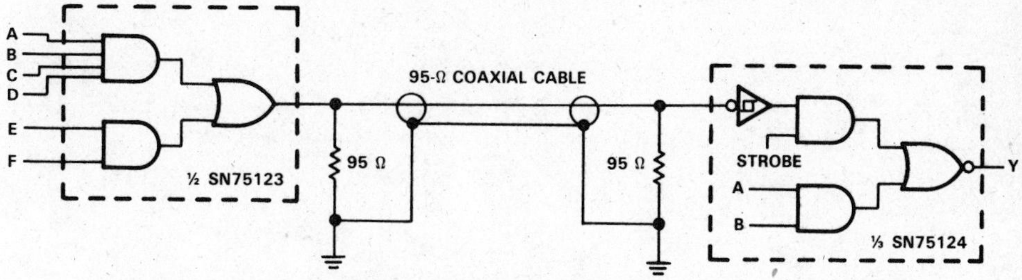
A
B
C
D
E
F
½ SN75123
95-Ω COAXIAL CABLE
95 Ω
95 Ω
STROBE
A
B
Y
⅓ SN75124

FIGURE 3. UNBALANCED LINE COMMUNICATION USING SN75123 AND SN75124

D2239, JANUARY 1977—REVISED SEPTEMBER 1986

- Meets IBM 360/370 I/O Specification
- Input Resistance . . . 7 kΩ to 20 kΩ
- Output Compatible with TTL
- Schottky-Clamped Transistors
- Operates from Single 5-V Supply
- High Speed . . . Low Propagation Delay
- Ratio Specification for Propagation Delay Time, Low-to-High/High-to-Low
- Seven Channels in One 16-Pin Package
- Standard V_{CC} and Ground Positioning on SN75127

SN75125 . . . D, J, OR N PACKAGE
(TOP VIEW)

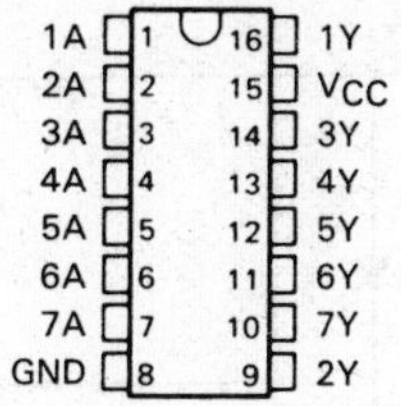
1A 1
2A 2
3A 3
4A 4
5A 5
6A 6
7A 7
GND 8
16 1Y
15 VCC
14 3Y
13 4Y
12 5Y
11 6Y
10 7Y
9 2Y

SN75127 . . . D, J, OR N PACKAGE
(TOP VIEW)

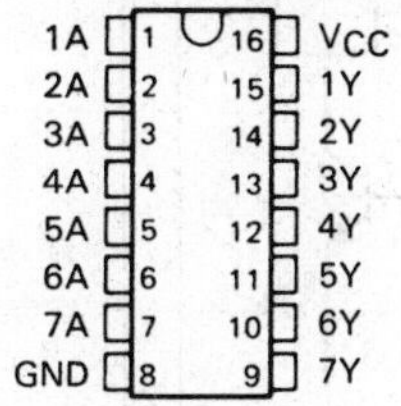
1A 1
2A 2
3A 3
4A 4
5A 5
6A 6
7A 7
GND 8
16 VCC
15 1Y
14 2Y
13 3Y
12 4Y
11 5Y
10 6Y
9 7Y

description

The SN75125 and SN75127 are monolithic seven-channel line receivers designed to satisfy the requirements of the IBM System 360/370 input/output interface specifications. Special low-power design and Schottky-clamped transistors allow for low supply-current requirements while maintaining fast switching speeds and high-current TTL outputs.

The SN75125 and SN75127 are characterized for operation from 0°C to 70°C.

logic symbols†

SN75125

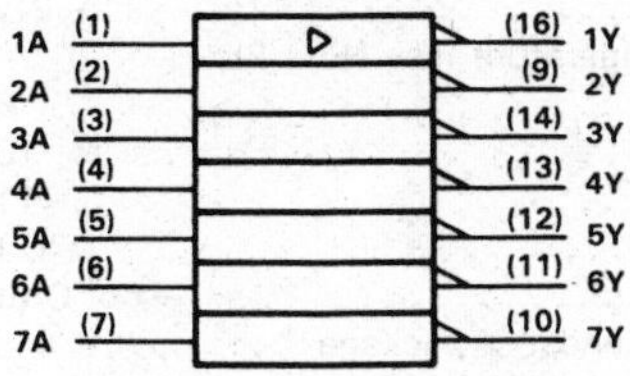
1A (1)
2A (2)
3A (3)
4A (4)
5A (5)
6A (6)
7A (7)
(16) 1Y
(9) 2Y
(14) 3Y
(13) 4Y
(12) 5Y
(11) 6Y
(10) 7Y

SN75127

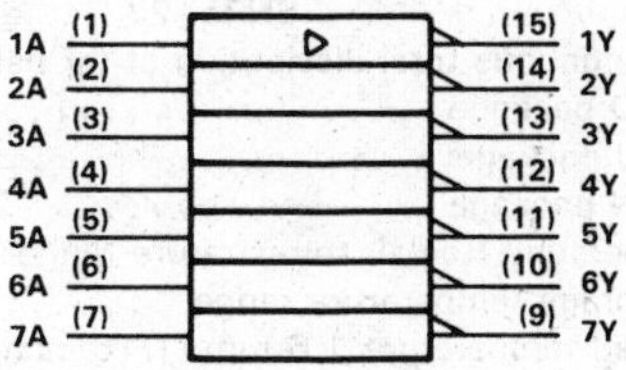
1A (1)
2A (2)
3A (3)
4A (4)
5A (5)
6A (6)
7A (7)
(15) 1Y
(14) 2Y
(13) 3Y
(12) 4Y
(11) 5Y
(10) 6Y
(9) 7Y

† These symbols are in accordance with ANSI/IEEE Std 91-1984 and IEC Publicaiton 617-12.

schematic (each receiver)

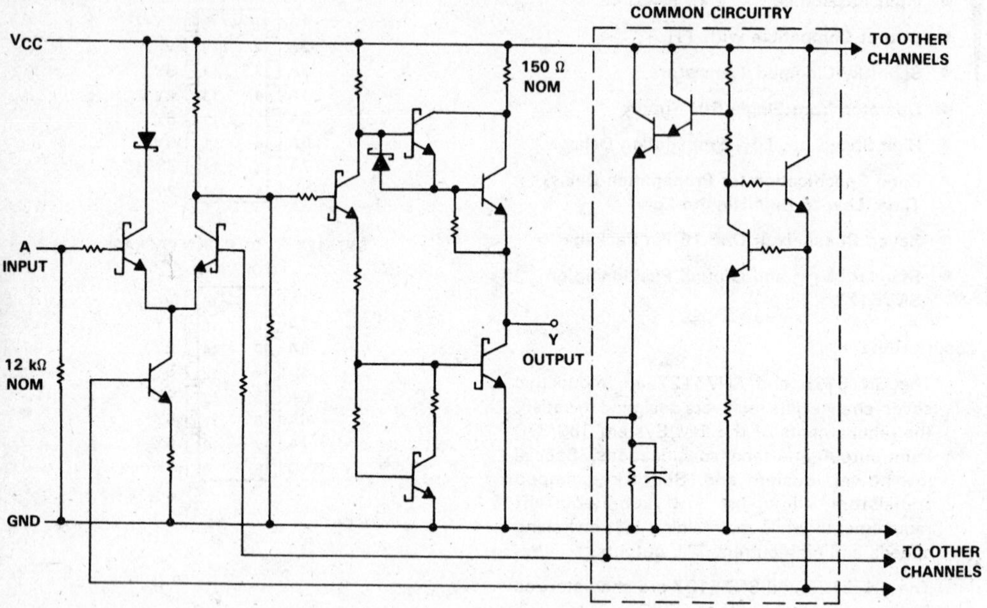
COMMON CIRCUITRY
VCC
TO OTHER CHANNELS
150 Ω NOM
A INPUT
12 kΩ NOM
Y OUTPUT
GND
TO OTHER CHANNELS

absolute maximum ratings over operating free-air temperature range (unless otherwise noted)

Supply voltage, V_{CC} (see Note 1)	7 V
Input voltage range: SN75125	−0.15 V to 7 V
SN75127	−2 V to 7 V
Continuous total dissipation at (or below) 25°C free-air temperature (see Note 2):	
D package	950 mW
J package	1025 mW
N package	1150 mW
Operating free-air temperature range	0°C to 70°C
Storage temperature range	−65°C to 150°C
Lead temperature 1,6 mm (1/16 inch) from case for 60 seconds: J package	300°C
Lead temperature 1,6 mm (1/16 inch) from case for 10 seconds: D or N package	260°C

NOTES: 1. All voltage values are with respect to network ground terminal.
2. For operation above 25°C free-air temperature, refer to the Dissipation Derating Curves in Appendix A. In the J package, SN75125 and SN75127 chips are glass mounted. For these devices in the N package, use the 9.2-mW/°C curve.

recommended operating conditions

	MIN	NOM	MAX	UNIT
Supply voltage, V_{CC}	4.5	5	5.5	V
High-level input voltage, V_{IH}	1.7			V
Low-level input voltage, V_{IL}			0.7	V
High-level output current, I_{OH}			−0.4	V
Low-level output current, I_{OL}			16	mA
Operating free-air temperature, T_A	0		70	°C

electrical characteristics over recommended operating free-air temperature range (unless otherwise noted)

PARAMETER		TEST CONDITIONS	MIN	TYP†	MAX	UNIT
V_{OH}	High-level output voltage	V_{CC} = 4.5 V, V_{IL} = 0.7 V, I_{OH} = −0.4 mA	2.4	3.1		V
V_{OL}	Low-level output voltage	V_{CC} = 4.5 V, V_{IH} = 1.7 V, I_{OL} = 16 mA		0.4	0.5	V
I_{IH}	High-level input current	V_{CC} = 5.5 V, V_I = 3.11 V		0.3	0.42	mA
I_{IL}	Low-level input current	V_{CC} = 5.5 V, V_I = 0.15 V			30	μA
I_{OS}	Short-circuit output current‡	V_{CC} = 5.5 V, V_O = 0	−18		−60	mA
r_i	Input resistance	V_{CC} = 4.5 V, 0 V, or open, ΔV_I = 0.15 V to 4.15 V	7		20	kΩ
I_{CC}	Supply current	V_{CC} = 5.5 V, I_{OH} = −0.4 mA, All inputs at 0.7 V		15	25	mA
		V_{CC} = 5.5 V, I_{OL} = 16 mA, All inputs at 4 V		28	47	mA

† All typical values are at V_{CC} = 5 V, T_A = 25°C.
‡ Not more than one output should be shorted at a time.

switching characteristics, V_{CC} = 5 V, T_A = 25°C

PARAMETER		TEST CONDITIONS	MIN	TYP	MAX	UNIT
t_{PLH}	Propagation delay time, low-to-high-level output	R_L = 400 Ω, C_L = 50 pF, See Figure 1	7	14	25	ns
t_{PHL}	Propagation delay time, high-to-low-level output		10	18	30	ns
$\frac{t_{PLH}}{t_{PHL}}$	Ratio of propagation delay times		0.5	0.8	1.3	
t_{TLH}	Transition time, low-to-high-level output		1	7	12	ns
t_{THL}	Transition time, high-to-low-level output		1	3	12	ns

PARAMETER MEASUREMENT INFORMATION

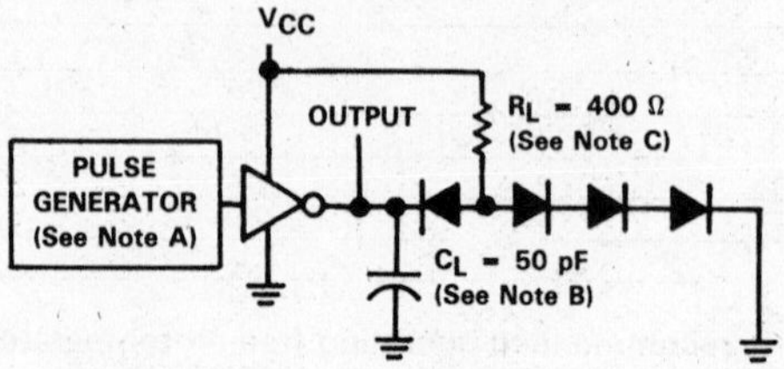
VCC
OUTPUT
RL = 400 Ω
(See Note C)
PULSE
GENERATOR
(See Note A)
CL = 50 pF
(See Note B)

TEST CIRCUIT

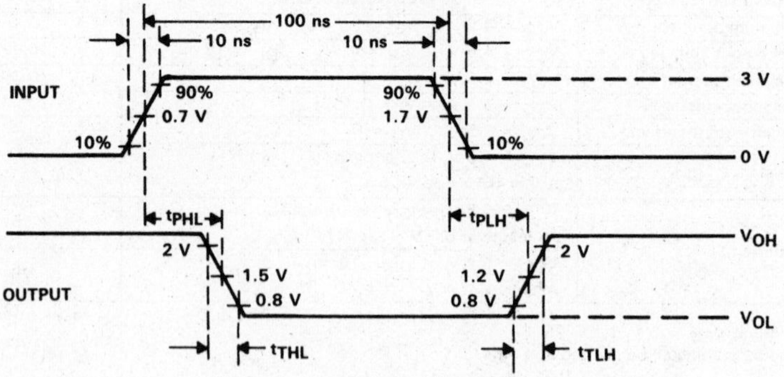
100 ns
10 ns
10 ns
INPUT
90%
0.7 V
10%
90%
1.7 V
10%
3 V
0 V
tPHL
tPLH
2 V
1.5 V
0.8 V
OUTPUT
1.2 V
0.8 V
2 V
VOH
VOL
tTHL
tTLH

VOLTAGE WAVEFORMS

NOTES: A. The pulse generator has the following characteristics: $Z_{out} \approx 50\ \Omega$, PRR $\leq$ 5 MHz.
B. C_L includes probe and jig capacitance.
C. All diodes are 1N3064 or equivalent.

FIGURE 1

TEXAS INSTRUMENTS
POST OFFICE BOX 655012 • DALLAS, TEXAS 75265

TYPICAL CHARACTERISTICS

VOLTAGE TRANSFER CHARACTERISTICS

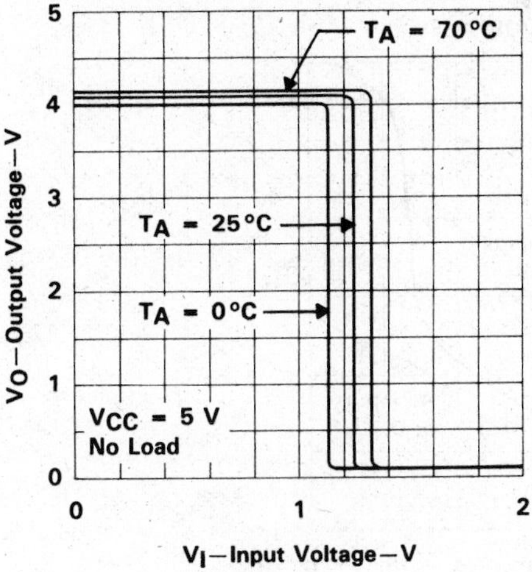
5
4
3
2
1
0
T_A = 70°C
T_A = 25°C
T_A = 0°C
V_{CC} = 5 V
No Load
V_O—Output Voltage—V
0
1
2
V_I—Input Voltage—V

FIGURE 2

VOLTAGE TRANSFER CHARACTERISTICS

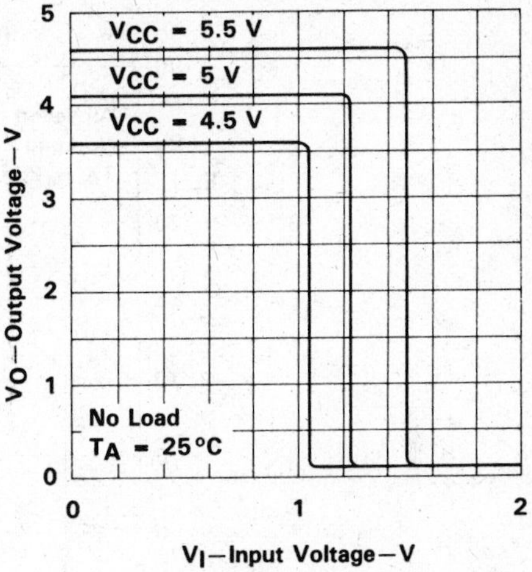
5
4
3
2
1
0
V_{CC} = 5.5 V
V_{CC} = 5 V
V_{CC} = 4.5 V
No Load
T_A = 25°C
V_O—Output Voltage—V
0
1
2
V_I—Input Voltage—V

FIGURE 3

INPUT CURRENT
vs
INPUT VOLTAGE

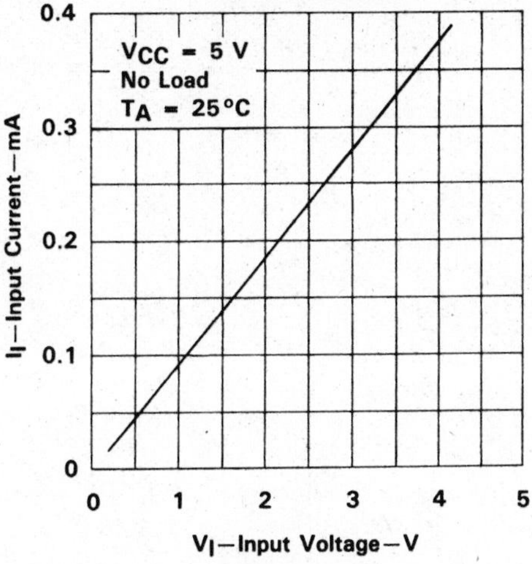
0.4
0.3
0.2
0.1
0
V_{CC} = 5 V
No Load
T_A = 25°C
I_I—Input Current—mA
0
1
2
3
4
5
V_I—Input Voltage—V

FIGURE 4

LOW-LEVEL OUTPUT VOLTAGE
vs
OUTPUT CURRENT

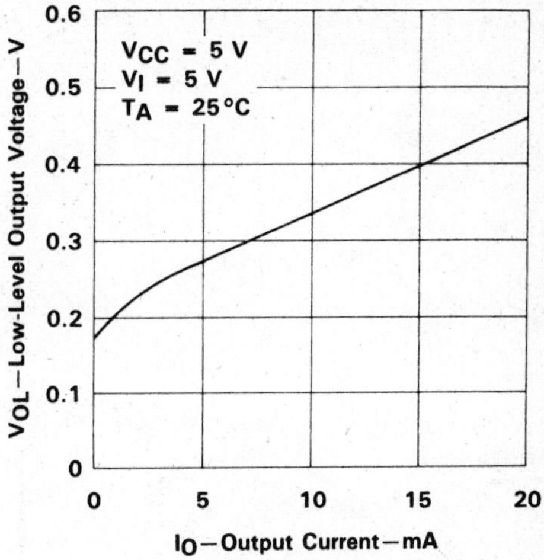
0.6
0.5
0.4
0.3
0.2
0.1
0
V_{CC} = 5 V
V_I = 5 V
T_A = 25°C
V_{OL}—Low-Level Output Voltage—V
0
5
10
15
20
I_O—Output Current—mA

FIGURE 5

TYPICAL CHARACTERISTICS

SUPPLY CURRENT
vs
SUPPLY VOLTAGE

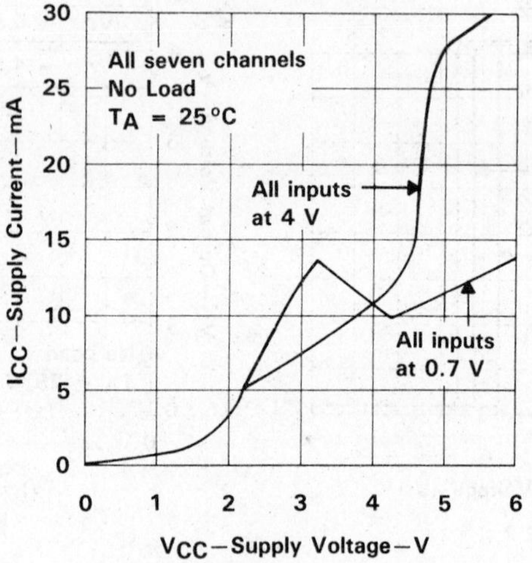
All seven channels
No Load
T_A = 25°C
All inputs at 4 V
All inputs at 0.7 V
I_{CC}—Supply Current—mA
V_{CC}—Supply Voltage—V
0
5
10
15
20
25
30
0
1
2
3
4
5
6

FIGURE 6

D2305, JANUARY 1977—REVISED SEPTEMBER 1986

- Meets IBM 360/370 I/O Specification
- Input Resistance . . . 7 kΩ to 20 kΩ
- Output Compatible with TTL
- Schottky-Clamped Transistors
- Operates from a Single 5-Volt Supply
- High Speed . . . Low Propagation Delay
- Ratio Specification . . . t_{PLH}/t_{THL}
- Common Strobe for Each Group of Four Receivers
- SN75128 . . . Active-High Strobes
  SN75129 . . . Active-Low Strobes

DW, J, OR N PACKAGE
(TOP VIEW)

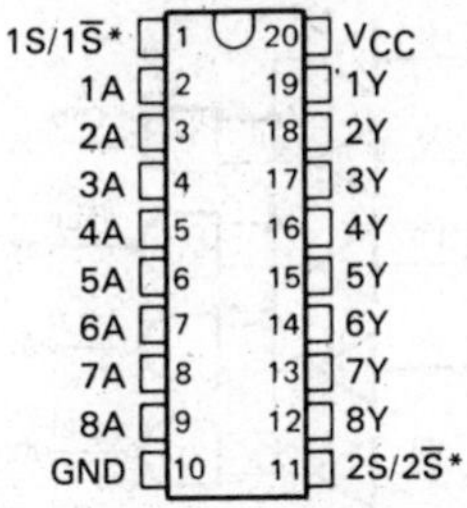
1S/1S̄* 1
1A 2
2A 3
3A 4
4A 5
5A 6
6A 7
7A 8
8A 9
GND 10
20 VCC
19 1Y
18 2Y
17 3Y
16 4Y
15 5Y
14 6Y
13 7Y
12 8Y
11 2S/2S̄*

*S and S̄ for SN75128 and SN75129, respectively

description

The SN75128 and SN75129 are eight-channel line receivers designed to satisfy the requirements of the input-output interface specification for IBM 360/370. Both devices feature common strobes for each group of four devices. The SN75128 has active-high strobes; the SN75129 has active-low strobes. Special low-power design and Schottky-diode-clamped transistors allow low supply-current requirements while maintaining fast switching speeds and high-current TTL outputs.

The SN75128 and SN75129 are characterized for operation from 0°C to 70°C.

logic symbols†

SN75128

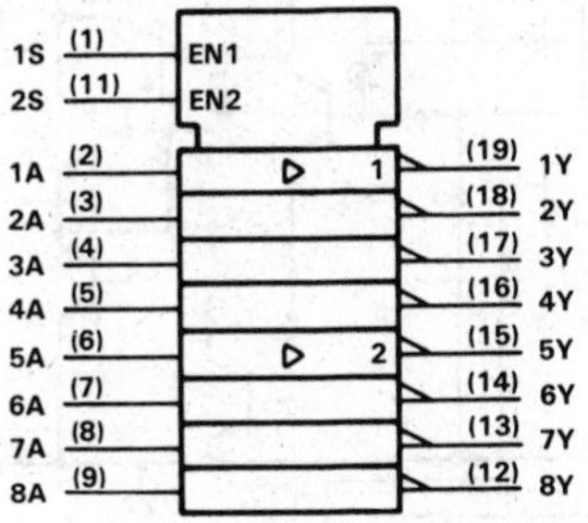
1S (1) EN1
2S (11) EN2
1A (2)
2A (3)
3A (4)
4A (5)
5A (6)
6A (7)
7A (8)
8A (9)
1
2
(19) 1Y
(18) 2Y
(17) 3Y
(16) 4Y
(15) 5Y
(14) 6Y
(13) 7Y
(12) 8Y

SN75129

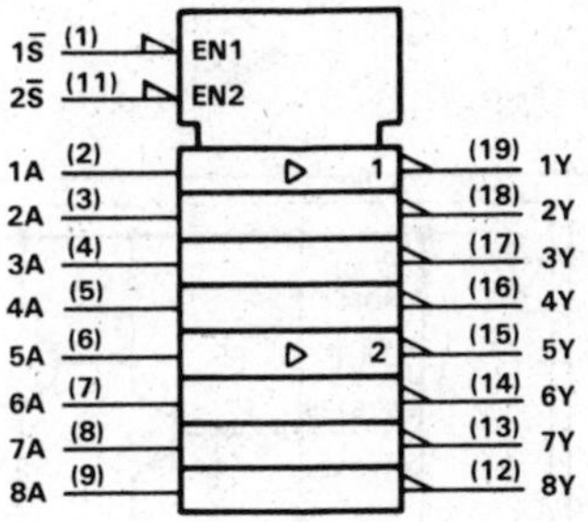
1S̄ (1) EN1
2S̄ (11) EN2
1A (2)
2A (3)
3A (4)
4A (5)
5A (6)
6A (7)
7A (8)
8A (9)
1
2
(19) 1Y
(18) 2Y
(17) 3Y
(16) 4Y
(15) 5Y
(14) 6Y
(13) 7Y
(12) 8Y

† These symbols are in accordance with ANSI/IEEE Std 91-1984 and IEC Publication 617-12.

logic diagrams (positive logic)

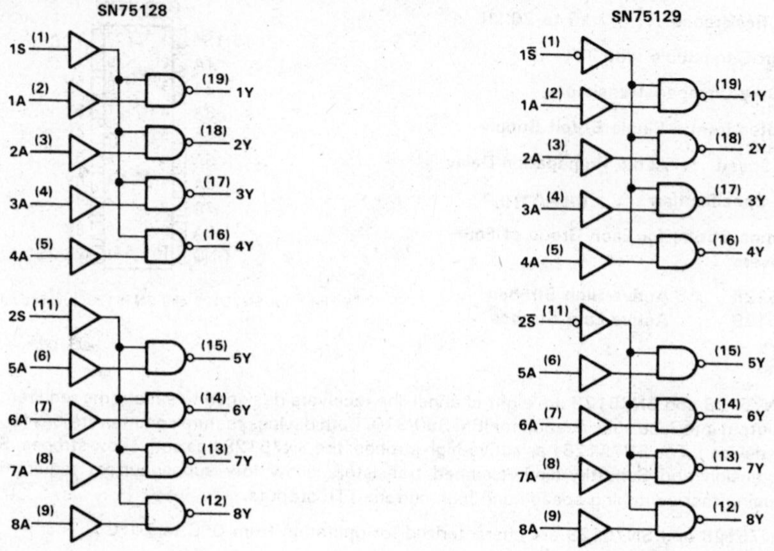
SN75128
1S (1)
1A (2)
2A (3)
3A (4)
4A (5)
(19) 1Y
(18) 2Y
(17) 3Y
(16) 4Y
2S (11)
5A (6)
6A (7)
7A (8)
8A (9)
(15) 5Y
(14) 6Y
(13) 7Y
(12) 8Y
SN75129
1S̄ (1)
1A (2)
2A (3)
3A (4)
4A (5)
(19) 1Y
(18) 2Y
(17) 3Y
(16) 4Y
2S̄ (11)
5A (6)
6A (7)
7A (8)
8A (9)
(15) 5Y
(14) 6Y
(13) 7Y
(12) 8Y

schematic (each driver)

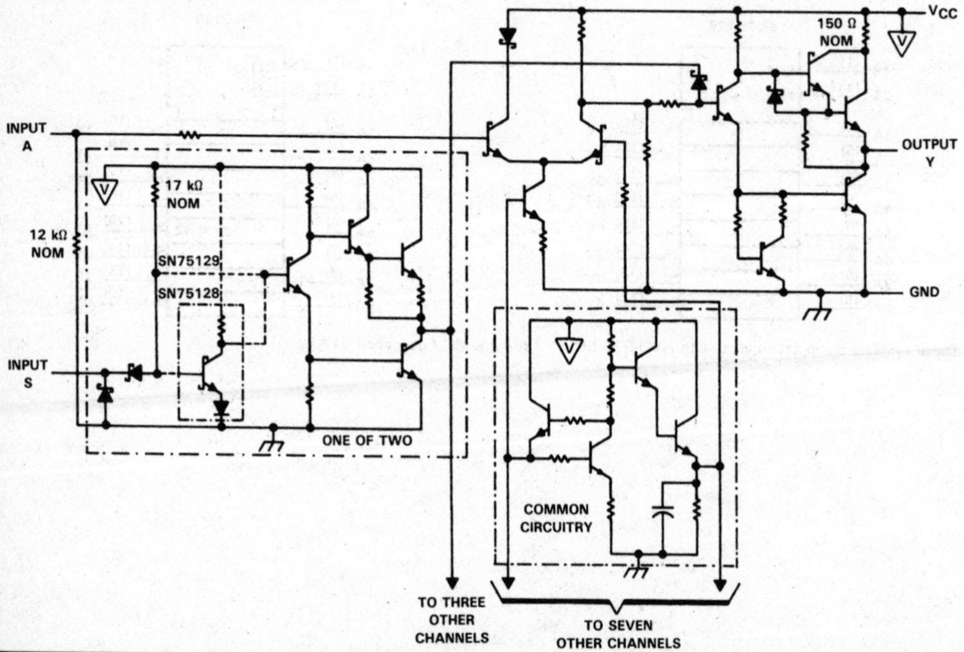
VCC
150 Ω NOM
INPUT A
17 kΩ NOM
12 kΩ NOM
SN75129
SN75128
INPUT S
ONE OF TWO
OUTPUT Y
GND
COMMON CIRCUITRY
TO THREE OTHER CHANNELS
TO SEVEN OTHER CHANNELS

absolute maximum ratings over operating free-air temperature range (unless otherwise noted)

Supply voltage, V_{CC} (see Note 1) . . . . 7 V
A input voltage range . . . . −0.15 V to 7 V
Strobe input voltage . . . . 7 V
Continuous total dissipation at (or below) 25 °C free-air temperature (see Note 2):
  DW package . . . . 1125 mW
  J package . . . . 1025 mW
  N package . . . . 1150 mW
Operating free-air temperature range . . . . 0 °C to 70 °C
Storage temperature range . . . . −65 °C to 150 °C
Lead temperature 1,6 mm (1/16 inch) from case for 10 seconds: DW or N package . . . . 260 °C
Lead temperature 1,6 mm (1/16 inch) from case for 60 seconds: J package . . . . 300 °C

NOTES: 1. All voltage values are with respect to network ground terminal.
2. For operation above 25 °C free-air temperature, refer to the Dissipation Derating Curves in Appendix A. In the J package, SN75128 amd SN75129 chips are glass mounted. For these devices in the N package, use the 9.2-mW/°C curve.

recommended operating conditions

		MIN	NOM	MAX	UNIT
Supply voltage, V_{CC}		4.5	5	5.5	V
High-level input voltage, V_{IH}	A	1.7			V
	S	2			V
Low-level input voltage, V_{IL}	A			0.7	V
	S			0.7	V
High-level output current, I_{OH}				−0.4	mA
Low-level output current, I_{OL}				16	mA
Operating free-air temperature, T_A		0		70	°C

electrical characteristics over recommended operating free-air temperature range (unless otherwise noted)

PARAMETER			TEST CONDITIONS	MIN	TYP†	MAX	UNIT
V_{OH}	High-level output voltage		V_{CC} = 4.5 V, V_{IL} = 0.7 V, I_{OH} = −0.4 mA	2.4	3.1		V
V_{OL}	Low-level output voltage		V_{CC} = 4.5 V, V_{IH} = 1.7 V, I_{OL} = 16 mA		0.4	0.5	V
V_{IK}	Input clamp voltage	S	V_{CC} = 4.5 V, I_I = −18 mA			−1.5	V
I_{IH}	High-level input current	A	V_{CC} = 5.5 V, V_I = 3.11 V		0.3	0.42	mA
		S	V_{CC} = 5.5 V, V_I = 2.7 V			20	µA
I_{IL}	Low-level input current	A	V_{CC} = 5.5 V, V_I = 0.15 V			30	µA
		S	V_{CC} = 5.5 V, V_I = 0.4 V			−0.4	mA
I_{OS}	Short-circuit output current‡		V_{CC} = 5.5 V, V_O = 0	−18		−60	mA
r_i	Input resistance		V_{CC} = 4.5 V, 0 V, or open; ΔV_I = 0.15 V to 4.15 V	7		20	kΩ
I_{CC}	Supply current	SN75128	V_{CC} = 5.5 V, Strobe at 2.4 V, All A inputs at 0.7 V		19	31	mA
		SN75129	V_{CC} = 5.5 V, Strobe at 0.4 V, All A inputs at 0.7 V		19	31	mA
		SN75128	V_{CC} = 5.5 V, Strobe at 2.4 V, All A inputs at 4 V		32	53	mA
		SN75129	V_{CC} = 5.5 V, Strobe at 0.4 V, All A inputs at 4 V		32	53	mA

† All typical values are at V_{CC} = 5 V, T_A = 25 °C.
‡ Not more than one output should be shorted at a time.

switching characteristics, V_{CC} = 5 V, T_A = 25°C

PARAMETER		FROM	TEST CONDITIONS	SN75128			SN75129			UNIT
				MIN	TYP	MAX	MIN	TYP	MAX	
t_{PLH}	Propagation delay time, low-to-high-level output	A	R_L = 400 Ω, C_L = 50 pF, See Figure 1	7	14	25	7	14	25	ns
t_{PHL}	Propagation delay time, high-to-low-level output	A		10	18	30	10	18	30	ns
t_{PLH}	Propagation delay time, low-to-high-level output	S			26	40		20	35	ns
t_{PHL}	Propagation delay time, high-to-low-level output	S			22	35		16	30	ns
$\frac{t_{PLH}}{t_{PHL}}$	Ratio of propagation delay times	A		0.5	0.8	1.3	0.5	0.8	1.3	
t_{TLH}	Transition time, low-to-high-level output			1	7	12	1	7	12	ns
t_{THL}	Transition time, high-to-low-level output			1	3	12	1	3	12	ns

PARAMETER MEASUREMENT INFORMATION

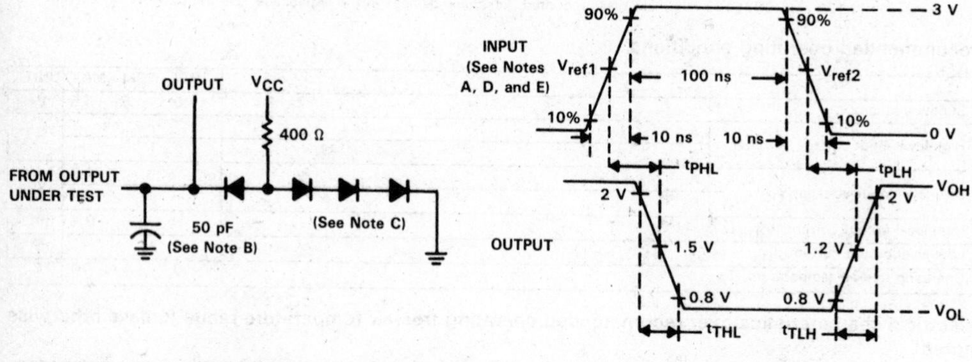
OUTPUT
VCC
400 Ω
FROM OUTPUT UNDER TEST
50 pF (See Note B)
(See Note C)
INPUT (See Notes A, D, and E)
90%
Vref1
10%
100 ns
10 ns
10 ns
Vref2
3 V
0 V
tPHL
tPLH
VOH
2 V
1.5 V
1.2 V
0.8 V
VOL
OUTPUT
tTHL
tTLH

LOAD CIRCUIT

VOLTAGE WAVEFORMS

NOTES:
A. Input pulses are supplied by a generator having the following characteristics: Z_O = 50 Ω, PRR ≤ 5 MHz.
B. Includes probe and jig capacitance.
C. All diodes are 1N3064 or equivalent.
D. The strobe inputs of SN75129 are in-phase with the output.
E. V_{ref1} = 0.7 V and V_{ref2} = 1.7 V for testing data (A) inputs, V_{ref1} = V_{ref2} = 1.3 V for strobe inputs.

FIGURE 1

TEXAS INSTRUMENTS
POST OFFICE BOX 655012 • DALLAS, TEXAS 75265

TYPICAL CHARACTERISTICS

VOLTAGE TRANSFER CHARACTERISTICS

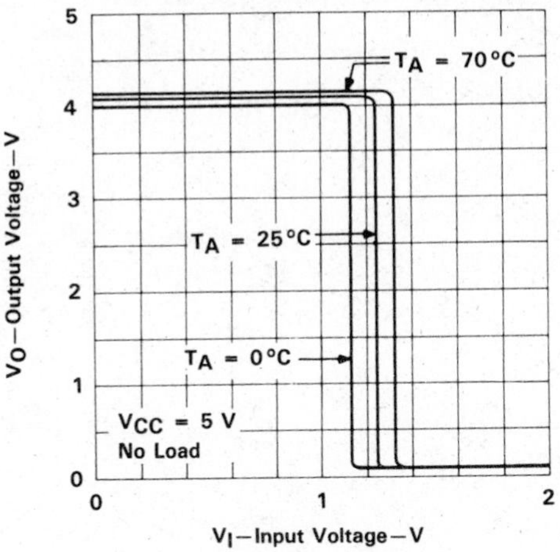
T_A = 70°C
T_A = 25°C
T_A = 0°C
V_{CC} = 5 V
No Load
V_O—Output Voltage—V
V_I—Input Voltage—V

FIGURE 2

VOLTAGE TRANSFER CHARACTERISTICS
FROM A INPUTS

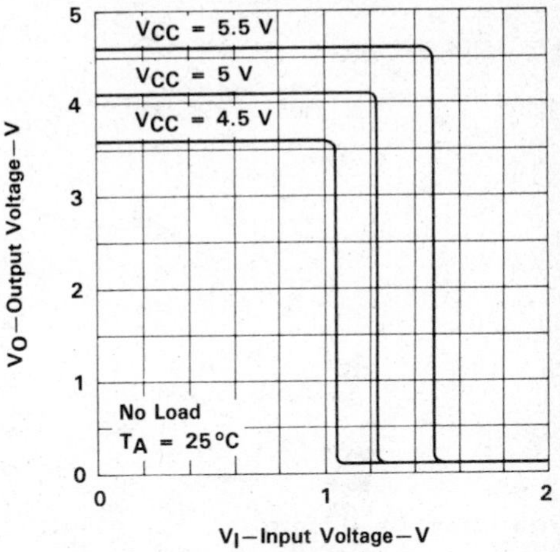
V_{CC} = 5.5 V
V_{CC} = 5 V
V_{CC} = 4.5 V
No Load
T_A = 25°C
V_O—Output Voltage—V
V_I—Input Voltage—V

FIGURE 3

INPUT CURRENT
vs
INPUT VOLTAGE

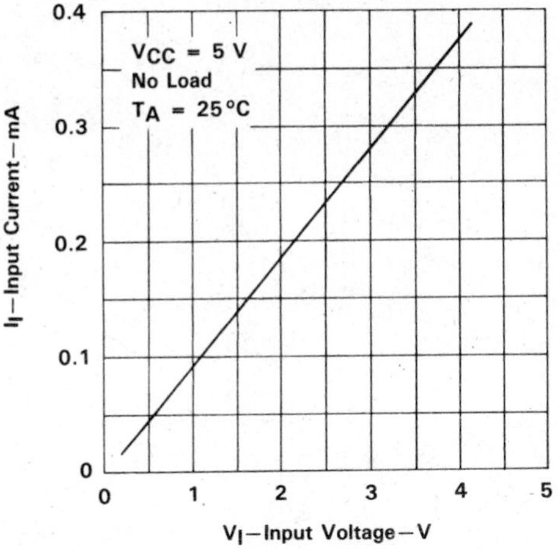
V_{CC} = 5 V
No Load
T_A = 25°C
I_I—Input Current—mA
V_I—Input Voltage—V

FIGURE 4

LOW-LEVEL OUTPUT VOLTAGE
vs
OUTPUT CURRENT

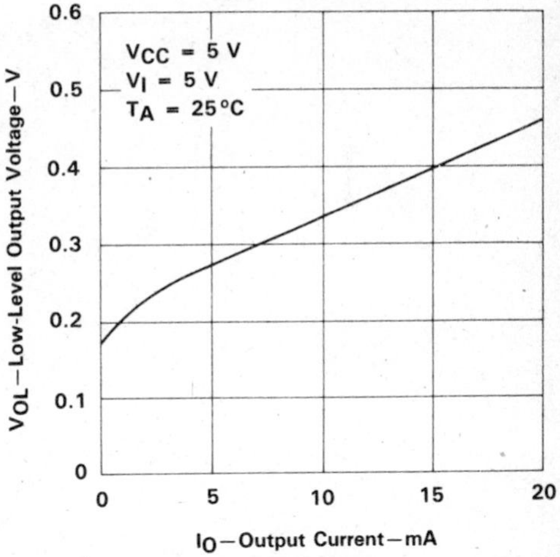
V_{CC} = 5 V
V_I = 5 V
T_A = 25°C
V_{OL}—Low-Level Output Voltage—V
I_O—Output Current—mA

FIGURE 5

D2291, JANUARY 1977—REVISED SEPTEMBER 1986

- P-N-P Inputs for Minimal Input Loading (200 μA Maximum)
- High-Speed Schottky Circuitry
- 3-State Outputs for Driver and Receiver
- Party-Line (Data-Bus) Operation
- Single 5-V Supply
- Driver has 40-mA Current Sink Capability
- Designed to be Functionally Interchangeable with Signetics N8T26, also Called 8T26

D, J, OR N PACKAGE
(TOP VIEW)

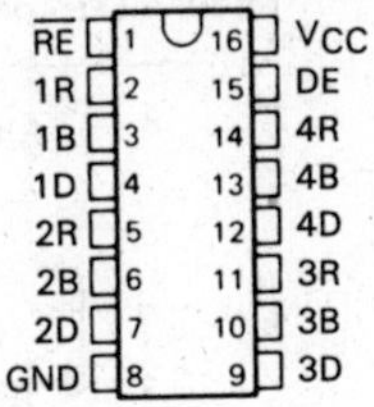
$\overline{RE}$ 1
1R 2
1B 3
1D 4
2R 5
2B 6
2D 7
GND 8
16 V_{CC}
15 DE
14 4R
13 4B
12 4D
11 3R
10 3B
9 3D

description

The SN75136 is a quadruple transceiver utilizing Schottky-diode-clamped transistors. Both the driver and receiver have three-state outputs. With p-n-p inputs, the input loading is reduced to a maximum input current of 200 microamperes.

The SN75136 is characterized for operation from 0°C to 70°C.

FUNCTION TABLE (DRIVER)

INPUT		OUTPUT
D	DE	B
L	H	H
H	H	L
X	L	Z

FUNCTION TABLE (RECEIVER)

INPUT		OUTPUT
B	$\overline{RE}$	R
L	L	H
H	L	L
X	H	Z

H = high level
L = low level
X = irrelevant
Z = high impedance

logic symbol†

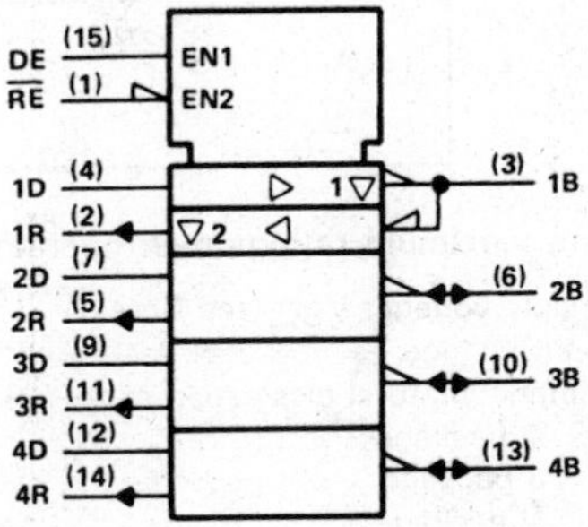
DE (15)
$\overline{RE}$ (1)
EN1
EN2
1D (4)
1R (2)
2D (7)
2R (5)
3D (9)
3R (11)
4D (12)
4R (14)
1∇
∇2
(3) 1B
(6) 2B
(10) 3B
(13) 4B

† This symbol is in accordance with ANSI/IEEE Std 91-1984 and IEC Publication 617-12.

logic diagram (positive logic)

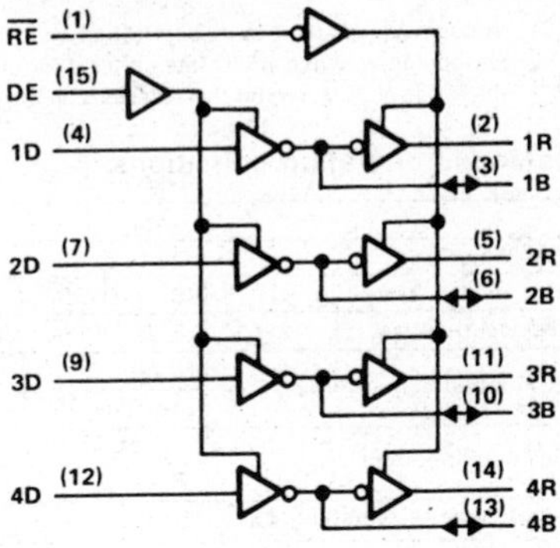
$\overline{RE}$ (1)
DE (15)
1D (4)
2D (7)
3D (9)
4D (12)
(2) 1R
(3) 1B
(5) 2R
(6) 2B
(11) 3R
(10) 3B
(14) 4R
(13) 4B

TEXAS INSTRUMENTS
POST OFFICE BOX 655012 • DALLAS, TEXAS 75265

schematics of inputs and outputs

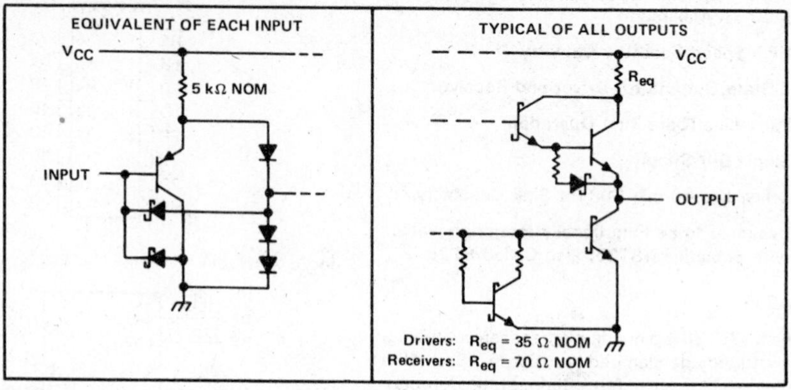
EQUIVALENT OF EACH INPUT
VCC
5 kΩ NOM
INPUT
TYPICAL OF ALL OUTPUTS
VCC
Req
OUTPUT
Drivers: Req = 35 Ω NOM
Receivers: Req = 70 Ω NOM

absolute maximum ratings over operating free-air temperature range (unless otherwise noted)

Supply voltage, V_{CC} (see Note 1) . . . . 7 V
Input voltage . . . . 5.5 V
Continuous total dissipation at (or below) 25°C free-air temperature (see Note 2):
D package . . . . 950 mW
J package . . . . 1025 mW
N package . . . . 1150 mW
Operating free-air temperature range . . . . 0°C to 70°C
Storage temperature range . . . . −65°C to 150°C
Lead temperature 1,6 mm (1/16 inch) from case for 60 seconds: J package . . . . 300°C
Lead temperature 1,6 mm (1/16 inch) from case for 10 seconds: D or N package . . . . 260°C

NOTES: 1. Voltage values are with respect to network ground terminal.
2. For operation above 25°C free-air temperature, refer to Dissipation Derating Curves in Appendix A. In the N package, use the 9.2-mW/°C curve for these devices. In the J package, SN75136 chips are glass mounted, and use the 8.25 mW/°C curve.

recommended operating conditions

		MIN	NOM	MAX	UNIT
Supply voltage, V_{CC}		4.75	5	5.25	V
High-level input voltage, V_{IH}	B, D, DE, $\overline{RE}$	2			V
Low-level input voltage, V_{IL}	B, D, DE, $\overline{RE}$			0.85	V
High-level output current, I_{OH}	Driver, B			−10	mA
	Receiver, R			−2	
Low-level output current, I_{OL}	Driver, B			40	mA
	Receiver, R			16	
Operating free-air temperature, T_A		0		70	°C

TEXAS INSTRUMENTS
POST OFFICE BOX 655012 • DALLAS, TEXAS 75265

electrical characteristics over recommended operating free-air temperature and supply voltage range (unless otherwise noted)

PARAMETER			TEST CONDITIONS	MIN	TYP†	MAX	UNIT
V_{IK}	Input clamp voltage	B,D,DE,$\overline{RE}$	$I_I = -5$ mA			−1	V
V_{OH}	High-level output voltage	B	$V_{IH} = 2$ V, $V_{IL} = 0.85$ V, $I_{OH} = -10$ mA	2.6	3.1		V
		R	$V_{IL} = 0.85$ V, $I_{OH} = -2$ mA	2.6	3.1		
V_{OL}	Low-level output voltage	B	$V_{IH} = 2$ V, $I_{OL} = 40$ mA			0.5	V
		R	$V_{IH} = 2$ V, $V_{IL} = 0.85$ V, $I_{OL} = 16$ mA			0.5	
I_{OZ}	Off-state (high-impedance state) output current	B,R	DE at 0.85 V, $\overline{RE}$ at 2 V, $V_O = 2.6$ V			100	μA
		R	$\overline{RE}$ at 2 V, $V_O = 0.5$ V			−100	
I_{IH}	High-level input current	D,DE,$\overline{RE}$	$V_I = 5.25$ V			25	μA
I_{IL}	Low-level input current	B,D,DE,$\overline{RE}$	$V_I = 0.4$ V			−200	μA
I_{OS}	Short-circuit output current‡	B	$V_{CC} = 5.25$ V	−50		−150	mA
		R		−30		−75	
I_{CC}	Supply current		$V_{CC} = 5.25$ V, No load			87	mA

switching characteristics, $V_{CC} = 5$ V, $T_A = 25°C$

PARAMETER		FROM	TO	TEST CONDITIONS	MIN	TYP	MAX	UNIT
t_{PLH}	Propagation delay time, low-to-high-level output	B	R	$C_L = 30$ pF, See Figure 1		8	18	ns
t_{PHL}	Propagation delay time, high-to-low-level output					7	14	
t_{PLH}	Propagation delay time, low-to-high-level output	D	B	$C_L = 300$ pF, See Figure 2		11	20	ns
t_{PHL}	Propagation delay time, high-to-low-level output					16	24	
t_{PLZ}	Output disable time from low level	$\overline{RE}$	R	$C_L = 30$ pF, See Figure 3		16	24	ns
t_{PZL}	Output enable time to low level					15	30	
t_{PLZ}	Output disable time from low level	DE	B	$C_L = 300$ pF, See Figure 4		9	24	ns
t_{PZL}	Output enable time to low level					31	38	

†All typical values are at $T_A = 25°C$ and $V_{CC} = 5$ V.
‡Only one output should be shorted to ground at a time, and duration of the short circuit should not exceed one second.

PARAMETER MEASUREMENT INFORMATION

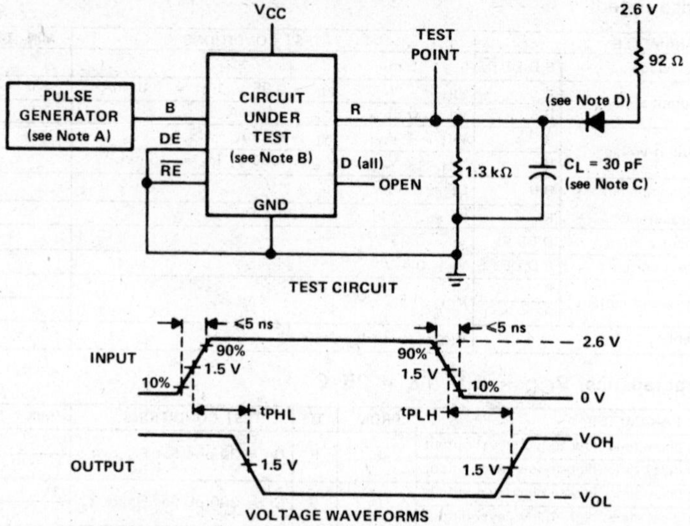
VCC
TEST POINT
2.6 V
92 Ω
PULSE GENERATOR (see Note A)
B
DE
RE
CIRCUIT UNDER TEST (see Note B)
R
D (all)
OPEN
GND
(see Note D)
1.3 kΩ
CL = 30 pF (see Note C)
TEST CIRCUIT
≤5 ns
INPUT
90%
1.5 V
10%
2.6 V
0 V
tPHL
tPLH
OUTPUT
VOH
VOL
VOLTAGE WAVEFORMS

FIGURE 1. PROPAGATION DELAY TIMES FROM BUS TO RECEIVER OUTPUT

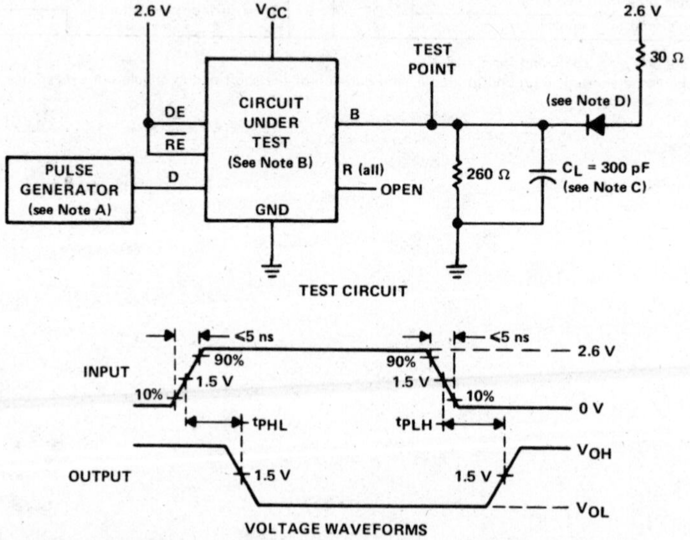
2.6 V
VCC
TEST POINT
30 Ω
DE
RE
CIRCUIT UNDER TEST (See Note B)
B
(see Note D)
PULSE GENERATOR (see Note A)
D
R (all)
OPEN
GND
260 Ω
CL = 300 pF (see Note C)
TEST CIRCUIT
≤5 ns
INPUT
90%
1.5 V
10%
2.6 V
0 V
tPHL
tPLH
OUTPUT
VOH
VOL
VOLTAGE WAVEFORMS

FIGURE 2. PROPAGATION DELAY TIMES FROM DRIVER INPUT TO BUS

NOTES: A. The pulse generator in Figures 1 and 2 has the following characteristics: PRR ≤ 10 MHz, duty cycle = 50%, Z_{out} ≈ 50 Ω.
B. All inputs and outputs not shown are open.
C. C_L includes probe and jig capacitance.
D. All diodes are 1N916 or 1N3064.

TEXAS INSTRUMENTS
POST OFFICE BOX 655012 • DALLAS, TEXAS 75265

PARAMETER MEASUREMENT INFORMATION

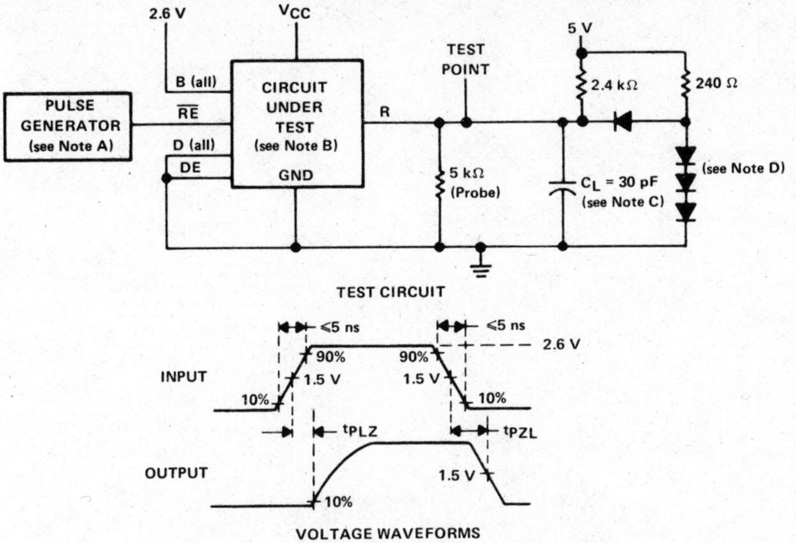
2.6 V
VCC
B (all)
RE
D (all)
DE
PULSE GENERATOR (see Note A)
CIRCUIT UNDER TEST (see Note B)
GND
R
TEST POINT
5 V
2.4 kΩ
240 Ω
5 kΩ (Probe)
CL = 30 pF (see Note C)
(see Note D)
TEST CIRCUIT
≤5 ns
2.6 V
90%
1.5 V
10%
INPUT
tPLZ
tPZL
OUTPUT
VOLTAGE WAVEFORMS

FIGURE 3. RECEIVER ENABLE AND DISABLE TIMES

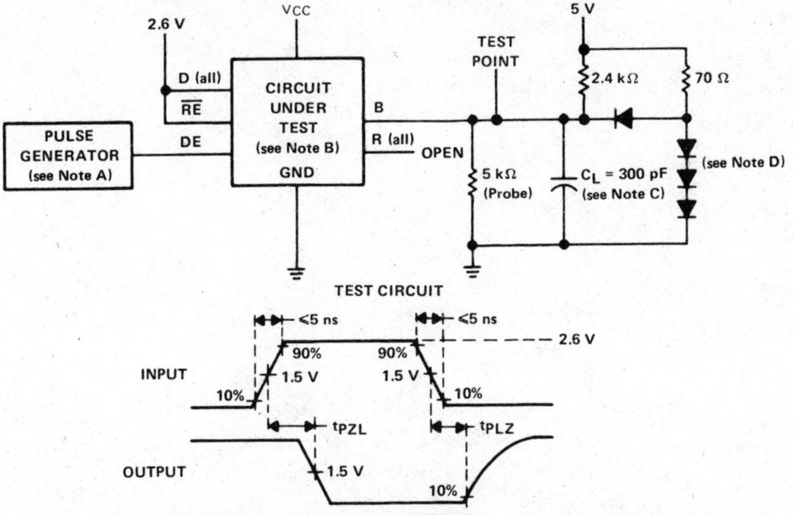
VCC
2.6 V
D (all)
RE
DE
PULSE GENERATOR (see Note A)
CIRCUIT UNDER TEST (see Note B)
GND
B
R (all)
OPEN
TEST POINT
5 V
2.4 kΩ
70 Ω
5 kΩ (Probe)
CL = 300 pF (see Note C)
(see Note D)
TEST CIRCUIT
≤5 ns
2.6 V
90%
1.5 V
10%
INPUT
tPZL
tPLZ
OUTPUT
VOLTAGE WAVEFORMS

FIGURE 4. DRIVER ENABLE AND DISABLE TIMES

NOTES:
A. The pulse generator in Figures 3 and 4 has the following characteristics: PRR ≤ 5 MHz, duty cycle = 50%, Z_{out} ≈ 50 Ω.
B. All inputs and outputs not shown are open.
C. C_L includes probe and jig capacitance.
D. All diodes are 1N916 or 1N3064.

D1663, SEPTEMBER 1973—REVISED SEPTEMBER 1986

- **Single 5-V Supply**
- **High-Input-Impedance, High-Threshold Receivers**
- **Common Driver Strobe**
- **TTL-Compatible Driver and Strobe Inputs with Clamp Diodes**
- **High-Speed Operation**
- **100-mA Open-Collector Driver Outputs**
- **Four Independent Channels**
- **TTL-Compatible Receiver Output**

SN55138 . . . J PACKAGE
SN75138 . . . D, J, OR N PACKAGE
(TOP VIEW)

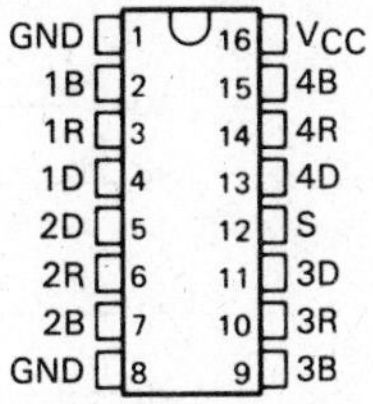
GND 1
1B 2
1R 3
1D 4
2D 5
2R 6
2B 7
GND 8
16 VCC
15 4B
14 4R
13 4D
12 S
11 3D
10 3R
9 3B

SN55138 . . . FK PACKAGE
(TOP VIEW)

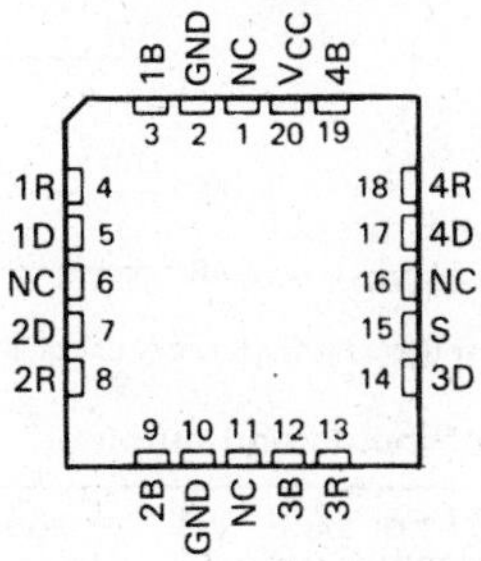
1B GND NC VCC 4B
3 2 1 20 19
1R 4
1D 5
NC 6
2D 7
2R 8
18 4R
17 4D
16 NC
15 S
14 3D
9 10 11 12 13
2B GND NC 3B 3R

NC—No internal connection

description

The SN55138 and SN75138 quad bus transceivers are designed for two-way data communication over single-ended transmission lines. Each of the four identical channels consists of a driver with TTL inputs and a receiver with a TTL output. The driver output is of the open-collector type, and is designed to handle loads of up to 100 milliamperes (50 ohms to 5 volts). The receiver input is internally connected to the driver output, and has a high impedance to minimize loading of the transmission line. Because of the high driver-output current and the high receiver-input impedance, a very large number (typically hundreds) of transceivers may be connected to a single data bus.

The receiver design also features a threshold of 2.3 volts (typical), providing a wider noise margin than would be possible with a receiver having the usual TTL threshold. A strobe turns off all drivers (high impedance) but does not affect receiver operation. These circuits are designed for operation from a single five-volt supply and include a provision to minimize loading of the data bus when the power-supply voltage is zero.

The SN55138 is characterized for operation over the full military temperature range of −55°C to 125°C; the SN75138 is characterized for operation from 0°C to 70°C.

FUNCTION TABLE
(TRANSMITTING)

INPUTS		OUTPUTS	
S	D	B	R
L	H	L	H
L	L	H	L

FUNCTION TABLE
(RECEIVING)

INPUTS			OUTPUT
S	B	D	R
H	H	X	L
H	L	X	H

H = high level, L = low level, X = irrelevant

logic symbol†

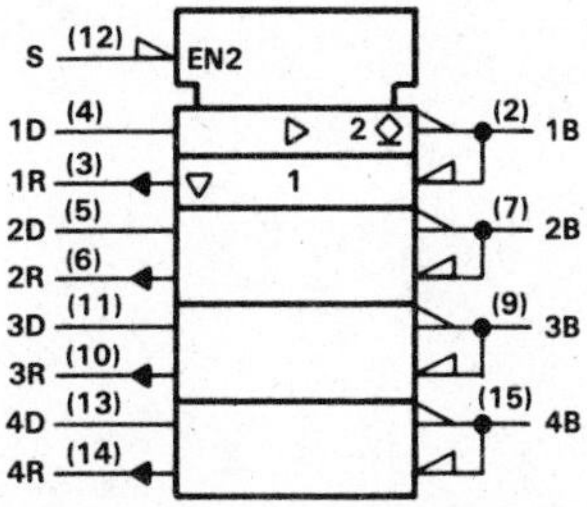
S (12)
EN2
1D (4)
1R (3)
2D (5)
2R (6)
3D (11)
3R (10)
4D (13)
4R (14)
2
1
(2) 1B
(7) 2B
(9) 3B
(15) 4B

† This symbol is in accordance with ANSI/IEEE Std 91-1984 and IEC Publication 617-12.
Pin numbers shown are for D, J, and N packages.

logic diagram (positive logic)

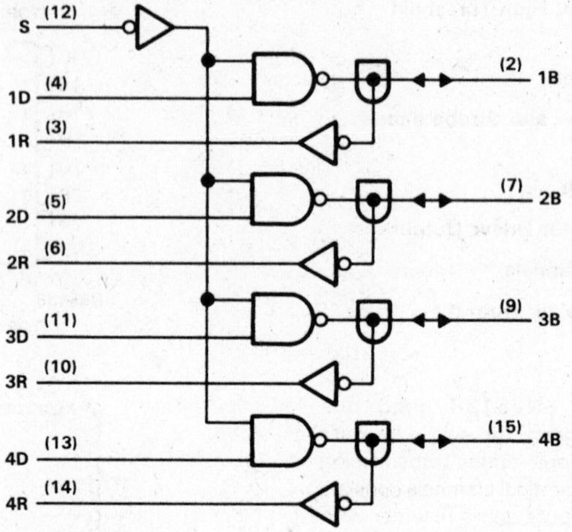
S (12)
1D (4)
1R (3)
2D (5)
2R (6)
3D (11)
3R (10)
4D (13)
4R (14)
(2) 1B
(7) 2B
(9) 3B
(15) 4B

Pin numbers showns are for D, J, and N packages.

schematics of inputs and outputs

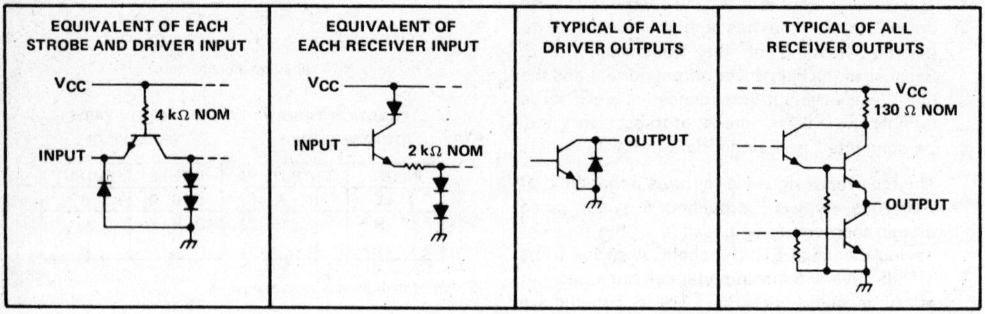
EQUIVALENT OF EACH STROBE AND DRIVER INPUT
VCC
4 kΩ NOM
INPUT
EQUIVALENT OF EACH RECEIVER INPUT
VCC
INPUT
2 kΩ NOM
TYPICAL OF ALL DRIVER OUTPUTS
OUTPUT
TYPICAL OF ALL RECEIVER OUTPUTS
VCC
130 Ω NOM
OUTPUT

TEXAS INSTRUMENTS
POST OFFICE BOX 655012 • DALLAS, TEXAS 75265

absolute maximum ratings over operating free-air temperature range (unless otherwise noted)

		SN55138	SN75138	UNIT
Supply voltage, V_{CC} (see Note 1)		7	7	V
Input voltage		5.5	5.5	V
Driver off-state output voltage		7	7	V
Low-level output current into the driver output		150	150	mA
Continuous total dissipation at (or below) 25°C free-air temperature (see Note 2)	D package		950	mW
	FK or J package	1375		
	J package		1025	
	N package		1150	
Operating free-air temperature range		−55 to 125	0 to 70	°C
Storage temperature range		−65 to 150	−65 to 150	°C
Lead temperature 1,6 mm (1/16 inch) from case for 60 seconds: J package		300	300	°C
Lead temperature 1,6 mm (1/16 inch) from case for 10 seconds: D or N package			260	°C
Case temperature for 60 seconds: FK package		260		°C

NOTES: 1. All voltage values are with respect to both ground terminals connected together.
2. For operation above 25°C free-air temperature, refer to the Dissipation Derating Curves in Appendix A. In the J package, SN55138 chips are alloy mounted and SN75138 chips are glass mounted. In the N package, use the 9.2-mW/°C curve for these devices.

recommended operating conditions

		SN55138			SN75138			UNIT
		MIN	NOM	MAX	MIN	NOM	MAX	
Supply voltage, V_{CC}		4.5	5	5.5	4.75	5	5.25	V
High-level input voltage, V_{IH}	Driver or strobe	2			2			V
	Receiver	3.2			2.9			
Low-level input voltage, V_{IL}	Driver or strobe			0.8			0.8	V
	Receiver			1.5			1.8	
High-level output current, I_{OH}	Receiver output			−400			−400	μA
Low-level output current, I_{OL}	Driver output			100			100	mA
	Receiver output			16			16	
Operating free-air temperature, T_A		−55		125	0		70	°C

electrical characteristics over recommended operating free-air temperature range (unless otherwise noted)

PARAMETER			TEST CONDITIONS†	SN55138			SN75138			UNIT
				MIN	TYP‡	MAX	MIN	TYP‡	MAX	
V_{IK}	Input clamp voltage	Driver or strobe	V_{CC} = MIN, I_I = −12 mA			−1.5			−1.5	V
V_{OH}	High-level output voltage	Receiver	V_{CC} = MIN, $V_{IH(S)}$ = 2 V, $V_{IL(R)}$ = V_{IL} max, I_{OH} = −400 μA	2.4	3.5		2.4	3.5		V
V_{OL}	Low-level output voltage	Driver	V_{CC} MIN, $V_{IH(D)}$ = 2 V, $V_{IL(S)}$ = 0.8 V, I_{OL} = 100 mA			0.45			0.45	V
		Receiver	V_{CC} = MIN, $V_{IH(R)}$ = V_{IH} min, $V_{IH(S)}$ = 2 V, I_{OL} = 16 mA			0.4			0.4	
I_I	Input current at maximum input voltage	Driver or strobe	V_{CC} = MAX, V_I = V_{CC}			1			1	mA
I_{IH}	High-level input current	Driver or strobe	V_{CC} = MAX, V_I = 2.4 V			40			40	μA
		Receiver	V_{CC} = 5 V, $V_{I(R)}$ = 4.5 V, $V_{I(S)}$ = 2 V		25	300		25	300	
I_{IL}	Low-level input current	Driver or strobe	V_{CC} = MAX, V_I = 0.4 V		−1	−1.6		−1	−1.6	mA
		Receiver	V_{CC} = MAX, $V_{I(R)}$ = 0.45 V, $V_{I(S)}$ = 2 V			−50			−50	μA
	Input current with power off	Receiver	V_{CC} = 0, V_I = 4.5 V		1.1	1.5		1.1	1.5	mA
I_{OS}	Short-circuit output current§	Receiver	V_{CC} = MAX,	−20		−55	−18		−55	mA
I_{CC}	Supply current	All driver outputs low	V_{CC} = MAX, $V_{I(D)}$ = 2 V, $V_{I(S)}$ = 0.8 V		50	65		50	65	mA
		All driver outputs high	V_{CC} = MAX, $V_{I(R)}$ = 3.5 V, $V_{I(S)}$ = 2 V, Receiver outputs open		42	55		42	55	

†For conditions shown as MIN or MAX, use the appropriate value specified under recommended operating conditions. Parenthetical letters D, R, and S used with V_I refer to the driver input, receiver input, and strobe input, respectively.
‡All typical values are at V_{CC} = 5 V, T_A = 25°C.
§Not more than one output should be shorted at a time.

switching characteristics, V_{CC} = 5 V, T_A = 25°C

PARAMETER†	FROM (INPUT)	TO (OUTPUT)	TEST CONDITIONS	MIN	TYP	MAX	UNIT
t_{PLH}	Driver	Driver	C_L = 50 pF, R_L = 50 Ω, See Figure 1		15	24	ns
t_{PHL}					14	24	
t_{PLH}	Strobe	Driver			18	28	ns
t_{PHL}					22	32	
t_{PLH}	Receiver	Receiver	C_L = 15 pF, R_L = 400 Ω, See Figure 2		7	15	ns
t_{PHL}					8	15	

†t_{PLH} ≡ propagation delay time, low-to-high-level output
t_{PHL} ≡ propagation delay time, high-to-low-level output

TEXAS INSTRUMENTS
POST OFFICE BOX 655012 • DALLAS, TEXAS 75265

PARAMETER MEASUREMENT INFORMATION

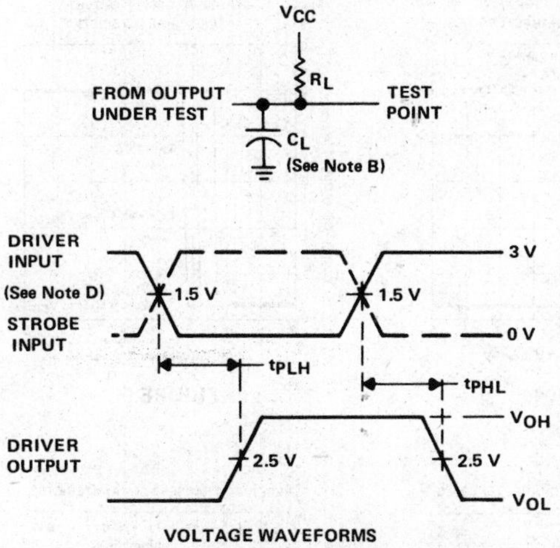
VCC
RL
FROM OUTPUT UNDER TEST
TEST POINT
CL
(See Note B)
DRIVER INPUT
(See Note D)
STROBE INPUT
1.5 V
1.5 V
3 V
0 V
tPLH
tPHL
VOH
DRIVER OUTPUT
2.5 V
2.5 V
VOL
VOLTAGE WAVEFORMS

FIGURE 1. PROPAGATION DELAY TIMES FROM DATA AND STROBE INPUTS

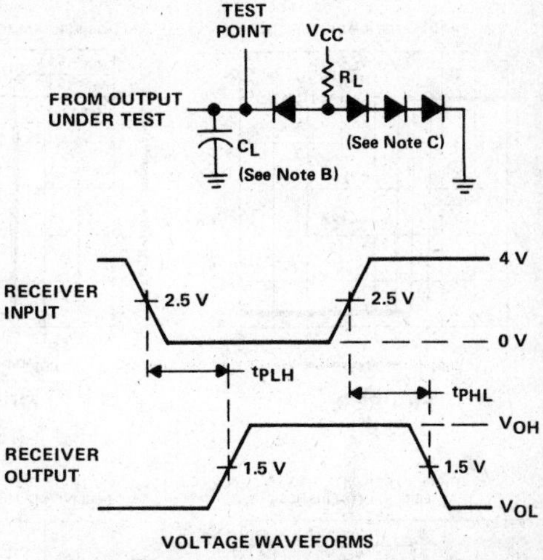
TEST POINT
VCC
RL
FROM OUTPUT UNDER TEST
CL
(See Note B)
(See Note C)
RECEIVER INPUT
2.5 V
2.5 V
4 V
0 V
tPLH
tPHL
VOH
RECEIVER OUTPUT
1.5 V
1.5 V
VOL
VOLTAGE WAVEFORMS

FIGURE 2. PROPAGATION DELAY TIMES FROM RECEIVER INPUT

NOTES:
A. Input pulses are supplied by generators having the following characteristics: t_W = 100 ns, PRR ≤ 1 MHz, t_r ≤10 ns, t_f ≤10 ns, Z_{out} ≈50 Ω.
B. C_L includes probe and jig capacitance.
C. All diodes are 1N916 or 1N3064.
D. When testing driver input (solid line) strobe must be low; when testing strobe input (dashed line) driver input must be high.

TYPICAL CHARACTERISTICS†

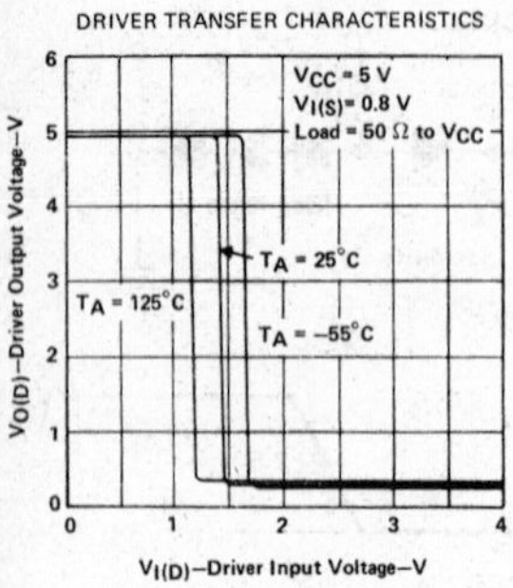
DRIVER TRANSFER CHARACTERISTICS
VCC = 5 V
VI(S) = 0.8 V
Load = 50 Ω to VCC
TA = 25°C
TA = 125°C
TA = –55°C
VO(D)–Driver Output Voltage–V
VI(D)–Driver Input Voltage–V

FIGURE 3

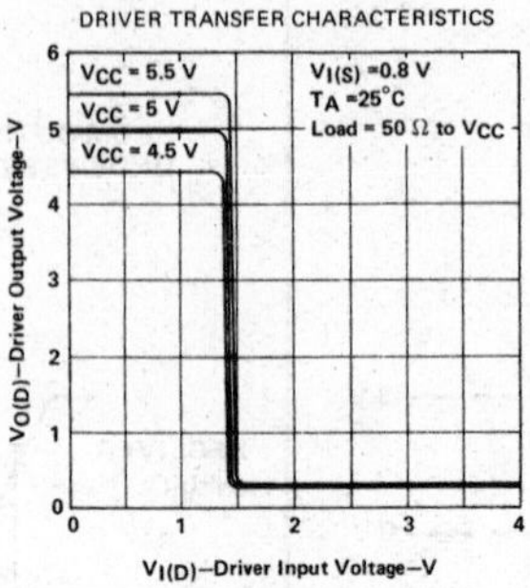
DRIVER TRANSFER CHARACTERISTICS
VCC = 5.5 V
VCC = 5 V
VCC = 4.5 V
VI(S) = 0.8 V
TA = 25°C
Load = 50 Ω to VCC
VO(D)–Driver Output Voltage–V
VI(D)–Driver Input Voltage–V

FIGURE 4

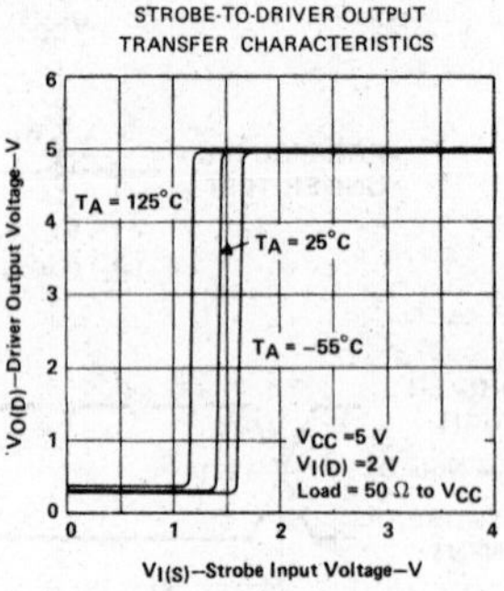
STROBE-TO-DRIVER OUTPUT
TRANSFER CHARACTERISTICS
TA = 125°C
TA = 25°C
TA = –55°C
VCC = 5 V
VI(D) = 2 V
Load = 50 Ω to VCC
VO(D)–Driver Output Voltage–V
VI(S)–Strobe Input Voltage–V

FIGURE 5

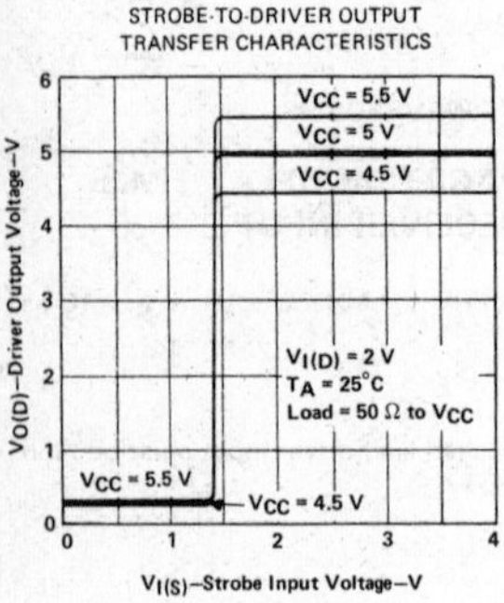
STROBE-TO-DRIVER OUTPUT
TRANSFER CHARACTERISTICS
VCC = 5.5 V
VCC = 5 V
VCC = 4.5 V
VI(D) = 2 V
TA = 25°C
Load = 50 Ω to VCC
VCC = 5.5 V
VCC = 4.5 V
VO(D)–Driver Output Voltage–V
VI(S)–Strobe Input Voltage–V

FIGURE 6

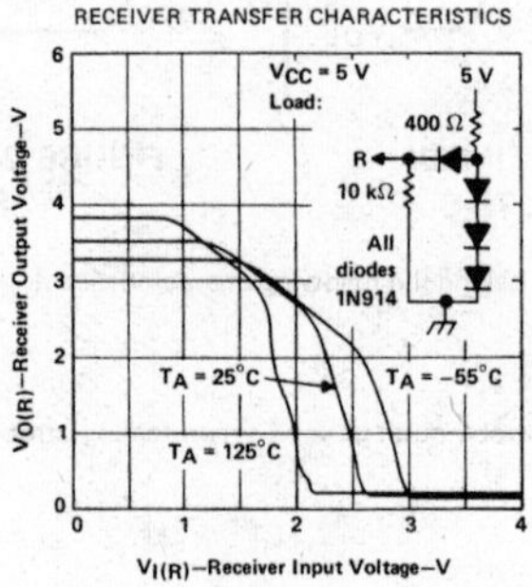
RECEIVER TRANSFER CHARACTERISTICS
VCC = 5 V
Load:
5 V
400 Ω
R
10 kΩ
All
diodes
1N914
TA = 25°C
TA = –55°C
TA = 125°C
VO(R)–Receiver Output Voltage–V
VI(R)–Receiver Input Voltage–V

FIGURE 7

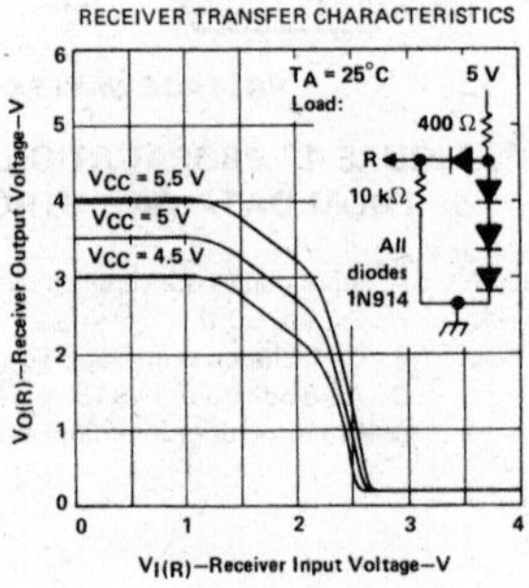
RECEIVER TRANSFER CHARACTERISTICS
TA = 25°C
Load:
5 V
400 Ω
R
10 kΩ
All
diodes
1N914
VCC = 5.5 V
VCC = 5 V
VCC = 4.5 V
VO(R)–Receiver Output Voltage–V
VI(R)–Receiver Input Voltage–V

FIGURE 8

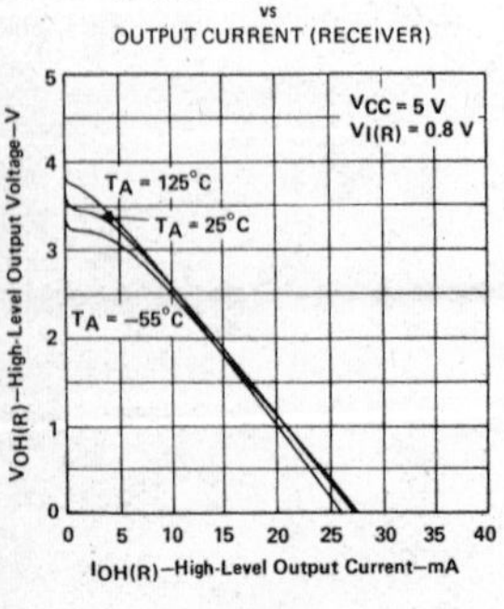
HIGH-LEVEL OUTPUT VOLTAGE
vs
OUTPUT CURRENT (RECEIVER)
VCC = 5 V
VI(R) = 0.8 V
TA = 125°C
TA = 25°C
TA = –55°C
VOH(R)–High-Level Output Voltage–V
IOH(R)–High-Level Output Current–mA

FIGURE 9

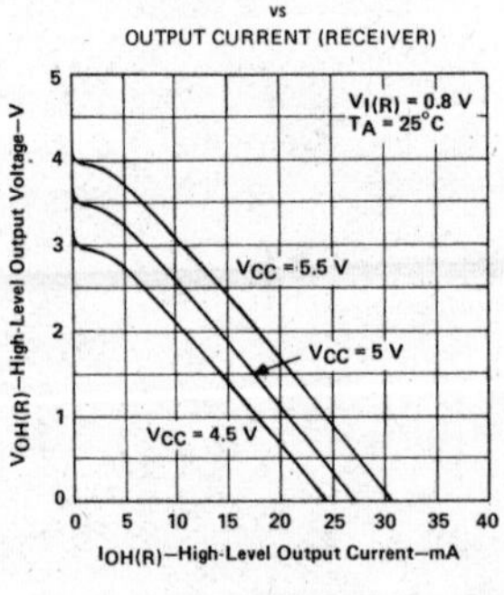
HIGH-LEVEL OUTPUT VOLTAGE
vs
OUTPUT CURRENT (RECEIVER)
VI(R) = 0.8 V
TA = 25°C
VCC = 5.5 V
VCC = 5 V
VCC = 4.5 V
VOH(R)–High-Level Output Voltage–V
IOH(R)–High-Level Output Current–mA

FIGURE 10

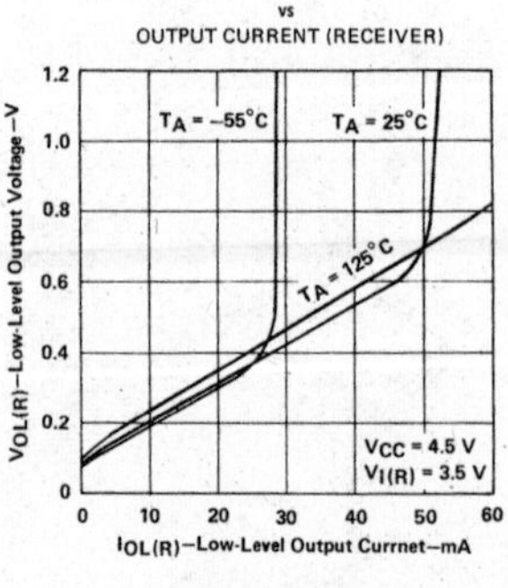
LOW-LEVEL OUTPUT VOLTAGE
vs
OUTPUT CURRENT (RECEIVER)
TA = –55°C
TA = 25°C
TA = 125°C
VCC = 4.5 V
VI(R) = 3.5 V
VOL(R)–Low-Level Output Voltage–V
IOL(R)–Low-Level Output Currnet–mA

FIGURE 11

†Data for temperatures below 0°C and above 70°C is applicable to SN55138 circuits only.

TEXAS INSTRUMENTS
POST OFFICE BOX 655012 • DALLAS, TEXAS 75265

TYPICAL CHARACTERISTICS†

LOW-LEVEL OUTPUT VOLTAGE vs OUTPUT CURRENT (DRIVER OUTPUT)

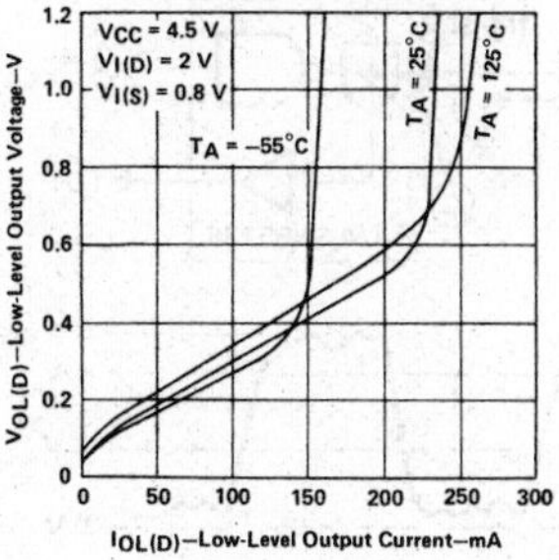
VCC = 4.5 V
VI(D) = 2 V
VI(S) = 0.8 V
TA = −55°C
TA = 25°C
TA = 125°C
VOL(D)—Low-Level Output Voltage—V
IOL(D)—Low-Level Output Current—mA

FIGURE 12

RECEIVER INPUT CURRENT vs RECEIVER INPUT VOLTAGE

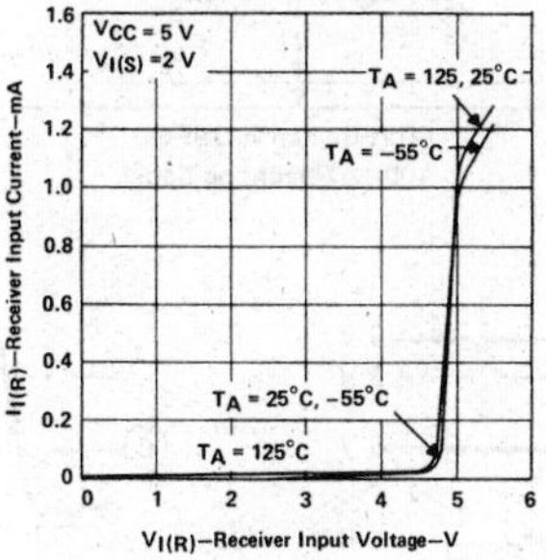
VCC = 5 V
VI(S) = 2 V
TA = 125, 25°C
TA = −55°C
TA = 25°C, −55°C
TA = 125°C
II(R)—Receiver Input Current—mA
VI(R)—Receiver Input Voltage—V

FIGURE 13

RECEIVER INPUT CURRENT vs RECEIVER INPUT VOLTAGE

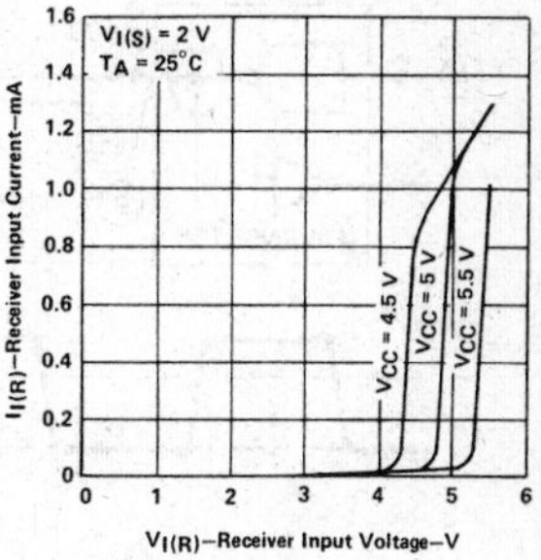
VI(S) = 2 V
TA = 25°C
VCC = 4.5 V
VCC = 5 V
VCC = 5.5 V
II(R)—Receiver Input Current—mA
VI(R)—Receiver Input Voltage—V

FIGURE 14

SUPPLY CURRENT vs SUPPLY VOLTAGE (ALL DRIVER OUTPUTS LOW)

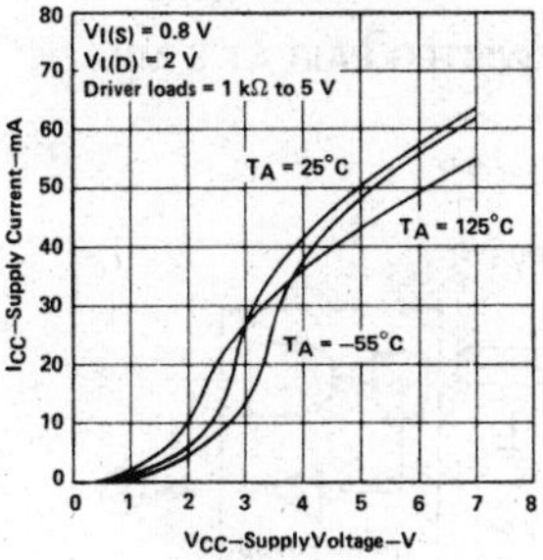
VI(S) = 0.8 V
VI(D) = 2 V
Driver loads = 1 kΩ to 5 V
TA = 25°C
TA = 125°C
TA = −55°C
ICC—Supply Current—mA
VCC—Supply Voltage—V

FIGURE 15

SUPPLY CURRENT vs SUPPLY VOLTAGE

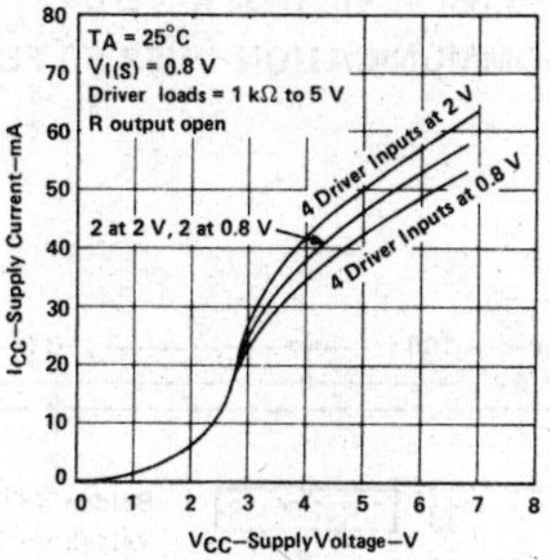
TA = 25°C
VI(S) = 0.8 V
Driver loads = 1 kΩ to 5 V
R output open
4 Driver Inputs at 2 V
2 at 2 V, 2 at 0.8 V
4 Driver Inputs at 0.8 V
ICC—Supply Current—mA
VCC—Supply Voltage—V

FIGURE 16

PROPAGATION DELAY TIMES vs FREE-AIR TEMPERATURE

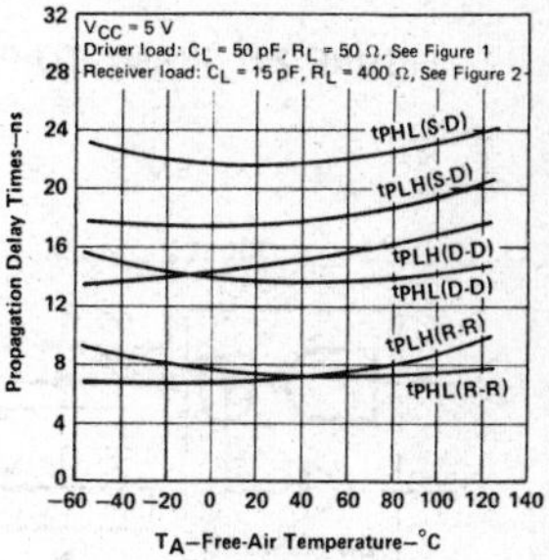
VCC = 5 V
Driver load: CL = 50 pF, RL = 50 Ω, See Figure 1
Receiver load: CL = 15 pF, RL = 400 Ω, See Figure 2
tPHL(S-D)
tPLH(S-D)
tPLH(D-D)
tPHL(D-D)
tPLH(R-R)
tPHL(R-R)
Propagation Delay Times—ns
TA—Free-Air Temperature—°C

FIGURE 17

PROPAGATION DELAY TIMES vs SUPPLY VOLTAGE

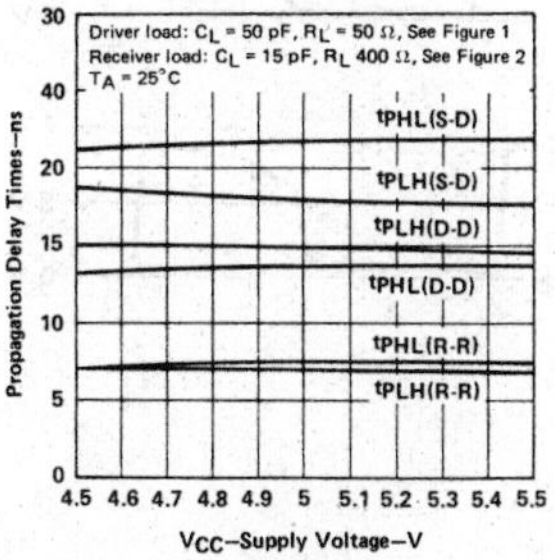
Driver load: CL = 50 pF, RL = 50 Ω, See Figure 1
Receiver load: CL = 15 pF, RL 400 Ω, See Figure 2
TA = 25°C
tPHL(S-D)
tPLH(S-D)
tPLH(D-D)
tPHL(D-D)
tPHL(R-R)
tPLH(R-R)
Propagation Delay Times—ns
VCC—Supply Voltage—V

FIGURE 18

DRIVER PROPAGATION DELAY TIMES vs LOAD CAPACITANCE

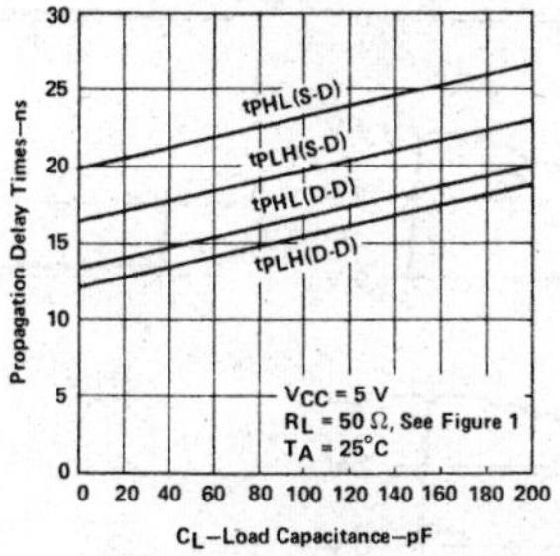
tPHL(S-D)
tPLH(S-D)
tPHL(D-D)
tPLH(D-D)
VCC = 5 V
RL = 50 Ω, See Figure 1
TA = 25°C
Propagation Delay Times—ns
CL—Load Capacitance—pF

FIGURE 19

RECEIVER PROPAGATION DELAY TIMES vs LOAD CAPACITANCE

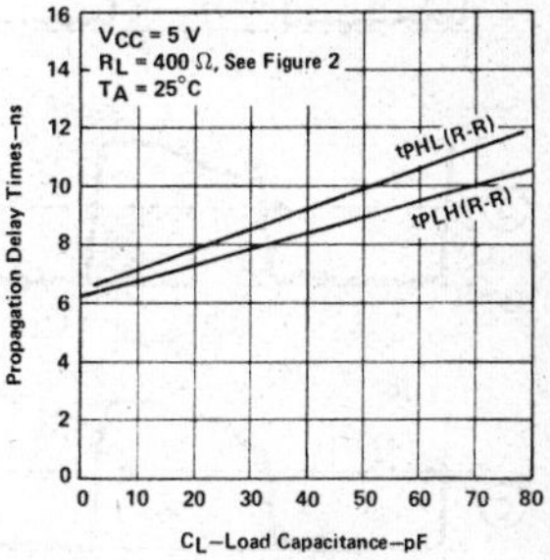
VCC = 5 V
RL = 400 Ω, See Figure 2
TA = 25°C
tPHL(R-R)
tPLH(R-R)
Propagation Delay Times—ns
CL—Load Capacitance—pF

FIGURE 20

†Data for temperatures below 0°C and above 70°C is applicable to SN55138 circuits only.

TYPICAL APPLICATION DATA

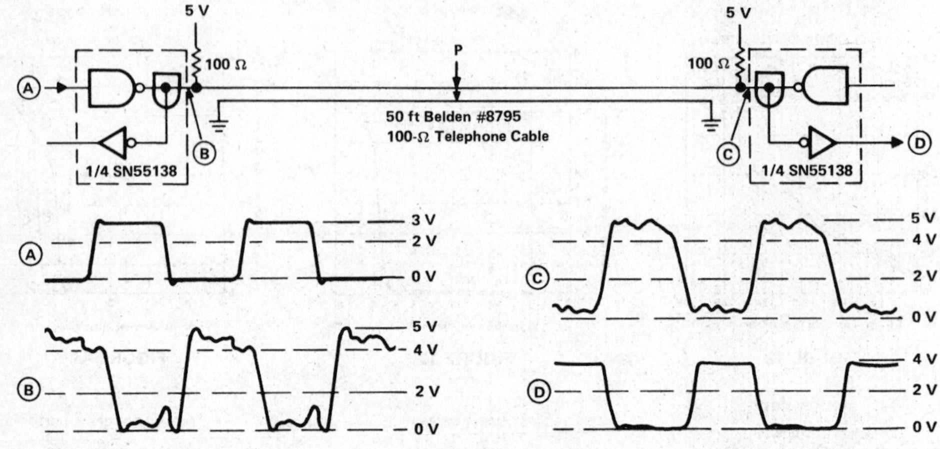
5 V
100 Ω
P
5 V
100 Ω
A
B
50 ft Belden #8795
100-Ω Telephone Cable
C
D
1/4 SN55138
1/4 SN55138
3 V
2 V
0 V
5 V
4 V
2 V
0 V
5 V
4 V
2 V
0 V
4 V
2 V
0 V

TYPICAL VOLTAGE WAVEFORMS

FIGURE 21. POINT-TO-POINT COMMUNICATION OVER 50 FEET OF TWISTED PAIR AT 5 MHz

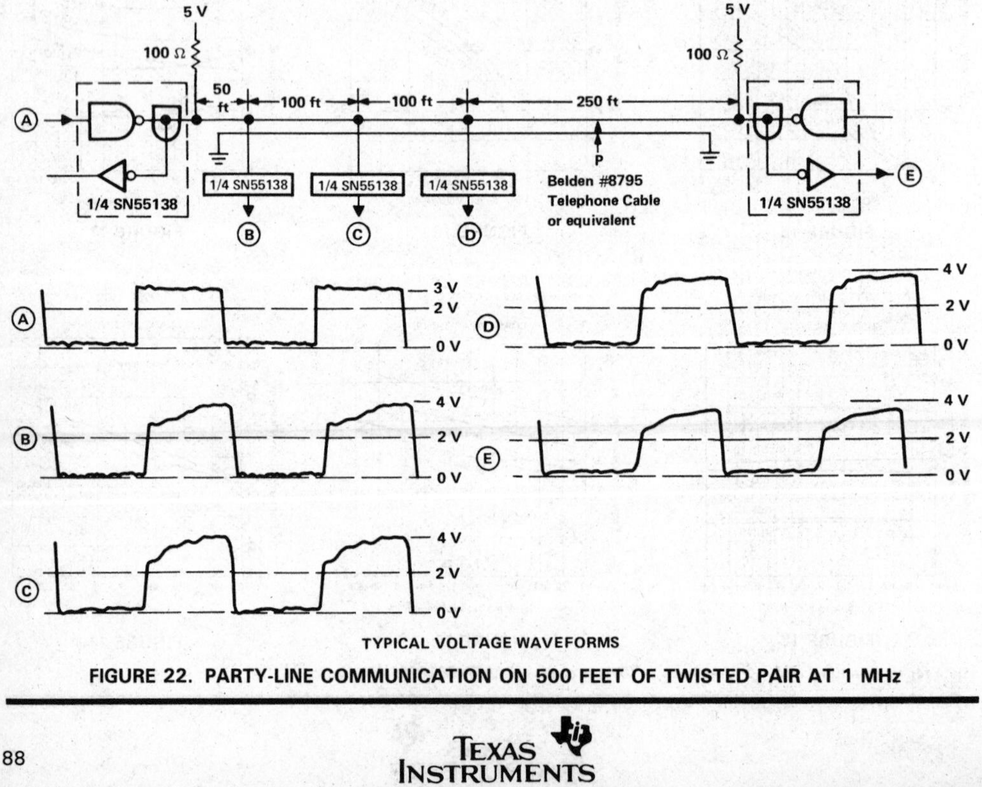
5 V
100 Ω
5 V
100 Ω
50 ft
100 ft
100 ft
250 ft
A
1/4 SN55138
1/4 SN55138
1/4 SN55138
1/4 SN55138
1/4 SN55138
B
C
D
E
P
Belden #8795
Telephone Cable
or equivalent
3 V
2 V
0 V
4 V
2 V
0 V

TYPICAL VOLTAGE WAVEFORMS

FIGURE 22. PARTY-LINE COMMUNICATION ON 500 FEET OF TWISTED PAIR AT 1 MHz

TYPICAL APPLICATION DATA

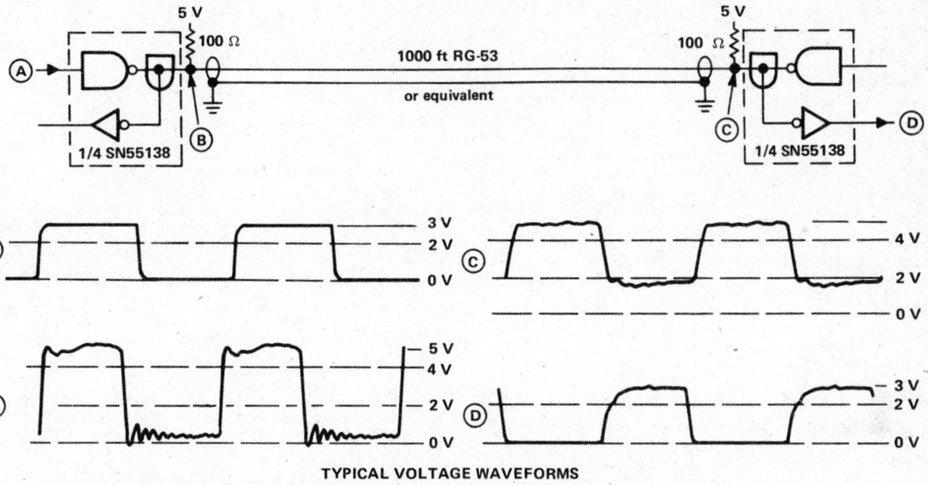
5 V
100 Ω
1000 ft RG-53
or equivalent
5 V
100 Ω
A
B
C
D
1/4 SN55138
1/4 SN55138
3 V
2 V
0 V
4 V
2 V
0 V
5 V
4 V
2 V
0 V
3 V
2 V
0 V
TYPICAL VOLTAGE WAVEFORMS

FIGURE 23. POINT-TO-POINT COMMUNICATION OVER 1000 FEET OF COAX AT 1 MHz

D2155, JANUARY 1977–REVISED OCTOBER 1986

- Single 5-V Supply
- ±100 mV Sensitivity
- For Application As:
  Single-Ended Line Receiver
  Gated Oscillator
  Level Comparator
- Adjustable Reference Voltage
- TTL Outputs
- TTL-Compatible Strobe
- Designed for Party-Line (Data-Bus) Applications
- Common Reference Pin
- Common Strobe
- '141 Has Diode-Protected Input Stage for Power-Off Condition

description

Each of these devices consists of a dual single-ended line receiver with TTL-compatible strobes and outputs. The reference voltage (switching threshold) is applied externally and can be adjusted from 1.5 volts to 3.5 volts, making it possible to optimize noise immunity for a given system design. Due to their low input current (less than 100 microamperes), they are ideally suited for party-line (bus-organized) systems.

The '140 has a common reference voltage pin and a common strobe. The '141 is the same as the '140 except that the input stage is diode protected.

The SN75140 and SN75141 are characterized for operation from 0°C to 70°C.

D, JG, OR P PACKAGE
(TOP VIEW)

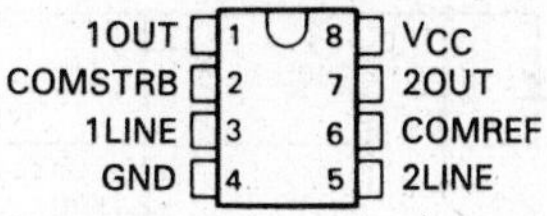
1OUT
COMSTRB
1LINE
GND
1
2
3
4
8
7
6
5
V_{CC}
2OUT
COMREF
2LINE

logic symbol†

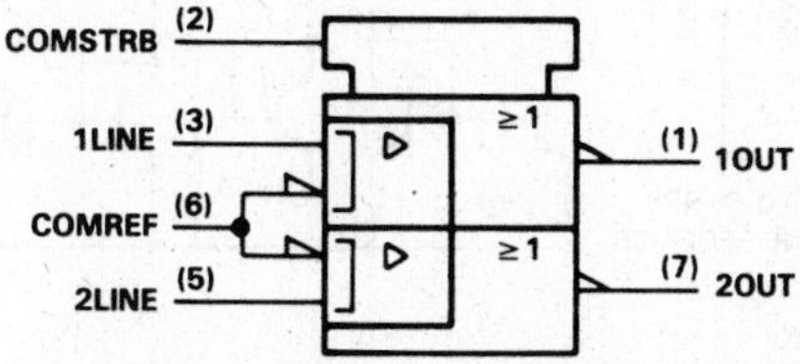
COMSTRB (2)
1LINE (3)
COMREF (6)
2LINE (5)
≥1
≥1
(1) 1OUT
(7) 2OUT

† This symbol is in accordance with ANSI/IEEE Std 91-1984 and IEC Publication 617-12.

logic diagram (positive logic)

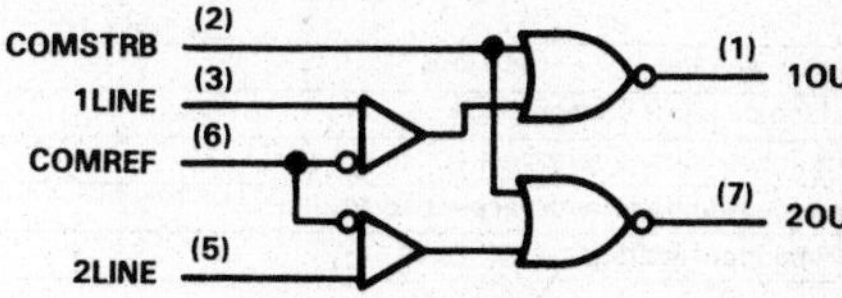
COMSTRB (2)
1LINE (3)
COMREF (6)
2LINE (5)
(1) 1OUT
(7) 2OUT

FUNCTION TABLE
(EACH RECEIVER)

LINE INPUT	STROBE	OUTPUT
≤ V_{ref} − 100 mV	L	H
≥ V_{ref} + 100 mV	X	L
X	H	L

H = high level, L = low level, X = irrelevant

schematic (each receiver)

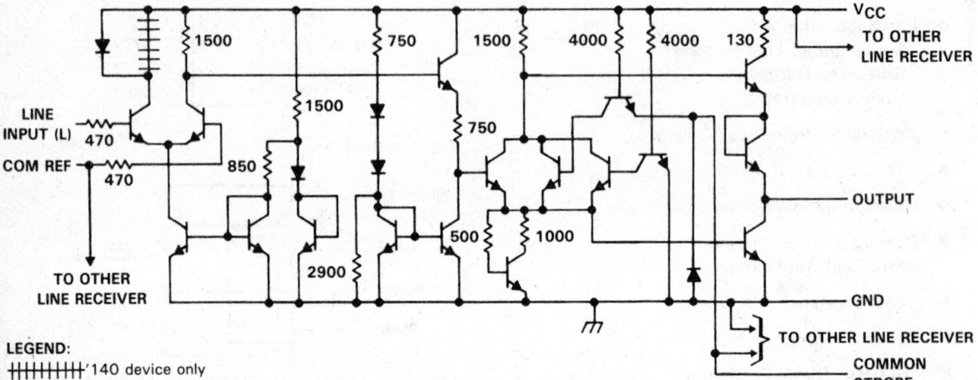
VCC
TO OTHER LINE RECEIVER
1500
750
1500
4000
4000
130
LINE INPUT (L)
470
COM REF
470
1500
750
850
OUTPUT
500
1000
2900
TO OTHER LINE RECEIVER
GND
TO OTHER LINE RECEIVER
COMMON STROBE

LEGEND:
'140 device only
Resistor values shown are nominal and in ohms.

absolute maximum ratings over operating free-air temperature range (unless otherwise noted)

			UNIT
Supply voltage, V_{CC} (see Note 1)		7	V
Reference input voltage, V_{ref}		5.5	V
Line input voltage with respect to ground		−2 to 5.5	V
Line input voltage with respect to V_{ref}		±5	V
Strobe input voltage		5.5	V
Continuous total dissipation at (or below) 25°C free-air temperature (see Note 2)	D package	725	mW
	JG package	825	mW
	P package	1000	mW
Operating free-air temperature range		0 to 70	°C
Storage temperature range		−65 to 150	°C
Lead temperature 1,6 mm (1/16 inch) from case for 60 seconds: JG package		300	°C
Lead temperature 1,6 mm (1/16 inch) from case for 10 seconds: D or P package		260	°C

NOTES: 1. Unless otherwise specified, voltage values are with respect to network terminal.
2. For operation above 25°C free-air temperature, refer to Dissipation Derating Curves in Appendix A. In the JG package, these chips are glass mounted. For SN75140 and SN75141 devices in the P package, use the 8.0-mW/°C curve.

recommended operating conditions

	MIN	NOM	MAX	UNIT
Supply voltage, V_{CC}	4.5	5	5.5	V
Reference input voltage, V_{ref}	1.5		3.5	V
High-level line input voltage, $V_{IH(L)}$	$V_{ref}+0.1$		$V_{CC}-1$	V
Low-level line input voltage, $V_{IL(L)}$	0		$V_{ref}-0.1$	V
High-level strobe input voltage, $V_{IH(S)}$	2		5.5	V
Low-level strobe input voltage, $V_{IL(S)}$	0		0.8	V

electrical characteristics over recommended operating free-air temperature range, V_{CC} = 5 V ± 10%, V_{ref} = 1.5 V to 3.5 V (unless otherwise noted)

PARAMETER			TEST CONDITIONS	MIN	TYP†	MAX	UNIT
$V_{IK(S)}$	Strobe input clamp voltage		$I_{I(S)}$ = −12 mA			−1.5	V
V_{OH}	High-level output voltage		$V_{IL(L)} = V_{ref}$ − 100 mV, $V_{IL(S)}$ = 0.8 V, I_{OH} = −400 μA	2.4			V
V_{OL}	Low-level output voltage		$V_{IH(L)} = V_{ref}$ + 100 mV, $V_{IL(S)}$ = 0.8 V, I_{OL} = 16 mA			0.4	V
			$V_{IL(L)} = V_{ref}$ − 100 mV, $V_{IH(S)}$ = 2 V, I_{OL} = 16 mA			0.4	
$I_{I(S)}$	Strobe input current at maximum input voltage	Strobe	$V_{I(S)}$ = 5.5 V			1	mA
		Com strb				2	
I_{IH}	High-level input current	Strobe	$V_{I(S)}$ = 2.4 V			40	μA
		Com strb				80	
		Line input	$V_{I(L)}$ = 3.5 V, V_{ref} = 1.5 V		35	100	
		Reference	$V_{I(L)}$ = 0 V, V_{ref} = 3.5 V		35	100	
		Com ref			70	200	
I_{IL}	Low-level input current	Strobe	$V_{I(S)}$ = 0.4 V			−1.6	mA
		Com strb				−3.2	
		Line input	$V_{I(L)}$ = 0 V, V_{ref} = 1.5 V			−10	μA
		Reference	$V_{I(L)}$ = 1.5 V, V_{ref} = 0 V			−10	
		Com ref				−20	
I_{OS}	Short-circuit output current‡		V_{CC} = 5.5 V	−18		−55	mA
I_{CCH}	Supply current, output high		$V_{I(S)}$ = 0 V, $V_{I(L)} = V_{ref}$ − 100 mV		18	30	mA
I_{CCL}	Supply current, output low		$V_{I(S)}$ = 0 V, $V_{I(L)} = V_{ref}$ + 100 mV		20	35	mA

† All typical values are at V_{CC} = 5 V, T_A = 25 °C.
‡ Only one output should be shorted at a time.

switching characteristics, V_{CC} = 5 V, V_{ref} = 2.5 V, T_A = 25 °C

PARAMETER		TEST CONDITIONS	MIN	TYP	MAX	UNIT
$t_{PLH(L)}$	Propagation delay time, low-to-high-level output from line input	C_L = 15 pF, R_L = 400 Ω, See Figure 1		22	35	ns
$t_{PHL(L)}$	Propagation delay time, high-to-low-level output from line input			22	30	
$t_{PLH(S)}$	Propagation delay time, low-to-high-level output from strobe input			12	22	ns
$t_{PHL(S)}$	Propagation delay time, high-to-low-level output from strobe input			8	15	

PARAMETER MEASUREMENT INFORMATION

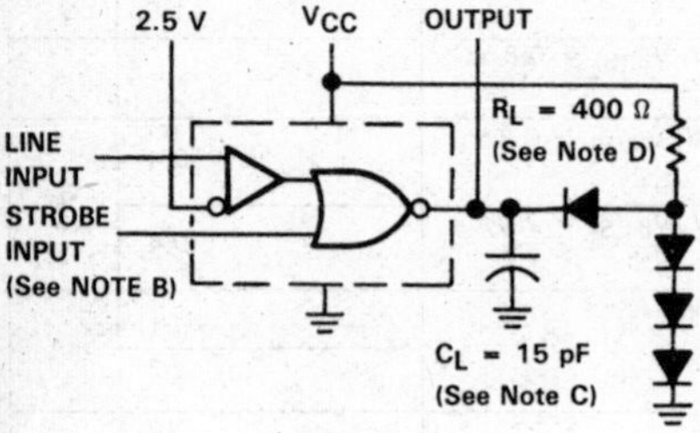
2.5 V
VCC
OUTPUT
LINE
INPUT
STROBE
INPUT
(See NOTE B)
RL = 400 Ω
(See Note D)
CL = 15 pF
(See Note C)

TEST CIRCUIT

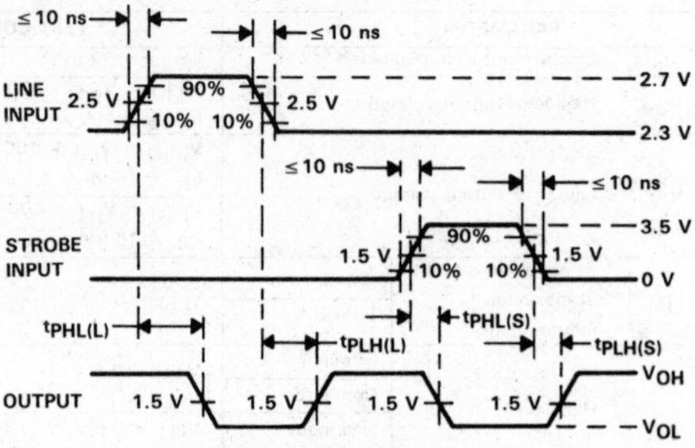
≤10 ns
≤10 ns
LINE
INPUT
2.5 V
90%
10%
10%
2.5 V
2.7 V
2.3 V
≤10 ns
≤10 ns
STROBE
INPUT
1.5 V
90%
10%
10%
1.5 V
3.5 V
0 V
tPHL(L)
tPLH(L)
tPHL(S)
tPLH(S)
OUTPUT
1.5 V
1.5 V
1.5 V
1.5 V
VOH
VOL

VOLTAGE WAVEFORMS

NOTES: A. Input pulses are supplied by generators having the following characteristics: PRR ≤ 1 MHz, duty cycle ≤50%, Z_{out} = 50 Ω.
B. Unused strobes are to be grounded.
C. C_L includes probe and jig capacitance.
D. All diodes are 1N3064.

FIGURE 1

TYPICAL CHARACTERISTICS

OUTPUT VOLTAGE
vs
LINE INPUT VOLTAGE

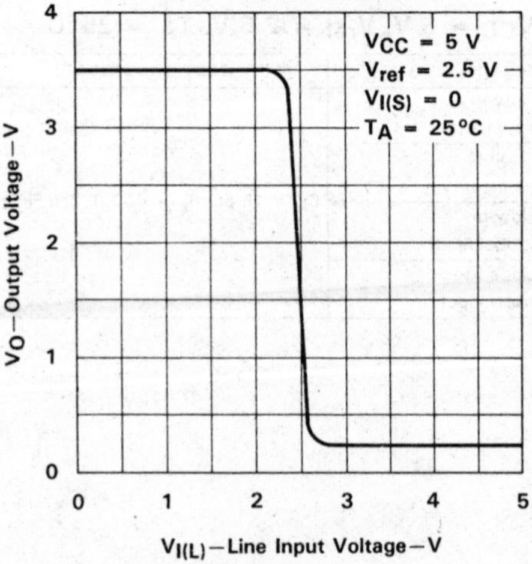
VCC = 5 V
Vref = 2.5 V
VI(S) = 0
TA = 25°C
VO—Output Voltage—V
VI(L)—Line Input Voltage—V
0
1
2
3
4
5

FIGURE 2

TEXAS INSTRUMENTS
POST OFFICE BOX 655012 • DALLAS, TEXAS 75265

TYPICAL APPLICATION DATA

line receiver

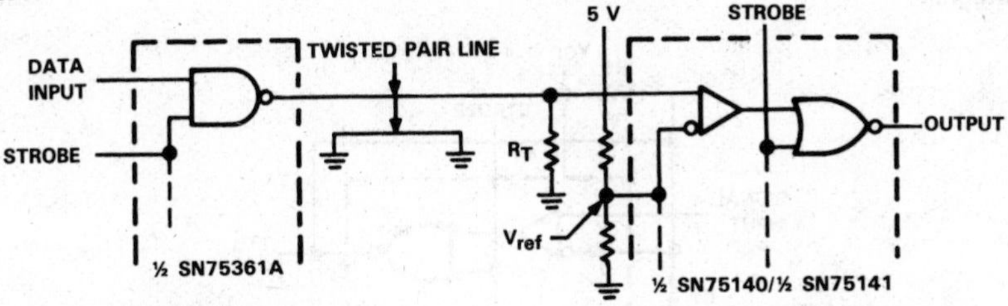
5 V
STROBE
TWISTED PAIR LINE
DATA
INPUT
OUTPUT
STROBE
R_T
V_{ref}
½ SN75361A
½ SN75140/½ SN75141

high fan-out from standard TTL gate

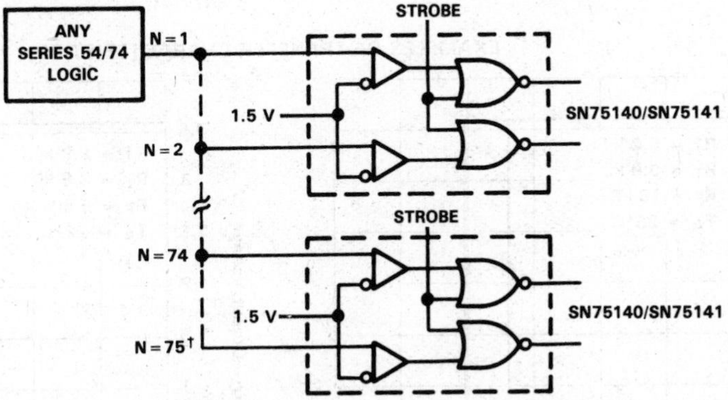
ANY
SERIES 54/74
LOGIC
N = 1
STROBE
1.5 V
SN75140/SN75141
N = 2
STROBE
N = 74
1.5 V
SN75140/SN75141
N = 75†

†Although most Series 54/74 circuits have a guaranteed 2.4-V output at 400 μA, they are typically capable of maintaining a 2.4-V output level under a load of 7.5 mA.

dual bus transceiver

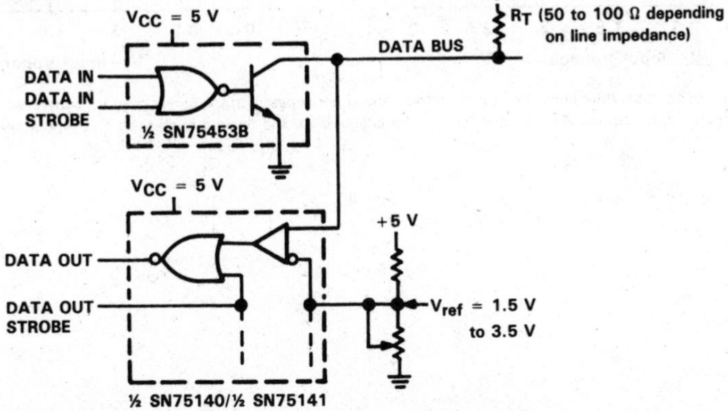
+5 V
R_T (50 to 100 Ω depending
on line impedance)
V_{CC} = 5 V
DATA BUS
DATA IN
DATA IN
STROBE
½ SN75453B
V_{CC} = 5 V
+5 V
DATA OUT
DATA OUT
STROBE
V_{ref} = 1.5 V
to 3.5 V
½ SN75140/½ SN75141

Using this arrangement, as many as 100 transceivers can be connected to a single data bus. The adjustable reference voltage feature allows the noise margin to be optimized for a given system. The complete dual bus transceiver (SN75453B driver and SN75140 receiver) can be assembled in approximately the same space required by a single 16-pin package and only one power supply is required (+5 V). Data In and Data Out terminals are TTL compatible.

TYPICAL APPLICATION DATA

schmitt trigger

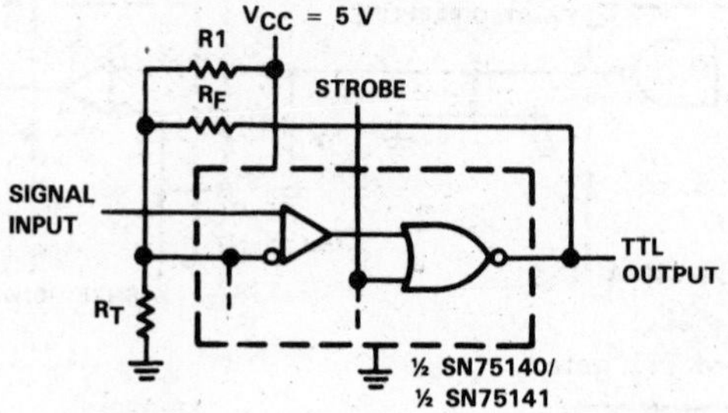
VCC = 5 V
R1
RF
STROBE
SIGNAL INPUT
TTL OUTPUT
RT
½ SN75140/
½ SN75141

EXAMPLES OF TRANSFER CHARACTERISTICS

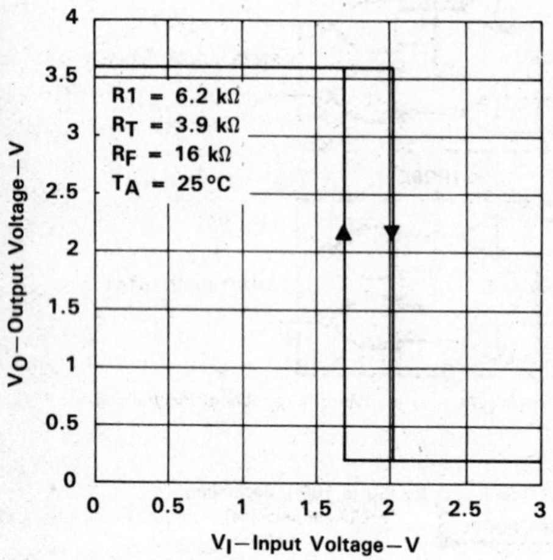
R1 = 6.2 kΩ
RT = 3.9 kΩ
RF = 16 kΩ
TA = 25°C
VO—Output Voltage—V
VI—Input Voltage—V
0
0.5
1
1.5
2
2.5
3
3.5
4

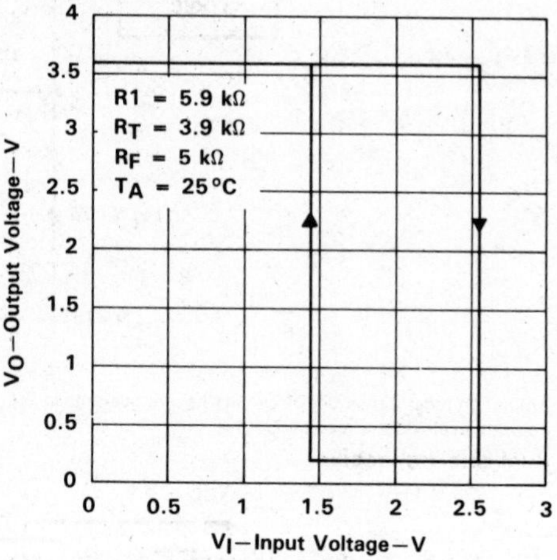
R1 = 5.9 kΩ
RT = 3.9 kΩ
RF = 5 kΩ
TA = 25°C
VO—Output Voltage—V
VI—Input Voltage—V
0
0.5
1
1.5
2
2.5
3
3.5
4

Slowly changing input levels from data lines, optical detectors, and other types of transducers may be converted to standard TTL signals with this Schmitt trigger circuit. R1, R_F, and R_T may be adjusted for the desired hysteresis and trigger levels.

TYPICAL APPLICATION DATA

gated oscillator

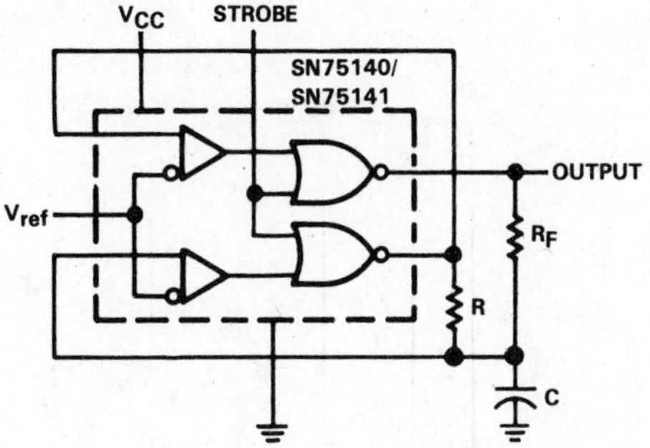
VCC
STROBE
SN75140/
SN75141
Vref
OUTPUT
RF
R
C

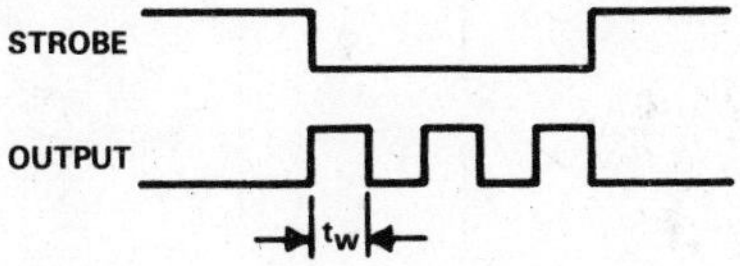
STROBE
OUTPUT
tw

OSCILLATOR FREQUENCY
vs
RC TIME CONSTANT

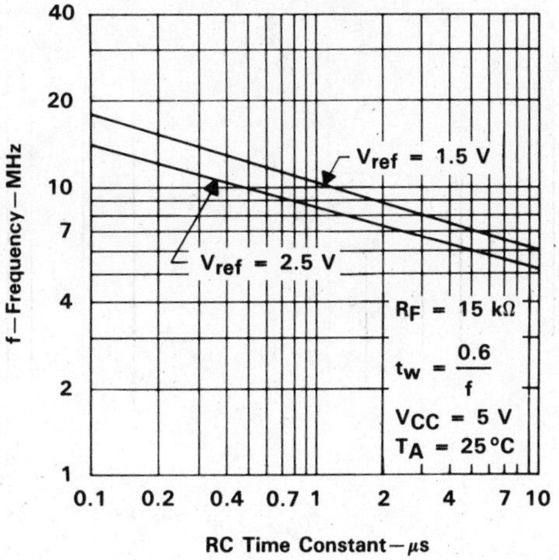
Vref = 1.5 V
Vref = 2.5 V
RF = 15 kΩ
tw = 0.6/f
VCC = 5 V
TA = 25°C
f—Frequency—MHz
RC Time Constant—μs

D2609, FEBRUARY 1986

- Meets EIA Standards RS-422-A and RS-423-A
- Meets EIA Standards RS-232 and CCITT V.28 with External Components
- Meets Federal Standards 1020 and 1030
- Built-in 5-MHz Low-Pass Filter
- Operates from Single 5-V Power Supply
- Wide Common-Mode Voltage Range
- High Input Impedance
- TTL-Compatible Outputs
- 8-Pin Dual-In-Line Package
- Pinout Compatible with the μA9637 and μA9639

D, JG, OR P PACKAGE
(TOP VIEW)

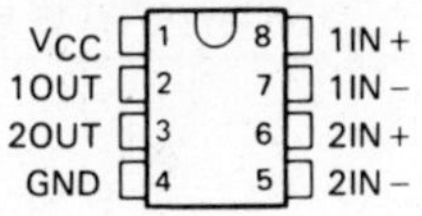
VCC
1OUT
2OUT
GND
1
2
3
4
8
7
6
5
1IN+
1IN−
2IN+
2IN−

logic symbol†

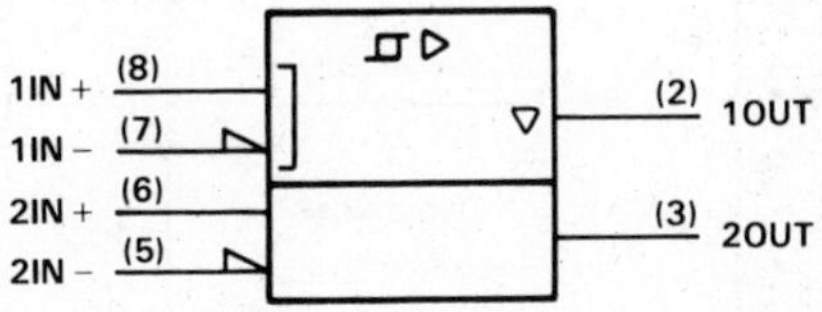
1IN+
1IN−
2IN+
2IN−
(8)
(7)
(6)
(5)
(2)
(3)
1OUT
2OUT

† This symbol is in accordance with ANSI/IEEE Std 91-1984 and IEC Publication 617-12.

logic diagram

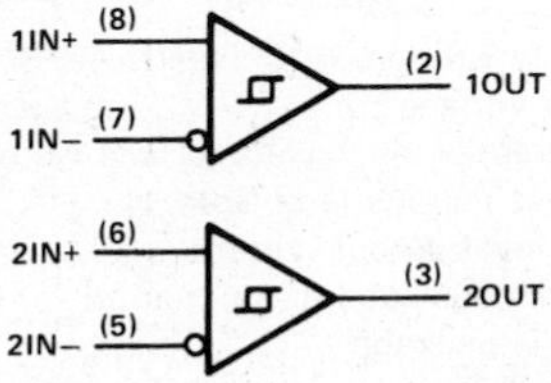
1IN+
1IN−
2IN+
2IN−
(8)
(7)
(6)
(5)
(2)
(3)
1OUT
2OUT

description

The SN75146 is a dual differential line receiver designed to meet EIA standards RS-422-A and RS-423-A. The receiver is designed to have a constant impedance with input voltages of ±3 volts to ±25 volts allowing it to meet the requirements of EIA standard RS-232-C and CCITT recommendation V.28 with the addition of an external bias resistor. This receiver is designed for low-speed operation below 355 kilohertz, and has a built-in 5-megahertz low-pass filter to attenuate high-frequency noise. The inputs are compatible with either a single-ended or a differential line system and the outputs are TTL compatible. This device operates from a single 5-volt power supply and is supplied in both the 8-pin dual-in-line and small outline packages.

The SN75146 is characterized for operation from 0°C to 70°C.

TEXAS INSTRUMENTS
POST OFFICE BOX 655012 • DALLAS, TEXAS 75265

schematics of inputs and outputs

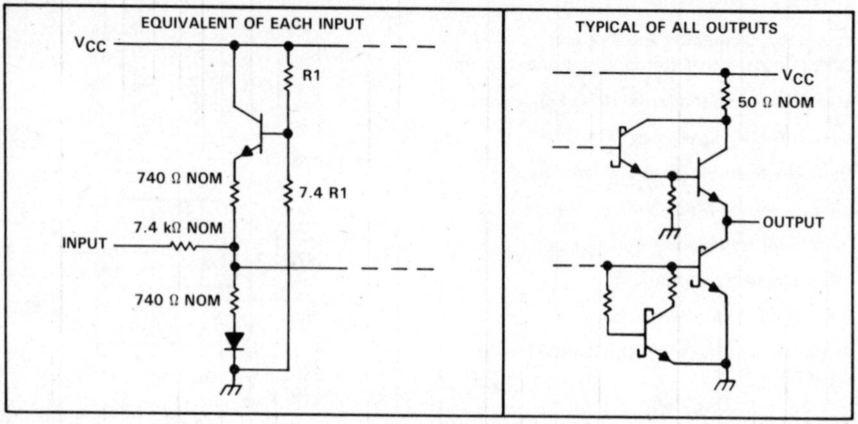
EQUIVALENT OF EACH INPUT
VCC
R1
740 Ω NOM
7.4 R1
7.4 kΩ NOM
INPUT
740 Ω NOM
TYPICAL OF ALL OUTPUTS
VCC
50 Ω NOM
OUTPUT

absolute maximum ratings over operating free-air temperature range (unless otherwise noted)

Supply voltage, V_{CC} (see Note 1) . . . . . −0.5 V to 7 V
Input voltage . . . . . ±25 V
Differential input voltage (see Note 2) . . . . . ±25 V
Output voltage (see Note 1) . . . . . −0.5 V to 5.5 V
Low-level output current . . . . . 50 mA
Continuous total dissipation at (or below) 25°C free-air temperature (see Note 3):
D package . . . . . 725 mW
JG package . . . . . 825 mW
P package . . . . . 1000 mW
Operating free-air temperature range . . . . . 0°C to 70°C
Storage temperature range . . . . . −65°C to 150°C
Lead temperature 1,6 mm (1/16 inch) from case for 60 seconds: JG package . . . . . 300°C
Lead temperature 1,6 mm (1/16 inch) from case for 10 seconds: D and P package . . . . . 260°C

NOTES: 1. All voltage values, except differential input voltage, are with respect to the network ground terminal.
2. Differential input voltage is measured at the noninverting input with respect to the corresponding inverting input.
3. For operation above 25°C free-air temperature, derate the JG package to 528 mW at 70°C at the rate of 6.6 mW/°C, the D package to 464 mW at 70°C at the rate of 5.8 mW/°C, and the P package to 640 mW at 70°C at the rate of 8 mW/°C. The SN75146 chips are glass mounted in the JG package.

recommended operating conditions

	MIN	NOM	MAX	UNIT
Supply voltage, V_{CC}	4.75	5	5.25	V
Common-mode input voltage, V_{IC}			±7	V
Operating free-air temperature, T_A	0	25	70	°C

TEXAS INSTRUMENTS
POST OFFICE BOX 655012 • DALLAS, TEXAS 75265

electrical characteristics over recommended ranges of supply voltage, common-mode input voltage, and operating free-air temperature (unless otherwise noted)

PARAMETER		TEST CONDITIONS		MIN	TYP†	MAX	UNIT
V_T	Threshold voltage (V_{T+} and V_{T-})			−0.2‡		0.2	V
		See Note 4		−0.4‡		0.4	
V_{hys}	Hysteresis ($V_{T+} - V_{T-}$)				70		mV
V_{IB}	Input bias voltage	$I_I = 0$		2		2.4	V
V_{OH}	High-level output voltage	$V_{ID} = 0.2$ V,	$I_O = -1$ mA	2.5	3.5		V
V_{OL}	Low-level output voltage	$V_{ID} = -0.2$ V,	$I_O = 20$ mA		0.35	0.5	V
r_i	Input resistance	See Note 5,	$V_I = 3$ V to 25 V or $V_I = -3$ V to -25 V	6	7.8	9.5	kΩ
I_I	Input current	$V_{CC} = 0$ to 5.5 V, See Note 6	$V_I = 10$ V		1.1	3.25	mA
			$V_I = -10$ V		−1.6	−3.25	
I_{OS}	Short-circuit output current§	$V_O = 0$,	$V_{ID} = 0.2$ V	−40	−75	−100	mA
I_{CC}	Supply current	$V_{ID} = -0.5$ V,	No load		35	50	mA

† All typical values are at $V_{CC} = 5$ V, $T_A = 25°C$.
‡ The algebraic convention, in which the less positive (more negative) limit is designated as minimum, is used in this data sheet for threshold levels only.
§ Only one output should be shorted at a time, and duration of the short-circuit should not exceed one second.
NOTES: 4. The expanded threshold parameter is tested with a 500-Ω resistor in series with each input.
5. r_i is defined by $\Delta V_I / \Delta I_I$.
6. The input not under test is grounded.

switching characteristics, $V_{CC} = 5$ V, $T_A = 25°C$

PARAMETER		TEST CONDITION	MIN	TYP	MAX	UNIT
t_{PLH}	Propagation delay time, low-to-high-level output	$C_L = 30$ pF, See Figure 1	100	150	300	ns
t_{PHL}	Propagation delay time, high-to-low-level output		100	150	300	ns

PARAMETER MEASUREMENT INFORMATION

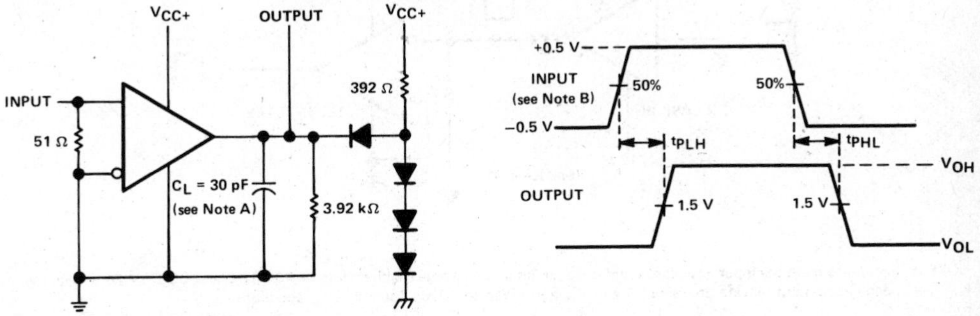
VCC+
OUTPUT
VCC+
INPUT
51 Ω
392 Ω
CL = 30 pF
(see Note A)
3.92 kΩ
+0.5 V
INPUT
(see Note B)
50%
50%
−0.5 V
tPLH
tPHL
VOH
OUTPUT
1.5 V
1.5 V
VOL

TEST CIRCUIT

VOLTAGE WAVEFORM

NOTES: A. C_L includes probe and jig capacitance.
B. The input pulse is supplied by a generator having the following characteristics: $t_r \leq 5$ ns, $t_f \leq 5$ ns, PRR ≤ 300 kHz, duty cycle = 50%.

FIGURE 1. TRANSITION TIMES

4

Line Drivers/Receivers

TYPICAL CHARACTERISTICS

OUTPUT VOLTAGE
vs
DIFFERENTIAL INPUT VOLTAGE

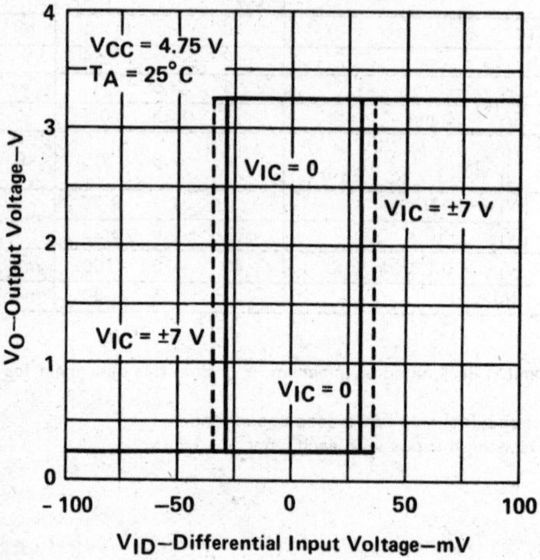

4
3
2
1
0
V_{CC} = 4.75 V
T_A = 25°C
V_{IC} = 0
V_{IC} = ±7 V
V_{IC} = ±7 V
V_{IC} = 0
V_O—Output Voltage—V
−100
−50
0
50
100
V_{ID}—Differential Input Voltage—mV

FIGURE 2

OUTPUT VOLTAGE
vs
DIFFERENTIAL INPUT VOLTAGE

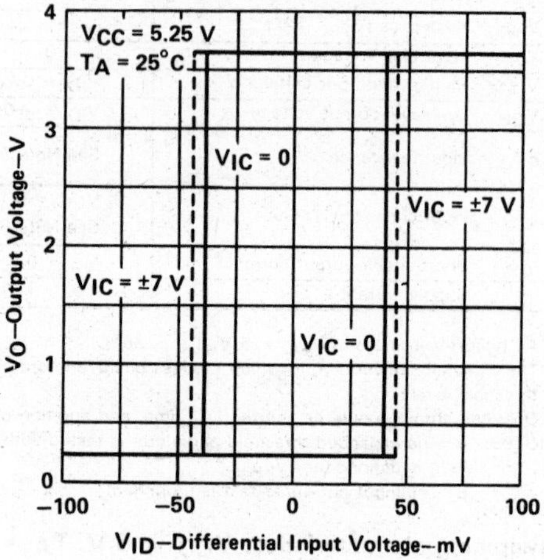

4
3
2
1
0
V_{CC} = 5.25 V
T_A = 25°C
V_{IC} = 0
V_{IC} = ±7 V
V_{IC} = ±7 V
V_{IC} = 0
V_O—Output Voltage—V
−100
−50
0
50
100
V_{ID}—Differential Input Voltage—mV

FIGURE 3

TYPICAL APPLICATION DATA

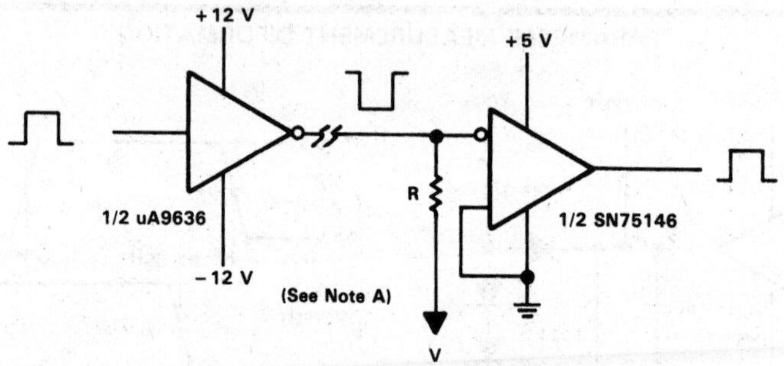

+12 V
−12 V
1/2 uA9636
+5 V
R
1/2 SN75146
(See Note A)
V

NOTE A: In order to meet the input-impedance and open-circuit-input voltage requirements of RS-232-C and CCITT V.28 and guarantee open-circuit-input failsafe operation, R and V are selected to satisfy the following equations:

$$V = -1.1 - 3.3 \frac{R}{r_i} \text{ volts}$$

$$3\ k\Omega \leq \frac{R(r_i)}{R + r_i} \leq 7\ k\Omega$$

FIGURE 4. RS-232-C SYSTEM APPLICATIONS

TYPICAL APPLICATION DATA

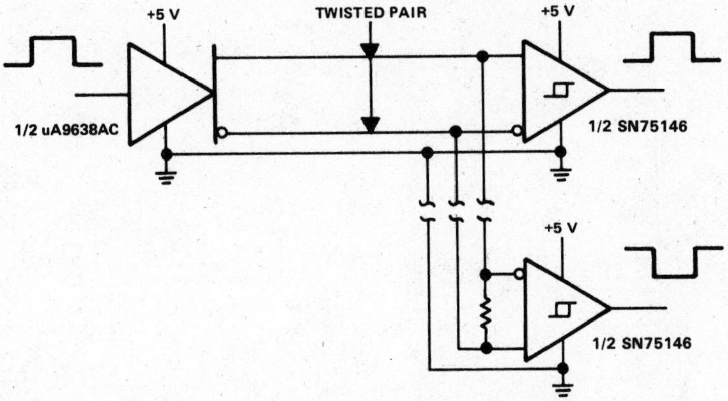
+5 V
TWISTED PAIR
+5 V
1/2 uA9638AC
1/2 SN75146
+5 V
1/2 SN75146

FIGURE 5. RS-422-A SYSTEM APPLICATIONS

D951, JANUARY 1971—REVISED SEPTEMBER 1986

- **Satisfies Requirements of EIA Standard RS-232-C**
- **Withstands Sustained Output Short-Circuit to any Low-Impedance Voltage Between −25 V and 25 V**
- **2 μs Max Transition Time Through the +3 V to −3 V Transition Region Under Full 2500-pF Load**
- **Inputs Compatible with Most TTL Families**
- **Common Strobe Input**
- **Inverting Output**
- **Slew Rate can be Controlled with an External Capacitor at the Output**
- **Standard Supply Voltages . . . ±12 V**

SN55150 . . . JG PACKAGE
SN75150 . . . D, JG, OR P PACKAGE
(TOP VIEW)

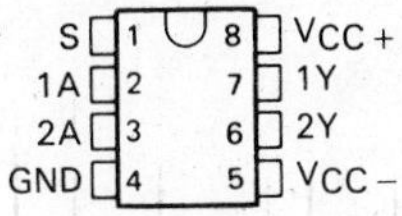
S 1
1A 2
2A 3
GND 4
8 VCC+
7 1Y
6 2Y
5 VCC−

SN55150 . . . FK PACKAGE
(TOP VIEW)

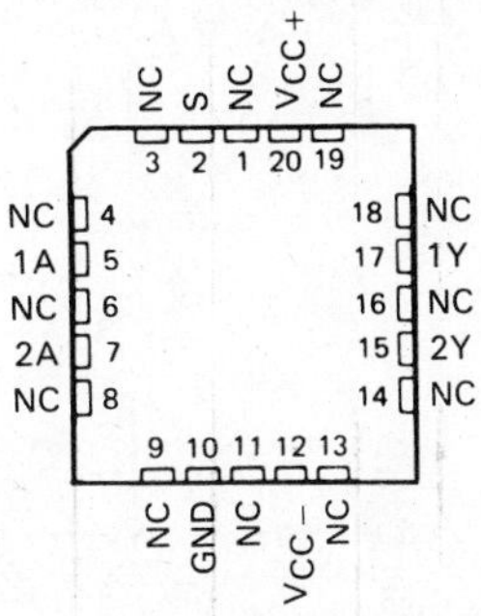
NC S NC VCC+ NC
3 2 1 20 19
NC 4
1A 5
NC 6
2A 7
NC 8
18 NC
17 1Y
16 NC
15 2Y
14 NC
9 10 11 12 13
NC GND NC VCC− NC

NC—No internal connection

description

The SN55150 and SN75150 are monolithic dual line drivers designed to satisfy the requirements of the standard interface between data terminal equipment and data communication equipment as defined by EIA Standard RS-232-C. A rate of 20,000 bits per second can be transmitted with a full 2500-pF load. Other applications are in data-transmission systems using relatively short single lines, in level translators, and for driving MOS devices. The logic input is compatible with most TTL families. Operation is from +12-volt and −12-volt power supplies.

The SN55150 is characterized for operation over the full military temperature range of −55°C to 125°C. The SN75150 is characterized for operation from 0°C to 70°C.

logic symbol†

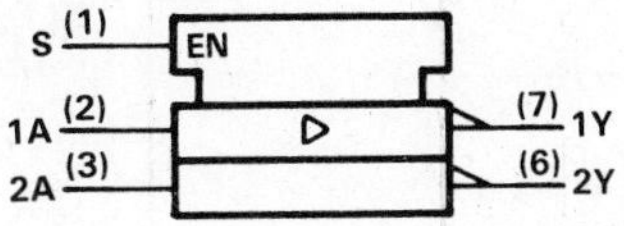
S (1)
EN
1A (2)
2A (3)
(7) 1Y
(6) 2Y

† This symbol is in accordance with ANSI/IEEE Std 91-1984 and IEC Publication 617-12.
Pin numbers shown are for D, JG, and P packages.

logic diagram (positive logic)

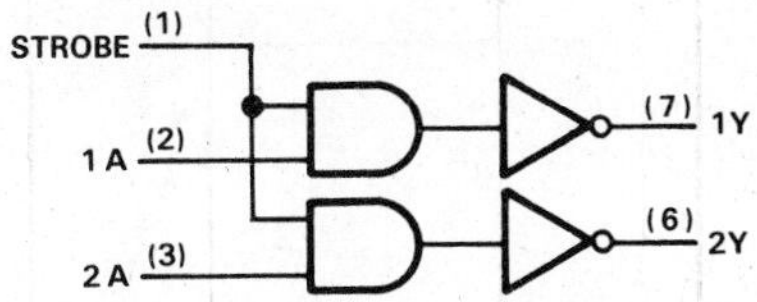
STROBE (1)
1 A (2)
2 A (3)
(7) 1Y
(6) 2Y

schematic (each line driver)

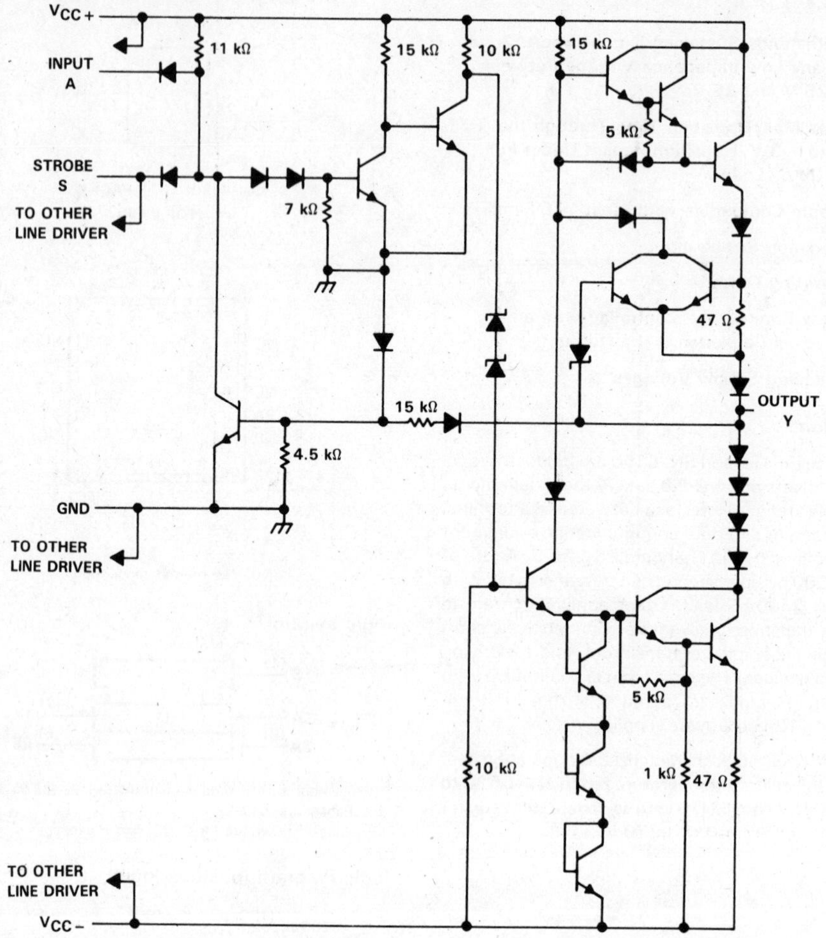
VCC+
INPUT A
11 kΩ
15 kΩ
10 kΩ
15 kΩ
5 kΩ
STROBE S
TO OTHER LINE DRIVER
7 kΩ
47 Ω
15 kΩ
OUTPUT Y
4.5 kΩ
GND
TO OTHER LINE DRIVER
5 kΩ
10 kΩ
1 kΩ
47 Ω
TO OTHER LINE DRIVER
VCC−

TEXAS INSTRUMENTS
POST OFFICE BOX 655012 • DALLAS, TEXAS 75265

absolute maximum ratings over operating free-air temperature range (unless otherwise noted)

		SN55150	SN75150	UNIT
Supply voltage V_{CC+} (see Note 1)		15	15	V
Supply voltage V_{CC-}		−15	−15	V
Input voltage		15	15	V
Applied output voltage		±25	±25	V
Continuous total dissipation at (or below) 25°C free-air temperature (see Note 2)	D package		725	mW
	FK package	1375		mW
	JG package	1050	825	mW
	P package		1000	mW
Operating free-air temperature range		−55 to 125	0 to 70	°C
Storage temperature range		−65 to 150	−65 to 150	°C
Case temperature for 60 seconds: FK package		260		°C
Lead temperature 1,6 mm (1/16 inch) from case for 60 seconds: JG package		300		°C
Lead temperature 1,6 mm (1/16 inch) from case for 10 seconds: D or P package			260	°C

NOTES: 1. Voltage values are with respect to network ground terminal.
2. For operation above 25°C free-air temperature, refer to Dissipation Derating Curves in Appendix A. In the JG package, SN55150 chips are alloy mounted and SN75150 chips are glass mounted. In the P package use the 8.0-mW/°C curve for these devices.

recommended operating conditions

	SN55150			SN75150			UNIT
	MIN	NOM	MAX	MIN	NOM	MAX	
Supply voltage, V_{CC+}	10.8	12	13.2	10.8	12	13.2	V
Supply voltage, V_{CC-}	−10.8	−12	−13.2	−10.8	−12	−13.2	V
High-level input voltage, V_{IH}	2		5.5	2		5.5	V
Low-level input voltage, V_{IL}	0		0.8	0		0.8	V
Applied output voltage, V_O			±15			±15	V
Operating free-air temperature, T_A	−55		125	0		70	°C

electrical characteristics over recommended operating free-air temperature range (unless otherwise noted)

PARAMETER		TEST CONDITIONS		MIN	TYP†	MAX	UNIT
V_{OH}	High-level output voltage	V_{CC+} = 10.8 V, V_{CC-} = −13.2 V, V_{IL} = 0.8 V, R_L = 3 kΩ to 7 kΩ		5	8		V
V_{OL}	Low-level output voltage (see Note 3)	V_{CC+} = 10.8 V, V_{CC-} = −10.8 V, V_{IH} = 2 V, R_L = 3 kΩ to 7 kΩ			−8	−5	V
I_{IH}	High-level input current	V_{CC+} = 13.2 V, V_{CC-} = −13.2 V, V_I = 2.4 V	Data input		1	10	μA
			Strobe input		2	20	
I_{IL}	Low-level input current	V_{CC+} = 13.2 V, V_{CC-} = −13.2 V, V_I = 0.4 V	Data input		−1	−1.6	mA
			Strobe input		−2	−3.2	
I_{OS}	Short-circuit output current‡	V_{CC+} = 13.2 V, V_{CC-} = −13.2 V	V_O = 25 V		2	8	mA
			V_O = −25 V		−3	−8	
			V_O = 0, V_I = 3 V	10	15	30	
			V_O = 0, V_I = 0	−10	−15	−30	
I_{CCH+}	Supply current from V_{CC+}, high-level output	V_{CC+} = 13.2 V, V_{CC-} = −13.2 V, V_I = 0, R_L = 3 kΩ, T_A = 25°C			10	22	mA
I_{CCH-}	Supply current from V_{CC-}, high-level output				−1	−10	
I_{CCL+}	Supply current from V_{CC+}, low-level output	V_{CC+} = 13.2 V, V_{CC-} = −13.2 V, V_I = 3 V, R_L = 3 kΩ, T_A = 25°C			8	17	mA
I_{CCL-}	Supply current from V_{CC-}, low-level output				−9	−20	

† All typical values are at V_{CC+} = 12 V, V_{CC-} = −12 V, T_A = 25°C.
‡ Not more than one output should be shorted at a time.
NOTE 3: The algebraic convention, in which the less positive (more negative) limit is designated as minimum, is used in this data sheet for logic levels only, e.g., when −5 V is the maximum, the typical value is a more negative voltage.

switching characteristics, V_{CC+} = 12 V, V_{CC-} = −12 V, T_A = 25°C (see Figure 1)

PARAMETER		TEST CONDITIONS	MIN	TYP	MAX	UNIT
t_{TLH}	Transition time, low-to-high-level output	C_L = 2500 pF, R_L = 3 kΩ to 7 kΩ	0.2	1.4	2	μs
t_{THL}	Transition time, high-to-low-level output		0.2	1.5	2	μs
t_{TLH}	Transition time, low-to-high-level output	C_L = 15 pF, R_L = 7 kΩ		40		ns
t_{THL}	Transition time, high-to-low-level output			20		ns
t_{PLH}	Propagation delay time, low-to-high-level output	C_L = 15 pF, R_L = 7 kΩ		60		ns
t_{PHL}	Propagation delay time, high-to-low-level output			45		ns

TEXAS INSTRUMENTS
POST OFFICE BOX 655012 • DALLAS, TEXAS 75265

PARAMETER MEASUREMENT INFORMATION

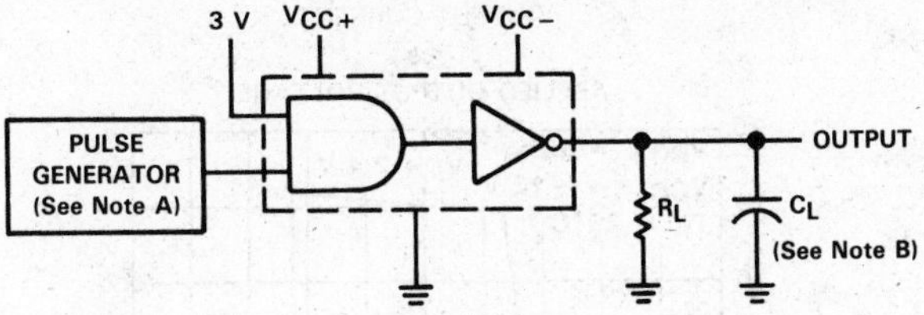
3 V
V_{CC+}
V_{CC-}
PULSE GENERATOR (See Note A)
OUTPUT
R_L
C_L
(See Note B)

TEST CIRCUIT

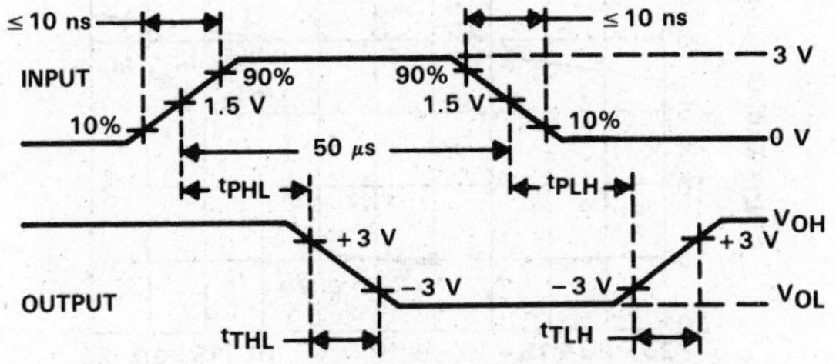
≤ 10 ns
≤ 10 ns
INPUT
90%
1.5 V
10%
3 V
0 V
50 µs
t_{PHL}
t_{PLH}
+3 V
−3 V
V_{OH}
V_{OL}
OUTPUT
t_{THL}
t_{TLH}

VOLTAGE WAVEFORMS

NOTES: A. The pulse generator has the following characteristics: duty cycle ≤ 50%, Z_{out} ≈ 50 Ω.
B. C_L includes probe and jig capacitance.

FIGURE 1. SWITCHING CHARACTERISTICS

TYPICAL CHARACTERISTICS

OUTPUT CURRENT
vs
APPLIED OUTPUT VOLTAGE

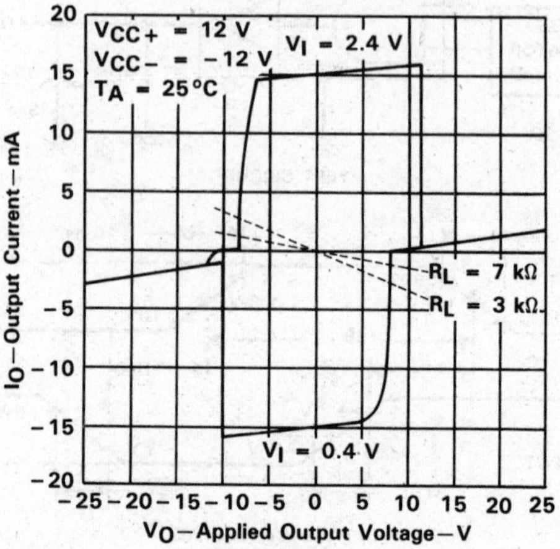
20
15
10
5
0
−5
−10
−15
−20
V_{CC+} = 12 V
V_{CC-} = −12 V
T_A = 25°C
V_I = 2.4 V
R_L = 7 kΩ
R_L = 3 kΩ
V_I = 0.4 V
I_O—Output Current—mA
−25 −20 −15 −10 −5 0 5 10 15 20 25
V_O—Applied Output Voltage—V

FIGURE 2

TYPICAL APPLICATION DATA

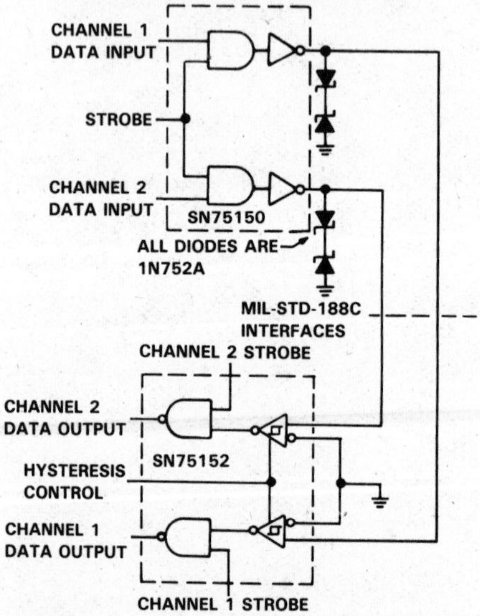
CHANNEL 1
DATA INPUT
STROBE
CHANNEL 2
DATA INPUT
SN75150
ALL DIODES ARE
1N752A
MIL-STD-188C
INTERFACES
CHANNEL 2 STROBE
CHANNEL 2
DATA OUTPUT
SN75152
HYSTERESIS
CONTROL
CHANNEL 1
DATA OUTPUT
CHANNEL 1 STROBE

FIGURE 3. DUAL-CHANNEL SINGLE-ENDED INTERFACE CIRCUIT MEETING MIL-STD-188C, PARAGRAPH 7.2.

TEXAS INSTRUMENTS
POST OFFICE BOX 655012 • DALLAS, TEXAS 75265

QUAD DIFFERENTIAL LINE DRIVERS WITH 3-STATE OUTPUTS

D2453, DECEMBER 1978—REVISED OCTOBER 1986

- **Meets EIA Standard RS-422-A**
- **High-Impedance Output State for Party-Line Operation**
- **High Output Impedance in Power-Off Condition**
- **Low Input Current to Minimize Loading**
- **Single 5-V Supply**
- **40-mA Sink- and Source-Current Capability**
- **High-Speed Schottky Circuitry**
- **Low Power Requirements**

SN75151
DW, J, OR N PACKAGE
(TOP VIEW)

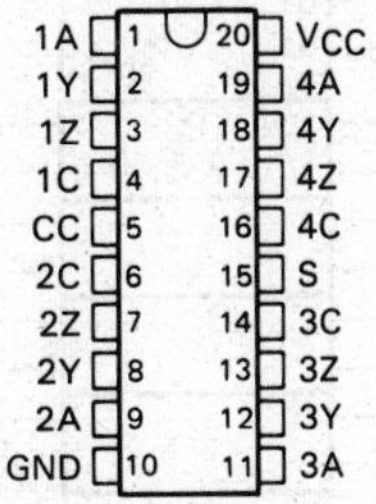
1A 1
1Y 2
1Z 3
1C 4
CC 5
2C 6
2Z 7
2Y 8
2A 9
GND 10
20 VCC
19 4A
18 4Y
17 4Z
16 4C
15 S
14 3C
13 3Z
12 3Y
11 3A

SN75153
J OR N DUAL-IN-LINE PACKAGE
(TOP VIEW)

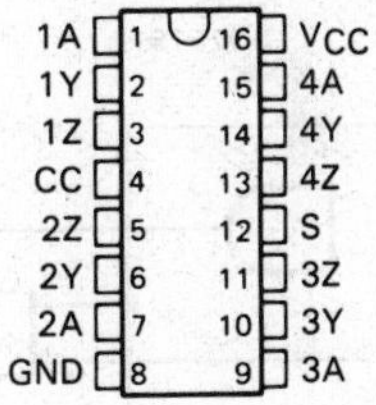
1A 1
1Y 2
1Z 3
CC 4
2Z 5
2Y 6
2A 7
GND 8
16 VCC
15 4A
14 4Y
13 4Z
12 S
11 3Z
10 3Y
9 3A

description

These line drivers are designed to provide differential signals with high current capability on balanced lines. These circuits provide strobe and enable inputs to control all four drivers, and the SN75151 provides an additional enable input for each driver. The output circuits have active pull-up and pull-down and are capable of sinking or sourcing 40 milliamperes.

The SN75151 and SN75153 meet all requirements of EIA Standard RS-422-A and Federal Standard 1020. They are characterized for operation from 0°C to 70°C.

FUNCTION TABLES

SN75151

INPUTS				OUTPUTS	
ENABLE CC	ENABLE C	STROBE S	DATA A	Y	Z
L	X	X	X	Z	Z
X	L	X	X	Z	Z
H	H	L	X	L	H
H	H	X	L	L	H
H	H	H	H	H	L

SN75153

INPUTS			OUTPUTS	
ENABLE CC	STROBE S	DATA A	Y	Z
L	X	X	Z	Z
H	L	X	L	H
H	X	L	L	H
H	H	H	H	L

logic symbols†

SN75151

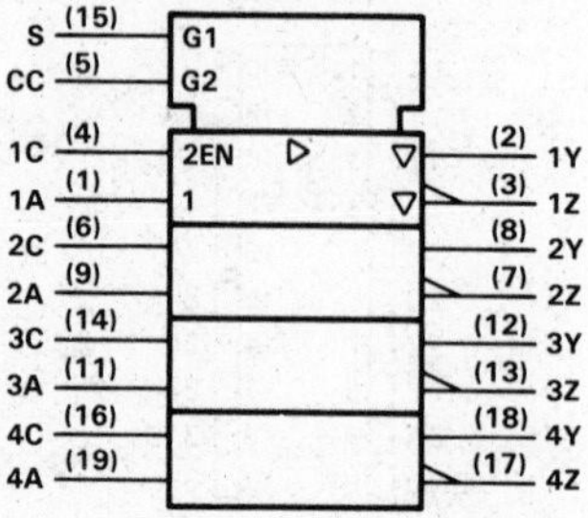
S (15)
CC (5)
1C (4)
1A (1)
2C (6)
2A (9)
3C (14)
3A (11)
4C (16)
4A (19)
G1
G2
2EN
1
(2) 1Y
(3) 1Z
(8) 2Y
(7) 2Z
(12) 3Y
(13) 3Z
(18) 4Y
(17) 4Z

SN75153

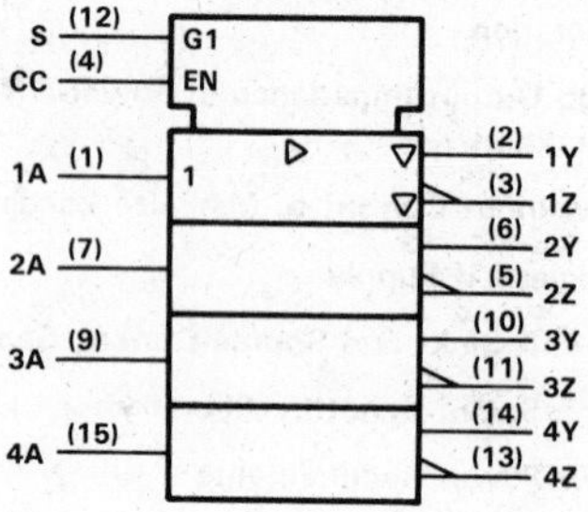
S (12)
CC (4)
1A (1)
2A (7)
3A (9)
4A (15)
G1
EN
1
(2) 1Y
(3) 1Z
(6) 2Y
(5) 2Z
(10) 3Y
(11) 3Z
(14) 4Y
(13) 4Z

†These symbols are in accordance with ANSI/IEEE Std 91-1984 and IEC Publication 617-12.

logic diagrams (positive logic)

SN75151

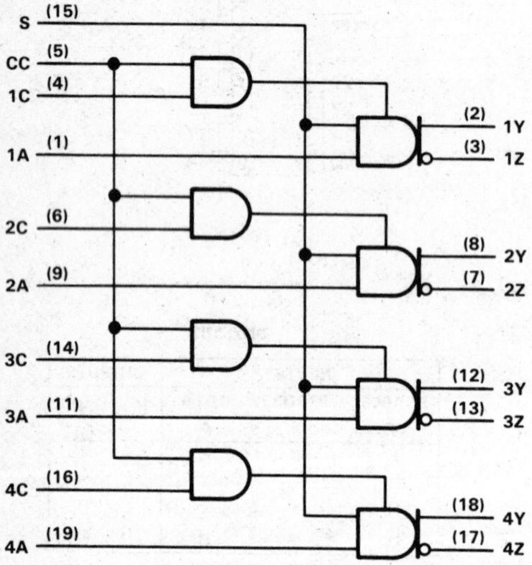
S (15)
CC (5)
1C (4)
1A (1)
2C (6)
2A (9)
3C (14)
3A (11)
4C (16)
4A (19)
(2) 1Y
(3) 1Z
(8) 2Y
(7) 2Z
(12) 3Y
(13) 3Z
(18) 4Y
(17) 4Z

SN75153

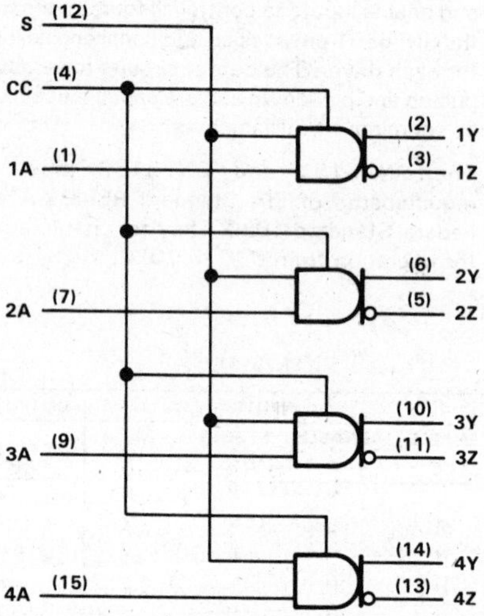
S (12)
CC (4)
1A (1)
2A (7)
3A (9)
4A (15)
(2) 1Y
(3) 1Z
(6) 2Y
(5) 2Z
(10) 3Y
(11) 3Z
(14) 4Y
(13) 4Z

TEXAS INSTRUMENTS
POST OFFICE BOX 655012 • DALLAS, TEXAS 75265

schematic

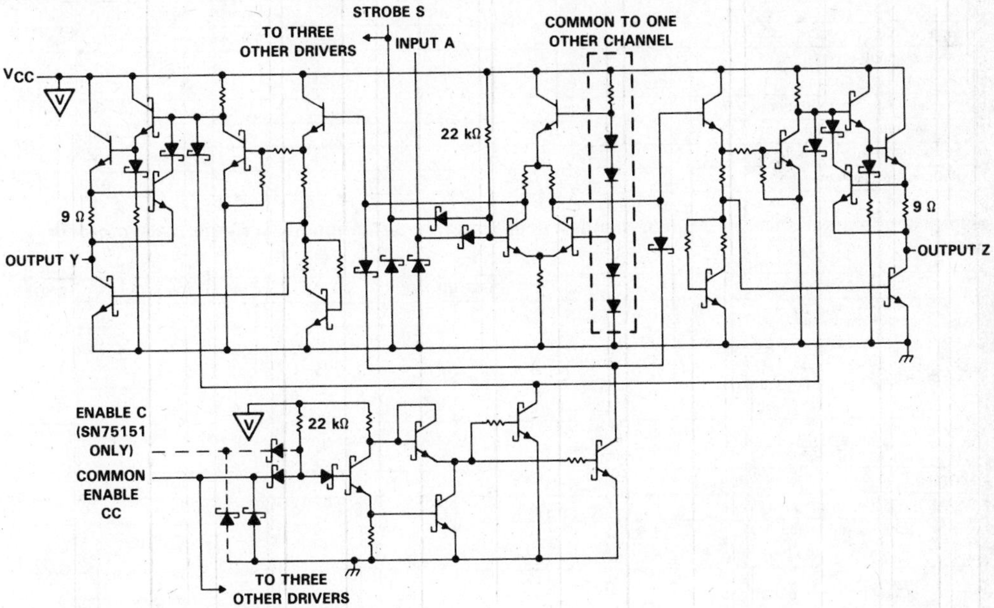
STROBE S
TO THREE OTHER DRIVERS
INPUT A
COMMON TO ONE OTHER CHANNEL
VCC
22 kΩ
9 Ω
OUTPUT Y
9 Ω
OUTPUT Z
ENABLE C (SN75151 ONLY)
COMMON ENABLE CC
22 kΩ
TO THREE OTHER DRIVERS

All resistor values shown are nominal.

absolute maximum ratings over operating free-air temperature range (unless otherwise noted)

Supply voltage, V_{CC} (see Note 1)	7 V
Input voltage	5.5 V
Continuous total dissipation at (or below) 25°C free-air temperature (see Note 2):	
DW package	1125 mW
J package	1025 mW
N package	1150 mW
Operating free-air temperature range	0°C to 70°C
Storage temperature range	−65°C to 150°C
Lead temperature 1,6 mm (1/16 inch) from case for 60 seconds: J package	300°C
Lead temperature 1,6 mm (1/16 inch) from case for 10 seconds: DW or N package	260°C

NOTES: 1. All voltage values, except differential output voltage V_{OD}, are with respect to network ground terminal.
2. For operation above 25°C free-air temperature, derate the DW package at the rate of 9 mW/°C, the J package at the rate of 8.2 mW/°C, and the N package at the rate of 9.2 mW/°C. In the J package, the chips are glass mounted.

recommended operating conditions

	MIN	NOM	MAX	UNIT
Supply voltage, V_{CC}	4.75	5	5.25	V
High-level input voltage, V_{IH}	2			V
Low-level input voltage, V_{IL}			0.8	V
Common-mode output voltage, V_{OC}	−0.25		6	V
High-level output current, I_{OH}			−40	mA
Low-level output current, I_{OL}			40	mA
Operating free-air temperature, T_A	0		70	°C

electrical characteristics over recommended operating free-air temperature range (unless otherwise noted)

PARAMETER		TEST CONDITIONS†		MIN	TYP‡	MAX	UNIT
V_{IK}	Input clamp voltage	V_{CC} = MIN, I_I = −12 mA	CC, S			−2	V
			All others		−0.9	−1.5	
V_{OH}	High-level output voltage	V_{CC} = MIN, V_{IL} = MAX, V_{IH} = 2 V	I_{OH} = −20 mA	2.5			V
			I_{OH} = −40 mA	2.4			
V_{OL}	Low-level output voltage	V_{CC} = MIN, V_{IH} = 2 V,	V_{IL} = MAX, I_{OL} = 40 mA			0.5	V
$\|V_{OD1}\|$	Differential output voltage	V_{CC} = MAX,	I_O = 0		3.4	$2V_{OD2}$	V
$\|V_{OD2}\|$	Differential output voltage	V_{CC} = MIN	R_L = 100 Ω, See Figure 1	2	2.8		V
$\Delta\|V_{OD}\|$	Change in magnitude of differential output voltage§	V_{CC} = MIN			±0.01	±0.4	V
V_{OC}	Common-mode output voltage¶	V_{CC} = MAX			1.8	3	V
		V_{CC} = MIN			1.6	3	
$\Delta\|V_{OC}\|$	Change in magnitude of common-mode output voltage§	V_{CC} = MIN or MAX			±0.02	±0.4	V
I_{OZ}	Off-state (high-impedance-state) output current	V_{CC} = MAX, Enable at 0.8 V	V_O = 0.5 V			−20	μA
			V_O = 2.5 V			20	
			V_O = V_{CC}			20	
I_O	Output current with power off	V_{CC} = 0	V_O = 6 V		0.1	100	μA
			V_O = −0.25 V		−0.1	−100	
			V_O = −0.25 V to 6 V			±100	
I_I	Input current at maximum input voltage	V_{CC} = MAX,	V_I = 5.5 V			0.1	mA
I_{IH}	High-level input current	V_{CC} = MAX, V_I = 2.4 V	C('151), A			20	μA
			CC, S			80	
I_{IL}	Low-level input current	V_{CC} = MAX, V_I = 0.4 V	C ('151), A			−0.36	mA
			CC,S			−1.6	
I_{OS}	Short-circuit output current#	V_{CC} = MAX		−50	−90	−150	mA
I_{CC}	Supply current (both drivers)	V_{CC} = MAX, No load	Outputs disabled		30	60	mA
			Outputs enabled		60	80	

† For conditions shown as MIN or MAX, use the appropriate value specified under recommended operating conditions.
‡ All typical values are at T_A = 25°C and V_{CC} = 5 V except for V_{OC}, for which V_{CC} is as stated under test conditions.
§ $\Delta|V_{OD}|$ and $\Delta|V_{OC}|$ are the changes in magnitudes of V_{OD} and V_{OC}, respectively, that occur when the input is changed from a high level to a low level.
¶ In EIA Standard RS-422-A, V_{OC}, which is the average of the two output voltages with respect to ground, is called output offset voltage, V_{OS}.
# Only one output should be shorted at a time, and duration of the short-circuit should not exceed one second.

switching characteristics, V_{CC} = 5 V, T_A = 0°C to 70°C (unless otherwise noted)

PARAMETER		TEST CONDITIONS	MIN	TYP†	MAX	UNIT
t_{PLH}	Propagation delay time, low-to-high-level output	C_L = 30 pF, R_L = 100 Ω, See Figure 2, Termination A		15	30	ns
t_{PHL}	Propagation delay time, high-to-low-level output			15	30	ns
t_{PLH}	Propagation delay time, low-to-high-level output	C_L = 30 pF, See Figure 2, Termination B		13	25	ns
t_{PHL}	Propagation delay time, high-to-low-level output			13	25	ns
t_{TLH}	Transition time, low-to-high-level output	C_L = 30 pF, R_L = 100 Ω, See Figure 2, Termination A		12	20	ns
t_{THL}	Transition time, high-to-low-level output			12	20	ns
t_{PZH}	Outut enable time to high level	C_L = 30 pF, R_L = 60 Ω, See Figure 3		18	35	ns
t_{PZL}	Output enable time to low level	C_L = 30 pF, R_L = 111 Ω, See Figure 4		20	35	ns
t_{PHZ}	Output disable time from high level	C_L = 30 pF, R_L = 60 Ω, See Figure 3		19	30	ns
t_{PLZ}	Output disable time from low level	C_L = 30 pF, R_L = 111 Ω, See Figure 4		13	30	ns
	Overshoot factor	R_L = 100 Ω, See Figure 2, Termination C			10	%

†All typical values are at T_A = 25°C.

PARAMETER MEASUREMENT INFORMATION

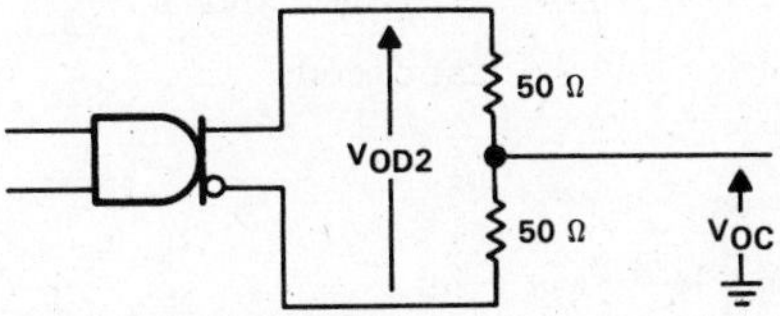
50 Ω
VOD2
50 Ω
VOC

FIGURE 1. DIFFERENTIAL AND COMMON-MODE OUTPUT VOLTAGES

PARAMETER MEASUREMENT INFORMATION

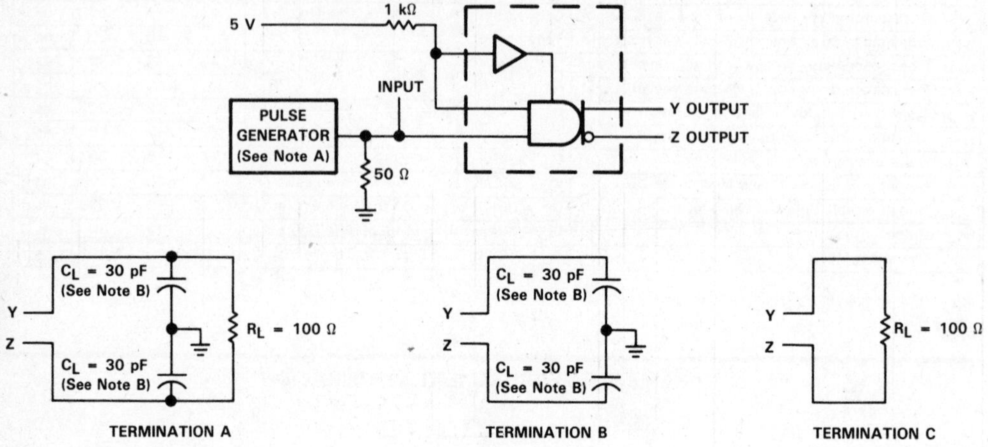
5 V
1 kΩ
INPUT
PULSE GENERATOR (See Note A)
50 Ω
Y OUTPUT
Z OUTPUT
Y
Z
C_L = 30 pF (See Note B)
C_L = 30 pF (See Note B)
R_L = 100 Ω
TERMINATION A
Y
Z
C_L = 30 pF (See Note B)
C_L = 30 pF (See Note B)
TERMINATION B
Y
Z
R_L = 100 Ω
TERMINATION C

TEST CIRCUITS

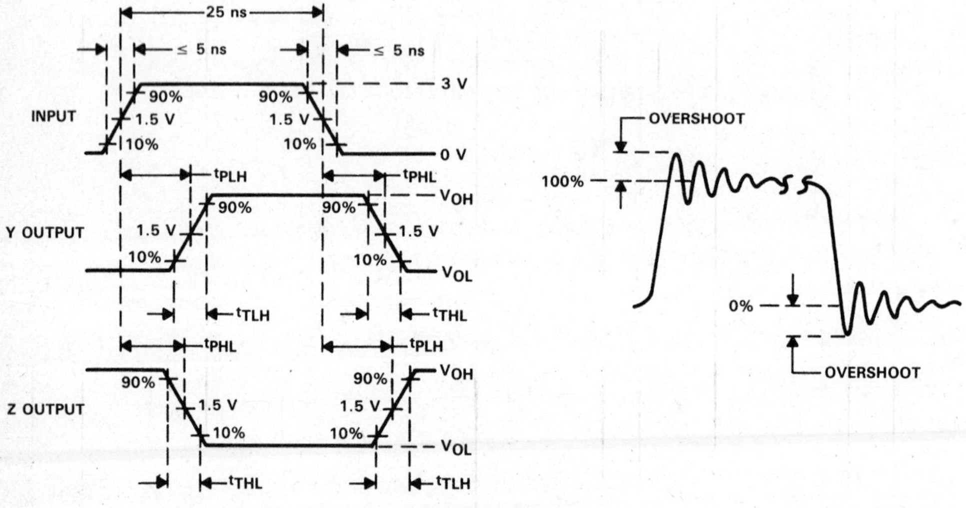
25 ns
≤ 5 ns
≤ 5 ns
3 V
INPUT
90%
1.5 V
10%
90%
1.5 V
10%
0 V
t_{PLH}
t_{PHL}
V_{OH}
Y OUTPUT
90%
1.5 V
10%
90%
1.5 V
10%
V_{OL}
t_{TLH}
t_{THL}
t_{PHL}
t_{PLH}
V_{OH}
Z OUTPUT
90%
1.5 V
10%
90%
1.5 V
10%
V_{OL}
t_{THL}
t_{TLH}
OVERSHOOT
100%
0%
OVERSHOOT

VOLTAGE WAVEFORMS

NOTES: A. The pulse generator has the following characteristics: Z_{out} = 50 Ω, PRR ≤ 10 MHz.
B. C_L includes probe and jig capacitance.

FIGURE 2. t_{PLH}, t_{PHL}, t_{TLH}, t_{THL}, AND OVERSHOOT FACTOR

PARAMETER MEASUREMENT INFORMATION

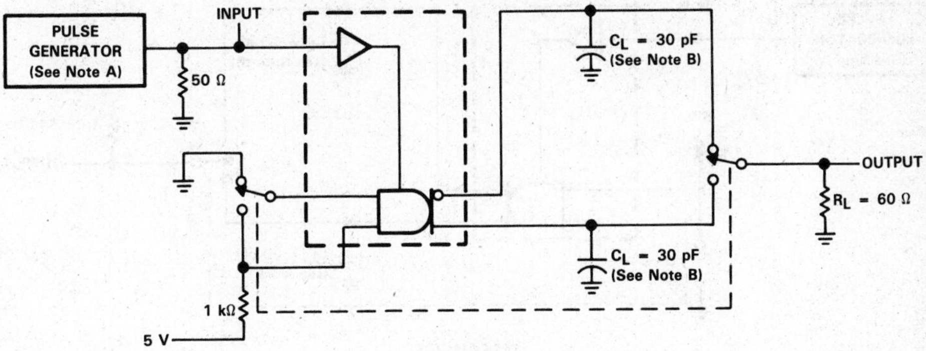
INPUT
PULSE GENERATOR (See Note A)
50 Ω
1 kΩ
5 V
C_L = 30 pF (See Note B)
C_L = 30 pF (See Note B)
OUTPUT
R_L = 60 Ω

TEST CIRCUIT

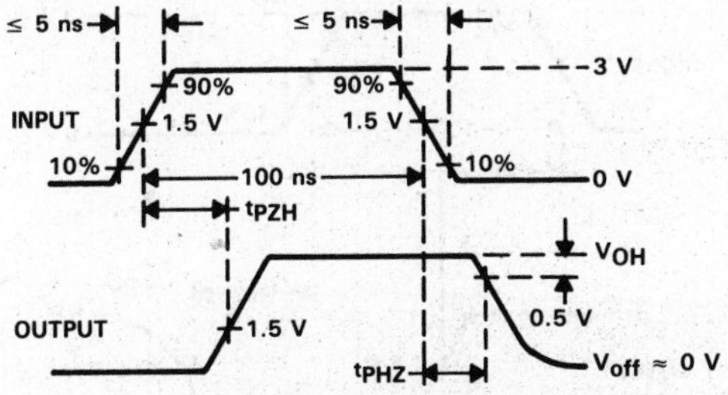
≤ 5 ns
≤ 5 ns
3 V
90%
90%
INPUT
1.5 V
1.5 V
10%
10%
0 V
100 ns
t_{PZH}
V_{OH}
0.5 V
OUTPUT
1.5 V
t_{PHZ}
V_{off} ≈ 0 V

VOLTAGE WAVEFORMS

FIGURE 3. t_{PZH} AND t_{PHZ}

NOTES: A. The pulse generators have the following characteristics: Z_{out} = 50 Ω, PRR ≤ 500 kHz.
B. C_L includes probe and jig capacitance.

PARAMETER MEASUREMENT INFORMATION

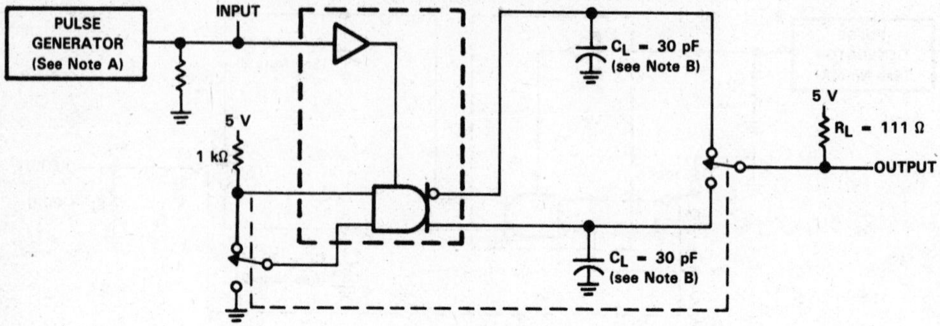
PULSE
GENERATOR
(See Note A)
INPUT
5 V
1 kΩ
C_L = 30 pF
(see Note B)
C_L = 30 pF
(see Note B)
5 V
R_L = 111 Ω
OUTPUT

TEST CIRCUIT

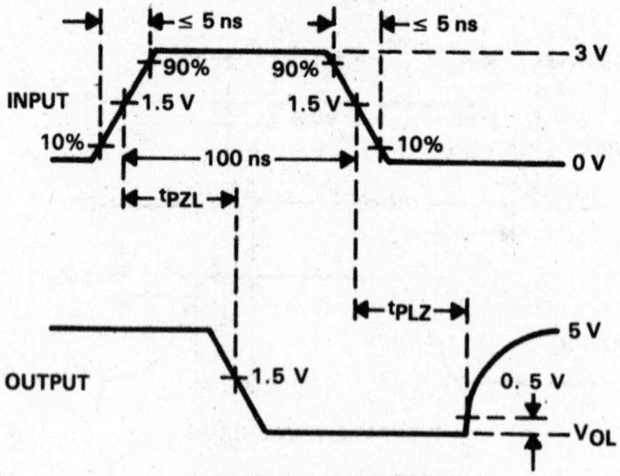
≤ 5 ns
≤ 5 ns
3 V
90%
90%
INPUT
1.5 V
1.5 V
10%
10%
100 ns
0 V
t_{PZL}
t_{PLZ}
5 V
OUTPUT
1.5 V
0. 5 V
V_{OL}

VOLTAGE WAVEFORMS

FIGURE 4. t_{PZL} AND t_{PLZ}

NOTES: A. The pulse generators have the following characteristics: Z_{out} = 50 Ω, PRR ≤ 500 kHz.
B. C_L includes probe and jig capacitance.

TEXAS INSTRUMENTS
POST OFFICE BOX 655012 • DALLAS, TEXAS 75265

TYPICAL CHARACTERISTICS

Y OUTPUT VOLTAGE
vs
DATA INPUT VOLTAGE

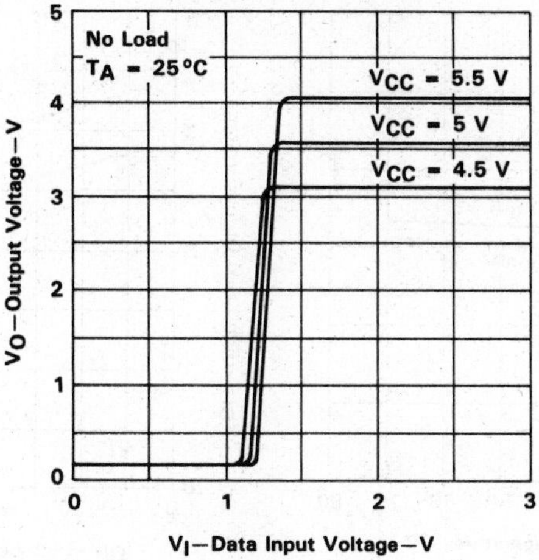
No Load
T_A = 25°C
V_{CC} = 5.5 V
V_{CC} = 5 V
V_{CC} = 4.5 V
V_O—Output Voltage—V
V_I—Data Input Voltage—V

FIGURE 5

Y OR Z OUTPUT VOLTAGE
vs
ENABLE INPUT VOLTAGE

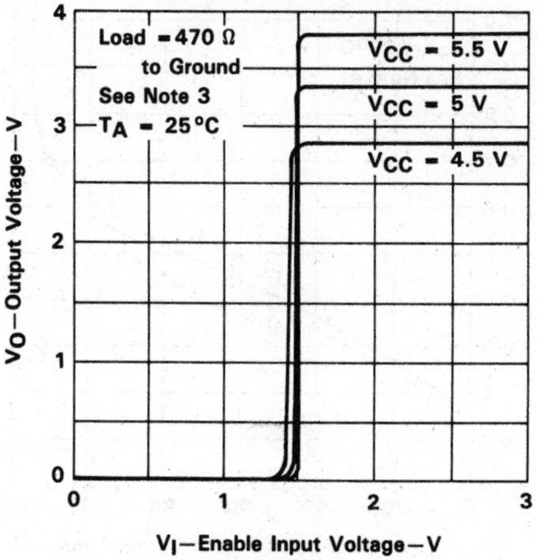
Load = 470 Ω to Ground
See Note 3
T_A = 25°C
V_{CC} = 5.5 V
V_{CC} = 5 V
V_{CC} = 4.5 V
V_O—Output Voltage—V
V_I—Enable Input Voltage—V

FIGURE 6

Y OR Z OUTPUT VOLTAGE
vs
ENABLE INPUT VOLTAGE

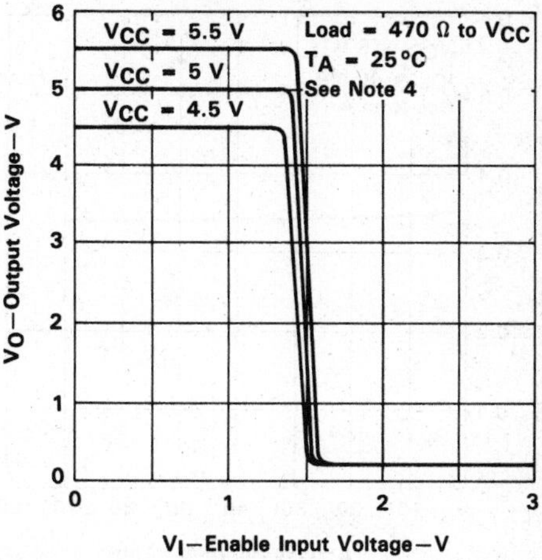
V_{CC} = 5.5 V
V_{CC} = 5 V
V_{CC} = 4.5 V
Load = 470 Ω to V_{CC}
T_A = 25°C
See Note 4
V_O—Output Voltage—V
V_I—Enable Input Voltage—V

FIGURE 7

NOTES: 3. The A input is connected to V_{CC} during the testing of the Y outputs and to ground during testing of the Z outputs.
4. The A input is connected to ground during the testing of the Y outputs and to V_{CC} during the testing of the Z outputs.

TYPICAL CHARACTERISTICS

HIGH-LEVEL OUTPUT VOLTAGE
vs
FREE-AIR TEMPERATURE

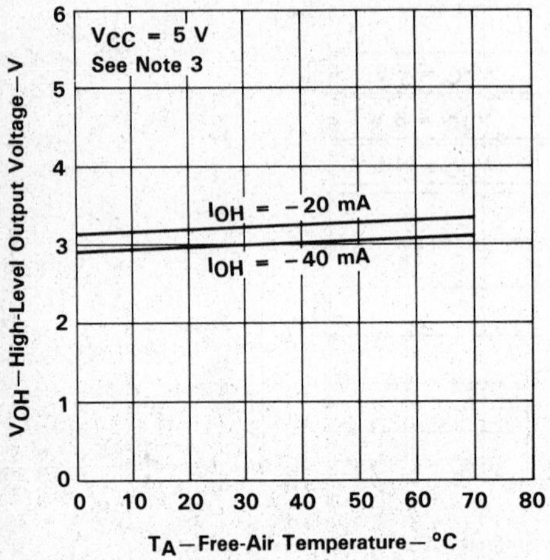
V_{CC} = 5 V
See Note 3
I_{OH} = −20 mA
I_{OH} = −40 mA
V_{OH}—High-Level Output Voltage—V
T_A—Free-Air Temperature—°C
0
1
2
3
4
5
6
10
20
30
40
50
60
70
80

FIGURE 8

HIGH-LEVEL OUTPUT VOLTAGE
vs
OUTPUT CURRENT

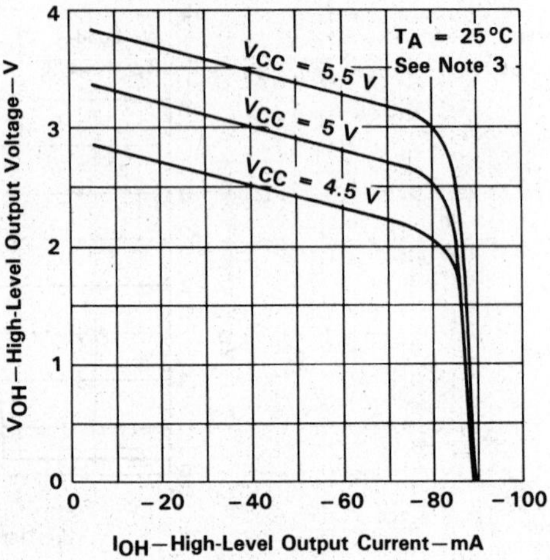
T_A = 25°C
See Note 3
V_{CC} = 5.5 V
V_{CC} = 5 V
V_{CC} = 4.5 V
V_{OH}—High-Level Output Voltage—V
I_{OH}—High-Level Output Current—mA
0
1
2
3
4
−20
−40
−60
−80
−100

FIGURE 9

LOW-LEVEL OUTPUT VOLTAGE
vs
FREE-AIR TEMPERATURE

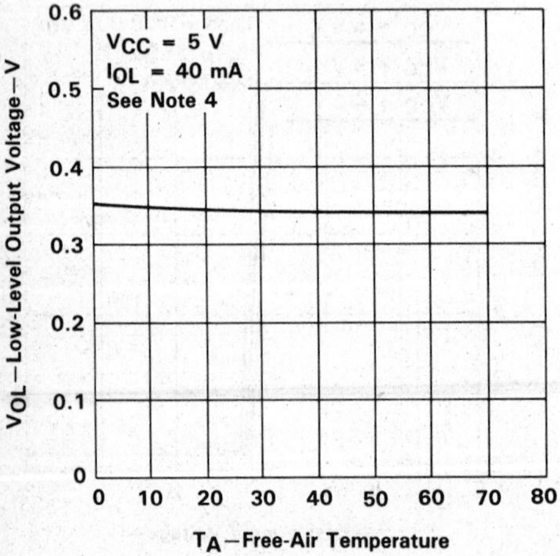
V_{CC} = 5 V
I_{OL} = 40 mA
See Note 4
V_{OL}—Low-Level Output Voltage—V
T_A—Free-Air Temperature
0
0.1
0.2
0.3
0.4
0.5
0.6
10
20
30
40
50
60
70
80

FIGURE 10

LOW-LEVEL OUTPUT VOLTAGE
vs
OUTPUT CURRENT

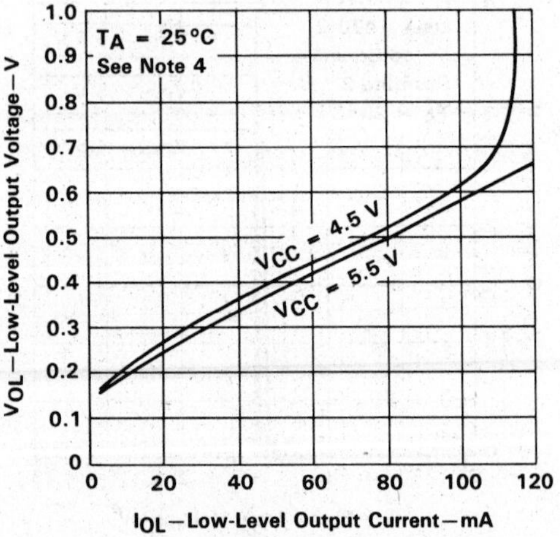
T_A = 25°C
See Note 4
V_{CC} = 4.5 V
V_{CC} = 5.5 V
V_{OL}—Low-Level Output Voltage—V
I_{OL}—Low-Level Output Current—mA
0
0.1
0.2
0.3
0.4
0.5
0.6
0.7
0.8
0.9
1.0
20
40
60
80
100
120

FIGURE 11

NOTES: 3. The A input is connected to V_{CC} during the testing of the Y outputs and to ground during testing of the Z outputs.
4. The A input is connected to ground during the testing of the Y outputs and to V_{CC} during the testing of the Z inputs.

TEXAS INSTRUMENTS
POST OFFICE BOX 655012 • DALLAS, TEXAS 75265

TYPICAL CHARACTERISTICS

SUPPLY CURRENT
vs
SUPPLY VOLTAGE

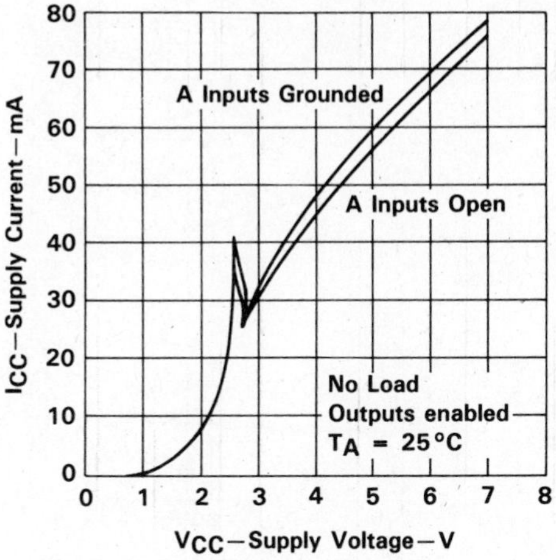
A Inputs Grounded
A Inputs Open
No Load
Outputs enabled
T_A = 25°C
I_{CC}—Supply Current—mA
V_{CC}—Supply Voltage—V
0
10
20
30
40
50
60
70
80
1
2
3
4
5
6
7
8

FIGURE 12

SUPPLY CURRENT
vs
SUPPLY VOLTAGE

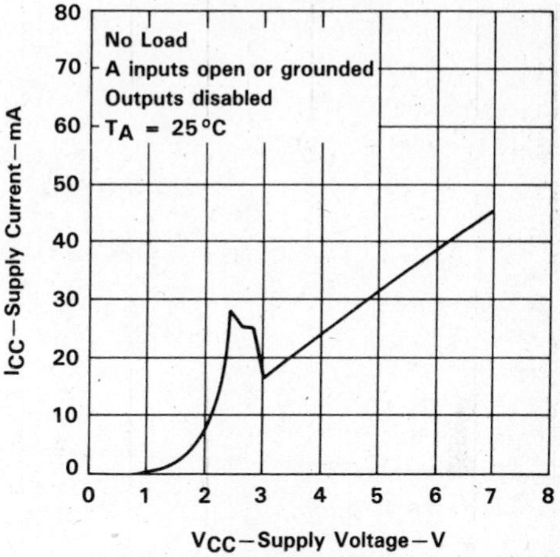
No Load
A inputs open or grounded
Outputs disabled
T_A = 25°C
I_{CC}—Supply Current—mA
V_{CC}—Supply Voltage—V
0
10
20
30
40
50
60
70
80
1
2
3
4
5
6
7
8

FIGURE 13

NOTES: 3. The A input is connected to V_{CC} during the testing of the Y outputs and to ground during testing of the Z outputs.
4. The A input is connected to ground during the testing of the Y outputs and to V_{CC} during the testing of the Z inputs.

- **Meets Specifications of EIA RS-232-C or MIL-STD-188C†**
- **Dual Differential Receiver with Independent Strobes**
- **Common-Mode Input Voltage Range . . . ±25 V**
- **Differential Input Capability with One Input Grounded . . . ±25 V**
- **Continuously Adjustable Hysteresis with External Resistors**
- **Standard Supply Voltages . . . +12 V and −12 V**
- **Input Hysteresis (Double Thresholds) Remain Approximately Fixed for Power Supply and/or Temperature Variations**

SN55152 . . . J PACKAGE
SN75152 . . . D, J, OR N PACKAGE
(TOP VIEW)

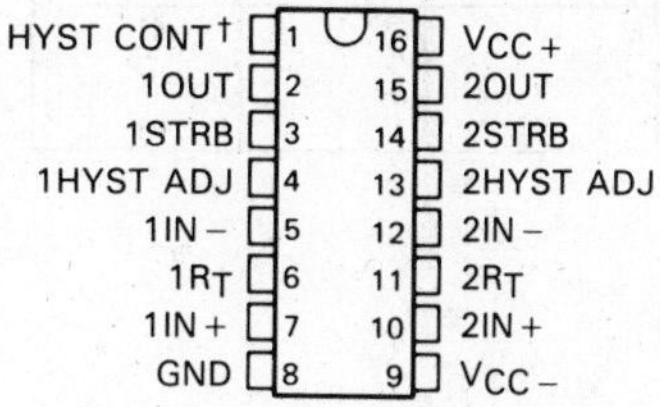
HYST CONT† 1
1OUT 2
1STRB 3
1HYST ADJ 4
1IN− 5
$1R_T$ 6
1IN+ 7
GND 8
16 V_{CC+}
15 2OUT
14 2STRB
13 2HYST ADJ
12 2IN−
11 $2R_T$
10 2IN+
9 V_{CC-}

SN55152 . . . FK PACKAGE
(TOP VIEW)

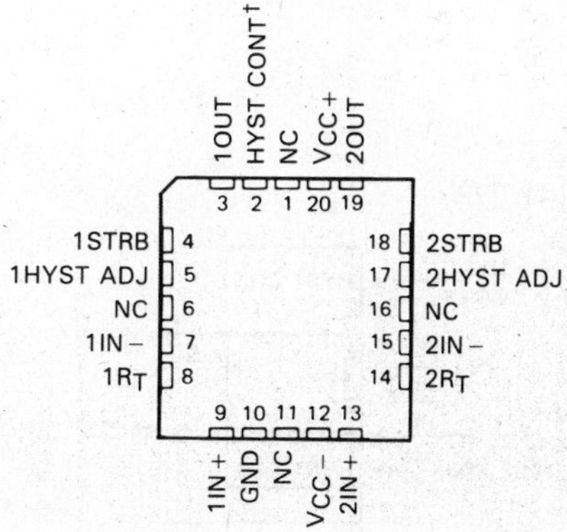
1OUT
HYST CONT†
NC
V_{CC+}
2OUT
3 2 1 20 19
1STRB 4
1HYST ADJ 5
NC 6
1IN− 7
$1R_T$ 8
18 2STRB
17 2HYST ADJ
16 NC
15 2IN−
14 $2R_T$
9 10 11 12 13
1IN+
GND
NC
V_{CC-}
2IN+

NC—No internal connection

description

The SN55152 and SN75152 are dual differential line receivers designed to meet the requirements of EIA Standard RS-232-C or MIL-STD-188 interfaces. A single control, HYST CONT, sets the input hysteresis for the required operation. An added feature is the capability of adjusting the hysteresis to any voltage between ±0.3 volt typical and ±5 volts typical by means of the hysteresis adjust terminals, 1HYST ADJ and 2HYST ADJ, making the SN55152 and SN75152 useful for a wide variety of line receiver and Schmitt trigger applications. The large common-mode input voltage range and differential input voltage (±25 volts) give the circuit added versatility. The SN55152 and SN75152 are designed for operation from standard ±12-volt supplies with ±10% variation. Each receiver has an output strobe that is TTL compatible.

The SN55152 is characterized for operation over the full military temperature range of −55°C to 125°C. The SN75152 is characterized for operation from 0°C to 70°C.

† To meet the specifications of EIA Standard RS-232-C, connect the hysteresis control pin, HYST CONT, to V_{CC-}. Also, connect termination resistor pin $1R_T$ to inverting input 1IN−, and termination resistor pin $2R_T$ to inverting input 2IN−. To meet the specifications of MIL-STD-188, leave HYST CONT, $1R_T$, and $2R_T$ open.

FUNCTION TABLE
(EACH RECEIVER)

LINE INPUT	STROBE	OUTPUT
H	H	H
L	H	L
X	L	H

Definition of logic levels:

For the strobe: H (high) is any voltage between V_{IH} min and V_{CC}.
L (low) is any voltage between ground and V_{IL} max.

For the line input: H (high) is any differential input voltage (V_{ID})‡ more positive than V_{T-}, once the level of V_{T+} has been reached.
L (low) is any differential input voltage (V_{ID})‡ more negative than V_{T+}, once the level of V_{T-} has been reached.
X (irrelevant) is any input voltage permitted by maximum ratings.

‡ Differential input voltages (V_T and V_{ID}) are at the noninverting input terminal IN+ with respect to the inverting input terminal IN−.

4

Line Drivers/Receivers

logic symbol†

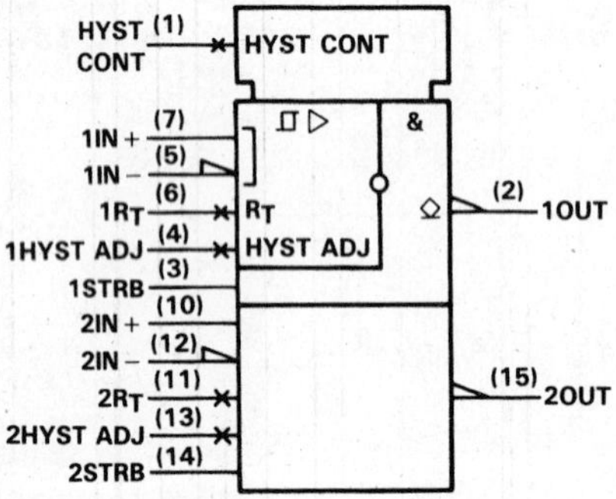
HYST CONT (1)
HYST CONT
1IN+ (7)
1IN− (5)
1RT (6)
RT
1HYST ADJ (4)
HYST ADJ
1STRB (3)
&
2IN+ (10)
2IN− (12)
2RT (11)
2HYST ADJ (13)
2STRB (14)
(2) 1OUT
(15) 2OUT

† This symbol is in accordance with ANSI/IEEE Std 91-1984 and IEC Publication 617-12.
Pin numbers shown are for D, J, and N packages.

logic diagram (positive logic)

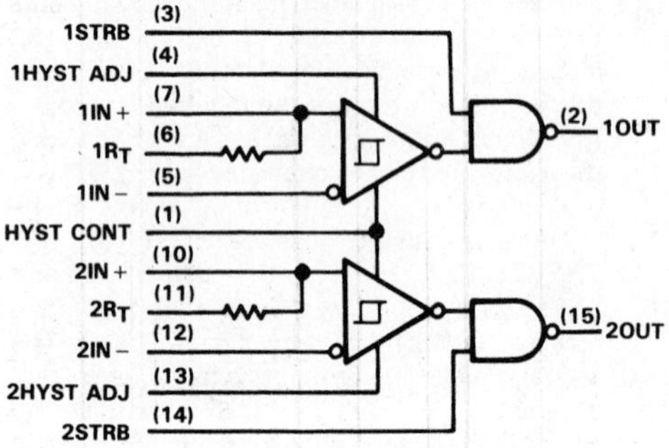
1STRB (3)
1HYST ADJ (4)
1IN+ (7)
1RT (6)
1IN− (5)
HYST CONT (1)
2IN+ (10)
2RT (11)
2IN− (12)
2HYST ADJ (13)
2STRB (14)
(2) 1OUT
(15) 2OUT

TEXAS INSTRUMENTS
POST OFFICE BOX 655012 • DALLAS, TEXAS 75265

schematic (each receiver)

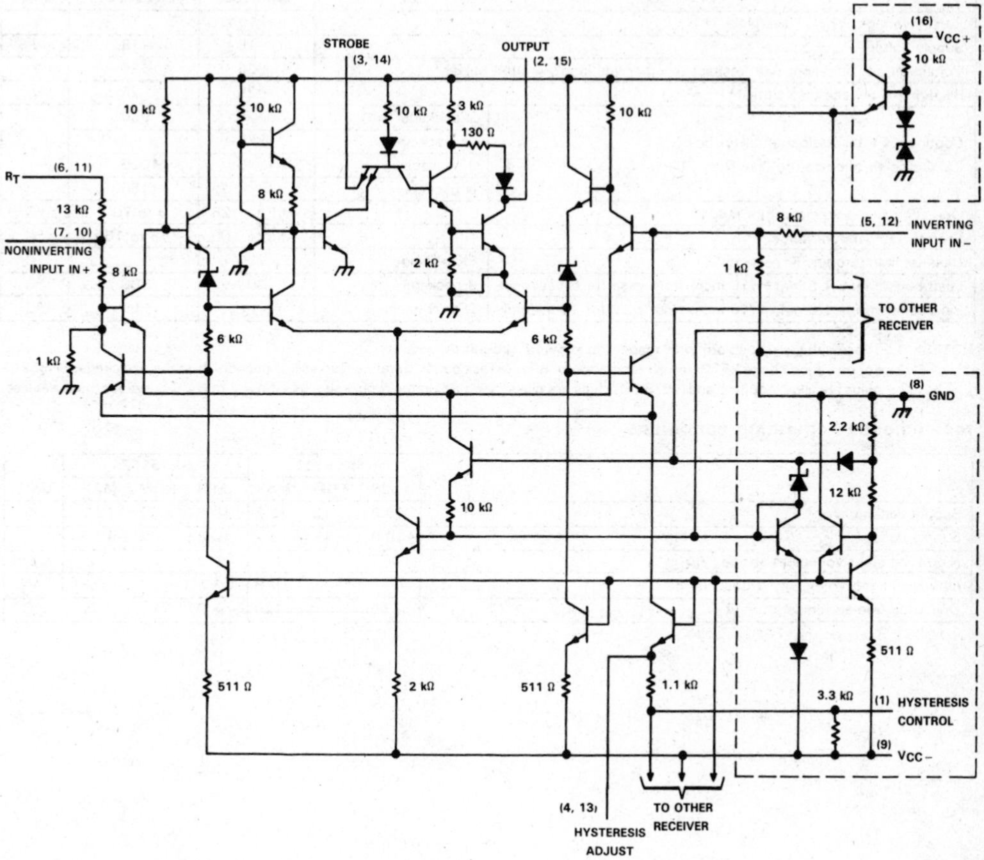
STROBE
(3, 14)
OUTPUT
(2, 15)
(16)
VCC+
10 kΩ
10 kΩ
10 kΩ
10 kΩ
3 kΩ
130 Ω
10 kΩ
RT
(6, 11)
13 kΩ
(7, 10)
NONINVERTING
INPUT IN+
8 kΩ
8 kΩ
8 kΩ
(5, 12)
INVERTING
INPUT IN−
2 kΩ
1 kΩ
TO OTHER
RECEIVER
1 kΩ
6 kΩ
6 kΩ
(8)
GND
2.2 kΩ
12 kΩ
10 kΩ
511 Ω
511 Ω
2 kΩ
511 Ω
1.1 kΩ
3.3 kΩ
(1)
HYSTERESIS
CONTROL
(9)
VCC−
(4, 13)
HYSTERESIS
ADJUST
TO OTHER
RECEIVER

Portions of circuit within dashed lines are common to both receivers.
Resistor values shown are nominal.

absolute maximum ratings over operating free-air temperature range (unless otherwise noted)

		SN55152	SN75152	UNIT
Supply voltage, V_{CC+} (see Note 1)		15	15	V
Supply voltage, V_{CC-} (see Note 1)		−15	−15	V
Voltage at any line input with respect to other line input, ground, or R_T		±25	±25	V
R_T terminal voltage (see Note 1)		±25	±25	V
Continuous total dissipation at (or below) 25°C free-air temperature (see Note 2)	D package		950	mW
	FK package	1375		mW
	J package	1375	1025	mW
	N package		1150	mW
Operating free-air temperature range		−55 to 125	0 to 70	°C
Storage temperature range		−65 to 150	−65 to 150	°C
Case temperature for 60 seconds	FK package	260		°C
Lead temperature 1,6 mm (1/16 inch) from case for 60 seconds	J package	300	300	°C
Lead temperature 1,6 mm (1/16 inch) from case for 10 seconds	D or N package		260	°C

NOTES: 1. These voltage values are with respect to network ground terminal.
2. For operation above 25°C free-air temperature, refer to Dissipation Derating Curves in Appendix A. In the J package, SN55152 chips are alloy mounted and SN75152 chips are glass mounted. In the N package, use the 9.2-mW/°C curve for these devices.

recommended operating conditions

	SN55152			SN75152			UNIT
	MIN	NOM	MAX	MIN	NOM	MAX	
Supply voltage, V_{CC+}	10.8	12	13.2	10.8	12	13.2	V
Supply voltage, V_{CC-}	−10.8	−12	−13.2	−10.8	−12	−13.2	V
High-level input voltage at strobe, $V_{IH(S)}$	2			2			V
Low-level input voltage at strobe, $V_{IL(S)}$			0.8			0.8	V
Operating free-air temperature, T_A	−55		125	0		70	°C

TEXAS INSTRUMENTS
POST OFFICE BOX 655012 • DALLAS, TEXAS 75265

electrical characteristics over operating free-air temperature range, $V_{CC+} = 12\ V \pm 10\%$, $V_{CC-} = -12\ V \pm 10\%$ (unless otherwise noted)

PARAMETER			TEST FIGURE	TEST CONDITIONS†		MIN	TYP‡	MAX	UNIT
						(SEE NOTE 3)			
V_{T+}	Positive-going threshold voltage	See Figure 8	1	MIL-STD-188 Conditions	'75152	0.1	0.3	0.5	V
					'55152	0.03	0.3	0.5	
V_{T-}	Negative-going threshold voltage				'75152	−0.5	−0.3	−0.1	V
					'55152	−0.5	−0.3	−0.03	
V_{T+}	Positive-going threshold voltage		2	EIA RS-232-C Conditions		1.5	2.2	3	V
V_{T-}	Negative-going threshold voltage					−3	−2.2	−1.5	
V_{OH}	High-level output voltage		1 and 2	$V_{ID} = V_{T+}$ max, $V_{I(strobe)} = 2\ V$, $I_{OH} = -500\ \mu A$		3	4.1	6	V
			1 and 2	$V_{ID} = V_{T-}$ min, $V_{I(strobe)} = 0.8\ V$, $I_{OH} = -500\ \mu A$		3	4.1	6	
V_{OL}	Low-level output voltage		1 and 2	$V_{ID} = V_{T-}$ min, $V_{I(strobe)} = 2\ V$, $I_{OL} = 6.4\ mA$		0	0.15	0.4	V
I_I	Input current into strobe at maximum strobe voltage		3	$V_{I(strobe)} = 5.5\ V$			0.1	1	mA
I_{IH}	High-level strobe current		3	$V_{I(strobe)} = 2.4\ V$			30	80	μA
I_{IL}	Low-level strobe current		3	$V_{I(strobe)} = 0.4\ V$			−0.5	−1.5	mA
r_I	Input resistance	MIL-STD-188	4	$\lvert V_{ID}\rvert = 0\ V$ to $25\ V$, R_T open, $T_A = 25°C$		6	9		kΩ
		EIA RS-232-C	4	$\lvert V_{ID}\rvert = 3\ V$ to $25\ V$, R_T connected to inverting line input, $T_A = 25°C$		3	5	7	
$V_{I(open)}$	Open-circuit input voltage		5				+1	±2	V
I_{OS}	Short-circuit output current		6	$V_{ID} = 3\ V$			−1.9	−4	mA
I_{CC+}	Supply current from V_{CC+}		1	$V_{ID} = -3\ V$, $V_{I(strobe)} = 2.4\ V$			10	16	mA
I_{CC-}	Supply current from V_{CC-}		1	$V_{ID} = -3\ V$, $V_{I(strobe)} = 2.4\ V$			−7	−13	mA

† Differential input voltages (V_T and V_{ID}) are at the noninverting line input terminal with respect to the inverting line input terminal.
‡ Typical values are at $V_{CC+} = 12\ V$, $V_{CC-} = -12\ V$, $T_A = 25°C$.
NOTE 3: The algebraic convention, in which the less positive (more negative) limit is designated as minimum, is used in this data sheet for threshold levels only, e.g., when −0.1 V is the maximum, the minimum limit is a more negative voltage.

switching characteristics, $V_{CC+} = 12\ V$, $V_{CC-} = -12\ V$, $T_A = 25°C$

PARAMETER		TEST FIGURE	TEST CONDITIONS	MIN	TYP	MAX	UNIT
t_{PLH}	Propagation delay time, low-to-high-level output	7	$C_L = 15\ pF$		40		ns
t_{PHL}	Propagation delay time, high-to-low-level output				60		

PARAMETER MEASUREMENT INFORMATION

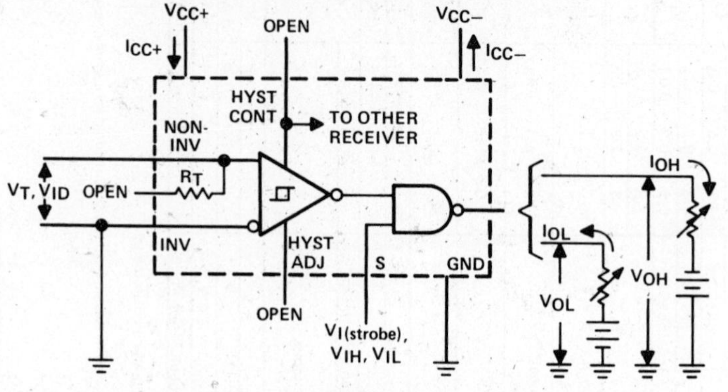
VCC+
ICC+
OPEN
VCC−
ICC−
HYST
CONT
TO OTHER
RECEIVER
NON-
INV
RT
VT, VID
OPEN
INV
HYST
ADJ
S
GND
OPEN
VI(strobe),
VIH, VIL
IOH
IOL
VOH
VOL

NOTE: Output is open for testing I_{CC+} and I_{CC-}

FIGURE 1. MIL-STD-188 CONDITION

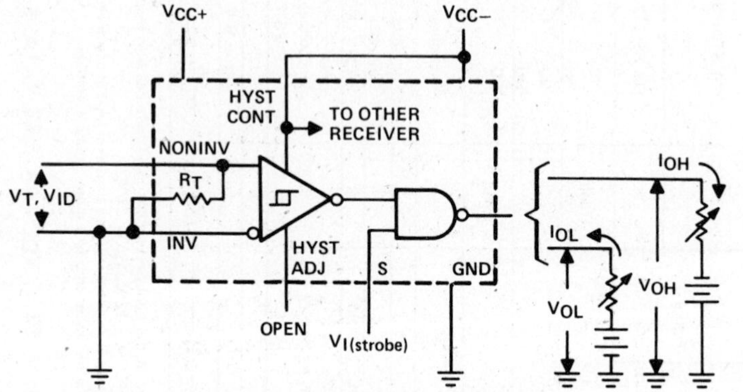
VCC+
VCC−
HYST
CONT
TO OTHER
RECEIVER
NONINV
RT
VT, VID
INV
HYST
ADJ
S
GND
OPEN
VI(strobe)
IOH
IOL
VOH
VOL

FIGURE 2. EIA RS-232-C CONDITION

TEXAS INSTRUMENTS
POST OFFICE BOX 655012 • DALLAS, TEXAS 75265

PARAMETER MEASUREMENT INFORMATION

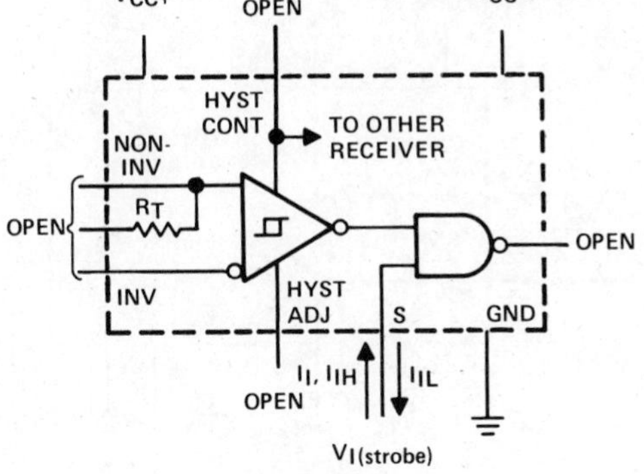
VCC+
OPEN
VCC–
HYST
CONT
TO OTHER
RECEIVER
NON-
INV
RT
OPEN
OPEN
INV
HYST
ADJ
S
GND
II, IIH
IIL
OPEN
VI(strobe)

FIGURE 3

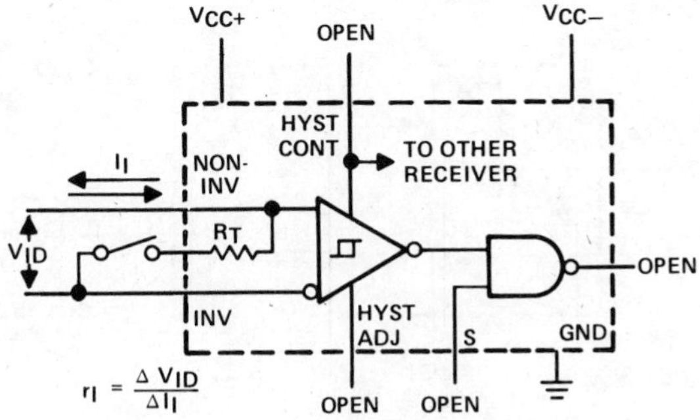
VCC+
OPEN
VCC–
HYST
CONT
TO OTHER
RECEIVER
II
NON-
INV
RT
VID
OPEN
INV
HYST
ADJ
S
GND
rI = Δ VID / Δ II
OPEN
OPEN

FIGURE 4

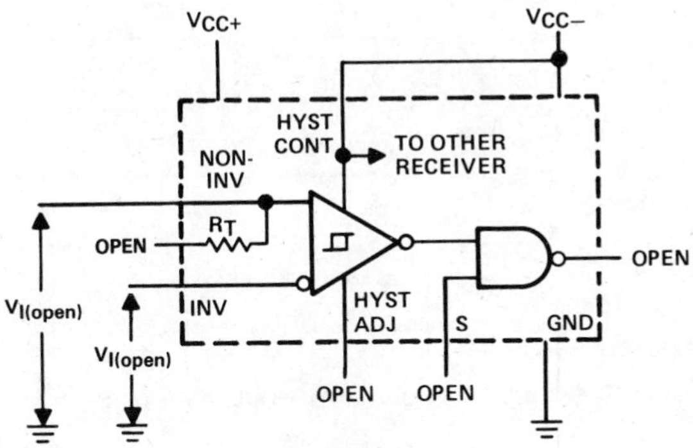
VCC+
VCC–
HYST
CONT
TO OTHER
RECEIVER
NON-
INV
RT
OPEN
OPEN
VI(open)
INV
VI(open)
HYST
ADJ
S
GND
OPEN
OPEN

FIGURE 5

TEXAS INSTRUMENTS
POST OFFICE BOX 655012 • DALLAS, TEXAS 75265

PARAMETER MEASUREMENT INFORMATION

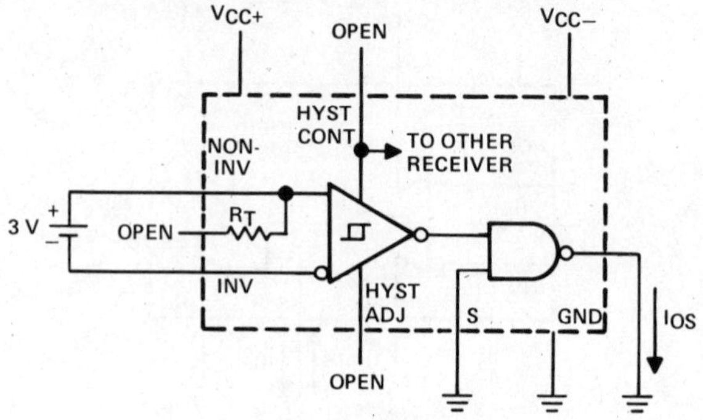
VCC+
OPEN
VCC−
HYST
CONT
TO OTHER
RECEIVER
NON-
INV
3 V
OPEN
RT
INV
HYST
ADJ
S
GND
IOS
OPEN

FIGURE 6

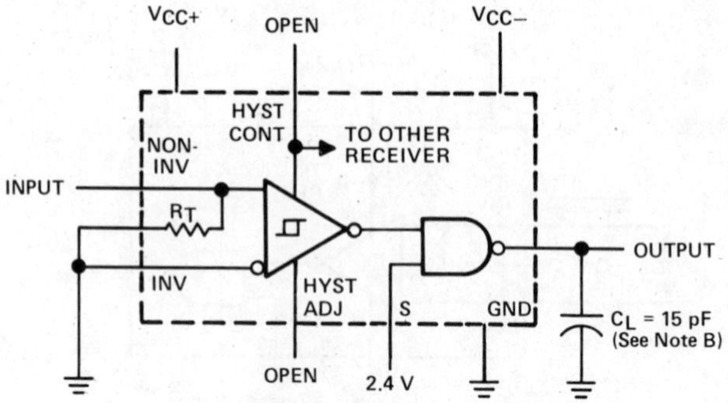
VCC+
OPEN
VCC−
HYST
CONT
TO OTHER
RECEIVER
NON-
INV
INPUT
RT
INV
HYST
ADJ
S
GND
OUTPUT
CL = 15 pF
(See Note B)
OPEN
2.4 V

TEST CIRCUIT

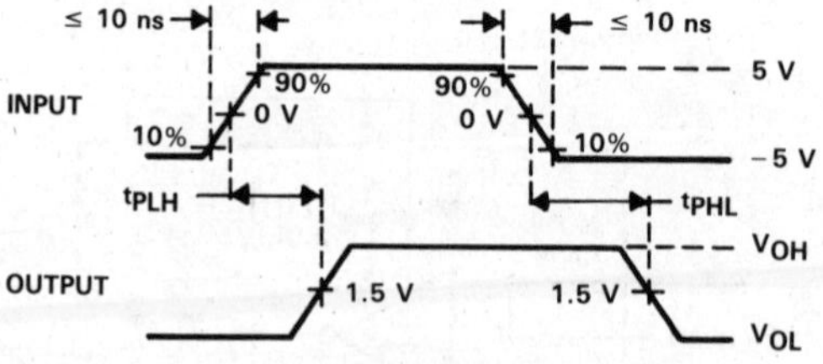
≤ 10 ns
≤ 10 ns
5 V
INPUT
90%
90%
0 V
0 V
10%
10%
−5 V
tPLH
tPHL
VOH
OUTPUT
1.5 V
1.5 V
VOL

VOLTAGE WAVEFORMS

NOTES: A. The input pulse is supplied by a generator having the following characteristics: PRR ≤ 1 MHz, duty cycle = 50%, $Z_{out} \approx 50\ \Omega$.
B. C_L includes probe and jig capacitance.

FIGURE 7. PROPAGATION DELAY TIMES

TYPICAL CHARACTERISTICS

OUTPUT VOLTAGE
vs
DIFFERENTIAL INPUT VOLTAGE

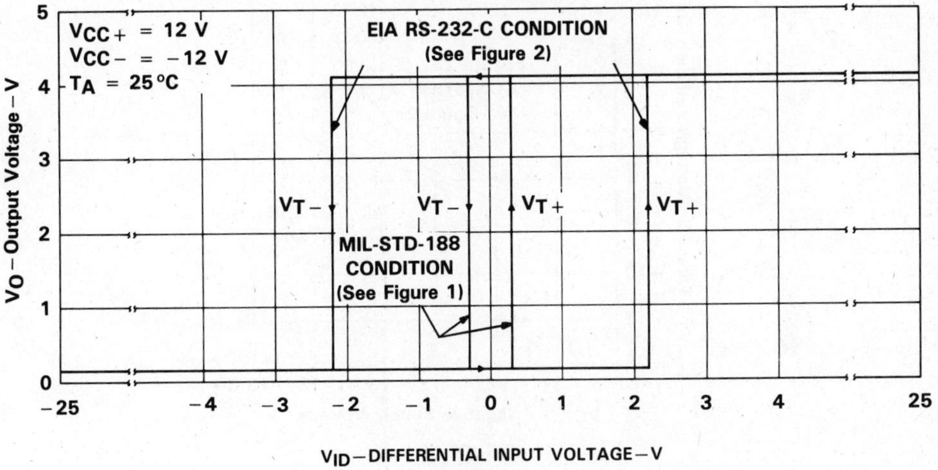
V_{CC+} = 12 V
V_{CC-} = −12 V
T_A = 25°C
EIA RS-232-C CONDITION
(See Figure 2)
V_{T-}
V_{T-}
V_{T+}
V_{T+}
MIL-STD-188
CONDITION
(See Figure 1)
V_O – Output Voltage – V
5
4
3
2
1
0
−25
−4
−3
−2
−1
0
1
2
3
4
25
V_{ID} – DIFFERENTIAL INPUT VOLTAGE – V

FIGURE 8

THRESHOLD VOLTAGE VARIATION
vs
POSITIVE SUPPLY VOLTAGE

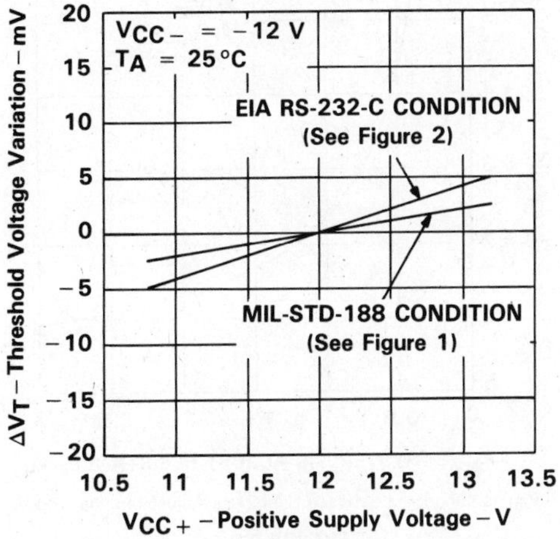
V_{CC-} = −12 V
T_A = 25°C
EIA RS-232-C CONDITION
(See Figure 2)
MIL-STD-188 CONDITION
(See Figure 1)
ΔV_T – Threshold Voltage Variation – mV
20
15
10
5
0
−5
−10
−15
−20
10.5
11
11.5
12
12.5
13
13.5
V_{CC+} – Positive Supply Voltage – V

FIGURE 9

TYPICAL CHARACTERISTICS

THRESHOLD VOLTAGE VARIATION
vs
NEGATIVE POWER SUPPLY

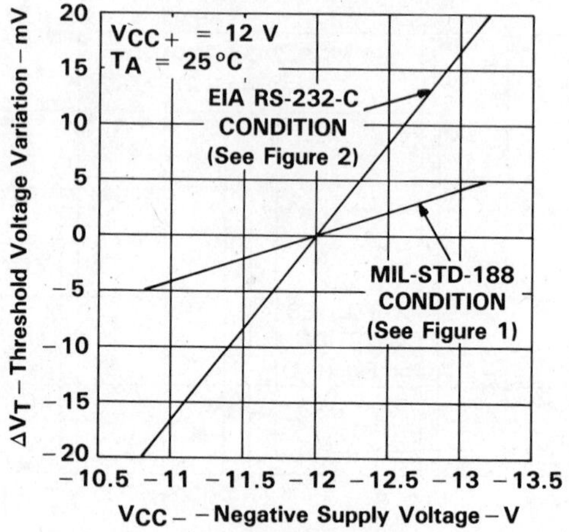
V_{CC+} = 12 V
T_A = 25°C
EIA RS-232-C CONDITION (See Figure 2)
MIL-STD-188 CONDITION (See Figure 1)
ΔV_T – Threshold Voltage Variation – mV
20
15
10
5
0
−5
−10
−15
−20
−10.5 −11 −11.5 −12 −12.5 −13 −13.5
V_{CC-} – Negative Supply Voltage – V

FIGURE 10

THRESHOLD VOLTAGE
vs
HYSTERESIS ADJUST RESISTANCE

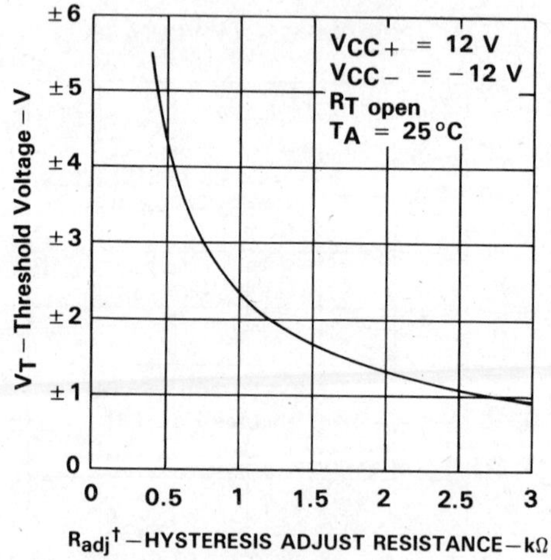
V_{CC+} = 12 V
V_{CC-} = −12 V
R_T open
T_A = 25°C
V_T – Threshold Voltage – V
±6
±5
±4
±3
±2
±1
0
0 0.5 1 1.5 2 2.5 3
R_{adj}† – HYSTERESIS ADJUST RESISTANCE – kΩ

† R_{adj} is connected between Hysteresis Adjust terminal and V_{CC-}.

FIGURE 11

TEXAS INSTRUMENTS
POST OFFICE BOX 655012 • DALLAS, TEXAS 75265

TYPICAL CHARACTERISTICS

PROPAGATION DELAY TIME
vs
FREE-AIR TEMPERATURE

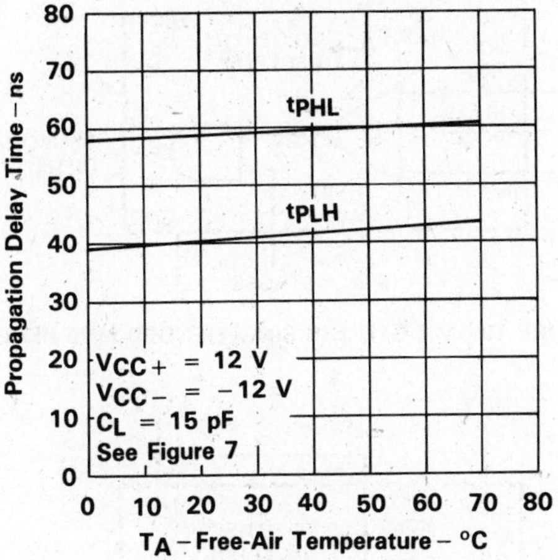
Propagation Delay Time – ns
80
70
60
50
40
30
20
10
0
tPHL
tPLH
VCC+ = 12 V
VCC– = –12 V
CL = 15 pF
See Figure 7
0 10 20 30 40 50 60 70 80
TA – Free-Air Temperature – °C

FIGURE 12

TYPICAL APPLICATIONS

Some typical applications of the SN55152 and SN75152 are as follows:

- MIL-STD-188 Interface Receiver
- EIA RS-232-C Interface Receiver
- Single-Ended Line Receiver
- Differential Line Receiver
- High-Noise-Immunity Line Receiver
- Schmitt Trigger
- High-Voltage-Logic-to-TTL Translator
- MOS-to-TTL Converter
- Pulse Generator
- Threshold Detector
- Pulse Shaper

TYPICAL APPLICATIONS

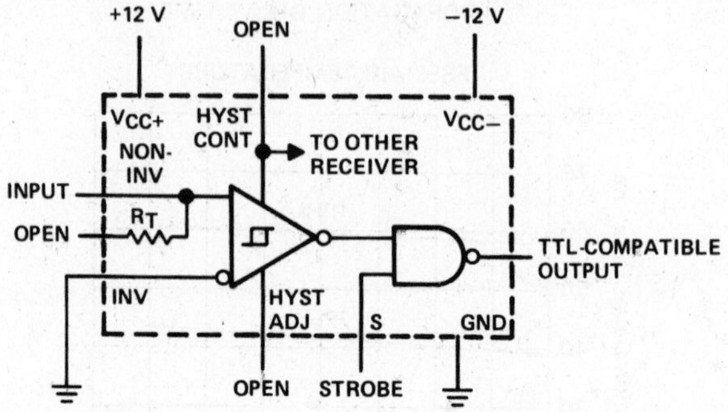
+12 V
OPEN
−12 V
VCC+
HYST
CONT
VCC−
NON-
INV
TO OTHER
RECEIVER
INPUT
RT
OPEN
TTL-COMPATIBLE
OUTPUT
INV
HYST
ADJ
S
GND
OPEN
STROBE

FIGURE 13. MIL-STD-188 SINGLE-ENDED LINE RECEIVER

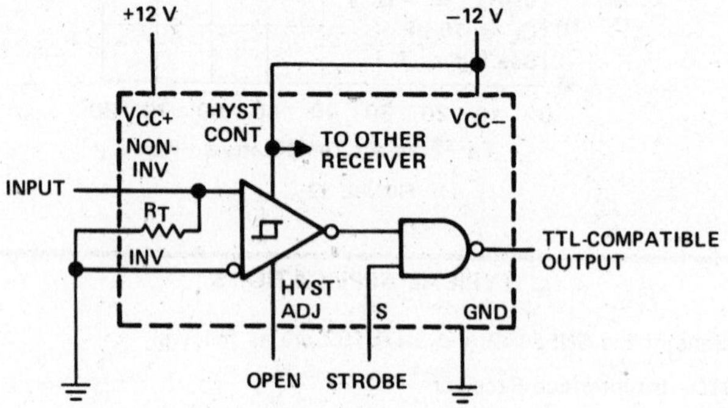
+12 V
−12 V
VCC+
HYST
CONT
VCC−
NON-
INV
TO OTHER
RECEIVER
INPUT
RT
INV
TTL-COMPATIBLE
OUTPUT
HYST
ADJ
S
GND
OPEN
STROBE

NORMAL OPERATION

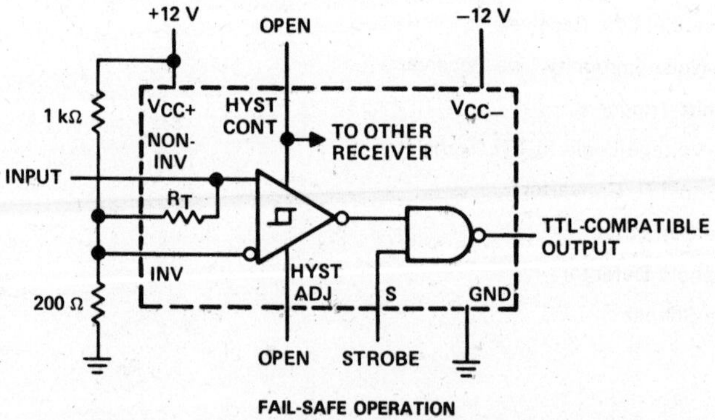
+12 V
OPEN
−12 V
1 kΩ
VCC+
HYST
CONT
VCC−
NON-
INV
TO OTHER
RECEIVER
INPUT
RT
INV
TTL-COMPATIBLE
OUTPUT
200 Ω
HYST
ADJ
S
GND
OPEN
STROBE

FAIL-SAFE OPERATION

FIGURE 14. EIA RS-232-C SINGLE-ENDED RECEIVER

TYPICAL APPLICATIONS

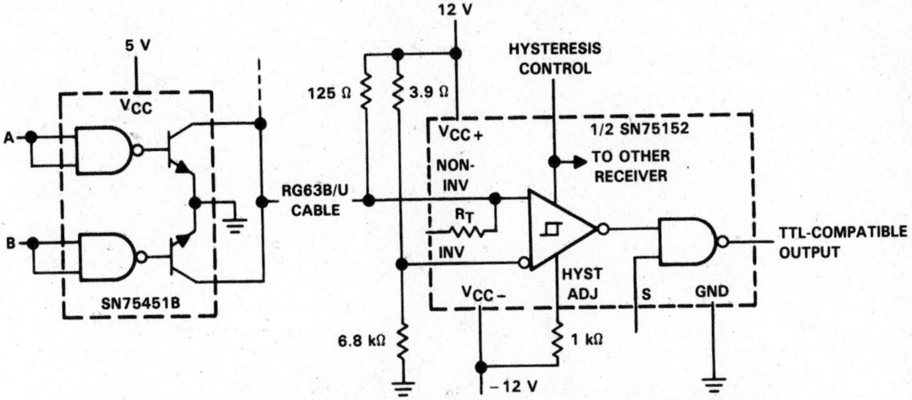
5 V
VCC
A
B
SN75451B
12 V
125 Ω
3.9 Ω
HYSTERESIS
CONTROL
RG63B/U
CABLE
VCC+
1/2 SN75152
TO OTHER
RECEIVER
NON-
INV
RT
INV
TTL-COMPATIBLE
OUTPUT
HYST
ADJ
VCC−
S
GND
6.8 kΩ
1 kΩ
−12 V

FIGURE 15. SINGLE-ENDED TRANSMITTER WITH DRIVER "OR" CAPABILITY AND RECEIVER WITH ADJUSTABLE NOISE IMMUNITY

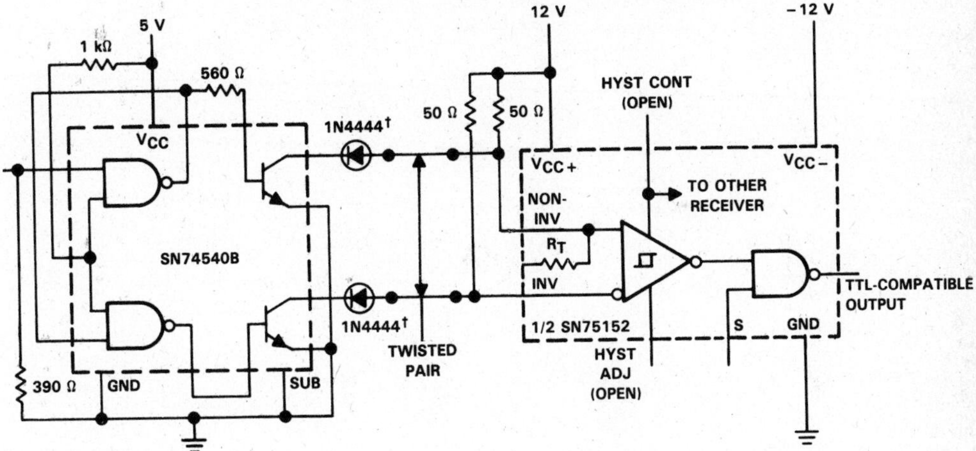
5 V
1 kΩ
560 Ω
VCC
SN74540B
390 Ω
GND
SUB
1N4444†
1N4444†
12 V
50 Ω
50 Ω
TWISTED
PAIR
HYST CONT
(OPEN)
−12 V
VCC+
VCC−
TO OTHER
RECEIVER
NON-
INV
RT
INV
1/2 SN75152
S
GND
HYST
ADJ
(OPEN)
TTL-COMPATIBLE
OUTPUT

Frequency to 0.5 MHz
Common-Mode Voltage . . . −12 V to +10 V

† The 1N4444 diodes are required only for negative common-mode protection at the driver outputs.

FIGURE 16. BALANCED LINE OPERATION WITH HIGH COMMON-MODE-VOLTAGE CAPABILITY

D899, NOVEMBER 1970—REVISED SEPTEMBER 1986

- Satisfies Requirements of EIA Standard RS-232-C
- Input Resistance . . . 3 kΩ to 7 kΩ over Full RS-232-C Voltage Range
- Input Threshold Adjustable to Meet "Fail-Safe" Requirements Without Using External Components
- Built-In Hysteresis for Increased Noise Immunity
- Inverting Output Compatible with TTL
- Output with Active Pull-Up for Symmetrical Switching Speeds
- Standard Supply Voltages . . . 5 V or 12 V

SN55154 . . . J PACKAGE
SN75154 . . . D, J, OR N PACKAGE
(TOP VIEW)

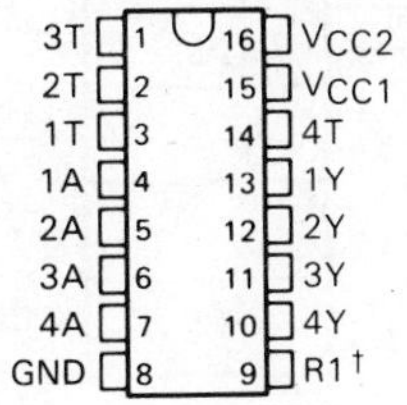
3T 1
2T 2
1T 3
1A 4
2A 5
3A 6
4A 7
GND 8
16 VCC2
15 VCC1
14 4T
13 1Y
12 2Y
11 3Y
10 4Y
9 R1†

SN55154 . . . FK PACKAGE
(TOP VIEW)

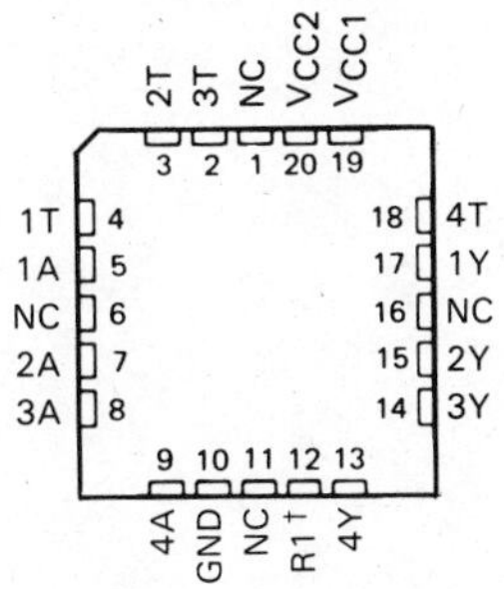
2T 3T NC VCC2 VCC1
3 2 1 20 19
1T 4
1A 5
NC 6
2A 7
3A 8
18 4T
17 1Y
16 NC
15 2Y
14 3Y
9 10 11 12 13
4A GND NC R1† 4Y

NC - No internal connection
†For function of R1, see schematic

description

The SN55154 and SN75154 are monolithic Low-Power Schottky line receivers designed to satisfy the requirements of the standard interface between data terminal equipment and data communication equipment as defined by EIA standard RS-232-C. Other applications are for relatively short, single-line, point-to-point data transmission and for level translators. Operation is normally from a single five-volt supply; however, a built-in option allows operation from a 12-volt supply without the use of additional components. The output is compatible with most TTL circuits when either supply voltage is used.

In normal operation, the threshold-control terminals are connected to the V_{CC1} terminal, even if power is being supplied via the alternate V_{CC2} terminal. This provides a wide hysteresis loop, which is the difference between the positive-going and negative-going threshold voltages. See typical characteristics. In this mode of operation, if the input voltage goes to zero, the output voltage will remain at the low or high level as determined by the previous input.

For fail-safe operation, the threshold-control terminals are open. This reduces the hysteresis loop by causing the negative-going threshold voltage to be above zero. The positive-going threshold voltage remains above zero as it is unaffected by the disposition of the threshold terminals. In the fail-safe mode, if the input voltage goes to zero or an open-circuit condition, the output will go to the high level regardless of the previous input condition.

The SN55154 is characterized for operation over the full military temperature range of −55°C to 125°C. The SN75154 is characterized for operation from 0°C to 70°C.

logic symbol†

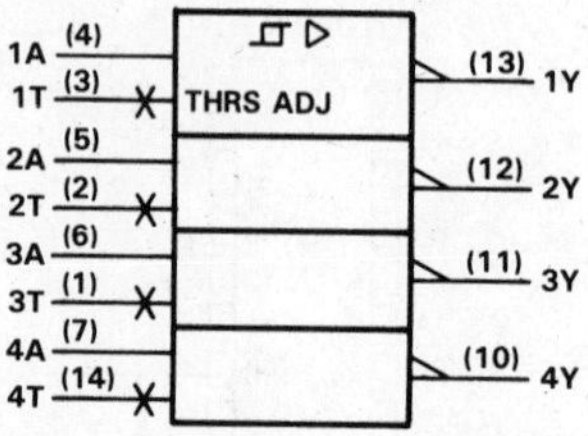
1A (4)
1T (3)
THRS ADJ
2A (5)
2T (2)
3A (6)
3T (1)
4A (7)
4T (14)
(13) 1Y
(12) 2Y
(11) 3Y
(10) 4Y

† This symbol is in accordance with ANSI/IEEE Std 91-1984 and IEC Publication 617-12.

Pin numbers shown are for D, J, and N packages.

logic diagram

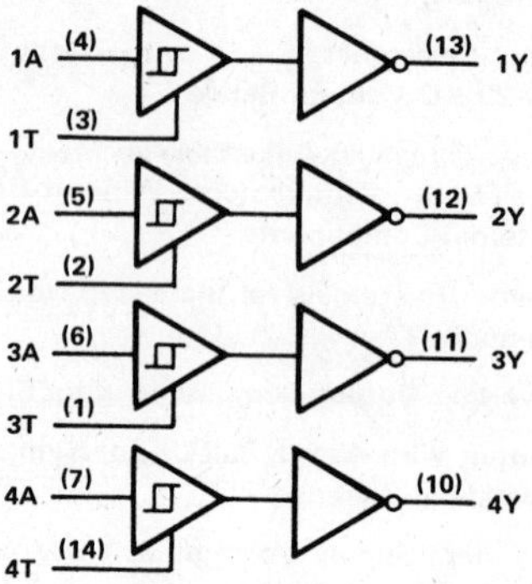
1A (4)
1T (3)
(13) 1Y
2A (5)
2T (2)
(12) 2Y
3A (6)
3T (1)
(11) 3Y
4A (7)
4T (14)
(10) 4Y

schematic

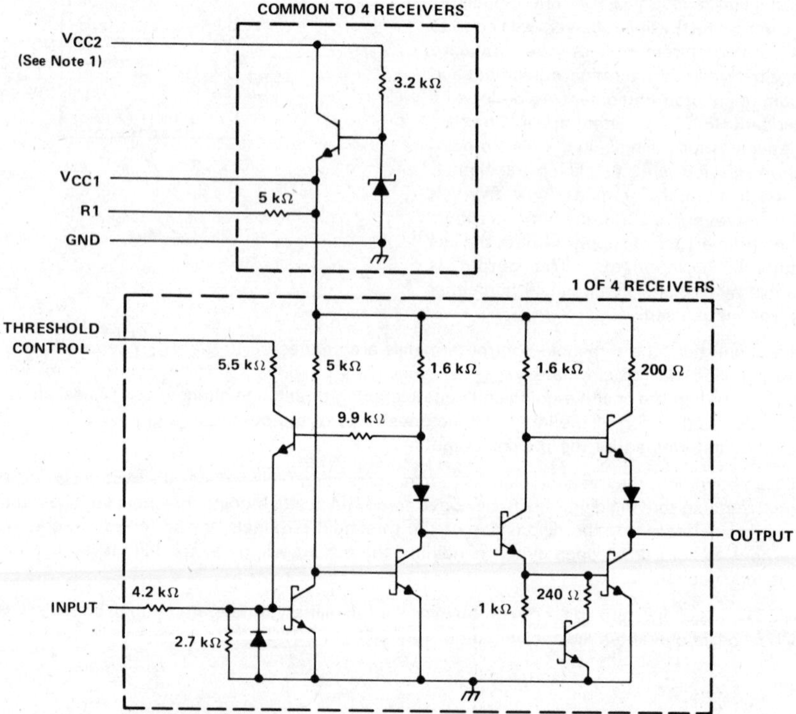
COMMON TO 4 RECEIVERS
VCC2
(See Note 1)
3.2 kΩ
VCC1
R1
5 kΩ
GND
1 OF 4 RECEIVERS
THRESHOLD
CONTROL
5.5 kΩ
5 kΩ
1.6 kΩ
1.6 kΩ
200 Ω
9.9 kΩ
OUTPUT
INPUT
4.2 kΩ
2.7 kΩ
240 Ω
1 kΩ

Component values shown are nominal.

⏚ . . . Substrate

NOTE 1: When V_{CC1} is used, V_{CC2} may be left open or shorted to V_{CC1}. When V_{CC2} is used, V_{CC1} must be left open or connected to the threshold control pins.

absolute maximum ratings over operating free-air temperature range (unless otherwise noted)

		SN55154	SN75154	UNIT
Normal supply voltage, V_{CC1} (see Note 2)		7	7	V
Alternate supply voltage, V_{CC2}		14	14	V
Input voltage		±25	±25	V
Continuous total dissipation at (or below) 25°C free-air temperature (see Note 3)	D package		950	mW
	FK package	1375		mW
	J package	1375	1025	mW
	N package		1150	mW
Operating free-air temperature range		−55 to 125	0 to 70	°C
Storage temperature range		−65 to 150	−65 to 150	°C
Case temperature for 60 seconds: FK package		260		°C
Lead temperature 1,6 mm (1/16 inch) from case for 60 seconds: J package		300	300	°C
Lead temperature 1,6 mm (1/16 inch) from case for 10 seconds: D or N package			260	°C

recommended operating conditions

	SN55154			SN75154			
	MIN	NOM	MAX	MIN	NOM	MAX	UNIT
Normal supply voltage, V_{CC1}	4.5	5	5.5	4.5	5	5.5	V
Alternate supply voltage, V_{CC2}	10.8	12	13.2	10.8	12	13.2	V
High-level input voltage, V_{IH} (see Note 4)	3		15	3		15	V
Low-level input voltage, V_{IL} (see Note 4)	−15		−3	−15		−3	V
High-level output current, I_{OH}			−400			−400	μA
Low-level output current, I_{OL}			16			16	mA
Operating free-air temperature, T_A	−55		125	0		70	°C

NOTES: 2. Voltage values are with respect to network ground terminal.
3. For operation above 25°C free-air temperature, refer to Dissipation Derating Curves in Appendix A. In the J package, SN55154 chips are alloy mounted and SN75154 chips are glass mounted. In the N package, use the 9.2-mW/°C curve for these devices.
4. The algebraic convention, where the less positive (more negative) limit is designated as minimum, is used in this data sheet for logic and threshold levels only, e.g., when 0 V is the maximum, the minimum limit is a more negative voltage.

electrical characteristics over recommended operating free-air temperature range (unless otherwise noted)

PARAMETER			TEST FIGURE	TEST CONDITIONS	MIN	TYP‡ (SEE NOTE 4)	MAX	UNIT
V_{T+}	Positive-going threshold voltage	Normal operation	1		0.8	2.2	3	V
		Fail-safe operation			0.8	2.2	3	
V_{T-}	Negative-going threshold voltage	Normal operation	1		−3	−1.1	0	V
		Fail-safe operation			0.8	1.4	3	
V_{hys}	Hysteresis ($V_{T+} - V_{T-}$)	Normal operation	1		0.8	3.3	6	V
		Fail-safe operation			0	0.8	2.2	
V_{OH}	High-level output voltage		1	$I_{OH} = -400\ \mu A$	2.4	3.5		V
V_{OL}	Low-level output voltage		1	$I_{OL} = 16$ mA		0.29	0.4	V
r_i	Input resistance		2	$\Delta V_I = -25$ V to -14 V	3	5	7	kΩ
				$\Delta V_I = -14$ V to -3 V	3	5	7	
				$\Delta V_I = -3$ V to 3 V	3	6	8	
				$\Delta V_I = 3$ V to 14 V	3	5	7	
				$\Delta V_I = 14$ V to 25 V	3	5	7	
$V_{I(open)}$	Open-circuit input voltage		3	$I_I = 0$	0	0.2	2	V
I_{OS}	Short-circuit output current†		4	$V_{CC1} = 5.5$ V, $V_I = -5$ V	−10	−20	−40	mA
I_{CC1}	Supply current from V_{CC1}		5	$V_{CC1} = 5.5$ V, $T_A = 25°C$		20	35	mA
I_{CC2}	Supply current from V_{CC2}			$V_{CC2} = 13.2$ V, $T_A = 25°C$		23	40	

†Not more than one output should be shorted at a time.
‡All typical values are at $V_{CC1} = 5$ V, $T_A = 25°C$.

switching characteristics, $V_{CC1} = 5$ V, $T_A = 25°C$, N = 10

PARAMETER		TEST FIGURE	TEST CONDITIONS	MIN	TYP	MAX	UNIT
t_{PLH}	Propagation delay time, low-to-high-level output	6	$C_L = 50$ pF, $R_L = 390\ \Omega$		11		ns
t_{PHL}	Propagation delay time, high-to-low-level output				8		ns
t_{TLH}	Transition time, low-to-high-level output				7		ns
t_{THL}	Transition time, high-to-low-level output				2.2		ns

TYPICAL CHARACTERISTICS

OUTPUT VOLTAGE vs INPUT VOLTAGE

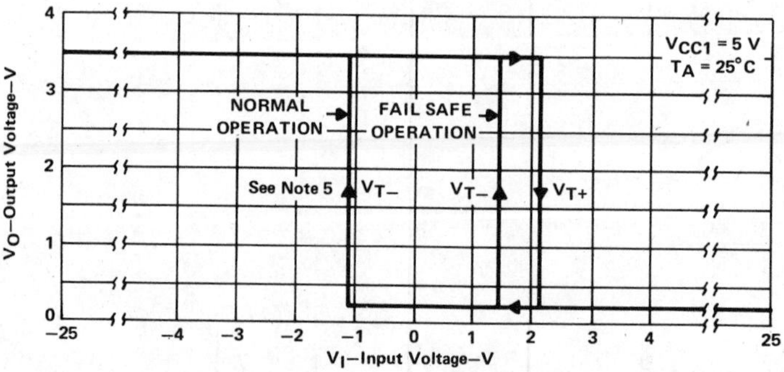
VCC1 = 5 V
TA = 25°C
NORMAL OPERATION
FAIL SAFE OPERATION
See Note 5
VT−
VT−
VT+
VO—Output Voltage—V
VI—Input Voltage—V
4
3
2
1
0
−25
−4
−3
−2
−1
0
1
2
3
4
25

NOTE 5: For normal operation, the threshold controls are connectd to V_{CC1}. For fail-safe operation, the threshold controls are open.

TEXAS INSTRUMENTS
POST OFFICE BOX 655012 • DALLAS, TEXAS 75265

PARAMETER MEASUREMENT INFORMATION

d-c test circuits†

TEST TABLE

TEST	MEASURE	A	T	Y	V_{CC1} (PIN 15)	V_{CC2} (PIN 16)
Open-circuit input	V_{OH}	Open	Open	I_{OH}	4.5 V	Open
(fail safe)	V_{OH}	Open	Open	I_{OH}	Open	10.8 V
V_{T+} min,	V_{OH}	0.8 V	Open	I_{OH}	5.5 V	Open
V_{T-} min (fail safe)	V_{OH}	0.8 V	Open	I_{OH}	Open	13.2 V
V_{T+} min (normal)	V_{OH}	Note A	Pin 15	I_{OH}	5.5 V and T	Open
	V_{OH}	Note A	Pin 15	I_{OH}	T	13.2 V
V_{IL} max,	V_{OH}	−3 V	Pin 15	I_{OH}	5.5 V and T	Open
V_{T-} min (normal)	V_{OH}	−3 V	Pin 15	I_{OH}	T	13.2 V
V_{IH} min, V_{T+} max,	V_{OL}	3 V	Open	I_{OL}	4.5 V	Open
V_{T-} max (fail safe)	V_{OL}	3 V	Open	I_{OL}	Open	10.8 V
V_{IH} min, V_{T+} max	V_{OL}	3 V	Pin 15	I_{OL}	4.5 V and T	Open
(normal)	V_{OL}	3 V	Pin 15	I_{OL}	T	10.8 V
V_{T-} max (normal)	V_{OL}	Note B	Pin 15	I_{OL}	5.5 V and T	Open
	V_{OL}	Note B	Pin 15	I_{OL}	T	13.2 V

NOTES: A. Momentarily apply −5 V, then 0.8 V.
B. Momentarily apply 5 V, then ground.

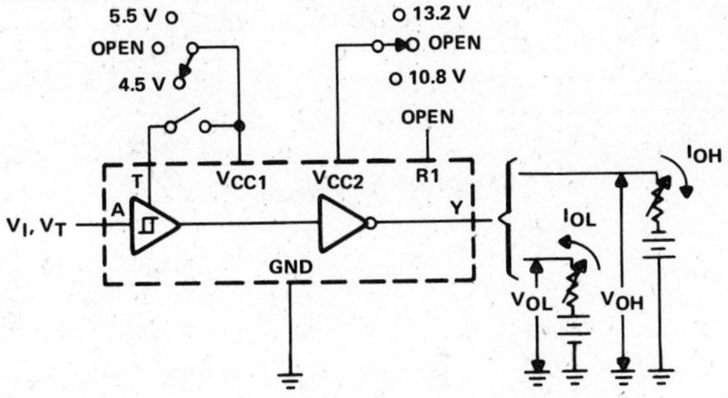
5.5 V
OPEN
4.5 V
13.2 V
OPEN
10.8 V
OPEN
T
VCC1
VCC2
R1
A
Y
VI, VT
GND
IOH
IOL
VOL
VOH

FIGURE 1. V_{IH}, V_{IL}, V_{T+}, V_{T-}, V_{OH}, V_{OL}

†Arrows indicate actual direction of current flow. Current into a terminal is a positive value.

PARAMETER MEASUREMENT INFORMATION

d-c test circuits† (continued)

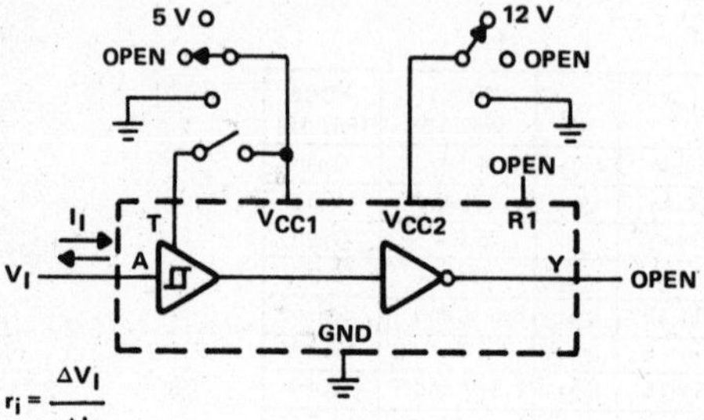
5 V
OPEN
12 V
OPEN
OPEN
T
V_{CC1}
V_{CC2}
R1
I_I
A
V_I
Y
OPEN
GND

$r_i = \frac{\Delta V_I}{\Delta I_I}$

TEST TABLE

T	V_{CC1} (PIN 15)	V_{CC2} (PIN 16)
Open	5 V	Open
Open	GND	Open
Open	Open	Open
Pin 15	T and 5 V	Open
GND	GND	Open
Open	Open	12 V
Open	Open	GND
Pin 15	T	12 V
Pin 15	T	GND
Pin 15	T	Open

FIGURE 2. r_i

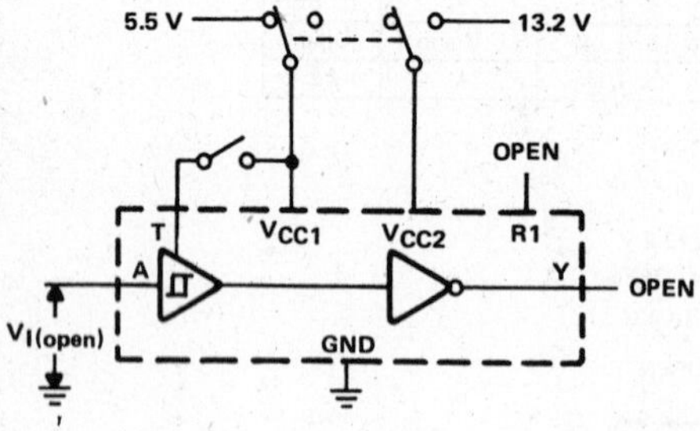
5.5 V
13.2 V
OPEN
T
V_{CC1}
V_{CC2}
R1
A
Y
OPEN
$V_{I(open)}$
GND

TEST TABLE

T	V_{CC1} (PIN 15)	V_{CC2} (PIN 16)
Open	5.5 V	Open
Pin 15	5.5 V	Open
Open	Open	13.2 V
Pin 15	T	13.2 V

FIGURE 3. $V_{I(open)}$

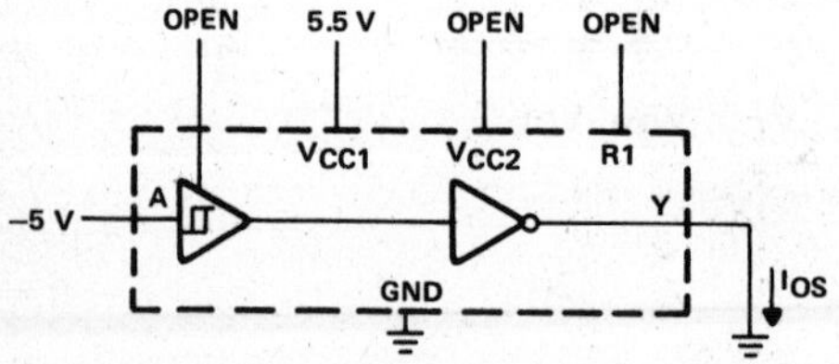
OPEN
5.5 V
OPEN
OPEN
V_{CC1}
V_{CC2}
R1
A
−5 V
Y
GND
I_{OS}

Each output is tested separately.

FIGURE 4. I_{OS}

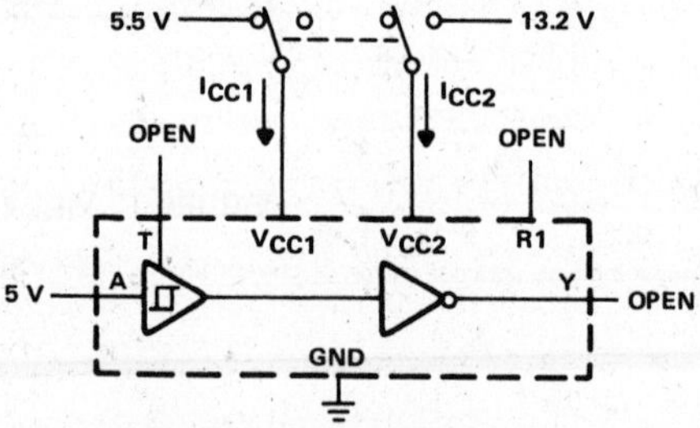
5.5 V
13.2 V
I_{CC1}
I_{CC2}
OPEN
OPEN
T
V_{CC1}
V_{CC2}
R1
A
5 V
Y
OPEN
GND

All four line receivers are tested simultaneously.

FIGURE 5. I_{CC}

†Arrows indicate actual direction of current flow. Current into a terminal is a positive value.

TEXAS INSTRUMENTS
POST OFFICE BOX 655012 • DALLAS, TEXAS 75265

PARAMETER MEASUREMENT INFORMATION

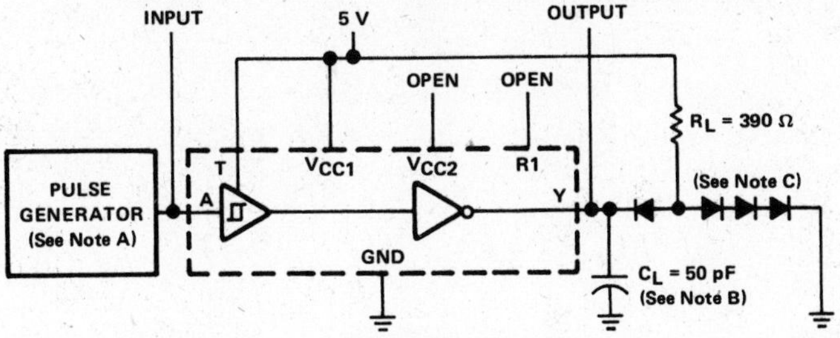
INPUT
5 V
OUTPUT
OPEN
OPEN
R_L = 390 Ω
PULSE
GENERATOR
(See Note A)
T
V_{CC1}
V_{CC2}
R1
(See Note C)
A
Y
GND
C_L = 50 pF
(See Note B)

TEST CIRCUIT

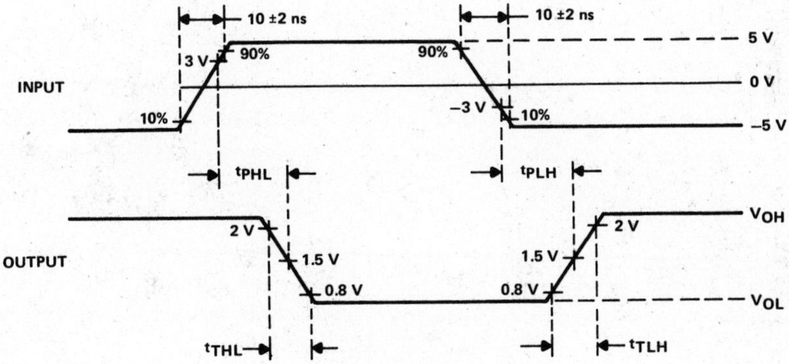
10 ±2 ns
10 ±2 ns
5 V
3 V
90%
90%
INPUT
0 V
10%
–3 V
10%
–5 V
t_{PHL}
t_{PLH}
V_{OH}
2 V
2 V
OUTPUT
1.5 V
1.5 V
0.8 V
0.8 V
V_{OL}
t_{THL}
t_{TLH}

VOLTAGE WAVEFORMS

NOTES: A. The pulse generator has the following characteristics: Z_{out} = 50 Ω, t_w = 200 ns, duty cycle ≤ 20%.
B. C_L includes probe and jig capacitance.
C. All diodes are 1N3064.

FIGURE 6. SWITCHING TIMES

SN75155
LINE DRIVER AND RECEIVER

D2951, JULY 1986

- Meets EIA Standard RS-232-C
- 10-mA Current Limited Output
- Wide Range of Supply Voltage . . . V_{CC} = 4.5 V to 15 V
- Low Power . . . 130 mW
- Built-In 5-Volt Regulator
- Response Control Provides:
  - Input Threshold Shifting
  - Input Noise Filtering
- Power-Off Output Resistance . . . 300 Ω Typ
- Driver Input TTL Compatible

D, JG, OR P PACKAGE
(TOP VIEW)

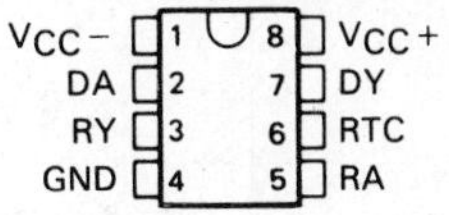
VCC−
DA
RY
GND
VCC+
DY
RTC
RA
1
2
3
4
5
6
7
8

logic symbol†

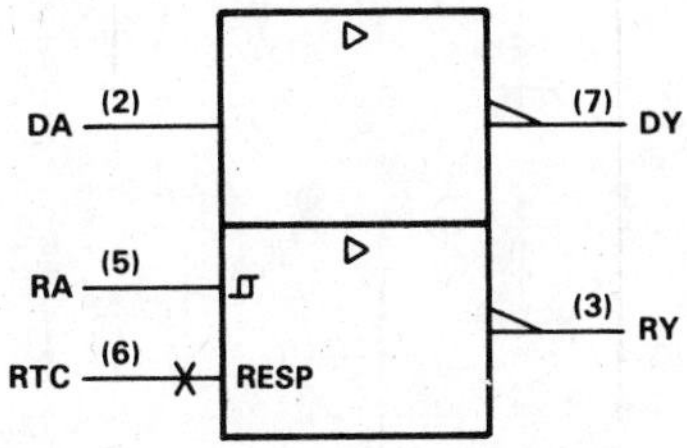
DA
(2)
(7)
DY
RA
(5)
RTC
(6)
RESP
(3)
RY

†This symbol is in accordance with ANSI/IEEE Std 91-1984 and IEC Publication 617-12

logic diagram

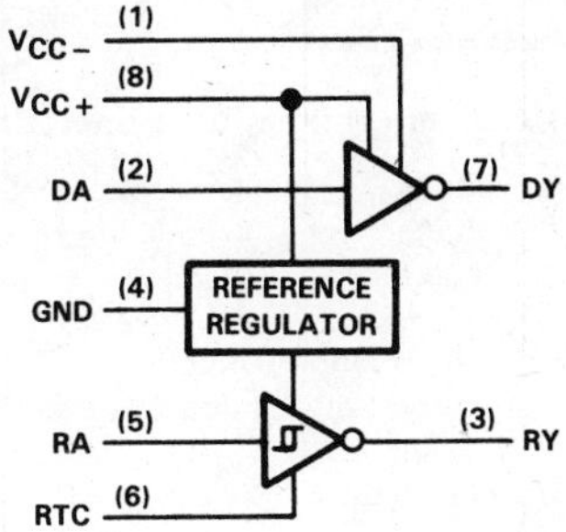
VCC−
(1)
VCC+
(8)
DA
(2)
(7)
DY
GND
(4)
REFERENCE
REGULATOR
RA
(5)
(3)
RY
RTC
(6)

description

The SN75155 is a monolithic line driver and receiver that is designed to satisfy the requirements of the standard interface between data terminal equipment and data communication equipment as defined by EIA standard RS-232-C. A Response Control input is provided for the receiver. A resistor or a resistor and a bias voltage can be connected between the response control input and ground to provide noise filtering. The driver used is similar to the SN75188. The receiver used is similar to the SN75189A.

The SN75155 is characterized for operation from 0°C to 70°C.

TEXAS INSTRUMENTS
POST OFFICE BOX 655012 • DALLAS, TEXAS 75265

schematic

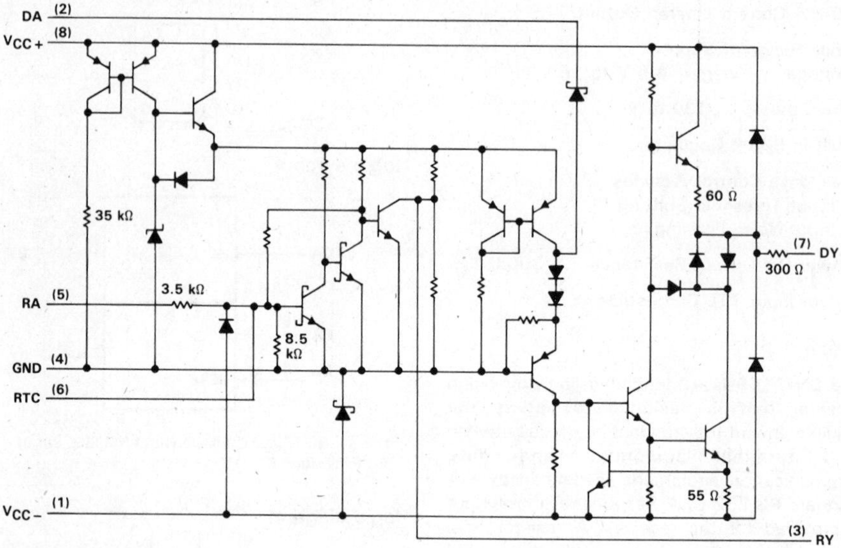
DA (2)
VCC+ (8)
35 kΩ
3.5 kΩ
RA (5)
8.5 kΩ
GND (4)
RTC (6)
VCC− (1)
60 Ω
(7) DY
300 Ω
55 Ω
(3) RY

All resistor values shown are nominal.

absolute maximum ratings over operating free-air temperature range (unless otherwise noted)

Supply voltage, V_{CC+} (see Note 1) .... 15 V
Supply voltage, V_{CC-} (see Note 1) .... −15 V
Input voltage range: Driver .... −15 V to 15 V
Receiver .... −30 V to 30 V
Output voltage range (Driver) .... −15 V to 15 V
Continuous total dissipation at (or below) 25°C free-air temperature (see Note 2):
D package .... 725 mW
JG package .... 825 mW
P package .... 1000 mW
Operating free-air temperature range .... 0°C to 70°C
Storage temperature range .... −65°C to 150°C
Lead temperature 1,6 mm (1/16 inch) from case for 60 seconds, JG package .... 300°C
Case temperature for 60 seconds, FK package .... 260°C
Lead temperature 1,6 mm (1/16 inch) from case for 10 seconds, D or P package .... 260°C

NOTES: 1. All voltage values are with respect to network ground terminal.
2. For operation above 25°C free-air temperature, refer to Dissipation Derating Table. In the JG package, SN75155 chips are glass mounted.

DISSIPATION DERATING TABLE

PACKAGE	T_A = 25°C POWER RATING	DERATING FACTOR	ABOVE T_A	T_A = 70°C POWER RATING
D	725 mW	5.8 mW/°C	25°C	464 mW
JG	825 mW	6.6 mW/°C	25°C	528 mW
P	1000 mW	8.0 mW/°C	25°C	640 mW

TEXAS INSTRUMENTS
POST OFFICE BOX 655012 • DALLAS, TEXAS 75265

recommended operating conditions

PARAMETERS	MIN	NOM	MAX	UNIT
Supply voltage, V_{CC+}	4.5	12	15	V
Supply voltage, V_{CC-}	−4.5	−12	−15	V
Input voltage, driver, $V_{I(D)}$			±15	V
Input voltage, receiver, $V_{I(R)}$	−25		25	V
High-level input voltage, driver, V_{IH}	2			V
Low-level input voltage, driver, V_{IL}			0.8	V
Response control current			±5.5	mA
Output current, receiver, $I_{O(R)}$			24	mA
Operating free-air temperature, T_A	0		70	°C

electrical characteristics over recommended operating free-air temperature range (unless otherwise noted)

total device

PARAMETERS		TEST CONDITIONS			MIN	TYP†	MAX	UNIT
I_{CCH+}	High-level supply current	V_{CC+} = 5 V,	V_{CC-} = −5 V,	$V_{I(D)}$ = 2 V,		6.3	8.1	mA
		V_{CC+} = 9 V,	V_{CC-} = −9 V,	$V_{I(R)}$ = 2.3 V,		9.1	11.9	
		V_{CC+} = 12 V,	V_{CC-} = −12 V,	Output open		10.4	14	
I_{CCL+}	Low-level supply current	V_{CC+} = 5 V,	V_{CC-} = −5 V,	$V_{I(D)}$ = 0.8 V,		2.5	3.4	mA
		V_{CC+} = 9 V,	V_{CC-} = −9 V,	$V_{I(R)}$ = 0.6 V,		3.7	5.1	
		V_{CC+} = 12 V,	V_{CC-} = −12 V,	Output open		4.1	5.6	
I_{CC+}	Supply current	V_{CC+} = 5 V,	V_{CC-} = 0,	$V_{I(R)}$ = 2.3 V,		4.8	6.4	mA
		V_{CC+} = 9 V,	V_{CC-} = 0,	$V_{I(D)}$ = 0		6.7	9.1	
I_{CCH-}	High-level supply current	V_{CC+} = 5 V,	V_{CC-} = −5 V,	$V_{I(D)}$ = 2 V,		−2.4	−3.1	mA
		V_{CC+} = 9 V,	V_{CC-} = −9 V,	$V_{I(R)}$ = 2.3 V,		−3.9	−4.9	
		V_{CC+} = 12 V,	V_{CC-} = −12 V,	Output open		−4.8	−6.1	
I_{CCL-}	Low-level supply current	V_{CC+} = 5 V,	V_{CC-} = −5 V,	$V_{I(D)}$ = 0.8 V,		−0.2	−0.35	mA
		V_{CC+} = 9 V,	V_{CC-} = −9 V,	$V_{I(R)}$ = 0.6 V,		−0.25	−0.4	
		V_{CC+} = 12 V,	V_{CC-} = −12 V,	Output open		−0.27	−0.45	

†All typical values are at T_A = 25°C.

electrical characteristics over recommended operating free-air temperature range, V_{CC+} = 12 V, V_{CC-} = −12 V (unless otherwise noted)

driver section

PARAMETER		TEST CONDITIONS		MIN	TYP†	MAX	UNIT
V_{OH}	High-level output voltage	V_{IL} = 0.8 V, R_L = 3 kΩ	V_{CC+} = 5 V, V_{CC-} = −5 V	3.2	3.7		V
			V_{CC+} = 9 V, V_{CC-} = −9 V	6.5	7.2		
			V_{CC+} = 12 V, V_{CC-} = −12 V	8.9	9.8		
V_{OL}	Low-level output voltage (see Note 3)	V_{IH} = 2 V, R_L = 3 kΩ	V_{CC+} = 5 V, V_{CC-} = −5 V		−3.6	−3.2	V
			V_{CC+} = 9 V, V_{CC-} = −9 V		−7.1	−6.4	
			V_{CC+} = 12 V, V_{CC-} = −12 V		−9.7	−8.8	
I_{IH}	High-level input current	V_I = 7 V				5	μA
I_{IL}	Low-level input current	V_I = 0			−0.73	−1.2	mA
I_{OSH}	High-level short-circuit output current	V_I = 0.8 V, V_O = 0		−7	−12	−14.5	mA
I_{OSL}	Low-level short-circuit output current	V_I = 2 V, V_O = 0		6.5	11.5	15	mA
R_O	Output resistance with power off	V_O = −2 V to 2 V			300		Ω

receiver section

PARAMETER		TEST CONDITIONS		MIN	TYP†	MAX	UNIT
V_{T+}	Positive-going threshold voltage			1.2	1.9	2.3	V
V_{T-}	Negative-going threshold voltage			0.6	0.95	1.2	V
V_{hys}	Hysteresis			0.6			V
V_{OH}	High-level output voltage	V_I = 0.6 V, I_{OH} = 10 μA	V_{CC+} = 5 V, V_{CC-} = −5 V	3.7	4.1	4.5	V
			V_{CC+} = 12 V, V_{CC-} = −12 V	4.4	4.7	5.2	
		V_I = 0.6 V, I_{OH} = 0.4 mA	V_{CC+} = 5 V, V_{CC-} = −5 V	3.1	3.4	3.8	
			V_{CC+} = 12 V, V_{CC-} = −12 V	3.6	4	4.5	
V_{OL}	Low-level output voltage	V_I = 2.3 V, I_{OL} = 24 mA			0.2	0.3	V
I_{IH}	High-level input current	V_I = 25 V		3.6	6.7	10	mA
		V_I = 3 V		0.43	0.67	1	mA
I_{IL}	Low-level input current	V_I = −25 V		−3.6	−6.7	−10	mA
		V_I = −3 V		−0.43	−0.67	−1	mA
I_{OS}	Short-circuit output current	V_I = 0.6 V			−2.8	−3.7	mA

†All typical values are at T_A = 25°C.

NOTE 3: The algebraic limit system, in which the more positive (less negative) limit is designated as maximum, is used in this data sheet for logic voltage levels only, e.g., if −8.8 V is the maximum, the typical value is a more negative value.

TEXAS INSTRUMENTS
POST OFFICE BOX 655012 • DALLAS, TEXAS 75265

switching characteristics over recommended operating free-air temperature range, V_{CC+} = 5 V, V_{CC-} = −5 V, C_L = 50 pF (unless otherwise noted)

driver section (see Figure 2)

PARAMETER		TEST CONDITIONS	MIN	TYP	MAX	UNIT
t_{PLH}	Propagation delay time, low-to-high-level output	R_L = 3 kΩ		250	480	ns
t_{PHL}	Propagation delay time high-to-low-level output			80	150	
t_r	Output rise time	R_L = 3 kΩ		67	180	ns
		R_L = 3 kΩ to 7 kΩ, C_L = 2500 pF		2.4	3	μs
t_f	Output fall time	R_L = 3 kΩ		48	160	ns
		R_L = 3 kΩ to 7 kΩ, C_L = 2500 pF		1.9	3	μs

receiver section (see Figure 3)

PARAMETER		TEST CONDITIONS	MIN	TYP	MAX	UNIT
t_{PLH}	Propagation delay time, low-to-high-level output	R_L = 400 Ω		175	245	ns
t_{PHL}	Propagation delay time, high-to-low-level output			37	100	
t_r	Output rise time	R_L = 400 Ω		255	360	ns
t_f	Output fall time	R_L = 400 Ω		23	50	ns

†All typical values are at T_A = 25°C.

PARAMETER MEASUREMENT INFORMATION

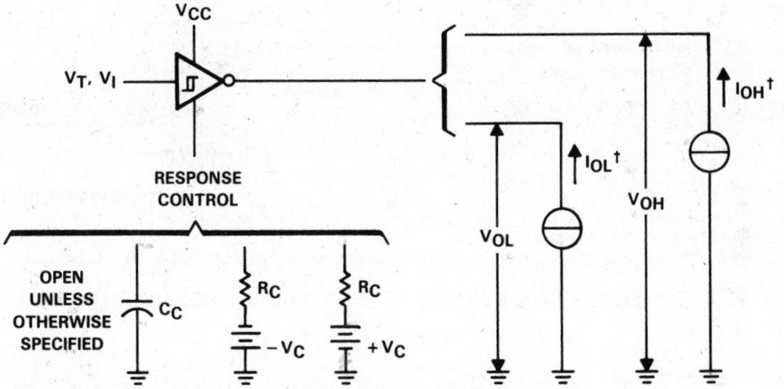
VCC
VT, VI
RESPONSE
CONTROL
OPEN
UNLESS
OTHERWISE
SPECIFIED
CC
RC
−VC
RC
+VC
IOL†
IOH†
VOL
VOH

†Arrows indicate actual direction of current flow. Current into a terminal is a positive value.

FIGURE 1. RECEIVER SECTION TEST CIRCUIT (V_{T+}, V_{T-}, V_{OH}, V_{OL})

PARAMETER MEASUREMENT INFORMATION

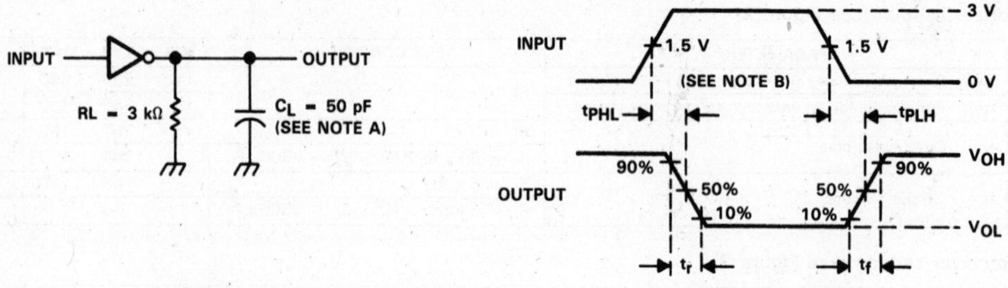
INPUT
OUTPUT
RL = 3 kΩ
CL = 50 pF
(SEE NOTE A)
INPUT
1.5 V
1.5 V
3 V
(SEE NOTE B)
0 V
tPHL
tPLH
VOH
90%
90%
OUTPUT
50%
50%
10%
10%
VOL
tr
tf

TEST CIRCUIT

VOLTAGE WAVEFORMS

NOTES: A. C_L includes probe and jig capacitance.
B. The input waveform is supplied by a generator with the following characteristics: $Z_{out} \approx 50\ \Omega$, $t_w = 1\ \mu s$, $t_r \leq 10$ ns, $t_f \leq 10$ ns.

FIGURE 2. DRIVER SECTION SWITCHING TEST CIRCUIT AND VOLTAGE WAVEFORMS

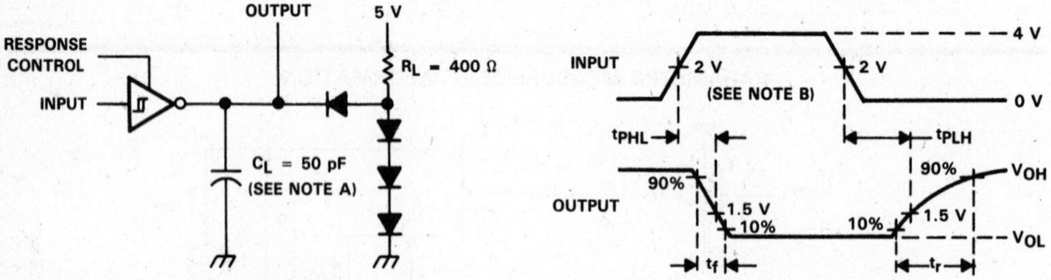
OUTPUT
5 V
RESPONSE
CONTROL
INPUT
RL = 400 Ω
CL = 50 pF
(SEE NOTE A)
INPUT
2 V
2 V
4 V
(SEE NOTE B)
0 V
tPHL
tPLH
90%
90%
VOH
OUTPUT
1.5 V
10%
10%
1.5 V
VOL
tf
tr

TEST CIRCUIT

VOLTAGE WAVEFORMS

NOTES: A. C_L includes probe and jig capacitance.
B. The input waveform is supplied by a generator with the following characteristics: $Z_{out} \approx 50\ \Omega$, $t_w = 1\ \mu s$, $t_r \leq 10$ ns, $t_f \leq 10$ ns.

FIGURE 3. RECEIVER SECTION SWITCHING TEST CIRCUIT AND VOLTAGE WAVEFORMS

TEXAS INSTRUMENTS
POST OFFICE BOX 655012 • DALLAS, TEXAS 75265

TYPICAL CHARACTERISTICS
(DRIVER)

VOLTAGE TRANSFER CHARACTERISTICS

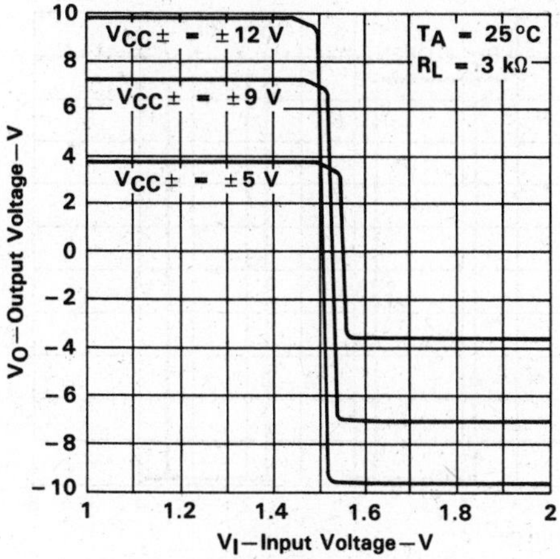
$V_{CC\pm}$ = ±12 V
$V_{CC\pm}$ = ±9 V
$V_{CC\pm}$ = ±5 V
T_A = 25°C
R_L = 3 kΩ
V_O—Output Voltage—V
V_I—Input Voltage—V
10 8 6 4 2 0 −2 −4 −6 −8 −10
1 1.2 1.4 1.6 1.8 2

FIGURE 4

OUTPUT CURRENT
vs
OUTPUT VOLTAGE

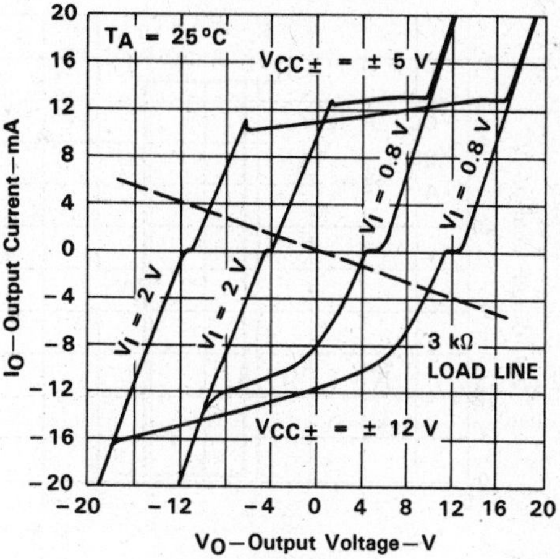
T_A = 25°C
$V_{CC\pm}$ = ± 5 V
$V_{CC\pm}$ = ± 12 V
V_I = 0.8 V
V_I = 0.8 V
V_I = 2 V
V_I = 2 V
3 kΩ
LOAD LINE
I_O—Output Current—mA
V_O—Output Voltage—V
20 16 12 8 4 0 −4 −8 −12 −16 −20
−20 −12 −4 0 4 8 12 16 20

FIGURE 5

SHORT-CIRCUIT OUTPUT CURRENT
vs
FREE-AIR TEMPERATURE

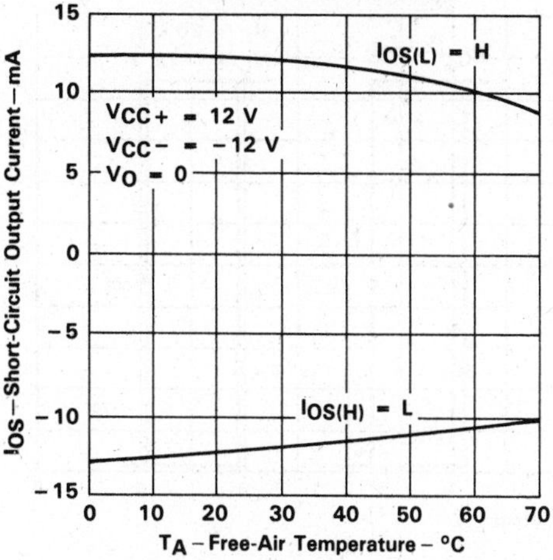
$I_{OS(L)}$ = H
V_{CC+} = 12 V
V_{CC-} = −12 V
V_O = 0
$I_{OS(H)}$ = L
I_{OS}—Short-Circuit Output Current—mA
T_A – Free-Air Temperature – °C
15 10 5 0 −5 −10 −15
0 10 20 30 40 50 60 70

FIGURE 6

SLEW RATE
vs
LOAD CAPACITANCE

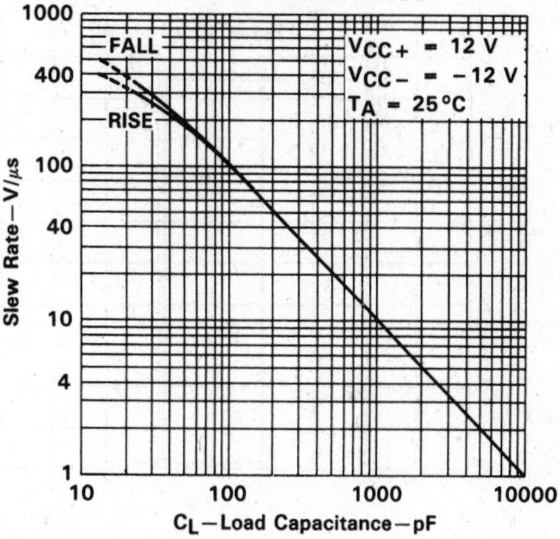
FALL
RISE
V_{CC+} = 12 V
V_{CC-} = −12 V
T_A = 25°C
Slew Rate—V/μs
C_L—Load Capacitance—pF
1000 400 100 40 10 4 1
10 100 1000 10000

FIGURE 7

4

Line Drivers/Receivers

ADVANCE INFORMATION

TYPICAL CHARACTERISTICS (RECEIVER)

OUTPUT VOLTAGE
vs
INPUT VOLTAGE

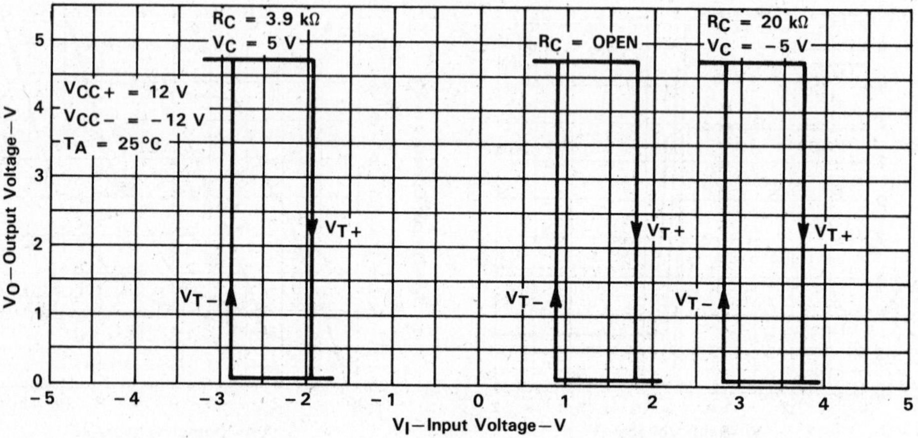
R_C = 3.9 kΩ
V_C = 5 V
R_C = OPEN
R_C = 20 kΩ
V_C = −5 V
V_{CC+} = 12 V
V_{CC-} = −12 V
T_A = 25°C
V_{T+}
V_{T-}
V_O—Output Voltage—V
V_I—Input Voltage—V

FIGURE 8

OUTPUT VOLTAGE
vs
INPUT VOLTAGE

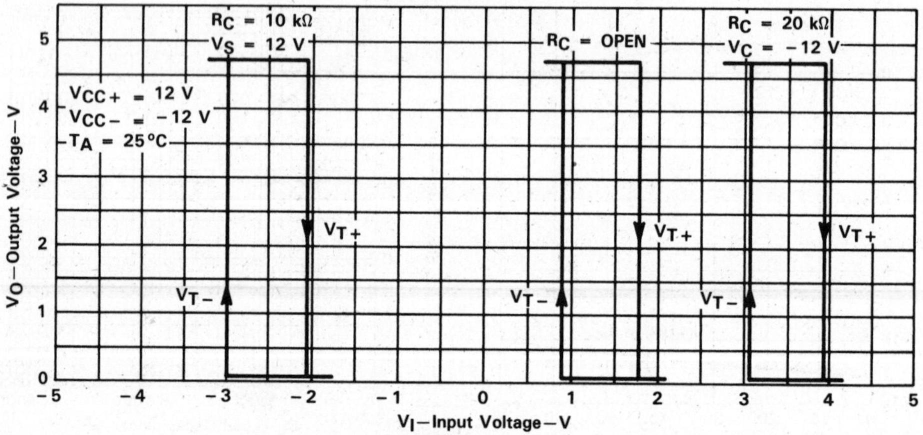
R_C = 10 kΩ
V_S = 12 V
R_C = OPEN
R_C = 20 kΩ
V_C = −12 V
V_{CC+} = 12 V
V_{CC-} = −12 V
T_A = 25°C
V_{T+}
V_{T-}
V_O—Output Voltage—V
V_I—Input Voltage—V

FIGURE 9

TEXAS INSTRUMENTS
POST OFFICE BOX 655012 • DALLAS, TEXAS 75265

TYPICAL CHARACTERISTICS
(RECEIVER)

INPUT THRESHOLD VOLTAGE
vs
FREE-AIR TEMPERATURE

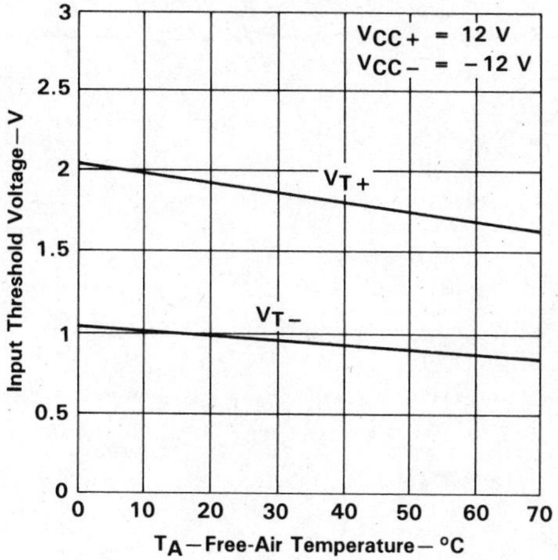
V_{CC+} = 12 V
V_{CC-} = −12 V
V_{T+}
V_{T-}
Input Threshold Voltage—V
T_A—Free-Air Temperature—°C

FIGURE 10

INPUT CURRENT
vs
INPUT VOLTAGE

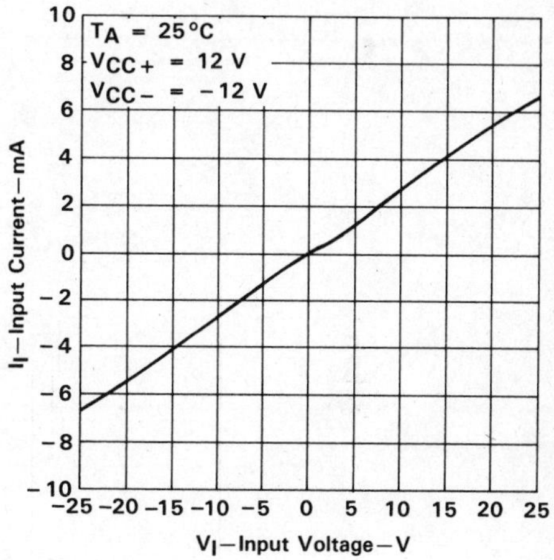
T_A = 25°C
V_{CC+} = 12 V
V_{CC-} = −12 V
I_I—Input Current—mA
V_I—Input Voltage—V

FIGURE 11

NOISE REJECTION

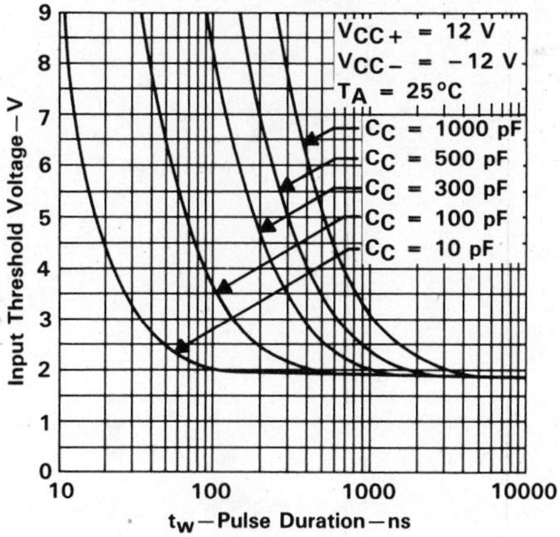
V_{CC+} = 12 V
V_{CC-} = −12 V
T_A = 25°C
C_C = 1000 pF
C_C = 500 pF
C_C = 300 pF
C_C = 100 pF
C_C = 10 pF
Input Threshold Voltage—V
t_w—Pulse Duration—ns

FIGURE 12

D2300, SEPTEMBER 1980—REVISED SEPTEMBER 1986

- Meets EIA Standards RS-422-A and RS-423-A
- Meets Federal Standards 1020 and 1030
- Operates from Single 5-V Power Supply
- Wide Common-Mode Voltage Range
- High Input Impedance
- TTL-Compatible Outputs
- High-Speed Schottky Circuitry
- 8-Pin Dual-In-Line Package
- Similar to uA9637AC except for Corner V_{CC} and Ground Pin Positions

SN55157 . . . JG PACKAGE
SN75157 . . . D, JG, OR P PACKAGE
(TOP VIEW)

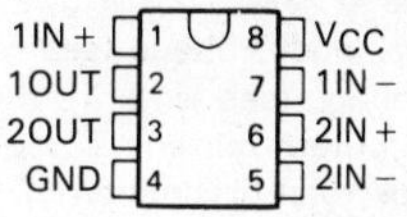
1IN+ 1
1OUT 2
2OUT 3
GND 4
8 VCC
7 1IN−
6 2IN+
5 2IN−

description

The SN75157 is a dual differential line receiver designed to meet EIA standards RS-422-A and RS-423-A and Federal Standards 1020 and 1030. It utilizes Schottky circuitry and has TTL-compatible outputs. The inputs are compatible with either a single-ended or a differential-line system. The device operates from a single 5-volt power supply and is supplied in an 8-pin dual-in-line package and small outline package.

The SN55157 is characterized over the full military temperature range of −55°C to 125°C. The SN75157 is characterized for operation from 0°C to 70°C.

logic symbol†

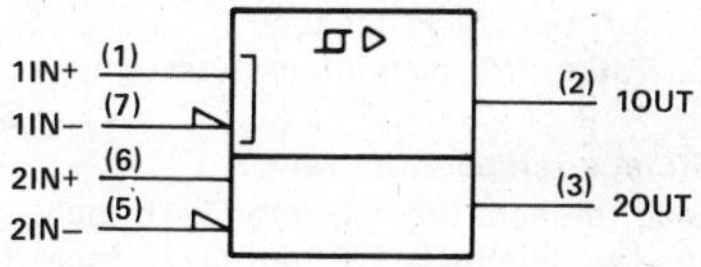
1IN+ (1)
1IN− (7)
2IN+ (6)
2IN− (5)
(2) 1OUT
(3) 2OUT

†This symbol is in accordance with ANSI/IEEE Std 91-1984 and IEC Publication 617-12.

logic diagram

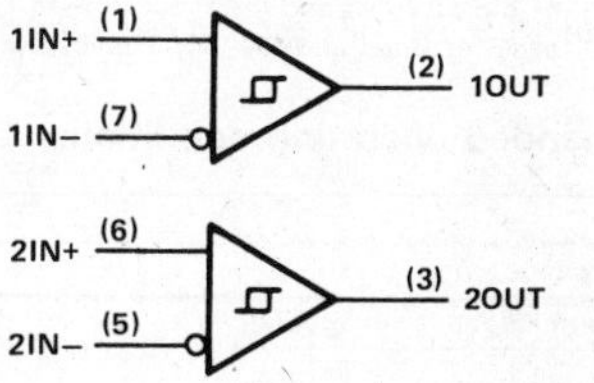
1IN+ (1)
1IN− (7)
(2) 1OUT
2IN+ (6)
2IN− (5)
(3) 2OUT

schematics of inputs and outputs

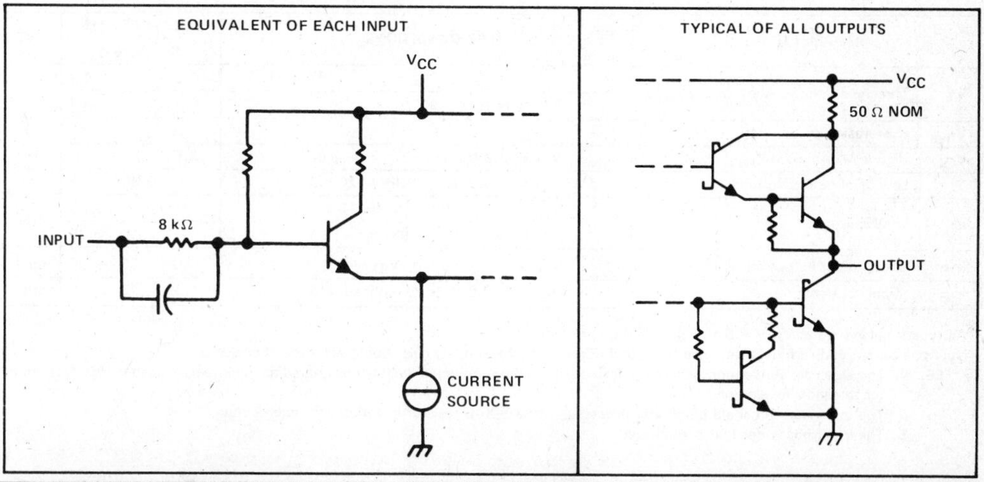
EQUIVALENT OF EACH INPUT
VCC
INPUT
8 kΩ
CURRENT SOURCE
TYPICAL OF ALL OUTPUTS
VCC
50 Ω NOM
OUTPUT

TEXAS INSTRUMENTS

POST OFFICE BOX 655012 • DALLAS, TEXAS 75265

absolute maximum ratings over operating free-air temperature range (unless otherwise noted)

Supply voltage, V_{CC} (see Note 1) . . . . . −0.5 V to 7 V
Input voltage . . . . . ±15 V
Differential input voltage (see Note 2) . . . . . ±15 V
Output voltage (see Note 1) . . . . . −0.5 V to 5.5 V
Low-level output current . . . . . 50 mA
Continuous total dissipation at (or below) 25°C free-air temperature (see Note 3):
  SN55157 JG package . . . . . 1050 mW
  SN75157 D package . . . . . 725 mW
    JG package . . . . . 825 mW
    P package . . . . . 1000 mW
Operating free-air temperature range: SN55157 . . . . . −55°C to 125°C
  SN75157 . . . . . 0°C to 70°C
Storage temperature range . . . . . −65°C to 150°C
Lead temperature 1,6 mm (1/16 inch) from case for 60 seconds JG package . . . . . 300°C
Lead temperature 1,6 mm (1/16 inch) from case for 10 seconds D or P package . . . . . 260°C

NOTES: 1. All voltage values, except differential input voltage, are with respect to the network ground terminal.
2. Differential input voltage is measured at the noninverting input with respect to the corresponding inverting input.
3. For operation above 25°C free-air temperature, derate the SN55157 JG package to 672 mW at 70°C at the rate of 8.4 mW/°C, the SN75157 JG package to 528 mW at 70°C at the rate of 6.6 mW/°C, the D package to 464 mW at 70°C at the rate of 5.8 mW/°C, and the P package to 640 mW at 70°C at the rate of 8.0 mW/°C. In the JG package, SN55157 chips are alloy mounted and SN75157 chips are glass mounted.

recommended operating conditions

		MIN	NOM	MAX	UNIT
Supply voltage, V_{CC}		4.75	5	5.25	V
Common-mode input voltage, V_{IC}				±7	V
Operating free-air temperature, T_A	SN55157	−55	25	125	°C
	SN75157	0	25	70	

electrical characteristics over recommended ranges of supply voltage, common-mode input voltage, and operating free-air temperature (unless otherwise noted)

PARAMETER		TEST CONDITIONS		MIN	TYP†	MAX (See Note 4)	UNIT
V_T	Threshold voltage (V_{T+} and V_{T-})			−0.2		0.2	V
		See Note 5		−0.4		0.4	
V_{hys}	Hysteresis ($V_{T+} - V_{T-}$)				70		mV
V_{OH}	High-level output voltage	V_{ID} = 0.2 V,	I_O = −1 mA	2.5	3.5		V
V_{OL}	Low-level output voltage	V_{ID} = −0.2 V,	I_O = 20 mA		0.35	0.5	V
I_I	Input current	V_{CC} = 0 to 5.5 V, See Note 6	V_I = 10 V		1.1	3.25	mA
			V_I = −10 V		−1.6	−3.25	
I_{OS}	Short-circuit output current‡	V_O = 0,	V_{ID} = 0.2 V	−40	−75	−100	mA
I_{CC}	Supply current	V_{ID} = −0.5 V,	No load		35	50	mA

†All typical values are at V_{CC} = 5 V, T_A = 25°C.
‡Only one output should be shorted at a time, and duration of the short-circuit should not exceed one second.
NOTES: 4. The algebraic convention, where the less-positive (more-negative) limit is designated as minimum, is used in this data sheet for threshold levels only.
5. The expanded threshold parameter is tested with a 500-Ω resistor in series with each input.
6. The input not under test is grounded.

switching characteristics, V_{CC} = 5 V, T_A = 25°C

PARAMETER		TEST CONDITION	MIN	TYP	MAX	UNIT
t_{PLH}	Propagation delay time, low-to-high-level output	C_L = 15 pF, See Figure 1		15	25	ns
t_{PHL}	Propagation delay time, high-to-low-level output			13	25	ns

PARAMETER MEASUREMENT INFORMATION

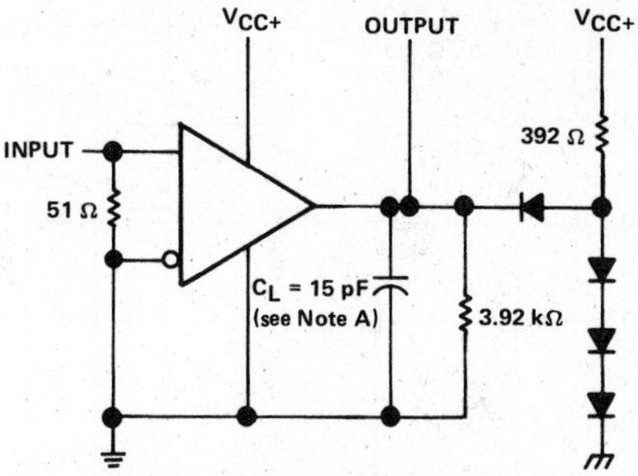
VCC+
OUTPUT
VCC+
INPUT
51 Ω
392 Ω
CL = 15 pF
(see Note A)
3.92 kΩ

TEST CIRCUIT

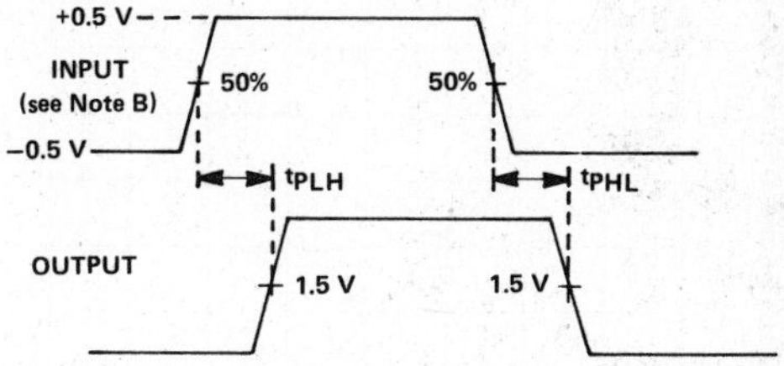
+0.5 V
INPUT
(see Note B)
−0.5 V
50%
50%
tPLH
tPHL
OUTPUT
1.5 V
1.5 V

VOLTAGE WAVEFORMS

NOTES: A. C_L includes probe and jig capacitance.
B. The input pulse is supplied by a generator having the following characteristics: $t_r \leq 5$ ns, $t_f \leq 5$ ns, PRR ≤ 5 MHz, duty cycle = 50%.

FIGURE 1. TRANSITION TIMES

TYPICAL CHARACTERISTICS

OUTPUT VOLTAGE
vs
DIFFERENTIAL INPUT VOLTAGE

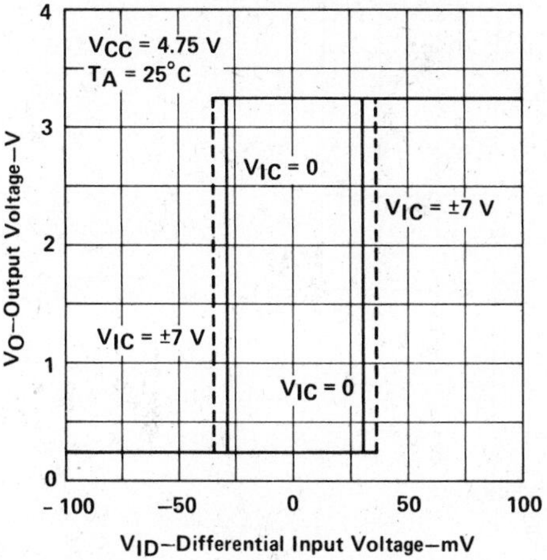
VCC = 4.75 V
TA = 25°C
VIC = 0
VIC = ±7 V
VIC = ±7 V
VIC = 0
VO—Output Voltage—V
VID—Differential Input Voltage—mV
4
3
2
1
0
−100
−50
0
50
100

FIGURE 2

OUTPUT VOLTAGE
vs
DIFFERENTIAL INPUT VOLTAGE

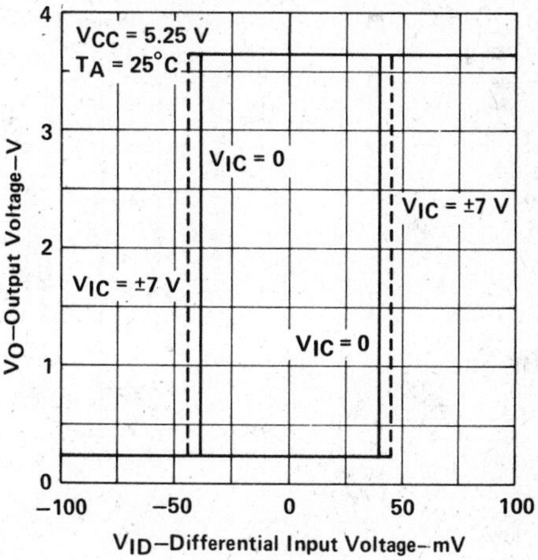
VCC = 5.25 V
TA = 25°C
VIC = 0
VIC = ±7 V
VIC = ±7 V
VIC = 0
VO—Output Voltage—V
VID—Differential Input Voltage—mV
4
3
2
1
0
−100
−50
0
50
100

FIGURE 3

4

Line Drivers/Receivers

TYPICAL CHARACTERISTICS

HIGH-LEVEL OUTPUT VOLTAGE
vs
HIGH-LEVEL OUTPUT CURRENT

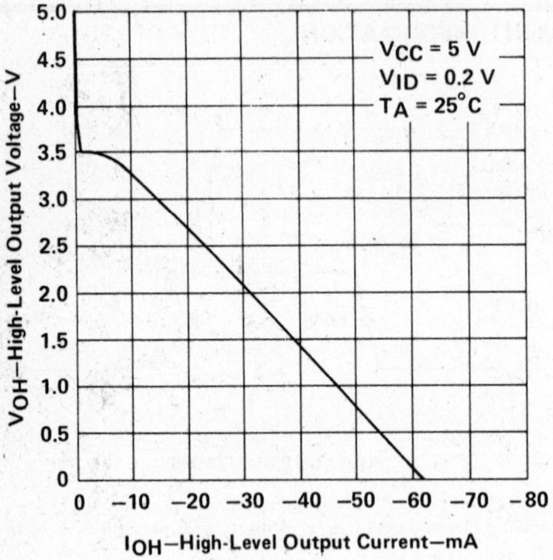
5.0
4.5
4.0
3.5
3.0
2.5
2.0
1.5
1.0
0.5
0
V_{OH}—High-Level Output Voltage—V
V_{CC} = 5 V
V_{ID} = 0.2 V
T_A = 25°C
0 −10 −20 −30 −40 −50 −60 −70 −80
I_{OH}—High-Level Output Current—mA

FIGURE 4

LOW-LEVEL OUTPUT VOLTAGE
vs
LOW-LEVEL OUTPUT CURRENT

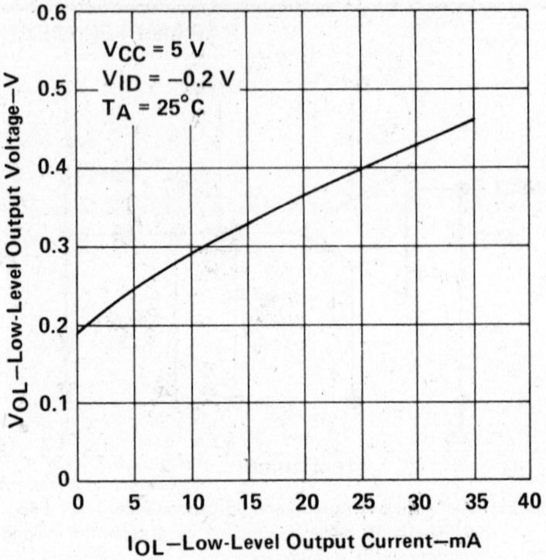
0.6
0.5
0.4
0.3
0.2
0.1
0
V_{OL}—Low-Level Output Voltage—V
V_{CC} = 5 V
V_{ID} = −0.2 V
T_A = 25°C
0 5 10 15 20 25 30 35 40
I_{OL}—Low-Level Output Current—mA

FIGURE 5

SUPPLY CURRENT
vs
SUPPLY VOLTAGE

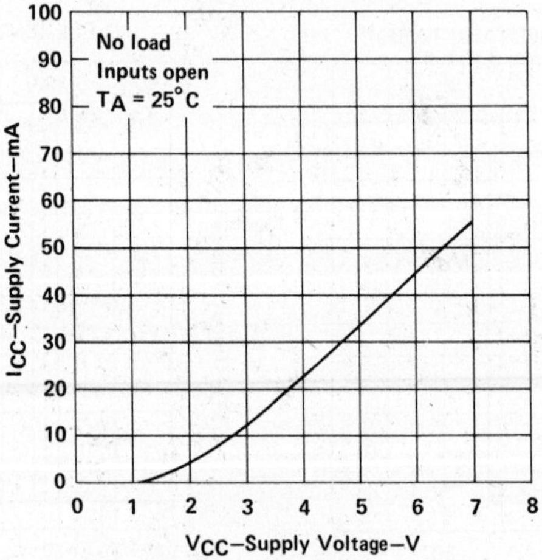
100
90
80
70
60
50
40
30
20
10
0
I_{CC}—Supply Current—mA
No load
Inputs open
T_A = 25°C
0 1 2 3 4 5 6 7 8
V_{CC}—Supply Voltage—V

FIGURE 6

TEXAS INSTRUMENTS
POST OFFICE BOX 655012 • DALLAS, TEXAS 75265

4 Line Drivers/Receivers

TYPICAL APPLICATION DATA

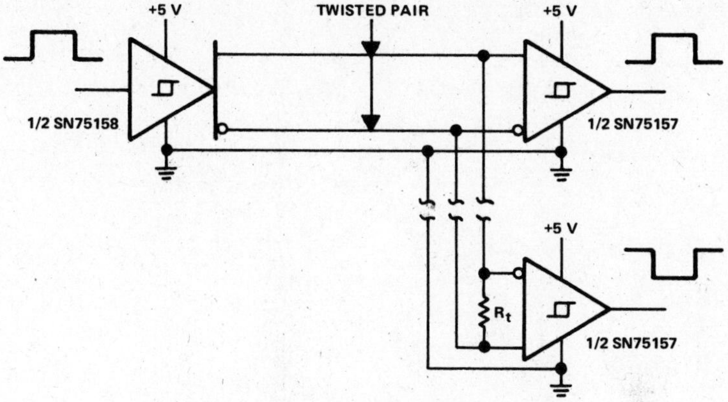
+5 V
TWISTED PAIR
+5 V
1/2 SN75158
1/2 SN75157
+5 V
R_t
1/2 SN75157

FIGURE 7. RS-422-A SYSTEM APPLICATIONS

DUAL DIFFERENTIAL LINE DRIVERS

D2292, JANUARY 1977—REVISED SEPTEMBER 1986

- Meets EIA Standard RS-422-A
- Single 5-V Supply
- Balanced-Line Operation
- TTL-Compatible
- High Output Impedance in Power-Off Condition
- High-Current Active-Pullup Outputs
- Short-Circuit Protection
- Dual Channels
- Input Clamp Diodes

SN55158 . . . JG PACKAGE
SN75158 . . . D, JG, OR P PACKAGE
(TOP VIEW)

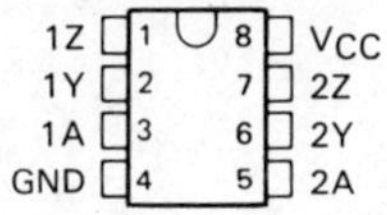
1Z 1 8 VCC
1Y 2 7 2Z
1A 3 6 2Y
GND 4 5 2A

description

The SN55158 and SN75158 are dual complementary-output line drivers designed to satisfy the requirements set by the EIA Standard RS-422-A interface specifications. The outputs provide complementary signals with high-current capability for driving balanced lines, such as twisted pair, at normal line impedance without high power dissipation. The output stages are TTL totem-pole outputs providing a high-impedance state in the power-off condition.

The SN55158 is characterized for operation over the full military temperature range of −55 °C to 125 °C. The SN75158 is characterized for operation from 0 °C to 70 °C.

logic symbol†

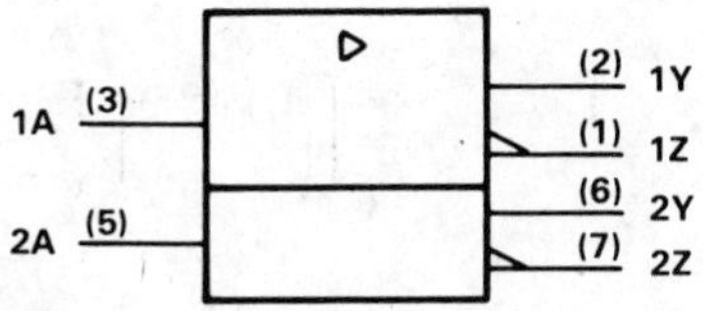
1A (3)
2A (5)
(2) 1Y
(1) 1Z
(6) 2Y
(7) 2Z

†This symbol is in accordance with ANSI/IEEE Std 91-1984 and IEC Publication 617-12.

logic diagram (positive logic)

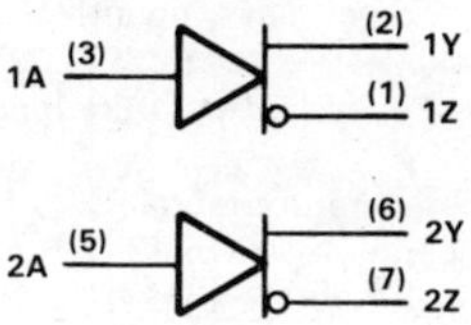
1A (3)
(2) 1Y
(1) 1Z
2A (5)
(6) 2Y
(7) 2Z

schematics of inputs and outputs

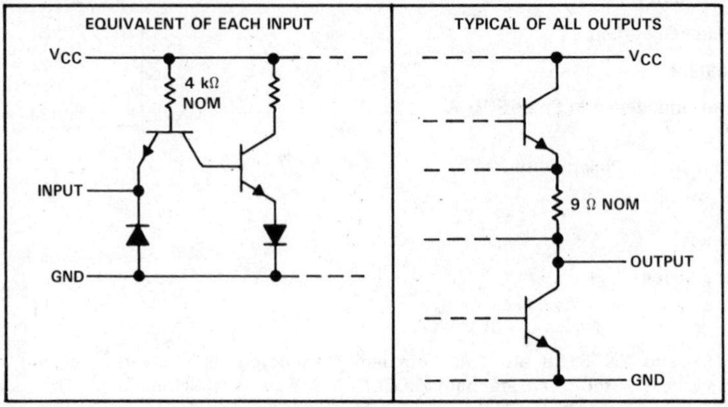
EQUIVALENT OF EACH INPUT
VCC
4 kΩ
NOM
INPUT
GND
TYPICAL OF ALL OUTPUTS
VCC
9 Ω NOM
OUTPUT
GND

absolute maximum ratings over operating free-air temperature range (unless otherwise noted)

Supply voltage, V_{CC} . . . . 7 V
Input voltage . . . . 5.5 V
Continuous total dissipation at (or below) 25 °C free-air temperature (see Note 2):
- D package . . . . 725 mW
- JG package (alloy mount), SN55158 . . . . 1050 mW
- JG package (glass mount), SN75158 . . . . 825 mW
- P package . . . . 1000 mW

Operating free-air temperature range: SN55158 . . . . −55 °C to 125 °C
SN75158 . . . . 0 °C to 70 °C
Storage temperature range . . . . −65 °C to 150 °C
Lead temperature 1,6 mm (1/16 inch) from case for 60 seconds: JG package . . . . 300 °C
Lead temperature 1,6 mm (1/16 inch) from case for 10 seconds: D or P package . . . . 260 °C

NOTES: 1. All voltage values except differential output voltage V_{OD} are with respect to network ground terminal. V_{OD} is at the Y output with respect to the Z output.
2. For operation above 25 °C free-air temperature, refer to Dissipation Derating Curves in Appendix A. In the JG package, SN55158 chips are alloy mounted and SN75158 chips are glass mounted. In the P package, use the 8.0-mW/°C curve for these devices.

recommended operating conditions

	SN55158			SN75158			UNIT
	MIN	NOM	MAX	MIN	NOM	MAX	
Supply voltage, V_{CC}	4.5	5	5.5	4.75	5	5.25	V
High-level input voltage, V_{IH}	2			2			V
Low-level input voltage, V_{IL}			0.8			0.8	V
High-level output current, I_{OH}			−40			−40	mA
Low-level output current, I_{OL}			40			40	mA
Operating free-air temperature, T_A	−55		125	0		70	°C

TEXAS INSTRUMENTS
POST OFFICE BOX 655012 • DALLAS, TEXAS 75265

electrical characteristics over operating free-air temperature range (unless otherwise noted)

PARAMETER		TEST CONDITIONS†		SN55158			SN75158			UNIT
				MIN	TYP‡	MAX	MIN	TYP‡	MAX	
V_{IK}	Input clamp voltage	V_{CC} = MIN, I_I = −12 mA			−0.9	−1.5		−0.9	−1.5	V
V_{OH}	High-level output voltage	V_{CC} = MIN, V_{IL} = 0.8 V, V_{IH} = 2 V, I_{OH} = −40 mA		2	3.0		2.4	3.0		V
V_{OL}	Low-level output voltage	V_{CC} = MIN, V_{IL} = 0.8 V, V_{IH} = 2 V, I_{OL} = 40 mA			0.2	0.4		0.2	0.4	V
$\|V_{OD1}\|$	Differential output voltage	V_{CC} = MAX, I_O = 0			3.5	$2V_{OD2}$		3.5	$2V_{OD2}$	V
$\|V_{OD2}\|$	Differential output voltage	V_{CC} = MIN	R_L = 100 Ω, See Figure 1	2	3.0		2	3.0		V
$\Delta\|V_{OD}\|$	Change in magnitude of differential output voltage§	V_{CC} = MIN	R_L = 100 Ω, See Figure 1		±0.02	±0.4		±0.02	±0.4	V
V_{OC}	Common-mode output voltage¶	V_{CC} = MAX	R_L = 100 Ω, See Figure 1		1.9	3		1.8	3	V
		V_{CC} = MIN	R_L = 100 Ω, See Figure 1		1.4	3		1.5	3	V
$\Delta\|V_{OC}\|$	Change in magnitude of common-mode output voltage§	V_{CC} = MIN or MAX	R_L = 100 Ω, See Figure 1		±0.01	±0.4		±0.01	±0.4	V
I_O	Output current with power off	V_{CC} = 0	V_O = 6 V		0.1	100		0.1	100	μA
			V_O = −0.25 V		−0.1	−100		−0.1	−100	μA
			V_O = −0.25 to 6 V			±100			±100	μA
I_I	Input current at maximum input voltage	V_{CC} = MAX, V_I = 5.5 V				1			1	mA
I_{IH}	High-level input current	V_{CC} = MAX, V_I = 2.4 V				40			40	μA
I_{IL}	Low-level input current	V_{CC} = MAX, V_I = 0.4 V			−1	−1.6		−1	−1.6	mA
I_{OS}	Short-circuit output current#	V_{CC} = MAX		−40	−90	−150	−40	−90	−150	mA
I_{CC}	Supply current (both drivers)	V_{CC} = MAX, Inputs grounded, No load, T_A = 25 °C			37	50		37	50	mA

† For conditions shown as MIN or MAX, use the appropriate value specified under recommended operating conditions.
‡ All typical values are at V_{CC} = 5 V and T_A = 25 °C except for V_{OC}, for which V_{CC} is as stated under test conditions.
§ $\Delta|V_{OD}|$ and $\Delta|V_{OC}|$ are the changes in magnitudes of V_{OD} and V_{OC}, respectively, that occur when the input is changed from a high level to a low level.
¶ In EIA Standard RS-422-A, V_{OC}, which is the average of the two output voltages with respect to ground, is called output offset voltage, V_{OS}.
# Only one output should be shorted at a time, and duration of the short-circuit should not exceed one second.

switching characteristics, V_{CC} = 5 V, T_A = 25 °C

PARAMETER		TEST CONDITIONS	SN55158			SN75158			UNIT
			MIN	TYP	MAX	MIN	TYP	MAX	
t_{PLH}	Propagation delay time, low-to-high-level output	See Figure 2, Termination A		16	25		16	25	ns
t_{PHL}	Propagation delay time, high-to-low-level output	See Figure 2, Termination A		10	20		10	20	ns
t_{PLH}	Propagation delay time, low-to-high-level output	See Figure 2, Termination B		13	20		13	20	ns
t_{PHL}	Propagation delay time, high-to-low-level output	See Figure 2, Termination B		9	15		9	15	ns
t_{TLH}	Transition time, low-to-high-level output	See Figure 2, Termination A		4	20		4	20	ns
t_{THL}	Transition time, high-to-low-level output	See Figure 2, Termination A		4	20		4	20	ns
	Overshoot factor	See Figure 2, Termination C			10			10	%

4

Line Drivers/Receivers

PARAMETER MEASUREMENT INFORMATION

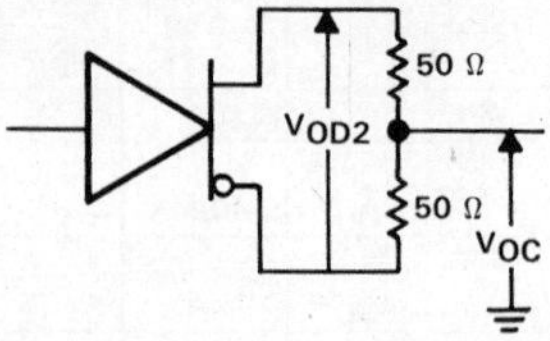
50 Ω
V_{OD2}
50 Ω
V_{OC}

FIGURE 1. DIFFERENTIAL AND COMMON-MODE OUTPUT VOLTAGES

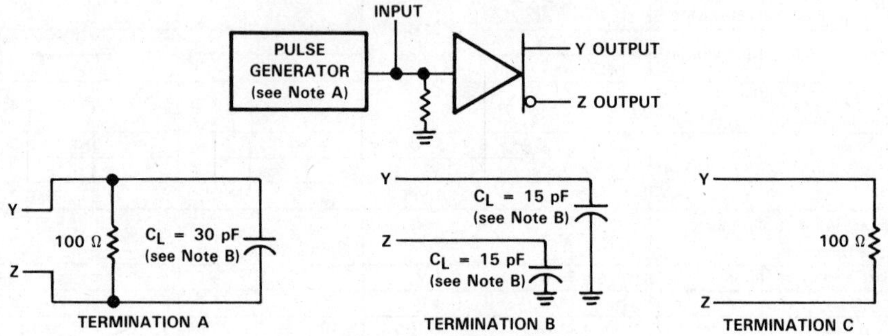
INPUT
PULSE GENERATOR (see Note A)
Y OUTPUT
Z OUTPUT
Y
Z
100 Ω
C_L = 30 pF (see Note B)
TERMINATION A
C_L = 15 pF (see Note B)
C_L = 15 pF (see Note B)
TERMINATION B
100 Ω
TERMINATION C

TEST CIRCUIT

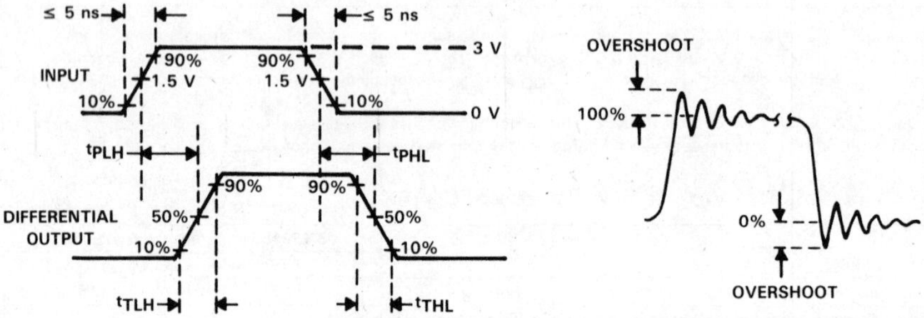
≤ 5 ns
≤ 5 ns
3 V
0 V
INPUT
90%
1.5 V
10%
t_{PLH}
t_{PHL}
DIFFERENTIAL OUTPUT
50%
10%
t_{TLH}
t_{THL}
OVERSHOOT
100%
0%
OVERSHOOT

VOLTAGE WAVEFORMS

NOTES: A. The pulse generator has the following characteristics: Z_{out} = 50 Ω, t_w = 25 ns, PRR ≤ 10 MHz.
B. C_L includes probe and jig capacitance.

FIGURE 2. SWITCHING TIMES

TYPICAL CHARACTERISTICS†

OUTPUT VOLTAGE
vs
DATA INPUT VOLTAGE

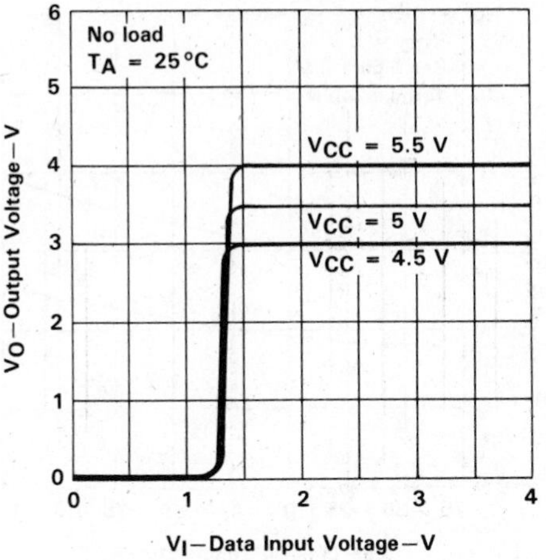
No load
T_A = 25°C
V_{CC} = 5.5 V
V_{CC} = 5 V
V_{CC} = 4.5 V
V_O—Output Voltage—V
V_I—Data Input Voltage—V
0
1
2
3
4
5
6

FIGURE 3

OUTPUT VOLTAGE
vs
DATA INPUT VOLTAGE

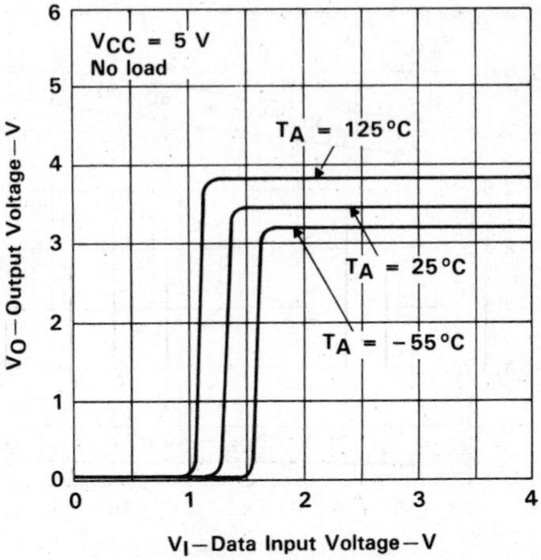
V_{CC} = 5 V
No load
T_A = 125°C
T_A = 25°C
T_A = −55°C
V_O—Output Voltage—V
V_I—Data Input Voltage—V

FIGURE 4

HIGH-LEVEL OUTPUT VOLTAGE
vs
OUTPUT CURRENT

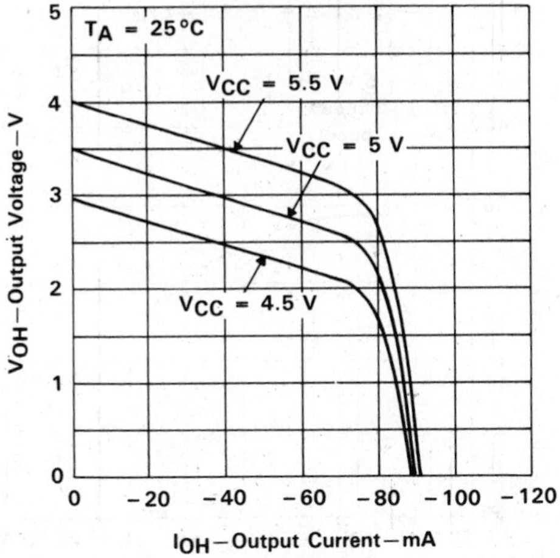
T_A = 25°C
V_{CC} = 5.5 V
V_{CC} = 5 V
V_{CC} = 4.5 V
V_{OH}—Output Voltage—V
I_{OH}—Output Current—mA
0
−20
−40
−60
−80
−100
−120

FIGURE 5

LOW-LEVEL OUTPUT VOLTAGE
vs
OUTPUT CURRENT

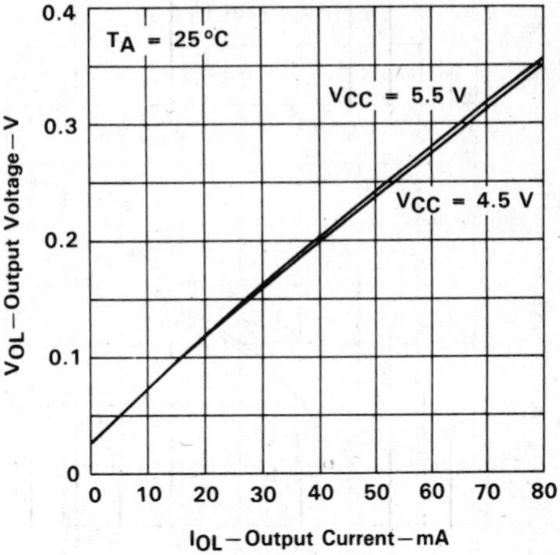
T_A = 25°C
V_{CC} = 5.5 V
V_{CC} = 4.5 V
V_{OL}—Output Voltage—V
I_{OL}—Output Current—mA
0
0.1
0.2
0.3
0.4
10
20
30
40
50
60
70
80

FIGURE 6

†Data for temperatures below 0°C and above 70°C are applicable to SN55158 circuits only.

TYPICAL CHARACTERISTICS†

OUTPUT VOLTAGE
vs
FREE-AIR TEMPERATURE

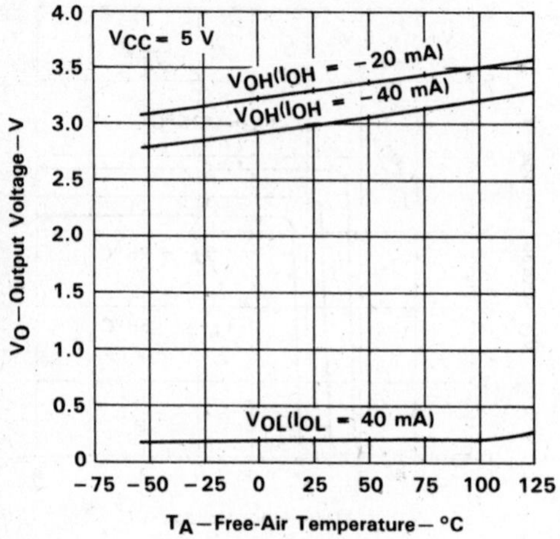
VCC = 5 V
VOH(IOH = −20 mA)
VOH(IOH = −40 mA)
VOL(IOL = 40 mA)
VO—Output Voltage—V
4.0
3.5
3.0
2.5
2.0
1.5
1.0
0.5
0
−75 −50 −25 0 25 50 75 100 125
TA—Free-Air Temperature—°C

FIGURE 7

PROPAGATION DELAY TIMES
vs
FREE-AIR TEMPERATURE

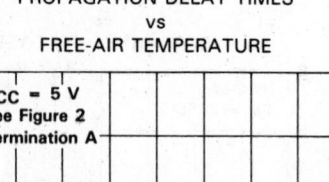

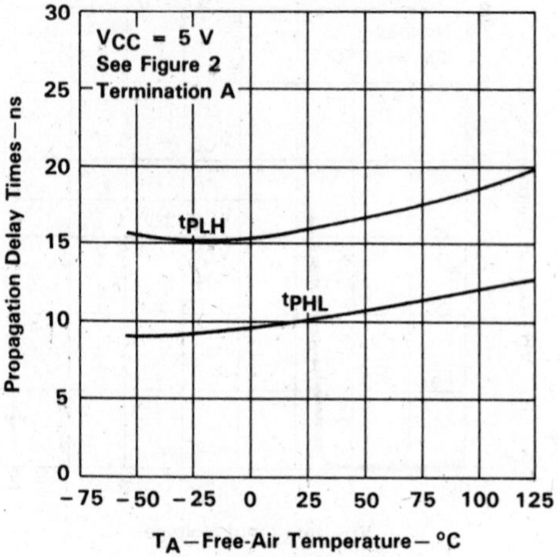
VCC = 5 V
See Figure 2
Termination A
tPLH
tPHL
Propagation Delay Times—ns
30
25
20
15
10
5
0
−75 −50 −25 0 25 50 75 100 125
TA—Free-Air Temperature—°C

FIGURE 8

SUPPLY CURRENT
(BOTH DRIVERS)
vs
SUPPLY VOLTAGE

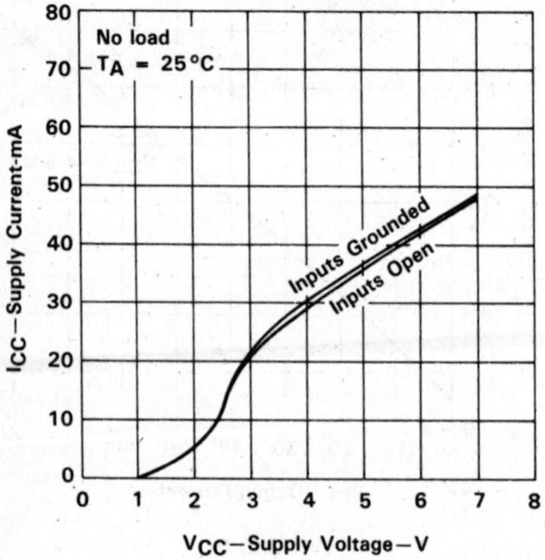
No load
TA = 25°C
Inputs Grounded
Inputs Open
ICC—Supply Current-mA
80
70
60
50
40
30
20
10
0
0 1 2 3 4 5 6 7 8
VCC—Supply Voltage—V

FIGURE 9

SUPPLY CURRENT
(BOTH DRIVERS)
vs
FREE-AIR TEMPERATURE

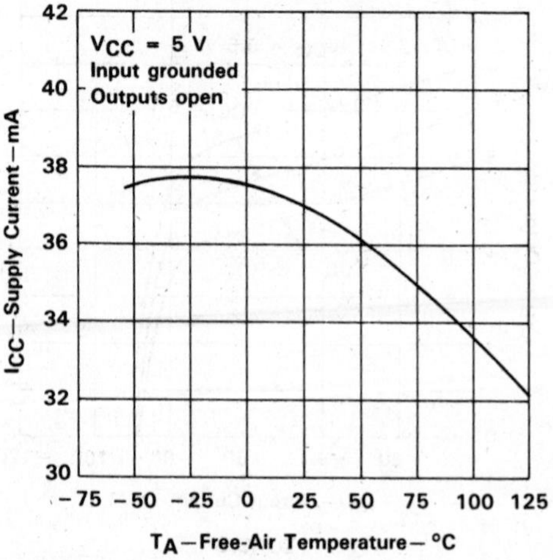
VCC = 5 V
Input grounded
Outputs open
ICC—Supply Current—mA
42
40
38
36
34
32
30
−75 −50 −25 0 25 50 75 100 125
TA—Free-Air Temperature—°C

FIGURE 10

†Data for temperatures below 0 °C and above 70 °C are applicable to SN55158 circuits only.

TYPICAL CHARACTERISTICS

SUPPLY CURRENT
(BOTH DRIVERS)
vs
FREQUENCY

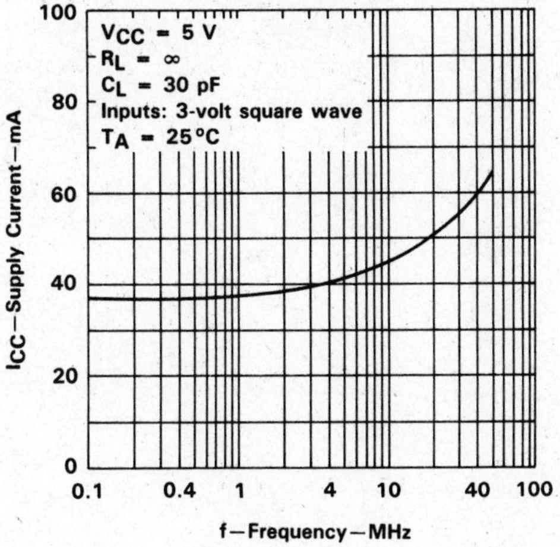
V_{CC} = 5 V
R_L = ∞
C_L = 30 pF
Inputs: 3-volt square wave
T_A = 25°C
100
80
60
40
20
0
I_{CC}—Supply Current—mA
0.1
0.4
1
4
10
40
100
f—Frequency—MHz

FIGURE 11

- Meets EIA Standard RS-422-A
- Single 5-V Supply
- Balanced Line Operation
- TTL-Compatible
- High-Impedance Output State for Party-Line Applications
- High-Current Active-Pull-Up Outputs
- Short-Circuit Protection
- Dual Channels
- Clamp Diodes at Inputs

D, J, OR N PACKAGE
(TOP VIEW)

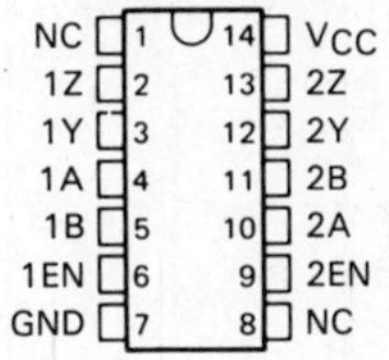
NC 1
1Z 2
1Y 3
1A 4
1B 5
1EN 6
GND 7
8 NC
9 2EN
10 2A
11 2B
12 2Y
13 2Z
14 V_{CC}

NC—No internal connection

description

The SN75159 dual differential line driver with three-state outputs is designed to provide all the features of the SN75158 line driver with the added feature of driver output controls. There is an individual control for each driver. When the output control is low, the associated outputs are in a high-impedance state and the outputs can neither drive nor load the bus. This permits many devices to be connected together on the same transmission line for party-line applications.

The SN75159 is characterized for operation from 0°C to 70°C.

logic symbol†

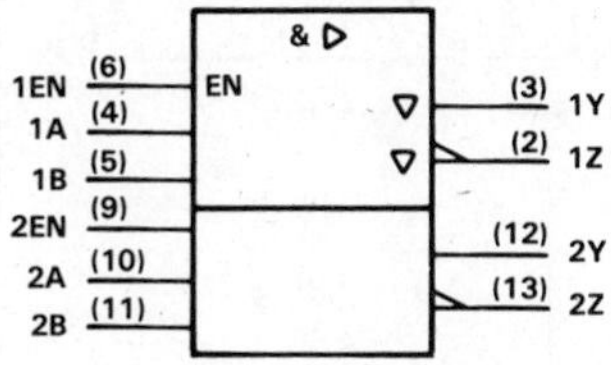
& ▷
EN
1EN (6)
1A (4)
1B (5)
2EN (9)
2A (10)
2B (11)
(3) 1Y
(2) 1Z
(12) 2Y
(13) 2Z

† This symbol is in accordance with ANSI/IEEE Std 91-1984 and IEC Publication 617-12.

logic diagram (positive logic)

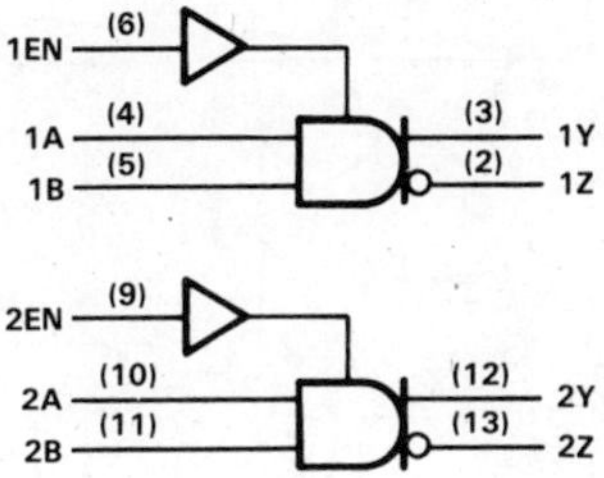
1EN (6)
1A (4)
1B (5)
(3) 1Y
(2) 1Z
2EN (9)
2A (10)
2B (11)
(12) 2Y
(13) 2Z

schematic (each driver)

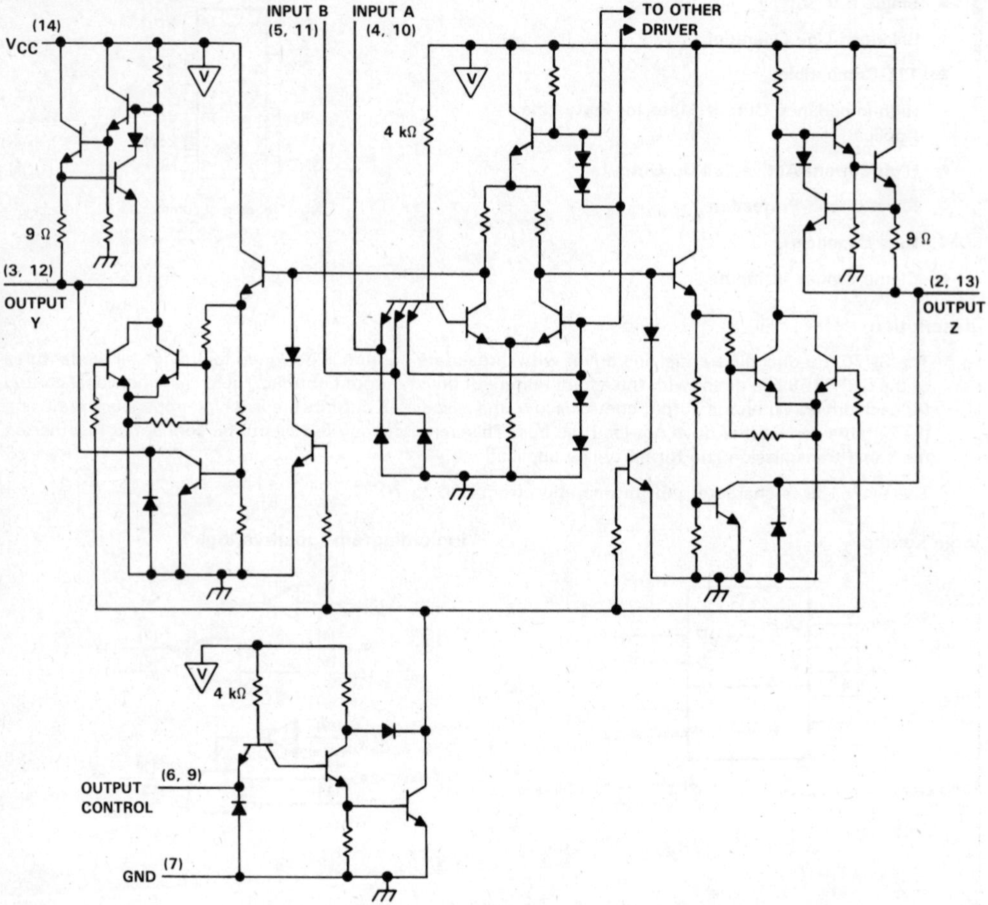
INPUT B
(5, 11)
INPUT A
(4, 10)
TO OTHER
DRIVER
(14)
V_{CC}
V
4 kΩ
9 Ω
(3, 12)
OUTPUT
Y
(2, 13)
OUTPUT
Z
(6, 9)
OUTPUT
CONTROL
(7)
GND

V . . . V_{CC} bus

Resistor values shown are nominal.

TEXAS INSTRUMENTS
POST OFFICE BOX 655012 • DALLAS, TEXAS 75265

absolute maximum ratings over operating free-air temperature range (unless otherwise noted)

Supply voltage, V_{CC} (see Note 1) . . . . 7 V
Input voltage . . . . 5.5 V
Off-state voltage applied to open-collector outputs . . . . 12 V
Continuous total dissipation at (or below) 25°C free-air temperature (see Note 2):
D package . . . . 950 mW
J package . . . . 1025 mW
N package . . . . 1150 mW
Operating free-air temperature range . . . . 0°C to 70°C
Storage temperature range . . . . −65°C to 150°C
Lead temperature 1,6 mm (1/16 inch) from case for 60 seconds: J package . . . . 300°C
Lead temperature 1,6 mm (1/16 inch) from case for 10 seconds: D or N package . . . . 260°C

NOTES: 1. All voltage values except differential output voltage V_{OD} are with respect to the network ground terminal. V_{OD} is at the Y output with respect to the Z output.
2. For operation above 25°C free-air temperature, derate the D package to 608 mW at 70°C at the rate of 7.6 mW/°C, the J package to 656 mW at 70°C at the rate of 8.2 mW/°C, and the N package to 736 mW at 70°C at the rate of 9.2 mW/°C. In the J package, SN75159 chips are glass mounted.

recommended operating conditions

	MIN	NOM	MAX	UNIT
Supply voltage, V_{CC}	4.75	5	5.25	V
High-level input voltage, V_{IH}	2			V
Low-level input voltage, V_{IL}			0.8	V
High-level output voltage, I_{OH}			−40	mA
Low-level output current, I_{OL}			40	mA
Operating free-air temperature, T_A	0		70	°C

electrical characteristics over operating free-air temperature range (unless otherwise noted)

PARAMETER		TEST CONDITIONS			MIN	TYP†	MAX	UNIT
V_{IK}	Input clamp voltage	V_{CC} = 4.75 V, I_I = −12 mA				−0.9	−1.5	V
V_{OH}	High-level output voltage	V_{CC} = 4.75 V, V_{IL} = 0.8 V, V_{IH} = 2 V, I_{OH} = −40 mA			2.4	3.0		V
V_{OL}	Low-level output voltage	V_{CC} = 4.75 V, V_{IL} = 0.8 V, V_{IH} = 2 V, I_{OL} = 40 mA				0.25	0.4	V
V_{OK}	Output clamp voltage	V_{CC} = 5.25 V, I_O = −40 mA				−1.1	−1.5	V
V_O	Output voltage	V_{CC} = 4.75 V to 5.25 V, I_O = 0			0		6	V
$\|V_{OD1}\|$	Differential output voltage	V_{CC} = 5.25 V, I_O = 0				3.5	$2V_{OD2}$	V
$\|V_{OD2}\|$	Differential output voltage	V_{CC} = 4.75 V			2	3.0		V
$\Delta\|V_{OD}\|$	Change in magnitude of differential output voltage‡	V_{CC} = 4.75 V	R_L = 100 Ω, See Figure 1			±0.02	±0.4	V
V_{OC}	Common-mode output voltage§	V_{CC} = 5.25 V	R_L = 100 Ω, See Figure 1			1.8	3	V
		V_{CC} = 4.75 V	R_L = 100 Ω, See Figure 1			1.5	3	V
$\Delta\|V_{OC}\|$	Change in magnitude of common-mode output voltage‡	V_{CC} = 4.75 V to 5.25 V	R_L = 100 Ω, See Figure 1			±0.01	±0.4	V
I_O	Output current with power off	V_{CC} = 0	V_O = 6 V			0.1	100	μA
			V_O = −0.25 V			−0.1	−100	
			V_O = −0.25 V to 6 V				±100	
I_{OZ}	Off-state (high impedance-state) output current	V_{CC} = 5.25 V, Output controls at 0.8 V	T_A = 25°C,	V_O = 0 to V_{CC}			±10	μA
			T_A = 70°C	V_O = 0			−20	
				V_O = 0.4 V			±20	
				V_O = 2.4 V			±20	
				V_O = V_{CC}			20	
I_I	Input current at maximum input voltage	V_{CC} = 5.25 V, V_I = 5.5 V					1	mA
I_{IH}	High-level input current	V_{CC} = 5.25 V, V_I = 2.4 V					40	μA
I_{IL}	Low-level input current	V_{CC} = 5.25 V, V_I = 0.4 V				−1	−1.6	mA
I_{OS}	Short-circuit output current¶	V_{CC} = 5.25 V			−40	−90	−150	mA
I_{CC}	Supply current (both drivers)	VCC = 5.25 V, Inputs grounded, No load, T_A = 25°C				47	65	mA

† All typical values are at V_{CC} = 5 V and T_A = 25°C except for V_{OC}, for which V_{CC} is as stated under test conditions.
‡ $\Delta|V_{OD}|$ and $\Delta|V_{OC}|$ are the changes in magnitudes of V_{OD} and V_{OC}, respectively, that occur when the input is changed from a high level to a low level.
§ In EIA Standard RS-422-A, V_{OC}, which is the average of the two output voltages with respect to ground, is called output offset voltage, V_{OS}.
¶ Only one output should be shorted at a time, and duration of the short-circuit should not exceed one second.

TEXAS INSTRUMENTS
POST OFFICE BOX 655012 • DALLAS, TEXAS 75265

switching characteristics over operating free-air temperature range, V_{CC} = 5 V (unless otherwise noted)

PARAMETER		TEST CONDITIONS	MIN	TYP†	MAX	UNIT
t_{PLH}	Propagation delay time, low-to-high-level output	C_L = 30 pF, R_L = 100 Ω, See Figure 2, Termination A		16	25	ns
t_{PHL}	Propagation delay time, high-to-low-level output			11	20	ns
t_{PLH}	Propagation delay time, low-to-high-level output	C_L = 15 pF, See Figure 2, Termination B		13	20	ns
t_{PHL}	Propagation delay time, high-to-low-level output			9	15	ns
t_{TLH}	Transition time, low-to-high-level output	C_L = 30 pF, R_L = 100 Ω, See Figure 2, Termination A		4	20	ns
t_{THL}	Transition time, high-to-low-level output			4	20	ns
t_{PZH}	Output enable time to high level	C_L = 30 pF, R_L = 180 Ω, See Figure 3		7	20	ns
t_{PZL}	Output enable time to low level	C_L = 30 pF, R_L = 250 Ω, See Figure 4		14	40	ns
t_{PHZ}	Output disable time from high level	C_L = 30 pF, R_L = 180 Ω, See Figure 3		10	30	ns
t_{PLZ}	Output disable time from low level	C_L = 30 pF, R_L = 250 Ω, See Figure 4		17	35	ns
	Overshoot factor	R_L = 100 Ω, See Figure 2, Termination C			10	%

† All typical values are at T_A = 25°C.

SYMBOL EQUIVALENTS

DATA SHEET PARAMETER	RS-422-A
V_O	V_{oa}, V_{ob}
$\|V_{OD1}\|$	V_o
$\|V_{OD2}\|$	V_t
$\Delta\|V_{OD}\|$	$\| \|V_t\| - \|\bar{V}_t\| \|$
V_{OC}	$\|V_{os}\|$
$\Delta\|V_{OC}\|$	$\| V_{os} - \bar{V}_{os} \|$
I_{OS}	$\|I_{sa}\|$, $\|I_{sb}\|$
I_O	$\|I_{xa}\|$, $\|I_{xb}\|$

PARAMETER MEASUREMENT INFORMATION

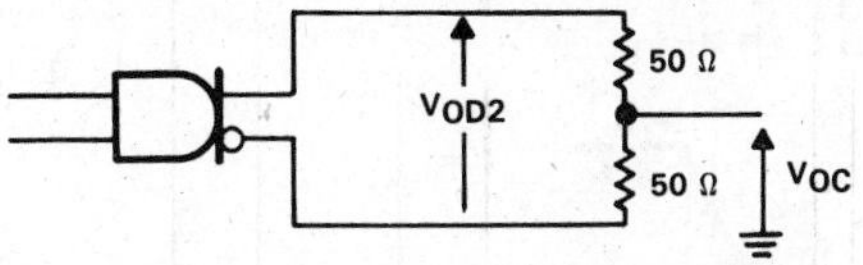
50 Ω
VOD2
50 Ω
VOC

FIGURE 1. DIFFERENTIAL AND COMMON-MODE OUTPUT VOLTAGES

4 Line Drivers/Receivers

PARAMETER MEASUREMENT INFORMATION

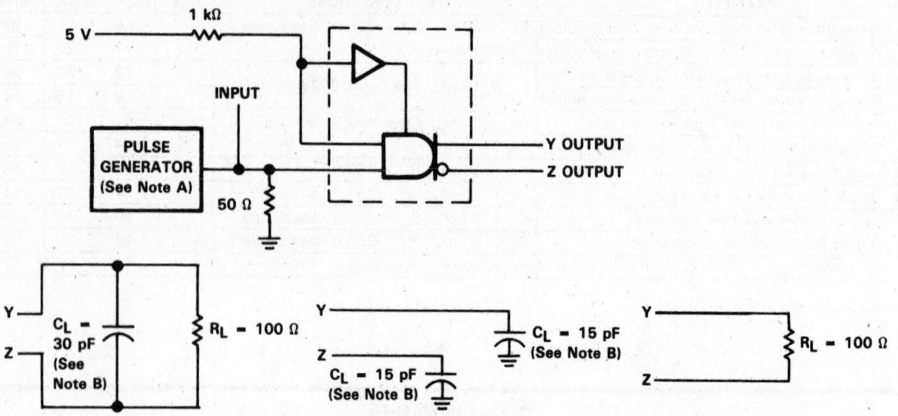
5 V
1 kΩ
INPUT
PULSE GENERATOR (See Note A)
50 Ω
Y OUTPUT
Z OUTPUT
Y
Z
CL = 30 pF (See Note B)
RL = 100 Ω
CL = 15 pF (See Note B)
CL = 15 pF (See Note B)
RL = 100 Ω

TERMINATION A　　TERMINATION B　　TERMINATION C

TEST CIRCUITS

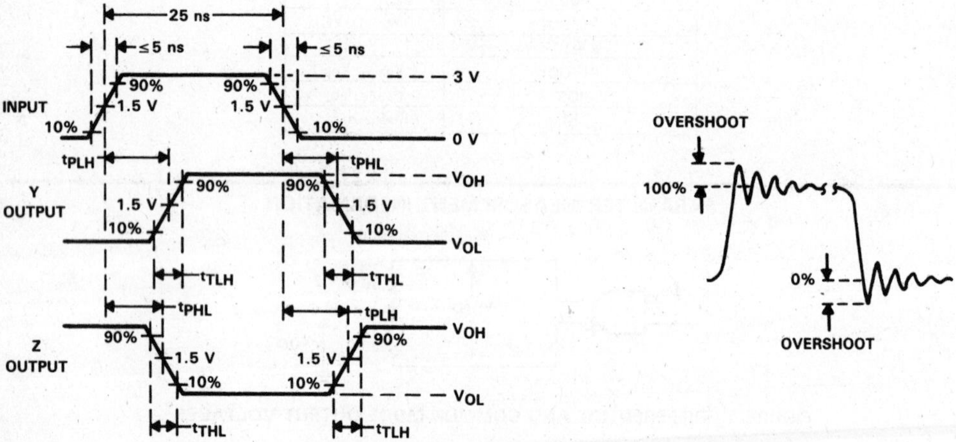
25 ns
≤5 ns
≤5 ns
3 V
0 V
INPUT
90%
1.5 V
10%
tPLH
tPHL
Y OUTPUT
VOH
VOL
tTLH
tTHL
Z OUTPUT
OVERSHOOT
100%
0%
OVERSHOOT

VOLTAGE WAVEFORMS

NOTES: A. The pulse generator has the following characteristics: Z_{out} = 50 Ω, PRR ≤ 10 MHz.
B. C_L includes probe and jig capacitance.

FIGURE 2. t_{PLH}, t_{PHL}, t_{TLH}, t_{THL}, AND OVERSHOOT FACTOR

PARAMETER MEASUREMENT INFORMATION

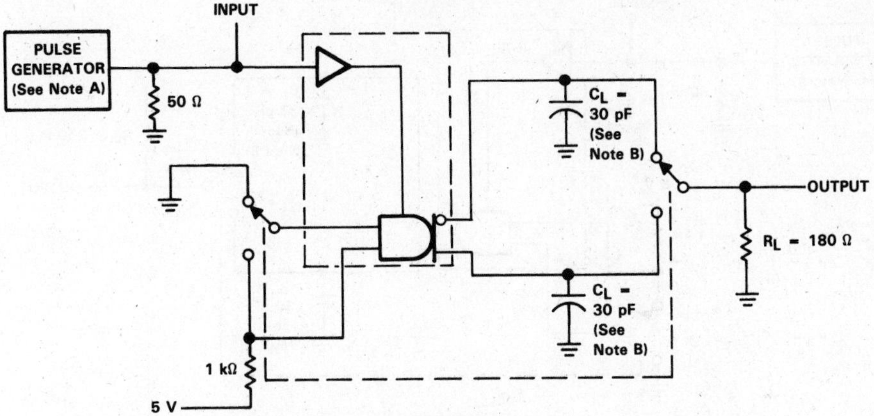
INPUT
PULSE GENERATOR (See Note A)
50 Ω
C_L = 30 pF (See Note B)
OUTPUT
R_L = 180 Ω
C_L = 30 pF (See Note B)
1 kΩ
5 V

TEST CIRCUIT

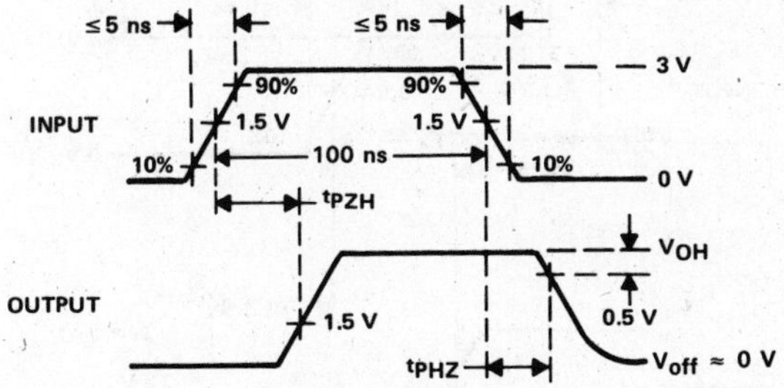
≤5 ns
≤5 ns
3 V
90%
90%
INPUT
1.5 V
1.5 V
10%
100 ns
10%
0 V
tPZH
VOH
OUTPUT
0.5 V
1.5 V
tPHZ
Voff ≈ 0 V

VOLTAGE WAVEFORMS

NOTES: A. The pulse generator has the following characteristics: Z_{out} = 50 Ω, PRR ≤ 500 kHz.
B. C_L includes probe and jig capacitance.

FIGURE 3. t_{PZH} AND t_{PHZ}

PARAMETER MEASUREMENT INFORMATION

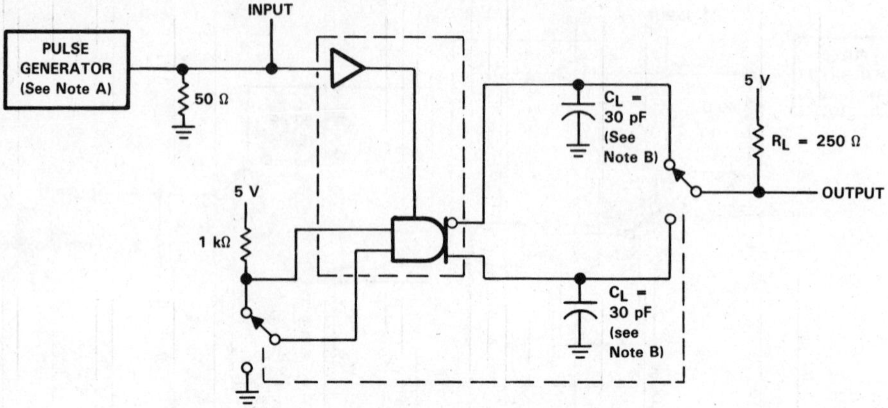
INPUT
PULSE GENERATOR (See Note A)
50 Ω
5 V
1 kΩ
C_L = 30 pF (See Note B)
5 V
R_L = 250 Ω
OUTPUT
C_L = 30 pF (see Note B)

TEST CIRCUIT

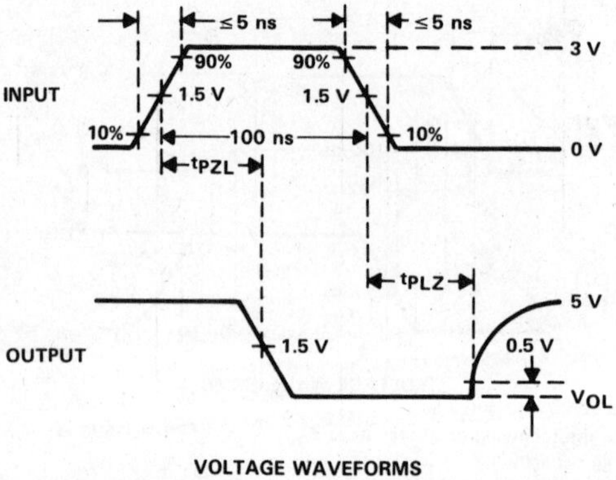
≤5 ns
≤5 ns
3 V
90%
90%
INPUT
1.5 V
1.5 V
10%
100 ns
10%
0 V
t_{PZL}
t_{PLZ}
5 V
1.5 V
0.5 V
OUTPUT
V_{OL}

VOLTAGE WAVEFORMS

NOTES: A. The pulse generator has the following characteristics: Z_{out} = 50 Ω, PRR ≤ 500 kHz.
C. C_L includes probe and jig capacitance.

FIGURE 4. t_{PZL} AND t_{PLZ}

TEXAS INSTRUMENTS
POST OFFICE BOX 655012 • DALLAS, TEXAS 75265

TYPICAL CHARACTERISTICS

OUTPUT VOLTAGE
vs
DATA INPUT VOLTAGE

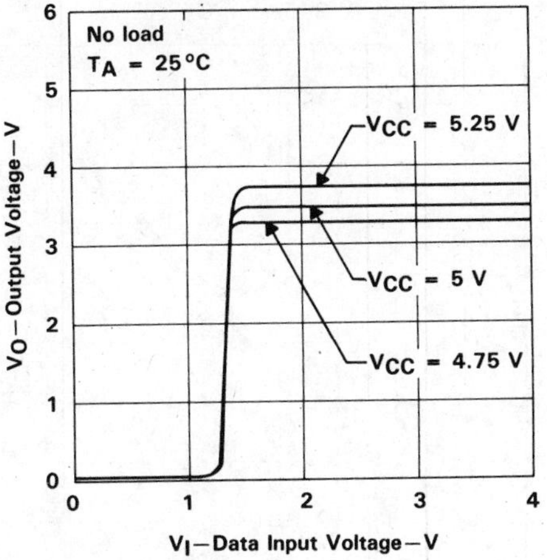
No load
TA = 25°C
VCC = 5.25 V
VCC = 5 V
VCC = 4.75 V
VO—Output Voltage—V
VI—Data Input Voltage—V

FIGURE 5

OUTPUT VOLTAGE
vs
DATA INPUT VOLTAGE

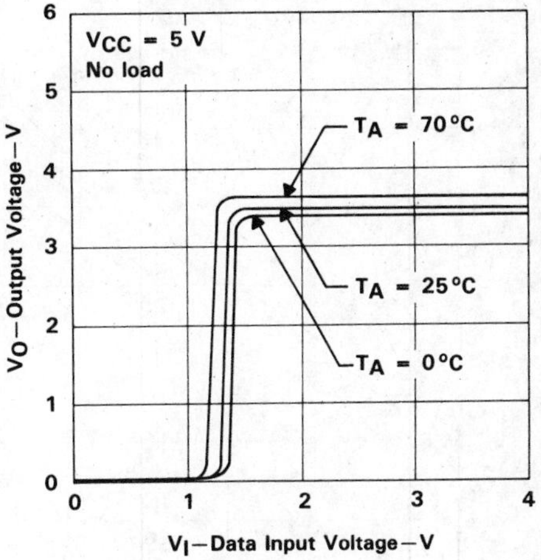
VCC = 5 V
No load
TA = 70°C
TA = 25°C
TA = 0°C
VO—Output Voltage—V
VI—Data Input Voltage—V

FIGURE 6

OUTPUT VOLTAGE
vs
FREE-AIR TEMPERATURE

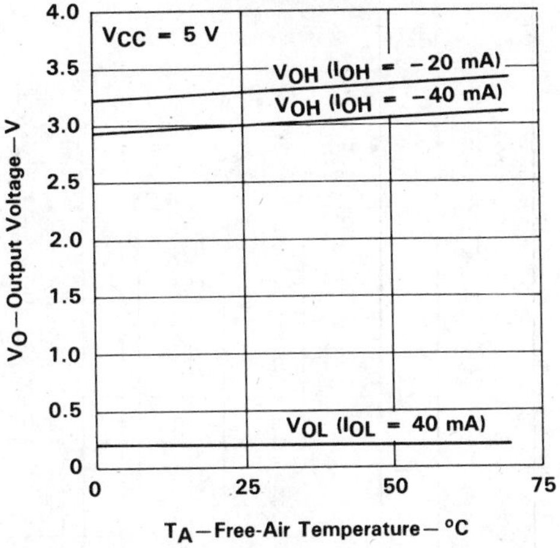
VCC = 5 V
VOH (IOH = −20 mA)
VOH (IOH = −40 mA)
VOL (IOL = 40 mA)
VO—Output Voltage—V
TA—Free-Air Temperature—°C

FIGURE 7

HIGH-LEVEL OUTPUT VOLTAGE
vs
OUTPUT CURRENT

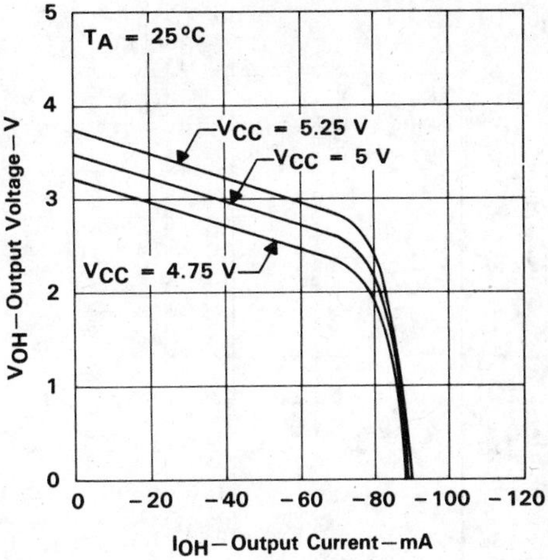
TA = 25°C
VCC = 5.25 V
VCC = 5 V
VCC = 4.75 V
VOH—Output Voltage—V
IOH—Output Current—mA

FIGURE 8

TYPICAL CHARACTERISTICS

LOW-LEVEL OUTPUT VOLTAGE
vs
OUTPUT CURRENT

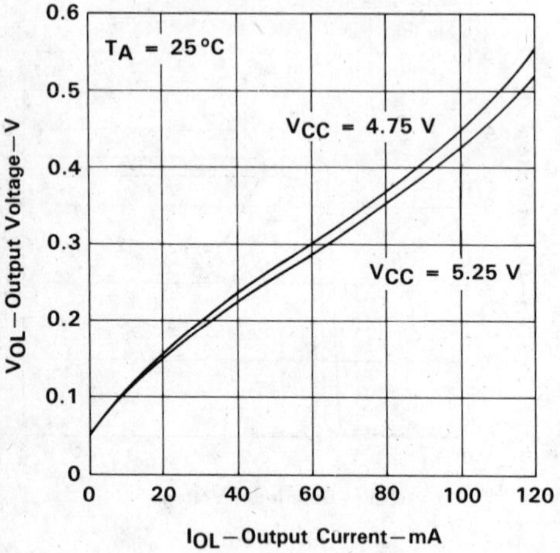
T_A = 25°C
V_{CC} = 4.75 V
V_{CC} = 5.25 V
V_{OL}—Output Voltage—V
I_{OL}—Output Current—mA

FIGURE 9

SUPPLY CURRENT
(BOTH DRIVERS)
vs
SUPPLY VOLTAGE

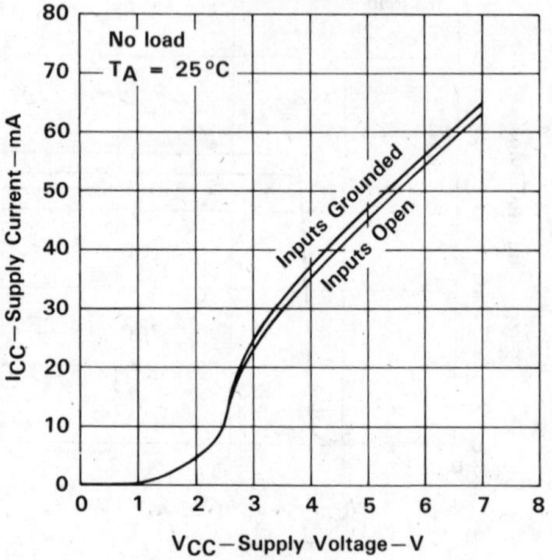
No load
T_A = 25°C
Inputs Grounded
Inputs Open
I_{CC}—Supply Current—mA
V_{CC}—Supply Voltage—V

FIGURE 10

SUPPLY CURRENT
(BOTH DRIVERS)
vs
FREE-AIR TEMPERATURE

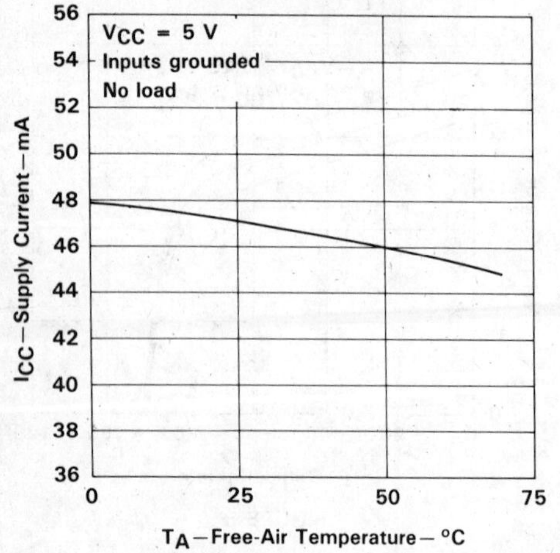
V_{CC} = 5 V
Inputs grounded
No load
I_{CC}—Supply Current—mA
T_A—Free-Air Temperature—°C

FIGURE 11

SUPPLY CURRENT
(BOTH DRIVERS)
vs
FREQUENCY

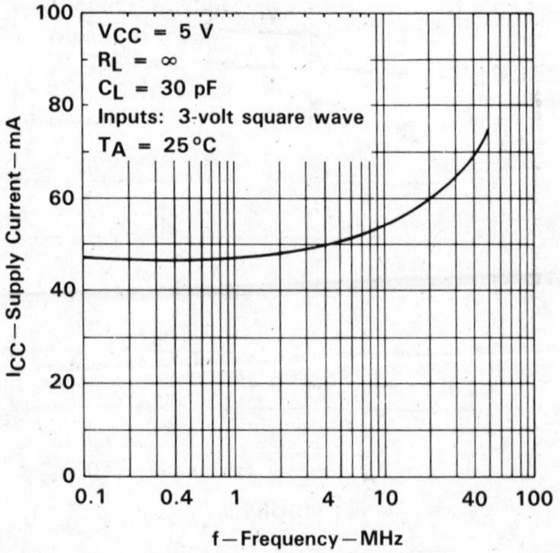
V_{CC} = 5 V
R_L = ∞
C_L = 30 pF
Inputs: 3-volt square wave
T_A = 25°C
I_{CC}—Supply Current—mA
f—Frequency—MHz

FIGURE 12

TYPICAL CHARACTERISTICS

PROPAGATION DELAY TIMES
FROM DATA INPUTS
vs
FREE-AIR TEMPERATURE

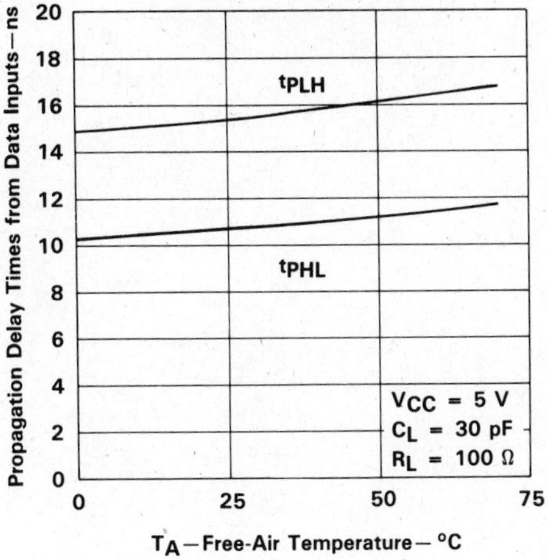
Propagation Delay Times from Data Inputs—ns
20
18
16
14
12
10
8
6
4
2
0
tPLH
tPHL
VCC = 5 V
CL = 30 pF
RL = 100 Ω
0
25
50
75
TA—Free-Air Temperature—°C

FIGURE 13

OUTPUT ENABLE AND DISABLE TIMES
vs
FREE-AIR TEMPERATURE

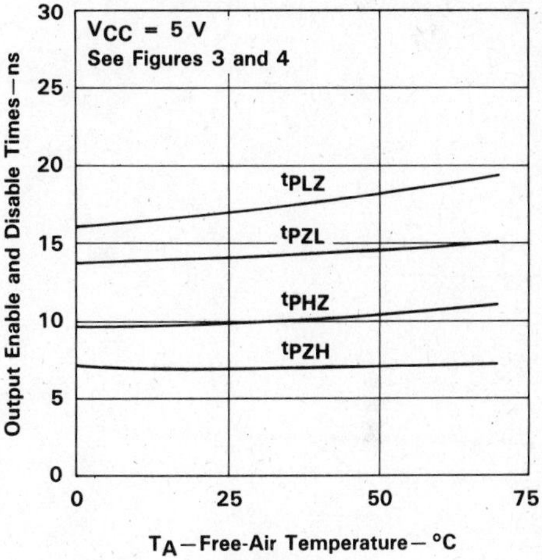
Output Enable and Disable Times—ns
30
25
20
15
10
5
0
VCC = 5 V
See Figures 3 and 4
tPLZ
tPZL
tPHZ
tPZH
0
25
50
75
TA—Free-Air Temperature—°C

FIGURE 14

SN75160B OCTAL GENERAL-PURPOSE INTERFACE BUS TRANSCEIVER

MEETS IEEE STANDARD 488-1978 (GPIB)

- 8-Channel Bidirectional Transceiver
- Power-Up/Power-Down Protection (Glitch-Free)
- High-Speed, Low-Power Schottky Circuitry
- Low-Power Dissipation . . . 72 mW Max per Channel
- Fast Propagation Times . . . 22 ns Max
- High-Impedance P-N-P Inputs
- Receiver Hysteresis . . . 650 mV Typ
- Open-Collector Driver Output Option
- No Loading of Bus When Device is Powered Down (V_{CC} = 0)

DW, J, OR N DUAL-IN-LINE PACKAGE
(TOP VIEW)

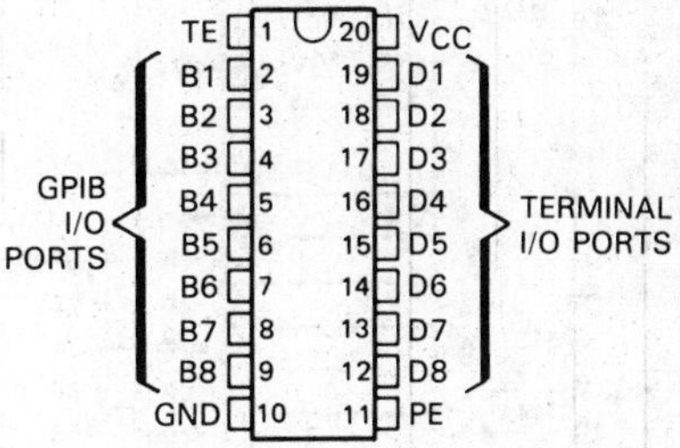
TE 1
B1 2
B2 3
B3 4
B4 5
B5 6
B6 7
B7 8
B8 9
GND 10
20 VCC
19 D1
18 D2
17 D3
16 D4
15 D5
14 D6
13 D7
12 D8
11 PE
GPIB I/O PORTS
TERMINAL I/O PORTS

FUNCTION TABLES

EACH DRIVER

INPUTS			OUTPUT
D	TE	PE	B
H	H	H	H
L	H	X	L
H	X	L	Z†
X	L	X	Z†

EACH RECEIVER

INPUTS			OUTPUT
B	TE	PE	D
L	L	X	L
H	L	X	H
X	H	X	Z

H = high level, L = low level, X = irrelevant, Z = High-impedance state.

† This is the high-impedance state of a normal 3-state output modified by the internal resistors to V_{CC} and ground.

description

The SN75160B 8-channel general-purpose interface bus transceiver is a monolithic, high-speed, low-power Schottky device designed for two-way data communications over single-ended transmission lines. It is designed to meet the requirements of IEEE Standard 488-1978. The transceiver features driver outputs that can be operated in either the passive-pullup or three-state mode. If Talk Enable (TE) is high, these ports have the characteristics of passive-pullup outputs when Pullup Enable (PE) is low, and of three-state outputs when PE is high. Taking TE low places these ports in the high-impedance state. The driver outputs are designed to handle loads up to 48 milliamperes of sink current.

Output glitches during power-up and power-down are eliminated by an internal circuit that disables both the bus and receiver outputs. The outputs do not load the bus when V_{CC} = 0 volts. When combined with the SN75161B or SN75162B management bus transceivers, the pair provides the complete 16-wire interface for the IEEE 488 bus.

The SN75160B is characterized for operation from 0°C to 70°C.

logic symbol†

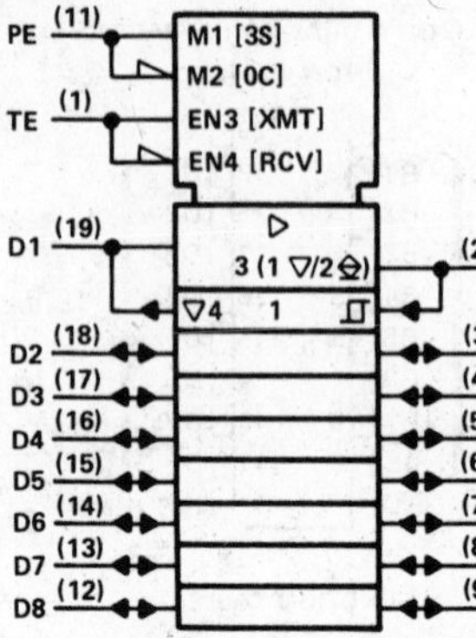
PE (11)
TE (1)
M1 [3S]
M2 [0C]
EN3 [XMT]
EN4 [RCV]
D1 (19)
3 (1 ▽/2 ⩟)
▽4 1
(2) B1
D2 (18)
D3 (17)
D4 (16)
D5 (15)
D6 (14)
D7 (13)
D8 (12)
(3) B2
(4) B3
(5) B4
(6) B5
(7) B6
(8) B7
(9) B8

† This symbol is in accordance with ANSI/IEEE Std 91-1984 and IEC Publication 617-12.
▽ Designates 3-state outputs.
⩟ Designates passive-pullup outputs.

logic diagram (positive logic)

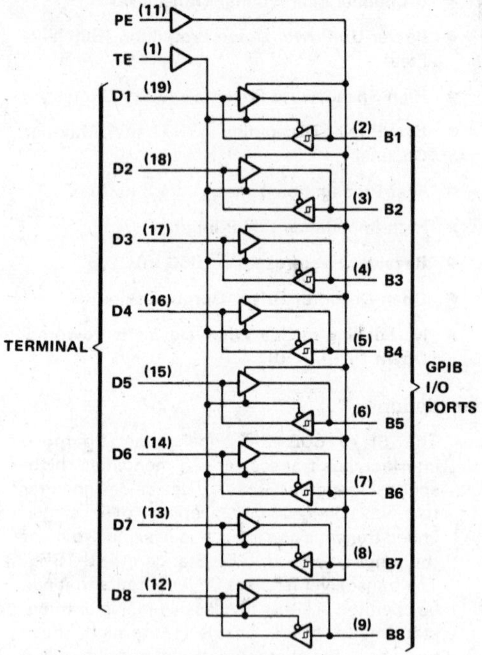
PE (11)
TE (1)
D1 (19)
D2 (18)
D3 (17)
D4 (16)
D5 (15)
D6 (14)
D7 (13)
D8 (12)
TERMINAL
(2) B1
(3) B2
(4) B3
(5) B4
(6) B5
(7) B6
(8) B7
(9) B8
GPIB I/O PORTS

schematics of inputs and outputs

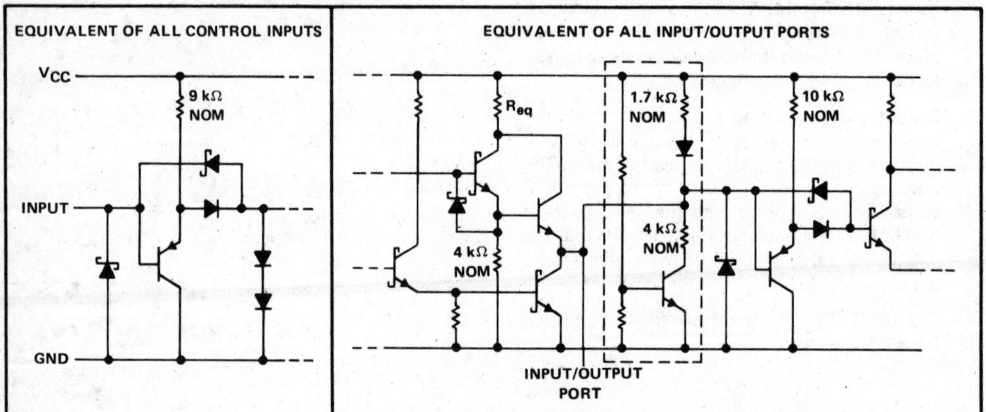
EQUIVALENT OF ALL CONTROL INPUTS
VCC
9 kΩ NOM
INPUT
GND
EQUIVALENT OF ALL INPUT/OUTPUT PORTS
Req
1.7 kΩ NOM
10 kΩ NOM
4 kΩ NOM
4 kΩ NOM
INPUT/OUTPUT PORT

Driver output R_{eq} = 30 Ω NOM
Receiver output R_{eq} = 110 Ω NOM
Circuit inside dashed lines is on the driver outputs only.

absolute maximum ratings over operating free-air temperature range (unless otherwise noted)

Supply voltage, V_{CC} (see Note 1) . . . . . 7 V
Input voltage . . . . . 5.5 V
Low-level driver output current . . . . . 100 mA
Continuous total dissipation at (or below) 25°C free-air temperature (see Note 2):
DW package . . . . . 1125 mW
J package . . . . . 1375 mW
N package . . . . . 1150 mW
Operating free-air temperature range . . . . . 0°C to 70°C
Storage temperature range . . . . . −65°C to 150°C
Lead temperature 1,6 mm (1/16 inch) from the case for 60 seconds: J package . . . . . 300°C
Lead temperature 1,6 mm (1/16 inch) from the case for 10 seconds: DW or N package . . . . 260°C

NOTES: 1. All voltage values are with respect to network ground terminal.
2. For operation above 25°C free-air temperature, derate the DW package at the rate of 9.0 mW/°C, the N package at the rate of 9.2 mW/°C, and the J package at the rate of 11.0 mW/°C. In the J package, SN75160B chips are alloy mounted.

recommended operating conditions

		MIN	NOM	MAX	UNIT
Supply voltage, V_{CC}		4.75	5	5.25	V
High-level input voltage, V_{IH}		2			V
Low-level input voltage, V_{IL}				0.8	V
High-level output current, I_{OH}	Bus ports with pull ups active			−5.2	mA
	Terminal ports			−800	μA
Low-level output current, I_{OL}	Bus ports			48	mA
	Terminal ports			16	
Operating free-air temperature, T_A		0		70	°C

electrical characteristics over recommended ranges of supply voltage and operating free-air temperature (unless otherwise noted)

PARAMETER			TEST CONDITIONS		MIN	TYP†	MAX	UNIT
V_{IK}	Input clamp voltage		I_I = −18 mA			−0.8	−1.5	V
V_{hys}	Hysteresis ($V_{T+} - V_{T-}$)	Bus			0.4	0.65		V
V_{OH}	High-level output voltage	Terminal	I_{OH} = −800 μA,	TE at 0.8 V	2.7	3.5		V
		Bus	I_{OH} = −5.2 mA,	PE and TE at 2 V	2.5	3.3		
V_{OL}	Low-level output voltage	Terminal	I_{OL} = 16 mA,	TE at 0.8 V		0.3	0.5	V
		Bus	I_{OL} = 48 mA,	TE at 2 V		0.35	0.5	
I_I	Input current at maximum input voltage	Terminal	V_I = 5.5 V			0.2	100	μA
I_{IH}	High-level input current	Terminal	V_I = 2.7 V			0.1	20	μA
I_{IL}	Low-level input current	Terminal	V_I = 0.5 V			−10	−100	μA
$V_{I/O(bus)}$	Voltage at bus port		Driver disabled	$I_{I(bus)}$ = 0	2.5	3.0	3.7	V
				$I_{I(bus)}$ = −12 mA			−1.5	
$I_{I/O(bus)}$	Current into bus port	Power on	Driver disabled	$V_{I(bus)}$ = −1.5 V to 0.4 V	−1.3			mA
				$V_{I(bus)}$ = 0.4 V to 2.5 V	0		−3.2	
				$V_{I(bus)}$ = 2.5 V to 3.7 V			+2.5 −3.2	
				$V_{I(bus)}$ = 3.7 V to 5 V	0		2.5	
				$V_{I(bus)}$ = 5 V to 5.5 V	0.7		2.5	
		Power off	V_{CC} = 0,	$V_{I(bus)}$ = 0 V to 2.5 V			−40	μA
I_{OS}	Short-circuit output current	Terminal			−15	−35	−75	mA
		Bus			−25	−50	−125	
I_{CC}	Supply current		No load	Receivers low and enabled		70	90	mA
				Drivers low and enabled		85	110	
$C_{i/o(bus)}$	Bus-port capacitance		V_{CC} = 5 V to 0 V, f = 1 MHz	$V_{I/O}$ = 0 to 2 V,		30		pF

†All typical values are at V_{CC} = 5 V, T_A = 25°C.

switching characteristics, V_{CC} = 5 V, C_L = 15 pF, T_A = 25°C (unless otherwise noted)

PARAMETER		FROM	TO	TEST CONDITIONS	MIN	TYP	MAX	UNIT
t_{PLH}	Propagation delay time, low-to-high-level output	Terminal	Bus	C_L = 30 pF, See Figure 1		14	20	ns
t_{PHL}	Propagation delay time, high-to-low-level output					14	20	
t_{PLH}	Propagation delay time, low-to-high-level output	Bus	Terminal	C_L = 30 pF, See Figure 2		10	20	ns
t_{PHL}	Propagation delay time, high-to-low-level output					15	22	
t_{PZH}	Output enable time to high level	TE	Bus	See Figure 3		25	35	ns
t_{PHZ}	Output disable time from high level					13	22	
t_{PZL}	Output enable time to low level					22	35	
t_{PLZ}	Output disable time from low level					22	32	
t_{PZH}	Output enable time to high level	TE	Terminal	See Figure 4		20	30	ns
t_{PHZ}	Output disable time from high level					12	20	
t_{PZL}	Output enable time to low level					23	32	
t_{PLZ}	Output disable time from low level					19	30	
t_{en}	Output pull-up enable time	PE	Bus	See Figure 5		15	22	ns
t_{dis}	Output pull-up disable time					13	20	

TEXAS INSTRUMENTS
POST OFFICE BOX 655012 • DALLAS, TEXAS 75265

PARAMETER MEASUREMENT INFORMATION

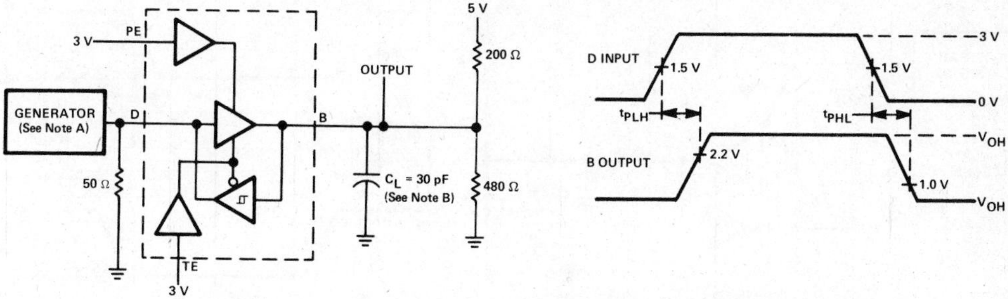
3 V
PE
GENERATOR
(See Note A)
D
50 Ω
TE
3 V
B
OUTPUT
5 V
200 Ω
C_L = 30 pF
(See Note B)
480 Ω
D INPUT
1.5 V
1.5 V
3 V
0 V
t_{PLH}
t_{PHL}
B OUTPUT
2.2 V
1.0 V
V_{OH}
V_{OH}

TEST CIRCUIT VOLTAGE WAVEFORMS

FIGURE 1. TERMINAL-TO-BUS PROPAGATION DELAY TIMES

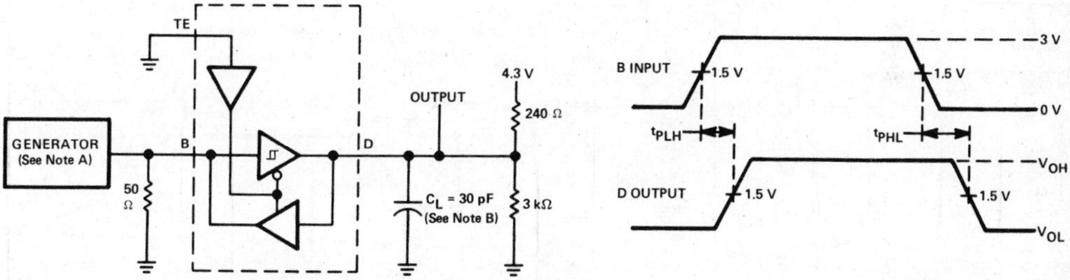
TE
GENERATOR
(See Note A)
B
50 Ω
D
OUTPUT
4.3 V
240 Ω
C_L = 30 pF
(See Note B)
3 kΩ
B INPUT
1.5 V
1.5 V
3 V
0 V
t_{PLH}
t_{PHL}
D OUTPUT
1.5 V
1.5 V
V_{OH}
V_{OL}

TEST CIRCUIT VOLTAGE WAVEFORMS

FIGURE 2. BUS-TO-TERMINAL PROPAGATION DELAY TIMES

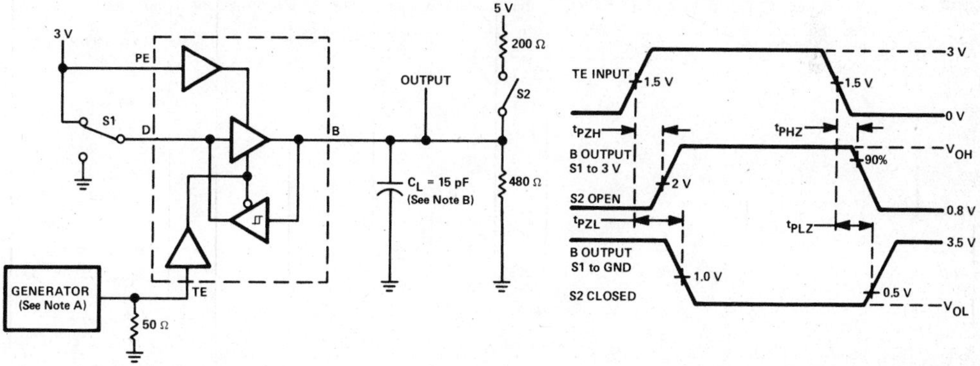
3 V
PE
S1
D
B
TE
GENERATOR
(See Note A)
50 Ω
OUTPUT
5 V
200 Ω
S2
C_L = 15 pF
(See Note B)
480 Ω
TE INPUT
1.5 V
1.5 V
3 V
0 V
t_{PZH}
t_{PHZ}
B OUTPUT
S1 to 3 V
S2 OPEN
2 V
90%
V_{OH}
0.8 V
t_{PZL}
t_{PLZ}
B OUTPUT
S1 to GND
S2 CLOSED
1.0 V
0.5 V
3.5 V
V_{OL}

TEST CIRCUIT VOLTAGE WAVEFORMS

FIGURE 3. TE-TO-BUS ENABLE AND DISABLE TIMES

NOTES: A. The input pulse is supplied by a generator having the following characteristics: PRR ≤ 1 MHz, 50% duty cycle, t_r ≤6 ns, t_f ≤ns, Z_{out} = 50 Ω.

B. C_L includes probe and jig capacitance.

PARAMETER MEASUREMENT INFORMATION

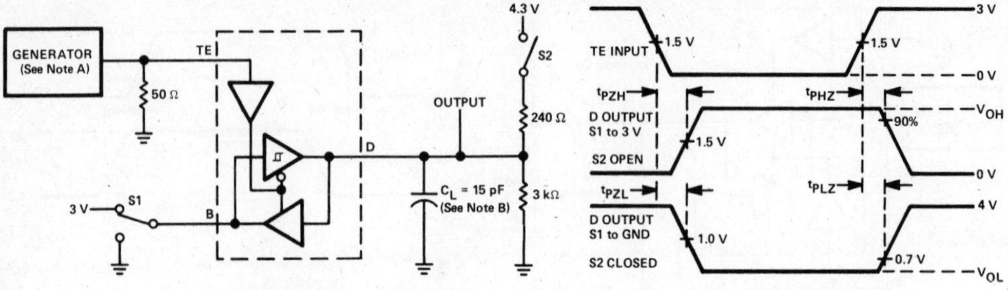
GENERATOR (See Note A)
TE
50 Ω
3 V
S1
B
D
OUTPUT
4.3 V
S2
240 Ω
C_L = 15 pF (See Note B)
3 kΩ
TE INPUT
1.5 V
1.5 V
3 V
0 V
t_{PZH}
t_{PHZ}
V_{OH}
D OUTPUT S1 to 3 V
1.5 V
90%
S2 OPEN
0 V
t_{PZL}
t_{PLZ}
4 V
D OUTPUT S1 to GND
1.0 V
S2 CLOSED
0.7 V
V_{OL}

TEST CIRCUIT

VOLTAGE WAVEFORMS

FIGURE 4. TE-TO-TERMINAL ENABLE AND DISABLE TIMES

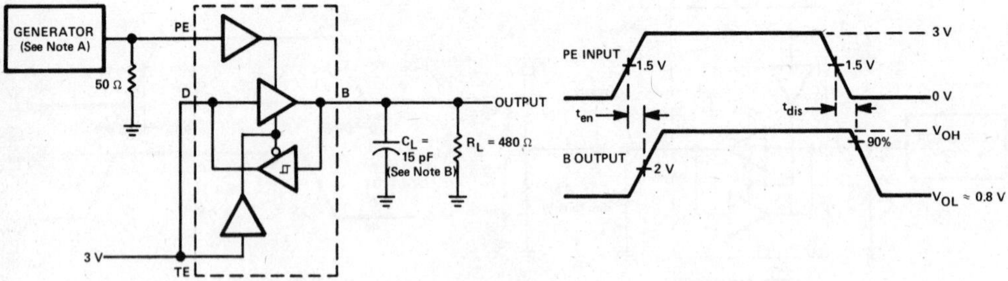
GENERATOR (See Note A)
PE
50 Ω
D
B
OUTPUT
C_L = 15 pF (See Note B)
R_L = 480 Ω
3 V
TE
PE INPUT
1.5 V
1.5 V
3 V
0 V
t_{en}
t_{dis}
V_{OH}
90%
B OUTPUT
2 V
V_{OL} ≈ 0.8 V

TEST CIRCUIT

VOLTAGE WAVEFORMS

FIGURE 5. PE-TO-BUS PULLUP ENABLE AND DISABLE TIMES

NOTES: A. The input pulse is supplied by a generator having the following characteristics: PRR ≤ 1 MHz, 50% duty cycle, t_r ≤6 ns, t_f ≤ns, Z_{out} = 50 Ω.

B. C_L includes probe and jig capacitance.

TYPICAL CHARACTERISTICS

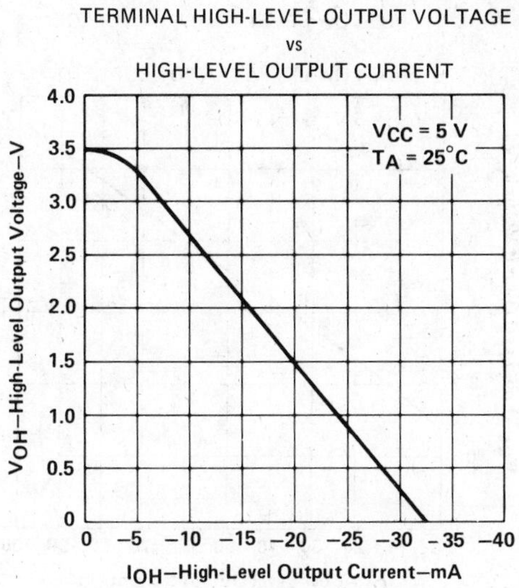
TERMINAL HIGH-LEVEL OUTPUT VOLTAGE
vs
HIGH-LEVEL OUTPUT CURRENT
V_{CC} = 5 V
T_A = 25°C
V_{OH}—High-Level Output Voltage—V
4.0
3.5
3.0
2.5
2.0
1.5
1.0
0.5
0
0
−5
−10
−15
−20
−25
−30
−35
−40
I_{OH}—High-Level Output Current—mA

FIGURE 6

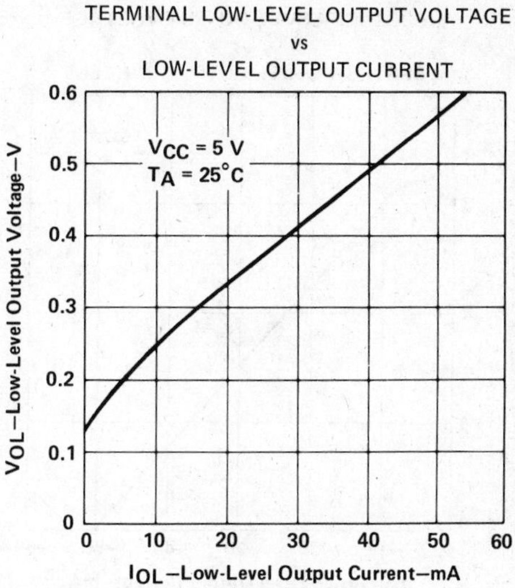
TERMINAL LOW-LEVEL OUTPUT VOLTAGE
vs
LOW-LEVEL OUTPUT CURRENT
V_{CC} = 5 V
T_A = 25°C
V_{OL}—Low-Level Output Voltage—V
0.6
0.5
0.4
0.3
0.2
0.1
0
0
10
20
30
40
50
60
I_{OL}—Low-Level Output Current—mA

FIGURE 7

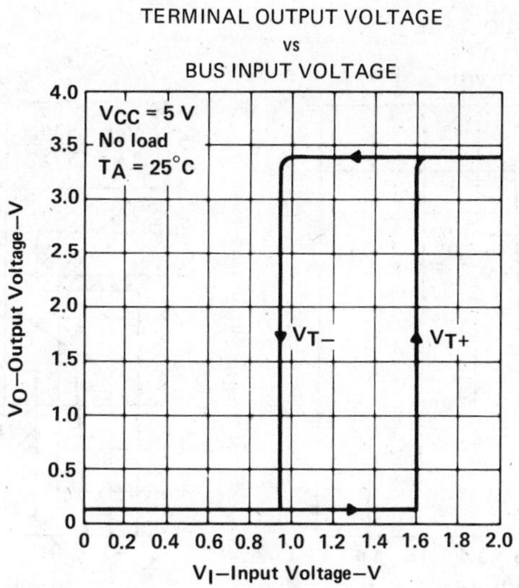
TERMINAL OUTPUT VOLTAGE
vs
BUS INPUT VOLTAGE
V_{CC} = 5 V
No load
T_A = 25°C
V_O—Output Voltage—V
4.0
3.5
3.0
2.5
2.0
1.5
1.0
0.5
0
V_{T-}
V_{T+}
0
0.2
0.4
0.6
0.8
1.0
1.2
1.4
1.6
1.8
2.0
V_I—Input Voltage—V

FIGURE 8

4
Line Drivers/Receivers

TYPICAL CHARACTERISTICS

BUS HIGH-LEVEL OUTPUT VOLTAGE
vs
BUS HIGH-LEVEL OUTPUT CURRENT

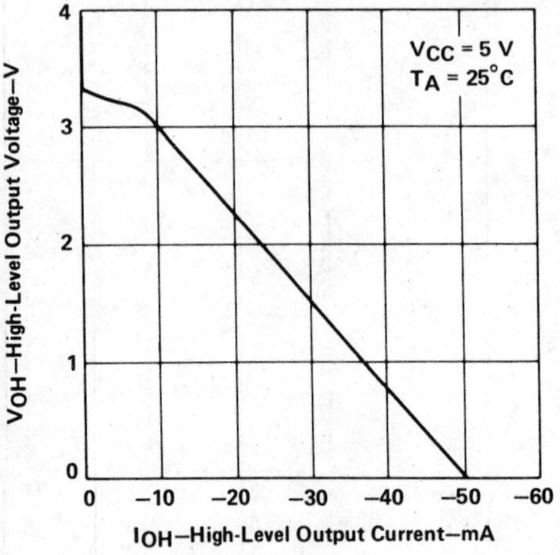
V_{CC} = 5 V
T_A = 25°C
V_{OH}—High-Level Output Voltage—V
0 1 2 3 4
0 −10 −20 −30 −40 −50 −60
I_{OH}—High-Level Output Current—mA

FIGURE 9

BUS LOW-LEVEL OUTPUT VOLTAGE
vs
BUS LOW-LEVEL OUTPUT CURRENT

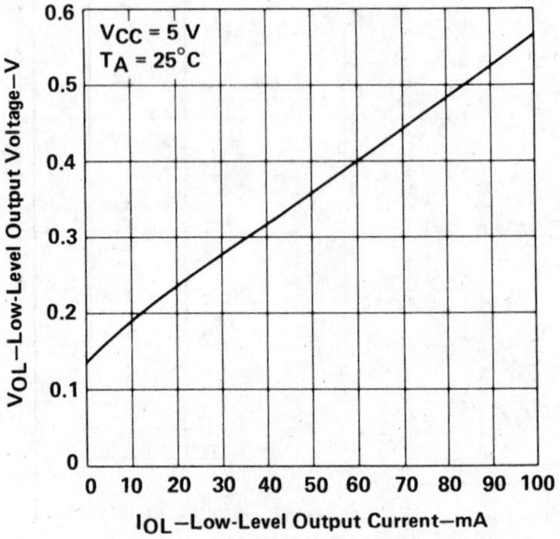
V_{CC} = 5 V
T_A = 25°C
V_{OL}—Low-Level Output Voltage—V
0 0.1 0.2 0.3 0.4 0.5 0.6
0 10 20 30 40 50 60 70 80 90 100
I_{OL}—Low-Level Output Current—mA

FIGURE 10

BUS OUTPUT VOLTAGE
vs
TERMINAL INPUT VOLTAGE

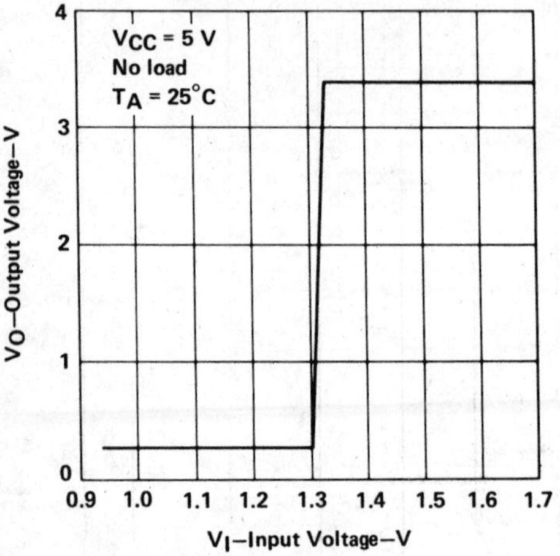
V_{CC} = 5 V
No load
T_A = 25°C
V_O—Output Voltage—V
0 1 2 3 4
0.9 1.0 1.1 1.2 1.3 1.4 1.5 1.6 1.7
V_I—Input Voltage—V

FIGURE 11

BUS CURRENT
vs
BUS VOLTAGE

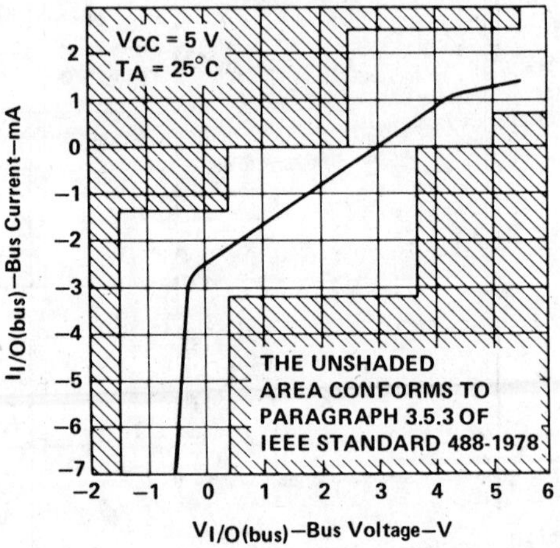
V_{CC} = 5 V
T_A = 25°C
THE UNSHADED AREA CONFORMS TO PARAGRAPH 3.5.3 OF IEEE STANDARD 488-1978
$I_{I/O(bus)}$—Bus Current—mA
−7 −6 −5 −4 −3 −2 −1 0 1 2
−2 −1 0 1 2 3 4 5 6
$V_{I/O(bus)}$—Bus Voltage—V

FIGURE 12

MEETS IEEE STANDARD 488-1978 (GPIB)

- 8-Channel Bidirectional Transceiver
- Power-Up/Power-Down Protection (Glitch-Free)
- Designed to Implement Control Bus Interface
- SN75161B Designed for Single Controller
- SN75162B Designed for Multi-Controllers
- High-Speed, Low-Power Schottky Circuitry
- Low-Power Dissipation . . . 72 mW Max Per Channel
- Fast Propagation Times . . . 22 ns Max
- High-Impedance P-N-P Inputs
- Receiver Hysteresis . . . 650 mV Typ
- Bus-Terminating Resistors Provided on Driver Outputs
- No Loading of Bus When Device is Powered Down (V_{CC} = 0)

SN75161B . . . DW, J, OR N DUAL-IN-LINE PACKAGE
(TOP VIEW)

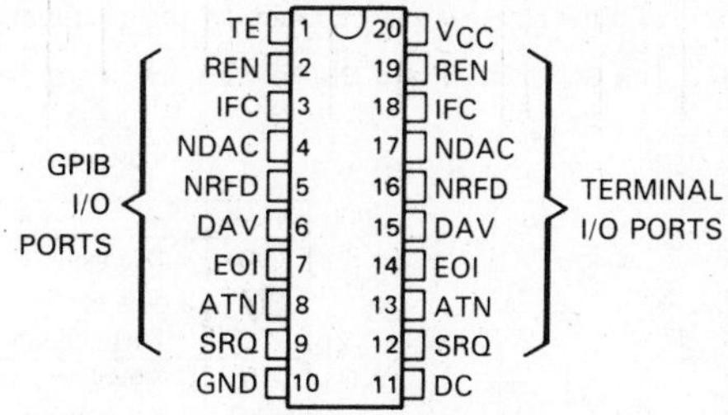
TE 1 20 VCC
REN 2 19 REN
IFC 3 18 IFC
NDAC 4 17 NDAC
NRFD 5 16 NRFD
DAV 6 15 DAV
EOI 7 14 EOI
ATN 8 13 ATN
SRQ 9 12 SRQ
GND 10 11 DC
GPIB I/O PORTS
TERMINAL I/O PORTS

SN75162B . . . DW DUAL-IN-LINE PACKAGE
(TOP VIEW)

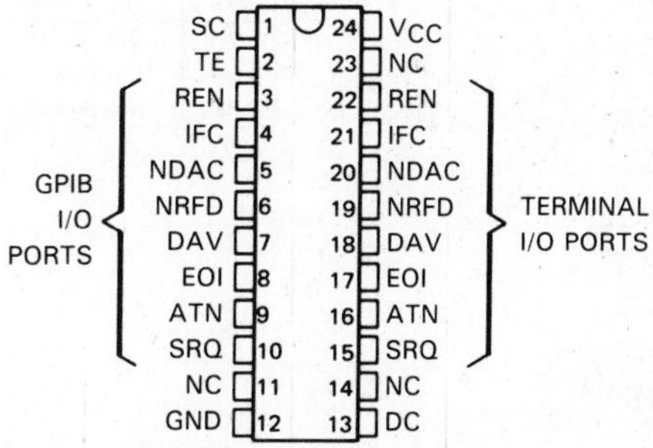
SC 1 24 VCC
TE 2 23 NC
REN 3 22 REN
IFC 4 21 IFC
NDAC 5 20 NDAC
NRFD 6 19 NRFD
DAV 7 18 DAV
EOI 8 17 EOI
ATN 9 16 ATN
SRQ 10 15 SRQ
NC 11 14 NC
GND 12 13 DC
GPIB I/O PORTS
TERMINAL I/O PORTS

SN75162B . . . N DUAL-IN-LINE PACKAGE
(TOP VIEW)

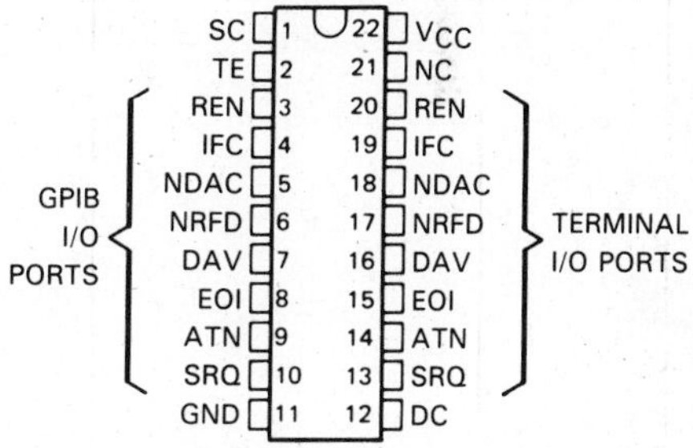
SC 1 22 VCC
TE 2 21 NC
REN 3 20 REN
IFC 4 19 IFC
NDAC 5 18 NDAC
NRFD 6 17 NRFD
DAV 7 16 DAV
EOI 8 15 EOI
ATN 9 14 ATN
SRQ 10 13 SRQ
GND 11 12 DC
GPIB I/O PORTS
TERMINAL I/O PORTS

NC—No internal connection.

description

The SN75161B and SN75162B eight-channel general-purpose interface bus transceivers are monolithic, high-speed, low-power Schottky devices designed to meet the requirements of IEEE Standard 488-1978. Each transceiver is designed to provide the bus-management and data-transfer signals between operating units of a single- or multiple-controller instrumentation system. When combined with the SN75160B octal bus transceiver, the SN75161B or SN75162B provides the complete 16-wire interface for the IEEE 488 bus.

The SN75161B and SN75162B each features eight driver-receiver pairs connected in a front-to-back configuration to form input/output (I/O) ports at both the bus and terminal sides. A power up/down disable circuit is included on all bus and receiver outputs. This provides glitch-free operation during V_{CC} power-up and power-down. The direction of data through these driver-receiver pairs is determined by the DC, TE, and SC (on SN75162B) enable signals. The SC input on the SN75162B allows the REN and IFC transceivers to be controlled independently.

The driver outputs (GPIB I/O ports) feature active bus-terminating resistor circuits designed to provide a high impedance to the bus when supply voltage V_{CC} is 0. The drivers are designed to handle loads up to 48 milliamperes of sink current. Each receiver features p-n-p transistor inputs for high input impedance and a guaranteed hysteresis of 400 millivolts for increased noise immunity. All receivers have 3-state outputs to present a high impedance to the terminal when disabled.

The SN75161B and SN75162B are characterized for operation from 0°C to 70°C.

CHANNEL IDENTIFICATION TABLE

NAME	IDENTITY	CLASS
DC	Direction Control	Control
TE	Talk Enable	
SC	System Control (SN75162B only)	
ATN	Attention	Bus Management
SRQ	Service Request	
REN	Remote Enable	
IFC	Interface Clear	
EOI	End or Identify	
DAV	Data Valid	Data Transfer
NDAC	Not Data Accepted	
NRFD	Not Ready for Data	

SN75161B logic symbol†

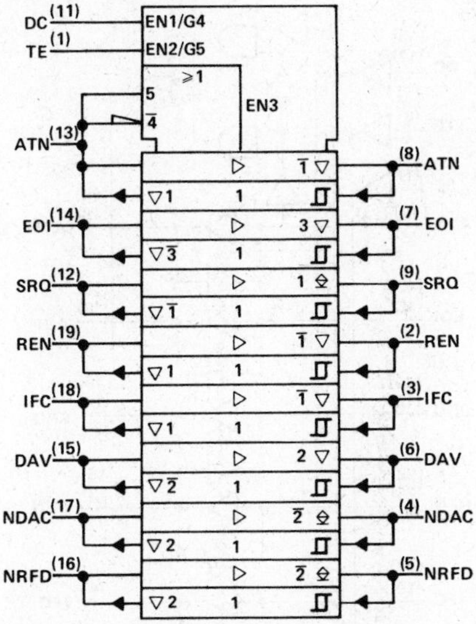
DC (11)
TE (1)
EN1/G4
EN2/G5
≥1
5
4
EN3
ATN (13)
EOI (14)
SRQ (12)
REN (19)
IFC (18)
DAV (15)
NDAC (17)
NRFD (16)
(8) ATN
(7) EOI
(9) SRQ
(2) REN
(3) IFC
(6) DAV
(4) NDAC
(5) NRFD

† This symbol is in accordance with IEEE Std 91-1984 and IEC publication 617-12.
▽designates 3-state output, ⩢designates passive-pullup outputs.

SN75161B logic diagram (positive logic)

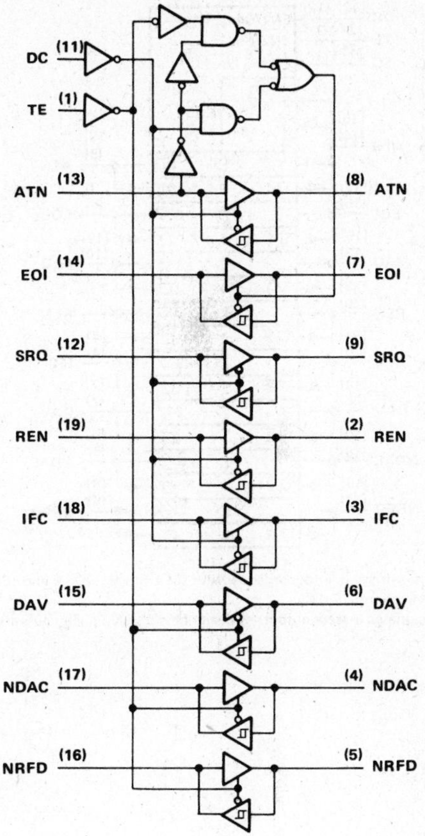
DC (11)
TE (1)
ATN (13)
EOI (14)
SRQ (12)
REN (19)
IFC (18)
DAV (15)
NDAC (17)
NRFD (16)
(8) ATN
(7) EOI
(9) SRQ
(2) REN
(3) IFC
(6) DAV
(4) NDAC
(5) NRFD

SN75162B logic symbol†

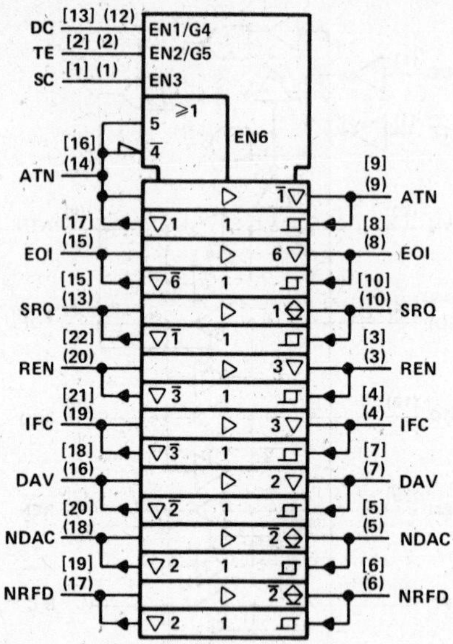
DC [13] (12) EN1/G4
TE [2] (2) EN2/G5
SC [1] (1) EN3
≥1
5
4
EN6
ATN [16] (14)
[9] (9) ATN
[17] (15) EOI
[8] (8) EOI
[15] (13) SRQ
[10] (10) SRQ
[22] (20) REN
[3] (3) REN
[21] (19) IFC
[4] (4) IFC
[18] (16) DAV
[7] (7) DAV
[20] (18) NDAC
[5] (5) NDAC
[19] (17) NRFD
[6] (6) NRFD

† This symbol is in accordance with IEEE Std 91-1984 and IEC publication 617-12.
▽designates 3-state output, ⬙designates passive-pullup outputs.

SN75162B logic diagram (positive logic)

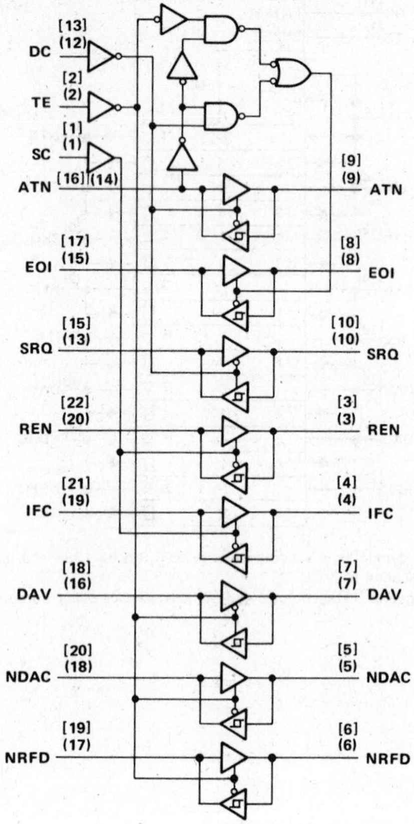
DC [13] (12)
TE [2] (2)
SC [1] (1)
ATN [16] (14)
[9] (9) ATN
EOI [17] (15)
[8] (8) EOI
SRQ [15] (13)
[10] (10) SRQ
REN [22] (20)
[3] (3) REN
IFC [21] (19)
[4] (4) IFC
DAV [18] (16)
[7] (7) DAV
NDAC [20] (18)
[5] (5) NDAC
NRFD [19] (17)
[6] (6) NRFD

[ ] Denotes pin numbers for DW package.
( ) Denotes pin numbers for N package.

TEXAS INSTRUMENTS
POST OFFICE BOX 655012 • DALLAS, TEXAS 75265

SN75161B
RECEIVE/TRANSMIT FUNCTION TABLE

<table>
<tr><th colspan="3">CONTROLS</th><th colspan="5">BUS-MANAGEMENT CHANNELS</th><th colspan="3">DATA-TRANSFER CHANNELS</th></tr>
<tr><th>DC</th><th>TE</th><th>ATN†</th><th>ATN†</th><th>SRQ</th><th>REN</th><th>IFC</th><th rowspan="2">EOI</th><th>DAV</th><th>NDAC</th><th>NRFD</th></tr>
<tr><th></th><th></th><th></th><th colspan="4">(Controlled by DC)</th><th colspan="3">(Controlled by TE)</th></tr>
<tr><td>H</td><td>H</td><td>H</td><td rowspan="2">R</td><td rowspan="2">T</td><td rowspan="2">R</td><td rowspan="2">R</td><td>T</td><td rowspan="2">T</td><td rowspan="2">R</td><td rowspan="2">R</td></tr>
<tr><td>H</td><td>H</td><td>L</td><td>R</td></tr>
<tr><td>L</td><td>L</td><td>H</td><td rowspan="2">T</td><td rowspan="2">R</td><td rowspan="2">T</td><td rowspan="2">T</td><td>R</td><td rowspan="2">R</td><td rowspan="2">T</td><td rowspan="2">T</td></tr>
<tr><td>L</td><td>L</td><td>L</td><td>T</td></tr>
<tr><td>H</td><td>L</td><td>X</td><td>R</td><td>T</td><td>R</td><td>R</td><td>R</td><td>R</td><td>T</td><td>T</td></tr>
<tr><td>L</td><td>H</td><td>X</td><td>T</td><td>R</td><td>T</td><td>T</td><td>T</td><td>T</td><td>R</td><td>R</td></tr>
</table>

SN75162B
RECEIVE/TRANSMIT FUNCTION TABLE

<table>
<tr><th colspan="4">CONTROLS</th><th colspan="5">BUS-MANAGEMENT CHANNELS</th><th colspan="3">DATA-TRANSFER CHANNELS</th></tr>
<tr><th rowspan="2">SC</th><th>DC</th><th>TE</th><th>ATN†</th><th>ATN†</th><th>SRQ</th><th>REN</th><th>IFC</th><th rowspan="2">EOI</th><th>DAV</th><th>NDAC</th><th>NRFD</th></tr>
<tr><th></th><th></th><th></th><th colspan="2">(Controlled by DC)</th><th colspan="2">(Controlled by SC)</th><th colspan="3">(Controlled by TE)</th></tr>
<tr><td rowspan="6"></td><td>H</td><td>H</td><td>H</td><td rowspan="2">R</td><td rowspan="2">T</td><td rowspan="6" colspan="2"></td><td>T</td><td rowspan="2">T</td><td rowspan="2">R</td><td rowspan="2">R</td></tr>
<tr><td>H</td><td>H</td><td>L</td><td>R</td></tr>
<tr><td>L</td><td>L</td><td>H</td><td rowspan="2">T</td><td rowspan="2">R</td><td>R</td><td rowspan="2">R</td><td rowspan="2">T</td><td rowspan="2">T</td></tr>
<tr><td>L</td><td>L</td><td>L</td><td>T</td></tr>
<tr><td>H</td><td>L</td><td>X</td><td>R</td><td>T</td><td>R</td><td>R</td><td>T</td><td>T</td></tr>
<tr><td>L</td><td>H</td><td>X</td><td>T</td><td>R</td><td>T</td><td>T</td><td>R</td><td>R</td></tr>
<tr><td>H</td><td rowspan="2" colspan="3"></td><td rowspan="2" colspan="2"></td><td>T</td><td>T</td><td rowspan="2"></td><td rowspan="2" colspan="3"></td></tr>
<tr><td>L</td><td>R</td><td>R</td></tr>
</table>

H = high level, L = low level, R = receive, T = transmit, X = irrelevant

Direction of data transmission is from the terminal side to the bus side, and the direction of data receiving is from the bus side to the terminal side. Data transfer is noninverting in both directions.

† ATN is a normal transceiver channel that functions additionally as an internal direction control or talk enable for EOI whenever the DC and TE inputs are in the same state. When DC and TE are in opposite states, the ATN channel functions as an independent transceiver only.

schematics of inputs and outputs

EQUIVALENT OF ALL CONTROL INPUTS

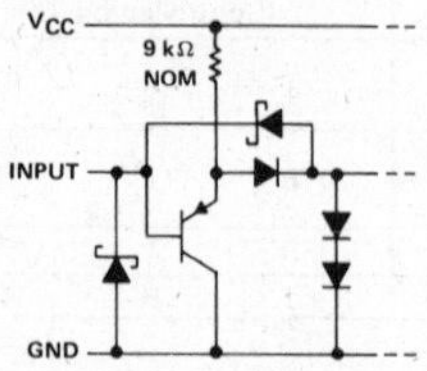
VCC
9 kΩ
NOM
INPUT
GND

TYPICAL OF SRQ, NDAC, and NRFD GPIB I/O PORT

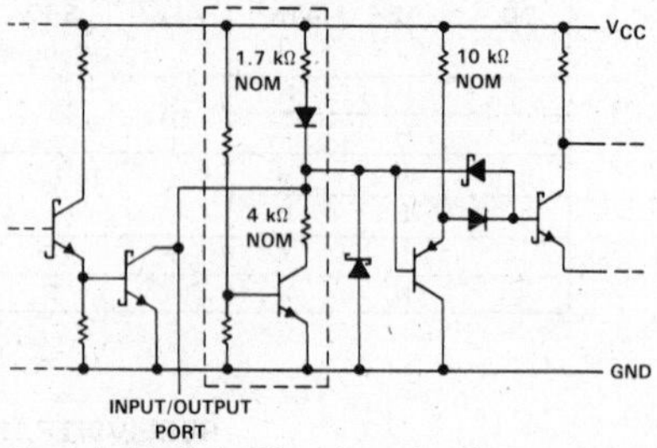
VCC
1.7 kΩ
NOM
10 kΩ
NOM
4 kΩ
NOM
GND
INPUT/OUTPUT
PORT

Circuit inside dashed lines is on the driver outputs only.

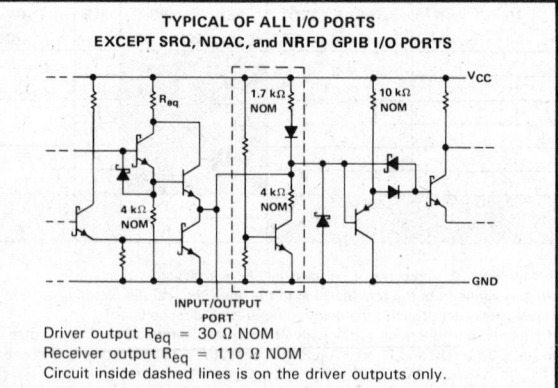

TYPICAL OF ALL I/O PORTS EXCEPT SRQ, NDAC, and NRFD GPIB I/O PORTS

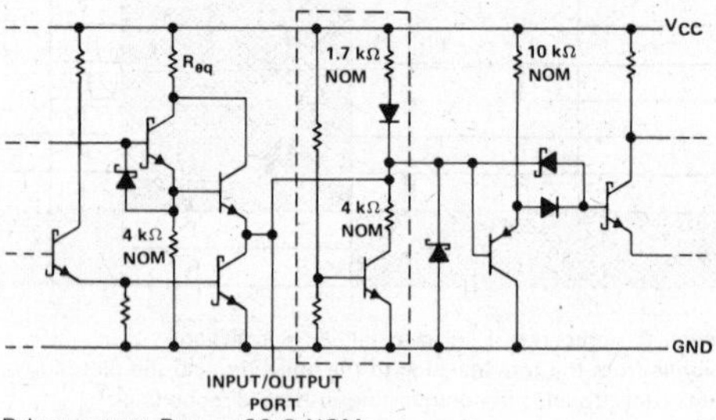
VCC
Req
1.7 kΩ
NOM
10 kΩ
NOM
4 kΩ
NOM
4 kΩ
NOM
GND
INPUT/OUTPUT
PORT

Driver output R_{eq} = 30 Ω NOM
Receiver output R_{eq} = 110 Ω NOM
Circuit inside dashed lines is on the driver outputs only.

absolute maximum ratings over operating free-air temperature range (unless otherwise noted)

Supply voltage, V_{CC} (see Note 1)	7 V
Input voltage	5.5 V
Low-level driver output current	100 mA
Continuous total dissipation at (or below) 25°C free-air temperature (see Note 2):	
DW package (20 pin)	1125 mW
DW package (24 pin)	1350 mW
J package	1375 mW
N package (20 pin)	1150 mW
N package (22 pin)	1700 mW
Operating free-air temperature range	0°C to 70°C
Storage temperature range	−65°C to 150°C
Lead temperature 1,6 mm (1/16) inch from the case for 60 seconds: J package	300°C
Lead temperature 1,6 mm (1/16) inch from the case for 10 seconds: DW or N package	260°C

NOTES: 1. All voltage values are with respect to network ground terminal.
2. For operation above 25°C free-air temperature, derate the 20-pin DW package at the rate of 9.0 mW/°C, the 24-pin DW package at the rate of 10.8 mW/°C, the 20-pin N package at the rate of 9.2 mW/°C, the 22-pin N package at the rate of 13.6 mW/°C, and the J package at the rate of 11.0 mW/°C. In the J package, SN75161B chips are alloy mounted.

recommended operating conditions

		MIN	NOM	MAX	UNIT
Supply voltage, V_{CC}		4.75	5	5.25	V
High-level input voltage, V_{IH}		2			V
Low-level input voltage, V_{IL}				0.8	V
High-level output current, I_{OH}	Bus ports with 3-state outputs			−5.2	mA
	Terminal ports			−800	μA
Low-level output current, I_{OL}	Bus ports			48	mA
	Terminal ports			16	mA
Operating free-air temperature, T_A		0		70	°C

electrical characteristics over recommended ranges of supply voltage and operating free-air temperature (unless otherwise noted)

PARAMETER			TEST CONDITIONS		MIN	TYP†	MAX	UNIT
V_{IK}	Input clamp voltage		$I_I = -18$ mA			−0.8	−1.5	V
V_{hys}	Hysteresis ($V_{T+} - V_{T-}$)	Bus			0.4	0.65		V
V_{OH}‡	High-level output voltage	Terminal	$I_{OH} = -800$ μA		2.7	3.5		V
		Bus	$I_{OH} = -5.2$ mA		2.5	3.3		V
V_{OL}	Low-level output voltage	Terminal	$I_{OL} = 16$ mA			0.3	0.5	V
		Bus	$I_{OL} = 48$ mA			0.35	0.5	V
I_I	Input current at maximum input voltage	Terminal	$V_I = 5.5$ V			0.2	100	μA
I_{IH}	High-level input current	Terminal and control inputs	$V_I = 2.7$ V			0.1	20	μA
I_{IL}	Low-level input current	Terminal and control inputs	$V_I = 0.5$ V			−10	−100	μA
$V_{I/O(bus)}$	Voltage at bus port		Driver disabled	$I_{I(bus)} = 0$	2.5	3.0	3.7	V
			Driver disabled	$I_{I(bus)} = -12$ mA			−1.5	V
$I_{I/O(bus)}$	Current into bus port	Power on	Driver disabled	$V_{I(bus)} = -1.5$ V to 0.4 V	−1.3			mA
		Power on	Driver disabled	$V_{I(bus)} = 0.4$ V to 2.5 V	0		−3.2	mA
		Power on	Driver disabled	$V_{I(bus)} = 2.5$ V to 3.7 V			+2.5 −3.2	mA
		Power on	Driver disabled	$V_{I(bus)} = 3.7$ V to 5 V	0		2.5	mA
		Power on	Driver disabled	$V_{I(bus)} = 5$ V to 5.5 V	0.7		2.5	mA
		Power off	$V_{CC} = 0$,	$V_{I(bus)} = 0$ V to 2.5 V			−40	μA
I_{OS}	Short-circuit output current	Terminal			−15	−35	−75	mA
		Bus			−25	−50	−125	mA
I_{CC}	Supply current		No load,	TE, DC, and SC low			110	mA
$C_{i/o(bus)}$	Bus-port capacitance		$V_{CC} = 5$ V to 0 V, $V_{I/O} = 0$ to 2 V, f = 1 MHz			30		pF

†All typical values are at $V_{CC} = 5$ V, $T_A = 25$ °C.
‡V_{OH} applies for three-state outputs only.

switching characteristics, V_{CC} = 5 V, C_L = 15 pF, T_A = 25°C (unless otherwise noted)

PARAMETER		FROM	TO	TEST CONDITIONS	MIN	TYP	MAX	UNIT
t_{PLH}	Propagation delay time, low-to-high-level output	Terminal	Bus	C_L = 30 pF, See Figure 1		14	20	ns
t_{PHL}	Propagation delay time, high-to-low-level output					14	20	
t_{PLH}	Propagation delay time, low-to-high-level output	Terminal	Bus (SRQ, NDAC NRFD)	C_L = 30 pF, See Figure 1		29	35	ns
t_{PLH}	Propagation delay time, low-to-high-level output	Bus	Terminal	C_L = 30 pF, See Figure 2		10	20	ns
t_{PHL}	Propagation delay time, high-to-low-level output					15	22	
t_{PZH}	Output enable time to high level	TE, DC, or SC	BUS (ATTN, EOI, REN, IFC, and DAV)	See Figure 3			60	ns
t_{PHZ}	Output disable time from high level						45	
t_{PZL}	Output enable time to low level						60	
t_{PLZ}	Output disable time from low level						55	
t_{PZH}	Output enable time to high level	TE, DC, or SC	Terminal	See Figure 4			55	ns
t_{PHZ}	Output disable time from high level						50	
t_{PZL}	Output enable time to low level						45	
t_{PLZ}	Output disable time from low level						55	

PARAMETER MEASUREMENT INFORMATION

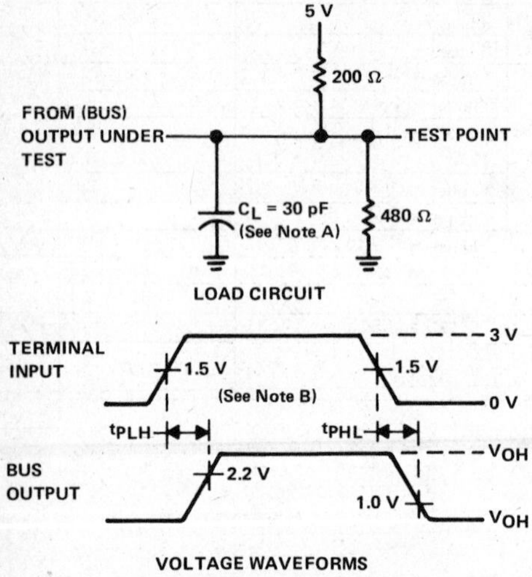
5 V
200 Ω
FROM (BUS)
OUTPUT UNDER
TEST
TEST POINT
C_L = 30 pF
(See Note A)
480 Ω
LOAD CIRCUIT
TERMINAL
INPUT
3 V
1.5 V
1.5 V
(See Note B)
0 V
tPLH
tPHL
VOH
BUS
OUTPUT
2.2 V
1.0 V
VOH
VOLTAGE WAVEFORMS

FIGURE 1. TERMINAL-TO-BUS PROPAGATION DELAY TIMES

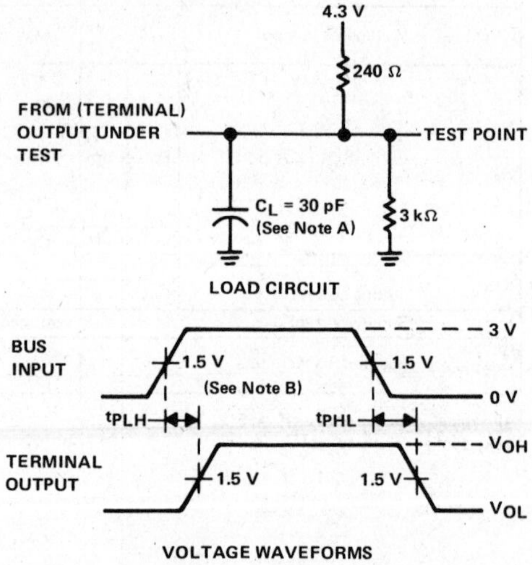
4.3 V
240 Ω
FROM (TERMINAL)
OUTPUT UNDER
TEST
TEST POINT
C_L = 30 pF
(See Note A)
3 kΩ
LOAD CIRCUIT
BUS
INPUT
3 V
1.5 V
1.5 V
(See Note B)
0 V
tPLH
tPHL
VOH
TERMINAL
OUTPUT
1.5 V
1.5 V
VOL
VOLTAGE WAVEFORMS

FIGURE 2. BUS-TO-TERMINAL PROPAGATION DELAY TIMES

NOTES: A. C_L includes probe and jig capacitance.

B. The input pulse is supplied by a generator having the following characteristics: PRR ≤ 1 MHz, 50% duty cycle, t_r ≤6 ns, t_f ≤6 ns, Z_{out} = 50 Ω.

PARAMETER MEASUREMENT INFORMATION

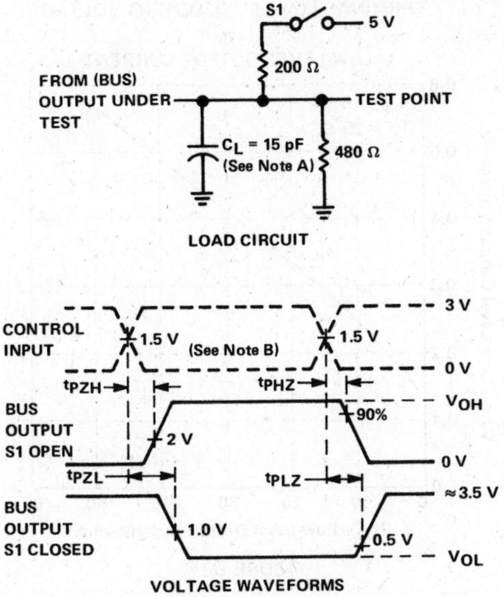
S1
5 V
200 Ω
FROM (BUS)
OUTPUT UNDER
TEST
TEST POINT
C_L = 15 pF
(See Note A)
480 Ω
LOAD CIRCUIT
CONTROL
INPUT
1.5 V
(See Note B)
1.5 V
3 V
0 V
t_{PZH}
t_{PHZ}
BUS
OUTPUT
S1 OPEN
2 V
90%
V_{OH}
0 V
t_{PZL}
t_{PLZ}
BUS
OUTPUT
S1 CLOSED
1.0 V
≈3.5 V
0.5 V
V_{OL}
VOLTAGE WAVEFORMS

FIGURE 3. BUS ENABLE AND DISABLE TIMES

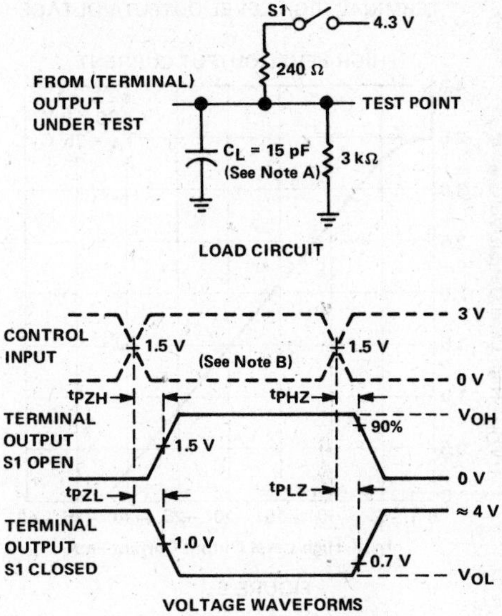
S1
4.3 V
240 Ω
FROM (TERMINAL)
OUTPUT
UNDER TEST
TEST POINT
C_L = 15 pF
(See Note A)
3 kΩ
LOAD CIRCUIT
CONTROL
INPUT
1.5 V
(See Note B)
1.5 V
3 V
0 V
t_{PZH}
t_{PHZ}
TERMINAL
OUTPUT
S1 OPEN
1.5 V
90%
V_{OH}
0 V
t_{PZL}
t_{PLZ}
TERMINAL
OUTPUT
S1 CLOSED
1.0 V
≈ 4 V
0.7 V
V_{OL}
VOLTAGE WAVEFORMS

FIGURE 4. TERMINAL ENABLE AND DISABLE TIMES

NOTES: A. C_L includes probe and jig capacitance.
B. The input pulse is supplied by a generator having the following characteristics: PRR ≤ 1 MHz, 50% duty cycle, t_r ≤6 ns, t_f ≤6 ns, Z_{out} = 50 Ω.

TYPICAL CHARACTERISTICS

TERMINAL HIGH-LEVEL OUTPUT VOLTAGE
vs
HIGH-LEVEL OUTPUT CURRENT

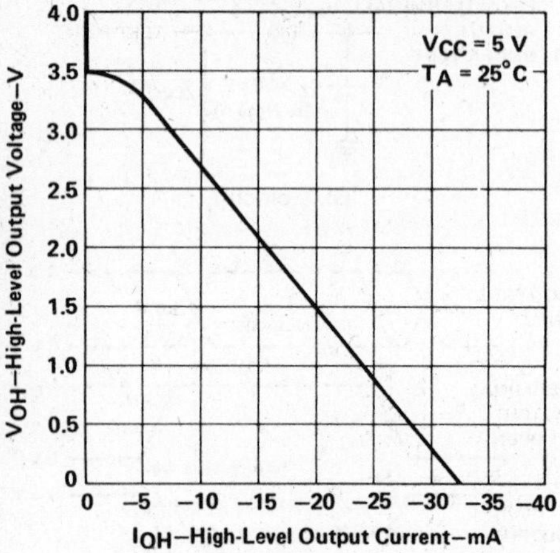
VCC = 5 V
TA = 25°C
VOH—High-Level Output Voltage—V
4.0 3.5 3.0 2.5 2.0 1.5 1.0 0.5 0
0 −5 −10 −15 −20 −25 −30 −35 −40
IOH—High-Level Output Current—mA

FIGURE 5

TERMINAL LOW-LEVEL OUTPUT VOLTAGE
vs
LOW-LEVEL OUTPUT CURRENT

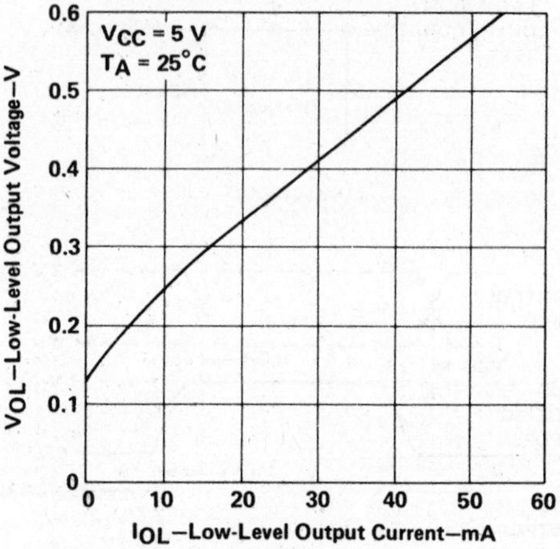
VCC = 5 V
TA = 25°C
VOL—Low-Level Output Voltage—V
0.6 0.5 0.4 0.3 0.2 0.1 0
0 10 20 30 40 50 60
IOL—Low-Level Output Current—mA

FIGURE 6

TERMINAL OUTPUT VOLTAGE
vs
BUS INPUT VOLTAGE

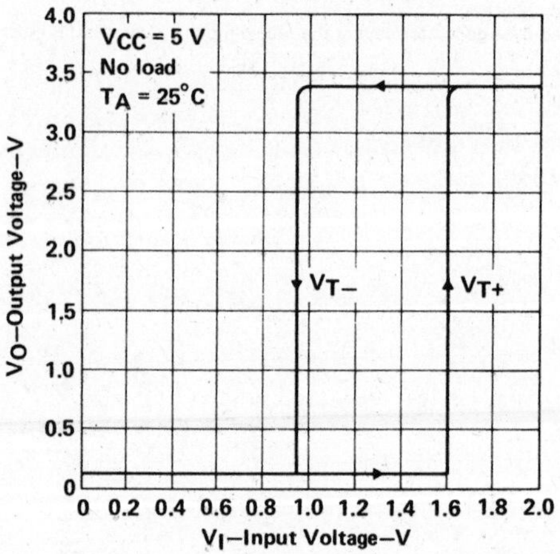
VCC = 5 V
No load
TA = 25°C
VO—Output Voltage—V
4.0 3.5 3.0 2.5 2.0 1.5 1.0 0.5 0
VT−
VT+
0 0.2 0.4 0.6 0.8 1.0 1.2 1.4 1.6 1.8 2.0
VI—Input Voltage—V

FIGURE 7

TEXAS INSTRUMENTS
POST OFFICE BOX 655012 • DALLAS, TEXAS 75265

TYPICAL CHARACTERISTICS

BUS HIGH-LEVEL OUTPUT VOLTAGE
vs
HIGH-LEVEL OUTPUT CURRENT

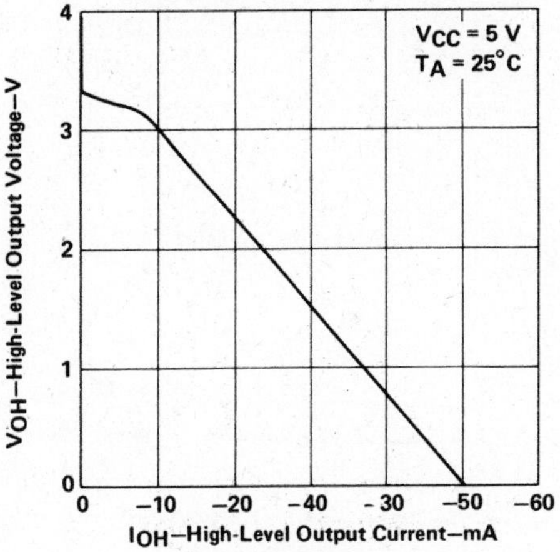
VCC = 5 V
TA = 25°C
VOH—High-Level Output Voltage—V
0 1 2 3 4
0 −10 −20 −40 −30 −50 −60
IOH—High-Level Output Current—mA

FIGURE 8

BUS-LOW LEVEL OUTPUT VOLTAGE
vs
LOW-LEVEL OUTPUT CURRENT

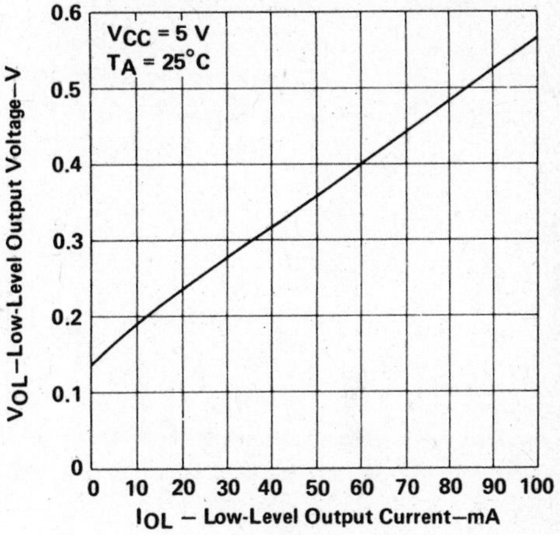
VCC = 5 V
TA = 25°C
VOL—Low-Level Output Voltage—V
0 0.1 0.2 0.3 0.4 0.5 0.6
0 10 20 30 40 50 60 70 80 90 100
IOL — Low-Level Output Current—mA

FIGURE 9

BUS OUTPUT VOLTAGE
vs
TERMINAL INPUT VOLTAGE

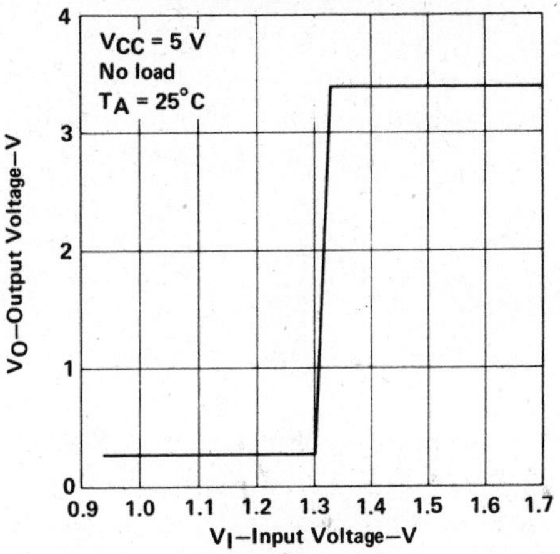
VCC = 5 V
No load
TA = 25°C
VO—Output Voltage—V
0 1 2 3 4
0.9 1.0 1.1 1.2 1.3 1.4 1.5 1.6 1.7
VI—Input Voltage—V

FIGURE 10

BUS CURRENT
vs
BUS VOLTAGE

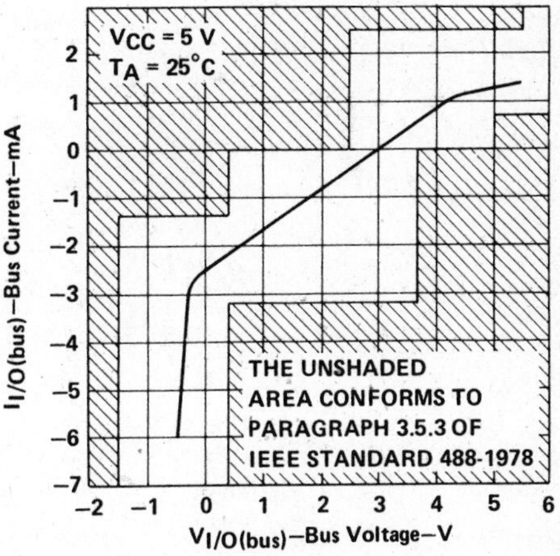
VCC = 5 V
TA = 25°C
THE UNSHADED AREA CONFORMS TO PARAGRAPH 3.5.3 OF IEEE STANDARD 488-1978
II/O(bus)—Bus Current—mA
2 1 0 −1 −2 −3 −4 −5 −6 −7
−2 −1 0 1 2 3 4 5 6
VI/O(bus)—Bus Voltage—V

FIGURE 11

D2611, OCTOBER 1985

- 8-Channel Bidirectional Transceivers
- Power-Up/Power-Down Protection (Glitch-Free)
- High-Speed Low-Power Schottky Circuitry
- Low Power Dissipation . . . 66 mW Max per Channel
- High-Impedance P-N-P Inputs
- Receiver Hysteresis . . . 650 mV Typ
- Open-Collector Driver Output Option
- No Loading of Bus When Device is Powered Down (V_{CC} = 0)

DW, J, OR N DUAL-IN-LINE PACKAGE
(TOP VIEW)

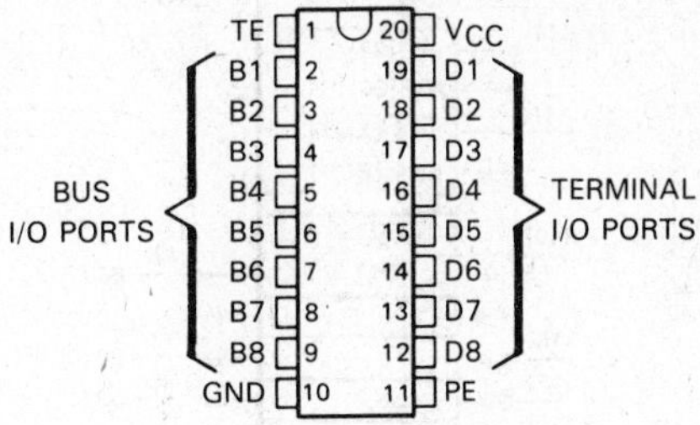
TE 1
B1 2
B2 3
B3 4
B4 5
B5 6
B6 7
B7 8
B8 9
GND 10
20 VCC
19 D1
18 D2
17 D3
16 D4
15 D5
14 D6
13 D7
12 D8
11 PE
BUS I/O PORTS
TERMINAL I/O PORTS

FUNCTION TABLES

EACH DRIVER

INPUTS			OUTPUT
D	TE	PE	B
H	H	H	H
L	H	H	L
H	X	L	Z
L	H	L	L
X	L	X	Z

EACH RECEIVER

INPUTS			OUTPUT
B	TE	PE	D
L	L	X	L
H	L	X	H
X	H	X	Z

H = high level, L = low level, X = irrelevant, Z = high-impedance state.

description

The SN75163B octal general-purpose interface bus transceiver is a monolithic, high-speed, low-power Schottky device. It is designed for two-way data communications over single-ended transmission lines. The transceiver features driver outputs that can be operated in either the open-collector or three-state modes. If Talk Enable (TE) is high, these outputs have the characteristics of open-collector outputs when Pullup Enable (PE) is low and of three-state outputs when PE is high. Taking TE low places the outputs in the high-impedance state. The driver outputs are designed to handle loads of up to 48 milliamperes of sink current. Each receiver features p-n-p transistor inputs for high input impedance and 400 millivolts of guaranteed hysteresis for increased noise immunity.

Output glitches during power-up and power-down are eliminated by an internal circuit that disables both the bus and receiver outputs. The outputs do not load the bus when V_{CC} = 0 volts.

The SN75163B is characterized for operation from 0°C to 70°C.

logic symbol†

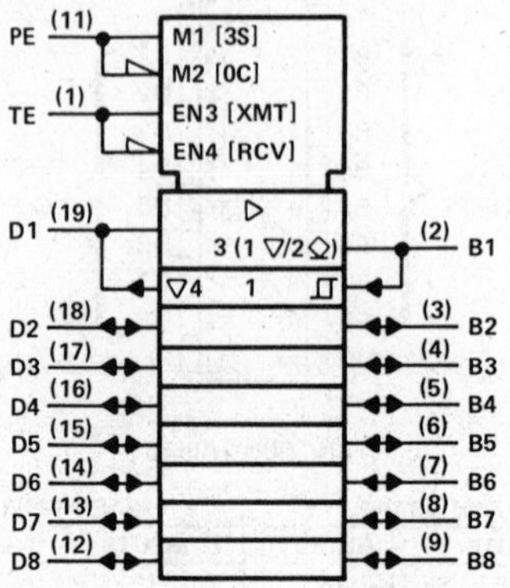
PE (11)
TE (1)
M1 [3S]
M2 [0C]
EN3 [XMT]
EN4 [RCV]
D1 (19)
3 (1 ▽/2 ◇)
▽4 1
(2) B1
D2 (18)
D3 (17)
D4 (16)
D5 (15)
D6 (14)
D7 (13)
D8 (12)
(3) B2
(4) B3
(5) B4
(6) B5
(7) B6
(8) B7
(9) B8

† This symbol is in accordance with ANSI/IEEE Std 91-1984 and IEC Publication 617-12.
▽ Designates 3-state outputs.
◇ Designates open-collector outputs.

logic diagram (positive logic)

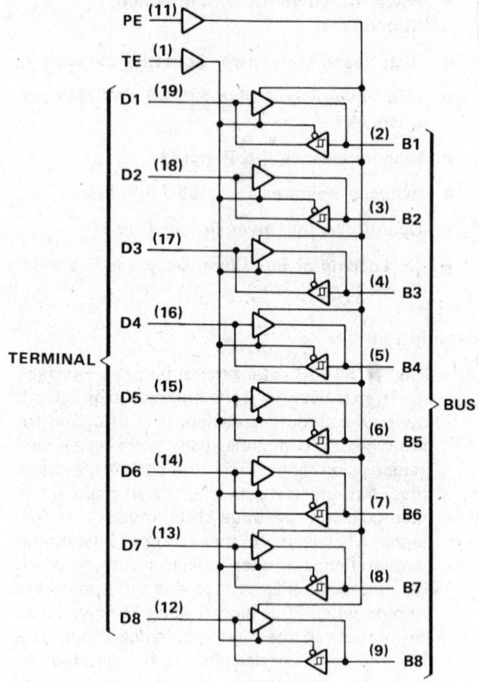
PE (11)
TE (1)
TERMINAL
D1 (19)
D2 (18)
D3 (17)
D4 (16)
D5 (15)
D6 (14)
D7 (13)
D8 (12)
(2) B1
(3) B2
(4) B3
(5) B4
(6) B5
(7) B6
(8) B7
(9) B8
BUS

schematics of inputs and outputs

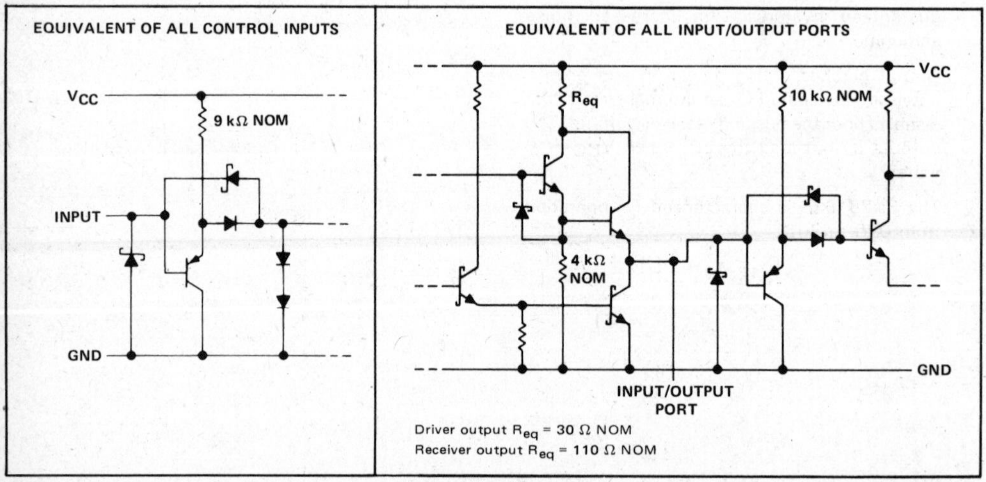
EQUIVALENT OF ALL CONTROL INPUTS
VCC
9 kΩ NOM
INPUT
GND
EQUIVALENT OF ALL INPUT/OUTPUT PORTS
Req
10 kΩ NOM
VCC
4 kΩ NOM
GND
INPUT/OUTPUT PORT
Driver output Req = 30 Ω NOM
Receiver output Req = 110 Ω NOM

TEXAS INSTRUMENTS
POST OFFICE BOX 655012 • DALLAS, TEXAS 75265

absolute maximum ratings over operating free-air temperature range (unless otherwise noted)

Supply voltage, V_{CC} (see Note 1)	7 V
Input voltage	5.5 V
Low-level driver output current	100 mA
Continuous total dissipation at (or below) 25°C free-air temperature (see Note 2):	
DW package	1125 mW
J package	1375 mW
N package	1150 mW
Operating free-air temperature range	0°C to 70°C
Storage temperature range	−65°C to 150°C
Lead temperature 1,6 mm (1/16) inch from the case for 60 seconds: J package	300°C
Lead temperature 1,6 mm (1/16 inch) from the case for 10 seconds: N package	260°C

NOTES: 1. All voltage values are with respect to network ground terminal.
2. For operation above 25°C free-air temperature, derate the DW package at the rate of 9.0 mW/°C, the N package at the rate of 9.2 mW/°C, and the J package at the rate of 11.0 mW/°C. In the J package, SN75163B chips are alloy mounted.

recommended operating conditions

		MIN	NOM	MAX	UNIT
Supply voltage, V_{CC}		4.75	5	5.25	V
High-level input voltage, V_{IH}		2			V
Low-level input voltage, V_{IL}				0.8	V
High-level output current, I_{OH}	Bus ports with pullups active			−10	mA
	Terminal ports			−800	μA
Low-level output current, I_{OL}	Bus ports			48	mA
	Terminal ports			16	
Operating free-air temperature range, T_A		0		70	°C

4

Line Drivers/Receivers

electrical characteristics over recommended ranges of supply voltage and operating free-air temperature (unless otherwise noted)

PARAMETER			TEST CONDITIONS		MIN	TYP†	MAX	UNIT
V_{IK}	Input clamp voltage		$I_I = -18$ mA			−0.8	−1.5	V
V_{hys}	Hysteresis ($V_{T+} - V_{T-}$)‡	Bus			0.4	0.65		V
V_{OH}	High level output voltage	Terminal	$I_{OH} = -800$ μA,	TE at 0.8 V	2.7	3.5		V
		Bus	$I_{OH} = -10$ mA,	PE and TE at 2 V	2.5	3.3		
V_{OL}	Low-level output voltage	Terminal	$I_{OL} = 16$ mA,	TE at 0.8 V		0.3	0.5	V
		Bus	$I_{OL} = 48$ mA,	PE and TE at 2 V		0.4	0.5	
I_{OH}	High-level output current (open-collector mode)	Bus	$V_O = 5.5$ V, D and TE at 2 V	PE at 0.8 V,			100	μA
I_{OZ}	Off-state output current (3-state mode)	Bus	PE at 2 V, TE at 0.8 V	$V_O = 2.7$ V			20	μA
				$V_O = 0.4$ V			−20	
I_I	Input current at maximum input voltage	Terminal	$V_I = 5.5$ V			0.2	100	μA
I_{IH}	High-level input current	Terminal	$V_I = 2.7$ V			0.1	20	μA
I_{IL}	Low-level input current	Terminal	$V_I = 0.5$ V			−10	−100	μA
I_{OS}	Short-circuit output current	Terminal			−15	−35	−75	mA
		Bus			−25	−50	−125	
I_{CC}	Supply current		No load	Receivers low and enabled			80	mA
				Drivers low and enabled			100	
$C_{i/o(bus)}$	Bus-port capacitance		$V_{CC} = 5$ V or 0 V, $V_{I/O} = 0$ to 2 V, f = 1 MHz			30		pF

†All typical values are at $V_{CC} = 5$, $T_A = 25°C$.
‡Hysteresis is the difference between the positive-going input threshold voltage, V_{T+}, and the negative-going input threshold voltage, V_{T-}.

switching characteristics, $V_{CC} = 5$ V, $C_L = 15$ pF, $T_A = 25°C$ (unless otherwise noted)

PARAMETER		FROM	TO	TEST CONDITIONS	MIN	TYP	MAX	UNIT
t_{PLH}	Propagation delay time, low-to-high-level output	Terminal	Bus	$C_L = 30$ pF, See Figure 1		14	20	ns
t_{PHL}	Propagation delay time, high-to-low-level output					14	20	
t_{PLH}	Propagation delay time, low-to-high-level output	Bus	Terminal	$C_L = 30$ pF, See Figure 2		10	20	ns
t_{PHL}	Propagation delay time, high-to-low-level output					15	22	
t_{PZH}	Output enable time to high level	TE	Bus	See Figure 3		25	35	ns
t_{PHZ}	Output disable time from high level					13	22	
t_{PZL}	Output enable time to low level					22	35	
t_{PLZ}	Output disable time from low level					22	32	
t_{PZH}	Output enable time to high level	TE	Terminal	See Figure 4		20	30	ns
t_{PHZ}	Output disable time from high level					12	20	
t_{PZL}	Output enable time to low level					23	32	
t_{PLZ}	Output disable time from low level					19	30	
t_{en}	Output pull-up enable time	PE	Terminal	See Figure 5		15	22	ns
t_{dis}	Output pull-up disable time					13	20	

TEXAS INSTRUMENTS
POST OFFICE BOX 655012 • DALLAS, TEXAS 75265

PARAMETER MEASUREMENT INFORMATION

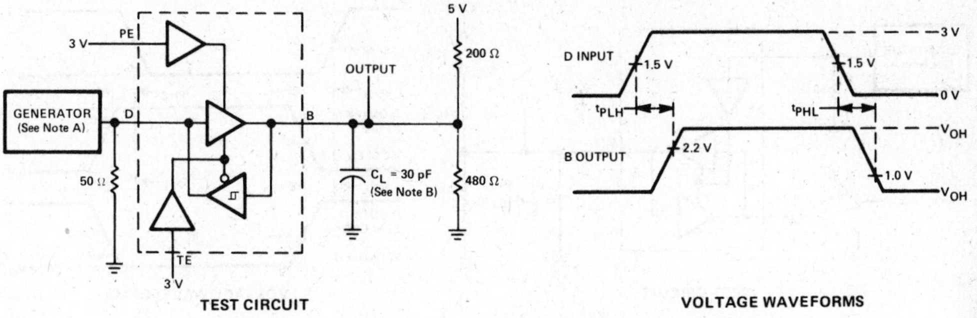
3 V
PE
GENERATOR (See Note A)
D
50 Ω
TE
3 V
B
OUTPUT
5 V
200 Ω
C_L = 30 pF (See Note B)
480 Ω
D INPUT
1.5 V
1.5 V
3 V
0 V
t_{PLH}
t_{PHL}
B OUTPUT
2.2 V
1.0 V
V_{OH}
V_{OH}
TEST CIRCUIT
VOLTAGE WAVEFORMS

FIGURE 1. TERMINAL-TO-BUS PROPAGATION DELAY TIMES

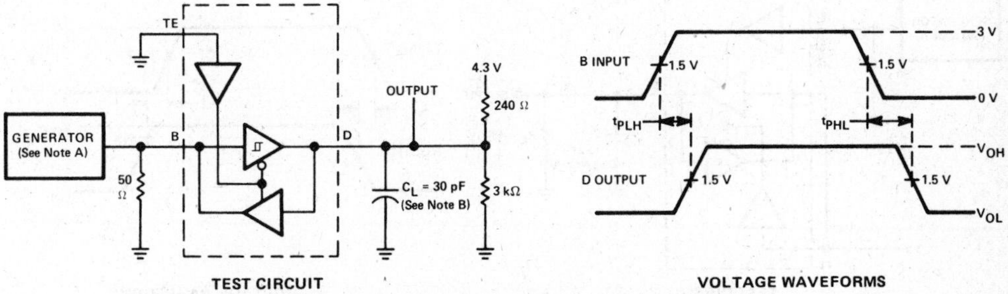
TE
GENERATOR (See Note A)
B
50 Ω
D
OUTPUT
4.3 V
240 Ω
C_L = 30 pF (See Note B)
3 kΩ
B INPUT
1.5 V
1.5 V
3 V
0 V
t_{PLH}
t_{PHL}
D OUTPUT
1.5 V
1.5 V
V_{OH}
V_{OL}
TEST CIRCUIT
VOLTAGE WAVEFORMS

FIGURE 2. BUS-TO-TERMINAL PROPAGATION DELAY TIMES

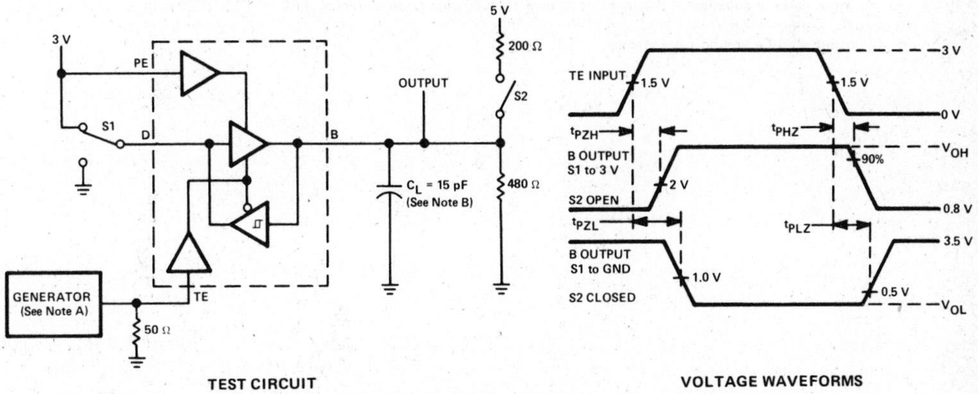
3 V
PE
S1
D
B
OUTPUT
5 V
200 Ω
S2
C_L = 15 pF (See Note B)
480 Ω
GENERATOR (See Note A)
TE
50 Ω
TE INPUT
1.5 V
1.5 V
3 V
0 V
t_{PZH}
t_{PHZ}
B OUTPUT S1 to 3 V
S2 OPEN
90%
2 V
V_{OH}
0.8 V
t_{PZL}
t_{PLZ}
3.5 V
B OUTPUT S1 to GND
S2 CLOSED
1.0 V
0.5 V
V_{OL}
TEST CIRCUIT
VOLTAGE WAVEFORMS

FIGURE 3. TE-TO-BUS ENABLE AND DISABLE TIMES

NOTES: A. The input pulse is supplied by a generator having the following characteristics: PRR ≤ 1 MHz, 50% duty cycle, t_r ≤6 ns, t_f ≤ns, Z_{out} = 50 Ω.

B. C_L includes probe and jig capacitance.

PARAMETER MEASUREMENT INFORMATION

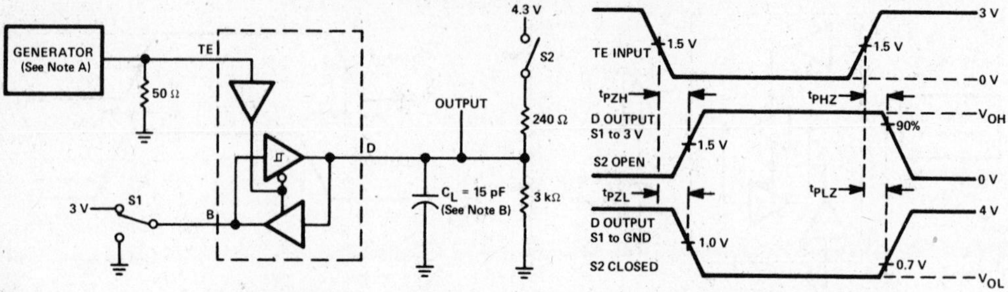
GENERATOR
(See Note A)
TE
50 Ω
3 V
S1
B
D
OUTPUT
4.3 V
S2
240 Ω
C_L = 15 pF
(See Note B)
3 kΩ
TE INPUT
1.5 V
1.5 V
3 V
0 V
t_{PZH}
t_{PHZ}
D OUTPUT
S1 to 3 V
S2 OPEN
1.5 V
90%
V_{OH}
0 V
t_{PZL}
t_{PLZ}
D OUTPUT
S1 to GND
S2 CLOSED
1.0 V
0.7 V
4 V
V_{OL}

TEST CIRCUIT

VOLTAGE WAVEFORMS

FIGURE 4. TE-TO-TERMINAL ENABLE AND DISABLE TIMES

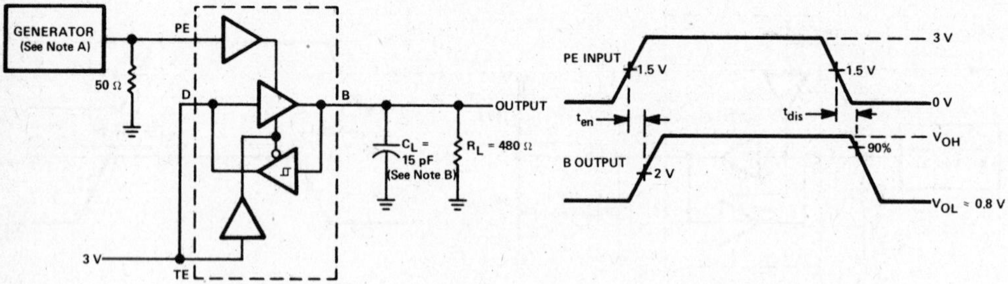
GENERATOR
(See Note A)
PE
50 Ω
D
B
OUTPUT
C_L =
15 pF
(See Note B)
R_L = 480 Ω
3 V
TE
PE INPUT
1.5 V
1.5 V
3 V
0 V
t_{en}
t_{dis}
B OUTPUT
2 V
90%
V_{OH}
V_{OL} ≈ 0.8 V

TEST CIRCUIT

VOLTAGE WAVEFORMS

FIGURE 5. PE-TO-BUS PULLUP ENABLE AND DISABLE TIMES

NOTES: A. The input pulse is supplied by a generator having the following characteristics: PRR ≤ 1 MHz, 50% duty cycle, t_r ≤6 ns, t_f ≤ns, Z_{out} = 50 Ω.

B. C_L includes probe and jig capacitance.

TEXAS INSTRUMENTS
POST OFFICE BOX 655012 • DALLAS, TEXAS 75265

TYPICAL CHARACTERISTICS

TERMINAL HIGH-LEVEL OUTPUT VOLTAGE
vs
HIGH-LEVEL OUTPUT CURRENT

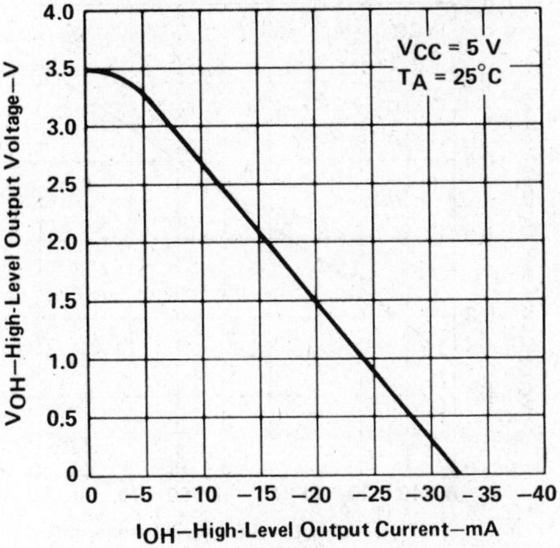
VCC = 5 V
TA = 25°C
VOH–High-Level Output Voltage–V
4.0
3.5
3.0
2.5
2.0
1.5
1.0
0.5
0
0 −5 −10 −15 −20 −25 −30 −35 −40
IOH–High-Level Output Current–mA

FIGURE 6

TERMINAL LOW-LEVEL OUTPUT VOLTAGE
vs
LOW-LEVEL OUTPUT CURRENT

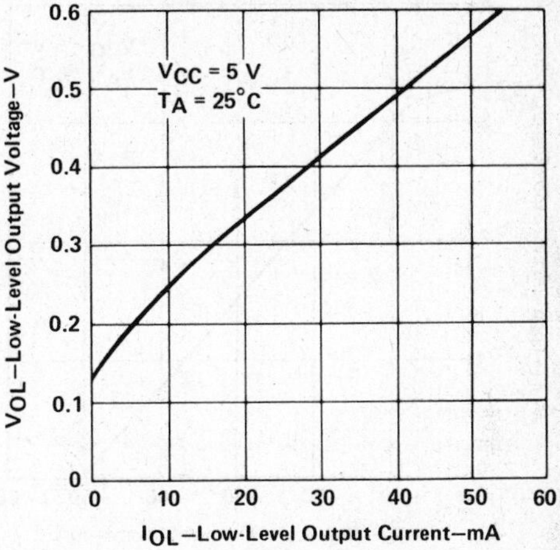
VCC = 5 V
TA = 25°C
VOL–Low-Level Output Voltage–V
0.6
0.5
0.4
0.3
0.2
0.1
0
0 10 20 30 40 50 60
IOL–Low-Level Output Current–mA

FIGURE 7

TERMINAL OUTPUT VOLTAGE
vs
BUS INPUT VOLTAGE

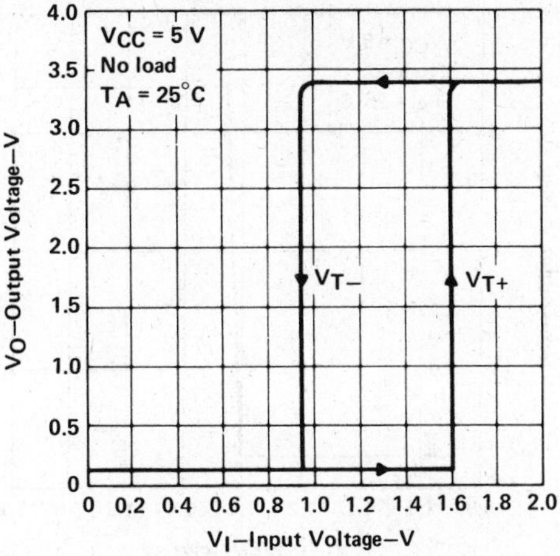
VCC = 5 V
No load
TA = 25°C
VT−
VT+
VO–Output Voltage–V
4.0
3.5
3.0
2.5
2.0
1.5
1.0
0.5
0
0 0.2 0.4 0.6 0.8 1.0 1.2 1.4 1.6 1.8 2.0
VI–Input Voltage–V

FIGURE 8

4

Line Drivers/Receivers

TYPICAL CHARACTERISTICS

BUS HIGH-LEVEL OUTPUT VOLTAGE
vs
BUS HIGH-LEVEL OUTPUT CURRENT

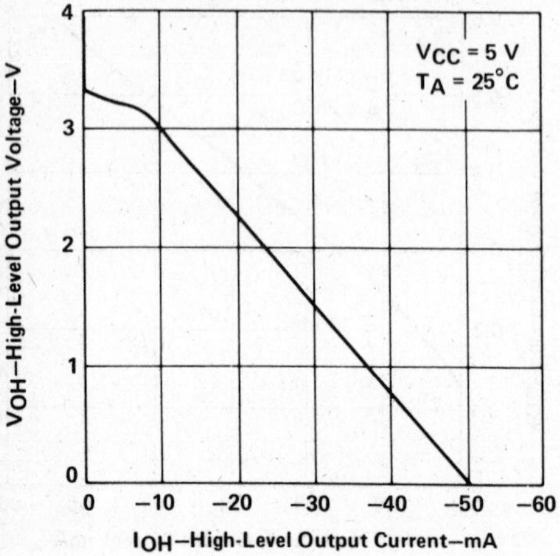
V_{CC} = 5 V
T_A = 25°C
V_{OH}—High-Level Output Voltage—V
I_{OH}—High-Level Output Current—mA
4
3
2
1
0
0 −10 −20 −30 −40 −50 −60

FIGURE 9

BUS LOW-LEVEL OUTPUT VOLTAGE
vs
BUS LOW-LEVEL OUTPUT CURRENT

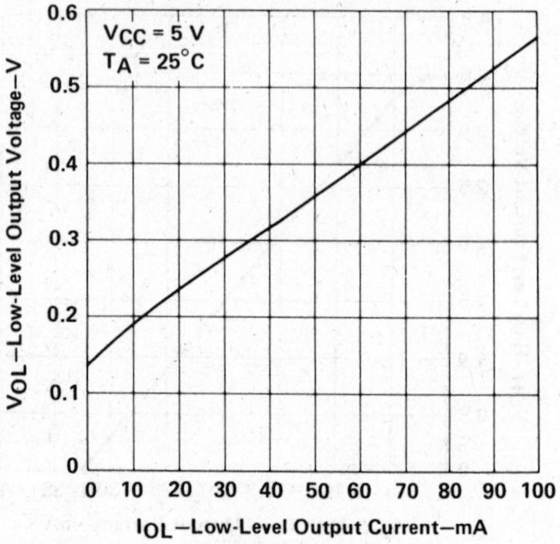
V_{CC} = 5 V
T_A = 25°C
V_{OL}—Low-Level Output Voltage—V
I_{OL}—Low-Level Output Current—mA
0.6
0.5
0.4
0.3
0.2
0.1
0
0 10 20 30 40 50 60 70 80 90 100

FIGURE 10

BUS OUTPUT VOLTAGE
vs
TERMINAL INPUT VOLTAGE

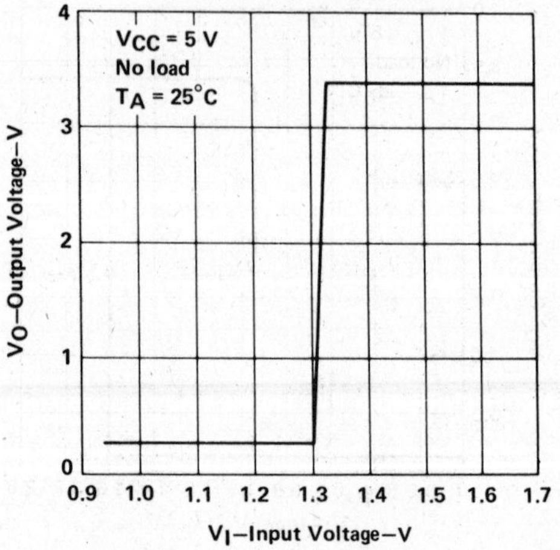
V_{CC} = 5 V
No load
T_A = 25°C
V_O—Output Voltage—V
V_I—Input Voltage—V
4
3
2
1
0
0.9 1.0 1.1 1.2 1.3 1.4 1.5 1.6 1.7

FIGURE 11

TEXAS INSTRUMENTS
POST OFFICE BOX 655012 • DALLAS, TEXAS 75265

D2908, OCTOBER 1985

- 8-Channel Bidirectional Transceiver
- Power-Up/Power-Down Protection (Glitch-Free)
- ATN + EOI (OR Function) Output to Simplify Board Layout
- Designed to Implement Control Bus Interface for Multi-Controllers
- Low-Power Dissipation . . . 72 mW Max Per Channel
- Fast Propagation Times . . . 22 ns Max
- High-Impedance P-N-P Inputs
- Receiver Hysteresis . . . 650 mV Typ
- Bus-Terminating Resistors Provided on Driver Outputs
- No Loading of Bus When Device is Powered Down (V_{CC} = 0)

DW SMALL OUTLINE PACKAGE
(TOP VIEW)

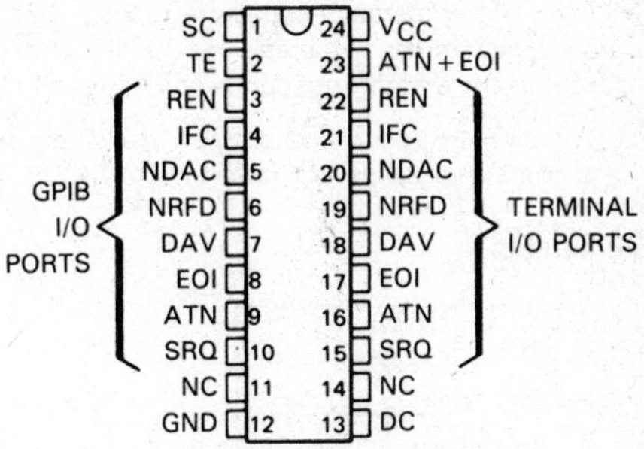
SC 1
TE 2
REN 3
IFC 4
NDAC 5
NRFD 6
DAV 7
EOI 8
ATN 9
SRQ 10
NC 11
GND 12
24 VCC
23 ATN + EOI
22 REN
21 IFC
20 NDAC
19 NRFD
18 DAV
17 EOI
16 ATN
15 SRQ
14 NC
13 DC
GPIB I/O PORTS
TERMINAL I/O PORTS

N DUAL-IN-LINE PACKAGE
(TOP VIEW)

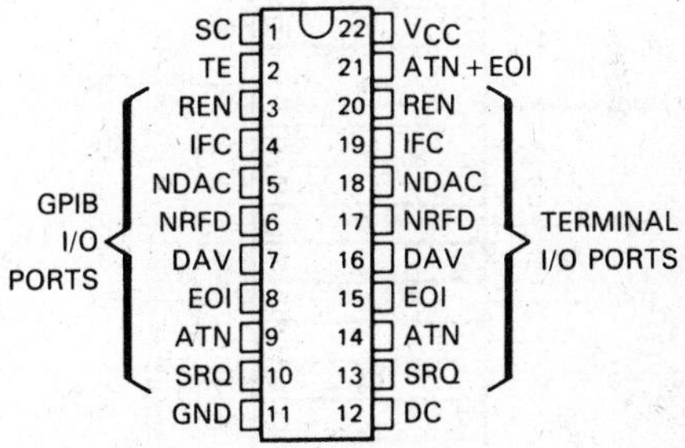
SC 1
TE 2
REN 3
IFC 4
NDAC 5
NRFD 6
DAV 7
EOI 8
ATN 9
SRQ 10
GND 11
22 VCC
21 ATN + EOI
20 REN
19 IFC
18 NDAC
17 NRFD
16 DAV
15 EOI
14 ATN
13 SRQ
12 DC
GPIB I/O PORTS
TERMINAL I/O PORTS

NC—No internal connection.

description

The SN75164B eight-channel general-purpose interface bus transceiver is a monolithic, high-speed, low-power Schottky device designed to meet the requirements of IEEE Standard 488-1978. Each transceiver is designed to provide the bus-management and data-transfer signals between operating units of a multiple-controller instrumentation system. When combined with the SN75160B octal bus transceiver, the SN75164B provides the complete 16-wire interface for the IEEE 488 bus.

The SN75164B features eight driver-receiver pairs connected in a front-to-back configuration to form input/output (I/O) ports at both the bus and terminal sides. All outputs are disabled (at a high-impedance state) during V_{CC} power-up and power-down transitions for glitch-free operation. The direction of data flow through these driver-receiver pairs is determined by the DC, TE, and SC enable signals. The SN75164B is identical to the SN75162B with the addition of an OR gate to help simplify board layouts in several popular applications. The ATN and EOI signals are ORed to pin 21, which is a standard totem-pole output.

CHANNEL IDENTIFICATION TABLE

NAME	IDENTITY	CLASS
DC TE SC	Direction Control Talk Enable System Control	Control
ATN SRQ REN IFC EOI	Attention Service Request Remote Enable Interface Clear End or Identify	Bus Management
ATN + EOI	ATN logical OR EOI	Logic
DAV NDAC NRFD	Data Valid Not Data Accepted Not Ready for Data	Data Transfer

The driver outputs (GPIB I/O ports) feature active bus-terminating resistor circuits designed to provide a high impedance to the bus when supply voltage V_{CC} is 0. The drivers are designed to handle loads up to 48 milliamperes of sink current. Each receiver features p-n-p transistor inputs for high input impedance and a guaranteed hysteresis of 400 millivolts for increased noise immunity. All receivers have 3-state outputs to present a high impedance to the terminal when disabled.

The SN75164B is manufactured in a 22-pin dual-in-line and 24-pin Small Outline package. The SN75164B is characterized for operation from 0°C to 70°C.

logic symbol†

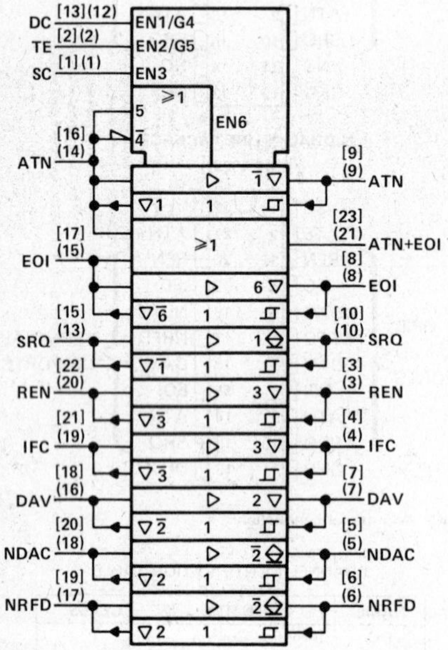
DC [13](12)
TE [2](2)
SC [1](1)
EN1/G4
EN2/G5
EN3
≥1
EN6
5
4
ATN [16](14)
[9](9) ATN
[23](21) ATN+EOI
EOI [17](15)
[8](8) EOI
SRQ [15](13)
[10](10) SRQ
REN [22](20)
[3](3) REN
IFC [21](19)
[4](4) IFC
DAV [18](16)
[7](7) DAV
NDAC [20](18)
[5](5) NDAC
NRFD [19](17)
[6](6) NRFD

†This symbol is in accordance with ANSI/IEEE Std 91-1984 and IEC Publication 617-12.

logic diagram (positive logic)

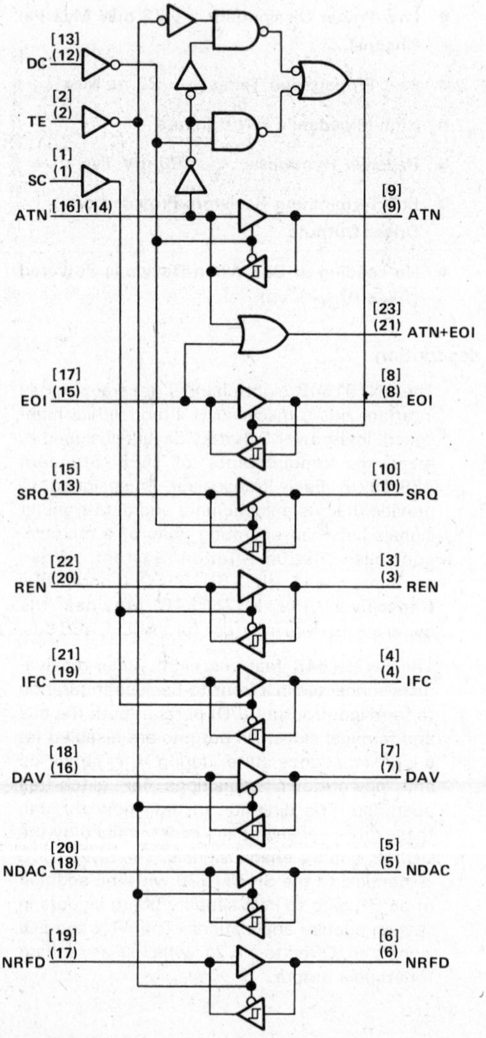
DC [13](12)
TE [2](2)
SC [1](1)
ATN [16](14)
[9](9) ATN
[23](21) ATN+EOI
EOI [17](15)
[8](8) EOI
SRQ [15](13)
[10](10) SRQ
REN [22](20)
[3](3) REN
IFC [21](19)
[4](4) IFC
DAV [18](16)
[7](7) DAV
NDAC [20](18)
[5](5) NDAC
NRFD [19](17)
[6](6) NRFD

[ ] Denotes pin numbers for DW package.
( ) Denotes pin numbers for N package.

TEXAS INSTRUMENTS
POST OFFICE BOX 655012 • DALLAS, TEXAS 75265

RECEIVE/TRANSMIT FUNCTION TABLE

CONTROLS				BUS-MANAGEMENT CHANNELS					DATA-TRANSFER CHANNELS		
SC	DC	TE	ATN†	ATN† (Controlled by DC)	SRQ (Controlled by DC)	REN (Controlled by SC)	IFC (Controlled by SC)	EOI	DAV (Controlled by TE)	NDAC (Controlled by TE)	NRFD (Controlled by TE)
	H	H	H	R	T			T	T	R	R
	H	H	L					R			
	L	L	H	T	R			R	R	T	T
	L	L	L					T			
	H	L	X	R	T			R	R	T	T
	L	H	X	T	R			T	T	R	R
H						T	T				
L						R	R				

H = high level, L = low level, R = receive, T = transmit, X = irrelevant
Direction of data transmission is from the terminal side to the bus side, and the direction of data receiving is from the bus side to the terminal side. Data transfer is noninverting in both directions.

† ATN is a normal transceiver channel that functions additionally as an internal direction control or talk enable for EOI whenever the DC and TE inputs are in the same state. When DC and TE are in opposite states, the ATN channel functions as an independent transceiver only.

ATN + EOI FUNCTION TABLE

INPUTS		OUTPUT
ATN	EOI	ATN + EOI
H	X	H
X	H	H
L	L	L

schematics of inputs and outputs

EQUIVALENT OF ALL CONTROL INPUTS

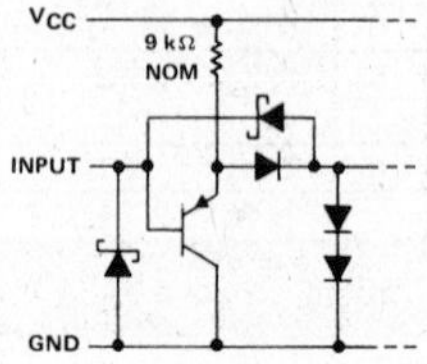
VCC
9 kΩ NOM
INPUT
GND

TYPICAL OF SRQ, NDAC, and NRFD GPIB I/O PORT

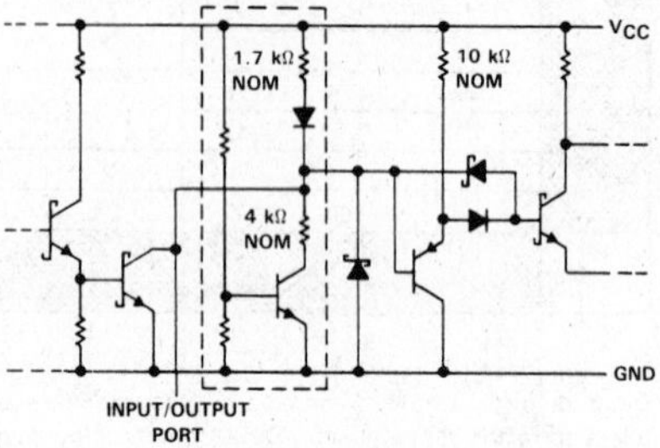
VCC
1.7 kΩ NOM
10 kΩ NOM
4 kΩ NOM
GND
INPUT/OUTPUT PORT

Circuit inside dashed lines is on GPIB I/O ports only.

TYPICAL OF ALL I/O PORTS EXCEPT SRQ, NDAC, and NRFD GPIB I/O PORTS

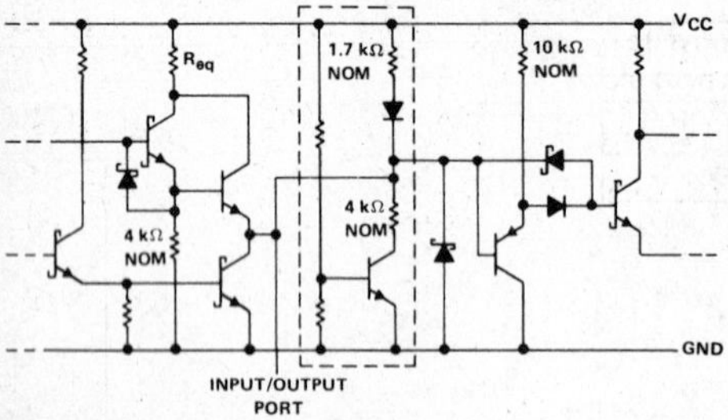
VCC
Req
1.7 kΩ NOM
10 kΩ NOM
4 kΩ NOM
4 kΩ NOM
GND
INPUT/OUTPUT PORT

Driver output R_{eq} = 30 Ω NOM
Receiver output R_{eq} = 110 Ω NOM
Circuit inside dashed lines is on GPIB I/O ports only.

ATN+EOI OUTPUT

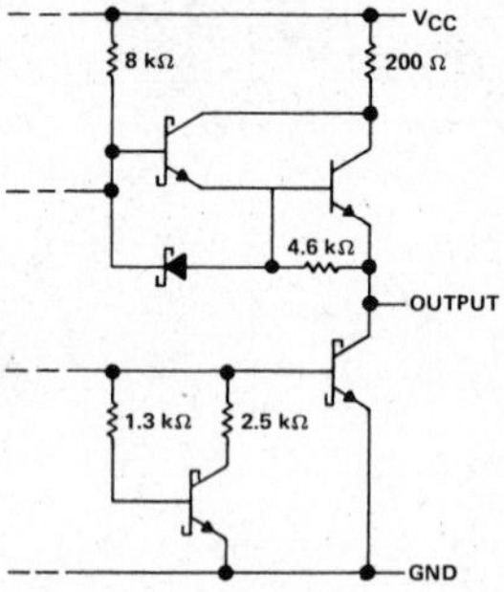
VCC
8 kΩ
200 Ω
4.6 kΩ
OUTPUT
1.3 kΩ
2.5 kΩ
GND

absolute maximum ratings over operating free-air temperature range (unless otherwise noted)

Supply voltage, V_{CC} (see Note 1) . . . . . 7 V
Input voltage . . . . . 5.5 V
Low-level driver output current . . . . . 100 mA
Continuous total dissipation at (or below) 25°C free-air temperature (see Note 2):
  DW package . . . . . 1350 mW
  N package . . . . . 1700 mW
Operating free-air temperature range . . . . . 0°C to 70°C
Storage temperature range . . . . . −65°C to 150°C
Lead temperature 1,6 mm (1/16) inch from the case for 10 seconds: DW or N package . . . . 260°C

NOTES: 1. All voltage values are with respect to network ground terminal.
2. For operation above 25°C free-air temperature, derate the DW package at the rate of 10.8 mW/°C, the N package at the rate of 13.6 mW/°C.

4 Line Drivers/Receivers

TEXAS INSTRUMENTS
POST OFFICE BOX 655012 • DALLAS, TEXAS 75265

recommended operating conditions

		MIN	NOM	MAX	UNIT
Supply voltage, V_{CC1}		4.75	5	5.25	V
High-level input voltage, V_{IH}		2			V
Low-level input voltage, V_{IL}				0.8	V
High-level output current, I_{OH}	Bus ports with 3-state outputs			−5.2	mA
High-level output current, I_{OH}	Terminal ports			−800	μA
High-level output current, I_{OH}	ATN+EOI			−400	μA
Low-level output current, I_{OL}	Bus ports			48	mA
Low-level output current, I_{OL}	Terminal ports			16	mA
Low-level output current, I_{OL}	ATN+EOI			4	mA
Operating free-air temperature, T_A		0		70	°C

electrical characteristics over recommended ranges of supply voltage and operating free-air temperature (unless otherwise noted)

PARAMETER			TEST CONDITIONS		MIN	TYP†	MAX	UNIT
V_{IK}	Input clamp voltage		I_I = −18 mA				−1.5	V
V_{hys}	Hysteresis ($V_{T+} - V_{T-}$)	Bus			0.4			V
V_{OH}‡	High-level output voltage	Terminal	I_{OH} = −800 μA		2.7			V
V_{OH}‡	High-level output voltage	Bus	I_{OH} = −5.2 mA		2.5			V
V_{OH}‡	High-level output voltage	ATN+EOI	I_{OH} = −400 μA		2.7			V
V_{OL}	Low-level output voltage	Terminal	I_{OL} = 16 mA				0.5	V
V_{OL}	Low-level output voltage	Bus	I_{OL} = 48 mA				0.5	V
V_{OL}	Low-level output voltage	ATN+EOI	I_{OL} = 4 mA				0.4	V
I_I	Input current at maximum input voltage	Terminal§	V_I = 5.5 V				100	μA
I_I	Input current at maximum input voltage	ATN, EOI	V_I = 5.5 V				200	μA
I_{IH}	High-level input current	Terminal, control	V_I = 2.7 V				20	μA
I_{IH}	High-level input current	ATN, EOI	V_I = 2.7 V				40	μA
I_{IL}	Low-level input current	Terminal, control	V_I = 0.5 V				−100	μA
I_{IL}	Low-level input current	ATN, EOI	V_I = 0.5 V				−500	μA
$V_{I/O(bus)}$	Voltage at bus port		Driver disabled	$I_{I(bus)}$ = 0	2.5		3.7	V
$V_{I/O(bus)}$	Voltage at bus port		Driver disabled	$I_{I(bus)}$ = −12 mA			−1.5	V
$I_{I/O(bus)}$	Current into bus port	Power on	Driver disabled	$V_{I(bus)}$ = −1.5 V to 0.4 V	−1.3			mA
$I_{I/O(bus)}$	Current into bus port	Power on	Driver disabled	$V_{I(bus)}$ = 0.4 V to 2.5 V	0		−3.2	mA
$I_{I/O(bus)}$	Current into bus port	Power on	Driver disabled	$V_{I(bus)}$ = 2.5 V to 3.7 V			+2.5 −3.2	mA
$I_{I/O(bus)}$	Current into bus port	Power on	Driver disabled	$V_{I(bus)}$ = 3.7 V to 5 V	0		2.5	mA
$I_{I/O(bus)}$	Current into bus port	Power on	Driver disabled	$V_{I(bus)}$ = 5 V to 5.5 V	0.7		2.5	mA
$I_{I/O(bus)}$	Current into bus port	Power off	V_{CC} = 0,	$V_{I(bus)}$ = 0 V to 2.5 V			−40	μA
I_{OS}	Short-circuit output current	Terminal			−15		−75	mA
I_{OS}	Short-circuit output current	Bus			−25		−125	mA
I_{OS}	Short-circuit output current	ATN+EOI			−10		−100	mA
I_{CC}	Supply current		No load,	TE, DC, and SC low			120	mA
$C_{i/o(bus)}$	Bus-port capacitance		V_{CC} = 5 V to 0 V, $V_{I/O}$ = 0 to 2 V, f = 1 MHz			30		pF

† All typical values are at V_{CC} = 5 V, T_A = 25 °C.
‡ V_{OH} applies for three-state outputs only.
§ Except ATN and EOI terminal pins.

switching characteristics, V_{CC} = 5 V, C_L = 15 pF, T_A = 25°C (unless otherwise noted)

PARAMETER		FROM	TO	TEST CONDITIONS	MIN	TYP	MAX	UNIT
t_{PLH}	Propagation delay time, low-to-high-level output	Terminal	Bus	C_L = 30 pF, See Figure 1		14	20	ns
t_{PHL}	Propagation delay time, high-to-low-level output	Terminal	Bus	C_L = 30 pF, See Figure 1		14	20	ns
t_{PLH}	Propagation delay time, low-to-high-level output	Terminal	Bus (SRQ, NDAC NRFD)	C_L = 30 pF, See Figure 1		29	35	ns
t_{PLH}	Propagation delay time low-to-high-level output	Bus	Terminal	C_L = 30 pF, See Figure 2		10	20	ns
t_{PHL}	Propagation delay time, high-to-low-level output	Bus	Terminal	C_L = 30 pF, See Figure 2		15	22	ns
t_{PLH}	Propagation delay time, low-to-high-level output	Terminal ATN or Terminal EOI	ATN + EOI	See Figure 3		14		ns
t_{PHL}	Propagation delay time, high-to-low-level output	Terminal ATN or Terminal EOI	ATN + EOI	See Figure 3		14		ns
t_{PZH}	Output enable time to high level	TE, DC, or SC	BUS (ATTN, EOI, REN, IFC, and DAV)	See Figure 4			60	ns
t_{PHZ}	Output disable time from high level	TE, DC, or SC	BUS (ATTN, EOI, REN, IFC, and DAV)	See Figure 4			45	ns
t_{PZL}	Output enable time to low level	TE, DC, or SC	BUS (ATTN, EOI, REN, IFC, and DAV)	See Figure 4			60	ns
t_{PLZ}	Output disable time from low level	TE, DC, or SC	BUS (ATTN, EOI, REN, IFC, and DAV)	See Figure 4			55	ns
t_{PZH}	Output enable time to high level	TE, DC, or SC	Terminal	See Figure 5			55	ns
t_{PHZ}	Output disable time from high level	TE, DC, or SC	Terminal	See Figure 5			50	ns
t_{PZL}	Output enable time to low level	TE, DC, or SC	Terminal	See Figure 5			45	ns
t_{PLZ}	Output disable time from low level	TE, DC, or SC	Terminal	See Figure 5			55	ns

TEXAS INSTRUMENTS
POST OFFICE BOX 655012 • DALLAS, TEXAS 75265

PARAMETER MEASUREMENT INFORMATION

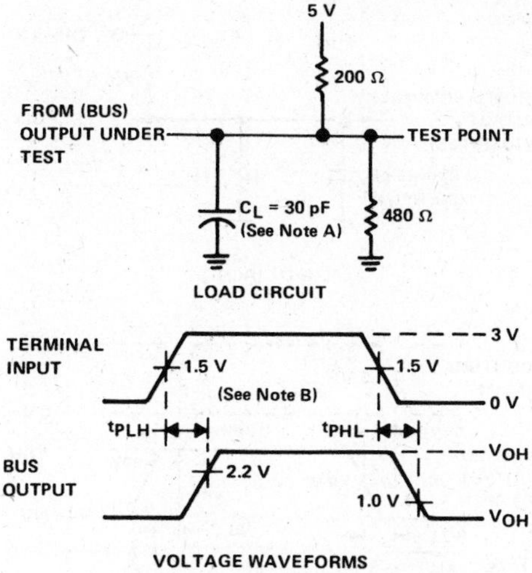
5 V
200 Ω
FROM (BUS)
OUTPUT UNDER
TEST
TEST POINT
C_L = 30 pF
(See Note A)
480 Ω
LOAD CIRCUIT
TERMINAL
INPUT
1.5 V
1.5 V
(See Note B)
3 V
0 V
t_{PLH}
t_{PHL}
BUS
OUTPUT
2.2 V
1.0 V
V_{OH}
V_{OH}
VOLTAGE WAVEFORMS

FIGURE 1. TERMINAL-TO-BUS PROPAGATION DELAY TIMES

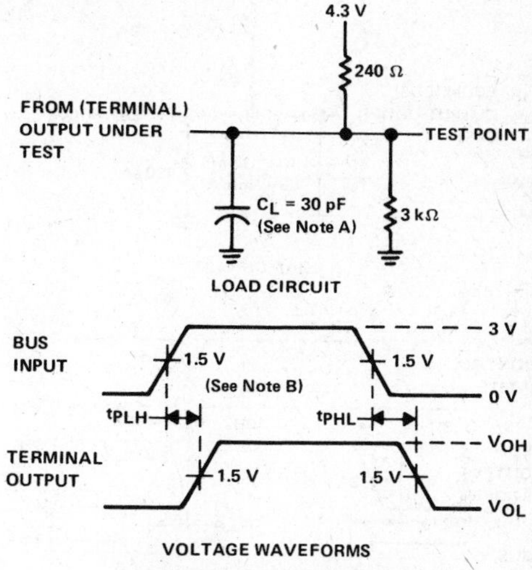
4.3 V
240 Ω
FROM (TERMINAL)
OUTPUT UNDER
TEST
TEST POINT
C_L = 30 pF
(See Note A)
3 kΩ
LOAD CIRCUIT
BUS
INPUT
1.5 V
1.5 V
(See Note B)
3 V
0 V
t_{PLH}
t_{PHL}
TERMINAL
OUTPUT
1.5 V
1.5 V
V_{OH}
V_{OL}
VOLTAGE WAVEFORMS

FIGURE 2. BUS-TO-TERMINAL PROPAGATION DELAY TIMES

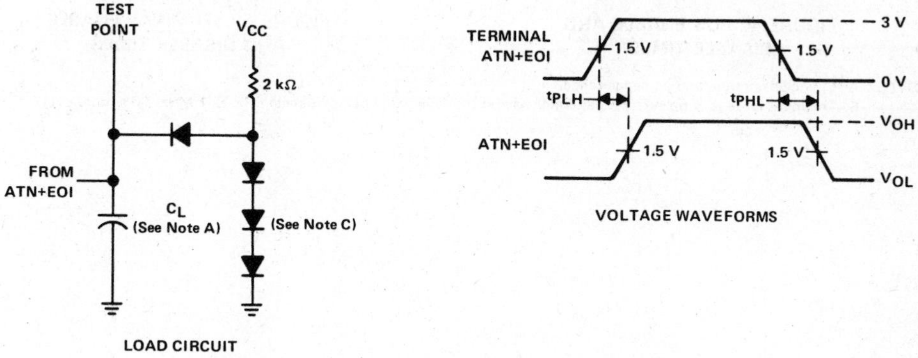
TEST
POINT
V_{CC}
2 kΩ
FROM
ATN+EOI
C_L
(See Note A)
(See Note C)
LOAD CIRCUIT
TERMINAL
ATN+EOI
1.5 V
1.5 V
3 V
0 V
t_{PLH}
t_{PHL}
ATN+EOI
1.5 V
1.5 V
V_{OH}
V_{OL}
VOLTAGE WAVEFORMS

FIGURE 3. ATN + EOI PROPAGATION DELAY TIMES

NOTES:
A. C_L includes probe and jig capacitance.
B. The input pulse is supplied by a generator having the following characteristics: PRR ≤ 1 MHz, 50% duty cycle, $t_r \leq 6$ ns, $t_f \leq 6$ ns, Z_{out} = 50 Ω.
C. All diodes are 1N916 or 1N3064.

PARAMETER MEASUREMENT INFORMATION

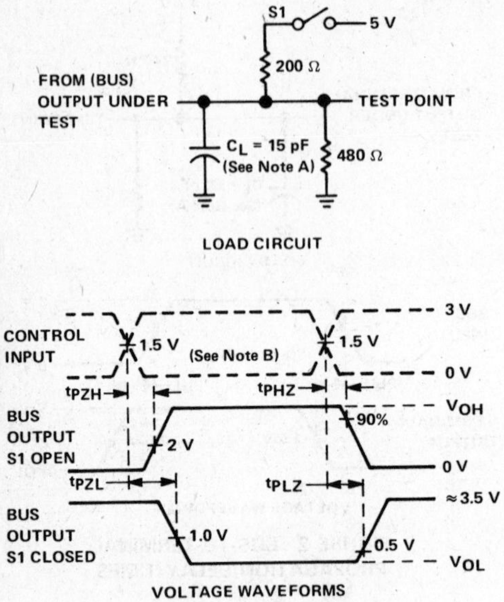
S1
5 V
200 Ω
FROM (BUS)
OUTPUT UNDER
TEST
TEST POINT
C_L = 15 pF
(See Note A)
480 Ω
LOAD CIRCUIT
CONTROL
INPUT
1.5 V
(See Note B)
3 V
0 V
t_{PZH}
t_{PHZ}
V_{OH}
BUS
OUTPUT
S1 OPEN
2 V
90%
t_{PZL}
t_{PLZ}
≈3.5 V
BUS
OUTPUT
S1 CLOSED
1.0 V
0.5 V
V_{OL}
VOLTAGE WAVEFORMS

FIGURE 4. BUS ENABLE AND DISABLE TIMES

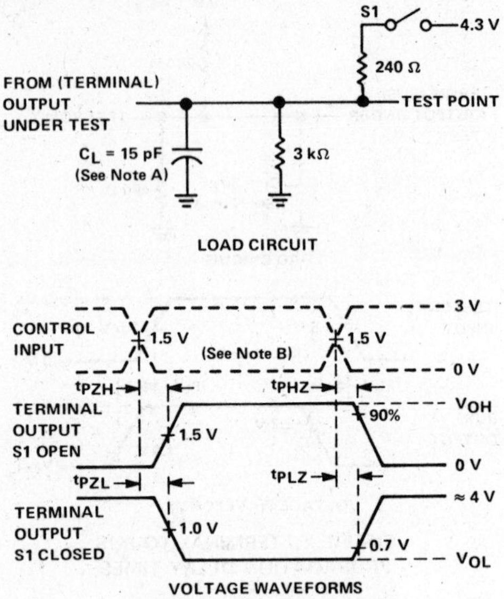
S1
4.3 V
240 Ω
FROM (TERMINAL)
OUTPUT
UNDER TEST
TEST POINT
C_L = 15 pF
(See Note A)
3 kΩ
LOAD CIRCUIT
CONTROL
INPUT
1.5 V
(See Note B)
3 V
0 V
t_{PZH}
t_{PHZ}
V_{OH}
TERMINAL
OUTPUT
S1 OPEN
1.5 V
90%
t_{PZL}
t_{PLZ}
≈ 4 V
TERMINAL
OUTPUT
S1 CLOSED
1.0 V
0.7 V
V_{OL}
VOLTAGE WAVEFORMS

FIGURE 5. TERMINAL ENABLE AND DISABLE TIMES

NOTES: A. C_L includes probe and jig capacitance.
B. The input pulse is supplied by a generator having the following characteristics: PRR ≤ 1 MHz, 50% duty cycle, t_r ≤6 ns, t_f ≤6 ns, Z_{out} = 50 Ω.

TEXAS INSTRUMENTS
POST OFFICE BOX 655012 • DALLAS, TEXAS 75265

TYPICAL CHARACTERISTICS

TERMINAL HIGH-LEVEL OUTPUT VOLTAGE
vs
HIGH-LEVEL OUTPUT CURRENT

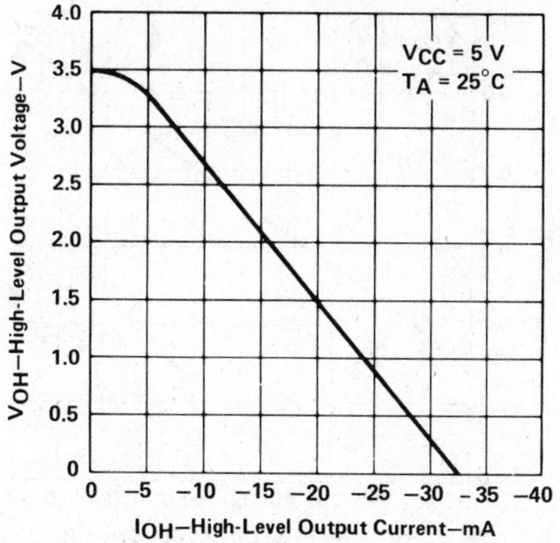
4.0
3.5
3.0
2.5
2.0
1.5
1.0
0.5
0
V_{OH}—High-Level Output Voltage—V
V_{CC} = 5 V
T_A = 25°C
0
−5
−10
−15
−20
−25
−30
−35
−40
I_{OH}—High-Level Output Current—mA

FIGURE 6

TERMINAL LOW-LEVEL OUTPUT VOLTAGE
vs
LOW-LEVEL OUTPUT CURRENT

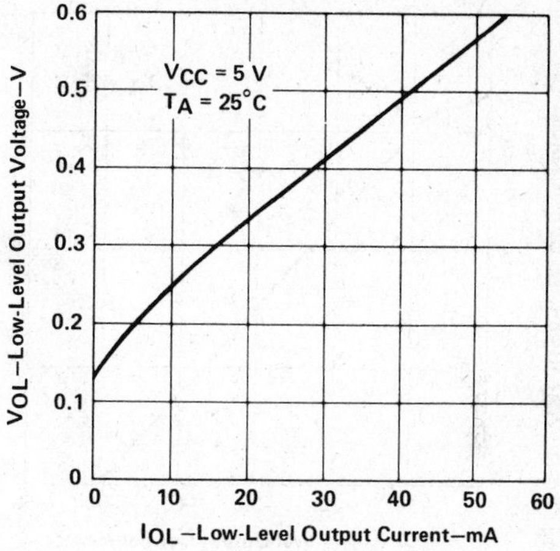
0.6
0.5
0.4
0.3
0.2
0.1
0
V_{OL}—Low-Level Output Voltage—V
V_{CC} = 5 V
T_A = 25°C
0
10
20
30
40
50
60
I_{OL}—Low-Level Output Current—mA

FIGURE 7

TERMINAL OUTPUT VOLTAGE
vs
BUS INPUT VOLTAGE

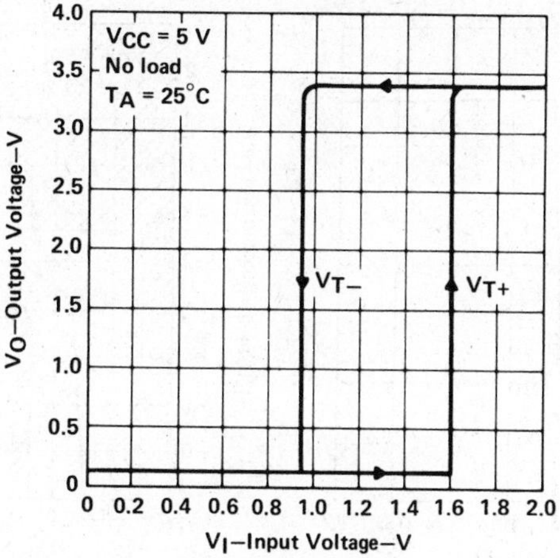
4.0
3.5
3.0
2.5
2.0
1.5
1.0
0.5
0
V_O—Output Voltage—V
V_{CC} = 5 V
No load
T_A = 25°C
V_{T-}
V_{T+}
0
0.2
0.4
0.6
0.8
1.0
1.2
1.4
1.6
1.8
2.0
V_I—Input Voltage—V

FIGURE 8

TYPICAL CHARACTERISTICS

BUS HIGH-LEVEL OUTPUT VOLTAGE
vs
HIGH-LEVEL OUTPUT CURRENT

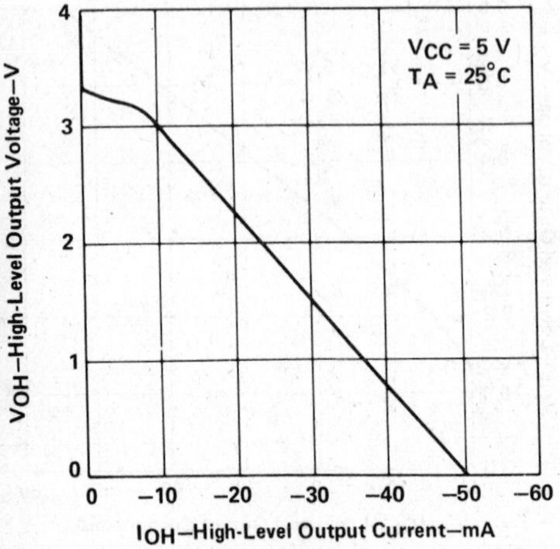
V_{CC} = 5 V
T_A = 25°C
V_{OH}—High-Level Output Voltage—V
4
3
2
1
0
0
−10
−20
−30
−40
−50
−60
I_{OH}—High-Level Output Current—mA

FIGURE 9

BUS LOW-LEVEL OUTPUT VOLTAGE
vs
LOW-LEVEL OUTPUT CURRENT

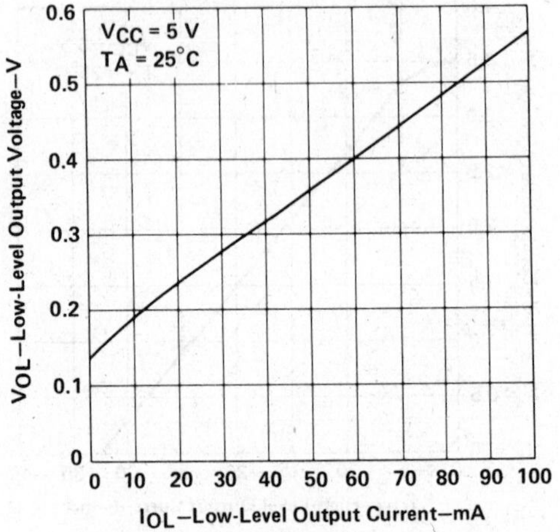
V_{CC} = 5 V
T_A = 25°C
V_{OL}—Low-Level Output Voltage—V
0.6
0.5
0.4
0.3
0.2
0.1
0
0 10 20 30 40 50 60 70 80 90 100
I_{OL}—Low-Level Output Current—mA

FIGURE 10

BUS OUTPUT VOLTAGE
vs
TERMINAL INPUT VOLTAGE

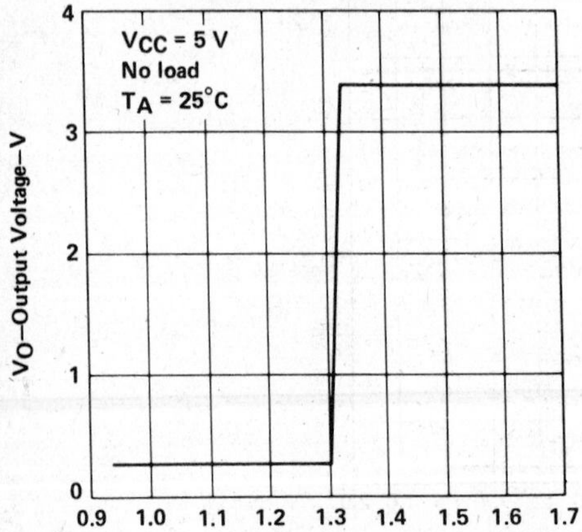
V_{CC} = 5 V
No load
T_A = 25°C
V_O—Output Voltage—V
4
3
2
1
0
0.9 1.0 1.1 1.2 1.3 1.4 1.5 1.6 1.7
V_I—Input Voltage—V

FIGURE 11

BUS CURRENT
vs
BUS VOLTAGE

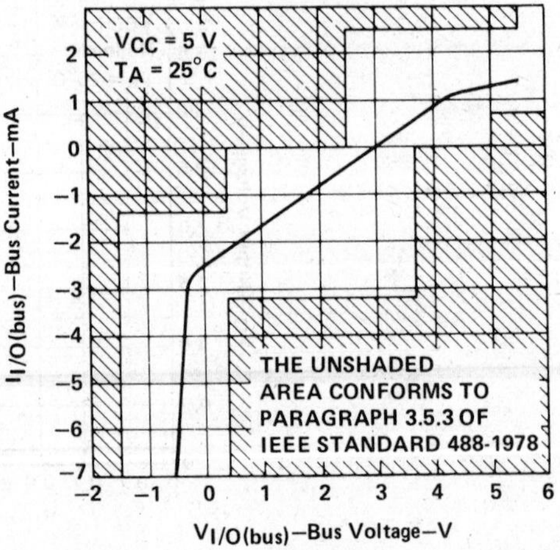
V_{CC} = 5 V
T_A = 25°C
THE UNSHADED AREA CONFORMS TO PARAGRAPH 3.5.3 OF IEEE STANDARD 488-1978
$I_{I/O(bus)}$—Bus Current—mA
2
1
0
−1
−2
−3
−4
−5
−6
−7
−2 −1 0 1 2 3 4 5 6
$V_{I/O(bus)}$—Bus Voltage—V

FIGURE 12

D2596, OCTOBER 1980—REVISED SEPTEMBER 1986

- Meets EIA Standards RS-422-A and RS-485
- Meets CCITT Recommendations V.11 and X.27
- Designed for Multipoint Transmission on Long Bus Lines in Noisy Environments
- 3-State Outputs
- Common-Mode Output Voltage Range of −7 V to 12 V
- Active-High and Active-Low Enables
- Thermal Shutdown Protection
- Positive- and Negative-Current Limiting
- Operates from Single 5-V Supply
- Low Power Requirements
- Functionally Interchangeable with AM26LS31

J OR N DUAL-IN-LINE PACKAGE
(TOP VIEW)

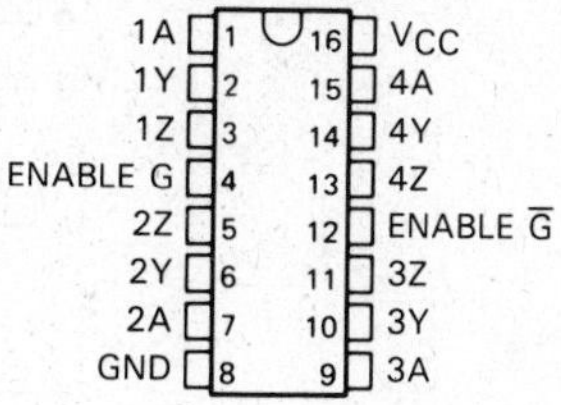
1A 1 16 VCC
1Y 2 15 4A
1Z 3 14 4Y
ENABLE G 4 13 4Z
2Z 5 12 ENABLE G̅
2Y 6 11 3Z
2A 7 10 3Y
GND 8 9 3A

description

The SN75172 is a monolithic quadruple differential line driver with three-state outputs. It is designed to meet the requirements of EIA Standards RS-422-A and RS-485 and CCITT Recommendations V.11 and X.27. The device is optimized for balanced multipoint bus transmission at rates of up to 4 megabaud. Each driver features wide positive and negative common-mode output voltage ranges making it suitable for party-line applications in noisy environments.

The SN75172 provides positive- and negative-current limiting and thermal shutdown for protection from line fault conditions on the transmission bus line. Shutdown occurs at a junction temperature of approximately 150°C. This device offers optimum performance when used with the SN75173 or SN75175 quadruple differential line receivers.

The SN75172 is characterized for operation from 0°C to 70°C.

logic symbol†

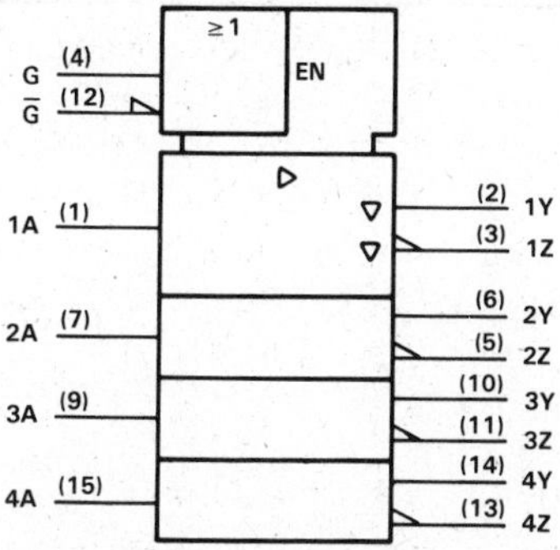
≥1
EN
G (4)
G̅ (12)
1A (1)
(2) 1Y
(3) 1Z
2A (7)
(6) 2Y
(5) 2Z
3A (9)
(10) 3Y
(11) 3Z
4A (15)
(14) 4Y
(13) 4Z

† This symbol is in accordance with ANSI/IEEE Std 91-1984 and IEC Publication 617-12.

FUNCTION TABLE (EACH DRIVER)

INPUT	ENABLES		OUTPUTS	
A	G	G̅	Y	Z
H	H	X	H	L
L	H	X	L	H
H	X	L	H	L
L	X	L	L	H
X	L	H	Z	Z

H = high level, L = low level
X = irrelevant, Z = high impedance (off)

logic diagram (positive logic)

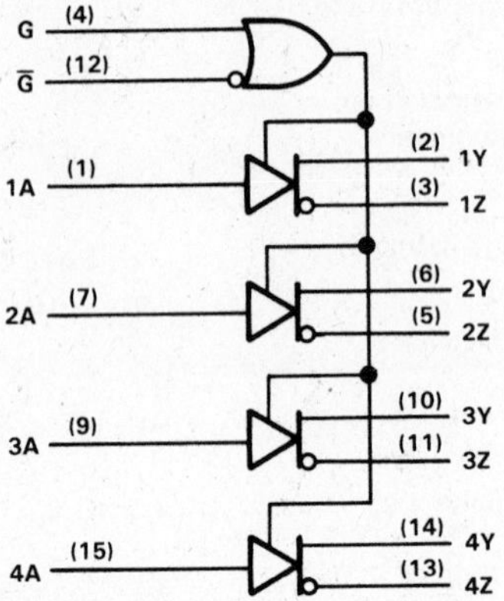
G
(4)
$\overline{G}$
(12)
1A
(1)
(2)
1Y
(3)
1Z
2A
(7)
(6)
2Y
(5)
2Z
3A
(9)
(10)
3Y
(11)
3Z
4A
(15)
(14)
4Y
(13)
4Z

schematics of inputs and outputs

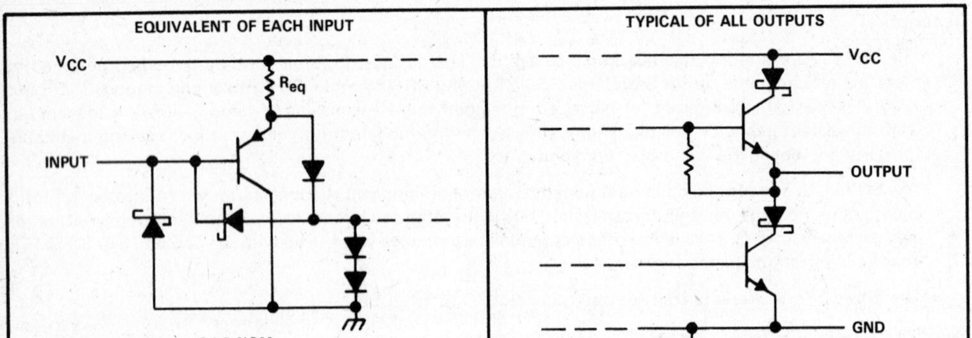
EQUIVALENT OF EACH INPUT
V_{CC}
R_{eq}
INPUT
TYPICAL OF ALL OUTPUTS
V_{CC}
OUTPUT
GND

Data inputs: R_{eq} = 3 kΩ NOM
Enable inputs: R_{eq} = 8 kΩ NOM

TEXAS INSTRUMENTS
POST OFFICE BOX 655012 • DALLAS, TEXAS 75265

absolute maximum ratings over operating free-air temperature (unless otherwise noted)

Supply voltage, V_{CC} . . . . . . 7 V
Input voltage . . . . . . 5.5 V
Continuous total dissipation at (or below)
25°C free-air temperature (see Note 2): J package . . . . . . 1375 mW
N package . . . . . . 1150 mW
Operating free-air temperature . . . . . . 0°C to 70°C
Storage temperature range . . . . . . −65°C to 150°C
Lead temperature 1,6 mm (1/16 inch) from case for 60 seconds: J package . . . . . . 300°C
Lead temperature 1,6 mm (1/16 inch) from case for 10 seconds: N package . . . . . . 260°C

NOTES: 1. All voltage values are with respect to the network ground terminal.
2. For operation above 25°C free-air temperature, derate the J package to 880 mW at 70°C at the rate of 11.0 mW/°C and the N package to 736 mW at 70°C at the rate of 9.2 mW/°C. In the J package, SN75172 chips are alloy mounted.

recommended operating conditions

	MIN	NOM	MAX	UNIT
Supply voltage, V_{CC}	4.75	5	5.25	V
High-level input voltage, V_{IH}	2			V
Low-level input voltage, V_{IL}			0.8	V
Common-mode output voltage, V_{OC}			−7 to 12	V
High-level output current, I_{OH}			−60	mA
Low-level output current, I_{OL}			60	mA
Operating free-air temperature, T_A	0		70	°C

electrical characteristics over recommended ranges of supply voltage and operating free-air temperature (unless otherwise noted)

PARAMETER		TEST CONDITIONS		MIN	TYP†	MAX	UNIT
V_{IK}	Input clamp voltage	$I_I = -18$ mA				−1.5	V
V_O	Output voltage	$I_O = 0$		0		6	V
$\|V_{OD1}\|$	Differential output voltage	$I_O = 0$		1.5		6	V
$\|V_{OD2}\|$	Differential output voltage	$R_L = 100\ \Omega$,	See Figure 1	½ V_{OD1} or 2			V
		$R_L = 54\ \Omega$,	See Figure 1	1.5	2.5	5	V
V_{OD3}	Differential output voltage	See Note 3		1.5		5	V
$\Delta\|V_{OD}\|$	Change in magnitude of differential output voltage‡	$R_L = 54\ \Omega$ or $100\ \Omega$, See Figure 1				±0.2	V
V_{OC}	Common-mode output voltage§					+3 −1	V
$\Delta\|V_{OC}\|$	Change in magnitude of common-mode output voltage‡					±0.2	V
I_O	Output current with power off	$V_{CC} = 0$,	$V_O = -7$ V to 12 V			±100	μA
I_{OZ}	High-impedance-state output current	$V_O = -7$ V to 12 V				±100	μA
I_{IH}	High-level input current	$V_I = 2.7$ V				20	μA
I_{IL}	Low-level input current	$V_I = 0.5$ V				−360	μA
I_{OS}	Short-circuit output current	$V_O = -7$ V				−180	mA
		$V_O = V_{CC}$				180	
		$V_O = 12$ V				500	
I_{CC}	Supply current (all drivers)	No load	Outputs enabled		38	60	mA
			Outputs disabled		18	40	

† All typical values are at $V_{CC} = 5$ V and $T_A = 25°C$.
‡ $\Delta|V_{OD}|$ and $\Delta|V_{OC}|$ are the changes in magnitude of V_{OD} and V_{OC}, respectively, that occur when the input is changed from a high level to a low level.
§ In EIA Standard RS-422-A, V_{OC}, which is the average of the two output voltages with respect to ground, is called output offset voltage, V_{OS}.
NOTE 3: See EIA Standard RS-485 Figure 3-5, Test Termination Measurement 2.

SYMBOL EQUIVALENTS

DATA SHEET PARAMETER	RS-422-A	RS-485
V_O	V_{oa}, V_{ob}	V_{oa}, V_{ob}
$\|V_{OD1}\|$	V_o	V_o
$\|V_{OD2}\|$	V_t ($R_L = 100\ \Omega$)	V_t ($R_L = 54\ \Omega$)
$\|V_{OD3}\|$		V_t (Test Termination Measurement 2)
$\Delta\|V_{OD}\|$	$\|\,\|V_t\| - \|\overline{V}_t\|\,\|$	$\|\,\|V_t\| - \|\overline{V}_t\|\,\|$
V_{OC}	$\|V_{os}\|$	$\|V_{os}\|$
$\Delta\|V_{OC}\|$	$\|V_{os} - \overline{V}_{os}\|$	$\|V_{os} - \overline{V}_{os}\|$
I_{OS}	$\|I_{sa}\|$, $\|I_{sb}\|$	
I_O	$\|I_{xa}\|$, $\|I_{xb}\|$	I_{ia}, I_{ib}

switching characteristics, V_{CC} = 5 V, T_A = 25°C

PARAMETER		TEST CONDITIONS	MIN	TYP	MAX	UNIT
t_{DD}	Differential-output delay time	R_L = 54 Ω, See Figure 2		45	65	ns
t_{TD}	Differential-output transition time			80	120	ns
t_{PZH}	Output enable time to high level	R_L = 110 Ω, See Figure 3		80	120	ns
t_{PZL}	Output enable time to low level	R_L = 110 Ω, See Figure 4		45	80	ns
t_{PHZ}	Output disable time from high level	R_L = 110 Ω, See Figure 3		78	115	ns
t_{PLZ}	Output disable time from low level	R_L = 110 Ω, See Figure 4		18	30	ns

PARAMETER MEASUREMENT INFORMATION

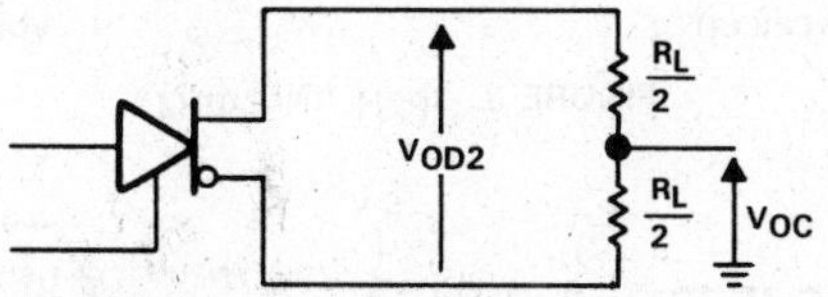
RL
2
VOD2
RL
2
VOC

FIGURE 1. DIFFERENTIAL AND COMMON-MODE OUTPUT VOLTAGES

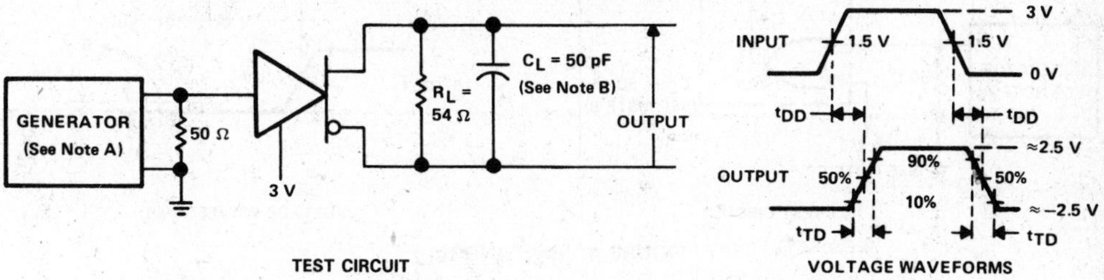
GENERATOR
(See Note A)
50 Ω
3 V
RL =
54 Ω
CL = 50 pF
(See Note B)
OUTPUT
TEST CIRCUIT
INPUT
1.5 V
1.5 V
3 V
0 V
tDD
tDD
OUTPUT
50%
90%
50%
10%
≈2.5 V
≈−2.5 V
tTD
tTD
VOLTAGE WAVEFORMS

FIGURE 2. DRIVER DIFFERENTIAL-OUTPUT DELAY AND TRANSITION TIMES

NOTES: A. The input pulse is supplied by a generator having the following characteristics: $t_r \le 5$ ns, $t_f \le 5$ ns, PRR ≤ 1 MHz, duty cycle = 50%, Z_O = 50 Ω.
B. C_L includes probe and stray capacitance.

PARAMETER MEASUREMENT INFORMATION

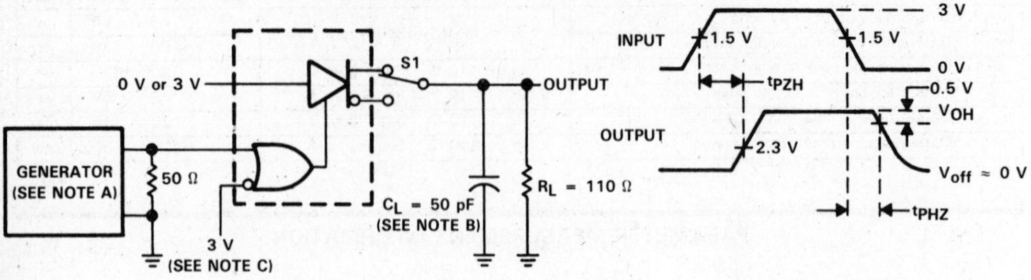
0 V or 3 V
S1
OUTPUT
GENERATOR (SEE NOTE A)
50 Ω
C_L = 50 pF (SEE NOTE B)
R_L = 110 Ω
3 V
(SEE NOTE C)
INPUT
1.5 V
1.5 V
3 V
0 V
t_{PZH}
0.5 V
V_{OH}
OUTPUT
2.3 V
V_{off} ≈ 0 V
t_{PHZ}

TEST CIRCUIT VOLTAGE WAVEFORMS

FIGURE 3. t_{PZH} AND t_{PHZ}

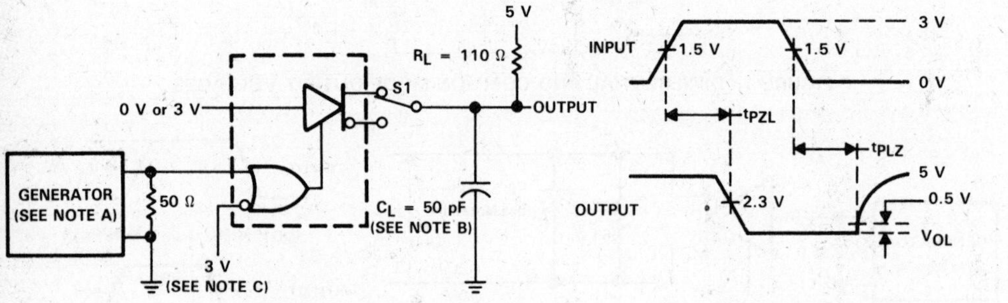
5 V
R_L = 110 Ω
0 V or 3 V
S1
OUTPUT
GENERATOR (SEE NOTE A)
50 Ω
C_L = 50 pF (SEE NOTE B)
3 V
(SEE NOTE C)
INPUT
1.5 V
1.5 V
3 V
0 V
t_{PZL}
t_{PLZ}
5 V
0.5 V
OUTPUT
2.3 V
V_{OL}

TEST CIRCUIT VOLTAGE WAVEFORMS

FIGURE 4. t_{PZL} AND t_{PLZ}

NOTES: A. The input pulse is supplied by a generator having the following characteristics: $t_f \leq 5$ ns, $t_f \leq 5$ ns, PRR ≤ 1 MHz, duty cycle = 50%, Z_{out} = 50 Ω.

B. C_L include probe and jig capacitance.

C. To test the active-low enable $\overline{G}$, ground G and apply an inverted waveform to $\overline{G}$.

TYPICAL CHARACTERISTICS

HIGH-LEVEL OUTPUT VOLTAGE
vs
HIGH-LEVEL OUTPUT CURRENT

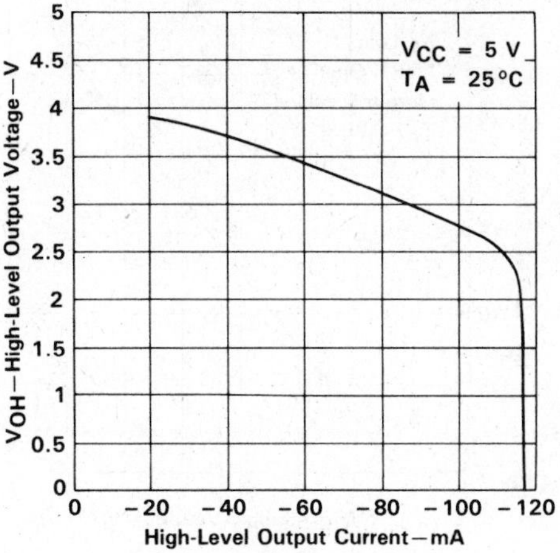
VCC = 5 V
TA = 25°C
VOH—High-Level Output Voltage—V
High-Level Output Current—mA
0 −20 −40 −60 −80 −100 −120
0 0.5 1 1.5 2 2.5 3 3.5 4 4.5 5

FIGURE 5

LOW-LEVEL OUTPUT VOLTAGE
vs
LOW-LEVEL OUTPUT CURRENT

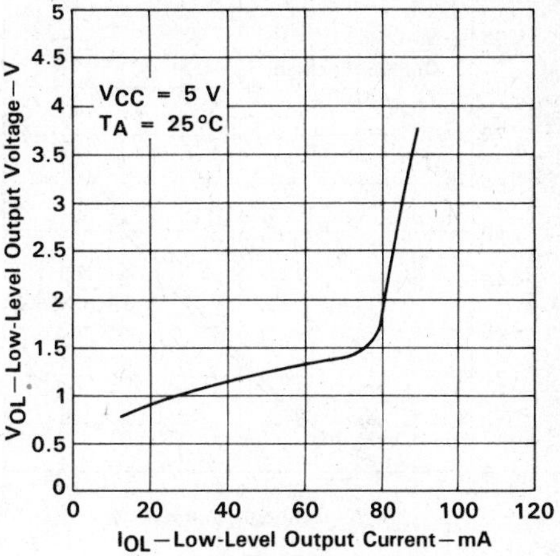
VCC = 5 V
TA = 25°C
VOL—Low-Level Output Voltage—V
IOL—Low-Level Output Current—mA
0 20 40 60 80 100 120
0 0.5 1 1.5 2 2.5 3 3.5 4 4.5 5

FIGURE 6

DIFFERENTIAL OUTPUT VOLTAGE
vs
OUTPUT CURRENT

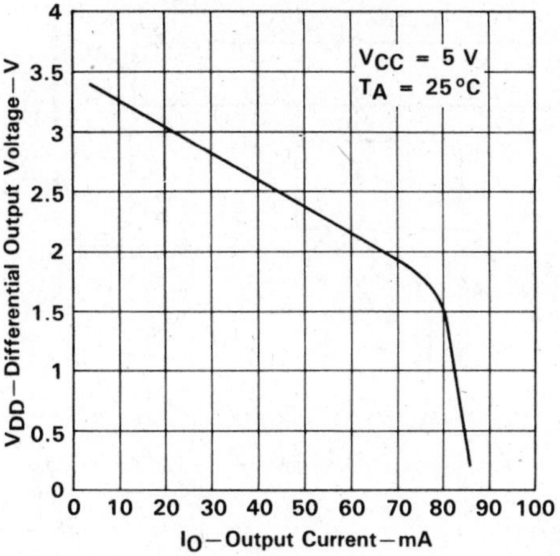
VCC = 5 V
TA = 25°C
VDD—Differential Output Voltage—V
IO—Output Current—mA
0 10 20 30 40 50 60 70 80 90 100
0 0.5 1 1.5 2 2.5 3 3.5 4

FIGURE 7

OUTPUT CURRENT
vs
OUTPUT VOLTAGE

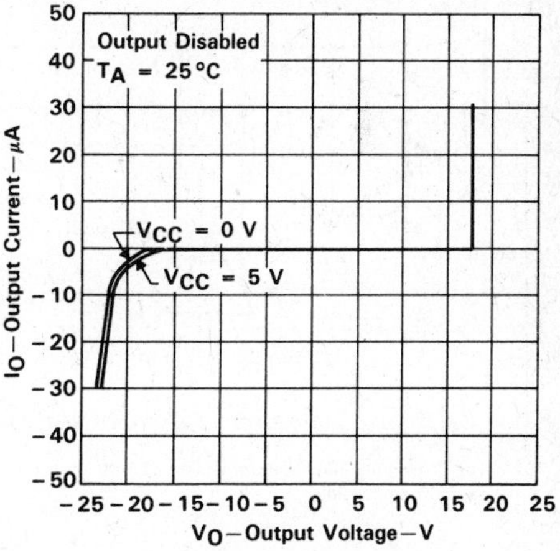
Output Disabled
TA = 25°C
VCC = 0 V
VCC = 5 V
IO—Output Current—μA
VO—Output Voltage—V
−25 −20 −15 −10 −5 0 5 10 15 20 25
−50 −40 −30 −20 −10 0 10 20 30 40 50

FIGURE 8

Line Drivers/Receivers 4

TYPICAL CHARACTERISTICS

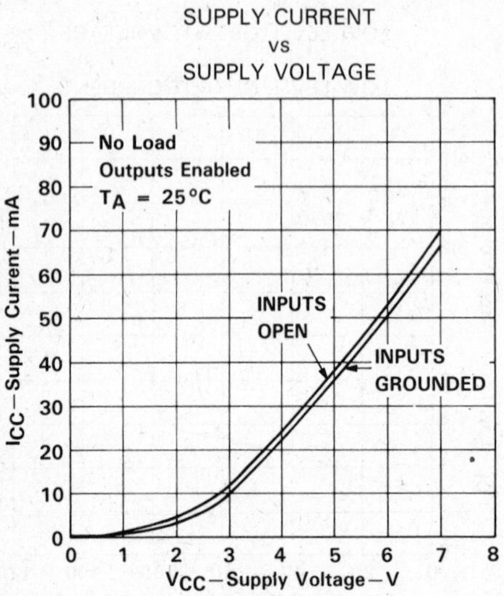
SUPPLY CURRENT
vs
SUPPLY VOLTAGE
No Load
Outputs Enabled
T_A = 25°C
INPUTS OPEN
INPUTS GROUNDED
I_{CC}—Supply Current—mA
V_{CC}—Supply Voltage—V
0
10
20
30
40
50
60
70
80
90
100
0
1
2
3
4
5
6
7
8

FIGURE 9

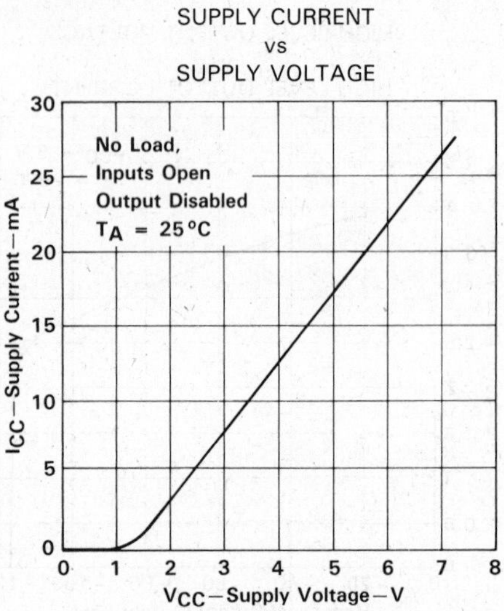
SUPPLY CURRENT
vs
SUPPLY VOLTAGE
No Load,
Inputs Open
Output Disabled
T_A = 25°C
I_{CC}—Supply Current—mA
V_{CC}—Supply Voltage—V
0
5
10
15
20
25
30
0
1
2
3
4
5
6
7
8

FIGURE 10

TYPICAL APPLICATION

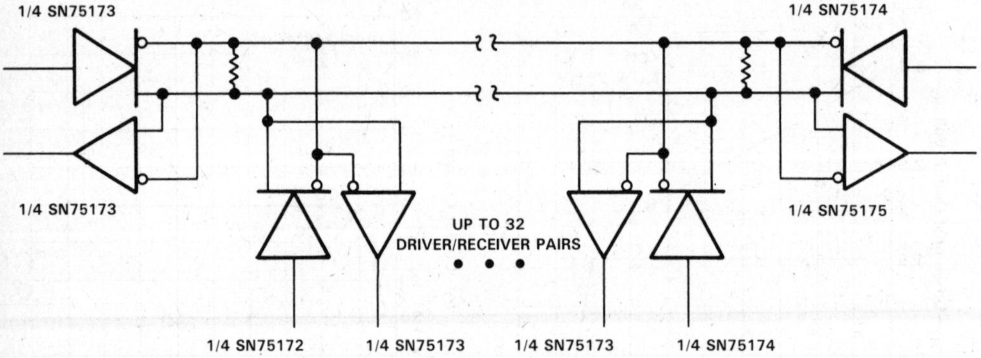
1/4 SN75173
1/4 SN75173
1/4 SN75174
1/4 SN75175
UP TO 32
DRIVER/RECEIVER PAIRS
1/4 SN75172
1/4 SN75173
1/4 SN75173
1/4 SN75174

NOTE A: The line length should be terminated at both ends in its characteristic impedance. Stub lengths off the main line should be kept as short as possible.

FIGURE 11

D2600, OCTOBER 1980—SEPTEMBER 1986

- **Meets EIA Standards RS-422-A, RS-423-A, and RS-485**
- **Meets CCITT Recommendations V.10, V.11, X.26, and X.27**
- **Designed for Multipoint Bus Transmission on Long Bus Lines in Noisy Environments**
- **3-State Outputs**
- **Common-Mode Input Voltage Range . . . −12 to 12 V**
- **Input Sensitivity . . . ±200 mV**
- **Input Hysteresis . . . 50 mV Typ**
- **High Input Impedance . . . 12 kΩ Min**
- **Operates from Single 5-Volt Supply**
- **Low Power Requirements**
- **Plug-In Replacement for AM26LS32**

**D, J, OR N
DUAL-IN-LINE PACKAGE
(TOP VIEW)**

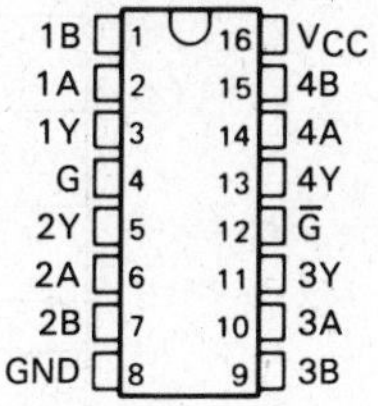
1B 1
1A 2
1Y 3
G 4
2Y 5
2A 6
2B 7
GND 8
16 VCC
15 4B
14 4A
13 4Y
12 G̅
11 3Y
10 3A
9 3B

description

The SN75173 is a monolithic quadruple differential line receiver with three-state outputs. It is designed to meet the requirements of EIA Standards RS-422-A, RS-423-A, and RS-485 and several CCITT recommendations. The device is optimized for balanced multipoint bus transmission at rates up to 10 megabits per second. Each of the two pairs of receivers has a common active-high enable. The device features high input impedance, input hysteresis for increased noise immunity, and input sensitivity of ±200 millivolts over a common-mode input voltage range of −12 to 12 volts. The SN75173 is designed for optimum performance when used with the SN75172 or SN75174 quadruple differential line drivers.

The SN75173 is characterized for operation from 0°C to 70°C.

logic symbol

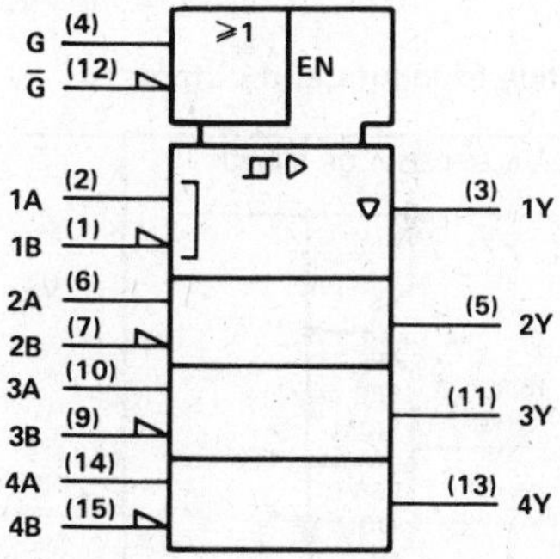
G (4)
G̅ (12)
≥1
EN
1A (2)
1B (1)
(3) 1Y
2A (6)
2B (7)
(5) 2Y
3A (10)
3B (9)
(11) 3Y
4A (14)
4B (15)
(13) 4Y

logic diagram (positive logic)

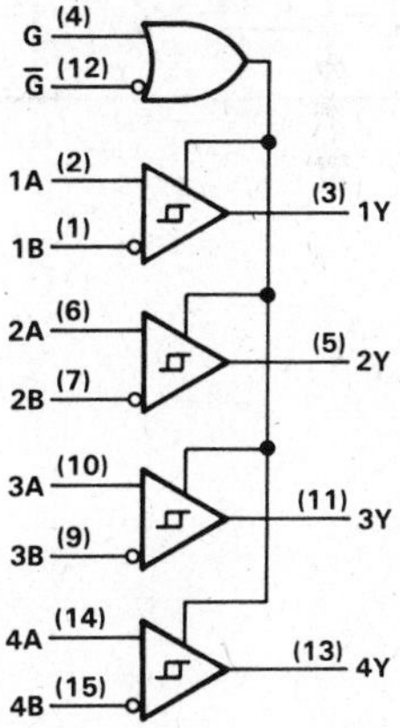
G (4)
G̅ (12)
1A (2)
1B (1)
(3) 1Y
2A (6)
2B (7)
(5) 2Y
3A (10)
3B (9)
(11) 3Y
4A (14)
4B (15)
(13) 4Y

TEXAS INSTRUMENTS
POST OFFICE BOX 655012 • DALLAS, TEXAS 75265

FUNCTION TABLE (EACH RECEIVER)

DIFFERENTIAL A−B	ENABLES G	ENABLES $\overline{G}$	OUTPUT Y
$V_{ID} \geq 0.2$ V	H	X	H
	X	L	H
-0.2 V $< V_{ID} <$ 0.2 V	H	X	?
	X	L	?
$V_{ID} \leq -0.2$ V	H	X	L
	X	L	L
X	L	H	Z

H = high level
L = low level
X = irrelevant
? = indeterminate
Z = high-impedance (off)

schematics of inputs and outputs

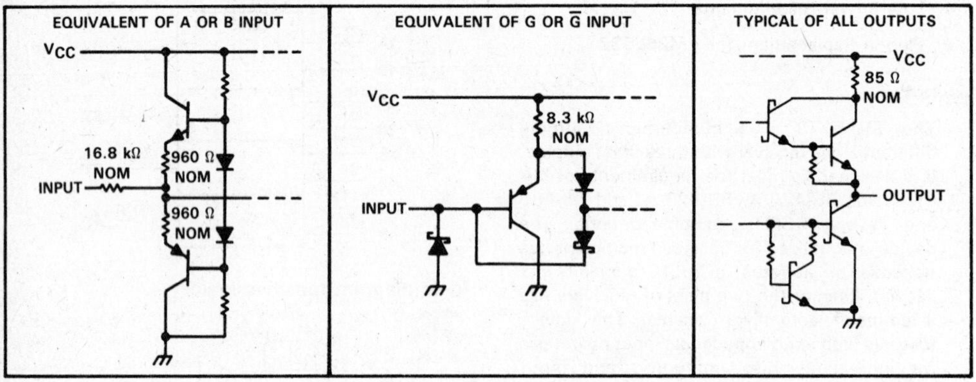
EQUIVALENT OF A OR B INPUT
VCC
16.8 kΩ
NOM
INPUT
960 Ω
NOM
960 Ω
NOM
EQUIVALENT OF G OR $\overline{G}$ INPUT
VCC
8.3 kΩ
NOM
INPUT
TYPICAL OF ALL OUTPUTS
VCC
85 Ω
NOM
OUTPUT

TEXAS INSTRUMENTS
POST OFFICE BOX 655012 • DALLAS, TEXAS 75265

absolute maximum ratings over operating free-air temperature range (unless otherwise noted)

Supply voltage, V_{CC} (see Note 1) . . . . 7 V
Input voltage, A or B inputs . . . . ±25 V
Differential input voltage (see Note 2) . . . . ±25 V
Enable input voltage . . . . 7 V
Low-level output current . . . . 50 mA
Continuous total dissipation at (or below) 25°C free-air temperature (see Note 3):
D Package . . . . 950 mW
J Package . . . . 1025 mW
N Package . . . . 1150 mW
Operating free-air temperature range . . . . 0°C to 70°C
Storage temperature range . . . . −65°C to 150°C
Lead temperature 1,6 mm (1/16 inch) from case for 60 seconds: J package . . . . 300°C
Lead temperature 1,6 mm (1/16 inch) from case for 10 seconds: D or N package . . . . 260°C

NOTES: 1. All voltage values, except differential input voltage, are with respect to network ground terminal.
2. Differential-input voltage is measured at the noninverting input with respect to the corresponding inverting input.
3. For operation above 25°C free-air temperature, derate the D package to 608 mW at 70°C at the rate of 7.6 mW/°C, the J package to 656 mW at 70°C at the rate of 8.2 mW/°C, and the N package to 736 mW at 70°C at the rate of 9.2 mW/°C. In the J package, SN75173 chips are glass mounted.

recommended operating conditions

	MIN	NOM	MAX	UNIT
Supply voltage, V_{CC}	4.75	5	5.25	V
Common-mode input voltage, V_{IC}			±12	V
Differential input voltage, V_{ID}			±12	V
High-level input voltage, V_{IH}	2			V
Low-level input voltage, V_{IL}			0.8	V
High-level output current, I_{OH}			−400	μA
Low-level output current, I_{OL}			16	mA
Operating free-air temperature, T_A	0		70	°C

electrical characteristics over recommended ranges of common-mode input voltage, supply voltage, and operating free-air temperature (unless otherwise noted)

PARAMETER		TEST CONDITIONS		MIN	TYP†	MAX	UNIT
V_{TH}	Differential-input high-threshold voltage	V_O = 2.7 V,	I_O = −0.4 mA			0.2	V
V_{TL}	Differential-input low-threshold voltage	V_O = 0.5 V,	I_O = 16 mA	−0.2‡			V
V_{hys}	Hysteresis§				50		mV
V_{IK}	Enable-input clamp voltage	I_I = −18 mA				−1.5	V
V_{OH}	High-level output voltage	V_{ID} = 200 mV,	I_{OH} = −400 μA	2.7			V
V_{OL}	Low-level output voltage	V_{ID} = −200 mV,	I_{OL} = 8 mA			0.45	V
			I_{OL} = 16 mA			0.5	
I_{OZ}	High-impedance-state output current	V_O = 0.4 V to 2.4 V				±20	μA
I_I	Line input current	Other input at 0 V, See Note 4	V_I = 12 V			1	mA
			V_I = −7 V			−0.8	
I_{IH}	High-level enable-input current	V_{IH} = 2.7 V				20	μA
I_{IL}	Low-level enable-input current	V_{IL} = 0.4 V				−100	μA
r_i	Input resistance			12			kΩ
I_{OS}	Short-circuit output current¶			−15		−85	mA
I_{CC}	Supply current	Outputs disabled				70	mA

† All typical values are at V_{CC} = 5 V, T_A = 25°C.
‡ The algebraic convention, in which the less positive (more negative) limit is designated minimum, is used in this data sheet for threshold voltage levels only.
§ Hysteresis is the difference between the positive-going input threshold voltage, V_{T+}, and the negative-going input threshold voltage, V_{T-}. See Figure 4.
¶ Not more than one output should be shorted at a time and the duration of the short-circuit should not exceed one second.
NOTE 4: Refer to EIA Standard RS-422-A and RS-423-A for exact conditions.

switching characteristics, V_{CC} = 5 V, T_A = 25°C

PARAMETER		TEST CONDITIONS		MIN	TYP	MAX	UNIT
t_{PLH}	Propagation delay time, low-to-high-level output	V_{ID} = −1.5 V to 1.5 V, C_L = 15 pF, See Figure 1			20	35	ns
t_{PHL}	Propagation delay time, high-to-low-level output				22	35	ns
t_{PZH}	Output enable time to high level	C_L = 15 pF,	See Figure 2		17	22	ns
t_{PZL}	Output enable time to low level	C_L = 15 pF,	See Figure 3		20	25	ns
t_{PHZ}	Output disable time from high level	C_L = 5 pF,	See Figure 2		21	30	ns
t_{PLZ}	Output disable time from low level	C_L = 5 pF,	See Figure 3		30	40	ns

PARAMETER MEASUREMENT INFORMATION

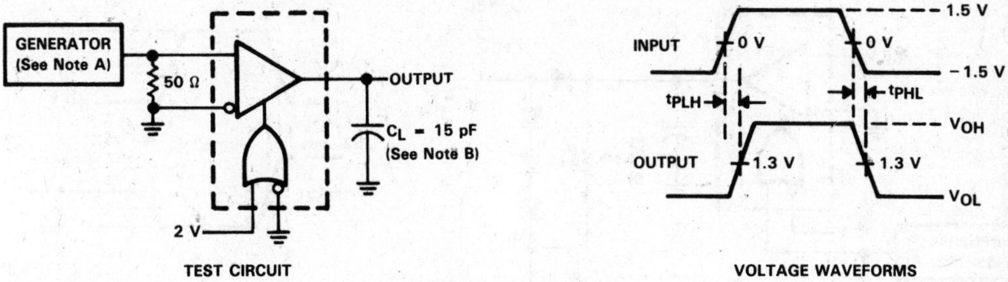
GENERATOR
(See Note A)
50 Ω
OUTPUT
C_L = 15 pF
(See Note B)
2 V
TEST CIRCUIT
1.5 V
INPUT
0 V
0 V
−1.5 V
t_{PLH}
t_{PHL}
V_{OH}
OUTPUT
1.3 V
1.3 V
V_{OL}
VOLTAGE WAVEFORMS

FIGURE 1. t_{PLH}, t_{PHL}

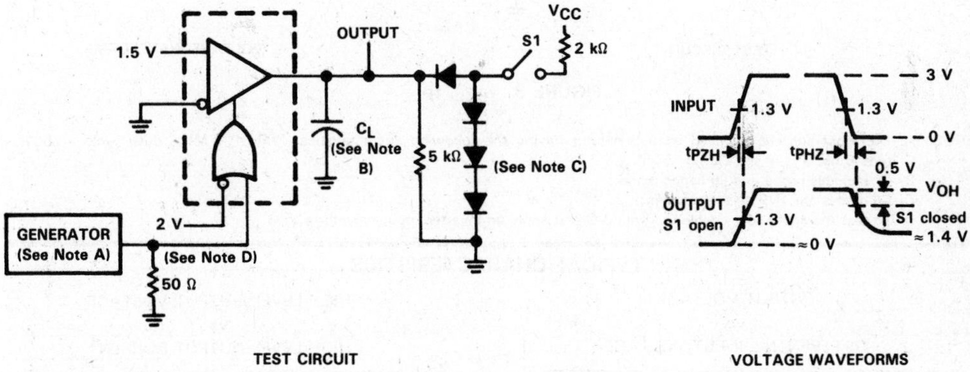
V_{CC}
OUTPUT
S1
2 kΩ
1.5 V
C_L
(See Note B)
5 kΩ
(See Note C)
2 V
GENERATOR
(See Note A)
(See Note D)
50 Ω
TEST CIRCUIT
3 V
INPUT
1.3 V
1.3 V
0 V
t_{PZH}
t_{PHZ}
0.5 V
V_{OH}
OUTPUT
S1 open
1.3 V
S1 closed
≈1.4 V
≈0 V
VOLTAGE WAVEFORMS

FIGURE 2. t_{PHZ}, t_{PZH}

NOTES: A. The input pulse is supplied by a generator having the following characteristics: PRR ≤ 1 MHz, duty cycle = 50%, $t_r \le 6$ ns, $t_f \le 6$ ns, Z_{out} = 50 Ω.

B. C_L includes probe and jig capacitance.

C. All diodes are 1N916 or equivalent.

D. To test the active-low enable $\overline{G}$, ground G and apply an inverted input waveform to $\overline{G}$.

PARAMETER MEASUREMENT INFORMATION

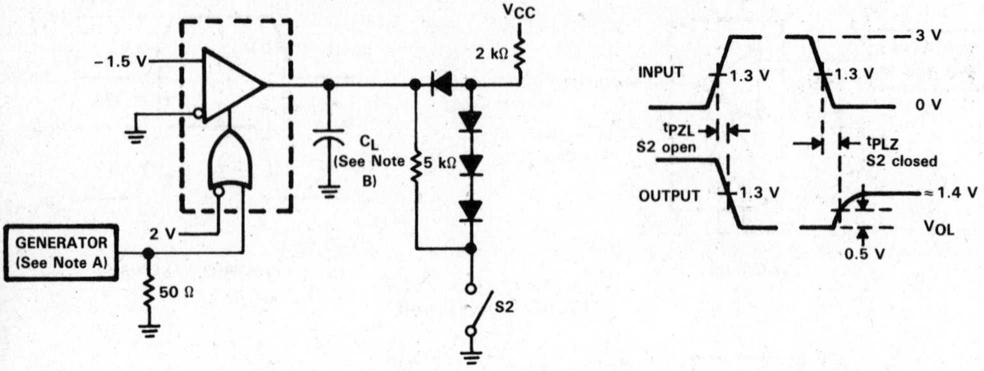
VCC
2 kΩ
−1.5 V
CL
(See Note B)
5 kΩ
GENERATOR
(See Note A)
2 V
50 Ω
S2
INPUT
1.3 V
1.3 V
3 V
0 V
tPZL
S2 open
tPLZ
S2 closed
OUTPUT
1.3 V
≈ 1.4 V
VOL
0.5 V

TEST CIRCUIT VOLTAGE WAVEFORMS

FIGURE 3. t_{PZL}, t_{PLZ}

NOTES: A. The input pulse is supplied by a generator having the following characteristics: PRR ≤ 1 MHz, duty cycle = 50%, $t_r \le 6$ ns, $t_f \le 6$ ns, $Z_{out} = 50\ \Omega$.
B. C_L includes probe and jig capacitance.
C. All diodes are 1N916 or equivalent.
D. To test the active-low enable $\overline{G}$, ground G and apply an inverted input waveform to $\overline{G}$.

TYPICAL CHARACTERISTICS

OUTPUT VOLTAGE
vs
DIFFERENTIAL INPUT VOLTAGE

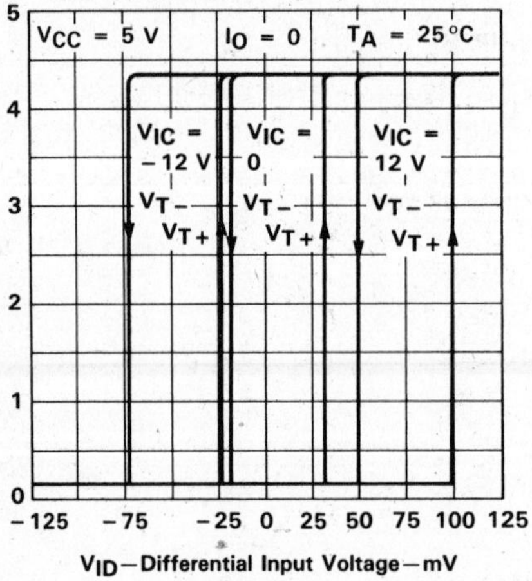
VCC = 5 V
IO = 0
TA = 25°C
VIC = −12 V
VIC = 0
VIC = 12 V
VT−
VT+
VT−
VT+
VT−
VT+
VO−Output Voltage−V
5
4
3
2
1
0
−125
−75
−25
0
25
50
75
100
125
VID−Differential Input Voltage−mV

FIGURE 4

HIGH-LEVEL OUTPUT VOLTAGE
vs
HIGH-LEVEL OUTPUT CURRENT

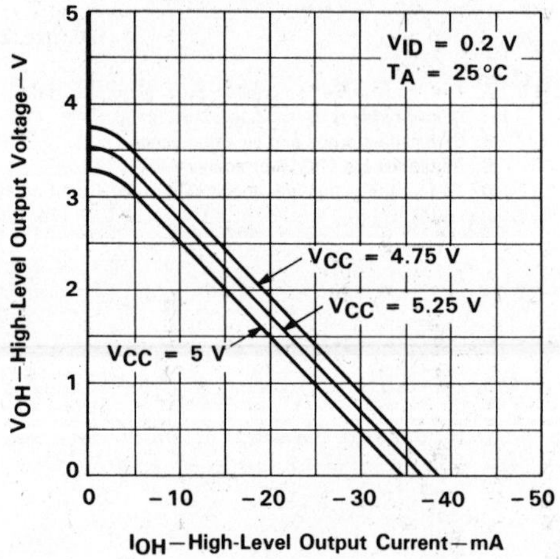
VID = 0.2 V
TA = 25°C
VCC = 4.75 V
VCC = 5.25 V
VCC = 5 V
VOH−High-Level Output Voltage−V
5
4
3
2
1
0
0
−10
−20
−30
−40
−50
IOH−High-Level Output Current−mA

FIGURE 5

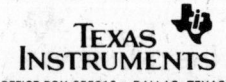

TYPICAL CHARACTERISTICS

HIGH-LEVEL OUTPUT VOLTAGE
vs
FREE-AIR TEMPERATURE

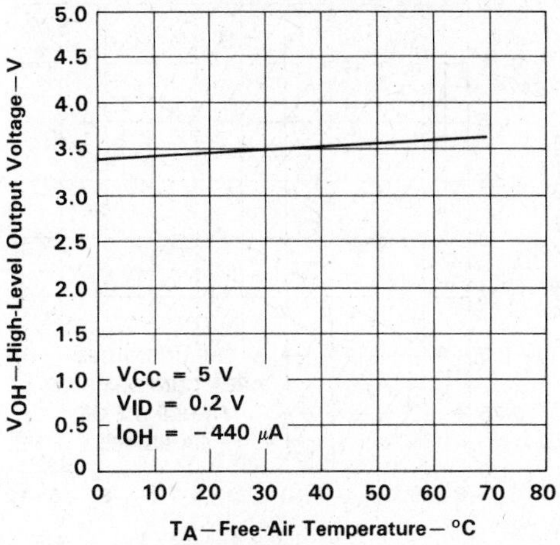
V_{OH}—High-Level Output Voltage—V
5.0
4.5
4.0
3.5
3.0
2.5
2.0
1.5
1.0
0.5
0
V_{CC} = 5 V
V_{ID} = 0.2 V
I_{OH} = −440 μA
0 10 20 30 40 50 60 70 80
T_A—Free-Air Temperature—°C

FIGURE 6

LOW-LEVEL OUTPUT VOLTAGE
vs
LOW-LEVEL OUTPUT CURRENT

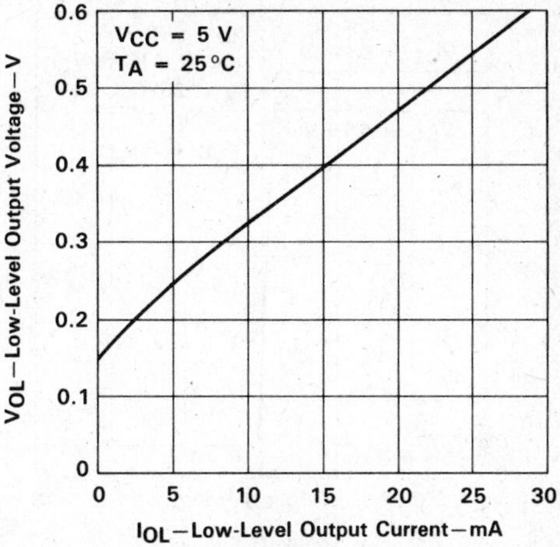
V_{OL}—Low-Level Output Voltage—V
0.6
0.5
0.4
0.3
0.2
0.1
0
V_{CC} = 5 V
T_A = 25°C
0 5 10 15 20 25 30
I_{OL}—Low-Level Output Current—mA

FIGURE 7

LOW-LEVEL OUTPUT VOLTAGE
vs
FREE-AIR TEMPERATURE

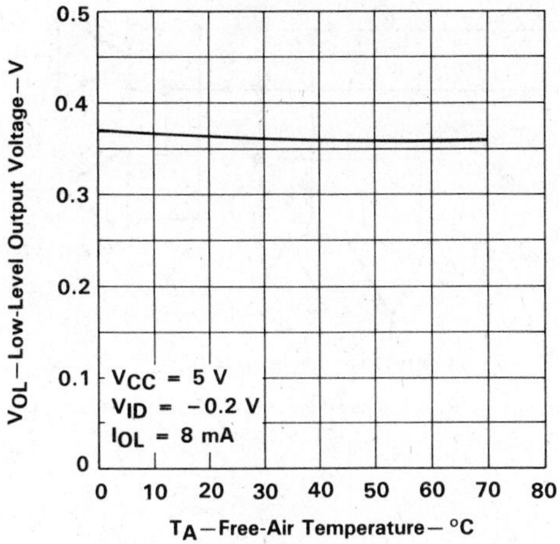
V_{OL}—Low-Level Output Voltage—V
0.5
0.4
0.3
0.2
0.1
0
V_{CC} = 5 V
V_{ID} = −0.2 V
I_{OL} = 8 mA
0 10 20 30 40 50 60 70 80
T_A—Free-Air Temperature—°C

FIGURE 8

OUTPUT VOLTAGE
vs
ENABLE G VOLTAGE

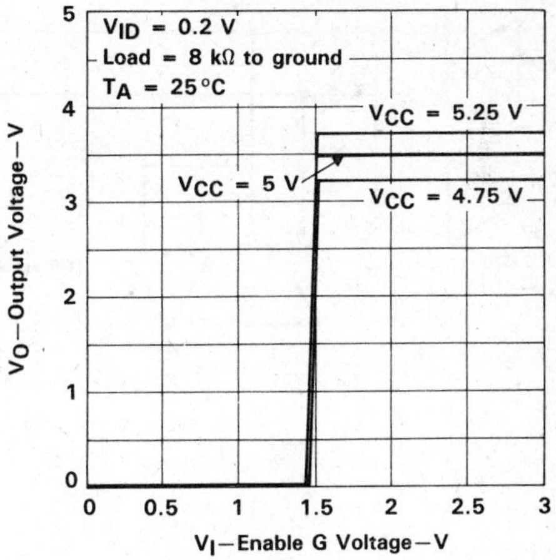
V_O—Output Voltage—V
5
4
3
2
1
0
V_{ID} = 0.2 V
Load = 8 kΩ to ground
T_A = 25°C
V_{CC} = 5.25 V
V_{CC} = 5 V
V_{CC} = 4.75 V
0 0.5 1 1.5 2 2.5 3
V_I—Enable G Voltage—V

FIGURE 9

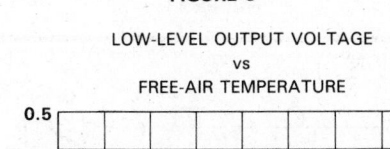

TYPICAL CHARACTERISTICS

OUTPUT VOLTAGE
vs
ENABLE G VOLTAGE

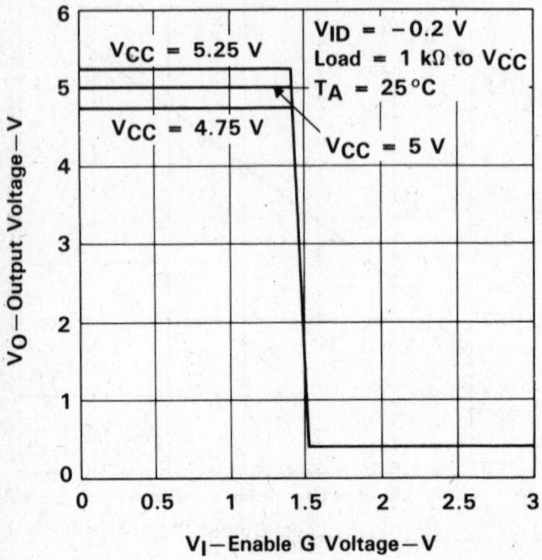
6
5
4
3
2
1
0
0
0.5
1
1.5
2
2.5
3
VCC = 5.25 V
VCC = 4.75 V
VID = −0.2 V
Load = 1 kΩ to VCC
TA = 25°C
VCC = 5 V
VO—Output Voltage—V
VI—Enable G Voltage—V

FIGURE 10

INPUT CURRENT
vs
INPUT VOLTAGE

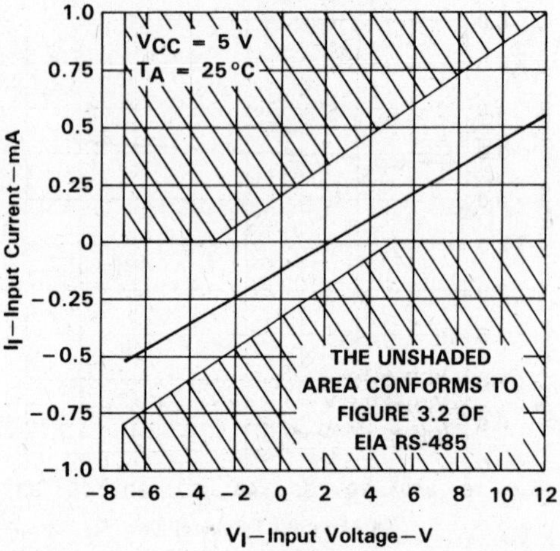
1.0
0.75
0.5
0.25
0
−0.25
−0.5
−0.75
−1.0
−8
−6
−4
−2
0
2
4
6
8
10
12
VCC = 5 V
TA = 25°C
THE UNSHADED
AREA CONFORMS TO
FIGURE 3.2 OF
EIA RS-485
II—Input Current—mA
VI—Input Voltage—V

FIGURE 11

TYPICAL APPLICATION

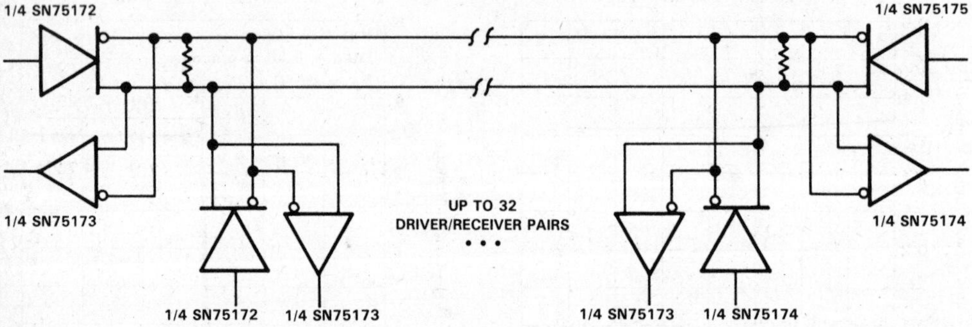
1/4 SN75172
1/4 SN75175
1/4 SN75173
1/4 SN75174
UP TO 32
DRIVER/RECEIVER PAIRS
• • •
1/4 SN75172
1/4 SN75173
1/4 SN75173
1/4 SN75174

NOTE 4: The line should be terminated at both ends in its characteristic impedance. Stub lengths off the main line should be kept as short as possible.

TEXAS INSTRUMENTS
POST OFFICE BOX 655012 • DALLAS, TEXAS 75265

D2601, OCTOBER 1980—REVISED OCTOBER 1986

- **Meets EIA Standards RS-422-A and RS-485 and CCITT Recommendations V.11 and X.27**
- **Designed for Multipoint Transmission on Long Bus Lines in Noisy Environments**
- **3-State Outputs**
- **Common-Mode Output Voltage Range of −7 V to 12 V**
- **Active-High Enable**
- **Thermal Shutdown Protection**
- **Positive- and Negative-Current Limiting**
- **Operates from Single 5-V Supply**
- **Low Power Requirements**
- **Functionally Interchangeable with MC3487**

J OR N DUAL-IN-LINE PACKAGE
(TOP VIEW)

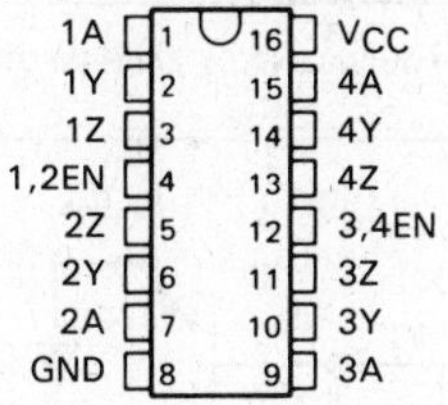
1A 1
1Y 2
1Z 3
1,2EN 4
2Z 5
2Y 6
2A 7
GND 8
16 VCC
15 4A
14 4Y
13 4Z
12 3,4EN
11 3Z
10 3Y
9 3A

FUNCTION TABLE (EACH DRIVER)

INPUT	ENABLE	OUTPUTS Y	OUTPUTS Z
H	H	H	L
L	H	L	H
X	L	Z	Z

H = TTL high level,
L = TTL low level,
X = irrelevant,
Z = High impedance (off)

description

The SN75174 is a monolithic quadruple differential line driver with three-state outputs. It is designed to meet the requirements of EIA Standards RS-422-A and RS-485 and CCITT Recommendations V.11 and X.27. The device is optimized for balanced multipoint bus transmission at rates up to 4 megabaud. Each driver features wide positive and negative common-mode output voltage ranges making it suitable for party-line applications in noisy environments.

The SN75174 provides positive- and negative-current limiting and thermal shutdown for protection from line fault conditions on the transmission bus line. Shutdown occurs at a junction temperature of approximately 150°C. This device offers optimum performance when used with the SN75173 or SN75175 quadruple differential line receivers.

The SN75174 is characterized for operation from 0°C to 70°C.

logic symbol†

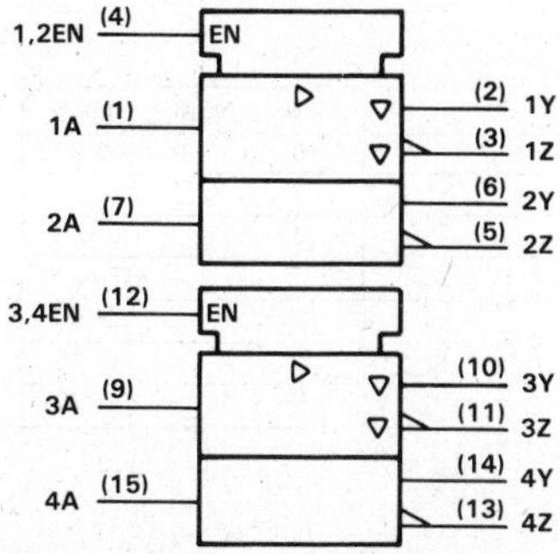
1,2EN (4)
EN
1A (1)
2A (7)
(2) 1Y
(3) 1Z
(6) 2Y
(5) 2Z
3,4EN (12)
EN
3A (9)
4A (15)
(10) 3Y
(11) 3Z
(14) 4Y
(13) 4Z

†This symbol is in accordance with ANSI/IEEE Std 91-1984 and IEC Publication 617-12.

logic diagram, each driver (positive logic)

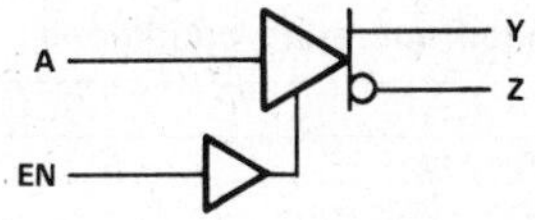
A
EN
Y
Z

schematics of inputs and outputs

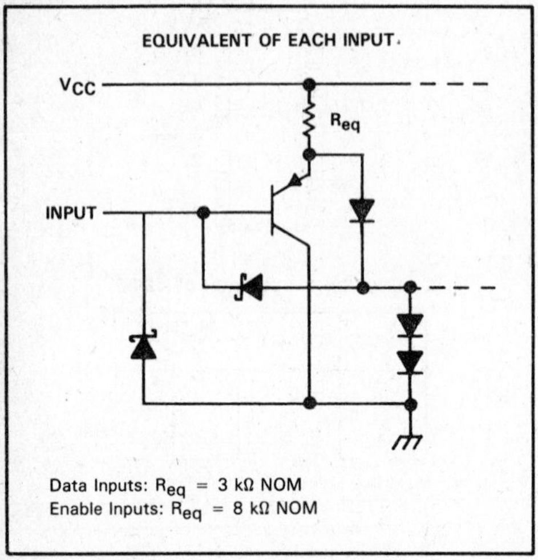
EQUIVALENT OF EACH INPUT
VCC
Req
INPUT
Data Inputs: Req = 3 kΩ NOM
Enable Inputs: Req = 8 kΩ NOM

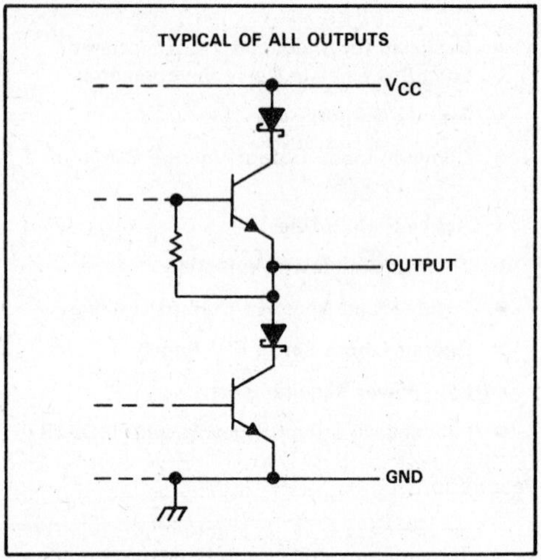
TYPICAL OF ALL OUTPUTS
VCC
OUTPUT
GND

absolute maximum ratings over operating free-air temperature (unless otherwise noted)

Supply voltage, V_{CC} (see Note 1) . . . . 7 V
Input voltage . . . . 5.5 V
Continuous total dissipation at (or below) 25°C free-air temperature (see Note 2):
- J package . . . . 1375 mW
- N package . . . . 1625 mW

Operating free-air temperature . . . . 0°C to 70°C
Storage temperature range . . . . −65°C to 150°C
Lead temperature 1,6 mm (1/16 inch) from case for 60 seconds: J package . . . . 300°C
Lead temperature 1,6 mm (1/16 inch) from case for 10 seconds: N package . . . . 260°C

NOTES: 1. All voltage values are with respect to the network terminal.
2. For operation above 25°C free-air temperature, derate the J package to 880 mW at 70°C at the rate of 11.0 mW/°C and the N package to 1040 mW at 70°C at the rate of 13.0 mW/°C.

recommended operating conditions

	MIN	NOM	MAX	UNIT
Supply voltage, V_{CC}	4.75	5	5.25	V
High-level input voltage, V_{IH}	2			V
Low-Level input voltage, V_{IL}			0.8	V
Common-mode output voltage, V_{OC}			−7 to 12	V
High-level output curent, I_{OH}			−60	mA
Low-level output current, I_{OL}			60	mA
Operating free-air temperature, T_A	0		70	°C

electrical characteristics over recommended ranges of supply voltage and operating free-air temperature (unless otherwise noted)

PARAMETER		TEST CONDITIONS		MIN	TYP†	MAX	UNIT
V_{IK}	Input clamp voltage	I_I = −18 mA				−1.5	V
V_{OH}	High-level output voltage	V_{IH} = 2 V, I_{OH} = −33 mA	V_{IL} = 0.8 V,		3.7		V
V_{OL}	Low-level output voltage	V_{IH} = 2 V, I_{OL} = 33 mA	V_{IL} = 0.8 V,		1.1		V
V_O	Output voltage	I_O = 0		0		6	V
$\|V_{OD1}\|$	Differential output voltage	I_O = 0		1.5		6	V
$\|V_{OD2}\|$	Differential output voltage	R_L = 100 Ω,	See Figure 1	½ V_{OD1} 2			V
		R_L = 54 Ω,	See Figure 1	1.5	2.5	5	V
V_{OD3}	Differential output voltage	See Note 3		1.5		5	V
$\Delta\|V_{OD}\|$	Change in magnitude of differential output voltage‡	R_L = 54 Ω or 100 Ω,	See Figure 1			±0.2	V
V_{OC}	Common mode output voltage					+3 −1	V
$\Delta\|V_{OC}\|$	Change in magnitude of common mode output voltage‡					±0.2	V
I_O	Output current with power off	V_{CC} = 0,	V_O = −7 V to 12 V			±100	μA
I_{OZ}	High-impedance-state output current	V_O = −7 V to 12 V				±100	μA
I_{IH}	High-level input current	V_I = 2.7 V				20	μA
I_{IL}	Low-level input current	V_I = 0.5 V				−360	μA
I_{OS}	Short-circuit output current	V_O = −7 V				−250	mA
		V_O = V_{CC}				180	
		V_O = 12 V				500	
I_{CC}	Supply current (all drivers)	No load	Outputs enabled		38	60	mA
			Oututs disabled		18	40	

† All typical values are at V_{CC} = 5 V and T_A = 25 °C.
‡ $\Delta|V_{OD}|$ and $\Delta|V_{OC}|$ are the changes in magnitude of V_{OD} and V_{OC}, respectively, that occur when the input is changed from a high level to a low level.
NOTE 3: See EIA Standard RS-485 Figure 3.5, Test Termination Measurement 2.

switching characteristics, V_{CC} = 5 V, T_A = 25 °C

PARAMETER		TEST CONDITIONS		MIN	TYP	MAX	UNIT
t_{DD}	Differential-output delay time	R_L = 54 Ω,	See Figure 2		45	65	ns
t_{TD}	Differential-output transition time				80	120	ns
t_{PZH}	Output enable time to high level	R_L = 110 Ω,	See Figure 3		80	120	ns
t_{PZL}	Output enable time to low level	R_L = 110 Ω,	See Figure 4		55	80	ns
t_{PHZ}	Outut disable time from high level	R_L = 110 Ω,	See Figure 3		75	115	ns
t_{PLZ}	Output disable time from low level	R_L = 110 Ω,	See Figure 4		18	30	ns

SYMBOL EQUIVALENTS

DATA SHEET PARAMETER	RS-422-A	RS-485
V_O	V_{oa}, V_{ob}	V_{oa}, V_{ob}
$\lvert V_{OD1} \rvert$	V_o	V_o
$\lvert V_{OD2} \rvert$	V_t ($R_L = 100\ \Omega$)	V_t ($R_L = 54\ \Omega$)
$\lvert V_{OD3} \rvert$		V_t (Test Termination Measurement 2)
$\Delta \lvert V_{OD} \rvert$	$\lvert \lvert V_t \rvert - \lvert \overline{V}_t \rvert \rvert$	$\lvert \lvert V_t \rvert - \lvert \overline{V}_t \rvert \rvert$
V_{OC}	$\lvert V_{os} \rvert$	$\lvert V_{os} \rvert$
$\Delta \lvert V_{OC} \rvert$	$\lvert V_{os} - \overline{V}_{os} \rvert$	$\lvert V_{os} - \overline{V}_{os} \rvert$
I_{OS}	$\lvert I_{sa} \rvert, \lvert I_{sb} \rvert$	
I_O	$\lvert I_{xa} \rvert, \lvert I_{xb} \rvert$	I_{ia}, I_{ib}

PARAMETER MEASUREMENT INFORMATION

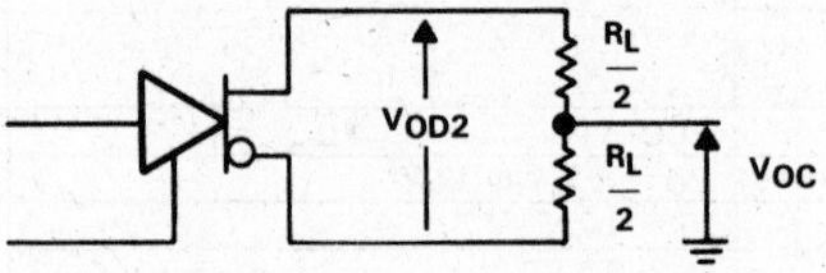
R_L/2
V_{OD2}
R_L/2
V_{OC}

FIGURE 1. DIFFERENTIAL AND COMMON-MODE OUTPUT VOLTAGES

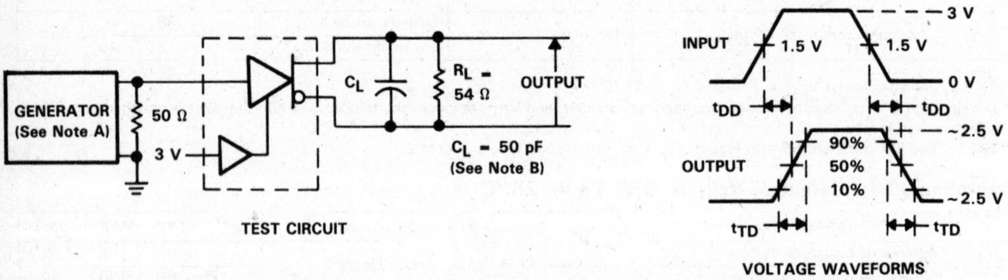
GENERATOR
(See Note A)
50 Ω
3 V
C_L
R_L = 54 Ω
OUTPUT
C_L = 50 pF
(See Note B)
TEST CIRCUIT
3 V
INPUT
1.5 V
1.5 V
0 V
t_{DD}
t_{DD}
~2.5 V
90%
OUTPUT
50%
10%
~2.5 V
t_{TD}
t_{TD}
VOLTAGE WAVEFORMS

NOTES: A. The input pulse is supplied by a generator having the following characteritics: $t_r \leq 5$ ns, $t_f \leq 5$ ns, PRR $\leq$ 1 MHz, duty cycle = 50%, Z_o = 50 Ω.

B. C_L includes probe and stray capacitance.

FIGURE 2. DIFFERENTIAL-OUTPUT DELAY AND TRANSITION TIMES

PARAMETER MEASUREMENT INFORMATION

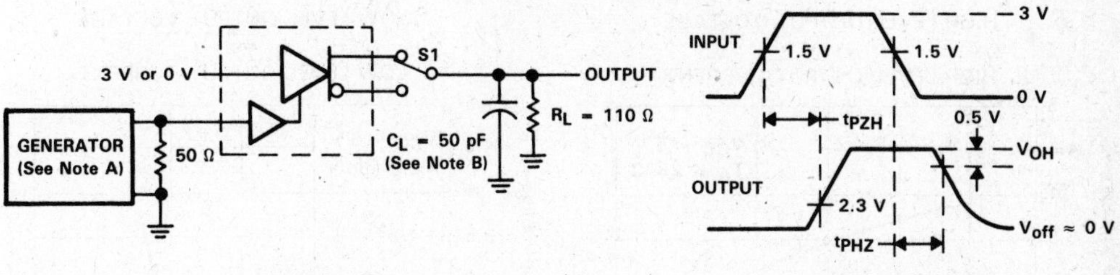
3 V or 0 V
S1
OUTPUT
GENERATOR
(See Note A)
50 Ω
R_L = 110 Ω
C_L = 50 pF
(See Note B)
INPUT
3 V
1.5 V
1.5 V
0 V
t_{PZH}
0.5 V
V_{OH}
OUTPUT
2.3 V
V_{off} ≈ 0 V
t_{PHZ}

TEST CIRCUIT VOLTAGE WAVEFORMS

FIGURE 3. t_{PZH} AND t_{PHZ}

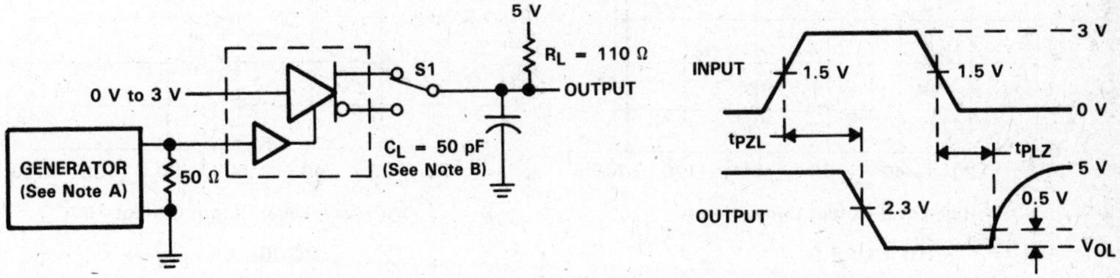
5 V
R_L = 110 Ω
S1
0 V to 3 V
OUTPUT
GENERATOR
(See Note A)
50 Ω
C_L = 50 pF
(See Note B)
INPUT
3 V
1.5 V
1.5 V
0 V
t_{PZL}
t_{PLZ}
5 V
0.5 V
OUTPUT
2.3 V
V_{OL}

TEST CIRCUIT VOLTAGE WAVEFORMS

FIGURE 4. t_{PZL} AND t_{PLZ}

NOTES: A. The input pulse is supplied by a generator having the following characteritics: PRR ≤ 1 MHz, duty cycle = 50%, t_r ≤ 5 ns, t_f ≤ 5 ns, Z_O = 50 Ω.
B. C_L includes probe and stray capacitance.

TYPICAL CHARACTERISTICS

HIGH-LEVEL OUTPUT VOLTAGE
vs
HIGH-LEVEL OUTPUT CURRENT

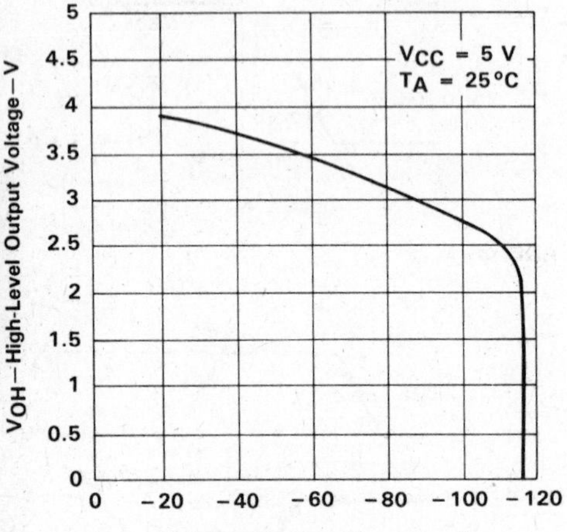
V_{CC} = 5 V
T_A = 25°C
V_{OH}—High-Level Output Voltage—V
5
4.5
4
3.5
3
2.5
2
1.5
1
0.5
0
0
−20
−40
−60
−80
−100
−120
High-Level Output Current—mA

FIGURE 5

LOW-LEVEL OUTPUT VOLTAGE
vs
LOW-LEVEL OUTPUT CURRENT

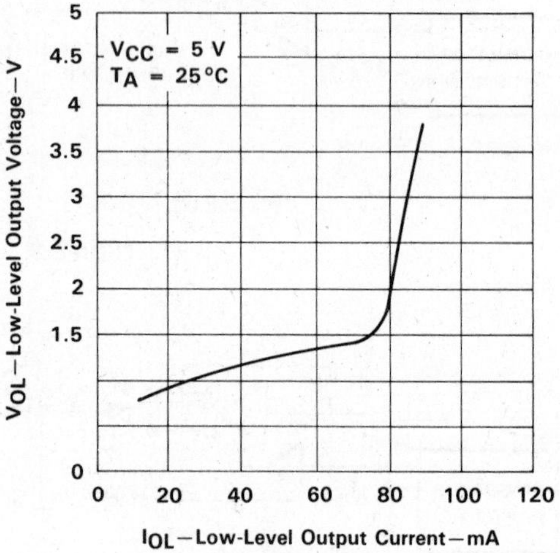
V_{CC} = 5 V
T_A = 25°C
V_{OL}—Low-Level Output Voltage—V
5
4.5
4
3.5
3
2.5
2
1.5
0
0
20
40
60
80
100
120
I_{OL}—Low-Level Output Current—mA

FIGURE 6

DIFFERENTIAL OUTPUT VOLTAGE
vs
OUTPUT CURRENT

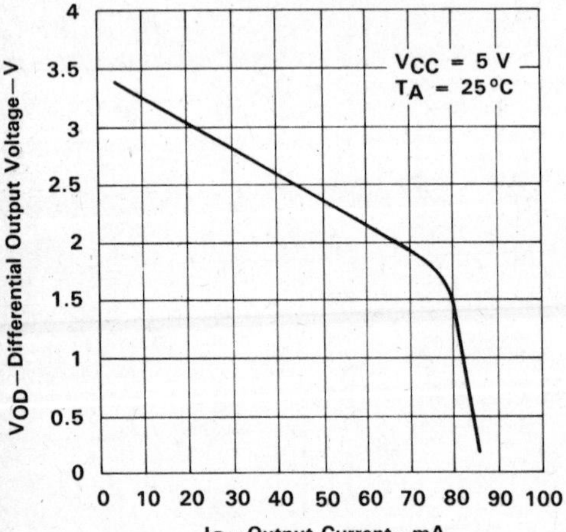
V_{CC} = 5 V
T_A = 25°C
V_{OD}—Differential Output Voltage—V
4
3.5
3
2.5
2
1.5
1
0.5
0
0
10
20
30
40
50
60
70
80
90
100
I_O—Output Current—mA

FIGURE 7

OUTPUT CURRENT
vs
OUTPUT VOLTAGE

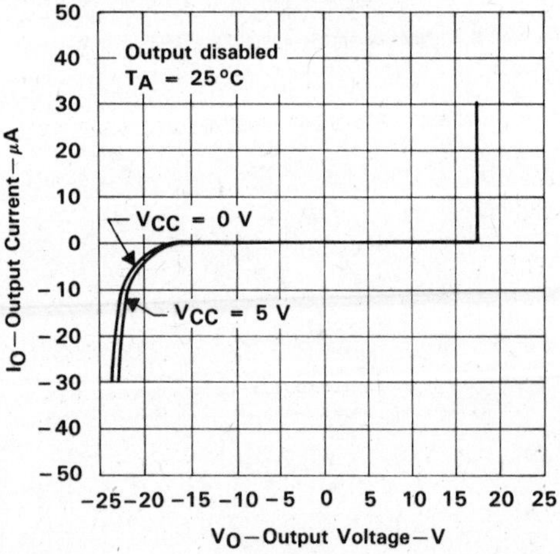
Output disabled
T_A = 25°C
V_{CC} = 0 V
V_{CC} = 5 V
I_O—Output Current—μA
50
40
30
20
10
0
−10
−20
−30
−40
−50
−25
−20
−15
−10
−5
0
5
10
15
20
25
V_O—Output Voltage—V

FIGURE 8

TEXAS INSTRUMENTS
POST OFFICE BOX 655012 • DALLAS, TEXAS 75265

TYPICAL CHARACTERISTICS

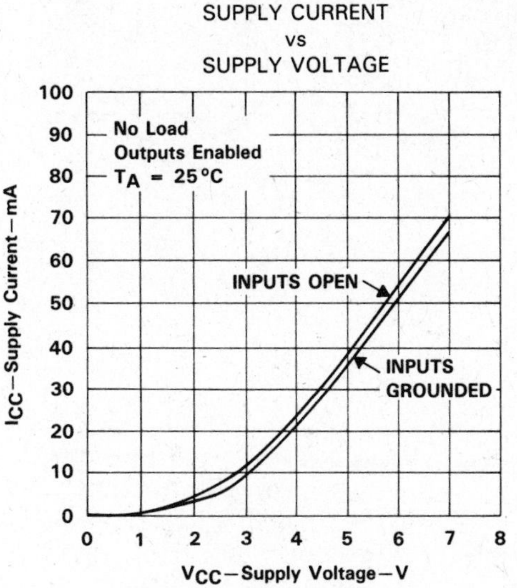
SUPPLY CURRENT
vs
SUPPLY VOLTAGE
No Load
Outputs Enabled
T_A = 25°C
INPUTS OPEN
INPUTS GROUNDED
I_{CC}—Supply Current—mA
V_{CC}—Supply Voltage—V
0
10
20
30
40
50
60
70
80
90
100
0 1 2 3 4 5 6 7 8

FIGURE 9

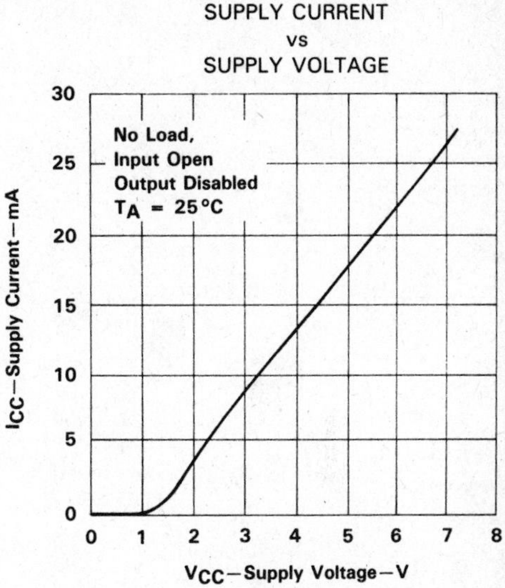
SUPPLY CURRENT
vs
SUPPLY VOLTAGE
No Load,
Input Open
Output Disabled
T_A = 25°C
I_{CC}—Supply Current—mA
V_{CC}—Supply Voltage—V
0
5
10
15
20
25
30
0 1 2 3 4 5 6 7 8

FIGURE 10

TYPICAL APPLICATION

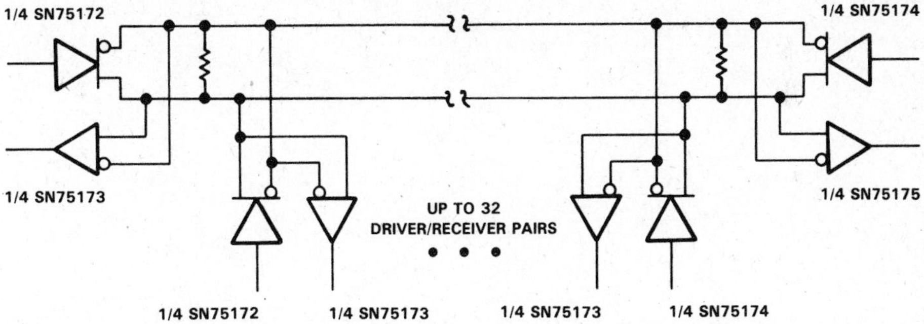
1/4 SN75172
1/4 SN75173
1/4 SN75174
1/4 SN75175
UP TO 32
DRIVER/RECEIVER PAIRS
1/4 SN75172
1/4 SN75173
1/4 SN75173
1/4 SN75174

NOTE: The line length should be terminated at both ends in its characteristic impedance. Stub lengths off the main line should be kept as short as possible.

FIGURE 11

D2602, OCTOBER 1980—REVISED SEPTEMBER 1986

- **Meets EIA Standards RS-422-A, RS-423-A, and RS-485**
- **Meets CCITT Recommendations V.10, V.11, X.26, and X.27**
- **Designed for Multipoint Bus Transmission on Long Bus Lines in Noisy Environments**
- **3-State Outputs**
- **Common-Mode Input Voltage Range −12 V to 12 V**
- **Input Sensitivity . . . ±200 mV**
- **Input Hysteresis . . . 50 mV Typ**
- **High Input Impedance . . . 12 kΩ Min**
- **Operates from Single 5-Volt Supply**
- **Low Power Requirements**
- **Plug-in Replacement for MC3486**

**D, J, OR N
DUAL-IN-LINE PACKAGE
(TOP VIEW)**

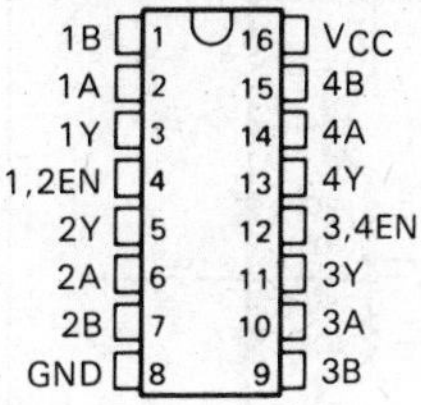
1B 1
1A 2
1Y 3
1,2EN 4
2Y 5
2A 6
2B 7
GND 8
16 VCC
15 4B
14 4A
13 4Y
12 3,4EN
11 3Y
10 3A
9 3B

description

The SN75175 is a monolithic quadruple differential line receiver with three-state outputs. It is designed to meet the requirements of EIA Standards RS-422-A, RS-423-A, and RS-485 and several CCITT recommendations. The device is optimized for balanced multipoint bus transmission at rates up to 10 megabits per second. Each of the two pairs of receivers has a common active-high enable.

The receivers feature high input impedance, input hysteresis for increased noise immunity, and input sensitivity of ±200 millivolts over a common-mode input voltage range of ±12 volts. The SN75175 is designed for optimum performance when used with the SN75172 or SN75174 quadruple differential line drivers.

The SN75175 is characterized for operation from 0°C to 70°C.

FUNCTION TABLE (EACH RECEIVER)

DIFFERENTIAL INPUTS A — B	ENABLE	OUTPUT Y
$V_{ID} \geq 0.2$ V	H	H
-0.2 V $< V_{ID} < 0.2$ V	H	?
$V_{ID} \geq -0.2$ V	H	L
X	L	Z

H = high level, L = low level, ? = indeterminate,
X = irrelevant, Z = high impedance (off)

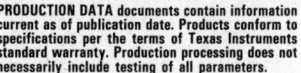

logic symbol†

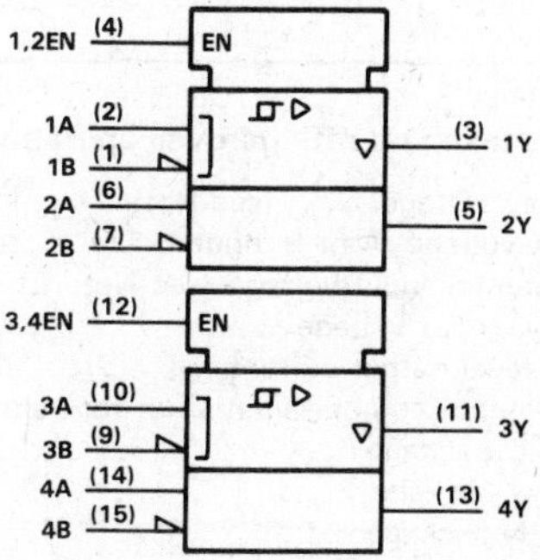
1,2EN (4)
EN
1A (2)
1B (1)
(3) 1Y
2A (6)
2B (7)
(5) 2Y
3,4EN (12)
EN
3A (10)
3B (9)
(11) 3Y
4A (14)
4B (15)
(13) 4Y

† This symbol is in accordance with ANSI/IEEE Std 91-1984 and IEC Publication 617-12.

logic diagram (positive logic)

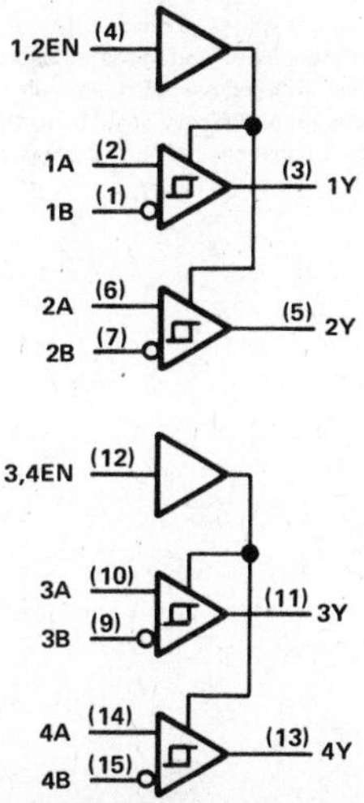
1,2EN (4)
1A (2)
1B (1)
(3) 1Y
2A (6)
2B (7)
(5) 2Y
3,4EN (12)
3A (10)
3B (9)
(11) 3Y
4A (14)
4B (15)
(13) 4Y

PRODUCTION DATA documents contain information current as of publication date. Products conform to specifications per the terms of Texas Instruments standard warranty. Production processing does not necessarily include testing of all parameters.

schematics of inputs and outputs

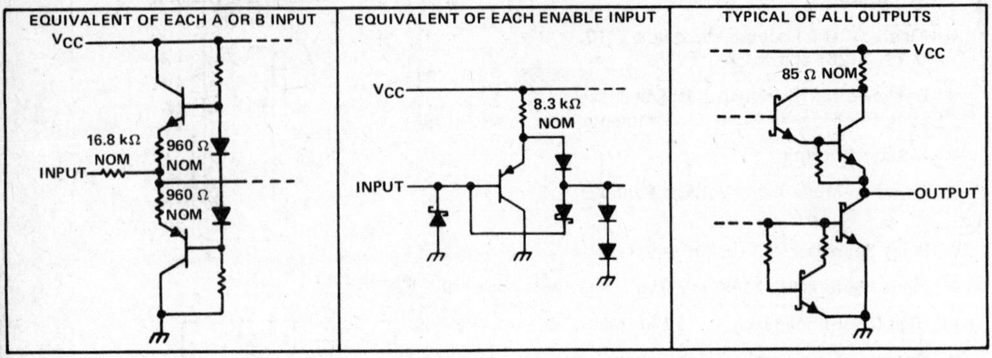
EQUIVALENT OF EACH A OR B INPUT
VCC
16.8 kΩ NOM
INPUT
960 Ω NOM
960 Ω NOM
EQUIVALENT OF EACH ENABLE INPUT
VCC
8.3 kΩ NOM
INPUT
TYPICAL OF ALL OUTPUTS
VCC
85 Ω NOM
OUTPUT

absolute maximum ratings over operating free-air temperature range (unless otherwise noted)

Supply voltage, V_{CC} (see Note 1)	7 V
Input voltage, A or B inputs	±25 V
Differential input voltage (see Note 2)	±25 V
Enable input voltage	7 V
Low-level output current	50 mA
Continuous total dissipation at (or below) 25°C free-air temperature (see Note 3):	
D package	950 mW
J package	1025 mW
N package	1150 mW
Operating free-air temperature range	0°C to 70°C
Storage temperature range	−65°C to 150°C
Lead temperature 1.6 mm (1/16 inch) from case for 60 seconds: J package	−300°C
Lead temperature 1.6 mm (1/16 inch) from case for 10 seconds: D or N package	−260°C

NOTES: 1. All voltage values, except differential input voltage, are with respect to network ground terminal.
2. Differential-input voltage is measured at the noninverting input with respect to the corresponding inverting input.
3. For operation above 25°C free-air temperature, derate the D package to 608 mW at 70°C at the rate of 7.6 mW/°C, the J package to 656 mW at 70°C at the rate of 8.2 mW/°C, and the N package to 736 mW at 70°C at the rate of 9.2 mW/°C. In the J package, SN75175 chips are glass mounted.

recommended operating conditions

	MIN	NOM	MAX	UNIT
Supply voltage, V_{CC}	4.75	5	5.25	V
Common-mode input voltage, V_{IC}			±12	V
Differential input voltage, V_{ID}			±12	V
High-level enable input voltage, V_{IH}	2			V
Low-level enable input voltage, V_{IL}			0.8	V
High-level output current, I_{OH}			−400	μA
Low-level output current, I_{OL}			16	mA
Operating free-air temperature, T_A	0		70	°C

electrical characteristics over recommended ranges of common-mode input voltage, supply voltage, and operating free-air temperature (unless otherwise noted)

PARAMETER		TEST CONDITIONS		MIN	TYP†	MAX	UNIT
V_{TH}	Differential-input high-threshold voltage	V_O = 2.7 V, I_O = −0.4 mA				0.2	V
V_{TL}	Differential-input low-threshold voltage	V_O = 0.5 V, I_O = 16 mA		−0.2‡			V
V_{hys}	Hysteresis§				50		mV
V_{IK}	Enable-input clamp voltage	I_I = −18 mA				−1.5	V
V_{OH}	High-level output voltage	V_{ID} = 200 mV, I_{OH} = −400 μA, See Figure 1		2.7			V
V_{OL}	Low-level output voltage	V_{ID} = −200 mV, See Figure 1	I_{OL} = 8 mA			0.45	V
			I_{OL} = 16 mA			0.5	V
I_{OZ}	High-impedance-state output current	V_O = 0.4 V to 2.4 V				±20	μA
I_I	Line input current	Other input at 0 V, See Note 4	V_I = 12 V			1	mA
			V_I = −7 V			−0.8	mA
I_{IH}	High-level enable-input current	V_{IH} = 2.7 V				20	μA
I_{IL}	Low-level enable-input current	V_{IL} = 0.4 V				−100	μA
r_i	Input resistance			12			kΩ
I_{OS}	Short-circuit output current¶			−15		−85	mA
I_{CC}	Supply current	Outputs disabled				70	mA

† All typical values are at V_{CC} = 5 V, T_A = 25°C.
‡ The algebraic convention, in which the less positive (more negative) limit is designated as minimum, is used in this data sheet for threshold voltage levels only.
§ Hysteresis is the difference between the positive-going input threshold voltage, V_{T+}, and the negative-going input threshold voltage, V_{T-}. See Figure 4.
¶ Not more than one output should be shorted at a time and the duration of the short-circuit should not exceed one second.
NOTE 4: Refer to EIA standards RS-422-A, RS-423-A, and RS-485 for exact conditions.

switching characteristics, V_{CC} = 5 V, T_A = 25°C

PARAMETER		TEST CONDITIONS	MIN	TYP	MAX	UNIT
t_{PLH}	Propagation delay time, low-to-high-level output	C_L = 15 pF, See Figure 2		22	35	ns
t_{PHL}	Propagation delay time, high-to-low-level output	C_L = 15 pF, See Figure 2		25	35	ns
t_{PZH}	Output enable time to high level	C_L = 15 pF, See Figure 3		13	30	ns
t_{PZL}	Output enable time to low level	C_L = 15 pF, See Figure 3		19	30	ns
t_{PHZ}	Output disable time from high level	C_L = 15 pF, See Figure 3		26	35	ns
t_{PLZ}	Output disable time from low level	C_L = 15 pF, See Figure 3		25	35	ns

PARAMETER MEASUREMENT INFORMATION

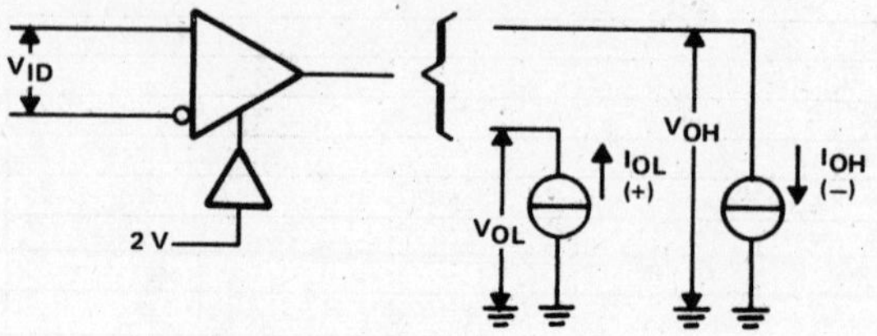
VID
2 V
VOH
IOL
(+)
IOH
(−)
VOL

FIGURE 1. V_{OH}, V_{OL}

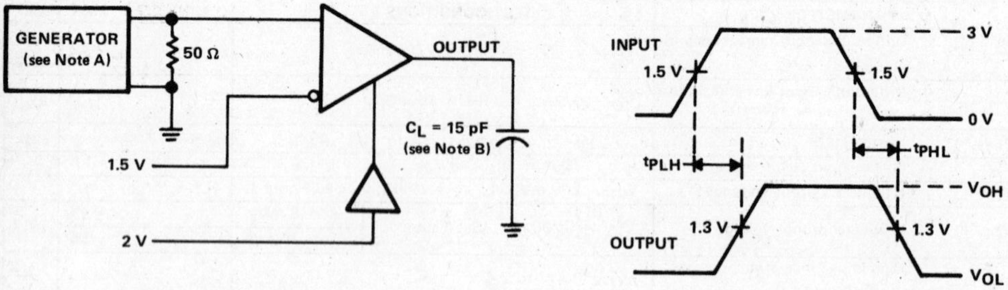
GENERATOR
(see Note A)
50 Ω
OUTPUT
CL = 15 pF
(see Note B)
1.5 V
2 V
INPUT
1.5 V
1.5 V
3 V
0 V
tPLH
tPHL
VOH
1.3 V
1.3 V
OUTPUT
VOL

TEST CIRCUIT

VOLTAGE WAVEFORMS

FIGURE 2. PROPAGATION DELAY TIMES

NOTES: A. The input pulse is supplied by a generator having tne following characteristics: PRR ≤ 1 MHz, duty cycle = 50%, $t_r \leq 6$ ns, $t_f \leq 6$ ns, $Z_{out} = 50\ \Omega$.

B. C_L includes probe and stray capacitance.

PARAMETER MEASUREMENT INFORMATION

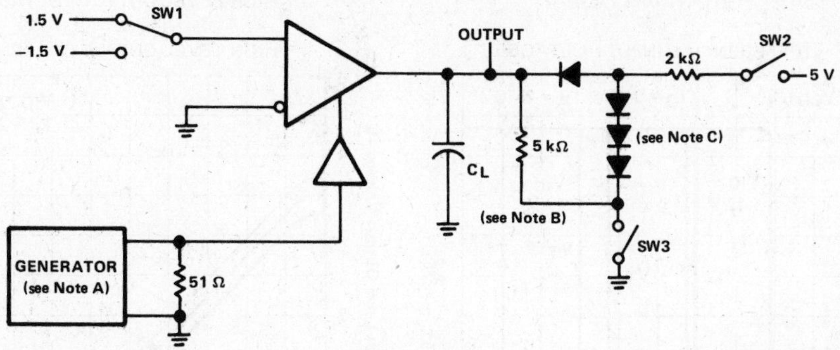
1.5 V
−1.5 V
SW1
OUTPUT
2 kΩ
SW2
5 V
(see Note C)
5 kΩ
C_L
(see Note B)
SW3
GENERATOR
(see Note A)
51 Ω

TEST CIRCUIT

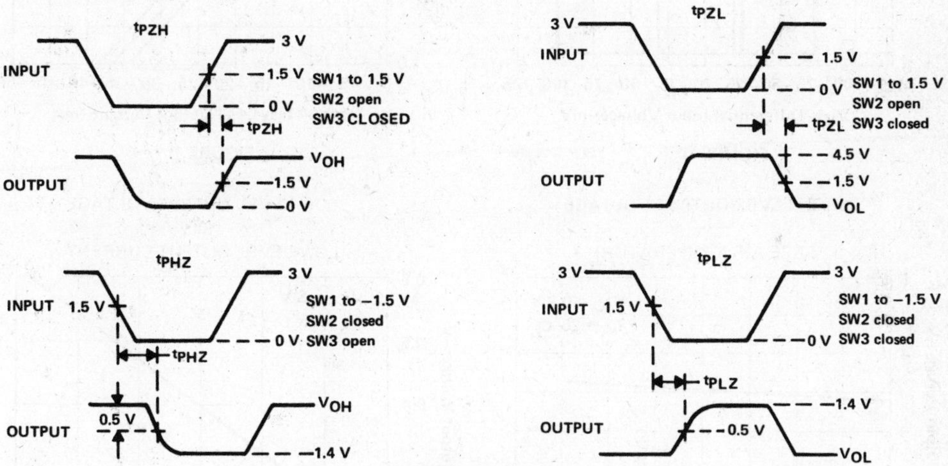
t_{PZH}
INPUT
3 V
1.5 V
0 V
SW1 to 1.5 V
SW2 open
SW3 CLOSED
t_{PZH}
OUTPUT
V_{OH}
1.5 V
0 V
t_{PZL}
3 V
INPUT
1.5 V
0 V
SW1 to 1.5 V
SW2 open
SW3 closed
t_{PZL}
OUTPUT
4.5 V
1.5 V
V_{OL}
t_{PHZ}
3 V
INPUT
1.5 V
SW1 to −1.5 V
SW2 closed
0 V
SW3 open
t_{PHZ}
OUTPUT
0.5 V
V_{OH}
1.4 V
t_{PLZ}
3 V
INPUT
1.5 V
3 V
SW1 to −1.5 V
SW2 closed
0 V
SW3 closed
t_{PLZ}
1.4 V
OUTPUT
0.5 V
V_{OL}

VOLTAGE WAVEFORMS

FIGURE 3. ENABLE AND DISABLE TIMES

NOTES: A. The input pulse is supplied by a generator having the following characteristics: PRR ≤ 1 MHz, duty cycle = 50%, t_f ≤ 6 ns, t_r ≤ 6 ns, Z_{out} = 50 Ω.

B. C_L includes probe and stray capacitance.

C. All diodes are 1N916 or equivalent.

TYPICAL CHARACTERISTICS

OUTPUT VOLTAGE
vs
DIFFERENTIAL INPUT VOLTAGE

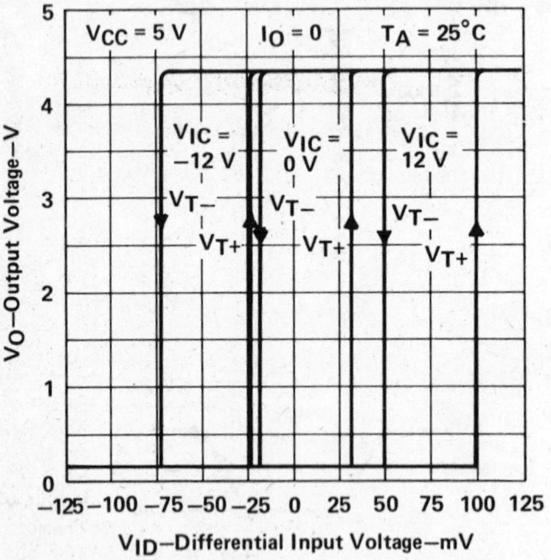
VCC = 5 V
IO = 0
TA = 25°C
VIC = −12 V
VIC = 0 V
VIC = 12 V
VT−
VT+
VO—Output Voltage—V
0 1 2 3 4 5
−125 −100 −75 −50 −25 0 25 50 75 100 125
VID—Differential Input Voltage—mV

FIGURE 4

HIGH-LEVEL OUTPUT VOLTAGE
vs
HIGH-LEVEL OUTPUT CURRENT

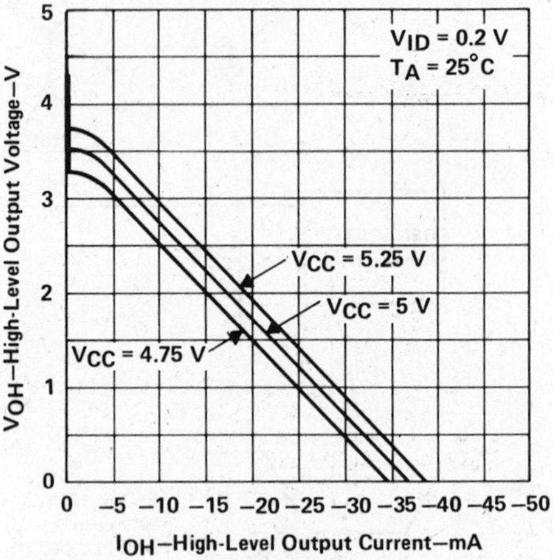
VID = 0.2 V
TA = 25°C
VCC = 5.25 V
VCC = 5 V
VCC = 4.75 V
VOH—High-Level Output Voltage—V
0 1 2 3 4 5
0 −5 −10 −15 −20 −25 −30 −35 −40 −45 −50
IOH—High-Level Output Current—mA

FIGURE 5

HIGH-LEVEL OUTPUT VOLTAGE
vs
FREE-AIR TEMPERATURE

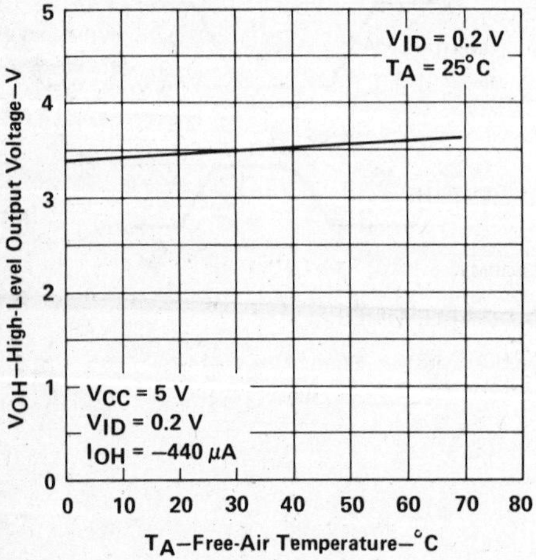
VID = 0.2 V
TA = 25°C
VCC = 5 V
VID = 0.2 V
IOH = −440 μA
VOH—High-Level Output Voltage—V
0 1 2 3 4 5
0 10 20 30 40 50 60 70 80
TA—Free-Air Temperature—°C

FIGURE 6

LOW-LEVEL OUTPUT VOLTAGE
vs
LOW-LEVEL OUTPUT CURRENT

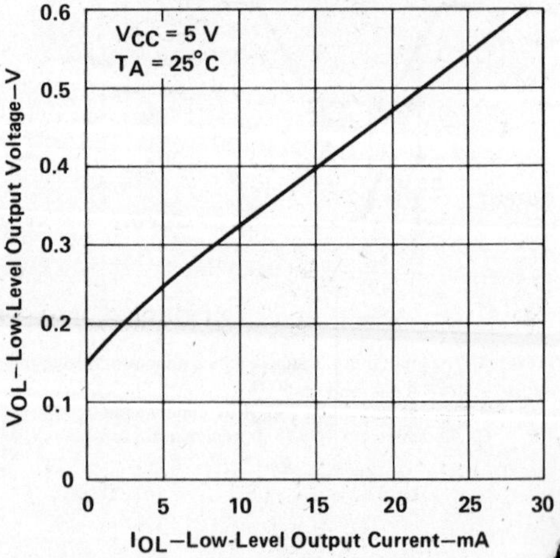
VCC = 5 V
TA = 25°C
VOL—Low-Level Output Voltage—V
0 0.1 0.2 0.3 0.4 0.5 0.6
0 5 10 15 20 25 30
IOL—Low-Level Output Current—mA

FIGURE 7

TYPICAL CHARACTERISTICS

LOW-LEVEL OUTPUT VOLTAGE
vs
FREE-AIR TEMPERATURE

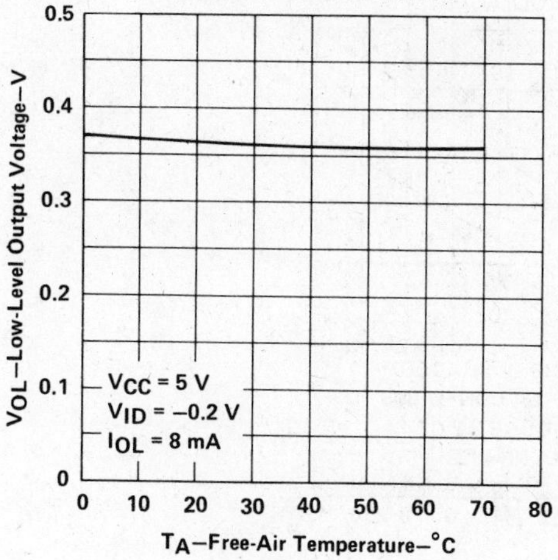
V_{CC} = 5 V
V_{ID} = −0.2 V
I_{OL} = 8 mA
V_{OL}—Low-Level Output Voltage—V
T_A—Free-Air Temperature—°C

FIGURE 8

OUTPUT VOLTAGE
vs
ENABLE VOLTAGE

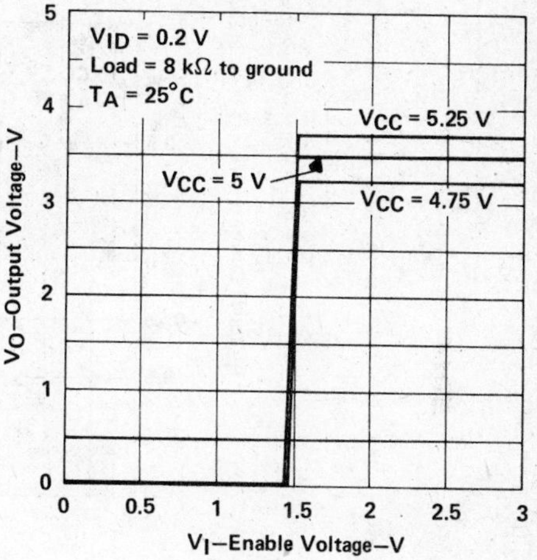
V_{ID} = 0.2 V
Load = 8 kΩ to ground
T_A = 25°C
V_{CC} = 5.25 V
V_{CC} = 5 V
V_{CC} = 4.75 V
V_O—Output Voltage—V
V_I—Enable Voltage—V

FIGURE 9

OUTPUT VOLTAGE
vs
ENABLE VOLTAGE

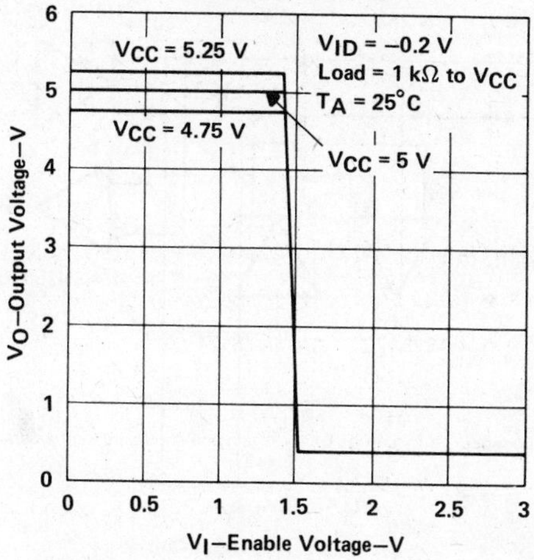
V_{CC} = 5.25 V
V_{ID} = −0.2 V
Load = 1 kΩ to V_{CC}
T_A = 25°C
V_{CC} = 4.75 V
V_{CC} = 5 V
V_O—Output Voltage—V
V_I—Enable Voltage—V

FIGURE 10

SUPPLY CURRENT
(ALL RECEIVERS)
vs
SUPPLY VOLTAGE

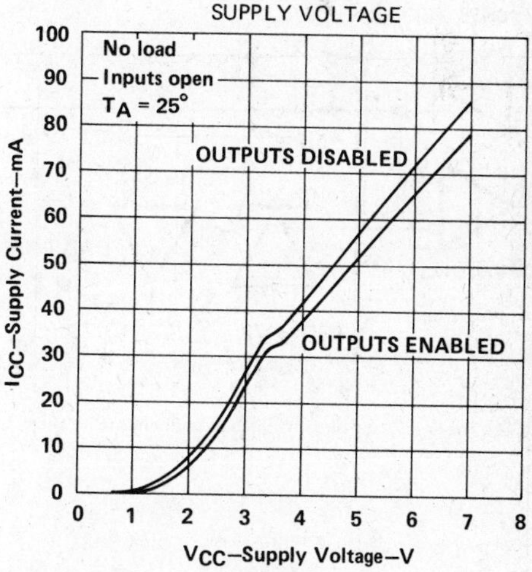
No load
Inputs open
T_A = 25°
OUTPUTS DISABLED
OUTPUTS ENABLED
I_{CC}—Supply Current—mA
V_{CC}—Supply Voltage—V

FIGURE 11

TYPICAL CHARACTERISTICS

INPUT CURRENT
vs
INPUT VOLTAGE

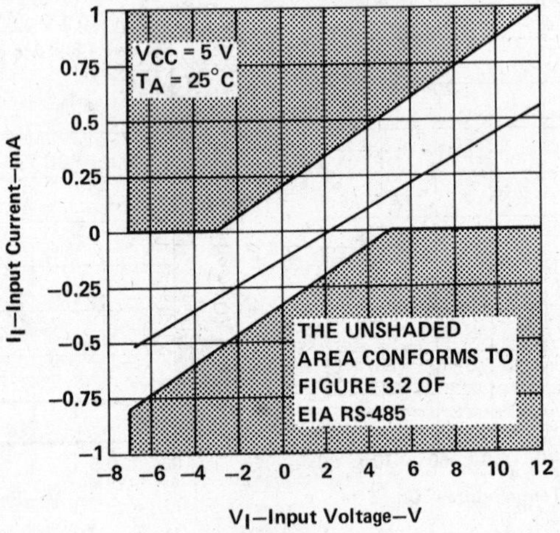
V_{CC} = 5 V
T_A = 25°C
THE UNSHADED AREA CONFORMS TO FIGURE 3.2 OF EIA RS-485
I_I—Input Current—mA
1
0.75
0.5
0.25
0
−0.25
−0.5
−0.75
−1
−8
−6
−4
−2
0
2
4
6
8
10
12
V_I—Input Voltage—V

FIGURE 12

TYPICAL APPLICATION

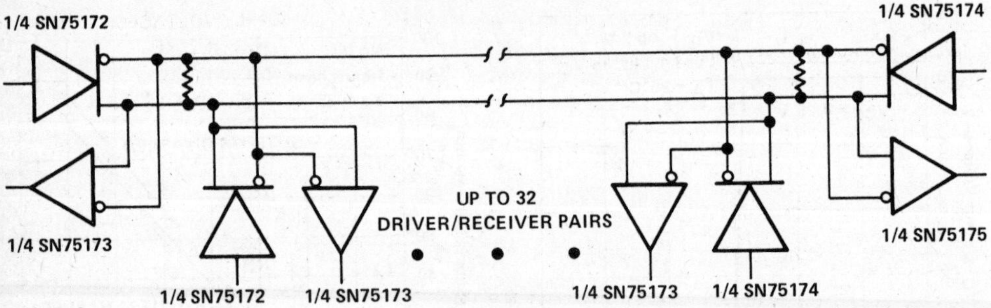
1/4 SN75172
1/4 SN75174
1/4 SN75173
1/4 SN75175
UP TO 32
DRIVER/RECEIVER PAIRS
1/4 SN75172
1/4 SN75173
1/4 SN75173
1/4 SN75174

FIGURE 13

NOTE: The line should be terminated at both ends in its characteristic impedance. Stub lengths off the main line should be kept as short as possible.

TEXAS INSTRUMENTS
POST OFFICE BOX 655012 • DALLAS, TEXAS 75265

ADVANCE INFORMATION

SN65176B, SN75176B DIFFERENTIAL BUS TRANSCEIVERS

D2619, JULY 1985—REVISED OCTOBER 1986

- Bidirectional Transceiver
- Meets EIA Standards RS-422-A and RS-485 and CCITT Recommendations V.11 and X.27
- Designed for Multipoint Transmission on Long Bus Lines in Noisy Environments
- 3-State Driver and Receiver Outputs
- Individual Driver and Receiver Enables
- Wide Positive and Negative Input/Output Bus Voltage Ranges
- Driver Output Capability. . . ±60 mA Max
- Thermal Shutdown Protection
- Driver Positive and Negative Current Limiting
- Receiver Input Impedance . . . 12 kΩ Min
- Receiver Input Sensitivity . . . ±200 mV
- Receiver Input Hysteresis . . . 50 mV Typ
- Operates from Single 5-Volt Supply
- Low Power Requirements

D, JG, OR P PACKAGE (TOP VIEW)

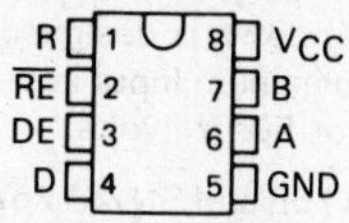
R 1
RE 2
DE 3
D 4
8 VCC
7 B
6 A
5 GND

FUNCTION TABLE (DRIVER)

INPUT D	ENABLE DE	OUTPUTS A	OUTPUTS B
H	H	H	L
L	H	L	H
X	L	Z	Z

FUNCTION TABLE (RECEIVER)

DIFFERENTIAL INPUTS A – B	ENABLE $\overline{RE}$	OUTPUT R
$V_{ID} \geq 0.2$ V	L	H
-0.2 V $< V_{ID} < 0.2$ V	L	?
$V_{ID} \leq -0.2$ V	L	L
X	H	Z

H = high level, L = low level, ? = indeterminate, X = irrelevant, Z = high impedance (off)

description

The SN65176B and SN75176B differential bus transceivers are monolithic integrated circuits designed for bidirectional data communication on multipoint bus transmission lines. They are designed for balanced transmission lines and meet EIA Standard RS-422-A and RS-485 and CCITT Recommendations V.11 and X.27.

The SN65176B and SN75176B combine a three-state differential line driver and a differential input line receiver both of which operate from a single 5-volt power supply. The driver and receiver have active-high and active-low enables, respectively, that can be externally connected together to function as a direction control. The driver differential outputs and the receiver differential inputs are connected internally to form differential input/output (I/O) bus ports that are designed to offer minimum loading to the bus whenever the driver is disabled or V_{CC} = 0 volts. These ports feature wide positive and negative common-mode voltage ranges making the device suitable for party-line applications.

logic symbol†

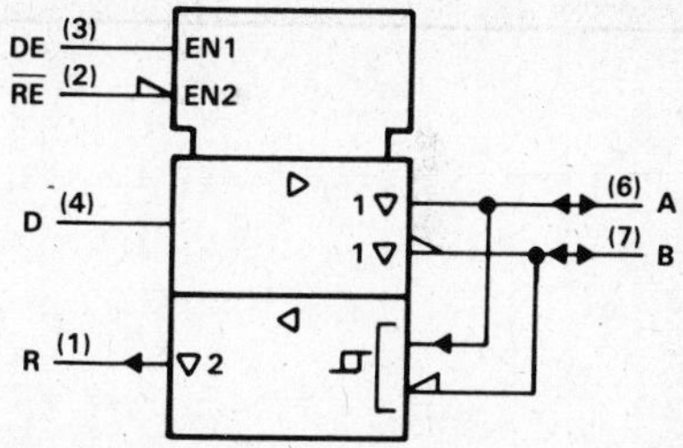
DE (3)
RE (2)
EN1
EN2
D (4)
1
1
(6) A
(7) B
R (1)
2

†This symbol is in accordance with ANSI/IEEE Std 91-1984 and IEC Publication 617-12.

logic diagram (positive logic)

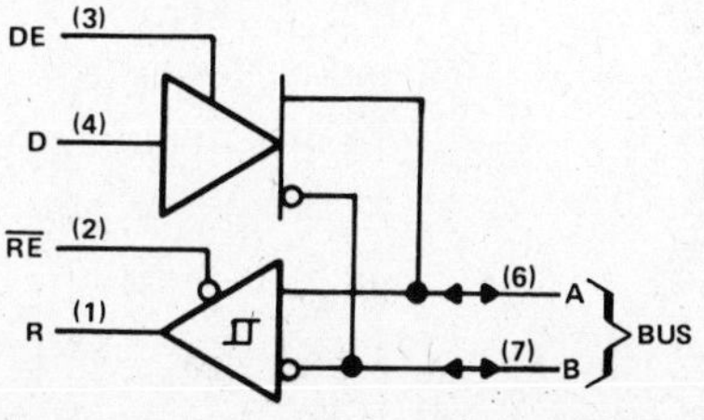
DE (3)
D (4)
RE (2)
R (1)
(6) A
(7) B
BUS

ADVANCE INFORMATION documents contain information on new product in the sampling or preproduction phase of development. Characteristic data and other specifications are subject to change without notice.

TEXAS INSTRUMENTS
POST OFFICE BOX 655012 • DALLAS, TEXAS 75265

The driver is designed to handle loads up to 60 milliamperes of sink or source current. The driver features positive- and negative-current limiting and thermal shutdown for protection from line fault conditions. Thermal shutdown is designed to occur at a junction temperature of approximately 150 °C. The receiver features a minimum input impedance of 12 kΩ, an input sensitivity of ±200 millivolts, and a typical input hysteresis of 50 millivolts.

The SN65176B and SN75176B can be used in transmission line applications employing the SN75172 and SN75174 quadruple differential line drivers and SN75173 and SN75175 quadruple differential line receivers.

The SN65176B is characterized for operation from −40 °C to 85 °C and the SN75176B is characterized for operation from 0 °C to 70 °C.

schematics of inputs and outputs

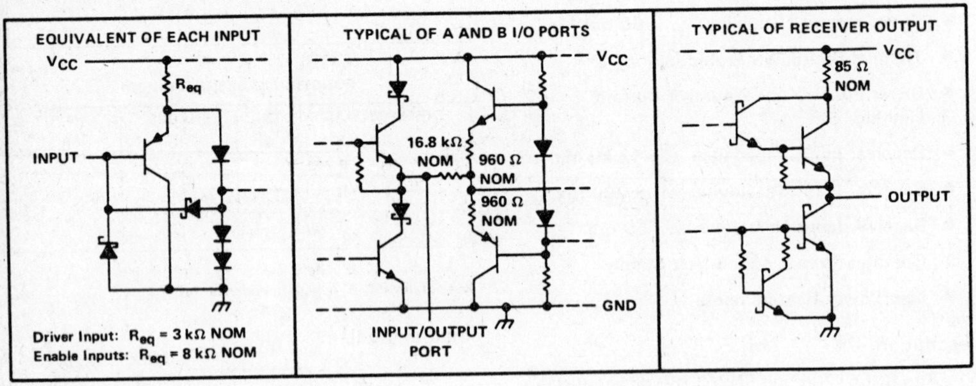
EQUIVALENT OF EACH INPUT
VCC
Req
INPUT
Driver Input: Req = 3 kΩ NOM
Enable Inputs: Req = 8 kΩ NOM
TYPICAL OF A AND B I/O PORTS
VCC
16.8 kΩ
NOM
960 Ω
NOM
960 Ω
NOM
GND
INPUT/OUTPUT
PORT
TYPICAL OF RECEIVER OUTPUT
VCC
85 Ω
NOM
OUTPUT

absolute maximum ratings over operating free-air temperature range (unless otherwise noted)

Supply voltage, V_{CC} (see Note 1) . . . . 7 V
Voltage at any bus terminal . . . . −10 V to 15 V
Enable input voltage . . . . 5.5 V
Continuous total dissipation at (or below) 25°C free-air temperature (see Note 2):
D package . . . . 725 mW
JG package . . . . 825 mW
P package . . . . 1100 mW
Operating free-air temperature range: SN65176B . . . . −40°C to 85°C
SN75176B . . . . 0°C to 70°C
Storage temperature range . . . . −65°C to 150°C
Lead temperature 1,6 mm (1/16 inch) from the case for 60 seconds: JG package . . . . 300°C
Lead temperature 1,6 mm (1/16 inch) from case for 10 seconds: D or P package . . . . 260°C

NOTES: 1. All voltage values, except differential input/output bus voltage, are with respect to network ground terminal.
2. For operation above 25°C free-air temperature, refer to Dissipation Derating Table. In the JG package, the chips are glass mounted.

DISSIPATION DERATING TABLE

PACKAGE	T_A = 25°C POWER RATING	DERATING FACTOR ABOVE T_A = 25°C	T_A = 70°C POWER RATING	T_A = 85°C POWER RATING
D	725 mW	5.8 mW/°C	464 mW	377 mW
JG	825 mW	6.6 mW/°C	528 mW	429 mW
P	1100 mW	8.8 mW/°C	702 mW	570 mW

recommended operating conditions

		MIN	TYP	MAX	UNIT
Supply voltage, V_{CC}		4.75	5	5.25	V
Voltage at any bus terminal (separately or common mode), V_I or V_{IC}				12	V
				−7	
High-level input voltage, V_{IH}	D, DE, and $\overline{RE}$	2			V
Low-level input voltage, V_{IL}	D, DE, and $\overline{RE}$			0.8	V
Differential input voltage, V_{ID} (see Note 3)				±12	V
High-level output current, I_{OH}	Driver			−60	mA
	Receiver			−400	μA
Low-level output current, I_{OL}	Driver			60	mA
	Receiver			8	
Operating free-air temperature, T_A	SN65176B	−40		85	°C
	SN75176B	0		70	

NOTE 3: Differential-input/output bus voltage is measured at the noninverting terminal A with respect to the inverting terminal B.

DRIVER SECTION

driver electrical characteristics over recommended ranges of supply voltage and operating free-air temperature (unless otherwise noted)

PARAMETER		TEST CONDITIONS†		MIN	TYP‡	MAX	UNIT
V_{IK}	Input clamp voltage	$I_I = -18$ mA				−1.5	V
V_O	Output voltage	$I_O = 0$		0		6	V
$\|V_{OD1}\|$	Differential output voltage	$I_O = 0$		1.5		6	V
$\|V_{OD2}\|$	Differential output voltage	$R_L = 100\ \Omega$,	See Figure 1	½ V_{OD1} or 2			V
		$R_L = 54\ \Omega$,	See Figure 1	1.5	2.5	5	V
V_{OD3}	Differential output voltage	See Note 4		1.5		5	V
$\Delta\|V_{OD}\|$	Change in magnitude of differential output voltage§	$R_L = 54\ \Omega$ or $100\ \Omega$,	See Figure 1			±0.2	V
V_{OC}	Common-mode output voltage					+3 / −1	V
$\Delta\|V_{OC}\|$	Change in magnitude of common-mode output voltage§					±0.2	V
I_O	Output current	Output disabled, See Note 5	$V_O = 12$ V			1	mA
			$V_O = -7$ V			−0.8	
I_{IH}	High-level input current	$V_I = 2.4$ V				20	μA
I_{IL}	Low-level input current	$V_I = 0.4$ V				−400	μA
I_{OS}	Short-circuit output current	$V_O = -7$ V				−250	mA
		$V_O = 0$				−150	
		$V_O = V_{CC}$				250	
		$V_O = 12$ V				250	
I_{CC}	Supply current (total package)	No load	Outputs enabled		42	55	mA
			Outputs disabled		26	35	

† The power-off measurement in EIA Standard RS-422-A applies to disabled outputs only and is not applied to combined inputs and outputs.

‡ All typical values are at $V_{CC} = 5$ V and $T_A = 25°C$.

§ $\Delta|V_{OD}|$ and $\Delta|V_{OC}|$ are the changes in magnitude of V_{OD} and V_{OC} respectively, that occur when the input is changed from a high level to a low level.

NOTES: 4. See EIA Standard RS-485 Figure 3.5, Test Termination Measurement 2.

5. This applies for both power on and off; refer to EIA Standard RS-485 for exact conditions. The RS-422-A limit does not apply for a combined driver and receiver terminal.

driver switching characteristics, $V_{CC} = 5$ V, $T_A = 25°C$

PARAMETER		TEST CONDITIONS		MIN	TYP	MAX	UNIT
t_{DD}	Differential-output delay time	$R_L = 54\ \Omega$,	See Figure 3		15	22	ns
t_{TD}	Differential-output transition time				20	30	ns
t_{PZH}	Output enable time to high level	$R_L = 110\ \Omega$,	See Figure 4		85	120	ns
t_{PZL}	Output enable time to low level	$R_L = 110\ \Omega$,	See Figure 5		40	60	ns
t_{PHZ}	Output disable time from high level	$R_L = 110\ \Omega$,	See Figure 4		150	250	ns
t_{PLZ}	Output disable time from low level	$R_L = 110\ \Omega$,	See Figure 5		20	30	ns

TEXAS INSTRUMENTS
POST OFFICE BOX 655012 • DALLAS, TEXAS 75265

SYMBOL EQUIVALENTS

DATA SHEET PARAMETER	RS-422-A	RS-485
V_O	V_{oa}, V_{ob}	V_{oa}, V_{ob}
$\|V_{OD1}\|$	V_o	V_o
$\|V_{OD2}\|$	V_t (R_L = 100 Ω)	V_t (R_L = 54 Ω)
$\|V_{OD3}\|$		V_t (Test Termination Measurement 2)
$\Delta\|V_{OD}\|$	$\|\|V_t\| - \|\bar{V}_t\|\|$	$\|\|V_t - \|\bar{V}_t\|\|$
V_{OC}	$\|V_{os}\|$	$\|V_{os}\|$
$\Delta\|V_{OC}\|$	$\| V_{os} - \bar{V}_{os} \|$	$\| V_{os} - \bar{V}_{os} \|$
I_{OS}	$\|I_{sa}\|, \|I_{sb}\|$	
I_O	$\|I_{xa}\|, \|I_{xb}\|$	I_{ia}, I_{ib}

RECEIVER SECTION

receiver electrical characteristics over recommended ranges of common-mode input voltage, supply voltage, and operating free-air temperature (unless otherwise noted)

PARAMETER		TEST CONDITIONS		MIN	TYP†	MAX	UNIT
V_{TH}	Differential-input high-threshold voltage	V_O = 2.7 V,	I_O = −0.4 mA			0.2	V
V_{TL}	Differential-input low-threshold voltage	V_O = 0.5 V,	I_O = 8 mA	−0.2‡			V
V_{hys}	Hysteresis§				50		mV
V_{IK}	Enable-input clamp voltage	I_I = −18 mA				−1.5	V
V_{OH}	High-level output voltage	V_{ID} = −200 mV, See Figure 2	I_{OH} = −400 μA,	2.7			V
V_{OL}	Low-level output voltage	V_{ID} = −200 mV, See Figure 2	I_{OL} = 8 mA,			0.45	V
I_{OZ}	High-impedance-state output current	V_O = 0.4 V to 2.4 V				±20	μA
I_I	Line input current	Other input = 0 V, See Note 6	V_I = 12 V			1	mA
			V_I = −7 V			−0.8	
I_{IH}	High-level enable-input current	V_{IH} = 2.7 V				20	μA
I_{IL}	Low-level enable-input current	V_{IL} = 0.4 V				−100	μA
r_i	Input resistance			12			kΩ
I_{OS}	Short-circuit output current			−15		−85	mA
I_{CC}	Supply current (total package)	No load	Outputs enabled		42	55	mA
			Outputs disabled		26	35	

† All typical values are at V_{CC} = 5 V, T_A = 25°C.
‡ The algebraic convention, in which the less-positive (more-negative) limit is designated minimum, is used in this data sheet for common-mode input voltage and threshold voltage levels only.
§ Hysteresis is the difference between the positive-going input threshold voltage, V_{T+}, and the negative-going input threshold voltage, V_{T-}. See Figure 4.
NOTE 6: This applies for both power on and power off. Refer to EIA Standard RS-485 for exact conditions.

receiver switching characteristics, V_{CC} = 5 V, T_A = 25°C

PARAMETER		TEST CONDITIONS		MIN	TYP	MAX	UNIT
t_{PLH}	Propagation delay time, low-to-high-level output	V_{ID} = 0 V to 3 V, C_L = 15 pF,	See Figure 6		21	35	ns
t_{PHL}	Propagation delay time, high-to-low-level output				23	35	ns
t_{PZH}	Output enable time to high level	C_L = 15 pF,	See Figure 7		10	20	ns
t_{PZL}	Output enable time to low level				12	20	ns
t_{PHZ}	Output disable time from high level	C_L = 15 pF,	See Figure 7		20	35	ns
t_{PLZ}	Output disable time from low level				17	25	ns

PARAMETER MEASUREMENT INFORMATION

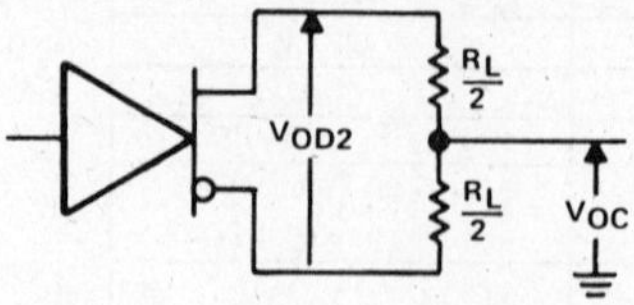
R_L/2
V_{OD2}
R_L/2
V_{OC}

FIGURE 1. DRIVER V_{OD} AND V_{OC}

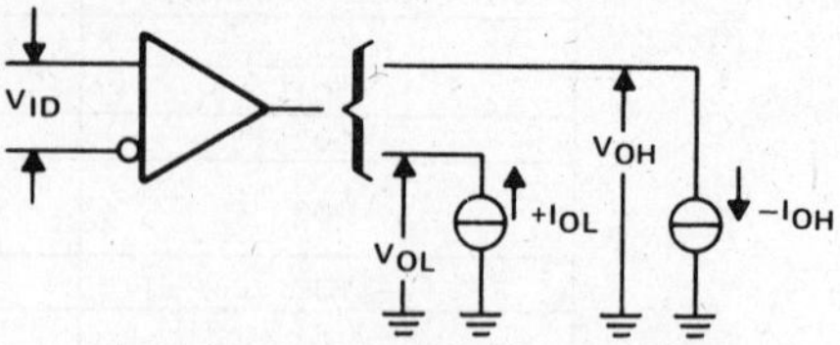
V_{ID}
V_{OH}
V_{OL}
$+I_{OL}$
$-I_{OH}$

FIGURE 2. RECEIVER V_{OH} AND V_{OL}

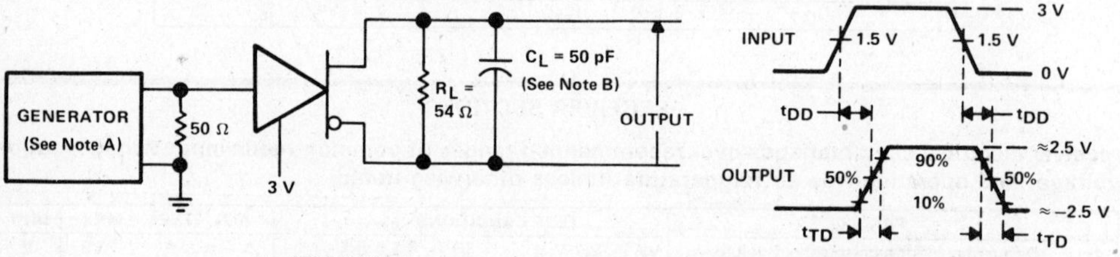
GENERATOR
(See Note A)
50 Ω
3 V
R_L =
54 Ω
C_L = 50 pF
(See Note B)
OUTPUT
INPUT
1.5 V
1.5 V
3 V
0 V
t_{DD}
t_{DD}
OUTPUT
50%
90%
10%
50%
≈2.5 V
≈ −2.5 V
t_{TD}
t_{TD}

TEST CIRCUIT

VOLTAGE WAVEFORMS

FIGURE 3. DRIVER DIFFERENTIAL-OUTPUT DELAY AND TRANSITION TIMES

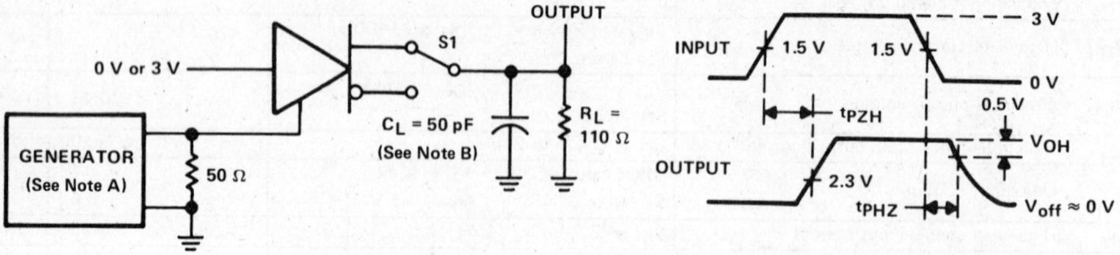
OUTPUT
0 V or 3 V
S1
GENERATOR
(See Note A)
50 Ω
C_L = 50 pF
(See Note B)
R_L =
110 Ω
INPUT
1.5 V
1.5 V
3 V
0 V
t_{PZH}
0.5 V
V_{OH}
OUTPUT
2.3 V
t_{PHZ}
V_{off} ≈ 0 V

TEST CIRCUIT

VOLTAGE WAVEFORMS

FIGURE 4. DRIVER ENABLE AND DISABLE TIMES

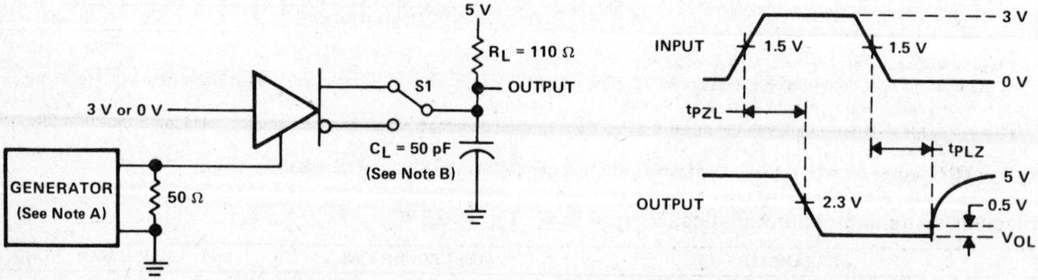
5 V
R_L = 110 Ω
OUTPUT
S1
3 V or 0 V
GENERATOR
(See Note A)
50 Ω
C_L = 50 pF
(See Note B)
INPUT
1.5 V
1.5 V
3 V
0 V
t_{PZL}
t_{PLZ}
5 V
OUTPUT
2.3 V
0.5 V
V_{OL}

TEST CIRCUIT

VOLTAGE WAVEFORMS

FIGURE 5. DRIVER ENABLE AND DISABLE TIMES

NOTES: A. The input pulse is supplied by a generator having the following characteristics: PRR ≤ 1 MHz, 50% duty cycle, $t_r \leq 6$ ns, $t_f \leq 6$ ns, Z_{out} = 50 Ω.
B. C_L includes probe and jig capacitance.

TEXAS INSTRUMENTS
POST OFFICE BOX 655012 • DALLAS, TEXAS 75265

PARAMETER MEASUREMENT INFORMATION

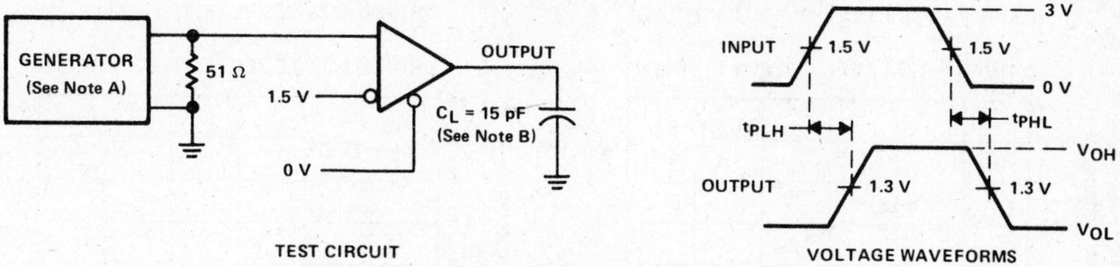
GENERATOR
(See Note A)
51 Ω
1.5 V
0 V
OUTPUT
C_L = 15 pF
(See Note B)
TEST CIRCUIT
INPUT
1.5 V
1.5 V
3 V
0 V
t_{PLH}
t_{PHL}
OUTPUT
1.3 V
1.3 V
V_{OH}
V_{OL}
VOLTAGE WAVEFORMS

FIGURE 6. RECEIVER PROPAGATION DELAY TIMES

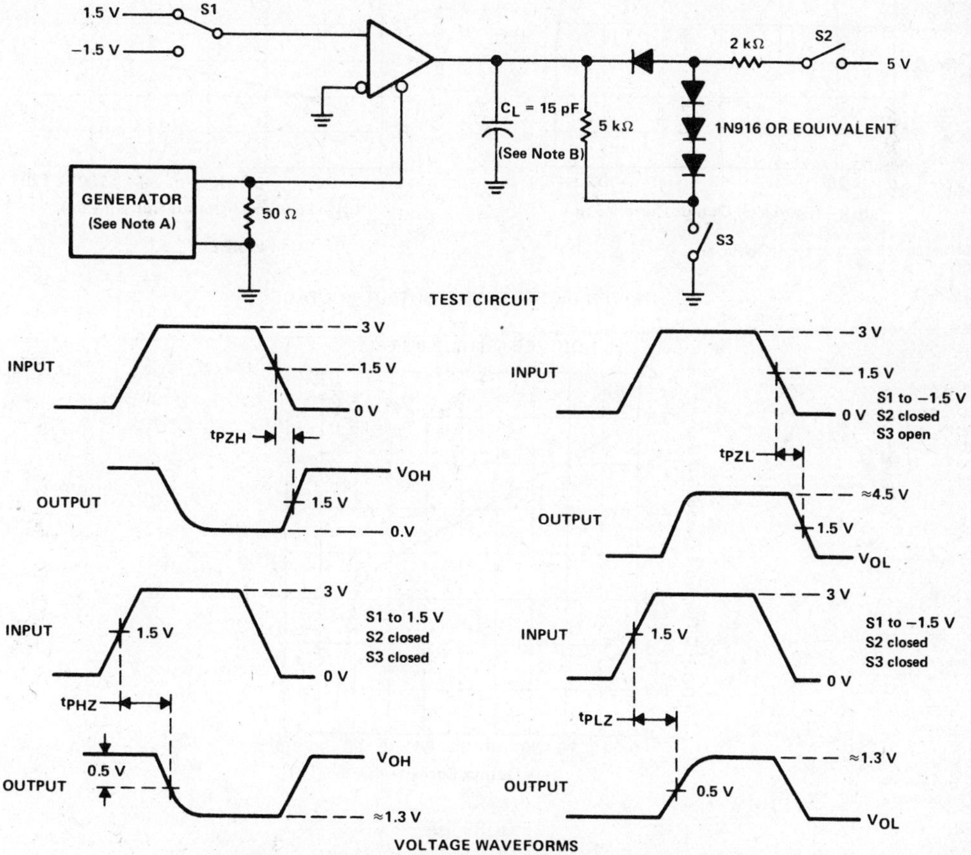
1.5 V
−1.5 V
S1
S2
2 kΩ
5 V
C_L = 15 pF
(See Note B)
5 kΩ
1N916 OR EQUIVALENT
GENERATOR
(See Note A)
50 Ω
S3
TEST CIRCUIT
INPUT
3 V
1.5 V
0 V
t_{PZH}
OUTPUT
V_{OH}
1.5 V
0 V
INPUT
3 V
1.5 V
0 V
S1 to −1.5 V
S2 closed
S3 open
t_{PZL}
OUTPUT
≈4.5 V
1.5 V
V_{OL}
INPUT
3 V
1.5 V
0 V
S1 to 1.5 V
S2 closed
S3 closed
t_{PHZ}
OUTPUT
0.5 V
V_{OH}
≈1.3 V
INPUT
3 V
1.5 V
0 V
S1 to −1.5 V
S2 closed
S3 closed
t_{PLZ}
OUTPUT
≈1.3 V
0.5 V
V_{OL}
VOLTAGE WAVEFORMS

FIGURE 7. RECEIVER OUTPUT ENABLE AND DISABLE TIMES

NOTES: A. The input pulse is supplied by a generator having the following characteristics: PRR ≤ 1 MHz, 50% duty cycle, t_r ≤ 6 ns, t_f ≤ 6 ns, Z_{out} = 50 Ω.

B. C_L includes probe and jig capacitance.

TYPICAL CHARACTERISTICS

DRIVER HIGH-LEVEL OUTPUT VOLTAGE
vs
DRIVER HIGH-LEVEL OUTPUT CURRENT

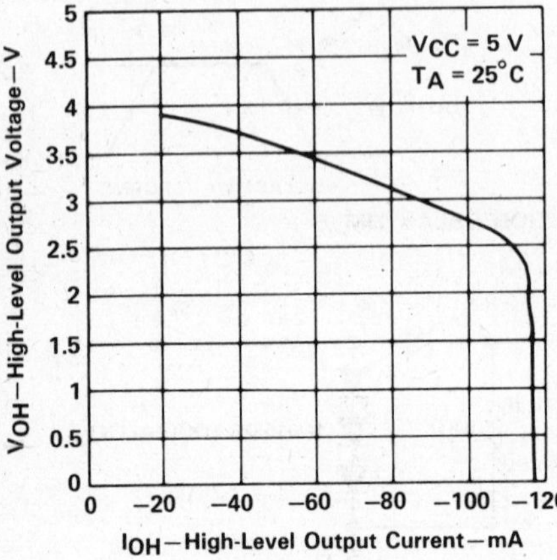
V_{CC} = 5 V
T_A = 25°C
V_{OH}—High-Level Output Voltage—V
I_{OH}—High-Level Output Current—mA
0 0.5 1 1.5 2 2.5 3 3.5 4 4.5 5
0 −20 −40 −60 −80 −100 −120

FIGURE 8

DRIVER LOW-LEVEL OUTPUT VOLTAGE
vs
DRIVER LOW-LEVEL OUTPUT CURRENT

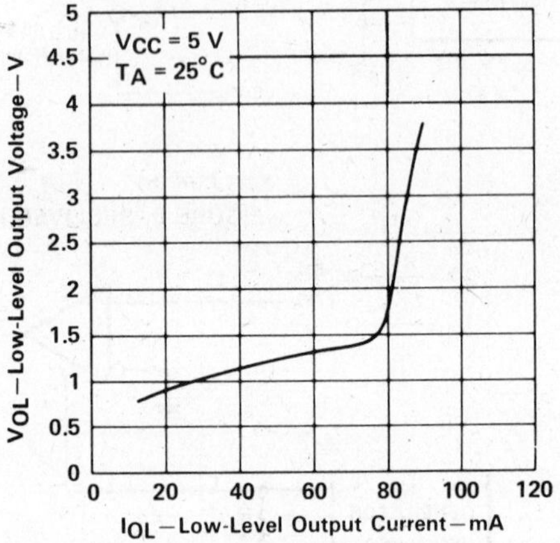
V_{CC} = 5 V
T_A = 25°C
V_{OL}—Low-Level Output Voltage—V
I_{OL}—Low-Level Output Current—mA
0 0.5 1 1.5 2 2.5 3 3.5 4 4.5 5
0 20 40 60 80 100 120

FIGURE 9

DRIVER DIFFERENTIAL OUTPUT VOLTAGE
vs
DRIVER OUTPUT CURRENT

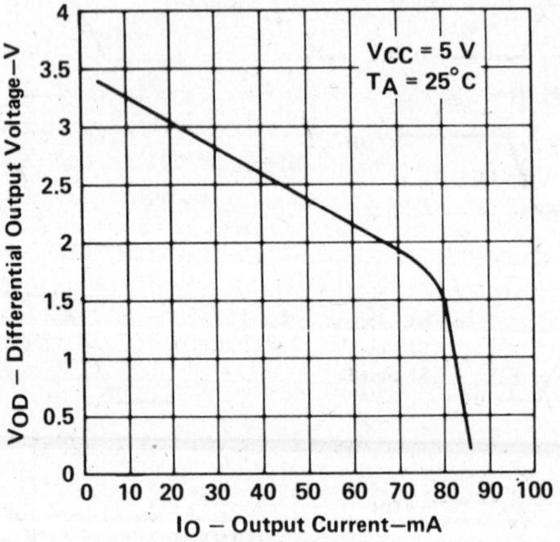
V_{CC} = 5 V
T_A = 25°C
V_{OD} – Differential Output Voltage—V
I_O – Output Current—mA
0 0.5 1 1.5 2 2.5 3 3.5 4
0 10 20 30 40 50 60 70 80 90 100

FIGURE 10

TEXAS INSTRUMENTS
POST OFFICE BOX 655012 • DALLAS, TEXAS 75265

TYPICAL CHARACTERISTICS

RECEIVER HIGH-LEVEL OUTPUT VOLTAGE
vs
HIGH-LEVEL OUTPUT CURRENT

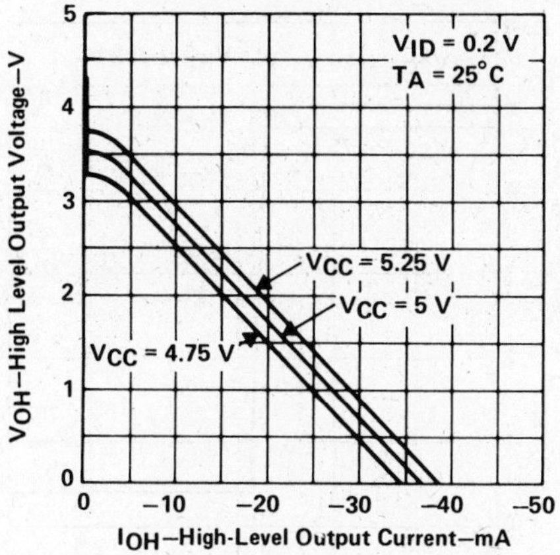
VID = 0.2 V
TA = 25°C
VCC = 5.25 V
VCC = 5 V
VCC = 4.75 V
VOH—High Level Output Voltage—V
IOH—High-Level Output Current—mA

FIGURE 11

RECEIVER HIGH-LEVEL OUTPUT
vs
FREE-AIR TEMPERATURE

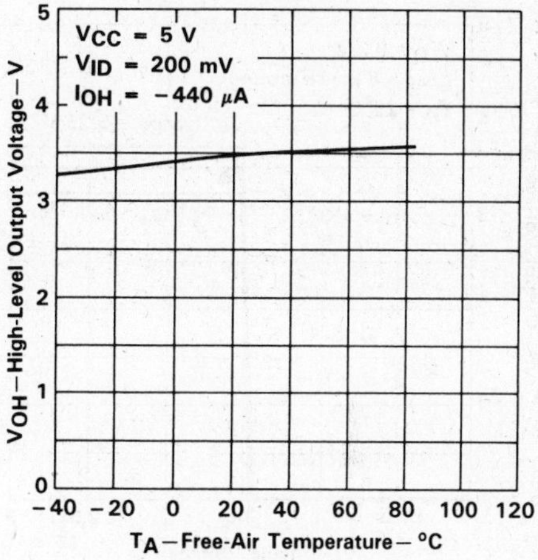
VCC = 5 V
VID = 200 mV
IOH = −440 μA
VOH—High-Level Output Voltage—V
TA—Free-Air Temperature—°C

FIGURE 12

RECEIVER LOW-LEVEL OUTPUT VOLTAGE
vs
RECEIVER LOW-LEVEL OUTPUT CURRENT

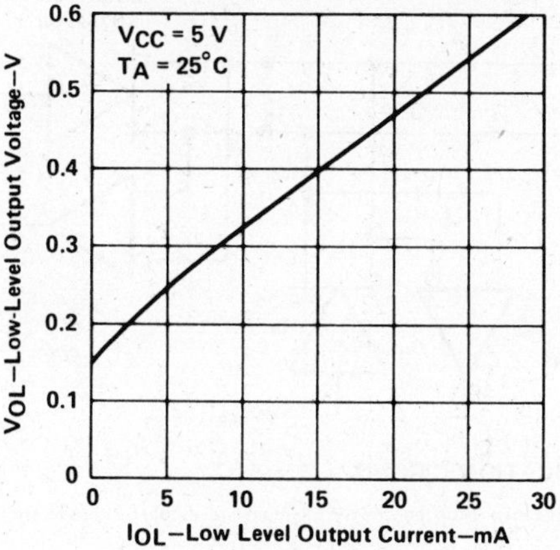
VCC = 5 V
TA = 25°C
VOL—Low-Level Output Voltage—V
IOL—Low Level Output Current—mA

FIGURE 13

RECEIVER LOW-LEVEL OUTPUT VOLTAGE
vs
FREE-AIR TEMPERATURE

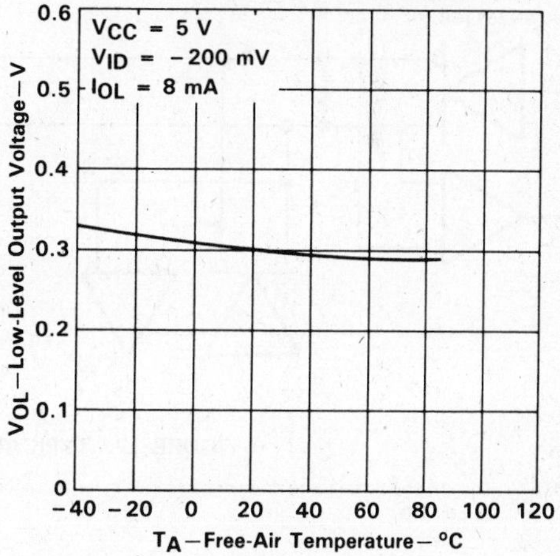
VCC = 5 V
VID = −200 mV
IOL = 8 mA
VOL—Low-Level Output Voltage—V
TA—Free-Air Temperature—°C

FIGURE 14

TYPICAL CHARACTERISTICS

RECEIVER OUTPUT VOLTAGE vs ENABLE VOLTAGE

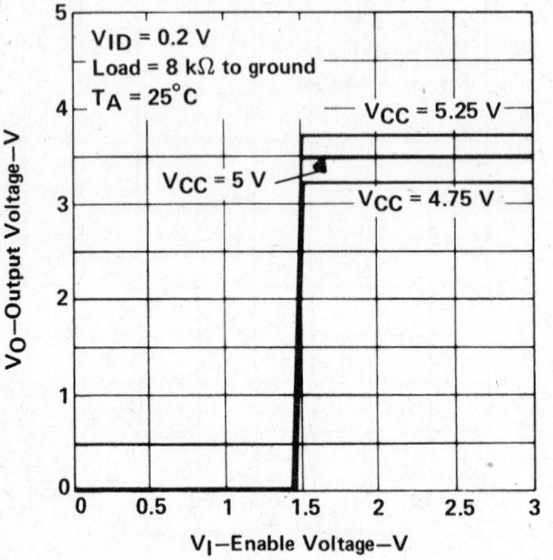
V_{ID} = 0.2 V
Load = 8 kΩ to ground
T_A = 25°C
V_{CC} = 5.25 V
V_{CC} = 5 V
V_{CC} = 4.75 V
V_O—Output Voltage—V
V_I—Enable Voltage—V
0
0.5
1
1.5
2
2.5
3
4
5

FIGURE 15

RECEIVER OUTPUT VOLTAGE vs ENABLE VOLTAGE

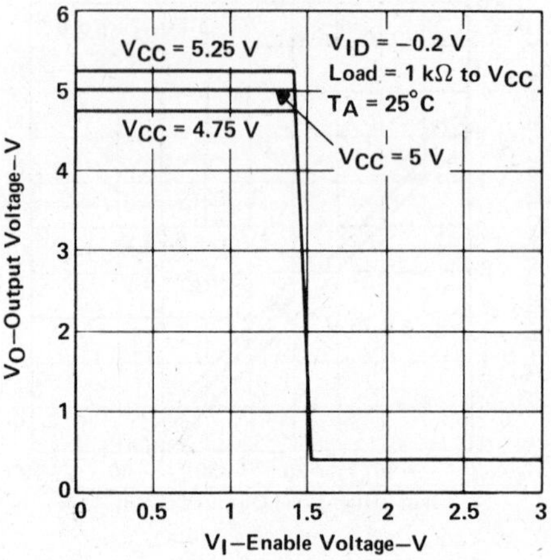
V_{CC} = 5.25 V
V_{ID} = −0.2 V
Load = 1 kΩ to V_{CC}
T_A = 25°C
V_{CC} = 4.75 V
V_{CC} = 5 V
V_O—Output Voltage—V
V_I—Enable Voltage—V
0
0.5
1
1.5
2
2.5
3
4
5
6

FIGURE 16

TYPICAL APPLICATION

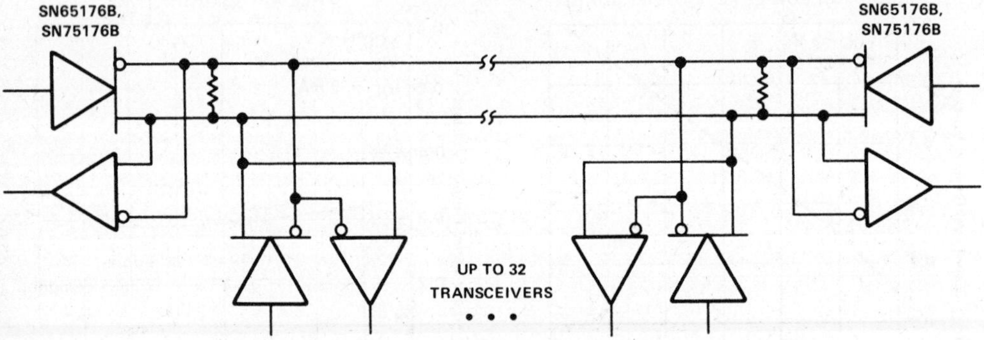
SN65176B, SN75176B
SN65176B, SN75176B
UP TO 32 TRANSCEIVERS

FIGURE 17. TYPICAL APPLICATION CIRCUIT

NOTE 7: The line should be terminated at both ends in its characteristic impedance. Stub lengths off the main line should be kept as short as possible.

TEXAS INSTRUMENTS
POST OFFICE BOX 655012 • DALLAS, TEXAS 75265

- Meets EIA Standards RS-422-A and RS-485 and CCITT Recommendations V.11 and X.27
- Designed for Multipoint Transmission on Long Bus Lines in Noisy Environments
- 3-State Outputs
- Bus Voltage Range . . . −7 V to 12 V
- Positive and Negative Current Limiting
- Driver Output Capability . . . 60 mA Max
- Driver Thermal Shutdown Protection
- Receiver Input Impedance . . . 12 kΩ Min
- Receiver Input Sensitivity . . . ±200 mV
- Receiver Input Hysteresis . . . 50 mV Typ
- Operates from Single 5-Volt Supply
- Low Power Requirements

SN75177B . . . D, JG, OR P
DUAL-IN-LINE PACKAGE
(TOP VIEW)

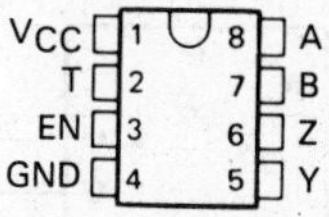
VCC 1
T 2
EN 3
GND 4
8 A
7 B
6 Z
5 Y

SN75178B . . . D, JG, OR P
DUAL-IN-LINE PACKAGE
(TOP VIEW)

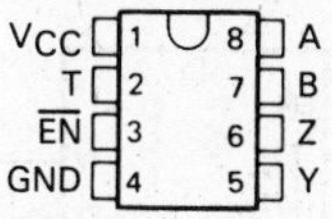
VCC 1
T 2
EN 3
GND 4
8 A
7 B
6 Z
5 Y

SN75177B FUNCTION TABLE

DIFFERENTIAL INPUTS A−B	ENABLE EN	OUTPUTS T	Y	Z
$V_{ID} \geq 0.2$ V	H	H	H	L
$-0.2\ V < V_{ID} < 0.2$ V	H	?	?	?
$V_{ID} \leq 0.2$ V	H	L	L	H
X	L	Z	Z	Z

SN75178B FUNCTION TABLE

DIFFERENTIAL INPUTS A−B	ENABLE $\overline{EN}$	OUTPUTS T	Y	Z
$V_{ID} \geq 0.2$ V	L	H	H	L
$-0.2\ V < V_{ID} < 0.2$ V	L	?	?	?
$V_{ID} \leq 0.2$ V	L	L	L	H
X	H	Z	Z	Z

H = high level, L = low level, ? = indeterminate,
X = irrelevant, Z = impedance (off)

description

The SN75177B and SN75178B differential bus repeaters are monolithic integrated devices each designed for one-way data communication on multipoint bus transmission lines. These devices are designed for balanced transmission bus line applications and meet EIA Standards RS-422-A and RS-485 and CCITT Recommendations V.11 and X.27. Each device is designed to improve the performance of the data communication over long bus lines. The SN75177B and SN75178B are identical except for the complementary enable inputs, which allow the devices to be used in pairs for bidirectional communication.

The SN75177B and SN75178B feature positive- and negative-current limiting three-state outputs for the receiver and driver. The receiver features high input impedance, input hysteresis for increased noise immunity, and input sensitivity of ±200 millivolts over a common-mode input voltage range of −7 volts to 12 volts. The driver features thermal shutdown for protection from line fault conditions. Thermal shutdown is designed to occur at a junction temperature of approximately 150°C. The driver is designed to drive current loads up to 60 milliamperes maximum.

The SN75177B and SN75178B are designed for optimum performance when used on transmission buses employing the SN75172 and SN75174 differential line drivers, SN75173 and SN75175 differential line receivers, or SN75176B bus transceivers.

logic symbols†

SN75177B

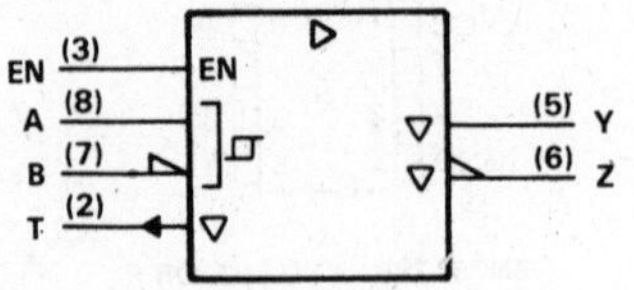
EN (3)
EN
A (8)
B (7)
T (2)
(5) Y
(6) Z

SN75178B

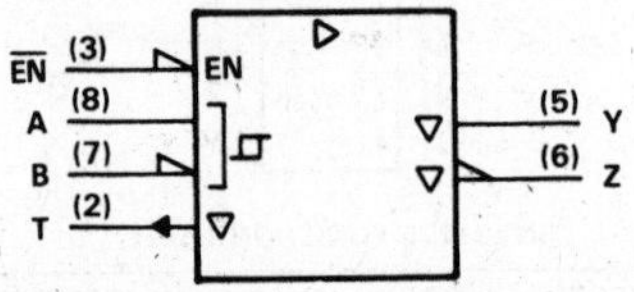
EN (3)
EN
A (8)
B (7)
T (2)
(5) Y
(6) Z

†These symbols are in accordance with ANSI/IEEE Std 91-1984 and IEC Publication 617-12.

logic diagrams (positive logic)

SN75177B

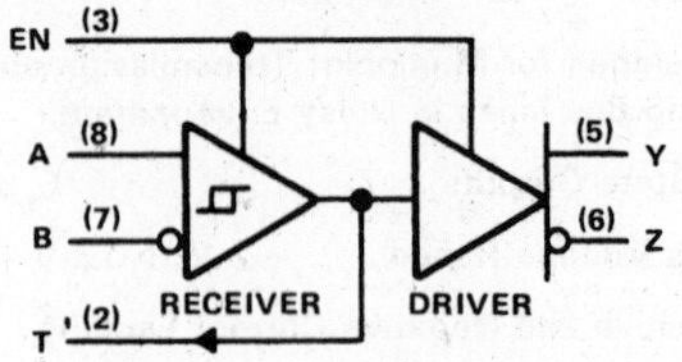
EN (3)
A (8)
B (7)
T (2)
RECEIVER
DRIVER
(5) Y
(6) Z

SN75178B

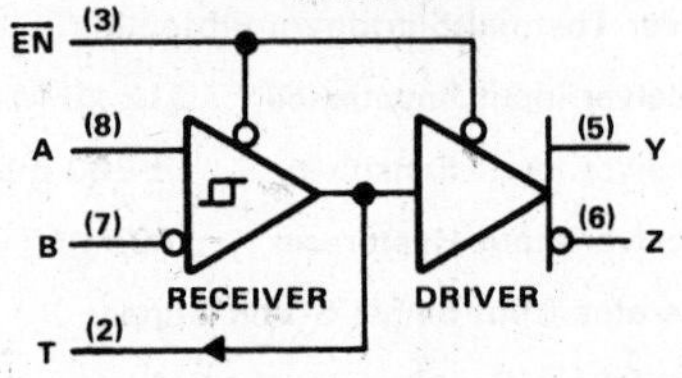
EN (3)
A (8)
B (7)
T (2)
RECEIVER
DRIVER
(5) Y
(6) Z

schematics of inputs and outputs

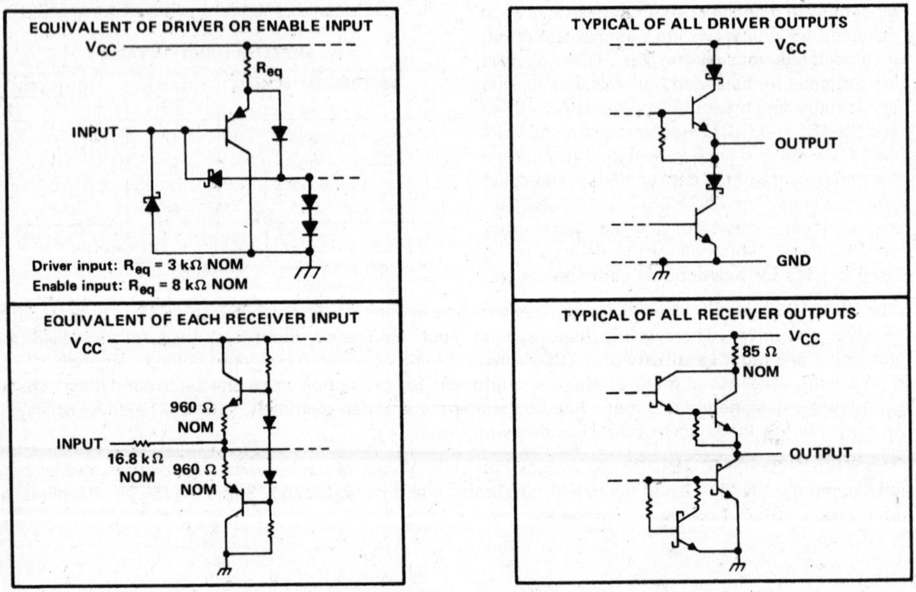
EQUIVALENT OF DRIVER OR ENABLE INPUT
VCC
Req
INPUT
Driver input: Req = 3 kΩ NOM
Enable input: Req = 8 kΩ NOM
TYPICAL OF ALL DRIVER OUTPUTS
VCC
OUTPUT
GND
EQUIVALENT OF EACH RECEIVER INPUT
VCC
960 Ω
NOM
INPUT
16.8 kΩ
NOM
960 Ω
NOM
TYPICAL OF ALL RECEIVER OUTPUTS
VCC
85 Ω
NOM
OUTPUT

TEXAS INSTRUMENTS
POST OFFICE BOX 655012 • DALLAS, TEXAS 75265

absolute maximum ratings over operating free-air temperature range (unless otherwise noted)

Supply voltage, V_{CC} (see Note 1) .... 7 V
Voltage at any bus terminal .... −10 V to 15 V
Differential input voltage (see Note 2) .... ±25 V
Enable input voltage .... 5.5 V
Continuous total dissipation at (or below) 25°C free-air temperature (see Note 3):
  D package .... 725 mW
  JG package .... 825 mW
  P package .... 1000 mW
Operating free-air temperature range .... 0°C to 70°C

NOTES: 1. All voltage values, except differential input voltage, are with respect to network ground terminal.
2. Differential input voltage is measured at the noninverting input with respect to the corresponding inverting input.
3. For operation above 25°C free-air temperature, derate the D package to 464 mW at 70°C at the rate of 5.8 mW/°C, the JG package to 528 mW at 70°C at the rate of 6.6 mW/°C and the P package to 640 mW at 70°C at the rate of 8.0 mW/°C. In the JG package, SN75177B and SN75178B chips are glass mounted.

recommended operating conditions

		MIN	NOM	MAX	UNIT
Supply voltage, V_{CC}		4.75	5	5.25	V
High-level input voltage, V_{IH}	EN or $\overline{EN}$	2			V
Low-level input voltage, V_{IL}	EN or $\overline{EN}$			0.8	V
Common-mode input voltage, V_{IC}		−7†		12	V
Differential input voltage, V_{ID}				±12	V
High-level output current, I_{OH}	Driver			−60	mA
	Receiver			−400	μA
Low-level output current, I_{OL}	Driver			60	mA
	Receiver			8	mA
Operating free-air temperature, T_A		0		70	°C

† The algebraic convention, where the less-positive (more-negative) limit is designated minimum, is used in this data sheet for common-mode input voltage and threshold voltage.

DRIVER SECTION

driver electrical characteristics over recommended ranges of supply voltage and operating free-air temperature (unless otherwise noted)

PARAMETER		TEST CONDITIONS		MIN	TYP†	MAX	UNIT
V_{IK}	Input clamp voltage	$I_I = -18$ mA				−1.5	V
V_O	Output voltage	$I_O = 0$		0		6	V
$\|V_{OD1}\|$	Differential output voltage	$I_O = 0$		1.5		6	V
$\|V_{OD2}\|$	Differential output voltage	$R_L = 100\ \Omega$, See Figure 1		½V_{OD1}			V
				2			V
		$R_L = 54\ \Omega$, See Figure 1		1.5	2.5	5	V
$\|V_{OD3}\|$	Differential output voltage	See Note 4		1.5		5	V
$\Delta\|V_{OD}\|$	Change in magnitude of differential output voltage‡	$R_L = 54\ \Omega$ or $100\ \Omega$, See Figure 1				±0.2	V
V_{OC}	Common-mode output voltage					3	V
						−1	
$\Delta\|V_{OC}\|$	Change in magnitude of common-mode output voltage‡					±0.2	V
I_O	Output current	$V_{CC} = 0$, $V_O = -7$ V to 12 V				±100	μA
I_{OZ}	High-impedance-state output current	$V_O = -7$ V to 12 V				±100	μA
I_{IH}	High-level input current	$V_I = 2.4$ V,				20	μA
I_{IL}	Low-level input current	$V_I = 0.4$ V,				−400	μA
I_{OS}	Short-circuit output current	$V_O = -7$ V				−250	mA
		$V_O = V_{CC}$				250	mA
		$V_O = 12$ V				250	mA
I_{CC}	Supply current (total package)	No load	Outputs enabled		57	70	mA
			Outputs disabled		26	35	mA

†All typical values are at $V_{CC} = 5$ V and $T_A = 25°C$.
‡$\Delta|V_{OD}|$ and $\Delta|V_{OC}|$ are the changes in magnitude of V_{OD} and V_{OC}, respectively, that occur when the input is changed from a high level to a low level.
NOTE 4: See EIA Standard RS-485 Figure 3.5, Test Termination Measurement 2.

driver switching characteristics, $V_{CC} = 5$ V, $T_A = 25°C$

PARAMETER		TEST CONDITIONS		MIN	TYP	MAX	UNIT
t_{DD}	Differential-output delay time	$R_L = 54\ \Omega$,	See Figure 3		15	22	ns
t_{TD}	Differential-output transition time				20	30	ns
t_{PZH}	Output enable time to high level	$R_L = 110\ \Omega$,	See Figure 4		85	120	ns
t_{PZL}	Output enable time to low level	$R_L = 110\ \Omega$,	See Figure 5		40	60	ns
t_{PHZ}	Output disable time from high level	$R_L = 110\ \Omega$,	See Figure 4		150	250	ns
t_{PLZ}	Output disable time from low level	$R_L = 110\ \Omega$,	See Figure 5		20	30	ns

TEXAS INSTRUMENTS
POST OFFICE BOX 655012 • DALLAS, TEXAS 75265

SYMBOL EQUIVALENTS

DATA SHEET PARAMETER	RS-422-A	RS-485
V_O	V_{oa}, V_{ob}	V_{oa}, V_{ob}
$\|V_{OD1}\|$	V_o	V_o
$\|V_{OD2}\|$	V_t (R_L = 100 Ω)	V_t (R_L = 54 Ω)
$\|V_{OD3}\|$		V_t (Test termination Measurement 2)
$\Delta\|V_{OD}\|$	$\| \|V_t\| - \|\bar{V}_t\| \|$	$\| \|V_t\| - \|\bar{V}_t\| \|$
V_{OC}	$\|V_{os}\|$	$\|V_{os}\|$
$\Delta\|V_{OC}\|$	$\| V_{os} - \bar{V}_{os} \|$	$\| V_{os} - \bar{V}_{os} \|$
I_{OS}	$\|I_{sa}\|$, $\|I_{sb}\|$	
I_O	$\|I_{xa}\|$, $\|I_{xb}\|$	I_{ia}, I_{ib}

RECEIVER SECTION

receiver electrical characteristics over recommended ranges of common-mode input voltage, supply voltage, and operating free-air temperature (unless otherwise noted)

PARAMETER		TEST CONDITIONS		MIN	TYP†	MAX	UNIT
V_{TH}	Differential-input high-threshold voltage	V_O = 2.7 V,	I_O = −0.4 mA			0.2	V
V_{TL}	Differential-input low-threshold voltage	V_O = 0.5 V,	I_O = 8 mA	−0.2‡			V
V_{hys}	Hysteresis§				50		mV
V_{IK}	Enable-input clamp voltage	I_I = −18 mA				−1.5	V
V_{OH}	High-level output voltage	V_{ID} = 200 mV, See Figure 2	I_{OH} = −400 μA,	2.7			V
V_{OL}	Low-level output voltage	V_{ID} = −200 mV, See Figure 2	I_{OL} = 8 mA,			0.45	V
I_{OZ}	High-impedance-state output current	V_O = 0.4 V to 2.4 V				20	μA
						−400	μA
I_I	Line input current	Other input at 0 V, See Note 5	V_I = 12 V			1	mA
			V_I = −7 V			−0.8	mA
I_{IH}	High-level enable-input current	V_{IH} = 2.7 V				20	μA
I_{IL}	Low-level enable-input current	V_{IL} = 0.4 V				−100	μA
r_i	Input resistance			12			kΩ
I_{OS}	Short-circuit output current			−15		−85	mA
I_{CC}	Supply current (total package)	No load	Outputs enabled		57	70	mA
			Outputs disabled		26	35	mA

† All typical values are at V_{CC} = 5 V, T_A = 25°C.
‡ The algebraic convention, where the less-positive (more-negative) limit is designated minimum, is used in this data sheet for common-mode input voltage and threshold voltage levels only.
§ Hysteresis is the difference between the positive-going input threshold voltage, V_{T+}, and the negative-going input threshold voltage, V_{T-}. See Figure 12.
NOTE 5: Refer to EIA Standard RS-422-A for exact conditions.

receiver switching characteristics, V_{CC} = 5 V, T_A = 25°C

PARAMETER		TEST CONDITIONS	MIN	TYP	MAX	UNIT
t_{PLH}	Propagation delay time, low-to-high-level output	V_{ID} = −1.5 V to 1.5 V, C_L = 15 pF, See Figure 6		19	35	ns
t_{PHL}	Propagation delay time, high-to-low-level output			30	40	ns
t_{PZH}	Output enable time to high level	C_L = 15 pF, See Figure 7		10	20	ns
t_{PZL}	Output enable time to low level			12	20	ns
t_{PHZ}	Output disable time from high level	C_L = 15 pF, See Figure 7		25	35	ns
t_{PLZ}	Output disable time from low level			17	25	ns

PARAMETER MEASUREMENT INFORMATION

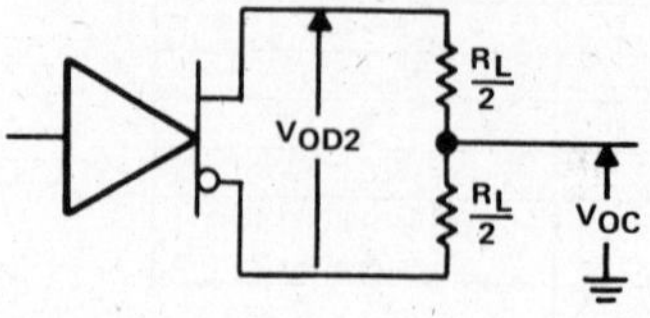
VOD2
RL/2
RL/2
VOC

FIGURE 1. DRIVER V_{OD} AND V_{OC}

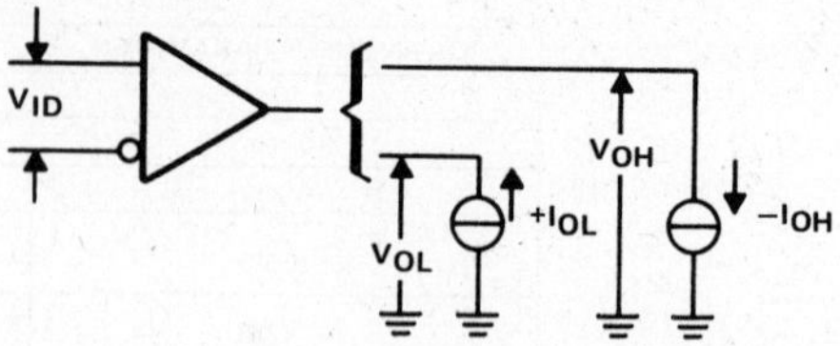
VID
VOH
+IOL
−IOH
VOL

FIGURE 2. RECEIVER V_{OH} AND V_{OL}

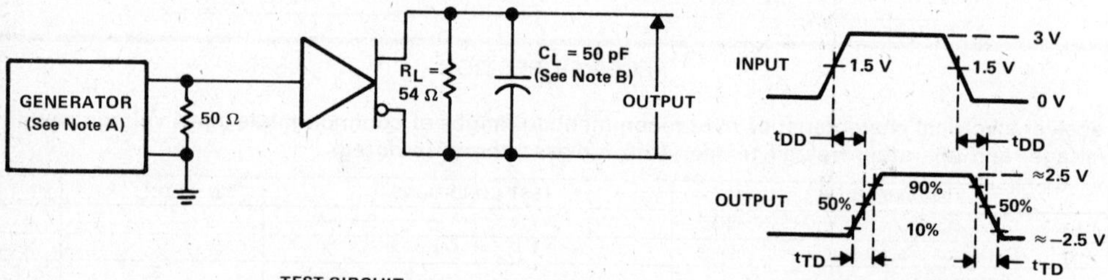
GENERATOR
(See Note A)
50 Ω
RL = 54 Ω
CL = 50 pF
(See Note B)
OUTPUT
INPUT
1.5 V
1.5 V
3 V
0 V
tDD
tDD
≈2.5 V
90%
OUTPUT
50%
50%
10%
≈−2.5 V
tTD
tTD

TEST CIRCUIT

VOLTAGE WAVEFORMS

FIGURE 3. DRIVER DIFFERENTIAL-OUTPUT DELAY AND TRANSITION TIMES

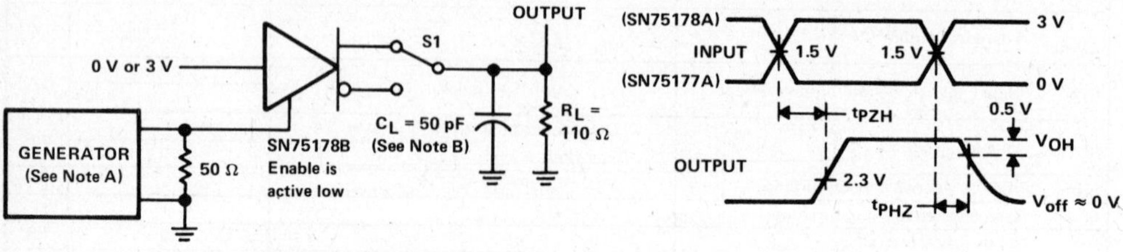
OUTPUT
S1
0 V or 3 V
GENERATOR
(See Note A)
50 Ω
SN75178B
Enable is
active low
CL = 50 pF
(See Note B)
RL =
110 Ω
(SN75178A)
INPUT
(SN75177A)
1.5 V
1.5 V
3 V
0 V
tPZH
0.5 V
VOH
OUTPUT
2.3 V
tPHZ
Voff ≈ 0 V

TEST CIRCUIT

VOLTAGE WAVEFORMS

FIGURE 4. DRIVER ENABLE AND DISABLE TIMES (t_{PZH}, t_{PHZ})

NOTES: A. The input pulse is supplied by a generator having the following characteristics: PRR ≤ 1 MHz, 50% duty cycle, t_r ≤6 ns, t_f ≤6 ns, Z_{out} = 50 Ω.

B. C_L includes probe and jig capacitance.

PARAMETER MEASUREMENT INFORMATION

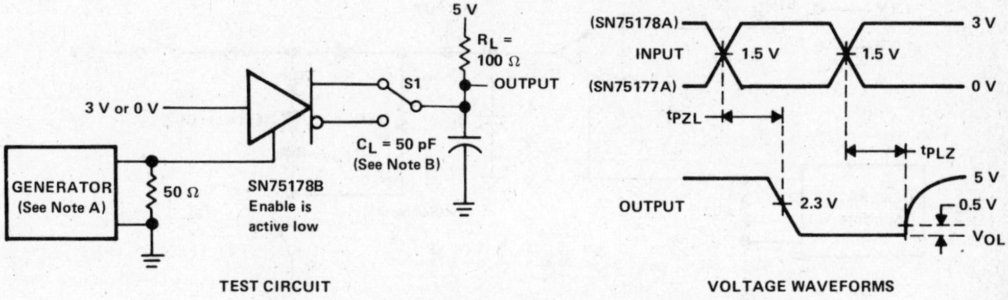
5 V
R_L =
100 Ω
S1
OUTPUT
3 V or 0 V
C_L = 50 pF
(See Note B)
GENERATOR
(See Note A)
50 Ω
SN75178B
Enable is
active low
TEST CIRCUIT
(SN75178A)
INPUT
(SN75177A)
1.5 V
1.5 V
3 V
0 V
t_{PZL}
t_{PLZ}
OUTPUT
2.3 V
5 V
0.5 V
V_{OL}
VOLTAGE WAVEFORMS

FIGURE 5. DRIVER ENABLE AND DISABLE TIMES (t_{PZL}, t_{PLZ})

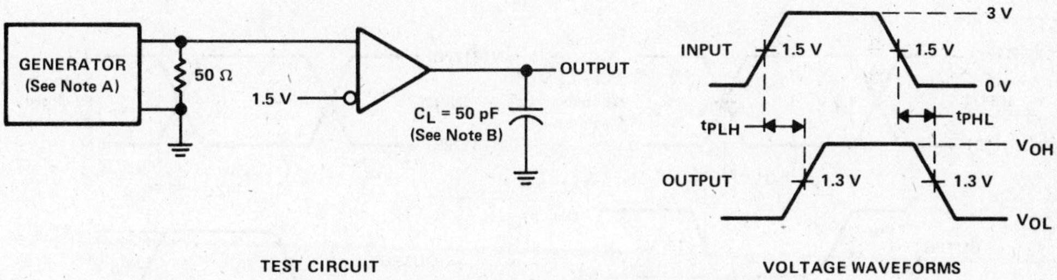
GENERATOR
(See Note A)
50 Ω
1.5 V
OUTPUT
C_L = 50 pF
(See Note B)
TEST CIRCUIT
3 V
INPUT
1.5 V
1.5 V
0 V
t_{PLH}
t_{PHL}
V_{OH}
OUTPUT
1.3 V
1.3 V
V_{OL}
VOLTAGE WAVEFORMS

FIGURE 6. RECEIVER PROPAGATION DELAY TIMES

NOTES: A. The input pulse is supplied by a generator having the following characteristics: PRR ≤ 1 MHz, 50% duty cycle, t_r ≤6 ns, t_f ≤6 ns, Z_{out} = 50 Ω.

B. C_L includes probe and jig capacitance.

PARAMETER MEASUREMENT INFORMATION

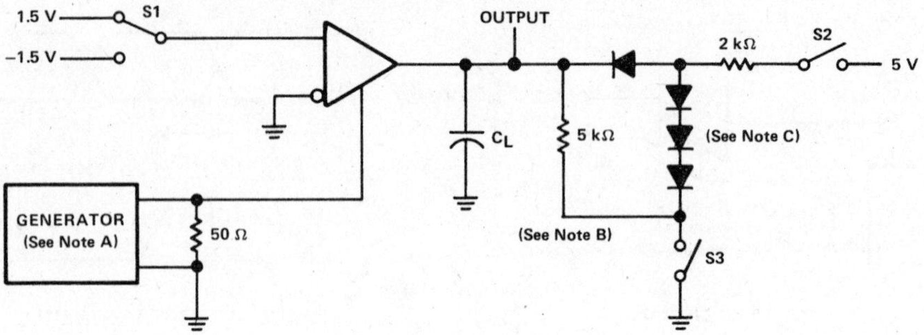
1.5 V
S1
−1.5 V
OUTPUT
2 kΩ
S2
5 V
C_L
5 kΩ
(See Note C)
GENERATOR
(See Note A)
50 Ω
(See Note B)
S3

TEST CIRCUIT

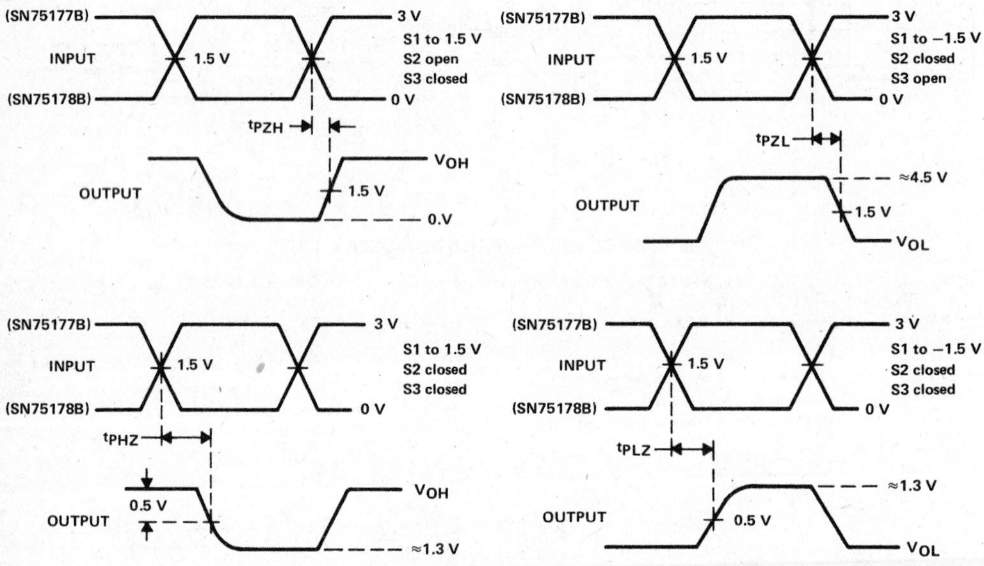
(SN75177B)
INPUT
(SN75178B)
1.5 V
3 V
0 V
S1 to 1.5 V
S2 open
S3 closed
t_{PZH}
OUTPUT
V_{OH}
1.5 V
0.V
S1 to −1.5 V
S2 closed
S3 open
t_{PZL}
≈4.5 V
V_{OL}
S1 to 1.5 V
S2 closed
S3 closed
t_{PHZ}
0.5 V
≈1.3 V
S1 to −1.5 V
S2 closed
S3 closed
t_{PLZ}

VOLTAGE WAVEFORMS

NOTES: A. The input pulse is supplied by a generator having the following characteristics: PRR ≤ 1 MHz, duty cycle ≈ 50%, t_r = t_f = 6 ns.
B. C_L includes probe and jig capacitance.
C. All diodes are 1N916 or equivalent.

FIGURE 7. RECEIVER ENABLE AND DISABLE TIMES

TYPICAL CHARACTERISTICS

DRIVER HIGH-LEVEL OUTPUT VOLTAGE
vs
HIGH-LEVEL OUTPUT CURRENT

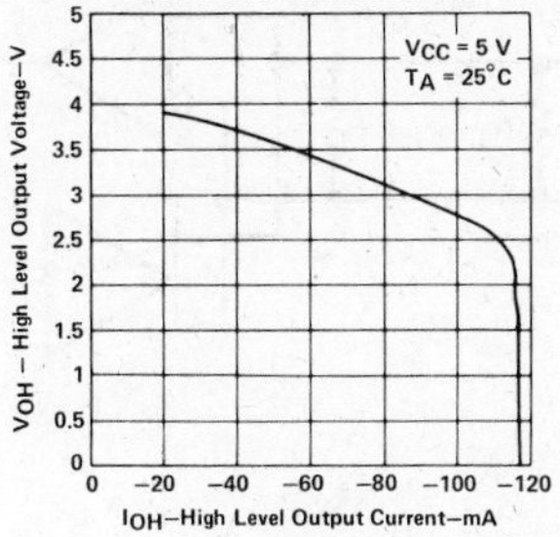
VCC = 5 V
TA = 25°C
VOH – High Level Output Voltage–V
IOH–High Level Output Current–mA

FIGURE 8

DRIVER LOW-LEVEL OUTPUT VOLTAGE
vs
LOW-LEVEL OUTPUT CURRENT

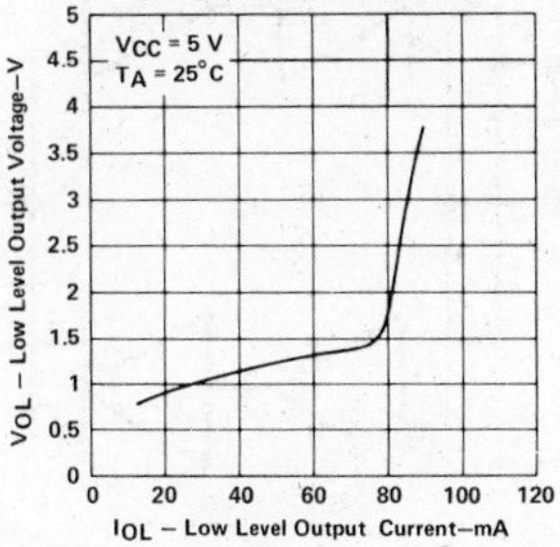
VCC = 5 V
TA = 25°C
VOL – Low Level Output Voltage–V
IOL – Low Level Output Current–mA

FIGURE 9

DRIVER DIFFERENTIAL OUTPUT VOLTAGE
vs
DRIVER OUTPUT CURRENT

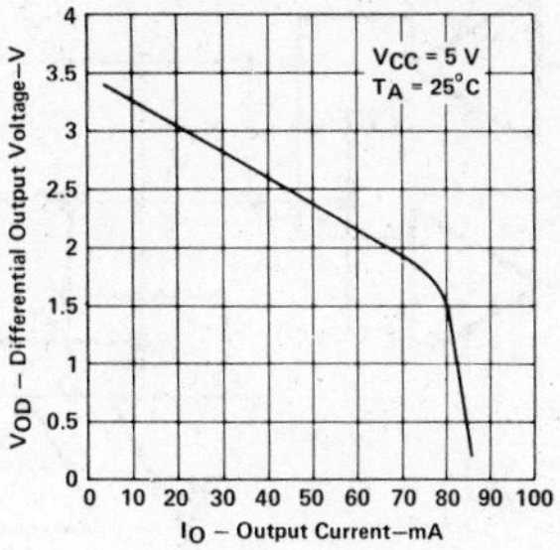
VCC = 5 V
TA = 25°C
VOD – Differential Output Voltage–V
IO – Output Current–mA

FIGURE 10

RECEIVER OUTPUT VOLTAGE
vs
DIFFERENTIAL INPUT VOLTAGE

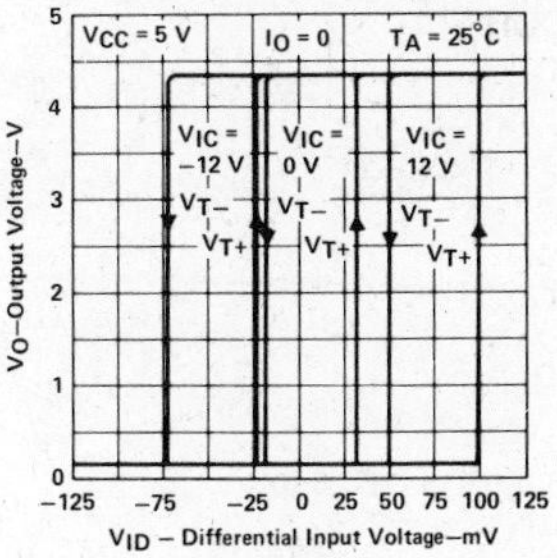
VCC = 5 V
IO = 0
TA = 25°C
VIC = –12 V
VIC = 0 V
VIC = 12 V
VT–
VT+
VO–Output Voltage–V
VID – Differential Input Voltage–mV

FIGURE 11

RECEIVER HIGH-LEVEL OUTPUT VOLTAGE
vs
HIGH-LEVEL OUTPUT CURRENT

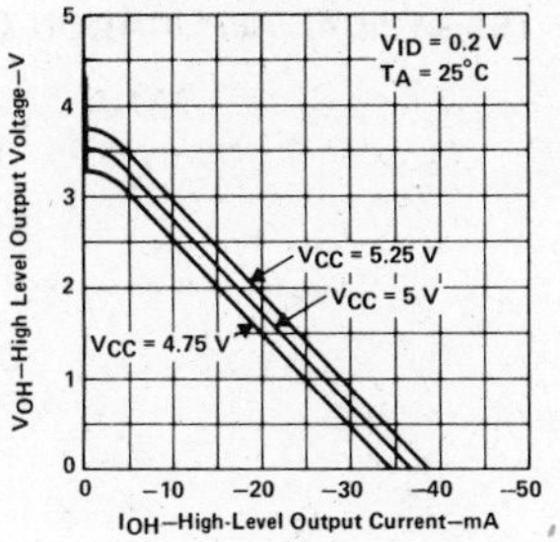
VID = 0.2 V
TA = 25°C
VCC = 5.25 V
VCC = 5 V
VCC = 4.75 V
VOH–High Level Output Voltage–V
IOH–High-Level Output Current–mA

FIGURE 12

RECEIVER HIGH-LEVEL OUTPUT VOLTAGE
vs
FREE-AIR TEMPERATURE

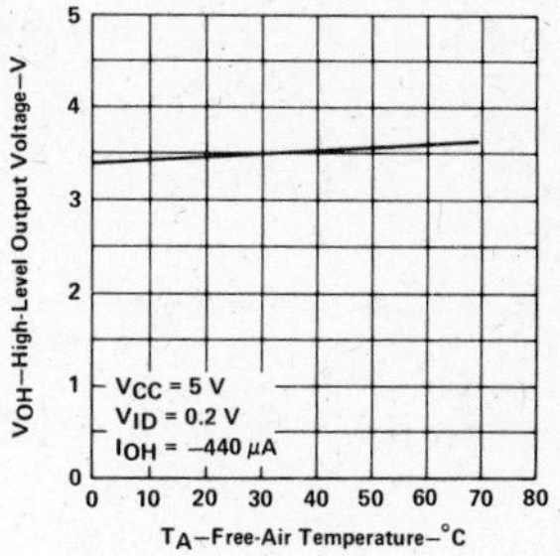
VCC = 5 V
VID = 0.2 V
IOH = –440 μA
VOH–High-Level Output Voltage–V
TA–Free-Air Temperature–°C

FIGURE 13

RECEIVER LOW-LEVEL OUTPUT VOLTAGE
vs
LOW-LEVEL OUTPUT CURRENT

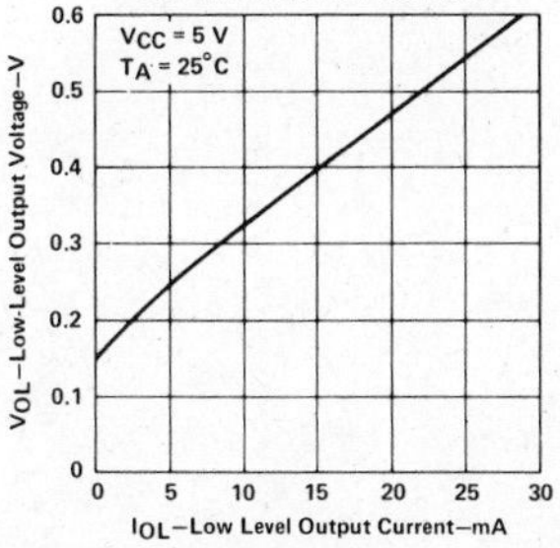
VCC = 5 V
TA = 25°C
VOL–Low-Level Output Voltage–V
IOL–Low Level Output Current–mA

FIGURE 14

RECEIVER LOW-LEVEL OUTPUT VOLTAGE
vs
FREE-AIR TEMPERATURE

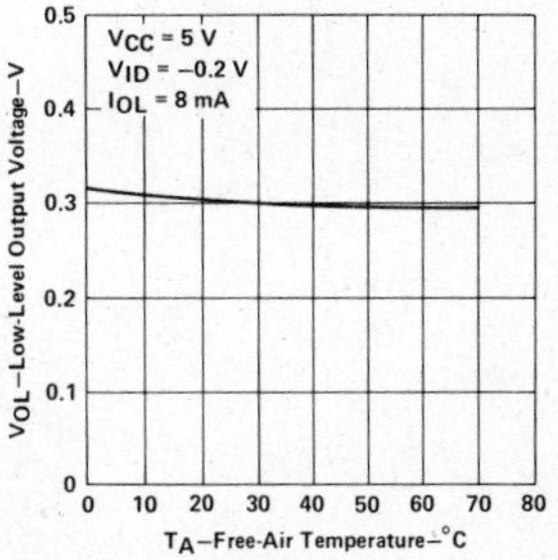
VCC = 5 V
VID = –0.2 V
IOL = 8 mA
VOL–Low-Level Output Voltage–V
TA–Free-Air Temperature–°C

FIGURE 15

TYPICAL APPLICATION

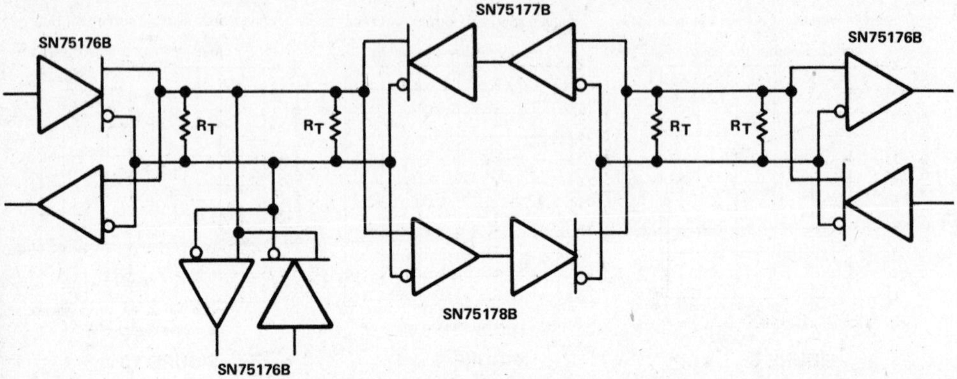
SN75177B
SN75176B
SN75176B
RT
RT
RT
RT
SN75178B
SN75176B

NOTE 6: The line should be terminated at both ends in its characteristic impedance. Stub lengths off the main line should be kept as short as possible.

FIGURE 16. TYPICAL APPLICATION CIRCUIT

TEXAS INSTRUMENTS
POST OFFICE BOX 655012 • DALLAS, TEXAS 75265

SN75179B
DIFFERENTIAL DRIVER AND RECEIVER PAIR

D2845, OCTOBER 1985

- Meets EIA Standards RS-422-A and RS-485 and CCITT Recommendations V.11 and X.27
- Bus Voltage Range . . . −7 V to 12 V
- Positive and Negative Current Limiting
- Driver Output Capability . . . 60 mA Max
- Driver Thermal Shutdown Protection
- Receiver Input Impedance . . . 12 kΩ Min
- Receiver Input Sensitivity . . . ±200 mV
- Receiver Input Hysteresis . . . 50 mV Typ
- Operates from Single 5-V Supply
- Low Power Requirements

D, JG, OR P DUAL-IN-LINE PACKAGE
(TOP VIEW)

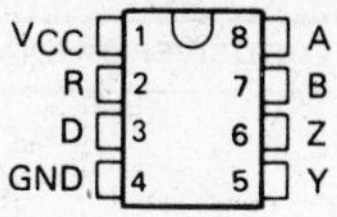
VCC 1 8 A
R 2 7 B
D 3 6 Z
GND 4 5 Y

FUNCTION TABLE (DRIVER)

INPUT	OUTPUTS	
D	Y	Z
H	H	L
L	L	H

FUNCTION TABLE (RECEIVER)

DIFFERENTIAL INPUTS A − B	OUTPUT R
$V_{ID} \geq 0.2$ V	H
-0.2 V $< V_{ID} <$ 0.2 V	?
$V_{ID} \leq -0.2$ V	L

H = high level, L = low level, ? = indeterminate

description

The SN75179B driver and bus receiver circuit is a monolithic integrated device designed for balanced transmission line applications and meets EIA Standards RS-422-A and RS-485 and CCITT Recommendations V.11 and X.27. It is designed to improve the performance of full-duplex data communications over long bus lines.

The SN75179B driver outputs provide limiting for both positive and negative currents. The receiver features high input impedance, input hysteresis for increased noise immunity, and input sensitivity of ±200 millivolts over a common-mode input voltage range of −12 volts to 12 volts. The driver provides thermal shutdown for protection from line fault conditions. Thermal shutdown is designed to occur at a junction temperature of approximately 150 °C. The device is designed to drive current loads of up to 60 milliamperes maximum.

The SN75179B is characterized for operation from 0 °C to 70 °C.

logic symbol†

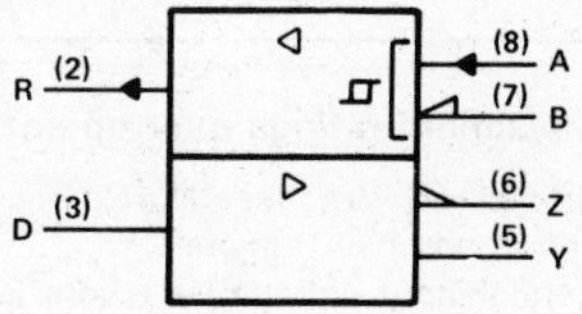
R (2)
D (3)
(8) A
(7) B
(6) Z
(5) Y

†This symbol is in accordance with ANSI/IEEE Std 91-1984 and IEC Publication 617-12.

logic diagram

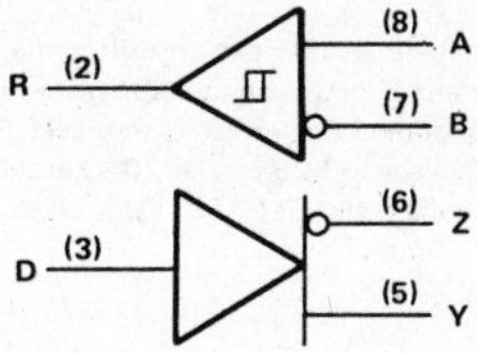
R (2)
D (3)
(8) A
(7) B
(6) Z
(5) Y

4

Line Drivers/Receivers

ADVANCE INFORMATION

ADVANCE INFORMATION documents contain information on new products in the sampling or preproduction phase of development. Characteristic data and other specifications are subject to change without notice.

TEXAS INSTRUMENTS
POST OFFICE BOX 655012 • DALLAS, TEXAS 75265

schematics of inputs and outputs

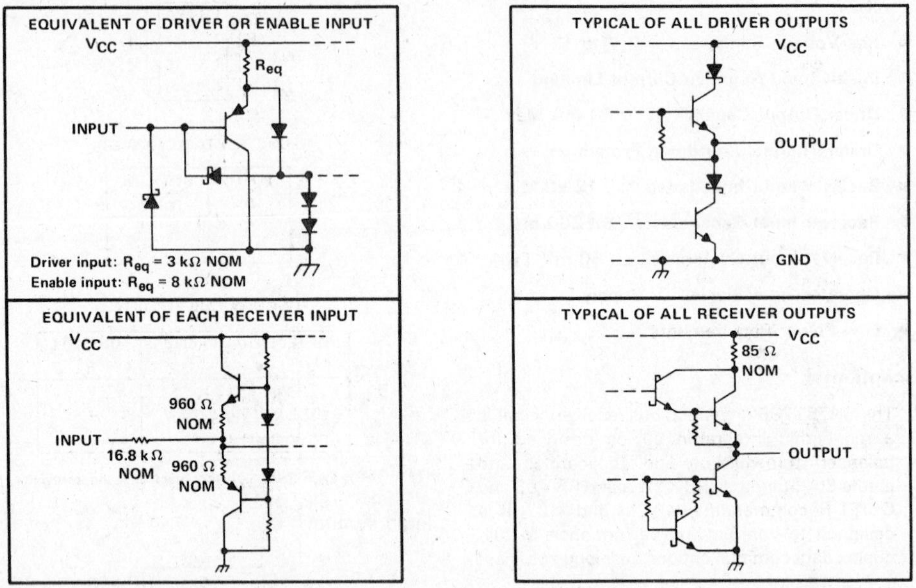
EQUIVALENT OF DRIVER OR ENABLE INPUT
VCC
Req
INPUT
Driver input: Req = 3 kΩ NOM
Enable input: Req = 8 kΩ NOM
TYPICAL OF ALL DRIVER OUTPUTS
VCC
OUTPUT
GND
EQUIVALENT OF EACH RECEIVER INPUT
VCC
960 Ω
NOM
INPUT
16.8 kΩ
NOM
960 Ω
NOM
TYPICAL OF ALL RECEIVER OUTPUTS
VCC
85 Ω
NOM
OUTPUT

absolute maximum ratings over operating free-air temperature range (unless otherwise noted)

Supply voltage, V_{CC} (see Note 1)	7 V
Voltage at any bus terminal	−10 V to 15 V
Differential input voltage (see Note 2)	±25 V
Continuous total dissipation at (or below) 25°C free-air temperature (see Note 3):	
D Package	725 mW
JG Package	825 mW
P Package	1000 mW
Operating free-air temperature range	0°C to 70°C

NOTES: 1. All voltage values, except differential input voltage, are with respect to network ground terminal.
2. Differential-input voltage is measured at the noninverting input with respect to the corresponding inverting input.
3. For operation above 25°C free-air temperature, derate the D package to 464 mW at 70°C at the rate of 5.8 mW/°C, the JG package to 528 mW at 70°C at the rate of 6.6 mW/°C and the P package to 640 mW at 70°C at the rate of 8.0 mW/°C. In the JG package SN75179B, chips are glass mounted.

TEXAS INSTRUMENTS
POST OFFICE BOX 655012 • DALLAS, TEXAS 75265

recommended operating conditions

		MIN	NOM	MAX	UNIT
Supply voltage, V_{CC}		4.75	5	5.25	V
High-level input voltage, V_{IH}	Driver	2			V
Low-level input voltage, V_{IL}	Driver			0.8	V
Common-mode input voltage, V_{IC}		−7†		12	V
Differential input voltage, V_{ID}				±12	V
High-level output current, I_{OH}	Driver			−60	mA
	Receiver			−400	μA
Low-level output current, I_{OL}	Driver			60	mA
	Receiver			8	mA
Operating free-air temperature, T_A		0		70	°C

† The algebraic convention, where the less-positive (more-negative) limit is designated minimum, is used in this data sheet for common-mode input voltage and threshold voltage.

DRIVER SECTION

driver electrical characteristics over recommended ranges of supply voltage and operating free-air temperature (unless otherwise noted)

PARAMETER		TEST CONDITIONS		MIN	TYP†	MAX	UNIT
V_{IK}	Input clamp voltage	$I_I = -18$ mA				−1.5	V
V_O	Output voltage	$I_O = 0$		0		6	V
$\|V_{OD1}\|$	Differential output voltage	$I_O = 0$		1.5		6	V
$\|V_{OD2}\|$	Differential output voltage	$R_L = 100\ \Omega$,	See Figure 1	½V_{OD1} or 2			V
		$R_L = 54\ \Omega$,	See Figure 1	1.5	2.5	5	V
$\|V_{OD3}\|$	Differential output voltage	See Note 4		1.5		5	V
$\Delta\|V_{OD}\|$	Change in magnitude of differential output voltage‡					±0.2	V
V_{OC}	Common-mode output voltage	$R_L = 54\ \Omega$ or $100\ \Omega$,	See Figure 1			+3 / −1	V
$\Delta\|V_{OC}\|$	Change in magnitude of common-mode output voltage‡					±0.2	V
I_O	Output current	$V_{CC} = 0$,	$V_O = -7$ V to 12 V			±100	μA
I_{IH}	High-level input current	$V_I = 2.4$ V				20	μA
I_{IL}	Low-level input current	$V_I = 0.4$ V				−200	μA
I_{OS}	Short-circuit output current	$V_O = -7$ V				−250	mA
		$V_O = V_{CC}$ or 12 V				250	mA
I_{CC}	Supply current (total package)	No load			57	70	mA

† All typical values are at $V_{CC} = 5$ V and $T_A = 25°C$.
‡ $\Delta|V_{OD}|$ and $\Delta|V_{OC}|$ are the changes in magnitude of V_{OD} and V_{OC}, respectively, that occur when the input is changed from a high level to a low level.
NOTE 4: See EIA Standard RS-485, Figure 3.5, Test Termination Measurement 2.

driver switching characteristics, $V_{CC} = 5$ V, $T_A = 25°C$

PARAMETER		TEST CONDITIONS		MIN	TYP	MAX	UNIT
t_{DD}	Differential-output delay time	$R_L = 54\ \Omega$,	See Figure 3		15	22	ns
t_{TD}	Differential-output transition time	$R_L = 54\ \Omega$,	See Figure 3		20	30	ns

SYMBOL EQUIVALENTS

DATA SHEET PARAMETER	RS-422-A	RS-485
V_O	V_{oa}, V_{ob}	V_{oa}, V_{ob}
$\|V_{OD1}\|$	V_o	V_o
$\|V_{OD2}\|$	V_t (R_L = 100 Ω)	V_t (R_L = 54 Ω)
$\|V_{OD3}\|$		V_t (Test termination Measurement 2)
$\Delta\|V_{OD}\|$	$\|\,\|V_t\| - \|\bar{V}_t\|\,\|$	$\|\,\|V_t\| - \|\bar{V}_t\|\,\|$
V_{OC}	$\|V_{os}\|$	$\|V_{os}\|$
$\Delta\|V_{OC}\|$	$\|V_{os} - \bar{V}_{os}\|$	$\|V_{os} - \bar{V}_{os}\|$
I_{OS}	$\|I_{sa}\|$, $\|I_{sb}\|$	
I_O	$\|I_{xa}\|$, $\|I_{xb}\|$	I_{ia}, I_{ib}

RECEIVER SECTION

receiver electrical characteristics over recommended ranges of common-mode input voltage, supply voltage, and operating free-air temperature (unless otherwise noted)

PARAMETER		TEST CONDITIONS		MIN	TYP†	MAX	UNIT
V_{TH}	Differential-input high-threshold voltage	V_O = 2.7 V,	I_O = −0.4 mA			0.2	V
V_{TL}	Differential-input low-threshold voltage	V_O = 0.5 V,	I_O = 8 mA	−0.2‡			V
V_{hys}	Hysteresis§				50		mV
V_{OH}	High-level output voltage	V_{ID} = 200 mV, See Figure 2	I_{OH} = −400 μA,	2.7			V
V_{OL}	Low-level output voltage	V_{ID} = −200 mV, See Figure 2	I_{OL} = 8 mA,			0.45	V
I_I	Line input current	Other input at 0 V, See Note 5	V_I = 12 V			1	mA
			V_I = −7 V			−0.8	mA
r_i	Input resistance			12			kΩ
I_{OS}	Short-circuit output current			−15		−85	mA
I_{CC}	Supply current (total package)	No load			57	70	mA

† All typical values are at V_{CC} = 5 V, T_A = 25 °C.
‡ The algebraic convention, where the less-positive (more-negative) limit is designated minimum, is used in this data sheet for common-mode input voltage and threshold voltage levels only.
§ Hysteresis is the difference between the positive-going input threshold voltage, V_{T+}, and the negative-going input threshold voltage, V_{T-}. See Figure 9.
NOTE 5: Refer to EIA Standard RS-422-A for exact conditions.

receiver switching characteristics, V_{CC} = 5 V, T_A = 25°C

PARAMETER		TEST CONDITIONS	MIN	TYP	MAX	UNIT
t_{PLH}	Propagation delay time, low-to-high-level output	V_{ID} = −1.5 V to 1.5 V,		19	35	ns
t_{PHL}	Propagation delay time, high-to-low-level output	C_L = 15 pF, See Figure 4		30	40	ns

TEXAS INSTRUMENTS
POST OFFICE BOX 655012 • DALLAS, TEXAS 75265

PARAMETER MESUREMENT INFORMATION

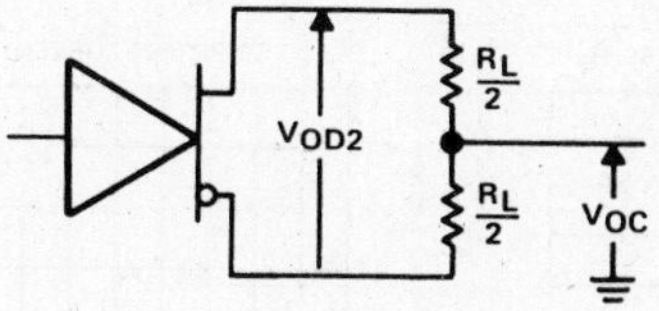
V_{OD2}
$\frac{R_L}{2}$
$\frac{R_L}{2}$
V_{OC}

FIGURE 1. DRIVER V_{OD} AND V_{OC}

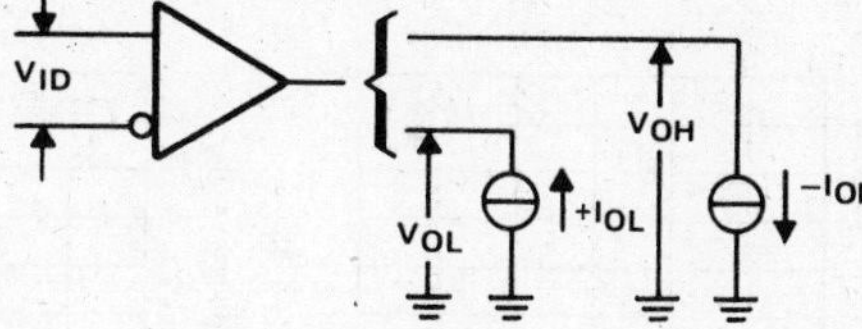
V_{ID}
V_{OH}
V_{OL}
$+I_{OL}$
$-I_{OH}$

FIGURE 2. RECEIVER V_{OH} AND V_{OL}

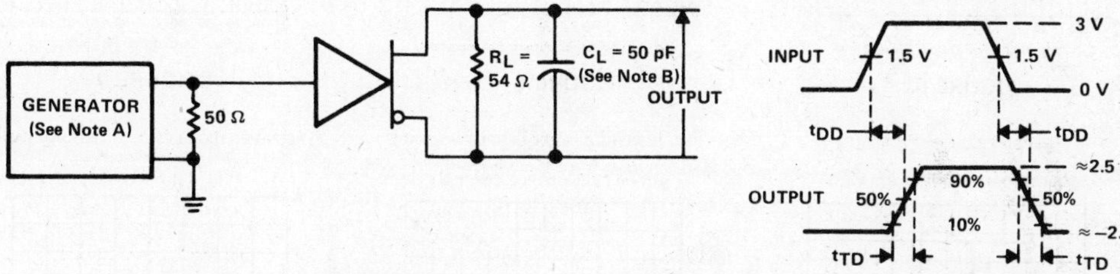
GENERATOR
(See Note A)
50 Ω
R_L =
54 Ω
C_L = 50 pF
(See Note B)
OUTPUT
INPUT
1.5 V
1.5 V
3 V
0 V
t_{DD}
t_{DD}
≈2.5 V
OUTPUT
50%
90%
50%
10%
≈ −2.5 V
t_{TD}
t_{TD}

TEST CIRCUIT

VOLTAGE WAVEFORMS

FIGURE 3. DRIVER DIFFERENTIAL-OUTPUT DELAY AND TRANSITION TIMES

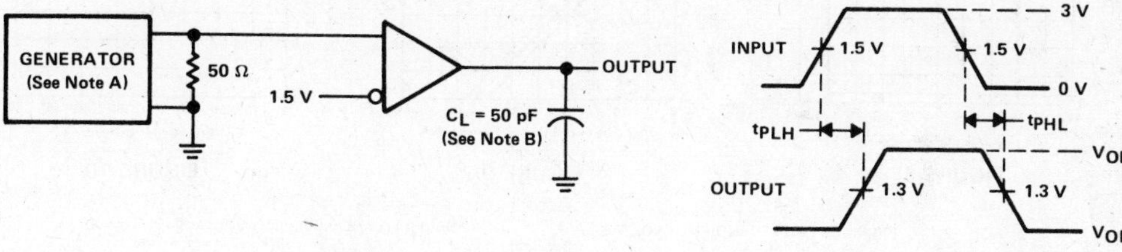
GENERATOR
(See Note A)
50 Ω
1.5 V
OUTPUT
C_L = 50 pF
(See Note B)
INPUT
1.5 V
1.5 V
3 V
0 V
t_{PLH}
t_{PHL}
V_{OH}
OUTPUT
1.3 V
1.3 V
V_{OL}

TEST CIRCUIT

VOLTAGE WAVEFORMS

FIGURE 4. RECEIVER PROPAGATION DELAY TIMES

NOTES: A. The input pulse is supplied by a generator having the following characteristics: PRR ≤ 1 MHz, 50% duty cycle, t_r ≤6 ns, t_f ≤6 ns, Z_{out} = 50 Ω.

B. C_L includes probe and jig capacitance.

TYPICAL CHARACTERISTICS

DRIVER HIGH-LEVEL OUTPUT VOLTAGE
vs
HIGH-LEVEL OUTPUT CURRENT

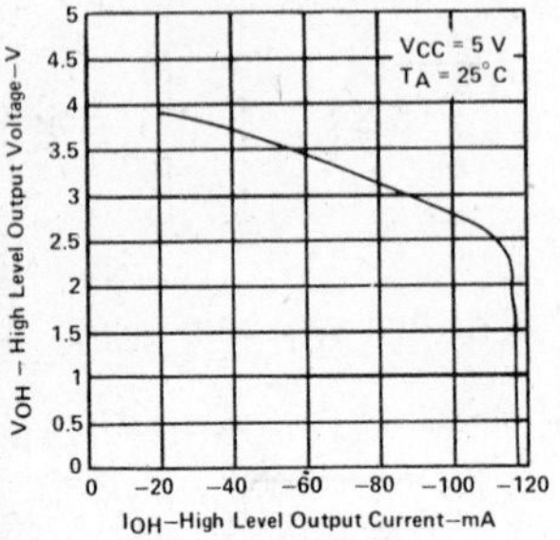
VCC = 5 V
TA = 25°C
VOH – High Level Output Voltage–V
IOH–High Level Output Current–mA

FIGURE 5

DRIVER LOW-LEVEL OUTPUT VOLTAGE
vs
LOW-LEVEL OUTPUT CURRENT

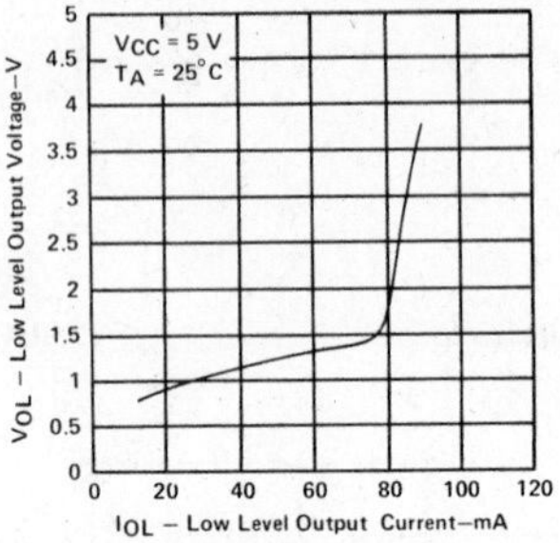
VCC = 5 V
TA = 25°C
VOL – Low Level Output Voltage–V
IOL – Low Level Output Current–mA

FIGURE 6

DRIVER DIFFERENTIAL OUTPUT VOLTAGE
vs
OUTPUT CURRENT

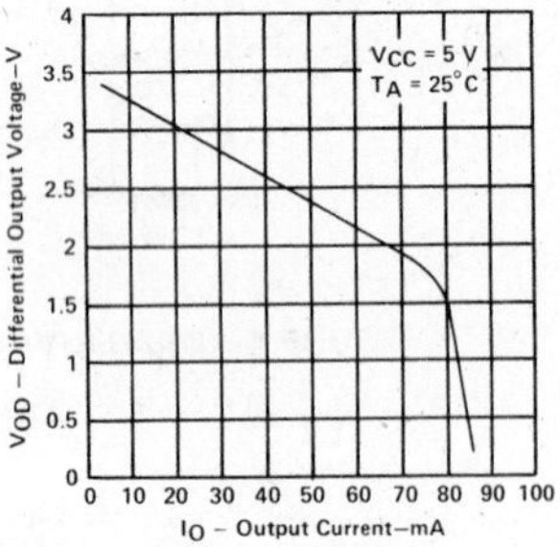
VCC = 5 V
TA = 25°C
VOD – Differential Output Voltage–V
IO – Output Current–mA

FIGURE 7

RECEIVER OUTPUT VOLTAGE
vs
DIFFERENTIAL INPUT VOLTAGE

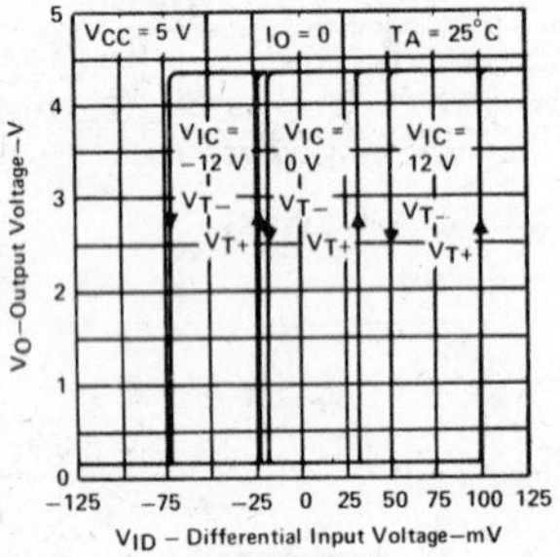
VCC = 5 V
IO = 0
TA = 25°C
VIC = −12 V
VIC = 0 V
VIC = 12 V
VT−
VT+
VO–Output Voltage–V
VID – Differential Input Voltage–mV

FIGURE 8

RECEIVER HIGH-LEVEL OUTPUT VOLTAGE
vs
HIGH-LEVEL OUTPUT CURRENT

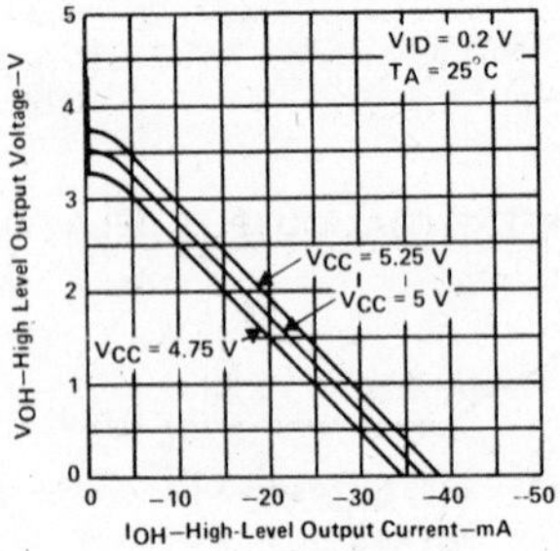
VID = 0.2 V
TA = 25°C
VCC = 5.25 V
VCC = 5 V
VCC = 4.75 V
VOH–High Level Output Voltage–V
IOH–High-Level Output Current–mA

FIGURE 9

RECEIVER HIGH-LEVEL OUTPUT VOLTAGE
vs
FREE-AIR TEMPERATURE

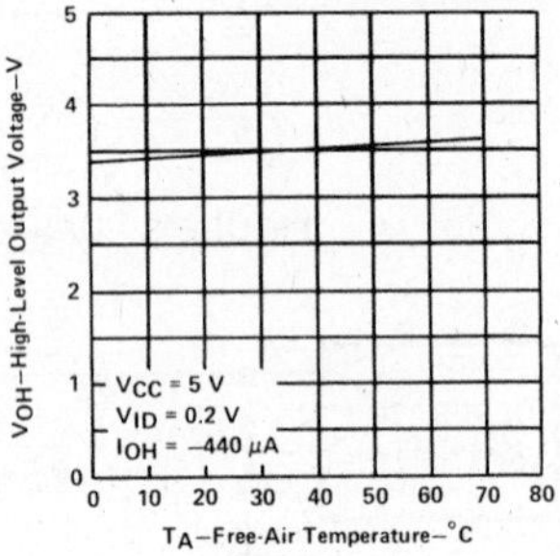
VCC = 5 V
VID = 0.2 V
IOH = −440 μA
VOH–High-Level Output Voltage–V
TA–Free-Air Temperature–°C

FIGURE 10

RECEIVER LOW-LEVEL OUTPUT VOLTAGE
vs
LOW-LEVEL OUTPUT CURRENT

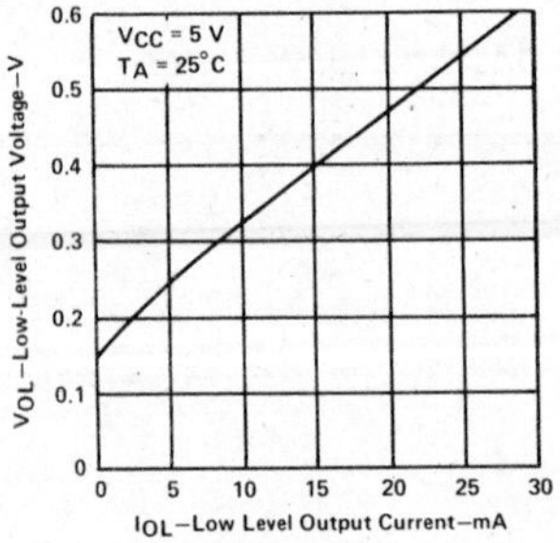
VCC = 5 V
TA = 25°C
VOL–Low-Level Output Voltage–V
IOL–Low Level Output Current–mA

FIGURE 11

RECEIVER LOW-LEVEL OUTPUT VOLTAGE
vs
FREE-AIR TEMPERATURE

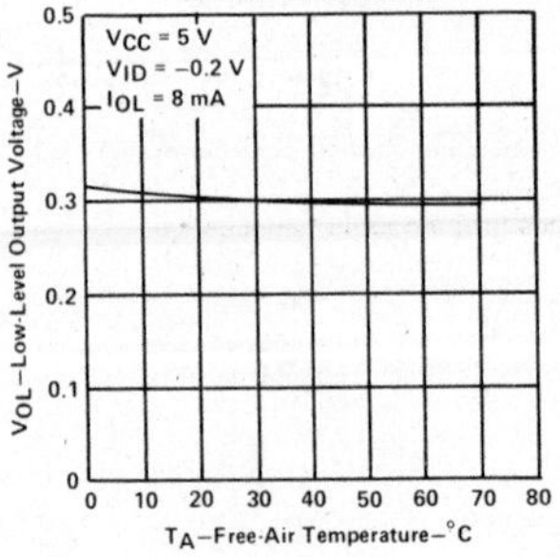
VCC = 5 V
VID = −0.2 V
IOL = 8 mA
VOL–Low-Level Output Voltage–V
TA–Free-Air Temperature–°C

FIGURE 12

TEXAS INSTRUMENTS
POST OFFICE BOX 655012 • DALLAS, TEXAS 75265

D1292, OCTOBER 1972—REVISED SEPTEMBER 1986

- **Single 5-V Supply**
- **Differential Line Operation**
- **Dual Channels**
- **TTL Compatibility**
- **±15 V Common-Mode Input Voltage Range**
- **±15 V Differential Input Voltage Range**
- **Individual Channel Strobes**
- **Built-In Optional Line-Termination Resistor**
- **Individual Frequency Response Controls**
- **Designed for Use with Dual Differential Drivers SN55183 and SN75183**
- **Designed to be Interchangeable with National Semiconductor DS7820A and DS8820A**

SN55182 . . . J PACKAGE
SN75182 . . . D, J OR N PACKAGE
(TOP VIEW)

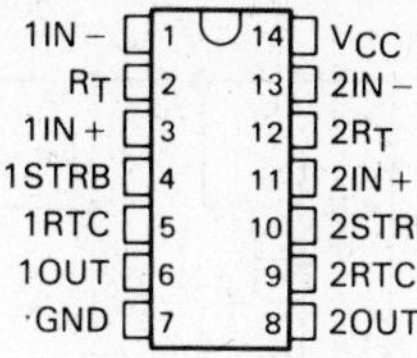
1IN−
R_T
1IN+
1STRB
1RTC
1OUT
GND
V_{CC}
2IN−
$2R_T$
2IN+
2STRB
2RTC
2OUT
1 2 3 4 5 6 7 8 9 10 11 12 13 14

SN55182 . . . FK PACKAGE
(TOP VIEW)

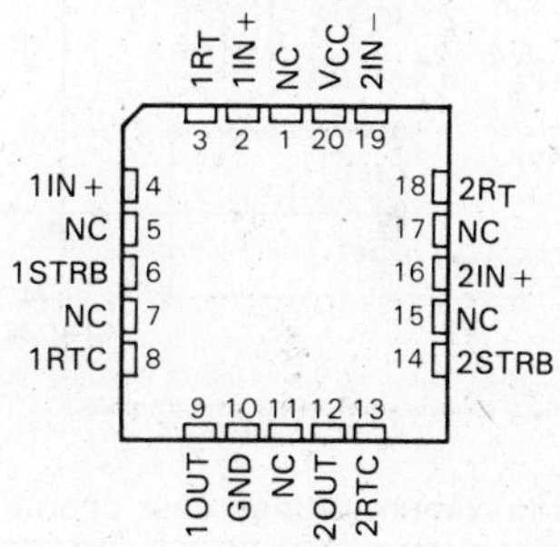
$1R_T$
1IN+
NC
V_{CC}
2IN−
3 2 1 20 19
1IN+ 4
NC 5
1STRB 6
NC 7
1RTC 8
18 $2R_T$
17 NC
16 2IN+
15 NC
14 2STRB
9 10 11 12 13
1OUT
GND
NC
2OUT
2RTC

NC—No internal connection.

description

The SN55182 and SN75182 dual differential line receivers are designed to sense small differential signals in the presence of large common-mode noise. These devices give TTL-compatible output signals as a function of the polarity of the differential input voltage. The frequency response of each channel may be easily controlled by a single external capacitor to provide immunity to differential noise spikes. The output goes to a high level when the inputs are open-circuited. A strobe input is provided which, when in the low level, disables the receiver and forces the output to a high level.

The receiver is of monolithic single-chip construction, and both halves of the dual circuits use common power supply and ground terminals.

The SN55182 is characterized for operation over the full military temperature range of −55 °C to 125 °C. The SN75182 is characterized for operation from 0 °C to 70 °C.

logic symbol†

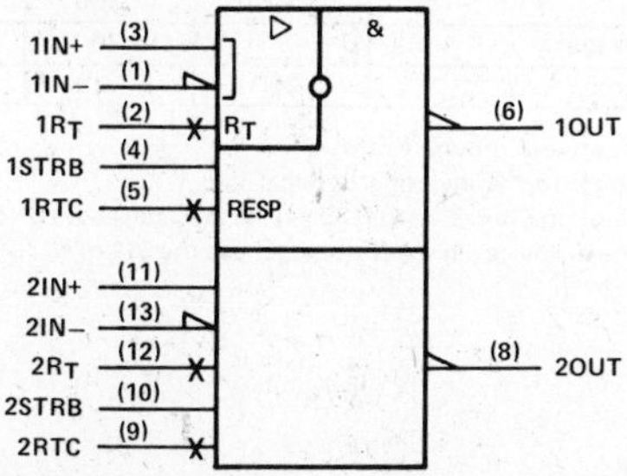
1IN+ (3)
1IN− (1)
$1R_T$ (2)
1STRB (4)
1RTC (5)
2IN+ (11)
2IN− (13)
$2R_T$ (12)
2STRB (10)
2RTC (9)
&
R_T
RESP
(6) 1OUT
(8) 2OUT

†This symbol is in accordance with ANSI/IEEE Std 91-1984 and IEC Publication 617-12.
Pin numbers shown are for D, J and N packages.

logic diagram (positive logic)

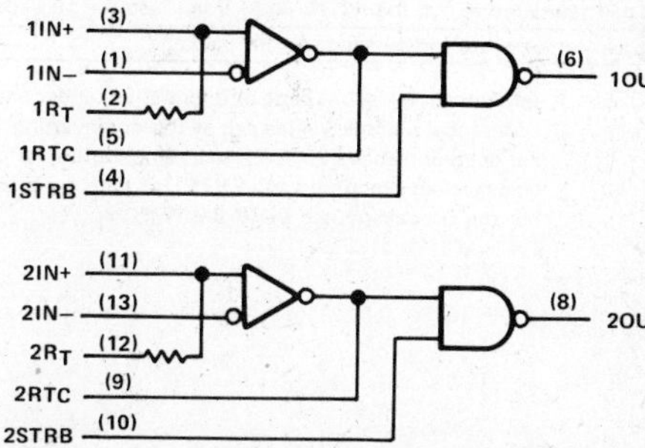
1IN+ (3)
1IN− (1)
$1R_T$ (2)
1RTC (5)
1STRB (4)
(6) 1OUT
2IN+ (11)
2IN− (13)
$2R_T$ (12)
2RTC (9)
2STRB (10)
(8) 2OUT

schematic (each receiver)

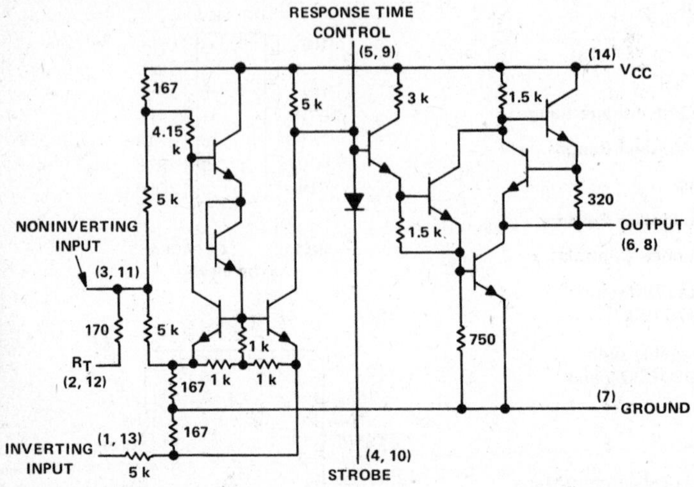
RESPONSE TIME
CONTROL
(5, 9)
(14)
VCC
167
5 k
3 k
1.5 k
4.15 k
5 k
320
NONINVERTING
INPUT
(3, 11)
1.5 k
OUTPUT
(6, 8)
170
5 k
1 k
750
RT
(2, 12)
167
1 k
1 k
(7)
GROUND
167
INVERTING
INPUT
(1, 13)
5 k
(4, 10)
STROBE

Resistor values shown are nominal and in ohms.

FUNCTION TABLE

STROBE	DIFF INPUT	OUTPUT
L	X	H
H	H	H
H	L	L

H = $V_I \geq V_{IH}$ min or V_{ID} more positive than V_{TH} max
L = $V_I \leq V_{IL}$ max or V_{ID} more negative than V_{TL} max
X = irrelevant

absolute maximum ratings over operating free-air temperature range (unless otherwise noted)

		SN55182	SN75182	UNIT
Supply voltage, V_{CC1} (see Note 1)		8	8	V
Common-mode input voltage		± 20	± 20	V
Differential input voltage (see Note 2)		± 20	± 20	V
Strobe input voltage		8	8	V
Output sink current		50	50	mA
Continuous total dissipation at (or below) 25°C free-air temperature (see Note 3)	D package		950	mW
	FK package	1375		
	J package	1375	1025	
	N package		1150	
Operating free-air temperature range		− 55 to 125	0 to 70	°C
Storage temperature range		− 65 to 150	− 65 to 150	°C
Lead temperature 1,6 mm (1/16 inch) from case for 60 seconds: J package		300	300	°C
Lead temperature 1,6 mm (1/16 inch) from case for 10 seconds: D or N package			260	°C
Case temperature for 60 seconds: FK package		260		°C

NOTES: 1. All voltage values, except differential voltages, are with respect to network ground terminal.
2. Differential voltage values are at the noninverting terminal with respect to the inverting terminal.
3. For operation above 25°C free-air temperature, refer to Dissipation Derating Curves in Appendix A. In the J package, SN55182 chips are alloy mounted and SN75182 chips are glass mounted. For these devices in the N package, use the 9.2-mW/°C curve. For the D package use the 7.6-mW/°C curve.

recommended operating conditions

	SN55182			SN75182			UNIT
	MIN	NOM	MAX	MIN	NOM	MAX	
Supply voltage, V_{CC}	4.5	5	5.5	4.5	5	5.5	V
Common-mode input voltage, V_{IC}			±15			±15	V
High-level strobe input voltage, $V_{IH(strobe)}$	2.1		5.5	2.1		5.5	V
Low-level strobe input voltage, $V_{IL(strobe)}$	0		0.9	0		0.9	V
High-level output current, I_{OH}			−400			−400	μA
Low-level output current, I_{OL}			16			16	mA
Operating free-air temperature, T_A	−55		125	0		70	°C

electrical characteristics over recommended ranges of V_{CC}, V_{IC}, and operating free-air temperature (unless otherwise noted)

PARAMETER			TEST CONDITIONS†		MIN	TYP‡	MAX	UNIT
V_{TH}	Differential input high-threshold voltage		V_O = 2.5 V, I_{OH} = −400 μA	V_{IC} = −3 V to 3 V			0.5	V
				V_{IC} = −15 V to 15 V			1	
V_{TL}	Differential input low-threshold voltage		V_O = 0.4 V, I_{OL} = 16 mA,	V_{IC} = −3 V to 3 V			−0.5	V
				V_{IC} = −15 V to 15 V			−1	
V_{OH}	High-level output voltage		V_{ID} = 1 V, I_{OH} = −400 μA	V_{strobe} = 2.1 V,	2.5	4.2	5.5	V
			V_{ID} = −1 V, I_{OH} = −400 μA	V_{strobe} = 0.4 V,	2.5	4.2	5.5	
V_{OL}	Low-level output voltage		V_{ID} = −1 V, I_{OL} = 16 mA	V_{strobe} = 2.1 V,		0.25	0.4	V
I_I	Input current	Inverting input	V_{IC} = 15 V			3	4.2	mA
			V_{IC} = 0			0	−0.5	
			V_{IC} = −15 V			−3	−4.2	
		Noninverting input	V_{IC} = 15 V			5	7	mA
			V_{IC} = 0			−1	−1.4	
			V_{IC} = −15 V			−7	−9.8	
I_{SH}	High-level strobe current		V_{strobe} = 5.5 V				5	μA
I_{SL}	Low-level strobe current		V_{strobe} = 0			−1	−1.4	mA
r_i	Input resistance	Inverting input			3.6	5		kΩ
		Noninverting input			1.8	2.5		kΩ
R_T	Line terminating resistnce		T_A = 25°C		120	170	250	Ω
I_{OS}	Short-circuit output current		V_{CC} = 5.5 V,	V_O = 0	−2.8	−4.5	−6.7	mA
I_{CC}	Supply current (average per receiver)		V_{IC} = 15 V,	V_{ID} = −1 V		4.2	6	mA
			V_{IC} = 0,	V_{ID} = −0.5 V		6.8	10.2	
			V_{IC} = −15 V,	V_{ID} = −1 V		9.4	14	

†Unless otherwise noted, $V_{strobe} \geq 2.1$ V or open.
‡All typical values are at V_{CC} = 5 V, V_{IC} = 0, and T_A = 25°C.

switching characteristics, V_{CC} = 5 V, T_A = 25°C

PARAMETER		TEST CONDITIONS	MIN	TYP	MAX	UNIT
$t_{PLH(D)}$	Propagation delay time, low-to-high-level output from differential input	R_L = 400 Ω, C_L = 15 pF, See Figure 1		18	40	ns
$t_{PHL(D)}$	Propagation delay time, high-to-low-level output from differential input			31	45	ns
$t_{PLH(S)}$	Propagation delay time, low-to-high-level output from strobe input			9	30	ns
$t_{PHL(S)}$	Propagation delay time, high-to-low-level output from strobe input			15	25	ns

PARAMETER MEASUREMENT INFORMATION

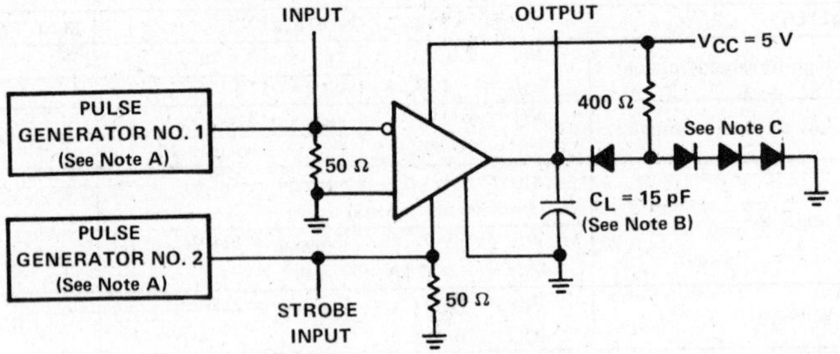
INPUT
OUTPUT
VCC = 5 V
PULSE GENERATOR NO. 1 (See Note A)
PULSE GENERATOR NO. 2 (See Note A)
50 Ω
400 Ω
See Note C
CL = 15 pF (See Note B)
50 Ω
STROBE INPUT

TEST CIRCUIT

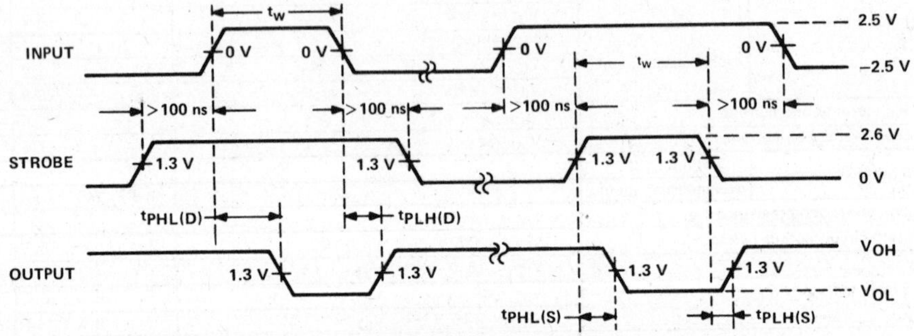
tw
INPUT
0 V
0 V
0 V
0 V
2.5 V
−2.5 V
>100 ns
>100 ns
>100 ns
>100 ns
STROBE
1.3 V
1.3 V
1.3 V
1.3 V
2.6 V
0 V
tPHL(D)
tPLH(D)
OUTPUT
1.3 V
1.3 V
1.3 V
1.3 V
VOH
VOL
tPHL(S)
tPLH(S)

VOLTAGE WAVEFORMS

NOTES: A. The pulse generators have the following characteristics: Z_O = 50 Ω, $t_r \leq$ 10 ns, $t_f \leq$ 10 ns, t_w = 0.5 ±0.1 μs, PRR ≤ 1 MHz.
B. C_L includes probe and jig capacitance.
C. All diodes are 1N3064 or equivalent.

FIGURE 1. PROPAGATION DELAY TIMES

TYPICAL CHARACTERISTICS†

DIFFERENTIAL INPUT THRESHOLD VOLTAGE
vs
SUPPLY VOLTAGE

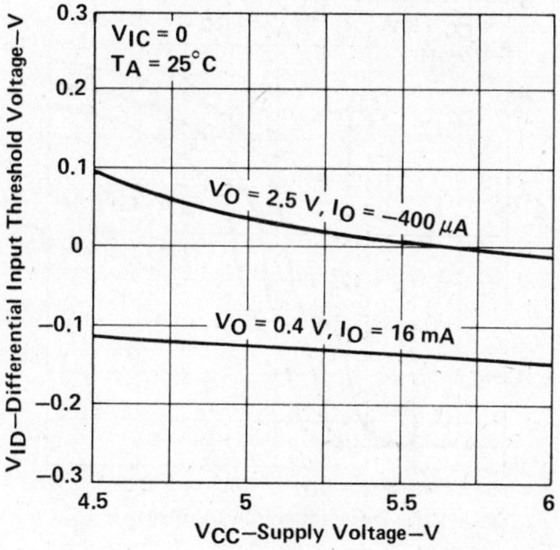
$V_{IC} = 0$
$T_A = 25°C$
$V_O = 2.5$ V, $I_O = -400$ μA
$V_O = 0.4$ V, $I_O = 16$ mA
0.3
0.2
0.1
0
−0.1
−0.2
−0.3
4.5
5
5.5
6
V_{ID}—Differential Input Threshold Voltage—V
V_{CC}—Supply Voltage—V

FIGURE 2.

DIFFERENTIAL INPUT THRESHOLD VOLTAGE
vs
COMMON-MODE VOLTAGE

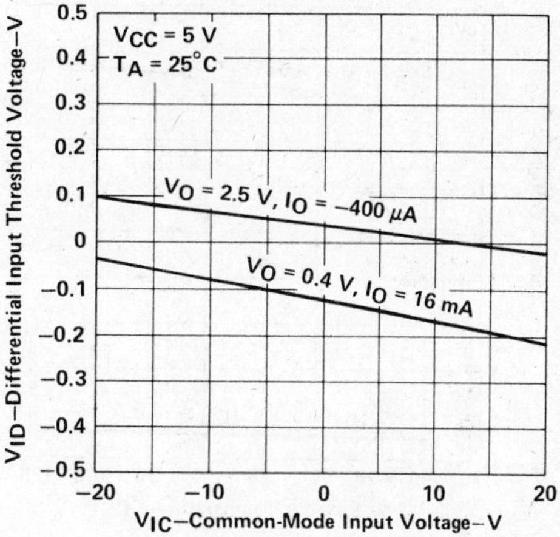
$V_{CC} = 5$ V
$T_A = 25°C$
$V_O = 2.5$ V, $I_O = -400$ μA
$V_O = 0.4$ V, $I_O = 16$ mA
0.5
0.4
0.3
0.2
0.1
0
−0.1
−0.2
−0.3
−0.4
−0.5
−20
−10
0
10
20
V_{ID}—Differential Input Threshold Voltage—V
V_{IC}—Common-Mode Input Voltage—V

FIGURE 3.

DIFFERENTIAL INPUT THRESHOLD VOLTAGE
vs
FREE-AIR TEMPERATURE

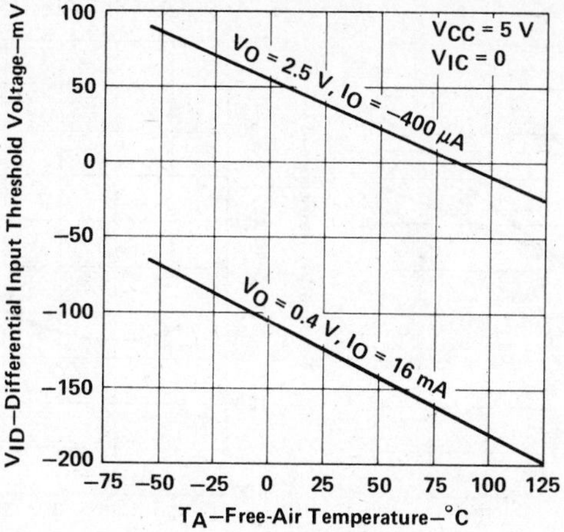
$V_{CC} = 5$ V
$V_{IC} = 0$
$V_O = 2.5$ V, $I_O = -400$ μA
$V_O = 0.4$ V, $I_O = 16$ mA
100
50
0
−50
−100
−150
−200
−75
−50
−25
0
25
50
75
100
125
V_{ID}—Differential Input Threshold Voltage—mV
T_A—Free-Air Temperature—°C

FIGURE 4.

†Data for temperatures below 0°C and above 70°C are applicable to SN55182 circuits only.

TYPICAL CHARACTERISTICS†

OUTPUT VOLTAGE
vs
FREE-AIR TEMPERATURE

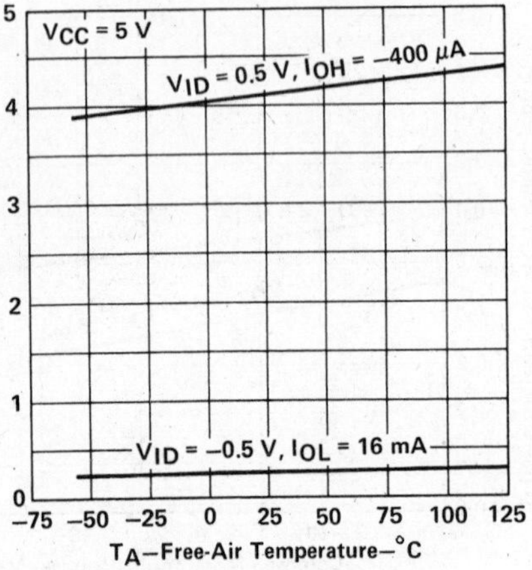
VCC = 5 V
VID = 0.5 V, IOH = −400 μA
VID = −0.5 V, IOL = 16 mA
VO—Output Voltage—V
TA—Free-Air Temperature—°C

FIGURE 5

VOLTAGE TRANSFRER CHARACTERISTICS

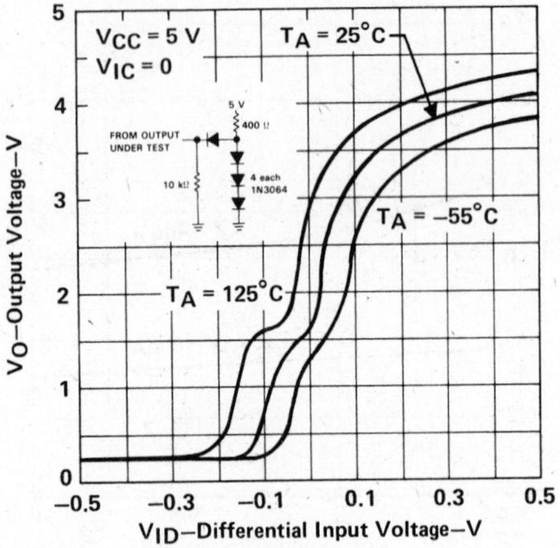
VCC = 5 V
VIC = 0
TA = 25°C
TA = −55°C
TA = 125°C
5 V
400 Ω
FROM OUTPUT UNDER TEST
10 kΩ
4 each 1N3064
VO—Output Voltage—V
VID—Differential Input Voltage—V

FIGURE 6

INPUT CURRENT
vs
INPUT VOLTAGE

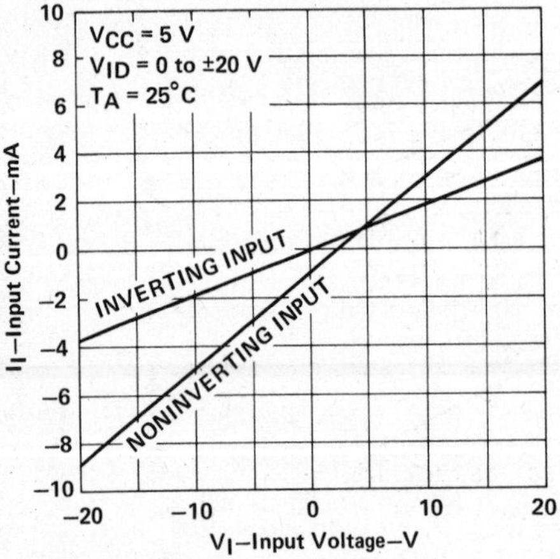
VCC = 5 V
VID = 0 to ±20 V
TA = 25°C
INVERTING INPUT
NONINVERTING INPUT
II—Input Current—mA
VI—Input Voltage—V

FIGURE 7

TERMINATING RESISTANCE
vs
FREE-AIR TEMPERATURE

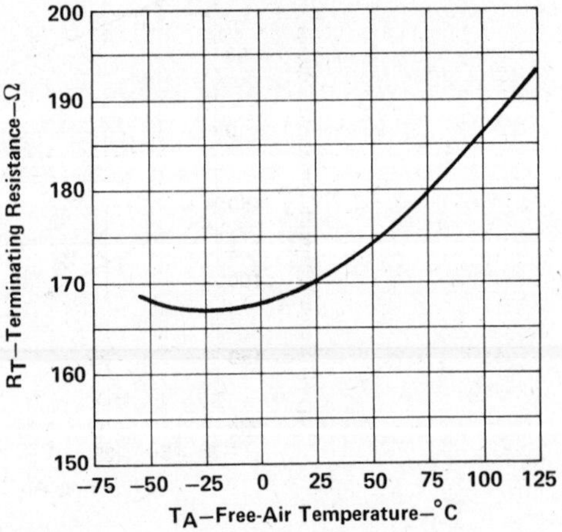
RT—Terminating Resistance—Ω
TA—Free-Air Temperature—°C

FIGURE 8

†Data for temperatures below 0°C and above 70°C are applicable to SN55182 circuits only.

TYPICAL CHARACTERISTICS†

SUPPLY CURRENT
(AVERAGE PER RECEIVER)
vs
COMMON-MODE INPUT VOLTAGE

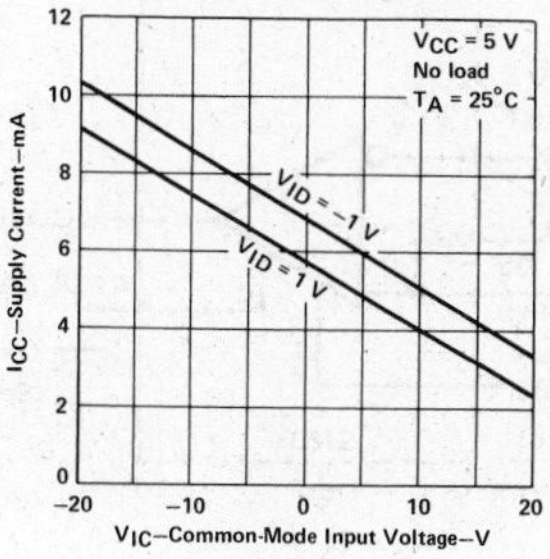

I_CC–Supply Current–mA
V_CC = 5 V
No load
T_A = 25°C
V_ID = –1 V
V_ID = 1 V
12
10
8
6
4
2
0
–20
–10
0
10
20
V_IC–Common-Mode Input Voltage–V

FIGURE 9

POWER DISSIPATION
(AVERAGE PER RECEIVER)
vs
COMMON-MODE INPUT VOLTAGE

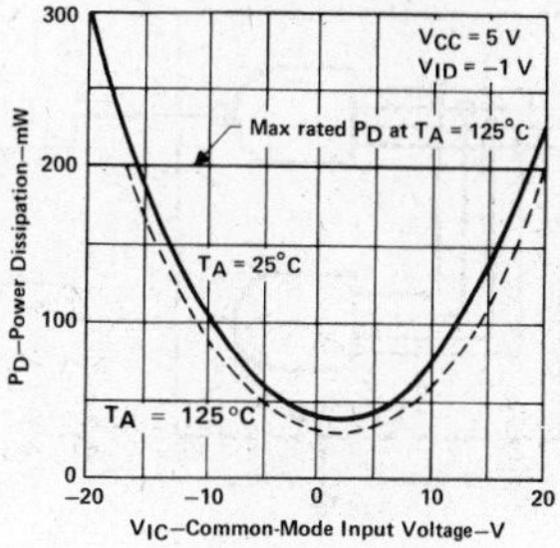

P_D–Power Dissipation–mW
V_CC = 5 V
V_ID = –1 V
Max rated P_D at T_A = 125°C
T_A = 25°C
T_A = 125°C
300
200
100
0
–20
–10
0
10
20
V_IC–Common-Mode Input Voltage–V

FIGURE 10

NOISE PULSE DURATION
vs
RESPONSE TIME-CONTROL CAPACITANCE

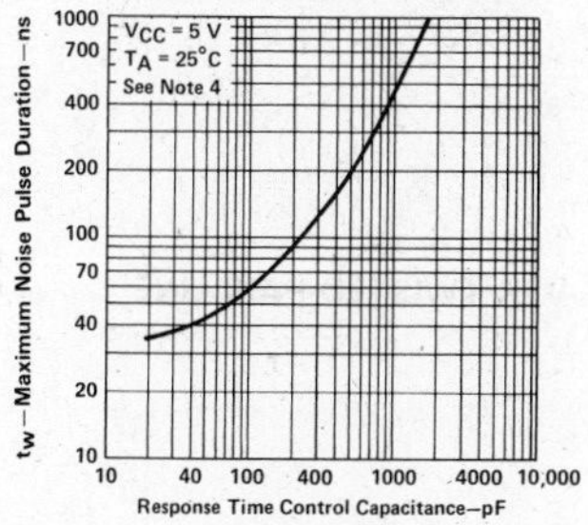

t_w–Maximum Noise Pulse Duration–ns
V_CC = 5 V
T_A = 25°C
See Note 4
1000
700
400
200
100
70
40
20
10
10
40
100
400
1000
4000
10,000
Response Time Control Capacitance–pF

FIGURE 11

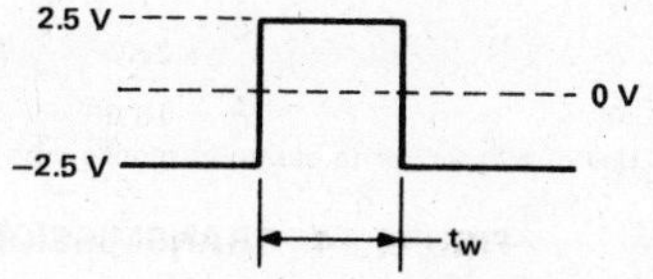

2.5 V
0 V
–2.5 V
t_w

INPUT PULSE FOR FIGURE 11

PROPAGATION DELAY TIMES FROM
DIFFERENTIAL INPUT
vs
FREE-AIR TEMPERATURE

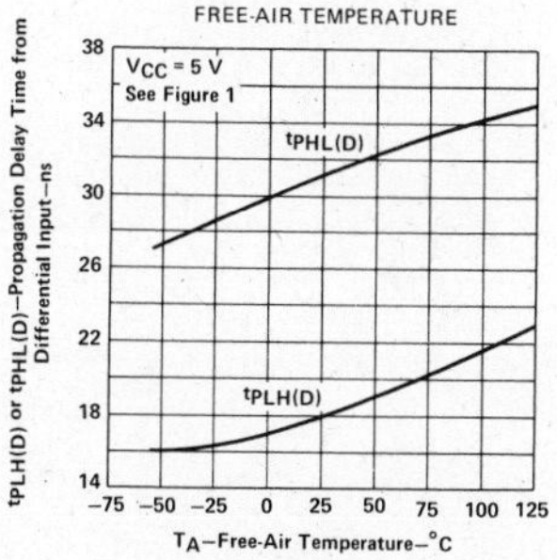

t_PLH(D) or t_PHL(D)–Propagation Delay Time from Differential Input–ns
V_CC = 5 V
See Figure 1
t_PHL(D)
t_PLH(D)
38
34
30
26
22
18
14
–75
–50
–25
0
25
50
75
100
125
T_A–Free-Air Temperature–°C

FIGURE 12

PROPAGATION DELAY TIMES FROM STROBE INPUT
vs
FREE-AIR TEMPERATURE

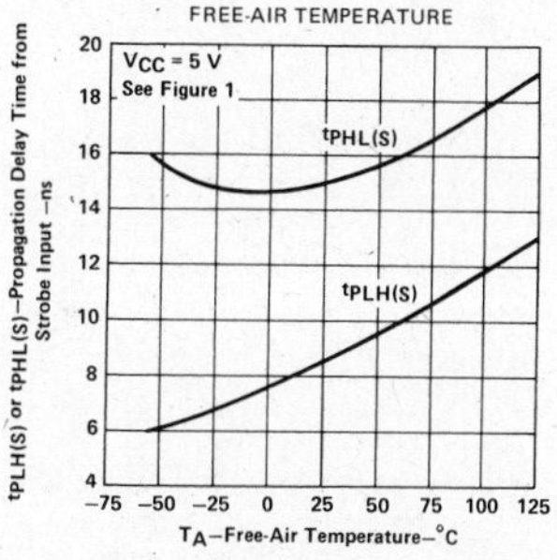

t_PLH(S) or t_PHL(S)–Propagation Delay Time from Strobe Input–ns
V_CC = 5 V
See Figure 1
t_PHL(S)
t_PLH(S)
20
18
16
14
12
10
8
6
4
–75
–50
–25
0
25
50
75
100
125
T_A–Free-Air Temperature–°C

FIGURE 13

†Data for temperatures below 0°C and above 70°C are applicable to SN55182 circuits only.

NOTE 4: Figure 11 shows the maximum duration of the illustrated pulse that can be applied differentially without the output changing from the low to high level.

TEXAS INSTRUMENTS
POST OFFICE BOX 655012 • DALLAS, TEXAS 75265

TYPICAL APPLICATION DATA

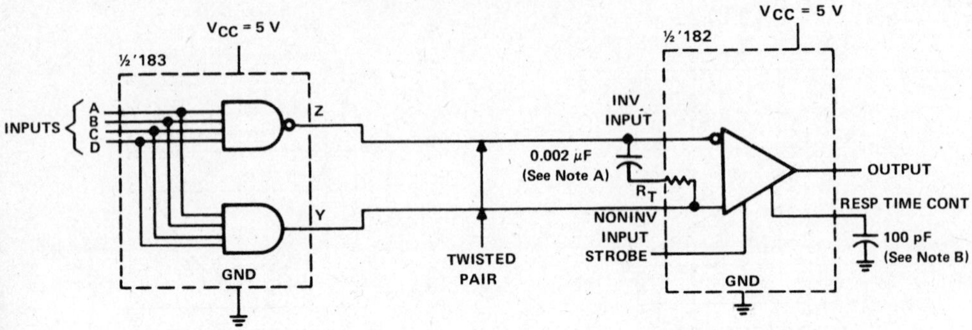
VCC = 5 V
½'183
INPUTS
A
B
C
D
Z
Y
GND
TWISTED
PAIR
VCC = 5 V
½'182
INV
INPUT
0.002 µF
(See Note A)
RT
NONINV
INPUT
STROBE
OUTPUT
RESP TIME CONT
100 pF
(See Note B)
GND

NOTES: A. When the inputs are open-circuited, the output will be high. A capacitor may be used for dc isolation of the line-terminating resistor. At the frequency of operation, the impedance of the capacitor should be relatively small.

Example: let $f = 5\ \text{MHz}$, $C = 0.002\ \mu F$

$$Z_C = \frac{1}{2\pi fC} = \frac{1}{2\pi\ (5 \times 10^6)\ (0.002 \times 10^{-6})}$$

$$Z_C \approx 16\ \Omega$$

B. Use of a capacitor to control response time is optional.

FIGURE 14. TRANSMISSION OF DIGITAL DATA OVER TWISTED-PAIR LINE

TEXAS INSTRUMENTS
POST OFFICE BOX 655012 • DALLAS, TEXAS 75265

D1292, OCTOBER 1972—REVISED SEPTEMBER 1986

- Single 5-V Supply
- Differential Line Operation
- Dual Channels
- TTL Compatibility
- Short-Circuit Protection of Outputs
- Output Clamp Diodes to Terminate Line Transients
- High-Current Outputs
- Quad Inputs
- Single-Ended or Differential AND/NAND Outputs
- Designed for Use with Dual Differential Drivers SN55182 and SN75182
- Designed to be Interchangeable with National Semiconductor DS7830 and DS8830

SN55183 . . . J PACKAGE
SN75183 . . . D, J OR N PACKAGE
(TOP VIEW)

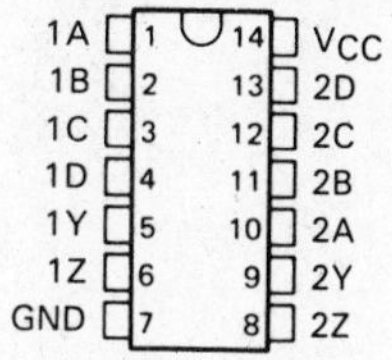
1A 1 14 VCC
1B 2 13 2D
1C 3 12 2C
1D 4 11 2B
1Y 5 10 2A
1Z 6 9 2Y
GND 7 8 2Z

SN55183 . . . FK PACKAGE
(TOP VIEW)

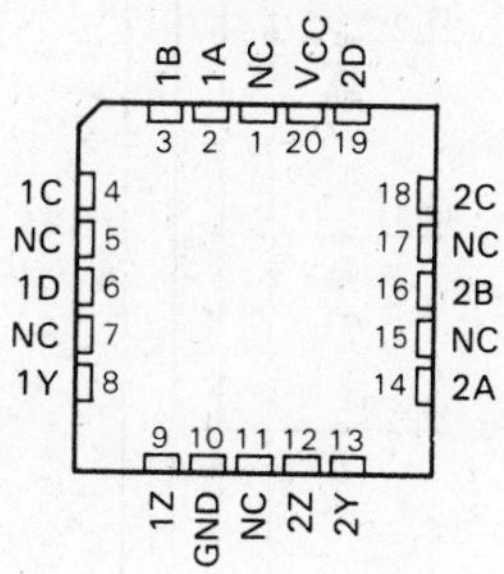
1B 1A NC VCC 2D
3 2 1 20 19
1C 4
NC 5
1D 6
NC 7
1Y 8
18 2C
17 NC
16 2B
15 NC
14 2A
9 10 11 12 13
1Z GND NC 2Z 2Y

NC—No internal connection.

description

The SN55183 and SN75183 dual differential line drivers are designed to provide differential output signals with high-current capability for driving balanced lines, such as twisted-pair, at normal line impedances without high power dissipation. These devices may be used as TTL expander/phase splitters, as the output stages are similar to TTL totem-pole outputs.

The driver is of monolithic single-chip construction, and both halves of the dual circuits use common power supply and ground terminals.

The SN55183 is characterized for operation over the full military temperature range of −55°C to 125°C. The SN75183 is characterized for operation from 0°C to 70°C.

logic symbol†

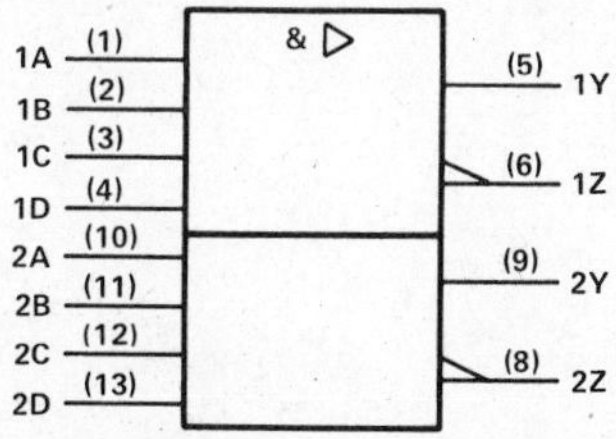
&
1A (1)
1B (2)
1C (3)
1D (4)
2A (10)
2B (11)
2C (12)
2D (13)
(5) 1Y
(6) 1Z
(9) 2Y
(8) 2Z

† This symbol is in accordance with ANSI/IEEE Std 91-1984 and IEC Publication 617-12.
Pin numbers shown are for D, J, and N packages.

logic diagram (positive logic)

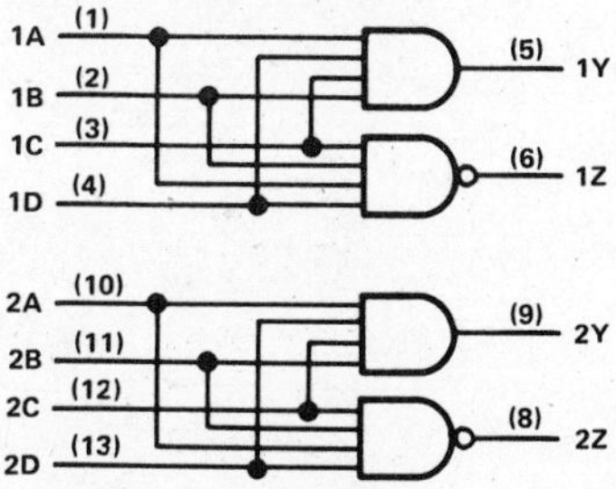
1A (1)
1B (2)
1C (3)
1D (4)
(5) 1Y
(6) 1Z
2A (10)
2B (11)
2C (12)
2D (13)
(9) 2Y
(8) 2Z

positive logic: $Y = ABCD$
$Z = \overline{ABCD}$

schematic (each driver)

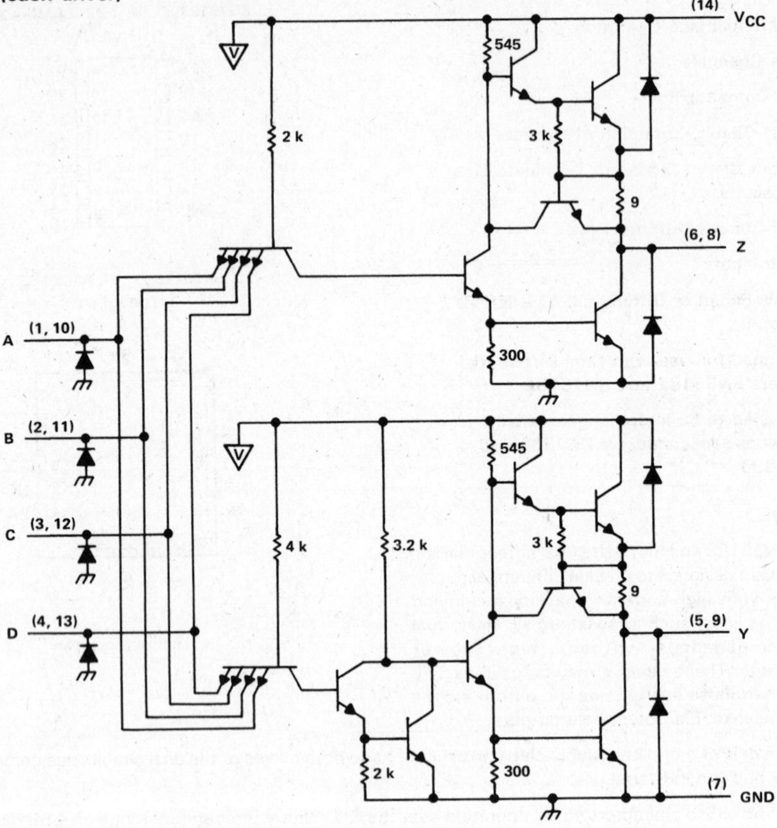
(14)
VCC
545
3 k
9
2 k
(6, 8)
Z
300
(1, 10)
A
(2, 11)
B
(3, 12)
C
(4, 13)
D
545
4 k
3.2 k
3 k
9
(5, 9)
Y
2 k
300
(7)
GND

Resistor values shown are nominal and in ohms.

absolute maximum ratings over operating free-air temperature range (unless otherwise noted)

		SN55183	SN75183	UNIT
Supply voltage, V_{CC} (see Note 1)		7	7	V
Input voltage		5.5	5.5	V
Duration of output short-circuit (see Note 2)		1	1	s
Continuous total dissipation at (or below) 25°C free-air temperature (see Note 3)	D package		950	mW
	FK package	1375		mW
	J package	1375	1025	mW
	N package		1150	mW
Operating free-air temperature range		−55 to 125	0 to 70	°C
Storage temperature range		−65 to 150	−65 to 150	°C
Case temperature for 60 seconds: FK package		260		°C
Lead temperature 1,6 mm (1/16 inch) from case for 60 seconds: J package		300	300	°C
Lead temperature 1,6 mm (1/16 inch) from case for 10 seconds: D or N package			260	°C

NOTES: 1. All voltage values, except differential voltages, are with respect to network ground terminal.
2. Not more than one output should be shorted to ground at a time.
3. For operation above 25°C free-air temperature, refer to Dissipation Derating Curves in Appendix A. In the J package, SN55183 chips are alloy mounted and SN75183 chips are glass mounted. For these devices in the N package, use the 9.2-mW/°C curve. For the D package, use the 7.6-mW/°C curve.

recommended operating conditions

	SN55183			SN75183			UNIT
	MIN	NOM	MAX	MIN	NOM	MAX	
Supply voltage, V_{CC}	4.5	5	5.5	4.75	5	5.25	V
High-level input voltage, V_{IH}	2			2			V
Low-level input voltage, V_{IL}			0.8			0.8	V
High-level output current, I_{OH}			−40			−40	mA
Low-level output current, I_{OL}			40			40	mA
Operating free-air temperature, T_A	−55		125	0		70	°C

electrical characteristics over recommended ranges of V_{CC} and operating free-air temperature (unless otherwise noted)

PARAMETER			TEST CONDITIONS		MIN	TYP†	MAX	UNIT
V_{OH}	High-level output voltage	Y (AND) OUTPUT	V_{IH} = 2 V,	I_{OH} = −0.8 mA	2.4			V
			V_{IH} = 2 V,	I_{OH} = −40 mA	1.8	3.3		
V_{OL}	Low-level output voltage		V_{IL} = 0.8 V,	I_{OL} = 32 mA		0.2		V
			V_{IL} = 0.8 V,	I_{OL} = 40 mA		0.22	0.4	
V_{OH}	High-level output voltage	Z (NAND) OUTPUT	V_{IL} = 0.8 V,	I_{OH} = −0.8 mA	2.4			V
			V_{IL} = 0.8 V,	I_{OH} = −40 mA	1.8	3.3		
V_{OL}	Low-levael output voltage		V_{IH} = 2 V,	I_{OL} = 32 mA		0.2		V
			V_{IH} = 2 V,	I_{OL} = 40 mA		0.22	0.4	
I_{IH}	High-level input current		V_{IH} = 2.4 V				120	μA
I_I	Input current at maximum input voltage		V_{IH} = 5.5 V				2	mA
I_{IL}	Low-level input current		V_{IL} = 0.4 V				−4.8	mA
I_{OS}	Short-circuit output current‡		V_{CC} = 5 V,	T_A = 125°C	−40	−100	−120	mA
I_{CC}	Supply current (average per driver)		V_{CC} = 5 V, No load	All inputs at 5 V,		10	18	mA

†All typical values are at V_{CC} = 5 V, T_A = 25°C.
‡Not more than one output should be shorted to ground at a time and duration of the short-circuit should not exceed one second.

switching characteristics, V_{CC} = 5 V, T_A = 25°C

PARAMETER			TEST CONDITIONS	MIN	TYP	MAX	UNIT
t_{PLH}	Propagation delay time, low-to-high-level Y output	AND gates	C_L = 15 pF, See Figure 1(a)		8	12	ns
t_{PHL}	Propagation delay time, high-to-low-level Y output				12	18	ns
t_{PLH}	Propagation delay time, low-to-high-level Z output	NAND gates			6	12	ns
t_{PHL}	Propagation delay time, high-to-low-level Z output				6	8	ns
t_{PLH}	Propagation delay time, low-to-high-level differential output	Y output with respect to Z output	Z_L = 100 Ω in series with 5000 pF, See Figure 1(b)		9	16	ns
t_{PHL}	Propagation delay time high-to-low-level differential output				8	16	ns

PARAMETER MEASUREMENT INFORMATION

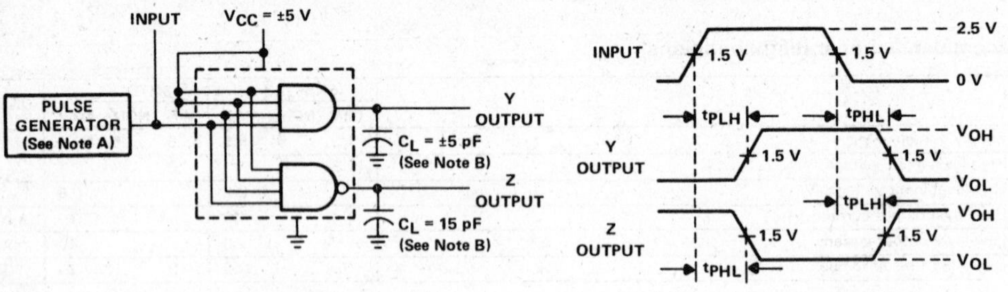
INPUT
V_{CC} = ±5 V
PULSE GENERATOR (See Note A)
Y OUTPUT
C_L = ±5 pF (See Note B)
Z OUTPUT
C_L = 15 pF (See Note B)
2.5 V
INPUT
1.5 V
1.5 V
0 V
t_{PLH}
t_{PHL}
V_{OH}
Y OUTPUT
V_{OL}
V_{OH}
Z OUTPUT
V_{OL}

TEST CIRCUIT

VOLTAGE WAVEFORMS

(a)—OUTPUTS Y AND Z

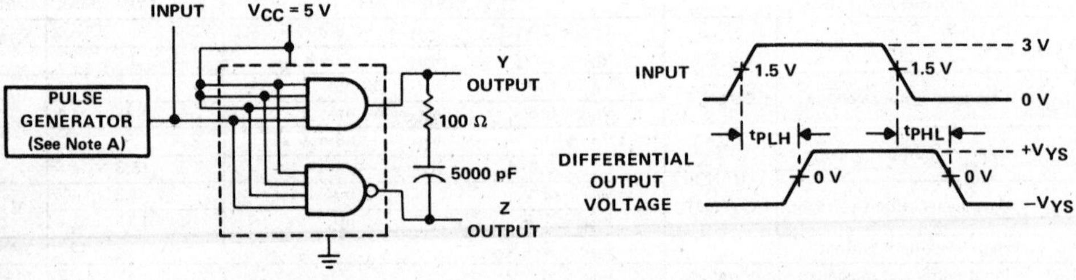
INPUT
V_{CC} = 5 V
PULSE GENERATOR (See Note A)
Y OUTPUT
100 Ω
5000 pF
Z OUTPUT
3 V
INPUT
1.5 V
1.5 V
0 V
t_{PLH}
t_{PHL}
+V_{YS}
DIFFERENTIAL OUTPUT VOLTAGE
0 V
0 V
−V_{YS}

TEST CIRCUIT

VOLTAGE WAVEFORMS

(b)—DIFFERENTIAL OUTPUT

NOTES: A. The pulse generators have the following characteristics: Z_O = 50 Ω, $t_r \le$ 10 ns, $t_f \le$ 10 ns, t_w = 0.5 μs, PRR ≤ 1 MHz.
B. C_L includes probe and jig capacitance.
C. Waveforms are monitored on an oscilloscope with $R_{in} \ge$ 1 MΩ.

FIGURE 1. PROPAGATION DELAY TIMES

TYPICAL CHARACTERISTICS†

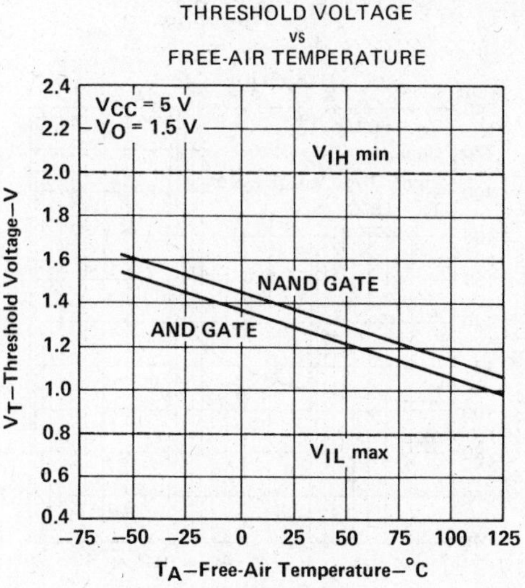
THRESHOLD VOLTAGE
vs
FREE-AIR TEMPERATURE
VCC = 5 V
VO = 1.5 V
VIH min
NAND GATE
AND GATE
VIL max
VT–Threshold Voltage–V
TA–Free-Air Temperature–°C
2.4
2.2
2.0
1.8
1.6
1.4
1.2
1.0
0.8
0.6
0.4
–75 –50 –25 0 25 50 75 100 125

FIGURE 2

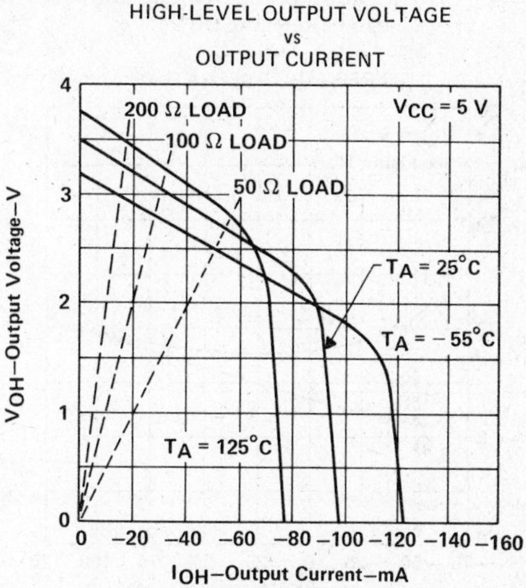
HIGH-LEVEL OUTPUT VOLTAGE
vs
OUTPUT CURRENT
200 Ω LOAD
100 Ω LOAD
50 Ω LOAD
VCC = 5 V
TA = 25°C
TA = – 55°C
TA = 125°C
VOH–Output Voltage–V
IOH–Output Current–mA
4
3
2
1
0
0 –20 –40 –60 –80 –100 –120 –140 –160

FIGURE 3

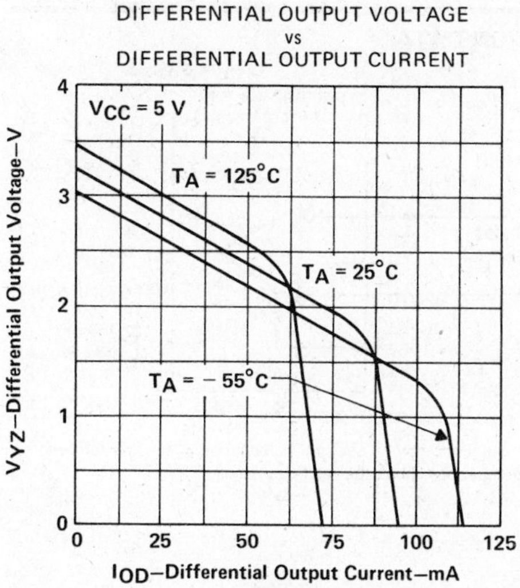
DIFFERENTIAL OUTPUT VOLTAGE
vs
DIFFERENTIAL OUTPUT CURRENT
VCC = 5 V
TA = 125°C
TA = 25°C
TA = – 55°C
VYZ–Differential Output Voltage–V
IOD–Differential Output Current–mA
4
3
2
1
0
0 25 50 75 100 125

FIGURE 4

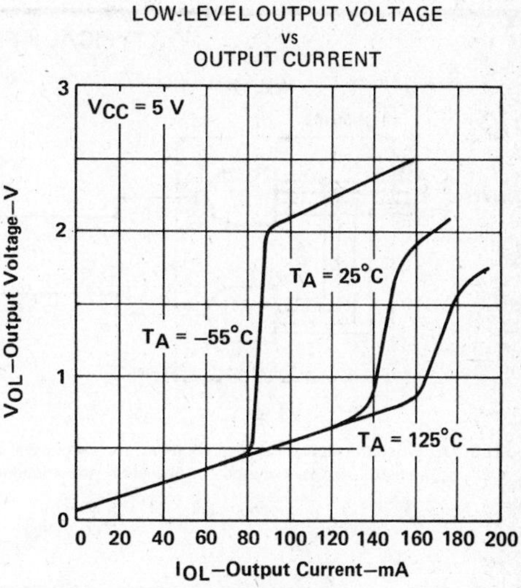
LOW-LEVEL OUTPUT VOLTAGE
vs
OUTPUT CURRENT
VCC = 5 V
TA = 25°C
TA = –55°C
TA = 125°C
VOL–Output Voltage–V
IOL–Output Current–mA
3
2
1
0
0 20 40 60 80 100 120 140 160 180 200

FIGURE 5

†Data for temperatures below 0°C and above 70°C are applicable to SN55183 circuits only.

TYPICAL CHARACTERISTICS†

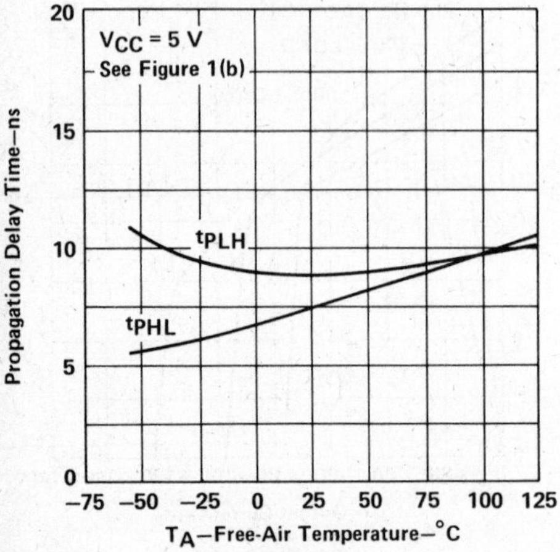
PROPAGATION DELAY TIME OF
DIFFERENTIAL OUTPUT
vs
FREE-AIR TEMPERATURE
VCC = 5 V
See Figure 1(b)
tPLH
tPHL
Propagation Delay Time—ns
TA—Free-Air Temperature—°C

FIGURE 6

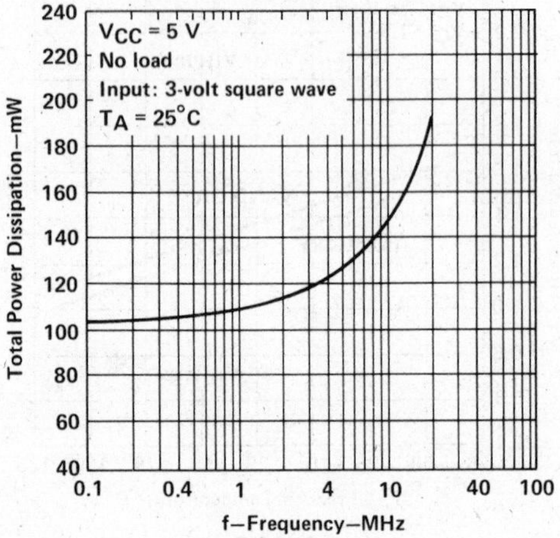
TOTAL POWER DISSIPATION
(BOTH DRIVERS)
vs
FREQUENCY
VCC = 5 V
No load
Input: 3-volt square wave
TA = 25°C
Total Power Dissipation—mW
f—Frequency—MHz

FIGURE 7

†Data for temperatures below 0°C and above 70°C are applicable to SN55183 circuits only.

TYPICAL APPLICATION DATA

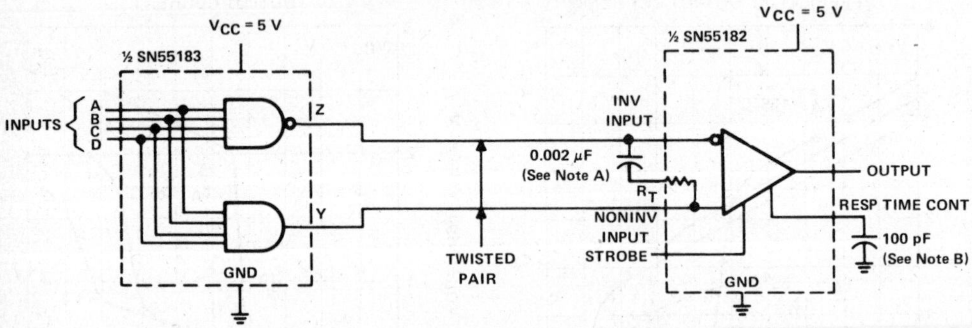
VCC = 5 V
½ SN55183
INPUTS
A
B
C
D
Z
Y
GND
TWISTED
PAIR
VCC = 5 V
½ SN55182
INV
INPUT
0.002 μF
(See Note A)
RT
NONINV
INPUT
STROBE
OUTPUT
RESP TIME CONT
100 pF
(See Note B)
GND

NOTES: A. When the inputs are open-circuited, the output will be high. A capacitor may be used for dc isolation of the line-terminating resistor. At the frequency of operation, the impedance of the capacitor should be relatively small.

Example: let $f = 5\ \text{MHz}$, $C = 0.002\ \mu F$

$$Z_C = \frac{1}{2\pi f C} = \frac{1}{2\pi\,(5 \times 10^6)\,(0.002 \times 10^{-6})}$$

$$Z_C \approx 16\ \Omega$$

B. Use of a capacitor to control response time is optional.

FIGURE 8. TRANSMISSION OF DIGITAL DATA OVER TWISTED-PAIR LINE

D1323, SEPTEMBER 1983—REVISED SEPTEMBER 1986

- **Meets Specifications of EIA RS-232-C**
- **Designed to be Interchangeable with Motorola MC1488**
- **Current-Limited Output: 10 mA Typ**
- **Power-Off Output Impedance: 300 Ω Min**
- **Slew Rate Control by Load Capacitor**
- **Flexible Supply Voltage Range**
- **Input Compatible with Most TTL Circuits**

SN55188 . . . J PACKAGE
SN75188 . . . D OR J PACKAGE
(TOP VIEW)

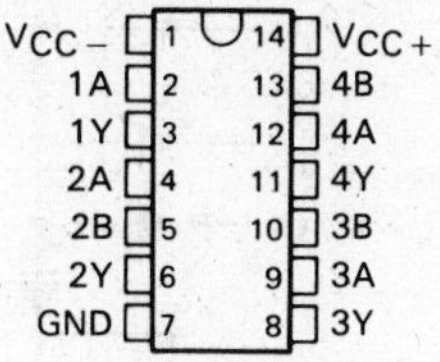
VCC− 1
1A 2
1Y 3
2A 4
2B 5
2Y 6
GND 7
14 VCC+
13 4B
12 4A
11 4Y
10 3B
9 3A
8 3Y

SN55188 . . . FK
CHIP CARRIER PACKAGE
(TOP VIEW)

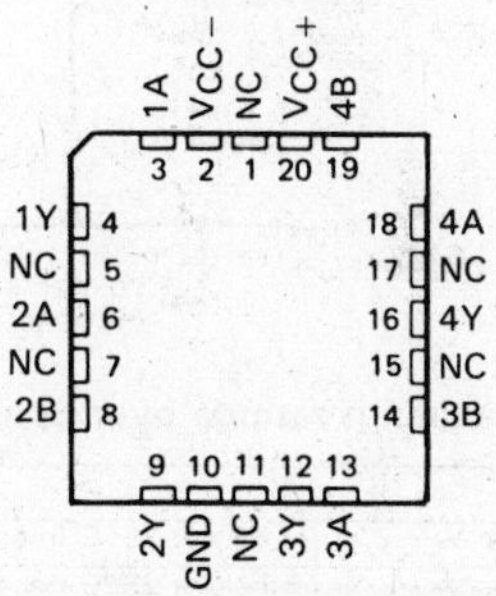
1A VCC− NC VCC+ 4B
3 2 1 20 19
1Y 4
NC 5
2A 6
NC 7
2B 8
18 4A
17 NC
16 4Y
15 NC
14 3B
9 10 11 12 13
2Y GND NC 3Y 3A

NC—No internal connection

description

The SN55188 and SN75188 are monolithic quadruple line drivers designed to interface data terminal equipment with data communications equipment in conformance with EIA Standard RS-232-C using a diode in series with each supply-voltage terminal as shown under typical applications.

The SN55188 is characterized for operation over the full military temperature range of −55°C to 125°C. The SN75188 is characterized for operation from 0°C to 70°C.

logic symbol†

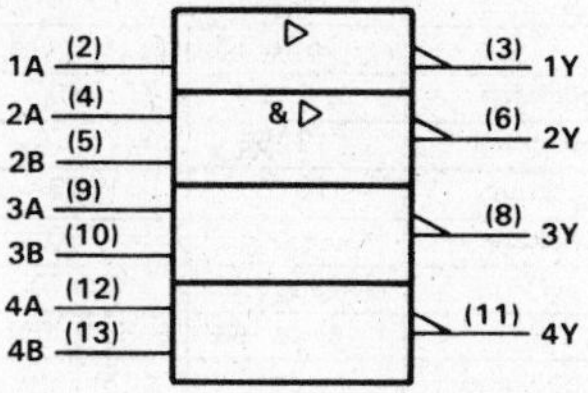
1A (2)
2A (4)
2B (5)
3A (9)
3B (10)
4A (12)
4B (13)
▷
& ▷
(3) 1Y
(6) 2Y
(8) 3Y
(11) 4Y

† This symbol is in accordance with ANSI/IEEE Std 91-1984 and IEC Publication 617-12.

FUNCTION TABLE
(DRIVERS 2 THRU 4)

A	B	Y
H	H	L
L	X	H
X	L	H

H = high level,
L = low level,
X = irrelevant

logic diagram (positive logic)

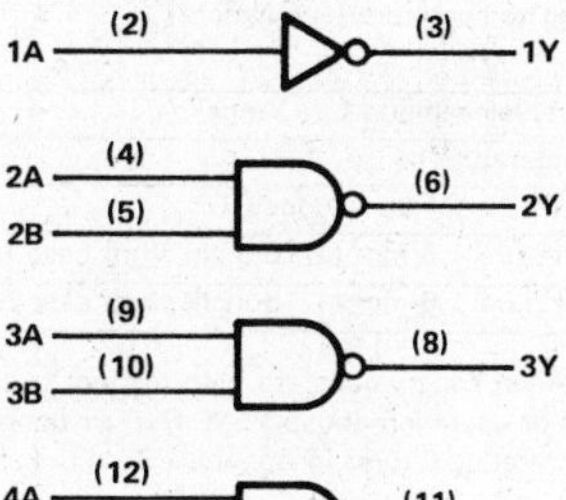
1A (2)
(3) 1Y
2A (4)
2B (5)
(6) 2Y
3A (9)
3B (10)
(8) 3Y
4A (12)
4B (13)
(11) 4Y

Positive logic
$Y = \overline{A}$ (driver 1)
$Y = \overline{AB}$ or $\overline{A} + \overline{B}$ (drivers 2 thru 4)

Pin numbers shown are for D, J, and N packages.

schematic (each driver)

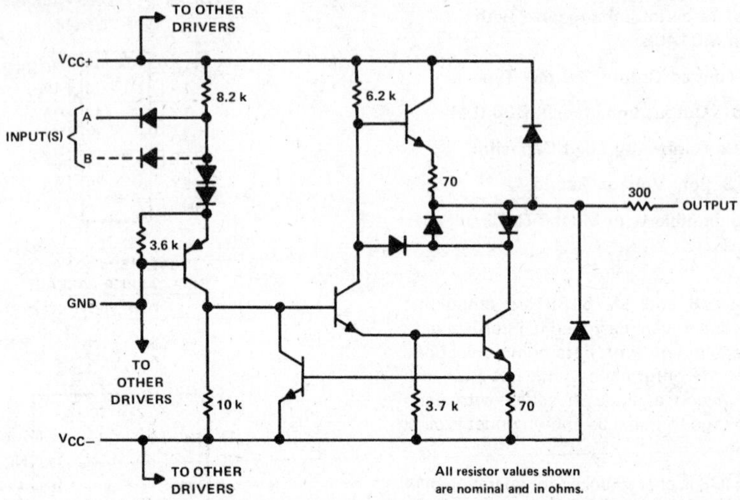
TO OTHER DRIVERS
VCC+
8.2 k
6.2 k
A
INPUT(S)
B
70
300
OUTPUT
3.6 k
GND
TO OTHER DRIVERS
10 k
3.7 k
70
VCC−
TO OTHER DRIVERS
All resistor values shown are nominal and in ohms.

absolute maximum ratings over operating free-air temperature range (unless otherwise noted)

		SN55188	SN75188	UNIT
Supply voltage V_{CC+} at (or below) 25°C free-air temperature (see Notes 1 and 2)		15	15	V
Supply voltage V_{CC-} at (or below) 25°C free-air temperature (see Notes 1 and 2)		−15	−15	V
Input voltage range		−15 to 7	−15 to 7	V
Output voltage range		−15 to 15	−15 to 15	V
Continuous total dissipation at (or below) 25°C free-air temperature (see Note 2)	D package		950	mW
	FK package	1375		
	J package	1375	1025	
	N package		1150	
Operating free-air temperature range		−55 to 125	0 to 70	°C
Storage temperature range		−65 to 150	−65 to 150	°C
Case temperature for 60 seconds	FK package	260		°C
Lead temperature 1,6 mm (1/16 inch) from case for 60 seconds	J package	300	300	
Lead temperature 1,6 mm (1/16 inch) from case for 10 seconds: D or N package			260	°C

NOTES: 1. All voltage values are with respect to the network ground terminal.
2. For operation above 25°C free-air temperature, refer to the Maximum Supply Voltage Curve, Figure 6, and the Dissipation Derating Curves in Appendix A. In the J package, SN55188 chips are alloy mounted and SN75188 chips are glass mounted.

recommended operating conditions

	SN55188			SN75188			UNIT
	MIN	NOM	MAX	MIN	NOM	MAX	
Supply voltage, V_{CC+}	7.5	9	15	7.5	9	15	V
Supply voltage, V_{CC-}	−7.5	−9	−15	−7.5	−9	−15	V
High-level input voltage, V_{IH}	1.9			1.9			V
Low-level input voltage, V_{IL}			0.8			0.8	V
Operating free-air temperature, T_A	−55		125	0		70	°C

electrical characteristics over operating free-air temperature range, $V_{CC+} = 9$ V, $V_{CC-} = -9$ V (unless otherwise noted)

PARAMETER		TEST CONDITIONS		SN55188 MIN	SN55188 TYP†	SN55188 MAX	SN75188 MIN	SN75188 TYP†	SN75188 MAX	UNIT
				(See Note 3)			(See Note 3)			
V_{OH}	High-level output voltage	$V_{IL} = 0.8$ V, $R_L = 3$ kΩ	$V_{CC+} = 9$ V, $V_{CC-} = -9$ V	6	7		6	7		V
			$V_{CC+} = 13.2$ V, $V_{CC-} = -13.2$ V	9	10.5		9	10.5		
V_{OL}	Low-level output voltage	$V_{IH} = 1.9$ V, $R_L = 3$ kΩ	$V_{CC+} = 9$ V, $V_{CC-} = -9$ V		−7	−6		−7	−6	V
			$V_{CC+} = 13.2$ V, $V_{CC-} = -13.2$ V		−10.5	−9		−10.5	−9	
I_{IH}	High-level input current	$V_I = 5$ V				10			10	μA
I_{IL}	Low-level input current	$V_I = 0$			−1	−1.6		−1	−1.6	mA
$I_{OS(H)}$	Short-circuit output current at high level‡	$V_I = 0.8$ V,	$V_O = 0$	−4.6	−9	−13.5	−6	−9	−12	mA
$I_{OS(L)}$	Short-circuit output current at low level‡	$V_I = 1.9$ V,	$V_O = 0$	4.6	9	13.5	6	9	12	mA
r_o	Output resistance, power off	$V_{CC+} = 0$, $V_O = -2$ V to 2 V	$V_{CC-} = 0$,	300			300			Ω
I_{CC+}	Supply current from V_{CC+}	$V_{CC+} = 9$ V, No load	All inputs at 1.9 V		15	20		15	20	mA
			All inputs at 0.8 V		4.5	6		4.5	6	
		$V_{CC+} = 12$ V, No load	All inputs at 1.9 V		19	25		19	25	
			All inputs at 0.8 V		5.5	7		5.5	7	
		$V_{CC+} = 15$ V, No load, $T_A = 25°C$	All inputs at 1.9 V			34			34	
			All inputs at 0.8 V			12			12	
I_{CC-}	Supply current from I_{CC-}	$V_{CC-} = -9$ V, No load	All inputs at 1.9 V		−13	−17		−13	−17	mA
			All inputs at 0.8 V			−0.5			−0.015	
		$V_{CC-} = -12$ V, No load	All inputs at 1.9 V		−18	−23		−18	−23	
			All inputs at 0.8 V			−0.5			−0.015	
		$V_{CC-} = -15$ V, No load, $T_A = 25°C$	All inputs at 1.9 V			−34			−34	
			All inputs at 0.8 V			−2.5			−2.5	
P_D	Total power dissipation	$V_{CC+} = 9$ V, No load	$V_{CC-} = -9$ V,			333			333	mW
		$V_{CC+} = 12$ V, No load	$V_{CC-} = -12$ V,			576			576	

† All typical values are at $T_A = 25°C$.
‡ Not more than one output should be shorted at a time.
NOTE 3: The algebraic convention in which the less positive (more negative) limit is designated as minimum, is used in this data sheet for logic voltage levels only, e.g., if −6 V is a maximum, the typical value is a more negative voltage.

switching characteristics, $V_{CC+} = 9$ V, $V_{CC-} = -9$ V, $T_A = 25°C$

PARAMETER		TEST CONDITIONS	MIN	TYP	MAX	UNIT
t_{PLH}	Propagation delay time, low-to-high-level output	$R_L = 3$ kΩ, $C_L = 15$ pF, See Figure 1		220	350	ns
t_{PHL}	Propagation delay time, high-to-low-level output			100	175	ns
t_{TLH}	Transition time, low-to-high-level output†			55	100	ns
t_{THL}	Transition time, high-to-low-level output†			45	75	ns
t_{TLH}	Transition time, low-to-high-level output‡	$R_L = 3$ kΩ to 7 kΩ, $C_L = 2500$ pF, See Figure 1		2.5		μs
t_{THL}	Transition time, high-to-low-level output‡			3.0		μs

† Measured between 10% and 90% points of output waveform.
‡ Measured between +3 V and −3 V points on the output waveform (EIA RS-232-C conditions)

PARAMETER MEASUREMENT INFORMATION

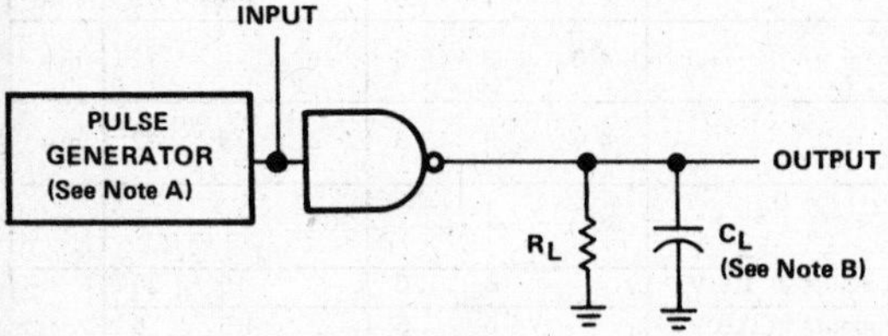
INPUT
PULSE GENERATOR (See Note A)
OUTPUT
RL
CL (See Note B)

TEST CIRCUIT

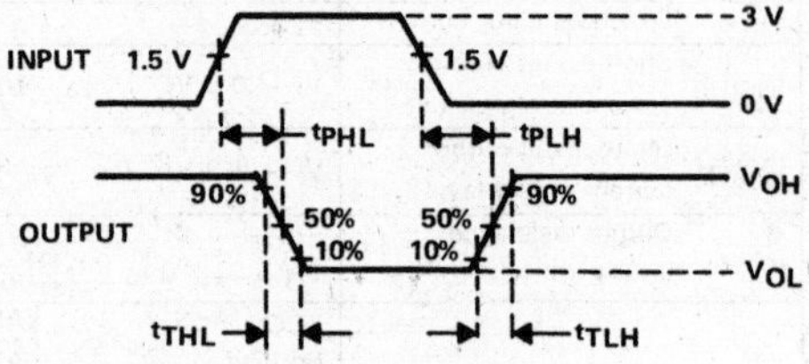
INPUT
1.5 V
1.5 V
3 V
0 V
tPHL
tPLH
OUTPUT
90%
90%
50%
50%
10%
10%
VOH
VOL
tTHL
tTLH

VOLTAGE WAVEFORMS

NOTES: A. The pulse generator has the following characteristics: $t_w = 0.5$ μs, PRR ≤ 1 MHz, $Z_O = 50$ Ω.
B. C_L includes probe and jig capacitance.

FIGURE 1. PROPAGATION AND TRANSITION TIMES

TEXAS INSTRUMENTS
POST OFFICE BOX 655012 • DALLAS, TEXAS 75265

TYPICAL CHARACTERISTICS†

VOLTAGE TRANSFER CHARACTERISTICS

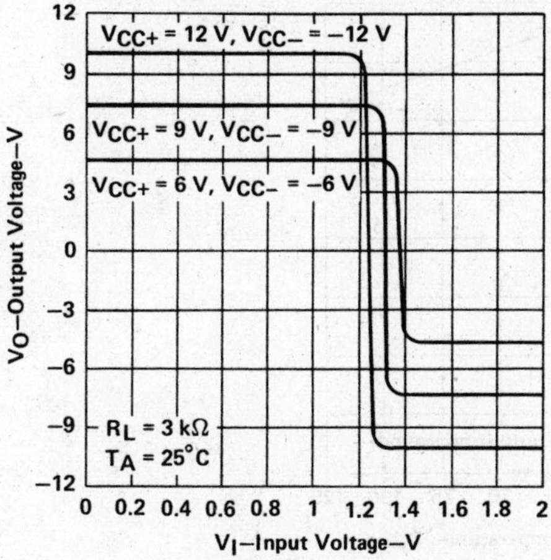
VCC+ = 12 V, VCC− = −12 V
VCC+ = 9 V, VCC− = −9 V
VCC+ = 6 V, VCC− = −6 V
RL = 3 kΩ
TA = 25°C
VO—Output Voltage—V
12
9
6
3
0
−3
−6
−9
−12
0 0.2 0.4 0.6 0.8 1 1.2 1.4 1.6 1.8 2
VI—Input Voltage—V

FIGURE 2

OUTPUT CURRENT
vs
OUTPUT VOLTAGE

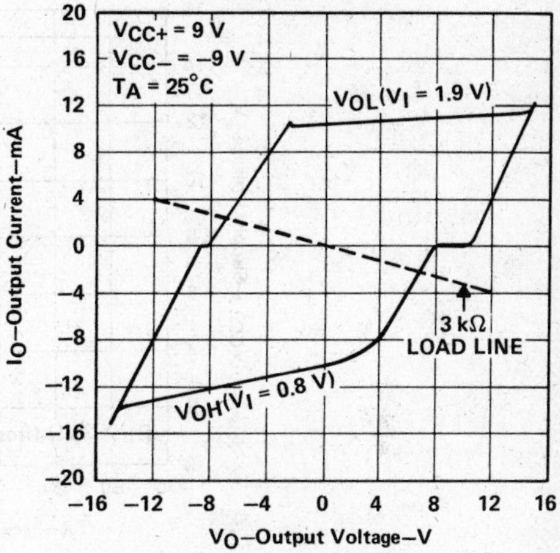
VCC+ = 9 V
VCC− = −9 V
TA = 25°C
VOL(VI = 1.9 V)
VOH(VI = 0.8 V)
3 kΩ
LOAD LINE
IO—Output Current—mA
20
16
12
8
4
0
−4
−8
−12
−16
−20
−16 −12 −8 −4 0 4 8 12 16
VO—Output Voltage—V

FIGURE 3

SHORT-CIRCUIT OUTPUT CURRENT
vs
FREE-AIR TEMPERATURE

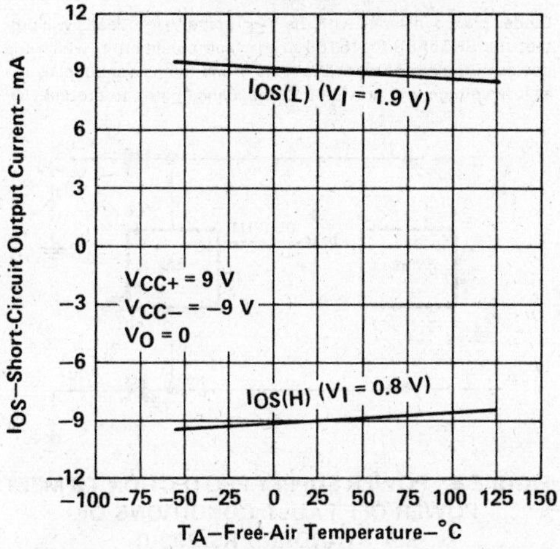
IOS(L) (VI = 1.9 V)
VCC+ = 9 V
VCC− = −9 V
VO = 0
IOS(H) (VI = 0.8 V)
IOS—Short-Circuit Output Current—mA
12
9
6
3
0
−3
−6
−9
−12
−100 −75 −50 −25 0 25 50 75 100 125 150
TA—Free-Air Temperature—°C

FIGURE 4

SLEW RATE
vs
LOAD CAPACITANCE

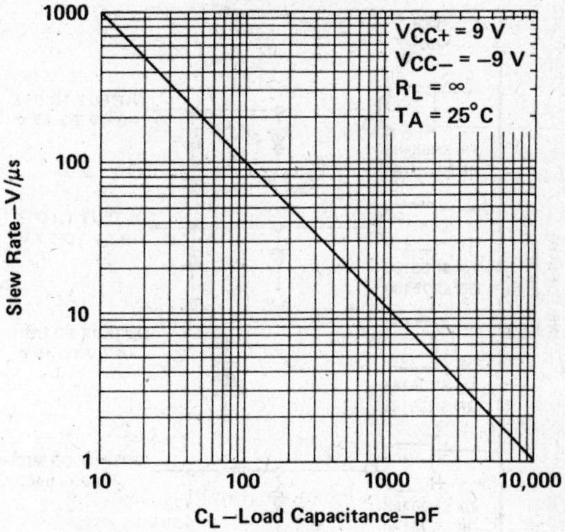
VCC+ = 9 V
VCC− = −9 V
RL = ∞
TA = 25°C
Slew Rate—V/µs
1000
100
10
1
10 100 1000 10,000
CL—Load Capacitance—pF

FIGURE 5

†Data for tempeatures below 0°C and above 70°C are applicable to SN55188 circuit only.

THERMAL INFORMATION†

MAXIMUM SUPPLY VOLTAGE
vs
FREE-AIR TEMPERATURE

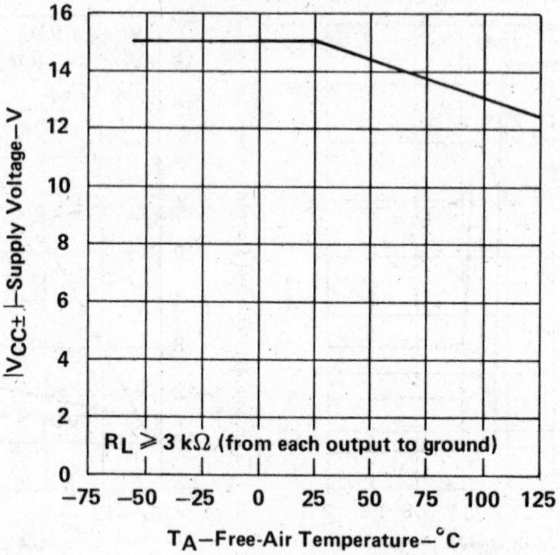
16
14
12
10
8
6
4
2
0
|VCC±|—Supply Voltage—V
RL ⩾ 3 kΩ (from each output to ground)
−75 −50 −25 0 25 50 75 100 125
TA—Free-Air Temperature—°C

FIGURE 6

†Data for temperatures below 0°C and above 70°C are applicable to SN55188 circuit only.

TYPICAL APPLICATION DATA

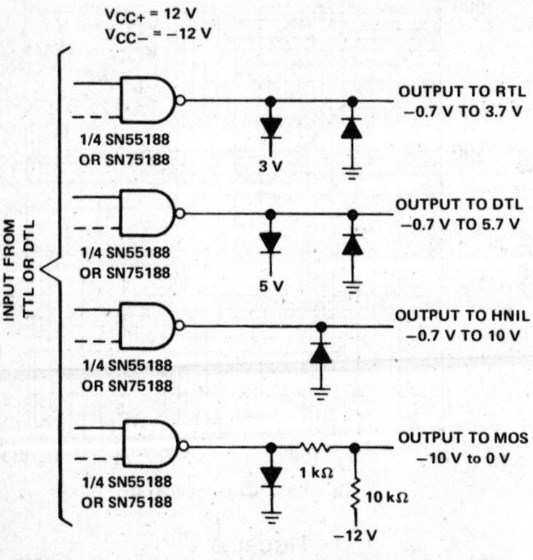
VCC+ = 12 V
VCC− = −12 V
INPUT FROM TTL OR DTL
1/4 SN55188 OR SN75188
OUTPUT TO RTL −0.7 V TO 3.7 V
3 V
1/4 SN55188 OR SN75188
OUTPUT TO DTL −0.7 V TO 5.7 V
5 V
1/4 SN55188 OR SN75188
OUTPUT TO HNIL −0.7 V TO 10 V
1/4 SN55188 OR SN75188
OUTPUT TO MOS −10 V to 0 V
1 kΩ
10 kΩ
−12 V

FIGURE 7. LOGIC TRANSLATOR APPLICATIONS

Diodes placed in series with the V_{CC+} and V_{CC-} leads will protect the SN55188/SN75188 in the fault condition in which the device outputs are shorted to ± 15 V and the power supplies are at low voltage and provide low-impedance paths to ground.

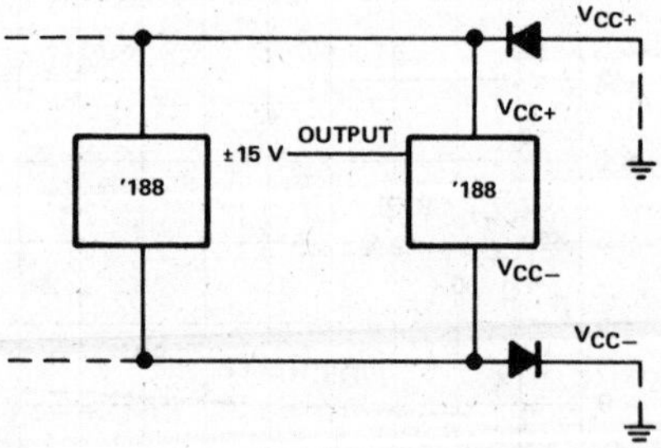
VCC+
VCC+
±15 V
OUTPUT
'188
'188
VCC−
VCC−

FIGURE 8. POWER SUPPLY PROTECTION TO MEET POWER-OFF FAULT CONDITIONS OF EIA STANDARD RS-232-C

D1619, SEPTEMBER 1973—REVISED SEPTEMBER 1986

- Input Resistance . . . 3 kΩ to 7 kΩ
- Input Signal Range . . . ±30 V
- Operates from Single 5-V Supply
- Built-in Input Hysteresis (Double Thresholds)
- Response Control Provides:
  - Input Threshold Shifting
  - Input Noise Filtering
- Satisfies Requirements of EIA RS-232-C
- Fully Interchangeable with Motorola MC1489, MC1489A

SN55189, SN55189A . . . J PACKAGE
SN75189, SN75189A . . . D, J, OR N PACKAGE
(TOP VIEW)

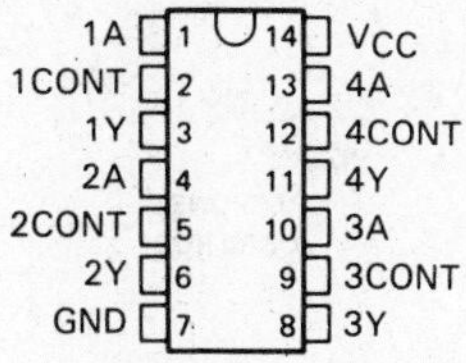
1A 1
1CONT 2
1Y 3
2A 4
2CONT 5
2Y 6
GND 7
14 VCC
13 4A
12 4CONT
11 4Y
10 3A
9 3CONT
8 3Y

SN55189, SN55189A . . . FK PACKAGE
(TOP VIEW)

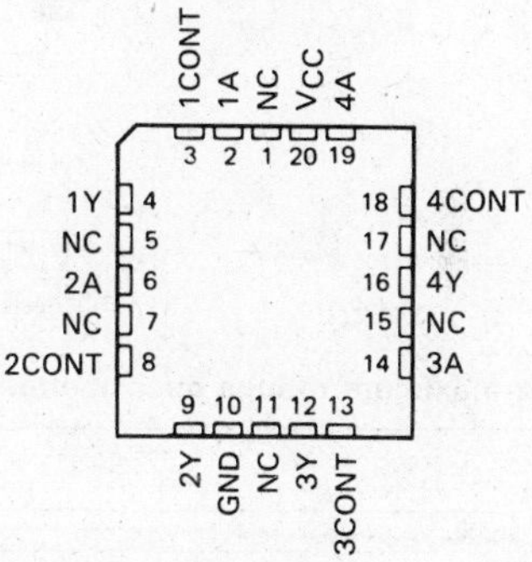
1CONT
1A
NC
VCC
4A
3 2 1 20 19
1Y 4
NC 5
2A 6
NC 7
2CONT 8
18 4CONT
17 NC
16 4Y
15 NC
14 3A
9 10 11 12 13
2Y
GND
NC
3Y
3CONT

NC—No internal connection

description

These devices are monolithic Low-power Schottky quadruple line receivers designed to satisfy the requirements of the standard interface between data terminal equipment and data communication equipment as defined by EIA Standard RS-232-C. A separate response control terminal is provided for each receiver. A resistor or a resistor and bias voltage source can be connected between this terminal and ground to shift the input threshold levels. An external capacitor can be connected between this terminal and ground to provide input noise filtering.

The SN55189 and SN55189A are characterized for operation over the full military temperature range of −55°C to 125°C. The SN75189 and SN75189A are characterized for operation from 0°C to 70°C.

logic diagram (each receiver)

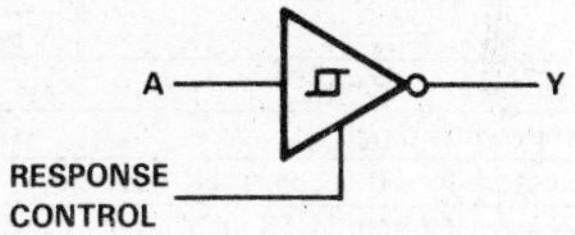
A
Y
RESPONSE
CONTROL

logic symbol†

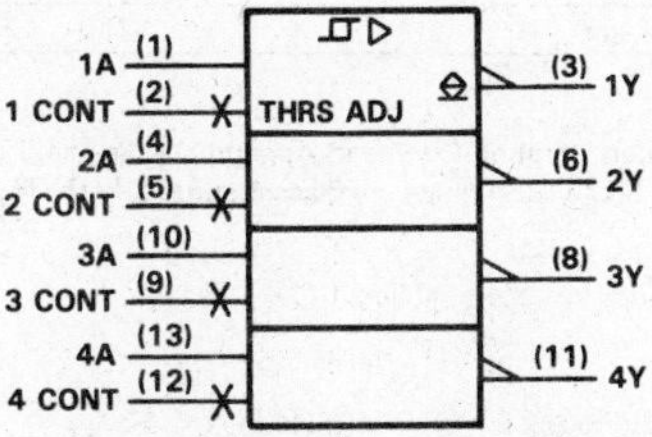
1A (1)
1 CONT (2)
THRS ADJ
(3) 1Y
2A (4)
2 CONT (5)
(6) 2Y
3A (10)
3 CONT (9)
(8) 3Y
4A (13)
4 CONT (12)
(11) 4Y

†This symbol is in accordance with ANSI/IEEE Std 91-1984 and IEC Publication 617-12.
Pin numbers shown are for D, J, and N packages.

schematic (each receiver)

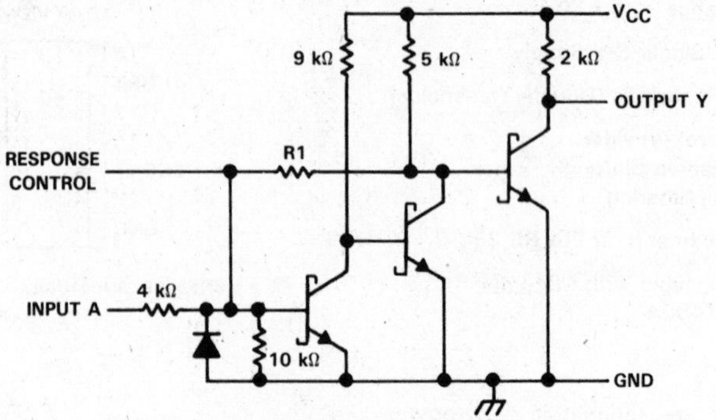
VCC
9 kΩ
5 kΩ
2 kΩ
OUTPUT Y
RESPONSE CONTROL
R1
4 kΩ
INPUT A
10 kΩ
GND

	SN55189 SN75189	SN55189A SN75189A
R1	10 kΩ	2 kΩ

Resistor values shown are nominal.

absolute maximum ratings over operating free-air temperature range (unless otherwise noted)

		SN55189 SN55189A	SN75189 SN75189A	UNIT
Supply voltage, V_{CC} (see Note 1)		10	10	V
Input voltage		±30	±30	V
Output current		20	20	mA
Continuous total dissipation at (or below) 25°C free-air temperature (see Note 2)	D package		950	mW
	FK or J package	1375		mW
	J package		1025	mW
	N package		1150	mW
Operating temperature range		−55 to 125	0 to 70	°C
Storage temperature range		−65 to 150	−65 to 150	°C
Case temperature for 60 seconds: FK package		260		°C
Lead temperature 1,6 mm (1/16 inch) from case for 60 seconds: J package		300	300	°C
Lead temperature 1,6 mm (1/16 inch) from case for 10 seconds: D or N package			260	°C

NOTES: 1. All voltage values are with respect to network ground terminals.

2. For operation above 25°C free-air temperature, refer to the Dissipation Derating Curves in Appendix A. In the J package, SN55189 and SN55189A chips are alloy mounted and SN75189 and SN75189A chips are glass mounted. In the N package, use the 9.2-mW/°C curve for these devices.

TEXAS INSTRUMENTS
POST OFFICE BOX 655012 • DALLAS, TEXAS 75265

electrical characteristics over operating free-air temperature range, V_{CC} = 5 V ± 1%, (unless otherwise noted)

PARAMETER		TEST FIGURE	TEST CONDITIONS†		SN55189 SN55189A MIN	TYP‡	MAX	SN75189 SN75189A MIN	TYP‡	MAX	UNIT
V_{T+}	Positive-going threshold voltage	1	'189	T_A = 25°C	1	1.3	1.5	1	1.3	1.5	V
				T_A = 0°C to 70°C				0.9		1.6	
				T_A = −55°C to 125°C	0.6		1.9				
			'189A	T_A = 25°C	1.75	1.9	2.25	1.75	1.9	2.25	
				T_A = 0°C to 70°C				1.55		2.25	
				T_A = −55°C to 125°C	1.30		2.65				
V_{T-}	Negative-going threshold voltage	1	'189, '189A	T_A = 25°C	0.75	1.0	1.25	0.75	1.0	1.25	V
				T_A = 0°C to 70°C				0.65		1.25	
				T_A = −55°C to 125°C	0.35		1.6				
V_{OH}	High-level output voltage	1	V_I = 0.75 V,	I_{OH} = −0.5 mA	2.6	4	5	2.6	4	5	V
			Input open,	I_{OH} = −0.5 mA	2.6	4	5	2.6	4	5	
V_{OL}	Low-level output voltage	1	V_I = 3 V,	I_{OL} = 10 mA		0.2	0.45		0.2	0.45	V
I_{IH}	High-level input current	2	V_I = 25 V		3.6		8.3	3.6		8.3	mA
			V_I = 3 V		0.43			0.43			
I_{IL}	Low-level input current	2	V_I = −25 V		−3.6		−8.3	−3.6		−8.3	mA
			V_I = −3 V		−0.43			−0.43			
I_{OS}	Short-circuit output current	3				−3			−3		mA
I_{CC}	Supply current	2	V_I = 5 V,	Outputs open		20	26		20	26	mA

† All characteristics are measured with the response control terminal open.
‡ All typical values are at V_{CC} = 5 V, T_A = 25°C.

switching characteristics, V_{CC} = 5 V, T_A = 25°C

PARAMETER		TEST FIGURE	TEST CONDITIONS	MIN	TYP	MAX	UNIT
t_{PLH}	Propagation delay time, low-to-high-level output	4	C_L = 15 pF, R_L = 3.9 kΩ		25	85	ns
t_{PHL}	Propagation delay time, high-to-low-level outut		C_L = 15 pF, R_L = 390 Ω		25	50	
t_{TLH}	Transition time, low-to-high-level output		C_L = 15 pF, R_L = 3.9 kΩ		120	175	ns
t_{THL}	Transition time, high-to-low-level output		C_L = 15 pF, R_L = 390 Ω		10	20	

PARAMETER MEASUREMENT INFORMATION†

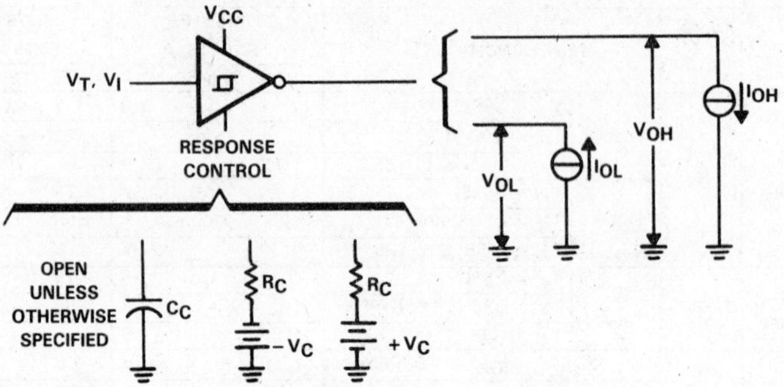
VCC
VT, VI
RESPONSE
CONTROL
IOH
VOH
IOL
VOL
OPEN
UNLESS
OTHERWISE
SPECIFIED
CC
RC
RC
−VC
+VC

FIGURE 1. V_{T+}, V_{T-}, V_{OH}, V_{OL}

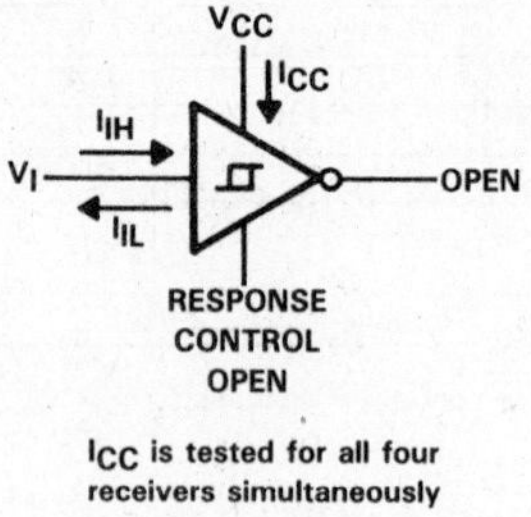
VCC
ICC
IIH
VI
IIL
OPEN
RESPONSE
CONTROL
OPEN
ICC is tested for all four
receivers simultaneously

FIGURE 2. I_{IH}, I_{IL}, I_{CC}

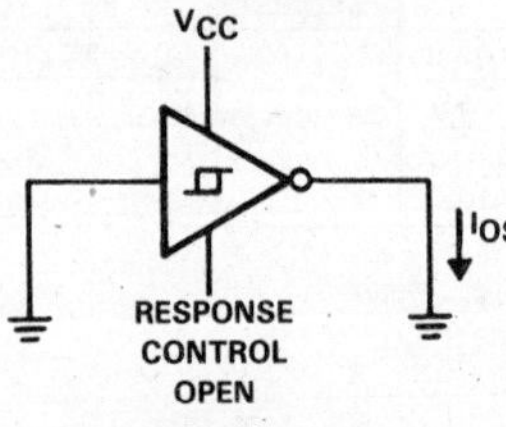
VCC
IOS
RESPONSE
CONTROL
OPEN

FIGURE 3. I_{OS}

†Arrows indicate actual direction of current flow. Current into a terminal is a positive value.

TEXAS INSTRUMENTS
POST OFFICE BOX 655012 • DALLAS, TEXAS 75265

PARAMETER MEASUREMENT INFORMATION†

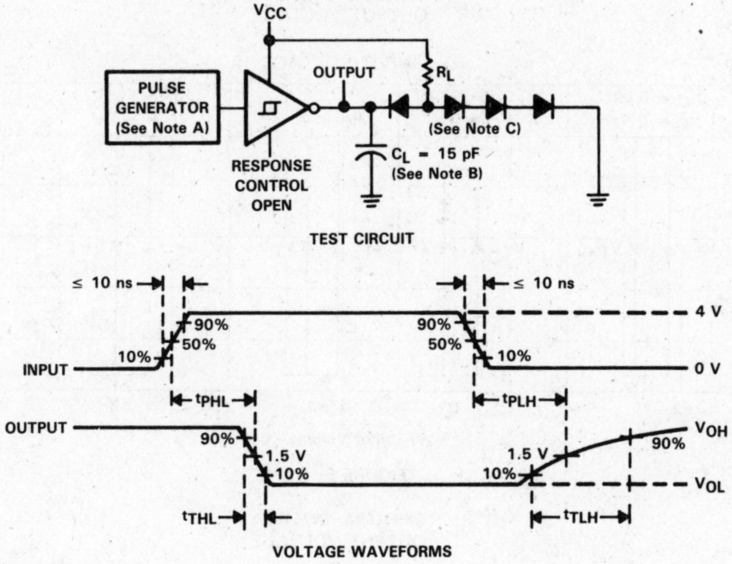
VCC
PULSE GENERATOR (See Note A)
OUTPUT
RL
(See Note C)
CL = 15 pF (See Note B)
RESPONSE CONTROL OPEN
TEST CIRCUIT
≤ 10 ns
≤ 10 ns
4 V
90%
50%
10%
INPUT
0 V
tPHL
tPLH
OUTPUT
VOH
1.5 V
VOL
tTHL
tTLH
VOLTAGE WAVEFORMS

NOTES: A. The pulse generator has the following characteristics: $Z_{out} \approx 50\ \Omega$, $t_w = 500$ ns.
B. C_L includes probe and jig capacitances.
C. All diodes are 1N3064 or equivalent.

FIGURE 4. SWITCHING TIMES

†Arrows indicate actual direction of current flow. Current into a terminal is a positive value.

TYPICAL CHARACTERISTICS†

SN55189, SN75189
OUTPUT VOLTAGE
vs
INPUT VOLTAGE

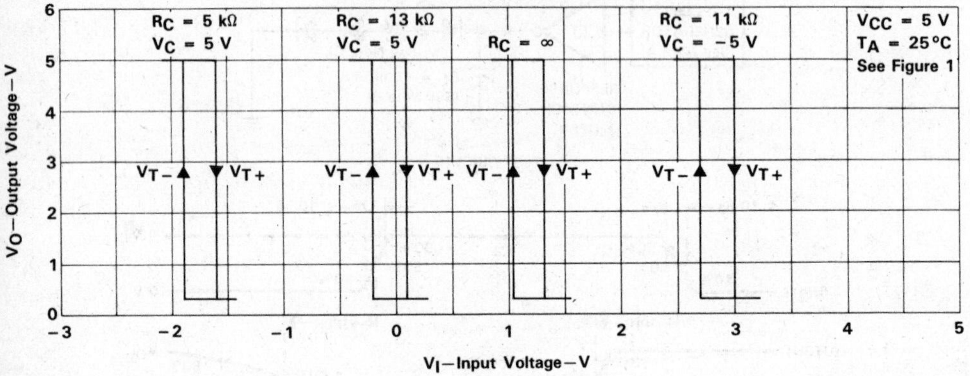
R_C = 5 kΩ
V_C = 5 V
R_C = 13 kΩ
V_C = 5 V
R_C = ∞
R_C = 11 kΩ
V_C = −5 V
V_{CC} = 5 V
T_A = 25°C
See Figure 1
V_{T-}
V_{T+}
V_O—Output Voltage—V
V_I—Input Voltage—V

FIGURE 5

SN55189A, SN75189A
OUTPUT VOLTAGE
vs
INPUT VOLTAGE

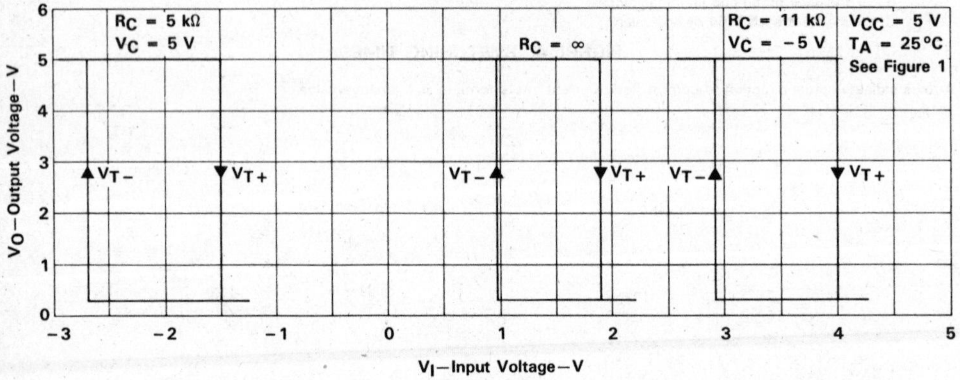
R_C = 5 kΩ
V_C = 5 V
R_C = ∞
R_C = 11 kΩ
V_C = −5 V
V_{CC} = 5 V
T_A = 25°C
See Figure 1
V_{T-}
V_{T+}
V_O—Output Voltage—V
V_I—Input Voltage—V

FIGURE 6

†Data for free-air temperatures below 0°C and above 70°C are applicable to SN55189 and SN55189A circuits only.

TYPICAL CHARACTERISTICS†

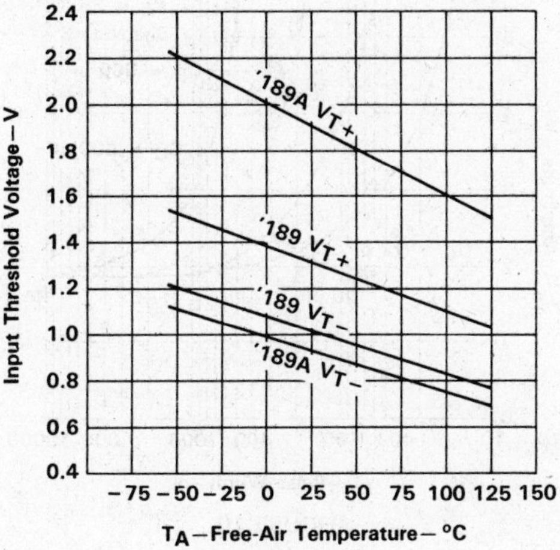
Input Threshold Voltage – V
2.4
2.2
2.0
1.8
1.6
1.4
1.2
1.0
0.8
0.6
0.4
'189A VT+
'189 VT+
'189 VT−
'189A VT−
−75 −50 −25 0 25 50 75 100 125 150
TA – Free-Air Temperature – °C

FIGURE 7

INPUT THRESHOLD VOLTAGE
vs
SUPPLY VOLTAGE

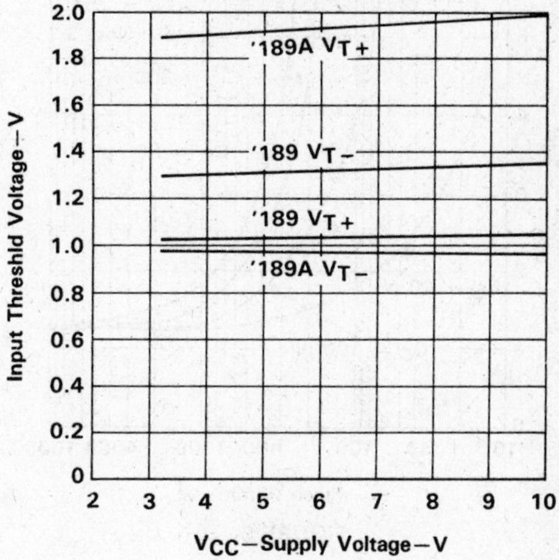
Input Threshold Voltage – V
2.0
1.8
1.6
1.4
1.2
1.0
0.8
0.6
0.4
0.2
0
'189A VT+
'189 VT−
'189 VT+
'189A VT−
2 3 4 5 6 7 8 9 10
VCC – Supply Voltage – V

FIGURE 8

†Data for free-air temperatures below 0°C and above 70°C are applicable to SN55189 and SN55189A circuits only.

TYPICAL CHARACTERISTICS

NOISE REJECTION

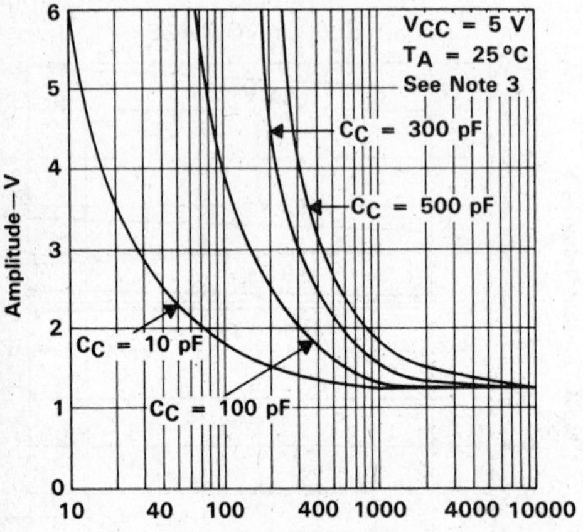
VCC = 5 V
TA = 25°C
See Note 3
CC = 300 pF
CC = 500 pF
CC = 10 pF
CC = 100 pF
Amplitude—V
tw—Pulse Width—ns

FIGURE 9

NOISE REJECTION

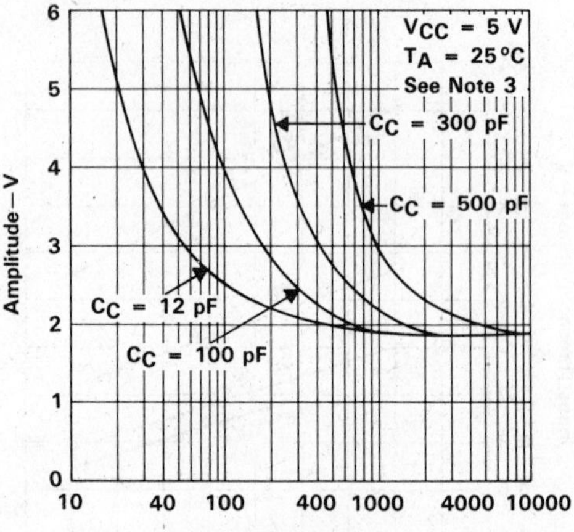
VCC = 5 V
TA = 25°C
See Note 3
CC = 300 pF
CC = 500 pF
CC = 12 pF
CC = 100 pF
Amplitude—V
tw—Pulse Width—ns

FIGURE 10

INPUT CURRENT
vs
INPUT VOLTAGE

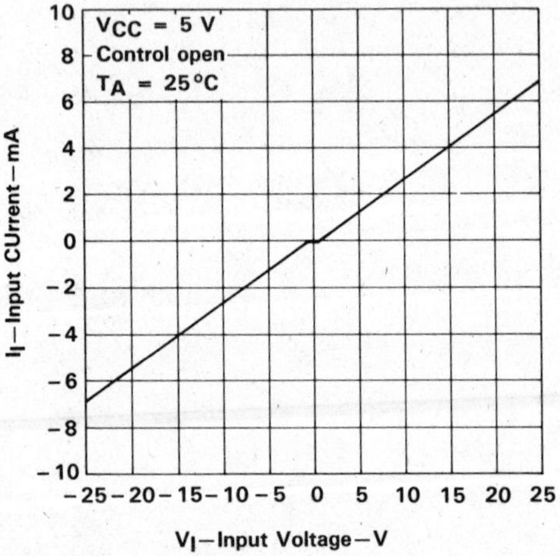
VCC = 5 V
Control open
TA = 25°C
II—Input CUrrent—mA
VI—Input Voltage—V

FIGURE 11

NOTE 3: This figure shows the maximum amplitude of a positive-going pulse that, starting from zero volts, will not cause a change of the output level.

TEXAS INSTRUMENTS
POST OFFICE BOX 655012 • DALLAS, TEXAS 75265

D1314, JULY 1973—REVISED SEPTEMBER 1986

- **Plug-in Replacement for SN75107A, SN75107B, SN75108A, SN75108B with Improved Characteristics**
- **±10 mV Guaranteed Input Sensitivity**
- **TTL Compatible**
- **Standard Supply Voltages . . . ±5 V**
- **Differential Input Common-Mode Voltage Range of ±3 V**
- **Strobe Inputs for Channel Selection**
- **'207 and '207B Have Totem-Pole Outputs**
- **'208 and '208B Have Open-Collector Outputs**
- **"B" Versions Have Diode-Protected Input Stage for Power-Off Condition**
- **Sense Amplifier for MOS Memories**
- **Dual Comparator**
- **High-Sensitivity Line Receiver**

D, J, OR N PACKAGE
(TOP VIEW)

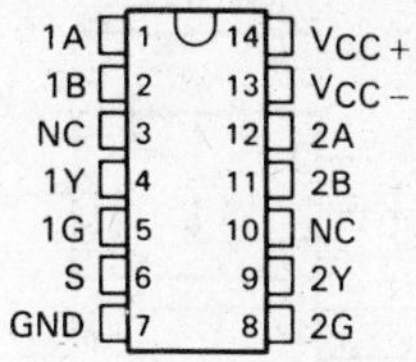
1A 1 14 VCC+
1B 2 13 VCC−
NC 3 12 2A
1Y 4 11 2B
1G 5 10 NC
S 6 9 2Y
GND 7 8 2G

NC—No internal connection

FUNCTION TABLE

DIFFERENTIAL INPUTS A-B	STROBES G	STROBES S	OUTPUT Y
$V_{ID} \geq 10$ mV	X	X	H
-10 mV $< V_{ID} < 10$ mV	X	L	H
	L	X	H
	H	H	Indeterminate
$V_{ID} \leq -10$ mV	X	L	H
	L	X	H
	H	H	L

H = high level, L = low level, X = irrelevant

description

The SN75207, SN75207B, SN75208, and SN75208B are pin-for-pin replacements for the SN75107A, SN75107B, SN75108A, and SN75108B, respectively. The improved input sensitivity makes them more suitable for MOS memory sense amplifiers and can result in faster memory cycles. Improved sensitivity also makes them more useful in line receiver applications by allowing use of longer transmission line lengths. The '207 and '207B each features a TTL-compatible active-pull-up output. The '208 and '208B each features an open-collector output that permits wired-AND logic connections with similar output configurations.

The essential difference between the unsuffixed and "B" versions can be seen in the schematics. Input-protection diodes are in series with the collectors of the differential-input transistors of the "B" versions. These diodes are useful in certain "party-line" systems that may have multiple V_{CC+} power supplies and may be operated with some of the V_{CC+} supplies turned off. In such a system, if a supply is turned off and allowed to go to ground, the equivalent input circuit connected to that supply would be as follows:

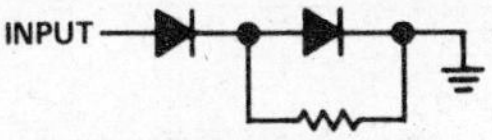
INPUT

UNSUFFIXED VERSION

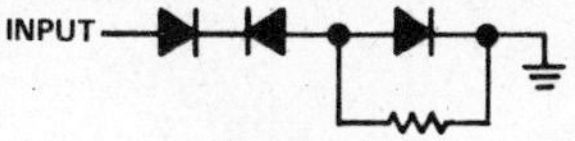
INPUT

"B" VERSION

This would be a problem in specific systems that might possibly have the transmission lines biased to some potential greater than 1.4 volts.

These devices are characterized for operation from 0°C to 70°C and are available in ceramic dual-in-line (J) package, plastic small outline (D) package, or plastic dual-in-line (N) package.

logic symbols†

SN75207

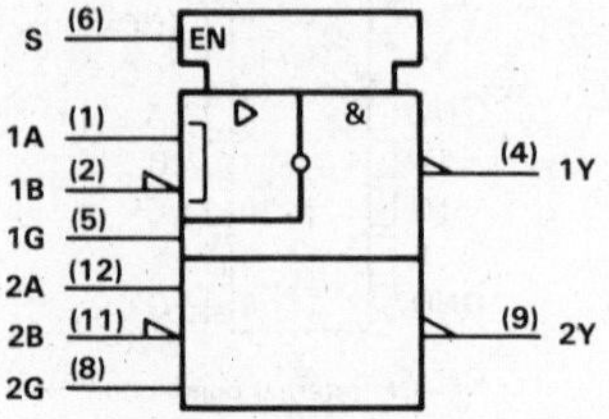
S (6)
EN
1A (1)
1B (2)
1G (5)
2A (12)
2B (11)
2G (8)
&
(4) 1Y
(9) 2Y

SN75208

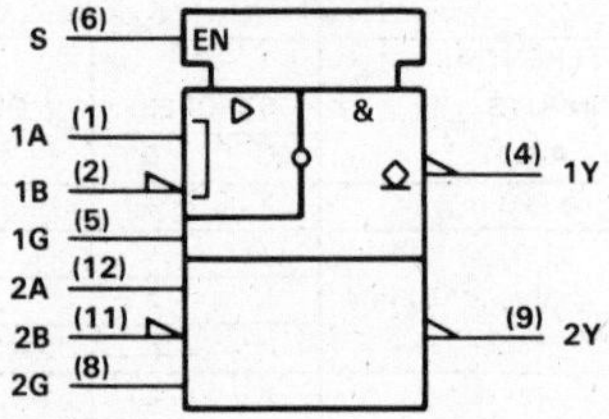
S (6)
EN
1A (1)
1B (2)
1G (5)
2A (12)
2B (11)
2G (8)
&
(4) 1Y
(9) 2Y

† These symbols are in accordance with ANSI/IEEE Std 91-1984 and IEC Publication 617-12.

logic diagram (positive logic)

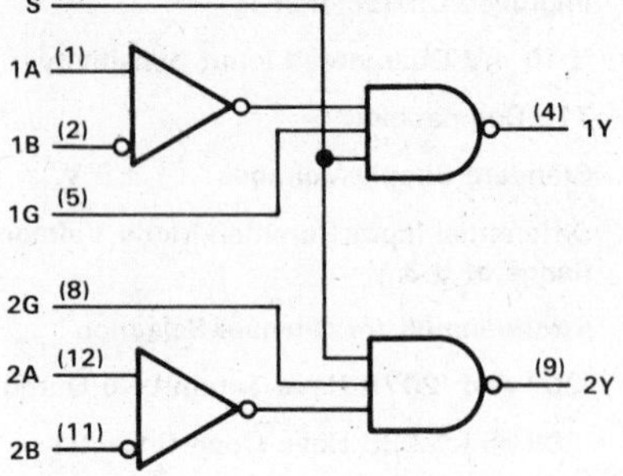
S (6)
1A (1)
1B (2)
1G (5)
2G (8)
2A (12)
2B (11)
(4) 1Y
(9) 2Y

schematic (each receiver)

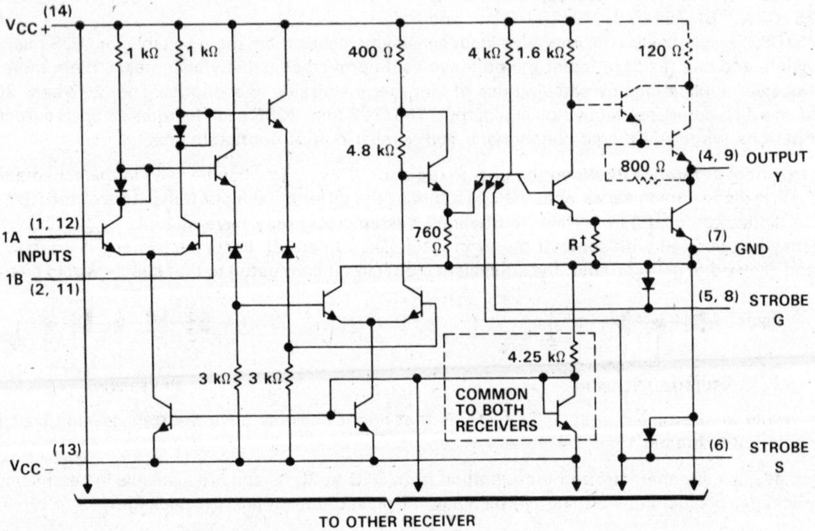
(14)
VCC+
1 kΩ
1 kΩ
400 Ω
4 kΩ
1.6 kΩ
120 Ω
4.8 kΩ
800 Ω
(4, 9) OUTPUT
Y
760 Ω
R†
(7) GND
(1, 12)
1A
INPUTS
1B
(2, 11)
(5, 8) STROBE
G
4.25 kΩ
3 kΩ
3 kΩ
COMMON
TO BOTH
RECEIVERS
(6) STROBE
S
(13)
VCC−
TO OTHER RECEIVER

†R = 1 kΩ for '207 and '207B, 750 Ω for '208 and '208B.

NOTES: A. Resistor values shown are nominal.

B. Components shown with dashed lines in the output circuitry are applicable to the '207 and '207B only. Diodes in series with the collectors of the differential input transistors are short-circuited on '207 and '208.

design characteristics

The '207, '207B, '208, and '208B line receivers/sense amplifiers are TTL-compatible dual circuits intended for use in high-speed data-transmission systems or MOS memory systems. They are designed to detect low-level differential signals in the presence of common-mode noise and variations of temperature and supplies. Dc specifications reflect worst-case conditions of temperature, supply voltages, and input voltages.

The input common-mode voltage range is ±3 volts. This is adequate for application in most systems. In systems with requirements for greater common-mode voltage range, input attenuators may be used to decrease the noise to an acceptable level at the receiver-input terminals.

The circuits feature individual strobe inputs for each channel and a strobe input common to both channels for logic versatility. The strobe inputs are tested to guarantee 400 millivolts of dc noise margin when interfaced with Series 54/74 TTL.

The circuits feature high input impedance and low input currents, which induce very little loading on the transmission line. This makes these devices especially useful in party-line systems. The excellent input sensitivity (3 millivolts typical) is particularly important when data is to be detected at the end of a long transmission line and the amplitude of the data has deteriorated due to cable losses. The circuits are designed to detect input signals of 10 millivolts (or greater) amplitude and convert the polarity of the signal into appropriate TTL-compatible output logic levels.

absolute maximum ratings over operating free-air temperature range (unless otherwise noted)

Supply voltage, V_{CC+} (see Note 1)	7 V
Supply voltage, V_{CC-}	−7 V
Differential input voltage (see Note 2)	±6 V
Common-mode input voltage (see Note 3)	±5 V
Strobe input voltage	5.5 V
Continuous total dissipation at (or below) 25°C free-air temperature: (see Note 4)	
D package	950 mW
J package	1025 mW
N package	1150 mW
Operating free-air temperature range	0°C to 70°C
Storage temperature range	−65°C to 150°C
Lead temperature 1,6 mm (1/16 inch) from case for 60 seconds:	
J package	300°C
Lead temperature 1,6 mm (1/16 inch) from case for 10 seconds:	
D or N package	260°C

NOTES: 1. All voltage values, except differential voltages, are with respect to ground terminal.
2. Differential input voltage values are at the noninverting (A) terminal with respect to the inverting (B) terminal.
3. Common-mode input voltage is the average of the voltages at the A and B inputs.
4. For operation above 25°C free-air temperature, derate linearly to 608 mW at 70°C at the rate of 7.6 mW/°C for the D package, 656 mW at 70°C at the rate of 8.2 mW/°C for the J package, and 736 mW at 70°C at the rate of 9.2 mW/°C for the N package.

recommended operating conditions (see Note 5)

	MIN	NOM	MAX	UNIT
Supply voltage, V_{CC+}	4.75	5	5.25	V
Supply voltage, V_{CC-}	−4.75	−5	−5.25	V
High-level differential input voltage V_{IDH} (see Note 6)	0.01		5	V
Low-level differential input voltage, V_{IDL}	−5†		−0.01	V
Common-mode input voltage, V_{IC} (see Notes 6 and 7)	−3†		3	V
Input voltage, any differential input to ground (see Note 6)	−5†		3	V
High-level input voltage at strobe inputs, $V_{IH(S)}$	2		5.5	V
Low-level input voltage at strobe inputs, $V_{IL(S)}$	0		0.8	V
Low-level output current, I_{OL}			−16	mA
Operating free-air temperature, T_A	0		70	°C

† The algebraic convention, in which the less positive (more negative) limit is designated as minimum, is used in this data sheet for logic voltage levels only.

NOTES: 5. When using only one channel of the line receiver, the strobe G of the unused channel should be grounded and at least one of the differential inputs of the unused receiver should be terminated at some voltage between −3 V and 3 V.
6. The recommended combinations of input voltages fall within the shaded area of the figure shown.
7. The common-mode voltage may be as low as −4 V provided that the more positive of the two inputs is not more negative than −3 V.

RECOMMENDED COMBINATIONS
OF INPUT VOLTAGES

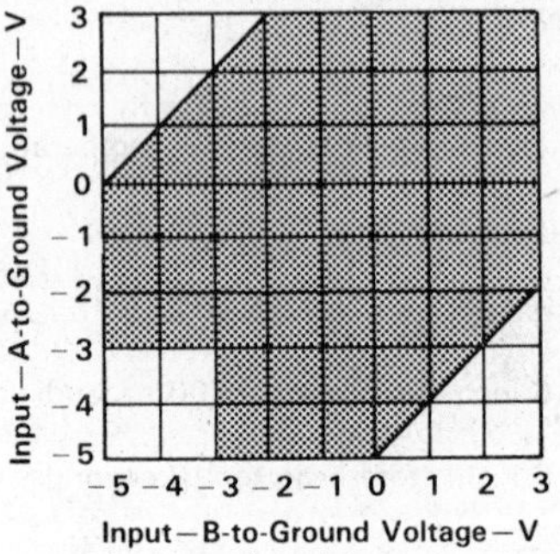
Input—A-to-Ground Voltage—V
3
2
1
0
−1
−2
−3
−4
−5
−5 −4 −3 −2 −1 0 1 2 3
Input—B-to-Ground Voltage—V

electrical characteristics over recommended free-air temperature range (unless otherwise noted)

PARAMETER			TEST CONDITIONS		'207, '207B			'208, '208B			UNIT
					MIN	TYP†	MAX	MIN	TYP†	MAX	
I_{IH}	High-level input current	A	$V_{CC\pm} = \pm 5.25$ V	$V_{ID} = 5$ V		30	75		30	75	μA
		B		$V_{ID} = -5$ V		30	75		30	75	
I_{IL}	Low-level input current	A	$V_{CC\pm} = \pm 5.25$ V	$V_{ID} = -5$ V			−10			−10	μA
		B		$V_{ID} = 5$ V			−10			−10	
I_{IH}	High-level input current into 1G or 2G		$V_{CC\pm} = \pm 5.25$ V, $V_{IH(S)} = 2.4$ V				40			40	μA
			$V_{CC\pm} = \pm 5.25$ V, $V_{IH(S)} = \pm 5.25$ V				1			1	mA
I_{IL}	Low-level input current into 1G or 2G		$V_{CC\pm} = \pm 5.25$ V, $V_{IL(S)} = 0.4$ V				−1.6			−1.6	mA
I_{IH}	High-level input current into S		$V_{CC\pm} = \pm 5.25$ V, $V_{IH(S)} = 2.4$ V				80			80	μA
			$V_{CC\pm} = \pm 5.25$ V, $V_{IH(S)} = \pm 5.25$ V				2			2	mA
I_{IL}	Low-level input current into S		$V_{CC\pm} = \pm 5.25$ V, $V_{IL(S)} = 0.4$ V				−3.2			−3.2	mA
V_{OH}	High-level output voltage		$V_{CC\pm} = \pm 4.75$ V, $V_{IL(S)} = 0.8$ V, $V_{IDH} = 10$ mV, $I_{OH} = -400$ μA, $V_{IC} = -3$ V to 3 V		2.4						V
V_{OL}	Low-level output voltage		$V_{CC\pm} = \pm 4.75$ V, $V_{IH(S)} = 2$ V, $V_{IDL} = -10$ mV, $I_{OL} = 16$ mA, $V_{IC} = -3$ V to 3 V				0.4			0.4	V
I_{OH}	High-level output current		$V_{CC\pm} = \pm 4.75$ V, $V_{OH} = 5.25$ V							250	μA
I_{OS}	Short-circuit output current‡		$V_{CC\pm} = \pm 5.25$ V		−18		−70				mA
I_{CCH+}	Supply current from V_{CC+}, outputs high		$V_{CC\pm} = \pm 5.25$ V, $T_A = 25$ °C			18	30		18	30	mA
I_{CCH-}	Supply current from V_{CC-}, outputs high		$V_{CC\pm} = \pm 5.25$ V, $T_A = 25$ °C			−8.4	−15		−8.4	15	mA

† All typical values are at $V_{CC+} = 5$ V, $V_{CC-} = -5$ V, $T_A = 25$ °C.
‡ Not more than one output should be shorted at a time.

switching characteristics, $V_{CC+} = 5$ V, $V_{CC-} = -5$ V, $T_A = 25$ °C

PARAMETER		TEST CONDITIONS	'207, '207B		'208, '208B		UNIT
			MIN	MAX	MIN	MAX	
$t_{PLH(D)}$	Propagation delay time, low-to-high-level output, from differential inputs A and B	$R_L = 470$ Ω, $C_L = 15$ pF, See Figure 1		35		35	ns
$t_{PHL(D)}$	Propagation delay time, high-to-low-level output, from differential inputs A and B			20		20	ns
$t_{PLH(S)}$	Propagation delay time, low-to-high-level output, from strobe input G or S			17		17	ns
$t_{PHL(S)}$	Propagation delay time, high-to-low-level output, from strobe input G or S			17		17	ns

PARAMETER MEASUREMENT INFORMATION

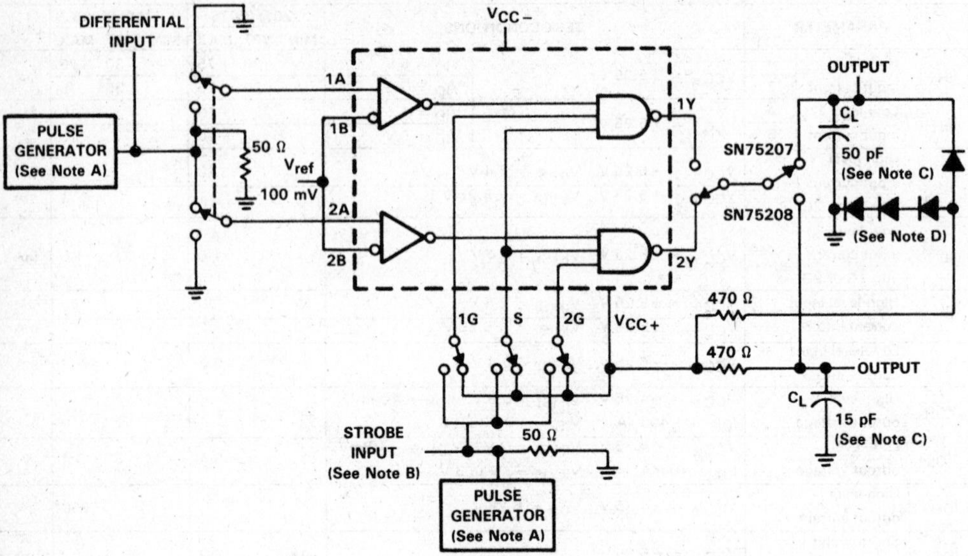
DIFFERENTIAL
INPUT
V_{CC-}
PULSE
GENERATOR
(See Note A)
50 Ω
V_{ref}
100 mV
1A
1B
2A
2B
1Y
2Y
OUTPUT
C_L
50 pF
(See Note C)
SN75207
SN75208
(See Note D)
470 Ω
470 Ω
OUTPUT
C_L
15 pF
(See Note C)
1G
S
2G
V_{CC+}
STROBE
INPUT
(See Note B)
50 Ω
PULSE
GENERATOR
(See Note A)

TEST CIRCUIT

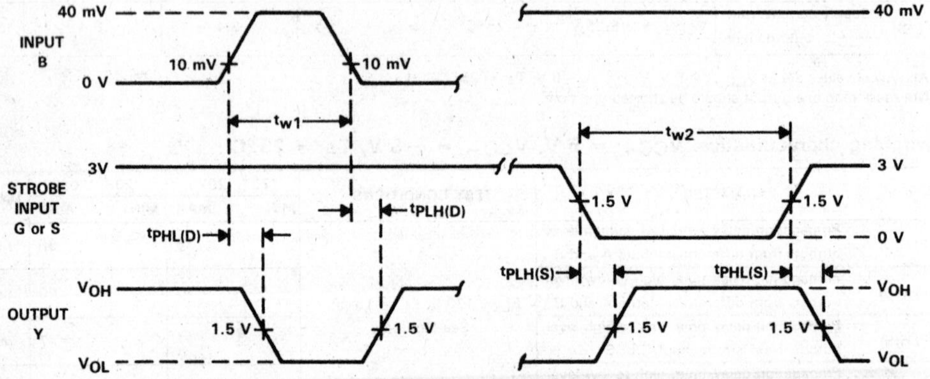
40 mV
INPUT
B
10 mV
10 mV
0 V
40 mV
t_{w1}
t_{w2}
3V
3 V
STROBE
INPUT
G or S
1.5 V
1.5 V
0 V
$t_{PLH(D)}$
$t_{PHL(D)}$
$t_{PLH(S)}$
$t_{PHL(S)}$
V_{OH}
V_{OH}
OUTPUT
Y
1.5 V
1.5 V
1.5 V
1.5 V
V_{OL}
V_{OL}

VOLTAGE WAVEFORMS

NOTES: A. The pulse generators have the following characteristics: Z_{out} = 50 Ω, $t_r \leq 5$ ns, $t_f \leq 5$ ns, t_{w1} = 500 ns with PRR ≤ 1 MHz, t_{w2} = 1 μs with PRR ≤ 500 kHz.

B. Strobe input pulse is applied to Strobe 1G when inputs 1A-1B are being tested, to Strobe S when inputs 1A-1B or 2A-2B are being tested, and to Strobe 2G when inputs 2A-2B are being tested.

C. C_L includes probe and jig capacitance.

D. All diodes are 1N916.

TEXAS INSTRUMENTS
POST OFFICE BOX 655012 • DALLAS, TEXAS 75265

TYPICAL APPLICATION DATA

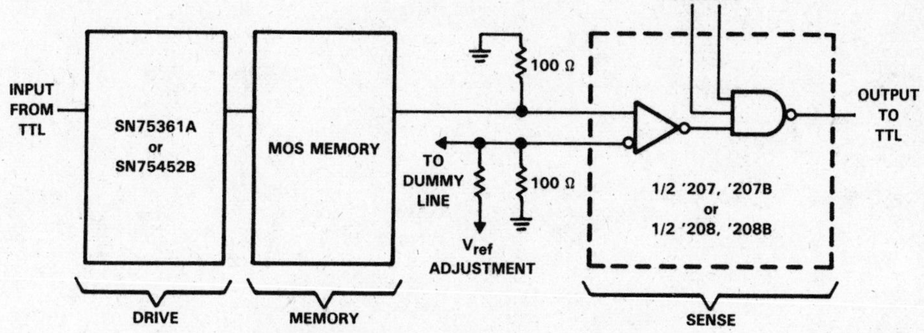
STROBES
100 Ω
INPUT FROM TTL
SN75361A or SN75452B
MOS MEMORY
OUTPUT TO TTL
TO DUMMY LINE
100 Ω
1/2 '207, '207B or 1/2 '208, '208B
Vref ADJUSTMENT
DRIVE
MEMORY
SENSE

FIGURE 2. MOS MEMORY SENSE AMPLIFIER

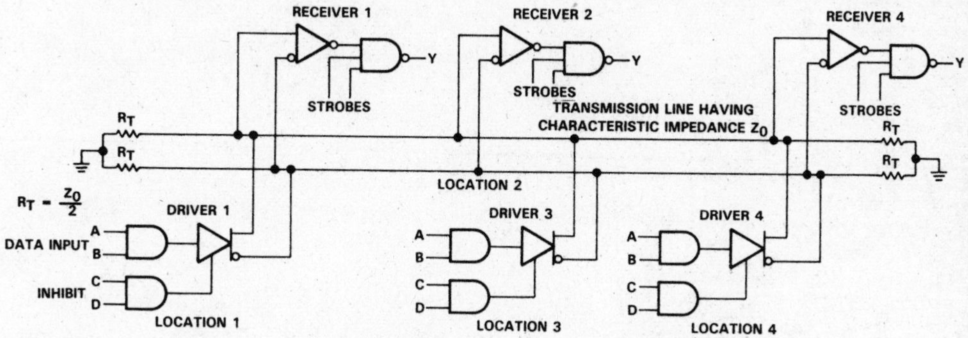
RECEIVER 1
RECEIVER 2
RECEIVER 4
Y
STROBES
TRANSMISSION LINE HAVING CHARACTERISTIC IMPEDANCE Z0
RT
LOCATION 2
RT = Z0/2
DRIVER 1
DRIVER 3
DRIVER 4
DATA INPUT
INHIBIT
A
B
C
D
LOCATION 1
LOCATION 3
LOCATION 4

Receivers are '207, '207B, or 208', or '208B; drivers are SN55109A, SN75109A, SN55110A, SN75110A, or SN75112.

FIGURE 3. DATA-BUS OR PARTY-LINE SYSTEM

PRECAUTIONS: When only one receiver in a package is being used, at least one of the differential inputs of the unused receiver should be terminated at some voltage between −3 volts and 3 volts, preferably at ground. Failure to do so will cause improper operation of the unit being used because of common bias circuitry for the current sources of the two receivers. Strobe G of the unused channel should be grounded.

D2299, FEBRUARY 1986—REVISED OCTOBER 1986

SN75ALS126 QUADRUPLE LINE DRIVER

- Meets IBM 360/370 I/O Interface Specification GA22-6974-3 (Also see SN75ALS130)
- Minimum Output Voltage of 3.11 V at I_{OH} = −60 mA
- Fault Flag Circuit Output Signals Driver Output Fault
- Fault-Detection Current Limit Circuit Minimizes Power Dissipation During a Fault Condition
- Advanced Low-Power Schottky Circuitry
- Dual Common Enable
- Individual Fault Flags
- Designed to be an Improved Replacement for the MC3481

SN75ALS126 . . . D, J, OR N PACKAGE
(TOP VIEW)

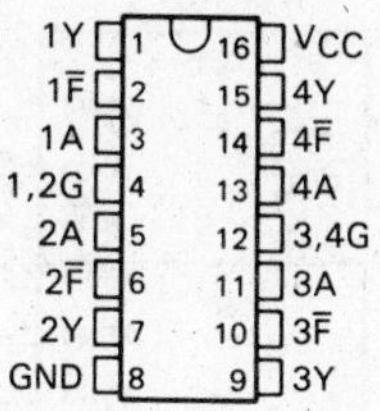
1Y 1
1F̄ 2
1A 3
1,2G 4
2A 5
2F̄ 6
2Y 7
GND 8
16 VCC
15 4Y
14 4F̄
13 4A
12 3,4G
11 3A
10 3F̄
9 3Y

FUNCTION TABLE

INPUTS		OUTPUTS	
G	A	Y	$\overline{F}$
L	X	L	H
H	H	H	H
H	H	S	L

H = high level, L = low level, X = irrelevant, S = shorted to ground

description

The SN75ALS126 quadruple line driver is designed to meet the IBM360/370 I/O specifications GA22-6974-3. The output voltage is 3.11 volts minimum (at I_{OH} = −59.3 milliamperes) over the recommended ranges of supply voltage (4.5 volts to 5.5 volts) and temperature (0 °C to 70 °C). Driver outputs use a fault-detection current-limit circuit to allow high drive current but still minimize power dissipation when the output is shorted to ground. The SN75ALS126 is compatible with standard TTL logic and supply voltages.

The SN75ALS126 employs the IMPACT™ process to achieve fast switching speeds and low power dissipation. Fault-flag circuitry is designed to sense and signal a line short on any Y line. Upon detecting an output fault condition, the fault-flag circuit forces the driver output into a low state and signals a fault condition by causing the fault-flag output to go low.

The SN75ALS126 will drive a 50-ohm load as required in the IBM GA22-6974-3 specification or a 90-ohm load as used in many I/O systems. Optimum performance can be achieved when the device is used with either the SN75125, SN75127, SN75128, or SN75129 line receivers.

The SN75ALS126 is characterized for operation from 0 °C to 70 °C.

IMPACT is a trademark of Texas Instruments Incorporated

logic symbol†

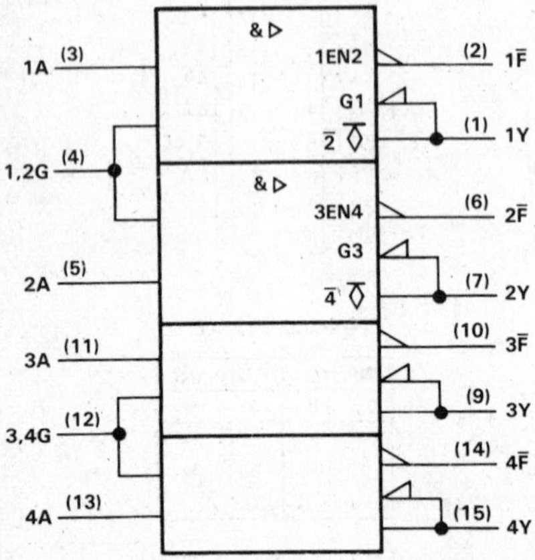
& ▷
1EN2
G1
2
& ▷
3EN4
G3
4
1A (3)
1,2G (4)
2A (5)
3A (11)
3,4G (12)
4A (13)
(2) 1F̄
(1) 1Y
(6) 2F̄
(7) 2Y
(10) 3F̄
(9) 3Y
(14) 4F̄
(15) 4Y

† This symbol is in accordance with ANSI/IEEE Std 91-1984 and IEC Publication 617-12.

logic diagram (positive logic)

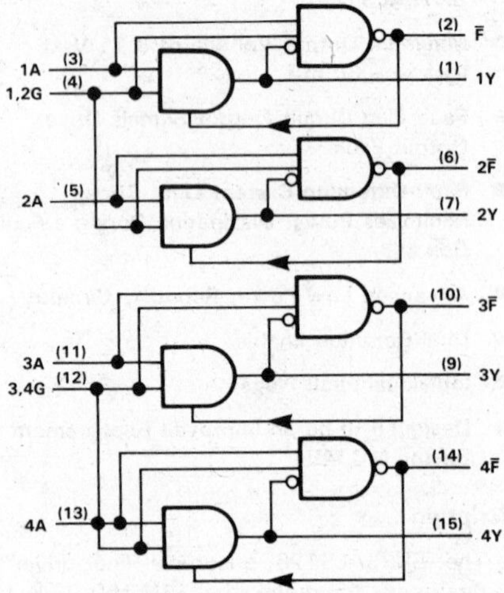
1A (3)
1,2G (4)
2A (5)
3A (11)
3,4G (12)
4A (13)
(2) F̄
(1) 1Y
(6) 2F̄
(7) 2Y
(10) 3F̄
(9) 3Y
(14) 4F̄
(15) 4Y

schematics of inputs and outputs

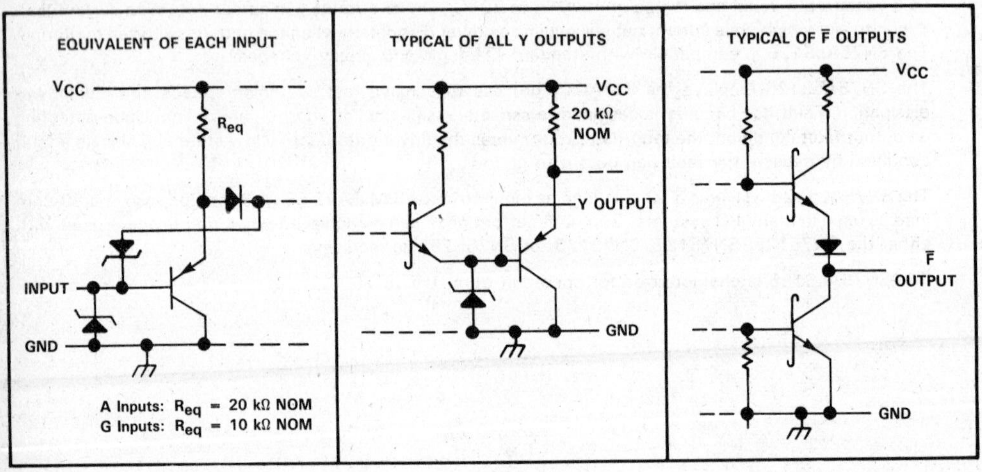
EQUIVALENT OF EACH INPUT
VCC
Req
INPUT
GND
A Inputs: Req = 20 kΩ NOM
G Inputs: Req = 10 kΩ NOM
TYPICAL OF ALL Y OUTPUTS
VCC
20 kΩ
NOM
Y OUTPUT
GND
TYPICAL OF F̄ OUTPUTS
VCC
F̄
OUTPUT
GND

absolute maximum ratings over operating free-air temperature range (unless otherwise noted)

Supply voltage, V_{CC} .... 7 V
Input voltage .... 7 V
Continuous total dissipation at (or below) 25 °C free-air temperature (see Note 1):
D package .... 950 mW
J package .... 1025 mW
N package .... 1150 mW
Operating free-air temperature range .... 0 °C to 70 °C
Storage temperature range .... −65 °C to 150 °C
Lead temperature 1,6 mm (1/16 inch) from case for 10 seconds: D, N package .... 260 °C
Lead temperature 1,6 mm (1/16 inch) from case for 60 seconds: J package .... 300 °C

NOTE 1: For operation above 25 °C free-air temperature, derate the D package to 608 mW at 70 °C at the rate of 7.6 mW/°C, the J package to 656 mW at 70 °C at the rate of 8.2 mW/°C, and the N package to 736 mW at 70 °C at the rate of 9.2 mW/°C.

recommended operating conditions

	MIN	NOM	MAX	UNIT
Supply voltage, V_{CC}	4.5	5	5.95	V
High-level input voltage, V_{IH}	2			V
Low-level input voltage, V_{IL}			0.8	V
High-level output current, I_{OH}			−59.3	mA
Operating free-air temperature, T_A	0		70	°C

electrical characteristics over recommended ranges of supply voltage and operating free-air temperature

PARAMETER			TEST CONDITIONS	MIN	MAX	UNIT
V_{IK}	Input clamp voltage	A,G	$I_I = -18$ mA		−1.5	V
V_{OH}	High-level output voltage	Y	$V_{CC} = 4.5$ V, $I_{OH} = -59.3$ mA, $V_{IH} = 2$ V	3.11		V
		Y	$V_{CC} = 5.25$ V, $I_{OH} = -41$ mA, $V_{IH} = 2$ V	3.9		
		$\overline{F}$	$V_{CC} = 4.5$ V, $I_{OH} = -400$ μA, $V_{IH} = 2$ V	2.5		
V_{OL}	Low-level output voltage	Y	$V_{CC} = 5.5$ V, $I_{OL} = -240$ μA, $V_{IL} = 0.8$ V		0.15	V
		Y	$V_{CC} = 5.95$ V, $I_{OL} = -1$ mA, $V_{IL} = 0.8$ V		0.15	
		$\overline{F}$	$V_{CC} = 4.5$ V, $I_{OL} = 8$ mA, Y at 0 V		0.5	
$I_{O(off)}$	Off-state output current	Y	$V_{CC} = 4.5$ V, $V_{IL} = 0$, $V_O = 3.11$ V		100	μA
		Y	$V_{CC} = 0$, $V_{IL} = 0$, $V_O = 3.11$ V		200	
I_I	Input current	A	$V_{CC} = 4.5$ V, $V_{IH} = 5.5$ V		100	μA
		G			400	
I_{IH}	High-level input current	A	$V_{CC} = 4.5$ V, $V_{IH} = 2.7$ V		20	μA
		G			80	
I_{IL}	Low-level input current	A	$V_{CC} = 5.95$ V, $V_{IL} = 0.4$ V		250	μA
		G			−1000	
I_{OS}	Short-circuit output	Y	$V_{CC} = 5.5$ V, $V_O = 0$		−5	mA
		$\overline{F}$		−15	−100	
		Y	$V_{CC} = 5.95$ V, $V_O = 0$		−5	
		$\overline{F}$		−15	−110	
I_{CCH}	Supply current, all outputs high		$V_{CC} = 5.5$ V, No load		25	mA
			$V_{CC} = 5.95$ V, No load		27	
I_{CCL}	Supply current, Y outputs low		$V_{CC} = 5.5$ V, No load		45	mA
			$V_{CC} = 5.95$ V, No load		47	

switching characteristics over recommended operating free-air temperature range

PARAMETER		FROM	TO	TEST CONDITIONS	MIN	MAX	UNIT
t_{PLH}	Propagation delay time, low-to-high-level output	A	Y	$V_{CC} = 4.5$ V to 5.5 V, $R_L = 50\ \Omega$, $C_L = 50$ pF, $V_{H(ref)} = 3.11$ V, See Figures 1 and 2		30	ns
t_{PHL}	Propagation delay time, high-to-low-level output					28	ns
$\frac{t_{PLH}}{t_{PHL}}$	Ratio of propagation delay times				0.3	3	
t_{PLH}	Propagation delay time, low-to-high-level output	A	Y	$V_{CC} = 5.25$ V to 5.95 V, $R_L = 90\ \Omega$, $C_L = 50$ pF, $V_{H(ref)} = 3.9$ V, See Figures 1 and 2		34	ns
t_{PHL}	Propagation delay time, high-to-low-level output					34	ns
t_{PLH}	Propagation delay time, low-to-high-level output	A	$\overline{F}$	$V_{CC} = 5$ V, $R_L = 2$ kΩ, $C_L = 15$ pF, See Figures 1 and 2		45	ns
t_{PHL}	Propagation delay time, high-to-low-level output					75	ns

TEXAS INSTRUMENTS
POST OFFICE BOX 655012 • DALLAS, TEXAS 75265

PARAMETER MEASUREMENT INFORMATION

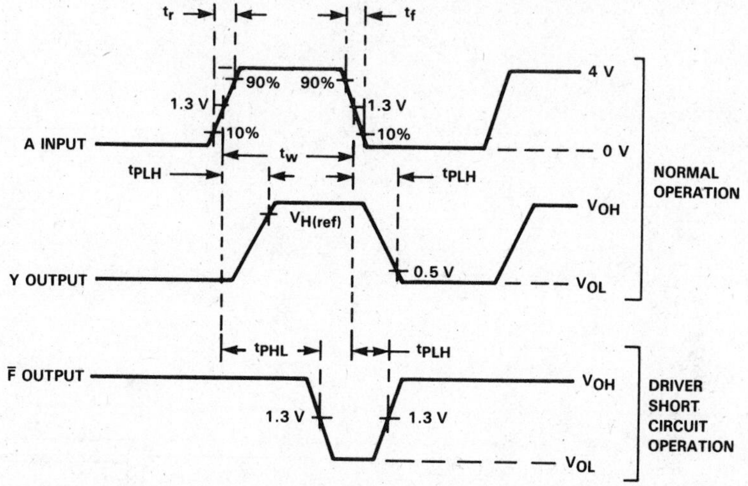
t_r
t_f
90%
90%
1.3 V
1.3 V
10%
10%
4 V
0 V
A INPUT
t_w
t_{PLH}
t_{PLH}
NORMAL OPERATION
$V_{H(ref)}$
V_{OH}
Y OUTPUT
0.5 V
V_{OL}
t_{PHL}
t_{PLH}
F̄ OUTPUT
V_{OH}
1.3 V
1.3 V
V_{OL}
DRIVER SHORT CIRCUIT OPERATION

NOTE A: The input pulse is supplied by a generator having the following characteristics: PRR = 1 MHz, duty cycle = 50%, $t_r \le 6$ ns, $t_f \le 6$ ns, $Z_{out} = 50\ \Omega$.

FIGURE 1. INPUT AND OUTPUT VOLTAGE WAVEFORMS

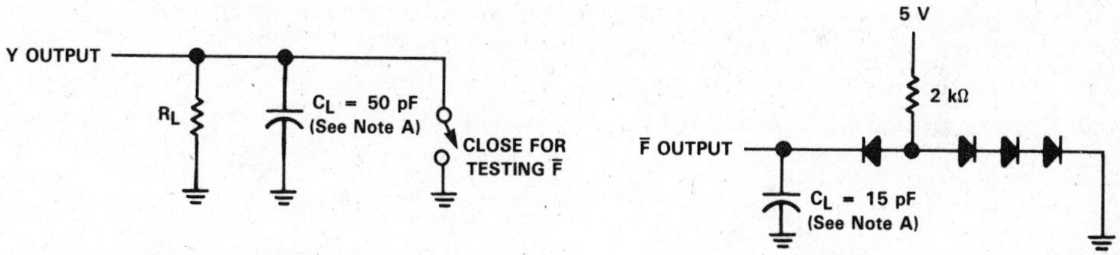
Y OUTPUT
R_L
C_L = 50 pF (See Note A)
CLOSE FOR TESTING F̄
5 V
2 kΩ
F̄ OUTPUT
C_L = 15 pF (See Note A)

NOTE A: C_L includes probe and stray capacitance.

FIGURE 2. SWITCHING CHARACTERISTICS LOAD CIRCUITS

D2299, FEBRUARY 1986

- Meets IBM 360/370 I/O Interface Specification GA22-6974-3 (Also see SN75ALS126)
- Minimum Output Voltage of 3.11 V at I_{OH} = −60 mA
- Fault-Flag Circuit Output Signals Driver Output Fault
- Fault-Detection Current Limit Circuit Minimizes Power Dissipation During a Fault Condition
- Advanced Low-Power Schottky Circuitry
- Common Enable and Common Fault Flag
- Designed to be an Improved Replacement for the MC3485

D, J, OR N PACKAGE
(TOP VIEW)

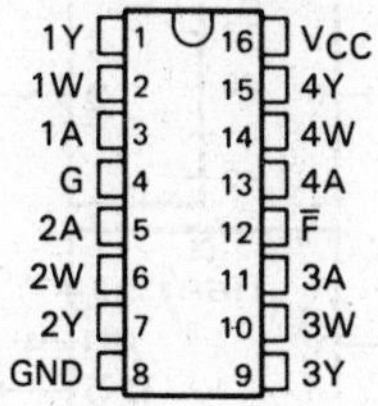
1Y 1
1W 2
1A 3
G 4
2A 5
2W 6
2Y 7
GND 8
16 VCC
15 4Y
14 4W
13 4A
12 F̄
11 3A
10 3W
9 3Y

FUNCTION TABLE

INPUTS		OUTPUTS		
G†	A	Y	F̄†	W
L	X	L	H	H
X	L	L	H	H
H	H	H	H	L
H	H	S	L	H

H = high level, L = low level, X = irrelevant, S = shorted to ground

† G and F̄ are common to the four drivers. If any of the four Y outputs is shorted, the Fault-Flag will respond.

description

The SN75ALS130 quadruple line driver is designed to meet the IBM 360/370 I/O specifications GA22-6974-3. The output voltage is 3.11 volts minimum (at I_{OH} = −59.3 milliamperes) over the recommended ranges of supply voltage (4.5 volts to 5.5 volts) and temperature (0°C to 70°C). Driver outputs use a fault-detection current limit circuit to allow high drive current but still minimize power dissipation when the output is shorted to ground. The SN75ALS130 is compatible with standard TTL logic and supply voltages.

The SN75ALS130 employs the IMPACT™ process to achieve fast switching speeds and low power dissipation. Fault-flag circuitry is designed to sense and signal a line short on any Y line. Upon detecting an output fault condition, the fault-flag circuit forces the driver output into the off (low) state and signals a fault condition by causing the fault-flag output to go low.

The SN75ALS130 will drive a 50-ohm load as required in the IBM GA22-6974-3 specification or a 90-ohm load as used in many I/O systems. Optimum performance can be achieved when the device is used with either the SN75125, SN75127, SN75128, or SN75129 line receivers.

The SN75ALS130 is characterized for operation from 0°C to 70°C.

IMPACT is a trademark of Texas Instruments Incorporated

TEXAS INSTRUMENTS

POST OFFICE BOX 655012 • DALLAS, TEXAS 75265

logic symbol†

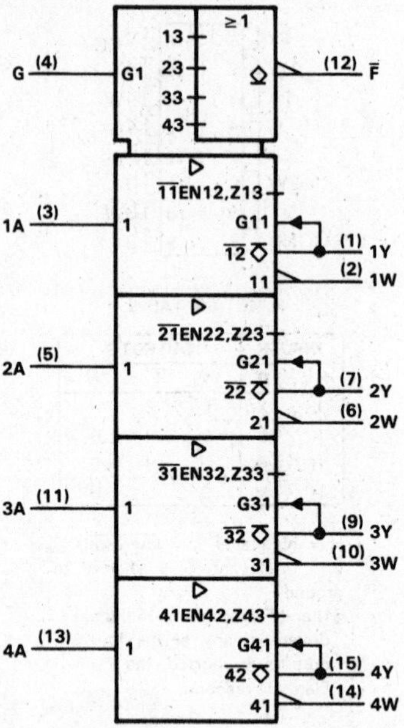
≥1
13
23
33
43
G (4)
G1
(12) F̄
▷
1̄1̄EN12,Z13
1A (3)
1
G11
(1) 1Y
12
(2) 1W
11
▷
2̄1̄EN22,Z23
2A (5)
1
G21
(7) 2Y
22
(6) 2W
21
▷
3̄1̄EN32,Z33
3A (11)
1
G31
(9) 3Y
32
(10) 3W
31
▷
41EN42,Z43
4A (13)
1
G41
(15) 4Y
42
(14) 4W
41

† This symbol is in accordance with ANSI/IEEE Std 91-1984 and IEC Publication 617-12.

logic diagram (positive logic)

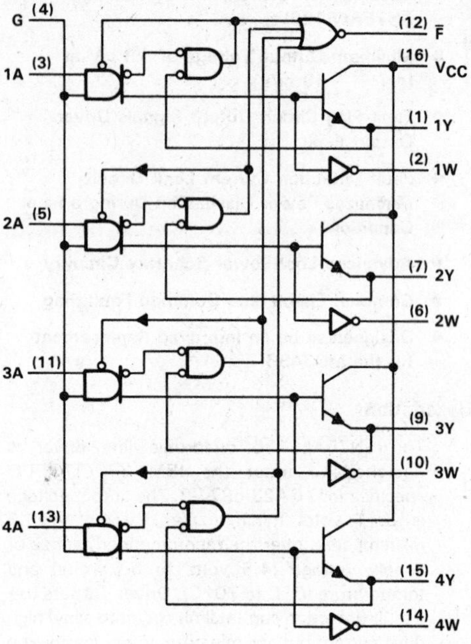
G (4)
1A (3)
2A (5)
3A (11)
4A (13)
(12) F̄
(16) VCC
(1) 1Y
(2) 1W
(7) 2Y
(6) 2W
(9) 3Y
(10) 3W
(15) 4Y
(14) 4W

TEXAS INSTRUMENTS
POST OFFICE BOX 655012 • DALLAS, TEXAS 75265

schematics of inputs and outputs

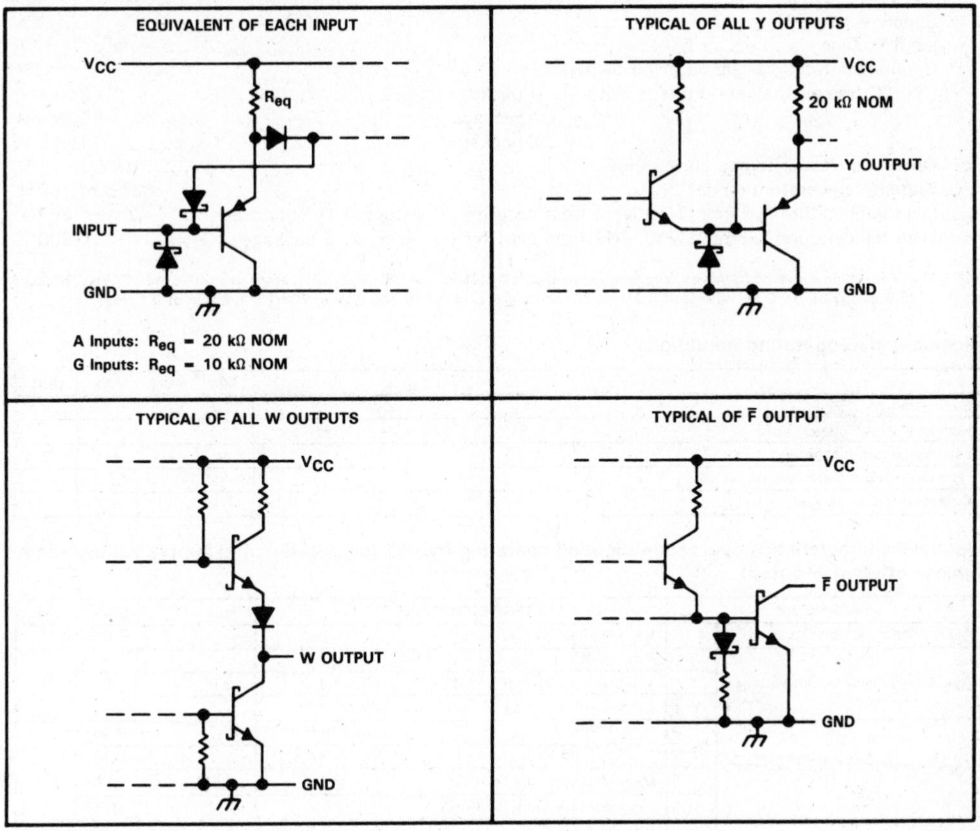
EQUIVALENT OF EACH INPUT
VCC
Req
INPUT
GND
A Inputs: Req = 20 kΩ NOM
G Inputs: Req = 10 kΩ NOM
TYPICAL OF ALL Y OUTPUTS
VCC
20 kΩ NOM
Y OUTPUT
GND
TYPICAL OF ALL W OUTPUTS
VCC
W OUTPUT
GND
TYPICAL OF F̄ OUTPUT
VCC
F̄ OUTPUT
GND

absolute maximum ratings over operating free-air temperature range (unless otherwise noted)

Supply voltage, V_{CC} . . . . 7 V
Input voltage . . . . 7 V
Continuous total dissipation at (or below)
  25 °C free-air temperature (see Note 1): D package . . . . 950 mW
  J package . . . . 1025 mW
  N package . . . . 1150 mW
Operating free-air temperature range . . . . 0 °C to 70 °C
Storage temperature range . . . . −65 °C to 150 °C
Lead temperature 1,6 mm (1/16 inch) from case for 10 seconds: D, N package . . . . 260 °C
Lead temperature 1,6 mm (1/16 inch) from case for 60 seconds: J package . . . . 300 °C

NOTE 1: For operation above 25 °C free-air temperature, derate the D package to 608 mW at 70 °C at the rate of 7.6 mW/°C, the J package to 656 mW at 70 °C at the rate of 8.2 mW/°C, and the N package to 736 mW at 70 °C at the rate of 9.2 mW/°C.

recommended operating conditions

	MIN	NOM	MAX	UNIT
Supply voltage, V_{CC}	4.5	5	5.95	V
High-level input voltage, V_{IH}	2			V
Low-level input voltage, V_{IL}			0.8	V
High-level output current, I_{OH}			−59.3	mA
Operating free-air temperature, T_A	0		70	°C

electrical characteristics over recommended operating free-air temperature and supply voltage range (unless otherwise noted)

PARAMETER			TEST CONDITIONS	MIN	MAX	UNIT
V_{IK}	Input clamp voltage	A,G	$I_I = -18$ mA		−1.5	V
V_{OH}	High-level output voltage	Y	$V_{CC} = 4.5$ V, $I_{OH} = -59.3$ mA, $V_{IH} = 2$ V	3.11		V
		Y	$V_{CC} = 5.25$ V, $I_{OH} = -41$ mA, $V_{IH} = 2$ V	3.9		
		W	$V_{CC} = 4.5$ V, $I_{OH} = -400$ µA, $V_{IH} = 2$ V	2.5		
V_{OL}	Low-level output voltage	Y	$V_{CC} = 5.5$ V, $I_{OL} = -240$ µA, $V_{IL} = 0.8$ V		0.15	V
		Y	$V_{CC} = 5.95$ V, $I_{OL} = -1$ mA, $V_{IL} = 0.8$ V		0.15	
		$\overline{F}$	$V_{CC} = 4.5$ V, $I_{OL} = 8$ mA, Y at 0 V		0.5	
		W	$V_{CC} = 4.5$ V, $I_{OL} = 8$ mA		0.5	
$I_{O(off)}$	Off-state output current	Y	$V_{CC} = 4.5$ V, $V_{IL} = 0$, $V_O = 3.11$ V		100	µA
		Y	$V_{CC} = 0$, $V_{IL} = 0$, $V_O = 3.11$ V		200	
I_{OH}	High-level output current	$\overline{F}$	$V_{CC} = 5.95$ V, $V_{OH} = 5.95$ V		100	µA
I_I	Input current	A	$V_{CC} = 4.5$ V, $V_{IH} = 5.5$ V		100	µA
		G			400	
I_{IH}	High-level input current	A	$V_{CC} = 4.5$ V, $V_{IH} = 2.7$ V		20	µA
		G			80	
I_{IL}	Low-level input current	A	$V_{CC} = 5.95$ V, $V_{IL} = 0.4$ V		250	µA
		G			−1000	
I_{OS}	Short-circuit output	Y	$V_{CC} = 5.5$ V, $V_O = 0$		−5	mA
		W		−15	−100	
		Y	$V_{CC} = 5.95$ V, $V_O = 0$		−5	
		W		−15	−110	
I_{CCH}	Supply current, all outputs high		$V_{CC} = 5.5$ V, No load		30	mA
			$V_{CC} = 5.95$ V, No load		32	
I_{CCL}	Supply current, Y outputs low		$V_{CC} = 5.5$ V, No load		45	mA
			$V_{CC} = 5.95$ V, No load		47	

TEXAS INSTRUMENTS
POST OFFICE BOX 655012 • DALLAS, TEXAS 75265

switching characteristics over recommended operating free-air temperature range

PARAMETER		FROM	TO	TEST CONDITIONS	MIN	MAX	UNIT
t_{PLH}	Propagation delay time, low-to-high-level output	A	Y	V_{CC} = 4.5 V to 5.5 V, R_L = 50 Ω, C_L = 50 pF, $V_{H(ref)}$ = 3.11 V, Input f = 1 MHz See Figures 1 and 2		30	ns
t_{PHL}	Propagation delay time, high-to-low-level output					28	ns
$\frac{t_{PLH}}{t_{PHL}}$	Ratio of propagation delay times				0.3	3	
t_{PLH}	Propagation delay time, low-to-high-level output	A	Y	V_{CC} = 5.25 V to 5.95 V, R_L = 90 Ω, C_L = 50 pF, $V_{H(ref)}$ = 3.9 V, Input f = 5 MHz See Figures 1 and 2		34	ns
t_{PHL}	Propagation delay time, high-to-low-level output					34	ns
t_{PLH}	Propagation delay time, low-to-high-level output	A	W	V_{CC} = 5 V, R_L = 2 kΩ, C_L = 15 pF, See Figures 1 and 2		34	ns
t_{PHL}	Propagation delay time, high-to-low-level output					21	ns
t_{PLH}	Propagation delay time, low-to-high-level output	A	$\overline{F}$	V_{CC} = 5 V, R_L = 2 kΩ, C_L = 15 pF, See Figures 1 and 2		45	ns
t_{PHL}	Propagation delay time, high-to-low-level output					75	ns

PARAMETER MEASUREMENT INFORMATION

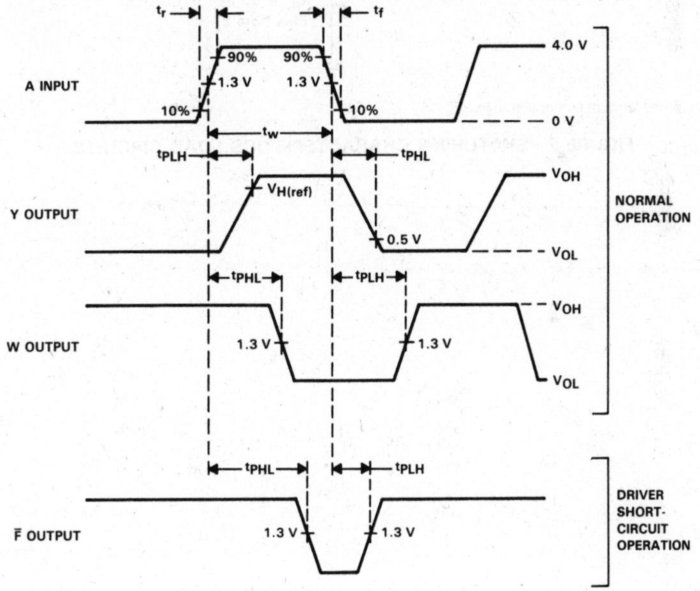

tr
tf
4.0 V
90%
90%
A INPUT
1.3 V
1.3 V
10%
10%
0 V
tw
tPLH
tPHL
VOH
VH(ref)
Y OUTPUT
NORMAL OPERATION
0.5 V
VOL
tPHL
tPLH
VOH
W OUTPUT
1.3 V
1.3 V
VOL
tPHL
tPLH
DRIVER SHORT-CIRCUIT OPERATION
F̄ OUTPUT
1.3 V
1.3 V

NOTE A: The input pulse is supplied by a generator having the following characteristics: PRR = 1 MHz, duty cycle = 50%, $t_r \leq 6$ ns, $t_f \leq 6$ ns, Z_{out} = 50 Ω.

FIGURE 1. INPUT AND OUTPUT VOLTAGE WAVEFORMS

PARAMETER MEASUREMENT INFORMATION

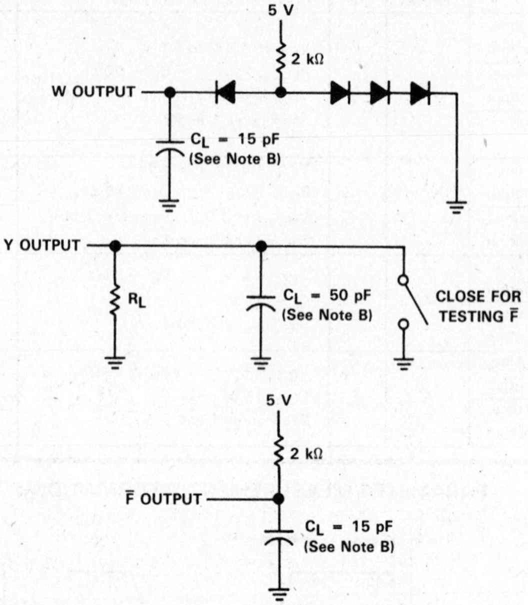
5 V
2 kΩ
W OUTPUT
C_L = 15 pF
(See Note B)
Y OUTPUT
R_L
C_L = 50 pF
(See Note B)
CLOSE FOR
TESTING $\overline{F}$
5 V
2 kΩ
$\overline{F}$ OUTPUT
C_L = 15 pF
(See Note B)

NOTE B: C_L includes probe and stray capacitance.

FIGURE 2. SWITCHING CHARACTERISTICS LOAD CIRCUITS

MEETS IEEE STANDARD 488-1978 (GPIB)

- 8-Channel Bidirectional Transceiver
- High-Speed Advanced Low-Power Schottky Circuitry
- Low Power Dissipation . . . 46 mW Max per Channel
- Fast Propagation Times . . . 20 ns Max
- High-Impedance P-N-P Inputs
- Receiver Hysteresis . . . 650 mV Typ
- Open-Collector Driver Output Option
- No Loading of Bus When Device is Powered Down (V_{CC} = 0)
- Power-Up/Power-Down Protection (Glitch-Free)

DW, J, OR N PACKAGE
(TOP VIEW)

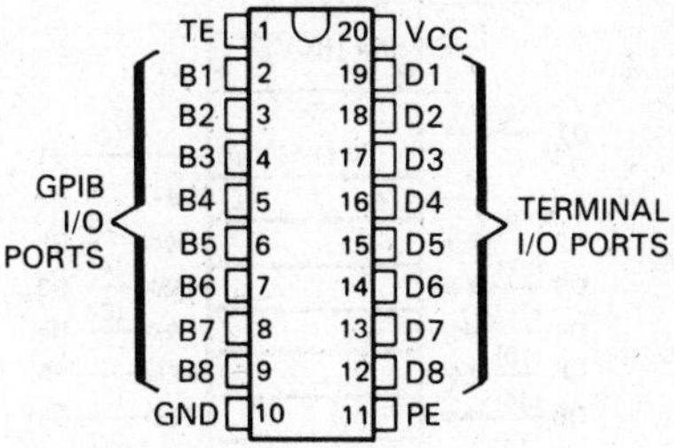
TE 1
B1 2
B2 3
B3 4
B4 5
B5 6
B6 7
B7 8
B8 9
GND 10
20 VCC
19 D1
18 D2
17 D3
16 D4
15 D5
14 D6
13 D7
12 D8
11 PE
GPIB I/O PORTS
TERMINAL I/O PORTS

FUNCTION TABLES

EACH DRIVER

INPUTS			OUTPUT
D	TE	PE	B
H	H	H	H
L	H	X	L
H	X	L	Z†
X	L	X	Z†

EACH RECEIVER

INPUTS			OUTPUT
B	TE	PE	D
L	L	X	L
H	L	X	H
X	H	X	Z

H = high level, L = low level, X = irrelevant,
Z = high-impedance state.

† This is the high-impedance state of a normal 3-state output modified by the internal resistors to V_{CC} and ground.

description

The SN75ALS160 eight-channel general-purpose interface bus transceiver is a monolithic, high-speed, Advanced Low-Power Schottky device designed for two-way data communications over single-ended transmission lines. It is designed to meet the requirements of IEEE Standard 488-1978. The transceiver features driver outputs that can be operated in either the passive-pullup or three-state mode. If Talk Enable (TE) is high, these ports have the characteristics of passive-pullup outputs when Pullup Enable (PE) is low, and of three-state outputs when PE is high. Taking TE low places these ports in the high-impedance state. The driver outputs are designed to handle loads up to 48 milliamperes of sink current.

An active turn-off feature has been incorporated into the bus-terminating resistors so that the device exhibits a high impedance to the bus when V_{CC} = 0. When combined with the SN75ALS161 or SN75ALS162 management bus transceiver, the pair provides the complete 16-wire interface for the IEEE 488 bus.

The SN75ALS160 is manufactured in a 20-pin package and is characterized for operation from 0°C to 70°C.

TEXAS INSTRUMENTS
POST OFFICE BOX 655012 • DALLAS, TEXAS 75265

logic symbol†

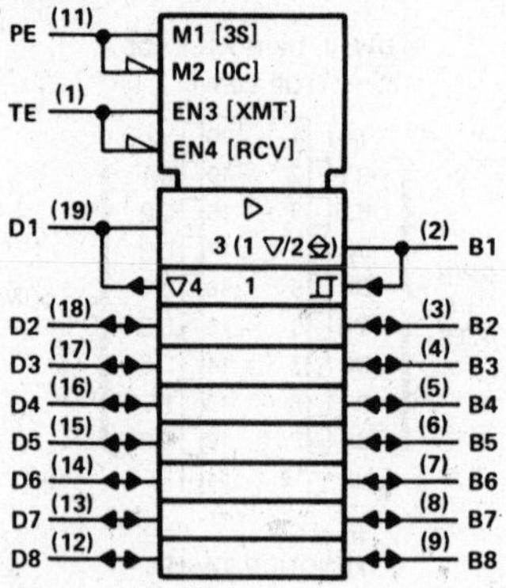
PE (11)
M1 [3S]
M2 [0C]
TE (1)
EN3 [XMT]
EN4 [RCV]
D1 (19)
3 (1 ▽/2 ⬙)
▽4 1
(2) B1
D2 (18)
D3 (17)
D4 (16)
D5 (15)
D6 (14)
D7 (13)
D8 (12)
(3) B2
(4) B3
(5) B4
(6) B5
(7) B6
(8) B7
(9) B8

† This symbol is in accordance with ANSI/IEEE Std 91-1984 and IEC Publication 617-12.
▽ Designates 3-state outputs.
⬙ Designates passive-pullup outputs.

logic diagram (positive logic)

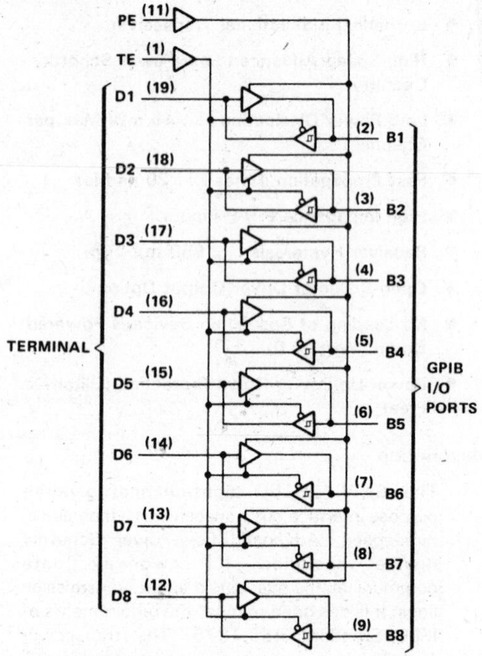
PE (11)
TE (1)
D1 (19)
D2 (18)
D3 (17)
D4 (16)
D5 (15)
D6 (14)
D7 (13)
D8 (12)
TERMINAL
(2) B1
(3) B2
(4) B3
(5) B4
(6) B5
(7) B6
(8) B7
(9) B8
GPIB I/O PORTS

schematics of inputs and outputs

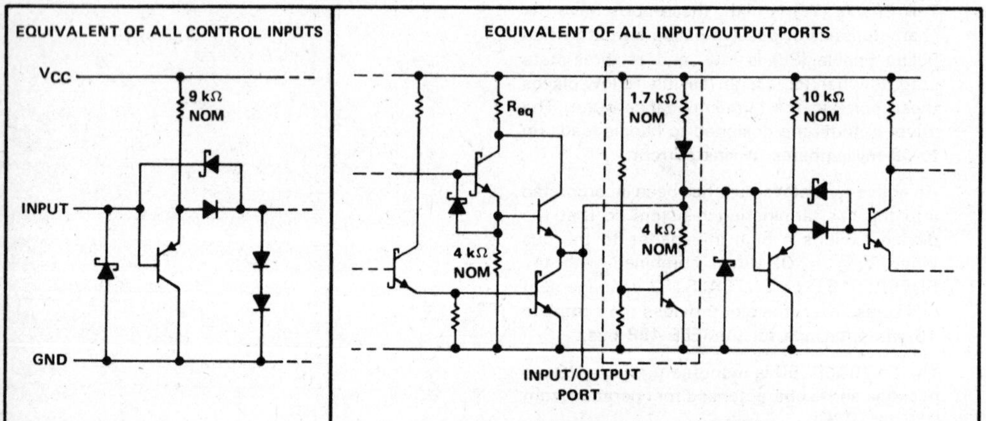
EQUIVALENT OF ALL CONTROL INPUTS
VCC
9 kΩ NOM
INPUT
GND
EQUIVALENT OF ALL INPUT/OUTPUT PORTS
Req
1.7 kΩ NOM
10 kΩ NOM
4 kΩ NOM
4 kΩ NOM
INPUT/OUTPUT PORT

Driver output R_{eq} = 30 Ω NOM
Receiver output R_{eq} = 110 Ω NOM
Circuit inside dashed lines is on the driver outputs only.

TEXAS INSTRUMENTS
POST OFFICE BOX 655012 • DALLAS, TEXAS 75265

absolute maximum ratings over operating free-air temperature range (unless otherwise noted)

Supply voltage, V_{CC} (see Note 1) .... 7 V
Input voltage .... 5.5 V
Low-level driver output current .... 100 mA
Continuous total dissipation at (or below) 25°C free-air temperature (see Note 2):
- DW package .... 1125 mW
- J package .... 1375 mW
- N package .... 1150 mW

Operating free-air temperature range .... 0°C to 70°C
Storage temperature range .... −65°C to 150°C
Lead temperature 1,6 mm (1/16 inch) from the case for 60 seconds: J package .... 300°C
Lead temperature 1,6 mm (1/16 inch) from the case for 10 seconds: DW or N package .... 260°C

NOTES: 1. All voltage values are with respect to network ground terminal.
2. For operation above 25°C free-air temperature, derate the DW package to 720 mW at 70°C at the rate of 9.0 mW/°C, derate the J package to 880 mW at 70°C at the rate of 11.0 mW/°C, and derate the N package to 736 mW at 70°C at the rate of 9.2 mW/°C.

recommended operating conditions

		MIN	NOM	MAX	UNIT
Supply voltage, V_{CC}		4.75	5	5.25	V
High-level input voltage, V_{IH}		2			V
Low-level input voltage, V_{IL}				0.8	V
High-level output current, I_{OH}	Bus ports with pullups active			−5.2	mA
	Terminal ports			−800	μA
Low-level output current, I_{OL}	Bus ports			48	mA
	Terminal ports			16	mA
Operating free-air temperature, T_A		0		70	°C

electrical characteristics over recommended ranges of supply voltage and operating free-air temperature (unless otherwise noted)

PARAMETER			TEST CONDITIONS		MIN	TYP†	MAX	UNIT
V_{IK}	Input clamp voltage		I_I = −18 mA			−0.8	−1.5	V
V_{hys}	Hysteresis ($V_{T+} - V_{T-}$)	Bus			0.4	0.65		V
V_{OH}‡	High-level output voltage	Terminal	I_{OH} = −800 μA,	TE at 0.8 V	2.7	3.5		V
		Bus	I_{OH} = −5.2 mA,	PE and TE at 2 V	2.5	3.3		
V_{OL}	Low-level output voltage	Terminal	I_{OL} = 16 mA,	TE at 0.8 V		0.3	0.5	V
		Bus	I_{OL} = 48 mA,	TE at 2 V		0.35	0.5	
I_I	Input current at maximum input voltage	Terminal	V_I = 5.5 V			0.2	100	μA
I_{IH}	High-level input current	Terminal, PE, or TE	V_I = 2.7 V			0.1	20	μA
I_{IL}	Low-level input current		V_I = 0.5 V			−10	−100	μA
$V_{I/O(bus)}$	Voltage at bus port		Driver disabled	$I_{I(bus)}$ = 0	2.5	3.0	3.7	V
				$I_{I(bus)}$ = −12 mA			−1.5	
$I_{I/O(bus)}$	Current into bus port	Power on	Driver disabled	$V_{I(bus)}$ = −1.5 V to 0.4 V	−1.3			mA
				$V_{I(bus)}$ = 0.4 V to 2.5 V	0		−3.2	
				$V_{I(bus)}$ = 2.5 V to 3.7 V			+2.5 −3.2	
				$V_{I(bus)}$ = 3.7 V to 5 V	0		2.5	
				$V_{I(bus)}$ = 5 V to 5.5 V	0.7		2.5	
		Power off	V_{CC} = 0,	$V_{I(bus)}$ = 0 V to 2.5 V			−40	μA
I_{OS}	Short-circuit output current	Terminal			−15	−35	−75	mA
		Bus			−25	−50	−125	
I_{CC}	Supply current		No load	Terminal outputs low and enabled		42	56	mA
				Bus outputs low and enabled		52	70	
$C_{i/o(bus)}$	Bus-port capacitance		V_{CC} = 5 V to 0 V, $V_{I/O}$ = 0 to 2 V, f = 1 MHz			30		pF

† All typical values are at V_{CC} = 5 V, T_A = 25°C.
‡ V_{OH} applies to three-state outputs only.

switching characteristics over recommended range of operating free-air temperature (unless otherwise noted), V_{CC} = 5 V

PARAMETER		FROM	TO	TEST CONDITIONS	MIN	TYP†	MAX	UNIT
t_{PLH}	Propagation delay time, low-to-high-level output	Terminal	Bus	C_L = 30 pF, See Figure 1		10	20	ns
t_{PHL}	Propagation delay time, high-to-low-level output					12	20	
t_{PLH}	Propagation delay time, low-to-high-level output	Bus	Terminal	C_L = 30 pF, See Figure 2		5	10	ns
t_{PHL}	Propagation delay time, high-to-low-level output					7	14	
t_{PZH}	Output enable time to high level	TE	Bus	C_L = 15 pF, See Figure 3		11	20	ns
t_{PHZ}	Output disable time from high level					3	10	
t_{PZL}	Output enable time to low level					18	35	
t_{PLZ}	Output disable time from low level					5	20	
t_{PZH}	Output enable time to high level	TE	Terminal	C_L = 15 pF, See Figure 4		5	20	ns
t_{PHZ}	Output disable time from high level					8	20	
t_{PZL}	Output enable time to low level					9	20	
t_{PLZ}	Output disable time from low level					8	20	
t_{en}	Output pull-up enable time	PE	Bus	C_L = 15 pF, See Figure 5		3	10	ns
t_{dis}	Output pull-up disable time					4	12	

†Typical values are at T_A = 25°C.

PARAMETER MEASUREMENT INFORMATION

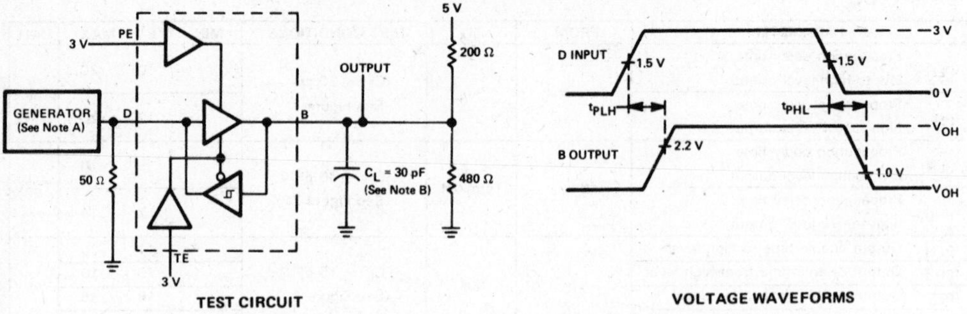
5 V
PE
3 V
200 Ω
OUTPUT
GENERATOR (See Note A)
D
B
50 Ω
C_L = 30 pF (See Note B)
480 Ω
TE
3 V
TEST CIRCUIT
D INPUT
1.5 V
3 V
0 V
t_{PLH}
t_{PHL}
B OUTPUT
2.2 V
1.0 V
V_{OH}
VOLTAGE WAVEFORMS

FIGURE 1. TERMINAL-TO-BUS PROPAGATION DELAY TIMES

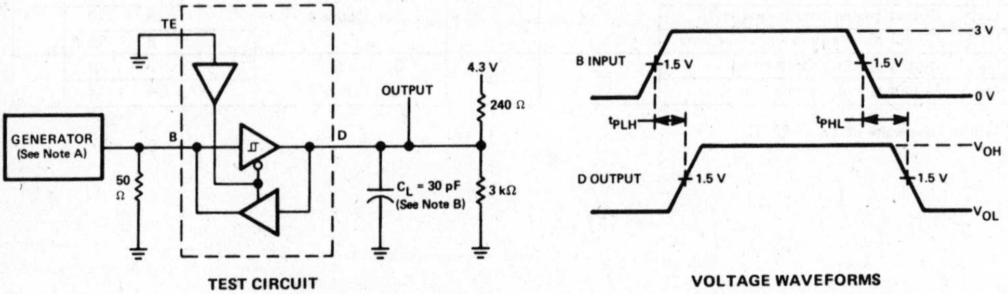
TE
4.3 V
OUTPUT
240 Ω
GENERATOR (See Note A)
B
D
50 Ω
C_L = 30 pF (See Note B)
3 kΩ
TEST CIRCUIT
B INPUT
1.5 V
3 V
0 V
t_{PLH}
t_{PHL}
D OUTPUT
V_{OH}
V_{OL}
VOLTAGE WAVEFORMS

FIGURE 2. BUS-TO-TERMINAL PROPAGATION DELAY TIMES

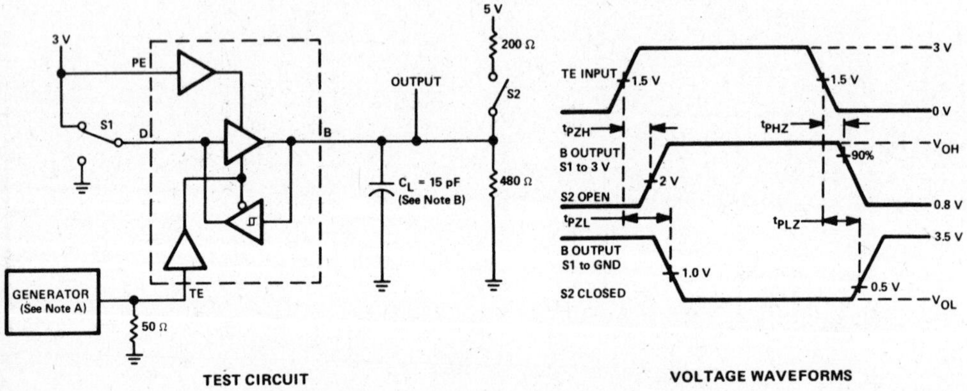
5 V
3 V
200 Ω
PE
OUTPUT
S2
S1
D
B
C_L = 15 pF (See Note B)
480 Ω
GENERATOR (See Note A)
TE
50 Ω
TEST CIRCUIT
TE INPUT
1.5 V
3 V
0 V
t_{PZH}
t_{PHZ}
B OUTPUT S1 to 3 V S2 OPEN
90%
V_{OH}
2 V
0.8 V
t_{PZL}
t_{PLZ}
B OUTPUT S1 to GND S2 CLOSED
3.5 V
1.0 V
0.5 V
V_{OL}
VOLTAGE WAVEFORMS

FIGURE 3. TE-TO-BUS ENABLE AND DISABLE TIMES

NOTES: A. The input pulse is supplied by a generator having the following characteristics: PRR ≤ 1 MHz, 50% duty cycle, $t_r \leq 6$ ns, $t_f \leq 6$ ns, Z_{out} = 50 Ω.
B. C_L includes probe and jig capacitance.

TEXAS INSTRUMENTS
POST OFFICE BOX 655012 • DALLAS, TEXAS 75265

PARAMETER MEASUREMENT INFORMATION

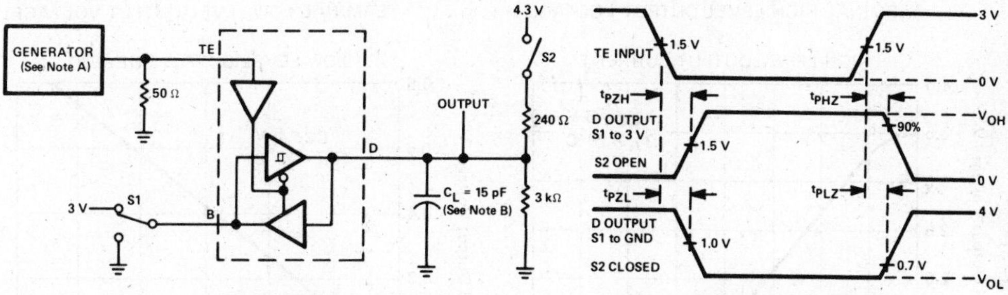
GENERATOR
(See Note A)
TE
50 Ω
3 V
S1
B
D
OUTPUT
4.3 V
S2
240 Ω
C_L = 15 pF
(See Note B)
3 kΩ
TE INPUT
1.5 V
1.5 V
3 V
0 V
t_{PZH}
t_{PHZ}
D OUTPUT
S1 to 3 V
S2 OPEN
1.5 V
90%
V_{OH}
0 V
t_{PZL}
t_{PLZ}
4 V
D OUTPUT
S1 to GND
S2 CLOSED
1.0 V
0.7 V
V_{OL}

TEST CIRCUIT — VOLTAGE WAVEFORMS

FIGURE 4. TE-TO-TERMINAL ENABLE AND DISABLE TIMES

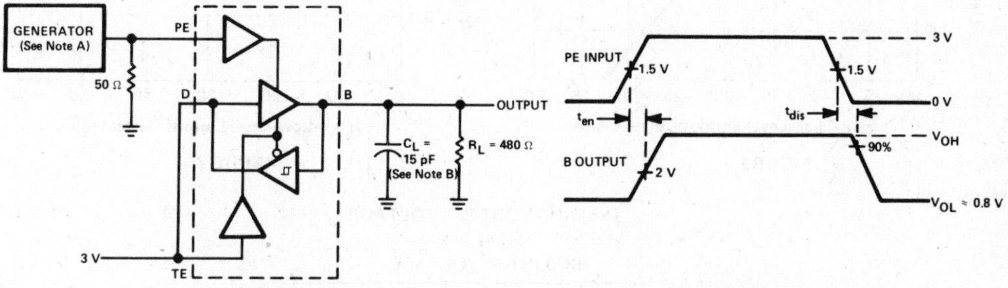
GENERATOR
(See Note A)
PE
50 Ω
D
B
OUTPUT
3 V
TE
C_L =
15 pF
(See Note B)
R_L = 480 Ω
PE INPUT
1.5 V
1.5 V
3 V
0 V
t_{en}
t_{dis}
B OUTPUT
2 V
90%
V_{OH}
V_{OL} ≈ 0.8 V

TEST CIRCUIT — VOLTAGE WAVEFORMS

FIGURE 5. PE-TO-BUS PULLUP ENABLE AND DISABLE TIMES

NOTES: A. The input pulse is supplied by a generator having the following characteristics: PRR ≤ 1 MHz, 50% duty cycle, t_r ≤6 ns, t_f ≤6 ns, Z_{out} = 50 Ω.

B. C_L includes probe and jig capacitance.

TYPICAL CHARACTERISTICS

TERMINAL HIGH-LEVEL OUTPUT VOLTAGE
vs
HIGH-LEVEL OUTPUT CURRENT

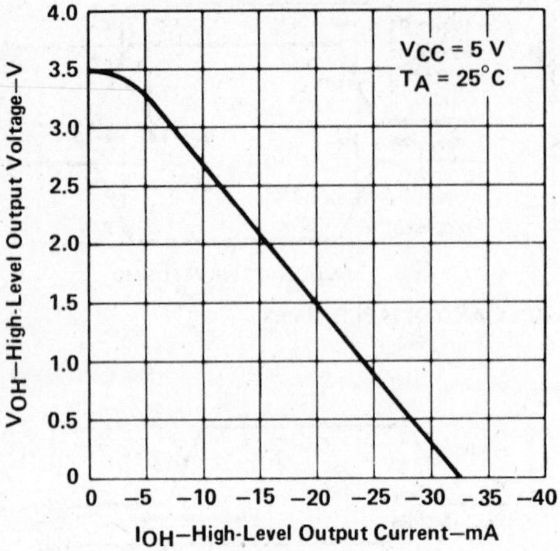
VCC = 5 V
TA = 25°C
4.0
3.5
3.0
2.5
2.0
1.5
1.0
0.5
0
VOH—High-Level Output Voltage—V
0
−5
−10
−15
−20
−25
−30
−35
−40
IOH—High-Level Output Current—mA

FIGURE 6

TERMINAL LOW-LEVEL OUTPUT VOLTAGE
vs
LOW-LEVEL OUTPUT CURRENT

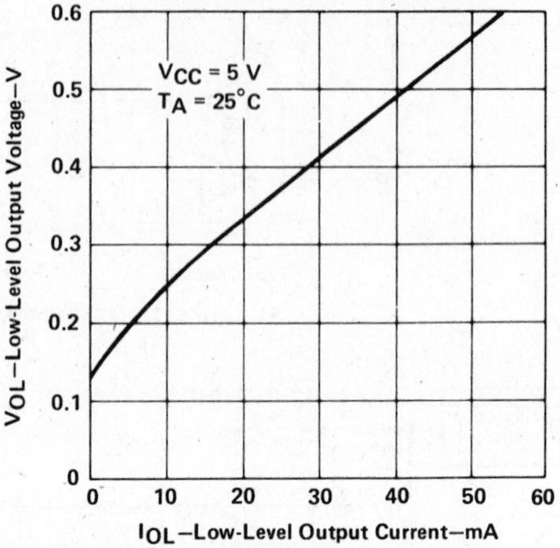
VCC = 5 V
TA = 25°C
0.6
0.5
0.4
0.3
0.2
0.1
0
VOL—Low-Level Output Voltage—V
0
10
20
30
40
50
60
IOL—Low-Level Output Current—mA

FIGURE 7

TERMINAL OUTPUT VOLTAGE
vs
BUS INPUT VOLTAGE

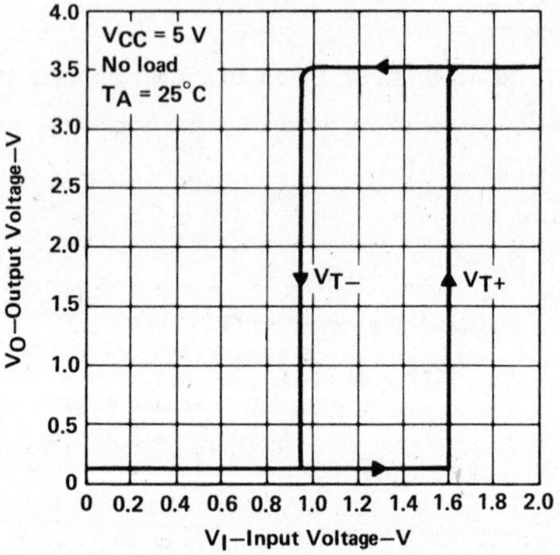
VCC = 5 V
No load
TA = 25°C
VT−
VT+
4.0
3.5
3.0
2.5
2.0
1.5
1.0
0.5
0
VO—Output Voltage—V
0
0.2
0.4
0.6
0.8
1.0
1.2
1.4
1.6
1.8
2.0
VI—Input Voltage—V

FIGURE 8

TEXAS INSTRUMENTS
POST OFFICE BOX 655012 • DALLAS, TEXAS 75265

TYPICAL CHARACTERISTICS

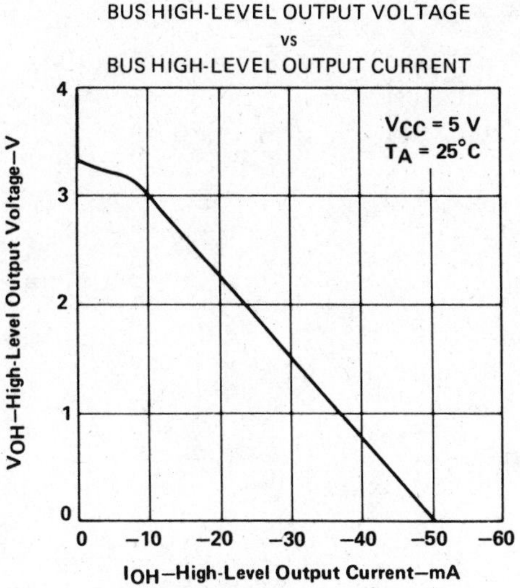
BUS HIGH-LEVEL OUTPUT VOLTAGE
vs
BUS HIGH-LEVEL OUTPUT CURRENT
VCC = 5 V
TA = 25°C
VOH—High-Level Output Voltage—V
IOH—High-Level Output Current—mA
0
1
2
3
4
0
−10
−20
−30
−40
−50
−60

FIGURE 9

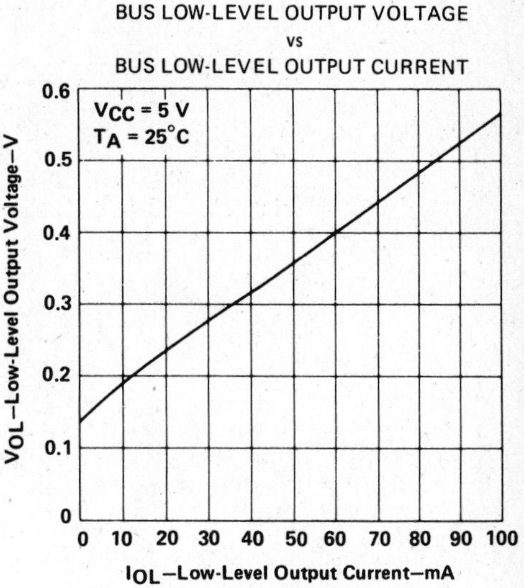
BUS LOW-LEVEL OUTPUT VOLTAGE
vs
BUS LOW-LEVEL OUTPUT CURRENT
VCC = 5 V
TA = 25°C
VOL—Low-Level Output Voltage—V
IOL—Low-Level Output Current—mA
0
0.1
0.2
0.3
0.4
0.5
0.6
0
10
20
30
40
50
60
70
80
90
100

FIGURE 10

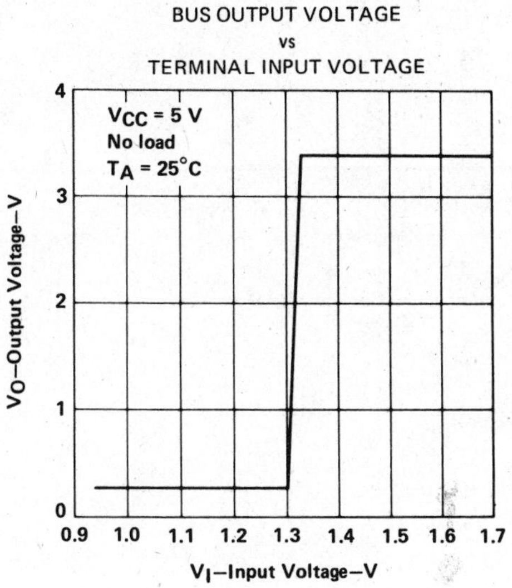
BUS OUTPUT VOLTAGE
vs
TERMINAL INPUT VOLTAGE
VCC = 5 V
No load
TA = 25°C
VO—Output Voltage—V
VI—Input Voltage—V
0
1
2
3
4
0.9
1.0
1.1
1.2
1.3
1.4
1.5
1.6
1.7

FIGURE 11

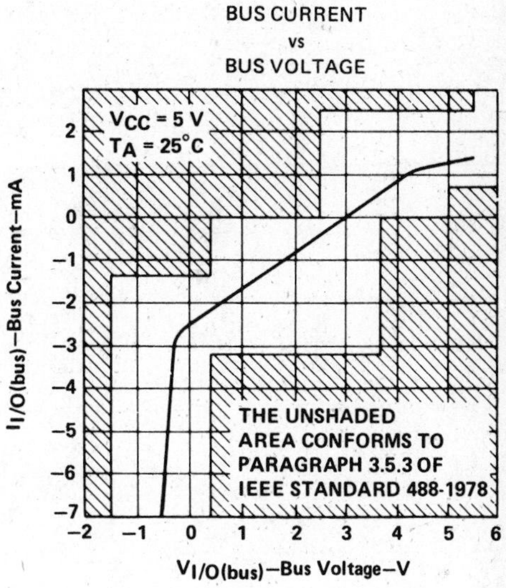
BUS CURRENT
vs
BUS VOLTAGE
VCC = 5 V
TA = 25°C
THE UNSHADED
AREA CONFORMS TO
PARAGRAPH 3.5.3 OF
IEEE STANDARD 488-1978
II/O(bus)—Bus Current—mA
VI/O(bus)—Bus Voltage—V
2
1
0
−1
−2
−3
−4
−5
−6
−7
−2
−1
0
1
2
3
4
5
6

FIGURE 12

TEXAS INSTRUMENTS
POST OFFICE BOX 655012 • DALLAS, TEXAS 75265

SN75ALS161 OCTAL GENERAL-PURPOSE INTERFACE BUS TRANSCEIVER

D2618, JUNE 1986

MEETS IEEE STANDARD 488-1978 (GPIB)

- 8-Channel Bidirectional Transceiver
- Designed to Implement Control Bus Interface
- Designed for Single Controller
- High-Speed Advanced Low-Power Schottky Circuitry
- Low Power Dissipation . . . 46 mW Max per Channel
- Fast Propagation Times . . . 20 ns Max
- High-Impedance P-N-P Inputs
- Receiver Hysteresis . . . 650 mV Typ
- Bus-Terminating Resistors Provided on Driver Outputs
- No Loading of Bus When Device is Powered Down (V_{CC} = 0)
- Power-Up/Power-Down Protection (Glitch-Free)

DW, J, OR N PACKAGE
(TOP VIEW)

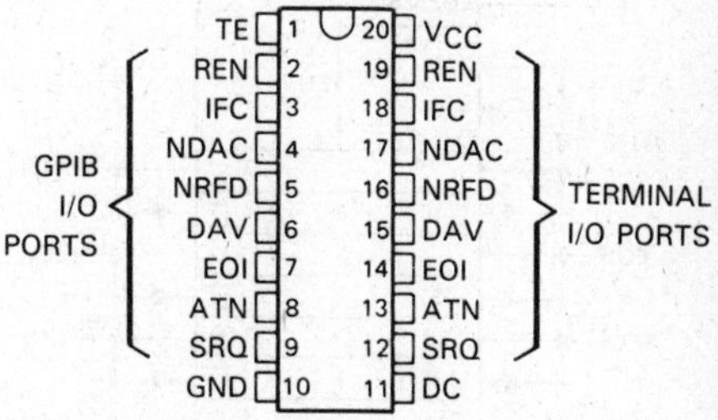
TE 1
REN 2
IFC 3
NDAC 4
NRFD 5
DAV 6
EOI 7
ATN 8
SRQ 9
GND 10
20 VCC
19 REN
18 IFC
17 NDAC
16 NRFD
15 DAV
14 EOI
13 ATN
12 SRQ
11 DC
GPIB I/O PORTS
TERMINAL I/O PORTS

CHANNEL IDENTIFICATION TABLE

NAME	IDENTITY	CLASS
DC	Direction Control	Control
TE	Talk Enable	
ATN	Attention	Bus Management
SRQ	Service Request	
REN	Remote Enable	
IFC	Interface Clear	
EOI	End or Identify	
DAV	Data Valid	Data Transfer
NDAC	Not Data Accepted	
NRFD	Not Ready for Data	

description

The SN75ALS161 eight-channel general-purpose interface bus transceiver is a monolithic, high-speed, Advanced Low-Power Schottky process device designed to provide the bus-management and data-transfer signals between operating units of a single controller instrumentation system. When combined with the SN75ALS160 octal bus transceiver, the SN75ALS161 provides the complete 16-wire interface for the IEEE 488 bus.

The SN75ALS161 features eight driver-receiver pairs connected in a front-to-back configuration to form input/output (I/O) ports at both the bus and terminal sides. The direction of data through these driver-receiver pairs is determined by the DC and TE enable signals.

The driver outputs (GPIB I/O ports) feature active bus-terminating resistor circuits designed to provide a high impedance to the bus when V_{CC} = 0. The drivers are designed to handle loads up to 48 milliamperes of sink current. Each receiver features p-n-p transistor inputs for high input impedance and a guaranteed hysteresis of 400 millivolts minimum for increased noise immunity. All receivers have 3-state outputs to present a high impedance to the terminal when disabled.

The SN75ALS161 is manufactured in a 20-pin package and is characterized for operation from 0 °C to 70 °C.

logic symbol†

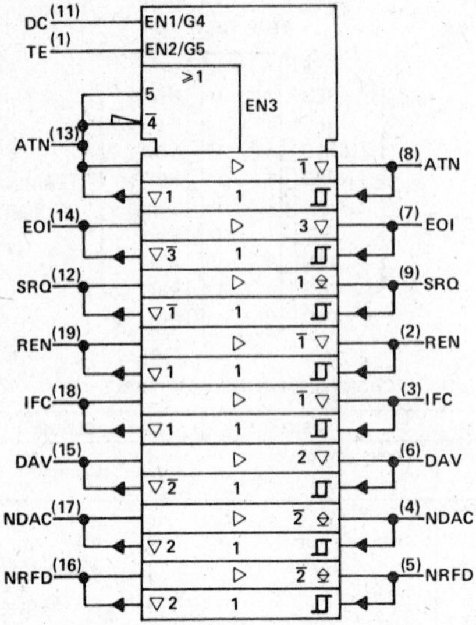
DC (11)
TE (1)
EN1/G4
EN2/G5
≥1
5
4
EN3
ATN (13)
EOI (14)
SRQ (12)
REN (19)
IFC (18)
DAV (15)
NDAC (17)
NRFD (16)
(8) ATN
(7) EOI
(9) SRQ
(2) REN
(3) IFC
(6) DAV
(4) NDAC
(5) NRFD

† This symbol is in accordance with ANSI/IEEE Std 91-1984 and IEC Publication 617-12.
▽ Designates 3-state outputs.
⩠ Designates passive-pullup outputs.

logic diagram (positive logic)

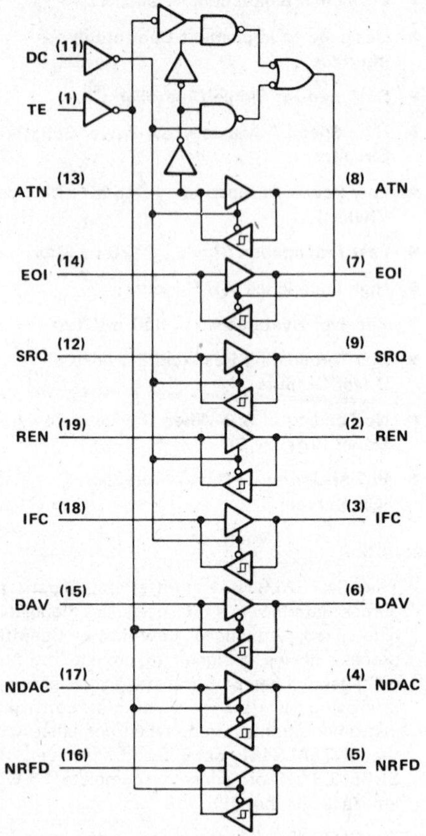
DC (11)
TE (1)
ATN (13)
EOI (14)
SRQ (12)
REN (19)
IFC (18)
DAV (15)
NDAC (17)
NRFD (16)
(8) ATN
(7) EOI
(9) SRQ
(2) REN
(3) IFC
(6) DAV
(4) NDAC
(5) NRFD

RECEIVE/TRANSMIT FUNCTION TABLE

CONTROLS			BUS-MANAGEMENT CHANNELS (Controlled by DC)					DATA-TRANSFER CHANNELS (Controlled by TE)		
DC	TE	ATN†	ATN†	SRQ	REN	IFC	EOI	DAV	NDAC	NRFD
H	H	H	R	T	R	R	T	T	R	R
H	H	L					R			
L	L	H	T	R	T	T	R	R	T	T
L	L	L					T			
H	L	X	R	T	R	R	R	R	T	T
L	H	X	T	R	T	T	T	T	R	R

H = high level, L = low level, R = receive, T = transmit, X = irrelevant
Direction of data transmission is from the terminal side to the bus side, and the direction of data receiving is from the bus side to the terminal side. Data transfer is noninverting in both directions.
†ATN is a normal transceiver channel that functions additionally as an internal direction control or talk enable for EOI whenever the DC and TE inputs are in the same state. When DC and TE are in opposite states, the ATN channel functions as an independent transceiver only.

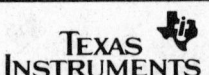

schematics of inputs and outputs

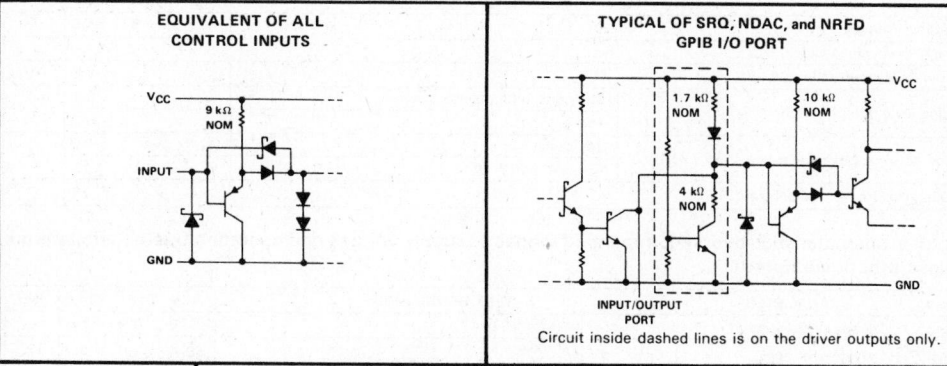

EQUIVALENT OF ALL
CONTROL INPUTS

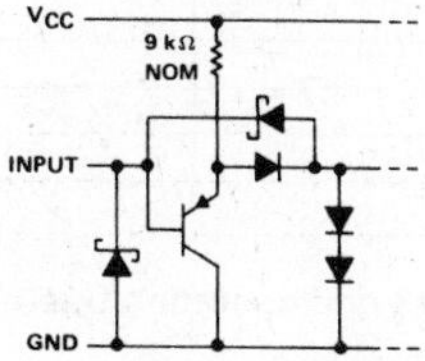
VCC
9 kΩ
NOM
INPUT
GND

TYPICAL OF SRQ, NDAC, and NRFD
GPIB I/O PORT

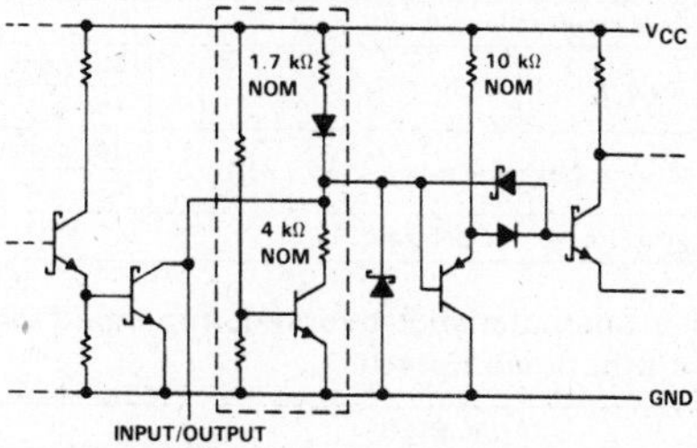
VCC
1.7 kΩ
NOM
10 kΩ
NOM
4 kΩ
NOM
GND
INPUT/OUTPUT
PORT

Circuit inside dashed lines is on the driver outputs only.

TYPICAL OF ALL I/O PORTS
EXCEPT SRQ, NDAC, and NRFD GPIB I/O PORTS

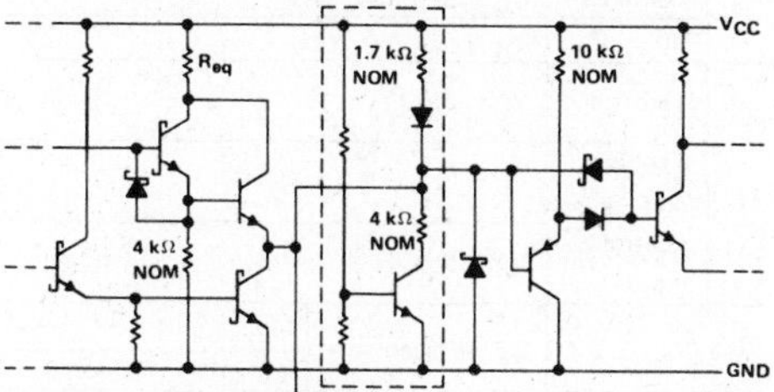
VCC
Req
1.7 kΩ
NOM
10 kΩ
NOM
4 kΩ
NOM
4 kΩ
NOM
GND
INPUT/OUTPUT
PORT

Driver output R_{eq} = 30 Ω NOM
Receiver output R_{eq} = 110 Ω NOM
Circuit inside dashed lines is on the driver outputs only.

absolute maximum ratings over operating free-air temperature range (unless otherwise noted)

Supply voltage, V_{CC} (see Note 1)	7 V
Input voltage	5.5 V
Low-level driver output current	100 mA
Continuous total dissipation at (or below) 25°C free-air temperature (see Note 2):	
DW package	1125 mW
J package	1375 mW
N package	1150 mW
Operating free-air temperature range	0°C to 70°C
Storage temperature range	−65°C to 150°C
Lead temperature 1,6 mm (1/16 inch) from the case for 60 seconds: J package	300°C
Lead temperature 1,6 mm (1/16 inch) from the case for 10 seconds: DW or N package	260°C

NOTES: 1. All voltage values are with respect to network ground terminal.
2. For operation above 25°C free-air temperature, derate the DW package to 720 mW at 70°C at the rate of 9.0 mW/°C, derate the J package to 880 mW at 70°C at the rate of 11.0 mW/°C, and derate the N package to 736 mW at 70°C at the rate of 9.2 mW/°C.

recommended operating conditions

		MIN	NOM	MAX	UNIT
Supply voltage, V_{CC}		4.75	5	5.25	V
High-level input voltage, V_{IH}		2			V
Low-level input voltage, V_{IL}				0.8	V
High-level output current, I_{OH}	Bus ports with pullups active			−5.2	mA
	Terminal ports			−800	μA
Low-level output current, I_{OL}	Bus ports			48	mA
	Terminal ports			16	mA
Operating free-air temperature, T_A		0		70	°C

electrical characteristics over recommended ranges of supply voltage and operating free-air temperature (unless otherwise noted)

PARAMETER			TEST CONDITIONS			MIN	TYP†	MAX	UNIT
V_{IK}	Input clamp voltage		$I_I = -18$ mA				−0.8	−1.5	V
V_{hys}	Hysteresis ($V_{T+} - V_{T-}$)	Bus				0.4	0.65		V
V_{OH}‡	High-level output voltage	Terminal	$I_{OH} = -800$ μA			2.7	3.5		V
		Bus	$I_{OH} = -5.2$ mA			2.5	3.3		V
V_{OL}	Low-level output voltage	Terminal	$I_{OL} = 16$ mA				0.3	0.5	V
		Bus	$I_{OL} = 48$ mA				0.35	0.5	V
I_I	Input current at maximum input voltage	Terminal	$V_I = 5.5$ V				0.2	100	μA
I_{IH}	High-level input current	Terminal and control inputs	$V_I = 2.7$ V				0.1	20	μA
I_{IL}	Low-level input current	Terminal and control inputs	$V_I = 0.5$ V				−10	−100	μA
$V_{I/O(bus)}$	Voltage at bus port		Driver disabled	$I_{I(bus)} = 0$		2.5	3.0	3.7	V
			Driver disabled	$I_{I(bus)} = -12$ mA				−1.5	V
$I_{I/O(bus)}$	Current into bus port	Power on	Driver disabled	$V_{I(bus)} = -1.5$ V to 0.4 V		−1.3			mA
		Power on	Driver disabled	$V_{I(bus)} = 0.4$ V to 2.5 V		0		−3.2	mA
		Power on	Driver disabled	$V_{I(bus)} = 2.5$ V to 3.7 V				+2.5 −3.2	mA
		Power on	Driver disabled	$V_{I(bus)} = 3.7$ V to 5 V		0		2.5	mA
		Power on	Driver disabled	$V_{I(bus)} = 5$ V to 5.5 V		0.7		2.5	mA
		Power off	$V_{CC} = 0$,	$V_{I(bus)} = 0$ V to 2.5 V				−40	μA
I_{OS}	Short-circuit output current	Terminal				−15	−35	−75	mA
		Bus				−25	−50	−125	mA
I_{CC}	Supply current		No load,	TE and DC low			55	75	mA
$C_{i/o(bus)}$	Bus-port capacitance		$V_{CC} = 5$ V to 0 V, $V_{I/O} = 0$ to 2 V, f = 1 MHz				30		pF

† All typical values are at $V_{CC} = 5$ V, $T_A = 25$°C.
‡ V_{OH} applies for three-state outputs only.

TEXAS INSTRUMENTS
POST OFFICE BOX 655012 • DALLAS, TEXAS 75265

switching characteristics over recommended range of operating free-air temperature (unless otherwise noted), V_{CC} = 5 V

PARAMETER		FROM	TO	TEST CONDITIONS	MIN	TYP†	MAX	UNIT
t_{PLH}	Propagation delay time, low-to-high-level output	Terminal	Bus	C_L = 30 pF, See Figure 1		10	20	ns
t_{PHL}	Propagation delay time, high-to-low-level output					12	20	
t_{PLH}	Propagation delay time, low-to-high-level output	Bus	Terminal	C_L = 30 pF, See Figure 2		5	10	ns
t_{PHL}	Propagation delay time, high-to-low-level output					7	14	
t_{PZH}	Output enable time to high level	TE or DC	BUS (ATTN, EOI, REN, IFC, and DAV)	C_L = 15 pF, See Figure 3			30	ns
t_{PHZ}	Output disable time from high level						20	
t_{PZL}	Output enable time to low level						45	
t_{PLZ}	Output disable time from low level						20	
t_{PZH}	Output enable time to high level	TE or DC	Terminal	C_L = 15 pF, See Figure 4			20	ns
t_{PHZ}	Output disable time from high level						25	
t_{PZL}	Output enable time to low level						30	
t_{PLZ}	Output disable time from low level						25	

†All typical values are at T_A = 25°C.

PARAMETER MEASUREMENT INFORMATION

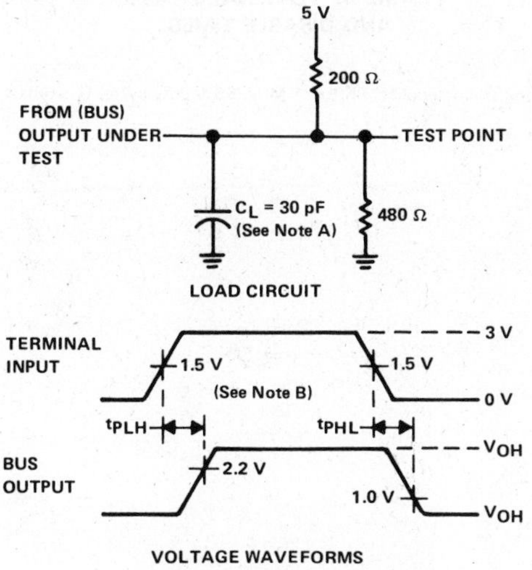
5 V
200 Ω
FROM (BUS) OUTPUT UNDER TEST
TEST POINT
C_L = 30 pF (See Note A)
480 Ω
LOAD CIRCUIT
TERMINAL INPUT
1.5 V
1.5 V
(See Note B)
3 V
0 V
t_{PLH}
t_{PHL}
BUS OUTPUT
2.2 V
1.0 V
V_{OH}
V_{OH}
VOLTAGE WAVEFORMS

FIGURE 1. TERMINAL-TO-BUS PROPAGATION DELAY TIMES

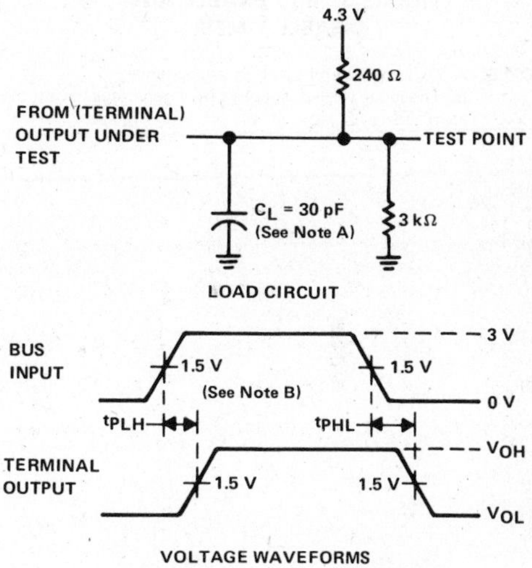
4.3 V
240 Ω
FROM (TERMINAL) OUTPUT UNDER TEST
TEST POINT
C_L = 30 pF (See Note A)
3 kΩ
LOAD CIRCUIT
BUS INPUT
1.5 V
1.5 V
(See Note B)
3 V
0 V
t_{PLH}
t_{PHL}
TERMINAL OUTPUT
1.5 V
1.5 V
V_{OH}
V_{OL}
VOLTAGE WAVEFORMS

FIGURE 2. BUS-TO-TERMINAL PROPAGATION DELAY TIMES

NOTES: A. C_L includes probe and jig capacitance.
B. The input pulse is supplied by a generator having the following characteristics: PRR ≤ 1 MHz, 50% duty cycle, t_r ≤ 6 ns, t_f ≤ 6 ns, Z_{out} = 50 Ω.

PARAMETER MEASUREMENT INFORMATION

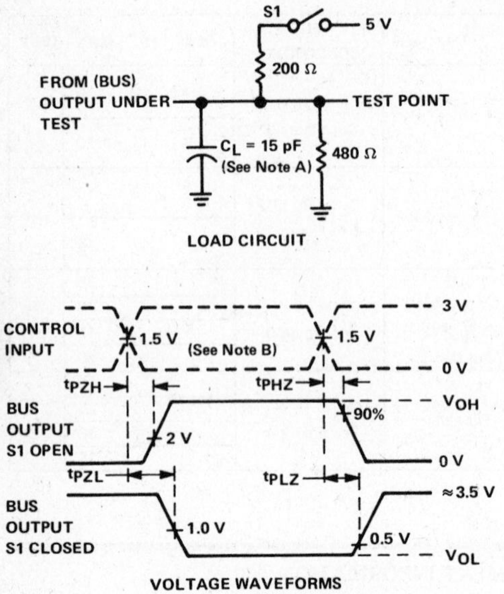
S1
5 V
200 Ω
FROM (BUS)
OUTPUT UNDER
TEST
TEST POINT
C_L = 15 pF
(See Note A)
480 Ω
LOAD CIRCUIT
CONTROL
INPUT
1.5 V
(See Note B)
3 V
0 V
t_{PZH}
t_{PHZ}
BUS
OUTPUT
S1 OPEN
V_{OH}
90%
2 V
t_{PZL}
t_{PLZ}
≈3.5 V
BUS
OUTPUT
S1 CLOSED
1.0 V
0.5 V
V_{OL}
VOLTAGE WAVEFORMS

FIGURE 3. BUS ENABLE AND DISABLE TIMES

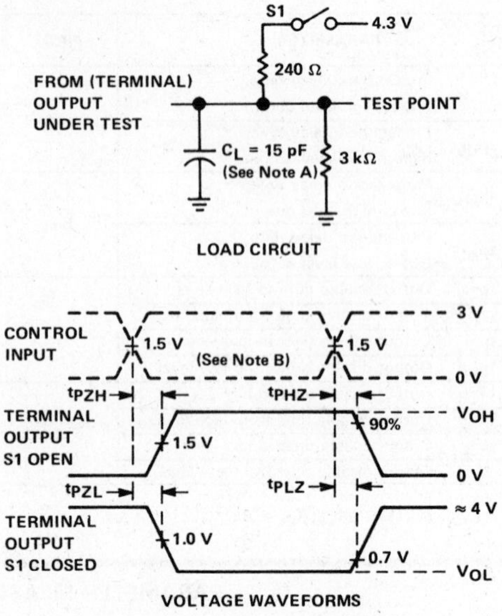
S1
4.3 V
240 Ω
FROM (TERMINAL)
OUTPUT
UNDER TEST
TEST POINT
C_L = 15 pF
(See Note A)
3 kΩ
LOAD CIRCUIT
CONTROL
INPUT
1.5 V
(See Note B)
3 V
0 V
t_{PZH}
t_{PHZ}
TERMINAL
OUTPUT
S1 OPEN
V_{OH}
90%
1.5 V
t_{PZL}
t_{PLZ}
≈ 4 V
TERMINAL
OUTPUT
S1 CLOSED
1.0 V
0.7 V
V_{OL}
VOLTAGE WAVEFORMS

FIGURE 4. TERMINAL ENABLE AND DISABLE TIMES

NOTES: A. C_L includes probe and jig capacitance.
B. The input pulse is supplied by a generator having the following characteristics: PRR ≤ 1 MHz, 50% duty cycle, t_r ≤6 ns, t_f ≤6 ns, Z_{out} = 50 Ω.

TEXAS INSTRUMENTS
POST OFFICE BOX 655012 • DALLAS, TEXAS 75265

TYPICAL CHARACTERISTICS

TERMINAL HIGH-LEVEL OUTPUT VOLTAGE
vs
HIGH-LEVEL OUTPUT CURRENT

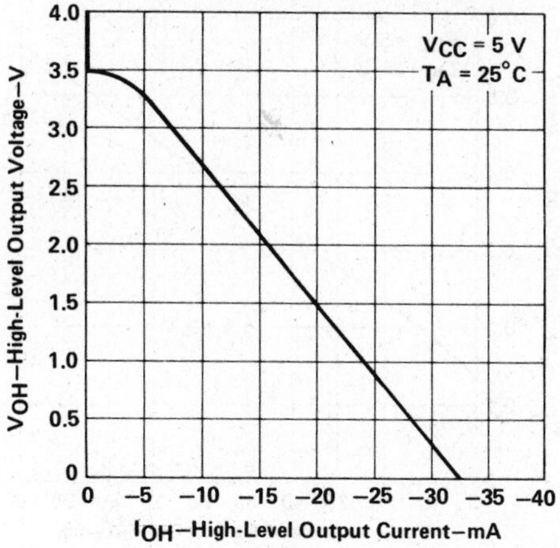
V_{CC} = 5 V
T_A = 25°C
V_{OH}—High-Level Output Voltage—V
4.0
3.5
3.0
2.5
2.0
1.5
1.0
0.5
0
0 −5 −10 −15 −20 −25 −30 −35 −40
I_{OH}—High-Level Output Current—mA

FIGURE 5

TERMINAL LOW-LEVEL OUTPUT VOLTAGE
vs
LOW-LEVEL OUTPUT CURRENT

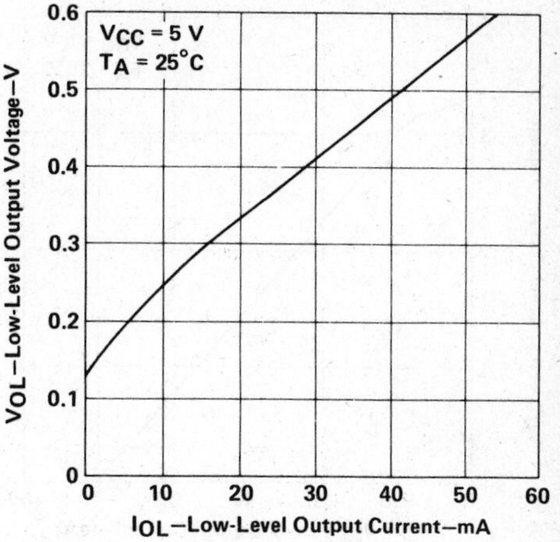
V_{CC} = 5 V
T_A = 25°C
V_{OL}—Low-Level Output Voltage—V
0.6
0.5
0.4
0.3
0.2
0.1
0
0 10 20 30 40 50 60
I_{OL}—Low-Level Output Current—mA

FIGURE 6

TERMINAL OUTPUT VOLTAGE
vs
BUS INPUT VOLTAGE

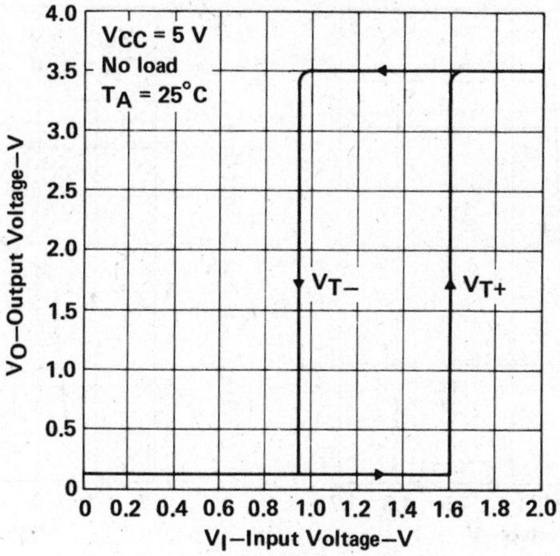
V_{CC} = 5 V
No load
T_A = 25°C
V_{T-}
V_{T+}
V_O—Output Voltage—V
4.0
3.5
3.0
2.5
2.0
1.5
1.0
0.5
0
0 0.2 0.4 0.6 0.8 1.0 1.2 1.4 1.6 1.8 2.0
V_I—Input Voltage—V

FIGURE 7

TYPICAL CHARACTERISTICS

BUS HIGH-LEVEL OUTPUT VOLTAGE
vs
HIGH-LEVEL OUTPUT CURRENT

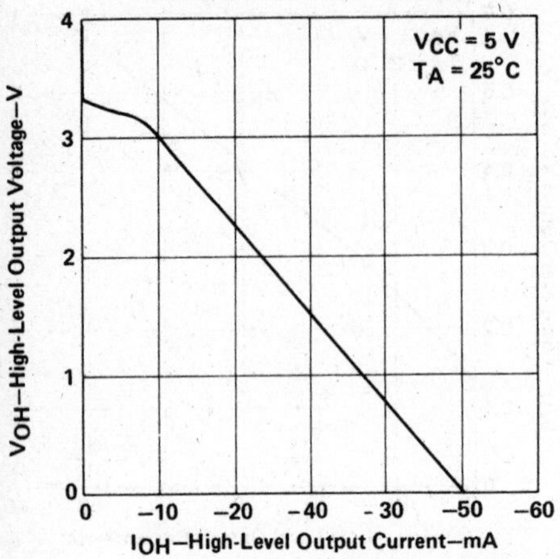
VCC = 5 V
TA = 25°C
VOH—High-Level Output Voltage—V
IOH—High-Level Output Current—mA
4
3
2
1
0
0
−10
−20
−40
- 30
−50
−60

FIGURE 8

BUS-LOW LEVEL OUTPUT VOLTAGE
vs
LOW-LEVEL OUTPUT CURRENT

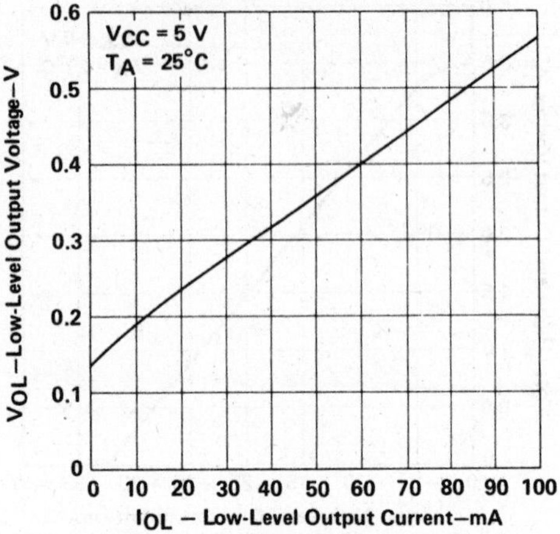
VCC = 5 V
TA = 25°C
VOL—Low-Level Output Voltage—V
IOL – Low-Level Output Current—mA
0.6
0.5
0.4
0.3
0.2
0.1
0
0 10 20 30 40 50 60 70 80 90 100

FIGURE 9

BUS OUTPUT VOLTAGE
vs
TERMINAL INPUT VOLTAGE

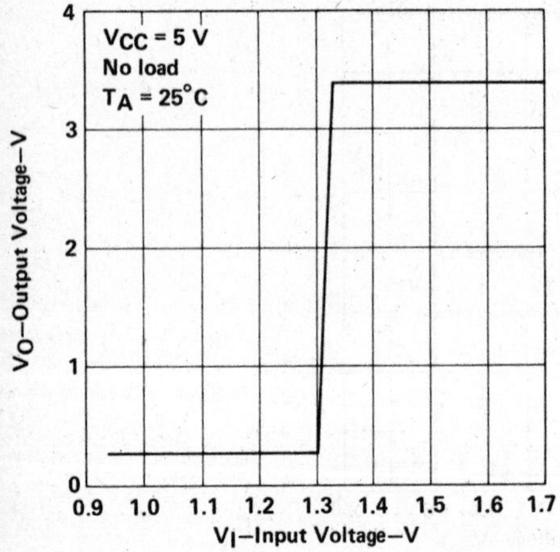
VCC = 5 V
No load
TA = 25°C
VO—Output Voltage—V
VI—Input Voltage—V
4
3
2
1
0
0.9 1.0 1.1 1.2 1.3 1.4 1.5 1.6 1.7

FIGURE 10

BUS CURRENT
vs
BUS VOLTAGE

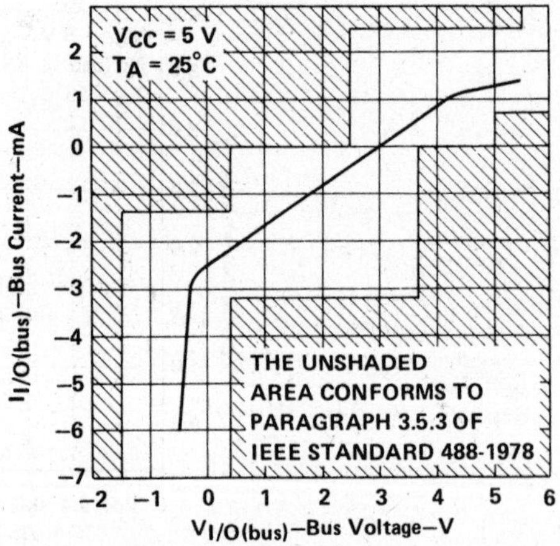
VCC = 5 V
TA = 25°C
THE UNSHADED
AREA CONFORMS TO
PARAGRAPH 3.5.3 OF
IEEE STANDARD 488-1978
II/O(bus)—Bus Current—mA
VI/O(bus)—Bus Voltage—V
2
1
0
−1
−2
−3
−4
−5
−6
−7
−2 −1 0 1 2 3 4 5 6

FIGURE 11

4

Line Drivers/Receivers

ADVANCE INFORMATION

SN75ALS162 OCTAL GENERAL-PURPOSE INTERFACE BUS TRANSCEIVER

D2618, JUNE 1986

MEETS IEEE STANDARD 488-1978 (GPIB)

- 8-Channel Bidirectional Transceiver
- Designed to Implement Control Bus Interface
- Designed for Multicontrollers
- High-Speed Advanced Low-Power Schottky Circuitry
- Low Power Dissipation . . . 46 mW Max per Channel
- Fast Propagation Times . . . 20 ns Max
- High-Impedance P-N-P Inputs
- Receiver Hysteresis . . . 650 mV Typ
- Bus-Terminating Resistors Provided on Driver Outputs
- No Loading of Bus When Device is Powered Down (V_{CC} = 0)
- Power-Up/Power-Down Protection (Glitch-Free)

DW PACKAGE
(TOP VIEW)

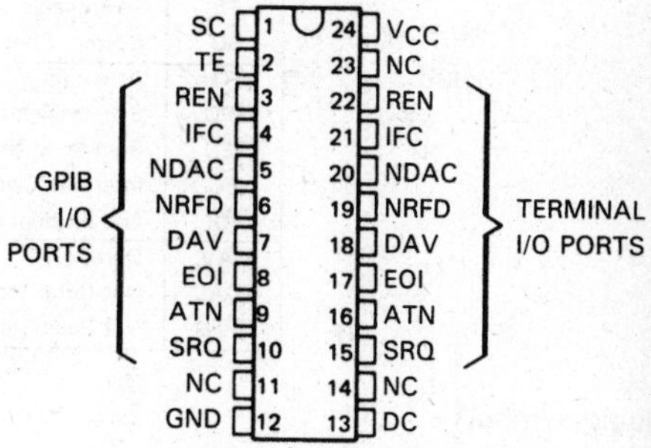
SC 1
TE 2
REN 3
IFC 4
NDAC 5
NRFD 6
DAV 7
EOI 8
ATN 9
SRQ 10
NC 11
GND 12
24 VCC
23 NC
22 REN
21 IFC
20 NDAC
19 NRFD
18 DAV
17 EOI
16 ATN
15 SRQ
14 NC
13 DC
GPIB I/O PORTS
TERMINAL I/O PORTS

N DUAL-IN-LINE PACKAGE
(TOP VIEW)

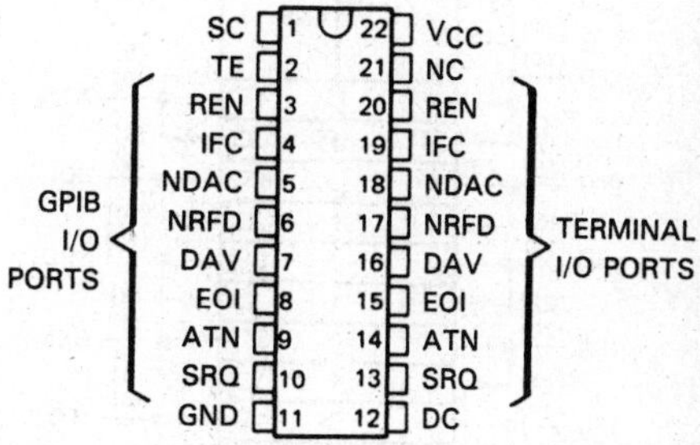
SC 1
TE 2
REN 3
IFC 4
NDAC 5
NRFD 6
DAV 7
EOI 8
ATN 9
SRQ 10
GND 11
22 VCC
21 NC
20 REN
19 IFC
18 NDAC
17 NRFD
16 DAV
15 EOI
14 ATN
13 SRQ
12 DC
GPIB I/O PORTS
TERMINAL I/O PORTS

NC—No internal connection.

description

The SN75ALS162 eight-channel general-purpose interface bus transceiver is a monolithic, high-speed, Advanced Low-Power Schottky process device designed to provide the bus-management and data-transfer signals between operating units of a multiple-controller instrumentation system. When combined with the SN75ALS160 octal bus transceiver, the SN75ALS162 provides the complete 16-wire interface for the IEEE 488 bus.

The SN75ALS162 features eight driver-receiver pairs connected in a front-to-back configuration to form input/output (I/O) ports at both the bus and terminal sides. The direction of data through these driver-receiver pairs is determined by the DC, TE, and SC enable signals. The SC input allows the REN and IFC transceivers to be controlled independently.

The driver outputs (GPIB I/O ports) feature active bus-terminating resistor circuits designed to provide a high impedance to the bus when V_{CC} = 0. The drivers are designed to handle loads up to 48 milliamperes of sink current. Each receiver features p-n-p transistor inputs for high input impedance and a guaranteed hysteresis of 400 millivolts minimum for increased noise immunity. All receivers have 3-state outputs to present a high impedance to the terminal when disabled.

The SN75ALS162 is manufactured in a 22-pin dual-in-line N package and in 24-pin DW package, and is characterized for operation from 0°C to 70°C.

CHANNEL IDENTIFICATION TABLE

NAME	IDENTITY	CLASS
DC	Direction Control	Control
TE	Talk Enable	
SC	System Control	
ATN	Attention	Bus Management
SRQ	Service Request	
REN	Remote Enable	
IFC	Interface Clear	
EOI	End or Identify	
DAV	Data Valid	Data Transfer
NDAC	Not Data Accepted	
NRFD	Not Ready for Data	

logic symbol†

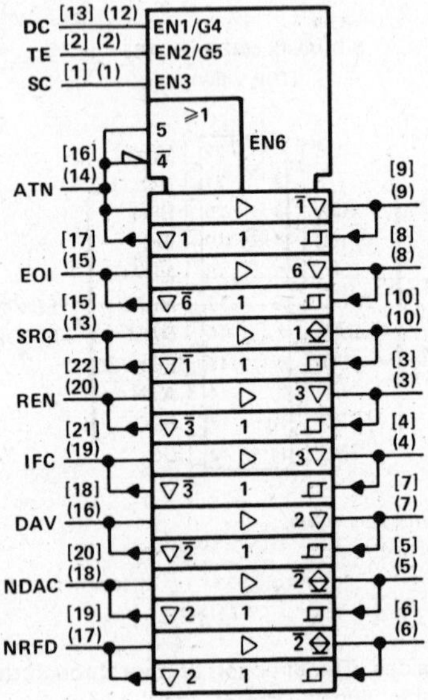
DC [13] (12) EN1/G4
TE [2] (2) EN2/G5
SC [1] (1) EN3
≥1
5
4
EN6
ATN [16] (14)
[9] (9) ATN
EOI [17] (15)
[8] (8) EOI
SRQ [15] (13)
[10] (10) SRQ
REN [22] (20)
[3] (3) REN
IFC [21] (19)
[4] (4) IFC
DAV [18] (16)
[7] (7) DAV
NDAC [20] (18)
[5] (5) NDAC
NRFD [19] (17)
[6] (6) NRFD

†This symbol is in accordance with ANSI/IEEE Std 91-1984 and IEC Publication 617-12.
▽ Designates 3-state outputs.
⬙ Designates passive-pullup outputs.

logic diagram (positive logic)

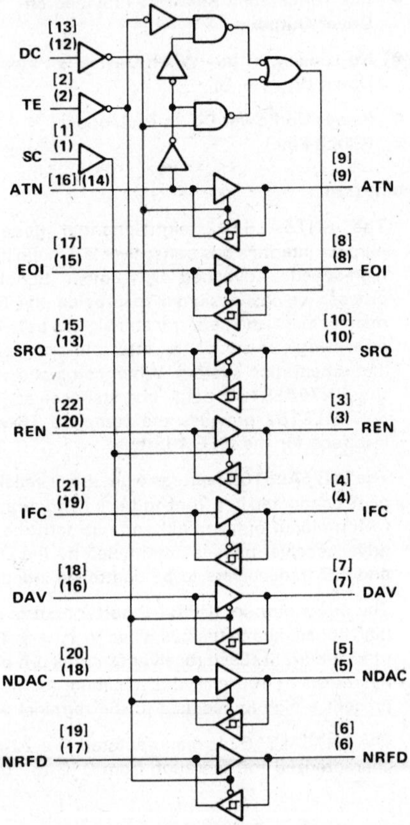
DC [13] (12)
TE [2] (2)
SC [1] (1)
ATN [16] (14)
[9] (9) ATN
EOI [17] (15)
[8] (8) EOI
SRQ [15] (13)
[10] (10) SRQ
REN [22] (20)
[3] (3) REN
IFC [21] (19)
[4] (4) IFC
DAV [18] (16)
[7] (7) DAV
NDAC [20] (18)
[5] (5) NDAC
NRFD [19] (17)
[6] (6) NRFD

[ ] Denotes pin numbers for DW package.
( ) Denotes pin numbers for N package.

TEXAS INSTRUMENTS
POST OFFICE BOX 655012 • DALLAS, TEXAS 75265

RECEIVE/TRANSMIT FUNCTION TABLE

CONTROLS				BUS-MANAGEMENT CHANNELS					DATA-TRANSFER CHANNELS		
SC	DC	TE	ATN†	ATN† (Controlled by DC)	SRQ (Controlled by DC)	REN (Controlled by SC)	IFC (Controlled by SC)	EOI	DAV (Controlled by TE)	NDAC (Controlled by TE)	NRFD (Controlled by TE)
	H	H	H	R	T			T	T	R	R
	H	H	L					R			
	L	L	H	T	R			R	R	T	T
	L	L	L					T			
	H	L	X	R	T			R	R	T	T
	L	H	X	T	R			T	T	R	R
H						T	T				
L						R	R				

H = high level, L = low level, R = receive, T = transmit, X = irrelevant
Direction of data transmission is from the terminal side to the bus side, and the direction of data receiving is from the bus side to the terminal side. Data transfer is noninverting in both directions.
†ATN is a normal transceiver channel that functions additionally as an internal direction control or talk enable for EOI whenever the DC and TE inputs are in the same state. When DC and TE are in opposite states, the ATN channel functions as an independent transceiver only.

schematics of inputs and outputs

EQUIVALENT OF ALL CONTROL INPUTS

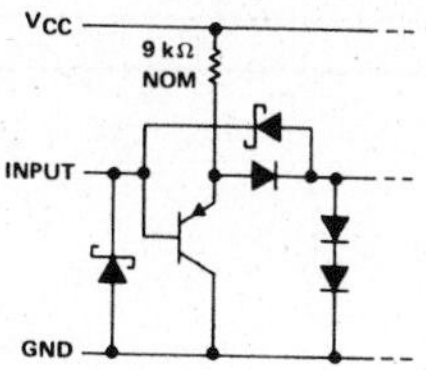
VCC
9 kΩ NOM
INPUT
GND

TYPICAL OF SRQ, NDAC, and NRFD GPIB I/O PORT

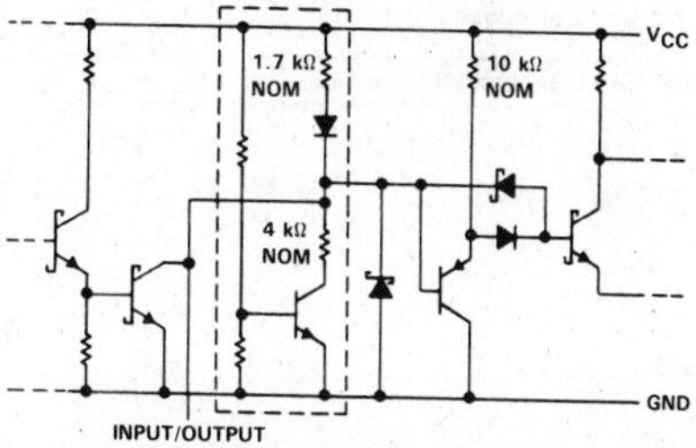
VCC
1.7 kΩ NOM
10 kΩ NOM
4 kΩ NOM
GND
INPUT/OUTPUT PORT

Circuit inside dashed lines is on the driver outputs only.

TYPICAL OF ALL I/O PORTS EXCEPT SRQ, NDAC, and NRFD GPIB I/O PORTS

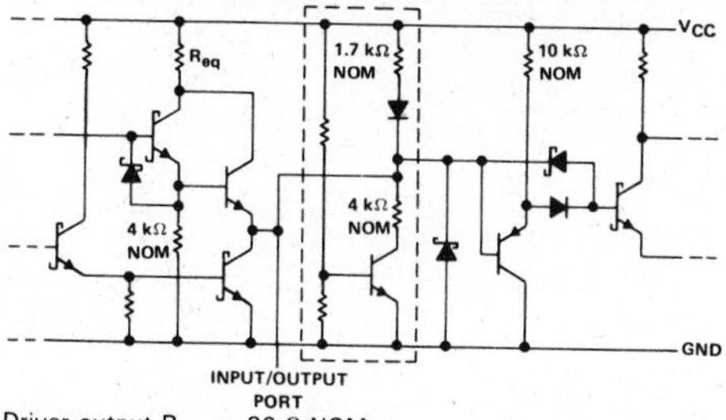
VCC
Req
1.7 kΩ NOM
10 kΩ NOM
4 kΩ NOM
4 kΩ NOM
GND
INPUT/OUTPUT PORT

Driver output R_{eq} = 30 Ω NOM
Receiver output R_{eq} = 110 Ω NOM
Circuit inside dashed lines is on the driver outputs only.

absolute maximum ratings over operating free-air temperature range (unless otherwise noted)

Supply voltage, V_{CC} (see Note 1) . . . . . . 7 V
Input voltage . . . . . . 5.5 V
Low-level driver output current . . . . . . 100 mA
Continuous total dissipation at (or below) 25°C free-air temperature (see Note 2):
DW package . . . . . . 1350 mW
N package . . . . . . 1700 mW
Operating free-air temperature range . . . . . . 0°C to 70°C
Storage temperature range . . . . . . −65°C to 150°C
Lead temperature 1,6 mm (1/16 inch) from the case for 10 seconds: DW or N package . . . . . . 260°C

NOTES: 1. All voltage values are with respect to network ground terminal.
2. For operation above 25°C free-air temperature, derate the DW package to 864 mW at 70°C at the rate of 10.8 mW/°C, and derate the N package to 1088 mW at 70°C at the rate of 13.6 mW/°C.

recommended operating conditions

		MIN	NOM	MAX	UNIT
Supply voltage, V_{CC}		4.75	5	5.25	V
High-level input voltage, V_{IH}		2			V
Low-level input voltage, V_{IL}				0.8	V
High-level output current, I_{OH}	Bus ports with 3-state outputs			−5.2	mA
	Terminal ports			−800	μA
Low-level output current, I_{OL}	Bus ports			48	mA
	Terminal ports			16	mA
Operating free-air temperature, T_A		0		70	°C

TEXAS INSTRUMENTS
POST OFFICE BOX 655012 • DALLAS, TEXAS 75265

electrical characteristics over recommended ranges of supply voltage and operating free-air temperature (unless otherwise noted)

PARAMETER			TEST CONDITIONS		MIN	TYP†	MAX	UNIT
V_{IK}	Input clamp voltage		$I_I = -18$ mA			−0.8	−1.5	V
V_{hys}	Hysteresis ($V_{T+} - V_{T-}$)	Bus			0.4	0.65		V
V_{OH}‡	High-level output voltage	Terminal	$I_{OH} = -800$ µA		2.7	3.5		V
		Bus	$I_{OH} = -5.2$ mA		2.5	3.3		V
V_{OL}	Low-level output voltage	Terminal	$I_{OL} = 16$ mA			0.3	0.5	V
		Bus	$I_{OL} = 48$ mA			0.35	0.5	V
I_I	Input current at maximum input voltage	Terminal	$V_I = 5.5$ V			0.2	100	µA
I_{IH}	High-level input current	Terminal and control inputs	$V_I = 2.7$ V			0.1	20	µA
I_{IL}	Low-level input current	Terminal and control inputs	$V_I = 0.5$ V			−10	−100	µA
$V_{I/O(bus)}$	Voltage at bus port		Driver disabled	$I_{I(bus)} = 0$	2.5	3.0	3.7	V
			Driver disabled	$I_{I(bus)} = -12$ mA			−1.5	V
$I_{I/O(bus)}$	Current into bus port	Power on	Driver disabled	$V_{I(bus)} = -1.5$ V to 0.4 V	−1.3			mA
		Power on	Driver disabled	$V_{I(bus)} = 0.4$ V to 2.5 V	0		−3.2	mA
		Power on	Driver disabled	$V_{I(bus)} = 2.5$ V to 3.7 V			+2.5 −3.2	mA
		Power on	Driver disabled	$V_{I(bus)} = 3.7$ V to 5 V	0		2.5	mA
		Power on	Driver disabled	$V_{I(bus)} = 5$ V to 5.5 V	0.7		2.5	mA
		Power off	$V_{CC} = 0$,	$V_{I(bus)} = 0$ V to 2.5 V			−40	µA
I_{OS}	Short-circuit output current	Terminal			−15	−35	−75	mA
		Bus			−25	−50	−125	mA
I_{CC}	Supply current		No load,	TE, DC, and SC low		55	75	mA
$C_{i/o(bus)}$	Bus-port capacitance		$V_{CC} = 5$ V to 0 V, $V_{I/O} = 0$ to 2 V, f = 1 MHz			30		pF

† All typical values are at $V_{CC} = 5$ V, $T_A = 25$°C.
‡ V_{OH} applies for three-state outputs only.

switching characteristics over recommended range of operating free-air temperature (unless otherwise noted), V_{CC} = 5 V

PARAMETER		FROM	TO	TEST CONDITIONS	MIN	TYP†	MAX	UNIT
t_{PLH}	Propagation delay time, low-to-high-level output	Terminal	Bus	C_L = 30 pF, See Figure 1		10	20	ns
t_{PHL}	Propagation delay time, high-to-low-level output					12	20	
t_{PLH}	Propagation delay time, low-to-high-level output	Bus	Terminal	C_L = 30 pF, See Figure 2		5	10	ns
t_{PHL}	Propagation delay time, high-to-low-level output					7	14	
t_{PZH}	Output enable time to high level	TE, DC, or SC	BUS (ATTN, EOI, REN, IFC, and DAV)	C_L = 15 pF, See Figure 3			30	ns
t_{PHZ}	Output disable time from high level						20	
t_{PZL}	Output enable time to low level						45	
t_{PLZ}	Output disable time from low level						20	
t_{PZH}	Output enable time to high level	TE, DC, or SC	Terminal	C_L = 15 pF, See Figure 4			20	ns
t_{PHZ}	Output disable time from high level						25	
t_{PZL}	Output enable time to low level						30	
t_{PLZ}	Output disable time from low level						25	

†All typical values are at T_A = 25°C.

PARAMETER MEASUREMENT INFORMATION

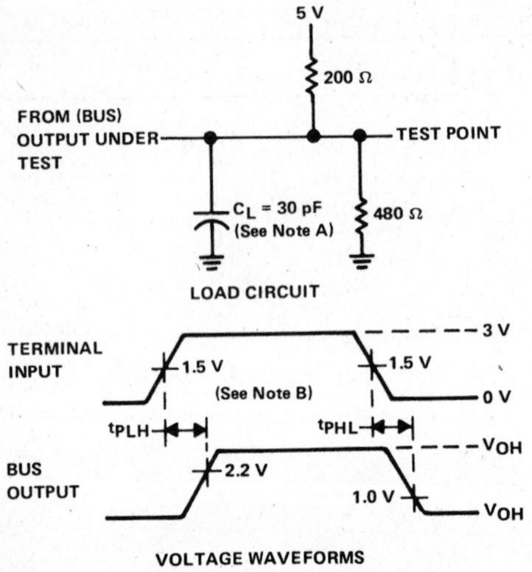
5 V
200 Ω
FROM (BUS) OUTPUT UNDER TEST
TEST POINT
C_L = 30 pF (See Note A)
480 Ω
LOAD CIRCUIT
TERMINAL INPUT
1.5 V
(See Note B)
1.5 V
3 V
0 V
t_{PLH}
t_{PHL}
BUS OUTPUT
2.2 V
1.0 V
V_{OH}
V_{OH}
VOLTAGE WAVEFORMS

FIGURE 1. TERMINAL-TO-BUS PROPAGATION DELAY TIMES

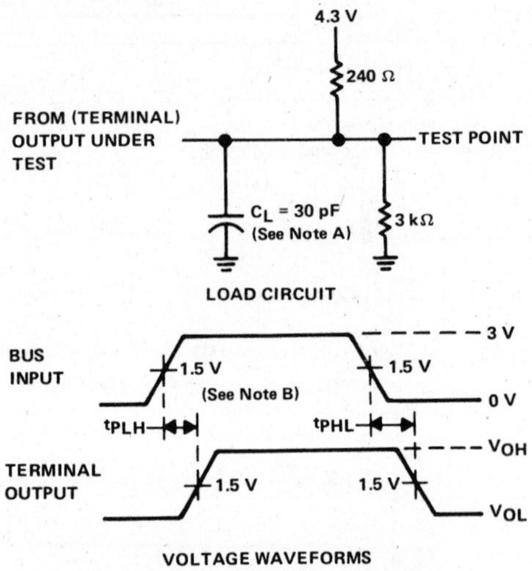
4.3 V
240 Ω
FROM (TERMINAL) OUTPUT UNDER TEST
TEST POINT
C_L = 30 pF (See Note A)
3 kΩ
LOAD CIRCUIT
BUS INPUT
1.5 V
(See Note B)
1.5 V
3 V
0 V
t_{PLH}
t_{PHL}
TERMINAL OUTPUT
1.5 V
1.5 V
V_{OH}
V_{OL}
VOLTAGE WAVEFORMS

FIGURE 2. BUS-TO-TERMINAL PROPAGATION DELAY TIMES

NOTES: A. C_L includes probe and jig capacitance.
B. The input pulse is supplied by a generator having the following characteristics: PRR ≤ 1 MHz, 50% duty cycle, t_r ≤6 ns, t_f ≤6 ns, Z_{out} = 50 Ω.

PARAMETER MEASUREMENT INFORMATION

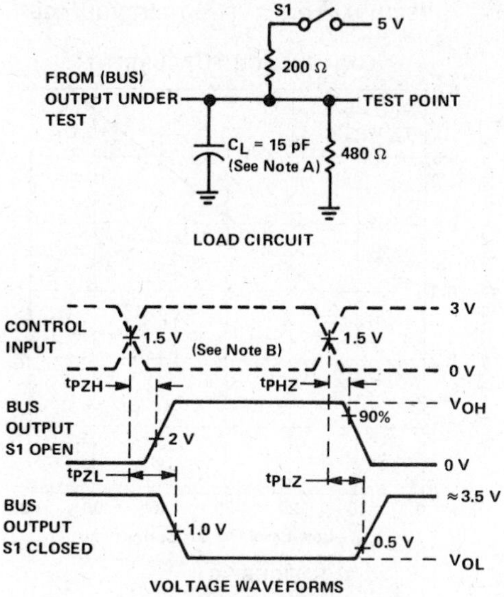
S1
5 V
200 Ω
FROM (BUS)
OUTPUT UNDER
TEST
TEST POINT
C_L = 15 pF
(See Note A)
480 Ω
LOAD CIRCUIT
CONTROL
INPUT
1.5 V
(See Note B)
1.5 V
3 V
0 V
t_{PZH}
t_{PHZ}
BUS
OUTPUT
S1 OPEN
V_{OH}
90%
2 V
0 V
t_{PZL}
t_{PLZ}
BUS
OUTPUT
S1 CLOSED
≈3.5 V
1.0 V
0.5 V
V_{OL}
VOLTAGE WAVEFORMS

FIGURE 3. BUS ENABLE AND DISABLE TIMES

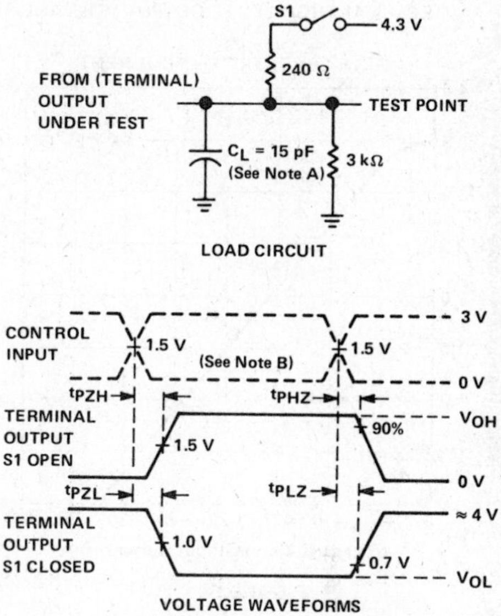
S1
4.3 V
240 Ω
FROM (TERMINAL)
OUTPUT
UNDER TEST
TEST POINT
C_L = 15 pF
(See Note A)
3 kΩ
LOAD CIRCUIT
CONTROL
INPUT
1.5 V
(See Note B)
1.5 V
3 V
0 V
t_{PZH}
t_{PHZ}
TERMINAL
OUTPUT
S1 OPEN
V_{OH}
90%
1.5 V
0 V
t_{PZL}
t_{PLZ}
TERMINAL
OUTPUT
S1 CLOSED
≈4 V
1.0 V
0.7 V
V_{OL}
VOLTAGE WAVEFORMS

FIGURE 4. TERMINAL ENABLE AND DISABLE TIMES

NOTES: A. C_L includes probe and jig capacitance.

B. The input pulse is supplied by a generator having the following characteristics: PRR ≤ 1 MHz, 50% duty cycle, t_r ≤6 ns, t_f ≤6 ns, Z_{out} = 50 Ω.

TYPICAL CHARACTERISTICS

TERMINAL HIGH-LEVEL OUTPUT VOLTAGE
vs
HIGH-LEVEL OUTPUT CURRENT

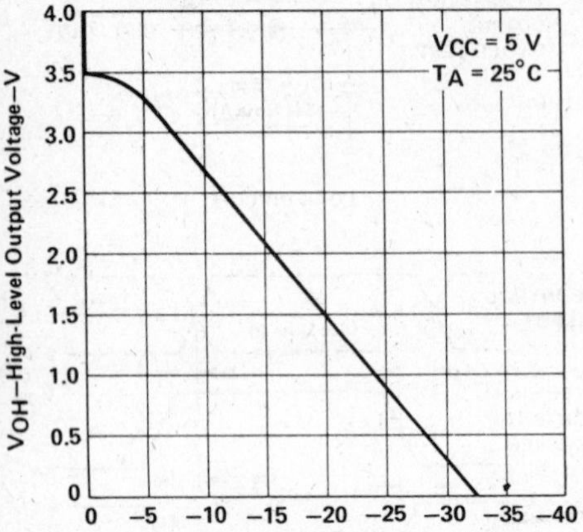
VCC = 5 V
TA = 25°C
4.0
3.5
3.0
2.5
2.0
1.5
1.0
0.5
0
VOH–High-Level Output Voltage–V
0 –5 –10 –15 –20 –25 –30 –35 –40
IOH–High-Level Output Current–mA

FIGURE 5

TERMINAL LOW-LEVEL OUTPUT VOLTAGE
vs
LOW-LEVEL OUTPUT CURRENT

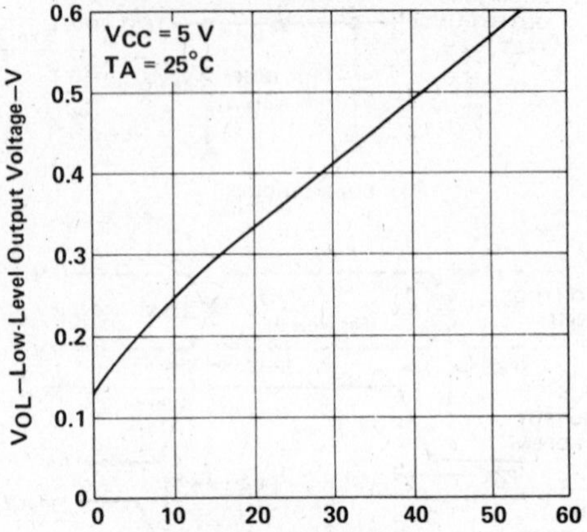
VCC = 5 V
TA = 25°C
0.6
0.5
0.4
0.3
0.2
0.1
0
VOL–Low-Level Output Voltage–V
0 10 20 30 40 50 60
IOL–Low-Level Output Current–mA

FIGURE 6

TERMINAL OUTPUT VOLTAGE
vs
BUS INPUT VOLTAGE

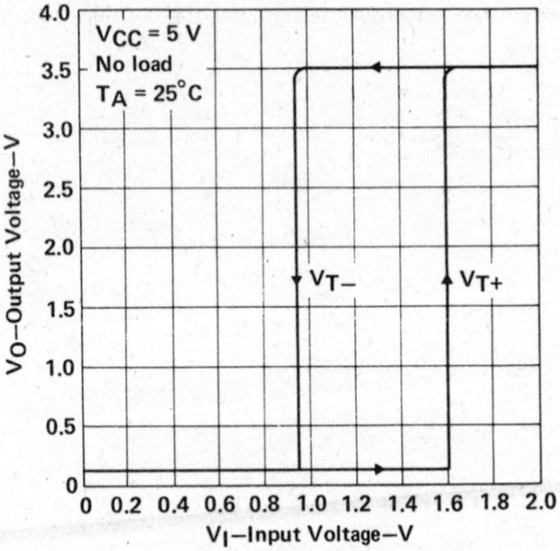
VCC = 5 V
No load
TA = 25°C
4.0
3.5
3.0
2.5
2.0
1.5
1.0
0.5
0
VO–Output Voltage–V
VT–
VT+
0 0.2 0.4 0.6 0.8 1.0 1.2 1.4 1.6 1.8 2.0
VI–Input Voltage–V

FIGURE 7

TEXAS INSTRUMENTS
POST OFFICE BOX 655012 • DALLAS, TEXAS 75265

TYPICAL CHARACTERISTICS

BUS HIGH-LEVEL OUTPUT VOLTAGE
vs
HIGH-LEVEL OUTPUT CURRENT

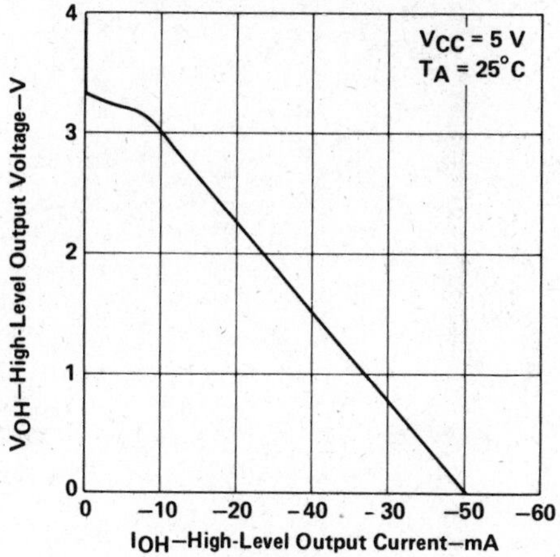
V_{CC} = 5 V
T_A = 25°C
V_{OH}—High-Level Output Voltage—V
4
3
2
1
0
0
−10
−20
−40
- 30
−50
−60
I_{OH}—High-Level Output Current—mA

FIGURE 8

BUS-LOW LEVEL OUTPUT VOLTAGE
vs
LOW-LEVEL OUTPUT CURRENT

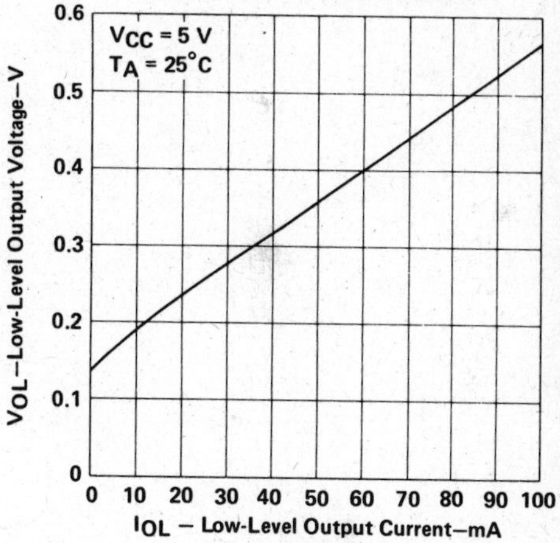
V_{CC} = 5 V
T_A = 25°C
V_{OL}—Low-Level Output Voltage—V
0.6
0.5
0.4
0.3
0.2
0.1
0
0 10 20 30 40 50 60 70 80 90 100
I_{OL} – Low-Level Output Current—mA

FIGURE 9

BUS OUTPUT VOLTAGE
vs
TERMINAL INPUT VOLTAGE

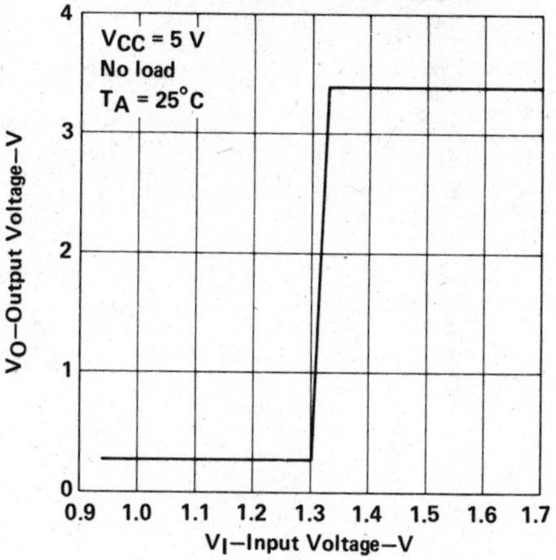
V_{CC} = 5 V
No load
T_A = 25°C
V_O—Output Voltage—V
4
3
2
1
0
0.9 1.0 1.1 1.2 1.3 1.4 1.5 1.6 1.7
V_I—Input Voltage—V

FIGURE 10

BUS CURRENT
vs
BUS VOLTAGE

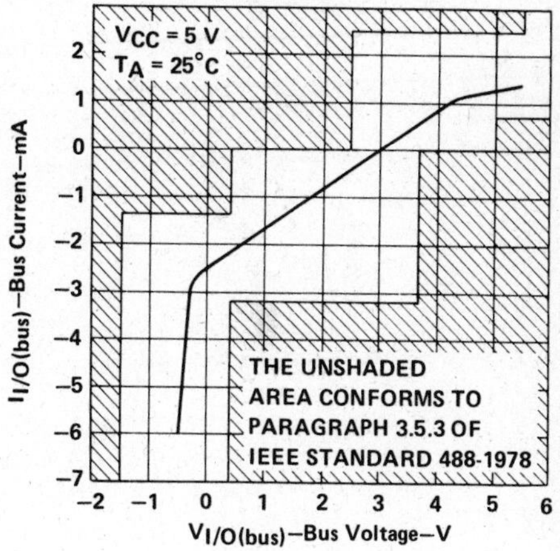
V_{CC} = 5 V
T_A = 25°C
THE UNSHADED AREA CONFORMS TO PARAGRAPH 3.5.3 OF IEEE STANDARD 488-1978
$I_{I/O(bus)}$—Bus Current—mA
2
1
0
−1
−2
−3
−4
−5
−6
−7
−2 −1 0 1 2 3 4 5 6
$V_{I/O(bus)}$—Bus Voltage—V

FIGURE 11

SN75ALS163
OCTAL GENERAL-PURPOSE INTERFACE BUS TRANSCEIVER

D2611, JUNE 1986

- 8-Channel Bidirectional Transceivers
- High-Speed Advanced Low-Power Schottky Circuitry
- Low Power Dissipation . . . 46 mW Max per Channel
- Fast Propagation Times . . . 20 ns Max
- High-Impedance P-N-P Inputs
- Receiver Hysteresis . . . 650 mV Typ
- Open-Collector Driver Output Option
- No Loading of Bus When Device is Powered Down ($V_{CC} = 0$)
- Power-Up/Power-Down Protection (Glitch-Free)

DW, J, OR N PACKAGE
(TOP VIEW)

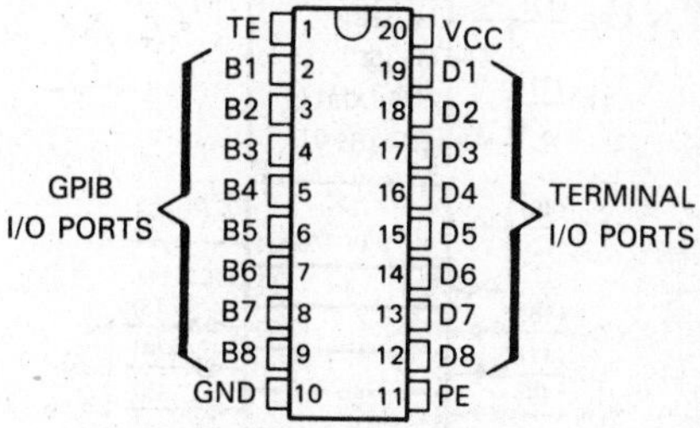
TE 1
B1 2
B2 3
B3 4
B4 5
B5 6
B6 7
B7 8
B8 9
GND 10
20 VCC
19 D1
18 D2
17 D3
16 D4
15 D5
14 D6
13 D7
12 D8
11 PE
GPIB I/O PORTS
TERMINAL I/O PORTS

FUNCTION TABLES

EACH DRIVER

INPUTS			OUTPUT
D	TE	PE	B
H	H	H	H
L	H	X	L
H	X	L	Z
X	L	X	Z

EACH RECEIVER

INPUTS			OUTPUT
B	TE	PE	D
L	L	X	L
H	L	X	H
X	H	X	Z

H = high level, L = low level, X = irrelevant, Z = High-impedance state.

description

The SN75ALS163 octal general-purpose interface bus transceiver is a monolithic, high-speed, Advanced Low-Power Schottky device. It is designed for two-way data communications over single-ended transmission lines. The transceiver features driver outputs that can be operated in either the open-collector or three-state mode. If Talk Enable (TE) is high, these outputs have the characteristics of open-collector outputs when Pullup Enable (PE) is low and of three-state outputs when PE is high. Taking TE low places the outputs in the high-impedance state. The driver outputs are designed to handle loads of up to 48 milliamperes of sink current. Each receiver features p-n-p transistor inputs for high input impedance and 400 millivolts minimum of guaranteed hysteresis for increased noise immunity.

Output glitches during power-up and power-down are eliminated by an internal circuit that disables both the bus and receiver outputs. The outputs do not load the bus when $V_{CC} = 0$.

The SN75ALS163 is characterized for operation from 0°C to 70°C.

logic symbol†

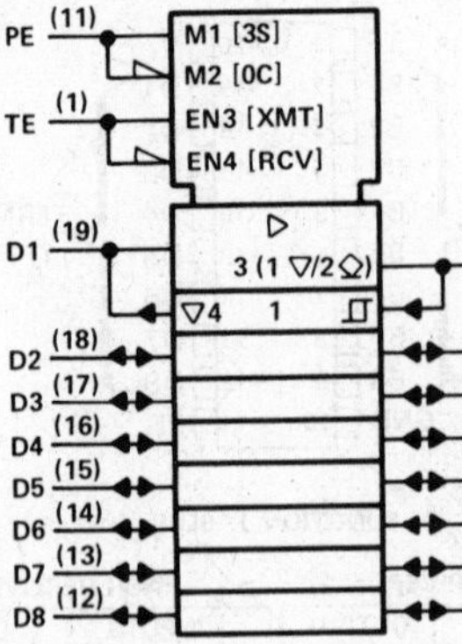
PE (11)
M1 [3S]
M2 [0C]
TE (1)
EN3 [XMT]
EN4 [RCV]
D1 (19)
3 (1 ▽/2 ◇)
▽4 1
(2) B1
D2 (18)
(3) B2
D3 (17)
(4) B3
D4 (16)
(5) B4
D5 (15)
(6) B5
D6 (14)
(7) B6
D7 (13)
(8) B7
D8 (12)
(9) B8

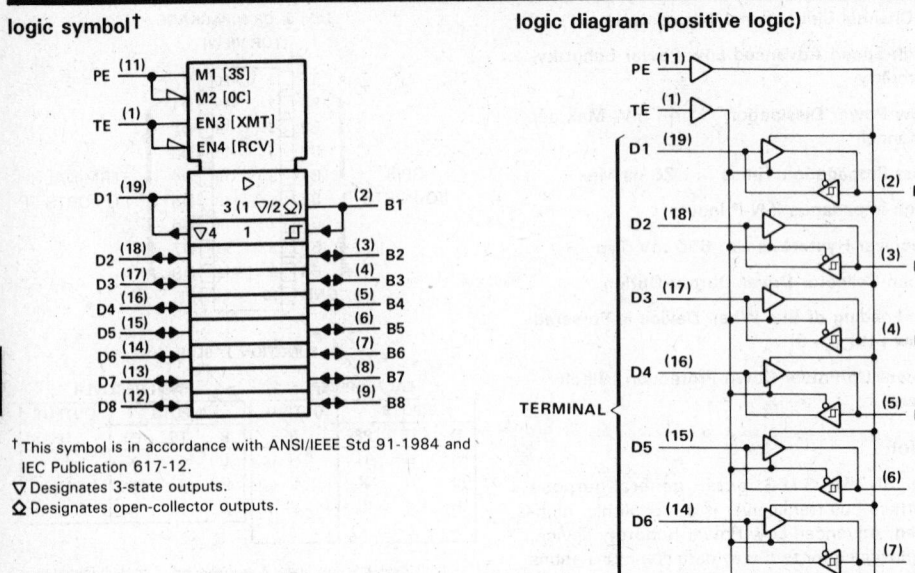

† This symbol is in accordance with ANSI/IEEE Std 91-1984 and IEC Publication 617-12.

▽ Designates 3-state outputs.

◇ Designates open-collector outputs.

logic diagram (positive logic)

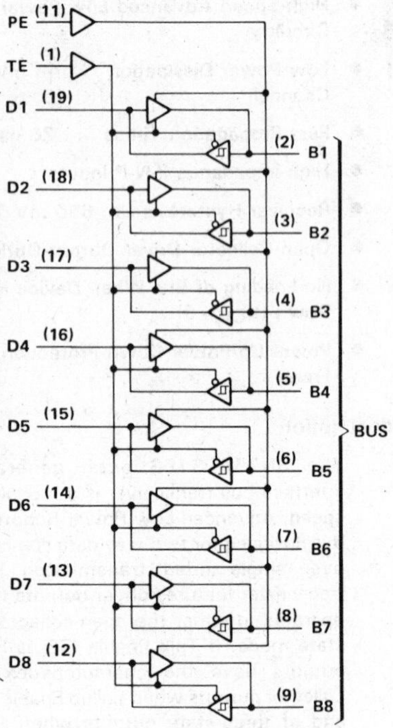
PE (11)
TE (1)
D1 (19)
(2) B1
D2 (18)
(3) B2
D3 (17)
(4) B3
D4 (16)
(5) B4
TERMINAL
D5 (15)
BUS
(6) B5
D6 (14)
(7) B6
D7 (13)
(8) B7
D8 (12)
(9) B8

schematics of inputs and outputs

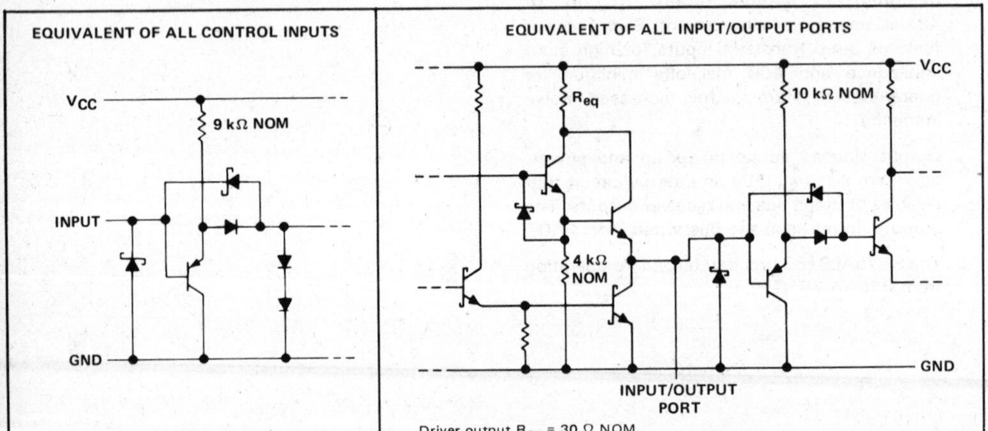
EQUIVALENT OF ALL CONTROL INPUTS
VCC
9 kΩ NOM
INPUT
GND
EQUIVALENT OF ALL INPUT/OUTPUT PORTS
VCC
Req
10 kΩ NOM
4 kΩ NOM
GND
INPUT/OUTPUT PORT

Driver output R_{eq} = 30 Ω NOM

Receiver output R_{eq} = 110 Ω NOM

TEXAS INSTRUMENTS
POST OFFICE BOX 655012 • DALLAS, TEXAS 75265

absolute maximum ratings over operating free-air temperature range (unless otherwise noted)

Supply voltage, V_{CC} (see Note 1) .... 7 V
Input voltage .... 5.5 V
Low-level driver output current .... 100 mA
Continuous total dissipation at (or below) 25 °C free-air temperature (see Note 2):
  DW package .... 1125 mW
  J package .... 1375 mW
  N package .... 1150 mW
Operating free-air temperature range .... 0 °C to 70 °C
Storage temperature range .... − 65 °C to 150 °C
Lead temperature 1,6 mm (1/16) inch from the case for 60 seconds: J package .... 300 °C
Lead temperature 1,6 mm (1/16 inch) from the case for 10 seconds: N package .... 260 °C

NOTES: 1. All voltage values are with respect to network ground terminal.
2. For operation above 25 °C free-air temperature, derate the DW package to 720 mW at 70 °C at the rate of 9.0 mW/ °C, derate the J package to 880 mW at 70 °C at the rate of 11.0 mW/ °C, and derate the N package to 736 mW at 70 °C at the rate of 9.2 mW/ °C.

recommended operating conditions

		MIN	NOM	MAX	UNIT
Supply voltage, V_{CC}		4.75	5	5.25	V
High-level input voltage, V_{IH}		2			V
Low-level input voltage, V_{IL}				0.8	V
High-level output current, I_{OH}	Bus ports with pullups active			−10	mA
	Terminal ports			−800	μA
Low-level output current, I_{OL}	Bus ports			48	mA
	Terminal ports			16	
Operating free-air temperature range, T_A		0		70	°C

4

Line Drivers/Receivers

ADVANCE INFORMATION

electrical characteristics over recommended ranges of supply voltage and operating free-air temperature (unless otherwise noted)

PARAMETER			TEST CONDITIONS		MIN	TYP†	MAX	UNIT
V_{IK}	Input clamp voltage		$I_I = -18$ mA			−0.8	−1.5	V
V_{hys}	Hysteresis ($V_{T+} - V_{T-}$)‡	Bus			0.4	0.65		V
V_{OH}	High level output voltage	Terminal	$I_{OH} = -800$ μA,	TE at 0.8 V	2.7	3.5		V
		Bus	$I_{OH} = -10$ mA,	PE and TE at 2 V	2.5	3.3		
V_{OL}	Low-level output voltage	Terminal	$I_{OL} = 16$ mA,	TE at 0.8 V		0.3	0.5	V
		Bus	$I_{OL} = 48$ mA,	PE and TE at 2 V		0.35	0.5	
I_{OH}	High-level output current (open-collector mode)	Bus	$V_O = 5.5$ V, D and TE at 2 V	PE at 0.8 V,			100	μA
I_{OZ}	Off-state output current (3-state mode)	Bus	PE at 2 V, TE at 0.8 V	$V_O = 2.7$ V			20	μA
				$V_O = 0.5$ V			−100	
I_I	Input current at maximum input voltage	Terminal	$V_I = 5.5$ V			0.2	100	μA
I_{IH}	High-level input current	Terminal PE or TE	$V_I = 2.7$ V			0.1	20	μA
I_{IL}	Low-level input current	Terminal PE or TE	$V_I = 0.5$ V			−10	−100	μA
I_{OS}	Short-circuit output current	Terminal			−15	−35	−75	mA
		Bus			−25	−50	−125	
I_{CC}	Supply current		No load	Terminal outputs low and enabled		42	56	mA
				Bus outputs low and enabled		52	70	
$C_{i/o(bus)}$	Bus-port capacitance		$V_{CC} = 5$ V or 0 V, $V_{I/O} = 0$ to 2 V, f = 1 MHz			30		pF

† All typical values are at $V_{CC} = 5$ V, $T_A = 25°C$.
‡ Hysteresis is the difference between the positive-going input threshold voltage, V_{T+}, and the negative-going input threshold voltage, V_{T-}.

switching characteristics over recommended range of operating free-air temperature (unless otherwise noted), $V_{CC} = 5$ V

PARAMETER		FROM	TO	TEST CONDITIONS	MIN	TYP†	MAX	UNIT
t_{PLH}	Propagation delay time, low-to-high-level output	Terminal	Bus	$C_L = 30$ pF, See Figure 1		10	20	ns
t_{PHL}	Propagation delay time, high-to-low-level output					12	20	
t_{PLH}	Propagation delay time, low-to-high-level output	Bus	Terminal	$C_L = 30$ pF, See Figure 2		5	10	ns
t_{PHL}	Propagation delay time, high-to-low-level output					7	14	
t_{PZH}	Output enable time to high level	TE	Bus	$C_L = 15$ pF, See Figure 3		11	20	ns
t_{PHZ}	Output disable time from high level					3	10	
t_{PZL}	Output enable time to low level					18	35	
t_{PLZ}	Output disable time from low level					5	20	
t_{PZH}	Output enable time to high level	TE	Terminal	$C_L = 15$ pF, See Figure 4		5	20	ns
t_{PHZ}	Output disable time from high level					8	20	
t_{PZL}	Output enable time to low level					9	20	
t_{PLZ}	Output disable time from low level					8	20	
t_{en}	Output pull-up enable time	PE	Bus	$C_L = 15$ pF, See Figure 5		3	10	ns
t_{dis}	Output pull-up disable time					4	12	

† All typical values are at $T_A = 25°C$.

PARAMETER MEASUREMENT INFORMATION

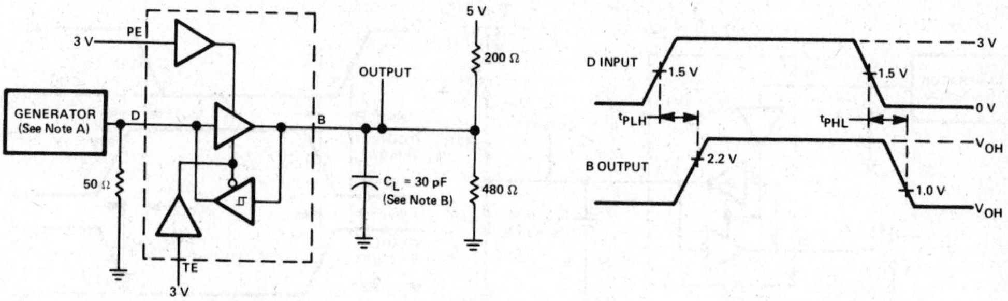
3 V
PE
GENERATOR (See Note A)
D
50 Ω
TE
3 V
B
OUTPUT
5 V
200 Ω
C_L = 30 pF (See Note B)
480 Ω
D INPUT
1.5 V
1.5 V
3 V
0 V
t_{PLH}
t_{PHL}
B OUTPUT
2.2 V
1.0 V
V_{OH}
V_{OH}

TEST CIRCUIT

VOLTAGE WAVEFORMS

FIGURE 1. TERMINAL-TO-BUS PROPAGATION DELAY TIMES

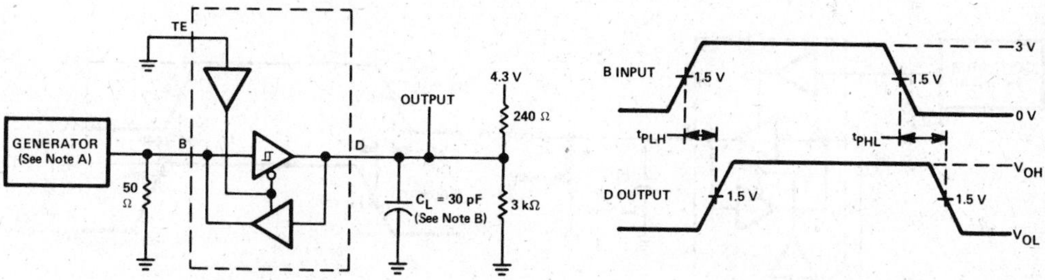
TE
GENERATOR (See Note A)
B
50 Ω
D
OUTPUT
4.3 V
240 Ω
C_L = 30 pF (See Note B)
3 kΩ
B INPUT
1.5 V
1.5 V
3 V
0 V
t_{PLH}
t_{PHL}
D OUTPUT
1.5 V
1.5 V
V_{OH}
V_{OL}

TEST CIRCUIT

VOLTAGE WAVEFORMS

FIGURE 2. BUS-TO-TERMINAL PROPAGATION DELAY TIMES

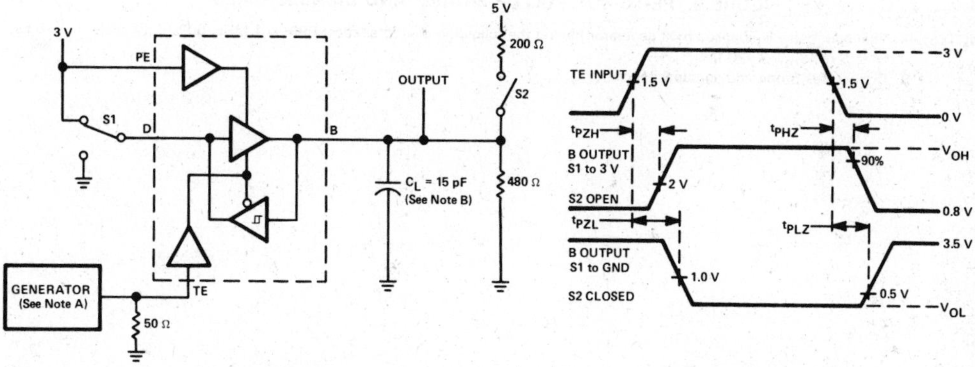
3 V
PE
S1
D
B
TE
GENERATOR (See Note A)
50 Ω
OUTPUT
5 V
200 Ω
S2
C_L = 15 pF (See Note B)
480 Ω
TE INPUT
1.5 V
1.5 V
3 V
0 V
t_{PZH}
t_{PHZ}
B OUTPUT S1 to 3 V S2 OPEN
2 V
90%
V_{OH}
0.8 V
t_{PZL}
t_{PLZ}
B OUTPUT S1 to GND S2 CLOSED
1.0 V
0.5 V
3.5 V
V_{OL}

TEST CIRCUIT

VOLTAGE WAVEFORMS

FIGURE 3. TE-TO-BUS ENABLE AND DISABLE TIMES

NOTES: A. The input pulse is supplied by a generator having the following characteristics: PRR ≤ 1 MHz, 50% duty cycle, $t_r \le 6$ ns, $t_f \le 6$ ns, $Z_{out} = 50\ \Omega$.
B. C_L includes probe and jig capacitance.

PARAMETER MEASUREMENT INFORMATION

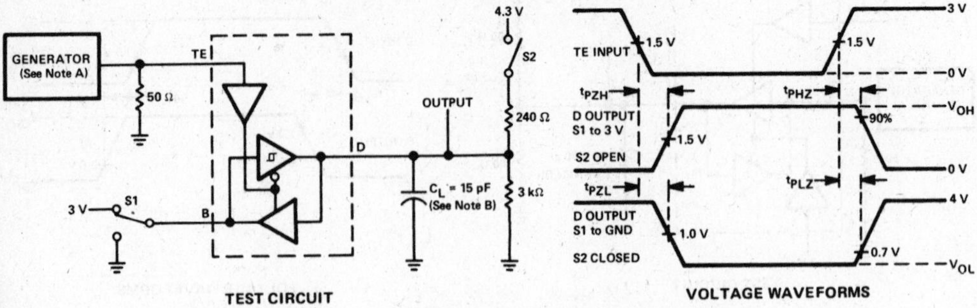
GENERATOR
(See Note A)
TE
50 Ω
S1
3 V
B
D
OUTPUT
C_L = 15 pF
(See Note B)
4.3 V
S2
240 Ω
3 kΩ
TEST CIRCUIT
TE INPUT
1.5 V
3 V
0 V
t_{PZH}
t_{PHZ}
D OUTPUT
S1 to 3 V
S2 OPEN
90%
V_{OH}
t_{PZL}
t_{PLZ}
D OUTPUT
S1 to GND
S2 CLOSED
1.0 V
0.7 V
4 V
V_{OL}
VOLTAGE WAVEFORMS

FIGURE 4. TE-TO-TERMINAL ENABLE AND DISABLE TIMES

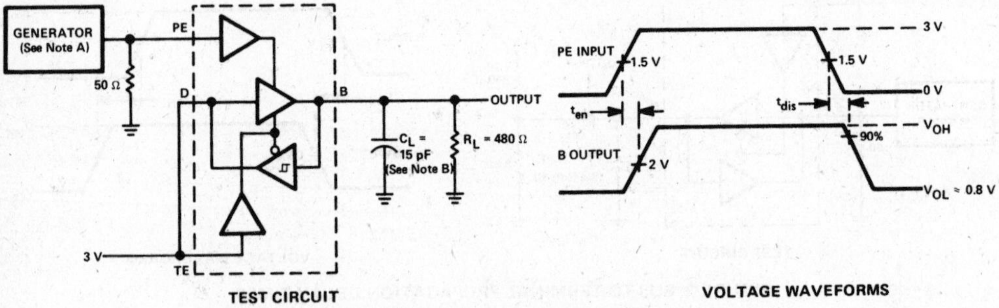
GENERATOR
(See Note A)
PE
50 Ω
D
B
OUTPUT
C_L = 15 pF
(See Note B)
R_L = 480 Ω
3 V
TE
TEST CIRCUIT
PE INPUT
1.5 V
3 V
0 V
t_{en}
t_{dis}
V_{OH}
90%
B OUTPUT
2 V
V_{OL} ≈ 0.8 V
VOLTAGE WAVEFORMS

FIGURE 5. PE-TO-BUS PULLUP ENABLE AND DISABLE TIMES

NOTES: A. The input pulse is supplied by a generator having the following characteristics: PRR ≤ 1 MHz, 50% duty cycle, $t_r \leq 6$ ns, $t_f \leq 6$ ns, Z_{out} = 50 Ω.
B. C_L includes probe and jig capacitance.

TEXAS INSTRUMENTS
POST OFFICE BOX 655012 • DALLAS, TEXAS 75265

TYPICAL CHARACTERISTICS

TERMINAL HIGH-LEVEL OUTPUT VOLTAGE
vs
HIGH-LEVEL OUTPUT CURRENT

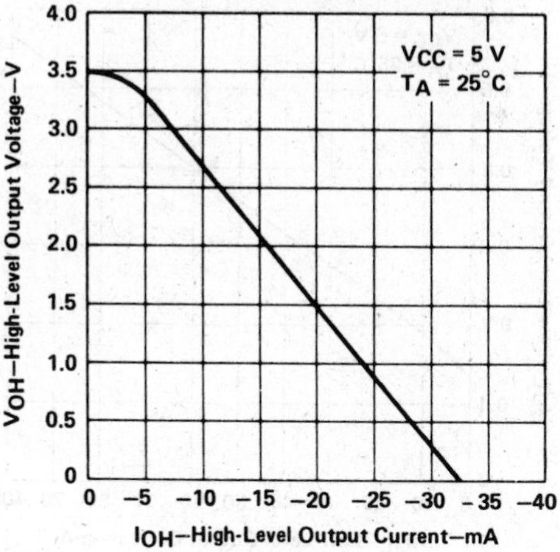
V_{CC} = 5 V
T_A = 25°C
V_{OH}—High-Level Output Voltage—V
4.0
3.5
3.0
2.5
2.0
1.5
1.0
0.5
0
0 −5 −10 −15 −20 −25 −30 −35 −40
I_{OH}—High-Level Output Current—mA

FIGURE 6

TERMINAL LOW-LEVEL OUTPUT VOLTAGE
vs
LOW-LEVEL OUTPUT CURRENT

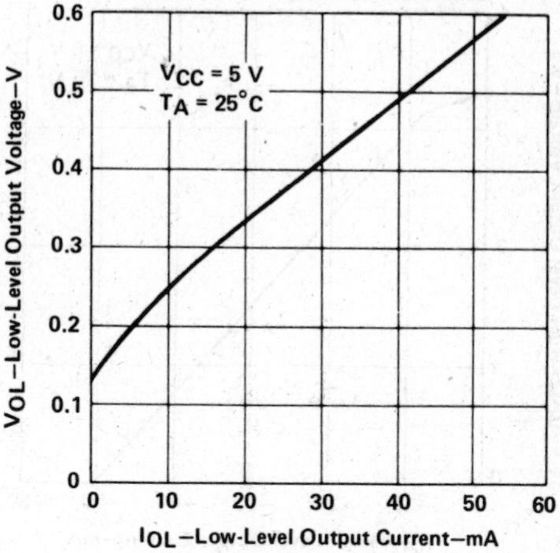
V_{CC} = 5 V
T_A = 25°C
V_{OL}—Low-Level Output Voltage—V
0.6
0.5
0.4
0.3
0.2
0.1
0
0 10 20 30 40 50 60
I_{OL}—Low-Level Output Current—mA

FIGURE 7

TERMINAL OUTPUT VOLTAGE
vs
BUS INPUT VOLTAGE

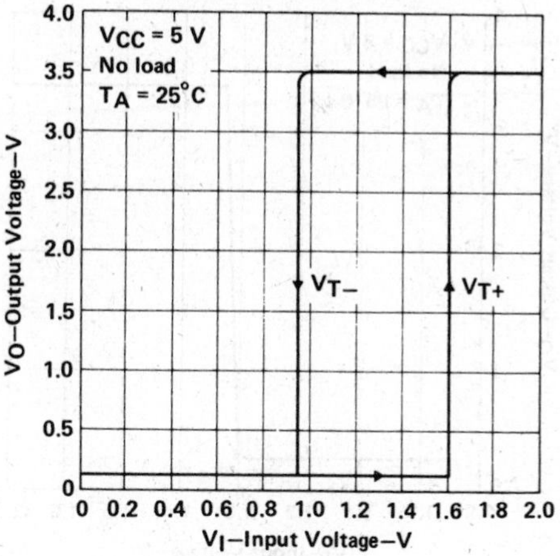
V_{CC} = 5 V
No load
T_A = 25°C
V_O—Output Voltage—V
4.0
3.5
3.0
2.5
2.0
1.5
1.0
0.5
0
V_{T-}
V_{T+}
0 0.2 0.4 0.6 0.8 1.0 1.2 1.4 1.6 1.8 2.0
V_I—Input Voltage—V

FIGURE 8

4

Line Drivers/Receivers

ADVANCE INFORMATION

TYPICAL CHARACTERISTICS

BUS HIGH-LEVEL OUTPUT VOLTAGE
vs
BUS HIGH-LEVEL OUTPUT CURRENT

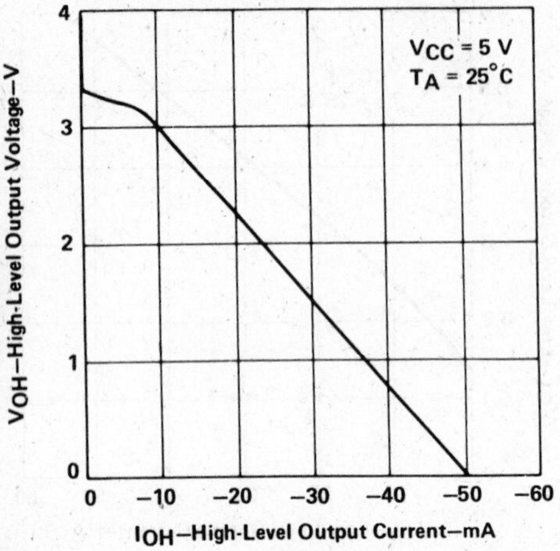
V_{CC} = 5 V
T_A = 25°C
V_{OH}—High-Level Output Voltage—V
0
1
2
3
4
0
−10
−20
−30
−40
−50
−60
I_{OH}—High-Level Output Current—mA

FIGURE 9

BUS LOW-LEVEL OUTPUT VOLTAGE
vs
BUS LOW-LEVEL OUTPUT CURRENT

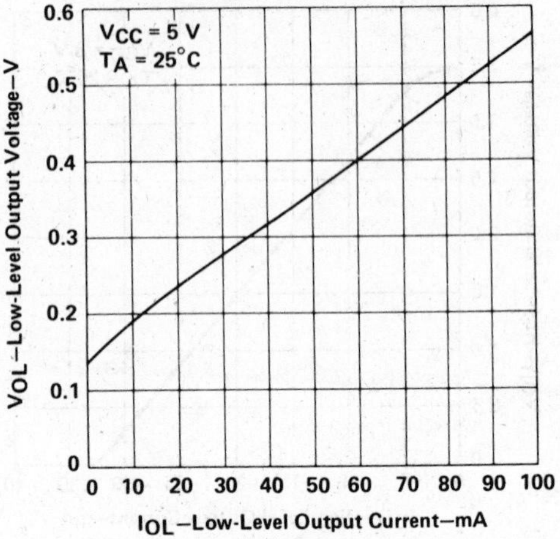
V_{CC} = 5 V
T_A = 25°C
V_{OL}—Low-Level Output Voltage—V
0
0.1
0.2
0.3
0.4
0.5
0.6
0
10
20
30
40
50
60
70
80
90
100
I_{OL}—Low-Level Output Current—mA

FIGURE 10

BUS OUTPUT VOLTAGE
vs
TERMINAL INPUT VOLTAGE

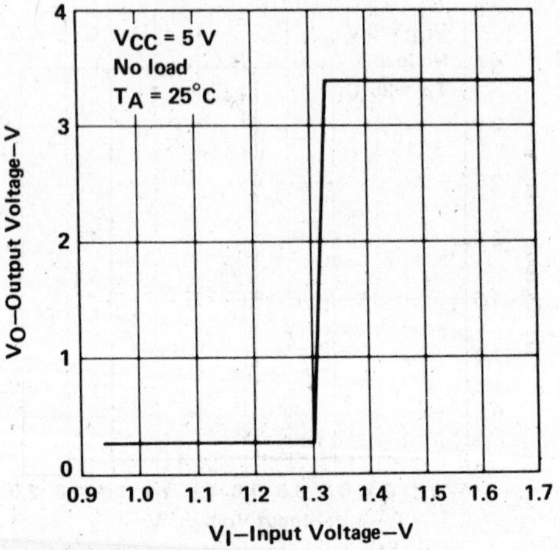
V_{CC} = 5 V
No load
T_A = 25°C
V_O—Output Voltage—V
0
1
2
3
4
0.9
1.0
1.1
1.2
1.3
1.4
1.5
1.6
1.7
V_I—Input Voltage—V

FIGURE 11

TEXAS INSTRUMENTS
POST OFFICE BOX 655012 • DALLAS, TEXAS 75265

SN75ALS164
OCTAL GENERAL-PURPOSE INTERFACE BUS TRANSCEIVER

D2908, JUNE 1986

- 8-Channel Bidirectional Transceiver
- Designed to Implement Control Bus Interface
- Designed for Multicontrollers
- High-Speed Advanced Low-Power Schottky Circuitry
- Low Power Dissipation . . . 46 mW Max per Channel
- Fast Propagation Times . . . 20 ns Max
- High-Impedance P-N-P Inputs
- Receiver Hysteresis . . . 650 mV Typ
- Bus-Terminating Resistors Provided on Driver Outputs
- No Loading of Bus When Device is Powered Down (V_{CC} = 0)
- Power-Up/Power-Down Protection (Glitch-Free)

DW PACKAGE
(TOP VIEW)

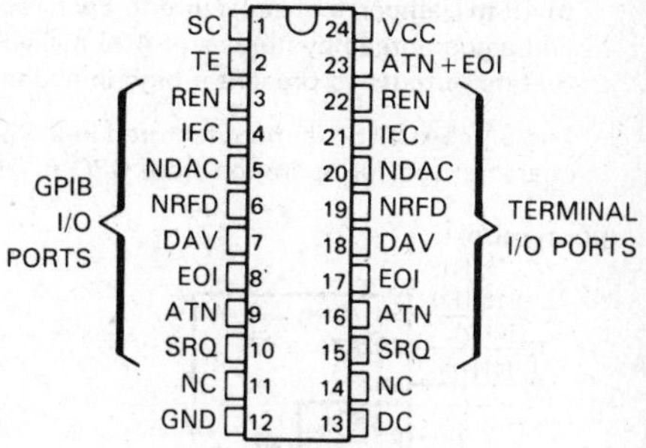
SC 1
TE 2
REN 3
IFC 4
NDAC 5
NRFD 6
DAV 7
EOI 8
ATN 9
SRQ 10
NC 11
GND 12
24 VCC
23 ATN + EOI
22 REN
21 IFC
20 NDAC
19 NRFD
18 DAV
17 EOI
16 ATN
15 SRQ
14 NC
13 DC
GPIB I/O PORTS
TERMINAL I/O PORTS

N DUAL-IN-LINE PACKAGE
(TOP VIEW)

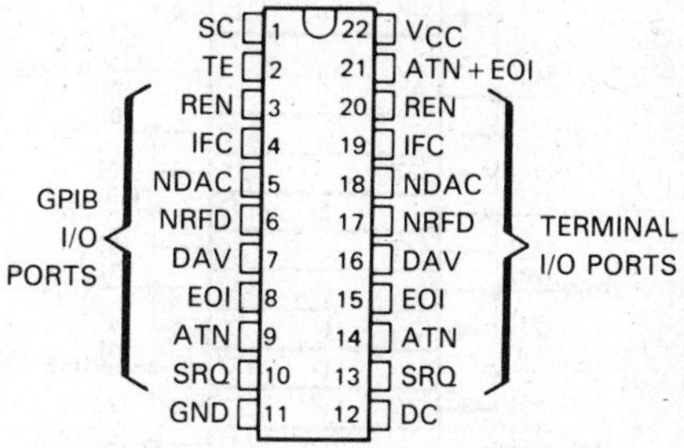
SC 1
TE 2
REN 3
IFC 4
NDAC 5
NRFD 6
DAV 7
EOI 8
ATN 9
SRQ 10
GND 11
22 VCC
21 ATN + EOI
20 REN
19 IFC
18 NDAC
17 NRFD
16 DAV
15 EOI
14 ATN
13 SRQ
12 DC
GPIB I/O PORTS
TERMINAL I/O PORTS

NC—No internal connection.

description

The SN75ALS164 eight-channel general-purpose interface bus transceiver is a monolithic, high-speed, Advanced Low-Power Schottky device designed to meet the requirements of IEEE Standard 488-1978. Each transceiver is designed to provide the bus-management and data-transfer signals between operating units of a multiple-controller instrumentation system. When combined with the SN75ALS160 octal bus transceiver, the SN75ALS164 provides the complete 16-wire interface for the IEEE 488 bus.

The SN75ALS164 features eight driver-receiver pairs connected in a front-to-back configuration to form input/output (I/O) ports at both the bus and terminal sides. All outputs are disabled (at a high-impedance state) during V_{CC} power-up and power-down transitions for glitch-free operation. The direction of data flow through these driver-receiver pairs is determined by the DC, TE, and SC enable signals. The SN75ALS164 is identical to the SN75ALS162 with the addition of an OR gate to help simplify board layouts in several popular applications. The ATN and EOI signals are ORed to pin 21, which is a standard totem-pole output.

CHANNEL IDENTIFICATION TABLE

NAME	IDENTITY	CLASS
DC TE SC	Direction Control Talk Enable System Control	Control
ATN SRQ REN IFC EOI	Attention Service Request Remote Enable Interface Clear End or Identify	Bus Management
ATN + EOI	ATN logical OR EOI	Logic
DAV NDAC NRFD	Data Valid Not Data Accepted Not Ready for Data	Data Transfer

ADVANCE INFORMATION documents contain information on new products in the sampling or preproduction phase of development. Characteristic data and other specifications are subject to change without notice.

TEXAS INSTRUMENTS
POST OFFICE BOX 655012 • DALLAS, TEXAS 75265

The driver outputs (GPIB I/O ports) feature active bus-terminating resistor circuits designed to provide a high impedance to the bus when supply voltage V_{CC} is 0. The drivers are designed to handle loads up to 48 milliamperes of sink current. Each receiver features p-n-p transistor inputs for high input impedance and a guaranteed hysteresis of 400 millivolts minimum for increased noise immunity. All receivers have 3-state outputs to present a high impedance to the terminal when disabled.

The SN75ALS164 is manufactured in a 22-pin dual-in-line N package and in 24-pin DW package, and is characterized for operation from 0°C to 70°C.

logic symbol†

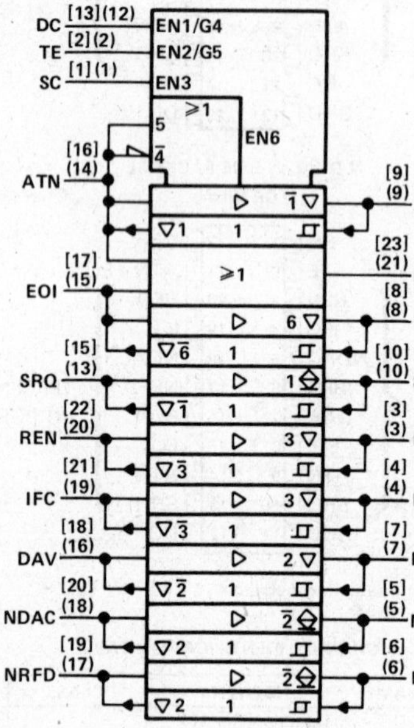
DC [13] (12)
TE [2] (2)
SC [1] (1)
EN1/G4
EN2/G5
EN3
≥1
5
4
EN6
ATN [16] (14)
[9] (9) ATN
[23] (21) ATN+EOI
EOI [17] (15)
[8] (8) EOI
SRQ [15] (13)
[10] (10) SRQ
REN [22] (20)
[3] (3) REN
IFC [21] (19)
[4] (4) IFC
DAV [18] (16)
[7] (7) DAV
NDAC [20] (18)
[5] (5) NDAC
NRFD [19] (17)
[6] (6) NRFD

†This symbol is in accordance with ANSI/IEEE Std 91-1984 and IEC Publication 617-12.
▽Designates 3-state outputs.
◊Designates passive-pullup outputs.

logic diagram (positive logic)

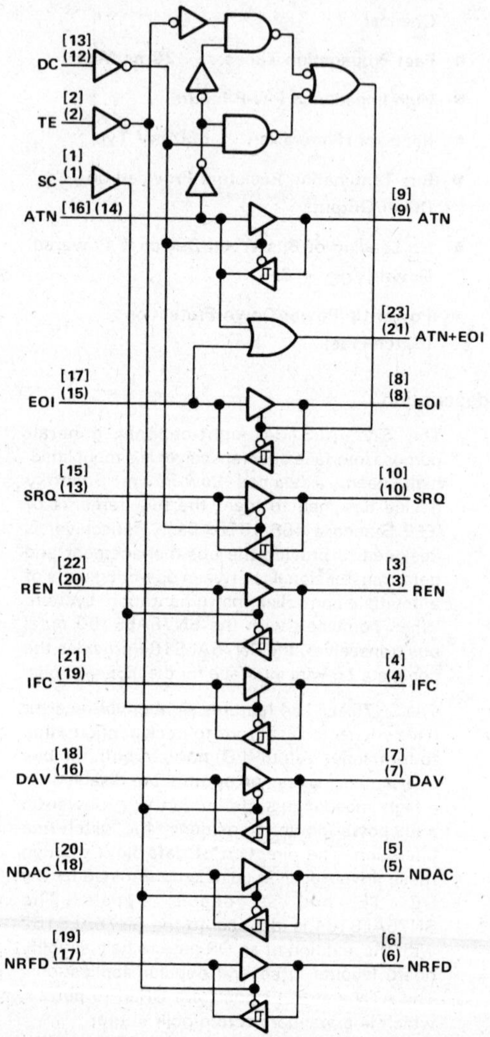
DC [13] (12)
TE [2] (2)
SC [1] (1)
ATN [16] (14)
[9] (9) ATN
[23] (21) ATN+EOI
EOI [17] (15)
[8] (8) EOI
SRQ [15] (13)
[10] (10) SRQ
REN [22] (20)
[3] (3) REN
IFC [21] (19)
[4] (4) IFC
DAV [18] (16)
[7] (7) DAV
NDAC [20] (18)
[5] (5) NDAC
NRFD [19] (17)
[6] (6) NRFD

[ ] Denotes pin numbers for DW package.
( ) Denotes pin numbers for N package.

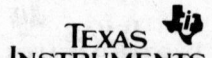

TEXAS INSTRUMENTS
POST OFFICE BOX 655012 • DALLAS, TEXAS 75265

RECEIVE/TRANSMIT FUNCTION TABLE

CONTROLS				BUS-MANAGEMENT CHANNELS					DATA-TRANSFER CHANNELS		
SC	DC	TE	ATN†	ATN† (Controlled by DC)	SRQ (Controlled by DC)	REN (Controlled by SC)	IFC (Controlled by SC)	EOI	DAV (Controlled by TE)	NDAC (Controlled by TE)	NRFD (Controlled by TE)
	H	H	H	R	T			T	T	R	R
	H	H	L	R	T			R	T	R	R
	L	L	H	T	R			R	R	T	T
	L	L	L	T	R			T	R	T	T
	H	L	X	R	T			R	R	T	T
	L	H	X	T	R			T	T	R	R
H						T	T				
L						R	R				

H = high level, L = low level, R = receive, T = transmit, X = irrelevant

Direction of data transmission is from the terminal side to the bus side, and the direction of data receiving is from the bus side to the terminal side. Data transfer is noninverting in both directions.

†ATN is a normal transceiver channel that functions additionally as an internal direction control or talk enable for EOI whenever the DC and TE inputs are in the same state. When DC and TE are in opposite states, the ATN channel functions as an independent transceiver only.

ATN + EOI FUNCTION TABLE

INPUTS		OUTPUT
ATN	EOI	ATN + EOI
H	X	H
X	H	H
L	L	L

schematics of inputs and outputs

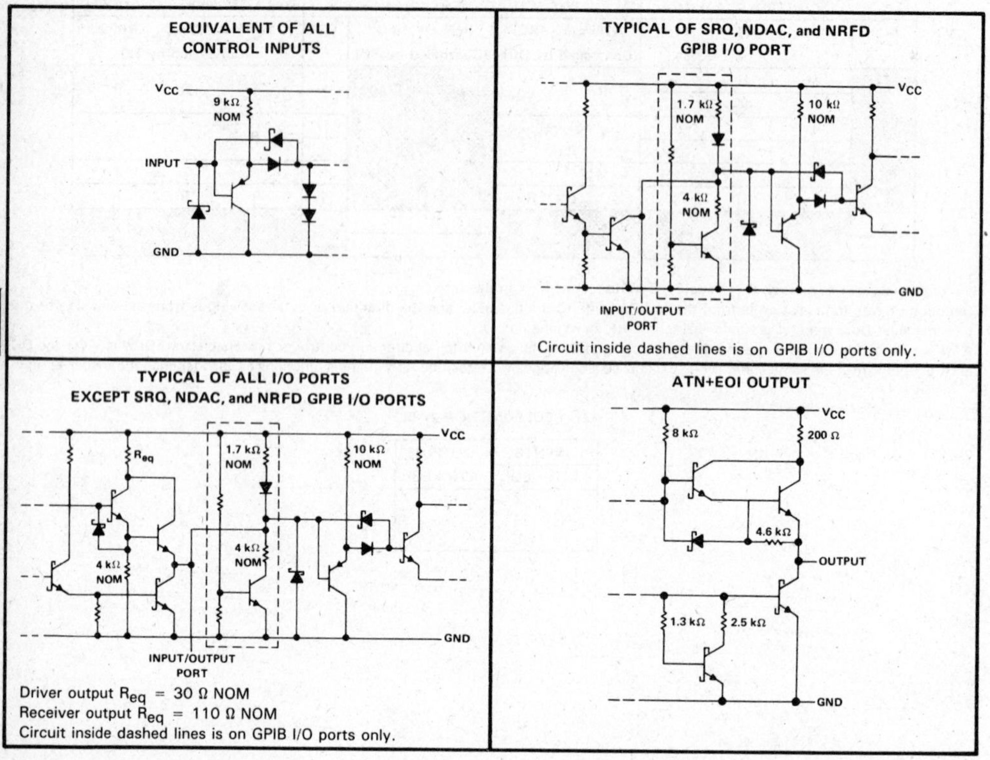
EQUIVALENT OF ALL CONTROL INPUTS
VCC
9 kΩ NOM
INPUT
GND
TYPICAL OF SRQ, NDAC, and NRFD GPIB I/O PORT
1.7 kΩ NOM
10 kΩ NOM
4 kΩ NOM
VCC
GND
INPUT/OUTPUT PORT
TYPICAL OF ALL I/O PORTS EXCEPT SRQ, NDAC, and NRFD GPIB I/O PORTS
Req
1.7 kΩ NOM
10 kΩ NOM
4 kΩ NOM
4 kΩ NOM
VCC
GND
INPUT/OUTPUT PORT
ATN+EOI OUTPUT
VCC
8 kΩ
200 Ω
4.6 kΩ
OUTPUT
1.3 kΩ
2.5 kΩ
GND

Circuit inside dashed lines is on GPIB I/O ports only.

Driver output R_{eq} = 30 Ω NOM
Receiver output R_{eq} = 110 Ω NOM
Circuit inside dashed lines is on GPIB I/O ports only.

absolute maximum ratings over operating free-air temperature range (unless otherwise noted)

Supply voltage, V_{CC} (see Note 1) . . . . . . 7 V
Input voltage . . . . . . 5.5 V
Low-level driver output current . . . . . . 100 mA
Continuous total dissipation at (or below) 25°C free-air temperature (see Note 2):
- DW package . . . . . . 1350 mW
- N package . . . . . . 1700 mW

Operating free-air temperature range . . . . . . 0°C to 70°C
Storage temperature range . . . . . . −65°C to 150°C
Lead temperature 1,6 mm (1/16) inch from the case for 10 seconds: DW or N package . . . . 260°C

NOTES: 1. All voltage values are with respect to network ground terminal.
2. For operation above 25°C free-air temperature, derate the DW package to 864 mW at 70°C at the rate of 10.8 mW/°C, and derate the N package to 1088 mW at 70°C at the rate of 13.6 mW/°C.

TEXAS INSTRUMENTS
POST OFFICE BOX 655012 • DALLAS, TEXAS 75265

recommended operating conditions

		MIN	NOM	MAX	UNIT
Supply voltage, V_{CC}		4.75	5	5.25	V
High-level input voltage, V_{IH}		2			V
Low-level input voltage, V_{IL}				0.8	V
High-level output current, I_{OH}	Bus ports with 3-state outputs			−5.2	mA
	Terminal ports			−800	μA
	ATN + EOI			−400	μA
Low-level output current, I_{OL}	Bus ports			48	mA
	Terminal ports			16	mA
	ATN + EOI			4	mA
Operating free-air temperature, T_A		0		70	°C

electrical characteristics over recommended ranges of supply voltage and operating free-air temperature (unless otherwise noted)

PARAMETER			TEST CONDITIONS	MIN	TYP†	MAX	UNIT
V_{IK}	Input clamp voltage		$I_I = -18$ mA		−0.8	−1.5	V
V_{hys}	Hysteresis ($V_{T+} - V_{T-}$)	Bus		0.4	0.65		V
V_{OH}‡	High-level output voltage	Terminal	$I_{OH} = -800$ μA	2.7	3.5		V
		Bus	$I_{OH} = -5.2$ mA	2.5	3.3		V
		ATN + EOI	$I_{OH} = -400$ μA	2.7			V
V_{OL}	Low-level output voltage	Terminal	$I_{OL} = 16$ mA		0.3	0.5	V
		Bus	$I_{OL} = 48$ mA		0.35	0.5	V
		ATN + EOI	$I_{OL} = 4$ mA			0.4	V
I_I	Input current at maximum input voltage	Terminal§	$V_I = 5.5$ V		0.2	100	μA
		ATN, EOI	$V_I = 5.5$ V			200	μA
I_{IH}	High-level input current	Terminal control	$V_I = 2.7$ V		0.1	20	μA
		ATN, EOI	$V_I = 2.7$ V			40	μA
I_{IL}	Low-level input current	Terminal, control	$V_I = 0.5$ V		−10	−100	μA
		ATN, EOI	$V_I = 0.5$ V			−500	μA
$V_{I/O(bus)}$	Voltage at bus port		Driver disabled, $I_{I(bus)} = 0$	2.5	3.0	3.7	V
			Driver disabled, $I_{I(bus)} = -12$ mA			−1.5	V
$I_{I/O(bus)}$	Current into bus port	Power on	Driver disabled, $V_{I(bus)} = -1.5$ V to 0.4 V	−1.3			mA
		Power on	Driver disabled, $V_{I(bus)} = 0.4$ V to 2.5 V	0		−3.2	mA
		Power on	Driver disabled, $V_{I(bus)} = 2.5$ V to 3.7 V			+2.5 −3.2	mA
		Power on	Driver disabled, $V_{I(bus)} = 3.7$ V to 5 V	0		2.5	mA
		Power on	Driver disabled, $V_{I(bus)} = 5$ V to 5.5 V	0.7		2.5	mA
		Power off	$V_{CC} = 0$, $V_{I(bus)} = 0$ V to 2.5 V			−40	μA
I_{OS}	Short-circuit output current	Terminal		−15	−35	−75	mA
		Bus		−25	−50	−125	mA
		ATN + EOI		−10		−100	mA
I_{CC}	Supply current		No load, TE, DC, and SC low		55	75	mA
$C_{i/o(bus)}$	Bus-port capacitance		$V_{CC} = 5$ V to 0 V, $V_{I/O} = 0$ to 2 V, f = 1 MHz		30		pF

† All typical values are at $V_{CC} = 5$ V, $T_A = 25°C$.
‡ V_{OH} applies for three-state outputs only.
§ Except ATN and EOI terminal pins.

switching characteristics over recommended range of operating free-air temperature (unless otherwise noted), V_{CC} = 5 V

PARAMETER		FROM	TO	TEST CONDITIONS	MIN	TYP	MAX	UNIT
t_{PLH}	Propagation delay time, low-to-high-level output	Terminal	Bus	C_L = 30 pF, See Figure 1		10	20	ns
t_{PHL}	Propagation delay time, high-to-low-level output					12	20	
t_{PLH}	Propagation delay time low-to-high-level output	Bus	Terminal	C_L = 30 pF, See Figure 2		5	10	ns
t_{PHL}	Propagation delay time, high-to-low-level output					7	14	
t_{PLH}	Propagation delay time, low-to-high-level output	Terminal ATN or Terminal EOI	ATN+EOI	C_L = 15 pF, See Figure 3		3.5	10	ns
t_{PHL}	Propagation delay time, high-to-low-level output	Terminal ATN or Terminal EOI	ATN+EOI	C_L = 15 pF, See Figure 3		7	15	ns
t_{PZH}	Output enable time to high level	TE, DC, or SC	BUS (ATTN, EOI, REN, IFC, and DAV)	C_L = 15 pF, See Figure 4			30	ns
t_{PHZ}	Output disable time from high level						20	
t_{PZL}	Output enable time to low level						45	
t_{PLZ}	Output disable time from low level						20	
t_{PZH}	Output enable time to high level	TE, DC, or SC	Terminal	C_L = 15 pF, See Figure 5			20	ns
t_{PHZ}	Output disable time from high level						25	
t_{PZL}	Output enable time to low level						30	
t_{PLZ}	Output disable time from low level						25	

TEXAS INSTRUMENTS
POST OFFICE BOX 655012 • DALLAS, TEXAS 75265

PARAMETER MEASUREMENT INFORMATION

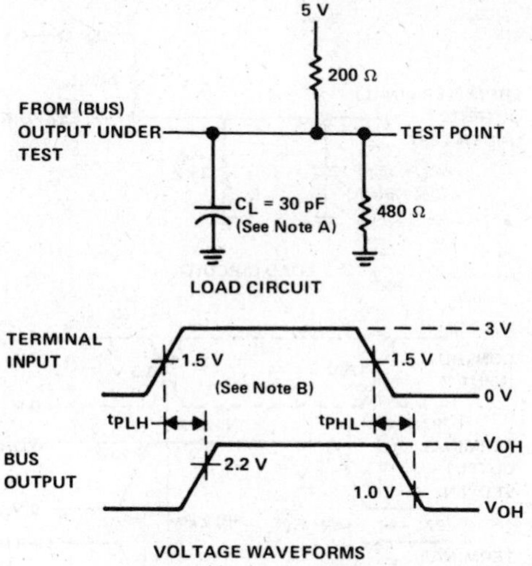
5 V
200 Ω
FROM (BUS) OUTPUT UNDER TEST
TEST POINT
C_L = 30 pF (See Note A)
480 Ω
LOAD CIRCUIT
TERMINAL INPUT
3 V
1.5 V
1.5 V
(See Note B)
0 V
t_{PLH}
t_{PHL}
V_{OH}
BUS OUTPUT
2.2 V
1.0 V
V_{OH}
VOLTAGE WAVEFORMS

FIGURE 1. TERMINAL-TO-BUS PROPAGATION DELAY TIMES

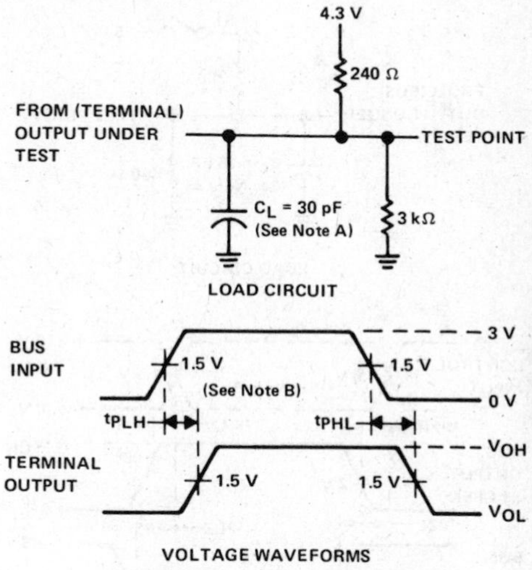
4.3 V
240 Ω
FROM (TERMINAL) OUTPUT UNDER TEST
TEST POINT
C_L = 30 pF (See Note A)
3 kΩ
LOAD CIRCUIT
BUS INPUT
3 V
1.5 V
1.5 V
(See Note B)
0 V
t_{PLH}
t_{PHL}
V_{OH}
TERMINAL OUTPUT
1.5 V
1.5 V
V_{OL}
VOLTAGE WAVEFORMS

FIGURE 2. BUS-TO-TERMINAL PROPAGATION DELAY TIMES

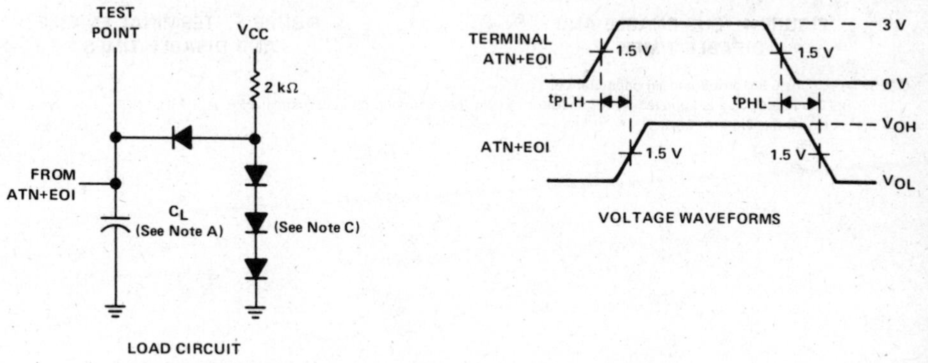
TEST POINT
V_{CC}
2 kΩ
FROM ATN+EOI
C_L (See Note A)
(See Note C)
LOAD CIRCUIT
TERMINAL ATN+EOI
3 V
1.5 V
1.5 V
0 V
t_{PLH}
t_{PHL}
V_{OH}
ATN+EOI
1.5 V
1.5 V
V_{OL}
VOLTAGE WAVEFORMS

FIGURE 3. ATN + EOI PROPAGATION DELAY TIMES

NOTES: A. C_L includes probe and jig capacitance.

B. The input pulse is supplied by a generator having the following characteristics: PRR ≤ 1 MHz, 50% duty cycle, t_r ≤6 ns, t_f ≤6 ns, Z_{out} = 50 Ω.

C. All diodes are 1N916 or 1N3064.

PARAMETER MEASUREMENT INFORMATION

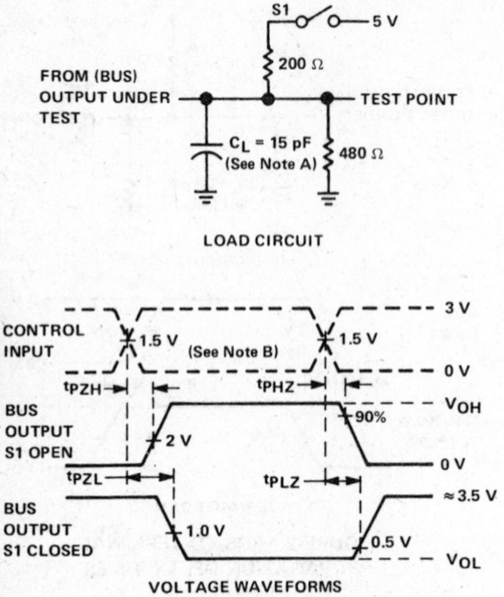
S1
5 V
200 Ω
FROM (BUS)
OUTPUT UNDER
TEST
TEST POINT
CL = 15 pF
(See Note A)
480 Ω
LOAD CIRCUIT
CONTROL
INPUT
1.5 V
(See Note B)
1.5 V
3 V
0 V
tPZH
tPHZ
BUS
OUTPUT
S1 OPEN
2 V
90%
VOH
0 V
tPZL
tPLZ
≈3.5 V
BUS
OUTPUT
S1 CLOSED
1.0 V
0.5 V
VOL
VOLTAGE WAVEFORMS

FIGURE 4. BUS ENABLE AND DISABLE TIMES

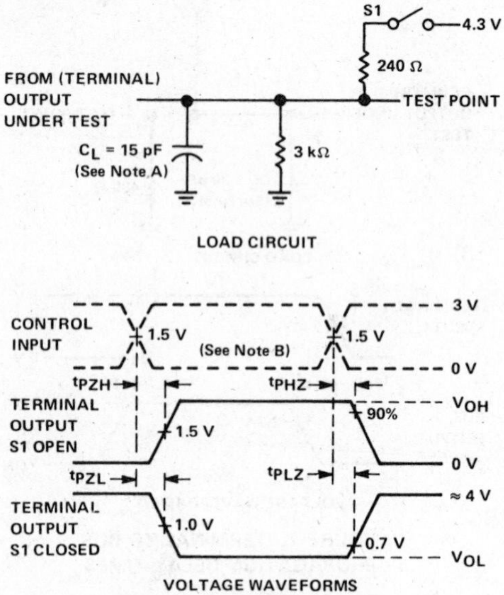
S1
4.3 V
240 Ω
FROM (TERMINAL)
OUTPUT
UNDER TEST
TEST POINT
CL = 15 pF
(See Note A)
3 kΩ
LOAD CIRCUIT
CONTROL
INPUT
1.5 V
(See Note B)
1.5 V
3 V
0 V
tPZH
tPHZ
TERMINAL
OUTPUT
S1 OPEN
1.5 V
90%
VOH
0 V
tPZL
tPLZ
≈ 4 V
TERMINAL
OUTPUT
S1 CLOSED
1.0 V
0.7 V
VOL
VOLTAGE WAVEFORMS

FIGURE 5. TERMINAL ENABLE AND DISABLE TIMES

NOTES: A. C_L includes probe and jig capacitance.
B. The input pulse is supplied by a generator having the following characteristics: PRR ≤ 1 MHz, 50% duty cycle, t_r ≤6 ns, t_f ≤6 ns, Z_{out} = 50 Ω.

TYPICAL CHARACTERISTICS

TERMINAL HIGH-LEVEL OUTPUT VOLTAGE
vs
HIGH-LEVEL OUTPUT CURRENT

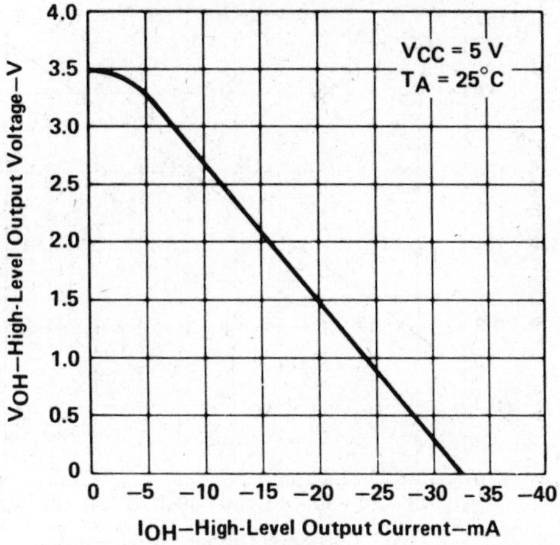
VCC = 5 V
TA = 25°C
VOH—High-Level Output Voltage—V
4.0
3.5
3.0
2.5
2.0
1.5
1.0
0.5
0
0 −5 −10 −15 −20 −25 −30 −35 −40
IOH—High-Level Output Current—mA

FIGURE 6

TERMINAL LOW-LEVEL OUTPUT VOLTAGE
vs
LOW-LEVEL OUTPUT CURRENT

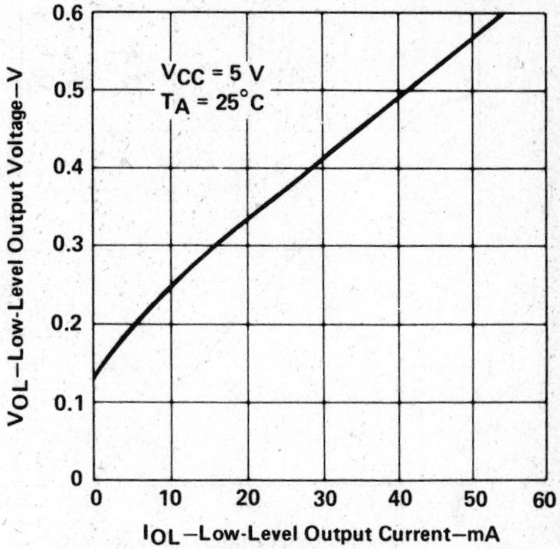
VCC = 5 V
TA = 25°C
VOL—Low-Level Output Voltage—V
0.6
0.5
0.4
0.3
0.2
0.1
0
0 10 20 30 40 50 60
IOL—Low-Level Output Current—mA

FIGURE 7

TERMINAL OUTPUT VOLTAGE
vs
BUS INPUT VOLTAGE

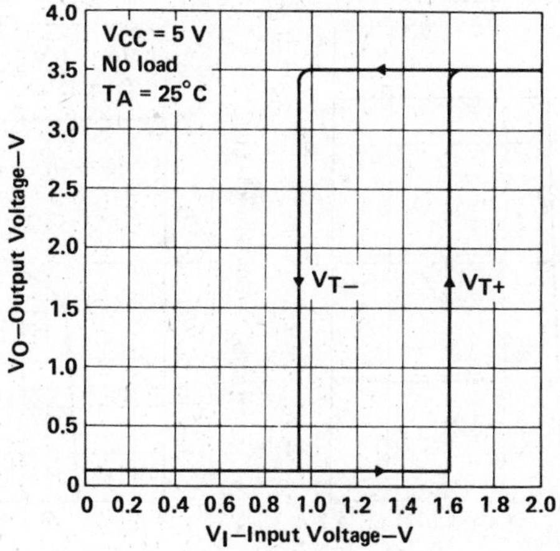
VCC = 5 V
No load
TA = 25°C
VT−
VT+
VO—Output Voltage—V
4.0
3.5
3.0
2.5
2.0
1.5
1.0
0.5
0
0 0.2 0.4 0.6 0.8 1.0 1.2 1.4 1.6 1.8 2.0
VI—Input Voltage—V

FIGURE 8

4

Line Drivers/Receivers

ADVANCE INFORMATION

TYPICAL CHARACTERISTICS

BUS HIGH-LEVEL OUTPUT VOLTAGE
vs
HIGH-LEVEL OUTPUT CURRENT

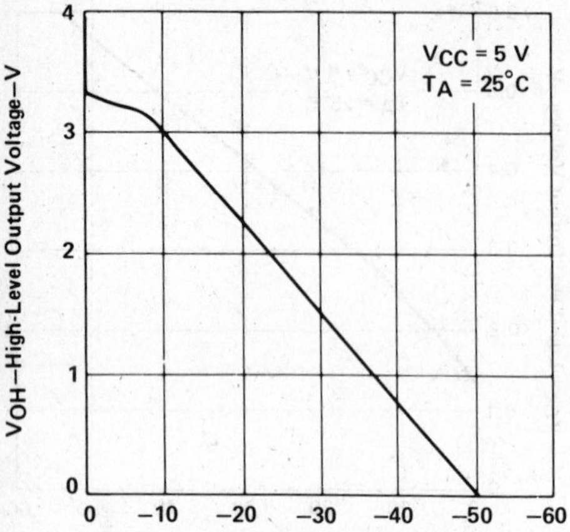
VCC = 5 V
TA = 25°C
VOH—High-Level Output Voltage—V
IOH—High-Level Output Current—mA
4
3
2
1
0
0
−10
−20
−30
−40
−50
−60

FIGURE 9

BUS LOW-LEVEL OUTPUT VOLTAGE
vs
LOW-LEVEL OUTPUT CURRENT

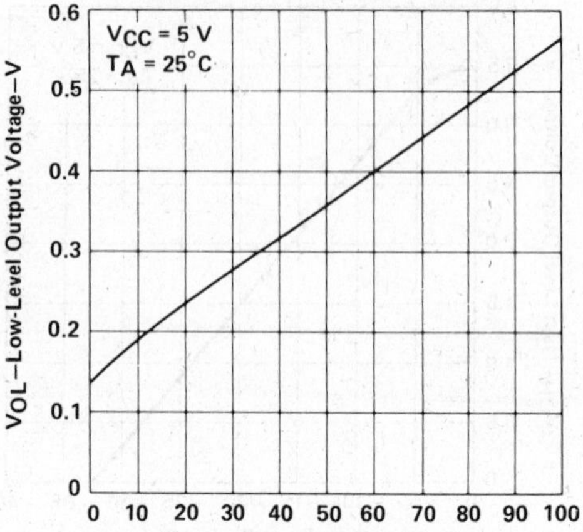
VCC = 5 V
TA = 25°C
VOL—Low-Level Output Voltage—V
IOL—Low-Level Output Current—mA
0.6
0.5
0.4
0.3
0.2
0.1
0
0
10
20
30
40
50
60
70
80
90
100

FIGURE 10

BUS OUTPUT VOLTAGE
vs
TERMINAL INPUT VOLTAGE

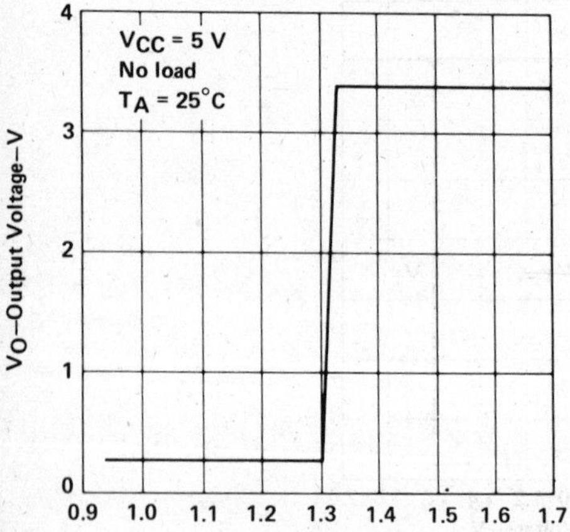
VCC = 5 V
No load
TA = 25°C
VO—Output Voltage—V
VI—Input Voltage—V
4
3
2
1
0
0.9
1.0
1.1
1.2
1.3
1.4
1.5
1.6
1.7

FIGURE 11

BUS CURRENT
vs
BUS VOLTAGE

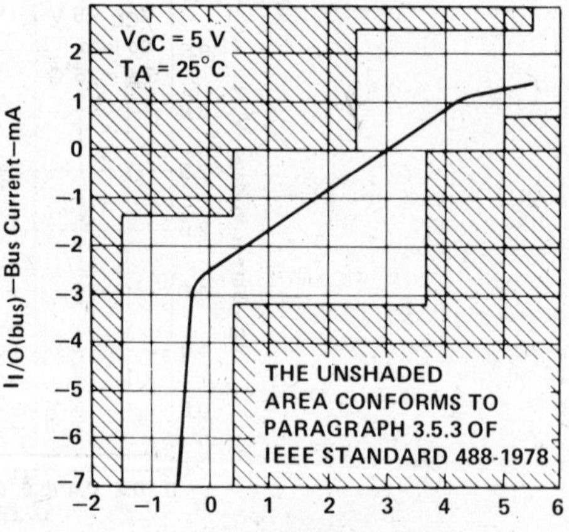
VCC = 5 V
TA = 25°C
THE UNSHADED AREA CONFORMS TO PARAGRAPH 3.5.3 OF IEEE STANDARD 488-1978
II/O(bus)—Bus Current—mA
VI/O(bus)—Bus Voltage—V
2
1
0
−1
−2
−3
−4
−5
−6
−7
−2
−1
0
1
2
3
4
5
6

FIGURE 12

4 Line Drivers/Receivers

ADVANCE INFORMATION

OCTAL GENERAL-PURPOSE INTERFACE BUS TRANSCEIVER

D3011, JUNE 1986

MEETS IEEE STANDARD 488-1978 (GPIB)

- 8-Channel Bidirectional Transceiver
- High-Speed Advanced Low-Power Schottky Circuitry
- Low Power Dissipation . . . 46 mW Max per Channel
- Fast Propagation Times . . . 20 ns Max
- High-Impedance P-N-P Inputs
- Receiver Hysteresis . . . 650 mV Typ
- No Loading of Bus When Device is Powered Down (V_{CC} = 0)
- Power-Up/Power-Down Protection (Glitch-Free)
- Driver and Receiver Can Be Disabled Simultaneously

DW, J, OR N PACKAGE

(TOP VIEW)

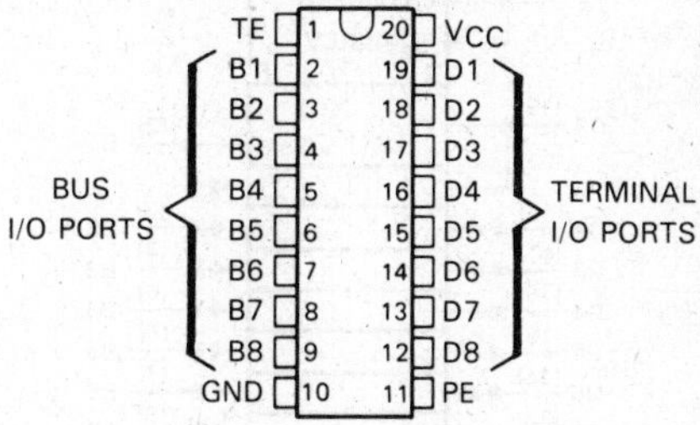
TE 1
B1 2
B2 3
B3 4
B4 5
B5 6
B6 7
B7 8
B8 9
GND 10
20 VCC
19 D1
18 D2
17 D3
16 D4
15 D5
14 D6
13 D7
12 D8
11 PE
BUS I/O PORTS
TERMINAL I/O PORTS

FUNCTION TABLES

EACH DRIVER

INPUTS			OUTPUT
D	TE	PE	B
H	H	H	H
L	H	X	L
H	X	L	Z†
X	L	X	Z†

EACH RECEIVER

INPUTS			OUTPUT
B	TE	PE	D
L	L	H	L
H	L	H	H
X	H	X	Z
X	X	L	Z

H = high level, L = low level, X = irrelevant, Z = high-impedance state.

†This is the high-impedance state of a normal 3-state output modified by the internal resistors to V_{CC} and ground.

description

The SN75ALS165 eight-channel general-purpose interface bus transceiver is a monolithic, high-speed, Advanced Low-Power Schottky device designed for two-way data communications over single-ended transmission lines. It is designed to meet the requirements of IEEE Standard 488-1978. The transceiver features driver outputs that can be operated in either the passive-pullup or three-state mode. If Talk Enable (TE) is high, these ports have the characteristics of passive-pullup outputs when Pullup Enable (PE) is low and of three-state outputs when PE is high. Taking TE low places these ports in the high-impedance state. Taking TE and PE low places both the drivers and receivers in the high-impedance state. The driver outputs are designed to handle loads up to 48 milliamperes of sink current.

An active turn-off feature has been incorporated into the bus-terminating resistors so that the device exhibits a high impedance to the bus when V_{CC} = 0. When combined with the SN75ALS161 or SN75ALS162 management bus transceiver, the pair provides the complete 16-wire interface for the IEEE 488 bus.

The SN75ALS165 is manufactured in a 20-pin package and is characterized for operation from 0°C to 70°C.

logic symbol†

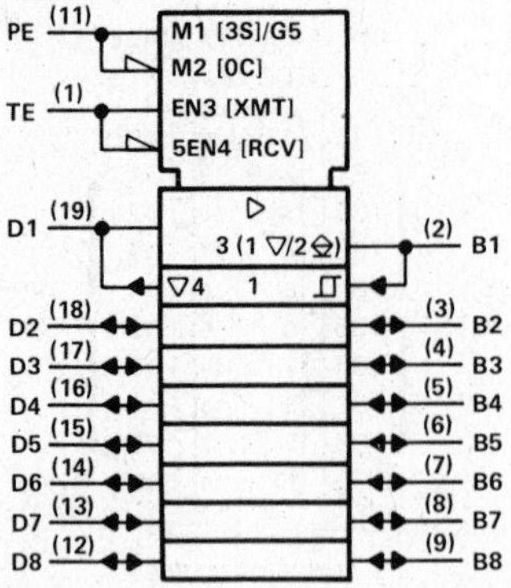
PE (11)
TE (1)
M1 [3S]/G5
M2 [0C]
EN3 [XMT]
5EN4 [RCV]
D1 (19)
3 (1 ▽/2 ⬙)
▽4 1
(2) B1
D2 (18)
D3 (17)
D4 (16)
D5 (15)
D6 (14)
D7 (13)
D8 (12)
(3) B2
(4) B3
(5) B4
(6) B5
(7) B6
(8) B7
(9) B8

† This symbol is in accordance with ANSI/IEEE Std 91-1984 and IEC Publication 617-12.
▽ Designates 3-state outputs.
⬙ Designates passive-pullup outputs.

logic diagram (positive logic)

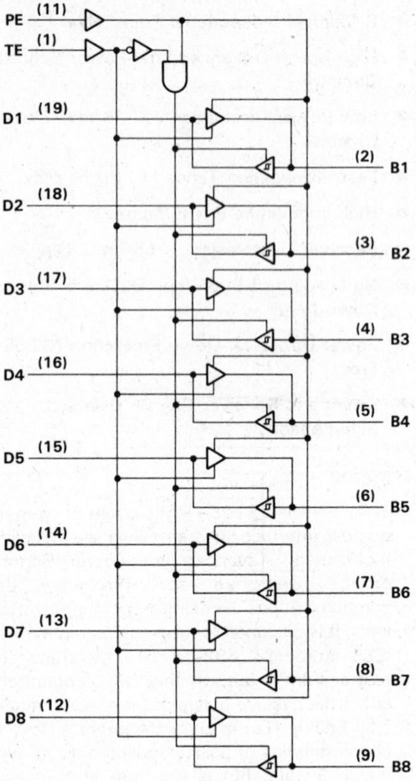
PE (11)
TE (1)
D1 (19)
(2) B1
D2 (18)
(3) B2
D3 (17)
(4) B3
D4 (16)
(5) B4
D5 (15)
(6) B5
D6 (14)
(7) B6
D7 (13)
(8) B7
D8 (12)
(9) B8

schematics of inputs and outputs

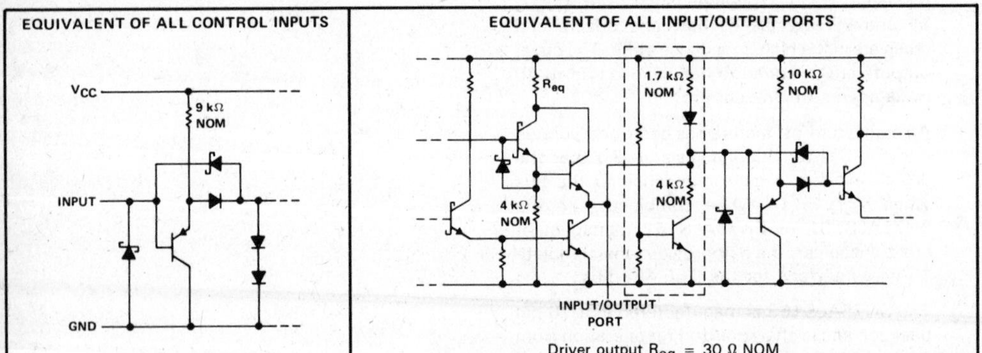
EQUIVALENT OF ALL CONTROL INPUTS
VCC
9 kΩ NOM
INPUT
GND
EQUIVALENT OF ALL INPUT/OUTPUT PORTS
Req
1.7 kΩ NOM
10 kΩ NOM
4 kΩ NOM
4 kΩ NOM
INPUT/OUTPUT PORT

Driver output R_{eq} = 30 Ω NOM
Receiver output R_{eq} = 110 Ω NOM
Circuit inside dashed lines is on the driver outputs only.

TEXAS INSTRUMENTS
POST OFFICE BOX 655012 • DALLAS, TEXAS 75265

absolute maximum ratings over operating free-air temperature range (unless otherwise noted)

Supply voltage, V_{CC} (see Note 1) . . . . 7 V
Input voltage . . . . 5.5 V
Low-level driver output current . . . . 100 mA
Continuous total dissipation at (or below) 25°C free-air temperature (see Note 2):
DW package . . . . 1125 mW
J package . . . . 1375 mW
N package . . . . 1150 mW
Operating free-air temperature range . . . . 0°C to 70°C
Storage temperature range . . . . −65°C to 150°C
Lead temperature 1,6 mm (1/16 inch) from the case for 60 seconds: J package . . . . 300°C
Lead temperature 1,6 mm (1/16 inch) from the case for 10 seconds: DW or N package . . . . 260°C

NOTES: 1. All voltage values are with respect to network ground terminal.
2. For operation above 25°C free-air temperature, derate the DW package to 720 mW at 70°C at the rate of 9.0 mW/°C, derate the J package to 880 mW at 70°C at the rate of 11.0 mW/°C, and derate the N package to 736 mW at 70°C at the rate of 9.2 mW/°C.

recommended operating conditions

		MIN	NOM	MAX	UNIT
Supply voltage, V_{CC}		4.75	5	5.25	V
High-level input voltage, V_{IH}		2			V
Low-level input voltage, V_{IL}				0.8	V
High-level output current, I_{OH}	Bus ports with pullups active			−5.2	mA
	Terminal ports			−800	μA
Low-level output current, I_{OL}	Bus ports			48	mA
	Terminal ports			16	
Operating free-air temperature, T_A		0		70	°C

electrical characteristics over recommended ranges of supply voltage and operating free-air temperature (unless otherwise noted)

PARAMETER			TEST CONDITIONS	MIN	TYP†	MAX	UNIT
V_{IK}	Input clamp voltage		$I_I = -18$ mA		−0.8	−1.5	V
V_{hys}	Hysteresis ($V_{T+} - V_{T-}$)	Bus		0.4	0.65		V
V_{OH}‡	High-level output voltage	Terminal	$I_{OH} = -800$ μA, TE at 0.8 V	2.7	3.5		V
		Bus	$I_{OH} = -5.2$ mA, PE and TE at 2 V	2.5	3.3		
V_{OL}	Low-level output voltage	Terminal	$I_{OL} = 16$ mA, TE at 0.8 V		0.3	0.5	V
		Bus	$I_{OL} = 48$ mA, TE at 2 V		0.35	0.5	
I_I	Input current at maximum input voltage	Terminal	$V_I = 5.5$ V		0.2	100	μA
I_{IH}	High-level input current	Terminal and control inputs	$V_I = 2.7$ V		0.1	20	μA
I_{IL}	Low-level input current		$V_I = 0.5$ V		−10	−100	μA
$V_{I/O(bus)}$	Voltage at bus port		Driver disabled, $I_{I(bus)} = 0$	2.5	3.0	3.7	V
			Driver disabled, $I_{I(bus)} = -12$ mA			−1.5	
$I_{I/O(bus)}$	Current into bus port	Power on	Driver disabled, $V_{I(bus)} = -1.5$ V to 0.4 V	−1.3			mA
			Driver disabled, $V_{I(bus)} = 0.4$ V to 2.5 V	0		−3.2	
			Driver disabled, $V_{I(bus)} = 2.5$ V to 3.7 V			+2.5 −3.2	
			Driver disabled, $V_{I(bus)} = 3.7$ V to 5 V	0		2.5	
			Driver disabled, $V_{I(bus)} = 5$ V to 5.5 V	0.7		2.5	
		Power off	$V_{CC} = 0$, $V_{I(bus)} = 0$ V to 2.5 V			−40	μA
I_{OS}	Short-circuit output current	Terminal		−15	−35	−75	mA
		Bus		−25	−50	−125	
I_{CC}	Supply current		No load, Terminal outputs low and enabled		42	56	mA
			No load, Bus outputs low and enabled		52	70	
$C_{i/o(bus)}$	Bus-port capacitance		$V_{CC} = 5$ V to 0 V, $V_{I/O} = 0$ to 2 V, f = 1 MHz		30		pF

† All typical values are at $V_{CC} = 5$ V, $T_A = 25$ °C.
‡ V_{OH} applies for three-state outputs only.

switching characteristics over recommended range of operating free-air temperature (unless otherwise noted), V_{CC} = 5 V

PARAMETER		FROM	TO	TEST CONDITIONS	MIN	TYP†	MAX	UNIT
t_{PLH}	Propagation delay time, low-to-high-level output	Terminal	Bus	C_L = 30 pF, See Figure 1		10	20	ns
t_{PHL}	Propagation delay time, high-to-low-level output					12	20	
t_{PLH}	Propagation delay time, low-to-high-level output	Bus	Terminal	C_L = 30 pF, See Figure 2		5	10	ns
t_{PHL}	Propagation delay time, high-to-low-level output					7	14	
t_{PZH}	Output enable time to high level	TE	Bus	C_L = 15 pF, See Figure 3		11	20	ns
t_{PHZ}	Output disable time from high level					3	10	
t_{PZL}	Output enable time to low level					18	35	
t_{PLZ}	Output disable time from low level					5	20	
t_{PZH}	Output enable time to high level	TE	Terminal	C_L = 15 pF, See Figure 4		5	20	ns
t_{PHZ}	Output disable time from high level					8	20	
t_{PZL}	Output enable time to low level					9	20	
t_{PLZ}	Output disable time from low level					8	20	
t_{en}	Output pull-up enable time	PE	Terminal	C_L = 15 pF, See Figure 5		3	10	ns
t_{dis}	Output pull-up disable time					4	12	

†All typical values are at T_A = 25°C.

PARAMETER MEASUREMENT INFORMATION

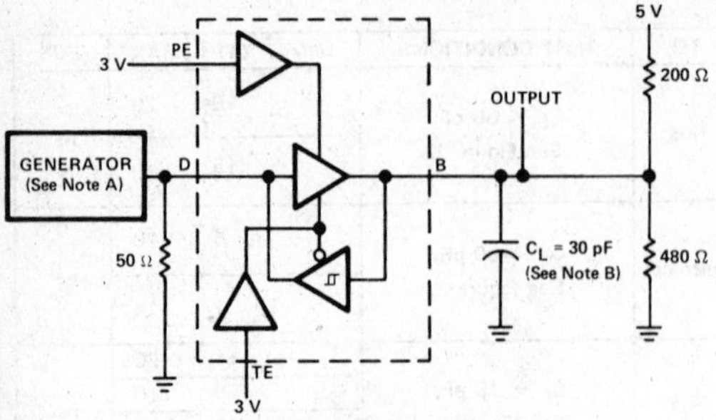
PE
3 V
GENERATOR
(See Note A)
D
50 Ω
TE
3 V
B
OUTPUT
5 V
200 Ω
C_L = 30 pF
(See Note B)
480 Ω

TEST CIRCUIT

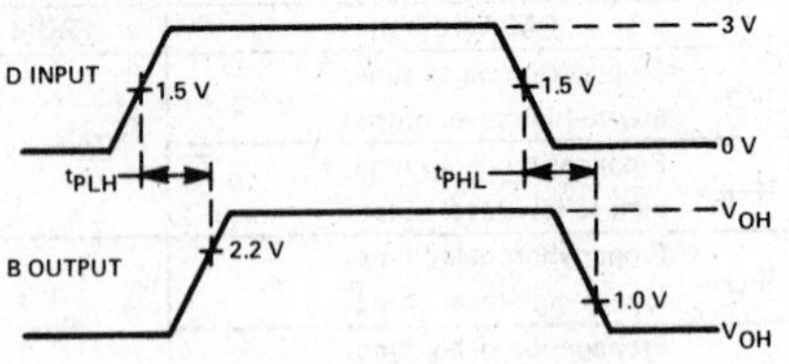
D INPUT
1.5 V
1.5 V
3 V
0 V
t_{PLH}
t_{PHL}
B OUTPUT
2.2 V
1.0 V
V_{OH}
V_{OH}

VOLTAGE WAVEFORMS

FIGURE 1. TERMINAL-TO-BUS PROPAGATION DELAY TIMES

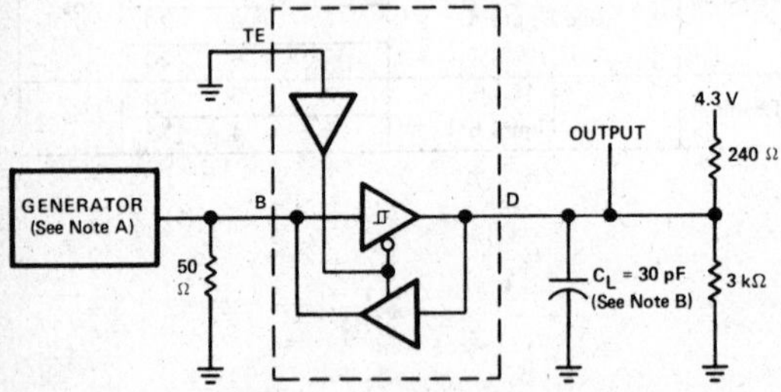
TE
GENERATOR
(See Note A)
B
50 Ω
D
OUTPUT
4.3 V
240 Ω
C_L = 30 pF
(See Note B)
3 kΩ

TEST CIRCUIT

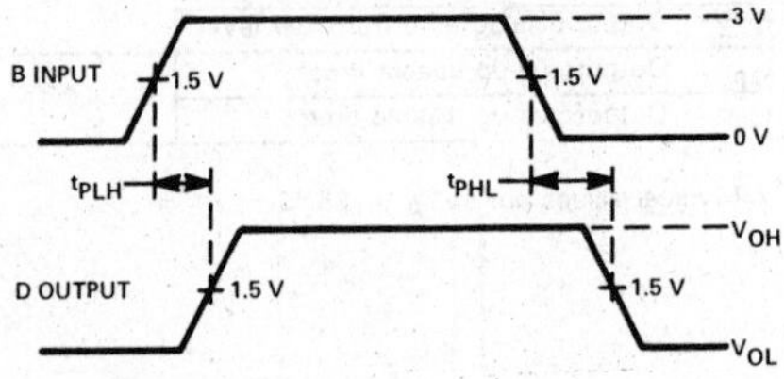
B INPUT
1.5 V
1.5 V
3 V
0 V
t_{PLH}
t_{PHL}
D OUTPUT
1.5 V
1.5 V
V_{OH}
V_{OL}

VOLTAGE WAVEFORMS

FIGURE 2. BUS-TO-TERMINAL PROPAGATION DELAY TIMES

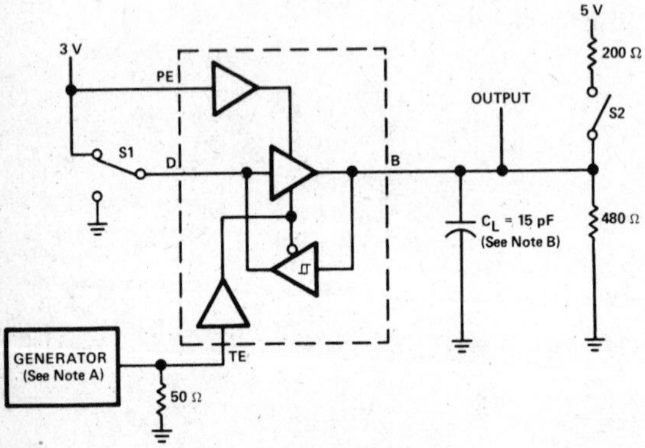
3 V
PE
S1
D
GENERATOR
(See Note A)
TE
50 Ω
B
OUTPUT
5 V
200 Ω
S2
C_L = 15 pF
(See Note B)
480 Ω

TEST CIRCUIT

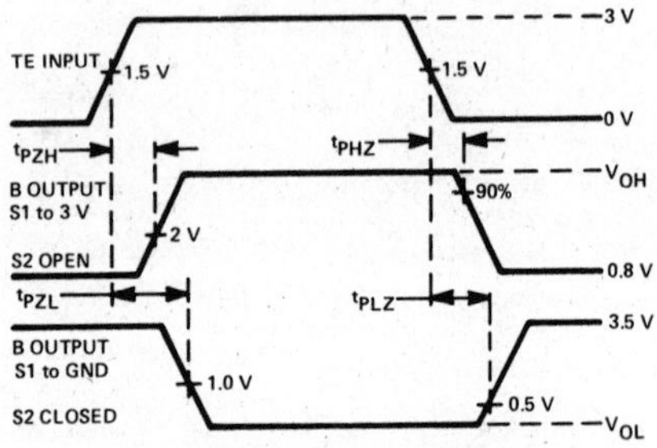
TE INPUT
1.5 V
1.5 V
3 V
0 V
t_{PZH}
t_{PHZ}
B OUTPUT
S1 to 3 V
S2 OPEN
2 V
90%
V_{OH}
0.8 V
t_{PZL}
t_{PLZ}
3.5 V
B OUTPUT
S1 to GND
S2 CLOSED
1.0 V
0.5 V
V_{OL}

VOLTAGE WAVEFORMS

FIGURE 3. TE-TO-BUS ENABLE AND DISABLE TIMES

NOTES: A. The input pulse is supplied by a generator having the following characteristics: PRR ≤ 1 MHz, 50% duty cycle, $t_r \leq 6$ ns, $t_f \leq 6$ ns, $Z_{out} = 50\ \Omega$.

B. C_L includes probe and jig capacitance.

TEXAS INSTRUMENTS
POST OFFICE BOX 655012 • DALLAS, TEXAS 75265

PARAMETER MEASUREMENT INFORMATION

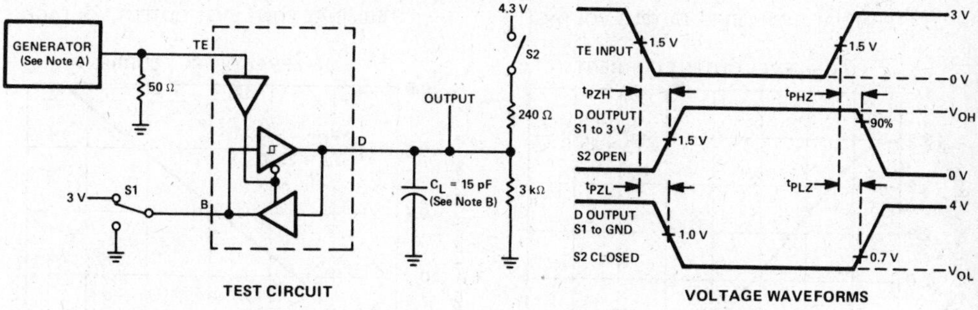
GENERATOR (See Note A)
TE
50 Ω
3 V
S1
B
D
OUTPUT
C_L = 15 pF (See Note B)
4.3 V
S2
240 Ω
3 kΩ
TE INPUT
1.5 V
1.5 V
3 V
0 V
t_{PZH}
t_{PHZ}
D OUTPUT S1 to 3 V
S2 OPEN
1.5 V
90%
V_{OH}
0 V
t_{PZL}
t_{PLZ}
D OUTPUT S1 to GND
S2 CLOSED
1.0 V
0.7 V
4 V
V_{OL}
TEST CIRCUIT
VOLTAGE WAVEFORMS

FIGURE 4. TE-TO-TERMINAL ENABLE AND DISABLE TIMES

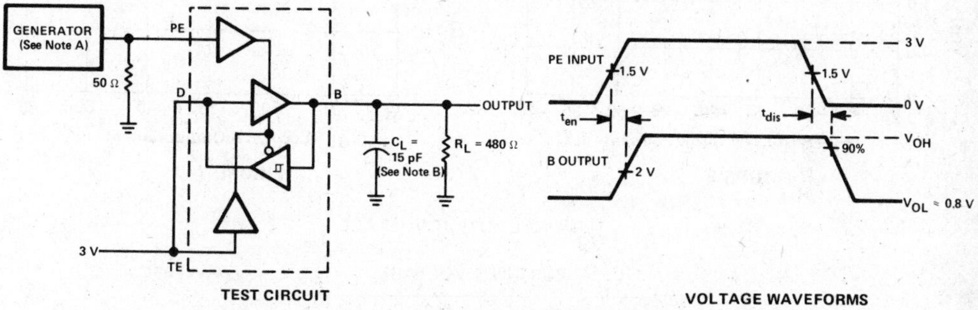
GENERATOR (See Note A)
PE
50 Ω
D
B
OUTPUT
C_L = 15 pF (See Note B)
R_L = 480 Ω
3 V
TE
PE INPUT
1.5 V
1.5 V
3 V
0 V
t_{en}
t_{dis}
B OUTPUT
2 V
90%
V_{OH}
V_{OL} ≈ 0.8 V
TEST CIRCUIT
VOLTAGE WAVEFORMS

FIGURE 5. PE-TO-BUS PULLUP ENABLE AND DISABLE TIMES

NOTES: A. The input pulse is supplied by a generator having the following characteristics: PRR ≤ 1 MHz, 50% duty cycle, $t_r \leq 6$ ns, $t_f \leq 6$ ns, Z_{out} = 50 Ω.

B. C_L includes probe and jig capacitance.

TYPICAL CHARACTERISTICS

TERMINAL HIGH-LEVEL OUTPUT VOLTAGE
vs
HIGH-LEVEL OUTPUT CURRENT

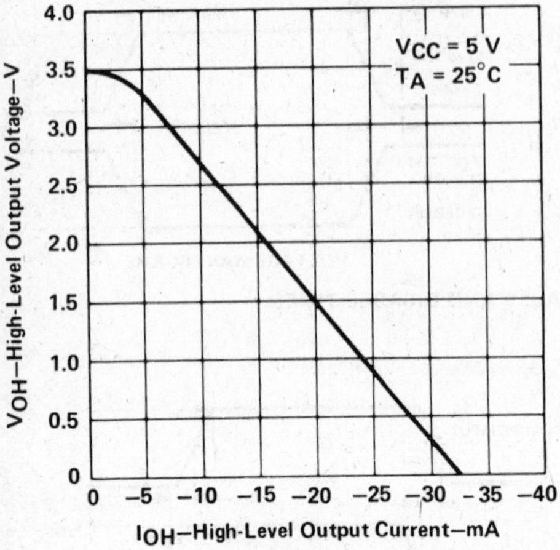
VCC = 5 V
TA = 25°C
4.0
3.5
3.0
2.5
2.0
1.5
1.0
0.5
0
0 −5 −10 −15 −20 −25 −30 −35 −40
VOH—High-Level Output Voltage—V
IOH—High-Level Output Current—mA

FIGURE 6

TERMINAL LOW-LEVEL OUTPUT VOLTAGE
vs
LOW-LEVEL OUTPUT CURRENT

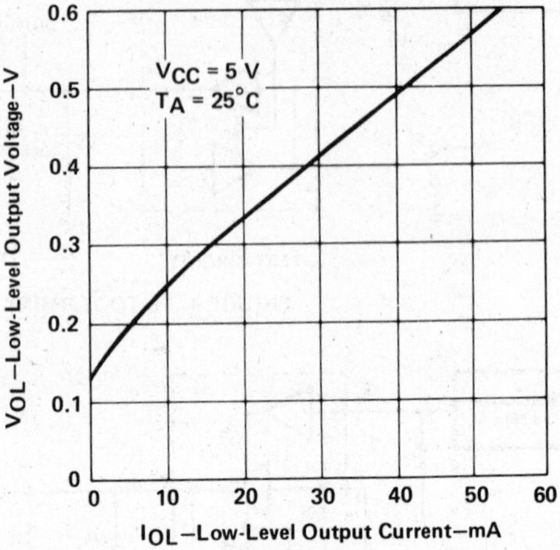
VCC = 5 V
TA = 25°C
0.6
0.5
0.4
0.3
0.2
0.1
0
0 10 20 30 40 50 60
VOL—Low-Level Output Voltage—V
IOL—Low-Level Output Current—mA

FIGURE 7

TERMINAL OUTPUT VOLTAGE
vs
BUS INPUT VOLTAGE

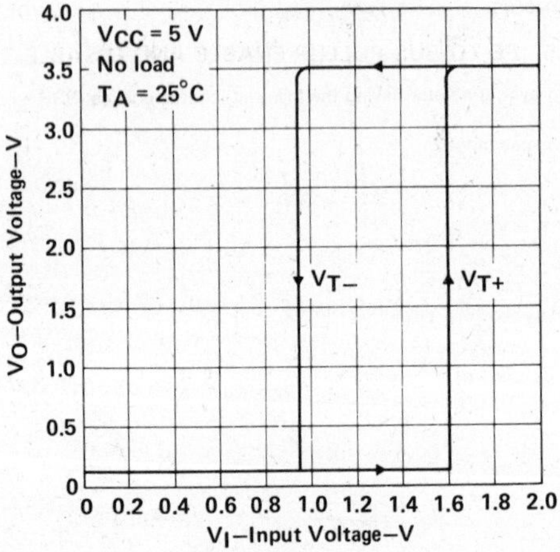
VCC = 5 V
No load
TA = 25°C
VT−
VT+
4.0
3.5
3.0
2.5
2.0
1.5
1.0
0.5
0
0 0.2 0.4 0.6 0.8 1.0 1.2 1.4 1.6 1.8 2.0
VO—Output Voltage—V
VI—Input Voltage—V

FIGURE 8

TYPICAL CHARACTERISTICS

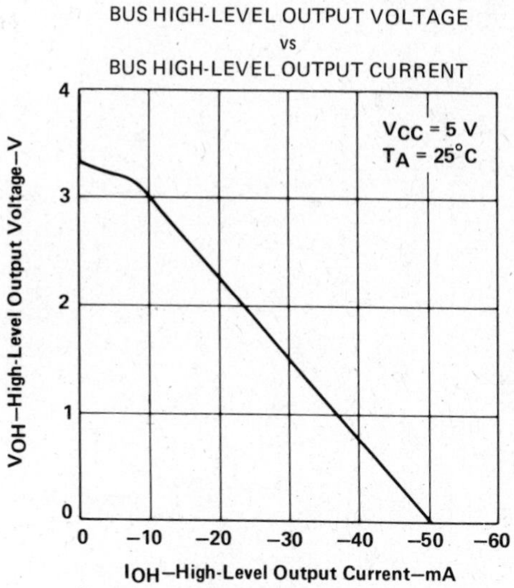
BUS HIGH-LEVEL OUTPUT VOLTAGE
vs
BUS HIGH-LEVEL OUTPUT CURRENT
VCC = 5 V
TA = 25°C
VOH—High-Level Output Voltage—V
4
3
2
1
0
0
−10
−20
−30
−40
−50
−60
IOH—High-Level Output Current—mA

FIGURE 9

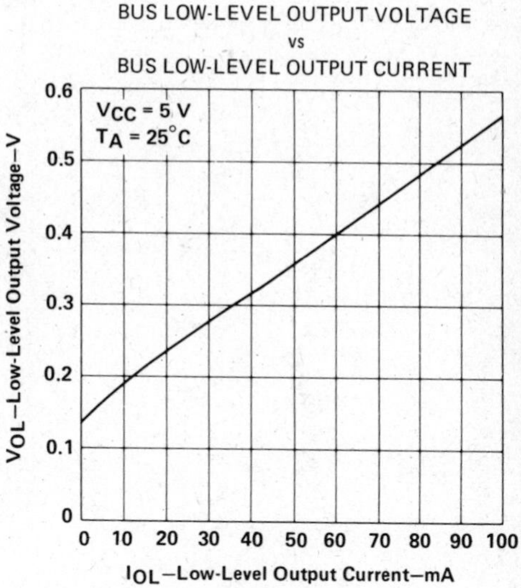
BUS LOW-LEVEL OUTPUT VOLTAGE
vs
BUS LOW-LEVEL OUTPUT CURRENT
VCC = 5 V
TA = 25°C
VOL—Low-Level Output Voltage—V
0.6
0.5
0.4
0.3
0.2
0.1
0
0 10 20 30 40 50 60 70 80 90 100
IOL—Low-Level Output Current—mA

FIGURE 10

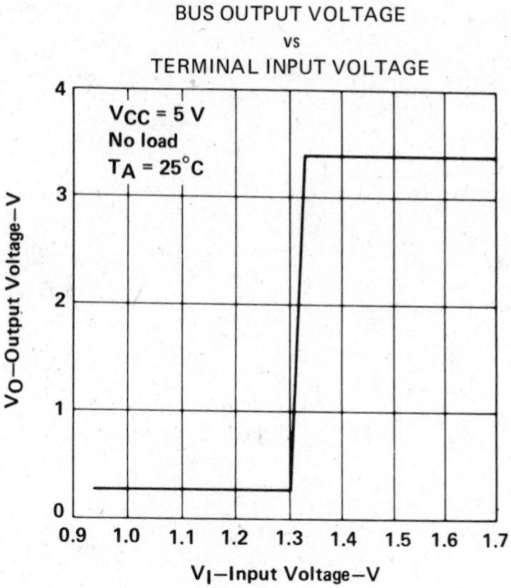
BUS OUTPUT VOLTAGE
vs
TERMINAL INPUT VOLTAGE
VCC = 5 V
No load
TA = 25°C
VO—Output Voltage—V
4
3
2
1
0
0.9 1.0 1.1 1.2 1.3 1.4 1.5 1.6 1.7
VI—Input Voltage—V

FIGURE 11

D2904, JULY 1985

- Meets EIA Standard RS-422-A
- High-Speed, Low-Power ALS Design
- 3-State TTL Compatible
- Single 5-V Supply Operation
- High Output Impedance in Power-Off Condition
- Complementary Output Enable Inputs
- Improved Replacement for the AM26LS31

SN75ALS192 . . . D, J, N DUAL-IN-LINE PACKAGE
(TOP VIEW)

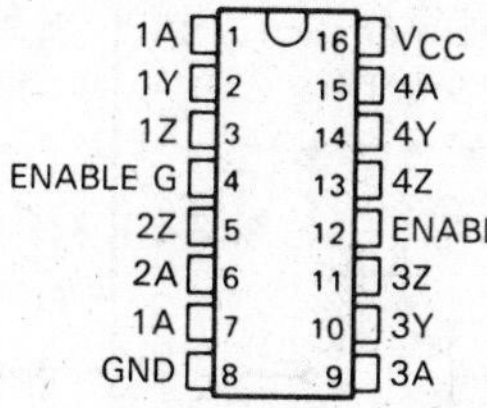
1A 1
1Y 2
1Z 3
ENABLE G 4
2Z 5
2A 6
1A 7
GND 8
16 VCC
15 4A
14 4Y
13 4Z
12 ENABLE $\overline{G}$
11 3Z
10 3Y
9 3A

description

This quadruple complementary-output line driver is designed for data transmission over twisted-pair or parallel-wire transmission lines. It meets the requirements of EIA Standard RS-422-A and is compatible with 3-state TTL circuits. Advanced Low-Power Schottky technology provides high speed without the usual power penalties. Standby supply current is typically only 26 milliamperes, while typical propagation delay time is less than 10 nanoseconds.

FUNCTION TABLE (EACH DRIVER)

INPUT	ENABLES		OUTPUTS	
A	G	$\overline{G}$	Y	Z
H	H	X	H	L
L	H	X	L	H
H	X	L	H	L
L	X	L	L	H
X	L	H	Z	Z

H = high level, L = low level,
Z = high impedance (off),
X = irrelevant

High-impedance inputs maintain input currents low, less than 1 microampere for a high level and less than 100 microamperes for a low level. Complementary control inputs, G and $\overline{G}$, allow these devices to be enabled at either a high input level or low input level. The SN75ALS192 is capable of data rates in excess of 20 megabits per second and is designed to operate with the SN75ALS193 quadruple line receiver.

The SN75ALS192 is characterized for operation from 0 °C to 70 °C.

4

Line Drivers/Receivers

logic symbol†

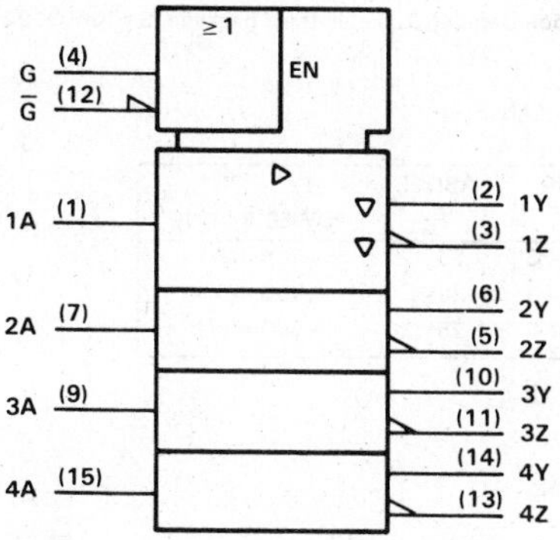
≥1
EN
G (4)
$\overline{G}$ (12)
1A (1)
2A (7)
3A (9)
4A (15)
(2) 1Y
(3) 1Z
(6) 2Y
(5) 2Z
(10) 3Y
(11) 3Z
(14) 4Y
(13) 4Z

†This symbol is in accordance with ANSI/IEEE Std 91-1984 and IEC Publication 617-12.

logic diagram (positive logic)

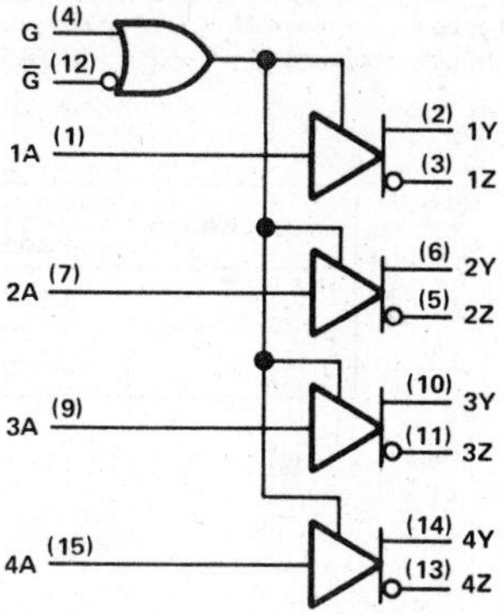
G (4)
$\overline{G}$ (12)
1A (1)
2A (7)
3A (9)
4A (15)
(2) 1Y
(3) 1Z
(6) 2Y
(5) 2Z
(10) 3Y
(11) 3Z
(14) 4Y
(13) 4Z

schematics of inputs and outputs

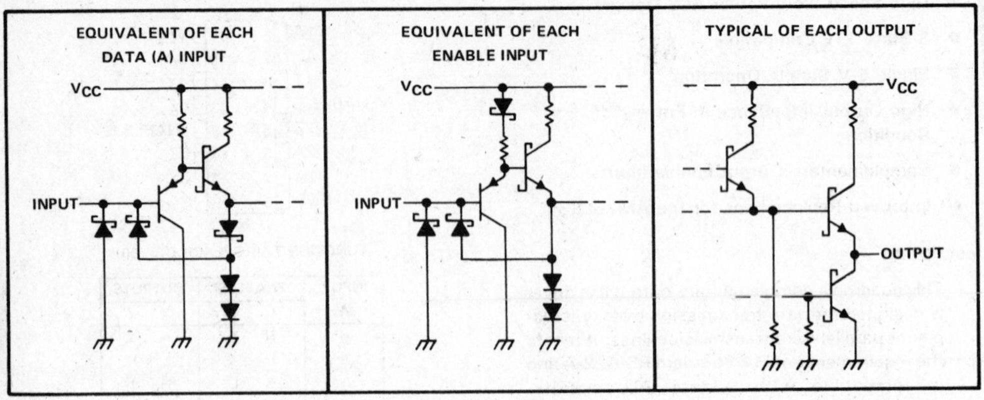
EQUIVALENT OF EACH
DATA (A) INPUT
V_{CC}
INPUT
EQUIVALENT OF EACH
ENABLE INPUT
V_{CC}
INPUT
TYPICAL OF EACH OUTPUT
V_{CC}
OUTPUT

absolute maximum ratings over operating free-air temperature range (unless otherwise noted)

Supply voltage, V_{CC} (see Note 1) .... 7 V
Input voltage, V_I .... 7 V
Output off-state voltage .... 5.5 V
Continuous total dissipation at (or below) 25 °C free-air temperature (see Note 2):
D package .... 950 mW
J package .... 1000 mW
N package .... 875 mW
Operating free-air temperature range .... 0 °C to 70 °C
Lead temperature 1,6 mm (1/16 inch) from case for 60 seconds: J package .... 300 °C
Lead temperature 1,6 mm (1/16 inch) from case for 10 seconds: D or N package .... 260 °C

NOTES: 1. All voltage values except differential output voltage V_{OD} are with respect to network ground terminal.
2. For operation above 25 °C free-air temperature, refer to the Dissipation Derating Table. In the J package, SN75ALS192 chips are glass mounted.

DISSIPATION DERATING TABLE

PACKAGE	T_A = 25 °C POWER RATING	DERATING FACTOR	ABOVE T_A	T_A = 70° POWER RATING
D	950 mW	7.6 mW/°C	25 °C	608 mW
J (Glass mount)	1000 mW	8.2 mW/°C	28 °C	656 mW
N	875 mW	7.0 mW/°C	25 °C	560 mW

recommended operating conditions

	MIN	NOM	MAX	UNIT
Supply voltage, V_{CC}	4.75	5	5.25	V
High-level input voltage, V_{IH}	2			V
Low-level input voltage, V_{IL}			0.8	V
High-level output current, I_{OH}			−20	mA
Low-level output current, I_{OL}			20	mA
Operating free-air temperature, T_A	0		70	°C

electrical characteristics over recommended operating free-air temperature range (unless otherwise noted)

PARAMETER		TEST CONDITIONS		MIN	TYP†	MAX	UNIT
V_{IK}	Input clamp voltage	V_{CC} = 4.75 V, I_I = −18 mA				−1.5	V
V_{OH}	High-level output voltage	V_{CC} = 4.75 V, I_{OH} = −20 mA		2.5			V
V_{OL}	Low-level output voltage	V_{CC} = 4.75 V, I_{OL} = 20 mA				0.5	V
I_{OZ}	Off-state (high-impedance state) output current	V_{CC} = 5.25 V	V_O = 0.5 V			−20	μA
			V_O = 2.5 V			20	
I_I	Input current at maximum input voltage	V_{CC} = 5.25 V, V_I = 7 V				0.1	mA
I_{IH}	High-level input current	V_{CC} = 5.25 V, V_I = 2.7 V				20	μA
I_{IL}	Low-level input current	V_{CC} = 5.25 V, V_I = 0.4 V				0.2	mA
I_{OS}	Short-circuit output current‡	V_{CC} = 5.25 V		−30		−150	mA
I_{CC}	Supply current (all drivers)	V_{CC} = 5.25 V, All outputs disabled			26	45	mA

†All typical values are at V_{CC} = 5 V, T_A = 25 °C.
‡Not more than one output should be shorted at a time, and duration of the short-circuit should not exceed one second.

switching characteristics, V_{CC} = 5 V, T_A = 25 °C (see Figure 1)

PARAMETER		TEST CONDTIONS	MIN	TYP	MAX	UNIT
t_{PLH}	Propagation delay time, low-to-high-level output	C_L = 30 pF, S1 and S2 open		6	13	ns
t_{PHL}	Propagation delay time, high-to-low-level output			9	14	ns
	Output-to-output skew			3	6	ns
t_{PZH}	Output enable time to high level	R_L = 75 Ω		11	15	ns
t_{PZL}	Output enable time to low level	R_L = 180 Ω		16	20	ns
t_{PHZ}	Output disable time from high level	C_L = 10 pF, S1 and S2 closed		8	15	ns
t_{PLZ}	Output disable time from low level			18	20	ns

PARAMETER MEASUREMENT INFORMATION

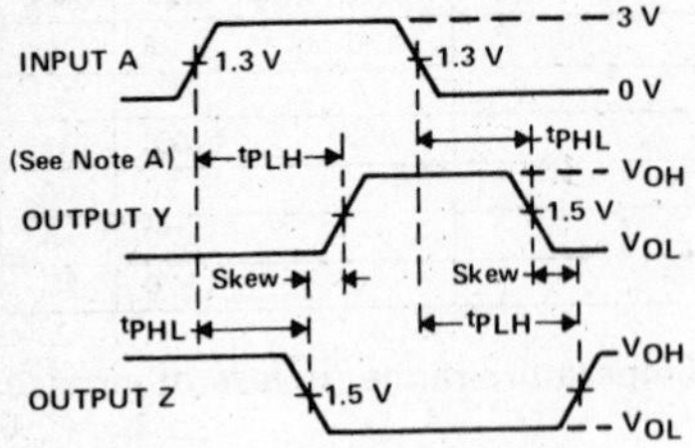
INPUT A
1.3 V
1.3 V
3 V
0 V
(See Note A)
tPLH
tPHL
VOH
OUTPUT Y
1.5 V
VOL
Skew
Skew
tPHL
tPLH
VOH
OUTPUT Z
1.5 V
VOL

PROPAGATION DELAY TIMES AND SKEW

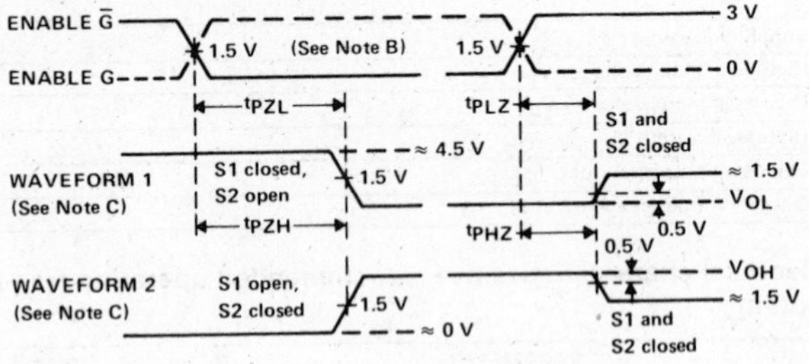
ENABLE G̅
ENABLE G
1.5 V
(See Note B)
1.5 V
3 V
0 V
tPZL
tPLZ
S1 and
S2 closed
≈ 4.5 V
WAVEFORM 1
(See Note C)
S1 closed,
S2 open
1.5 V
≈ 1.5 V
VOL
0.5 V
tPZH
tPHZ
0.5 V
VOH
WAVEFORM 2
(See Note C)
S1 open,
S2 closed
1.5 V
≈ 1.5 V
≈ 0 V
S1 and
S2 closed

ENABLE AND DISABLE TIMES

VOLTAGE WAVEFORMS

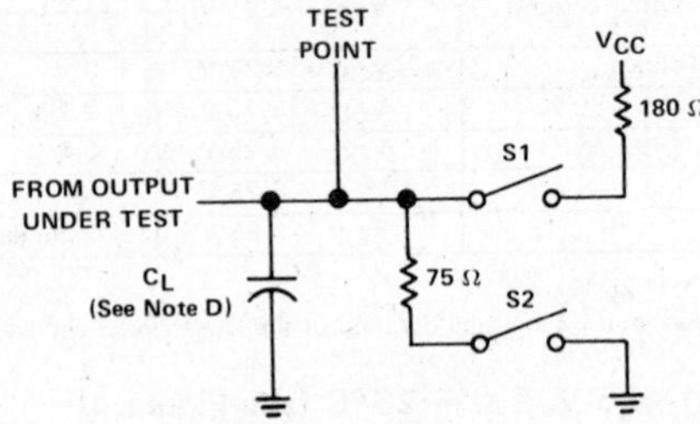
TEST
POINT
VCC
180 Ω
S1
FROM OUTPUT
UNDER TEST
CL
(See Note D)
75 Ω
S2

TEST CIRCUIT

NOTES: A. When measuring propagation delay times and skew, switches S1 and S2 are open.
B. Each enable is tested separately.
C. Waveform 1 is for an output with internal conditions such that the output is low except when disabled by the enable inputs. Waveform 2 is for an output with internal conditions such that the output is high except when disabled by the enable inputs.
D. C_L includes probe and jig capacitance.
E. All input pulses are supplied by generators having the following characteristics: PRR $\leq$ 1 MHz, $Z_{out} \approx 50\ \Omega$, $t_r \leq 15$ ns, and $t_f \leq 6$ ns.

FIGURE 1. SWITCHING TIMES

TYPICAL CHARACTERISTICS

Y OUTPUT VOLTAGE
vs
DATA INPUT VOLTAGE

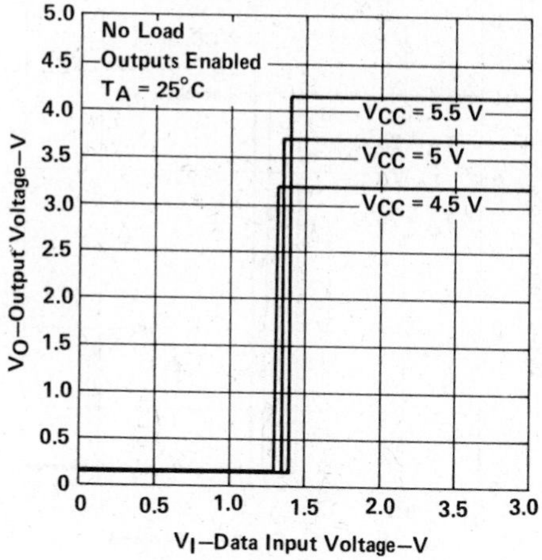
No Load
Outputs Enabled
T_A = 25°C
V_{CC} = 5.5 V
V_{CC} = 5 V
V_{CC} = 4.5 V
V_O—Output Voltage—V
5.0
4.5
4.0
3.5
3.0
2.5
2.0
1.5
1.0
0.5
0
0
0.5
1.0
1.5
2.0
3.5
3.0
V_I—Data Input Voltage—V

FIGURE 2

Y OUTPUT VOLTAGE
vs
DATA INPUT VOLTAGE

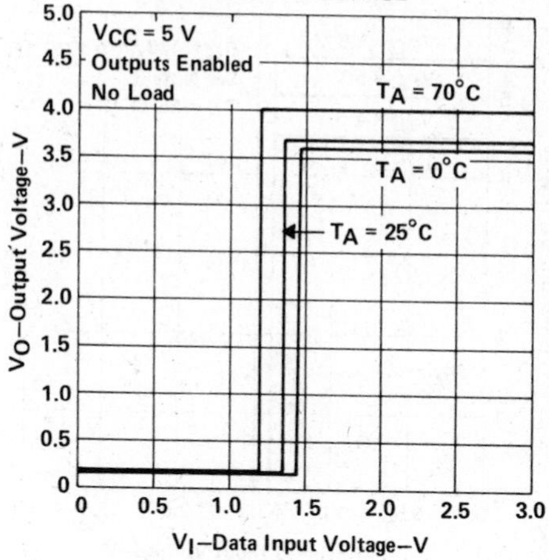
V_{CC} = 5 V
Outputs Enabled
No Load
T_A = 70°C
T_A = 0°C
T_A = 25°C
V_O—Output Voltage—V
5.0
4.5
4.0
3.5
3.0
2.5
2.0
1.5
1.0
0.5
0
0
0.5
1.0
1.5
2.0
2.5
3.0
V_I—Data Input Voltage—V

FIGURE 3

Y OUTPUT VOLTAGE
vs
ENABLE G INPUT VOLTAGE

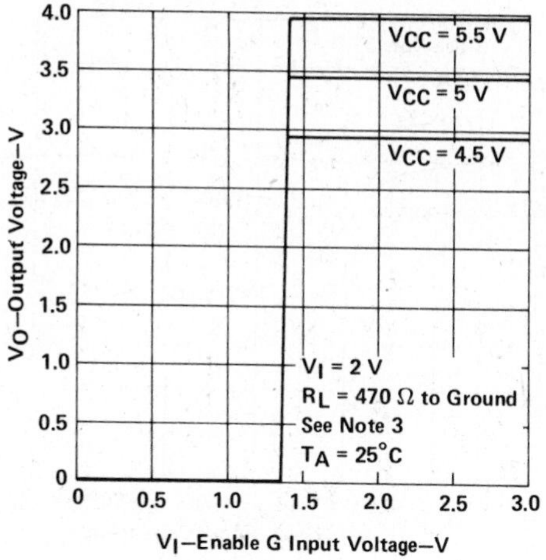
V_{CC} = 5.5 V
V_{CC} = 5 V
V_{CC} = 4.5 V
V_I = 2 V
R_L = 470 Ω to Ground
See Note 3
T_A = 25°C
V_O—Output Voltage—V
4.0
3.5
3.0
2.5
2.0
1.5
1.0
0.5
0
0
0.5
1.0
1.5
2.0
2.5
3.0
V_I—Enable G Input Voltage—V

FIGURE 4

Y OUTPUT VOLTAGE
vs
ENABLE G INPUT VOLTAGE

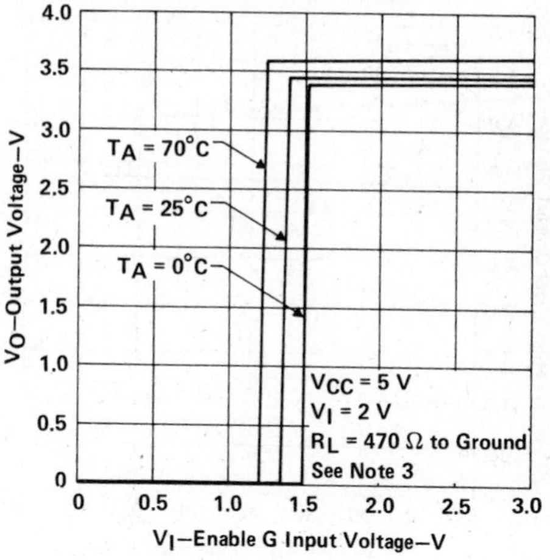
T_A = 70°C
T_A = 25°C
T_A = 0°C
V_{CC} = 5 V
V_I = 2 V
R_L = 470 Ω to Ground
See Note 3
V_O—Output Voltage—V
4.0
3.5
3.0
2.5
2.0
1.5
1.0
0.5
0
0
0.5
1.0
1.5
2.0
2.5
3.0
V_I—Enable G Input Voltage—V

FIGURE 5

NOTE 3: The A input is connected to V_{CC} during the testing of the Y outputs and to ground during the testing of the Z outputs.

TYPICAL CHARACTERISTICS

Z OUTPUT VOLTAGE
vs
ENABLE G INPUT VOLTAGE

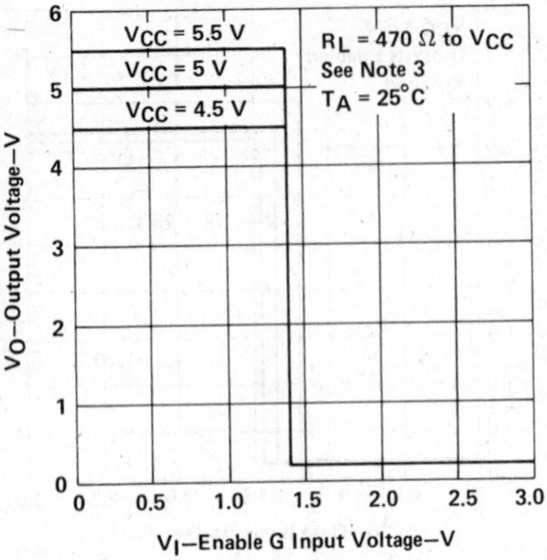
V_{CC} = 5.5 V
V_{CC} = 5 V
V_{CC} = 4.5 V
R_L = 470 Ω to V_{CC}
See Note 3
T_A = 25°C
V_O—Output Voltage—V
V_I—Enable G Input Voltage—V
0
1
2
3
4
5
6
0.5
1.0
1.5
2.0
2.5
3.0

FIGURE 6

Z OUTPUT VOLTAGE
vs
ENABLE G INPUT VOLTAGE

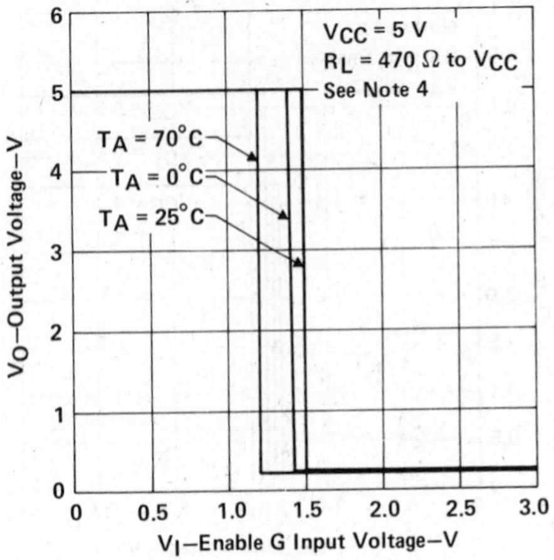
V_{CC} = 5 V
R_L = 470 Ω to V_{CC}
See Note 4
T_A = 70°C
T_A = 0°C
T_A = 25°C
V_O—Output Voltage—V
V_I—Enable G Input Voltage—V
0
1
2
3
4
5
6
0.5
1.0
1.5
2.0
2.5
3.0

FIGURE 7

HIGH-LEVEL OUTPUT VOLTAGE
vs
FREE-AIR TEMPERATURE

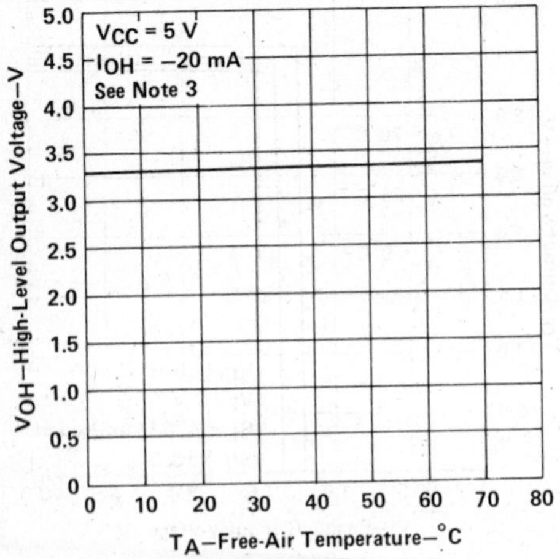
V_{CC} = 5 V
I_{OH} = −20 mA
See Note 3
V_{OH}—High-Level Output Voltage—V
T_A—Free-Air Temperature—°C
0
0.5
1.0
1.5
2.0
2.5
3.0
3.5
4.0
4.5
5.0
10
20
30
40
50
60
70
80

FIGURE 8

HIGH-LEVEL OUTPUT VOLTAGE
vs
OUTPUT CURRENT

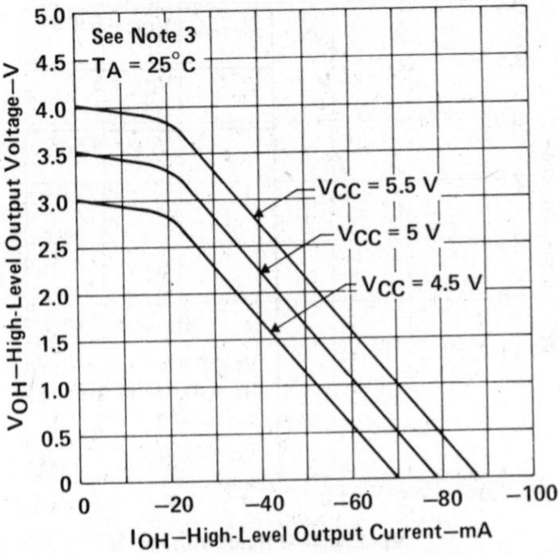
See Note 3
T_A = 25°C
V_{CC} = 5.5 V
V_{CC} = 5 V
V_{CC} = 4.5 V
V_{OH}—High-Level Output Voltage—V
I_{OH}—High-Level Output Current—mA
0
0.5
1.0
1.5
2.0
2.5
3.0
3.5
4.0
4.5
5.0
−20
−40
−60
−80
−100

FIGURE 9

NOTES: 3. The A input is connected to V_{CC} during the testing of the Y outputs and to ground during the testing of the Z outputs.
4. The A input is connected to ground during the testing of the Y outputs and to V_{CC} during the testing of the Z outputs.

TEXAS INSTRUMENTS
POST OFFICE BOX 655012 • DALLAS, TEXAS 75265

TYPICAL CHARACTERISTICS

LOW-LEVEL OUTPUT VOLTAGE
vs
FREE-AIR TEMPERATURE

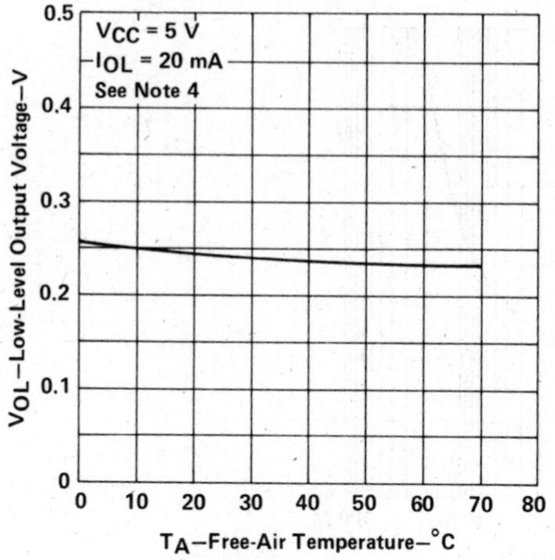
V_{CC} = 5 V
I_{OL} = 20 mA
See Note 4
V_{OL}—Low-Level Output Voltage—V
T_A—Free-Air Temperature—°C

FIGURE 10

LOW-LEVEL OUTPUT VOLTAGE
vs
OUTPUT CURRENT

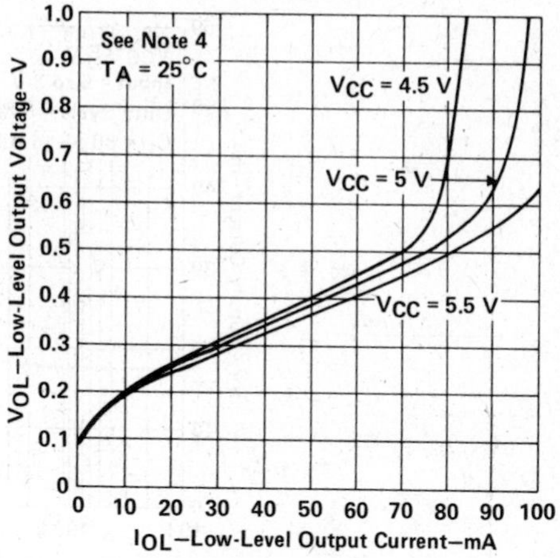
See Note 4
T_A = 25°C
V_{CC} = 4.5 V
V_{CC} = 5 V
V_{CC} = 5.5 V
V_{OL}—Low-Level Output Voltage—V
I_{OL}—Low-Level Output Current—mA

FIGURE 11

SUPPLY CURRENT
vs
SUPPLY VOLTAGE

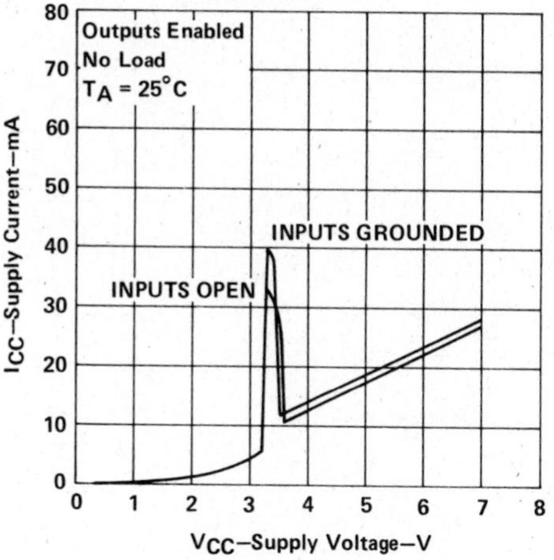
Outputs Enabled
No Load
T_A = 25°C
INPUTS GROUNDED
INPUTS OPEN
I_{CC}—Supply Current—mA
V_{CC}—Supply Voltage—V

FIGURE 12

SUPPLY CURRENT
vs
SUPPLY VOLTAGE

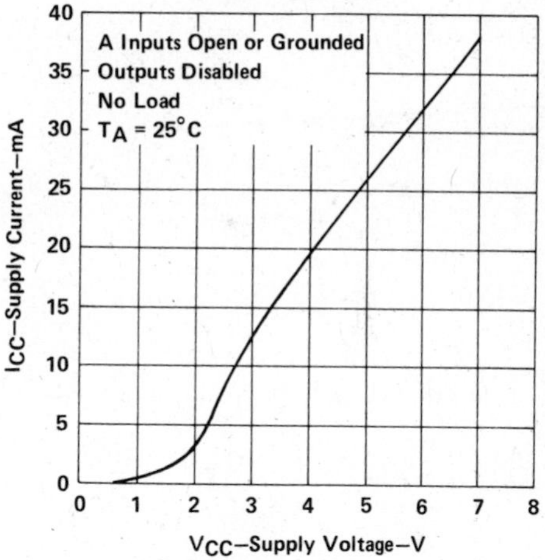
A Inputs Open or Grounded
Outputs Disabled
No Load
T_A = 25°C
I_{CC}—Supply Current—mA
V_{CC}—Supply Voltage—V

FIGURE 13

NOTES: 3. The A input is connected to V_{CC} during the testing of the Y outputs and to ground during the testing of the Z outputs.
4. The A input is connected to ground during the testing of the Y outputs and to V_{CC} during the testing of the Z outputs.

TYPICAL CHARACTERISTICS

SUPPLY CURRENT
vs
FREQUENCY

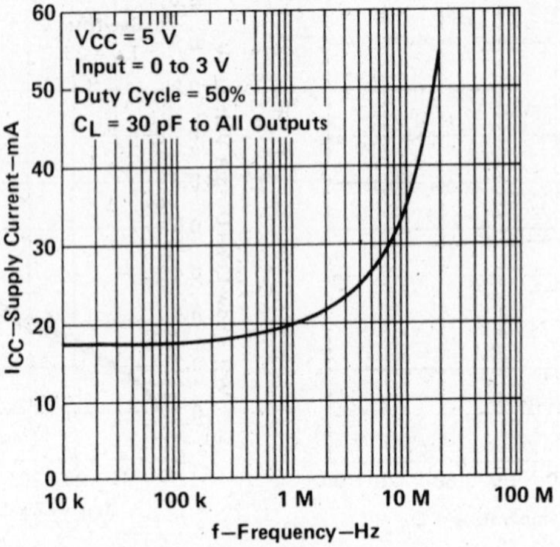
V_{CC} = 5 V
Input = 0 to 3 V
Duty Cycle = 50%
C_L = 30 pF to All Outputs
I_{CC}–Supply Current–mA
60
50
40
30
20
10
0
10 k
100 k
1 M
10 M
100 M
f–Frequency–Hz

FIGURE 14

ADVANCE INFORMATION

SN75ALS193 QUADRUPLE DIFFERENTIAL LINE RECEIVER WITH 3-STATE OUTPUTS

D2931, JUNE 1986—REVISED OCTOBER 1986

- **Meets EIA Standards RS-422-A and RS-423-A**
- **Meets CCITT Recommendations V.10, V.11, X.26, and X.27**
- **Designed for Multipoint Bus Transmission on Long Bus Lines in Noisy Environments**
- **3-State Outputs**
- **Common-Mode Input Voltage Range . . . −7 V to 7 V**
- **Input Sensitivity . . . ±200 mV**
- **Input Hysteresis . . . 120 mV Typ**
- **High Input Impedance . . . 12 kΩ Min**
- **Operates from Single 5-Volt Supply**
- **Low I_{CC} Requirements: I_{CC} . . . 35 mA Max**
- **Improved Speed and Power Consumption Compared to AM26LS32A**

J DUAL-IN-LINE PACKAGE (TOP VIEW)

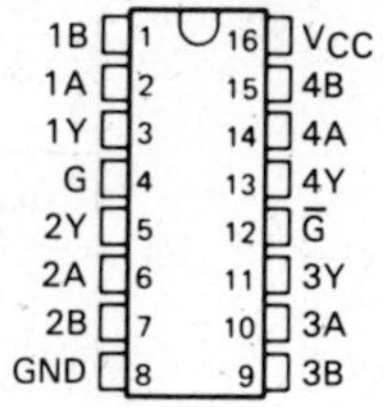
1B 1
1A 2
1Y 3
G 4
2Y 5
2A 6
2B 7
GND 8
16 VCC
15 4B
14 4A
13 4Y
12 G̅
11 3Y
10 3A
9 3B

description

The SN75ALS193 is a monolithic quadruple line receiver with three-state outputs designed using Advanced Low-Power Schottky technology. This technology provides combined improvements in bar design, tooling production, and wafer fabrication. This, in turn, provides significantly less power requirements and permits much higher data throughput than other designs. The device meets the specifications of EIA Standards RS-422-A and RS-423-A. It features three-state outputs that permit direct connection to a bus-organized system with a Fail-Safe design that ensures the outputs will always be high if the inputs are open.

The device is optimized for balanced multipoint bus transmission at rates up to 10 megabits per second. The input features high input impedance, input hysteresis for increased noise immunity, and an input sensitivity of ±200 millivolts over a common-mode input voltage range of −7 to 7 volts. It also features active-high and active-low enable functions that are common to the four channels. The SN75ALS193 is designed for optimum performance when used with the SN75ALS192 quadruple differential line driver.

The SN75ALS193 is characterized for operation from 0°C to 70°C.

logic symbol†

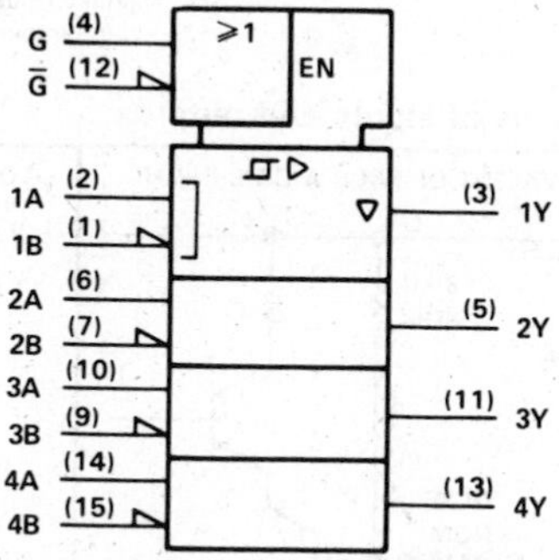
G (4)
G̅ (12)
≥1
EN
1A (2)
1B (1)
(3) 1Y
2A (6)
2B (7)
(5) 2Y
3A (10)
3B (9)
(11) 3Y
4A (14)
4B (15)
(13) 4Y

†This symbol is in accordance with ANSI/IEEE Std 91-1984 and IEC Publication 617-12.

logic diagram (positive logic)

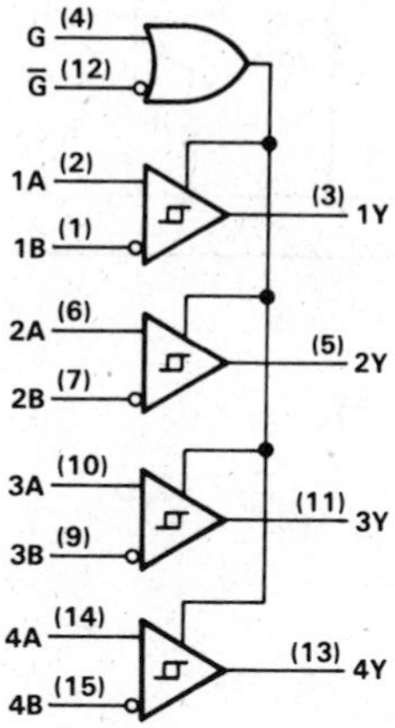
G (4)
G̅ (12)
1A (2)
1B (1)
(3) 1Y
2A (6)
2B (7)
(5) 2Y
3A (10)
3B (9)
(11) 3Y
4A (14)
4B (15)
(13) 4Y

FUNCTION TABLE (EACH RECEIVER)

DIFFERENTIAL A−B	ENABLES G	ENABLES $\overline{G}$	OUTPUT Y
$V_{ID} \geq 0.2$ V	H	X	H
	X	L	H
$-0.2 \text{ V} < V_{ID} < 0.2$ V	H	X	?
	X	L	?
$V_{ID} \leq -0.2$ V	H	X	L
	X	L	L
X	L	H	Z

H = high level
L = low level
X = irrelevant
? = indeterminate
Z = high-impedance (off)

schematics of inputs and outputs

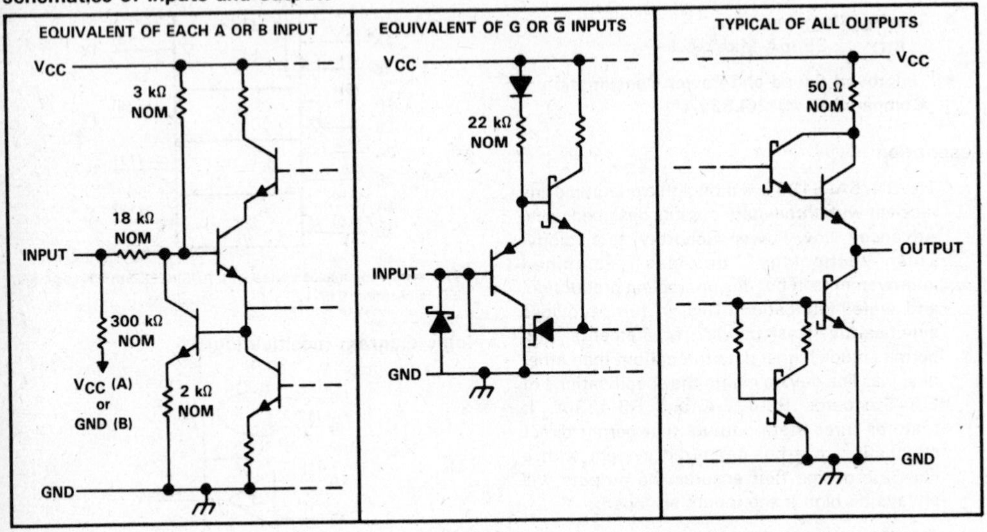
EQUIVALENT OF EACH A OR B INPUT
VCC
3 kΩ NOM
18 kΩ NOM
INPUT
300 kΩ NOM
VCC (A) or GND (B)
2 kΩ NOM
GND
EQUIVALENT OF G OR G̅ INPUTS
VCC
22 kΩ NOM
INPUT
GND
TYPICAL OF ALL OUTPUTS
VCC
50 Ω NOM
OUTPUT
GND

TEXAS INSTRUMENTS
POST OFFICE BOX 655012 • DALLAS, TEXAS 75265

absolute maximum ratings over operating free-air temperature range (unless otherwise noted)

Supply voltage, V_{CC} (see Note 1)	7 V
Input voltage, A or B inputs	±15 V
Differential input voltage (see Note 2)	±15 V
Enable input voltage	7 V
Low-level output current	50 mA
Continuous total dissipation at (or below) 25°C free-air temperature (see Note 3)	1025 mW
Operating free-air temperature range	0°C to 70°C
Storage temperature range	−65°C to 150°C
Lead temperature 1,6 mm (1/16 inch) from case for 60 seconds	300°C

NOTES: 1. All voltage values, except differential input voltage, are with respect to network ground terminal.
2. Differential-input voltage is measured at the noninverting input with respect to the corresponding inverting input.
3. For operation above 25°C free-air temperature, derate to 656 mW at 70°C at the rate of 8.2 mW/°C. In the J package, SN75ALS193 chips are glass mounted.

recommended operating conditions

	MIN	NOM	MAX	UNIT
Supply voltage, V_{CC}	4.75	5	5.25	V
Common-mode input voltage, V_{IC}			±7	V
Differential input voltage, V_{ID}			±12	V
High-level input voltage, V_{IH}	2			V
Low-level input voltage, V_{IL}			0.8	V
High-level output current, I_{OH}			−400	μA
Low-level output current, I_{OL}			16	mA
Operating free-air temperature, T_A	0		70	°C

electrical characteristics over recommended range of common-mode input voltage, supply voltage, and operating free-air temperature (unless otherwise noted)

PARAMETER		TEST CONDITIONS		MIN	TYP†	MAX	UNIT
V_{T+}	Positive-going threshold voltage					200	mV
V_{T-}	Negative-going threshold voltage			−200‡			mV
V_{hys}	Hysteresis§				120		mV
V_{IK}	Enable-input clamp voltage	I_I = −18 mA				−1.5	V
V_{OH}	High-level output voltage	V_{ID} = 200 mV,	I_{OH} = −400 μA	2.7	3.6		V
V_{OL}	Low-level output voltage	V_{ID} = −200 mV	I_{OL} = 8 mA			0.45	V
			I_{OL} = 16 mA			0.5	
I_{OZ}	High-impedance-state output current	V_{CC} = 5.25 V	V_O = 2.4 V			20	μA
			V_O = 0.4 V			−20	
I_I	Line input current	Other input at 0 V, See Note 4	V_I = 15 V		0.7	1.2	mA
			V_I = −15 V		−1.0	−1.7	
I_{IH}	High-level enable-input current		V_{IH} = 2.7 V			20	μA
			V_{IH} = 5.25 V			100	
I_{IL}	Low-level enable-input current	V_{IL} = 0.4 V				−100	μA
	Input resistance			12	18		kΩ
I_{OS}	Short-circuit output current	V_{ID} = 3 V, See Note 5	V_O = 0,	−15	−78	−130	mA
I_{CC}	Supply current	Outputs disabled			22	35	mA

† All typical values are at V_{CC} = 5 V, T_A = 25°C.
‡ The algebraic convention, in which the less positive limit is designated minimum, is used in this data sheet for threshold voltage levels only.
§ Hysteresis is the difference between the positive-going input threshold voltage, V_{T+}, and the negative-going input threshold voltage, V_{T-}.

NOTES: 4. Refer to EIA Standard RS-422-A and RS-423-A for exact conditions.
5. Not more than one output should be shorted at a time and the duration of the short-circuit should not exceed one second.

switching characteristics, V_{CC} = 5 V, T_A = 25°C

PARAMETER		TEST CONDITIONS	MIN	TYP	MAX	UNIT
t_{PLH}	Propagation delay time, low-to-high-level output	V_{ID} = −2.5 V to 2.5 V, C_L = 15 pF, See Figure 2		15	22	ns
t_{PHL}	Propagation delay time, high-to-low-level output			15	22	ns
t_{PZH}	Output enable time to high level	C_L = 15 pF, See Figure 3		13	25	ns
t_{PZL}	Output enable time to low level			11	25	
t_{PHZ}	Output disable time from high level	C_L = 15 pF, See Figure 3		13	25	ns
t_{PLZ}	Output disable time from low level			15	22	

PARAMETER MEASUREMENT INFORMATION

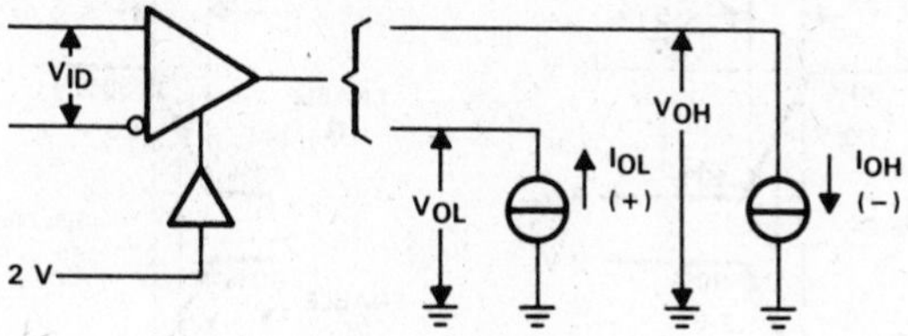
V_{ID}
V_{OH}
I_{OL} (+)
I_{OH} (−)
V_{OL}
2 V

FIGURE 1. V_{OH}, V_{OL}

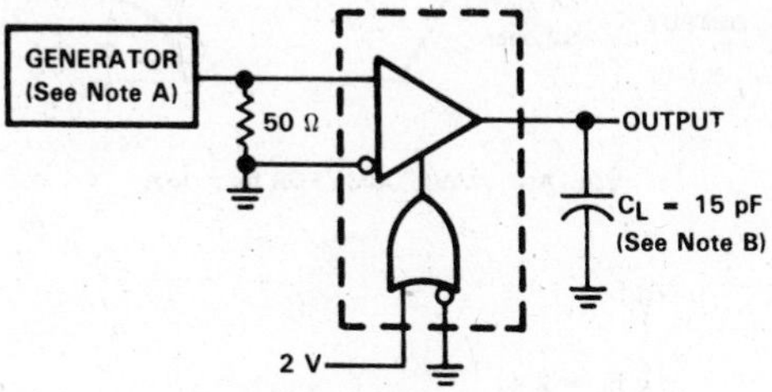
GENERATOR (See Note A)
50 Ω
OUTPUT
C_L = 15 pF (See Note B)
2 V

TEST CIRCUIT

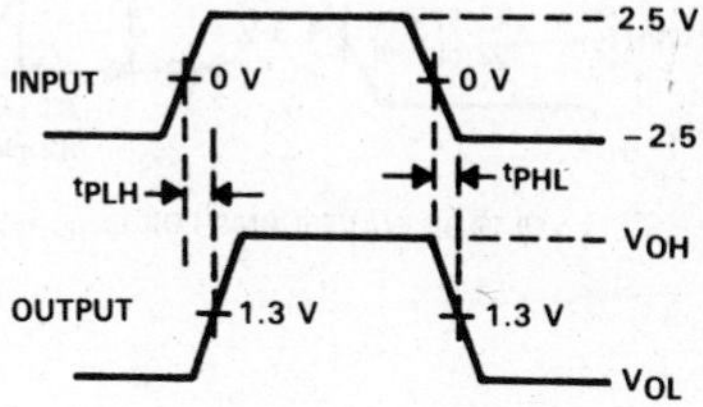
2.5 V
INPUT
0 V
0 V
−2.5 V
t_{PLH}
t_{PHL}
V_{OH}
OUTPUT
1.3 V
1.3 V
V_{OL}

VOLTAGE WAVEFORMS

NOTES: A. The input pulse is supplied by a generator having the following characteristics: PRR ≤ 1 MHz, duty cycle ≤ 50%, Z_{out} = 50 Ω, t_r ≤ 6 ns, t_f ≤ 6 ns.

B. C_L includes probe and jig capacitance.

FIGURE 2. t_{PLH}, t_{PHL}

PARAMETER MEASUREMENT INFORMATION

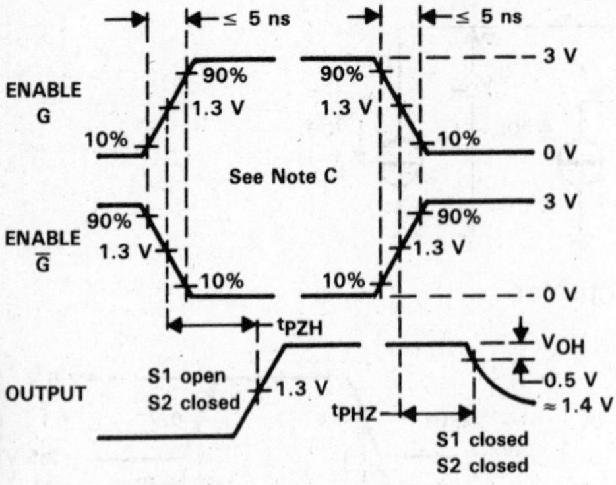
≤ 5 ns
≤ 5 ns
3 V
ENABLE G
90%
1.3 V
10%
0 V
See Note C
ENABLE $\overline{G}$
3 V
0 V
tPZH
tPHZ
OUTPUT
S1 open
S2 closed
VOH
0.5 V
≈ 1.4 V
S1 closed
S2 closed

VOLTAGE WAVEFORMS FOR t_{PHZ}, t_{PZH}

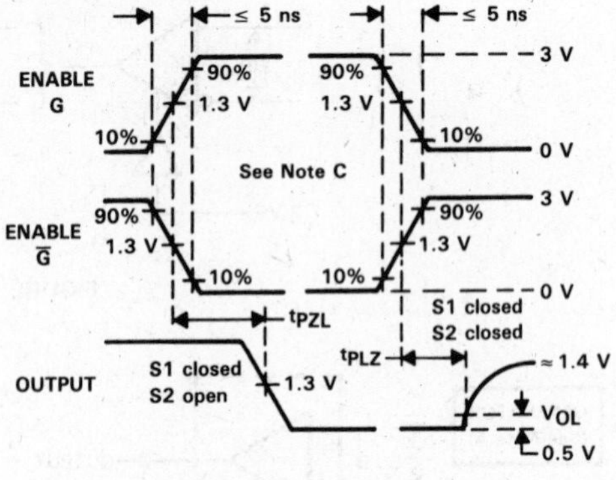
≤ 5 ns
≤ 5 ns
3 V
ENABLE G
90%
1.3 V
10%
0 V
See Note C
ENABLE $\overline{G}$
3 V
0 V
S1 closed
S2 closed
tPZL
tPLZ
OUTPUT
S1 closed
S2 open
≈ 1.4 V
VOL
0.5 V

VOLTAGE WAVEFORMS FOR t_{PLZ}, t_{PZL}

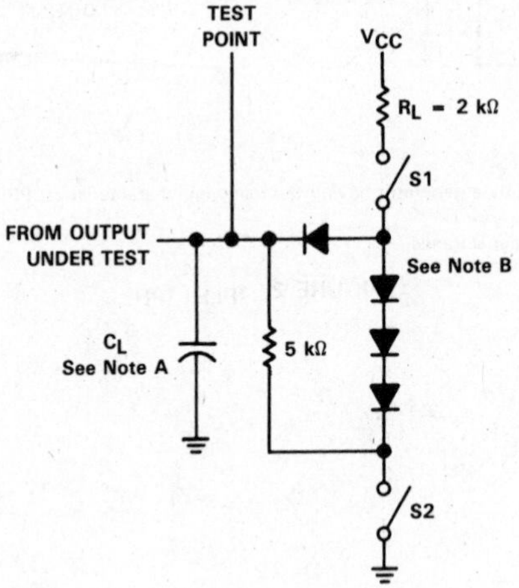
TEST
POINT
VCC
RL = 2 kΩ
S1
FROM OUTPUT
UNDER TEST
See Note B
CL
See Note A
5 kΩ
S2

LOAD CIRCUIT

NOTES: A. C_L includes probe and jig capacitance.
B. All diodes are 1N3064 or equivalent.
C. Enable G is tested with $\overline{G}$ high; $\overline{G}$ is tested with G low.

FIGURE 3. t_{PHZ}, t_{PZH}, t_{PLZ}, t_{PZL}

TYPICAL CHARACTERISTICS

OUTPUT VOLTAGE
vs
ENABLE VOLTAGE

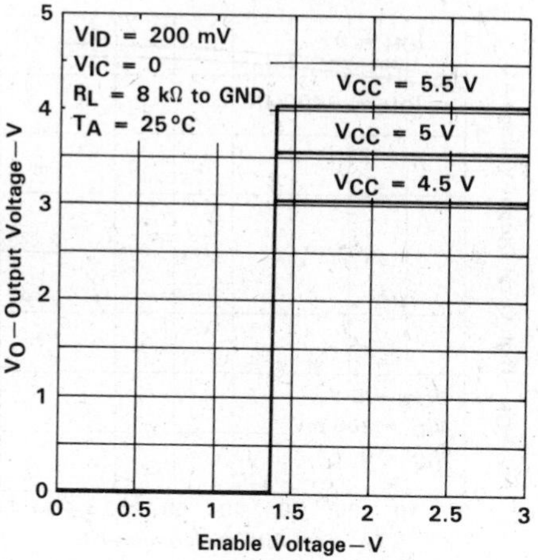
V_{ID} = 200 mV
V_{IC} = 0
R_L = 8 kΩ to GND
T_A = 25°C
V_{CC} = 5.5 V
V_{CC} = 5 V
V_{CC} = 4.5 V
V_O—Output Voltage—V
Enable Voltage—V

FIGURE 4

OUTPUT VOLTAGE
vs
ENABLE VOLTAGE

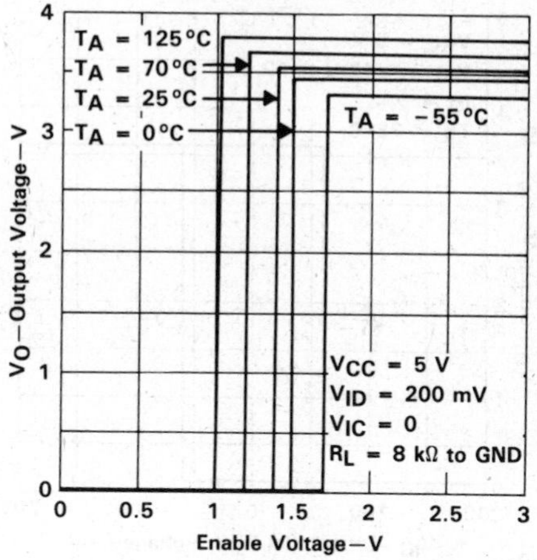
T_A = 125°C
T_A = 70°C
T_A = 25°C
T_A = 0°C
T_A = −55°C
V_{CC} = 5 V
V_{ID} = 200 mV
V_{IC} = 0
R_L = 8 kΩ to GND
V_O—Output Voltage—V
Enable Voltage—V

FIGURE 5

OUTPUT VOLTAGE
vs
ENABLE VOLTAGE

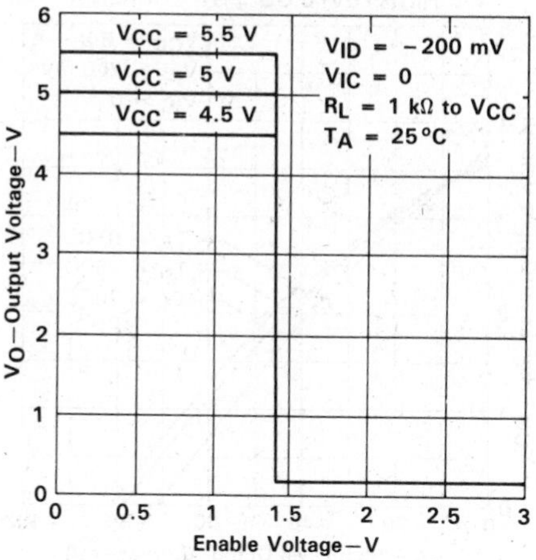
V_{CC} = 5.5 V
V_{CC} = 5 V
V_{CC} = 4.5 V
V_{ID} = −200 mV
V_{IC} = 0
R_L = 1 kΩ to V_{CC}
T_A = 25°C
V_O—Output Voltage—V
Enable Voltage—V

FIGURE 6

OUTPUT VOLTAGE
vs
ENABLE VOLTAGE

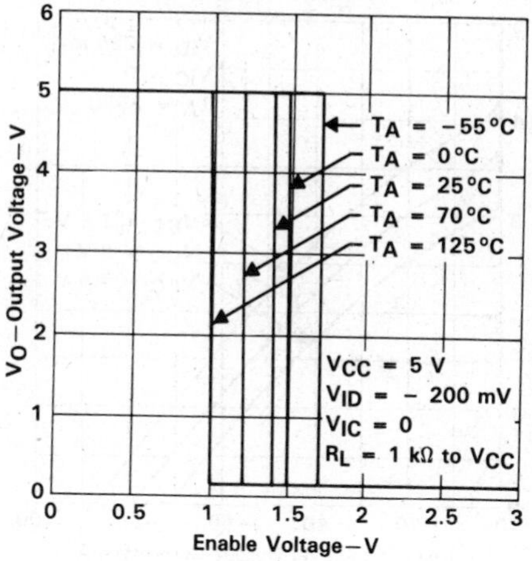
T_A = −55°C
T_A = 0°C
T_A = 25°C
T_A = 70°C
T_A = 125°C
V_{CC} = 5 V
V_{ID} = − 200 mV
V_{IC} = 0
R_L = 1 kΩ to V_{CC}
V_O—Output Voltage—V
Enable Voltage—V

FIGURE 7

TYPICAL CHARACTERISTICS

OUTPUT VOLTAGE
vs
DIFFERENTIAL INPUT VOLTAGE

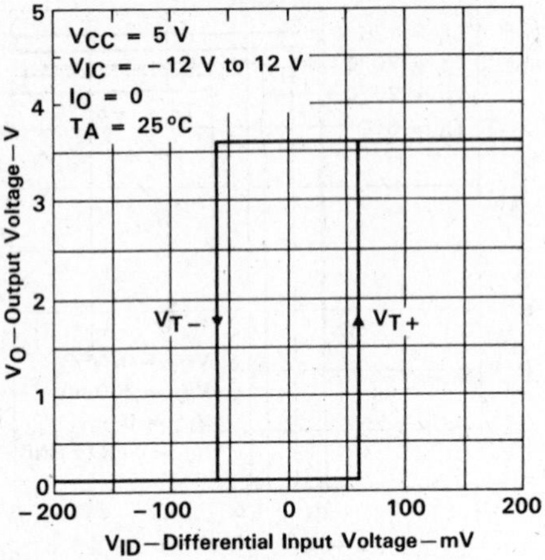
VCC = 5 V
VIC = −12 V to 12 V
IO = 0
TA = 25°C
VT−
VT+
VO—Output Voltage—V
VID—Differential Input Voltage—mV
−200
−100
0
100
200
0
1
2
3
4
5

FIGURE 8

HIGH-LEVEL OUTPUT VOLTAGE
vs
FREE-AIR TEMPERATURE

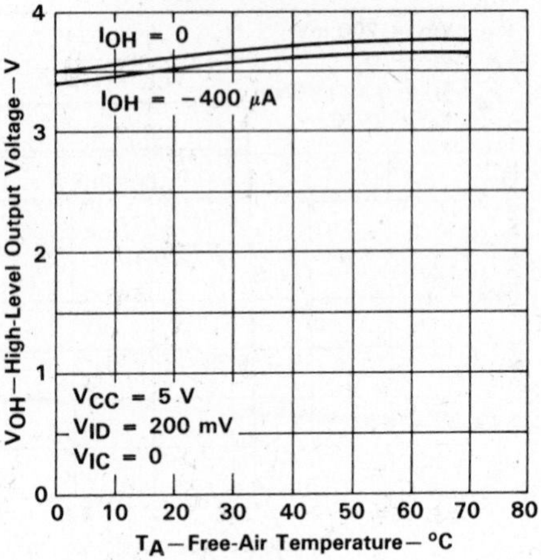
IOH = 0
IOH = −400 μA
VCC = 5 V
VID = 200 mV
VIC = 0
VOH—High-Level Output Voltage—V
TA—Free-Air Temperature—°C
0
10
20
30
40
50
60
70
80
0
1
2
3
4

FIGURE 9

HIGH-LEVEL OUTPUT VOLTAGE
vs
HIGH-LEVEL OUTPUT CURRENT

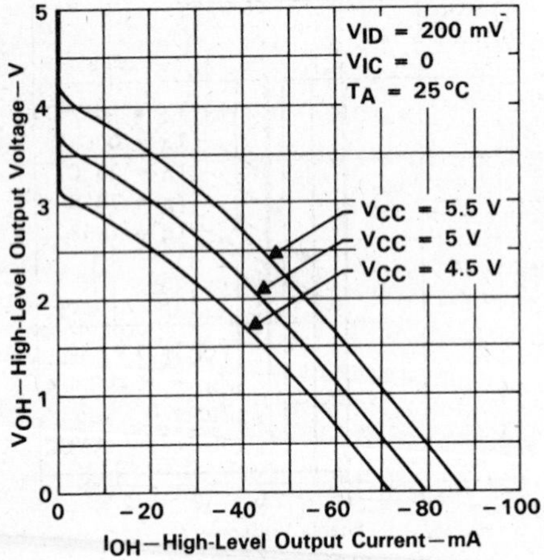
VID = 200 mV
VIC = 0
TA = 25°C
VCC = 5.5 V
VCC = 5 V
VCC = 4.5 V
VOH—High-Level Output Voltage—V
IOH—High-Level Output Current—mA
0
−20
−40
−60
−80
−100
0
1
2
3
4
5

FIGURE 10

HIGH-LEVEL OUTPUT VOLTAGE
vs
HIGH-LEVEL OUTPUT CURRENT

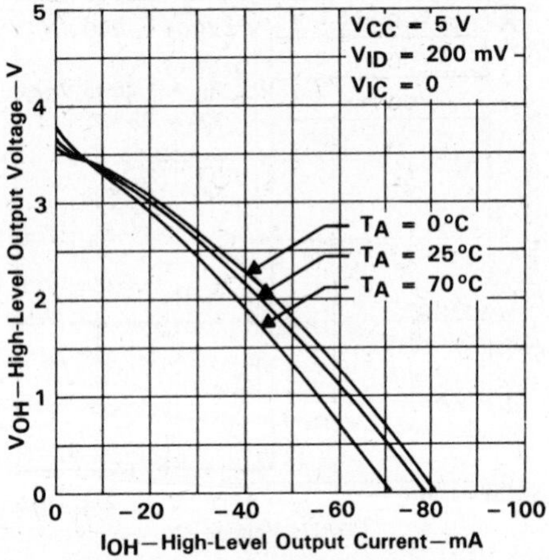
VCC = 5 V
VID = 200 mV
VIC = 0
TA = 0°C
TA = 25°C
TA = 70°C
VOH—High-Level Output Voltage—V
IOH—High-Level Output Current—mA
0
−20
−40
−60
−80
−100
0
1
2
3
4
5

FIGURE 11

TEXAS INSTRUMENTS
POST OFFICE BOX 655012 • DALLAS, TEXAS 75265

TYPICAL CHARACTERISTICS

LOW-LEVEL OUTPUT VOLTAGE
vs
FREE-AIR TEMPERATURE

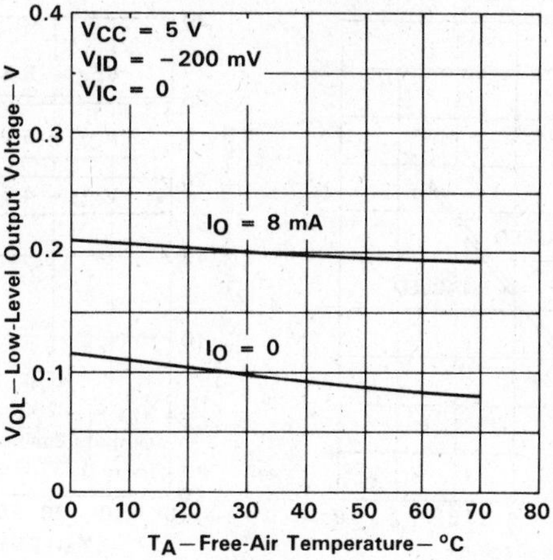
V_{CC} = 5 V
V_{ID} = −200 mV
V_{IC} = 0
I_O = 8 mA
I_O = 0
V_{OL}—Low-Level Output Voltage—V
T_A—Free-Air Temperature—°C

FIGURE 12

LOW-LEVEL OUTPUT VOLTAGE
vs
LOW-LEVEL OUTPUT CURRENT

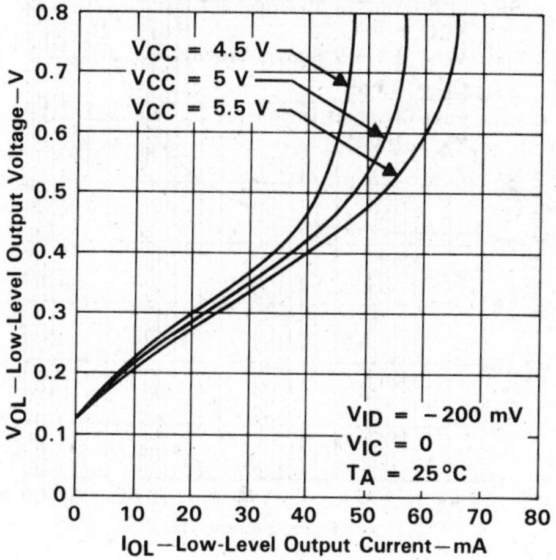
V_{CC} = 4.5 V
V_{CC} = 5 V
V_{CC} = 5.5 V
V_{ID} = −200 mV
V_{IC} = 0
T_A = 25°C
V_{OL}—Low-Level Output Voltage—V
I_{OL}—Low-Level Output Current—mA

FIGURE 13

LOW-LEVEL OUTPUT VOLTAGE
vs
LOW-LEVEL OUTPUT CURRENT

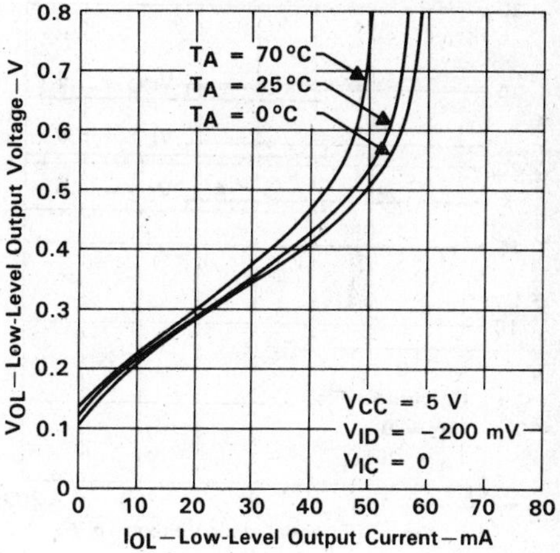
T_A = 70°C
T_A = 25°C
T_A = 0°C
V_{CC} = 5 V
V_{ID} = −200 mV
V_{IC} = 0
V_{OL}—Low-Level Output Voltage—V
I_{OL}—Low-Level Output Current—mA

FIGURE 14

TYPICAL CHARACTERISTICS

SUPPLY CURRENT
vs
SUPPLY VOLTAGE

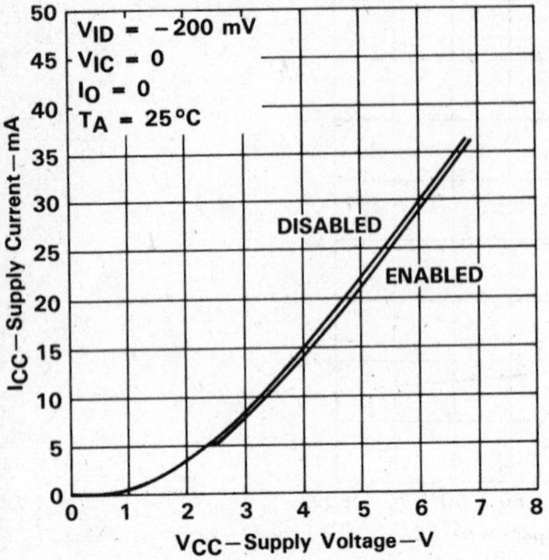
V_{ID} = −200 mV
V_{IC} = 0
I_O = 0
T_A = 25°C
DISABLED
ENABLED
I_{CC}—Supply Current—mA
V_{CC}—Supply Voltage—V

FIGURE 15

SUPPLY CURRENT
vs
FREE-AIR TEMPERATURE

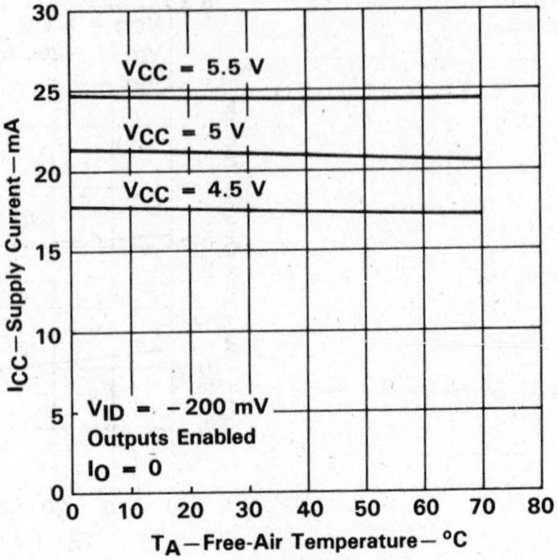
V_{CC} = 5.5 V
V_{CC} = 5 V
V_{CC} = 4.5 V
V_{ID} = −200 mV
Outputs Enabled
I_O = 0
I_{CC}—Supply Current—mA
T_A—Free-Air Temperature—°C

FIGURE 16

SUPPLY CURRENT
vs
DIFFERENTIAL INPUT VOLTAGE

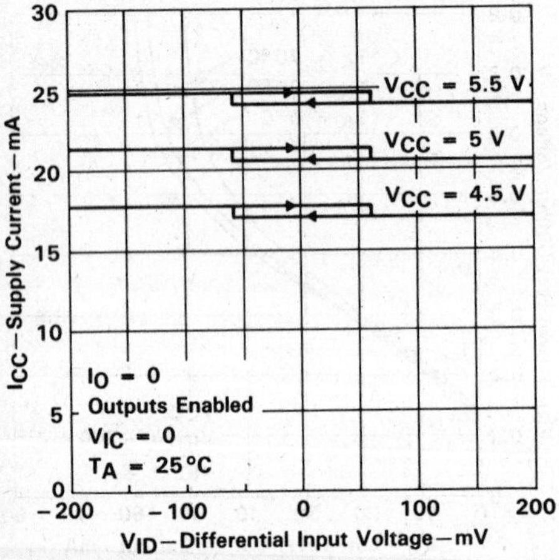
V_{CC} = 5.5 V
V_{CC} = 5 V
V_{CC} = 4.5 V
I_O = 0
Outputs Enabled
V_{IC} = 0
T_A = 25°C
I_{CC}—Supply Current—mA
V_{ID}—Differential Input Voltage—mV

FIGURE 17

SUPPLY CURRENT
vs
FREQUENCY

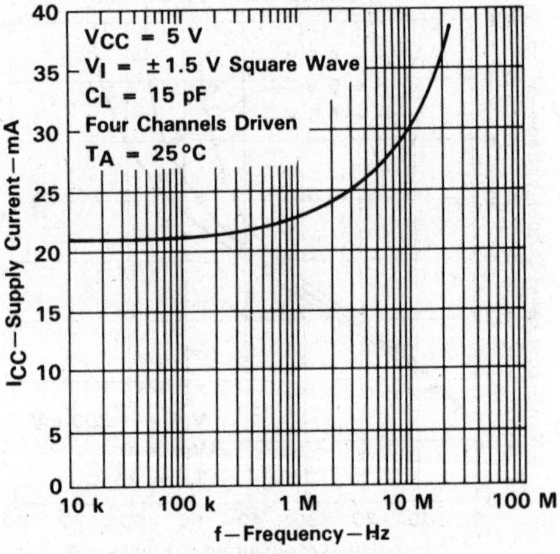
V_{CC} = 5 V
V_I = ±1.5 V Square Wave
C_L = 15 pF
Four Channels Driven
T_A = 25°C
I_{CC}—Supply Current—mA
f—Frequency—Hz

FIGURE 18

4 Line Drivers/Receivers

TEXAS INSTRUMENTS
POST OFFICE BOX 655012 • DALLAS, TEXAS 75265

TYPICAL CHARACTERISTICS

INPUT RESISTANCE
vs
FREE-AIR TEMPERATURE

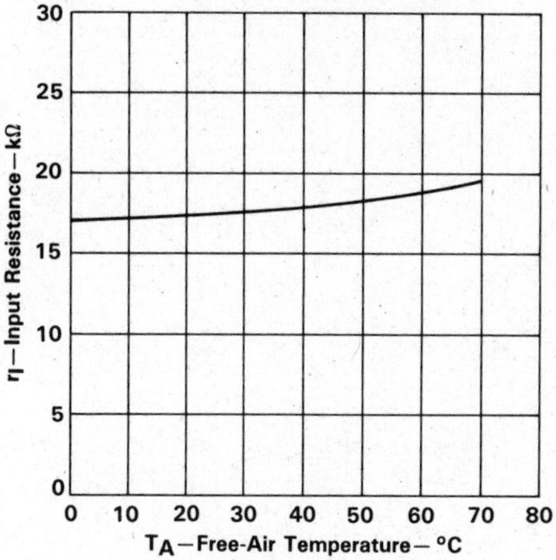
30
25
20
15
10
5
0
0 10 20 30 40 50 60 70 80
r_I—Input Resistance—kΩ
T_A—Free-Air Temperature—°C

FIGURE 19

INPUT CURRENT
vs
INPUT VOLTAGE

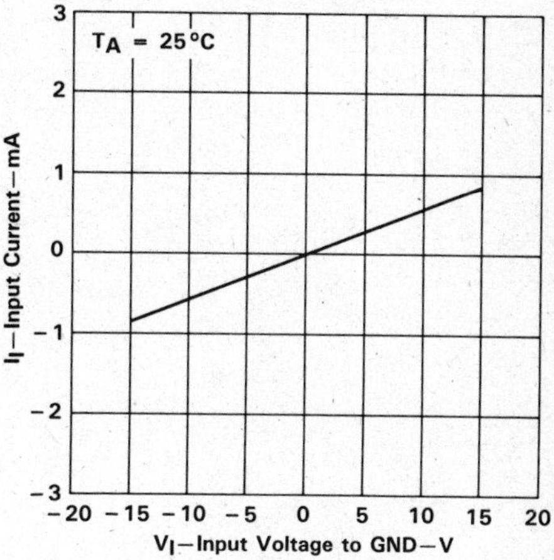
T_A = 25°C
3
2
1
0
−1
−2
−3
−20 −15 −10 −5 0 5 10 15 20
I_I—Input Current—mA
V_I—Input Voltage to GND—V

FIGURE 20

SWITCHING CHARACTERISTICS
vs
FREE-AIR TEMPERATURE

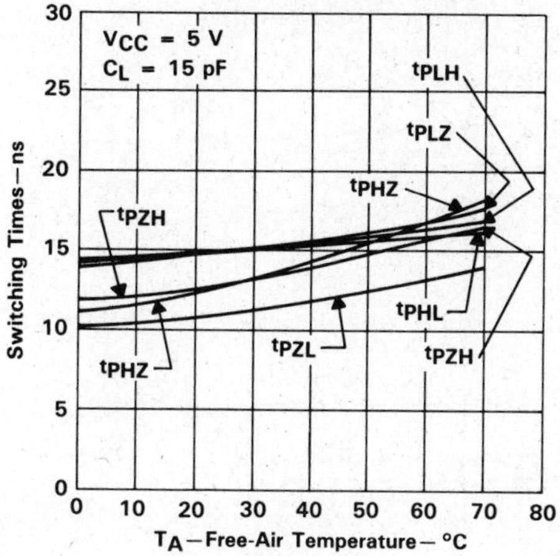
V_{CC} = 5 V
C_L = 15 pF
t_{PLH}
t_{PLZ}
t_{PHZ}
t_{PZH}
t_{PHL}
t_{PZL}
t_{PZH}
t_{PHZ}
30
25
20
15
10
5
0
0 10 20 30 40 50 60 70 80
Switching Times—ns
T_A—Free-Air Temperature—°C

FIGURE 21

PROPAGATION DELAY TIME
vs
SUPPLY VOLTAGE

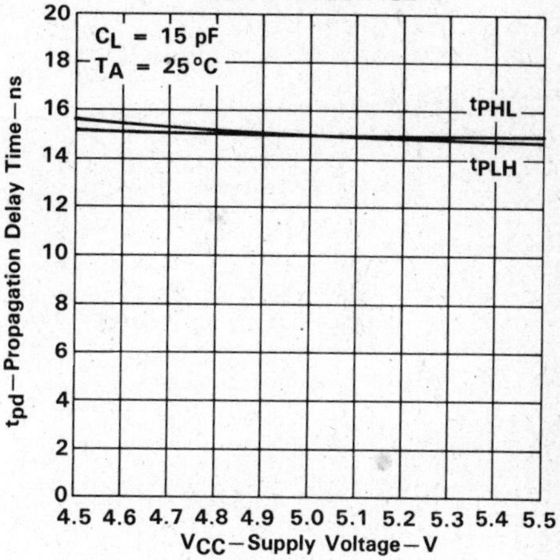
C_L = 15 pF
T_A = 25°C
t_{PHL}
t_{PLH}
20
18
16
14
12
10
8
6
4
2
0
4.5 4.6 4.7 4.8 4.9 5.0 5.1 5.2 5.3 5.4 5.5
t_{pd}—Propagation Delay Time—ns
V_{CC}—Supply Voltage—V

FIGURE 22

4

Line Drivers/Receivers

QUADRUPLE DIFFERENTIAL LINE DRIVER WITH 3-STATE OUTPUTS

D2917, OCTOBER 1985—REVISED OCTOBER 1986

- Meets EIA Standard RS-422-A
- High-Speed ALS Design
- 3-State TTL-Compatible
- Single 5-V Supply Operation
- High Output Impedance in Power-Off Condition
- Two Pairs of Drivers Independently Enabled
- Designed as a Replacement for the MC3487 with Improvements: I_{CC} 50% Lower, Switching Speed 30% Faster

D, J, N PACKAGE
(TOP VIEW)

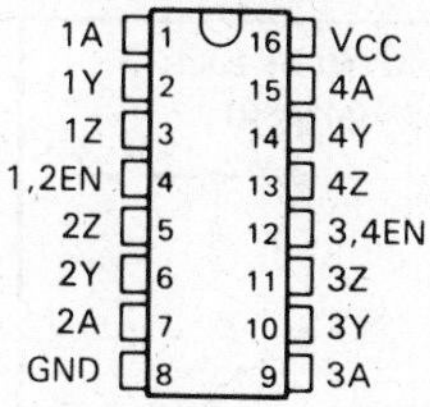
1A 1
1Y 2
1Z 3
1,2EN 4
2Z 5
2Y 6
2A 7
GND 8
16 VCC
15 4A
14 4Y
13 4Z
12 3,4EN
11 3Z
10 3Y
9 3A

FUNCTION TABLE (EACH DRIVER)

INPUT	OUTPUT ENABLE	OUTPUTS Y	OUTPUTS Z
H	H	H	L
L	H	L	H
X	L	High-Impedance	High-Impedance

H = TTL high level, L = TTL low level, X = irrelevant

description

This quadruple complementary-output line driver is designed for data transmission over twisted-pair or parallel-wire transmission lines. It meets the requirements of EIA Standard RS-422-A and is compatible with 3-state TTL circuits. Advanced Low-Power Schottky technology provides high speed without the usual power penalty. Standby supply current is typically only 26 milliamperes, while typical propagation delay time is less than 10 nanoseconds and enable/disable times are typically less than 16 nanoseconds.

High-impedance inputs keep input currents low, less than 1 microampere for a high level and less than 100 microamperes for a low level. The driver circuits can be enabled in pairs by separate active-high enable inputs. The SN75ALS194 is capable of data rates in excess of 10 megabits per second and is designed to operate with the SN75ALS195 quadruple line receiver.

The SN75ALS194 is characterized for operation from 0 °C to 70 °C.

logic symbol†

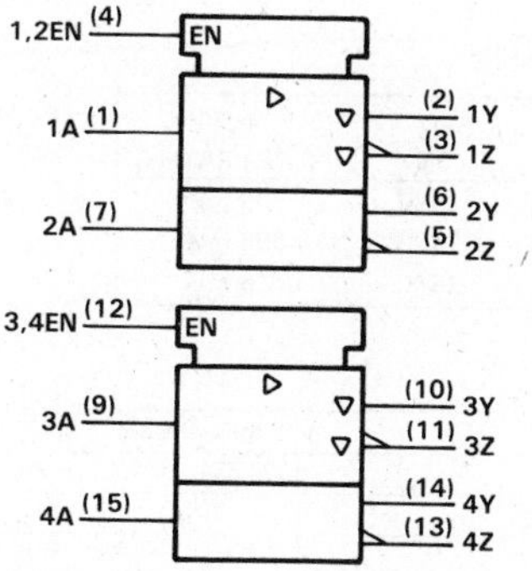
1,2EN (4)
EN
1A (1)
(2) 1Y
(3) 1Z
2A (7)
(6) 2Y
(5) 2Z
3,4EN (12)
EN
3A (9)
(10) 3Y
(11) 3Z
4A (15)
(14) 4Y
(13) 4Z

† This symbol is in accordance with ANSI/IEEE Std 91-1984 and IEC Publication 617-12.

logic diagram (positive logic)

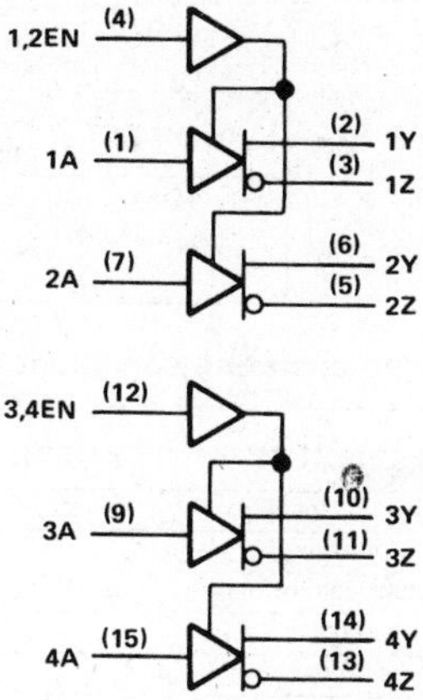
1,2EN (4)
1A (1)
(2) 1Y
(3) 1Z
2A (7)
(6) 2Y
(5) 2Z
3,4EN (12)
3A (9)
(10) 3Y
(11) 3Z
4A (15)
(14) 4Y
(13) 4Z

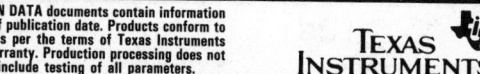

TEXAS INSTRUMENTS
POST OFFICE BOX 655012 • DALLAS, TEXAS 75265

schematics of inputs and outputs

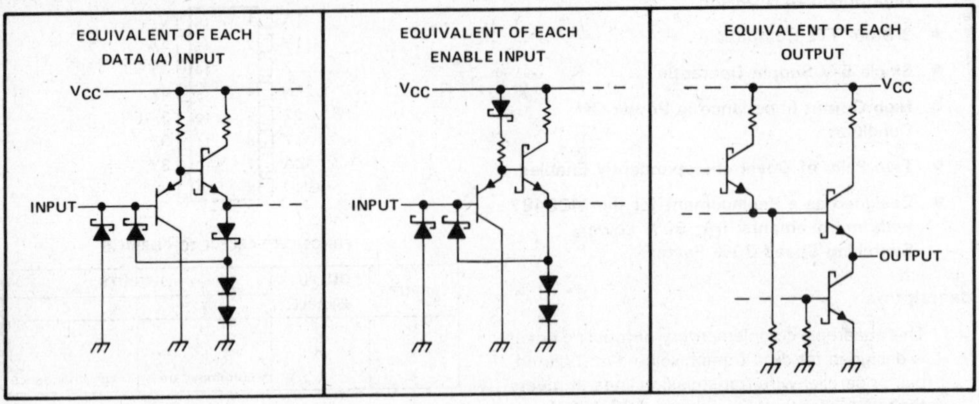
EQUIVALENT OF EACH
DATA (A) INPUT
VCC
INPUT
EQUIVALENT OF EACH
ENABLE INPUT
VCC
INPUT
EQUIVALENT OF EACH
OUTPUT
VCC
OUTPUT

absolute maximum ratings over operating free-air temperature range (unless otherwise noted)

Supply voltage, V_{CC} (see Note 1) . . . . . 7 V
Input voltage, V_I . . . . . 5.5 V
Continuous total dissipation at (or below) 25 °C free-air temperature (see Note 2):
- D package . . . . . 950 mW
- J package . . . . . 1025 mW
- N package . . . . . 1150 mW

Operating free-air temperature range . . . . . 0 °C to 70 °C
Lead temperature 1,6 mm (1/16 inch) from case for 60 seconds: J package . . . . . 300 °C
Lead temperature 1,6 mm (1/16 inch) from case for 10 seconds: D or N package . . . . . 260 °C

NOTES: 1. All voltage values are with respect to network ground terminal.
2. For operation above 25 °C free-air temperature, refer to the Dissipation Derating Table.

DISSIPATION DERATING TABLE

PACKAGE	T_A = 25 °C POWER RATING	DERATING FACTOR	ABOVE T_A	T_A = 70 °C POWER RATING
D	950 mW	7.6 mW/°C	25 °C	608 mW
J (Glass mount)	1025 mW	8.2 mW/°C	25 °C	656 mW
N	1150 mW	9.2 mW/°C	25 °C	736 mW

recommended operating conditions

	MIN	NOM	MAX	UNIT
Supply voltage, V_{CC}	4.75	5	5.25	V
High-level input voltage, V_{IH}	2			V
Low-level input voltage, V_{IL}			0.8	V
High-level output current, I_{OH}			−20	mA
Low-level output current, I_{OL}			48	mA
Operating free-air temperature, T_A	0		70	°C

TEXAS INSTRUMENTS
POST OFFICE BOX 655012 • DALLAS, TEXAS 75265

electrical characteristics over recommended ranges of supply voltage and operating free-air temperature range (unless otherwise noted)

PARAMETER		TEST CONDITIONS		MIN	TYP†	MAX	UNIT
V_{IK}	Input clamp voltage	$I_I = -18$ mA				−1.5	V
V_{OH}	High-level output voltage	$I_{OH} = -20$ mA		2.5			V
V_{OL}	Low-level output voltage	$I_{OL} = 48$ mA				0.5	V
V_O	Output voltage	$I_O = 0$		0		6	V
$\|V_{OD1}\|$	Differential output voltage	$I_O = 0$		2		6	V
$\|V_{OD2}\|$	Differential output voltage	$R_L = 100\ \Omega$, See Figure 1		½ V_{OD1}			V
				2			
$\Delta\|V_{OD}\|$	Change in magnitude of differential output voltage‡					±0.4	V
V_{OC}	Common-mode output voltage					±3	V
$\Delta\|V_{OC}\|$	Change in magnitude of common-mode output voltage‡					±0.4	V
I_O	Output current with power off	$V_{CC} = 0$	$V_O = 6$ V			100	μA
			$V_O = -0.25$ V			−100	
I_{OZ}	High-impedance state output current	Output enables at 0.8 V	$V_O = 2.7$ V			100	μA
			$V_O = 0.5$ V			−100	
I_I	Input current at maximum input voltage	$V_I = 5.5$ V				100	μA
I_{IH}	High-level input current	$V_I = 2.7$ V				50	μA
I_{IL}	Low-level input current	$V_I = 0.5$ V				−200	μA
I_{OS}	Short-circuit output current§	$V_I = 2$ V		−40		−140	mA
I_{CC}	Supply current (all drivers)	$V_{CC} = 5.25$ V,	All outputs disabled		26	45	mA

† All typical values are at $V_{CC} = 5$ V, $T_A = 25°C$.
‡ $\Delta|V_{OD}|$ and $\Delta|V_{OC}|$ are the changes in magnitude of V_{OD} and V_{OC}, respectively, that occur when the input is changed from a high level to a low level.
§ Not more than one output should be shorted at a time, and duration of the short-circuit should not exceed one second.

switching characteristics, $V_{CC} = 5$ V, $T_A = 25°C$

PARAMETER		TEST CONDTIONS	MIN	TYP	MAX	UNIT
t_{PLH}	Propagation delay time, low-to-high-level output	$C_L = 15$ pF, See Figure 1		6	13	ns
t_{PHL}	Propagation delay time, high-to-low-level output			9	14	ns
	Output-to-output skew			3.5	6	ns
t_{TD}	Differential-output transition time	$C_L = 15$ pF, See Figure 2		8	14	ns
t_{PZH}	Output enable time to high level	$C_L = 50$ pF, See Figure 3		9	12	ns
t_{PZL}	Output enable time to low level			12	20	ns
t_{PHZ}	Output disable time from high level			9	14	ns
t_{PLZ}	Output disable time from low level			12	15	ns

SYMBOL EQUIVALENTS

DATA SHEET PARAMETER	RS-422-A
V_O	V_{oa}, V_{ob}
$\|V_{OD1}\|$	V_o
$\|V_{OD2}\|$	V_t ($R_L = 100\ \Omega$)
$\Delta\|V_{OD}\|$	$\|\ \|V_t\| - \|\bar{V}_t\|\ \|$
V_{OC}	$\|V_{os}\|$
$\Delta\|V_{OC}\|$	$\|\ V_{os} - \bar{V}_{os}\ \|$
I_{OS}	$\|I_{sa}\|$, $\|I_{sb}\|$
I_O	$\|I_{xa}\|$, $\|I_{xb}\|$

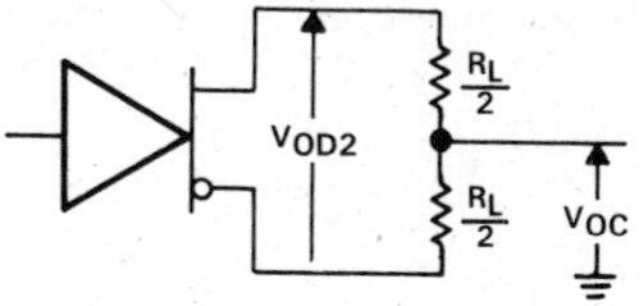
V_{OD2}
R_L/2
R_L/2
V_{OC}

FIGURE 1. DRIVER V_{OD} AND V_{OC}

4 Line Drivers/Receivers

PARAMETER MEASUREMENT INFORMATION

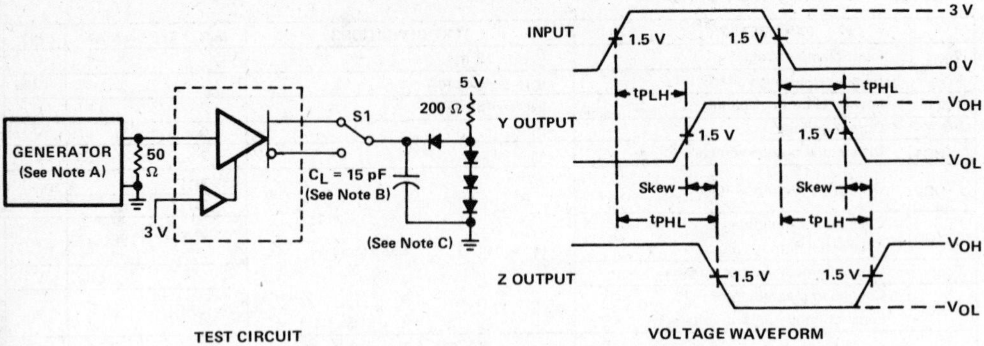
GENERATOR
(See Note A)
50 Ω
3 V
5 V
200 Ω
S1
C_L = 15 pF
(See Note B)
(See Note C)
TEST CIRCUIT
INPUT
1.5 V
1.5 V
3 V
0 V
t_{PLH}
t_{PHL}
Y OUTPUT
V_{OH}
V_{OL}
Skew
Skew
t_{PHL}
t_{PLH}
Z OUTPUT
VOLTAGE WAVEFORM

FIGURE 2. PROPAGATION DELAY TIMES

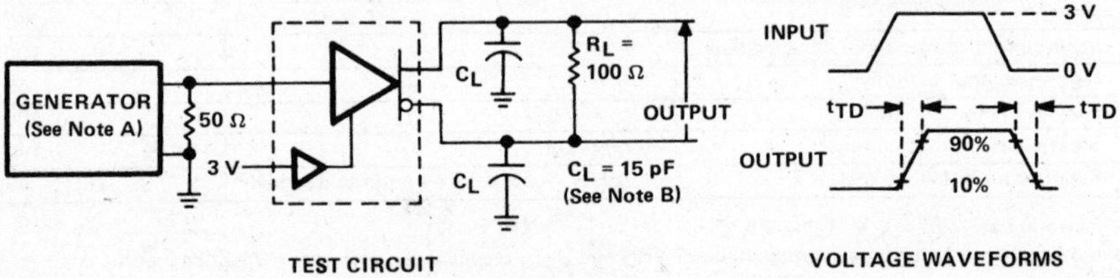
GENERATOR
(See Note A)
50 Ω
3 V
C_L
R_L = 100 Ω
OUTPUT
C_L = 15 pF
(See Note B)
TEST CIRCUIT
INPUT
3 V
0 V
t_{TD}
t_{TD}
OUTPUT
90%
10%
VOLTAGE WAVEFORMS

FIGURE 3. DIFFERENTIAL-OUTPUT TRANSITION TIMES

NOTES: A. The input pulse is supplied by a generator having the following characteristics: $t_r \leq 5$ ns, $t_f \leq 5$ ns, PRR ≤ 1 MHz, duty cycle = 50%, Z_O = 50 Ω.
B. C_L includes probe and stray capacitance.
C. All diodes are 1N916 or 1N3064.

PARAMETER MEASUREMENT INFORMATION

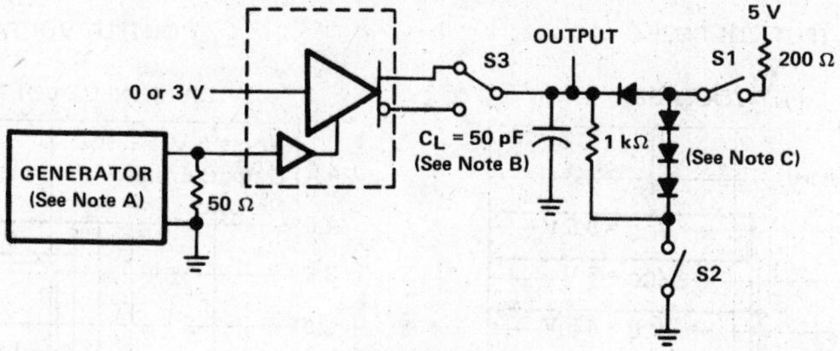
0 or 3 V
GENERATOR
(See Note A)
50 Ω
S3
OUTPUT
5 V
S1
200 Ω
C_L = 50 pF
(See Note B)
1 kΩ
(See Note C)
S2

TEST CIRCUIT

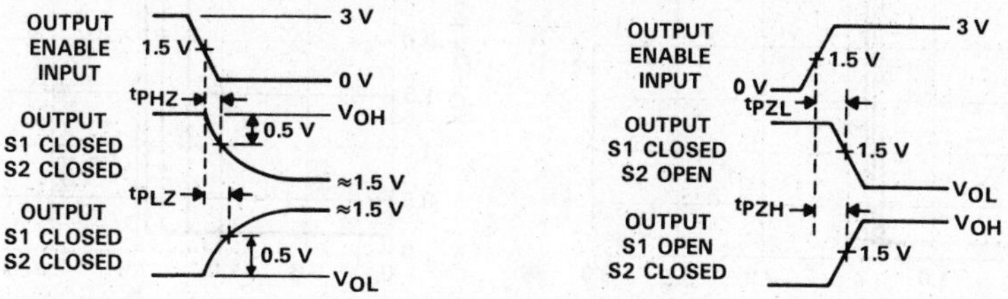
OUTPUT
ENABLE
INPUT
1.5 V
3 V
0 V
t_{PHZ}
OUTPUT
S1 CLOSED
S2 CLOSED
0.5 V
V_{OH}
≈1.5 V
t_{PLZ}
≈1.5 V
OUTPUT
S1 CLOSED
S2 CLOSED
0.5 V
V_{OL}
OUTPUT
ENABLE
INPUT
0 V
3 V
1.5 V
t_{PZL}
OUTPUT
S1 CLOSED
S2 OPEN
1.5 V
V_{OL}
t_{PZH}
OUTPUT
S1 OPEN
S2 CLOSED
V_{OH}
1.5 V

VOLTAGE WAVEFORMS

NOTES: A. The input pulse is supplied by a generator having the following characteristics: $t_r \leq 5$ ns, $t_f \leq 5$ ns, PRR $\leq$ 1 MHz, duty cycle = 50%, Z_O = 50 Ω.
B. C_L includes probe and stray capacitance.
C. All diodes are 1N916 or 1N3064.

FIGURE 4. DRIVER ENABLE AND DISABLE TIMES

TYPICAL CHARACTERISTICS

Y OUTPUT VOLTAGE
vs
DATA INPUT VOLTAGE

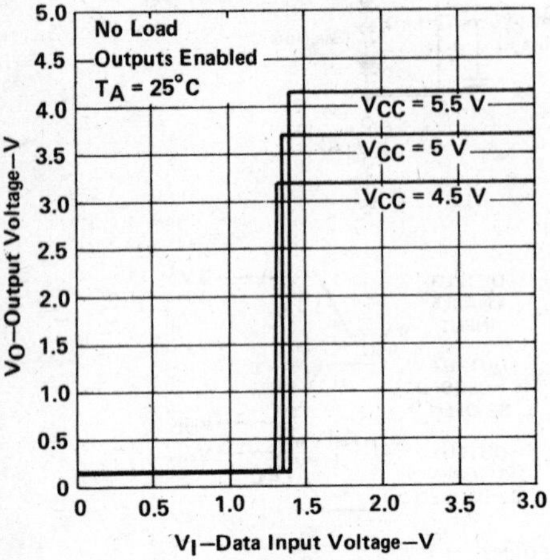
No Load
Outputs Enabled
T_A = 25°C
V_{CC} = 5.5 V
V_{CC} = 5 V
V_{CC} = 4.5 V
V_O—Output Voltage—V
V_I—Data Input Voltage—V

FIGURE 5

Y OUTPUT VOLTAGE
vs
DATA INPUT VOLTAGE

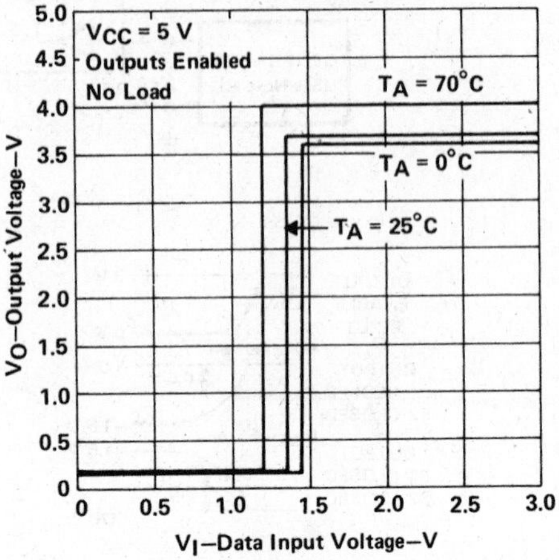
V_{CC} = 5 V
Outputs Enabled
No Load
T_A = 70°C
T_A = 0°C
T_A = 25°C
V_O—Output Voltage—V
V_I—Data Input Voltage—V

FIGURE 6

Y OUTPUT VOLTAGE
vs
ENABLE G INPUT VOLTAGE

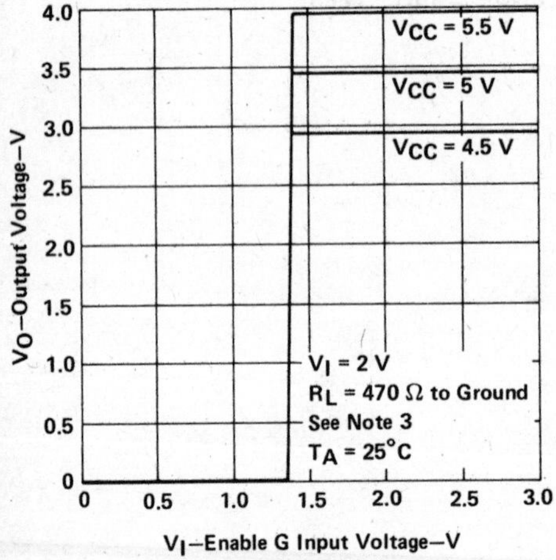
V_{CC} = 5.5 V
V_{CC} = 5 V
V_{CC} = 4.5 V
V_I = 2 V
R_L = 470 Ω to Ground
See Note 3
T_A = 25°C
V_O—Output Voltage—V
V_I—Enable G Input Voltage—V

FIGURE 7

Y OUTPUT VOLTAGE
vs
ENABLE G INPUT VOLTAGE

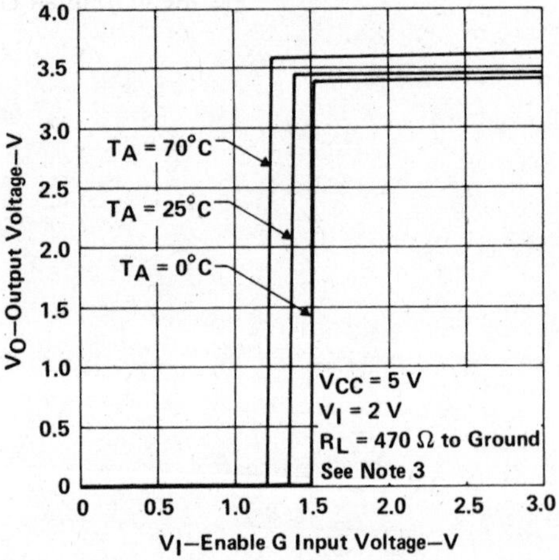
T_A = 70°C
T_A = 25°C
T_A = 0°C
V_{CC} = 5 V
V_I = 2 V
R_L = 470 Ω to Ground
See Note 3
V_O—Output Voltage—V
V_I—Enable G Input Voltage—V

FIGURE 8

NOTE 3: The A input is connected to V_{CC} during the testing of the Y outputs and to ground during the testing of the Z outputs.

TYPICAL CHARACTERISTICS

Z OUTPUT VOLTAGE
vs
ENABLE G INPUT VOLTAGE

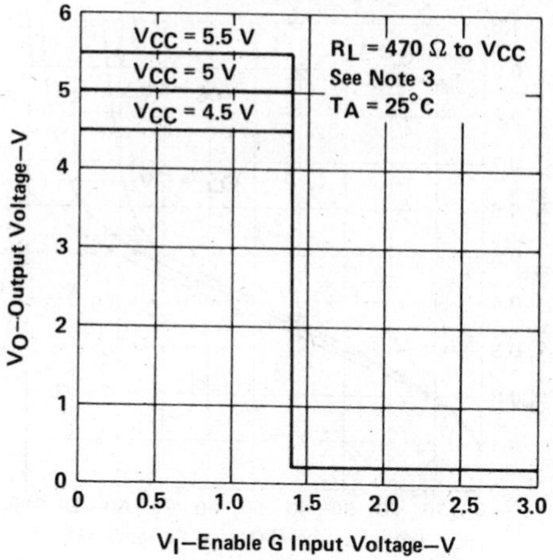
VCC = 5.5 V
VCC = 5 V
VCC = 4.5 V
RL = 470 Ω to VCC
See Note 3
TA = 25°C
VO—Output Voltage—V
VI—Enable G Input Voltage—V
0
0.5
1.0
1.5
2.0
2.5
3.0
1
2
3
4
5
6

FIGURE 9

Z OUTPUT VOLTAGE
vs
ENABLE G INPUT VOLTAGE

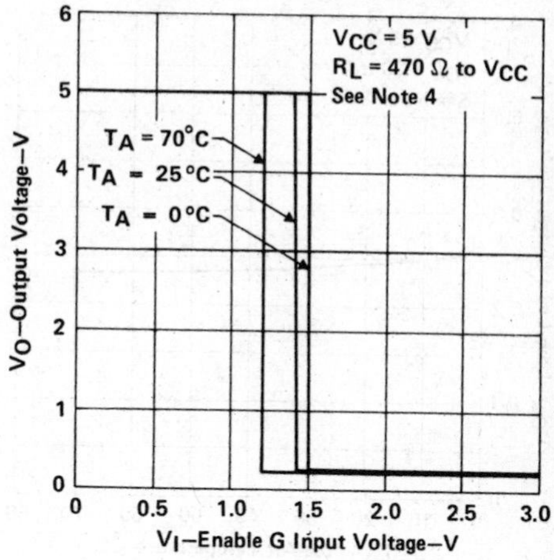
VCC = 5 V
RL = 470 Ω to VCC
See Note 4
TA = 70°C
TA = 25°C
TA = 0°C
VO—Output Voltage—V
VI—Enable G Input Voltage—V
0
0.5
1.0
1.5
2.0
2.5
3.0
1
2
3
4
5
6

FIGURE 10

HIGH-LEVEL OUTPUT VOLTAGE
vs
FREE-AIR TEMPERATURE

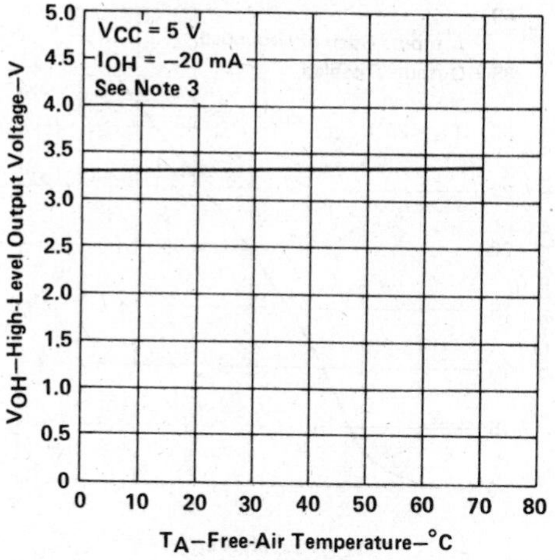
VCC = 5 V
IOH = −20 mA
See Note 3
VOH—High-Level Output Voltage—V
TA—Free-Air Temperature—°C
0
10
20
30
40
50
60
70
80
0.5
1.0
1.5
2.0
2.5
3.0
3.5
4.0
4.5
5.0

FIGURE 11

HIGH-LEVEL OUTPUT VOLTAGE
vs
OUTPUT CURRENT

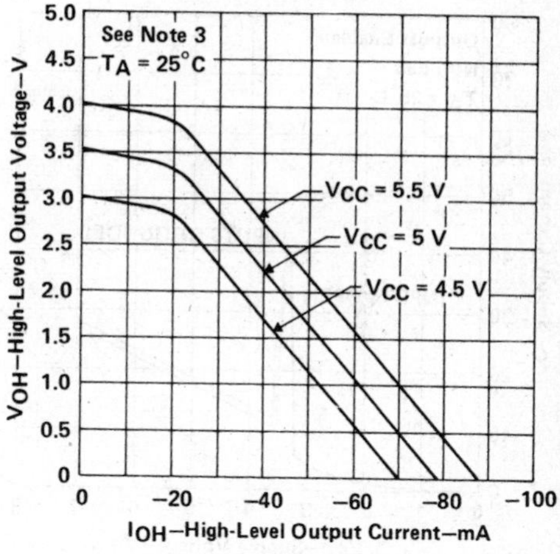
See Note 3
TA = 25°C
VCC = 5.5 V
VCC = 5 V
VCC = 4.5 V
VOH—High-Level Output Voltage—V
IOH—High-Level Output Current—mA
0
−20
−40
−60
−80
−100
0.5
1.0
1.5
2.0
2.5
3.0
3.5
4.0
4.5
5.0

FIGURE 12

NOTES: 3. The A input is connected to V_{CC} during the testing of the Y outputs and to ground during the testing of the Z outputs.
4. The A input is connected to ground during the testing of the Y outputs and to V_{CC} during the testing of the Z outputs.

TYPICAL CHARACTERISTICS

LOW-LEVEL OUTPUT VOLTAGE
vs
FREE-AIR TEMPERATURE

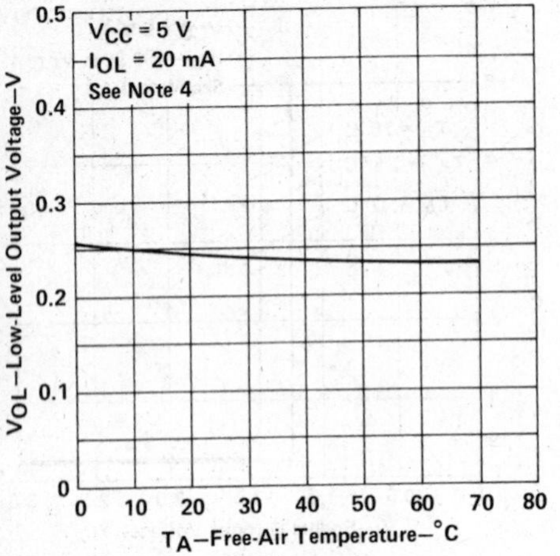
VCC = 5 V
IOL = 20 mA
See Note 4
VOL–Low-Level Output Voltage–V
TA–Free-Air Temperature–°C
0
0.1
0.2
0.3
0.4
0.5
0 10 20 30 40 50 60 70 80

FIGURE 13

LOW-LEVEL OUTPUT VOLTAGE
vs
OUTPUT CURRENT

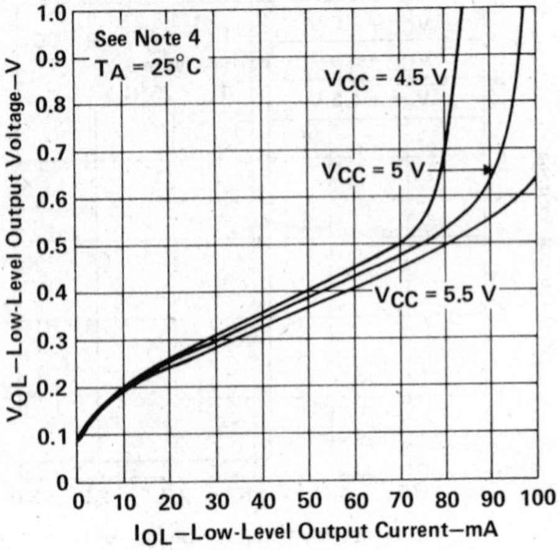
See Note 4
TA = 25°C
VCC = 4.5 V
VCC = 5 V
VCC = 5.5 V
VOL–Low-Level Output Voltage–V
IOL–Low-Level Output Current–mA
0
0.1
0.2
0.3
0.4
0.5
0.6
0.7
0.8
0.9
1.0
0 10 20 30 40 50 60 70 80 90 100

FIGURE 14

SUPPLY CURRENT
vs
SUPPLY VOLTAGE

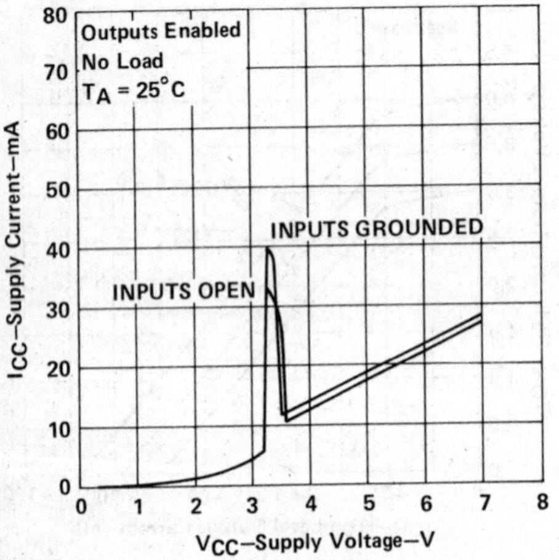
Outputs Enabled
No Load
TA = 25°C
INPUTS GROUNDED
INPUTS OPEN
ICC–Supply Current–mA
VCC–Supply Voltage–V
0
10
20
30
40
50
60
70
80
0 1 2 3 4 5 6 7 8

FIGURE 15

SUPPLY CURRENT
vs
SUPPLY VOLTAGE

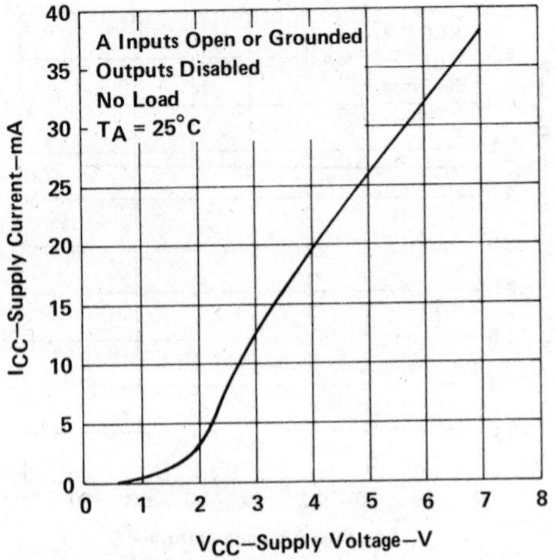
A Inputs Open or Grounded
Outputs Disabled
No Load
TA = 25°C
ICC–Supply Current–mA
VCC–Supply Voltage–V
0
5
10
15
20
25
30
35
40
0 1 2 3 4 5 6 7 8

FIGURE 16

NOTE 4: The A input is connected to ground during the testing of the Y outputs and to V_{CC} during the testing of the Z outputs.

TEXAS INSTRUMENTS
POST OFFICE BOX 655012 • DALLAS, TEXAS 75265

TYPICAL CHARACTERISTICS

SUPPLY CURRENT
vs
FREQUENCY

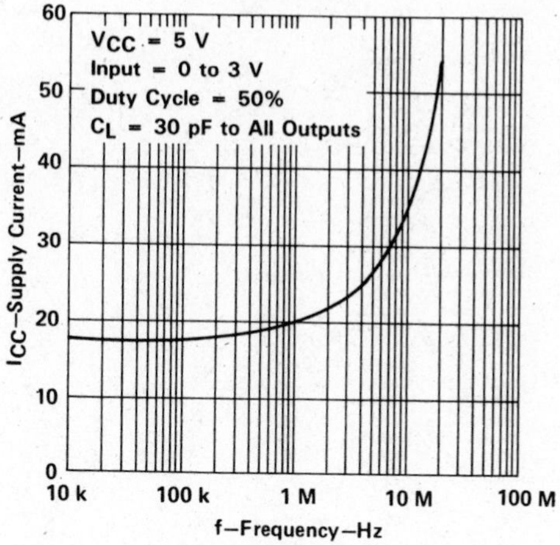
V_{CC} = 5 V
Input = 0 to 3 V
Duty Cycle = 50%
C_L = 30 pF to All Outputs
60
50
40
30
20
10
0
I_{CC}—Supply Current—mA
10 k
100 k
1 M
10 M
100 M
f—Frequency—Hz

FIGURE 17

SN75ALS195
QUADRUPLE DIFFERENTIAL LINE RECEIVER WITH 3-STATE OUTPUTS

D2928; JUNE 1986—REVISED OCTOBER 1986

- Meets EIA Standards RS-422-A and RS-423-A
- Meets CCITT Recommendations V.10, V.11, X.26, and X.27
- −7 V to 7 V Common-Mode Range with 200-mV Sensitivity
- 3-State TTL-Compatible Outputs
- High Input Impedance . . . 12 kΩ Min
- Input Hysteresis . . . 120 mV Typ
- Single 5-V Supply Operation
- Low Supply Current Requirement . . . 35 mA Max
- Improved Speed and Power Consumption Compared to MC3486

J PACKAGE
(TOP VIEW)

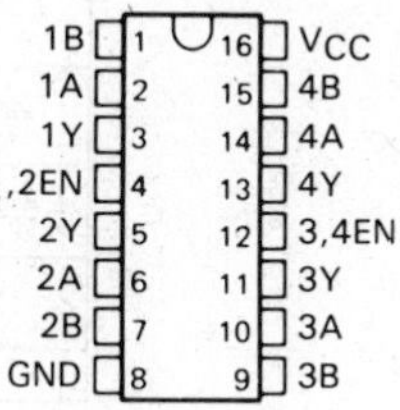
1B 1
1A 2
1Y 3
1,2EN 4
2Y 5
2A 6
2B 7
GND 8
16 VCC
15 4B
14 4A
13 4Y
12 3,4EN
11 3Y
10 3A
9 3B

description

The SN75ALS195 is a monolithic quadruple line receiver with three-state outputs designed using Advanced Low-Power Schottky technology. This technology provides combined improvements in bar design, tooling production, and wafer fabrication, providing significantly less power consumption and permitting much higher data throughput than other designs. The device meets the specifications of EIA Standards RS-422-A and RS-423-A.

The SN75ALS195 features three-state outputs that permit direct connection to a bus-organized system with a fail-safe design that ensures the outputs will always be high if the inputs are open. The device is optimized for balanced multipoint bus transmission at rates up to 10 megabits per second. The input features high input impedance, input hysteresis for increased noise immunity, and an input sensitivity of ±200 millivolts over a common-mode input voltage range of ±7 volts. It also features an active-high enable function for each of two receiver pairs. The SN75ALS195 is designed for optimum performance when used with the SN75ALS194 quadruple differential line driver.

The SN75ALS195 is characterized for operation from 0°C to 70°C.

logic symbol†

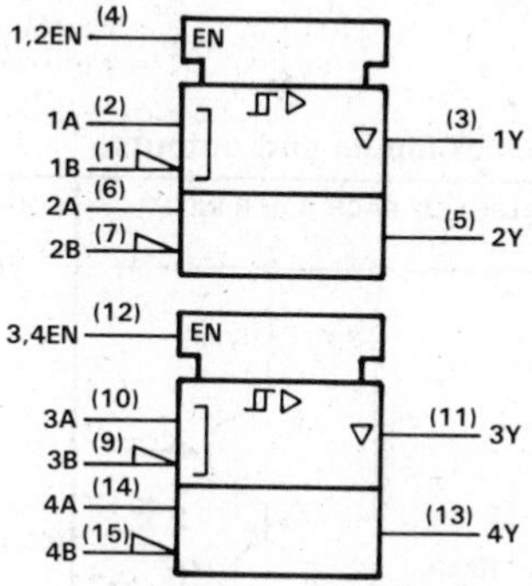
1,2EN (4)
EN
1A (2)
1B (1)
2A (6)
2B (7)
(3) 1Y
(5) 2Y
3,4EN (12)
EN
3A (10)
3B (9)
4A (14)
4B (15)
(11) 3Y
(13) 4Y

†This symbol is in accordance with ANSI/IEEE Std 91-1984 and IEC Publication 617-12.

logic diagram

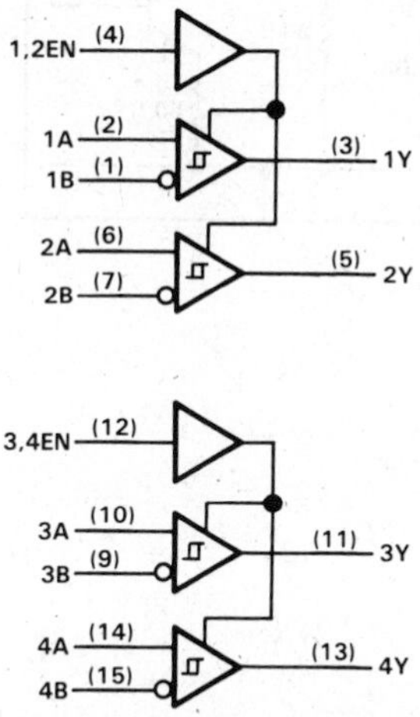
1,2EN (4)
1A (2)
1B (1)
(3) 1Y
2A (6)
2B (7)
(5) 2Y
3,4EN (12)
3A (10)
3B (9)
(11) 3Y
4A (14)
4B (15)
(13) 4Y

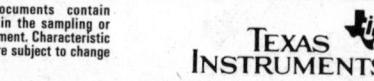

ADVANCE INFORMATION documents contain information on new products in the sampling or preproduction phase of development. Characteristic data and other specifications are subject to change without notice.

TEXAS INSTRUMENTS
POST OFFICE BOX 655012 • DALLAS, TEXAS 75265

FUNCTION TABLE (EACH RECEIVER)

DIFFERENTIAL A−B	ENABLES G	ENABLES $\overline{G}$	OUTPUT Y
$V_{ID} \geq 0.2$ V	H	X	H
	X	L	H
-0.2 V $< V_{ID} <$ 0.2 V	H	X	?
	X	L	?
$V_{ID} \leq -0.2$ V	H	X	L
	X	L	L
X	L	H	Z

H = high level
L = low level
X = irrelevant
? = indeterminate
Z = high-impedance (off)

schematics of inputs and outputs

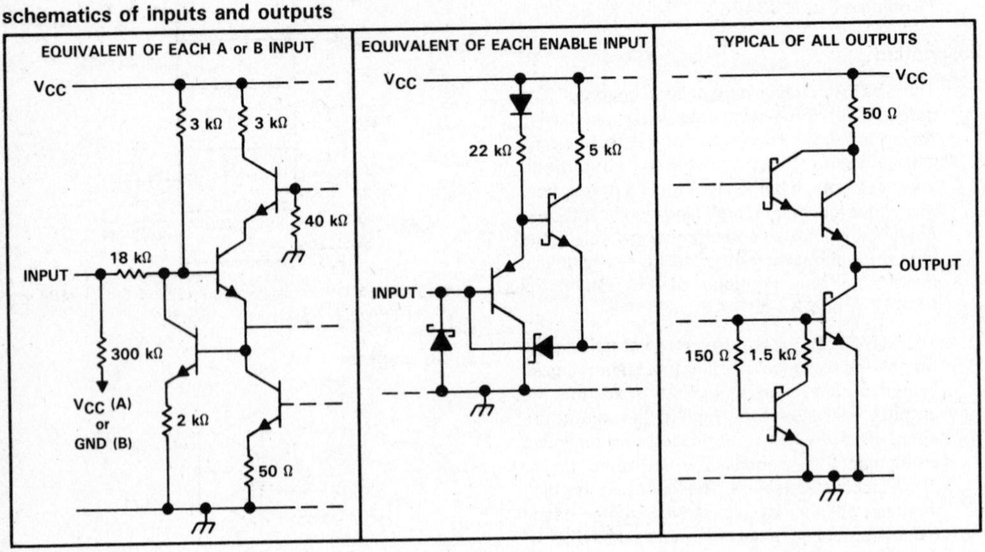
EQUIVALENT OF EACH A or B INPUT
VCC
3 kΩ
3 kΩ
40 kΩ
18 kΩ
INPUT
300 kΩ
VCC (A)
or
GND (B)
2 kΩ
50 Ω
EQUIVALENT OF EACH ENABLE INPUT
VCC
22 kΩ
5 kΩ
INPUT
TYPICAL OF ALL OUTPUTS
VCC
50 Ω
OUTPUT
150 Ω
1.5 kΩ

TEXAS INSTRUMENTS
POST OFFICE BOX 655012 • DALLAS, TEXAS 75265

absolute maximum ratings over operating free-air temperature range (unless otherwise noted)

Supply voltage, V_{CC} (see Note 1) .... 7 V
Input voltage, A or B inputs, V_I .... ±15 V
Differential input voltage (see Note 2) .... ±15 V
Enable input voltage .... 7 V
Low-level output current .... 50 mA
Continuous total dissipation at (or below) 25 °C free-air temperature (see Note 3) .... 1025 mW
Operating free-air temperature range .... 0 °C to 70 °C
Storage temperature range .... −65 °C to 150 °C
Lead temperature 1,6 mm (1/16 inch) from case for 60 seconds: J package .... 300 °C

NOTES: 1. All voltage values, except differential input voltage, are with respect to network ground terminal.
2. Differential-input voltage is measured at the noninverting input with respect to the corresponding inverting input.
3. For operating above 25 °C free-air temperature, derate the J package to to 656 mW at 70 °C at the rate of 8.2 mW/°C. In the J package, SN75ALS195 chips are glass mounted.

recommended operating conditions

	MIN	NOM	MAX	UNIT
Supply voltage, V_{CC}	4.75	5	5.25	V
Common-mode input voltage, V_{IC}			±7	V
Differential input voltage, V_{ID}			±12	V
High-level input voltage, V_{IH}	2			V
Low-level input voltage, V_{IL}			0.8	V
High-level output current, I_{OH}			−400	μA
Low-level output current, I_{OL}			16	mA
Operating free-air temperature, T_A	0		70	°C

electrical characteristics over recommended ranges of common-mode input voltage, supply voltage, and operating free-air temperature (unless otherwise noted)

PARAMETER		TEST CONDITIONS		MIN	TYP†	MAX	UNIT
V_{T+}	Positive-going threshold voltage					200	mV
V_{T-}	Negative-going threshold voltage			−200‡			mV
V_{hys}	Hysteresis§				120		mV
V_{IK}	Enable-input clamp voltage	I_I = −18 mA				−1.5	V
V_{OH}	High-level output voltage	V_{ID} = 200 mV,	I_{OH} = −400 μA	2.7	3.6		V
V_{OL}	Low-level output voltage	V_{ID} = −200 mV	I_{OL} = 8 mA			0.45	V
			I_{OL} = 16 mA			0.5	
I_{OZ}	High-impedance state output current	V_{IL} = 0.8 V, V_O = 2.7 V	V_{ID} = −3 V,			20	μA
		V_{IL} = 0.8 V, V_O = 0.5 V	V_{ID} = 3 V,			−20	
I_I	Line input current	Other input at 0 V, See Note 4	V_I = 15 V		0.7	1.2	mA
			V_I = −15 V		−1.0	−1.7	
I_{IH}	High-level enable-input current		V_{IH} = 2.7 V			20	μA
			V_{IH} = 5.25 V			100	
I_{IL}	Low-level enable-input current	V_{IL} = 0.4 V				−100	μA
	Input resistance			12	18		kΩ
I_{OS}	Short-circuit output current	V_{ID} = 3 V, See Note 5	V_O = 0,	−15	−78	−130	mA
I_{CC}	Supply current	Outputs disabled			22	35	mA

† All typical values are at V_{CC} = 5 V, T_A = 25°C.
‡ The algebraic convention, in which the less positive limit is designated minimum, is used in this data sheet for threshold voltage levels only.
§ Hysteresis is the difference between the positive-going input threshold voltage, V_{T+}, and the negative-going input threshold voltage, V_{T-}.

NOTES: 4. Refer to EIA Standard RS-422-A and RS-423-A for exact conditions.
5. Not more than one output should be shorted at a time and the duration of the short-circuit should not exceed one second.

switching characteristics, V_{CC} = 5 V, T_A = 25°C

PARAMETER		TEST CONDITIONS		MIN	TYP	MAX	UNIT
t_{PLH}	Propagation delay time, low-to-high-level output	V_{ID} = 0 V to 3 V, See Figure 2	C_L = 15 pF,		15	22	ns
t_{PHL}	Propagation delay time, high-to-low-level output				15	22	ns
t_{PZH}	Output enable time to high level	C_L = 15 pF,	See Figure 3		13	25	ns
t_{PZL}	Output enable time to low level				11	25	
t_{PHZ}	Output disable time from high level	C_L = 15 pF,	See Figure 3		13	25	ns
t_{PLZ}	Output disable time from low level				15	22	

PARAMETER MEASUREMENT INFORMATION

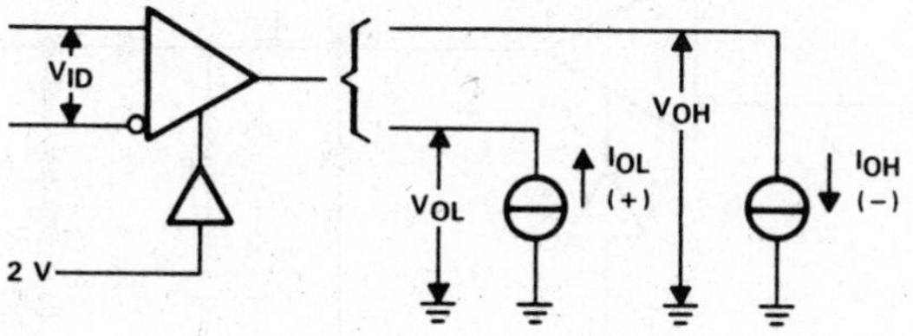
V_{ID}
V_{OH}
V_{OL}
I_{OL} (+)
I_{OH} (−)
2 V

FIGURE 1. V_{OH}, V_{OL}

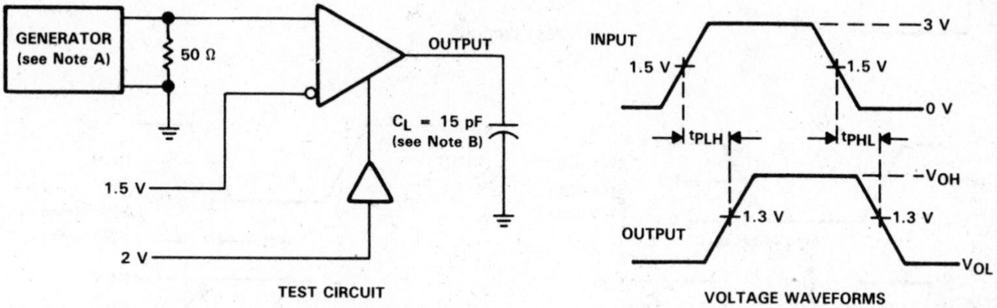
GENERATOR (see Note A)
50 Ω
OUTPUT
C_L = 15 pF (see Note B)
1.5 V
2 V
TEST CIRCUIT
INPUT
3 V
1.5 V
1.5 V
0 V
t_{PLH}
t_{PHL}
V_{OH}
1.3 V
1.3 V
OUTPUT
V_{OL}
VOLTAGE WAVEFORMS

NOTES: A. The input pulse is supplied by a generator having the following characteristics: PRR ≤ 1 MHz, duty cycle ≤ 50%, Z_{out} = 50 Ω, t_r ≤ 6 ns, t_f ≤ 6 ns.

B. C_L includes probe and jig capacitance.

FIGURE 2. PROPAGATION DELAY TIMES

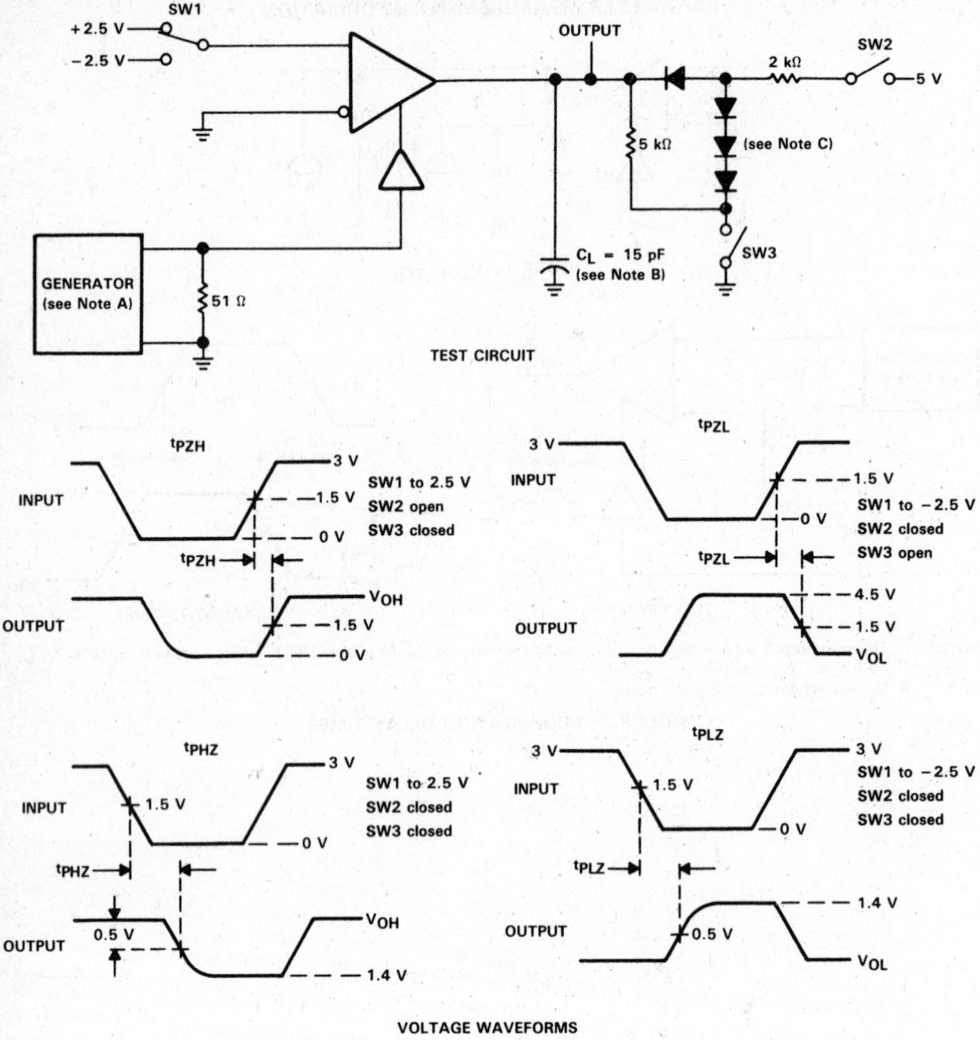
SW1
+2.5 V
−2.5 V
OUTPUT
SW2
2 kΩ
5 V
5 kΩ
(see Note C)
SW3
C_L = 15 pF
(see Note B)
GENERATOR
(see Note A)
51 Ω
TEST CIRCUIT
t_{PZH}
INPUT
3 V
1.5 V
0 V
SW1 to 2.5 V
SW2 open
SW3 closed
OUTPUT
V_{OH}
t_{PZL}
SW1 to −2.5 V
SW2 closed
SW3 open
4.5 V
V_{OL}
t_{PHZ}
SW2 closed
SW3 closed
0.5 V
1.4 V
t_{PLZ}
VOLTAGE WAVEFORMS

NOTES: A. The input pulse is supplied by a generator having the following characteristics: PRR ≤ 1 MHz, duty cycle ≤ 50%, Z_{out} = 50 Ω, t_r ≤ 6 ns, t_f ≤ 6 ns.

B. C_L includes probe and jig capacitance.

C. All diodes are 1N3064 or equivalent.

FIGURE 3. ENABLE AND DISABLE TIMES

TEXAS INSTRUMENTS
POST OFFICE BOX 655012 • DALLAS, TEXAS 75265

TYPICAL CHARACTERISTICS

OUTPUT VOLTAGE
vs
ENABLE VOLTAGE

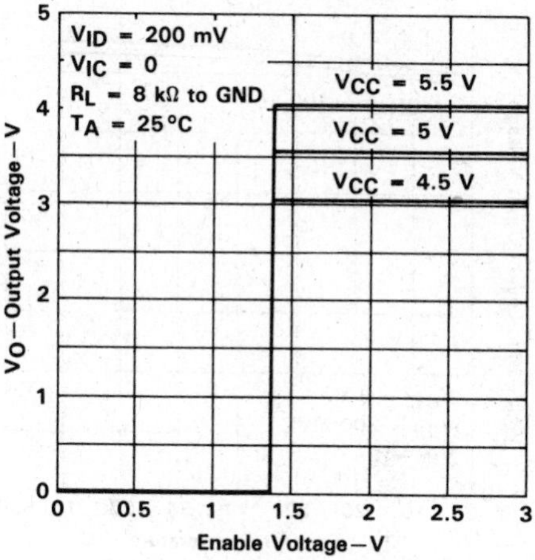
V_{ID} = 200 mV
V_{IC} = 0
R_L = 8 kΩ to GND
T_A = 25°C
V_{CC} = 5.5 V
V_{CC} = 5 V
V_{CC} = 4.5 V
V_O—Output Voltage—V
Enable Voltage—V

FIGURE 4

OUTPUT VOLTAGE
vs
ENABLE VOLTAGE

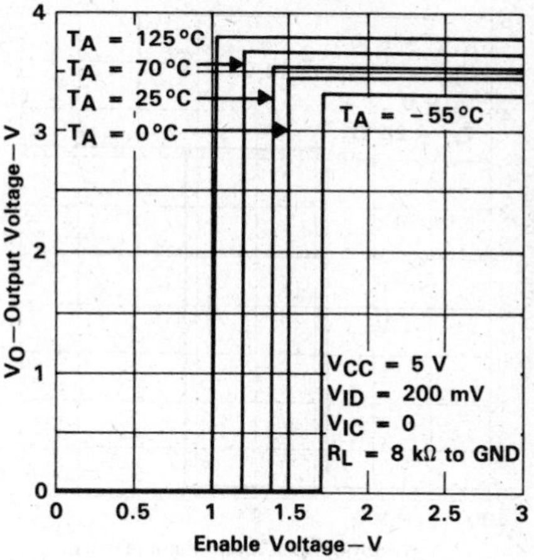
T_A = 125°C
T_A = 70°C
T_A = 25°C
T_A = 0°C
T_A = −55°C
V_{CC} = 5 V
V_{ID} = 200 mV
V_{IC} = 0
R_L = 8 kΩ to GND
V_O—Output Voltage—V
Enable Voltage—V

FIGURE 5

OUTPUT VOLTAGE
vs
ENABLE VOLTAGE

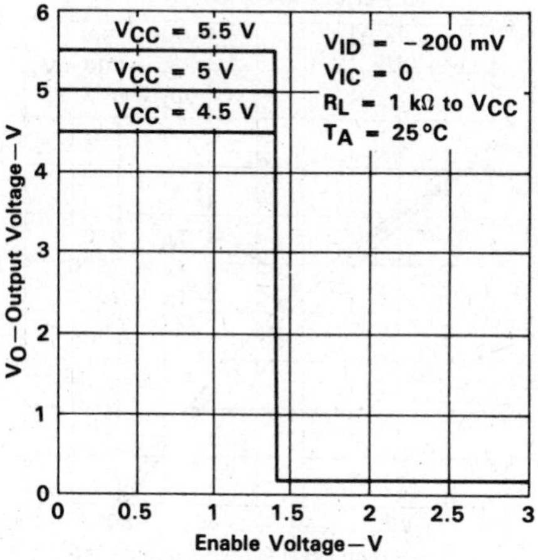
V_{CC} = 5.5 V
V_{CC} = 5 V
V_{CC} = 4.5 V
V_{ID} = −200 mV
V_{IC} = 0
R_L = 1 kΩ to V_{CC}
T_A = 25°C
V_O—Output Voltage—V
Enable Voltage—V

FIGURE 6

OUTPUT VOLTAGE
vs
ENABLE VOLTAGE

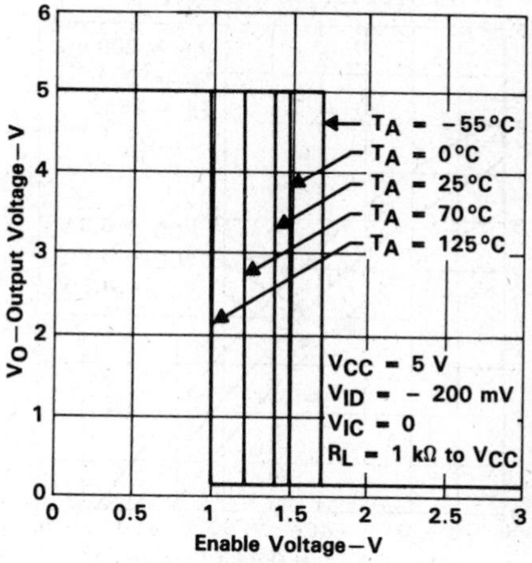
T_A = −55°C
T_A = 0°C
T_A = 25°C
T_A = 70°C
T_A = 125°C
V_{CC} = 5 V
V_{ID} = − 200 mV
V_{IC} = 0
R_L = 1 kΩ to V_{CC}
V_O—Output Voltage—V
Enable Voltage—V

FIGURE 7

TYPICAL CHARACTERISTICS

OUTPUT VOLTAGE
vs
DIFFERENTIAL INPUT VOLTAGE

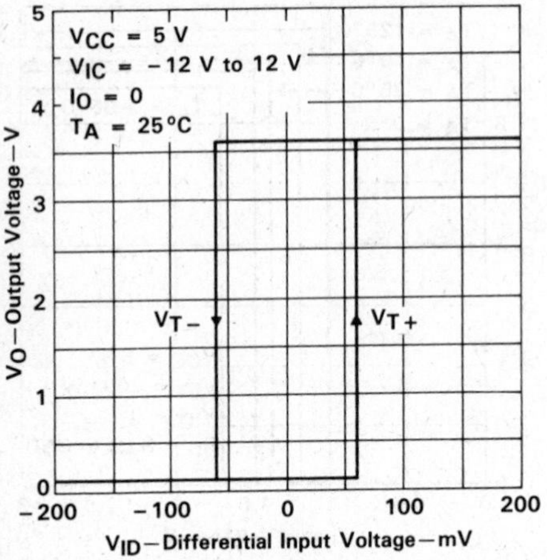
V_{CC} = 5 V
V_{IC} = −12 V to 12 V
I_O = 0
T_A = 25°C
V_{T-}
V_{T+}
V_O—Output Voltage—V
V_{ID}—Differential Input Voltage—mV
5
4
3
2
1
0
−200
−100
0
100
200

FIGURE 8

HIGH-LEVEL OUTPUT VOLTAGE
vs
FREE-AIR TEMPERATURE

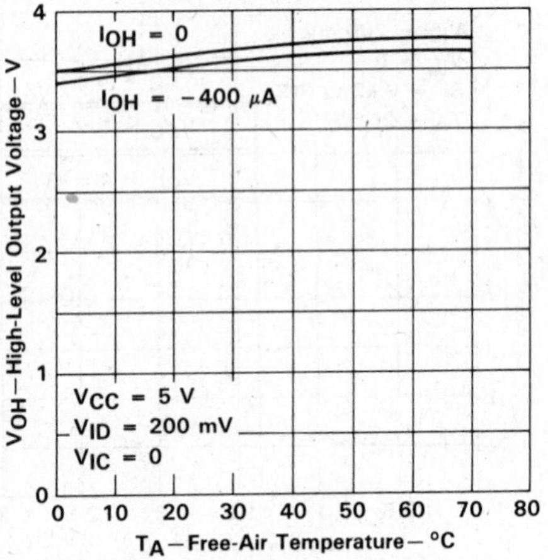
I_{OH} = 0
I_{OH} = −400 μA
V_{CC} = 5 V
V_{ID} = 200 mV
V_{IC} = 0
V_{OH}—High-Level Output Voltage—V
T_A—Free-Air Temperature—°C
4
3
2
1
0
0
10
20
30
40
50
60
70
80

FIGURE 9

HIGH-LEVEL OUTPUT VOLTAGE
vs
HIGH-LEVEL OUTPUT CURRENT

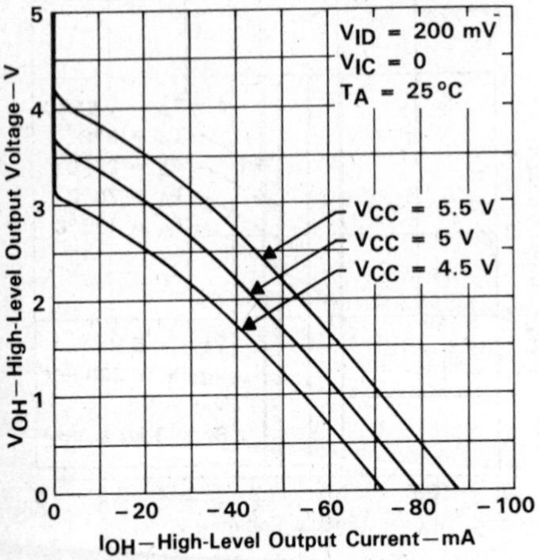
V_{ID} = 200 mV
V_{IC} = 0
T_A = 25°C
V_{CC} = 5.5 V
V_{CC} = 5 V
V_{CC} = 4.5 V
V_{OH}—High-Level Output Voltage—V
I_{OH}—High-Level Output Current—mA
5
4
3
2
1
0
0
−20
−40
−60
−80
−100

FIGURE 10

HIGH-LEVEL OUTPUT VOLTAGE
vs
HIGH-LEVEL OUTPUT CURRENT

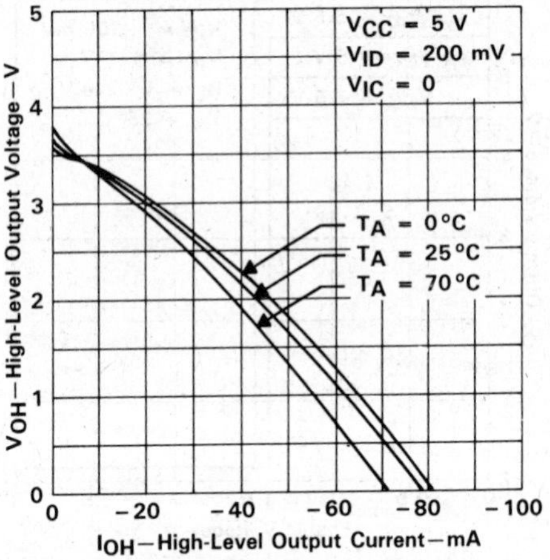
V_{CC} = 5 V
V_{ID} = 200 mV
V_{IC} = 0
T_A = 0°C
T_A = 25°C
T_A = 70°C
V_{OH}—High-Level Output Voltage—V
I_{OH}—High-Level Output Current—mA
5
4
3
2
1
0
0
−20
−40
−60
−80
−100

FIGURE 11

TEXAS INSTRUMENTS
POST OFFICE BOX 655012 • DALLAS, TEXAS 75265

TYPICAL CHARACTERISTICS

LOW-LEVEL OUTPUT VOLTAGE
vs
FREE-AIR TEMPERATURE

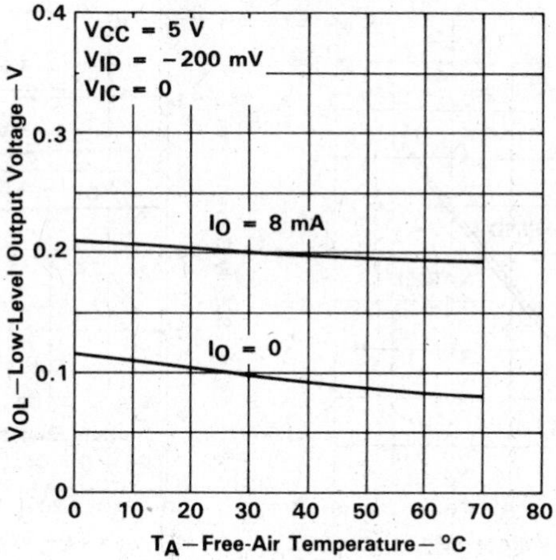
V_{CC} = 5 V
V_{ID} = −200 mV
V_{IC} = 0
I_O = 8 mA
I_O = 0
V_{OL}—Low-Level Output Voltage—V
0.4
0.3
0.2
0.1
0
0 10 20 30 40 50 60 70 80
T_A—Free-Air Temperature—°C

FIGURE 12

LOW-LEVEL OUTPUT VOLTAGE
vs
LOW-LEVEL OUTPUT CURRENT

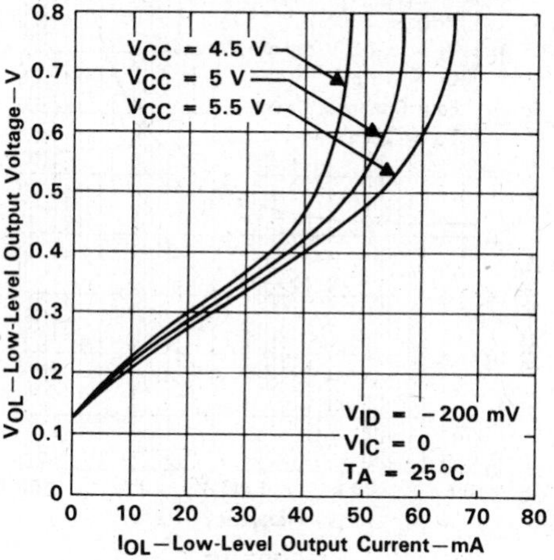
V_{CC} = 4.5 V
V_{CC} = 5 V
V_{CC} = 5.5 V
V_{ID} = −200 mV
V_{IC} = 0
T_A = 25°C
V_{OL}—Low-Level Output Voltage—V
0.8
0.7
0.6
0.5
0.4
0.3
0.2
0.1
0
0 10 20 30 40 50 60 70 80
I_{OL}—Low-Level Output Current—mA

FIGURE 13

LOW-LEVEL OUTPUT VOLTAGE
vs
LOW-LEVEL OUTPUT CURRENT

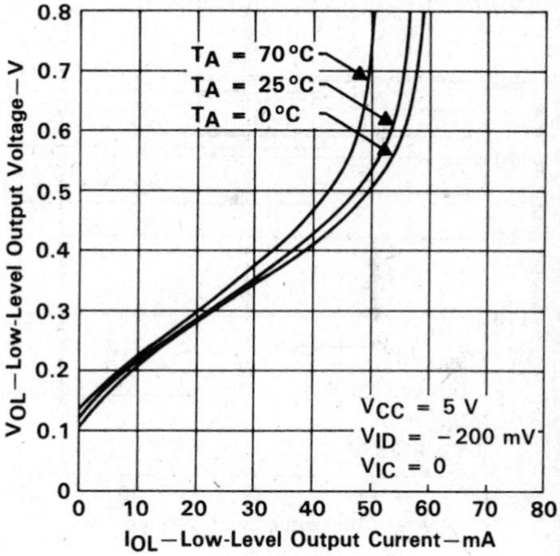
T_A = 70°C
T_A = 25°C
T_A = 0°C
V_{CC} = 5 V
V_{ID} = −200 mV
V_{IC} = 0
V_{OL}—Low-Level Output Voltage—V
0.8
0.7
0.6
0.5
0.4
0.3
0.2
0.1
0
0 10 20 30 40 50 60 70 80
I_{OL}—Low-Level Output Current—mA

FIGURE 14

TYPICAL CHARACTERISTICS

SUPPLY CURRENT
vs
SUPPLY VOLTAGE

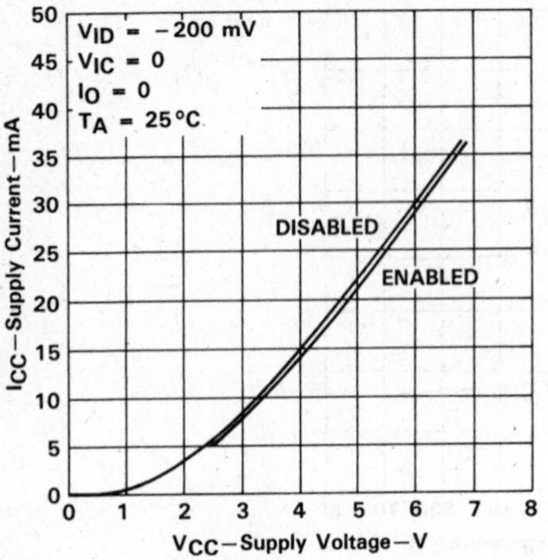
V_{ID} = −200 mV
V_{IC} = 0
I_O = 0
T_A = 25°C
DISABLED
ENABLED
I_{CC}—Supply Current—mA
V_{CC}—Supply Voltage—V
0
1
2
3
4
5
6
7
8
5
10
15
20
25
30
35
40
45
50

FIGURE 15

SUPPLY CURRENT
vs
FREE-AIR TEMPERATURE

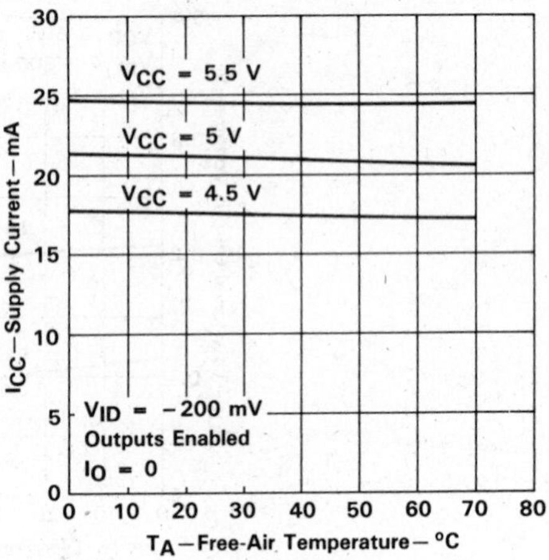
V_{CC} = 5.5 V
V_{CC} = 5 V
V_{CC} = 4.5 V
V_{ID} = −200 mV
Outputs Enabled
I_O = 0
I_{CC}—Supply Current—mA
T_A—Free-Air Temperature—°C
0
10
20
30
40
50
60
70
80
5
15
25

FIGURE 16

SUPPLY CURRENT
vs
DIFFERENTIAL INPUT VOLTAGE

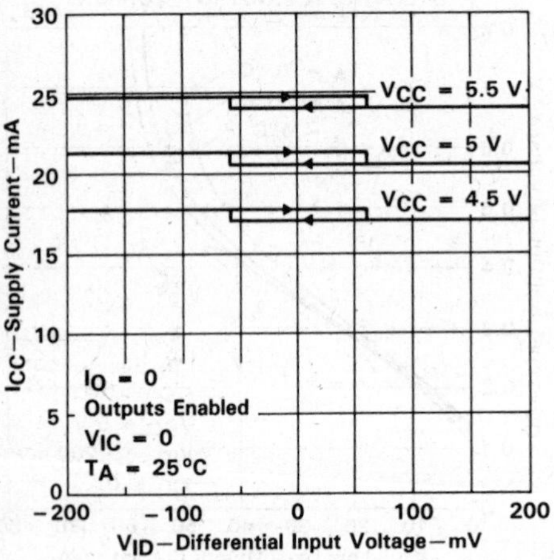
V_{CC} = 5.5 V
V_{CC} = 5 V
V_{CC} = 4.5 V
I_O = 0
Outputs Enabled
V_{IC} = 0
T_A = 25°C
I_{CC}—Supply Current—mA
V_{ID}—Differential Input Voltage—mV
−200
−100
0
100
200
5
10
15
20
25
30

FIGURE 17

SUPPLY CURRENT
vs
FREQUENCY

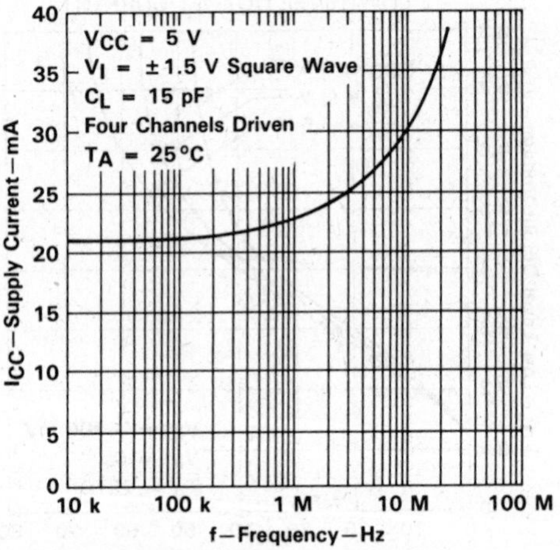
V_{CC} = 5 V
V_I = ±1.5 V Square Wave
C_L = 15 pF
Four Channels Driven
T_A = 25°C
I_{CC}—Supply Current—mA
f—Frequency—Hz
10 k
100 k
1 M
10 M
100 M
0
5
10
15
20
25
30
35
40

FIGURE 18

4

Line Drivers/Receivers

ADVANCE INFORMATION

TEXAS INSTRUMENTS

POST OFFICE BOX 655012 • DALLAS, TEXAS 75265

TYPICAL CHARACTERISTICS

INPUT RESISTANCE
vs
FREE-AIR TEMPERATURE

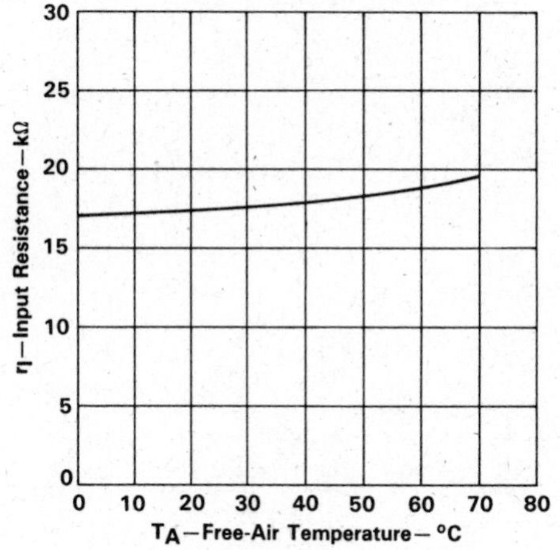
30
25
20
15
10
5
0
0 10 20 30 40 50 60 70 80
r_I—Input Resistance—kΩ
T_A—Free-Air Temperature—°C

FIGURE 19

INPUT CURRENT
vs
INPUT VOLTAGE

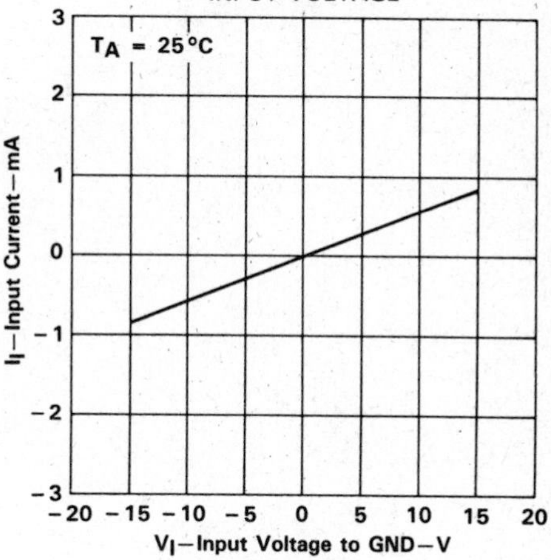
T_A = 25°C
3
2
1
0
−1
−2
−3
−20 −15 −10 −5 0 5 10 15 20
I_I—Input Current—mA
V_I—Input Voltage to GND—V

FIGURE 20

SWITCHING CHARACTERISTICS
vs
FREE-AIR TEMPERATURE

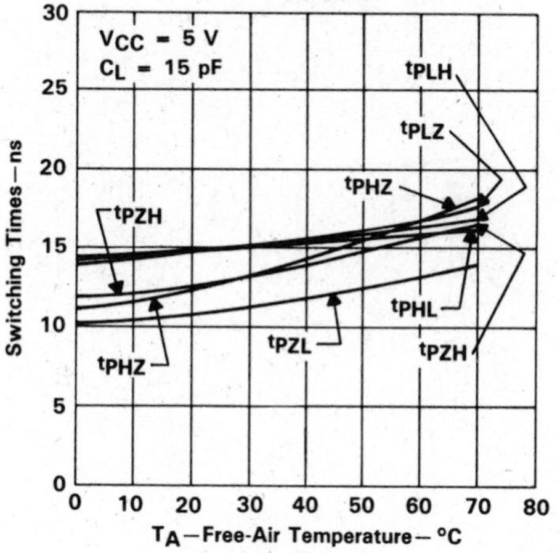
V_{CC} = 5 V
C_L = 15 pF
t_{PLH}
t_{PLZ}
t_{PHZ}
t_{PZH}
t_{PHL}
t_{PZL}
t_{PHZ}
t_{PZH}
30
25
20
15
10
5
0
0 10 20 30 40 50 60 70 80
Switching Times—ns
T_A—Free-Air Temperature—°C

FIGURE 21

PROPAGATION DELAY TIME
vs
SUPPLY VOLTAGE

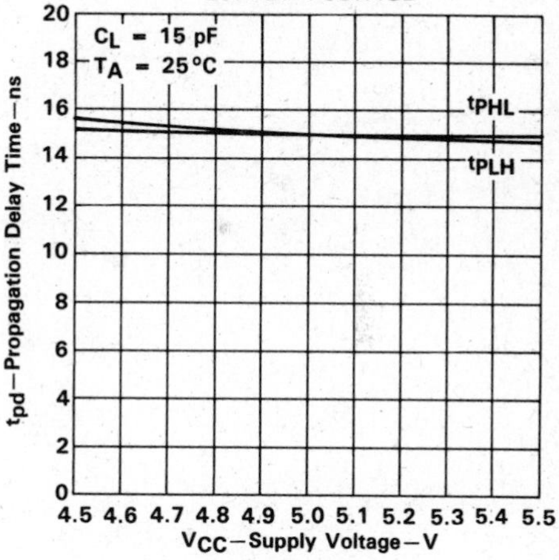
C_L = 15 pF
T_A = 25°C
t_{PHL}
t_{PLH}
20
18
16
14
12
10
8
6
4
2
0
4.5 4.6 4.7 4.8 4.9 5.0 5.1 5.2 5.3 5.4 5.5
t_{pd}—Propagation Delay Time—ns
V_{CC}—Supply Voltage—V

FIGURE 22

DUAL LINE DRIVERS WITH ADJUSTABLE SLEW RATE

D2608, OCTOBER 1980—REVISED SEPTEMBER 1986

- Meets EIA Standards RS-423-A and RS-232-C and Federal Standard 1030
- Slew Rate Control
- Output Short-Circuit-Current Limiting
- Wide Supply Voltage Range
- 8-Pin Dual-In-Line Package
- Designed to be Interchangeable with Fairchild 9636A

D, JG, OR P DUAL-IN-LINE PACKAGE
(TOP VIEW)

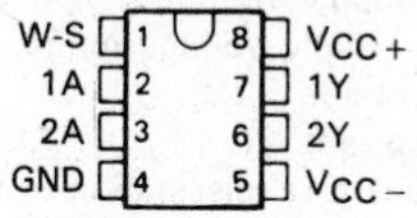
W-S
1A
2A
GND
VCC+
1Y
2Y
VCC−

logic symbol†

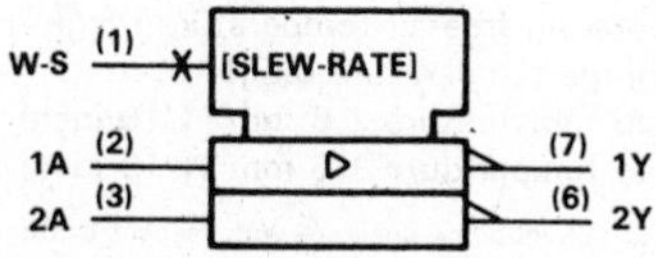
W-S
(1)
[SLEW-RATE]
1A
(2)
2A
(3)
(7)
1Y
(6)
2Y

† This symbol is in accordance with ANSI/IEEE Std 91-1984 and IEC Publication 617-12.

logic diagram

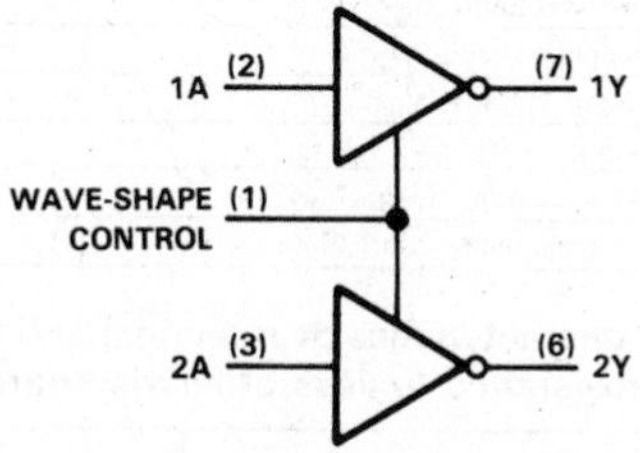
1A
(2)
(7)
1Y
WAVE-SHAPE (1)
CONTROL
2A
(3)
(6)
2Y

description

The uA9636AC is a dual single-ended line driver designed to meet EIA Standards RS-423-A and RS-232-C and Federal Standard 1030. The slew rates of both amplifiers are controlled by a single external resistor, R_{WS}, connected between the wave-shape-control terminal and ground. Output current limiting is provided. Inputs are compatible with TTL and CMOS and are diode-protected against negative transients. This device operates from ± 12 volts and is supplied in an 8-pin dual-in-line package.

The uA9636AC is characterized for operation from 0°C to 70°C.

schematics of inputs and outputs

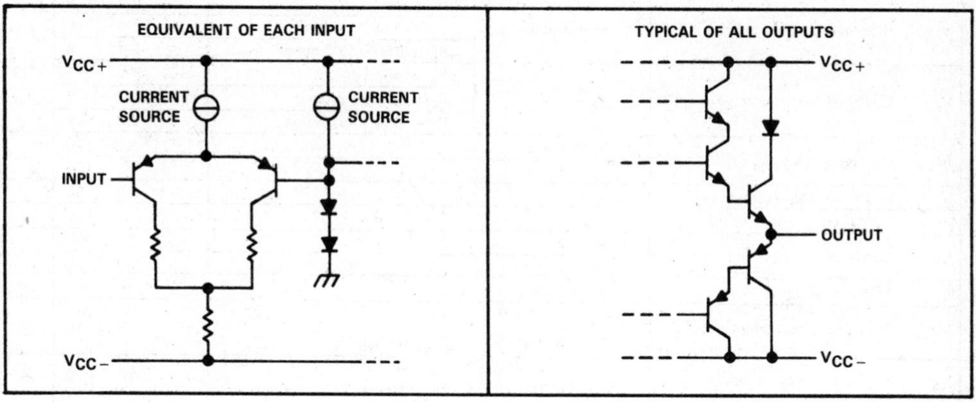
EQUIVALENT OF EACH INPUT
VCC+
CURRENT SOURCE
CURRENT SOURCE
INPUT
VCC−
TYPICAL OF ALL OUTPUTS
VCC+
OUTPUT
VCC−

TEXAS INSTRUMENTS
POST OFFICE BOX 655012 • DALLAS, TEXAS 75265

absolute maximum ratings over operating free-air temperature range (unless otherwise noted)

Positive supply voltage range, V_{CC+} (see Note 1) .... V_{CC-} to 15 V
Negative supply voltage range, V_{CC-} .... 0.5 V to −15 V
Output voltage .... ±15 V
Output current .... ±150 mA
Continuous total dissipation at (or below) 25°C free-air temperature (see Note 2):
- D package .... 725 mW
- JG package .... 825 mW
- P package .... 1000 mW

Operating free-air temperature range .... 0°C to 70°C
Storage temperature range .... −65°C to 150°C
Lead temperature 1,6 mm (1/16 inch) from case for 60 seconds: JG package .... 300°C
Lead temperature 1,6 mm (1/16 inch) from case for 10 seconds: D and P packages .... 260°C

NOTES: 1. All voltage values are with respect to the network ground terminal.
2. For operation above 25°C free-air temperature, refer to Dissipation Derating Curves in Appendix A. In the JG package, uA9636AC chips are glass mounted. In the P package, use the 8.0 mW/°C curve for these devices.

recommended operating conditions

	MIN	NOM	MAX	UNIT
Positive supply voltage, V_{CC+}	10.8	12	13.2	V
Negative supply voltage, V_{CC-}	−10.8	−12	−13.2	V
High-level input voltage, V_{IH}	2			V
Low-level input voltage, V_{IL}			0.8	V
Wave-shaping resistor, R_{WS}	10		1000	kΩ
Operating free-air temperature, T_A	0		70	°C

electrical characteristics over recommended range of free-air temperature, supply voltage, and wave-shaping resistance (unless otherwise noted)

PARAMETER		TEST CONDITIONS		MIN	TYP†	MAX (See Note 3)	UNIT
V_{IK}	Input clamp voltage	I_I = −15 mA			−1.1	−1.5	V
V_{OH}	High-level output voltage	V_I = 0.8 V	R_L = ∞	5	5.6	6	V
			R_L = 3 kΩ to ground	5	5.6	6	
			R_L = 450 Ω to ground	4	5.4	6	
V_{OL}	Low-level output voltage	V_I = 2 V	R_L = ∞	−6	−5.7	−5	V
			R_L = 3 kΩ to ground	−6	−5.6	−5	
			R_L = 450 Ω to ground	−6	−5.4	−4	
I_{IH}	High-level input current	V_I = 2.4 V				10	μA
		V_I = 5.5 V				100	
I_{IL}	Low-level input current	V_I = 0.4 V			−20	−80	μA
I_O	Output current (power off)	$V_{CC±}$ = 0,	V_O = ±6 V			±100	μA
I_{OS}	Short-circuit output current‡	V_I = 2 V		15	25	150	mA
		V_I = 0		−15	−40	−150	
r_o	Output resistance	R_L = 450 Ω			25	50	Ω
I_{CC+}	Positive supply current	V_{CC} = ±12 V, R_{WS} = 100 kΩ,	V_I = 0, Output open		13	18	mA
I_{CC-}	Negative supply current	V_{CC} = ±12 V, R_{WS} = 100 kΩ,	V_I = 0, Output open		−13	−18	mA

†All typical values are at V_{CC} ±12 V, T_A = 25°C.
‡Not more than one output should be shorted to ground at a time.
NOTE 3: The algebraic convention, in which the less-positive (more-negative) limit is designated as minimum, is used in this data sheet for logic voltage levels, e.g., when −5 V is the maximum, the minimum is a more-negative voltage.

switching characteristics, $V_{CC\pm}$ = 12 V, T_A = 25°C, see Figure 1

PARAMETER		TEST CONDITIONS		MIN	TYP	MAX	UNIT
t_{TLH}	Transition time, low-to-high-level output	R_L = 450 Ω, C_L = 30 pF	R_{WS} = 10 kΩ	0.8	1.1	1.4	μs
			R_{WS} = 100 kΩ	8	11	14	
			R_{WS} = 500 kΩ	40	55	70	
			R_{WS} = 1 MΩ	80	110	140	
t_{THL}	Transition time, high-to-low-level output	R_L = 450 Ω, C_L = 30 pF	R_{WS} = 10 kΩ	0.8	1.1	1.4	μs
			R_{WS} = 100 kΩ	8	11	14	
			R_{WS} = 500 kΩ	40	55	70	
			R_{WS} = 1 mΩ	80	110	140	

PARAMETER MEASUREMENT INFORMATION

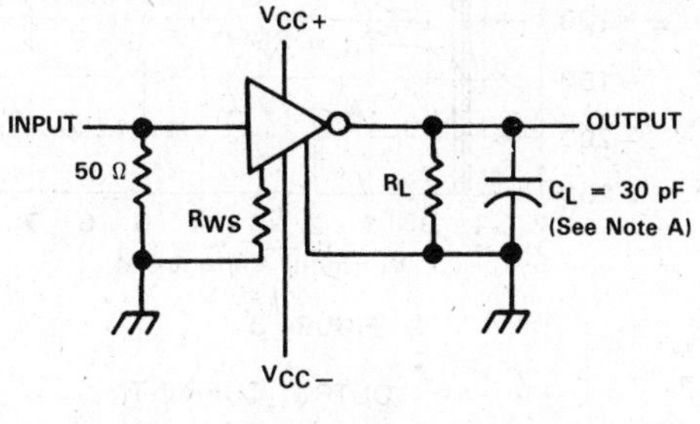
VCC+
INPUT
50 Ω
RWS
RL
OUTPUT
CL = 30 pF
(See Note A)
VCC−

TEST CIRCUIT

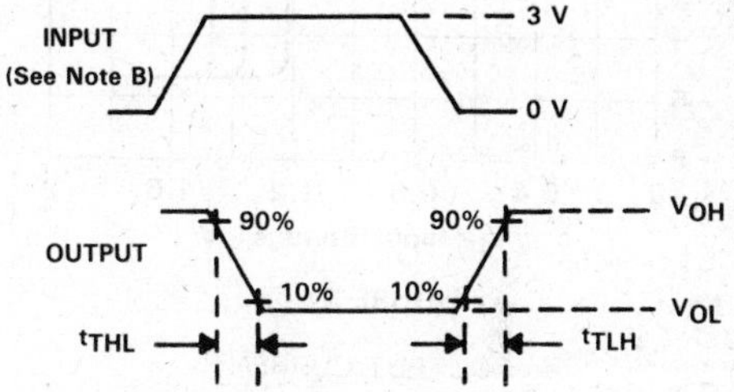
INPUT
(See Note B)
3 V
0 V
OUTPUT
90%
90%
10%
10%
VOH
VOL
tTHL
tTLH

VOLTAGE WAVEFORMS

NOTES: A. C_L includes probe and jig capacitance.
B. The input pulse is supplied by a generator having the following characteristics: t_r ≤ 10 ns, t_f ≤ 10 ns, Z_{out} = 50 Ω, PRR ≤ 1 kHz, duty cycle = 50%.

FIGURE 1. TRANSITION TIMES

TYPICAL CHARACTERISTICS

OUTPUT VOLTAGE
vs
INPUT VOLTAGE

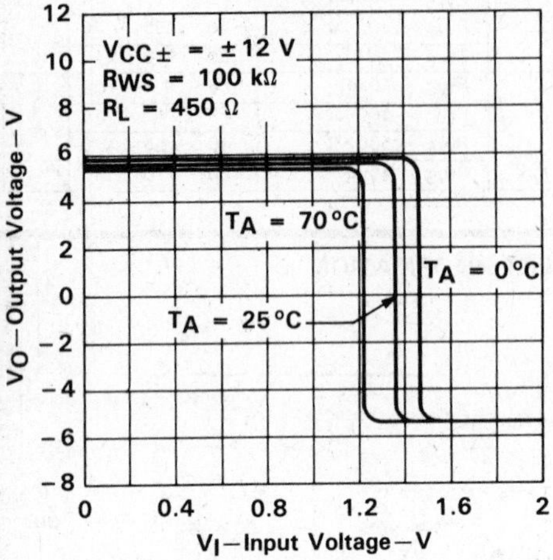
VCC± = ±12 V
RWS = 100 kΩ
RL = 450 Ω
TA = 70°C
TA = 0°C
TA = 25°C
12
10
8
6
4
2
0
−2
−4
−6
−8
0
0.4
0.8
1.2
1.6
2
VO—Output Voltage—V
VI—Input Voltage—V

FIGURE 2

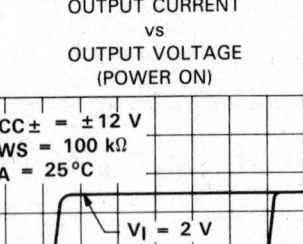

INPUT CURRENT
vs
INPUT VOLTAGE

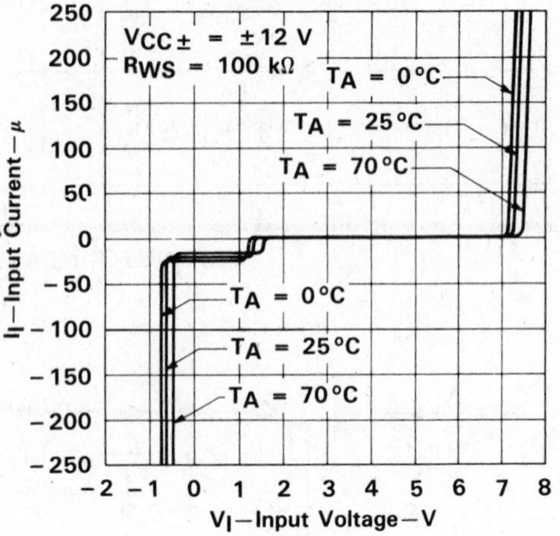
VCC± = ±12 V
RWS = 100 kΩ
TA = 0°C
TA = 25°C
TA = 70°C
TA = 0°C
TA = 25°C
TA = 70°C
250
200
150
100
50
0
−50
−100
−150
−200
−250
−2
−1
0
1
2
3
4
5
6
7
8
II—Input Current—μ
VI—Input Voltage—V

FIGURE 3

OUTPUT CURRENT
vs
OUTPUT VOLTAGE
(POWER ON)

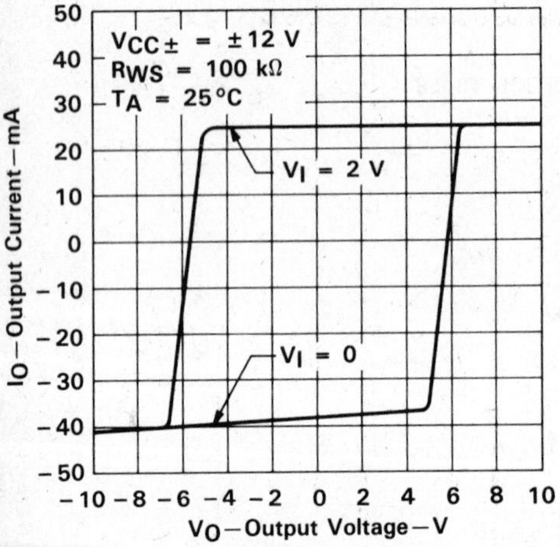
VCC± = ±12 V
RWS = 100 kΩ
TA = 25°C
VI = 2 V
VI = 0
50
40
30
20
10
0
−10
−20
−30
−40
−50
−10
−8
−6
−4
−2
0
2
4
6
8
10
IO—Output Current—mA
VO—Output Voltage—V

FIGURE 4

OUTPUT CURRENT
vs
OUTPUT VOLTAGE
(POWER OFF)

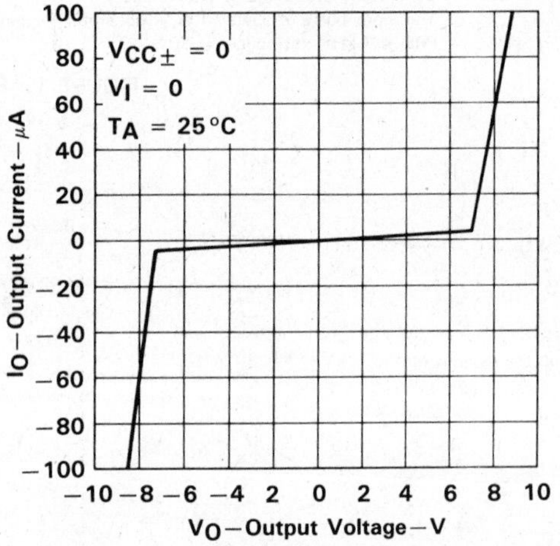
VCC± = 0
VI = 0
TA = 25°C
100
80
60
40
20
0
−20
−40
−60
−80
−100
−10
−8
−6
−4
2
0
2
4
6
8
10
IO—Output Current—μA
VO—Output Voltage—V

FIGURE 5

TYPICAL CHARACTERISTICS

TRANSITION TIMES
vs
WAVESHAPING RESISTANCE

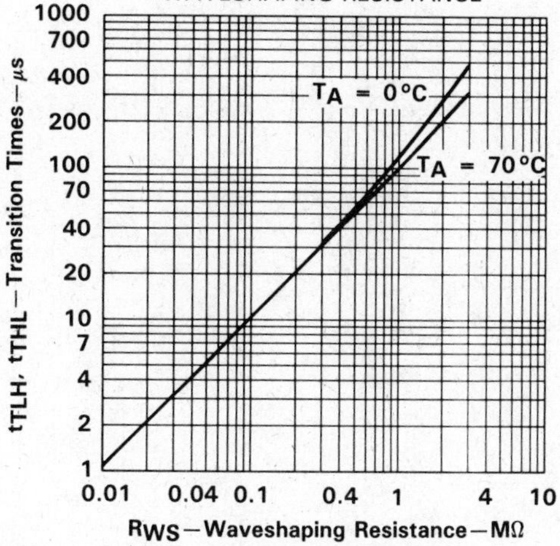
1000
700
400
200
100
70
40
20
10
7
4
2
1
tTLH, tTHL – Transition Times – μs
TA = 0°C
TA = 70°C
0.01
0.04
0.1
0.4
1
4
10
RWS – Waveshaping Resistance – MΩ

FIGURE 6

TYPICAL APPLICATION DATA

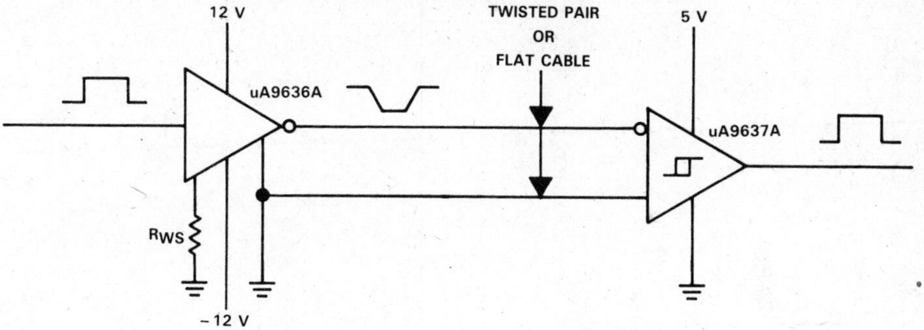
12 V
uA9636A
TWISTED PAIR
OR
FLAT CABLE
5 V
uA9637A
RWS
–12 V

FIGURE 7. RS-423-A SYSTEM APPLICATION

4
Line Drivers/Receivers

- Meets EIA Standards RS-422-A and RS-423-A
- Meets Federal Standards 1020 and 1030
- Operates from Single 5-V Power Supply
- Wide Common-Mode Voltage Range
- High Input Impedance
- TTL-Compatible Outputs
- High-Speed Schottky Circuitry
- 8-Pin Dual-In-Line and "Small Outline" Packages
- Similar to SN75157 except for Corner V_{CC} and Ground Pin Positions
- Designed to Be Interchangeable with Fairchild μA9637A

uA9637M . . . JG PACKAGE
uA9637C . . . D, JG, OR P PACKAGE
(TOP VIEW)

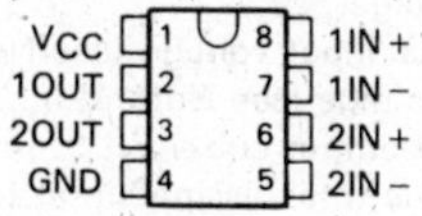
VCC
1OUT
2OUT
GND
1
2
3
4
8
7
6
5
1IN+
1IN−
2IN+
2IN−

logic symbol†

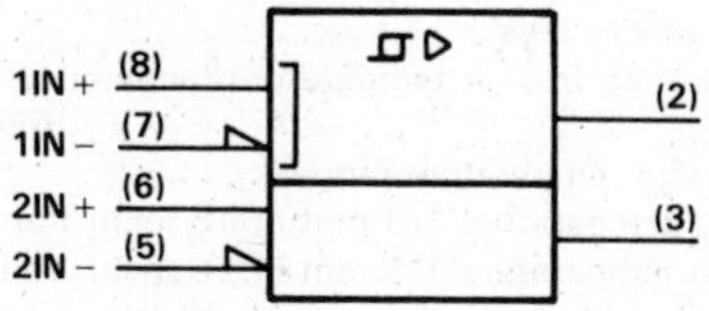
1IN+
(8)
1IN−
(7)
2IN+
(6)
2IN−
(5)
(2)
1OUT
(3)
2OUT

† This symbol is in accordance with ANSI/IEEE Std 91-1984 and IEC Publication 617-12.

logic diagram

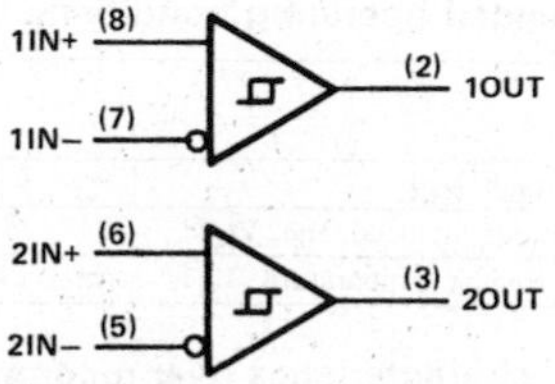
1IN+
(8)
1IN−
(7)
(2)
1OUT
2IN+
(6)
2IN−
(5)
(3)
2OUT

description

The uA9637AC is a dual differential line receiver designed to meet EIA standards RS-422-A and RS-423-A and Federal Standards 1020 and 1030. It utilizes Schottky circuitry and has TTL-compatible outputs. The inputs are compatible with either a single-ended or a differential-line system. This device operates from a single 5-volt power supply and is supplied in an 8-pin dual-in-line package and small outline package.

The uA9637AM is characterized over the full military temperature range of −55°C to 125°C. The uA9637AC is characterized for operation from 0°C to 70°C.

schematics of inputs and outputs

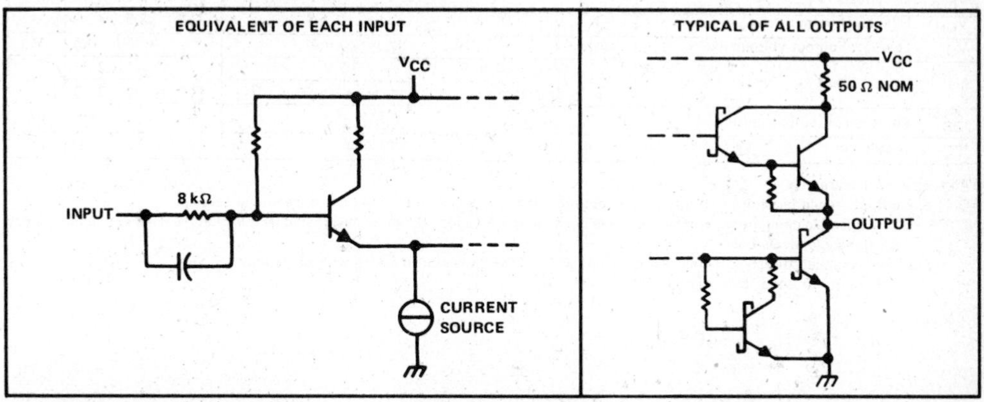
EQUIVALENT OF EACH INPUT
VCC
8 kΩ
INPUT
CURRENT
SOURCE
TYPICAL OF ALL OUTPUTS
VCC
50 Ω NOM
OUTPUT

absolute maximum ratings over operating free-air temperature range (unless otherwise noted)

Supply voltage, V_{CC} (see Note 1) . . . . −0.5 V to 7 V
Input voltage . . . . ±15 V
Differential input voltage (see Note 2) . . . . ±15 V
Output voltage (see Note 1) . . . . −0.5 V to 5.5 V
Low-level output current . . . . 50 mA
Continuous total dissipation at (or below) 25°C free-air temperature (see Note 3):
  D package . . . . 725 mW
  JG package: uA9637AM . . . . 1050 mW
  uA9637AC . . . . 825 mW
  P package . . . . 1000 mW
Operating free-air temperature range: uA9637AM . . . . −55°C to 125°C
  uA9637AC . . . . 0°C to 70°C
Storage temperature range . . . . −65°C to 150°C
Lead temperature 1,6 mm (1/16 inch) from case for 60 seconds: JG package . . . . 300°C
Lead temperature 1,6 mm (1/16 inch) from case for 10 seconds: D or P package . . . . 260°C

NOTES: 1. All voltage values, except differential input voltage, are with respect to the network ground terminal.
2. Differential input voltage is measured at the noninverting input with respect to the corresponding inverting input.
3. For operation above 25°C free-air temperature, derate linearly at the following rates: 5.8 mW/°C for the D package, 8.4 mW/°C for uA9637AM in the JG package, 6.6 mW/°C for uA9637AC in the JG package, and 8.0 mW/°C for the P package.

recommended operating conditions

	uA9637AM			uA9637AC			UNIT
	MIN	NOM	MAX	MIN	NOM	MAX	
Supply voltage, V_{CC}	4.5	5	5.5	4.75	5	5.25	V
Common-mode input voltage, V_{IC}			±7			±7	V
Operating free-air temperature, T_A	−55		125	0		70	°C

electrical characteristics over recommended ranges of supply voltage, common-mode input voltage, and operating free-air temperature (unless otherwise noted)

PARAMETER		TEST CONDITIONS		MIN	TYP†	MAX (See Note 4)	UNIT
V_T	Threshold voltage (V_{T+} and V_{T-})			−0.2		0.2	V
		See Note 5		−0.4		0.4	
V_{hys}	Hysteresis ($V_{T+} - V_{T-}$)				70		mV
V_{OH}	High-level output voltage	V_{ID} = 0.2 V,	I_O = −1 mA	2.5	3.5		V
V_{OL}	Low-level output voltage	V_{ID} = −0.2 V,	I_O = 20 mA		0.35	0.5	V
I_I	Input current	V_{CC} = 0 to 5.5 V, See Note 6	V_I = 10 V		1.1	3.25	mA
			V_I = −10 V		−1.6	−3.25	
I_{OS}	Short-circuit output current‡	V_O = 0,	V_{ID} = 0.2 V	−40	−75	−100	mA
I_{CC}	Supply current	V_{ID} = −0.5 V,	No load		35	50	mA

†All typical values are at V_{CC} = 5 V, T_A = 25°C.
‡Only one output should be shorted at a time, and duration of the short-circuit should not exceed one second.
NOTES: 4. The algebraic convention, in which the less positive (more negative) limit is designated as minimum, is used in this data sheet for threshold levels only.
5. The expanded threshold parameter is tested with a 500-Ω resistor in series with each input.
6. The input not under test is grounded.

switching characteristics, V_{CC} = 5 V, T_A = 25°C

PARAMETER		TEST CONDITION	MIN	TYP	MAX	UNIT
t_{PLH}	Propagation delay time, low-to-high-level output	C_L = 30 pF, See Figure 1		15	25	ns
t_{PHL}	Propagation delay time, high-to-low-level output			13	25	ns

PARAMETER MEASUREMENT INFORMATION

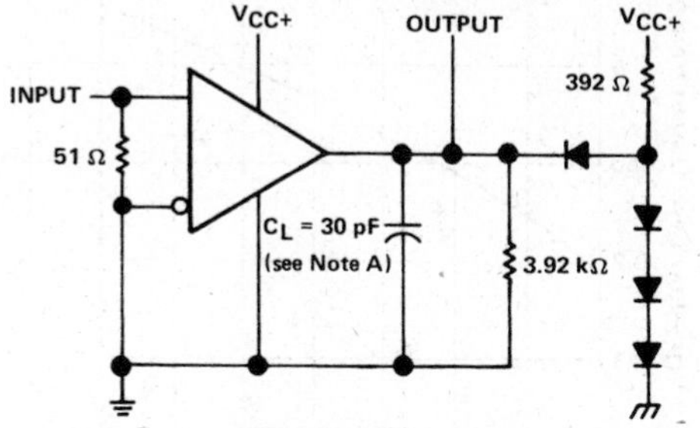
V_{CC+}
OUTPUT
V_{CC+}
INPUT
392 Ω
51 Ω
C_L = 30 pF
(see Note A)
3.92 kΩ

TEST CIRCUIT

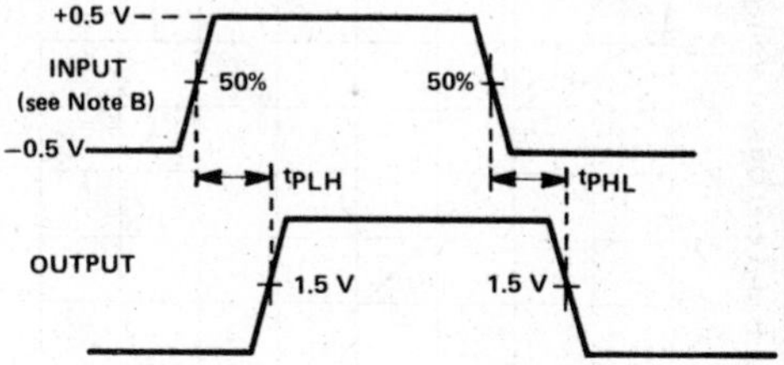
+0.5 V
INPUT
(see Note B)
50%
50%
−0.5 V
t_{PLH}
t_{PHL}
OUTPUT
1.5 V
1.5 V

VOLTAGE WAVEFORM

NOTES: A. C_L includes probe and jig capacitance.
B. The input pulse is supplied by a generator having the following characteristics: t_r ≤ 5 ns, t_f ≤ 5 ns, PRR ≤ 5 MHz, duty cycle = 50%.

FIGURE 1. TRANSITION TIMES

TYPICAL CHARACTERISTICS

OUTPUT VOLTAGE
vs
DIFFERENTIAL INPUT VOLTAGE

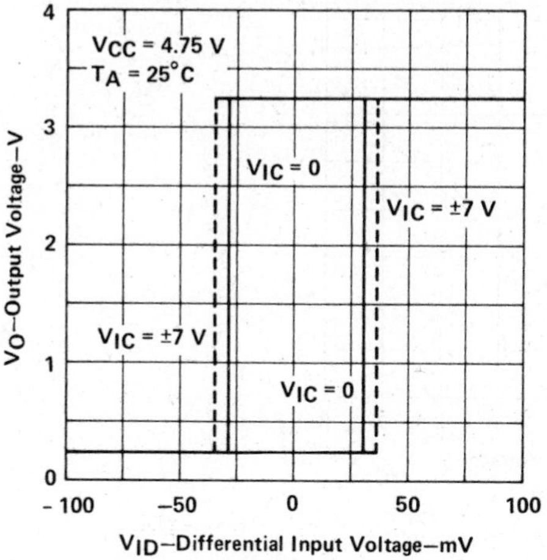
V_{CC} = 4.75 V
T_A = 25°C
V_{IC} = 0
V_{IC} = ±7 V
V_{IC} = ±7 V
V_{IC} = 0
V_O—Output Voltage—V
4
3
2
1
0
− 100
−50
0
50
100
V_{ID}—Differential Input Voltage—mV

FIGURE 2

OUTPUT VOLTAGE
vs
DIFFERENTIAL INPUT VOLTAGE

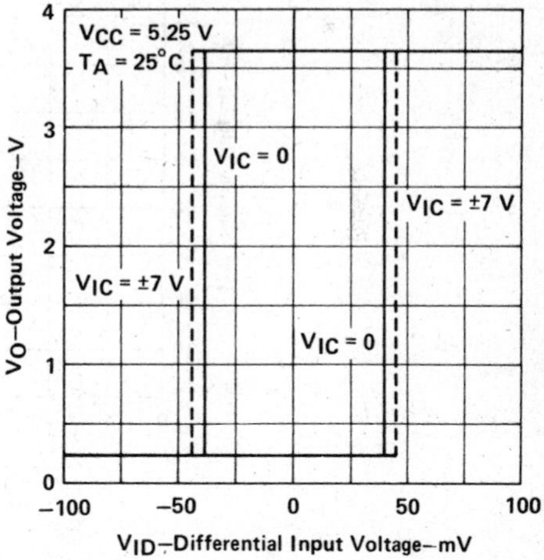
V_{CC} = 5.25 V
T_A = 25°C
V_{IC} = 0
V_{IC} = ±7 V
V_{IC} = ±7 V
V_{IC} = 0
V_O—Output Voltage—V
4
3
2
1
0
−100
−50
0
50
100
V_{ID}—Differential Input Voltage—mV

FIGURE 3

TYPICAL CHARACTERISTICS

HIGH-LEVEL OUTPUT VOLTAGE
vs
HIGH-LEVEL OUTPUT CURRENT

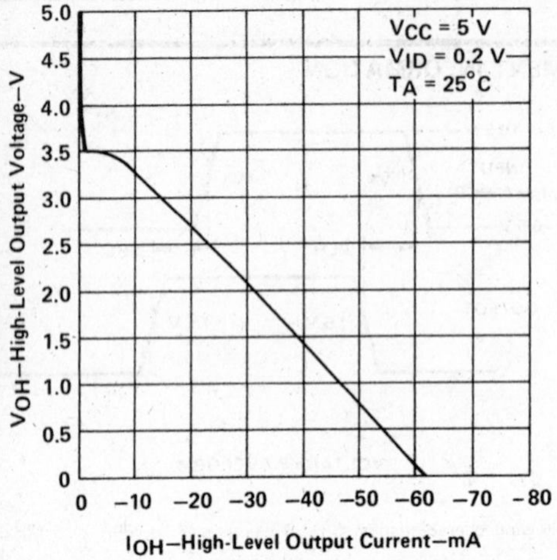
V_{CC} = 5 V
V_{ID} = 0.2 V
T_A = 25°C
V_{OH}—High-Level Output Voltage—V
5.0
4.5
4.0
3.5
3.0
2.5
2.0
1.5
1.0
0.5
0
0 −10 −20 −30 −40 −50 −60 −70 −80
I_{OH}—High-Level Output Current—mA

FIGURE 4

LOW-LEVEL OUTPUT VOLTAGE
vs
LOW-LEVEL OUTPUT CURRENT

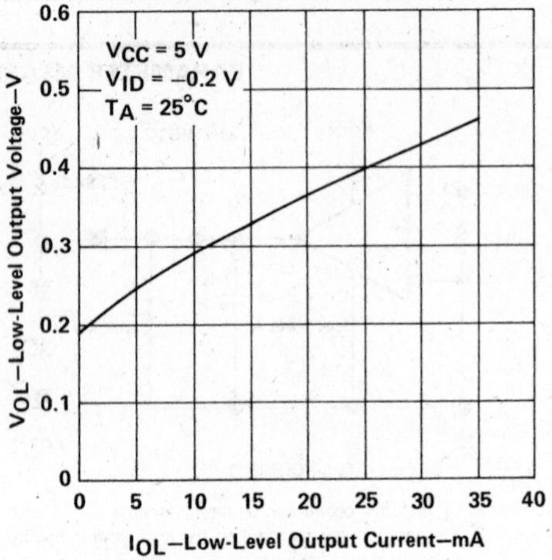
V_{CC} = 5 V
V_{ID} = −0.2 V
T_A = 25°C
V_{OL}—Low-Level Output Voltage—V
0.6
0.5
0.4
0.3
0.2
0.1
0
0 5 10 15 20 25 30 35 40
I_{OL}—Low-Level Output Current—mA

FIGURE 5

SUPPLY CURRENT
vs
SUPPLY VOLTAGE

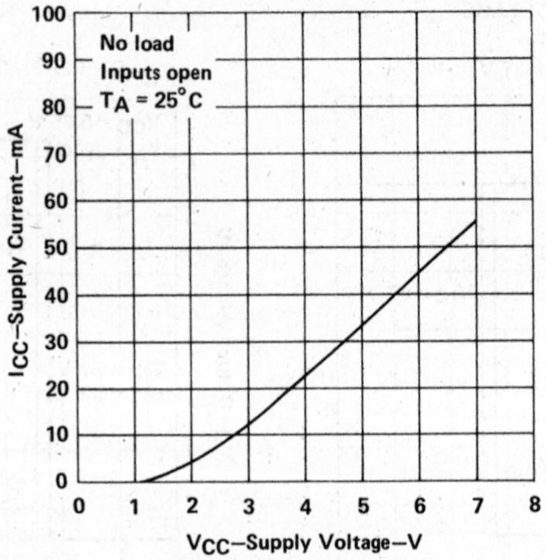
No load
Inputs open
T_A = 25°C
I_{CC}—Supply Current—mA
100
90
80
70
60
50
40
30
20
10
0
0 1 2 3 4 5 6 7 8
V_{CC}—Supply Voltage—V

FIGURE 6

4 Line Drivers/Receivers

TYPICAL APPLICATION DATA

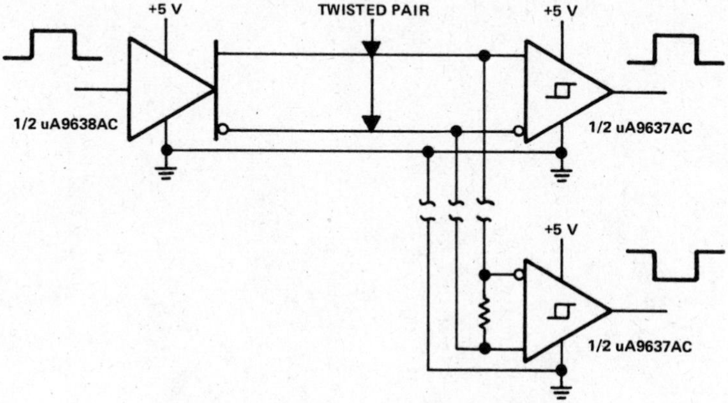
+5 V
TWISTED PAIR
+5 V
1/2 uA9638AC
1/2 uA9637AC
+5 V
1/2 uA9637AC

FIGURE 7. RS-422-A SYSTEM APPLICATIONS

D2612, OCTOBER 1980—REVISED SEPTEMBER 1986

- Meets EIA Standard RS-422-A
- Operates From a Single 5-V Supply
- TTL-and CMOS-Input Compatibility
- Output Short-Circuit Protection
- Schottky Circuitry
- Designed to be Interchangeable with Fairchild 9638

D, JG, OR P DUAL-IN-LINE PACKAGE
(TOP VIEW)

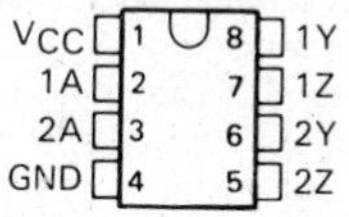
VCC 1
1A 2
2A 3
GND 4
8 1Y
7 1Z
6 2Y
5 2Z

description

The uA9638C is a dual high-speed differential line driver designed to meet EIA Standard RS-422-A. The inputs are TTL- and CMOS-compatible and have input clamp diodes. Schottky-diode-clamped transistors are used to minimize propagation delay time. This device operates from a single 5-volt power supply and is supplied in an 8-pin dual-in-line package.

The uA9638C is characterized for operation from 0°C to 70°C.

logic symbol†

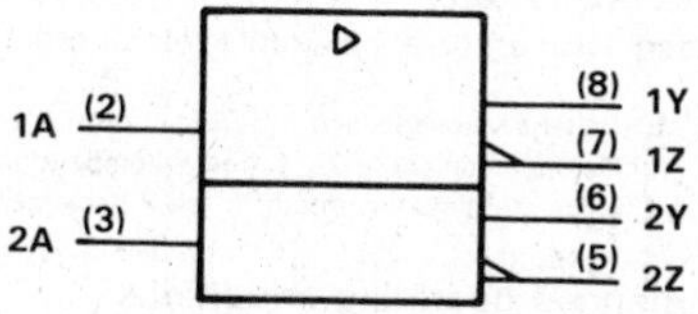
1A (2)
2A (3)
(8) 1Y
(7) 1Z
(6) 2Y
(5) 2Z

logic diagram

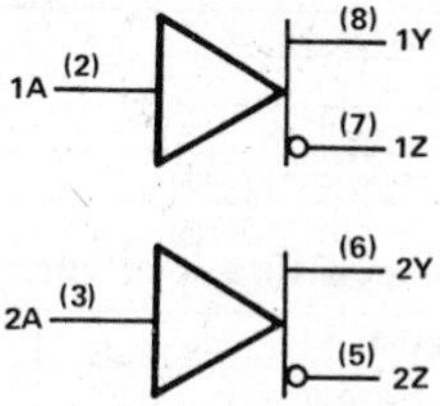
1A (2)
(8) 1Y
(7) 1Z
2A (3)
(6) 2Y
(5) 2Z

†This symbol is in accordance with ANSI/IEEE Std 91-1984 and IEC Publication 617-12.

schematics of inputs and outputs

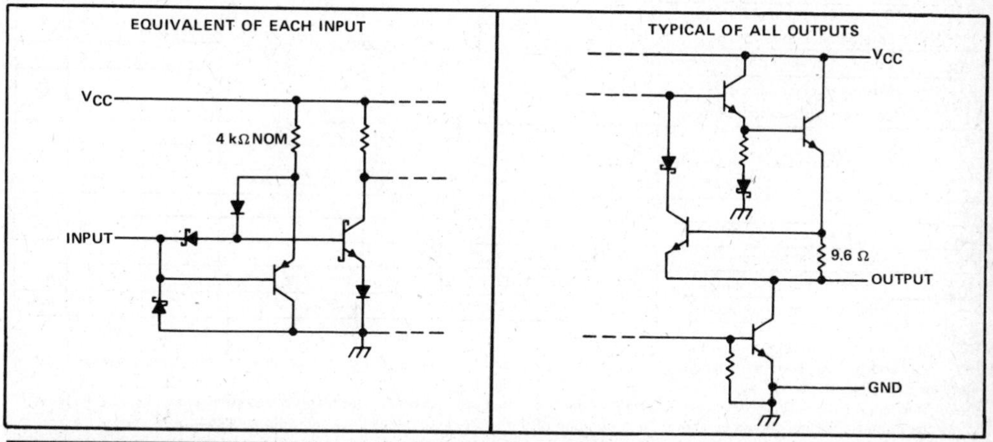
EQUIVALENT OF EACH INPUT
VCC
4 kΩ NOM
INPUT
TYPICAL OF ALL OUTPUTS
VCC
9.6 Ω
OUTPUT
GND

absolute maximum ratings over operating free-air temperature range (unless otherwise noted)

Supply voltage range, V_{CC} (see Note 1) .... −0.5 V to 7 V
Input voltage range .... −0.5 V to 7 V
Continuous total dissipation at (or below) 25°C free-air temperature (see Note 2):
  D package .... 725 mW
  JG package .... 825 mW
  P package .... 1000 mW
Operating free-air temperature range .... 0°C to 70°C
Storage temperature range .... −65°C to 150°C
Lead temperature 1,6 mm (1/16 inch) from case for 60 seconds: JG package .... 300°C
Lead temperature 1,6 mm (1/16 inch) from 10 seconds: D and P package .... 260°C

NOTES: 1. Voltage values except differential output voltages are with respect to network ground terminal.
2. For operation above 25°C free-air temperature, refer to Dissipation Derating Curves in Appendix A. In the JG package, uA9638C chips are glass mounted. In the P package, use the 8.0-mW/°C curve for these devices.

recommended operating conditions

	MIN	NOM	MAX	UNIT
Supply voltage, V_{CC}	4.75	5	5.25	V
High-level input voltage, V_{IH}	2			V
Low-level input voltage, V_{IL}			0.8	V
High-level output current, I_{OH}			−50	mA
Low-level output current, I_{OL}			50	mA
Operating free-air temperature, T_A	0		70	°C

electrical characteristics over operating free-air temperature range (unless otherwise noted)

PARAMETER		TEST CONDITIONS		MIN	TYP†	MAX	UNIT
V_{IK}	Input clamp voltage	V_{CC} = 4.75 V, I_I = −18 mA			−1	−1.2	V
V_{OH}	High-level output voltage	V_{CC} = 4.75 V, V_{IH} = 2 V, V_{IL} = 0.8 V	I_{OH} = −10 mA	2.5	3.5		V
			I_{OH} = −40 mA	2			V
V_{OL}	Low-level output voltage	V_{CC} = 4.75 V, V_{IH} = 2 V, V_{IL} = 0.8 V, I_{OL} = 40 mA				0.5	V
$\|V_{OD1}\|$	Differential output voltage	V_{CC} = 5.25 V, I_O = 0				$2V_{OD2}$	V
$\|V_{OD2}\|$	Differential output voltage	V_{CC} = 4.75 V to 5.25 V, R_L = 100 Ω, See Figure 1		2			V
$\Delta\|V_{OD}\|$	Change in magnitude of‡ differential output voltage					±0.4	V
V_{OC}	Common-mode output voltage§					3	V
$\Delta\|V_{OC}\|$	Change in magnitude of‡ common-mode output voltage					±0.4	V
I_O	Output current with power off	V_{CC} = 0,	V_O = 6 V		0.1	100	μA
			V_O = −0.25 V		−0.1	−100	μA
			V_O = −0.25 V to 6 V			±100	μA
I_I	Input current	V_{CC} = 5.25 V, V_I = 5.5 V				50	μA
I_{IH}	High-level input current	V_{CC} = 5.25 V, V_I = 2.7 V				25	μA
I_{IL}	Low-level input current	V_{CC} = 5.25 V, V_I = 0.5 V				−200	μA
I_{OS}	Short-circuit output current¶	V_{CC} = 5.25 V, V_O = 0		−50		−150	mA
I_{CC}	Supply current (all drivers)	V_{CC} = 5.25 V, No load,	All inputs at 0 V		45	65	mA

†All typical values are at V_{CC} = 5 V and T_A = 25°C.
‡$\Delta|V_{OD}|$ and $\Delta|V_{OC}|$ are the changes in magnitude of V_{OD} and V_{OC}, respectively, that occur when the input is changed from a high level to a low level.
§In EIA Standard RS-422-A, V_{OC}, which is the average of the two output voltages with respect to ground, is called output offset voltage, V_{OS}.
¶Only one output at a time should be shorted and duration of the short-circuit should not exceed one second.

switching characteristics, V_{CC} = 5 V, T_A = 25°C

PARAMETER		TEST CONDITION	MIN	TYP	MAX	UNIT
t_{DD}	Differential-output delay time	C_L = 15 pF, R_L = 100 Ω, See Figure 2		10	15	ns
t_{TD}	Differential-output transition time			10	15	ns
	Skew			1		ns

PARAMETER MEASUREMENT INFORMATION

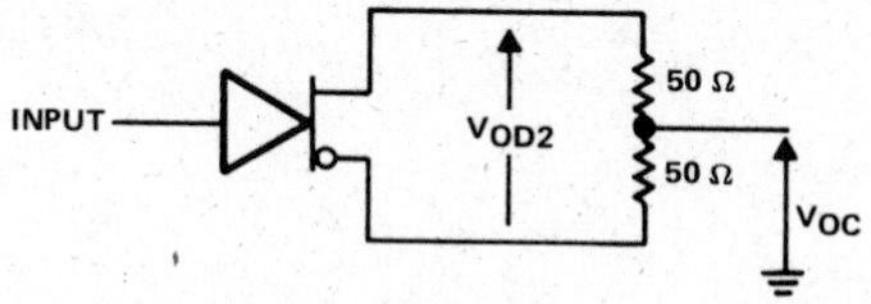
INPUT
V_{OD2}
50 Ω
50 Ω
V_{OC}

FIGURE 1. DIFFERENTIAL AND COMMON-MODE OUTPUT VOLTAGES

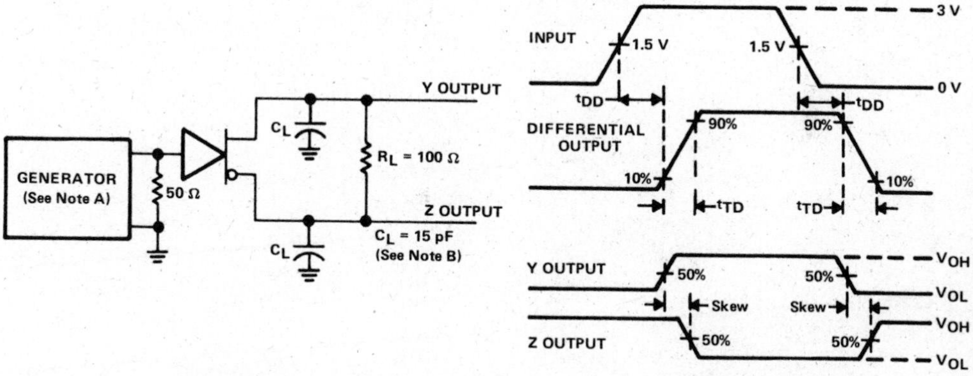
GENERATOR
(See Note A)
50 Ω
C_L
R_L = 100 Ω
Y OUTPUT
Z OUTPUT
C_L = 15 pF
(See Note B)
INPUT
1.5 V
1.5 V
3 V
0 V
t_{DD}
t_{DD}
DIFFERENTIAL
OUTPUT
90%
90%
10%
10%
t_{TD}
t_{TD}
Y OUTPUT
50%
50%
Skew
Skew
Z OUTPUT
50%
50%
V_{OH}
V_{OL}
V_{OH}
V_{OL}

TEST CIRCUITS

VOLTAGE WAVEFORMS

NOTES: A. The input pulse generator has the following characteristics: Z_{OUT} = 50 Ω, PRR ≤ 500 kHz, t_w = 100 ns, t_r = ≤ 5 ns.
B. C_L includes probe and jig capacitance.

FIGURE 2. SWITCHING TIMES

D3009, OCTOBER 1986

- Meets EIA Standards RS-422-A and RS-423-A
- Meets Federal Standards 1020 and 1030
- Operates from Single 5-V Power Supply
- Wide Common-Mode Voltage Range
- High Input Impedance
- TTL-Compatible Outputs
- High-Speed Schottky Circuitry
- 8-Pin Dual-In-Line and "Small Outline" Packages
- Designed to be Interchangeable with Fairchild μA9639AC

D, JG, OR P PACKAGE
(TOP VIEW)

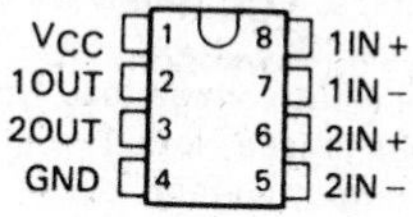
VCC 1
1OUT 2
2OUT 3
GND 4
8 1IN+
7 1IN−
6 2IN+
5 2IN−

logic symbol†

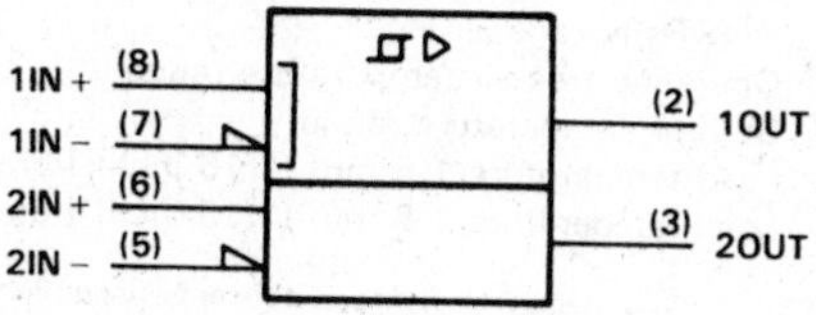
1IN+ (8)
1IN− (7)
2IN+ (6)
2IN− (5)
(2) 1OUT
(3) 2OUT

† This symbol is in accordance with ANSI/IEEE Std 91-1984 and IEC Publication 617-12.

description

The uA9639C is a dual differential line receiver designed to meet EIA standards RS-422-A and RS-423-A and Federal Standards 1020 and 1030. It utilizes Schottky circuitry and has TTL-compatible outputs. The inputs are compatible with either a single-ended or a differential-line system. This device operates from a single 5-volt power supply and is supplied in an 8-pin dual-in-line package and "small outline" package.

The uA9639C is characterized for operation from 0°C to 70°C.

logic diagram

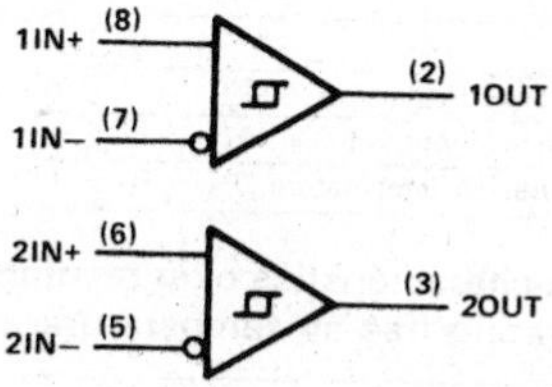
1IN+ (8)
1IN− (7)
(2) 1OUT
2IN+ (6)
2IN− (5)
(3) 2OUT

schematics of inputs and outputs

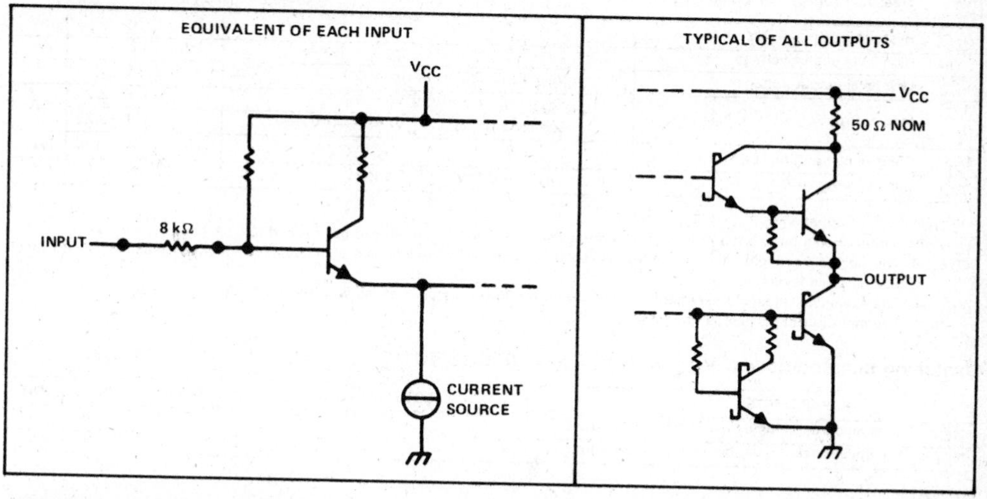
EQUIVALENT OF EACH INPUT
VCC
INPUT
8 kΩ
CURRENT SOURCE
TYPICAL OF ALL OUTPUTS
VCC
50 Ω NOM
OUTPUT

absolute maximum ratings over operating free-air temperature range (unless otherwise noted)

Supply voltage, V_{CC} (see Note 1) . . . . −0.5 V to 7 V
Input voltage . . . . ±15 V
Differential input voltage (see Note 2) . . . . ±15 V
Output voltage (see Note 1) . . . . −0.5 V to 5.5 V
Low-level output current . . . . 50 mA
Continuous total dissipation at (or below) 25°C free-air temperature (see Note 3):
- D package . . . . 725 mW
- JG package . . . . 825 mW
- P package . . . . 1000 mW

Operating free-air temperature range . . . . 0°C to 70°C
Storage temperature range . . . . −65°C to 150°C
Lead temperature 1,6 mm (1/16 inch) from case for 60 seconds: JG package . . . . 300°C
Lead temperature 1,6 mm (1/16 inch) from case for 10 seconds: D and P package . . . . 260°C

NOTES: 1. All voltage values, except differential input voltage, are with respect to the network ground terminal.
2. Differential input voltage is measured at the noninverting input with respect to the corresponding inverting input.
3. For operation above 25°C free-air temperature, derate the D package to 464 mW at 70°C at the rate of 5.8 mW/°C, the JG package to 528 mW at 70°C at the rate of 6.6 mW/°C, and the P package to 640 mW at 70°C at the rate of 8.0 mW/°C.

recommended operating conditions

	MIN	NOM	MAX	UNIT
Supply voltage, V_{CC}	4.75	5	5.25	V
Common-mode input voltage, V_{IC}			±7	V
Operating free-air temperature, T_A	0		70	°C

electrical characteristics over recommended ranges of supply voltage, common-mode input voltage, and operating free-air temperature (unless othewise noted)

PARAMETER		TEST CONDITIONS		MIN (See Note 4)	TYP†	MAX	UNIT
V_T	Threshold voltage (V_{T+} and V_{T-})			−0.2		0.2	V
		See Note 5		−0.4		0.4	
V_{hys}	Hysteresis ($V_{T+} - V_{T-}$)				70		mV
V_{OH}	High-level output voltage	V_{ID} = 0.2 V,	I_O = −1 mA	2.5	3.5		V
V_{OL}	Low-level output voltage	V_{ID} = −0.2 V,	I_O = 20 mA		0.35	0.5	V
I_I	Input current	V_{CC} = 0 to 5.5 V, See Note 6	V_I = 10 V		1.1	3.25	mA
			V_I = −10 V		−1.6	−3.25	
I_{OS}	Short-circuit output current‡	V_O = 0,	V_{ID} = 0.2 V	−40	−75	−100	mA
I_{CC}	Supply current	V_{ID} = −0.5 V,	No load		35	50	mA

†All typical values are at V_{CC} = 5 V, T_A = 25°C.
‡Only one output should be shorted at a time, and duration of the short-circuit should not exceed one second.
NOTES: 4. The algebraic convention, in which the less positive (more negative) limit is designated as minimum, is used in this data sheet for threshold levels only.
5. The expanded threshold parameter is tested with a 500-Ω resistor in series with each input.
6. The input not under test is grounded.

switching characteristics, V_{CC} = 5 V, T_A = 0°C to 70°C

PARAMETER		TEST CONDITION	MIN	MAX	UNIT
t_{PLH}	Propagation delay time, low-to-high-level output	C_L = 30 pF, See Figure 1		85	ns
t_{PHL}	Propagation delay time, high-to-low-level output			85	ns

TEXAS INSTRUMENTS
POST OFFICE BOX 655012 • DALLAS, TEXAS 75265

PARAMETER MEASUREMENT INFORMATION

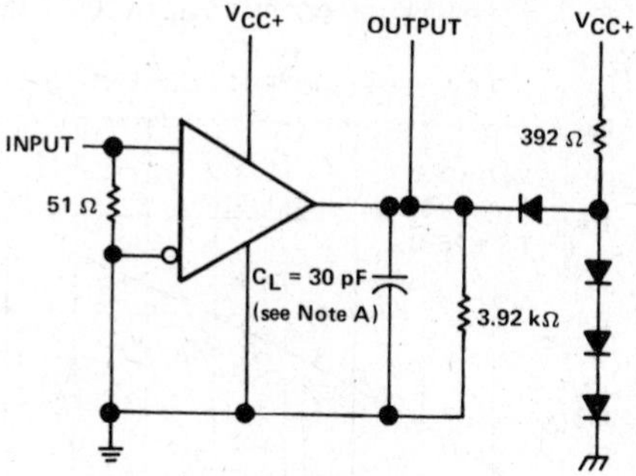
VCC+
OUTPUT
VCC+
INPUT
51 Ω
392 Ω
CL = 30 pF
(see Note A)
3.92 kΩ

TEST CIRCUIT

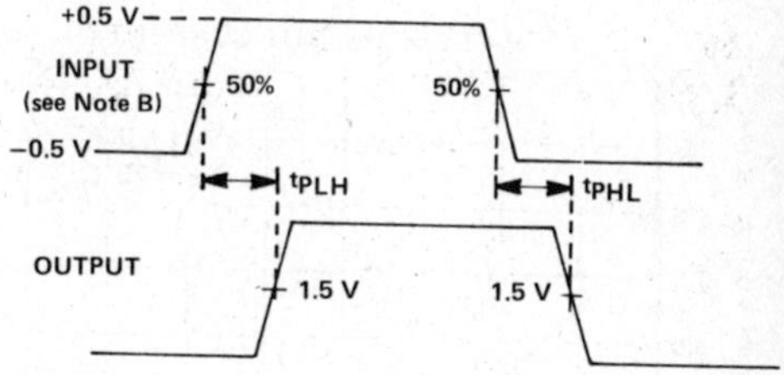
+0.5 V
INPUT
(see Note B)
50%
50%
−0.5 V
tPLH
tPHL
OUTPUT
1.5 V
1.5 V

VOLTAGE WAVEFORM

NOTES: A. C_L includes probe and jig capacitance.
B. The input pulse is supplied by a generator having the following characteristics: $t_r \leq 5$ ns, $t_f \leq 5$ ns, PRR ≤ 5 MHz, duty cycle = 50%.

FIGURE 1. TRANSITION TIMES

TYPICAL CHARACTERISTICS

OUTPUT VOLTAGE
vs
DIFFERENTIAL INPUT VOLTAGE

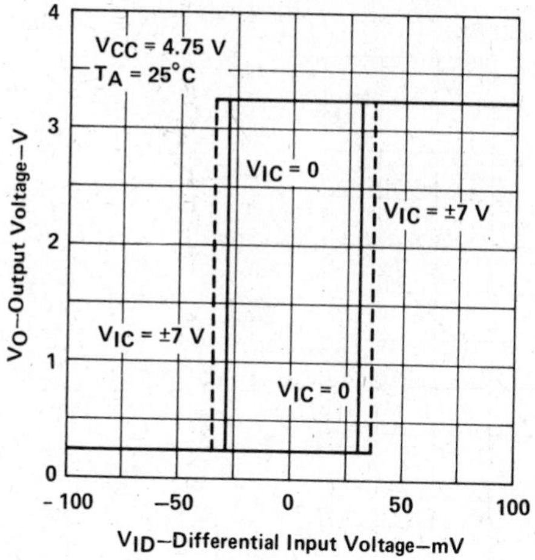
VCC = 4.75 V
TA = 25°C
VIC = 0
VIC = ±7 V
VIC = ±7 V
VIC = 0
VO—Output Voltage—V
VID—Differential Input Voltage—mV

FIGURE 2

OUTPUT VOLTAGE
vs
DIFFERENTIAL INPUT VOLTAGE

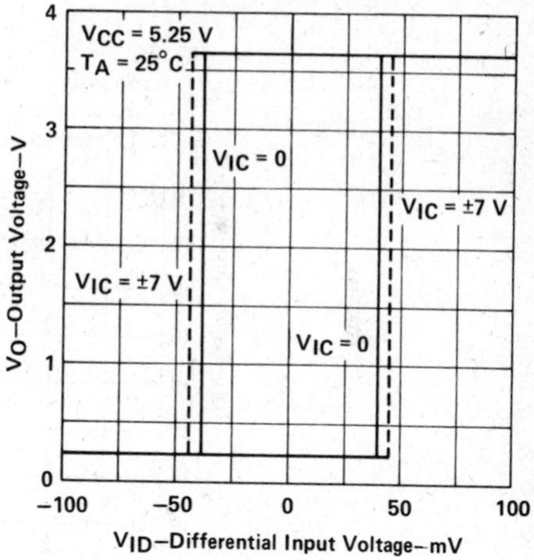
VCC = 5.25 V
TA = 25°C
VIC = 0
VIC = ±7 V
VIC = ±7 V
VIC = 0
VO—Output Voltage—V
VID—Differential Input Voltage—mV

FIGURE 3

TYPICAL CHARACTERISTICS

HIGH-LEVEL OUTPUT VOLTAGE
vs
HIGH-LEVEL OUTPUT CURRENT

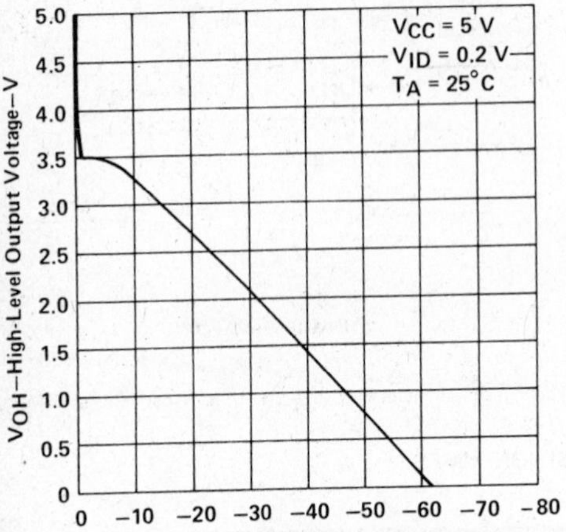
VCC = 5 V
VID = 0.2 V
TA = 25°C
VOH–High-Level Output Voltage–V
5.0
4.5
4.0
3.5
3.0
2.5
2.0
1.5
1.0
0.5
0
0
–10
–20
–30
–40
–50
–60
–70
–80
IOH–High-Level Output Current–mA

FIGURE 4

LOW-LEVEL OUTPUT VOLTAGE
vs
LOW-LEVEL OUTPUT CURRENT

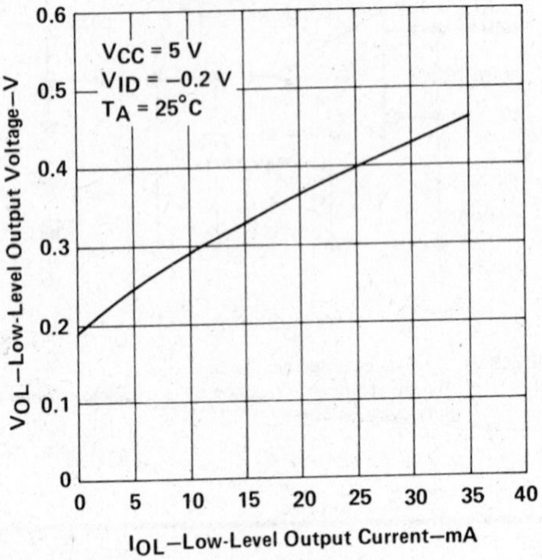
VCC = 5 V
VID = –0.2 V
TA = 25°C
VOL–Low-Level Output Voltage–V
0.6
0.5
0.4
0.3
0.2
0.1
0
0
5
10
15
20
25
30
35
40
IOL–Low-Level Output Current–mA

FIGURE 5

SUPPLY CURRENT
vs
SUPPLY VOLTAGE

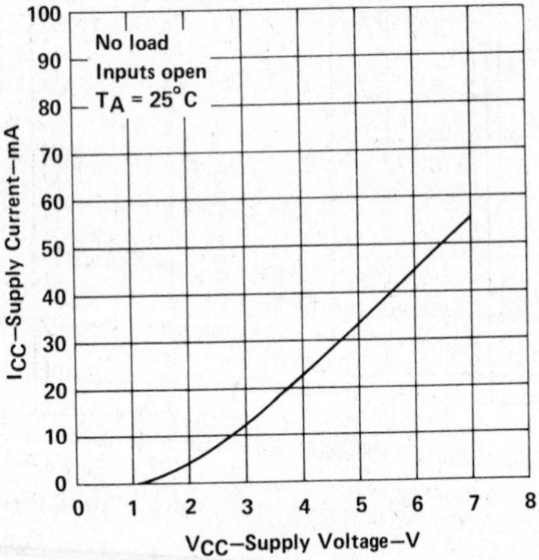
No load
Inputs open
TA = 25°C
ICC–Supply Current–mA
100
90
80
70
60
50
40
30
20
10
0
0
1
2
3
4
5
6
7
8
VCC–Supply Voltage–V

FIGURE 6

TEXAS INSTRUMENTS
POST OFFICE BOX 655012 • DALLAS, TEXAS 75265

TYPICAL APPLICATION DATA

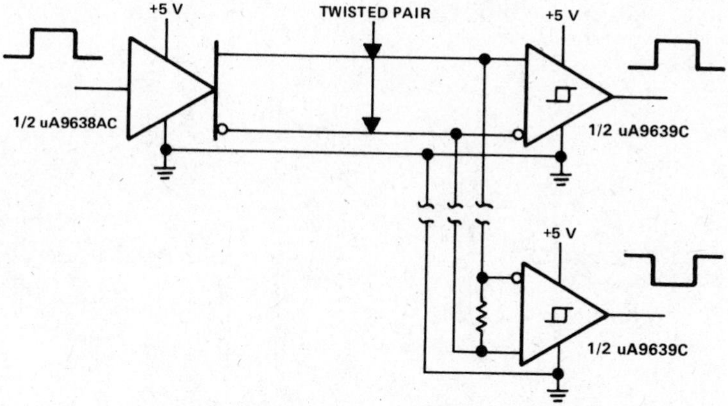
+5 V
TWISTED PAIR
+5 V
1/2 uA9638AC
1/2 uA9639C
+5 V
1/2 uA9639C

FIGURE 7. RS-422-A SYSTEM APPLICATIONS

General Information — 1
Alphanumeric Index
Selection Guide

Data Acquisition Circuits — 2
Cross-Reference Guide
Data Sheets

Display Drivers — 3
Data Sheets

Line Drivers and Receivers — 4
Cross-Reference Guide
Data Sheets

Peripheral Drivers/Actuators — 5
Cross-Reference Guide
Data Sheets

Memory Interface Circuits — 6
Data Sheets

Speech Synthesis Circuits — 7
Data Sheets

Appendix A Power Derating Curves — A

Appendix B Ordering Instructions, Mechanical Data, IC Sockets — B

Appendix C Explanation of Logic Symbols — C

CROSS-REFERENCE GUIDE
(manfacturers arranged alphabetically)

Replacements were based on similarity of electrical and mechanical characteristics as shown in currently published data. Interchangeability in particular applications is not guaranteed. Before using a device as a substitute, the user should compare the specifications of the substitute device with the specifications of the original.

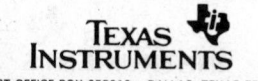

Texas Instruments makes no warranty as to the information furnished and the buyer assumes all risk in the use thereof. No liability is assumed for damages resulting from the use of the information contained in this list.

FAIRCHILD	SUGGESTED TI REPLACEMENT	PAGE NO.
μA75451	SN75451B	5-81
μA75452	SN75452B	5-81
μA75453	SN75453B	5-81
μA75454	SN75454B	5-81
μA75461	SN75461	5-93
μA75462	SN75462	5-93
MC1412	ULN2002A	5-173
MC1413	ULN2003A	5-173
μA3680	DS36805-173	
μA9665	ULN2001A	5-173
μA9666	ULN2002A	5-173
μA9667	ULN2003A	5-173
μA9668	ULN2004A	5-173

MOTOROLA	SUGGESTED TI REPLACEMENT	PAGE NO.
MC1411	ULN2001A	5-173
MC1412	ULN2002A	5-173
MC1413	ULN2003A	5-173
MC1413T	SN75468	5-101
MC1416	ULN2004A	5-173
MC1471	SN75476	5-117
MC1473	SN75478	5-117
MC1474	SN75479	5-117
SN75451B	SN75451B	5-81
SN75452B	SN75452B	5-81
SN75453B	SN75453B	5-81
SN75454B	SN75454B	5-81
UDN2841	UDN2841	5-169
UDN2845	UDN2845	5-169
ULN2001	ULN2001A	5-173
ULN2002	ULN2002A	5-173
ULN2003	ULN2003A	5-173
ULN2004	ULN2004A	5-173
ULN2064	ULN2064	5-181
ULN2065	ULN2065	5-181
ULN2066	ULN2066	5-181
ULN2067	ULN2067	5-181
ULN2068	ULN2068	5-187
ULN2069	ULN2069	5-187
ULN2074	ULN2074	5-193
ULN2075	ULN2075	5-193

NATIONAL	SUGGESTED TI REPLACEMENT	PAGE NO.
DS3611	SN75471	5-109
DS3612	SN75472	5-109
DS3613	SN75473	5-109
DS3658	SN75437A	5-63
DS3668	SN75435	5-57
DS3669	SN75440	5-69
DS3680	DS3680	5-5
DS75361	SN75372	5-33
DS75365	SN75374	5-43
DS75451	SN75451B	5-81
DS75452	SN75452B	5-81
DS75453	SN75453B	5-81
DS75454	SN75454B	5-81
DS75461	SN75461	5-93
DS75462	SN75462	5-93
DS75463	SN75463	5-93
LM3611	SN75471	5-109
LM3612	SN75472	5-109
LM3613	SN75473	5-109
LM75453	SN75453B	5-81

RIFA	SUGGESTED TI REPLACEMENT	PAGE NO.
PBD352301	ULN2001A	5-173
PBD352302	ULN2004A	5-173
PBD352303	ULN2003A	5-173
PBD352304	ULN2002A	5-173
PBD352311	SN75466	5-101
PBD352312	SN75469	5-101
PBD352313	SN75468	5-101
PBD352314	SN75467	5-101
UC3717	PBL3717A	5-19

SGS-ATES	SUGGESTED TI REPLACEMENT	PAGE NO.
L293	L293	5-9
L293	SN754411*	5-159
L293D	L293D	5-13
L293D	SN754410*	5-153
L298	L298	5-19
L201	ULN2001A	5-173
L202	ULN2002A	5-173
L203	ULN2003A	5-173
L204	ULN2004A	5-173
ULN2001	ULN2001A	5-173
ULN2002	ULN2002A	5-173
ULN2003	ULN2003A	5-173
ULN2004	ULN2004A	5-173
ULN2064	ULN2064	5-181
ULN2065	ULN2065	5-181
ULN2066	ULN2066	5-181
ULN2067	ULN2067	5-181
ULN2068	ULN2068	5-187
ULN2069	ULN2069	5-187
ULN2074	ULN2074	5-193
ULN2075	ULN2075	5-193
PBL3717A	PBL3717A	5-19

SILICON GENERAL	SUGGESTED TI REPLACEMENT	PAGE NO.
SG2001	ULN2001A	5-173
SG2002	ULN2002A	5-173
SG2003	ULN2003A	5-173
SG2004	ULN2004A	5-173
SG2022	SN75467	5-101
SG2023	SN75468	5-101
SG2024	SN75469	5-101
SG75451	SN75451B	5-81
SG75452	SN75452B	5-81
SG75453	SN75453B	5-81
SG75454	SN75454B	5-81
SG75461	SN75461	5-93
SG75462	SN75462	5-93
SG75463	SN75463	5-93
SG75473	SN75473	5-109

SPRAGUE	SUGGESTED TI REPLACEMENT	PAGE NO.
UDM-5732	SN75407*	5-53
UDN-2541	SN75437A	5-63
UDN-2841	UDN2841	5-169
UDN-2845	UDN2845	5-169
UDN-2949	SN75605*	5-123
UDN-2950	SN75605*	5-123
UDN-3611	SN75471	5-109
UDN-3612	SN75472	5-109
UDN-3613	SN75473	5-109
UDN-5711	SN75476	5-117
UDN-5713	SN75478	5-117
UDN-5714	SN75479	5-117
UDN-5722	SN75477	5-117
ULN-2001	ULN2001A	5-173
ULN-2002	ULN2002A	5-173
ULN-2003	ULN2003A	5-173
ULN-2004	ULN2004A	5-173
ULN-2005	ULN2005A	5-173
ULN-2021	SN75266	5-101
ULN-2022	SN75267	5-101
ULN-2023	SN75268	5-101
ULN-2024	SN75269	5-101
ULN-2025	SN75265	5-101
ULN-2064	ULN2064	5-181
ULN-2065	ULN2065	5-181
ULN-2066	ULN2066	5-181
ULN-2067	ULN2067	5-181
ULN-2068	ULN2068	5-187
ULN-2069	ULN2069	5-187
ULN-2074	ULN2074	5-193
ULN-2075	ULN2075	5-193

UNITRODE	SUGGESTED TI REPLACEMENT	PAGE NO.
L293	L293	5-9
L293	SN754411*	5-159
L293D	L293D	5-13
L293D	SN754410*	5-153
L298	L298	5-17
PBL3717A	PBL3717A	5-19

*Consult product data sheet for possible slight product differences.

TEXAS INSTRUMENTS
POST OFFICE BOX 655012 • DALLAS, TEXAS 75265

DS3680
QUAD TELEPHONE RELAY DRIVER

D2758, MARCH 1986

- Designed for −52-V Battery Operation
- 50-mA Output Current Capability
- Input Compatible with TTL and CMOS
- High Common-Mode Input Voltage Range
- Very Low Input Current
- Fail-Safe Disconnect Feature
- Built-In Output Clamp Diode
- Direct Replacement for National DS3680 and Fairchild μA3680

D, J OR N PACKAGE
(TOP VIEW)

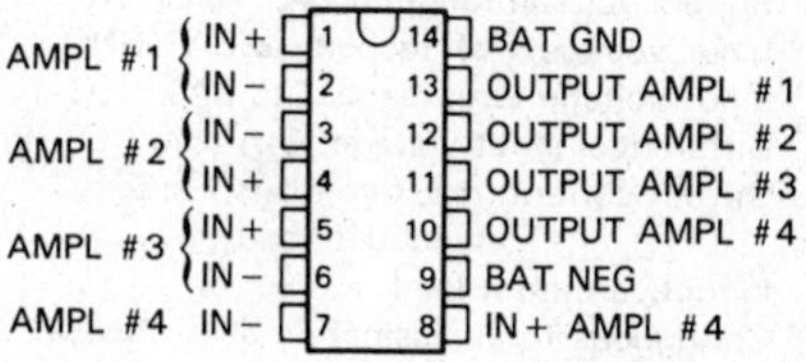
AMPL #1
IN+ 1
IN− 2
AMPL #2
IN− 3
IN+ 4
AMPL #3
IN+ 5
IN− 6
AMPL #4 IN− 7
14 BAT GND
13 OUTPUT AMPL #1
12 OUTPUT AMPL #2
11 OUTPUT AMPL #3
10 OUTPUT AMPL #4
9 BAT NEG
8 IN+ AMPL #4

description

The DS3680 telephone relay driver is a monolithic integrated circuit designed to interface −48-volt relay systems to TTL or other systems in telephone applications. It is capable of sourcing up to 50 milliamperes from standard −52-volt battery power. To reduce the effects of noise and IR drop between logic ground and battery ground, these drivers are designed to operate with a common-mode input range of ±20 volts referenced to battery ground. The common-mode input voltages for the four drivers can be different, so a wide range of input elements can be accommodated. The high-impedance inputs are compatible with positive TTL and CMOS levels or negative logic levels. A clamp network is included in the driver outputs to limit high-voltage transients generated by the relay coil during switching. The complementary inputs ensure that the driver output will be "off" as a fail-safe condition when either output is open.

The DS3680 is characterized for operation from −25°C to 85°C.

symbol (each driver)

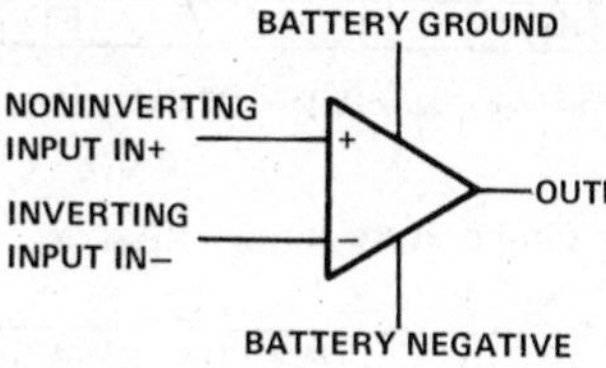
BATTERY GROUND
NONINVERTING
INPUT IN+
INVERTING
INPUT IN−
+
−
OUTPUT
BATTERY NEGATIVE

schematic diagram (each driver)

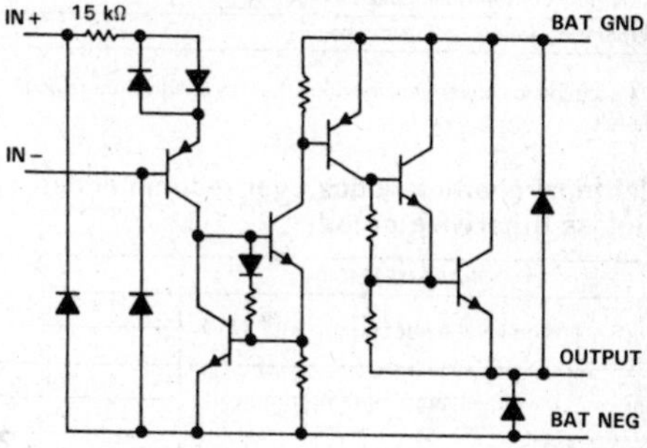
IN+
15 kΩ
BAT GND
IN−
OUTPUT
BAT NEG

absolute maximum ratings over operating free-air temperature range (unless otherwise noted)

Supply voltage range at BAT NEG, V_{B-} . . . . −70 V to 0.5 V
Input voltage with respect to BAT GND . . . . −70 V to 20 V
Input voltage with respect to BAT NEG . . . . −0.5 V to 70 V
Differential input voltage, V_{ID} (see Note 2) . . . . ±20 V
Output current: resistive load . . . . −100 mA
inductive load . . . . −50 mA
Inductive output load . . . . 5 H
Continuous total dissipation at (or below) 25°C free-air temperature (see Note 3):
D package . . . . 900 mW
J package . . . . 1025 mW
N package . . . . 1650 mW
Operating free-air temperature range . . . . −25°C to 85°C
Storage temperature range . . . . −65°C to 150°C
Lead temperature 1,6 mm (1/16 inch) from case for 60 seconds: J package . . . . 300°C
N package . . . . 260°C

NOTES: 1. All voltages are with respect to the BAT GND terminal, unless otherwise specified.
2. Differential input voltages are at the noninverting input terminal IN+ with respect to the inverting input terminal IN−.
3. For operation above 25°C free-air temperature, derate linearly at the rate of 7.2 mW/°C for the D package, 8.2 mW/°C for the J package, and 13.2 mW/°C for the N package.

recommended operating conditions

	MIN	MAX	UNIT
Supply voltage, V_{B-}	−10	−60	V
Input voltage, either input	−20†	20	V
High-level differential input voltage, V_{IDH}	2	20	V
Low-level differential input voltage, V_{IDL}	−20†	0.8	V
Operating free-air temperature, T_A	−25	85	°C

†The algebraic convention, in which the less positive (more negative) limit is designated minimum, is used in this data sheet for input voltage levels.

electrical characteristics over recommended operating free-air temperature range, $V_{B-} = -52$ V (unless otherwise noted)

PARAMETER		TEST CONDITIONS		MIN	TYP‡	MAX	UNIT
I_{IH}	High-level input current (into IN+)	V_{ID} = 2 V			40	100	μA
		V_{ID} = 7 V			375	1000	
I_{IL}	Low-level input current (into IN+)	V_{ID} = 0.4 V			0.01	5	μA
		V_{ID} = −7 V			−1	−100	
$V_{O(on)}$	On-state output voltage	I_O = 50 mA,	V_{ID} = 2 V		−1.6	−2.1	V
$I_{O(off)}$	Off-state output current	$V_O = V_{B-}$	V_{ID} = 0.8 V		−2	−100	μA
			Inputs open		−2	−100	
I_R	Clamp diode reverse current	V_O = 0			2	100	μA
V_{OK}	Output clamp voltage	I_O = 50 mA			0.9	1.2	V
		I_O = −50 mA,	V_{B-} = 0		−0.9	−1.2	
$I_{B(on)}$	On-state battery current	All drivers on			−2	−4.4	mA
$I_{B(off)}$	Off-state battery current	All drivers off			−1	−100	μA

‡All typical values are at T_A = 25°C.

switching characteristics V_{B-} = −52 V, T_A = 25°C

PARAMETER		TEST CONDITIONS	MIN	TYP	MAX	UNIT
t_{on}	Turn-on time	V_{ID} = 3-V pulse, R_L = 1 kΩ, L = 1 H, See Figure 1		1	10	μs
t_{off}	Turn-off time			1	10	μs

PARAMETER MEASUREMENT INFORMATION

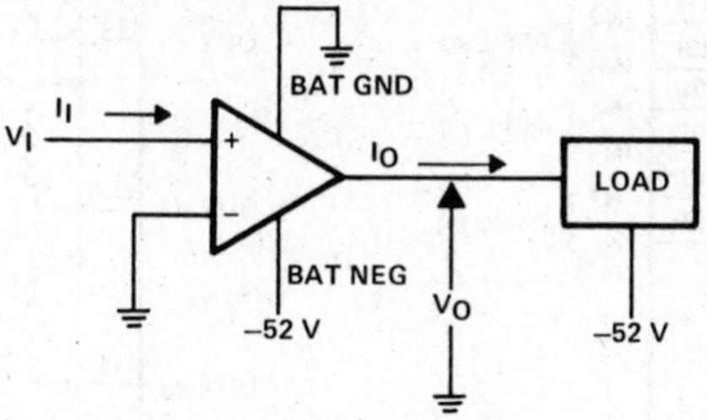
BAT GND
I_I
V_I
+
−
I_O
LOAD
BAT NEG
−52 V
V_O
−52 V

FIGURE 1. GENERALIZED TEST CIRCUIT, EACH DRIVER

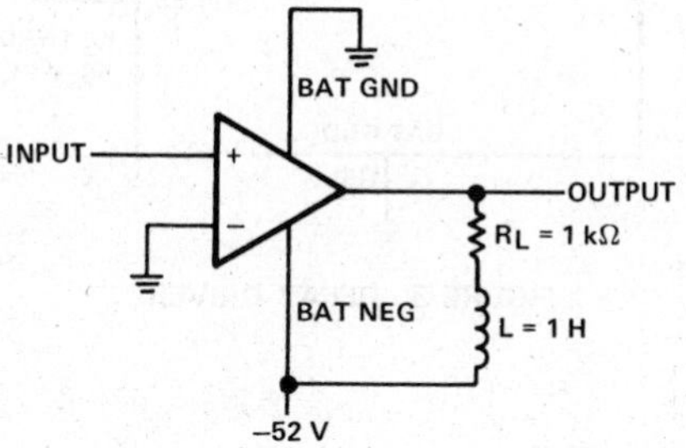
BAT GND
INPUT
+
−
OUTPUT
R_L = 1 kΩ
BAT NEG
L = 1 H
−52 V

TEST CIRCUIT

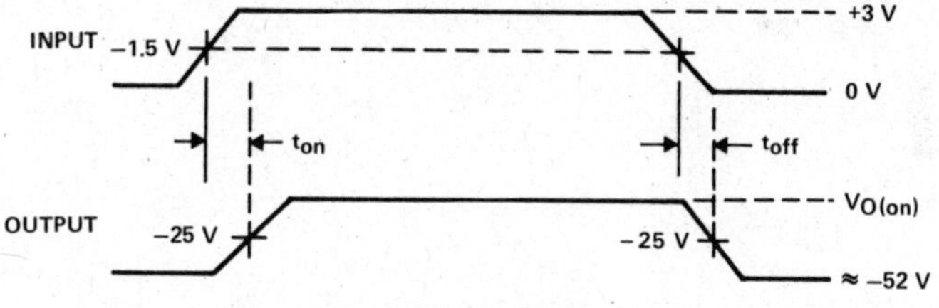
INPUT
1.5 V
+3 V
0 V
t_{on}
t_{off}
OUTPUT
−25 V
−25 V
$V_{O(on)}$
≈ −52 V

VOLTAGE WAVEFORMS

FIGURE 2. SWITCHING CHARACTERISTICS, EACH DRIVER

TYPICAL APPLICATION DATA

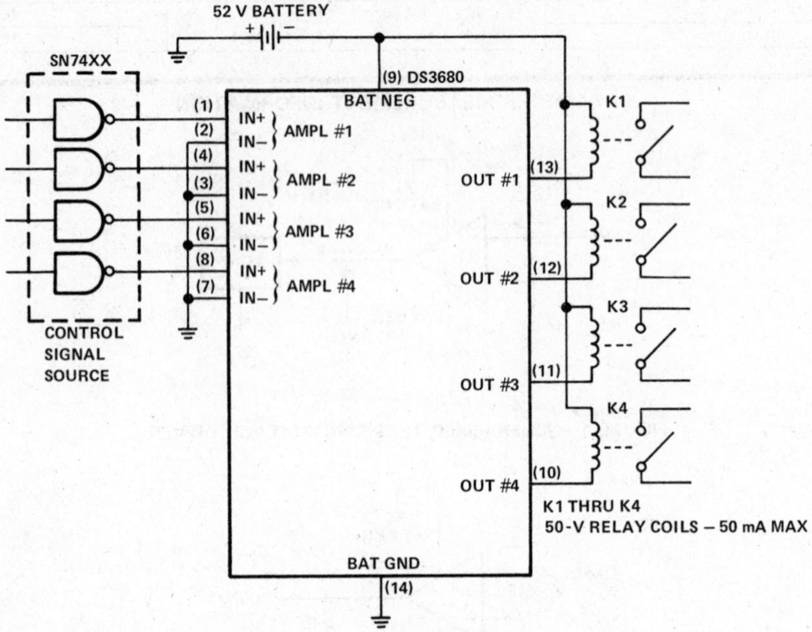
52 V BATTERY
+
−
SN74XX
(9) DS3680
BAT NEG
(1)
(2)
(4)
(3)
(5)
(6)
(8)
(7)
IN+
IN−
AMPL #1
IN+
IN−
AMPL #2
IN+
IN−
AMPL #3
IN+
IN−
AMPL #4
CONTROL
SIGNAL
SOURCE
OUT #1
(13)
OUT #2
(12)
OUT #3
(11)
OUT #4
(10)
K1
K2
K3
K4
K1 THRU K4
50-V RELAY COILS – 50 mA MAX
BAT GND
(14)

FIGURE 3. RELAY DRIVER

L293
QUADRUPLE HALF-H DRIVER

D2942, SEPTEMBER 1986

- 1-A Output Current Capability per Channel
- Wide Supply Voltage Range: 4.5 V to 36 V
- Separate Input-Logic Supply
- Thermal Shutdown
- Internal ESD Protection
- High-Noise-Immunity Inputs
- Designed to be Interchangeable with SGS L293

NE DUAL-IN-LINE PACKAGE
(TOP VIEW)

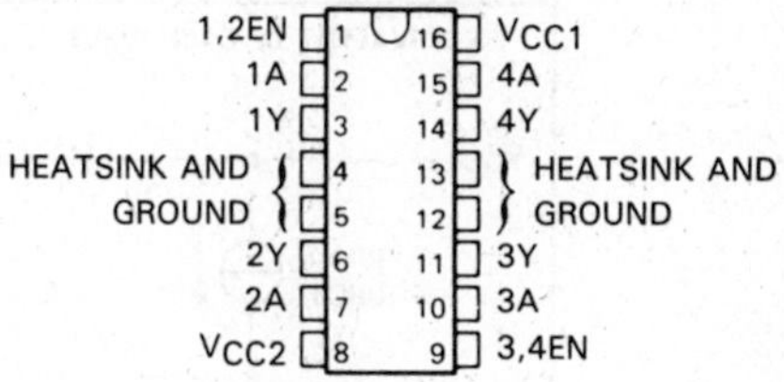
1,2EN 1
1A 2
1Y 3
HEATSINK AND GROUND 4
5
2Y 6
2A 7
VCC2 8
16 VCC1
15 4A
14 4Y
13 HEATSINK AND GROUND
12
11 3Y
10 3A
9 3,4EN

FUNCTION TABLE
(EACH CHANNEL)

INPUTS		OUTPUT
A	EN	Y
H	H	H
L	H	L
X	L	Z

H = high-level
L = low-level
X = irrelevant
Z = high-impedance (off)

description

The L293 is a quadruple high-current half-H driver designed to provide bidirectional drive currents of up to one ampere at voltages from 4.5 volts to 36 volts. It is designed to drive inductive loads such as relays, solenoids, dc and stepping motors, as well as other high-current/high-voltage loads in positive-supply applications.

All inputs are TTL-compatible. Each output is a complete totem-pole drive circuit with a Darlington transistor sink and a psuedo-Darlington source. Channels are enabled in pairs with channels 1 and 2 enabled by 1,2EN and channels 3 and 4 enabled by 3,4EN. When an enable input is high, the associated channels are enabled and their outputs are active and in phase with their inputs. When the enable input is low, those channels are disabled and their outputs are off and in a high-impedance state. With the proper data inputs, each pair of drivers form a full-H (or bridge) reversible drive suitable for solenoid or motor applications.

External high-speed output clamp diodes should be used for inductive transient suppression. A V_{CC1} terminal , separate from V_{CC2}, is provided for the logic inputs to minimize device power dissipation.

The L293 is designed for operation from 0°C to 70°C.

logic symbol†

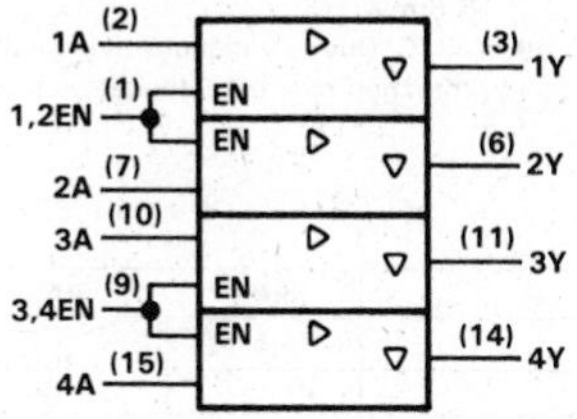
1A (2)
1,2EN (1)
EN
2A (7)
3A (10)
3,4EN (9)
4A (15)
(3) 1Y
(6) 2Y
(11) 3Y
(14) 4Y

† This symbol is in accordance with ANSI/IEEE Std 91-1984 and IEC Publication 617-12.

logic diagram

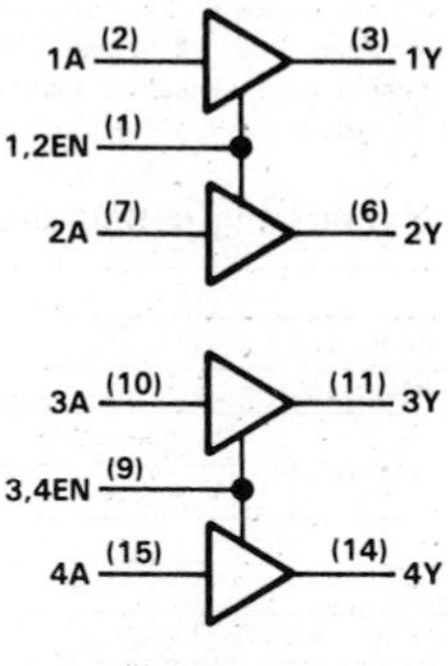
1A (2)
1,2EN (1)
2A (7)
(3) 1Y
(6) 2Y
3A (10)
3,4EN (9)
4A (15)
(11) 3Y
(14) 4Y

ADVANCE INFORMATION documents contain information on new products in the sampling or preproduction phase of development. Characteristic data and other specifications are subject to change without notice.

schematics of inputs and outputs

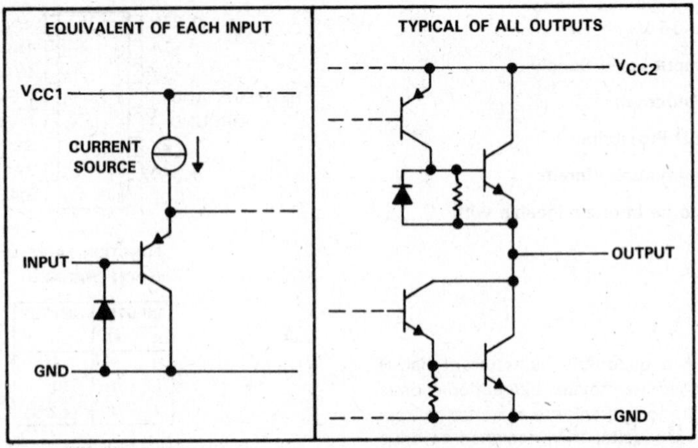
EQUIVALENT OF EACH INPUT
VCC1
CURRENT SOURCE
INPUT
GND
TYPICAL OF ALL OUTPUTS
VCC2
OUTPUT
GND

absolute maximum ratings over operating free-air temperature range (unless otherwise noted)

Logic supply voltage, V_{CC1} (see Note 1)	36 V
Output supply voltage, V_{CC2}	36 V
Input voltage	7 V
Output voltage range	−3 V to V_{CC2} + 3 V
Peak output current (nonrepetitive, $t \leq 5$ ms)	±2 A
Continuous output current	±1 A
Continuous total dissipation at (or below) 25°C free-air temperature (see Notes 2 and 3)	2075 mW
Continuous total dissipation at (or below) 25°C case temperature (see Note 3)	7375 mW
Continuous total dissipation at 80°C case temperature (see Note 3)	4130 mW
Operating case or virtual junction temperature range	−40°C to 150°C
Storage temperature range	−65°C to 150°C
Lead temperature 1,6 mm (1/16 inch) from case for 10 seconds	260°C

NOTES: 1. All voltage values are with respect to the network ground terminal.
2. For operation above 25°C free-air temperature, derate linearly at the rate of 16.6 mW/°C.
3. For operation above 25°C case temperature, derate linearly at the rate of 59 mW/°C. Due to variations in individual device electrical characteristics and thermal resistance, the built-in thermal overload protection may be activated at power levels slightly above or below the rated dissipation.

recommended operating conditions

		MIN	MAX	UNIT
Logic supply voltage, V_{CC1}		4.5	7	V
Output supply voltage, V_{CC2}		V_{CC1}	36	V
High-level input voltage, V_{IH}	$V_{CC1} \leq 7$ V	2.3	V_{CC1}	V
	$V_{CC1} \geq 7$ V	2.3	7	
Low-level input voltage, V_{IL}		−0.3†	1.5	V
Output current, I_O			±1	A
Operating free-air temperature, T_A		0	70	°C

†The algebraic convention, in which the least positive (most negative) designated minimum, is used in this data sheet for logic voltage levels.

TEXAS INSTRUMENTS
POST OFFICE BOX 655012 • DALLAS, TEXAS 75265

electrical characteristics, V_{CC1} = 5 V, V_{CC2} = 24 V, T_A = 25°C

PARAMETER			TEST CONDITIONS		MIN	TYP	MAX	UNIT
V_{OH}	High-level output voltage		I_{OH} = −1 A		V_{CC2}−1.8	V_{CC2}−1.4		V
V_{OL}	Low-level output voltage		I_{OL} = 1 A			1.2	1.8	V
I_{IH}	High-level input current	A	V_I = 7 V			0.2	100	μA
		EN				0.2	±10	
I_{IL}	Low-level input current	A	V_I = 0			−3	−10	μA
		EN				−2	−100	
I_{CC1}	Logic supply current		I_O = 0	All outputs at high level			22	mA
				All outputs at low level			60	
				All outputs at high impedance			24	
I_{CC2}	Output supply current		I_O = 0	All outputs at high level			24	mA
				All outputs at low level			6	
				All outputs at high impedance			4	

switching characteristics, V_{CC1} = 5 V, V_{CC2} = 24 V, T_A = 25°C

PARAMETER		TEST CONDITIONS	MIN	TYP	MAX	UNIT
t_{PLH}	Propagation delay time, low-to-high-level output from A input	C_L = 30 pF, See Figure 1		800		ns
t_{PHL}	Propagation delay time, high-to-low-level output from A input			400		ns
t_{TLH}	Transition time, low-to-high-level output			300		ns
t_{THL}	Transition time, high-to-low-level output			300		ns

PARAMETER MEASUREMENT INFORMATION

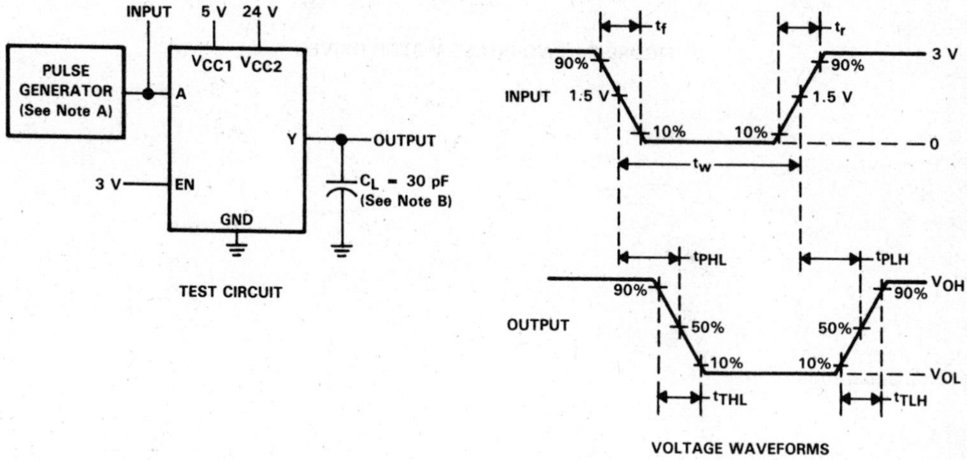
INPUT
5 V
24 V
PULSE GENERATOR (See Note A)
VCC1
VCC2
A
Y
OUTPUT
3 V
EN
GND
CL = 30 pF (See Note B)
TEST CIRCUIT
tf
tr
90%
3 V
INPUT
1.5 V
10%
0
tw
tPHL
tPLH
OUTPUT
50%
VOH
VOL
tTHL
tTLH
VOLTAGE WAVEFORMS

NOTES: A. The pulse generator has the following characteristics: $t_r \leq 10$ ns, $t_f \leq 10$ ns, t_w = 10 μs, PRR = 5 kHz, Z_{out} = 50 Ω.
B. C_L includes probe and jig capacitance.

FIGURE 1. SWITCHING TIMES

TYPICAL APPLICATION DATA

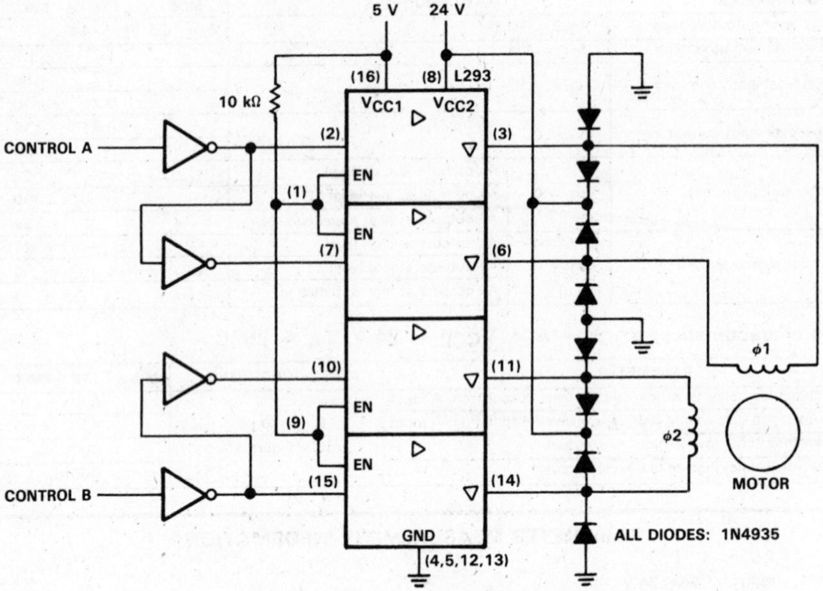
5 V
24 V
(16)
(8)
L293
10 kΩ
V_{CC1}
V_{CC2}
CONTROL A
(2)
(3)
EN
(1)
EN
(7)
(6)
$\phi 1$
(10)
(11)
EN
(9)
$\phi 2$
MOTOR
EN
(15)
(14)
CONTROL B
GND
(4,5,12,13)
ALL DIODES: 1N4935

FIGURE 2. TWO-PHASE MOTOR DRIVER

L293D
QUADRUPLE HALF-H DRIVER

D2942, SEPTEMBER 1986

- 600-mA Output Current Capability per Channel
- Output Clamp Diodes for Inductive Transient Suppression
- Wide Supply Voltage Range: 4.5 V to 36 V
- Separate Input-Logic Supply
- Thermal Shutdown
- Internal ESD Protection
- High-Noise-Immunity Inputs
- Designed to be Interchangeable with SGS L293D

NE DUAL-IN-LINE PACKAGE
(TOP VIEW)

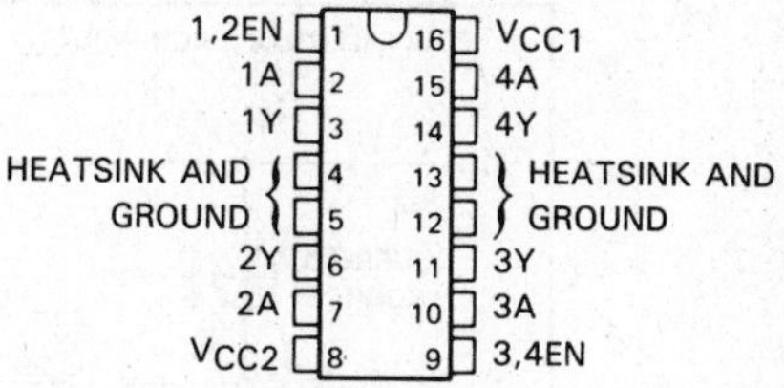
1,2EN
1A
1Y
HEATSINK AND GROUND
2Y
2A
VCC2
VCC1
4A
4Y
HEATSINK AND GROUND
3Y
3A
3,4EN

FUNCTION TABLE
(EACH CHANNEL)

INPUTS		OUTPUT
A	EN	Y
H	H	H
L	H	L
X	L	Z

H = high-level
L = low-level
X = irrelevant
Z = high-impedance (off)

description

The L293D is a quadruple high-current half-H driver designed to provide bidirectional drive currents of up to 600 milliamperes at voltages from 4.5 volts to 36 volts. It is designed to drive inductive loads such as relays, solenoids, dc and stepping motors, as well as other high-current/high-voltage loads in positive-supply applications.

All inputs are TTL-compatible. Each output is a complete totem-pole drive circuit with a Darlington transistor sink and a psuedo-Darlington source. Channels are enabled in pairs with channels 1 and 2 enabled by 1,2EN and channels 3 and 4 enabled by 3,4EN. When an enable input is high,the associated channels are enabled and their outputs are active and in phase with their inputs. When the enable input is low, those channels are disabled and their outputs are off and in a high-impedance state. With the proper data inputs, each pair of drivers form a full-H (or bridge) reversible drive suitable for solenoid or motor applications.

A V_{CC1} terminal, separate from V_{CC2}, is provided for the logic inputs to minimize device power dissipation.

The L293D is designed for operation from 0°C to 70°C.

logic symbol†

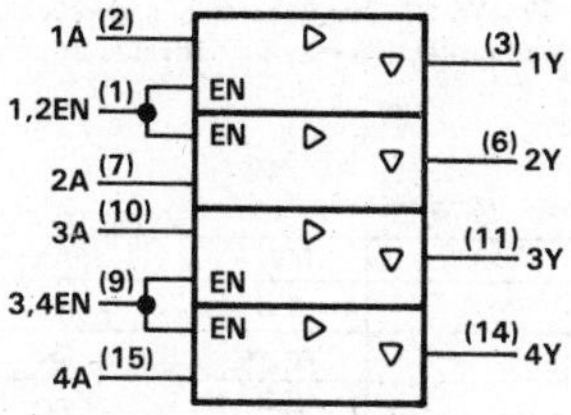
1A (2)
1,2EN (1)
2A (7)
3A (10)
3,4EN (9)
4A (15)
EN
EN
EN
EN
(3) 1Y
(6) 2Y
(11) 3Y
(14) 4Y

† This symbol is in accordance with ANSI/IEEE Std 91-1984 and IEC Publication 617-12.

logic diagram

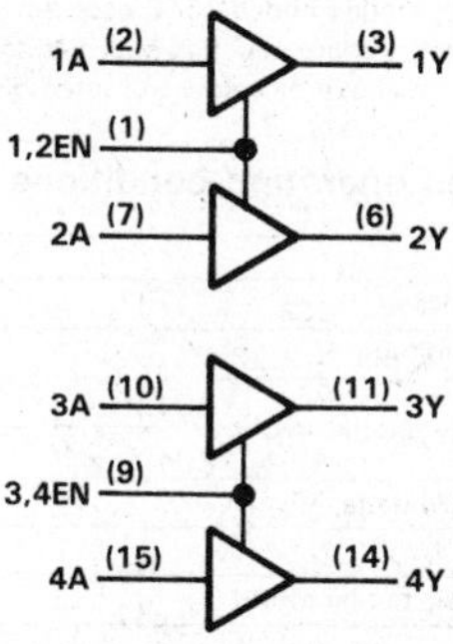
1A (2)
(3) 1Y
1,2EN (1)
2A (7)
(6) 2Y
3A (10)
(11) 3Y
3,4EN (9)
4A (15)
(14) 4Y

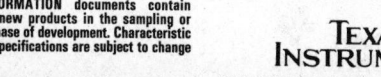

TEXAS INSTRUMENTS
POST OFFICE BOX 655012 • DALLAS, TEXAS 75265

schematics of inputs and outputs

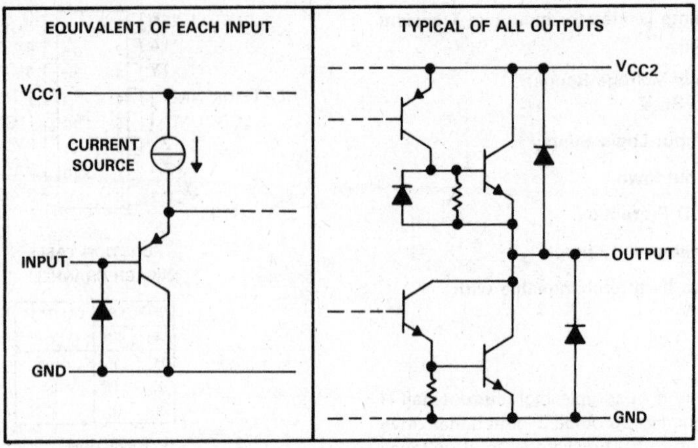
EQUIVALENT OF EACH INPUT
TYPICAL OF ALL OUTPUTS
V_{CC1}
CURRENT
SOURCE
INPUT
GND
V_{CC2}
OUTPUT
GND

absolute maximum ratings over operating free-air temperature range (unless otherwise noted)

Logic supply voltage, V_{CC1} (see Note 1) .... 36 V
Output supply voltage, V_{CC2} .... 36 V
Input voltage .... 7 V
Output voltage range .... -3 V to $V_{CC2}+3$ V
Peak output current (nonrepetitive, $t \leq 100\ \mu s$) .... ±1.2 A
Continuous output current .... ±600 mA
Continuous total dissipation at (or below) 25°C free-air temperature (see Notes 2 and 3) .... 2075 mW
Continuous total dissipation at (or below) 25°C case temperature (see Note 3) .... 7375 mW
Continuous total dissipation at 80°C case temperature (see Note 3) .... 4130 mW
Operating case or virtual junction temperature range .... −40°C to 150°C
Storage temperature range .... −65°C to 150°C
Lead temperature 1,6 mm (1/16 inch) from case for 10 seconds .... 260°C

NOTES: 1. All voltage values are with respect to the network ground terminal.
2. For operation above 25°C free-air temperature, derate linearly at the rate of 16.6 mW/°C.
3. For operation above 25°C case temperature, derate linearly at the rate of 59 mW/°C. Due to variations in individual device electrical characteristics and thermal resistance, the built-in thermal overload protection may be activated at power levels slightly above or below the rated dissipation.

recommended operating conditions

		MIN	MAX	UNIT
Logic supply voltage, V_{CC1}		4.5	7	V
Output supply voltage, V_{CC2}		V_{CC1}	36	V
High-level input voltage, V_{IH}	$V_{CC1} \leq 7$ V	2.3	V_{CC1}	V
	$V_{CC1} \geq 7$ V	2.3	7	
Low-level input voltage, V_{IL}		−0.3†	1.5	V
Output current, I_O			±600	mA
Operating free-air temperature, T_A		0	70	°C

† The algebraic convention, in which the least positive (most negative) limit is designated minimum, is used in this data sheet for logic voltage levels.

TEXAS INSTRUMENTS
POST OFFICE BOX 655012 • DALLAS, TEXAS 75265

electrical characteristics, V_{CC1} = 5 V, V_{CC2} = 24 V, T_A = 25°C

PARAMETER			TEST CONDITIONS		MIN	TYP	MAX	UNIT
V_{OH}	High-level output voltage		I_{OH} = −0.6 A		$V_{CC2}-1.8$	$V_{CC2}-1.4$		V
V_{OL}	Low-level output voltage		I_{OL} = 0.6 A			1.2	1.8	V
V_{OKH}	High-level output clamp voltage		I_{OK} = 0.6 A			$V_{CC2}+1.3$		V
V_{OKL}	Low-level output clamp voltage		I_{OK} = −0.6 A			1.3		V
I_{IH}	High-level input current	A	V_I = 7 V			0.2	100	μA
		EN				0.2	±10	
I_{IL}	Low-level input current	A	V_I = 0			−3	−10	μA
		EN				−2	−100	
I_{CC1}	Logic supply current		I_O = 0	All outputs at high level			22	mA
				All outputs at low level			60	
				All outputs at high impedance			24	
I_{CC2}	Output supply current		I_O = 0	All outputs at high level			24	mA
				All outputs at low level			6	
				All outputs at high impedance			4	

switching characteristics, V_{CC1} = 5 V, V_{CC2} = 24 V, T_A = 25°C

PARAMETER		TEST CONDITIONS	MIN	TYP	MAX	UNIT
t_{PLH}	Propagation delay time, low-to-high-level output from A input	C_L = 30 pF, See Figure 1		800		ns
t_{PHL}	Propagation delay time, high-to-low-level output from A input			400		ns
t_{TLH}	Transition time, low-to-high-level output			300		ns
t_{THL}	Transition time, high-to-low-level output			300		ns

PARAMETER MEASUREMENT INFORMATION

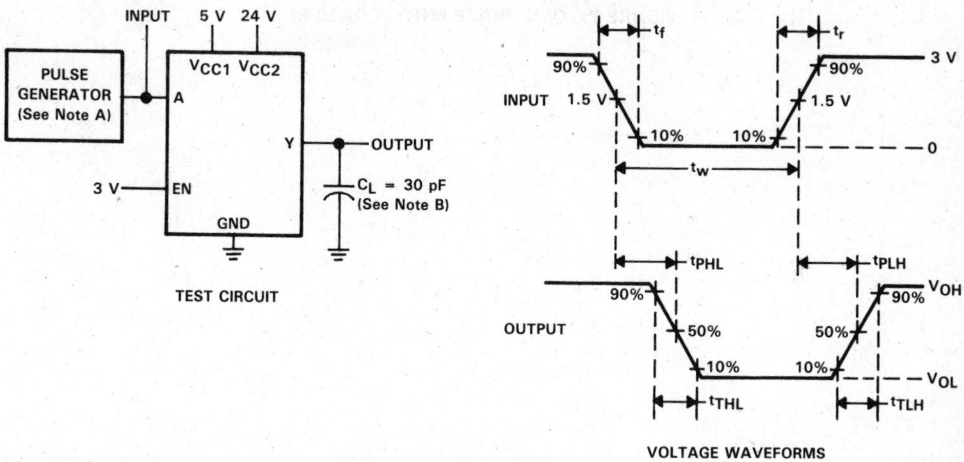
INPUT
5 V
24 V
PULSE GENERATOR (See Note A)
VCC1 VCC2
A
Y
OUTPUT
3 V
EN
CL = 30 pF (See Note B)
GND
TEST CIRCUIT
tf
tr
3 V
90%
90%
INPUT
1.5 V
1.5 V
10%
10%
0
tw
tPHL
tPLH
VOH
90%
90%
OUTPUT
50%
50%
10%
10%
VOL
tTHL
tTLH
VOLTAGE WAVEFORMS

NOTES: A. The pulse generator has the following characteristics: $t_r \le 10$ ns, $t_f \le 10$ ns, $t_w = 10$ μs, PRR = 5 kHz, $Z_{out} = 50\ \Omega$.
B. C_L includes probe and jig capacitance.

FIGURE 1. SWITCHING TIMES

TYPICAL APPLICATION DATA

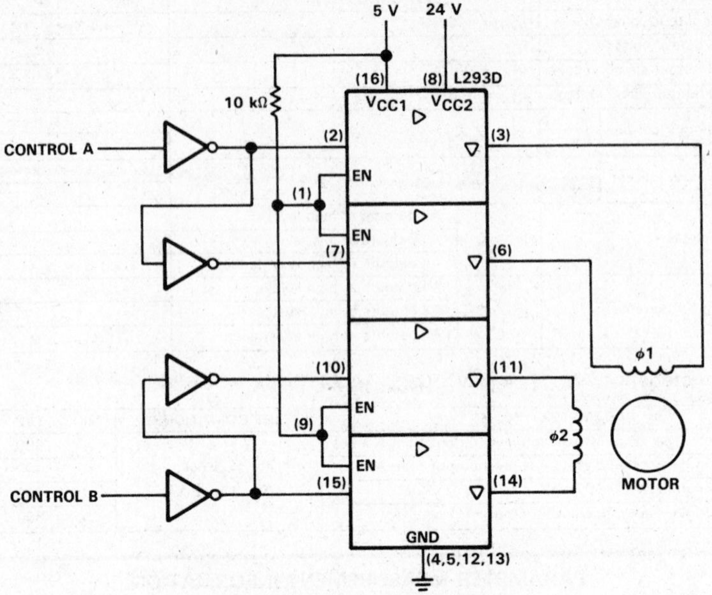
5 V
24 V
(16)
(8)
L293D
10 kΩ
V_{CC1}
V_{CC2}
CONTROL A
(2)
(3)
EN
(1)
EN
(7)
(6)
φ1
(10)
(11)
EN
(9)
φ2
EN
MOTOR
CONTROL B
(15)
(14)
GND
(4,5,12,13)

FIGURE 2. TWO-PHASE MOTOR DRIVER

TEXAS INSTRUMENTS
POST OFFICE BOX 655012 • DALLAS, TEXAS 75265

L298
DUAL FULL-H DRIVER

D2942, OCTOBER 1986

- 2 A Per Channel Output Capability
- Wide Range of Output Supply Voltage . . . 5 V to 46 V
- Separate Input-Logic Supply Voltage
- Thermal Shutdown
- Internal Electrostatic Discharge Protection
- High Noise Immunity
- Direct Replacement for SGS L298

KV PACKAGE
(TOP VIEW)

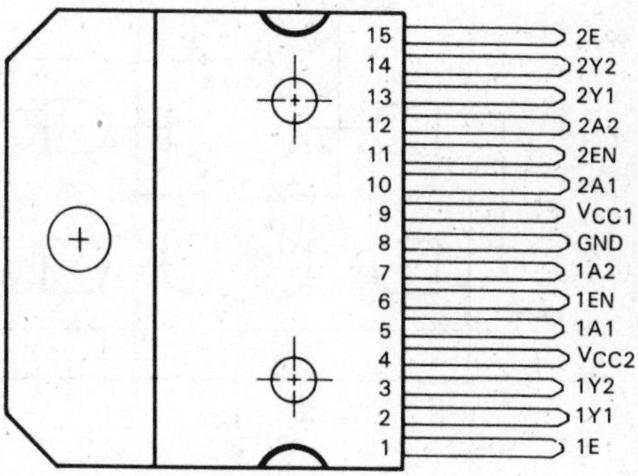

15
14
13
12
11
10
9
8
7
6
5
4
3
2
1
2E
2Y2
2Y1
2A2
2EN
2A1
V_{CC1}
GND
1A2
1EN
1A1
V_{CC2}
1Y2
1Y1
1E

The tab is electrically connected to pin 8.

logic symbol†

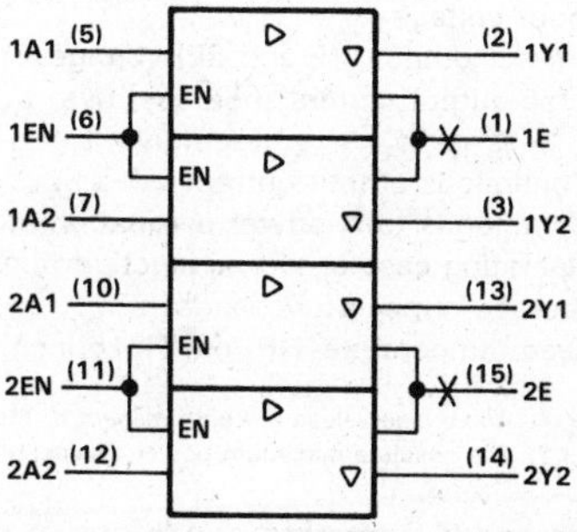

1A1 (5)
1EN (6)
1A2 (7)
2A1 (10)
2EN (11)
2A2 (12)
EN
EN
EN
EN
(2) 1Y1
(1) 1E
(3) 1Y2
(13) 2Y1
(15) 2E
(14) 2Y2

† This symbol is in accordance with ANSI/IEEE Std 91-1984 and IEC Publication 617-12.

description

The L298 is a dual high-current full-H driver designed to provide bidirectional drive currents of up to two amperes at voltages from 5 volts to 46 volts. It is designed to drive inductive loads such as relays, solenoids, dc motors, stepping motors, and other high-current or high-voltage loads in positive-supply applications. All inputs are TTL compatible. Each output (Y) is a complete totem-pole drive with a Darlington transistor sink and a psuedo-Darlington source. Each full-H driver is enabled separately. Outputs 1Y1 and 1Y2 are enabled by 1EN and outputs 2Y1 and 2Y2 are enabled by 2EN. When an EN input is high, the associated channels are active. When an EN input is low, the associated channels are off (i.e., in the high-impedance state).

Each half of the device forms a full-H reversible driver suitable for solenoid or motor applications. The current in each full-H driver can be monitored by connecting a resistor between the sense output terminal 1E and ground and another resistor between sense output terminal 2E and ground.

External high-speed output-clamp diodes should be used for inductive transient suppression. To minimize device power dissipation, a V_{CC1} supply voltage, separate from V_{CC2}, is provided for the logic inputs.

logic diagram (positive logic)

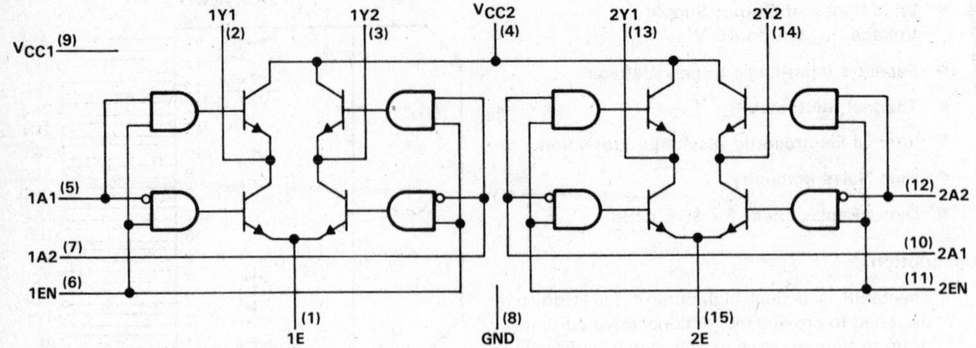
1Y1
(2)
1Y2
(3)
VCC2
(4)
2Y1
(13)
2Y2
(14)
VCC1 (9)
1A1 (5)
1A2 (7)
1EN (6)
(12) 2A2
(10) 2A1
(11) 2EN
(1)
1E
(8)
GND
(15)
2E

absolute maximum ratings over operating temperature (unless otherwise noted)

Logic supply voltage, V_{CC1}, (see Note 1) .... 7 V
Output supply voltage, V_{CC2} .... 50 V
Input voltage .... 7 V
Emitter output (1E and 2E) voltage .... −1 to 2.3 V
Peak output current (nonrepetitive, $t_w \leq 0.1$ ms) .... ±3 A
(repetitive, $t_w \leq 10$ ms, duty cycle ≤ 80%) .... ±2.5 A
Continuous output current .... ±2 A
Continuous total power dissipation at 75°C case temperature (see Note 2) .... 25 W
Operating case or virtual junction temperature range .... −40°C to 150°C
Storage temperature range .... −65°C to 150°C
Lead temperature 1,6 mm (1/16 inch) from the case for 10 seconds .... 260°C

NOTES: 1. All voltage values are with respect to the network ground terminal.
2. The absolute maximum power dissipation ratings are design goals. For further information contact the factory.

recommended operating conditions

		MIN	MAX	UNIT
V_{CC1}	Logic supply voltage	4.5	7	V
V_{CC2}	Output supply voltage	V_{IH} + 2.5	46	V
V_{IH}	High-level input voltage	2.3	V_{CC1}	V
V_{IL}	Low-level input voltage	−0.3†	1.5	V
I_O	Output current		±2	A
T_A	Operating free-air temperature	0	70	°C

†The algebraic convention, in which the least positive (most negative designated minimum), is used in this data sheet for logic voltage levels.

D2985, FEBRUARY 1987

- Three Operating Modes . . . Full Step, Half Step, and Quarter Step
- Both Digital and Analog Control of Output Current
- 1-Ampere Bidirectional Output Current Capability
- Chop-Mode Current Regulation
- Wide Output Supply Voltage Range . . . 10 V to 46 V
- Separate Input-Logic Supply
- Thermal Shutdown Protection
- Output Clamp Diodes for Inductive Transient Protection
- Internal ESD Protection
- Direct Replacement for SGS PBL3717A

NE PACKAGE
(TOP VIEW)

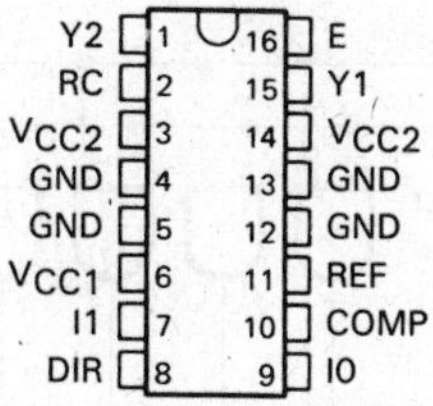
Y2 1
RC 2
VCC2 3
GND 4
GND 5
VCC1 6
I1 7
DIR 8
16 E
15 Y1
14 VCC2
13 GND
12 GND
11 REF
10 COMP
9 I0

FUNCTION TABLE

LOGIC INPUTS			OUTPUTS		
DIR	I1	I0	Y1	Y2	LEVEL
H	L	L	Source	Sink	High Current
L			Sink	Source	
H	L	H	Source	Sink	Medium Current
L			Sink	Source	
H	H	L	Source	Sink	Low Current
L			Sink	Source	
X	H	H			Off

description

The PBL3717A is a high-current, high-voltage full-H reversible driver designed to control and drive one phase of a bipolar stepper motor. It is designed to provide bidirectional drive currents of up to one ampere at voltages from 10 volts to 46 volts in positive-supply applications. Two PBL3717A devices, with a few external components, form a complete two-phase bipolar stepper motor driver. All inputs are TTL-compatible. Each output (Y) forms a complete totem-pole drive with a Darlington transistor sink and a psuedo-Darlington source. Logic input pin DIR selects the direction of current flow in a load connected between outputs Y1 and Y2. A high level at the DIR input causes the load current to flow from output Y1 (source) to output Y2 (sink). A low level at the DIR input causes the load current to flow from output Y2 (source) to output Y1 (sink). When logic inputs I0 and I1 are both high, the Y1 and Y2 outputs are disabled and in the high-impedance state.

The current in the full-H driver load can be monitored by connecting a resistor between the sense output terminal E and ground. Voltage feedback from terminal E to the COMP input pin provides output current regulation via chop-mode operation of the sink output transistors. Three levels of output current can be selected by programming two logic inputs, I0 and I1, as shown in the function table. These inputs are internally decoded to enable one of three comparators to set the output current level to low, medium, or high. The precise level of output current is set by the comparator selected, the comparator reference voltage applied to the REF pin, the value of the sense resistor, and the sense output voltage fed back to the COMP input. When chop-mode current regulation is used, an internal monostable circuit, programmed by an external RC network at the RC pin, sets the current decay time.

The device contains built-in high-speed output clamp diodes for inductive transient protection. A separate supply voltage (V_{CC1}) is provided for the logic input circuits to minimize device power dissipation. Supply voltage V_{CC2} is used for the output circuits. Both V_{CC2} supply pins should be connected together as close to the package as possible.

The PBL3717A is characterized for operation from 0°C to 70°C.

logic diagram (position logic)

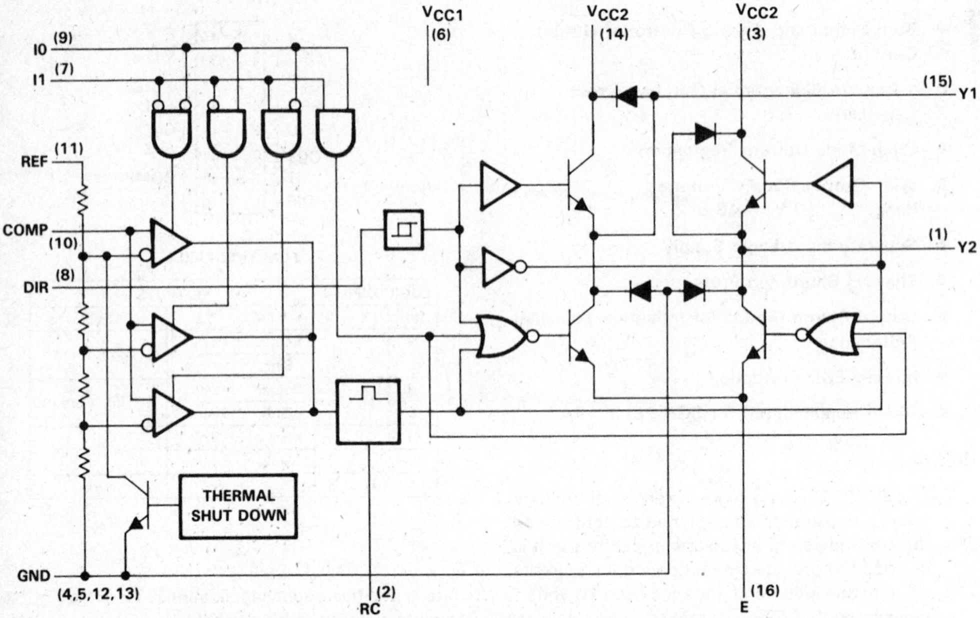
VCC1 (6)
VCC2 (14)
VCC2 (3)
I0 (9)
I1 (7)
(15) Y1
REF (11)
COMP (10)
(1) Y2
DIR (8)
THERMAL SHUT DOWN
GND (4,5,12,13)
(2) RC
(16) E

absolute maximum ratings over operating free-air temperature range (unless otherwise noted)

Output supply voltage, V_{CC2} (see Notes 1 and 2)	50 V
Logic supply voltage, V_{CC1}	7 V
Logic input voltage	6 V
Comparator input voltage	V_{CC1}
Reference voltage, V_{ref}	15 V
Continuous output current	±1.2 A
Continuous total dissipation at (or below) 25 °C free-air temperature (see Note 3)	2075 mW
Continuous total dissipation at (or below) 25 °C case temperature (see Note 3)	7375 mW
Operating case or virtual junction temperature	0 °C to 150 °C
Storage temperature range	−65 °C to 150 °C
Lead temperature 1,6 mm (1/16 inch) from the case for 10 seconds	260 °C

NOTES: 1. All voltage values are with respect to the GND terminal.
2. Both V_{CC2} pins must be connected together as close to the package as possible for optimum testing and operation of the device.
3. For operation above 25 °C free-air temperature, derate linearly at the rate of 16.6 mW/°C. For operation above 25 °C case temperature, derate linearly at the rate of 59 mW/°C. To avoid exceeding the design maximum virtual junction temperature, these ratings should not be exceeded. Due to variations in individual device electrical characteristics and thermal resistance, the built-in thermal overload protection may be activated at power levels slightly above or below the rated dissipation.

recommended operating conditions

	MIN	MAX	UNIT
Output supply voltage, V_{CC2}	10	46	V
Logic supply voltage, V_{CC1}	4.75	5.25	V
High-level input voltage, V_{IH}	2	V_{CC1}	V
Low-level input voltage, V_{IL}		0.8	V
Output current, I_O		±1	A
Operating free-air temperature, T_A	0	70	°C

D2620, FEBRUARY 1981—REVISED SEPTEMBER 1986

- Output Collector Current . . . 1.5 A Max
- 2-W Dissipation Rating
- High Output Voltage Capability
- Outputs Diode-Clamped for Inductive Loads
- Common-Emitter Circuit for Current Sink
- SN75064 and SN75065 Have TTL-Compatible Inputs
- SN75066 and SN75067 Have CMOS- and PMOS-Compatible Inputs
- Functionally Interchangeable with ULN2064 thru ULN2067, Respectively

NE
DUAL-IN-LINE PACKAGE
(TOP VIEW)

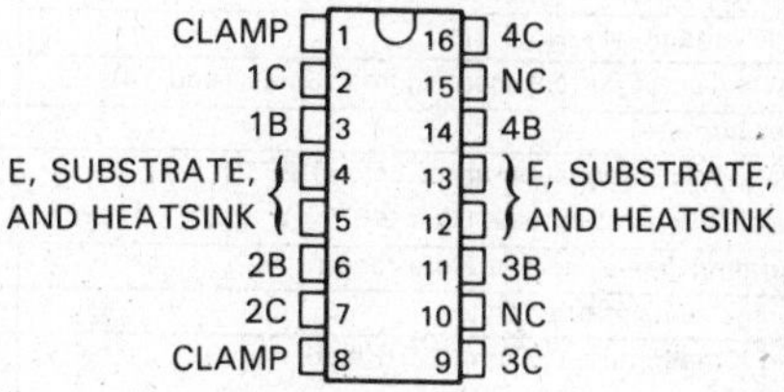
CLAMP 1
1C 2
1B 3
E, SUBSTRATE, AND HEATSINK 4
5
2B 6
2C 7
CLAMP 8
9 3C
10 NC
11 3B
12 AND HEATSINK
13 E, SUBSTRATE,
14 4B
15 NC
16 4C

NC—No internal connection

description

The SN75064, SN75065, SN75066, and SN75067 are monolithic high-voltage, high-current Darlington transistor switches. Each comprises four n-p-n Darlington pairs. All units feature high-voltage outputs with common-cathode clamp diodes for switching inductive loads. Outputs and inputs may each be paralleled for higher current capability. Applications include relay drivers, hammer drivers, lamp drivers, display drivers (LED and gas discharge), line drivers, and logic buffers. These common-emitter circuits are designed to operate as current sinks to the load.

The SN75064 and SN75065 are intended for use with TTL and 5-volt MOS logic. The SN75066 and SN75067 are intended for use with PMOS and higher voltage CMOS logic.

The SN75064 thru SN75067 are characterized for operation from 0°C to 70°C.

schematic (each Darlington pair)

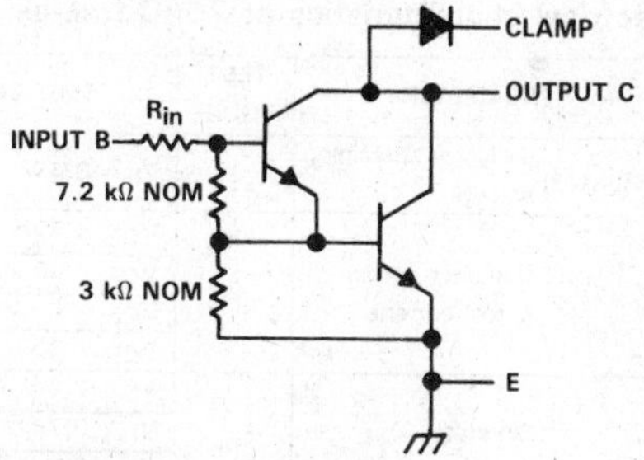
CLAMP
OUTPUT C
R_{in}
INPUT B
7.2 kΩ NOM
3 kΩ NOM
E

SN75064, SN75065: R_{in} = 350 Ω NOM
SN75066, SN75067: R_{in} = 3 kΩ NOM

logic symbol†

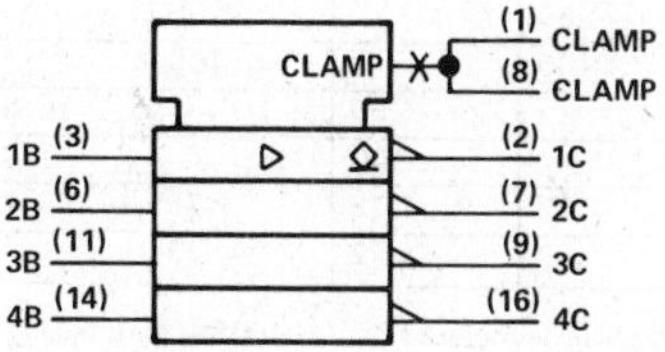
CLAMP
(1) CLAMP
(8) CLAMP
1B (3)
2B (6)
3B (11)
4B (14)
(2) 1C
(7) 2C
(9) 3C
(16) 4C

†This symbol is in accordance with ANSI/IEEE Std 91-1984 and IEC Publication 617-12.

logic diagram

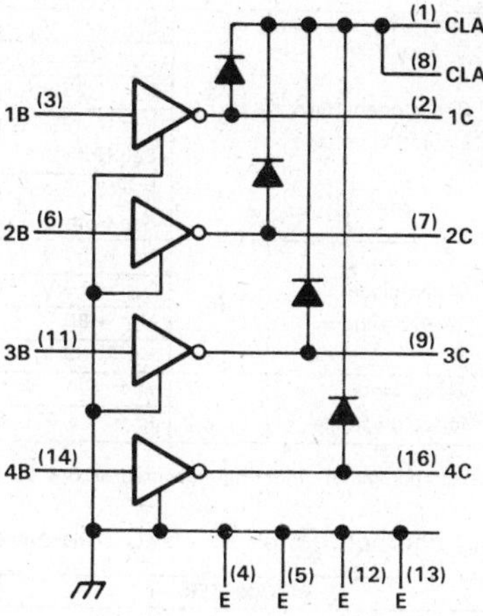
(1) CLAMP
(8) CLAMP
1B (3)
(2) 1C
2B (6)
(7) 2C
3B (11)
(9) 3C
4B (14)
(16) 4C
(4) E
(5) E
(12) E
(13) E

absolute maximum ratings at 25 °C free-air temperature for each switch (unless otherwise noted)

	SN75064	SN75065	SN75066	SN75067	UNIT
Collector-emitter voltage	50	80	50	80	V
Input voltage (see Note 1)	15	15	30	30	V
Peak collector current (see Figures 12, 13, and 14)	1.5	1.5	1.5	1.5	A
Input current	25	25	25	25	mA
Total power dissipation at (or below) 25 °C free-air temperature (see Note 2)	2075	2075	2075	2075	mW
Operating free-air temperature range	0 to 70	0 to 70	0 to 70	0 to 70	°C
Storage temperature range	−55 to 150	−55 to 150	−55 to 150	−55 to 150	°C
Lead temperature 1,6 mm (1/16 inch) from the case for 10 seconds	260	260	260	260	°C

NOTES: 1. All voltage values (unless otherwise noted) are with respect to the emitter/substrate terminal E.
2. For operation above 25 °C free-air temperature, derate to 1328 mW at 70 °C at the rate of 16.6 mW/°C.

electrical characteristics at 25 °C free-air temperature (unless otherwise noted)

PARAMETER		TEST FIGURE	TEST CONDITIONS	SN75064		SN75065		SN75066		SN75067		UNIT
				MIN	MAX	MIN	MAX	MIN	MAX	MIN	MAX	
$V_{CEX(sus)}$	Collector sustaining voltage	1	$V_I = 0.4$ V, $I_C = 100$ mA	35		50		35		50		V
I_{CEX}	Collector output cutoff current	2	$V_{CE} = 50$ V		100				100			µA
			$V_{CE} = 50$ V, $T_A = 70$ °C		500				500			
			$V_{CE} = 80$ V				100				100	
			$V_{CE} = 80$ V, $T_A = 70$ °C				500				500	
$I_{I(on)}$	On-state input current	3	$V_I = 2.4$ V	2	4.3	2	4.3					mA
			$V_I = 3.75$ V	4.5	9.6	4.5	9.6					
			$V_I = 5$ V					0.9	1.8	0.9	1.8	
			$V_I = 12$ V					2.75	5.2	2.75	5.2	
$V_{I(on)}$	On-state input voltage	4	$V_{CE} = 2$ V, $I_C = 1$ A		2		2		6.5		6.5	V
			$V_{CE} = 2$ V, $I_C = 1.5$ A, See Note 3		2.5		2.5		10		10	
$V_{CE(sat)}$	Collector-emitter saturation voltage	5	$I_I = 625$ µA, $I_C = 500$ mA		1.13		1.13		1.13		1.13	V
			$I_I = 935$ µA, $I_C = 750$ mA		1.25		1.25		1.25		1.25	
			$I_I = 1.25$ mA, $I_C = 1$ A		1.4		1.4		1.4		1.4	
			$I_I = 2$ mA, $I_C = 1.25$ A, See Note 3		1.6				1.6			
			$I_I = 2.25$ mA, $I_C = 1.5$ A, See Note 3				1.7				1.7	
I_R	Clamp-diode reverse current	6	$V_R = 50$ V		50				50			µA
			$V_R = 50$ V, $T_A = 70$ °C		100				100			
			$V_R = 80$ V				50				50	
			$V_R = 80$ V, $T_A = 70$ °C				100				100	
V_F	Clamp-diode forward voltage	7	$I_F = 1$ A		1.75		1.75		1.75		1.75	V
			$I_F = 1.5$ A, See Note 3		2		2		2		2	

NOTE 3: These parameters must be measured on one output at a time using pulse techniques, $t_w = 10$ ms, duty cycle ≤ 10%.

switching characteristics at 25 °C free-air temperature, $V_{CC} = 5$ V

PARAMETER		TEST CONDITIONS	MIN	TYP	MAX	UNIT
t_{PLH}	Propagation delay time, low-to-high-level output	See Figure 8			1	µs
t_{PHL}	Propagation delay time, high-to-low-level output				1.5	µs

PARAMETER MEASUREMENT INFORMATION

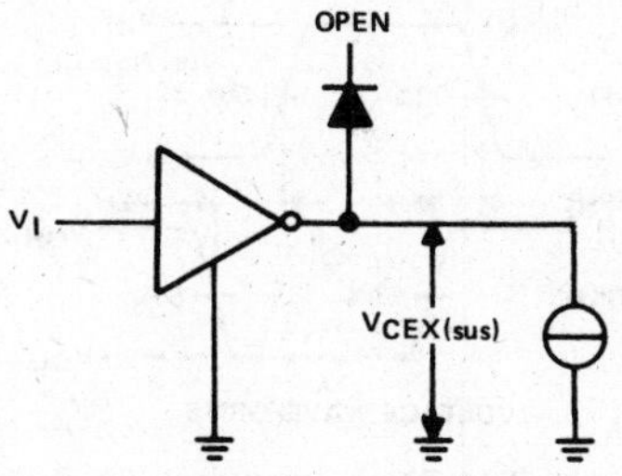
OPEN
V_I
$V_{CEX(sus)}$
I_C

FIGURE 1. $V_{CEX(sus)}$

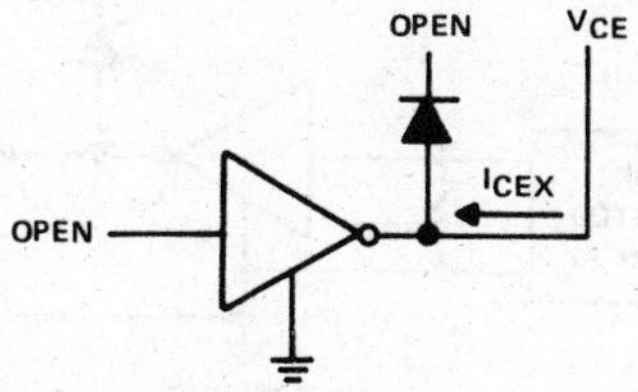
OPEN
V_{CE}
I_{CEX}
OPEN

FIGURE 2. I_{CEX}

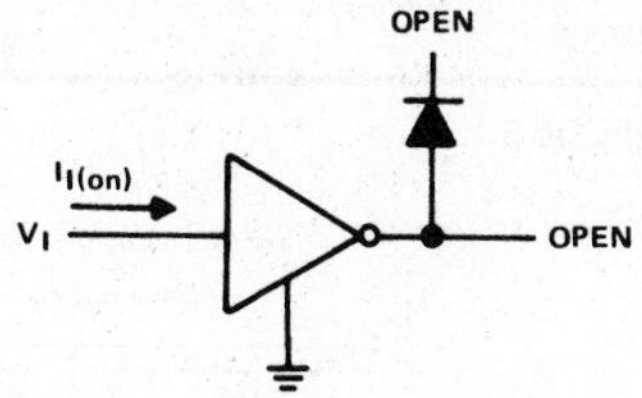
OPEN
$I_{I(on)}$
V_I
OPEN

FIGURE 3. $I_{I(on)}$

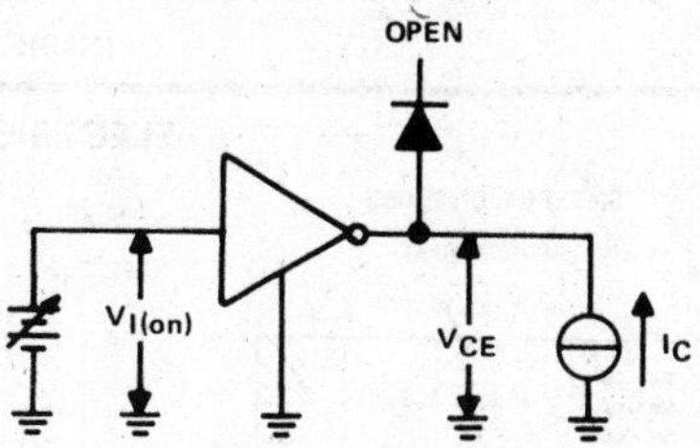
OPEN
$V_{I(on)}$
V_{CE}
I_C

FIGURE 4. $V_{I(on)}$

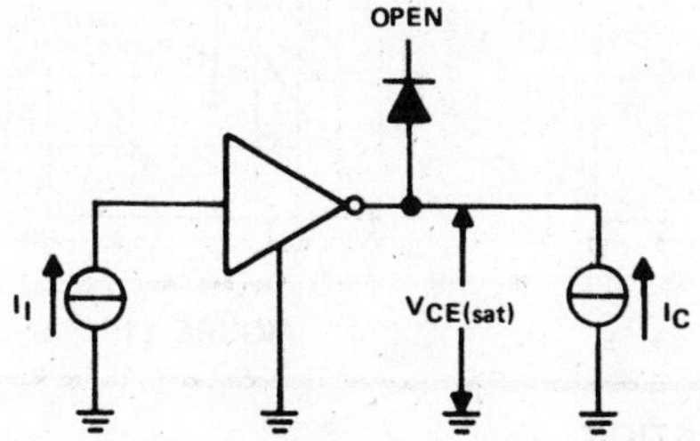
OPEN
I_I
$V_{CE(sat)}$
I_C

FIGURE 5. $V_{CE(sat)}$

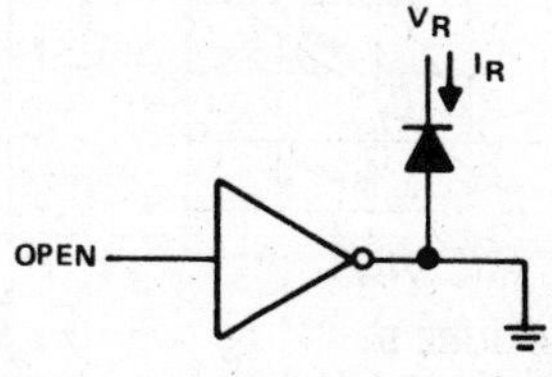
V_R
I_R
OPEN

FIGURE 6. I_R

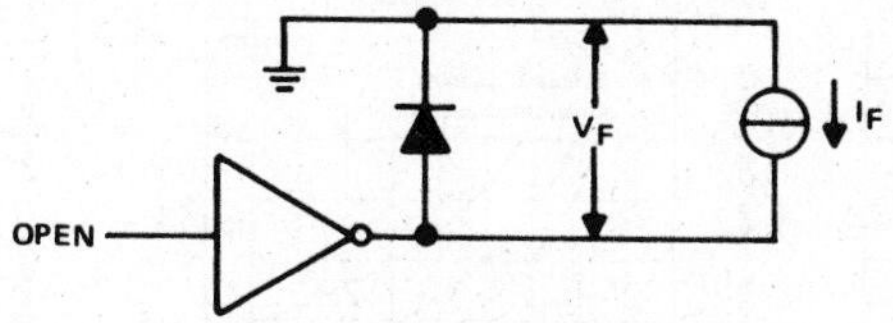
V_F
I_F
OPEN

FIGURE 7. V_F

PARAMETER MEASUREMENT INFORMATION

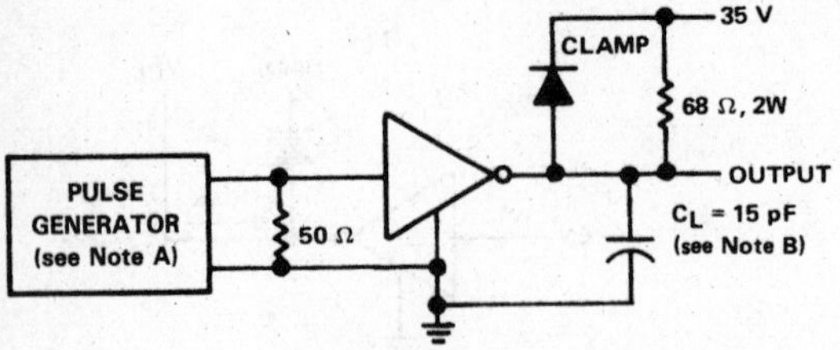
35 V
CLAMP
68 Ω, 2W
PULSE GENERATOR (see Note A)
50 Ω
OUTPUT
C_L = 15 pF (see Note B)

TEST CIRCUIT

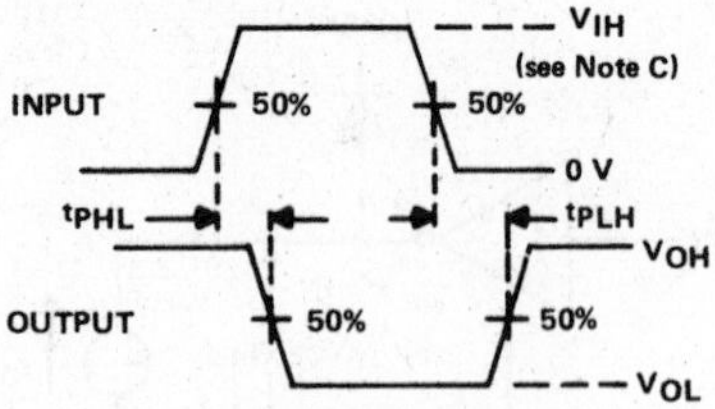
V_{IH} (see Note C)
INPUT
50%
50%
0 V
t_{PHL}
t_{PLH}
V_{OH}
OUTPUT
50%
50%
V_{OL}

VOLTAGE WAVEFORMS

NOTES: A. The input pulse is supplied by a generator having the following characteristics: PRR = 50 kHz, duty cycle = 10%, Z_o = 50 Ω.
B. C_L includes all probe and stray capacitance.
C. V_{IH} = 2.5 V for SN75064 and SN75065. V_{IH} = 10 V for SN75066 and SN75067.

FIGURE 8. SWITCHING TIMES

ELECTRICAL CHARACTERISTICS

SN75064, SN75065

INPUT CURRENT
vs
INPUT VOLTAGE

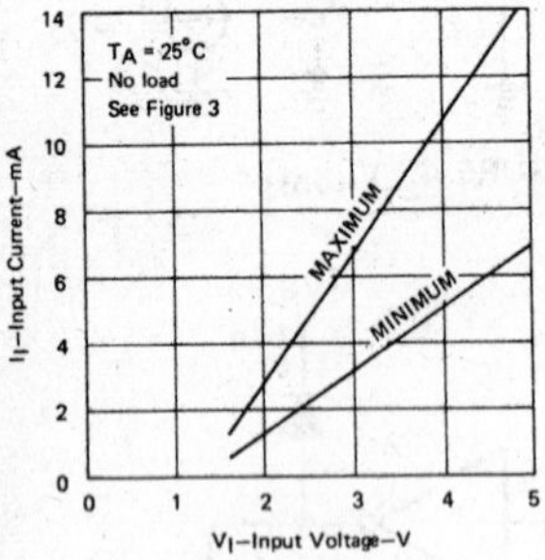
T_A = 25°C
No load
See Figure 3
MAXIMUM
MINIMUM
I_I—Input Current—mA
V_I—Input Voltage—V

FIGURE 9

SN75066, SN75067

INPUT CURRENT
vs
INPUT VOLTAGE

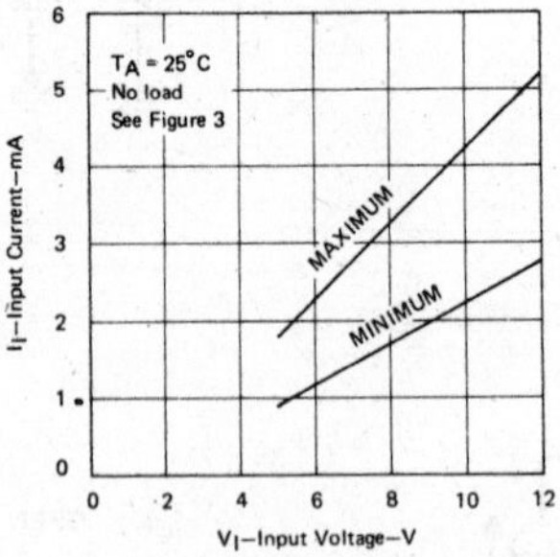
T_A = 25°C
No load
See Figure 3
MAXIMUM
MINIMUM
I_I—Input Current—mA
V_I—Input Voltage—V

FIGURE 10

TYPICAL COLLECTOR CURRENT
vs
BASE CURRENT

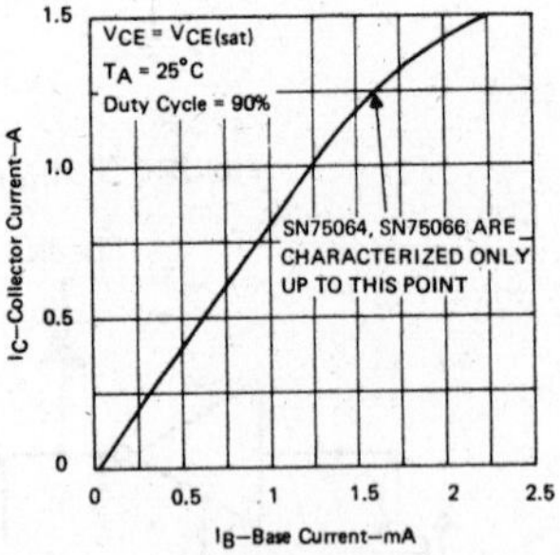
V_{CE} = $V_{CE(sat)}$
T_A = 25°C
Duty Cycle = 90%
SN75064, SN75066 ARE
CHARACTERIZED ONLY
UP TO THIS POINT
I_C—Collector Current—A
I_B—Base Current—mA

FIGURE 11

THERMAL INFORMATION

MAXIMUM COLLECTOR CURRENT
vs
DUTY CYCLE

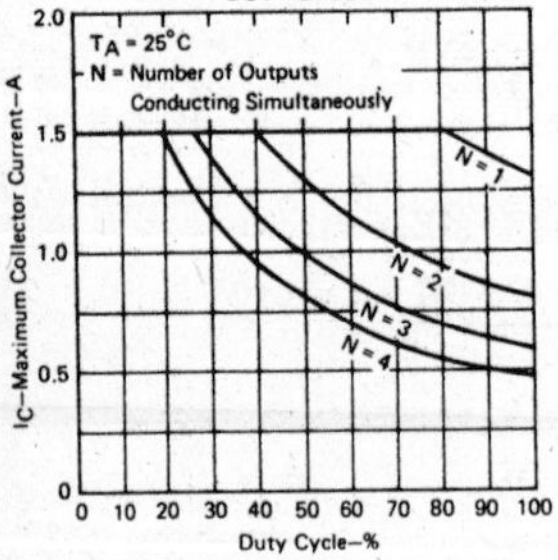
T_A = 25°C
N = Number of Outputs
Conducting Simultaneously
N = 1
N = 2
N = 3
N = 4
I_C—Maximum Collector Current—A
Duty Cycle—%

FIGURE 12

MAXIMUM COLLECTOR CURRENT
vs
DUTY CYCLE

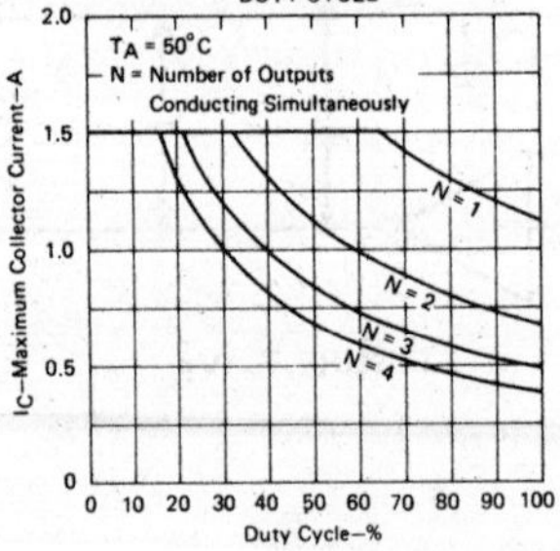
T_A = 50°C
N = Number of Outputs
Conducting Simultaneously
N = 1
N = 2
N = 3
N = 4
I_C—Maximum Collector Current—A
Duty Cycle—%

FIGURE 13

MAXIMUM COLLECTOR CURRENT
vs
DUTY CYCLE

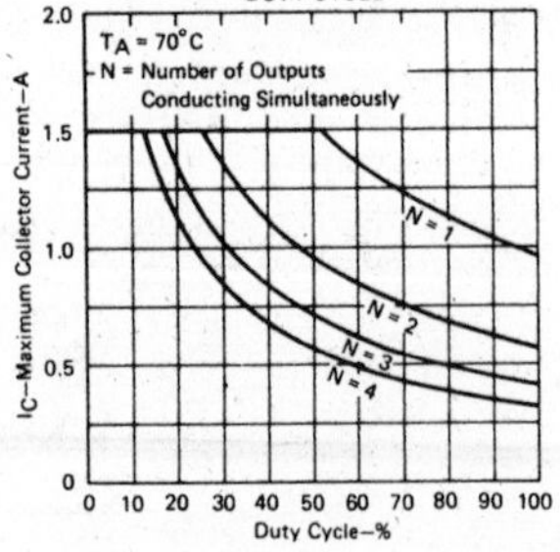
T_A = 70°C
N = Number of Outputs
Conducting Simultaneously
N = 1
N = 2
N = 3
N = 4
I_C—Maximum Collector Current—A
Duty Cycle—%

FIGURE 14

TYPICAL APPLICATION DATA

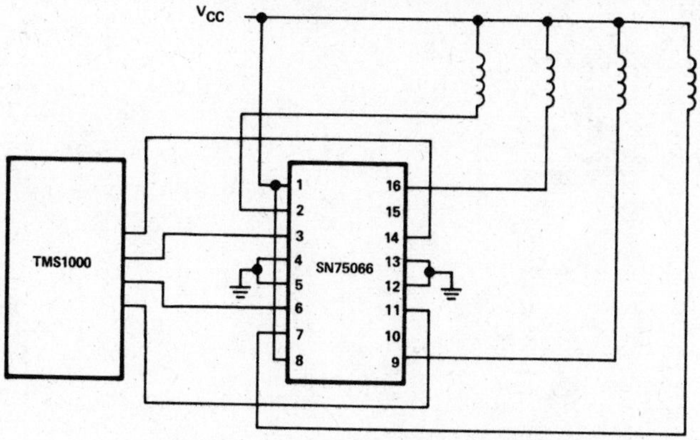
VCC
TMS1000
SN75066
1
2
3
4
5
6
7
8
16
15
14
13
12
11
10
9

FIGURE 15. RELAY DRIVER INTERFACE

D2621, DECEMBER 1979—REVISED FEBRUARY 1987

- Output Collector Current . . . 1.5 A Max
- 2-W Dissipation Rating
- High Output-Voltage Capability
- Preamp for High Current Gain
- Outputs Diode-Clamped for Inductive Loads
- Common-Emitter Circuit for Current Sink
- Inputs Compatible with TTL and 5-Volt CMOS
- Functionally Interchangeable with ULN2068 and ULN2069

NE
DUAL-IN-LINE PACKAGE
(TOP VIEW)

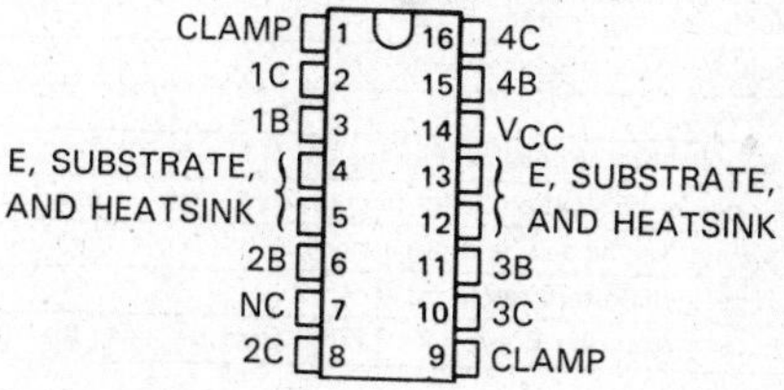
CLAMP 1 16 4C
1C 2 15 4B
1B 3 14 VCC
E, SUBSTRATE, AND HEATSINK 4 13 E, SUBSTRATE, AND HEATSINK
5 12
2B 6 11 3B
NC 7 10 3C
2C 8 9 CLAMP

NC—No internal connection

description

The SN75068 and SN75069 are monolithic integrated circuits each consisting of four high-voltage, high-current n-p-n cascaded transistor switches. Each switch includes a first stage compatible with both TTL and 5-volt CMOS signal levels. The second and third stages form uncommitted-collector outputs with common-cathode clamp diodes for switching inductive loads.

The SN75068 and SN75069 can sink up to 1.5 amperes per switch. Applications include logic buffers, MOS drivers, memory drivers, line drivers, relay drivers, hammer drivers, lamp drivers, and display drivers (LED and gas discharge).

The SN75068 and SN75069 are characterized for operation from 0°C to 70°C.

schematic (each switch)

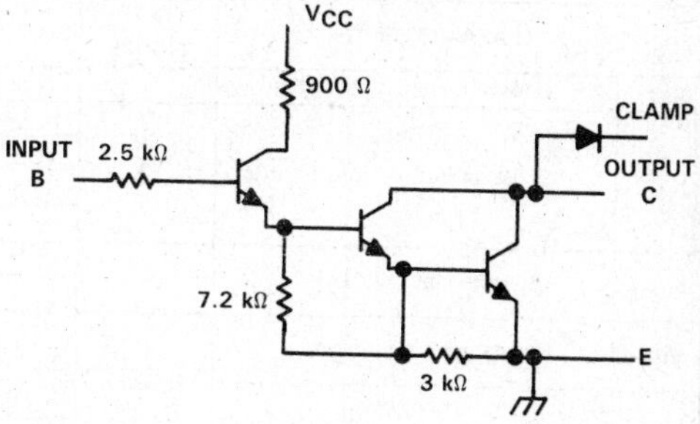
VCC
900 Ω
CLAMP
INPUT B
2.5 kΩ
OUTPUT C
7.2 kΩ
3 kΩ
E

Resistor values shown are nominal.

logic symbol†

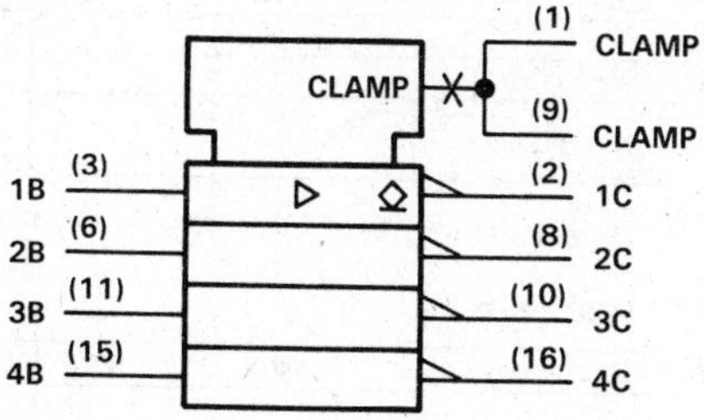
CLAMP
(1) CLAMP
(9) CLAMP
1B (3)
(2) 1C
2B (6)
(8) 2C
3B (11)
(10) 3C
4B (15)
(16) 4C

†This symbol is in accordance with ANSI/IEEE Std 91-1984 and IEC Publication 617-12.

logic diagram

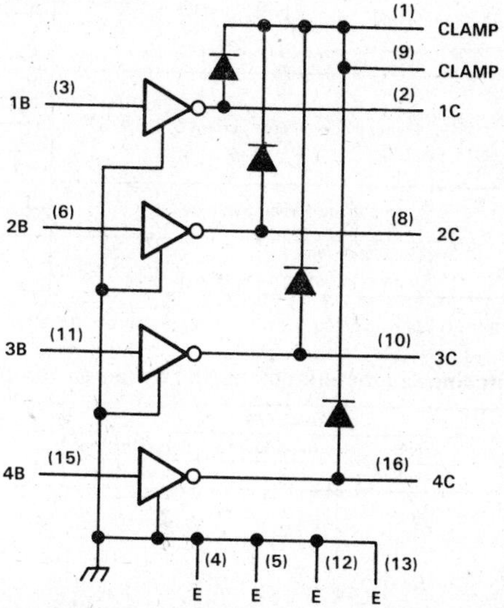
(1) CLAMP
(9) CLAMP
1B (3)
(2) 1C
2B (6)
(8) 2C
3B (11)
(10) 3C
4B (15)
(16) 4C
(4) E
(5) E
(12) E
(13) E

TEXAS INSTRUMENTS
POST OFFICE BOX 655012 • DALLAS, TEXAS 75265

absolute maximum ratings at 25°C free-air temperature for each switch (unless otherwise noted)

	SN75068	SN75069	UNIT
Collector-emitter voltage	50	80	V
Supply voltage, V_{CC} (see Note 1)	10	10	V
Input voltage	15	15	V
Peak collector current (see Figures 10, 11, and 12)	1.5	1.5	A
Total power dissipation at (or below) 25°C free-air temperature (see Note 2)	2075	2075	mW
Operating free-air temperature range	0 to 70	0 to 70	°C
Storage temperature range	−55 to 150	−55 to 150	°C
Lead temperature 1,6 mm (1/16 inch) from the case for 10 seconds	260	260	°C

NOTES: 1. All voltage values (unless otherwise noted) are with respect to the emitter/substrate terminal E.
2. For operation above 25°C free-air temperature, derate total power to 1328 mW at 70°C at the rate of 16.6 mW/°C.

electrical characteristics at 25°C free-air temperature, V_{CC} = 5 V (unless otherwise noted)

PARAMETER		TEST FIGURE	TEST CONDITIONS	SN75068		SN75069		UNIT
				MIN	MAX	MIN	MAX	
$V_{CEX(sus)}$	Collector sustaining voltage	1	V_I = 0.4 V, I_C = 100 mA	35		50		V
I_{CEX}	Collector output cutoff current	2	V_{CE} = 50 V		100			μA
			V_{CE} = 50 V, T_A = 70°C		500			
			V_{CE} = 80 V				100	
			V_{CE} = 80 V, T_A = 70°C				500	
$I_{I(on)}$	On-state input current	3	V_I = 2.4 V		250		250	μA
			V_I = 3.75 V		1000		1000	
$V_{I(on)}$	On-state input voltage	4	V_{CE} = 2 V, I_C = 1.5 A, See Note 3		2.4		2.4	V
$V_{CE(sat)}$	Collector-emitter saturation voltage	5	V_I = 2.4 V, I_C = 500 mA		1.13		1.13	V
			V_I = 2.4 V, I_C = 750 mA		1.25		1.25	
			V_I = 2.4 V, I_C = 1 A		1.4		1.4	
			V_I = 2.4 V, I_C = 1.25 V, See Note 3		1.6			
			V_I = 2.4 V, I_C = 1.5 A, See Note 3				1.7	
I_R	Clamp-diode reverse current	6	V_R = 50 V		50			μA
			V_R = 50 V, T_A = 70°C		100			
			V_R = 80 V				50	
			V_R = 80 V, T_A = 70°C				100	
V_F	Clamp-diode forward voltage	7	I_F = 1 A		1.75		1.75	V
			I_F = 1.5 A, See Note 3		2		2	
I_{CC}	Supply current (only one switch conducting)	8	V_I = 2.4 V, I_C = 500 mA		6		6	mA

NOTE 3: These parameters must be measured on one output at a time using pulse techniques, t_w = 10 ms, duty cycle ≤ 10%.

switching characteristics at 25°C free-air temperature, V_{CC} = 5 V

PARAMETER		TEST CONDITIONS	MIN	TYP	MAX	UNIT
t_{PLH}	Propagation delay time, low-to-high-level output	See Figure 9			1	μs
t_{PHL}	Propagation delay time, high-to-low-level output				1.5	μs

PARAMETER MEASUREMENT INFORMATION

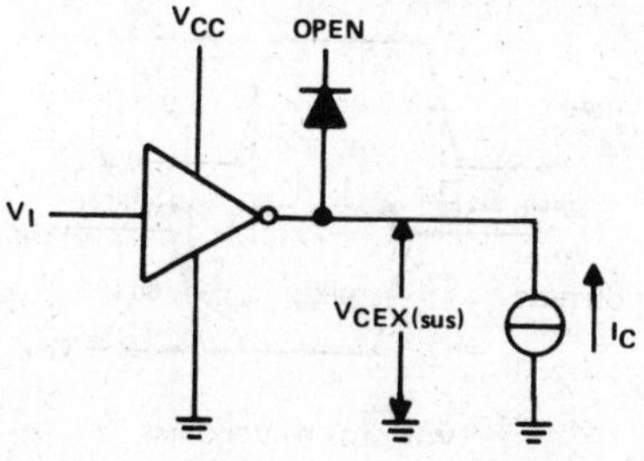
V_{CC}
OPEN
V_I
$V_{CEX(sus)}$
I_C

FIGURE 1. $V_{CEX(sus)}$

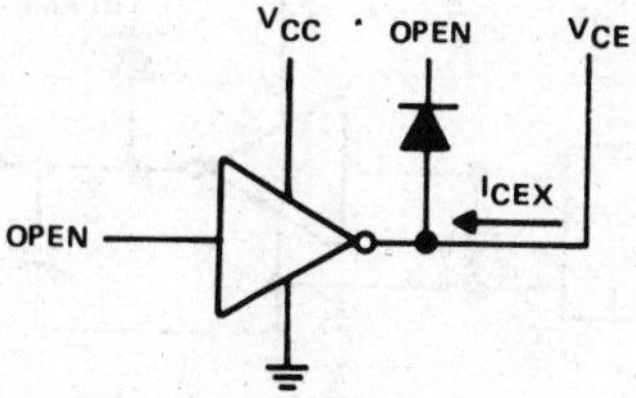
V_{CC}
OPEN
V_{CE}
I_{CEX}
OPEN

FIGURE 2. I_{CEX}

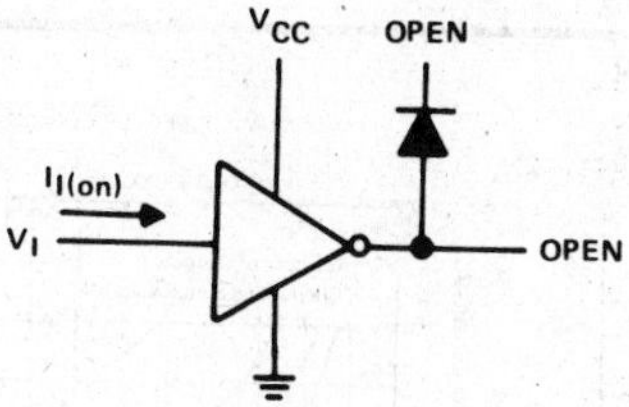
V_{CC}
OPEN
$I_{I(on)}$
V_I
OPEN

FIGURE 3. $I_{I(on)}$

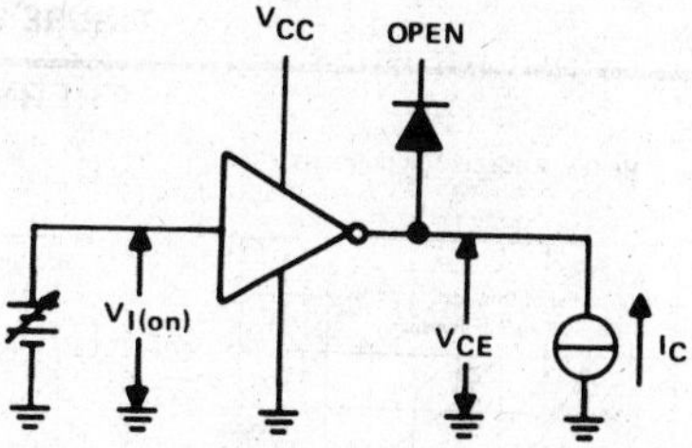
V_{CC}
OPEN
$V_{I(on)}$
V_{CE}
I_C

FIGURE 4. $V_{I(on)}$

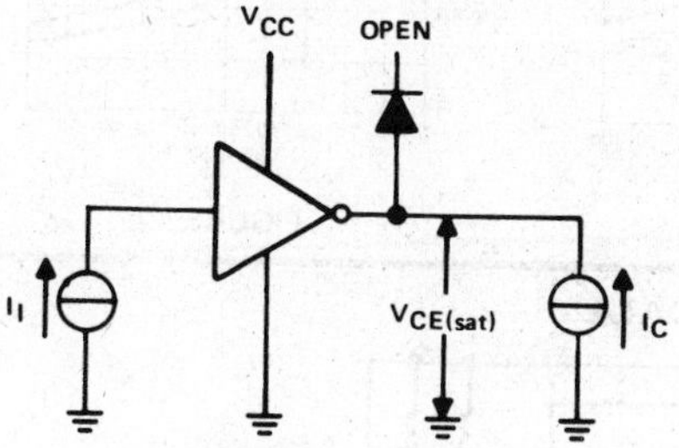
V_{CC}
OPEN
I_I
$V_{CE(sat)}$
I_C

FIGURE 5. $V_{CE(sat)}$

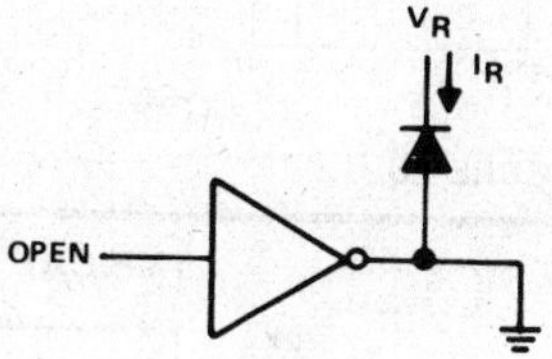
V_R
I_R
OPEN

FIGURE 6. I_R

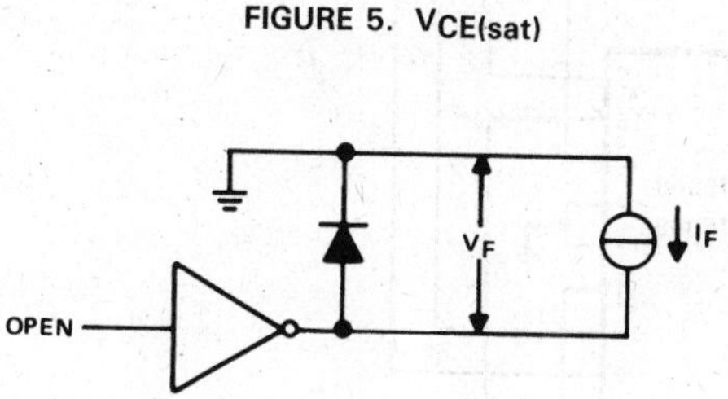
V_F
I_F
OPEN

FIGURE 7. V_F

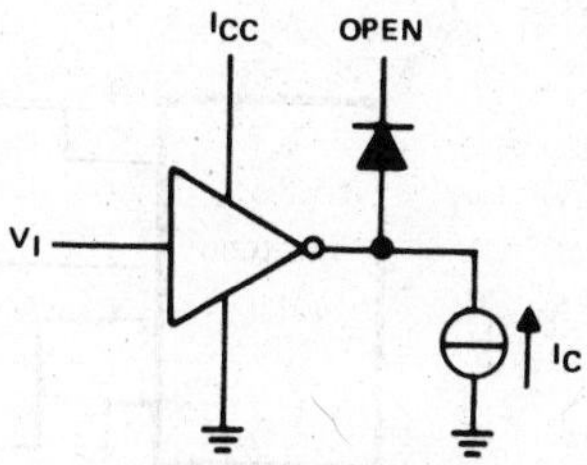
I_{CC}
OPEN
V_I
I_C

FIGURE 8. I_{CC}

PARAMETER MEASUREMENT INFORMATION

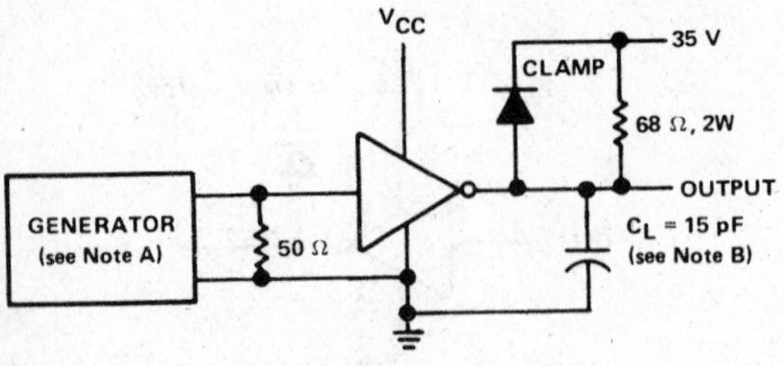
VCC
CLAMP
35 V
68 Ω, 2W
OUTPUT
GENERATOR
(see Note A)
50 Ω
CL = 15 pF
(see Note B)

TEST CIRCUIT

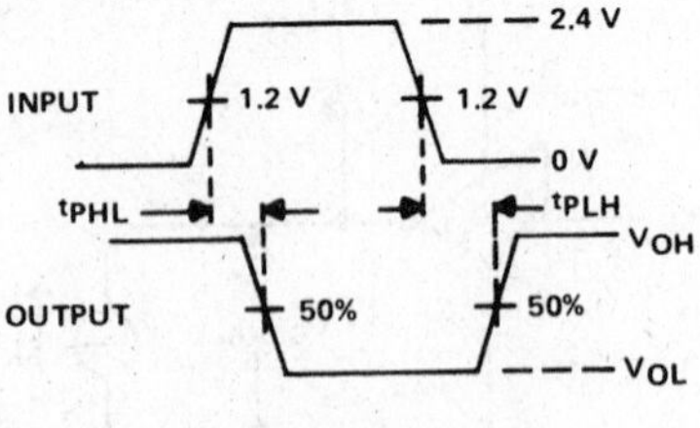
2.4 V
INPUT
1.2 V
1.2 V
0 V
tPHL
tPLH
VOH
OUTPUT
50%
50%
VOL

VOLTAGE WAVEFORMS

NOTES: A. The input pulse is supplied by a generator having the following characteristics: PRR = 50 kHz, duty cycle = 10%, Z_o = 50 Ω.
B. C_L includes all probe and stray capacitance.

FIGURE 9. SWITCHING TIMES

THERMAL INFORMATION

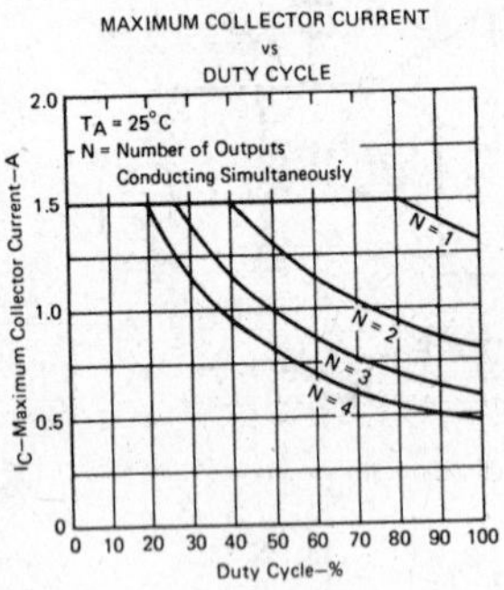
MAXIMUM COLLECTOR CURRENT
vs
DUTY CYCLE
TA = 25°C
N = Number of Outputs
Conducting Simultaneously
N = 1
N = 2
N = 3
N = 4
IC—Maximum Collector Current—A
Duty Cycle—%

FIGURE 10

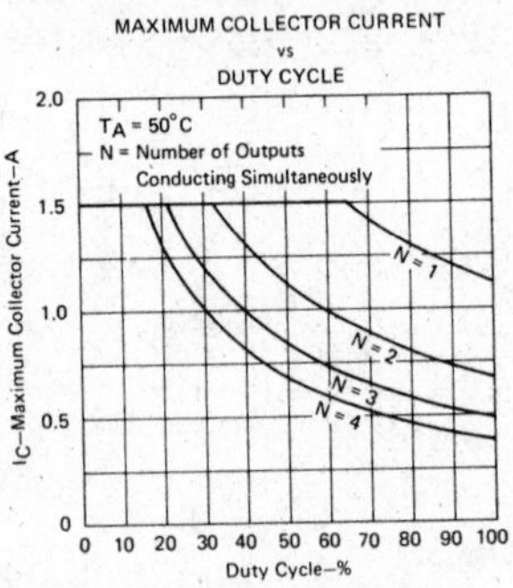
MAXIMUM COLLECTOR CURRENT
vs
DUTY CYCLE
TA = 50°C
N = Number of Outputs
Conducting Simultaneously
N = 1
N = 2
N = 3
N = 4
IC—Maximum Collector Current—A
Duty Cycle—%

FIGURE 11

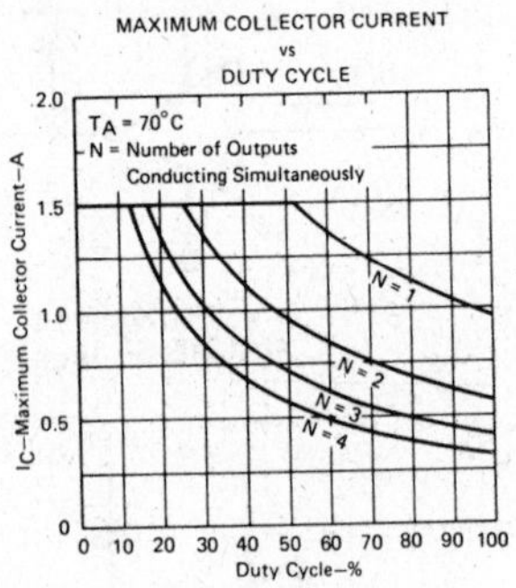
MAXIMUM COLLECTOR CURRENT
vs
DUTY CYCLE
TA = 70°C
N = Number of Outputs
Conducting Simultaneously
N = 1
N = 2
N = 3
N = 4
IC—Maximum Collector Current—A
Duty Cycle—%

FIGURE 12

TYPICAL APPLICATION DATA

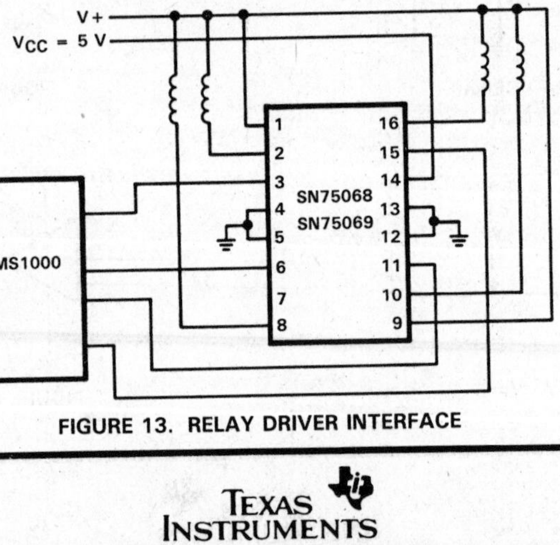
V+
VCC = 5 V
TMS1000
SN75068
SN75069

FIGURE 13. RELAY DRIVER INTERFACE

TEXAS INSTRUMENTS
POST OFFICE BOX 655012 • DALLAS, TEXAS 75265

D3004, JULY 1986

- Dual Circuits Capable of Driving High-Capacitance Loads at High Speeds
- Output Supply Voltage Range Up to 24 V
- Low Standby Power Dissipation

D OR P PACKAGE
(TOP VIEW)

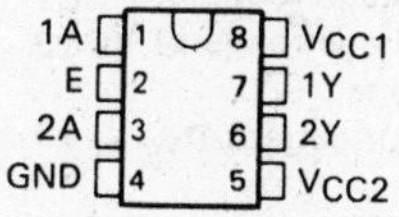
1A
E
2A
GND
1
2
3
4
8
7
6
5
VCC1
1Y
2Y
VCC2

description

The SN75372 is a dual NAND gate interface circuit designed to drive power MOSFETs from TTL inputs. It provides high current and voltage levels necessary to drive large capacitive loads at high speeds. The device operates from a V_{CC1} of 5 volts, and a V_{CC2} of up to 24 volts.

The SN75372 is characterized for operation from 0°C to 70°C.

logic symbol†

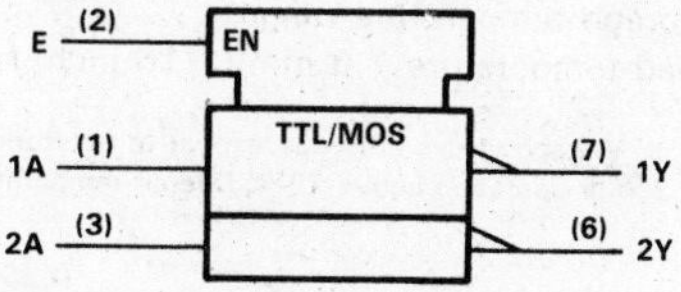
E
(2)
EN
TTL/MOS
1A
(1)
(7)
1Y
2A
(3)
(6)
2Y

†This symbol is in accordance with ANSI/IEEE Std 91-1984 and IEC Publication 617-12.

logic diagram (positive logic)

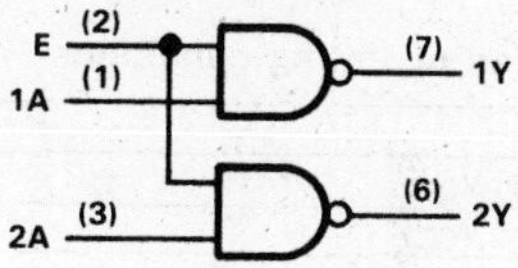
E
(2)
1A
(1)
(7)
1Y
2A
(3)
(6)
2Y

schematic (each driver)

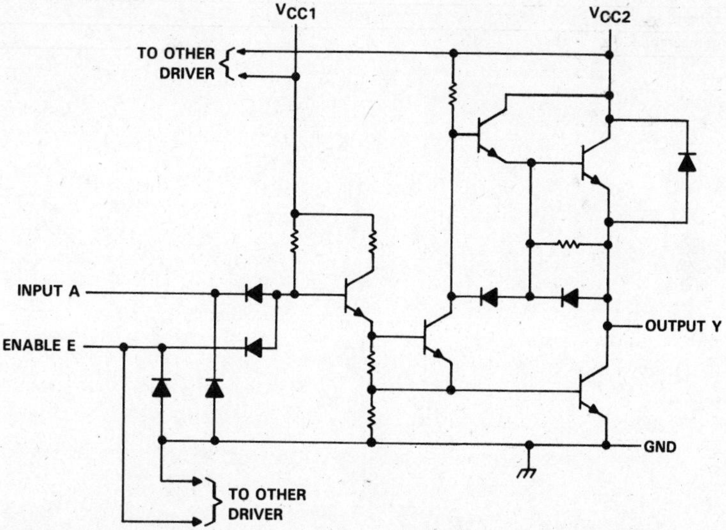
VCC1
VCC2
TO OTHER
DRIVER
INPUT A
ENABLE E
OUTPUT Y
GND
TO OTHER
DRIVER

TEXAS INSTRUMENTS
POST OFFICE BOX 655012 • DALLAS, TEXAS 75265

absolute maximum ratings over operating free-air temperature range (unless otherwise noted)

Supply voltage range of V_{CC1} (see Note 1) . . . . −0.5 V to 7 V
Supply voltage range of V_{CC2} . . . . −0.5 V to 25 V
Input voltage . . . . 5.5 V
Peak output current (t_w < 10 ms, duty cycle < 50%): Sink . . . . 500 mA
Source . . . . 500 mA
Continuous total dissipation at (or below) 25°C free-air temperature (see Note 2):
D package . . . . 725 mW
P package . . . . 1200 mW
Storage temperature range . . . . −65°C to 150°C
Lead temperature 1,6 mm (1/16 inch) from case for 10 seconds . . . . 260°C

NOTES: 1. Voltage values are with respect to network ground terminal.
2. For operation above 25°C free-air temperature, see the Dissipation Derating Table.

DISSIPATION DERATING TABLE

PACKAGE	POWER RATING	DERATING FACTOR	ABOVE T_A
D	725 mW	5.8 mW/°C	25°C
P	1200 mW	9.6 mW/°C	25°C

recommended operating conditions

	MIN	NOM	MAX	UNIT
Supply voltage, V_{CC1}	4.75	5	5.25	V
Supply voltage, V_{CC2}	4.75	20	24	V
High-level input voltage, V_{IH}	2			V
Low-level input voltage, V_{IL}			0.8	V
High-level output current, I_{OH}			−10	mA
Low-level output current, I_{OL}			40	mA
Operating free-air temperature, T_A	0		70	°C

electrical characteristics over recommended ranges of V_{CC1}, V_{CC2}, and operating free-air temperature (unless otherwise noted)

PARAMETER			TEST CONDITIONS		MIN	TYP†	MAX	UNIT
V_{IK}	Input clamp voltage		$I_I = -12$ mA				−1.5	V
V_{OH}	High-level output voltage		$V_{IL} = 0.8$ V,	$I_{OH} = -50\ \mu A$	$V_{CC2}-1.3$	$V_{CC2}-0.8$		V
			$V_{IL} = 0.8$ V,	$I_{OH} = -10$ mA	$V_{CC2}-2.5$	$V_{CC2}-1.8$		
V_{OL}	Low-level output voltage		$V_{IH} = 2$ V,	$I_{OL} = 10$ mA		0.15	0.3	V
			$V_{CC2} = 15$ V to 24 V, $I_{OL} = 40$ mA	$V_{IH} = 2$ V,		0.25	0.5	
V_F	Output clamp diode forward voltage		$V_I = 0$,	$I_F = 20$ mA			1.5	V
I_I	Input current at maximum input voltage		$V_I = 5.5$ V				1	mA
I_{IH}	High-level input current	Any A	$V_I = 2.4$ V				40	μA
		Any E					80	
I_{IL}	Low-level input current	Any A	$V_I = 0.4$ V			−1	−1.6	mA
		Any E				−2	−3.2	
$I_{CC1(H)}$	Supply current from V_{CC1}, both outputs high		$V_{CC1} = 5.25$ V, All inputs at 0 V,	$V_{CC2} = 24$ V, No load		2	4	mA
$I_{CC2(H)}$	Supply current from V_{CC2}, both outputs high						0.5	mA
$I_{CC1(L)}$	Supply current from V_{CC1}, both outputs low		$V_{CC1} = 5.25$ V, All inputs at 5 V,	$V_{CC2} = 24$ V, No load		16	24	mA
$I_{CC2(L)}$	Supply current from V_{CC2}, both outputs low					7	13	mA
$I_{CC2(S)}$	Supply current from V_{CC2}, standby condition		$V_{CC1} = 0$, All inputs at 5 V,	$V_{CC2} = 24$ V, No load			0.5	mA

†All typical values are at $V_{CC1} = 5$ V, $V_{CC2} = 20$ V, and $T_A = 25°C$.

switching characteristics, $V_{CC1} = 5$ V, $V_{CC2} = 20$ V, $T_A = 25°C$

PARAMETER		TEST CONDITIONS	MIN	TYP	MAX	UNIT
t_{DLH}	Delay time, low-to-high-level output	$C_L = 390$ pF, $R_D = 10\ \Omega$, See Figure 1		20	35	ns
t_{DHL}	Delay time, high-to-low-level output			10	20	ns
t_{TLH}	Transition time, low-to-high-level output			20	30	ns
t_{THL}	Transition time, high-to-low-level output			20	30	ns
t_{PLH}	Propagation delay time, low-to-high-level output		10	40	65	ns
t_{PHL}	Propagation delay time, high-to-low-level output		10	30	50	ns

PARAMETER MEASUREMENT INFORMATION

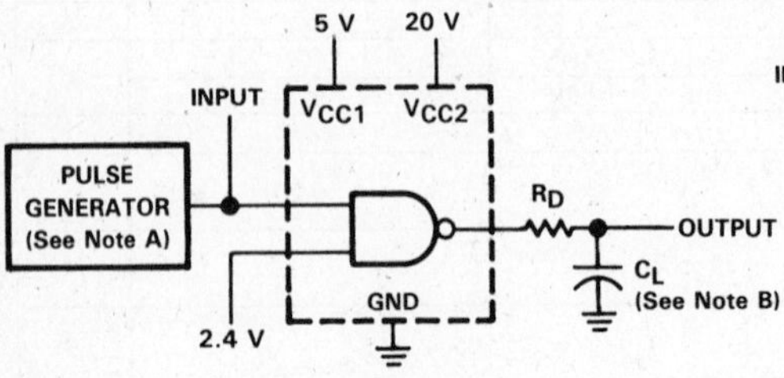
5 V
20 V
INPUT
V_{CC1}
V_{CC2}
PULSE GENERATOR (See Note A)
R_D
OUTPUT
C_L (See Note B)
GND
2.4 V

TEST CIRCUIT

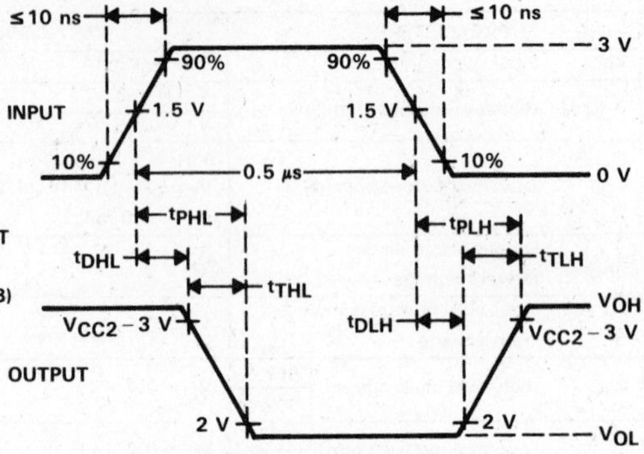
≤10 ns
≤10 ns
3 V
90%
90%
INPUT
1.5 V
1.5 V
10%
10%
0.5 μs
0 V
t_{PHL}
t_{PLH}
t_{DHL}
t_{TLH}
t_{THL}
V_{OH}
V_{CC2} − 3 V
t_{DLH}
V_{CC2} − 3 V
OUTPUT
2 V
2 V
V_{OL}

VOLTAGE WAVEFORMS

NOTES: A. The pulse generator has the following characteristics: PRR = 1 MHz, $Z_{out} \approx 50\ \Omega$.
B. C_L includes probe and jig capacitance.

FIGURE 1. SWITCHING TIMES, EACH DRIVER

TYPICAL CHARACTERISTICS

HIGH-LEVEL OUTPUT VOLTAGE
vs
OUTPUT CURRENT

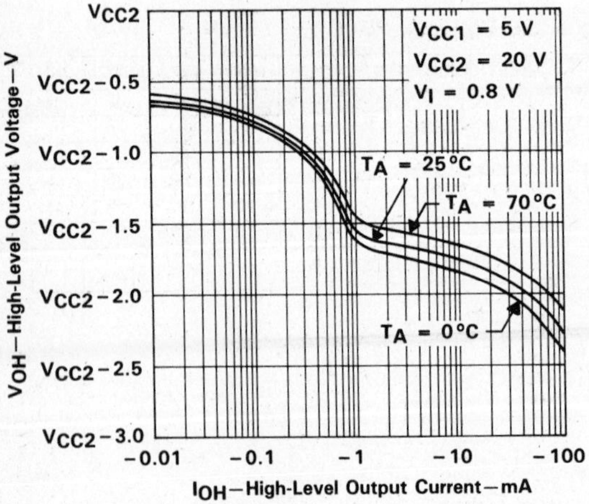
V_{CC1} = 5 V
V_{CC2} = 20 V
V_I = 0.8 V
T_A = 25°C
T_A = 70°C
T_A = 0°C
V_{CC2}
V_{CC2} − 0.5
V_{CC2} − 1.0
V_{CC2} − 1.5
V_{CC2} − 2.0
V_{CC2} − 2.5
V_{CC2} − 3.0
−0.01
−0.1
−1
−10
−100
V_{OH}—High-Level Output Voltage—V
I_{OH}—High-Level Output Current—mA

FIGURE 2

LOW-LEVEL OUTPUT VOLTAGE
vs
OUTPUT CURRENT

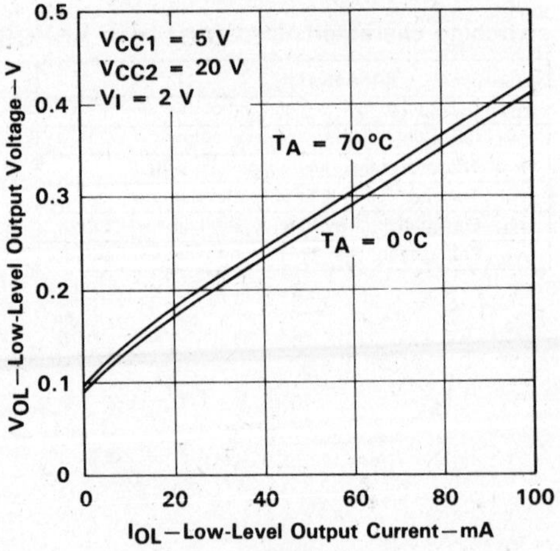
V_{CC1} = 5 V
V_{CC2} = 20 V
V_I = 2 V
T_A = 70°C
T_A = 0°C
0.5
0.4
0.3
0.2
0.1
0
0
20
40
60
80
100
V_{OL}—Low-Level Output Voltage—V
I_{OL}—Low-Level Output Current—mA

FIGURE 3

TYPICAL CHARACTERISTICS

VOLTAGE TRANSFER CHARACTERISTICS

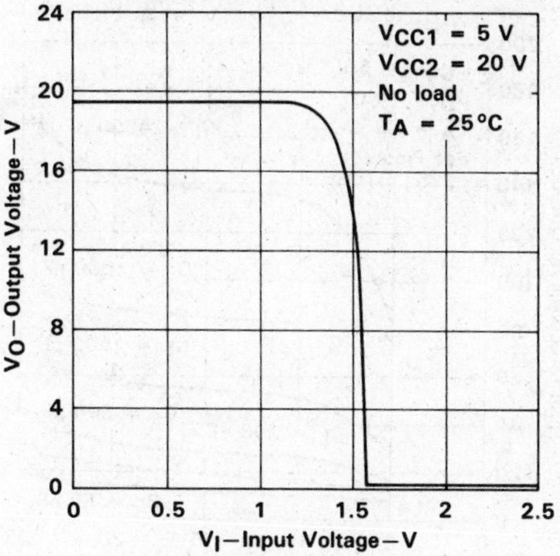
VCC1 = 5 V
VCC2 = 20 V
No load
TA = 25°C
VO—Output Voltage—V
VI—Input Voltage—V
0
4
8
12
16
20
24
0.5
1
1.5
2
2.5

FIGURE 4

POWER DISSIPATION (BOTH DRIVERS) vs FREQUENCY

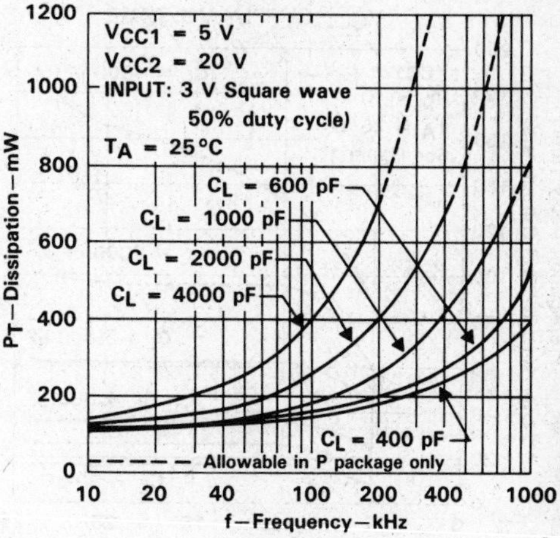
VCC1 = 5 V
VCC2 = 20 V
INPUT: 3 V Square wave
50% duty cycle)
TA = 25°C
CL = 600 pF
CL = 1000 pF
CL = 2000 pF
CL = 4000 pF
CL = 400 pF
Allowable in P package only
PT—Dissipation—mW
f—Frequency—kHz
0
200
400
600
800
1000
1200
10
20
40
100
200
400
1000

FIGURE 5

PROPAGATION DELAY TIME, LOW-TO-HIGH-LEVEL OUTPUT vs FREE-AIR TEMPERATURE

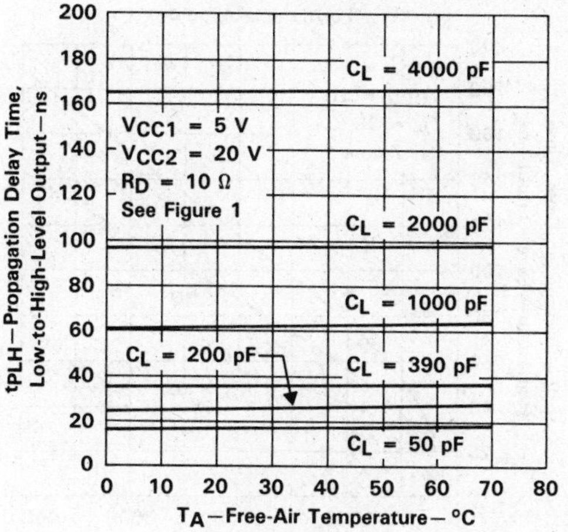
CL = 4000 pF
VCC1 = 5 V
VCC2 = 20 V
RD = 10 Ω
See Figure 1
CL = 2000 pF
CL = 1000 pF
CL = 200 pF
CL = 390 pF
CL = 50 pF
tPLH—Propagation Delay Time, Low-to-High-Level Output—ns
TA—Free-Air Temperature—°C
0
20
40
60
80
100
120
140
160
180
200
10
30
50
70

FIGURE 6

PROPAGATION DELAY TIME, HIGH-TO-LOW-LEVEL OUTPUT vs FREE-AIR TEMPERATURE

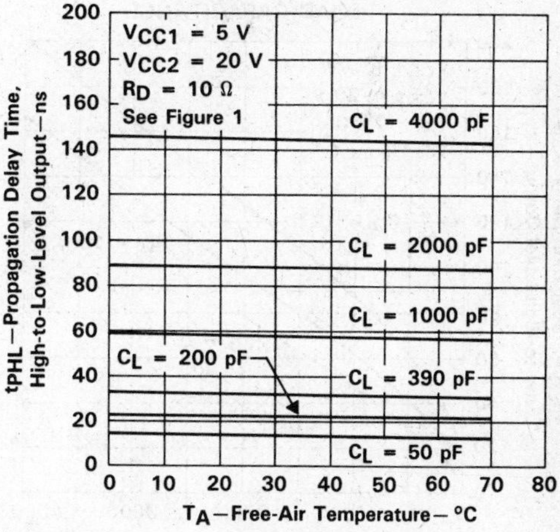
VCC1 = 5 V
VCC2 = 20 V
RD = 10 Ω
See Figure 1
CL = 4000 pF
CL = 2000 pF
CL = 1000 pF
CL = 200 pF
CL = 390 pF
CL = 50 pF
tPHL—Propagation Delay Time, High-to-Low-Level Output—ns
TA—Free-Air Temperature—°C
0
20
40
60
80
100
120
140
160
180
200
10
30
50
70

FIGURE 7

TYPICAL CHARACTERISTICS

PROPAGATION DELAY TIME,
LOW-TO-HIGH-LEVEL OUTPUT
vs
V_{CC2} SUPPLY VOLTAGE

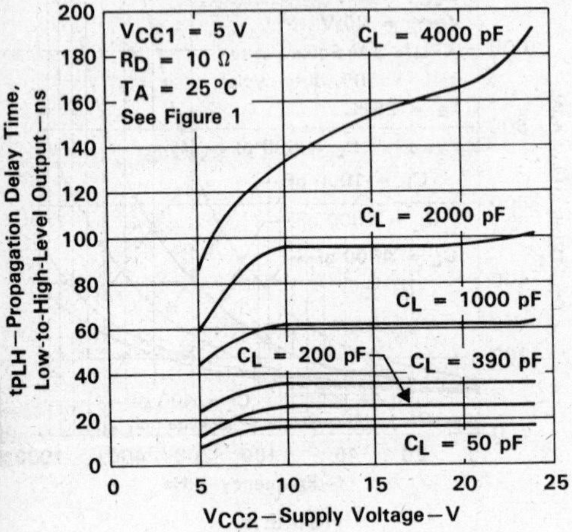
VCC1 = 5 V
RD = 10 Ω
TA = 25°C
See Figure 1
CL = 4000 pF
CL = 2000 pF
CL = 1000 pF
CL = 200 pF
CL = 390 pF
CL = 50 pF
tPLH—Propagation Delay Time, Low-to-High-Level Output—ns
VCC2—Supply Voltage—V
0
20
40
60
80
100
120
140
160
180
200
5
10
15
20
25

FIGURE 8

PROPAGATION DELAY TIME,
HIGH-TO-LOW-LEVEL OUTPUT
vs
V_{CC2} SUPPLY VOLTAGE

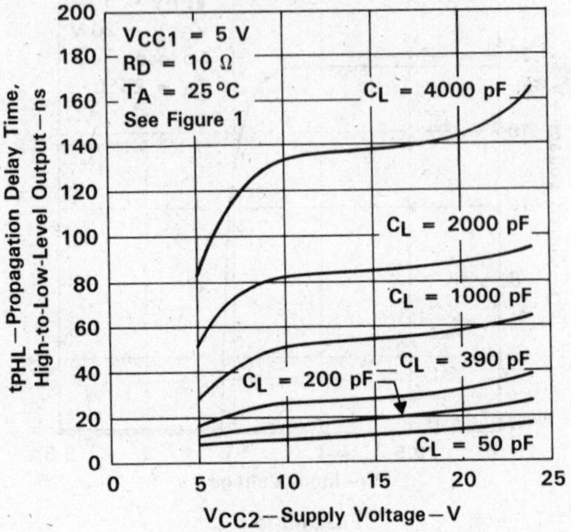
VCC1 = 5 V
RD = 10 Ω
TA = 25°C
See Figure 1
CL = 4000 pF
CL = 2000 pF
CL = 1000 pF
CL = 200 pF
CL = 390 pF
CL = 50 pF
tPHL—Propagation Delay Time, High-to-Low-Level Output—ns
VCC2—Supply Voltage—V
0
20
40
60
80
100
120
140
160
180
200
5
10
15
20
25

FIGURE 9

PROPAGATION DELAY TIME,
LOW-TO-HIGH-LEVEL OUTPUT
vs
LOAD CAPACITANCE

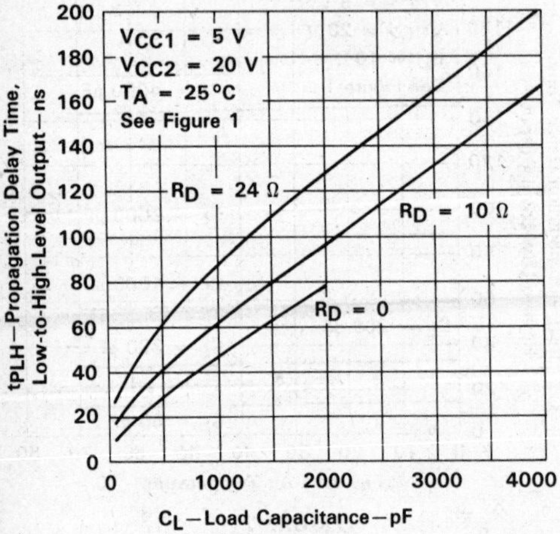
VCC1 = 5 V
VCC2 = 20 V
TA = 25°C
See Figure 1
RD = 24 Ω
RD = 10 Ω
RD = 0
tPLH—Propagation Delay Time, Low-to-High-Level Output—ns
CL—Load Capacitance—pF
0
20
40
60
80
100
120
140
160
180
200
1000
2000
3000
4000

FIGURE 10

PROPAGATION DELAY TIME,
HIGH-TO-LOW-LEVEL OUTPUT
vs
LOAD CAPACITANCE

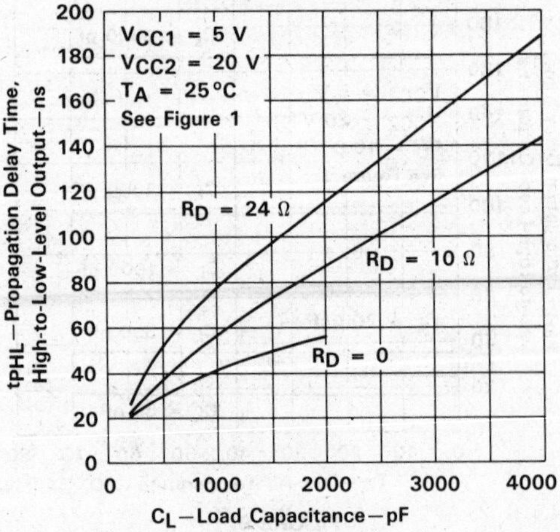
VCC1 = 5 V
VCC2 = 20 V
TA = 25°C
See Figure 1
RD = 24 Ω
RD = 10 Ω
RD = 0
tPHL—Propagation Delay Time, High-to-Low-Level Output—ns
CL—Load Capacitance—pF
0
20
40
60
80
100
120
140
160
180
200
1000
2000
3000
4000

FIGURE 11

NOTE: For $R_D = 0$, operation with $C_L > 2000$ pF violates absolute maximum current rating.

TEXAS INSTRUMENTS
POST OFFICE BOX 655012 • DALLAS, TEXAS 75265

APPLICATIONS INFORMATION

driving power MOSFETs

The drive requirements of power MOSFETs are much lower than comparable bipolar power transistors. The input impedance of a FET consists of a reverse biased PN junction that can be described as a large capacitance in parallel with a very high resistance. For this reason, the commonly used open-collector driver with a pull-up resistor is not satisfactory for high-speed applications. In Figure 12(a), an IRF151 power MOSFET switching an inductive load is driven by an open-collector transistor driver with a 470 Ω pull-up resistor. The input capacitance (C_{iss}) specification for an IRF151 is 4000 pF maximum. The resulting long turn-on time due to the combination of C_{iss} and the pull-up resistor is shown in Figure 12(b).

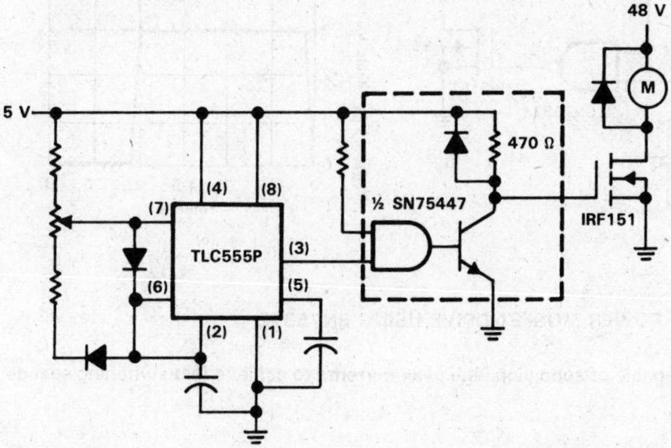
48 V
M
5 V
470 Ω
½ SN75447
IRF151
(4)
(8)
(7)
TLC555P
(3)
(6)
(5)
(2)
(1)

(a)

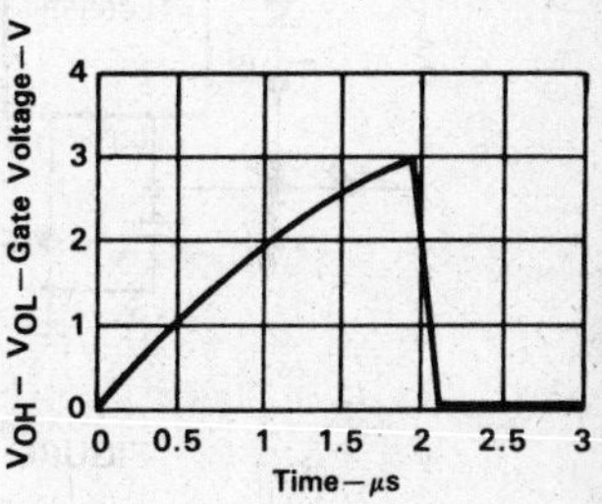
VOH − VOL − Gate Voltage − V
4
3
2
1
0
0
0.5
1
1.5
2
2.5
3
Time − μs

(b)

FIGURE 12. POWER MOSFET DRIVE USING SN75447

APPLICATIONS INFORMATION

A faster, more efficient drive circuit uses an active pull-up as well as an active pull-down output configuration, referred to as a totem-pole output. The SN75372 driver provides the high speed, totem-pole drive desired in an application of this type, see Figure 13(a). The resulting faster switching speeds are shown in Figure 13(b).

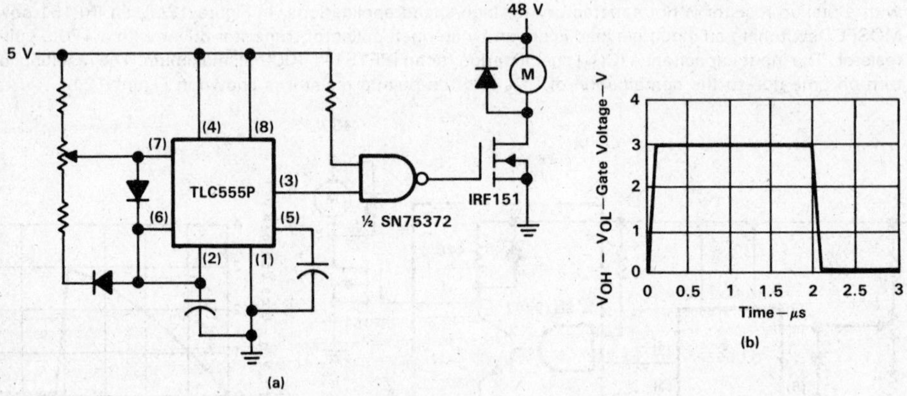
48 V
5 V
M
(4)
(8)
(7)
TLC555P
(3)
(6)
(5)
(2)
(1)
½ SN75372
IRF151
(a)
$V_{OH} - V_{OL}$—Gate Voltage—V
0
1
2
3
4
0 0.5 1 1.5 2 2.5 3
Time—μs
(b)

FIGURE 13. POWER MOSFET DRIVE USING SN75372

Power MOSFET drivers must be capable of supplying high peak currents to achieve fast switching speeds as shown by the equation

$$I_{pk} = \frac{VC}{t_r}$$

where C is the capacitive load, and t_r is the desired rise time. V is the voltage that the capacitance is charged to. In the circuit shown in Figure 13(a), V is found by the equation

$$V = V_{OH} - V_{OL}$$

Peak current required to maintain a rise time of 100 ns in the circuit of Figure 13(a) is

$$I_{PK} = \frac{(3\text{-}0)4(10^9)}{100(10^9)} = 120 \text{ mA}$$

Circuit capacitance can be ignored because it is very small compared to the input capacitance of the IRF151. With a V_{CC} of 5 V, and assuming worst-case conditions, the gate drive voltage is 3 V.

For applications in which the full voltage of V_{CC2} must be supplied to the MOSFET gate, the SN75374 QUAD MOSFET driver should be used.

TEXAS INSTRUMENTS
POST OFFICE BOX 655012 • DALLAS, TEXAS 75265

THERMAL INFORMATION

power dissipation precautions

Significant power may be dissipated in the SN75372 driver when charging and discharging high-capacitance loads over a wide voltage range at high frequencies. Figure 5 shows the power dissipated in a typical SN75372 as a function of load capacitance and frequency. Average power dissipated by this driver is derived from the equation

$$P_{T(AV)} = P_{DC(AV)} + P_{C(AV)} + P_{S(AV)}$$

where $P_{DC(AV)}$ is the steady-state power dissipation with the output high or low, $P_{C(AV)}$ is the power level during charging or discharging of the load capacitance, and $P_{S(AV)}$ is the power dissipation during switching between the low and high levels. None of these include energy transferred to the load and all are averaged over a full cycle.

The power components per driver channel are

$$P_{DC(AV)} = \frac{P_H t_H + P_L t_L}{T}$$

$$P_{C(AV)} \approx C\,V_C^2\,f$$

$$P_{S(AV)} = \frac{P_{LH} t_{LH} + P_{HL} t_{HL}}{T}$$

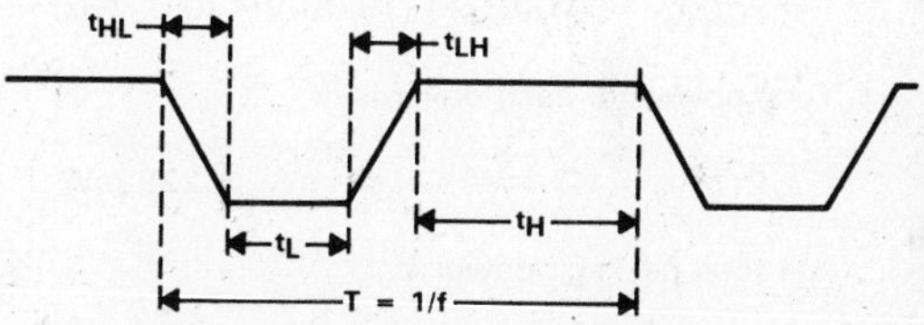
tHL
tLH
tL
tH
T = 1/f

FIGURE 14. OUTPUT VOLTAGE WAVEFORM

where the times are as defined in Figure 14.

P_L, P_H, P_{LH}, and P_{HL} are the respective instantaneous levels of power dissipation, C is the load capacitance. V_C is the voltage across the load capacitance during the charge cycle shown by the equation

$$V_C = V_{OH} - V_{OL}$$

$P_{S(AV)}$ may be ignored for power calculations at low frequencies.

THERMAL INFORMATION

In the following power calculation, both channels are operating under identical conditions: V_{OH} = 19.2 volts and V_{OL} = 0.15 volts with V_{CC1} = 5 volts, V_{CC2} = 20 volts, V_C = 19.05 volts, C = 1000 picofarads, and the duty cycle = 60%. At 0.5 MHz, $P_{S(AV)}$ is negligible and can be ignored. When the output voltage is high, I_{CC2} is negligible and can be ignored.

On a per-channel basis using data sheet values

$$P_{DC(AV)} = \left[(5\ V)\left(\frac{2\ mA}{2}\right) + (20\ V)\left(\frac{0\ mA}{2}\right)\right](0.6) + \left[(5\ V)\left(\frac{16\ mA}{2}\right) + (20\ V)\left(\frac{7\ mA}{2}\right)\right](0.4)$$

$$P_{DC(AV} = 47\ mW \text{ per channel}$$

Power during the charging time of the load capacitance is

$$P_{C(AV)} = (1000\ pF)\ (19.05\ V)^2\ (0.5\ MHz) = 182\ mW \text{ per channel}$$

Total power for each driver is

$$P_{T(AV)} = 47\ mW + 182\ mW = 229\ mW$$

and total package power is

$$P_{T(AV)} = (229)\ (2) = 458\ mW.$$

- Quadruple Circuits Capable of Driving High-Capacitance Loads at High Speeds
- Output Supply Voltage Range from 5 V to 24 V
- Low Standby Power Dissipation
- V_{CC3} Supply Maximizes Output Source Voltage

D OR N PACKAGE
(TOP VIEW)

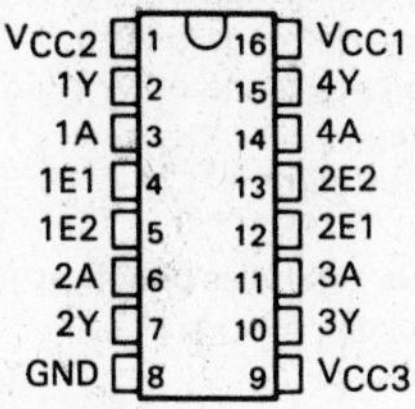
VCC2
1
16
VCC1
1Y
2
15
4Y
1A
3
14
4A
1E1
4
13
2E2
1E2
5
12
2E1
2A
6
11
3A
2Y
7
10
3Y
GND
8
9
VCC3

description

The SN75374 is a quadruple NAND interface circuit designed to drive power MOSFETs from TTL inputs. It provides the high current and voltage necessary to drive large capacitive loads at high speeds.

The outputs can be switched very close to the V_{CC2} supply rail when V_{CC3} is about 3 volts higher than V_{CC2}. The V_{CC3} pin can also be tied directly to V_{CC2} when the source voltage requirements are lower.

The SN75374 is characterized for operation from 0°C to 70°C.

schematic (each driver)

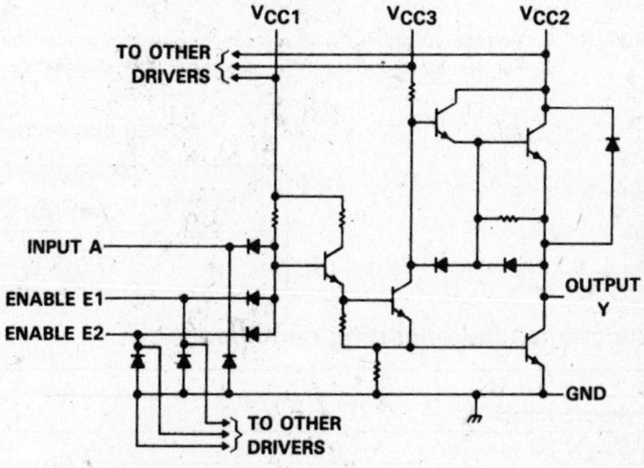
VCC1
VCC3
VCC2
TO OTHER DRIVERS
INPUT A
ENABLE E1
ENABLE E2
OUTPUT Y
GND
TO OTHER DRIVERS

logic symbol†

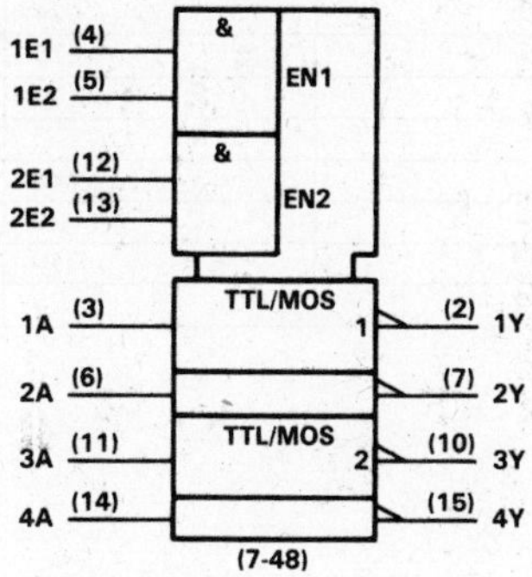
1E1
(4)
1E2
(5)
&
EN1
2E1
(12)
2E2
(13)
&
EN2
1A
(3)
TTL/MOS
1
(2)
1Y
2A
(6)
(7)
2Y
3A
(11)
TTL/MOS
2
(10)
3Y
4A
(14)
(15)
4Y
(7-48)

†This symbol is in accordance with ANSI/IEEE Std 91-1984 and IEC Publication 617-12.

logic diagram (positive logic)

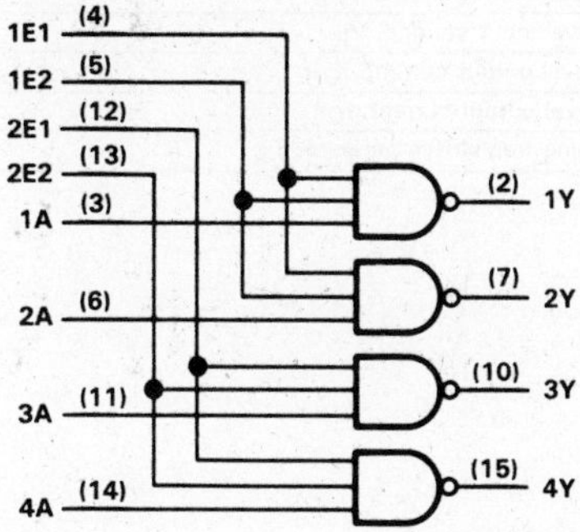
1E1
(4)
1E2
(5)
2E1
(12)
2E2
(13)
1A
(3)
(2)
1Y
2A
(6)
(7)
2Y
3A
(11)
(10)
3Y
4A
(14)
(15)
4Y

absolute maximum ratings over operating free-air temperature range (unless otherwise noted)

Supply voltage range of V_{CC1}, (see Note 1)	−0.5 V to 7 V
Supply voltage range of V_{CC2}	−0.5 V to 25 V
Supply voltage range of V_{CC3}	−0.5 V to 30 V
Input voltage	5.5 V
Peak output current (t_w < 10 ms, duty cycle < 50%): Sink	500 mA
Source	500 mA
Continuous total dissipation at (or below) 25°C free-air temperature (see Note 2):	
D package	1025 mW
N package	1650 mW
Operating free-air temperature range	0°C to 70°C
Storage temperature range	−65°C to 150°C
Lead temperature 1,6 mm (1/16 inch) from case for 10 seconds	260°C

NOTES: 1. Voltage values are with respect to network ground terminal.
2. For operation above 25°C, see the Power Dissipation Derating Table.

POWER DISSIPATION DERATING TABLE

PACKAGE	POWER RATING	DERATING FACTOR	ABOVE T_A
D	1025 mW	8.2 mW/°C	25°C
N	1650 mW	13.2 mW/°C	25°C

recommended operating conditions

	MIN	NOM	MAX	UNIT
Supply voltage, V_{CC1}	4.75	5	5.25	V
Supply voltage, V_{CC2}	4.75	20	24	V
Supply voltage, V_{CC3}	V_{CC2}	24	28	V
Voltage difference between supply voltages: $V_{CC3} - V_{CC2}$	0	4	10	V
High-level input voltage, V_{IH}	2			V
Low-level input voltage, V_{IL}			0.8	V
High-level output current, I_{OH}			−10	mA
Low-level output current, I_{OL}			40	mA
Operating free-air temperature, T_A	0		70	°C

electrical characteristics over recommended ranges of V_{CC1}, V_{CC2}, V_{CC3}, and operating free-air temperature (unless otherwise noted)

PARAMETER			TEST CONDITIONS	MIN	TYP†	MAX	UNIT
V_{IK}	Input clamp voltage		I_I = −12 mA			−1.5	V
V_{OH}	High-level output voltage		$V_{CC3} = V_{CC2} + 3$ V, V_{IL} = 0.8 V, I_{OH} = −100 μA	$V_{CC2}-0.3$	$V_{CC2}-0.1$		V
			$V_{CC3} = V_{CC2} + 3$ V, V_{IL} = 0.8 V, I_{OH} = −10 mA	$V_{CC2}-1.3$	$V_{CC2}-0.9$		
			$V_{CC3} = V_{CC2}$, V_{IL} = 0.8 V, I_{OH} = −50 μA	$V_{CC2}-1$	$V_{CC2}-0.7$		
			$V_{CC3} = V_{CC2}$, V_{IL} = 0.8 V, I_{OH} = −10 mA	$V_{CC2}-2.5$	$V_{CC2}-1.8$		
V_{OL}	Low-level output voltage		V_{IH} = 2 V, I_{OL} = 10 mA		0.15	0.3	V
			V_{CC2} = 15 V to 28 V, V_{IH} = 2 V, I_{OL} = 40 mA		0.25	0.5	
V_F	Output clamp diode forward voltage		V_I = 0, I_F = 20 mA			1.5	V
I_I	Input current at maximum input voltage		V_I = 5.5 V			1	mA
I_{IH}	High-level input current	Any A	V_I = 2.4 V			40	μA
		Any E				80	
I_{IL}	Low-level input current	Any A	V_I = 0.4 V		−1	−1.6	mA
		Any E			−2	−3.2	
$I_{CC1(H)}$	Supply current from V_{CC1}, all outputs high		V_{CC1} = 5.25 V, V_{CC2} = 24 V, V_{CC3} = 28 V, All inputs at 0 V, No load		4	8	mA
$I_{CC2(H)}$	Supply current from V_{CC2}, all outputs high				−2.2	0.25	
$I_{CC3(H)}$	Supply current from V_{CC3}, all outputs high				2.2	3.5	
$I_{CC1(L)}$	Supply current from V_{CC1}, all outputs low		V_{CC1} = 5.25 V, V_{CC2} = 24 V, V_{CC3} = 28 V, All inputs at 5 V, No load		31	47	mA
$I_{CC2(L)}$	Supply current from V_{CC2}, all outputs low					2	
$I_{CC3(L)}$	Supply current from V_{CC3}, all outputs low				16	27	
$I_{CC2(H)}$	Supply current from V_{CC2}, all outputs high		V_{CC1} = 5.25 V, V_{CC2} = 24 V, V_{CC3} = 24 V, All inputs at 0 V, No load			0.25	mA
$I_{CC3(H)}$	Supply current from V_{CC3}, all outputs high					0.5	
$I_{CC2(S)}$	Supply current from V_{CC2}, standby condition		V_{CC1} = 0, V_{CC2} = 24 V, V_{CC3} = 24 V, All inputs at 0 V, No load			0.25	mA
$I_{CC3(S)}$	Supply current from V_{CC3}, standby condition					0.5	

†All typical values are at V_{CC1} = 5 V, V_{CC2} = 20 V, V_{CC3} = 24 V, and T_A = 25°C except for V_{OH} for which V_{CC2} and V_{CC3} are as stated under test conditions.

switching characteristics, V_{CC1} = 5 V, V_{CC2} = 20 V, V_{CC3} = 24 V, T_A = 25°C

PARAMETER		TEST CONDITIONS	MIN	TYP	MAX	UNIT
t_{DLH}	Delay time, low-to-high-level output	C_L = 200 pF, R_D = 24 Ω, See Figure 1		20	30	ns
t_{DHL}	Delay time, high-to-low-level output			10	20	ns
t_{TLH}	Transition time, low-to-high-level output			20	30	ns
t_{THL}	Transition time, high-to-low-level output			20	30	ns
t_{PLH}	Propagation delay time, low-to-high-level output		10	40	60	ns
t_{PHL}	Propagation delay time, high-to-low-level output		10	30	50	ns

PARAMETER MEASUREMENT INFORMATION

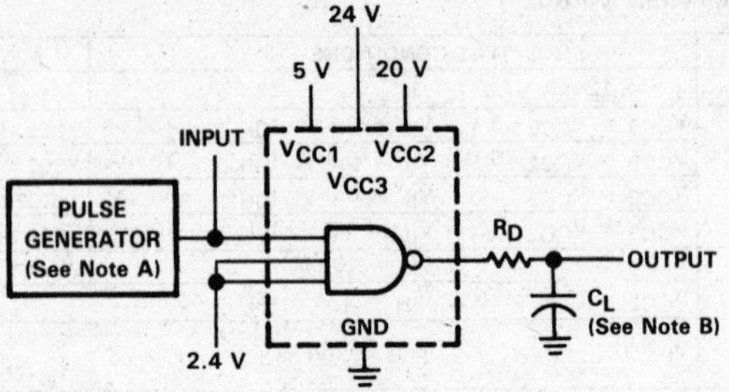
24 V
5 V
20 V
INPUT
V_{CC1}
V_{CC2}
V_{CC3}
PULSE GENERATOR (See Note A)
R_D
OUTPUT
C_L (See Note B)
GND
2.4 V

TEST CIRCUIT

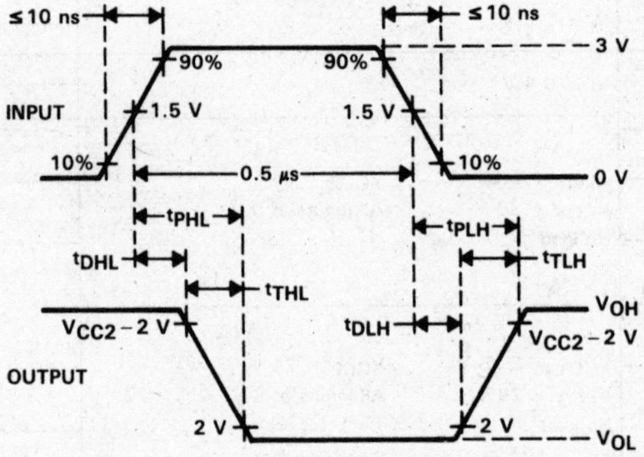
≤10 ns
≤10 ns
3 V
90%
90%
INPUT
1.5 V
1.5 V
10%
10%
0.5 µs
0 V
t_{PHL}
t_{PLH}
t_{DHL}
t_{TLH}
t_{THL}
V_{OH}
$V_{CC2} - 2$ V
t_{DLH}
$V_{CC2} - 2$ V
OUTPUT
2 V
2 V
V_{OL}

VOLTAGE WAVEFORMS

NOTES: A. The pulse generator has the following characteristics: PRR = 1 MHz, $Z_{out} \approx 50\ \Omega$.
B. C_L includes probe and jig capacitance.

FIGURE 1. SWITCHING TIMES, EACH DRIVER

TYPICAL CHARACTERISTICS

HIGH-LEVEL OUTPUT VOLTAGE
vs
OUTPUT CURRENT

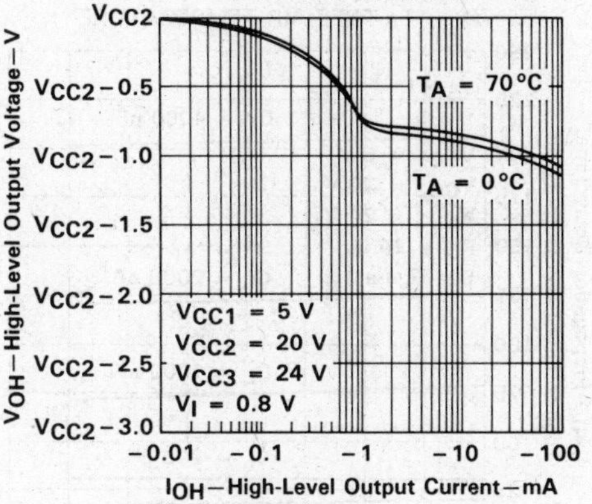
V_{CC2}
$V_{CC2}-0.5$
$V_{CC2}-1.0$
$V_{CC2}-1.5$
$V_{CC2}-2.0$
$V_{CC2}-2.5$
$V_{CC2}-3.0$
V_{OH}—High-Level Output Voltage—V
T_A = 70°C
T_A = 0°C
V_{CC1} = 5 V
V_{CC2} = 20 V
V_{CC3} = 24 V
V_I = 0.8 V
−0.01
−0.1
−1
−10
−100
I_{OH}—High-Level Output Current—mA

FIGURE 2

HIGH-LEVEL OUTPUT VOLTAGE
vs
OUTPUT CURRENT

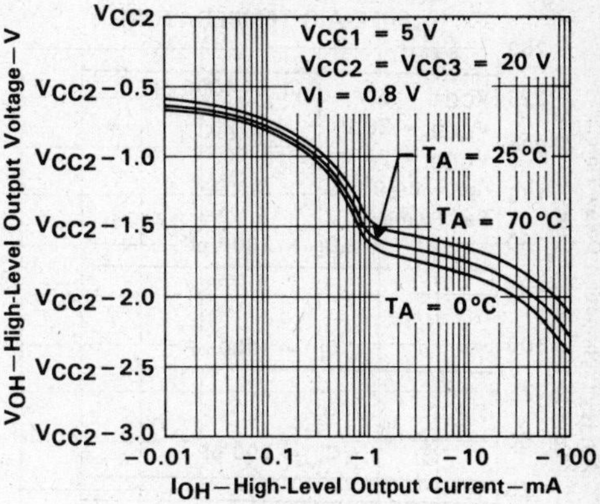
V_{CC2}
$V_{CC2}-0.5$
$V_{CC2}-1.0$
$V_{CC2}-1.5$
$V_{CC2}-2.0$
$V_{CC2}-2.5$
$V_{CC2}-3.0$
V_{OH}—High-Level Output Voltage—V
V_{CC1} = 5 V
V_{CC2} = V_{CC3} = 20 V
V_I = 0.8 V
T_A = 25°C
T_A = 70°C
T_A = 0°C
−0.01
−0.1
−1
−10
−100
I_{OH}—High-Level Output Current—mA

FIGURE 3

LOW-LEVEL OUTPUT VOLTAGE
vs
OUTPUT CURRENT

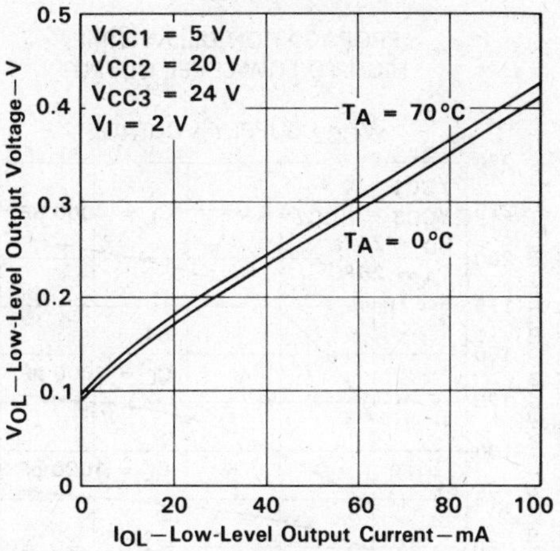
0.5
0.4
0.3
0.2
0.1
0
V_{OL}—Low-Level Output Voltage—V
V_{CC1} = 5 V
V_{CC2} = 20 V
V_{CC3} = 24 V
V_I = 2 V
T_A = 70°C
T_A = 0°C
0
20
40
60
80
100
I_{OL}—Low-Level Output Current—mA

FIGURE 4

VOLTAGE TRANSFER CHARACTERISTICS

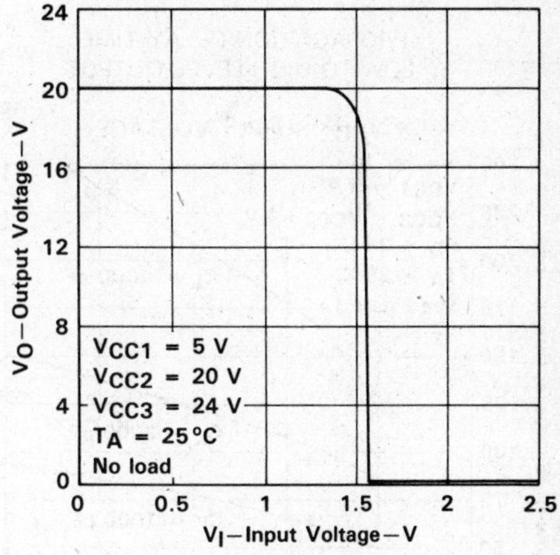
24
20
16
12
8
4
0
V_O—Output Voltage—V
V_{CC1} = 5 V
V_{CC2} = 20 V
V_{CC3} = 24 V
T_A = 25°C
No load
0
0.5
1
1.5
2
2.5
V_I—Input Voltage—V

FIGURE 5

TYPICAL CHARACTERISTICS

PROPAGATION DELAY TIME,
LOW-TO-HIGH-LEVEL OUTPUT
vs
FREE-AIR TEMPERATURE

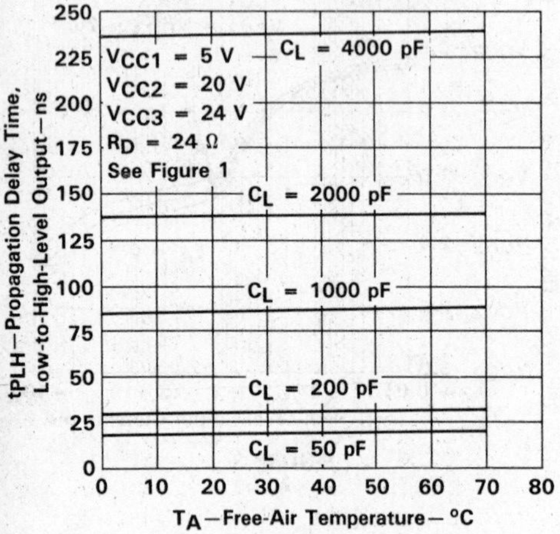
tPLH—Propagation Delay Time, Low-to-High-Level Output—ns
250
225
200
175
150
125
100
75
50
25
0
VCC1 = 5 V
VCC2 = 20 V
VCC3 = 24 V
RD = 24 Ω
See Figure 1
CL = 4000 pF
CL = 2000 pF
CL = 1000 pF
CL = 200 pF
CL = 50 pF
0 10 20 30 40 50 60 70 80
TA—Free-Air Temperature—°C

FIGURE 6

PROPAGATION DELAY TIME,
HIGH-TO-LOW-LEVEL OUTPUT
vs
T_A-FREE-AIR TEMPERATURE

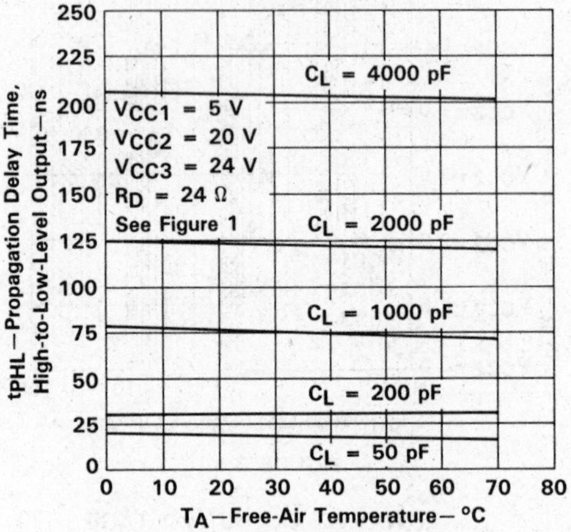
tPHL—Propagation Delay Time, High-to-Low-Level Output—ns
250
225
200
175
150
125
100
75
50
25
0
CL = 4000 pF
VCC1 = 5 V
VCC2 = 20 V
VCC3 = 24 V
RD = 24 Ω
See Figure 1
CL = 2000 pF
CL = 1000 pF
CL = 200 pF
CL = 50 pF
0 10 20 30 40 50 60 70 80
TA—Free-Air Temperature—°C

FIGURE 7

PROPAGATION DELAY TIME,
LOW-TO-HIGH-LEVEL OUTPUT
vs
V_{CC2} SUPPLY VOLTAGE

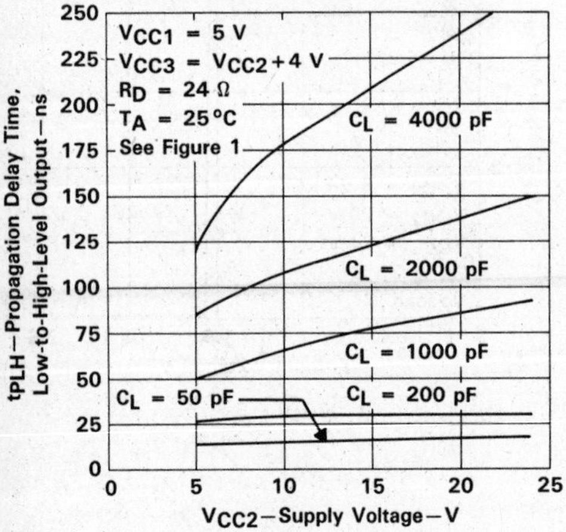
tPLH—Propagation Delay Time, Low-to-High-Level Output—ns
250
225
200
175
150
125
100
75
50
25
0
VCC1 = 5 V
VCC3 = VCC2+4 V
RD = 24 Ω
TA = 25°C
See Figure 1
CL = 4000 pF
CL = 2000 pF
CL = 1000 pF
CL = 50 pF
CL = 200 pF
0 5 10 15 20 25
VCC2—Supply Voltage—V

FIGURE 8

PROPAGATION DELAY TIME,
HIGH-TO-LOW-LEVEL OUTPUT
vs
V_{CC2} SUPPLY VOLTAGE

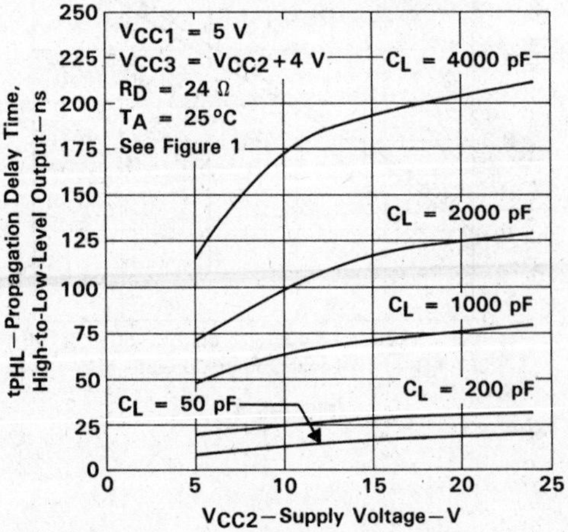
tPHL—Propagation Delay Time, High-to-Low-Level Output—ns
250
225
200
175
150
125
100
75
50
25
0
VCC1 = 5 V
VCC3 = VCC2+4 V
RD = 24 Ω
TA = 25°C
See Figure 1
CL = 4000 pF
CL = 2000 pF
CL = 1000 pF
CL = 200 pF
CL = 50 pF
0 5 10 15 20 25
VCC2—Supply Voltage—V

FIGURE 9

TEXAS INSTRUMENTS
POST OFFICE BOX 655012 • DALLAS, TEXAS 75265

TYPICAL CHARACTERISTICS

PROPAGATION DELAY TIME,
LOW-TO-HIGH-LEVEL OUTPUT
vs
LOAD CAPACITANCE

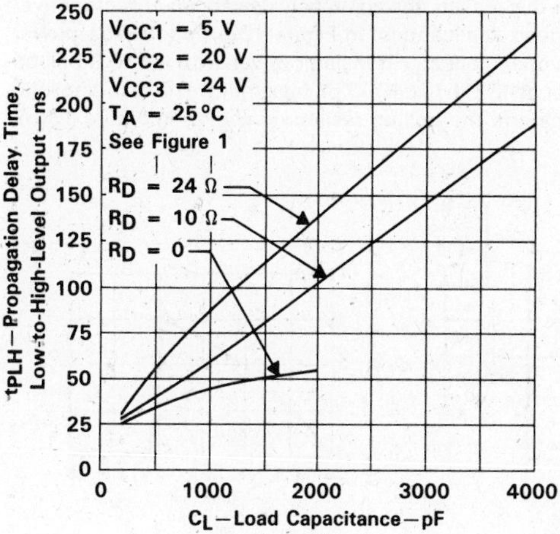
VCC1 = 5 V
VCC2 = 20 V
VCC3 = 24 V
TA = 25°C
See Figure 1
RD = 24 Ω
RD = 10 Ω
RD = 0
tPLH—Propagation Delay Time, Low-to-High-Level Output—ns
0
25
50
75
100
125
150
175
200
225
250
0
1000
2000
3000
4000
CL—Load Capacitance—pF

FIGURE 10

PROPAGATION DELAY TIME,
HIGH-TO-LOW-LEVEL OUTPUT
vs
LOAD CAPACITANCE

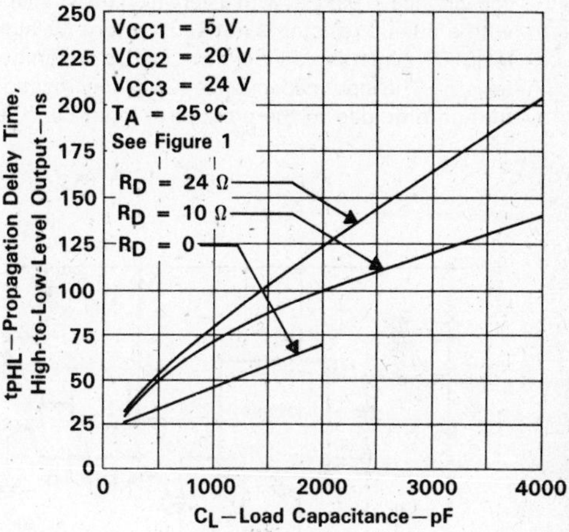
VCC1 = 5 V
VCC2 = 20 V
VCC3 = 24 V
TA = 25°C
See Figure 1
RD = 24 Ω
RD = 10 Ω
RD = 0
tPHL—Propagation Delay Time, High-to-Low-Level Output—ns
0
25
50
75
100
125
150
175
200
225
250
0
1000
2000
3000
4000
CL—Load Capacitance—pF

FIGURE 11

POWER DISSIPATION (ALL DRIVERS)
vs
FREQUENCY

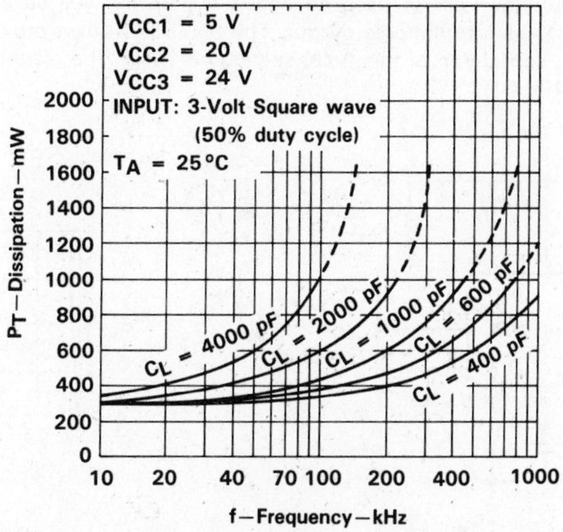
VCC1 = 5 V
VCC2 = 20 V
VCC3 = 24 V
INPUT: 3-Volt Square wave
(50% duty cycle)
TA = 25°C
CL = 4000 pF
CL = 2000 pF
CL = 1000 pF
CL = 600 pF
CL = 400 pF
PT—Dissipation—mW
0
200
400
600
800
1000
1200
1400
1600
1800
2000
10
20
40
70
100
200
400
1000
f—Frequency—kHz

FIGURE 12

NOTE: For $R_D = 0$, operation with $C_L > 2000$ pF violates absolute maximum current rating.

APPLICATIONS INFORMATION

driving power MOSFETs

The drive requirements of power MOSFETs are much lower than comparable bipolar power transistors. The input impedance of a FET consists of a reverse biased PN junction that can be described as a large capacitance in parallel with a very high resistance. For this reason, the commonly used open-collector driver with a pull-up resistor is not satisfactory for high-speed applications. In Figure 13(a), an IRF151 power MOSFET switching an inductive load is driven by an open-collector transistor driver with a 470 Ω pull-up resistor. The input capacitance (C_{iss}) specification for an IRF151 is 4000 pF maximum. The resulting long turn-on time due to the product of input capacitance and the pull-up resistor is shown in Figure 13(b).

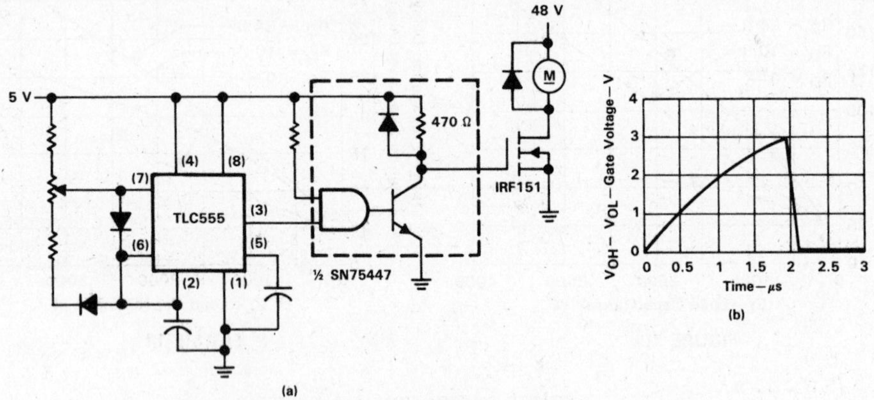
48 V
5 V
470 Ω
M
(4)
(8)
(7)
IRF151
TLC555
(3)
(6)
(5)
(2)
(1)
½ SN75447
VOH − VOL − Gate Voltage − V
0
1
2
3
4
0 0.5 1 1.5 2 2.5 3
Time − μs
(b)
(a)

FIGURE 13. POWER MOSFET DRIVE USING SN75447

A faster, more efficient drive circuit uses an active pull-up as well as an active pull-down output configuration, referred to as a totem-pole output. The SN75374 driver provides the high-speed totem-pole drive desired in an application of this type, see Figure 14(a). The resulting faster switching speeds are shown in Figure 14(b).

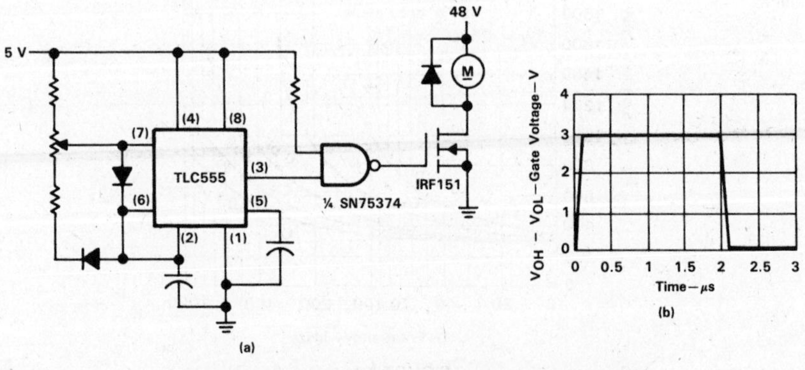
48 V
5 V
M
(4)
(8)
(7)
TLC555
(3)
IRF151
(6)
(5)
¼ SN75374
(2)
(1)
VOH − VOL − Gate Voltage − V
0
1
2
3
4
0 0.5 1 1.5 2 2.5 3
Time − μs
(b)
(a)

FIGURE 14. POWER MOSFET DRIVE USING SN75374

APPLICATIONS INFORMATION

Power MOSFET drivers must be capable of supplying high peak currents to achieve fast switching speeds as shown by the equation

$$I_{pk} = \frac{VC}{t_r}$$

where C is the capacitive load, and t_r is the desired rise time. V is the voltage that the capacitance is charged to. In the circuit shown in Figure 14(a), V is found by the equation

$$V = V_{OH} - V_{OL}$$

Peak current required to maintain a rise time of 100 ns in the circuit of Figure 14(a) is

$$I_{PK} = \frac{(3\text{-}0)4(10^9)}{100(10^9)} = 120 \text{ mA}$$

Circuit capacitance can be ignored because it is very small compared to the input capacitance of the IRF151. With a V_{CC} of 5 V, and assuming worst-case conditions, the gate drive voltage is 3 V.

For applications in which the full voltage of V_{CC2} must be supplied to the MOSFET gate, V_{CC3} should be at least 3 volts higher than V_{CC2}.

THERMAL INFORMATION

power dissipation precautions

Significant power may be dissipated in the SN75374 driver when charging and discharging high-capacitance loads over a wide voltage range at high frequencies. Figure 12 shows the power dissipated in a typical SN75374 as a function of frequency and load capacitance. Average power dissipated by this driver is derived from the equation

$$P_{T(AV)} = P_{DC(AV)} + P_{C(AV)} + P_{S(AV)}$$

where $P_{DC(AV)}$ is the steady-state power dissipation with the output high or low, $P_{C(AV)}$ is the power level during charging or discharging of the load capacitance, and $P_{S(AV)}$ is the power dissipation during switching between the low and high levels. None of these include energy transferred to the load and all are averaged over a full cycle.

The power components per driver channel are

$$P_{DC(AV)} = \frac{P_H t_H + P_L t_L}{T}$$

$$P_{C(AV)} \approx C\, V^2{}_C\, f$$

$$P_{S(AV)} = \frac{p_{LH} t_{LH} + P_{HL} t_{HL}}{T}$$

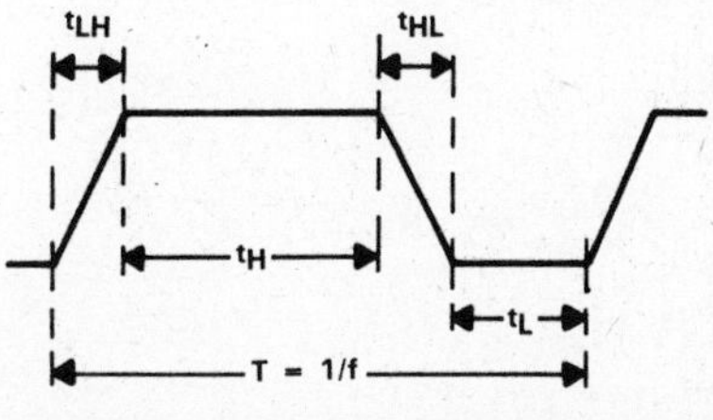
t_{LH}
t_{HL}
t_H
t_L
T = 1/f

FIGURE 15. OUTPUT VOLTAGE WAVEFORM

where the times are as defined in Figure 15.

THERMAL INFORMATION

P_L, P_H, P_{LH}, and P_{HL} are the respective instantaneous levels of power dissipation, C is the load capacitance. V_C is the voltage across the load capacitance during the charge cycle shown by the equation

$$V_C = V_{OH} - V_{OL}$$

$P_{S(AV)}$ may be ignored for power calculations at low frequencies.

$P_{S(AV)}$ may be ignored for power calculations at low frequencies.

In the following power calculation, all four channels are operating under identical conditions: f = 0.2 MHz, V_{OH} = 19.9 volts and V_{OL} = 0.15 volts with V_{CC1} = 5 volts, V_{CC2} = 20 volts, V_{CC3} = 24 volts, V_C = 19.75 volts, C = 1000 picofarads, and the duty cycle = 60%. At 0.2 MHz for $C_L < 2000$ pF, $P_{S(AV)}$ is negligible and can be ignored. When the output voltage is low, I_{CC2} is negligible and can be ignored.

On a per-channel basis using data sheet values

$$P_{DC(AV)} = \left[(5\ V)\left(\frac{4\ mA}{4}\right) + (20\ V)\left(\frac{-2.2\ mA}{4}\right) + (24\ V)\left(\frac{2.2\ mA}{4}\right)\right](0.6) + \left[(5\ V)\left(\frac{31\ mA}{4}\right) + (20\ V)\left(\frac{0\ mA}{4}\right) + (24\ V)\left(\frac{16\ mA}{4}\right)\right](0.4)$$

$$P_{DC(AV} = 58.2\ mW\ \text{per channel}$$

Power during the charging time of the load capacitance is

$$P_{C(AV)} = (1000\ pF)\ (19.75\ V)^2\ (0.2\ MHz) = 78\ mW\ \text{per channel}$$

Total power for each driver is

$$P_{T(AV)} = 58.2\ mW + 78\ mW = 136.2\ mW$$

The total package power is

$$P_{T(AV)} = (136.2)\ (4) = 544.8\ mW$$

TEXAS INSTRUMENTS
POST OFFICE BOX 655012 • DALLAS, TEXAS 75265

D2829, SEPTEMBER 1986

- Characterized for Use to 500 mA
- No Output Latch-Up at 50 V
- Very Low Quiescent Power . . . 100 mW Typical
- Very Low Input Current . . . 1 μA Typical
- Output Clamp Diodes for Transient Suppression
- TTL- or MOS-Compatible Diode-Clamped Inputs
- Standard 5-V Supply Voltage
- New Plastic DIP (P) with Copper Lead Frame Provides Cooler Operation and Improved Reliability

D OR P PACKAGE
(TOP VIEW)

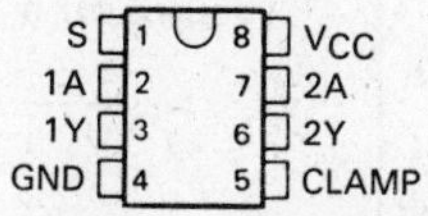
S
1A
1Y
GND
1
2
3
4
8
7
6
5
V_{CC}
2A
2Y
CLAMP

FUNCTION TABLES

SN75407
(EACH NAND DRIVER)

INPUTS		OUTPUT
A	S	Y
H	H	L
L	X	H
X	L	H

SN75408
(EACH OR DRIVER)

INPUTS		OUTPUT
A	S	Y
H	X	H
X	H	H
L	L	L

description

The SN75407 and SN75408 dual peripheral drivers are designed for use in systems that require high current, high voltage, and fast switching outputs. These devices have diode-clamped inputs as well as high-current, high-voltage output clamp diodes for switching inductive loads. Special circuits enable these devices to feature very low quiescent power and minimal input current requirements. Applications include logic buffers, hammer drivers, dc motor drivers, and dc relay/solenoid drivers.

The SN75407 and SN75408 are characterized for operation from 0°C to 70°C.

logic symbols

SN75407

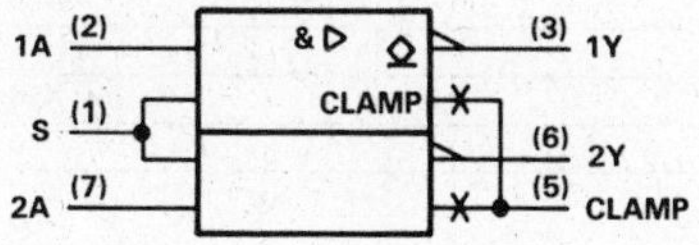
1A
(2)
S
(1)
2A
(7)
&
CLAMP
(3)
1Y
(6)
2Y
(5)
CLAMP

SN75408

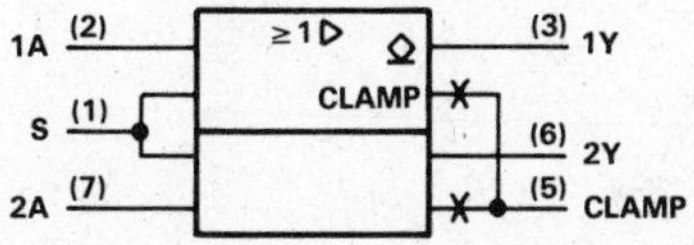
1A
(2)
S
(1)
2A
(7)
≥1
CLAMP
(3)
1Y
(6)
2Y
(5)
CLAMP

functional block diagrams (positive logic)

SN75407

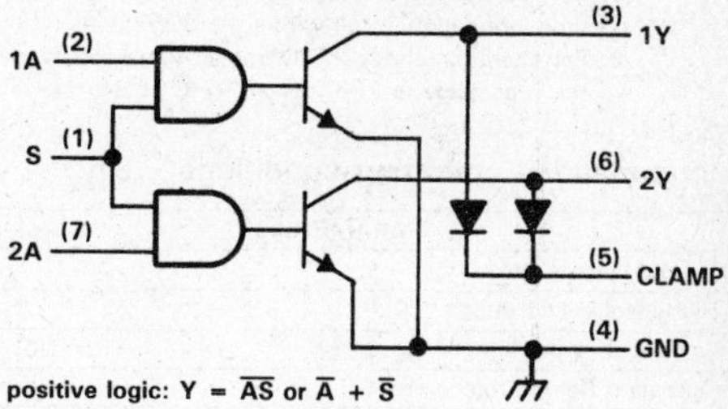
1A
(2)
S
(1)
2A
(7)
(3)
1Y
(6)
2Y
(5)
CLAMP
(4)
GND

positive logic: $Y = \overline{AS}$ or $\overline{A} + \overline{S}$

SN75408

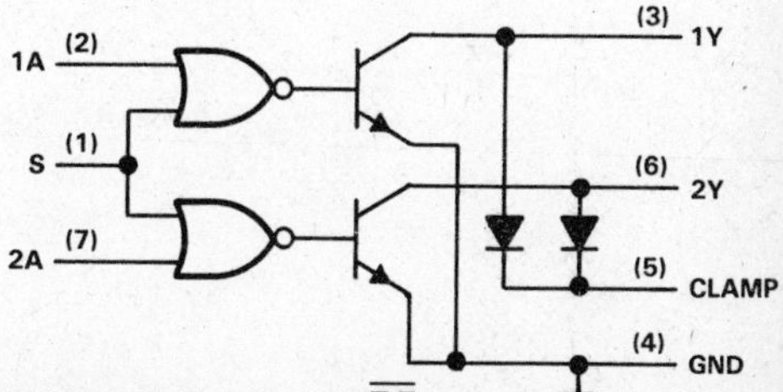
1A
(2)
S
(1)
2A
(7)
(3)
1Y
(6)
2Y
(5)
CLAMP
(4)
GND

positive logic: $Y = A + S$ or $\overline{\overline{A}\,\overline{S}}$

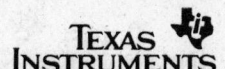

schematics of inputs and outputs

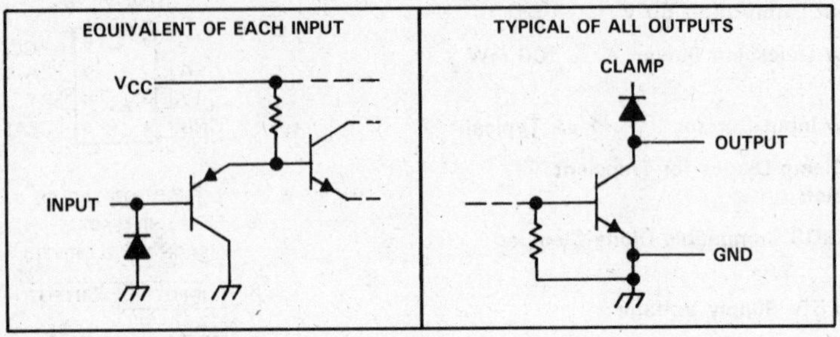
EQUIVALENT OF EACH INPUT
V_{CC}
INPUT
TYPICAL OF ALL OUTPUTS
CLAMP
OUTPUT
GND

absolute maximum ratings over operating free-air temperature (unless otherwise noted)

Supply voltage, V_{CC} (see Note 1) .... 7 V
Input voltage .... 5.5 V
Output current (see Note 2) .... 550 mA
Output clamp diode current .... 550 mA
Continuous total dissipation at (or below) 25°C free-air temperature (see Note 3):
  D package .... 725 mW
  P package .... 1200 mW
Operating free-air temperature range .... 0°C to 70°C
Storage temperature range .... −65°C to 150°C
Lead temperature 1,6 mm (1/16 inch) from case for 10 seconds .... 260°C

NOTES: 1. All voltage values are with respect to the network ground terminal.
2. Both halves of this dual circuit may conduct rated current simultaneously; however, power dissipation averaged over a short time interval must fall within the continuous dissipation ratings.
3. For operation above 25°C free-air temperature, derate the D package to 464 mW at 70°C at the rate of 5.8 mW/°C and the P package to 768 mW at 70°C at the rate of 9.6 mW/°C.

recommended operating conditions

PARAMETERS	MIN	NOM	MAX	UNIT
Supply voltage, V_{CC}	4.75	5	5.25	V
High-level input voltage, V_{IH}	2			V
Low-level input voltage, V_{IL}			0.8	V
Operating free-air temperature, T_A	0		70	°C

electrical characteristics over recommended operating free-air temperature range (unless otherwise noted)

PARAMETER			TEST CONDITIONS		MIN	TYP†	MAX	UNIT
V_{IK}	Input clamp voltage		$I_I = -12$ mA			−0.9	−1.5	V
I_{OH}	High-level output current		$V_{CC} = 4.75$ V, $V_{IL} = 0.8$ V,	$V_{IH} = 2$ V, $V_{OH} = 70$ V		1	100	μA
V_{OL}	Low-level output voltage		$V_{CC} = 4.75$ V, $V_{IH} = 2$ V, $V_{IL} = 0.8$ V	$I_{OL} = 100$ mA		0.10	0.3	V
				$I_{OL} = 200$ mA		0.22	0.45	
				$I_{OL} = 300$ mA		0.45	0.65	
				$I_{OL} = 500$ mA		0.8	1	
$V_{(BR)O}$	Output breakdown voltage		$V_{CC} = 4.75$ V,	$I_{OH} = 100$ μA	70	100		V
$V_{R(K)}$	Output clamp diode reverse voltage		$V_{CC} = 4.75$ V,	$I_R = 100$ μA	70	100		V
$V_{F(K)}$	Output clamp diode forward voltage		$V_{CC} = 4.75$ V,	$I_F = 500$ mA	0.6	1.2	2	V
I_{IH}	High-level input current		$V_{CC} = 5.25$ V,	$V_I = 5.25$ V		0.01	10	μA
I_{IL}	Low-level input current	A input	$V_{CC} = 5.25$ V,	$V_I = 0.8$ V		−0.5	−10	μA
		Strobe S				−1	−20	
I_{CCH}	Supply current, outputs high	SN75407	$V_{CC} = 5.25$ V	$V_I = 0$		20	30	mA
		SN75408		$V_I = 5$ V		20	30	
I_{CCL}	Supply current, outputs low	SN75407	$V_{CC} = 5.25$ V	$V_I = 5$ V		20	30	mA
		SN75408		$V_I = 0$		20	30	

switching characteristics, $V_{CC} = 5$ V, $T_A = 25°C$

PARAMETER		TEST CONDITIONS	MIN	TYP†	MAX	UNIT
t_{PLH}	Propagation delay time, low-to-high-level output	$C_L = 15$ pF, $R_L = 100\ \Omega$, See Figures 1 and 3		0.5	1	μs
t_{PHL}	Propagation delay time, high-to-low-level output			0.4	0.8	μs
t_{TLH}	Transition time, low-to-high-level output			0.1	0.2	μs
t_{THL}	Transition time, high-to-low-level output			0.1	0.2	μs
V_{OH}	High-level output voltage after switching	$V_S = 50\ \Omega$, $R_L = 100\ \Omega$, See Figures 2 and 3	$V_S - 10$			mV

† All typical values are at $V_{CC} = 5$ V, $T_A = 25°C$.

PARAMETER MEASUREMENT INFORMATION

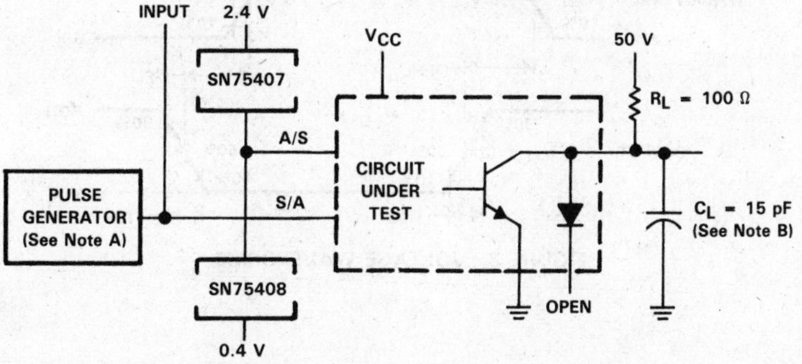
INPUT
2.4 V
SN75407
A/S
S/A
SN75408
0.4 V
PULSE GENERATOR (See Note A)
VCC
CIRCUIT UNDER TEST
50 V
RL = 100 Ω
CL = 15 pF (See Note B)
OPEN

NOTES A. The pulse generator has the following characteristics; $t_w = 5$ μs, PRR = 100 kHz, $Z_{out} = 50\ \Omega$.
B. C_L includes probe and jig capacitance.

FIGURE 1. TEST CIRCUIT FOR SWITCHING TIMES

PARAMETER MEASUREMENT INFORMATION (continued)

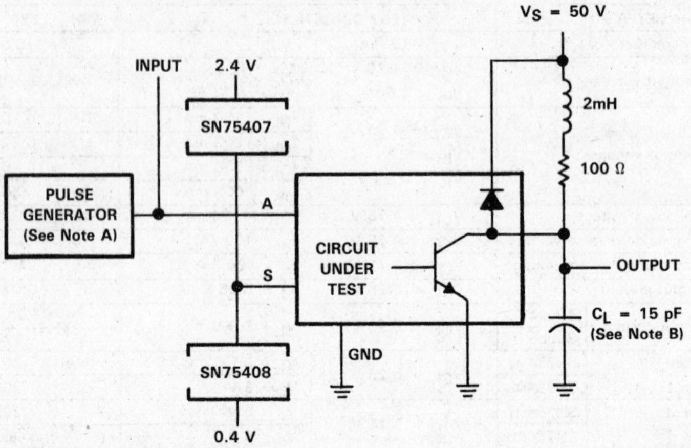
V_S = 50 V
INPUT
2.4 V
SN75407
2mH
100 Ω
PULSE GENERATOR (See Note A)
A
CIRCUIT UNDER TEST
S
OUTPUT
C_L = 15 pF (See Note B)
GND
SN75408
0.4 V

NOTES A. The pulse generator has the following characteristics; t_w = 40 μs, PRR = 12.5 kHz, Z_{out} = 50 Ω.
B. C_L includes probe and jig capacitance.

FIGURE 2. TEST CIRCUIT FOR HIGH-LEVEL OUTPUT VOLTAGE AFTER SWITCHING

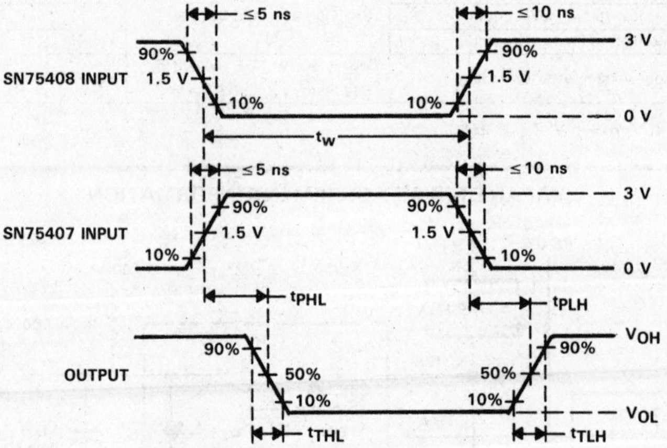
≤5 ns
≤10 ns
3 V
90%
SN75408 INPUT
1.5 V
10%
0 V
t_w
SN75407 INPUT
t_{PHL}
t_{PLH}
V_{OH}
OUTPUT
50%
V_{OL}
t_{THL}
t_{TLH}

FIGURE 3. VOLTAGE WAVEFORMS

TEXAS INSTRUMENTS
POST OFFICE BOX 655012 • DALLAS, TEXAS 75265

QUADRUPLE PERIPHERAL DRIVER WITH OUTPUT FAULT PROTECTION

D2848, FEBRUARY 1985

- **Saturating Outputs With Low On Resistance**
- **Very Low Standby Power . . . 53 mW Max**
- **High-Impedance MOS- or TTL-Compatible Inputs**
- **Standard 5-V Supply Voltage**
- **No Output Glitch During Power-Up or Power-Down**
- **Output Clamp Diodes for Transient Suppression**
- **2-W Power Package . . . 60 °C/W $R_{\theta JA}$**
- **600-mA Output Current**
- **35-V Switching Voltage**

NE DUAL-IN-LINE PACKAGE
(TOP VIEW)

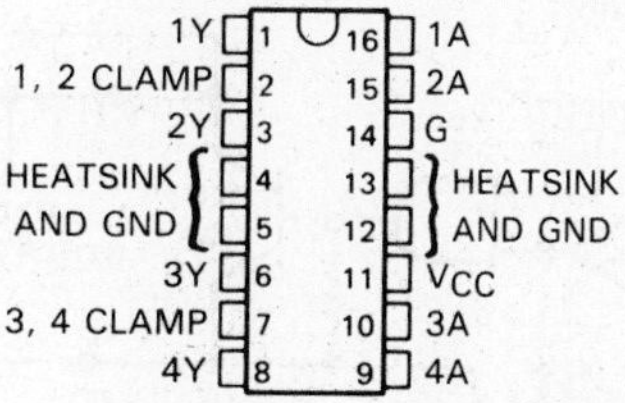
1Y
1, 2 CLAMP
2Y
HEATSINK
AND GND
3Y
3, 4 CLAMP
4Y
1
2
3
4
5
6
7
8
16
15
14
13
12
11
10
9
1A
2A
G
HEATSINK
AND GND
VCC
3A
4A

FUNCTION TABLE
(EACH NAND DRIVER)

INPUTS		OUTPUT
A	G	Y
L	X	H
X	L	H
H	H	L

H = high level, L = low level
X = irrelevant

description

The SN75435 quadruple peripheral driver is designed for use in systems requiring high current, high voltage, and high load power. It features four inverting open-collector drivers with a common enable input that, when taken low, disables all four outputs. Each driver is protected against load shorts with its own latching over-current shutdown circuitry, which will turn the output off when a load short is detected. A short on one load will not affect operation of the other three drivers. The latch for the shutdown will hold the output off until the input or enable pin is taken low and then high again. A delay circuit is incorporated in the over-current shutdown to allow load capacitance of up to 5 nF at 35 volts.

Applications include relay drivers, lamp drivers, solenoid drivers, motor drivers, LED drivers, line drivers, logic buffers, hammer drivers, and memory drivers.

The SN75435 is characterized for operation from 0 °C to 70 °C.

logic symbol†

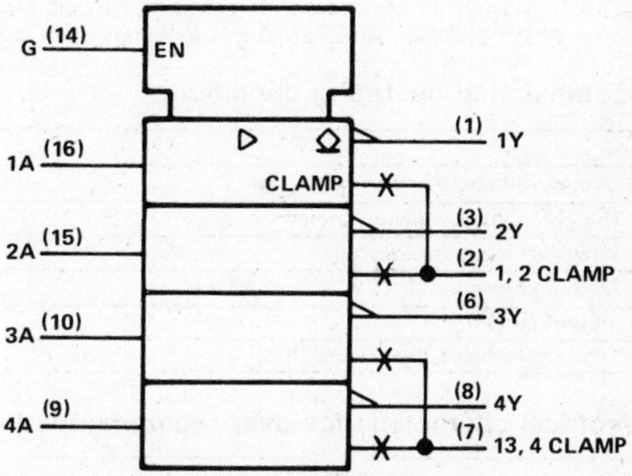
G (14)
EN
1A (16)
2A (15)
3A (10)
4A (9)
CLAMP
(1) 1Y
(3) 2Y
(2) 1, 2 CLAMP
(6) 3Y
(8) 4Y
(7) 13, 4 CLAMP

†This symbol is in accordance with ANSI/IEEE Std 91-1984 and IEC Publication 617-12.

logic diagram (positive logic)

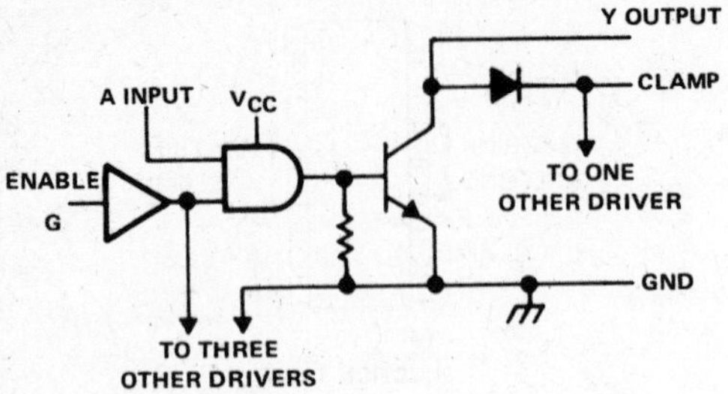
Y OUTPUT
CLAMP
A INPUT
V_{CC}
ENABLE
G
TO ONE
OTHER DRIVER
GND
TO THREE
OTHER DRIVERS

schematic of inputs

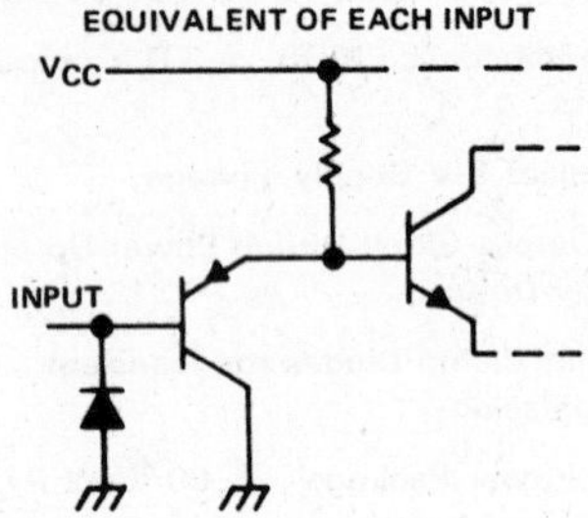
EQUIVALENT OF EACH INPUT
V_{CC}
INPUT

absolute maximum ratings over operating free-air temperature range (unless otherwise noted)

Supply voltage range of V_{CC} (see Note 1) . . . . 7 V
Input voltage . . . . 5.5 V
Output supply voltage . . . . 70 V
Output diode clamp current . . . . 1 A
Continuous total dissipation at (or below) 25 °C free-air temperature (see Note 2) . . . . 2075 mW
Operating free-air temperature range . . . . 0 °C to 70 °C
Storage temperature range . . . . −65 °C to 150 °C
Lead temperature 1,6 mm (1/16 inch) from case for 10 seconds . . . . 260 °C

NOTES: 1. All voltage values are with respect to network ground terminal.
2. For operation above 25 °C free-air temperature, derate linearly at the rate of 16.6 mW/°C.

recommended operating conditions

	MIN	NOM	MAX	UNIT
Supply voltage, V_{CC}	4.75	5	5.25	V
High-level input voltage, V_{IH}	2			V
Low-level input voltage, V_{IL}			0.8	V
Output voltage			35	V
Output current			600	mA
Load capacitance (See Figure 3)			35	nF

electrical characteristics over recommended operating free-air temperature range

PARAMETER		TEST CONDITIONS		MIN	TYP†	MAX	UNIT
V_{IK}	Input clamp voltage	V_{CC} = 4.75 V,	I_I = −12 mA		−0.9	−1.5	V
I_{OH}	High-level output current	V_{CC} = 4.75 V, V_{IL} = 0.8 V,	V_{IH} = 2 V, V_{OH} = 70 V			100	μA
V_{OL}	Low-level output voltage	V_{CC} = 4.75 V, V_{IH} = 2 V	I_{OL} = 300 mA		0.25	0.5	V
			I_{OL} = 600 mA		0.55	1	
V_R	Output clamp diode reverse voltage	V_{CC} = 4.75 V,	I_R = 100 μA	70	100		V
V_F	Output clamp diode forward voltage	I_F = 600 mA			1.2	1.6	V
I_{IH}	High-level input current	V_{CC} = 5.25 V,	V_I = 5.25 V		0.01	10	μA
I_{IL}	Low-level input current	V_{CC} = 5.25 V,	V_I = 0.8 V		−0.5	−10	μA
	Over-current shutdown current	V_{CC} = 4.75 V to 5.25 V		650	850		mA
I_{CCH}	Supply current, outputs high	V_{CC} = 5.25 V,	V_I = 0		6	10	mA
I_{CCL}	Supply current, outputs low	V_{CC} = 5.25 V,	V_I = 5 V		55	75	mA

†All typical values are at V_{CC} = 5 V, T_A = 25 °C.

TEXAS INSTRUMENTS
POST OFFICE BOX 655012 • DALLAS, TEXAS 75265

switching characteristics, V_{CC} = 5 V, T_A = 25°C

PARAMETER		TEST CONDITIONS	MIN	TYP	MAX	UNIT
t_{PLH}	Propogation delay time, low-to-high-level output	C_L = 30 pF, R_L = 60 Ω, See Figure 1		750		ns
t_{PHL}	Propagation delay time, high-to-low-level output			750		ns
t_{TLH}	Transition time, low-to-high-level output			200		ns
t_{THL}	Transition time, high-to-low-level output			200		ns
V_{OH}	High-level output voltage after switching	See Figure 2	$V_S - 10$			mV

PARAMETER MEASUREMENT INFORMATION

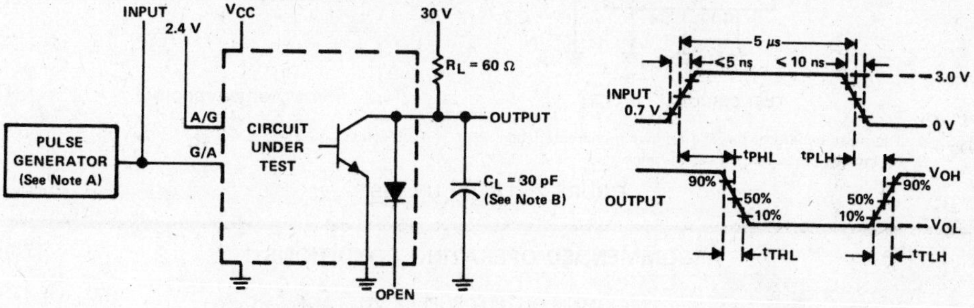
INPUT
2.4 V
VCC
30 V
RL = 60 Ω
A/G
G/A
PULSE GENERATOR (See Note A)
CIRCUIT UNDER TEST
OUTPUT
CL = 30 pF (See Note B)
OPEN
5 µs
≤5 ns
≤ 10 ns
3.0 V
INPUT
0.7 V
0 V
tPHL
tPLH
90%
VOH
90%
OUTPUT
50%
10%
50%
10%
VOL
tTHL
tTLH

TEST CIRCUIT

VOLTAGE WAVEFORMS

NOTES: A. The pulse generator has the following characteristics: PRR = 100 kHz, Z_{out} = 50 Ω.
B. C_L includes probe and jig capacitance.

FIGURE 1. SWITCHING CHARACTERISTICS

PARAMETER MEASUREMENT INFORMATION

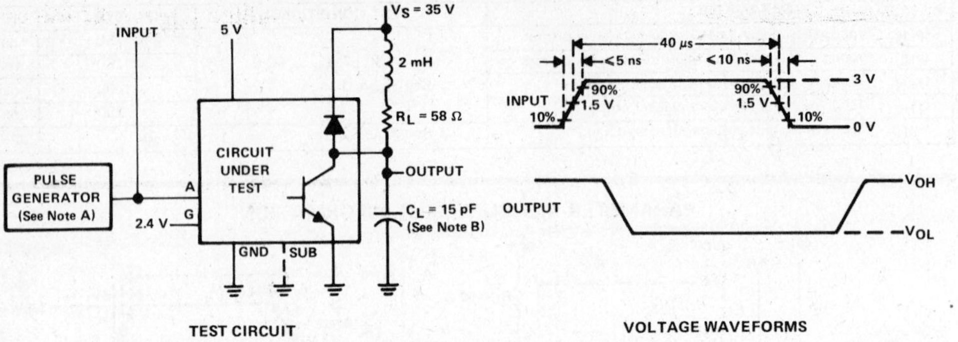
INPUT
5 V
V_S = 35 V
2 mH
R_L = 58 Ω
CIRCUIT UNDER TEST
PULSE GENERATOR (See Note A)
A
G
2.4 V
OUTPUT
C_L = 15 pF (See Note B)
GND
SUB
TEST CIRCUIT
40 μs
<5 ns
<10 ns
90%
1.5 V
10%
3 V
0 V
INPUT
OUTPUT
V_{OH}
V_{OL}
VOLTAGE WAVEFORMS

NOTES: A. The pulse generator has the following characteristics: PRR = 12.5 kHz, Z_{out} = 50 Ω.
B. C_L include probe and jig capacitance.

FIGURE 2. LATCH-UP TEST

RECOMMENDED OPERATING CONDITIONS

MAXIMUM OUTPUT SUPPLY VOLTAGE
vs
LOAD CAPACITANCE

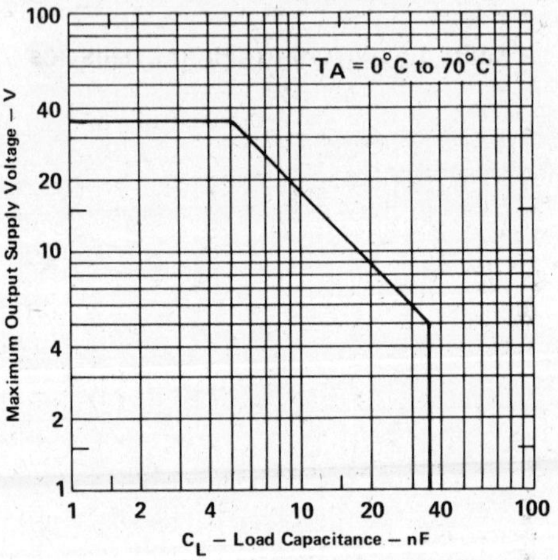
100
40
20
10
4
2
1
1
2
4
10
20
40
100
T_A = 0°C to 70°C
Maximum Output Supply Voltage – V
C_L – Load Capacitance – nF

FIGURE 3

TYPICAL APPLICATION DATA

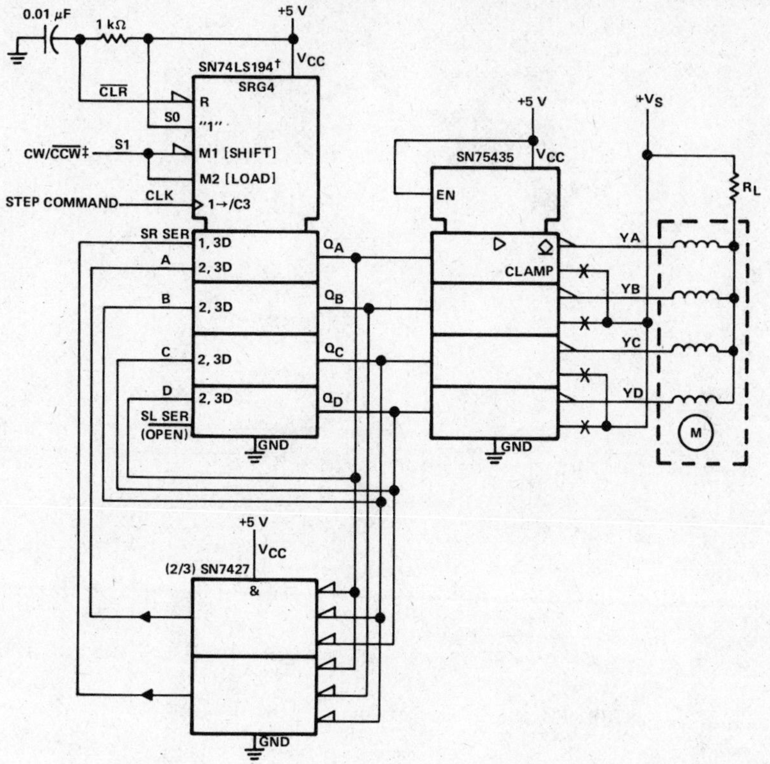
0.01 μF
1 kΩ
+5 V
VCC
SN74LS194†
SRG4
CLR
R
S0
"1"
CW/CCW‡
S1
M1 [SHIFT]
M2 [LOAD]
STEP COMMAND
CLK
1→/C3
SR SER
1, 3D
A
2, 3D
B
2, 3D
C
2, 3D
D
2, 3D
SL SER
(OPEN)
GND
QA
QB
QC
QD
+5 V
VCC
SN75435
EN
CLAMP
GND
+VS
RL
YA
YB
YC
YD
M
+5 V
VCC
(2/3) SN7427
&
GND

4-WINDING STEPPER MOTOR CONTROL CIRCUIT

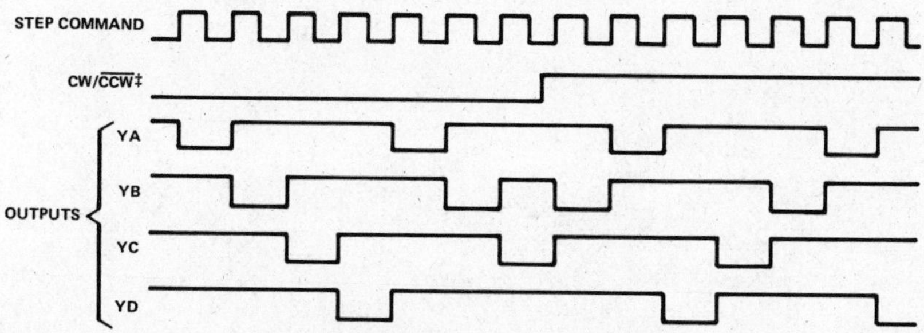
STEP COMMAND
CW/CCW‡
YA
YB
YC
YD
OUTPUTS

TIMING DIAGRAM FOR MOTOR CONTROL CIRCUIT

†The SN74LS194 is a universal shift register with both shift-right and shift-left capability. In this application S0 (pin 9) is wired high and only the shift-right and parallel-load modes are utilized. The logic symbol shown above has been simplified to show only the utilized modes.

‡This signal is CW/$\overline{\text{CCW}}$ or $\overline{\text{CW}}$/CCW depending on motor winding.

- **Saturating Outputs with Low On-State Resistance**
- **High-Impedance Inputs Compatible with CMOS, MOS, and TTL Levels**
- **Very Low Standby Power . . . 21 mW Maximum**
- **High-Voltage Outputs . . . 70 V Min**
- **No Output Glitch During Power Up or Power Down**
- **No Latch-Up Within Recommended Operating Conditions**
- **Output Clamp Diodes for Transient Suppression**
- **2-Watt Power Package**

NE DUAL-IN-LINE PACKAGE (TOP VIEW)

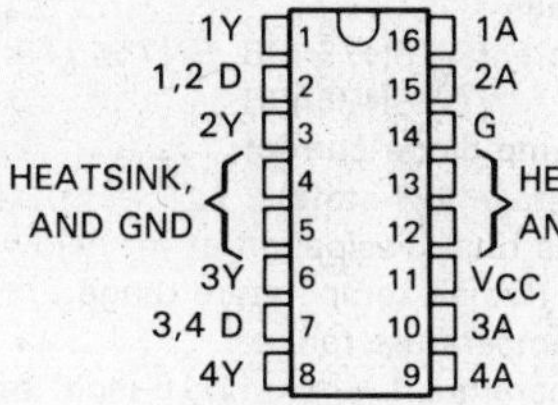
1Y
1,2 D
2Y
HEATSINK, AND GND
3Y
3,4 D
4Y
1A
2A
G
HEATSINK, AND GND
VCC
3A
4A
1
2
3
4
5
6
7
8
16
15
14
13
12
11
10
9

FUNCTION TABLE (each NAND driver)

INPUTS		OUTPUT
A	G	Y
H	H	L
L	X	H
X	L	H

H = high level,
L = low level,
X = irrelevant

description

The SN75436, SN75437A, and SN75438 quadruple peripheral drivers are designed for use in systems requiring high current, high voltage, and high load power. Each device features four inverting open-collector outputs with a common enable input that, when taken low, disables all four outputs. The envelope of I-V characteristics exceeds the specifications sufficiently to avoid high-current latch up. Applications include driving relays, lamps, solenoids, motors, LED's, transmission lines, hammers, and other high-power-demand devices.

equivalent schematic of each input

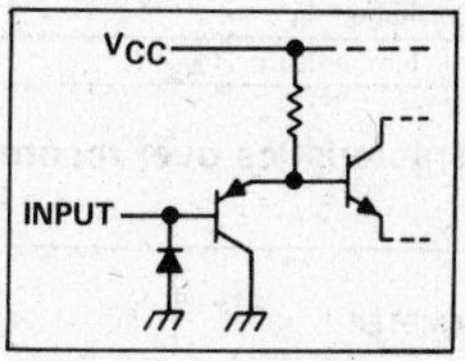
VCC
INPUT

logic symbol†

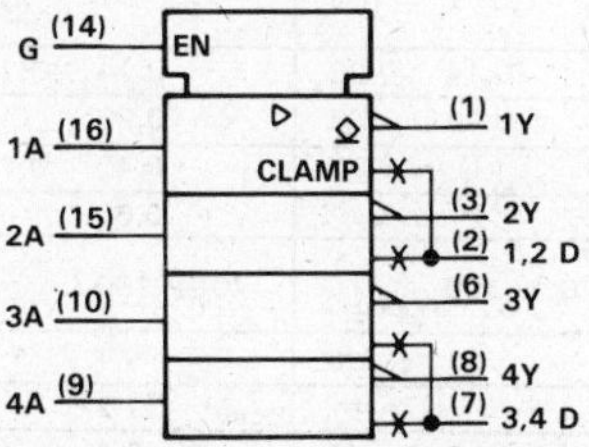
G
(14)
EN
1A
(16)
CLAMP
(1)
1Y
2A
(15)
(3)
2Y
(2)
1,2 D
3A
(10)
(6)
3Y
4A
(9)
(8)
4Y
(7)
3,4 D

† This symbol is in accordance with ANSI/IEEE Std 91-1984 and IEC Publication 617-12.

logic diagram (positive logic, each driver)

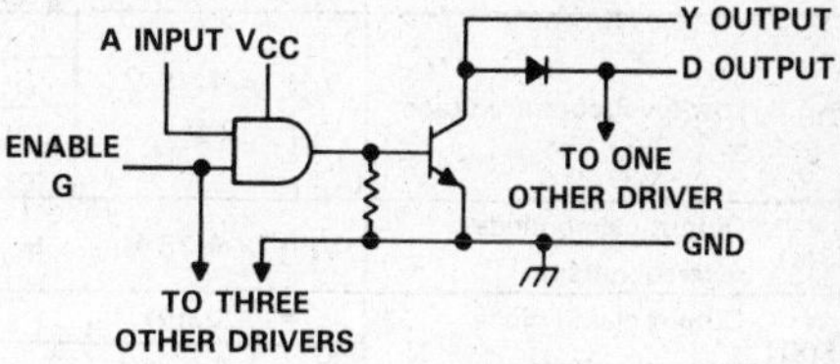
A INPUT VCC
ENABLE
G
TO THREE
OTHER DRIVERS
Y OUTPUT
D OUTPUT
TO ONE
OTHER DRIVER
GND

SELECTION GUIDE

FEATURE	SN75436	SN75437A	SN75438	UNIT
Maximum recommended output current	0.5	0.5	1	A
Maximum V_{OL} at maximum I_{OL}	0.5	0.5	1	V
Maximum recommended output supply voltage in an inductive switching circuit, V_S	50	35	35	V

absolute maximum ratings (unless otherwise noted)

Supply voltage, V_{CC} . . . . 7 V
Input voltage . . . . 30 V
Output current: SN75436, SN75437A (see Note 1) . . . . 0.75 A
SN75438 . . . . 1.25 A
Output clamp diode current . . . . 1.25 A
Output voltage (off-state) . . . . 70 V
Continuous total dissipation at (or below) 25°C free-air temperature (see Note 2) . . . . 2075 mW
Operating free-air temperature range . . . . 0°C to 70°C
Storage temperature range . . . . −65°C to 150°C
Lead temperature 1,6 mm (1/16-inch) from case for 10 seconds . . . . 260°C

NOTES: 1. All four sections of these circuits may conduct rated current simultaneously; however, power dissipation average over a short time interval must fall within the continuous dissipation ratings.
2. For operation above 25°C free-air temperature, derate linearly to 1328 mW at 70°C at the rate of 16.6 mW/°C.

recommended operating conditions

PARAMETER	SN75436			SN75437A			SN75438			UNIT
	MIN	NOM	MAX	MIN	NOM	MAX	MIN	NOM	MAX	
Supply voltage, V_{CC}	4.75	5	5.25	4.75	5	5.25	4.75	5	5.25	V
Output current, I_{OL}			0.5			0.5			1	A
Output supply voltage in inductive switching circuit (see Figure 2), V_S			50			35			35	V
High-level input voltage, V_{IH}	2			2			2			V
Low-level input voltage, V_{IL}			0.8			0.8			0.8	V
Operating free-air temperature, T_A	0		70	0		70	0		70	°C

electrical characteristics over recommended operating free-air temperature range (unless otherwise noted)

PARAMETER		TEST CONDITIONS		SN75436 SN75437A			SN75438			UNIT
				MIN	TYP†	MAX	MIN	TYP†	MAX	
V_{IK}	Input clamp	V_{CC} = 4.75 V,	I_I = −12 mA		−0.9	−1.5		−0.9	−1.5	V
I_{OH}	High-level output current	V_{CC} = 4.75 V, V_{IL} = 0.8 V,	V_{IH} = 2 V, V_{OH} = 70 V		1	100		1	100	μA
V_{OL}	Low-level output voltage	V_{CC} = 4.75 V, V_{IH} = 2 V	I_{OL} = 250 mA		0.14	0.25		0.14	0.25	V
			I_{OL} = 500 mA		0.28	0.5		0.28	0.5	
			I_{OL} = 750 mA					0.42	0.75	
			I_{OL} = 1 A					0.60	1	
$V_{R(K)}$	Output clamp diode reverse voltage	V_{CC} = 4.75 V,	I_R = 100 μA	70	100		70	100		V
$V_{F(K)}$	Output clamp diode forward voltage	I_F = 500 mA			1	1.6		1	1.6	V
		I_F = 1 A						1.2	2	
I_{IH}	High-level input current	V_{CC} = 5.25 V,	V_I = 5.25 V		0.1	10		0.1	10	μA
I_{IL}	Low-level input current	V_{CC} = 5.25 V,	V_I = 0.8 V		−0.25	−10		−0.25	−10	μA
I_{CCH}	Supply current, outputs high	V_{CC} = 5.25 V,	V_I = 0 V		1	4		1	4	mA
I_{CCL}	Supply current, outputs low	V_{CC} = 5.25 V,	V_I = 5 V		45	65		45	65	mA

† All typical values are at V_{CC} = 5 V, T_A = 25°C.

switching characteristics, V_{CC} = 5 V, T_A = 25°C

PARAMETER			TEST CONDITIONS		MIN	TYP	MAX	UNIT
t_{PLH}	Propagation delay time, low-to-high-level output		C_L = 30 pF, See Figure 1	R_L = 60 Ω		1950	5000	ns
t_{PHL}	Propagation delay time, high-to-low-level output					150	500	ns
t_{TLH}	Transition time, low-to-high-level output					40		ns
t_{THL}	Transition time, high-to-low-level output					36		ns
V_{OH}	High-level output voltage, after switching	SN75436	V_S = 50 V, R_L = 100 Ω,	I_O ≈ 500 mA, See Figure 2	V_S − 10			mV
		SN75437A	V_S = 35 V, R_L = 70 Ω,	I_O ≈ 500 mA, See Figure 2	V_S − 10			mV
		SN75438	V_S = 35 V, R_L = 35 Ω,	I_O ≈ 1 A, See Figure 2	V_S − 10			mV

PARAMETER MEASUREMENT INFORMATION

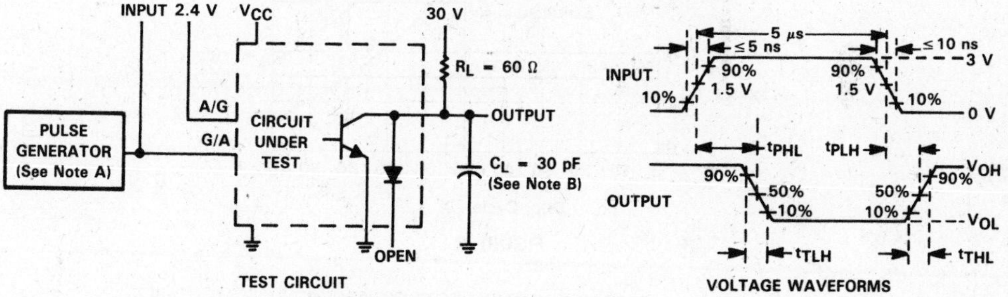
INPUT 2.4 V
V_{CC}
30 V
R_L = 60 Ω
A/G
G/A
PULSE GENERATOR (See Note A)
CIRCUIT UNDER TEST
OUTPUT
C_L = 30 pF (See Note B)
OPEN
TEST CIRCUIT
5 μs
≤5 ns
≤10 ns
3 V
INPUT
90%
1.5 V
10%
0 V
t_{PHL}
t_{PLH}
V_{OH}
OUTPUT
50%
V_{OL}
t_{TLH}
t_{THL}
VOLTAGE WAVEFORMS

NOTES: A. The pulse generator has the following characteristics: PRR = 100 kHz, Z_{out} = 50 Ω.
B. C_L includes probe and jig capacitance.

FIGURE 1. SWITCHING CHARACTERISTICS

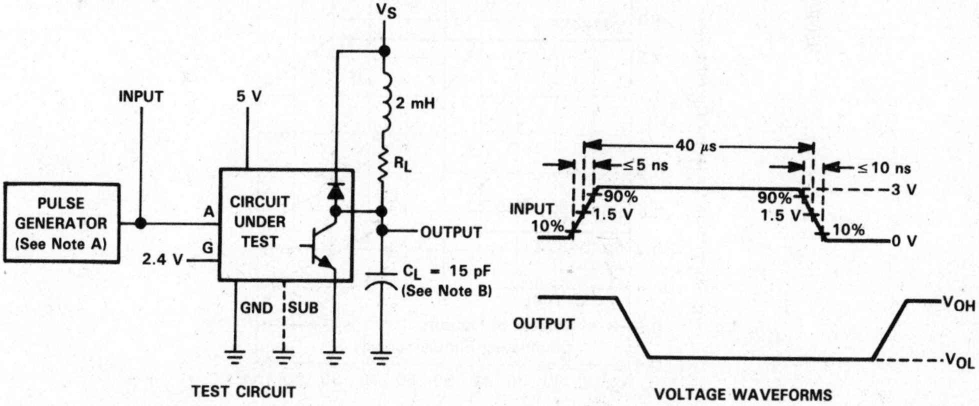
V_S
INPUT
5 V
2 mH
R_L
PULSE GENERATOR (See Note A)
A
G
2.4 V
CIRCUIT UNDER TEST
OUTPUT
C_L = 15 pF (See Note B)
GND
SUB
TEST CIRCUIT
40 μs
≤5 ns
≤10 ns
3 V
INPUT
90%
1.5 V
10%
0 V
OUTPUT
V_{OH}
V_{OL}
VOLTAGE WAVEFORMS

NOTES: A. The pulse generator has the following characteristics: PRR = 12.5 kHz, Z_{out} = 50 Ω.
B. C_L includes probe and jig capacitance.

FIGURE 2. LATCH-UP TEST

PARAMETER MEASUREMENT INFORMATION

MAXIMUM COLLECTOR CURRENT
vs
DUTY CYCLE

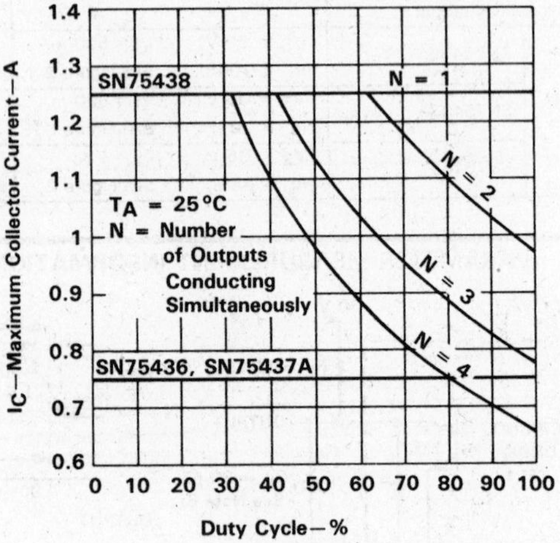
1.4
1.3
1.2
1.1
1
0.9
0.8
0.7
0.6
SN75438
N = 1
N = 2
N = 3
N = 4
TA = 25°C
N = Number of Outputs Conducting Simultaneously
SN75436, SN75437A
IC—Maximum Collector Current—A
0 10 20 30 40 50 60 70 80 90 100
Duty Cycle—%

FIGURE 3

MAXIMUM COLLECTOR CURRENT
vs
DUTY CYCLE

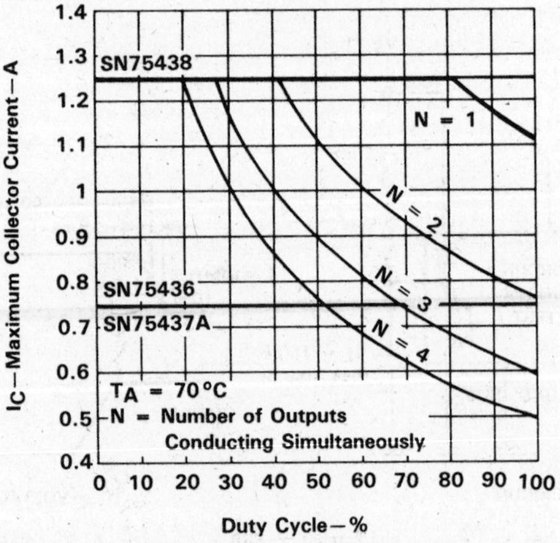
1.4
1.3
1.2
1.1
1
0.9
0.8
0.7
0.6
0.5
0.4
SN75438
N = 1
N = 2
N = 3
N = 4
SN75436
SN75437A
TA = 70°C
N = Number of Outputs Conducting Simultaneously
IC—Maximum Collector Current—A
0 10 20 30 40 50 60 70 80 90 100
Duty Cycle—%

FIGURE 4

TYPICAL APPLICATION DATA

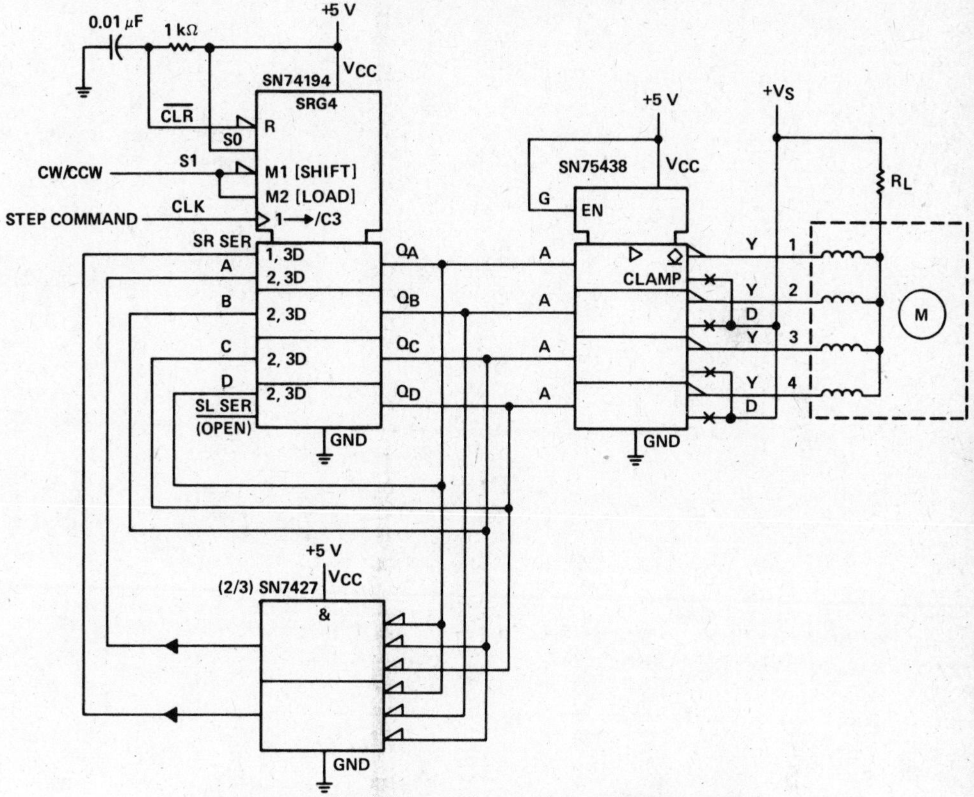
+5 V
0.01 μF
1 kΩ
VCC
SN74194
SRG4
CLR
R
S0
S1
CW/CCW
M1 [SHIFT]
M2 [LOAD]
CLK
STEP COMMAND
1→/C3
SR SER
1, 3D
A
2, 3D
B
2, 3D
C
2, 3D
D
2, 3D
SL SER
(OPEN)
GND
QA
QB
QC
QD
+5 V
+VS
SN75438
VCC
G
EN
CLAMP
Y
D
1
2
3
4
RL
M
GND
+5 V
VCC
(2/3) SN7427
&
GND

FIGURE 5. 4-WINDING STEPPER MOTOR CONTROL CIRCUIT

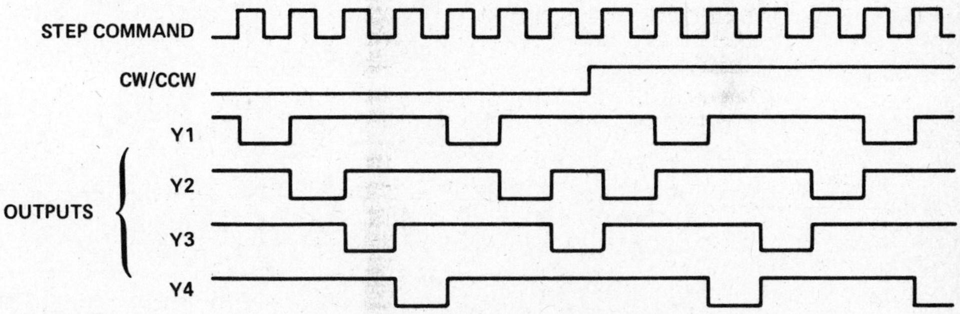
STEP COMMAND
CW/CCW
Y1
Y2
Y3
Y4
OUTPUTS

FIGURE 6. TIMING DIAGRAM

D2872, JANUARY 1985—REVISED SEPTEMBER 1986

- **Saturating Outputs with Low On-State Resistance**
- **High-Impedance Inputs Compatible with CMOS, MOS, and TTL Levels**
- **Very Low Standby Power . . . 21 mW Maximum**
- **High-Voltage Outputs . . . 70 V Min**
- **No Output Glitch During Power Up or Power Down**
- **No Latch-Up Within Recommended Operating Conditions**
- **Output Clamp Diodes for Transient Suppression**
- **2-Watt Power Packages**
- **Direct Replacement for National Semiconductor DS3669**

NE DUAL-IN-LINE PACKAGE
(TOP VIEW)

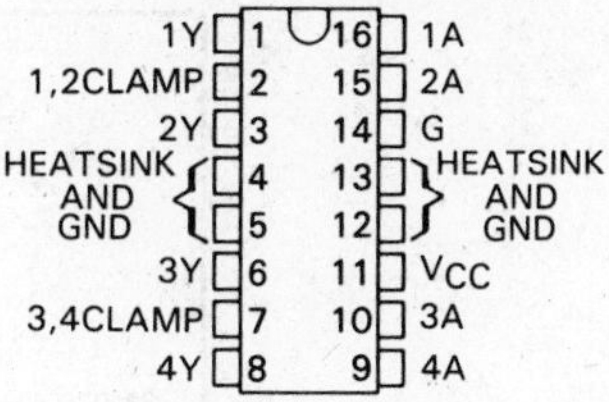
1Y 1
1,2CLAMP 2
2Y 3
HEATSINK AND GND 4
5
3Y 6
3,4CLAMP 7
4Y 8
16 1A
15 2A
14 G
13 HEATSINK AND GND
12
11 V_{CC}
10 3A
9 4A

FUNCTION TABLE

INPUTS		OUTPUT
A	G	Y
L	H	L
H	X	H
X	L	H

H = high-level
L = low-level
X = irrelevant

description

The SN75440 quadruple peripheral driver is designed for use in systems requiring high current, high voltage, and high load power. Each device features four noninverting open-collector outputs with a common enable input that, when taken low, disables all four outputs. The envelope of I-V characteristics exceeds the specifications sufficiently to avoid high-current latch up. Applications include driving relays, lamps, solenoids, motors, LEDs, transmission lines, hammers, and other high-power-demand devices.

logic symbols (alternatives)†

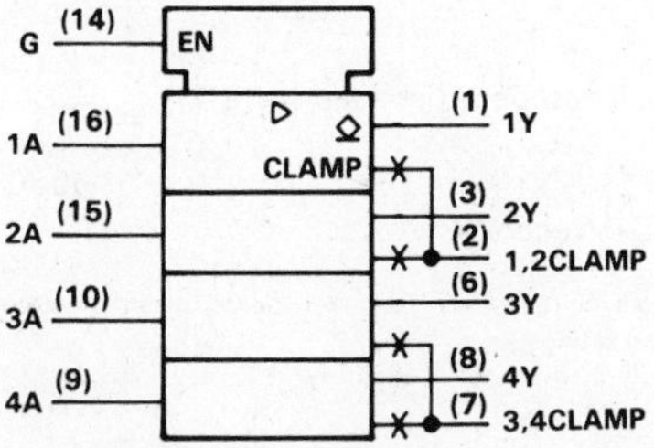
G (14)
EN
1A (16)
CLAMP
(1) 1Y
2A (15)
(3) 2Y
(2) 1,2CLAMP
3A (10)
(6) 3Y
4A (9)
(8) 4Y
(7) 3,4CLAMP

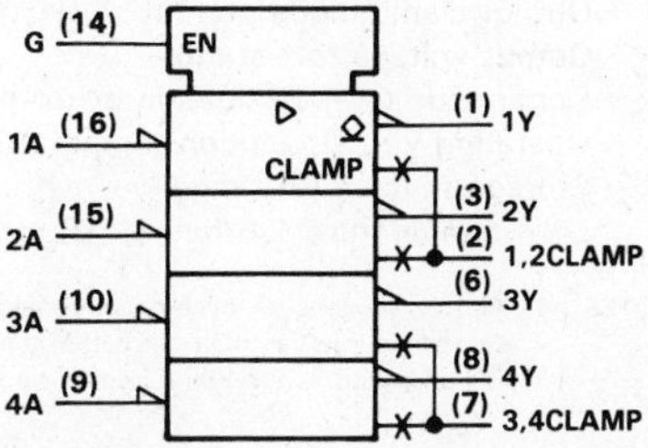
G (14)
EN
1A (16)
CLAMP
(1) 1Y
2A (15)
(3) 2Y
(2) 1,2CLAMP
3A (10)
(6) 3Y
4A (9)
(8) 4Y
(7) 3,4CLAMP

† These symbols are in accordance with ANSI/IEEE Std 91-1984 and IEC Publication 617-12.

equivalent schematic of each input

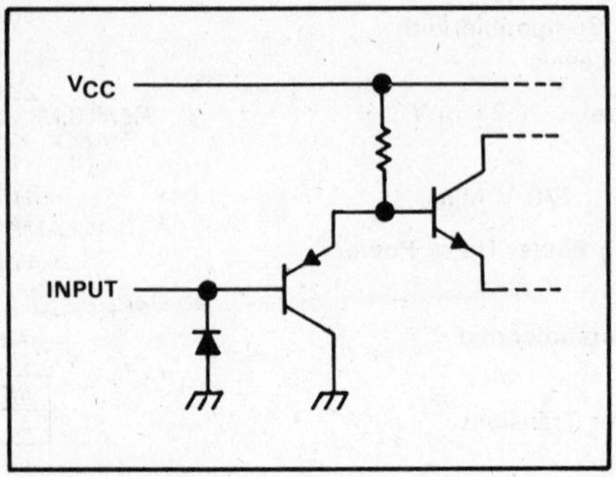
VCC
INPUT

logic diagram (each driver, positive logic)

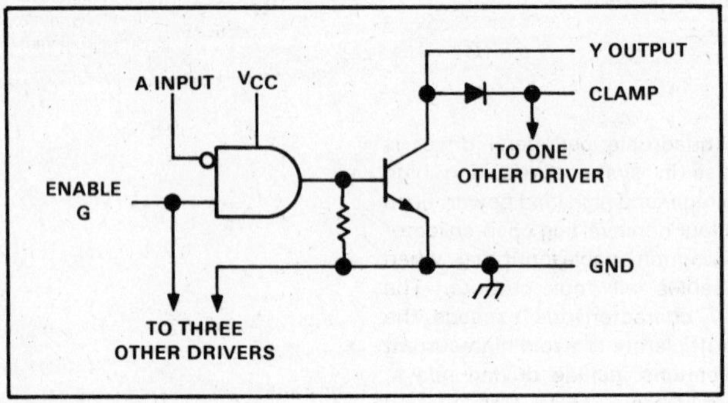
Y OUTPUT
A INPUT
VCC
CLAMP
TO ONE
OTHER DRIVER
ENABLE
G
GND
TO THREE
OTHER DRIVERS

absolute maximum ratings (unless otherwise noted)

Supply voltage, V_{CC}	7 V
Input voltage	30 V
Output current (see Note 1)	0.75 A
Output clamp diode current	1 A
Output voltage (off-state)	70 V
Continuous total dissipation at (or below) 25°C free-air temperature (see Note 2)	2075 mW
Operating virtual junction temperature	150°C
Storage temperature range	−65°C to 150°C
Lead temperature 1.6 mm (1/16 inch) from case for 10 seconds	260°C

NOTES: 1. All four sections of these circuits may conduct rated current simultaneously; however, power dissipation averaged over a short time interval must fall within the continuous dissipation ratings.
2. For operation above 25°C free-air temperature, derate linearly at the rate of 16.6 mW/°C.

TEXAS INSTRUMENTS
POST OFFICE BOX 655012 • DALLAS, TEXAS 75265

recommended operating conditions

PARAMETER	MIN	NOM	MAX	UNIT
Supply Voltage, V_{CC}	4.75	5	5.25	V
Output current, I_{OL}			600	mA
Output supply voltage in Figure 2 (Inductive switching circuit, V_S			35	V
High-level input voltage, V_{IH}	2			V
Low-level input voltage, V_{IL}			0.8	V
Operating free-air temperature, T_A	0		70	°C

electrical characteristics over recommended operating free-air temperature range (unless otherwise noted)

PARAMETER		TEST CONDITIONS		MIN	TYP†	MAX	UNIT
V_{IK}	Input clamp voltage	V_{CC} = 4.75 V,	I_I = −12 mA		−0.9	−1.5	V
I_{OH}	High-level output current	V_{CC} = 4.75 V, V_{IL} = 0.8 V,	V_{IH} = 2 V, V_{OH} = 70 V		1	100	μA
V_{OL}	Low-level output voltage	V_{CC} = 4.75 V, V_{IH} = 2 V, V_{IL} = 0.8 V	I_{OL} = 300 mA			0.4	V
			I_{OL} = 600 mA			0.7	V
$V_{R(D)}$	Output clamp diode reverse voltage	V_{CC} = 4.75 V,	I_R = 100 μA	70	100		V
$V_{F(D)}$	Output clamp diode forward voltage	I_F = 800 mA			1	1.6	V
I_{IH}	High-level input current	V_{CC} = 5.25 V,	V_I = 5.25 V		0.1	10	μA
I_{IL}	Low-level input current	V_{CC} = 5.25 V,	V_I = 0.4 V		−0.25	−10	μA
I_{CCH}	Supply current, outputs high	V_{CC} = 5.25 V,	V_{IH} = 2 V		1	4	mA
I_{CCL}	Supply current, outputs low	V_{CC} = 5.25 V, V_{IL} = 0 V	V_{IH} = 2 V,		50	65	mA

† All typical values are at V_{CC} = 5 V, T_A = 25°C.

switching characteristics, V_{CC} = 5 V, T_A = 25°C

PARAMETER		TEST CONDITIONS		MIN	TYP	MAX	UNIT
t_{PLH}	Propagation delay time, low-to-high-level output	R_L = 60 Ω,	A Input		1.4	5	μs
			G Input		1.5	5	μs
t_{PHL}	Propagation delay time, high-to-low-level output	C_L = 30 pF,	A Input		0.1	0.5	μs
			G Input		2.5	5	μs
t_{TLH}	Transition time, low-to-high-level output	See Figure 1			200		ns
t_{THL}	Transition time, high-to-low-level output	See Figure 1			50		ns
V_{OH}	High-level output voltage, after switching	V_S = 35 V, R_L = 70 Ω,	I_O ≈ 500 mA, See Figure 2	V_S −10			mV

PARAMETER MEASUREMENT INFORMATION

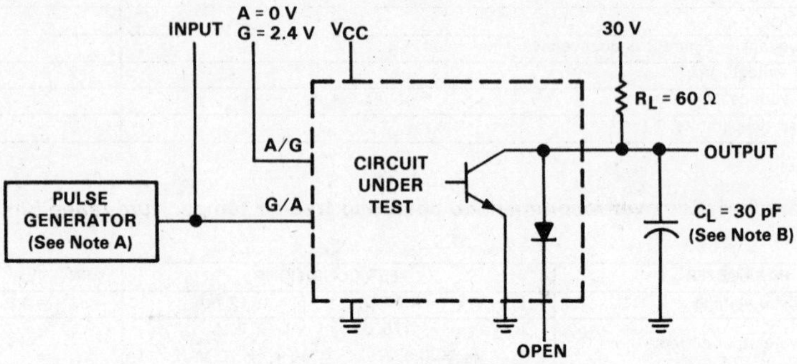
INPUT
A = 0 V
G = 2.4 V
VCC
30 V
RL = 60 Ω
A/G
G/A
CIRCUIT
UNDER
TEST
OUTPUT
PULSE
GENERATOR
(See Note A)
CL = 30 pF
(See Note B)
OPEN

TEST CIRCUIT

NOTES: A. The pulse generator has the following characteristics: PRR ≤ 100 kHz, Z_{out} = 50 Ω.
B. C_L includes probe and jig capacitance.

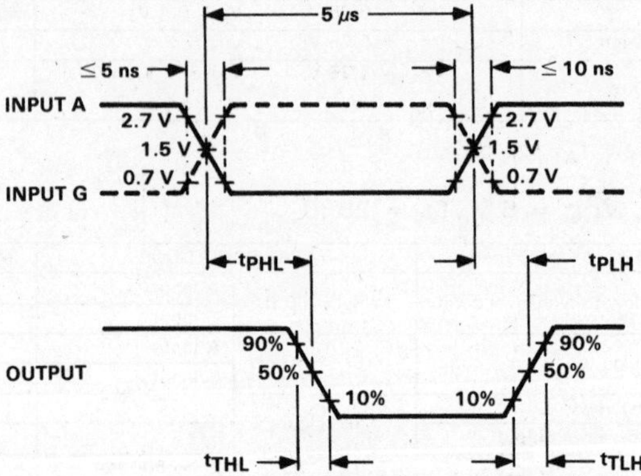
5 μs
≤ 5 ns
≤ 10 ns
INPUT A
2.7 V
1.5 V
0.7 V
INPUT G
2.7 V
1.5 V
0.7 V
0 V
tPHL
tPLH
OUTPUT
90%
50%
10%
10%
90%
50%
tTHL
tTLH

VOLTAGE WAVEFORMS

FIGURE 1. SWITCHING CHARACTERISTICS

TEXAS INSTRUMENTS
POST OFFICE BOX 655012 • DALLAS, TEXAS 75265

PARAMETER MEASUREMENT INFORMATION

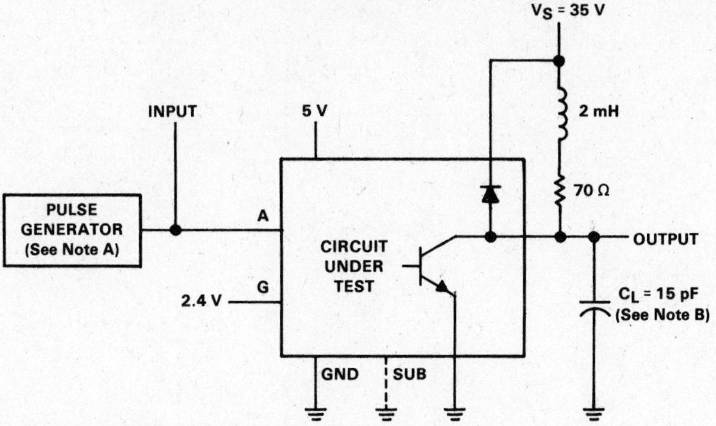
VS = 35 V
2 mH
70 Ω
INPUT
5 V
PULSE GENERATOR (See Note A)
A
CIRCUIT UNDER TEST
G
2.4 V
OUTPUT
CL = 15 pF (See Note B)
GND
SUB

NOTES: A. The pulse generator has the following characteristics: PRR ≤ 12.5 kHz, Z_{out} = 50 Ω.
B. C_L includes probe and jig capacitance.

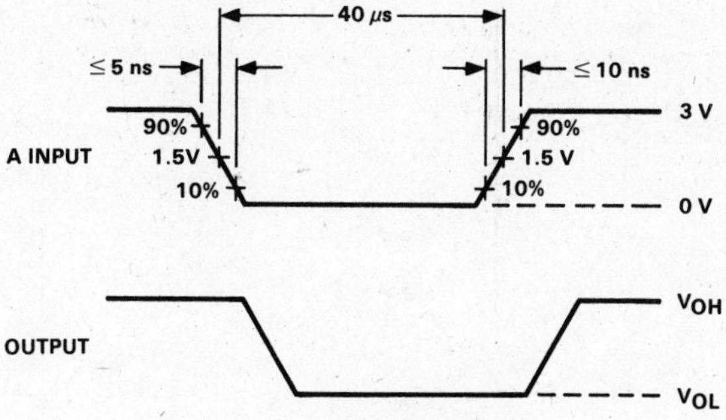
40 μs
≤ 5 ns
≤ 10 ns
A INPUT
90%
1.5V
10%
90%
1.5 V
10%
3 V
0 V
OUTPUT
VOH
VOL

VOLTAGE WAVEFORMS

FIGURE 2. LATCH-UP TEST

D2481, DECEMBER 1978—REVISED FEBRUARY 1987

- **Very Low Power Requirements**
- **Very Low Input Current**
- **Characterized for Use to 350 mA**
- **No Output Latch-Up at 50 V (After Conducting 300 mA)**
- **High-Voltage Outputs (70 V Min)**
- **Output Clamp Diodes for Transient Suppression (350 mA, 70 V)**
- **TTL- or MOS-Compatible Diode-Clamped Inputs**
- **Standard Supply Voltage**
- **Suitable for Hammer-Driver Applications**

SN75446, SN75447, SN75448, SN75449 . . . D OR P PACKAGE
(TOP VIEW)

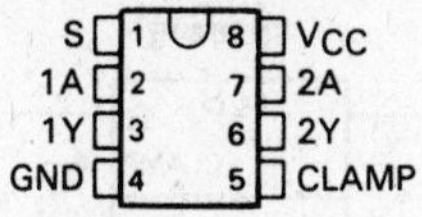
S 1
1A 2
1Y 3
GND 4
8 VCC
7 2A
6 2Y
5 CLAMP

description

Series 75446 dual peripheral drivers are designed for use in systems that require high current, high voltage, and fast switching times. The SN75446, SN75447, SN75448, and SN75449 provide AND, NAND, OR, and NOR drivers, respectively. These devices have diode-clamped inputs as well as high-current, high-voltage inductive-clamp diodes on the outputs.

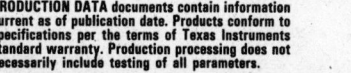

Series 75446 drivers are characterized for operation from 0°C to 70°C.

FUNCTION TABLES

SN75446
(EACH AND DRIVER)

INPUTS		OUTPUT
A	S	Y
H	H	H
L	X	L
X	L	L

SN75447
(EACH NAND DRIVER)

INPUTS		OUTPUT
A	S	Y
H	H	L
L	X	H
X	L	H

SN75448
(EACH OR DRIVER)

INPUTS		OUTPUT
A	S	Y
H	X	H
X	H	H
L	L	L

SN75449
(EACH NOR DRIVER)

INPUTS		OUTPUT
A	S	Y
H	X	L
X	H	L
L	L	H

H = high level
L = low level
X = irrelevant

schematics of inputs and outputs

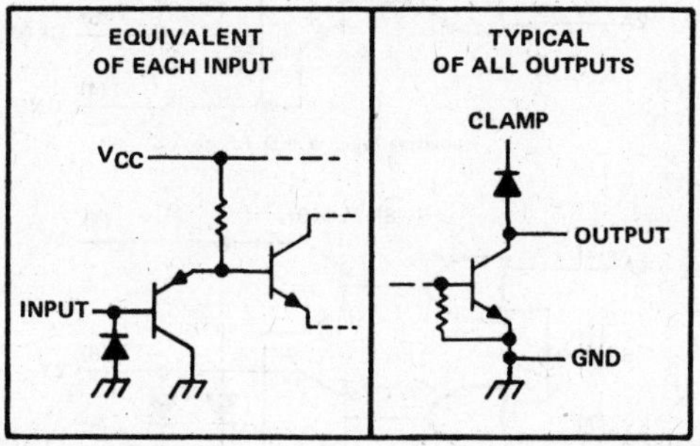
EQUIVALENT OF EACH INPUT
VCC
INPUT
TYPICAL OF ALL OUTPUTS
CLAMP
OUTPUT
GND

logic symbols†

SN75446

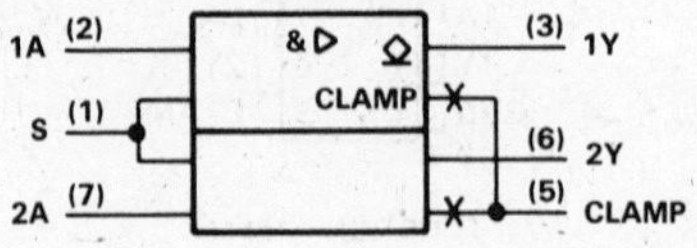
1A
(2)
S
(1)
2A
(7)
&
CLAMP
(3)
1Y
(6)
2Y
(5)
CLAMP

SN75447

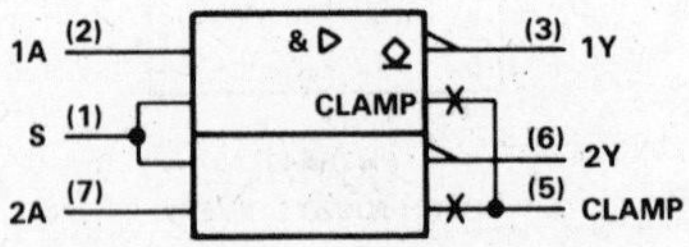
1A
(2)
S
(1)
2A
(7)
&
CLAMP
(3)
1Y
(6)
2Y
(5)
CLAMP

SN75448

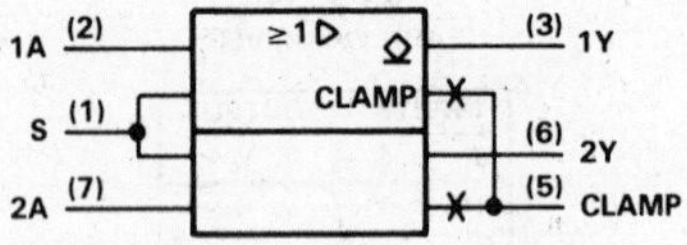
1A
(2)
S
(1)
2A
(7)
≥1
CLAMP
(3)
1Y
(6)
2Y
(5)
CLAMP

SN75449

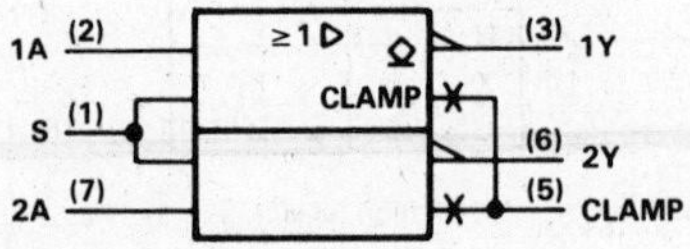
1A
(2)
S
(1)
2A
(7)
≥1
CLAMP
(3)
1Y
(6)
2Y
(5)
CLAMP

† These symbols are in accordance with ANSI/IEEE Std 91-1984 and IEC Publication 617-12.

logic diagrams (positive logic)

SN75446

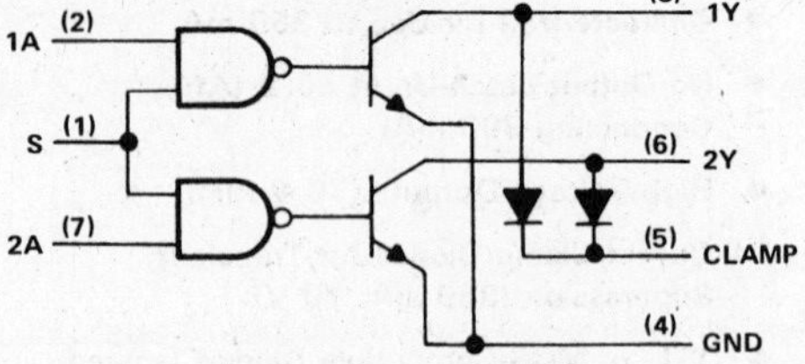
1A
(2)
S
(1)
2A
(7)
(3)
1Y
(6)
2Y
(5)
CLAMP
(4)
GND

positive logic: Y = AS or $\overline{\overline{A}+\overline{S}}$

SN75447

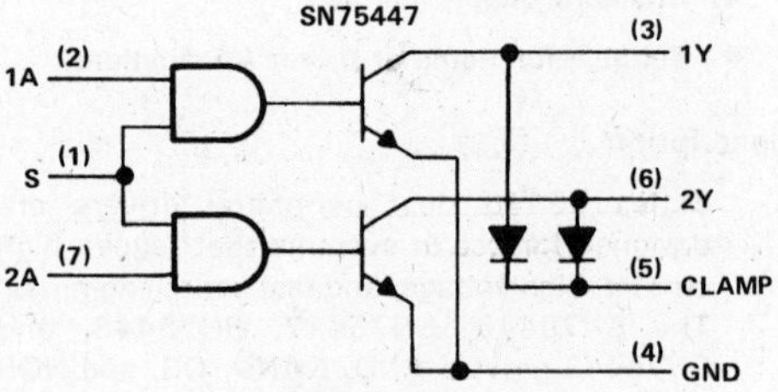
1A
(2)
S
(1)
2A
(7)
(3)
1Y
(6)
2Y
(5)
CLAMP
(4)
GND

positive logic: Y = $\overline{AS}$ or $\overline{A}+\overline{S}$

SN75448

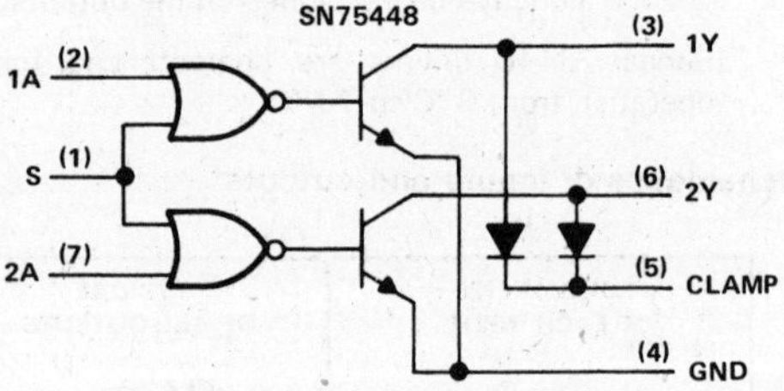
1A
(2)
S
(1)
2A
(7)
(3)
1Y
(6)
2Y
(5)
CLAMP
(4)
GND

positive logic: Y = A + S or $\overline{\overline{A}\,\overline{S}}$

SN75449

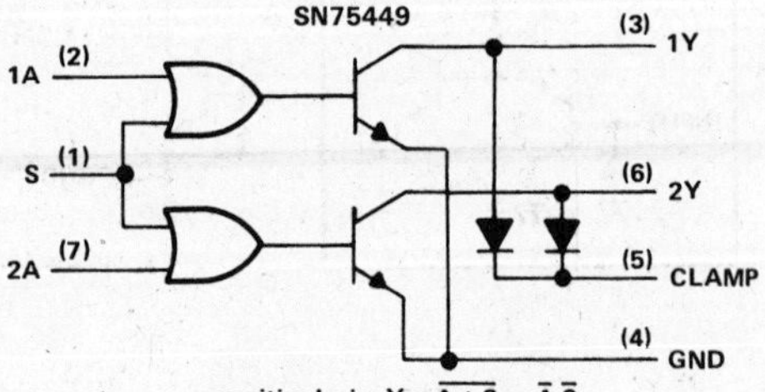
1A
(2)
S
(1)
2A
(7)
(3)
1Y
(6)
2Y
(5)
CLAMP
(4)
GND

positive logic: Y = $\overline{A+S}$ or $\overline{A}\,\overline{S}$

absolute maximum ratings over operating free-air temperature range (unless otherwise noted)

Supply voltage, V_{CC} (see Note 1) ... 7 V
Input voltage ... 5.5 V
Output current (see Note 2) ... 400 mA
Output clamp diode current ... 400 mA
Continuous total dissipation at (or below)
25°C free-air temperature (see Note 3): D package ... 725 mW
P package ... 1200 mW
Storage temperature range ... −65°C to 150°C
Lead temperature 1,6 mm (1/16 inch) from case for 10 seconds ... 260°C

NOTES: 1. Voltage values are with respect to network ground terminal.
2. Both halves of this dual circuit may conduct rated current simultaneously; however, power dissipation averaged over a short time interval must fall within the continuous dissipation ratings.
3. For operation above 25°C free-air temperature, derate the D package at the rate of 5.8 mW/°C and the P package at the rate of 9.6 mW/°C.

recommended operating conditions

	MIN	NOM	MAX	UNIT
Supply voltage, V_{CC}	4.75	5	5.25	V
High-level input voltage, V_{IH}	2			V
Low-level input voltage, V_{IL}			0.8	V
Operating free-air temperature, T_A	0		70	°C

electrical characteristics over recommended operating free-air temperature range (unless otherwise noted)

PARAMETER			TEST CONDITIONS		MIN	TYP†	MAX	UNIT
V_{IK}	Input clamp voltage		$I_I = -12$ mA			−0.9	−1.5	V
I_{OH}	High-level output current		$V_{CC} = 4.75$ V, $V_{IL} = 0.8$ V,	$V_{IH} = 2$ V, $V_{OH} = 70$ V		1	100	μA
V_{OL}	Low-level output voltage		$V_{CC} = 4.75$ V, $V_{IH} = 2$ V, $V_{IL} = 0.8$ V	$I_{OL} = 100$ mA		0.10	0.3	V
				$I_{OL} = 200$ mA		0.22	0.45	
				$I_{OL} = 300$ mA		0.45	0.65	
				$I_{OL} = 350$ mA		0.55	0.75	
$V_{(BR)O}$	Output breakdown voltage		$V_{CC} = 4.75$ V,	$I_{OH} = 100$ μA	70	100		V
$V_{R(K)}$	Output clamp diode reverse voltage		$V_{CC} = 4.75$ V,	$I_R = 100$ μA	70	100		V
$V_{F(K)}$	Output clamp diode forward voltage		$V_{CC} = 4.75$ V,	$I_F = 350$ mA	0.6	1.2	1.6	V
I_{IH}	High-level input current		$V_{CC} = 5.25$ V,	$V_I = 5.25$ V		0.01	10	μA
I_{IL}	Low-level input current	A input	$V_{CC} = 5.25$ V,	$V_I = 0.8$ V		−0.5	−10	μA
		Strobe S				−1	−20	
I_{CCH}	Supply current, outputs high	SN75446	$V_{CC} = 5.25$ V	$V_I = 5$ V		11	18	mA
		SN75447		$V_I = 0$		11	18	
		SN75448		$V_I = 5$ V		18	25	
		SN75449		$V_I = 0$		18	25	
I_{CCL}	Supply current, outputs low	SN75446	$V_{CC} = 5.25$ V	$V_I = 0$		11	18	mA
		SN75447		$V_I = 5$ V		11	18	
		SN75448		$V_I = 0$		18	25	
		SN75449		$V_I = 5$ V		18	25	

† All typical values are at $V_{CC} = 5$ V, $T_A = 25$°C.

switching characteristics, V_{CC} = 5 V, T_A = 25°C

PARAMETER		TEST CONDITIONS	MIN	TYP	MAX	UNIT
t_{PLH}	Propagation delay time, low-to-high-level output	C_L = 15 pF, R_L = 100 Ω, See Figure 1		300	750	ns
t_{PHL}	Propagation delay time, high-to-low-level output			200	500	ns
t_{TLH}	Transition time, low-to-high-level output			50	100	ns
t_{THL}	Transition time, high-to-low-level output			50	100	ns
V_{OH}	High-level output voltage after switching	V_S = 55 V, I_O ≈ 300 mA, See Figure 2	V_S − 0.018			V

PARAMETER MEASUREMENT INFORMATION

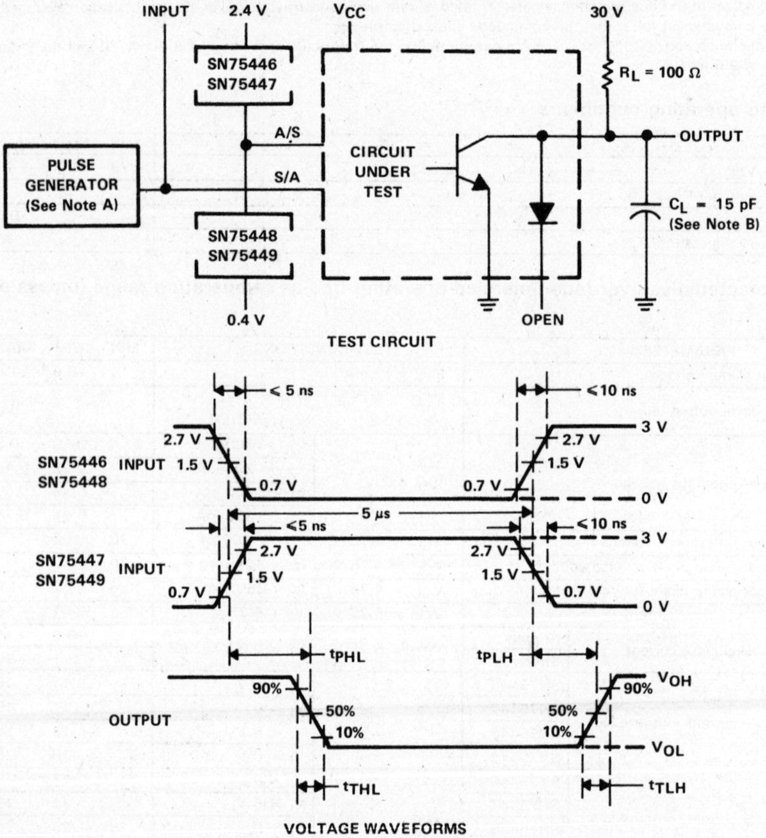
INPUT
2.4 V
VCC
30 V
SN75446
SN75447
RL = 100 Ω
A/S
OUTPUT
PULSE
GENERATOR
(See Note A)
CIRCUIT
UNDER
TEST
S/A
CL = 15 pF
(See Note B)
SN75448
SN75449
0.4 V
OPEN
TEST CIRCUIT
≤ 5 ns
≤10 ns
3 V
2.7 V
SN75446
SN75448
INPUT
1.5 V
0.7 V
0 V
5 µs
≤5 ns
≤10 ns
SN75447
SN75449
tPHL
tPLH
90%
VOH
OUTPUT
50%
10%
VOL
tTHL
tTLH
VOLTAGE WAVEFORMS

NOTES: A. The pulse generator has the following characteristics: PRR = 100 kHz, Z_{out} = 50 Ω.
B. C_L includes probe and jig capacitance.

FIGURE 1. SWITCHING CHARACTERISTICS

TEXAS INSTRUMENTS
POST OFFICE BOX 655012 • DALLAS, TEXAS 75265

PARAMETER MEASUREMENT INFORMATION

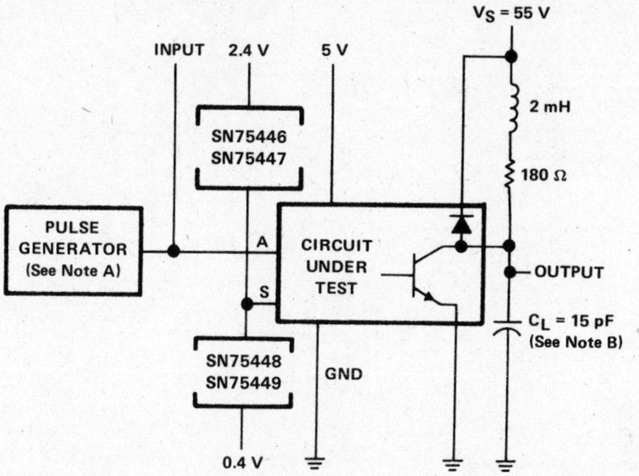
VS = 55 V
INPUT
2.4 V
5 V
SN75446
SN75447
2 mH
180 Ω
PULSE
GENERATOR
(See Note A)
A
CIRCUIT
UNDER
TEST
S
OUTPUT
CL = 15 pF
(See Note B)
SN75448
SN75449
GND
0.4 V

TEST CIRCUIT

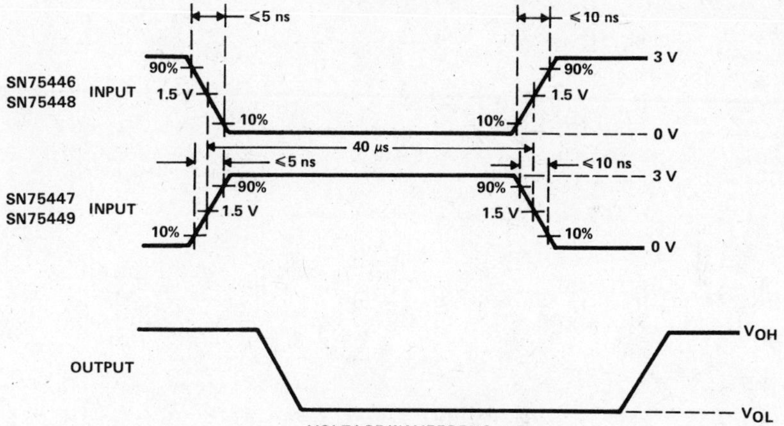
≤5 ns
≤10 ns
SN75446
SN75448
INPUT
90%
1.5 V
10%
3 V
0 V
40 μs
≤5 ns
≤10 ns
SN75447
SN75449
INPUT
OUTPUT
VOH
VOL

VOLTAGE WAVEFORMS

NOTES: A. The pulse generator has the following characteristics: PRR = 12.5 kHz, Z_{out} = 50 Ω.
B. C_L includes probe and jig capacitance.

FIGURE 2. LATCH-UP TEST

D2217, DECEMBER 1976—REVISED SEPTEMBER 1986

PERIPHERAL DRIVERS FOR HIGH-CURRENT SWITCHING AT VERY HIGH SPEEDS

- Characterized for Use to 300 mA
- High-Voltage Outputs
- No Output Latch-Up at 20 V (After Conducting 300 mA)
- High-Speed Switching
- Circuit Flexibility for Varied Applications
- TTL-Compatible Diode-Clamped Inputs
- Standard Supply Voltages
- New Plastic DIP (P) with Copper Lead Frame Provides Cooler Operation and Improved Reliability
- Package Options Include Plastic "Small Outline" Packages, Ceramic Chip Carriers, and Standard Plastic and Ceramic 300-mil DIPs

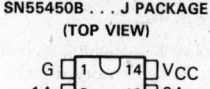

SN55450B . . . J PACKAGE
(TOP VIEW)

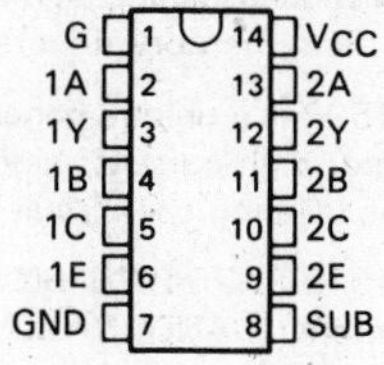
G 1
1A 2
1Y 3
1B 4
1C 5
1E 6
GND 7
14 VCC
13 2A
12 2Y
11 2B
10 2C
9 2E
8 SUB

SN55450B . . . FK PACKAGE
(TOP VIEW)

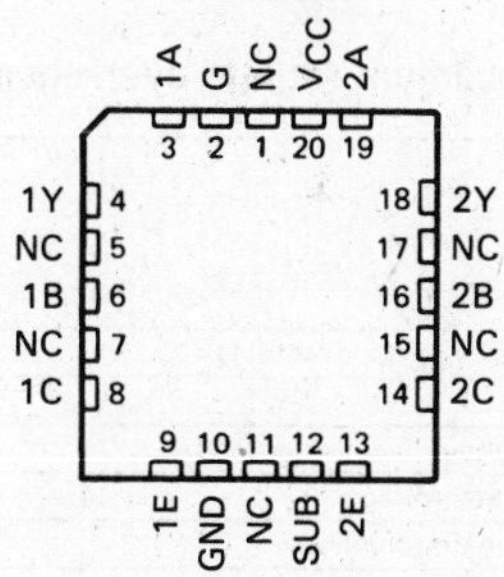
1A G NC VCC 2A
3 2 1 20 19
1Y 4
NC 5
1B 6
NC 7
1C 8
18 2Y
17 NC
16 2B
15 NC
14 2C
9 10 11 12 13
1E GND NC SUB 2E

SUMMARY OF SERIES 55450B/75451B

DEVICE	LOGIC OF COMPLETE CIRCUIT	PACKAGES
SN55450B	AND†	FK,J
SN55451B	AND	FK,JG
SN55452B	NAND	FK,JG
SN55453B	OR	FK,JG
SN55454B	NOR	FK,JG
SN75451B	AND	D,P
SN75452B	NAND	D,P
SN75453B	OR	D,P
SN75454B	NOR	D,P

†With output transistor base connected externally to output of gate.

SN55451B, SN55452B,
SN55453B, SN55454B . . . JG PACKAGE
SN75451B, SN75452B,
SN75453B, SN75454B . . . D OR P PACKAGE
(TOP VIEW)

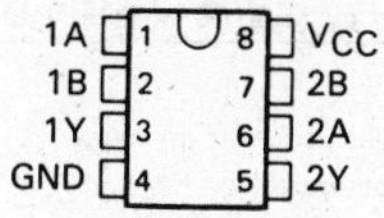
1A 1
1B 2
1Y 3
GND 4
8 VCC
7 2B
6 2A
5 2Y

description

Series 55450B/75451B dual peripheral drivers are a family of versatile devices designed for use in systems that employ TTL logic. This family is functionally interchangeable with and replaces the 75450 family and the 75450A family devices manufactured previously. The speed of the 55450B/75451B family is equal to that of the 75450 family, and the parts have been designed to ensure freedom from latch-up. Diode-clamped inputs simplify circuit design.

SN55451B, SN55452B,
SN55453B, SN55454B, . . . FK PACKAGE
(TOP VIEW)

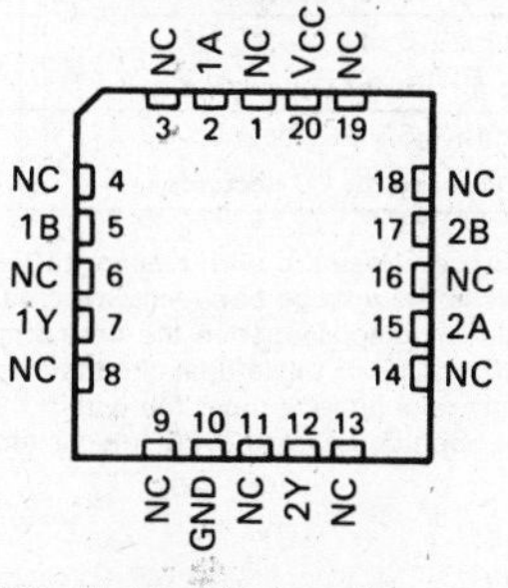
NC 1A NC VCC NC
3 2 1 20 19
NC 4
1B 5
NC 6
1Y 7
NC 8
18 NC
17 2B
16 NC
15 2A
14 NC
9 10 11 12 13
NC GND NC 2Y NC

NC—No internal connection

5

Peripheral Drivers/Actuators

TEXAS INSTRUMENTS
POST OFFICE BOX 655012 • DALLAS, TEXAS 75265

description (cont'd)

Typical applications include high-speed logic buffers, power drivers, relay drivers, lamp drivers. MOS drivers, line drivers, and memory drivers.

The SN55450B is a unique general-purpose device, featuring two standard Series 54 TTL gates and two uncommitted, high-current, high-voltage n-p-n transistors. The device offers the system designer the flexibility of tailoring the circuit to the application.

The SN55451B/SN75451B, SN55452B/SN75452B, SN55453B/SN75453B, and SN55454B/SN75454B are dual peripheral AND, NAND, OR, and NOR drivers, respectively, (assuming positive logic), with the output of the logic gates internally connected to the bases of the n-p-n output transistors.

Series 55450B drivers are characterized for operation over the full military range of −55°C to 125°C. Series 75451B drivers are characterized for operation from 0°C to 70°C.

absolute maximum ratings over operating free-air temperature range (unless otherwise noted)

		SN55450B	SN55451B SN55452B SN55453B SN55454B	SN75451B SN75452B SN75453B SN75454B	UNIT
Supply voltage, V_{CC} (see Note 1)		7	7	7	V
Input voltage		5.5	5.5	5.5	V
Interemitter voltage (see Note 2)		5.5	5.5	5.5	V
V_{CC}-to-substrate voltage		35			V
Collector-to-substrate voltage		35			V
Collector-base voltage		35			V
Collector-emitter voltage (see Note 3)		30			V
Emitter-base voltage		5			V
Off-state output voltage			30	30	V
Continuous collector or output current (see Note 4)		400	400	400	mA
Peak collector or output current ($t_w \leq$ 10 ms, duty cycle ≤ 50%, (see Note 4)		500	500	500	mA
Continuous total dissipation at (or below) 25°C free-air temperature (see Note 5)	D package			725	mW
	FK package	1375	1375		mW
	J package	1375			mW
	JG package		1050		mW
	P package			1200	mW
Operating free-air temperature range		−55 to 125	−55 to 125	0 to 70	°C
Storage temperature range		−65 to 150	−65 to 150	−65 to 150	°C
Case temperature for 60 seconds	FK package	260	260		°C
Lead temperature 1,6 mm (1/16 inch) from case for 60 seconds	J or JG package	300	300		°C
Lead temperature 1,6 mm (1/16 inch) from case for 10 seconds	D or P package			260	°C

NOTES: 1. Voltage values are with respect to the network ground terminal unless otherwise specified.
2. This is the voltage between two emitters of a multiple-emitter transistor.
3. This value applies when the base-emitter resistance (R_{BE}) is equal to or less than 500 Ω.
4. Both halves of these dual circuits may conduct rated current simultaneously; however, power dissipation averaged over a short time interval must fall within the continuous dissipation rating.
5. For operation above 25°C free-air temperature, refer to the Dissipation Derating Table.

TEXAS INSTRUMENTS
POST OFFICE BOX 655012 • DALLAS, TEXAS 75265

DISSIPATION DERATING TABLE

PACKAGE	POWER RATING	DERATING FACTOR	ABOVE T_A
D	725 mW	5.8 mW/°C	25°C
FK	1375 mW	11.0 mW/°C	25°C
J	1375 mW	11.0 mW/°C	25°C
JG	1050 mW	8.4 mW/°C	25°C
P	1200 mW	9.6 mW/°C	25°C

recommended operating conditions (see Note 6)

	SERIES 55450B			SERIES 75451B			UNIT
	MIN	NOM	MAX	MIN	NOM	MAX	
Supply voltage, V_{CC}	4.5	5	5.5	4.75	5	5.25	V
High-level input voltage, V_{IH}	2.2			2			V
Low-level input voltage, V_{IL}			0.8			0.8	V
Operating free-air temperature, T_A	−55		125	0		70	°C

NOTE 6: For the SN55450B only, the substrate (pin 8) must always be at the most negative device voltage for proper operation.

logic symbol†

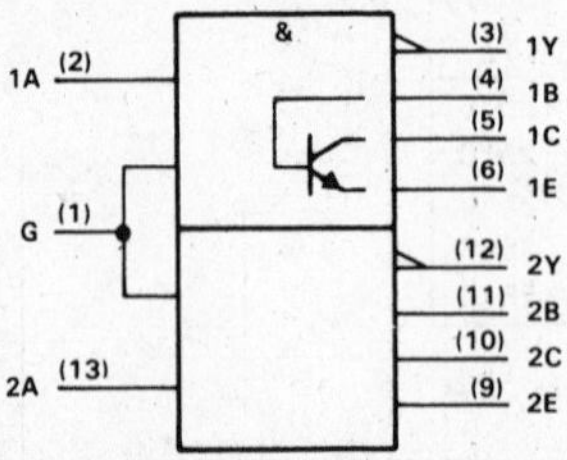
&
1A
(2)
G
(1)
2A
(13)
(3) 1Y
(4) 1B
(5) 1C
(6) 1E
(12) 2Y
(11) 2B
(10) 2C
(9) 2E

†This symbol is in accordance with ANSI/IEEE Std 91-1984 and IEC Publication 617-12.

logic diagram (positive logic)

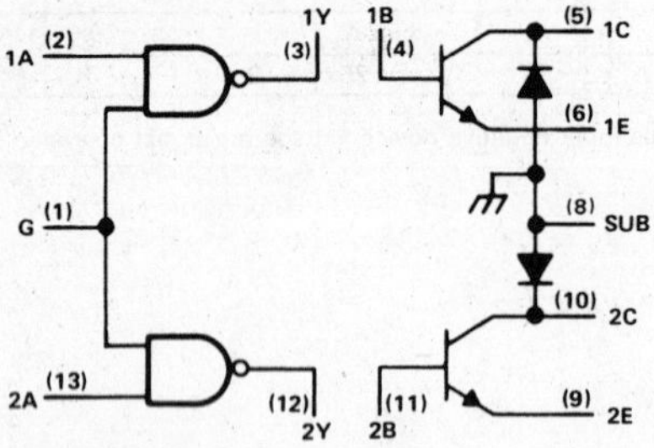
1A (2)
G (1)
2A (13)
1Y (3)
1B (4)
(5) 1C
(6) 1E
(8) SUB
(10) 2C
(12) 2Y
(11) 2B
(9) 2E

positive logic:

Y = $\overline{AG}$ or $\overline{A}+\overline{G}$ (gate only)

C = AG or $\overline{\overline{A}+\overline{G}}$ (gate and transistor)

Pin numbers shown are for the J package.

schematic

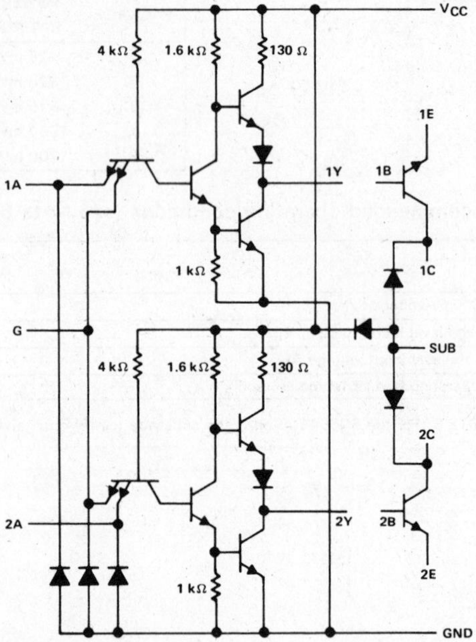
VCC
4 kΩ
1.6 kΩ
130 Ω
1A
1Y
1B
1E
1C
1 kΩ
G
SUB
2C
2A
2Y
2B
2E
GND

Resistor values shown are nominal.

electrical characteristics over recommended operating free-air temperature range (unless otherwise noted)

TTL gates

	PARAMETER		TEST CONDITIONS‡		SN55450B MIN	SN55450B TYP§	SN55450B MAX	UNIT
V_{IK}	Input clamp voltage		V_{CC} = MIN,	I_I = −12 mA		−1.2	−1.5	V
V_{OH}	High-level output voltage		V_{CC} = MIN, I_{OH} = −400 μA	V_{IL} = 0.8 V,	2.4	3.3		V
V_{OL}	Low-level output voltage		V_{CC} = MIN, I_{OL} = 16 mA	V_{IH} = MIN,		0.25	0.5	V
I_I	Input current at maximum input voltage	input A	V_{CC} = MAX,	V_I = 5.5 V			1	mA
		input G					2	
I_{IH}	High-level input current	input A	V_{CC} = MAX,	V_I = 2.4 V			40	μA
		input G					80	
I_{IL}	Low-level input current	input A	V_{CC} = MAX,	V_I = 0.4 V			−1.6	mA
		input G					−3.2	
I_{OS}	Short-circuit output current¶		V_{CC} = MAX,	V_O = 0	−18	−35	−55	mA
I_{CCH}	Supply current, outputs high		V_{CC} = MAX,	V_I = 0		2.8	4	mA
I_{CCL}	Supply current, outputs low		V_{CC} = MAX,	V_I = 5 V		7	11	mA

‡For conditions shown as MIN or MAX, use the appropriate value specified under recommended operating conditions.

§All typical values are at V_{CC} = 5 V, T_A = 25°C.

¶Not more than one output should be shorted at a time.

TEXAS INSTRUMENTS
POST OFFICE BOX 655012 • DALLAS, TEXAS 75265

electrical characteristics over recommended operating free-air temperature range (unless otherwise noted)

output transistors

PARAMETER		TEST CONDITIONS†		SN55450B MIN	TYP‡	MAX	UNIT
$V_{(BR)CBO}$	Collector-base breakdown voltage	I_C = 100 μA, I_E = 0		35			V
$V_{(BR)CER}$	Collector-emitter breakdown voltage	I_C = 100 μA, R_{BE} = 500 Ω		30			V
$V_{(BR)EBO}$	Emitter-base breakdown voltage	I_E = 100 μA, I_C = 0		5			V
h_{FE}	Static forward current transfer ratio	V_{CE} = 3 V, T_A = 25°C See Note 7	I_C = 100 mA	25			
			I_C = 300 mA	30			
		V_{CE} = 3 V, T_A = MIN, See Note 7	I_C = 100 mA	10			
			I_C = 300 mA	15			
V_{BE}	Base-emitter voltage	I_B = 10 mA, I_C = 100 mA, See Note 7			0.85	1.2	V
		I_B = 30 mA, I_C = 300 mA, See Note 7			1	1.4	
$V_{CE(sat)}$	Collector-emitter saturation voltage	I_B = 10 mA, I_C = 100 mA, See Note 7			0.25	0.5	V
		I_B = 30 mA, I_C = 300 mA, See Note 7			0.45	0.8	

† For conditions shown as MIN or MAX, use the appropriate value specified under recommended operating conditions.
‡ All typical values are at V_{CC} = 5 V, T_A = 25°C.
NOTE 7: These parameters must be measured using pulse techniques. t_w = 300 μs, duty cycle ≤ 2%.

switching characteristics, V_{CC} = 5 V, T_A = 25°C

TTL gates

PARAMETER		TEST CONDITIONS	MIN	TYP	MAX	UNIT
t_{PLH}	Propagation delay time, low-to-high-level output	C_L = 15 pF, R_L = 400 Ω, See Figure 1		12		ns
t_{PHL}	Propagation delay time, high-to-low-level output			8		ns

output transistors

PARAMETER		TEST CONDITIONS§	MIN	TYP	MAX	UNIT
t_d	Delay time	I_C = 200 mA, $I_{B(1)}$ = 20 mA, $I_{B(2)}$ = −40 mA, $V_{BE(off)}$ = −1 V, C_L = 15 pF, R_L = 50 Ω, See Figure 2		8		ns
t_r	Rise time			12		ns
t_s	Storage time			7		ns
t_f	Fall time			6		ns

§ Voltage and current values shown are nominal; exact values vary slightly with transistor parameters.

gate and transistors combined

PARAMETER		TEST CONDITIONS	MIN	TYP	MAX	UNIT
t_{PLH}	Propagation delay time, low-to-high-level output	I_C ≈ 200 mA, C_L = 15 pF, R_L = 50 Ω, See Figure 3		20	30	ns
t_{PHL}	Propagation delay time, high-to-low-level output			20	30	ns
t_{TLH}	Transition time, low-to-high-level output			7	12	ns
t_{THL}	Transition time, high-to-low-level output			9	15	ns
V_{OH}	High-level output voltage after switching	V_S = 20 V, I_C ≈ 300 mA, R_{BE} = 500 Ω, See Figure 4	V_S−6.5			mV

logic symbol†

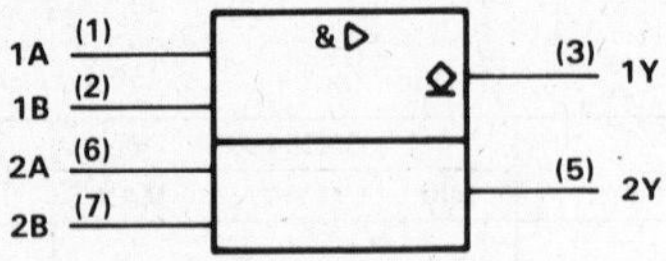
1A
1B
2A
2B
(1)
(2)
(6)
(7)
&
(3)
1Y
(5)
2Y

†This symbol is in accordance with ANSI/IEEE STD 91-1984 and IEC Publication 617-12.

logic diagram (positive logic)

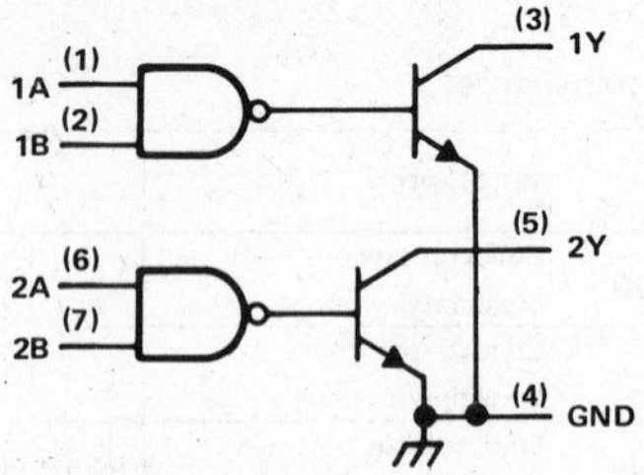
1A
(1)
(2)
1B
(3)
1Y
(5)
2Y
2A
(6)
(7)
2B
(4)
GND

FUNCTION TABLE
(EACH DRIVER)

A	B	Y
L	L	L (on state)
L	H	L (on state)
H	L	L (on state)
H	H	H (off state)

positive logic:
$Y = \overline{AB}$ or $\overline{A} + \overline{B}$

schematic (each driver)

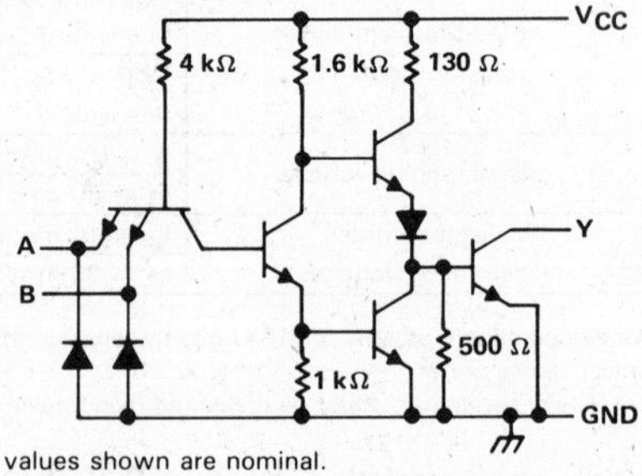
VCC
4 kΩ
1.6 kΩ
130 Ω
A
B
Y
500 Ω
1 kΩ
GND

Pin numbers shown are for D, JG, and P packages.

Resistor values shown are nominal.

electrical characteristics over recommended operating free-air temperature range (unless otherwise noted)

PARAMETER		TEST CONDITIONS‡		SN55451B			SN75451B			UNIT
				MIN	TYP§	MAX	MIN	TYP§	MAX	
V_{IK}	Input clamp voltage	V_{CC} = MIN,	I_I = −12 mA		−1.2	−1.5		−1.2	−1.5	V
I_{OH}	High-level output current	V_{CC} = MIN, V_{OH} = 30 V	V_{IH} = MIN,			300			100	μA
V_{OL}	Low-level output voltage	V_{CC} = MIN, I_{OL} = 100 mA	V_{IL} = 0.8 V,		0.25	0.5		0.25	0.4	V
		V_{CC} = MIN, I_{OL} = 300 mA	V_{IL} = 0.8 V,		0.5	0.8		0.5	0.7	
I_I	Input current at maximum input voltage	V_{CC} = MAX,	V_I = 5.5 V			1			1	mA
I_{IH}	High-level input current	V_{CC} = MAX,	V_I = 2.4 V			40			40	μA
I_{IL}	Low-level input current	V_{CC} = MAX,	V_I = 0.4 V		−1	−1.6		−1	−1.6	mA
I_{CCH}	Supply current, outputs high	V_{CC} = MAX,	V_I = 5 V		7	11		7	11	mA
I_{CCL}	Supply current, outputs low	V_{CC} = MAX,	V_I = 0		52	65		52	65	mA

‡For conditions shown as MIN or MAX, use the appropriate value specified under recommended operating conditons.
§All typical values are at V_{CC} = 5 V, T_A = 25°C.

switching characteristics, V_{CC} = 5 V, T_A = 25°C

PARAMETER		TEST CONDITIONS		MIN	TYP	MAX	UNIT
t_{PLH}	Propagation delay time, low-to-high-level output	$I_O \approx$ 200 mA, R_L = 50 Ω,	C_L = 15 pF, See Figure 3		18	25	ns
t_{PHL}	Propagation delay time, high-to-low-level output				18	25	ns
t_{TLH}	Transition time, low-to-high-level output				5	8	ns
t_{THL}	Transition time, high-to-low-level output				7	12	ns
V_{OH}	High-level output voltage after switching	V_S = 20 V, See Figure 4	$I_O \approx$ 300 mA,	V_S − 6.5			mV

TEXAS INSTRUMENTS
POST OFFICE BOX 655012 • DALLAS, TEXAS 75265

logic symbol†

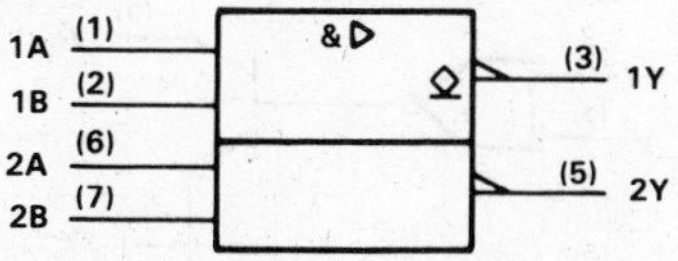
1A
(1)
1B
(2)
2A
(6)
2B
(7)
&▷
(3)
1Y
(5)
2Y

†This symbol is in accordance with ANSI/IEEE STD 91-1984 and IEC Publication 617-12.

FUNCTION TABLE (EACH DRIVER)

A	B	Y
L	L	H (off state)
L	H	H (off state)
H	L	H (off state)
H	H	L (on state)

positive logic:
$Y = \overline{AB}$ or $\overline{A} + \overline{B}$

Pin numbers shown are for D, JG, and P packages.

logic diagram (positive logic)

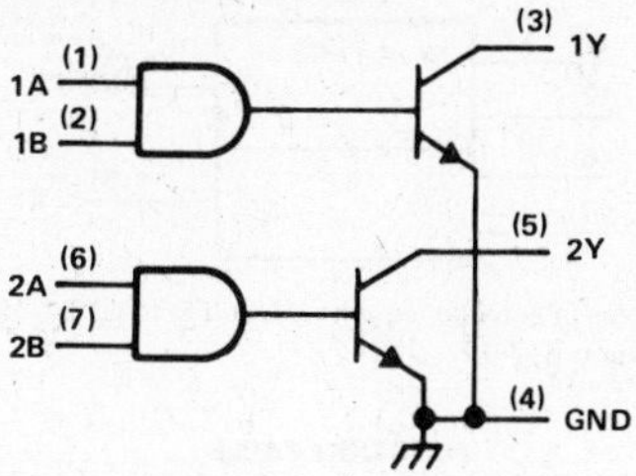
1A
(1)
1B
(2)
(3)
1Y
2A
(6)
2B
(7)
(5)
2Y
(4)
GND

schematic (each driver)

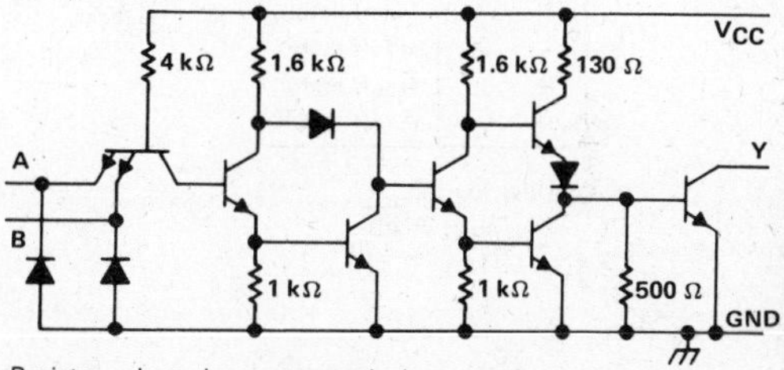
VCC
4 kΩ
1.6 kΩ
1.6 kΩ
130 Ω
A
B
Y
1 kΩ
1 kΩ
500 Ω
GND

Resistor values shown are nominal.

electrical characteristics over recommended operating free-air temperature range (unless otherwise noted)

PARAMETER		TEST CONDITIONS‡		SN55452B MIN	SN55452B TYP§	SN55452B MAX	SN75452B MIN	SN75452B TYP§	SN75452B MAX	UNIT
V_{IK}	Input clamp voltage	V_{CC} = MIN,	I_I = −12 mA		−1.2	−1.5		−1.2	−1.5	V
I_{OH}	High-level output current	V_{CC} = MIN, V_{OH} = 30 V	V_{IL} = 0.8 V,			300			100	μA
V_{OL}	Low-level output voltage	V_{CC} = MIN, I_{OL} = 100 mA	V_{IH} = MIN,		0.25	0.5		0.25	0.4	V
		V_{CC} = MIN, I_{OL} = 300 mA	V_{IH} = MIN,		0.5	0.8		0.5	0.7	
I_I	Input current at maximum input voltage	V_{CC} = MAX,	V_I = 5.5 V			1			1	mA
I_{IH}	High-level input current	V_{CC} = MAX,	V_I = 2.4 V			40			40	μA
I_{IL}	Low-level input current	V_{CC} = MAX,	V_I = 0.4 V		−1.1	−1.6		−1.1	−1.6	mA
I_{CCH}	Supply current, outputs high	V_{CC} = MAX,	V_I = 0		11	14		11	14	mA
I_{CCL}	Supply current, outputs low	V_{CC} = MAX,	V_I = 5 V		56	71		56	71	mA

‡For conditions shown as MIN or MAX, use the appropriate value specified under recommended operating conditions.
§All typical values are at V_{CC} = 5 V, T_A = 25°C.

switching characteristics, V_{CC} = 5 V, T_A = 25°C

PARAMETER		TEST CONDITIONS		MIN	TYP	MAX	UNIT
t_{PLH}	Propagation delay time, low-to-high-level output	I_O ≈ 200 mA, R_L = 50 Ω,	C_L = 15 pF, See Figure 3		26	35	ns
t_{PHL}	Propagation delay time, high-to-low-level output				24	35	ns
t_{TLH}	Transition time, low-to-high-level output				5	8	ns
t_{THL}	Transition time, high-to-low-level output				7	12	ns
V_{OH}	High-level output voltage after switching	V_S = 20 V, See Figure 4	I_O ≈ 300 mA,	V_S−6.5			mV

logic symbol†

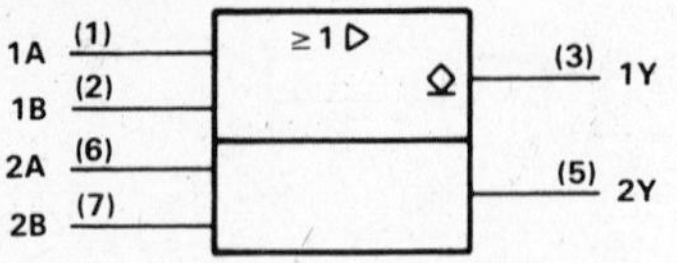
≥1
1A
1B
2A
2B
(1)
(2)
(6)
(7)
(3)
(5)
1Y
2Y

†This symbol is in accordance with ANSI/IEEE STD 91-1984 and IEC Publication 617-12.

logic diagram (positive logic)

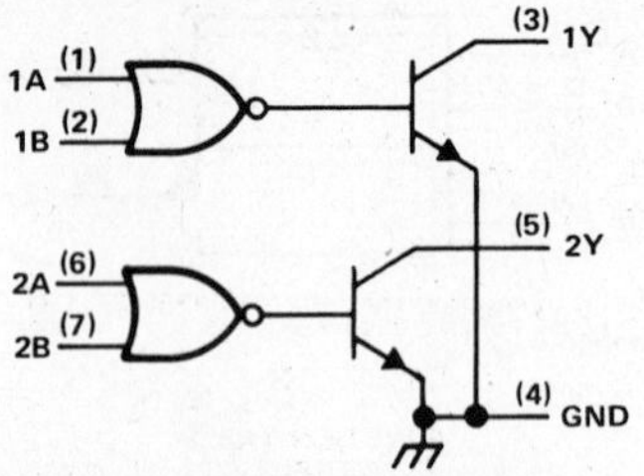
1A
1B
2A
2B
(1)
(2)
(6)
(7)
(3) 1Y
(5) 2Y
(4) GND

FUNCTION TABLE
(EACH DRIVER)

A	B	Y
L	L	L (on state)
L	H	H (off state)
H	L	H (off state)
H	H	H (off state)

positive logic:
$Y = A + B$ or $\overline{\overline{A}\overline{B}}$

schematic (each driver)

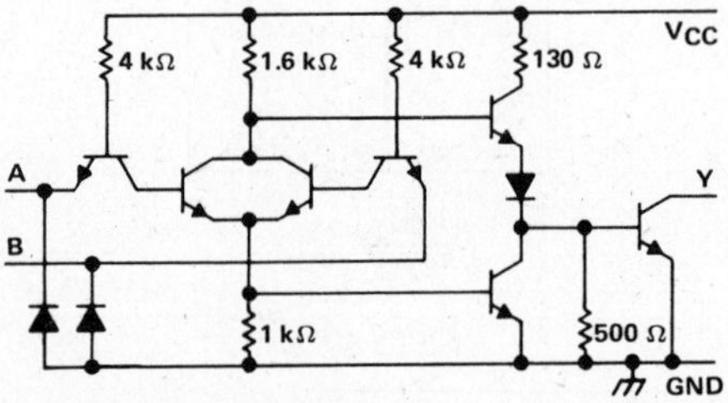
4 kΩ
1.6 kΩ
4 kΩ
130 Ω
VCC
A
B
Y
1 kΩ
500 Ω
GND

Pin numbers shown are for D, JG, and P packages.

Resistor values shown are nominal.

electrical characteristics over recommended operating free-air temperature range (unless otherwise noted)

PARAMETER		TEST CONDITIONS‡	SN55453B			SN75453B			UNIT
			MIN	TYP§	MAX	MIN	TYP§	MAX	
V_{IK}	Input clamp voltage	V_{CC} = MIN, I_I = −12 mA		−1.2	−1.5		−1.2	−1.5	V
I_{OH}	High-level output current	V_{CC} = MIN, V_{IH} = MIN, V_{OH} = 30 V			300			100	μA
V_{OL}	Low-level output voltage	V_{CC} = MIN, V_{IL} = 0.8 V, I_{OL} = 100 mA		0.25	0.5		0.25	0.4	V
		V_{CC} = MIN, V_{IL} = 0.8 V, I_{OL} = 300 mA		0.5	0.8		0.5	0.7	
I_I	Input current at maximum input voltage	V_{CC} = MAX, V_I = 5.5 V			1			1	mA
I_{IH}	High-level input current	V_{CC} = MAX, V_I = 2.4 V			40			40	μA
I_{IL}	Low-level input current	V_{CC} = MAX, V_I = 0.4 V		−1	−1.6		−1	−1.6	mA
I_{CCH}	Supply current, outputs high	V_{CC} = MAX, V_I = 5 V		8	11		8	11	mA
I_{CCL}	Supply current, outputs low	V_{CC} = MAX, V_I = 0		54	68		54	68	mA

‡For conditions shown as MIN or MAX, use the appropriate value specified under recommended operating conditions.
§All typical values are at V_{CC} = 5 V, T_A = 25°C.

switching characteristics, V_{CC} = 5 V, T_A = 25°C

PARAMETER		TEST CONDITIONS	MIN	TYP	MAX	UNIT
t_{PLH}	Propagation delay time, low-to-high-level output	$I_O \approx$ 200 mA, C_L = 15 pF, R_L = 50 Ω, See Figure 3		18	25	ns
t_{PHL}	Propagation delay time, high-to-low-level output			16	25	ns
t_{TLH}	Transition time, low-to-high-level output			5	8	ns
t_{THL}	Transition time, high-to-low-level output			7	12	ns
V_{OH}	High-level output voltage after switching	V_S = 20 V, $I_O \approx$ 300 mA, See Figure 4	V_S−6.5			mV

TEXAS INSTRUMENTS
POST OFFICE BOX 655012 • DALLAS, TEXAS 75265

logic symbol†

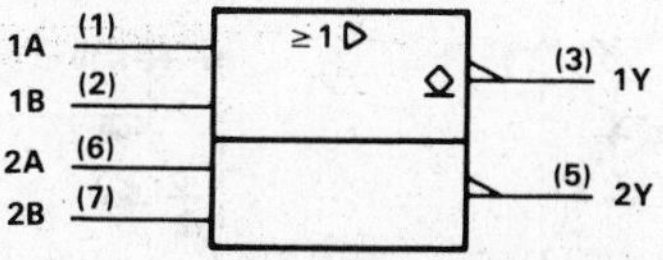
≥1
1A (1)
1B (2)
2A (6)
2B (7)
(3) 1Y
(5) 2Y

†This symbol is in accordance with ANSI/IEEE STD 91-1984 and IEC Publication 617-12.

logic diagram (positive logic)

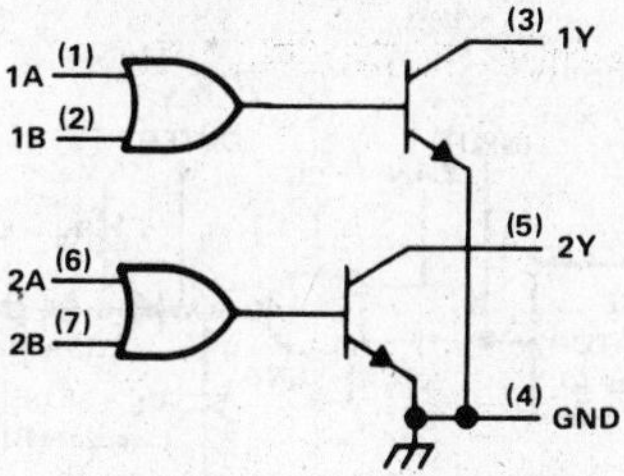
1A (1)
1B (2)
(3) 1Y
(5) 2Y
2A (6)
2B (7)
(4) GND

FUNCTION TABLE (EACH DRIVER)

A	B	Y
L	L	H (off state)
L	H	L (on state)
H	L	L (on state)
H	H	L (on state)

positive logic:
$Y = \overline{A+B}$ or $\overline{A}\,\overline{B}$

Pin numbers shown are for D, JG, and P packages.

schematic (each driver)

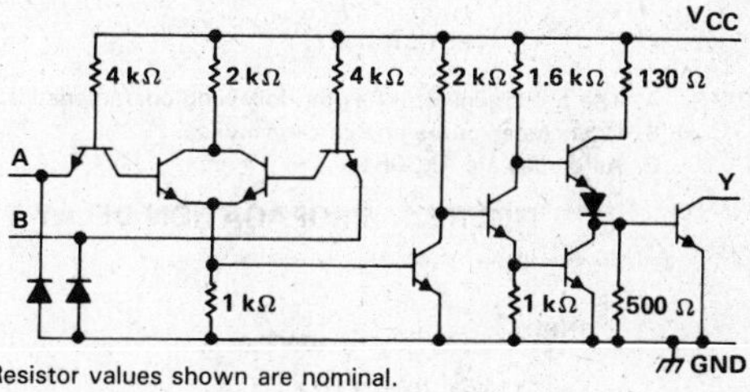
VCC
4 kΩ
2 kΩ
4 kΩ
2 kΩ
1.6 kΩ
130 Ω
A
B
Y
1 kΩ
1 kΩ
500 Ω
GND

Resistor values shown are nominal.

electrical characteristics over recommended operating free-air temperature range (unless otherwise noted)

PARAMETER		TEST CONDITIONS‡	SN55454B MIN	SN55454B TYP§	SN55454B MAX	SN75454B MIN	SN75454B TYP§	SN75454B MAX	UNIT
V_{IK}	Input clamp voltage	V_{CC} = MIN, I_I = −12 mA		−1.2	−1.5		−1.2	−1.5	V
I_{OH}	High-level output current	V_{CC} = MIN, V_{IL} = 0.8 V, V_{OH} = 30 V			300			100	µA
V_{OL}	Low-level output voltage	V_{CC} = MIN, V_{IH} = MIN, I_{OL} = 100 mA		0.25	0.5		0.25	0.4	V
		V_{CC} = MIN, V_{IH} = MIN, I_{OL} = 300 mA		0.5	0.8		0.5	0.7	
I_I	Input current at maximum input voltage	V_{CC} = MAX, V_I = 5.5 V			1			1	mA
I_{IH}	High-level input current	V_{CC} = MAX, V_I = 2.4 V			40			40	µA
I_{IL}	Low-level input current	V_{CC} = MAX, V_I = 0.4 V		−1	−1.6		−1	−1.6	mA
I_{CCH}	Supply current, outputs high	V_{CC} = MAX, V_I = 0		13	17		13	17	mA
I_{CCL}	Supply current, outputs low	V_{CC} = MAX, V_I = 5 V		61	79		61	79	mA

‡For conditions shown as MIN or MAX, use the appropriate value specified under recommended operating conditions.
§All typical values are at V_{CC} = 5 V, T_A = 25°C.

switching characteristics, V_{CC} = 5 V, T_A = 25°C

PARAMETER		TEST CONDITIONS	MIN	TYP	MAX	UNIT
t_{PLH}	Propagation delay time, low-to-high-level output	I_O ≈ 200 mA, C_L = 15 pF, R_L = 50 Ω, See Figure 3		27	35	ns
t_{PHL}	Propagation delay time, high-to-low-level output			24	35	ns
t_{TLH}	Transition time, low-to-high-level output			5	8	ns
t_{THL}	Transition time, high-to-low-level output			7	12	ns
V_{OH}	High-level output voltage after switching	V_S = 20 V, I_O ≈ 300 mA, See Figure 4	V_S−6.5			mV

PARAMETER MEASUREMENT INFORMATION

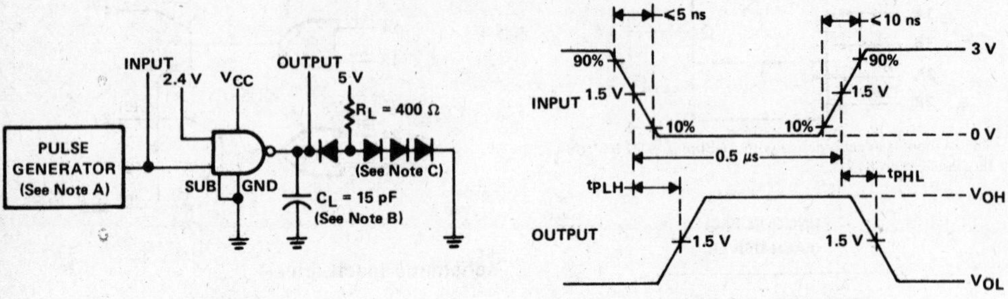
INPUT
2.4 V
VCC
OUTPUT
5 V
RL = 400 Ω
PULSE GENERATOR (See Note A)
SUB
GND
(See Note C)
CL = 15 pF (See Note B)
<5 ns
<10 ns
3 V
90%
1.5 V
10%
0 V
0.5 μs
tPLH
tPHL
VOH
VOL
OUTPUT

TEST CIRCUIT

VOLTAGE WAVEFORMS

NOTES: A. The pulse generator has the following characteristics: PRR ≤ 1 MHz, Z_{out} ≈ 50 Ω.
B. C_L includes probe and jig capacitance.
C. All diodes are 1N3064.

FIGURE 1. PROPAGATION DELAY TIMES, EACH GATE (SN55450B ONLY)

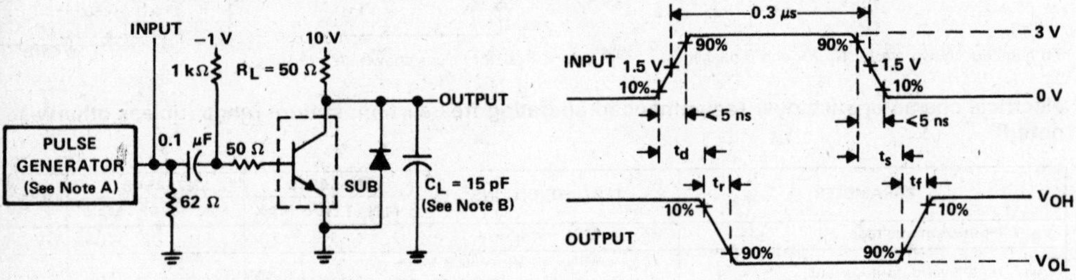
INPUT
−1 V
10 V
1 kΩ
RL = 50 Ω
OUTPUT
PULSE GENERATOR (See Note A)
0.1 μF
50 Ω
62 Ω
SUB
CL = 15 pF (See Note B)
0.3 μs
3 V
90%
1.5 V
10%
0 V
<5 ns
td
ts
tr
tf
VOH
VOL

TEST CIRCUIT

VOLTAGE WAVEFORMS

NOTES: A. The pulse generator has the following characteristics: duty cycle ≤ 1%, Z_{out} ≈ 50 Ω.
B. C_L includes probe and jig capacitance.

FIGURE 2. SWITCHING TIMES, EACH TRANSISTOR (SN55450B ONLY)

TEXAS INSTRUMENTS
POST OFFICE BOX 655012 • DALLAS, TEXAS 75265

PARAMETER MEASUREMENT INFORMATION

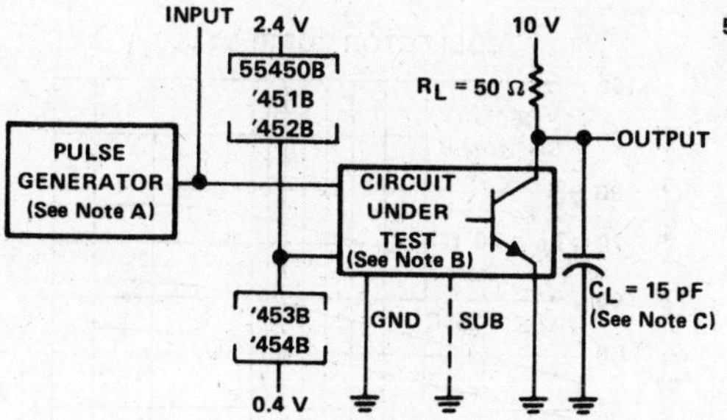
INPUT
2.4 V
55450B
'451B
'452B
PULSE
GENERATOR
(See Note A)
10 V
R_L = 50 Ω
OUTPUT
CIRCUIT
UNDER
TEST
(See Note B)
'453B
'454B
0.4 V
GND
SUB
C_L = 15 pF
(See Note C)

TEST CIRCUIT

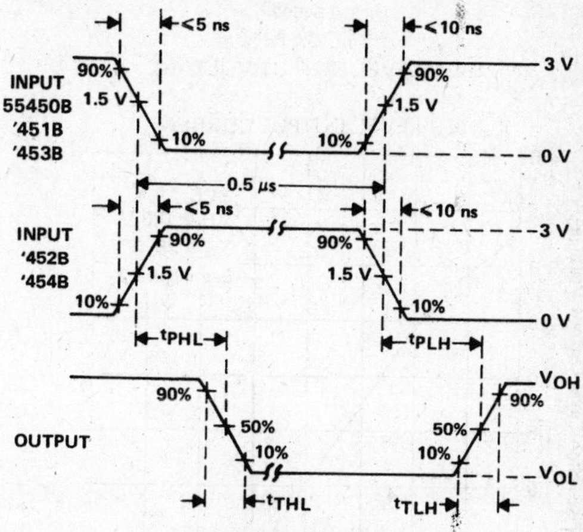
<5 ns
<10 ns
INPUT
55450B
'451B
'453B
90%
1.5 V
10%
3 V
0 V
0.5 μs
INPUT
'452B
'454B
tPHL
tPLH
OUTPUT
50%
VOH
VOL
tTHL
tTLH

VOLTAGE WAVEFORMS

NOTES: A. The pulse generator has the following characteristics: PRR ≤ 1 MHZ, Z_{out} ≈ 50 Ω.
B. When testing SN55450B, connect output Y to transistor base and ground the substrate terminal.
C. C_L includes probe and jig capacitance.

FIGURE 3. SWITCHING TIMES OF COMPLETE DRIVERS

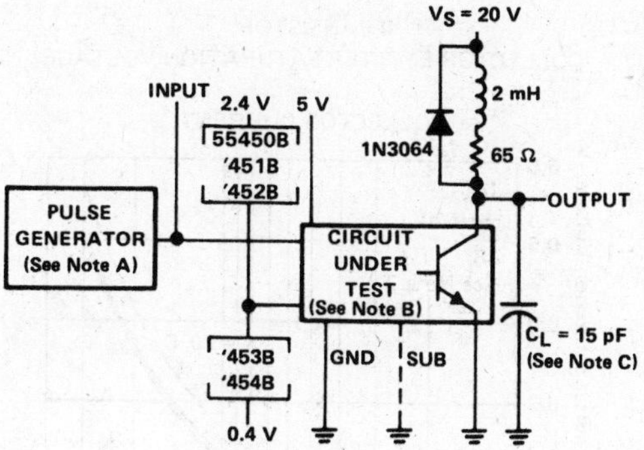
V_S = 20 V
INPUT
2.4 V
5 V
55450B
'451B
'452B
2 mH
1N3064
65 Ω
OUTPUT
PULSE
GENERATOR
(See Note A)
CIRCUIT
UNDER
TEST
(See Note B)
'453B
'454B
0.4 V
GND
SUB
C_L = 15 pF
(See Note C)

TEST CIRCUIT

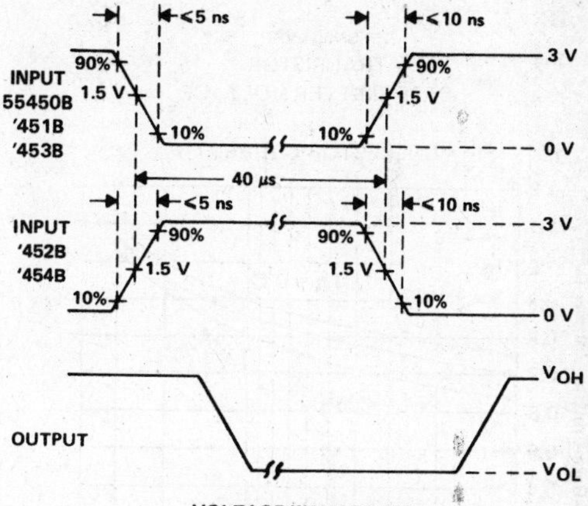
<5 ns
<10 ns
INPUT
55450B
'451B
'453B
90%
1.5 V
10%
3 V
0 V
40 μs
INPUT
'452B
'454B
OUTPUT
VOH
VOL

VOLTAGE WAVEFORMS

NOTES: A. The pulse generator has the following characteristics: PRR ≤ 12.5 kHz, Z_{out} = 50 Ω.
B. When testing SN55450B, connect output Y to transistor base with a 500-Ω resistor from there to ground, and ground the substrate terminal.
C. C_L includes probe and jig capacitance.

FIGURE 4. LATCH-UP TEST OF COMPLETE DRIVERS

TYPICAL CHARACTERISTICS

SN55450B
TTL GATE
HIGH-LEVEL OUTPUT VOLTAGE
vs
HIGH-LEVEL OUTPUT CURRENT

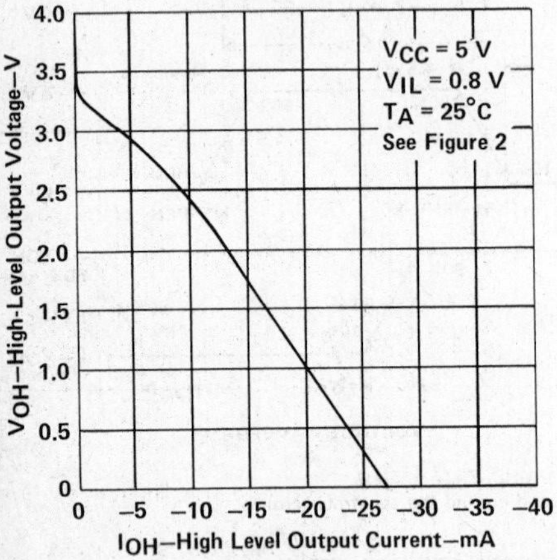
VCC = 5 V
VIL = 0.8 V
TA = 25°C
See Figure 2
VOH–High-Level Output Voltage–V
4.0
3.5
3.0
2.5
2.0
1.5
1.0
0.5
0
0 –5 –10 –15 –20 –25 –30 –35 –40
IOH–High Level Output Current–mA

FIGURE 5

SN55450B
TRANSISTOR
STATIC FORWARD CURRENT TRANSFER RATIO
vs
COLLECTOR CURRENT

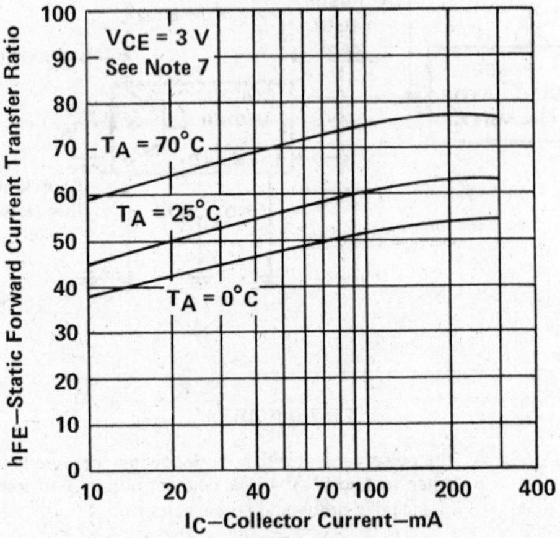
VCE = 3 V
See Note 7
TA = 70°C
TA = 25°C
TA = 0°C
hFE–Static Forward Current Transfer Ratio
100
90
80
70
60
50
40
30
20
10
0
10 20 40 70 100 200 400
IC–Collector Current–mA

FIGURE 6

SN55450B
TRANSISTOR
BASE-EMITTER VOLTAGE
vs
COLLECTOR CURRENT

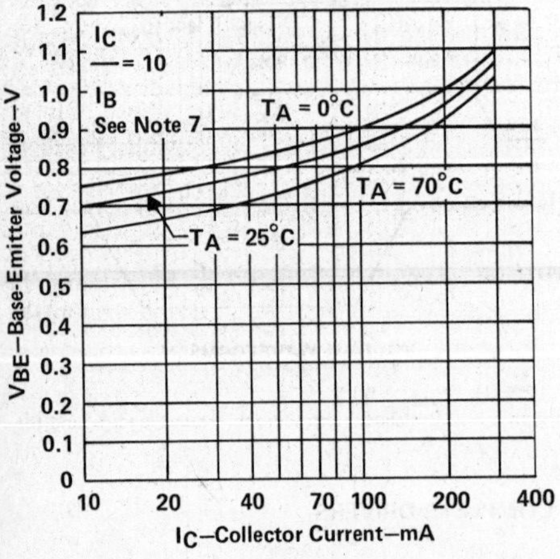
IC/IB = 10
See Note 7
TA = 0°C
TA = 70°C
TA = 25°C
VBE–Base-Emitter Voltage–V
1.2
1.1
1.0
0.9
0.8
0.7
0.6
0.5
0.4
0.3
0.2
0.1
0
10 20 40 70 100 200 400
IC–Collector Current–mA

FIGURE 7

TRANSISTOR
COLLECTOR-EMITTER SATURATION VOLTAGE
vs
COLLECTOR CURRENT

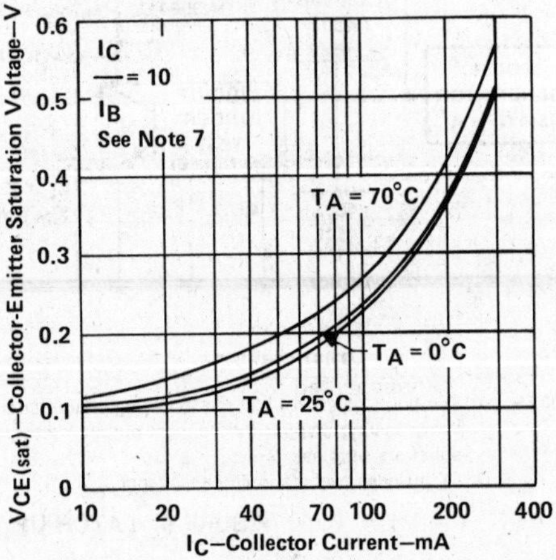
IC/IB = 10
See Note 7
TA = 70°C
TA = 0°C
TA = 25°C
VCE(sat)–Collector-Emitter Saturation Voltage–V
0.6
0.5
0.4
0.3
0.2
0.1
0
10 20 40 70 100 200 400
IC–Collector Current–mA

FIGURE 8

NOTE 7: These parameters must be measured using pulse techniques, t_W = 300 μs, duty cycle ≤2%.

D2218, DECEMBER 1976—REVISED SEPTEMBER 1986

PERIPHERAL DRIVERS FOR HIGH-VOLTAGE, HIGH-CURRENT DRIVER APPLICATIONS

- Characterized for Use to 300 mA
- High-Voltage Outputs
- No Output Latch-Up at 30 V (After Conducting 300 mA)
- Medium-Speed Switching
- Circuit Flexibility for Varied Applications and Choice of Logic Function
- TTL-Compatible Diode-Clamped Inputs
- Standard Supply Voltages
- New Plastic DIP (P) with Copper Lead Frame for Cooler Operation and Improved Reliability
- Package Options Include Plastic "Small Outline" Packages, Ceramic Chip Carriers, and Standard Plastic and Ceramic 300-mil DIPs

SN55461, SN55462,
SN55463, SN55464 . . . JG PACKAGE
SN75461, SN75462,
SN75463 . . . D OR P PACKAGE
(TOP VIEW)

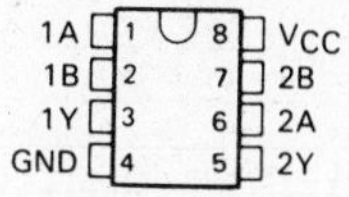
1A
1B
1Y
GND
1
2
3
4
8
7
6
5
VCC
2B
2A
2Y

SN55461, SN55462,
SN55463, SN55464, . . . FK PACKAGE
(TOP VIEW)

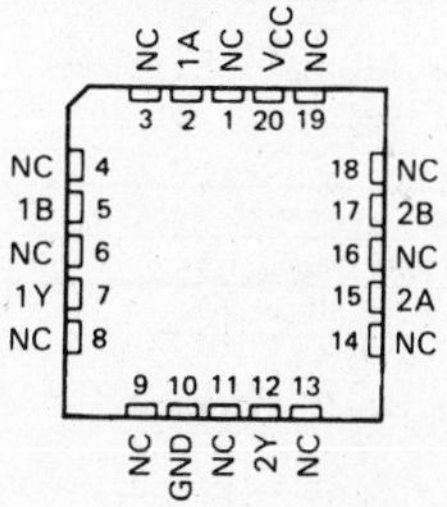
NC
1A
NC
VCC
NC
3 2 1 20 19
NC 4
1B 5
NC 6
1Y 7
NC 8
18 NC
17 2B
16 NC
15 2A
14 NC
9 10 11 12 13
NC
GND
NC
2Y
NC

NC—No internal connection

SUMMARY OF SERIES 55461/75461

DEVICE	LOGIC	PACKAGES
SN55461	AND	FK,JG
SN55462	NAND	FK,JG
SN55463	OR	FK,JG
SN55464	NOR	FK,JG
SN75461	AND	D,P
SN75462	NAND	D,P
SN75463	OR	D,P

description

These dual peripheral drivers are functionally interchangeable with SN55451B through SN55454B and SN75451B through SN75453B peripheral drivers, but are designed for use in systems that require higher breakdown voltages than those devices can provide at the expense of slightly slower switching speeds. Typical applications include logic buffers, power drivers, relay drivers, lamp drivers, MOS drivers, line drivers, and memory drivers.

The SN55461/SN75461, SN55462/SN75462, SN55463/SN75463, and SN55464 are dual peripheral AND, NAND, OR, and NOR drivers, respectively, (assuming positive logic), with the output of the gates internally connected to the bases of the n-p-n output transistors.

Series 55461 drivers are characterized for operation over the full military temperature range of −55°C to 125°C; Series 75461 drivers are characterized for operation from 0°C to 70°C.

absolute maximum ratings over operating free-air temperature range (unless otherwise noted)

		SN55461 SN55462 SN55463 SN55464	SN75461 SN75462 SN75463	UNIT
Supply voltage, V_{CC} (see Note 1)		7	7	V
Input voltage		5.5	5.5	V
Interemitter voltage (see Note 2)		5.5	5.5	V
Off-state output voltage		35	35	V
Continuous collector or output current (see Note 3)		400	400	mA
Peak collector or output current ($t_w \leq$ 10 ms, duty cycle $\leq$ 50%, see Note 3)		500	500	mA
Continuous total dissipation at (or below) 25°C free-air temperature (see Note 4)	D package		725	mW
	FK package	1375		mW
	JG package	1050		mW
	P package		1200	mW
Operating free-air temperature range		−55 to 125	0 to 70	°C
Storage temperature range		−65 to 150	−65 to 150	°C
Case temperature for 60 seconds	FK package	260		°C
Lead temperature 1,6 mm (1/16 inch) from case for 60 seconds	JG package	300		°C
Lead temperature 1,6 mm (1/16 inch) from case for 10 seconds	D or P package		260	°C

NOTES: 1. Voltage values are with respect to network ground terminal unless otherwise specified.
2. This is the voltage between two emitters of a multiple-emitter transistor.
3. Both halves of these dual circuits may conduct rated current simultaneously; however, power dissipation averaged over a short time interval must fall within the continuous dissipation rating.
4. For operation above 25°C free-air temperature, refer to the Dissipation Derating Table.

DISSIPATION DERATING TABLE

PACKAGE	POWER RATING	DERATING FACTOR	ABOVE T_A
D	725 mW	5.8 mW/°C	25°C
FK	1375 mW	11 mW/°C	25°C
JG	1050 mW	8.4 mW/°C	25°C
P	1200 mW	9.6 mW/°C	25°C

recommended operating conditions

	SN55461 THRU SN55464			SN75461 THRU SN75463			UNIT
	MIN	NOM	MAX	MIN	NOM	MAX	
Supply voltage, V_{CC}	4.5	5	5.5	4.75	5	5.25	V
High-level input voltage, V_{IH}	2.2			2			V
Low-level input voltage, V_{IL}			0.8			0.8	V
Operating free-air temperature, T_A	−55		125	0		70	°C

logic symbol†

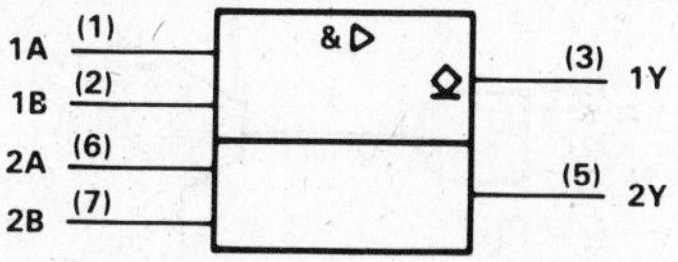
1A
(1)
1B
(2)
2A
(6)
2B
(7)
&
(3)
1Y
(5)
2Y

†This symbol is in accordance with ANSI/IEEE STD 91-1984 and IEC Publication 617-12.

logic diagram (positive logic)

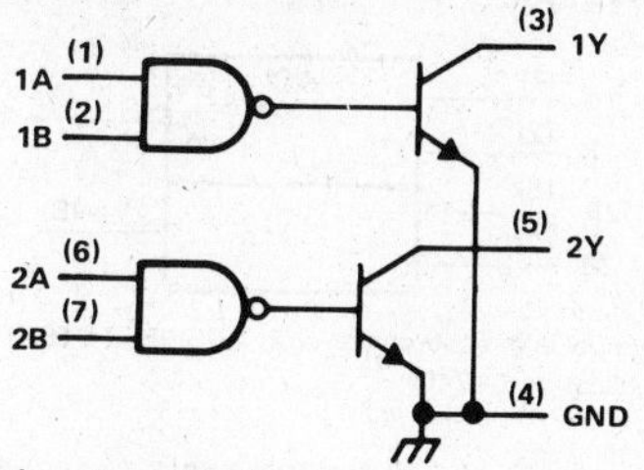
1A
(1)
1B
(2)
2A
(6)
2B
(7)
(3)
1Y
(5)
2Y
(4)
GND

FUNCTION TABLE (EACH DRIVER)

A	B	Y
L	L	L (on state)
L	H	L (on state)
H	L	L (on state)
H	H	H (off state)

positive logic:
$Y = \overline{AB}$ or $\overline{A} + \overline{B}$

schematic (each driver)

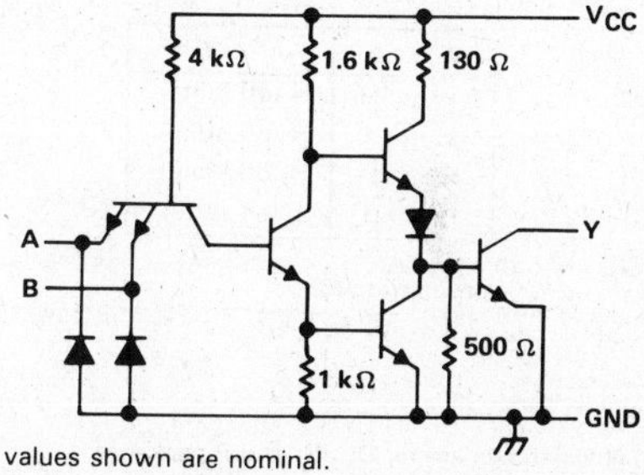
VCC
4 kΩ
1.6 kΩ
130 Ω
A
B
Y
500 Ω
1 kΩ
GND

Pin numbers shown are for D, JG, and P packages.

Resistor values shown are nominal.

electrical characteristics over recommended operating free-air temperature range (unless otherwise noted)

PARAMETER		TEST CONDITIONS†	SN55461 MIN	SN55461 TYP‡	SN55461 MAX	SN75461 MIN	SN75461 TYP‡	SN75461 MAX	UNIT
V_{IK}	Input clamp voltage	V_{CC} = MIN, I_I = −12 mA		−1.2	−1.5		−1.2	−1.5	V
I_{OH}	High-level output current	V_{CC} = MIN, V_{IH} = MIN, V_{OH} = 35 V			300			100	μA
V_{OL}	Low-level output voltage	V_{CC} = MIN, V_{IL} = 0.8 V, I_{OL} = 100 mA		0.25	0.5		0.25	0.4	V
		V_{CC} = MIN, V_{IL} = 0.8 V, I_{OL} = 300 mA		0.5	0.8		0.5	0.7	
I_I	Input current at maximum input voltage	V_{CC} = MAX, V_I = 5.5 V			1			1	mA
I_{IH}	High-level input current	V_{CC} = MAX, V_I = 2.4 V			40			40	μA
I_{IL}	Low-level input current	V_{CC} = MAX, V_I = 0.4 V		−1	−1.6		−1	−1.6	mA
I_{CCH}	Supply current, outputs high	V_{CC} = MAX, V_I = 5 V		8	11		8	11	mA
I_{CCL}	Supply current, outputs low	V_{CC} = MAX, V_I = 0		56	76		56	76	mA

†For conditions shown as MIN or MAX, use the appropriate value specified under recommended operating conditions.
‡All typical values are at V_{CC} = 5 V, T_A = 25°C.

switching characteristics, V_{CC} = 5 V, T_A = 25°C

PARAMETER		TEST CONDITIONS	MIN	TYP	MAX	UNIT
t_{PLH}	Propagation delay time, low-to-high-level output	$I_O \approx$ 200 mA, C_L = 15 pF, R_L = 50 Ω, See Figure 1		30	55	ns
t_{PHL}	Propagation delay time, high-to-low-level output			25	40	ns
t_{TLH}	Transition time, low-to-high-level output			8	20	ns
t_{THL}	Transition time, high-to-low-level output			10	20	ns
V_{OH}	High-level output voltage after switching	V_S = 30 V, $I_O \approx$ 300 mA, See Figure 2	$V_S - 10$			mV

logic symbol†

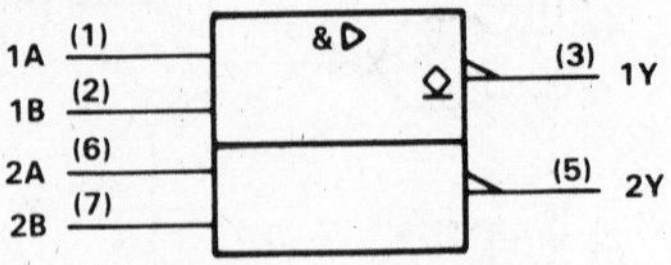
&D
1A
(1)
1B
(2)
2A
(6)
2B
(7)
(3)
1Y
(5)
2Y

†This symbol is in accordance with ANSI/IEEE STD 91-1984 and IEC Publication 617-12.

logic diagram (positive logic)

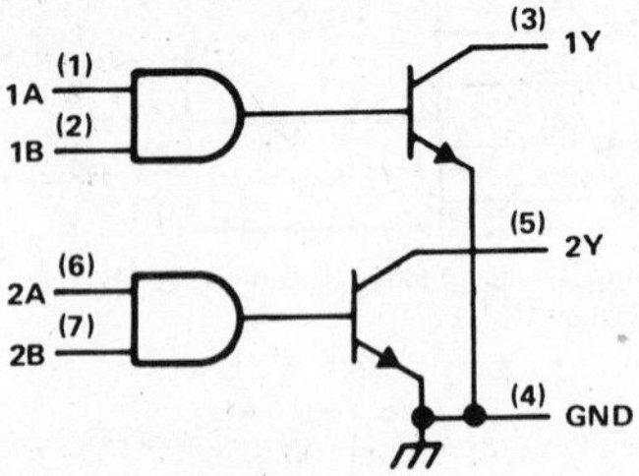
1A
(1)
1B
(2)
2A
(6)
2B
(7)
(3)
1Y
(5)
2Y
(4)
GND

FUNCTION TABLE
(EACH DRIVER)

A	B	Y
L	L	H (off state)
L	H	H (off state)
H	L	H (off state)
H	H	L (on state)

positive logic:
$Y = \overline{AB}$ or $\overline{A} + \overline{B}$

schematic (each driver)

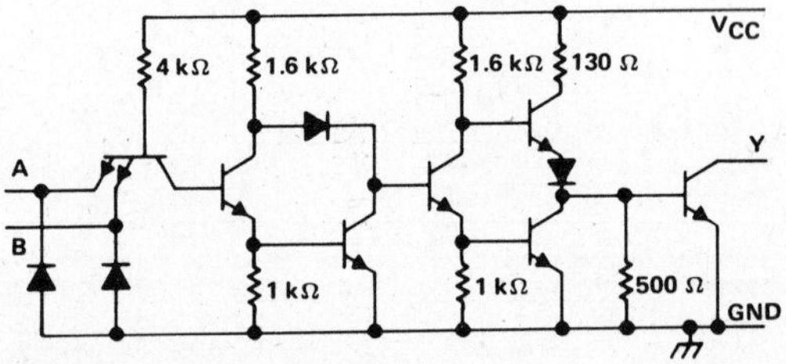
V_{CC}
4 kΩ
1.6 kΩ
1.6 kΩ
130 Ω
A
B
Y
1 kΩ
1 kΩ
500 Ω
GND

Pin numbers shown are for D, JG, and P packages.

Resistor values shown are nominal.

electrical characteristics over recommended operating free-air temperature range (unless otherwise noted)

PARAMETER		TEST CONDITIONS†	SN55462 MIN	SN55462 TYP‡	SN55462 MAX	SN75462 MIN	SN75462 TYP‡	SN75462 MAX	UNIT
V_{IK}	Input clamp voltage	V_{CC} = MIN, $I_I = -12$ mA		−1.2	−1.5		−1.2	−1.5	V
I_{OH}	High-level output current	V_{CC} = MIN, V_{IL} = 0.8 V, V_{OH} = 35 V			300			100	µA
V_{OL}	Low-level output voltage	V_{CC} = MIN, V_{IH} = MIN, I_{OL} = 100 mA		0.25	0.5		0.25	0.4	V
		V_{CC} = MIN, V_{IH} = MIN, I_{OL} = 300 mA		0.5	0.8		0.5	0.7	
I_I	Input current at maximum input voltage	V_{CC} = MAX, V_I = 5.5 V			1			1	mA
I_{IH}	High-level input current	V_{CC} = MAX, V_I = 2.4 V			40			40	µA
I_{IL}	Low-level input current	V_{CC} = MAX, V_I = 0.4 V		−1.1	−1.6		−1.1	−1.6	mA
I_{CCH}	Supply current, outputs high	V_{CC} = MAX, V_I = 0		13	17		13	17	mA
I_{CCL}	Supply current, outputs low	V_{CC} = MAX, V_I = 5 V		61	76		61	76	mA

†For conditions shown as MIN or MAX, use the appropriate value specified under recommended operating conditions.
‡All typical values are at V_{CC} = 5 V, T_A = 25 °C.

switching characteristics, V_{CC} = 5 V, T_A = 25 °C

PARAMETER		TEST CONDITIONS	MIN	TYP	MAX	UNIT
t_{PLH}	Propagation delay time, low-to-high-level output	$I_O \approx 200$ mA, C_L = 15 pF, R_L = 50 Ω, See Figure 1		45	65	ns
t_{PHL}	Propagation delay time, high-to-low-level output			30	50	ns
t_{TLH}	Transition time, low-to-high-level output			13	25	ns
t_{THL}	Transition time, high-to-low-level output			10	20	ns
V_{OH}	High-level output voltage after switching	V_S = 30 V, $I_O \approx 300$ mA, See Figure 2	$V_S - 10$			mV

logic symbol†

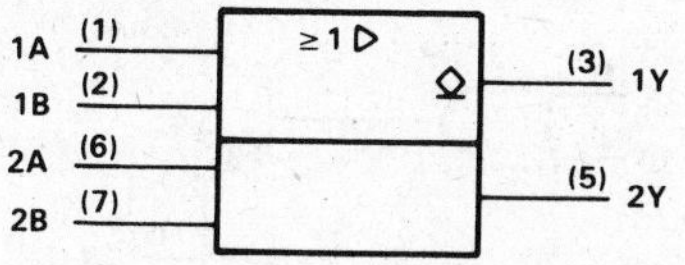
≥1
1A (1)
1B (2)
2A (6)
2B (7)
(3) 1Y
(5) 2Y

†This symbol is in accordance with ANSI/IEEE STD 91-1984 and IEC Publication 617-12.

logic diagram (positive logic)

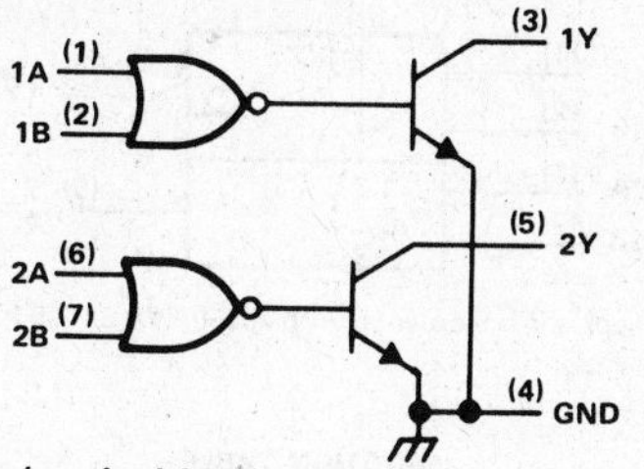
1A (1)
1B (2)
2A (6)
2B (7)
(3) 1Y
(5) 2Y
(4) GND

FUNCTION TABLE
(EACH DRIVER)

A	B	Y
L	L	L (on state)
L	H	H (off state)
H	L	H (off state)
H	H	H (off state)

positive logic:
$Y = A + B$ or $\overline{\overline{A}\overline{B}}$

schematic (each driver)

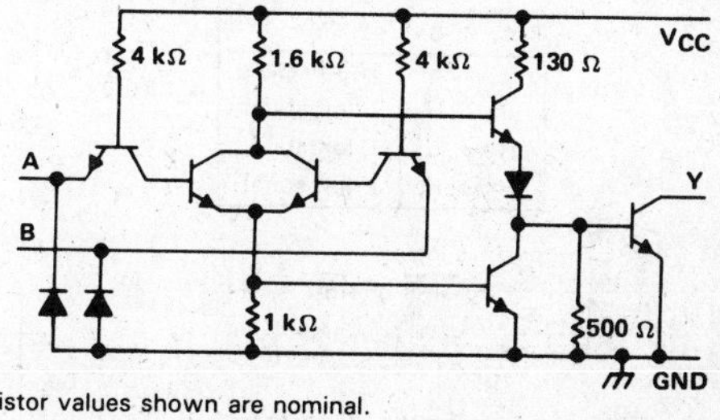
VCC
4 kΩ
1.6 kΩ
4 kΩ
130 Ω
A
B
Y
1 kΩ
500 Ω
GND

Pin numbers shown are for D, JG, and P packages.

Resistor values shown are nominal.

electrical characteristics over recommended operating free-air temperature range (unless otherwise noted)

PARAMETER		TEST CONDITIONS†		SN55463			SN75463			UNIT
				MIN	TYP‡	MAX	MIN	TYP‡	TYP	
V_{IK}	Input clamp voltage	V_{CC} = MIN,	I_I = −12 mA		−1.2	−1.5		−1.2	−1.5	V
I_{OH}	High-level output current	V_{CC} = MIN, V_{OH} = 35 V	V_{IH} = MIN,			300			100	μA
V_{OL}	Low-level output voltage	V_{CC} = MIN, I_{OL} = 100 mA	V_{IL} = 0.8 V,		0.25	0.5		0.25	0.4	V
		V_{CC} = MIN, I_{OL} = 300 mA	V_{IL} = 0.8 V,		0.5	0.8		0.5	0.7	
I_I	Input current at maximum input voltage	V_{CC} = MAX,	V_I = 5.5 V			1			1	mA
I_{IH}	High-level input current	V_{CC} = MAX,	V_I = 2.4 V			40			40	μA
I_{IL}	Low-level input current	V_{CC} = MAX,	V_I = 0.4 V		−1	−1.6		−1	−1.6	mA
I_{CCH}	Supply current, outputs high	V_{CC} = MAX,	V_I = 5 V		8	11		8	11	mA
I_{CCL}	Supply current, outputs low	V_{CC} = MAX,	V_I = 0		58	76		58	76	mA

†For conditions shown as MIN or MAX, use the appropriate value specified under recommended operating conditions.
‡All typical values are at V_{CC} = 5 V, T_A = 25°C.

switching characteristics, V_{CC} = 5 V, T_A = 25°C

PARAMETER		TEST CONDITIONS		MIN	TYP	MAX	UNIT
t_{PLH}	Propagation delay time, low-to-high-level output	$I_O \approx$ 200 mA, R_L = 50 Ω,	C_L = 15 pF, See Figure 1		30	55	ns
t_{PHL}	Propagation delay time, high-to-low-level output				25	40	ns
t_{TLH}	Transition time, low-to-high-level output				8	25	ns
t_{THL}	Transition time, high-to-low-level output				10	25	ns
V_{OH}	High-level output voltage after switching	V_S = 30 V, See Figure 2	$I_O \approx$ 300 mA,	$V_S - 10$			mV

logic symbol†

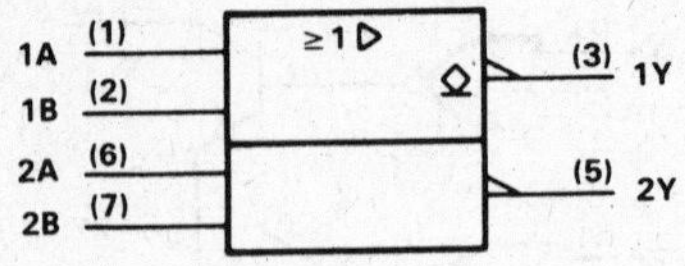
≥1
1A (1)
1B (2)
2A (6)
2B (7)
(3) 1Y
(5) 2Y

†This symbol is in accordance with ANSI/IEEE STD 91-1984 and IEC Publication 617-12.

FUNCTION TABLE
(EACH DRIVER)

A	B	Y
L	L	H (off state)
L	H	L (on state)
H	L	L (on state)
H	H	L (on state)

positive logic:
$Y = \overline{A+B}$ or $\overline{A}\,\overline{B}$

Pin numbers shown are for the JG package.

logic diagram (positive logic)

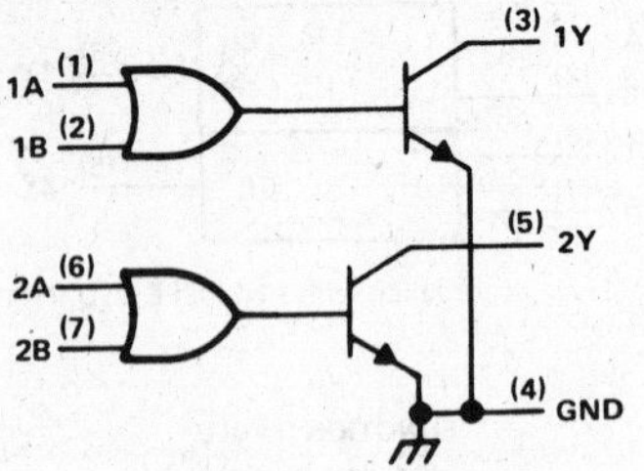
1A (1)
1B (2)
(3) 1Y
(5) 2Y
2A (6)
2B (7)
(4) GND

schematic (each driver)

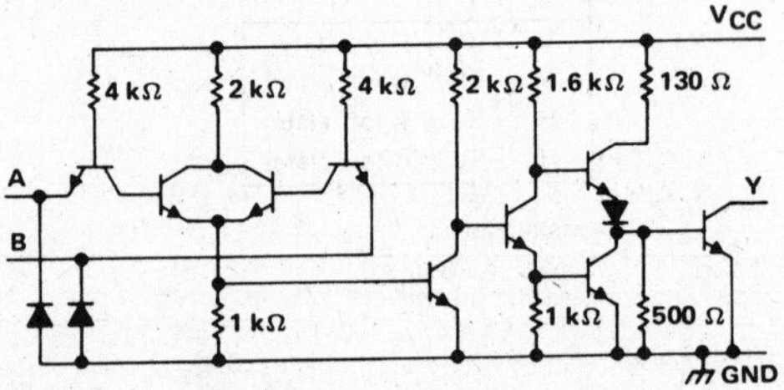
VCC
4 kΩ
2 kΩ
4 kΩ
2 kΩ
1.6 kΩ
130 Ω
A
B
Y
1 kΩ
1 kΩ
500 Ω
GND

Resistor values shown are nominal.

electrical characteristics over recommended operating free-air temperature range (unless otherwise noted)

PARAMETER		TEST CONDITIONS†	SN55464 MIN	SN55464 TYP‡	SN55464 MAX	UNIT
V_{IK}	Input clamp voltage	V_{CC} = MIN, I_I = −12 mA		−1.2	−1.5	V
I_{OH}	High-level output current	V_{CC} = MIN, V_{IL} = 0.8 V, V_{OH} = 35 V			300	μA
V_{OL}	Low-level output voltage	V_{CC} = MIN, V_{IH} = MIN, I_{OL} = 100 mA		0.25	0.5	V
		V_{CC} = MIN, V_{IH} = MIN, I_{OL} = 300 mA		0.5	0.8	
I_I	Input current at maximum input voltage	V_{CC} = MAX, V_I = 5.5 V			1	mA
I_{IH}	High-level input current	V_{CC} = MAX, V_I = 2.4 V			40	μA
I_{IL}	Low-level input current	V_{CC} = MAX, V_I = 0.4 V		−1	−1.6	mA
I_{CCH}	Supply current, outputs high	V_{CC} = MAX, V_I = 0		14	19	mA
I_{CCL}	Supply current, outputs low	V_{CC} = MAX, V_I = 5 V		67	85	mA

† For conditions shown as MIN or MAX, use the appropriate value specified under recommended operating conditions.
‡ All typical values are at V_{CC} = 5 V, T_A = 25°C.

switching characteristics, V_{CC} = 5 V, T_A = 25°C

PARAMETER		TEST CONDITIONS	MIN	TYP	MAX	UNIT
t_{PLH}	Propagation delay time, low-to-high-level output	$I_O \approx$ 200 mA, C_L = 15 pF, R_L = 50 Ω, See Figure 1		40	65	ns
t_{PHL}	Propagation delay time, high-to-low-level output			30	50	ns
t_{TLH}	Transition time, low-to-high-level output			8	20	ns
t_{THL}	Transition time, high-to-low-level output			10	20	ns
V_{OH}	High-level output voltage after switching	V_S = 30 V, $I_O \approx$ 300 mA, See Figure 2	$V_S - 10$			mV

TEXAS INSTRUMENTS
POST OFFICE BOX 655012 • DALLAS, TEXAS 75265

PARAMETER MEASUREMENT INFORMATION

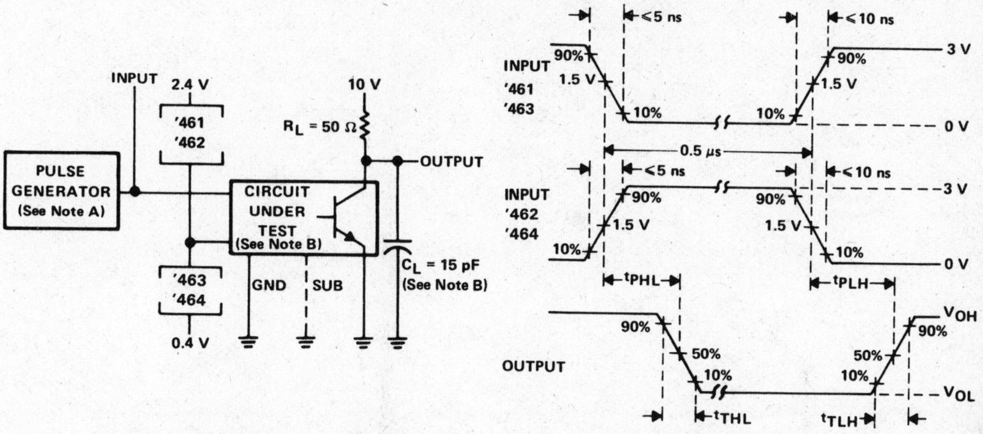
INPUT
2.4 V
'461
'462
10 V
R_L = 50 Ω
OUTPUT
PULSE GENERATOR (See Note A)
CIRCUIT UNDER TEST (See Note B)
'463
'464
GND
SUB
C_L = 15 pF (See Note B)
0.4 V
≤5 ns
≤10 ns
INPUT '461 '463
90%
1.5 V
10%
3 V
0 V
0.5 μs
INPUT '462 '464
t_{PHL}
t_{PLH}
OUTPUT
50%
V_{OH}
V_{OL}
t_{THL}
t_{TLH}

TEST CIRCUIT   VOLTAGE WAVEFORMS

NOTES: A. The pulse generator has the following characteristics: PRR ≤ 1 MHZ, Z_{out} ≈ 50 Ω.
B. C_L includes probe and jig capacitance.

FIGURE 1. SWITCHING TIMES

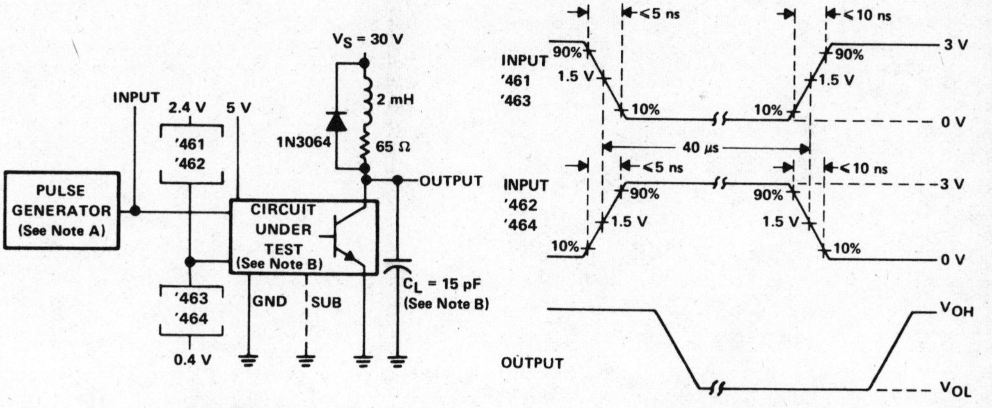
V_S = 30 V
INPUT
2.4 V
5 V
'461
'462
2 mH
1N3064
65 Ω
OUTPUT
PULSE GENERATOR (See Note A)
CIRCUIT UNDER TEST (See Note B)
'463
'464
GND
SUB
C_L = 15 pF (See Note B)
0.4 V
≤5 ns
≤10 ns
INPUT '461 '463
90%
1.5 V
10%
3 V
0 V
40 μs
INPUT '462 '464
OUTPUT
V_{OH}
V_{OL}

TEST CIRCUIT   VOLTAGE WAVEFORMS

NOTES: A. The pulse generator has the following characteristics: PRR ≤ 12.5 kHz, Z_{out} = 50 Ω.
B. C_L includes probe and jig capacitance.

FIGURE 2. LATCH-UP TEST

D2625, DECEMBER 1976—REVISED SEPTEMBER 1986

HIGH-VOLTAGE HIGH-CURRENT DARLINGTON TRANSISTOR ARRAYS

- **500 mA Rated Collector Current (Single Output)**
- **High-Voltage Outputs . . . 100 V**
- **Output Clamp Diodes**
- **Inputs Compatible with Various Types of Logic**
- **Relay Driver Applications**
- **Higher-Voltage Versions of ULN2005A, ULN2001A, ULN2002A, ULN2003A, and ULN2004A, Respectively, for Commercial Temperature Range**

D OR N PACKAGE
(TOP VIEW)

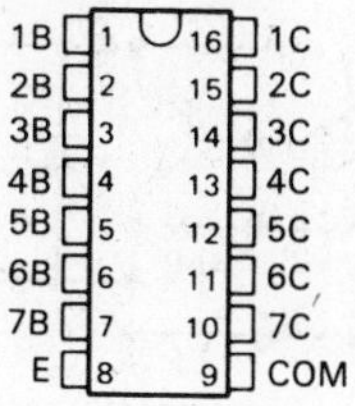
1B 1
2B 2
3B 3
4B 4
5B 5
6B 6
7B 7
E 8
16 1C
15 2C
14 3C
13 4C
12 5C
11 6C
10 7C
9 COM

description

The SN75465, SN75466, SN75467, SN75468, and SN75469 are monolithic high-voltage, high-current Darlington transistor arrays. Each consists of seven n-p-n Darlington pairs that feature high-voltage outputs with common-cathode clamp diodes for switching inductive loads. The collector-current rating of each Darlington pair is 500 milliamperes. The Darlington pairs may be paralleled for higher current capability. Applications include relay drivers, hammer drivers, lamp drivers, display drivers (LED and gas discharge), line drivers, and logic buffers.

The SN75465 has a 1050-ohm series base resistor and is especially designed for use with TTL where higher current is required and loading of the driving source is not a concern. The SN75466 is a general-purpose array and may be used with TTL, P-MOS, CMOS, and other MOS technologies. The SN75467 is specifically designed for use with 14- to 25-volt P-MOS devices and each input has a zener diode and resistor in series to limit the input current to a safe limit. The SN75468 has a 2700-ohm series base resistor for each Darlington pair for operation directly with TTL or 5-volt CMOS. The SN75469 has a 10.5-kilohm series base resistor to allow its operation directly from CMOS or P-MOS that use supply voltages of 6 to 15 volts. The required input current is below that of the SN75468 and the required voltage is less than that required by the SN75467.

logic symbol†

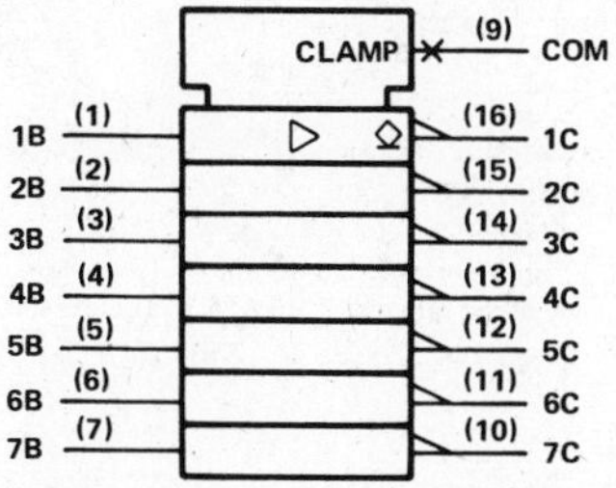
CLAMP
(9) COM
1B (1)
2B (2)
3B (3)
4B (4)
5B (5)
6B (6)
7B (7)
(16) 1C
(15) 2C
(14) 3C
(13) 4C
(12) 5C
(11) 6C
(10) 7C

†This symbol is in accordance with ANSI/IEEE Std 91-1984 and IEC Publication 617-12.

logic diagram

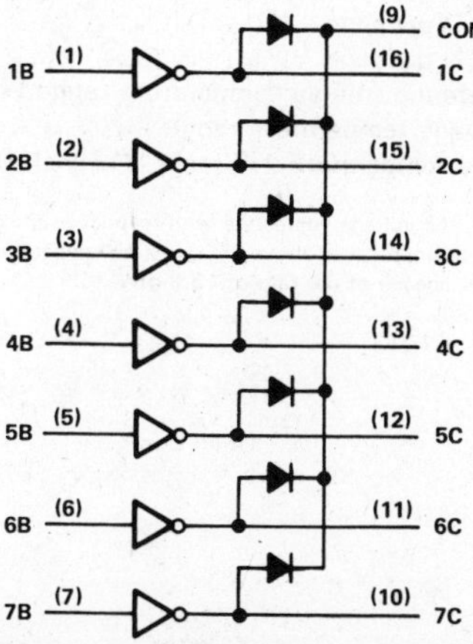
(9) COM
1B (1)
(16) 1C
2B (2)
(15) 2C
3B (3)
(14) 3C
4B (4)
(13) 4C
5B (5)
(12) 5C
6B (6)
(11) 6C
7B (7)
(10) 7C

5

Peripheral Drivers/Actuators

TEXAS INSTRUMENTS
POST OFFICE BOX 655012 • DALLAS, TEXAS 75265

schematics (each Darlington pair)

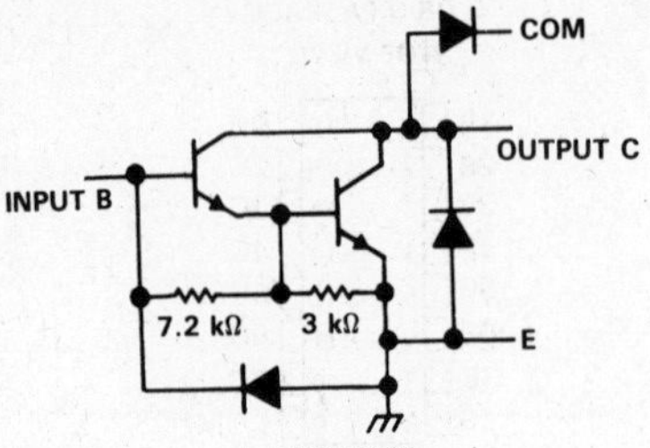
COM
OUTPUT C
INPUT B
7.2 kΩ
3 kΩ
E

SN75466

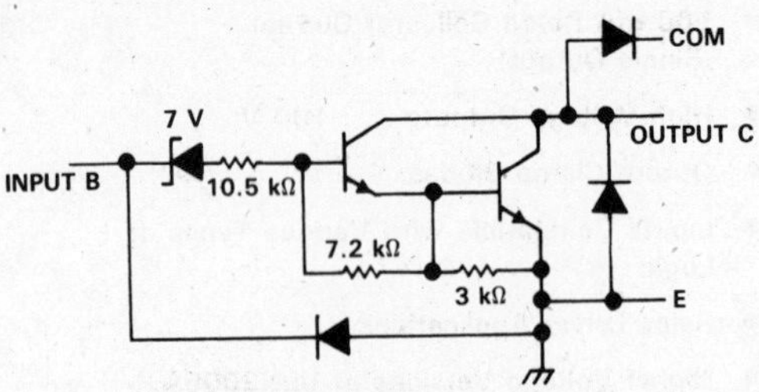
COM
7 V
OUTPUT C
INPUT B
10.5 kΩ
7.2 kΩ
3 kΩ
E

SN75467

SN75465: R_B = 1.05 kΩ
SN75468: R_B = 2.7 kΩ
SN75469: R_B = 10.5 kΩ

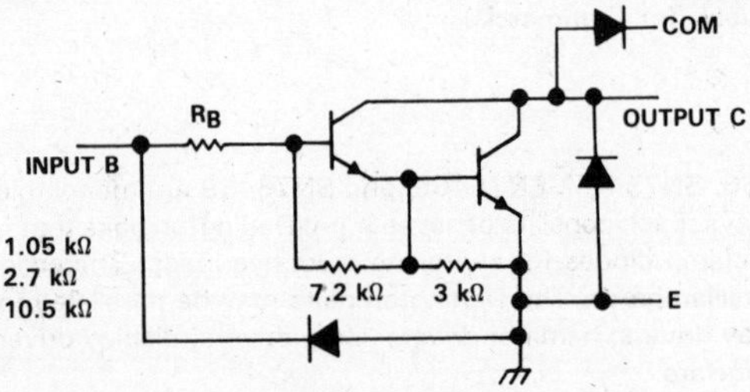
COM
R_B
OUTPUT C
INPUT B
7.2 kΩ
3 kΩ
E

SN75465, SN75468, SN75469

All resistor values shown are nominal.

absolute maximum ratings at 25°C free-air temperature (unless otherwise noted)

Collector-emitter voltage	100 V
Input voltage (see Note 1): SN75465	15 V
SN75467, SN75468, SN75469	30 V
Peak collector current (see Figures 14 and 15)	500 mA
Output clamp diode current	500 mA
Total emitter-terminal current	−2.5 A
Continuous dissipation (total package) at (or below) 25°C free-air temperature (see Note 2):	
D package	950 mW
N package	1650 mW
Operating free-air temperature range	0°C to 70°C
Storage temperature range	−65°C to 150°C
Lead temperature 1,6 mm (1/16 inch) from case for 10 seconds	260°C

NOTES: 1. All voltage values are with respect to the emitter/substrate terminal, E, unless otherwise noted.
2. For operation above 25°C free-air temperature, derate the D package linearly at the rate of 7.6 mW/°C and derate the N package linearly at the rate of 13.2 mW/°C.

electrical characteristics at 25 °C free-air temperature (unless otherwise noted)

PARAMETER		TEST FIGURE	TEST CONDITIONS	SN75465			UNIT
				MIN	TYP	MAX	
I_{CEX}	Collector cutoff current	1	V_{CE} = 100 V, I_I = 0			50	µA
			V_{CE} = 100 V, I_I = 0, T_A = 70 °C			100	
$I_{I(off)}$	Off-state input current	3	V_{CE} = 100 V, I_C = 500 µA, T_A = 70 °C	50	65		µA
I_I	Input current	4	V_I = 3 V		1.5	2.4	mA
$V_{I(on)}$	On-state input voltage	5	V_{CE} = 2 V, I_C = 350 mA			2.4	V
$V_{CE(sat)}$	Collector-emitter saturation voltage	6	I_I = 250 µA, I_C = 100 mA		0.9	1.1	V
			I_I = 350 µA, I_C = 200 mA		1.0	1.3	
			I_I = 500 µA, I_C = 350 mA		1.2	1.6	
I_R	Clamp diode reverse current	7	V_R = 100 V			50	µA
			V_R = 100 V, T_A = 70 °C			100	
V_F	Clamp diode forward voltage	8	I_F = 350 mA		1.7	2	V
C_i	Input capacitance		V_I = 0 V, f = 1 MHz		15	25	pF

electrical characteristics at 25 °C free-air temperature (unless otherwise noted)

PARAMETER		TEST FIGURE	TEST CONDITIONS	SN75466			SN75467			UNIT
				MIN	TYP	MAX	MIN	TYP	MAX	
I_{CEX}	Collector cutoff current	1	V_{CE} = 100 V, I_I = 0			50			50	µA
			V_{CE} = 100 V, I_I = 0, T_A = 70 °C			100			100	
		2	T_A = 70 °C, V_I = 6 V						500	
$I_{I(off)}$	Off-state input current	3	V_{CE} = 50 V, I_C = 500 µA, T_A = 70 °C	50	65		50	65		µA
I_I	Input current	4	V_I = 17 V					0.82	1.25	mA
h_{FE}	Static forward current transfer ratio	6	V_{CE} = 2 V, I_C = 350 mA	1000						
$V_{I(on)}$	On-state input voltage	5	V_{CE} = 2 V, I_C = 300 mA						13	V
$V_{CE(sat)}$	Collector-emitter saturation voltage	6	I_I = 250 µA, I_C = 100 mA		0.9	1.1		0.9	1.1	V
			I_I = 350 µA, I_C = 200 mA		1.0	1.3		1.0	1.3	
			I_I = 500 µA, I_C = 350 mA		1.2	1.6		1.2	1.6	
I_R	Clamp diode reverse current	7	V_R = 100 V			50			50	µA
			V_R = 100 V, T_A = 70 °C			100			100	
V_F	Clamp diode forward voltage	8	I_F = 350 mA		1.7	2		1.7	2	V
C_i	Input capacitance		V_I = 0 V, f = 1 MHz		15	25		15	25	pF

electrical characteristics at 25°C free-air temperature (unless otherwise noted)

PARAMETER		TEST FIGURE	TEST CONDITIONS		SN75468 MIN	SN75468 TYP	SN75468 MAX	SN75469 MIN	SN75469 TYP	SN75469 MAX	UNIT
I_{CEX}	Collector cutoff current	1	V_{CE} = 100 V, I_I = 0				50			50	μA
		1	V_{CE} = 100 V, T_A = 70°C	I_I = 0			100			100	
		2		V_I = 1 V						500	
$I_{I(off)}$	Off-state input current	3	V_{CE} = 50 V, I_C = 500 μA, T_A = 70°C		50	65		50	65		μA
I_I	Input current	4	V_I = 3.85 V			0.93	1.35				mA
			V_I = 5 V						0.35	0.5	
			V_I = 12 V						1.0	1.45	
$V_{I(on)}$	On-state input voltage	5	V_{CE} = 2 V	I_C = 125 mA						5	V
				I_C = 200 mA			2.4			6	
				I_C = 250 mA			2.7				
				I_C = 275 mA						7	
				I_C = 300 mA			3				
				I_C = 350 mA						8	
$V_{CE(sat)}$	Collector-emitter saturation voltage	6	I_I = 250 μA, I_C = 100 mA			0.9	1.1		0.9	1.1	V
			I_I = 350 μA, I_C = 200 mA			1.0	1.3		1.0	1.3	
			I_I = 500 μA, I_C = 350 mA			1.2	1.6		1.2	1.6	
I_R	Clamp diode reverse current	7	V_R = 100 V				50			50	μA
			V_R = 100 V, T_A = 70°C				100			100	
V_F	Clamp diode forward voltage	8	I_F = 350 mA			1.7	2		1.7	2	V
C_i	Input capacitance		V_I = 0 V, f = 1 MHz			15	25		15	25	pF

switching characteristics at 25°C free-air temperature

PARAMETER		TEST CONDITIONS	MIN	TYP	MAX	UNIT
t_{PLH}	Propagation delay time, low-to-high-level output	V_S = 50 V, R_L = 163 Ω, C_L = 15 pF, See Figure 9		0.25	1	μs
t_{PHL}	Propagation delay time, high-to-low-level output			0.25	1	μs
V_{OH}	High-level output voltage after switching	V_S = 50 V, I_O ≈ 300 mA, See Figure 10	$V_S - 20$			mV

PARAMETER MEASUREMENT INFORMATION

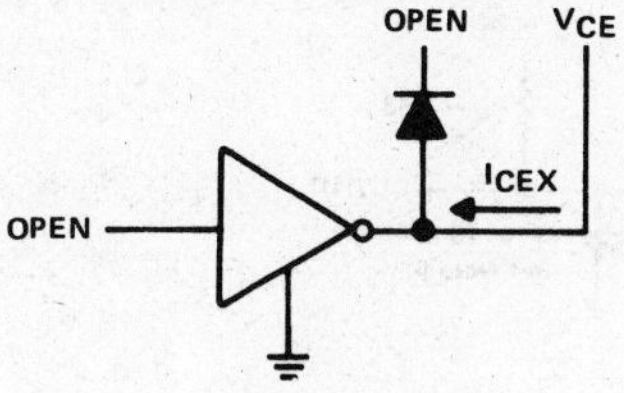
OPEN
VCE
ICEX
OPEN

FIGURE 1. I_{CEX}

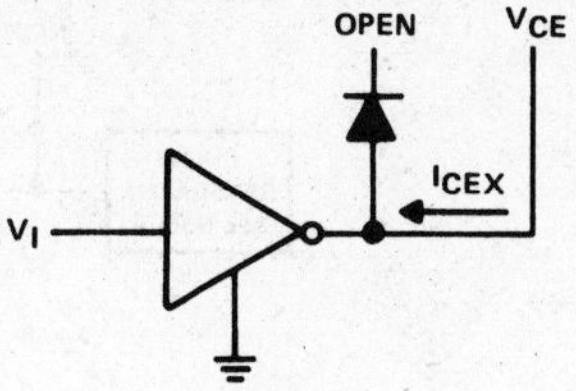
OPEN
VCE
ICEX
VI

FIGURE 2. I_{CEX}

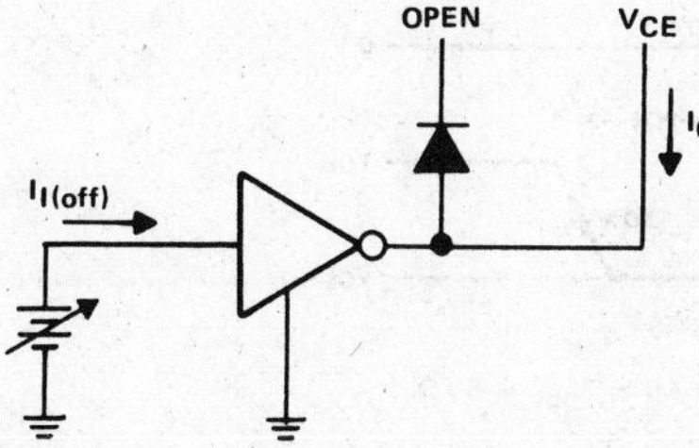
OPEN
VCE
IC
II(off)

FIGURE 3. $I_{I(off)}$

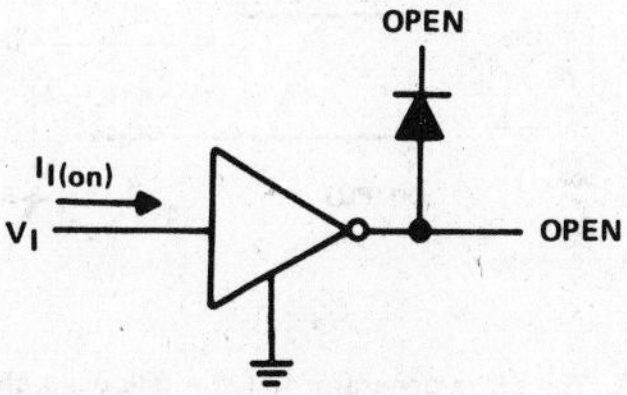
OPEN
II(on)
VI
OPEN

FIGURE 4. I_I

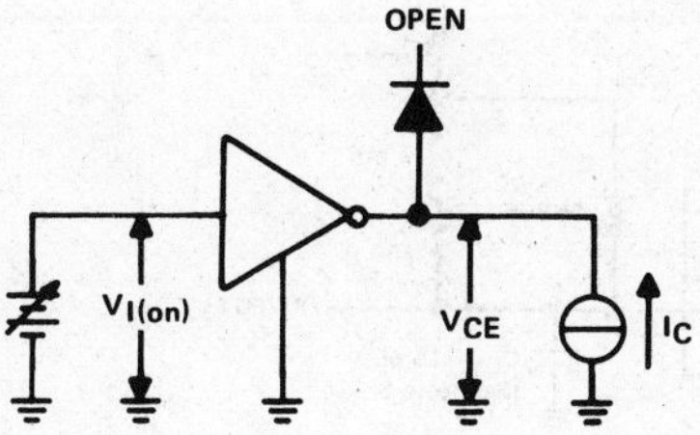
OPEN
VI(on)
VCE
IC

FIGURE 5. $V_{I(on)}$

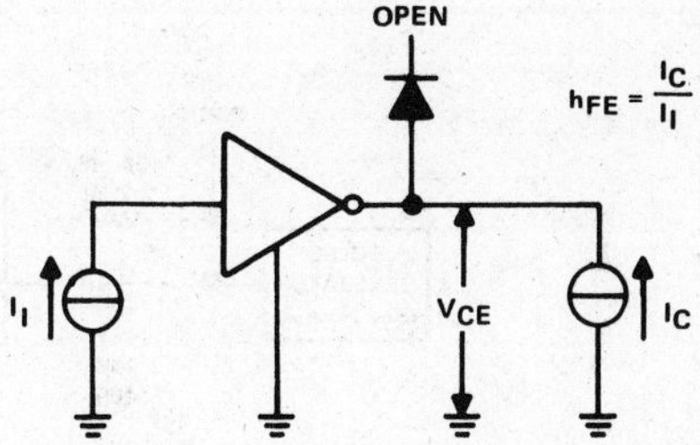
OPEN
hFE = IC / II
II
VCE
IC

NOTE: I_I is fixed for measuring $V_{CE(sat)}$, variable for measuring h_{FE}.

FIGURE 6. h_{FE}, $V_{CE(sat)}$

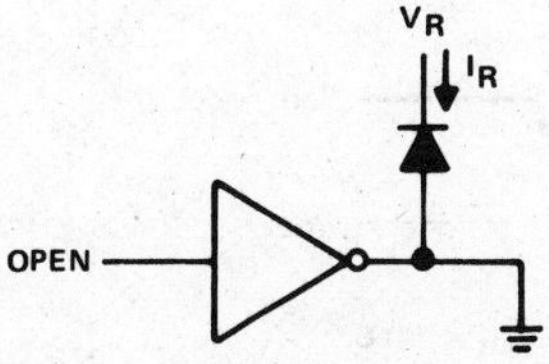
VR
IR
OPEN

FIGURE 7. I_R

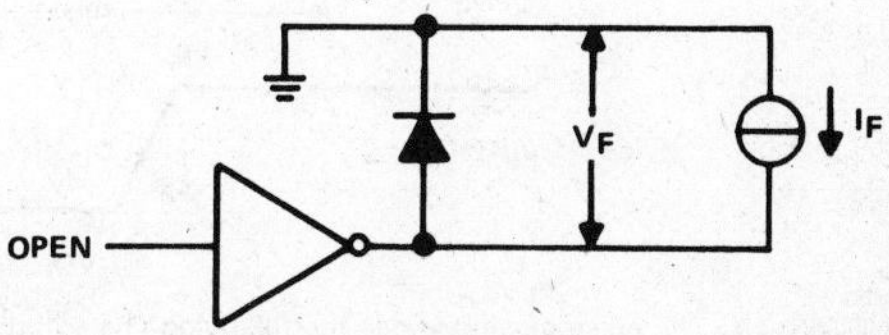
VF
IF
OPEN

FIGURE 8. V_F

PARAMETER MEASUREMENT INFORMATION

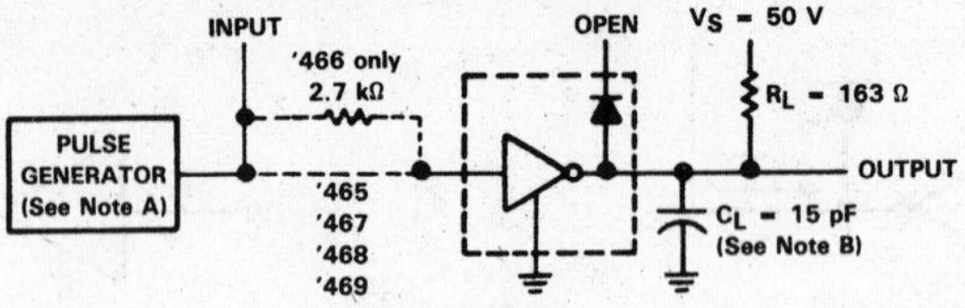
INPUT
'466 only
2.7 kΩ
OPEN
V_S = 50 V
R_L = 163 Ω
PULSE
GENERATOR
(See Note A)
OUTPUT
'465
'467
'468
'469
C_L = 15 pF
(See Note B)

TEST CIRCUIT

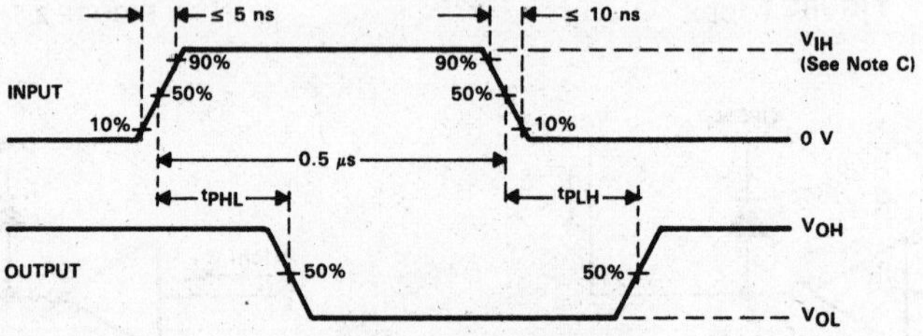
≤ 5 ns
≤ 10 ns
V_{IH}
(See Note C)
INPUT
90%
50%
10%
0 V
0.5 μs
t_{PHL}
t_{PLH}
V_{OH}
OUTPUT
50%
V_{OL}

VOLTAGE WAVEFORMS

NOTES: A. The pulse generator has the following characteristics: PRR = 1 MHz, Z_{out} ≈ 50 Ω.
B. C_L includes probe and jig capacitance.
C. For testing the '465, '466, and '468, V_{IH} = 3 V; for the '467, V_{IH} = 13 V; for the '469, V_{IH} = 8 V.

FIGURE 9. PROPAGATION DELAY TIMES

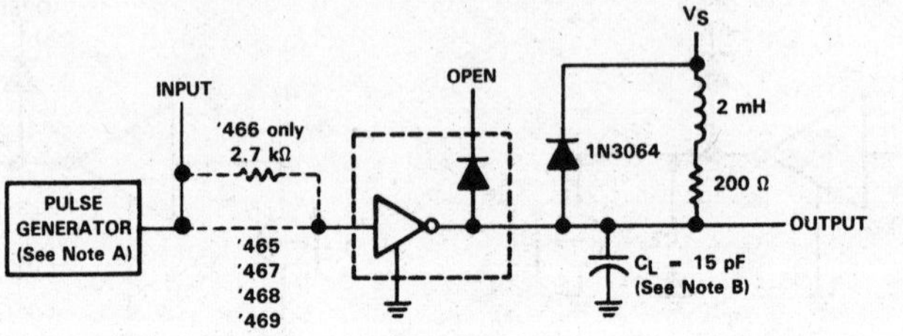
V_S
INPUT
OPEN
2 mH
'466 only
2.7 kΩ
1N3064
200 Ω
PULSE
GENERATOR
(See Note A)
OUTPUT
'465
'467
'468
'469
C_L = 15 pF
(See Note B)

TEST CIRCUIT

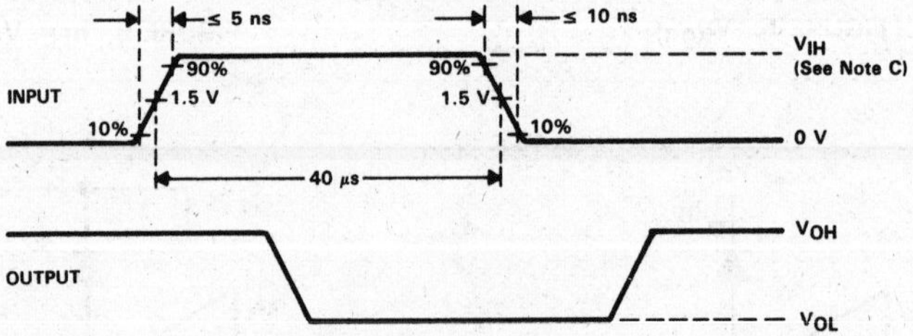
≤ 5 ns
≤ 10 ns
V_{IH}
(See Note C)
INPUT
90%
1.5 V
10%
0 V
40 μs
V_{OH}
OUTPUT
V_{OL}

VOLTAGE WAVEFORMS

NOTES: A. The pulse generator has the following characteristics: PRR = 12.5 kHz, Z_{out} = 50 Ω.
B. C_L includes probe and jig capacitance.
C. For testing the '465, '466, and '468, V_{IH} = 3 V; for the '467, V_{IH} = 13 V; for the '469, V_{IH} = 8 V.

FIGURE 10. LATCH-UP TEST

TEXAS INSTRUMENTS
POST OFFICE BOX 655012 • DALLAS, TEXAS 75265

TYPICAL CHARACTERISTICS

COLLECTOR-EMITTER SATURATION VOLTAGE vs COLLECTOR CURRENT (ONE DARLINGTON)

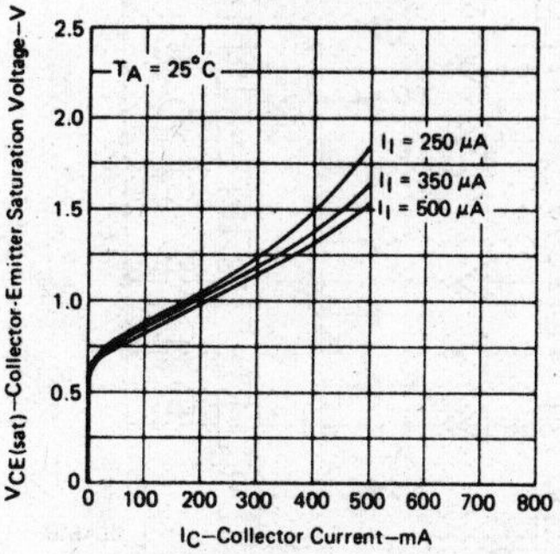
T_A = 25°C
I_I = 250 μA
I_I = 350 μA
I_I = 500 μA
VCE(sat)—Collector-Emitter Saturation Voltage—V
IC—Collector Current—mA
0
0.5
1.0
1.5
2.0
2.5
0 100 200 300 400 500 600 700 800

FIGURE 11

COLLECTOR-EMITTER SATURATION VOLTAGE vs COLLECTOR CURRENT (TWO DARLINGTONS PARALLELED)

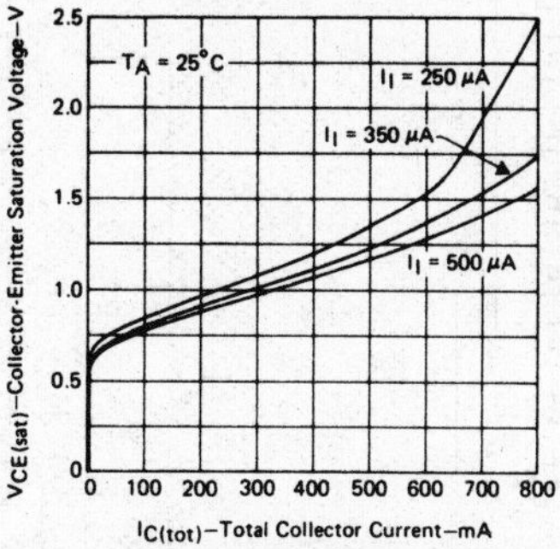
T_A = 25°C
I_I = 250 μA
I_I = 350 μA
I_I = 500 μA
VCE(sat)—Collector-Emitter Saturation Voltage—V
IC(tot)—Total Collector Current—mA
0
0.5
1.0
1.5
2.0
2.5
0 100 200 300 400 500 600 700 800

FIGURE 12

COLLECTOR CURRENT vs INPUT CURRENT

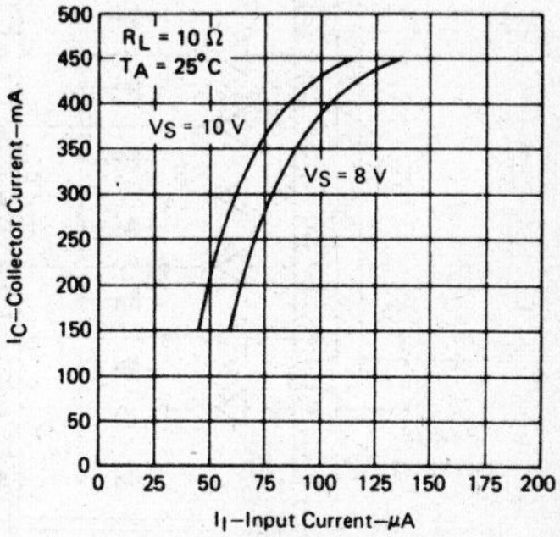
R_L = 10 Ω
T_A = 25°C
V_S = 10 V
V_S = 8 V
IC—Collector Current—mA
II—Input Current—μA
0
50
100
150
200
250
300
350
400
450
500
0 25 50 75 100 125 150 175 200

FIGURE 13

THERMAL INFORMATION

D PACKAGE
MAXIMUM COLLECTOR CURRENT
vs
DUTY CYCLE

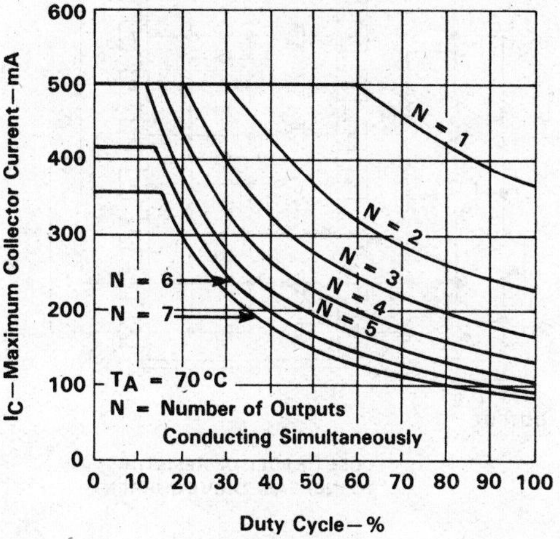
N = 1
N = 2
N = 3
N = 4
N = 5
N = 6
N = 7
T_A = 70°C
N = Number of Outputs
Conducting Simultaneously
IC—Maximum Collector Current—mA
Duty Cycle—%
0
100
200
300
400
500
600
0 10 20 30 40 50 60 70 80 90 100

FIGURE 14

N PACKAGE
MAXIMUM COLLECTOR CURRENT
vs
DUTY CYCLE

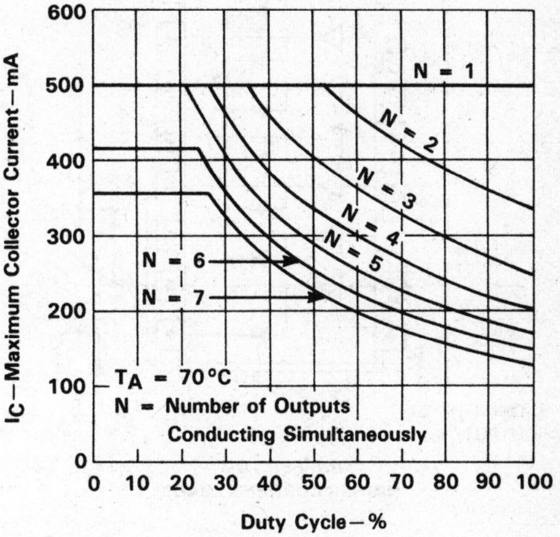
N = 1
N = 2
N = 3
N = 4
N = 5
N = 6
N = 7
T_A = 70°C
N = Number of Outputs
Conducting Simultaneously
IC—Maximum Collector Current—mA
Duty Cycle—%
0
100
200
300
400
500
600
0 10 20 30 40 50 60 70 80 90 100

FIGURE 15

TYPICAL APPLICATION DATA

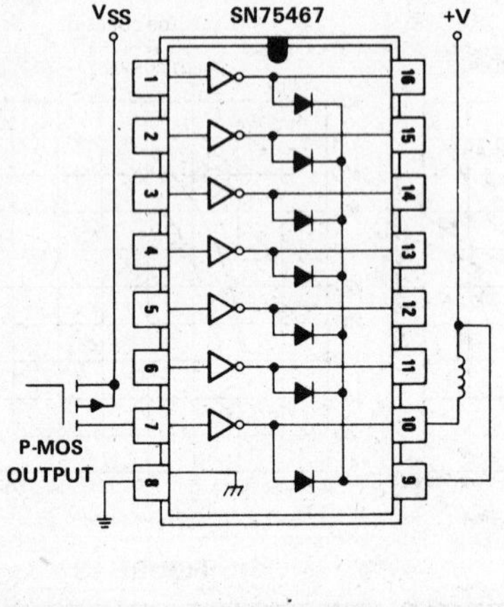
VSS
SN75467
+V
P-MOS
OUTPUT

P-MOS TO LOAD

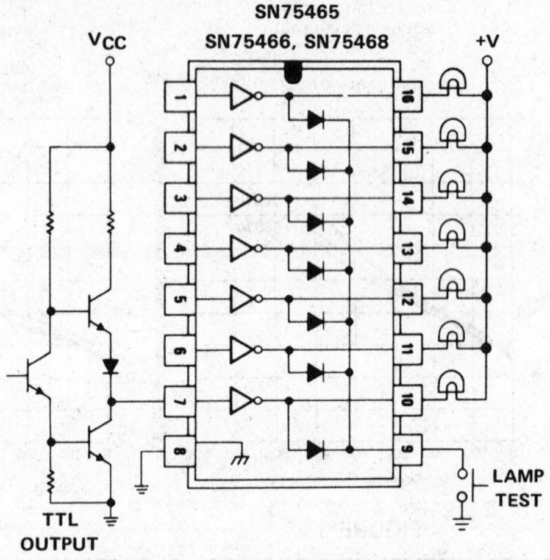
VCC
SN75465
SN75466, SN75468
+V
TTL
OUTPUT
LAMP
TEST

TTL TO LOAD

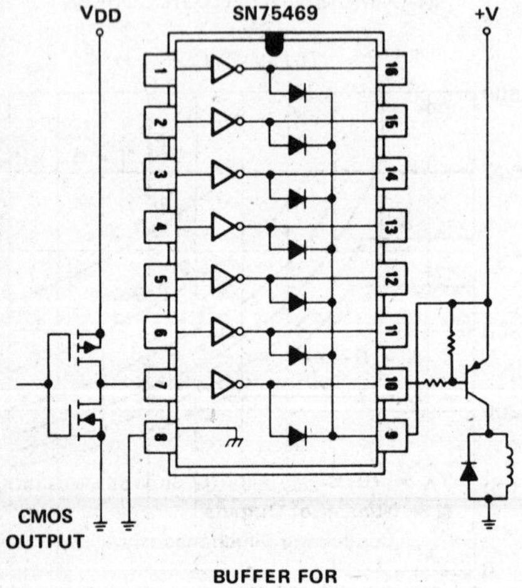
VDD
SN75469
+V
CMOS
OUTPUT

BUFFER FOR
HIGHER CURRENT LOADS

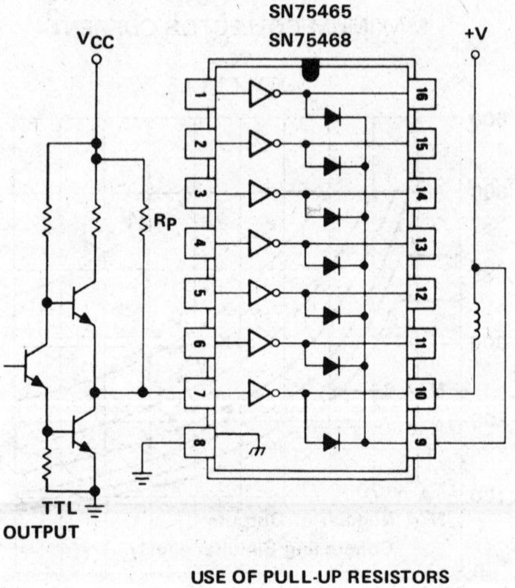
VCC
RP
TTL
OUTPUT
SN75465
SN75468
+V

USE OF PULL-UP RESISTORS
TO INCREASE DRIVE CURRENT

TEXAS INSTRUMENTS
POST OFFICE BOX 655012 • DALLAS, TEXAS 75265

D2130, DECEMBER 1976—REVISED SEPTEMBER 1986

PERIPHERAL DRIVERS FOR HIGH-VOLTAGE, HIGH-CURRENT DRIVER APPLICATIONS

- Characterized for Use to 300 mA
- High-Voltage Outputs
- No Output Latch-Up at 55 V (After Conducting 300 mA)
- Medium-Speed Switching
- Circuit Flexibility for Varied Applications and Choice of Logic Function
- TTL-Compatible Diode-Clamped Inputs
- Standard Supply Voltages
- New Plastic DIP (P) with Copper Lead Frame Provides Cooler Operation and Improved Reliability
- Package Options Include Plastic ''Small Outline'' Packages, Ceramic Chip Carriers, and Standard Plastic and Ceramic 300-mil DIPs

SN55471, SN55472,
SN55473, SN55474 . . . JG PACKAGE
SN75471, SN75472,
SN75473 . . . D OR P PACKAGE
(TOP VIEW)

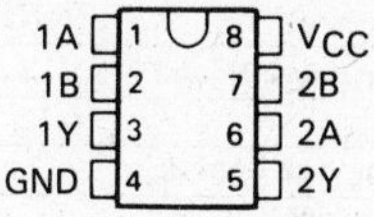
1A 1
1B 2
1Y 3
GND 4
8 VCC
7 2B
6 2A
5 2Y

SN55471, SN55472,
SN55473, SN55474 . . . FK PACKAGE
(TOP VIEW)

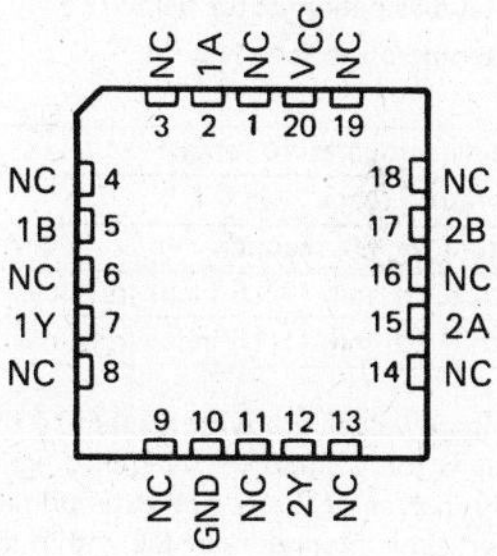
NC 1A NC VCC NC
3 2 1 20 19
NC 4
1B 5
NC 6
1Y 7
NC 8
18 NC
17 2B
16 NC
15 2A
14 NC
9 10 11 12 13
NC GND NC 2Y NC

NC—No internal connection

SUMMARY OF SERIES 55471/75471

DEVICE	LOGIC OF COMPLETE CIRCUIT	PACKAGES
SN55471	AND	FK,JG
SN55472	NAND	FK,JG
SN55473	OR	FK,JG
SN55474	NOR	FK,JG
SN75471	AND	D,P
SN75472	NAND	D,P
SN75473	OR	D,P

description

Series 55471/75471 dual peripheral drivers are functionally interchangeable with Series 55451B/75451B and Series 55461/75461 peripheral drivers, but are designed for use in systems that require higher breakdown voltages than either of those series can provide at the expense of slightly slower switching speeds than Series 55451B/75451B (limits are the same as Series 55461/75461). Typical applications include logic buffers, power drivers, relay drivers, lamp drivers, MOS drivers, line drivers, and memory drivers.

The SN55471/SN75471, SN55472/SN75472, SN55473/SN75473, and SN55474 are dual peripheral AND, NAND, OR, and NOR drivers, respectively, (assuming positive logic) with the output of the logic gates internally connected to the bases of the n-p-n output transistors.

Series 55471 drivers are characterized for operation over the full military temperature range of −55°C to 125°C. Series 75471 drivers are characterized for operation from 0°C to 70°C.

absolute maximum ratings over operating free-air temperature range (unless otherwise noted)

		SN55471 SN55472 SN55473 SN55474	SN75471 SN75472 SN75473	UNIT
Supply voltage, V_{CC} (see Note 1)		7	7	V
Input voltage		5.5	5.5	V
Interemitter voltage (see Note 2)		5.5	5.5	V
Off-state output voltage		70	70	V
Continuous collector or output current (see Note 3)		400	400	mA
Peak collector or output current ($t_w \leq$ 10 ms, duty cycle $\leq$ 50%, see Note 3)		500	500	mA
Continuous total dissipation at (or below) 25°C free-air temperature (see Note 4)	D package		725	mW
	FK package	1375		mW
	JG package	1050		mW
	P package		1200	mW
Operating free-air temperature range		−55 to 125	0 to 70	°C
Storage temperature range		−65 to 150	−65 to 150	°C
Case temperature for 60 seconds	FK package	260		°C
Lead temperature 1,6 mm (1/16 inch) from case for 60 seconds	JG package	300		°C
Lead temperature 1,6 mm (1/16 inch) from case for 10 seconds	D or P package		260	°C

NOTES:
1. Voltage values are with respect to the network ground terminal unless otherwise specified.
2. This is the voltage between two emitters of a multiple-emitter transistor.
3. Both halves of these dual circuits may conduct rated current simultaneously; however, power dissipation averaged over a short time interval must fall within the continuous dissipation rating.
4. For operation above 25°C free-air temperature, refer to the Dissipation Derating Table.

DISSIPATION DERATING TABLE

PACKAGE	POWER RATING	DERATING FACTOR	ABOVE T_A
D	725 mW	5.8 mW/°C	25°C
FK	1375 mW	11.0 mW/°C	25°C
JG	1050 mW	8.4 mW/°C	25°C
P	1200 mW	9.6 mW/°C	25°C

recommended operating conditions

	SN55471 SN55472 SN55473 SN55474			SN75471 SN75472 SN75473			UNIT
	MIN	NOM	MAX	MIN	NOM	MAX	
Supply voltage, V_{CC}	4.5	5	5.5	4.75	5	5.25	V
High-level input voltage, V_{IH}	2.2			2			V
Low-level input voltage, V_{IL}			0.8			0.8	V
Operating free-air temperature, T_A	−55		125	0		70	°C

logic symbol†

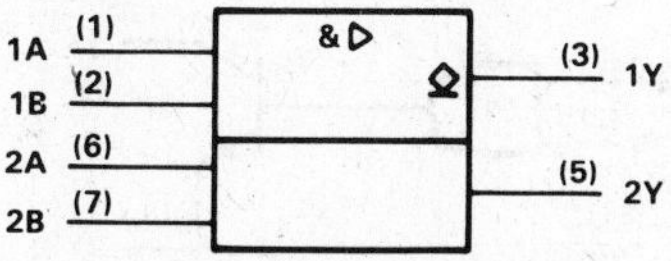
1A
(1)
1B
(2)
2A
(6)
2B
(7)
&▷
(3)
1Y
(5)
2Y

†This symbol is in accordance with ANSI/IEEE STD 91-1984 and IEC Publication 617-12.

logic diagram (positive logic)

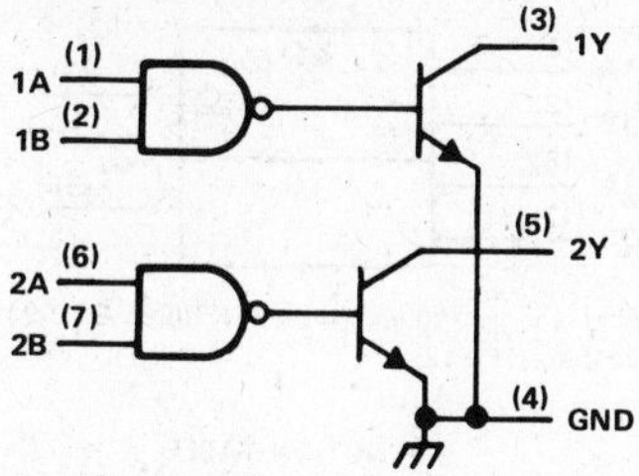
1A
(1)
1B
(2)
(3)
1Y
2A
(6)
2B
(7)
(5)
2Y
(4)
GND

FUNCTION TABLE (EACH DRIVER)

A	B	Y
L	L	L (on state)
L	H	L (on state)
H	L	L (on state)
H	H	H (off state)

positive logic:
$Y = AB$ or $\overline{\overline{A}+\overline{B}}$

schematic (each driver)

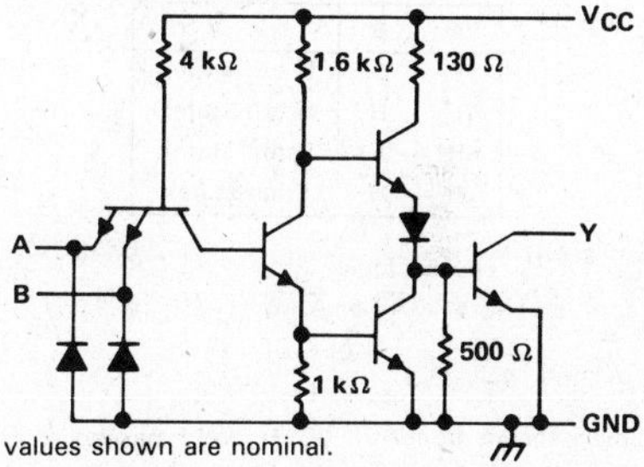
V_CC
4 kΩ
1.6 kΩ
130 Ω
A
B
Y
500 Ω
1 kΩ
GND

Pin numbers shown are for the JG, D, and P packages.

Resistor values shown are nominal.

electrical characteristics over recommended operating free-air temperature range (unless otherwise noted)

PARAMETER		TEST CONDITIONS‡		SN55471			SN75471			UNIT
				MIN	TYP§	MAX	MIN	TYP§	MAX	
V_{IK}	Input clamp voltage	V_{CC} = MIN,	$I_I = -12$ mA		−1.2	−1.5		−1.2	−1.5	V
I_{OH}	High-level output current	V_{CC} = MIN, V_{OH} = 70 V	V_{IH} = MIN,			300			100	μA
V_{OL}	Low-level output voltage	V_{CC} = MIN, I_{OL} = 100 mA	V_{IL} = 0.8 V,		0.25	0.5		0.25	0.4	V
		V_{CC} = MIN, I_{OL} = 300 mA	V_{IL} = 0.8 V,		0.5	0.8		0.5	0.7	
I_I	Input current at maximum input voltage	V_{CC} = MAX,	V_I = 5.5 V			1			1	mA
I_{IH}	High-level input current	V_{CC} = MAX,	V_I = 2.4 V			40			40	μA
I_{IL}	Low-level input current	V_{CC} = MAX,	V_I = 0.4 V		−1	−1.6		−1	−1.6	mA
I_{CCH}	Supply current, outputs high	V_{CC} = MAX,	V_I = 5 V		8	11		8	11	mA
I_{CCL}	Supply current, outputs low	V_{CC} = MAX,	V_I = 0		56	76		56	76	mA

‡For conditions shown as MIN or MAX, use the appropriate value specified under recommended operating conditions.
§All typical values are at V_{CC} = 5 V, T_A = 25°C.

switching characteristics, V_{CC} = 5 V, T_A = 25°C

PARAMETER		TEST CONDITIONS		MIN	TYP	MAX	UNIT
t_{PLH}	Propagation delay time, low-to-high-level output	$I_O \approx$ 200 mA, R_L = 50 Ω,	C_L = 15 pF, See Figure 1		30	55	ns
t_{PHL}	Propagation delay time, high-to-low-level output				25	40	ns
t_{TLH}	Transition time, low-to-high-level output				8	20	ns
t_{THL}	Transition time, high-to-low-level output				10	20	ns
V_{OH}	High-level output voltage after switching	V_S = 55 V, See Figure 2	$I_O \approx$ 300 mA,	$V_S - 18$			mV

logic symbol†

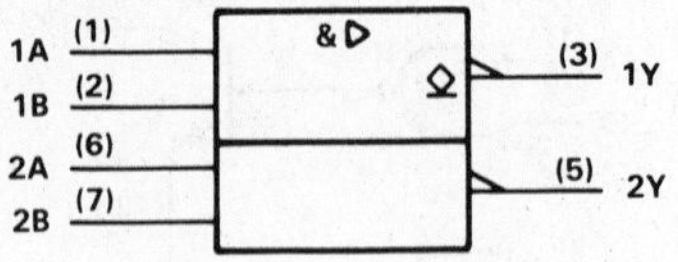
1A
(1)
1B
(2)
2A
(6)
2B
(7)
&
(3)
1Y
(5)
2Y

†This symbol is in accordance with ANSI/IEEE STD 91-1984 and IEC Publication 617-12.

logic diagram (positive logic)

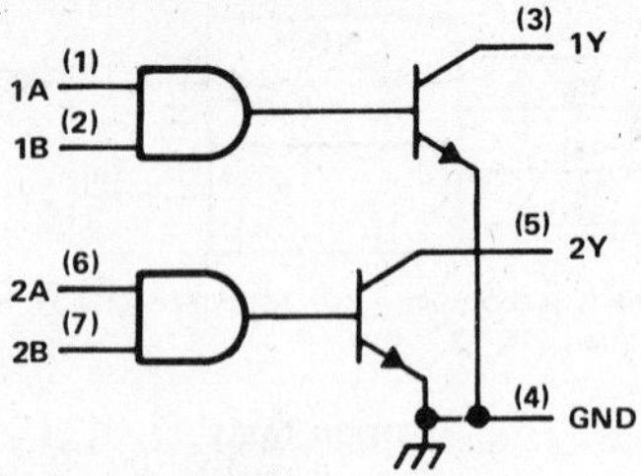
(1)
1A
(2)
1B
(6)
2A
(7)
2B
(3)
1Y
(5)
2Y
(4)
GND

FUNCTION TABLE (EACH DRIVER)

A	B	Y
L	L	H (off state)
L	H	H (off state)
H	L	H (off state)
H	H	L (on state)

positive logic:

$Y = \overline{AB}$ or $\overline{A} + \overline{B}$

Pin numbers shown are for the JG, D, and P packages.

schematic (each driver)

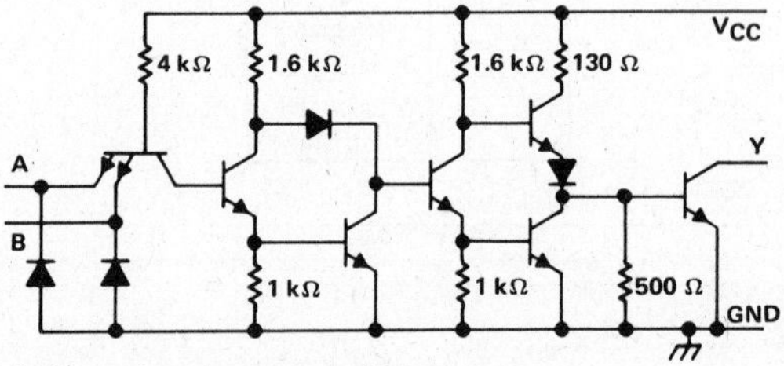
4 kΩ
1.6 kΩ
1.6 kΩ
130 Ω
VCC
A
B
Y
1 kΩ
1 kΩ
500 Ω
GND

Resistor values shown are nominal.

electrical characteristics over recommended operating free-air temperature range (unless otherwise noted)

PARAMETER		TEST CONDITIONS‡	SN55472 MIN	SN55472 TYP§	SN55472 MAX	SN75472 MIN	SN75472 TYP§	SN75472 MAX	UNIT
V_{IK}	Input clamp voltage	V_{CC} = MIN, $I_I = -12$ mA		−1.2	−1.5		−1.2	−1.5	V
I_{OH}	High-level output current	V_{CC} = MIN, V_{IL} = 0.8 V, V_{OH} = 70 V			300			100	μA
V_{OL}	Low-level output voltage	V_{CC} = MIN, V_{IH} = MIN, I_{OL} = 100 mA		0.25	0.5		0.25	0.4	V
		V_{CC} = MIN, V_{IH} = MIN, I_{OL} = 300 mA		0.5	0.8		0.5	0.7	V
I_I	Input current at maximum input voltage	V_{CC} = MAX, V_I = 5.5 V			1			1	mA
I_{IH}	High-level input current	V_{CC} = MAX, V_I = 2.4 V			40			40	μA
I_{IL}	Low-level input current	V_{CC} = MAX, V_I = 0.4 V		−1.1	−1.6		−1.1	−1.6	mA
I_{CCH}	Supply current, outputs high	V_{CC} = MAX, V_I = 0		13	17		13	17	mA
I_{CCL}	Supply current, outputs low	V_{CC} = MAX, V_I = 5 V		61	76		61	76	mA

‡For conditions shown as MIN or MAX, use the appropriate value specified under recommended operating conditions.
§All typical values are at V_{CC} = 5 V, T_A = 25°C.

switching characteristics, V_{CC} = 5 V, T_A = 25°C

PARAMETER		TEST CONDITIONS	MIN	TYP	MAX	UNIT
t_{PLH}	Propagation delay time, low-to-high-level output	$I_O \approx 200$ mA, C_L = 15 pF, R_L = 50 Ω, See Figure 1		45	65	ns
t_{PHL}	Propagation delay time, high-to-low-level output			30	50	ns
t_{TLH}	Transition time, low-to-high-level output			13	25	ns
t_{THL}	Transition time, high-to-low-level output			10	20	ns
V_{OH}	High-level output voltage after switching	V_S = 55 V, $I_O \approx 300$ mA, See Figure 2	$V_S - 18$			mV

logic symbol†

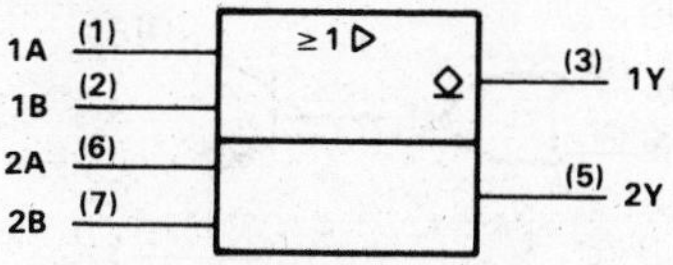
1A (1)
1B (2)
2A (6)
2B (7)
≥1
(3) 1Y
(5) 2Y

†This symbol is in accordance with ANSI/IEEE STD 91-1984 and IEC Publication 617-12.

FUNCTION TABLE
(EACH DRIVER)

A	B	Y
L	L	L (on state)
L	H	H (off state)
H	L	H (off state)
H	H	H (off state)

positive logic:
$Y = A + B$ or $\overline{\overline{A}\,\overline{B}}$

Pin numbers shown are for the JG, D, and P packages.

logic diagram (positive logic)

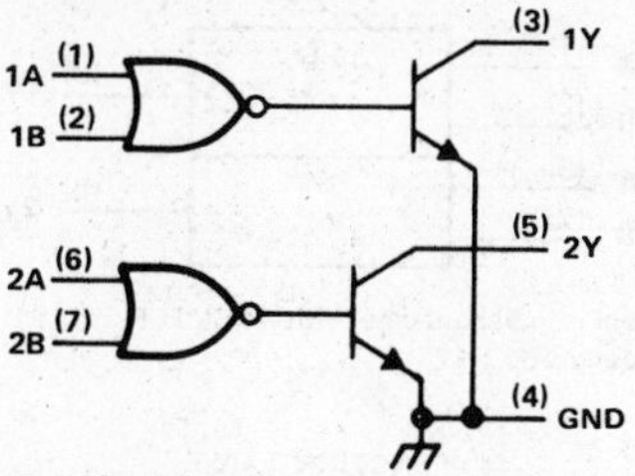
1A (1)
1B (2)
(3) 1Y
2A (6)
2B (7)
(5) 2Y
(4) GND

schematic (each driver)

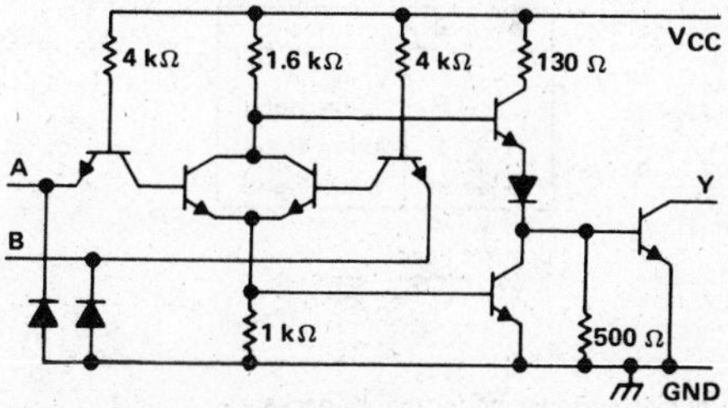
4 kΩ
1.6 kΩ
4 kΩ
130 Ω
VCC
A
B
Y
1 kΩ
500 Ω
GND

Resistor values shown are nominal.

electrical characteristics over recommended operating free-air temperature range (unless otherwise noted)

PARAMETER		TEST CONDITIONS‡	SN55473			SN75473			UNIT
			MIN	TYP§	MAX	MIN	TYP§	TYP	
V_{IK}	Input clamp voltage	V_{CC} = MIN, I_I = −12 mA		−1.2	−1.5		−1.2	−1.5	V
I_{OH}	High-level output current	V_{CC} = MIN, V_{IH} = MIN, V_{OH} = 70 V			300			100	μA
V_{OL}	Low-level output voltage	V_{CC} = MIN, V_{IL} = 0.8 V, I_{OL} = 100 mA		0.25	0.5		0.25	0.4	V
		V_{CC} = MIN, V_{IL} = 0.8 V, I_{OL} = 300 mA		0.5	0.8		0.5	0.7	
I_I	Input current at maximum input voltage	V_{CC} = MAX, V_I = 5.5 V			1			1	mA
I_{IH}	High-level input current	V_{CC} = MAX, V_I = 2.4 V			40			40	μA
I_{IL}	Low-level input current	V_{CC} = MAX, V_I = 0.4 V		−1	−1.6		−1	−1.6	mA
I_{CCH}	Supply current, outputs high	V_{CC} = MAX, V_I = 5 V		8	11		8	11	mA
I_{CCL}	Supply current, outputs low	V_{CC} = MAX, V_I = 0		58	76		58	76	mA

‡For conditions shown as MIN or MAX, use the appropriate value specified under recommended operating conditions.
§All typical values are at V_{CC} = 5 V, T_A = 25°C.

switching characteristics, V_{CC} = 5 V, T_A = 25°C

PARAMETER		TEST CONDITIONS	MIN	TYP	MAX	UNIT
t_{PLH}	Propagation delay time, low-to-high-level output	$I_O \approx$ 200 mA, C_L = 15 pF, R_L = 50 Ω, See Figure 1		30	55	ns
t_{PHL}	Propagation delay time, high-to-low-level output			25	40	ns
t_{TLH}	Transition time, low-to-high-level output			8	25	ns
t_{THL}	Transition time, high-to-low-level output			10	25	ns
V_{OH}	High-level output voltage after switching	V_S = 55 V, $I_O \approx$ 300 mA, See Figure 2	$V_S - 18$			mV

logic symbol†

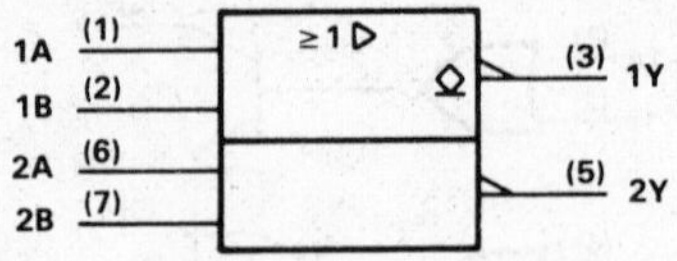
1A (1)
1B (2)
2A (6)
2B (7)
≥1
(3) 1Y
(5) 2Y

†This symbol is in accordance with ANSI/IEEE STD 91-1984 and IEC Publication 617-12.

logic diagram (positive logic)

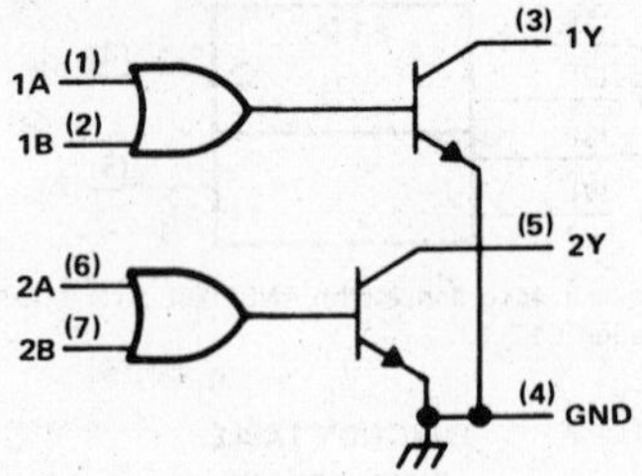
1A (1)
1B (2)
2A (6)
2B (7)
(3) 1Y
(5) 2Y
(4) GND

FUNCTION TABLE
(EACH DRIVER)

A	B	Y
L	L	H (off state)
L	H	L (on state)
H	L	L (on state)
H	H	L (on state)

positive logic:
$Y = \overline{A+B}$ or $\overline{A}\,\overline{B}$

schematic (each driver)

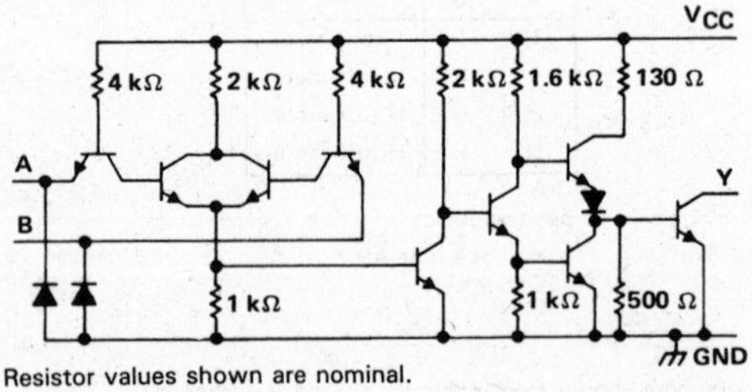
VCC
4 kΩ
2 kΩ
4 kΩ
2 kΩ
1.6 kΩ
130 Ω
A
B
Y
1 kΩ
1 kΩ
500 Ω
GND

Pin numbers shown are for the JG package.

Resistor values shown are nominal.

electrical characteristics over recommended operating free-air temperature range (unless otherwise noted)

PARAMETER		TEST CONDITIONS‡		SN55474 MIN	TYP§	MAX	UNIT
V_{IK}	Input clamp voltage	V_{CC} = MIN,	I_I = −12 mA		−1.2	−1.5	V
I_{OH}	High-level output current	V_{CC} = MIN, V_{OH} = 70 V	V_{IL} = 0.8 V,			300	μA
V_{OL}	Low-level output voltage	V_{CC} = MIN, I_{OL} = 100 mA	V_{IH} = MIN,		0.25	0.5	V
		V_{CC} = MIN, I_{OL} = 300 mA	V_{IH} = 2 V,		0.5	0.8	
I_I	Input current at maximum input voltage	V_{CC} = MAX,	V_I = 5.5 V			1	mA
I_{IH}	High-level input current	V_{CC} = MAX,	V_I = 2.4 V			40	μA
I_{IL}	Low-level input current	V_{CC} = MAX,	V_I = 0.4 V		−1	−1.6	mA
I_{CCH}	Supply current, outputs high	V_{CC} = MAX,	V_I = 0		14	19	mA
I_{CCL}	Supply current, outputs low	V_{CC} = MAX,	V_I = 5 V		67	85	mA

‡For conditions shown as MIN or MAX, use the appropriate value specified under recommended operating conditions.
§All typical values are at V_{CC} = 5 V, T_A = 25°C.

switching characteristics, V_{CC} = 5 V, T_A = 25°C

PARAMETER		TEST CONDITIONS		MIN	TYP	MAX	UNIT
t_{PLH}	Propagation delay time, low-to-high-level output	$I_O \approx$ 200 mA, R_L = 50 Ω,	C_L = 15 pF, See Figure 1		40	65	ns
t_{PHL}	Propagation delay time, high-to-low-level output				30	50	ns
t_{TLH}	Transition time, low-to-high-level output				8	20	ns
t_{THL}	Transition time, high-to-low-level output				10	20	ns
V_{OH}	High-level output voltage after switching	V_S = 55 V, See Figure 2	$I_O \approx$ 300 mA,		V_S−18		mV

PARAMETER MEASUREMENT INFORMATION

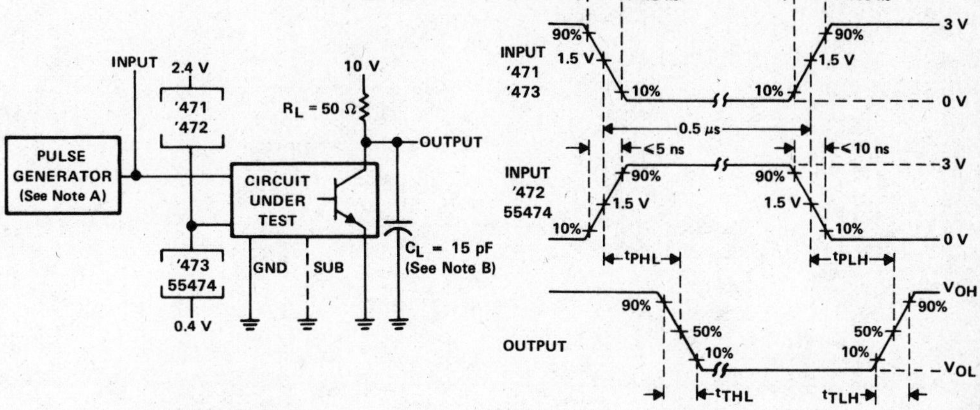
INPUT
2.4 V
'471
'472
10 V
R_L = 50 Ω
OUTPUT
PULSE
GENERATOR
(See Note A)
CIRCUIT
UNDER
TEST
'473
55474
GND
SUB
C_L = 15 pF
(See Note B)
0.4 V
<5 ns
<10 ns
3 V
90%
INPUT
'471
'473
1.5 V
10%
0 V
0.5 μs
INPUT
'472
55474
t_{PHL}
t_{PLH}
V_{OH}
50%
OUTPUT
V_{OL}
t_{THL}
t_{TLH}

TEST CIRCUIT

VOLTAGE WAVEFORMS

NOTES: A. The pulse generator has the following characteristics: PRR ≤ 1 MHz, Z_{out} ≈ 50 Ω.
B. C_L includes probe and jig capacitance.

FIGURE 1. SWITCHING TIMES

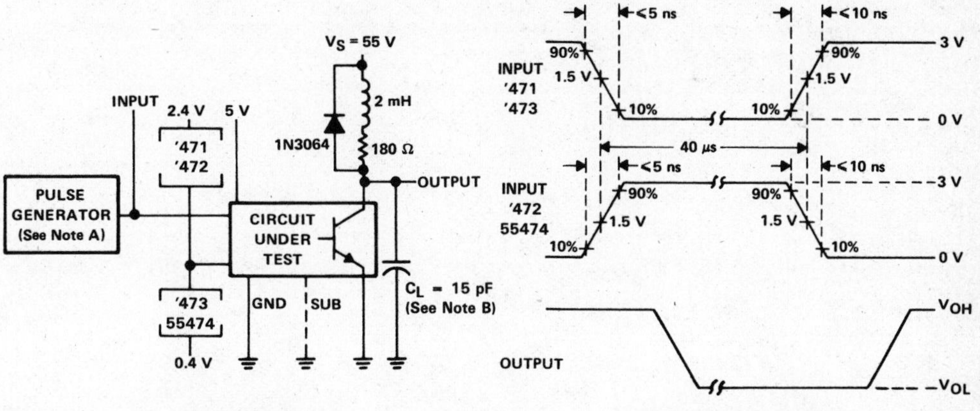
V_S = 55 V
2 mH
1N3064
180 Ω
INPUT
2.4 V
5 V
'471
'472
OUTPUT
PULSE
GENERATOR
(See Note A)
CIRCUIT
UNDER
TEST
'473
55474
GND
SUB
C_L = 15 pF
(See Note B)
0.4 V
<5 ns
<10 ns
3 V
90%
INPUT
'471
'473
1.5 V
10%
0 V
40 μs
INPUT
'472
55474
V_{OH}
OUTPUT
V_{OL}

TEST CIRCUIT

VOLTAGE WAVEFORMS

NOTES: A. The pulse generator has the following characteristics: PRR ≤ 12.5 kHz, Z_{out} ≈ 50 Ω.
B. C_L includes probe and jig capacitance.

FIGURE 2. LATCH-UP TEST

D2284, DECEMBER 1976—REVISED SEPTEMBER 1986

- **Characterized for Use to 300 mA**
- **No Output Latch-Up at 55 V (After Conducting 300 mA)**
- **High-Voltage Outputs (100 V Typical)**
- **Output Clamp Diodes for Transient Suppression (300 mA, 70 V)**
- **TTL- or MOS-Compatible Diode-Clamped Inputs**
- **P-N-P Inputs Reduce Input Current**
- **Standard Supply Voltage**
- **Suitable for Hammer-Driver Applications**
- **New Plastic DIP (P) with Copper Lead Frame Provides Cooler Operation and Improved Reliability**

D OR P PACKAGE
(TOP VIEW)

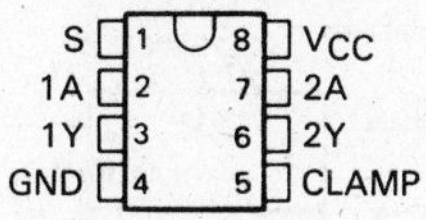
S 1
1A 2
1Y 3
GND 4
8 VCC
7 2A
6 2Y
5 CLAMP

description

Series 75476 dual peripheral drivers are designed for use in systems that require high current, high voltage, and fast switching times. The SN75476, SN75477, SN75478, and SN75479 provide AND, NAND, OR, and NOR drivers, respectively. These devices have diode-clamped inputs as well as high-current, high-voltage clamp diodes on the outputs for inductive transient protection.

The SN75476, SN75477, SN75478, and SN75479 drivers are characterized for operation from 0°C to 70°C.

FUNCTION TABLES

SN75476
(EACH AND DRIVER)

INPUTS		OUTPUT
A	S	Y
H	H	H
L	X	L
X	L	L

SN75477
(EACH NAND DRIVER)

INPUTS		OUTPUT
A	S	Y
H	H	L
L	X	H
X	L	H

SN75478
(EACH OR DRIVER)

INPUTS		OUTPUT
A	S	Y
H	X	H
X	H	H
L	L	L

SN75479
(EACH NOR DRIVER)

INPUTS		OUTPUT
A	S	Y
H	X	L
X	H	L
L	L	H

H = high level
L = low level
X = irrelevant

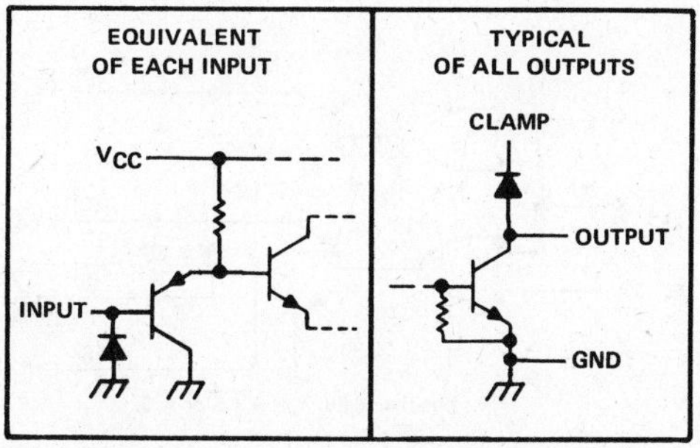
EQUIVALENT
OF EACH INPUT
VCC
INPUT
TYPICAL
OF ALL OUTPUTS
CLAMP
OUTPUT
GND

TEXAS INSTRUMENTS
POST OFFICE BOX 655012 • DALLAS, TEXAS 75265

logic symbols†

SN75476

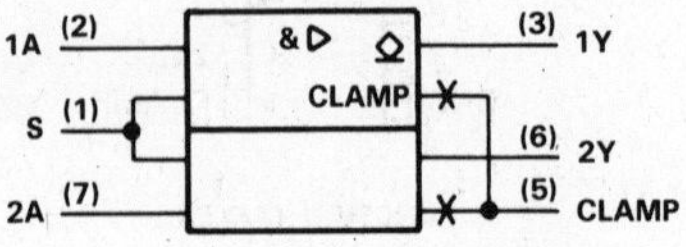
1A (2)
S (1)
2A (7)
&
CLAMP
(3) 1Y
(6) 2Y
(5) CLAMP

SN75477

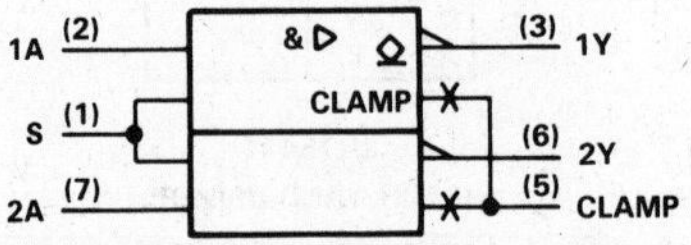
1A (2)
S (1)
2A (7)
&
CLAMP
(3) 1Y
(6) 2Y
(5) CLAMP

SN75478

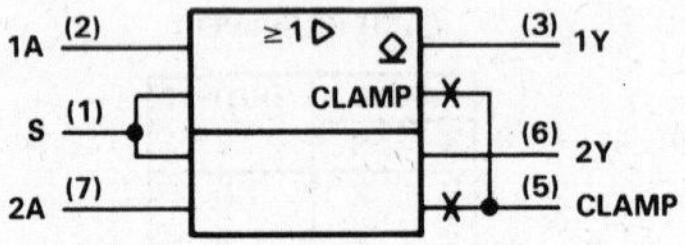
1A (2)
S (1)
2A (7)
≥1
CLAMP
(3) 1Y
(6) 2Y
(5) CLAMP

SN75479

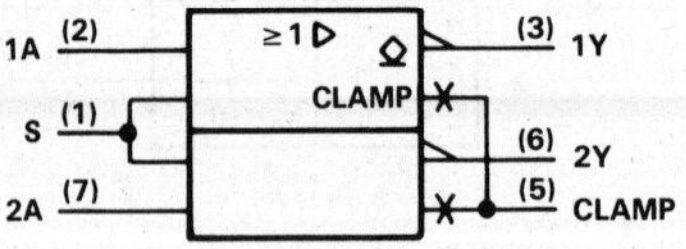
1A (2)
S (1)
2A (7)
≥1
CLAMP
(3) 1Y
(6) 2Y
(5) CLAMP

logic diagrams (positive logic)

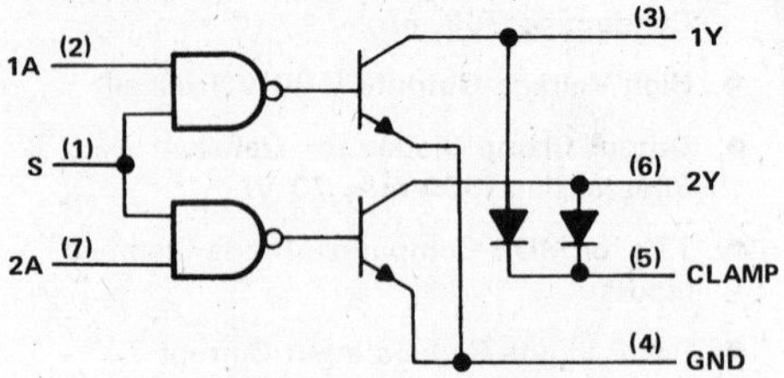
1A (2)
S (1)
2A (7)
(3) 1Y
(6) 2Y
(5) CLAMP
(4) GND

positive logic: Y = AS or $\overline{\overline{A}+\overline{S}}$

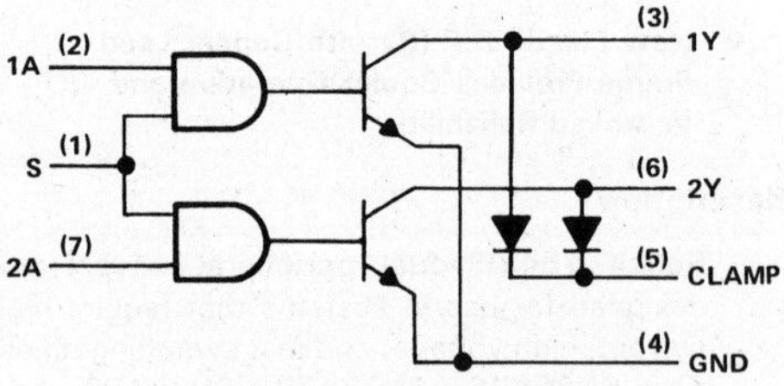
1A (2)
S (1)
2A (7)
(3) 1Y
(6) 2Y
(5) CLAMP
(4) GND

positive logic: Y = $\overline{AS}$ or $\overline{A}+\overline{S}$

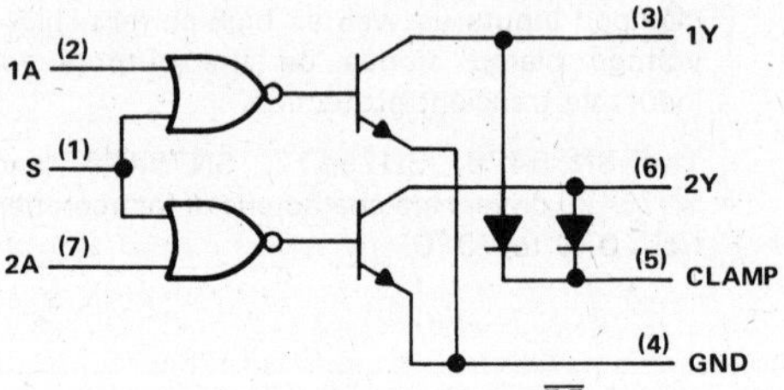
1A (2)
S (1)
2A (7)
(3) 1Y
(6) 2Y
(5) CLAMP
(4) GND

positive logic: Y = A + S or $\overline{\overline{A}\,\overline{S}}$

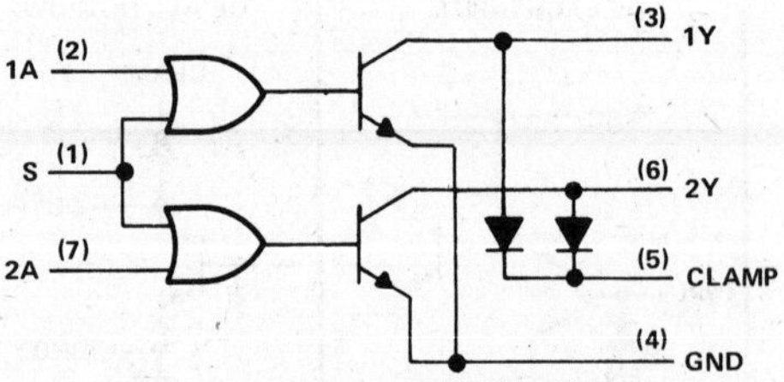
1A (2)
S (1)
2A (7)
(3) 1Y
(6) 2Y
(5) CLAMP
(4) GND

positive logic: Y = $\overline{A+S}$ or $\overline{A}\,\overline{S}$

† These symbols are in accordance with ANSI/IEEE Std 91-1984 and IEC Publication 617-12.

5

Peripheral Drivers/Actuators

absolute maximum ratings over operating free-air temperature range (unless otherwise noted)

Supply voltage, V_{CC} (see Note 1) . . . 7 V
Input voltage . . . 5.5 V
Continuous output current (see Note 2) . . . 400 mA
Peak output current: $t_w \leq 10$ ms, duty cycle $\leq 50\%$ . . . 500 mA
$t_w \leq 30$ ns, duty cycle $\leq 0.002\%$ . . . 3 A
Output clamp diode current . . . 400 mA
Continuous total dissipation at (or below) 25°C free-air temperature (see Note 3):
D package . . . 725 mW
P package . . . 1200 mW
Storage temperature range . . . −65°C to 150°C
Lead temperature 1,6 mm (1/16 inch) from case for 10 seconds . . . 260°C

NOTES: 1. Voltage values are with respect to network ground terminal.
2. Both halves of this dual circuit may conduct rated current simultaneously; however, power dissipation averaged over a short time interval must fall within the continuous dissipation ratings.
3. For operation above 25°C free-air temperature, derate the D package at the rate of 5.8 mW/°C and the P package at the rate of 9.6 mW/°C.

recommended operating conditions

	MIN	NOM	MAX	UNIT
Supply voltage, V_{CC}	4.5	5	5.5	V
High-level input voltage, V_{IH}	2			V
Low-level input voltage, V_{IL}			0.8	V
Operating free-air temperature, T_A	0		70	°C

electrical characteristics over recommended operating free-air temperature range (unless otherwise noted)

PARAMETER			TEST CONDITIONS		MIN	TYP†	MAX	UNIT
V_{IK}	Input clamp voltage		$I_I = -12$ mA			−0.95	−1.5	V
I_{OH}	High-level output current		$V_{CC} = 4.5$ V, $V_{IL} = 0.8$ V,	$V_{IH} = 2$ V, $V_{OH} = 70$ V		1	100	μA
V_{OL}	Low-level output voltage		$V_{CC} = 4.5$ V, $V_{IH} = 2$ V, $V_{IL} = 0.8$ V	$I_{OL} = 100$ mA		0.16	0.3	V
				$I_{OL} = 175$ mA		0.22	0.5	
				$I_{OL} = 300$ mA		0.33	0.6	
$V_{(BR)O}$	Output breakdown voltage		$V_{CC} = 4.5$ V,	$I_{OH} = 100$ μA	70	100		V
$V_{R(K)}$	Output clamp diode reverse voltage		$V_{CC} = 4.5$ V,	$I_R = 100$ μA	70	100		V
$V_{F(K)}$	Output clamp diode forward voltage		$V_{CC} = 4.5$ V,	$I_F = 300$ mA	0.8	1.15	1.6	V
I_{IH}	High-level input current		$V_{CC} = 5.5$ V,	$V_I = 5.5$ V		0.01	10	μA
I_{IL}	Low-level input current	A input	$V_{CC} = 5.5$ V,	$V_I = 0.8$ V		−80	−110	μA
		Strobe S				−160	−220	
I_{CCH}	Supply current, outputs high	SN75476	$V_{CC} = 5.5$ V	$V_I = 5$ V		10	17	mA
		SN75477		$V_I = 0$		10	17	
		SN75478		$V_I = 5$ V		10	17	
		SN75479		$V_I = 0$		10	17	
I_{CCL}	Supply current, outputs low	SN75476	$V_{CC} = 5.5$ V	$V_I = 0$		54	75	mA
		SN75477		$V_I = 5$ V		54	75	
		SN75478		$V_I = 0$		54	75	
		SN75479		$V_I = 5$ V		54	75	

†All typical values are at $V_{CC} = 5$ V, $T_A = 25$°C.

switching characteristics, V_{CC} = 5 V, T_A = 25°C

PARAMETER		TEST CONDITIONS	MIN	TYP	MAX	UNIT
t_{PLH}	Propagation delay time, low-to-high-level output	C_L = 15 pF, R_L = 100 Ω, See Figure 1		200	350	ns
t_{PHL}	Propagation delay time, high-to-low-level output			200	350	ns
t_{TLH}	Transition time, low-to-high-level output			50	125	ns
t_{THL}	Transition time, high-to-low-level output			90	125	ns
V_{OH}	High-level output voltage after switching	V_S = 55 V, I_O ≈ 300 mA, See Figure 2	$V_S - 18$			mV

PARAMETER MEASUREMENT INFORMATION

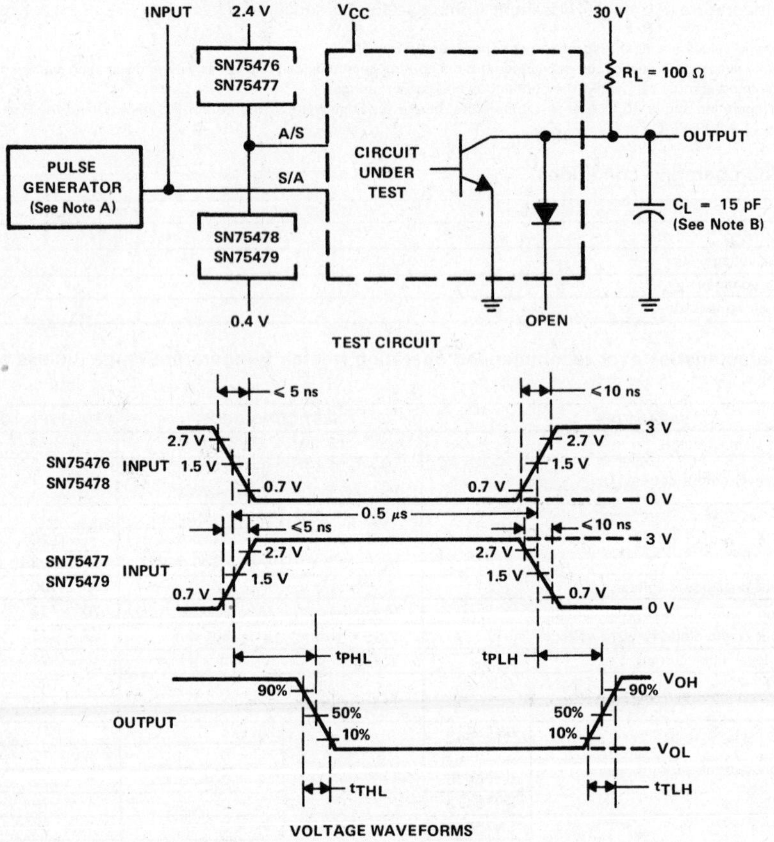
INPUT
2.4 V
VCC
30 V
SN75476
SN75477
RL = 100 Ω
A/S
S/A
OUTPUT
CIRCUIT
UNDER
TEST
PULSE
GENERATOR
(See Note A)
CL = 15 pF
(See Note B)
SN75478
SN75479
0.4 V
OPEN
TEST CIRCUIT
≤ 5 ns
≤10 ns
3 V
2.7 V
SN75476 INPUT 1.5 V
SN75478
0.7 V
0 V
0.5 μs
<5 ns
SN75477
SN75479
INPUT
1.5 V
tPHL
tPLH
VOH
90%
OUTPUT
50%
10%
VOL
tTHL
tTLH
VOLTAGE WAVEFORMS

NOTES: A. The pulse generator has the following characteristics: PRR = 1 MHz, Z_{out} = 50 Ω.
B. C_L includes probe and jig capacitance.

FIGURE 1. SWITCHING CHARACTERISTICS

PARAMETER MEASUREMENT INFORMATION

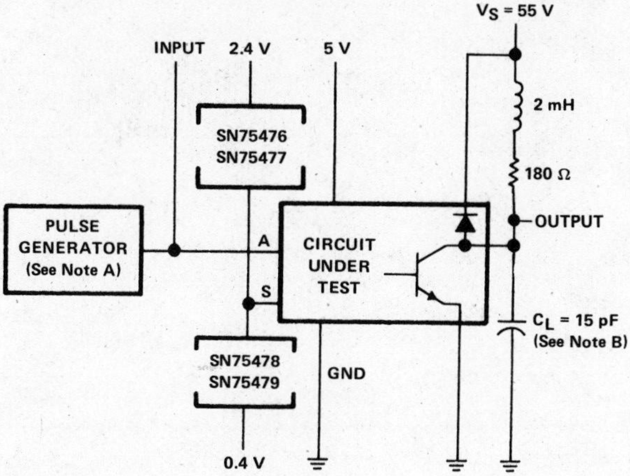
VS = 55 V
INPUT
2.4 V
5 V
SN75476
SN75477
2 mH
180 Ω
OUTPUT
PULSE
GENERATOR
(See Note A)
A
CIRCUIT
UNDER
TEST
S
CL = 15 pF
(See Note B)
SN75478
SN75479
GND
0.4 V

TEST CIRCUIT

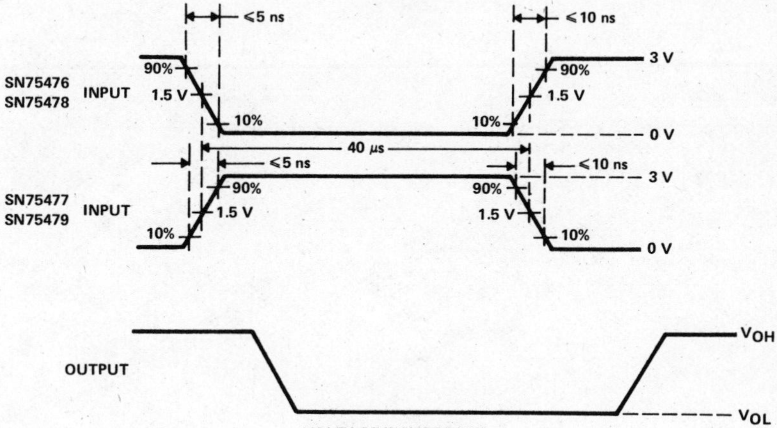
≤5 ns
≤10 ns
3 V
90%
SN75476
SN75478
INPUT
1.5 V
10%
0 V
40 μs
≤5 ns
≤10 ns
3 V
90%
SN75477
SN75479
INPUT
1.5 V
10%
0 V
VOH
OUTPUT
VOL

VOLTAGE WAVEFORMS

NOTES: A. The pulse generator has the following characteristics: PRR = 12.5 kHz, Z_{out} = 50 Ω.
B. C_L includes probe and jig capacitance.

FIGURE 2. LATCH-UP TEST

D2832, MARCH 1984—REVISED OCTOBER 1986

- Three-State Outputs
- Continuous Output Current of ±2 A
- Outputs Can Switch 40 V
- Transient Suppression
- Thermal Shutdown
- Inputs Compatible with TTL and 5-V CMOS
- V_{CC} Range: 8 V to 40 V

KC, KH, AND KV PACKAGES

(TOP VIEW)

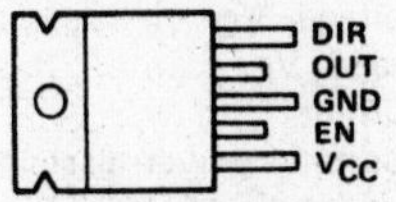
DIR
OUT
GND
EN
VCC

GROUND TERMINAL IS IN ELECTRICAL CONTACT WITH MOUNTING BASE

description

The SN75603, SN75604, and SN75605 are high-current half-H drivers designed for high-current switching of bidirectional loads at voltages from 8 V to 40 V. The devices are ideal for the switching of bidirectional dc and stepping motors.

FUNCTION TABLE

INPUTS		OUTPUT		
EN	DIR	SN75603	SN75604	SN75605
L	L	Z	Z	L
L	H	Z	Z	Z
H	L	L	H	Z
H	H	H	L	H

The SN75603 and SN75604 are designed to be used together in pairs, which eliminates the need for additional control logic. The SN75605 is a functional replacement for Sprague UDN2949. By controlling the enable and direction inputs, these devices may be placed in the high-impedance output state.

logic symbols

SN75603

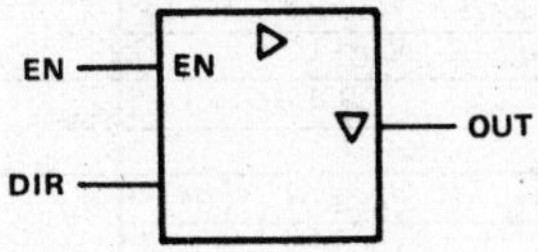
EN
EN
OUT
DIR

SN75604

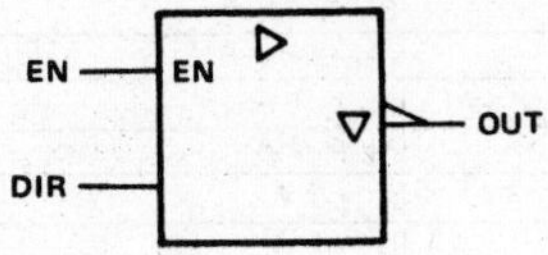
EN
EN
OUT
DIR

SN75605

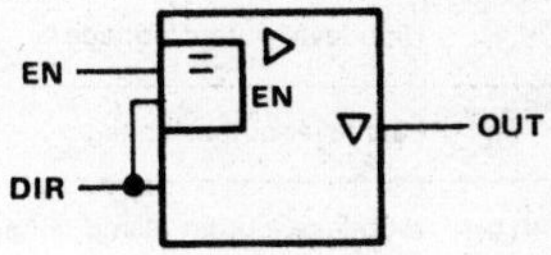
EN
=
EN
OUT
DIR

logic diagrams (positive logic)

SN75603

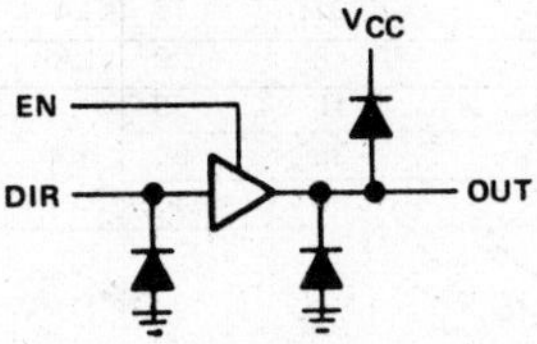
VCC
EN
DIR
OUT

SN75604

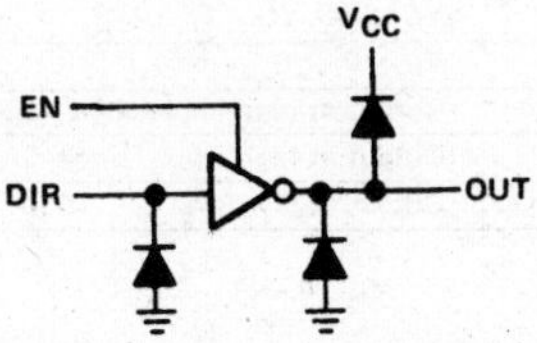
VCC
EN
DIR
OUT

SN75605

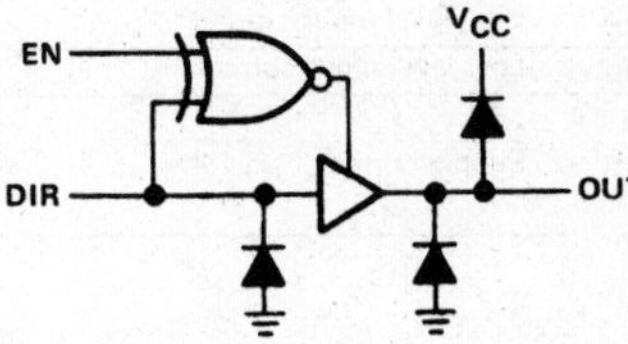
VCC
EN
DIR
OUT

absolute maximum ratings over operating free-air temperature range (unless otherwise noted)

Supply voltage, V_{CC} .... 40 V
Output voltage, V_O .... 42 V
Input voltage, V_I .... 40 V
Output current, I_O .... ±2.5 A
Continuous total power dissipation at (or below) 110°C case temperature (see Note 1) .... 10 W
Continuous total power dissipation at (or below) 25°C free-air temperature (see Note 2) .... 2 W
Operating case or virtual junction temperature range .... −40°C to 150°C
Storage temperature range .... −65°C to 150°C
Lead temperature 1,6 mm (1/16 inch) from case for 10 seconds .... 260°C

NOTES: 1. For operation above 110°C case temperature, derate linearly at the rate of 250 mW/°C.
2. For operation above 25°C free-air temperature, derate linearly at the rate of 16 mW/°C.

recommended operating conditions

		MIN	NOM	MAX	UNIT
V_{CC}	Supply voltage	8		40	V
V_{IH}	High-level input voltage	2			V
V_{IL}	Low-level input voltage			0.8	V
T_C	Case temperature	−40		125	°C
T_J	Junction temperature	−40		150	°C

electrical characteristics, V_{CC} = 8 V to 40 V, T_A = 25°C

PARAMETER		TEST CONDITIONS	MIN	TYP	MAX	UNIT
V_{IK}	Input clamp voltage	I_I = 12 mA		−0.9	−1.5	V
V_{OH}	High-level output voltage	I_{OH} = −1 A	V_{CC}−1.5	V_{CC}−0.9		V
		I_{OH} = −2 A	V_{CC}−2	V_{CC}−1.2		
V_{OL}	Low-level output voltage	I_{OL} = 1 A		0.9	1.5	V
		I_{OL} = 2 A		1.1	2	
V_{OKH}	High-level output clamp voltage	I_O = 1 A		V_{CC}+1.2	V_{CC}+1.5	V
		I_O = 2 A		V_{CC}+1.4	V_{CC}+2	
V_{OKL}	Low-level output clamp voltage	I_O = −1 A		−1.2	−1.5	V
		I_O = −2 A		−1.4	−2	
I_{OZ}	High-impedance-state output current	V_O = 40 V		0.1	100	μA
		V_O = 0		−0.1	−100	
I_{IH}	High-level input current	V_I = 5.5 V		0.01	10	μA
I_{IL}	Low-level input current	V_I = 0		−8	−20	μA
I_{CC}	Supply current	Output at high impedance		16	30	mA
		Output at high level		35	50	
		Output at low level		30	40	

switching characteristics, T_A = 25°C

PARAMETER		TEST CONDITIONS	MIN	TYP	MAX	UNIT
V_{OZH}	High-impedance-state output voltage after switching with high-level voltage applied	V_{CC} = 40 V, $I_{OL} \approx$ 2 A, L = 2 mH, R = 19 Ω, C_L = 15 pF, See Figure 1	$V_{CC} - 10$			mV
V_{OZL}	High-impedance-state output voltage after switching with low-level voltage applied	V_{CC} = 40 V, $I_{OH} \approx$ −2 A, L = 2 mH, R = 19 Ω, C_L = 15 pF, See Figure 2			10	mV
t_{PZH}	Enable time to the high level	V_{CC} = 25 V, V_{CC} = 25 Ω, C_L = 15 pF, See Figure 3		1.3		μs
t_{PHZ}	Disable time from the high level			1.8		μs
t_{TZH}	Enable transition time to the high level			70		ns
t_{THZ}	Disable transition time from the high level			500		ns
t_{PZL}	Enable time to the low level	V_{CC} = 25 V, R_L = 25 Ω, C_L = 15 pF, See Figure 4		1		μs
t_{PLZ}	Disable time from the low level			2.5		μs
t_{TZL}	Enable transition time to the low level			100		ns
t_{TLZ}	Disable transition time from the low level			100		ns

PARAMETER MEASUREMENT INFORMATION

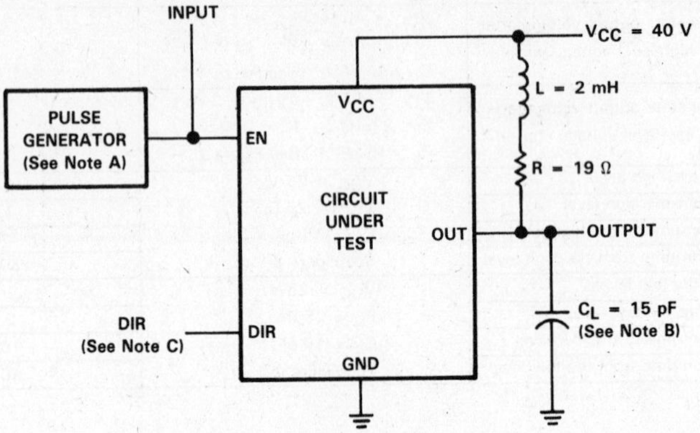
INPUT
V_{CC} = 40 V
L = 2 mH
PULSE GENERATOR (See Note A)
V_{CC}
EN
R = 19 Ω
CIRCUIT UNDER TEST
OUT
OUTPUT
C_L = 15 pF (See Note B)
DIR (See Note C)
DIR
GND

TEST CIRCUIT

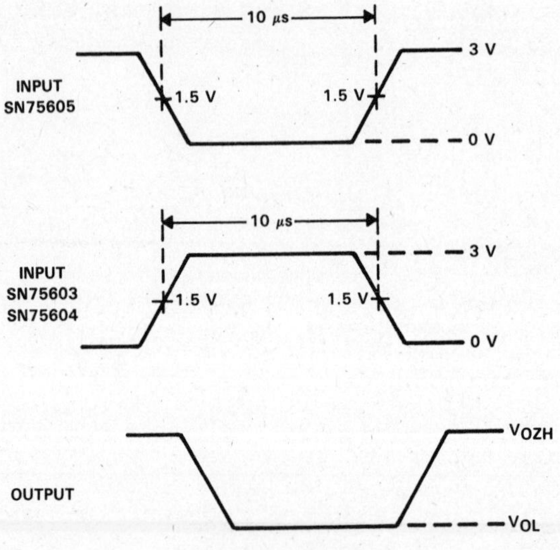
10 μs
3 V
INPUT SN75605
1.5 V
1.5 V
0 V
10 μs
3 V
INPUT SN75603 SN75604
1.5 V
1.5 V
0 V
V_{OZH}
OUTPUT
V_{OL}

VOLTAGE WAVEFORMS

NOTES: A. The pulse generator has the following characteristics: PRR = 50 kHz, Z_{out} = 50 Ω, $t_r \le 5$ ns, $t_f \le 5$ ns.
B. C_L includes probe and jig capacitance.
C. DIR is low for SN75603 and SN75605, and high for SN75604.

FIGURE 1. LATCH-UP TEST

PARAMETER MEASUREMENT INFORMATION

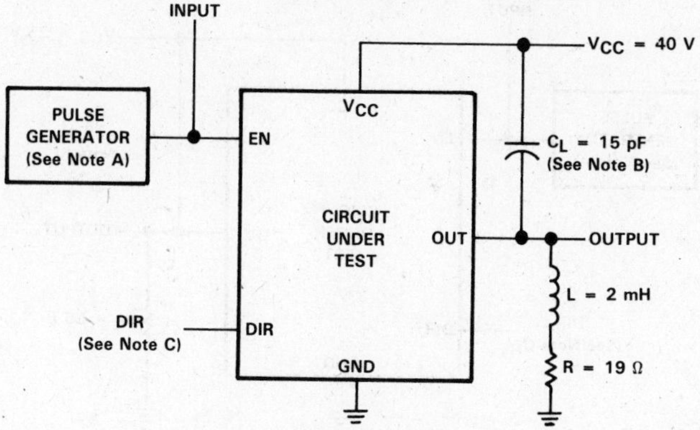
INPUT
PULSE
GENERATOR
(See Note A)
EN
VCC
VCC = 40 V
CL = 15 pF
(See Note B)
CIRCUIT
UNDER
TEST
OUT
OUTPUT
L = 2 mH
DIR
(See Note C)
DIR
GND
R = 19 Ω

TEST CIRCUIT

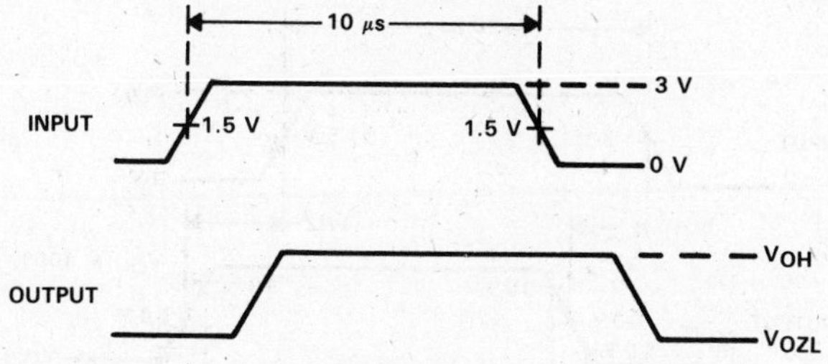
10 μs
3 V
INPUT
1.5 V
1.5 V
0 V
VOH
OUTPUT
VOZL

VOLTAGE WAVEFORMS

NOTES: A. The pulse generator has the following characteristics: PRR = 50 kHz, Z_{out} = 50 Ω, $t_r \leq 5$ ns, $t_f \leq 5$ ns.
B. C_L includes probe and jig capacitance.
C. DIR is low for SN75603 and SN75605, and high for SN75604.

FIGURE 2. LATCH-UP TEST

PARAMETER MEASUREMENT INFORMATION

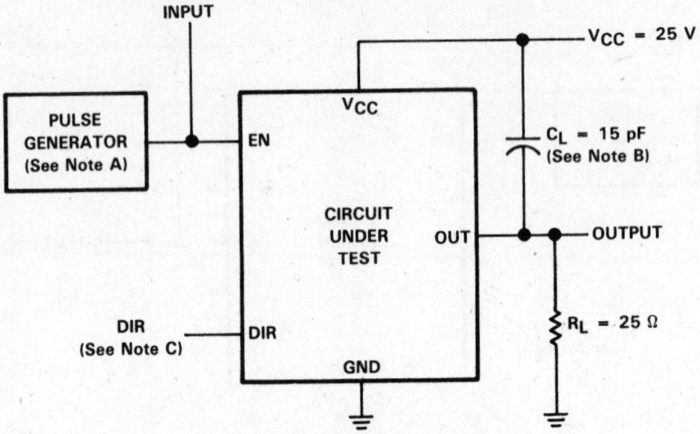
INPUT
VCC = 25 V
PULSE
GENERATOR
(See Note A)
EN
VCC
CL = 15 pF
(See Note B)
CIRCUIT
UNDER
TEST
OUT
OUTPUT
DIR
(See Note C)
DIR
RL = 25 Ω
GND

TEST CIRCUIT

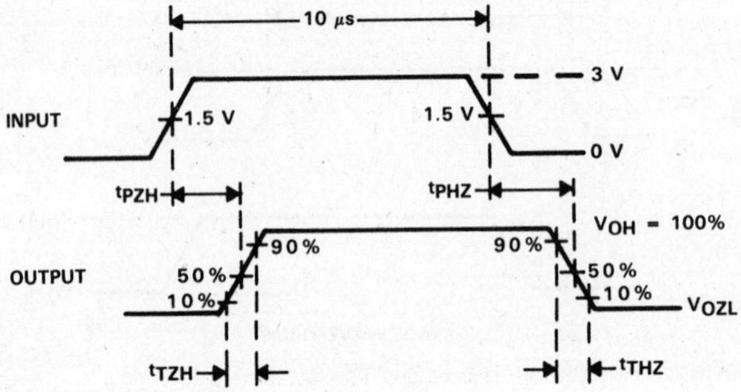
10 μs
3 V
INPUT
1.5 V
1.5 V
0 V
tPZH
tPHZ
VOH = 100%
90%
90%
OUTPUT
50%
10%
50%
10%
VOZL
tTZH
tTHZ

VOLTAGE WAVEFORMS

NOTES: A. The pulse generator has the following characteristics: PRR = 50 kHz, Z_{out} 50 Ω, $t_r \leq 5$ ns, $t_f \leq 5$ ns.
B. C_L includes probe and jig capacitance.
C. DIR is high for SN75603 and SN75605, and low for SN75604.

FIGURE 3. SWITCHING TIMES, ENABLE TIME TO HIGH-LEVEL AND DISABLE TIME FROM HIGH-LEVEL

PARAMETER MEASUREMENT INFORMATION

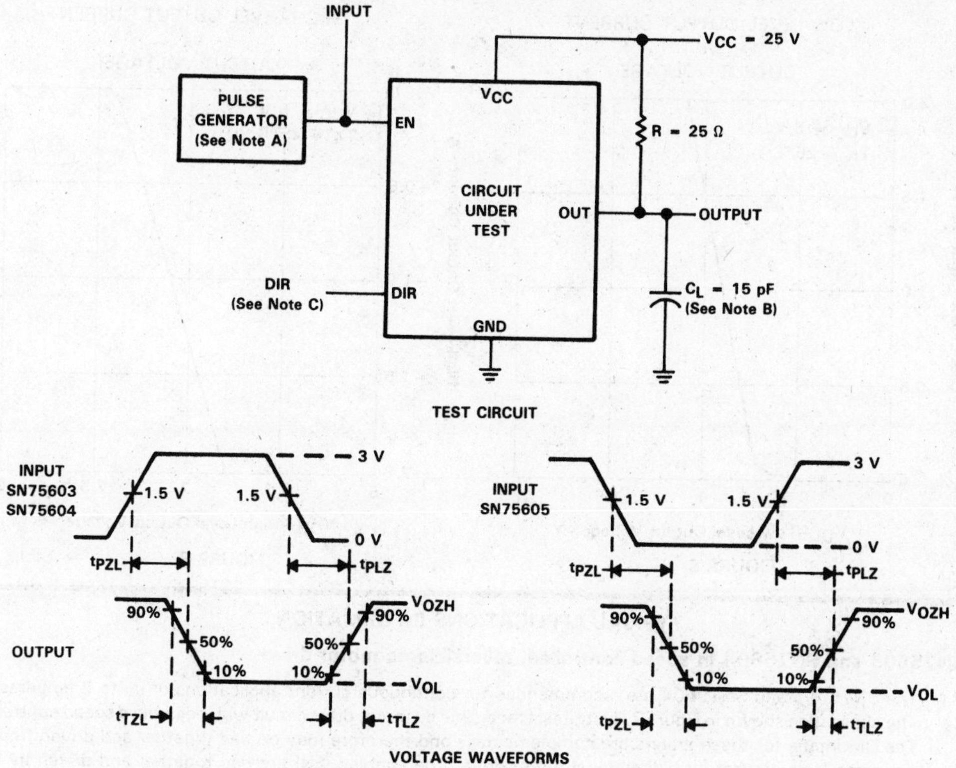
INPUT
PULSE GENERATOR (See Note A)
EN
VCC
VCC = 25 V
R = 25 Ω
CIRCUIT UNDER TEST
OUT
OUTPUT
DIR (See Note C)
DIR
GND
CL = 15 pF (See Note B)
TEST CIRCUIT
INPUT SN75603 SN75604
1.5 V
1.5 V
3 V
0 V
tPZL
tPLZ
90%
50%
10%
VOZH
VOL
OUTPUT
tTZL
tTLZ
INPUT SN75605
VOLTAGE WAVEFORMS

NOTES: A. The pulse generator has the following characteristics: PRR = 50 kHz, Z_{out} = 50 Ω, $t_r \leq 5$ ns, $t_f \leq 5$ ns.
B. C_L includes probe and jig capacitance.
C. DIR is low for SN75603 and SN75605, and high for SN75604.

FIGURE 4. SWITCHING TIMES, ENABLE TIME TO LOW LEVEL, AND DISABLE TIME FROM LOW LEVEL

TYPICAL CHARACTERISTICS

LOW-LEVEL OUTPUT CURRENT
vs
OUTPUT VOLTAGE

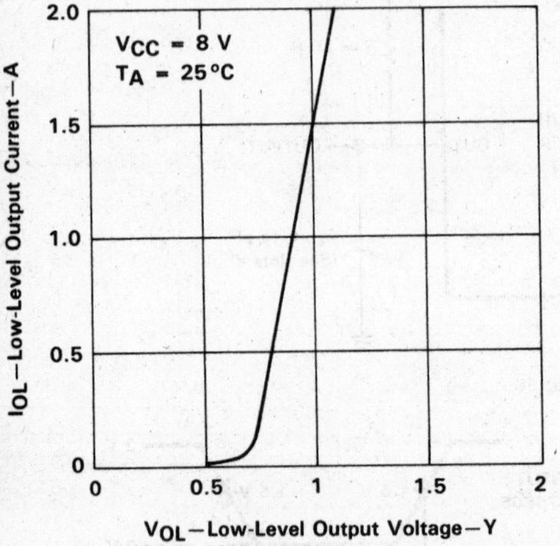
2.0
1.5
1.0
0.5
0
V_{CC} = 8 V
T_A = 25°C
I_{OL}—Low-Level Output Current—A
0
0.5
1
1.5
2
V_{OL}—Low-Level Output Voltage—Y

FIGURE 5

HIGH-LEVEL OUTPUT CURRENT
vs
OUTPUT VOLTAGE

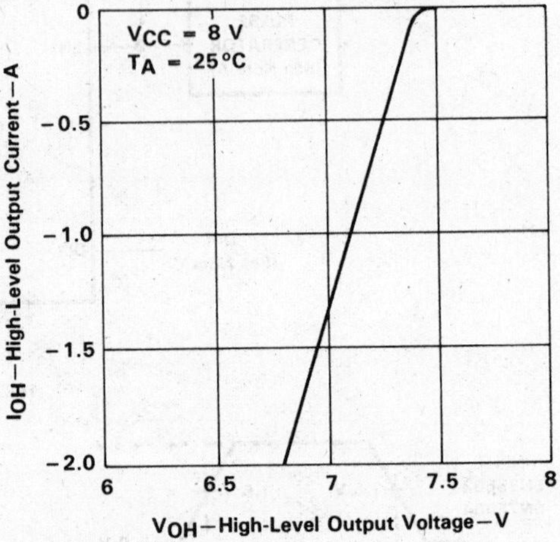
0
−0.5
−1.0
−1.5
−2.0
V_{CC} = 8 V
T_A = 25°C
I_{OH}—High-Level Output Current—A
6
6.5
7
7.5
8
V_{OH}—High-Level Output Voltage—V

FIGURE 6

TYPICAL APPLICATIONS INFORMATION

SN75603 and SN75604 in speed-controlled, reversible dc motor drive

The SN75603 and SN75604 are recommended for continuous current applications of up to 2 amperes. The application shown in Figure 7 illustrates a reversible dc motor drive circuit with adjustable speed control. The DIR inputs for these drivers are complementary and therefore may be tied together and driven from the same logic control for bidirectional motor drive. The enables (EN) are tied together and driven by a pulse-width-modulated generator providing ''on'' duty cycles of 10% to 90% for speed control. A separate enable control is provided through a SN7409 logic gate.

TEXAS INSTRUMENTS
POST OFFICE BOX 655012 • DALLAS, TEXAS 75265

TYPICAL APPLICATIONS INFORMATION

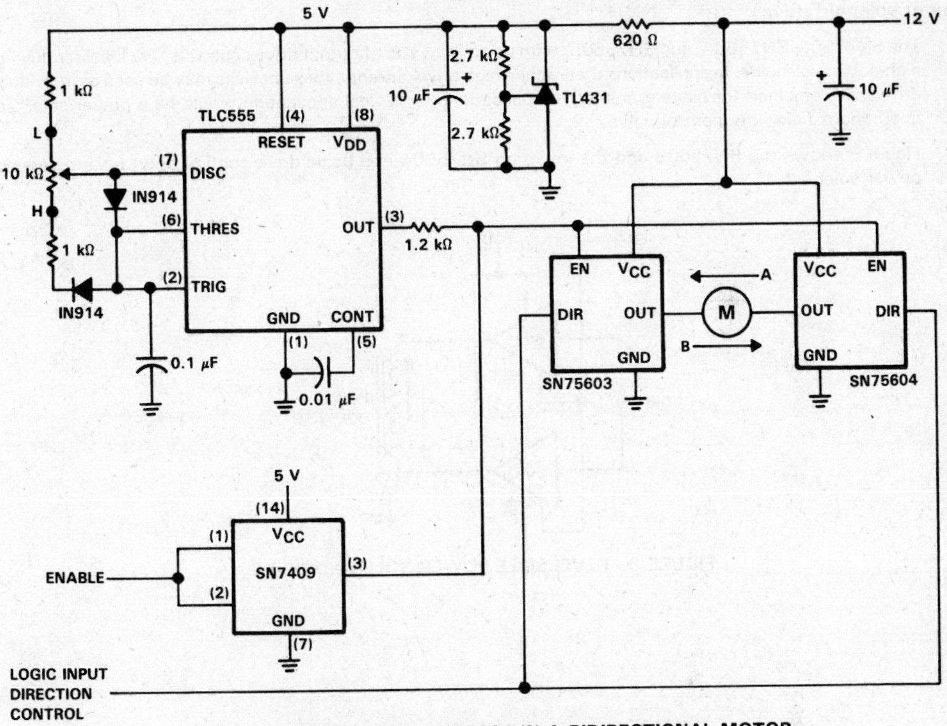
5 V
12 V
620 Ω
2.7 kΩ
10 µF
TL431
2.7 kΩ
10 µF
1 kΩ
L
TLC555
(4)
(8)
RESET
VDD
(7)
10 kΩ
DISC
IN914
H
(6)
THRES
OUT
(3)
1.2 kΩ
1 kΩ
(2)
TRIG
IN914
GND
CONT
(1)
(5)
0.1 µF
0.01 µF
EN
VCC
A
VCC
EN
DIR
OUT
M
OUT
DIR
B
GND
GND
SN75603
SN75604
5 V
(14)
(1)
VCC
ENABLE
SN7409
(3)
(2)
GND
(7)
LOGIC INPUT
DIRECTION
CONTROL

FIGURE 7. SN75603 AND SN75604 IN A BIDIRECTIONAL MOTOR CONTROL APPLICATION WITH SPEED CONTROL

FUNCTIONAL TABLE FOR MOTOR CONTROL CIRCUIT

EN	DC	SPC	MOTOR DIRECTION	MOTOR SPEED
L	X	X	OFF	OFF
H	L	N	A	SLOW
H	L	W	A	FAST
H	H	N	B	SLOW
H	H	W	B	FAST

DEFINITION OF TERMS USED IN FUNCTION TABLE

EN. . . . . .Enable
DC . . . . .Direction control
SPC . . . .Speed control
A . . . . . .Direction of current—right to left
B . . . . . .Direction of current—left to right
H . . . . . .High logic level
L. . . . . . .Low logic level
N . . . . . .Speed control set for narrow pulse width
W . . . . . .Speed control set for wide pulse width
X . . . . . .Irrelevant

TYPICAL APPLICATIONS INFORMATION

power solenoid drive

The SN75603, SN75604, and SN75605, with up to 70 watts of output drive capability, are ideal for driving high-power solenoids. In applications that require high drive currents, these devices may be used as individual drivers or combined for bidirectional drive applications. A typical application would be a power solenoid operating a fluid-flow-control valve.

Figure 8 shows the SN75603 and the inverting SN75604 in a basic drive configuration for a reversing power solenoid.

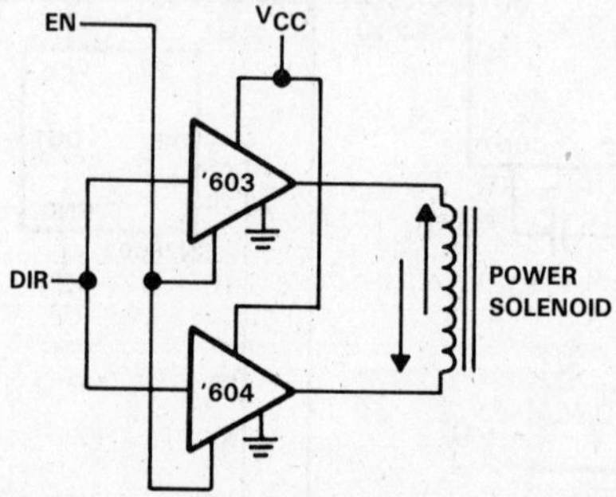
EN
V_{CC}
'603
DIR
'604
POWER
SOLENOID

FIGURE 8. REVERSIBLE POWER SOLENOID DRIVE

TEXAS INSTRUMENTS
POST OFFICE BOX 655012 • DALLAS, TEXAS 75265

SN75608
DUAL FLUX-REGULATING ACTUATOR

D2955, DECEMBER 1986

- 2.5-A Current Capability per Channel
- For Split-Supply Applications
- Wide Differential Supply Voltage Range . . . 30 V to 60 V
- High-Impedance Clamped Inputs Compatible with TTL or CMOS Devices
- Output Clamp Diodes for Inductive Transient Suppression
- Thermal Shutdown
- Internal ESD Protection
- Short-Circuit Protection on Sink Outputs
- Input Hysteresis Improves Noise Immunity
- No Output Glitch During Power-Up or Power-Down
- 15-Pin SIP Power Package

KV SINGLE-IN-LINE PACKAGE
(TOP VIEW)

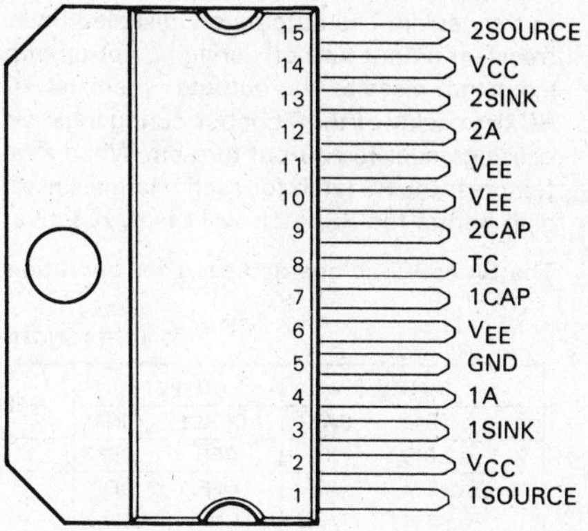
15 2SOURCE
14 V_{CC}
13 2SINK
12 2A
11 V_{EE}
10 V_{EE}
9 2CAP
8 TC
7 1CAP
6 V_{EE}
5 GND
4 1A
3 1SINK
2 V_{CC}
1 1SOURCE

The tab is electrically connected to the V_{EE} pins.

description

The SN75608 is a high-current dual flux-regulating actuator designed for switching double-ended loads with currents up to 2.5 amperes at differential supply voltages from 30 volts to 60 volts. It is designed to drive and control the electromagnetic flux in printheads, solenoids, relays, and other loads whose inductance value varies during operation.

The SN75608 performs the function of flux regulation under control of standard TTL or CMOS input signals for two independent channels. Flux is proportional to the integral of the inductive-load voltage. It is a function of the total amount of current in the load and is the magnetic field maintained in the load. With flux regulation, the load current will vary to compensate for core saturation, temperature changes, and other variations of load inductance during operation while maintaining controlled, relatively constant flux in the load.

Each channel has separate sink and source driver outputs for driving each end of the inductive load. Internal feedback, consisting of an integrator and voltage comparator, provides flux regulation via chop-mode operation of the source output. The integrator circuit provides current to the capacitor terminal (CAP) proportional to the differential voltage between the sink and source outputs for each channel. The integrator requires an external capacitor connected between the CAP and V_{EE} terminals. The voltage at the CAP terminal, referenced to V_{EE}, is proportional to the integral of the source to sink (load) voltage.

The feedback path is completed by a differential voltage comparator that controls the state of the source output. The inverting comparator input is connected to the CAP terminal, and the noninverting input is connected to an analog voltage, $V_{ref(TC)}$, which is referenced to V_{EE}. $V_{ref(TC)}$ is proportional to the Threshold Control (TC) voltage, which is referenced to ground. The comparator hysteresis controls the charge and discharge voltage excursions at the CAP terminal and thus controls the on and off time of the source output chopper.

TEXAS INSTRUMENTS
POST OFFICE BOX 655012 • DALLAS, TEXAS 75265

description (continued)

The SN75608 features built-in thermal protection and a sink output over-current sensor to prevent damage to the device. The outputs are disabled under low-V_{CC} or low-V_{EE} supply-voltage conditions to prevent transient output turn-on during power-up or power-down. The TC input is a combined threshold and logic input that disables the outputs when the TC input voltage is less than 0.8 V. This permits an external RC time delay at the TC input during logic system power-up to allow logic at input A to stabilize without causing undesired output turn-on. When a fault condition is detected by any one of these five protection features, the RS latch for each channel is set. The fault condition must be removed and the A input taken high before the RS latch will reset, reactivating the channel.

The SN75608 is characterized for operation from −20°C to 85°C.

DRIVER FUNCTION TABLE (EACH CHANNEL)

INPUTS			OUTPUTS		OPERATING MODE	COMMENTS
A	TC	CAP	SOURCE	SINK		
X	≤0.8 V	X	OFF	OFF	Disabled	TC acts as a digital input referenced to GND
H	≥2 V	X	OFF	OFF	Active	TC acts an an analog input referenced to GND (See Note 1)
L	≥2 V	$<V_{T+}$	ON	ON		
L	≥2 V	$>V_{T-}$	OFF	ON		

INTEGRATOR FUNCTION TABLE (EACH CHANNEL)

VOLTAGE INPUTS		CAP VOLTAGE	DIFFERENTIAL VOLTAGE $V_{O(SOURCE)} - V_{O(SINK)}$	CAP TERMINAL (See schematic)	INTEGRATOR MODE OF OPERATION
A	TC				
X	≤0.8 V	X	X	Q1 Sinking	Reset (Disabled)
H	≥2 V	X	X	Q1 Sinking	Reset
L	≥2 V	X	≥300 mV	Q2 Sourcing	Charge
L	≥2 V	$>V_{T-}$	≤−300 mV	Q3 Sinking	Discharge

H = high level; L = low level; X = irrelevant

NOTE 1: The TC input has an operating range from 0 V to 6 V, but its effect on the CAP terminal is linear from approximately 2 V to 6 V. The best linearity is achieved within the recommended operating linear range.

logic diagram (each channel, positive logic)

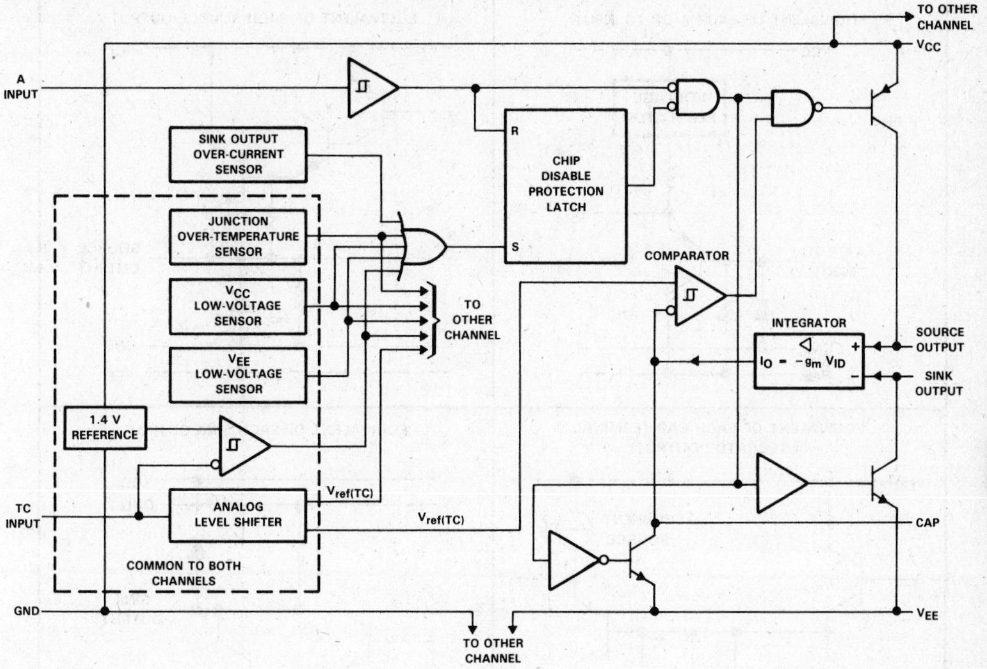
TO OTHER CHANNEL
V_{CC}
A INPUT
SINK OUTPUT OVER-CURRENT SENSOR
JUNCTION OVER-TEMPERATURE SENSOR
V_{CC} LOW-VOLTAGE SENSOR
V_{EE} LOW-VOLTAGE SENSOR
1.4 V REFERENCE
TC INPUT
ANALOG LEVEL SHIFTER
COMMON TO BOTH CHANNELS
$V_{ref(TC)}$
$V_{ref(TC)}$
R
S
CHIP DISABLE PROTECTION LATCH
TO OTHER CHANNEL
COMPARATOR
INTEGRATOR
$I_O = -g_m V_{ID}$
SOURCE OUTPUT
SINK OUTPUT
CAP
GND
V_{EE}
TO OTHER CHANNEL

schematics of inputs and outputs

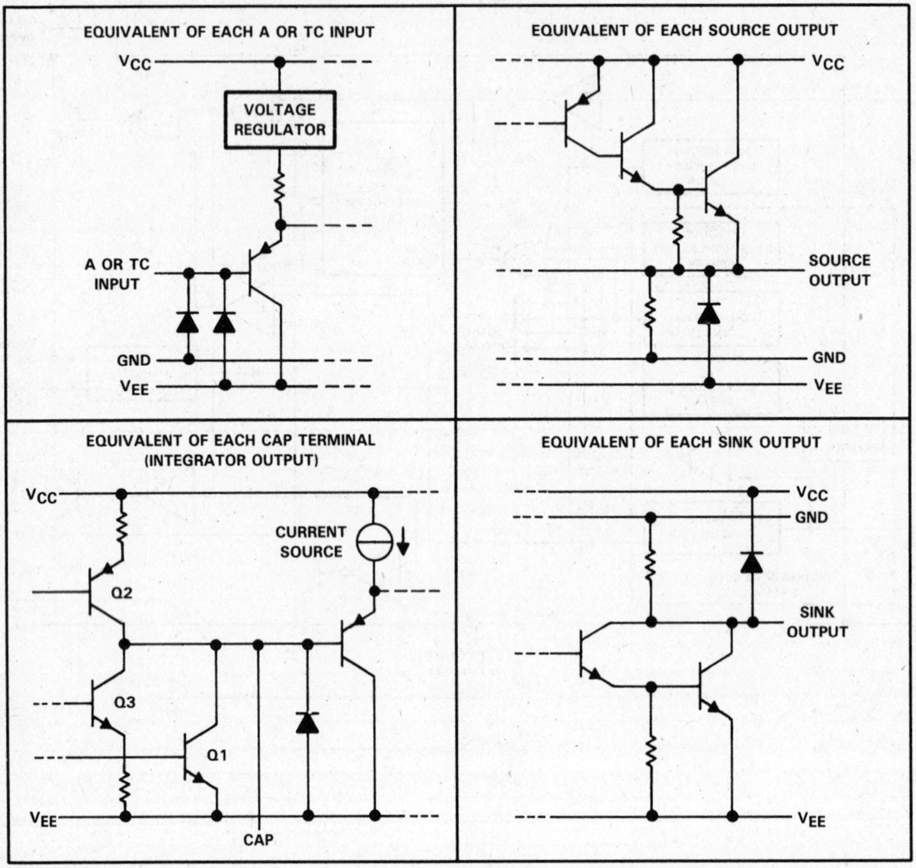
EQUIVALENT OF EACH A OR TC INPUT
VCC
VOLTAGE REGULATOR
A OR TC INPUT
GND
VEE
EQUIVALENT OF EACH SOURCE OUTPUT
VCC
SOURCE OUTPUT
GND
VEE
EQUIVALENT OF EACH CAP TERMINAL (INTEGRATOR OUTPUT)
VCC
CURRENT SOURCE
Q2
Q3
Q1
VEE
CAP
EQUIVALENT OF EACH SINK OUTPUT
VCC
GND
SINK OUTPUT
VEE

absolute maximum ratings over operating temperature range (unless otherwise noted)

Supply voltage range, V_{CC} (see Notes 2 and 3) . . . . −0.3 V to 60 V
Supply voltage range, V_{EE} (see Note 3) . . . . −60 V to 0.3 V
Voltage difference range between supply voltages, $V_{CC} - V_{EE}$ . . . . −0.3 V to 60 V
Input voltage range, V_I, A and TC inputs (see Note 4) . . . . −1.6 V to V_{EE}+60 V
CAP terminal range, $V_{(CAP)}$ . . . . V_{EE}−0.3 V to V_{CC}
Source output voltage range, $V_{O(SRC)}$ (see Note 4) . . . . V_{EE}−3 V to V_{CC}+0.3 V
Sink output voltage range, $V_{O(SNK)}$ (see Note 4) . . . . V_{EE}−0.3 V to V_{CC}+4 V
Input current, I_I, A and TC inputs . . . . −15 mA
Peak output current, source and sink outputs (nonrepetitive, $t_w \leq 100\ \mu s$), I_{OM} . . . . ±3 A
Continuous dissipation at (or below) 90°C case temperature (see Note 5) . . . . 20 W
Continuous dissipation at (or below) 25°C free-air temperature (see Note 5) . . . . 3.575 W
Operating case or virtual junction temperature range . . . . −20°C to 150°C
Storage temperature range . . . . −65°C to 150°C
Lead temperature 1,6 mm (1/16 inch) from the case for 10 seconds . . . . 260°C

NOTES: 2. All voltage values, except differential voltages, are with respect to the network ground terminal.
3. Both V_{CC} pins must be connected together as close to the package as possible for optimum testing and operation of the device. V_{EE} pins are handled in a like manner.
4. The maximum current limitation at this terminal generally occurs at a voltage of lower magnitude than the voltage limit. Neither the maximum current nor the maximum voltage for this terminal should be exceeded.
5. For operation above 25°C free-air temperature, derate linearly at the rate of 28.6 mW/°C. For operation above 90°C case temperature, derate linearly at the rate of 333 mW/°C. To avoid exceeding the design maximum virtual junction temperature, these ratings should not be exceeded. Due to variations in individual device electrical characteristics and thermal resistance, the built-in thermal overload protection may be activated at power levels slightly above or below the rated dissipation.

recommended operating conditions

	MIN	MAX	UNIT
Supply voltage, V_{CC}	15	30	V
Supply voltage, V_{EE}	−15	−30	V
Voltage difference between supply voltages, $V_{CC} - V_{EE}$	30	60	V
High-level input voltage at A, V_{IH}	2	7	V
Low-level input voltage at A, V_{IL}	−0.3†	0.8	V
Linear-range TC input voltage (see Note 1)	2	6	V
Continuous output current, I_O		±1	A
Peak output current at 40% duty cycle, I_O		±2.5	A
Operating virtual junction temperature, T_J	−20	125	°C

† The algebraic convention, in which the least positive (most negative) designated minimum, is used in this data sheet for logic voltage levels.
NOTE 1: The TC input has a operating range from 0 V to 6 V, but its effect on the CAP terminal is linear from approximately 2 V to 6 V. The best linearity is achieved within the recommended operating linear range.

electrical characteristics over recommended ranges of V_{CC}, V_{EE}, and virtual junction operating temperature (unless otherwise noted)

PARAMETER			TEST CONDITIONS		MIN	TYP†	MAX	UNIT
V_{IK}	Input clamp voltage at A or TC		$I_I = -12$ mA			−0.9	−1.5	V
$V_{O(on)}$	On-state output voltage	Source output	$I_O = -1$ A	See Note 6	$V_{CC}-2.8$	$V_{CC}-1.9$		V
			$I_O = -2.5$ A		$V_{CC}-3.7$	$V_{CC}-2.5$		
		Sink output	$I_O = 1$ A		$V_{EE}+1.8$	$V_{EE}+1.2$		
			$I_O = 2.5$ A		$V_{EE}+3$	$V_{EE}+2$		
$V_{(CAP)}$	CAP terminal output voltage		$I_{(CAP)} = 1$ mA,	Reset mode	$V_{EE}+0.4$	$V_{EE}+0.2$		V
$V_{OK(SRC)}$	Source output clamp voltage		$I_O = -1$ A	See Note 6		$V_{EE}-1.1$	$V_{EE}-1.6$	V
			$I_O = -2.5$ A			$V_{EE}-1.5$	$V_{EE}-2.2$	
$V_{OK(SNK)}$	Sink output clamp voltage		$I_O = 1$ A	See Note 6		$V_{CC}+1.4$	$V_{CC}+2.1$	
			$I_O = 2.5$ A			$V_{CC}+2.2$	$V_{CC}+3.3$	
$I_{O(off)}$	Off-state output current	Source output	$V_O = V_{EE}$			−0.4	−2	mA
		Sink output	$V_O = V_{CC}$			0.3	1	
I_{IH}	High-level input current at A		$V_I = 5.5$ V				10	μA
I_{IL}	Low-level input current at A		$V_I = 0$				−10	μA
I_I	Input current at TC		$V_I = 0$ to 6 V				±10	μA
$I_{(CAP)}$	Current at CAP terminal	Reset mode	$V_{(CAP)} = V_{EE} + 1$ V to $V_{EE} + 6$ V		1	4		mA
		Charge mode	$V_{I(TC)} = 5$ V, $V_{(CAP)} = V_{EE} + 2$ V, $R_L = 1$ kΩ from Source to Sink			−115		μA
		Discharge mode	$V_{I(TC)} = 2$ V, $V_{(CAP)} = V_{EE} + 5$ V, $I_{O(SRC)} = -10$ mA, $I_{O(SNK)} = 10$ mA			5		
V_{T+}	Positive-going threshold voltage at CAP terminal (charge mode)		$R_L = 1$ kΩ from Source to Sink, $V_{I(TC)} = 2$ V to 6 V	See Notes 7 and 8		$V_{I(TC)}$		
V_{T-}	Negative-going threshold voltage at CAP terminal (discharge mode)					0.95 $V_{I(TC)}$		
$\frac{V_{hys}}{V_{T+}}$	Normalized hysteresis at CAP Terminal $[(V_{T+} - V_{T-})/V_{T+}]$			See Notes 7, 8, and 9		0.05		
g_m‡	Transconductance of integrator	Charge mode	$R_L = 1$ kΩ from Source to Sink, $V_{(CAP)} = V_{EE}$ to $V_{T+} - 0.1$ V			3		μS
		Discharge mode	$I_{O(SRC)} = -10$ mA, $I_{O(SINK)} = 10$ mA, $V_{CAP} = V_{T+} + 0.1$ V to $V_{T-} + 0.1$ V			3		

† All typical values are at $V_{CC} = 20$ V, $V_{EE} = -20$ V, $T_J = 25°C$.

‡ Transconductance (g_m) of the integrator is: $I_{(CAP)}/[V_{(SRC)} - V_{(SNK)}]$. The ratio of $V_{T+}/V_{I(TC)}$ is factory adjusted to compensate for variances in g_m. This causes the integration time to be more constant from unit to unit.

NOTES: 6. These parameters must be measured on one output at a time using pulse techniques, $t_w = 1$ ms, duty cycle < 10%.

7. Threshold values are those voltage levels at the CAP terminal at which the source output changes state. A level more positive than V_{T+} causes the source output to go to the off state, and a level more negative than V_{T-} causes the source output to go to the on state. Both V_{T+} and V_{T-} are variable values that are dependent on the voltage level at the TC input.

8. V_{T+} and V_{T-} are measured differentially with CAP terminal voltage referenced to the V_{EE} terminal.

9. Both V_{T+} and V_{T-} must be measured at the same junction temperature using the same TC voltage.

electrical characteristics over recommended ranges of V_{CC}, V_{EE}, and virtual junction operating temperature (unless otherwise noted) (continued)

PARAMETER			TEST CONDITIONS		MIN	TYP†	MAX	UNIT
I_{CC}	Supply current from V_{CC}	Disabled	V_{CC} = 25 V, V_{EE} = −25 V, No load	$V_{I(A)}$ = 0, $V_{I(TC)}$ = 0		8	14	mA
I_{EE}	Supply current from V_{EE}					−13	−23	mA
I_{CC}	Supply current from V_{CC}	All source and sink outputs off		$V_{I(A)}$ = 5 V, $V_{I(TC)}$ = 5 V		10	18	mA
I_{EE}	Supply current from V_{EE}					−16	−28	mA
I_{CC}	Supply current from V_{CC}	All source and sink outputs on		$V_{I(A)}$ = 0, $V_{I(TC)}$ = 5 V, $V_{(CAP)}$ = V_{EE} + 2 V		16	28	mA
I_{EE}	Supply current from V_{EE}					−25	−45	mA
I_{CC}	Supply current from V_{CC}	All source outputs off and all sink outputs on		$V_{I(A)}$ = 0, $V_{I(TC)}$ = 3 V, $V_{(CAP)}$ = V_{EE} + 5 V		10	18	mA
I_{EE}	Supply current from V_{EE}					−20	−35	mA

† All typical values are at V_{CC} = 20 V, V_{EE} = −20 V, T_J = 25°C.

switching characteristics, T_A = 25°C

PARAMETER		TEST CONDITIONS	MIN	TYP	MAX	UNIT
t_{on}	Source output turn-on time from A input	V_{CC} = 20 V, V_{EE} = −20 V, C_L = 30 pF, See Figure 1		700		ns
t_{off}	Source output turn-off time from A input			900		ns
t_{rv}	Source output voltage rise time (turning on)			200		ns
t_{fv}	Source output voltage fall time (turning off)			150		ns
t_{on}	Sink output turn-on time from A input			600		ns
t_{off}	Sink output turn off time from A input			900		ns
t_{fv}	Sink output voltage fall time (turning on)			100		ns
t_{rv}	Sink output voltage rise time (turning off)			150		ns

PARAMETER MEASUREMENT INFORMATION

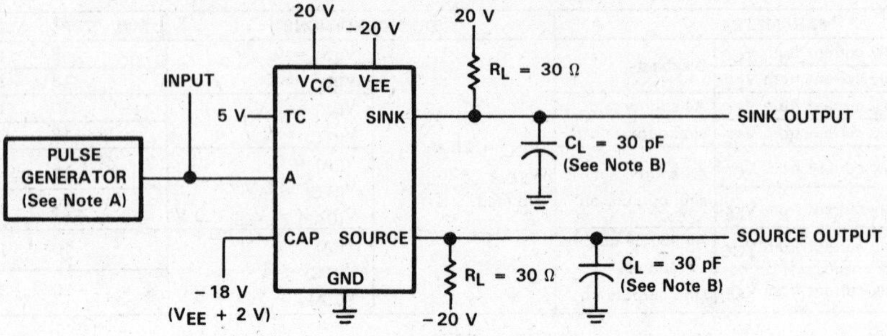
20 V
−20 V
20 V
INPUT
V_{CC}
V_{EE}
R_L = 30 Ω
5 V
TC
SINK
SINK OUTPUT
PULSE
GENERATOR
(See Note A)
A
C_L = 30 pF
(See Note B)
CAP
SOURCE
SOURCE OUTPUT
GND
R_L = 30 Ω
C_L = 30 pF
(See Note B)
−18 V
(V_{EE} + 2 V)
−20 V

TEST CIRCUIT

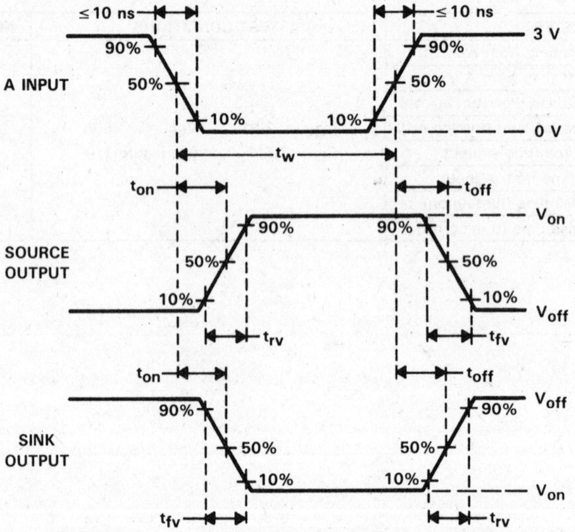
≤10 ns
≤10 ns
3 V
90%
90%
A INPUT
50%
50%
10%
10%
0 V
t_w
t_{on}
t_{off}
V_{on}
90%
90%
SOURCE
OUTPUT
50%
50%
10%
10%
V_{off}
t_{rv}
t_{fv}
t_{on}
t_{off}
V_{off}
90%
90%
SINK
OUTPUT
50%
50%
10%
10%
V_{on}
t_{fv}
t_{rv}

VOLTAGE WAVEFORMS

NOTES: A. The pulse generator has the following characteristics: PRR ≤ 5 kHz, t_w = 10 μs, Z_O = 50 Ω.
B. C_L includes probe and jig capacitance.

FIGURE 1. SWITCHING TIMES FROM A INPUTS

TYPICAL APPLICATION DATA

A typical application of the SN75608 Dual Flux-Regulating Actuator driving inductive loads is shown in Figure 2. Figure 3 illustrates representative waveforms that occur with the circuit connected as shown in Figure 2. The waveforms illustrate adjustment of output current in the load to compensate for a change in the load inductance while maintaining a constant CAP voltage waveform and thus constant electromagnetic flux in the load.

For optimum operation, both V_{CC} pins must be connected together, close to the package, as well as all three V_{EE} pins. A low-impedance bypass capacitor, 10 microfarads or larger, should be connected between V_{CC} and V_{EE}, also close to the package. The value of the integration capacitor connected between the CAP terminal and V_{EE} is dependent on the load characteristics and the performance desired. The analog voltage on the TC terminal may be varied between 2 volts and 6 volts for fine adjustment of integrator timing characteristics.

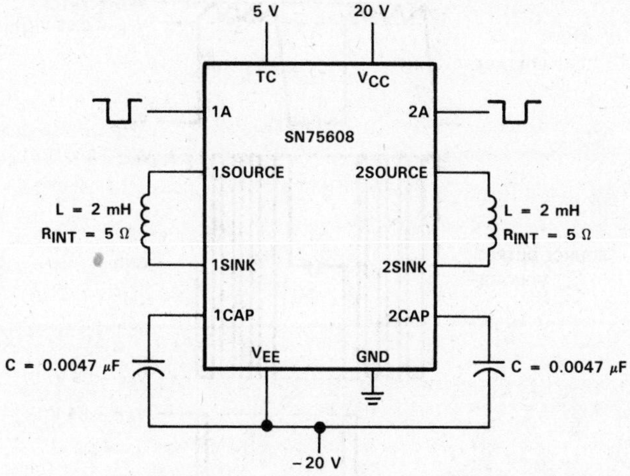
5 V
20 V
TC
VCC
1A
2A
SN75608
1SOURCE
2SOURCE
L = 2 mH
RINT = 5 Ω
L = 2 mH
RINT = 5 Ω
1SINK
2SINK
1CAP
2CAP
C = 0.0047 μF
VEE
GND
C = 0.0047 μF
−20 V

FIGURE 2. TYPICAL DOT-MATRIX PRINTHEAD-DRIVER APPLICATION

TYPICAL APPLICATION DATA

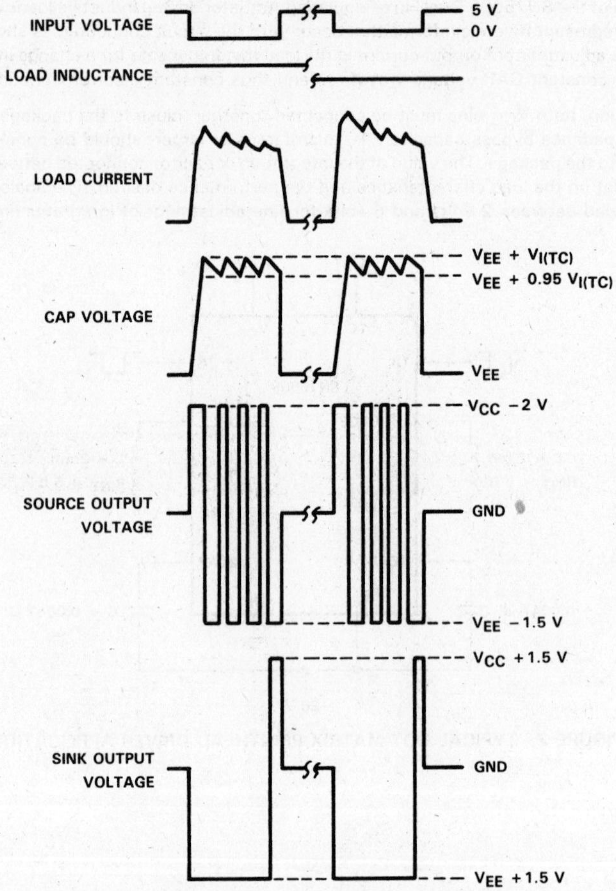
INPUT VOLTAGE
5 V
0 V
LOAD INDUCTANCE
LOAD CURRENT
CAP VOLTAGE
VEE + VI(TC)
VEE + 0.95 VI(TC)
VEE
VCC −2 V
SOURCE OUTPUT
VOLTAGE
GND
VEE −1.5 V
VCC +1.5 V
SINK OUTPUT
VOLTAGE
GND
VEE +1.5 V

FIGURE 3. REPRESENTATIVE WAVEFORMS

D2973, DECEMBER 1986

- 2.5-A Current Capability per Channel
- For Positive Supply Applications
- Wide Supply Voltage Range . . . 30 V to 60 V
- High-Impedance Clamped Inputs Compatible with TTL or CMOS Devices
- Output Clamp Diodes for Inductive Transient Suppression
- Thermal Shutdown
- Internal ESD Protection
- Short-Circuit Protection on Sink Outputs
- Input Hysteresis Improves Noise Immunity
- No Output Glitch During Power-Up or Power-Down
- 15-Pin SIP Power Package

KV SINGLE-IN-LINE PACKAGE
(TOP VIEW)

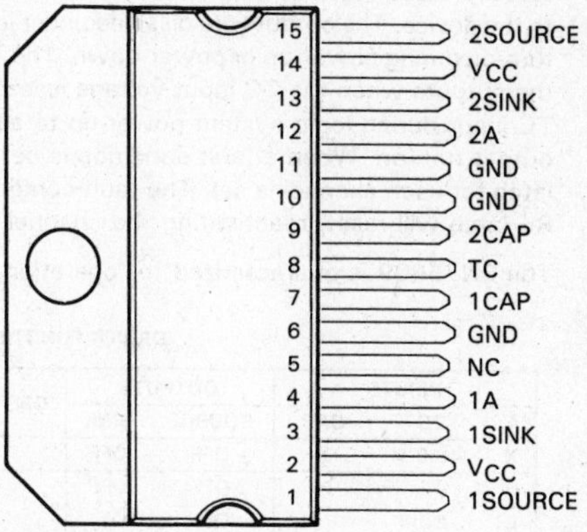
15 2SOURCE
14 V_{CC}
13 2SINK
12 2A
11 GND
10 GND
9 2CAP
8 TC
7 1CAP
6 GND
5 NC
4 1A
3 1SINK
2 V_{CC}
1 1SOURCE

NC—No internal connection
The tab is electrically connected to the GND pins.

description

The SN75609 is a high-current dual flux-regulating actuator designed for switching double-ended loads with currents up to 2.5 amperes at supply voltages from 30 volts to 60 volts. It is designed to drive and control the electromagnetic flux in printheads, solenoids, relays, and other loads whose inductance value varies during operation.

The SN75609 performs the function of flux regulation under control of standard TTL or CMOS input signals for two independent channels. Flux is proportional to the integral of the inductive-load voltage. It is a function of the total amount of current in the load and is the magnetic field maintained in the load. With flux regulation, the load current will vary to compensate for core saturation, temperature changes, and other variations of load inductance during operation while maintaining controlled, relatively constant flux in the load.

Each channel has separate sink and source driver outputs for driving each end of the inductive load. Internal feedback, consisting of an integrator and voltage comparator, provides flux regulation via chop-mode operation of the source output. The integrator circuit provides current to the capacitor terminal (CAP) proportional to the differential voltage between the sink and source outputs for each channel. The integrator requires an external capacitor connected between the CAP and GND terminals. The voltage at the CAP terminal, referenced to GND, is proportional to the integral of the source-to-sink (load) voltage.

The feedback path is completed by a differential voltage comparator that controls the state of the source output. The inverting comparator input is connected to the CAP terminal, and the noninverting input is connected to an analog voltage, $V_{ref(TC)}$, which is referenced to GND. $V_{ref(TC)}$ is proportional to the Threshold Control (TC) voltage, which is referenced to ground. The comparator hysteresis controls the charge and discharge voltage excursions at the CAP terminal and thus controls the on and off time of the source output chopper.

TEXAS INSTRUMENTS
POST OFFICE BOX 655012 • DALLAS, TEXAS 75265

description (continued)

The SN75609 features built-in thermal protection and a sink output over-current sensor to prevent damage to the device. The outputs are disabled under low-V_{CC} supply-voltage conditions to prevent transient output turn-on during power-up or power-down. The TC input is a combined threshold and logic input that disables the outputs when the TC input voltage is less than 0.8 V. This permits an external RC time delay at the TC input during logic system power-up to allow logic at input A to stabilize without causing undesired output turn-on. When a fault condition is detected by any one of these four protection features, the RS latch for each channel is set. The fault condition must be removed and the A input taken high before the RS latch will reset, reactivating the channel.

The SN75609 is characterized for operation from −20°C to 85°C.

DRIVER FUNCTION TABLE (EACH CHANNEL)

INPUTS			OUTPUTS		OPERATING MODE	COMMENTS
A	TC	CAP	SOURCE	SINK		
X	≤0.8 V	X	OFF	OFF	Disabled	TC acts as a digital input referenced to GND
H	≥2 V	X	OFF	OFF	Active	TC acts as an analog input referenced to GND (See Note 1)
L	≥2 V	$<V_{T+}$	ON	ON		
L	≥2 V	$>V_{T-}$	OFF	ON		

INTEGRATOR FUNCTION TABLE (EACH CHANNEL)

VOLTAGE INPUTS		CAP VOLTAGE	DIFFERENTIAL VOLTAGE $V_{O(SOURCE)} - V_{O(SINK)}$	CAP TERMINAL (See schematic)	INTEGRATOR MODE OF OPERATION
A	TC				
X	≤0.8 V	X	X	Q1 Sinking	Reset (Disabled)
H	≥2 V	X	X	Q1 Sinking	Reset
L	≥2 V	X	≥300 mV	Q2 Sourcing	Charge
L	≥2 V	$>V_{T-}$	≤−300 mV	Q3 Sinking	Discharge

H = high level; L = low level; X = irrelevant

NOTE 1: The TC input has an operating range from 0 to 6 volts, but its effect on the CAP terminal is linear from approximately 2 V to 6 V. Best linearity is achieved within the recommended operating linear range.

logic diagram (each channel, positive logic)

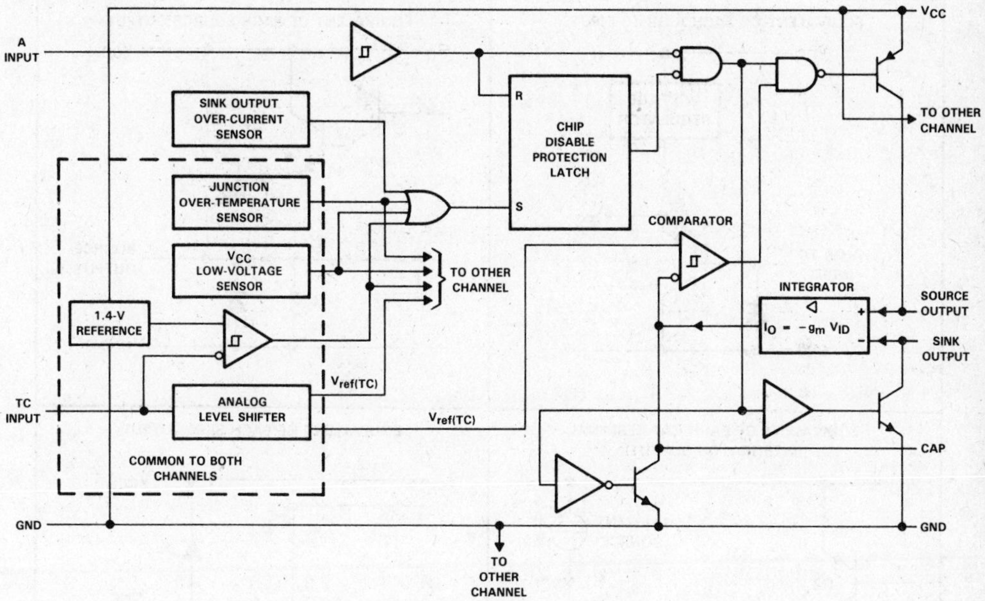
A
INPUT
R
CHIP
DISABLE
PROTECTION
LATCH
S
VCC
TO OTHER
CHANNEL
SINK OUTPUT
OVER-CURRENT
SENSOR
JUNCTION
OVER-TEMPERATURE
SENSOR
VCC
LOW-VOLTAGE
SENSOR
TO OTHER
CHANNEL
COMPARATOR
INTEGRATOR
IO = −gm VID
SOURCE
OUTPUT
SINK
OUTPUT
1.4-V
REFERENCE
Vref(TC)
TC
INPUT
ANALOG
LEVEL SHIFTER
Vref(TC)
CAP
COMMON TO BOTH
CHANNELS
GND
GND
TO
OTHER
CHANNEL

schematics of inputs and outputs

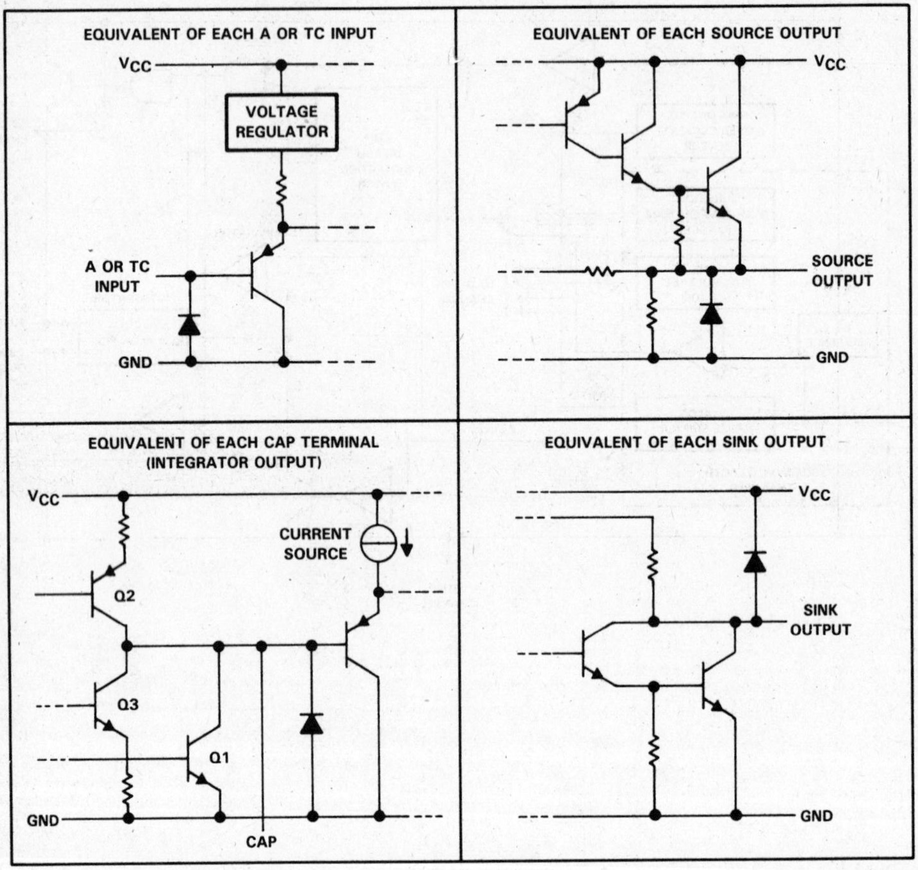
EQUIVALENT OF EACH A OR TC INPUT
V_{CC}
VOLTAGE
REGULATOR
A OR TC
INPUT
GND
EQUIVALENT OF EACH SOURCE OUTPUT
V_{CC}
SOURCE
OUTPUT
GND
EQUIVALENT OF EACH CAP TERMINAL
(INTEGRATOR OUTPUT)
V_{CC}
CURRENT
SOURCE
Q2
Q3
Q1
GND
CAP
EQUIVALENT OF EACH SINK OUTPUT
V_{CC}
SINK
OUTPUT
GND

absolute maximum ratings over operating temperature range (unless otherwise noted)

Supply voltage range, V_{CC} (see Notes 2 and 3)	−0.3 V to 60 V
Input voltage range, V_I, A and TC inputs (see Note 4)	−1.6 V to 60 V
CAP terminal range, $V_{(CAP)}$	−0.3 V to V_{CC}
Source output voltage range, $V_{O(SRC)}$ (see Note 4)	−3 V to V_{CC}+0.3 V
Sink output voltage range, $V_{O(SNK)}$ (see Note 4)	−0.3 V to V_{CC}+4 V
Input current, I_I, A and TC inputs	−15 mA
Peak output current, source and sink outputs (nonrepetitive, $t_w \leq 100\ \mu s$), I_{OM}	±3 A
Continuous dissipation at (or below) 90°C case temperature (see Note 5)	20 W
Continuous dissipation at (or below) 25°C free-air temperature (see Note 5)	3.575 W
Operating case or virtual junction temperature range	−20°C to 150°C
Storage temperature range	−65°C to 150°C
Lead temperature 1,6 mm (1/16 inch) from the case for 10 seconds	260°C

NOTES: 2. All voltage values, except differential voltages, are with respect to the network ground terminal.
3. Both V_{CC} pins must be connected together as close to the package as possible for optimum testing and operation of the device. GND pins are handled in a like manner.
4. The maximum current limitation at this terminal generally occurs at a voltage of lower magnitude than the voltage limit. Neither the maximum current nor the maximum voltage for this terminal should be exceeded.
5. For operation above 25°C free-air temperature, derate linearly at the rate of 28.6 mW/°C. For operation above 90°C case temperature, derate linearly at the rate of 333 mW/°C. To avoid exceeding the design maximum virtual junction temperature, these ratings should not be exceeded. Due to variations in individual device electrical characteristics and thermal resistance, the built-in thermal overload protection may be activated at power levels slightly above or below the rated dissipation.

recommended operating conditions

	MIN	MAX	UNIT
Supply voltage, V_{CC}	30	60	V
High-level input voltage at A, V_{IH}	2	7	V
Low-level input voltage at A, V_{IL}	−0.3†	0.8	V
Linear-range TC input voltage (see Note 1)	3	6	V
Continuous output current, I_O		±1	A
Peak output current at 40% duty cycle, I_O		±2.5	A
Operating virtual junction temperature, T_J	−20	125	°C

†The algebraic convention, in which the least positive (most negative) designated minimum, is used in this data sheet for logic voltage levels.

NOTE 1: The TC input has an operating range from 0 to 6 volts, but its effect on the CAP terminal is linear from approximately 2 V to 6 V. Best linearity is achieved within the recommended operating linear range.

electrical characteristics over recommended ranges of V_{CC} and virtual junction operating temperature (unless otherwise noted)

PARAMETER			TEST CONDITIONS		MIN	TYP†	MAX	UNIT
V_{IK}	Input clamp voltage at A or TC		$I_I = -12$ mA			−0.9	−1.5	V
$V_{O(on)}$	On-state output voltage	Source output	$I_O = -1$ A	See Note 6	$V_{CC}-2.8$	$V_{CC}-1.9$		V
			$I_O = -2.5$ A		$V_{CC}-3.7$	$V_{CC}-2.5$		
		Sink output	$I_O = 1$ A			1.2	1.8	
			$I_O = 2.5$ A			2	3	
$V_{(CAP)}$	CAP terminal output voltage		$I_{(CAP)} = 1$ mA,	Reset mode		0.2	0.4	V
$V_{OK(SRC)}$	Source output clamp voltage		$I_O = -1$ A	See Note 6		−1.1	−1.6	V
			$I_O = -2.5$ A			−1.5	−2.2	
$V_{OK(SNK)}$	Sink output clamp voltage		$I_O = 1$ A	See Note 6		$V_{CC}+1.4$	$V_{CC}+2.1$	
			$I_O = 2.5$ A			$V_{CC}+2.2$	$V_{CC}+3.3$	
$I_{O(off)}$	Off-state output current	Source output	$V_O = 0$ V			−0.4	−2	mA
		Sink output	$V_O = V_{CC}$			0.3	1	
I_{IH}	High-level input current at A		$V_I = 5.5$ V				10	μA
I_{IL}	Low-level input current at A		$V_I = 0$				−10	μA
I_I	Input current at TC		$V_I = 0$ to 6 V				±10	μA
$I_{(CAP)}$	Current at CAP terminal	Reset mode	$V_{(CAP)} = 1$ V to 6 V		1	4		mA
		Charge mode	$V_{I(TC)} = 5$ V, $V_{(CAP)} = 2$ V, $R_L = 1$ kΩ from Source to Sink			−115		μA
		Discharge mode	$V_{I(TC)} = 2$ V, $V_{(CAP)} = 5$ V, $I_{O(SRC)} = -10$ mA, $I_{O(SNK)} = 10$ mA			5		
V_{T+}	Positive-going threshold voltage at CAP terminal (charge mode)		$R_L = 1$ kΩ from Source to Sink, $V_{I(TC)} = 3$ V to 6 V	See Notes 7 and 8		$V_{I(TC)}$		
V_{T-}	Negative-going threshold voltage at CAP terminal (discharge mode)					0.95 $V_{I(TC)}$		
$\frac{V_{hys}}{V_{T+}}$	Normalized hysteresis at CAP terminal $[(V_{T+} - V_{T-})/V_{T+}]$			See Notes 7, 8, and 9		0.05		
g_m‡	Transconductance of integrator	Charge mode	$R_L = 1$ kΩ from Source to Sink, $V_{(CAP)} = 0$ V to $V_{T+} - 0.1$ V			3		μS
		Discharge mode	$I_{O(SRC)} = -10$ mA, $I_{O(SINK)} = 10$ mA, $V_{CAP} = V_{T+} + 0.1$ V to $V_{T-} + 0.1$ V			3.		

† All typical values are at $V_{CC} = 40$ V, $T_A = 25°C$.

‡ Transconductance (g_m) of the integrator is: $I_{(CAP)}/[V_{(SRC)} - V_{(SNK)}]$. The ratio of $V_{T+}/V_{I(TC)}$ is factory adjusted to compensate for variances in g_m. This causes the integration time to be more constant from unit to unit.

NOTES: 6. These parameters must be measured on one output at a time using pulse techniques, $t_w = 1$ ms, duty cycle $< 10\%$.

7. Threshold values are those voltage levels at the CAP terminal at which the source output changes state. A level more positive than V_{T+} causes the source output to go to the off state, and a level more negative than V_{T-} causes the source output to go to the on state. Both V_{T+} and V_{T-} are variable values that are dependent on the voltage level at the TC input.

8. V_{T+} and V_{T-} are measured differentially with CAP terminal voltage referenced to the GND terminal.

9. Both V_{T+} and V_{T-} must be measured at the same junction temperature using the same TC voltage.

electrical characteristics over recommended ranges of V_{CC} and virtual junction operating temperature (unless otherwise noted) (continued)

PARAMETER			TEST CONDITIONS		MIN	TYP†	MAX	UNIT
I_{CC}	Supply current	Disabled	V_{CC} = 50 V, No load	$V_{I(A)}$ = 0, $V_{I(TC)}$ = 0		19	34	mA
		All source and sink outputs off		$V_{I(A)}$ = 5 V, $V_{I(TC)}$ = 5 V		26	37	
		All source and sink outputs on		$V_{I(A)}$ = 0, $V_{I(TC)}$ = 5 V, $V_{(CAP)}$ = 2 V		26	46	
		All source outputs off and all sink outputs on		$V_{I(A)}$ = 0, $V_{I(TC)}$ = 3 V, $V_{(CAP)}$ = 5 V		26	46	

† All typical values are at V_{CC} = 40 V, T_A = 25°C.

switching characteristics, T_A = 25°C

PARAMETER		TEST CONDITIONS	MIN	TYP	MAX	UNIT
t_{on}	Source output turn-on time from A input	V_{CC} = 40 V, C_L = 30 pF, See Figure 1		700		ns
t_{off}	Source output turn-off time from A input			900		ns
t_{rv}	Source output voltage rise time (turning on)			200		ns
t_{fv}	Source output voltage fall time (turning off)			150		ns
t_{on}	Sink output turn-on time from A input			600		ns
t_{off}	Sink output turn off time from A input			900		ns
t_{fv}	Sink output voltage fall time (turning on)			100		ns
t_{rv}	Sink output voltage rise time (turning off)			150		ns

PARAMETER MEASUREMENT INFORMATION

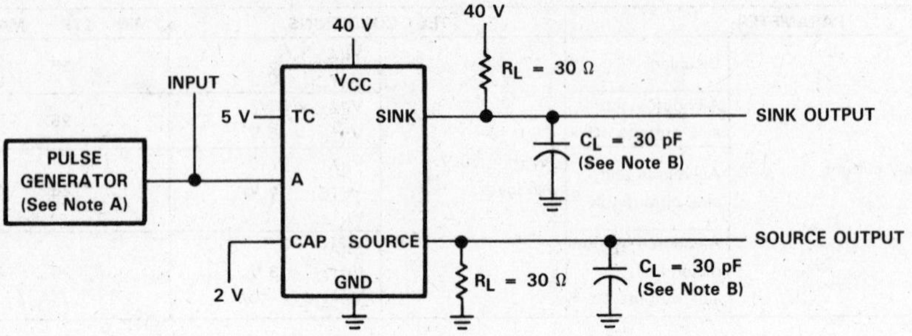
40 V
40 V
R_L = 30 Ω
INPUT
V_{CC}
5 V
TC
SINK
SINK OUTPUT
C_L = 30 pF
(See Note B)
PULSE
GENERATOR
(See Note A)
A
CAP
SOURCE
SOURCE OUTPUT
R_L = 30 Ω
C_L = 30 pF
(See Note B)
2 V
GND

TEST CIRCUIT

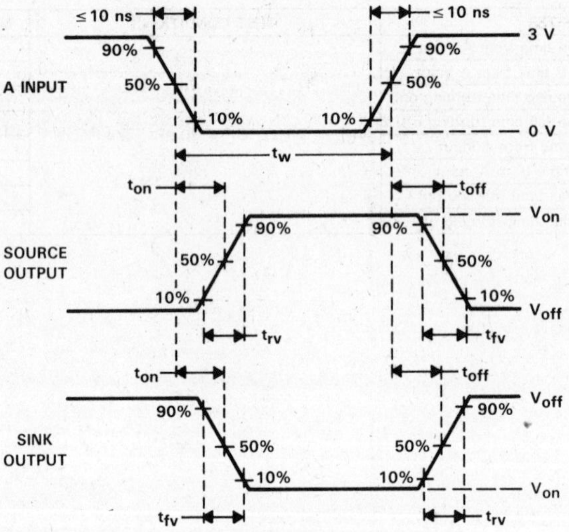
≤ 10 ns
≤ 10 ns
3 V
90%
90%
A INPUT
50%
50%
10%
10%
0 V
t_w
t_{on}
t_{off}
V_{on}
90%
90%
SOURCE
OUTPUT
50%
50%
10%
10%
V_{off}
t_{rv}
t_{fv}
t_{on}
t_{off}
V_{off}
90%
90%
SINK
OUTPUT
50%
50%
10%
10%
V_{on}
t_{fv}
t_{rv}

VOLTAGE WAVEFORMS

NOTES: A. The pulse generator has the following characteristics: PRR ≤ 5 kHz, t_W = 10 μs, Z_O = 50 Ω.
B. C_L includes probe and jig capacitance.

FIGURE 1. SWITCHING TIMES FROM A INPUTS

TYPICAL APPLICATION DATA

A typical application of the SN75609 Dual Flux-Regulating Actuator driving inductive loads is shown in Figure 2. Figure 3 illustrates representative waveforms that occur with the circuit connected as shown in Figure 2. The waveforms illustrate adjustment of output current in the load to compensate for a change in the load inductance while maintaining a constant CAP voltage waveform and thus constant electromagnetic flux in the load.

For optimum operation, both V_{CC} pins must be connected together, close to the package, as well as all three GND pins. A low-impedance bypass capacitor, 10 microfarads or larger, should be connected between V_{CC} and GND, also close to the package. The value of the integration capacitor connected between the CAP terminal and GND is dependent on the load characteristics and the performance desired. The analog voltage on the TC terminal may be varied between 3 volts and 6 volts for fine adjustment of integrator timing characteristics.

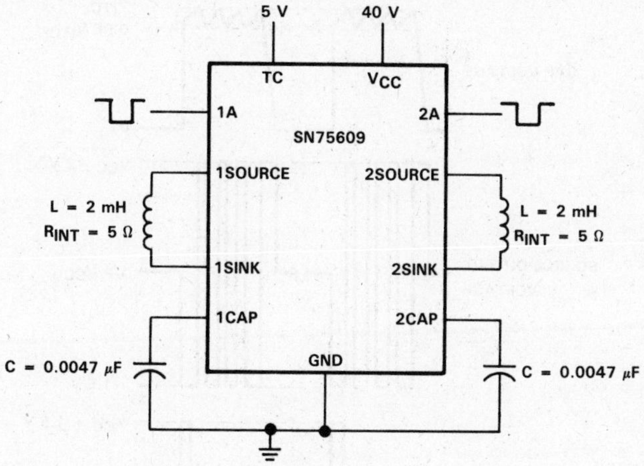
5 V
40 V
TC
VCC
1A
2A
SN75609
1SOURCE
2SOURCE
L = 2 mH
RINT = 5 Ω
L = 2 mH
RINT = 5 Ω
1SINK
2SINK
1CAP
2CAP
GND
C = 0.0047 μF
C = 0.0047 μF

FIGURE 2. TYPICAL DOT-MATRIX PRINTHEAD-DRIVER APPLICATION

TYPICAL APPLICATION DATA

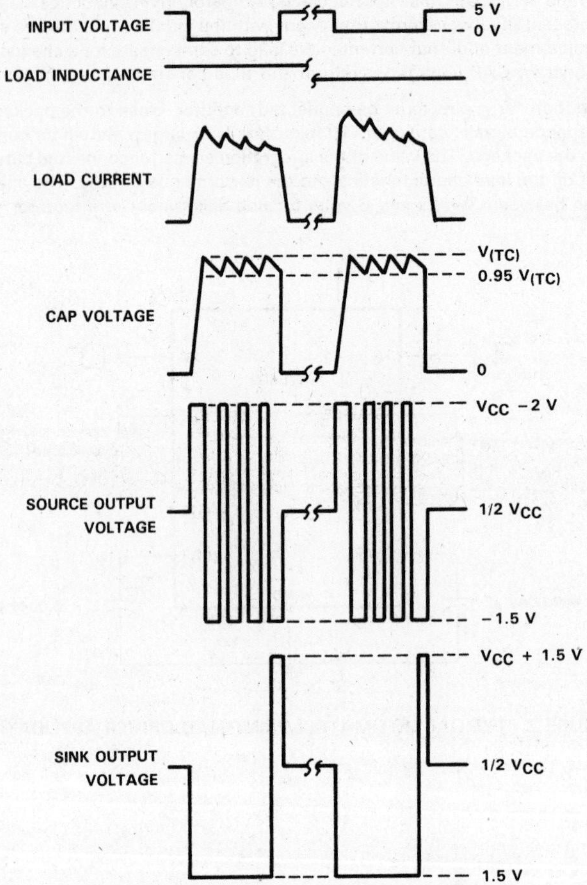
INPUT VOLTAGE
5 V
0 V
LOAD INDUCTANCE
LOAD CURRENT
V(TC)
0.95 V(TC)
CAP VOLTAGE
0
VCC −2 V
SOURCE OUTPUT VOLTAGE
1/2 VCC
−1.5 V
VCC + 1.5 V
SINK OUTPUT VOLTAGE
1/2 VCC
1.5 V

FIGURE 3. REPRESENTATIVE WAVEFORMS

SN754410
QUADRUPLE HALF-H DRIVER

D2942, NOVEMBER 1986

- 1-A Output Current Capability per Channel
- Output Clamp Diodes for Inductive Transient Suppression
- Applications Include Half-H and Full-H Solenoid Drivers and Motor Drivers
- Designed for Positive-Supply Applications
- Wide Supply Voltage Range: 4.5 V to 36 V
- TTL- and CMOS-Compatible High-Impedance Diode-Clamped Inputs
- Separate Input-Logic Supply
- Thermal Shutdown
- Internal ESD Protection
- Input Hysteresis Improves Noise Immunity
- Three-State Outputs
- Minimized Power Dissipation
- Sink/Source Interlock Circuitry Prevents Simultaneous Conduction
- No Output "Glitch" During Power-Up or Power-Down
- Improved Functional Replacement for the SGS L293D

NE DUAL-IN-LINE PACKAGE
(TOP VIEW)

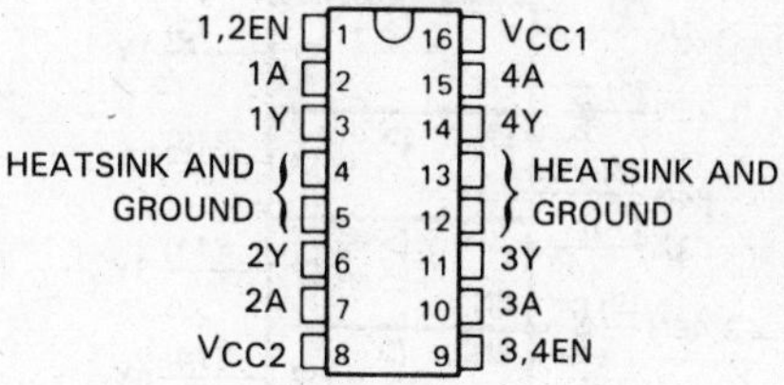
1,2EN 1
1A 2
1Y 3
HEATSINK AND GROUND 4
5
2Y 6
2A 7
VCC2 8
16 VCC1
15 4A
14 4Y
13 HEATSINK AND GROUND
12
11 3Y
10 3A
9 3,4EN

FUNCTION TABLE
(EACH CHANNEL)

INPUTS		OUTPUT
A	EN	Y
H	H	H
L	H	L
X	L	Z

H = high-level
L = low-level
X = irrelevant
Z = high-impedance (off)

description

The SN754410 is a quadruple high-current half-H driver designed to provide bidirectional drive currents of up to one ampere at voltages from 4.5 volts to 36 volts. It is designed to drive inductive loads such as relays, solenoids, dc and stepping motors, as well as other high-current/high-voltage loads in positive-supply applications.

All inputs are compatible with TTL and low-level CMOS logic. Each output (Y) is a complete totem-pole driver with a Darlington transistor sink and a psuedo-Darlington source. Channels are enabled in pairs with channels 1 and 2 enabled by 1,2EN and channels 3 and 4 enabled by 3,4EN. When an enable input is high, the associated channels are enabled and their outputs become active and in phase with their inputs. When the enable input is low, those channels are disabled and their outputs are off and in a high-impedance state. With the proper data inputs, each pair of drivers form a full-H (or bridge) reversible drive suitable for solenoid or motor applications.

A separate supply voltage (V_{CC1}) is provided for the logic input circuits to minimize device power dissipation. Supply voltage (V_{CC2}) is used for the output circuits.

The SN754410 is designed for operation from −40°C to 85°C.

logic symbol†

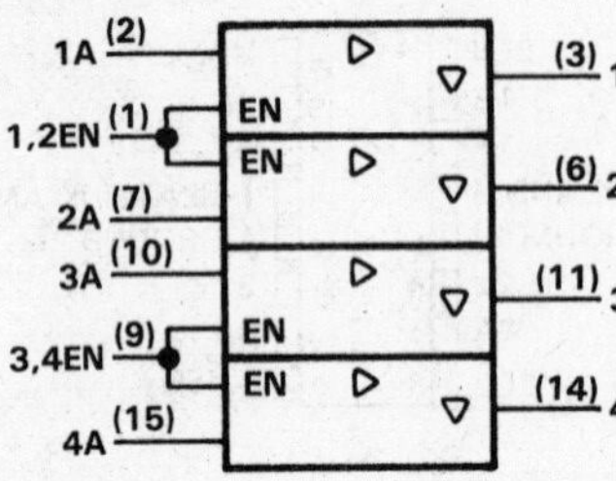
1A (2)
1,2EN (1)
2A (7)
3A (10)
3,4EN (9)
4A (15)
EN
EN
EN
EN
(3) 1Y
(6) 2Y
(11) 3Y
(14) 4Y

† This symbol is in accordance with ANSI/IEEE Std 91-1984 and IEC Publication 617-12.

logic diagram

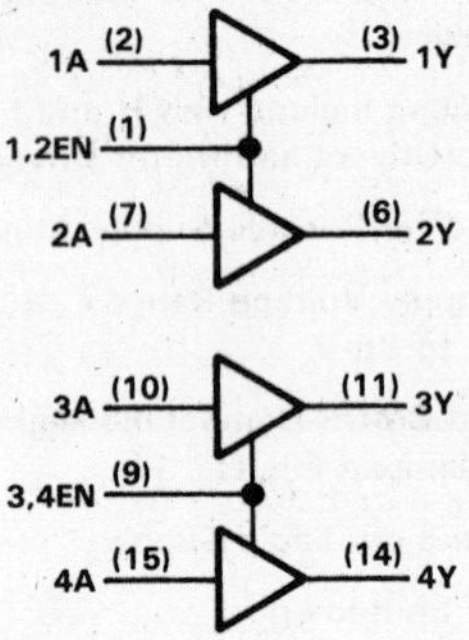
1A (2)
(3) 1Y
1,2EN (1)
2A (7)
(6) 2Y
3A (10)
(11) 3Y
3,4EN (9)
4A (15)
(14) 4Y

schematics of inputs and outputs

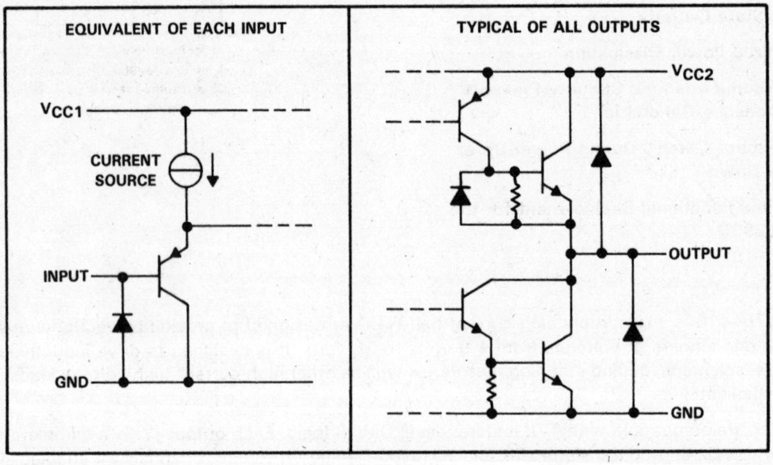
EQUIVALENT OF EACH INPUT
VCC1
CURRENT SOURCE
INPUT
GND
TYPICAL OF ALL OUTPUTS
VCC2
OUTPUT
GND

absolute maximum ratings over operating free-air temperature range (unless otherwise noted)

Logic supply voltage range, V_{CC1} (see Note 1)	−0.5 V to 36 V
Output supply voltage range, V_{CC2}	−0.5 V to 36 V
Input voltage	36 V
Output voltage range, V_O	−3 V to V_{CC2}+3 V
Peak output current (nonrepetitive, $t_w \leq 5$ ms), I_{PK}	±2 A
Continuous output current, I_O	±1.1 A
Continuous total dissipation at (or below) 25 °C free-air temperature (see Note 2)	2075 mW
Continuous total dissipation at (or below) 25 °C case temperature (see Note 2)	7375 mW
Operating case or virtual junction temperature range	−40 °C to 150 °C
Storage temperature range	−65 °C to 150 °C
Lead temperature 1,6 mm (1/16 inch) from case for 10 seconds	260 °C

NOTES: 1. All voltage values are with respect to the network ground terminal.
2. For operation above 25 °C free-air temperature, derate linearly at the rate of 16.6 mW/°C. For operation above 25 °C case temperature, derate linearly at the rate of 59 mW/°C. To avoid exceeding the design maximum virtual junction temperature, these ratings should not be exceeded. Due to variations in individual device electrical characteristics and thermal resistance, the built-in thermal overload protection may be activated at power levels slightly above or below the rated dissipation.

recommended operating conditions

	MIN	MAX	UNIT
Logic supply voltage, V_{CC1}	4.5	5.5	V
Output supply voltage, V_{CC2}	4.5	36	V
High-level input voltage, V_{IH}	2	5.5	V
Low-level input voltage, V_{IL}	−0.3†	0.8	V
Output current, I_O		±1	A
Operating virtual junction temperature, T_J	−40	125	°C
Operating free-air temperature, T_A	−40	85	°C

†The algebraic convention, in which the least positive (most negative designated minimum), is used in this data sheet for logic voltage levels.

electrical characteristics over recommended ranges of V_{CC1}, V_{CC2}, and operating virtual junction temperature (unless otherwise noted)

PARAMETER		TEST CONDITIONS		MIN	TYP†	MAX	UNIT
V_{IK}	Input clamp voltage	$I_I = -12$ mA			−0.9	−1.5	V
V_{OH}	High-level output voltage	$I_{OH} = -0.5$ A		$V_{CC2}-1.5$	$V_{CC2}-1.1$		V
		$I_{OH} = -1$ A		$V_{CC2}-2$			
		$I_{OH} = -1$ A, $T_J = 25°C$		$V_{CC2}-1.8$	$V_{CC2}-1.4$		
V_{OL}	Low-level output voltage	$I_{OL} = 0.5$ A			1	1.4	V
		$I_{OL} = 1$ A				2	
		$I_{OL} = 1$ A, $T_J = 25°C$			1.2	1.8	
V_{OKH}	High-level output clamp voltage	$I_{OK} = 0.5$ A			$V_{CC2}+1.4$	$V_{CC2}+2$	V
		$I_{OK} = 1$ A			$V_{CC2}+1.9$	$V_{CC2}+2.5$	
V_{OKL}	Low-level output clamp voltage	$I_{OK} = -0.5$ A			−1.1	−2	V
		$I_{OK} = -1$ A			−1.3	−2.5	
I_{OZ}	Off-state (high impedance-state) output current	$V_O = V_{CC2}$				500	μA
		$V_O = 0$				−500	
I_{IH}	High-level input current	$V_I = 5.5$ V				10	μA
I_{IL}	Low-level input current	$V_I = 0$				−10	μA
I_{CC1}	Logic supply current	$I_O = 0$	All outputs at high level			38	mA
			All outputs at low level			70	
			All outputs at high impedance			25	
I_{CC2}	Output supply current	$I_O = 0$	All outputs at high level			33	mA
			All outputs at low level			20	
			All outputs at high impedance			5	

†All typical values are at $V_{CC1} = 5$ V, $V_{CC2} = 24$ V, $T_A = 25°C$.

switching characteristics, $V_{CC1} = 5$ V, $V_{CC2} = 24$ V, $T_A = 25°C$

PARAMETER		TEST CONDITIONS	MIN	TYP	MAX	UNIT
t_{DLH}	Delay time, low-to-high-level output from A input	$C_L = 30$ pF, See Figure 1		800		ns
t_{DHL}	Delay time, high-to-low-level output from A input			400		ns
t_{TLH}	Transition time, low-to-high-level output			300		ns
t_{THL}	Transition time, high-to-low-level output			300		ns
t_{PZH}	Enable time to the high level	$C_L = 30$ pF, See Figure 2		700		ns
t_{PZL}	Enable time to the low level			400		ns
t_{PHZ}	Disable time from the high level			900		ns
t_{PLZ}	Disable time from the low level			600		ns

PARAMETER MEASUREMENT INFORMATION

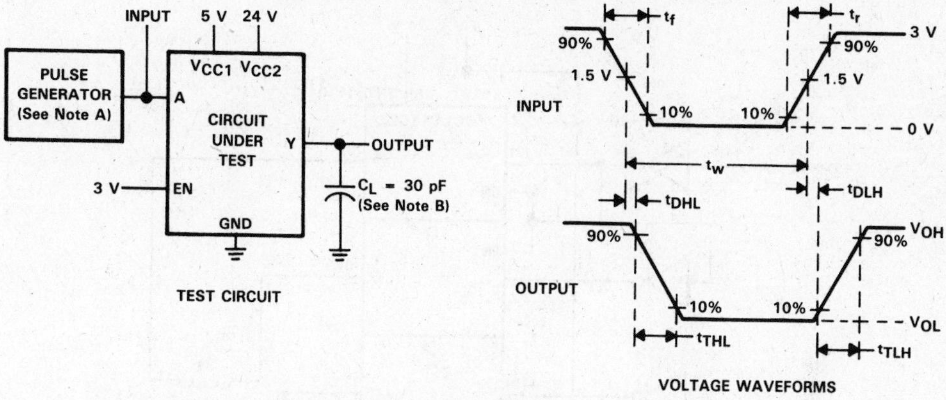
INPUT
5 V
24 V
PULSE GENERATOR (See Note A)
VCC1 VCC2
A
CIRCUIT UNDER TEST
Y
OUTPUT
3 V
EN
CL = 30 pF (See Note B)
GND
TEST CIRCUIT
tf
tr
90%
1.5 V
INPUT
10%
3 V
0 V
tw
tDHL
tDLH
VOH
VOL
OUTPUT
tTHL
tTLH
VOLTAGE WAVEFORMS

NOTES: A. The pulse generator has the following characteristics: $t_r \leq 10$ ns, $t_f \leq 10$ ns, $t_w = 10\ \mu s$, PRR = 5 kHz, $Z_{out} = 50\ \Omega$.
B. C_L includes probe and jig capacitance.

FIGURE 1. SWITCHING TIMES FROM DATA INPUTS

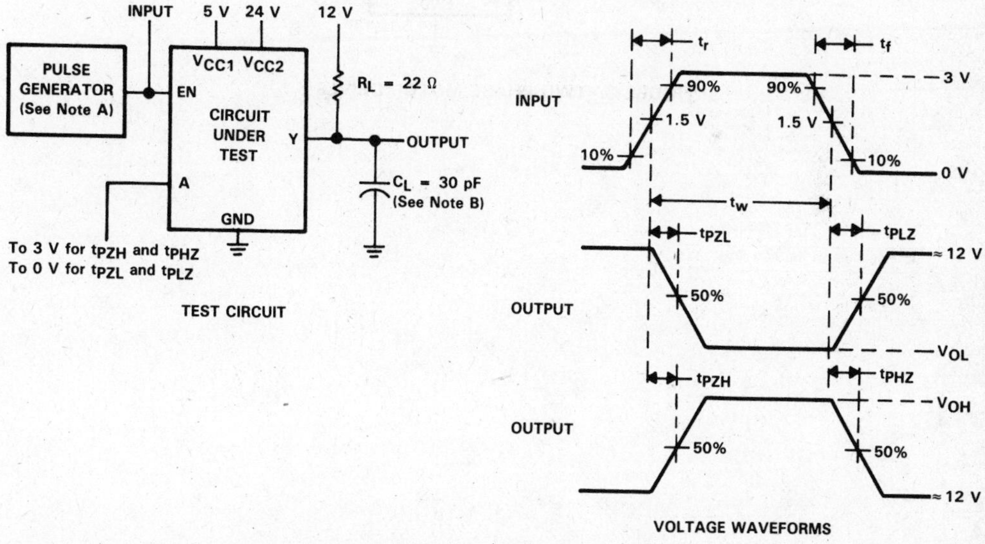
INPUT
5 V
24 V
12 V
PULSE GENERATOR (See Note A)
VCC1 VCC2
EN
RL = 22 Ω
CIRCUIT UNDER TEST
Y
OUTPUT
A
CL = 30 pF (See Note B)
GND
To 3 V for tPZH and tPHZ
To 0 V for tPZL and tPLZ
TEST CIRCUIT
tr
tf
INPUT
90%
1.5 V
10%
3 V
0 V
tw
tPZL
tPLZ
≈ 12 V
50%
OUTPUT
VOL
tPZH
tPHZ
VOH
≈ 12 V
VOLTAGE WAVEFORMS

NOTES: A. The pulse generator has the following characteristics: $t_r \leq 10$ ns, $t_f \leq 10$ ns, $t_w = 10\ \mu s$, PRR = 5 kHz, $Z_{out} = 50\ \Omega$.
B. C_L includes probe and jig capacitance.

FIGURE 2. SWITCHING TIMES FROM ENABLE INPUTS

TYPICAL APPLICATION DATA

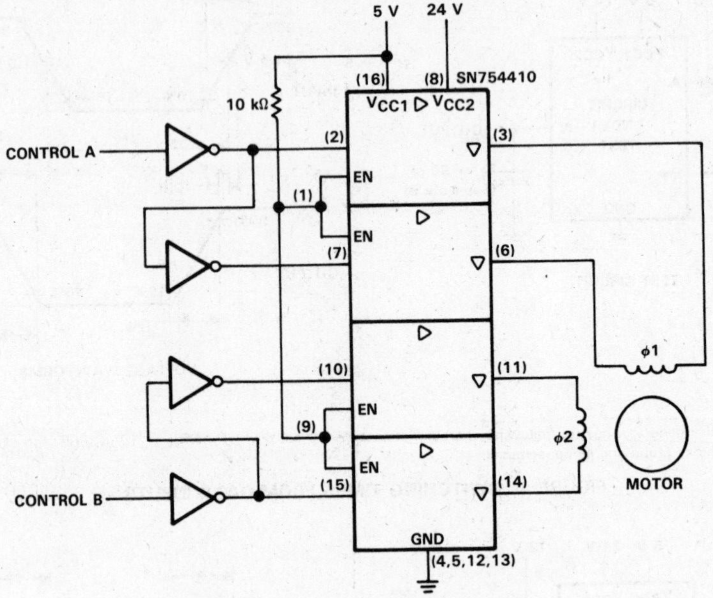
5 V
24 V
10 kΩ
(16)
(8)
SN754410
V_{CC1}
V_{CC2}
CONTROL A
(2)
(3)
EN
(1)
EN
(7)
(6)
φ1
(10)
(11)
EN
(9)
φ2
MOTOR
EN
(15)
(14)
CONTROL B
GND
(4,5,12,13)

FIGURE 3. TWO-PHASE MOTOR DRIVER

SN754411
QUADRUPLE HALF-H DRIVER

D2942, NOVEMBER 1986

- 1-A Output Current Capability per Channel
- Applications Include Half-H and Full-H Solenoid Drivers and Motor Drivers
- Designed for Positive-Supply Applications
- Wide Supply Voltage Range: 4.5 V to 36 V
- TTL- and CMOS-Compatible High-Impedance Diode-Clamped Inputs
- Separate Input-Logic Supply
- Thermal Shutdown
- Internal ESD Protection
- Input Hysteresis Improves Noise Immunity
- Three-State Outputs
- Minimized Power Dissipation
- Sink/Source Interlock Circuitry Prevents Simultaneous Conduction
- No Output "Glitch" During Power-Up or Power-Down
- Improved Functional Replacement for the SGS L293

NE DUAL-IN-LINE PACKAGE
(TOP VIEW)

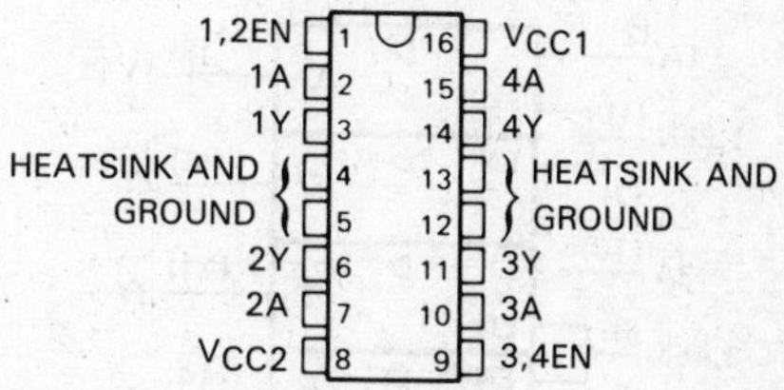
1,2EN
1A
1Y
HEATSINK AND GROUND
2Y
2A
V_{CC2}
1
2
3
4
5
6
7
8
16
15
14
13
12
11
10
9
V_{CC1}
4A
4Y
HEATSINK AND GROUND
3Y
3A
3,4EN

FUNCTION TABLE
(EACH CHANNEL)

INPUTS		OUTPUT
A	EN	Y
H	H	H
L	H	L
X	L	Z

H = high-level
L = low-level
X = irrelevant
Z = high-impedance (off)

description

The SN754411 is a quadruple high-current half-H driver designed to provide bidirectional drive currents of up to one ampere at voltages from 4.5 volts to 36 volts. It is designed to drive inductive loads such as relays, solenoids, dc and stepping motors, as well as other high-current/high-voltage loads in positive-supply applications.

All inputs are compatible with TTL and low-level CMOS logic. Each output (Y) is a complete totem-pole driver with a Darlington transistor sink and a psuedo-Darlington source. Channels are enabled in pairs with channels 1 and 2 enabled by 1,2EN and channels 3 and 4 enabled by 3,4EN. When an enable input is high, the associated channels are enabled and their outputs become active and in phase with their inputs. When the enable input is low, those channels are disabled and their outputs are off and in a high-impedance state. With the proper data inputs, each pair of drivers form a full-H (or bridge) reversible drive suitable for solenoid or motor applications.

External high-speed output clamp diodes should be used for inductive-transient suppression. A separate supply voltage (V_{CC1}) is provided for the logic input circuits to minimize device power dissipation. Supply voltage (V_{CC2}) is used for the output circuits.

The SN754411 is designed for operation from −40°C to 85°C.

TEXAS INSTRUMENTS
POST OFFICE BOX 655012 • DALLAS, TEXAS 75265

logic symbol†

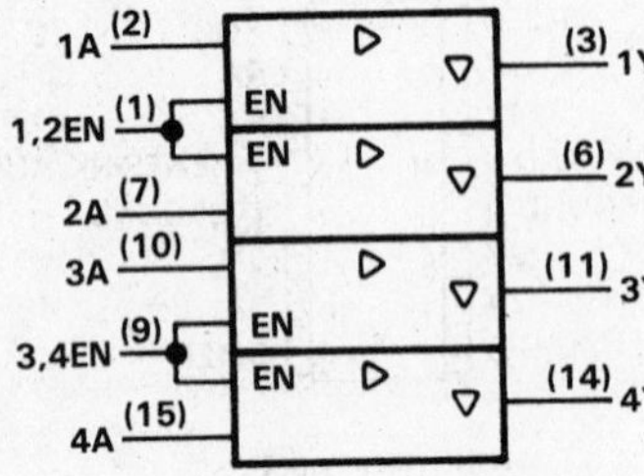
1A
(2)
1,2EN
(1)
EN
EN
2A
(7)
3A
(10)
3,4EN
(9)
EN
EN
4A
(15)
(3)
1Y
(6)
2Y
(11)
3Y
(14)
4Y

† This symbol is in accordance with ANSI/IEEE Std 91-1984 and IEC Publication 617-12.

logic diagram

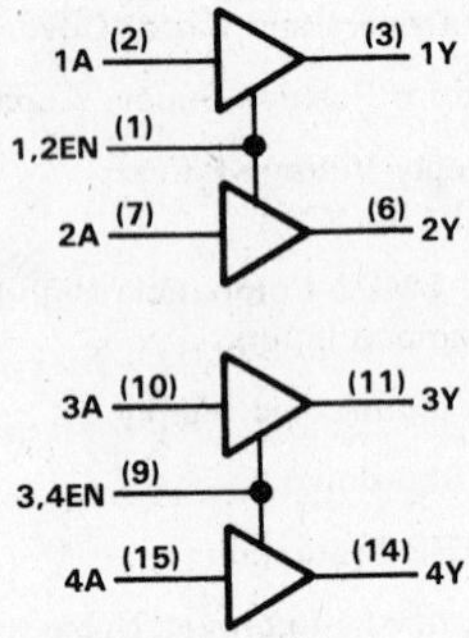
1A
(2)
(3)
1Y
1,2EN
(1)
2A
(7)
(6)
2Y
3A
(10)
(11)
3Y
3,4EN
(9)
4A
(15)
(14)
4Y

schematics of inputs and outputs

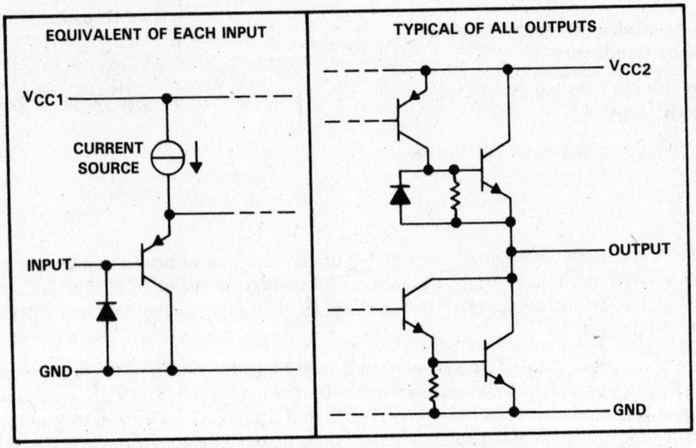
EQUIVALENT OF EACH INPUT
VCC1
CURRENT
SOURCE
INPUT
GND
TYPICAL OF ALL OUTPUTS
VCC2
OUTPUT
GND

TEXAS INSTRUMENTS
POST OFFICE BOX 655012 • DALLAS, TEXAS 75265

absolute maximum ratings over operating free-air temperature range (unless otherwise noted)

Logic supply voltage range, V_{CC1} (see Note 1)	−0.5 V to 36 V
Output supply voltage range, V_{CC2}	−0.5 V to 36 V
Input voltage	36 V
Output voltage range, V_O	−3 V to V_{CC2} + 3 V
Peak output current (nonrepetitive, $t_w \leq 5$ ms), I_{PK}	±2 A
Continuous output current, I_O	±1.1 A
Continuous total dissipation at (or below) 25°C free-air temperature (see Note 2)	2075 mW
Continuous total dissipation at (or below) 25°C case temperature (see Note 2)	7375 mW
Operating case or virtual junction temperature range	−40°C to 150°C
Storage temperature range	−65°C to 150°C
Lead temperature 1,6 mm (1/16 inch) from case for 10 seconds	260°C

NOTES: 1. All voltage values are with respect to the network ground terminal.
2. For operation above 25°C free-air temperature, derate linearly at the rate of 16.6 mW/°C. For operation above 25°C case temperature, derate linearly at the rate of 59 mW/°C. To avoid exceeding the design maximum virtual junction temperature, these ratings should not be exceeded. Due to variations in individual device electrical characteristics and thermal resistance, the built-in thermal overload protection may be activated at power levels slightly above or below the rated dissipation.

recommended operating conditions

	MIN	MAX	UNIT
Logic supply voltage, V_{CC1}	4.5	5.5	V
Output supply voltage, V_{CC2}	4.5	36	V
High-level input voltage, V_{IH}	2	5.5	V
Low-level input voltage, V_{IL}	−0.3†	0.8	V
Output current, I_O		±1	A
Operating virtual junction temperature, T_J	−40	125	°C
Operating free-air temperature, T_A	−40	85	°C

†The algebraic convention, in which the least positive (most negative designated minimum), is used in this data sheet for logic voltage levels.

electrical characteristics over recommended ranges of V_{CC1}, V_{CC2}, and operating virtual junction temperature (unless otherwise noted)

PARAMETER		TEST CONDITIONS		MIN	TYP†	MAX	UNIT
V_{IK}	Input clamp voltage	$I_I = -12$ mA			−0.9	−1.5	V
V_{OH}	High-level output voltage	$I_{OH} = -0.5$ A		$V_{CC2}-1.5$	$V_{CC2}-1.1$		V
		$I_{OH} = -1$ A		$V_{CC2}-2$			
		$I_{OH} = -1$ A, $T_J = 25°C$		$V_{CC2}-1.8$	$V_{CC2}-1.4$		
V_{OL}	Low-level output voltage	$I_{OL} = 0.5$ A			1	1.4	V
		$I_{OL} = 1$ A				2	
		$I_{OL} = 1$ A, $T_J = 25°C$			1.2	1.8	
I_{OZ}	Off-state (high impedance-state) output current	$V_O = V_{CC2}$				500	μA
		$V_O = 0$				−500	
I_{IH}	High-level input current	$V_I = 5.5$ V				10	μA
I_{IL}	Low-level input current	$V_I = 0$				−10	μA
I_{CC1}	Logic supply current	$I_O = 0$	All outputs at high level			38	mA
			All outputs at low level			70	
			All outputs at high impedance			25	
I_{CC2}	Output supply current	$I_O = 0$	All outputs at high level			33	mA
			All outputs at low level			20	
			All outputs at high impedance			5	

†All typical values are at $V_{CC1} = 5$ V, $V_{CC2} = 24$ V, $T_A = 25°C$.

switching characteristics, $V_{CC1} = 5$ V, $V_{CC2} = 24$ V, $T_A = 25°C$

PARAMETER		TEST CONDITIONS	MIN	TYP	MAX	UNIT
t_{DLH}	Delay time, low-to-high-level output from A input	$C_L = 30$ pF, See Figure 1		800		ns
t_{DHL}	Delay time, high-to-low-level output from A input			400		ns
t_{TLH}	Transition time, low-to-high-level output			300		ns
t_{THL}	Transition time, high-to-low-level output			300		ns
t_{PZH}	Enable time to the high level	$C_L = 30$ pF, See Figure 2		700		ns
t_{PZL}	Enable time to the low level			400		ns
t_{PHZ}	Disable time from the high level			900		ns
t_{PLZ}	Disable time from the low level			600		ns

PARAMETER MEASUREMENT INFORMATION

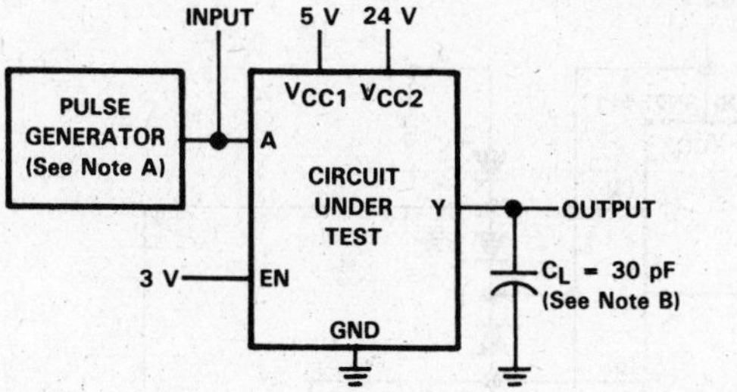
INPUT
5 V
24 V
PULSE GENERATOR (See Note A)
VCC1 VCC2
A
CIRCUIT UNDER TEST
Y
OUTPUT
3 V
EN
CL = 30 pF (See Note B)
GND

TEST CIRCUIT

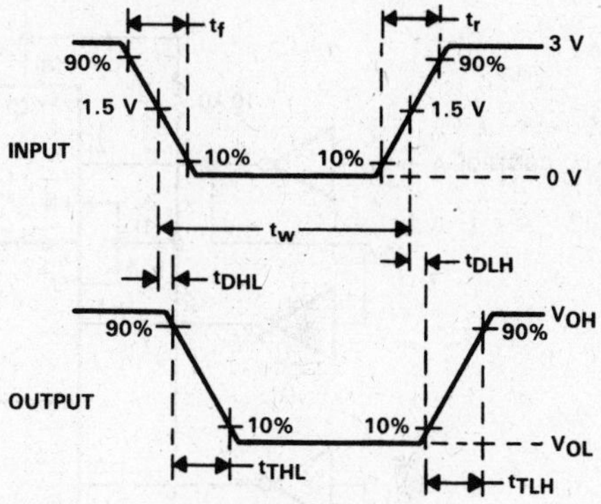
tf
tr
3 V
90%
1.5 V
INPUT
10%
0 V
tw
tDHL
tDLH
VOH
OUTPUT
VOL
tTHL
tTLH

VOLTAGE WAVEFORMS

NOTES: A. The pulse generator has the following characteristics: $t_r \leq 10$ ns, $t_f \leq 10$ ns, $t_w = 10\ \mu s$, PRR = 5 kHz, $Z_{out} = 50\ \Omega$.
B. C_L includes probe and jig capacitance.

FIGURE 1. SWITCHING TIMES FROM DATA INPUTS

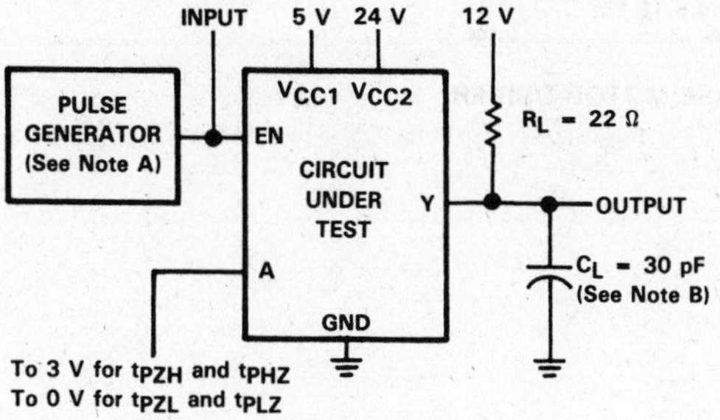
INPUT
5 V
24 V
12 V
PULSE GENERATOR (See Note A)
VCC1 VCC2
EN
RL = 22 Ω
CIRCUIT UNDER TEST
Y
OUTPUT
A
CL = 30 pF (See Note B)
GND
To 3 V for tPZH and tPHZ
To 0 V for tPZL and tPLZ

TEST CIRCUIT

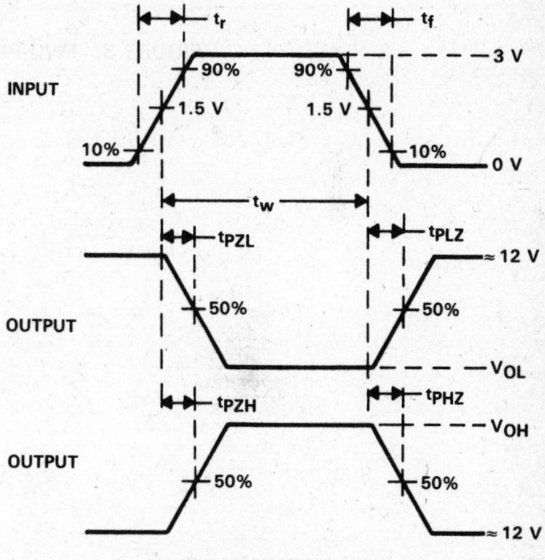
tr
tf
3 V
90%
INPUT
1.5 V
10%
0 V
tw
tPZL
tPLZ
≈ 12 V
50%
OUTPUT
VOL
tPZH
tPHZ
VOH
≈ 12 V

VOLTAGE WAVEFORMS

NOTES: A. The pulse generator has the following characteristics: $t_r \leq 10$ ns, $t_f \leq 10$ ns, $t_w = 10\ \mu s$, PRR = 5 kHz, $Z_{out} = 50\ \Omega$.
B. C_L includes probe and jig capacitance.

FIGURE 2. SWITCHING TIMES FROM ENABLE INPUTS

TYPICAL APPLICATION DATA

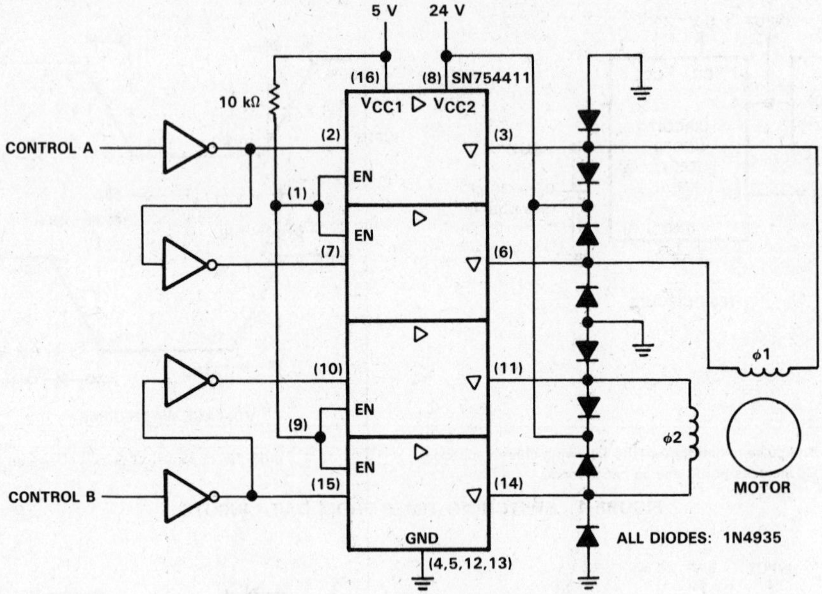
5 V
24 V
(16)
(8)
SN754411
10 kΩ
V_{CC1}
V_{CC2}
CONTROL A
(2)
(3)
EN
(1)
EN
(7)
(6)
(10)
(11)
EN
(9)
EN
(15)
(14)
CONTROL B
GND
(4,5,12,13)
φ1
φ2
MOTOR
ALL DIODES: 1N4935

FIGURE 3. TWO-PHASE MOTOR DRIVER

D2570, DECEMBER 1979

- Three Independent Inverting Stepper-Motor Driver Circuits
- High Output Source Current . . . 500 mA Typ
- High Output Sink Current . . . 500 mA Typ
- Inputs Are Compatible with Bipolar and MOS
- Wide Supply Voltage Range . . . 4 V to 18 V
- Threshold Voltage Range is Approximately One-Half V_{CC}
- Active Pull-Down on Each Input
- Low Standby Power Dissipation
- 14-Pin NE Power Package

NE DUAL-IN-LINE PACKAGE
(TOP VIEW)

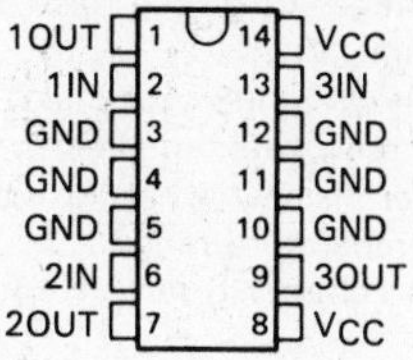
1OUT 1
1IN 2
GND 3
GND 4
GND 5
2IN 6
2OUT 7
14 VCC
13 3IN
12 GND
11 GND
10 GND
9 3OUT
8 VCC

logic symbol†

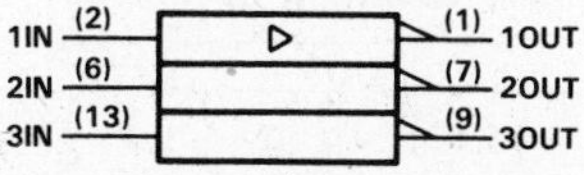
1IN (2)
2IN (6)
3IN (13)
(1) 1OUT
(7) 2OUT
(9) 3OUT

†This symbol is in accordance with ANSI/IEEE Std 91-1984 and IEC Publication 617-12.

description

The TL376C is a monolithic bipolar three-channel stepper-motor driver. The input signal is inverted through the device and drives a totem-pole output section. Each output can source or sink up to 500 milliamperes. The wide supply-voltage range coupled with a threshold voltage level of approximately one-half V_{CC} allows this device to interface with MOS as well as bipolar outputs. An active-pull-down circuit is included on each input. In typical operation, a microprocessor supplies a three-phase signal to the device, which then drives a two-winding stepper-motor.

The TL376C is characterized for operation from 0°C to 70°C.

schematic (each driver)

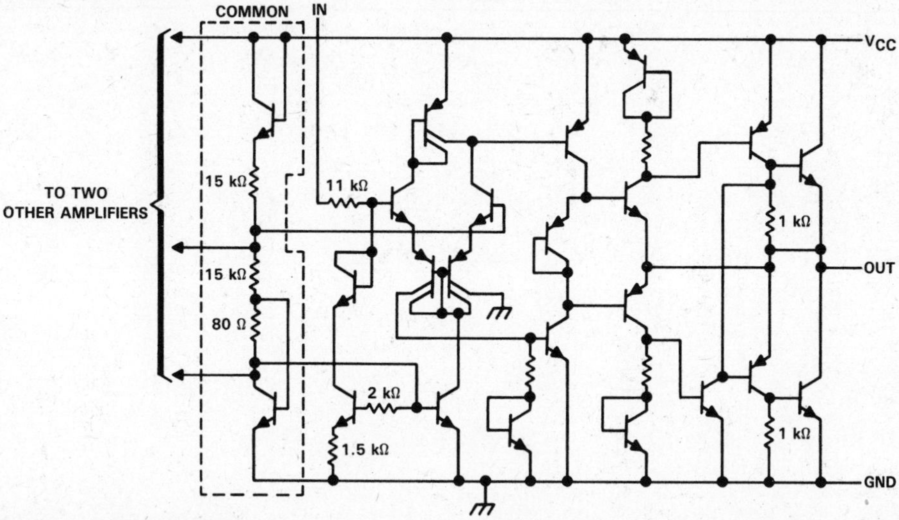
COMMON
IN
VCC
TO TWO
OTHER AMPLIFIERS
15 kΩ
11 kΩ
15 kΩ
80 Ω
2 kΩ
1.5 kΩ
1 kΩ
1 kΩ
OUT
GND

Resistor values shown are nominal.

absolute maximum ratings over operating free-air temperature (unless otherwise noted)

Supply voltage, V_{CC} (see Note 1) .... 22 V
Input voltage, V_I .... V_{CC}
Output voltage range .... −0.9 V to V_{CC}+1 V
Output current, each amplifier .... 550 mA
Total power dissipation at (or below) 25 °C free-air temperature (see Note 2) .... 2075 mW
Storage temperature range .... −65 °C to 150 °C
Lead temperature 1,6 mm (1/16 inch) from case for 10 seconds .... 260 °C

NOTES: 1. Voltage values are with respect to the network ground terminal.
2. For operation above 25 °C free-air temperature, derate linearly at the rate of 16.6 mW/°C.

recommended operating conditions

	MIN	NOM	MAX	UNIT
High-level input voltage, V_{IH}	$\frac{V_{CC}}{2}+0.8$		V_{CC}	V
Low-level input voltage, V_{IL}			$\frac{V_{CC}}{2}-0.2$	V
Supply voltage range, V_{CC}	4	11	18	V
Operating free-air temperature, T_A	0		70	°C

electrical characteristics over recommended ranges of supply voltage and operating free-air temperature (unless otherwise noted)

PARAMETER		TEST CONDITIONS		MIN	TYP†	MAX	UNIT
V_{OL}	Low-level output voltage	I_{OL} = 500 mA,	V_I = V_{IH} min			1.5	V
V_{OH}	High-level output voltage	I_{OH} = −500 mA,	V_I = V_{IL} max	V_{CC} − 1.5			V
I_I	Input current	V_I = V_{CC}				100	μA
		V_I = 1.8 V		5			μA
I_{CC}	Supply current	Inputs open, Outputs open, V_{CC} = 18 V			0.7	2	mA

†Typical values are measured at V_{CC} = 15 V, T_A = 25 °C.

TEXAS INSTRUMENTS
POST OFFICE BOX 655012 • DALLAS, TEXAS 75265

TYPICAL CHARACTERISTICS

INPUT CURRENT
vs
INPUT VOLTAGE

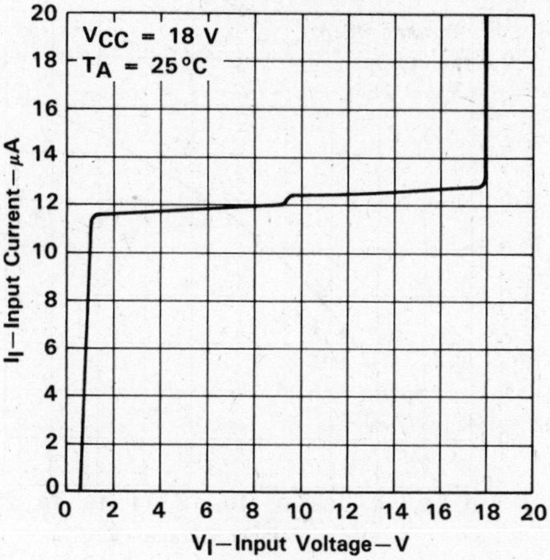
V_{CC} = 18 V
T_A = 25°C
I_I—Input Current—μA
V_I—Input Voltage—V
0 2 4 6 8 10 12 14 16 18 20

FIGURE 1

OUTPUT VOLTAGE
vs
INPUT VOLTAGE

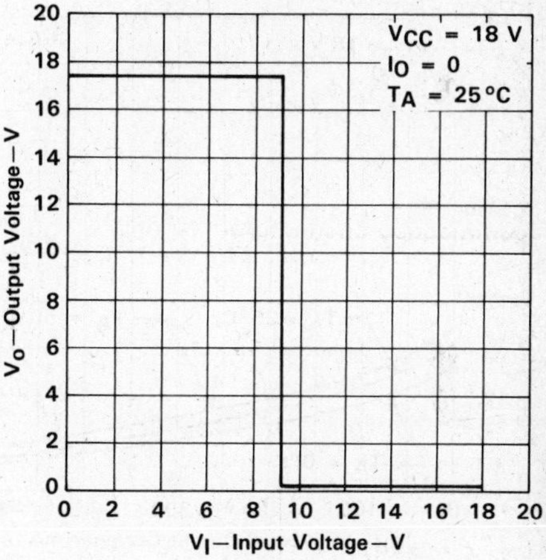
V_{CC} = 18 V
I_O = 0
T_A = 25°C
V_O—Output Voltage—V
V_I—Input Voltage—V
0 2 4 6 8 10 12 14 16 18 20

FIGURE 2

LOW-LEVEL OUTPUT VOLTAGE
vs
LOW-LEVEL OUTPUT CURRENT

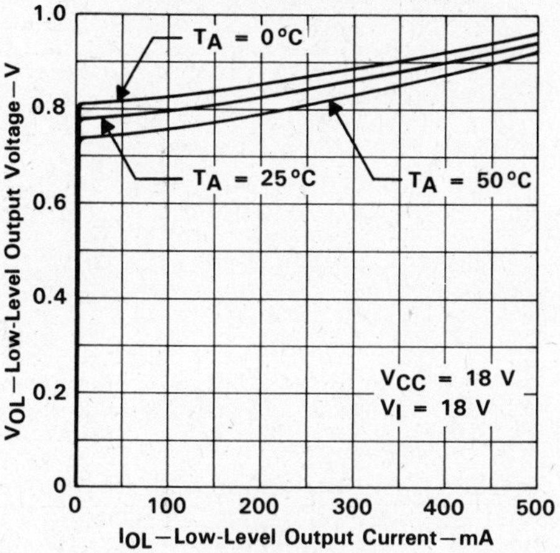
T_A = 0°C
T_A = 25°C
T_A = 50°C
V_{CC} = 18 V
V_I = 18 V
V_{OL}—Low-Level Output Voltage—V
0 0.2 0.4 0.6 0.8 1.0
0 100 200 300 400 500
I_{OL}—Low-Level Output Current—mA

FIGURE 3

TYPICAL CHARACTERISTICS

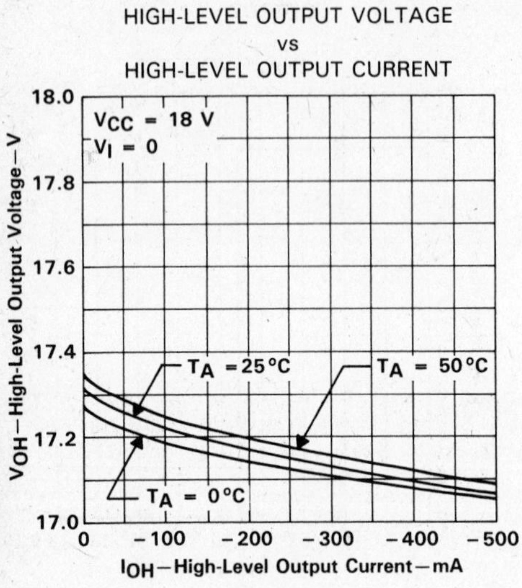
HIGH-LEVEL OUTPUT VOLTAGE
vs
HIGH-LEVEL OUTPUT CURRENT
VCC = 18 V
VI = 0
TA = 25°C
TA = 50°C
TA = 0°C
VOH—High-Level Output Voltage—V
18.0
17.8
17.6
17.4
17.2
17.0
0
−100
−200
−300
−400
−500
IOH—High-Level Output Current—mA

FIGURE 4

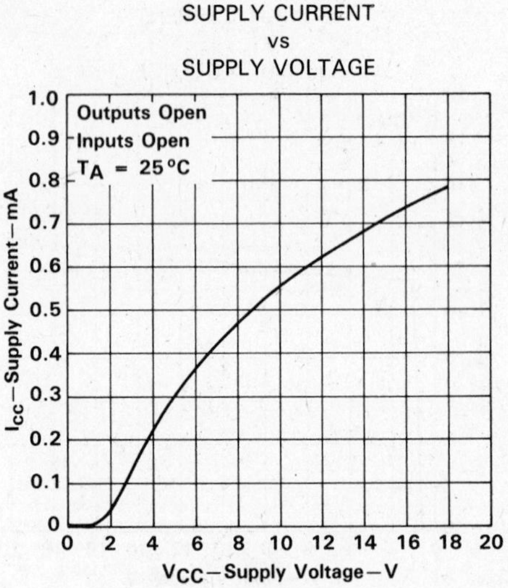
SUPPLY CURRENT
vs
SUPPLY VOLTAGE
Outputs Open
Inputs Open
TA = 25°C
ICC—Supply Current—mA
1.0
0.9
0.8
0.7
0.6
0.5
0.4
0.3
0.2
0.1
0
0
2
4
6
8
10
12
14
16
18
20
VCC—Supply Voltage—V

FIGURE 5

D2507, DECEMBER 1980—REVISED AUGUST 1986

UDN2841, UDN2845 QUADRUPLE HIGH-CURRENT DARLINGTON DRIVERS

- For Use with Negative Supplies
- Current Sink . . . UDN2841
- Sink or Source Combination . . . UDN2845
- Output Current Capability . . . 1.5 A
- High Output-Voltage Capability . . . 50 V
- Preamplifier for High Current Gain
- Inputs Compatible with TTL and 5-V CMOS
- Reliable Monolithic Construction
- Designed to be Interchangeable with Sprague UDN2841 and UDN2845

NE DUAL-IN-LINE PACKAGE
(TOP VIEW)

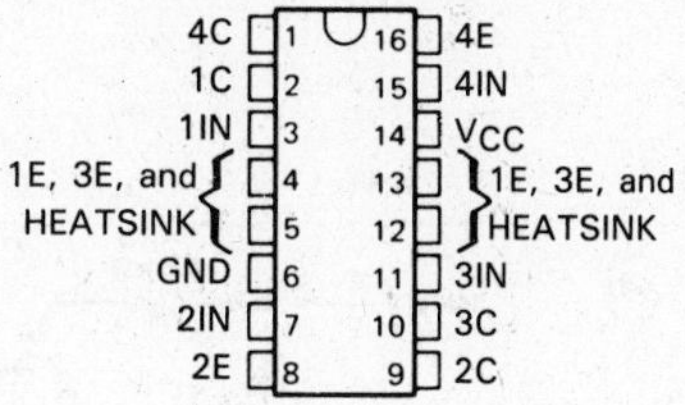
4C 1
1C 2
1IN 3
1E, 3E, and HEATSINK 4, 5
GND 6
2IN 7
2E 8
16 4E
15 4IN
14 VCC
13, 12 1E, 3E, and HEATSINK
11 3IN
10 3C
9 2C

description

These quadruple Darlington switches are monolithic bipolar devices especially designed for high-current, high-voltage peripheral driver applications. The devices are designed to offer solutions to interface problems involving electronic-discharge printers, bipolar and unipolar dc motor drivers, telephone relays, LEDs, PIN diodes, and other high-current loads operating from negative power supplies.

The UDN2841 is intended for current-sink applications with the load connected to ground and the device switching the negative supply. The UDN2845 is a sink and source combination for use in bipolar switching applications where both ends of the load are floating. The UDN2841 and UDN2845 each feature inputs that are compatible with standard TTL and 5-volt CMOS signals. The p-n-p input transistor serves as a level translator and the first n-p-n transistor stage is designed to provide sufficient current gain to drive the output Darlington-connected pair.

Driver channels 2 and 4 have uncommitted collectors and emitters while 1 and 3 have emitters internally connected to the substrate. For proper operation, the substrate must be connected to the most-negative supply voltage.

The UDN2841 and UDN2845 are characterized for operation from −20°C to 85°C.

logic symbol†

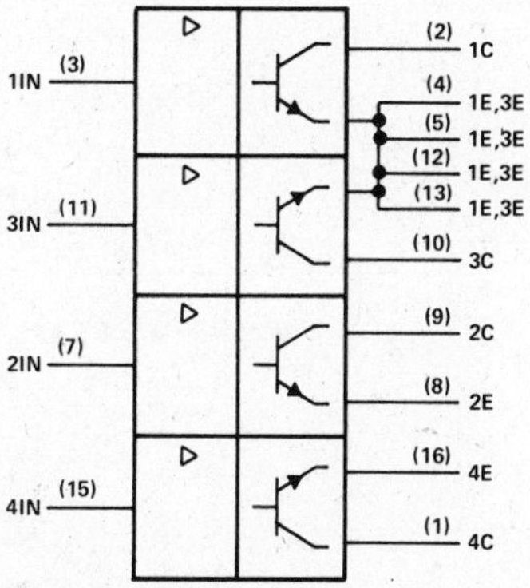
1IN (3)
3IN (11)
2IN (7)
4IN (15)
(2) 1C
(4) 1E,3E
(5) 1E,3E
(12) 1E,3E
(13) 1E,3E
(10) 3C
(9) 2C
(8) 2E
(16) 4E
(1) 4C

logic diagram

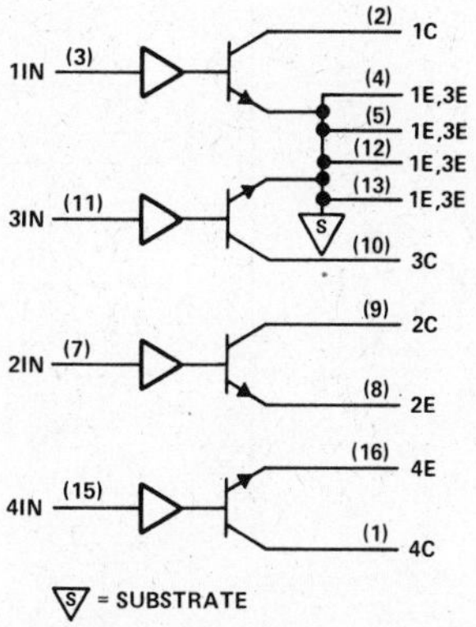
1IN (3)
3IN (11)
2IN (7)
4IN (15)
(2) 1C
(4) 1E,3E
(5) 1E,3E
(12) 1E,3E
(13) 1E,3E
(10) 3C
(9) 2C
(8) 2E
(16) 4E
(1) 4C
S = SUBSTRATE

†This symbol is in accordance with ANSI/IEEE Std 91-1984 and IEC Publication 617-12.

schematic diagram (each driver)

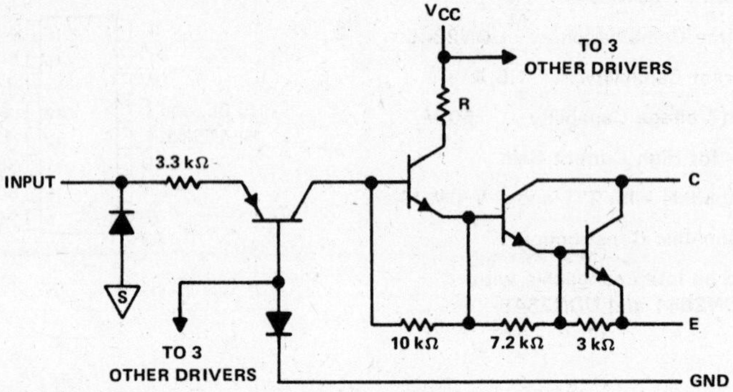
VCC
TO 3
OTHER DRIVERS
R
3.3 kΩ
INPUT
C
S
TO 3
OTHER DRIVERS
10 kΩ
7.2 kΩ
3 kΩ
E
GND

S = Substrate

UDN2841: R = 15 kΩ each channel

UDN2845: R = 15 kΩ channels 1 and 3, R = 1 kΩ, channels 2 and 4.

Resistor values shown are nominal.

absolute maximum ratings at 25°C free-air temperature (unless otherwise noted)

Collector-emitter voltage	50 V
Supply voltage, V_{CC} (see Note 1)	10 V
Input voltage	10 V
Substrate voltage	−50 V
Peak output current	1.75 A
Total power dissipation at (or below) 25°C free-air temperature (see Note 2)	2075 mW
Operating free-air temperature range	−20°C to 85°C
Storage temperature range	−55°C to 150°C
Lead temperature 1,6 mm (1/16 inch) from case for 10 seconds	260°C

NOTES: 1. All voltage values, except collector-emitter voltage, are with respect to the network ground terminal.
2. For operation above 25°C free-air temperature, derate total power linearly to 1079 mW at 85°C at the rate of 16.6 mW/°C.

electrical characteristics at 25°C free-air temperature (unless otherwise noted), V_{CC} = 5 V, see Figures 1 and 2

PARAMETER		TEST CONDITIONS		UDN2841 MIN	UDN2841 TYP	UDN2841 MAX	UDN2845 MIN	UDN2845 TYP	UDN2845 MAX	UNIT
$V_{CEX(sus)}$	Collector sustaining voltage	V_{EE} = −50 V, V_I = 0.4 V, I_O = 100 mA		35	50		35	50		V
I_{CEX}	Collector output cutoff current	V_{EE} = −50 V, V_I = 0.4 V				100			100	μA
		V_{EE} = −50 V, V_I = 0.4 V, T_A = 70°C				500			500	
$I_{I(on)}$	On-state input current	I_O = 0.5 A	Drivers 1 and 3		300	500		300	500	μA
			Drivers 2 and 4		300	500		350	500	
$V_{I(on)}$	On-state input voltage	I_O = 1.5 A, See Note 3				2.4			2.4	V
$V_{CE(sat)}$	Collector-emitter saturation voltage	V_I = 2.4 V, See Note 3	I_O = 0.5 A			1.1			1.1	V
			I_O = 1 A			1.4			1.4	
			I_O = 1.5 A			1.6			1.6	
I_{CC}	Supply current (each driver)	I_O = 0.5 A, See Note 3	Drivers 1 and 3		2.5	3.75		2.5	3.75	mA
			Drivers 2 and 4		2.5	3.75		3.75	7.5	

NOTE 3: These parameters must be measured on one output at a time using pulse techniques, t_w = 10 ms, duty cycle ≤10%.

switching characteristics at V_{EE} = −40 V, R_L = 39 Ω, C_L = 15 pF, T_A = 25°C

PARAMETER		TEST CONDITIONS	UDN2841 MIN	UDN2841 TYP	UDN2841 MAX	UDN2845 MIN	UDN2845 TYP	UDN2845 MAX	UNIT
t_{on}	Turn-on time	See Figure 3			2			2	μs
t_{off}	Turn-off time	See Figure 3			5			5	μs

PARAMETER MEASUREMENT INFORMATION

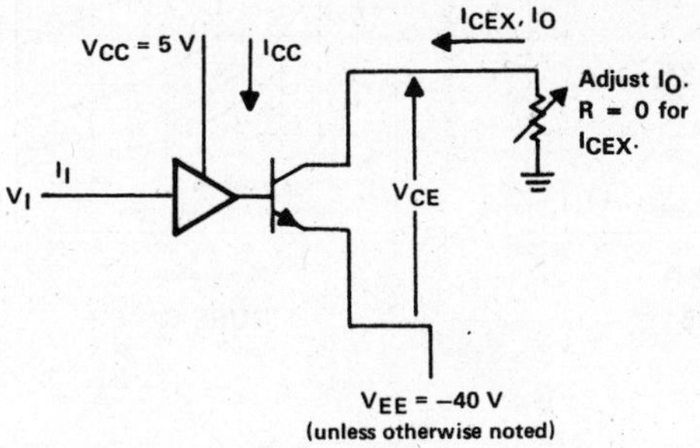
V_{CC} = 5 V
I_{CC}
I_{CEX}, I_O
Adjust I_O.
R = 0 for
I_{CEX}.
I_I
V_I
V_{CE}
V_{EE} = −40 V
(unless otherwise noted)

NOTE: UDN2841 driver channels 1 through 4 and UDN2845 driver channels 1 and 3 only.

TEST CIRCUIT

FIGURE 1. SINK-CURRENT DRIVER

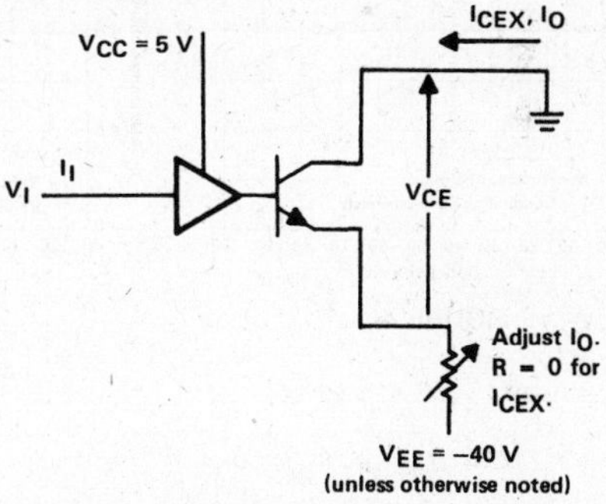
V_{CC} = 5 V
I_{CEX}, I_O
I_I
V_I
V_{CE}
Adjust I_O.
R = 0 for
I_{CEX}.
V_{EE} = −40 V
(unless otherwise noted)

NOTE: UDN2845 driver channels 2 and 4 only.

TEST CIRCUIT

FIGURE 2. SOURCE-CURRENT DRIVER

PARAMETER MEASUREMENT INFORMATION

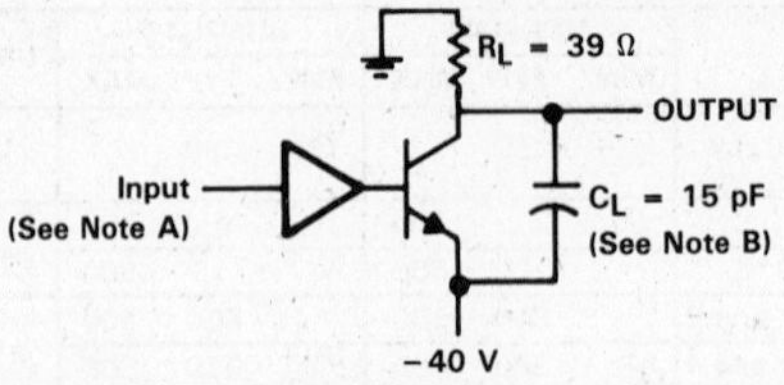
R_L = 39 Ω
OUTPUT
Input
(See Note A)
C_L = 15 pF
(See Note B)
−40 V

TEST CIRCUIT

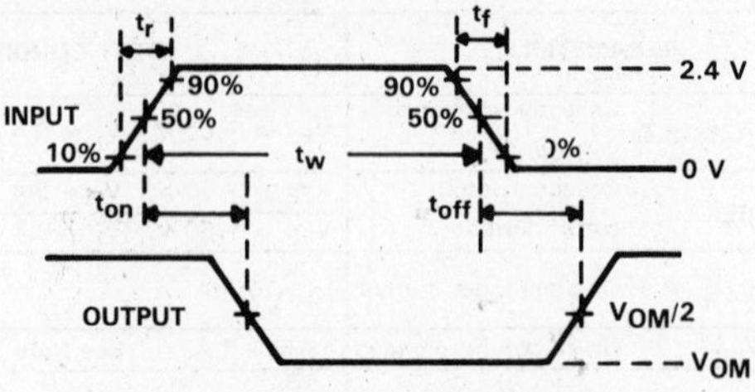
t_r
t_f
2.4 V
90%
90%
INPUT
50%
50%
10%
t_w
0 V
t_{on}
t_{off}
OUTPUT
$V_{OM}/2$
V_{OM}

VOLTAGE WAVEFORMS

NOTES: A. The input pulse is supplied by a generator with the following characteristics: PRR = 50 kHz, t_w = 10 μs, $t_r \leq 5$ ns, $t_f \leq 5$ ns, Z_O = 50 Ω.
B. C_L includes probe and jig capacitance.

FIGURE 3. SWITCHING CHARACTERISTICS

THERMAL INFORMATION

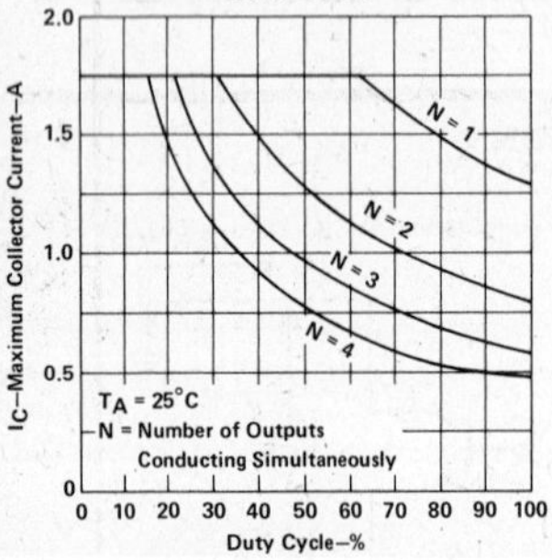
MAXIMUM COLLECTOR CURRENT
vs
DUTY CYCLE
I_C—Maximum Collector Current—A
N = 1
N = 2
N = 3
N = 4
T_A = 25°C
N = Number of Outputs
Conducting Simultaneously
Duty Cycle—%

FIGURE 4

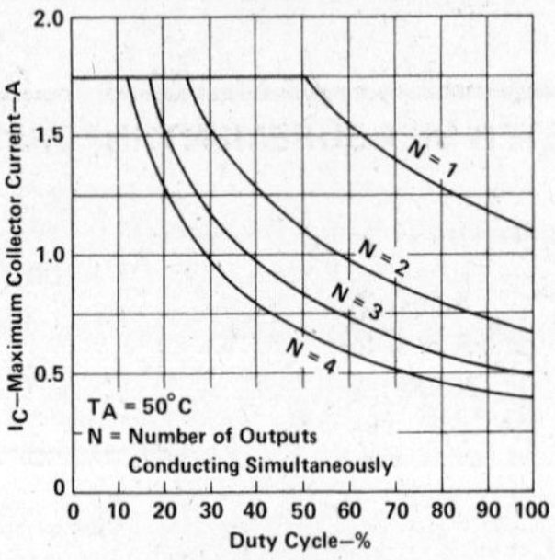
MAXIMUM COLLECTOR CURRENT
vs
DUTY CYCLE
I_C—Maximum Collector Current—A
N = 1
N = 2
N = 3
N = 4
T_A = 50°C
N = Number of Outputs
Conducting Simultaneously
Duty Cycle—%

FIGURE 5

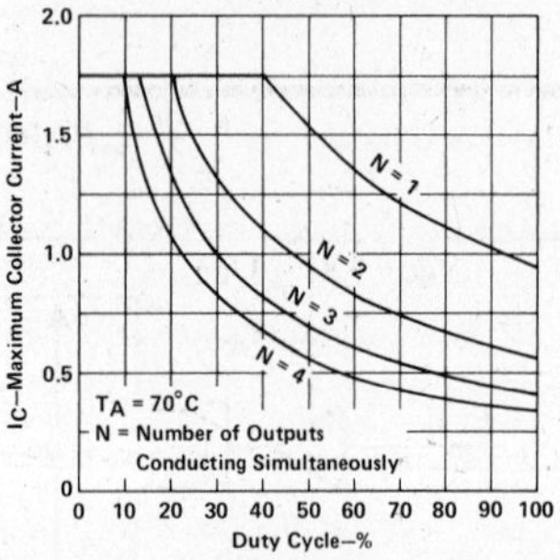
MAXIMUM COLLECTOR CURRENT
vs
DUTY CYCLE
I_C—Maximum Collector Current—A
N = 1
N = 2
N = 3
N = 4
T_A = 70°C
N = Number of Outputs
Conducting Simultaneously
Duty Cycle—%

FIGURE 6

D2624, DECEMBER 1976—REVISED SEPTEMBER 1986

HIGH-VOLTAGE HIGH-CURRENT DARLINGTON TRANSISTOR ARRAYS

- **500 mA Rated Collector Current (Single Output)**
- **High-Voltage Outputs . . . 50 V**
- **Output Clamp Diodes**
- **Inputs Compatible with Various Types of Logic**
- **Relay Driver Applications**
- **Designed to be Interchangeable with Sprague ULN2001A Series**

D OR N PACKAGE
(TOP VIEW)

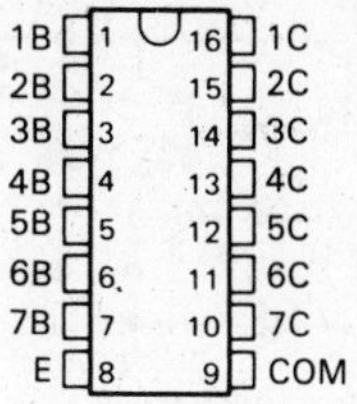
1B 1 16 1C
2B 2 15 2C
3B 3 14 3C
4B 4 13 4C
5B 5 12 5C
6B 6 11 6C
7B 7 10 7C
E 8 9 COM

description

The ULN2001A, ULN2002A, ULN2003A, ULN2004A, and ULN2005A are monolithic high-voltage, high-current Darlington transistor arrays. Each consists of seven n-p-n Darlington pairs that feature high-voltage outputs with common-cathode clamp diodes for switching inductive loads. The collector-current rating of a single Darlington pair is 500 milliamperes. The Darlington pairs may be paralleled for higher current capability. Applications include relay drivers, hammer drivers, lamp drivers, display drivers (LED and gas discharge), line drivers, and logic buffers. For 100-volt (otherwise interchangeable) versions, see the SN75465 through SN75469.

The ULN2001A is a general-purpose array and may be used with TTL, P-MOS, CMOS, and other MOS technologies. The ULN2002A is specifically designed for use with 14- to 25-volt P-MOS devices and each input has a zener diode and resistor in series to limit the input current to a safe limit. The ULN2003A has a 2700-ohm series base resistor for each Darlington pair for operation directly with TTL or 5-volt CMOS. The ULN2004A has a 10.5-kilohm series base resistor to allow its operation directly from CMOS or P-MOS that use supply voltages of 6 to 15 volts. The required input current is below that of the ULN2003A and the required voltage is less than that required by the ULN2002A. The ULN2005A has a 1050-ohm series base resistor and is especially designed for use with TTL where higher output current is required and loading of the driving source is not a concern.

logic symbol†

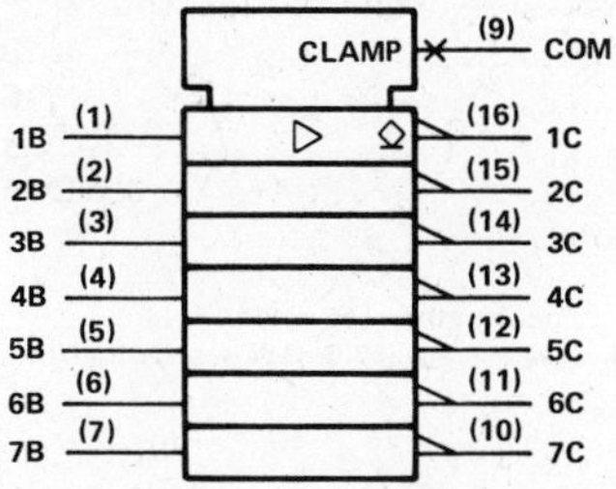
CLAMP
(9) COM
1B (1)
(16) 1C
2B (2)
(15) 2C
3B (3)
(14) 3C
4B (4)
(13) 4C
5B (5)
(12) 5C
6B (6)
(11) 6C
7B (7)
(10) 7C

†This symbol is in accordance with ANSI/IEEE Std 91-1984 and IEC Publication 617-12.

logic diagram

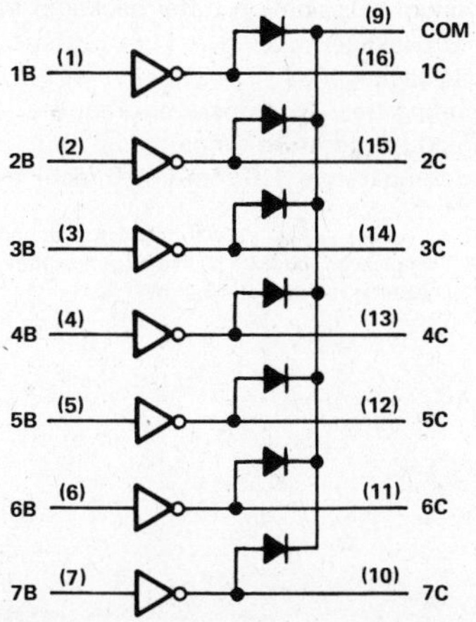
(9) COM
1B (1)
(16) 1C
2B (2)
(15) 2C
3B (3)
(14) 3C
4B (4)
(13) 4C
5B (5)
(12) 5C
6B (6)
(11) 6C
7B (7)
(10) 7C

schematics (each Darlington pair)

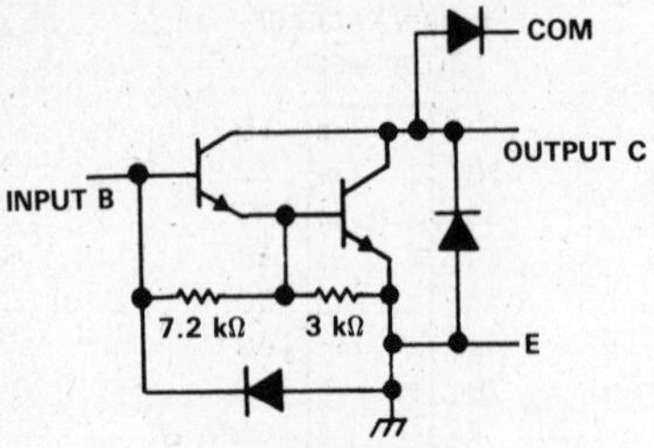
COM
OUTPUT C
INPUT B
7.2 kΩ
3 kΩ
E

ULN2001A

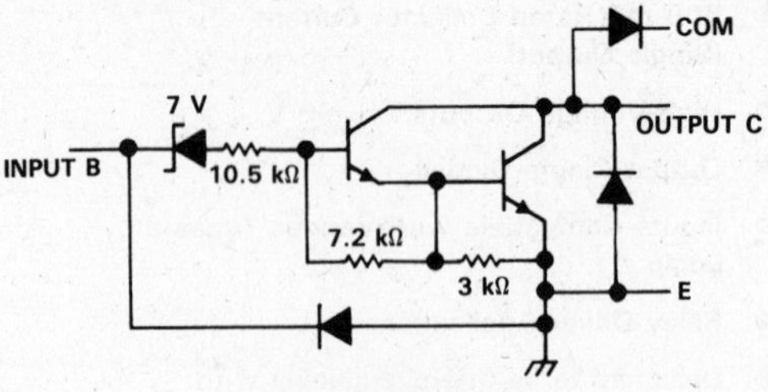
COM
7 V
OUTPUT C
INPUT B
10.5 kΩ
7.2 kΩ
3 kΩ
E

ULN2002A

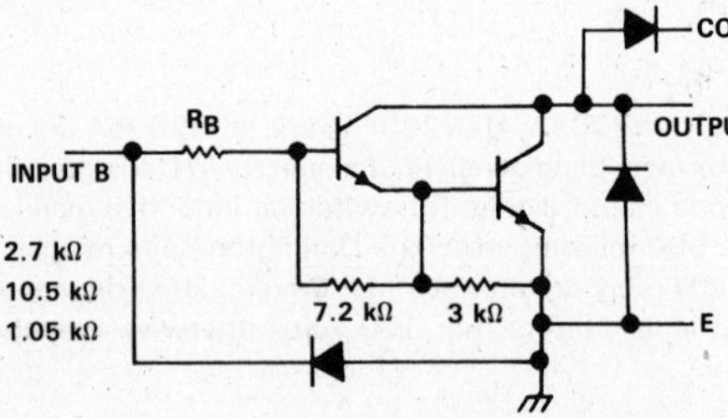
COM
R$_B$
OUTPUT C
INPUT B
7.2 kΩ
3 kΩ
E

ULN2003A: R_B = 2.7 kΩ
ULN2004A: R_B = 10.5 kΩ
ULN2005A: R_B = 1.05 kΩ

ULN2003A, ULN2004A, ULN2005A

All resistor values shown are nominal.

absolute maximum ratings at 25°C free-air temperature (unless otherwise noted)

Collector-emitter voltage	50 V
Input voltage (see Note 1): ULN2002A, ULN2003A, ULN2004A	30 V
ULN2005A	15 V
Peak collector current (see Figures 14 and 15)	500 mA
Output clamp diode current	500 mA
Total emitter-terminal current	−2.5 A
Continuous dissipation (total package) at (or below) 25°C free-air temperature (see Note 2):	
D package	950 mW
N package	1650 mW
Operating free-air temperature range	−20°C to 85°C
Storage temperature range	−65°C to 150°C
Lead temperature 1,6 mm (1/16 inch) from case for 10 seconds	260°C

NOTES: 1. All voltage values are with respect to the emitter/substrate terminal, E, unless otherwise noted.
2. For operation above 25°C free-air temperature, derate the D package linearly at the rate of 7.6 mW/°C and derate the N package linearly at the rate of 13.2 mW/°C.

electrical characteristics at 25°C free-air temperature (unless otherwise noted)

PARAMETER		TEST FIGURE	TEST CONDITIONS	ULN2001A MIN	ULN2001A TYP	ULN2001A MAX	ULN2002A MIN	ULN2002A TYP	ULN2002A MAX	UNIT
I_{CEX}	Collector cutoff current	1	V_{CE} = 50 V, I_I = 0			50			50	μA
		1	V_{CE} = 50 V, T_A = 70°C, I_I = 0			100			100	
		2	V_{CE} = 50 V, T_A = 70°C, V_I = 6 V						500	
$I_{I(off)}$	Off-state input current	3	V_{CE} = 50 V, I_C = 500 μA, T_A = 70°C	50	65		50	65		μA
I_I	Input current	4	V_I = 17 V					0.82	1.25	mA
h_{FE}	Static forward current transfer ratio	5	V_{CE} = 2 V, I_C = 350 mA	1000						
$V_{I(on)}$	On-state input voltage	6	V_{CE} = 2 V, I_C = 300 mA						13	V
$V_{CE(sat)}$	Collector-emitter saturation voltage	5	I_I = 250 μA, I_C = 100 mA		0.9	1.1		0.9	1.1	V
			I_I = 350 μA, I_C = 200 mA		1.0	1.3		1.0	1.3	
			I_I = 500 μA, I_C = 350 mA		1.2	1.6		1.2	1.6	
I_R	Clamp diode reverse current	7	V_R = 50 V			50			50	μA
			V_R = 50 V, T_A = 70°C			100			100	
V_F	Clamp diode forward voltage	8	I_F = 350 mA		1.7	2		1.7	2	V
C_i	Input capacitance		V_I = 0 V, f = 1 MHz		15	25		15	25	pF

electrical characteristics at 25°C free-air temperature (unless otherwise noted)

PARAMETER		TEST FIGURE	TEST CONDITIONS	ULN2003A MIN	ULN2003A TYP	ULN2003A MAX	ULN2004A MIN	ULN2004A TYP	ULN2004A MAX	UNIT
I_{CEX}	Collector cutoff current	1	V_{CE} = 50 V, I_I = 0			50			50	μA
		1	V_{CE} = 50 V, T_A = 70°C, I_I = 0			100			100	
		2	V_{CE} = 50 V, T_A = 70°C, V_I = 1 V						500	
$I_{I(off)}$	Off-state input current	3	V_{CE} = 50 V, I_C = 500 μA, T_A = 70°C	50	65		50	65		μA
I_I	Input current	4	V_I = 3.85 V		0.93	1.35				mA
			V_I = 5 V					0.35	0.5	
			V_I = 12 V					1.0	1.45	
$V_{I(on)}$	On-state input voltage	6	V_{CE} = 2 V, I_C = 125 mA						5	V
			V_{CE} = 2 V, I_C = 200 mA			2.4			6	
			V_{CE} = 2 V, I_C = 250 mA			2.7				
			V_{CE} = 2 V, I_C = 275 mA						7	
			V_{CE} = 2 V, I_C = 300 mA			3				
			V_{CE} = 2 V, I_C = 350 mA						8	
$V_{CE(sat)}$	Collector-emitter saturation voltage	5	I_I = 250 μA, I_C = 100 mA		0.9	1.1		0.9	1.1	V
			I_I = 350 μA, I_C = 200 mA		1.0	1.3		1.0	1.3	
			I_I = 500 μA, I_C = 350 mA		1.2	1.6		1.2	1.6	
I_R	Clamp diode reverse current	7	V_R = 50 V			50			50	μA
			V_R = 50 V, T_A = 70°C			100			100	
V_F	Clamp diode forward voltage	8	I_F = 350 mA		1.7	2		1.7	2	V
C_i	Input capacitance		V_I = 0 V, f = 1 MHz		15	25		15	25	pF

electrical characteristics at 25°C free-air temperature (unless otherwise noted)

PARAMETER		TEST FIGURE	TEST CONDITIONS	ULN2005A MIN	ULN2005A TYP	ULN2005A MAX	UNIT
I_{CEX}	Collector cutoff current	1	$V_{CE} = 50$ V, $I_I = 0$			50	μA
			$V_{CE} = 50$ V, $I_I = 0$, $T_A = 70°C$			100	
$I_{I(off)}$	Off-state input current	3	$V_{CE} = 50$ V, $I_C = 500$ μA, $T_A = 70°C$	50	65		μA
I_I	Input current	4	$V_I = 3$ V		1.5	2.4	mA
$V_{I(on)}$	On-state input voltage	6	$V_{CE} = 2$ V, $I_C = 350$ mA			2.4	V
$V_{CE(sat)}$	Collector-emitter saturation voltage	5	$I_I = 250$ μA, $I_C = 100$ mA		0.9	1.1	V
			$I_I = 350$ μA, $I_C = 200$ mA		1.0	1.3	
			$I_I = 500$ μA, $I_C = 350$ mA		1.2	1.6	
I_R	Clamp diode reverse current	7	$V_R = 50$ V			50	μA
			$V_R = 50$ V, $T_A = 70°C$			100	
V_F	Clamp diode forward voltage	8	$I_F = 350$ mA		1.7	2	V
C_i	Input capacitance		$V_I = 0$ V, f = 1 MHz		15	25	pF

switching characteristics at 25°C free-air temperature

PARAMETER		TEST CONDITIONS	MIN	TYP	MAX	UNIT
t_{PLH}	Propagation delay time, low-to-high-level output	See Figure 9		0.25	1	μs
t_{PHL}	Propagation delay time, high-to-low-level output			0.25	1	μs
V_{OH}	High-level output voltage after switching	$V_S = 50$ V, $I_O \approx 300$ mA, See Figure 10	$V_S - 20$			mV

TEXAS INSTRUMENTS
POST OFFICE BOX 655012 • DALLAS, TEXAS 75265

PARAMETER MEASUREMENT INFORMATION

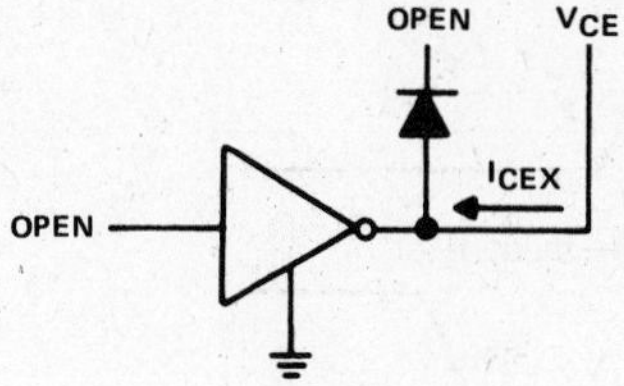
OPEN
VCE
ICEX
OPEN

FIGURE 1. I_{CEX}

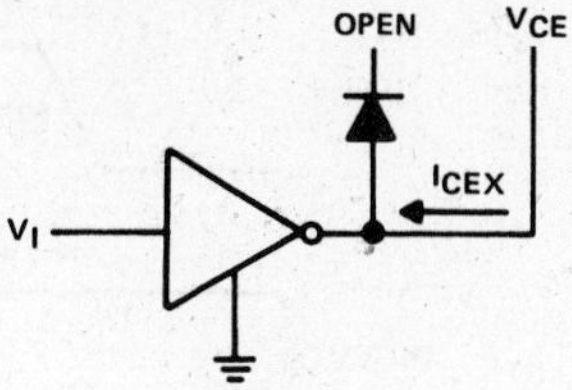
OPEN
VCE
ICEX
VI

FIGURE 2. I_{CEX}

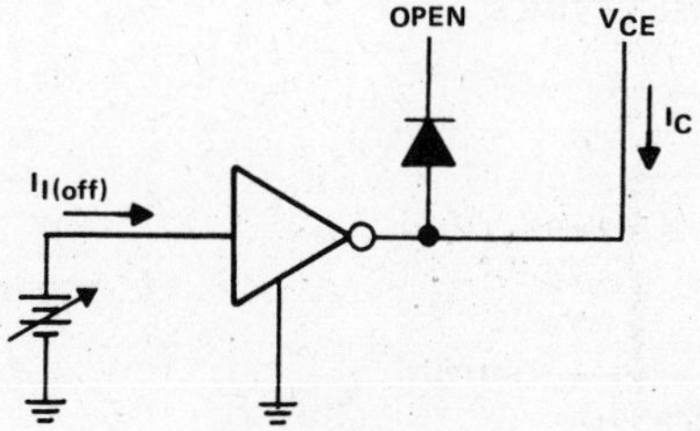
OPEN
VCE
IC
II(off)

FIGURE 3. $I_{I(off)}$

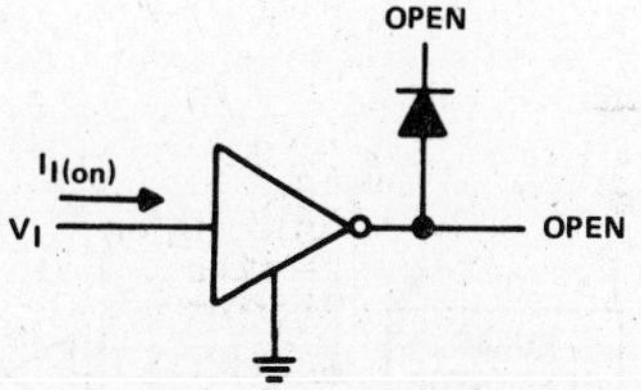
OPEN
II(on)
VI
OPEN

FIGURE 4. I_I

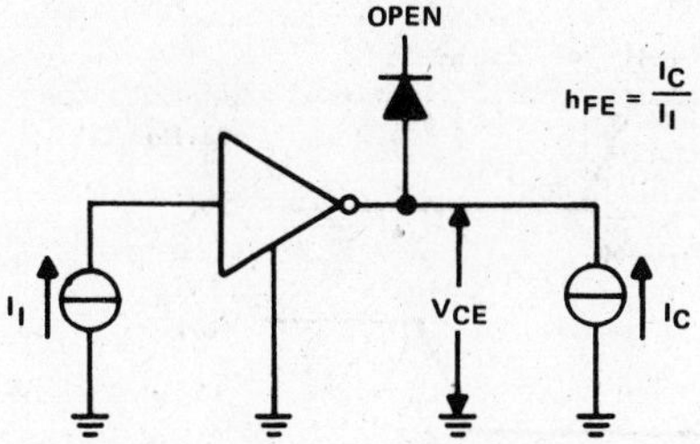
OPEN
hFE = IC / II
II
VCE
IC

NOTE: I_I is fixed for measuring $V_{CE(sat)}$, variable for measuring h_{FE}.

FIGURE 5. h_{FE}, $V_{CE(sat)}$

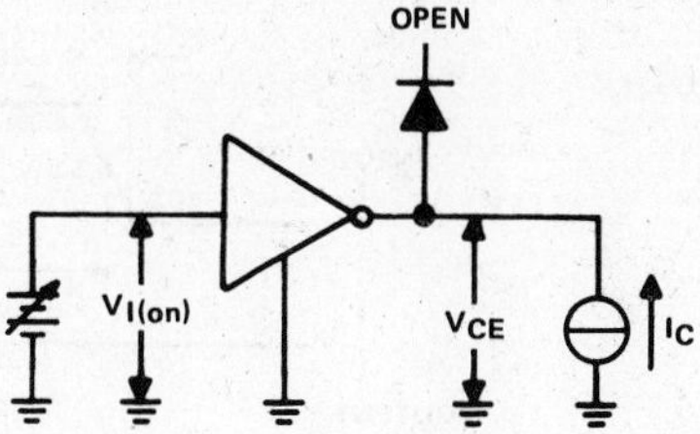
OPEN
VI(on)
VCE
IC

FIGURE 6. $V_{I(on)}$

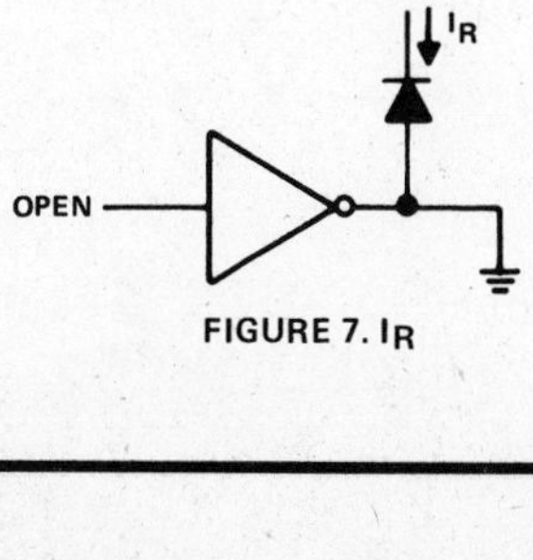
VR
IR
OPEN

FIGURE 7. I_R

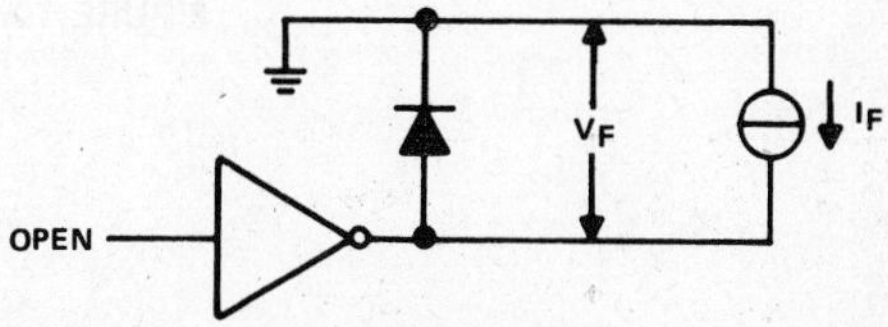
VF
IF
OPEN

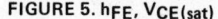

FIGURE 8. V_F

PARAMETER MEASUREMENT INFORMATION

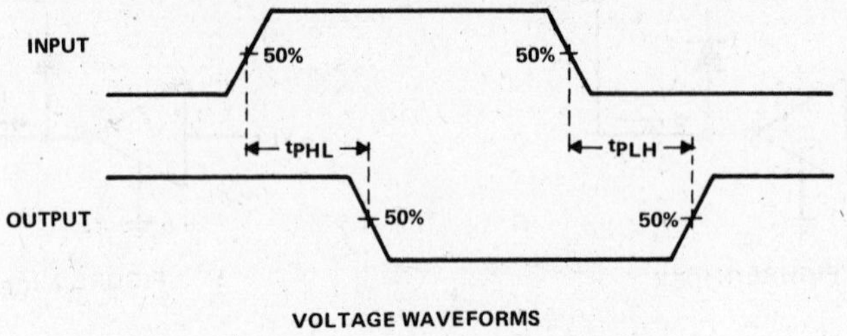
INPUT
50%
50%
tPHL
tPLH
OUTPUT
50%
50%
VOLTAGE WAVEFORMS

FIGURE 9. PROPAGATION DELAY TIMES

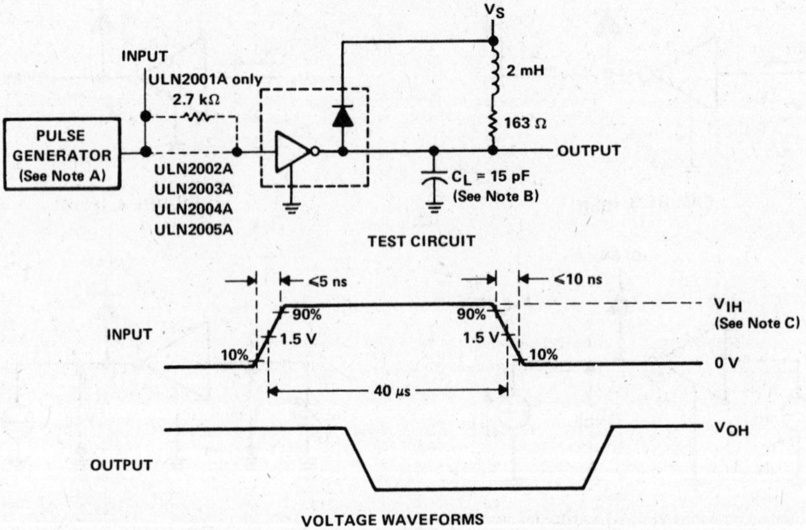
VS
INPUT
ULN2001A only
2.7 kΩ
2 mH
163 Ω
PULSE GENERATOR (See Note A)
OUTPUT
ULN2002A
ULN2003A
ULN2004A
ULN2005A
CL = 15 pF
(See Note B)
TEST CIRCUIT
≤5 ns
≤10 ns
VIH
(See Note C)
90%
90%
1.5 V
1.5 V
10%
10%
0 V
INPUT
40 μs
VOH
OUTPUT
VOLTAGE WAVEFORMS

NOTES: A. The pulse generator has the following characteristics: PRR = 12.5 kHz, Z_{out} = 50 Ω.
B. C_L includes probe and jig capacitance.
C. For testing the ULN2001A, ULN2003A, and the ULN2005A, V_{IH} = 3 V; for the ULN2002A, V_{IH} = 13 V; for the ULN2004A, V_{IH} = 8 V.

FIGURE 10. LATCH-UP TEST

TEXAS INSTRUMENTS
POST OFFICE BOX 655012 • DALLAS, TEXAS 75265

TYPICAL CHARACTERISTICS

COLLECTOR-EMITTER SATURATION VOLTAGE vs COLLECTOR CURRENT (ONE DARLINGTON)

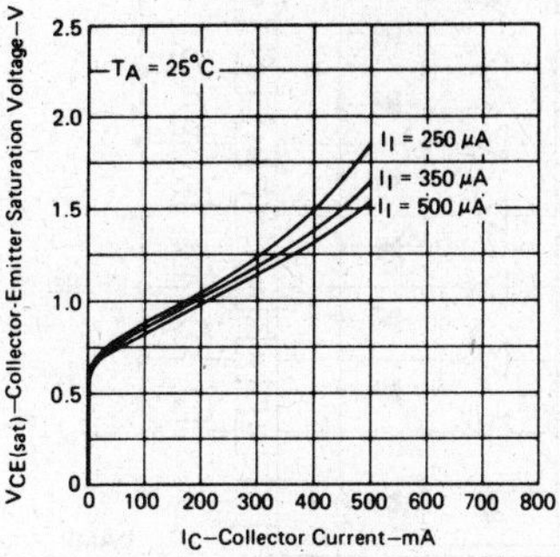
T_A = 25°C
I_I = 250 μA
I_I = 350 μA
I_I = 500 μA
VCE(sat)—Collector-Emitter Saturation Voltage—V
IC—Collector Current—mA

FIGURE 11

COLLECTOR-EMITTER SATURATION VOLTAGE vs COLLECTOR CURRENT (TWO DARLINGTONS PARALLELED)

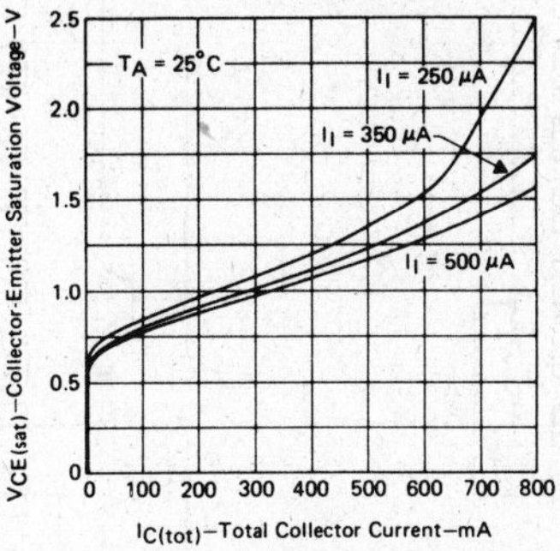
T_A = 25°C
I_I = 250 μA
I_I = 350 μA
I_I = 500 μA
VCE(sat)—Collector-Emitter Saturation Voltage—V
IC(tot)—Total Collector Current—mA

FIGURE 12

COLLECTOR CURRENT vs INPUT CURRENT

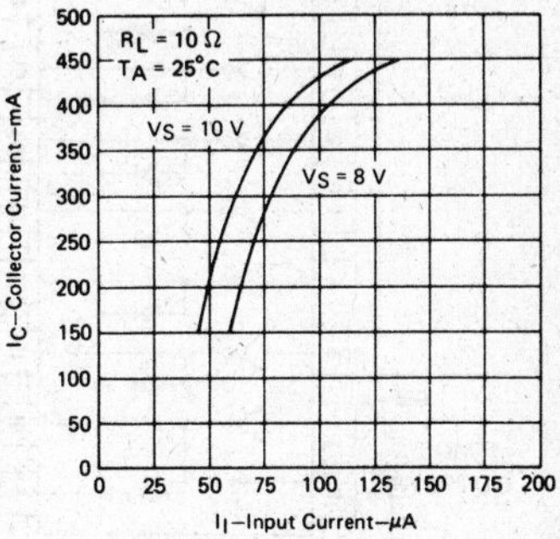
R_L = 10 Ω
T_A = 25°C
V_S = 10 V
V_S = 8 V
IC—Collector Current—mA
II—Input Current—μA

FIGURE 13

THERMAL INFORMATION

D PACKAGE
MAXIMUM COLLECTOR CURRENT
vs
DUTY CYCLE

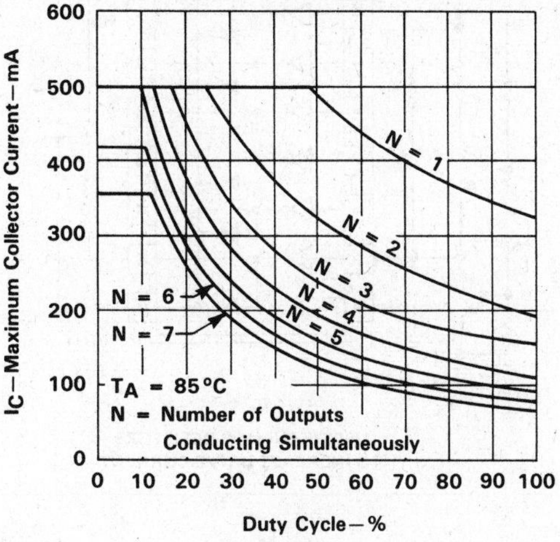
N = 1
N = 2
N = 3
N = 4
N = 5
N = 6
N = 7
T_A = 85°C
N = Number of Outputs Conducting Simultaneously
IC—Maximum Collector Current—mA
Duty Cycle—%

FIGURE 14

N PACKAGE
MAXIMUM COLLECTOR CURRENT
vs
DUTY CYCLE

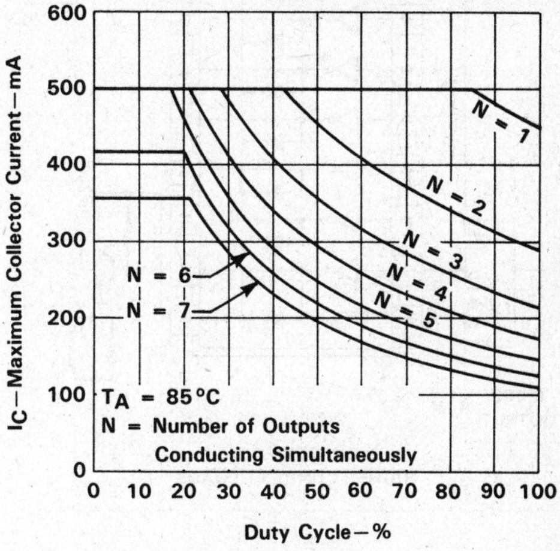
N = 1
N = 2
N = 3
N = 4
N = 5
N = 6
N = 7
T_A = 85°C
N = Number of Outputs Conducting Simultaneously
IC—Maximum Collector Current—mA
Duty Cycle—%

FIGURE 15

TYPICAL APPLICATION DATA

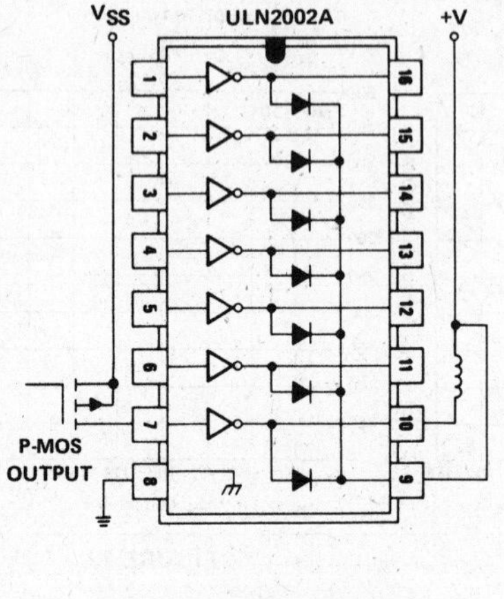
VSS
ULN2002A
+V
P-MOS
OUTPUT

P-MOS TO LOAD

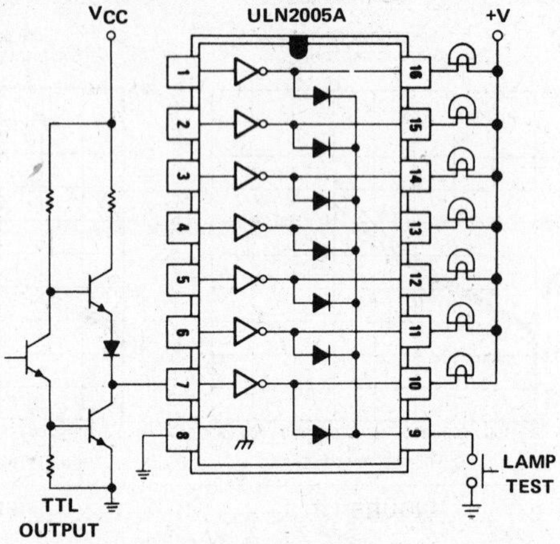
VCC
ULN2003A
ULN2005A
+V
LAMP
TEST
TTL
OUTPUT

TTL TO LOAD

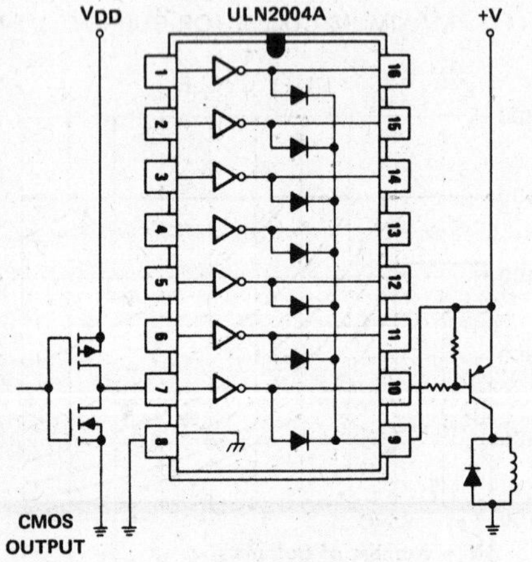
VDD
ULN2004A
+V
CMOS
OUTPUT

BUFFER FOR
HIGHER CURRENT LOADS

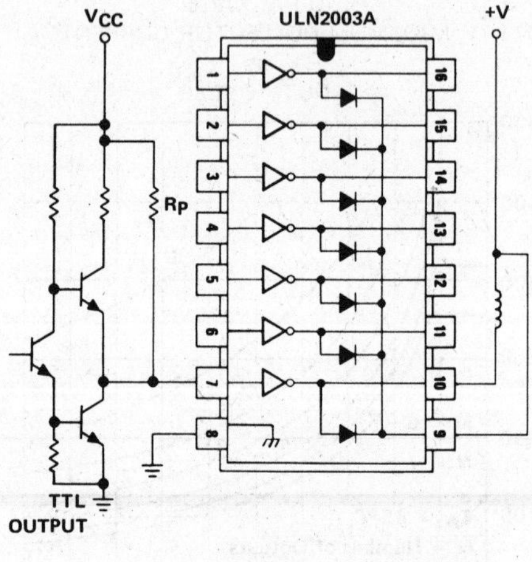
VCC
ULN2003A
+V
RP
TTL
OUTPUT

USE OF PULL-UP RESISTORS
TO INCREASE DRIVE CURRENT

TEXAS INSTRUMENTS
POST OFFICE BOX 655012 • DALLAS, TEXAS 75265

D2528, DECEMBER 1979—REVISED SEPTEMBER 1986

- Output Collector Current . . . 1.5 A Max
- 2-W Dissipation Rating
- High Output-Voltage Capability
- Outputs Diode-Clamped for Inductive Loads
- Common-Emitter Circuit for Current Sink
- ULN2064 and ULN2065 Have TTL Compatible Inputs
- ULN2066 and ULN2067 Have CMOS- and PMOS-Compatible Inputs
- Designed for Interchangeability with Sprague ULN2064 thru ULN2067, Respectively

NE DUAL-IN-LINE PACKAGE
(TOP VIEW)

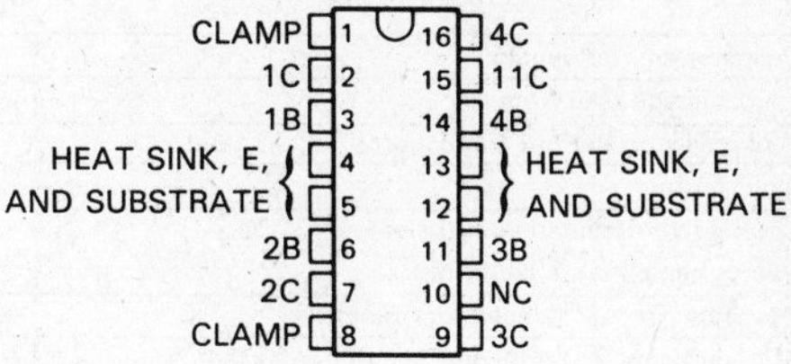
CLAMP 1
1C 2
1B 3
HEAT SINK, E, AND SUBSTRATE 4
5
2B 6
2C 7
CLAMP 8
16 4C
15 11C
14 4B
13 HEAT SINK, E, AND SUBSTRATE
12
11 3B
10 NC
9 3C

NC—No internal connection

description

The ULN2064, ULN2065, ULN2066, and ULN2067 are monolithic high-voltage, high-current darlington transistor switches. Each comprises four n-p-n darlington pairs. All units feature high-voltage outputs with common-cathode clamp diodes for switching inductive loads. Outputs and inputs may each be paralleled for higher current capability. Applications include relay drivers, hammer drivers, lamp drivers, display drivers (LED and gas discharge), line drivers, and logic buffers. These common-emitter circuits are designed to operate as current sinks to the load.

The ULN2064 and ULN2065 are intended for use with TTL and 5-volt MOS logic. The ULN2066 and ULN2067 are intended for use with PMOS and higher-voltage CMOS logic.

The ULN2064, ULN2065, ULN2066, and ULN2067 are characterized for operation from −20°C to 85°C.

schematic (each darlington pair)

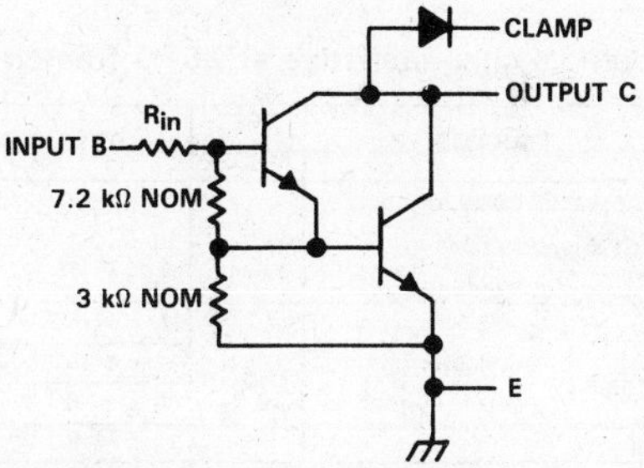
CLAMP
OUTPUT C
R_{in}
INPUT B
7.2 kΩ NOM
3 kΩ NOM
E

ULN2064, ULN2065: R_{in} = 350 Ω NOM
ULN2066, ULN2067: R_{in} = 3 kΩ NOM

logic symbol†

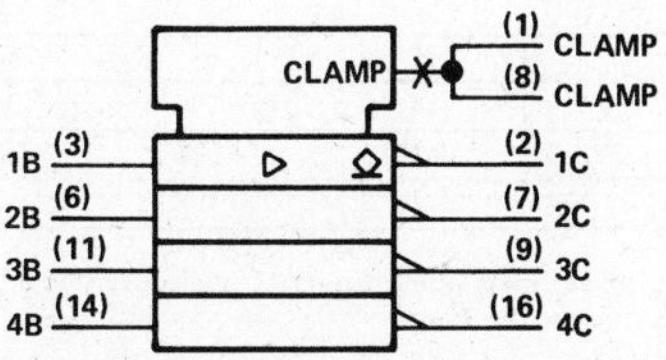
CLAMP
(1) CLAMP
(8) CLAMP
1B (3)
2B (6)
3B (11)
4B (14)
(2) 1C
(7) 2C
(9) 3C
(16) 4C

† This symbol is in accordance with ANSI/IEEE Std 91-1984 and IEC Publication 617-12.

logic diagram

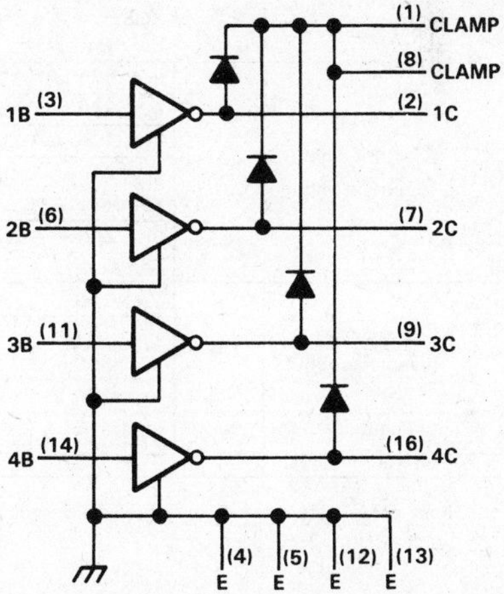
(1) CLAMP
(8) CLAMP
1B (3)
(2) 1C
2B (6)
(7) 2C
3B (11)
(9) 3C
4B (14)
(16) 4C
(4) E
(5) E
(12) E
(13) E

absolute maximum ratings at 25°C free-air temperature for each switch (unless otherwise noted)

	ULN2064	ULN2065	ULN2066	ULN2067	UNIT
Collector-emitter voltage	50	80	50	80	V
Input voltage (see Note 1)	15	15	30	30	V
Peak collector current (see Figures 12, 13, and 14)	1.5	1.5	1.5	1.5	A
Input current	25	25	25	25	mA
Total power dissipation at (or below) 25°C free-air temperature (see Note 2)	2075	2075	2075	2075	mW
Operating free-air temperature range	−20 to 85	−20 to 85	−20 to 85	−20 to 85	°C
Storage temperature range	−55 to 150	−55 to 150	−55 to 150	−55 to 150	°C
Lead temperature 1,6 mm (1/16 inch) from the case for 10 seconds	260	260	260	260	°C

NOTES: 1. All voltage values (unless otherwise noted) are with respect to the emitter/substrate terminal E.
2. For operation above 25°C free-air temperature, derate total power linearly to 1079 mW at 85°C at the rate of 16.6 mW/°C.

electrical characteristics at 25°C free-air temperature (unless otherwise noted)

PARAMETER		TEST FIGURE	TEST CONDITIONS	ULN2064 MIN	ULN2064 MAX	ULN2065 MIN	ULN2065 MAX	ULN2066 MIN	ULN2066 MAX	ULN2067 MIN	ULN2067 MAX	UNIT
$V_{CEX(sus)}$	Collector sustaining voltage	1	V_I = 0.4 V, I_C = 100 mA	35		50		35		50		V
I_{CEX}	Collector output cutoff current	2	V_{CE} = 50 V		100				100			μA
			V_{CE} = 50 V, T_A = 70°C		500				500			
			V_{CE} = 80 V				100				100	
			V_{CE} = 80 V, T_A = 70°C				500				500	
$I_{I(on)}$	On-state input current	3	V_I = 2.4 V	1.4	4.3	1.4	4.3					mA
			V_I = 3.75 V	3.3	9.6	3.3	9.6					
			V_I = 5 V					0.6	1.8	0.6	1.8	
			V_I = 12 V					1.7	5.2	1.7	5.2	
$V_{I(on)}$	On-state input voltage	4	V_{CE} = 2 V, I_C = 1 A		2		2		6.5		6.5	V
			V_{CE} = 2 V, I_C = 1.5 A, See Note 3		2.5		2.5		10		10	
$V_{CE(sat)}$	Collector-emitter saturation voltage	5	I_I = 625 μA, I_C = 500 mA		1.1		1.1		1.1		1.1	V
			I_I = 935 μA, I_C = 750 mA		1.2		1.2		1.2		1.2	
			I_I = 1.25 mA, I_C = 1 A		1.3		1.3		1.3		1.3	
			I_I = 2 mA, I_C = 1.25 A, See Note 3		1.4				1.4			
			I_I = 2.25 mA, I_C = 1.5 A, See Note 3				1.5				1.5	
I_R	Clamp-diode reverse current	6	V_R = 50 V		50				50			μA
			V_R = 50 V, T_A = 70°C		100				100			
			V_R = 80 V				50				50	
			V_R = 80 V, T_A = 70°C				100				100	
V_F	Clamp-diode forward voltage	7	I_F = 1 A		1.75		1.75		1.75		1.75	V
			I_F = 1.5 A, See Note 3		2		2		2		2	

NOTE 3: These parameters must be measured on one output at a time using pulse techniques, t_w = 10 ms, duty cycle ≤ 10%.

switching characteristics at 25°C free-air temperature, V_{CC} = 5 V

PARAMETER		TEST CONDITIONS	MIN	TYP	MAX	UNIT
t_{PLH}	Propagation delay time, low-to-high-level output	See Figure 8			1	μs
t_{PHL}	propagation delay time, high-to-low-level output				1.5	μs

PARAMETER MEASUREMENT INFORMATION

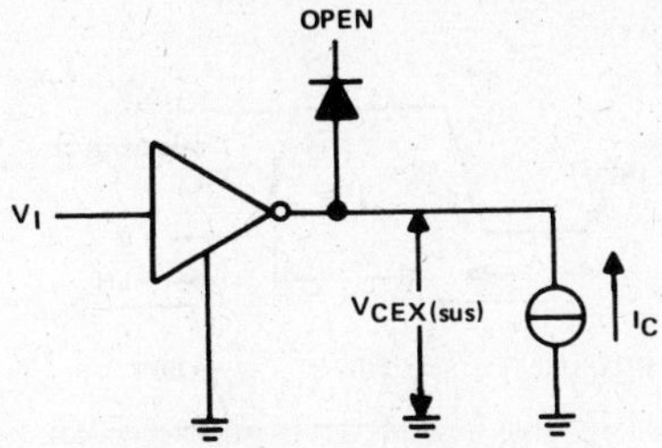
OPEN
V_I
V_CEX(sus)
I_C

FIGURE 1. $V_{CEX(sus)}$

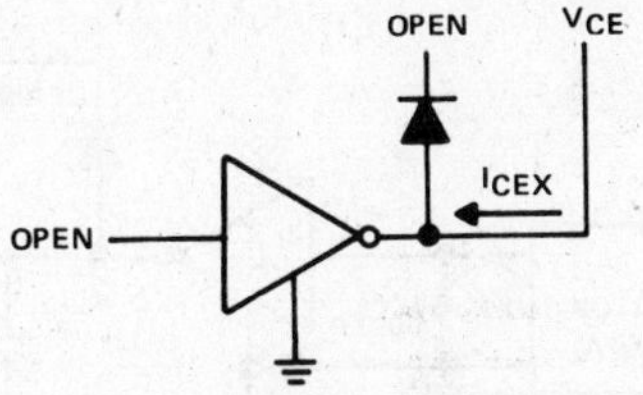
OPEN
V_CE
I_CEX
OPEN

FIGURE 2. I_{CEX}

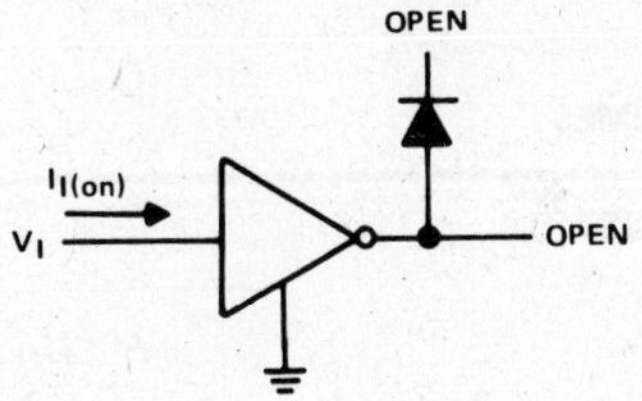
OPEN
I_I(on)
V_I
OPEN

FIGURE 3. $I_{I(on)}$

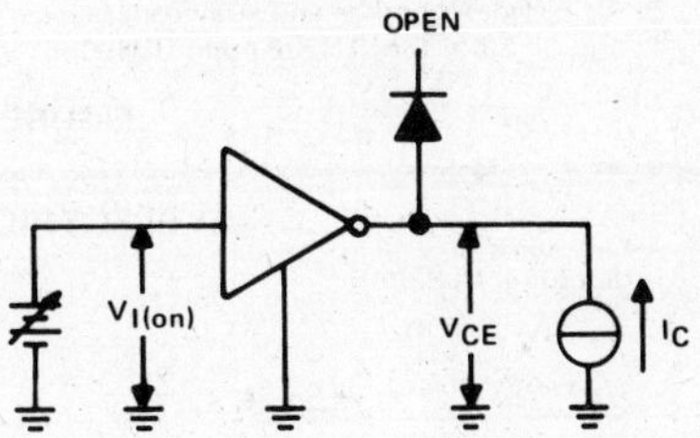
OPEN
V_I(on)
V_CE
I_C

FIGURE 4. $V_{I(on)}$

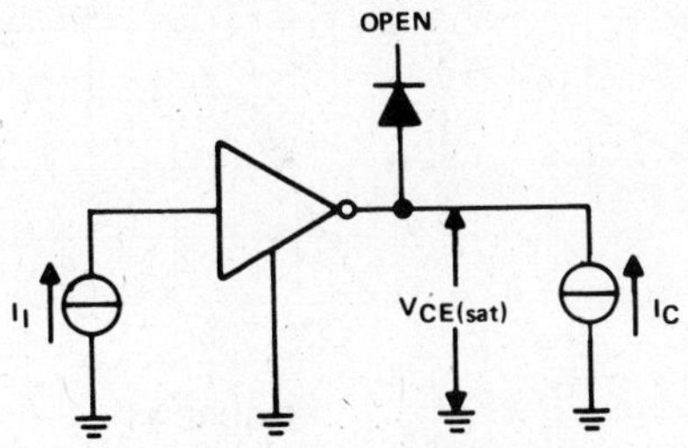
OPEN
I_I
V_CE(sat)
I_C

FIGURE 5. $V_{CE(sat)}$

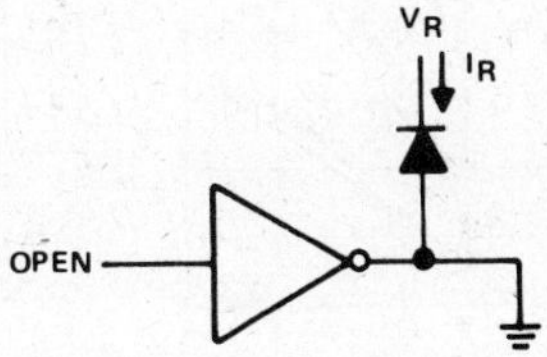
V_R
I_R
OPEN

FIGURE 6. I_R

PARAMETER MEASUREMENT INFORMATION

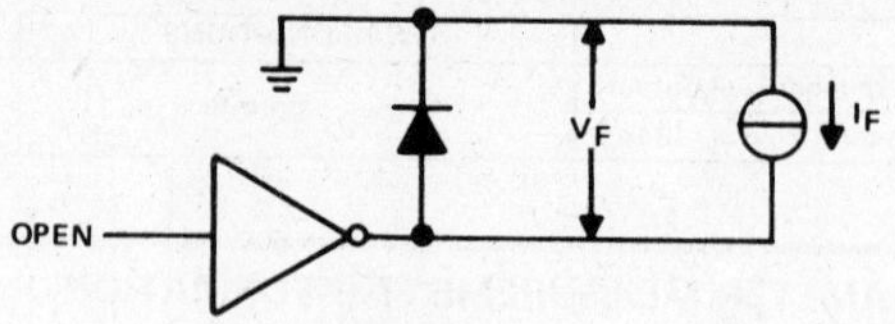
OPEN
V_F
I_F

FIGURE 7. V_F

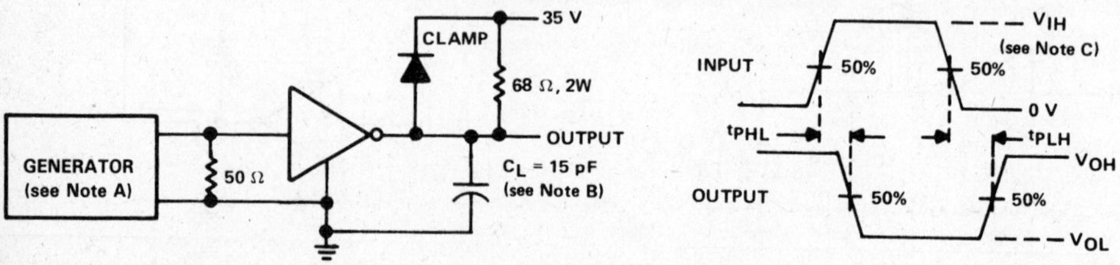
CLAMP
35 V
68 Ω, 2W
OUTPUT
GENERATOR
(see Note A)
50 Ω
C_L = 15 pF
(see Note B)
INPUT
50%
50%
V_{IH}
(see Note C)
0 V
t_{PHL}
t_{PLH}
V_{OH}
OUTPUT
50%
50%
V_{OL}

NOTES: A. The input pulse is supplied by a generator having the following characteristics: PRR = 50 kHz, duty cycle = 10%, Z_o = 50 Ω.
B. C_L includes all probe and stray capacitance.
C. V_{IH} = 2.5 V for ULN2064 and ULN2065. V_{IH} = 10 V for ULN2065 and ULN2067.

FIGURE 8. SWITCHING TIMES

ELECTRICAL CHARACTERISTICS

ULN2064, ULN2065

INPUT CURRENT
vs
INPUT VOLTAGE

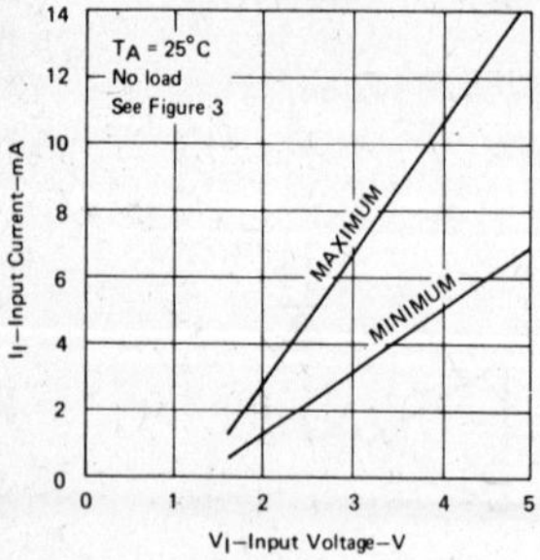
T_A = 25°C
No load
See Figure 3
MAXIMUM
MINIMUM
I_I—Input Current—mA
V_I—Input Voltage—V

FIGURE 9

ULN2066, ULN2067

INPUT CURRENT
vs
INPUT VOLTAGE

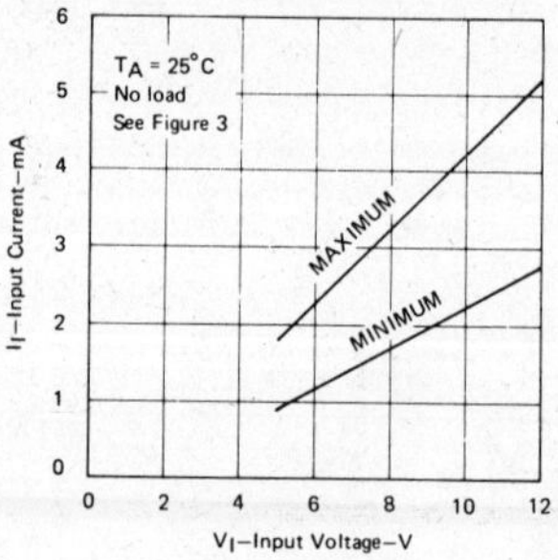
T_A = 25°C
No load
See Figure 3
MAXIMUM
MINIMUM
I_I—Input Current—mA
V_I—Input Voltage—V

FIGURE 10

COLLECTOR CURRENT
vs
BASE CURRENT

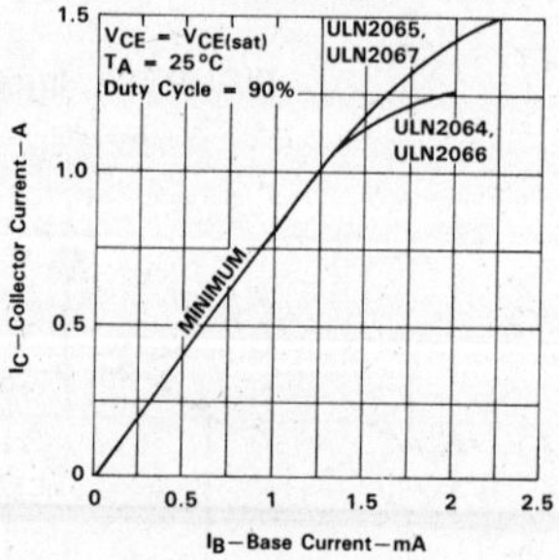
V_{CE} = $V_{CE(sat)}$
T_A = 25°C
Duty Cycle = 90%
ULN2065,
ULN2067
ULN2064,
ULN2066
MINIMUM
I_C—Collector Current—A
I_B—Base Current—mA

FIGURE 11

TEXAS INSTRUMENTS
POST OFFICE BOX 655012 • DALLAS, TEXAS 75265

THERMAL INFORMATION

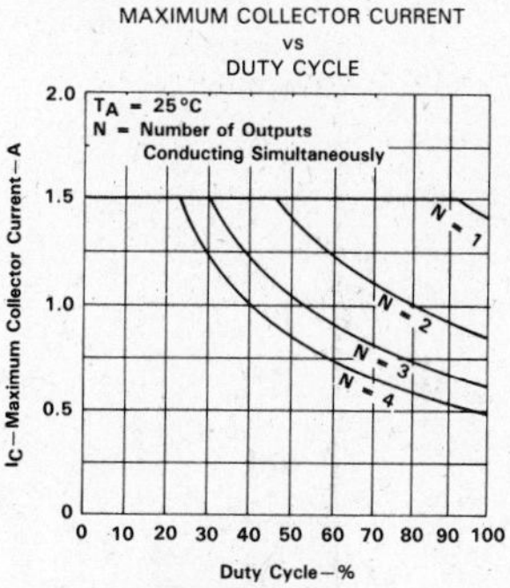
MAXIMUM COLLECTOR CURRENT
vs
DUTY CYCLE
T_A = 25°C
N = Number of Outputs Conducting Simultaneously
I_C—Maximum Collector Current—A
Duty Cycle—%
N = 1
N = 2
N = 3
N = 4

FIGURE 12

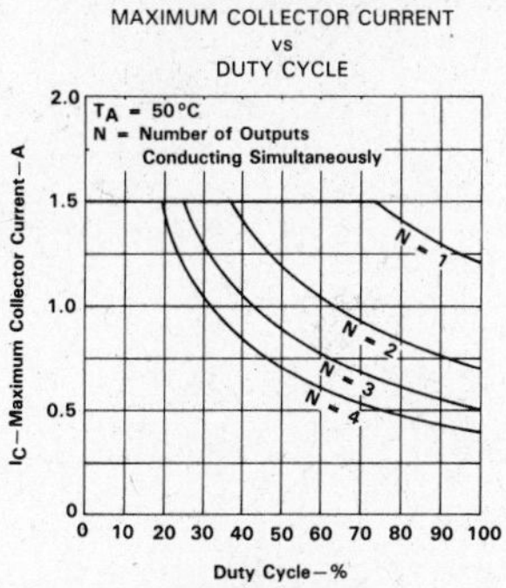
MAXIMUM COLLECTOR CURRENT
vs
DUTY CYCLE
T_A = 50°C
N = Number of Outputs Conducting Simultaneously
I_C—Maximum Collector Current—A
Duty Cycle—%
N = 1
N = 2
N = 3
N = 4

FIGURE 13

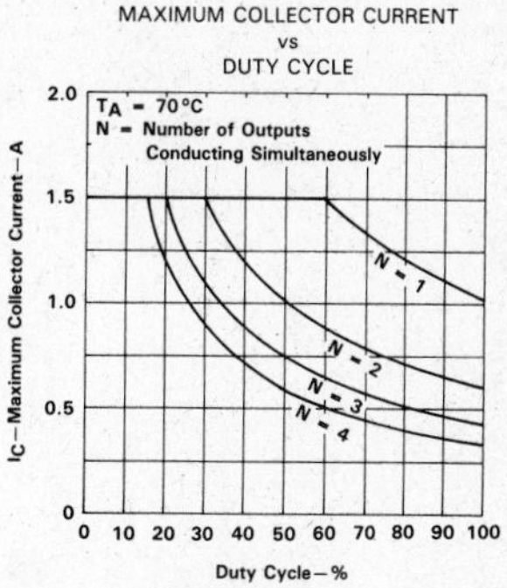
MAXIMUM COLLECTOR CURRENT
vs
DUTY CYCLE
T_A = 70°C
N = Number of Outputs Conducting Simultaneously
I_C—Maximum Collector Current—A
Duty Cycle—%
N = 1
N = 2
N = 3
N = 4

FIGURE 14

TYPICAL APPLICATION DATA

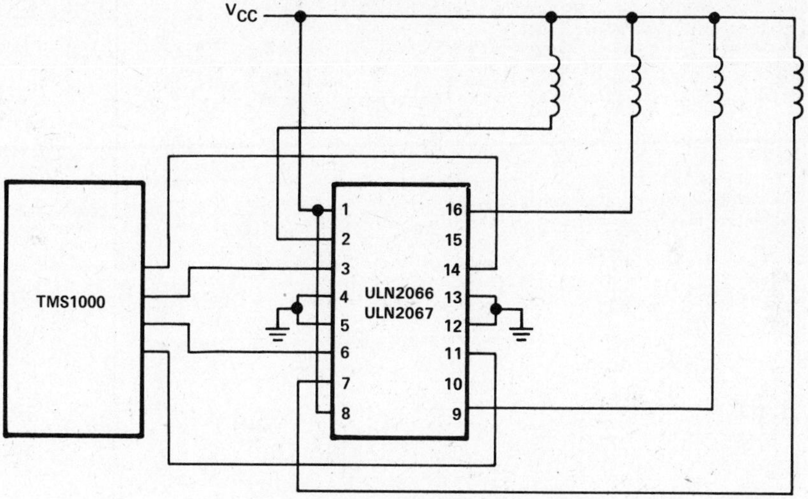
V_{CC}
TMS1000
ULN2066
ULN2067
1
2
3
4
5
6
7
8
16
15
14
13
12
11
10
9

FIGURE 15. RELAY DRIVER INTERFACE

D2579, MAY 1980—REVISED SEPTEMBER 1986

ULN2068, ULN2069 QUADRUPLE HIGH-CURRENT DARLINGTON SWITCHES

- Output Collector Current . . . 1.5 A Max
- 2-W Dissipation Rating
- High Output-Voltage Capability
- Preamp for High Current Gain
- Outputs Diode-Clamped for Inductive Loads
- Common-Emitter Circuit for Current Sink
- Inputs Compatible with TTL and 5-Volt CMOS
- Designed for Interchangeability with Sprague ULN2068 and ULN2069

NE DUAL-IN-LINE PACKAGE
(TOP VIEW)

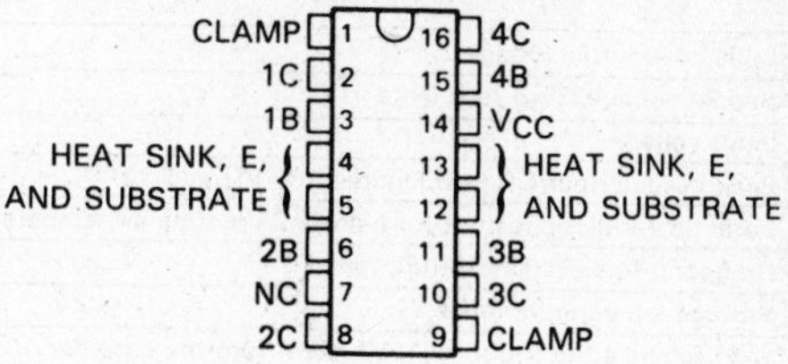
CLAMP 1
1C 2
1B 3
HEAT SINK, E, AND SUBSTRATE 4
5
2B 6
NC 7
2C 8
16 4C
15 4B
14 V_{CC}
13 HEAT SINK, E, AND SUBSTRATE
12
11 3B
10 3C
9 CLAMP

NC—No internal connection

description

The ULN2068 and ULN2069 are monolithic integrated circuits each consisting of four high-voltage, high-current n-p-n cascaded transistor switches. Each switch includes a first stage compatible with both TTL and 5-volt CMOS signal levels. The second and third stages form uncommitted-collector outputs with common-cathode clamp diodes for switching inductive loads.

The ULN2068 and ULN2069 can sink up to 1.5 amperes per switch. Applications include logic buffers, MOS drivers, memory drivers, line drivers, relay drivers, hammer drivers, lamp drivers, and display drivers (LED and gas discharge).

The ULN2068 and ULN2069 are characterized for operation from −20°C to 85°C.

schematic (each switch)

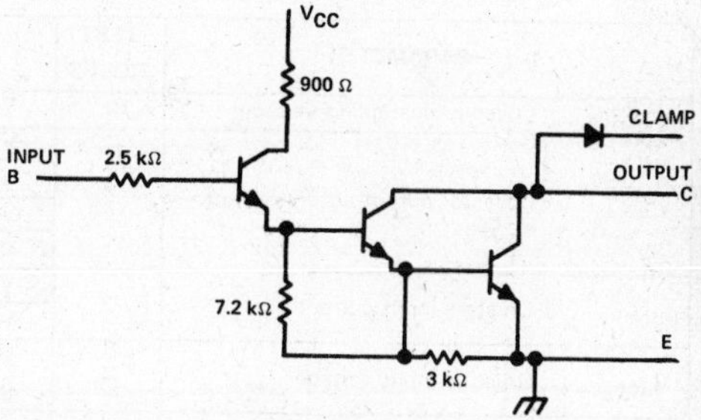
V_{CC}
900 Ω
CLAMP
INPUT B
2.5 kΩ
OUTPUT C
7.2 kΩ
E
3 kΩ

Resistor values shown are nominal.

logic symbol†

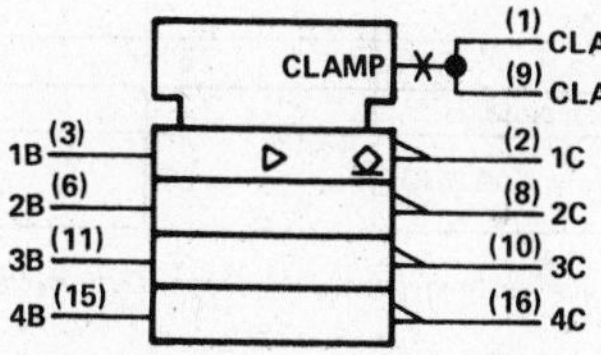
CLAMP
(1) CLAMP
(9) CLAMP
1B (3)
2B (6)
3B (11)
4B (15)
(2) 1C
(8) 2C
(10) 3C
(16) 4C

†This symbol is in accordance with ANSI/IEEE Std 91-1984 and IEC Publication 617-12.

logic diagram (positive logic)

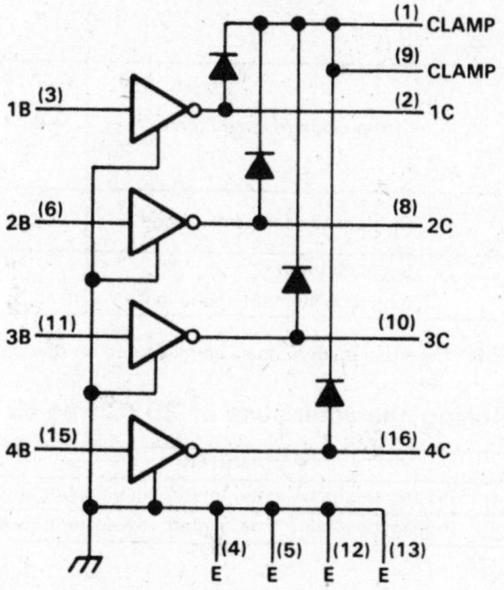
(1) CLAMP
(9) CLAMP
1B (3)
(2) 1C
2B (6)
(8) 2C
3B (11)
(10) 3C
4B (15)
(16) 4C
(4) E
(5) E
(12) E
(13) E

absolute maximum ratings at 25°C free-air temperature for each switch (unless otherwise noted)

	ULN2068	ULN2069	UNIT
Collector-emitter voltage	50	80	V
Supply voltage, V_{CC} (see Note 1)	10	10	V
Input voltage	15	15	V
Peak collector current (see Figures 10, 11, and 12)	1.5	1.5	A
Total power dissipation at (or below) 25°C free-air temperature (see Note 2)	2075	2075	mW
Operating free-air temperature range	−20 to 85	−20 to 85	°C
Storage temperature range	−55 to 150	−55 to 150	°C
Lead temperature 1,6 mm (1/16 inch) from the case for 10 seconds	260	260	°C

NOTES: 1. All voltage values (unless otherwise noted) are with respect to the emitter/substrate terminal E.
2. For operation above 25°C free-air temperature, derate total power linearly to 1079 mW at 85°C at the rate of 16.6 mW/°C.

electrical characteristics at 25°C free-air temperature, V_{CC} = 5 V (unless otherwise noted)

PARAMETER		TEST FIGURE	TEST CONDITIONS	ULN2068 MIN	ULN2068 MAX	ULN2069 MIN	ULN2069 MAX	UNIT
$V_{CEX(sus)}$	Collector sustaining voltage	1	V_I = 0.4 V, I_C = 100 mA	35		50		V
I_{CEX}	Collector output cutoff current	2	V_{CE} = 50 V		100			μA
			V_{CE} = 50 V, T_A = 70°C		500			
			V_{CE} = 80 V				100	
			V_{CE} = 80 V, T_A = 70°C				500	
$I_{I(on)}$	On-state input current	3	V_I = 2.4 V		250		250	μA
			V_I = 3.75 V		1000		1000	
$V_{I(on)}$	On-state input voltage	4	V_{CE} = 2 V, I_C = 1.5 A, See Note 3		2.4		2.4	V
$V_{CE(sat)}$	Collector-emitter saturation voltage	5	V_I = 2.4 V, I_C = 500 mA		1.1		1.1	V
			V_I = 2.4 V, I_C = 750 mA		1.2		1.2	
			V_I = 2.4 V, I_C = 1 A		1.3		1.3	
			V_I = 2.4 V, I_C = 1.25 A, See Note 3		1.4			
			V_I = 2.4 V, I_C = 1.5 A, See Note 3				1.5	
I_R	Clamp-diode reverse current	6	V_R = 50 V		50			μA
			V_R = 50 V, T_A = 70°C		100			
			V_R = 80 V				50	
			V_R = 80 V, T_A = 70°C				100	
V_F	Clamp-diode forward voltage	7	I_F = 1 A		1.75		1.75	V
			I_F = 1.5 V, See Note 3		2		2	
I_{CC}	Supply current (only one switch conducting)	8	V_I = 2.4 V, I_C = 500 mA		6		6	mA

NOTE 3: These parameters must be measured on one output at a time using pulse techniques, t_w = 10 ms, Duty cycle ≤ 10%.

switching characteristics at 25°C free-air temperature, V_{CC} = 5 V

PARAMETER		TEST CONDITIONS	MIN	TYP	MAX	UNIT
t_{PLH}	Propagation delay time, low-to-high-level output	See Figure 9			1	μs
t_{PHL}	Propagation delay time, high-to-low-level output				1.5	μs

PARAMETER MEASUREMENT INFORMATION

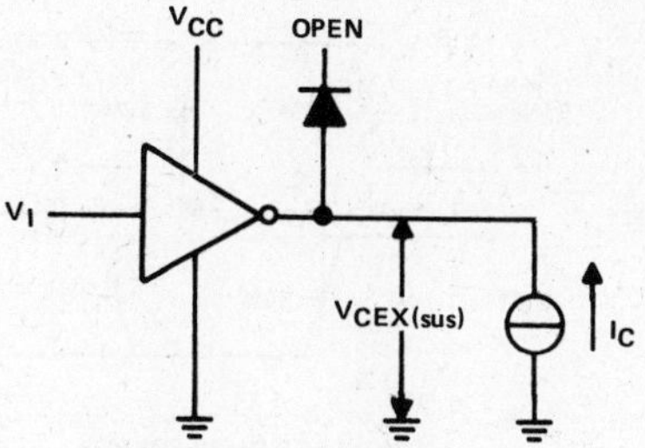
V$_{CC}$
OPEN
V$_I$
V$_{CEX(sus)}$
I$_C$

FIGURE 1. $V_{CEX(sus)}$

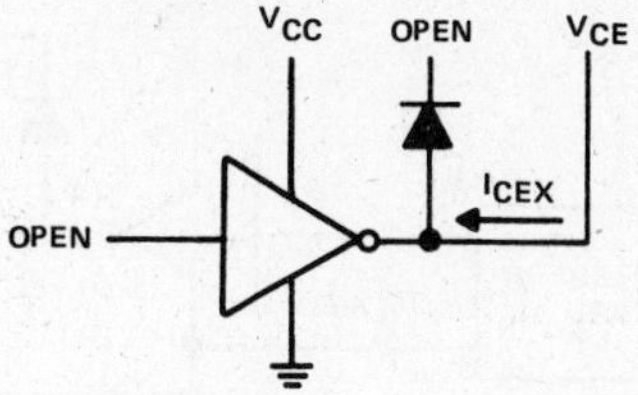
V$_{CC}$
OPEN
V$_{CE}$
I$_{CEX}$
OPEN

FIGURE 2. I_{CEX}

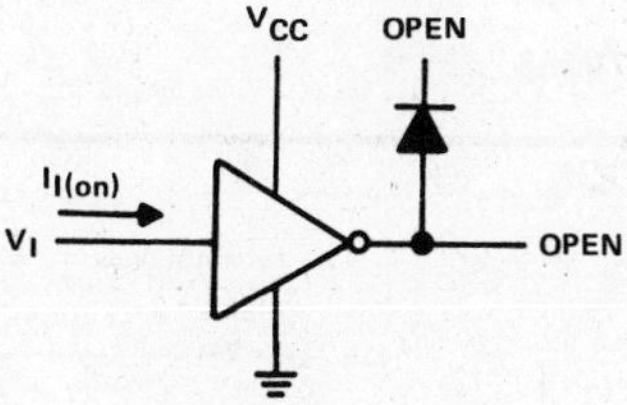
V$_{CC}$
OPEN
I$_{I(on)}$
V$_I$
OPEN

FIGURE 3. $I_{I(on)}$

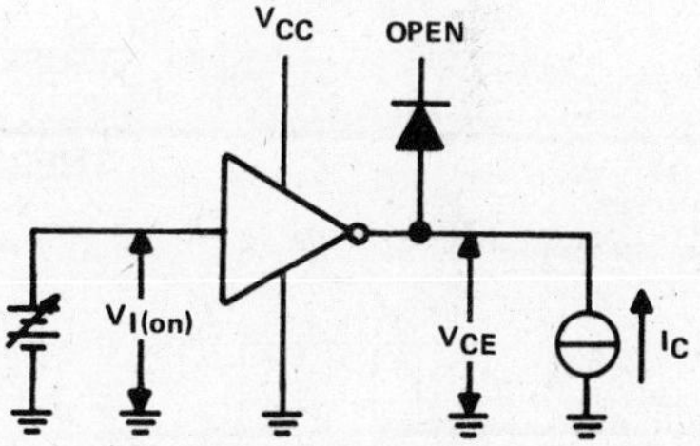
V$_{CC}$
OPEN
V$_{I(on)}$
V$_{CE}$
I$_C$

FIGURE 4. $V_{I(on)}$

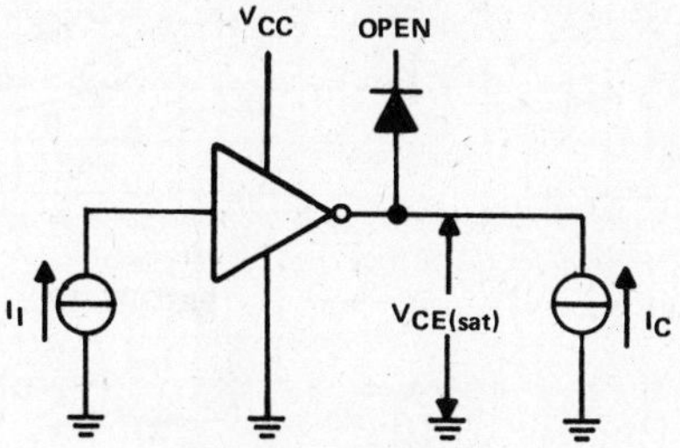
V$_{CC}$
OPEN
I$_I$
V$_{CE(sat)}$
I$_C$

FIGURE 5. $V_{CE(sat)}$

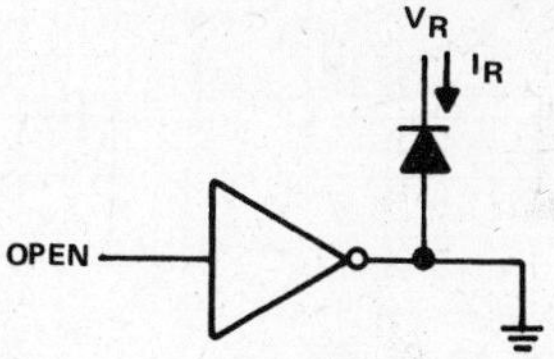
V$_R$
I$_R$
OPEN

FIGURE 6. I_R

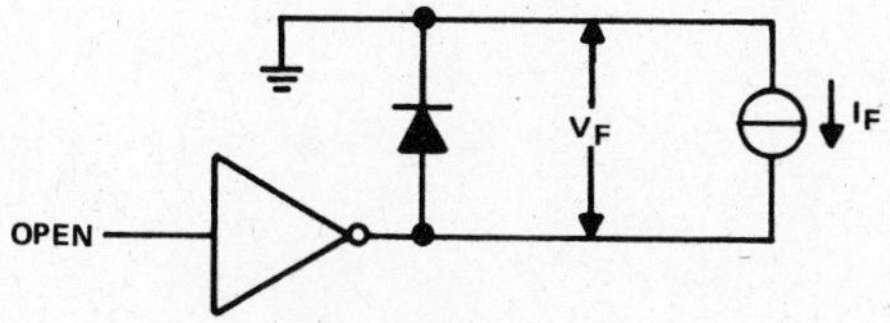
V$_F$
I$_F$
OPEN

FIGURE 7. V_F

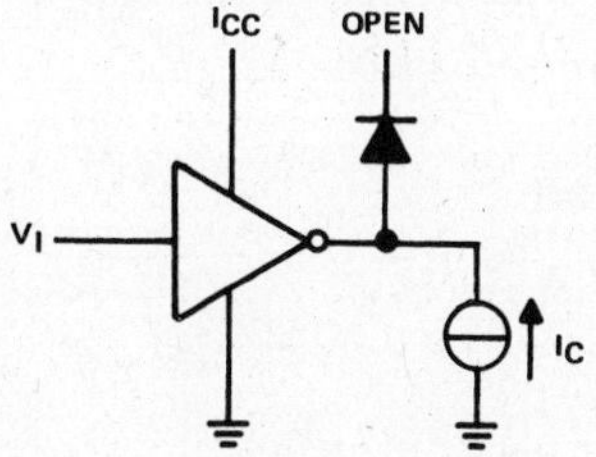
I$_{CC}$
OPEN
V$_I$
I$_C$

FIGURE 8. I_{CC}

PARAMETER MEASUREMENT INFORMATION

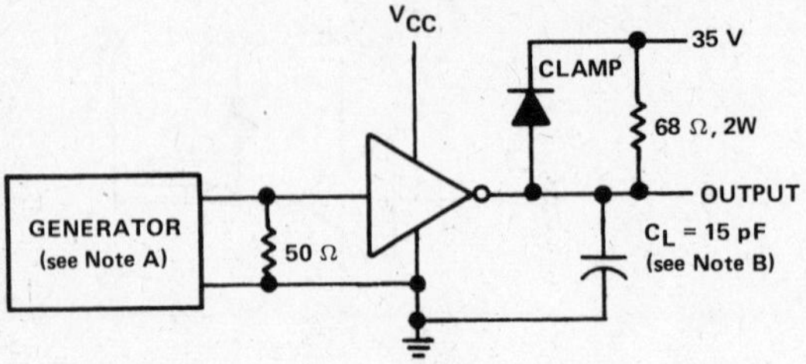
VCC
35 V
CLAMP
68 Ω, 2W
OUTPUT
GENERATOR (see Note A)
50 Ω
CL = 15 pF (see Note B)

TEST CIRCUIT

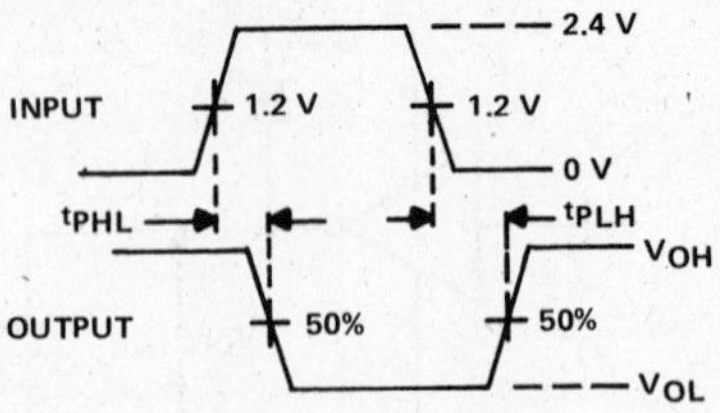
INPUT
2.4 V
1.2 V
1.2 V
0 V
tPHL
tPLH
VOH
OUTPUT
50%
50%
VOL

VOLTAGE WAVEFORMS

NOTES: A. The input pulse is supplied by a generator having the following characteristics: PRR = 50 kHz, duty cycle = 10%, Z_O = 50 Ω.
B. C_L includes all probe and stray capacitance.

FIGURE 9. SWITCHING TIMES

THERMAL INFORMATION

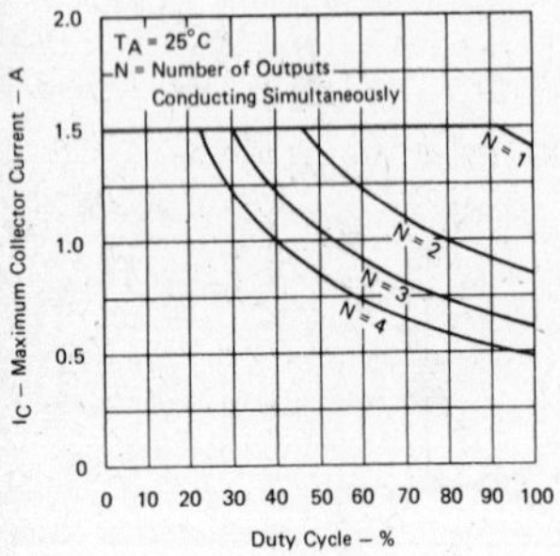
MAXIMUM COLLECTOR CURRENT vs DUTY CYCLE
TA = 25°C
N = Number of Outputs Conducting Simultaneously
N = 1
N = 2
N = 3
N = 4
IC – Maximum Collector Current – A
Duty Cycle – %

FIGURE 10

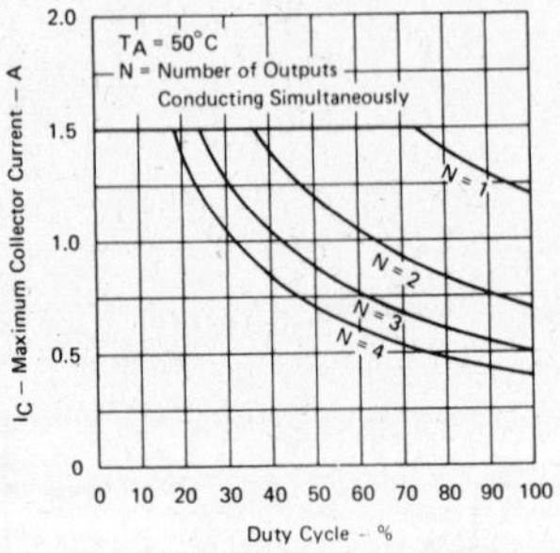
MAXIMUM COLLECTOR CURRENT vs DUTY CYCLE
TA = 50°C
N = Number of Outputs Conducting Simultaneously
N = 1
N = 2
N = 3
N = 4
IC – Maximum Collector Current – A
Duty Cycle – %

FIGURE 11

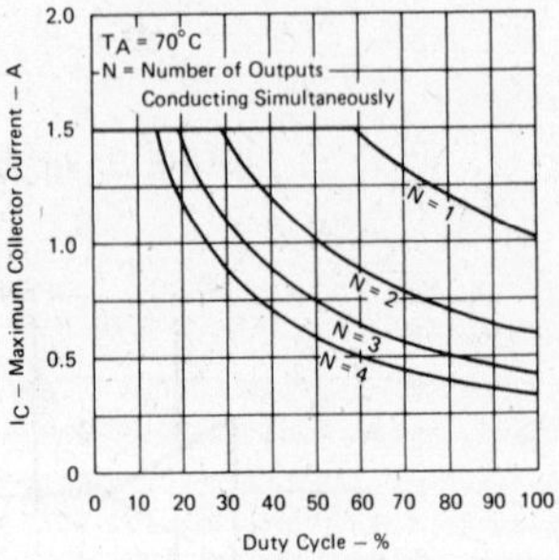
MAXIMUM COLLECTOR CURRENT vs DUTY CYCLE
TA = 70°C
N = Number of Outputs Conducting Simultaneously
N = 1
N = 2
N = 3
N = 4
IC – Maximum Collector Current – A
Duty Cycle – %

FIGURE 12

TEXAS INSTRUMENTS
POST OFFICE BOX 655012 • DALLAS, TEXAS 75265

TYPICAL APPLICATION DATA

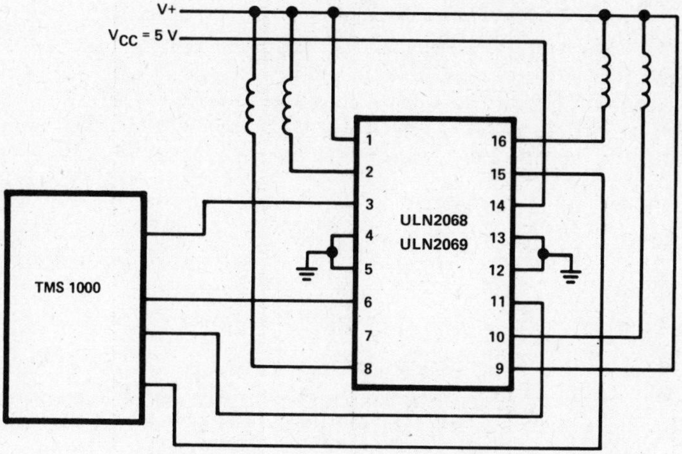
V+
V_{CC} = 5 V
TMS 1000
ULN2068
ULN2069
1
2
3
4
5
6
7
8
16
15
14
13
12
11
10
9

FIGURE 13. RELAY DRIVER INTERFACE

D2580, MAY 1980—REVISED SEPTEMBER 1986

- **Output Collector Current . . . 1.5 A Max**
- **2-W Dissipation Rating**
- **High Output-Voltage Capability**
- **Output Sink- or Source-Current Capabilities**
- **Input Compatible with TTL or 5-V CMOS**
- **Designed for Interchangeability with Sprague ULN2074 and ULN2075**

NE
DUAL-IN-LINE PACKAGE
(TOP VIEW)

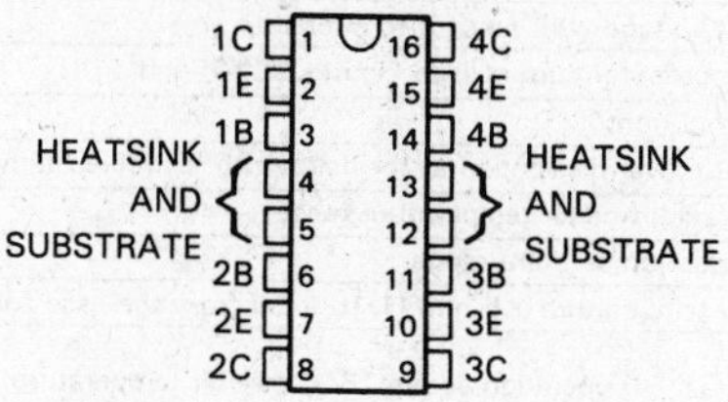
1C
1E
1B
HEATSINK
AND
SUBSTRATE
2B
2E
2C
1
2
3
4
5
6
7
8
16
15
14
13
12
11
10
9
4C
4E
4B
HEATSINK
AND
SUBSTRATE
3B
3E
3C

description

The ULN2074 and ULN2075 are monolithic, quadruple, high-voltage, high-current n-p-n darlington-transistor amplifier devices. They feature high-voltage outputs with collector-current ratings of 1.5 amperes for each darlington pair.

The ULN2074 and ULN2075 are unique general-purpose devices, each featuring uncommitted collectors and emitters to allow for either sinking or sourcing the output current. These devices offer the system designer the flexibility of tailoring the circuit to the application. Typical applications include logic buffers, relay drivers, lamp drivers, and hammer drivers.

For proper operation, the substrate must be connected to the most negative voltage.

The ULN2074 and ULN2075 are characterized for operation from −20 °C to 85 °C.

schematic (each switch)

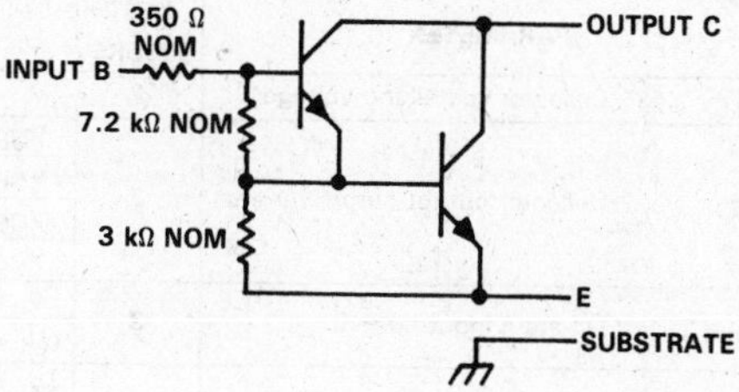
350 Ω
NOM
INPUT B
7.2 kΩ NOM
3 kΩ NOM
OUTPUT C
E
SUBSTRATE

logic symbol†

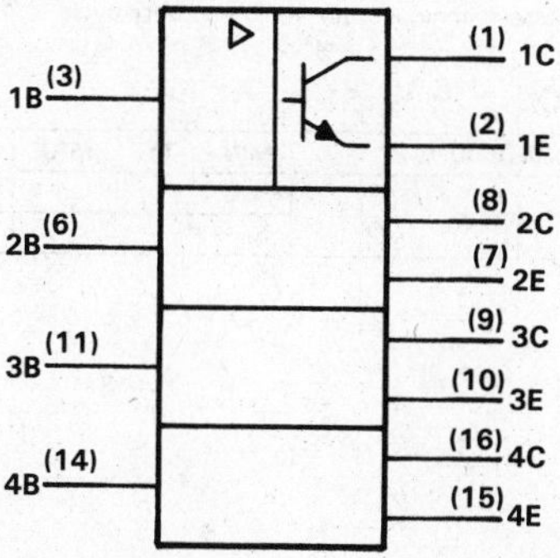
1B
(3)
2B
(6)
3B
(11)
4B
(14)
(1)
1C
(2)
1E
(8)
2C
(7)
2E
(9)
3C
(10)
3E
(16)
4C
(15)
4E

‡This symbol is in accordance with ANSI/IEEE Std 91-1984 and IEC Publication 617-12.

TEXAS INSTRUMENTS
POST OFFICE BOX 655012 • DALLAS, TEXAS 75265

absolute maximum ratings at 25°C free-air temperature for each switch (unless otherwise noted)

	ULN2074	ULN2075	UNIT
Collector-emitter voltage	50	80	V
Input voltage with respect to substrate	30	60	V
Peak collector current (see Figures 9, 10, and 11)	1.5	1.5	A
Input current	25	25	mA
Total power dissipation at (or below) 25°C free-air temperature (see Note 1)	2075	2075	mW
Operating free-air temperature range	−20 to 85	−20 to 85	°C
Storage temperature range	−55 to 150	−55 to 150	°C
Lead temperature 1,6 mm (1/16 inch) from the case for 10 seconds	260	260	°C

NOTE 1: For operation above 25°C free-air temperature, derate total power linearly to 1079 mW at 85°C at the rate of 16.6 mW/°C.

electrical characteristics at 25°C free-air temperature (unless otherwise noted)

PARAMETER		TEST FIGURE	TEST CONDITIONS	ULN2074		ULN2075		UNIT
				MIN	MAX	MIN	MAX	
$V_{CEX(sus)}$	Collector sustaining voltage	1	V_I = 0.4 V, I_C = 100 mA	35		50		V
I_{CEX}	Collector output cutoff current	2	V_{CE} = 50 V		100			μA
			V_{CE} = 50 V, T_A = 70°C		500			
			V_{CE} = 80 V				100	
			V_{CE} = 80 V, T_A = 70°C				500	
$I_{I(on)}$	On-state input current	3	V_I = 2.4 V	2	4.3	2	4.3	mA
			V_I = 3.75 V	4.5	9.6	4.5	9.6	
$V_{I(on)}$	On-state input voltage	4	V_{CE} = 2 V, I_C = 1 A		2		2	V
			V_{CE} = 2 V, I_C = 1.5 A, See Note 2		2.5		2.5	
$V_{CE(sat)}$	Collector-emitter saturation voltage	5	I_I = 625 μA, I_C = 500 mA		1.1		1.1	V
			I_I = 935 μA, I_C = 750 mA		1.2		1.2	
			I_I = 1.25 mA, I_C = 1 A		1.3		1.3	
			I_I = 2 mA, I_C = 1.25 A, See Note 2		1.4			
			I_I = 2.25 mA, I_C = 1.5 A, See Note 2				1.5	

NOTE 2: These parameters must be measured on one output at a time using pulse techniques, t_w = 10 ms, duty cycle ≤ 10%.

switching characteristics at 25°C free-air temperature, V_{CC} = 5 V

PARAMETER		TEST CONDITIONS	MIN	TYP	MAX	UNIT
t_{PLH}	Propagation delay time, low-to-high-level output	See Figure 6			1	μs
t_{PHL}	Propagation delay time, high-to-low-level output				1.5	μs

TEXAS INSTRUMENTS
POST OFFICE BOX 655012 • DALLAS, TEXAS 75265

PARAMETER MEASUREMENT INFORMATION

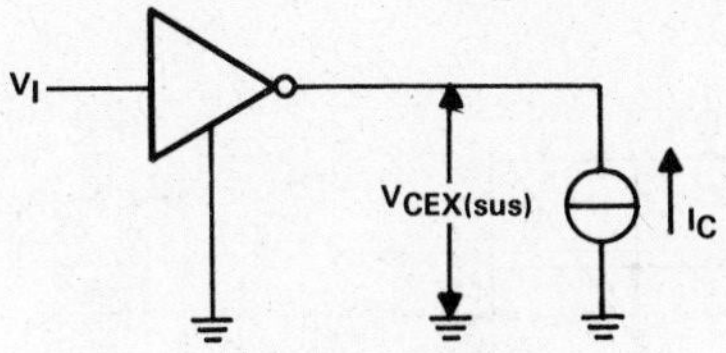
V_I
V_CEX(sus)
I_C

FIGURE 1. $V_{CEX(sus)}$

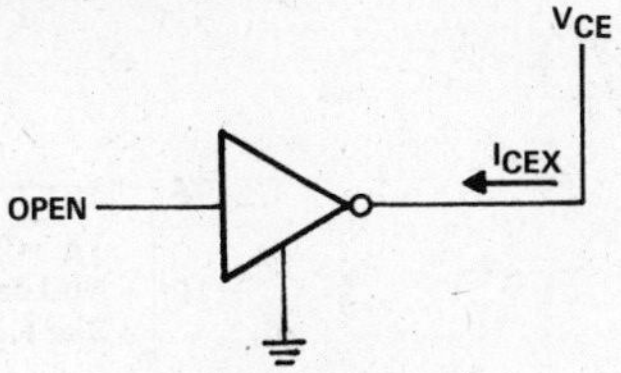
V_CE
I_CEX
OPEN

FIGURE 2. I_{CEX}

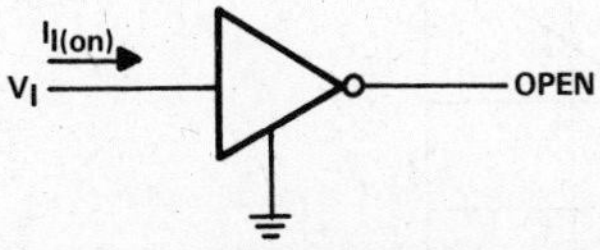
I_I(on)
V_I
OPEN

FIGURE 3. $I_{I(on)}$

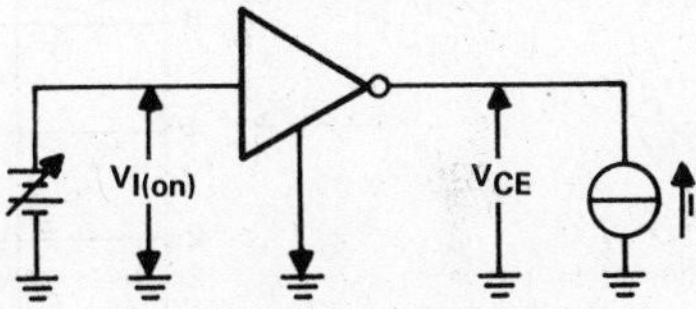
V_I(on)
V_CE
I_C

FIGURE 4. $V_{I(on)}$

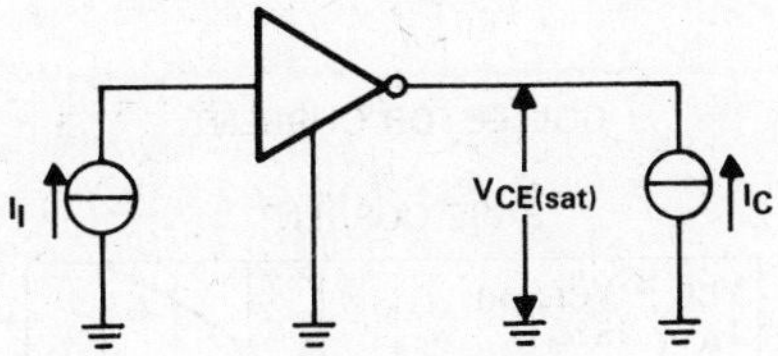
I_I
V_CE(sat)
I_C

FIGURE 5. $V_{CE(sat)}$

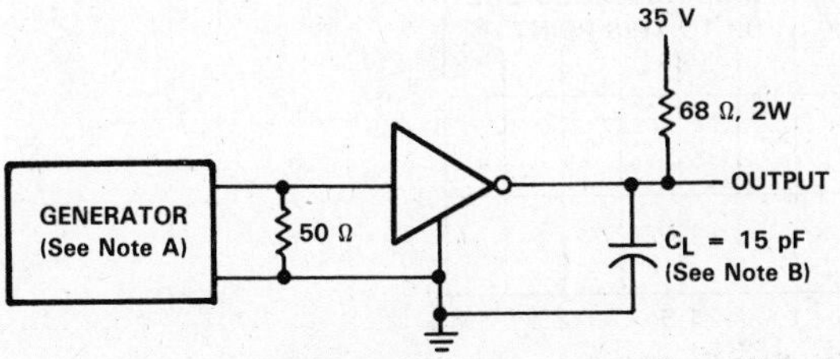
35 V
68 Ω, 2W
OUTPUT
GENERATOR
(See Note A)
50 Ω
C_L = 15 pF
(See Note B)

TEST CIRCUITS

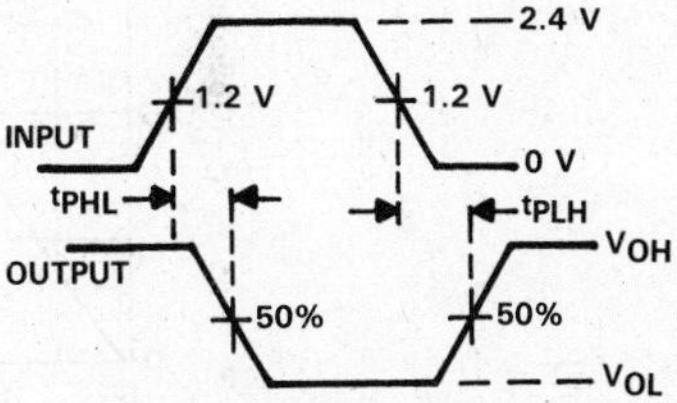
2.4 V
1.2 V
1.2 V
INPUT
0 V
t_PHL
t_PLH
V_OH
OUTPUT
50%
50%
V_OL

VOLTAGE WAVEFORMS

NOTES: A. The input pulse is supplied by a generator having the following characteristics: PRR = 50 kHz, duty cycle = 10%, Z_o = 50 Ω.
B. C_L includes all probe and stray capacitance.

FIGURE 6. SWITCHING CHARACTERISTICS

ELECTRICAL CHARACTERISTICS

INPUT CURRENT
vs
INPUT VOLTAGE

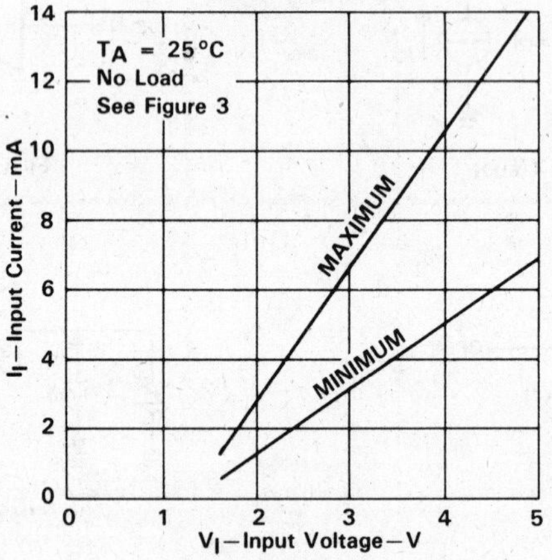
14
12
10
8
6
4
2
0
T_A = 25°C
No Load
See Figure 3
MAXIMUM
MINIMUM
I_I—Input Current—mA
0
1
2
3
4
5
V_I—Input Voltage—V

FIGURE 7

COLLECTOR CURRENT
vs
BASE CURRENT

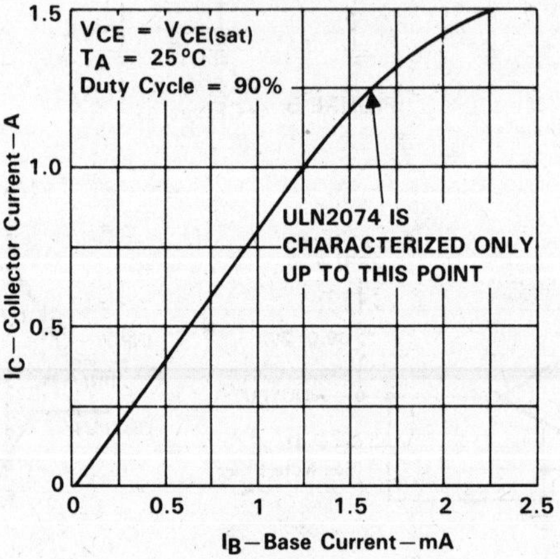
1.5
1.0
0.5
0
V_{CE} = $V_{CE(sat)}$
T_A = 25°C
Duty Cycle = 90%
ULN2074 IS
CHARACTERIZED ONLY
UP TO THIS POINT
I_C—Collector Current—A
0
0.5
1
1.5
2
2.5
I_B—Base Current—mA

FIGURE 8

THERMAL INFORMATION

MAXIMUM COLLECTOR CURRENT
vs
DUTY CYCLE

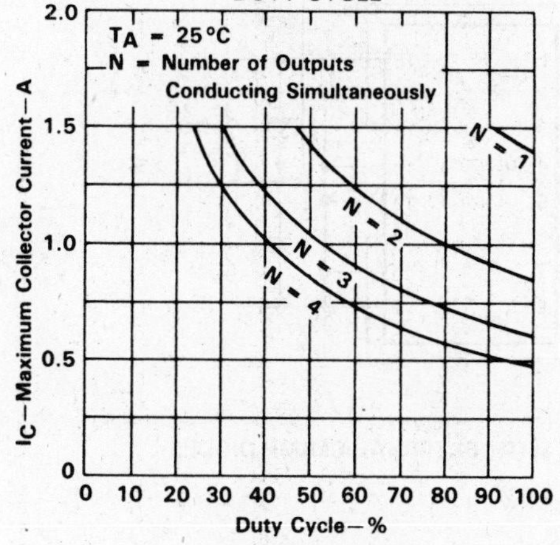
T_A = 25°C
N = Number of Outputs Conducting Simultaneously
N = 1
N = 2
N = 3
N = 4
2.0
1.5
1.0
0.5
0
0 10 20 30 40 50 60 70 80 90 100
I_C—Maximum Collector Current—A
Duty Cycle—%

FIGURE 9

MAXIMUM COLLECTOR CURRENT
vs
DUTY CYCLE

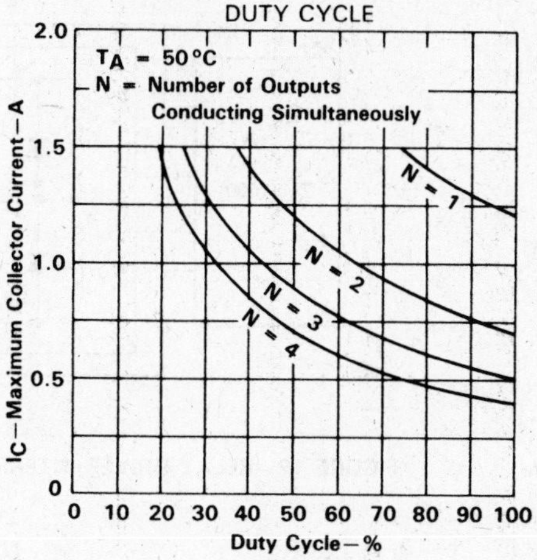
T_A = 50°C
N = Number of Outputs Conducting Simultaneously
N = 1
N = 2
N = 3
N = 4
2.0
1.5
1.0
0.5
0
0 10 20 30 40 50 60 70 80 90 100
I_C—Maximum Collector Current—A
Duty Cycle—%

FIGURE 10

MAXIMUM COLLECTOR CURRENT
vs
DUTY CYCLE

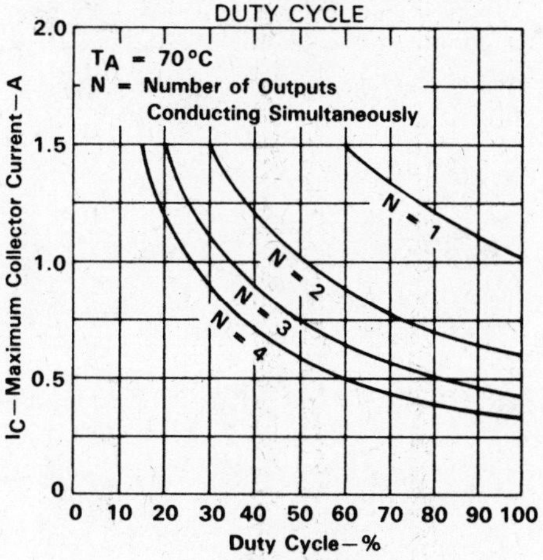
T_A = 70°C
N = Number of Outputs Conducting Simultaneously
N = 1
N = 2
N = 3
N = 4
2.0
1.5
1.0
0.5
0
0 10 20 30 40 50 60 70 80 90 100
I_C—Maximum Collector Current—A
Duty Cycle—%

FIGURE 11

TYPICAL APPLICATION DATA

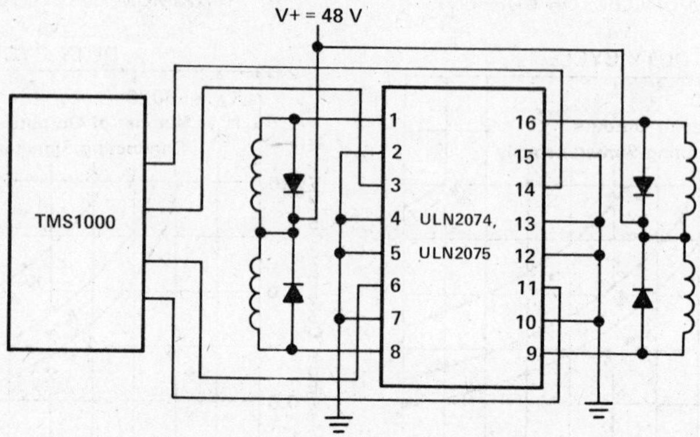
V+ = 48 V
TMS1000
ULN2074,
ULN2075
1
2
3
4
5
6
7
8
16
15
14
13
12
11
10
9

FIGURE 12. RELAY DRIVER INTERFACE WITH EXTERNAL CLAMP DIODES

TEXAS INSTRUMENTS
POST OFFICE BOX 655012 • DALLAS, TEXAS 75265

General Information — 1
Alphanumeric Index
Selection Guide

Data Acquisition Circuits — 2
Cross-Reference Guide
Data Sheets

Display Drivers — 3
Data Sheets

Line Drivers and Receivers — 4
Cross-Reference Guide
Data Sheets

Peripheral Drivers/Actuators — 5
Cross-Reference Guide
Data Sheets

Memory Interface Circuits — 6
Data Sheets

Speech Synthesis Circuits — 7
Data Sheets

Appendix A — A
Power Derating Curves

Appendix B — B
Ordering Instructions
Mechanical Data
IC Sockets

Appendix C — C
Explanation of
Logic Symbols

DUAL-CHANNEL SENSE AMPLIFIER WITH COMPLEMENTARY OUTPUTS

D2338, DECEMBER 1986

- High Speed and Fast Recovery Time
- Time and Amplitude Signal Discrimination
- Adjustable Input Threshold Voltage Levels
- Narrow Window of Threshold Voltage Uncertainty
- Multiple Differential-Input Preamplifiers
- High DC Noise Margin . . . 1 V Typ
- Good Fanout Capability
- TTL Drive Capability
- Standard Logic Supply Voltages
- Plug-in Configuration Ideal for Flow-Soldering Techniques

J PACKAGE
(TOP VIEW)

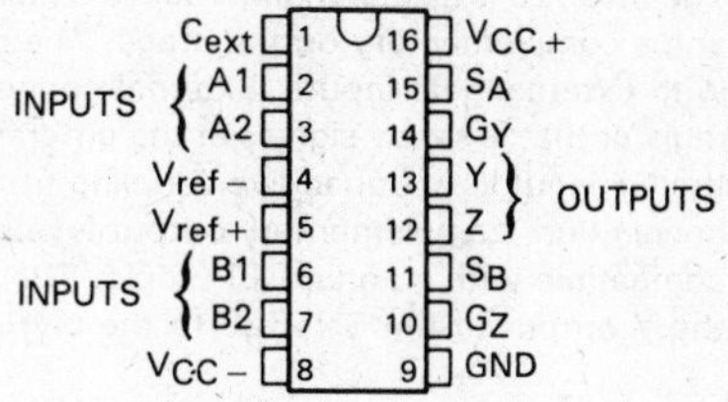

Cext 1
16 VCC+
INPUTS
A1 2
15 SA
A2 3
14 GY
Vref− 4
13 Y
OUTPUTS
Vref+ 5
12 Z
INPUTS
B1 6
11 SB
B2 7
10 GZ
VCC− 8
9 GND

FUNCTION TABLE

INPUTS						OUTPUTS	
A	B	G_Y	G_Z	S_A	S_B	Y	Z
X	X	L	X	X	X	H	$\overline{G}_Z$
H	X	X	X	H	X	H	$\overline{G}_Z$
X	H	X	X	X	H	H	$\overline{G}_Z$
L	L	H	X	X	X	L	H
L	X	H	X	X	L	L	H
X	L	H	X	L	X	L	H
X	X	H	X	L	L	L	H
X	X	X	L	X	X	X	H

positive logic: $Y = \overline{G}_Y + A \cdot S_A + B \cdot S_B$

$Z = \overline{G}_Z + \overline{Y}$

$Z = \overline{G}_Z + G_Y(\overline{A}+\overline{S}_A)(\overline{B}+\overline{S}_B)$

DEFINITION OF LOGIC LEVELS

INPUT	H	L	X
A or B†	$V_{ID} \geq V_T$ max	$V_{ID} \leq V_T$ min	Irrelevant
Any G or S	$V_I \geq V_{IH}$ min	$V_I \leq V_{IL}$ max	Irrelevant

† A and B are differential voltages (V_{ID}) between A1 and A2 or B1 and B2, respectively. For these circuits, V_{ID} is considered positive regardless of which terminal of each pair is positive with respect to the other.

description

The SN5520 monolithic sense amplifier is designed for use with high-speed memory systems. This sense amplifier detects bipolar differential-input signals from the memory and provides the necessary interface circuitry between the memory and the logic section. Low-level pulses originating in the memory are transformed into logic levels compatible with standard TTL circuits.

The SN5520 sense amplifier features multiple differential-input preamplifiers and versatile gating and output circuits, permitting a significant reduction in the circuitry required to accomplish the sensing function. A unique circuit design provides an inherent stability of the input threshold level over a wide range of supply voltage levels and temperature ranges. Independent strobing of each of the dual sense-input channels ensures maximum versatility and permits detection to occur when the signal-to-noise ratio is at a maximum. The gate and strobe inputs and the outputs are compatible with standard TTL logic circuits.

With the Z output connected to the G_Y input, the SN5520 may be used to perform the functions of a flip-flop or register that responds to the sense and strobe input conditions.

The SN5520 is available in the J ceramic dual-in-line package and is characterized for operation over the full military temperature range of −55°C to 125°C.

TEXAS INSTRUMENTS
POST OFFICE BOX 655012 • DALLAS, TEXAS 75265

circuit operation

The SN5520 is a dual-channel sense amplifier with the preamplifiers connected to a common output stage and a complementary output stage. The output circuit is composed of two cascaded NAND gates, each with external gate inputs. External connection of the Z output and the G_Y input results in a flip-flop or register that is set by signals at the differential-input terminals. Reset of the register is performed by taking the G_Z input low. Capacitive coupling from output Z to G_Y results in output pulse stretching. With either connection, complementary output levels are available. The gate and strobe inputs and the outputs are compatible with standard TTL logic. The input function of the SN5520 can be expanded by connecting the Y output of the SN5522 to the G_Y input of the circuit being expanded.

logic symbol†

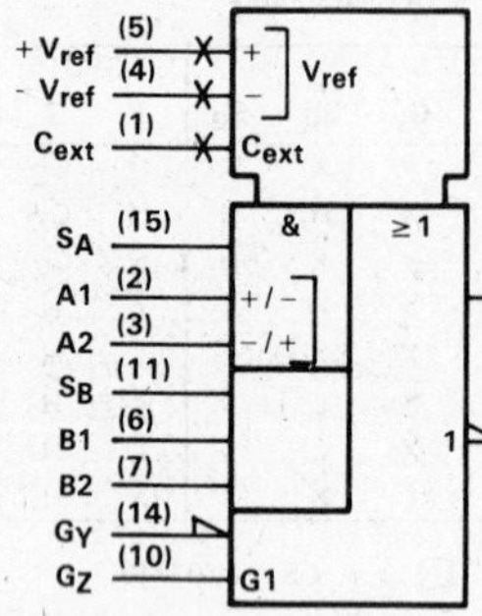
+Vref
- Vref
Cext
SA
A1
A2
SB
B1
B2
GY
GZ
(5)
(4)
(1)
(15)
(2)
(3)
(11)
(6)
(7)
(14)
(10)
Vref
Cext
&
≥1
+/−
−/+
G1
(13)
Y
(12)
Z
1

†This symbol is in accordance with ANSI/IEEE Std 91-1984 and IEC Publication 617-12.

logic diagram

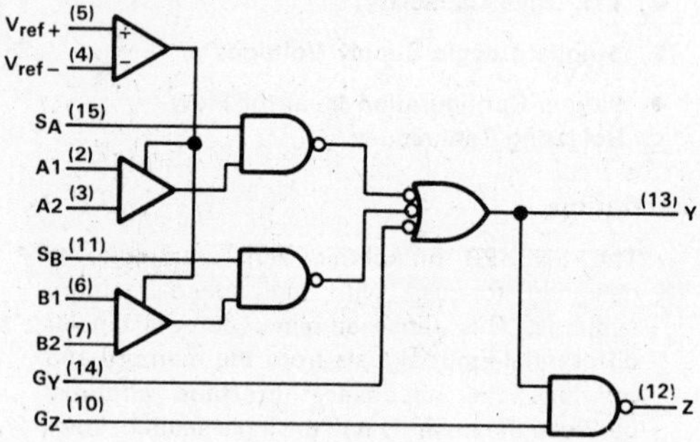
Vref+
Vref−
SA
A1
A2
SB
B1
B2
GY
GZ
(5)
(4)
(15)
(2)
(3)
(11)
(6)
(7)
(14)
(10)
(13)
Y
(12)
Z

TEXAS INSTRUMENTS
POST OFFICE BOX 655012 • DALLAS, TEXAS 75265

schematic

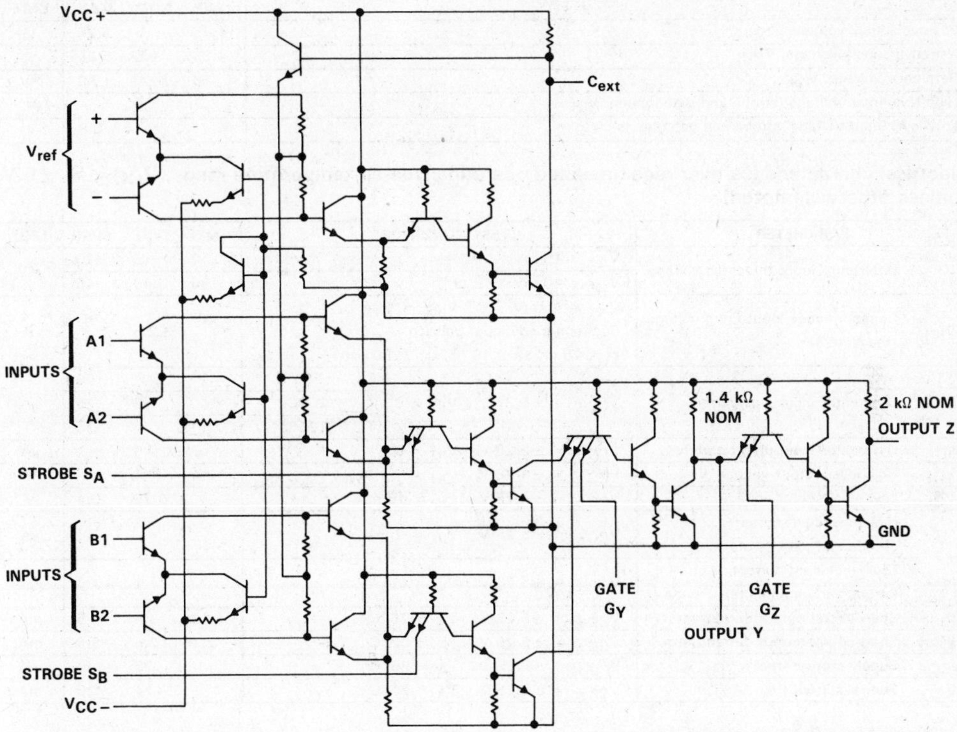
VCC+
Cext
Vref
+
−
INPUTS
A1
A2
STROBE SA
INPUTS
B1
B2
STROBE SB
VCC−
1.4 kΩ
NOM
2 kΩ NOM
OUTPUT Z
GND
GATE
GY
GATE
GZ
OUTPUT Y

absolute maximum ratings over operating free-air temperature range (unless otherwise noted)

Supply voltage, V_{CC+} (see Note 1)	7 V
Supply voltage, V_{CC-}	−7 V
Differential input voltage, V_{ID} or V_{ref}	±5 V
Voltage from any input to ground (see Note 2)	5.5 V
Continuous total power dissipation at (or below) 25 °C free-air temperature (see Note 3)	1375 mW
Operating free-air temperature range	−55 °C to 125 °C
Storage temperature range	−65 °C to 150 °C
Lead temperature 1,6 mm (1/16 inch) from case for 60 seconds	300 °C

NOTES: 1. Voltage values, except differential voltages, are with respect to the network ground terminal.
2. Strobe and gate input voltages must be zero or positive with respect to the network ground terminal.
3. For operating above 25 °C free-air temperature, derate total power linearly at the rate of 11.0 mW/°C.

recommended operating conditions

	MIN	NOM	MAX	UNIT
Positive supply voltage, V_{CC+}	4.75	5	5.25	V
Negative supply voltage, V_{CC-}	−4.75	−5	−5.25	V
Reference voltage, V_{ref}	15		40	mV
High-level input voltage (strobe and gate inputs), V_{IH}	2			V
Low-level input voltage (strobe and gate inputs), V_{IL}			0.8	V

electrical characteristics over recommended operating free-air temperature range, $V_{CC\pm} = \pm 5$ V, (unless otherwise noted)

PARAMETER		TEST CONDITIONS		MIN	TYP†	MAX	UNIT
V_T	Differential-input threshold voltage	$V_{ref} = 15$ mV		10	15	20	mV
		$V_{ref} = 40$ mV		35	40	45	
V_{ICF}	Common-mode input firing voltage (see Note 4)	$V_{ref} = 40$ mV, $V_{I(S)} = V_{IH}$ Common-mode input pulse: $t_r \le 15$ ns, $t_f \le 15$ ns, $t_w = 50$ ns			±2.5		V
I_{IB}	Differential-input bias current	$V_{CC\pm} = \pm 5.25$ V, $V_{ID} = 0$	$T_A = -55$°C			100	μA
			$T_A = 25$°C		30	75	
			$T_A = 125$°C			75	
I_{IO}	Differential-input offset current	$V_{CC\pm} = \pm 5.25$ V, $V_{ID} = 0$			0.5		μA
V_{OH}	High-level output voltage	$V_{CC\pm} = \pm 4.75$ V, $I_{OH} = -400$ μA		2.4	4		V
V_{OL}	Low-level output voltage	$V_{CC\pm} = \pm 4.75$ V, $I_{OL} = 16$ mA			0.25	0.4	V
I_{IH}	High-level input current (strobe and gate inputs)	$V_{CC\pm} = \pm 5.25$ V, $V_{IH} = 2.4$ V				40	μA
I_{IL}	Low-level input current (strobe and gate inputs)	$V_{CC\pm} = \pm 5.25$ V, $V_{IL} = 0.4$ V			−1	−1.6	mA
$I_{OS(Y)}$	Short-circuit output current into Y	$V_{CC\pm} = \pm 5.25$ V, $T_A = 25$°C		−3	−3.8	−5	mA
$I_{OS(Z)}$	Short-circuit output current into Z	$V_{CC\pm} = \pm 5.25$ V, $T_A = 25$°C		−2.1	−2.6	−3.5	mA
I_{CC+}	Supply current from V_{CC+}	$V_{CC\pm} = \pm 5.25$ V, $T_A = 25$°C			28	40	mA
I_{CC-}	Supply current from V_{CC-}	$V_{CC\pm} = \pm 5.25$ V, $T_A = 25$°C			−14	−20	mA

†Typical values are at $V_{CC\pm} = \pm 5$ V, $T_A = 25$°C.

NOTE 4: Common-mode input firing voltage is the minimum common-mode voltage that will exceed the dynamic range of the input at the specified conditions and cause the logic output to switch. The common-mode input signal is applied when the strobe is high.

switching characteristics, $V_{CC\pm}$ = ±5 V, T_A = 25°C

PARAMETER	FROM (INPUT)	TO (OUTPUT)	TEST FIGURE	TEST CONDITIONS	MIN	TYP	MAX	UNIT
$t_{PLH(DY)}$	A1-A2 or B1-B2	Y	1	C_L = 15 pF, C_{ext} ≥ 100 pF, R_L = 288 Ω		25	40	ns
$t_{PHL(DY)}$						20		
$t_{PLH(DZ)}$	A1-A2 or B1-B2	Z	1			30		ns
$t_{PHL(DZ)}$						35	55	
$t_{PLH(SY)}$	STROBE A or B	Y	1			15	30	ns
$t_{PHL(SY)}$						20		
$t_{PLH(SZ)}$	STROBE A or B	Z	1			30		ns
$t_{PHL(SZ)}$						35	55	
$t_{PLH(GY, Y)}$	GATE G_Y	Y	2			15	25	ns
$t_{PHL(GY, Y)}$						10		
$t_{PLH(GY, Z)}$	GATE G_Y	Z	2			15		ns
$t_{PHL(GY,Z)}$						20	30	
$t_{PLH(GZ, Z)}$	GATE G_Z	Z	3			15		ns
$t_{PHL(GZ, Z)}$						10	20	

recovery and cycle times, $V_{CC\pm}$ = ±5 V, C_{ext} ≥ 100 pF, T_A = 25°C

PARAMETER		TEST CONDITIONS	MIN	TYP	MAX	UNIT
t_{orD}	Differential-input overload recovery time (see Note 5)	Differential Input Pulse: V_{ID} = 2 V, t_r = t_f = 20 ns		20		ns
t_{orC}	Common-mode input overload recovery time (see Note 6)	Common-mode Input Pulse: V_{IC} = ±2 V, t_r = t_f = 20 ns		20		ns
$t_{cyc(min)}$	Minimum cycle time			200		ns

NOTES: 5. Differential-input overload recovery time is the time necessary for the device to recover from the specified differential-input overload signal prior to the strobe-enable signal.

6. Common-mode-input overload recovery time is the time necessary for the device to recover from the specified common-mode-input overload signal prior to the strobe-enable signal.

PARAMETER MEASUREMENT INFORMATION

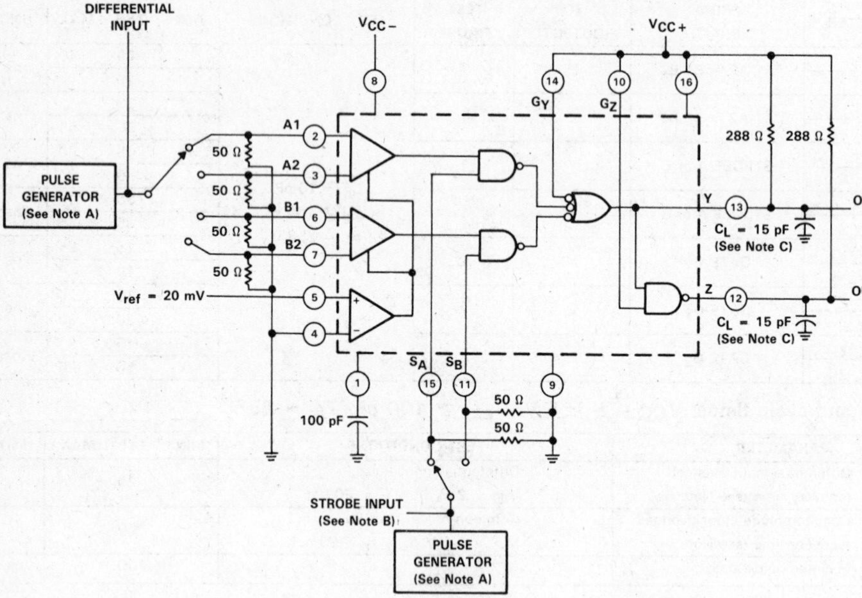
DIFFERENTIAL INPUT
VCC−
VCC+
PULSE GENERATOR (See Note A)
A1
A2
B1
B2
50 Ω
50 Ω
50 Ω
50 Ω
Vref = 20 mV
GY
GZ
288 Ω
288 Ω
Y
OUTPUT Y
CL = 15 pF (See Note C)
Z
OUTPUT Z
CL = 15 pF (See Note C)
SA
SB
100 pF
50 Ω
50 Ω
STROBE INPUT (See Note B)
PULSE GENERATOR (See Note A)

TEST CIRCUIT

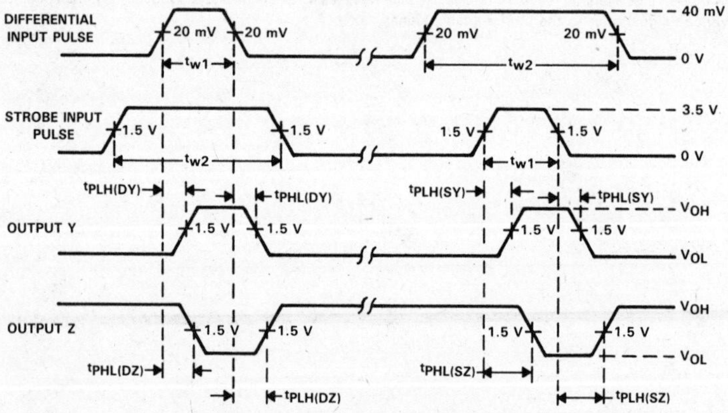
DIFFERENTIAL INPUT PULSE
20 mV
20 mV
40 mV
0 V
tw1
tw2
STROBE INPUT PULSE
1.5 V
3.5 V
tPLH(DY)
tPHL(DY)
tPLH(SY)
tPHL(SY)
OUTPUT Y
VOH
VOL
OUTPUT Z
tPHL(DZ)
tPLH(DZ)
tPHL(SZ)
tPLH(SZ)

VOLTAGE WAVEFORMS

NOTES: A. The pulse generators have the following characteristics: Z_{out} = 50 Ω, t_r = 15 ±5 ns, t_f = 15 ±5 ns, t_{w1} = 100 ns, t_{w2} = 300 ns, and PRR ≤ 1 MHz.

B. The strobe input pulse is applied to Strobe S_A when inputs A1-A2 are being tested and to Strobe S_B when inputs B1-B2 are being tested.

C. C_L includes probe and jig capacitance.

FIGURE 1. PROPAGATION DELAY TIMES FROM DIFFERENTIAL AND STROBE INPUTS

PARAMETER MEASUREMENT INFORMATION

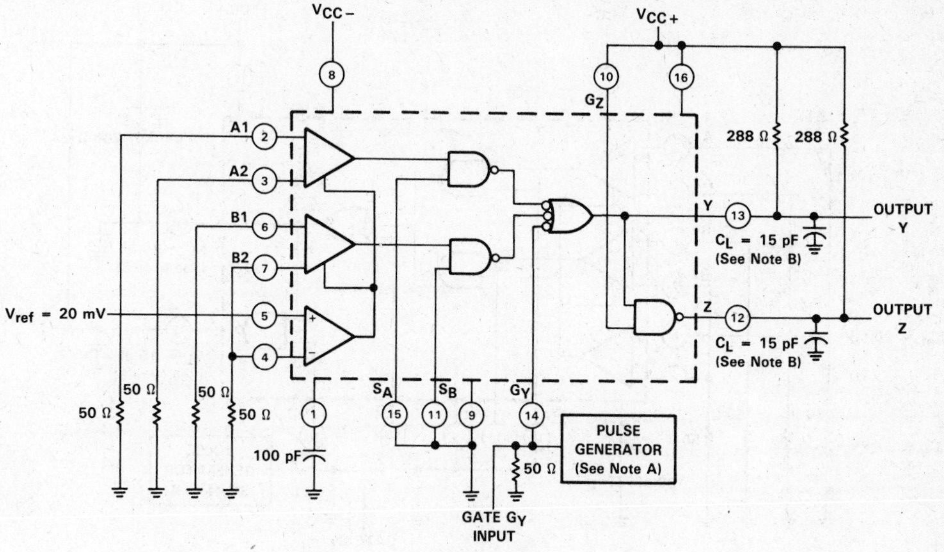
V_{CC-}
V_{CC+}
8
10
16
G_Z
A1
2
A2
3
B1
6
B2
7
V_{ref} = 20 mV
5
4
288 Ω
288 Ω
Y
13
OUTPUT Y
C_L = 15 pF (See Note B)
Z
12
OUTPUT Z
C_L = 15 pF (See Note B)
50 Ω
50 Ω
50 Ω
50 Ω
1
100 pF
S_A
S_B
G_Y
15
11
9
14
50 Ω
PULSE GENERATOR (See Note A)
GATE G_Y INPUT

TEST CIRCUIT

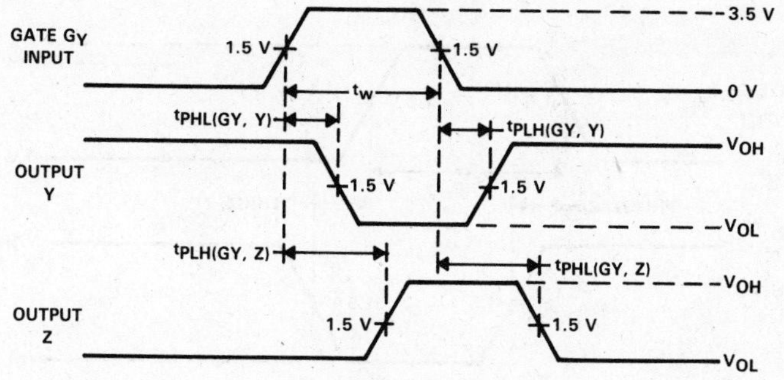
GATE G_Y INPUT
1.5 V
1.5 V
3.5 V
0 V
t_w
$t_{PHL(GY, Y)}$
$t_{PLH(GY, Y)}$
V_{OH}
OUTPUT Y
1.5 V
1.5 V
V_{OL}
$t_{PLH(GY, Z)}$
$t_{PHL(GY, Z)}$
V_{OH}
OUTPUT Z
1.5 V
1.5 V
V_{OL}

VOLTAGE WAVEFORMS

NOTES: A. The pulse generator has the following characteristics: Z_{out} = 50 Ω, t_r = 15 ± 5 ns, t_f = 15 ± 5 ns, t_w = 100 ns, and PRR ≤ 1 MHz.
B. C_L includes probe and jig capacitance.

FIGURE 2. PROPAGATION DELAY TIMES FROM GATE G_Y

PARAMETER MEASUREMENT INFORMATION

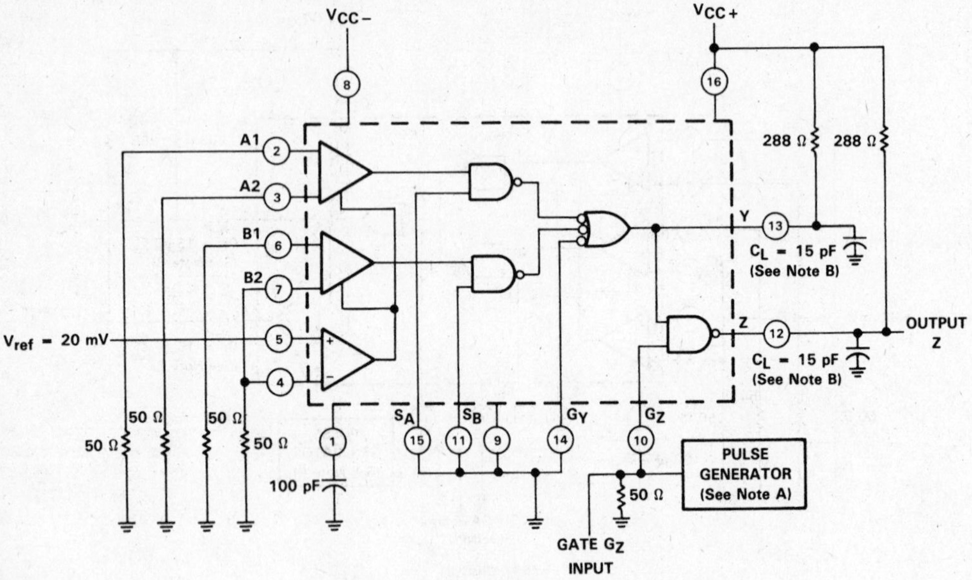
VCC−
VCC+
8
16
A1
2
A2
3
B1
6
B2
7
Vref = 20 mV
5
4
288 Ω
288 Ω
Y
13
CL = 15 pF
(See Note B)
Z
12
OUTPUT
Z
CL = 15 pF
(See Note B)
50 Ω
50 Ω
50 Ω
50 Ω
1
100 pF
SA
SB
GY
GZ
15
11
9
14
10
50 Ω
PULSE
GENERATOR
(See Note A)
GATE GZ
INPUT

TEST CIRCUIT

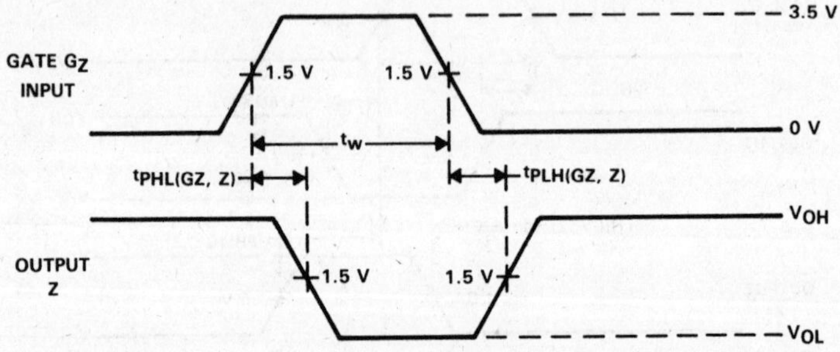
3.5 V
GATE GZ
INPUT
1.5 V
1.5 V
0 V
tw
tPHL(GZ, Z)
tPLH(GZ, Z)
VOH
OUTPUT
Z
1.5 V
1.5 V
VOL

VOLTAGE WAVEFORMS

NOTES: A. The pulse generator has the following characteristics: Z_{out} = 50 Ω, t_r = 15 ±5 ns, t_f = 15 ±5 ns, t_w = 100 ns, and PRR ≤ 1 MHz.
B. C_L includes probe and jig capacitance.

FIGURE 3. PROPAGATION DELAY TIMES FROM GATE G_Z

TEXAS INSTRUMENTS
POST OFFICE BOX 655012 • DALLAS, TEXAS 75265

TYPICAL CHARACTERISTICS

THRESHOLD VOLTAGE
vs
REFERENCE VOLTAGE

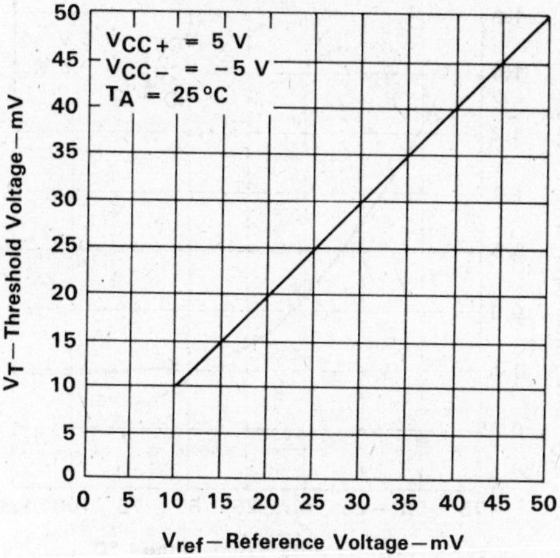
VCC+ = 5 V
VCC− = −5 V
TA = 25°C
VT—Threshold Voltage—mV
Vref—Reference Voltage—mV
0 5 10 15 20 25 30 35 40 45 50

FIGURE 4

THRESHOLD VOLTAGE
vs
SUPPLY VOLTAGE

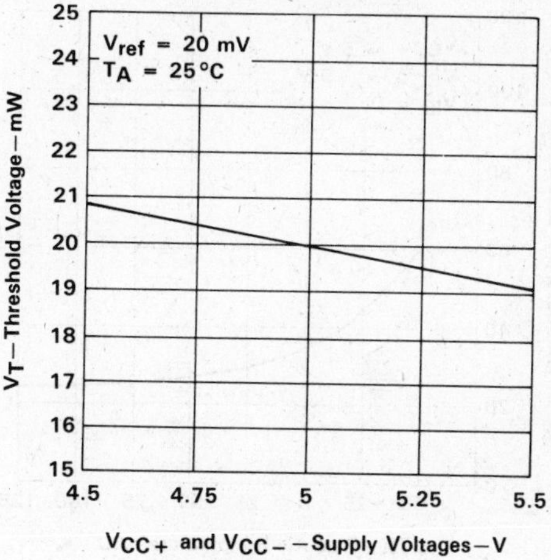
Vref = 20 mV
TA = 25°C
VT—Threshold Voltage—mW
15 16 17 18 19 20 21 22 23 24 25
4.5 4.75 5 5.25 5.5
VCC+ and VCC−—Supply Voltages—V

FIGURE 5

NORMALIZED THRESHOLD VOLTAGE
vs
PULSE REPETITION RATE

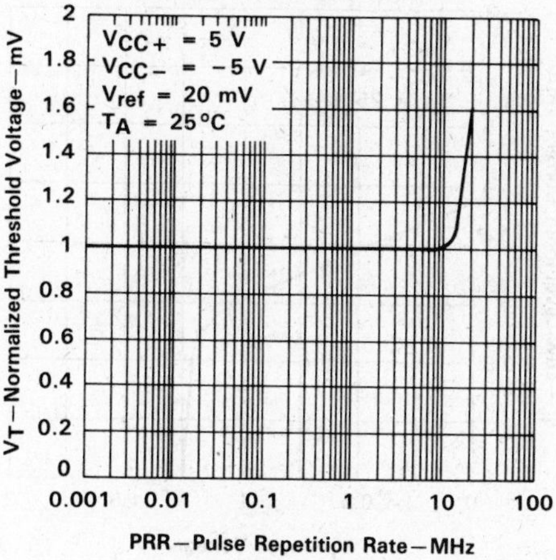
VCC+ = 5 V
VCC− = −5 V
Vref = 20 mV
TA = 25°C
VT—Normalized Threshold Voltage—mV
0 0.2 0.4 0.6 0.8 1 1.2 1.4 1.6 1.8 2
0.001 0.01 0.1 1 10 100
PRR—Pulse Repetition Rate—MHz

FIGURE 6

COMMON-MODE FIRING VOLTAGE
vs
FREE-AIR TEMPERATURE

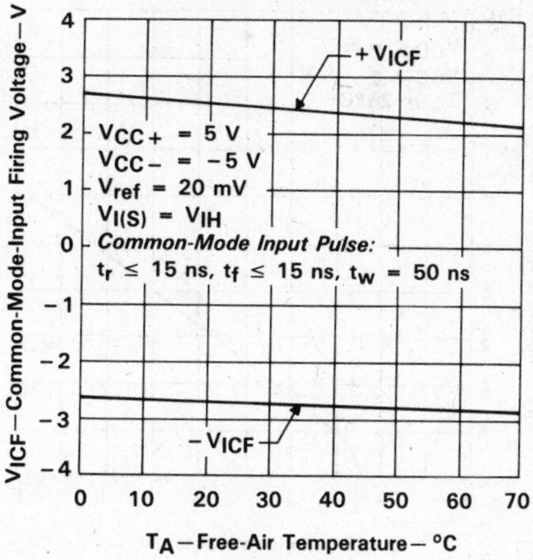
+VICF
−VICF
VCC+ = 5 V
VCC− = −5 V
Vref = 20 mV
VI(S) = VIH
Common-Mode Input Pulse:
tr ≤ 15 ns, tf ≤ 15 ns, tw = 50 ns
VICF—Common-Mode-Input Firing Voltage—V
−4 −3 −2 −1 0 1 2 3 4
0 10 20 30 40 50 60 70
TA—Free-Air Temperature—°C

FIGURE 7

TYPICAL CHARACTERISTICS

DIFFERENTIAL-INPUT BIAS CURRENT
vs
FREE-AIR TEMPERATURE

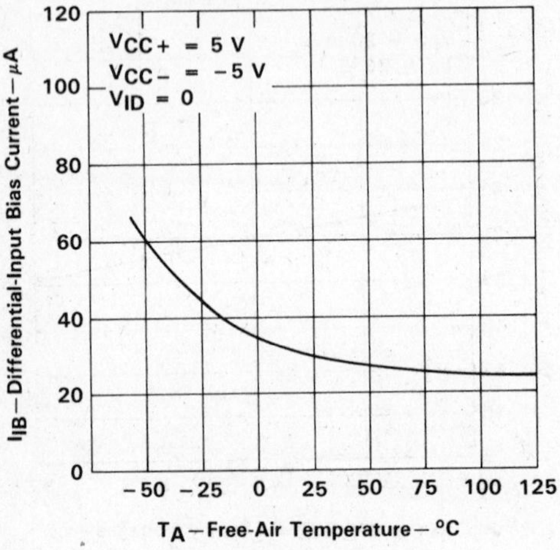
V_{CC+} = 5 V
V_{CC-} = −5 V
V_{ID} = 0
I_{IB}—Differential-Input Bias Current—μA
120
100
80
60
40
20
0
−50 −25 0 25 50 75 100 125
T_A—Free-Air Temperature—°C

FIGURE 8

DIFFERENTIAL-INPUT OFFSET CURRENT
vs
FREE-AIR TEMPERATURE

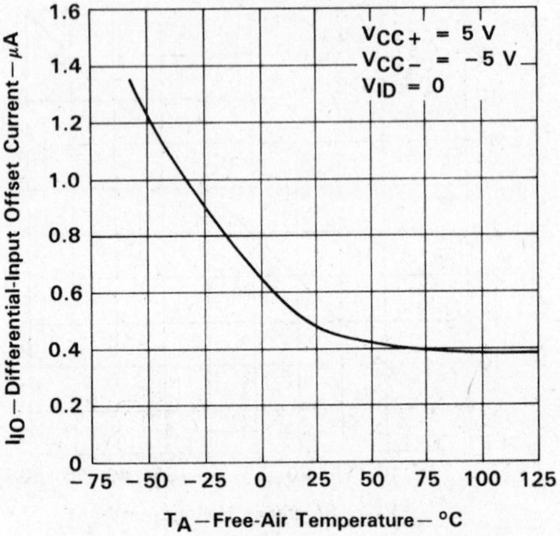
V_{CC+} = 5 V
V_{CC-} = −5 V
V_{ID} = 0
I_{IO}—Differential-Input Offset Current—μA
1.6
1.4
1.2
1.0
0.8
0.6
0.4
0.2
0
−75 −50 −25 0 25 50 75 100 125
T_A—Free-Air Temperature—°C

FIGURE 9

HIGH-LEVEL INPUT CURRENT
vs
INPUT VOLTAGE

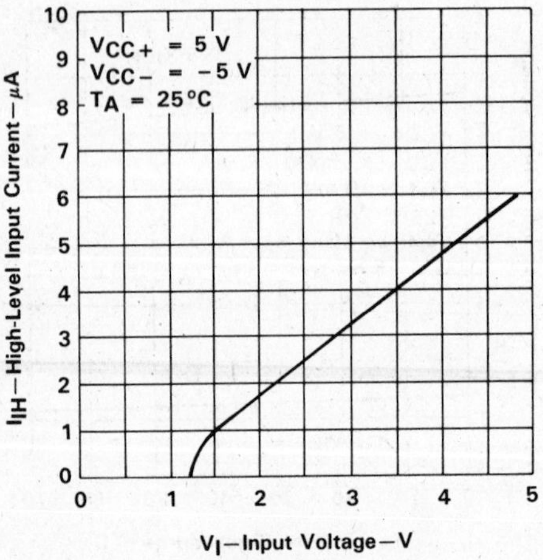
V_{CC+} = 5 V
V_{CC-} = −5 V
T_A = 25°C
I_{IH}—High-Level Input Current—μA
10
9
8
7
6
5
4
3
2
1
0
0 1 2 3 4 5
V_I—Input Voltage—V

FIGURE 10

LOW-LEVEL INPUT CURRENT
vs
INPUT VOLTAGE

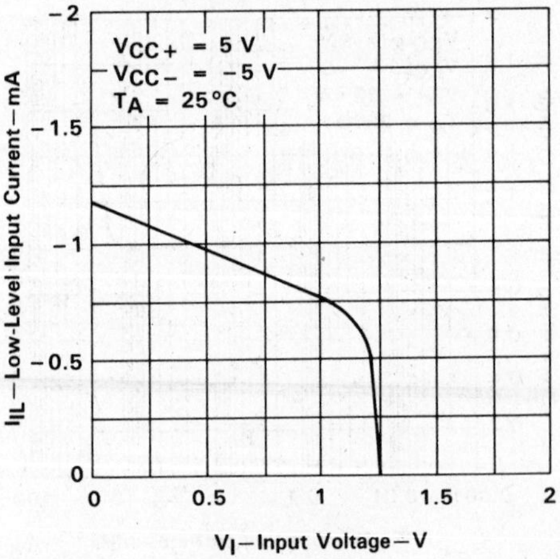
V_{CC+} = 5 V
V_{CC-} = −5 V
T_A = 25°C
I_{IL}—Low-Level Input Current—mA
−2
−1.5
−1
−0.5
0
0 0.5 1 1.5 2
V_I—Input Voltage—V

FIGURE 11

TEXAS INSTRUMENTS
POST OFFICE BOX 655012 • DALLAS, TEXAS 75265

TYPICAL CHARACTERISTICS

Y OUTPUT VOLTAGE
vs
DIFFERENTIAL-INPUT VOLTAGE

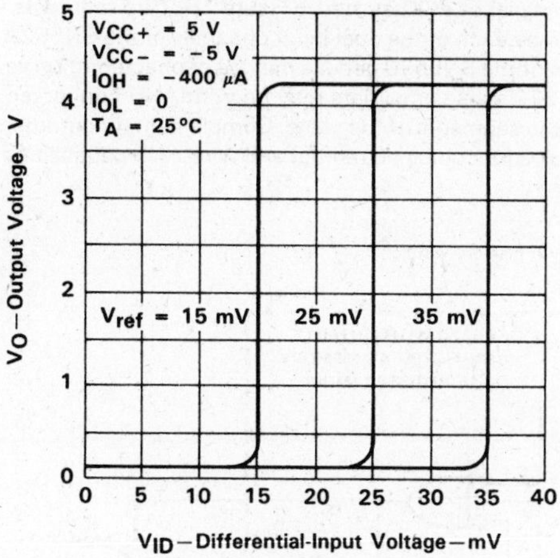
V_{CC+} = 5 V
V_{CC-} = −5 V
I_{OH} = −400 μA
I_{OL} = 0
T_A = 25°C
V_{ref} = 15 mV
25 mV
35 mV
V_O—Output Voltage—V
V_{ID}—Differential-Input Voltage—mV
0
1
2
3
4
5
0
5
10
15
20
25
30
35
40

FIGURE 12

Z OUTPUT VOLTAGE
vs
DIFFERENTIAL-INPUT VOLTAGE

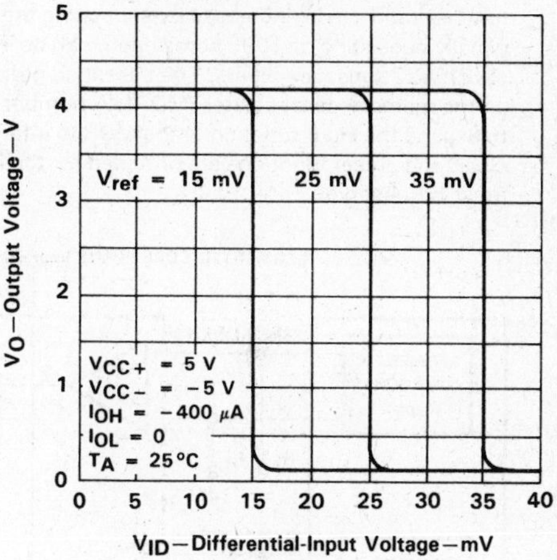
V_{ref} = 15 mV
25 mV
35 mV
V_{CC+} = 5 V
V_{CC-} = −5 V
I_{OH} = −400 μA
I_{OL} = 0
T_A = 25°C
V_O—Output Voltage—V
V_{ID}—Differential-Input Voltage—mV
0
1
2
3
4
5
0
5
10
15
20
25
30
35
40

FIGURE 13

HIGH-LEVEL OUTPUT VOLTAGE
vs
HIGH-LEVEL OUTPUT CURRENT

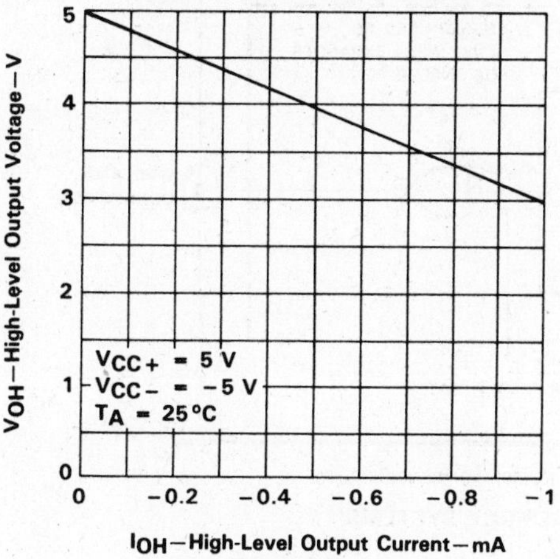
V_{CC+} = 5 V
V_{CC-} = −5 V
T_A = 25°C
V_{OH}—High-Level Output Voltage—V
I_{OH}—High-Level Output Current—mA
0
1
2
3
4
5
0
−0.2
−0.4
−0.6
−0.8
−1

FIGURE 14

LOW-LEVEL OUTPUT VOLTAGE
vs
LOW-LEVEL OUTPUT CURRENT

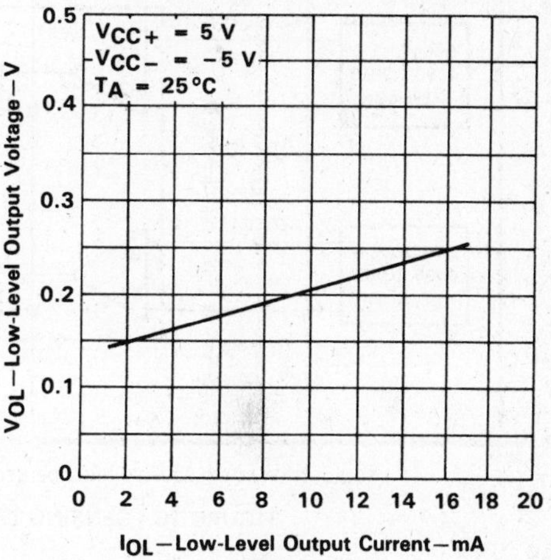
V_{CC+} = 5 V
V_{CC-} = −5 V
T_A = 25°C
V_{OL}—Low-Level Output Voltage—V
I_{OL}—Low-Level Output Current—mA
0
0.1
0.2
0.3
0.4
0.5
0
2
4
6
8
10
12
14
16
18
20

FIGURE 15

TYPICAL APPLICATIONS

large memory systems

This application demonstrates an improved method of sensing data from large memory systems. The signal-to-noise ratio can be increased by sectioning the large core planes as illustrated in Figure 16. Two segments, usually consisting of 4096 cores each, can be interfaced by each of the dual-input channels of the SN5520 or SN5522 sense amplifiers. The cascaded output gates of the SN5520 circuits may be connected to serve as the memory data register (MDR). A number of SN5522 sense amplifiers may be wire-AND connected to expand the input function of the MDR to interface all the segments of the plane. Complementary outputs, clear, and preset functions are provided for the MDR. Rules for combined fan-out and wire-AND capabilities must be observed.

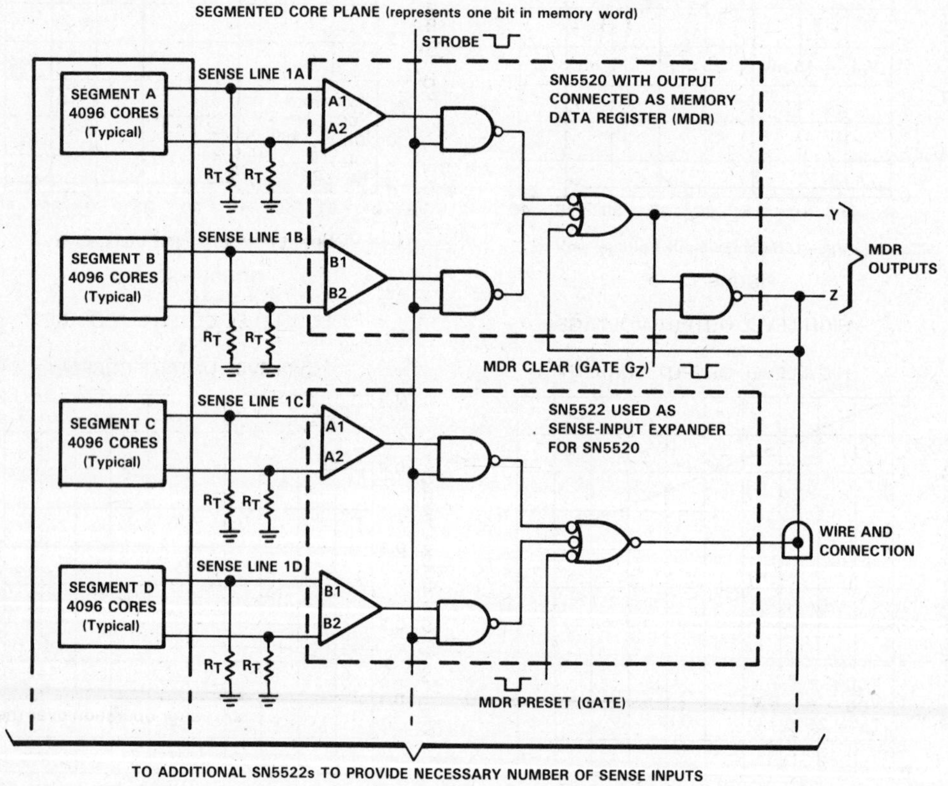
SEGMENTED CORE PLANE (represents one bit in memory word)
STROBE
SENSE LINE 1A
SEGMENT A
4096 CORES
(Typical)
A1
A2
SN5520 WITH OUTPUT
CONNECTED AS MEMORY
DATA REGISTER (MDR)
RT
SENSE LINE 1B
SEGMENT B
4096 CORES
(Typical)
B1
B2
Y
Z
MDR
OUTPUTS
MDR CLEAR (GATE GZ)
SENSE LINE 1C
SEGMENT C
4096 CORES
(Typical)
SN5522 USED AS
SENSE-INPUT EXPANDER
FOR SN5520
WIRE AND
CONNECTION
SENSE LINE 1D
SEGMENT D
4096 CORES
(Typical)
MDR PRESET (GATE)
TO ADDITIONAL SN5522s TO PROVIDE NECESSARY NUMBER OF SENSE INPUTS

FIGURE 16. SENSING LARGE MEMORY SYSTEMS

DUAL-CHANNEL SENSE AMPLIFIER WITH OPEN-COLLECTOR OUTPUTS

D2338, DECEMBER 1986

- **High Speed and Fast Recovery Time**
- **Time and Amplitude Signal Discrimination**
- **Adjustable Input Threshold Voltage Levels**
- **Narrow Window of Threshold Voltage Uncertainty**
- **Multiple Differential-Input Preamplifiers**
- **High DC Noise Margin . . . 1 V Typ**
- **Good Fanout Capability**
- **TTL Drive Capability**
- **Standard Logic Supply Voltages**
- **Plug-in Configuration Ideal for Flow-Soldering Techniques**

J PACKAGE
(TOP VIEW)

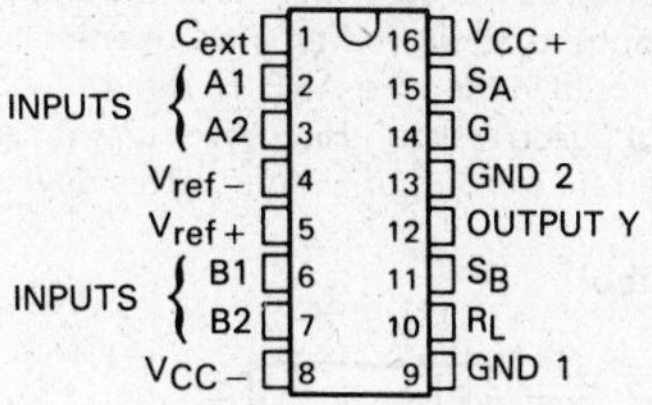
Cext 1
INPUTS A1 2
A2 3
Vref− 4
Vref+ 5
INPUTS B1 6
B2 7
VCC− 8
16 VCC+
15 SA
14 G
13 GND 2
12 OUTPUT Y
11 SB
10 RL
9 GND 1

FUNCTION TABLE

INPUTS					OUTPUT
A	B	G	S_A	S_B	Y
L	L	H	X	X	H
L	X	H	X	L	H
X	L	H	L	X	H
X	X	H	L	L	H
X	X	L	X	X	L
H	X	X	H	X	L
X	H	X	X	H	L

positive logic: $Y = G(\overline{A} + \overline{S}_A)(\overline{B} + \overline{S}_B)$

DEFINITION OF LOGIC LEVELS

INPUT	H	L	X
A or B†	$V_{ID} \geq V_T$ max	$V_{ID} \leq V_T$ min	Irrelevant
Any G or S	$V_I \geq V_{IH}$ min	$V_I \leq V_{IL}$ max	Irrelevant

†A and B are differential voltages (V_{ID}) between A1 and A2 or B1 and B2, respectively. For these circuits, V_{ID} is considered positive regardless of which terminal of each pair is positive with respect to the other.

description

The SN5522 monolithic sense amplifier is designed for use with high-speed memory systems. This sense amplifier detects bipolar differential-input signals from the memory and provides the necessary interface circuitry between the memory and the logic section. Low-level pulses originating in the memory are transformed into logic levels compatible with standard TTL circuits.

The SN5522 sense amplifier features multiple differential-input preamplifiers and versatile gating and output circuits, permitting a significant reduction in the circuitry required to accomplish the sensing function. A unique circuit design provides an inherent stability of the input threshold level over a wide range of supply voltage levels and temperature ranges. Independent strobing of each of the dual sense-input channels ensures maximum versatility and permits detection to occur when the signal-to-noise ratio is at a maximum. The gate and strobe inputs and the outputs are compatible with standard TTL logic circuits.

The circuit features a high-fanout, single-ended, open-collector output stage. In addition, it may be used to expand the inputs to a SN5520 circuit or to perform the wire-AND function.

The SN5522 is available in the J ceramic dual-in-line package and is characterized for operation over the full military temperature range of −55°C to 125°C.

circuit operation

The SN5522 is a dual-channel sense amplifier with the preamplifiers connected to a common output stage. The output circuit features an open-collector output that permits two or more of these outputs to be connected in the wire-AND configuration. Each package includes a load resistor that may be used as the output pullup resistor. High sink-current capability is a feature of this design, and a separate ground terminal is used for the output circuitry. This device can also be used as an input expander for the SN5520 circuit.

logic symbol†

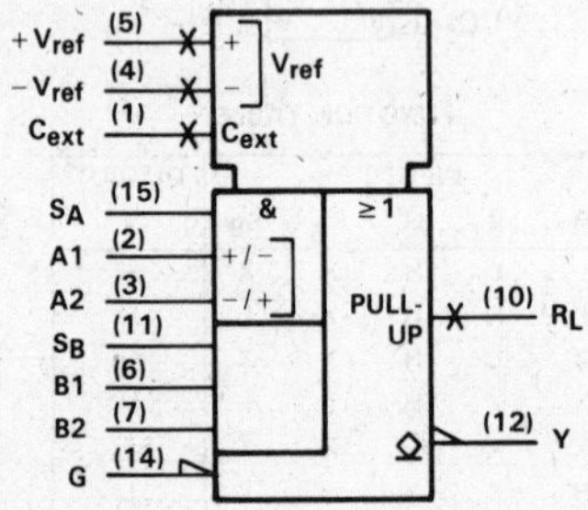
+Vref (5)
−Vref (4)
Cext (1)
Vref
Cext
SA (15)
A1 (2)
A2 (3)
SB (11)
B1 (6)
B2 (7)
G (14)
&
≥1
+/−
−/+
PULL-UP
(10) RL
(12) Y

†This symbol is in accordance with ANSI/IEEE Std 91-1984 and IEC Publication 617-12.

logic diagram

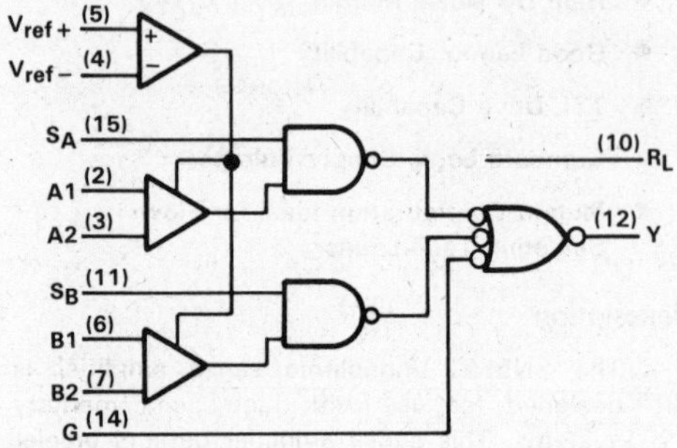
Vref+ (5)
Vref− (4)
SA (15)
A1 (2)
A2 (3)
SB (11)
B1 (6)
B2 (7)
G (14)
(10) RL
(12) Y

schematic

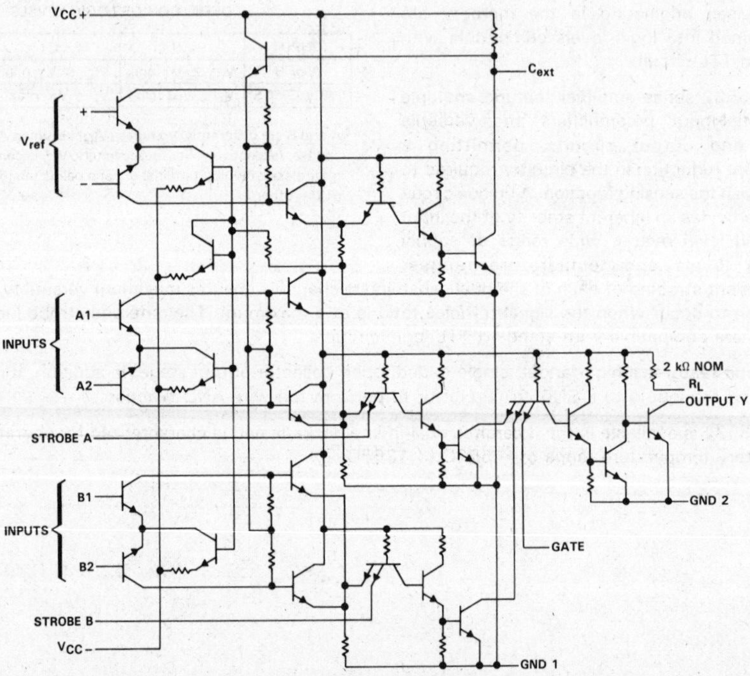
VCC+
Cext
Vref
+
−
A1
INPUTS
A2
STROBE A
B1
INPUTS
B2
STROBE B
VCC−
2 kΩ NOM
RL
OUTPUT Y
GND 2
GATE
GND 1

TEXAS INSTRUMENTS
POST OFFICE BOX 655012 • DALLAS, TEXAS 75265

absolute maximum ratings over operating free-air temperature range (unless otherwise noted)

Supply voltage, V_{CC+} (see Note 1) .... 7 V
Supply voltage, V_{CC-} .... −7 V
Differential input voltage, V_{ID} or V_{ref} .... ±5 V
Voltage from any input to ground (see Note 2) .... 5.5 V
Off-state voltage applied to open-collector outputs .... 5.5 V
Continuous total power dissipation at (or below) 25 °C free-air temperature (see Note 3) .... 1375 mW
Operating free-air temperature range .... −55 °C to 125 °C
Storage temperature range .... −65 °C to 150 °C
Lead temperature 1,6 mm (1/16 inch) from case for 60 seconds .... 300 °C

NOTES: 1. Voltage values, except differential voltages, are with respect to the network ground terminal.
2. Strobe and gate input voltages must be zero or positive with respect to the network ground terminal.
3. For operating above 25 °C free-air temperature, derate total power linearly at the rate of 11.0 mW/°C.

recommended operating conditions

	MIN	NOM	MAX	UNIT
Positive supply voltage, V_{CC+}	4.75	5	5.25	V
Negative supply voltage, V_{CC-}	−4.75	−5	−5.25	V
Reference voltage, V_{ref}	15		40	mV
High-level input voltage (strobe and gate inputs), V_{IH}	2			V
Low-level input voltage (strobe and gate inputs), V_{IL}			0.8	V

electrical characteristics over recommended operating free-air temperature range, $V_{CC\pm} = \pm 5$ V, R_L connected to output, (unless otherwise noted)

PARAMETER		TEST CONDITIONS	MIN	TYP†	MAX	UNIT
V_T	Differential-input threshold voltage	$V_{ref} = 15$ mV	10	15	20	mV
		$V_{ref} = 40$ mV	35	40	45	
V_{ICF}	Common-mode input firing voltage (see Note 4)	$V_{ref} = 40$ mV, $V_{I(S)} = V_{IH}$; Common-mode input pulse: $t_r \le 15$ ns, $t_f \le 15$ ns, $t_w = 50$ ns		±2.5		V
I_{IB}	Differential-input bias current	$V_{CC\pm} = \pm 5.25$ V, $V_{ID} = 0$, $T_A = -55$°C			100	µA
		$V_{CC\pm} = \pm 5.25$ V, $V_{ID} = 0$, $T_A = 25$°C		30	75	
		$V_{CC\pm} = \pm 5.25$ V, $V_{ID} = 0$, $T_A = 125$°C			75	
I_{IO}	Differential-input offset current	$V_{CC\pm} = \pm 5.25$ V, $V_{ID} = 0$		0.5		µA
V_{OH}	High-level output voltage	$V_{CC\pm} = \pm 4.75$ V, $I_{OH} = -400$ µA	2.4	4		V
V_{OL}	Low-level output voltage	$V_{CC\pm} = \pm 4.75$ V, $I_{OL} = 16$ mA		0.25	0.4	V
I_{IH}	High-level input current (strobe and gate inputs)	$V_{CC\pm} = \pm 5.25$ V, $V_{IH} = 2.4$ V			40	µA
		$V_{CC\pm} = \pm 5.25$ V, $V_{IH} = 5.25$ V			1	mA
I_{IL}	Low-level input current (strobe and gate inputs)	$V_{CC\pm} = \pm 5.25$ V, $V_{IL} = 0.4$ V		−1	−1.6	mA
I_{OH}	High-level output current	$V_{CC\pm} = \pm 4.75$ V, $V_O = 5.25$ V			250	µA
I_{OS}	Short-circuit output current	$V_{CC\pm} = \pm 5.25$ V, $T_A = 25$°C	−2.1	−2.6	−3.5	mA
I_{CC+}	Supply current from V_{CC+}	$V_{CC\pm} = \pm 5.25$ V, $T_A = 25$°C		27	40	mA
I_{CC-}	Supply current from V_{CC-}	$V_{CC\pm} = \pm 5.25$ V, $T_A = 25$°C		−15	−20	mA

†Typical values are at $V_{CC\pm} = \pm 5$ V, $T_A = 25$°C.
NOTE 4: Common-mode input firing voltage is the minimum common-mode voltage that will exceed the dynamic range of the input at the specified conditions and cause the logic output to switch. The common-mode input signal is applied when the strobe is high.

switching characteristics, $V_{CC\pm} = \pm 5$ V, $T_A = 25°C$

PARAMETER	FROM (INPUT)	TO (OUTPUT)	TEST FIGURE	TEST CONDITIONS	MIN	TYP	MAX	UNIT
$t_{PLH(D)}$	A1-A2 or B1-B2	Y	1	$C_L = 15$ pF, $C_{ext} \geq 100$ pF, $R_L = 288\ \Omega$		20		ns
$t_{PHL(D)}$						30	45	
$t_{PLH(S)}$	STROBE A or B	Y	1			20		ns
$t_{PHL(S)}$						20	40	
$t_{PLH(G)}$	GATE	Y	2			10		ns
$t_{PHL(G)}$						15	25	

recovery and cycle times, $V_{CC\pm} = \pm 5$ V, $C_{ext} \geq 100$ pF, $T_A = 25°C$

PARAMETER		TEST CONDITIONS	MIN	TYP	MAX	UNIT
t_{orD}	Differential-input overload recovery time (see Note 5)	Differential Input Pulse: $V_{ID} = 2$ V, $t_r = t_f = 20$ ns		20		ns
t_{orC}	Common-mode input overload recovery time (see Note 6)	Common-mode Input Pulse: $V_{IC} = \pm 2$ V, $t_r = t_f = 20$ ns		20		ns
$t_{cyc(min)}$	Minimum cycle time			200		ns

NOTES: 5. Differential-input overload recovery time is the time necessary for the device to recover from the specified differential-input overload signal prior to the strobe-enable signal.
6. Common-mode-input overload recovery time is the time necessary for the device to recover from the specified common-mode-input overload signal prior to the strobe-enable signal.

driving SN54XXX loads and combining outputs

The table below provides minimum and maximum resistor values for driving one to ten SN54XXX loads and wire-AND connecting two to seven parallel outputs. Each value shown for one wire-AND output is determined by the fanout plus the cutoff current of a single output transistor. Extension beyond seven wire-AND connections is permitted with fanouts of seven or less if a valid minimum and maximum R_L is possible. When fanning out to ten SN54XXX loads, the calculation for the minimum value of R_L indicates that an infinite resistance should be used; however, the use of a 4-kΩ resistor in this case will satisfy the high-level condition and limit the low level to less than 0.43 volt.

FANOUT TO TTL LOADS	WIRE-AND OUTPUTS							
	1	2	3	4	5	6	7	1 to 7
1	8965	4814	3291	2500	2015	1688	1452	319
2	7878	4482	3132	2407	1954	1645	1420	359
3	7027	4193	2988	2321	1897	1604	1390	410
4	6341	3939	2857	2241	1843	1566	1361	479
5	5777	3714	2736	2166	1793	1529	1333	575
6	5306	3513	2626	2096	1744	1494	1306	718
7	4905	3333	2524	2031	1699	1460	1280	958
8	4561	3170	2429	1969	1656	X	X	1437
9	4262	3023	X	X	X	X	X	2875
10	4000	X	X	X	X	X	X	4000
	MAXIMUM							MIN
	LOAD RESISTOR VALUE IN OHMS							

‡All values shown in the table are based on:
High-level conditions: $V_{CC} = 5$ V, V_{OH} min = 2.4 V
Low-level conditions: $V_{CC} = 5$ V, V_{OL} max = 0.4 V
X — Not recommended or not possible.
See explanation in text.

TEXAS INSTRUMENTS
POST OFFICE BOX 655012 • DALLAS, TEXAS 75265

PARAMETER MEASUREMENT INFORMATION

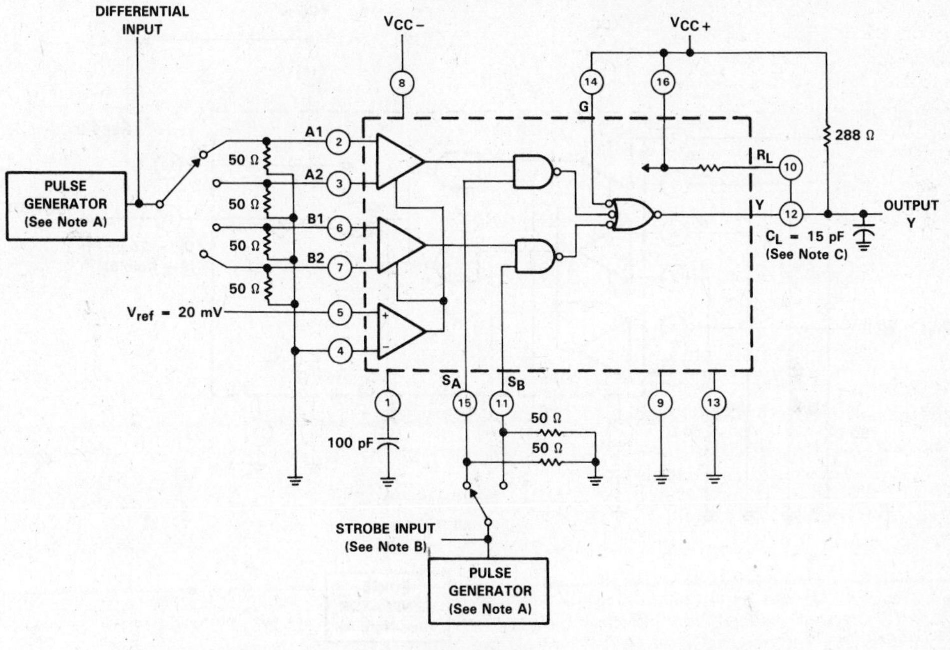
DIFFERENTIAL INPUT
VCC−
VCC+
PULSE GENERATOR (See Note A)
50 Ω
A1
A2
B1
B2
Vref = 20 mV
G
288 Ω
RL
Y
OUTPUT Y
CL = 15 pF (See Note C)
SA
SB
100 pF
STROBE INPUT (See Note B)
PULSE GENERATOR (See Note A)

TEST CIRCUIT

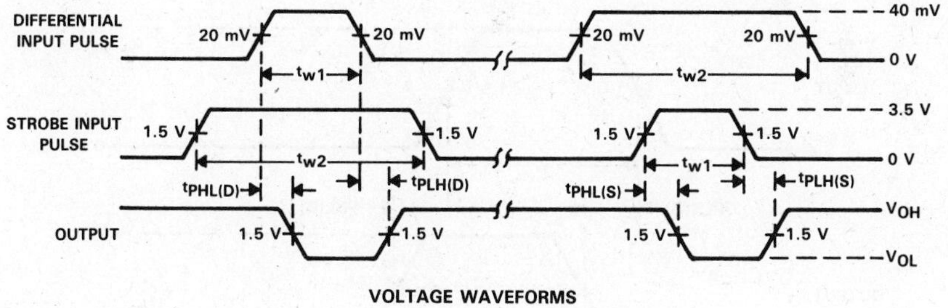
DIFFERENTIAL INPUT PULSE
20 mV
40 mV
0 V
tw1
tw2
STROBE INPUT PULSE
1.5 V
3.5 V
tPHL(D)
tPLH(D)
tPHL(S)
tPLH(S)
OUTPUT
VOH
VOL

VOLTAGE WAVEFORMS

NOTES: A. The pulse generators have the following characteristics: Z_{out} = 50 Ω, $t_r = t_f$ = 15 ±5 ns, t_{w1} = 100 ns, t_{w2} = 300 ns, and PRR ≤ 1 MHz.

B. The strobe input pulse is applied to Strobe S_A when testing inputs A1-A2 and to Strobe S_B when testing inputs B1-B2.

C. C_L includes probe and jig capacitance.

FIGURE 1. PROPAGATION DELAY TIMES FROM DIFFERENTIAL AND STROBE INPUTS

PARAMETER MEASUREMENT INFORMATION

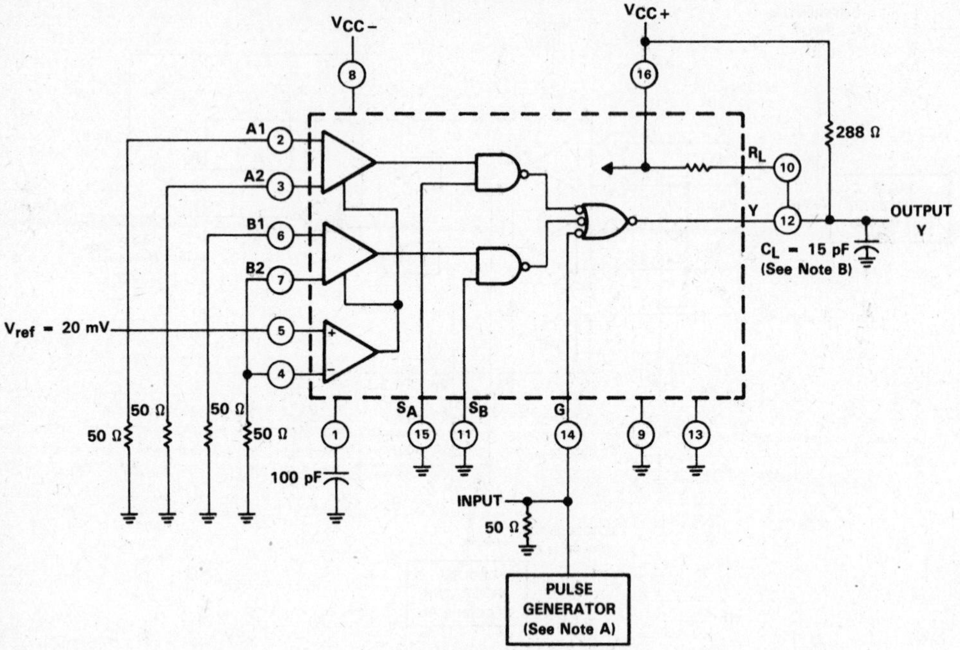
VCC−
VCC+
8
16
A1
A2
B1
B2
2
3
6
7
5
4
288 Ω
RL
10
Y
12
OUTPUT
Y
CL = 15 pF
(See Note B)
Vref = 20 mV
50 Ω
50 Ω
50 Ω
50 Ω
1
100 pF
SA
SB
G
15
11
14
9
13
INPUT
50 Ω
PULSE
GENERATOR
(See Note A)

TEST CIRCUIT

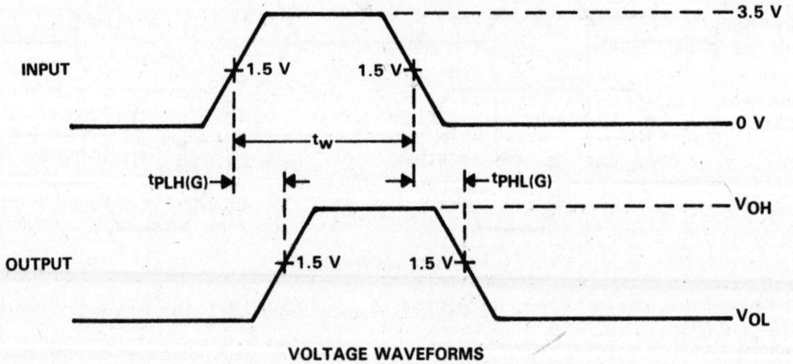
3.5 V
INPUT
1.5 V
1.5 V
0 V
tw
tPLH(G)
tPHL(G)
VOH
OUTPUT
1.5 V
1.5 V
VOL

VOLTAGE WAVEFORMS

NOTES: A. The pulse generator has the following characteristics: Z_{out} = 50 Ω, t_r = 15 ±5 ns, t_f = 15 ±5 ns, t_w = 100 ns, and PRR ≤ 1 MHz.
B. C_L includes probe and jig capacitance.

FIGURE 2. PROPAGATION DELAY TIMES FROM GATE INPUT

TEXAS INSTRUMENTS
POST OFFICE BOX 655012 • DALLAS, TEXAS 75265

TYPICAL CHARACTERISTICS

THRESHOLD VOLTAGE
vs
REFERENCE VOLTAGE

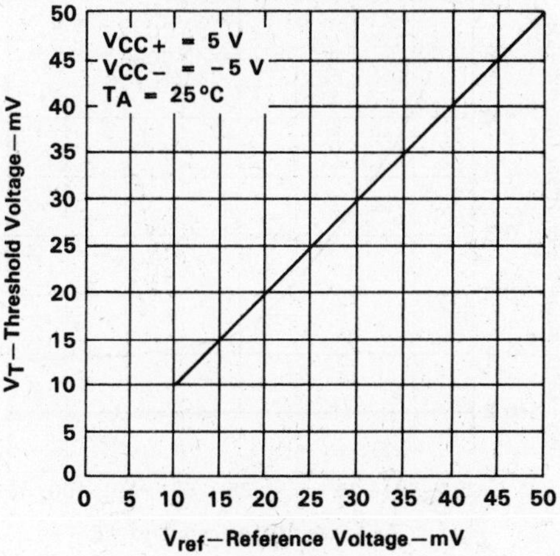
V_{CC+} = 5 V
V_{CC-} = −5 V
T_A = 25°C
V_T—Threshold Voltage—mV
V_{ref}—Reference Voltage—mV

FIGURE 3

THRESHOLD VOLTAGE
vs
SUPPLY VOLTAGE

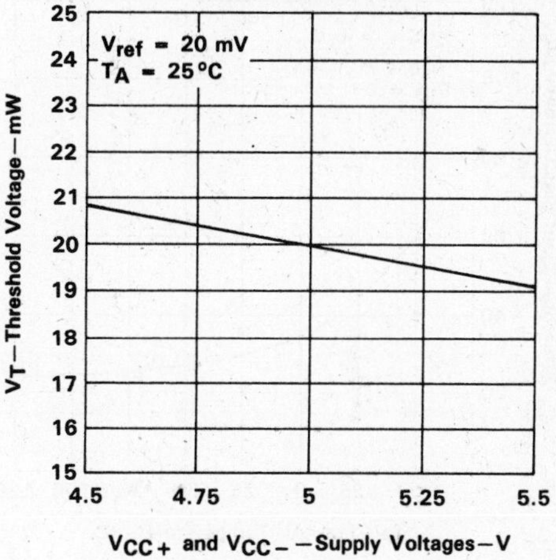
V_{ref} = 20 mV
T_A = 25°C
V_T—Threshold Voltage—mW
V_{CC+} and V_{CC-}—Supply Voltages—V

FIGURE 4

NORMALIZED THRESHOLD VOLTAGE
vs
PULSE REPETITION RATE

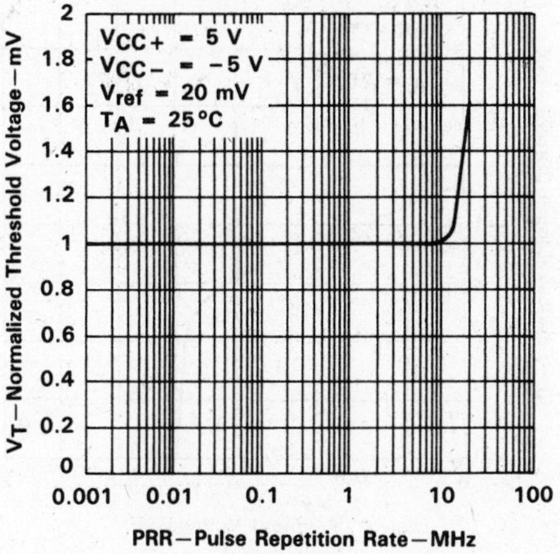
V_{CC+} = 5 V
V_{CC-} = −5 V
V_{ref} = 20 mV
T_A = 25°C
V_T—Normalized Threshold Voltage—mV
PRR—Pulse Repetition Rate—MHz

FIGURE 5

COMMON-MODE FIRING VOLTAGE
vs
FREE-AIR TEMPERATURE

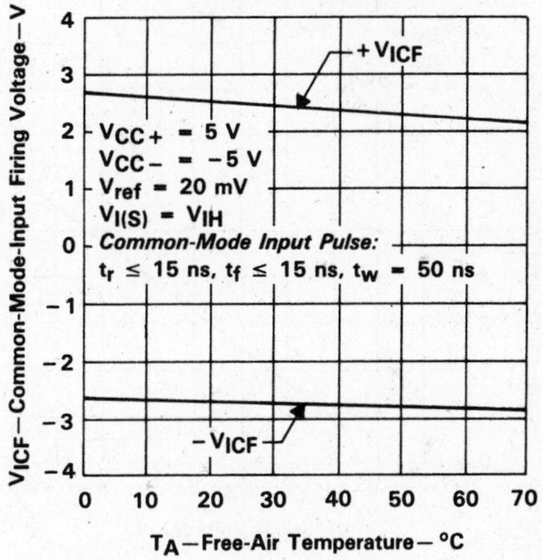
$+V_{ICF}$
V_{CC+} = 5 V
V_{CC-} = −5 V
V_{ref} = 20 mV
$V_{I(S)}$ = V_{IH}
Common-Mode Input Pulse:
$t_r \leq 15$ ns, $t_f \leq 15$ ns, $t_w = 50$ ns
$-V_{ICF}$
V_{ICF}—Common-Mode-Input Firing Voltage—V
T_A—Free-Air Temperature—°C

FIGURE 6

TYPICAL CHARACTERISTICS

DIFFERENTIAL-INPUT BIAS CURRENT
vs
FREE-AIR TEMPERATURE

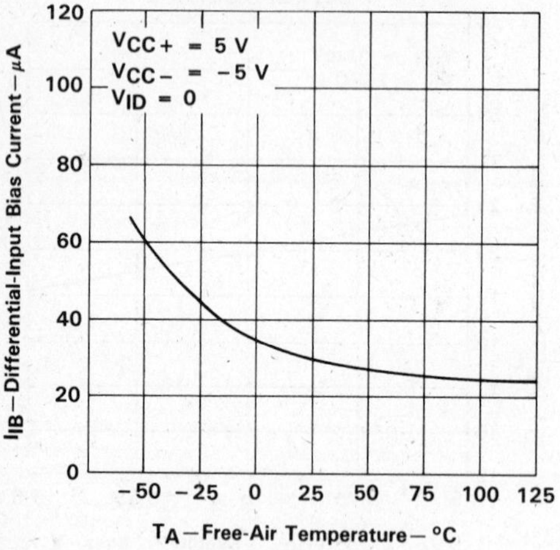
VCC+ = 5 V
VCC− = −5 V
VID = 0
IB—Differential-Input Bias Current—μA
0
20
40
60
80
100
120
−50
−25
0
25
50
75
100
125
TA—Free-Air Temperature—°C

FIGURE 7

DIFFERENTIAL-INPUT OFFSET CURRENT
vs
FREE-AIR TEMPERATURE

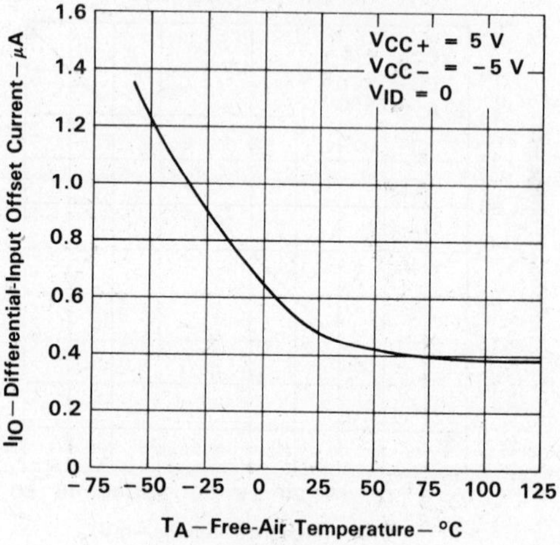
VCC+ = 5 V
VCC− = −5 V
VID = 0
IIO—Differential-Input Offset Current—μA
0
0.2
0.4
0.6
0.8
1.0
1.2
1.4
1.6
−75
−50
−25
0
25
50
75
100
125
TA—Free-Air Temperature—°C

FIGURE 8

HIGH-LEVEL INPUT CURRENT
vs
INPUT VOLTAGE

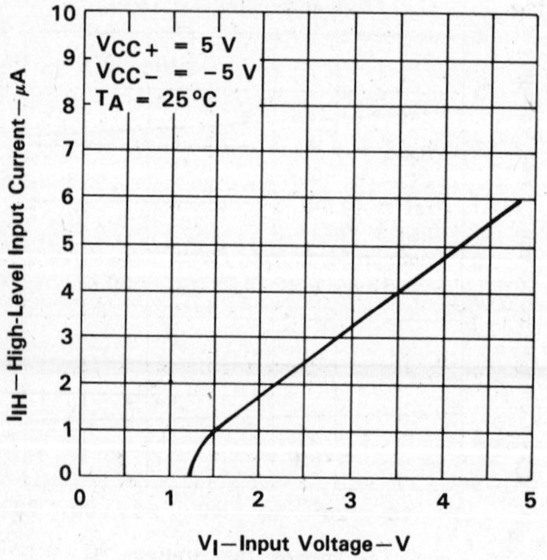
VCC+ = 5 V
VCC− = −5 V
TA = 25°C
IIH—High-Level Input Current—μA
0
1
2
3
4
5
6
7
8
9
10
0
1
2
3
4
5
VI—Input Voltage—V

FIGURE 9

LOW-LEVEL INPUT CURRENT
vs
INPUT VOLTAGE

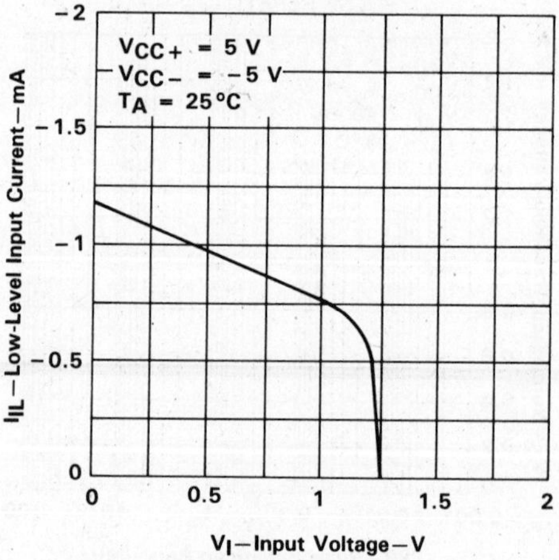
VCC+ = 5 V
VCC− = −5 V
TA = 25°C
IIL—Low-Level Input Current—mA
0
−0.5
−1
−1.5
−2
0
0.5
1
1.5
2
VI—Input Voltage—V

FIGURE 10

TYPICAL CHARACTERISTICS

OUTPUT VOLTAGE
vs
DIFFERENTIAL-INPUT VOLTAGE

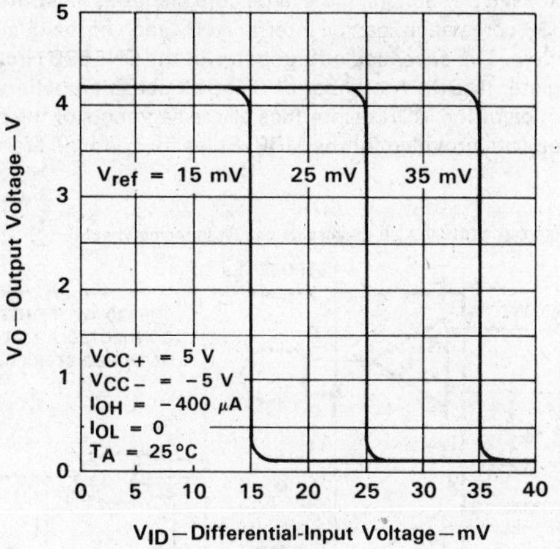
5
4
3
2
1
0
V_O—Output Voltage—V
V_{ref} = 15 mV
25 mV
35 mV
V_{CC+} = 5 V
V_{CC-} = −5 V
I_{OH} = −400 μA
I_{OL} = 0
T_A = 25°C
0 5 10 15 20 25 30 35 40
V_{ID}—Differential-Input Voltage—mV

FIGURE 11

HIGH-LEVEL OUTPUT VOLTAGE
vs
HIGH-LEVEL OUTPUT CURRENT

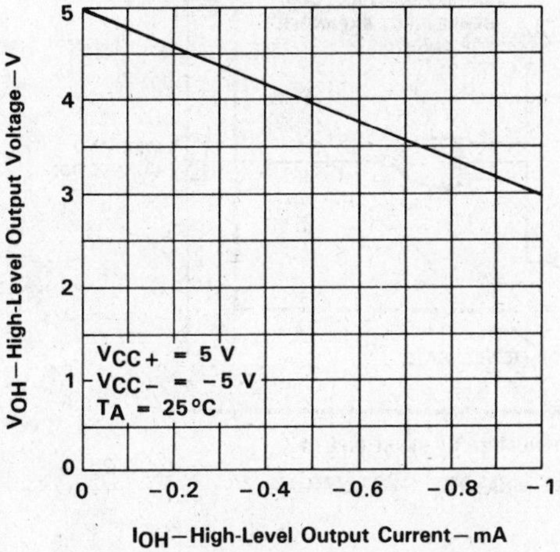
5
4
3
2
1
0
V_{OH}—High-Level Output Voltage—V
V_{CC+} = 5 V
V_{CC-} = −5 V
T_A = 25°C
0 −0.2 −0.4 −0.6 −0.8 −1
I_{OH}—High-Level Output Current—mA

FIGURE 12

LOW-LEVEL OUTPUT VOLTAGE
vs
LOW-LEVEL OUTPUT CURRENT

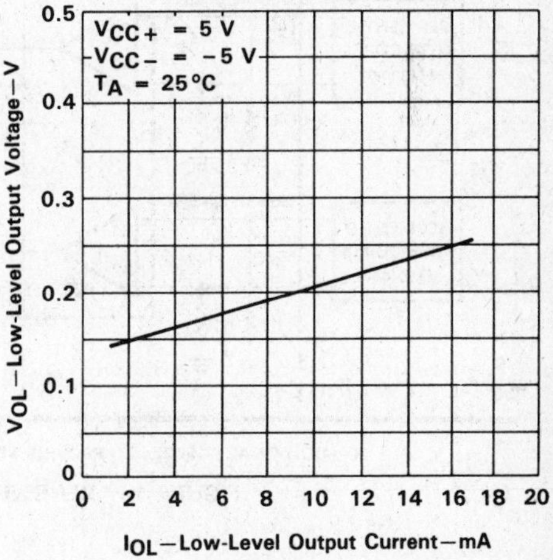
0.5
0.4
0.3
0.2
0.1
0
V_{OL}—Low-Level Output Voltage—V
V_{CC+} = 5 V
V_{CC-} = −5 V
T_A = 25°C
0 2 4 6 8 10 12 14 16 18 20
I_{OL}—Low-Level Output Current—mA

FIGURE 13

APPLICATION DATA

large memory systems

This application demonstrates an improved method of sensing data from large memory systems. The signal-to-noise ratio can be increased by sectioning the large core planes as illustrated in Figure 14. Two segments, usually consisting of 4096 cores each, can be interfaced by each of the dual-input channels of the SN5520 or SN5522 sense amplifiers. The cascaded output gates of the SN5520 circuits may be connected to serve as the memory data register (MDR). A number of SN5522 sense amplifiers may be wire-AND connected to expand the input function of the MDR to interface all the segments of the plane. Complementary outputs, clear, and preset functions are provided for the MDR. Rules for combined fan-out and wire-AND capabilities must be observed.

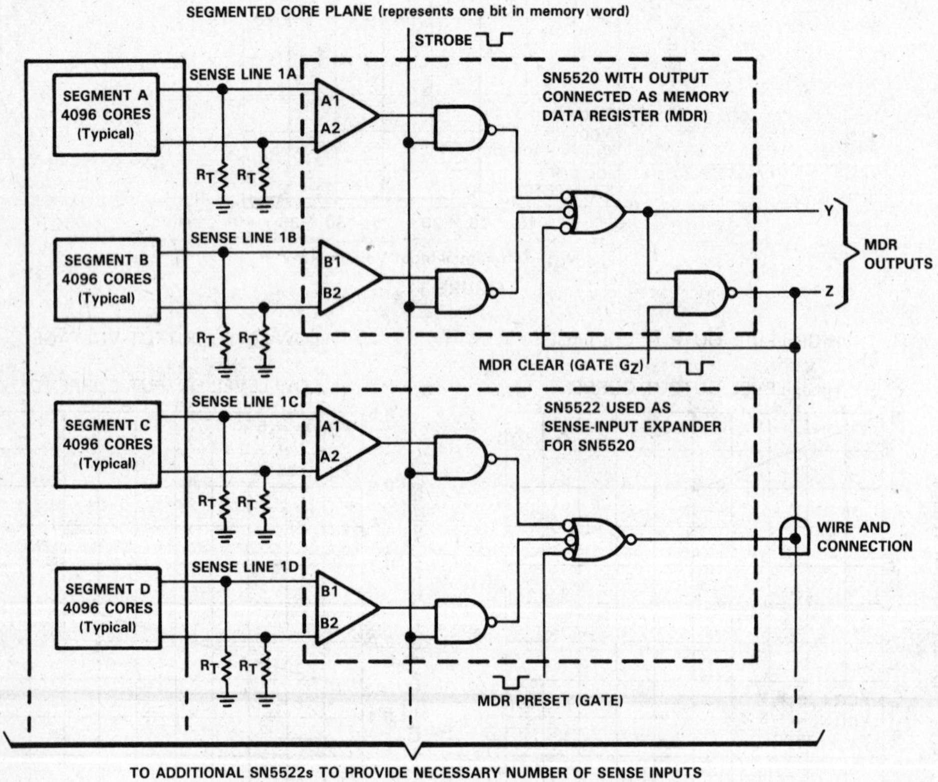
SEGMENTED CORE PLANE (represents one bit in memory word)
STROBE
SENSE LINE 1A
SEGMENT A
4096 CORES
(Typical)
A1
A2
SN5520 WITH OUTPUT
CONNECTED AS MEMORY
DATA REGISTER (MDR)
RT
RT
Y
MDR
OUTPUTS
Z
SENSE LINE 1B
SEGMENT B
4096 CORES
(Typical)
B1
B2
MDR CLEAR (GATE GZ)
SENSE LINE 1C
SEGMENT C
4096 CORES
(Typical)
SN5522 USED AS
SENSE-INPUT EXPANDER
FOR SN5520
WIRE AND
CONNECTION
SENSE LINE 1D
SEGMENT D
4096 CORES
(Typical)
MDR PRESET (GATE)
TO ADDITIONAL SN5522s TO PROVIDE NECESSARY NUMBER OF SENSE INPUTS

FIGURE 14. SENSING LARGE MEMORY SYSTEMS

6
Memory Interface Circuits

- **High Speed and Fast Recovery Time**
- **Time and Amplitude Signal Discrimination**
- **Adjustable Input Threshold Voltage Levels**
- **Narrow Window of Threshold Voltage Uncertainty**
- **Multiple Differential-Input Preamplifiers**
- **High DC Noise Margin . . . 1 V Typ**
- **Good Fanout Capability**
- **TTL Drive Capability**
- **Standard Logic Supply Voltages**
- **Plug-in Configuration Ideal for Flow-Soldering Techniques**

J PACKAGE
(TOP VIEW)

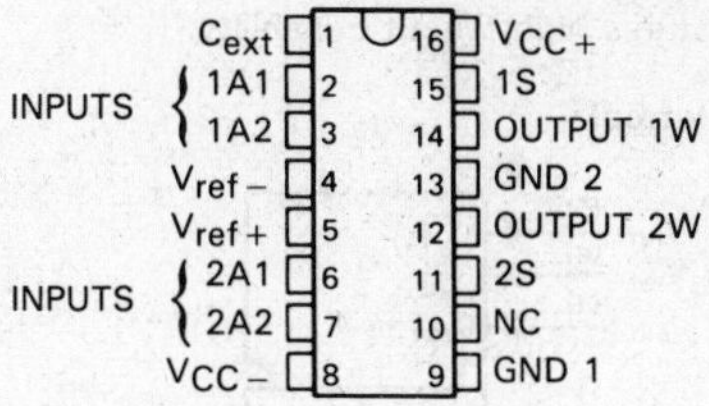
Cext 1
INPUTS 1A1 2
1A2 3
Vref− 4
Vref+ 5
INPUTS 2A1 6
2A2 7
VCC− 8
16 VCC+
15 1S
14 OUTPUT 1W
13 GND 2
12 OUTPUT 2W
11 2S
10 NC
9 GND 1

NC—No internal connection

FUNCTION TABLE

INPUTS A	INPUTS S	OUTPUT W
H	H	H
L	X	L
X	L	L

positive logic: W = AS

DEFINITION OF LOGIC LEVELS

INPUT	H	L	X
A†	$V_{ID} \geq V_T$ max	$V_{ID} \leq V_T$ min	Irrelevant
S	$V_I \geq V_{IH}$ min	$V_I \leq V_{IL}$ max	Irrelevant

†A is a differential voltage (V_{ID}) between A1 and A2. For these circuits, V_{ID} is considered positive regardless of which terminal is positive with respect to the other.

description

The SN5524 monolithic sense amplifier is designed for use with high-speed memory systems. This sense amplifier detects bipolar differential-input signals from the memory and provides the necessary interface circuitry between the memory and the logic section. Low-level pulses originating in the memory are transformed into logic levels compatible with standard TTL circuits.

The SN5524 sense amplifier features multiple differential-input preamplifiers and versatile gating and output circuits, permitting a significant reduction in the circuitry required to accomplish the sensing function. A unique circuit design provides an inherent stability of the input threshold level over a wide range of supply voltage levels and temperature ranges. Independent strobing of each of the dual sense-input channels ensures maximum versatility and permits detection to occur when the signal-to-noise ratio is at a maximum. The strobe inputs and the outputs are compatible with standard TTL logic circuits.

The circuit provides for independent, dual-channel sensing with separate outputs.

The SN5524 is available in the J ceramic dual-in-line package and is characterized for operation over the full military temperature range of −55°C to 125°C.

circuit operation

The SN5524 circuit features two completely independent sense amplifiers in a single package. Each amplifier features high fanout capability.

logic symbol†

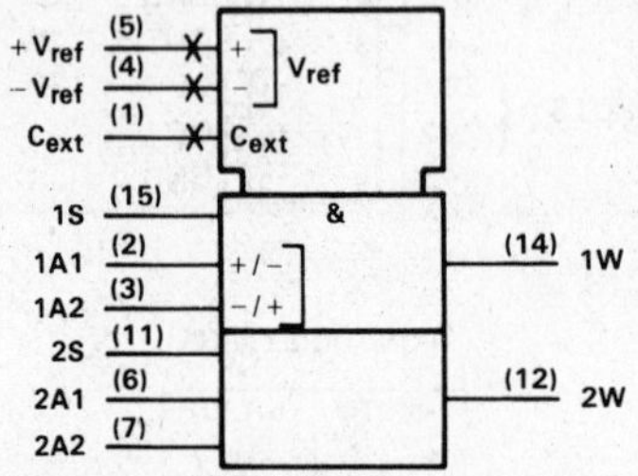
+Vref (5)
−Vref (4)
Cext (1)
1S (15)
1A1 (2)
1A2 (3)
2S (11)
2A1 (6)
2A2 (7)
Vref
Cext
&
+/−
−/+
(14) 1W
(12) 2W

†This symbol is in accordance with ANSI/IEEE Std 91-1984 and IEC Publication 617-12.

logic diagram

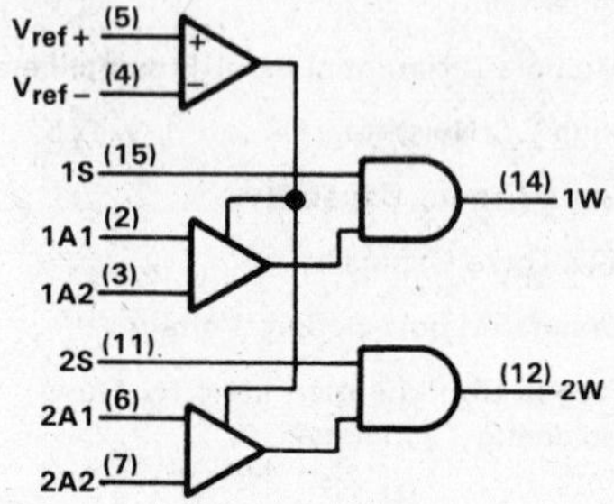
Vref+ (5)
Vref− (4)
1S (15)
1A1 (2)
1A2 (3)
2S (11)
2A1 (6)
2A2 (7)
(14) 1W
(12) 2W

schematic

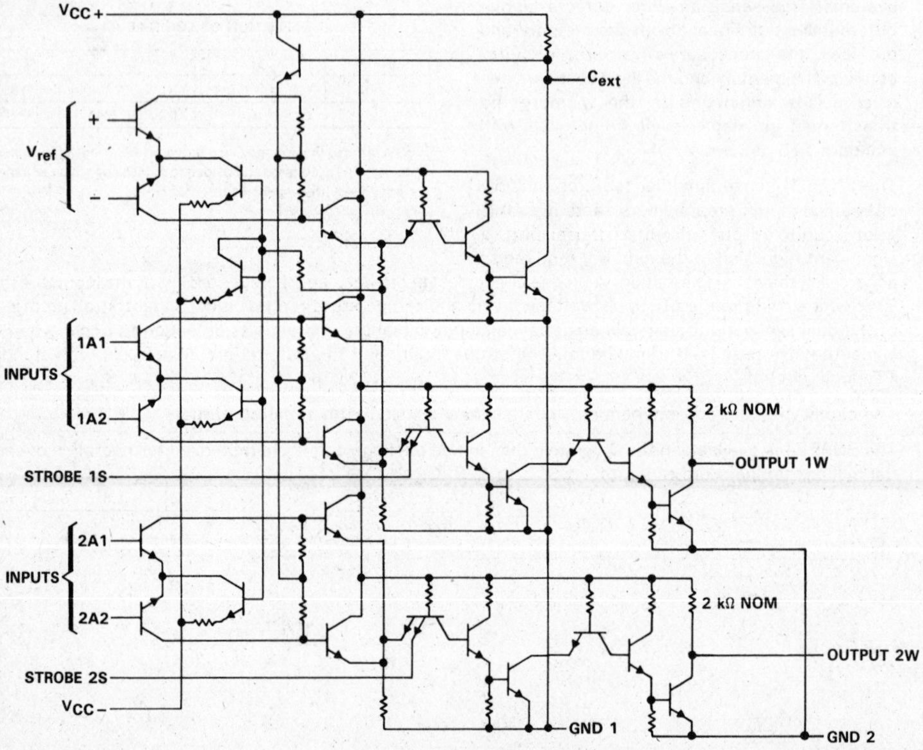
VCC+
Cext
Vref
+
−
1A1
INPUTS
1A2
STROBE 1S
2A1
INPUTS
2A2
STROBE 2S
VCC−
2 kΩ NOM
OUTPUT 1W
2 kΩ NOM
OUTPUT 2W
GND 1
GND 2

TEXAS INSTRUMENTS
POST OFFICE BOX 655012 • DALLAS, TEXAS 75265

absolute maximum ratings over operating free-air temperature range (unless otherwise noted)

Supply voltage, V_{CC+} (see Note 1) .... 7 V
Supply voltage, V_{CC-} .... −7 V
Differential input voltage, V_{ID} or V_{ref} .... ±5 V
Voltage from any input to ground (see Note 2) .... 5.5 V
Continuous total power dissipation at (or below) 25°C free-air temperature (see Note 3) .... 1375 mW
Operating free-air temperature range .... −55°C to 125°C
Storage temperature range .... −65°C to 150°C
Lead temperature 1,6 mm (1/16 inch) from case for 60 seconds .... 300°C

NOTES: 1. Voltage values, except differential voltages, are with respect to the network ground terminal.
2. Strobe input voltages must be zero or positive with respect to the network ground terminal.
3. For operating above 25°C free-air temperature, derate total power linearly at the rate of 11.0 mW/°C.

recommended operating conditions

	MIN	NOM	MAX	UNIT
Positive supply voltage, V_{CC+}	4.75	5	5.25	V
Negative supply voltage, V_{CC-}	−4.75	−5	−5.25	V
Reference voltage, V_{ref}	15		40	mV
High-level input voltage (strobe inputs), V_{IH}	2			V
Low-level input voltage (strobe inputs), V_{IL}			0.8	V

electrical characteristics over recommended operating free-air temperature range, $V_{CC\pm}$ = ±5 V, (unless otherwise noted)

PARAMETER		TEST CONDITIONS	MIN	TYP†	MAX	UNIT
V_T	Differential-input threshold voltage	V_{ref} = 15 mV	10	15	20	mV
		V_{ref} = 40 mV	35	40	45	
V_{ICF}	Common-mode input firing voltage (see Note 4)	V_{ref} = 40 mV, $V_{I(S)} = V_{IH}$; Common-mode input pulse: $t_r \le 15$ ns, $t_f \le 15$ ns, t_w = 50 ns		±2.5		V
I_{IB}	Differential-input bias current	$V_{CC\pm}$ = ±5.25 V, V_{ID} = 0, T_A = −55°C			100	μA
		$V_{CC\pm}$ = ±5.25 V, V_{ID} = 0, T_A = 25°C		30	75	
		$V_{CC\pm}$ = ±5.25 V, V_{ID} = 0, T_A = 125°C			75	
I_{IO}	Differential-input offset current	$V_{CC\pm}$ = ±5.25 V, V_{ID} = 0		0.5		μA
V_{OH}	High-level output voltage	$V_{CC\pm}$ = ±4.75 V, I_{OH} = −400 μA	2.4	4		V
V_{OL}	Low-level output voltage	$V_{CC\pm}$ = ±4.75 V, I_{OL} = 16 mA		0.25	0.4	V
I_{IH}	High-level input current (strobe inputs)	$V_{CC\pm}$ = ±5.25 V, V_{IH} = 2.4 V			40	μA
		$V_{CC\pm}$ = ±5.25 V, V_{IH} = 5.25 V			1	mA
I_{IL}	Low-level input current (strobe inputs)	$V_{CC\pm}$ = ±5.25 V, V_{IL} = 0.4 V		−1	−1.6	mA
I_{OS}	Short-circuit output current	$V_{CC\pm}$ = ±5.25 V, T_A = 25°C	−2.1	−2.6	−3.5	mA
I_{CC+}	Supply current from V_{CC+}	$V_{CC\pm}$ = ±5.25 V, T_A = 25°C		25	40	mA
I_{CC-}	Supply current from V_{CC-}	$V_{CC\pm}$ = ±5.25 V, T_A = 25°C		−15	−20	mA

†Typical values are at $V_{CC\pm}$ = ±5 V, T_A = 25°C.
NOTE 4: Common-mode input firing voltage is the minimum common-mode voltage that will exceed the dynamic range of the input at the specified conditions and cause the logic output to switch. The common-mode input signal is applied when the strobe is high.

switching characteristics, $V_{CC\pm}$ = ±5 V, T_A = 25°C

PARAMETER	FROM (INPUT)	TO (OUTPUT)	TEST FIGURE	TEST CONDITIONS	MIN	TYP	MAX	UNIT
$t_{PLH(D)}$	A1-A2	W	1	C_L = 15 pF, C_{ext} ≥ 100 pF, R_L = 288 Ω		25	40	ns
$t_{PHL(D)}$						20		
$t_{PLH(S)}$	STROBE	W	1			15	30	ns
$t_{PHL(S)}$						20		

recovery and cycle times, $V_{CC\pm}$ = ±5 V, C_{ext} ≥ 100 pF, T_A = 25°C

PARAMETER		TEST CONDITIONS	MIN	TYP	MAX	UNIT
t_{orD}	Differential-input overload recovery time (see Note 5)	Differential Input Pulse: V_{ID} = 2 V, t_r = t_f = 20 ns		20		ns
t_{orC}	Common-mode input overload recovery time (see Note 6)	Common-mode Input Pulse: V_{IC} = ±2 V, t_r = t_f = 20 ns		20		ns
$t_{cyc(min)}$	Minimum cycle time			200		ns

NOTES: 5. Differential-input overload recovery time is the time necessary for the device to recover from the specified differential-input overload signal prior to the strobe-enable signal.
6. Common-mode-input overload recovery time is the time necessary for the device to recover from the specified common-mode-input overload signal prior to the strobe-enable signal.

TEXAS INSTRUMENTS
POST OFFICE BOX 655012 • DALLAS, TEXAS 75265

PARAMETER MEASUREMENT INFORMATION

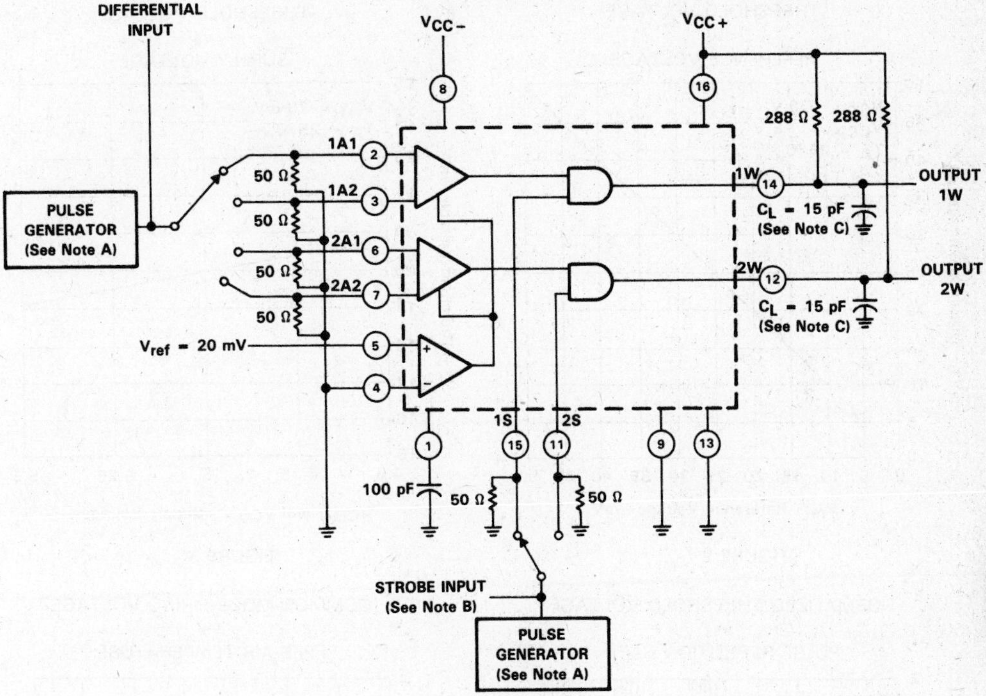
DIFFERENTIAL INPUT
PULSE GENERATOR (See Note A)
VCC−
VCC+
8
16
288 Ω
288 Ω
1A1
1A2
2A1
2A2
50 Ω
50 Ω
50 Ω
50 Ω
2
3
6
7
5
4
Vref = 20 mV
1W
14
2W
12
OUTPUT 1W
OUTPUT 2W
CL = 15 pF (See Note C)
CL = 15 pF (See Note C)
1S
2S
1
15
11
9
13
100 pF
50 Ω
50 Ω
STROBE INPUT (See Note B)
PULSE GENERATOR (See Note A)

TEST CIRCUIT

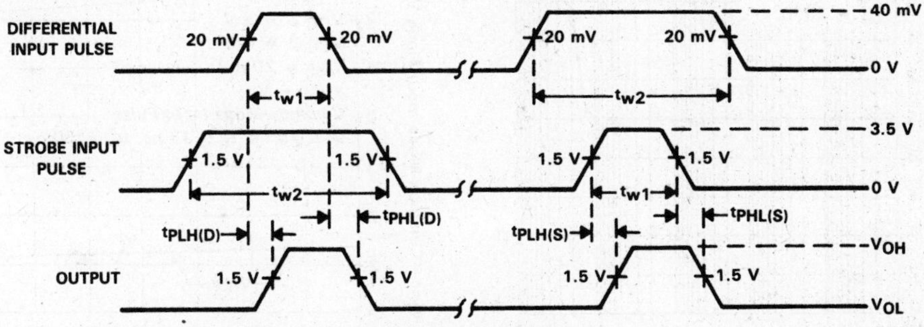
DIFFERENTIAL INPUT PULSE
20 mV
20 mV
20 mV
20 mV
40 mV
0 V
tw1
tw2
STROBE INPUT PULSE
1.5 V
1.5 V
1.5 V
1.5 V
3.5 V
0 V
tw2
tw1
tPLH(D)
tPHL(D)
tPLH(S)
tPHL(S)
OUTPUT
1.5 V
1.5 V
1.5 V
1.5 V
VOH
VOL

VOLTAGE WAVEFORMS

NOTES: A. The pulse generators have the following characteristics: Z_{out} = 50 Ω, t_r = 15 ± 5 ns, t_f = 15 ±5 ns, t_{w1} = 100 ns, t_{w2} = 300 ns, and PRR ≤ 1 MHz.

B. The strobe input pulse is applied to Strobe 1S when inputs 1A1-1A2 are being tested and to Strobe 2S when inputs 2A1-2A2 are being tested.

C. C_L includes probe and jig capacitance.

FIGURE 1. PROPAGATION DELAY TIMES

TYPICAL CHARACTERISTICS

THRESHOLD VOLTAGE
vs
REFERENCE VOLTAGE

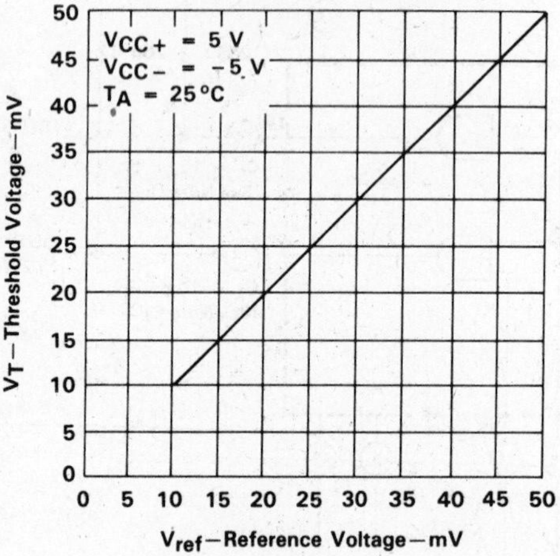
VCC+ = 5 V
VCC− = −5 V
TA = 25°C
VT—Threshold Voltage—mV
0 5 10 15 20 25 30 35 40 45 50
Vref—Reference Voltage—mV

FIGURE 2

THRESHOLD VOLTAGE
vs
SUPPLY VOLTAGE

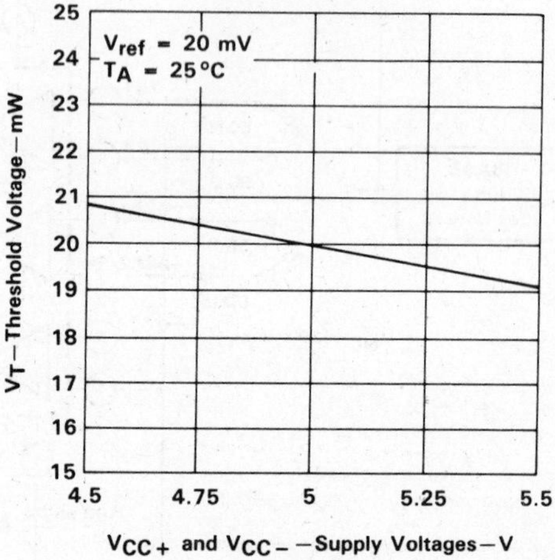
Vref = 20 mV
TA = 25°C
VT—Threshold Voltage—mW
15 16 17 18 19 20 21 22 23 24 25
4.5 4.75 5 5.25 5.5
VCC+ and VCC− —Supply Voltages—V

FIGURE 3

NORMALIZED THRESHOLD VOLTAGE
vs
PULSE REPETITION RATE

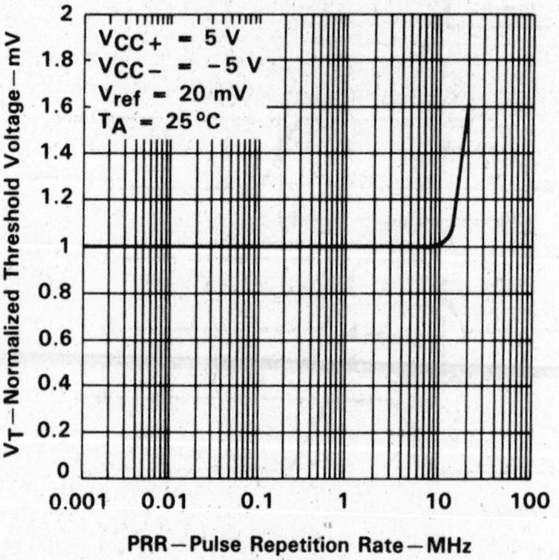
VCC+ = 5 V
VCC− = −5 V
Vref = 20 mV
TA = 25°C
VT—Normalized Threshold Voltage—mV
0 0.2 0.4 0.6 0.8 1 1.2 1.4 1.6 1.8 2
0.001 0.01 0.1 1 10 100
PRR—Pulse Repetition Rate—MHz

FIGURE 4

COMMON-MODE FIRING VOLTAGE
vs
FREE-AIR TEMPERATURE

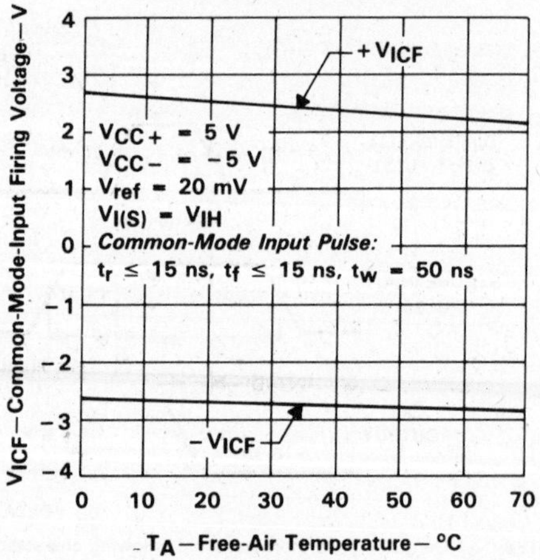
+VICF
−VICF
VCC+ = 5 V
VCC− = −5 V
Vref = 20 mV
VI(S) = VIH
Common-Mode Input Pulse:
tr ≤ 15 ns, tf ≤ 15 ns, tw = 50 ns
VICF—Common-Mode-Input Firing Voltage—V
−4 −3 −2 −1 0 1 2 3 4
0 10 20 30 40 50 60 70
TA—Free-Air Temperature—°C

FIGURE 5

TEXAS INSTRUMENTS
POST OFFICE BOX 655012 • DALLAS, TEXAS 75265

TYPICAL CHARACTERISTICS

DIFFERENTIAL-INPUT BIAS CURRENT
vs
FREE-AIR TEMPERATURE

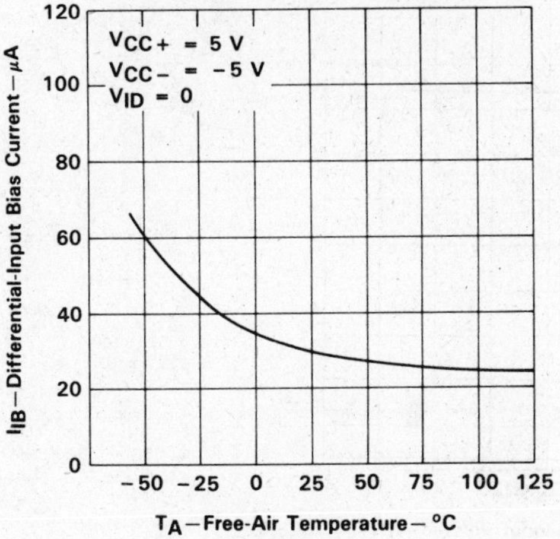
V_{CC+} = 5 V
V_{CC-} = −5 V
V_{ID} = 0
I_{IB}—Differential-Input Bias Current—μA
120
100
80
60
40
20
0
−50 −25 0 25 50 75 100 125
T_A—Free-Air Temperature—°C

FIGURE 6

DIFFERENTIAL-INPUT OFFSET CURRENT
vs
FREE-AIR TEMPERATURE

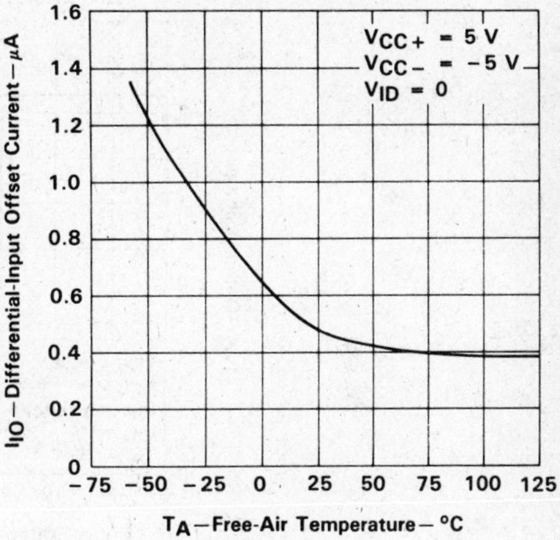
V_{CC+} = 5 V
V_{CC-} = −5 V
V_{ID} = 0
I_{IO}—Differential-Input Offset Current—μA
1.6
1.4
1.2
1.0
0.8
0.6
0.4
0.2
0
−75 −50 −25 0 25 50 75 100 125
T_A—Free-Air Temperature—°C

FIGURE 7

HIGH-LEVEL INPUT CURRENT
vs
INPUT VOLTAGE

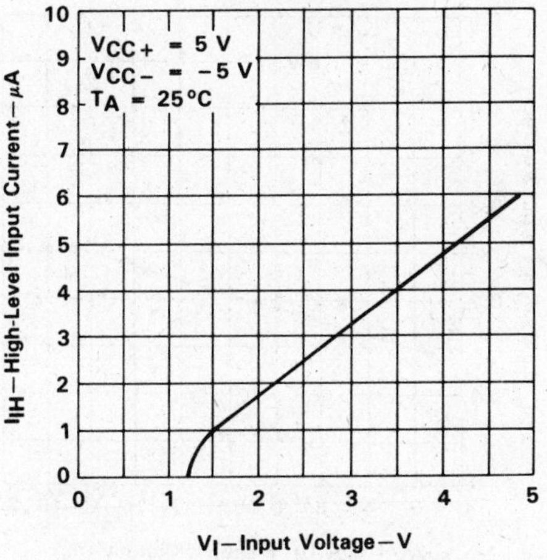
V_{CC+} = 5 V
V_{CC-} = −5 V
T_A = 25°C
I_{IH}—High-Level Input Current—μA
10
9
8
7
6
5
4
3
2
1
0
0 1 2 3 4 5
V_I—Input Voltage—V

FIGURE 8

LOW-LEVEL INPUT CURRENT
vs
INPUT VOLTAGE

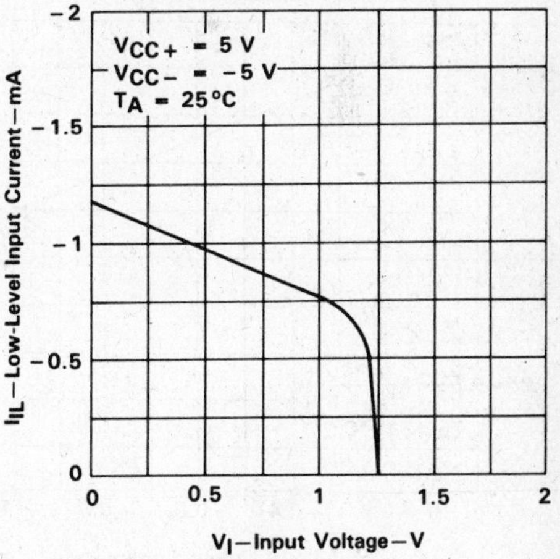
V_{CC+} = 5 V
V_{CC-} = −5 V
T_A = 25°C
I_{IL}—Low-Level Input Current—mA
−2
−1.5
−1
−0.5
0
0 0.5 1 1.5 2
V_I—Input Voltage—V

FIGURE 9

6

Memory Interface Circuits

TYPICAL CHARACTERISTICS

OUTPUT VOLTAGE
vs
DIFFERENTIAL-INPUT VOLTAGE

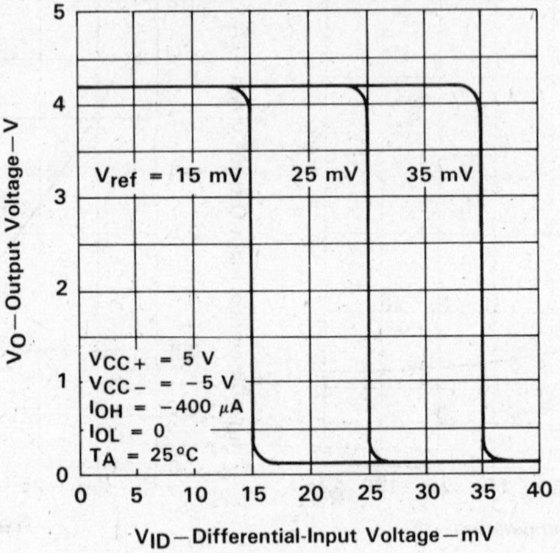
5
4
3
2
1
0
V_O—Output Voltage—V
V_{ref} = 15 mV
25 mV
35 mV
V_{CC+} = 5 V
V_{CC-} = −5 V
I_{OH} = −400 μA
I_{OL} = 0
T_A = 25°C
0 5 10 15 20 25 30 35 40
V_{ID}—Differential-Input Voltage—mV

FIGURE 10

HIGH-LEVEL OUTPUT VOLTAGE
vs
HIGH-LEVEL OUTPUT CURRENT

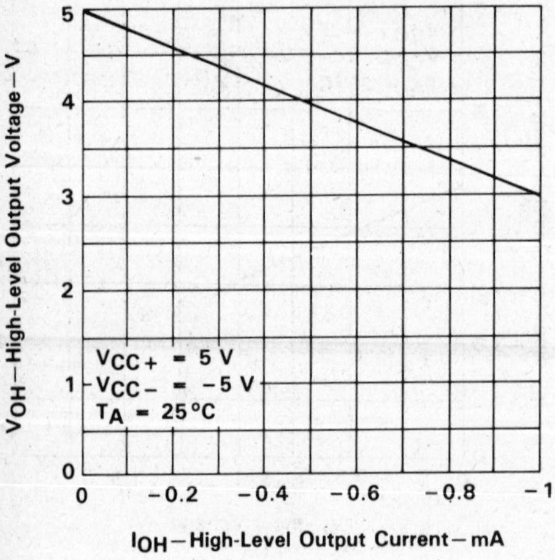
5
4
3
2
1
0
V_{OH}—High-Level Output Voltage—V
V_{CC+} = 5 V
V_{CC-} = −5 V
T_A = 25°C
0 −0.2 −0.4 −0.6 −0.8 −1
I_{OH}—High-Level Output Current—mA

FIGURE 11

LOW-LEVEL OUTPUT VOLTAGE
vs
LOW-LEVEL OUTPUT CURRENT

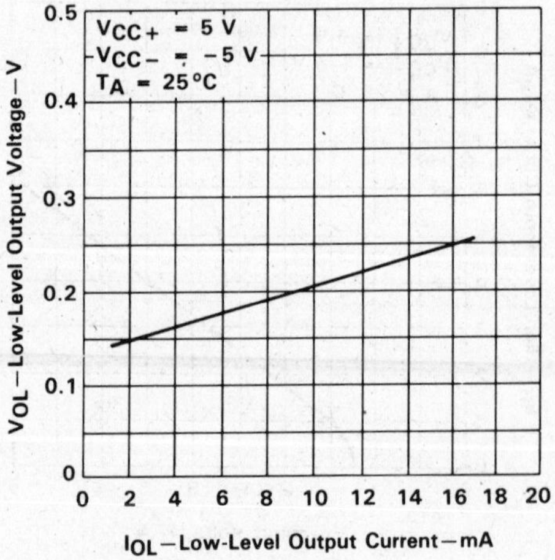
0.5
0.4
0.3
0.2
0.1
0
V_{OL}—Low-Level Output Voltage—V
V_{CC+} = 5 V
V_{CC-} = −5 V
T_A = 25°C
0 2 4 6 8 10 12 14 16 18 20
I_{OL}—Low-Level Output Current—mA

FIGURE 12

6

Memory Interface Circuits

TYPICAL APPLICATIONS

small memory systems

This application demonstrates an improved method of sensing data from small memory systems. Two individual core planes, usually consisting of 4096 cores each, can be interfaced by each of the dual-channel SN5524 sense amplifiers, see Figure 13. Standard TTL integrated circuits, driven directly from the compatible sense-amplifier outputs, may be selected to serve as the memory data register (MDR).

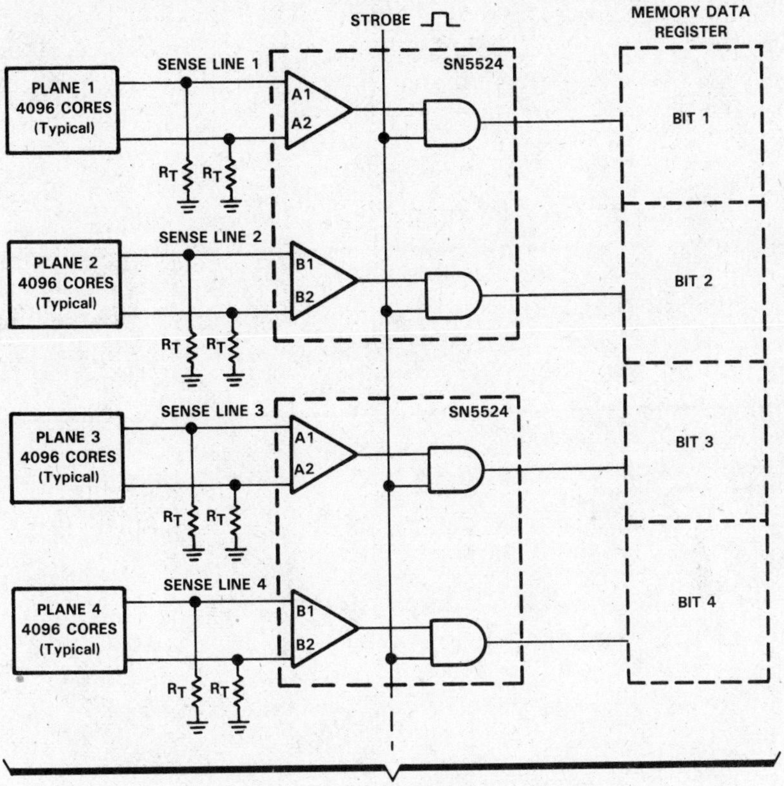
STROBE
MEMORY DATA REGISTER
SENSE LINE 1
SN5524
PLANE 1 4096 CORES (Typical)
A1
A2
BIT 1
RT
RT
SENSE LINE 2
PLANE 2 4096 CORES (Typical)
B1
B2
BIT 2
SENSE LINE 3
SN5524
PLANE 3 4096 CORES (Typical)
BIT 3
SENSE LINE 4
PLANE 4 4096 CORES (Typical)
BIT 4
TO ADDITIONAL PLANES AND SN5524s AS NECESSARY FOR COMPLETE MEMORY WORD

FIGURE 13. SENSING SMALL MEMORY SYSTEMS

D2338, DECEMBER 1986

- High Speed and Fast Recovery Time
- Time and Amplitude Signal Discrimination
- Adjustable Input Threshold Voltage Levels
- Narrow Window of Threshold Voltage Uncertainty
- Multiple Differential-Input Preamplifiers
- High DC Noise Margin . . . 1 V Typ
- Good Fanout Capability
- TTL Drive Capability
- Standard Logic Supply Voltages
- Plug-in Configuration Ideal for Flow-Soldering Techniques

J PACKAGE
(TOP VIEW)

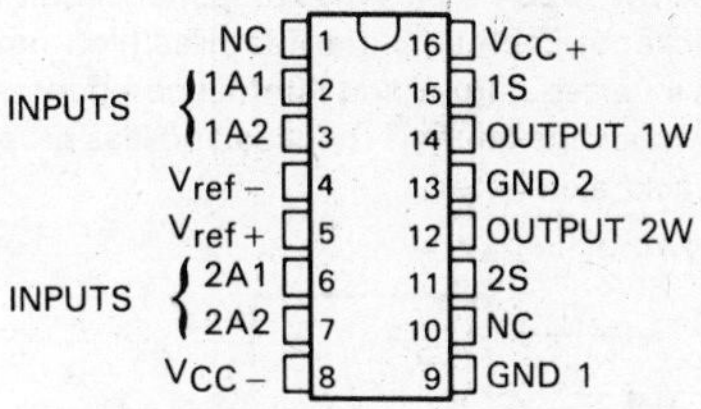
NC 1
INPUTS 1A1 2
1A2 3
Vref − 4
Vref + 5
INPUTS 2A1 6
2A2 7
VCC − 8
16 VCC +
15 1S
14 OUTPUT 1W
13 GND 2
12 OUTPUT 2W
11 2S
10 NC
9 GND 1

NC — No internal connection

FUNCTION TABLE

INPUTS A	INPUTS S	OUTPUT W
H	H	L
L	X	H
X	L	H

positive logic: $W = \overline{AS}$

DEFINTION OF LOGIC LEVELS

INPUT	H	L	X
A†	$V_{ID} \geq V_T$ max	$V_{ID} \leq V_T$ min	Irrelevant
S	$V_I \geq V_{IH}$ min	$V_I \leq V_{IL}$ max	Irrelevant

†A is a differential voltage (V_{ID}) between A1 and A2. For these circuits, V_{ID} is considered positive regardless of which terminal is positive with respect to the other.

description

The SN55234 monolithic sense amplifier is designed for use with high-speed memory systems. This sense amplifier detects bipolar differential-input signals from the memory and provides the necessary interface circuitry between the memory and the logic section. Low-level pulses originating in the memory are transformed into logic levels compatible with standard TTL circuits.

The SN55234 sense amplifier features multiple differential-input preamplifiers and versatile gating and output circuits, permitting a significant reduction in the circuitry required to accomplish the sensing function. A unique circuit design provides an inherent stability of the input threshold level over a wide range of supply voltage levels and temperature ranges. Independent strobing of each of the dual sense-input channels ensures maximum versatility and permits detection to occur when the signal-to-noise ratio is at a maximum. The strobe inputs and the outputs are compatible with standard TTL logic circuits.

The SN55234 sense amplifier is available in the J ceramic dual-in-line package and is characterized for operation over the full military temperature range of −55°C to 125°C.

TEXAS INSTRUMENTS
POST OFFICE BOX 655012 • DALLAS, TEXAS 75265

circuit operation

The SN55234 dual sense amplifier circuit features two completely independent sense amplifiers in a single package. Each amplifier features high fanout capability. An additional stage in the output gate provides an inverted output and internal compensation has been added. By not using a separate gate for inversion, the package count is reduced and less propagation delay is added. The need for an external roll-off capacitor is also eliminated.

logic symbol†

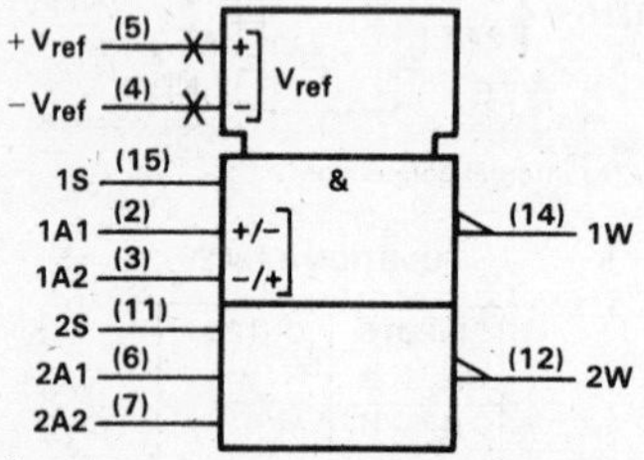
+Vref (5)
−Vref (4)
1S (15)
1A1 (2)
1A2 (3)
2S (11)
2A1 (6)
2A2 (7)
Vref
&
+/−
−/+
(14) 1W
(12) 2W

†This symbol is in accordance with ANSI/IEEE Std 91-1984 and IEC Publication 617-12.

logic diagram

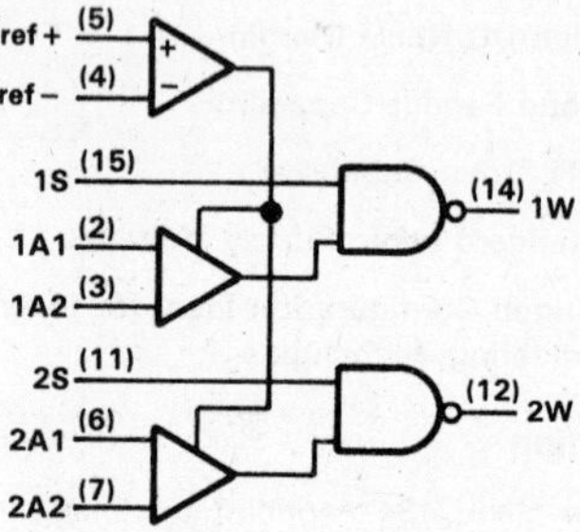
Vref+ (5)
Vref− (4)
1S (15)
1A1 (2)
1A2 (3)
2S (11)
2A1 (6)
2A2 (7)
(14) 1W
(12) 2W

schematic

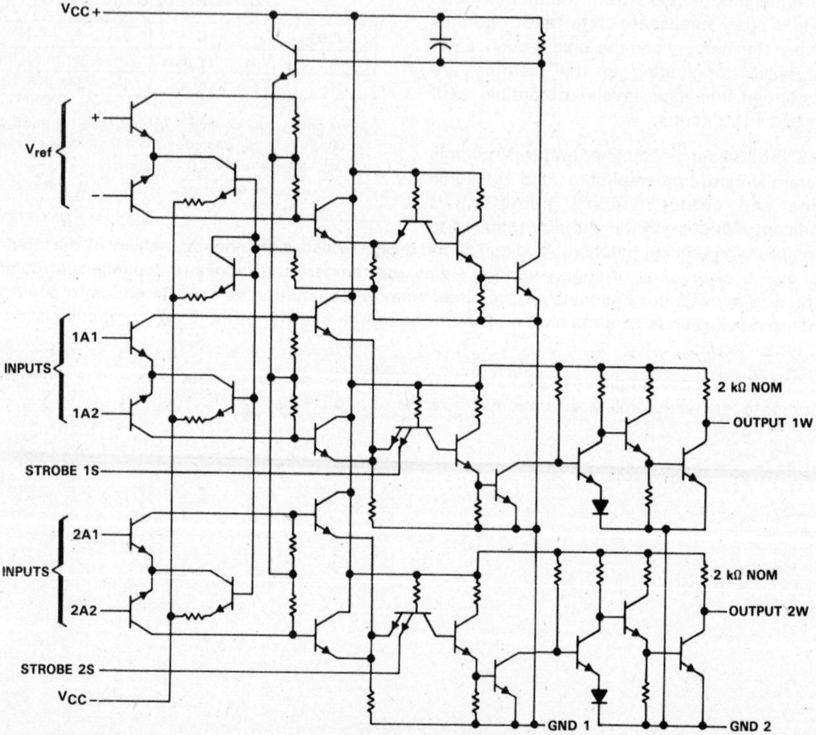
VCC+
Vref
+
−
1A1
INPUTS
1A2
STROBE 1S
2A1
INPUTS
2A2
STROBE 2S
VCC−
2 kΩ NOM
OUTPUT 1W
2 kΩ NOM
OUTPUT 2W
GND 1
GND 2

TEXAS INSTRUMENTS
POST OFFICE BOX 655012 • DALLAS, TEXAS 75265

absolute maximum ratings over operating free-air temperature range (unless otherwise noted)

Supply voltage, V_{CC+} (see Note 1) . . . . . 7 V
Supply voltage, V_{CC-} . . . . . −7 V
Differential input voltage, V_{ID} or V_{ref} . . . . . ±5 V
Voltage from any input to ground (see Note 2) . . . . . 5.5 V
Continuous total power dissipation at (or below) 25°C free-air temperature (see Note 3) . . . . . 1375 mW
Operating free-air temperature range . . . . . −55°C to 125°C
Storage temperature range . . . . . −65°C to 150°C
Lead temperature 1,6 mm (1/16 inch) from case for 60 seconds . . . . . 300°C

NOTES: 1. Voltage values, except differential voltages, are with respect to the network ground terminal.
2. Strobe input voltages must be zero or positive with respect to the network ground terminal.
3. For operating above 25°C free-air temperature, derate total power linearly at the rate of 11.0 mW/°C.

recommended operating conditions

	MIN	NOM	MAX	UNIT
Positive supply voltage, V_{CC+}	4.75	5	5.25	V
Negative supply voltage, V_{CC-}	−4.75	−5	−5.25	V
Reference voltage, V_{ref}	15		40	mV
High-level input voltage (strobe inputs), V_{IH}	2			V
Low-level input voltage (strobe inputs), V_{IL}			0.8	V

electrical characteristics over recommended operating free-air temperature range, $V_{CC\pm}$ = ±5 V, (unless otherwise noted)

PARAMETER		TEST CONDITIONS	MIN	TYP†	MAX	UNIT
V_T	Differential-input threshold voltage	V_{ref} = 15 mV	7	15	20	mV
		V_{ref} = 40 mV	32	40	48	mV
V_{ICF}	Common-mode input firing voltage (see Note 4)	V_{ref} = 40 mV, $V_{I(S)} = V_{IH}$, Common-mode input pulse: $t_r \le$ 15 ns, $t_f \le$ 15 ns, t_w = 50 ns		±2.5		V
I_{IB}	Differential-input bias current	$V_{CC\pm}$ = ±5.25 V, V_{ID} = 0, T_A = −55°C			100	μA
		$V_{CC\pm}$ = ±5.25 V, V_{ID} = 0, T_A = 25°C		30	75	μA
		$V_{CC\pm}$ = ±5.25 V, V_{ID} = 0, T_A = 125°C			75	μA
I_{IO}	Differential-input offset current	$V_{CC\pm}$ = ±5.25 V, V_{ID} = 0		0.5		μA
V_{OH}	High-level output voltage	$V_{CC\pm}$ = ±4.75 V, I_{OH} = −400 μA	2.4	4		V
V_{OL}	Low-level output voltage	$V_{CC\pm}$ = ±4.75 V, I_{OL} = 16 mA		0.25	0.4	V
I_{IH}	High-level input current (strobe inputs)	$V_{CC\pm}$ = ±5.25 V, V_{IH} = 2.4 V			40	μA
		$V_{CC\pm}$ = ±5.25 V, V_{IH} = 5.25 V			1	mA
I_{IL}	Low-level input current (strobe inputs)	$V_{CC\pm}$ = ±5.25 V, V_{IL} = 0.4 V		−1	−1.6	mA
I_{OS}	Short-circuit output current	$V_{CC\pm}$ = ±5.25 V, T_A = 25°C	−2.1	−2.6	−3.5	mA
I_{CC+}	Supply current from V_{CC+}	$V_{CC\pm}$ = ±5.25 V, T_A = 25°C		25	40	mA
I_{CC-}	Supply current from V_{CC-}	$V_{CC\pm}$ = ±5.25 V, T_A = 25°C		−15	−20	mA

†Typical values are at $V_{CC\pm}$ = ±5 V, T_A = 25°C.
NOTE 4: Common-mode input firing voltage is the minimum common-mode voltage that will exceed the dynamic range of the input at the specified conditions and cause the logic output to switch. The common-mode input signal is applied when the strobe is high.

switching characteristics, $V_{CC\pm}$ = ±5 V, T_A = 25°C

PARAMETER	FROM (INPUT)	TO (OUTPUT)	TEST FIGURE	TEST CONDITIONS	MIN	TYP	MAX	UNIT
$t_{PLH(D)}$	A1-A2	W	1	C_L = 15 pF, R_L = 288 Ω		25		ns
$t_{PHL(D)}$						25	40	
$t_{PLH(S)}$	STROBE	W	1			25		ns
$t_{PHL(S)}$						15	30	

recovery and cycle times, $V_{CC\pm}$ = ±5 V, T_A = 25°C

PARAMETER		TEST CONDITIONS	MIN	TYP	MAX	UNIT
t_{orD}	Differential-input overload recovery time (see Note 5)	Differential Input Pulse: V_{ID} = 2 V, t_r = t_f = 20 ns		20		ns
t_{orC}	Common-mode input overload recovery time (see Note 6)	Common-mode Input Pulse: V_{IC} = ±2 V, t_r = t_f = 20 ns		20		ns
$t_{cyc(min)}$	Minimum cycle time			200		ns

NOTES: 5. Differential-input overload recovery time is the time necessary for the device to recover from the specified differential-input overload signal prior to the strobe-enable signal.

6. Common-mode-input overload recovery time is the time necessary for the device to recover from the specified common-mode-input overload signal prior to the strobe-enable signal.

TEXAS INSTRUMENTS
POST OFFICE BOX 655012 • DALLAS, TEXAS 75265

PARAMETER MEASUREMENT INFORMATION

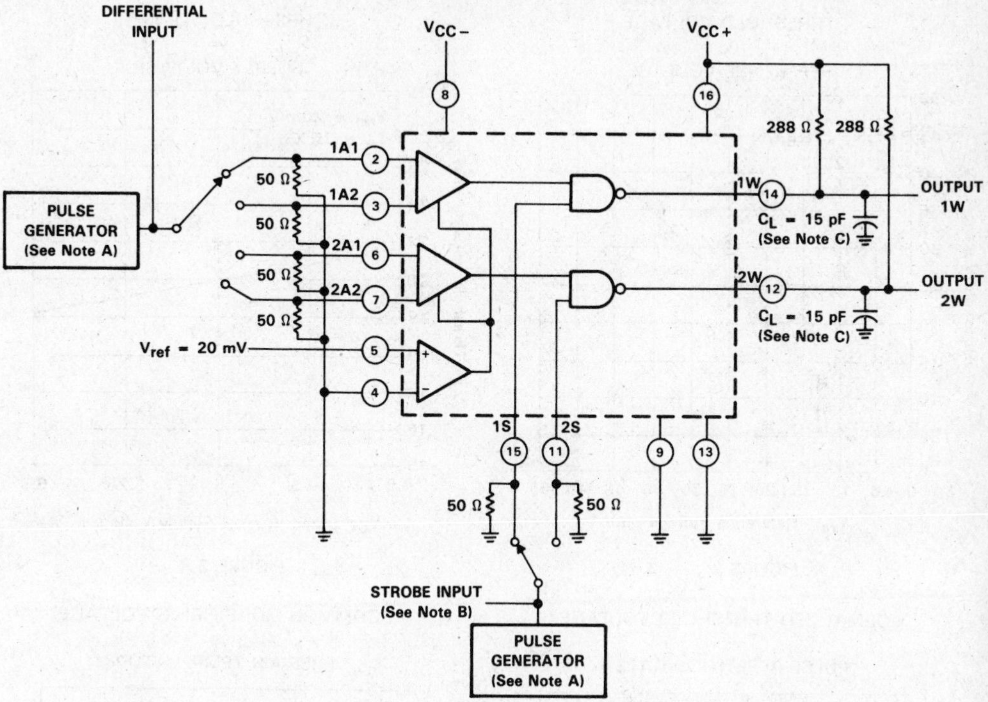
DIFFERENTIAL
INPUT
PULSE
GENERATOR
(See Note A)
VCC−
VCC+
8
16
288 Ω
288 Ω
1A1
1A2
2A1
2A2
50 Ω
50 Ω
50 Ω
50 Ω
2
3
6
7
5
4
Vref = 20 mV
1W
14
2W
12
OUTPUT
1W
OUTPUT
2W
CL = 15 pF
(See Note C)
CL = 15 pF
(See Note C)
1S
2S
15
11
9
13
50 Ω
50 Ω
STROBE INPUT
(See Note B)
PULSE
GENERATOR
(See Note A)

TEST CIRCUIT

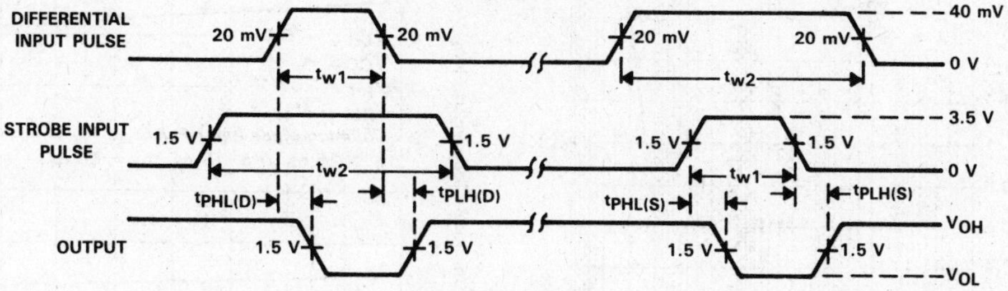
DIFFERENTIAL
INPUT PULSE
20 mV
20 mV
20 mV
20 mV
40 mV
0 V
tw1
tw2
STROBE INPUT
PULSE
1.5 V
1.5 V
1.5 V
1.5 V
3.5 V
0 V
tw2
tw1
tPHL(D)
tPLH(D)
tPHL(S)
tPLH(S)
OUTPUT
1.5 V
1.5 V
1.5 V
1.5 V
VOH
VOL

VOLTAGE WAVEFORMS

NOTES: A. The pulse generators have the following characteristics: Z_{out} = 50 Ω, t_r = 15 ±5 ns, t_f = 15 ±5 ns, t_{w1} = 100 ns, t_{w2} = 300 ns, and PRR ≤ 1 MHz.

B. The strobe input pulse is applied to Strobe 1S when inputs 1A1-1A2 are being tested and to Strobe 2S when inputs 2A1-2A2 are being tested.

C. C_L includes probe and jig capacitance.

FIGURE 1. PROPAGATION DELAY TIMES

6

Memory Interface Circuits

TYPICAL CHARACTERISTICS

THRESHOLD VOLTAGE
vs
REFERENCE VOLTAGE

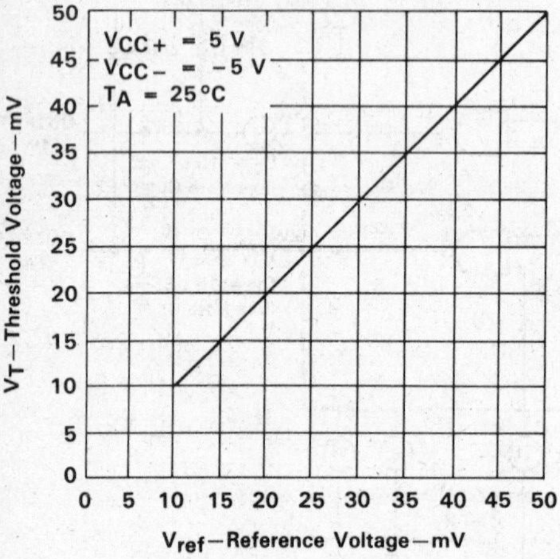
V_{CC+} = 5 V
V_{CC-} = −5 V
T_A = 25°C
V_T—Threshold Voltage—mV
V_{ref}—Reference Voltage—mV

FIGURE 2

THRESHOLD VOLTAGE
vs
SUPPLY VOLTAGE

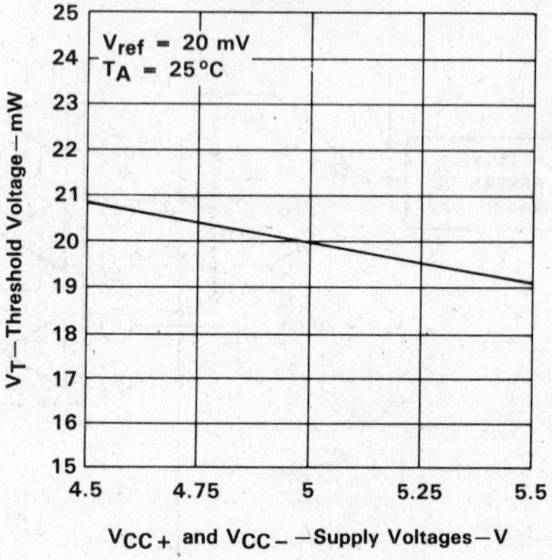
V_{ref} = 20 mV
T_A = 25°C
V_T—Threshold Voltage—mW
V_{CC+} and V_{CC-}—Supply Voltages—V

FIGURE 3

NORMALIZED THRESHOLD VOLTAGE
vs
PULSE REPETITION RATE

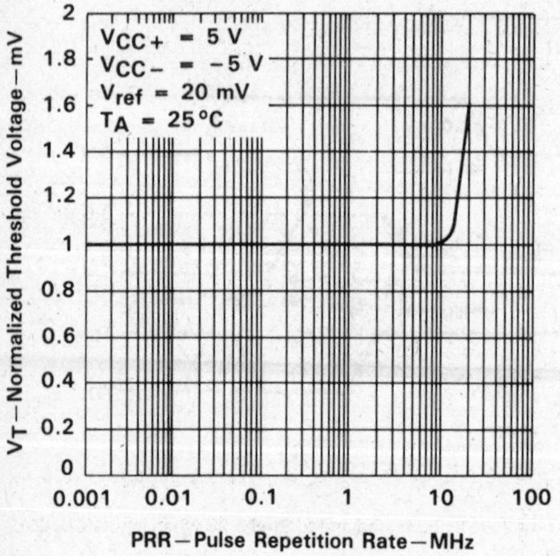
V_{CC+} = 5 V
V_{CC-} = −5 V
V_{ref} = 20 mV
T_A = 25°C
V_T—Normalized Threshold Voltage—mV
PRR—Pulse Repetition Rate—MHz

FIGURE 4

COMMON-MODE FIRING VOLTAGE
vs
FREE-AIR TEMPERATURE

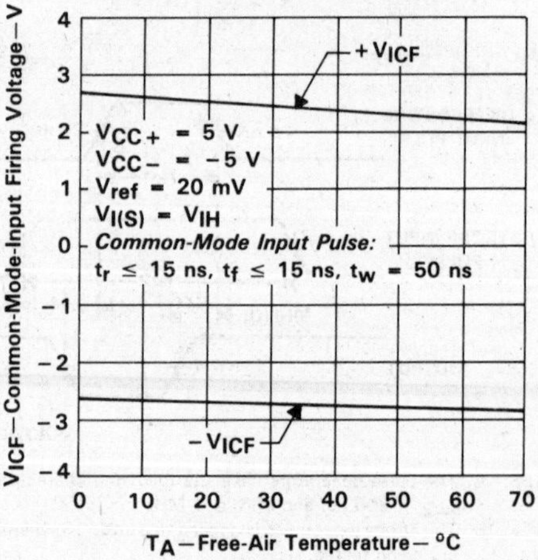
$+V_{ICF}$
V_{CC+} = 5 V
V_{CC-} = −5 V
V_{ref} = 20 mV
$V_{I(S)}$ = V_{IH}
Common-Mode Input Pulse:
$t_r \le 15$ ns, $t_f \le 15$ ns, $t_w = 50$ ns
$-V_{ICF}$
V_{ICF}—Common-Mode-Input Firing Voltage—V
T_A—Free-Air Temperature—°C

FIGURE 5

TYPICAL CHARACTERISTICS

DIFFERENTIAL-INPUT BIAS CURRENT
vs
FREE-AIR TEMPERATURE

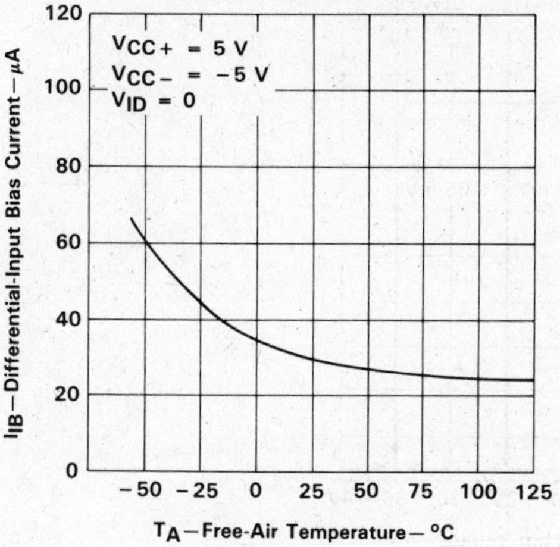
V_{CC+} = 5 V
V_{CC-} = −5 V
V_{ID} = 0
I_{IB} — Differential-Input Bias Current — μA
120
100
80
60
40
20
0
−50
−25
0
25
50
75
100
125
T_A — Free-Air Temperature — °C

FIGURE 6

DIFFERENTIAL-INPUT OFFSET CURRENT
vs
FREE-AIR TEMPERATURE

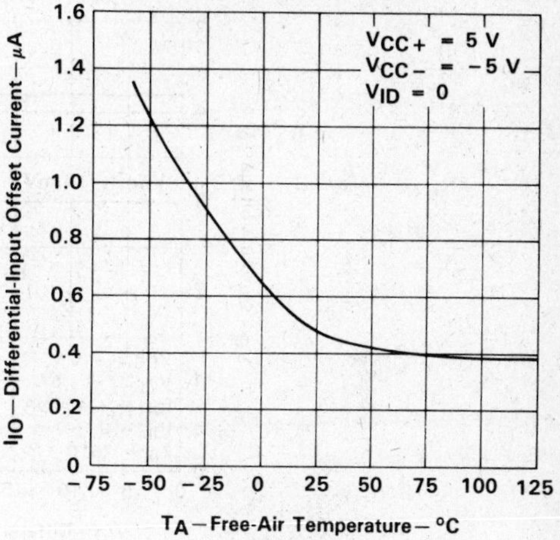
V_{CC+} = 5 V
V_{CC-} = −5 V
V_{ID} = 0
I_{IO} — Differential-Input Offset Current — μA
1.6
1.4
1.2
1.0
0.8
0.6
0.4
0.2
0
−75
−50
−25
0
25
50
75
100
125
T_A — Free-Air Temperature — °C

FIGURE 7

HIGH-LEVEL INPUT CURRENT
vs
INPUT VOLTAGE

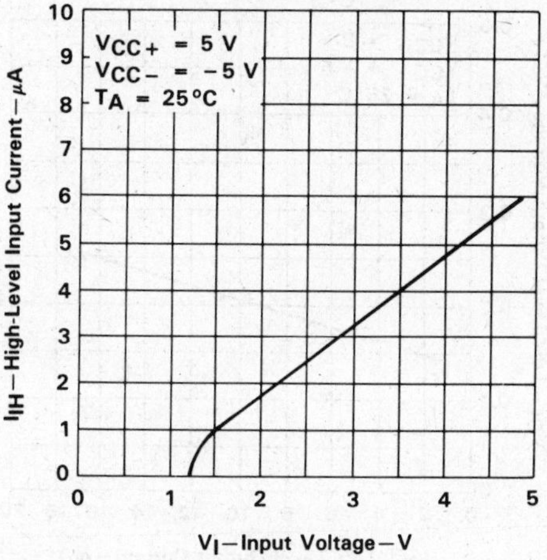
V_{CC+} = 5 V
V_{CC-} = −5 V
T_A = 25°C
I_{IH} — High-Level Input Current — μA
10
9
8
7
6
5
4
3
2
1
0
0
1
2
3
4
5
V_I — Input Voltage — V

FIGURE 8

LOW-LEVEL INPUT CURRENT
vs
INPUT VOLTAGE

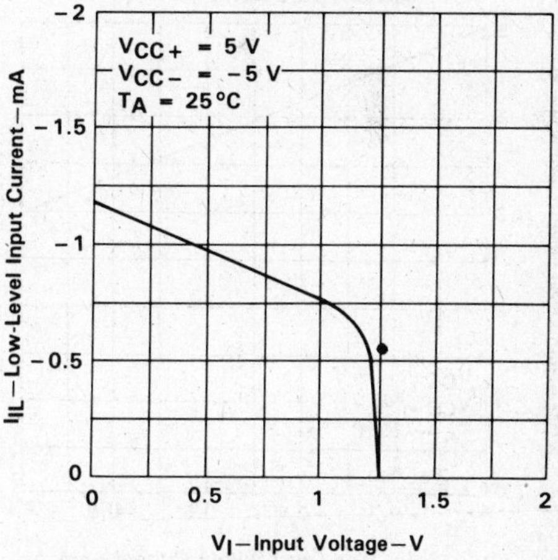
V_{CC+} = 5 V
V_{CC-} = −5 V
T_A = 25°C
I_{IL} — Low-Level Input Current — mA
−2
−1.5
−1
−0.5
0
0
0.5
1
1.5
2
V_I — Input Voltage — V

FIGURE 9

6
Memory Interface Circuits

TYPICAL CHARACTERISTICS

Z OUTPUT VOLTAGE
vs
DIFFERENTIAL-INPUT VOLTAGE

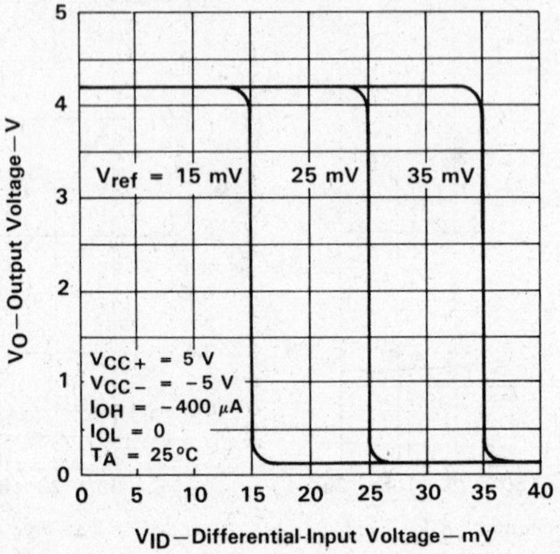
V_{ref} = 15 mV
25 mV
35 mV
V_{CC+} = 5 V
V_{CC-} = −5 V
I_{OH} = −400 μA
I_{OL} = 0
T_A = 25°C
V_O—Output Voltage—V
V_{ID}—Differential-Input Voltage—mV

FIGURE 10

HIGH-LEVEL OUTPUT VOLTAGE
vs
HIGH-LEVEL OUTPUT CURRENT

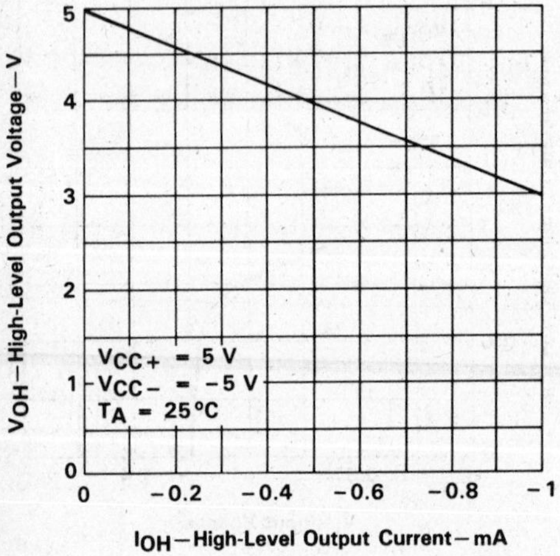
V_{CC+} = 5 V
V_{CC-} = −5 V
T_A = 25°C
V_{OH}—High-Level Output Voltage—V
I_{OH}—High-Level Output Current—mA

FIGURE 11

LOW-LEVEL OUTPUT VOLTAGE
vs
LOW-LEVEL OUTPUT CURRENT

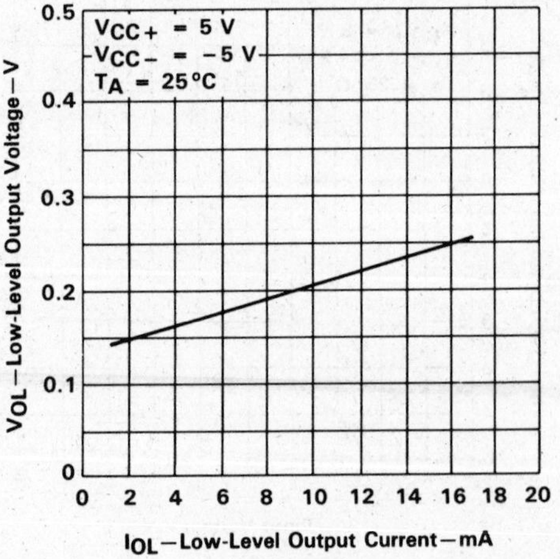
V_{CC+} = 5 V
V_{CC-} = −5 V
T_A = 25°C
V_{OL}—Low-Level Output Voltage—V
I_{OL}—Low-Level Output Current—mA

FIGURE 12

TEXAS INSTRUMENTS
POST OFFICE BOX 655012 • DALLAS, TEXAS 75265

TYPICAL APPLICATIONS

small memory systems

This application demonstrates an improved method of sensing data from small memory systems. Two individual core planes, usually consisting of 4096 cores each, can be interfaced by each of the dual-channel SN55234 sense amplifiers, see Figure 13. Standard TTL integrated circuits, driven from the compatible sense-amplifier outputs, may be selected to serve as the memory data register (MDR).

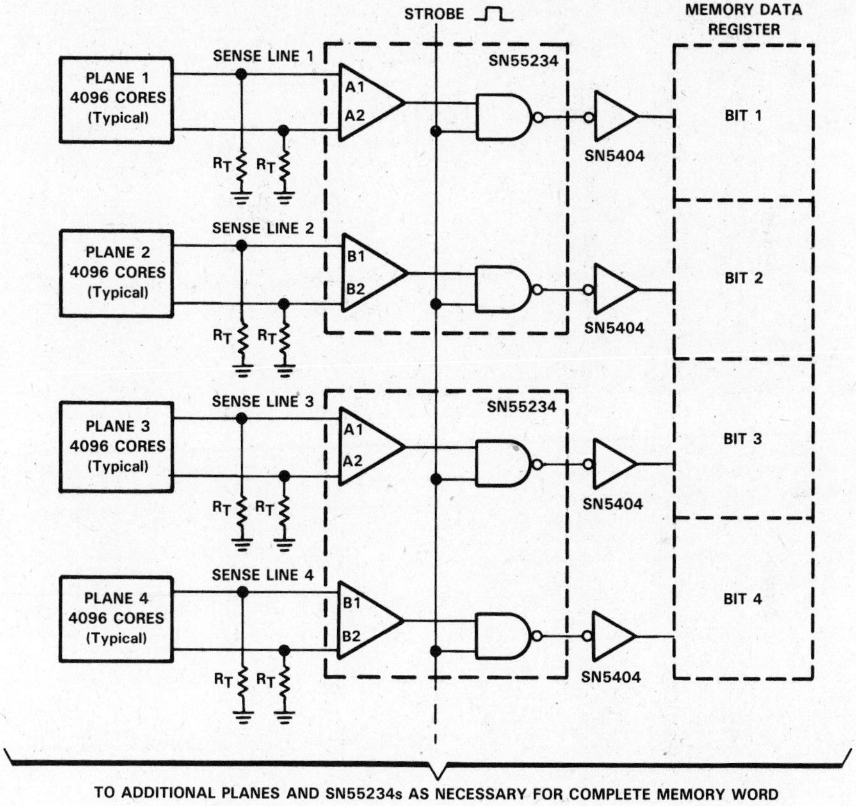
STROBE
MEMORY DATA REGISTER
SENSE LINE 1
SENSE LINE 2
SENSE LINE 3
SENSE LINE 4
PLANE 1 4096 CORES (Typical)
PLANE 2 4096 CORES (Typical)
PLANE 3 4096 CORES (Typical)
PLANE 4 4096 CORES (Typical)
SN55234
SN55234
A1
A2
B1
B2
RT
SN5404
BIT 1
BIT 2
BIT 3
BIT 4
TO ADDITIONAL PLANES AND SN55234s AS NECESSARY FOR COMPLETE MEMORY WORD

FIGURE 14. SENSING SMALL MEMORY SYSTEMS

D969, MARCH 1971—REVISED SEPTEMBER 1986

- 600-mA Output Capability
- Fast Switching Times
- Output Transient-Voltage Protection
- Dual Sink and Dual Source Outputs
- Minimum Time Skew Between Address and Output Current Rise
- 24-Volt Output Capability
- Source Base Drive Externally Adjustable
- TTL Compatibility
- Input Clamping Diodes
- Transformer Coupling Eliminated
- Reliability Increased
- Drive-Line Lengths Reduced
- Use of External Components Minimized

J PACKAGE
(TOP VIEW)

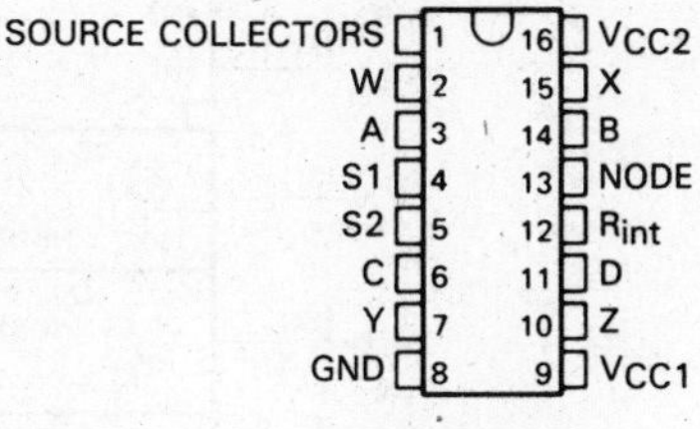
SOURCE COLLECTORS 1
W 2
A 3
S1 4
S2 5
C 6
Y 7
GND 8
16 VCC2
15 X
14 B
13 NODE R
12 Rint
11 D
10 Z
9 VCC1

FK PACKAGE
(TOP VIEW)

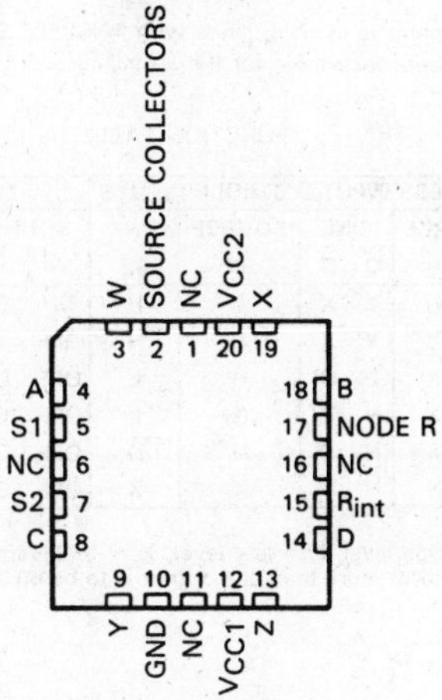
W
SOURCE COLLECTORS
NC
VCC2
X
3 2 1 20 19
A 4
S1 5
NC 6
S2 7
C 8
18 B
17 NODE R
16 NC
15 Rint
14 D
9 10 11 12 13
Y
GND
NC
VCC1
Z

NC—No internal connection

description

The SN55325 monolithic integrated circuit memory driver is designed for use with magnetic memories.

The device contains two 600-milliampere source switches and two 600-milliampere sink switches. Source selection is determined by one of two logic inputs, and source turn-on is determined by the source strobe. Likewise, sink selection is determined by one of two logic inputs, and sink turn-on is determined by the sink strobe. This arrangement allows selection of one of the four switches and its subsequent turn-on with minimum time skew of the output current rise.

When R_{int} and node R are connected together, the amount of base drive available for the source-1 or source-2 output transistor is set internally by a 575-ohm resistor. This method provides adequate base drive for source currents up to 375 mA with a V_{CC2} voltage of 15 volts or 600 mA with a V_{CC2} voltage of 24 volts.

When source currents greater than 375 mA are required, it is recommended that a resistor of the appropriate value be connected between V_{CC2} and node R and R_{int} must remain open. By using this method the source base current may usually be regulated within ±5%. An advantage of this method of setting the base drive is that the power dissipated by this resistor is external to the package and allows the integrated circuit to operate at higher source currents for a given junction temperature.

Each sink-output collector has an internal pull-up resistor in parallel with a clamping diode connected to V_{CC2}. This arrangement provides protection from voltage surges associated with switching inductive loads.

The SN55325 is characterized for operation over the full military temperature range of −55°C to 125°C.

logic symbol†

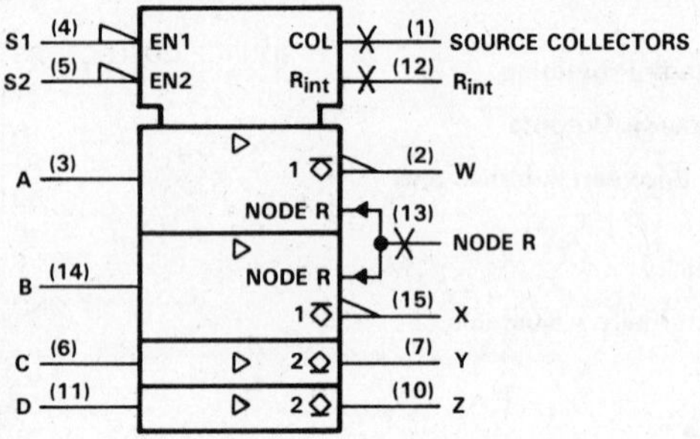
S1
(4)
EN1
S2
(5)
EN2
COL
(1)
SOURCE COLLECTORS
R_{int}
(12)
R_{int}
A
(3)
1
(2)
W
NODE R
(13)
NODE R
B
(14)
NODE R
1
(15)
X
C
(6)
2
(7)
Y
D
(11)
2
(10)
Z

†This symbol is in accordance with ANSI/IEEE Std 91-1984 and IEC Publication 617-12.
Pin numbers shown are for the J package.

FUNCTION TABLE

ADDRESS INPUTS				STROBE INPUTS		OUTPUTS			
SOURCE		SINK		SOURCE	SINK	SOURCE		SINK	
A	B	C	D	S1	S2	W	X	Y	Z
L	H	X	X	L	H	ON	OFF	OFF	OFF
H	L	X	X	L	H	OFF	ON	OFF	OFF
X	X	L	H	H	L	OFF	OFF	ON	OFF
X	X	H	L	H	L	OFF	OFF	OFF	ON
X	X	X	X	H	H	OFF	OFF	OFF	OFF
H	H	H	H	X	X	OFF	OFF	OFF	OFF

H = high level, L = low level, X = irrelevant
NOTE: Not more than one output is to be on at any one time.

logic diagram (positive logic)

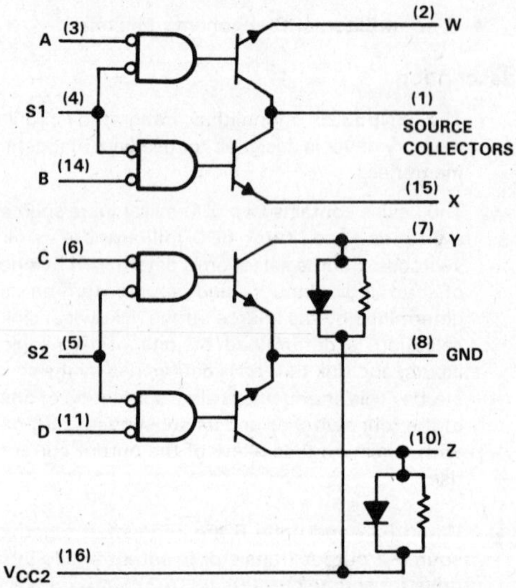
A
(3)
(2)
W
S1
(4)
(1)
SOURCE
COLLECTORS
B
(14)
(15)
X
C
(6)
(7)
Y
S2
(5)
(8)
GND
D
(11)
(10)
Z
V_{CC2}
(16)

absolute maximum ratings over operating free-air temperature range (unless otherwise noted)

Supply voltage V_{CC1} (see Note 1) . . . . 7 V
Supply voltage V_{CC2} (see Note 1) . . . . 25 V
Input voltage (any address or strobe input) . . . . 5.5 V
Continuous total dissipation at (or below) 25°C free-air temperature (see Note 2) . . . . . . 1375 mW
Operating free-air temperature range . . . . −55 to 125°C
Storage temperature range . . . . −65 to 150°C
Case temperature for 10 seconds: FK package . . . . 260°C
Lead temperature 1,6 mm (1/16 inch) from case for 60 seconds: J package . . . . 300°C

NOTES: 1. Voltage values are with respect to the network ground terminal.
2. For operation above 25°C free-air temperature, derate linearly at the rate of 11.0 mW/°C for both packages.

TEXAS INSTRUMENTS
POST OFFICE BOX 655012 • DALLAS, TEXAS 75265

recommended operating conditions

	MIN	NOM	MAX	UNIT
Supply voltage, V_{CC1}	4.5	5	5.5	V
Supply voltage, V_{CC2}	4.5		24	V
High-level input voltage, V_{IH}	2			V
Low-level input voltage, V_{IL}			0.8	V
Operating free-air temperatuare, T_A	−55		125	°C

electrical characteristics over rated operating free-air temperature range (unless otherwise noted)

PARAMETER			TEST CONDITIONS		MIN	TYP†	MAX	UNIT
V_{IK}	Input clamp voltage		V_{CC1} = 4.5 V, I_I = −10 mA,	V_{CC2} = 24 V, T_A = 25°C		−1.3	−1.7	V
$I_{(off)}$	Source-collectors terminal off-state current		V_{CC1} = 4.5 V, V_{CC2} = 24 V	T_A = −55°C to 125°C			500	μA
				T_A = 25°C		3	150	μA
V_{OH}	High-level sink output voltage		V_{CC1} = 4.5 V, I_O = 0	V_{CC2} = 24 V,	19	23		V
$V_{(sat)}$	Saturation voltage	Source outputs	V_{CC1} = 4.5 V, V_{CC2} = 15 V, R_L = 24 Ω to V_{CC2}, $I_{(source)} \approx -600$ mA‡, See Note 3	T_A = −55°C to 125°C			0.9	V
				T_A = 25°C		0.43	0.7	
		Sink outputs	V_{CC1} = 4.5 V, V_{CC2} = 15 V, R_L = 24 Ω to V_{CC2}, $I_{(sink)} \approx 600$ mA‡, See Note 3	T_A = −55°C to 125°C			0.9	
				T_A = 25°C		0.43	0.7	
I_I	Input current at maximum input voltage	Address inputs	V_{CC1} = 5.5 V, V_I = 5.5 V	V_{CC2} = 24 V,			1	mA
		Strobe inputs					2	
I_{IH}	High-level input current	Address inputs	V_{CC1} = 5.5 V, V_I = 2.4 V	V_{CC2} = 24 V,		3	40	μA
		Strobe inputs				6	80	
I_{IL}	Low-level input current	Address inputs	V_{CC1} = 5.5 V, V_I = 0.4 V	V_{CC2} = 24 V,		−1	−1.6	mA
		Strobe inputs				−2	−3.2	
$I_{CC(off)}$	Supply current, all sources and sinks off	From V_{CC1}	V_{CC1} = 5.5 V, T_A = 25°C	V_{CC2} = 24 V,		14	22	mA
		From V_{CC2}				7.5	20	
I_{CC1}	Supply current from V_{CC1} either sink on		V_{CC1} = 5.5 V, $I_{(sink)}$ = 50 mA,	V_{CC2} = 24 V, T_A = 25°C		55	70	mA
I_{CC2}	Supply current from V_{CC2}, either source on		V_{CC1} = 5.5 V, $I_{(source)}$ = −50 mA, See Note 3	V_{CC2} = 24 V, T_A = 25°C,		32	50	mA

† All typical values are at T_A = 25°C.
‡ Under these conditions, not more than one output is to be on at any one time.
NOTE 3: These parameters must be measured using pulse techniques, t_w = 200 μs, duty cycle ≤2%.

switching characteristics, V_{CC1} = 5 V, T_A = 25°C

PARAMETER†	TO (OUTPUT)	TEST CONDITIONS	MIN	TYP	MAX	UNIT
t_{PLH}	Source collectors	V_{CC2} = 15 V, R_L = 24 Ω, C_L = 25 pF		45	50	ns
t_{PHL}				45	50	
t_{TLH}	Source outputs	V_{CC2} = 20 V, R_L = 1 kΩ, C_L = 25 pF		55		ns
t_{THL}				7		
t_{PLH}	Sink outputs	V_{CC2} = 15 V, R_L = 24 Ω, C_L = 25 pF		25	45	ns
t_{PHL}				25	45	
t_{TLH}	Sink outputs	V_{CC2} = 15 V, R_L = 24 Ω, C_L = 25 pF		10	15	ns
t_{THL}				15	20	
t_s	Sink outputs	V_{CC2} = 15 V, R_L = 24 Ω, C_L = 25 pF		20	30	ns

†t_{PLH} ≡ propagation delay time, low-to-high-level output
t_{PHL} ≡ propagation delay time, high-to-low-level output
t_{TLH} ≡ transition time, low-to-high-level output
t_{THL} ≡ transition time, high-to-low-level output
t_s ≡ storage time

schematic

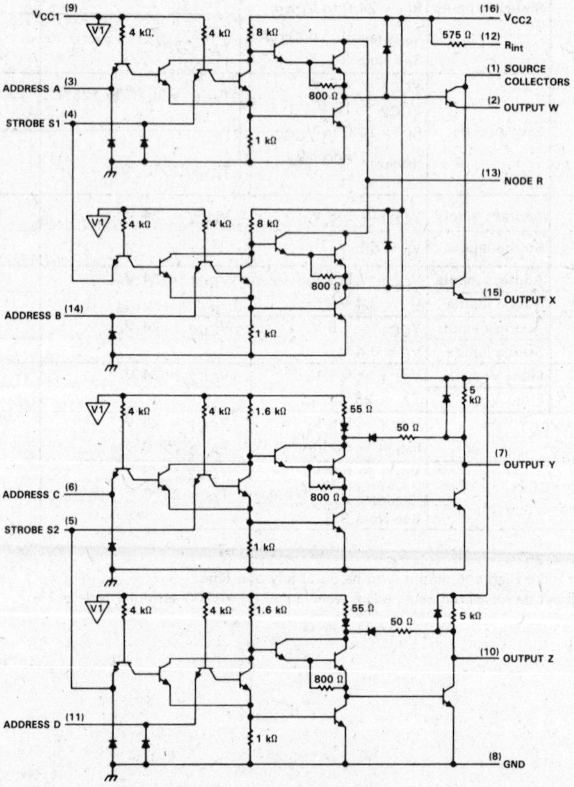
VCC1 (9)
V1
4 kΩ
8 kΩ
1.6 kΩ
800 Ω
1 kΩ
55 Ω
50 Ω
5 kΩ
575 Ω
ADDRESS A (3)
STROBE S1 (4)
ADDRESS B (14)
ADDRESS C (6)
STROBE S2 (5)
ADDRESS D (11)
(16) VCC2
(12) Rint
(1) SOURCE COLLECTORS
(2) OUTPUT W
(13) NODE R
(15) OUTPUT X
(7) OUTPUT Y
(10) OUTPUT Z
(8) GND

Component values shown are nominal. Pin numbers shown are for the J package.

TEXAS INSTRUMENTS
POST OFFICE BOX 655012 • DALLAS, TEXAS 75265

TYPICAL APPLICATION DATA

balanced bipolar logic-line driver

The circuit shown in Figure 1 converts standard TTL logic to bipolar logic. Bipolar logic is primarily used in transmitting data or clock pulses over long lines. This line-driver may be operated from a single 5-volt supply; however, the output drive may be increased by raising the supply voltage to the source collectors. The circuit features a three-state output that is off during the absence of data, thus not dissipating high power. It provides a balanced drive circuit giving maximum noise immunity when used with the proper line receiver. Large drive levels can be used to further increase noise immunity. The circuit is capable of driving twisted-pair lines of several thousand feet in length or low-impedance coaxial lines.

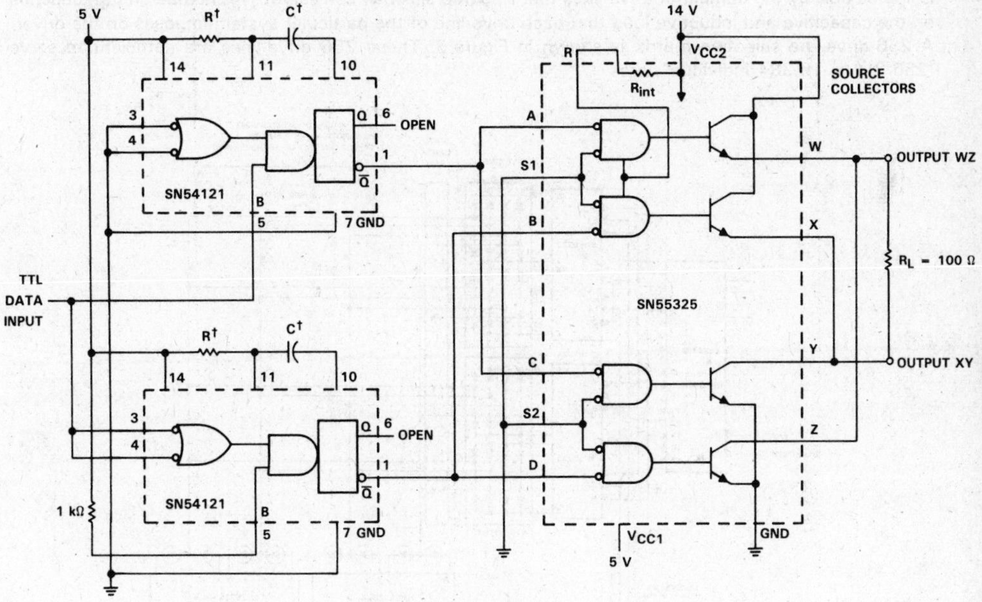
5 V
R†
C†
14
11
10
3
4
6
OPEN
1
SN54121
B
5
7 GND
TTL
DATA
INPUT
1 kΩ
14 V
R
VCC2
Rint
SOURCE
COLLECTORS
A
S1
B
C
S2
D
W
X
Y
Z
OUTPUT WZ
RL = 100 Ω
OUTPUT XY
SN55325
VCC1
5 V
GND

TEST CIRCUIT

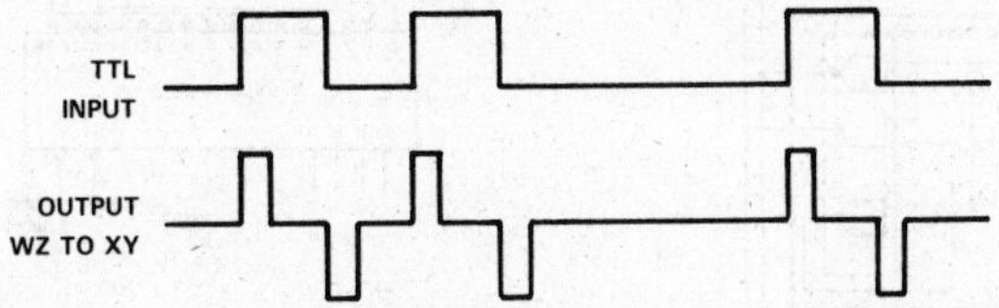
TTL
INPUT
OUTPUT
WZ TO XY

VOLTAGE WAVEFORMS

†R and C are adjusted to give the desired bipolar output pulse width.

FIGURE 1. BALANCED BIPOLAR LOGIC-LINE DRIVER

TYPICAL APPLICATION DATA

balanced bipolar logic-line driver

In memory-drive applications, the SN55325 can be connected in any of several ways. Typically, however, sources and sinks are arranged in pairs from which many drive-lines branch off as shown in Figure 2. Here each drive-line is served by a unique combination of two source/sink pairs so that a selection matrix is formed. To select drive-line 13, SN54154 No. 1 must be set to 3 (with mode select high), enabling source X of SN55325 No. 2 to drive lines 12 through 15, and SN54154 No. 2 must be set to 2, providing a sink at Y of SN55325 No. 4 for drive-line 13 only. Alternatively, to drive current in drive-line 13 in the opposite direction, only the mode-select voltage would be changed from high to low. The size of such a matrix is limited only by the number of drive-lines that a source/sink pair can serve. This number in turn depends on the capacitive and inductive load that each drive-line of the particular system imposes on the driver. A 256-drive-line selection matrix is shown in Figure 3. These 256 drive-lines are sufficient to serve $(256/2)^2 = 16{,}384$ individual cores.

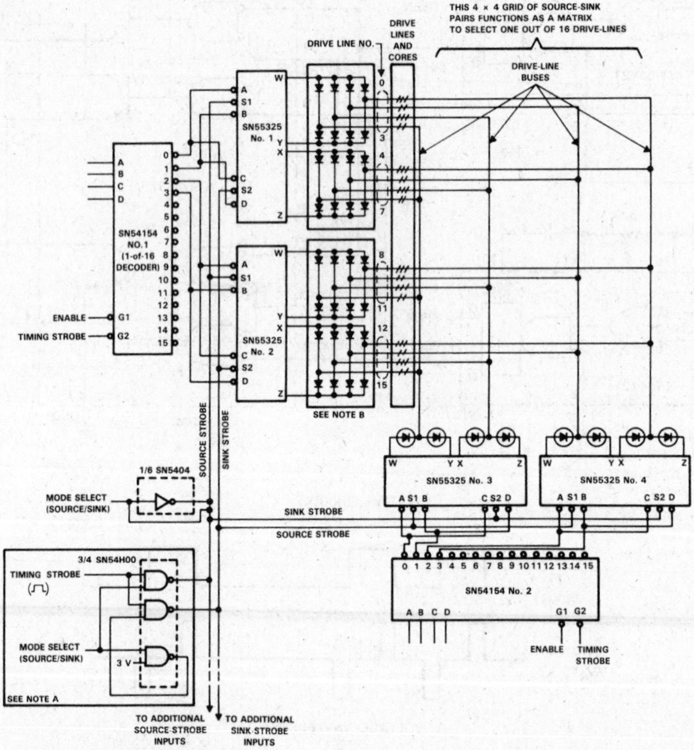
THIS 4 × 4 GRID OF SOURCE-SINK PAIRS FUNCTIONS AS A MATRIX TO SELECT ONE OUT OF 16 DRIVE-LINES
DRIVE LINE NO.
DRIVE LINES AND CORES
DRIVE-LINE BUSES
SN55325 No. 1
SN55325 No. 2
SN54154 NO.1 (1-of-16 DECODER)
ENABLE
TIMING STROBE
SEE NOTE B
SOURCE STROBE
SINK STROBE
1/6 SN5404
MODE SELECT (SOURCE/SINK)
SN55325 No. 3
SN55325 No. 4
SN54154 No. 2
3/4 SN54H00
3 V
SEE NOTE A
TO ADDITIONAL SOURCE-STROBE INPUTS
TO ADDITIONAL SINK-STROBE INPUTS

NOTE A: This optional mode-select and timing-strobe technique can be used in place of the SN5404 mode-select and SN54154 timing-strobe when minimum time skew is desired.

B. All diodes are IN4607.

FIGURE 2. SN55325 USED AS A MEMORY DRIVER TO SELECT ONE OF SIXTEEN DRIVE LINES

TYPICAL APPLICATION DATA

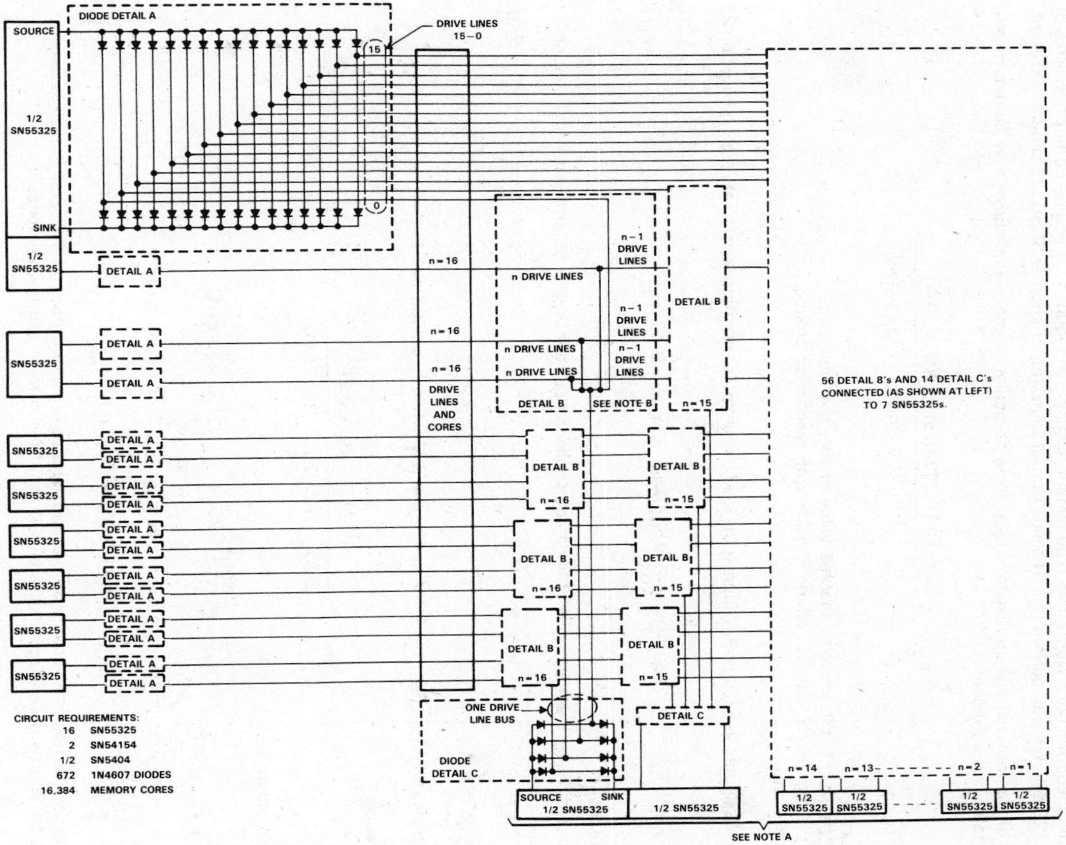
DIODE DETAIL A
SOURCE
1/2 SN55325
SINK
DRIVE LINES 15–0
15
0
1/2 SN55325
DETAIL A
SN55325
n = 16
DRIVE LINES AND CORES
n DRIVE LINES
n – 1 DRIVE LINES
DETAIL B
SEE NOTE B
n = 15
56 DETAIL B's AND 14 DETAIL C's CONNECTED (AS SHOWN AT LEFT) TO 7 SN55325s.
ONE DRIVE LINE BUS
DIODE DETAIL C
DETAIL C
SOURCE
SINK
1/2 SN55325
n = 14
n = 13
n = 2
n = 1
SEE NOTE A
CIRCUIT REQUIREMENTS:
16 SN55325
2 SN54154
1/2 SN5404
672 1N4607 DIODES
16,384 MEMORY CORES

NOTES: A. Outputs from one SN54154 decoder are connected to each SN55325 as shown in Figure 2. Source strobe and sink strobe from an SN5404 are connected to each SN55325 as shown in Figure 2.
B. The division of the drive-line bus into four segments reduces the capacitive load on the SN55325 driver.
C. All diodes are IN4607.

FIGURE 3. SN55325 SERVING 256 DRIVE LINES IN A MAGNETIC MEMORY

TYPICAL APPLICATION DATA

external resistor calculation

A typical magnetic-memory word-drive requirement is shown in Figure 4. A source-output transistor of one SN55325 delivers load current (I_L). The sink-output transistor of another SN55325 sinks this current.

The value of the external pull-up resistor (R_{ext}) for a particular memory application may be determined using the following equation:

$$R_{ext} = \frac{16\,[V_{CC2(min)} - V_S - 2.2]}{I_L - 1.6\,[V_{CC2(min)} - V_S - 2.9]} \quad (1)$$

where: R_{ext} is in kΩ,
$V_{CC2(min)}$ is the lowest expected value of V_{CC2} in volts,
V_S is the source output voltage in volts with respect to ground,
I_L is in mA.

The power dissipated in resistor R_{ext} during the load current pulse duration is calculated using Equation 2,

$$P_{Rext} \approx \frac{I_L}{16}\,[V_{CC2(min)} - V_S - 2] \quad (2)$$

where: P_{Rext} is in mW.

After solving for R_{ext}, the magnitude of the source collector current (I_{CS}) is determined from Equation 3,

$$I_{CS} \approx 0.94\,I_L \quad (3)$$

where: I_{CS} in in mA.

As an example, let $V_{CC2(min)} = 20$ V and $V_L = 3$ V while I_L of 500 mA flows.

Using Equation 1,

$$R_{ext} = \frac{16\,(20 - 3 - 2.2)}{500 - 1.6\,(20 - 3 - 2.9)} = 0.5\ \text{k}\Omega$$

and from Equation 2,

$$P_{Rext} \approx \frac{500}{16}\,[20 - 3 - 2] \approx 470\ \text{mW}$$

The amount of the memory system current source (I_{CS}) from Equation 3 is:

$$I_{CS} \approx 0.94\,(500) \approx 470\ \text{mA}$$

In this example, the regulated source-output transistor base current through the external pull-up resistor (R_{ext}) and the source gate is approximately 30 mA. This current and I_{CS} comprise I_L.

TEXAS INSTRUMENTS
POST OFFICE BOX 655012 • DALLAS, TEXAS 75265

TYPICAL APPLICATION DATA

external resistor calculation (continued)

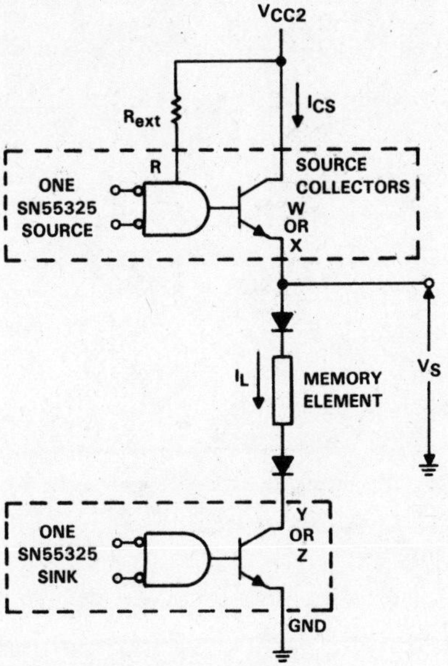
VCC2
ICS
Rext
R
ONE
SN55325
SOURCE
SOURCE
COLLECTORS
W
OR
X
IL
MEMORY
ELEMENT
VS
ONE
SN55325
SINK
Y
OR
Z
GND

NOTES: A. For clarity, partial logic diagrams of two SN55325s are shown.
B. Source and sink shown are in different packages.

FIGURE 4

D1496, SEPTEMBER 1973—REVISED SEPTEMBER 1986

Common Features

- Inputs Compatible with TTL Logic Levels
- Minimum Time Skew Between Strobe and Output-Current Rise
- Compatible with High-Speed Magnetic Core Memories

SN55326 Features

- Quad Positive-OR Sink Driver
- 600-mA Output Current Sink Capability
- 24-V Output Capability
- Output Clamp Voltage Variable to 24 V

SN55327 Features

- Quad Positive-OR Source Driver
- 600-mA Output Source Capability
- V_{CC2} Voltage Variable to 24 V
- Output Capable of Swinging Between V_{CC2} and Ground

SN55326 . . . J PACKAGE
(TOP VIEW)

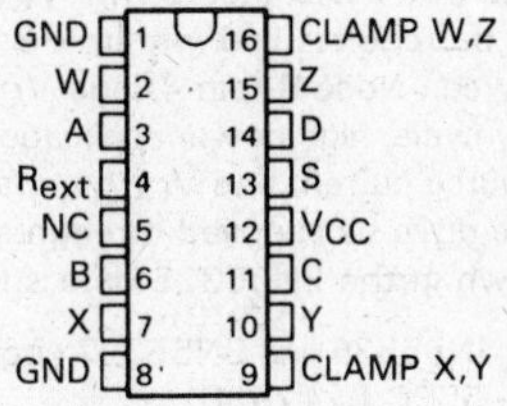
GND 1
W 2
A 3
Rext 4
NC 5
B 6
X 7
GND 8
16 CLAMP W,Z
15 Z
14 D
13 S
12 VCC
11 C
10 Y
9 CLAMP X,Y

SN55327 . . . J PACKAGE
(TOP VIEW)

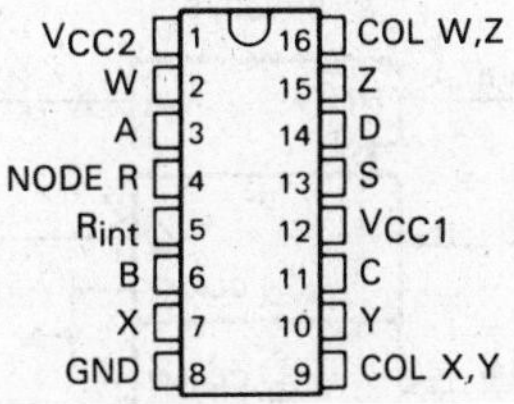
VCC2 1
W 2
A 3
NODE R 4
Rint 5
B 6
X 7
GND 8
16 COL W,Z
15 Z
14 D
13 S
12 VCC1
11 C
10 Y
9 COL X,Y

NC—No internal connection

FUNCTION TABLE

INPUTS					OUTPUTS			
ADDRESS				STROBE				
A	B	C	D	S	W	X	Y	Z
L	H	H	H	L	ON	OFF	OFF	OFF
H	L	H	H	L	OFF	ON	OFF	OFF
H	H	L	H	L	OFF	OFF	ON	OFF
H	H	H	L	L	OFF	OFF	OFF	ON
H	H	H	H	X	OFF	OFF	OFF	OFF
X	X	X	X	H	OFF	OFF	OFF	OFF

H = high level, L = low level, X = irrelevant
NOTE: Not more than one output is to be on at any one time.

description

The SN55326 and SN55327 are monolithic integrated circuit quadruple memory core drivers. These devices accept standard TTL decoder input signals and provide high-current and high-voltage output levels suitable for driving magnetic memory elements. Output transistor selection is determined by using one of the four address inputs and the common timing strobe.

The SN55326 memory core driver can sink up to 600 milliamperes and operate from a single 5-volt supply. Each driver is similar to the sink drivers of the SN55325. The four output transistors share a common base-drive resistor and it is recommended that only one of the four driver gates be selected at a time. Output-transistor base current may be increased by connecting an external resistor between R_{ext} (pin 4) and V_{CC}. Each output collector is protected from voltage surges during inductive switching by a clamping diode in parallel with its internal pull-up resistor. The two clamp pins may be returned to a power supply of from 4.5 volts to 24 volts.

The SN55327 memory core switch can source or sink up to 600 milliamperes and operate from two supplies; one of five volts and the other from 4.5 volts to 24 volts. Each switch is similar to the source drivers of the SN55325. They can function as either sink drivers or source drivers since the voltages at the output transistor terminals are capable of swinging between V_{CC2} and ground. The four output transistors share a common base-drive resistor and it is recommended that only one of the four outputs be selected at a time. An internal base-drive resistor is available on the chip and can be used by connecting Node R (pin 4) to R_{int} (pin 5). This resistor provides adequate base current to the output transistors for output sink currents

description (continued)

up to 375 milliamperes with V_{CC2} at 15 volts or 600 milliamperes with V_{CC2} at 24 volts. Base current can be regulated to within ±5 percent by substituting for this resistor an external resistor connected between Node R (pin 4) and V_{CC2} with R_{int} (pin 5) remaining open. This method is preferable in high-duty-cycle, high-power applications since the power dissipated in this resistor is outside the package. When a source current and V_{CC2} voltage other than the above values are required, it is recommended that the base drive be supplied through an external resistor of the appropriate value calculated using Equation 1 shown in the SN55325 data sheet.

The SN55326 and SN55327 circuits are characterized for operation over the full military temperature range of −55°C to 125°C.

logic symbols†

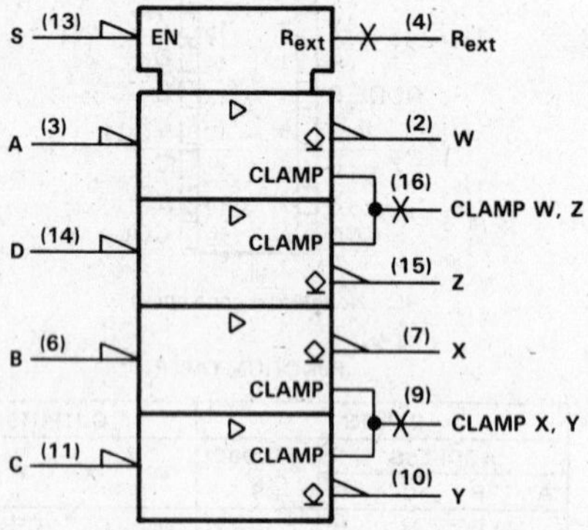
SN55326
S (13) EN
Rext (4) Rext
A (3)
(2) W
CLAMP
(16) CLAMP W, Z
D (14)
CLAMP
(15) Z
B (6)
(7) X
CLAMP
(9) CLAMP X, Y
C (11)
CLAMP
(10) Y

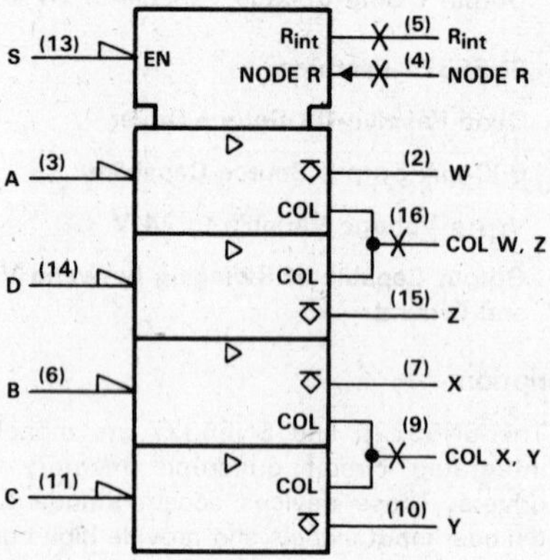
SN55327
S (13) EN
Rint (5) Rint
NODE R (4) NODE R
A (3)
(2) W
COL
(16) COL W, Z
D (14)
COL
(15) Z
B (6)
(7) X
COL
(9) COL X, Y
C (11)
COL
(10) Y

†These symbols are in accordance with ANSI/IEEE Std 91-1984 and IEC Publication 617-12.

TEXAS INSTRUMENTS
POST OFFICE BOX 655012 • DALLAS, TEXAS 75265

logic diagrams (positive logic)

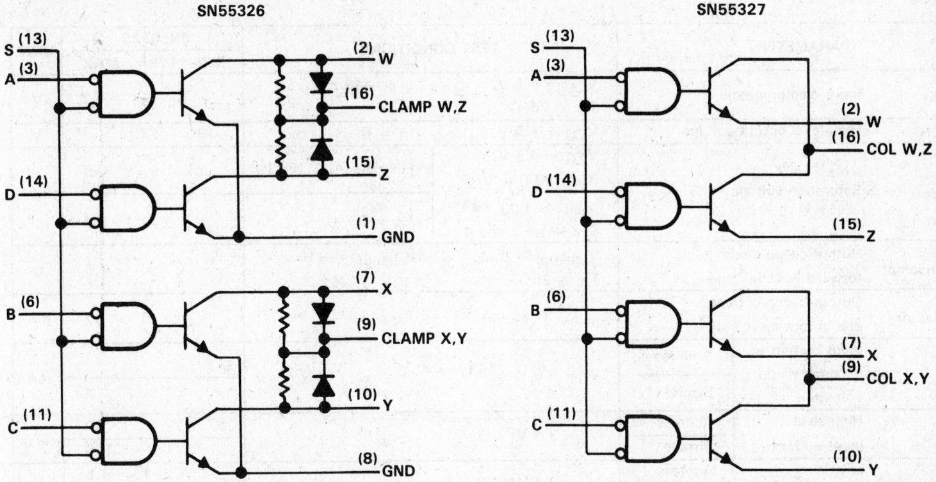
SN55326
S (13)
A (3)
(2) W
(16) CLAMP W,Z
(15) Z
D (14)
(1) GND
(7) X
B (6)
(9) CLAMP X,Y
(10) Y
C (11)
(8) GND
SN55327
S (13)
A (3)
(2) W
(16) COL W,Z
D (14)
(15) Z
B (6)
(7) X
(9) COL X,Y
C (11)
(10) Y

absolute maximum ratings over operating free-air temperature range (unless otherwise noted)

	SN55326	SN55327	UNIT
Supply voltage, V_{CC} or V_{CC1} (see Note 1)	7	7	V
Supply voltage, V_{CC2}		25	V
Input voltage, any address or strobe	5.5	5.5	V
Output collector voltage	25	25	V
Output clamp voltage	25		V
Output collector current	750	750	mA
Continuous total dissipation at (or below) 25°C free-air temperature (see Note 2)	1375	1375	mW
Operating free-air temperature range	−55 to 125	−55 to 125	°C
Storage temperature range	−65 to 150	−65 to 150	°C
Lead temperature 1,6 mm (1/16 inch) from case for 60 seconds	300	300	°C

NOTES: 1. Voltage values are with respect to the network ground terminal(s).
2. For operation above 25°C free-air temperature, derate linearly at the rate of 11.0 mW/°C.

recommended operating conditions

	SN55326			SN55327			UNIT
	MIN	NOM	MAX	MIN	NOM	MAX	
Supply voltage, V_{CC} or V_{CC1}	4.5	5	5.5	4.5	5	5.5	V
Supply voltage, V_{CC2}				4.5		24	V
High-level input voltage, V_{IH}	2			2			V
Low-level input voltage, V_{IL}			0.8			0.8	V
Output collector voltage			24			24	V
Output clamp voltage, $V_{(clamp)}$	4.5		24				V
Output collector current			600			600	mA
Operating free-air temperature, T_A	−55		125	−55		125	°C

electrical characteristics over recommended operating free-air temperature range (unless otherwise noted)

PARAMETER			TEST CONDITIONS†		SN55326 MIN	TYP‡	MAX	UNIT
V_{IK}	Input clamp voltage		V_{CC} = 4.5 V, T_A = 25°C	I_I = −10 mA,		−1	−1.7	V
V_{OH}	High-level output voltage		V_{CC} = 4.5 V,	I_O = 0	19	23		V
$V_{(sat)}$	Saturation voltage		V_{CC} = 4.5 V, R_L = 23 Ω to V_{CC}, $I_{(sink)}$ ≈ 600 mA§, See Notes 3 and 4	Full range			0.9	V
				T_A = 25°C		0.43	0.7	
$V_{F(clamp)}$	Output-clamp-diode forward voltage		$V_{(clamp)}$ = 0, T_A = 25°C	$I_{(clamp)}$ = −10 mA,			1.5	V
$I_{(clamp)}$	Output-clamp-current, one output on		$I_{(sink)}$ = 50 mA,	T_A = 25°C		5	7	mA
I_I	Input current at maximum input voltage	Address	V_I = 5.5 V				1	mA
		Strobe					4	
I_{IH}	High-level input current	Address	V_I = 2.4 V				40	μA
		Strobe					160	
I_{IL}	Low-level input current	Address	V_I = 0.4 V			−1	−1.6	mA
		Strobe				−4	−6.4	
$I_{CC(off)}$	Supply current, all outputs off		All inputs at 5 V,	T_A = 25°C		18	25	mA
$I_{CC(on)}$	Supply current, one output on		$I_{(sink)}$ = 50 mA,	T_A = 25°C		58	75	mA

switching characteristics, V_{CC} = 5 V, T_A = 25°C

PARAMETER¶	TO (OUTPUT)	TEST CONDITIONS§	MIN	TYP	MAX	UNIT
t_{PLH}	W, X, Y, or Z	V_S = $V_{(clamp)}$ = 15 V, R_L = 24 Ω, C_L = 25 pF, See Figure 3		40	50	ns
t_{PHL}				35	50	
t_{TLH}	W, X, Y, or Z			10	15	ns
t_{THL}				15	20	
t_s	W, X, Y, or Z			30	35	ns
V_{OH}	W, X, Y, or Z	V_S = $V_{(clamp)}$ = 24 V, R_L = 47 Ω, C_L = 25 pF, $I_{(sink)}$ ≈ 500 mA, See Figure 3		$V_S - 1$		mV

† Unless otherwise noted, V_{CC} = 5.5 V, $V_{(clamp)}$ = 24 V. See Figure 1.
‡ All typical values are at T_A = 25°C.
§ Under these conditions, not more than one output is to be on at any one time.
¶ t_{PLH} ≡ propagation delay time, low-to-high-level output
t_{PHL} ≡ propagation delay time, high-to-low-level output
t_{TLH} ≡ transition time, low-to-high-level output
t_{THL} ≡ transition time, high-to-low-level output
t_s ≡ storage time
V_{OH} ≡ high-level output voltage (after switching)

NOTES: 3. These parameters must be measured using pulse techniques. t_w = 200 μs, duty cycle ≤ 2%.
4. R_{ext} is connected to V_{CC} through a 40-Ω resistor.

TEXAS INSTRUMENTS
POST OFFICE BOX 655012 • DALLAS, TEXAS 75265

electrical characteristics over recommended operating free-air temperature range (unless otherwise noted)

PARAMETER			TEST CONDITIONS†		SN55327 MIN	TYP‡	MAX	UNIT
V_{IK}	Input clamp voltage		V_{CC} = 4.5 V, T_A = 25°C	I_I = −10 mA,		−1	−1.7	V
$I_{(off)}$	Collectors terminal off-state current		V_{CC1} = 4.5 V, $V_{(col)}$ = 24 V	Full range			500	μA
				T_A = 25°C			150	
$V_{(sat)}$	Saturation voltage		V_{CC1} = 4.5 V, V_O = 0, R_L = 25 Ω to 15 V, $I_{(source)}$ ≈ −600 mA§, See Notes 3 and 5	Full range			0.9	V
				T_A = 25°C		0.43	0.7	
I_I	Input current at maximum input voltage	Address	V_I = 5.5 V				1	mA
		Strobe					4	
I_{IH}	High-level input current	Address	V_I = 2.4 V				40	μA
		Strobe					160	
I_{IL}	Low-level input current	Address	V_I = 0.4 V			−1	−1.6	mA
		Strobe				−4	−6.4	
$I_{CC(off)}$	Supply current, all outputs off	From V_{CC1}	All inputs at 5 V,	T_A = 25°C		7	10	mA
		From V_{CC2}				13	20	
$I_{CC(on)}$	Supply current, one output on	From V_{CC1}	$V_{(col)}$ = 6 V, T_A = 25°C,	$I_{(source)}$ = −50 mA, See Note 3		8	12	mA
		From V_{CC2}				36	55	

switching characteristics, V_{CC1} = 5 V, T_A = 25°C

PARAMETER¶	TO (OUTPUT)	TEST CONDITIONS§	MIN	TYP	MAX	UNIT
t_{PLH}	Collectors W, Z or X, Y	V_S = V_{CC2} = 15 V, R_L = 24 Ω, C_L = 25 pF, See Figure 3 and Note 5		35	55	ns
t_{PHL}				30	55	
t_{TLH}	W, X, Y, or Z	$V_{(col)}$ = V_{CC2} = 20 V, R_L = 100 Ω, C_L = 25 pF, See Figure 4 and Note 5		30		ns
t_{THL}				10		
V_{OH}	Collectors W, Z or X, Y	V_S = V_{CC2} = 24 V, $I_{(sink)}$ ≈ 500 mA, R_L = 47 Ω, C_L = 25 pF, See Figure 3 and Note 5		$V_S - 1$		mV

† Unless otherwise noted, V_{CC1} = 5.5 V, V_{CC2} = 24 V. See Figure 2.
‡ All typical values are at T_A = 25°C.
§ Under these conditions, not more than one output is to be on at any one time.
¶ t_{PLH} ≡ propagation delay time, low-to-high-level output
t_{PHL} ≡ propagation delay time, high-to-low-level output
t_{TLH} ≡ transition time, low-to-high-level output
t_{THL} ≡ transition time, high-to-low-level output
V_{OH} ≡ high-level output voltage (after switching)

NOTES: 3. These parameters must be measured using pulse techniques. t_w = 200 μs, duty cycle ≤ 2%.
5. A 350-Ω resistor is connected between node R (pin 4) and V_{CC2} (pin 1) with R_{int} (pin 5) open.

6 Memory Interface Circuits

schematic

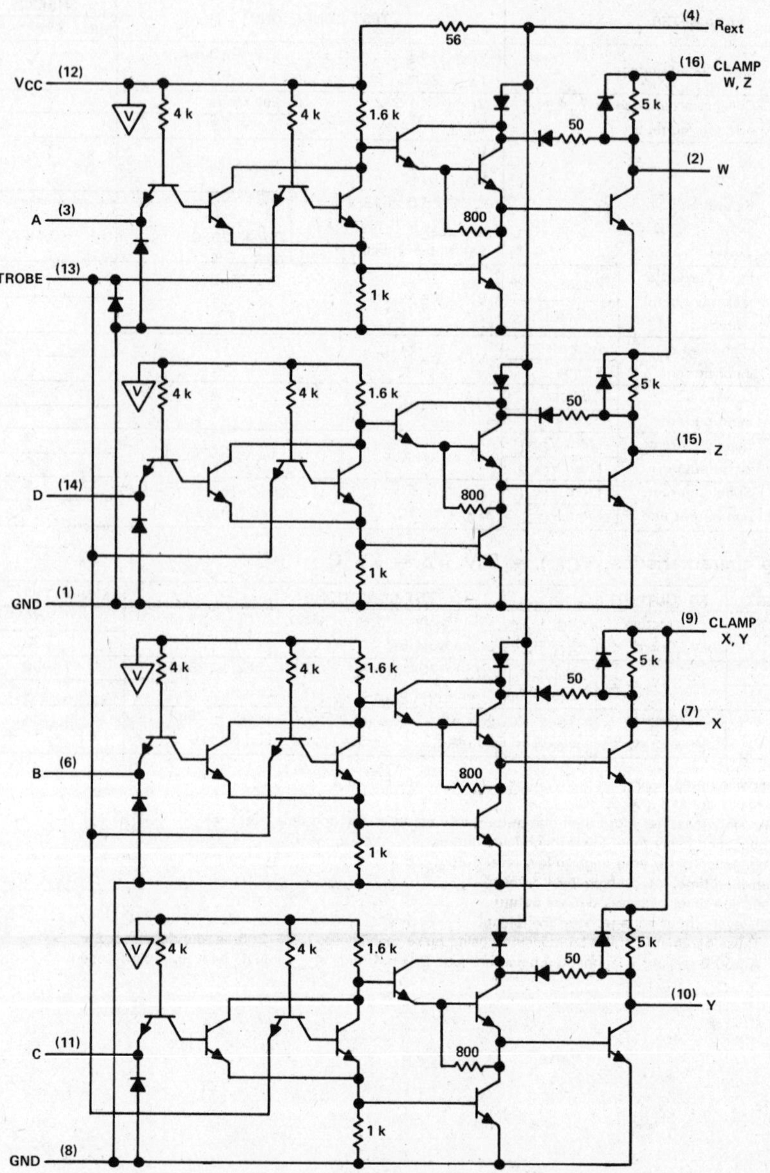
SN55326
(4) Rext
56
VCC (12)
(16) CLAMP W, Z
4 k
4 k
1.6 k
5 k
50
(2) W
A (3)
800
STROBE (13)
1 k
(15) Z
D (14)
GND (1)
(9) CLAMP X, Y
(7) X
B (6)
(10) Y
C (11)
GND (8)
V

Resistor values shown are nominal and in ohms.

TEXAS INSTRUMENTS
POST OFFICE BOX 655012 • DALLAS, TEXAS 75265

schematic

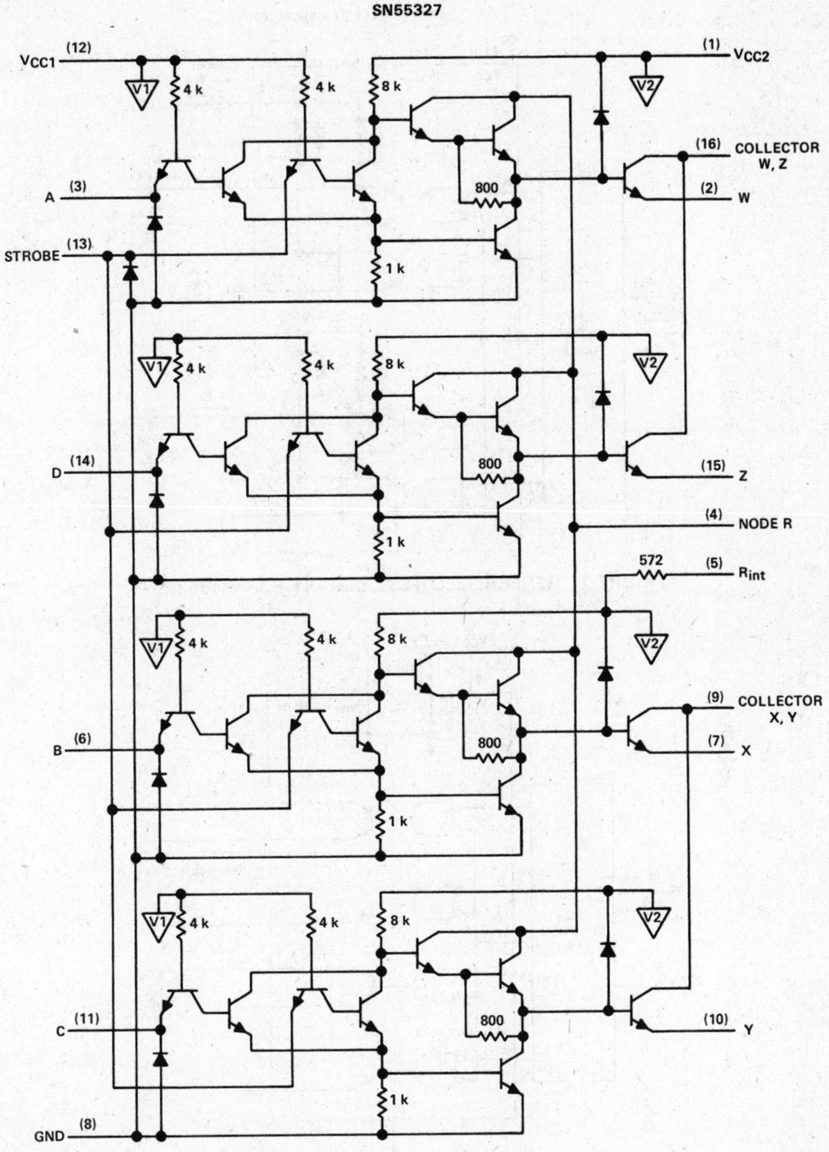
SN55327
VCC1
(12)
(1)
VCC2
V1
V2
4 k
8 k
800
1 k
A
(3)
STROBE
(13)
(16)
COLLECTOR
W, Z
(2)
W
D
(14)
(15)
Z
(4)
NODE R
572
(5)
Rint
B
(6)
(9)
COLLECTOR
X, Y
(7)
X
C
(11)
(10)
Y
GND
(8)

Resistor values shown are nominal and in ohms.

PARAMETER MEASUREMENT INFORMATION

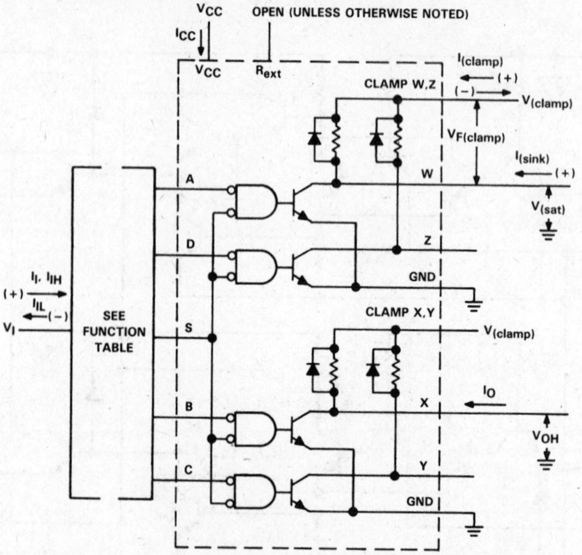
VCC
OPEN (UNLESS OTHERWISE NOTED)
ICC
VCC
Rext
I(clamp)
(+)
CLAMP W,Z
(−)
V(clamp)
VF(clamp)
I(sink)
(+)
W
V(sat)
A
D
Z
GND
II, IIH
(+)
IIL
(−)
VI
SEE
FUNCTION
TABLE
S
CLAMP X,Y
V(clamp)
IO
X
B
VOH
Y
C
GND

FIGURE 1. GENERALIZED TEST CIRCUIT FOR SN55326

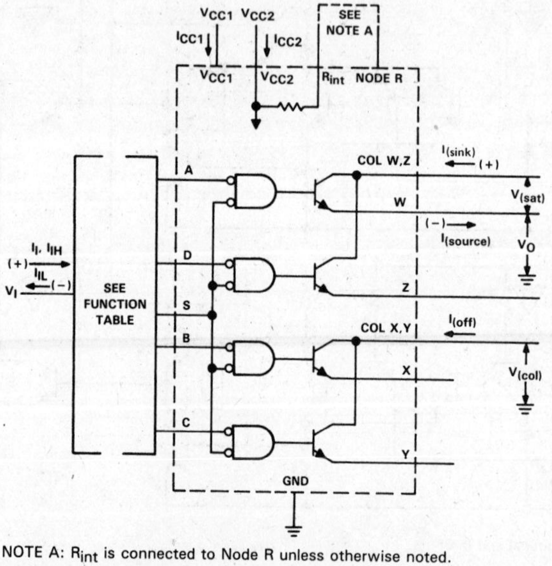
VCC1
VCC2
SEE
NOTE A
ICC1
ICC2
VCC1
VCC2
Rint
NODE R
I(sink)
(+)
COL W,Z
A
V(sat)
W
(−)
I(source)
VO
II, IIH
(+)
IIL
(−)
VI
D
SEE
FUNCTION
TABLE
S
Z
COL X,Y
I(off)
B
X
V(col)
C
Y
GND

NOTE A: R_{int} is connected to Node R unless otherwise noted.

FIGURE 2. GENERALIZED TEST CIRCUIT FOR SN55327

PARAMETER MEASUREMENT INFORMATION

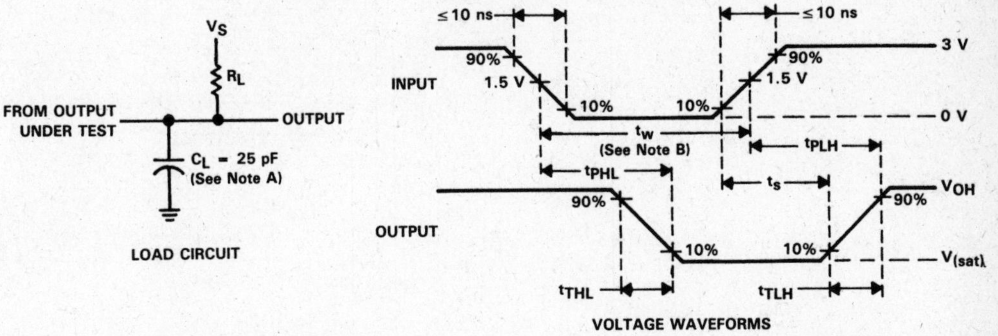
VS
RL
FROM OUTPUT UNDER TEST
OUTPUT
CL = 25 pF
(See Note A)
LOAD CIRCUIT
≤ 10 ns
≤ 10 ns
INPUT
90%
1.5 V
10%
10%
1.5 V
90%
3 V
0 V
tw
(See Note B)
tPHL
tPLH
ts
OUTPUT
90%
10%
10%
90%
VOH
V(sat)
tTHL
tTLH
VOLTAGE WAVEFORMS

NOTES: A. C_L includes probe and jig capacitance.

B. Input pulses are supplied by generators having the following characteristics: $Z_{out} \approx 50\ \Omega$. For testing V_{OH} (after switching), $t_w = 40\ \mu s$, PRR ≤ 12.5 kHz. For all other tests, $t_w = 200$ ns, duty cycle ≤ 1%.

FIGURE 3. SWITCHING TIMES

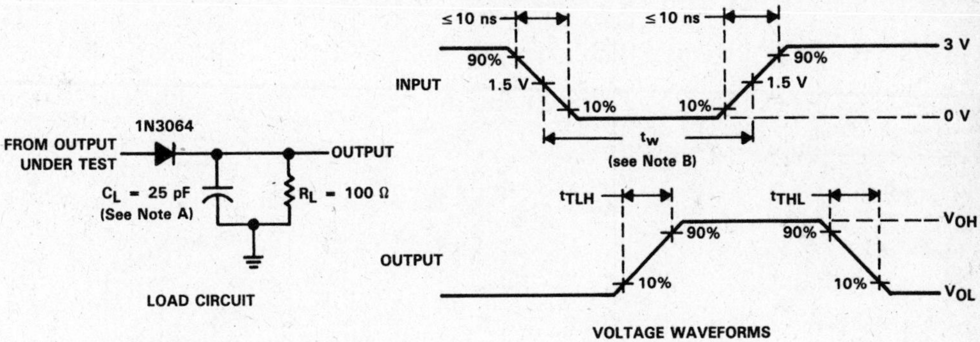
1N3064
FROM OUTPUT UNDER TEST
OUTPUT
CL = 25 pF
(See Note A)
RL = 100 Ω
LOAD CIRCUIT
≤ 10 ns
≤ 10 ns
INPUT
90%
1.5 V
10%
10%
1.5 V
90%
3 V
0 V
tw
(see Note B)
tTLH
tTHL
OUTPUT
90%
90%
10%
10%
VOH
VOL
VOLTAGE WAVEFORMS

NOTES: A. C_L includes probe and jig capacitance.

B. Input pulses are supplied by generators having the following characteristics: $Z_{out} \approx 50\ \Omega$, $t_w = 200$ ns, duty cycle ≤ 1%.

FIGURE 4. SWITCHING TIMES

6

Memory Interface Circuits

General Information 1
Alphanumeric Index
Selection Guide

Data Acquisition Circuits 2
Cross-Reference Guide
Data Sheets

Display Drivers 3
Data Sheets

Line Drivers and Receivers 4
Cross-Reference Guide
Data Sheets

Peripheral Drivers/Actuators 5
Cross-Reference Guide
Data Sheets

Memory Interface Circuits 6
Data Sheets

Speech Synthesis Circuits 7
Data Sheets

Appendix A Power Derating Curves A

Appendix B Ordering Instructions, Mechanical Data, IC Sockets B

Appendix C Explanation of Logic Symbols C

PRODUCT SUMMARY

This data is excerpted from the *TSP50C40A Speech Synthesizer Data Manual* intended for publication in early 1987.

- **LPC-10 Synthesizer**
- **8-Bit Microprocessor**
- **8K-Byte ROM**
- **32-Byte RAM**
- **18 Digital I/O Lines**
- **4-V to 6-V CMOS Technology for Low Power Dissipation**
- **High-Efficiency Push-Pull PWM (Pulse-Width Modulation) Digital-to-Analog Output Can Drive a Speaker Directly**
- **10- or 8-kHz Speech Sample Rate**

N PACKAGE
(TOP VIEW)

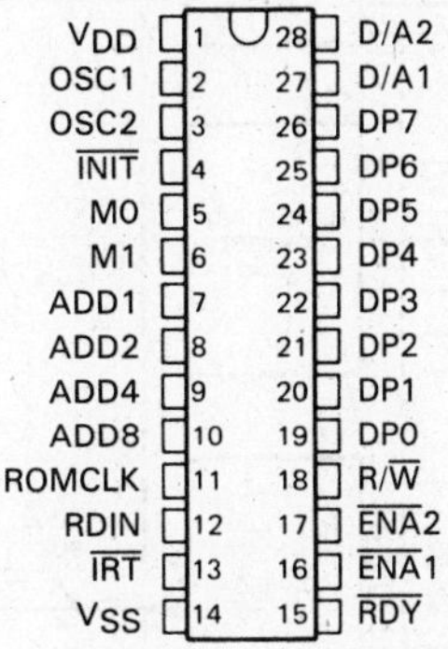
V_{DD} 1 28 D/A2
OSC1 2 27 D/A1
OSC2 3 26 DP7
$\overline{INIT}$ 4 25 DP6
M0 5 24 DP5
M1 6 23 DP4
ADD1 7 22 DP3
ADD2 8 21 DP2
ADD4 9 20 DP1
ADD8 10 19 DP0
ROMCLK 11 18 R/$\overline{W}$
RDIN 12 17 $\overline{ENA2}$
$\overline{IRT}$ 13 16 $\overline{ENA1}$
V_{SS} 14 15 $\overline{RDY}$

description

The TSP50C40A combines an 8-bit microprocessor, a speech synthesizer, a 32-byte RAM, and a 64K-bit ROM onto one chip.

The TSP50C40A implements an LPC-10 speech synthesis algorithm and a 10-pole latice filter to generate speech at a low data rate. The TSP50C40A internal processor accesses speech data from an internal and an external ROM (TSP60C20), decodes the speech data, and sends the decoded data to the synthesizer at a predetermined frame data rate. The internal processor also performs interpolations to smooth the transitions between frames. The output of the synthesizer can be used to drive a small speaker directly or can be used with an external low-power filter and amplifier.

The TSP50C40A is characterized for operation from 0°C to 70°C.

functional block diagram

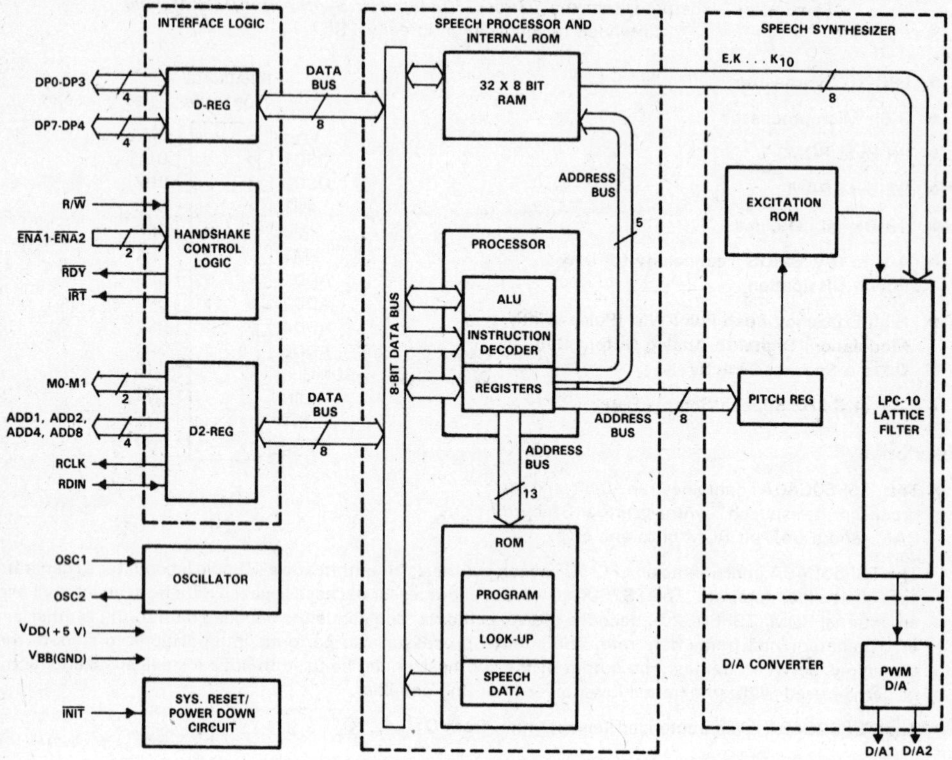
INTERFACE LOGIC
DP0-DP3
DP7-DP4
D-REG
R/W
ENA1-ENA2
HANDSHAKE CONTROL LOGIC
RDY
IRT
M0-M1
ADD1, ADD2, ADD4, ADD8
D2-REG
RCLK
RDIN
DATA BUS
OSC1
OSC2
OSCILLATOR
VDD(+5 V)
VBB(GND)
INIT
SYS. RESET/ POWER DOWN CIRCUIT
SPEECH PROCESSOR AND INTERNAL ROM
8-BIT DATA BUS
32 X 8 BIT RAM
ADDRESS BUS
PROCESSOR
ALU
INSTRUCTION DECODER
REGISTERS
ROM
PROGRAM
LOOK-UP
SPEECH DATA
SPEECH SYNTHESIZER
E,K . . . K10
EXCITATION ROM
PITCH REG
LPC-10 LATTICE FILTER
D/A CONVERTER
PWM D/A
D/A1
D/A2

TEXAS INSTRUMENTS
POST OFFICE BOX 655012 • DALLAS, TEXAS 75265

PIN FUNCTION TABLE

PIN NAME	PIN NO.	I/O	DESCRIPTION
ADD1	7	O	Address weight 1 output
ADD2	8	O	Address weight 2 output
ADD4	9	O	Address weight 4 output
ADD8	10	O	Address weight 8 output
DP0	19	I/O	Data bit 0 (LSB)
DP1	20	I/O	Data bit 1
DP2	21	I/O	Data bit 2
DP3	22	I/O	Data bit 3
DP4	23	I/O	Data bit 4
DP5	24	I/O	Data bit 5
DP6	25	I/O	Data bit 6
DP7	26	I/O	Data bit 7 (MSB)
D/A1	27	O	Noninverting digital-to-analog converter output (PWM)
D/A2	28	O	Inverting digital-to-analog converter output (PWM)
$\overline{ENA}$1	16	I	Input/output control of data bus (DP7–DP0).
$\overline{ENA}$2	17	I	Read mode (R/$\overline{W}$ high) $\overline{ENA}$1: Most significant nibble of data register is sent to the data bus (DP7–DP4) while $\overline{ENA}$1 is low. $\overline{ENA}$2: Least significant nibble of data register is output to the data bus (DP3←DP0) while $\overline{ENA}$2 is low. When $\overline{ENA}$1 goes low, $\overline{IRT}$ goes high. Write mode (R/$\overline{W}$) low) ENA1: Data already set on data bus DP7–DP4 are strobed into most significant nibble of data register toggling $\overline{ENA}$1 (high-low-high). Data is latched as $\overline{ENA}$1 goes high. ENA2: Data already set on data bus DP3–DP0 are strobed into least significant nibble of data register toggling $\overline{ENA}$2 (high-low-high). Data is latched as $\overline{ENA}$2 goes high.
$\overline{INIT}$	4	I	Initialize input. When low, device is initialized and goes into low-power mode.
$\overline{IRT}$	13	O	Ready for data output. $\overline{IRT}$ goes high as data in the data register is read on data bus DP4 through DP7. $\overline{IRT}$ can also be masked programmed to act as an event counter input.
M0	5	O	ROM mode control output
M1	6	O	ROM mode control output
OSC1	2	I	Clock Input
OSC2	3	O	Oscillator output. 2.86 MHz for speech data of 8-kHz sampling and 3.58 MHz for speech data of 10-kHz sampling. Can be used with crystal, ceramic resonator, or external input to OSC2.
RDIN	12	I/O	ROM data input
$\overline{RDY}$	15	O	Ready for data input. $\overline{RDY}$ goes high as data on data bus DP0 through DP3 is written into the data register. $\overline{RDY}$ is reset to a low level by internal command.
ROMCLK	11	O	Clock output to a ROM
R/$\overline{W}$	18	I	Read/Write input. Determines direction of data bus R/$\overline{W}$: When high, data transfers from data register (read). R/$\overline{W}$: When low, data transfers into data register (write).
V_{DD}	1	I	5-volt nominal supply voltage
V_{SS}	14	I	Substrate voltage (ground pin)

absolute maximum ratings over free-air temperature range

Supply voltage†, V_{DD} .... −0.3 V to 7 V
Input voltage, V_I .... −0.3 V to V_{DD} + 0.3 V
Output voltage, V_O .... −0.3 V to V_{DD} + 0.3 V
Storage temperature range .... −30°C to 125°C
Lead temperature 1,6 mm (1/16 inch) from case for 10 seconds .... 260°C

† Unless otherwise noted, all voltages are with respect to V_{SS}.

recommended operating conditions

			MIN	NOM	MAX	UNIT
V_{DD}	Supply voltage		4	5	6	V
V_{IH}	High-level input voltage	V_{DD} = 6 V	V_{DD} − 1.2		V_{DD}	V
		V_{DD} = 5 V	V_{DD} − 1		V_{DD}	
		V_{DD} = 4 V	V_{DD} − 0.7		V_{DD}	
V_{IL}	Low-level input voltage	V_{DD} = 6 V	V_{SS}		1.2	V
		V_{DD} = 5 V	V_{SS}		1	
		V_{DD} = 4 V	V_{SS}		0.7	
f_{osc}	Oscillator frequency		2	3.6	4	MHz
T_A	Operating free-air temperature		0		70	°C

electrical characteristics over recommended operating free-air temperature range (unless otherwise noted)

PARAMETER				MIN	TYP‡	MAX	UNIT
V_{OH}	High-level output voltage	V_{DD} = 5 V,	I_{OH} = −1 mA	V_{DD} − 2.5	V_{DD} − 0.5	V_{DD}	V
		V_{DD} = 5 V,	I_{OH} = 0.3 mA	V_{DD} − 0.5	V_{DD} − 0.2	V_{DD}	
V_{OL}	Low-level output voltage	V_{DD} = 5 V,	I_{OL} = 1.7 mA	V_{SS}	0.3	0.4	V
I_I	Input current	V_{IH} = V_{DD}				5	μA
		V_{IL} = V_{SS}		−5			
I_{OH}	High-level output current	V_{DD} = 6 V,	V_{OH} = V_{DD} − 0.5 V	−0.35	−1.5		mA
		V_{DD} = 5 V,	V_{OH} = V_{DD} − 0.5 V	−0.3	−1.2		
		V_{DD} = 4 V,	V_{OH} = V_{DD} − 0.5 V	−0.2	−0.8		
I_{OL}	Low-level output current	V_{DD} = 6 V,	V_{OL} = 0.4 V	2	2.8		mA
		V_{DD} = 5 V,	V_{OL} = 0.4 V	1.7	2.4		
		V_{DD} = 4 V,	V_{OL} = 0.4 V	1.2	1.8		
I_{DD}	Supply current	V_{DD} = 5 V, All outputs open	f_{osc} = 3.6 MHz,			2	mA

‡All typical values are at V_{DD} = 5 V, T_A = 25°C.

D2978, DECEMBER 1986

PRODUCT SUMMARY

This data is excerpted from the *TSP50C50 Speech Processor Data Manual* intended for publication in early 1987.

- **High-Quality Speech**
- **LPC12 or LPC10 Synthesis Algorithm**
- **10-kHz or 8-kHz Sample Rate**
- **Pitch Accuracy Better Than 0.1% at 1 kHz**
- **Parameter Interpolation Pitch Synchronous and Asynchronous**
- **On-Chip A-Law D/A Converter, 13-Bit Dynamic Range**
- **On-Chip 6th-Order Low-Pass Filter**
- **Phonetic Coding and Music Capabilities**
- **Standard 8- or 4-Bit Microprocessor Interface**
- **Serial Interface to TSP60C20 Speech ROM**
- **Standard PCM Interface-to-Digital Network**
- **4.32 MHz Crystal Oscillator**
- **28-Pin Plastic/Ceramic Dual-in-Line Package**
- **Silicon Gate CMOS Technology**
- **Single 5-V Power Supply**
- **Low-Power Standby Mode**

J OR N PACKAGE
(TOP VIEW)

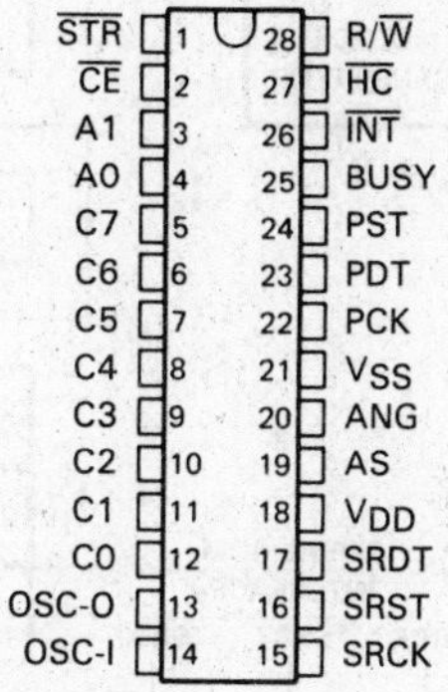
$\overline{STR}$ 1
$\overline{CE}$ 2
A1 3
A0 4
C7 5
C6 6
C5 7
C4 8
C3 9
C2 10
C1 11
C0 12
OSC-O 13
OSC-I 14
28 R/$\overline{W}$
27 $\overline{HC}$
26 $\overline{INT}$
25 BUSY
24 PST
23 PDT
22 PCK
21 V_{SS}
20 ANG
19 AS
18 V_{DD}
17 SRDT
16 SRST
15 SRCK

description

The TSP50C50 is a silicon-gate CMOS LSI device consisting of a speech processor, speech acquisition logic, PCM interface logic, analog output circuits, and microprocessor controlled interface. These functional areas contain features and controls that produce band-limited high-quality speech and extends the capability of the TSP50C50 allowing its use with a wide range of speech products. It offers specific controls for constructive speech synthesis and musical and telecommunication network applications.

The TSP50C50 is characterized for operation from 0°C to 70°C.

functional block diagram

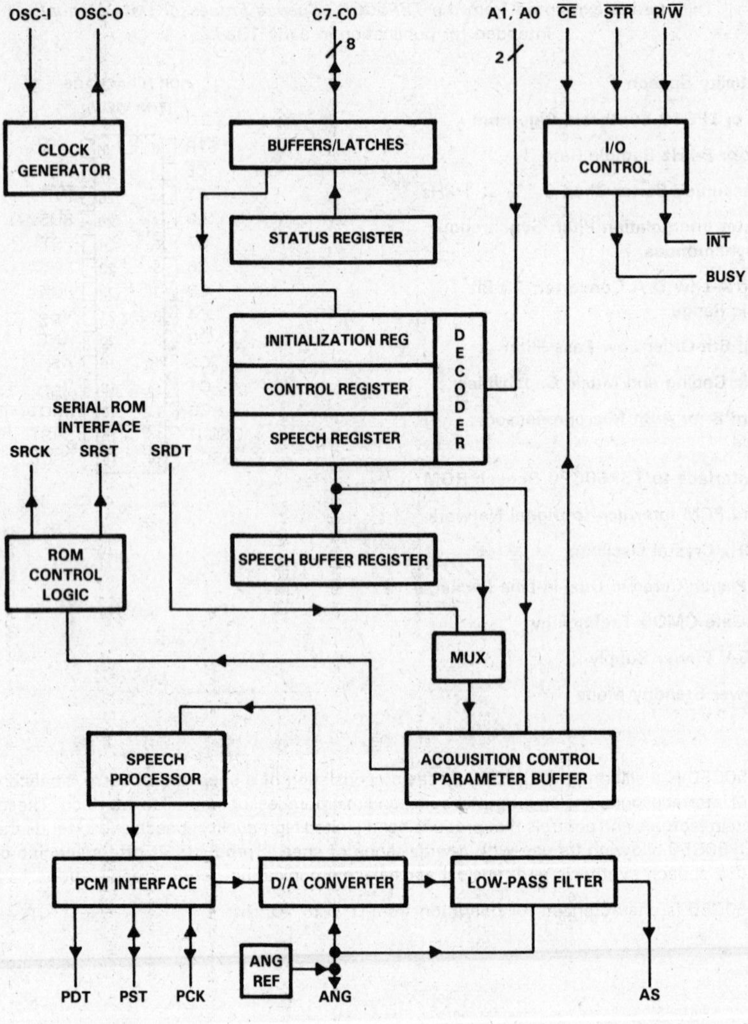
OSC-I
OSC-O
C7-C0
8
A1, A0
2
CE
STR
R/W
CLOCK GENERATOR
BUFFERS/LATCHES
I/O CONTROL
STATUS REGISTER
INT
BUSY
INITIALIZATION REG
CONTROL REGISTER
SPEECH REGISTER
DECODER
SERIAL ROM INTERFACE
SRCK
SRST
SRDT
ROM CONTROL LOGIC
SPEECH BUFFER REGISTER
MUX
SPEECH PROCESSOR
ACQUISITION CONTROL PARAMETER BUFFER
PCM INTERFACE
D/A CONVERTER
LOW-PASS FILTER
ANG REF
PDT
PST
PCK
ANG
AS

PIN FUNCTION TABLE

PIN NAME	PIN NO.	I/O	DESCRIPTION
ANG	20		2.5 V analog ground. May be used to polarize an external speaker or to filter parasitic ground noise.
AS	19	O	Analog speech (filtered audio) output. Output is suitable for driving a low-power audio amplifier or a low-power speaker.
A0	4	I	Register address inputs. Used in conjunction with $R/\overline{W}$ to define the function of the registers.
A1	3	I	
BUSY	25	O	Busy output. When high, indicates that the interface is busy and a write strobe cannot be accepted.
$\overline{CE}$	2	I	Chip enable input. When high, $\overline{CE}$ disables read and write. When low and strobe is low, $\overline{CE}$ enables read and write.
C0	12	I/O	Data input/output
C1	11	I/O	
C2	10	I/O	
C3	9	I/O	
C4	8	I/O	
C5	7	I/O	
C6	6	I/O	
C7	5	I/O	
$\overline{HC}$	27	I	Hardware clear input. When high, disables the low-power standby mode and enables the crystal oscillator. When low, stops the crystal oscillator, sets the register address to nonoperating state, sets the INIT register to the all-high default value, and enables the low-power standby mode.
$\overline{INT}$	26	O	Interrupt output. When low, indicates that a condition exists in the TSP50C50 that requires immediate action by the microprocessor.
OSC-0	13	O	Oscillator crystal connection
OSC-1	14	I	Oscillator crystal connection or input to TSP50C50 from external oscillator
PCK	22	I/O	Clock input/output. A free-running data-rate clock at 2.16 MHz.
PDT	23	O	Serial PCM/binary two's compliment speech data output
PST	24	I/O	Sample-rate input/output strobe (8 kHz or 10 kHz)
$R/\overline{W}$	28	I	Read/write select input. When high, selects read status register and, when low, selects write to address register.
SRCK	15	O	720-kHz clock output
SRDT	17	I	External serial speech data input to the multiplexer
SRST	16	O	Single-bit speech data request strobe output to external speech ROM — 360 kHz maximum
$\overline{STR}$	1	I	Strobe input. When high, disables read and write. When low and $\overline{CE}$ is low, enables read and write.
V_{DD}	18		5-volt power supply
V_{SS}	21		Digital ground

absolute maximum ratings over free-air temperature range (unless otherwise noted)

Supply voltage, V_{DD} (see Note 1) . . . . −0.3 V to 7 V
All input voltages (see Note 1) . . . . −0.3 V to V_{DD}+0.3 V
All output voltages (see Note 1) . . . . −0.3 V to V_{DD}+0.3 V
Operating free-air temperature range . . . . −55°C to 150°C
Lead temperature 1,6 mm (1/16 inch) from case for 10 seconds . . . . 260°C

NOTE 1: All voltages are with respect to V_{SS}.

recommended operating conditions

PARAMETER			MIN	TYP	MAX	UNIT
V_{DD}	Supply voltage		4.5	5	5.5	V
V_{IH}	High-level input voltage	OSC/input	2.8		V_{DD}	V
		All others	2		V_{DD}	
V_{IL}	Low-level input voltage	OSC/input	V_{SS}		0.6	V
		All others	V_{SS}		0.8	
f_{osc}	Crystal oscillator frequency			4.32		MHz
T_A	Operating free-air temperature		0		70	°C

electrical characteristics over full range of recommended operating conditions

PARAMETER			TEST CONDITIONS	MIN	TYP	MAX	UNIT
V_{OH}	High-level output voltage		I_{OH} = −400 μA	V_{DD}−0.5			V
V_{OL}	Low-level output voltage		I_{OL} = 1.6 mA			0.4	V
I_I	Input current	OSC/I input	V_I = V_{SS} to V_{DD}		8	20	μA
		C0-C7 inputs			10		
		All other inputs			2		
I_{DD}	Supply current		V_{DD} = 5 V, All outputs open, f_{osc} = 4.32 MHz		5		mA
$I_{DD(stby)}$	Standby current		V_I = V_{SS}−0.5 V to V_{SS}+0.5 V		50		μA

PRODUCT SUMMARY

This data is excerpted from the *TSP5110A Voice Synthesis Processor Data Manual* intended for publication in early 1987

- **LPC-10 Pitch Asynchronous Synthesis Algorithm**
- **On-Chip Oscillator and Clock Generation**
- **Direct Drive for 50-Ω Speaker**
- **Simple Control Interface for 4-Bit Microcomputers**

N PACKAGE
(TOP VIEW)

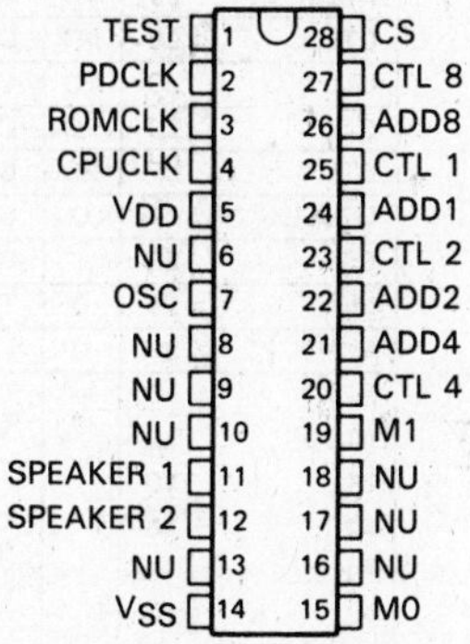
TEST 1
PDCLK 2
ROMCLK 3
CPUCLK 4
VDD 5
NU 6
OSC 7
NU 8
NU 9
NU 10
SPEAKER 1 11
SPEAKER 2 12
NU 13
VSS 14
28 CS
27 CTL 8
26 ADD8
25 CTL 1
24 ADD1
23 CTL 2
22 ADD2
21 ADD4
20 CTL 4
19 M1
18 NU
17 NU
16 NU
15 MO

NU—Make no external connection

description

The TSP5110A is a PMOS Voice Synthesis Processor. Speech is synthesized by processing an externally provided variable data-rate bit stream of encoded speech data and converting the result to an audible output with an on-chip eight-bit digital-to-analog converter and push-pull amplifier. The TSP5110A outputs all control signals necessary for direct control of the TSP6100 vocabulary ROM. Control of the TSP5110A is provided by an external device (e.g., TMS1000) through four control pins and a command clock.

The TSP5110A is characterized for operation from 0 °C to 70 °C

functional block diagram

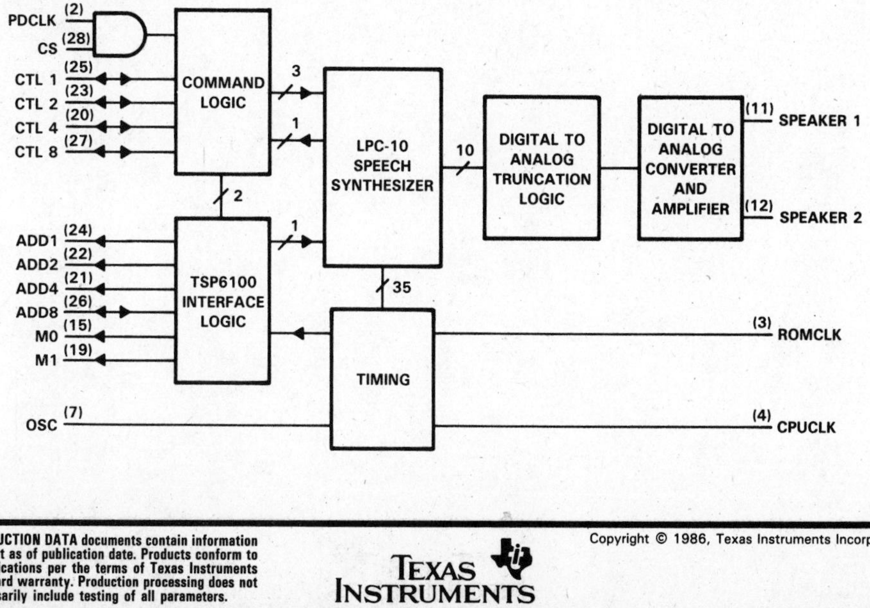
PDCLK (2)
CS (28)
CTL 1 (25)
CTL 2 (23)
CTL 4 (20)
CTL 8 (27)
COMMAND LOGIC
3
1
2
LPC-10 SPEECH SYNTHESIZER
10
DIGITAL TO ANALOG TRUNCATION LOGIC
DIGITAL TO ANALOG CONVERTER AND AMPLIFIER
(11) SPEAKER 1
(12) SPEAKER 2
ADD1 (24)
ADD2 (22)
ADD4 (21)
ADD8 (26)
M0 (15)
M1 (19)
TSP6100 INTERFACE LOGIC
1
35
TIMING
(3) ROMCLK
OSC (7)
(4) CPUCLK

PIN FUNCTION TABLE

PIN NAME	PIN NO.	I/O	DESCRIPTION
ADD1	24	O	Address weight 1 output to TSP6100
ADD2	22	O	Address weight 2 output to TSP6100
ADD4	21	O	Address weight 4 output to TSP6100
ADD8	26	I/O	Address weight 8 output to TSP6100 or serial data input
CTL 1	25	I/O	Control output to TSP6100 or read data bus input (LSB)
CTL 2	23	I/O	Control output to TSP6100 or read data bus input
CTL 4	20	I/O	Control output to TSP6100 or read data bus input
CTL 8	27	I/O	Control output to TSP6100 or read data bus input (MSB)
CPUCLK	4	O	Clock output for CPU (320 kHz)
CS	28	I	Chip select input
M0	15	O	Transfers data to TSP6100 control input
M1	19	O	Load address to TSP6100 control input
NU	6		Make no external connection. Used for factory testing only
NU	9		
NU	10		
NU	13		
NU	16		
NU	17		
NU	18		
NU	8		
OSC	7	I	Optional oscillator input (640 kHz)
PDCLK	2	I	Processor data clock
ROMCLK	3	O	Clock to control TSP6100 ROM (160 kHz)
SPEAKER 1	11	O	Analog speech out
SPEAKER 2	12	O	Analog speech out
TEST	1		For factory testing only
V_{DD}	5	I	−9 V nominal supply voltage
V_{SS}	14	I	Substrate supply voltage (0 V nominal)

TEXAS INSTRUMENTS
POST OFFICE BOX 655012 • DALLAS, TEXAS 75265

absolute maximum ratings over free-air temperature range

Supply voltage range, V_{DD} (see Note 1) . . . . −15 V to 0.3 V
Voltage applied to any device pin . . . . −15 V to 0.3 V
Operating free-air temperature range . . . . 0°C to 70°C
Storage temperature range . . . . −30°C to 125°C
Lead temperature 1,6 mm (1/16 inch) from case for 10 seconds . . . . 260°C

NOTE 1: Voltage values are with respect to V_{SS}.

recommended operating conditions

			MIN†	NOM	MAX†	UNIT
V_{DD}	Drain supply voltage		−8.3	−9	−9.7	V
V_{SS}	Substrate supply voltage			0		V
V_{IH}	High-level input voltage	ADD8	−0.7		0	V
		CTL 1 thru CTL 2, PDC, CS	−0.95		0	
V_{IL}	Low-level input voltage		V_{DD}		−4	V
T_A	Operating free-air temperature		0		70	°C

electrical characteristics over recommended operating free-air temperature range, V_{DD} = −9 V

PARAMETER			TEST CONDITIONS	MIN†	TYP‡	MAX†	UNIT
V_{OH}	High-level output voltage	ROM CLK, CPU CLK, I0, I1, ADD1 thru ADD8	I_{OH} = −100 μA	−0.5			V
		CTL 1 thru CTL 8	I_{OH} = −400 μA	−0.7		0	
V_{OL}	Low-level output voltage		I_{OL} = 100 μA	V_{DD}		−5	V
I_{IH}	High-level input current		V_I = 0			100	μA
I_{IL}	Low-level input current		V_I = −12 V			−50	μA
I_{DD}	Supply current		All inputs and outputs open		−20	−35	mA
P_o	Power output to speaker		100 Ω speaker load, 50 Ω each output	30			mW

†The algebraic convention, in which the more negative (less positive) limit is designated as minimum, is used in the data sheet for logic voltage levels only.
‡All typical values are at T_A = 25°C.

7

Speech Synthesis Circuits

D2960, SEPTEMBER 1986

TSP5220C VOICE SYNTHESIS PROCESSOR

PRODUCT SUMMARY

This data is excerpted from the *TSP5220C Voice Synthesis Processor Data Manual* intended for publication in early 1987

- **Pitch-Excited LPC-10 Synthesis Algorithm**
- **Choice of 4-kHz or 5-kHz Voice Input Bandwidth**
- **Low Data-Rate Range . . . 1000 to 1700 bps**
- **On-Chip 8-Bit Digital-to-Analog Converter**
- **TTL-Compatible Inputs and Outputs**
- **Compressed Voice Data Input Through Either 8-Bit Control/Data Bus and FIFO for Controller-Supplied Speech Data or Serial Interface for Use with TSP6100 Custom-Masked Speech ROMs**

N PACKAGE
(TOP VIEW)

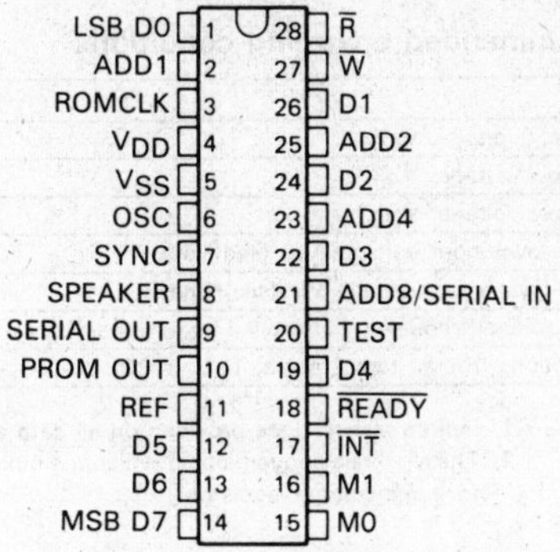
LSB D0 1
ADD1 2
ROMCLK 3
VDD 4
VSS 5
OSC 6
SYNC 7
SPEAKER 8
SERIAL OUT 9
PROM OUT 10
REF 11
D5 12
D6 13
MSB D7 14
28 R̄
27 W̄
26 D1
25 ADD2
24 D2
23 ADD4
22 D3
21 ADD8/SERIAL IN
20 TEST
19 D4
18 READY
17 INT
16 M1
15 M0

description

The TSP5220C is an LPC-10 voice synthesis function on a chip. A flexible interface structure allows a choice of storage media for the compressed model data and simple microprocessor selection of the spoken phrase or transfer of the data representing that phrase. For best performance, an external low-pass filter should be used between the TSP5220C and the audio amplifier or speaker.

The TSP5220C is characterized for operation from 0 °C to 70 °C.

functional block diagram

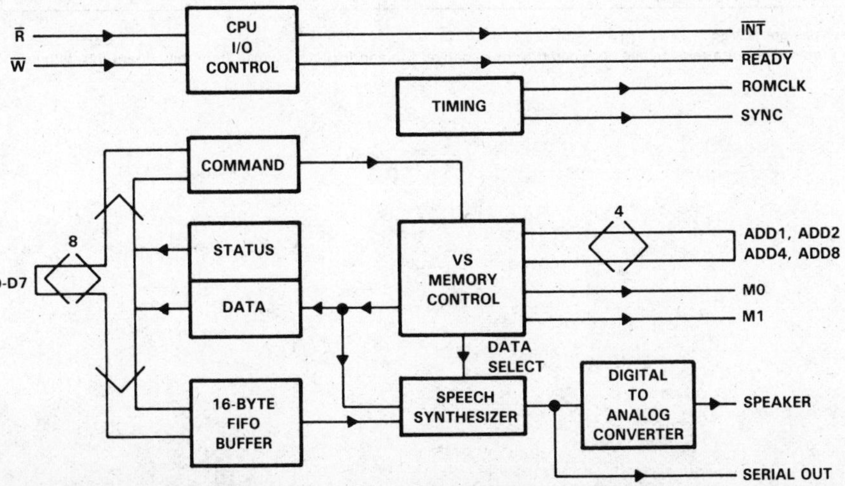
R̄
W̄
CPU I/O CONTROL
INT
READY
TIMING
ROMCLK
SYNC
COMMAND
8
D0-D7
STATUS
DATA
VS MEMORY CONTROL
4
ADD1, ADD2
ADD4, ADD8
M0
M1
DATA SELECT
16-BYTE FIFO BUFFER
SPEECH SYNTHESIZER
DIGITAL TO ANALOG CONVERTER
SPEAKER
SERIAL OUT

absolute maximum ratings over operating free-air temperature range (unless otherwise noted)

Any pin with respect to V_{SS} .... −15 V to 0.3 V
Continuous total dissipation .... 600 mW
Operating free-air temperature .... 0°C to 70°C
Storage temperature range .... −40°C to 125°C
Lead temperature 1,6 mm (1/16 inch) from case for 10 seconds .... 260°C

recommended operating conditions

	MIN	NOM	MAX	UNIT
Supply voltage, V_{SS}	4.75	5	5.25	V
Supply voltage, V_{ref}		0		V
Supply voltage, V_{DD}	−4.75	−5	−5.25	V
High-level input voltage, V_{IH} (see Note 1)	$V_{SS}-0.6$		V_{SS}	V
Low-level input voltage, V_{IL} (see Note 2)	V_{DD}		$V_{SS}-4$	V
Operational frequency (controlled by external RC)	620		825	kHz
Operating free-air temperature, T_A	0		70	°C

NOTES: 1. Pull-up resistors are provided on all data and select inputs. This permits direct drive from TTL-compatible devices.
2. The algebraic convention, in which the more positive (less negative) limit is designated as maximum, is used in this data sheet for logic voltage levels only.

electrical characteristics over full range of recommended operating conditions

PARAMETER			TEST CONDITIONS	MIN	TYP†	MAX	UNIT
V_{OH}	High-level output voltage	D0-D7, $\overline{W}$, $\overline{R}$, $\overline{INT}$	I_{OH} = 0.4 mA	2.4		V_{SS}	V
		ROMCLK, ADD1-ADD8, M0, M1	I_{OH} = 100 μA	$V_{SS}-0.5$		V_{SS}	
V_{OL}	Low-level output voltage	D0-D7, $\overline{W}$, $\overline{R}$, $\overline{INT}$	I_{OL} = 1.6 mA	$V_{ref}-0.5$		$V_{ref}+0.5$	V
		ROMCLK, ADD1-ADD8, M0, M1	I_{OL} = 100 μA	$V_{SS}-0.5$		V_{SS}	
I_{ref}	Supply current from V_{ref} (see Note 3)				−3	−5	mA
I_{DD}	Supply current from V_{DD} (see Note 3)				−10	−35	mA
C_i	Input capacitance, except data bus				15		pF
C_o	Output capacitance, except data bus				15		pF
C_L	Load capacitance, data bus			25		300	pF

†Typical values are at V_{SS} = 5 V, V_{DD} = −5 V, T_A = 25°C.
NOTE 3: Currents out of a terminal are considered to be negative. I_{ref} and I_{DD} are sourced from the current into terminal V_{SS} (I_{SS}).

TEXAS INSTRUMENTS
POST OFFICE BOX 655012 • DALLAS, TEXAS 75265

TYPICAL APPLICATION DATA

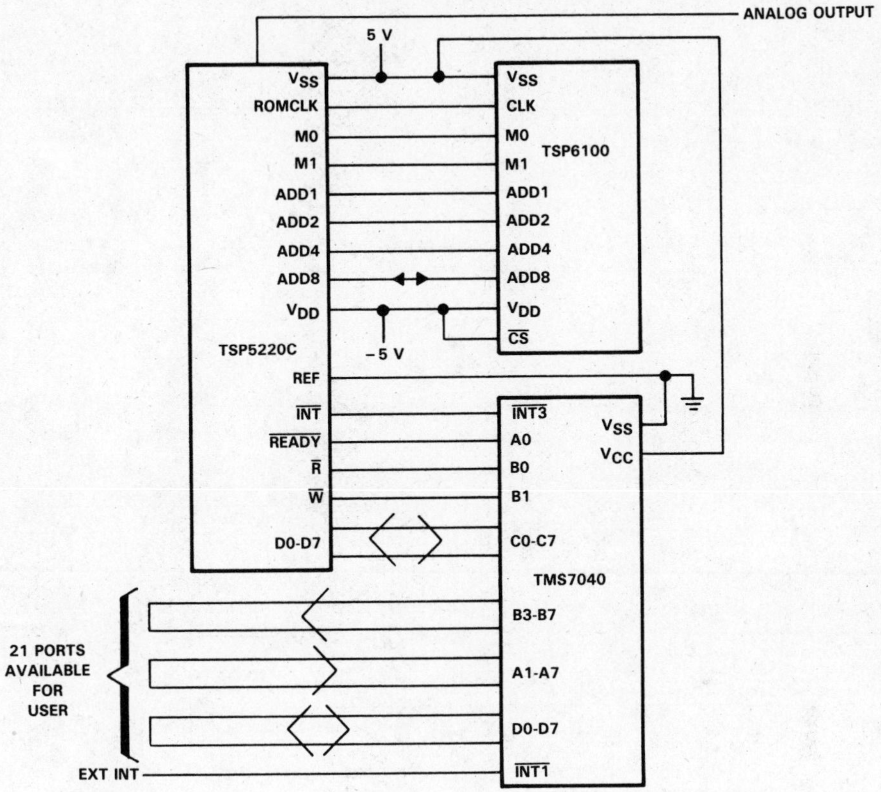
ANALOG OUTPUT
5 V
V_{SS}
ROMCLK
M0
M1
ADD1
ADD2
ADD4
ADD8
V_{DD}
TSP5220C
REF
$\overline{INT}$
$\overline{READY}$
$\overline{R}$
$\overline{W}$
D0-D7
V_{SS}
CLK
M0
M1
TSP6100
ADD1
ADD2
ADD4
ADD8
V_{DD}
$\overline{CS}$
−5 V
$\overline{INT3}$
A0
B0
B1
V_{SS}
V_{CC}
C0-C7
TMS7040
B3-B7
A1-A7
D0-D7
$\overline{INT1}$
21 PORTS AVAILABLE FOR USER
EXT INT

FIGURE 1. TSP5220C SYSTEM USING TSP6100 STORAGE FOR SPEECH DATA

D2998, JANUARY 1987

PRODUCT SUMMARY

This data is excerpted from the *TSP60C20 Read Only Memory Data Manual* intended for publication in early 1987.

- **256K-Bit ROM (Internal Organization 16K × 16K Bits)**
- **On-Chip Address Register with Automatic Incrementing Feature**
- **Parallel Port (4- or 8-Bit Format, User Controlled)**
- **Address/Data/Control Compatible with Memory-Mapped Systems**
- **Serial Port for Interfacing with Speech Synthesizers or Other Application Requiring Decoding of Variable Length Codes or Data**
- **Standby Mode for Very Low Power Consumption**
- **On-Chip Table Lookup of Address (Indirect Addressing)**
- **TTL and MOS Compatabile**

N PACKAGE
(TOP VIEW)

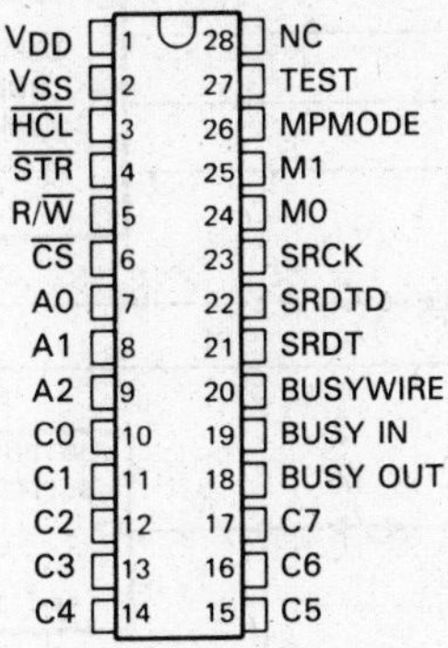
VDD 1
VSS 2
HCL 3
STR 4
R/W 5
CS 6
A0 7
A1 8
A2 9
C0 10
C1 11
C2 12
C3 13
C4 14
28 NC
27 TEST
26 MPMODE
25 M1
24 M0
23 SRCK
22 SRDTD
21 SRDT
20 BUSYWIRE
19 BUSY IN
18 BUSY OUT
17 C7
16 C6
15 C5

NC—No internal connection

description

The TMS60C20 is a 256K-bit ROM fabricated in CMOS technology for low operating and standby power consumption. The design is optimized for the data storage requirements of synthetic speech systems. Low cost and high density are maximized by increasing access time.

The TMS60C20 is characterized for operation from −40°C to 80°C.

functional block diagram

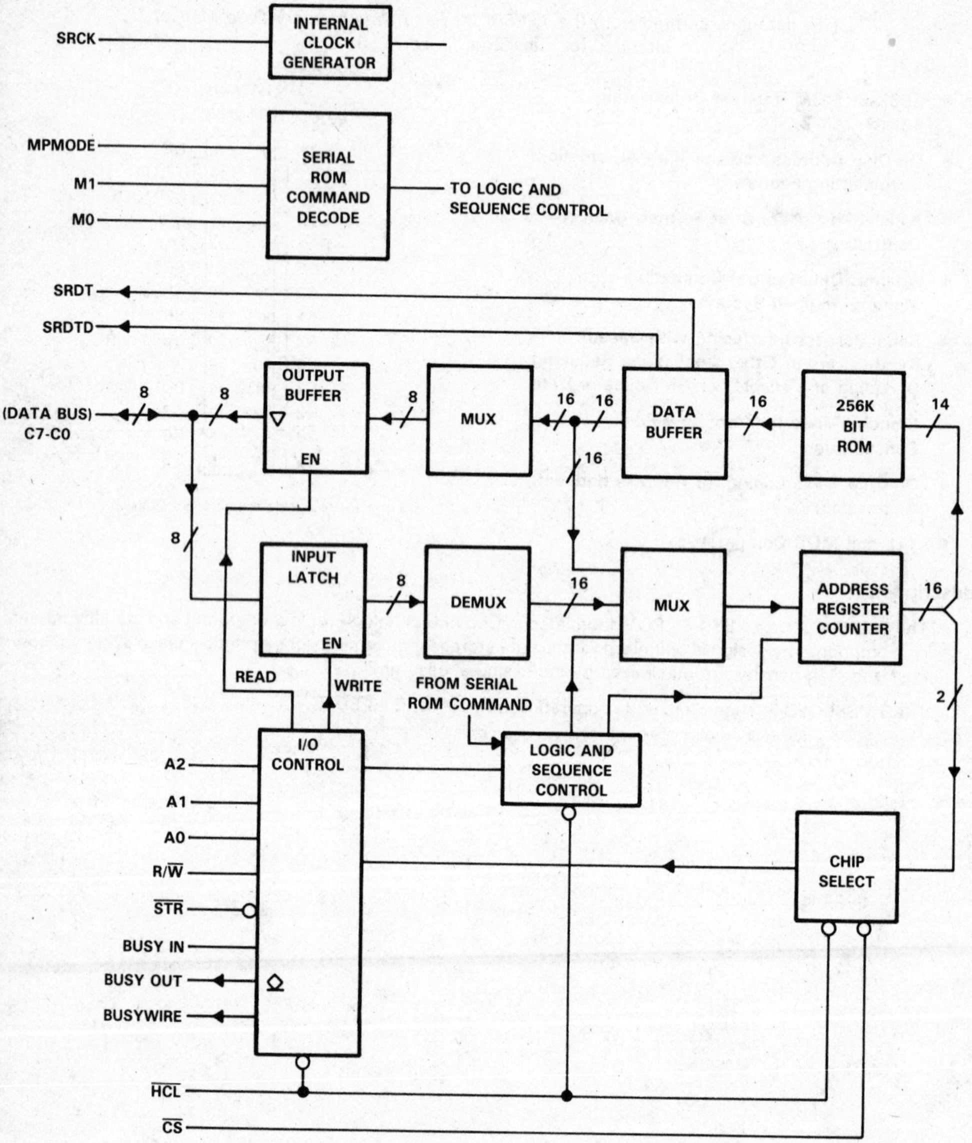
SRCK
INTERNAL
CLOCK
GENERATOR
MPMODE
M1
M0
SERIAL
ROM
COMMAND
DECODE
TO LOGIC AND
SEQUENCE CONTROL
SRDT
SRDTD
(DATA BUS)
C7-C0
8
8
OUTPUT
BUFFER
EN
8
MUX
16
16
DATA
BUFFER
16
256K
BIT
ROM
14
16
8
INPUT
LATCH
EN
8
DEMUX
16
MUX
ADDRESS
REGISTER
COUNTER
16
READ
WRITE
FROM SERIAL
ROM COMMAND
2
I/O
CONTROL
LOGIC AND
SEQUENCE
CONTROL
A2
A1
A0
R/W̄
STR̄
BUSY IN
BUSY OUT
BUSYWIRE
CHIP
SELECT
HCL̄
CS̄

PIN FUNCTIONAL DESCRIPTION

PIN NAME	PIN NO.	I/O	DESCRIPTION
A0	7	I	Register address inputs
A1	8	I	
A2	9	I	
BUSY OUT	18	O	Busy output
BUSY IN	19	I	Busy input
BUSYWIRE	20	O	Busy output — Open drain
$\overline{CS}$	6	I	Chip select
CO	10	I/O	Data input/output
C1	11	I/O	
C2	12	I/O	
C3	13	I/O	
C4	14	I/O	
C5	15	I/O	
C6	16	I/O	
C7	17	I/O	
$\overline{HCL}$	3	I	Hardware clear input
MO	24	I	Command bit 0
M1	25	I	Command bit 1
MPMODE	26	I	Microprocessor mode (M1)
R/$\overline{W}$	5	I	Read/Write input
$\overline{STR}$	4	I	Input strobe
SRDT	21	I	Serial data input
SRDTD	22	I	Serial data input delayed
SRCK	23	I	Clock input
TEST	27	I	Test enable
V_{DD}	1		5-V power supply
V_{SS}	2		Ground

absolute maximum ratings over free-air temperature range

Supply voltage, V_{DD} (see Note 1) . . . . −0.3 V to 7 V
Input voltage range . . . . −0.3 V to V_{DD} + 0.3 V
Output voltage range . . . . −0.3 V to V_{DD} + 0.3 V
Operating free-air temperature range . . . . −40 °C to 80 °C
Storage temperature range . . . . −55 °C to 150 °C
Lead temperature 1,6 mm (1/16 inch) from case for 10 seconds . . . . 260 °C

NOTE 1: All voltages are with respect to V_{SS} terminal.

recommended operating conditions

	MIN	TYP	MAX	UNIT
Supply voltage, V_{DD}	4.5	5	5.5	V
High-level input voltage, V_{IH}	2			V
Low-level input voltage, V_{IL}			0.8	V
Clock frequency, f_{clock}	100		900	kHz
Operating free-air temperature, T_A	−40		80	°C

electrical characteristics over recommended ranges of supply voltage and operating free-air temperature

PARAMETER			TEST CONDITIONS	MIN	MAX	UNIT
V_{OH}	High-level output voltage		I_{OH} = −400 μA	2.4	V_{DD}	V
			I_{OH} = −100 μA	$V_{DD}-0.5$	V_{DD}	
V_{OL}	Low-level output voltage		V_{DD} = 4.5 V to 5.5 V, I_{OL} = 1.6 mA		0.4	V
I_{CC}	Supply current	Standby	BUSYWIRE open, See Note 2		0.2	mA
		Operating			1	
C_i	Input capacitance				10	pF
C_o	Output capacitance				10	pF

NOTE 2: These limits apply with all inputs between 0 and 0.5 V or V_{DD} − 0.5 V and V_{DD}. For TTL levels, standby supply current will increase significantly in the input buffers.

timing requirements

		MIN	MAX	UNIT
t_w	Pulse duration, strobe low	100		ns
t_{su1}	Setup time, C7-C0 high or low before strobe↓	60		ns
t_{su2}	Setup time, A2-A0 high or low before strobe↓	20		ns
t_{su3}	Setup time, M1 high or low before strobe	20		ns
t_{su4}	Setup time, R/$\overline{W}$ high before strobe↓	50		ns
t_{su5}	Setup time, R/$\overline{W}$ low before strobe ↓	20		ns
t_{h1}	Hold time, data valid after strobe	20		ns
t_{h2}	Hold time, R/$\overline{W}$ high after strobe↑	20		ns
t_{h3}	Hold time, R/W low after strobe↑	20		ns
t_t	Transition time (see Note 3)	30		ns

NOTE 3: Transition time is defined from the 10% point to the 67% point for the rise time and from the 90% point to the 33% point for the fall time.

D2974, NOVEMBER 1986

PRODUCT SUMMARY

This data is excerpted from the *TSP6100 Customed Masked ROM Data Manual* intended for publication in early 1987

- **128K Bits of Serially Accessed Speech Data**
- **One-of-16 Chip Select Decode Plus Optional External Chip Select**
- **Auto Incrementing Address Counter (Presettable by TSP5220C or TSP5110A)**
- **Built-In Indirect Addressing Logic**

N PACKAGE
(TOP VIEW)

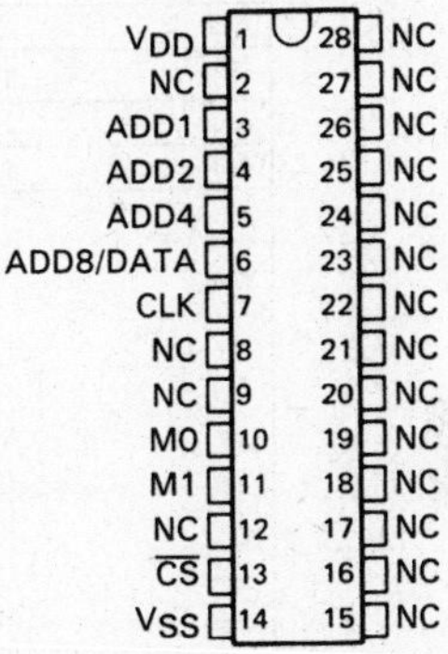
VDD 1
NC 2
ADD1 3
ADD2 4
ADD4 5
ADD8/DATA 6
CLK 7
NC 8
NC 9
M0 10
M1 11
NC 12
CS 13
VSS 14
28 NC
27 NC
26 NC
25 NC
24 NC
23 NC
22 NC
21 NC
20 NC
19 NC
18 NC
17 NC
16 NC
15 NC

NC—No internal connection

description

The TSP6100 is a 128K-bit serial-interface masked ROM that provides economical speech data storage for TSP5220C- or TSP5110A-based speech systems when production volume warrants masking. The TSP6100 is designed for direct connection to the TSP5220C or TSP5110A ROM ports and takes advantage of several special ROM setup instructions provided in these devices. No connection of the TSP6100 to the rest of the speech control system is needed.

The TSP6100 is characterized for operation from 0°C to 70°C.

functional block diagram

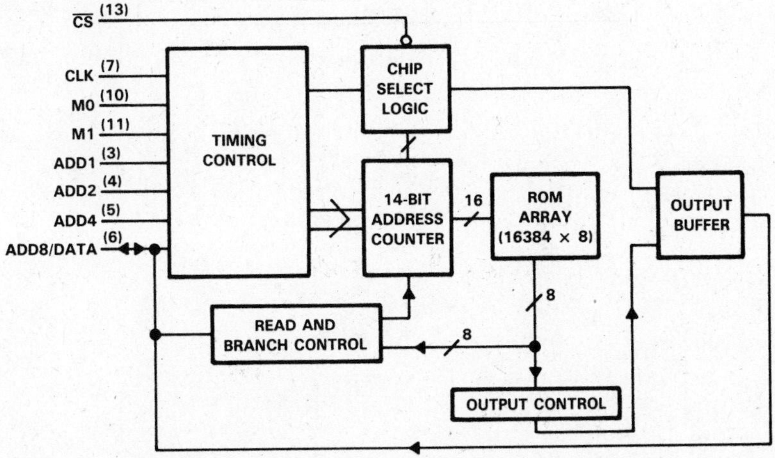
CS (13)
CLK (7)
M0 (10)
M1 (11)
ADD1 (3)
ADD2 (4)
ADD4 (5)
ADD8/DATA (6)
TIMING CONTROL
CHIP SELECT LOGIC
14-BIT ADDRESS COUNTER
16
ROM ARRAY (16384 × 8)
OUTPUT BUFFER
8
READ AND BRANCH CONTROL
8
OUTPUT CONTROL

TEXAS INSTRUMENTS
POST OFFICE BOX 655012 • DALLAS, TEXAS 75265

PIN FUNCTION TABLE

PIN NAME	NO.	I/O	DESCRIPTION
ADD1	3	I	Address 1 input
ADD2	4	I	Address 2 input
ADD4	5	I	Address 4 input
ADD8/DATA	6	I/O	Address 8 input or serial data output
CLK	7	I	Clock input
$\overline{CS}$	13	I	Chip Select input
M0	10	I	Mode Select 0 input
M1	11	I	Mode Select 1 input
NC	2, 8, 9, 12, 15, 16, 17, 18, 19, 20, 21, 22, 23, 24, 25, 26, 27, 28		No internal connection
V_{DD}	1		−10-V supply voltage
V_{SS}	14		Ground

TEXAS INSTRUMENTS
POST OFFICE BOX 655012 • DALLAS, TEXAS 75265

absolute maximum ratings over operating free-air temperature range

Voltage applied to any pin (see Note 1) .... −15 V to 0.3 V
Supply voltage range, V_{DD} .... −15 V to 0.3 V
Operating free-air temperature range .... 0°C to 70°C
Storage temperature range .... −30°C to 125°C
Lead temperature 1,6 mm (1/16 inch) from case for 10 seconds .... 260°C

NOTE 1: Voltage values are with respect to V_{SS}.

recommended operating characteristics

		MIN†	NOM	MAX	UNIT
V_{DD}	Drain supply voltage	−8.3		−10.5	V
V_{SS}	Substrate supply voltage		0		V
V_{IH}	High-level input voltage	−1		V_{SS}	V
V_{IL}	Low-level input voltage	V_{DD}		−4	V
T_A	Operating free-air temperature	0		70	°C

†The algebraic convention, in which the more negative (less positive) limit is designated as minimum, is used in the data sheet for logic voltage levels only.

electrical characteristics over recommended operating free-air temperature range, V_{DD} = −10 V

PARAMETER		TEST CONDITIONS	MIN	NOM	MAX	UNIT
V_{OH}	High-level output voltage	I_{OH} = −100 μA	−0.6			V
V_{OL}	Low-level output voltage	I_{OL} = 100 μA			−4.2	V
I_{IH}	High-level input current	V_{IH} = −0.6 V			10	μA
I_{IL}	Low-level input current	V_{IL} = −4.2 V			−10	μA
I_O	Output current	V_I = V_{SS} to V_{DD}			±10	μA
I_{DD}	Drain supply current				−10	mA

General Information 1

Alphanumeric Index
Selection Guide

Data Acquisition Circuits 2

Cross-Reference Guide
Data Sheets

Display Drivers 3

Data Sheets

Line Drivers and Receivers 4

Cross-Reference Guide
Data Sheets

Peripheral Drivers/Actuators 5

Cross-Reference Guide
Data Sheets

Memory Interface Circuits 6

Data Sheets

Speech Synthesis Circuits 7

Data Sheets

Appendix A Power Derating Curves A

Appendix B Ordering Instructions, Mechanical Data, IC Sockets B

Appendix C Explanation of Logic Symbols C

plastic "small outline" packages

These curves are for use with the continuous dissipation ratings specified on the individual data sheets. Those ratings apply up to the temperature at which the rated level intersects the appropriate derating curve or the maximum operating free-air temperature.

DISSIPATION DERATING CURVES

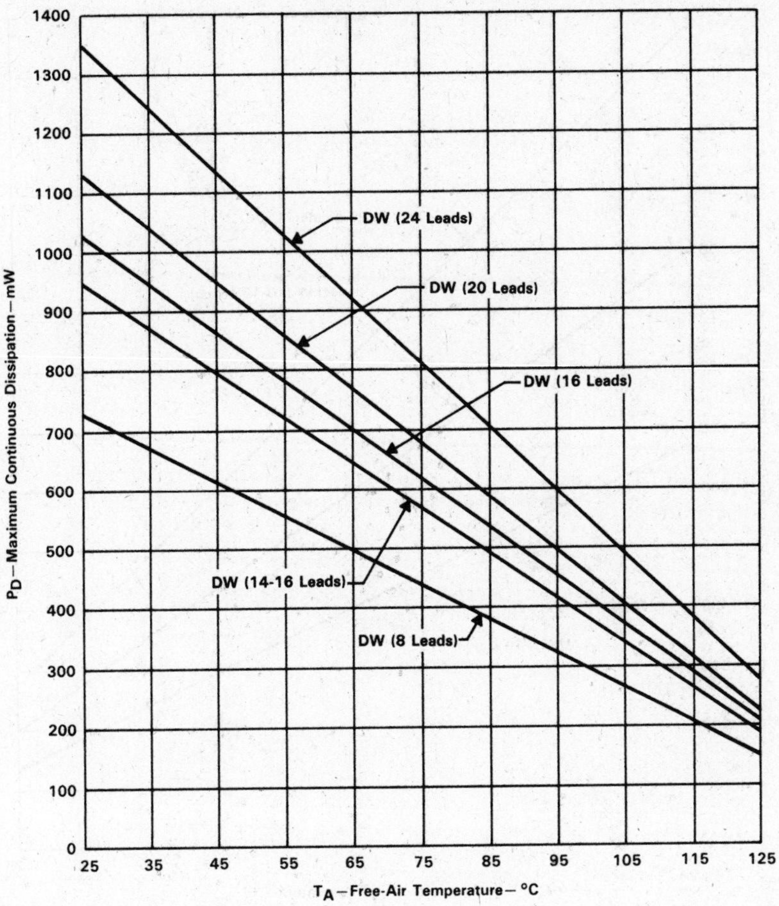
DW (24 Leads)
DW (20 Leads)
DW (16 Leads)
DW (14-16 Leads)
DW (8 Leads)
1400
1300
1200
1100
1000
900
800
700
600
500
400
300
200
100
0
25
35
45
55
65
75
85
95
105
115
125
P_D—Maximum Continuous Dissipation—mW
T_A—Free-Air Temperature—°C

leadless ceramic chip carrier and flat packages

These curves are for use with the continuous dissipation ratings specified on the individual data sheets. Those ratings apply up to the temperature at which the rated level intersects the appropriate derating curve or the maximum operating free-air temperature.

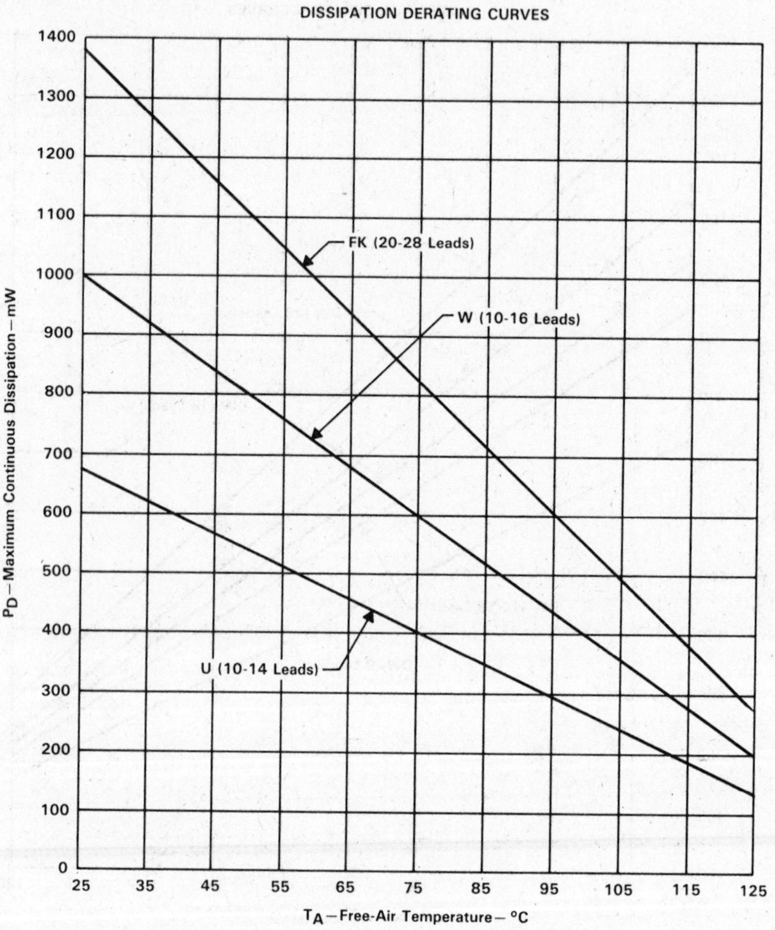
DISSIPATION DERATING CURVES
FK (20-28 Leads)
W (10-16 Leads)
U (10-14 Leads)
PD—Maximum Continuous Dissipation—mW
1400
1300
1200
1100
1000
900
800
700
600
500
400
300
200
100
0
25
35
45
55
65
75
85
95
105
115
125
TA—Free-Air Temperature—°C

A
Derating Curves

TEXAS INSTRUMENTS
POST OFFICE BOX 655012 • DALLAS, TEXAS 75265

plastic chip-carrier packages

These curves are for use with the continuous dissipation ratings specified on the individual data sheets. Those ratings apply up to the temperature at which the rated level intersects the appropriate derating curve or the maximum operating free-air temperature.

DISSIPATION DERATING CURVES

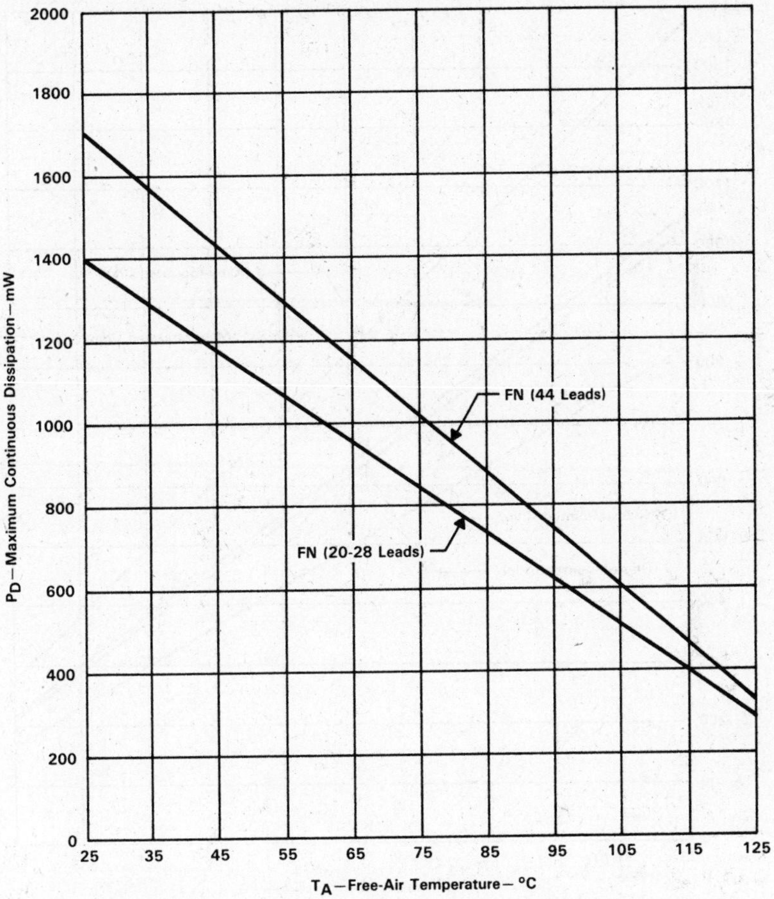
2000
1800
1600
1400
1200
1000
800
600
400
200
0
25
35
45
55
65
75
85
95
105
115
125
FN (44 Leads)
FN (20-28 Leads)
P_D—Maximum Continuous Dissipation—mW
T_A—Free-Air Temperature—°C

ceramic dual-in-line packages

These curves are for use with the continuous dissipation ratings specified on the individual data sheets. Those ratings apply up to the temperature at which the rated level intersects the appropriate derating curve or the maximum operating free-air temperature.

DISSIPATION DERATING CURVES

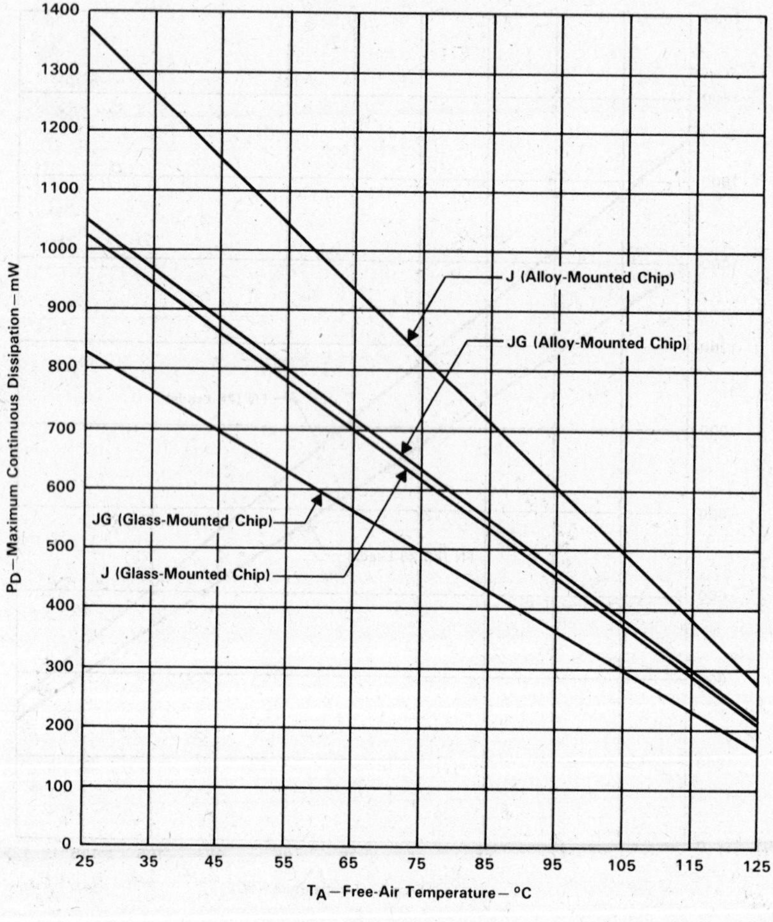

1400
1300
1200
1100
1000
900
800
700
600
500
400
300
200
100
0
P_D—Maximum Continuous Dissipation—mW
J (Alloy-Mounted Chip)
JG (Alloy-Mounted Chip)
JG (Glass-Mounted Chip)
J (Glass-Mounted Chip)
25
35
45
55
65
75
85
95
105
115
125
T_A—Free-Air Temperature—°C

TEXAS INSTRUMENTS
POST OFFICE BOX 655012 • DALLAS, TEXAS 75265

ceramic packages — side-braze

These curves are for use with the continuous dissipation ratings specified on the individual data sheets. Those ratings apply up to the temperature at which the rated level intersects the appropriate derating curve or the maximum operating free-air temperature.

DISSIPATION DERATING CURVES

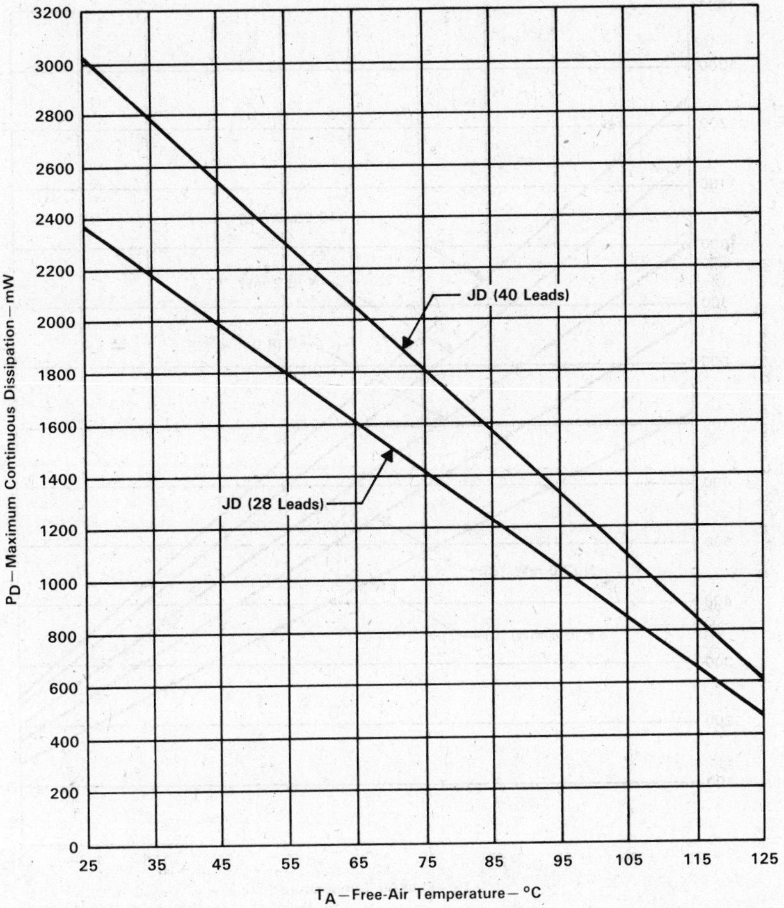
3200
3000
2800
2600
2400
2200
2000
1800
1600
1400
1200
1000
800
600
400
200
0
P_D—Maximum Continuous Dissipation—mW
JD (40 Leads)
JD (28 Leads)
25
35
45
55
65
75
85
95
105
115
125
T_A—Free-Air Temperature—°C

TEXAS INSTRUMENTS
POST OFFICE BOX 655012 • DALLAS, TEXAS 75265

plastic dual-in-line packages

These curves are for use with the continuous dissipation ratings specified on the individual data sheets. Those ratings apply up to the temperature at which the rated level intersects the appropriate derating curve or the maximum operating free-air temperature.

DISSIPATION DERATING CURVES

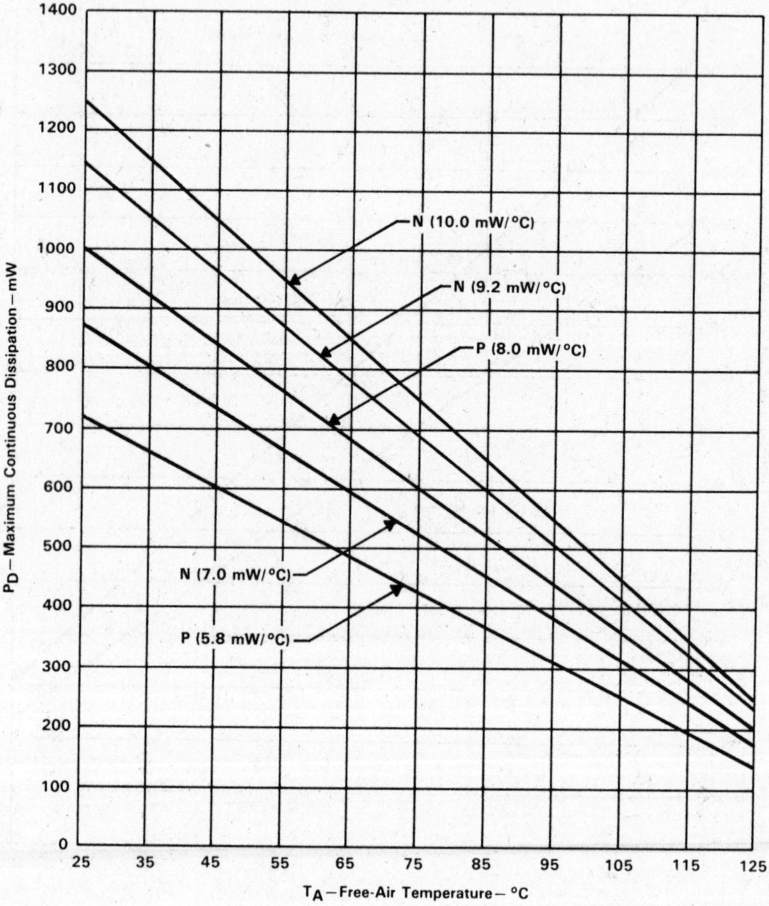
1400
1300
1200
1100
1000
900
800
700
600
500
400
300
200
100
0
N (10.0 mW/°C)
N (9.2 mW/°C)
P (8.0 mW/°C)
N (7.0 mW/°C)
P (5.8 mW/°C)
P_D — Maximum Continuous Dissipation — mW
25
35
45
55
65
75
85
95
105
115
125
T_A — Free-Air Temperature — °C

plastic dual-in-line packages (continued)

These curves are for use with the continuous dissipation ratings specified on the individual data sheets. Those ratings apply up to the temperature at which the rated level intersects the appropriate derating curve or the maximum operating free-air temperature.

DISSIPATION DERATING CURVES

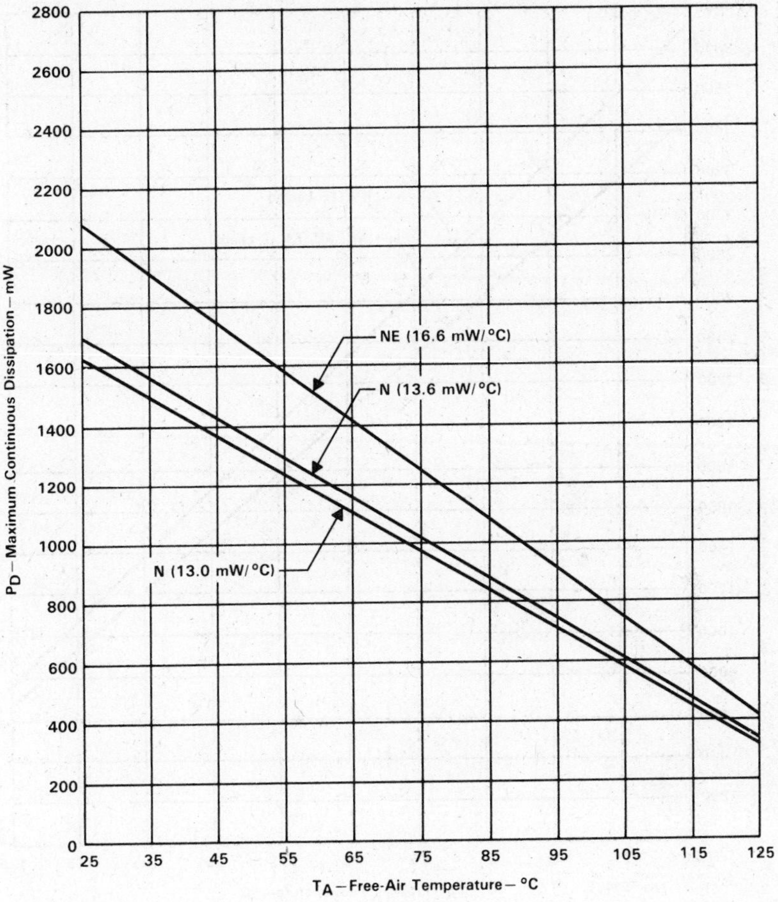

2800
2600
2400
2200
2000
1800
1600
1400
1200
1000
800
600
400
200
0
NE (16.6 mW/°C)
N (13.6 mW/°C)
N (13.0 mW/°C)
25
35
45
55
65
75
85
95
105
115
125
P_D – Maximum Continuous Dissipation – mW
T_A – Free-Air Temperature – °C

plastic power tab packages

These curves are for use with the continuous dissipation ratings specified on the individual data sheets. Those ratings apply up to the temperature at which the rated level intersects the appropriate derating curve or the maximum operating free-air temperature.

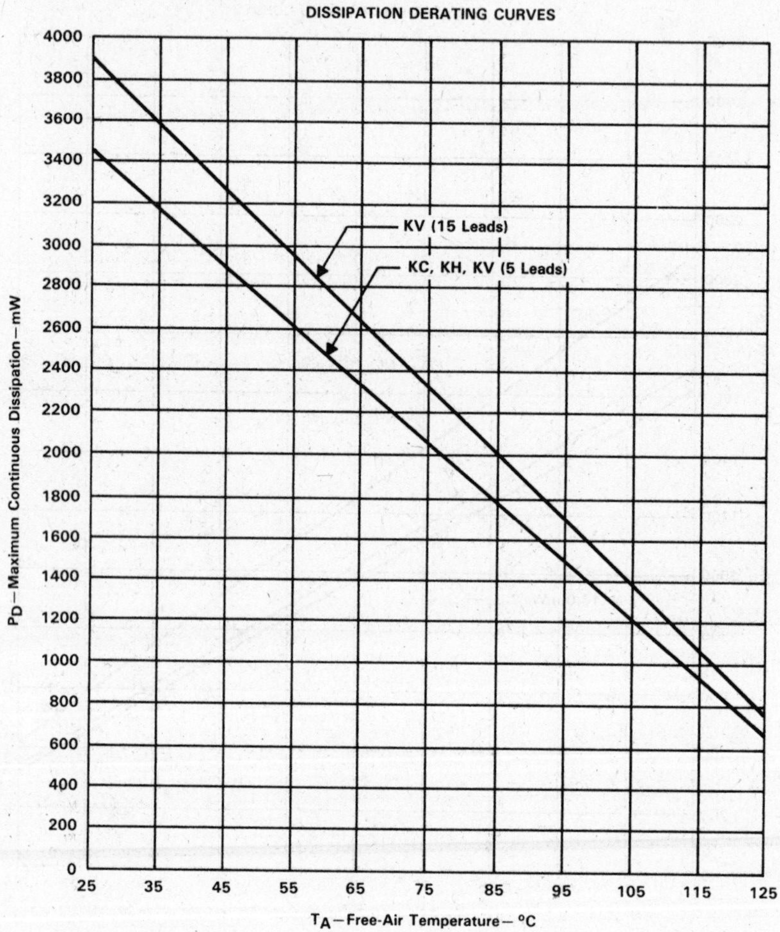
DISSIPATION DERATING CURVES
KV (15 Leads)
KC, KH, KV (5 Leads)
P_D—Maximum Continuous Dissipation—mW
4000
3800
3600
3400
3200
3000
2800
2600
2400
2200
2000
1800
1600
1400
1200
1000
800
600
400
200
0
25
35
45
55
65
75
85
95
105
115
125
T_A—Free-Air Temperature—°C

TEXAS INSTRUMENTS
POST OFFICE BOX 655012 • DALLAS, TEXAS 75265

General Information 1

Alphanumeric Index
Selection Guide

Data Acquisition Circuits 2

Cross-Reference Guide
Data Sheets

Display Drivers 3

Data Sheets

Line Drivers and Receivers 4

Cross-Reference Guide
Data Sheets

Peripheral Drivers/Actuators 5

Cross-Reference Guide
Data Sheets

Memory Interface Circuits 6

Data Sheets

Speech Synthesis Circuits 7

Data Sheets

Appendix A Power Derating Curves A

Appendix B Ordering Instructions
Mechanical Data
IC Sockets B

Appendix C Explanation of
Logic Symbols C

ORDERING INSTRUCTIONS

Electrical characteristics presented in this data book, unless otherwise noted, apply for the circuit type(s) listed in the page heading regardless of package. The availability of a circuit function in a particular package is denoted by an alphabetical reference above the pin-correction diagram(s). These alphabetical references refer to mechanical outline drawings shown in this section.

Factory orders for circuits described in this data book should include a four-part type number as explained in the following example.

EXAMPLE: TL 4066M J /883B

1. **Prefix**

MUST CONTAIN TWO OR THREE LETTERS

SN .... TI Special Function or Interface Products
TL .... TI Linear Products (excluding Interface)
TLC .... TI Linear Silicon-Gate CMOS Products
TSP .... Speech Products

STANDARD SECOND-SOURCE PREFIXES

ADC	Analog Devices	N8T	Signetics
AM	ADM	µA	Fairchild
DS	National	UCN	Sprague
L	SGS	UDN	Sprague
MC	Motorola	ULN	Sprague

NOTE: Due to size limitations of the 8-pin D package, some devices in this data book use abbreviated symbolization. Example: The TLC549CD becomes '549CD.

2. **Unique Circuit Designator Including Temperature Range (If not already specified by the prefix)**

MUST CONTAIN ONE TO EIGHT CHARACTERS
(From Individual Data Sheets)

Examples: 4, 293, 607M, 533AM, 65554, 75ALS194

3. **Package**

MUST CONTAIN ONE OR TWO LETTERS

D, DW, FD, FH, FJ, FK, FN, J, JD, JG, KC, KH, KV, N, P, U, W
(From Pin-Connection Diagram on Individual Data Sheet)

4. **MIL-STD-883B, Method 5004, Class B**

OMIT/883B WHEN NOT APPLICABLE

Circuits are shipped in one of the carriers below. Unless a specific method of shipment is specified by the customer (with possible additional costs), circuits will be shipped on the most practical carrier.

(D,DW,J,JD,JG,N,P)
—Slide Magazines
—A-Channel Plastic Tubing
—Barnes Carrier
—Sectioned Cardboard Box
—Individual Cardboard Box

Chip Carriers (FD,FH,FJ,FK,FN)
—Anti-Static Plastic Tubing

Flat (U,W)
—Barnes Carrier
—Milton Ross Carrier

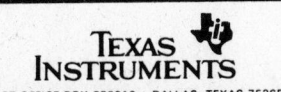

Power Tab (KC,KH,KV)
—Sleeves

ORDERING INSTRUCTIONS

Mechanical Data

B

D plastic ''small outline'' packages

Each of these ''small outline'' packages consists of a circuit mounted on a lead frame and encapsulated within a plastic compound. The compound will withstand soldering temperature with no deformation, and circuit performance characteristics will remain stable when operated in high-humidity conditions. Leads require no additional cleaning or processing when used in soldered assembly.

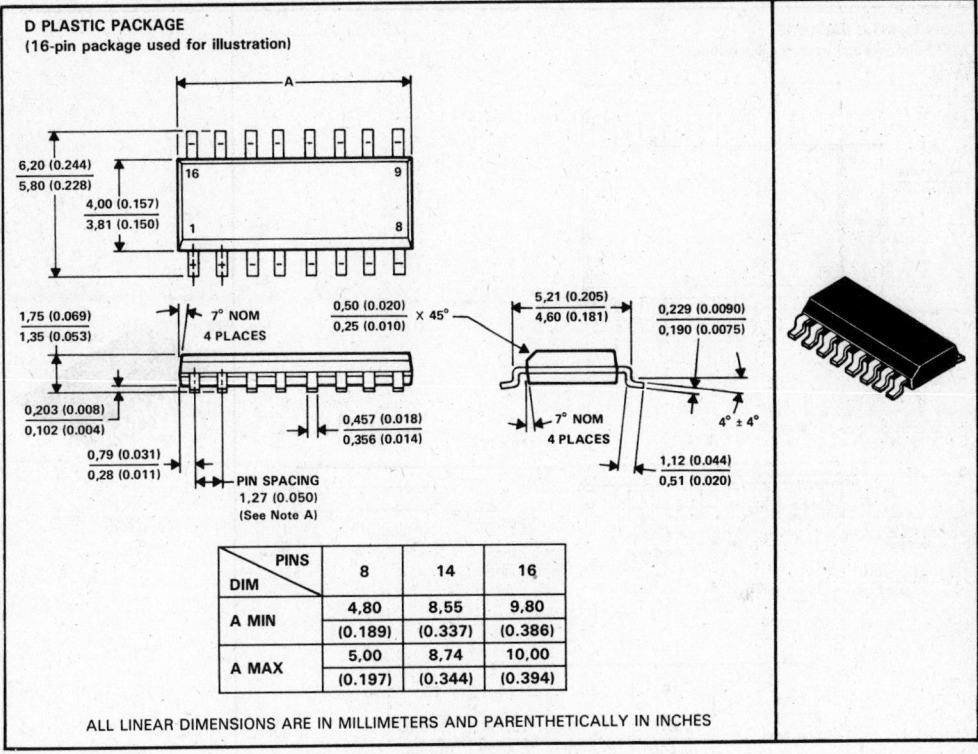

D PLASTIC PACKAGE
(16-pin package used for illustration)

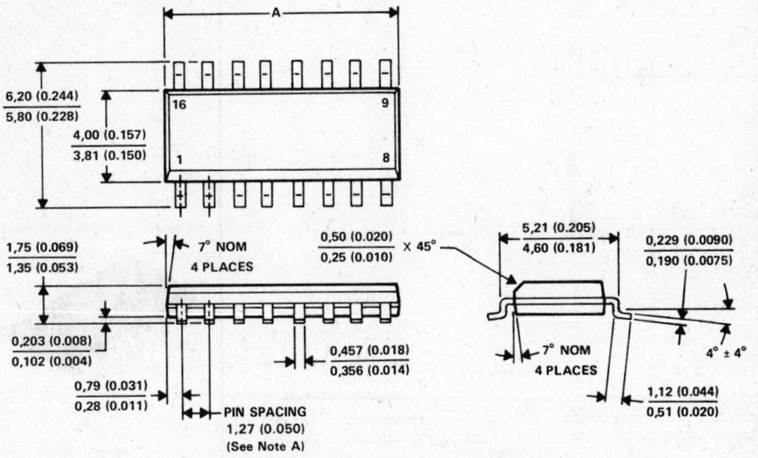
A
6,20 (0.244)
5,80 (0.228)
4,00 (0.157)
3,81 (0.150)
16
9
1
8
1,75 (0.069)
1,35 (0.053)
7° NOM
4 PLACES
0,50 (0.020)
0,25 (0.010)
X 45°
5,21 (0.205)
4,60 (0.181)
0,229 (0.0090)
0,190 (0.0075)
0,203 (0.008)
0,102 (0.004)
0,457 (0.018)
0,356 (0.014)
7° NOM
4 PLACES
4° ± 4°
0,79 (0.031)
0,28 (0.011)
PIN SPACING
1,27 (0.050)
(See Note A)
1,12 (0.044)
0,51 (0.020)

DIM \ PINS	8	14	16
A MIN	4,80 (0.189)	8,55 (0.337)	9,80 (0.386)
A MAX	5,00 (0.197)	8,74 (0.344)	10,00 (0.394)

ALL LINEAR DIMENSIONS ARE IN MILLIMETERS AND PARENTHETICALLY IN INCHES

NOTES:
A. Leads are within 0,25 (0.010) radius of true position at maximum material dimension.
B. Body dimensions do not include mold flash or protrusion.
C. Mold flash or protrusion shall not exceed 0,15 (0.006).
D. Lead tips to be planar within ±0,051 (0.002) exclusive of solder.

DW plastic "small outline" packages

Each of these "small outline" packages consists of a circuit mounted on a lead frame and encapsulated within a plastic compound. The compound will withstand soldering temperature with no deformation, and circuit performance characteristics will remain stable when operated in high-humidity conditions. Leads require no additional cleaning or processing when used in soldered assembly.

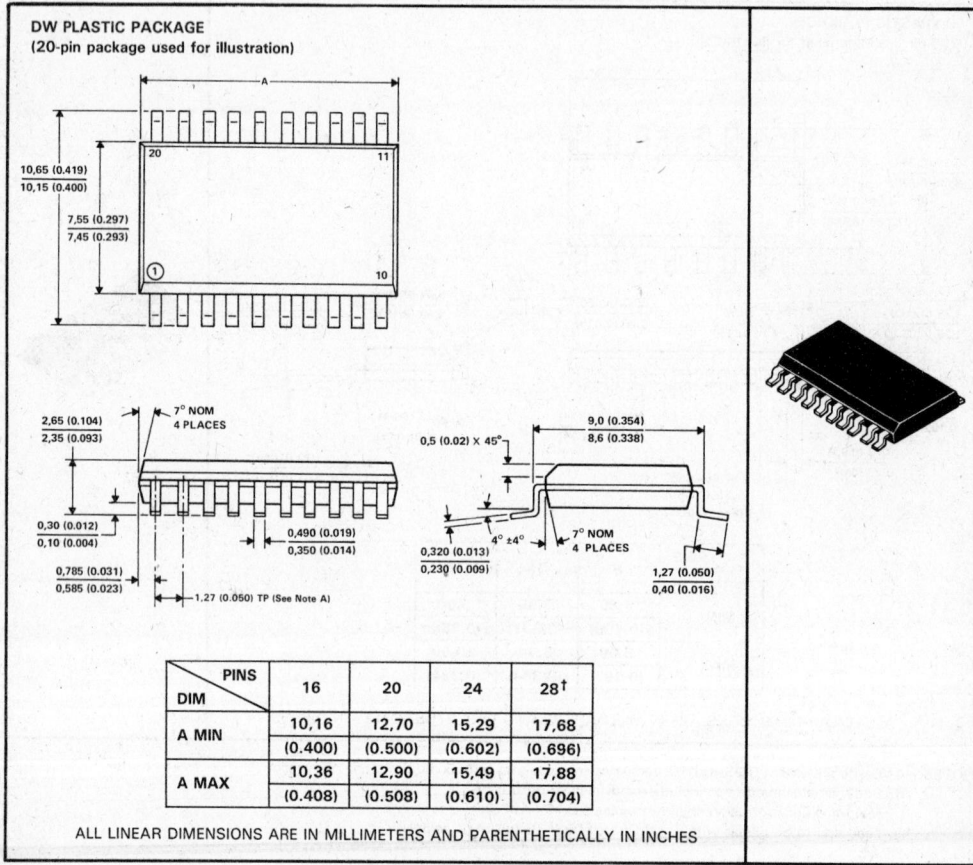

DW PLASTIC PACKAGE
(20-pin package used for illustration)

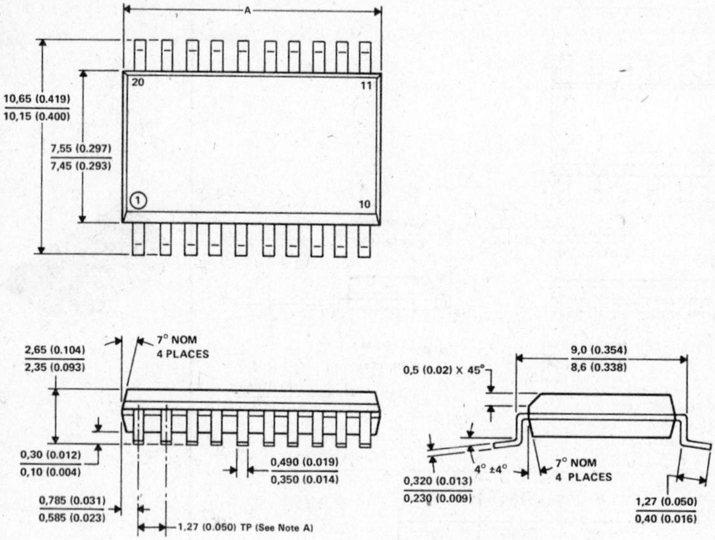
A
20
11
10
10,65 (0.419)
10,15 (0.400)
7,55 (0.297)
7,45 (0.293)
2,65 (0.104)
2,35 (0.093)
7° NOM
4 PLACES
0,30 (0.012)
0,10 (0.004)
0,490 (0.019)
0,350 (0.014)
0,785 (0.031)
0,585 (0.023)
1,27 (0.050) TP (See Note A)
0,5 (0.02) X 45°
9,0 (0.354)
8,6 (0.338)
4° ±4°
7° NOM
4 PLACES
0,320 (0.013)
0,230 (0.009)
1,27 (0.050)
0,40 (0.016)

PINS / DIM		16	20	24	28†
A MIN		10,16	12,70	15,29	17,68
		(0.400)	(0.500)	(0.602)	(0.696)
A MAX		10,36	12,90	15,49	17,88
		(0.408)	(0.508)	(0.610)	(0.704)

ALL LINEAR DIMENSIONS ARE IN MILLIMETERS AND PARENTHETICALLY IN INCHES

†The 28-pin package drawing is presently classified as Advance Information.

NOTES: A. Leads are within 0,25 (0.010) radius of true position at maximum material dimension.
B. Body dimensions do not include mold flash or protrusion.
C. Mold flash or protrusion shall not exceed 0,15 (0.006).
D. Lead tips to be planar within ±0,051 (0.002) exclusive of solder.

FD and FK leadless ceramic chip carrier packages

Each of these hermetically sealed chip carrier packages has a three-layer ceramic base with a metal lid and braze seal. The packages are intended for surface mounting on solder lands on 1,27 (0.050-inch) centers. Terminals require no additional cleaning or processing when used in soldered assembly.

FK package terminal assignments conform to JEDEC standards 1, 2, and 11.

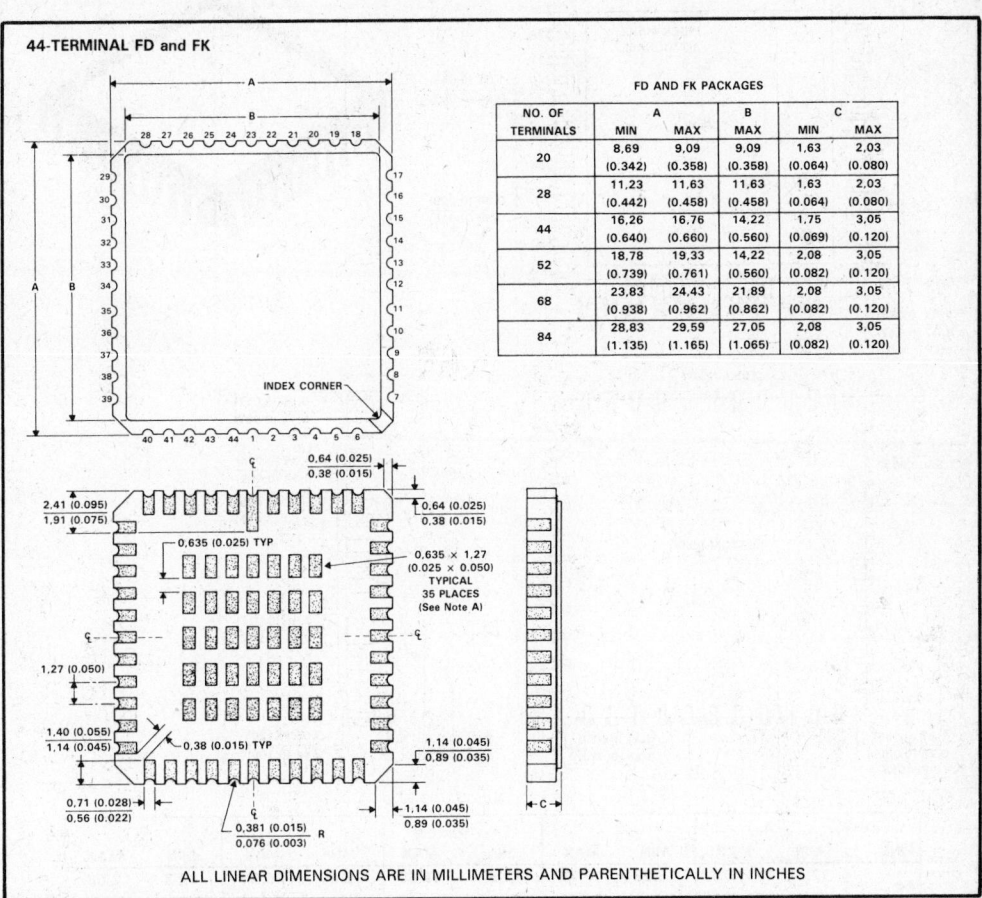

44-TERMINAL FD and FK

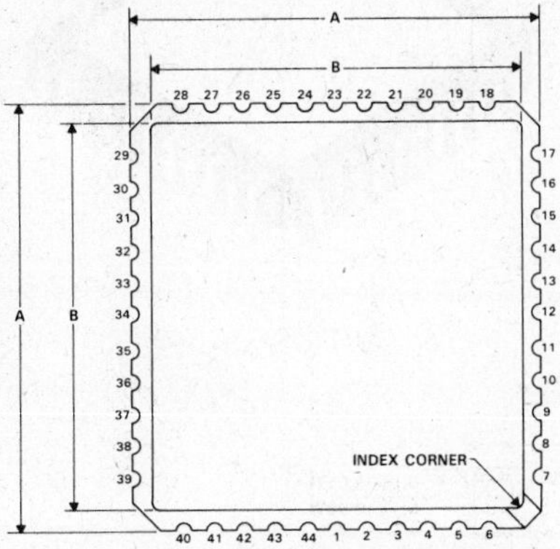
A
B
28 27 26 25 24 23 22 21 20 19 18
29 30 31 32 33 34 35 36 37 38 39
17 16 15 14 13 12 11 10 9 8 7
40 41 42 43 44 1 2 3 4 5 6
INDEX CORNER

FD AND FK PACKAGES

NO. OF TERMINALS	A MIN	A MAX	B MAX	C MIN	C MAX
20	8,69 (0.342)	9,09 (0.358)	9,09 (0.358)	1,63 (0.064)	2,03 (0.080)
28	11,23 (0.442)	11,63 (0.458)	11,63 (0.458)	1,63 (0.064)	2,03 (0.080)
44	16,26 (0.640)	16,76 (0.660)	14,22 (0.560)	1,75 (0.069)	3,05 (0.120)
52	18,78 (0.739)	19,33 (0.761)	14,22 (0.560)	2,08 (0.082)	3,05 (0.120)
68	23,83 (0.938)	24,43 (0.962)	21,89 (0.862)	2,08 (0.082)	3,05 (0.120)
84	28,83 (1.135)	29,59 (1.165)	27,05 (1.065)	2,08 (0.082)	3,05 (0.120)

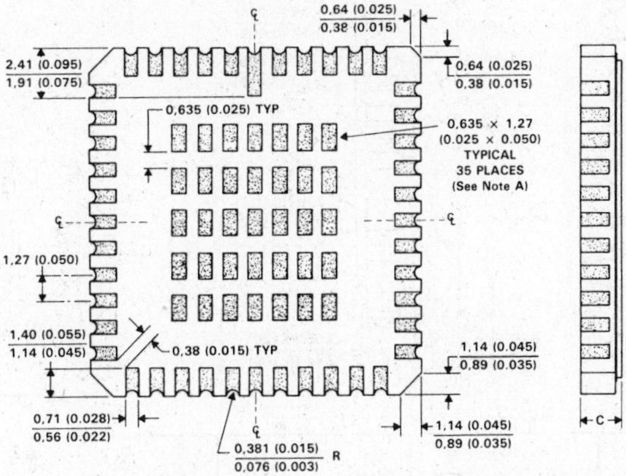
0,64 (0.025)
0,38 (0.015)
2,41 (0.095)
1,91 (0.075)
0,64 (0.025)
0,38 (0.015)
0,635 (0.025) TYP
0,635 × 1,27
(0.025 × 0.050)
TYPICAL
35 PLACES
(See Note A)
1,27 (0.050)
1,40 (0.055)
1,14 (0.045)
0,38 (0.015) TYP
1,14 (0.045)
0,89 (0.035)
0,71 (0.028)
0,56 (0.022)
0,381 (0.015)
0,076 (0.003)
R
1,14 (0.045)
0,89 (0.035)
C

ALL LINEAR DIMENSIONS ARE IN MILLIMETERS AND PARENTHETICALLY IN INCHES

NOTE A: The checkerboard pattern is aligned vertically with the contact pads and is symmetrical horizontally as shown; it is applicable to some 44-terminal packages only.

Mechanical Data

B

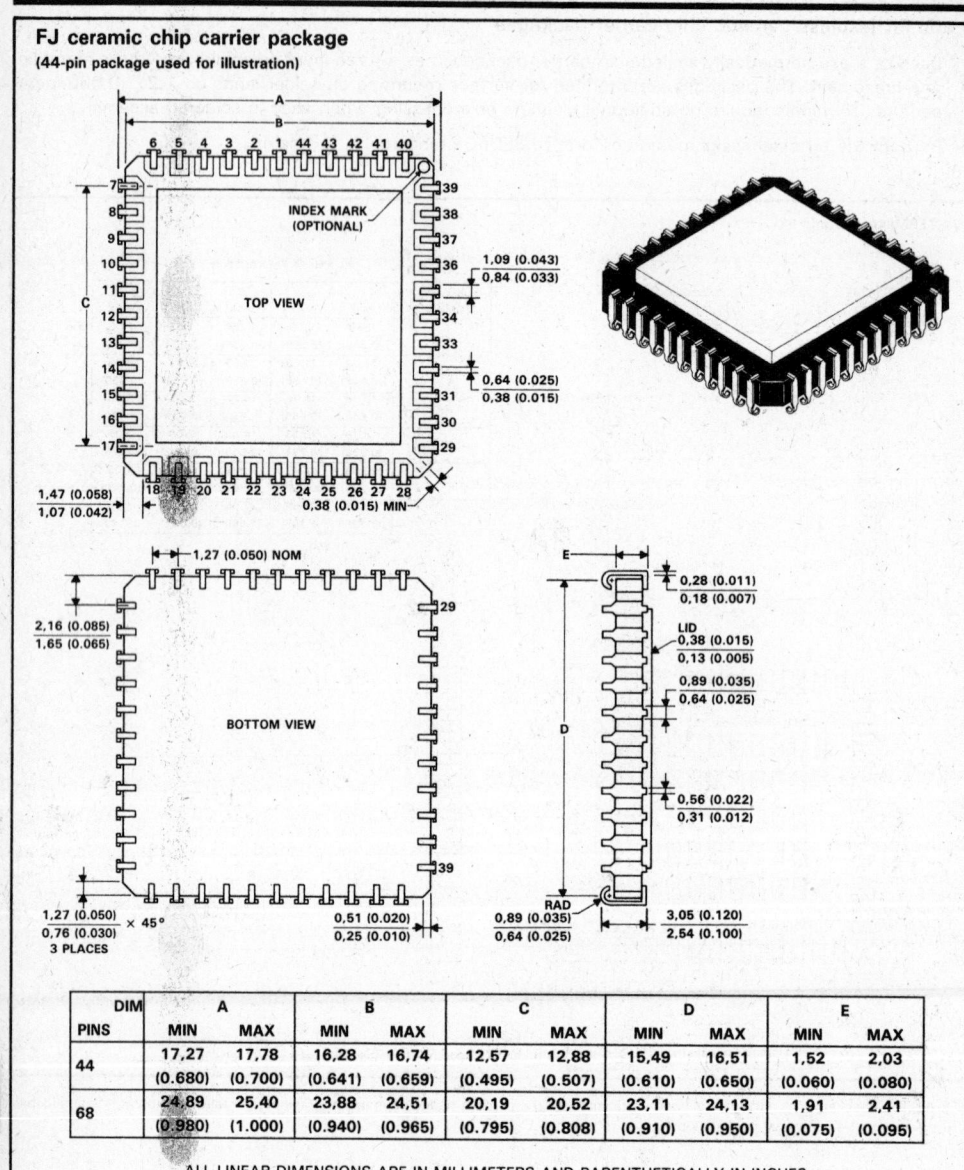

FJ ceramic chip carrier package

(44-pin package used for illustration)

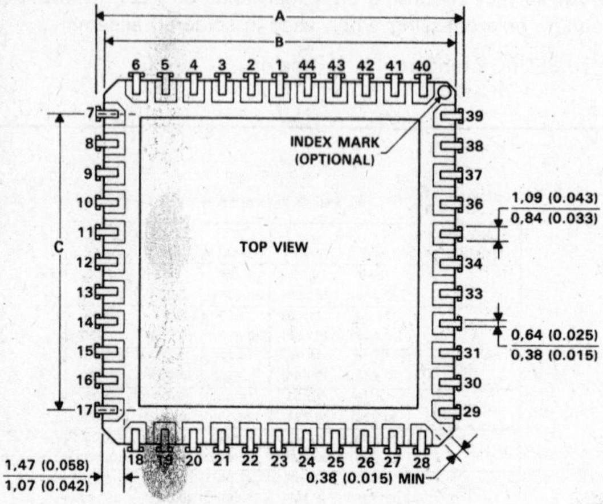
A
B
6 5 4 3 2 1 44 43 42 41 40
7
8
9
10
11
C
12
13
14
15
16
17
INDEX MARK
(OPTIONAL)
TOP VIEW
39
38
37
36
34
33
31
30
29
1,09 (0.043)
0,84 (0.033)
0,64 (0.025)
0,38 (0.015)
18 19 20 21 22 23 24 25 26 27 28
0,38 (0.015) MIN
1,47 (0.058)
1,07 (0.042)

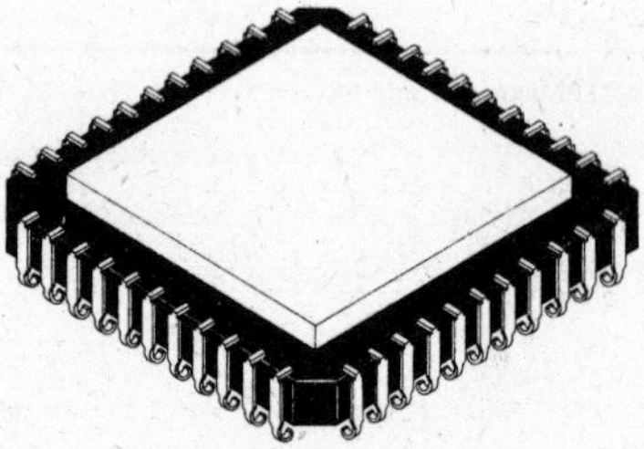

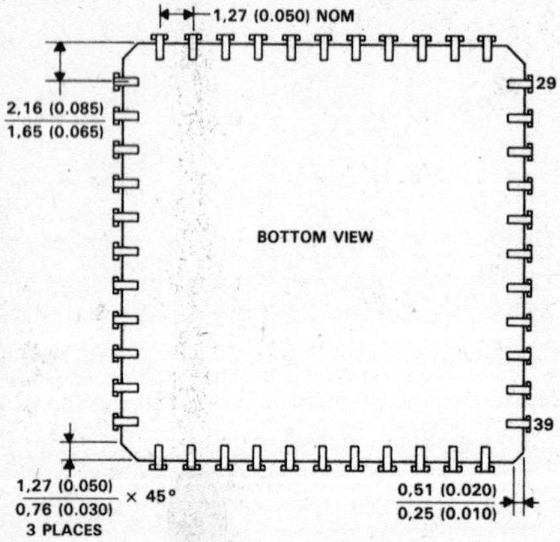
1,27 (0.050) NOM
29
2,16 (0.085)
1,65 (0.065)
BOTTOM VIEW
39
1,27 (0.050)
0,76 (0.030)
× 45°
3 PLACES
0,51 (0.020)
0,25 (0.010)

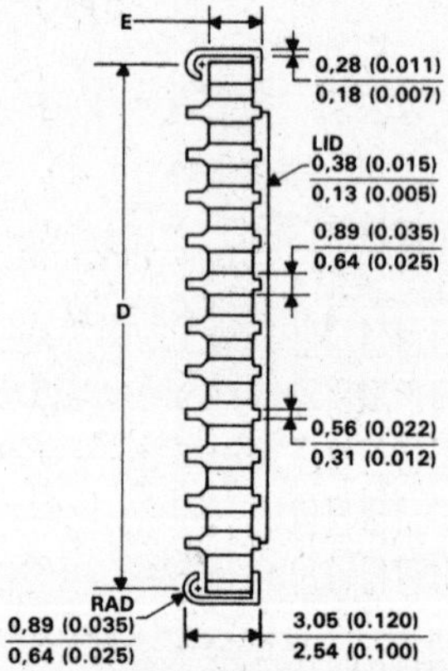
E
0,28 (0.011)
0,18 (0.007)
LID
0,38 (0.015)
0,13 (0.005)
0,89 (0.035)
0,64 (0.025)
D
0,56 (0.022)
0,31 (0.012)
RAD
0,89 (0.035)
0,64 (0.025)
3,05 (0.120)
2,54 (0.100)

PINS \ DIM	A MIN	A MAX	B MIN	B MAX	C MIN	C MAX	D MIN	D MAX	E MIN	E MAX
44	17,27 (0.680)	17,78 (0.700)	16,28 (0.641)	16,74 (0.659)	12,57 (0.495)	12,88 (0.507)	15,49 (0.610)	16,51 (0.650)	1,52 (0.060)	2,03 (0.080)
68	24,89 (0.980)	25,40 (1.000)	23,88 (0.940)	24,51 (0.965)	20,19 (0.795)	20,52 (0.808)	23,11 (0.910)	24,13 (0.950)	1,91 (0.075)	2,41 (0.095)

ALL LINEAR DIMENSIONS ARE IN MILLIMETERS AND PARENTHETICALLY IN INCHES

TEXAS INSTRUMENTS

POST OFFICE BOX 655012 • DALLAS, TEXAS 75265

FN plastic chip carrier package

Each of these chip carrier packages consists of a circuit mounted on a lead frame and encapsulated within an electrically nonconductive plastic compound. The compound withstands soldering temperatures with no deformation, and circuit performance characteristics remain stable when the devices are operated in high-humidity conditions. The packages are intended for surface mounting on solder lands on 1,27 (0.050) centers. Leads require no additional cleaning or processing when used in soldered assembly.

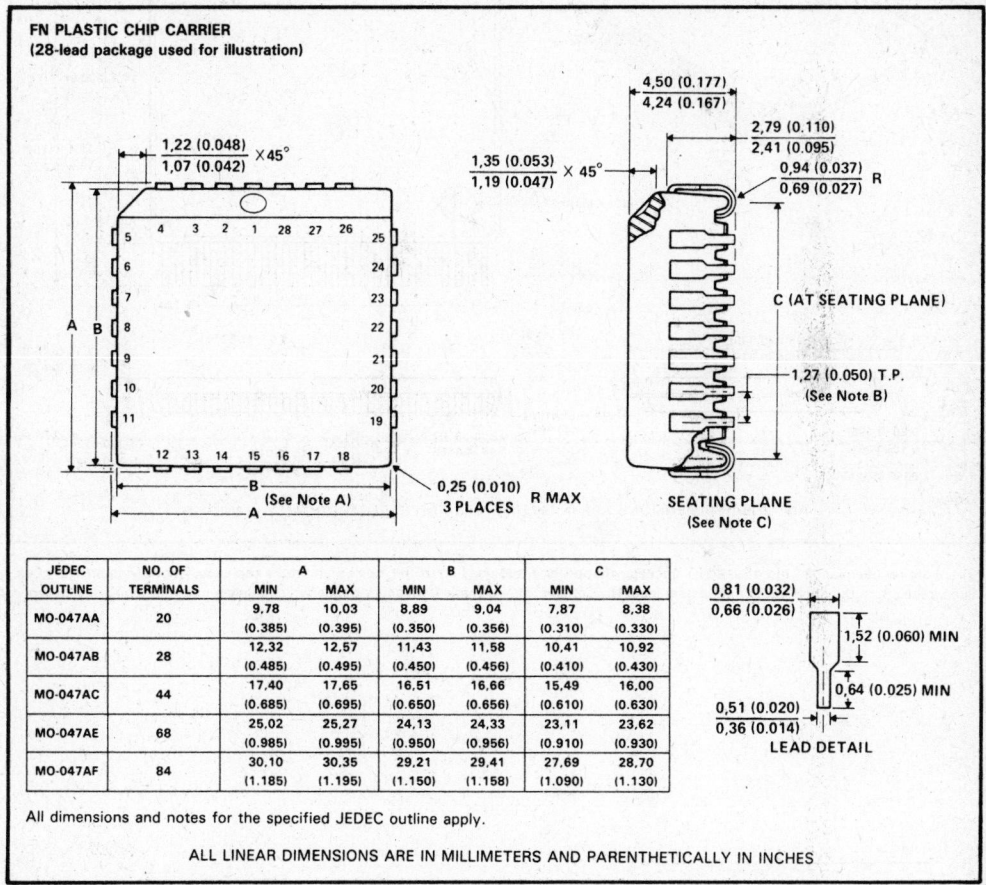

FN PLASTIC CHIP CARRIER
(28-lead package used for illustration)

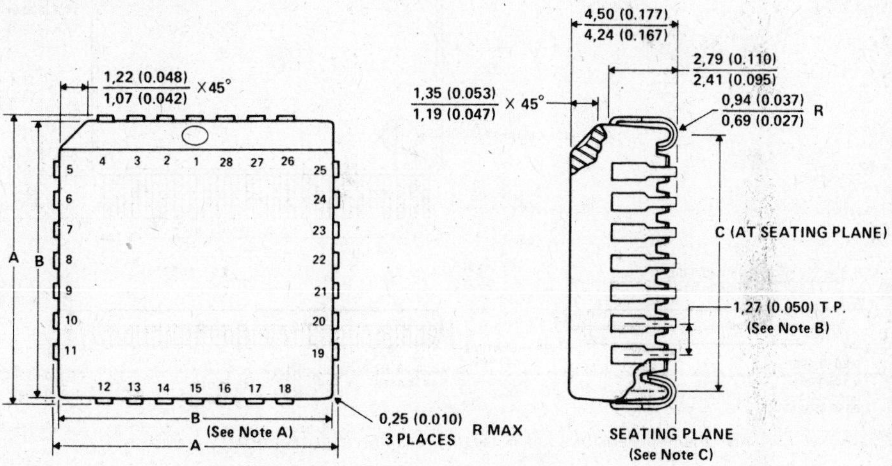
4,50 (0.177)
4,24 (0.167)
2,79 (0.110)
2,41 (0.095)
1,22 (0.048)
1,07 (0.042)
X 45°
1,35 (0.053)
1,19 (0.047)
X 45°
0,94 (0.037)
0,69 (0.027)
R
4 3 2 1 28 27 26
5 6 7 8 9 10 11
25 24 23 22 21 20 19
12 13 14 15 16 17 18
A
B
C (AT SEATING PLANE)
1,27 (0.050) T.P.
(See Note B)
B
(See Note A)
A
0,25 (0.010)
3 PLACES
R MAX
SEATING PLANE
(See Note C)

JEDEC OUTLINE	NO. OF TERMINALS	A MIN	A MAX	B MIN	B MAX	C MIN	C MAX
MO-047AA	20	9,78 (0.385)	10,03 (0.395)	8,89 (0.350)	9,04 (0.356)	7,87 (0.310)	8,38 (0.330)
MO-047AB	28	12,32 (0.485)	12,57 (0.495)	11,43 (0.450)	11,58 (0.456)	10,41 (0.410)	10,92 (0.430)
MO-047AC	44	17,40 (0.685)	17,65 (0.695)	16,51 (0.650)	16,66 (0.656)	15,49 (0.610)	16,00 (0.630)
MO-047AE	68	25,02 (0.985)	25,27 (0.995)	24,13 (0.950)	24,33 (0.956)	23,11 (0.910)	23,62 (0.930)
MO-047AF	84	30,10 (1.185)	30,35 (1.195)	29,21 (1.150)	29,41 (1.158)	27,69 (1.090)	28,70 (1.130)

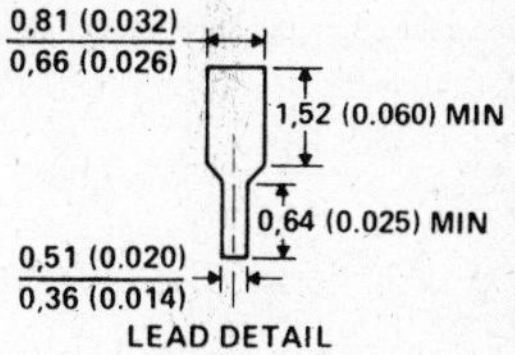
0,81 (0.032)
0,66 (0.026)
1,52 (0.060) MIN
0,64 (0.025) MIN
0,51 (0.020)
0,36 (0.014)

LEAD DETAIL

All dimensions and notes for the specified JEDEC outline apply.

ALL LINEAR DIMENSIONS ARE IN MILLIMETERS AND PARENTHETICALLY IN INCHES

NOTES: A. Centerline of center lead on each side is within 0,10 (0.004) of package centerline as determined by dimension B.
B. Location of each lead within 0,127 (0.005) of true position with respect to center lead on each side.
C. The lead contact points are planar within 0,10 (0.004).

FT CERAMIC FLAT PACKAGE
(48-LEAD FT)

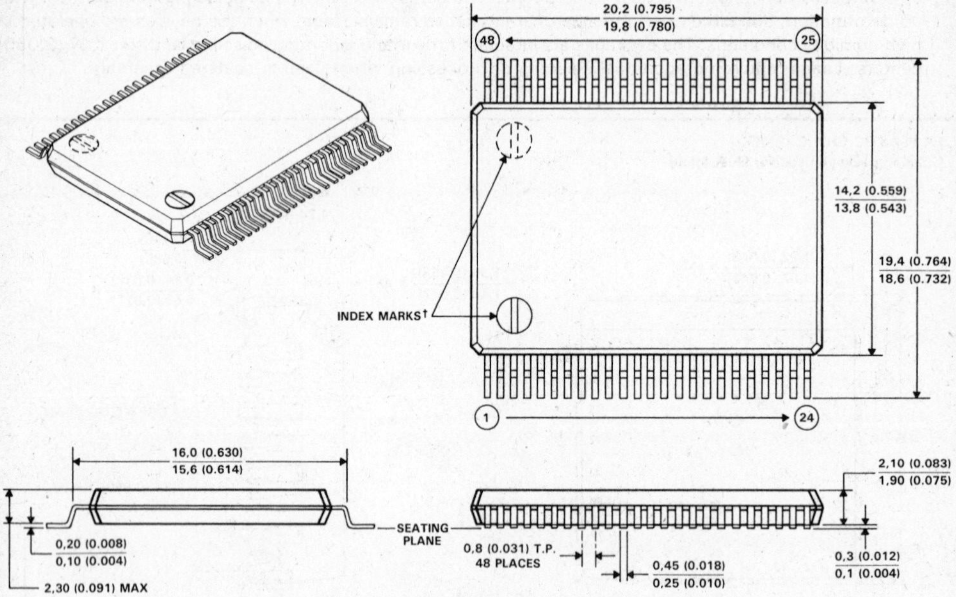
20,2 (0.795)
19,8 (0.780)
48
25
14,2 (0.559)
13,8 (0.543)
19,4 (0.764)
18,6 (0.732)
INDEX MARKS†
1
24
16,0 (0.630)
15,6 (0.614)
SEATING PLANE
0,20 (0.008)
0,10 (0.004)
2,30 (0.091) MAX
0,8 (0.031) T.P.
48 PLACES
0,45 (0.018)
0,25 (0.010)
2,10 (0.083)
1,90 (0.075)
0,3 (0.012)
0,1 (0.004)

ALL LINEAR DIMENSIONS ARE IN MILLIMETERS AND PARENTHETICALLY IN INCHES

†There are two versions of the 48-lead FT package that differ in the position of the index mark in the top view. In one version, the mark is near lead 3, in the other version, it is near lead 46. Consult the individual data sheet to see which applies for a particular device type.

J ceramic dual-in-line package

This hermetically sealed dual-in-line package consists of a ceramic base, ceramic cap, and a lead frame. Hermetic sealing is accomplished with glass. The package is intended for insertion in mounting-hole rows on 7,62 (0.300) centers. Once the leads are compressed and inserted, sufficient tension is provided to secure the package in the board during soldering. Tin-plated (''bright-dipped'') leads require no additional cleaning or processing when used in soldered assembly.

14-PIN J CERAMIC

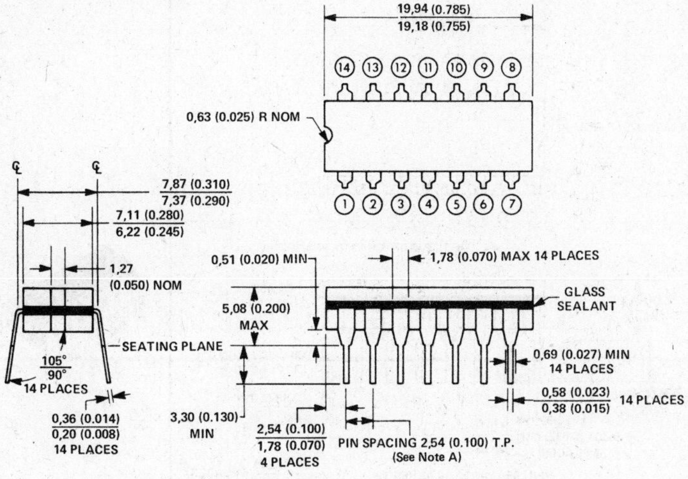
19,94 (0.785)
19,18 (0.755)
14 13 12 11 10 9 8
0,63 (0.025) R NOM
1 2 3 4 5 6 7
7,87 (0.310)
7,37 (0.290)
7,11 (0.280)
6,22 (0.245)
1,27
(0.050) NOM
0,51 (0.020) MIN
1,78 (0.070) MAX 14 PLACES
GLASS
SEALANT
5,08 (0.200)
MAX
SEATING PLANE
105°
90°
14 PLACES
0,69 (0.027) MIN
14 PLACES
0,58 (0.023)
0,38 (0.015)
14 PLACES
0,36 (0.014)
0,20 (0.008)
14 PLACES
3,30 (0.130)
MIN
2,54 (0.100)
1,78 (0.070)
4 PLACES
PIN SPACING 2,54 (0.100) T.P.
(See Note A)

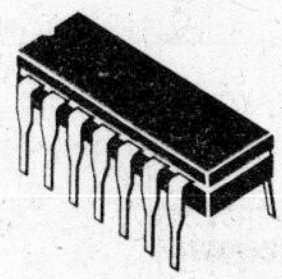

Falls Within JEDEC TO-116 and EIA MO-001AA Dimensions

ALL LINEAR DIMENSIONS ARE IN MILLIMETERS AND PARENTHETICALLY IN INCHES

NOTE A: Each pin centerline is located within 0,25 (0.010) of its true longitudinal position.

J ceramic dual-in-line package

This hermetically sealed dual-in-line package consists of a ceramic base, ceramic cap, and a lead frame. Hermetic sealing is accomplished with glass. The package is intended for insertion in mounting-hole rows on 7,62 (0.300) centers. Once the leads are compressed and inserted, sufficient tension is provided to secure the package in the board during soldering. Tin-plated ("bright-dipped") leads require no additional cleaning or processing when used in soldered assembly.

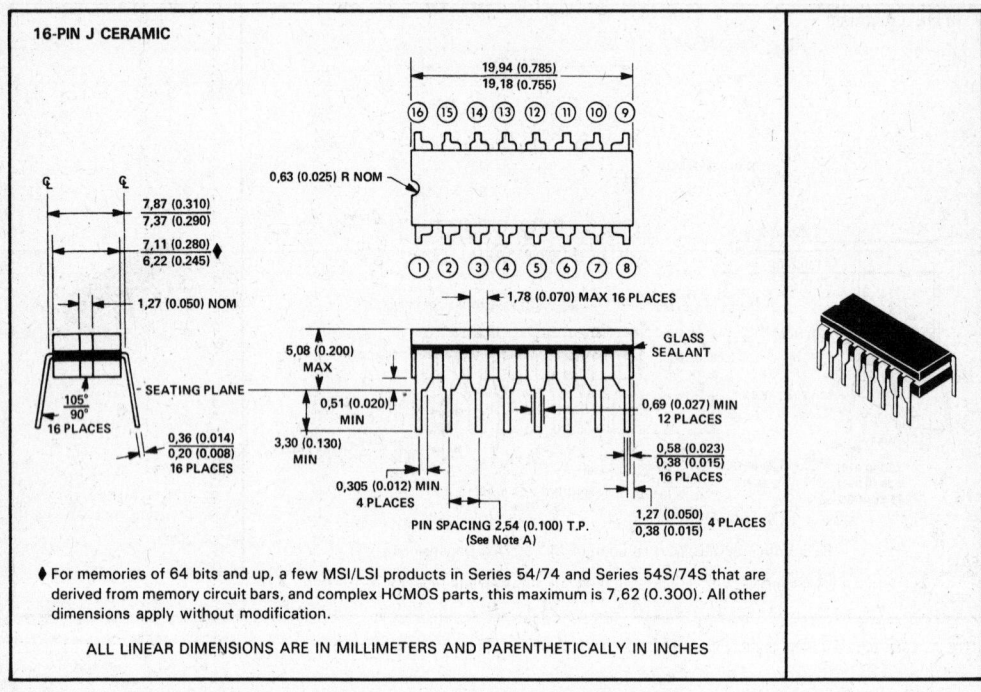

16-PIN J CERAMIC

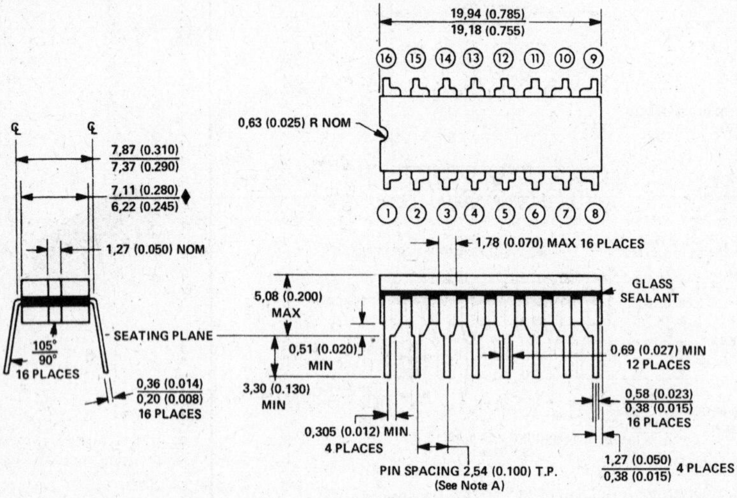
19,94 (0.785)
19,18 (0.755)
16 15 14 13 12 11 10 9
0,63 (0.025) R NOM
1 2 3 4 5 6 7 8
7,87 (0.310)
7,37 (0.290)
7,11 (0.280) ♦
6,22 (0.245)
1,27 (0.050) NOM
1,78 (0.070) MAX 16 PLACES
5,08 (0.200) MAX
GLASS SEALANT
SEATING PLANE
105°
90°
16 PLACES
0,51 (0.020) MIN
0,69 (0.027) MIN 12 PLACES
0,36 (0.014)
0,20 (0.008)
16 PLACES
3,30 (0.130) MIN
0,58 (0.023)
0,38 (0.015)
16 PLACES
0,305 (0.012) MIN 4 PLACES
PIN SPACING 2,54 (0.100) T.P. (See Note A)
1,27 (0.050)
0,38 (0.015)
4 PLACES

♦ For memories of 64 bits and up, a few MSI/LSI products in Series 54/74 and Series 54S/74S that are derived from memory circuit bars, and complex HCMOS parts, this maximum is 7,62 (0.300). All other dimensions apply without modification.

ALL LINEAR DIMENSIONS ARE IN MILLIMETERS AND PARENTHETICALLY IN INCHES

NOTE A: Each pin centerline is located within 0,25 (0.010) of its true longitudinal position.

J ceramic dual-in-line package

This hermetically sealed dual-in-line package consists of a ceramic base, ceramic cap, and a lead frame. Hermetic sealing is accomplished with glass. The package is intended for insertion in mounting-hole rows on 7,62 (0.300) centers. Once the leads are compressed and inserted, sufficient tension is provided to secure the package in the board during soldering. Tin-plated ("bright-dipped") leads require no additional cleaning or processing when used in soldered assembly.

18-PIN J CERAMIC

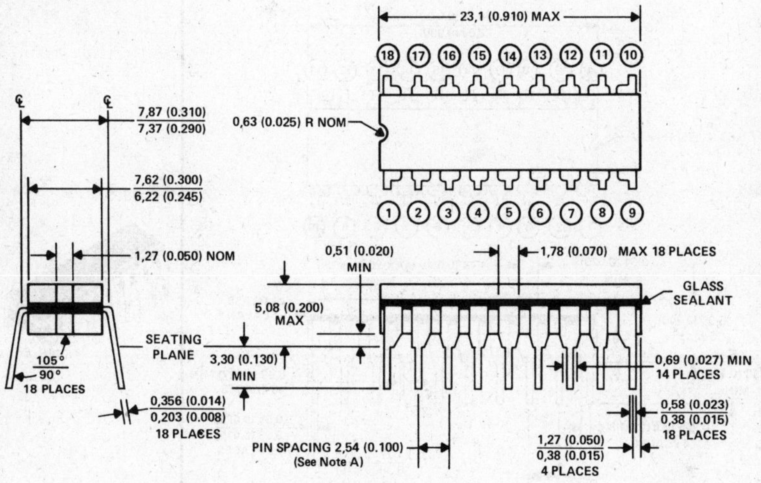
23,1 (0.910) MAX
18 17 16 15 14 13 12 11 10
1 2 3 4 5 6 7 8 9
7,87 (0.310)
7,37 (0.290)
0,63 (0.025) R NOM
7,62 (0.300)
6,22 (0.245)
1,27 (0.050) NOM
0,51 (0.020) MIN
1,78 (0.070) MAX 18 PLACES
GLASS SEALANT
5,08 (0.200) MAX
SEATING PLANE
3,30 (0.130) MIN
105°
90°
18 PLACES
0,69 (0.027) MIN
14 PLACES
0,356 (0.014)
0,203 (0.008)
18 PLACES
0,58 (0.023)
0,38 (0.015)
18 PLACES
PIN SPACING 2,54 (0.100)
(See Note A)
1,27 (0.050)
0,38 (0.015)
4 PLACES

ALL LINEAR DIMENSIONS ARE IN MILLIMETERS AND PARENTHETICALLY IN INCHES

NOTE A: Each pin centerline is located within 0,25 (0.010) of its true longitudinal position.

J ceramic dual-in-line package

This hermetically sealed dual-in-line package consists of a ceramic base, ceramic cap, and a lead frame. Hermetic sealing is accomplished with glass. The package is intended for insertion in mounting-hole rows on 7,62 (0.300) centers. Once the leads are compressed and inserted, sufficient tension is provided to secure the package in the board during soldering. Tin-plated ("bright-dipped") leads require no additional cleaning or processing when used in soldered assembly.

20-PIN J CERAMIC

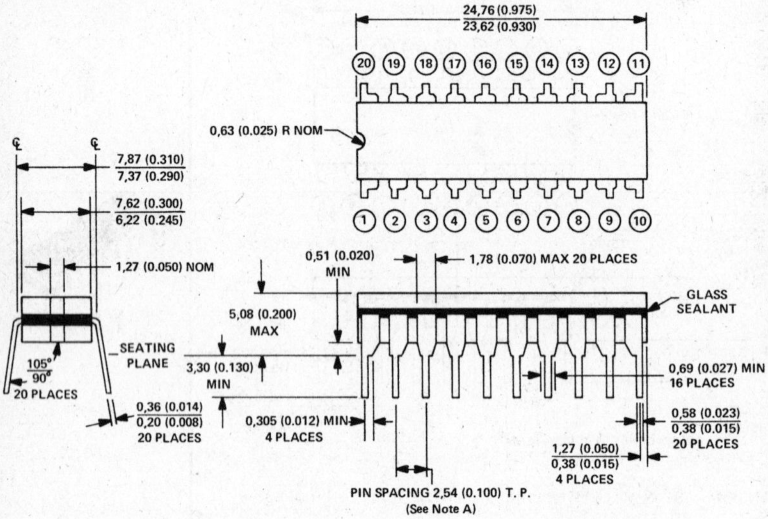

24,76 (0.975)
23,62 (0.930)
20 19 18 17 16 15 14 13 12 11
0,63 (0.025) R NOM
1 2 3 4 5 6 7 8 9 10
℄ ℄
7,87 (0.310)
7,37 (0.290)
7,62 (0.300)
6,22 (0.245)
1,27 (0.050) NOM
0,51 (0.020)
MIN
1,78 (0.070) MAX 20 PLACES
GLASS
SEALANT
5,08 (0.200)
MAX
SEATING
PLANE
3,30 (0.130)
MIN
105°
90°
20 PLACES
0,69 (0.027) MIN
16 PLACES
0,36 (0.014)
0,20 (0.008)
20 PLACES
0,305 (0.012) MIN
4 PLACES
0,58 (0.023)
0,38 (0.015)
20 PLACES
1,27 (0.050)
0,38 (0.015)
4 PLACES
PIN SPACING 2,54 (0.100) T. P.
(See Note A)

ALL LINEAR DIMENSIONS ARE IN MILLIMETERS AND PARENTHETICALLY IN INCHES

NOTE A: Each pin centerline is located within 0,25 (0.010) of its true longitudinal position.

Mechanical Data

B

TEXAS INSTRUMENTS

POST OFFICE BOX 655012 • DALLAS, TEXAS 75265

JD ceramic side-brade dual-in-line packages

This is a hermetically sealed ceramic package with a metal cap and side-brazed tin-plated leads.

JD CERAMIC—SIDE-BRAZE

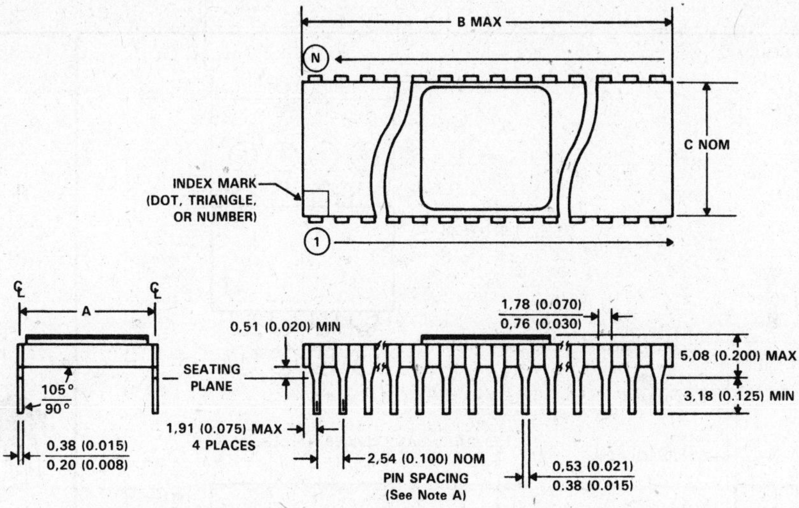
B MAX
N
C NOM
INDEX MARK
(DOT, TRIANGLE,
OR NUMBER)
1
A
1,78 (0.070)
0,76 (0.030)
0,51 (0.020) MIN
5,08 (0.200) MAX
SEATING
PLANE
3,18 (0.125) MIN
105°
90°
1,91 (0.075) MAX
4 PLACES
0,38 (0.015)
0,20 (0.008)
2,54 (0.100) NOM
PIN SPACING
(See Note A)
0,53 (0.021)
0.38 (0.015)

DIM \ PINS (N)	16	18	20	22	24
A +0,51 (+0.020) −0,25 (−0.010)	7,62 (0.300)	7,62 (0.300)	7,62 (0.300)	10,16 (0.400)	7,62 (0.300)
B (MAX)	20,57 (0.810)	23,11 (0.910)	25,65 (1.010)	27,94 (1.100)	30,86 (1.215)
C (NOM)	7,37 (0.290)	7,37 (0.290)	7,37 (0.290)	9,91 (0.390)	7,37 (0.290)

DIM \ PINS (N)	24	28	40	48	52	64
A +0,51 (+0.020) −0,25 (−0.010)	15,24 (0,600)	15,24 (0.600)	15,24 (0.600)	15,24 (0.600)	15,24 (0.600)	22,86 (0.900)
B (MAX)	31,8 (1.250)	36,8 (1,450)	52,1 (2.050)	62,2 (2.450)	67,3 (2.650)	82,6 (3.250)
C (NOM)	15,0 (0.590)	15,0 (0.590)	15,0 (0.590)	15,0 (0.590)	15,0 (0.590)	22,6 (0.890)

ALL LINEAR DIMENSIONS ARE IN MILLIMETERS AND PARENTHETICALLY IN INCHES

NOTE A: Each pin centerline is located within 0,25 (0.010) of its true longitudinal position.

JG ceramic dual-in-line package

This hermetically sealed dual-in-line package consists of a ceramic base, ceramic cap, and an 8-pin lead frame. The package is intended for insertion in mounting-hole rows 7,62 (0.300) centers (see Note A). Once the leads are compressed and inserted, sufficient tension is provided to secure the package in the board during soldering.

8-PIN JG CERAMIC

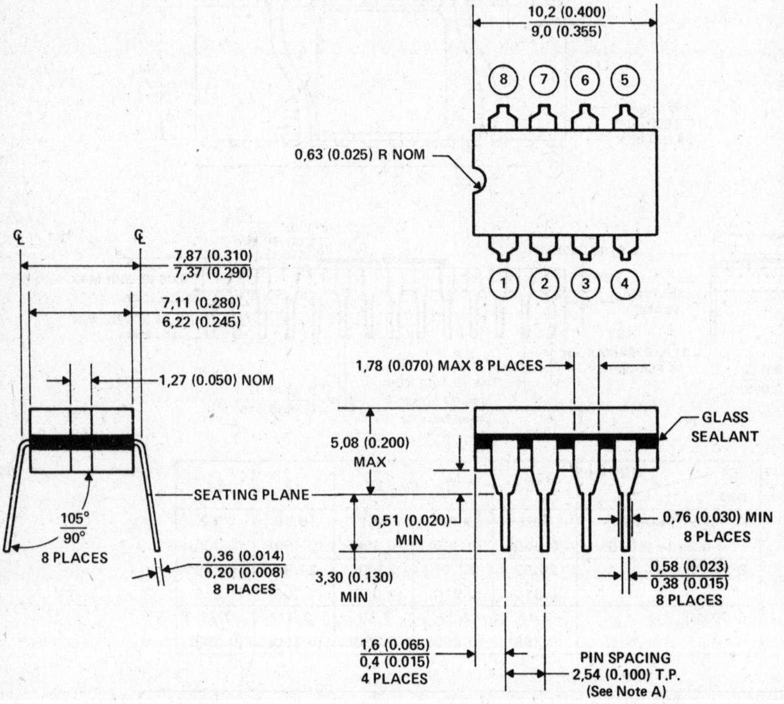
10,2 (0.400)
9,0 (0.355)
8
7
6
5
0,63 (0.025) R NOM
1
2
3
4
7,87 (0.310)
7,37 (0.290)
7,11 (0.280)
6,22 (0.245)
1,27 (0.050) NOM
105°
90°
8 PLACES
SEATING PLANE
0,36 (0.014)
0,20 (0.008)
8 PLACES
1,78 (0.070) MAX 8 PLACES
5,08 (0.200)
MAX
GLASS
SEALANT
0,51 (0.020)
MIN
3,30 (0.130)
MIN
0,76 (0.030) MIN
8 PLACES
0,58 (0.023)
0,38 (0.015)
8 PLACES
1,6 (0.065)
0,4 (0.015)
4 PLACES
PIN SPACING
2,54 (0.100) T.P.
(See Note A)

ALL LINEAR DIMENSIONS ARE IN MILLIMETERS AND PARENTHETICALLY IN INCHES

NOTE A: Each pin centerline is located within 0,25 (0.010) of its true longitudinal position.

TEXAS INSTRUMENTS
POST OFFICE BOX 655012 • DALLAS, TEXAS 75265

KC plastic package

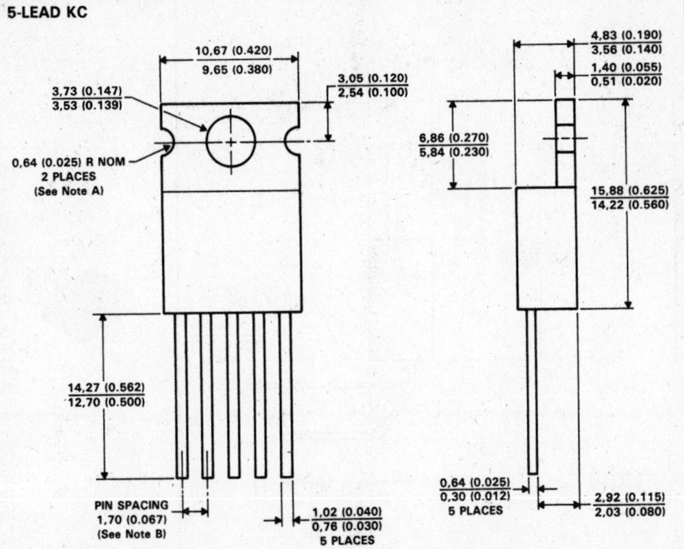
5-LEAD KC
10,67 (0.420)
9,65 (0.380)
3,73 (0.147)
3,53 (0.139)
3,05 (0.120)
2,54 (0.100)
0,64 (0.025) R NOM
2 PLACES
(See Note A)
14,27 (0.562)
12,70 (0.500)
PIN SPACING
1,70 (0.067)
(See Note B)
1,02 (0.040)
0,76 (0.030)
5 PLACES
4,83 (0.190)
3,56 (0.140)
1,40 (0.055)
0,51 (0.020)
6,86 (0.270)
5,84 (0.230)
15,88 (0.625)
14,22 (0.560)
0,64 (0.025)
0,30 (0.012)
5 PLACES
2,92 (0.115)
2,03 (0.080)

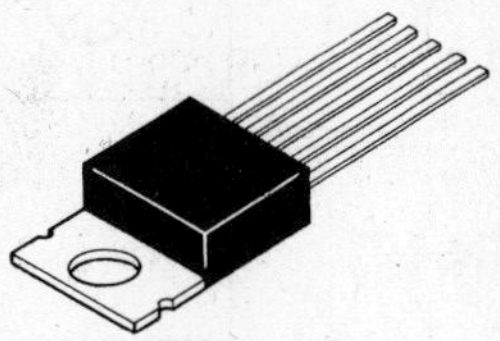

ALL LINEAR DIMENSIONS ARE IN MILLIMETERS AND PARENTHETICALLY IN INCHES

NOTES: A. Notches may or may not be present.
B. Leads are within 0,13 (0.005) radius of true position (T.P.) at maximum material conditions.

TEXAS INSTRUMENTS
POST OFFICE BOX 655012 • DALLAS, TEXAS 75265

KH plastic package

5-LEAD KH

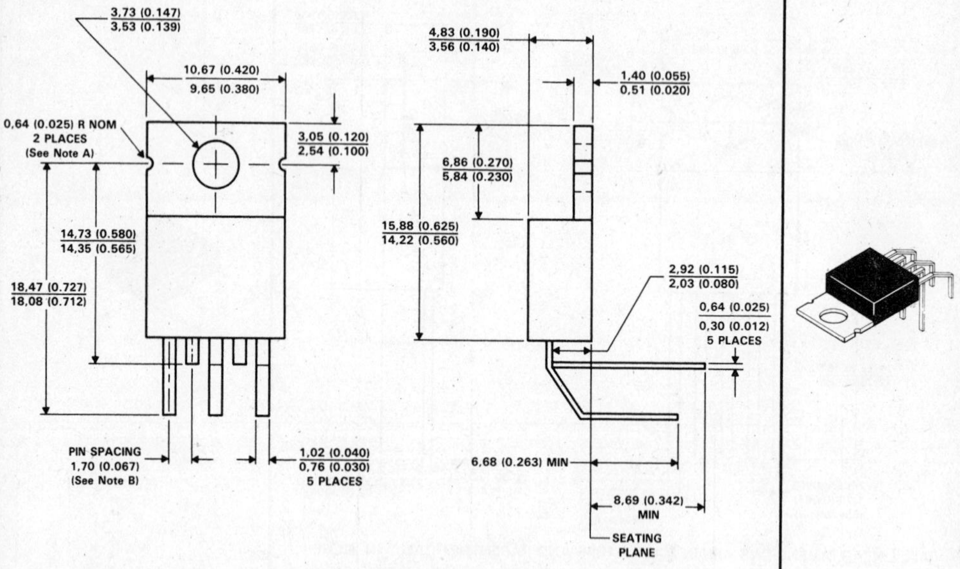
3,73 (0.147)
3,53 (0.139)
10,67 (0.420)
9,65 (0.380)
0,64 (0.025) R NOM
2 PLACES
(See Note A)
3,05 (0.120)
2,54 (0.100)
14,73 (0.580)
14,35 (0.565)
18,47 (0.727)
18,08 (0.712)
PIN SPACING
1,70 (0.067)
(See Note B)
1,02 (0.040)
0,76 (0.030)
5 PLACES
4,83 (0.190)
3,56 (0.140)
1,40 (0.055)
0,51 (0.020)
6,86 (0.270)
5,84 (0.230)
15,88 (0.625)
14,22 (0.560)
2,92 (0.115)
2,03 (0.080)
0,64 (0.025)
0,30 (0.012)
5 PLACES
6,68 (0.263) MIN
8,69 (0.342)
MIN
SEATING
PLANE

ALL LINEAR DIMENSIONS ARE IN MILLIMETERS AND PARENTHETICALLY IN INCHES

NOTES: A. Notches may or may not be present.
B. Leads are within 0,13 (0.005) radius of true position (T.P.) at maximum material conditions.

KV plastic package

5-LEAD KV

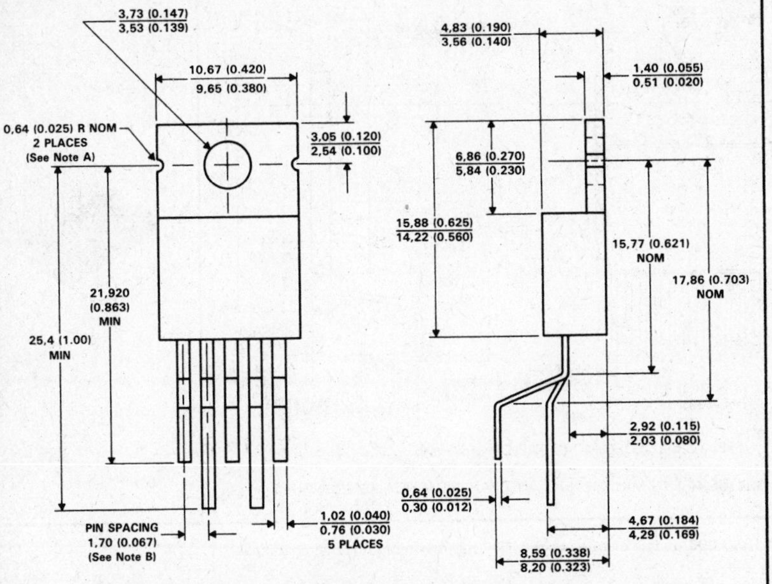
3,73 (0.147)
3,53 (0.139)
10,67 (0.420)
9,65 (0.380)
0,64 (0.025) R NOM
2 PLACES
(See Note A)
3,05 (0.120)
2,54 (0.100)
21,920
(0.863)
MIN
25,4 (1.00)
MIN
PIN SPACING
1,70 (0.067)
(See Note B)
1,02 (0.040)
0,76 (0.030)
5 PLACES
4,83 (0.190)
3,56 (0.140)
1,40 (0.055)
0,51 (0.020)
6,86 (0.270)
5,84 (0.230)
15,88 (0.625)
14,22 (0.560)
15,77 (0.621)
NOM
17,86 (0.703)
NOM
2,92 (0.115)
2,03 (0.080)
0,64 (0.025)
0,30 (0.012)
4,67 (0.184)
4,29 (0.169)
8,59 (0.338)
8,20 (0.323)

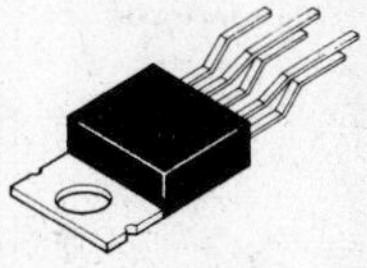

ALL LINEAR DIMENSIONS ARE IN MILLIMETERS AND PARENTHETICALLY IN INCHES

NOTES: A. Notches may or may not be present.
B. Leads are within 0,13 (0.005) radius of true position (T.P.) at maximum material conditions.

KV plastic package

15-LEAD KV

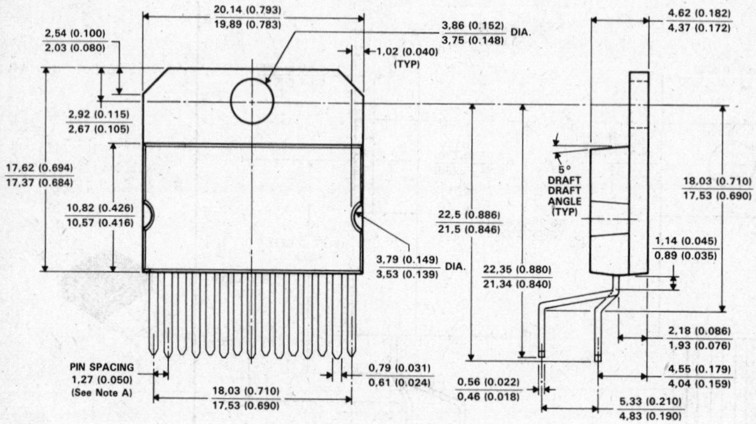
20,14 (0.793)
19,89 (0.783)
2,54 (0.100)
2,03 (0.080)
3,86 (0.152)
3,75 (0.148) DIA.
1,02 (0.040) (TYP)
4,62 (0.182)
4,37 (0.172)
2,92 (0.115)
2,67 (0.105)
17,62 (0.694)
17,37 (0.684)
5°
DRAFT
DRAFT
ANGLE
(TYP)
18,03 (0.710)
17,53 (0.690)
10,82 (0.426)
10,57 (0.416)
22,5 (0.886)
21,5 (0.846)
1,14 (0.045)
0,89 (0.035)
3,79 (0.149)
3,53 (0.139) DIA.
22,35 (0.880)
21,34 (0.840)
2,18 (0.086)
1,93 (0.076)
PIN SPACING
1,27 (0.050)
(See Note A)
0,79 (0.031)
0,61 (0.024)
4,55 (0.179)
4,04 (0.159)
18,03 (0.710)
17,53 (0.690)
0,56 (0.022)
0,46 (0.018)
5,33 (0.210)
4,83 (0.190)

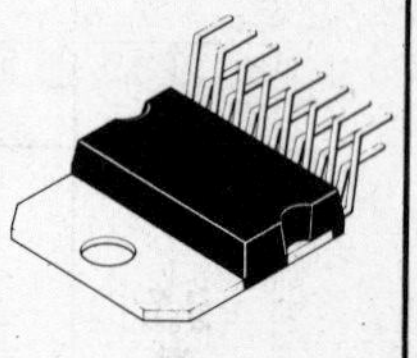

Falls Within JEDEC MO-048AB Dimensions

ALL LINEAR DIMENSIONS ARE IN MILLIMETERS AND PARENTHETICALLY IN INCHES

NOTE A: Leads are within 0,13 (0.005) radius of true position (T.P.) at maximum material conditions.

N plastic dual-in-line package

This dual-in-line package consists of a circuit mounted on a lead frame and encapsulated within an electrically nonconductive plastic compound. The compound will withstand soldering temperature with no deformation, and circuit performance characteristics will remain stable when operated in high-humidity conditions. The package is intended for insertion in mounting-hole rows on 7,62 (0.300) centers. Once the leads are compressed and inserted, sufficient tension is provided to secure the package in the board during soldering. Leads require no additional cleaning or processing when used in soldered assembly.

14-PIN N PLASTIC

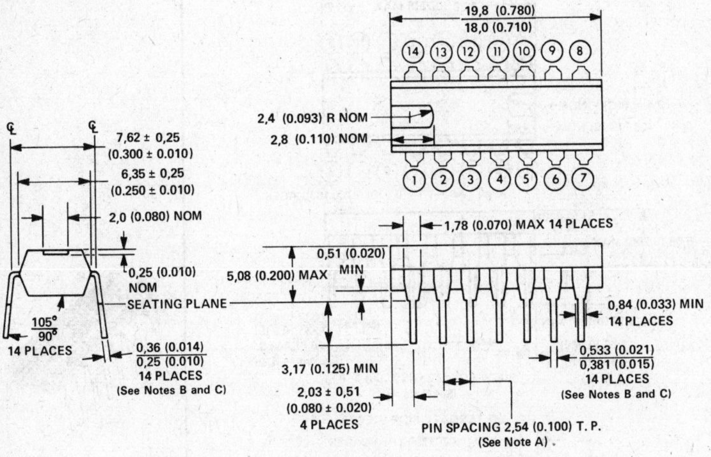
19,8 (0.780)
18,0 (0.710)
14 13 12 11 10 9 8
2,4 (0.093) R NOM
2,8 (0.110) NOM
1 2 3 4 5 6 7
7,62 ± 0,25
(0.300 ± 0.010)
6,35 ± 0,25
(0.250 ± 0.010)
2,0 (0.080) NOM
1,78 (0.070) MAX 14 PLACES
0,51 (0.020) MIN
5,08 (0.200) MAX
0,25 (0.010) NOM
SEATING PLANE
0,84 (0.033) MIN
14 PLACES
105°
90°
14 PLACES
0,36 (0.014)
0,25 (0.010)
14 PLACES
(See Notes B and C)
3,17 (0.125) MIN
0,533 (0.021)
0,381 (0.015)
14 PLACES
(See Notes B and C)
2,03 ± 0,51
(0.080 ± 0.020)
4 PLACES
PIN SPACING 2,54 (0.100) T. P.
(See Note A)

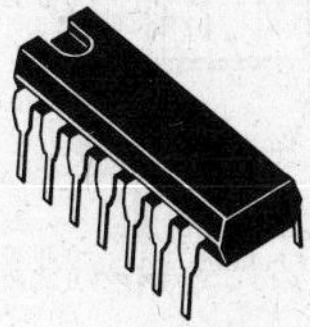

Falls Within JEDEC TO-116 and EIA MO-001AA Dimensions

ALL LINEAR DIMENSIONS ARE IN MILLIMETERS AND PARENTHETICALLY IN INCHES

NOTES: A. Each pin centerline is located within 0,25 (0.010) of its true longitudinal position.
B. This dimension does not apply for solder-dipped leads.
C. When solder-dipped leads are specified, dipped area of the lead extends from the lead tip to at least 0,51 (0.020) above seating plane.

TEXAS INSTRUMENTS
POST OFFICE BOX 655012 • DALLAS, TEXAS 75265

N plastic dual-in-line package

This dual-in-line package consists of a circuit mounted on a lead frame and encapsulated within an electrically nonconductive plastic compound. The compound will withstand soldering temperature with no deformation, and circuit performance characteristics will remain stable when operated in high-humidity conditions. The package is intended for insertion in mounting-hole rows on 7,62 (0.300) centers. Once the leads are compressed and inserted, sufficient tension is provided to secure the package in the board during soldering. Leads require no additional cleaning or processing when used in soldered assembly.

16-PIN N PLASTIC

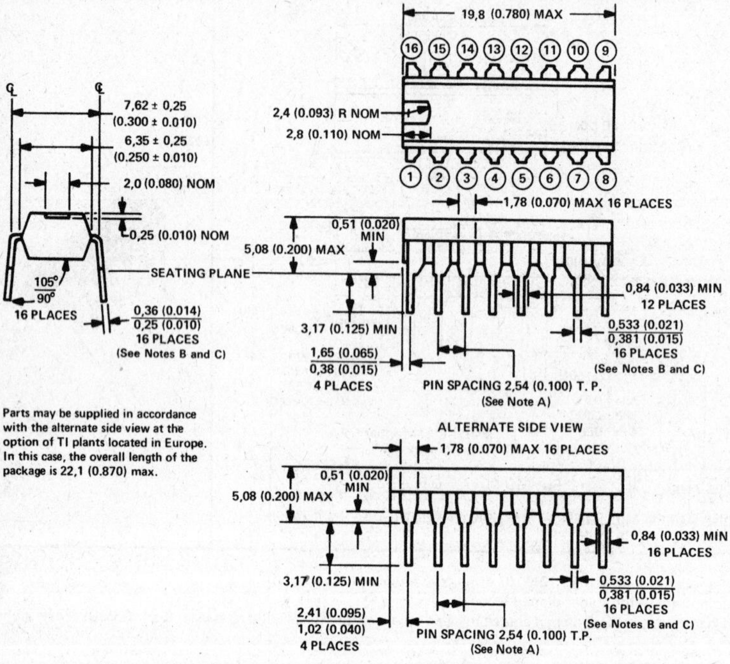
19,8 (0.780) MAX
16 15 14 13 12 11 10 9
1 2 3 4 5 6 7 8
2,4 (0.093) R NOM
2,8 (0.110) NOM
7,62 ± 0,25 (0.300 ± 0.010)
6,35 ± 0,25 (0.250 ± 0.010)
2,0 (0.080) NOM
0,25 (0.010) NOM
SEATING PLANE
105° / 90°
16 PLACES
0,36 (0.014) / 0,25 (0.010) 16 PLACES (See Notes B and C)
1,78 (0.070) MAX 16 PLACES
0,51 (0.020) MIN
5,08 (0.200) MAX
0,84 (0.033) MIN 12 PLACES
3,17 (0.125) MIN
0,533 (0.021) / 0,381 (0.015) 16 PLACES (See Notes B and C)
1,65 (0.065) / 0,38 (0.015) 4 PLACES
PIN SPACING 2,54 (0.100) T. P. (See Note A)
ALTERNATE SIDE VIEW
1,78 (0.070) MAX 16 PLACES
0,51 (0.020) MIN
5,08 (0.200) MAX
0,84 (0.033) MIN 16 PLACES
3,17 (0.125) MIN
0,533 (0.021) / 0,381 (0.015) 16 PLACES (See Notes B and C)
2,41 (0.095) / 1,02 (0.040) 4 PLACES
PIN SPACING 2,54 (0.100) T.P. (See Note A)

Parts may be supplied in accordance with the alternate side view at the option of TI plants located in Europe. In this case, the overall length of the package is 22,1 (0.870) max.

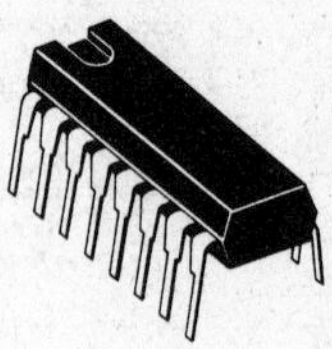

ALL LINEAR DIMENSIONS ARE IN MILLIMETERS AND PARENTHETICALLY IN INCHES

NOTES: A. Each pin centerline is located within 0,25 (0.010) of its true longitudinal position.
B. This dimension does not apply for solder-dipped leads.
C. When solder-dipped leads are specified, dipped area of the lead extends from the lead tip to at least 0,51 (0.020) above seating plane.

N plastic dual-in-line package

This dual-in-line package consists of a circuit mounted on a lead frame and encapsulated within an electrically nonconductive plastic compound. The compound will withstand soldering temperature with no deformation, and circuit performance characteristics will remain stable when operated in high-humidity conditions. The package is intended for insertion in mounting-hole rows on 7,62 (0.300) centers. Once the leads are compressed and inserted, sufficient tension is provided to secure the package in the board during soldering. Leads require no additional cleaning or processing when used in soldered assembly.

18-PIN N PLASTIC

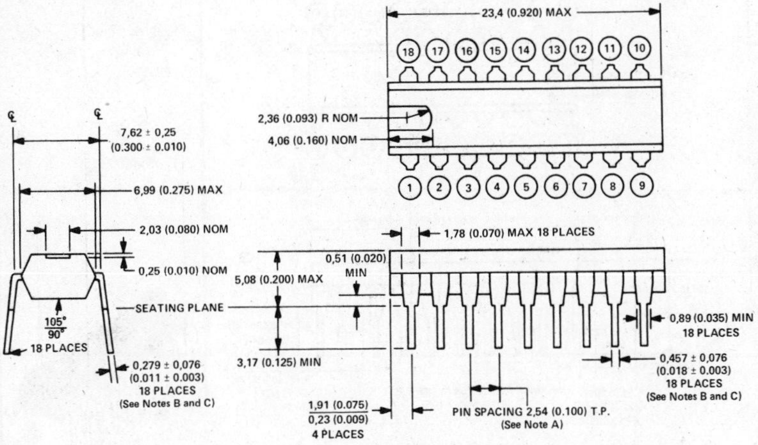
23,4 (0.920) MAX
18
17
16
15
14
13
12
11
10
2,36 (0.093) R NOM
4,06 (0.160) NOM
1
2
3
4
5
6
7
8
9
7,62 ± 0,25
(0.300 ± 0.010)
6,99 (0.275) MAX
2,03 (0.080) NOM
0,25 (0.010) NOM
SEATING PLANE
105°
90°
18 PLACES
0,279 ± 0,076
(0.011 ± 0.003)
18 PLACES
(See Notes B and C)
1,78 (0.070) MAX 18 PLACES
0,51 (0.020)
MIN
5,08 (0.200) MAX
3,17 (0.125) MIN
1,91 (0.075)
0,23 (0.009)
4 PLACES
PIN SPACING 2,54 (0.100) T.P.
(See Note A)
0,89 (0.035) MIN
18 PLACES
0,457 ± 0,076
(0.018 ± 0.003)
18 PLACES
(See Notes B and C)

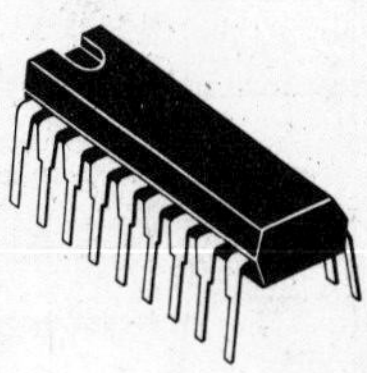

ALL LINEAR DIMENSIONS ARE IN MILLIMETERS AND PARENTHETICALLY IN INCHES

NOTES:
- A. Each pin centerline is located within 0,25 (0.010) of its true longitudinal position.
- B. This dimension does not apply for solder-dipped leads.
- C. When solder-dipped leads are specified, dipped area of the lead extends from the lead tip to at least 0,51 (0.020) above seating plane.

N plastic dual-in-line package

This dual-in-line package consists of a circuit mounted on a lead frame and encapsulated within an electrically nonconductive plastic compound. The compound will withstand soldering temperature with no deformation, and circuit performance characteristics will remain stable when operated in high-humidity conditions. The package is intended for insertion in mounting-hole rows on 7,62 (0.300) centers. Once the leads are compressed and inserted, sufficient tension is provided to secure the package in the board during soldering. Leads require no additional cleaning or processing when used in soldered assembly.

20-PIN N PLASTIC

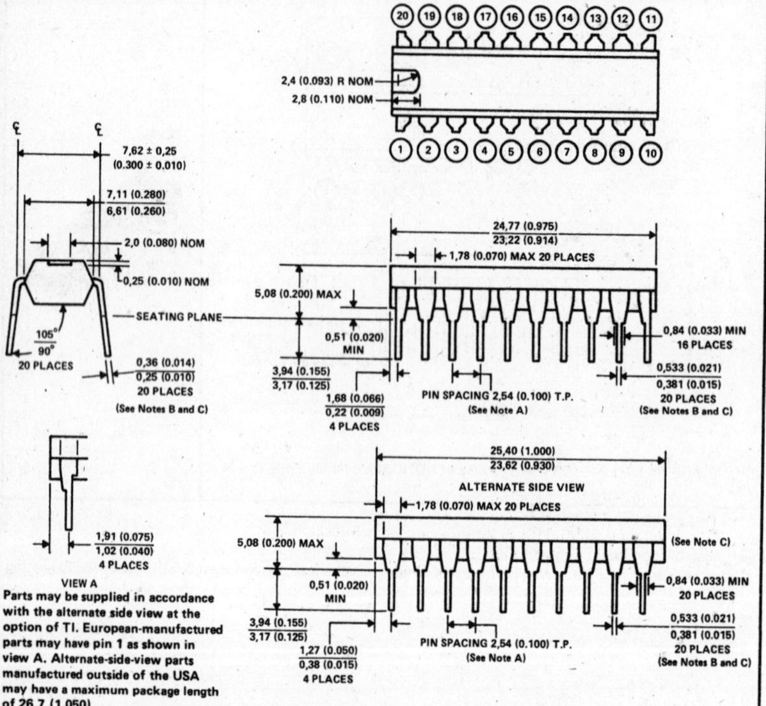
20 19 18 17 16 15 14 13 12 11
2,4 (0.093) R NOM
2,8 (0.110) NOM
1 2 3 4 5 6 7 8 9 10
7,62 ± 0,25 (0.300 ± 0.010)
7,11 (0.280) / 6,61 (0.260)
2,0 (0.080) NOM
0,25 (0.010) NOM
SEATING PLANE
105° / 90° 20 PLACES
0,36 (0.014) / 0,25 (0.010) 20 PLACES (See Notes B and C)
24,77 (0.975) / 23,22 (0.914)
1,78 (0.070) MAX 20 PLACES
5,08 (0.200) MAX
0,51 (0.020) MIN
0,84 (0.033) MIN 16 PLACES
3,94 (0.155) / 3,17 (0.125)
1,68 (0.066) / 0,22 (0.009) 4 PLACES
PIN SPACING 2,54 (0.100) T.P. (See Note A)
0,533 (0.021) / 0,381 (0.015) 20 PLACES (See Notes B and C)
1,91 (0.075) / 1,02 (0.040) 4 PLACES
VIEW A
25,40 (1.000) / 23,62 (0.930)
ALTERNATE SIDE VIEW
1,78 (0.070) MAX 20 PLACES
5,08 (0.200) MAX
(See Note C)
0,51 (0.020) MIN
0,84 (0.033) MIN 20 PLACES
3,94 (0.155) / 3,17 (0.125)
1,27 (0.050) / 0,38 (0.015) 4 PLACES
PIN SPACING 2,54 (0.100) T.P. (See Note A)
0,533 (0.021) / 0,381 (0.015) 20 PLACES (See Notes B and C)

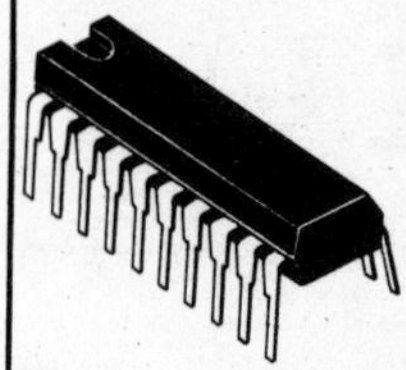

Parts may be supplied in accordance with the alternate side view at the option of TI. European-manufactured parts may have pin 1 as shown in view A. Alternate-side-view parts manufactured outside of the USA may have a maximum package length of 26,7 (1.050).

ALL LINEAR DIMENSIONS ARE IN MILLIMETERS AND PARENTHETICALLY IN INCHES

NOTES: A. Each pin centerline is located within 0,25 (0.010) of its true longitudinal position.
B. This dimension does not apply for solder-dipped leads.
C. When solder-dipped leads are specified, dipped area of the lead extends from the lead tip to at least 0,51 (0.020) above seating plane.

TEXAS INSTRUMENTS
POST OFFICE BOX 655012 • DALLAS, TEXAS 75265

N plastic dual-in-line package

This dual-in-line package consists of a circuit mounted on a lead frame and encapsulated within an electrically nonconductive plastic compound. The compound will withstand soldering temperature with no deformation, and circuit performance characteristics will remain stable when operated in high-humidity conditions. The package is intended for insertion in mountng-hole rows on 7,62 (0.300) centers. Once the leads are compressed and inserted, sufficient tension is provided to secure the package in the board during soldering. Leads require no additional cleaning or processing when used in soldered assembly.

22-PIN N PLASTIC

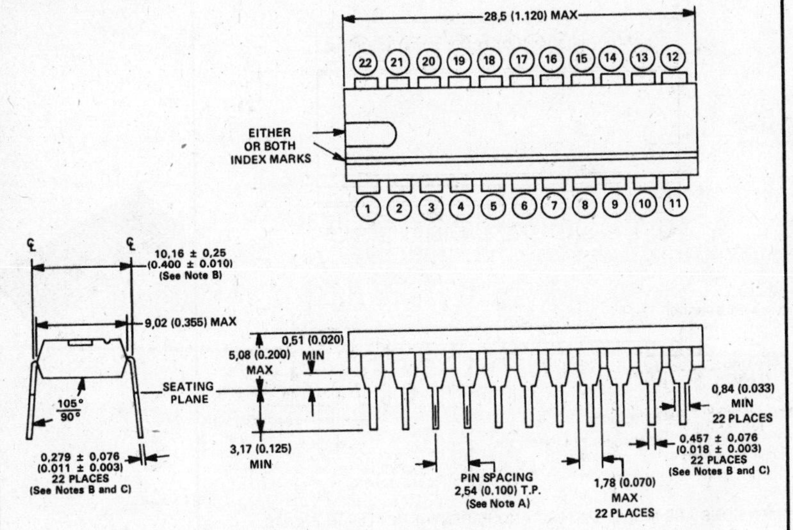
28,5 (1.120) MAX
22
21
20
19
18
17
16
15
14
13
12
EITHER OR BOTH INDEX MARKS
1
2
3
4
5
6
7
8
9
10
11
10,16 ± 0,25 (0.400 ± 0.010) (See Note B)
9,02 (0.355) MAX
0,51 (0.020) MIN
5,08 (0.200) MAX
SEATING PLANE
105°
90°
3,17 (0.125) MIN
0,279 ± 0,076 (0.011 ± 0.003) 22 PLACES (See Notes B and C)
PIN SPACING 2,54 (0.100) T.P. (See Note A)
1,78 (0.070) MAX 22 PLACES
0,84 (0.033) MIN 22 PLACES
0,457 ± 0,076 (0.018 ± 0.003) 22 PLACES (See Notes B and C)

ALL LINEAR DIMENSIONS ARE IN MILLIMETERS AND PARENTHETICALLY IN INCHES

NOTES: A. Each pin centerline is located within 0,25 (0.010) of its true longitudinal position.
B. This dimension does not apply for solder-dipped leads.
C. When solder-dipped leads are specified, dipped area of the lead extends from the lead tip to at least 0,51 (0.020) above seating plane.

N plastic dual-in-line package

This dual-in-line package consists of a circuit mounted on a lead frame and encapsulated within an electrically nonconductive plastic compound. The compound will withstand soldering temperature with no deformation, and circuit performance characteristics will remain stable when operated in high-humidity conditions. The package is intended for insertion in mounting-hole rows on 15,24 (0.600) centers. Once the leads are compressed and inserted, sufficient tension is provided to secure the package in the board during soldering. Leads require no additional cleaning or processing when used in soldered assembly.

28-PIN N PLASTIC

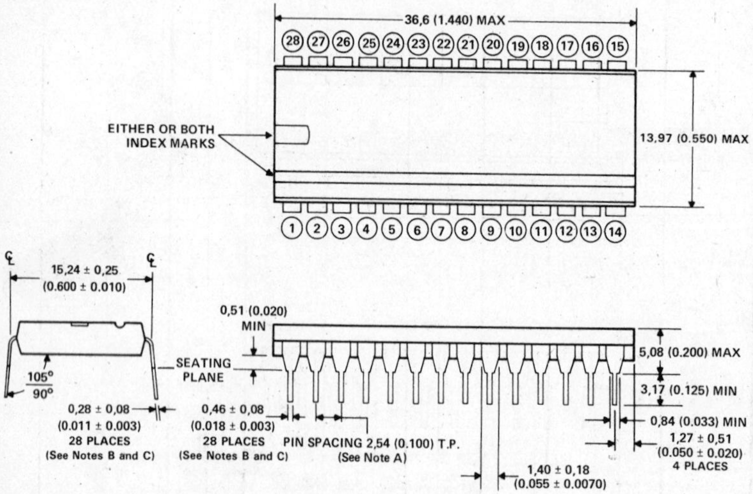
36,6 (1.440) MAX
28 27 26 25 24 23 22 21 20 19 18 17 16 15
EITHER OR BOTH INDEX MARKS
13.97 (0.550) MAX
1 2 3 4 5 6 7 8 9 10 11 12 13 14
15,24 ± 0,25 (0.600 ± 0.010)
0,51 (0.020) MIN
5,08 (0.200) MAX
SEATING PLANE
3,17 (0.125) MIN
105°
90°
0,28 ± 0,08 (0.011 ± 0.003) 28 PLACES (See Notes B and C)
0,46 ± 0,08 (0.018 ± 0.003) 28 PLACES (See Notes B and C)
PIN SPACING 2,54 (0.100) T.P. (See Note A)
0,84 (0.033) MIN
1,27 ± 0,51 (0.050 ± 0.020) 4 PLACES
1,40 ± 0,18 (0.055 ± 0.0070)

ALL LINEAR DIMENSIONS ARE IN MILLIMETERS AND PARENTHETICALLY IN INCHES

NOTES: A. Each pin centerline is located within 0,25 (0.010) of its true longitudinal position.
B. This dimension does not apply for solder-dipped leads.
C. When solder-dipped leads are specified, dipped area of the lead extends from the lead tip to at least 0,51 (0.020) above seating plane.

Mechanical Data B

N plastic dual-in-line package

This dual-in-line package consists of a circuit mounted on a lead frame and encapsulated within an electrically nonconductive plastic compound. The compound will withstand soldering temperature with no deformation, and circuit performance characteristics will remain stable when operated in high-humidity conditions. The package is intended for insertion in mounting-hole rows on 15,24 (0.600) centers. Once the leads are compressed and inserted, sufficient tension is provided to secure the package in the board during soldering. Leads require no additional cleaning or processing when used in soldered assembly.

40-PIN N PLASTIC

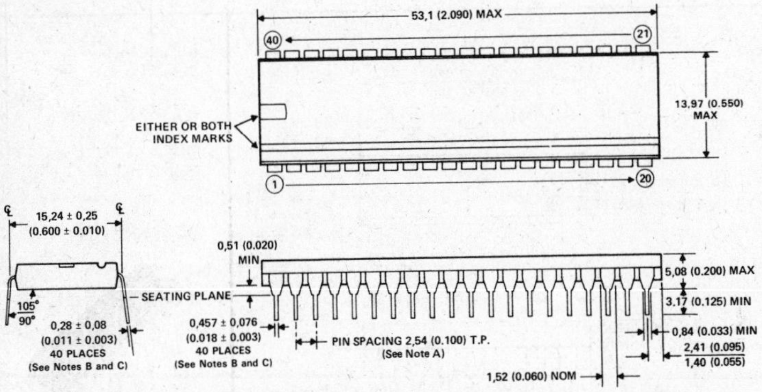
53,1 (2.090) MAX
40
21
13,97 (0.550) MAX
EITHER OR BOTH INDEX MARKS
1
20
15,24 ± 0,25
(0.600 ± 0.010)
0,51 (0.020) MIN
5,08 (0.200) MAX
SEATING PLANE
3,17 (0.125) MIN
105°
90°
0,28 ± 0,08
(0.011 ± 0.003)
40 PLACES
(See Notes B and C)
0,457 ± 0,076
(0.018 ± 0.003)
40 PLACES
(See Notes B and C)
PIN SPACING 2,54 (0.100) T.P.
(See Note A)
0,84 (0.033) MIN
2,41 (0.095)
1,40 (0.055)
1,52 (0.060) NOM

ALL LINEAR DIMENSIONS ARE IN MILLIMETERS AND PARENTHETICALLY IN INCHES

NOTES: A. Each pin centerline is located within 0,25 (0.010) of its true longitudinal position.
B. This dimension does not apply for solder-dipped leads.
C. When solder-dipped leads are specified, dipped area of the lead extends from the lead tip to at least 0,51 (0.020) above seating plane.

NE plastic dual-in-line packages

This dual-in-line package consists of a circuit mounted on a 14-pin lead frame and encapsulated within an electrically nonconductive plastic compound. The compound will withstand soldering temperature with no deformation, and circuit performance characteristics will remain stable when operated in high-humidity conditions. For better heat dissipation there are internal tabs connecting the three central leads on each side of the 14-pin package. The package is intended for insertion in mounting-hole rows on 7,62 (0.300) centers. Once the leads are compressed and inserted, sufficient tension is provided to secure the package in the board during soldering. Leads require no additional cleaning or processing when used in soldered assembly.

14-PIN NE PLASTIC

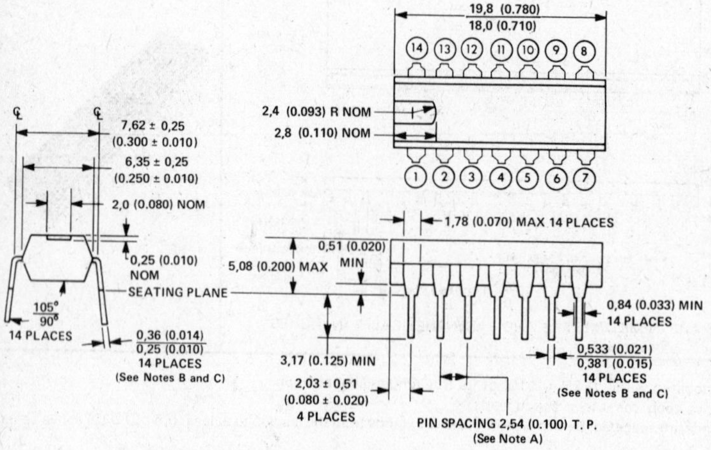

19,8 (0.780)
18,0 (0.710)
14 13 12 11 10 9 8
1 2 3 4 5 6 7
2,4 (0.093) R NOM
2,8 (0.110) NOM
7,62 ± 0,25
(0.300 ± 0.010)
6,35 ± 0,25
(0.250 ± 0.010)
2,0 (0.080) NOM
1,78 (0.070) MAX 14 PLACES
0,51 (0.020)
MIN
5,08 (0.200) MAX
0,25 (0.010)
NOM
SEATING PLANE
105°
90°
14 PLACES
0,84 (0.033) MIN
14 PLACES
0,36 (0.014)
0,25 (0.010)
14 PLACES
(See Notes B and C)
3,17 (0.125) MIN
0,533 (0.021)
0,381 (0.015)
14 PLACES
(See Notes B and C)
2,03 ± 0,51
(0.080 ± 0.020)
4 PLACES
PIN SPACING 2,54 (0.100) T. P.
(See Note A)

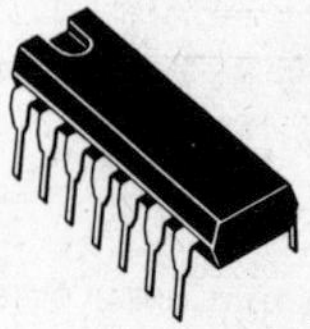

Falls Within JEDEC TO-116 and EIA MO-001AA Dimensions

ALL LINEAR DIMENSIONS ARE IN MILLIMETERS AND PARENTHETICALLY IN INCHES

NOTES: A. Each pin centerline is located within 0,25 (0.010) of its true longitudinal position.
B. This dimension does not apply for solder-dipped leads.
C. When solder-dipped leads are specified, dipped area of the lead extends from the lead tip to at least 0,51 (0.020) above seating plane.

NE plastic dual-in-line packages

This dual-in-line package consists of a circuit mounted on a 16-pin lead frame and encapsulated within an electrically nonconductive plastic compound. The compound will withstand soldering temperature with no deformation, and circuit performance characteristics will remain stable when operated in high-humidity conditions. For better heat dissipation there are internal tabs connecting the two central leads on each side of the 16-pin package. The package is intended for insertion in mounting-hole rows on 7,62 (0.300) centers. Once the leads are compressed and inserted, sufficient tension is provided to secure the package in the board during soldering. Leads require no additional cleaning or processing when used in soldered assembly.

16-PIN NE PLASTIC

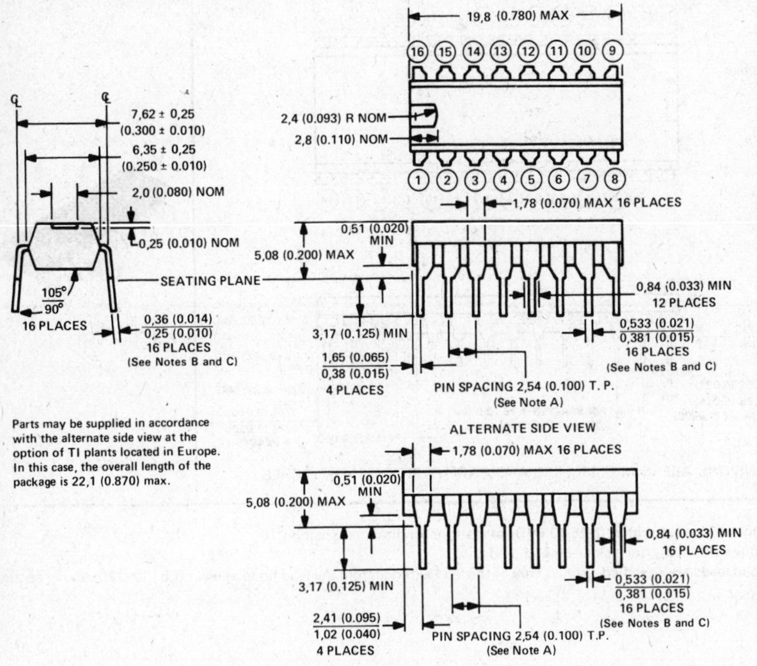
19,8 (0.780) MAX
16 15 14 13 12 11 10 9
1 2 3 4 5 6 7 8
2,4 (0.093) R NOM
2,8 (0.110) NOM
7,62 ± 0,25 (0.300 ± 0.010)
6,35 ± 0,25 (0.250 ± 0.010)
2,0 (0.080) NOM
0,25 (0.010) NOM
SEATING PLANE
105° / 90° 16 PLACES
0,36 (0.014) / 0,25 (0.010) 16 PLACES (See Notes B and C)
1,78 (0.070) MAX 16 PLACES
0,51 (0.020) MIN
5,08 (0.200) MAX
3,17 (0.125) MIN
1,65 (0.065) / 0,38 (0.015) 4 PLACES
0,84 (0.033) MIN 12 PLACES
0,533 (0.021) / 0,381 (0.015) 16 PLACES (See Notes B and C)
PIN SPACING 2,54 (0.100) T. P. (See Note A)
Parts may be supplied in accordance with the alternate side view at the option of TI plants located in Europe. In this case, the overall length of the package is 22,1 (0.870) max.
ALTERNATE SIDE VIEW
1,78 (0.070) MAX 16 PLACES
0,51 (0.020) MIN
5,08 (0.200) MAX
0,84 (0.033) MIN 16 PLACES
3,17 (0.125) MIN
0,533 (0.021) / 0,381 (0.015) 16 PLACES (See Notes B and C)
2,41 (0.095) / 1,02 (0.040) 4 PLACES
PIN SPACING 2,54 (0.100) T.P. (See Note A)

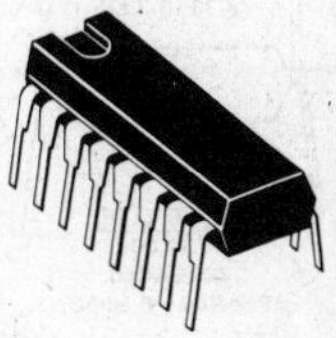

ALL LINEAR DIMENSIONS ARE IN MILLIMETERS AND PARENTHETICALLY IN INCHES

NOTES: A. Each pin centerline is located within 0,25 (0.010) of its true longitudinal position.
B. This dimension does not apply for solder-dipped leads.
C. When solder-dipped leads are specified, dipped area of the lead extends from the lead tip to at least 0,51 (0.020) above seating plane.

NF plastic dual-in-line packages

This dual-in-line package consists of a circuit mounted on a 28-pin lead frame and encapsulated within a plastic compound. The compound will withstand soldering temperature with no deformation, and circuit performance characteristics will remain stable when operated in high-humidity conditions. The package is intended for insertion in mounting-hole rows on 10,16 (0.400) centers. Pin spacing within the rows is 1,78 (0.070). Once the leads are compressed and inserted, sufficient tension is provided to secure the package in the board during soldering. Solder-plated leads require no additional cleaning or processing when used in soldered assembly.

28-PIN NF PLASTIC

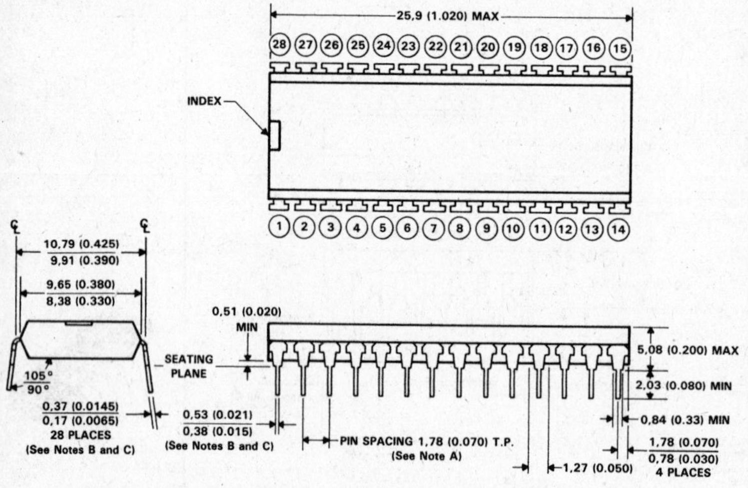
25,9 (1.020) MAX
28 27 26 25 24 23 22 21 20 19 18 17 16 15
INDEX
1 2 3 4 5 6 7 8 9 10 11 12 13 14
10,79 (0.425)
9,91 (0.390)
9,65 (0.380)
8,38 (0.330)
0,51 (0.020) MIN
SEATING PLANE
105°
90°
5,08 (0.200) MAX
2,03 (0.080) MIN
0,37 (0.0145)
0,17 (0.0065)
28 PLACES
(See Notes B and C)
0,53 (0.021)
0,38 (0.015)
(See Notes B and C)
0,84 (0.33) MIN
PIN SPACING 1,78 (0.070) T.P.
(See Note A)
1,78 (0.070)
0,78 (0.030)
4 PLACES
1,27 (0.050)

ALL LINEAR DIMENSIONS ARE IN MILLIMETERS AND PARENTHETICALLY IN INCHES

NOTES:
A. Each pin centerline is located within 0,25 (0.010) of its true longitudinal position.
B. This dimension does not apply for solder-dipped leads.
C. When solder-dipped leads are specified, dipped area of the lead extends from the lead tip to at least 0,51 (0.020) above seating plane.

NT plastic dual-in-line package

This dual-in-line package consists of a circuit mounted on a lead frame and encapsulated within an electrically nonconductive plastic compound. The compound will withstand soldering temperature with no deformation, and circuit performance characteristics will remain stable when operated in high-humidity conditions. The package is intended for insertion in mounting-hole rows on 7,62 (0.300) centers. Once the leads are compressed and inserted, sufficient tension is provided to secure the package in the board during soldering. Leads require no additional cleaning or processing when used in soldered assembly.

NOTE: For all except 24-pin packages, the letter N is used by itself since only the 24-pin package is available in more than one row-spacing. For the 24-pin package, the 7,62 (0.300) version is designated NT; the 15,24 (0.600) version is designated NW. If no second letter or row-spacing is specified, the package is assumed to have 15,24 (0.600) row-spacing.

24-PIN NT PLASTIC

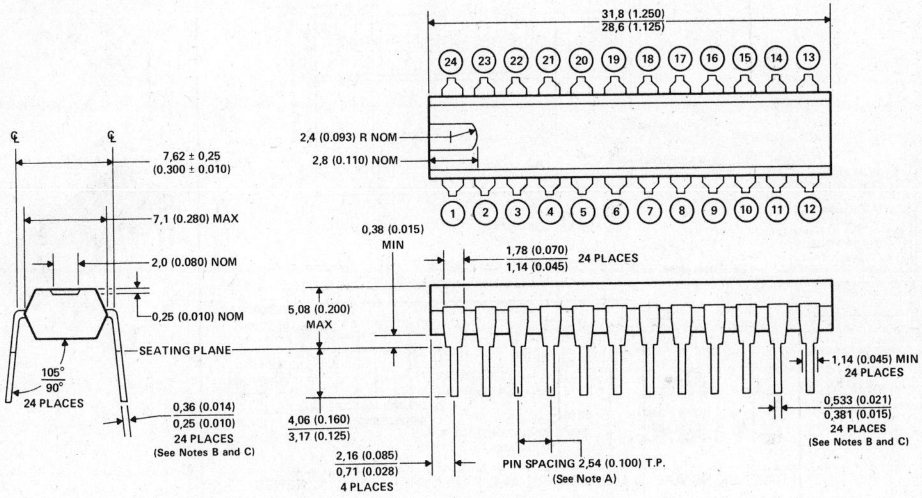
31,8 (1.250)
28,6 (1.125)
2,4 (0.093) R NOM
2,8 (0.110) NOM
7,62 ± 0,25
(0.300 ± 0.010)
7,1 (0.280) MAX
2,0 (0.080) NOM
0,25 (0.010) NOM
SEATING PLANE
105°
90°
24 PLACES
0,36 (0.014)
0,25 (0.010)
24 PLACES
(See Notes B and C)
0,38 (0.015)
MIN
5,08 (0.200)
MAX
4,06 (0.160)
3,17 (0.125)
2,16 (0.085)
0,71 (0.028)
4 PLACES
1,78 (0.070)
1,14 (0.045)
24 PLACES
1,14 (0.045) MIN
24 PLACES
0,533 (0.021)
0,381 (0.015)
24 PLACES
(See Notes B and C)
PIN SPACING 2,54 (0.100) T.P.
(See Note A)

ALL LINEAR DIMENSIONS ARE IN MILLIMETERS AND PARENTHETICALLY IN INCHES

NOTES:
A. Each pin centerline is located within 0,25 (0.010) of its true longitudinal position.
B. This dimension does not apply for solder-dipped leads.
C. When solder-dipped leads are specified, dipped area of the lead extends from the lead tip to at least 0,51 (0.020) above seating plane.

NW plastic dual-in-line package

This dual-in-line package consists of a circuit mounted on a lead frame and encapsulated within an electrically nonconductive plastic compound. The compound will withstand soldering temperature with no deformation, and circuit performance characteristics will remain stable when operated in high-humidity conditions. The package is intended for insertion in mounting-hole rows on 15,24 (0.600) centers. Once the leads are compressed and inserted, sufficient tension is provided to secure the package in the board during soldering. Leads require no additional cleaning or processing when used in soldered assembly.

NOTE: For all except 24-pin packages, the letter N is used by itself since only the 24-pin package is available in more than one row-spacing. For the 24-pin package, the 7,62 (0.300) version is designated NT; the 15,24 (0.600) version is designated NW. If no second letter or row-spacing is specified, the package is assumed to have 15,24 (0.600) row-spacing.

24-PIN NW PLASTIC

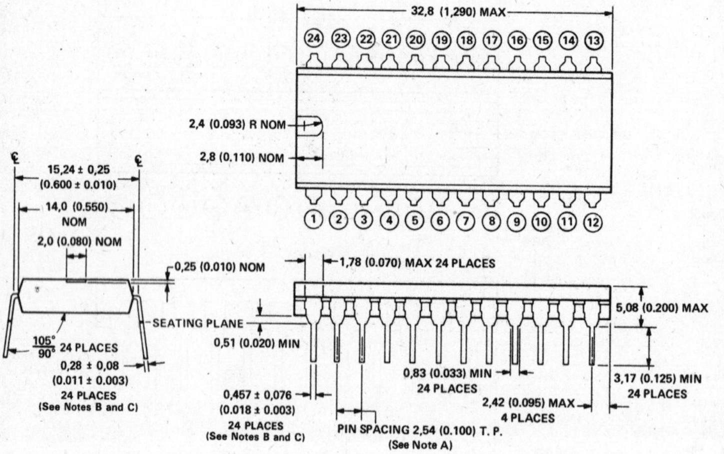
32,8 (1,290) MAX
24 23 22 21 20 19 18 17 16 15 14 13
2,4 (0.093) R NOM
2,8 (0.110) NOM
1 2 3 4 5 6 7 8 9 10 11 12
15,24 ± 0,25
(0.600 ± 0.010)
14,0 (0.550)
NOM
2,0 (0.080) NOM
0,25 (0.010) NOM
SEATING PLANE
105° / 90° 24 PLACES
0,28 ± 0,08
(0.011 ± 0.003)
24 PLACES
(See Notes B and C)
1,78 (0.070) MAX 24 PLACES
0,51 (0.020) MIN
5,08 (0.200) MAX
0,83 (0.033) MIN
24 PLACES
3,17 (0.125) MIN
24 PLACES
0,457 ± 0,076
(0.018 ± 0.003)
24 PLACES
(See Notes B and C)
2,42 (0.095) MAX
4 PLACES
PIN SPACING 2,54 (0.100) T. P.
(See Note A)

ALL LINEAR DIMENSIONS ARE IN MILLIMETERS AND PARENTHETICALLY IN INCHES

NOTES:
A. Each pin centerline is located within 0,25 (0.010) of its true longitudinal position.
B. This dimension does not apply for solder-dipped leads.
C. When solder-dipped leads are specified, dipped area of the lead extends from the lead tip to at least 0,51 (0.020) above seating plane.

P dual-in-line plastic package

This package consists of a circuit mounted on an 8-pin lead frame and encapsulated within a plastic compound. The compound will withstand soldering temperature with no deformation, and circuit performance characteristics will remain stable when operated in high-humidity conditions. The package is intended for insertion in mounting-hole rows on 7,62 (0.300) centers (See Note A). Once the leads are compressed and inserted, sufficient tension is provided to secure the package in the board during soldering. Solder-plated leads require no additional cleaning or processing when used in soldered assembly.

8-PIN P PLASTIC

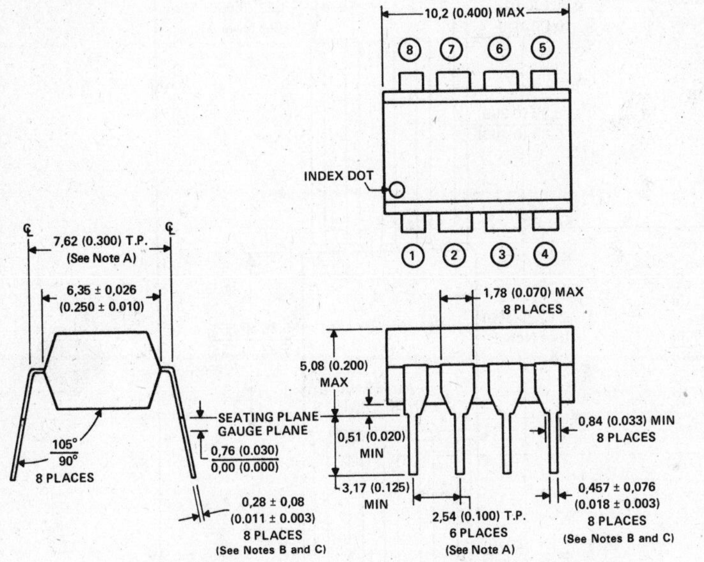
10,2 (0.400) MAX
8
7
6
5
INDEX DOT
1
2
3
4
7,62 (0.300) T.P.
(See Note A)
6,35 ± 0,026
(0.250 ± 0.010)
1,78 (0.070) MAX
8 PLACES
5,08 (0.200)
MAX
SEATING PLANE
GAUGE PLANE
0,76 (0.030)
0,00 (0.000)
0,51 (0.020)
MIN
0,84 (0.033) MIN
8 PLACES
105°
90°
8 PLACES
3,17 (0.125)
MIN
0,457 ± 0,076
(0.018 ± 0.003)
8 PLACES
(See Notes B and C)
0,28 ± 0,08
(0.011 ± 0.003)
8 PLACES
(See Notes B and C)
2,54 (0.100) T.P.
6 PLACES
(See Note A)

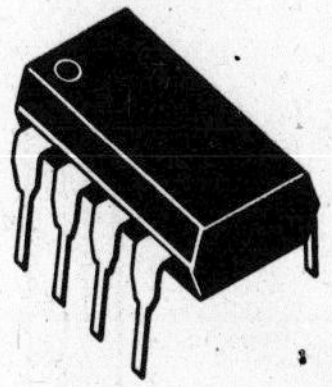

ALL LINEAR DIMENSIONS ARE IN MILLIMETERS AND PARENTHETICALLY IN INCHES

NOTES:
A. Each pin centerline is located within 0,25 (0.010) of its true longitudinal position.
B. This dimension does not apply for solder-dipped leads.
C. When solder-dipped leads are specified, dipped area of the lead extends from the lead tip to at least 0,51 (0.020) above seating plane.

U ceramic flat package

This flat package consists of a ceramic base, ceramic cap, and lead frame. Circuit bars are alloy mounted. Hermetic sealing is accomplished with glass. Leads require no additional cleaning or processing when used in soldered assembly.

10-PIN U CERAMIC

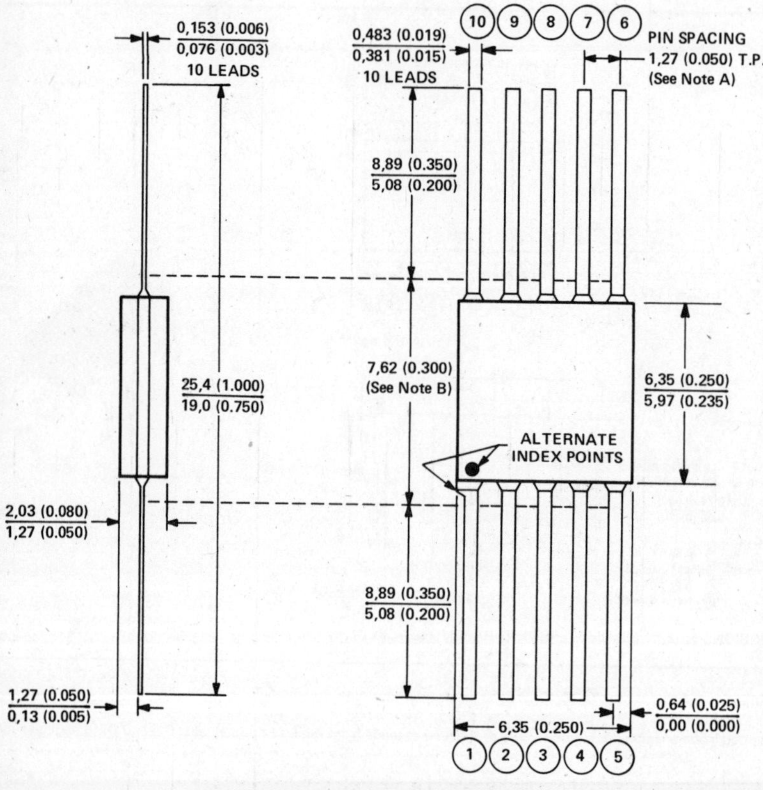
0,153 (0.006)
0,076 (0.003)
10 LEADS
0,483 (0.019)
0,381 (0.015)
10 LEADS
10
9
8
7
6
PIN SPACING
1,27 (0.050) T.P.
(See Note A)
8,89 (0.350)
5,08 (0.200)
25,4 (1.000)
19,0 (0.750)
7,62 (0.300)
(See Note B)
6,35 (0.250)
5,97 (0.235)
ALTERNATE
INDEX POINTS
2,03 (0.080)
1,27 (0.050)
8,89 (0.350)
5,08 (0.200)
1,27 (0.050)
0,13 (0.005)
6,35 (0.250)
0,64 (0.025)
0,00 (0.000)
1
2
3
4
5

Fall Within JEDEC MO-004AE Dimensions

ALL LINEAR DIMENSIONS ARE IN MILLIMETERS AND PARENTHETICALLY IN INCHES

NOTES: A. Leads are within 0,13 (0.005) radius of true position (T.P.) at maximum material conditions.
B. This dimension determines a zone within which all body and lead irregularities lie.

W ceramic flat packages

Each of these hermetically sealed flat packages consists of an electrically nonconductive ceramic base and cap and a lead frame. Hermetic sealing is accomplished with glass. Leads require no additional cleaning or processing when used in soldered assembly.

14-PIN W CERAMIC

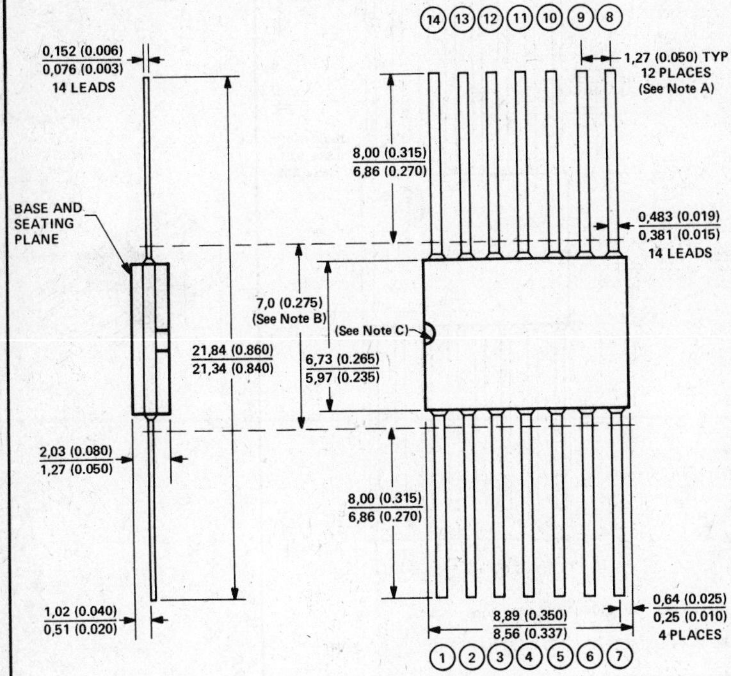
14 13 12 11 10 9 8
0,152 (0.006)
0,076 (0.003)
14 LEADS
1,27 (0.050) TYP
12 PLACES
(See Note A)
8,00 (0.315)
6,86 (0.270)
BASE AND SEATING PLANE
0,483 (0.019)
0,381 (0.015)
14 LEADS
7,0 (0.275)
(See Note B)
(See Note C)
21,84 (0.860)
21,34 (0.840)
6,73 (0.265)
5,97 (0.235)
2,03 (0.080)
1,27 (0.050)
8,00 (0.315)
6,86 (0.270)
1,02 (0.040)
0,51 (0.020)
8,89 (0.350)
8,56 (0.337)
0,64 (0.025)
0,25 (0.010)
4 PLACES
1 2 3 4 5 6 7

Fall Within JEDEC MO-004AA Dimensions

ALL LINEAR DIMENSIONS ARE IN MILLIMETERS AND PARENTHETICALLY IN INCHES

NOTES: A. Leads are within 0,13 (0.005) radius of true position (T.P.) at maximum material condition.
B. This dimension determines a zone within which all body and lead irregularities lie.
C. Index point is provided on cap for terminal identification only.

W ceramic flat packages

16-PIN W CERAMIC

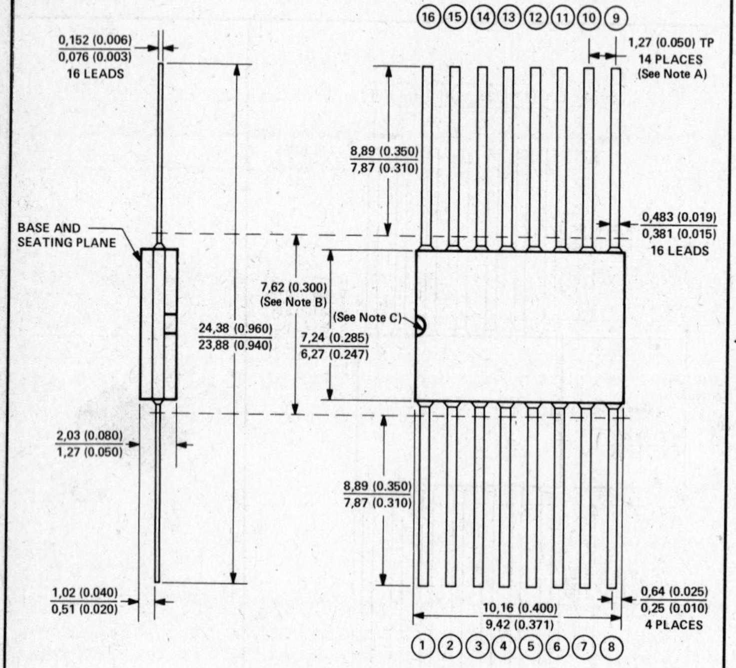
0,152 (0.006)
0,076 (0.003)
16 LEADS
16 15 14 13 12 11 10 9
1,27 (0.050) TP
14 PLACES
(See Note A)
8,89 (0.350)
7,87 (0.310)
BASE AND
SEATING PLANE
0,483 (0.019)
0,381 (0.015)
16 LEADS
7,62 (0.300)
(See Note B)
(See Note C)
24,38 (0.960)
23,88 (0.940)
7,24 (0.285)
6,27 (0.247)
2,03 (0.080)
1,27 (0.050)
8,89 (0.350)
7,87 (0.310)
1,02 (0.040)
0,51 (0.020)
10,16 (0.400)
9,42 (0.371)
0,64 (0.025)
0,25 (0.010)
4 PLACES
1 2 3 4 5 6 7 8

Falls Within JEDEC MO-004-AA Dimensions

ALL LINEAR DIMENSIONS ARE IN MILLIMETERS AND PARENTHETICALLY IN INCHES

NOTES: A. Leads are within 0,13 (0.005) radius of true position (T.P.) at maximum material condition.
B. This dimension determines a zone within which all body and lead irregularities lie.
C. Index point is provided on cap for terminal identification only.

W ceramic flat packages

20-PIN W CERAMIC

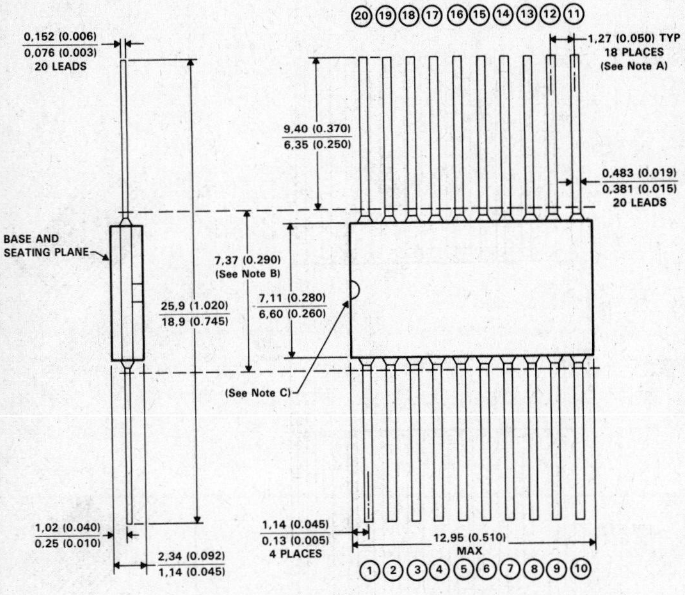
20
19
18
17
16
15
14
13
12
11
0,152 (0.006)
0,076 (0.003)
20 LEADS
1,27 (0.050) TYP
18 PLACES
(See Note A)
9,40 (0.370)
6,35 (0.250)
0,483 (0.019)
0,381 (0.015)
20 LEADS
BASE AND
SEATING PLANE
7,37 (0.290)
(See Note B)
25,9 (1.020)
18,9 (0.745)
7,11 (0.280)
6,60 (0.260)
(See Note C)
1,02 (0.040)
0,25 (0.010)
1,14 (0.045)
0,13 (0.005)
4 PLACES
12,95 (0.510)
MAX
2,34 (0.092)
1,14 (0.045)
1
2
3
4
5
6
7
8
9
10

ALL LINEAR DIMENSIONS ARE IN MILLIMETERS AND PARENTHETICALLY IN INCHES

NOTES: A. Leads are within 0,13 (0.005) radius of true position (T.P.) at maximum material condition.
B. This dimension determines a zone within which all body and lead irregularities lie.
C. Index point is provided on cap for terminal identification only.

W ceramic flat packages

24-PIN W CERAMIC

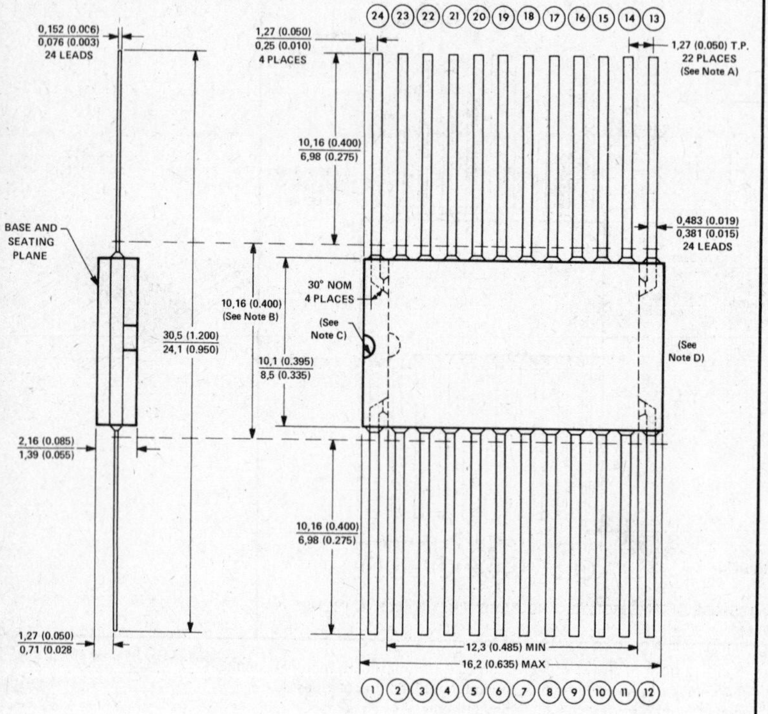
0,152 (0.006)
0,076 (0.003)
24 LEADS
1,27 (0.050)
0,25 (0.010)
4 PLACES
24 23 22 21 20 19 18 17 16 15 14 13
1,27 (0.050) T.P.
22 PLACES
(See Note A)
10,16 (0.400)
6,98 (0.275)
BASE AND
SEATING
PLANE
0,483 (0.019)
0,381 (0.015)
24 LEADS
30° NOM
4 PLACES
10,16 (0.400)
(See Note B)
(See
Note C)
30,5 (1.200)
24,1 (0.950)
10,1 (0.395)
8,5 (0.335)
(See
Note D)
2,16 (0.085)
1,39 (0.055)
10,16 (0.400)
6,98 (0.275)
1,27 (0.050)
0,71 (0.028)
12,3 (0.485) MIN
16,2 (0.635) MAX
1 2 3 4 5 6 7 8 9 10 11 12

Falls Within JEDEC MO-019AA Dimensions

ALL LINEAR DIMENSIONS ARE IN MILLIMETERS AND PARENTHETICALLY IN INCHES

NOTES:
A. Leads are within 0,13 (0.005) radius of true position (T.P.) at maximum material condition.
B. This dimension determines a zone within which all body and lead irregularities lie.
C. Index point is provided on cap for terminal identification only.
D. End configuration of 24-pin package is at the option of TI.

Mechanical Data

B

TEXAS INSTRUMENTS
POST OFFICE BOX 655012 • DALLAS, TEXAS 75265

WA ceramic flat package

14-PIN WA CERAMIC

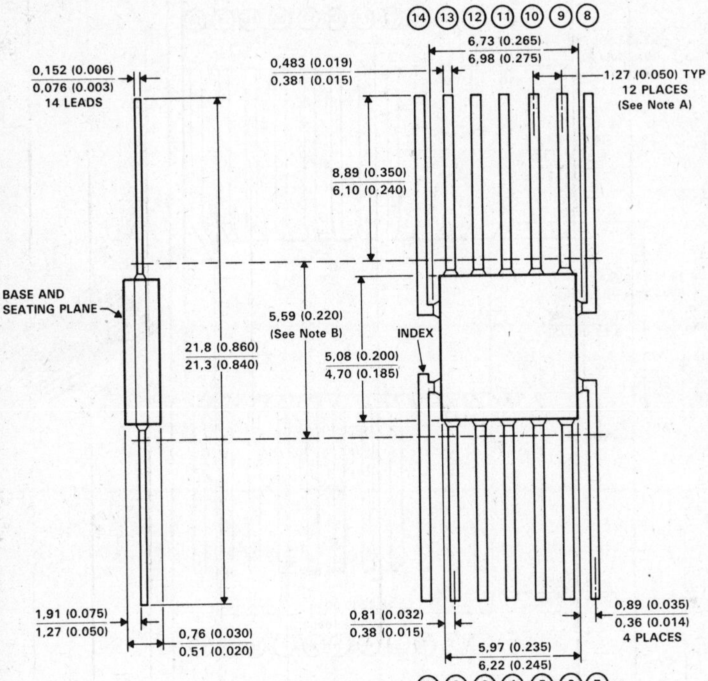
14 13 12 11 10 9 8
6,73 (0.265)
6,98 (0.275)
0,152 (0.006)
0,076 (0.003)
14 LEADS
0,483 (0.019)
0,381 (0.015)
1,27 (0.050) TYP
12 PLACES
(See Note A)
8,89 (0.350)
6,10 (0.240)
BASE AND
SEATING PLANE
5,59 (0.220)
(See Note B)
INDEX
21,8 (0.860)
21,3 (0.840)
5,08 (0.200)
4,70 (0.185)
1,91 (0.075)
1,27 (0.050)
0,76 (0.030)
0,51 (0.020)
0,81 (0.032)
0,38 (0.015)
0,89 (0.035)
0,36 (0.014)
4 PLACES
5,97 (0.235)
6,22 (0.245)
1 2 3 4 5 6 7

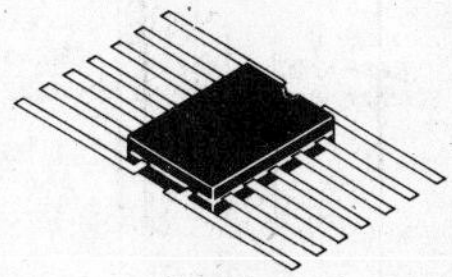

ALL LINEAR DIMENSIONS ARE IN MILLIMETERS AND PARENTHETICALLY IN INCHES

NOTES: A. Leads are within 0,13 (0.005) radius of true position (T.P.) at maximum material condition.
B. This dimension determines a zone within which all body and lead irregularities lie.
C. Index point is provided on cap for terminal identification only.
D. End configuration of 24-pin package is at the option of TI.

WC ceramic flat package

24-PIN WC CERAMIC

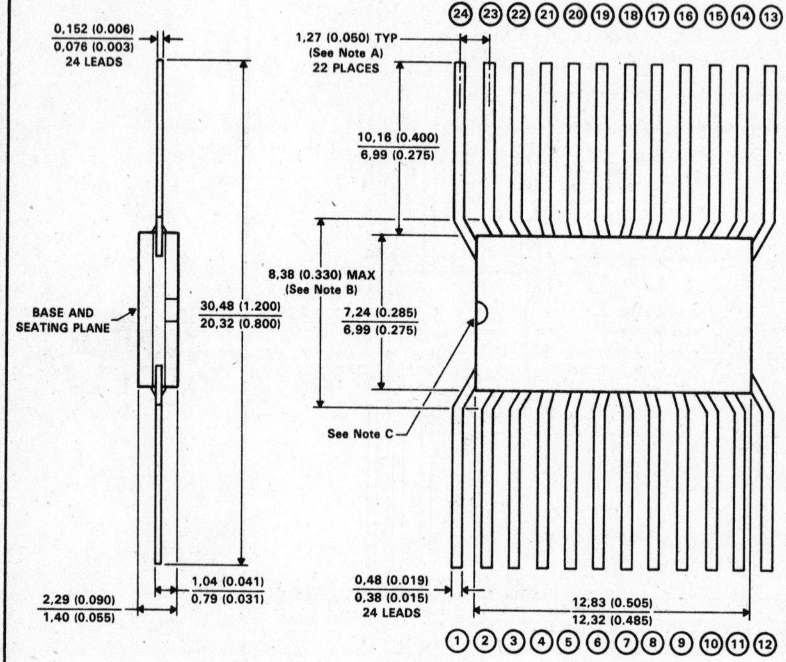
0,152 (0.006)
0,076 (0.003)
24 LEADS
BASE AND
SEATING PLANE
30,48 (1.200)
20,32 (0.800)
1,04 (0.041)
0,79 (0.031)
2,29 (0.090)
1,40 (0.055)
1,27 (0.050) TYP
(See Note A)
22 PLACES
10,16 (0.400)
6,99 (0.275)
8,38 (0.330) MAX
(See Note B)
7,24 (0.285)
6,99 (0.275)
See Note C
0,48 (0.019)
0,38 (0.015)
24 LEADS
12,83 (0.505)
12,32 (0.485)
24 23 22 21 20 19 18 17 16 15 14 13
1 2 3 4 5 6 7 8 9 10 11 12

ALL LINEAR DIMENSIONS ARE IN MILLIMETERS AND PARENTHETICALLY IN INCHES

NOTES: A. Leads are within 0,13 (0.005) radius of true position (T.P.) at maximum material condition.
B. This dimension determines a zone within which all body and lead irregularities lie.
C. Index point is provided on cap for terminal identification only.

INTRODUCTION

Texas Instruments has developed solutions for today's high density packaging needs. The TI facility at Attleboro, Massachusetts (one of the world's largest suppliers of multimetal systems) provides leading-edge technology which, combined with reliable, high-volume, off-the-shelf interconnection products, allows TI to quickly meet volume commercial applications.

During the last decade, TI has produced one of the largest IC socket families. TI's sockets include every type and size socket in common use today and are available in a wide choice of contact materials and designs.

Our sockets are designed for:

- ergonomical efficiencies in assembly
- compatibility with automatic assembly equipment
- maximum performance and board density

This section of the data book provides information on the following types of IC socket products.

PRODUCTION SOCKETS	TYPE
Plastic Leaded Chip Carrier	PLCC
Single-in-Line Packages	SIP
Pin-Grid Arrays	PGA
Dual In-Line	DIP
Dual In-Line 0.070-in spacing	Shrink Pack
Quad In-Line	QUIP
BURN-IN/TEST SOCKETS	**TYPE**
Plastic Leaded Chip Carrier	PLCC
Pin Grid Array	PGA
Dual In-Line	DIP
Dual In-Line 0.070-in spacing	Shrink Pack
Small Outline	Flatpack
Quad	Flatpack

Specially formulated alloys give the TI contact springs:

- Low Contact Resistance
- High Contact Strength (to stand up to repetitive insertions and withdrawals)
- High normal forces, assure gas tight reliability

A full line of reliable, readily available, low-cost interconnection systems means premium performance at an economical price.

Additional information, on these and other TI products, including pricing and delivery quotations may be obtained from your nearest TI Distributor, TI Sales Representative or:

Texas Instruments Incorporated
Connector Systems Department MS 14-3
Attleboro, Massachusetts 02703

Telephone: (617) 699-5242/5269
TELEX: 92-7708

PERFORMANCE SPECIFICATIONS

Mechanical

Recommended PCB thickness range: 0.062 in to 0.092 in
Recommended PCB hole size range: 0.032 in to 0.042 in
Vibration: 15 G
Shock: 100 G
Solderability: Per MIL-STD 202, Method 208
Insertion force: 0.59 lbs per position
Withdrawal force: 0.25 lbs per position
Normal force: 200 g min, 450 g typ
Wipe: 0.075 in min
Durability: 5 cycles min
Contact retention: 1.5 lbs min

Electrical

Current carrying capacity: 1 A
Insulation resistance: 5000 MΩ min
Dielectric withstanding voltage: 1000 V ac rms min
Capacitance: 1.0 pF max

Environmental

Operating temperature:
Operating: − 40 °C to 85 °C
Storage: −40 °C to 95 °C
Temperature cycling with humidity: will conform to final EIA specifications
Shelf life: 1 year min

MATERIALS

Body — Ryton R-4 (40% glass) U/L 94-VO rating
Contacts — CDA 510 spring temper
Contact finish — 90/10 tin (200 μin −400 μin) over 40 μin copper

Contact factory for detailed information

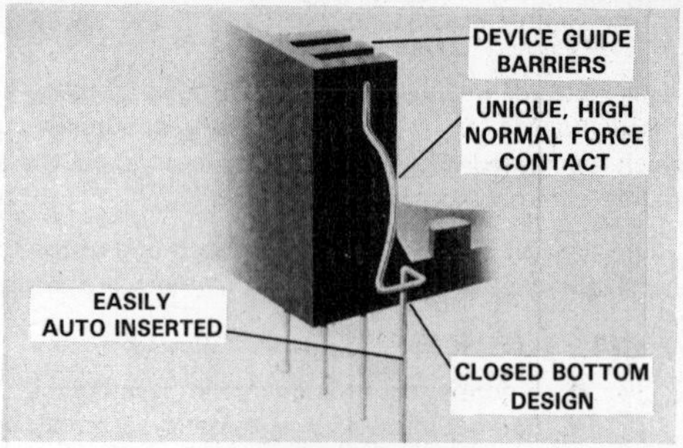
DEVICE GUIDE BARRIERS
UNIQUE, HIGH NORMAL FORCE CONTACT
EASILY AUTO INSERTED
CLOSED BOTTOM DESIGN

PART NUMBER SYSTEM

CPR PH XXX − X − X − O

- CPR — TI socket Series Plastic leaded chip carrier
- PH — Plated thru hole, solder tail
- XXX — Number of pos (044, 052, 068, 084)
- X — Contact spacing 1 — 0.050 in
- X — Contact surface 1 — tin lead plating

PLASTIC LEADER CHIP CARRIER CPR SERIES

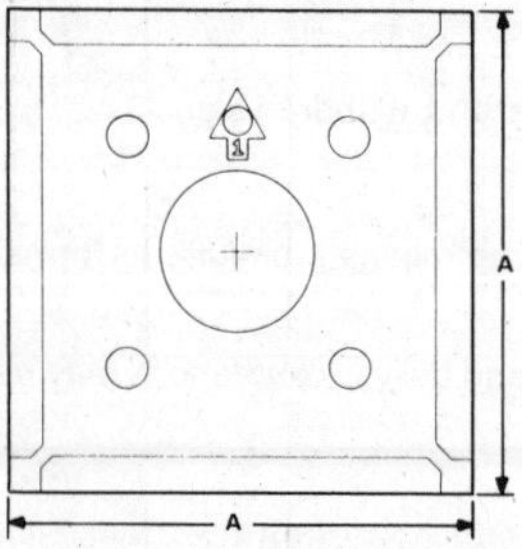
A
A

Device guide barriers not shown

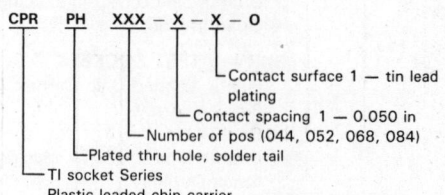

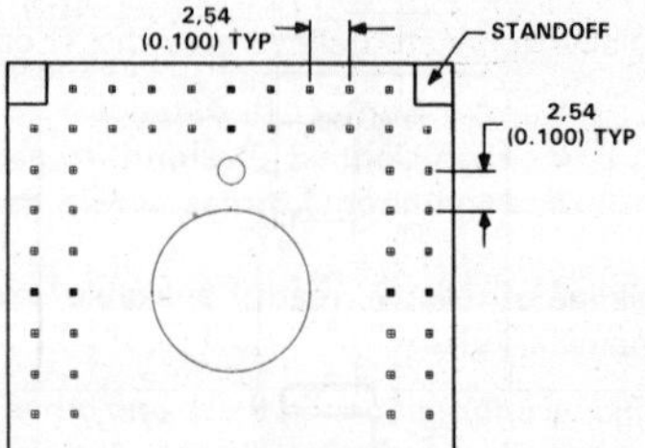
2,54
(0.100) TYP
STANDOFF
2,54
(0.100) TYP

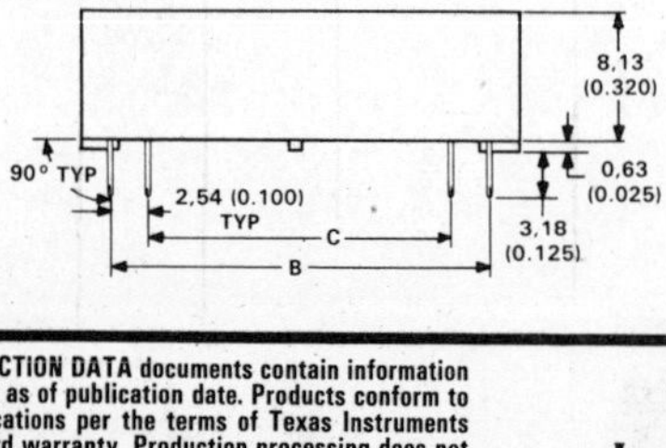
8,13
(0.320)
0,63
(0.025)
90° TYP
2,54 (0.100)
TYP
C
3,18
(0.125)
B

Pos	A	B	C
44	21,43 (0.844	17,78 (0.700)	12,70 (0.500)
52	23,98 (0.944)	20,32 (0.800)	15,24 (0.600)
68	29,06 (1.144)	25,40 (1.000)	20,32 (0.800)
84	34,14 (1.344)	30,48 (1.200)	25,40 (1.000)

Extraction tool available, consult factory.

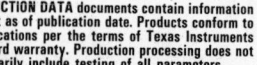

TEXAS INSTRUMENTS
34 Forest Street • Attleboro, Massachusetts 02703

PRODUCT FEATURES

Can be loaded by top actuated insertion or press-in insertion, either manually or automatically
High reliability due to high pressure contact point
Open body and high stand-off design provide high efficiency in heat dissipation
High durability up to 10,000 cycles
Compact design

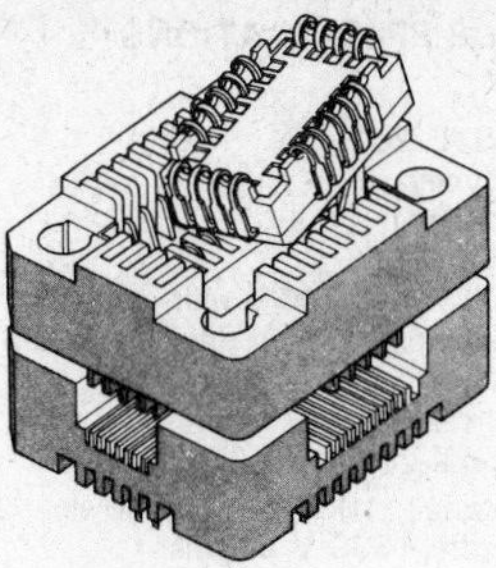

PERFORMANCE SPECIFICATIONS

Mechanical

Durability: 10,000 cycles
Operating Temperature: 180 °C max

Electrical

Contact rating: 1.0 A per contact
Contact resistance: 30 mΩ max
Insulation resistance: 1000 MΩ min
Dielectric withstanding voltage: 500 V ac rms min

MATERIALS

Body — ultem glass filled (U/L 94 VO)
Contact — copper alloy
Plating — overall gold plate

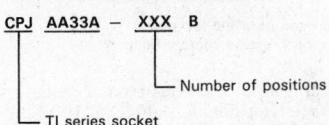

PART NUMBER SYSTEM

CPJ AA33A – XXX B

XXX — Number of positions
CPJ — TI series socket

PLCC BURN-IN/TEST SOCKETS CPJ SERIES

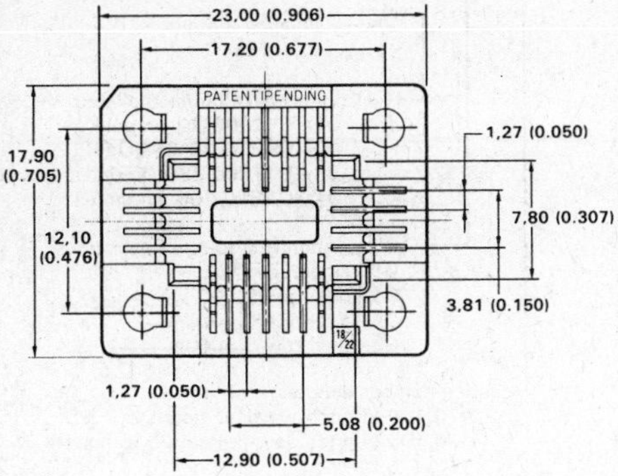
23,00 (0.906)
17,20 (0.677)
PATENTIPENDING
1,27 (0.050)
17,90 (0.705)
12,10 (0.476)
7,80 (0.307)
3,81 (0.150)
18/22
1,27 (0.050)
5,08 (0.200)
12,90 (0.507)

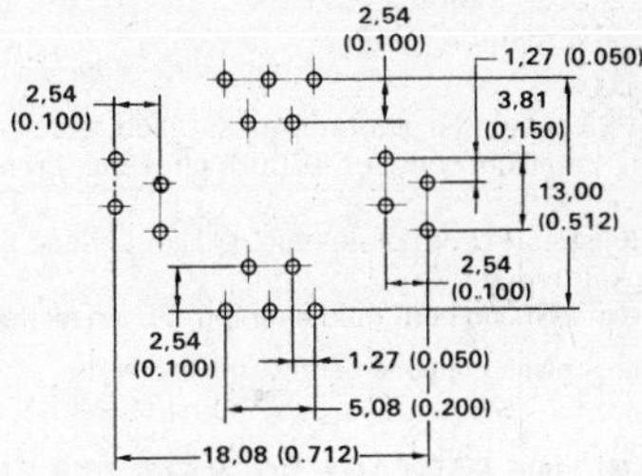
2,54 (0.100)
1,27 (0.050)
2,54 (0.100)
3,81 (0.150)
13,00 (0.512)
2,54 (0.100)
2,54 (0.100)
1,27 (0.050)
5,08 (0.200)
18,08 (0.712)

SIZES: 18 PIN
22 PIN

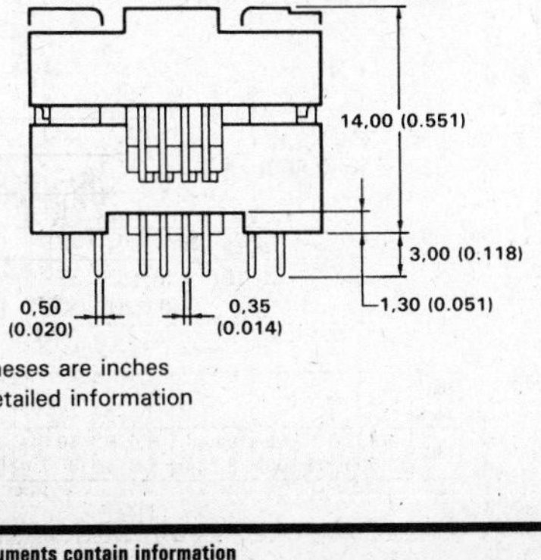
14,00 (0.551)
3,00 (0.118)
0,50 (0.020)
0,35 (0.014)
1,30 (0.051)

Dimensions in parentheses are inches
Contact factory for detailed information

TEXAS INSTRUMENTS
34 Forest Street • Attleboro, Massachusetts 02703

PERFORMANCE SPECIFICATIONS†

Mechanical
Vibration: MIL-STD-202
Durability: 30 cycles
Insertion force: Og
Withdrawal force: Og‡
Contact (normal) force: 200 g min
Contact retention force: 2 lbs per circuit min

Electrical
Contact rating: 1 A
Contact resistance: 30 mΩ max initial
Insulation resistance: 1000 MΩ at 500 dc
Dielectric strength: 1500 V ac rms
Capacitance: 2 pF max

†Values may vary due to test sequence and SIP module configuration
‡After module is unlocked from the receptacle
For a complete test report, please contact factory)

Environmental
(20 mΩ max contact resistance change after all tests)
Operating and storage temperature: −40°C to 100°C
Humidity: MIL-STD 202, Method 106D, 10 days
Temperature soak: 85°C for 160 hours
Thermal Shock: 5 cycles, −40°C to 85°C per MIL-STD 202, Method 107E
Shelf life: 12 months min

MATERIALS
Body — PES polyether sulfone, glass filled, black, 94 VO
Contact — Beryllium copper C17000; phosphor bronze alloy CA510
Contact finishes — Post plate min 200 μin tin/lead over min 50μin nickel overall
Post plate min 30 μin hard gold over min 75 μin nickel overall
For additional plating options contact the factory.

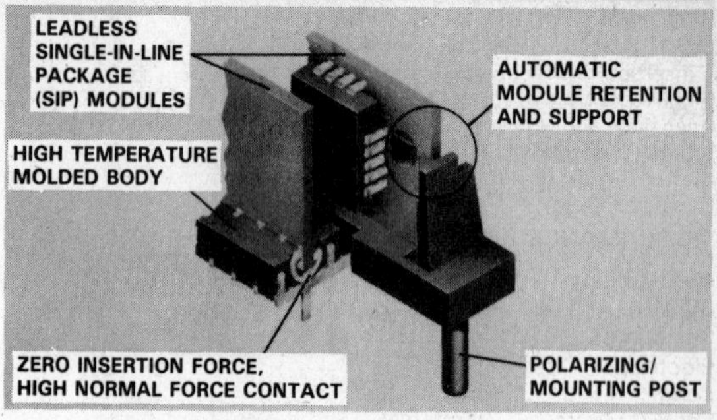
LEADLESS SINGLE-IN-LINE PACKAGE (SIP) MODULES
AUTOMATIC MODULE RETENTION AND SUPPORT
HIGH TEMPERATURE MOLDED BODY
ZERO INSERTION FORCE, HIGH NORMAL FORCE CONTACT
POLARIZING/ MOUNTING POST

PART NUMBER SYSTEM

TS8X XX XX X −XX − XX

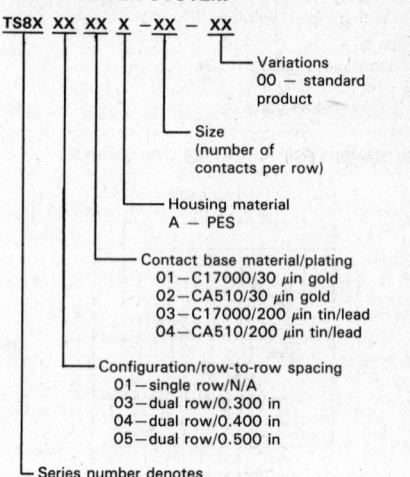

- XX (last): Variations
  00 — standard product
- −XX: Size (number of contacts per row)
- X: Housing material
  A — PES
- XX (third group): Contact base material/plating
  01—C17000/30 μin gold
  02—CA510/30 μin gold
  03—C17000/200 μin tin/lead
  04—CA510/200 μin tin/lead
- XX (second group): Configuration/row-to-row spacing
  01—single row/N/A
  03—dual row/0.300 in
  04—dual row/0.400 in
  05—dual row/0.500 in
- TS8X: Series number denotes
  0—0.100 in pitch, vertical mount
  1—0.100 in pitch, low-profile (25°) mount

Consult factory for availability of configurations, materials, and sizes.

SINGLE-IN-LINE PACKAGE SOCKETS TS8 SERIES

DUAL ROW VERTICAL

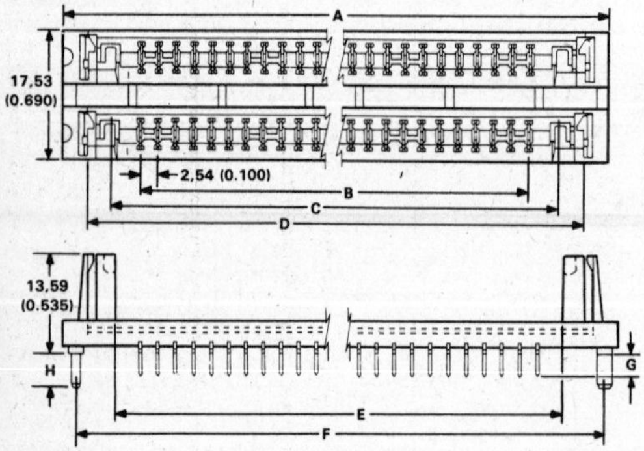
A
17,53
(0.690)
2,54 (0.100)
B
C
D
13,59
(0.535)
H
G
E
F

SINGLE ROW LOW PROFILE

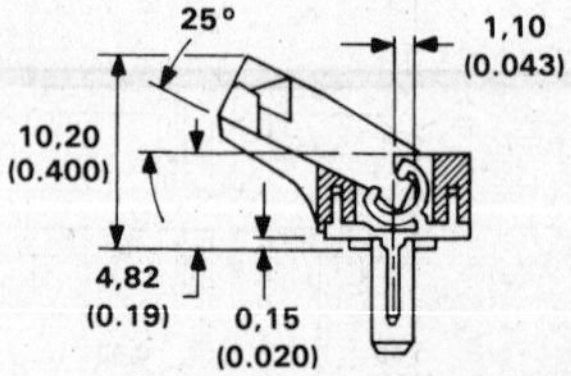
25°
1,10
(0.043)
10,20
(0.400)
4,82
(0.19)
0,15
(0.020)

Ckt. Size	A	B	C	D	E	F	G	H
30	96,52 (3.800)	73,66 (2.900)	82,14 (3.234)	89,28 (3.515)	80,52 (3.170)	92,71 (3.650)	2,79 (0.110)	3,86 (0.152)

Contact factory for detailed information

PERFORMANCE SPECIFICATIONS

Mechanical

Accommodates IC leads 0.015 in to 0.021 in diameter
Recommended PCB thickness range: 0.062 in to 0.092 in
Recommended PCB hole size range: 0.032 in to 0.042 in
Recommended hole grid pattern: 0.100 in ± 0.002 in each direction
Vibration: 15 G, 10-2000 Hz per MIL-STD 1344A, Method 2005.1 Test Condition III
Shock: 100 G, sawtooth waveform, 2 shocks each direction per MIL-STD 202, Method 213, Test Condition I
Durability: 5 cycles, 10 mΩ max contact resistance change per MIL-STD 1344, Method 2016
Solderability: per MIL-STD 202, Method 208
Insertion force: 3.6 oz (102 g) per pin typ using 0.018 in diameter test pin
Withdrawal force: 0.5 oz (14 g) per pin min using 0.018 in diameter test pin

Electrical

Contact rating: 1.0 A per contact
Contact resistance: 20 mΩ max initial
Insulation resistance: 1000 MΩ at 500 V dc per MIL-STD 1344, Method 3003.1
Dielectric withstanding voltage: 1000 V ac rms per MIL-STD 1344, Method 3001.1
Capacitance: 1.0 pF max per MIL-STD 202, Method 305

Environmental

Operating temperature: −65°C to 125°C, gold; −40°C to 100°C, tin
Corrosive atmosphere: 10 mΩ max contact resistance change when exposed to 22% ammonium sulfide for 4 hours
Gas tight: 10 mΩ max contact resistance change when exposed to nitric acid vapor for 1 hour
Temperature soak: 10 mΩ max contact resistance change when exposed to 105°C temperature for 48 hours
Shell life: 12 months min

MATERIALS

Body — PBT polyester U/L94-VO rating
On request, G10/FR4 or Mylar film
Outer sleeve — Machined Brass (QQ-B-626)
Inner contact — Beryllium copper (QQ-C-530) heat treated
Plating: (specified by part number)

PIN GRID ARRAY

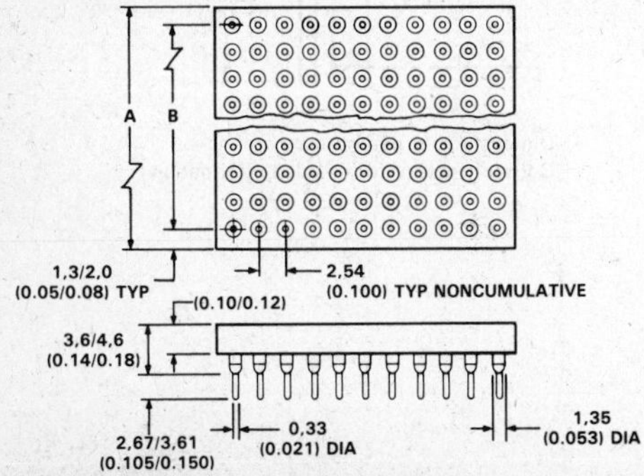
A
B
1,3/2,0
(0.05/0.08) TYP
2,54
(0.100) TYP NONCUMULATIVE
(0.10/0.12)
3,6/4,6
(0.14/0.18)
2,67/3,61
(0.105/0.150)
0,33
(0.021) DIA
1,35
(0.053) DIA

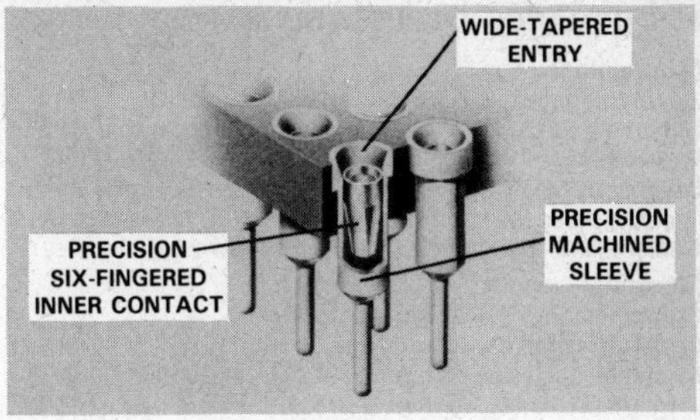
WIDE-TAPERED ENTRY
PRECISION SIX-FINGERED INNER CONTACT
PRECISION MACHINED SLEEVE

Inner contact — 30 μin gold over 50 μin nickel or 100 μin tin/lead over 50 μin nickel
Outer sleeve — 10 μin gold over 50 μin nickel or 50 μin tin/lead over 50 μin nickel

PART NUMBER SYSTEM

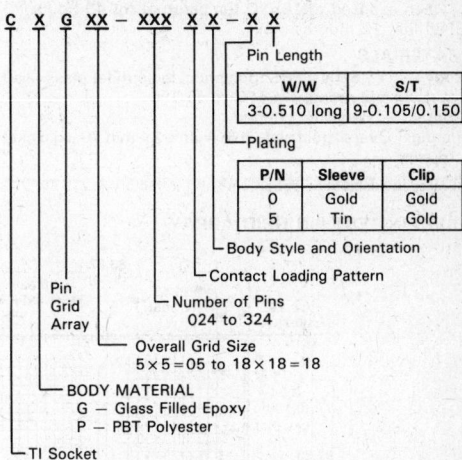

C X G XX – XXX X X – X X

Pin Length

W/W	S/T
3-0.510 long	9-0.105/0.150

Plating

P/N	Sleeve	Clip
0	Gold	Gold
5	Tin	Gold

Body Style and Orientation
Contact Loading Pattern
Number of Pins 024 to 324
Pin Grid Array
Overall Grid Size 5×5=05 to 18×18=18
BODY MATERIAL
G — Glass Filled Epoxy
P — PBT Polyester
TI Socket

Insulator Size	A ±0.010	B ±0.005†
9×9	(0.950) 24,13	(0.800) 20,32
10×10	(1.050) 26,67	(0.900) 22,86
11×11	(1.150) 29,21	(1.000) 25,40
12×12	(1.250) 31,75	(1.100) 27,94
13×13	(1.350) 34,29	(1.200) 30,48
14×14	(1.450) 36,83	(1.300) 33,02
15×15	(1.550) 39,37	(1.400) 35,56
16×16	(1.650) 41,91	(1.500) 38,10
17×17	(1.750) 44,45	(1.600) 40,64
18×18	(1.850) 46,99	(1.700) 43,18

†Noncumulative
Dimensions in parentheses are inches
Consult factory for detailed information

PERFORMANCE SPECIFICATIONS

Mechanical

Accommodates IC leads per specific IC device
Recommended PCB thickness range: 0.062 in to 0.092 in
Recommended PCB hole size range: 0.032 in to 0.042 in
Durability: 5000 cycles, 10 mΩ max contact resistance change per MIL-STD 1344, Method 2016
Solderability: per MIL-STD 202, Method 208

Electrical

Contact rating: 1.0 A per contact
Contact resistance: 20 mΩ max initial
Insulation resistance: 1.0 MΩ at 500 V dc per MIL-STD 1344, Method 3003.1
Dielectric withstanding voltage: 700 V ac rms per MIL-STD 1344, Method 3001.1
Capacitance: 1.0 pF max per MIL-STD 202, Method 305

Environmental

Operating temperature: −65°C to 170°C
Humidity: 10 mΩ max contact resistance change when tested per MIL-STD 202, Method 103B
Temperature soak; 10 mΩ max contact resistance change when exposed to 105°C temperature for 48 hours
Shelf life: 12 months max

MATERIALS

Body — CZF Series: PPS (polyphenylen sufide) glass filled U/L 94 VO rating, −65°C to 170°C
Contact — Beryllium copper
Plating:† Overall gold plate min 4 μin over min 70 μin nickel plating

†For additional plating option consult the factory.

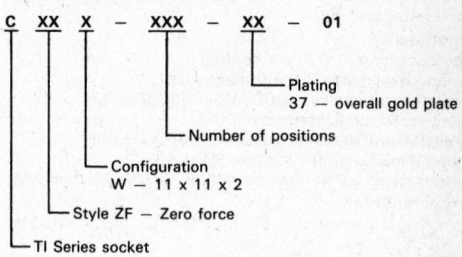

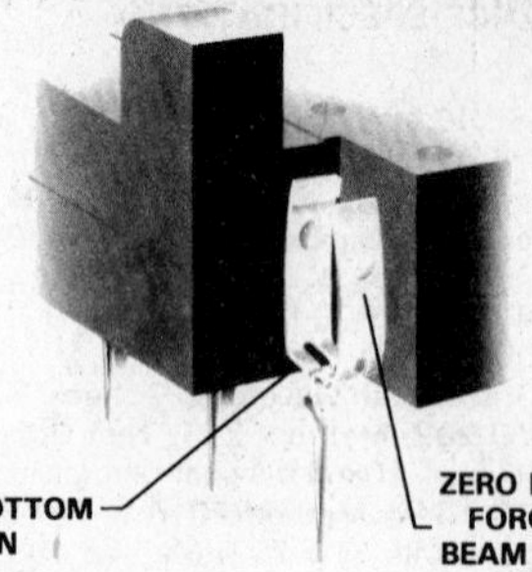
CLOSED BOTTOM DESIGN
ZERO INSERTION FORCE DUAL BEAM CONTACT SYSTEM

PART NUMBER SYSTEM

C XX X – XXX – XX – 01

- C — TI Series socket
- XX — Style ZF — Zero force
- X — Configuration: W — 11 x 11 x 2
- XXX — Number of positions
- XX — Plating: 37 — overall gold plate

BURN-IN TEST PIN GRID ARRAY

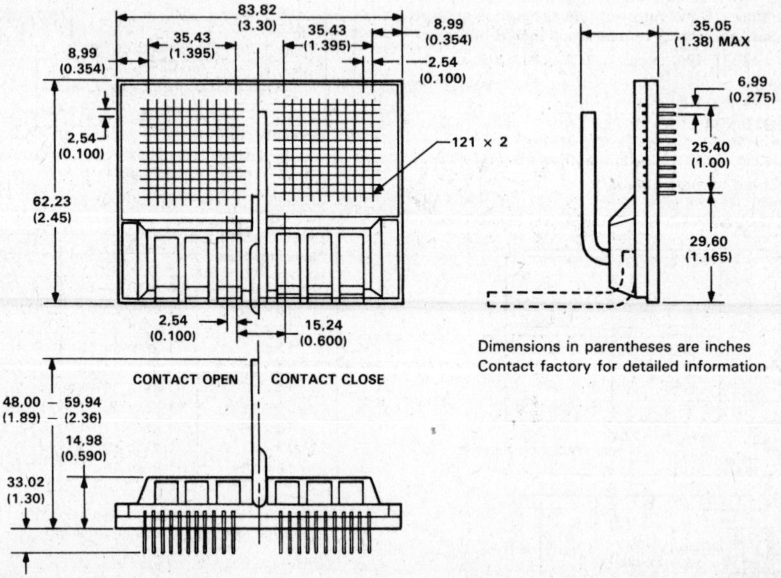
83,82
(3.30)
35,43
(1.395)
35,43
(1.395)
8,99
(0.354)
8,99
(0.354)
2,54
(0.100)
2,54
(0.100)
121 × 2
62,23
(2.45)
2,54
(0.100)
15,24
(0.600)
CONTACT OPEN
CONTACT CLOSE
48,00 – 59,94
(1.89) – (2.36)
14,98
(0.590)
33,02
(1.30)
35,05
(1.38) MAX
6,99
(0.275)
25,40
(1.00)
29,60
(1.165)

Dimensions in parentheses are inches
Contact factory for detailed information

TEXAS INSTRUMENTS
34 Forest Street • Attleboro, Massachusetts 02703

PERFORMANCE SPECIFICATIONS

Mechanical

Accommodates IC leads 0.011 ± 0.003 in by 0.018 ± 0.003
Recommended PCB thickness range: 0.062 in to 0.092 in
Recommended PCB hole size range: 0.032 in to 0.042 in
Recommended hole grid pattern: 0.100 in ± 0.003 in each direction
Vibration: 15 G, 10-2000 Hz per MIL-STD 1344A, Method 2005.1 Test Condition III.
Shock: 100 G, sawtooth waveform, 2 shocks each direction per MIL-STD 202, Method 213, Test Condition I
Durability: 5 cycles, 10 mΩ max contact resistance change per MIL-STD 1344, Method 2016
Solderability: per MIL-STD 202, Method 208
Insertion force (C7X and C86): 16 oz (454 g) per pin max
Insertion force (C50): 12 oz per pin max
Withdrawal force: (40 g) per pin min

Electrical

Contact rating: 1.0 A per contact
Contact resistance: 20 mΩ max initial
Insulation resistance: 1000 MΩ at 500 V dc per MIL-STD 1344, Method 3003
Dielectric withstanding voltage: 1000 V ac rms per MIL-STD 1344, Method 3001.1
Capacitance: 1.0 pF max per MIL-STD 202, Method 305

Environmental

Operating temperature: −55 °C to 125 °C, gold; −40 °C to 100 °C, tin
Corrosive atmosphere: 10 mΩ max contact resistance change when exposed to 22% ammonium sulfide for 4 hours
Gas tight: 10 mΩ max contact resistance change when exposed to nitric acid vapor for 1 hour
Temperature soak: 10 mΩ max contact resistance change when exposed to 105 °C temperature for 48 hours
Shelf life: 12 months min

Materials (C7X, C50, and C86)

Body — PBT polyester U/L 94 VO rating
C7X & C50 Contacts — Outer sleeve: brass
Clip: BECU or PHBR
Contact finish — clip 30 µin gold over 50 µin nickel or 50 µin tin/lead over 50 µin nickel
Specified by Part Number — sleeve 10 µin gold over 50 µin nickel or 50 µin tin/lead over 50 µin nickel
C86 Contacts — Phosphor bronze base metal
C86 Contact-finish — Tin plate 200 µin over copper flash

C7X SERIES — SCREW MACHINE

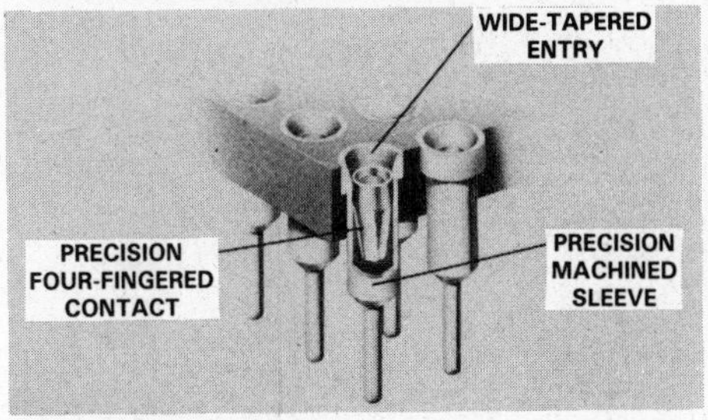
WIDE-TAPERED ENTRY
PRECISION FOUR-FINGERED CONTACT
PRECISION MACHINED SLEEVE

C7X SERIES PART NUMBER SYSTEM

C7X (X) XX − X X

- X — Variations
  - Solder Tail
    - 9 — Pin length 0.105/0.150
  - Wire Wrap
    - 3 — Pin length 0.510
- X — Plating (Sleeve/Clip)
  - 0 — Gold/Gold
  - 5 — Tin/Gold
- XX — Number of Positions
- (X) — S — Single-in-line package (where applicable)
- C7X — Screw Machine Socket
  - 1 — wire wrap
  - 2 — solder tail

C86 SERIES — STAMPED AND FORMED

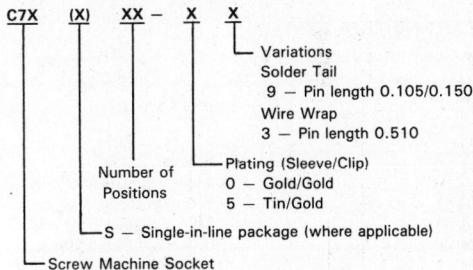

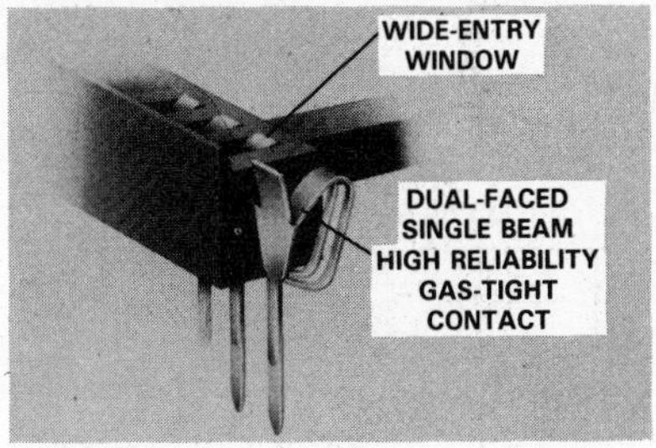
WIDE-ENTRY WINDOW
DUAL-FACED SINGLE BEAM HIGH RELIABILITY GAS-TIGHT CONTACT

C86 SERIES PART NUMBER SYSTEM

C 86 XX − 01

- 01 — Variation
  - 01 — Standard product
- XX — Number of positions
- 86 — Tin Dual Face Wipe Single Beam
- C — TI Socket Series

Mechanical Data

B

C50—UNICON

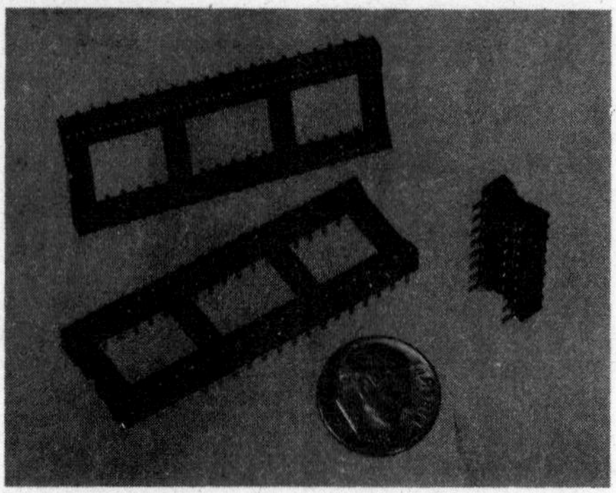

PART NUMBER SYSTEM

C50 – XX – XXX

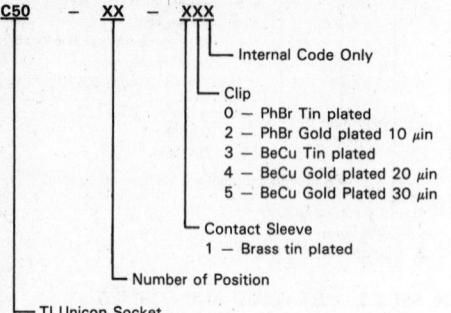

- XXX (third digit) — Internal Code Only
- XXX (second digit) — Clip
  - 0 – PhBr Tin plated
  - 2 – PhBr Gold plated 10 μin
  - 3 – BeCu Tin plated
  - 4 – BeCu Gold plated 20 μin
  - 5 – BeCu Gold Plated 30 μin
- XXX (first digit) — Contact Sleeve
  - 1 – Brass tin plated
- XX — Number of Position
- C50 — TI Unicon Socket

DUAL-IN-LINE C50, C7X AND C86 SERIES

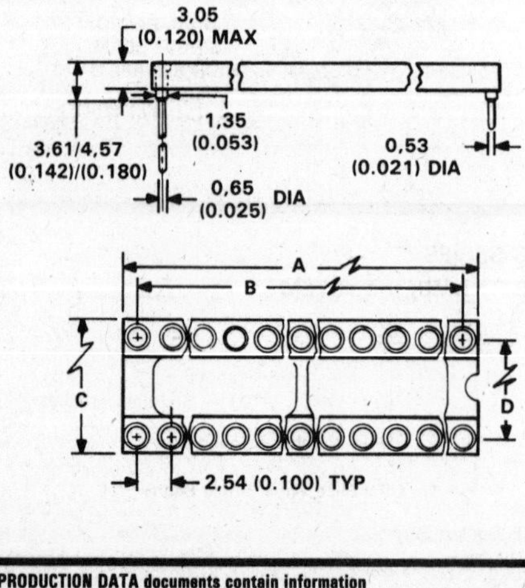
3,05
(0.120) MAX
1,35
(0.053)
3,61/4,57
(0.142)/(0.180)
0,53
(0.021) DIA
0,65
(0.025) DIA
A
B
C
D
2,54 (0.100) TYP

DIPS

Positions	Dim A Max	Dim B ±0.005	Dim C Max	Dim D ±0.005	Positions	Dim A Max	Dim B ±0.005	Dim C Max	Dim D ±0.005
6	7,62 (0.300)	5,08 (0.200)	10,16 (0.400)	7,62 (0.300)	†24	30,48 (1.200)	27,94 (1.100)	12,76 (0.500)	10,16 (0.400)
8	10,16 (0.400)	7,62 (0.300)	10,16 (0.400)	7,62 (0.300)	28	35,56 (1.400)	33,02 (1.300)	17,78 (0.700)	15,24 (0.600)
14	17,78 (0.700)	15,24 (0.600)	10,16 (0.400)	7,62 (0.300)	32	40,64 (1.600)	38,10 (1.500)	17,78 (0.700)	15,24 (0.600)
16	20,32 (0.800)	17,78 (0.700)	10,16 (0.400)	7,62 (0.300)	34	45,72 (1.800)	43,18 (1.700)	17,78 (0.700)	15,24 (0.600)
18	22,86 (0.900)	20,32 (0.800)	10,16 (0.400)	7,62 (0.300)	40	50,80 (2.000)	48,26 (1.900)	17,78 (0.700)	15,24 (0.600)
20	25,40 (1.000)	22,86 (0.900)	10,16 (0.400)	7,62 (0.300)	48	60,96 (2.400)	58,42 (2.300)	17,78 (0.700)	15,24 (0.600)
22	27,94 (1.100)	25,40 (1.000)	12,76 (0.500)	10,16 (0.400)	50	63,50 (2.500)	60,96 (2.400)	25,40 (1.000)	7,62 (0.900)
24	30,48 (1.200)	27,94 (1.100)	17,78 (0.700)	15,24 (0.600)	64	81,28 (3.200)	78,74 (3.100)	25,40 (1.000)	22,86 (0.900)
†24	30,48 (1.200)	27,94 (1.100)	10,16 (0.400)	7,62 (0.300)					

†Nonstandard sizes
Not all sizes available in each series

C7X SERIES

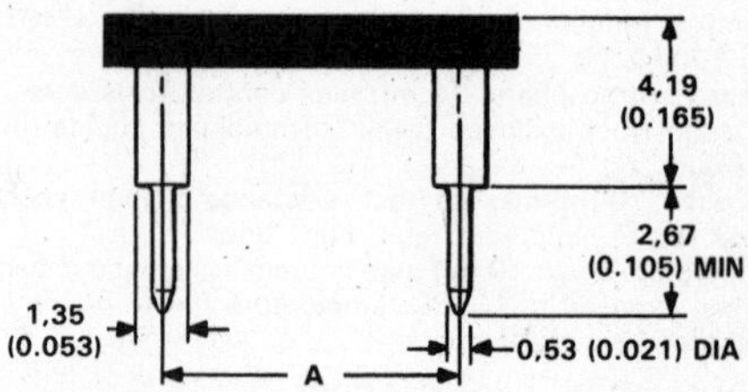
4,19
(0.165)
2,67
(0.105) MIN
1,35
(0.053)
0,53 (0.021) DIA
A

C86 SERIES

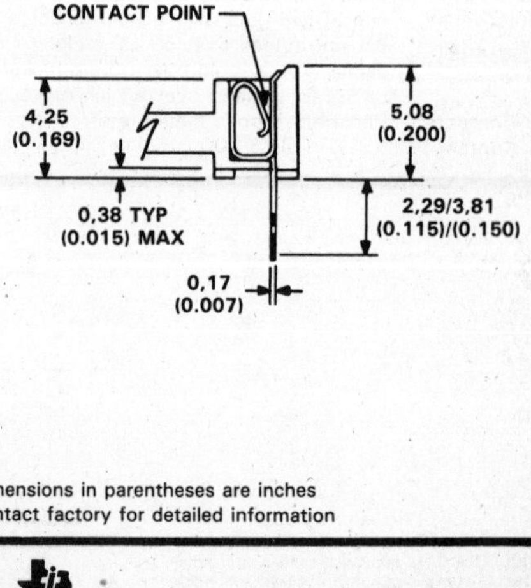
CONTACT POINT
4,25
(0.169)
5,08
(0.200)
0,38 TYP
(0.015) MAX
2,29/3,81
(0.115)/(0.150)
0,17
(0.007)

Dimensions in parentheses are inches
Contact factory for detailed information

Mechanical Data

B

TEXAS INSTRUMENTS
34 Forest Street • Attleboro, Massachusetts 02703

PERFORMANCE SPECIFICATIONS

Mechanical
Accommodates IC leads 0.011 in by 0.018 in NOM
Recommended PCB thickness range: 0.062 in to 0.092 in
Recommended PCB hold size range: 0.032 in to 0.042 in
Durability: 10K cycles — CM Series, 5K cycles — CP/CQ
Solderability: per MIL-STD 202, Method 208

Electrical
Contact rating: 1.0 A per contact
Contact resistance: 20 mΩ max initial
Insulation resistance: 1000 MΩ at 500 V dc
Dielectric withstanding voltage: 1000 V ac rms
Capacitance: 1.0 pF max per MIL-STD 202, Method 305

Environmental
Operating temperature: −65 °C to 170 °C — CP/CM Series, −65 °C to 150 °C — CQ Series
Humidity: 10 mΩ max contact resistance
Temperature Soak: 10 mΩ max contact resistance change

MATERIALS
Body — PPS (polyphenylen sulfide) glass filled U/L 94 VO
Contacts — Higher performance copper nickel alloy
Plating:† 4 μin of gold min over 100 μin of nickel min

†For additional plating options consult the factory

BURN-IN/TEST DIP SOCKETS

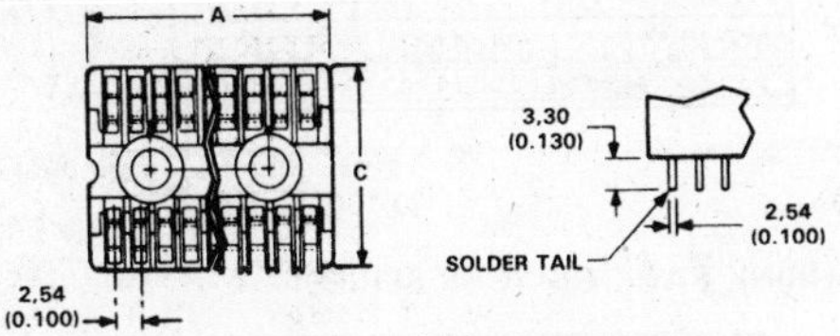
A
C
3,30
(0.130)
2,54
(0.100)
SOLDER TAIL
2,54
(0.100)

CQ37 SERIES

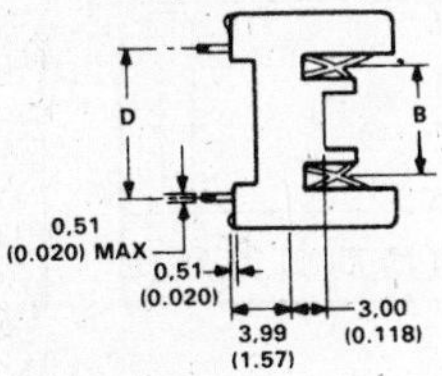
D
B
0,51
(0.020) MAX
0,51
(0.020)
3,00
(0.118)
3,99
(1.57)

CP37 SERIES

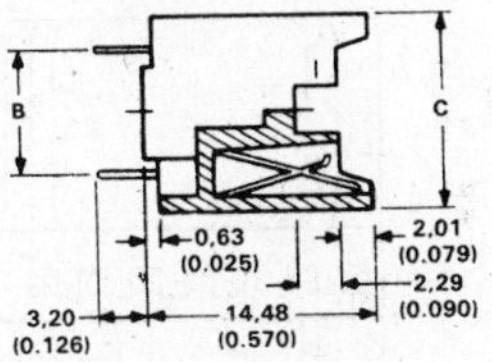
B
C
0,63
(0.025)
2,01
(0.079)
2,29
(0.090)
3,20
(0.126)
14,48
(0.570)

CM37 SERIES

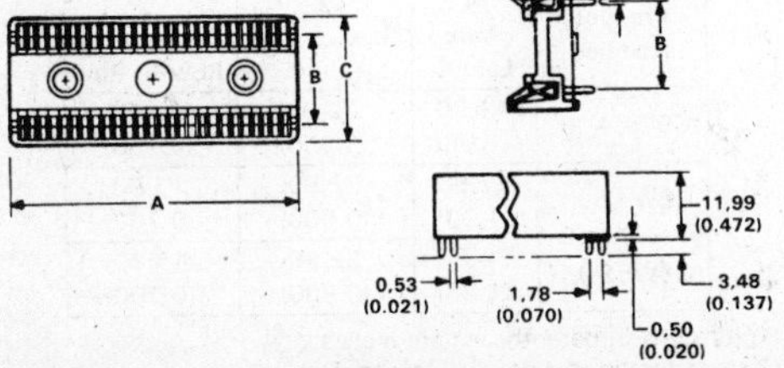
B
C
A
6,50
(0.256)
0,51
(0.020)
B
11,99
(0.472)
0,53
(0.021)
1,78
(0.070)
3,48
(0.137)
0,50
(0.020)

PART NUMBER SYSTEM

C X 37 XX — 22 S X

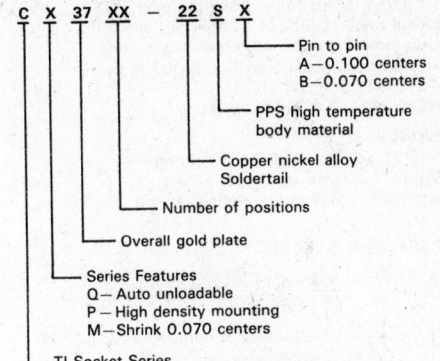

- X (last) — Pin to pin: A—0.100 centers; B—0.070 centers
- S — PPS high temperature body material
- 22 — Copper nickel alloy Soldertail
- XX — Number of positions
- 37 — Overall gold plate
- X — Series Features: Q—Auto unloadable; P—High density mounting; M—Shrink 0.070 centers
- C — TI Socket Series

CQ37 SERIES

Number of Positions	A ±0.01 Length	D ±0.02	C ±0.01 Width	B ±0.01 Contact
14	20,32 (0.800)	12,70 (0.500)	15,24 (0.600)	7,62 (0.300)
16	22,35 (0.880)			
18	24,89 (0.980)			
20	27,43 (1.080)			
24	32,51 (1,280)	19,05 (0.750)	22,86 (0.900)	15,24 (0.600)
28	37,59 (1.480)			
40	52,83 (2.080)			
42	55,37 (2.180)			

CP37 SERIES

Number of Positions	A max Length	B ±0.02	C max Width
8	11,68 (0.460)	7,62 (0.300)	12,70 (0.500)
14	17,78 (0.700)		
16	20,32 (0.800)		
18	22,86 (0.900)		
20	25,40 (1.000)		
24	30,48 (1.200)	15,24 (0.600)	20,32 (0.800)
28	35,56 (1.400)		
40	50,80 (2.000)		

CM37 SERIES

Number of Positions	A ±0.016 Length	B ±0.02	C ±0.016 Width
28	27,18 (1.070)	10,67 (0.420)	17,20 (0.677)
40	37,85 (1.490)	16,51 (0.650)	23,11 (0.910)
42	39,62 (1.560)		
54	50,29 (1.980)		
64	59,18 (2.330)	20,32 (0.800)	26,92 (1.060)

Dimensions in parentheses are inches
Contact factory for detailed information

TEXAS INSTRUMENTS
34 Forest Street • Attleboro, Massachusetts 02703

PERFORMANCE SPECIFICATIONS

Insertion force: 16 oz (454 g) per pin max
Withdrawal force: 1.5 oz (42 g) per pin min
Operating temperature: −40°C to 100°C, tin
Accommodates IC leads 0.011 ± 0.003 in by 0.018 ± 0.003 in
Contact rating: 1.0 A per contact

MATERIALS

Body — PBT polyester U/L 94 VO rating
C4S & CxW Contacts — Copper alloy
Contact finish — Reflow tin plating, 40 μin min

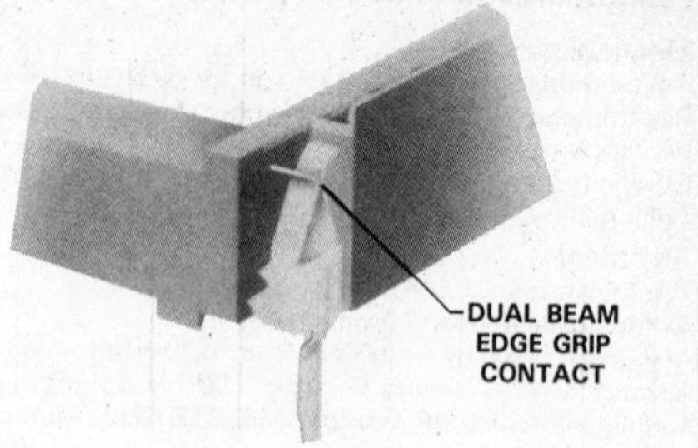
DUAL BEAM
EDGE GRIP
CONTACT

PART NUMBER SYSTEM

C X W XX − 11

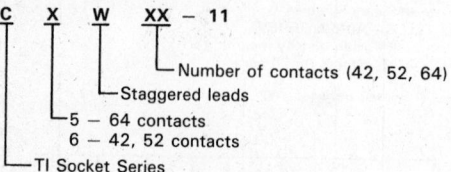

- XX — Number of contacts (42, 52, 64)
- W — Staggered leads
- X — 5 – 64 contacts; 6 – 42, 52 contacts
- C — TI Socket Series

PART NUMBER SYSTEM†

C 4 S XX − 02

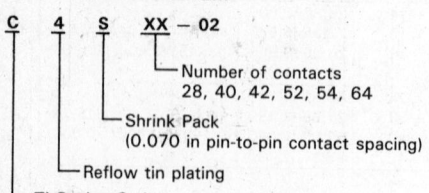

- XX — Number of contacts 28, 40, 42, 52, 54, 64
- S — Shrink Pack (0.070 in pin-to-pin contact spacing)
- 4 — Reflow tin plating
- C — TI Socket Series

†Also available in screw machine contacts

QUAD-IN-LINE (CxW SERIES)

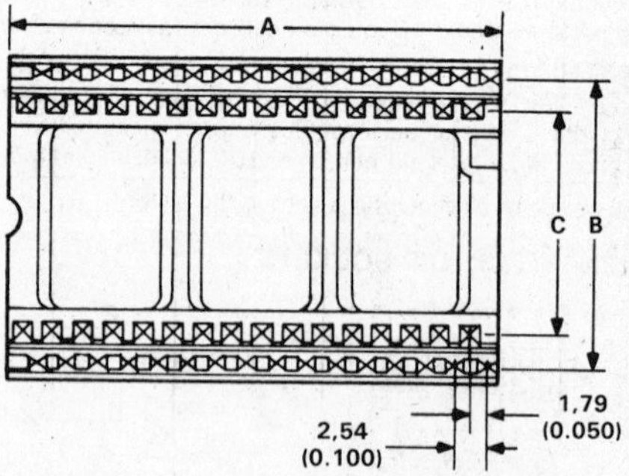
A
C
B
2,54
(0.100)
1,79
(0.050)

SHRINK PACK DIP (C4S SERIES)

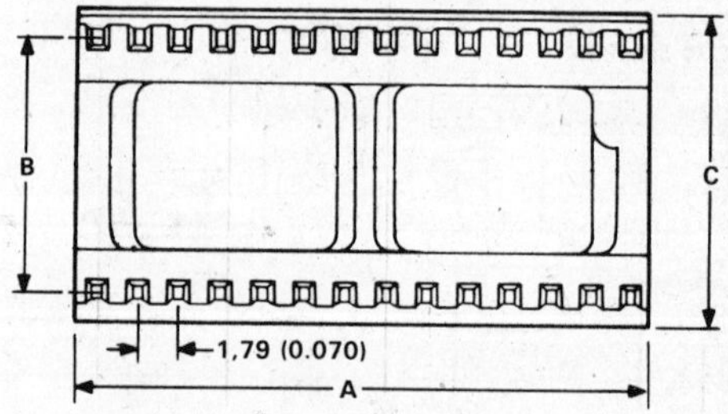
B
C
1,79 (0.070)
A

C4S SERIES

Positions	A max Length	B Row to Row	C max Width
28	25,02 (0.985)	10,16 (0.400)	13,00 (0.512)
40	35,69 (1.405)	15,24 (0.600)	17,98 (0.708)
64	57,07 (2.247)	19,05 (0.750)	21,62 (0.851)

Dimension in parentheses are inches

C4S SERIES

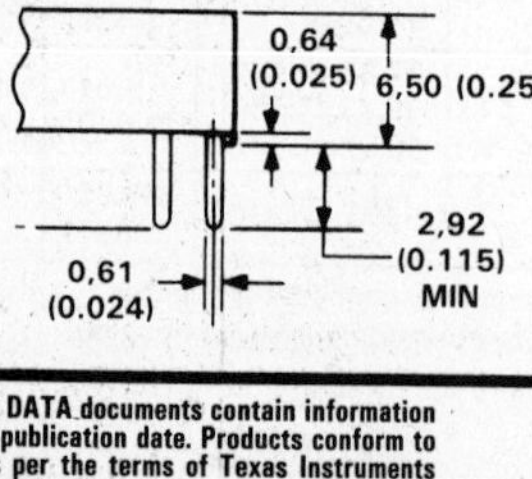
0,64
(0.025)
6,50 (0.256)
2,92
(0.115)
MIN
0,61
(0.024)

QUAD-IN-LINE (CxW SERIES)

Product Number	A Max Length	B Row to Row	C Max Row to Row
C5W64-11	41,90 (1.65)	22,90 (0.950)	19,05 (0.750)
C6W42-11	27,90 (1.10)	22,90 (0.900)	17,80 (0.700)
C6W52-11	34,30 (1.35)	22,90 (0.900)	17,80 (0.700)

Dimensions in parentheses are inches
Contact factory for detailed information

TEXAS INSTRUMENTS
34 Forest Street • Attleboro, Massachusetts 02703

PERFORMANCE SPECIFICATIONS

Mechanical

Accommodates IC leads per specific IC device
Recommended PCB thickness range: 0.062 in to 0.092 in
Recommended PCB hole size range: 0.032 in to 0.042 in
Durability: 5000 cycles, 10 mΩ max contact resistance change per MIL-STD 1344, Method 2016
Solderability: per MIL-STD 202, Method 208

Electrical

Contact rating: 1.0 A per contact
Contact resistance: 20 mΩ max initial
Insulation resistance: 1.0 MΩ at 500 V dc per MIL-STD 1344, Method 3003.1
Dielectric withstanding voltage: 700 V ac rms per MIL-STD 1344, Method 3001.1
Capacitance: 1.0 pF max per MIL-STD 202, Method 305

Environmental

Operating temperature: −65°C to 170°C
Humidity: 10 mΩ max contact resistance change when tested per MIL-STD 202, Method 103B
Temperature soak: 10 mΩ max contact resistance change when exposed to 105°C temperature for 48 hours
Shelf life: 12 months min

MATERIALS

Body — CFP Series — PES (polyether sulfone) glass filled U/L 94 VO rating
Temperature: −65°C to 170°C
Contact — Beryllium copper
Plating:† Overall gold plate min 4 μin over min 70 μin nickel plating

†For additional plating option consult the factory.
Dimensional drawings available from factory.

SMALL OUT LINE FLAT PACK (CFPH/K SERIES)

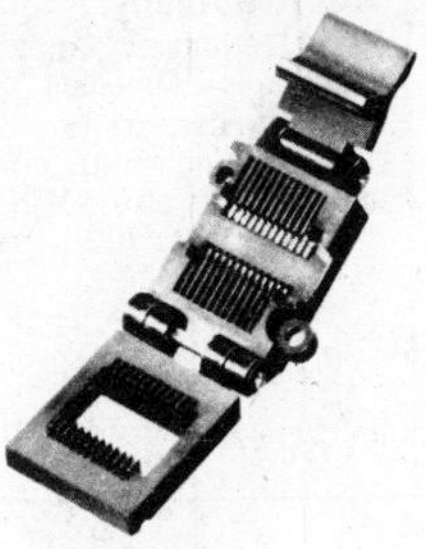

PART NUMBER SYSTEM

C XX X − XXX − XX − 01

- C — TI Series socket
- XX — Style FP — Flat pack
- X — Configuration: H — 14, 16, 18, 20 Positions; K — 24, 28 Positions
- XXX — Number of positions
- XX — Plating: 37 — overall gold plate

QUAD FLAT PACK (CFPM SERIES)

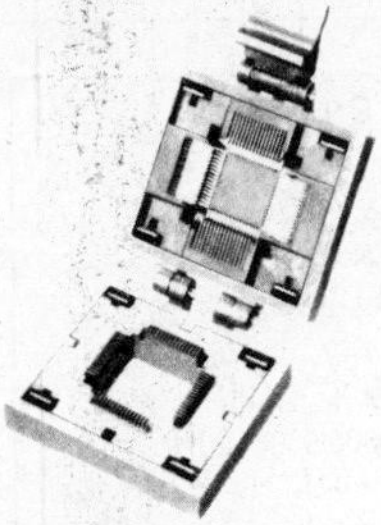

PART NUMBER SYSTEM

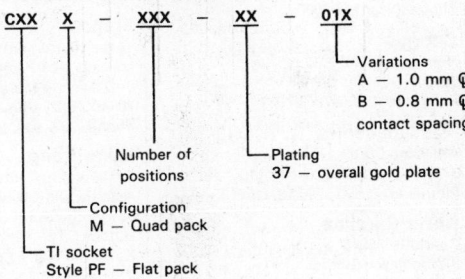

CXX X − XXX − XX − 01X

- CXX — TI socket, Style PF — Flat pack
- X — Configuration: M — Quad pack
- XXX — Number of positions
- XX — Plating: 37 — overall gold plate
- 01X — Variations: A — 1.0 mm ℄; B — 0.8 mm ℄ contact spacing

SMALL OUTLINE (J-LEADED)

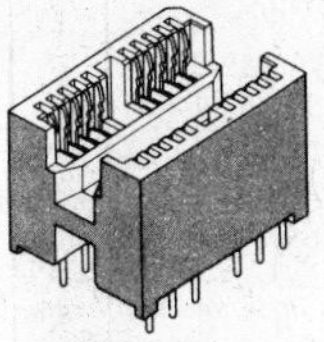

PART NUMBER SYSTEM

CSJT XXX − 37 − 02

- CSJT — TI series
- XXX — Number of positions

AVAILABLE SIZES

CFPH Series 14, 16, 18, 20 CFPK Series 24, 28	Flat Pack
CFPM Series 64, 80	Quad Pack
CSJT Series 20, 26	Small Outline J-Leaded

Contact factory for detailed information

Mechanical Data
B

TEXAS INSTRUMENTS
34 Forest Street • Attleboro, Massachusetts 02703

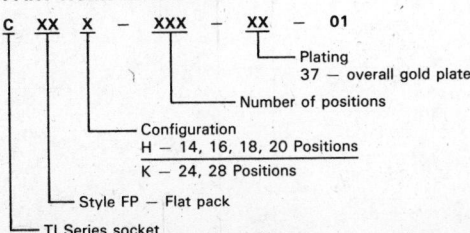

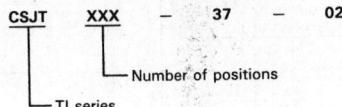

For more information contact your local distributor or contact TI directly:

Texas Instruments Incorporated
CSD Marketing, MS 14-1
Attleboro, MA 02703

(617) 699-5242/5269

Field Sales Offices

UNITED STATES

California
Irvine 91714
17891 Cartwright Road
Phone: (714) 660-8111

San Diego 92123
4333 View Ridge Ave., Suite B
Phone (619) 278-9600/9603

Torrence 90502
9505 Hamilton St.
Bldg. A, Suite One
Phone: (213) 217-7000

Georgia
Norcross 30092
5515 Spaulding Drive
Phone: (404) 662-7861/7931

Massachusetts
Attleboro 02703
34 Forest Street, MS 10-6/MS 14-3
Phone: (617) 699-5206/1278/5213

North Carolina
Charlotte 28210
8 Woodlawn Green
Suite 100
Phone: (704) 527-0930

Texas
Richardson 75081
1001 E. Campbell Rd., MS 328
Phone: (214) 680-5284/5268/5267

Texas Instruments provides customer assistance in varied technical areas. Since TI does not possess full access to data concerning all of the uses and applications of customers' products, responsibility is assumed by TI neither for customer product design nor for any infringement of patents or rights of others, which may result from TI assistance.

INTERNATIONAL

Australia
Texas Instruments Australia, Ltd.
P.O. Box 63
Elizabeth, South Australia 5112
Phone: 61-8-255-2066

England
Texas Instruments, Ltd.
Beffordia House
Prebend Stsreet
Bedford MK41 7PA
Phone: (0234) 63211, Ext. 1

France
Texas Instruments, Ltd.
Metallurgical Materials Division
8-10 Avenue Morane Saulnier
78140 Velizy-Villacoublay, Paris
Phone: 333. 946. 9712

Hong Kong
Texas Instruments Asia, Ltd.
Asia Pacific Division
8th Floor, World Shipping Centre
Harbor City 7, Canton Road
Kowloon, Hong Kong
Phone: 852-3-722-1223

Italy
Texas Instruments Italia SPA
Viale Europa, 40
I-20093 Cologno Monzese
Milano
Phone: 011-39-2-25.300.1

TEXAS INSTRUMENTS

Japan
Texas Instruments Japan, Ltd.
305 Tanagasnira
Oyama-Cho
Suntoh-Gun, Shizuoka-Ken
Japan 410-13
Phone: (81) 550-81211

Mexico
Texas Instruments de Mexico, SA
Av. Reforma No. 450-10 Piso
Col. Juarez
Delegacion: Cuauhtemoc
Mexico City, D.F.
Mexico 06600
Phone: 52-5-514-3583

Singapore
Texas Instruments Asia
#02-08, 12 Lorong Bakar Batu
Kolam Ayer Industrial Estate
Singapore 1334
Republic of Singapore
Phone: 65-747-2255

Taiwan
Texas Instruments Supply Co.
Taiwan Branch
Bank Tower
Room 903, 205 Tun Hwa N. Road
Taipei, Taiwan
Phone: 886-2-713-9311

West Germany
Texas Instruments Deutschland GMBH
Metallurgical Materials Div.
Rosenkavalierplatz 15
D-8000 Muenchen 81
Phone: 011-49-89-915081

General Information — 1

Alphanumeric Index
Selection Guide

Data Acquisition Circuits — 2

Cross-Reference Guide
Data Sheets

Display Drivers — 3

Data Sheets

Line Drivers and Receivers — 4

Cross-Reference Guide
Data Sheets

Peripheral Drivers/Actuators — 5

Cross-Reference Guide
Data Sheets

Memory Interface Circuits — 6

Data Sheets

Speech Synthesis Circuits — 7

Data Sheets

Appendix A — Power Derating Curves — A

Appendix B — Ordering Instructions, Mechanical Data, IC Sockets — B

Appendix C — Explanation of Logic Symbols — C

Explanation of Logic Symbols

by F.A. Mann

Contents

Section		Page
1	**Introduction**	C-5
2	**Symbol Composition**	C-5
3	**Qualifying Symbols**	C-6
3.1	General Qualifying Symbols	C-6
3.2	Qualifying Symbols for Inputs and Outputs	C-6
3.3	Symbols Inside the Outline	C-7
3.4	Combinations of Outlines and Internal Connections	C-8
4	**Dependency Notation**	C-9
4.1	General Explanation	C-9
4.2	G, AND	C-10
4.3	Conventions for the Application of Dependency Notation in General	C-11
4.4	V, OR	C-11
4.5	N, Negate (Exclusive-OR)	C-12
4.6	Z, Interconnection	C-12
4.7	X, Transmission	C-13
4.8	C, Control	C-13
4.9	EN, Enable	C-14
4.10	M, Mode	C-14
5	**Bistable Elements**	C-16
6	**Examples of Actual Device Symbols**	C-17
6.1	UDN2841 High-Current Darlington Drivers	C-17
6.2	SN75437 Quadruple Peripheral Driver	C-17
6.3	TL607 Analog Switch with Enable	C-18
6.4	SN75128 8-Channel Line Receiver	C-18
6.5	SN75122 Triple Line Receivers	C-19
6.6	DS8831 Quad Single-Ended or Dual Differential Line Drivers	C-19
6.7	SN75113 Differential Line Drivers with Split 3-State Outputs	C-20
6.8	SN75163B Octal General-Purpose Interface Bus Transceiver	C-20
6.9	SN75161B Octal IEEE Std 488 Interface Bus Transceiver	C-21
6.10	SN5520 Dual-Channel Sense Amplifier with Complementary Outputs	C-22
6.11	SN75500E AC Plasma Display Driver with CMOS-Compatible Inputs	C-23
6.12	SN75551 Electroluminescent Row Driver with CMOS-Compatible Inputs	C-24

List of Tables

Table		Page
1	General Qualifying Symbols	C-6
2	Qualifying Symbols for Inputs and Outputs	C-7
3	Symbols Inside the Outline	C-8
4	Symbols for Internal Connections	C-9
5	Summary of Dependency Notation	C-16

List of Illustrations

Figure		Page
1	Symbol Composition	C-5
2	Common-Control Block	C-8
3	G Dependency Between Inputs	C-10
4	G Dependency Between Outputs and Inputs	C-10
5	G Dependency with a Dynamic Input	C-10
6	ORed Affecting Inputs	C-11
7	V (OR) Dependency	C-11
8	N (Negate/Exclusive-OR) Dependency	C-12
9	Z (Interconnection) Dependency	C-12
10	X (Transmission) Dependency	C-13
11	Analog Data Selector (Multiplexer/Demultiplexer)	C-13
12	C (Control) Dependency	C-14
13	EN (Enable) Dependency	C-14
14	M (Mode) Dependency Affecting Inputs	C-15
15	Type of Output Determined by Mode	C-15
16	Latches and Flip-Flops	C-16

If you have questions on this Explanation of Logic Symbols, please contact:

F.A. Mann, MS 49
Texas Instruments Incorporated
P.O. Box 225012
Dallas, Texas 75265

Telephone (214) 995-2659

IEEE Standards may be purchased from:

Institute of Electrical and Electronics Engineers, Inc.
IEEE Standards Office
345 East 47th Street
New York, N.Y. 10017

International Electrotechnical Commission (IEC) publications may be purchased from:

American National Standards Institute, Inc.
1430 Broadway
New York, N.Y. 10018

1 Introduction

The International Electrotechnical Commission (IEC) has been developing a very powerful symbolic language that can show the relationship of each input of a digital logic circuit to each output without showing explicitly the internal logic. At the heart of the system is dependency notation, which will be explained in Section 4.

The system was introduced in the USA in a rudimentary form in IEEE/ANSI Standard Y32.14-1973. Lacking at that time a complete development of dependency notation, it offered little more than a substitution of rectangular shapes for the familiar distinctive shapes for representing the basic functions of AND, OR, negation, etc. This is no longer the case.

Internationally, Working Group 2 of IEC Technical Committee TC-3 has prepared a new document (Publication 617-12) that consolidates the original work started in the mid 1960's and published in 1972 (Publication 117-15) and the amendments and supplements that have followed. Similarly for the USA, IEEE Committee SCC 11.9 has revised the publication IEEE Std 91/ANSI Y32.14. Now numbered simply ANSI/IEEE Std 91-1984, the IEEE standard contains all of the IEC work that has been approved, and also a small amount of material still under international consideration. Texas Instruments is participating in the work of both organizations and this document introduces new logic symbols in accordance with the new standards. When changes are made as the standards develop, future editions will take those changes into account.

The following explanation of the new symbolic language is necessarily brief and greatly condensed from what the standards publications now contain. This is not intended to be sufficient for those people who will be developing symbols for new devices. It is primarily intended to make possible the understanding of the symbols used in this data book and is somewhat briefer than the explanation that appears in several of TI's data books on digital logic. However, it includes a new section (6.0) that explains several symbols for actual devices in detail. This has proven to be a powerful learning aid.

2 Symbol Composition

A symbol comprises an outline or a combination of outlines together with one or more qualifying symbols. The shape of the symbols is not significant. As shown in Figure 1, general qualifying symbols are used to tell exactly what logical operation is performed by the elements. Table 1 shows general qualifying symbols defined in the new standards. Input lines are placed on the left and output lines are placed on the right. When an exception is made to that convention, the direction of signal flow is indicated by an arrow as shown in Figure 9.

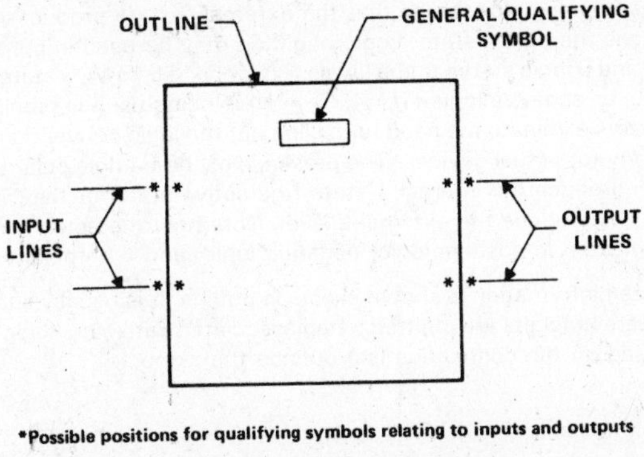
OUTLINE
GENERAL QUALIFYING SYMBOL
INPUT LINES
OUTPUT LINES
*Possible positions for qualifying symbols relating to inputs and outputs

Figure 1. Symbol Composition

Explanation of Logic Symbols

C

3 Qualifying Symbols

3.1 General Qualifying Symbols

Table 1 shows general qualifying symbols defined by ANSI/IEEE Standard 91. These characters are placed near the top center or the geometric center of a symbol or symbol element to define the basic function of the device represented by the symbol or of the element.

X/Y is the general qualifying symbol for identifying coders, code converters, and level converters. X and Y may be used in their own right to stand for some code or either or both may be replaced by some other indication of the code or level such as BCD or TTL. As might be expected, interface circuits make frequent use of this set of qualifying symbols.

Table 1. General Qualifying Symbols

SYMBOL	DESCRIPTION	
&	AND gate or function	
≥1	OR gate or function. The symbol was chosen to indicate that at least one active input is needed to activate the output.	
=1	Exclusive OR. One and only one input must be active to activate the output.	
1	A simple 1-input gate or element	
▷ or ◁	A buffer or element with more than usual output capability (symbol is oriented in the direction of signal flow).	
⎎	Schmitt trigger; element with hysteresis	
X/Y	Coder, code converter, level converter	
	The following are examples of subsets of this general class of qualifying symbol used in this book.	
	BCD/7-SEG	BCD to 7-segment display driver
	TTL/MOS	TTL to MOS level converter
	CMOS/PLASMA DISP	Plasma-display driver with CMOS-compatible inputs
	MOS/LED	Light-emitting-diode driver with MOS-compatible inputs
	CMOS/VAC FLUOR DISP	Vacuum-fluorescent display driver with CMOS-compatible inputs
	CMOS/EL DISP	Electroluminescent display driver with CMOS-compatible inputs
	TTL/GAS DISCH DISPLAY	Gas-discharge display driver with TTL-compatible inputs
SRGm	Shift register. m = number of bits.	

3.2 Qualifying Symbols for Inputs and Outputs

Qualifying symbols for inputs and outputs are shown in Table 2 and many will be familiar to most users, a likely exception being the logic polarity symbol for directly indicating active-low inputs and outputs. The older logic negation indicator means that the external 0 state produces the internal 1 state. The internal 1 state means the active state. Logic negation may be used in pure logic diagrams; in order to tie the external 1 and 0 logic states to the levels H (high) and L (low), a statement of whether positive logic (1 = H, 0 = L) or negative logic (1 = L, 0 = H) is being used is required or must be assumed. Logic polarity indicators eliminate the need for calling out the logic convention and are used in this data book in the symbology for actual devices. The presence of the triangle polarity indicator indicates that the L logic level will produce the internal 1 state (the active state) or that, in the case of an output, the internal 1 state will produce the external L level. Note how the active direction of transition for a dynamic input is indicated in positive logic, negative logic, and with polarity indication.

When nonstandardized information is shown inside an outline, it is usually enclosed in square brackets [like these]. The square brackets are omitted when associated with a nonlogic input, which is indicated by an X superimposed on the connection line outside the symbol.

Table 2. Qualifying Symbols for Inputs and Outputs

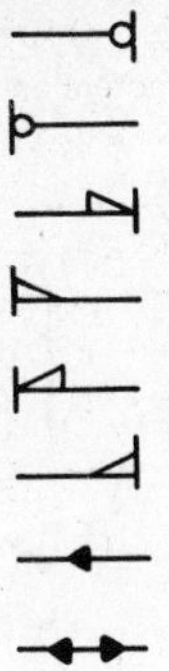

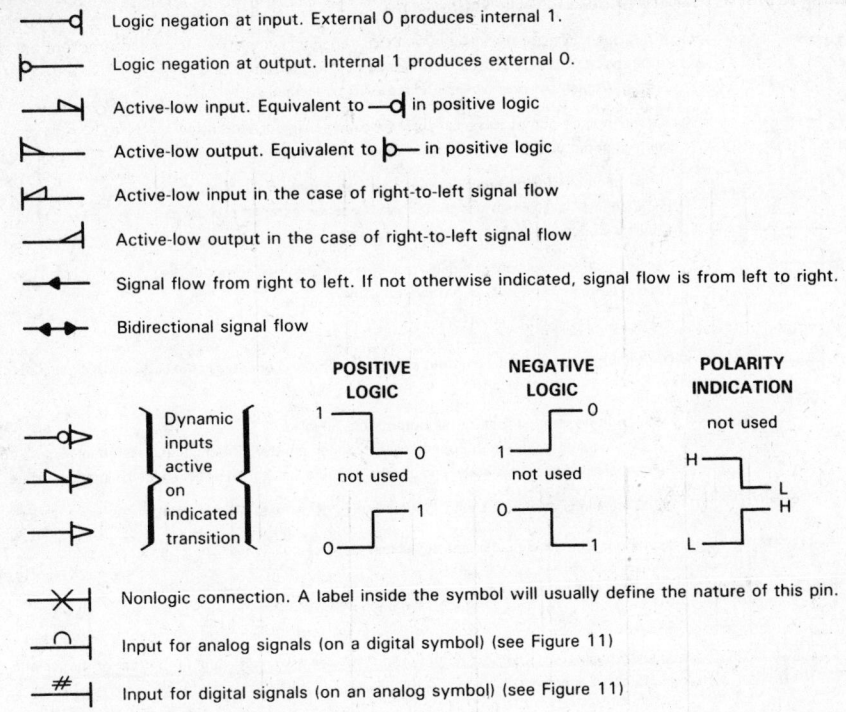

Logic negation at input. External 0 produces internal 1.

Logic negation at output. Internal 1 produces external 0.

Active-low input. Equivalent to —○ in positive logic

Active-low output. Equivalent to ○— in positive logic

Active-low input in the case of right-to-left signal flow

Active-low output in the case of right-to-left signal flow

Signal flow from right to left. If not otherwise indicated, signal flow is from left to right.

Bidirectional signal flow

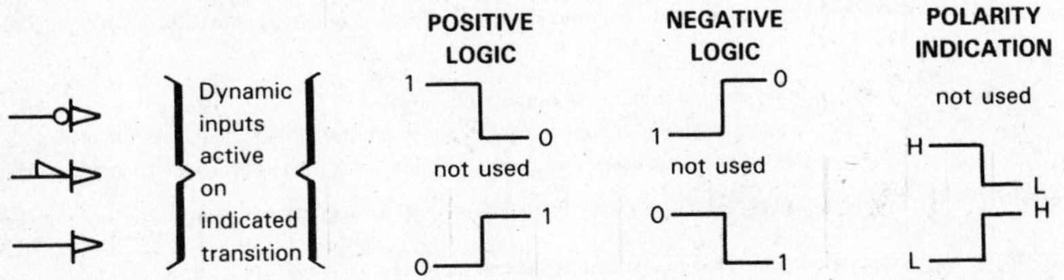
POSITIVE LOGIC
NEGATIVE LOGIC
POLARITY INDICATION
Dynamic inputs active on indicated transition
1
0
not used
0
1
1
0
not used
not used
0
1
1
0
H
L
H
L

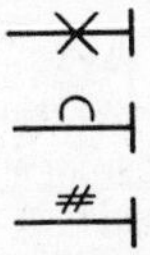

Nonlogic connection. A label inside the symbol will usually define the nature of this pin.

Input for analog signals (on a digital symbol) (see Figure 11)

Input for digital signals (on an analog symbol) (see Figure 11)

3.3 Symbols Inside the Outline

Table 3 shows some symbols used inside the outline. Note particularly that open-collector (open-drain), open-emitter (open-source), and three-state outputs have distinctive symbols. Also note that an EN input affects all of the outputs of the element and has no effect on inputs. An EN input affects all the external outputs of the element in which it is placed, plus the external outputs of any elements shown to be influenced by that element. It has no effect on inputs. When an enable input affects only certain outputs, affects outputs located outside the indicated influence of the element in which the enable input is placed, and/or affects one or more inputs, a form of dependency notation will indicate this (see 4.9). The effects of the EN input on the various types of outputs are shown.

It is particularly important to note that a D input is always the data input of a storage element. At its internal 1 state, the D input sets the storage element to its 1 state, and at its internal 0 state it resets the storage element to its 0 state.

The binary grouping symbol will be explained more fully in Section 6.11. Binary-weighted inputs are arranged in order and the binary weights of the least significant and the most significant lines are indicated by numbers. In this document weights of input and output lines will be represented by powers of two usually only when the binary grouping symbol is used, otherwise decimal numbers will be used. The grouped inputs generate an internal number on which a mathematical function can be performed or that can be an identifying number for dependency notation. This number is the sum of the weights ($1, 2, 4 \ldots 2^n$) of those input standing at their 1 states. A frequent use is in addresses for memories.

Reversed in direction, the binary grouping symbol can be used with outputs. The concept is analogous to that for the inputs and the weighted outputs will indicate the internal number assumed to be developed within the circuit.

Table 3. Symbols Inside the Outline

Symbol	Description
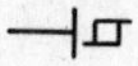	Bithreshold input (input with hysteresis)
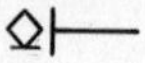	N-P-N open-collector or similar output that can supply a relatively low-impedance L level when not turned off. Requires external pull-up. Capable of positive-logic wired-AND connection. 
	Passive-pull-up output is similar to N-P-N open-collector output but is supplemented with a built-in passive pull-up.
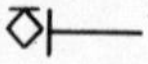	N-P-N open-emitter or similar output that can supply a relatively low-impedance H level when not turned off. Requires external pull-down. Capable of positive-logic wired-OR connection. 
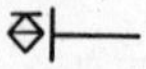	Passive-pull-down output is similar to N-P-N open-emitter output but is supplemented with a built-in passive pull-down.
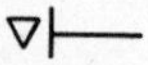	3-state output
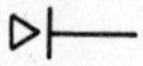	Output with more than usual output capability (symbol is oriented in the direction of signal flow).
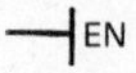 EN	Enable input When at its internal 1-state, all outputs are enabled. When at its internal 0-state, open-collector, open-emitter outputs, and three-state outputs at external high-impedance state, and all other outputs (i.e., totem-poles) are at the internal 0-state.
J, K, R, S, T	Usual meanings associated with flip-flops (e.g., R = reset, T = toggle)
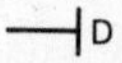 D	Data input to a storage element equivalent to: 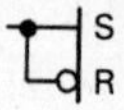 S R
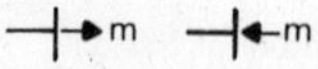 m m	Shift right (left) inputs, m = 1, 2, 3, etc. If m = 1, it is usually not shown.
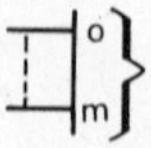 0 m	Binary grouping. m is highest power of 2. Produces a number equal to the sum of the weights of the active inputs
	Input line grouping . . . indicates two or more terminals used to implement a single logic input. e.g., differential inputs.

3.4 Combinations of Outlines and Internal Connections

When a circuit has one or more inputs that are common to more than one element of the circuit, the common-control block may be used. This is the only distinctively shaped outline used in the IEC system. Figure 2 shows that unless otherwise qualified by dependency notation, an input to the common-control block is an input to each of the elements below the common-control block.

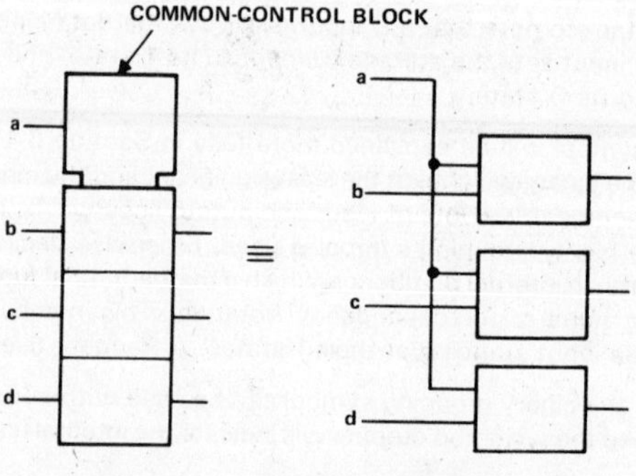
COMMON-CONTROL BLOCK
a
b
c
d
≡
a
b
c
d

Figure 2. Common-Control Block

The outlines of elements may be embedded within one another or abutted to form complex elements, in which case the following rules apply. There is no logic connection between elements when the line common to their outlines is in the direction of signal flow. There is at least one logic connection when the line common to two outlines is perpendicular to the direction of signal flow. If no indications are shown on either side of the common line, it is assumed that there is only one logic connection. If more than one internal connection exists between adjacent elements, the number of connections will be clarified by the use of one or more of the internal connection symbols from Table 4 and/or appropriate qualifying symbols or dependency notation.

Table 4. Symbols for Internal Connections

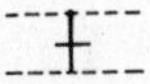	Internal connection. 1 state on left produces 1 state on right.
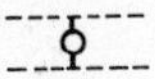	Negated internal connection. 1 state on left produces 0 state on right.
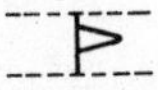	Dynamic internal connection. Transition from 0 to 1 on left produces transitory 1 state on right.
	Dynamic internal connection. Transition from 1 to 0 on left produces transitory 1 state on right.

Table 4 shows symbols that are used to represent internal connection with specific characteristics. The first is a simple noninverting connection, the second is inverting, the third is dynamic. As with this symbol and an external input line, the transition from 0 to 1 on the left produces a momentary 1-state on the right. The fourth symbol is similar except that the active transition on the left is from 1 to 0.

Only logic states, not levels, exist inside symbols. The negation symbol (O) is used internally even when direct polarity indication (▷) is used externally.

In an array of elements, if the same general qualifying symbol and the same qualifying symbols associated with inputs and outputs would appear inside each of the elements of the array, these qualifying symbols are usually shown only in the first element. This is done to reduce clutter and to save time in recognition. Similarly, large identical elements that are subdivided into smaller elements may each be represented by an unsubdivided outline. The SN75163B symbol (see 6.8) illustrates this principle.

4 Dependency Notation

Some readers will find it more to their liking to skip this section and proceed to the explanation of the symbols for a few actual devices in 6.0. Reference will be made there to various parts of this section as it is needed. If this procedure is followed, it is recommended that 5.0 be read after 6.0 and then all of 4.0 be reread.

4.1 General Explanation

Dependency notation is the powerful tool that sets the IEC symbols apart from previous systems and makes compact, meaningful, symbols possible. It provides the means of denoting the relationship between inputs, outputs, or inputs and outputs without actually showing all the elements and interconnections involved. The information provided by dependency notation supplements that provided by the qualifying symbols for an element's function.

In the convention for the dependency notation, use will be made of the terms "affecting" and "affected." In cases where it is not evident which inputs must be considered as being the affecting or the affected ones (e.g., if they stand in an AND relationship), the choice may be made in any convenient way.

So far, eleven types of dependency have been defined but only the eight used in this book are explained. They are listed below in the order in which they are presented and are summarized in Table 5 following 4.10.2.

Section	**Dependency Type or Other Subject**
4.2	G, AND
4.3	General Rules for Dependency Notation
4.4	V, OR
4.5	N, Negate (Exclusive-OR)
4.6	Z, Interconnection
4.7	X, Transmission
4.8	C, Control
4.9	EN, Enable
4.10	M, Mode

4.2 G (AND) Dependency

A common relationship between two signals is to have them ANDed together. This has traditionally been shown by explicitly drawing an AND gate with the signals connected to the inputs of the gate. The 1972 IEC publication and the 1973 IEEE/ANSI standard showed several ways to show this AND relationship using dependency notation. While ten other forms of dependency have since been defined, the ways to invoke AND dependency are now reduced to one.

In Figure 3 input **b** is ANDed with input **a** and the complement of **b** is ANDed with **c**. The letter G has been chosen to indicate AND relationships and is placed at input **b**, inside the symbol. A number considered appropriate by the symbol designer (1 has been used here) is placed after the letter G and also at each affected input. Note the bar over the 1 at input **c**.

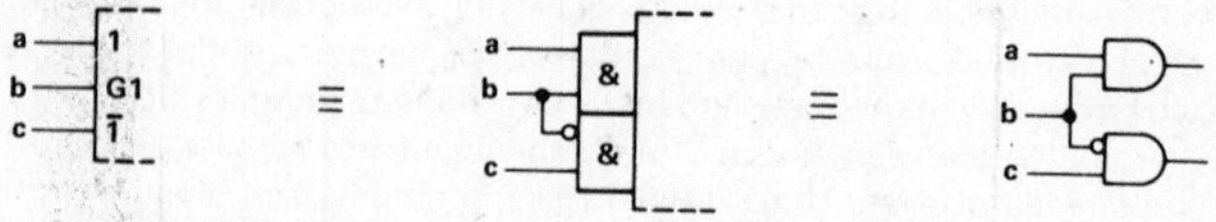
a
b
c
1
G1
1̄
≡
&
&
≡

Figure 3. G Dependency Between Inputs

In Figure 4, output **b** affects input **a** with an AND relationship. The lower example shows that it is the internal logic state of **b**, unaffected by the negation sign, that is ANDed. Figure 5 shows input **a** to be ANDed with a dynamic input **b**.

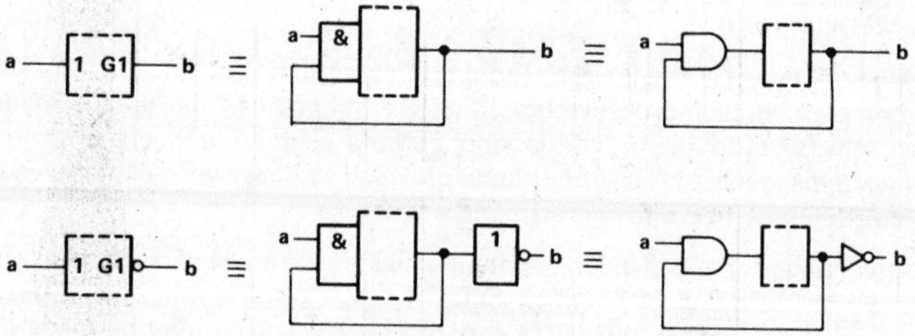
a
1 G1
b
≡
&
1
≡

Figure 4. G Dependency Between Outputs and Inputs

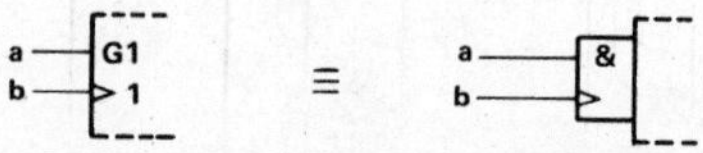
a
b
G1
1
≡
&

Figure 5. G Dependency with a Dynamic Input

The rules for G dependency can be summarized thus:

When a G*m* input or output (*m* is a number) stands at its internal 1 state, all inputs and outputs affected by G*m* stand at their normally defined internal logic states. When the G*m* input or output stands at its 0 state, all inputs and outputs affected by G*m* stand at their internal 0 states.

4.3 Conventions for the Application of Dependency Notation in General

The rules for applying dependency relationships in general follow the same pattern as was illustrated for G dependency.

Application of dependency notation is accomplished by:

1. Labeling the input (or output) *affecting* other inputs or outputs with the letter symbol indicating the relationship involved (e.g., G for AND) followed by an identifying number, appropriately chosen, and
2. Labeling each input or output *affected* by that affecting input (or output) with that same number.

If it is the complement of the internal logic state of the affecting input or output that does the affecting, then a bar is placed over the identifying numbers at the affected inputs or outputs (Figure 3).

If two affecting inputs or outputs have the same letter and same identifying number, they stand in an OR relationship to each other (Figure 6).

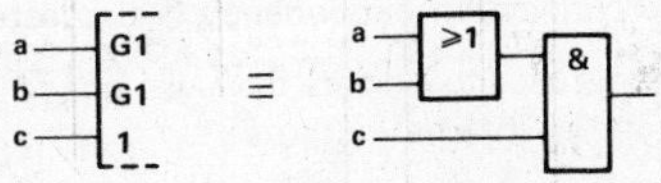

a
b
c
G1
G1
1
≡
a
b
c
≥1
&

Figure 6. ORed Affecting Inputs

If the affected input or output requires a label to denote its function (e.g., "D"), this label will be *prefixed* by the identifying number of the affecting input (Figure 12).

If an input or output is affected by more than one affecting input, the identifying numbers of each of the affecting inputs will appear in the label of the affected one, separated by commas. The normal reading order of these numbers is the same as the sequence of the affecting relationships (Figure 12).

4.4 V (OR) Dependency

The symbol denoting OR dependency is the letter V (Figure 7).

When a V*m* input or output stands at its internal 1 state, all inputs and outputs affected by V*m* stand at their internal 1 states. When the V*m* input or output stands at its internal 0 state, all inputs and outputs affected by V*m* stand at their normally defined internal logic states.

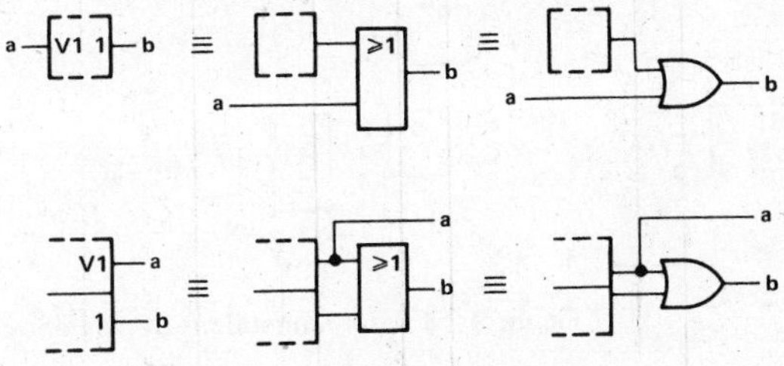

a
V1 1
b
≡
≥1
b
a
≡
a
b
V1
a
1
b
≡
a
≥1
b
≡
a
b

Figure 7. V (OR) Dependency

4.5 N (Negate) (Exclusive-OR) Dependency

The symbol denoting negate dependency is the letter N (Figure 8). Each input or output affected by an N*m* input or output stands in an Exclusive-OR relationship with the N*m* input or output.

When an N*m* input or output stands at its internal 1 state, the internal logic state of each input and each output affected by N*m* is the complement of what it would otherwise be. When an N*m* input or output stands at its internal 0 state, all inputs and outputs affected by N*m* stand at their normally defined internal logic states.

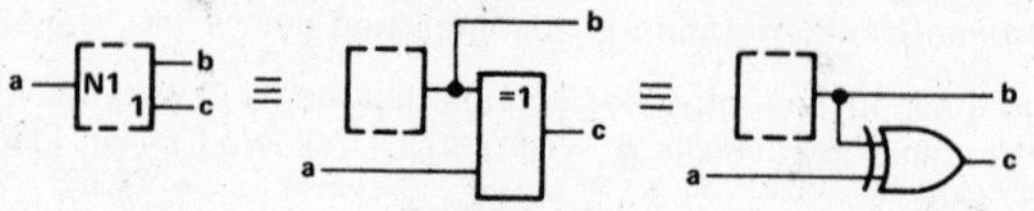

a
N1
1
b
c
≡
=1

If a = 0, then c = b
If a = 1, then c = $\overline{b}$

Figure 8. N (Negate) (Exclusive-OR) Dependency

4.6 Z (Interconnection) Dependency

The symbol denoting interconnection dependency is the letter Z.

Interconnection dependency is used to indicate the existence of internal logic connections between inputs, outputs, internal inputs, and/or internal outputs.

The internal logic state of an input or output affected by a Z*m* input or output will be the same as the internal logic state of the Z*m* input or output, unless modified by additional dependency notation (Figure 9).

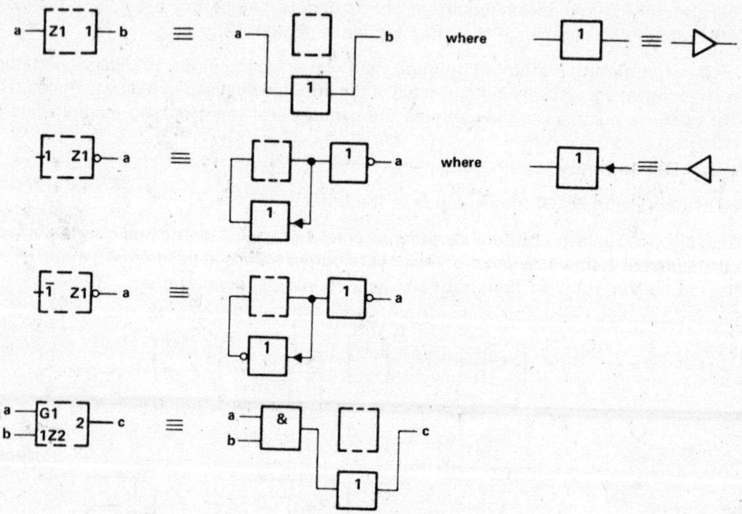

Z1
where
G1
1Z2
&

Figure 9. Z (Interconnection) Dependency

4.7 X (Transmission) Dependency

The symbol denoting transmission dependency is the letter X.

Transmission dependency is used to indicate controlled bidirectional connections between affected input/output ports (Figure 10).

When an X*m* input or output stands at its internal 1 state, all input-output ports affected by this X*m* input or output are bidirectionally connected together and stand at the same internal logic state or analog signal level. When an X*m* input or output stands at its internal 0 state, the connection associated with this set of dependency notation does not exist.

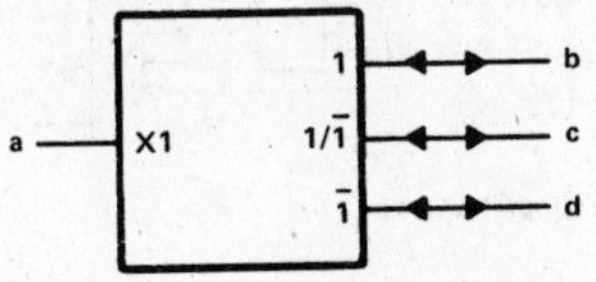

a
X1
1
1/$\bar{1}$
$\bar{1}$
b
c
d

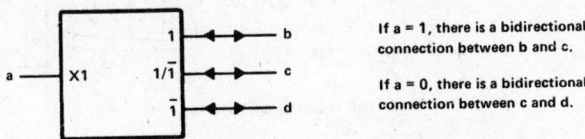

If a = 1, there is a bidirectional connection between b and c.

If a = 0, there is a bidirectional connection between c and d.

Figure 10. X (Transmission) Dependency

Although the transmission paths represented by X dependency are inherently bidirectional, use is not always made of this property. This is analogous to a piece of wire, which may be constrained to carry current in only one direction. If this is the case in a particular application, then the directional arrows shown in Figures 10 and 11 would be omitted.

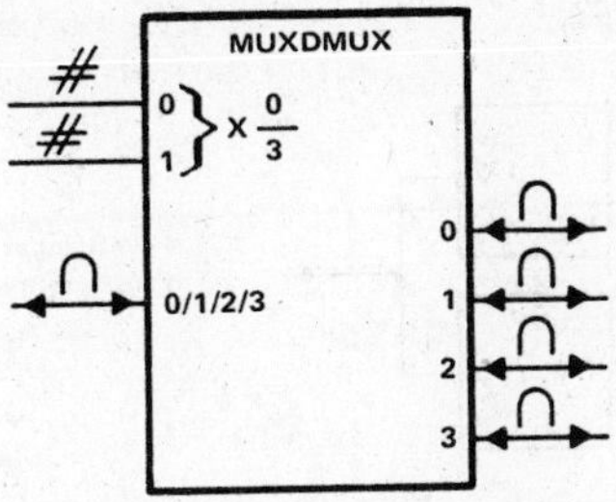

MUXDMUX
0
1
X $\frac{0}{3}$
0/1/2/3
0
1
2
3

Figure 11. Analog Data Selector (Multiplexer/Demultiplexer)

4.8 C (Control) Dependency

The symbol denoting control dependency is the letter C.

Control inputs are usually used to enable or disable the data (D, J, K, R, or S) inputs of storage elements. They may take on their internal 1 states (be active) either statically or dynamically. In the latter case the dynamic input symbol is used as shown in the second example of Figure 12.

When a C*m* input or output stands at its internal 1 state, the inputs affected by C*m* have their normally defined effect on the function of the element, i.e., these inputs are enabled. When a C*m* input or output stands at its internal 0 state, the inputs affected by C*m* are disabled and have no effect on the function of the element.

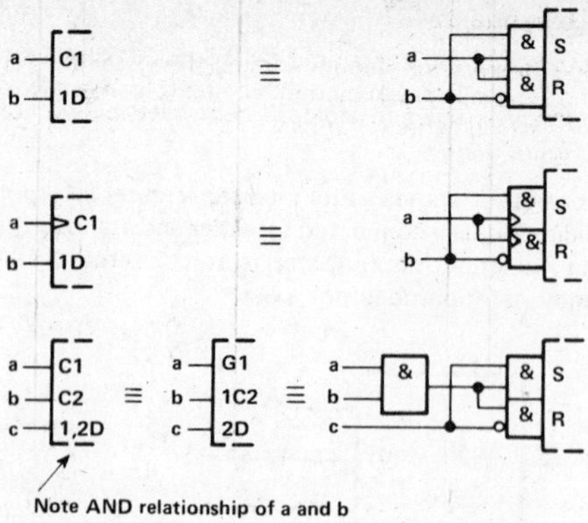
a
b
C1
1D
a
b
&
S
&
R
a
b
C1
1D
a
b
&
S
&
R
a
b
c
C1
C2
1,2D
a
b
c
G1
1C2
2D
a
b
c
&
&
S
&
R
Note AND relationship of a and b

Figure 12. C (Control) Dependency

4.9 EN (Enable) Dependency

The symbol denoting enable dependency is the combination of letters EN.

An EN*m* input has the same effect on outputs as an EN input, see 3.3, but it affects only those outputs labeled with the identifying number *m*. It also affects those inputs labeled with the identifying number *m*. By contrast, an EN input affects all outputs and no inputs. The effect of an EN*m* input on an affected input is identical to that of a C*m* input (Figure 13).

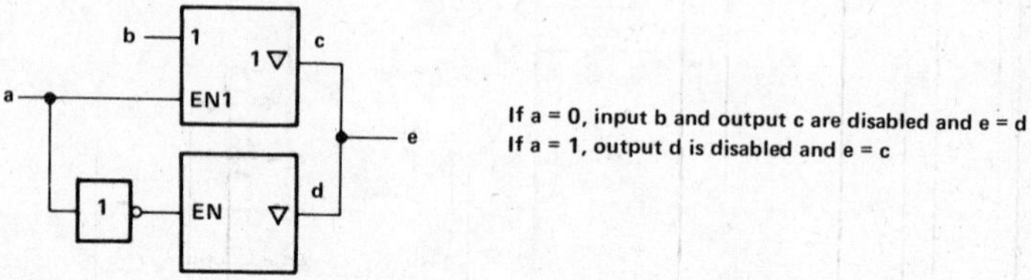
b
1
1▽
c
a
EN1
e
1
EN
▽
d
If a = 0, input b and output c are disabled and e = d
If a = 1, output d is disabled and e = c

Figure 13. EN (Enable) Dependency

When an EN*m* input stands at its internal 1 state, the inputs affected by EN*m* have their normally defined effect on the function of the element and the outputs affected by this input stand at their normally defined internal logic states, i.e., these inputs and outputs are enabled.

When an EN*m* input stands at its internal 0 state, the inputs affected by EN*m* are disabled and have no effect on the function of the element, and the outputs affected by EN*m* are also disabled. Open-collector outputs are turned off, three-state outputs stand at their high-impedance state, and all other outputs (e.g., totem-pole outputs) stand at their internal 0 states.

4.10 M (MODE) Dependency

The symbol denoting mode dependency is the letter M.

Mode dependency is used to indicate that the effects of particular inputs and outputs of an element depend on the mode in which the element is operating.

If an input or output has the same effect in different modes of operation, the identifying numbers of the relevant affecting M*m* inputs will appear in the label of that affected input or output between parentheses and separated by solidi, e.g., (1/2)CT=0 ≡ 1CT=0/2CT=0 where 1 and 2 refer to M1 and M2.

4.10.1 M Dependency Affecting Inputs

M dependency affects inputs the same as C dependency. When an M*m* input or M*m* output stands at its internal 1 state, the inputs affected by this M*m* input or M*m* output have their normally defined effect on the function of the element, i.e., the inputs are enabled.

When an M*m* input or M*m* output stands at its internal 0 state, the inputs affected by this M*m* input or M*m* output have no effect on the function of the element. When an affected input has several sets of labels separated by solidi (e.g., C4/2→/3+), any set in which the identifying number of the M*m* input or M*m* output appears has no effect and is to be ignored. This represents disabling of some of the functions of a multifunction input.

The circuit in Figure 14 has two inputs, **b** and **c**, that control which one of four modes (0, 1, 2, or 3) will exist at any time. Inputs **d**, **e**, and **f** are D inputs subject to dynamic control (clocking) by the **a** input. The numbers 1 and 2 are in the series chosen to indicate the modes so inputs **e** and **f** are only enabled in mode 1 (for parallel loading) and input **d** is only enabled in mode 2 (for serial loading). Note that input **a** has three functions. It is the clock for entering data. In mode 2, it causes right shifting of data, which means a shift away from the control block. In mode 3, it causes the contents of the register to be incremented by one count.

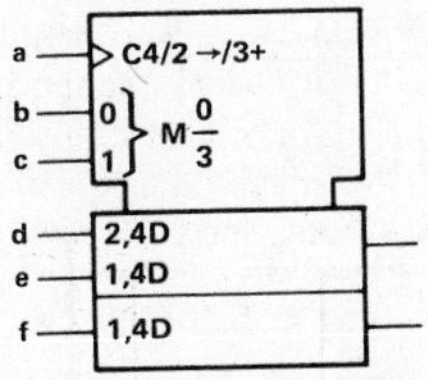
a
b
c
d
e
f
C4/2→/3+
0
1
M 0/3
2,4D
1,4D
1,4D

Note that all operations are synchronous.

In MODE 0 (b = 0, c = 0), the outputs remain at their existing states as none of the inputs has an effect.

In MODE 1 (b = 1, c = 0), parallel loading takes place thru inputs e and f.

In MODE 2 (b = 0, c = 1), shifting down and serial loading thru input d take place.

In MODE 3 (b = c = 1), counting up by increment of 1 per clock pulse takes place.

Figure 14. M (Mode) Dependency Affecting Inputs

4.10.2 M Dependency Affecting Outputs

When an M*m* input or M*m* output stands at its internal 1 state, the affected outputs stand at their normally defined internal logic states, i.e., the outputs are enabled.

When an M*m* input or M*m* output stands at its internal 0 state, at each affected output any set of labels containing the identifying number of that M*m* input or M*m* output has no effect and is to be ignored. When an output has several different sets of labels separated by solidi (e.g., 2,4/3,5), only those sets in which the identifying number of this M*m* input or M*m* output appears are to be ignored.

Figure 15 shows a symbol for a device whose output can behave like either a 3-state output or an open-collector output depending on the signal applied to input **a**. Mode 1 exists when input **a** stands at its internal 1 state and, in that case, the three-state symbol applies and the open-element symbol has no effect. When **a** = 0, mode 1 does not exist so the three-state symbol has no effect and the open-element symbol applies.

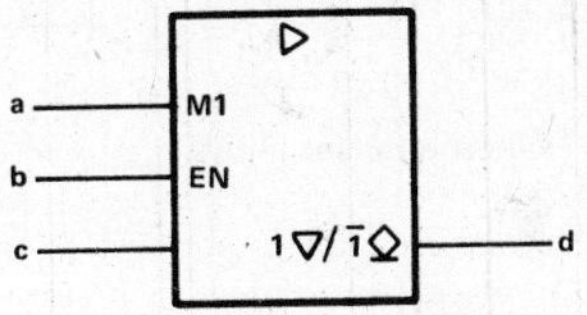
a
b
c
d
M1
EN

Figure 15. Type of Output Determined by Mode

Table 5. Summary of Dependency Notation

TYPE OF DEPENDENCY	LETTER SYMBOL*	AFFECTING INPUT AT ITS 1-STATE	AFFECTING INPUT AT ITS 0-STATE
Control	C	Permits action	Prevents action
Enable	EN	Permits action	Prevents action of inputs ◇ outputs turned off ▽ outputs at external high impedance Other outputs at internal 0 state
AND	G	Permits action	Imposes 0 state
Mode	M	Permits action (mode selected)	Prevents action (mode not selected)
Negate (Ex-NOR)	N	Complements state	No effect
OR	V	Imposes 1 state	Permits action
Transmission	X	Bidirectional connection exists	Bidirectional connection does not exist
Interconnection	Z	Imposes 1 state	Imposes 0 state

* These letter symbols appear at the AFFECTING input (or output) and are followed by a number. Each input (or output) AFFECTED by that input is labeled with that same number.

5 Bistable Elements

The dynamic input symbol and dependency notation provide the tools to identify different types of bistable elements and make synchronous and asynchronous inputs easily recognizable (Figure 16).

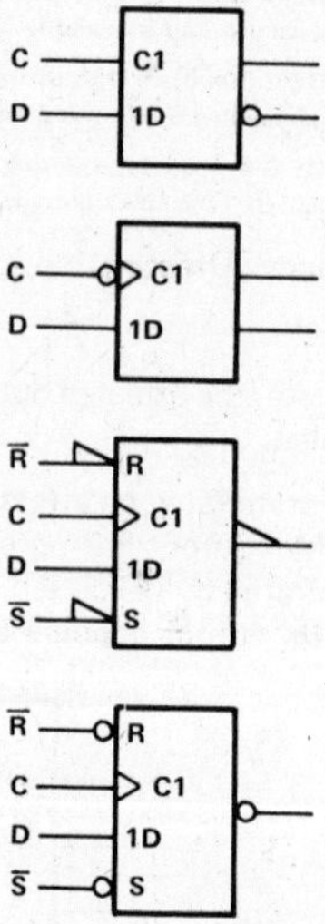
C
D
C1
1D
C
D
C1
1D
R̄
C
D
S̄
R
C1
1D
S
R̄
C
D
S̄
R
C1
1D
S

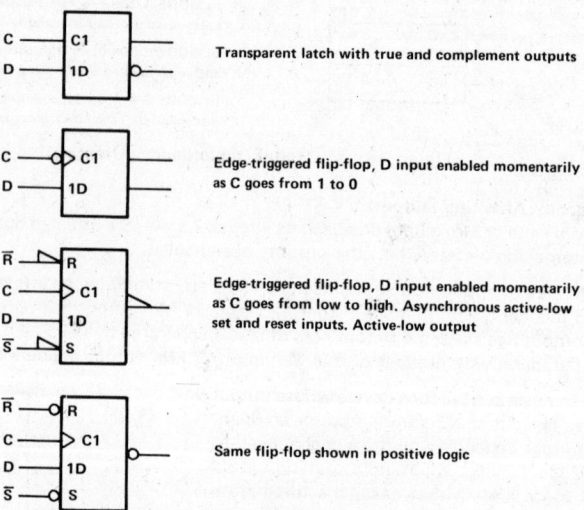

Transparent latch with true and complement outputs

Edge-triggered flip-flop, D input enabled momentarily as C goes from 1 to 0

Edge-triggered flip-flop, D input enabled momentarily as C goes from low to high. Asynchronous active-low set and reset inputs. Active-low output

Same flip-flop shown in positive logic

Figure 16. Latches and Flip-Flops

Transparent latches have a level-operated control input. The D input is active as long as the C input is at its internal 1 state. The outputs respond immediately. Edge-triggered elements accept data from D, J, K, R, or S inputs on the active transition of C.

Notice that synchronous inputs can be readily recognized by their dependency labels (a number preceding the functional label, 1D in these examples) compared to the asynchronous inputs (S and R), which are not dependent on the C inputs. Of course if the set and reset inputs were dependent on the C inputs, their labels would be similarly modified (e.g., 1S, 1R).

6 Examples of Actual Device Symbols

The symbols explained in this section include some of the most complex in this book. These were chosen, not to discourage the reader, but to illustrate the amount of information that can be conveyed. It is likely that if one reads these explanations and follows them reasonably well, most of the other symbols will seem simple indeed. The explanations are intended to be independent of each other so they may seem somewhat repetitious. However each illustrates new principles. They are arranged more or less in the order of complexity.

6.1 UDN2841 High-Current Darlington Drivers

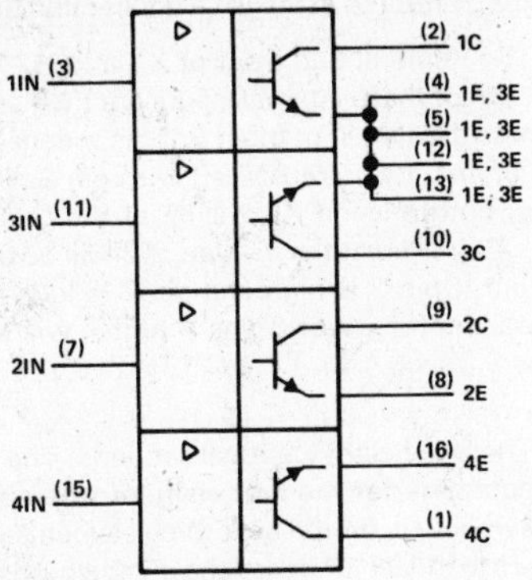
1IN (3)
3IN (11)
2IN (7)
4IN (15)
(2) 1C
(4) 1E, 3E
(5) 1E, 3E
(12) 1E, 3E
(13) 1E, 3E
(10) 3C
(9) 2C
(8) 2E
(16) 4E
(1) 4C

There are four identical sections. The emitters of the output transistors of the elements numbered 1 and 3 are connected together and share pins 4, 5, 12, and 13. The triangular qualifying symbol (▷) indicates amplification, the principal function of the device.

An extension of symbology used for analog devices has been used to show the output transistors. The emitter and collector terminals are lined up with the terminals to which they are connected.

6.2 SN75437 Quadruple Peripheral Driver

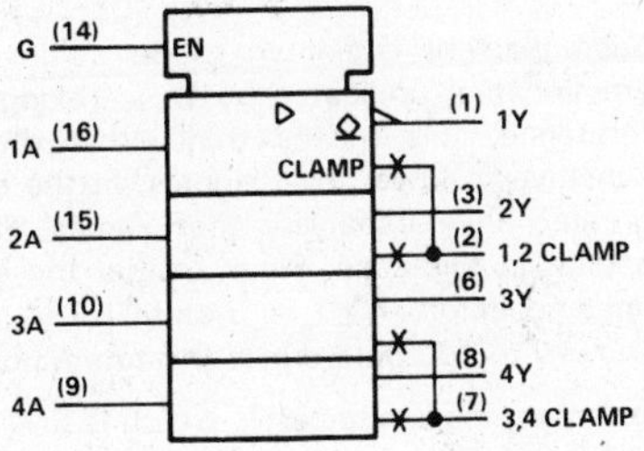
G (14)
EN
1A (16)
2A (15)
3A (10)
4A (9)
CLAMP
(1) 1Y
(3) 2Y
(2) 1,2 CLAMP
(6) 3Y
(8) 4Y
(7) 3,4 CLAMP

There are four identical sections. The symbology is complete for the first element; the absence of any symbology for the other elements indicates they are identical. The top two elements share a common output clamp, pin 2. This is shown to be a nonlogic connection by the superimposed X on the line. The function for this type of connection is indicated briefly and not necessarily exactly by a small amount of text within the symbol. The bottom two elements likewise share a common clamp.

Each element is shown to be an inverter with amplification (indicated by ▷). Taking TTL as a reference, this means that either the input is sensitive to lower level signals, or the output has greater drive capability than usual. The latter applies in this case. The output is shown by ◇ to be open collector.

All the outputs share a common EN input, pin 14. See Figure 2 for an explanation of the common control block. When EN = 0 (pin 14 is low), the outputs, being open-collector types, are turned off and would be pulled high by an external pullup resistor.

6.3 TL607 Analog Switch with Enable

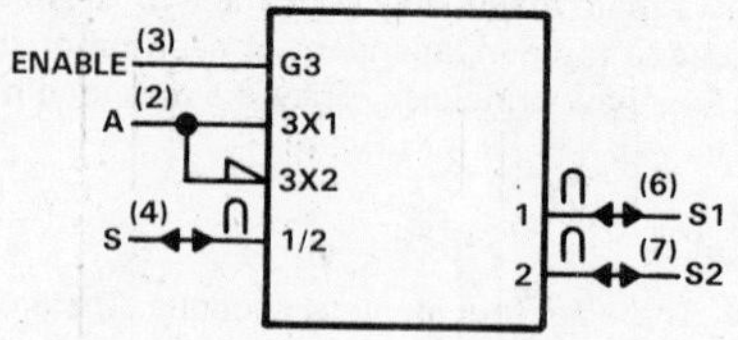

ENABLE
(3)
G3
A
(2)
3X1
3X2
S
(4)
1/2
1
2
(6)
S1
(7)
S2

This device is basically two single-pole, single-throw (SPST) switches connected between pins 4 and 6 and pins 4 and 7. This is indicated using X (Transmission) dependency, which is explained in 4.7. When the internal input X2 stands at its 1-state, a bilateral connection exists between those points affected by X2: pins 4 and 7. The 1 at pins 4 and 6 means they are affected by X1; X1 must stand at the 1-state to establish a connection between them.

The numeral 3 in front of X1 and X2 indicates that both of these internal inputs are themselves affected by affecting input number 3, which is G3. This is coincidently pin number 3. This means that both the active-high branch of pin 2 (X1) and the active-low branch (X2) are ANDed with pin 3. See 4.2 for an explanation of G (AND) dependency. If pin 3 is low, both X1 and X2 will be at the 0-state and both switches will be off. If pin 3 is high and pin 2 is high, X1 will be at the 1-state, X2 will be at the 0-state, and only the switch between pins 4 and 6 will be on. If pin 3 is high and pin 2 is low, X1 will be at the 0-state, X2 will be at the 1-state, and only the switch between pins 4 and 7 will be on.

6.4 SN75128 8-Channel Line Receiver

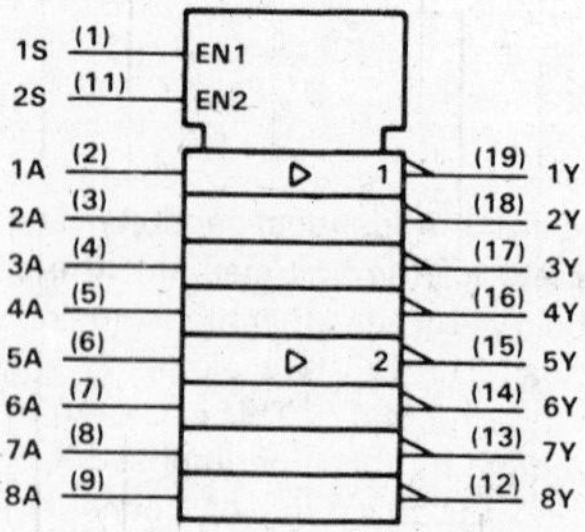

1S (1)
2S (11)
EN1
EN2
1A (2)
2A (3)
3A (4)
4A (5)
5A (6)
6A (7)
7A (8)
8A (9)
1
2
(19) 1Y
(18) 2Y
(17) 3Y
(16) 4Y
(15) 5Y
(14) 6Y
(13) 7Y
(12) 8Y

There are eight identical sections. The symbology is complete for the first element; the absence of any symbology for the next three elements indicates they are identical. Likewise the symbology is complete for the fifth element; the absence of any symbology for the next three elements indicates they are identical to the fifth.

Each element is shown to be an inverter with amplification (indicated by ▷). Taking TTL as a reference, this means that either the input is sensitive to lower level signals, or the output has greater drive capability than usual. The former applies in this case. Since neither the symbol for open-collector (⬦) or 3-state (▽) outputs is shown, the outputs are of the totem-pole type.

The top four outputs are shown to be affected by affecting input number 1, which is EN1, meaning they will be enabled if EN1 = 1 (pin 1 is high). See 4.9 for an explanation of EN dependency. If pin 1 is low, EN1 = 0 and the affected outputs will go to their inactive (high) levels. Similarly, the lower four outputs are controlled by pin 11.

6.5 SN75122 Triple Line Receivers

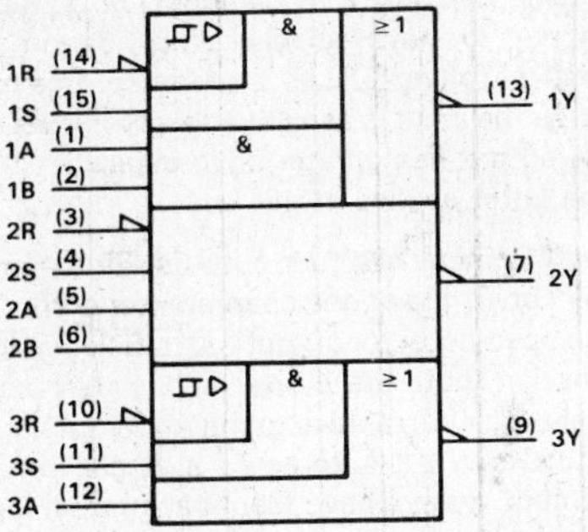

&
≥1
&
&
≥1
1R
1S
1A
1B
2R
2S
2A
2B
3R
3S
3A
(14)
(15)
(1)
(2)
(3)
(4)
(5)
(6)
(10)
(11)
(12)
(13)
1Y
(7)
2Y
(9)
3Y

There are two identical sections. The symbology is complete for the first section; the absence of any symbology for the next section indicates it is identical. Likewise the symbology is complete for the third section, which is similar, but not identical, to the first and second.

The top section may be considered to be an OR element (≥1) with two embedded ANDs (&), one of which has an active-low amplified input (▷) with hysteresis (⎍), pin 14. This is ANDed with pin 15 and the result is ORed with the AND of pins 1 and 2. The output of the OR, pin 13, is active-low.

The third section is identical to the first except that pin 12 has no input ANDed with it. Since neither the symbol for open-collector (◇) or 3-state (▽) outputs is shown, the outputs are of the totem-pole type.

6.6 DS8831 Quad Single-Ended or Dual Differential Line Drivers

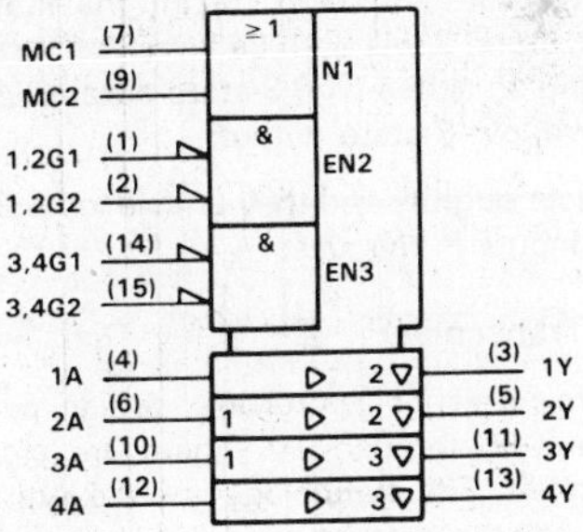

≥1
N1
&
EN2
&
EN3
MC1
MC2
1,2G1
1,2G2
3,4G1
3,4G2
(7)
(9)
(1)
(2)
(14)
(15)
1A
2A
3A
4A
(4)
(6)
(10)
(12)
1
1
2▽
2▽
3▽
3▽
(3)
(5)
(11)
(13)
1Y
2Y
3Y
4Y

There are four similar elements in the array. Each element is shown to be noninverting with amplification (indicated by ▷). Taking TTL as a reference, this means that either the input is sensitive to lower level signals or the output has greater drive capability than usual. The latter applies in this case. The outputs are shown by ▽ to be of the 3-state type.

The top two outputs are shown to be affected by affecting input number 2, which is EN2, meaning they will be enabled if EN2 = 1. See 4.9 for an explanation of EN dependency. If EN2 = 0, the affected outputs will go to their high-impedance (off) states. EN2 is the output of an AND gate (indicated by &) whose active-low inputs are pins 1 and 2. Both pins 1 and 2 must be low to enable pins 3 and 5. Likewise both pins 14 and 15 must be low to enable pins 11 and 13 through EN3.

Input pins 6 and 10 are shown to be affected by affecting input number 1, which is N1, meaning they will be negated if N1 = 1. See 4.5 for an explanation of N (negate or exclusive-OR) dependency. If N1 = 0, the input signals are not negated. N1 is the output of an OR gate (indicated by ≥1) whose active-high inputs are pins 7 and 9. Thus if either of these pins are high, then the second and third elements become inverters.

6.7 SN75113 Differential Line Drivers with Split 3-State Outputs

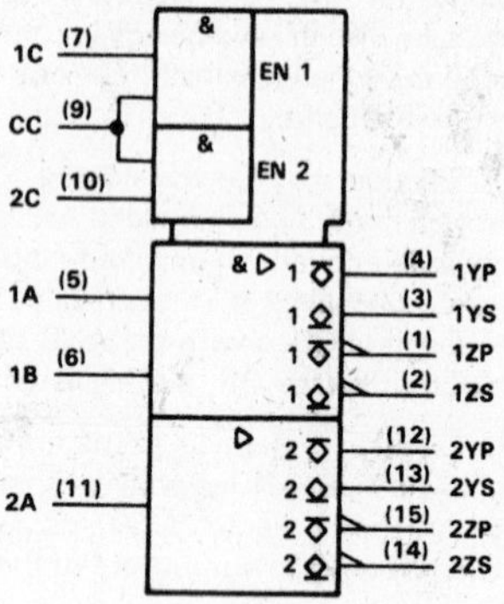

1C (7)
CC (9)
2C (10)
&
EN 1
&
EN 2
1A (5)
1B (6)
2A (11)
& ▷
▷
(4) 1YP
(3) 1YS
(1) 1ZP
(2) 1ZS
(12) 2YP
(13) 2YS
(15) 2ZP
(14) 2ZS

There are two similar elements in the array. The first is a 2-input AND element (indicated by &); the second has only a single input. Both elements are shown to have special amplification (indicated by ▷). Taking TTL as a reference, this means that either the input is sensitive to lower level signals, or the output has greater drive capability than usual. The latter applies in this case.

Each element has four outputs. Pins 4 and 3 are a pair consisting of one open-emitter output (◇̄) and one open-collector output (◇̲). Relative to the AND function, both are active high. Pins 1 and 2 are a similar pair but relative to the AND function, both are active low. All outputs of a single, unsubdivided element always have identical internal logic states determined by the function of the element except when otherwise indicated by an associated symbol or label inside the element. Here there is no such contrary indication. All four outputs are shown to be affected by affecting input number 1, which is EN1, meaning they will all be enabled if EN1 = 1. See 4.9 for an explanation of EN dependency. If EN1 = 0, all the affected outputs will be turned off. EN1 is the output of an AND gate (indicated by &) whose active-high inputs are pins 7 and 9. Both pins 7 and 9 must be high to enable the outputs of the top element. Assuming they are enabled and that pins 5 and 6 are both high, the internal state of all four outputs will be a 1. Pins 4 and 3 will both be high, pins 1 and 2 will both be low. The part is designed so that pins 3 and 4 may be connected together creating an active-high 3-state output. Likewise pins 1 and 2 may be connected together to create an active-low 3-state output.

All that has been said about the first element regarding its outputs and their enable inputs also applies to the second element. Pins 9 and 10 are the enable inputs in this case.

6.8 SN75163B Octal General-Purpose Interface Bus Transceiver

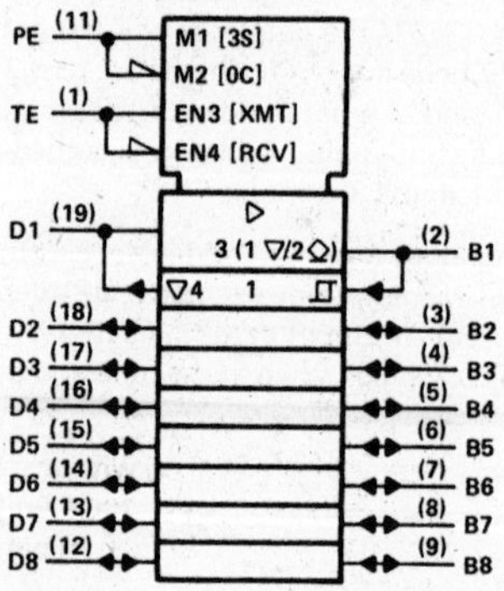

PE (11)
TE (1)
M1 [3S]
M2 [OC]
EN3 [XMT]
EN4 [RCV]
D1 (19)
▷
3 (1 ▽/2◇̲)
▽4 1 ⎍
(2) B1
D2 (18)
D3 (17)
D4 (16)
D5 (15)
D6 (14)
D7 (13)
D8 (12)
(3) B2
(4) B3
(5) B4
(6) B5
(7) B6
(8) B7
(9) B8

There are eight I/O ports on each side, pins 2 through 9 and 12 through 19. There are eight identical channels. The symbology is complete for the first channel; the absence of any symbology for the other channels indicates they are identical. The eight bidirectional channels each have amplification from left to right, that is, the outputs on the right have increased drive capability (indicated by ▷), and the inputs on the right all have hysteresis (indicated by ⎍).

The outputs on the left are shown to be 3-state outputs by the ▽. They are also shown to be affected by affecting input number 4, which is EN4, meaning they will be enabled if EN4 = 1 (pin 1 is low). See 4.9 for an explanation of EN dependency. If EN4 = 0 (pin 1 is high), the affected outputs will go to their high-impedance (off) states.

The labeling at pin 2, which applies to all the outputs on the right, is unusual because the outputs themselves have an unusual feature. The label includes both the symbol for a 3-state output (▽) and for an open-collector output (◇̲), separated by a slash indicating that these are alternatives.

The symbol for the 3-state output is shown to be affected by affecting input number 1, which is M1, meaning the ∇ label is valid when M1 = 1 (pin 11 is high), but is to be ignored when M1 = 0 (pin 11 is low). See 4.10 for an explanation of M (mode) dependency. Likewise the symbol for the open-collector output is shown to be affected by affecting input number 2, which is M2, meaning the ⍚ label is valid when M2 = 1 (pin 11 is low), but is to be ignored when M2 = 0 (pin 11 is high). These labels are enclosed in parentheses (used as in algebra); the numeral 3 indicates that in either case the output is affected by EN3. Thus the right-hand outputs will be off if pin 1 is low. It can now be seen that pin 1 is the direction control and pin 11 is used to determine whether the outputs are of the 3-state or open-collector variety.

6.9 SN75161B Octal IEEE Std 488 Interface Bus Transceiver

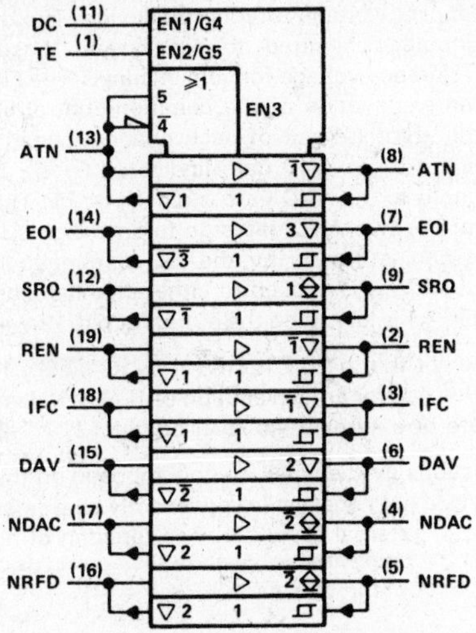

DC (11)
TE (1)
EN1/G4
EN2/G5
≥1
EN3
ATN (13)
EOI (14)
SRQ (12)
REN (19)
IFC (18)
DAV (15)
NDAC (17)
NRFD (16)
(8) ATN
(7) EOI
(9) SRQ
(2) REN
(3) IFC
(6) DAV
(4) NDAC
(5) NRFD

There are eight I/O ports on each side, pins 2 through 9 and 12 through 19. Pin 13 is not only an I/O port; the line running into the common-control block (see Figure 2) indicates that it also has control functions. Pins 1 and 11 are also controls. The eight bidirectional channels each have amplification from left to right, that is, the outputs on the right have increased drive capability (indicated by ▷), and the inputs on the right all have hysteresis (indicated by ⎍). All of the outputs are shown to be of the 3-state type by the ∇ symbol except for the outputs at pins 9, 4, and 5, which are shown to have passive pullups by the ⍚ symbol.

Starting with a typical I/O port, pin 18, the output portion is identified by an arrow indicating right-to-left signal flow and the three-state output symbol (∇). This output is shown to be affected by affecting input number 1, which is EN1, meaning it will be enabled as an output if EN1 = 1 (pin 11 is high). See 4.9 for an explanation of EN dependency. If pin 11 is low, EN1 = 0 and the output at pin 18 will be in its high-impedance (off) state. This also applies to the 3-state outputs at pins 13 and 19 and to the passive-pullup output at pin 9. On the other hand, the outputs at pins 8, 2, 3, and 12 all are affected by the complement of EN1. This is indicated by the bar over the 1 at each of those outputs. They are enabled only when pin 11 is low. Thus one function of pin 11 is to serve as direction control for the first, third, fourth, and fifth channels.

Similarly it can be seen that pin 1 serves as direction control for the sixth, seventh, and eighth channels. If pin 1 is high, transmission will be from left to right in the sixth channel, right to left in the seventh and eighth. These transmissions are reversed if pin 1 is low.

The direction control for the second channel, EN3, is more complex. EN3 is the output of an OR (≥ 1) function. One of the inputs to this OR is the active-high signal on pin 13. This signal is shown to be affected at the input to the OR gate by affecting input number 5, which is G5, meaning that pin 13 is ANDed with pin 1 before entering the OR gate. See 4.2 for an explanation of G (AND) dependency. The other input to the OR is the active-low signal on pin 13. This signal is ANDed with the complement of pin 11 before entering the OR gate. This is indicated by the G4 at pin 1 and the 4 with a bar over it at pin 13. Thus for EN3 to stand at the 1 state, which would enable transmission from pin 14 to pin 7, both pins 13 and 1 must be high or both pins 13 and 11 must be low.

6.10 SN5520 Dual-Channel Sense Amplifier with Complementary Outputs

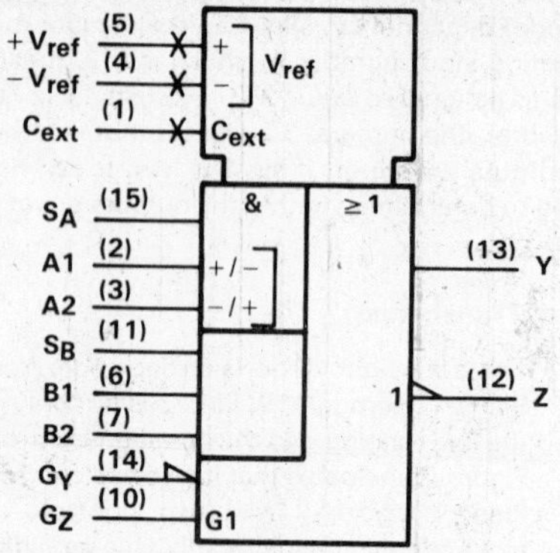
+Vref
(5)
−Vref
(4)
Cext
(1)
+
−
Vref
Cext
SA
(15)
A1
(2)
A2
(3)
SB
(11)
B1
(6)
B2
(7)
GY
(14)
GZ
(10)
&
+/−
−/+
≥1
G1
1
(13)
Y
(12)
Z

There are two input channels. The symbology is complete for the first channel; the absense of any symbology for the second channel indicates it is identical.

The square bracket around the input lines connected to pins 2 and 3 shows they constitute a single input pair. The +/− by one line and the −/+ by the other indicate that these are differential inputs, and in this case, only the magnitude and not the polarity of the applied voltage affects the device. The differential inputs connected to pins 5 and 4 supply the reference voltage for both channels; see Figure 2 for an explanation of the common-control block. The differential input of each channel constitutes one input of an AND gate (indicated by &). The other input to the AND gate is pin 15 or 11. The outputs of the two AND gates go to an OR gate (indicated by ≥1). Pin 14 is also an input to this OR gate and is shown by the ⊳ symbol to be active when low. Thus if a voltage greater in magnitude than V_{ref} is applied to either channel along with a high-level voltage to the respective "S" input, or if pin 14 is low, then the output at pin 13 will be high.

Pin 12 is shown to be the complement of pin 13 except that pin 12 is shown by its label "1" to also be affected by pin 10 (G1) in an AND relationship. See 4.2 for an explanation of G (AND) dependency. Thus if pin 10 is low, it imposes the internal 0 state or external high level on pin 12.

Pins 5, 4, and 1 are all shown to be nonlogic connections by the small X superimposed on those lines. The function of such a connection is indicated briefly and not necessarily very exactly by a small amount of text within the symbol. Pin 1 is a connection for an external capacitor. The function of pins 5 and 4 was explained above.

6.11 SN75500E AC Plasma Display Driver with CMOS-Compatible Inputs

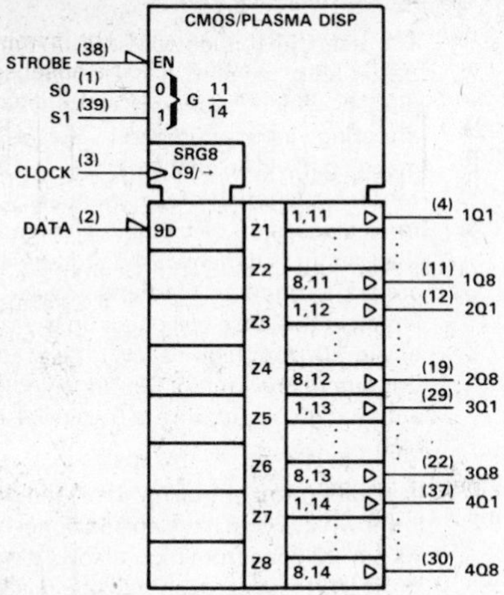
CMOS/PLASMA DISP
STROBE
(38)
EN
S0
(1)
S1
(39)
0
1
G
11
14
SRG8
CLOCK
(3)
C9/→
DATA
(2)
9D
Z1
Z2
Z3
Z4
Z5
Z6
Z7
Z8
1,11
8,11
1,12
8,12
1,13
8,13
1,14
8,14
(4)
1Q1
(11)
1Q8
(12)
2Q1
(19)
2Q8
(29)
3Q1
(22)
3Q8
(37)
4Q1
(30)
4Q8

The heart of this device and its symbol is an 8-bit shift register. It has a single D input, pin 2, which is shown to be affected by affecting input number 9, which is C9, meaning it will be enabled if C9 = 1. See 4.8 for an explanation of C dependency and 5.0 for a discussion of bistable elements. Since the C input is dynamic, the storage elements are edge-triggered flip-flops. While C9 = 1, which in this case will occur on the transition of pin 3 from low to high, the state of the D input will be stored. Pin 2 is shown to be active low so to store a 1, pin 2 must be low.

In addition to controlling the D input, pin 3 is shown by /→ to have an additional function. As pin 3 goes from low to high, data stored in the shift register is shifted one position. The right-pointing arrow means that the data is shifted away from the control block (down).

On the right side of the symbol an abbreviation technique has been used that is practical only when the internal labels and the pin numbers are both consecutive. Thus it should be clear that the input of the element whose output is pin 5 is affected by affecting input number 2, just as the input of the element whose output is pin 4 is affected by affecting input number 1. Affecting inputs 1 through 8 are Z inputs (Z1 through Z8), which means their signals are tranferred directly to the output elements. See 4.6 for an explanation of Z dependency.

The inputs of the 32 implicitly shown output elements are also shown to be affected by affecting inputs numbers 11, 12, 13, and 14 in four blocks of eight each. These inputs will be found in the common control block preceded by a letter G and a brace. The brace is called the binary grouping symbol. It is equivalent to a decoder with outputs in this case driving four G inputs (G11, G12, G13, and G14). The weights of the inputs to the coder are shown to be 2^0 and 2^1 for pins 1 and 39, respectively. The decoder has four outputs corresponding to the four possible sums of the weights of the activated decoder inputs. If pins 1 and 39 are both low, the sum of the weights = 0 and G11 = 1. If pin 1 is low while pin 39 is high, the sum = 2 and G13 = 1 and so forth. G indicates AND dependency, see 4.2. Only one of the four affecting G inputs at a time can take on the 1 state. The block of eight output elements affected by that G input are enabled; the 0 state is imposed on the other 24 output elements and externally those output pins are low.

Because of their high-current, high-voltage characteristics, the outputs are labeled with the amplification symbol ▷. All the outputs share a common EN input, pin 38. See Figure 2 for an explanation of the common control block. When EN = 0 (pin 38 is high), the outputs take on their internal 0 states. Being active high, that means they are forced low.

6.12 SN75551 Electroluminescent Row Driver with CMOS-Compatible Inputs

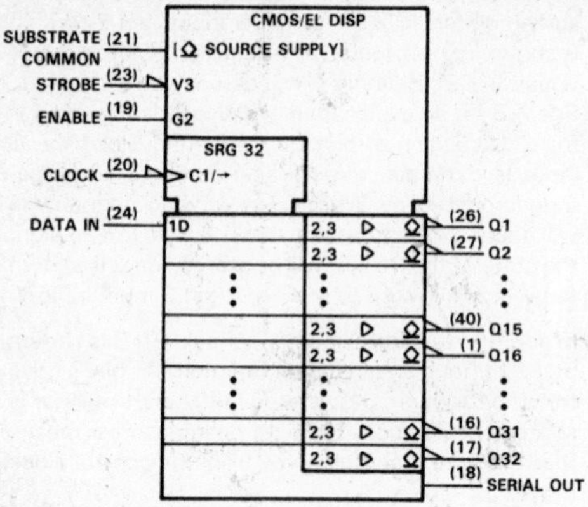
CMOS/EL DISP
SUBSTRATE COMMON (21)
[◇ SOURCE SUPPLY]
STROBE (23)
V3
ENABLE (19)
G2
SRG 32
CLOCK (20)
C1/→
DATA IN (24)
1D
2,3
(26) Q1
(27) Q2
(40) Q15
(1) Q16
(16) Q31
(17) Q32
(18) SERIAL OUT

The heart of this device and its symbol is a 32-bit shift register. It has a single D input, pin 24, which is shown to be affected by affecting input number 1, which is C1, meaning it will be enabled if C1 = 1. See 4.8 for an explanation of C dependency and 5.0 for a discussion of bistable elements. Since the C input is dynamic, the storage elements are edge-triggered flip-flops. While C1 = 1, which in this case will occur on the transition of pin 20 from high to low, the state of the D input will be stored. Pin 24 is shown to be active high so to store a 1, pin 24 must be high.

In addition to controlling the D input, pin 20 is shown by /→ to have an additional function. As pin 20 goes from high to low, data stored in the shift register is shifted one position. The right-pointing arrow means that the data is shifted away from the control block (down). The internal inputs of the output buffers are all shown to be affected by affecting inputs 2 and 3. Affecting input 2 is G2, meaning that pin 19 is ANDed with each of the internal register outputs, which are the buffer inputs. If pin 19 is high, the affected buffer inputs are enabled. If pin 19 is low, the 0 state is imposed on the affected buffer inputs. See 4.2 for an explanation of G (AND) dependency. Affecting input 3 is V3, meaning that pin 23 (active low) is ORed with each of the internal register outputs. If pin 23 is high, V3 = 0 and the affected buffer inputs are enabled. If pin 23 is low, V3 = 1 and the 1 state is imposed on the affected buffer inputs. See 4.4 for an explanation of V (OR) dependency. The effect of V3 is taken into account after that of G2 because of the order in which the labels appear. This means that the imposition of the 1 state on the internal buffer inputs by pin 23 would take precedence over the imposition of the 0 state by pin 19 in case both inputs were active. Pin 18 is shown to be an output directly from the thirty-second stage of the shift register. Pins 19 and 23 do not affect this output.

An abbreviation technique has been used for the shift register elements and associated the output lines. This technique is practical only when the pin numbers and pin names are both consecutive.

The symbol ◇ designates an n-p-n open-collector or similar output. In this device, the outputs are actually open-drain n-channel field-effect transistors. Instead of being grounded, the sources of these transistors are all connected to pin 21. This pin is used as an input to control the output voltage.

TI Sales Offices

ALABAMA: Huntsville (205) 837-7530.

ARIZONA: Phoenix (602) 995-1007;
Tucson (602) 624-3276.

CALIFORNIA: Irvine (714) 660-8187;
Sacramento (916) 929-1521;
San Diego (619) 278-9601;
Santa Clara (408) 980-9000;
Torrance (213) 217-7010;
Woodland Hills (818) 704-7759.

COLORADO: Aurora (303) 368-8000.

CONNECTICUT: Wallingford (203) 269-0074.

FLORIDA: Ft. Lauderdale (305) 973-8502;
Maitland (305) 660-4600; **Tampa** (813) 870-6420.

GEORGIA: Norcross (404) 662-7900.

ILLINOIS: Arlington Heights (312) 640-2925.

INDIANA: Ft. Wayne (219) 424-5174;
Indianapolis (317) 248-8555.

IOWA: Cedar Rapids (319) 395-9550.

MARYLAND: Baltimore (301) 944-8600.

MASSACHUSETTS: Waltham (617) 895-9100.

MICHIGAN: Farmington Hills (313) 553-1500;
Grand Rapids (616) 957-4200.

MINNESOTA: Eden Prairie (612) 828-9300.

MISSOURI: Kansas City (816) 523-2500;
St. Louis (314) 569-7600.

NEW JERSEY: Iselin (201) 750-1050.

NEW MEXICO: Albuquerque (505) 345-2555.

NEW YORK: East Syracuse (315) 463-9291;
Melville (516) 454-6600; **Pittsford** (716) 385-6770;
Poughkeepsie (914) 473-2900.

NORTH CAROLINA: Charlotte (704) 527-0930;
Raleigh (919) 876-2725.

OHIO: Beachwood (216) 464-6100;
Dayton (513) 258-3877.

OREGON: Beaverton (503) 643-6758.

PENNSYLVANIA: Ft. Washington (215) 643-6450.

PUERTO RICO: Hato Rey (809) 753-8700

TEXAS: Austin (512) 250-7655;
Houston (713) 778-6592; **Richardson** (214) 680-5082;
San Antonio (512) 496-1779.

UTAH: Murray (801) 266-8972.

VIRGINIA: Fairfax (703) 849-1400.

WASHINGTON: Redmond (206) 881-3080.

WISCONSIN: Brookfield (414) 785-7140.

CANADA: Nepean, Ontario (613) 726-1970;
Richmond Hill, Ontario (416) 884-9181;
St. Laurent, Quebec (514) 335-8392.

TI Regional Technology Centers

CALIFORNIA: Irvine (714) 660-8140,
Santa Clara (408) 748-2220.

GEORGIA: Norcross (404) 662-7945.

ILLINOIS: Arlington Heights (312) 640-2909.

MASSACHUSETTS: Waltham (617) 895-9197.

TEXAS: Richardson (214) 680-5066.

CANADA: Nepean, Ontario (613) 726-1970

Customer Response Center

TOLL FREE: (800) 232-3200

OUTSIDE USA: (214) 995-6611
(8:00 a.m. — 5:00 p.m. CST)

TI Distributors

TI AUTHORIZED DISTRIBUTORS IN USA
- **Arrow Electronics**
- **General Radio Supply Company**
- **Graham Electronics**
- **Hall-Mark Electronics**
- **Kierulff Electronics**
- **Marshall Industries**
- **Milgray Electronics**
- **Newark Electronics**
- **Schweber Electronics**
- **Time Electronics**
- **Wyle Laboratories**
- **Zeus Component, Inc. (Military)**

TI AUTHORIZED DISTRIBUTORS IN CANADA
- **Arrow Electronics Canada**
- **Future Electronics**

TI AUTHORIZED DISTRIBUTORS IN USA
—OBSOLETE PRODUCT ONLY—
Rochester Electronics, Inc.
Newburyport, Massachusetts
(617) 462-9332

ALABAMA: Arrow (205) 837-6955;
Hall-Mark (205) 837-8700; Kierulff (205) 883-6070;
Marshall (205) 881-9235; Schweber (205) 895-0480.

ARIZONA: Arrow (602) 968-4800;
Hall-Mark (602) 437-1200; Kierulff (602) 437-0750;
Marshall (602) 968-6181; Schweber (602) 997-4874;
Wyle (602) 866-2888.

CALIFORNIA: Los Angeles/Orange County:
Arrow (818) 701-7500, (714) 838-5422;
Hall-Mark (818) 716-7300, (714) 669-4700,
(213) 217-8400; Kierulff (213) 725-0325, (714) 731-5711,
(714) 220-6300; (818) 407-2500;
Marshall (818) 407-0101, (818) 442-7204,
(714) 458-5395; Schweber (818) 999-4702;
(714) 863-0200; (213) 327-8409; Wyle (213) 322-8100,
(818) 880-9001, (714) 863-9953; Zeus (714) 632-6880;
Sacramento: Hall-Mark (916) 722-8600;
Marshall (916) 635-9700; Schweber (916) 929-9732;
Wyle (916) 638-5282;
San Diego: Arrow (619) 565-4800;
Hall-Mark (619) 268-1201; Kierulff (619) 278-2112;
Marshall (619) 578-9600; Schweber (619) 450-0454;
Wyle (619) 565-9171;
San Francisco Bay Area: Arrow (408) 745-6600;
(415) 487-4600; Hall-Mark (408) 946-0900;
Kierulff (408) 971-2600; Marshall (408) 943-4600;
Schweber (408) 946-7171; Wyle (408) 727-2500;
Zeus (408) 998-5121.

COLORADO: Arrow (303) 696-1111;
Hall-Mark (303) 790-1662; Kierulff (303) 790-4444;
Marshall (303) 451-8444; Schweber (303) 799-0258;
Wyle (303) 457-9953.

CONNECTICUT: Arrow (203) 265-7741;
Hall-Mark (203) 269-0100; Kierulff (203) 265-1115;
Marshall (203) 265-3822; Milgray (203) 795-0714;
Schweber (203) 748-7080.

FLORIDA: Ft. Lauderdale: Arrow (305) 429-8200;
Hall-Mark (305) 971-9280; Kierulff (305) 486-4004;
Marshall (305) 928-0661; Schweber (305) 977-7511;
Orlando: Arrow (305) 725-1480;
Hall-Mark (305) 855-4020; Marshall (305) 841-1878;
Milgray (305) 647-5747; Schweber (305) 331-7555;
Zeus (305) 365-3000;
Tampa: Arrow (813) 576-8995;
Hall-Mark (813) 530-4543; Kierulff (813) 576-1966.

GEORGIA: Arrow (404) 449-8252;
Hall-Mark (404) 447-8000; Kierulff (404) 447-5252;
Marshall (404) 923-5750; Schweber (404) 449-9170.

ILLINOIS: Arrow (312) 397-3440;
Hall-Mark (312) 860-3800; Kierulff (312) 250-0500;
Marshall (312) 490-0155; Newark (312) 784-5100;
Schweber (312) 364-3750.

INDIANA: Indianapolis: Arrow (317) 243-9353;
Graham (317) 634-8202; Hall-Mark (317) 872-8875;
Marshall (317) 297-0483;
Ft. Wayne: Graham (219) 423-3422.

IOWA: Arrow (319) 395-7230;
Schweber (319) 373-1417.

KANSAS: Kansas City: Arrow (913) 541-9542;
Hall-Mark (913) 888-4747; Marshall (913) 492-3121;
Schweber (913) 492-2921.

MARYLAND: Arrow (301) 995-0003;
Hall-Mark (301) 988-9800; Kierulff (301) 840-1155;
Milgray (301) 995-6169; Marshall (301) 840-9450;
Schweber (301) 840-5900; Zeus (301) 997-1118.

MASSACHUSETTS: Arrow (617) 933-8130;
Hall-Mark (617) 667-0902; Kierulff (617) 667-8331;
Marshall (617) 658-0810; Schweber (617) 275-5100,
(617) 657-0760; Time (617) 532-6200;
Zeus (617) 863-8800.

MICHIGAN: Detroit: Arrow (313) 971-8220;
Marshall (313) 525-5850; Newark (313) 967-0600;
Schweber (313) 525-8100;
Grand Rapids: Arrow (616) 243-0912.

MINNESOTA: Arrow (612) 830-1800;
Hall-Mark (612) 941-2600; Kierulff (612) 941-7500;
Marshall (612) 559-2211; Schweber (612) 941-5280.

MISSOURI: St. Louis: Arrow (314) 567-6888;
Hall-Mark (314) 291-5350; Kierulff (314) 997-4956;
Schweber (314) 739-0526.

NEW HAMPSHIRE: Arrow (603) 668-6968;
Schweber (603) 625-2250.

NEW JERSEY: Arrow (201) 575-5300,
(609) 596-8000; General Radio (609) 964-8560;
Hall-Mark (201) 575-4415, (609) 235-1900;
Kierulff (201) 575-6750, (609) 235-1444;
Marshall (201) 882-0320, (609) 234-9100;
Milgray (609) 983-5010; Schweber (201) 227-7880.

NEW MEXICO: Arrow (505) 243-4566.

NEW YORK: Long Island: Arrow (516) 231-1000;
Hall-Mark (516) 737-0600; Marshall (516) 273-2053;
Milgray (516) 420-9800; Schweber (516) 334-7474;
Zeus (914) 937-7400;
Rochester: Arrow (716) 427-0300;
Marshall (716) 235-7620; Schweber (716) 424-2222.
Syracuse: Marshall (607) 798-1611.

NORTH CAROLINA: Arrow (919) 876-3132,
(919) 725-8711; Hall-Mark (919) 872-0712;
Kierulff (919) 872-8410; Marshall (919) 878-9882;
Schweber (919) 876-0000.

OHIO: Cleveland: Arrow (216) 248-3990;
Hall-Mark (216) 349-4632; Kierulff (216) 831-5222;
Marshall (216) 248-1788; Schweber (216) 464-2970.
Columbus: Arrow (614) 885-8362;
Hall-Mark (614) 888-3313;
Dayton: Arrow (513) 435-5563;
Graham (513) 435-8660; Kierulff (513) 439-0045;
Marshall (513) 236-8088; Schweber (513) 439-1800.

OKLAHOMA: Arrow (918) 665-7700;
Kierulff (918) 252-7537; Schweber (918) 622-8000.

OREGON: Arrow (503) 684-1690;
Kierulff (503) 641-9153; Wyle (503) 640-6000;
Marshall (503) 644-5050.

PENNSYLVANIA: Arrow (412) 856-7000,
(215) 928-1800; General Radio (215) 922-7037;
Schweber (215) 441-0600, (412) 782-1600.

TEXAS: Austin: Arrow (512) 835-4180;
Hall-Mark (512) 258-8848; Kierulff (512) 835-2090;
Marshall (512) 837-1991; Schweber (512) 458-8253;
Wyle (512) 834-9957;
Dallas: Arrow (214) 380-6464;
Hall-Mark (214) 553-4300; Kierulff (214) 343-2400;
Marshall (214) 233-5200; Schweber (214) 661-5010;
Wyle (214) 235-9953; Zeus (214) 783-7010;
Houston: Arrow (713) 530-4700;
Hall-Mark (713) 781-6100; Kierulff (713) 530-7030;
Marshall (713) 895-9200; Schweber (713) 784-3600;
Wyle (713) 879-9953.

UTAH: Arrow (801) 972-0404;
Hall-Mark (801) 972-1008; Kierulff (801) 973-6913;
Marshall (801) 485-1551; Wyle (801) 974-9953;

WASHINGTON: Arrow (206) 643-4800;
Kierulff (206) 575-4420; Wyle (206) 453-8300;
Marshall (206) 747-9100.

WISCONSIN: Arrow (414) 792-0150;
Hall-Mark (414) 797-7844; Kierulff (414) 784-8160;
Marshall (414) 797-8400; Schweber (414) 784-9020.

CANADA: Calgary: Future (403) 235-5325;
Edmonton: Future (403) 438-2858;
Montreal: Arrow Canada (514) 735-5511;
Future (514) 694-7710;
Ottawa: Arrow Canada (613) 226-6903;
Future (613) 820-8313;
Quebec City: Arrow Canada (418) 687-4231;
Toronto: Arrow Canada (416) 661-0220;
Future (416) 638-4771;
Vancouver: Future (604) 294-1166
Winnipeg: Future (204) 339-0554

TEXAS INSTRUMENTS

BR

TI Worldwide Sales Offices

ALABAMA: Huntsville: 500 Wynn Drive, Suite 514, Huntsville, AL 35805, (205) 837-7530.

ARIZONA: Phoenix: 8825 N. 23rd Ave., Phoenix, AZ 85021, (602) 995-1007.

CALIFORNIA: Irvine: 17891 Cartwright Rd., Irvine, CA 92714, (714) 660-8187; **Sacramento:** 1900 Point West Way, Suite 171, Sacramento, CA 95815, (916) 929-1521; **San Diego:** 4333 View Ridge Ave., Suite B., San Diego, CA 92123, (619) 278-9601; **Santa Clara:** 5353 Betsy Ross Dr., Santa Clara, CA 95054, (408) 980-9000; **Torrance:** 690 Knox St., Torrance, CA 90502, (213) 217-7010; **Woodland Hills:** 21220 Erwin St., Woodland Hills, CA 91367, (818) 704-7759.

COLORADO: Aurora: 1400 S. Potomac Ave., Suite 101, Aurora, CO 80012, (303) 368-8000.

CONNECTICUT: Wallingford: 9 Barnes Industrial Park Rd., Barnes Industrial Park, Wallingford, CT 06492, (203) 269-0074.

FLORIDA: Ft. Lauderdale: 2765 N.W. 62nd St., Ft. Lauderdale, FL 33309, (305) 973-8502; **Maitland:** 2601 Maitland Center Parkway, Maitland, FL 32751, (305) 660-4600; **Tampa:** 5010 W. Kennedy Blvd., Suite 101, Tampa, FL 33609, (813) 870-6420.

GEORGIA: Norcross: 5515 Spalding Drive, Norcross, GA 30092, (404) 662-7900

ILLINOIS: Arlington Heights: 515 W. Algonquin, Arlington Heights, IL 60005, (312) 640-2925.

INDIANA: Ft. Wayne: 2020 Inwood Dr., Ft. Wayne, IN 46815, (219) 424-5174; **Indianapolis:** 2346 S. Lynhurst, Suite J-400, Indianapolis, IN 46241, (317) 248-8555.

IOWA: Cedar Rapids: 373 Collins Rd. NE, Suite 200, Cedar Rapids, IA 52402, (319) 395-9550.

MARYLAND: Baltimore: 1 Rutherford Pl., 7133 Rutherford Rd., Baltimore, MD 21207, (301) 944-8600.

MASSACHUSETTS: Waltham: 504 Totten Pond Rd., Waltham, MA 02154, (617) 895-9100.

MICHIGAN: Farmington Hills: 33737 W. 12 Mile Rd., Farmington Hills, MI 48018, (313) 553-1500.

MINNESOTA: Eden Prairie: 11000 W. 78th St., Eden Prairie, MN 55344 (612) 828-9300.

MISSOURI: Kansas City: 8080 Ward Pkwy., Kansas City, MO 64114, (816) 523-2500; **St. Louis:** 11816 Borman Drive, St. Louis, MO 63146, (314) 569-7600.

NEW JERSEY: Iselin: 485E U.S. Route 1 South, Parkway Towers, Iselin, NJ 08830 (201) 750-1050

NEW MEXICO: Albuquerque: 2820-D Broadbent Pkwy NE, Albuquerque, NM 87107, (505) 345-2555.

NEW YORK: East Syracuse: 6365 Collamer Dr., East Syracuse, NY 13057, (315) 463-9291; **Endicott:** 112 Nanticoke Ave., P.O. Box 618, Endicott, NY 13760, (607) 754-3900; **Melville:** 1 Huntington Quadrangle, Suite 3C10, P.O. Box 2936, Melville, NY 11747, (516) 454-6600; **Pittsford:** 2851 Clover St., Pittsford, NY 14534, (716) 385-6770; **Poughkeepsie:** 385 South Rd., Poughkeepsie, NY 12601, (914) 473-2900.

NORTH CAROLINA: Charlotte: 8 Woodlawn Green, Woodlawn Rd., Charlotte, NC 28210, (704) 527-0930; **Raleigh:** 2809 Highwoods Blvd., Suite 100, Raleigh, NC 27625, (919) 876-2725.

OHIO: Beachwood: 23408 Commerce Park Rd., Beachwood, OH 44122, (216) 464-6100; **Dayton:** Kingsley Bldg., 4124 Linden Ave., Dayton, OH 45432, (513) 258-3877.

OREGON: Beaverton: 6700 SW 105th St., Suite 110, Beaverton, OR 97005, (503) 643-6758.

PENNSYLVANIA: Ft. Washington: 260 New York Dr., Ft. Washington, PA 19034, (215) 643-6450; **Coraopolis:** 420 Rouser Rd., 3 Airport Office Park, Coraopolis, PA 15108, (412) 771-8550.

PUERTO RICO: Hato Rey: Mercantil Plaza Bldg., Suite 505, Hato Rey, PR 00919, (809) 753-8700.

TEXAS: Austin: P.O. Box 2909, Austin, TX 78769, (512) 250-7655; **Richardson:** 1001 E. Campbell Rd., Richardson, TX 75080, (214) 680-5082; **Houston:** 9100 Southwest Frwy., Suite 237, Houston, TX 77036, (713) 778-6592; **San Antonio:** 1000 Central Parkway South, San Antonio, TX 78232, (512) 496-1779.

UTAH: Murray: 5201 South Green SE, Suite 200, Murray, UT 84107, (801) 266-8972.

VIRGINIA: Fairfax: 2750 Prosperity, Fairfax, VA 22031, (703) 849-1400.

WASHINGTON: Redmond: 5010 148th NE, Bldg B, Suite 107, Redmond, WA 98052, (206) 881-3080.

WISCONSIN: Brookfield: 450 N. Sunny Slope, Suite 150, Brookfield, WI 53005, (414) 785-7140.

CANADA: Nepean: 301 Moodie Drive, Mallorn Center, Nepean, Ontario, Canada, K2H9C4, (613) 726-1970. **Richmond Hill:** 280 Centre St. E., Richmond Hill L4C1B1, Ontario, Canada (416) 884-9181; **St. Laurent:** Ville St. Laurent Quebec, 9460 Trans Canada Hwy., St. Laurent, Quebec, Canada H4S1R7, (514) 335-8392.

ARGENTINA: Texas Instruments Argentina S.A.I.C.F.: Esmeralda 130, 15th Floor, 1035 Buenos Aires, Argentina, 1 + 394-3008.

AUSTRALIA (& NEW ZEALAND): Texas Instruments Australia Ltd.: 6-10 Talavera Rd., North Ryde (Sydney), New South Wales, Australia 2113, 2 + 887-1122; 5th Floor, 418 St. Kilda Road, Melbourne, Victoria, Australia 3004, 3 + 267-4677; 171 Philip Highway, Elizabeth, South Australia 5112, 8 + 255-2066.

AUSTRIA: Texas Instruments Ges.m.b.H.: Industriestrabe B/16, A-2345 Brunn/Gebirge, 2236-846210.

BELGIUM: Texas Instruments N.V. Belgium S.A.: Mercure Centre, Raketstraat 100, Rue de la Fusee, 1130 Brussels, Belgium, 2/720.80.00.

BRAZIL: Texas Instruments Electronicos do Brasil Ltda.: Rua Paes Leme, 524-7 Andar Pinheiros, 05424 Sao Paulo, Brazil, 0815-6166.

DENMARK: Texas Instruments A/S, Mairelundvej 46E, DK-2730 Herlev, Denmark, 2 - 91 74 00.

FINLAND: Texas Instruments Finland OY: Teollisuuskatu 19D 00511 Helsinki 51, Finland, (90) 701-3133.

FRANCE: Texas Instruments France: Headquarters and Prod. Plant, BP 05, 06270 Villeneuve-Loubet, (93) 20-01-01; Paris Office, BP 67 8-10 Avenue Morane-Saulnier, 78141 Velizy-Villacoublay, (3) 946-97-12; Lyon Sales Office, L'Oree D'Ecully, Batiment B, Chemin de la Forestiere, 69130 Ecully, (7) 833-04-40; Strasbourg Sales Office, Le Sebastopol 3, Quai Kleber, 67055 Strasbourg Cedex, (88) 22-12-66; Rennes, 23-25 Rue du Puits Mauger, 35100 Rennes, (99) 31-54-86; Toulouse Sales Office, Le Peripole—2, Chemin du Pigeonnier de la Cepiere, 31100 Toulouse, (61) 44-18-19; Marseille Sales Office, Noilly Paradis—146 Rue Paradis, 13006 Marseille, (91) 37-25-30.

TEXAS INSTRUMENTS

GERMANY (Fed. Republic of Germany): Texas Instruments Deutschland GmbH: Haggertystrasse 1, D-8050 Freising, 8161 + 80-4591; Kurfuerstendamm 195/196, D-1000 Berlin 15, 30 + 882-7365; III, Hagen 43/Kibbelstrasse, .19, D-4300 Essen, 201-24250; Frankfurter Allee 6-8, D-6236 Eschborm 1, 06196 + 8070; Hamburgerstrasse 11, D-2000 Hamburg 76, 040 + 220-1154, Kirchhorsterstrasse 2, D-3000 Hannover 51, 511 + 648021; Maybachstrabe 11, D-7302 Ostfildern 2-Nelingen, 711 + 547001; Mixikoring 19, D-2000 Hamburg 60, 40 + 637 + 0061; Postfach 1309, Roonstrasse 16, D-5400 Koblenz, 261 + 35044.

HONG KONG (+ PEOPLES REPUBLIC OF CHINA): Texas Instruments Asia Ltd., 8th Floor, World Shipping Ctr., Harbour City, 7 Canton Rd., Kowloon, Hong Kong, 3 + 722-1223.

IRELAND: Texas Instruments (Ireland) Limited: Brewery Rd., Stillorgan, County Dublin, Eire, 1 831311.

ITALY: Texas Instruments Semiconduttori Italia Spa: Viale Delle Scienze, 1, 02015 Cittaducale (Rieti), Italy, 746 694.1; Via Salaria KM 24 (Palazzo Cosma), Monterotondo Scalo (Rome), Italy, 6 + 9003241; Viale Europa, 38-44, 20093 Cologno Monzese (Milano), 2 2532541; Corso Svizzera, 185, 10100 Torino, Italy, 11 774545; Via J. Barozzi 6, 40100 Bologna, Italy, 51 355851.

JAPAN: Texas Instruments Asia Ltd.: 4F Aoyama Fuji Bldg., 6-12, Kita Aoyama 3-Chome, Minato-ku, Tokyo, Japan 107, 3-498-2111; Osaka Branch, 5F, Nissho Iwai Bldg., 30 Imabashi 3- Chome, Higashi-ku, Osaka, Japan 541, 06-204-1881; Nagoya Branch, 7F Daini Toyota West Bldg., 10-27, Meieki 4-Chome, Nakamura-ku Nagoya, Japan 450, 52-583-8691.

KOREA: Texas Instruments Supply Co.: 3rd Floor, Samon Bldg., Yuksam-Dong, Gangnam-ku, 135 Seoul, Korea, 2 + 462-8001.

MEXICO: Texas Instruments de Mexico S.A.: Mexico City, AV Reforma No. 450 — 10th Floor, Mexico, D.F., 06600, 5 + 514-3003.

MIDDLE EAST: Texas Instruments: No. 13, 1st Floor Mannai Bldg., Diplomatic Area, P.O. Box 26335, Manama Bahrain, Arabian Gulf, 973 + 274681.

NETHERLANDS: Texas Instruments Holland B.V., P.O. Box 12995, (Bullewijk) 1100 CB Amsterdam, Zuid-Oost, Holland 20 + 5602911.

NORWAY: Texas Instruments Norway A/S: PB106, Refstad 131, Oslo 1, Norway, (2) 155090.

PHILIPPINES: Texas Instruments Asia Ltd.: 14th Floor, Ba- Lepanto Bldg., 8747 Paseo de Roxas, Makati, Metro Manila, Philippines, 2 + 8188987.

PORTUGAL: Texas Instruments Equipamento Electronico (Portugal), Lda.: Rua Eng. Frederico Ulrich, 2650 Moreira Da Maia, 4470 Maia, Portugal, 2-948-1003.

SINGAPORE (+ INDIA, INDONESIA, MALAYSIA, THAILAND): Texas Instruments Asia Ltd.: 12 Lorong Bakar Batu, Unit 01-02, Kolam Ayer Industrial Estate, Republic of Singapore, 747-2255.

SPAIN: Texas Instruments Espana, S.A.: C/Jose Lazaro Galdiano No. 6, Madrid 16, 1/458.14.58.

SWEDEN: Texas Instruments International Trade Corporation (Sverigefilialen): Box 39103, 10054 Stockholm, Sweden, 8 - 235480.

SWITZERLAND: Texas Instruments, Inc., Reidstrasse 6, CH-8953 Dietikon (Zuerich) Switzerland, 1-740 2220.

TAIWAN: Texas Instruments Supply Co.: Room 903, 205 Tun Hwan Rd., 71 Sung-Kiang Road, Taipei, Taiwan, Republic of China, 2 + 521-9321.

UNITED KINGDOM: Texas Instruments Limited: Manton Lane, Bedford, MK41 7PA, England, 0234 67466; St. James House, Wellington Road North, Stockport, SK4 2RT, England, 61 + 442-7162.

BM

To correct your address or add an associate to TI's Linear Interface products mailing list, complete and return this card.

NAME

TITLE

COMPANY

ADDRESS & M/S

CITY STATE ZIP -

PHONE - EXTENSION

SSA03IIY705M

Expires 10/31/88

To correct your address or add an associate to TI's Linear Interface products mailing list, complete and return this card.

NAME

TITLE

COMPANY

ADDRESS & M/S

CITY STATE ZIP -

PHONE - EXTENSION

SSA03IIY705M

Expires 10/31/88

Copyright © 1987, Texas Instruments Incorporated

PLACE
STAMP
HERE

Texas Instruments
Literature Response Center
P.O. Box 809066
Dallas, Texas 75380-9066

PLACE
STAMP
HERE

Texas Instruments
Literature Response Center
P.O. Box 809066
Dallas, Texas 75380-9066